Handbook of Heat and Mass Transfer

Volume 1: Heat Transfer Operations

Gulf Publishing Company
Book Division
Houston, London, Paris, Tokyo

Handbook of Heat and Mass Transfer

Volume 1: Heat Transfer Operations

Nicholas P. Cheremisinoff, Editor

in collaboration with:

R. L. Adams
N. Afzal
K. Banerjee
H. H. Bau
R. Blaszczyk
H. de Bruijn
M. Bruxelmane
P. N. Cheremisinoff
R. Cole
A. Cordoba
U. Cywinska
H. Desplanches
R. Echigo
W. E. Genetti
N. S. Grewal
C. P. Gupta
S. Hasegawa
K. Hashizume
R. H. Heist
B. H. Hjertager
J. Iciek
B. M. Johnson
I. Kataoka
Y. Katto
M. Kielkiewicz
B. Korenic
M. Kosinski
T. Kumada
J. Lach
S. L. Lee
W. Leidenfrost
P. K. Nag
R. A. Nelson
J. M. Phelps
G. P. Purohit
O. Ratsep
A. Rey de Luna
E. Ruckenstein
A. Serizawa
W. T. Sha
M. M. Shah
A. V. Shenoy
A. Shimizu
S. Sieniutycz
A. Steiff
B. Sudhoff
K. Suzuki
T. N. Tandon
A. K. S. Thakur
H. K. Varma
P. M. Weinspach
W. J. Yang
M. Zamora
W. Zijl

Handbook of Heat and Mass Transfer

Volume 1: Heat Transfer Operations

ISBN 0-87201-411-8

Library of Congress Cataloging in Publication Data
Main entry under title:

Handbook of heat and mass transfer.

Includes indexes.
Contents: v. 1 Heat transfer operations-v.2. Mass transfer and reactor design.
1. Heat-Transmission-Handbooks, manuals, etc. 2. Mass transfer-Handbooks, manuals, etc.
I. Cheremisinoff, Nicholas P.
TJ260.H36 1986 621.402′2 84-25338
ISBN 0-87201-411-8 (v. 1)
ISBN 0-87201-412-6 (v. 2)

CONTENTS

Section I: Heat Transfer Mechanisms

Section II: Industrial Operations and Design

VOLUME 2: MASS TRANSFER AND REACTOR DESIGN

CONTENTS

Section I: Mass Transfer Principles

Section II: Distillation and Extraction

CONTRIBUTORS TO THIS VOLUME

R. L. Adams, Dept. of Mechanical Engineering, Oregon State University, Corvallis, Oregon, USA.

N. Afzal, Dept. of Mechanical Engineering, Aligarh Muslim University, Aligarh, India.

K. Banerjee, Dept. of Civil and Environmental Engineering, New Jersey Institute of Technology, Newark, New Jersey, USA.

H. H. Bau, Dept. of Mechanical Engineering and Applied Mechanics, University of Pennsylvania, Philadelphia, Pennsylvania, USA.

R. Blaszczyk, Institute of Chemical Engineering, Lodz Technical University, Lodz, Poland.

H. de Bruijn, Group of Seismics and Acoustics, Department of Applied Physics, Delft University of Technology, The Netherlands.

M. Bruxelmane, Faculte Polytechnique de Mons, B-7000 Mons, Belgium.

N. P. Cheremisinoff, Exxon Chemical Co., Linden, New Jersey, USA.

P. N. Cheremisinoff, Dept. of Civil and Environmental Engineering, New Jersey Institute of Technology, Newark, New Jersey, USA.

R. Cole, Dept. of Chemical Engineering and Institute of Colloid and Surface Science, Clarkson University, Potsdam, New York, USA.

A. Cordoba, Departmento de Termologia, Universidad de Sevilla, Spain.

U. Cywinska, Institute of Chemical Engineering, Lodz Technical University, Lodz, Poland.

H. Desplanches, Ecole Supérieure de Chimie, Faculte de Saint Jerome, F-13397 Marseille Cedex 4, France.

R. Echigo, Dept. of Nuclear Engineering, Kyushu University, Fukuoka 812, Japan.

W. E. Genetti, Chemical Engineering Dept., University of Mississippi, University, Mississippi, USA.

N. S. Grewal, Mechanical Engineering Dept., University of North Dakota, Grand Forks, North Dakota, USA.

C. P. Gupta, Dept. of Mechanical Engineering, University of Roorkee, India.

S. Hasegawa, Dept. of Nuclear Engineering, Kyushu University, Fukuoka 812, Japan.

K. Hashizume, Energy Science and Technology Laboratory, R&D Center, Toshiba Corp., Ukishimacho 4-1, Kawasaki, Japan.

R. H. Heist, Dept. of Chemical Engineering, University of Rochester, Rochester, New York, USA.

B. H. Hjertager, Chr. Michelson Institute, Dept. of Science and Technology, N-5036, Fantoft, Bergen, Norway.

J. Iciek, Institute of Chemical Engineering, Lodz Technical University, Lodz, Poland.

B. M. Johnson, Pacific Northwest Laboratory, Richland, Washington, USA.

I. Kataoka, Institute of Atomic Energy, Kyoto University, Uji, Kyoto 611, Japan.

Y. Katto, Dept. of Mechanical Engineering, University of Tokyo, Hongo, Bunkyo-ku, Tokyo, Japan.

M. Kielkiewicz, Warsaw Technical University, Institute of Heat Engineering, Warsaw, Poland.

B. Korenic, Baltimore Aircoil Co., P.O. Box 7322, Baltimore, Maryland, USA.

M. Kosinski, Warsaw Technical University, Institute of Heat Engineering, Warsaw, Poland.

T. Kumada, Hokkaido University, North 13, West 8, Sapporo 060, Japan.

J. Lach, Institute of Atomic Energy, Dept. of Reactor Engineering, Otwock-Swierk, Poland.

S. L. Lee, Dept. of Power Mechanical Engineering, National Tsing Hua University, Hsinchu, Taiwan 300, Republic of China.

W. Leidenfrost, School of Mechanical Engineering, Purdue University, West Lafayette, Indiana, USA.

P. K. Nag, Dept. of Mechanical Engineering, Indian Institute of Technology, Kharagpur, India.

R. A. Nelson, Los Alamos National Laboratory, Los Alamos, New Mexico, USA.

J. M. Phelps, Bowen Engineering Inc., Somerville, New Jersey, USA.

G. P. Purohit, Consulting Engineer, Parkview Lane, Irvine, California, USA.

O. Ratsep, Northeast Protech Inc. Boonton, New Jersey, USA.

A. Rey de Luna, Departamento de Termologia, Universidad de Sevilla, Spain.

E. Ruckenstein, Dept. of Chemical Engineering, State University of New York at Buffalo, Amherst, New York, USA.

A. Serizawa, Dept. of Nuclear Engineering, Kyoto University, Sakyo-ku, Kyoto 606, Japan.

W. T. Sha, Argonne National Laboratory, 9700 South Cass Avenue, Argonne, Illinois, USA.

M. M. Shah, Consulting Engineer, Port Jefferson Station, New York, USA.

A. V. Shenoy, Chemical Engineering Division, National Chemical Laboratory, Pune 411 008, India.

A. Shimizu, Dept. of Nuclear Engineering, Kyushu University, Fukuoka 812, Japan.

S. Sieniutycz, Institute of Chemical Engineering, Warsaw Technical University, 00-645 Warsaw, Warynskiego 1, Poland.

A. Steiff, University of Dortmund, Federal Republic of Germany.

B. Sudhoff, University of Dortmund, Federal Republic of Germany.

K. Suzuki, Dept. of Mechanical Engineering, Kyoto University, Kyoto 606, Japan.

T. N. Tandon, Dept. of Mechanical Engineering, M. M. M. Engineering College, Gorakhpur, India.

A. K. S. Thakur, Dept. of Physics, University of Sokoto, Sokoto, Nigeria.

H. K. Varma, Dept. of Mechanical Engineering, University of Roorkee, India.

P. M. Weinspach, University of Dortmund, Federal Republic of Germany.

W. J. Yang, Dept. of Mechanical Engineering and Applied Mechanics, The University of Michigan, Ann Arbor, Michigan, USA.

M. Zamora, Departamento de Termologia, Universidad de Sevilla, Spain.

W. Zijl, Research & Development Group, Institute for Applied Geosciences TNO, 2600 AG Delft, The Netherlands.

ABOUT THE EDITOR

Nicholas P. Cheremisinoff heads the product development group in the Elastomers Technology Division of Exxon Chemical Company. Previously, he led the Reactor and Fluid Dynamics Modeling Group at Exxon Research and Engineering Company. He received his B.S., M.S., and Ph.D. degrees in chemical engineering from Clarkson College of Technology, and he is also a member of a number of professional societies including AIChE, Tau Beta Pi, and Sigma Xi.

PREFACE

Transport phenomena involve the exchange of momentum, energy, and mass within and across defined system boundaries. Understanding these mechanisms and describing them in the universal language of mathematics enable the forces of nature to be harnessed for mankind's benefit. This is essentially accomplished every day in countless manufacturing operations that produce an endless stream of products and life-sustaining articles throughout the world.

Despite industry's success, our understanding of transport phenomena is limited. Most process industries handle complex flows of mixtures, process streams involving phase changes, and streams with multiple chemical reactions, often requiring energy to be transferred between equipment and process streams. Because the most up-to-date data are fragmented, scale-up is difficult, and all too often results in operations that perform below expectations resulting in costly upsets and shutdowns. This is largely because the literature on heat and mass transfer is so diverse and often contradictory for systems involving phase changes and chemical reactions. Clearly, a more unified approach to designing process operations can only be gained through better understanding of the physical mechanisms of mass and energy transport as well as the chemical kinetics.

This handbook is aimed at unifying heat and mass transport concepts for the practicing engineer. The overall theme of this work is to provide fundamental understanding of the physical processes and detailed guidance in applying principles to designing complex unit operations and chemical reactors involving multiphase flows. Emphasis is placed on more advanced industrial problems, although ample references are given to provide the user with reviews of basic principles.

The work is divided into two volumes.

Volume 1: Heat Transfer Operations, contains two sections. The first section, "Heat Transfer Mechanisms," devotes 20 chapters to discussions of principles and applications of transport phenomena. Momentum transport and fluid dynamics are related to various subjects of heat transfer. In addition, several chapters are devoted to more recent areas of research. The second section, "Industrial Operations and Design," provides 23 chapters on critical unit operations involving heat exchangers, drying operations, combustion, and heat transfer problems in specialized reactors.

Volume 2: Mass Transfer and Reactor Design, contains four sections. Section I, "Mass Transfer Principles," emphasizes multicomponent mass transfer in nine chapters. Discussions also cover mass transfer in reacting flows. Section II, "Distillation and Extraction," is composed of eight chapters, with emphasis given to design-related problems. Again, the references provide further information on the fundamental concepts. Section III, "Multiphase Reactor Systems," devotes 15 chapters to various types of reactors of industrial importance. Many of these chapters relate to reaction kinetics principles. Section IV, "Special Applications and Reactor Topics," provides five chapters on the subjects of industry crystallization, thermal diffusion columns, parametric pumping, and some special topics of interest.

To ensure the highest degree of reliability, the services of more than one hundred specialists were enlisted. This work presents the efforts of these experts. In addition, it incorporates the experience and opinions of scores of engineers and researchers who aided with advice and suggestions in reviewing and refereeing the material presented. Each contributor is to be regarded as responsible for the statements and recommendations in his/her chapter. These individuals are to be congratulated for devoting their time and efforts to producing these volumes. Without their efforts this work could not have been a reality. Special thanks is also expressed to Gulf Publishing Company and, in particular, to Melissa Lewis for her assistance in copyediting and styling.

Nicholas P. Cheremisinoff

SECTION I

HEAT TRANSFER MECHANISMS

CONTENTS

CHAPTER 1

FUNDAMENTALS OF MOMENTUM AND HEAT TRANSFER

Nicholas P. Cheremisinoff

Exxon Chemical Co.
Linden, New Jersey

CONTENTS

INTRODUCTION

This chapter is presented as a fundamental review of momentum and heat transfer mechanisms. Basic concepts and formulations important to momentum transfer and transport phenomena in general are reviewed. Since emphasis in the first section of the handbook is on heat transfer, discussions here are limited to governing statements concerning conduction, convection, and radiation. The chapter is presented mainly for the reader's orientation and to clarify terminology used in later chapters.

PRINCIPLES OF FLUID MECHANICS AND MOMENTUM TRANSFER

For fluids in motion, any property of the flow, w, that changes with respect to time and space (e.g., temperature, pressure, density, concentration, or velocity components w_x, w_y, and w_z) can be expressed in terms of its fluctuations at any moment at a fixed point in space (x, y, z). A stationary observer would see that the rate of change in w over time is constant.

The rate of change of w may be described by a partial derivative $\partial w/\partial \tau$ and hence, the variation in w over a time interval $d\tau$ is $(\partial w/\partial \tau)\,d\tau$. This is the *local* variation of the given variable and at steady state is zero.

When the observer moves along with the flow, w is the sum of resultant vectors. For a fluid particle displaced from point A (x,y,z) to point B (x + dx), (y + dy), (z + dz) over time $d\tau$ the corresponding path of projections dx, dy, and dz is $(\partial w/\partial x)\,dx$, $(\partial w/\partial y)\,dy$, and $(\partial w/\partial z)\,dz$. At steady state w changes by:

$$dw = \frac{\partial w}{\partial x}\,dx + \frac{\partial w}{\partial y}\,dy + \frac{\partial w}{\partial z}\,dz \tag{1}$$

Equation 1 characterizes the *convective* change of parameter w.

For unsteady phenomenon where $w = f(x,y,z,\tau)$, changes occur over time interval $d\tau$ by $(\partial w/\partial \tau)\,d\tau$. The *total* change in w is the sum of the local and convective variations:

$$dw = \frac{\partial w}{\partial \tau}\,d\tau + \frac{\partial w}{\partial x}\,dx + \frac{\partial w}{\partial y}\,dy + \frac{\partial w}{\partial z}\,dz \tag{2a}$$

or

$$\frac{dw}{d\tau} = \frac{\partial w}{\partial \tau} + \frac{\partial w}{\partial x}\frac{dx}{d\tau} + \frac{\partial w}{\partial y}\frac{dy}{\partial \tau} + \frac{\partial w}{\partial z}\frac{dz}{\partial \tau} \tag{2b}$$

Defining w as fluid velocity, then

$$\frac{dw}{d\tau} = \frac{\partial w}{\partial \tau} + \frac{\partial w}{\partial x}\,w_x + \frac{\partial w}{\partial y}\,w_y + \frac{\partial w}{\partial z}\,w_z \tag{3a}$$

When $\partial w/\partial \tau = 0$ (steady state)

$$\frac{dw}{d\tau} = \frac{\partial w}{\partial x}\,w_x + \frac{\partial w}{\partial y}\,w_y + \frac{\partial w}{\partial z}\,w_z \tag{3b}$$

Equations 3a and 3b are substantial derivatives which characterize changes in both time and space.

For flow through a constant cross-sectional area, the superficial mass velocity is:

$$G = W/A,\ kg/m^2\text{-}hr \tag{4}$$

and the volumetric flow rate is:

$$V = \frac{W}{\rho},\ m^3/hr \tag{5}$$

The mean linear velocity of the flow is:

$$\bar{w} = \frac{V}{A} \tag{6}$$

The local linear fluid velocity depends on the influence of solid boundaries. Fluid elements close to the pipe axis move faster than those in the region of the wall. Hence, information on how the velocity changes over the cross-section is important. For a symmetrical velocity distribution about the pipe axis, w is a function of the distance normal to the wall:

$$w = f(r) \tag{7}$$

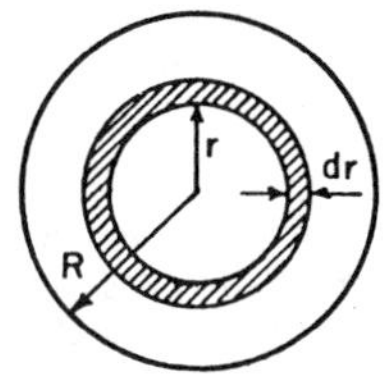

Figure 1. Front view of fluid flowing through a pipe.

For a circular flow element of radius r and thickness dr, moving along a pipe of radius R shown in Figure 1, the local volumetric flow is (where $dA = 2\pi rdr$):

$$dV = w2\pi rdr$$

Hence, the volumetric rate over the total cross-section is:

$$V = 2\pi \int_0^R wrdr \qquad (8)$$

From the $w = f(r)$ relationship, the integral can be evaluated for the volumetric flow rate.

The average linear velocity for flow through a circular section is:

$$w = \int_0^1 wd\left(\frac{r}{R}\right)^2 \qquad (9)$$

The local velocity w in Equation 8 can be stated in terms of the fractional pipe radius (r/R). By integrating this relationship over the limits of r/R from 0 to 1, the average velocity of the flow is obtained.

For a continuous flow, the mass rate is the same at any section within the system.

$$W_1 = W_2 = W_3 = \ldots \qquad (10)$$

From Equation 4, the continuitiy of flow is stated as:

$$G_1A_1 = G_2A_2 = G_3A_3 \ldots \qquad (11a)$$

or

$$\bar{w}_1\gamma_1A_1 = \bar{w}_2\gamma_2A_2 = \bar{w}_3\gamma_3A_3 = \ldots \qquad (11b)$$

Equations 10 and 11a, and b represent the material balance for any flow system. Equation 11b has special relevance to compressible flows, in which pressure, temperature, and specific gravity may vary at different cross-sections, resulting in different linear velocities. Equations 11b establishes a relationship between these velocities and their specific gravities. For liquid flow in a pipe of a constant cross-sectional area the balance equation is considerably simplified:

$$G_1 = G_2 = G_3 = \ldots = \bar{w}_1\rho_1 = \bar{w}_2\rho_2 = \bar{w}_3\rho_3 \qquad (12)$$

Equation 12 states that for any cross-section the mass rates of fluid will be the same even if the temperatures and pressures vary. Applying Equation 12 to a liquid flow undergoing a small temperature change, we may assume that the specific density of the liquid is approximately constant. Consequently, the equation of continuity for liquid flow is:

$$\bar{w}_1A_1 = \bar{w}_2A_2 = \bar{w}_3A_3 \qquad (13)$$

And its volumetric rate is:

$$V_1 = V_2 = V_3 = \ldots \qquad (14)$$

Hence, the average linear velocities for liquid flow are inversely proportional to the cross-sectional areas of flow. For a constant cross-section the linear velocities are the same:

$$\bar{w}_1 = \bar{w}_2 = \bar{w}_3 = \ldots \tag{15}$$

The fluid's ideal *residence time* in a system can be obtained by assuming that its mass rate W does not change over the cross-section and that there are no stagnant regions of flow. If dx is the distance along the flow direction, and if the flow travels over dx in time $d\tau$, then the average linear velocity is:

$$\bar{w} = \frac{dx}{d\tau} \tag{16}$$

or

$$W = \gamma A \frac{dx}{d\tau} \tag{17}$$

where A is the cross-sectional area normal to the flow. Note that the elemental volume of the pipe or piece of equipment is $dV_0 = Adx$. When γ is constant, on integration of Equation 17 the residence time is:

$$\tau = \frac{\gamma}{W} V_0 = \frac{V_0}{V} \tag{18}$$

Hence, theoretical residence time depends only on the system volume V_0 and the volumetric flow.

From the previous fundamental definitions, we can now summarize a rigorous description of flow through any system. Figure 2 shows a differential fluid element of volume $dV = dxdydz$, where the velocity component along the x-axis is w_x. The input per unit time, $d\tau$, in the x-direction through the left face of the elements ($dS = dydz$) is:

$$M_x = \rho w_x dydzd\tau \tag{19}$$

where ρ = fluid density at the left face of the element.

The velocity and density at the parallel face (located at x + dx) are $(w_x + [\partial(\rho w_x)/\partial x]\,dx)$ and $(\rho + (\partial\rho/\partial x)\,dx)$, respectively. Hence, the output at the right face of the element in the same time interval $d\tau$ is:

$$M_{x+dx} = \left[\rho w_x + \frac{\partial(\rho w_x)}{\partial x}dx\right] dydzd\tau \tag{20}$$

The mass differential in the element along the x-axis is:

$$dM_x = M_x - M_{x+dx} = -\frac{\partial(\rho w_x)}{\partial x} dxdydzd\tau \tag{21}$$

The total mass differential in elemental volume dV over time $d\tau$ is:

$$dM = -\left[\frac{\partial(\rho w_x)}{\partial x} + \frac{\partial(\rho w_y)}{\partial y} + \frac{\partial(\rho w_z)}{\partial z}\right] dxdydzd\tau \tag{22}$$

For a mass differential to occur in the fluid element, density also must change with respect to time.

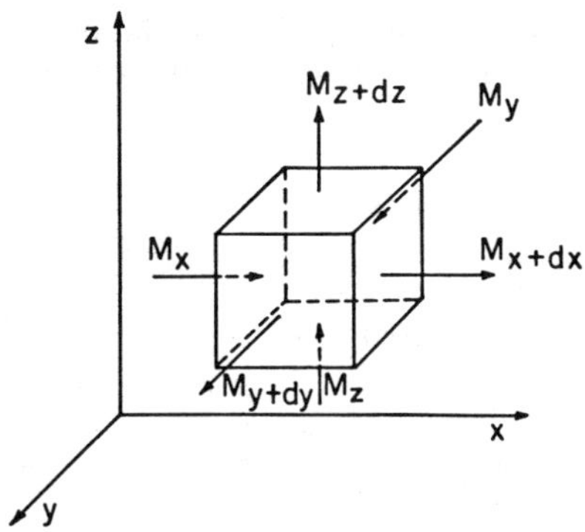

Figure 2. Fluid element showing mass transfer through defined boundaries.

$$dM = \frac{\partial \rho}{\partial \tau} dxdydzd\tau \tag{23}$$

Equating these last two expressions through dM and rearranging terms:

$$\frac{\partial \rho}{\partial \tau} + \frac{\partial (\rho w_x)}{\partial x} + \frac{\partial (\rho w_y)}{\partial y} + \frac{\partial (\rho w_z)}{\partial z} = 0 \tag{24}$$

Equation 24 is the differential equation of continuity for unsteady flow of a compressible fluid. The equation may be written in another form by differentiating the product ρw:

$$\frac{\partial \rho}{\partial \tau} + \frac{\partial \rho}{\partial x} w_x + \frac{\partial \rho}{\partial y} w_y + \frac{\partial \rho}{\partial z} w_z + \frac{\partial w_x}{\partial x} \rho + \frac{\partial w_y}{\partial y} \rho + \frac{\partial w_z}{\partial z} \rho = 0 \tag{25}$$

or

$$\frac{1}{\rho}\frac{d\rho}{d\tau} + \frac{\partial w_x}{\partial x} + \frac{\partial w_y}{\partial y} + \frac{\partial w_z}{\partial z} = 0 \tag{26}$$

where $\partial \rho / \partial \tau$ is a substantial derivative.

In steady state flows, density ρ does not change with time (i.e., $\partial \rho / \partial \tau = 0$), and Equation 24 simplifies to:

$$\frac{\partial (\rho w_x)}{\partial x} + \frac{\partial (\rho w_y)}{\partial y} + \frac{\partial (\rho w_z)}{\partial z} = 0 \tag{27}$$

For incompressible fluids:

$$\frac{\partial w_x}{\partial x} + \frac{\partial w_y}{\partial y} + \frac{\partial w_z}{\partial z} = 0 \tag{28}$$

Equation 28 is the differential continuity equation for the flow of incompressible fluids.

An ideal fluid is one having no consistency. For the fluid element in Figure 2 the projections of pressure and gravity forces along the coordinate axes acting on the elemental volume dV = dxdydz are:

$$-\frac{\partial P}{\partial x}\,dxdydz \qquad \text{(x-axis)}$$

$$-\frac{\partial P}{\partial y}\,dxdydz \qquad \text{(y-axis)}$$

$$-\left(\rho g+\frac{\partial P}{\partial z}\right)dxdydz \qquad \text{(z-axis)}$$

The sum of the projection of forces acting on the elemental volume in motion is equal to the product of mass and acceleration. Hence, for the x, y, and z axis projections:

$$\rho dxdydz\,\frac{dw_x}{d\tau}=-\frac{\partial P}{\partial x}\,dxdydz$$

$$\rho dxdydz\,\frac{dw_y}{\partial\tau}=-\frac{\partial P}{\partial y}\,dxdydz \qquad (29)$$

$$\rho dxdydz\,\frac{w_z}{d\tau}=-\frac{\partial P}{\partial z}\,dxdydz$$

After simplifying:

$$\rho\,\frac{dw_x}{d\tau}=-\frac{\partial P}{\partial x}$$

$$\rho\,\frac{dw_y}{d\tau}=-\frac{\partial P}{\partial y} \qquad (30)$$

$$\rho\,\frac{dw_z}{d\tau}=-\rho g-\frac{\partial P}{\partial z}$$

The substantial derivatives of the velocity components are:

$$\frac{dw_x}{d\tau}=\frac{\partial w_x}{\partial x}\,w_x+\frac{\partial w_x}{\partial y}\,w_y+\frac{\partial w_x}{\partial z}\,w_z$$

$$\frac{dw_y}{d\tau}=\frac{\partial w_y}{\partial x}\,w_x+\frac{\partial w_y}{\partial y}\,w_y+\frac{\partial w_y}{\partial z}\,w_z \qquad (31)$$

$$\frac{dw_z}{d\tau}=\frac{\partial w_z}{\partial x}\,w_x+\frac{\partial w_z}{\partial y}\,w_y+\frac{\partial w_z}{\partial z}\,w_z$$

The prior system of equations are Euler's differential equations of motion for an ideal fluid under steady-flow. In unsteady flow, velocity changes not only when fluid particles

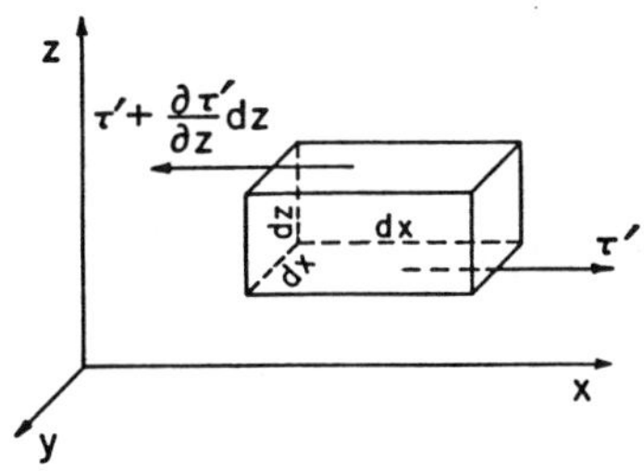

Figure 3. Forces acting on a volume element of moving fluid.

move from one point in space to another, but also with time. Therefore, the components of acceleration in Equation 30 must be expressed in the following form:

$$\frac{dw_x}{d\tau} = \frac{\partial w_x}{\partial \tau} + \frac{\partial w_x}{\partial x} w_x + \frac{\partial w_x}{\partial y} w_y + \frac{\partial w_x}{\partial z} w_z$$

$$\frac{dw_y}{d\tau} = \frac{\partial w_y}{\partial \tau} + \frac{\partial w_y}{\partial x} w_x + \frac{\partial w_y}{\partial y} w_y + \frac{\partial w_y}{\partial z} w_z \quad (32)$$

$$\frac{dw_z}{d\tau} = \frac{\partial w_z}{\partial \tau} + \frac{\partial w_z}{\partial x} w_x + \frac{\partial w_z}{\partial y} w_y + \frac{\partial w_z}{\partial z} w_z$$

The systems of equations in Equation 30 and 32 represent Euler's differential equations of motion for unsteady flow.

The motion of real or viscous fluids is influenced by frictional forces in addition to pressure and gravity. Frictional forces T′ acting on a volume element dV = dxdydz shown in Figure 3 result in shear stresses τ′ forming on the faces. For flow only in the x direction the velocity projection w_x depends only on distance z above some horizontal reference plane. In this case the shear stresses on the lower and upper faces of the volume element dA = dxdy are τ' and $\tau' + (\partial\tau'/\partial z)\,dz$, respectively. The partial derivative $(\partial\tau'/\partial z)$ expresses the change in shear stress along the z-axis at points located on the lower face of the volume element, and $(\partial\tau'/\partial z)\,dz$ is the change of shear stress along the total length of the element's edge dz. The projection of the resultant shear force in the x direction is:

$$\tau' dxdy - \left(\tau' + \frac{\partial \tau'}{\partial z} dz\right) dxdy = -\frac{\partial \tau'}{\partial z} dxdydz \quad (33)$$

From Newton's law, $\tau' = -\mu \frac{\partial w_x}{\partial z}$, and hence–

$$\mu \frac{\partial\left(\frac{\partial w_x}{\partial z}\right)}{\partial z} dxdydz = \mu \frac{\partial^2 w_x}{\partial z^2} dxdydz \quad (34)$$

For three-dimensional flow, component w_x changes along three coordinate axes, and the projection of the resultant shear forces in the x direction is:

$$\mu \left(\frac{\partial^2 w_x}{\partial z^2} + \frac{\partial w_x^2}{\partial y^2} + \frac{\partial^2 w_z}{\partial x^2}\right) dxdydz \quad (35a)$$

or

$$\frac{\partial^2 w_x}{\partial x^2} + \frac{\partial^2 w_x}{\partial y^2} + \frac{\partial^2 w_x}{\partial z^2} = \nabla^2 w_x \tag{35b}$$

where ∇^2 is the Laplacian operator:

$$\nabla^2 = \frac{\partial^2}{\partial x^2} + \frac{\partial^2}{\partial y^2} + \frac{\partial^2}{\partial z^2} \tag{36}$$

Hence, the resultant projections of gravity, pressure, and shear forces on respective coordinate axes are:

$$\left(-\frac{\partial P}{\partial x} + \mu \nabla^2 w_x\right) dxdydz \qquad \text{(x axis)}$$

$$\left(-\frac{\partial P}{\partial x} + \mu \nabla^2 w_y\right) dxdydz \qquad \text{(y axis)}$$

$$\left(-\rho g - \frac{\partial P}{\partial x} + \mu \nabla^2 w_z\right) dxdydz \qquad \text{(z axis)}$$

Where the gravity and pressure force projections are derived from Euler's equations.

The sum of the projection of forces equals the product of the fluid mass $\rho dxdydz$ in the elemental volume and the projections of acceleration. Applying this principle and after simplification, the Navier-Stokes equations of motion for viscous incompressible fluids are obtained in Cartesian coordinates:

$$\begin{aligned} \rho \frac{dw_x}{d\tau} &= -\frac{\partial P}{\partial x} + \mu \nabla^2 w_x \\ \rho \frac{dw_y}{d\tau} &= -\frac{\partial P}{\partial y} + \mu \nabla^2 w_y \\ \rho \frac{dw_z}{d\tau} &= -\rho g - \frac{\partial P}{\partial z} + \mu \nabla^2 w_z \end{aligned} \tag{37}$$

The Navier-Stokes equations for compressible fluids are as follows:

$$\begin{aligned} \rho \frac{dw_x}{d\tau} &= -\frac{\partial P}{\partial x} + \mu \left(\nabla^2 w_x + \frac{1}{3}\frac{\partial \theta}{\partial x}\right) \\ \rho \frac{dw_y}{d\tau} &= -\frac{\partial P}{\partial y} + \mu \left(\nabla^2 w_y + \frac{1}{3}\frac{\partial \theta}{\partial y}\right) \\ \rho \frac{dw_z}{d\tau} &= -\rho g - \frac{\partial P}{\partial z} + \mu \left(\nabla^2 w_z + \frac{1}{3}\frac{\partial \theta}{\partial z}\right) \end{aligned} \tag{38}$$

The partial derivatives $(\partial\theta/\partial x)$, $(\partial\theta/\partial y)$, and $(\partial\theta/\partial z)$ represent changes in velocities along the x, y and z axes and are associated with compression and tension forces acting on the fluid.

The terms of LHS of Equations 37 and 38 represents the rate of inertial forces for the moving fluid. Terms on the RHS describe three types of forces:

1. ρg denotes gravitational forces acting on the fluid volume element
2. The partial derivatives $(\partial P/\partial x)$, $(\partial P/\partial y)$, and $(\partial P/\partial z)$ denote pressure forces
3. The product of viscosity and the sum of the second derivatives of the velocity projections denote the viscous forces acting on the moving fluid element.

All terms in these two sets of equations have dimensions of force per unit volume. For ideal fluids the expressions reduce to Euler's equations.

The Navier-Stokes equations represent a system of nonlinear partial expressions having no universal solution. In fact, these equations are often of a higher order because viscosity can be a function of temperature as well as velocity. Two general approaches may be used in applying the Navier-Stokes equations to analyzing flow systems. The expressions may be reduced to specific solutions either by applications of a series of simplifying assumptions or by transformations techniques based on similarity theory. Principles of the latter method are reviewed herein.

Following Sedov [1] and Gukhman [2], the following scaling factors are defined:

$$\frac{\rho}{\rho_0} = P'; \frac{P}{P_0} = \mathscr{P}; \frac{\mu}{\mu_0} = M; \frac{\tau'}{\tau_0'} = T'; \frac{w}{w_0} = \tilde{W}; \frac{\ell}{\ell_0} = \frac{x}{x_0} = X; \frac{y}{y_0} = Y; \frac{z}{z_0} = Z; \frac{w_{x,y,z}}{w_{x,y,z0}} = W_{x,y,z0}$$

Rearranging Equation 38:

$$\rho\left[\frac{\partial w_x}{\partial \tau} + w_x\frac{\partial w_x}{\partial x} + w_y\frac{\partial w_y}{\partial y} + w_z\frac{\partial w_z}{\partial z}\right] = \rho g - \frac{\partial P}{\partial x} + \mu\left(\nabla^2 w_x + \frac{1}{3}\frac{\partial \theta}{\partial x}\right) \tag{39}$$

Substituting for scaling factors, we obtain:

$$P\rho_0\left[\frac{\partial W_x}{\partial T}\cdot\frac{w_0}{\tau_0} + W_x\frac{\partial W_x}{\partial x}\cdot\frac{w_0}{\ell_0} + w_0 W_y\frac{\partial W_y}{\partial y}\cdot\frac{w_0}{\ell_0} + w_0 W_z\frac{\partial W_z}{\partial z}\cdot\frac{w_0}{\ell_0}\right]$$
$$= P\rho_0 g - \frac{\partial \mathscr{P}}{\partial x}\frac{P_0}{\ell_0} + \mu M\left(\nabla^2 W_x\frac{w_0}{\ell_0^2} + \frac{1}{3}\frac{\partial \theta}{\partial x}\cdot\frac{w_o}{\ell_0^2}\right) \tag{40}$$

where

$$0 = \frac{\partial W_x}{\partial x} + \frac{\partial W_y}{\partial y} + \frac{\partial W_z}{\partial z} = \text{div}\,w$$

$$\nabla^2 W_x = \frac{\partial^2 W_x}{\partial x^2} + \frac{\partial^2 W_y}{\partial y^2} + \frac{\partial^2 W_z}{\partial z^2}$$

As ℓ_0, w_0, and ρ_0 are constants, they may be combined into a single coefficient: $\frac{\ell_0 w_0^2}{\rho_0}$

Diving both sides by $\ell_0 w_0^2/\rho_0$ we obtain:

$$P\left[\frac{\partial W_x}{\partial T}\left(\frac{\ell_0}{w_0\tau_0}\right) + W_x\frac{\partial W_x}{\partial x} + W_y\frac{\partial W_x}{\partial y} + W_z\frac{\partial W_x}{\partial z}\right]$$
$$= \frac{g_0\ell_0}{w_0^2}P - \left(\frac{P_0}{\rho_0 w_0^2}\right)\frac{\partial \mathscr{P}}{\partial x} + \left(\frac{\mu_0}{\ell_0 w_0 \rho_0}\right)M\left[\nabla^2 W_x + \frac{1}{3}\frac{\partial \theta}{\partial x}\right] \tag{41}$$

Equation 41 contains similarity invariants which are the following dimensionless groups:

$$\frac{w\tau}{\ell} = \text{Ho} = \text{homochronity number}$$

$$\frac{w^2}{g\ell} = \text{Fr} = \text{Froude number}$$

$$\frac{\ell w\rho}{\mu} = \text{Re} = \text{Reynolds number}$$

$$\frac{P}{\rho w^2} = \text{Eu} = \text{Euler number}$$

Hence, the generalized equations of motion can be written as:

$$P\left[\frac{\partial w_x}{\partial T}\cdot\frac{1}{\text{Ho}} + w_x\frac{\partial w_x}{\partial x} + w_y\frac{\partial w_y}{\partial y} + w_z\frac{\partial w_z}{\partial z}\right]$$
$$= \frac{1}{\text{Fr}_0}P - \text{Eu}_0\frac{\partial \mathscr{P}}{\partial x} + \frac{1}{\text{Re}_0}M\left[\nabla^2 W_x + \frac{1}{3}\frac{\partial\theta}{\partial x}\right] \tag{42}$$

Equation 42 illustrates the importance of understanding the physics of phenomenon when analyzing flow systems. For example, if gravity forces are unimportant then the Froude number may be neglected and the conditions of identity of a system of the Navier-Stokes equations can be determined strictly by the homochronity, Reynolds, and Euler numbers.

The Euler number should be considered as a nondetermining dimensionless group. Dimensionless groups of this type are unique functions of other determining groups. For this reason, the Navier-Stokes equations may be stated simply as:

$$\text{Eu} = \psi(\text{Ho},\text{Fr},\text{Re}) \tag{43}$$

The equations of motion alone are not sufficient to describe a system. A second constitutive equation needed is the continuity equation. Continuity depends only on the homochronity number:

$$\frac{1}{\text{Ho}}\frac{\partial P}{\partial T'} + \frac{\partial(\rho w_x)}{\partial x} + \frac{\partial(\rho w_y)}{\partial y} + \frac{\partial(\rho w_z)}{\partial z} = 0 \tag{44}$$

At steady-state, Ho vanishes and similarity in the absence of volume forces (i.e., the Froude number) is provided by identifying the Reynolds number. The Euler number in this case is a unique function of the Reynolds number whence Equation 43 reduces to:

$$\text{Eu} = \psi(\text{Re}) \tag{45}$$

This last statement reflects only the kinematic and dynamic similarities of flows. For total similarity between flow systems it is necessary to maintain geometric similarity. For flow in round pipes the simplex of geometric similarity is the ratio of ℓ/d. Denoting this simplex number as $\tilde{\Gamma} = \ell/d$, Equation 45 is rigorously stated as:

$$\text{Eu} = \psi(\text{Re},\tilde{\Gamma}) \tag{46}$$

The nature of fluid motion and the interactions that take place between fluid particles which ultimately impact on transport mechanisms can be described by the laminar and turbulent regimes. The *laminar regime* occurs at relatively low fluid velocities where the flow may be visualized as layers that slide over each other providing smooth flow patterns and no macroscopic mixing of fluid particles. In the *turbulent regime,* fluid velocities are higher and an unstable pattern within the bulk flow is observed in which eddies move at all angles. The thin fluid near the wall where the fluid motion retains a laminar identify is the *laminar boundary layer*.

The transition from laminar to turbulent flow in tubes is a function of the fluid velocity, density, and viscosity, and the tube diameter, i.e., the *Reynolds number*.

$$Re = \frac{wD}{\nu} = \frac{wD\rho}{\mu} = \frac{WD}{\mu} \tag{47}$$

where ν = kinematic viscosity
ρ = density
μ = dynamic viscosity
W = mass velocity
D = tube diameter (or some characteristic size of the system through which flow occurs)

The *critical Reynolds number* denotes the transition of one flow regime to another.

In a straight circular pipe, when the Reynolds number is less than 2,100, the flow is laminar. When the Reynolds number exceeds 4,000, the flow is turbulent, except for some very special cases. The flow between these two values is referred to as the *transition region,* where the motion may be either laminar or turbulent.

For steady-state laminar flow of a constant density fluid through a tube of length L and radius R, the motion of fluid layers results from a difference in pressure forces P_1 and P_2 at both ends of the fluid cylinder (ignoring end effects):

$$P_1 - P_2 = (p_1 - p_2)\pi r^2 \tag{48}$$

P_1 and P_2 are the hydrostatic pressures at two sections.

The motion of the fluid cylinder is resisted by a friction force:

$$T' = -\mu A \frac{dw_r}{dr} \tag{49}$$

where w_r = fluid velocity along the x axis at a distance r from the tube centerline
A = $2\pi r\ell$ = the cylinder's outside surface area
μ = fluid viscosity

The minus sign in Equation 49 denotes that the fluid velocity decreases as radius r increases. That is, as r approaches the radius of the tube, w_r becomes smaller, and at r = R, $w_r = 0$. The pressure gradient $P_1 - P_2$ represents the driving force for the flow and must be sufficiently large to overcome frictional force T′, i.e.:

$$(P_1 - P_2)\pi r^2 = -\mu 2\pi r\ell \frac{dw_r}{dr}$$

or

$$\frac{P_1 - P_2}{2\mu\ell} rdr = -dw_r \tag{50}$$

This differential equation may be integrated over the limits of r = 0 to R and w = w_r to 0:

$$\int_r^R \frac{P_1 - P_2}{2\mu\ell} rdr = -\int_{w_r}^0 dw_r \tag{51}$$

Hence,

$$\frac{P_1 - P_2}{2\mu\ell}\left(\frac{R^2}{2} - \frac{r^2}{2}\right) = w_r$$

or

$$w_r = \frac{P_1 - P_2}{4\mu\ell}(R^2 - \gamma^2) \tag{51a}$$

The maximum velocity occurs at r = 0 and therefore may be written as:

$$w_{max} = \frac{P_1 - P_2}{4\mu\ell} R^2 \tag{51b}$$

Combining equations we obtain:

$$w_r = w_{max}\frac{R^2 - r^2}{R} = w_{max}\left(1 - \frac{r^2}{R^2}\right) \tag{52}$$

Equation 52 is known as *Stokes law for the parabolic velocity distribution* describing laminar flow in a pipe of constant cross-section.

To evaluate the volumetric flow rate, consider a small hollow cylinder of fluid having an inside radius r and outer radius r + dr. The cross-sectional area of the annular fluid ring is $dS = 2\pi rdr$, and the volume rate passing through this section is:

$$dV_{sec} = w_r dS = w_r 2\pi rdr \tag{53}$$

Combining Equations 53 and 51a:

$$dV_{sec} = \frac{P_1 - P_2}{4\mu\ell}\int_0^R (R^2 - r^2)\,2\pi rdr = \frac{P_1 - P_2}{4\mu\ell}\left(2\pi R^2\int_0^R rdr - 2\pi\int_0^R r^3dr\right) = \frac{P_1 - P_2}{8\mu\ell}\pi R^4 \tag{54a}$$

Denoting the pipe diameter as D = 2R and the pressure gradient $(p_1 - p_2) = \Delta P$:

$$V_{sec} = \frac{\pi D^4 \Delta P}{128\mu\ell} \tag{54b}$$

Equation 54 expresses the volumetric rate for laminar flow in a straight circular pipe and is known as *Poiseuille's equation.*

The relationship between average velocity w and maximum velocity w_{max} is obtained by comparing the following expressions:

$$V_{sec} = wS = w\pi R^2 \text{ and } \pi R^2 w = \frac{P_1 - P_2}{8\mu\ell}\pi R^4$$

Hence,

$$w = \frac{P_1 - P_2}{8\mu\ell}R^2$$

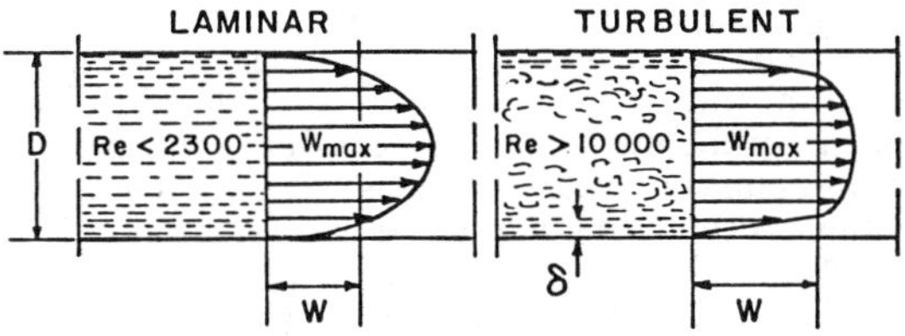

Figure 4. Velocity profiles for laminar and turbulent flows in a pipe.

Combining Equations 54b and 55 we obtain:

$$w = \frac{w_{max}}{2} \tag{56}$$

Thus, the average velocity for laminar flow in a circular pipe is equal to half the maximum velocity at the pipe axis.

Similarly, the parabolic velocity distribution for pipe flow Equation 52 may be rewritten as:

$$w_r = 2w\left(1 - \frac{r^2}{R^2}\right) \tag{57}$$

The previous example illustrates the ability to rigorously describe laminar flow for simple geometries. The phenomenon of turbulence, however, has no theoretically rigorous expressions. In turbulent flow, chaotic motion of fluid particles causes a smoothing of velocity streamlines in the bulk flow. In pipe flow, this results in a velocity distribution that is very different from the parabolic profile obtained with laminar flows. A comparison of laminar and turbulent velocity profiles for pipe flow is shown in Figure 4. Note that the turbulent distribution does not show any significantly wider apex. Experiments have shown that unlike laminar flow, the average velocity is not equal to half the maximum. In fact, the average velocity is observed to be considerably greater and is a function of the Reynolds number i.e., $w/w_{max} = f(Re)$. As a general rule, for turbulent Reynolds numbers up to 10,000 velocity $w \simeq 0.8\,w_{max}$ and at $Re > 10^4$, $w \simeq 0.9\,w_{max}$.

Present understanding of turbulence does not enable velocity profiles to be computed a priori as with laminar flow. The turbulent profile illustrated in Figure 4 does not represent an instantaneous distribution but rather one that is time-averaged. The instantaneous velocity at a point varies with time in both magnitude and direction. Therefore, it is an irregularly oscillating function that can be defined by a time-smoothed velocity $\bar{w}_2$ in the x direction (i.e., by taking a time-average of w_x over some time interval τ). The time interval must be large with respect to the time of turbulent oscillation but small with respect to actual time changes:

$$\bar{w}_x = \frac{\int_0^\tau w_x d\tau}{\tau} \tag{58}$$

where w_x is the instantaneous velocity (a function of time τ). Note that w_x is the sum of the time-smoothed velocity $\bar{w}_x$ and a velocity fluctuation Δw.

$$w_x = \bar{w}_x \pm \Delta w \tag{59}$$

The value of the time-averaged velocity $\bar{w}_x$ over a sufficiently large time interval is constant. As the frequency of the velocity fluctuation is extremely large, the time interval need be only a few seconds or even fractions of seconds. Consequently, we may consider a time-independent change of the velocities averaged over the cross-section of piping rather than the change of instantaneous values. That is, we shall consider local flow to be at quasi-stationary motion. Altough the instantaneous fluctuations are small, they have a dramatic effect on the flow characteristics and, hence, transport mechanisms. For example, the transverse movement of a fluid particle from a faster-moving region to a slower-moving one has the effect of increasing the velocity in the slower region and thus acts as an equivalent shear stress. This turbulent shear stress may be hundreds of times greater than the laminar stress due to the sliding of one fluid layer over another.

To evaluate the importance of fluctuating velocities relative to their average, turbulence is characterized by an intensity factor:

$$I_T = \frac{\overline{\Delta w}}{\bar{w}} \tag{60}$$

where $\overline{\Delta w}$ is the root-average square of fluctuating velocities. Hence,

$$I_T = \frac{\sqrt{\frac{1}{3}(\overline{\Delta w_x^2} + \overline{\Delta w_y^2} + \overline{\Delta w_z^2})}}{\bar{w}_x} \tag{61}$$

The intensity of turbulence is *an estimate of the fluctuations at a given point within the flow*. For turbulent flow in a pipe $I_T \simeq 0.01$–0.1.

The case in which the three average-square fluctuating velocities are equal is referred to as *isotropic turbulence*. Equation 61 simplifies to:

$$I_T = \sqrt{\overline{\Delta w_x^2}} \tag{62}$$

In a practical sense, turbulent flows almost never reach isotropic conditions. In pipe flow, for example, the isotropic state is only approached at the flow's axis. Greater deviation from isotropic turbulence occurs in fluid regions removed from the axis, particularly near the wall.

The closer two fluid particles are to each other, the closer their actual or instantaneous velocities. In contrast, for two fluid particles relatively removed from one another no relation exists between their velocities. Hence, we can assume that the fluid particles belong to different groups of fluid elements, which tend to move together and form some unified aggregates. These aggregates are called *eddies*. The size of eddies or the depth of their penetration before collapsing (which may be identified approximately by the distance between two close particles not belonging to the same eddy) depends on the rate of turbulence. This rate of turbulence is known as the *scale*, where a correlation coefficient is defined as:

$$R_c(y) = \frac{\overline{\Delta w_{x1} \Delta w_{x2}}}{\sqrt{\overline{\Delta w_{x1}^2}}\sqrt{\overline{\Delta w_{x2}^2}}} \tag{63}$$

For flow in the x direction, $\Delta\bar{w}_{x1}$ and $\Delta\bar{w}_{x2}$ are the fluctuating velocities occurring at the same time τ_1 at points 1 and 2, separated by distance y.

The turbulence scale is based on the correlation coefficient $R_c(y)$ expressed as a function between two points:

$$L_T = \int_0^\infty R_c(y)\,dy \tag{64}$$

Eddy motion also may exist in laminar flow because it is characterized simply by velocity differences over the flow area. The distinction between turbulent and laminar flows therefore is not denoted by eddy motion but by the presence of chaotic fluctuations of velocities at different points throughout the flow area. These fluctuations produce particle displacements in a direction normal to the axis of flow.

These chaotic fluctuations are characterized by the eddy viscosity. Considering two fluid particles flowing in the x direction parallel to the pipe axis where y is the distance between these particles in the direction normal to the pipe axis, the velocity components are $\bar{w}_{x1}$ and $\bar{w}_{x2}$ which differ from each other by $d\bar{w}_x$. Hence

$$\tau_N' = -\mu \frac{d\bar{w}_x}{dy} = -\rho\nu \frac{d\bar{w}_x}{dy} \tag{65}$$

where μ and ν are the dynamic and kinematic viscosities, respectively; ρ is density and subscript "N" denotes that we are dealing with a Newtonian fluid. Equation 65 is a general expression that only rigorously applies to laminar flow. In laminar flow, τ_N' is the only stress existing between fluid layers located distance dy apart. In turbulent flow however, fluid particles move relative to each other, not only in the longitudinal direction (i.e., together with the flow) but also in the transverse direction. This creates an additional shear stress τ_T' (subscript T refers to turbulent) which by analogy with τ_N' may be expressed as follows:

$$\tau_T = -E_v \frac{d\bar{w}_x}{dy} \tag{66}$$

Parameter E_v in Equation 66 is the *eddy viscosity* and is analogous to dynamic viscosity μ. The quantity $\varepsilon_m = E_v/\rho$ is the *eddy diffusivity of momentum* which is analogous to kinematic viscosity. These two fluid properties (E_v and ε_m) depend on the fluid velocity and the geometry of the flow system. That is, they are functions of all factors that influence the nature of turbulence and the deviation of velocities. In particular, they are sensitive to location within the turbulent field (i.e., distance from the wall, etc.). Thus, the total shear stress in a turbulent flow is the sum of viscous stresses and turbulent stresses.

$$\tau' = \tau_N' + \tau_T' = -\rho(\nu + \nu_T) \frac{d\bar{w}_x}{dy} \tag{67}$$

Examining the turbulent velocity profile again in Figure 4, note that the time-averaged velocities compose a relatively flat distribution in the bulk flow. It is only in the vicinity of the wall that velocities decrease rapidly and eventually become zero at the solid boundary. In the immediate vicinity of the wall, the fluid motion becomes less turbulent and more laminar. The solid boundary has the effect of dampening turbulent fluctuations in the transverse direction. Thus, turbulent motion never really exists in pure form; rather, it is accompanied by laminar motion.

Fluid motion is divided conditionally into two zones: the central or *bulk flow,* where the motion is turbulent, and the *hydrodynamic boundary layer* (the zone near the wall where turbulent flow changes into laminar). Within the hydrodynamic boundary layer a thin sublayer exists very close to the pipe wall. In this region viscous forces have a predominant influence on the fluid motion. The velocity gradient in the *laminar boundary sublayer* is very high. In turbulent flow the laminar sublayer is very thin (sometimes fractions of a millimeter in thickness) which decreases with increasing turbulence. Regardless of the thickness, the sublayer has a dramatic influence not only on the hydraulic resistance of fluid motion, but on the mechanisms of heat and mass transfer.

Between the bulk flow and the laminar sublayer exists the buffer zone. The thickness of this layer is determined by shear stresses that exist between fluid particles caused by viscous forces and turbulent fluctuations. Hence, in this region the magnitude of ν and ν_T are comparable.

CONSERVATION OF ENERGY

The equations of motion and continuity alone are not sufficient for describing a flow system. The additional constitutive equation is the energy balance. For a steady continuous-flow system as shown in Figure 5, the overall energy balance is composed of two portions, input and output. The energy input is the sum of the kinetic (e_{k1}), potential (E_{p1}), volumetric (E_{v1}), and internal (E_1) energies at section 1; the heat (Q) added through the exchanger; and the mechanical work (W′) performed on the fluid by the pump. The energy output comprises kinetic (E_{k2}), potential (E_{p2}), volumetric (E_{v2}) and internal (E_{i2}) energies at section 2. Equating the energy input to the system output, the following balance is obtained:

$$E_{k1} + E_{p1} + E_{v1} + E_1 + \dot{Q} + W' = E_{k2} + E_{p2} + E_{v2} + E_2 \qquad (68)$$

The importance of each term is as follows. The potential energy is the product of weight (1 kg) and the fluid's elevation above some specified datum plane.

$$E_{p1} = Z_1;\ E_{p2} = Z_2 \qquad (69)$$

where Z_1 and Z_2 are elevations at sections 1 and 2 relative to a reference.

The volumetric energy under pressure P is equivalent to the work expended to forming volume ν at this pressure. The volumetric energies of 1 kg of fluid at the two sections are:

$$E_{v1} = P_1\nu_1;\ E_{v2} = P_2\nu_2 \qquad (70)$$

Kinetic energy is the product of mass and one half the square of the linear velocity of the fluid. In this case, the weight of the fluid is equal to one kilogram-force. The mass of the fluid expressed in gravitational units is $1/g \times kg \times sec^2/m^4$. If the fluid flows with velocity w, then its kinetic energy is $w^2/2g$ and for sections 1 and 2:

$$E_{k1} = \frac{\bar{w}_1^2}{2g\alpha_1};\ E_{k2} = \frac{\bar{w}_2^2}{2g\alpha_2} \qquad (71)$$

(for turbulent flows, $\alpha = 1$).

The fourth term in Equation 68 is the system's internal energy, a thermodynamic property. It is defined relative to a specified reference state such as standard conditions. It is convenient to introduce a value that expresses the change of the internal energy of 1 kg of fluid as it passes through the system ($E_2 - E_1$). Denoting W′ as the work performed in pumping 1 kg of fluid and Q as the heat added in units of work, Equation 68 may be stated as follows:

$$P_1\nu_1 + \frac{\bar{w}_1^2}{2g\alpha_1} + z_1 + \dot{Q} + W' = P_2\nu_2 + \frac{\bar{w}_2^2}{2g\alpha_2} + z_2 + (E_2 - E_1) \qquad (72)$$

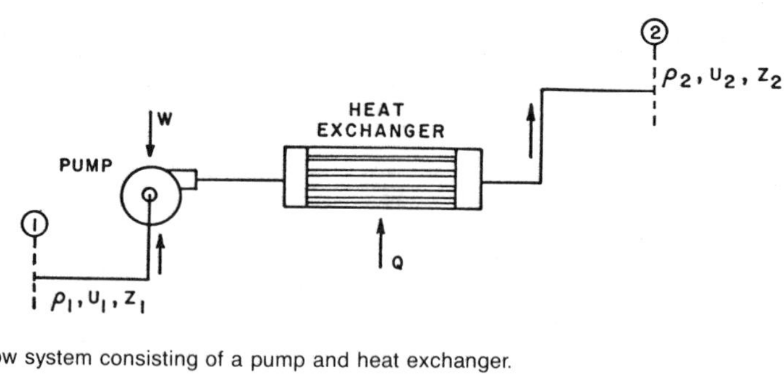

Figure 5. Flow system consisting of a pump and heat exchanger.

Equation 72 is the law of conservation of energy and may be applied both to compressible and incompressible flows. Each term has units of length, or more precisely, units of energy per unit weight.

For engineering calculations, the internal energy terms may be approximated by enthalpy.

$$i_1 = E_1 + P_1 v_1;\ i_2 = E_2 + P_2 v_2 \tag{73}$$

Substituting for enthalpy into Equation 72 gives:

$$\frac{\bar{w}_1^2}{2g\alpha_1} + z_1 + \dot{Q} + W' = \frac{\bar{w}_2^2}{2g\alpha_2} + z_2 + (i_2 - i_1) \tag{74}$$

For ideal gases the change in enthalpy may be determined from the product of heat capacity at constant pressure and the system's temperature differential $C_p(t_2 - t_1)$.

A special case of the total energy balance is derived by the integration of Euler's equations of motion for staedy flow, namely, the Bernoulli equation. Multiplying both sides of the system of equations (Equation 30 by dx, dy and dz, respectively, and dividing by density, ρ, we obtain:

$$\frac{dx}{d\tau} dw_x = -\frac{1}{\rho}\frac{\partial P}{\partial x} dx$$

$$\frac{dy}{d\tau} dw_y = -\frac{1}{\rho}\frac{\partial P}{\partial y} dy$$

$$\frac{dz}{d\tau} dw_z = -gdz - \frac{1}{\rho}\frac{\partial P}{\partial z} dz$$

Summing these equations and noting that derivatives $dx/d\tau$, $dy/d\tau$, and $dz/d\tau$ are the velocity components w_x, w_y, and w_z.

$$w_x dw_x + w_y dw_y + w_z dw_z = -gdz - \frac{1}{\rho}\left(\frac{\partial P}{\partial x} dx + \frac{\partial P}{\partial y} dy + \frac{\partial P}{\partial z} dz\right) \tag{75}$$

The terms on the LHS are:

$$w_x dw_x = d\left(\frac{w_x^2}{2}\right);\ w_y dw_y = d\left(\frac{w_y^2}{2}\right);\ w_z dw_z = d\left(\frac{w_z^2}{2}\right)$$

Where their sum is:

$$d\left(\frac{w_x^2}{2}\right) + d\left(\frac{w_y^2}{2}\right) + d\left(\frac{w_z^2}{2}\right) = d\left(\frac{w_x^2 + w_y^2 + w_z^2}{2}\right) = d\left(\frac{w^2}{2}\right) \tag{76}$$

where $w = |\vec{w}|$ is the velocity vector with components along coordinate axes w_x, w_y, and w_z.

The sum of the terms in parantheses on the RHS of Equation 75 is the total pressure differential dP. Since only spacial coordinates are considered

$$d\left(\frac{w^2}{2}\right) = -\frac{dP}{\rho} - dz \tag{77}$$

Dividing both sides by g and specifying incompressible homogeneous flow, i.e., ρ is constant:

$$d\left(\frac{w^2}{2g}\right) + \frac{dP}{\rho g} + dz = 0 \tag{78}$$

The sum of the differential then is substituted by the differential of the sum:

$$d\left(z + \frac{P}{\rho g} + \frac{w^2}{2g}\right) = 0$$

Hence,

$$z + \frac{P}{\rho g} + \frac{w^2}{2g} = \text{constant} \tag{79}$$

This expression may be related to any two cross-sections of a flow element:

$$z_1 + \frac{P_1}{\rho g} + \frac{w_1^2}{2\alpha g} = z_2 + \frac{P_2}{\rho g} + \frac{w_2^2}{2\alpha g} \tag{80}$$

Equation 80 is related not to the cross-section as a whole, but to any pair of compatible points in these sections. To compare congruent values over total cross-sections it must be assumed that the terms in Equation 65 may be approximated by Equation 79 which represents the point form of the *Bernoulli equation for ideal fluids*. It follows that the *hydrodynamic head* $(z + (P/\rho g) + (w^2/2g))$ is constant for all cross-sections of ideal steady flow. The first two terms of the hydrodynamic head, z and $P/\rho g$, are the *geometric head* which represents a specific potential energy at a given point within the cross-section. $P/\rho g$ is the static or piezometric head characterizing the specific potential energy of pressure at a given point. Both terms may be expressed in units either of length or specific energy (i.e., energy per unit weight of fluid).

$w^2/2g$ is the *velocity or dynamic head* which characterizes a specific kinetic energy at a given point within the cross-section.

The Bernoulli equation states that the sum of the sum of the static and dynamic heads and level do not change from one cross-section of flow to another. This statement is inclusive only of steady-state flow of an ideal fluid. It further follows that the sum of the potential and kinetic energies is constant.

A conversion of energy occurs when the flow area changes. This change in energy is reflected in the fluid velocity. Reducing a pipe's cross-sectional area causes part of the potential energy of pressure to be converted into kinetic energy and vice versa. When the flow area is increased, part of the kinetic energy is converted into potential, however, the total energy is unchanged. Hence, we may conclude that the amount of energy entering the initial pipe cross-section is equal to that leaving the pipe. For this reason, Equation 80 is a special application of the principle of conservation of energy.

Multiplying Equation 8 by the specific weight $\gamma = \rho g$, we obtain:

$$\rho g z_1 + P_1 + \frac{\rho w_1^2}{2} = \rho g z_2 + P_2 + \frac{\rho w_2^2}{2} \tag{81}$$

where each term expresses the specific energy of the flow per unit volume at a given point.

For a horizontal system, $z_1 = z_2$ and Equation 81 simplifies to:

$$\frac{P_1}{\rho g} + \frac{w_1^2}{2g} = \frac{P_2}{\rho g} + \frac{w_2^2}{2g} \tag{82}$$

For real fluids, both shear and friction forces play important roles. These latter forces exert resistance to fluid motion. A portion of the flow energy must therefore be devoted to overcome this *hydraulic resistance*. The total energy decreases continuously in the direction of flow as a portion is converted from potential energy into *lost energy* (the energy expended for friction). This conversion is irreversible and is lost in the form of heat dissipation to the surroundings.

The Bernoulli equation may be corrected for frictional losses by adding the term h_ℓ to the RHS of Equation 80. Hence, the Bernoulli equation for real fluids is:

$$z_1 + \frac{P_1}{\rho g} + \frac{w_1^2}{2g} = z_2 + \frac{P_2}{\rho g} + \frac{w_2^2}{2g} + h_\ell \tag{83}$$

where the lost head, h_ℓ, characterizes the *specific energy spent for overcoming hydraulic resistances*.

Another form of this expression can be obtained by multiplying both sides by ρg:

$$\rho g z_1 + \frac{\rho w_1^2}{2\alpha} + P_1 = \rho g z_2 + P_2 + \frac{\rho w_2^2}{2\alpha} + \Delta P_\ell \tag{84}$$

ΔP_ℓ is the lost pressure drop, defined as:

$$\Delta P_\ell = \rho g h_\ell \tag{85}$$

HYDRAULIC RESISTANCES TO FLOW

Frictional resistance arises from the motion of real fluids and is influenced by the flow regime. In turbulent flow, resistance are characterized not only by the fluid's consistency, but by the eddy viscosity, which depends on the hydrodynamic conditions. Such resistances result in additional energy losses.

Local resistances arise due to variations in fluid velocity caused by changes in flow area or direction. Among the many local resistances are friction losses caused by entrances and exits from piping, sudden contractions, and expansions, losses in fittings and valves, etc. The loss of head h_ℓ, therefore may be expressed by the sum of two terms:

$$h_\ell = h_{fr} + h_{\ell r} \tag{86}$$

where h_{fr} and $h_{\ell r}$ are head losses due to friction and local resistances, respectively.

For a horizontal pipe of constant cross-section, Bernoulli's equation is applied to determine the head loss due to friction:

$$\frac{p_1 - p_2}{\rho g} = \frac{\Delta P}{\rho g} = h_{fr}$$

Substituting $\Delta p = \rho g h_{fr}$ into Equation 54b and noting that V_{sec} may be estimated by the product of average velocity and the pipe's cross-sectional area $\pi d^2/4$ we obtain:

$$w = \pi d^2/4 = \frac{\pi d^4 \rho g h_{fr}}{128 \mu \ell}$$

where ℓ and d are the length and diameter of the pipe.

Solving for the head loss $h_{fr} = 32 w \mu \ell / \mu g d^2$ and multiplying the numerator and denominator of the RHS of this expression by 2w, we obtain:

$$h_{fr} = \frac{64\mu}{wd\rho} \frac{\ell}{d} \frac{w^2}{2g}$$

or

$$h_{fr} = \frac{64}{Re}\frac{\ell}{d}\frac{w^2}{2g} \tag{87}$$

Thus the head losses due to friction are expressed in terms of the velocity head $w^2/2g$. The quantity 64/Re in Equation 87 is the *friction factor* for laminar flow (λ).

The friction resistance coefficient is defined as:

$$\xi = \lambda \frac{\ell}{d} \tag{88}$$

Using this notation, Equation 87 is expressed as:

$$h_{fr} = \xi \frac{w^2}{2g} = \lambda \frac{\ell}{d}\frac{w^2}{2g} \tag{89}$$

And for pressure losses ($\Delta P_{fr} = \rho g h_{fr}$)

$$\Delta P_{fr} = \lambda \frac{\ell}{d}\frac{\rho w^2}{2} \tag{90}$$

Equation 89 describes the case of laminar flow through a pipe ($Re < 2{,}100$), where the friction factor is independent of the conduit's wall roughness.

For flow through noncircular cross-sections, an equivalent diameter is used for the characteristic system size. The equivalent diameter (d_{eq}) is defined as four times the hydraulic radius, where the hydraulic radius is the cross-sectional area of flow divided by the wetted perimeter. In applying Equation 89, the laminar friction factor is:

$$\lambda = \frac{B}{Re} \tag{91}$$

where B = 57 for a square cross-section, and B = 96 for an annular cross-section. Equation 89 also applies to determining friction losses in the turbulent regime (see Hodson [3], Rouse [4], and Davies [5]. However, in this case the friction factor relationship cannot be derived theoretically. Therefore, design equations for λ in turbulent flow are obtained from similarity theory, where the general expressions is:

$$Eu = A\, Re^m \left(\frac{\ell}{d}\right)^q \tag{92}$$

For flow through smooth-walled piping ($4{,}000 < Re < 100{,}000$), values of the coefficients are A = 0.158, m = − 0.25, q = 1; or

$$Eu = 0.158\, Re^{-0.25} \left(\frac{\ell}{d}\right) \tag{93}$$

and

$$h_{fr} = 0.316 \frac{\ell}{d}\frac{w^2}{2g} \tag{94}$$

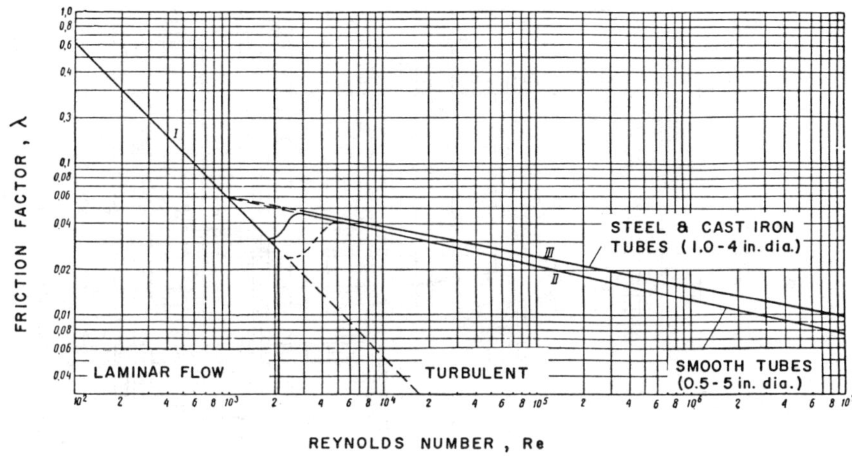

Figure 6. Friction factor vs. Reynolds number chart: curve 1 = laminar flow for smooth and rough pipes; curve 2 = turbulent flow for smooth pipes made from copper, lead, and glass; curve 3 = turbulent flow for rough pipes made from steel and cast iron.

From Equations 94 and 89 the friction factor is:

$$\lambda = 0.316\,\mathrm{Re}^{-0.25} \tag{95}$$

The friction factor depends not only on the Reynolds number but also on the roughness of solid boundaries. Roughness is characterized by the effective height of the protrusions, e. Figure 6 shows a plot of friction factor versus Reynolds number for different piping materials.

To normalize the effect of roughness, Moody [6] developed the generalized friction factor plot shown in Figure 7. The ordinate on the right of Figure 7 is the inverse of relative roughness, where $\varepsilon = e/d$.

In laminar flow through a pipe in which e/d is usually somewhat less than 0.01, the influence of wall roughness is negligible because the fluid fills the spaces between the protrusions and the inner fluid layers slide smoothly over a pipe of effective diameter $d - 2e$. In some initial range of turbulence, the wall roughness also can be neglected if it is smaller in height than the thickness of the viscous sublayer. In this case, the pipe is hydraulically smooth and λ then can be computed from Equation 95.

As the Reynolds number increases, three flow regimes are observed. The first is referred to as the zone of *smooth friction,* where $\lambda = f(\mathrm{Re})$. The second is the zone of *mixed friction,* where λ is both a function of Re and roughness. Finally, there is the self-modeling zone, where λ becomes practically independent of the Reynolds number and is primarily a function of wall roughness.

The critical Reynold numbers, $\mathrm{Re}_{cr,1}$, where roughness begins to influence λ, as well as critical values, and $\mathrm{Re}_{cr,2}$, when λ is only a function of pipe roughness (refer to dashed line in Figure 7), depend on the *relative roughness*:

$$\mathrm{Re}_{cr,1} \simeq \frac{23}{\varepsilon} \tag{96}$$

and

$$\mathrm{Re}_{cr,2} \simeq 220\varepsilon^{-9/8} \tag{97}$$

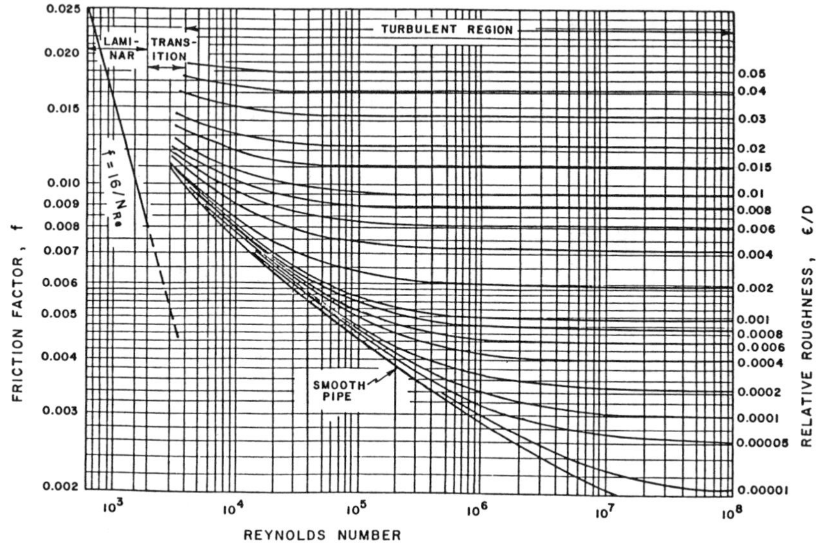

Figure 7. Reynolds number-friction factor chart as a function of relative roughness.

Hodson [3] gives the following correlation for all three zones of turbulent motion:

$$\frac{1}{\sqrt{\lambda}} = -2\log\left[\frac{\varepsilon}{3.7} + \left(\frac{6.81}{\text{Re}}\right)^{0.9}\right] \tag{98}$$

For the zone of smooth friction, λ can be calculated from Equation 95 or 98. In the latter equation, the first term inside the parantheses is neglected.

$$\frac{1}{\sqrt{\lambda}} = -2\log\left(\frac{6.81}{\text{Re}}\right)^{0.9} = 1.8\log\text{Re} - 1.5 \tag{99}$$

For the self-modeling zone, where λ is independent of Re, Equation 99 becomes:

$$\frac{1}{\sqrt{\lambda}} = 2\log\frac{3.7}{\varepsilon} \tag{100}$$

Care should be exercised in applying friction factors as there are several definitions. The Blasius friction factor is defined as four times the Fanning friction factor. So, for laminar flow, $\lambda_F = 16/\text{Re}$. Figure 7 is based on the Blasius friction factor. Some textbooks define a friction factor that is twice the value of the Fanning friction factor.

The friction factor expressions are based on isothermal flows. When the fluid is either heated or cooled through the pipe wall, physical property changes occur over the cross-section. Thus, the velocity field is modified and λ varies, especially in the laminar region where the temperature gradients are higher than in turbulent bulk flow.

For all the friction factor expressions given (excluding Equation 100), correction factors must be applied. These corrections are based on the assumption that the fluid temperature can be represented by a mean bulk temperature (defined as the arithmetic average of the inlet and outlet temperature). The correction factors, denoted by ψ, are:

For Re > 2,100:

$$\psi = \left(\frac{\mu}{\mu_w}\right)^{0.17} \quad \text{for heating}$$

$$= \left(\frac{\mu}{\mu_w}\right)^{0.11} \quad \text{for cooling}$$

For Re < 2,100:

$$\psi = \left(\frac{\mu}{\mu_w}\right)^{0.38} \quad \text{for heating}$$

$$= \left(\frac{\mu}{\mu_w}\right)^{0.23} \quad \text{for cooling}$$

where μ = viscosity of fluid at the average bulk temperature
μ_w = viscosity at the temperature of the pipe wall

To obtain the friction factor corresponding to the fluid's average bulk temperature, divide the isothermal λ value by ψ.

Variable Flow Areas

When the flow undergoes a sudden change in cross-sectional area, additional head losses result. For example, in a sudden expansion the head loss is due to shocks as the flow the smaller cross-section impacts against a slower moving fluid in the larger conduit.

The force of the rapid flow PA_2 acts against the plane A_2 in the larger section and in the opposite direction to the force from the slower-moving flow, P_1A_2. The momentum change $d\psi$ with time dt is the following:

$$d\psi = (p_1A_2 - p_2A_2)\, d\tau \tag{101}$$

The fluid momentum over time $d\tau$ is the product of mass and velocity, and the change in momentum is:

$$(w_2 - w_1)\, dM = \left(w_2 \frac{\gamma}{\rho} w_1\right) - (A_2w_2)\, d\tau \tag{102}$$

where A_2w_2 is the volumetric flow rate.

Equating these expressions:

$$\frac{P_1 - P_2}{\gamma} = \frac{w_2(w_2 - w_1)}{g} \tag{103}$$

This shows that the total pressure change is due to resistance and changes in kinetic energy.

For turbulent flow, the Bernoulli equation is applied to the section ahead of the expansion:

$$\frac{p_1}{\gamma} + \frac{w_1^2}{2g} = \frac{p_2}{\gamma} + \frac{w_2^2}{2g} + h_\ell \tag{104}$$

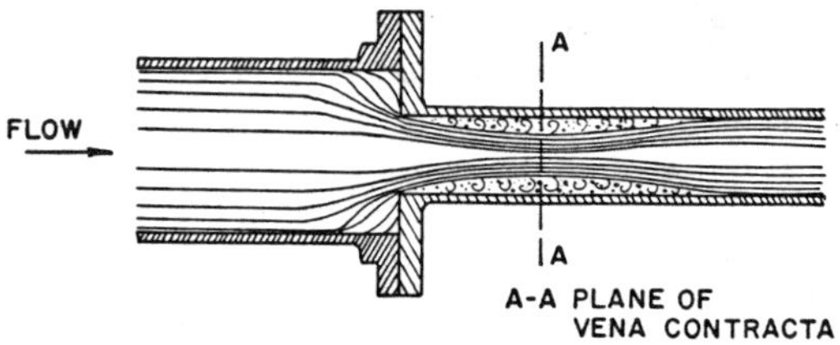

Figure 8. Flow through a sudden contraction.

Solving for resistance h_ℓ

$$h_\ell = \frac{(w_1 - w_2)^2}{2g} \tag{105}$$

The effects of sudden contractions on flows are discussed by Mott [7], Cheremisinoff [8, 9], and Simpson [10]. Figure 8 illustrates flow through a sudden contraction.

As shown, the flow's cross-section first contracts and then expands to fill the cross-section of smaller piping with energy losses or arising resistances. These losses can be estimated from the Weisbach equation:

$$h_\ell = \left(0.04 + \left(\frac{1}{a} - 1\right)^2\right)\frac{w_2^2}{2g} = \psi\,\frac{w_2^2}{2g} \tag{106}$$

where a is the contraction loss coefficient defined as the ratio of the minimum flow cross-section to the cross-section of the smaller pipes, and w_2 is the average velocity in the smaller section.

The coefficients a and ψ depend on the ratio of the pipe's cross-sections, A_2/A_1. Typical values are given in Table 1. Coefficient ψ also may be calculated from the following equation:

$$\psi = \frac{1.5\left(1 - \frac{A_2}{A_1}\right)}{3 - \frac{A_2}{A_1}} \tag{107}$$

These equations may be used to calculate resistances encountered in liquid discharge from large vessels through orifices or for liquid discharges from vessels into piping. For the latter case, A_2/A_1 in Equation 107 can be neglected. Equation 106 expresses only the head loss due to the change in flow cross-section. The total head loss also accounts for the change in kinetic energy in accordance with the Bernoulli equation.

Similar losses occur when flows undergo gradual changes in cross-section. For a smooth conical expansion where the cone's apex angle is $\beta < 10°$, head losses may be estimated from the following equation:

$$h_\ell = \left(\frac{A_2}{A_1} - 1\right)^2 \sin\beta \left(\frac{w_1^2}{2g}\right) \tag{108}$$

For $\beta > 10°$, head losses may be estimated from the expressions for a sudden expansion.

Table 1
Coefficients of Contraction for Use in Equation 106

A_1/A_2	a	ψ
0.01	0.6	0.5
0.1	0.61	0.46
0.2	0.62	0.42
0.4	0.65	0.33
0.6	0.7	0.23
0.8	0.77	0.13
1.0	1.0	0.0

Table 2
Data for Estimating Pressure Losses

$\frac{L}{D}$ Re	$\frac{p_0 - p_1}{\gamma \frac{w^2}{2g}}$
0.005	2.1
0.01	2.6
0.02	3.4
0.03	4.1
0.04	4.7
0.05	5.3
0.06	6.0

D = pipe diameter; L = pipe length; γ = liquid specific gravity.

For a gradual expansion, where $7° < \beta < 35°$, head losses may be estimeted from:

$$h_\ell = 0.35 \left(\log \frac{\beta}{2}\right)^{1.22} \frac{(w_1 - w_2)^2}{2g} \tag{109}$$

For gradual expansions with $\beta < 7°$, head losses are determined by integration of the Bernoulli and Darcy-Weisbach equations.

At $\beta > 40°$, head losses may become very high and even exceed those encountered in sudden expansions. Note also that the flow may be turbulent in the small cross-section but laminar in the larger section.

For gradual contractions head losses are small (especially for smooth contradictions and large Reynolds numbers). Equation 106 may be used with $h_\ell = 0.05$, independent of the A_2/A_1 ratio for turbulent flow in the narrrow section. If the flow in the contracted section is laminar (as in the case of discharge from a vessel), even when the resistance is small a pressure decrease occurs that does not correspond to Poiseuille's law. This decrease occurs over a length equivalent to 0.065 ReD (the entrance region of a pipe). The pressure gradient, $p_0 - p_1$, for laminar flow at the entrance of a pipe of length L may be calculated from the data given in Table 2.

Flow Through Fittings, Valves, and Components

Flow through fittings such as bends and elbows result in additional head losses that are described in terms of equivalent lengths of straight pipe of pipe diameters having the same head loss as the fittings. Head losses in fittings and valves may be expressed by a relationship of the form:

$$h_\ell = K_e \frac{w_1^2}{2} \tag{110}$$

Table 3
Friction loss Coefficients
for Turbulent Flow Through Valves and Fittings

Type Fitting or Valve	Head Loss Coefficient K_e	L_{eq}/D
Elbow, 45°	0.35	17
Elbow, 90°	0.75	35
Tee	1	50
Return bend	1.5	75
Coupling	0.04	2
Union	0.04	2
Gate valve		
Wide open	0.17	9
Half open	4.5	225
Globe valve		
Wide open	6.0	300
Half open	9.5	475
Angle valve		
Wide open	2.0	100
Check valve		
Ball	70.0	3,500
Swing	2.0	100
Water meter, disk	7.0	350

Table 4
Values for Parameter in Equation 112

R/r	X
2	4
4	1.2
6	1.5
8	1.7
10	2.2
20	4.8
32	7.6

From Kanakov [11].

where K_e is the head loss coefficient, which is a function of the ratio of the fittings' equivalent length to diameter ratio, L_{eq}/D. Table 3 gives typical values of K_e for different valves and fittings. Note that w_1 is the velocity in the pipe immediately upstream of the fitting. Application of Equation 110 is limited to turbulent flows.

To evaluate total head loss for a piping system, equivalent lengths, L_{eq}, of fittings and components are added to straight lengths of pipe in the system. That is, the total piping over which losses occur = L_{eq} + length of straight pipe. The Darcy equation then can be applied to obtain total head losses for a system:

$$\hat{F}_f = \frac{\Delta P_f}{\rho} = 4\lambda_F \frac{\Delta L_e}{D} \frac{w^2}{2} \tag{111}$$

where λ_F is the Fanning friction factor.

For turbulent flow through copper bends, Konakov [11] gives the following formula:

$$\frac{L_{eq}}{D} = 0.0202 X a^{1.10} Re^{0.032} \tag{112}$$

where a is the bending angle in grades, and X is an empirical function of the ratio of bending radius R to pipe radius r. Values for X are given in Table 4.

Table 5
Critical Reynolds Numbers for Coils

D/d	Re_{cr}
15.5	7,600
18.7	7,100
50	6,000
2,050	2,270

For flow through coils, the friction factor is a function of the ratio of coil diameter, D, to pipe diameter d, Typical critical Reynolds numbers for different D/d ratios are given in Table 5. The critical Re of 2,100 marking the transition between laminar and turbulent flows through pipes is greatly exceeded for coils. The resistance of laminar flow in coils may be estimated from the Darcy-Weisbach equation, ($h_\ell = -(\Delta P/\gamma) = \lambda(L/D)(w^2/2g)$) using the following friction factor expression:

$$\lambda = C\frac{64}{Re} \tag{113}$$

Coefficient C may be computed from the following formula:

$$C = \frac{1}{1 - \left[1 - \left(\frac{11.6}{Re\sqrt{d/D}}\right)^{0.45}\right]^{1/0.45}} \tag{114}$$

For turbulent flow through coils of $D/d > 500$, resistances are comparable to those in straight pipes. At very high Reynolds numbers ($Re > 110{,}000$), the resistance coefficient, λ, is independent of Re:

$$\lambda = 0.0238 + 0.0891\,d/D \tag{115}$$

Hydraulic resistances through coils can be estimated by applying Equation 112 where the coil is considered to be a return bend. This provides an effective L_e/D value. Until now, the analyses presented have focused on evaluating flow resistances through piping of uniform or round cross-sections. In practice, however, other configurations are encountered frequently. One example is an annular cross-section in a heat exchanger, where the fluid must flow between two concentric pipes. The formulae and analyses presented are still valid for other configurations; however, the Reynolds number and, consequently, the friction factor, are redefined in terms of an equivalent diameter.

By redefining Re in terms of r_h, head losses for turbulent flows may be estimated form Equation 111. There are two approaches to evaluating resistances for laminar flows through irregular configurations. In the first method, the Reynolds number is evaluated from an equivalent diameter, where the friction factor is determined from an expression of the form:

$$\lambda = \frac{a}{Re} \tag{116}$$

The value of a depends on the flow geometry. Values for different configurations along with formulae for the equivalent diameter are given in Table 6 [12].

The second method for estimating losses in the laminar regime for noncircular cross-sections is based on derivations of Poiseuille-type equations. A common flow

Table 6
Equivalent Diameters and Values for Parameter "a" in Equation 116

Cross-section	h/b	D_{eq}	a
Circular with diameter D	–	D	64
Square with side h	–	h	57
Triangle with side h	–	0.58h	53
Ring with width b	–	2b	96
Ellipse with axes h and b	0.7	1.17h	65
(h = small axis)	0.5	1.30h	68
	0.3	1.44h	73
	0.2	1.50h	76
	0.1	1.55h	78
Rectangle with sides h and b	1/∞	2h	96
(h = small side)	0.1	1.82h	85
	0.2	1.67h	76
	0.25	1.60h	73
	0.33	1.50h	69
	0.50	1.33h	62

Table 7
Values of n in Equation 118 for a Rectangular Cross-Section

a/b	n
0.1	0.03
0.2	0.06
0.3	0.08
0.4	0.10
0.5	0.11
0.6	0.133
0.7	0.136
0.8	0.138
0.9	0.139
1.0	0.140

configuration is that of an annular cross-section, where derivation from Poiseuille's law, results in an expression for the resistance per unit length of channel:

$$-\frac{dP}{dL} = \frac{32\mu\bar{w}}{D^2 + d^2 - \dfrac{D^2 - d^2}{\ln(D/d)}} \tag{117}$$

where $\bar{w}$ is the average superficial velocity. For a rectangular section of sides a and b, Poiseuille's equation takes the form:

$$-\frac{dP}{dL} = \frac{4\bar{w}\mu}{abn} \tag{118}$$

where n is a function of a/b; typical values are given in Table 7.

For an elliptical cross-section, the head loss is:

$$-\frac{dP}{dL} = \frac{4\bar{w}\mu(a^2 + b^2)}{a^2b^2} \tag{119}$$

where a and b are the semiaxes of an ellipse.

Finally, for two indefinitely wide plates separated by distance of 2b Poiseuille's equation is:

$$-\frac{dP}{dL} = \frac{3\mu\bar{w}}{b^2} \tag{120}$$

The velocity distribution between the two parallel plates is:

$$w = \frac{3}{4}\frac{\Gamma(b^2 - y^2)}{b^3\gamma} \tag{121}$$

where Γ = weight rate per unit plate width
y = distance to the plane of symmetry (i.e., the plane in the middle between the plates)
γ = specific gravity.

Flow normal to a bank of tubes is encountered in a variety of equipment, such as tubular heat exchangers and condensers. The magnitude of the flow resistance in such equipment is due to the contraction and expansion of the flow. This resistance also depends on the tube layout (e.g., staggered or unstaggered) relative to the flow direction. The critical Reynolds number for these flows is:

$$Re_{cr} = \frac{t' w_{max}\gamma}{\mu g} \simeq 40 \tag{122}$$

where t′ is the tube separation, and w_{max} is the maximum fluid velocity (in the narrow section).

For turbulent flow, i.e., Re > 40, flow resistance may be estimated from:

$$h_r = \lambda N' \frac{w_{max}^2}{2g} \tag{123}$$

where $h_r = \frac{\Delta P}{\gamma}$ = resistance of the liquid column
N' = number of tube rows in the direction of flow
λ = resistance coefficient depending on Re and the tube layout

For a single row of tube layout, λ can be computed from:

$$\lambda = \left[0.175 + \frac{0.32\ell/d}{(t'/d)^n}\right] Re_d \tag{124}$$

where e, d, t′ are defined in Figure 9, and $Re_d = \frac{t' w_{max}\gamma}{\mu g} = \frac{d w_{max}\gamma}{\mu g}$.

At $Re_d < 10{,}000$ (still turbulent), λ is constant. Note that when e = 2d, t′ = d, Equation 124 simplifies to:

$$\lambda = 1.82\, Re^{-0.15} \tag{125}$$

where $Re = Re_d$. For rows of staggered tubes in the Reynold number range of 2,000 to 4,000, the following equation is recommended:

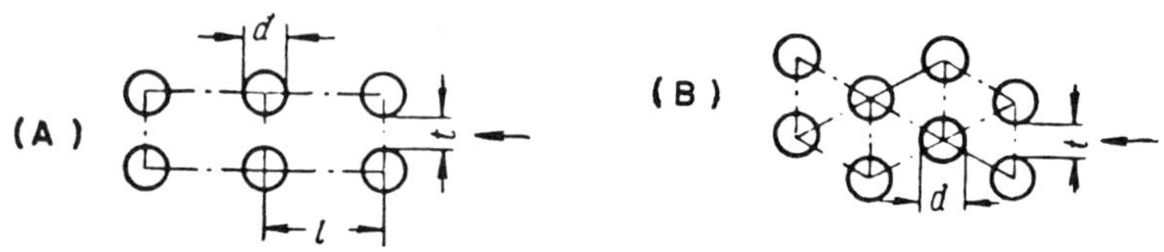

Figure 9. Flow normal to a bank of tubes. Flow resistance is a function of the tube layout: (A) an unstaggered layout; (B) a staggered layout.

$$\lambda = \left[0.92 + \frac{0.44}{(t'/d)^{1.08}}\right] Re_d^{-0.15} \tag{126}$$

And for t' = d,

$$\lambda = 1.40\, Re^{-0.15} \tag{127}$$

Equations 124 through 127 provide good estimates when the number of rows, N', exceeds 10. At $N' < 10$, actual head losses are somewhat higher than those predicted. At $N' = 4$, the error in applying these correlations is 7%; at $N' = 3$, 15%; and at $N' = 2$, up to 30%.

For flow around a single row of tubes ($N' = 1$), Table 8 gives values of λ for different t'/d ratios at Re = 10,000 (where λ is constant at $Re < 10{,}000$ for a specific t'/d).

For laminar flow around tubes ($Re < 40$), the resistance may be calculated from an expression analogous to the Darcy equation [14]:

$$h_r = \lambda \frac{\ell}{d_{eq}} N' \frac{w_{max}^2}{2g} \tag{128}$$

where $h_r = \Delta P/\gamma$ = the flow resistance
ℓ = distance between rows in the direction of flow
w_{max} = maximum velocity in the intertubular space
d_{eq} = equivalent diameter equal to 4 times hydraulic radius

For two rows of tubes, we note that $\ell(t' + d) - (\pi d^2/4)$, where $\pi d^2/4$ is the part of the volume occupied by corresponding segments of tubes. The tube surface is πd, therefore, the equivalent diameter is:

$$d_{eq} = 4\frac{\ell(t' + d) - (\pi d^2/4)}{\pi d} = \frac{4\ell(t' + d)}{\pi d} - d \tag{129}$$

The equivalent Reynolds number is:

$$Re_{eq} = \frac{d_{eq} w_{max} \gamma}{\mu g} \tag{130}$$

For a row of staggered tubes where $Re_{eq} = 1 \sim 100$:

$$\lambda = \frac{106}{Re_{eq}} \tag{131}$$

Table 8
Resistance Coefficients for Banks of Tubes

t'/d	λ
0.2	0.64
0.5	0.50
1.0	0.40
1.5	0.36
2.0	0.30
5.0	0.18
10.0	0.10

Substituting this expressions for λ into Equation 129, we obtain:

$$h_r = \frac{53\mu N'\ell w_{max}}{\gamma d_{eq}^2} \tag{132}$$

For a row of unstaggered tubes, turbulent λ values should be multiplied by 1.5 [15]. When the flow is accompanied by heat transfer, the tube surface has a temperature different from the average or bulk temperature of the fluid. The effect of wall temperature is considered in Siedel's method, in which λ_{is} corresponding to the flow temperature is estimated first and the true coefficient, λ, is corrected as follows:

$$\lambda = \frac{\lambda_{is}}{a\left(\frac{\mu}{\mu_w}\right)^n} \tag{133}$$

where μ is the viscosity at the temperature of the flow and μ_w is the viscosity at the temperature of the wall surface.

For laminar flow, a = 1.1 and n = 0.25; for turbulent flow, a = 1.0 and n = 0.14.

Closure

At this point we shall end our review of momentum transfer and basic fluid mechanics. More detailed discussions and basic calculations are given in References 8 and 9. Attention is now directed to principles and data of energy transfer. The concepts and descriptions of transport mechanisms and the analytical means of describing transfer will all be covered throughout this volume. The remainder of this chapter summarizes fundamental concepts and general information on standard heat transfer priciples.

PRINCIPLES OF CONDUCTION

One-Dimensional Conduction

The differential equation describing unsteady-state heat conduction is known as Fourier's law:

$$\frac{\partial t}{\partial \tau} = \frac{k}{c\rho}\left(\frac{\partial^2 t}{\partial x^2} + \frac{\partial^2 t}{\partial y^2} + \frac{\partial^2 t}{\partial z^2}\right) \tag{134}$$

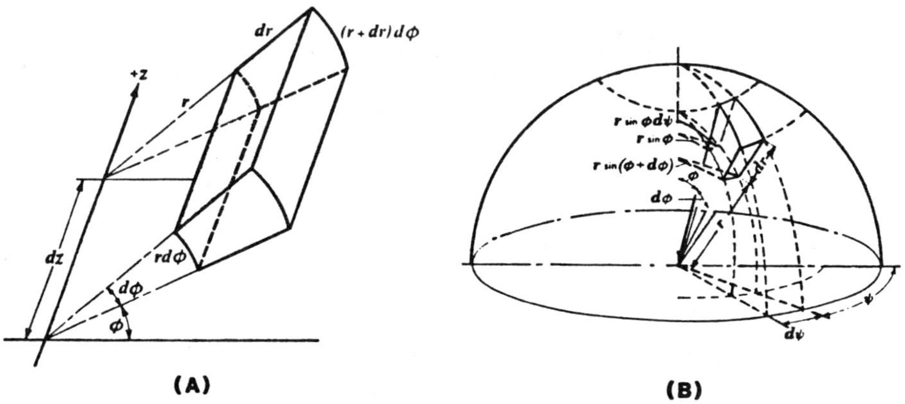

Figure 10. (A) Cylindrical coordinates; (B) spherical coordinates.

where $\alpha = k/c\rho$, is thermal diffusivity having units of:

$$\frac{k}{c\rho} = \left(\frac{\text{kcal/m} - \text{hr} - {}^\circ\text{C}}{\text{kcal/kg} - {}^\circ\text{C} \cdot \text{kg/m}^3}\right) = \left(\frac{\text{m}^2}{\text{hr}}\right)$$

The polynomial in this equation is the Laplace operator ($\nabla^2 t$):

$$\frac{\partial t}{\partial \tau} = \alpha \Delta^2 t \tag{135}$$

The heat conduction equation in cylindrical coordinates is (see Figure 10a):

$$\frac{\partial^2 t}{\partial r^2} + \frac{1}{r}\frac{\partial t}{\partial r} + \frac{1}{r^2}\frac{\partial^2 t}{\partial \varphi^2} + \frac{\partial^2 t}{\partial z^2} = \frac{1}{\alpha}\frac{\partial t}{\partial \tau} \tag{136}$$

and in spherical coordinates (Figure 10b):

$$\frac{1}{r^2}\frac{\partial}{\partial r}\left(r^2 \frac{\partial t}{\partial r}\right) + \frac{1}{r^2 \sin\varphi}\frac{\partial}{\partial \varphi}\left(\sin\varphi \frac{\partial t}{\partial \varphi}\right) + \frac{1}{r^2 \sin^2\varphi}\frac{\partial^2 t}{\partial \varphi^2} = \frac{1}{\alpha}\frac{\partial t}{\partial \tau} \tag{137}$$

Fourier's equation describes the propagation of heat through a body via conduction in either steady- or unsteady-state systems. The heating surfaces of interest in process equipment are usually well-defined geometries, e.g., flat, cylindrical, spherical. Therefore, the solutions to heat transfer problems involving conduction through standard geometries are of practical interest. We shall summarize the steady-state solutions to some of the frequently encountered geometries. Consider the steady-state conduction through a flat surface where the temperature distribution is one-dimensional (refer to Figure 11).

$$\frac{d^2 t}{\partial y^2} = \frac{\partial^2 t}{\partial z^2} = 0 \tag{138}$$

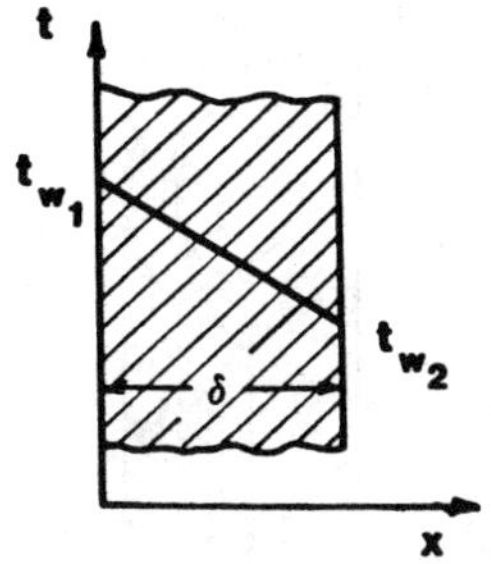

Figure 11. Heat conduction through a flat wall.

For this system, Fourier's equation is:

$$\frac{d^2t}{dx^2} = 0 \quad (139)$$

And integrating:

$$\frac{dt}{dx} = C_1$$

$$t = C_1x + C_2$$

The constants of integration are determined from the boundary conditions at $x = 0$ and $x = \delta$:

$$C_2 = t_{w1}$$

$$C_1 = \frac{dt}{dx} = \frac{t_{w2} - t_{w1}}{\delta}$$

or

$$t = \frac{t_{w2} - t_{w1}}{\delta} x + t_{w2} \quad (140)$$

Hence, the temperature along the thickness of a flat wall changes linearly at steady-state. One-dimentionsional heat flow for the flat wall case is:

$$dQ = k \frac{t_{w1} - t_{w2}}{\delta} dA d\tau$$

or

$$Q = \frac{k}{\delta} (t_{w1} - t_{w2}) A\tau \quad (141)$$

where k/δ = *thermal conductivity* (the inverse of this ratio is the *thermal resistance of the wall*) and A = heat transfer area.

Consider now the cylinder in Figure 12 which has length ℓ, inside radius r_1, and outside radius r_2, and a constant conductivity. At steady-state the inside and outside surface temperatures are constant, where $t_{w1} > t_{w2}$. For a one-dimensional analysis, the temperature changes only in the direction of the radius.

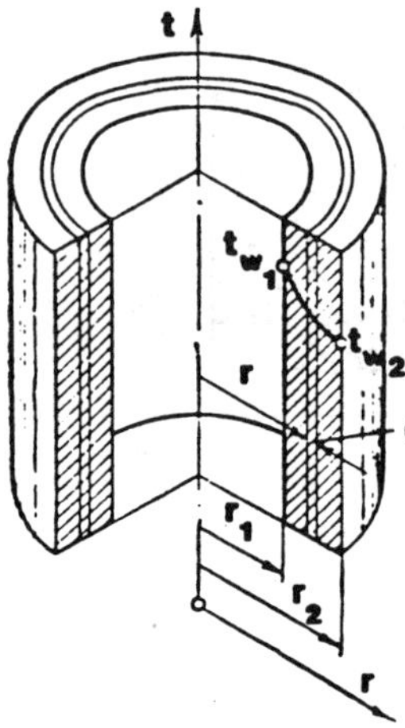

Figure 12. Heat conduction through a cylindrical wall.

The amount of heat passing through the circular layer inside the wall of thickness, dr, and radius r is:

$$dQ = -k\frac{\partial t}{\partial n}Adt = -k2\pi r\ell\frac{dt}{dr}d\tau \qquad (142)$$

For a steady-state thermal regime:

$$Q = -k\pi r\ell\frac{dt}{dr}\tau$$

or

$$dt = -\frac{Q}{2\pi\ell k\tau}\cdot\frac{dr}{r} \qquad (143)$$

Integrating over the limits from r_1 to r_2 and from t_{w1} to t_{w2} results in the one-dimensional conductivity equation at steady-state heat transfer:

$$t_{w2} - t_{w1} = -\frac{Q}{2\pi\ell k\tau}\,2.3\log\frac{r_2}{r_1}$$

or

$$Q = \frac{2\pi\ell}{\frac{1}{k}2.3\log\frac{r_2}{r_1}}(t_{w1} - t_{w2})\tau \qquad (144)$$

For the sphere shown in Figure 13, assume the interior and exterior surface temperatures are uniform and constant. The sphere has an inside radius r_1 and outer radius r_2. For a one-dimensional analysis, the temperature changes only along the direction of the radius.

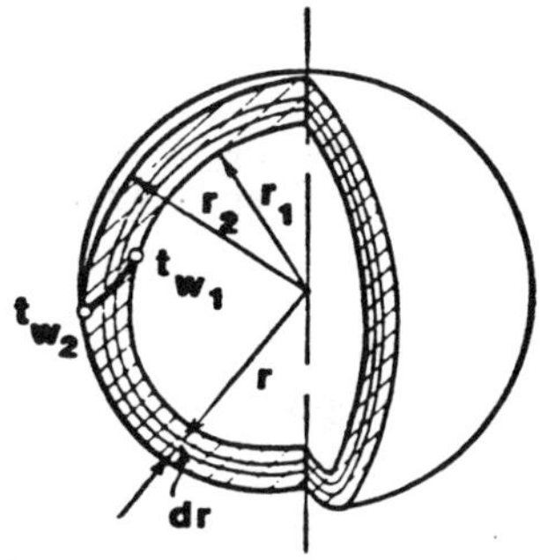

Figure 13. Heat conduction through a spherical wall.

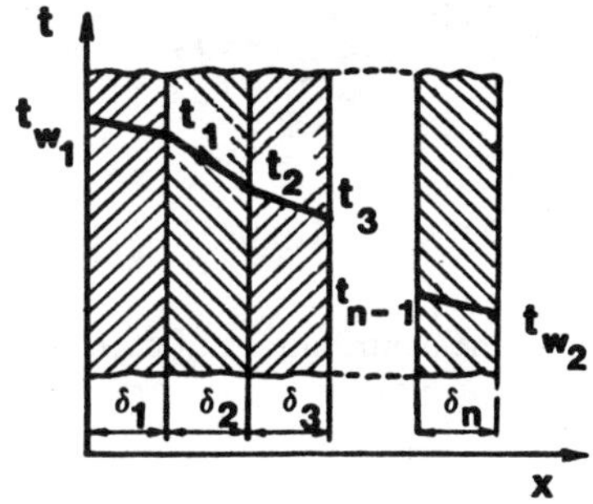

Figure 14. Conduction in a multilayer flat wall.

From Fourier's law, the amount of heat passing through a spherical layer of thickness dr and radius r inside the wall is:

$$Q = k\,\frac{dt}{dn}\,A\tau = -k4\pi r^2\,\frac{dt}{dr}\,\tau \tag{145}$$

or

$$dt = -\frac{Q}{4\pi k\tau}\,\frac{dr}{r^2} \tag{146}$$

Integrating over the limits from r_1 to r_2 from t_{w1} to t_{w2} gives:

$$t_{w2} - t_{w1} = -\frac{Q}{4\pi k\tau}\left(\frac{1}{r_1} - \frac{1}{r_2}\right)$$

or

$$Q = \frac{4\pi k}{\dfrac{1}{r_1} - \dfrac{1}{r_2}}\,(t_{w1} - t_{w2})\,\tau \tag{147}$$

The walls of process equipment are often constructed of several layers of different materials of varying thickness (refer to Figure 14). The most general one-dimensional case is a flat multilayer wall comprising n layers of thicknesses δ_1, $\delta_2 \ldots \delta_n$ and conductivities k_1, $k_2 \ldots k_n$.

The temperatures on the opposite sides of the multilayer walls are t_{w1} and t_{w2} (where $t_{w1} > t_{w2}$), and temperatures on the boundaries between layers are t_1, t_2, … t_{n-1}. At steady-state the heat flow is the same for all layers. Fourier's law can be written for each of these layers.

Layer 1:

$$Q = \frac{k_1}{\delta_1} A\tau(t_{w1} - t_1) \text{ or } \frac{Q}{A\tau}\frac{\delta_1}{k_1} = (t_{w1} - t_1)$$

Layer 2:

$$A = \frac{k_2}{\delta_2} A\tau(t_1 - t_2) \text{ or } \frac{Q}{A\tau}\frac{\delta_2}{k_2}(t_1 - t_2)$$

Layer n:

$$A = \frac{k_n}{\delta_n} A\tau(t_{n-1} - t_{w2}) \text{ or } \frac{Q}{A\tau}\frac{\delta_n}{k_n}(t_{n-1} - t_{w2})$$

Summing the contributions of each layer gives:

$$\frac{Q}{A\tau}\left(\frac{1}{k_1} + \frac{2}{k_2} + \cdots \frac{n}{k_n}\right) = (t_{w1} - t_{w2})$$

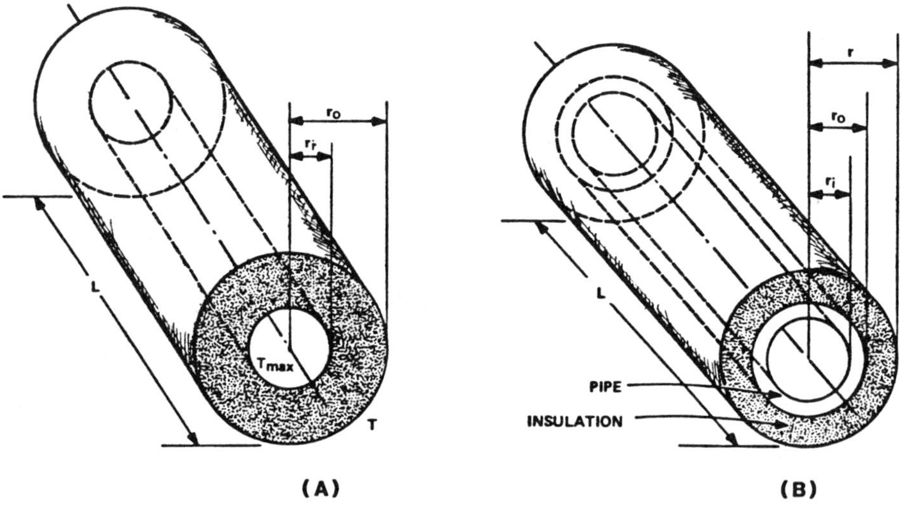

Figure 15. (A) Heat losses through single-wall pipe; (B) heat losses through insulated pipe.

or

$$Q = \frac{1}{\sum_{i=1}^{n} \frac{\delta_1}{k_1}} A (t_{w1} - t_{w2}) \tau \qquad (148)$$

where 1 = number of a layer;
n = number of layers

The equation states that the overall thermal resistance of a multilayer flat wall is the sum of the individual thermal resistances. For a multilayer cylindrical wall undergoing steady-state heat conduction, the following applies:

$$Q = \frac{2\pi\ell}{\sum_{i=1}^{n} \frac{2.3}{k_1} \log \frac{r_i + 1}{r_i}} (t_{w1} - t_{wz}) \tau \qquad (149)$$

The thermal conduction through a multilayer spherical wall at steady-state is:

$$Q = \frac{4\pi}{\sum_{i=1}^{n} \frac{1}{k_i} \left(\frac{1}{r_1} - \frac{1}{r_{i+1}}\right)} (t_{w1} - t_{w2}) \tau \qquad (150)$$

Two practical problems of interest are the heat losses from insulated piping, and losses through furnace walls. In both cases, we shall review the one-dimensional approximation.

As a first approximation, the radial heat flow through a cylindrical pipe can be considered as one-dimensional, steady-state conduction, where axial heat flow is ignored. The system is shown in Figure 15.

The heat flow through a solid pipe wall is:

$$q = \frac{T_{max} - T}{\sum R_T} \qquad (151)$$

where q = heat transfer rate, Btu/hr
T_{max} = maximum temperature in system, °F
T = temperature of surface toward which heat is flowing, °F
R_T = thermal resistance of solid per unit thickness, °F/Btu/(hr) (sq ft)

Thermal resistance is analogous to electrical resistance. The reciprocal of electrical resistance is conductance; similarly, the reciprocal of thermal resistance is thermal conductance. Thermal resistance or conductance is a characteristic of the piping material.

For cylindrical pipes, thermal resistance R_T is:

$$R_T = \frac{1}{2\pi kL} \ln\left(\frac{r_0}{r_i}\right) \qquad (152)$$

where k = thermal conductivity, Btu/(hr) (sq ft) (°F/ft)
L = length of pipe, ft
r_0 = outside pipe radius, ft
r_i = inside pipe radius, ft

Substitution of Equation 152 for R_T into Equation 151 gives:

$$q = \frac{2\pi kL (T_{max} - T)}{\ln \frac{r_0}{r_i}} \qquad (153)$$

This equation can be more conveniently expressed in terms of the surface areas of the inner and outer surfaces of the pipe of which there are several definitions:

Arithmetic mean area

$$A_A = \frac{A_i + A_0}{2} \tag{154}$$

where $A_i = 2\pi r_i L$ = inside surface area of pipe, sq ft
$A_0 = 2\pi r_0 L$ = outer surface of pipe, sq ft

Geometric mean area

$$A_G = \sqrt{A_i A_0} \tag{155}$$

Logarithmic mean area

$$A_L = \frac{A_0 - A_i}{\ln\left(\frac{A_0}{A_i}\right)} \tag{156}$$

The heat transfer rate for a composite pipe wall and/or when a combination of conduction and convection controls heat transfer, can be expressed in terms of an overall heat-transfer coefficient U.

$$q_T = UA_s\Delta t = UA_s(T_{max} - T) \tag{157}$$

where q_T = total rate of heat transfer, Btu/hr
U = overall heat-transfer coefficient, Btu/(hr) (sq ft) (°F)
A_s = surface area available to heat transfer, sq ft
Δt = overall temperature difference between hot and cold fluids, °F

Setting Equations 151 and 157 equal and solving for U gives:

$$U = \frac{1}{A_s \sum R} \tag{158}$$

Hence, for a pipe without insulation:

$$U = \frac{1}{A_s\left[\frac{1}{A_i h_i} + \frac{\ln(r_0/r_i)}{2\pi k_0 L} + \frac{1}{A_0 h_0}\right]} \tag{159}$$

where A_s = surface area (most commonly expressed in terms of the inner or outer surface area of the pipe), sq ft
A_i, A_0 = inside and outside areas, respectively, of pipe, sq ft
h_i, h_0 = convective heat transfer coefficients of the fluids inside and outside the pipe, respectively, Btu/(hr) (sq ft) (°F)

Equations 158 and 159 show that the overall coefficient U depends on the thermal conductivities of the pipe wall and insulating material, the thickness of the pipe wall and insulation, and the convective heat transfer coefficients of the fluid and surroundings.

We now briefly examine the problem of heat losses through a furnace wall. The heat generated in an industrial furnace is transmitted to the charge, the surrounding walls, and up the stack; however, we shall consider only the heat transferred by conduction through the wall under steady conditions. That is, no attention will be given to the effects of infiltration, furnace pressure, open joints, doors, etc.

Heat is transmitted through refractory walls mainly by conduction, and is then transmitted to the surrounding air and neighboring structures by radiation and convection from the outer wall surface. The drop in temperature through the wall depends upon the resistance offered by the wall to the flow of heat.

The quantity of heat transmitted by conduction through the wall per unit area varies directly with the temperature difference between the outer and inner surfaces of the wall, and the thermal conductivity of the refractory and insulation, and inversely with the thickness. Thermal conductivities of refractory and insulating materials are usually given as a function of mean or average temperatures.

Table 9 gives typical thermal conductivity values for different refractory materials. The heat which passes through the wall is dissipated to the surrounding air and objects. The rate of heat transfer depends on various factors, such as the temperature of the surface and the surroundings, the rate of circulation of air over the wall, and the geometry of the surface. The total rate of heat transfer through the wall depends upon the thermal resistance of the wall, as well as the resistance to heat dissipated from the surface of the wall. Therefore, before the actual transfer through the wall can be determined, it is necessary to solve for the simultaneous flow of heat through the wall and the surface. The true value of the heat flow is obtained when the solution gives the same heat flow through the wall (by conduction) and from its outer surface (by radiation and convection).

An approximate procedure for calculating the heat flow through furnace walls based on equilibrium conditions is to assume a temperature for the outer wall surface and from an assumed mean temperature of the various wall components and their conductivities compute the heat flow through the walls. The calculated heat flow through the wall is then compared with the heat loss from the outer surface by radiation and convection, which would occur when the outer surface temperature was the same as the assumed value. If the two losses are equal and the mean temperature for the wall components agree, the approximation is valid.

Table 9
Approximate Thermal Conductivity for Furnace Insulation
(Btu per ft² per hr per °F per in. thickness)

Insulation Material	Mean Temperature (°F)													
	100	200	400	600	800	1000	1200	1400	1600	1800	2000	2200	2400	2600
Fireclay brick	5.9	6.2	6.6	7.0	7.5	7.9	8.3	8.8	9.2	9.7	10.1	10.5	11.0	11.4
Common brick	4.6	4.8	5.3	5.8	6.3	6.8	7.4	7.9	8.4					
Magnesite brick	38.1	37.2	35.4	33.3	31.5	30.0	28.3	26.9	25.9	25.0	24.0	23.5	23.2	23.0
Chrome brick	9.3	9.5	10.0	10.5	11.0	11.5	12.0	12.5	13.0	13.5	14.0	14.5	15.0	15.5
Silica brick	6.4	6.8	7.5	8.3	9.0	9.8	10.5	11.3	12.0	12.8	13.5	14.3	15.0	15.8
Silicon carbide						104.0	105.0	106.0	106.5	107.0	108.0	109.0	109.5	110.0
APG plastic firebrick				3.0	3.3	3.6	4.0	4.4	4.8	5.2	5.5	5.9	6.3	
APG KAST-SET or HYDROSET				3.8	3.9	4.0	4.1	4.2	4.2	4.3	4.4	4.5		
APG KAST-O-LITE		1.6	1.7	1.8	1.9	2.0	2.2	2.4	2.6	2.9	3.1			
APG castable insulation No. 20		1.44	1.45	1.48	1.49	1.53	1.57	1.62	1.67	1.73				
APG block insulation		.42	.50	.60	.74	.89	1.08	1.30						
APG insulating comment		.65	.69	.76	.84	.94	1.05	1.19						
APG castable block mix		.53	.56	.60	.67	.75	.85	1.0						
APG loose wool	.28	.36	.54	.77	1.01									
APG G-26 I.F.B.		1.7	1.8	1.9	2.0	2.1	2.2	2.4	2.5	2.7	2.8			
APG G-23 I.F.B.		1.5	1.6	1.7	1.8	1.9	2.0	2.2	2.3	2.4	2.6			
Group 28 insulating firebrick		2.38	2.5	2.62	2.75	2.85	3.0	3.13	3.29	3.44	3.60			
Group 20 insulating firebrick		1.16	1.23	1.3	1.4	1.5	1.6	1.73	1.87	2.0				
Natural diatomaceous earth brick rt. angles to strata		.56	.61	.66	.71	.77	.82	.87						
Natural diatomaceous earth brick parallel to strata		.92	.97	1.02	1.07	1.12	1.17	1.22						
Magnesia 85%	.47	.49	.55											
Asbestos millboard	.85													
Concrete (stone)	6.5													
Sand (dry)	2.3													

Data extracted from A. P. Green Fire Brick Co. – Technical Note 3105-B – Dec. 6, 1948.

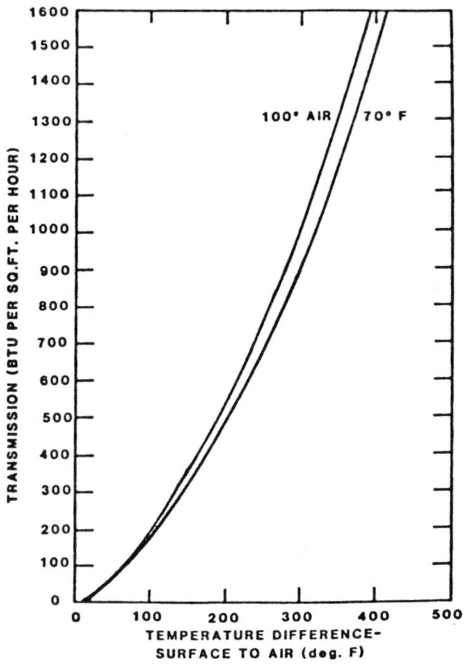

Figure 16. Heat transmission from vertical surfaces.

To faciliate calculations of the surface heat losses, the radiation and convection formulas have been combined and the results of calculations for various temperature differences between the wall surface and surrounding air are shown in Figure 16. The plot is based on heat losses from a vertical surface exposed to ambient air temperatures of 70° and 100°F., and with an emissivity or radiation constant of 0.95. Reference 16 gives example problems on the one-dimensional heat loss analysis for furnace walls.

Two-Dimensional Conduction

We now consider two-dimensional heat flow with application of the steady-state Laplace equation, assuming constant thermal conductivity:

$$\frac{\partial^2 t}{\partial x^2} + \frac{\partial^2 t}{\partial y^2} = 0 \tag{160}$$

Since this is a linear differential equation the principle of superposition of solutions applies. The solution of Equation 160 provides the temperature field in a two-dimensional body as a function of the x and y coordinates.

The heat flow in the x and y directions may be calculated from the Fourier equation:

$$q_x = -\,kA_x \frac{\partial t}{\partial x}$$
$$q_y = -\,kA_y \frac{\partial t}{\partial y} \tag{161}$$

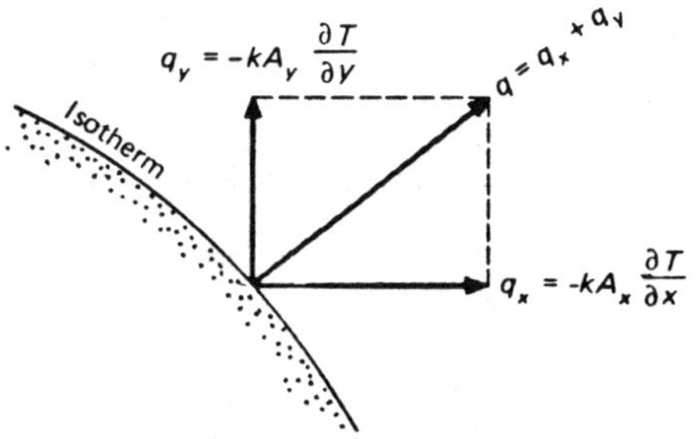

Figure 17. Heat flow in two dimensions.

The total heat flow at any point is the resultant of q_x and q_y at that point and directed perpendicular to the isotherm surface. This is shown in Figure 17.

Assume that the temperature distribution can be expressed by the product of two functions where X(x) is a function of x only, and Y(y) is a function of y only. Hence, temperature T is:

$$T = X(x) \cdot Y(y) \tag{162}$$

Then Fourier's equation becomes:

$$-\frac{1}{X}\frac{d^2X}{dx^2} = \frac{1}{Y}\frac{d^2Y}{dy^2} \tag{163}$$

Both sides of this expression contain only independent variable, and the equality is valid only if they are both equivalent to the same constant. That is:

$$-\frac{1}{X}\frac{d^2X}{dx^2} = \frac{1}{Y}\frac{d^2Y}{dy^2} = \Upsilon^2 \tag{164}$$

Equation 164 can now be expressed as the sum of two ordinary differential equations:

$$\frac{d^2X}{dx^2} + \Upsilon^2 = 0 \tag{165}$$

$$\frac{d^2Y}{dy^2} - \Upsilon^2 Y = 0 \tag{166}$$

The general solutions of these equations are:

$$Y = B_1 \sinh(\Upsilon y) + B_2 \cosh(\Upsilon y) \tag{167}$$

$$X = B_3 \sin(\Upsilon x) + B_4 \cos(\Upsilon x) \tag{168}$$

Substituting for X and Y into the assumed form of the temperature distribution (Equation 162) gives the following general solution to the two-dimensional, steady-state Laplace equation (Equation 160):

$$t = [B_1 \sinh(\Upsilon y) + B_2 \cosh(\Upsilon y)]\ [B_3 \sin(\Upsilon x) + B_4 \cos(\Upsilon x)] \tag{169}$$

where Υ and B_i are constants evaluated from specified boundary conditions.

We shall now summarize the two-dimensional steady-state solutions of several common systems. The first system for considerations is a rectangular plate having a specified temperature distribution on one edge while the remaining edges have zero temperature. The system is defined in Figure 18 as a rectangular plate of finite width L along the x coordinate direction and width W in the y coordinate direction.

The temperatures of the edges of the plate at $x = 0$, $x = L$, and $y = 0$ are zero, whereas at $y = W$ the temperature varies along that edge and can be described by the function $f(x)$ (when $0 < x < L$).

Hence, the boundary conditions that are to be applied to Equation 169 are:

1. At $x = 0$, $t = 0$.
2. At $x = L$, $t = 0$.
3. At $y = 0$, $t = 0$.
4. At $y = W$, $t = f(x)$.

We wish to determine the function $T = g(x,y)$ that gives the distribution of the temperature in the plate. Applying the third boundary condition gives $B_2 = 0$, and Equation 169 becomes:

$$t = \sinh(\Upsilon y)\,[B_3 \sin(\Upsilon x) + B_4 \cos(\Upsilon x)]$$

To satisfy the first boundary condition, $B_4 = 0$, then:

$$T = \sinh(\Upsilon y) \cdot B \sin(\Upsilon x) \tag{170}$$

where B replaces the product B_1B_3. Substituting in boundary condition 2 leads to:

$$0 = \sinh(\Upsilon y) \cdot B \sin(\Upsilon L)$$

For this statement to be satisfied for all values of y:

$$\sin \Upsilon L = 0$$

And this expressions is satisfied for $\Upsilon = 0, \pi/L, 2\pi/L, \ldots$, or

$$\Upsilon_n = \frac{n\pi}{L},\ n = 0, 1, 2, 3, \ldots \tag{171}$$

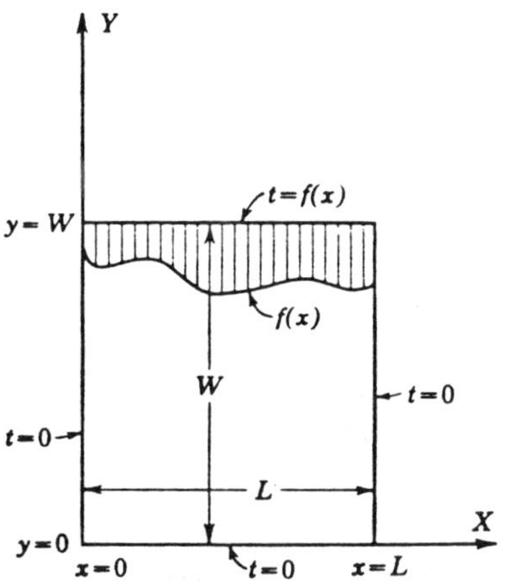

Figure 18. A rectangular plate with a specified temperature distribution on one edge.

Each of the Υ's provides a separate solution of Equation 169. Note that the general solutions is the sum of these individual solutions:

$$t = \sum_{n=0}^{\infty} B_n \sinh(\Upsilon_n y) \sin(\Upsilon_n x)$$

where B_n denotes the constant B for each of the solutions. Since $\Upsilon_n = 0$ for $n = 0$, no contributions is made by the first term, and

$$t = \sum_{n=1}^{\infty} B_n \sinh(\Upsilon_n y) \sin(\Upsilon_n x) \tag{172}$$

Applying boundary condition 4 for $y = W$:

$$f(x) = \sum_{n=1}^{\infty} B_n \sinh(\Upsilon_n W) \sin(\Upsilon_n x)$$

$$\Upsilon_n = \frac{n\pi}{L}, \; n = 1, 2, 3, \ldots \tag{173}$$

$$0 \leqslant x \leqslant L$$

It can be shown from orthogonality considerations that:

$$B_n \sinh(\Upsilon_n W) = C_n = \frac{2}{L} \int_0^L f(x) \sin(\Upsilon_n x)$$

where C_n is a constant.

Since $f(x)$ is a known boundary condition, then the previous integration can be performed where the final general solutions is:

$$t = \frac{2}{L} \sum_{n=1}^{\infty} \left(\frac{\sinh \frac{n\pi y}{L}}{\sinh \frac{n\pi W}{L}} \right) \sin \frac{n\pi x}{L} \int_0^L f(x) \sin \frac{n\pi x}{L} \, dx \tag{174}$$

For the case in which the edge at $y = W$ is at a constant temperature, i.e., $f(x) = T_1$, Equation 174 becomes:

$$\frac{t}{T} = 2 \sum_{n=1}^{\infty} \frac{[1 - (-1)^n]}{n\pi} \left(\frac{\sinh \frac{n\pi y}{L}}{\sinh \frac{n\pi W}{L}} \right) \sin \frac{n\pi x}{L} \tag{175}$$

This relation results from the evaluation of

$$\int_0^L T \sin \frac{n\pi x}{L} \, dx$$

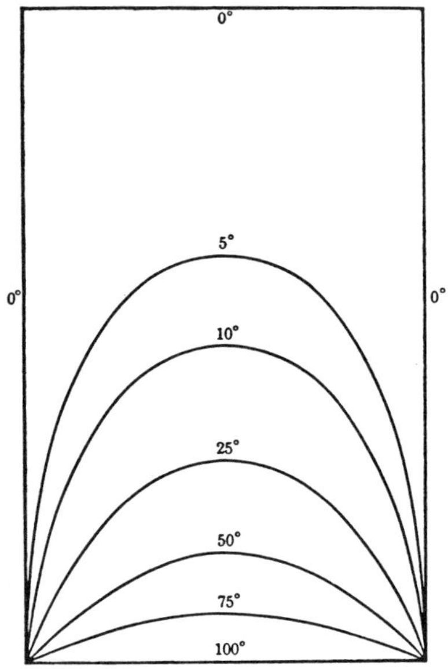

Figure 19. Shows isotherms in a rectangular plate heated on one edge.

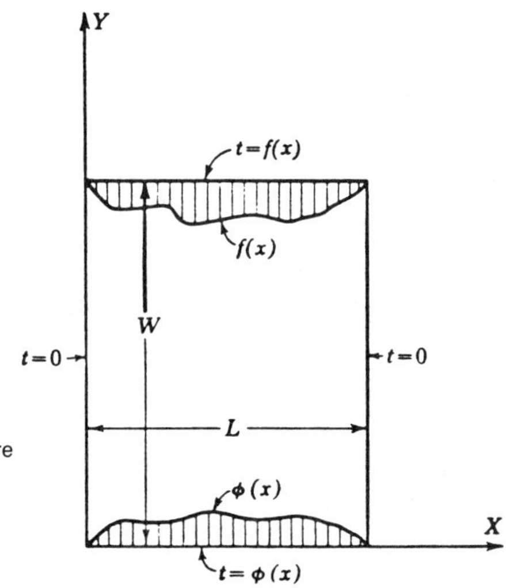

Figure 20. Shows a plate with a specified temperature distribution on more than one edge.

Equation 175 enables the temperature at any desired point in the plate to be computed. This is illustrated in Figure 19 which shows a plate, 10 by 6, with one edge held at 100° and all others held at 0°. The isothermal lines within the plate were generated from Equation 175.

A rectangular plate with specified temperature functions on more than one edge can be reduced to the same problem by employing the principle of superposition. Figure 20 shows a plate that has a distribution $t = f(x)$ specified at $y = W$ whereas $t = \psi(x)$ is imposed at y = 0. The other edges are maintained at t = 0. The equation to solve along with the boundary conditions is:

$$\frac{\partial^2 t}{\partial x^2} + \frac{\partial^2 t}{\partial y^2} = 0$$

$$\begin{array}{ll} t = f(x) & \text{at } y = W \\ t = \psi(x) & \text{at } y = 0 \\ t = 0 & \text{at } x = 0 \\ t = 0 & \text{at } x = L \end{array} \tag{176}$$

Defining the temperature field in terms of two independent variables:
$t = u + \nu$

where u and ν denote the solutions to the following two systems:

$$\frac{\partial^2 u}{\partial x^2} + \frac{\partial^2 u}{\partial y^2} = 0$$

$$\begin{array}{ll} u = f(x) & \text{at } y = W \\ u = 0 & \text{at } y = 0 \\ u = 0 & \text{at } x = 0 \\ u = 0 & \text{at } x = L \end{array} \tag{177a}$$

$$\frac{\partial^2 \nu}{\partial x^2} + \frac{\partial^2 \nu}{\partial y^2} = 0$$

$$\begin{array}{ll} \nu = 0 & \text{at } y = W \\ \nu = \varphi(x) & \text{at } y = 0 \\ \nu = 0 & \text{at } x = 0 \\ \nu = 0 & \text{at } x = L \end{array} \tag{177b}$$

The solution to Equation 177a is the same as that expressed in Equation 174 with u replacing t. By a change of variable the solution given in Equation 174 can be used to determining ν from Equation 177b.

$$\nu = \frac{2}{L} \sum_{n=1}^{\infty} \frac{\sinh\left(\frac{n\pi(W-y)}{L}\right)}{\sinh\left(\frac{n\pi W}{L}\right)} \sin\frac{n\pi x}{L} \int_0^L \varphi(x) \sin\frac{n\pi x}{L}\, dx \tag{178}$$

From $t = u + \nu$, the solution to Equation 176 is the sum of Equations 174 and 178.

A second example is that of steady conduction in a circular cylinder of finite length, shown in Figure 21. R is the outer radius of the cylinder and L is its length. In this case, we choose boundary conditions so that axial symmetry exists. The general conduction equation in cylindrical coordinates (where the temperature distribution is assumed to depend only on the coordinates) is:

$$\frac{\partial^2 t}{\partial r^2} + \frac{1}{r}\frac{\partial t}{\partial r} + \frac{\partial^2 t}{\partial z^2} = 0 \tag{179}$$

Note that the imposed boundary conditions must be independent of the θ coordinate. Following the above procedure, a solution of the following form is sought:

$$t = R(r) \cdot Z(z) \tag{180}$$

where $R(r)$ and $Z(z)$ are functions of r and z only, respectively.

Substituting Equation 180 into 179 gives:

$$\frac{1}{R}\frac{d^2R}{dr^2} + \frac{1}{r}\frac{1}{R}\frac{dR}{dr} = -\frac{1}{Z}\frac{d^2Z}{dz^2}$$

Note that the RHS is a function of z alone and LHS is a function of r alone.

As before, this expression can be written as two ordinary differential equations:

$$\frac{d^2R}{dr^2} + \frac{1}{r}\frac{dR}{dr} + \Upsilon^2 R = 0 \tag{181a}$$

$$\frac{d^2Z}{dz^2} - \Upsilon^2 Z = 0 \tag{181b}$$

where Υ^2 is a constant.

Equation 181a is Bessel's equation of zero order and Equation 181b leads to the hyperbolic functions. Thus, applying the definition of R and Z, the solution of the differential equation in Equation 179 is:

$$t = [B_1 J_0(\Upsilon r) + B_2 Y_0(\Upsilon r)]\,[B_3 \sinh(\Upsilon z) + B_4 \cosh(\Upsilon z)] \tag{182}$$

We now impose the following boundary conditions:

1. At $z = 0$, $t = 0$
2. At $z = L$, $t = f(r)$ (183)
3. At $r = R$, $t = 0$

Also, the additional condition that the temperature is finite at $r = 0$ is imposed. Knowing that $Y_0(\Upsilon r) \to \infty$ as $(\Upsilon r) \to 0$ then $B_2 = 0$. Boundary condition 1 sets $B_4 = 0$:

$$t = B \sinh(\Upsilon z)\, J_0(\Upsilon r)$$

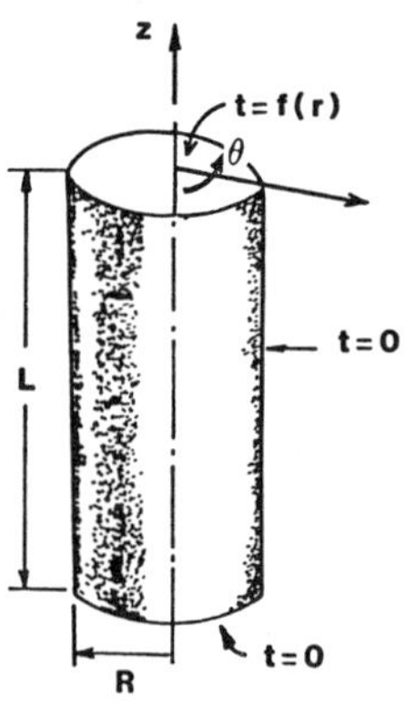

Figure 21. Steady conduction in a circular cylinder of finite length.

where B replaces B_1B_3. Boundary condition 3 produces:

$$B \sinh(\Upsilon z) J_0(\Upsilon R) = 0$$

For this condition to be satisfied for all Z values between 0 and L:

$$J_0(\Upsilon R) = 0$$

In this example there are an infinite number of Υ's satisfying the defining relation $J_0(\Upsilon_n R) = 0$. The first five are: $\Upsilon_1 R = 2.4048$, $\Upsilon_2 R = 5.5201$, $\Upsilon_3 R = 8.6537$, $\Upsilon_4 R = 11.7915$, $\Upsilon_5 R = 14.9309$. The general solution is the sum of all the solutions corresponding to each of the Υ_n's:

$$t = \sum_{n=1}^{\infty} (B_n \sinh(\Upsilon_n z)) J_0(\Upsilon_n r)$$

Applying boundary condition 2 determines the unknown B_n's:

$$f(r) = \sum_{n=1}^{\infty} (B_n \sinh(\Upsilon_n L)) J_0(\Upsilon_n r)$$

The constant coefficients $(B_n \sinh(\Upsilon_n L))$ are given by:

$$B_n \sinh \Upsilon_n L = \frac{\int_0^R r f(r) J_0(\Upsilon_n r)\, dr}{\dfrac{R^2}{2} J_1^2(\Upsilon_n R)}$$

The solution to the problem giving the distribution of the temperature through the cylinder is:

$$t = \frac{2}{R^2} \sum_{n=1}^{\infty} \frac{\sinh(\Upsilon_n z)}{\sinh(\Upsilon_n L)} \frac{J_0(\Upsilon_n r)}{J_1^2(\Upsilon_n R)} \int_0^R r f(r) J_0(\Upsilon_n r)\, dr. \tag{184}$$

For the case where $f(r) = T$ (a constant), the solution reduces to:

$$\frac{t}{T} = 2 \sum_{n=1}^{\infty} \frac{1}{\Upsilon_n R} \frac{\sinh(\Upsilon_n z)}{\sinh(\Upsilon_n L)} \frac{J_0(\Upsilon_n r)}{J_1(\Upsilon_n R)} \tag{185}$$

Equation 185 describes the temperature distribution in the cylinder where all the surface except the one end are zero.

Note that the temperature of the other surfaces may be at some constant value other than zero. In such cases T is simply interpreted as the difference between the temperature at the end $z = L$ and this constant value.

The principle of superposition can be applied to address other situations. Figure 22 provides a plot of the axial temperature distribution $(r = 0)$ for a solid cylinder heated to a constant temperature on one end. Also, the case of a cylinder heated on both ends which is derived by superposition from the first case is shown.

Unsteady-State Conduction

Unsteady-state heat conduction is important to many heat transfer applications involving start-ups, operating, and instability transients which may affect the rate at which

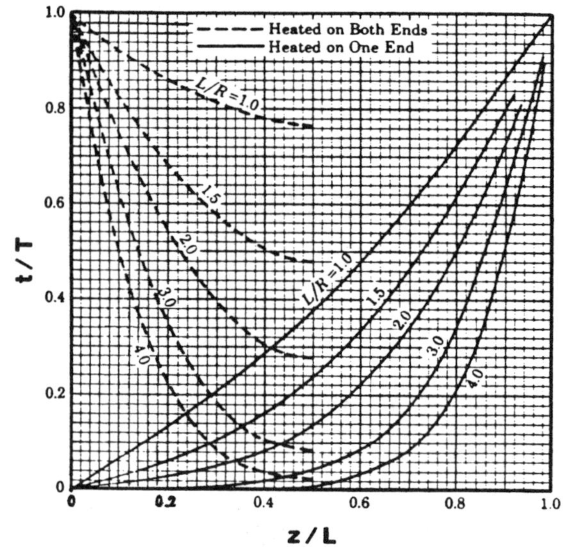

Figure 22. Plot of axial temperature distribution in solid cylinders of finite length.

process equipment attains stable operation. It is also important in defining the temperature limitations imposed on materials of construction for equipment in terms of heat transfer characteristics, thermal stresses, and temperature distortions. The general differential equation for unsteady-state conduction is derived from the basic Fourier equation.

To illustrate development of the transient heat conduction analysis, consider an infinitesimal parallel piped of sides dx, dy, and dz where the rate at which heat enters along the x, y, and z coordinates is:

$$\frac{dQ_{in}}{d\tau} = -\left(k_x dydz\,\frac{dt}{dx} + k_y dzdx\,\frac{dt}{dy} + k_z dxdy\,\frac{dt}{dz}\right) \tag{186}$$

Similarly, the rate of which heat exits along the x, y, and z axes is:

$$\frac{dQ_{out}}{d\tau} = -\left[k_x\frac{\partial t}{\partial x} + \frac{\partial}{\partial x}\left(k_x\frac{\partial t}{\partial x}\right)dx\right]dydz - \left[k_y\frac{\partial t}{\partial y} + \frac{\partial}{\partial y}\left(k_y\frac{\partial t}{\partial y}\right)dy\right]dzdx$$
$$-\left[k_z\frac{\partial t}{\partial z} + \frac{\partial}{\partial z}\left(k_z\frac{\partial t}{\partial z}\right)dz\right]dxdy \tag{187}$$

It then follows that the net rate of accumulation is:

$$\frac{dQ_{net}}{d\tau} = \frac{dQ_{in}}{d\tau} - \frac{dQ_{out}}{d\tau} = \left[\frac{\partial}{\partial x}\left(k_x\frac{\partial t}{\partial x}\right) + \frac{\partial}{\partial y}\left(k_y\frac{\partial t}{\partial y}\right)\right.$$
$$\left. + \frac{\partial}{\partial z}\left(k_z\frac{\partial t}{\partial z}\right)\right]dxdydz \tag{188}$$

The rate of heat accumulation in the element is equal to the rate of increase in enthalpy, which is related to the rate of temperature rise of the medium:

$$\frac{dQ}{d\tau} = \frac{dH}{d\tau} = \left(\gamma C_p \frac{\partial t}{\partial \tau}\right) dxdydz \tag{189}$$

That is, Equation 189 and 188 are equivalent. Equation these expressions results in the general differential equation for unsteady-state conduction:

$$\frac{1}{\gamma C_p}\left[\frac{\partial}{\partial x}\left(k_x \frac{\partial t}{\partial x}\right) + \frac{\partial}{\partial y}\left(k_y \frac{\partial t}{\partial y}\right) + \frac{\partial}{\partial z}\left(k_z \frac{\partial t}{\partial z}\right)\right] = \frac{\partial t}{\partial \tau} \tag{190}$$

For a homogeneous, isotropic medium, Equation 190 can be written as:

$$\frac{k}{\gamma C_p}\left[\left(\frac{\partial^2 t}{\partial x^2}\right) + \left(\frac{\partial^2 t}{\partial y^2}\right) + \left(\frac{\partial^2 t}{\partial z^2}\right)\right] = \frac{\partial t}{\partial \tau} \tag{191}$$

where we have assumed k to be constant over the temperature range of interest. Note that $k/C_p\gamma$ is the thermal diffusivity of the medium (denoted by symbol α). The sum of second-order derivatives on the LHS of Equation 191 is the Laplace operator ($\nabla^2 t$). Hence, Equation 191 may be written in notation form as:

$$\alpha \nabla^2 t = \frac{\partial t}{\partial \tau} \tag{192}$$

A classical solution to Equation 192 for a simple system will now be developed. The system for consideration is that of the heating/cooling of an infinite plate. The surface area of the plate is large in comparison to its thickness, hence, the problem can be considered one-dimensional where the heat transmission is across thickness x. Equation 191 thus reduces to:

$$\frac{k}{C_p r}\left(\frac{\partial^2 t}{\partial x^2}\right) = \frac{\partial t}{\partial \tau} \tag{193}$$

Initially, the plate's temperature is a constant t_l and is uniform. At some later time both of its surfaces are cooled to a new uniform temperature t_0. To define a convenient coordinate system, assume that the heat flow is from the inside of the plate to the surfaces for this one-dimensional problem. A dimensionless ratio is then defined to described the local temperature distribution at any point in time:

$$\Delta = \frac{t - t_0}{t_l - t_0} \tag{194}$$

where Δ varies between the limits of 0 and 1.

Introducing Δ into Equation 193 gives:

$$\alpha \frac{\partial^2 \Delta}{\partial x^2} = \frac{\partial \Delta}{\partial \tau} \tag{195}$$

Distance x is measured from one surface (at $x = 0$) across to the other surface ($x = 2R$, where 2R is the plate thickness). At time $\tau = 0$ the temperature at each point is t_l (i.e., $\Delta = 1$); and at steady-state (at some infinite time later) the temperature at each t_0, or $\Delta = 0$. That is, on the both sides of the plate, (i.e., at $x = 0$ and $x = 2R$), the temperature at any time is equal to t_0 and $\Delta = 0$.

We can therefore write the following four boundary conditions:

$$\begin{aligned} &1.\ \tau = 0, \quad \text{arbitrary } x, \Delta = 1 \\ &2.\ \tau = \infty, \text{ arbitrary } x, \Delta = 0 \\ &3.\ x = 0, \quad \text{arbitrary } \tau, \Delta = 0 \\ &4.\ x = 2R, \text{arbitrary } \tau, \Delta = 0 \end{aligned} \tag{196}$$

Applying these boundary conditions to Equation 195 and integrating provides the temperature at any point and time within the plate. Not that the function $\Delta = f(x, \tau)$ can be expressed as the product of two independent functions:

$$\Delta = f(x, y) = y(x)\psi(\tau) \tag{197}$$

Let $y(x) = y$ and $\psi(\tau) = \psi$; Equation 195 can be stated as:

$$\frac{1}{y}\frac{\partial^2 y}{\partial x^2} = \frac{1}{\alpha\psi}\frac{\partial\psi}{\partial\tau} \tag{198}$$

For this statement to be valid, y and ψ must both be equal to some constant, A^2. Setting both sides of Equation 198 equal to a^2 results in two total-differential equations, one of which has the following form:

$$\frac{1}{\alpha\psi}\left(\frac{d\psi}{d\tau}\right) = -A^2 \tag{199}$$

The general solution to this equation is:

$$\psi(\theta) = C_1 e^{-A2\alpha\tau} \tag{200}$$

where C_1 is a constant of integration.
The second equation has the form:

$$\frac{1}{y}\frac{d^2y}{dx^2} = -A^2 \tag{201}$$

Denoting $dy/dx = p$, then

$$\frac{d^2y}{dx^2} = \frac{dp}{dx} = \left(\frac{dp}{dy}\right)\left(\frac{dy}{dx}\right) = p\frac{dp}{dy} \tag{202}$$

And substituting for $\frac{d^2y}{dx^2}$ in Equation 201 gives:

$$\frac{p}{y}\cdot\frac{dp}{dy} = -A^2 \tag{203}$$

where the general solution is:

$$p = \sqrt{C_2 - A^2y^2} = \frac{dy}{dx} \tag{204}$$

And C_2 is a constant of integration.

A second integration of this expression produces:

$$\frac{1}{A}\sin^{-1}\left(\frac{Ay}{\sqrt{C_2}}\right) = x + C_3 \tag{205}$$

where C_3 is another integration constant.
Equation 205 is more conveniently written as:

$$y = \frac{\sqrt{C_2}}{A}\sin(Ax + AC_3) \tag{206}$$

After trigonometric transformation and introducing constants C_4 and C_5, the following function is obtained:

$$y(x) = C_4 \cos Ax + C_5 \sin Ax \tag{207}$$

From the functions $\psi(\tau)$ and y (x) (Equations 200 and 207), the Δ function (Equation 197) can be expressed as:

$$\Delta = C_1 e^{-A2\alpha\tau}(C_4 \cos Ax + C_5 \sin Ax) \tag{208}$$

We may now apply the boundary conditions of Equation 196 to evaluate the constants of integration. Boundary condition 2 ($\tau \rightarrow \infty$, $\Delta = 0$) is fulfilled at any value of x. Boundary condition 3 ($x = 0$, $\Delta = 0$) is fulfilled when $C_4 = 0$. Denoting $C_1 \cdot C_5 = C_6$ we obtain an expression that accounts for both conditions:

$$\Delta = C_6 e^{-A2\alpha\tau}\sin Ax \tag{209}$$

Applying bc 4 ($\Delta = 0$ at $x = 2R$) and at arbitrary τ we obtain:

$$0 = C_6 \sin 2RA \tag{210}$$

Note that the sine function is zero for $n\pi$ values where n = 1, 2, 3, 4. Consequently, the product 2 RA also should be multiples of π. Therefore, the constant A is evaluated as:

$$A = \frac{n\pi}{2R} \tag{211}$$

Substituting for A into Equation 209 gives:

$$A = C_6(e^{-\alpha n 2\pi/4R2})\sin\frac{n\pi x}{2R} \tag{212}$$

At n = 1, 2, 3, ... the detailed solution of Equation 195 in terms of first-order derivatives only is obtained. From the additivity principle, the general solution of Equation 212 is:

$$\Delta = S_1(e^{-\frac{\alpha\pi 2}{4R2}\tau})\sin\frac{\tau x}{2R} + S_2(e^{-\frac{4\alpha\pi 2}{4R2}})\sin\frac{2\pi x}{2R} + S_3(e^{-\frac{9\alpha\pi 2}{4R2}\tau})\sin\frac{3\pi x}{2R} + \ldots \tag{213}$$

where S_1, S_2, S_3 ... are constants corresponding to different C_6 values in Equation 212.
We now apply the first boundary condition (Equation 196 – $\Delta = 1$ at $\tau = 0$) for all values of x. Substituting these values in Equation 213 gives:

$$1 = S_1 \sin\left(\frac{\pi x}{2R}\right) + S_2 \sin\left(\frac{2\pi x}{2R}\right) + S_3 \sin\left(\frac{3\pi x}{2R}\right) + \ldots \tag{214}$$

Expanding the function f(z) = 1 in a Fourier series:

$$f(z) = \frac{4}{\pi}\left(\sin z + \frac{1}{3}\sin 3z + \frac{1}{5}\sin 5z + \ldots\right) = 1 \tag{215}$$

Comparison of these last two expressions gives the values of the constants, where the condition of identity for both equations is:

$$S_1 = \frac{4}{\pi};\ S_2 = 0;\ S_3 = \frac{4}{3}\pi;\ S_4 = 0,\ S_5 = \frac{4}{5}\pi \ldots$$

Those constants with even-number indices will be equal to zero, and those with uneven indices are $\frac{4}{\pi}, \frac{4}{3}\pi, \frac{4}{5}\pi \ldots$ Hence, the general form for n-constants is:

$$S_n = [1 - (-1)^n]\frac{2}{n\pi} \tag{216}$$

Incorporating the appropriate S_n-values into Equation 213 results in the final expression describing the temperature field in an infinite plate:

$$\frac{t - t_0}{t_1 - t_0} = \frac{4}{\pi}\left[\left(e^{-\left(\frac{\pi}{2}\right)^2 \frac{\alpha\tau}{R^2}} \sin\frac{\pi x}{2R}\right) + \left(\frac{1}{3}e^{-9\left(\frac{\pi}{2}\right)^2 \frac{\alpha\tau}{R^2}} \sin\frac{3\pi x}{2R}\right) + \left(\frac{1}{5}e^{-25\left(\frac{\pi}{2}\right)^2 \frac{\alpha\tau}{R^2}} \sin\frac{5\pi x}{2R}\right) + \ldots\right] \tag{217}$$

Equation 217 can be applied to evaluating the temperature distribution in a plate at different times $- \tau_1, \tau_2, \tau_3 \ldots$ Figure 23 qualitatively shows the temperature field for different times that can be computed from Equation 217. Note, that since the heat flow is in the direction of lower temperatures there is no heat flow along the plane of symmetry and hence, the temperature curves show maxima. If the symmetry plane is replaced by a perfectly insulating boundary (ignore half of the plate), the temperature distribution would remain unaltered. In other words, Equation 217 may also be applied to evaluating the time-temperature field in a plate that is heated or cooled from one side and insulated on the other. In this case R denote the total plate thickness.

Once the temperature distribution is known, the thermal capacity of the plate can be evaluated. That is, we may now evaluate the rate at which heat flows to or from the system during heating or cooling. The system's heat capacity initially (at temperature t_1) is:

$$Q_1 = C_p(t_1 - t_0)\gamma A 2R \tag{218}$$

where C_p = heat capacity
γ = specific weight
A = surface area of the plate

The heat flow per unit time over thickness dx is:

$$dQ = C_p(t - t_0)\gamma A dx \tag{219}$$

and the total heat capacity is:

$$Q = C_p\gamma A \int_0^{2R} (t - t_0)\, dx \tag{220}$$

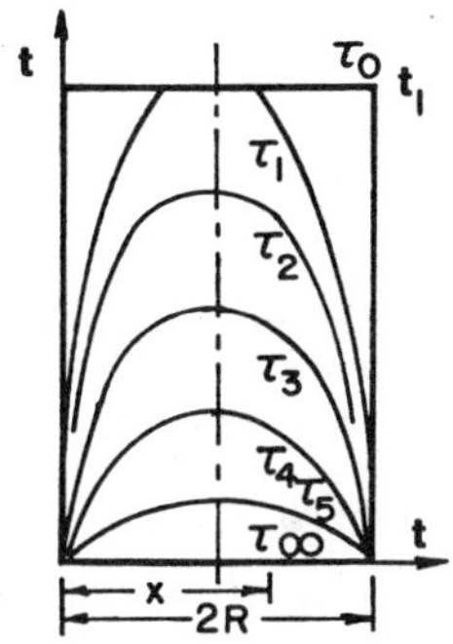

Figure 23. Distribution of temperatures vs. time in a slab.

Dividing through Q_1 we obtain:

$$\frac{Q}{Q_1} = \frac{1}{2R} \int_0^{2R} \Delta dx \tag{221}$$

where Δ is evaluated from Equation 217.

After integration:

$$\frac{Q}{Q_1} = \frac{8}{\pi^2}\left[e^{-\left(\frac{\pi}{2}\right)^2 \frac{\alpha\tau}{R^2}} + \frac{1}{9} e^{-9\left(\frac{\pi}{2}\right)^2 \frac{\alpha\tau}{R^2}} + \frac{1}{25} e^{-25\left(\frac{\pi}{2}\right)^2 \frac{\alpha\tau}{R^2}} + \ldots\right] \tag{222}$$

Equation 222 provides the heat rate Q in the plate for any time τ. For the limiting case of a short process time (i.e., $(d\tau/R^2) > (1/2)$), the relationship reduces to:

$$\ln \frac{Q}{Q_1} = \ln \frac{8}{\pi^2} - \left(\frac{\pi}{2}\right)^2 \frac{\alpha\tau}{R^2} \tag{223}$$

When Equation 223 is plotted on semi-logarithmic coordinates as $\ln Q$ vs, τ, a straight line results.

The prior analysis can also be used to develop similar expressions of the distribution for other geometries such as an infinite cylinder, a sphere or for a body limited only by one plane. The latter geometry is the most straightforward, with the temperature field reducing to the Gaussian "error integral."

$$\frac{t - t_0}{t_1 - t_0} = \frac{2}{\sqrt{\pi}} \int_0^b e^{-z^2} dz \tag{224}$$

Where the upper limit on the integral is:

$$b = \frac{1}{2\sqrt{\alpha\tau/R^2}} \tag{225}$$

R is defined as the distance from a given point on a plane (of a body limited from one side).

In general, the equations describing transient heat conduction, even for simple geometries, require tedious computations. To assist in more practical calculations, these equations have been integrated for various boundary conditions and have been prepared in graphical forms [17, 18]. The dimensionless temperature ratio in these equations is a function of two dimensionless groups, namely the:

Fourier modulus (dimensionless time):

$$Fo = \frac{\alpha\tau}{R^2} = \frac{k\tau}{\gamma c_p R^2} \tag{226}$$

and

Dimensionless size:

$$n = \frac{r}{R} \tag{227}$$

where r is the distance from a given point to the plane of symmetry for a slab as the symmetry axis for a cylinder.

Defining x as measured from one surface, and the cylinder radius or half of the slab thickness R, we obtain t as an absolute value from R to x. That is, the temperature distribution is a function of two parameters:

$$\Delta = f(Fo, n) \tag{228}$$

Equation 228 applies to simple bodies such as infinite cylinders, infinite plates, and spheres. For an infinite slab and cylinder, a family of lines in the coordinate system Δ-F_0 is obtained where each line corresponds to a definite n-value. For a body limited from one side, only one curve is obtained. The body surface is assumed to have temperature t_0 at all times. At very high flows past the surface, heat transfer by convection can be viewed as corresponding to an infinitely great heat transfer coefficient for the slab. An example is when heating metal with saturated steam. In the case of heating (or cooling) with gases of low heat transfer coefficients, the surface temperature of the body will not be equal to that of the gas t_0. Even for a high flow rate its temperature t_0 will be constant but the surface temperature of the body will change. As a different heat flow rate passes between the surface and liquid in the unsteady-state regime and if a heat transfer coefficient and one of the temperatures are constant, the other temperature should change, i.e., the surface temperature of the wall.

The thickness of an equivalent layer is equal to the ratio of the material conduction (cylinder, sphere, slab) to the heat transfer coefficient:

$$\delta_z = \frac{k}{h_k} \tag{229}$$

If interest is directed only at the temperature distribution along the slab thickness from a moment when at distance R the temperature is t_1, then the ratio of the slab's (cylinder, sphere) cross-section to the thickness of an equivalent layer δ_z is important and is expressed by the dimensionless Biot number.

$$Bi = \frac{\bar{h}R}{k} \tag{230}$$

Accounting for convection between the slab and a heating or cooling flow, then instead of Δ being a function of two variables, it is:

$$\Delta = f(Fo, Bi, n) \tag{231}$$

Derivation of this function for an infinite slab, infinite cylinder, and semi-limited body are given by Gurney [17]. Figure 24 shows the function Δ vs, Fo, n, and the inverse of the Biot number $m = 1/Bi$.

Each family of curves is determined by the boundary conditions, i.e., whether or not a convective resistance (represented by a heat transfer coefficient h) exists at the surface. Below these curves are located a family of curves at $m = 0$, which relates to ideal conditions of convection. At $m = \infty$ $Bi = 0$, i.e., at infinitely slow convection, the temperature of the slab will always be equal to t_1 or $\Delta = 1$ at any value n and Fo (space and time). Each of the curves in the family is related to a definite location in the slab (to a definite n-value).

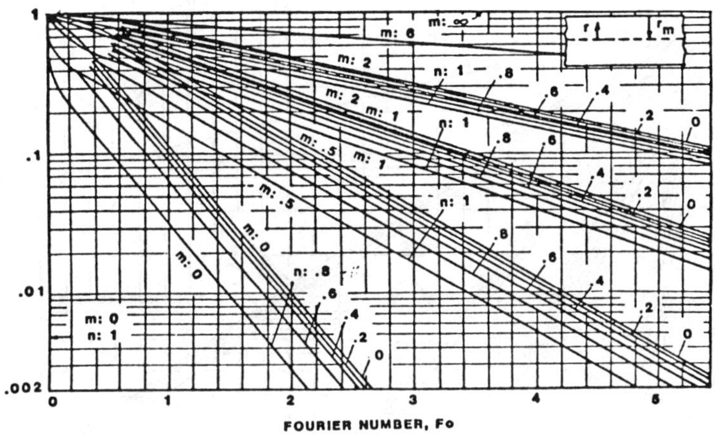

Figure 24. Dimensionless groups for a solid unlimited slab [19].

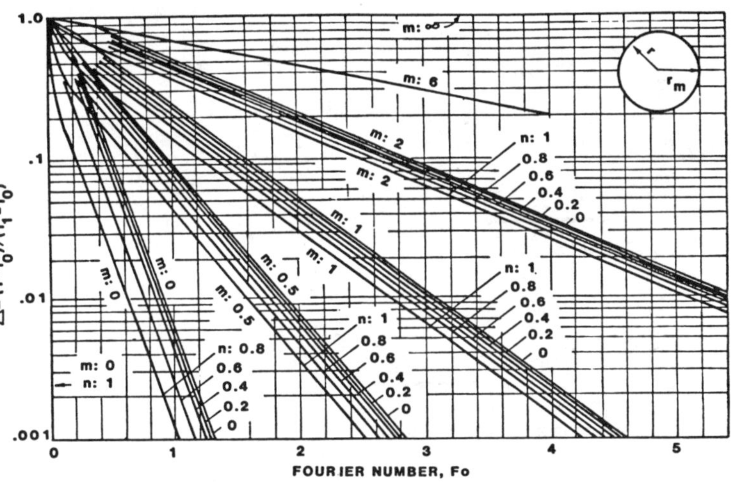

Figure 25. Function Δ vs. Fo (Fourier number) and inverse Biot number for an infinite cylinder.

At m = 0 (ideal convection), the temperature of the slab surface is equal to t_0, therefore, the value Δ at n = 0 is not given here because the corresponding line will coincide with the vertical axis of the system (at Fo = 0).

The case of an infinite cylinder is shown in Figure 25. Note that each curve corresponds to a definite value m or Bi, i.e., to a definite regime of convection. Using these charts one can evaluate the temperature in the respective geometric bodies depending on the location (n), time (Fo), and convection intensity (m). The case of a sphere is shown in Figure 26. Sample calculations on the use of these curves are given in Reference 16.

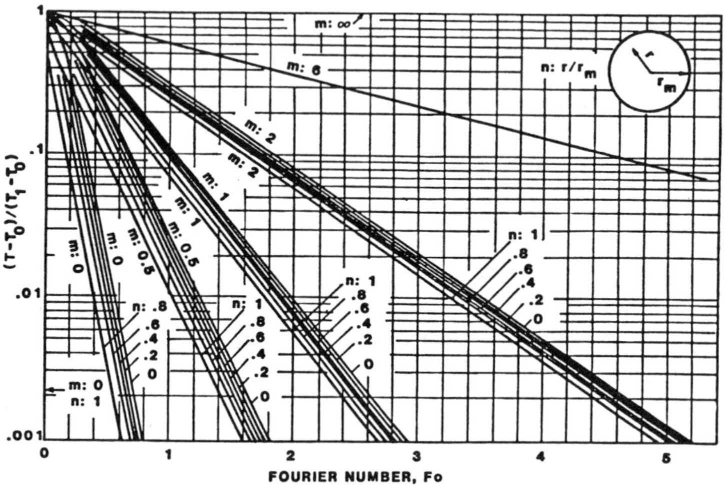

Figure 26. Dimensionless groups for a solid sphere [19].

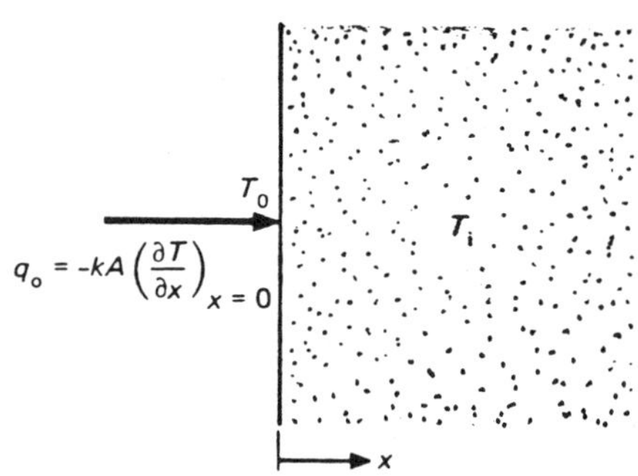

Figure 27. Transient heat flow in a semi-infinite solid.

We now direct attention to unsteady-state conduction in semi-infinite bodies. This class of problems deals with materials of relatively low thermal diffusivity, in which an external temperature change results in immediate and equal surface temperature changes.

Figure 27 defines a system where the surface is maintained at some initial temperature T_i. At some time later, the surface temperature is immediately lowered to T_0. This temperature distribution may subsequently be used to calculate heat flow at any x position in the solid as a function of time. Assuming constant properties, the temperature distribution $T(x, \tau)$ is

$$\frac{\partial^2 T}{\partial x^2} = \frac{1}{\alpha}\frac{\partial T}{\partial \tau} \tag{232}$$

The boundary and initial conditions are

$T(x, 0) = T_i$

$T(0, \tau) = T_0$ for $\tau > 0$

The problem may be solved by the Laplace transform technique, where the solution is

$$\frac{T(x, \tau) - T_0}{T_i - T_0} = \operatorname{erf} \frac{x}{2\sqrt{\alpha\tau}} \tag{233}$$

$$\operatorname{erf} \frac{x}{2\sqrt{\alpha\tau}} = \frac{2}{\sqrt{\pi}} \int_0^{x/2\sqrt{\alpha\tau}} e^{-\eta^2} d\eta \tag{234}$$

η is a dummy variable and the integral is a function of its upper limit. Introducing the error function:

$$\frac{T(x,\tau) - T_0}{T_i - T_0} = \frac{2}{\sqrt{\pi}} \int_0^{x/2\sqrt{\alpha\tau}} e^{-\eta^2} d\eta \tag{235}$$

The heat flow at any x position may be obtained from:

$$q_x = -kA \frac{\partial T}{\partial x}$$

And performing partial differentiation:

$$\begin{aligned} \frac{\partial T}{\partial x} &= (T_i - T_0) \frac{2}{\sqrt{\pi}} e^{-x^2/4\alpha\tau} \frac{\partial}{\partial x} \left(\frac{x}{2\sqrt{\alpha\tau}} \right) \\ &= \frac{T_i - T_0}{\sqrt{\pi\alpha\tau}} e^{-x^2/4\alpha\tau} \end{aligned} \tag{236}$$

At the surface the heat flow is:

$$q_0 = \frac{kA (T_0 - T_i)}{\sqrt{\pi\alpha\tau}} \tag{237}$$

The surface heat flux is determined by evaluating the temperature gradient at $x = 0$ from Equation 236. A plot of the temperature distribution for the semi-infinite solid is given in Figure 28.

If for the same uniform initial temperature distribution the surface is suddenly exposed to a constant surface heat flux q_0/A, the boundary conditions become:

$T(x, 0) = T_i$

$$\frac{q_0}{A} = -k \left. \frac{\partial T}{\partial x} \right|_{x=0} \quad \text{for } \tau > 0$$

The solution to Equation 232 for this case is:

$$T - T_i = \frac{2q_0\sqrt{\alpha\tau/\pi}}{kA} \exp\left(\frac{-x^2}{4\alpha\tau}\right) - \frac{q_0 x}{kA}\left(1 - \operatorname{erf}\frac{x}{2\sqrt{\alpha\tau}}\right) \tag{238}$$

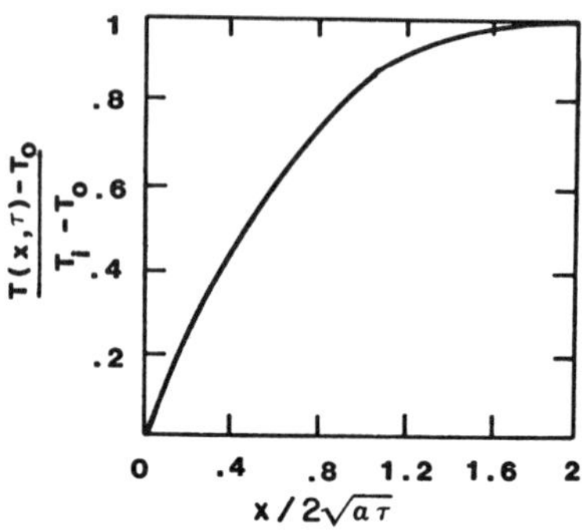

Figure 28. Shows the temperature distribution in a semi-infinite solid.

A very practical boundary condition problem is that connected with a convection boundary at the surface of the solid. The boundary conditions for the differential equation must be modified to take into account this convection heat transfer at the surface. For the semi-infinite-solid problem this is expressed by:

$$hA\,(T_h - T)_{x=0} = -\,kA\frac{\partial T}{\partial x}\bigg|_{x=0} \tag{239}$$

The solution for this problem is:

$$\frac{T - T_i}{T_\infty - T_i} = 1 - \mathrm{erf}X - \left[\exp\left(\frac{hx}{k} + \frac{h^2\alpha\tau}{k^2}\right)\right]\left[1 - \mathrm{erf}\left(X + \frac{h\sqrt{\alpha\tau}}{k}\right)\right] \tag{240}$$

where $X = x/2\sqrt{\alpha\tau})$
T_i = initial temperature of solid
T_∞ = environment temperature

A graphical solution is shown in Figure 29.

We will close this section with a short discussion of transient conduction in two- and three-dimensional bodies.

The basis for the analyses presented thus far has been the neglection of end effect (to which the prior simplifications of one-dimensional conduction is equivalent). This simplification is not always possible, and may lead to gross inaccuracies and wrong conclusions in engineering designs.

Evaluating the heating and/or cooling times of complex shapes is of practical interest to a variety of industrial processes. For example, in paper manufacturing, drying times and heat input on rotating drum driers depend on the temperature distribution of thousands of feets of sheets discharging from the dryer press. Or in a lumber yard, the manufacturer needs to know the time required for logs to air dry before he can begin cutting and processing. Many drying times of manufactured items can be approximated by combinations of three limiting geometries; namely, an unlimited cylinder, an unlimited slab, and a semi-limited body.

The transient temperature distribution over space is very difficult to evaluate for such systems due to the different paths that heat flows in. The problems of multidimensional heat flow satisfy a more complex differential equation; however, limiting solutions can be obtained by combining one-dimensional solutions. In addition, it is not always necessary to know the temperature distribution in each section of the body; that is, it is often necessary that only the distribution in localized regions be known, e.g., along the axis or center of symmetry. The functional temperature change in complex shapes may be

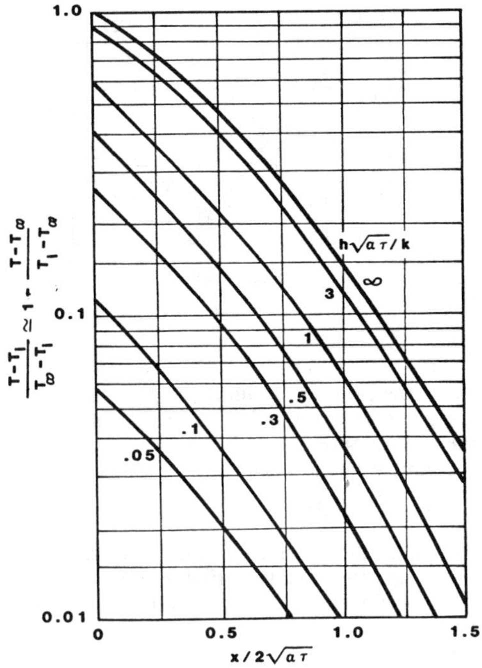

Figure 29. Temperature distribution in the semi-infinite solid with convection boundary condition.

determined by use of functions describing simple geometric shapes whose combinations approximate the system of interest. Thus, when the values of $\Delta = (T_s - T)/(T_s - T_0)$ for simple forms (e.g., symmetry plane of an unlimited slab, the axis of an unlimited cylinder, etc.) are Δ_1, Δ_2, $\Delta_3 \ldots$, the true system may by simulated as:

$$\Delta = \frac{t - t_0}{t_1 - t_0} = \Delta_1\Delta_2\Delta_3 \ldots \quad (241)$$

Figures 22, 24, through 26, and 29 provide graphical solutions to the most common geometries and boundary conditions. Depending on the location of interest in the actual body, the temperature distribution may be a function of only time or the Fourier modulus. Note that in most of the solutions presented the outer surface of the body was assumed to be at a constant temperature t_0; that is, heat transfer by convection has been ignored.

To illustrate the concepts, consider several bodies of complex shape, each of which can be approximated by simpler geometries. Let $\Delta P(a)$ denote the Δ-value for the plane of symmetry of an infinite slab of thickness a; $\Delta W(d)$ is the Δ-value for the axis of an infinite cylinder of diameter d, and $\Delta C(x)$ denotes the Δ-value for a semilimited body at a distance x from its surface. Figure 30 gives the Δ-values for these geometries. Various geometries are shown in Figure 31.

For a semiinfinite cylinder (Figure 31A) the temperature along the axis at a distance x from the outer surface is:

$$\Delta = \frac{t - t_0}{t_1 - t_0} = \Delta W(r)\,\Delta C(x) \quad (242)$$

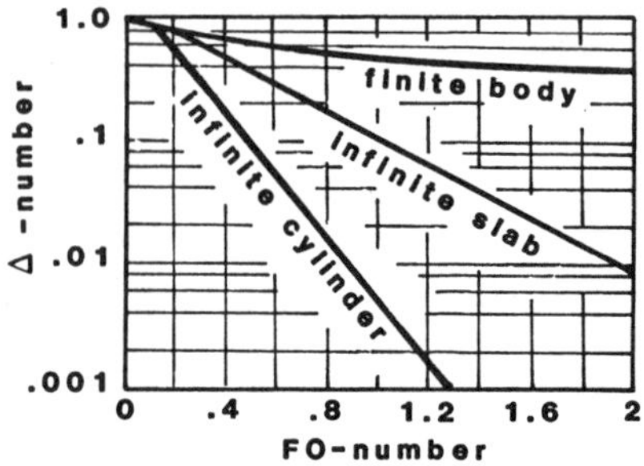

Figure 30. Δ-function for three basic geometries.

From the Fourier modulus the plot given in Figure 30 can be used to evaluate ΔC. For a cylinder of radius r, ΔW is obtained from the infinite cylinder curve in Figure 30. This allows the temperature along the axis at a distance x from the plane to be evaluated. Hence, for the limited cylinder shown in Figure 31B:

$$\Delta = \Delta W(r)\,\Delta P(a) \tag{243}$$

ΔP(a) for the specified Fourier modulus can be evaluted with respect to half the plate thickness (refer again to Figure 30).

Now consider a system where the region of interest is over one-fourth the edge of a limited body (Figure 31C). This system is formed by the intersection of two semilimited bodies:

$$\Delta = \Delta C(x_i)\,\Delta C(x_2) \tag{244}$$

where x_1 and x_2 are the distances from a given point from both planes.

For 1/8 of an infinite body (see Figure 31D):

$$\Delta = \Delta C(x_1)\Delta C(x_2)\,\Delta C(x_3) \tag{245}$$

For a semi-infinite plate limited by one-third of the perpendicular plane (Figure 31E):

$$\Delta = \Delta P(a)\,\Delta C(x) \tag{246}$$

For one fourth of an infinite plate limited by two perpendicular planes (Figure 31F):

$$\Delta = \Delta P(a)\,\Delta C(x_1)\,\Delta C(x_2) \tag{247}$$

The infinite rectangular bar (Figure 31G) is formed by the intersection of two infinite slabs. The temperature distribution along its axis of symmetry is:

$$\Delta = \Delta P(a)\,\Delta P(b) \tag{248}$$

where a and b are the sides of the bar section.

For a semilimited cylinder shown in Figure 31H:

$$\Delta = \Delta P(a)\,\Delta P(b)\,\Delta C(x) \tag{249}$$

where x is the distance from a given point along the axis of symmetry from the limiting plane.

Finally, for the center of symmetry of a parallelepiped (Figure 31I):

$$\Delta = \Delta P(a)\,\Delta P(b)\,\Delta P(c) \tag{250}$$

where a, b, c = sides of the parallelepiped.

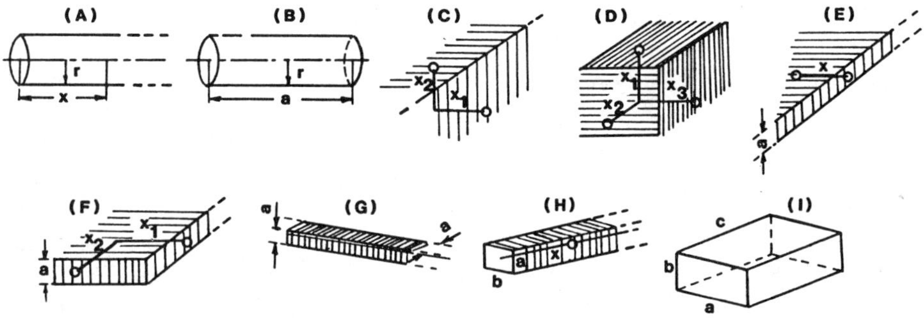

Figure 31. Typical shapes of solid bodies.

An alternate approach to evaluating Δ for bodies formed by the intersection of an infinite cylinder with a plate stems from the fact that for large values of the Fourier modulus the Δ relationships are linear (see Figure 30 for Fo > 0.2). Hence, Δ can be expressed as an exponential function which in its expanded form is:

$$\Delta = je^{-2.303\left(\frac{\alpha\tau}{R^2 t}\right)} \tag{251}$$

where R denotes the radius for a cylinder or sphere, or half the wall thickness for a plate. Physically, j is the extrapolated intersection of the lines and t is the inverse slope. Values of j and τ are: j = 1.273, t = 0.933 for an infinite slab; j = 1.602, t = 0.398 for an infinite cylinder; j = 2.000, t = 0.233 for a sphere.

Hence, for a complex-geometry approximated by the combination of simpler shapes we may write the following general solution:

$$\Delta = \Delta_1\Delta_2\Delta_3 = (j_1 j_2 j_3)\, e^{-2.303\,\alpha\tau\left[\frac{1}{R_1^2 t_1} + \frac{1}{R_2^2 t_2} + \frac{1}{R_3^2 t_3}\right]}$$

or

$$\Delta = j' e^{-2.303\tau/f} \tag{252}$$

where $j' = j_1 j_2 j_3$ and $1/f$ is defined as:

$$\frac{1}{f} = \left(\frac{1}{R_1^2 t_1} + \frac{1}{R_2^2 t_2} + \frac{1}{R_3^2 t_3}\right)\alpha \tag{253}$$

Note that our analysis assumes that the surface of the body is heated or cooled at a constant temperature t_0. This is the same as saying that the convection coefficient between the body's surface and the surrounding environment is very large. When the coefficient is small, however, the surface temperature will vary and the problem is much more complex.

We note, however, that conceptually the exponent f comprises both a conduction and a convection term (i.e., $f = f_{tc} + f_c$). Treatment of this problem is given by Rosenow et al. [20] along with additional references.

More detailed readings and problem solving can be found in References 21 through 39.

PRINCIPLES OF CONVECTIVE HEAT TRANSFER

Basic Concepts and Similarity Analysis

The convective heat transfer coefficient (also called Newton's formula) can be stated as follows:

$$dQ = \bar{h} dA (t_w - t_f) d\tau \tag{254}$$

This states that the heat transfer rate dQ per unit time $d\tau$ is directly proportional to the temperature difference between the wall t_w and fluid t_f and to the available heat transfer area dA. For a continuous process over the entire solid body surface:

$$Q = \bar{h} A (t_w - t_f) \tag{255}$$

where $\bar{h}$ = convective heat transfer coefficient where conventional units are J/m^2-s-°C or Wt/m^2-°C or kcal/m^2-hr-°C.

Coefficient $\bar{h}$ depends on the fluid velocity u, density ρ, viscosity μ, specific heat C_p, conductivity k, coefficient of thermal expansion β, system diameter d, length L, and wall roughness ε. It is essentially a measure of the amount of heat transferred from a unit area of wall surface to the surrounding fluid per unit time and over the temperature difference between the wall and fluid.

The most general differential equation of convective heat transfer which expresses the temperature distribution in a flowing fluid is the Fourier–Kirchhoff equation:

$$\frac{\partial \tau}{\partial \tau} + \frac{\partial t}{\partial x} w_x + \frac{\partial t}{\partial y} w_y + \frac{\partial t}{\partial z} w_z = \alpha \nabla^2 t \tag{256}$$

where $\alpha = k/C\rho$, thermal diffusivity.

For steady-state processes, $\partial t/\partial t = 0$; and for solid bodies we note that $u_x = u_y = u_z$ from whence Equation 256 reduces to the differential equation of conduction.

Figure 32 shows flow past a solid boundary. The temperature distribution in the fluid during convective heat transfer results in large gradients within the boundary layer. Hence, the thermal resistance in this region is established mainly by the intensity of heat transfer.

In practical applications, the convective heat transfer coefficient $\bar{h}$ is best expressed in terms of dimensionless groups. Definitions of the principal ones are summarized below:

- Nusselt number

$$Nu = \frac{\bar{h}\ell}{k} \tag{257}$$

Equality of Nusselt numbers characterizes the similarity of heat transfer at the interface between the wall and fluids in different systems. Nu is basically a measure of the ratio of the boundary layer thickness δ and the system geometric dimension ℓ (e.g., vessel diameter).

- Fourier number

$$Fo = \frac{\alpha \tau}{\ell^2} \tag{258}$$

Equality of Fourier numbers at compatible points of thermal flows is a necessary condition of similarity in unsteady-state heat transfer processes.

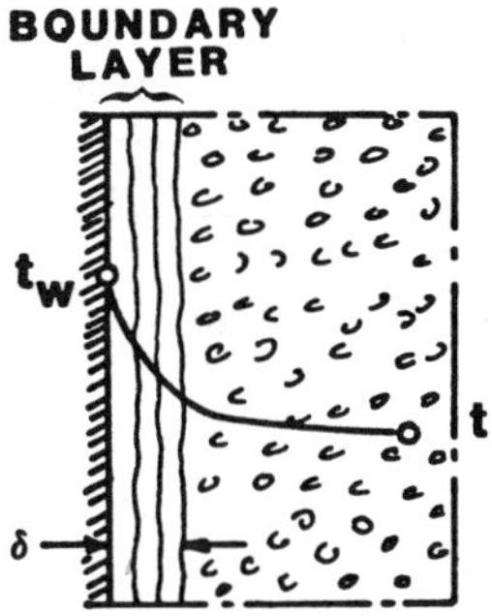

Figure 32. Shows temperature distribution for fluid flowing past a vertical surface.

- Peclet number

$$Pe = \frac{w\ell}{\alpha} \tag{259}$$

The Peclet number reflects the relative amounts of heat transferred by convection and conduction during convective heat transfer. It can be expressed as the product of the Reynolds and Prandtl numbers:

$$Pe = \frac{u\ell}{\nu} \cdot \frac{\nu}{\alpha} = \frac{u\ell\rho}{\mu} \cdot \frac{\mu C_p}{k} = RePr \tag{260}$$

- Prandtl number

$$Pr = \frac{\nu}{\alpha} = \frac{\mu C_p}{k} \tag{261}$$

The Prandtl number contains information on the physical properties of the fluid and characterizes the similarity of physical properties of heat carriers in the convective heat transfer process. For liquids, Pr varies between $3 \sim 300$; for gases Pr typically varies between $0.7 \sim 1.0$.

A summary of the important dimensionless groups is given in Table 10.

With the definitions of these dimensionless groups in hand, we shall now attempt to better define the convective heat transfer coefficient $\bar{h}$. As implied from the discussions thus far, a theoretically rigorous formula for $\bar{h}$ is not apparent. Therefore, the principles of similarity theory must be invoked to obtain working formulas. Given the complexity of the convective heat transfer coefficient, we state its general function as being one which depends on several system parameters and properties:

$$\bar{h} = f(u, \mu, \rho, c_p, k, \beta, d, L, \varepsilon) \tag{262}$$

As testified by many investigators, a generalized equation of $\bar{h}$ applicable to all situations is not possible. We can, however, normalize experimental results to obtain general expressions that apply to heat transfer situations, where $\bar{h}$ is expressed in terms of dimensionless groups important to the particular application. It is first necessary to evaluate the temperature distribution of the fluid in the vicinity of the wall. The general law of the temperature distribution in the fluid is expressed by the differential equation of convective heat transfer which is now developed.

Table 10
Important Dimensionless Groups

Symbol	Name	Group
Bi	Biot Number	$h\ell/k$
Fo	Fourier Number	$k\theta/\rho C\ell^2$
Gz	Graetz Number	wC/kL
Gr	Grashof Number	$D^3\rho^2 g\beta\Delta t/\mu^2$
Nu	Nusselt Number	hD/k
Pe	Peclet Number	DGC/k
Pr	Prandtl Number	$C\mu/k$
Re	Reynolds Number	DG/μ, $Dw\rho/\mu$
Sc	Schmidt Number	$\mu/\rho k_d$
St	Stanton Number	hC/G

Figure 33 shows an element volume of fluid in steady-steady motion. The properties of the fluid are assumed to be constant and its velocity components along the x, y, and z axes are w_x, w_y, and w_z, respectively. Temperature, t, varies along the fluid element's sides. We shall assume that all the heat introduced to the element is used in raising the systems enthalpy; then the heat transfer comprises conduction and convection mechanisms. Examining the element's face dydx, the amounts of heat entering (Q_x) and exiting on the opposite side (Q_{x+dx}) by convection are:

$$Q_x = \rho w_x dydzC_p td\tau$$

$$Q_{x+dx} = Q + dQ_x = \rho w_x dydzC_p td\tau +$$

$$C_p\left[\frac{\partial(\rho w_x t)}{\partial x}dx\right]dydzd\tau = \tag{263}$$

$$\rho w_x dydzC_p td\tau + C_p\left[t\frac{\partial(\rho w_x)}{\partial x} + \rho w_x\frac{\partial t}{\partial x}\right]dxdydzd\tau$$

Hence, the net rate of heat accumulation in the fluid volume element in the x direction is:

$$dQ_x = Q_x - Q_{x+dx} = -C_p\left[t\frac{\partial(\rho w_x)}{\partial x} + \rho w_x\frac{\partial t}{\partial x}\right]dxdydzd\tau \tag{264}$$

Similar expressions can be written for the y- and z-directions, and by summing the components we obtain the total amount of heat by convection for the element

$$dQ_{conv} = dQ_x + dQ_y + Q_z$$

$$= -C_p\left\{t\left[\frac{\partial(\rho w_x)}{\partial x} + \frac{\partial(\rho w_y)}{\partial y} + \frac{\partial(\rho w_z)}{\partial z}\right] + \rho w_x\frac{\partial t}{\partial x} + \rho w_y\frac{\partial t}{\partial y} + \rho w_x\frac{\partial t}{\partial z}\right\}dxdydzd\tau \tag{265}$$

From continuity and for an incompressible fluid (constant ρ), the term in the brackets on the LHS of this expression is zero and we note that $dV = dxdydz$. The convective contribution of heat flow thus appears as:

$$dQ_{conv} = -\rho C_p\left(\frac{\partial t}{\partial x}w_x + \frac{\partial t}{\partial y}w_y + \frac{\partial t}{\partial z}w_z\right)dVd\tau \tag{266}$$

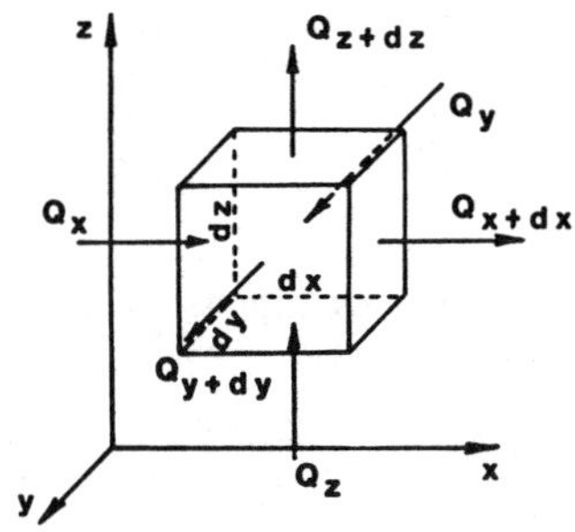

Figure 33. Elemental volume of fluid at steady-state flow.

The conduction term for the volume over time $d\tau$ is:

$$dQ_{cond} = k\left(\frac{\partial^2 t}{\partial x^2} + \frac{\partial^2 t}{\partial y^2} + \frac{\partial^2 t}{\partial z^2}\right)dVd\tau \tag{267}$$

Hence, the total amount of heat is:

$$dQ_{conv} + dQ_{cond} = -\rho C_p\left(\frac{\partial t}{\partial x}w_x + \frac{\partial t}{\partial y}w_y + \frac{\partial t}{\partial z}w_z\right)dVd\tau + k\left(\frac{\partial^2 t}{\partial x^2} + \frac{\partial^2 t}{\partial y^2} + \frac{\partial^2 t}{\partial z^2}\right)dVd\tau \tag{268}$$

Since it is assumed that the heat is used to raise the fluid volume's enthalpy, then $dQ = C_p\rho dV(dt/d\tau)\,d\tau$, which upon equating to Equation 268 and applying some simple transformation gives:

$$\frac{\partial t}{\partial \tau} + \frac{\partial t}{\partial x}w_x + \frac{\partial t}{\partial y}w_y + \frac{\partial t}{\partial z}w_z = \alpha\left(\frac{\partial^2 t}{\partial x^2} + \frac{\partial^2 t}{\partial y^2} + \frac{\partial^2 t}{\partial z^2}\right) \tag{269a}$$

$$\frac{\partial t}{\partial \tau} + \frac{\partial t}{\partial x}w_x + \frac{\partial t}{\partial y}w_y + \frac{\partial t}{\partial z}w_z = \alpha\nabla^2 t \tag{269b}$$

This last expression is the Fourier–Kirchoff equation (written earlier as Equation 256) and is the general temperature distribution law for a fluid in motion.

We will now transform this equation to dimensionless form by use of similarity boundary conditions. Heat transfer in the wall region takes place by conduction through the thermal sublayer, normal to the flow direction, and can be described by Fourier's law:

$$dQ = -k\frac{\partial t}{\partial \delta}dAd\tau \tag{270}$$

Similarly, the heat flow from the wall to the bulk flow is described by Newton's law:

$$dQ = \bar{h}(t_w - t_f)\,dAd\tau \tag{271}$$

For steady-state conditions, these two heat flows are equivalent and hence, conditions at the boundary may be written as:

$$- k\frac{\partial t}{\partial \delta} = \bar{h}(t_w - t_f) = \bar{h}\Delta t \qquad (272)$$

Dividing this expression through by $k\partial t/\partial \delta$ gives the following complex dimensionless group:

$$\frac{\bar{h}(t_w - t_f)}{k\partial t} = \frac{\bar{h}\Delta t \partial \delta}{k\partial t}$$

The expression immediately simplifies to the simplex dimensionless group, the Nusselt number (where δ is replaced by ℓ to denote a nonoriented linear size:

$$\mathrm{Nu} = \frac{\bar{h}\ell}{k}$$

The physical significance of this group was explained earlier. From an engineering viewpoint, problems in convective heat transfer are aimed at evaluating h, which is obtained from determining the Nusselt number.

We now direct attention to the similarity conditions for the bulk flow by invoking a transformation of Equation 269b. The LHS of the Fourier–Kirchhoff equation reflects the influence of the hydrodynamics on heat transfer. Intuitively, we can state this influence as:

$$\left(\frac{\partial t}{\partial x}w_x + \frac{\partial t}{\partial y}w_y + \frac{\partial t}{\partial z}w_z\right) \sim \frac{t}{2}w$$

where ℓ is a characteristic system size.

The RHS of Equation 269b characterizes heat transfer by conduction and can be represented by:

$$\alpha\left(\frac{\partial^2 t}{\partial x^2} + \frac{\partial^2 t}{\partial y^2} + \frac{\partial^2 t}{\partial z^2}\right) \sim \frac{\alpha t}{\ell^2}$$

And for the transient term:

$$\frac{\partial t}{\partial \tau} \sim t/\tau$$

Dividing t/τ by $\alpha t/\ell^2$ we obtain the dimensionless complex $\ell^2/\alpha\tau$ which is recognized as the inverse of the Fourier number:

$$\frac{\alpha\tau}{\ell^2} = \mathrm{Fo}$$

The Fourier number is analogous to the homochronity number Ho derived in the first section of this chapter. Finally, dividing the convective term $(t/\ell)w$ by $\alpha t/\ell^2$ results in the Peclet number

$$\frac{w\ell}{\alpha} = \mathrm{Pe}$$

Recall that the criteria for hydrodynamic similarity between systems is fulfilled by equality of the homochronity, Reynolds, and Froude numbers at compatible points, along with geometric similarity, by maintaining constant ratios of basic geometric tolerances. It follows then that the similarity of heat transfer processes is only possible when dynamic and geometric similarities are maintained. Hence, the convective heat transfer process may be written as a general function where the dependent variable of interest, $\bar{h}$, is expressed in terms of the Nusselt number:

$$Nu = f'\left(Fo, Pe, Ho, Re, Fr, \frac{L_1}{L_0}, \frac{L_2}{L_0} \ldots \frac{L_n}{L_0}\right) \tag{273}$$

Note also that the Peclet number is the product of the Reynolds and Prandtl numbers, where Pr reflects physical properties of the fluid and characterizes the similarity of physical properties in the convective heat transfer process.

At steady state Fo and Ho vanish from the function. Furthermore, during forced motion where gravity forces are small, the Froude number vanishes, and the function reduces to:

$$Nu = f''\left(Re, Pr, \frac{L_1}{L_0}, \frac{L_2}{L_0} \ldots \frac{L_n}{L_0}\right) \tag{274}$$

The function can be written in exponential form with the exponents evaluated from correlating experimental data. Note that it is also possible to rederive the previous expression in terms of the Stanton number, where the general working formulas have the forms:

$$Nu = C_1 (Re)^{n1} (Pr)^{n2} \left(\frac{d_w}{L}\right)^{n3}$$

$$St = C_2 (Re)^{n4} (Pr)^{n5} \left(\frac{d_w}{L}\right)^{n6} \tag{275}$$

where C_1, C_2 are constants and $n_1 \ldots n_6$ are exponents obtained from a regression of the data. Both equations are equivalent; each having certain advantages.

In terms of the definitions of Pr and Re the equations for tube flow become:

$$Nu_m = \frac{d_w^2}{4L} \frac{GC_p}{k} \frac{T_{b2} - T_{b1}}{(T_w - T_b)_{lm}} \tag{276}$$

$$St_m = \frac{d_w}{4L} \frac{T_{b2} - T_{b1}}{(T_w - T_b)_{lm}} \tag{277}$$

where T_{b1}, T_{b2} = inlet and outlet bulk temperatures, respectively.

For fully-developed turbulent flow, the Dittus–Boelter equation [40] is applicable for moderate temperature differences:

$$\left(\frac{h_m d_w}{k}\right)_0 = 0.023 \left(\frac{d_w G}{\mu}\right)_b^{0.8} \left(\frac{C_p \mu}{k}\right)_b^n \tag{278}$$

Equation 278 is based on fluid properties evaluated at the arithmetic mean bulk temperature; where exponent n = 0.4 for heating; n = 0.3 for cooling. The correlation is applicable over the following ranges of dimensionless groups:

$Re > 10{,}000;\ 0.7 < Pr < 100;\ L/d_w > 60$

For turbulent flow heat transfer in annuli the following correlation [41] applies:

$$\left(\frac{h_m d_e}{k}\right)_b = 0.023\left(\frac{d_e G}{\mu}\right)_b^{0.8}\left(\frac{C_p\mu}{k}\right)_b^{0.4}\left(\frac{d_2}{d_1}\right)^{0.45} \tag{279}$$

The correlation applies for $Re > 10^4$ and $d_e = (d_2 - d_1)$. Fluid properties are based on bulk temperatures.

For flow in noncircular smooth ducts (e.g., air flows in square, rectangular, and triangular ducts) the following correlation is useful [42]:

$$\left(\frac{h_m d_e}{k}\right) = 0.023\left[\frac{d_e G}{\mu}\right]^{0.08}\left[\frac{C_p\mu}{k}\right]^{0.4} \tag{280}$$

For gas flow in coiled tubes the following is applicable (for small tubes up to 1-1/4 in. steel pipe) for turbulent Reynolds numbers up to 150,000:

$$\frac{hD}{k} = \left(0.039 + \frac{0.138D}{D_c}\right)\left(\frac{DC_pG}{k}\right)^{0.76} \tag{281}$$

where D/D_c = ratio of the diameter of the tube to that of the helix.

For high velocity gas flows (Mach numbers in the range of 0.2 ~ 1.0) through a long straight tube, h can be computed from the following dimensionless correlation:

$$\left(\frac{h}{C_pG}\right)\left(\frac{C_p\mu}{k}\right)^{2/3} = \frac{0.027}{(DG/\mu_m)^{0.23}} \tag{282}$$

where μ_m = mean fluid viscosity.

For low viscosity liquids inside tubes ($Re > 2{,}100$) the following dimensionless correlation can be used [43]:

$$\frac{h}{C_pG}\left(\frac{C_p\mu_f}{k}\right)^{2/3} = 0.023\left(\frac{DG}{\mu_f}\right)^{-0.2}\frac{f}{2} \tag{283}$$

For high viscosity liquids in streamline flow (e.g., viscous oils in laminar flow; $Re < 2{,}100$), the Sieder–Tate equation can be used:

$$\frac{h_{am}D}{k} = 2.0\left(\frac{WC_p}{kL}\right)^{1/3}\left(\frac{\mu}{\mu_s}\right)^{0.14} \tag{284}$$

where W is the mass flow rate per tube of heated length and h_{am} is based on the arithmetic mean of the terminal temperature differences.

Equation 284 can also be applied for $Re > 2{,}100$, but calculations must be based on properties evaluated from the logarithmic mean temperature.

For cases where the Grashof number ($Gr = D^3\rho^2 gL\beta\Delta t/\mu^2$, evaluated at the stream bulk temperature) is large and Re is small, the right-hand side of Equation 284 should be multiplied by the following dimensionless group for horizontal tube flow:

$$C' - \frac{2.3(1 + 0.01\,Gr^{1/3})}{\log_{10}Re} \tag{285}$$

For liquid flow in vertical pipes:

$$\frac{h_m}{[k_f^2\rho_f^2C_pg\beta\Delta t/\mu_f]^{1/3}} = 0.13 \tag{286}$$

Equation 286 is based on water flow data;

where k = thermal conductivity
C_p = specific heat at constant pressure
ρ = density
β = coefficient of expansion
Δt = temperature difference between wall and fluid
g = gravitational constant
μ = viscosity

For air flow at right angles to the axes of single cylinders ranging in diameter from 0.001 ~ 3.75 in., and for gas temperatures up to 1800°F, the following correlations apply:

$$\frac{h_mD_0k_f}{(C_p\mu_f/k_f)^{0.3}} = 0.35 + 0.47\left(\frac{D_0G}{\mu_f}\right)^{0.52}; \ 0.1 < Re < 1{,}000 \tag{287}$$

$$\frac{h_mD_0/k_f}{(C_p\mu_f/k_f)^{0.3}} = 0.26\left(\frac{D_0G}{\mu_f}\right)^{0.6}; \ 1{,}000 < Re < 50{,}000 \tag{288}$$

For gases at moderate temperature at 1,000 < Re < 50,000:

$$h = \frac{0.3C_pG^{0.6}}{(D_0')^{0.4}} \tag{289}$$

For gases flowing at right angles to a bank of staggered tubes (3,000 < Re < 40,000):

$$\left(\frac{h_m}{C_pG_{max}}\right)\left(\frac{C_p\mu_f}{k}\right)^{2/3} = 0.33\left(\frac{D_0G_{max}}{\mu_f}\right)^{-0.4} \tag{290}$$

For free or natural convection of heat transfer between a surface and fluid, without a change in state, the following dimensionless correlation in terms of the Grashof number (which can be related to Re) applies:

$$\frac{h_cD}{k_f} = \frac{1}{a}\left(\frac{D^3\rho_f^2g\beta\Delta t}{\mu_f}\right)^n\left(\frac{C_p\mu_f}{k_f}\right)^m \tag{291}$$

For pipes having large length-to-diameter ratios, use the following values:

a = 2.12, n = 0.27, m = 0.25

Finally, for laminar flow heat transfer in annuli, the correlation of Chen et al. [44] can be used. The correlation is based on heat transfer data from the inner wall in four annuli with diameter ratios d_2/d_1 in the range of 1.09 ~ 2.

$$Nu_{am} = 1.02\,(Re)^{0.45}(Pr)^{0.5}\left(\frac{\mu_b}{\mu_1}\right)_{am}^{0.14}\left(\frac{d_e}{L}\right)^{0.4}\left(\frac{d_2}{d_1}\right)^{0.8}(Gr)^{0.05} \tag{292}$$

Equation 292 is applicable for $200 < Re < 2,000$, where Re is based on the equivalent diameter d_e. Note that μ_b = viscosity at the arithmetic mean bulk temperature; μ_1 = viscosity at temperature T_1 of the inside wall of the annulus. Other physical properties are based on the fluid's bulk temperature.

Analogies Between Heat Transfer and Momentum Transfer

In later chapters various authors apply analogical treatments to describing/modeling transport processes. This is a useful approach in that it forms a theoretical basis for correlating both heat and mass transfer data. A brief review of the analogies between heat transfer and momentum transfer is therefore warranted.

Consider the limiting case of isothermal flow of a fluid in turbulent motion inside a long, straight tube (where the system is well-developed incompressible flow). A radial investigation of the local velocity reveals three distinct zones, namely, the laminar sublayer, the buffer layer, and the turbulent core. The laminar sublayer heat transfer has already been described as molecular transport (conduction). In the buffer layer and turbulent core the warmer portions of the fluid are mechanically mixed with colder portions by eddy diffusion (convection).

Analogous to the transfer of momentum described in the first section Equation 67, we may define an eddy diffusivity for heat transfer, ε_H, where the rate of heat flux is:

$$q = -(k + \rho C_p \varepsilon_H)\frac{dt}{dy} \tag{293}$$

For fully-developed flow, the local shear stress is approximately linear with respect to the radius, that is:

$$\tau'/\tau'_w = r/r_w$$

As an approximation it may be assumed that a similar relation applies to the heat flux:

$$\frac{q}{q_w} = \frac{r}{r_w}$$

where subscript w refers to conditions at the wall. From these observations the momentum transport and heat flux expressions can be written as:

$$\frac{\tau w'}{\rho}(1 - y/r_w) = \left(\frac{\mu}{\rho} + \varepsilon_M\right)\frac{dw}{dy} \tag{294}$$

$$\frac{q_w}{\rho C_p}(1 - y/r_w) = -\left(\frac{k}{\rho C_p} + \varepsilon_H\right)\frac{dt}{dy} \tag{295}$$

As noted in the first section of this chapter, the term in the brackets on the RHS of Equation 294 is the sum of the molecular and eddy diffusivities of momentum. Analogously, in the RHS of Equation 295, the term in parentheses is the sum of the molecular and eddy diffusivities of heat.

We can now state three well-known analogies; namely that of Reynolds, von Karman, and Martinelli. In the Reynolds analogy, it is assumed that the sum of the thermal diffusivities equals the sum of the momentum diffusivities. By integrating the ratio of Equations 294 and 295 over the limits from $w = 0$ to $w = \bar{w}$ and from $t = T_w$ to $t = T_b$, we obtain:

$$q_w \bar{w}/C_p \tau'_w = T_w - T_b$$

Note that average properties are denoted by the overbar, and bulk properties are signified by subscript b. Since by definition

$h \equiv q/(T_w - T_b)$,

$\tau_w' = f\rho\bar{w}^2/2$

the following relation is obtained:

$$\frac{h}{C_p\bar{w}\rho} = \frac{f}{2} \tag{296}$$

Equation 296 is known as Reynolds analogy and is found to be a good approximation for common gases (i.e., where $Pr \simeq 1$) at moderate temperature differences. This of course is a very limited analogy, and does not incorporate sufficient detail of the velocity distribution.

A more representative analogy is that of von Karman [45, 46], who considered the resistance to heat transfer as being composed of separate contributions from the laminar sublayer, the buffer zone, and the turbulent core. Note that in the region of a flow field where the molecular viscosity can be neglected compared with the eddy-viscosity, we can write:

$$\tau' = \rho\ell^2\left(\frac{dw}{dy}\right)^2 \tag{297}$$

where ℓ is the Prandtl mixing length and is proportional to the distance, y, from the wall, i.e.

$$\ell = k^*y \tag{298}$$

von Karman proposed another expression for ℓ which is:

$$\ell = k^*\frac{(dw/dy)}{(d^2w/dy^2)} \tag{299}$$

This expression is equivalent to any eddy viscosity of the form:

$$\varepsilon = k^*\frac{(dw/dy)^3}{(d^2w/dy^2)^2} \tag{300}$$

By integration of Equation 67, these mixing length models lead to a generalized velocity distribution of the form:

$$w^+ = \frac{w}{w^*} = \frac{1}{k^*}\ln y^+ + c \tag{301}$$

where

$w^* = \sqrt{\frac{\tau w'}{\rho}}$ the friction velocity

$y^+ = yw^*\mu/\rho$ dimensionless distance

Fitting velocity data for the core of the tube ($y^+ > 30$) to Equation 301, by assuming laminar flow in the sublayer region, and by matching the results empirically in the buffer layer where the other expression are valid, the following universal velocity profile is obtained:

$$w^+ = y^+, \; y^+ < 5 \tag{302a}$$

$$w^+ = -3.05 + 5 \ln y^+, \; 5 < y^+ < 30 \tag{302b}$$

$$w^+ = 5.5 + 2.5 \ln y^+ \tag{302c}$$

Several similar expressions have been developed for the eddy viscosity such as that of Deissler [47]:

$$\varepsilon_M = n^2 wy (1 - e^{-n^2 wy\rho/\mu}); \; 0 < y^+ < 30 \tag{303}$$

where $n = 0.10$.

The von Karman analogy uses equations for the laminar sublayer ($y^+ < 5$, where the molecular thermal diffusivity and kinematic viscosity are considered but the eddy terms are ignored). This results in an expression for the temperature drop over this layer. For the buffer zone ($5 < y^+ < 30$), both the molecular and eddy transport terms are included, and Equation 302b is used to define the eddy viscosity which is then assumed equal to the eddy thermal diffusivity in the heat flux expression. By integrating this equation which includes both eddy and molecular thermal diffusivities, an expression for the temperature drop over the buffer region is obtained. The final step involves application of the Reynolds analogy to the turbulent core ($y^+ > 30$) which produces an equation for the temperature change associated with this resistance. The sum of the three temperature-drop expressions results in an equation describing the overall temperature variation. Dividing the heat flux q by the overall temperature drop gives the following relation for the heat transfer coefficient:

$$h = \frac{f\bar{w}\rho C_p/2}{1 + 5\sqrt{f/2}\,\{Pr - 1 + \ln[(1 + 5Pr)/6]\}} \tag{304}$$

Note that for a Prandtl number of one, this reduces to the Reynolds analogy (Equation 296).

The last analogy we shall consider is that of Martinelli [48] for the case of uniform heat flux. The principal difference between the Martinelli and von Karman analogies lies in the treatment of the turbulent core. In this model, the derivative of the velocity field (Equation 302c) is combined with Equation 294 to give:

$$\frac{\mu}{\rho} + \varepsilon_M = \frac{w^+}{2.5}(1 - y/r_w)\,y \tag{305}$$

This is rearranged to:

$$1 + \frac{\varepsilon_M}{\mu/\rho} = \sqrt{\frac{f}{2}}\,\frac{Re}{5}\left(1 - \frac{y}{r_w}\right)\frac{y}{r_w} \tag{306}$$

Combining this with the heat flux expression (Equation 295) results in the Martinelli equation:

$$h = \frac{\bar{w}\rho C_p\sqrt{f/2}}{5\theta\left[Pr + \ln(1 + 5Pr) + 0.5\xi' \ln\frac{Re}{60}\sqrt{\frac{f}{2}}\right]} \tag{307}$$

where $\xi' = \dfrac{\varepsilon_H}{\varepsilon_H + \dfrac{k}{\rho C_p}}$, $\theta = \dfrac{T_w - T_b}{T_w - T_c}$ and T_c is the temperature at the center of the tube. Tables 11 and 12 give values of ξ' and θ as functions of the Reynolds, Peclet, and Prandtl numbers.

There are a variety of other analogies, most of which are proposed modifications to these three. Some modifications include terms which account for differences between bulk velocity and centerline velocity, while others attempt to correct for the variation of τ'/q with radial position. The reader can refer to References 49 through 57 for further details.

PRINCIPLES OF RADIATION

In this last section we shall review only some basic terminology important to thermal radiation. Thermal radiation exhibits the same optical properties as visible light, and hence, may be qualitatively described by wave theory.

Planck proposed the concept that radiation is emitted not in a continuous energy state but in discrete quantities (called quanta). The intensity of radiation from a black body is:

$$I_{b,\lambda} = \frac{2C^2h\lambda^{-5}}{\exp\left(\dfrac{Ch}{\varkappa\lambda T}\right) - 1} \qquad (308)$$

$I_{b,\lambda}$ = radiation of intensity from a black body between wavelengths λ and $\lambda + d\lambda$
C = speed of light
h = Planck's constant
$\varkappa$ = Boltzmann constant
T = absolute temperature

Table 11
Values of ω' as a Function of the Reynolds and Peclet Numbers

Peclet Number	Reynolds Number 10^4	10^5	10^6
10^2	0.18	0.098	0.052
10^3	0.55	0.45	0.29
10^4	0.92	0.83	0.65
10^5	0.99	0.985	0.980
10^6	1.00	1.00	1.00

Table 12
Values of θ as a Function of Prandtl and Reynolds Numbers

Prandtl Number	Reynolds Number 10^4	10^5	10^6	10^7
0	0.564	0.558	0.553	0.550
10^{-4}	0.568	0.560	0.565	0.617
10^{-3}	0.570	0.572	0.627	0.728
10^{-2}	0.589	0.639	0.738	0.813
10^{-1}	0.692	0.761	0.823	0.864
1.0	0.865	0.877	0.897	0.912
10	0.958	0.962	0.963	0.966
10^2	0.992	0.993	0.993	0.994
10^3	1.00	1.00	1.00	1.00

The total emissive power between wavelengths λ and $\lambda + d\lambda$ is:

$$E_{b,\lambda} = \frac{2\pi C^2 h \lambda^{-5}}{\exp\left(\frac{Ch}{\varkappa \lambda T}\right) - 1} \tag{309}$$

Planck's law of radiation can be integrated over wavelength from 0 to ∞ to determine the total emissive power:

$$E_b = \int_0^\infty E_{b,\lambda} d\lambda = \frac{2\pi^5 \varkappa^4 T^4}{15 C^2 h^3} = \sigma T^4 \tag{310}$$

Equation 310 is known as the Stefan–Boltzmann law and σ = Stefan–Boltzmann constant (0.1714×10^{-8} Btu/hr-ft^2-°R^4).

Emissivity and absorptivity are the controlling properties in heat transmission via radiation. For blackbody radiation Equation 310 applies. For actual surfaces we note that:

$$E = \varepsilon E_b \tag{311}$$

where ε = emissivity of a surface

Emissivity varies with wavelengths as well as the direction of emission.

Several generalizations concerning the emissivity of surfaces are:

- Highly polished metallic surfaces usually have very low emissivities.
- ε depends on surface conditions.
- ε increases with temperature for all metallic surfaces.
- The formation of oxide layers and surface roughness increase ε.
- ε of nonmetallic surfaces are much higher than for metallic surfaces and show a decrease as temperature increases.

Absorptivity α of surfaces depends on the quality of the incident radiation. For most materials, in the temperature range of ambient to 2000°F, a good approximation is $\alpha = \varepsilon$.

If the incident radiation is derived from a high-temperature source (e.g., solar radiation ~ 10,000°F), emissivity and absorptivity of ordinary surfaces differ greatly. White metal oxides typically have an ε-value (and α) of about 0.95 at ordinary temperatures, but α drops sharply to 0.15 if the oxides are exposed to solar radiation. Freshly polished metallic surfaces have an ε-value of about 0.05.

A gray surface refers to a surface where the total average absorptivity is independent of the spectral-energy distribution of the incident radiation. This means that ε may be used in place of α, even though the temperature of the incident radiation and the receiver are not the same. Examples of good approximations of a gray surface are slate, tar board, and dark linoleum. Table 13 lists emissivities of different meterials at various temperatures.

For radiation between the surfaces separated by a non-absorbing medium, the net loss of energy from a body at room temperature T_1 in black surroundings T_2 is:

$$q_{1,net} = 0.173\, A_1 \left[\varepsilon_1 \left(\frac{T_1}{100}\right)^4 - \alpha_{1,2} \left(\frac{T_2}{100}\right)^4 \right] \tag{312}$$

where $q_{1,net}$ is in Btu/hr.

For $\alpha_{1,2} = \varepsilon_1$:

$$q_{1,net} = 0.173\, A_1 \varepsilon_1 \left[\left(\frac{T_1}{100}\right)^4 - \left(\frac{T_2}{100}\right)^4 \right] \tag{313}$$

Table 13
Normal Total Emissivities of Different Surfaces

Surface	t(°F)	Emissivity
Metals		
Aluminium (highly polished, 98.3% pure)	440–1070	0.039~0.057
Brass (highly polished)		
73.2% Cu, 26.7% Zn	476–674	0.028~0.031
82.9% Cu, 17.0% Zn	530	0.030
Copper		
Polished	242	0.023
Plate heated a 1110°F	390–1110	0.57
Molten-state	1970–2330	0.16~0.13
Gold	440~1160	0.018~0.035
Iron and steel		
Polished, electrolytic iron	350~440	0.052~0.064
Polished iron	800~1800	0.144~0.377
Sheet iron	1650~1900	0.55~0.60
Cast iron	1620~1810	0.60~0.70
Lead (unoxidized)	260~440	0.057~0.075
Mercury	32~212	0.09~0.12
Nickel (technically pure, polished)	440~710	0.07~0.087
Platinum (pure)	440~1160	0.054~0.104
Silver (pure)	440~1160	0.0198~0.0324
Refractories and Miscellaneous Materials		
Asbestos	74~700	0.93–0.96
Brick, red	70	0.93
Carbon		
Filament	1900~2560	0.526
Candle Soot	206~520	0.952
Lampblack	100~700	0.945
Glass	72	0.937
Gypsum	70	0.903
Plaster	50~190	0.91
Porcelain, glazed	72	0.924
Rubber	75	0.86~0.95
Water	32~212	0.95~0.963

The intensity of thermal radiation from a black body, measured at a specified distance and direction from the body, is identical with that which would be emitted from any other black body at the same temperature. The elements of the body's perimeter as viewed from the measuring point must be identical in direction to the corresponding elements of the original black body.

The general equation describing direct radiant-heat exchange between two black surfaces is:

$$q_{net} = (W_{B1} - W_{B2})AF' = 0.173\left[\left(\frac{T_1}{100}\right)^4 - \left(\frac{T_2}{100}\right)^4\right]AF' \tag{314}$$

where A = area of one of the surfaces
F′ = geometrical factor dependent on the shape and relative orientation of the two surfaces
W = dq/dA

Values of F′ as reported by Hottel [58] for various surface arrangements are given Figure 34. F′-values are based on the assumptions that ε_θ is constant, independent of θ (this is exact for black bodies and a good approximation for most metallic or oxidized or rough metal surfaces). Figure 34B gives F′ values for adjacent rectangles in perpendicular planes. For opposed parallel rectangles and disks of equal sizes use lines 1 ~ 4 of Figure 34C; for an infinite plane parallel to a system of parallel tubes use lines 1 and 3 of Figure 34D.

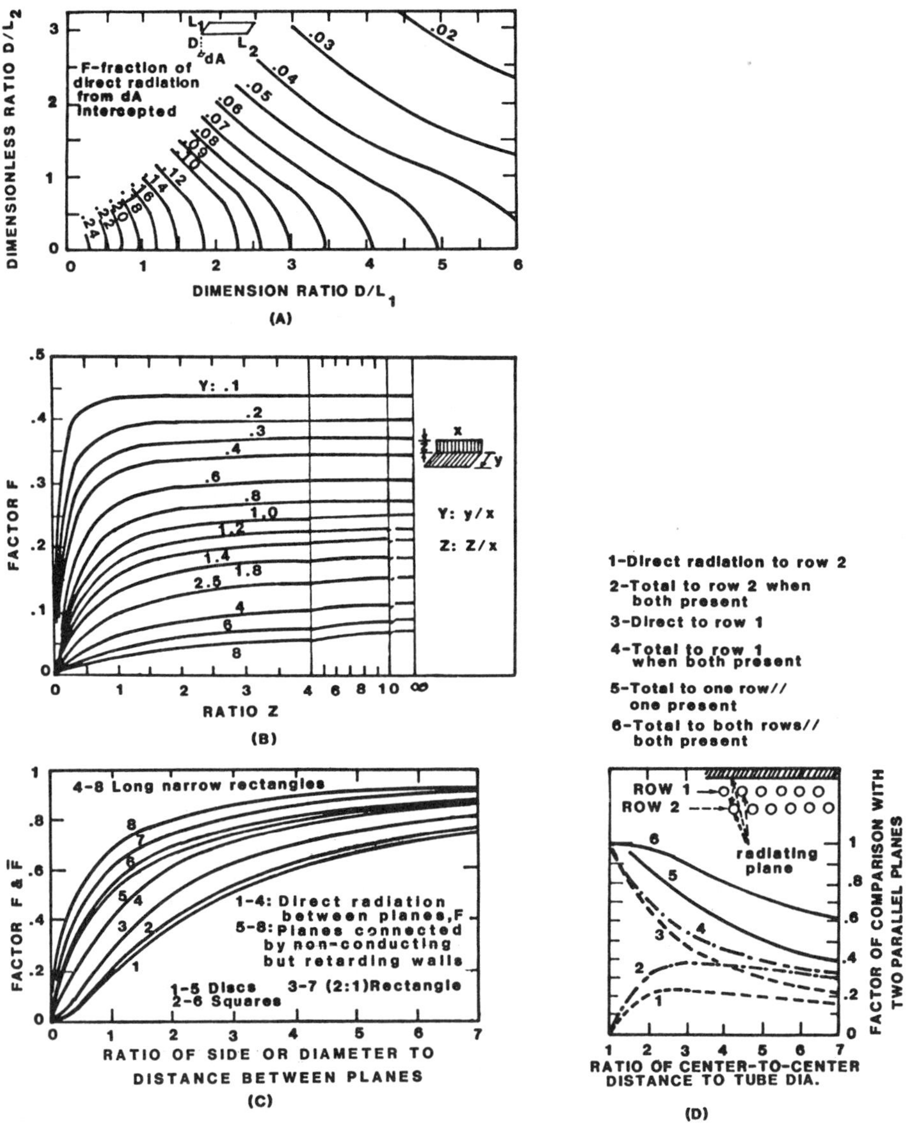

Figure 34. (A) Geometrical factor F′ for direct radiation between element dA and a parallel rectangle, in terms of the dimension ratios; (B) case of radiation between adjacent rectangles in perpendicular planes; (C) parallel planes, directly opposed; (D) radiation between a plane and rows of tubes parallel to the plane.

The net interchange of radiation between two black surfaces is:

$$q = \sigma A_1 F_{12}(T_1^4 - T_2^4) \tag{315}$$

$$q = \sigma A_2 F_{21}(T_1^4 - T_2^4) \tag{316}$$

A practical problem is that of radiation to refractory surfaces. An idealized case is an enclosure composed in part of black heat sources and sinks $A_1, A_2, \ldots$ and refractory surfaces $A_R, A_S, \ldots$ There is not net radiant heat flux in this case. Unknown refractory surface temperatures can be eliminated from heat balances to give an expression for the net flux from A_1 to A_2:

$$q_{1 \rightleftarrows 2} = A_1 \bar{F}_{12}\sigma(T_1^4 - T_2^4) \equiv A_2 F_{21}\sigma(T_1^4 - T_2^4) \tag{317}$$

Factor $\bar{F}$ is given for several surface arrangements in Figure 34C; F_{21} is given by line 5 or 6 of Figure 34D).

For gray surfaces, the net radiant interchange between A_1 and A_2 (caused by direct radiation, retardation from refractory surfaces and multiple reflections inside the enclosure)

$$q_{1 \rightleftarrows 2} = A_1 F_{12}\sigma(T_1^4 - T_2^4) \equiv A_2 F_{21}\sigma(T_1^4 - T_2^4) \tag{318}$$

For two nonrefractory surfaces A_1, A_2, with refractory zones, the F-factor can be evaluated from:

$$F_{12} = \frac{1}{\dfrac{1}{\bar{F}_{12}} + \left(\dfrac{1}{\varepsilon_1} - 1\right) + \dfrac{A_1}{A_2}\left(\dfrac{1}{\varepsilon_2} - 1\right)} \tag{319}$$

Table 14
Transmittance of Metal Foils (Monochromatic Radiation)

Wavelength λ	Silver	(Thickness in Microns)* Gold	Platinum
2.5	–	0.0272	0.0352
2.0	–	0.0238	0.0331
1.5	0.0222	0.0244	0.0309
1.0	0.0230	0.0266	0.0284
0.8	0.0236	0.0281	0.0274
0.7	0.0233	0.0311	0.0268

* Thickness refers to distance to which the intensity of radiation decreases to 1/10 of original value.

Data extracted from E. Hagen and H. Rubens, *Ann. Phys.*, 8: 432 (1902).

Table 15
Transmittance of Electric Nonconductors for Incident Black Radiation (95 °C)

Material	Thickness (mm)*
Glass	0.3
Writing paper	0.07
Tracing paper	0.09
Mica	0.25

* Thickness refers to distance to which the intensity of incident radiation decreases to 1/10 of originale value.

Data extracted from E. Eckert, *Forschung Geb. Ingenieurw.*, 9: 251 (1938).

Table 16
Geometric Mean Beam Lengths for Gas Radiation

Geometry	Characterizing Dimension	Beam Length (Fraction of Dimension)
Sphere	Diameter	2/3
Infinite cylinder	Diameter	1
Rectangular Parallelpipeds:		
1:1:1 (cube)	Edge	2/3
1:1:4, radiating to 1 × 4 face	Edge	0.90
radiating to 1 × 1 face	Shortest edge	0.86
radiating to all faces	Shortest edge	0.89
1:∞:∞ (infinite parallel planes)	Distance between planes	2
Space outside infinite bank of tubes with centers on equilateral triangles	Clearance (= tube diameters)	3.4

From Gukhman [2].

Detailed discussions of principles of radiation can be found in References 20, 58–59. Tables 14 through 16 contain useful data. Table 14 gives the transmittance of metal foils for monochromatic radiation. Table 15 gives transmittance values of electric nonconductors for incident black radiation. Table 16 lists geometric mean beam lengths for gas radiation.

NOTATION

- A — cross-sectional area; constant
- a — contraction loss coefficient in Equation 106
- B — constant in Equation 91
- b — function defined by Equation 225
- Bi — Biot number
- C — friction factor coefficient in Equation 113; speed of light in radiation discussions
- C, C_p — specific heat
- D, d — diameter
- E_k, E_p — kinetic and potential energies, respectively
- E_v — eddy viscosity
- $E_{bi\lambda}$ — emissive power
- Eu — Euler number
- e — height of protrusions on rough surface
- F′ — radiation geometric factor
- $\hat{F}_f$ — total frictional head loss
- Fo — Fourier number
- Fr — Froude number
- f — friction factor
- G — superficial mass velocity
- Gr — Grashof number
- Gz — Graetz number
- g — gravitational acceleration
- Ho — homochronity number
- h — convective heat transfer coefficient; Planck's constant in radiation discussions
- h_{fr} — frictional head loss
- h_ℓ — head loss
- I_T — turbulence intensity defined by Equation 61
- $I_{b,\lambda}$ — radiation intensity defined by Equation 308
- i — enthalpy
- j — heat transfer factor; see Equation 251
- K_e — head loss coefficient
- k — thermal conductivity; Boltzmann constant in radiation discussion
- L, ℓ — length
- L_e — equivalent length
- L_T — turbulence scale; see Equation 64
- M — mass
- m — inverse of Biot number; exponent in Equation 92
- N′ — number of tube rows
- Nu — Nusselt number
- n — number of layers
- P, p — pressure
- P′ — dimensionless length

Pe	Peclet number
Pr	Prandtl number
Q	heat flow
q	heat transfer rate
R, r	tube radius
R_T	thermal resistance
Re	Reynolds number
$R_c(y)$	correlation coefficient; see Equation 63
Sc	Schmidt number
St	Stanton number
T	absolute temperature
t	temperature
t′	tube separation
U	overall heat transfer coefficient
u	dummy variable
V	volumetric flow rate
W	mass flow rate
W′	work
w	dummy variable or velocity
w*	friction velocity
X	empirical function in Equation 112
x, y, z	coordinate axes
y*	dimensionless distance
Z	dimensionless distance
z	elevation

Greek Symbols

α	flow regime coefficient; thermal diffusivity
β	angle; coefficient of expansion
γ	specific weight
Γ	weight rate per unit plate width
$\tilde{G}$	dimensionless system size
δ_n	insulation layer thickness or boundary layer thickness
Δ	temperature function; see Equation 197
ε	wall roughness; emissivity in radiation discussions
ε_M	eddy viscosity of momentum
ε_H	eddy diffusivity of heat transfer
$\varkappa$	Boltzmann constant
λ	friction factor; wavelength in radiation discussions
λ_F	Fanning friction factor
μ	viscosity
ν	kinematic viscosity
ω	friction resistance coefficient
ρ	density
σ	Stephan–Boltzmann constant
φ	angle
τ	time
τ_T'	turbulent shear stress
$\tau_{N'}$	Newtonian shear stress
τ'	shear stress
T′	friction force
Υ	temperature distribution function defined by Equation 164
ψ	frictional correction factor; coefficient defined by Equation 107; or function (see Equation 199)

REFERENCES

1. Sedov, L. E., *Similarity and Dimensional Methods in Mechanics,* Academic Press, Inc., New York (1959).
2. Gukhman, A. A., *Introduction to the Theory of Similarity,* Academic Press, Inc., New York (1965).
3. Hodson, J. L., *Trans. ASME,* 51: 303 (1939).
4. Rouse, H., *Elementary Mechanics of Fluids,* John Wiley & Sons, Inc., New York (1948).
5. Davies, J. T., *Turbulence Phenomena,* Academic Press, Inc., New York (1972).
6. Moody, L. F., *Trans. ASME,* 66: 671 (1944).
7. Mott, R. L., *Applied Fluid Mechanics,* Charles E. Mezzill Pub. Co., Columbus, Ohio (1972).
8. Cheremisinoff, N. P., *Fluid Flow: Pumps, Pipes and Channels,* Ann Arbor Science Pub., Ann Arbor, Michigan (1981).
9. Azbel, D., and Cheremisinoff, N. P., *Fluid Mechanics and Unit Operations,* Ann Arbor Science Pub., Ann Arbor, Michigan (1983).
10. Simpson, L. L., *Chem. Eng.,* 76 (8): 167–181 (1969).
11. Kanakov, P. K., *Izv. an SSSR,* OTN, 7: 1029, USSR (1949).

12. Othmer, D. F., *Ind. Eng. Chem.*, 40 (5): 783 (1948).
13. Boucher, D. F., *Chem. Eng. Progr.*, 44 (10): 527, 601 (1947).
14. Chilton, T. H., *Trans. Am. Inst. Chem. Eng.*, 29: 161 (1938).
15. Bergelin, O. P., *Chem. Eng.*, 56 (5): 104–106 (1949).
16. Cheremisinoff, N. P., *Heat Transfer Pocket Handbook*, Gulf Publishing Co., Houston, Texas (1984).
17. McAdams, W. H., *Heat Transmission*, McGraw–Hill Book Co., New York (1942).
18. Perry, J. H. (Ed.), *Chemical Engineers Handbook*, 5th ed., McGraw–Hill Book Co., New York (1963).
19. Gurney, H. P., and Lurie, J., *Ind. Eng. Chem.*, 15: 1170 (1923).
20. Rosenow, W. M., and Hartnett, J. P. (Eds.), *Handbook of Heat Transfer*, McGraw–Hill Book Co., New York (1973).
21. Comings, E. W., *Ind. Eng. Chem.*, 39: 964 (1947).
22. Stolyarov, E. A., et al., *Zh. F. Kh.*, 24: 166, 279 (1950).
23. Gamson, B. W., *Chem. Eng. Progr.*, 45: 154 (1949).
24. Hirschfelder, J. O., Proc. Joint Conf. Thermodynamic and Transport Properties of Fluids, The Inst. Mech. Eng. (London), p. 133 (1958).
25. Buddenberg, J. W., *Ind. Eng. Chem.*, 41 (7): 1345 (1949).
26. Barratt, T., *Int. Crit. Tables*, 5: 227 (1929).
27. Jakob, M., *Heat Transmission*, McGraw–Hill, New York (1952).
28. Eckert, E. R. G., and Drake, R. M., Jr., *Analysis of Heat and Mass Transfer*, McGraw–Hill Book Co., New York (1972).
29. Holman, J. P., *Heat Transfer*, McGraw–Hill Book Co., New York (1976).
30. Chapman, A. J., *Heat Transfer*, MacMillan Co., New York (1967).
31. Jahnke, E., and Emde, F., *Tables of Functions*, 4th ed., Dover Publications, New York (1945).
32. Bennett, C. O., and Myers, J. E., *Momentum, Heat, and Mass Transfer*, McGraw–Hill Book Co., New York (1982).
33. G. G. Brown, et al., *Unit Operations*, John Wiley, New York (1950).
34. Gurney, H. P., *Ind. Eng. Chem.*, 15: 1170 (1923).
35. Schneider P. J., *Conduction Heat Transfer*, Addison–Wesley Pub. Co., Reading, MA (1955).
36. Heisler, M. P., *Trans. ASME*, 69: 227 (1947).
37. Neuman, A. B., *Trans. Am. Inst. Chem. Eng.*, 27: 203 (1931).
38. Olson, F. S., et al., *Ind. Eng. Chem.*, 34: 336, 874 (1942).
39. Mesill, D. G., *Ind. Eng. Chem.*, 40: 2263 (1948).
40. Dittus, F. W., and Boelter, L. M. K., Univ. Calif. (Berkeley), *Pubs. Eng.*, 2: 443 (1930).
41. Wiegand, J. H., *Trans. AIChE*, 41: 147 (1945).
42. Lowdermilk, W. H., Weiland, W. F., and Livingood, J. N. B., NACA RM E53J07 (1954).
43. Colburn, A. P., *Trans. AIChE*, 29: 174 (1933).
44. Chen, C. Y., Hawkins, G. A., and Solberg, H. L., *Trans. ASME*, 68: 99 (1946).
45. von Karman, T., *Eng.*, 148: 210 (1939).
46. von Karman, T., *Trans. ASME*, 705 (1939).
47. Deissler, R. E., NACA-TN-2138 (1952); see also NACA-TN-2129 (1950) and NACA-TN-3145 (1959).
48. Martinelli, R. C., *Trans. ASME*, 69: 947 (1947).
49. Prandtl, L., *Phys. Zeit*, 10: 1072 (1910).
50. Taylor, G. E., British Advisory Comm. Aeronaut., Report. Mem. 272 (31): 423 (1961).
51. Hoffmann, E., *Forsch. Gebiete Ingenieurw.*, 11: 159 (1940).
52. Boelter, L. M. K., Martinelli, R. C., and Jonassen, F., *Trans. ASME*, 63: 447 (1941).
53. Lyon, R. N., *Chem. Eng. Progress*, 47 (2): 75 (1951).
54. Deissler, R. G., *Trans. ASME*, 73: 101 (1951).
55. Friend, W. L., and Metzner, A. B., *AIChE J.*, 4 (4): 393 (1958).

CHAPTER 2

SCALING IN LAMINAR AND TURBULENT HEAT AND MASS TRANSFER

Eli Ruckenstein

Department of Chemical Engineering
State University of New York at Buffalo
Amherst, NY

CONTENTS

Introduction

Information regarding the rate of heat or mass transfer can be obtained by dimensional analysis, by similarity considerations applied to the transport equations, or by solving the transport equations for the appropriate boundary conditions. Of course, only the last mentioned procedure can provide complete analytical expressions for the heat and mass transfer rates. The other two provide a number of dimensionless groups which must be combined with experiment to give rise to useful correlations. Dimensional analysis requires physical intuition to identify the physical quantities upon which the transfer coefficients depend and uses dimensional considerations to group these quantities in a number of dimensionless groups. The number of dimensionless groups is equal to the number of physical quantities involved (including the mass transfer coefficient) less the number of dimensions. The similarity method extends the concept of similitude from geometry to the basic transport equations. By using scaling transformations for all the quantities involved, the conditions for two temperature or concentration fields to be similar are derived [1]. The results are the same dimensionless groups obtained by dimensional analysis. The most complete procedure for predicting the rate of mass transfer is by solving the transport equations for the appropriate boundary conditions. For laminar motion the transport equations can be formulated easily and difficulties may lie only in their solution. On the other hand, for turbulent motion the formulation of the basic equations for the time-averaged local quantities constitutes a major physical difficulty. Recently, new ideas concerning turbulence are being developed, in which turbulence (chaos) is considered a deterministic process predictable from the time-dependent transport equations. The latter point of view will be only qualitatively discussed in the last section of this chapter, due to the high degree of difficulty encountered in its application. For the time being procedures based on physical models will be used for obtaining some useful results concerning the turbulent mass or heat transfer.

The scope of this chapter is to emphasize some simple procedures which can be used to extract information concerning the heat and mass transfer coefficients in laminar and turbulent motions, without actually solving the transport equations, or by solving them only in some simpler cases. In the first section, a new kind of dimensional analysis is developed which, in contrast to the conventional one, enables us to identify not only various dimensionless groups, but also to provide some physical significance and some bounds to the exponent of the Reynolds number in the expression of the Sherwood number. Two length scales and two time scales are used in the "dimensional scaling" of the mass transfer coefficients: the macro-scales for the macroscopic quantities and the micro-scales for the molecular quantities. In the second section, a simple algebraic procedure is suggested to predict the rate of heat or mass transfer in complex situations. It involves replacement of the terms in the convective diffusion equation by algebraic expressions containing scaling quantities but multiplied by appropriate constants. The constants are determined from the solutions available for simpler limiting cases. To emphasize the difference between the kind of scaling involved in this algebraic procedure and that in the dimensional analysis, it is appropriate to call the former "physical scaling." The third and last section of the chapter deals with turbulent heat and mass transfer. A physical model is used to describe turbulence near a boundary, the motion being scaled by a succession of short laminar paths. The turbulent fluctuations bring elements of liquid in the proximity of the wall. These elements of liquid proceed short distances along the wall in laminar motion, after which they dissolve into the bulk, being replaced by other elements and so on. This description of turbulence involves a double scaling:

1. The scaling of turbulence by a succession of short laminar motions.
2. The scaling of the laminar motion itself.

Thus the problem of turbulence is reduced to the solution of a simple laminar motion. The path length of the laminar motion, as well as the velocity of the liquid element when it is brought close to the wall by turbulent fluctuations, is obtained on the basis of dimensional considerations.

The algebraic method is applied to a large number of problems, such as: laminar free convection over the entire range of Prandtl numbers; combined free and forced convection, for assisting flows, along a vertical plate; combined convection between two parallel vertical plates; penetration theory with chemical reaction; steady laminar flow along a plate with homogeneous chemical reaction; heat transfer in wedge flow with suction; mass transfer to a continuous phase from single bubbles at small and large Reynolds numbers; unsteady heat transfer to a liquid in steady laminar flow along a plate; and turbulent mass transfer with a first order chemical reaction treated by using either the turbulent diffusivity or the renewal models.

The multiple scaling in turbulent flow is applied to the flow of Newtonian and non-Newtonian fluids through a tube; to turbulent heat of mass transfer in a stirred vessel; to the heat transfer between a fluid flowing through a fixed or fluidized bed and the wall; to the heat transfer between a bubbling bed and the wall; to the boiling heat transfer and to separated flows. The quasi-steady laminar model previously proposed by the author is further developed to include the transient behaviour of the elements of liquid which are brought close to the wall by velocity fluctuations. The algebraic method is used to derive expressions for the rate in the short path laminar motion.

The main message of this chapter is that simple scaling procedures can provide information concerning the heat or mass transfer even in complex problems such as mixed convection or turbulent flow, without incurring the loss of understanding that invariably arises from numerical or series solutions.

DIMENSIONAL ANALYSIS BASED ON MACRO- AND MICRO-LENGTHS AND TIME SCALES

The conventional dimensional analysis, which makes use of single length and time scales, leads to the following correlations for the mass or heat transfer coefficients:

$$\mathrm{Sh} = \mathrm{A}\mathrm{Re}^{m}\mathrm{Sc}^{n} \tag{1}$$

and

$$\mathrm{Nu} = \mathrm{A}\mathrm{Re}^{m}\mathrm{Pr}^{n}, \tag{2}$$

where

$$\mathrm{Sh} \equiv \frac{kd}{D} \qquad \mathrm{Nu} \equiv \frac{hd}{k'} \tag{3}$$

$$\mathrm{Re} \equiv \frac{Ud}{\nu} \qquad \mathrm{Sc} \equiv \frac{\nu}{D} \qquad \mathrm{Pr} \equiv \frac{\nu}{a} \tag{4}$$

Here Sh is the Sherwood number, Nu is the Nusselt number, Re is the Reynolds number, Sc is the Schmidt number, Pr is the Prandtl number, k is the mass transfer coefficient, d is a characteristic length of the system (e.g. the diameter of the tube), U is a characteristic velocity, ν is the kinematic viscosity, D is the diffusion coefficient, h is the heat transfer coefficient, k′ is the thermal conductivity, a is the thermal diffusivity, and A, m, and n are constants.

The starting point of the present treatment is the observation that, for both laminar and turbulent flows, two kinds of length and time scales characterize the process: the micro- and macro-scales. The micro-length ℓ and micro-time t are associated with molecular phenomena and therefore they can be used to construct the dimensions of the molecular diffusion coefficient, of the thermal diffusivity, and of the kinematic viscosity. The macro-length L and macro-time T can be used to express, from a dimensional point of view, the macro-dimensions and macro-velocity of the system, such as the diameter and length of the tube and the velocity of the fluid. Since the mass transfer coefficient depends on both micro and macro processes, it is natural to include both these kinds of scales in its dimensions. Denoting the exponent of the macro-length by α and that of the macro-time by β, one can write the dimensions of the mass transfer coefficient k as

$$[k] = \frac{L^{\alpha}\ell^{1-\alpha}}{T^{\beta}t^{1-\beta}} \tag{5}$$

The dimensions of the other relevant quantities are:

$$[d] = L,\ [u] = LT^{-1},\ [D] = \ell^2t^{-1},\ \text{and}\ [\nu] = \ell^2t^{-1} \tag{6}$$

Writing, as usual,

$$k = AU^xd^y\nu^zD^v \tag{7}$$

then by use of dimensional analysis in which the macro- and micro-scales are treated separately,

$$\frac{kd}{D} = A\left(\frac{Ud}{\nu}\right)^{\beta}\left(\frac{\nu}{D}\right)^{1-v} \tag{8}$$

where

$$\beta = (1 + \alpha)/2. \tag{9}$$

Equation 8 has the same form as Equation 1 which was derived by using the conventional dimensional analysis. However, the exponent m of the Reynolds number

$$m = \beta = (1 + \alpha)/2 \tag{10}$$

has acquired some physical significance, because β is a measure of the participation of the macro-time scale and α is a measure of the participation of the macro-length scale in the dimensional composition of the mass transfer coefficient. Since the participation of the macro-scales should be greater in turbulent than in laminar motion, the value of m should be larger in the former case.

In the limiting case of $Re \to \infty$ (strong turbulence), no micro-participation is expected to occur because the molecular processes do not play any part in the transport rate. In this case, $\alpha = \beta = 1$ (see Equation 5). In the other limiting case, such as the vanishingly small Reynolds number flows around a sphere, the Sherwood number is constant. In this case, the micro-participation is dominant and $\alpha = -1$ and $\beta = 0$. In turbulent flows involving sufficiently large Reynolds numbers, some micro-participation will occur because the molecular quantities play a role in the rate of the process. However, the macro-participation remains greater than the micro-participation. At the other extreme, in the laminar flow past a sphere involving not too large Reynolds numbers, the micro-participation is expected to decrease in comparison to that for $Re \to 0$, but still remaining greater than that of the macro-participation. Extrapolation in both directions, a plausible assumption is that the macro-participation is greater than or equal to the micro-participation for the turbulent motion and that the opposite is valid for the laminar flow. Thus, for the turbulent case,

$$\alpha \geqslant 1 - \alpha \text{ and } \beta > 1 - \beta \tag{11}$$

The second equation does not contain the equality sign because α and β are not independent quantities, being related via Equation 9. Therefore, the previous two inequalities are simultaneously satisfied only if

$$0.5 \leqslant \alpha < 1 \text{ and } 0.75 \leqslant \beta < 1 \tag{12}$$

Similarly, for the laminar case,

$$1 - \alpha > \alpha \text{ and } 1 - \beta \geqslant \beta \tag{13}$$

These inequalities are satisfied without contradicting Equation 9, if

$$\beta \leqslant 0.5 \text{ and } \alpha \leqslant 0 \tag{14}$$

Equations 12 and 14 show that the exponent of the Reynolds number is greater or equal to 0.75 for turbulent flow and smaller or equal to 0.5 for laminar flow. It is, indeed, well known that in turbulent flow, the exponent of the Reynolds number is equal or larger than 0.75 [2]. In laminar flow along a plate or around a sphere, the exponent of the Reynolds number is 1/2 at sufficiently large Reynolds numbers; it is 1/3 for Stokes flow around a sphere; and it is zero for vanishingly small Reynolds number flows around a sphere, or for laminar motion in a long tube [2].

In addition, Equation 8 indicates that the exponent of the Schmidt number is independent of the exponents of the macro- and micro-scales. In other words, the exponent of ν/D should be the same for both laminar and turbulent flows. It is, indeed, known from experiment that the value of 1/3 for this exponent is valid for many laminar and turbulent flows involving solid interfaces and that the value of 1/2 is valid for laminar and turbulent motions involving fluid-fluid interfaces. It is worth noting that there is a jump from 0.5 to 0.75 in the value of the bound of the exponent m with the transition from laminar to turbulent flow, a result which is in agreement with what is observed experimentally [2].

ALGEBRAIC INTERPOLATION BASED ON PHYSICAL SCALING

Introducing the hydrodynamic boundary layer concept, Prandtl [1–5] suggested an order-of-magnitude evaluation of the terms in the Navier–Stokes equations. This leads to an expression for the thickness of the boundary layer in which only a proportionality constant remains undetermined. The interpolation procedure presented here for complex cases [6] replaces each term in the convective-diffusion equation by its corresponding expression in terms of scaling quantities (as done by Prandtl) but multiplied by an appropriate constant. Thus the convective diffusion equation is replaced by an algebraic equation containing a number of constants, for the thickness of the diffussion boundary layer. The undetermined constants are then obtained from solutions which are available for some limiting cases. Thus, Prandtl's evaluation procedure has been transformed into an interpolation method.

A Correlation for Laminar Free Convection from a Vertical Plate Over the Entire Range of Prandtl Numbers

The basic equations for laminar free convection from a vertical plate are

$$\frac{\partial u}{\partial x} + \frac{\partial v}{\partial y} = 0 \tag{15}$$

$$u\frac{\partial u}{\partial x} + v\frac{\partial u}{\partial y} = g\beta(T - T_\infty) + \nu\frac{\partial^2 u}{\partial y^2} \tag{16}$$

and

$$u\frac{\partial T}{\partial x} + v\frac{\partial T}{\partial y} = a\frac{\partial^2 T}{\partial y^2} \tag{17}$$

where u und v = the x and y components of the velocity
x = the distance up the plate
y = the distance to the plate
T = the temperature, T_∞ = the ambient temperature
g = the acceleration due to gravity
β = the volumetric coefficient of expansion

Since the velocity field is generated in free convection by the temperature field, the thickness δ of the thermal boundary layer can be used as the single length scale for both the temperature and velocity fields. The velocity scale in the x direction is denoted by u_0. The continuity equation (Equation 15) shows that the velocity scale v_0 in the y direction is of the order of $u_0\delta/x$.

Each of the terms in Equations 16 and 17 is evaluated by replacing u and ∂u by u_0, ∂x by x, ∂y by δ, ∂T and $T - T_\infty$ by $\Delta T = T_w - T_\infty$ (where T_w is the wall temperature). Consequently,

$$u\frac{\partial u}{\partial x} + v\frac{\partial u}{\partial y} \sim \frac{u_0^2}{x} \tag{18a}$$

$$\nu\frac{\partial^2 u}{\partial y^2} \sim \frac{\nu u_0}{\delta^2},\; a\frac{\partial^2 T}{\partial y^2} \sim \frac{a\Delta T}{\delta^2} \tag{18b}$$

and

$$u\frac{\partial T}{\partial x} + v\frac{\partial T}{\partial y} \sim \frac{u_0\Delta T}{x} \tag{18c}$$

Replacing each of the terms by their order of manitude evaluations multiplied by constants, Equations 16 and 19 become

$$A'\frac{u_0^2}{x} + B'\frac{\nu u_0}{\delta^2} = g\beta\Delta T \tag{19}$$

and

$$C'\frac{u_0}{x} = \frac{a}{\delta^2} \tag{20}$$

where A′, B′, and C′ are constants.

The velocity scale u_0 can be eliminated between Equations 19 and 20, to obtain

$$\delta^4 = \frac{a^2 x\left(A + B\frac{\nu}{a}\right)}{g\beta\Delta T} \tag{21}$$

The local heat transfer coefficient h_x is given by:

$$h_x = k'/\delta$$

and thus Equation 21 becomes:

$$Nu_x \equiv \frac{h_x x}{k'} = \left(\frac{g\beta\Delta T x^3}{\nu a}\right)^{1/4} \left(\frac{\nu}{a}\right)^{1/4} \left(A + B\frac{\nu}{a}\right)^{-1/4}$$

$$\equiv Ra_x^{1/4} Pr^{1/4} (A + B\,Pr)^{-1/4} \tag{22}$$

where $Ra_x \equiv g\beta\Delta T x^3/\nu a$ is the Rayleigh number and A and B are constants. For the two limiting cases $Pr \rightarrow 0$ and $Pr \rightarrow \infty$, Equations 22 acquires the forms of the exact relations given in the following. The constants A and B determined from these exact relations.

For an isothermal plate, the limiting solutions derived by LeFevre [8] are:

$$Nu_x = 0.6004\,(Ra_x Pr)^{1/4} \qquad \text{for } Pr \rightarrow 0 \tag{23a}$$

and

$$Nu_x = 0.5027\,Ra_x^{1/4} \qquad \text{for } Pr \rightarrow \infty. \tag{23b}$$

For uniform heat flux, the asymptotic solutions derived by Churchill and Ozoe [9] have the form:

$$Nu_x = 0.692\,(Ra_x Pr)^{1/4} \qquad \text{for } Pr \rightarrow 0 \tag{24a}$$

and

$$Nu_x = 0.563\,Ra_x^{1/4} \qquad \text{for } Pr \rightarrow \infty \tag{24b}$$

Using these asymptotic solutions, Equation (22) becomes [7]:

$$Nu_x = Ra_x^{1/4} Pr^{1/4} (7.6955 + 15.659\,Pr)^{-1/4} \tag{25}$$

for uniform wall temperature, and

$$Nu_x = Ra_x^{1/4} Pr^{1/4} (4.361 + 9.953\,Pr)^{-1/4} \tag{26}$$

for uniform heat flux.

Equations 25 and 26 are in good agreement with the empirical correlations used by Churchill and Ozoe [9] to represent the results of their numerical solution of the governing transport equations. See Tables 1 and 2 of References 7 for the comparison.

Combined Laminar Convection from Vertical Surfaces for Assisting Flows

Equation (15), (16) and (17) again constitute the appropriate boundary layer equations. However, the boundary conditions assume the form:

$$u = v = 0,\ T = T_w \text{ at } y = 0 \text{ and all } x \tag{27 a}$$

$$u = u_\infty,\ T = T_\infty \text{ at } y \rightarrow \infty \text{ and all } x \tag{27b}$$

$$u = u_\infty,\ T = T_\infty \text{ at } x = 0 \text{ and all } y \tag{27c}$$

In general, it is difficult to solve the system of Equations 15, 16 and 17 for these boundary conditions. If one is interested in cases where the Prandtl number is large, certain simplifying assumptions can be made. As the Prandtl number approaches large values, the thermal boundary layer becomes very thin. Consequently, only the fluid layer near the plate contributes significantly to the heat transfer resistance. The fluid velocities in Equation 17 are thus well approximated by the first term of their Taylor series expansions in terms of y. In addition, because the nonlinear inertial terms are negligible near the wall, one can further assume that the combined forced and free convection velocity is approximately equal to the sum of the velocities that would exist when these effects act independently. Therefore, for assisting flows at large Prandtl numbers (theoretically for $Pr \rightarrow \infty$), Equation 17 can be rewritten as

$$(u_F + u_N)\frac{\partial T}{\partial x} + (v_F + v_N)\frac{\partial T}{\partial y} = a\frac{\partial^2 T}{\partial y^2} \tag{28}$$

where subscripts F and N refer to forced and natural convection, respectively. The velocity components of Equation 28 are given by [11, 5]

$$u_F = 0.332 u_\infty^{3/2} \nu^{-1/2} x^{-1/2} y$$

$$v_F = 0.083 u_\infty^{3/2} \nu^{-1/2} x^{-3/2} y^2 \tag{29a}$$

and by [5]:

$$u_N = 0.679\, Pr^{-1/4} (g\beta\Delta T/\nu^2)^{3/4} \nu x^{1/4} y \text{ for } Pr \rightarrow \infty$$

and (29b)

$$v_N = -\,0.084\, Pr^{-1/4} (g\beta\Delta T/\nu^2)^{3/4} \nu x^{-3/4} y^2 \text{ for } Pr \rightarrow \infty$$

Using the order-of-magnitude analysis, one obtains from Equation 28 the expression

$$A\, Re_x^{3/2} Pr + B\, Ra_x^{3/4} = Nu_x^3 \tag{30}$$

where $Re_x \equiv \frac{u_\infty x}{\nu}$

The values of the constants A and B can be obtained from the limiting expressions where either free convection or forced convection is absent. In the case of isothermal plates [2, 5]

$$Nu_{x,F} = 0.339\, Re_x^{1/2} Pr^{1/3} \tag{31a}$$

and [9]

$$Nu_{x,N} = 0.503\, Ra_x^{1/4} \tag{31b}$$

When the wall heat flux is constant [2]

$$Nu_{x,F} = 0.464\, Re_x^{1/2} Pr^{1/3} \tag{32a}$$

and [9]

$$Nu_{x,N} = 0.563\, Ra_x^{1/4} \tag{32b}$$

The constants in Equation 30 can be obtained easily from the limiting cases provided by Equations 31 and 32. However, it is of more interest to note that, in terms of $Nu_{x,F}$ and $Nu_{x,N}$, Equation 30 can be rewritten as [10]:

$$Nu_x^3 = Nu_{x,F}^3 + Nu_{x,N}^3 \tag{33}$$

providing a derivation of Churchill's suggestion [12]. A similar expression can be derived for spheres or cylinders. The coordinate x is replaced in those cases by the diameter [10]. Acrivos [13], and Sparrow and Gregg [14] have presented a rigorous treatment of mixed convection.

Combined Convection in Internal Flows

In electrochemical reactors, where the externally imposed flow is often moderate or low, natural convection can exert a substantial influence. As an example, consider a vertical parallel plate reactor where the electrodes are separated by a distance d; the electrodes are assumed to be placed sufficiently far away from the entrance to the reactor so that the forced laminar flow can be considered to be fully developed. Since the reaction occurs only at the electrodes, the concentration profile begins to develop at the leading edges of the electrodes. The thickness of the concentration boundary layer throughout the length of the electrode is assumed to be much smaller than d—a condition that is usually satisfied in experiments. The velocity profile of the fully developed forced flow is given by

$$u_F = 6u_{av}\left(\frac{y}{d} - \frac{y^2}{d^2}\right)$$

which, close to the electrode, can be approximated by

$$u_F = (6u_{av}/d)y; \; v_F = 0, \tag{34}$$

where u_{av} is the average forced flow velocity through the reactor. The natural convection velocity is given by Equation 29b. The order-of-magnitude analysis gives the following equation for the Sherwood number

$$\left(\frac{k_x x}{D}\right)^3 \equiv Sh_x^3 = A\,Re_x\,Sc\frac{x}{d} \pm B\,Ra_x^{3/4} \tag{35}$$

The negative sign in Equation 35 applies when the external flow and the buoyancy-induced flow oppose each other.

Here $Ra_x \equiv \frac{g\Delta\rho x^3}{\rho\nu a}$ = the Rayleigh number for mass transfer

ρ = the density of the fluid

$\Delta\rho$ = the density difference between the fluid at the wall and at a large distance from the wall

k_x = the local mass transfer coefficient

$$Re_x \equiv \frac{u_{av}x}{\nu}$$

When pure natural convection is dominant, Equation 35 must take the form of Equation 23b rewritten for mass transfer. One thus obtains B = 0.1273

The first part of Equation 35 is just the cube of the Levèque [15] solution given by

$$Sh_{x,F} = 0.978\left[Re_x Sc\frac{x}{d}\right]^{1/3}$$

Consequently,

$$Sh_x = 0.978\,[Re_x\,Sc\,(x/d)]^{1/3}\left[1 \pm 0.136\frac{Ra_x^{3/4}}{Re_x Sc\,(x/d)}\right]^{1/3} \tag{36}$$

Experimentally one often measures the average value of the mass transfer coefficient, which is defined as

$$k_{av} = \frac{\int_0^A k_x dA}{A} = \frac{\int_0^L k_x W dx}{WL}$$

where A = WL
W = the width
L = the length of the plates

Integrating Equation 36 by parts, one can derive an expression for $Sh_{av} \equiv k_{av}d/D$. The integration leads to a power series in terms of x. Retaining the first two terms, one obtains [10]:

$$Sh_{av} = 1.47\,Gz^{1/3}\left[1 \pm 0.0989\,\frac{[Ra\,(d/L)]^{3/4}}{Gz}\right]^{1/3} \tag{37}$$

where $Gz = Re\,Sc\,(d/L)$ = the Graetz number

$$Re = \frac{u'_{av}d}{\nu}$$

$$Ra = \frac{g\Delta\rho d^3}{\rho\nu a}.$$

Penetration Theory for the Rate of Diffusion Accompanied by a Chemical Reaction

The problem of absorption accompanied by a chemical reaction in a liquid film flowing along a vertical wall can be treated within the framework of the penetration theory [16—20]. This theory has additional applications, since it constitutes the basis of the renewal models of turbulence [19, 21, 22].

In the framework of the penetration theory, the mass transport equation takes the form

$$\frac{\partial c}{\partial t} = D\frac{\partial^2 c}{\partial y^2} - K_1 c \tag{38}$$

where c is the concentration, t is the time and K_1 is the reaction rate constant. This equation is solved for the initial and boundary conditions:

$$c = 0 \qquad \text{for } t = 0 \text{ and all } y$$

$$c = c^* \qquad \text{for } y = 0 \text{ and all } t \tag{39}$$

and

$$c = 0 \qquad \text{for } y \rightarrow \infty \text{ and all } t.$$

As in the previous examples, the terms of Equation 38 are evaluated by replacing ∂t by t, ∂y by the thickness δ of the diffusion boundary layer, and c and ∂c by Δc. One thus obtains:

$$\frac{\partial c}{\partial t} \sim \frac{\Delta c}{t} \tag{40a}$$

and

$$D\frac{\partial^2 c}{\partial y^2} \sim \frac{D\Delta c}{\delta^2} \tag{40b}$$

Introducing these expressions multiplied by constants in Equation 38 yields the following algebraic equation:

$$\frac{A}{t} + B K_1 = \frac{D}{\delta^2}, \tag{41}$$

where A and B are constants. The rate N of absorption per unit area of the interface and unit time results from

$$N = \frac{c^* D}{\delta} \equiv c^* \left(\frac{D(A + BK_1t)}{t}\right)^{1/2} \tag{42}$$

The constants A and B can be obtained from the solutions available for the limiting cases in which either $K_1 = 0$ or $\frac{\partial c}{\partial t} = 0$. In the first limiting case,

$$\frac{\partial c}{\partial t} = D\frac{\partial^2 c}{\partial y^2}$$

and the rate is

$$N_I = \left(\frac{D}{\pi t}\right)^{1/2} c^* \equiv k_I c^* \tag{43}$$

where k_I is a mass transfer coefficient in the absence of reaction. In the second limiting case,

$$D\frac{d^2c}{dy^2} = K_1 c \tag{44}$$

and the rate of mass transfer is given by

$$N_{II} = c^* (K_1 D)^{1/2} \equiv k_{II} c^* \tag{45}$$

where k_{II} is also a mass transfer coefficient, but now for steady diffusion and reaction. Since Equation 42 must reduce to Equation 43 for $K_1 = 0$ and to Equation 45 for $t \rightarrow \infty$, one obtains

$$A = \frac{1}{\pi} \text{ and } B = 1 \tag{46}$$

Consequently

$$N = c^* \left[\frac{D}{\pi t}(1 + \pi K_1 t)\right]^{1/2} \equiv kc^* \tag{47}$$

and

$$k^2 = k_I^2 + k_{II}^2 \tag{48}$$

or

$$Sh^2 = Sh_I^2 + Sh_{II}^2 \tag{49}$$

The Sherwood numbers have been obtained by multiplying Equation 48 by x/D, where x is a length scale. The discrepancy between the simple equation 47 and the lengthy exact solution [16, 17] is less than about 10%.

Steady Laminar Flow Along a Plate with Homogeneous Reaction

The convective-diffusion equation has, in this case, the form

$$u\frac{\partial c}{\partial x} + v\frac{\partial c}{\partial y} = D\frac{\partial^2 c}{\partial y^2} - K_1 c \tag{50}$$

and the appropriate boundary conditions are

$$c = c_s \text{ at } y = 0 \text{ and for all } x$$

$$c = 0 \text{ for } y \to \infty \text{ and for all } x \tag{51}$$

and

$$c = 0 \text{ for } x = 0 \text{ and for all } y$$

For large Schmidt numbers, the depth of penetration by diffusion is very small. Therefore, only the region near the wall affects the mass or heat transfer rate. Consequently, one can retain only the first term in the Blasius expansions for the velocity components u and v

$$u \propto \frac{u_\infty^{3/2} y}{\nu^{1/2} x^{1/2}}, \quad v \propto \frac{u_\infty^{3/2} y^2}{\nu^{1/2} x^{3/2}} \tag{52}$$

The sign $\propto$ has the meaning of proportional to.

The terms of Equation 50 can be evaluated as in the previous examples, to obtain

$$A\,Re_x^{3/2} Sc + B\frac{K_1 x^2}{D}\frac{x}{\delta} = \left(\frac{x}{\delta}\right)^3 \tag{53}$$

where

$$Re_x = \frac{u_\infty x}{\nu}.$$

Since the local mass transfer coefficient k_x and the local Sherwood number are defined by

$$k_x = \frac{D}{\delta} \text{ and } Sh_x = \frac{k_x x}{D} = \frac{x}{\delta} \tag{54}$$

Equation 53 can be rewritten as

$$A\,Re_x^{3/2}\,Sc + B\frac{K_1 x^2}{D} Sh_x = Sh_x^3 \tag{55}$$

The constants A and B can be determined from the limiting cases, $K_1 = 0$ and $u_\infty = 0$. The first limiting case corresponds to the dissolution of a planar wall in a semi-infinite liquid in steady flow (Blasius flow). The second limit corresponds to the steady-state diffusion with chemical reaction in a stagnant fluid. For Blasius flow and large Schmidt numbers [2]

$$Sh_{x,I} = 0.339\,Re_x^{1/2}\,Sc^{1/3} \tag{56}$$

whereas for $u_\infty = 0$, one obtains (see the previous example)

$$Sh_{x,II} = \left(\frac{K_1 x^2}{D}\right)^{1/2} \equiv Da_x^{1/2} \tag{57}$$

where Da_x is the Damköhler number. Consequently, Equation 55 becomes

$$Sh_{x,I}^3 + (Sh_{x,II})^2\,Sh_x = Sh_x^3 \tag{58}$$

Rate of Heat Transfer in Wedge Flows with Suction at the Surface

The group of problems called "wedge flows," or Falkner-Skan flows [2, 4, 23], is characterized by free stream velocities (outside the momentum boundary layer) that vary as a power of the distance x as measured along the wedge from the leading edge $U(x) = U_0 x^m$. Near the wall the velocity components are given by

$$u = A(m)\,U(x)^{3/2}\nu^{-1/2}x^{-1/2}y \tag{59}$$

and

$$v = v_0(x) + B_1(m)\,U(x)^{3/2}\nu^{-1/2}x^{-3/2}y^2 \tag{60}$$

where $A(m)$ and $B_1(m)$ are functions of m, and the first term on the right hand side of Equation 60 is the suction velocity $v_0(x)$ at the wall. Replacing the terms of Equation 17 by their corresponding evaluations multiplied by constants, leads to

$$A\,Re_x^{3/2}Pr + B\,\frac{v_0 x}{a}\,Nu_x^2 = Nu_x^3 \tag{61}$$

where

$$Re_x = U(x)\,x/\nu \text{ and } Nu_x \equiv h_x x/k' = \frac{x}{\delta} \tag{62}$$

In the case of no suction and $m = 0$,

$$Nu_{x,I} = 0.339\,Re_x^{1/2}\,Pr^{1/3} \tag{63}$$

whereas for the limiting case in which $U(x) = 0$,

$$v_0(x)\,\frac{dT}{dy} = a\,\frac{d^2T}{dy^2}. \tag{64}$$

Using the boundary conditions

$$T = T_w \text{ at } y = 0 \text{ and } T = T_\infty \text{ as } y \to \infty, \tag{65}$$

the solution of Equation 64 is

$$T = e^{v_0 y/a}\,[T_\infty(e^{-v_0 y/a} - 1) + T_w]. \tag{66}$$

Equation 66 satisfies the boundary condition at $y \to \infty$ only if $v_0 < 0$, i.e. for suction. In the latter case, one obtains

$$Nu_{x,II} = -\,(v_0 x/a). \tag{67}$$

Using Equations 63 and 67, Equation 61 can be rewritten for $m = 0$ as [24]

$$Nu_{x,I}^3 + Nu_{x,II}\,Nu_x^2 = Nu_x^3 \tag{68}$$

One can show that Equation 68 remains valid for $m \neq 0$. A comparison between Equation 68 and the exact numerical solutions for Stewart and Prober [23] shows excellent agreement. Surprisingly, Equation 68 is accurate even of Prandtl numbers of 0 (1).

Mass Transfer to a Continuous Phase from a Single Spherical Drop

For sufficiently large Schmidt numbers, the depth of penetration by diffusion is small compared to the droplet radius R. Therefore, the convective diffusion equation can be approximated by

$$v_r \frac{\partial c}{\partial y} + \frac{v_\theta}{R} \frac{\partial c}{\partial \theta} = D \frac{\partial^2 c}{\partial y^2} \tag{69}$$

where y is the distance at the surface of the bubble, θ is the polar angle, and v_r and v_θ are the radial and tangential velocity components, respectively.

For low Reynolds numbers and large Schmidt numbers, the components of the velocity can be approximated by expanding in series the equations derived by Hadamard and Rybczynski [25, 26] and retaining the first two terms:

$$v_r = -2\left[u_0 \frac{y}{R} + \frac{1}{2}\left(\frac{y}{R}\right)^2 \left(\frac{3}{2}U - 2u_0\right)\right]\cos\theta \tag{70}$$

and

$$v_\theta = \left[u_0 + \frac{y}{R}\left(\frac{3}{2}U - 2u_0\right)\right]\sin\theta. \tag{71}$$

In these expressions, U is the velocity of the drop and u_0 is the velocity at the interface with respect to the center of the drop for $\theta = \pi/2$. For solid-like behavior, as occurs in the presence of surface active agents, u_0 becomes zero, while $u_0 \neq 0$ in systems in which they are either absent or in extremely small amounts [5, 27]. For a solid-like behavior of the interface:

$$Sh_{II} \equiv \frac{2Rk_{II}}{D} = 0.998\left(\frac{2UR}{D}\right)^{1/3} \equiv 0.998\,Pe^{1/3} \tag{72}$$

while for a liquid-like behavior of the interface [5, 27]

$$Sh_I \equiv \frac{2Rk_I}{D} = 0.923\left(\frac{u_0}{U}\right)^{1/2} Pe^{1/2} \tag{73}$$

where Pe is the Peclet number.

An interpolation equation between the two extreme cases, based on the order-of-magnitude evaluation, has the form [27]:

$$0.849\frac{u_0}{U} Pe\,Sh + 0.662\left(\frac{3}{2} - 2\frac{u_0}{U}\right)Pe = Sh^3 \tag{74}$$

which can be rewritten as

$$Sh_I^2\,Sh + Sh_{II}^3\left(1 - \frac{4}{3}\frac{u_0}{U}\right) = Sh^3 \tag{75}$$

For large Reynolds numbers, the following interpolation equation could be obtained [27]:

$$0.849\frac{u_0}{U} Sh^{-2} Re + 0.278\left(\frac{3}{2} - \frac{u_0}{U}\right)\left(\frac{3}{2} + \frac{u_0}{U}\right)^{1/2} Re^{3/2} Sh^{-3} = Sc^{-1} \tag{76}$$

where $Re \equiv \frac{2RU}{\nu}$. Equation 76 can be rearranged as

$$\frac{2}{3}\frac{u_0}{U} Sh\, Sh_I^2 + \left(1 - \frac{2}{3}\frac{u_0}{U}\right)\left(1 + \frac{2}{3}\frac{u_0}{U}\right)^{1/2} Sh_{II}^3 = Sh^3 \tag{77}$$

Here

$$Sh_I = 1.13\, Pe^{1/2} \tag{78}$$

is valid for potential flow $\left(\frac{u_0}{U} = \frac{3}{2}\right)$ [28], whereas

$$Sh_{II} = 0.8\, Re^{1/2}\, Sc^{1/3} \tag{79}$$

is valid when the interface behaves like a solid boundary ($u_0 = 0$) [29–32].

Unsteady Heat Transfer to a Liquid in Steady Laminar Flow Along a Plate

In this case the transport equation has the form

$$\frac{\partial T}{\partial t} + u\frac{\partial T}{\partial x} + v\frac{\partial T}{\partial y} = a\frac{\partial^2 T}{\partial y^2} \tag{80}$$

and the appropriate initial and boundary conditions are:

$$\begin{aligned} T &= T_0 \quad \text{for } t = 0 \text{ and for all x and y} \\ T &= T_w \quad \text{for } y = 0,\ t > 0 \text{ and all x} \\ T &= T_0 \quad \text{for } y \rightarrow \infty \text{ and all t and x} \end{aligned} \tag{81}$$

Assuming again large Prandtl numbers, one can use Equation 29a for the velocity components u and v. Replacing each of the terms of Equation 80 by their evaluations multiplied by constants, one obtains:

$$\frac{A}{t} + B\frac{u_\infty^{3/2}\delta}{\nu^{1/2}x^{3/2}} = \frac{a}{\delta^2}, \tag{82}$$

which can be rearranged as

$$A\frac{x^2}{at}\frac{x}{\delta} + B\, Re_x^{3/2}\, Pr = \left(\frac{x}{\delta}\right)^3 \tag{83}$$

The two limiting cases which permit one to calculate the values of A and B are

1. The steady state heating of the flowing liquid.
2. The unsteady heating of the stagnant liquid.

For the former case,

$$Nu_{x,I} \equiv \frac{h_{x,I}x}{k'} = 0.339\, Re_x^{1/2}\, Pr^{1/3} \tag{84}$$

whereas for the latter case

$$\mathrm{Nu}_{x,\mathrm{II}} \equiv \frac{h_{x,\mathrm{II}}x}{k'} = \left(\frac{x^2}{\pi a t}\right)^{1/2} \tag{85}$$

One thus obtains:

$$\left(\frac{x^2}{\pi a t}\right)\mathrm{Nu}_x + (0.339\,\mathrm{Re}_x^{1/2}\,\mathrm{Pr}^{1/3})^3 = \mathrm{Nu}_x^3 \tag{86}$$

which, using Equations 84 and 85, becomes [6]:

$$(\mathrm{Nu}_{x,\mathrm{I}})^3 + (\mathrm{Nu}_{x,\mathrm{II}})^2\,\mathrm{Nu}_x = \mathrm{Nu}_x^3 \tag{87}$$

This expression compares well with the more exact solution of Cess [32].

Turbulent Mass Transfer with a First-Order Chemical Reaction

Consider a tube through which a turbulent fluid is flowing. The wall of the tube is dissolving in the liquid, and the diffusison of the solute is accompanied by a first-order chemical reaction. The turbulent transport is represented by means of the turbulent diffusivity. Only the region near the wall is of interest, whose thickness is small compared to the radius of the tube. The effect of curvature can therefore be neglected and the transport equation can be written as:

$$\frac{d}{dy}\left[(D + \varepsilon)\frac{dc}{dy}\right] - K_1 c = 0, \tag{88}$$

where ε is the turbulent diffusivity and y the distance to the wall. The boundary conditions are

$$c = c_s \quad \text{for } y = 0$$
$$c = 0 \quad \text{for } y \to \infty \tag{89}$$

Since the concentration c decays to zero in a region near the wall whose thickness is extremely small, one can use for the turbulent diffusivity the expression

$$\varepsilon = \gamma_0 \left(\frac{y\,(\tau_0/\rho)^{1/2}}{\nu}\right)^3 \equiv \gamma y^3 \tag{90}$$

which is valid in that region [33, 34]. Here τ_0 is the shear stress at the wall, ρ is the density of the liquid, and γ_0 is a constant. The terms in Equation 88 can be evaluated as

$$\frac{d}{dy}D\frac{dc}{dy} \sim D\frac{c_s}{\delta^2} \tag{91a}$$

$$\frac{d}{dy}\varepsilon\frac{dc}{dy} \sim \gamma\delta c_s \tag{91b}$$

$$K_1 c \sim K_1 c_s \tag{91c}$$

Replacing each term in Equation (88) by their evaluations multiplied by constants, one obtains

$$A\frac{D}{\delta^2} + B\gamma\delta = K_1 \tag{92}$$

In the limiting case in which $\gamma = 0$, one easily obtains $K_1\delta^2/D = 1$. Consequently, $A = 1$. Since in the absence of a chemical reaction

$$\frac{D}{\delta_I^2} + B\gamma\delta_I = 0 \tag{93}$$

one can eliminate B between Equations 92 and 93, to obtain

$$\left(\frac{\delta_I}{\delta}\right)^2 - \left(\frac{\delta}{\delta_I}\right) = \frac{K_1\delta_I^2}{D} \tag{94}$$

Introducing the mass transfer coefficient, the latter equation becomes

$$\left(\frac{k}{k_I}\right)^2 - \frac{k_I}{k} = \frac{K_1 D}{k_I^2} \equiv M \tag{95}$$

One can show that there is excellent agreement between the "exact" numerical solution of Equation 88 obtained by Vieth, et al. [35], and Equation 95.

The Algebraic Method and the Renewal Models

Instead of the approach based on turbulent diffusivity, the turbulent mass transfer near a wall can be described by scaling the turbulence with a succession of short, steady, laminar motions along a plate. While the scaling of turbulence by a succession of short laminar path-lengths is examined in more detail in the next section of this chapter, here the emphasis is on the use of the algebraic method for the problem presented previously.

The length scale of the laminar path is denoted by x_0. Along each path of length x_0 we assume that there is a quasi-steady laminar flow of a semifinite fluid along a plate. This means that the hydrodynamic and diffusion boundary layers are assumed smaller than the thickness of the fluid elements brought by turbulent fluctuations near the wall. In addition, because in liquids the diffusion coefficient is small, the depth of penetration by diffusion in the element of liquid is also small and therefore one can use for the velocity components the first terms in the Taylor expansion of the Blasius expressions. Hence the rate of mass transfer in the laminar micro-structure can be obtained starting with the equation

$$u\frac{\partial c}{\partial x} + v\frac{\partial c}{\partial y} = D\frac{\partial^2 c}{\partial y^2} - K_1 c \tag{96}$$

where

$$u \propto \frac{u_0 y}{\left(\frac{x\nu}{u_0}\right)^{1/2}}, \quad v \propto u_0\frac{y^2}{x\left(\frac{x\nu}{u_0}\right)^{1/2}} \tag{97}$$

and u_0 is the velocity in the element of liquid which has just arrived at the wall because of a turbulent fluctuation. This element travels a path length x_0 along the wall in laminar motion, after which it dissolves into the bulk fluid, and is replaced by another element of fluid, and so on. There are now two length scales: the thickness δ of the diffusion boundary layer and the length x_0 of the laminar path length. The terms of Equation 96 can be evaluated as follows:

$$u\frac{\partial c}{\partial x} \sim v\frac{\partial c}{\partial y} \sim \frac{u_0 \delta \Delta c}{x_0\left(\dfrac{x_0 \nu}{u_0}\right)^{1/2}}$$

$$D\frac{\partial^2 c}{\partial y^2} \sim D\frac{\Delta c}{\delta^2}$$

$$K_1 c \sim K_1 \Delta c \tag{98}$$

where δ is an average thickness of the diffusion boundary layer over the path length x_0. Using these expressions, Equation 96 becomes

$$A\frac{D}{\delta^2} + B\frac{u_0\delta}{x_0\left(\dfrac{x_0 \nu}{u_0}\right)^{1/2}} = K_1 \tag{99}$$

From the limiting case in which convection is ignored, one easily obtains $A = 1$, while for the limiting case which is free of reaction one can write

$$\frac{D}{\delta_I^2} + B\frac{u_0\delta_I}{x_0\left(\dfrac{x_0 \nu}{u_0}\right)^{1/2}} = 0 \tag{100}$$

Eliminating **B** between Equations 99 and 100, one again finds Equations 94 and 95. The mass transfer coefficient for the case with reaction can thus be expressed in terms of the transfer coefficient without reaction and the dimensionless number $\mathbf{M} \equiv (K_1 D)/(k_I^2)$.

Discussion

In the preceding examples, expressions for the rate of transfer in complex situations have been derived by interpolating between two of their limiting cases. Although the method is extremely simple, its results are generally within 10% of the more exact (and more complicated) solutions. As expected, the form of the interpolation equation depends upon the particular physical problem. Indeed, in terms of the Nusselt (Sherwood) numbers for the two extreme cases between which the interpolation is carried out, the algebraic equation is quadratic in the case of penetration theory with chemical reaction, and cubic for mixed (free plus forced) convection. This difference can be associated with different kinds of velocity profiles; the tangential velocity component is independent of the distance from the interface in the penetration model, but it varies linearly with distance in the mixed convection case. In the other cases treated, convection with reaction, convection with suction, and steady convection with unsteady temperature field, the interpolation equation is also cubic, but one of the terms consists of mixed Sherwood (or Nusselt) numbers of the form $Sh_{x,II}^2 Sh_x$ or $Sh_{x,II} Sh_x^2$. When a chemical reaction or an unsteady concentration (or temperature) field is involved, the mixed term is of the form $Sh_{x,II}^2 Sh_x$; when suction is involved, the mixed term has the form $Sh_{x,II} Sh_x^2$.

MULTIPLE SCALING IN TURBULENT FLOW

In all the theoretical approaches developed thus far for predicting the heat or mass transfer to turbulent fluids flowing in a tube, the main difficulties were linked to the description of turbulence near the wall. Now, there is unanimous agreement that turbulence decays at the wall and not at a certain distance from the wall. The turbulent diffusivity ε decays to zero at the wall as [5, 33, 34, 36]

$$\frac{\varepsilon}{\nu} \propto \left\{ \frac{y\sqrt{\frac{\tau_0}{\rho}}}{\nu} \right\}^n \tag{101}$$

where values between 2 and 4 have been suggested for the exponent n. Measurement of the mass transfer coefficient [37] from the dissolution of the wall of a tube into a turbulent liquid with very large Schmidt numbers (up to $Sc = 10^5$) indicates a value of 3 for n. However, the representation of turbulence by means of the turbulent diffusivity is not entirely satisfactory from a conceptual point of view. Indeed, the turbulent diffusivity ε is given by

$$\varepsilon = \overline{v'\ell} \tag{102}$$

where v′ is the normal component of the velocity fluctuation and ℓ is the mixing length. The bar indicates a temporal averaging. In order to obtain Equation 101 one assumes that $\ell \propto y$ and that the tangential component of the fluctuation $u' \propto \bar{u} \propto y$, where $\bar{u}$ is the average local velocity of the main flow. The continuity equation

$$\frac{\partial u'}{\partial x} + \frac{\partial v'}{\partial y} = 0 \tag{103}$$

leads to

$$v' \propto y^2 \tag{104}$$

and consequently to

$$\varepsilon \propto y^3 \tag{105}$$

The turbulent diffusivity involves a local (point) description of the transfer process. Such a description is physically reasonable, as long as the mixing length is sufficiently small in comparison to the distance to the wall. However, near the wall (but not in its immediate vicinity), $\ell = 0.4y$ and therefore, the preceding condition is not satisfied. In addition, because $\bar{u}$ depends only on y, $\partial u'/\partial x$ differs from zero only if the proportionality constant between u′ and $\bar{u}$ depends on x. Since the experimental values of ε are independent of x, the dependence on x is probably repeated quasi-periodically over small successive paths. Consequently, ε might be considered a result of both a temporal and spatial averaging. The first comment suggests that a nonlocal description of turbulence near the wall might be more appropriate than a local one, whereas the second suggests that the hydrodynamic process repeats itself over short, successive paths. In fact, Fage and Townsend [38], following the motion of small suspended particles by means of an ultramicroscope, have observed large lateral displacements very near the wall and that the motions of the particles are almost rectilinear along the wall, in the interval between two such displacements. Both comments concerning turbulent diffusivity and previous experimental observations suggest that one should represent turbulence near a solid by a succession of short laminar motions, the quasi-local turbulent quantities being averages over the short path. In other words, this model considers that elements of liquid, which have been brought to the wall by turbulent fluctuations, are moving through short laminar paths along the wall, after which they dissolve into the bulk of fluid, being replaced by other elements of liquid and so on [39]. This model, proposed by Ruckenstein [39], borrows the renewal idea from theories formulated by Higbie [18] and Danckwerts [19] for the mass transfer near a liquid-fluid interface, but accounts for the micro-hydrodynamic structure of the elements of liquid instead of considering them to be solid-like. Further, in contrast to the latter models, which provide only the dependence of the mass transfer coefficient on the diffusion coefficient, [39] relates the mass transfer coefficient to both the diffusion coefficient as well as hydrodynamics.

More recent experiments [40] have shown a three dimensional structure for turbulence near the wall. In a plane normal to the mean flow, counterrotating eddy pairs are involved, whereas in the direction of the mean flow, the motion is quasi-periodic (as just described). Since the wave length along the mean flow is much larger than along the perimeter of the tube, a simplified bidimensional model may account only for what happens in the plane normal to the mean flow. The steady laminar motion along a plate can be used to represent the elementary phenomenon occuring along each path length, in the direction of the mean flow, whereas the steady stagnation flow pattern appears to be a more adequate simplifying decription, for the motion in the plane normal to the mean flow. However, as shown in the following, the final expression for the turbulent mass transfer is the same in both cases.

Model Based on Steady Laminar Motion Along a Plate

In this model, the velocity and concentration fields are described by the equations valid for the steady laminar motion of a semi-infinite fluid along a plate:

$$u\frac{\partial u}{\partial x} + v\frac{\partial u}{\partial y} = \nu\frac{\partial^2 u}{\partial y^2}, \tag{106}$$

$$\frac{\partial u}{\partial x} + \frac{\partial v}{\partial y} = 0 \tag{107}$$

$$u\frac{\partial c}{\partial x} + v\frac{\partial c}{\partial y} = D\frac{\partial^2 c}{\partial y^2} \tag{108}$$

This means that the length x_0 of the laminar path is so short that the thickness of the hydrodynamic boundary layer and the depth of penetration by diffusion remain smaller than the thickness of the liquid element. For other possibilities see Ref. 39. Denoting by u_0 the uniform velocity of an element of fluid which has just arrived at the wall and by δ_h the thickness of the hydrodynamic boundary layer, one can evaluate the terms in Equations 106 and 107 by

$$u\frac{\partial u}{\partial x} \sim \frac{u_0^2}{x}$$

$$v\frac{\partial u}{\partial y} \sim \frac{u_0\delta_h}{x}\frac{u_0}{\delta_h} \sim \frac{u_0^2}{x}$$

$$\nu\frac{\partial^2 u}{\partial y^2} \sim \frac{\nu u_0}{\delta_h^2} \tag{109}$$

Replacing in Equation 106 each term by its evaluation, one obtains

$$\delta_h^2 \propto \frac{\nu x}{u_0} \tag{110}$$

At sufficiently large Schmidt numbers, hence in liquids, where the diffusion coefficient is very small ($D \sim 10^{-5}\,cm^2/sec$), only the region very near the wall is affected by diffusion. In that region the velocity distribution can be approximated by the first term in its Taylor expansion:

$$\frac{u}{u_0} \propto \frac{y}{\delta_h} \equiv y\left(\frac{u_0}{\nu x}\right)^{1/2} \tag{111}$$

Using the continuity equation, one can write

$$\frac{v}{u_0} \propto \frac{y^2 u_0^{1/2}}{\nu^{1/2} x^{3/2}} \tag{112}$$

The terms of Equation 108 can be evaluated as follows:

$$u\frac{\partial c}{\partial x} \sim v\frac{\partial c}{\partial y} \sim \frac{\delta u_0^{3/2}}{\nu^{1/2} x^{1/2}}\frac{\Delta c}{x} \tag{113a}$$

and

$$D\frac{\partial^2 c}{\partial y^2} \sim D\frac{\Delta c}{\delta^2} \tag{113b}$$

where $\delta = \delta(x)$ is the thickness of the diffusion boundary layer. Consequently, Equation 106 can be replaced by the algebraic expression

$$\delta^3 \propto \frac{D\nu^{1/2} x^{3/2}}{u_0^{3/2}} \tag{114}$$

which rewritten in terms of the mass transfer coefficient $k_x = D/\delta$ becomes:

$$k_x \propto \left(\frac{u_0 \nu}{x}\right)^{1/2}\left(\frac{D}{\nu}\right)^{2/3} \tag{115}$$

The local turbulent mass transfer coefficient can be defined by the average

$$k = \frac{1}{x_0}\int_0^{x0} k_x dx \tag{116}$$

where x_0 is the length of the laminar path. Consequently,

$$k \propto \left(\frac{u_0 \nu}{x_0}\right)^{1/2}\left(\frac{D}{\nu}\right)^{2/3} \tag{117}$$

The path length x_0 can be expressed in terms of the shear stress at the wall and the velocity u_0 as follows. Indeed, the shear stress at the wall $\tau_{0,x}$ is defined as

$$\frac{\tau_{0,x}}{\rho} = \nu\left(\frac{\partial u}{\partial y}\right)_{y=0} \propto \frac{\nu u_0}{\delta_h} \tag{118}$$

The average τ_0 of $\tau_{0,x}$ over the path length x_0 can be considered as the local turbulent shear stress at the wall:

$$\tau_0 = \frac{1}{x_0}\int_0^{x0} \tau_{0,x} dx \propto \rho\,\frac{\nu^{1/2} u_0^{3/2}}{x_0^{1/2}} \tag{119}$$

from where it results

$$x_0 \propto \nu u_0^3 (\tau_0/\rho)^{-2} \tag{120}$$

Eliminating x_0 between Equations 117 and 120 one obtains

$$k = \left(\frac{D}{\nu}\right)^{2/3} \frac{\tau_0}{\rho} \frac{1}{u_0} \tag{121}$$

The proportionality constant is unity because the numerical coefficients in Equations 117 and 120, which can be obtained on the basis of Equations 29a and 31a, approach such a value. The only supplementary assumptions which are necessary to complete the theory concern u_0. The universal law of velocity distribution near the solid surface has the form:

$$\frac{u}{\left(\frac{\tau_0}{\rho}\right)^{1/2}} = F\left(\frac{y\sqrt{\frac{\tau_0}{\rho}}}{\nu}\right)$$

The region of origin of the eddies which penetrate up to the wall, must be at a critical distance y_{crit}, corresponding to a critical Reynolds number $\frac{y_{crit}\sqrt{\frac{\tau_0}{\rho}}}{\nu} = \text{const}$. Consequently,

$$u_0 \propto \left(\frac{\tau_0}{\rho}\right)^{1/2} \tag{122}$$

Combining Equations 121 and 122 one finally obtains:

$$k \propto \left(\frac{\tau_0}{\rho}\right)^{1/2} \left(\frac{D}{\nu}\right)^{2/3} \tag{123}$$

Model Based on Steady Stagnation Laminar Flow

As already noted, in the plane normal to the main flow, turbulence can be described by the stagnation flow. For steady laminar stagnation flow, $u = \alpha_0 x$ and $v = -\alpha_0 y$ far from the plane $y = 0$, and [2]:

$$k \propto \left(\frac{\nu}{D}\right)^{1/3} \frac{D}{\left(\frac{\nu}{\alpha_0}\right)^{1/2}} \tag{124}$$

where α_0 is a dimensional constant with the dimensions of $\sec^{-1}$. The state of turbulence near the wall can be characterized by τ_0, ρ, and η. Therefore:

$$\alpha_0 = \psi(\tau_0, \rho, \eta)$$

Since the number of quantities is 4 while the number of dimensions involved is 3, a single dimensionless group can be formed, which must be a constant. Consequently,

$$\alpha_0 \propto \frac{\tau_0}{\rho \nu} \tag{125}$$

Eliminating α_0 between Equations 124 and 125, Equation 123 is again obtained.

One may note that, two kinds of scaling are involved in this treatment of turbulent mass transfer near a solid boundary. Firstly, the local turbulent behaviour is scaled by laminar motions over short path lengths and secondly, the laminar microstructure is scaled in terms of its characteristics.

Unsteady Laminar Motion Used to Describe Turbulent Transport

Pinczewski and Sideman [41] suggest the introduction of additional transient terms in the model of Ruckenstein. Instead of Equations 106 and 107 they write the unsteady equations:

$$\frac{\partial u}{\partial t} + u\frac{\partial u}{\partial x} + v\frac{\partial u}{\partial y} = \nu\frac{\partial^2 u}{\partial y^2}, \tag{126}$$

$$\frac{\partial u}{\partial x} + \frac{\partial v}{\partial y} = 0 \tag{127}$$

$$\frac{\partial c}{\partial t} + u\frac{\partial c}{\partial x} + v\frac{\partial c}{\partial y} = D\frac{\partial^2 c}{\partial y^2} \tag{128}$$

The boundary conditions are:

$$\begin{aligned}
&u = u_0 \text{ for } t = 0 \text{ and all y and x}\\
&u = u_0 \text{ for } y \to \infty \text{ and all t and x}\\
&u = 0 \text{ for } y = 0 \text{ and } t > 0 \text{ and all x}\\
&c = c_0 \text{ for } t = 0 \text{ and all y and x}\\
&c = c_0 \text{ for } y \to \infty \text{ and all t and x}\\
&c = c_i \text{ for } y = 0 \text{ and } t > 0 \text{ and all x}
\end{aligned} \tag{129}$$

One may note that these boundary conditions, hence the model, involve the assumption that an element of liquid with a uniform velocity u_0 is brought at time $t = 0$ in a given small region of the surface and that fluid is flowing in this region along the wall during a time Δ. It is shown here that such a model does not lead to the right dependence of the mass transfer coefficient on the Schmidt number and that the values predicted by it are much too high. In contrast, the quasi-steady model leads to the correct dependence on the Schmidt number and to values which are only somewhat lower than the experimental ones. Since a simple analytical solution of this problem is not possible*, the algebraic approximation is used, in what follows, to obtain an equation for the mass transfer coefficient. The terms of Equation 126 can be evaluated as follows

$$\frac{\partial u}{\partial t} \sim \frac{u_0}{t}, \; u\frac{\partial u}{\partial x} \sim \frac{u_0^2}{x}, \; v\frac{\partial u}{\partial y} \sim u\frac{\partial u}{\partial x} \sim \frac{u_0^2}{x}$$

*The approximate solution which is available is incorrect since it involves the quasi-steady approximation at short times and the transient approximation at long times.

and

$$\nu \frac{\partial^2 u}{\partial y^2} \sim \frac{\nu u_0}{\delta_h^2} \tag{130}$$

Replacing each term in Equation 126 by its corresponding evaluation multiplied by a constant, one obtains

$$\frac{A}{t} + \frac{B u_0}{x} = \frac{\nu}{\delta_h^2} \tag{131}$$

The constants A and B can be obtained from the solutions available for the limiting cases of $u_0 = 0$ and $t \rightarrow \infty$. In the first,

$$\frac{\partial u}{\partial t} = \nu \frac{\partial^2 u}{\partial y^2} \tag{132}$$

The solution of Equation 132 for the appropriate initial and boundary conditions is

$$u = \frac{2}{\sqrt{\pi}} u_0 \int_0^{\frac{y}{2\sqrt{\nu t}}} e^{-\mu^2} d\mu \tag{133}$$

The velocity u becomes almost equal to u_0 when

$$\frac{\delta_h}{2\sqrt{\nu t}} \approx 1.5 \tag{134}$$

One thus obtains

$$A = \frac{1}{9}$$

In the other limiting case [2, 4]

$$\delta_h = 5.2 \left(\frac{\nu x}{u_0}\right)^{1/2} \tag{135}$$

which leads to

$$B = \frac{1}{27}$$

Equation 131 can be therefore rewritten as

$$\delta_h^2 = \frac{\nu}{\dfrac{1}{9t} + \dfrac{u_0}{27x}} \tag{136}$$

Using Equation 29a, it is easy to show that the first term in the Taylor expansion of u has the form

$$\frac{u}{u_0} = 1.7 \frac{y}{\delta_h} \tag{137}$$

Note that, on the basis of Equations 133 and 134, one also obtains $\frac{u}{u_0} = 1.7\frac{y}{\delta}$. In fact, the value of the constant in Equation 134 was chosen in such a way as to obtain in both limiting cases the same value for the coefficient which multiplies the ratio y/δ_h. From the continuity equation, one obtains:

$$\frac{v}{u_0} = \frac{1.7}{4x27}\frac{u_0 y^2}{\nu^{1/2}x^2}\left(\frac{1}{9t} + \frac{u_0}{27x}\right)^{-1/2} \tag{138}$$

The terms of Equation 128 can be evaluated as

$$\frac{\partial c}{\partial t} \sim \frac{\Delta c}{t}$$

$$u\frac{\partial c}{\partial x} \sim \frac{\delta u_0}{\nu^{1/2}x}\left(\frac{1}{9t} + \frac{u_0}{27x}\right)^{1/2}\Delta c$$

$$v\frac{\partial c}{\partial y} \sim \frac{u_0^2\delta}{\nu^{1/2}x^2}\left(\frac{1}{9t} + \frac{u_0}{27x}\right)^{-1/2}\Delta c$$

$$D\frac{\partial^2 c}{\partial y^2} \sim \frac{D}{\delta^2}\Delta c \tag{139}$$

Equation 139 permits one to replace Equation 128 by the algebraic expression:

$$\frac{A}{t} + B\frac{\delta u_0}{\nu^{1/2}x}\left(\frac{1}{9t} + \frac{u_0}{27x_0}\right)^{1/2} + C\frac{u_0^2\delta}{\nu^{1/2}x^2}\left(\frac{1}{9t} + \frac{u_0}{27x}\right)^{-1/2} = \frac{D}{\delta^2} \tag{140}$$

where A, B, and C are constants. Unfortunately the constants B and C cannot be determined independently. However, since Equation 140 is an interpolation equation, it is reasonable to consider that a similar expression could constitute a good approximation for the average mass transfer coefficient. The expression for the average mass transfer coefficient is obtained by replacing t by Δ, x by x_0 and δ by $\delta_m\left(=\frac{D}{k}\right)$. One thus obtains

$$\frac{A}{\Delta} + B\frac{\delta_m u_0}{\nu^{1/2}x_0}\left(\frac{1}{9\Delta} + \frac{u_0}{27x_0}\right)^{1/2} + C\frac{u_0^2\delta_m}{\nu^{1/2}x_0^2}\left(\frac{1}{9\Delta} + \frac{u_0}{27x_0}\right)^{-1/2} = \frac{D}{\delta_m^2} \tag{141}$$

Anticipating the considerations which lead to Equations 148, one has

$$\Delta \propto \frac{x_0}{u_0} \tag{142}$$

and, therefore, Equation 141 can be rewritten as:

$$\frac{A}{\Delta} + B'\frac{\delta_m}{\nu^{1/2}}\left(\frac{u_0}{x_0}\right)^{3/2} = \frac{D}{\delta_m^2} \tag{143}$$

where B′ is a new constant. In the extreme case of no convection, the mass transfer coefficient is given by the expression:

$$k_I = \left(\frac{4}{\pi}\right)^{1/2} \left(\frac{D}{\Delta}\right)^{1/2} \tag{144}$$

whereas in the extreme case of steady state, by the expression:

$$k_{II} = 0.68 \left(\frac{u_0}{x_0}\right)^{1/2} D^{2/3} \nu^{-1/6} \tag{145}$$

In terms of the mass transfer coefficient $k \equiv \frac{D}{\delta_m}$, Equation 143 can be rearranged in the form

$$k^3 = Ak\frac{D}{\Delta} + B' \frac{u_0^{3/2} D^2}{\nu^{1/2} x_0^{3/2}} \tag{146}$$

where, comparing with Equations 144 and 145, A = 1.27 and B′ = 0.314. Consequently,

$$k^3 = 1.27k \frac{D}{\Delta} + 0.314 \left(\frac{u_0}{x_0}\right)^{3/2} \frac{D^2}{\nu^{1/2}} \tag{147}$$

The state of turbulence near the wall can be characterized by τ_0, ρ, and η. Hence Δ, x_0 and u_0 are functions of the preceding quantities. Since only single dimensionless groups can be formed, they must be constant. Consequently,

$$\Delta \propto \frac{\nu}{\tau_0/\rho}, \quad u_0 \propto (\tau_0/\rho)^{1/2} \text{ and } x_0 \propto \frac{\nu}{(\tau_0/\rho)^{1/2}} \tag{148}$$

Combining Equations 146 and 148, one obtains

$$k^3 = \alpha k\, Sc^{-1} \frac{\tau_0}{\rho} + \beta\, Sc^{-2} \left(\frac{\tau_0}{\rho}\right)^{3/2} \tag{149}$$

From experiment [42]

$$\Delta = 243 \frac{\nu}{\tau_0/\rho} \tag{150}$$

On the other hand, for the flow in a tube, the wall shear stress τ_0 is given by the expression

$$\frac{\tau_0}{\rho} = \frac{f}{2} u_m^2 \tag{151}$$

where f is the friction factor and u_m is the average velocity. The friction factor f can be calculated using the Blasius expression:

$$\frac{f}{2} = 0.023\, Re^{-0.2} \tag{152}$$

where $Re \equiv (u_m 2R)/\nu$, u_m being the average velocity and R the radius of the tube. If one assumes $x_0/u_0 = \Delta$, one obtains

$$\left(\frac{k}{u_m}\right)^3 = 3 \cdot 10^{-7} Re^{-0.3} Sc^{-2} + 1.2 \cdot 10^{-4} \frac{k}{u_m} Sc^{-1} Re^{-0.2} \tag{153}$$

The first term on the right-hand side of Equation 153 is identical to that provided by the quasi-steady model, whereas the second represents the contribution of the transient process. Measurements of the mass transfer coefficient from the dissolution of the wall of a tube into a turbulent liquid having Schmidt numbers as large as 10^5 could be correlated with the expression [37]:

$$Sh \equiv \frac{2kR}{D} = 0.0096\, Re^{0.913}\, Sc^{0.346} \tag{154}$$

A comparison of the experimental expression (Equation 154) with the theoretical one (Equation 153) shows that the quasi-steady term leads to the same dependence on the Sc number, $Sh \sim Sc^{1/3}$, as the experimental one does, whereas the transient term leads to a too strong dependence, $Sh \sim Sc^{1/2}$. In addition, one can easily verify that for sufficiently large values of the Schmidt number the transient term has the dominant contribution to the value of the mass transfer coefficient and that the values predicted are much too large.

Using the quasi-steady model and writing that

$$\frac{x_0}{u_0} = 243 \frac{\nu}{\tau_0/\rho} \tag{150a}$$

one obtains

$$Sh = 0.0067\, Re^{0.9}\, Sc^{1/3} \tag{155}$$

The exponent of the Schmidt number in Equation 155 is equal to the experimental one. It is true that the coefficient 0.0067 is somewhat low. However, one must not take very seriously the value of 243 of the coefficient in Equation 150a. A statistical distribution of the path length x_0 and the associated averaging to obtain the average value of the mass transfer coefficient with respect to this distribution, might also constitute an explanation for the slight discrepancy.

In reality, the turbulent process is more complex than described by any of the previous models. The transient model exaggerates the importance of the transient process, while the quasi-steady model neglects transient effects. It is easier to understand the shortcomings of the former model than the compensatory effects which cause the latter model to be in better agreement with experiment. Indeed as already noted, the boundary conditions (Equation 129) involve the assumption that at time zero we have an element of liquid with the velocity u_0 in a given fixed region near the wall and that during a time Δ, the fluid is flowing in that region along the wall. However, an eddy that has arrived at the wall will move along it, being entrained by the flow of liquid. Therefore, the mentioned boundary conditions are not entirely realistic.

More complex models involving several length and velocity scales have been examined in Reference 39.

Turbulent Mass or Heat Transfer Coefficient in a Tube for Sc or Pr > 1

When the Schmidt (or Prandtl) number is sufficiently large, the transfer coefficient near the wall, as calculated in the previous section, coincides with the overall transfer coefficient. For smaller values of the Schmidt number, the resistance existing in the core region also has to be taken into account. For the latter region, Reynolds analogy leads to

$$\frac{c - c_0}{u - u_0} = \frac{N_0}{\tau_0/\rho} \tag{156}$$

where N_0 is the wall mass flux and u_0 and c_0 are the values of u and c at the outer limit of the wall layer. Since the region near the wall is very thin compared to the fully turbulent one, the average concentration is calculated by using only Equation 156 to obtain:

$$c_m - c_0 = \frac{N_0}{\tau_0/\rho}\left[\frac{(u'_m)^2}{u_m} - u_0\right] \tag{157}$$

where

$$c_m = \frac{1}{\pi R^2 u_m}\int_0^R 2\pi rucdr \tag{158}$$

and

$$(u'_m)^2 = \frac{1}{\pi R^2}\int_0^R 2\pi ru^2dr \tag{159}$$

r being the distance from the center of the tube. Since x_0 is short, we assume that the thickness of the hydrodynamic boundary layer is much smaller than the thickness of the element of fluid. In addition, if the Schmidt number is not too small, we still use the first terms in the expansion of the velocity components. Therefore, one can write for the wall region that:

$$N_0 = \beta^{1/3}\,Sc^{-2/3}\left(\frac{\tau_0}{\rho}\right)^{1/2}(c_i - c_0) \tag{160}$$

where c_i is the concentration at the interface. Since the overall mass transfer coefficient is defined by

$$N_0 = k\,(c_i - c_m) \tag{161}$$

one finally obtains [43]:

$$\frac{k}{u_m} = \frac{\frac{f}{2}}{\left(\frac{u'_m}{u_m}\right)^2 + \beta'\left(\frac{f}{2}\right)^{1/2}\left[\left(\frac{\nu}{D}\right)^{2/3} - 1\right]} \qquad (\beta' \equiv \beta^{-1/3}) \tag{162}$$

For heat transfer, good agreement with experiment is obtained with the equation:

$$\frac{h}{\rho c_p u_m} = \frac{\frac{f}{2}}{1.10 + 12.8\left(\frac{f}{2}\right)^{1/2}\left[\left(\frac{\nu}{a}\right)^{2/3} - 1\right]} \tag{163}$$

where c_p is the specific heat.

These expressions can be easily extended to non-Newtonian fluids by using, for the region near the wall, the corresponding expressions for the heat or mass transfer along a plate [44]. For a power law fluid we have

$$\tau = K\left(\frac{du}{dy}\right)^n \tag{164}$$

and one can still use Equations 162 and 163 by replacing the kinematic viscosity ν by the quantity ν' defined by

$$\nu' = \frac{\tau_0}{\rho} \Big/ \frac{du}{dy} = \left(\frac{\tau_0}{\rho}\right)^{\frac{n-1}{n}} \left(\frac{K}{\rho}\right)^{1/n} \tag{165}$$

Turbulent Heat or Mass Transfer in a Stirred Vessel

The motion which occurs in this case is a result of the dissipation of the mechanical energy supplied by the stirrer. Denoting by P the mechanical energy supplied per unit time and by V the volume of the vessel, the energy ξ_m dissipated on the average per unit volume is given by:

$$P = V\xi_m \tag{166}$$

Even though the energy dissipated per unit volume is not uniform and the nonuniformity depends upon the details of the stirring device, in what follows we will consider, for the sake of simplicity, that the energy dissipated per unit volume is uniform. The following problems are of interest with regard to the stirred vessel:

1. Heat transfer between the liquid and either the wall of the vessel or the cooling or heating coil introduced into the vessel
2. Rate of dissolution of large unsuspended crystals in the liquid
3. Rate of dissolution of suspended crystallites in the liquid.
4. Rate of absorption at the free surface of the liquid.

In case 1 turbulence can be described as in a pipe, differing only in that the stirring ensures uniformity of temperature in the core of the fluid vessel. Near the wall, the process can be described by the quasi-steady model. Therefore,

$$\frac{h}{\rho c_p} = \beta^{1/3} Pr^{-2/3} \left(\frac{\tau_0}{\rho}\right)^{1/2} \tag{167}$$

where, since

$$\frac{\tau_0}{\rho} = \nu\left(\frac{du}{dy}\right)_{y=0} \quad \text{and} \quad \frac{\xi_0}{\rho} = \nu\left(\frac{du}{dy}\right)^2_{y=0} \tag{168}$$

one can write

$$\frac{\tau_0}{\rho} = \left(\frac{\xi_0 \nu}{\rho}\right)^{1/2} \tag{169}$$

Consequently, Equation 167 can be rewritten as [45]:

$$\frac{h}{\rho c_p} = \beta^{1/3} Pr^{-2/3} \left(\frac{\xi_0 \nu}{\rho}\right)^{1/4} \tag{170}$$

Concerning the constant $\beta^{1/3}$, Equation 163 suggests a value of 0.078. However, the mechanical energy consumed per unit time by a blade stirrer can be evaluated by using expressions of the form [46a]:

$$P = d^5 n^3 \rho F\left(\frac{nd^2}{\nu}, \frac{gd^3}{\nu^2}, \ldots\right) \tag{171}$$

where d is the diameter of the circle described by the end points of the stirrer blades, n is the number of revolutions per unit time, and the friction coefficient F is a function of the Reynolds number $(nd^2)/\nu$, Galileo number $(gd^3)/\nu^2$ and various geometric ratios. Equation 170 can be rearranged in the form [45]:

$$\mathrm{Nu} \equiv \frac{d^{1/4}V^{1/4}h}{k'} = \beta^{1/3}F^{1/4}\mathrm{Re}^{3/4}\mathrm{Pr}^{1/3} \tag{172}$$

Since [46a]

$$F \propto \mathrm{Re}^{-m} \text{ with } 0.20 < m < 0.40 \tag{173}$$

Equation 172 leads to

$$\mathrm{Nu} \propto \mathrm{Re}^{0.7}\,\mathrm{Pr}^{1/3} \text{ for } m = 0.20$$

$$\mathrm{Nu} \propto \mathrm{Re}^{0.65}\,\mathrm{Pr}^{1/3} \text{ for } m = 0.40 \tag{174}$$

Experiment shows that for the heat transfer between the liquid and the wall [46b]:

$$\frac{hD}{k'} = 0.36\,\mathrm{Re}^{0.67}\,\mathrm{Pr}^{1/3}$$

where D is the diameter of the vessel. These expressions are also in agreement with the correlations obtained for the dissolution of a fixed crystal in a stirred liquid [47].

For suspended particles, the relative velocity is zero and turbulence brings elements of liquid into contact with the particles, elements which are renewed after a time Δ. In this case, one can use Equation 144:

$$k = 1.13\left(\frac{D}{\Delta}\right)^{1/2} \tag{175}$$

Since the state of turbulence is characterized by ξ_0, ρ, and η, dimensional considerations lead to

$$\Delta \propto \left(\frac{\xi_0}{\rho\nu}\right)^{-1/2} \tag{176}$$

and Equation 175 becomes [22]:

$$k \propto \left(\frac{D}{\nu}\right)^{1/2}\left(\frac{\xi_0\nu}{\rho}\right)^{1/4} \tag{177}$$

Equation 177 is of the same form as Equation 170 only the exponent of the Schmidt number is somewhat different. Now, Equation 172 is replaced by [22]:

$$\mathrm{Sh} \equiv \frac{d^{1/4}V^{1/4}k}{D} \propto F^{1/4}\,\mathrm{Re}^{3/4}\,\mathrm{Sc}^{1/2} \tag{178}$$

In the case of absorption through the free interface, the problem is somewhat more complicated. The elements of liquid brought to the interface by the turbulent fluctuations stay at the interface for a length of time Δ without experiencing deformations, after which they dissolve into the bulk, being replaced by other elements of liquid and so on. Equation 175 can be applied to such a situation. However, intuition now suggests that the time of contact depends not only on ξ_0, ρ, and η but also on σ, σ being the surface tension. Since the number of quantities is 5 and the number of dimensions involved is 3, two dimensionless groups can be formed and therefore, in contrast to the previous cases, dimensional analysis no longer provides an expression for Δ. In this case, dimensional analysis leads to

$$\varphi\left(\frac{\Delta}{\left(\frac{\rho\nu}{\xi_0}\right)^{1/2}}, \frac{\Delta}{\left(\frac{\rho}{\xi_0}\right)^{3/5}\left(\frac{\sigma}{\rho}\right)^{2/5}}\right) = 0 \tag{179}$$

In order to obtain explicit expressions for Δ, two extreme cases are considered. In one, Δ is assumed independent of σ, whereas in the other, Δ is assumed independent of ν. In the first case, one obtains

$$\Delta \propto \left(\frac{\xi_0}{\rho\nu}\right)^{-1/2} \text{ and } k \propto \left(\frac{D}{\nu}\right)^{1/2}\left(\frac{\nu\xi_0}{\rho}\right)^{1/4}, \tag{180}$$

whereas in the other case one obtains

$$\Delta \propto \left(\frac{\xi_0}{\rho}\right)^{-3/5}\left(\frac{\sigma}{\rho}\right)^{2/5} \text{ and } k \propto D^{1/2}\left(\frac{\xi_0}{\rho}\right)^{3/10}\left(\frac{\sigma}{\rho}\right)^{-1/5} \tag{181}$$

Substituting $\xi_0 = \frac{P}{V}$ and using Equation 171 for P, Equations 180 and 181 become, respectively,

$$k \propto D^{1/2}\nu^{-1/4}n^{3/4}F^{1/4}d^{5/4}V^{-1/4} \tag{182}$$

and

$$k \propto D^{1/2}n^{9/10}F^{3/10}d^{3/2}V^{-3/10}\left(\frac{\sigma}{\rho}\right)^{-1/5} \tag{183}$$

The dependence on n is stronger in the latter than in the former case. The experimental results of Davies, et al. [48] yield

$$k \propto n^{0.62} \tag{184}$$

which, when compared to Equations 182 and 183 indicates that the effect of viscosity is more important than that of the surface tension.

Heat Transfer Between a Fluid Flowing Through a Fixed or Fluidized Bed and a Wall

The quasi-steady laminar model is used to describe the motion near the wall. Therefore, Equation 170 is applied to the heat transfer coefficient. Note that while the shear stress at the wall can be related easily to the pressure drop for the flow in a tube, such a relation is more difficult to derive in a packed or fluidized bed. However, for the flow in a tube the dissipated energy is not uniform over the section of the tube. Such an assumption is more plausible for the flow through a packed bed which could be modeled as a bundle of tubes in each of which the behavior of the fluid is similar. The dissipated energy is given in this case by the expression

$$\xi = \frac{U\Delta p}{zS\varepsilon'} \tag{185}$$

where U = the volumetric flow rate
Δp = the pressure drop
z = the height of the bed
S = the cross-sectional area of the tube
ε' = free volume fraction of the bed.

However, for a fixed bed [46a]:

$$\Delta p = \lambda \frac{z}{d_e} \frac{v^2}{2} \rho \varepsilon' \tag{186}$$

where D_e = the equivalent diameter of the voids ($D_e = \frac{4\varepsilon'}{\sigma'}$, σ' being here the surface area of the particles per unit volume and ε' the porosity of the bed)
$v = U/S$
λ = the friction coefficient

The friction coefficient is related to the Reynolds number via the expressions [46a]:

$$\lambda = 140\left(\frac{vD_e}{\nu}\right)^{-1} \text{ for } \frac{vD_e}{\nu} < 40$$

$$\lambda = 16\left(\frac{vD_e}{\nu}\right)^{-1/5} \text{ for } \frac{vD_e}{\nu} > 40 \tag{187}$$

Combining Equations 170, 185, 186, and 187, yields

$$\frac{hD_e}{k'} \propto \left(\frac{vD_e}{\nu}\right)^{0.5} \mathrm{Pr}^{1/3} \text{ for } \frac{vD_e}{\nu} < 40$$

$$\frac{hD_e}{k'} \propto \left(\frac{vD_e}{\nu}\right)^{0.7} \mathrm{Pr}^{1/3} \text{ for } \frac{vD_e}{\nu} > 40$$

These expressions compare well with those obtained by experiment [49]:

$$\frac{hD_e}{k'} \propto \left(\frac{vD_e}{\nu}\right)^{0.5} \text{ for } \frac{vD_e}{\nu} < 40$$

$$\frac{hD_e}{k'} \propto \left(\frac{vD_e}{\nu}\right)^{0.8} \text{ for } \frac{vD_e}{\nu} > 40 \tag{188b}$$

A similar approach can be used for liquid fluidized beds. In this case

$$\frac{\Delta p}{z} = (1 - \varepsilon') g (\rho_s - \rho) \tag{189}$$

and

$$\frac{\xi}{\rho} = \frac{v(1 - \varepsilon') g (\rho_s - \rho)}{\rho \varepsilon'} \tag{190}$$

where ρ_s is the density of the particles.
Consequently,

$$\frac{h}{k'} = \beta^{1/3}\mathrm{Pr}^{1/3}\left(\frac{v(1-\varepsilon')vg(\rho_s-\rho)}{\varepsilon' v^3 \rho}\right)^{1/4} \tag{191}$$

Denoting by v_d the sedimentation velocity of the particle, the following expression relates v to ε' [50]

$$\frac{v}{v_d} = \varepsilon'^{4.6} \tag{192}$$

Equation 191 shows that h has a maximum for $\varepsilon' = 0.8$.

Equation 191 cannot be used for gas fluidized beds because in the latter case, the fluidized bed contains a large number of bubbles. The rate of heat transfer between the bed and wall is determined in the latter case by the heat transfer in the packets (clusters) of solid particles (through which the gas flows at the minimum fluidization velocity). The packets are exchanged, because of the bubbling, between the wall and the bulk of the fluidized bed. The heat transfer coefficient is given in the latter case by an expression similar to Equation 175

$$h = \left(\frac{4}{\pi}\rho'c'k'/\Delta\right)^{1/2} \tag{193}$$

where ρ', c', and k' are the density, specific heat, and thermal conductivity in the packets. In the packets, the void fraction is considered to be equal to that corresponding to the minimum fluidization velocity and the thermal conductivity to that of the corresponding fixed bed. The specific heat c' accounts for the heat capacity of the particles which constitute the packet.

The two-phase model of fluidized beds was used [52] to obtain an expression for Δ. In that model, the flow of gas in excess to that corresponding to the minimum fluidization velocity traverses the fluidized bed as bubbles. On this basis, one can write an equation for the volume fraction ψ occupied by the bubbles:

$$(1-\psi)(1-\varepsilon'_{min}) = 1-\varepsilon' \tag{194}$$

where ε' is the volume fraction occupied by the gas and ε'_{min} its value for the minimum fluidization velocity. For the velocity of the bubbles one can use the equation of Davis and Taylor [53]

$$v_b = 0.711\, g^{1/2} D_b^{1/2} \tag{195}$$

where D_b is the bubble diameter. On the basis of the two-phase model, one can relate v_b to ε' and v via the expression derived by Teoreanu [54]:

$$v_b - v_{min} = \frac{1-\varepsilon'_{min}}{\varepsilon' - \varepsilon'_{min}}(v - v_{min}) \tag{196}$$

where v is the superficial gas velocity.

Let us consider that the bubbles are uniformly distributed into the bed. The assumption of a simple cubical lattice for the arrangement of the bubbles in the bed leads to the following expression for the distance L between bubbles:

$$L = \left(\frac{\pi}{6}\right)^{1/3}\psi^{-1/3}D_b \tag{197}$$

If one considers that the renewal frequency Δ^{-1} is given by the renewal frequency of the bubbles in the vicinity of the wall, one can write

$$\Delta = L/v_b \tag{198}$$

Consequently, Equations 194 to 198 lead to [52]:

$$\frac{1}{\Delta} = 0.64\,g \frac{(\varepsilon' - \varepsilon'_{min})}{(1 - \varepsilon'_{min})^{1/3}\left[v_{min} + \dfrac{1 - \varepsilon'_{min}}{\varepsilon' - \varepsilon'_{min}}(v - v_{min})\right]} \tag{199}$$

This equation predicts for Δ^{-1} a value of about 6 sec^{-1} for the experimental conditions used by Mickley and Fairbanks, and is in satisfactory agreement with the experimental value.

Heat Transfer Between a Bubbling Bed and a Vertical Wall

Gas bubbling through a liquid generates two kinds of motions:

1. A general motion of circulation caused by the ensemble of bubbles
2. Local micro-motions generated by the individual bubbles.

The presence of the wall causes the number of bubbles there to be smaller than in the central region. This density gradient determines an ascending motion in the central region and a descending one near the wall. The rate of heat transfer is determined by the process which takes place near the wall. It is highly likely that the descending motion in this region is turbulent. Therefore, we use the quasi-steady laminar model to write

$$h \propto k' \left(\frac{u_0}{\nu x_0}\right)^{1/2} Pr^{1/3} \tag{200}$$

Dimensional analysis can provide expressions for the velocity u_0 and for the laminar path length x_0. Indeed, the circulatory motion is induced by the buoyancy force $g\Delta\rho$, where $\Delta\rho$ is the difference between the density near the wall (assumed to be equal to that of the liquid) and the density of the bubble bed

$$\Delta\rho = \rho - [\rho(1 - \varepsilon') + \rho'\varepsilon'] \simeq \varepsilon'\rho \tag{201}$$

where ρ and ρ' are the densities of the liquid and gas, respectively, and ε' is the gas volume fraction. Intuition suggests that u_0 and x_0 depend on $g\rho\varepsilon'$ and on the viscosity η and density ρ of the liquid. Dimensional analysis leads to

$$x_0 \propto \nu^{2/3}(g\,\varepsilon')^{-1/3} \text{ and } u_0 \propto \nu^{1/3}(g\varepsilon')^{1/3} \tag{202}$$

Consequently, Equation 200 becomes [55]:

$$h = Ck'\left(\frac{g\varepsilon}{\nu^2}\right)^{1/3}\left(\frac{\nu}{a}\right)^{1/3} \tag{203}$$

Experiment [55] provides the value $C = 0.28$ for the constant. Equation 203 is in good agreement with the results of experiments in References 55 to 58.

Boiling Heat Transfer from a Horizontal Surface in the Regime of Isolated Bubbles

Experiment [59, 60] has shown that, in the neighborhood of a bubble growing at an active site and detaching from it, strong forward and backward motions of the liquid take place.

As the number of active centers and the frequency of bubble generation increase, the bubble columns entrain the liquid almost continuously. This ascending motion of the liquid is compensated by a descending motion in the region of the space between the bubble columns and it is continued in the neighborhood of the heating surface with a movement quasi-parallel to the latter. Around every active center, a domain having a square form of area L^2 is ascribed, L being the distance between two such active centers [61]. Obviously, the rate of heat transfer is determined by the hydrodynamic process which takes place in the vicinity of the heating surface. Let us first assume that the motion along the surface is laminar. Though the motion of the liquid is more complex, we will assume for the sake of simplicity that it may be considered as the motion of a liquid along a plate, the distance covered by the liquid being taken as $1/2(L - R_0)$, where $R_0/2$ represents the average between the zero-size bubble and the size R_0 of the detaching bubble. For the heat transfer coefficient, evaluated as that along a plate, one obtains

$$h \propto k'\left(\frac{u_0}{\nu(L - R_0)}\right)^{1/2}\left(\frac{\nu}{a}\right)^{1/3} \tag{204}$$

where u_0 is the velocity of the descending liquid arrived at the plate. A factor $1 - (\pi/4)(R_0/L)^2$ should be included to account for the average surface area covered by bubbles.

The circulatory motion, just described is due to the buoyant force $g\Delta\rho$, where $\Delta\rho = \rho - [\rho(1 - \varepsilon') + \rho'\varepsilon'] = (\rho - \rho')\varepsilon'$, ρ being the density of the liquid, ρ' that of the vapor, and ε' the volume fraction of the vapor in the neighborhood of the surface. The velocity u_0 of the liquid should depend on $g\Delta\rho$, the kinematic viscosity ν, and density ρ. Since the number of physical quantities is four, while that of the dimensions involved is three, it follows that a single dimensionless group can be formed, which must be a constant. Consequently,

$$u_0 \propto \left(\frac{\rho - \rho'}{\rho} g\varepsilon'\nu\right)^{1/3} \tag{205}$$

Let us now consider that the motion of the liquid is turbulent and use the quasi-steady laminar model to describe turbulence. In this case, the heat transfer coefficient is given by the expression

$$h \propto k'\left(\frac{u_0}{\nu x_0}\right)^{1/2} \mathrm{Pr}^{1/3} \tag{206}$$

where x_0 is the length of the laminar path and u_0 is the velocity of the liquid at the start of the path. Note that, for both the laminar and turbulent cases the same equation is used. The difference lies in the fact that the laminar flow path is determined in the laminar case by the bubble spacing and in the turbulent case by the quantities which characterize the state of turbulence of the liquid. Now, both u_0 and x_0 depend on $g(\rho - \rho')\varepsilon'$, ρ, and η. Dimensional considerations lead to

$$u_0 \propto \left(\frac{\rho - \rho'}{\rho} g\nu\varepsilon'\right)^{1/3} \tag{207}$$

and

$$x_0 \propto \left(\frac{\rho - \rho'}{\rho} g\varepsilon'\right)^{-1/3} \nu^{2/3} \tag{208}$$

Combining Equations 206, 207, and 208 yields

$$h \propto k' \left(\frac{\rho - \rho'}{\rho} \frac{g\varepsilon'}{\nu^2}\right)^{1/3} \left(\frac{\nu}{a}\right)^{1/3} \tag{209}$$

A factor accounting for the contribution of the heating surface occupied by the bubbles should be included in Equation 209. Equation 209 was proposed by Zuber [62] on the basis of an analogy with turbulent free convection. The latter author has shown that Equation 209 is in good agreement with experiment. The previous discussion of boiling heat transfer emphasizes only the features of the problem related to turbulence and its treatment by scaling.

Mass Transfer to Separated Flows

In order to emphasize the versatility of the quasi-steady model, let us examine a problem which is somewhat different from the previous ones in the sense that the equation for a semi-infinite fluid flowing along a plate can no longer be used. For the turbulent motion in a tube, the shear stress $\tau \approx \tau_0 =$ constant near the wall, whereas, for turbulent separated flows, the shear stress is small at the wall near the separation point (becoming zero at this point) and depends upon the distance to the wall. In the last case, it is possible to write under certain conditions that [63]:

$$\tau = \tau_0 + y\alpha \tag{210}$$

where α can be assumed approximately equal to the pressure gradient $\partial p/\partial x$. The problem is extremely complicated because α is an unknown function of x. To gain some insight into the problem, let us consider only the region very close to the separation point and assume

$$\tau = \alpha y \tag{211}$$

with $\alpha =$ const.

Introducing the latter expression in

$$\tau = \eta \frac{du}{dy} \tag{212}$$

one obtains

$$u = \frac{\alpha}{2\eta} y^2 \tag{213}$$

The convective diffusion equation therefore takes the form

$$\frac{\alpha}{2\eta} y^2 \frac{\partial c}{\partial x} = D \frac{\partial^2 c}{\partial y^2} \tag{214}$$

This equation can be easily integrated exactly by using a similarity transformation. The scaling approach is, however, extremely simple and will be used in what follows. The terms of Equation 214 can be evaluated by

$$\frac{\alpha}{2\eta} y^2 \frac{\partial c}{\partial x} \sim \frac{\alpha}{2\eta} \delta^2 \frac{\Delta c}{x}$$

and (215)

$$D \frac{\partial^2 c}{\partial y^2} \sim D \frac{\Delta c}{\delta^2}$$

Therefore

$$\delta^4 \propto \frac{\eta x}{\alpha} \text{ and } k = \frac{D}{\delta} \propto D^{3/4}\left(\frac{\alpha}{\eta}\right)^{1/4} x_0^{-1/4} \tag{216}$$

The path length x_0 is a function of α, η, and ρ. Dimensional considerations lead to

$$x_0 \propto \nu^{1/3}(\alpha/\eta)^{-1/3} \tag{217}$$

and, consequently,

$$k \propto D^{3/4}\nu^{-5/12}(\alpha/\rho)^{1/3} \tag{218}$$

In the vicinity of the separation point

$$\alpha \approx \frac{\partial p}{\partial x} \propto U^2 \tag{219}$$

Equation 218 becomes

$$k \propto D^{3/4}\nu^{-5/12}U^{2/3} \tag{220}$$

It is of interest to note that, Richardson [64] has obtained by experiment, a value of 2/3 for the exponent of the velocity, while studying the heat and mass transfer in turbulent separated flows.

Discussion

In the conclusion of this section one can emphasize that the double scaling procedure can provide useful information, for a large number of problems. The fact that the micro-structure is characterized by a length scale which is small in comparison to any characteristic dimension of the system ensures some universality to the procedure in the sense that the same equation is valid in a variety of situations. The model based on a succession of short laminar motions was applied to predict the rate of transfer for the flow in a tube, for stirred vessels, for fixed and fluidized beds, for bubble beds, for boiling heat transfer, and for separated flows. In the last case, the form of the equation for the laminar path had to be changed because of the y directional dependence of the shear stress near a separation point. In the case of a fluid-liquid interface also, the form of the equation is different because the shear stress is zero. The essence of the procedure, however, remains the same.

The shortcoming of the renewal model approach to turbulence arises from the fact that the renewal assumption, while it is plausible, is still not a result of the equations of motion. However, for the mass transfer through a liquid-fluid interface the time dependent convective-diffusion equation was solved by assuming a periodical, time dependent, roll cell velocity field [65]. In this way, one of the results is the same as that obtained on the basis of the renewal assumption. However, the procedure no longer retains any essential element of that model, since renewal becomes a natural consequence of the assumed velocity distribution. Statistical elements (disorder) could be introduced in the previously ordered approach, by assuming a statistical distribution for the amplitude and/or for the wave number which appear in the velocity distribution. The overall mass transfer coefficient is obtained by averaging over the considered ensemble. Let us also note that, deterministic time-dependent velocity distributions can also have the disordered features of the turbulent motion and their use can avoid the need of a statistical distribution [66]. Indeed, an almost periodical function such as

$$u = B_1 \sin w_1 t + B_2 \sin w_2 t, \tag{221}$$

in which the ratio w_1/w_2 is an incommensurable number, has a disordered time dependence somewhat similar to that caused by turbulent fluctuations. The above function is surely too simple to simulate the "chaotic" unsteadiness of turbulence and therefore some other functions should be considered. Since the time dependence of the mass transfer coefficient is due to that of the velocity field, the former has the disordered features of the latter. This means that there is no simple periodicity in the time dependence of k on t and therefore its time average has to be defined by the limit:

$$\langle k\rangle = \frac{1}{T}\int_0^T k\,dt \qquad T\to\infty \tag{222}$$

Let us divide T into a large number of intervals and let us consider that the number of intervals in $d\tau$ is $\psi(\tau)d\tau$ and that, for each interval τ, there is a corresponding value of $k = k(\tau)$. Equation 222 can be therefore rewritten as

$$\langle k\rangle = \frac{\int_0^\infty k(\tau)\,\psi(\tau)\,d\tau}{\int_0^\infty \psi(\tau)\,d\tau} \tag{223}$$

which has the form of a statistical average, similar to that proposed in the renewal model of Danckwerts [19].

The fact that a deterministic function of the form of Equation 221 can simulate the chaotic time dependence of the turbulent velocity field suggests that there are deterministic solutions of the equations of motion which describe the turbulent phenomenon. Consequently, the most appropriate procedure to solve the problem of turbulent heat or mass transfer is to obtain the turbulent velocity distribution as the deterministic solution of the appropriate equations of motion and to use these velocity components in the convective diffusion equation. However, the computation of the deterministic turbulent velocity is a formidable problem. Progress has been achieved recently in the understanding of a more simple problem, namely, that of the wave motion of a thin film flowing along a vertical wall [67]. The unsteady equations of motions have been solved to explain, in a deterministic way, the chaotic dependence of the film thickness on time. The mass transfer problem is now being solved in this laboratory for the same situation.

NOTATION

A	constant
a	thermal diffusivity
B′	constant in Equation 143
C	constant in Equation 203
c	concentration
c_p	specific heat
c′	specific heat in fluidized-bed case
D	diffusion coefficient; vessel diameter in heat transfer discussion
D_b	bubble diameter
D_e	equivalent diameter
Da	Damkohler number
d	diameter; distance; characteristic dimension of system
f	friction factor
Gz	Graetz number
g	gravitational acceleration
h	heat transfer coefficient
k	consistency index in Equation 164
k_1, k_I	constants
k, k_{II}	mass transfer coefficient
k′	thermal conductivity
L, ℓ	length
M	dimensionless mass transfer group
m	constant
N	mass transfer rate or flux
Nu	Nusselt number
n	number of revolutions per unit time in Equation 171; exponent
P	mechanical energy per unit time and volume
Pe	Peclet number
Pr	Prandtl number
p	pressure

R radius
Ra Rayleigh number
Re Reynolds number
r distance from center of tube
S cross-sectional area
Sc Schmidt number
Sh Sherwood Number
T temperature; time
t time
U velocity; volumetric flow rate
u_m average velocity
u_0 local velocity
$\bar{u}$ average local velocity
u, v x and y velocity components, respectively
V volume
v superficial gas velocity
v_b bubble rise velocity
v_d particle settling velocity
v' normal component of velocity component
W plate width
x distance
y distance normal from wall
z bed height

Greek Symbols

α exponent; parameter approximated by pressure gradient in Equation 210
α_0 dimensionless constant in Equation 124
β volumetric coefficient of expansion; exponent
γ_0 constant in Equation 90
Δ function defined in Equation 148; time
δ boundary layer thickness
ε turbulent diffusivity
ε' void fraction of bed
η apparent viscosity
θ angle
λ friction factor
ν kinematic viscosity
ν' power law kinematic viscosity defined by Equation 165
ξ_m energy dissipated in agitation
ρ density
ρ_s particle density
σ surface tension
σ' surface area of particles per unit volume
τ time interval; shear stress
τ_0 wall shear stress
ψ volume fraction occupied by bubbles

REFERENCES

1. Landau, L. D., and Lifshitz, E. M., *Fluid Mechanics,* Pergamon Press, London, 1959.
2. Kays, W. M., and Crawford, M. E., *Convective Heat and Mass Transfer,* 2nd ed., McGraw-Hill, New York, 1980.
3. Prandtl, L., *Verhandlungen des dritten internationalen Mathematiker-Kongress* (Heidelberg 1904), Leipzig, p. 484, 1905.
4. Schlichting, H., *Boundary Layer Theory,* 6th ed., McGraw-Hill, New York, 1968.
5. Levich, V. G., *Physicochemical Hydrodynamics,* Prentice Hall, Inc., Englewoood Cliffs, N. J., 1962.
6. Ruckenstein, E., *Rev. (Roumaine) de Physique.* 7: 153 (1962); *Chem. Eng. Sci.* Vol. 37, p. 1505 (1982).
7. Ruckenstein, E., *A. I. Ch. E. J.,* 26: 830 (1980).
8. LeFevre, E. J., Proc. 9th Intern. Congr. Appl. Mech., Vol. 4, p. 168, Belgium, 1957.
9. Churchill, S. W., and Ozoe, H., *Journal Heat Transfer,* Trans. ASME, Series C, 95: 540 (1973).
10. Ruckenstein, E., and Rajagopalan, R., *Chem. Eng. Commun.,* 4: 15 (1980).
11. Blasius, H., *Z. Math. Phys.,* 56: 1 (1908).
12. Churchill, S. W., *A. I. Ch. E. J.,* 23: 10 (1977).
13. Acrivos, A., *Chem. Eng. Sci.,* 21: 343 (1966).
14. Sparrow, E. M., and Gregg, J. L., *J. Appl. Mech.,* 26: Trans. ASME, 81E, 133 (1959).

15. Levèque, M. A., *Ann. Mines,* 13: 201, 305, 381 (1928).
16. Astarita, G., *Mass Transfer with Chemical Reaction,* Elsevier, 1967.
17. Danckwerts, P. V., *Gas-Liquid Reactions,* McGraw-Hill, 1970.
18. Higbie, R., *Trans. A. I. Ch. E. J.,* 31: 365 (1935).
19. Danckwerts, P. V., *Ind. Eng. Chem.* 43: 1468 (1951).
20. Kramers, H., and Beek, W. J., Accad. Naz. Lincei, Fasci, *Alta Technol. Chim. V° Corso Estivo di Chimica,* Varese, p. 25, 1960.
21. Toor, H. L., and Marchello, J. M., *A. I. Ch. E. J.,* 4: 97 (1958).
22. Ruckenstein, E., *Rev. Chim. (Bucharest),* 8: 749 (1957); 9: 689 (1958).
23. Stewart, W. E., and Prober, R., *Int. J. Heat Mass Transfer,* 5: 1149 (1962).
24. Ruckenstein, E., and Rajagopalan, R., *Indian Chem. Eng.,* Recent Advances in the Engineering Analysis of Chemically Reacting Systems, L. Doraiswamy (Ed.), Wiley Eastern (1984).
25. Hadamard, J., *C. R. Acad. Sci. Paris,* 152: 1735 (1911).
26. Rybczynsky, W., Bull. Acad. Sci. Cracovie, A-40 (1911).
27. Ruckenstein, E., *Chem. Eng. Sci.,* 19: 131 (1964).
28. Boussinesq, J., *Ann. Chim. Phys.,* 29: 364 (1913).
29. Frössling, N., *Lunds Univ. Arrsskr. F. F. AVD,* 36: 4 (1940).
30. Akselrud, G. A., *Zh. Fiz. Khim. SSSR,* 24: 1446 (1953).
31. Garney, F. H., and Keey, R. B., *Chem. Eng. Sci.* 9: 119 (1958).
32. Cess, R. D., *Journal Heat Transfer,* Trans. ASME, Series C, 83: 274 (1961).
33. Murphree, E. V., *Ind. Eng. Chem.,* 24: 726 (1932).
34. Reichardt, H., *Z. A. M. M.* 31: 208 (1951); Lin, C. S., Moulton, R. W., and Putnam, G. L., *Ind. Eng. Chem.* 45: 636 (1953).
35. Vieth, W. R., Porter, J. H., and Sherwood, T. K., *Ind. Eng. Chem. Fundls.* 3: 1 (1963).
36. Frank-Kamenetsky, D. A., *Diffusion and Heat Transfer in Chemical Kinetics,* Plenum Press, New York, 1969.
37. Harriot, P., and Hamilton, R. M., *Chem. Eng. Sci.* 20: 1073 (1965).
38. Fage, A., and Townsend, H. C. H., *Proc. R. Soc.,* A135: 656 (1932).
39. Ruckenstein, E., *Chem. Eng. Sci.,* 7: 265 (1958); 18: 233 (1963).
40. Kline, S. J., and Rundstadler, P. W., *J. Appl. Mech.,* 26: 166 (1959); Bakewell, H. P., Jr., and Lumley, J. L., *Phys. Fluids,* 10: 1880 (1967); Kline, S. J., Reynolds, W. D., Schraub, F. A., and Rundstadler, P. W., *J. Fluid Mech.,* 30: 741 (1967).
41. Pinczewski, W. V., and Sideman, S., *Chem. Eng. Sci.,* 29: 1969 (1974).
42. Corino, E. R., and Brodkey, R. S., *J. Fluid Mech.,* 37: 1 (1969).
43. Ruckenstein, E., *Studii si Cercetari Fizica (Bucharest),* 9: 190 (1958).
44. Ruckenstein, E., *Chem. Eng. Sci.,* 27: 947 (1972).
45. Ruckenstein, E., *Bul. Sci. Ser. Mat.-Fiz.,* Acad. R. P. R., 6: 449 (1954).
46. (a) Bratu, E. A., *Operations and Equipments in Chemical Engineering,* Vol. I, 2nd ed., Ed. Technica, Bucharest, 1969.
    (b) Bratu, E. A., *Operations and Equipments in Chemical Engineering,* Vol. II, 2nd ed., Ed. Technica, Bucharest, 1961.
47. Hickson, A. W., and Baum, I. S., *Ind. Eng. Chem.,* 33: 1433 (1941); *Ind. Eng. Chem.,* 33: 1478 (1941).
48. Davies, J. T., Kilner, A. A., and Ratcliff, G. A., *Chem. Eng. Sci.,* 19: 583 (1964).
49. Yagi, S., and Wakao, N., *A. I. Ch. E. J.,* 5: 79 (1959).
50. Richardson, J. F., and Zaki, W. N., *Trans. Inst. Chem. Eng. (London),* 32: 35 (1954).
51. Mickley, H. S., and Fairbanks, D. F., *A. I. Ch. E. J.,* 1: 374 (1955).
52. Ruckenstein, E., *Ind. Eng. Chem. Fundamentals,* 5: 139 (1966).
53. Davidson, J. F., and Harrison, D., *Fluidized Particles,* Cambridge University Press, 1963.
54. Teoreanu, I., *Rev. Chim. Bucharest,* 11: 691 (1960).
55. Ruckenstein, E., and Smigelschi, O., *Trans. Inst. Chem. Engrs. (London),* 43: 334 (1965).

56. Kast, W., *Int. J. Heat Mass Transfer,* 5: 329 (1962); *Chem. Ing. Tech.*, 35: 785 (1963).
57. Sokolov, V. N., and Salamanin, A. D., *Zh. Prikl. Khim.*, 35: 1022 (1962).
58. Sokolov, V. N., and Bushkov, M. D., *Zh. Prikl. Khim.*, 37: 639 (1964).
59. Jakob, M., and Linke, W., *Phys. Z.,* 36: 367 (1935).
60. Jakob, M., and Fritz, W., *Forsch. Ver. Dt. Ing.,* 2: 435 (1931).
61. Ruckenstein, E., *Int. J. Heat Mass Transfer,* 7: 191 (1964).
62. Zuber, N., *Int. J. Heat Mass Transfer,* 6: 53 (1963).
63. Townsend, A. A., *J. Fluid Mech.,* 11: 97 (1961).
64. Richardson, P. D., *Chem. Eng. Sci.*, 18: 149 (1963).
65. Ruckenstein, E., and Berbente, C., *Chem. Eng. Sci.,* 25: 475 (1970).
66. Ruckenstein, E., *Chem. Eng. Journal,* 2: 1 (1971).
67. Sivashinsky, G. I., and Michelson, D. M., *Prog. Theor. Phys.* 63: 2112 (1980); Shlang, T., and Sivashinsky, G. I., *J. Physics,* 43: 459 (1982).

CHAPTER 3

HEAT FLUX IN THE BENARD-RAYLEIGH PROBLEM

M. Zamora, A. Cordoba, and A. Rey de Luna

Departamento de Termologia
Universidad de Sevilla, Spain

CONTENTS

INTRODUCTION

Ever since Rumford [1] established that convection as an energy transport mechanism was very different from conduction, innumerable scientific and technical contributions on these mechanisms and the effects of natural convection have followed. These contributions were at first either descriptive or excessively specific, making it difficult to justify structured theoretical knowledge which would enable prediction and understanding of different convective behavior.

As of 1900, the situation changed radically due to Bénard's work [2], which offered for the first time a reproducible and controlled systematic study of thermal convection in the laboratory. This radical change is also due to Lord Rayleigh's theoretical interpretation [3] in which the origin of the movement is attributed to the forces of buoyancy counterbalanced with dissipative forces. About 1930 activity in this field began increasing and still shows no signs of abating.

Interest in this subject ranges from the initial cause to the discovery of the mechanisms and the laws that govern convective heat transfer, to the fact that thermal convection allows us to study in detail the movement of fluid in a nonlinear regime. At the same time, the diversity of the phenomenology found through deeper study has made thermal convection and particularly the Bénard–Rayleigh problem appear as a generic case. In the simplest case there is a broad range of microscopic problems related to physical symmetry and break-down particularly in systems far removed from equilibrium, which lead continuously and controllably to chaotic turbulence.

In this chapter, we shall restrict ourselves to the description of heat flux and energetic behavior in the Bénard–Rayleigh problem, which is basically characterized by a shallow liquid layer heated from below and contained between two horizontal, parallel, solid surfaces that are much better heat conductors than the fluid itself.

The Bénard–Rayleigh problem has produced an extensive bibliography considering many different aspects of the phenomenon. See, for example, References 4 through 14 which provide an overview of the progress made in this problem.

THEORETICAL ASPECTS

Basic Equations

To simplify all of the mathematical development, we shall make use of the Oberbeck–Boussinesq [15–16] approximation for equations of movement. Density is considered constant except with respect to the force of gravity, and is then assumed to be linearly dependent, for a pure fluid, exclusively on temperature:

$$\rho = \rho_0[1 - \alpha(T - T_0)] \tag{1}$$

Assuming the characteristics of the fluid to be constant, the equations for conservation of mass and movement, under this approximation are:

$$\vec{\nabla}' \cdot \vec{v}' = 0 \tag{2}$$

$$\frac{\partial \vec{v}'}{\partial t'} + (\vec{v}' \cdot \vec{\nabla}')\vec{v}' = -\frac{1}{\rho_0}\vec{\nabla}' p + \frac{\rho}{\rho_0}\vec{g} + \nu \nabla'^2 \vec{v}' \tag{3}$$

and neglecting the term for viscous dissipation, the equation for conservation of energy is:

$$\frac{\partial T}{\partial t'} + (\vec{v}' \cdot \vec{\nabla}')T = \varkappa \nabla'^2 T \tag{4}$$

where ρ = density
ρ_0 = the density at the reference temperature T_0
T = temperature
$\vec{v}'$ = velocity
t' = time
p = pressure
$\vec{g}$ = gravity acceleration vector
ν = kinematic viscosity
α = expansion coefficient
$\varkappa$ = thermal diffusivity

The temperature and pressure fields can be expressed:

$$T = T_0 - \frac{\Delta T}{h} z' + \theta' \text{ and } p = p_s + \pi' \tag{5}$$

where ΔT is the temperature difference between the lower and upper boundaries of the fluid layer, h is its thickness, and z' denotes the vertical coordinate measured positively upwards, taking its origin in the center of the fluid, which when at rest yields $\theta' = 0$ and $\pi' = 0$, obtaining from Equations 3 and 5:

$$\frac{\partial p_s}{\partial z'} = -g\rho_0\left(1 + \frac{\alpha \Delta T}{h} z'\right)$$

with p_s corresponding to static pressure.

Changing the variables:

$$(x',y',z') = h\,(x,y,z);\; t' = \frac{h^2 t}{\varkappa};\; \theta' = \Delta T\theta;\; \pi' = \frac{\varkappa\nu\rho_0}{h^3}\pi;\; \vec{v}' = \frac{\varkappa}{h}\vec{v} \tag{6}$$

we can formulate the problem in a dimensionless way, with the results:

$$\vec{\nabla}\cdot\vec{v} = 0 \tag{7}$$

$$P^{-1}\left(\frac{\partial\vec{v}}{\partial t} + \vec{v}\cdot\vec{\nabla}\vec{v}\right) = -\vec{\nabla}\pi + R\theta\vec{k} + \nabla^2\vec{v} \tag{8}$$

$$\frac{\partial\theta}{\partial t} + \vec{v}\cdot\vec{\nabla}\theta = \nabla^2\theta + w \tag{9}$$

where $\vec{k}$ is the unitary upwards vertical vector, w the component z of velocity and P and R are dimensionless Prandtl and Rayleigh numbers given by:

$$P = \frac{\nu}{\varkappa} \text{ and } R = \frac{\alpha g h^3}{\varkappa\nu}\Delta T \tag{10}$$

Applying the curl operator to Equation 8, we manage to eliminate the dependent variable pressure, leaving only $(\vec{v}, \theta)$ as unknown functions.

The group of Equations 7 through 9 reflect the local behavior of the fluid and should be solved under appropriate boundary conditions. Two different situations may be considered among the boundary conditions for velocity: rigid walls and stress-free walls. If v_i is the normal component of the wall, it can be deduced that:

$$v_i = \frac{\partial v_i}{\partial x_i} = 0 \qquad \text{at rigid walls}$$

$$v_i = \frac{\partial^2 v_i}{\partial x_i^2} = 0 \qquad \text{at stress-free walls}$$

With respect to the boundary conditions for temperature, the horizontal plates are usually assumed to be perfect conductors ($\theta = 0$).

As we can see, the characteristic quantities of this problem are the fields of temperature and velocity. However, there are global quantities, typical of the system as a whole, which characterize convective patterns. The total heat flux through the fluid layer is a quantity of energy transfer inherent in every stationary state. Another global quantity typical of the fluid's stationary state is the energy contained therein.

Let us now look at the theoretical expression of these two quantities. In a stationary regime, Equations 2 and 4 tell us that the vectorial field $\varkappa\vec{\nabla}'T - \vec{v}'T$ has no scalar source nor sink in the fluid. Thus, the independent quantity of the coordinate z′

$$q = \int_{z' = \text{constant}} \rho c(-\varkappa\vec{\nabla}'T + \vec{v}'T)\cdot d\vec{S}_{z'} = \frac{1}{h}\int_{V'} \rho_o c\left(w'T - \varkappa\frac{\partial T}{\partial z'}\right)dV' \tag{11}$$

will represent the heat per unit of time that passes through the fluid. Substituting the expression of T given in Equation 5 and allowing for the conservation of mass and the boundary conditions of temperature, we arrive at

$$q = \frac{KS}{h}\Delta T + \frac{\rho_0 c}{h}\int w'\theta' \, dV' \tag{12}$$

where K is the thermal conductivity of the fluid.

The first term of the previous equation represents the heat flux transferred by conduction, to which must be added that transferred by convection, the second term, to obtain the total heat flux.

Defining the dimensionless Nusselt number as the relationship between total heat flux and heat transferred by conduction, we arrive at:

$$Nu = 1 + \frac{1}{\varkappa S \Delta T}\int w'\theta' \, dV' \tag{13}$$

and if we make use of the transformation equations (Equation 6), we obtain:

$$Nu = 1 + < w\theta > \tag{14}$$

where the bracket express the average value of the function considered for the whole fluid.

The second global quantity typical of every stationary state is structural energy. For each stationary situation of a thermodynamic system, it is possible to assign an energy, called contained energy, as an extension of the concept of internal energy [17–18]. Like internal energy, structural energy has no natural origin, so that we must choose a state of reference. Applying the following assumptions (1) the state of reference is that of the fluid, while maintaining the same temperature difference (2) the fluid is heated from above at the same temperature differential, then the contained energy will only be due to the convective structure, and it is referred to as structural energy.

The energies in question in our problem are kinetic, gravitational, and thermal in nature. A simple evaluation of these energies leads us to the conclusion that the one which stands out is the thermal type. Thus, the structural energy will be:

$$E = \int c\,(\rho T - \rho_r T_r) \, dV'$$

where the subindex r lets us known that we are dealing with the state of reference, that is

$$\rho_r = \rho_0[1 - \alpha(T_r - T_0)];\; T_r = T_0 - \frac{\Delta T}{h} z' \tag{15}$$

so that, taking into account Equations 1, 5 and 15, we arrive at:

$$E = \rho_0 c \int (1 - \alpha T_0)\,\theta' \, dV' - \rho_0 c \int \theta' \left(\theta' - 2\,\frac{\Delta T}{h} z'\right) dV' \tag{16}$$

The first integral is zero, given that the total fluid mass is the same for the reference situation as for that of convection:

$$0 = \int (\rho - \rho_r)\, dV' = -\,\alpha \int \theta' \, dV' \tag{17}$$

Making use of the transformation equations (Equation 6), the result is:

$$Z = \frac{E}{Mc\alpha\Delta T^2} = 2 < \theta z > - < \theta^2 > \tag{18}$$

where M is the total fluid mass. A new dimensionless number has been defined to express the structural energy, which we have designated Z.

The group of local quantities such as θ and $\vec{v}$, and global ones such as Nu and Z as functions of the independent parameters R and P give a complete view of the Bénard–Rayleigh problem.

Linear Analysis

Initially Rayleigh [3] made a linear analysis assuming stress-free horizontal walls, and this study was followed by Jeffreys [19] and Low [20]. But Pellew and Southwell [21] were the first to perform a rigorous linear analysis. They (and also Low, by a different method) found an exact solution for rigid horizontal walls, which is generally a more difficult case (albeit a more realistic one) to deal with than the stress-free walls. We owe the definitive treatment of the problem to Reid and Harris [22]. A very complete study of the linear problem is found in Chandrasekhar's book [4].

In linear analysis of stability, perturbations are considered to be small, and thus the nonlinear terms $\vec{v} \cdot \vec{\nabla} v$ and $\vec{v} \cdot \vec{\nabla} \theta$ are negligible in Equations 8 through 9. These equations are reduced to:

$$P^{-1} \frac{\partial \vec{v}}{\partial t} = - \vec{\nabla} \pi + R\theta\vec{k} + \nabla^2 \vec{v} \tag{19}$$

$$\frac{\partial \theta}{\partial t} = \nabla^2 \theta + w \tag{20}$$

Stability means stability with respect to any infinitesimal arbitrary perturbation. Any arbitrary perturbation is expressed as a superposition of certain basic modes and stability is analyzed according to these modes. Instability with respect to any one of them provides a sufficient condition of instability in general. If $A(\vec{r}, t; \alpha)$ is the amplitude of a perturbation (in velocity or temperature), with α being the wave number of that perturbation, the time dependence can be separated thus:

$$A(\vec{r}, t; \alpha) = A(\vec{r}, \alpha)\, e^{\sigma(\alpha) t} \tag{21}$$

where $\sigma(\alpha)$ is a constant that must be determined. The condition of stability implies that the real part of σ must be negative, so that any disturbance is lessened with time. This is why the condition $\sigma(\alpha) = 0$ determines the marginal state, from which point on instability is produced. It can be demonstrated that in the Bénard–Rayleigh problem σ is always real (the sometimes called "principle of exchange of stability"), in such a way that the stable state corresponding to the onset of convection is stationary. This is why the only relevant parameter for determining whether the quiescent state is stable or not is the Rayleigh number, and its value for when convection begins is called the critical Rayleigh number, R_c. The stationary solution of Equations 7, 19 and 20 takes on the form:

$$A = \sum_n c_n e^{i\vec{k}_n \cdot \vec{r}} f(z, \alpha) \tag{22}$$

where $\vec{r}$ is the vector of position and $|\vec{k}_n| = \alpha$ and $\vec{k}_n \cdot \vec{k} = 0$ for all n. The function $f(z, \alpha)$ is determined by an ordinary differential equation with α as a parameter. For stress-free horizontal walls, $f(z, \alpha)$ does not depend on α and the results are

$$f(z, \alpha) = \sqrt{2} \cos \pi z; \; R(\alpha) = \frac{(\pi^2 + \alpha^2)^3}{\alpha^2} \tag{23}$$

The minimum value of the function $R(\alpha)$ produces R_c and α_c

$$R_c = \frac{27\pi^4}{4}; \; \alpha_c = \frac{\pi}{\sqrt{2}} \tag{24}$$

In the most realistic case of rigid horizontal walls, the numerical calculations lead to:

$$R_c = 1707.76 \qquad \alpha_c = 3.117 \tag{25}$$

Linear analysis allows us to determine the values of R_c and α_c, but not the selected convective structure. There is a double degeneration of the eigenvalue R_c. On one hand, the infinite extension of the layer originates a degeneration of translation and orientation on the horizontal plane. When a finite horizontal extension is imposed, the continuous function $R(\alpha)$ will break into a group of discrete values, restricting degeneration, but is also necessary to specify the geometric configuration of the lateral walls. Thus, this degeneration cannot be saved without considering the effect of these walls.

On the other hand, there is the degeneration associated with the arbitrary choice of the coefficients c_n in Equation 22. So that Equation 22 can represent a real expression, the following conditions are imposed:

$$c_{-n} = c_n^+; \; \vec{k}_{-n} = -\vec{k}_n \tag{26}$$

The $^+$ sign denotes a conjugate complex, and assuming that there are N different vectors $\vec{k}_n$, $(N \to \infty)$ (Equation 22) can be expanded. We must also impose the normalization condition

$$\sum_{-N}^{N} |c_n|^2 = 1 \tag{27}$$

But the conditions of Equations 26 and 27 do not determine the unique choice of the coefficients c_n. Although part of this arbitrariness reflects the already mentioned degeneration of orientation and translation, the freedom of choice also reflects the wide variety of different forms of convection movements. This part of the degeneration problem can be solved considering the nonlinear problem and the stability of its solutions.

Nonlinear Developments

The only systematical method of analyzing the solutions of the nonlinear Equations 8 through 9 consists of carrying out a perturbation analysis as a function of a small parameter that gives a measurement of the extent of the perturbations. Obviously, development in perturbations is of limited worth at values of R not much greater than R_c. Thus, for high values of R, it is necessary to fall back on numerical methods.

The perturbation method was first used, independently of each other, by Gorkov [23], and Malkus and Veronis [24]. Using a perturbative outline slightly different from the previous ones, Kuo [25] obtained a good approximation for bidimensional solutions up to approximately $R = 8R_c$. Later, a systematic analysis of the variety of solutions to Equation 22 was carried out by Schluter, Lortz, and Busse [26]. While Malkus and Veronis researched the case as an infinite horizontal layer with free horizontal walls and two particular cases of tridimensional convection patterns, with rectangular and hexagonal cells, Schluter, Lortz, and Busse extended the method considering an arbitrary tridimensional flux with stress-free and rigid boundary conditions and infinite horizontal extension.

The applied perturbative method consits of introducing the expansion in series:

$$\begin{aligned} \vec{v} &= \varepsilon(\vec{v}^{(0)} + \varepsilon\vec{v}^{(1)} + \varepsilon^2\vec{v}^{(2)} + \ldots) \\ R &= R_c + \varepsilon R^{(1)} + \varepsilon^2 R^{(2)} + \ldots \end{aligned} \tag{28}$$

and analogous expressions for θ and π in Equations 7 through 9, proceeding to solve the linear equations that correspond to each power of ε. In a stationary regime, the term $\partial/\partial t$ is not considered and for ε^1 the result is the just-analyzed linear solution. For the order ε^2

nonhomogeneous linear equations are obtained, with the left-hand side terms given as a function of the solutions (Equation 22) with the conditions of Equation 26.

A solution of the nonhomogeneous system of equations exists if and only if the first member is orthogonal to all the solutions to the adjoining homogeneous problem (Fredholm alternative). It is easy to prove that this problem is self-adjoining.

The condition of solvability thus imposed as the nonhomogeneous part of the equations determines the coefficients $R^{(n)}$ and provides restrictions on the values of the coefficients c_n. The condition of solvability of the order ε^2 is trivially satisfied with $R^{(1)} = 0$. In the order ε^3, the following system of equations is obtained:

$$R^{(2)} |c_1|^2 - \sum_{n=-N}^{N} A(\vec{k}_1 \cdot \vec{k}_n) |c_n|^2 |c_1|^2 = 0 \text{ for } 1 = 1, \ldots, N \quad (29)$$

In the case where each vector $\vec{k}_n$ assumes the same set of angles with all the other vectors $\vec{k}$, it can be easily seen that:

$$|c_1|^2 = \ldots = |c_N|^2 = \frac{1}{2N} \quad (30)$$

represents a solution of Equation 29 along with the condition of Equation 27.

Only two of the solutions (Equation 30) are of physical importance. The simplest solution is given by N = 1, and corresponds to convection in the form of bidimensional rolls. For N = 3, they are hexagonal cells.

By solving Equation 29 convective heat flux may be calculated. In the case of free horizontal walls for bidimensional rolls, we obtain

$$H = (Nu - 1) R = 2(R - R_c) \quad (31)$$

and for the case of rigid walls:

$$\frac{H}{R - R_c} = \frac{1}{0.69942 - 0.00472\ P^{-1} + 0.00832\ P^{-2}} \text{ (for rolls)}$$

$$\frac{H}{R - R_c} = \frac{1}{0.77890 + 0.03996\ P^{-1} + 0.06363\ P^{-2}} \text{ (for square cells)}$$

$$\frac{H}{R - R_c} = \frac{1}{0.89360 + 0.04959\ P^{-1} + 0.06787\ P^{-2}} \text{ (for hexagons)} \quad (32)$$

For Equations 31 and 32 we can see the formula for the initial slope of the convective heat flux curve for a given pattern. On the other hand, when P is very small, Equation 32 yields the result that thermal conduction dominates convection. Really, when $P \to 0$ the radius of convergence of the expansion becomes very small and we must be cautious in interpreting the results.

The condition of Equation 29 significantly restricts the number of possible solutions, but many solutions still exist. An analysis of their stability is then necessary to see which of them are physically feasible. Schluter, Lortz, and Busse have demonstrated that only bidimensional rolls are stable for a range immediately above R_c and even among these, are stable only for a determinate range of wave numbers. Several modes may be stable for one R, and they are physically feasible depending on the initial conditions, as experiments confirm.

Another type of stability analysis is provided by the so-called energy method, which gives a global criteria of stability and which is linked to Lyapunov's [27] theory of stability.

Details on this method applied to the Bénard–Rayleigh problem can be found in Joseph's book [28].

The Finite Box Problem

The first rigorous study of a totally confined fluid was that of Davis [29], who considered containers of rectangular horizontal section having perfectly conducting walls. Davis takes advantage of the fact that any tridimensional movement can be generated by superimposing a sufficient number of bidimensional movements. His analysis of linear stability, using Galerkin's method, demonstrated that convection had to begin in the form of finite rolls, parallel to the shortest lateral wall of the container. It is necessary to emphasize the tridimensional nature of the finite rolls, which have no velocity component in the axial direction but depend on the three spacial coordinates, contrary to what happens with rolls in layers which are not laterally delimited whose velocity does not vary in the axial direction. Davis's previsions agree with what Stork and Muller [30] observed in their experiments. When the thickness of the layer is the smallest dimension, finite rolls of approximately square geometry are predicted. Otherwise, the rolls are narrower. Finally, it is worth noting that R_c decreases rapidly to 1,708 when the horizontal dimension increases, so that for an aspect ratio of several units, we can anticipate that $R_c \sim 1{,}708$. Later studies by Segel [31], Davies– Jones [32] and Catton [33–35] obtained similar results. Drazin [36] analytically developed some of Davis's ideas for a simplified bidimensional model with rigid lateral walls and stress-free horizontal ones. The same model has been studied by Daniels [37], who, following Segel, developed the solution by means of the multiple-scales method. Daniels found that if heat flux exists through the lateral walls, the bifurcation is replaced by a smooth transition to convection to finite amplitude. This coincides with Ahlers's experimental results [38]. Similar studies have been carried out by Brown and Stewartson [39], Hall and Walton [40], Cross et al. [41–42] and Kidachi [43]. Notable studies of cylindrical containers includes that of Charlson and Sani [44–45], Catton and Edwards [46], Brown and Stewartson [47–48], Shaumeyer et al. [49] and Rosenblat [50].

Heat Flux Calculations

All of these studies basically consider the problem of stability and selection of the structure, but most do not obtain values for heat flux against the Rayleigh number. This has been carried out largely by numerical procedures using Galerkin's method or by finite differences. Fittings of empirical laws to experimental data have also been suggested by different authors (see, for example, References 51 through 53), but we will not analyze them here.

Besides Equations 31 and 32 by Schluter, Lortz, and Busse and those of Malkus and Veronis, expansions in truncated series have allowed for expressions of Nu as a function of $R - R_c$ by Kuo [25], Kuo and Platzman [54] and Veronis [55]. The expression:

$$\left\{\frac{R}{R_c}\,\mathrm{Nu} - 1\right\} \propto \left\{\frac{R}{R_c} - 1\right\}^{\beta} \tag{33}$$

with $\beta = 1.24$ in the range $R < 8R_c$ has been obtained by Kuo, and with $\beta = 1.26$ in the range $R < 50R_c$, by Veronis (he used Fourier's expansion).

In spite of the obvious advantages of an analytical approximation to the problem by means of expansion within a small parameter, this method is only applicable to small separations of the second order in amplitude. This produces satisfactory results up to two or three times the value of R_c. Even calculations of stationary movement up to the eighth order within a small parameter [25] for two-dimensional rolls produce results that are valid up to $R \sim 8R_c$.

Several authors, especially Busse, have carried out systematic heat flux studies for the supercritical region for different ranges of the Prandtl number. The analysis of the different structures formed and their ranges of stability in the $R - P - \alpha$ space (α = wave number) can be seen in References 12 and 14. For an infinite Prandtl number, Busse [56] has found that for R = 22,600 three-dimensional flux begins (bimodal transition) and the following approximate law is obtained.

$$\frac{H}{R - R_c} = \left\{\frac{R - R_c}{645.3}\right\}^{0.250} \tag{34}$$

for convective heat flux, H.

Clever and Busse [57] have calculated Nu against R for two-dimensional rolls, different wavelengths of the structure, and a wide range of Prandtl numbers, finding that Nu is significantly affected by α for $P = 0.025$, and the influence is much smaller for $P = 0.71$ and higher values. Heat flux is greater for α_c and decreases as α decreases. This influence of α at high P has been experimentally determined by Moreno et al. [58], who have shown that it is small, although detectable. However, for a given α the heat flux is greater for greater values of P, with the differences being more noticeable for small $R - R_c$, although the magnitude $(Nu - 1)/P^2(R - R_c)$ increases as P decreases, and for small values of $R - R_c$ it is approximately constant. Convective heat flux is relatively independent of P at R greater than a critical value on the order of 5×10^3 [59] within the range $0.001 < P < 0.71$. Above $R \simeq 7 \times 10^3$ convective heat flux approaches the law $Nu = 0.175\ R^{0.25}$ [60].

Frick and Clever [61] and Frick [62] have considered the influence of the factor A = height/width in heat flux, covering the range $0.01 < A < 100$ and $R_c < R < 50\ R_c$ and several values of α ($A = 0$ corresponds to the ideal Bénard–Rayleigh case, and $A \to \infty$ corresponds to the so-called Hele–Shaw convection). For high values of R, differences in Nu for different values of A become small, and grow larger as R decreases. Convection begins for higher R as A increases, but with a higher slope in the Nu(R) curve, and as R increases, the Nu values come closer to each other. For high values of A, heat flux is approximately proportional to $R^{0.3}$.

More recently, Frick et al. [63] have considered the stability of three-dimensional structures and heat flux for infinite P. A result of their analysis is that the pattern of convection associated with maximum heat flux is unstable, thus confirming that maximum heat flux bears no relation to stability. They have calculated Nu against α_1 and α_2 (wave numbers characteristic of bimodal patterns) for R up to 70×10^3. Bimodal convection presents a heat flux in excess of the bidimensional solution. The change in slope of the $Nu - R$ curve in the bimodal transition depends on the values of α_1 and α_2. It is more noticeable when α_2 differs greatly from α_1 (for $\alpha_1 = 3.117$ and $\alpha_2 = 5$), but when α_2 comes closer to α_1 the change in slope is very slight (for $\alpha_1 = 3.117$ and $\alpha_2 = 4$).

A number of authors have carried out calculations by means of finite differences (see References 64 through 77) and have obtained results for the Nusselt number against R with diverse boundary conditions and range of parameters. These results are most often presented in the form of a numerical table, so it is difficult to come up with a simple, general Nu(R) law. Comparing experimental data, the results differ in many cases and the calculations often over estimate Nu with respect to the experimental value. It is also very common, because of the simplicity of the calculations, to consider two-dimensional regimes, and the results are therefore not applicable to ranges of R in which convection is three-dimensional. For the two-dimensional case, an expression $(Nu - 1) \sim P^2(R - R_c)$ has been obtained for $P \leqslant 0.1$ in the case of stress-free walls. For P on the order of one, $\alpha = \alpha_c$, and for high values of R, the approximate laws

$$Nu = 1.8\,(R/R_c)^{0.365};\quad Nu = 1.56\,(R/R_c)^{0.296} \tag{35}$$

have been obtained for stress-free and rigid walls, respectively. We believe that all of the information thus far provided by these calculations should be taken with reservations, and there is still room for study in this field.

Finally, we would like to point out that to obtain asymptotic Nu (R) laws for very high Rayleigh numbers mean field approximations have been made [78–80] in which nonlinear terms have been retained only to the extent that they modify the average temperature gradient. The stability of the boundary layer has also been analyzed [81–83], with the asymptotic relationships obtained being:

$$\begin{aligned} &\mathrm{Nu} = 0.325\,\mathrm{R}^{1/3} && \text{stress-free boundaries} \\ &\mathrm{Nu} = 0.24\,\mathrm{R}^{3/2}\ln \mathrm{R}^{3/2} && \text{rigid boundaries} \end{aligned} \tag{36}$$

Variations of the Fluid Properties

Until now, the Oberbeck–Boussinesq hypothesis is considered to have been strictly followed. However, the properties of the fluid depend in general on temperature and to a much lesser degree on pressure. The range of validity of the Oberbeck–Boussinesq approximation has been analyzed by several authors [84–87]. When density is at a maximum, the so-called penetrative convection takes place in which the convective layer is less thick than the fluid layer [88–95]. The separation of the Oberbeck–Boussinesq hypothesis produces, as its primary effect, a modification of the value of R_c, since variations in the properties of the fluid give rise to stabilizing or destabilizing effects and its greater or lesser importance also depends on the boundary conditions [96–107]. A second matter has to do with the convective pattern that arises and consequently with heat flux. Palm [96] was the first to analyze this subject and found for the case of variations in viscosity with temperature that a hexagonal pattern, resulting from three suitable superposed rolls, was stable for $R > R_c$. Palm's initial study was later completed by Segel and Stuart [97], Palm and Oiam [99], Segel [100] and Palm et al. [103], and, separately, by Busse [101–102]. When R continues to increase, these hexagons turn into rolls [103]. The first experimental evidence of these facts was found by Hoard, Robertson, and Acrivos [108]. For other experiences in which the heat flux is determined, see the next section of this chapter. From Busse's analysis [102] what is outstanding is that there is a zone of values of R where rolls and hexagons can be stable, which could provoke hysteresis in the heat flux and give rise to subcritical instabilities, since convection in the form of hexagons could begin before R reaches the value of R_c. For great variations of viscosity numerical studies have been carried out by Torrance and Turcotte [109]. They studied only two-dimensional cases, which makes their predictions doubtful since, as we said, the hexagonal pattern may be stable.

EXPERIMENTAL ASPECTS

The study of macroscopic systems far from equilibrium presents two essentially different points of view, as much for their objectives as for their theoretical and experimental techniques. We shall call the first one the "global" point of view (the term "macroscopic" is not suitable here, as the second point of view is not of the microscopic type), and its purpose is to study the system as a whole in parametric situations from the energetic interaction between the system and its immediate exterior surroundings. The second point of view, however, is "local" and studies the behavior of the system, investigating its properties at all points, macroscopically speaking, of the space it occupies.

The global study of systems far from equilibrium can be considered for its methods and terminology as an extension of "classical thermodynamics" (thermostatics), while the local study belongs to the terms of "thermodynamics of irreversible processes".

Situations which are globally describable for a set of parameters (external variables) are sometimes not uniquely determined by this set. In such circumstances, internal constraints arise in the system, creating structures characterized by new parameters of the type known as "of order" (internal variables).

The Bénard–Rayleigh problem can be situated within either a local or a global context. Interest in the global study has been superseded by the more attractive local study of this problem. However, from the thermodynamic viewpoint this is the simplest system that can be found far from equilibrium, as it only has two external thermal sources. Besides, mechanical powers of convection appear clearly differentiated from heat fluxes by conduction-convection.

In the light of all of this, experimental research of the Bénard–Rayleigh problem is proposed considering the following aspects:

1. Parametrization of situations accesible to the fluid.
- External variables: Rayleigh number; Prandtl number; geometric parameters,
- Internal variables: Parameters characteristic of the structure (for example, the wave number).
2. Global study of the problem.
- Heat flux measurement through the fluid layer: Nusselt number.
- Measurement of structural energy.
3. Local study of the problem.
- Measurement of the distribution of velocity in the fluid.
- Measurement of the temperature field.

A new dimension of the problem is generated when the fluid is not only ordered in space, but also in time, giving rise to oscillations characteristic of the system.

From the theoretical and experimental point of view, the simple Bénard–Rayleigh system quite adequately accounts for the different behavioral patterns of systems far from equilibrium, shedding more and more light on these phenomena.

General Techniques

Heat flux measurement in the Bénard–Rayleigh problem has been carried out by means of two basically different experimental set-ups. The first one consists of providing a certain heat power for the lower plate and determining the heat flux thus transferred, from the heat exchanges between the horizontal metal plates. The second one fixes the temperature of the metallic blocks and determines the heat flux that passes through the system by inserting a detection element between the blocks and the fluid. It must be pointed out that these two set-ups use very different means of thermally binding the fluid: while the first one imposes the heat flux which passes through the fluid, the second one fixes the temperatures in the plates, letting the fluid choose the heat flux to be transferred. From here on, for this reason, we shall call the first one the flux method, and the second one, the temperature method. It is usually this second method which is used when other information apart from that related to heat flux is sought from the Bénard–Rayleigh problem.

The major problem in heat flux measurement is that of heat losses, which is worsened in the Bénard–Rayleigh problem when the intention is to simultaneously observe the convective patterns, for which observation windows are required. These losses, especially the lateral ones, can play an important role in the formation of convective patterns in the fluid, as Zamora and Rey de Luna [110] have recently shown. In any case, the evaluation or control of these heat losses determine the degree of accuracy of the heat flux measurements.

Let us analyze each of the methods in detail and describe their most noticeable characteristics:

The Flux Method

The experimental set-up can be briefly described considering a shallow layer of fluid contained between two metallic blocks. The lower one may be heated by a built-in electrical coil, while the upper one maintains a constant temperature. In this case, in addition to the lateral heat losses, we must take into account those of the base of the lower piece. For this reason, if the heat flux through the fluid is to be measured from the electrical power supplied, which is frequently the case, special precautions must be taken to eliminate heat losses through the base, using a passive or programmable adiabatic wall or a system with a high vacuum. When heat flux is measured from the energy absorbed by the refrigeration liquid of the upper block, it becomes imperative to very precisely determine the temperature increase of the refrigeration fluid between the entrance and exit, as well as the flow rate of this fluid.

A negative aspect of the experimental apparatus is that the lower temperature of the fluid may fluctuate throughout the experiments, depending on the pattern formed. This means that the flux method is hardly satisfactory for studying the behavior of fluid in oscillating situations. In fact, when the fluid is weakly bound thermally, new degrees of freedom are added to those typical of the physical problem, making the latter even more complicated. Another negative aspect must be taken into account when considering a range of temperature differences applied to the fluid. Since the upper temperature is kept constant, the results are up to a point heterogeneous, as the average temperature is continuously modified. All in all, the method is based on very simple technical principles which are only complicated by the necessary auxiliary systems.

The Temperature Method

In this case, the upper and lower blocks are thermostated by means of an externally circulating liquid, with a heat-flux detecting device inserted between them and the fluid. The measuring principle is simple; knowing the thermal resistance of the inserted device, and measuring the temperature difference between the two parts of the block containing it, the heat flux that passes through can be measured by merely applying Fourier's law. For reasons of accuracy, the material introduced is generally a poor conductor; however, this makes the interior temperature of the block fluctuate more, and lengthens the characteristic time of the whole experimental system. For these reasons, the most suitable arrangement is the use of a measuring device with low thermal resistance and a short characteristic time of thermal effects, which permits a high resolution measurement of the temperature increase that its presence originates. This has been achieved thanks to the heat fluxmeter, which provides excellent results, minimizing the previously exposed difficulties. The heat fluxmeter is essentially made up of a set of thermocouples placed in electrical series whose threads go in the same direction as the heat conduction, and thus after calibration allows us to determine heat flux from the electrical response produced by the Seebeck effect. As the material used for its construction is essentially metallic, its thermal resistance is small and the thermal relaxation time can be studied a priori to obtain the best one for each experiment. The incorporation of the heat fluxmeter in the upper and lower blocks allows us to determine lateral heat losses in the fluid and eventually, as will be seen later, to also determine other energetic aspects of convective structures. The incorporation of heat fluxmeters in the study of thermal convection has made available a clean and simple technique for solving important aspects of this problem.

Experimental Observations

The first studies of rates of heat transfer by convection in fluid confined between two horizontal plates and heated from below served as cornerstones for engineering calculations. Thus, Jacob [111] reports the work of Mull and Reiher [112], who studied the heat transfer of a flat air layer for high temperature increments ($R > 0.7 \times 10^4$), i.e. above the onset of convection.

The earliest experiments for the detection of the onset of thermal instability and the determination of the critical Rayleigh number appear to be those of Schmidt and Milverton

[113]. In their experiments, heat flux was used to detect the onset of convection, and the method turned out to be so direct and simple, that it has served as the basis for many later experiments in this field. The basic ideas of Schmidt and Milverton were very ingenious. In a steady state, when the heat flux through the fluid layer is entirely by conduction, it clearly increases linearly with the temperature difference applied. This linear relation must break down at the onset of thermal instability, when other modes of heat transfer become operative. The experimental technique used corresponds to the flux method, but it disregards heat flux and measures only the electrical intensity of the current in the heating coils. Although their experiments were not designed for accurate heat flux measurement, they obtained qualitative success. Their most important result was the critical Rayleigh number, $R_c = 1770 \pm 140$, obtained for water with four different layer depths.

Schmidt and Saunders [114] essentially repeated the previous experiment in order to attain more precision and to extend the scope of the earlier research. The experimental arrangement used presented serious disadvantages because the two pieces, upper and lower, were submerged in a big container of fluid, then the fluid outside the test section contributed significantly to the heat flux of the layer. Under these circumstances, the values of heat flux in convection were at least 25 percent higher than those of Mull and Reiher, but Schmidt and Saunders observed a second break in the transfer curve for about R = 45,000, at which point turbulent motions are said to begin.

Extensive measurements of heat flux and fluid velocities were made by Malkus [115] with an experimental device based on the temperature method with passive external insulation. The values of the depth of the fluid ranged from 0.127 to 20.32 cm. The heat flux measurements were made by heat transfer through the fluid in a quasi-steady regime, i.e. the data were taken during the temperature relaxation period of the insulated blocks. The range of Rayleigh numbers studied was from 1.7×10^3 to 10^6, and in this interval six discrete transitions were obtained. On the basis of the results of these experiments, Malkus and Veronis [24] established a theoretical method that attributed to each discontinuity in the slope the induction of a new transfer mode. A criterion for relative stability was deduced, which was the maximum heat transfer for the onset of the second mode.

The former experiments were repeated by Silveston [116] to achieve greater precision. Silveston's apparatus was the most sophisticated used hitherto; its experimental arrangement was designed on the flux method, and its heat losses were carefully avoided. Silveston used five fluids, in which physical properties were also measured. The depth of the fluid layer confined could be varied from 1.45 to 13 mm. The experimental error estimated in the Nusselt number was about 3%, and its values were determined, for the five fluids, up to $R \simeq 4 \times 10^5$. Using the sudden increase of thermal conductivity of fluid after the onset of convection, Silveston found the value of the critical Rayleigh number to be $R_c = 1{,}700 \pm 51$. Another aspect of Silveston's experiments was the relationship between heat flux and the planform of motion.

Globe and Dropkin [117] made ambitious experiments to obtain data and develop a mathematical relationship between heat transfer and the Prandtl number. They used the flux method and wrapped the test apparatus in mineral wool. Their paper does not mention the estimated error in the measurements. Several fluids were studied covering a range of Prandtl numbers from 0.02 to 8,750. Using two different heights for the container, 3.53 and 6.66 cm, they obtain the expression

$$Nu = 0.069\ R^{1/3}\ P^{0.074} \tag{37}$$

valid for their experimental range, $1.51 \times 10^5 < R < 6.76 \times 10^8$. The authors noted that turbulent fluctuations of temperature in both plates were observed in the experiments on mercury above $R = 10^5$. These fluctuations increased in magnitude as the Rayleigh number increased and attained their highest value at 3×10^6. The authors also pointed out that the size of the containers used had no bearing on the heat-transfer rates obtained.

O'Toole and Silveston [118] basing their work on available data up to 1,960, proposed the following correlation for the Nusselt number:

$$Nu = 0.104\ R^{0.305}\ P^{0.084} \tag{38}$$

Willis and Deardorff [119], unlike the previous authors, used a convection chamber with square horizontal surfaces. They tried to confirm the existence of the discrete transitions observed by Malkus for square geometry and other values of the Prandtl number. Their experimental arrangement was designed to follow the temperature method, but they introduced test elements in the chamber. Thus, they estimated the error to be up to 10% in Nu − 1. Their results, and those of Malkus, are listed in Table 1.

The effects of the lateral boundaries were reported by Soberman [120], but the first systematic experimental study was made by Catton and Edwards [121]. They, using the temperature method, employed an apparatus based on that described by Malkus [115] but which was more carefully isolated. They used solutions of water and silicon oils, and to obtain high values of the cell aspect ratio, they introduced two alternate honeycomb cores of different materials, fiberglass-phenolic and aluminum in the chamber test. The range of aspect ratios investigated was from 0.4 to 33.33; the values of the critical Rayleigh number estimated from the fittings of the published diagrams are listed in Table 2.

In 1968 Rossby [122] published an experimental study to determine the relationship between the Nusselt number, the Rayleigh number, and the Taylor number (dimensionless

Table 1
Experimental Values of Rayleigh Number at Transition Points

Transition	Malkus	Willis and Deardorff	Linear Analysis for Rigids Surfaces
1	1,700	1,750	1,708
2	18,000	8,200	17,610
3		24,000	75,710
4	55,000	56,000	219,885
5	170,000	180,000	509,658
6	425,000	410,000	1,202,509
7	860,000	830,000	1,843,582
8	1,700,000	1,400,000	3,085,829
9		2,250,000	4,870,124

Table 2
Critical Rayleigh Number as Function of Aspect Ratio

Aspect Ratio	Insulating Walls	Conducting Walls
33.33	1,700	1,740
2	1,770	1,800
1	3,600	6,600
0.66	10,000	26,000
0.5	22,000	66,000
0.4	52,000	160,000

Table 3
Some Results of Rossby's Experiments

Fluid	Aspect Ratio	R_c	Range	Relationship
Mercury	152	1,680	$R > 20{,}000$	$Nu = 0.147\ R^{0.257}$
Water	211	1,760	$R > 34{,}000$	$Nu = 0.131\ R^{0.30}$
Silicon oil	211	1,810	$R > 4{,}000$	$Nu = 0.184\ R^{0.281}$

rate of rotation about the vertical axis). For this purpose he used an accurate calorimeter based on the temperature method with circular geometry and the possibility of rotation. The fluids studied were mercury, water, and silicon oil ($P = 100$), with several aspect ratios. The particular case of nonrotation was studied and discussed as a function of results of the earlier research. Some results are listed in Table 3. It is pertinent to point out here some of Rossby's qualitative comments. He found a larger dispersion in the results near the critical Rayleigh number, as Koshmieder had noted previously [123]. In comparing his results with those of Silveston, he concluded that the Nusselt number at high Rayleigh numbers differs by 16% at $R = 10^6$, which shows that one cannot expect to condense all results into a single formula, $Nu = f(R) \cdot g(P)$, capable of correctly describing the role of both the Rayleigh and the Prandtl numbers.

Golstein and Chu made two studies on thermal convection in air and water. In the first [124], they used absolute interferometry to obtain the temperature distribution in turbulent convection ($6 \times 10^5 < R < 1.2 \times 10^8$) and the heat flux, which was determined for the temperature gradient in the horizontal walls. Thus, they could study a rectangular layer of air with aspect ratios from 1.5 to 7. Their results of the Nusselt number were fitted by

$$Nu = 0.123\ R^{0.294} \tag{39}$$

In the second work [125] the flux method was used to measure heat flux in a rectangular water layer which had an aspect ratio of 20. The resulting heat transfer correlation was

$$Nu = 0.183\ R^{0.278} \tag{40}$$

for $2.76 \times 10^5 < R < 1.05 \times 10^8$.

A new proof of the discontinuous transitions detected by Malkus and Willis and Deardorff (both by quasi-steady measurements), the determination of the changes in slope of the heat flux lines, and the detection of the transition from a two-dimensional plane to a three-dimensional one, predicted theoretically by Busse [56] were made by Krishnamurti [126] by repeating careful experiments for several fluids ($0.71 < P < 8{,}500$) and different aspect ratios ($10 < L/h < 51$) in rectangular geometry. The calorimeter used by Krishnamurti was based on the temperature method. The experiments demonstrate that there is a second transition for $R_{II} = 12\,R_c$, with a discrete change of slope in the heat flux curve and evident hysteresis as R is increased from below or decreased from above. The change from a two-dimensional plane to a three-dimensional flow in this second transition, or bimodal transition, was confirmed.

In a second paper, Krishnamurti [127] studied the plane and heat flux in a horizontal convective layer of fluid for $10^3 < R < 10^6$ and $1 < P < 10^4$. The data obtained could be fitted by straight line segments of different slopes. As R increases, four transitions occur; the first or critical one in R_c, the second around $R_{II} = 12\,R_c$, the third, $R_{III} = 35\,R_c$, was found to be dependent upon the Prandtl number, and the fourth, $R_{IV} = 105\,R_c$, was studied less intensively. Two new results were found about the third transition: a hysteresis loop in the heat curve as R is increased from below or decreased from above, and an oscillatory behavior in the fluid layer for external steady conditions for R values higher than R_{III} detected with thermocouples within the convecting fluid. At Rayleigh numbers greater than R_{IV} a double frequency of oscillations began to appear, but this was only confirmed for $P < 10^2$. Lastly, when Krishnamurti and Howard [128] studied the transition to turbulence, they discovered the appearance of transient bubbles or plumes, which had been confirmed by Dubois [129] by means of laser anemometry. At still higher R, the plumes drift in one direction and a larger scale horizontal flow appears.

Velocity and heat transfer measurements in a cylindrical water layer were made by Garon and Goldstein [130], for $7 \times 10^5 < R < 1.2 \times 10^8$. For the heat flux measurements, they used the flux method with passive external isolation. The data were fitted by the expression

$$Nu = 0.130\,R^{0.293} \tag{41}$$

and four apparent discontinuities in the slope of the heat flux were observed at Rayleigh numbers of 5.9×10^7 and 1.3, 2.2, and 4.0×10^8.

A crucial study on this subjet was carried out by Koschmieder and Pallas [131], who measured heat flux in a circular liquid layer for a well-defined plane as a function of the wavelength of convective motion, the Prandtl number ($P > 50$), and the Rayleigh number ($R < 3 \times 10^5$). The experimental apparatus was based on the flux method, and the heat flux was carefully measured with a thermopile [132] to determine the temperature differences between the incoming and outcoming water cooling the upper plate. This plate was formed by a colorless monocrystal sapphire, so the fluid motion could be observed simultaneously with the heat flux measurements. Their results are important, qualitatively as well as quantitatively, thus, for the range $1 < R/R_c < 6$, the wavelength of motion seems to be a linear function of the Rayleigh number, with a slope:

$$\frac{\Delta\lambda}{\Delta(R/R_c)} = 0.44 \pm 6\% \qquad (42)$$

The initial slope of the convective heat flux was found to be independent of the Prandtl number, which confirmed the theoretical value found by Schlüter, Lortz, and Busse [26]. But the most important result was the experimental demonstration, for high Prandtl numbers, of the continuity of the heat flux curve at supercritical Rayleigh numbers (up to 170 R_c). (These results were obtained by other investigators, too [134–139]). On the previous results, Koschmieder and Pallas said "... the slope changes observed by other investigators may originate from time-dependent operations of the apparatus or, as seems more likely, from the scatter of the data which permits to draw linear sections of the heat transfer curve."

Booker [133] measured heat transfer rates in a cylindrical container for a fluid, Polybutene 8, with a strong temperature-dependent viscosity. The flux method was employed with passive isolation. Nevertheless the rough estimation of the heat losses and the limited number of measurements led him to find a Nusselt number for this fluid lower than that of a constant-viscosity fluid at the Rayleigh number. The ratio of both Rayleigh numbers was fitted, for $\nu_{max} / \nu_{min} > 10$, as:

$$Nu / Nu_0 = 1.033 - 0.0694 \log_{10}(\nu_{max} / \nu_{min}) \qquad (43)$$

where ν_{max} and ν_{min} are the extreme values of kinematic viscosity of fluid in a experience. Lastly he proposed [106] that this decrease can be entirely predicted from the effect of the variable viscosity on the critical Rayleigh number.

A collection of papers by Ahlers, Behringer, and Walden [38, 134–141] are of special interest. In these papers, low temperature techniques were applied to the study of the Bénard–Rayleigh problem in helium. The experimental apparatus and the procedure were carefully described in Behringer and Ahlers [141]. The principle used was the flux method, but here the system was suspended in a vacuum, in such a way that the heat losses and the extraneous heat transport parallel to the conveying fluid were very small. Another favorable aspect was the low heat capacity of copper in the upper and lower plates at working temperatures together with its high thermal conductivity. Both properties assured the absence of lateral temperature gradients and minimized the damping of temperature fluctuations provoked by the time evolution of the convective state. Poor thermal conductivity of the lateral walls and high resolution in the temperature changes completes a positive assessment of the experiments. Only the relatively high errors in the fluid properties present negatives aspects, particularly since their influence on the Rayleigh number is important because the mean working temperature is variable. The values of the critical Rayleigh number are obtained for the aspect ratios 2.08 and 4.72. While the temperature increments measured were $676.2 \pm 8\,\mu K$ and $599.6 \pm 4\,\mu K$, the critical Rayleigh numbers assigned were $1{,}694 \pm 250$ and $1{,}599 \pm 240$, respectively.

The most important results obtained in cylindrical geometry with three aspect ratios 2.08, 4.72, and 57 were the following:

- *Boussinesq fluid layer (temperature independent properties)*. The onset of convective flow was essentially sharp for the aspect ratios 2.08 and 4.72, while it was rounded for aspect ratio 57. The last result can be explained by the great relative variation of the thickness of the layer. The supercritical Nusselt number was measured as a function of Rayleigh number and aspect ratio. For the aspect ratio 2.08 it revealed two discontinuities with hysteresis at 3 R_c and 10 R_c. For the aspect ratio 4.72, one transition to a turbulent state was found at 2 R_c. For the largest aspect ratio a nonperiodic (turbulent) state was obtained just above R_c.

- *Non-Boussinesq fluid layer (expansion coefficient dependent on temperature)*. The non-Boussinesq fluids are characterized by the extent, Q, of departures from the Oberbeck–Boussinesq approximation:

$$Q = \sum_{i=0}^{4} \gamma_i \cdot P_i \tag{44}$$

where γ_i is the relative increases of the lower and upper values of the density, thermal expansion coefficient, kinematic viscosity, thermal diffusivity, and specific heat of fluid, and P_i is functions of the Prandtl number [102]. In these experiences, R_c is demonstrated to be a linear function of $|Q|$, as

$$R_c = R_{oc}(1 - 0.0047\,|Q|) \tag{45}$$

for $Q^2 \ll R_c$.

The two possible hysteretic transitions, [102] were only experimentally resolved for larger Q-values. It was impossible to observe both transitions separately for $|Q| < 7.8$. The initial slope of the Nusselt number as a function of the Rayleigh number was studied. The absence of a Q-dependence and values lower than the theoretical predictions were obtained. Finally, several influences of the initial conditions of heating that cause three distinct convective states were studied.

- *Penetrative convection*. This occurs when a fluid with a maximum density at T_0, is heated from below at $T_1 > T_0$ and cooled from above at $T_2 < T_0$. Then, two different situations appear in the fluid: a stable upper layer and an unstable lower layer. The extent of the penetration of the velocity field into the upper layer can be expressed by the penetration parameter, λ:

$$\lambda = \frac{T_1 - T_2}{T_1 - T_0} \tag{46}$$

The definition of the Rayleigh number was modified, using only the depth and the temperature difference of the unstable fluid layer. We will call this new number Ru. Measurements of the critical value of Ru and the Nusselt number were made for high penetration parameters. New hysteretic transitions were found, but only the critical loop was studied, because the time-dependent behavior of the fluid flow made the other measurements unreliable. For ΔT near ΔT_c, quasi-static changes of heat flux appeared in a series of transitions both hysteretic and reproducible, closely spaced in ΔT. This apparent multistability was justified by the discrete expulsion of one convective cell from the lateral finite system. Time-dependent flows were observed much closer to R_c, but they were not analyzed.

Recently, Lucas, Pfotenhauer, and Donnelly [142], using an experimental device similar to the one designed by previous authors, presented the first one of a promised series of papers about heat transport in rotating helium layers. This work does not enter in the objectives of this study.

Richter, Nataf, and Daly [143] have measured heat flux and horizontally average temperatures in non-Boussinesq fluids with large viscosity variations. The calorimeter used was based on the flux method, with evaluated heat losses. The precision of the calorimeter was confirmed by measuring heat flux in Boussinesq fluid and comparing the results with those of Rossby. The measurements with polibutene oil and golden syrup show the existence of a subcritical Rayleigh number. The supercritical values of relative Nusselt numbers do not coincide with Booker's results. Researchers do not all agree on this subject.

Moreno et al. [144] introduced a new technique for heat measurement in the Bénard–Rayleigh problem. The calorimeter was based on the temperature method, but the heat flux transducer was a heat fluxmeter (see Chapter 20 "Heat Flux Measurements"). The first study [58] was on heat transport for different wavelengths in a box with an aspect ratio of 10, filled with silicon oil. By different initial and boundary conditions, they induced several wavelengths in the fluid. Heat flux was a continuous function of the Rayleigh number and changed slightly between different structures and for bimodal transition. On the same basis, Zamora and Rey de Luna [110] have demonstrated the possibility of measuring the structural energy for each planform of convective motion. The method used was the energy balance for the inward and the outward heat flux in the fluid layer when it turn 180° in space, passing from a stable situation (with the fluid heated from above) to an unstable state (with the fluid heated from below) for the same steady external conditions.

CONCLUDING REMARKS

In spite of the tremendous amount of work gathered in the earlier parts of this paper, it is obvious that the results on heat flux show discrepancies when theoretical and experimental data are compared, as much with each other as among themselves. If these differences are usually quantitative, they are sometimes also qualitative, as with the changes in heat transfer when the convective pattern is changed. These differences cannot be entirely attributed to the theoretical techniques or instruments used but are rather related to deeper causes inherent in the problem itself. This can be clearly appreciated from theoretical analysis of stability. Thus, given a set of boundary conditions and a certain convective structure, its stability can be determinated. But if several stable structures exist, we lack criteria to judge which of them the system really selects. This indicates that the parametrization of the problem is incomplete, so it is necessary to define suitable additional parameters to discriminate among stable convective structures. From an experimental point of view, the lack of well-defined parametrization means that experiments which are apparently similar may correspond to different, incomparable real situations.

Because of all of this, on top of the above-mentioned difficulties in the solution of thermohydrodynamic equations due to their nonlinear character, we lack a phenomenological theory that, after the preliminar definition of a complete set of parameters, could establish the necessary selection criteria. Thus, the broadened space of these parameters would be determinist, in the sense that each point would correspond to one single real physical situation, avoiding the current indetermination.

Today's most creative ideas are basically oriented in two directions: transitory situations and conditions that lead to turbulence. At a local level, along with theoretical analysis in which very abstract concepts often predominate, very notable experimental contributions have been made. At a global level, new energy-measuring techniques can make important and immediate contributions, although the subject still remains open.

Along with the basic problem described up to now for pure Newtonian fluids with fairly idealized boundary conditions, we must not forget the continuous development of other aspects of the problem, such as the case of more realistic walls, mixtures of fluids, crystal-liquids, porous materials, systems submitted to rotation or force fields different from the gravitational field, etc. These subjects are not dealt with in this chapter for reasons of space.

NOTATION

A	amplitude of perturbation	M	mass
c	specific heat	Nu	Nusselt number
C_n	coefficients	P	Prandtl number
E	energy	P_s	static pressure
$\vec{g}$	gravity acceleration vector	Q	extent of departures
H	convective heat flux	q	heat flux
h	heat transfer coefficient, fluid layer thickness	R	Rayleigh number
		T	temperature
K	thermal conductivity	$\vec{v}'$	velocity
$\vec{k}$	upwards vertical vector	z′	vertical distance

Greek Symbols

α	coefficient of expansion; wave number	$\varkappa$	thermal diffusivity
		ν	kinematic viscosity
γ_i	relative increases of lower and upper values of physical properties	ρ	density

REFERENCES

1. Rumford, Count, *Complete Works,* Vol. 1, American Academy of Arts and Sciences, Boston, 1870, p. 239.
2. Bénard, H., *Revue Général des Sciences Pures et Appliquees,* 11: 1261, 1309 (1900); *Annales de Chimie et de Physique,* 23: 62 (1901).
3. Rayleigh, J. W. S., "On Convention Current in a Horizontal Layer of Fluid, when the Higher Temperature is on the Under Side," *Phil. Mag.,* 32: 529 (1916).
4. Chandrasekhar, S., *Hydrodynamics and Hydromagnetics Stability,* Clarendon, Oxford, 1961.
5. Segel, L. A., "Non-Linear Hydrodynamic Stability Theory and its Applications to Thermal Convection and Curved Flows," in *Non-equilibrium Thermodynamics, Variational Techniques and Stability,* R. J. Donnelly, R. Herman and I. Prigogine (Eds.), Univ. Chicago Press, Chicago, 1966, p. 165.
6. Koshmieder, E. L., "Bénard Convection," *Adv. Chem. Phys.,* 26: 177 (1974).
7. Palm, E., "Non-Linear Thermal Convection," *Ann. Rev. Fluid Mech.,* 7: 39 (1975).
8. Gershuni, G. Z., and Zhukhovitskii, E. M., *Convective Stability of Incompressible Fluids,* Keter Pub. Jerusalem, 1976.
9. Velarde, M. G., "Hydrodynamic Instabilities (in isotropic fluids)," in *Fluid Dynamics* ("Les Houches, 1973"), R. Balian and I. L. Peube (Eds.), Gordon and Breach, New York, 1976, p. 469.
10. Normand, C., Pomeau, Y., and Velarde M. G., "Convective Instability: A Physicist's Approach," *Rev. Mod. Phys.,* 49: 581 (1977).
11. Whithehead, J. A., "A Survey of Hydrodynamic Instabilities," in *Fluctuations, Instabilities, and Phase Transitions,* T. Riste (Ed.), Plenum Press, New York, 1975, p. 153.
12. Busse, F. H., "Non-Linear Properties of Convection," *Rep. Prog. Phys.,* 41: 1929 (1978).

13. Azouni, M. A., "Survey of Thermoconvective Instabilities of Confined Fluids," *J. Non-equilibrium Thermodynamics,* 4: 321 (1979).
14. Busse, F. H., "Transition to Turbulence in Rayleigh–Bénard Convection," *Topics in Applied Physics,* 45: 97 (1981).
15. Oberbeck, A., "Über die Wärmeleitung der Flüssigkeiten bei der Berücksichtigung der Strömungen infolge von Temperaturdifferenzen," *Ann. Phys. Chem.* (Wied. Ann.), 7: 271 (1879).
16. Boussinesq, J., *Theorie Analytique de la Chaleur,* Vol. 2, Gautier–Villars, Paris, 1903.
17. Tykodi, R. J., *Thermodynamics of Steady States,* Macmillan, New York, 1967.
18. Benofy, S. J., and Quay, P. M., "The Thermodynamics of Systems in Steady State," *J. Chem. Phys.,* 78 (6): 3177 (1983).
19. Jeffrey, H., "The Stability of a Layer of Fluid Heated from Below," *Phil. Mag.,* 2: 833 (1926).
20. Low, A. R., "On the Criterion for Stability for a Layer of Viscous Fluid Heated from Below," *Proc. Roy. Soc.,* A 125, 180 (1929).
21. Pellew, A., and Southwell, R. V., "On Maintained Convective Motion in a Fluid Heated from Below," *Proc. Roy. Soc.,* A 176: 312 (1940).
22. Reid, W. H., and Harris, D. L., "Some Further Results on the Bénard Problem," *Phys. Fluids,* 1: 102 (1958).
23. Gorkov, L. P., "Stationary Convection in a Plane Liquid Layer near the Critical Heat Transfer Point," *Sov. Phys.*-JETP, 6: 311 (1958).
24. Malkus, W. V. R., and Veronis, G., "Finite Amplitude Convection," *J. Fluid Mech.,* 4: 225 (1958).
25. Kuo, H. L., "Solution of the Non-Linear Equations of Cellular Convection and Heat Transport," *J. Fluid Mech.,* 10: 611 (1961).
26. Schluter, A., Lortz, D., and Busse, F. H., "On the Stability of Steady Finite Amplitude Convection," *J. Fluid Mech.,* 23: 129 (1965).
27. Lasalle, J. P., and Leshetz, S., *Stability by Liapunov's Direct Methods with Applications,* Academic Press, New York, 1961.
28. Joseph, D. D., *Stability of Fluid Motions,* Vols. 1 and 2, Springer, Berlin, 1976.
29. Davis, S. H., "Convection in a Box: Linear Theory," *J. Fluid Mech.,* 30: 465 (1967).
30. Stork, K., and Muller, U., "Convection in Boxes: Experiments," *J. Fluid Mech.,* 54: 599 (1972).
31. Segel, L. A., "Distant Side-walls Cause Slow Amplitude Modulation of Cellular Convection," *J. Fluid Mech.,* 38: 203 (1969).
32. Davies-Jones, R. P., "Thermal Convection in an Infinite Channel with No-Slip Sidewall," *J. Fluid Mech.,* 44: 695 (1970).
33. Catton, I., "Convection in a Closed Rectangular Region: the Onset of Motion," *J. Heat Transfer,* 92 C: 186 (1970).
34. Catton, I., "Effect of Wall Conduction on the Stability of a Fluid in a Rectangular Region Heated from Below," *J. Heat Transfer,* 94 C: 446 (1972).
35. Catton, I., "The Effect of Insulating Vertical Walls on the Onset of Motion in a Fluid Heated from Below," *Int. J. Heat Mass Transfer,* 15: 665 (1972).
36. Drazin, P. G., "On the Effects of Side-Walls on Bénard Convection," *J. Applied Math. and Physics* (ZAMP), 26: 239 (1975).
37. Daniels, P. G., "The Effect of Distant Sidewalls on the Transition to Finite Amplitude Bénard Convection," *Proc. Roy. Soc. London,* A 358: 173 (1977).
38. Ahlers, G., "Low Temperature Studies of the Rayleigh–Bénard Instability and Turbulence," in *Fluctuations, Instabilities, and Phase Transitions,* T. Riste (Ed.), Plenum Press, New York, 1975, p. 181.
39. Brown, S. N., and Stewartson, K., "On Thermal Conduction in a Large Box," *Studies in Applied Mathematics,* 57: 187 (1977).
40. Hall, P., and Walton, I. C., "The Smooth Transition to a Convective Regime in a 2-Dimensional Box," *Proc. Roy. Soc. London,* A 358: 199 (1977).

41. Cross, M. C., Daniels, P. G., Hohemberg, P. C., and Siggia, E. D., "Effect of Distant Sidewalls on Wavenumber Selection in Rayleigh–Bénard Convection," *Phys. Rev. Lett.*, 45: 898 (1980).
42. Cross, M. C., Daniels, P. G., Hohemberg, P. C., and Siggia, E. D., "Phase-Winding Solutions in a Finite Container above the Convective Threshold," *J. Fluid Mech.*, 127: 155 (1983).
43. Kidachi, H., "Side Wall Effect on the Pattern Formation of the Rayleigh–Bénard Convection," *Prog. Theor. Phys.*, 68: 49 (1982).
44. Charlson, G. S., and Sani, R. L., "Thermoconvective Instability in a Bounded Cylindrical Fluid Layer," *Int. J. Heat Mass Transfer*, 13: 1479 (1970).
45. Charlson, G. S., and Sani, R. L., "Finite Amplitude Axisymmetric Thermoconvective Flows in a Bounded Cylindrical Layer of Fluid," *J. Fluid Mech.*, 71: 209 (1975).
46. Catton, I., and Edwards, D. K., "Initiation of Thermal Convection in Finite Right Circular Cylinders," *AIChE Journal*, 16: 594 (1970).
47. Brown, S. N., and Stewartson, K., "On Finite Amplitude Bénard Convection in a Cylindrical Container," *Proc. Roy. Soc. London*, A 360: 455 (1978).
48. Brown, S. N., and Stewartson, K., "On Finite Amplitude Bénard Convection in a Cylindrical Container. Part II," *SIAM J. Applied Mathematics*, 36: 573 (1979).
49. Shaumeyer, J. N., Behringer, R. P., and Baierlein, R., "Linear Growth Rates for the Rayleigh–Bénard Instability in Cylindrical Geometry," *J. Fluid Mech.*, 109: 339 (1981).
50. Rosenblat, S., "Thermal Convection in a Vertical Circular Cylinder," *J. Fluid Mech.*, 122: 395 (1982).
51. Hollands, K. G. T., Raithby, G. D., and Konicek, L., "Correlation Equations for Free Convection Heat Transfer in Horizontal Layers of Air and Water," *Int. J. Heat Mass Transfer*, 18: 879 (1975).
52. Edwards, D. K., Arnold, J. N., and Wu, P. S., "Correlations for Natural Convection through High L/D Rectangular Cells," *J. Heat Transfer*, 101: 741 (1979).
53. Denton, R. A., and Wood, I. R., "Turbulent Convection Between Two Horizontal Plates," *Int. J. Heat Mass Transfer*, 22: 1339 (1979).
54. Kuo, H. L., and Platzmann, G. W., "A Normal Mode Nonlinear Solution of the Rayleigh Convection Problem," *Beitr. Phys. Atmos. Bd.*, 33 S: 137 (1961).
55. Veronis, G., "Large-Amplitude Bénard Convection," *J. Fluid Mech.*, 26: 49 (1966).
56. Busse, F. H., "On the Stability of Two-Dimensional Convection in a Layer Heated from Below," *J. Math. Phys.*, 46: 140 (1967).
57. Clever, R. M., and Busse, F. H., "Transition to Time-Dependent Convection," *J. Fluid Mech.*, 65: 625 (1974).
58. Zamora, M., Córdoba, A., and Moreno, J., "The Influence of Wavelength and Bimodal Transition on Heat Flux in the Bénard–Rayleigh Problem," *Physicochem. Hydrodynam.*, 4: 65 (1983).
59. Clever, R. M., and Busse, F. H., "Low-Prandtl-Number Convection in a Layer Heated from Below," *J. Fluid Mech.*, 102: 61 (1981).
60. Busse, F. H., and Clever, R. M., "An Asymptotic Model of Two-Dimensional Convection in the Limit of Low Prandtl Number," *J. Fluid Mech.*, 102: 75 (1981).
61. Frick, H., and Clever, R. M., "The Influence of Side Walls on Finite Amplitude Convection in a Layer Heated from Below," *J. Fluid Mech.*, 114: 467 (1982).
62. Frick, H., "The Effect of Thermal Boundary Conditions on the Heat Transport in Vertical Channels Heated from Below," *Int. J. Heat Mass Transfer*, 26: 681 (1983).
63. Frick, H., Busse, F. H., and Clever, R. M., "Steady Three-Dimensional Convection at High Prandtl Numbers," *J. Fluid Mech.*, 127: 141 (1983).
64. Deardorff, J. W., "A Numerical Study of Pseudo Three-Dimensional Parallel-Plate Convection," *J. Atm. Sci.*, 22: 419 (1965).
65. Fromm, J. E., "Numerical Solutions of the Non-Linear Equations for a Heated Fluid Layer," *Phys. Fluid*, 8: 1757 (1965).

66. Lipps, F. B., and Sommerville, R. C. J., "Dynamics of Variable Wavelength in Finite-Amplitude Bénard Convection," *Phys. Fluid,* 14: 759 (1965).
67. Deardorff, J. W., and Willis, G. E., "The Effect of Two-Dimensionality on the Suppression of Thermal Turbulence," *J. Fluid Mech.,* 23: 337 (1965).
68. Samuels, M. R., and Churchill, S. W., "Stability of a Fluid in a Rectangular Region Heated from Below," *AIChE J.,* 13: 77 (1967).
69. Aziz, K., and Hellum, J. D., "Numerical Solution of the Three-Dimensional Equations of Motion for Laminar Convection," *Phys. Fluids,* 10: 314 (1967).
70. Liang, S. F., Vidal, A., and Acrivos, A., "Buoyancy-Driven Convection in Cylindrical Geometries," *J. Fluid Mech.,* 36: 239 (1969).
71. Cabelli, A., and Vahl, Davis, G. de, "A Numerical Study of the Bénard Cell," *J. Fluid Mech.,* 45: 805 (1971).
72. Moore, D. R., and Weiss, N. O., "Two-Dimensional Rayleigh–Bénard Convection," *J. Fluid Mech.,* 58: 289 (1973).
73. Lipps, F. B., "Numerical Simulation of Three-Dimensional Bénard Convection in Air," *J. Fluid Mech.,* 75: 113 (1976).
74. Ozoe, H., et al., "Three-Dimensional Numerical Analysis of Laminar Natural Convection in a Confined Fluid Heated From Below," *J. Heat Transfer,* 98 C: 202 (1976).
75. Ozoe, H., et al., "Long Rolls Generated by Natural Convection in an Inclined, Rectangular Enclosure," *Int. J. Heat Mass Transfer,* 26: 1427 (1983).
76. McDonough, J. M., and Catton, I., "Accuracy of the Mean Field Approximation and the Physical Effects of Prandtl Number in Bénard Convection," *Phys. Fluid,* 25: 1502 (1982).
77. Grotzbach, G., "Direct Numerical Simulation of Laminar and Turbulent Bénard Convection," *J. Fluid Mech.,* 119: 27 (1982).
78. Herring, J. R., "Investigation of Problems in Thermal Convection," *J. Atmos. Sci.,* 20: 325 (1963).
79. Herring, J. R., "Investigation of Problem in Thermal Convection: Rigid Boundaries," *J. Atmos. Sci.,* 21: 277 (1964).
80. Elder, J. W., *J. Fluid Mech.,* 35: 417 (1969).
81. Howard, L. N., Woods Hole Oceanographic Inst. G. F. O. Notes No 65–51, I, (1965) p. 124.
82. Roberts, P. H., "On Non-Linear Bénard Convection," in *Non-Equilibrium Thermodynamics, Variational Techniques and Stability*, R. J. Donnelly et al. (Eds.), Univ. Chicago Press, Chicago, 1966, p. 125.
83. Stewartson, K., "Asymptotic Theory in the Limit $R \rightarrow \infty$," in *Non-Equilibrium Thermodynamics, Variational Techniques and Stability,* R. J. Donnelly et al. (Eds.), Univ. Chicago Press, Chicago, 1966, p. 158.
84. Mihaljan, J. A., "A rigorous Exposition of the Boussinesq Approximation Applicable to a Thin Fluid Layer," *Astrophys. J.,* 136: 1126 (1962).
85. Pérez-Cordon, R., and Velarde, M. G., "On the (Non-Linear) Foundations of Boussinesq Approximation Applicable to a Thin Layer of Fluid," *J. de Physique,* 36: 59 (1975).
86. Velarde, M. G., and Pérez-Cordon, R., "On the (Non-Linear) Foundations of Boussinesq Approximation Applicable to a Thin Layer of Fluid. (II) Viscous Dissipation and Large Cell Gap Effects," *J. de Physique,* 37: 431 (1976).
87. Gray, D. D., and Giorgini, A., "The Validity of the Boussinesq Approximation for Liquid and Gases," *Int. J. Heat Mass Transfer,* 19: 545 (1976).
88. Veronis, G., "Penetrative Convection," *Astrophys. J.,* 137: 641 (1963).
89. Sun, Z. S., Tien, C., and Yen, Y. C., "Thermal Instability of a Horizontal Layer of Liquid with Maximum Density," *AIChE J.,* 15: 910 (1969).
90. Moore, D. R., and Weiss, N. O., "Non-Linear Penetrative Convection," *J. Fluid Mech.,* 58: 289 (1973).

91. Legros, J. C., Longree, D., and Tomaes, G., "Bénard Problem in Water Near 4° C," *Physica,* 12: 410 (1974).
92. Seki, N., Fukusako, S., and Sugawara, M., "A Criterion of Onset of Free Convection in a Horizontal Melted Water Layer with Free Surface," *J. Heat Transfer,* 99: 92 (1977).
93. Dubois, M., Bergé, P., and Wesfreid, E., "Non Boussinesq Convective Structures in Water Near 4° C," *J. de Physique,* 39: 1253 (1978).
94. Gebhart, B., and Mollendorf, J. C., "Buoyancy-Induced Flows in Water Under Conditions in Which Density Extrema May Arise," *J. Fluid Mech.,* 89: 673 (1978).
95. Merker, G. P., Waas, P., and Grigull, U., "Onset of Convection in a Horizontal Water Layer with Maximum Density Effects," *Int. J. Heat Mass Transfer,* 22: 505 (1979).
96. Palm, E., "On the Tendency Towards Hexagonal Cells in Steady Convection," *J. Fluid Mech.,* 8: 183 (1960).
97. Segel, L. A., and Stuart, J. T., "On the Question of the Preferred Mode in Cellular Thermal Convection," *J. Fluid Mech.,* 13: 289 (1962).
98. Jensen, O., "Note on the Influence of Variable Viscosity on the Critical Rayleigh Number," *Acta Polytech. Scand.,* 24: 1 (1963).
99. Palm, E., and Øiann, H., "Contribution to the Theory of Cellular Thermal Convection," *J. Fluid Mech.,* 19: 353 (1964).
100. Segel, L. A., "The Nonlinear Interaction of a Finite Number of Disturbances to a Layer of Fluid Heated from Below," *J. Fluid Mech.,* 21: 359 (1965).
101. Busse, F. H., Dissertation, Univ. of Munich, English Translation by S. H. Davis, Rand Report LT-66-19, Rand Corporation, Santa Monica, California, 1962.
102. Busse, F. H., "The Stability of Finite Amplitude Cellular Convection and its Relation to an Extremum Principle," *J. Fluid Mech.,* 30: 625 (1967).
103. Palm, E., Ellingsen, T., and Gjevik, B., "On the Occurrence of Cellular Motion in Bénard Convection," *J. Fluid Mech.,* 30: 651 (1967).
104. Davis, S. H., and Segel, L. A., "Effects of Surface Curvature and Property Variation on Cellular Convection," *Phys. Fluids,* 11: 470 (1968).
105. Roja, D. S., and Fynlayson, B. A., "The Effect of Property Variations on the Convective Instability of Gases," *AIChE J.,* 16: 876 (1970).
106. Booker, J. R., and Stengel, K. C., "Further Thoughts on Convective Heat Transport in a Variable-Viscosity Fluid," *J. Fluid Mech.,* 86: 289 (1978).
107. Córdoba, A., "Effects of Variable Fluid Properties and Boundary Conditions on Thermal Convection," *Int. Comm. Heat Mass Transfer,* 10: 147 (1983).
108. Hoard, C. Q., Robertson, C. R., and Acrivos, A., "Experiments on the Cellular Structure in Bénard Convection," *Int. J. Heat Mass Transfer,* 13: 849 (1970).
109. Torrance, K. E., and Turcotte, D. L., "Thermal Convection with Large Viscosity Variations," *J. Fluid Mech.,* 47: 113 (1971).
110. Zamora, M., and Rey de Luna, A., "Energy in the Bénard–Rayleigh Problem," in *Cellular Structures in Instabilities,* J. E. Wesfreid and S. Zaleski (Eds.), Lectures notes in Physics, Springer-Verlag, Berlin, 1984.
111. Jakob, M., *Heat Transfer,* Vol. 1, Wiley, New York, 1949, p. 534.
112. Mull, W., and Reiher, H., "Der Wärmeschutz von Luftschichten," *Beih. Gesundh. Ing.,* Series I, 28: 1 (1930).
113. Schmidt, R. J., and Milverton, S. W., "On the Instability of a Fluid when Heated from Below," *Proc. Roy. Soc.,* A 152: 586 (1935).
114. Schmidt, R. J., and Saunders, O. A., "On the Motion of a Fluid Heated from Below," *Proc. Roy. Soc.,* A 165: 216 (1938).
115. Malkus, W. V. R., "Discrete Transitions in Turbulence Convection," *Proc. Roy. Soc.,* A 225: 185 (1954).
116. Silveston, P. L., "Wärmedurchgang in Waagrechten Flüssigkeitsschichten," Part. 1, *Forsch. Ing. Wes.,* 24: 29 and 59 (1958).

117. Globe, S., and Dropkin, D., "Natural-Convection Heat Transfer in Liquids Confined by Two Horizontal Plates and Heated from Below," *J. Heat Transfer,* 81: 24 (1959).
118. O'Toole, J. L., and Silveston, P. L., "Correlation of Convective Heat Transfer in Confined Horizontal Layers," *Chem. Eng. Progr. Symp.*, 57 (32): 81 (1961).
119. Willis, G. E., and Deardorff, J. W., "Confirmation and Renumbering of the Discrete Heat Flux Transitions of Malkus," *Phys. Fluids,* 10: 1861 (1967).
120. Soberman, R. K., "Effects of Lateral Boundaries on Natural Convection," *J. Appl. Phys.*, 29: 872 (1958).
121. Catton, I., and Edwards, D. K., "Effect of Side Walls on Natural Convection Between Horizontal Plates Heated from Below," *J. Heat Transfer,* 89 (4): 295 (1967).
122. Rossby, H. T., "A Study of Bénard Convection with and without Rotation," *J. Fluid Mech.*, 36 (2): 309 (1969).
123. Koshmieder, E. L., "On Convection on a Uniformly Heated Plane," *Beitr. Phys. Atmos.*, 39: 1 (1966).
124. Goldstein, R. J., and Chu, T. Y., "Thermal Convection in a Horizontal Layer of Air," *Prog. Heat Mass Transfer,* 2: 55 (1969).
125. Chu, T. Y., and Goldstein, R. J., "Turbulent Convection in a Horizontal Layer of Water," *J. Fluid Mech.*, 60 (1): 141 (1973).
126. Krishnamurti, R., "On the Transition to Turbulent Convection. Part 1. The Transition from Two- to Three-Dimensional Flow," *J. Fluid Mech.*, 42 (2): 295 (1970).
127. Krishnamurti, R., "On the Transition to Turbulent Convection. Part 2. The Transition to Time-Dependent Flow," *J. Fluid Mech.*, 42 (2): 309 (1970).
128. Krishnamurti, R., and Howard, L. N., "Large-Scale Flow Generation in Turbulent Convection," *Proc. Natl. Acad. Sci.*, 78 (4): 1981 (1981).
129. Dubois, M., "Approach of the Turbulence in Hydrodynamics Instabilities," in *Symmetries and Broken Symmetries in Condensed Matter Physics,* N. Boccara (Ed.), IDSET, Paris, 1981, p. 81.
130. Garon, A. M., and Goldstein, R. J., "Velocity and Heat Transfer Measurements in Thermal Convection," *Phys. Fluids,* 16: 1818 (1973).
131. Koschmieder, E. L., and Pallas, S. G., "Heat Transfer Through a Shallow, Horizontal Convecting Fluid Layer," *Int. J. Heat Mass Transfer,* 17: 991 (1974).
132. Koschmieder, E. L., and Pallas, S. G., "A Sensor for Heat Transfer Measurements," *Rev. Sci. Instrum.*, 45 (9): 1164 (1974).
133. Booker, J. R., "Thermal Convection with Strongly Temperature-Dependent Viscosity," *J. Fluid Mech.*, 76 (4): 741 (1976).
134. Ahlers, G., "Low Temperature Studies of the Rayleigh–Bénard Instability and Turbulence," *Phys. Rev. Lett.*, 33: 1185 (1974).
135. Ahlers, G., and Behringer, R. P., "The Rayleigh–Bénard Instability and the Evolution of Turbulence," *Prog. Theor. Phys. Suppl.*, 64: 186 (1978).
136. Ahlers, G., and Behringer, R. P., "Evolution of Turbulence from the Rayleigh–Bénard Instability," *Phys. Rev. Lett.*, 40: 712 (1978).
137. Ahlers, G., "Effects of Departures from the Oberbeck–Boussinesq Approximation on the Heat Transport of Horizontal Convecting Layers," *J. Fluid Mech.*, 98 (1): 137 (1980).
138. Ahlers, G., "Onset of Convection and Turbulence in a Cylindrical Container," in *Systems Far From Equilibrium,* L. Garrido (Ed.), Springer, 1980, p. 143.
139. Ahlers, G., and Walden, R. W., "Turbulence Near Onset of Convection," *Phys. Rev. Lett.*, 44: 445 (1980).
140. Walden, R. W., and Ahlers, G., "Non-Boussinesq and Penetrative Convection in a Cylindrical Cell," *J. Fluid Mech.*, 109: 89 (1981).
141. Behringer, R. P., and Ahlers, G., "Heat Transport and Temporal Evolution of Fluid Flow Near the Rayleigh–Bénard Instability in Cylindrical Container," *J. Fluid Mech.*, 125: 219 (1982).

142. Lucas, P. G. J., Pfotenhauer, J. M., and Donnelly, R. J., "Stability and Heat Transfer of Rotating Cryogens. Part 1. Influence of Rotation on the Onset of Convection in Liquid ^{4}He," *J. Fluid Mech.*, 129: 251 (1983).
143. Richter, F. M., Nataf, H. C., and Daly, R. J., "Heat Transfer and Horizontally Averaged Temperature of Convection with Large Viscosity Variations," *J. Fluid Mech.*, 129: 173 (1983).
144. Moreno, J., et al., "New Experimental Apparatus for the Study of the Bénard–Rayleigh problem," *Rev. Sci. Instrum*, 51 (1): 82 (1980).

CHAPTER 4

HYDRODYNAMICS OF FREE-LIQUID JETS AND THEIR INFLUENCE ON HEAT TRANSFER

Jan Iciek, Urszula Cywinska, and Roman Blaszczyk

Institute of Chemical Engineering
Lodz Technical University
Lodz, Poland

CONTENTS

INTRODUCTION

In every process of heat transfer the conditions under which heat-exchanging media are flowing, generally called process hydrodynamics, have a great influence on the heat transfer rate. The effect of hydrodynamics in generalized heat transfer equations is reflected by dimensionless numbers, system geometry, and transfer criteria. It is especially important to know direct contact heat transfer hydrodynamics for which it is difficult not only to determine heat transfer coefficients but also to establish one of the main parameters, i.e. the heat transfer surface.

The concept of direct contact heat transfer includes all processes which occur without getting in contact with the apparatus walls. In practice this is usually the case of steam condensation in cold fluid characterized by high values of heat transfer coefficients. Generally, two ways of such heat transfer can be distinguished. The first one, well-known and used more frequently, consists in submitting steam to a bubbler dipped in a mixer (Figure 1) (e.g. distillation with steam). In the other one steam is condensed on thin jets formed on a perforated plate (Figure 2). We shall consider the latter type of heat transfer which is not well-known as yet and is poorly and sometimes contradictorily presented in the literature.

So far the main interest of both practicians and theoreticians has been directed to maximum breakup of liquid jets either for diffusional mass transfer (e.g. absorption, cooling of hot air) applied in chemical technology or in order to form fine liquid drop

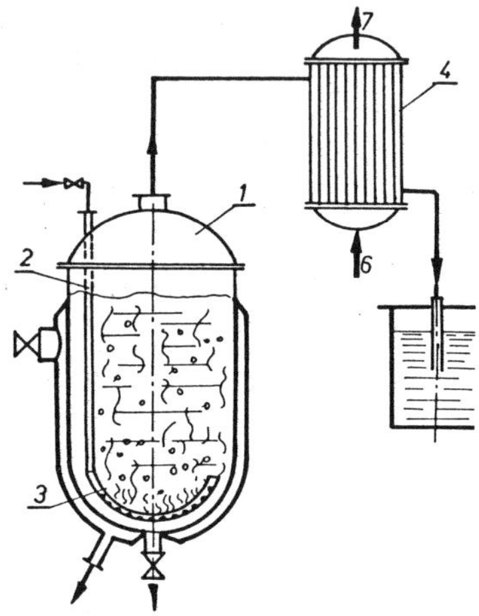

Figure 1. Steam distillation apparatus: 1—distiller; 2—steam supply; 3—bubbler; 4—condenser; 5—solvent separator; 6, 7—water inflow and outflow; 8, 9—outflow of distillate and condensation water.

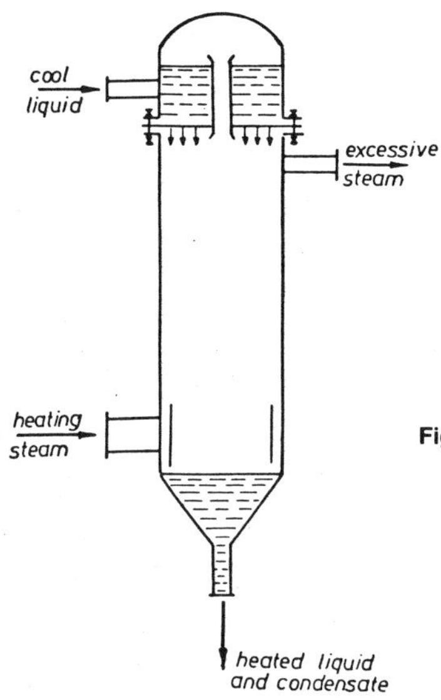

Figure 2. Direct contact jet heat exchanger.

suspensions in the air (e.g. spraying) used in power industry and agriculture. There are many theoretical and practical studies [1–10] concerned with a detailed description of the way the jet can be broken up into drops of a determined diameter, while a few publications present the data concerning the method for maintaining a continuous jet of determined turbulence [3, 11–16]. This information gap referring mainly to jet hydrodynamics considerably affects heat transfer correlations derived for special conditions (e.g. for vacuums [16–19], for fluids of high viscosity [3, 20–22], and for such cases which cannot be generalized due to slight differences in jet hydrodynamics). That was the reason why it was

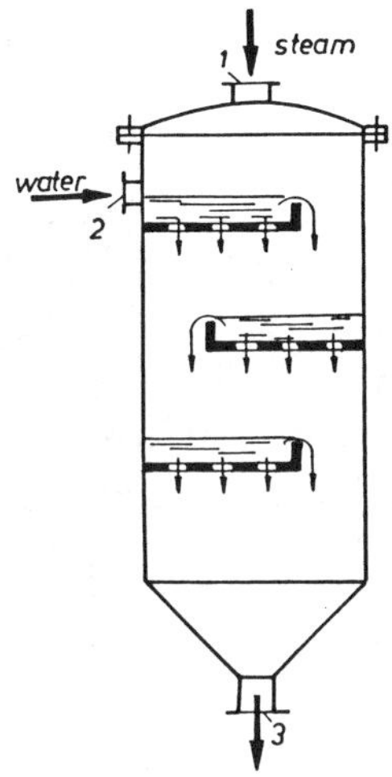

Figure 3. Barometric condenser: 1—steam supply; 2—liquid supply; 3—gas, water and condensate outflow.

considered necessary to explain in the study devoted basically to heat transfer, the problems connected with jet hydrodynamics in relation to the jet geometry (e.g. cylindrical, rectangular), its state (e.g. continuous, broken up), and character (e.g. laminar, turbulent). Various attempts have been made [14, 23] to choose the orifice shape which would enable the best possible conditions for heat transfer. We established that only the cylindrical shape was of practical importance because other shapes are difficult to produce and use and because sharp edges become rounded or fouled so that after some time the performance begins to differ significantly. Our considerations therefore deal mainly with jets formed on cylindrical orifices.

Direct contact heat transfer has been used for many years in some industrial apparatus [24–29], such as barometric condensers (Figure 3) and thermal degasifiers (Figure 4 through 8). In barometric condensers economically useless steam is condensed on cold water jets, while in thermal degasifiers cold water jets being in direct contact with steam are degasified. Recently there has been a growing interest in direct contact heat transfer on continuous, laminar or turbulent liquid jets. It appeared that although in broken up systems heat transfer was more intensive, the cost of subsequent phase separation exceeded the previously obtained profits. Direct contact heat transfer has been used already in sea-water desalting by multiple distillation [30] and in the sterilization of liquid substrates in food technology [31–34]. The system for multiphase water evaporation from brine is presented in Figure 9. A universal system for direct-contact sterilization or pasteurization and for cooling of aqueous solutions and suspensions is presented in Figures 10 and 11. This system operates faultlesly in a wide range of thermal conditions in various applications due to the elimination of the wall in heat exchangers and due to the application of natural motion of fluids in the system.

In this type of heat exchanger relatively high heat transfer coefficients may be obtained. They are in the range of $10^4 W/m^2K$[35] or even $10^7 W/m^2K$[36], and the difference between steam and jet temperatures is of great significance. The presence of inert gases, mainly of air, causes a significant decrease in the heat transfer coefficient [37–39] which is shown in Figure 12 [39]. As can be seen in the chart, when 1% of air is contained in the vapor, the heat transfer coefficient α' is only 40% of the value determined for the condensation of pure steam.

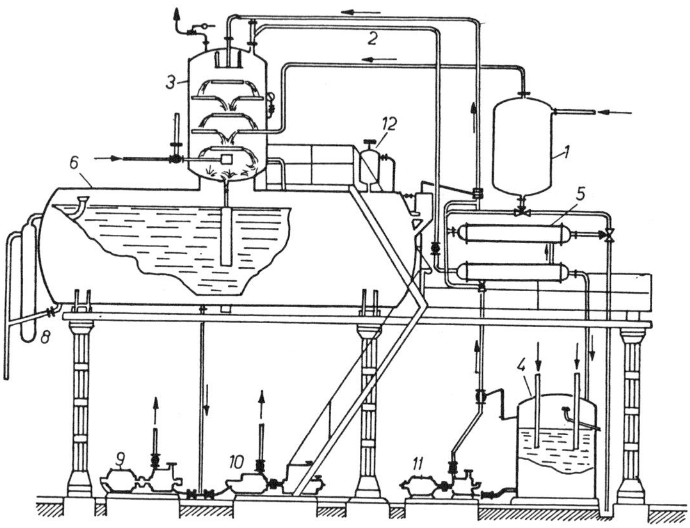

Figure 4. Thermal degasifier: 1—expander; 2—vapors; 3—degasifier; 4—intermediate tank; 5—heat exchanger; 6—feed water tank; 7—batch feeder; 8—syphon overflow; 9, 10—feeding pumps; 11—overflow pump; 12—boiler surface blow-off water [26].

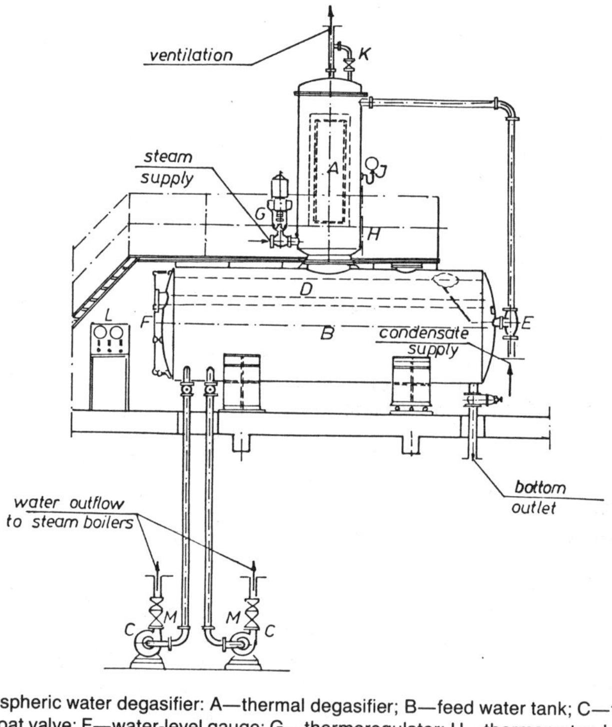

Figure 5. Atmospheric water degasifier: A—thermal degasifier; B—feed water tank; C—feed water pump; D—anode; E—float valve; F—water-level gauge; G—thermoregulator; H—thermometer; I—vacuum gauge; K—safety valve; L—switch board; M—check valves [27].

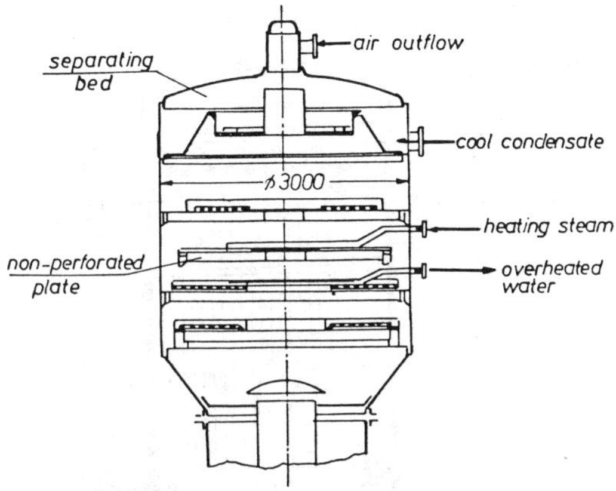

Figure 6. System of dripping plates in thermal degasifier [27].

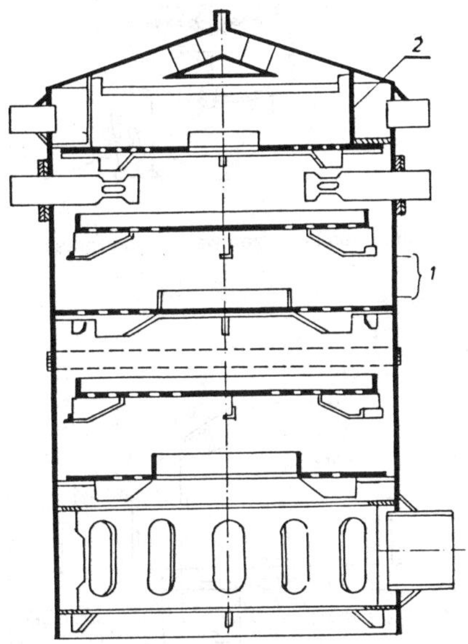

Figure 7. Mixing-type atmospheric degasifier: 1—part of the casing to be dismantled for control or plate exchange; 2—annular part of degasifier head [28].

JET HYDRODYNAMICS

From the point of view of heat transfer the only jets which are considered useful are those that maintain their shape, heat transfer surface, and inner energy state (turbulence level) during a long period of time irrespective of disturbances which may appear under industrial conditions. Thus, the process should be performed in conditions that would enable a

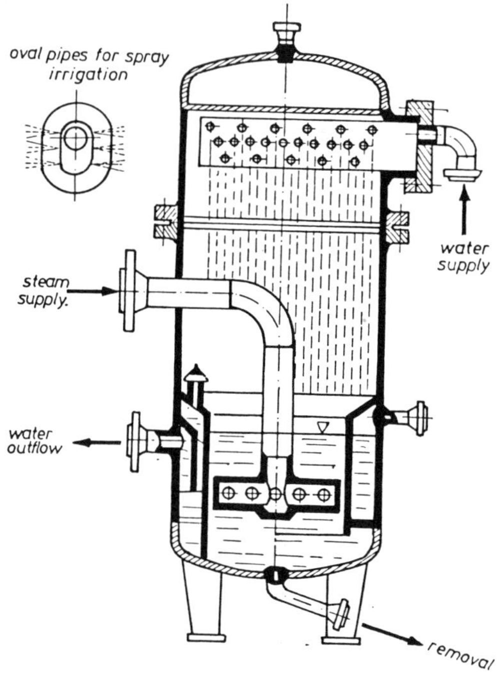

Figure 8. Thermal degasification of demineralized water under pressure through spray irrigation [29].

Figure 9. System for multi-stage water evaporation from brine [30].

possibly intensive heat transfer (turbulent jet), but at the same time its state should be sufficiently distant from the stability limit of the jet.

Selection of desired heat transfer conditions on free liquid jets requires the knowledge of basic hydrodynamic data, which include the conditions leading to jet breakup, conditions affecting the jet character, and conditions which determine the length of continuous portion of the jet. The outflow velocity, physical properties of the liquid, orifice geometry, and physical properties of the fluid into which the jet enters are the factors of basic importance for proper hydrodynamic conditions. The existing hydrodynamic conditions are described

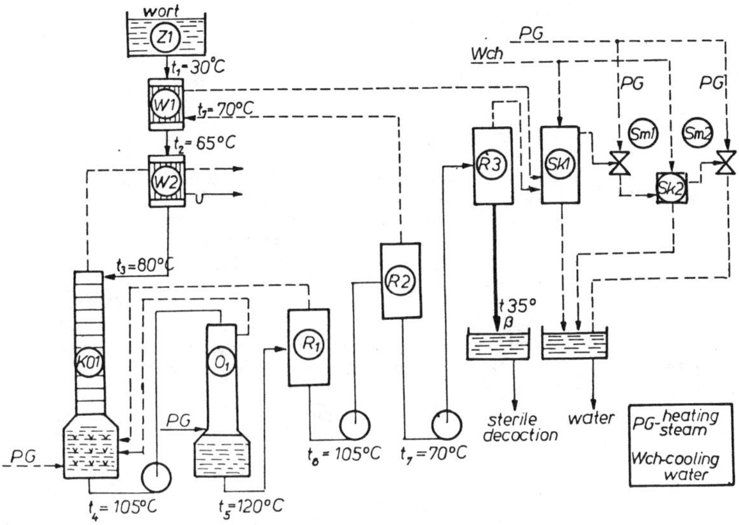

Figure 10. Installation for wort sterilization in spirits-producing plants: Z1—wort tank; W1,W2—heat exchangers; R1,R2,R3—expanders; O1—heater; KO1—stripping column; Sk1,Sk2—condensers; Sm1,Sm2—steam injectors.

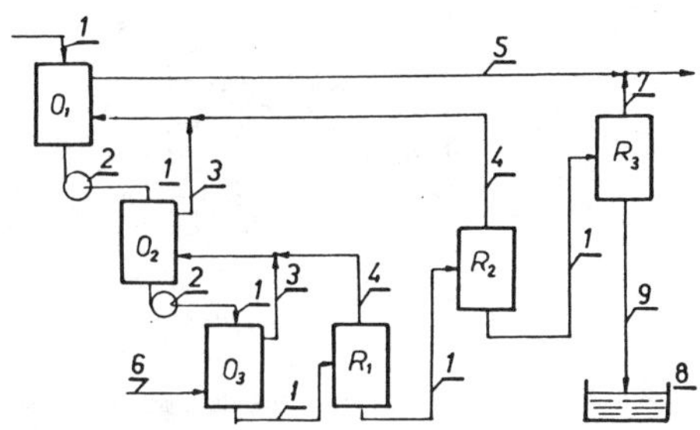

Figure 11. Universal system for direct contact sterilization or pasteurization and cooling of water solutions: O_1,O_2, O_3—heaters; R_1,R_2,R_3—expanders; 1—conduit supplying liquid from pumps; 4,5,7—vapor supplying conduit; 6—heating steam supplying conduit; 8—tank for sterilized and cool liquid; 9—barometric tail-type conduit.

by some state vector in which the previously mentioned parameters are grouped. Unfortunately, the present state of the art does not allow coefficients of this vector to be selected explicity, and in the present study the effects of particular parameter are discussed separately though their interrelations have been stated.

Types of Cylindrical Liquid Jet Breakup

Fluid velocity in the orifice determines the type of breakup. Usually three characteristic breakup types resulting from axisymmetrical, asymmetrical waves as well as aerodynamic forces are distinguished [1, 3] (Figure 13). The previously mentioned breakup types occur

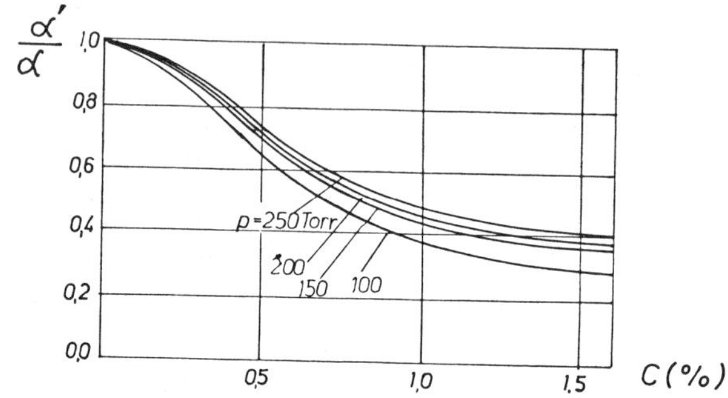

Figure 12. The effect of air content in vapor on condensation intensity: α'—heat transfer coefficient determined in the presence of air; C—content of air in the steam (after Kopp [39]).

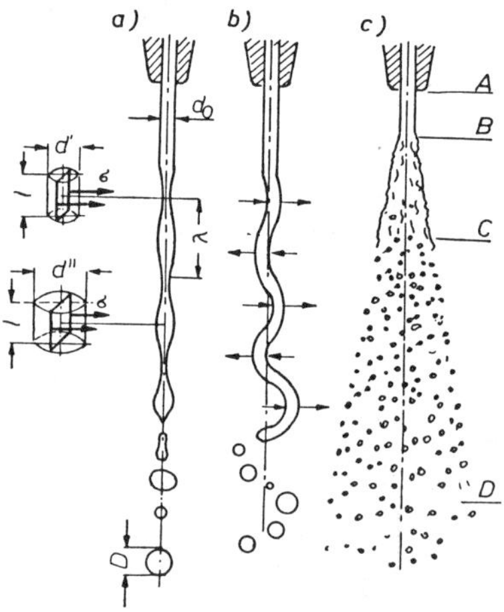

Figure 13. Cylindrical jet breakup caused by: (a) axisymmetrical waves; (b) asymmetrical waves; (c) aerodynamic forces.

at velocities approximately equal to 1 m/s, 10 m/s, and 100 m/s, respectively (in air at atmospheric pressure).

It is obvious that free jet outflow velocity does not exceed 10 m/s and that these jets can be broken up only due to either axisymmetrical or asymmetrical waves. The breakup caused by aerodynamic forces is characteristic of liquids sprayed from various types of nozzles and shall not be discussed here.

The problem of the breakup length of a stable jet under vacuum at low velocities occuring due to axisymmetrical waves was solved theoretically for the first time by Rayleigh in 1878–79 according to Gibasiewicz and Krzyżanowski [40]. During the breakup caused by axisymmetrical waves the jet becomes thinner in some places due to random internal disturbances. In these places liquid pressure is higher than in thick portions, which follows from the condition of equilibrium between pressure and surface tension forces

$$d'Lp_1 = 2L\sigma$$

$$d''Lp_2 = 2L\sigma$$

Hence,

$$p_l = \frac{d''}{d'} p_2$$

i.e., $p_l > p_2$, since $d'' > d'$. Therefore, fluid is forced from the contracted to the thickened parts, which results in drop formation.

Drops are formed in this case only due to surface tension forces (capillary forces). The volume of the drop of diameter D is equal to the volume of jet portion of diameter d_0 and length λ (Figure 13).

$$\frac{\pi D^3}{6} = \frac{\pi d_0^2}{4} \lambda$$

On the other hand it is known that free surface energy is a product of surface tension σ and the surface. Thus a jet is more stable when it has less energy (i.e., a smaller surface). Hence, for the drop and for the jet portion one may write

$$\pi D^2 \sigma \leqslant \pi d_0 \lambda \sigma$$

Therefore,

$$D \geqslant 1.5 d_0$$

is obtained.

Under real conditions drops have different diameters as they are formed of jet portions of various levels of contraction. The effect of viscosity is revealed in vibration damping, which causes a delay or lack of jet breakup. This may be illustrated by high-viscosity liquid jets obtained in polymer fiber production.

Since jet breakup into drops is a result not only of fluid velocity and physical properties of fluid and gas but also of given orifice geometry [1], there are no strict interrelationships that would enable the initiation of axisymmetrical breakup to be determined. Moreover, we have established that apart from the previously mentioned reasons, the breakup process of laminar jets is accelerated by even the slightest external disturbances, such as vibrations of the apparatus, and it is impossible to establish objective conditions for its appearance. Similar observations apply to the change from axisymmetrical to asymmetrical jet breakup. For instance, Fenn and Middleman [16] stated that for $We_a < 5.3$ the laminar jet was broken up due to axisymmetrical waves and for $We_a > 5.3$ due to asymmetrical waves. This was also confirmed by Phinney [41, 42]. From our observations it follows that this criterion can be considered valid only for laminar jets.

The conditions of jet breakup have been also presented by W. Ohnesorge according to Borovin [43] in the system of dimensionless numbers $\lg Lp^{-1/2}$ and $\lg Re$. Various types of fluids which were flowing to the plenum at atmospheric pressure through cylindrical orifices of various diameters were used in the investigations. As seen in Figure 14 particular zones are divided by straight lines. Other realtionships are given by Lyshevskiy [44] who selected in an experimental way the coefficients K, n, m of a dimensionless equation in the form

$$We_c = K \cdot L_p^n M^m; \; Lp = \frac{\rho_l \sigma d_0}{\eta_l^2}; \; M = \frac{\rho_g}{\rho_l}$$

and obtained for each breakup type different values of three coefficients (Table 1). Boundary lines corresponding to these equations are presented in Figure 15.

Results obtained by Lyshevskiy [44] have not been fully confirmed by other studies, and precise and verifiable criteria of this breakup are still being investigated.

Table 1
Coefficients K, n, m of a Dimensionless Equation in the Form $We_c = KLp^nM^m$ for Different Breakup Types

	Breakup type	K	n	m
I	Axisymmetrical	10	− 0.455	− 1.08
II	Asymmetrical	16.6	− 0.302	− 1.05
III	Spraying	266	− 0.133	− 0.8

The Jet Character

As has already been mentioned, the most useful for heat transfer are the turbulent continuous jets enabling the highest heat transfer coefficients to be obtained at a minimum energy input for jet formation and liquid dispersion removal.

The character of a free, liquid jet depends, first of all, on the fluid itself (the flow character inside the orifice) and, to some extent, on external conditions (effects on the inlet, coarseness of pipe surface). Most researchers [3, 6] attempt to determine jet character by the Reynolds number, but it can be easily observed that the effect of external conditions such as orifice geometry and the medium which the jet enters on jet character is significantly stronger than that on the character of flow of fluid in a long pipe. This is the reason why significant discrepancies are found when the jet character is defined only by the Reynolds number, since, for example, according to Haenklein after Grant and Middleman [3] a critical value of the Reynolds number is only 240, although the author doubts such a low value, while Hooper after Grant and Middleman [3] gives a value of 10,000.

The jet character may be determined from the Ohnesorge number for the determination of the critical Reynolds number as proposed by Grant and Middleman [3] who obtained the following correlation:

$$Re_c = 325 \cdot Z^{-0.28} \tag{1}$$

It should be stressed that Grant and Middleman [3] used jets produced by long nozzles, $\ell/d > 95$.

Similar relationships but slightly different values of coefficients in Equation 1 were obtained by Fenn and Middleman [16] and Phinney [15].

Sande and Smith [23, 46] who investigated the effect of orifice geometry on jet character stated that the shape of the orifice outlet may be neglected, while the inlet shape and nozzle length are significant. They proposed the following correlation to calculate the critical value of Reynolds number:

$$Re_c = 12{,}000\,(\ell/d)^{-0.3} \tag{2}$$

for water flowing through orifices 2, 3, and 5 mm in diameter and with bevelled edges at the inlet.

The problem becomes even more complex when the resistance of the medium into which the liquid enters, is significant since in this case the disturbances occurring at the orifice outlet do not cause axisymmetrical breakup at quite long continuous portions of the jet. However, these disturbances grow along the jet length and cause breakup due to jet spraying.

In the literature no correlations can be found describing the critical value of the Reynolds number which would take into account the effect of the medium into which the liquid enters. There are, however, a few publications [16–19] in which the effect of the medium on the length of the continuous jet portion is presented. This problem will be considered later.

Our own investigations suggested that such significant discrepancies in the determination of the criterion of jet character are due to very strong effects of orifice geometry, and

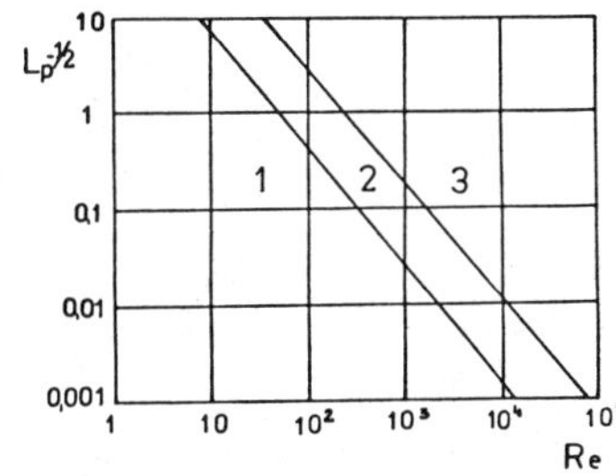

Figure 14. Cylindrical jet breakup regions: 1—axisymmetrical breakup; 2—asymmetrical breakup; 3—spraying.

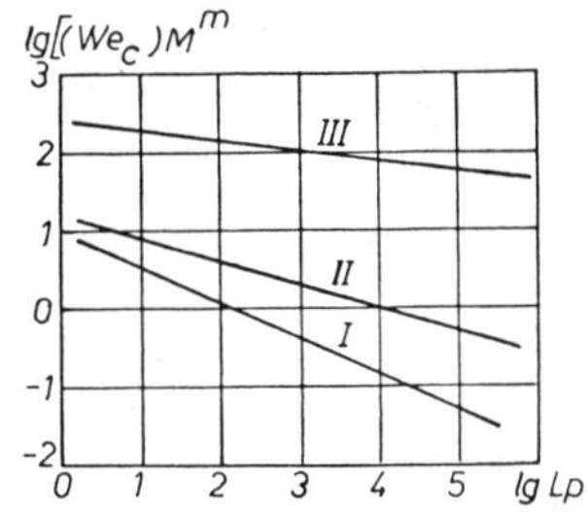

Figure 15. Boundary lines corresponding to different breakup types (Table 1).

external perturbations of various types. It should be stressed that the disturbance effect is crucial in the case of laminar jets, i.e., it determines the instance of transition from laminar to turbulent regime. In the case of turbulent jets the disturbances are not so important.

In the following we present our own results [47] for two significant elements of orifice geometry, i.e. the shape of the orifice inlet and its depth.

The Effect of Orifice Inlet Shape and Depth on Jet Character

Figure 16 presents three typical orifice inlet shapes with (a) sharp, (b) bevelled, and (c) rounded edges. It was found that the jet lost its laminar character first at the sharp inlet edge and last at the rounded edge. The edge is considered rounded when $2r/d > 0.30$ after Perry [48], where r is the rounding radius.

Such significant differences in jet shapes and character caused by the orifice inlet shape may be attributed to the abrupt change of cross-section. The jet contraction does not occur during laminar flow irrespective of the inlet shape and in flowing through bevelled and rounded edges irrespective of the flow character. However, it also appears that turbulence caused by a sharp inlet edge can be damped in some cases by the proper choice of the orifice depth.

During the outflow through a sharp-edged orifice ($\ell/d \approx 0$) and when the inertial forces exceed the forces of internal friction ($Re > ab. 2 \cdot 10^3$) the jet is broken up due to contraction, but the lack of wall friction and negligible effect of air friction do not cause the disturbances to develop further. The jet looks like a typical laminar jet, and only its continuous portion is shorter than in the outflow through a rounded-edge orifice (Figure 16 A–D) which is the net result of the reduced initial diameter of the jet. The fact that the outflow through the sharp-edged orifice allows much-longer-than-expected continuous portions of the jet to be obtained was observed by Gavis and Modan [12] and Van de Sande [23].

In the other extreme case when the sharp-edged nozzle is sufficiently long $\ell/d > 5$ the jet turbulence appears very early ($Re_c \sim 3{,}000$), and further increase of the nozzle length does

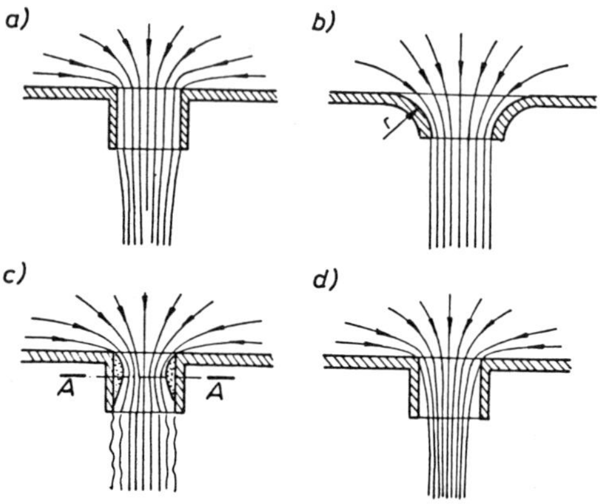

Figure 16. Streamlines in nozzles for different inlet shapes ratios of inertial to viscous forces and nozzle lengths.

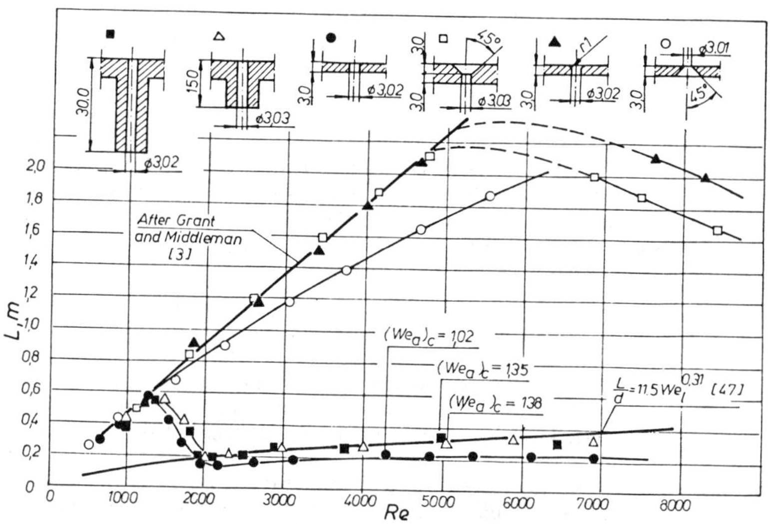

Figure 17. Influence of nozzle inlet shape and nozzle length on jet character (44% per weight water solution of ethylene glycol, d ≅ 3.0 mm).

not affect practically the jet hydrodynamics (Figures 17 and 18). If the nozzle inlet is bevelled the same effect is obtained at $\ell/d > 1.5$ and $Re_c > 4{,}000$.

While investigating hydrodynamics of the jet flowing from sharp-edged orifices we observed an interesting phenomenon called jet hysteresis. This phenomenon is presented below.

Jet Hysteresis

As has been mentioned, the flow through a short nozzle with sharp edges causes jet contraction. The conditions necessary for jet hysteresis (especially the outflow velocity)

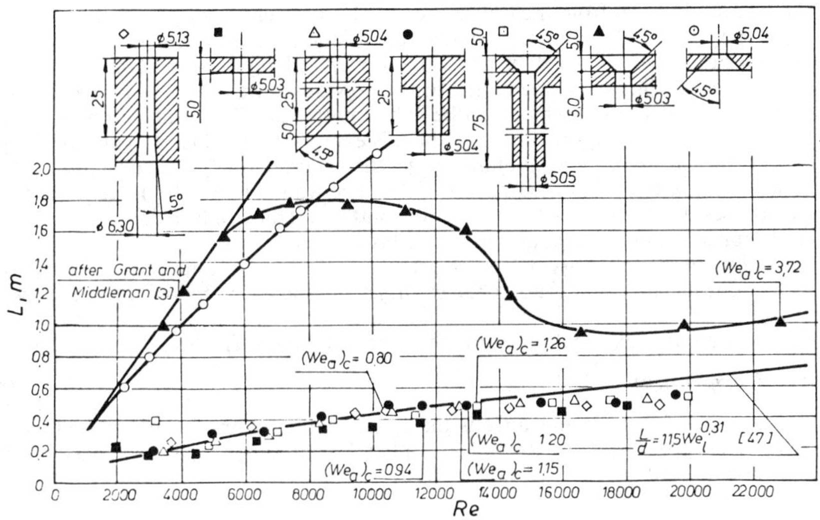

Figure 18. Influence of nozzle geometry on jet character (d ≅ 5.0 mm, water).

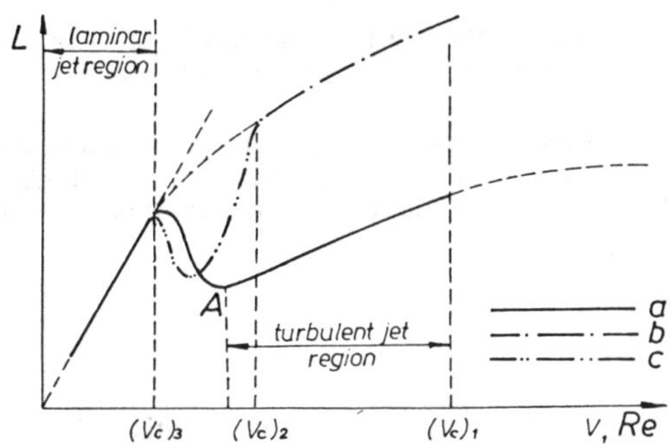

Figure 19. Dependence of breakup length on jet velocity for a short nozzle ($0.4 < \ell/d < 2$) with sharp inlet edges: (a) full cross-section (Figure 20A or 20B); (b) flow without wetting of inner nozzle walls (Figure 20D); (c) flow with partial wetting of inner nozzle walls.

were determined, and jet behavior during the increase and then decrease of outflow velocity was investigated by Iciek [49]. The results are presented in Figure 19.

In the velocity range $0 < v < v_{c3}$ the flow through a sharp-edged orifice is laminar and the jet is similar to the one outflowing either through rounded or bevelled orifices. At velocity v_{c3} contraction is taking place, and the jet loses laminarity, which is reflected by a decrease in the length of the continuous portion of the jet. An increase in outflow velocity causes a further temporary decrease in the length of the continuous portion of the jet up to the moment when turbulent flow is formed. After that (point A in Figure 19) there is an increase of the continuous portion length up to the critical velocity v_{c1} above which there is an abrupt and irreversible change in the outflow character and a significant increase of the continuous portion length of jet. This phenomenon is explained by separation of the jet from the nozzle wall (liquid does not wet the wall) (Figure 20), because when the dead space (shown in Figure 16 C as a dotted area) reaches the end of the nozzle the air enters the dead space. The separation of jet from the wall causes an abrupt decrease of the jet diameter and subsequently the jet becomes laminar, and therefore its continuous portion length increases.

A decrease of outflow velocity does not cause a return along the same continuous line, but as can be seen in Figure 19, the return to the previous state occurs at a much longer continuous portion of the jet, which can be called hysteresis. The wall becomes wetted significantly as the velocity approaches v_{c2} which is expressed by a fast decrease of the jet continuous portion length with a simultaneous jet distortion, namely, as a deviation of the jet from the axis towards the wetted part of the orifice wall (Figure 20 E). Regaining the velocity v_{c3} causes a return to full cross-section outflow. Thus, only below that velocity hysteresis disappears, and both continuous and broken lines overlap.

Determination of the velocity v_{c1} is of great practical importance for defining the boundary conditions for an abrupt change in jet character with loss of orifice wall wettability.

Velocity v_{c1} is a complex function of the following process variables

$$v_{c1} = f(d, \ell, \rho, \eta, \Theta, \sigma, g) \tag{3}$$

but it appears that of basic importance is the nozzle diameter and length as well as the process parameters grouped in the Reynolds number. Due to this, for practical reasons, the following simple dependence

$$\frac{\ell}{d} = f(Re)$$

may be used. It allows the determination for a given ℓ/d ratio of a critical value of the Reynolds number at which there is a change from full cross-section flow to unwetted cross-section flow.

If for given process conditions the ratio $(\ell/d)_{calc}$ is calculated, and if this value is smaller than the designed ℓ/d, there will be a full cross-section outflow. If this value is higher the wall will not be wetted during the outflow. We proposed [49] two forms of this correlation. The simpler one is

$$\frac{\ell}{d} = 0.0354\, Re^{0.34} \tag{4}$$

The other one is of more complex form but has better statistical estimation

$$\lg \frac{d}{\ell} = 5.42\left(\frac{1}{\lg Re}\right) - 1.25 \tag{5}$$

Each of them is practically sufficient to determine extreme parameters necessary for changing the jet flow character if only inertial forces prevail over the forces of internal friction ($Re > 2{,}800$)

The Length of Continuous Jet Portion

This is defined as the part between the orifice outlet and the point where the jet is broken up into droplets. It has been distinguished as an important process parameter, since, as mentioned at the beginning, the continuous jet part is the main surface of heat exchange and it determines the size of the direct-contact heat exchanger being designed. It has been already discussed that every jet, irrespective of its character, is broken up due to axisymmetrical or asymmetrical waves or the aerodynamic forces (spraying) occuring in it. The length of continuous jet portion is determined by several process variables, namely the orifice diameter, outflow velocity, orifice geometry, jet character, physical properties of the

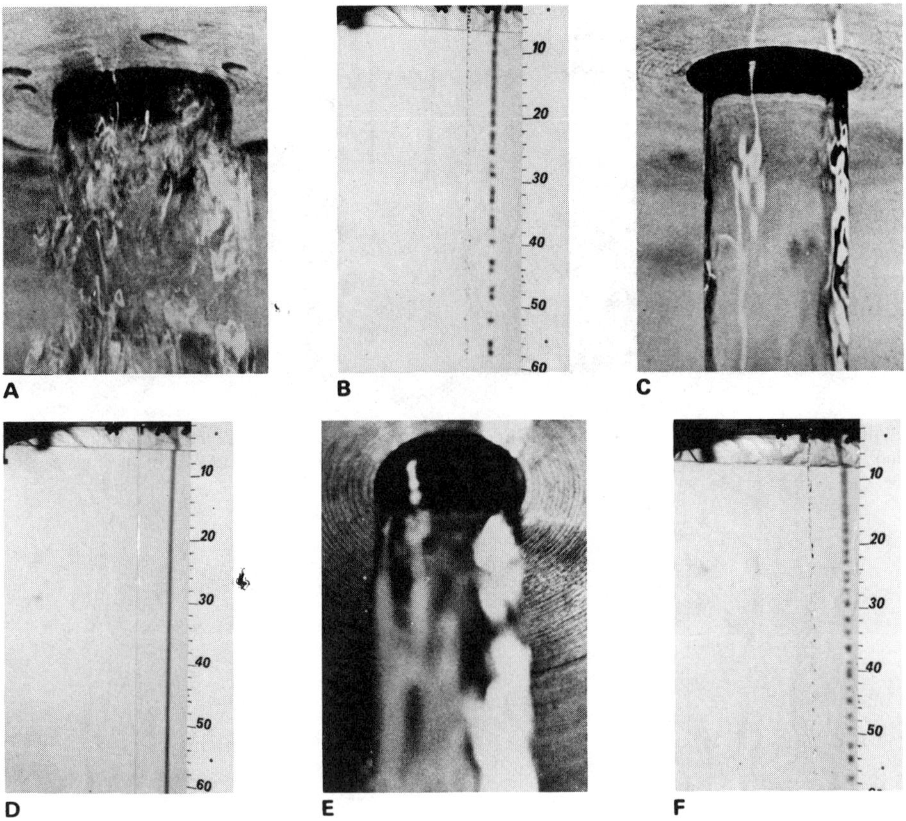

Figure 20. Photograph of a water jet (8°C) outflowing from a sharp-edged brass orifice (d = 5.03 mm, ℓ = 4.00 mm); (A,B) full cross-section flow; H = 0.400 m; Re = 8,230; (C,D) flow without wetting of inner walls; H = 0.400 m; Re = 6,300; (E,F) flow with partial wetting of inner walls; H = 0.103 m; Re = 3,300.

fluid, i.e. the medium into which the jet enters and the breakup type occurring under given conditions.

The effect of jet diameter on the length of its continuous part was analyzed by many authors and all of them concluded that with the increase in jet diameter there is an increase in the length of the continuous jet portion (at constant values of other variables [1]).

A much more complicated effect is exerted by the outflow velocity (Figure 22). For a laminar jet broken up by axisymmetrical waves (section BC) Weber [50] gave the following dependence which determines the length of continuous jet portion

$$\frac{L}{d} = \ln \frac{d}{2\delta} \sqrt{We_j}(1 + 3Z) \tag{6}$$

where the value of parameter ln (d/2δ) must be determined experimentally. On the basis of Heanklein's data Weber stated that (d/2δ) ≈ 12, which has been confirmed by other authors [14]. Slightly different values of ln (d/2δ) have been obtained by Grant and Middleman [3],

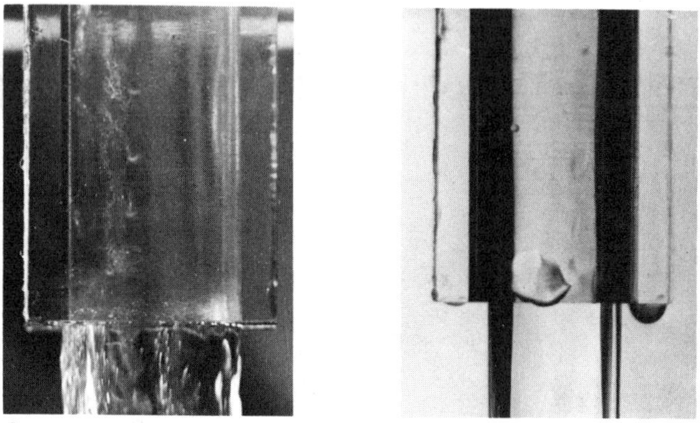

Figure 21. Photograph of a water jet (18°C) outflowing from a sharp-edged perspex nozzle (d = 17.0 mm, ℓ = 43.0 mm); a) flow without wetting of inner walls, H = 0.21 m, Re = 19,700; b) full cross-section flow, H = 0.21 m, Re = 25,700.

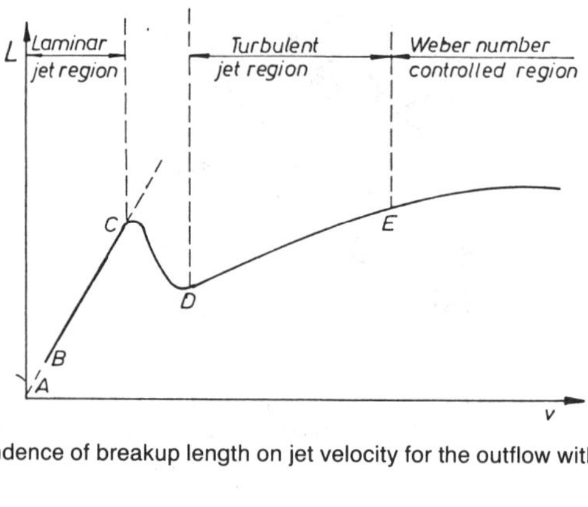

Figure 22. The dependence of breakup length on jet velocity for the outflow with a developed velocity profile.

Fenn and Middleman [16], and Phinney [15]. Different equations are given by Tyler and Wathin according to McCarthy and Molloy [14], and especially by Grant and Middleman [3] who stated that the value of parameter ln (d/2δ) in Equation 6 depends on the Ohnesorge number of the form

$$\ln \frac{d}{2\delta} = -2.66 \ln Z + 7.68 \tag{7}$$

Fenn and Middleman [16] and Phinney [15] confirmed the dependence of parameter ln (d/2δ) on the Ohnesorge number, although they obtained slightly different coefficients in Equation 7.

We stated [47] that for a given orifice and physical properties of the liquid the value of parameter ln (d/2δ) was constant in the whole range of laminar flow, and generally it depends on the Ohnesorge number in the way given by Grant and Middleman.

It was also stated that for a cylindrical laminar jet the length of its continuous portion does not depend on the orifice geometry. This was confirmed by McCarthy and Molloy [14] and Phinney [15].

For turbulent jets (above point D in Figure 22), however, there is no conformity of results obtained by various authors [2, 3, 11, 35]. For instance

Grant and Middleman [3]

$$\frac{L}{d} = 8.51\,(We)^{0.32} \tag{8a}$$

Phinney [11]

$$\frac{L}{d} = 55 + 1.085\,\sqrt{We_j} \tag{8b}$$

We found for our data the correlation that took into account the dependence $L/d = f(We_a)$, but the obtained statistical estimations were weak. It can be stated only that for satisfactorily formed turbulence (if for the bevelled-edge inlet $\ell/d > 15$ and $Re > 4{,}000$ and if for the sharp-edge inlet $\ell/d > 5$ and $Re > 3{,}000$) the jet breaks up due to axisymmetrical waves if $We_a < 1.2$.

The results obtained in the case of outflow through a sharp-edged orifice ($\ell/d \approx 0$) are also worth mentioning. According to Phinney [15] and Van de Sande [23] the jet flowing through such an orifice remains laminar despite high values of the Reynolds numbers and thus its continuous portion remains long. This was also confirmed by our studies [47]. We also proved that in that case the orifice wall ceased to be wetted. Such outflow is called pseudolaminar.

The preceding results are concerned with the outflow of the liquid to the space occupied by gas or freely circulating steam. Any transverse flow in relation to the direction of jet outflow causes a significant decrease of jet continuous length when the jet is out of plumb. In the literature, however, there is no quantitative determination of this flow, but some empirical data can be found in Brodowicz [51] and Kitamura and Takahaschi [52] (cf. Figure 23).

HEAT TRANSFER ON LIQUID JETS

There is a general interrelationship between hydrodynamics and heat transfer, which in the case of direct steam condensation allows quite different dimensionless dependences for laminar, turbulent, and drop-like flows of the jet to be expected.

The process of heat transfer on laminar jets is well known and described theoretically [53–55]. This is obvious since in this case the effect of orifice geometry is negligible and the most significant parameter appears to be the volumetric flow rate [53]. The correlations determining the heat transfer rate presented in the literature are usually consistent both as far as the function form and numerical factors are concerned. Unfortunately the values of heat transfer coefficients are low and this type of heat exchange is of little practical importance.

Condensation on turbulent jets is of basic importance in practice, as has been mentioned earlier [56]. The existing correlations are often inconsistent and their applicability is limited strictly to the conditions given by the authors. From the most recent investigations it follows that these discrepancies are caused by the effect of the geometry of the orifices forming the jet.

The third case mentioned previously, i.e., condensation on a drop-like segment of the jet is most complicated both from the theoretical and practical points of view. The amount and dimensions of droplets determining the heat transfer surface can be estimated only

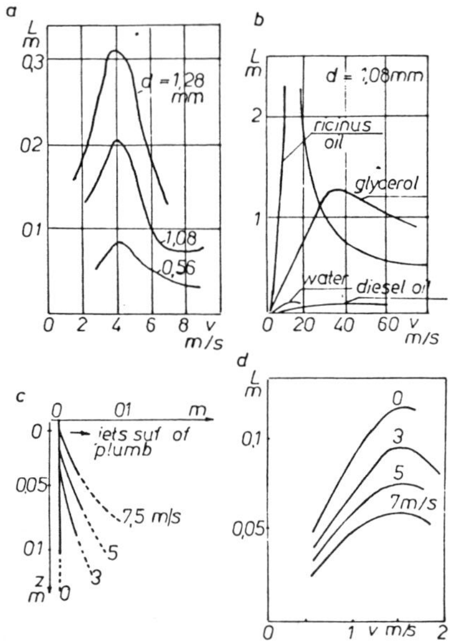

Figure 23. Results of experimental investigations on vertical jet length: (a,b) jet in the air without aerodynamic interaction (a—water; b—various liquids (d = 1.08 mm)), (c,d) deformation of water jet by transverse flow from a nozzle, d = 1.24 mm, (c—for v = 1.237 m/s, d—effect of velocity v at various jet flow velocities) (After Brodowicz[51]).

statistically and in the same way the heat transfer rate may be estimated [57–62]. Undoubtedly this method of heat exchange is most intensive [58, 60] but, simultaneously, it requires a significant power input and costly apparatus to disperse and then to separate droplets.

Heat Transfer on Laminar Jets

Steam condensation on a laminar liquid jet, especially on its smooth segment, is the simplest analytical case of direct contact heat transfer. The heat transfer rate for this case, determined by the mean heat transfer coefficient, is easily obtained from the balance equation

$$Q = G_{sL} \cdot r = \bar{\alpha} \cdot \overline{\Delta T} \cdot A \tag{9}$$

where G_{sL} = steam consumption on jet length (Figure 24)

$\bar{\alpha}$ = mean heat transfer coefficient on length L

$\overline{\Delta T}$ = mean temperature difference defined as

$$\overline{\Delta T} = \frac{T_L - T_0}{\ln \dfrac{T_s - T_0}{T_s - T_L}} \tag{10}$$

A = heat transfer surface, which, although being variable, can be established approximately as an external surface of the cylinder with a nozzle diameter from which the liquid flows $A = \pi L d$.

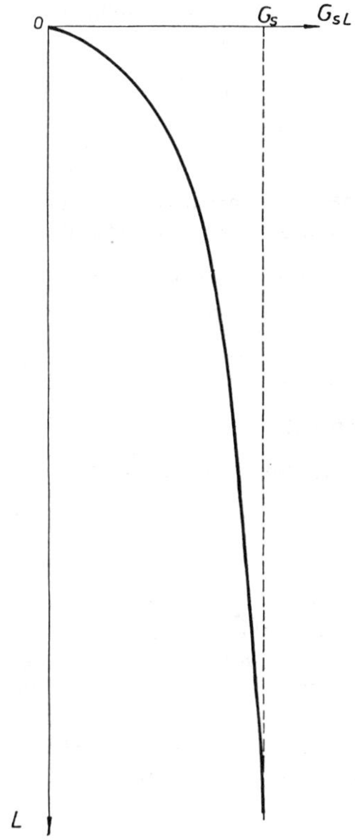

Figure 24. Steam output on jet length.

The balance of heat given up by steam and absorbed by cold liquid also holds

$$G_{sL} \cdot r + G_{sL} \cdot c_l (T_s - T_L) = G_0 \cdot c_l (T_L - T_0) \tag{11}$$

If the previously mentioned equations are combined together, taking account of

$$G_0 = \rho_l \cdot v_0 \frac{\pi d^2}{4} \tag{12}$$

the following simple dimensionless relation in the form

$$4\,\overline{St} = \frac{d}{L} \ln \frac{T_s - T_0}{T_s - T_L} \left[1 + \frac{1}{K} \left(\frac{T_s - T_L}{T_s - T_0} \right) \right]^{-1} \tag{13}$$

is obtained,

where $\overline{St} = \dfrac{\bar{\alpha}}{\rho_l \cdot c_l \cdot v_0}$

Such an approach to heat transfer on a laminar jet can be found in Trub and Litvin [19] and Gibasiewicz and Krzyżanowski [40], where it was confirmed experimentally.

However, it should be stressed that conformity of correlations derived from balance equations with the experimental data is valid first of all for the smooth segment of the jet. Axisymmetrical vibrations occurring in the jet cause rippling and then breakup into drops, which changes significantly both the surface and the heat transfer coefficient.

In our investigation of the effect of orifice geometry on the heat transfer rate between condensing steam and the jet [63] very important differences in heat transfer rates were observed in the smooth, rippled, and drop-like segments of the jet. The comparison was accomplished by constructing for each region the functions of dependence between the temperature and length of the jet. We obtained [63]

For the smooth segment of the jet:

$$T_L = 42.2 + 3.22 \ln L_j$$

For the rippled segment of the jet:

$$T_L = 60.6 + 12.8 \ln L_j$$

For the drop-like segment of the jet:

$$T_L = 79.1 + 28.8 \ln L_j$$

The obtained form of the relationship for laminar jets appeared to be the same as for pseudolaminar jets [49] formed during the outflow through short, sharp-edged nozzles without wetting the nozzle walls, at Reynolds numbers corresponding to apparently turbulent flow. Pseudolaminar jets, despite high power input, needed to achieve a very high velocity, to produce the same thermal effects as the laminar jets. Therefore, such a type of jet formation should be avoided when constructing direct-contact heat exchangers.

It should also be stressed that the laminar and pseudolaminar jets are extremely sensitive to the slightest external disturbances which cause a change of the jet direction and its breakup.

Heat Transfer on Turbulent Jets

The problem of heat transfer from condensing steam to a turbulent water jet is much more difficult to solve explicity, first of all, because of the lack of a uniform turbulence criterion which would take into account properly the effect of all process parameters. This lack has been pointed out earlier and is easily observed in the correlations describing the heat exchange rate. Table 2 presents the equations available in the literature. In Figure 25 a comparison of these relationships in a coordinate system L_j, T_L is shown.

As can be seen discrepancies are considerable and in our opinion they are caused by not paying enough attention to the influence of the nozzle geometry. Such a conclusion follows from Figure 26, which describes the rate of heating a jet outflowing from orifices of the same diameter but having different shapes, at Reynolds number equal to 4,100. Three distinct ranges for pseudolaminar flow can be distinguished: that corresponding to (a) a smooth segment of the jet, (b) a rippled segment, and (c) a drop-like flow. Two ranges of relationships for a turbulent flow on the cylindrical orifice with the ratio $\ell/d = 1$ in which (d) a continuous jet and (e) a drop-like jet can also be distinguished.

In orifices of smaller diameter the influence of geometry on heat transfer intensity is smaller (Figure 27). For nozzles having sharp inlet edges and length at which the full cross-section outflow should take place, the influence of length on heat transfer intensity is very small and the main heat exchange occurs on the drop-like segment of the jet (Figure 28).

As seen in the presented figures, in the case of cylindrical orifices the ratio L_j/d has great influence on the heat transfer rate. The outflow conditions are not significantly affected by

Table 2
Proposed Descriptions of the Process of Direct Contact Heating of the Cylindrical, Free-Falling Turbulent Jet

Author	Equation
Kutateladze [65]	$\lg \frac{T_s - T_0}{T_s - T_L} = 0.16 + 10.1 \left\{\frac{a L_j}{vd^2} + \frac{\varepsilon^* v^2}{5\varphi^{2.5} g d}\left[\left(1 + \frac{2\varphi^2 g L_j}{v^2}\right)^{5/4} - 1\right]\right\}$
	where $\varepsilon^* = 5 \cdot 10^{-4}$
Isachenko [60]	for $We_a \leqslant 2.7$
	$4St = 0.133 \left(\frac{L_j}{d}\right)^{-0.41} \cdot Re^{-0.18} \cdot Pr^{-0.05} \cdot K^{0.11} \cdot \exp(0.16\, We_a)$
Vasilev according to Kutateladze [35]	$\lg \frac{T_s - T_0}{T_s - T_L} = 0.029 \left(\frac{gd}{v^2}\right)^{0.2} \left(\frac{L_j}{d}\right)^{0.7}$
Benedek [69]	$St = 0.00286 \left(\frac{S}{A}\right)^{0.06} \cdot K^{0.084}$

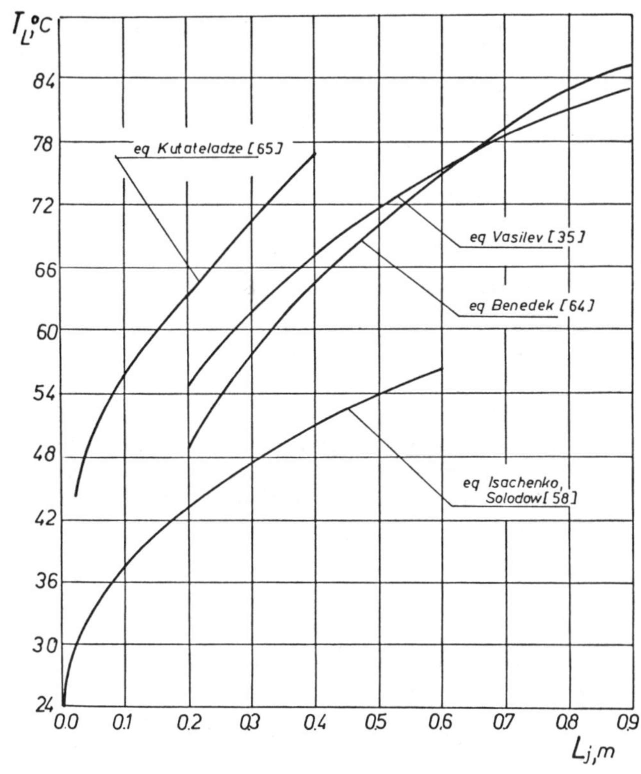

Figure 25. Comparison of results obtained from equations presented in Table 2.

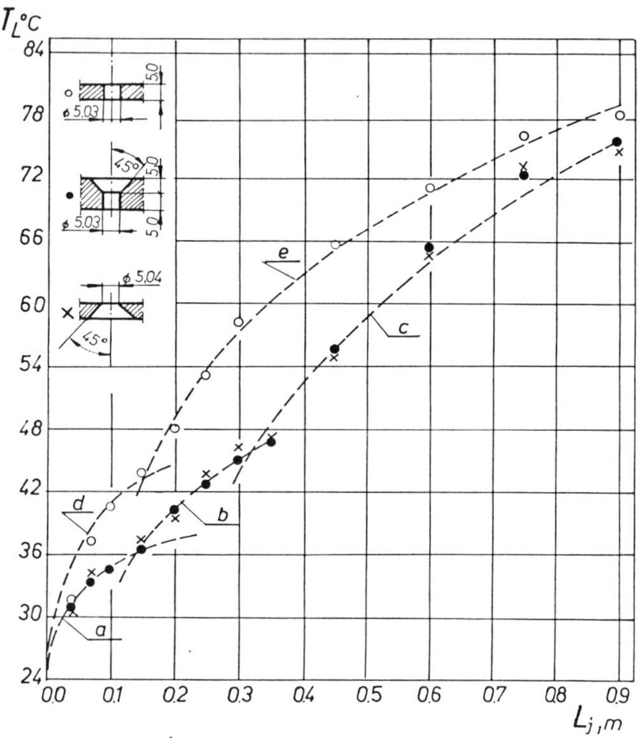

Figure 26. Influence of nozzle geometry on jet heating (T_0 = 24°C, Re ≅ 4,100).

the Froude number. The correlation, we obtained has a form statistically justified in our measurements

$$\lg \frac{T_s - T_0}{T_s - T_L} = f(L_j/d, Fr) \tag{14}$$

which is identical with that given by Vasilev according to Kutateladze [35], although coefficients achieved as a result of the identification are different (Figures 29 and 30) (cf. Iciek [63]). A detailed form of the correlation we used, is presented in the last section of this chapter.

Heat Transfer on Drop-Like Liquid Jets

While discussing heat transfer on a turbulent liquid jet we stated that it was necessary to consider separately a so-called "drop-like region" where the heating rate was significantly higher than on the continuous jet segment. The process is intensified when the jet is broken up directly after the nozzle (spraying). Then the heat transfer surface is abruptly enlarged, and thus the heat transfer rate increases. It is assumed [60] that two main stages of heat transfer at steam condensation on liquid drops can be distinguished. The first, defined as a "disequilibrium state," occurs immediately and consists in heating up the drop surface to

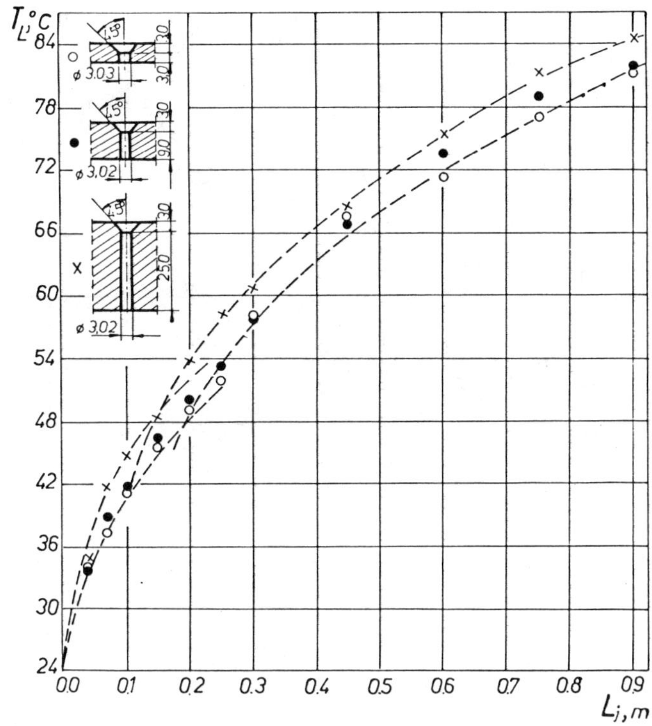

Figure 27. Influence of the nozzle length with conically tapered inlet edges on jet heating (T_0 = 24°C, Re ≅ 4,200).

the equilibrium temperature. In this stage the process rate is limited only by heat transfer resistance of order $10^{-7} m^2K/W$[58]. In the second stage called a "pseudoequilibrium state" the condensation occurs at a rate equal to the rate of heat diffusion into the droplets. Next the drops temperature is equalized and thus the heat transfer rate is reduced. This stage can be reflected by a simple mathematical model if we assume that the heat flux q does not change with the surface increase, i.e. that the main heat transfer resistance comes from the liquid.

$$dQ = r \cdot \rho_l \cdot 4\pi R^2 dR = q \cdot 4\pi R^2 d\tau = -\lambda_l \left(\frac{\partial T}{\partial \xi}\right)_{\xi = R} \cdot 4\pi R^2 d\tau$$

Hence

$$\frac{dR}{d\tau} = -\frac{\lambda_l}{r \cdot \rho_l}\left(\frac{\partial \vartheta}{\partial \xi}\right)_{\xi = R} = -\frac{\lambda_l \vartheta_0}{r\rho_l R_0}\left(\frac{\partial \Theta}{\partial \varepsilon}\right)_{\varepsilon = 1} \tag{15}$$

where $\vartheta_0 = T_s - T_0$

$$\Theta = \frac{\vartheta}{\vartheta_0}; \ \varepsilon = \frac{\xi}{R}$$

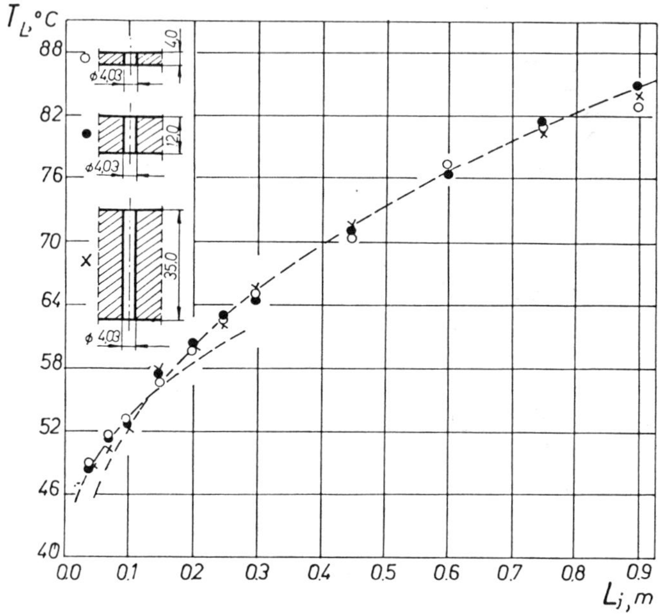

Figure 28. Influence of length of the nozzle with sharp inlet edges on jet heating (T_0 = 40°C, Re ≅ 6,900)

In the first approximation the value of $(\partial\Theta/\partial\varepsilon)_{\varepsilon=1}$ can be determined as for thermal conductivity in heating a liquid sphere at an assumed surface temperature

$$\left(\frac{\partial\vartheta}{\partial\varepsilon}\right)_{\varepsilon=1} = -2\sum_{i=1}^{\infty}\exp(-\varepsilon_i^2\mathrm{Fo}) \tag{16}$$

where $\varepsilon_i = i\pi$

$$\mathrm{Fo} = \frac{a_l\tau}{R^2}$$

Equations 15 and 16 lead to the following dependence

$$\frac{dR}{d\tau} = 2\,\frac{\lambda_l\vartheta}{r\rho_l R}\sum_{i=1}^{\infty}\exp[-(i\pi)^2\mathrm{Fo}]$$

The series is convergent and at Fo > 0.25 it can be approximated only by the first term [58]

$$\frac{dR}{d\tau} = 2\,\frac{\lambda_l\vartheta}{r\rho_l R}\exp\left(-\pi^2\frac{a_l\tau}{R^2}\right) \tag{17}$$

The sphere radius must satisfy, of course, the heat balance equation of the form

$$R^3cT = R_0^3c_lT_0 + (R^3 - R_0^3)(cT_s + r) \tag{18}$$

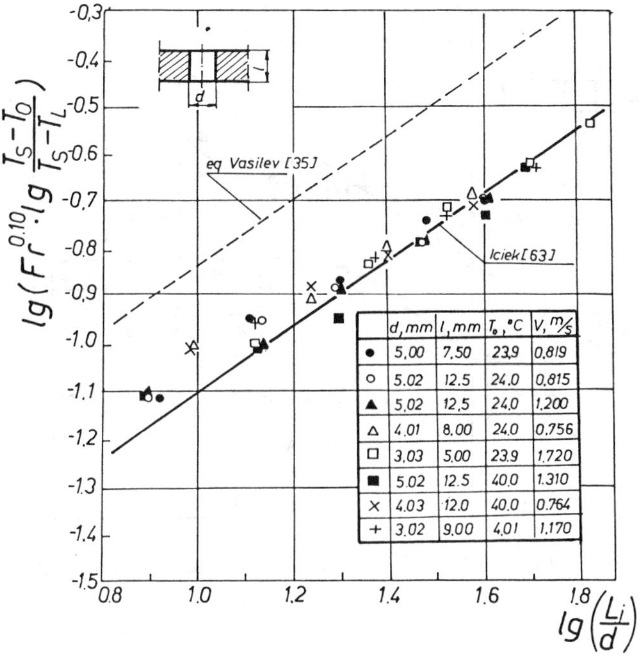

Figure 29. Turbulent jet temperature change vs. jet lengths (direct heating in the region of continuous jet part).

where T = average temperature of growing drops
r = heat of vaporization
c_l = specific heat of liquid

or in the dimensionless form

$$P = \sqrt[3]{(K + 1)/(K + \Theta)} \tag{19}$$

where $\Theta = \frac{T_s - T}{T_s - T_0}$

$$P = \frac{R}{R_0}$$

This mathematical model is confirmed by Hamieleck and Johnson [66] and Volynski and Lipatov [67], especially for low liquid velocity. It was stated that the drop growth was not large. For instance, for steam condensation on water drops at $T_R = 373\,K$ and $T_s - T_0 = 40\,K$, the radius increased by about 2.5%.

In a more detailed analysis we should take into account liquid circulation inside the drops induced by uneven distribution of forces affecting the drop upon its outflow from the nozzle. This refers especially to large drops [67] since in small drops if only $\eta_l / \eta_s > 10$ which occurs practically in most cases of condensation, the effect of convention inside the drops is negligible.

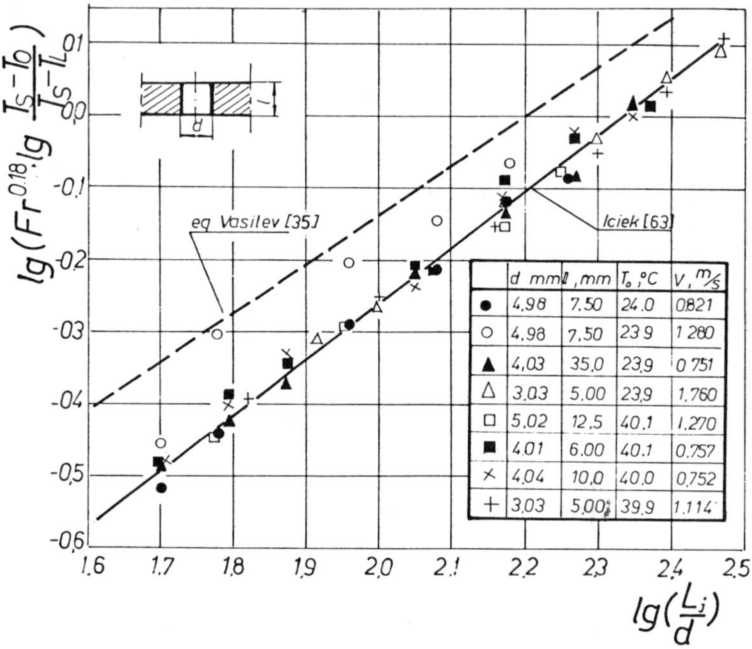

Figure 30. Turbulent jet temperature change vs. jet lengths (direct heating in the region of drop-like jet part).

Bigger drops, accounting for a small percent in diameter distribution, are deformed. This leads to intensified convection and an increase of heat transfer surface. The deformation coefficient is defined as

$$k_d = \frac{\varphi_d \cdot F_d}{\varphi_k \cdot F_k} \tag{20}$$

where φ, F are aerodynamic resistance coefficient and average cross-section of deformed and spherical drops, respectively
The deformation coefficient depends on the Weber number [68] in the way described by

$$k_d = \exp(0.03\, We^{1.5}) \tag{21}$$

In real systems the heat transfer surface is a typical statistical parameter as it depends directly on the number and size of droplets formed. Therefore, we may find in the literature [60, 66, 68, 69] attempts at approximating the phenomenon by properly chosen statistical models in the form of equation

$$Q_V = \int_V q_V dV = 4\pi \int_V \int_R r\rho_c R^2 \frac{dR}{d\tau} \frac{dN}{dR} dR\, dV \tag{22}$$

where $\frac{dR}{d\tau}$ = linear velocity of a single drop growth

$\frac{dN}{dR}$ = differential function of drops distribution according to their size

dN = the participation of drops of a given fraction $(R, R + dR)$

The application of these model types requires rather complicated calculations [58], and the obtained solutions are of importance only for the cases investigated experimentally for which initial conditions necessary to solve the equations are determined.

It should be stressed that heat transfer on broken up jets is extremely intensive. According to Isachenko [60] the complete heating up of the liquid takes place 0.02 m from the nozzle outlet if spraying occurs in pure steam and up to 0.1 m if it occurs in steam containing up to 0.8% by weight of nitrogen.

In spite of this the application of direct-contact heat transfer on liquid being sprayed encounters significant technical problems because the formed drops are usually very small, their elimination is troublesome (even at a small amount of uncondensed gases), and requires special demisters.

CONCLUSIONS

Direct-contact steam condensation on cold water jets is a very intensive process which is easily realized. It will be generally applied, especially in these processes where deposits decreasing heat transfer coefficients are formed. The process may be carried out on a simple apparatus in a reproducible and easily predictable way, assuming that proper outflow conditions will be maintained.

It has been found that the most suitable for this aim are continuous turbulent jets formed on cylindrical orifices in perforated plates with $\ell/d > (\ell/d)_{calc}$. The phenomena of hysteresis and pseudolaminar outflow are then prevented. Jets are identical and vertical and both continuous length and temperature can be calculated for them. For instance, when a direct contact heat exchanger for condensation of steam at 80° C on cold water jets at 20° C is designed, then for a nozzle of diameter d = 6 mm at outflow velocity v = 2 m/s the main calculations are as follows:

1. From Equation 5 the thickness of the plate is calculated

$$\lg \frac{d}{\ell} = 5.42\left(\frac{1}{\lg Re}\right) - 1.25$$

$$Re = \frac{vd\rho}{\eta} = \frac{2 \cdot 0.006 \cdot 998.2}{10^{-3}} = 11978.4 > Re_c$$

$$\ell = 5\,mm$$

2. From $\frac{L}{d} = 11.5\,We^{0.31}$ [47] continuous jet length is

$$We = \frac{\rho v^2 d}{\sigma} = \frac{998.2 \cdot 2^2 \cdot 0.006}{72.75 \cdot 10^{-3}} = 329.3$$

$$L = 0.416\,m$$

3. From Equation 23 the temperature of liquid jet for $L_j > L$ (for example, $L_j = 2\,m$) is calculated

$$\lg \frac{T_s - T_0}{T_s - T_{Lj}} = 0.0152\left(\frac{L_j}{d}\right)^{0.78} Fr^{-0.18} \tag{23}$$

$$Fr = \frac{v^2}{dg} = \frac{2^2}{0.006 \cdot 9.81} = 67.96$$

$$T_{Lj} = 70°C$$

If the parameters L and T_{Lj} are known the main parameters needed for designing a direct contact heat exchanger for an assumed heat duty can be determined.

NOTATION

A	heat transfer area (m^2)	n	coefficient in a dimensionless equation
c	specific heat (J/kg K)	N	number of drops
C	content of air in the steam (%)	m	coefficient in a dimensionless equation
d	nozzle diameter (m)	p	pressure of condensation (Torr)
d_0	jet segment diameter (m)	p_1	liquid pressure in contracted jet part
d′	diameter of contracted jet segment (m)	p_2	liquid pressure in thickened jet part
d″	diameter of thickened jet segment (m)	q	heat flux (W/m^2)
D	drop diameter (m)	Q	heat transfer rate (W)
g	acceleration of gravity (m^2/s)	r	latent heat of vaporization (J/kg)
G	liquid flow rate (kg/s)	R	drop radius (m)
H	height of hydrostatic head (m)	S	the total cross-section of water jet at the outflow (m^2)
k_d	deformation coefficient	T	temperature (°C)
K	coefficient in a dimensionless equation	v	velocity (m/s)
ℓ	nozzle length (m)		
L	jet breakup length (m)		

Greek Symbols

α	heat transfer coefficient (W/m^2K)	ξ	current drop radius (m)
δ	initial amplitude of a disturbance (m)	ρ	density (kg/m^3)
η	dynamic viscosity (Pa s)	σ	surface tension (N/m)
Θ	wetting angle (o)	τ	time (s)
λ	jet segment length (m)	φ	coefficient of discharge
λ_l	thermal conductivity (w/m K)		

Dimensionless Numbers

Fr	v^2/dg, Froude number	Z	$\eta/\sqrt{\rho d \sigma}$
K	$r/c(T_s - T_0)$, Kutateladze number	Re	$vd\rho/\eta$, Reynolds number
Lp	$\rho\sigma d/\eta_l^2$	St	$\alpha/\rho c v$, Stanton number
M	ρ_g/ρ_l	We	$\rho v^2 d/\sigma$, Weber number

Subscripts

a	ambient	l	liquid
c	critical	L	length
D	drop	s	steam
g	gas	0	initial
j	jet	V	volumetric

REFERENCES

1. Orzechowski, Z., *Atomization of Liquids,* Warsaw: Wydawnictwa Naukowo-Techniczne, 1976 (in Polish).
2. Chen, T. F., and Davis, J. R., "Disintegration of a Turbulent Water Jets," *Proc. Am. Soc. Civil Engng. J. Hydr. Div.,* Vol. 90 (1964), pp. 175–206.
3. Grant, R. P., and Middleman, S., "Newtonian Jet Stability," *AIChE Journal,* Vol. 12, No. 4 (1966), pp. 669–678.
4. Dosoudil, M., "Erzeugung gleich großer Tropfen nach dem Abtropfverfahren," *Chem. Ing. Tech.,* Vol. 43, No. 21 (1971), pp. 1172–1176.
5. Pazhi, D. G., and Galustor, V., *Liquid Atomizers.* Moskva: Khimia, 1976 (in Russian).
6. Panchenkov, T. M., and Mamleev, R. A., "Theories of Axisymmetrical Liquid Jets Spraying," *Zhurnal Fizicheskoy Khimii,* No. 1 (1976), pp. 187–190 (in Russian).
7. Telperin, N. I., et al., "Spraying of Liquids by Mechanical Sprayers," *Teoreticheskiye Osnovy Khimicheskoy Tekhnologii,* Vol. 8, No. 3 (1974), pp. 463–467 (in Russian).
8. Loparev, V. P., "An Experimental Investigation of Liquid Drop Breakup under the Condition of Gradual Growing of Inner Forces," *Mekhanika Zhidkosti i Gaza,* Izd. SSSR, No. 3 (1975), pp. 174–178 (in Russian).
9. Vivdenko, N. J., and Shablin, K. N., "The Mechanism of Jet Breakup into Large Drops," *Inzhenerno-Fizicheskiy Zhurnal,* Vol. 8, No. 4 (1965), pp. 445–446 (in Russian).
10. Goedde, E. F., and Yuen, H. C., "Experiments on Liquid Jet Instability," *J. of Fluid Mechanics,* Vol. 40 (1970), pp. 495–511.
11. Ito, R. et al. "Experimental Study on Behaviour of Turbulence in a Fully Developed Pipe Flow," *J. Chem. Engng. Japan.,* Vol. 10 (1977), pp. 194–199.
12. Gavis, J., and Modan, M., "Expansion and Contraction of Jets of Newtonian Liquids in Air: Effect of Tube Length," *Phys. Fluids,* Vol. 10 (1967), pp. 487–497.
13. Tikhonenko, L. K., Kevorkov, L. P., and Lutovinov, S. Z., "Critical Rates of Hot Water Free Outflow from Tubes," *Teploenergetika,* Vol. 5 (1979), pp. 32–36.
14. McCarthy, M. J., and Molloy, N. A., "Review of Stability of Liquid Jets and the Influence of Nozzle Design," *Chem. Engng. J.,* Vol. 7 (1974), pp. 1–20.
15. Phinney, R. E., "Stability of a Laminar Viscous Jet–the Influence of the Initial Disturbance Level," *AIChE J.,* Vol. 18 (1972), pp. 432–434.
16. Fenn, R. W., and Middleman, S., "Newtonian Jet Stability: the Role of Air Resistance," *AIChE J.,* Vol. 15 (1969), pp. 379–383.
17. Phinney, R. E., "Breakup of a Turbulent Liquid Jet in a Low Pressure Atmosphere," *AIChE J.,* Vol. 22, No. 5 (1975), pp. 996–999.
18. Shklover, G. G., and Rodivilin, M. D., "Generalization of the Experimental Data of Steam Condensation on Vertical Water Jets under Vacuum," *Teploenergetika,* No. 10 (1970), pp. 27–29 (in Russian).
19. Trub, J. A., and Litvin, O. P., "Heating of Free Falling Water Jets by Steam under Vacuum," *Energomashinostroenie,* No. 4 (1968), pp. 25–26 (in Russian).
20. Caulk, D. F., and Naghdi, P. M., "On the Onset of Breakup in Inviscid and Viscous Jets," *Journal of Applied Mechanics,* Vol. 46 (1979), pp. 291–297.
21. Vulis, M. A., and Kashkarov, V. P., *Theory of the Viscous Liquid Jet.* Izd. Nauka, 1965 (in Russian).

22. Novikov, A. A., "Non-Linear, Capillary Waves on the Surface of Viscous Liquid," *Mekhanika Zhidkosti i Gaza,* Izd. AN SSSR, No. 2 (1972), pp. 179–181 (in Russian).
23. Van de Sande, E., *Air Entrainment by Plunging Water Jets,* Ph. D. thesis, Delft, 1974.
24. Braynes, J. M., *Processes and Apparatus in Chemical Industry,* Goskhimizdat, 1947 (in Russian).
25. Jakubik, A., *Thermal Degasification of Water,* Warsaw: WNT, 1967 (in Polish).
26. Wróblewski, T., Sikorski, W., and Rzepa, K., *Steam Boilers,* Warsaw: WNT, 1973 (in Polish).
27. Olczakowski, W., and Ficki, Z., *Water in Industrial Plants,* Warsaw: PWN, 1955 (in Polish).
28. Bunkin, W. J., *Operation of Steam Turbines,* Warsaw: PWN, 1956 (in Polish).
29. Chojnowski, A., *Water and Sewage Technology,* Warsaw: PWN, 1972 (in Polish).
30. Sideman, S., and Moalem-Maron, D., *Advances in Heat Transfer,* Vol. 15 (1982), pp. 227–281.
31. Iciek, J., and Ciesielski, M., "Direct Contact Heat Exchangers and Their Application in Industry," *Przemysł Fermentacyjny i Rolny,* Vol. 11 (1977), pp. 13–15 (in Polish).
32. Strumiłło, C., et al., System for Pasteurization or Sterilization and Cooling of Water Solutions and Suspensions, *Polish Patent* No. P-107813, 1980 (in Polish).
33. Iciek, J., and Ciesielski, M., "Thermal Sterilization of Water Solution and Slurries," *Przemysł Fermentacyjny,* No. 3 (1978), pp. 15–17 (in Polish).
34. Burton, H., "The Direct Heating Process for the VHT Sterilization of Milk," *The Chemical Engineer,* No. 32, 1977.
35. Kutateladze, S. S., *Problems of Heat Transfer during Change of State of Substances,* Moskva: Gosenergoizdat, 1953 (in Russian).
36. Abramovich, G. A., *Theory of Turbulent Jets,* Moskva: Gosenergoizdat, 1960 (in Russian).
37. Taitel, Y., and Tamir, A., "Condensation in the Pressure of a Noncondensable Gas in Direct Contact," *Int. J. Heat Mass Transfer,* Vol. 12 (1969), pp. 1157–1169.
38. Hasson, D., Luss, D., and Navon, U., "An Experimental Study of Steam Condensation on a Laminar Water Sheet," *Heat Mass Transfer,* Vol. 7 (1967), pp. 983–1001.
39. Kopp, J. H., "Wärme- und Stoffaustausch bei Mischkondensation," *Berennst-Wärme-Kraft,* Vol. 18, No. 3 (1966), pp. 128–129.
40. Gibasiewicz, B., and Krzyżanowski, Z., "Problems of Fuel Apparatus Quality in Marine Diesel Engines," *Silniki Spalinowe,* No. 1 (1968) (in Polish).
41. Phinney, R. E., and Humphries, W., "Stability of a Laminar Jet of Viscous Liquid Influence of Nozzle Shape," *AIChE J.,* Vol. 19, No. 3 (1973), pp. 655–657.
42. Phinney, R. E., "Breakup of a Turbulent Liquid Jet in a Gaseous Atmosphere," *Fluid Mechanics,* Vol. 60 (1973), pp. 689–701.
43. Borovin, V. A., et al., *Atomization of Liquid,* Moskva: Izd. Mashinostroenie, 1967 (in Russian).
44. Lyshevskiy, A. S., *Fuel Injection in Diesel Engines.* Moskva, 1963 (in Russian).
45. Eckert, R. E. G., and Drake, R. M., *Analysis of Heat and Mass Transfer,* New York: McGraw–Hill, 1972.
46. Van de Sande, E., and Smith, J. M., "Jet Breakup and Air Entrainment by Low Velocity Turbulent Water Jets," *Chem. Eng. Sci.,* Vol. 31 (1976), pp. 219–224.
47. Iciek, J., "The Hydrodynamics of a Free Liquid Jet and Their Influence on Direct Contact Heat Transfer–I. Hydrodynamics of a Free Cylindrical Liquid Jet," *Int. J. Multiphase Flow,* Vol. 8, No. 3 (1982), pp. 239–249.
48. Perry, J. H., *Chemical Engineering Handbook.* New York: McGraw–Hill, 1963.
49. Iciek, J., "The Hydrodynamics of a Free Liquid Jet and Their Influence on Direct Contact Heat Transfer–II. Conditions of Change of Liquid Outflow Type through Sharp Inlet Edged Orifice," *Int. J. Multiphase Flow,* Vol. 8, No. 3 (1982), pp. 251–260.
50. Weber, C., "Zum Zerfall eines Flüssigkeitsstrahles," *Z. Angew. Math. Mech.,* Vol. 11 (1931), pp. 136–154.

51. Brodowicz, K., *Theory of Heat and Mass Exchangers,* Warsaw: PWN, 1982 (in Polish).
52. Kitamura, Y., and Takahaschi, T., "Stability of a Liquid Jet in Air Flow Normal to the Jet Axis," *J. Chem. Eng. Jap.,* Vol. 9, No. 4 (1976), pp. 282–286.
53. Hasson, D., Luss, D., and Peck, R., "Theoretical Analyses of Vapour Condensation on Laminar Liquid Jets," *Int. J. Heat Mass Transfer.,* Vol. 7 (1964), pp. 969–981.
54. Murty, N. S., and Sastri, V. M. K., "Direct Contact Heating of Laminar Falling Liquid Jets," *Int. J. Heat Mass Transfer,* Vol. 19 (1976), pp. 115–117.
55. Isachenko, V. P., Sotskov, S. A., and Yakusheva, A. V., "Heat Exchange During Steam Condensation on a Laminar Cylindrical Jet," *Teploenergetika,* No. 8 (1976), pp. 72–74 (in Russian).
56. Isachenko, V. P., et al., "Heat Exchange during Steam Condensation on Turbulent Liquid Jets," *Teploenergetika,* Izd. Energia, No. 2 (1971), pp. 7–10 (in Russian).
57. Lehic, A., Bajramovic, R., and Ford, J. D., "Droplet Size Distribution: An Improved Method for Fitting Experimental Data," *The Canadian Journal of Chemical Engineering,* Vol. 54, No. 5 (1976), pp. 399–402.
58. Isachenko, V. P., *Heat Exchange at Condensation.* Moskva: Energia, 1977 (in Russian).
59. Kulic, E., Rhodes, E., and Sullivan, G., "Heat Transfer Rate Predictions in Condensation on Droplets from Air-Steam Mixtures," *Can. J. Chem. Eng.,* Vol. 53, No. 3 (1975), pp. 252–258.
60. Isachenko, V. P., and Solodov, A. P., "Heat Exchange at Condensation of Vapour in Continuous and Dispersed Liquid Jets," *Teploenergetika,* No. 9 (1972), pp. 24–27 (in Russian).
61. Kaplan, R. D., and Shendalman, L. H., "Heat Transfer to a Cylindrical Laminar Liquid Jet Ejecting into a Gas," *Int. J. Heat Mass Transfer,* Vol. 16, No. 6 (1973), pp. 1231–1244.
62. Solodov, A. M., "Steam Condensation on a Laminar Flat Liquid Jet," *Teploenergetika,* No. 4 (1971), pp. 50–52 (in Russian).
63. Iciek, J., "The Hydrodynamics of a Free, Liquid Jet and Their Influence on Direct Contact Heat Transfer–III. Direct Contact Heating of a Cylindrical Free Falling Liquid Jet," *Int. J. Multiphase Flow,* Vol. 9, No. 2 (1983), pp. 167–179.
64. Benedek, S., "Heat Transfer at the Condensation of Steam on Turbulent Water Jet," *Int. J. Heat Mass Transfer,* Vol. 19 (1976), pp. 448–450.
65. Kutateladze, S. S., *Fundamentals of Heat Transfer,* Moskva: Atomizdat, 1979 (in Russian).
66. Isachenko, V. P., et al., "Proceedings of the Conference on Scientific Research in 1968–1969. On Heat Transfer Calculation during Condensation on Drop-Like Liquid Jets. Part 2," Moskva, 1969, pp. 59–70 (in Russian).
67. Hamieleck, A. E., and Johnson, A. J., "Viscous Flow around Fluid Spheres at Intermediate Reynolds Number," *Canadian Journal of Chemical Engineering,* Vol. 40, No. 2 (1962), pp. 41–45.
68. Volynski, M. S., and Lipatov, A. S., "Deformation and Rippling of Drops in a Gas Stream," *I. F. Zh.,* Vol. 18, No. 5 (1970), pp. 838–843 (in Russian).
69. Isachenko, V. P., Kushnyrev, V. I., and Solodov, A. P., "Heat Exchange during Steam Condensation from the Mixture of Waste Gases on Liquid Droplike Jet. Mass Transfer and Hydrodynamics of One- and Two-Phase Heating Media," *Trudy MEI,* Vol. 81 (1971), pp. 42–50 (in Russian).

CHAPTER 5

NATURAL CONVECTION HEAT TRANSFER TO POWER-LAW FLUIDS

A. V. Shenoy

Chemical Engineering Division
National Chemical Laboratory
Pune 411 008, India

CONTENTS

INTRODUCTION

Major research efforts on natural convection heat transfer to non-Newtonian fluids have emerged during the last two decades due to the obvious pragmatic importance of these types of fluids in industrial practice. Such fluids are commonly handled by the food, paint, polymer, and pharmaceutical industries. Most slurries, suspensions and dispersions, polymer solutions, and melts exhibit complex rheological behavior which is not experienced when handling ordinary low-molecular-weight Newtonian fluids such as air, water, silicon oils, etc. A study of the various transport processes in non-Newtonian fluids becomes vital for the design of equipment and for tackling problems related to the handling and processing of such fluids.

In any heat transfer circumstance, density differences are bound to arise near the heated surface thus resulting in the fluid near the surface becoming lighter and rising. The fluid from the extensive region far from the surface begins to continuously flow into this region near the surface to replace the rising material and thus sets up natural convection currents. Though in practice, forced convection is encountered more often, it is impossible to avoid the effects of natural convection during heat transport phenomena. The combined effect of forced and natural convection is, in general, much more complicated to analyze in comparison with the individual cases. Between the two, natural convection is far more difficult for theoretical analyses due to the coupling of the momentum and energy equations as will be shown later. When heat transfer by natural convection to non-Newtonian fluids is considered, the task of analysis becomes formidable due to the added complexity of the rheological behavior of such fluids.

In dealing with the complexities of non-Newtonian flow, methods that allow the description, interpretation, and correlation of flow properties are required. To this end, a number of mathematical models and techniques have been developed to describe the rheological behavior of such fluids [1, 2]. Models which have good theoretical foundations [3] are known to lack the flexibility to fit a wide range of non-Newtonian materials. On the other hand, some models though empirical have been found to correlate viscosity data adequately well for various types of materials through a limited number of meaningful parameters. Despite the trend to develop constitutive theories through the application of continuum theories [1], simple models for describing non-Newtonian behavior have found useful applications in industry [4]. Ideally, a simple model should give an accurate fit of experimental data, have a minimum number of independent constants which can be easily evaluated and have some physical basis. Such a model which has been most widely used for non-Newtonian fluids is the empirical power-law Ostwald-de Waele model [5–7].

In the present chapter, we will deal with the natural convection heat transfer to inelastic non-Newtonian fluids whose flow behavior can be described by the power-law model. Laminar as well as turbulent natural convection in both external and internal flow situations will be reviewed giving the theoretical developments together with the experimental efforts of various research workers. A separate section will be allocated for understanding the effects of natural convection on forced convection heat transfer to power-law fluids (which would be of high pragmatic value as such mixed-convection flows are truly present in all real heat transfer circumstances).

GOVERNING EQUATIONS

The governing equations for natural convection heat transfer are akin to those that are conventionally used for analyzing momentum and energy transport. However, in the natural convection case, the density appearing in the momentum equation is dependent on a temperature leading to a coupling of the momentum and energy equation. This complicates the solution of natural convection problems unlike forced convection cases wherein the velocity profile is determined independently from the momentum and continuity equation and then used in the energy equation to obtain the temperature profile. The exact solution of the coupled momentum and energy equation in the natural convection thus becomes a very formidable task. Certain approximations and simplifications are to be thus introduced in order to render the equations in a form that can be easily handled.

It is thus assumed in all cases that

- The natural convection flow is steady,
- All physical properties such as thermal conductivity, viscosity do not vary with temperature.
- The density parameter appearing in the buoyancy term alone varies with temperature and its variation is such that the Boussinesq approximation is valid.

- Boundary layer simplifications can be applied since the entire natural convection phenomena takes place very near the heated solid surface.

Thus the simplified forms of the governing equations of the conservation of mass, momentum, and energy for two-dimensional natural convection flow can be written as:

$$\frac{\partial u}{\partial x} + \frac{\partial v}{\partial y} = 0 \tag{1}$$

$$u\frac{\partial u}{\partial x} + v\frac{\partial u}{\partial y} = g\beta_0(T - T_\infty)\sin\varepsilon + \frac{1}{\rho}\frac{\partial \tau_{xy}}{\partial y} + \frac{1}{\rho}\frac{\partial}{\partial x}(\tau_{xx} - \tau_{yy}) \tag{2}$$

$$u\frac{\partial T}{\partial x} + v\frac{\partial T}{\partial y} = \frac{k}{\rho c_p}\frac{\partial^2 T}{\partial y^2} \tag{3}$$

The details of how the prior equations have been arrived at are available elsewhere [8] and hence are not repeated here.

The boundary conditions on the velocity and temperature are

$$u(x, o) = v(x, o) = 0$$

$$u(x,\delta) = v(x,\delta) = 0$$

$$T(x, 0) = T_w \text{ (constant temperature surface)}$$

$$= T_w(x) \text{ (constant heat flux and variable temperature surface)}$$

$$T(x, \delta_T) = T_\infty$$

$\sin\varepsilon$ is an arbitrary function introduced for characterizing the surface geometry. The left-hand side term in Equation 2 represents the inertial term which is generally neglected when dealing with natural convection in non-Newtonian fluids [8–12] in comparison to the terms on the right-hand side. The first term on the right-hand side of Equation 2 is the buoyancy term, the second the viscous stress, and third the elastic stress. For Newtonian and purely viscous non-Newtonian fluids such as those represented by the power-law fluid model, $\tau_{xx} - \tau_{yy}$ is identically equal to zero. The expression for the viscous stress in the Ostwald-de Waele power-law model is given by the following expression:

$$\tau_{xy} = K\left(\frac{\partial u}{\partial y}\right)^n \tag{4}$$

where K is the consistency index and n is the power-law index that allows the model to describe Newtonian ($n = 1$), shear thinning ($n < 1$), and shear thickening ($n > 1$) flow behavior. This relation is quite useful over a select portion in the medium shear rate range of viscosity curves of a number of non-Newtonian fluids.

Thus, for the power-law fluid, the momentum equation can be written in the simplified form as

$$0 = g\beta_0(T - T_\infty)\sin\varepsilon + K\frac{\partial}{\partial y}\left(\frac{\partial u}{\partial y}\right)^n \tag{5}$$

Equations 1, 3 and 5 thus form the set of governing equations that need to be solved in order to analyze natural convection heat transfer to power-law fluids.

LAMINAR NATURAL CONVECTION

Vertical Flat Plate

The simplest situation that can be envisaged when studying any transport process is that of the vertical flat plate. Laminar natural convection flow of a power-law fluid past a vertical flat plate has thus been studied in the greatest detail. All possible cases of surface temperature, namely, the constant temperature, and constant heat flux as well as the variable temperature instances have been theoretically tackled and experimentally investigated.

Constant Temperature Case

Acrivos [9] obtained the exact asymptotic solution of the appropriate laminar boundary layer equations as given earlier and obtained an expression for the local Nusselt number as follows:

$$Nu = -\theta'(0)\left(\frac{2n+1}{3n+1}\right)^{\frac{n}{3n+1}} Gr_{Lt}^{\frac{1}{2(n+1)}} Pr_{Lt}^{\frac{n}{3n+1}} \tag{6}$$

where

$$Gr_{Lt} = \frac{\rho^2 L^{n+2}[g\beta_0(T_w - T_\infty)]^{2-n}}{K^2} \tag{7}$$

$$Pr_{Lt} = \frac{\rho C_p}{k}\left(\frac{K}{\rho}\right)^{\frac{2}{n+1}} L^{\frac{n-1}{2(n+1)}}[g\beta_0(T_w - T_\infty)]^{\frac{3(n-1)}{2(n+1)}} \tag{8}$$

The plot depicting the dependence of $-\theta'(0)$ on the power-law index n as shown in Figure 1 indicates the insensitivity of the parameter in the range of $0 < n < \frac{3}{2}$. There is no available explanation for the minimum that is seen in the curve and can be assumed to be possibly only a mathematical artifact.

Using an approximate integral method, Shenoy and Ulbrecht [11] obtained a similarity solution under the appropriate assumptions of unequal boundary layer thicknesses and a specific choice of velocity and temperature profiles that satisfied correct boundary and compatibility conditions. The importance of these assumptions have been discussed by Shenoy and Mashelkar [10] in great details. The local and average Nusselt numbers as derived by Shenoy and Ulbrecht [11] are as follows:

$$Nu_x = 2\left\{\frac{f(\alpha_N)}{2\alpha_N(10n\alpha_N)^{\frac{1}{n}}}\right\}^{\frac{n}{3n+1}}\left(\frac{2n+1}{3n+1}\right)^{\frac{n}{3n+1}} Gr_{xt}^{\frac{1}{2(n+1)}} Pr_{xt}^{\frac{n}{3n+1}} \tag{9}$$

$$Nu = 2\left\{\frac{f(\alpha_N)}{2\alpha_N(10n\alpha_N)^{1/n}}\right\}^{\frac{n}{3n+1}}\left(\frac{3n+1}{2n+1}\right)^{\frac{2n+1}{3n+1}} Gr_{Lt}^{\frac{1}{2(n+1)}} Pr_{Lt}^{\frac{n}{3n+1}} \tag{10}$$

where

$$\alpha_N = \frac{1}{3n} \tag{11}$$

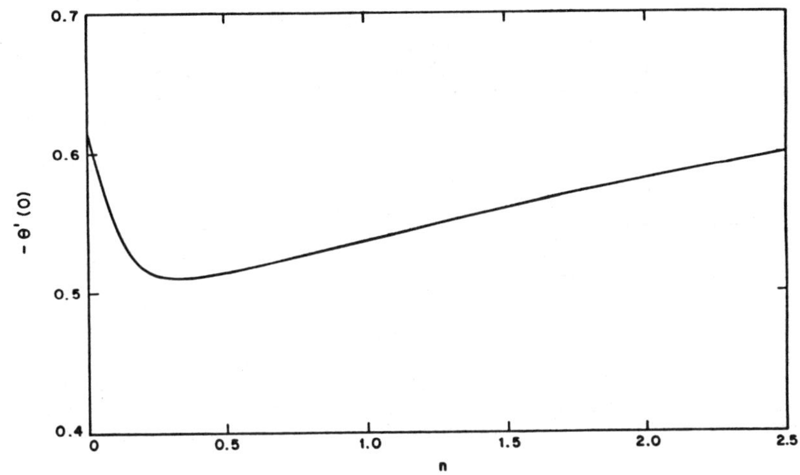

Figure 1. Dependence of $-\theta'(0)$ on the pseudoplasticity index n[9].

$$f(\alpha_N) = \frac{1}{15}\alpha_N - \frac{5}{42}\alpha_N^2 + \frac{3}{28}\alpha_N^3 - \frac{1}{18}\alpha_N^4 + \frac{1}{63}\alpha_N^5 - \frac{3}{1540}\alpha_N^6 \tag{12}$$

$$Gr_{xt} = \frac{\rho^2 x^{n+2}[g\beta_0(T_w - T_\infty)]^{2-n}}{K^2} \tag{13}$$

$$Pr_{xt} = \frac{\rho C_p}{k}\left(\frac{K}{\rho}\right)^{\frac{n-1}{2(n+1)}} x^{\frac{n-1}{2(n+1)}} [g\beta_0(T_w - T_\infty)]^{\frac{3(n-1)}{2(n+1)}} \tag{14}$$

Tien [13] also obtained an approximate integral solution for the vertical flat plate but assumed a velocity profile that did not attain a zero value at a well-defined momentum boundary layer thickness. This was a result of not using the appropriate boundary and compatibility conditions [10] when making the choice of the velocity and temperature profiles. The average Nusselt numbers predicted by Tien [3] are, however, correct as can be seen from Table 1 showing the comparison of the results of the three theoretical approaches.

An experimental study of the natural convection heat transfer to power-law fluids has been done by Reilly et al. [14], who found that the experimental values obtained for the average Nusselt numbers using 0.5% and 1% Carbopol solutions were only 5% to 10% below the theoretical predictions of Acrivos [9]. For pseudoplastic (shear thinning) fluids, the theoretical predictions of Shenoy and Ulbrecht [11] as well as Tien [13] are about 3–5% lower than those of Acrovos [9] as can be seen from Table 1 and can be expected to fit the data of Reilly et al. [14] more closely. Comparison between theory and experiments is bound to produce excellent agreement when average Nusselt numbers are compared. A true test of the theoretical prediction would be seen by comparing the actual velocity and temperature profiles obtained through controlled natural convection experiments. There is no experimental data available yet on the velocity profile for natural convection heat transfer to power-law fluid past an isothermal vertical flat plate. However, temperature

Table 1
A Comparison of the Average Nusselt Numbers for the Constant Temperature Vertical Flat Plate as Predicted by Different Approaches

n	$Nu/Gr_L^{\frac{1}{2(n+1)}} Pr_L^{\frac{n}{3n+1}}$		
	Shenoy and Ulbrecht's Solution [11]	Acrivos's Solution [9]	Tien's Solution [13]
0.5	0.5957	0.63	0.6098
1.0	0.6775	0.67	0.6838
1.5	0.7194	0.71	0.7229

profiles have been experimentally determined [11] and compared with the theoretical predictions as shown in Figure 2 and found to be in good agreement. The theoretical expressions for the velocity and temperature profiles can be obtained from Shenoy and Ulbrecht [11].

Constant Heat Flux Case

Chen and Wollersheim [15] used the fourth-order Runge–Kutta numerical scheme to obtain the solution of the governing equations for the vertical flat plate held under constant heat flux conditions. The mean Nusselt number obtained was of the following form:

$$Nu = \left(\frac{2}{3n+2}\right)^{\frac{n}{3n+2}} \frac{Gr_{Lh}^{\frac{1}{n+4}} Pr_{Lh}^{\frac{n}{3n+2}}}{\theta(0)} \tag{15}$$

where

$$Gr_{Lh} = \frac{\rho^2 L^4}{K^2}\left(\frac{g\beta_0 q}{k}\right) \tag{16}$$

$$Pr_{Lh} = \frac{\rho C_p}{k}\left(\frac{K}{\rho}\right)^{\frac{5}{n+4}} L^{\frac{2(n-1)}{n+4}} \left(\frac{g\beta_0 q}{k}\right)^{\frac{3(n-1)}{n+4}} \tag{17}$$

The dimensionless temperature difference $[\theta(0)]$ is given in Reference 15 for three different typical values of n.

A similarity solution was obtained by Shenoy [16] for the vertical flat plate at constant heat flux following the approximate integral method used earlier [11] for the constant temperature case. The local and average Nusselt numbers as obtained by Shenoy [16] can be written as

$$Nu_x = 2\left\{\frac{f(\alpha_N)}{2\alpha_N(20n\alpha_N)^{1/n}}\right\}^{\frac{n}{3n+2}} \left(\frac{2n+2}{3n+2}\right)^{\frac{n}{3n+2}} Gr_{xh}^{\frac{1}{n+4}} Pr_{xh}^{\frac{n}{3n+2}} \tag{18}$$

$$Nu = 2\left\{\frac{f(\alpha_N)}{2\alpha_N(20n\alpha_N)^{1/n}}\right\}^{\frac{n}{3n+2}} \left(\frac{3n+2}{2n+2}\right)^{\frac{2n+2}{3n+2}} Gr_{Lh}^{\frac{1}{n+4}} Pr_{Lh}^{\frac{n}{3n+2}} \tag{19}$$

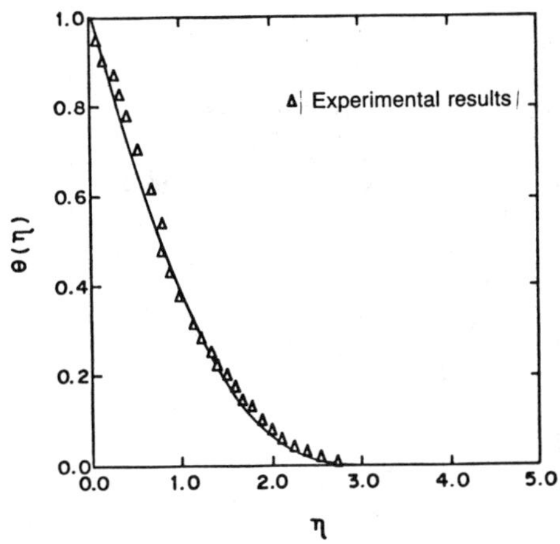

Figure 2. Comparison between the theoretical predictions for the dimensionless temperature distribution and the experimental results (Δ) in the case of an isothermal plate (T_w = 47°C) for a 0.05% CMC solution with a power-law index n = 0.958 [16]. Note that $\theta = (T - T_\infty)/(T_w - T_\infty)$ and $\eta = (y/x)\,(3n + 1)^{-n/(3n+1)}\,Gr_{xt}^{1/2(n+1)}\,Pr_{xt}^{n/(3n+1)}$.

where

$$Gr_{xh} = \frac{\rho^2 x^4}{K^2}\left(\frac{g\beta_0 q}{k}\right)^{2-n} \tag{20}$$

$$Pr_{xh} = \frac{\rho C_p}{k}\left(\frac{K}{\rho}\right)^{\frac{5}{n+4}} x^{\frac{2(n-1)}{n+4}}\left(\frac{g\beta_0 q}{k}\right)^{\frac{3(n-1)}{n+4}} \tag{21}$$

α_N, $f(\alpha_N)$, Gr_{Lh} and Pr_{Lh} have the same definitions as given by Equations 11, 12, 16, and 17, respectively.

Experimental data for laminar natural convection heat transfer from a vertical flat plate at constant heat flux condition has been provided by Dale and Emery [17]. The local Nusselt number obtained [17] for heat transfer to a 0.05% CMC solution with n = 0.888 was found to correlate by the following expression:

$$Nu_x = 0.6\left(Gr_{xh}^{\frac{3n+2}{n+4}}\,Pr_{xh}^{n}\right)^{0.2101} \tag{22}$$

The theoretical prediction as given by Shenoy [16] when rearranged appropriately and solved for n = 0.888 gives the following

$$Nu_x = 0.593\left(Gr_{xh}^{\frac{3n+2}{n+4}}\,Pr_{xh}^{n}\right)^{0.2144} \tag{23}$$

Thus, showing a very close agreement between the experimental data and the theoretical estimates.

As in the constant temperature case, it is of interest to see how well the exact procedure of Chen and Wollersheim [15] and approximate procedure of Shenoy [16] predicts the experimental temperature and velocity distributions. Figures 3 and 4 show that the

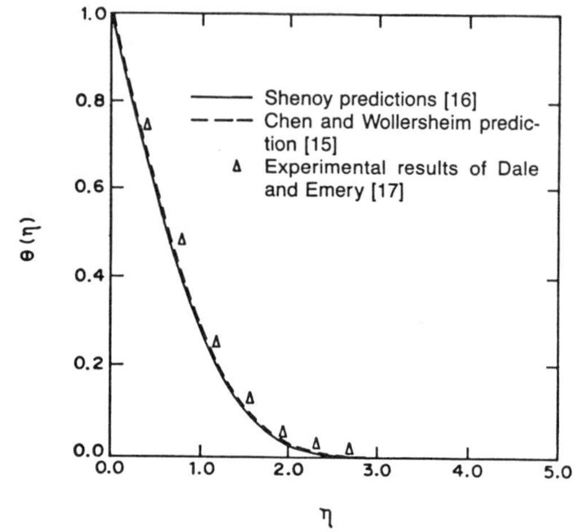

Figure 3. Comparison between the theoretical predictions for the dimensionless temperature distributions for a constant heat flux plate predicted by Shenoy [16] and Chen and Wollersheim [15], and the experimental results of Dale and Emery [17] for a 0.05% CMC solution with a power-law index n = 0.888 [16]. Note that

$$\theta(\eta) = \frac{(T - T_\infty)\, Gr_{xh}^{1/(n+4)}\, Pr_{xh}^{n/(3n+1)}}{(qL)/k\,[(3n+2)(x/L)]^{n/(3n+2)}}$$

and $\eta = (y/x)(3n+2)^{n/(3n+2)}\, Gr_{xh}^{1/(n+4)}\, Pr_{xh}^{n/(3n+2)}$

temperature distribution is predicted well by both the exact numerical as well as the approximate integral solution whereas the theoretical velocity profiles are far from actual experimental observations. No physical explanation can be provided except for the hypothesis that there might be the effect of slip at the wall probably present, leading to an apparent discrepancy. The effect of slip during the transport phenomena of structured fluids has been shown recently to explain a number of conflicting observations [18–20]. Slip velocity has never been considered in the solution of natural convection problems and its neglect is probably the reason for the discrepancy between theory and experiment.

Variable Temperature Case

The only available theoretical analysis for the variable temperature case is that of Shenoy [16] who provided the following expressions for the local and average Nusselt numbers for plate temperature varying as $[T_w(x_1) - T_\infty] = [T_{wL} - T_\infty]\, x_1^p$. Thus

$$Nu_x = 2\left\{\frac{f(\alpha_N)}{2\alpha_N(10n\alpha_N)^{1/n}}\right\}^{\frac{n}{3n+1}} \left\{\frac{2n+1+p}{3n+1}\right\}^{\frac{n}{3n+1}} Gr_{xv}^{\frac{1}{2(n+1)}+\frac{P}{(n+2)(3n+1)}}\, Pr_{xv}^{\frac{n}{3n+1}} \tag{24}$$

$$Nu = 2\left\{\frac{f(\alpha_N)}{2\alpha_N(10n\alpha_N)^{1/n}}\right\}^{\frac{n}{3n+1}} \left\{\frac{3n+1}{2n+1+p}\right\}^{\frac{2n+1}{3n+1}} Gr_{LV}^{\frac{1}{2(n+1)}+\frac{P}{(n+2)(3n+1)}}\, Pr_{LV}^{\frac{n}{3n+1}} \tag{25}$$

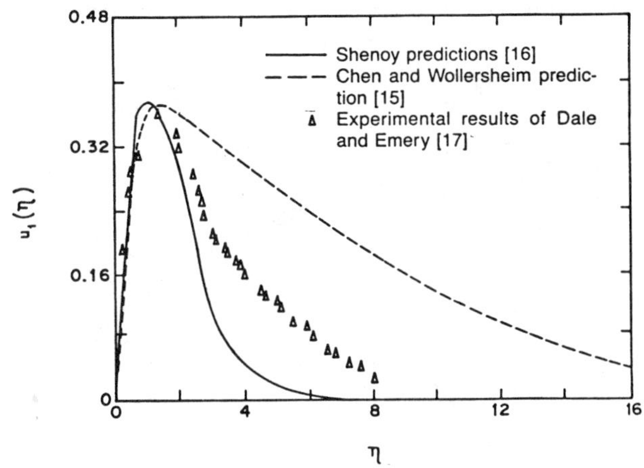

Figure 4. Comparison between the dimensionless velocity distributions for a constant heat flux plate predicted by Shenoy [16], and Chen and Wollersheim, and the experimental results of Dale and Emery [17] for the same system as in Figure 3 [16]. Note that

$$u_1(\eta) = \left(\frac{Pr_{xh}}{3n + 2}\right)^{(n+2)/(3n+2)} \frac{ux^{n/(2-n)}\,(K/\rho)^{1/(2-n)}}{Gr_{xh}^{(n+1)/(n+4)(2-n)}}$$

where

$$Gr_{xv} = \rho^2 x^{n+2} \frac{[g\beta_0(T_{wL} - T_\infty)]^{2-n}}{K^2} \tag{26}$$

$$Pr_{xv} = \frac{\rho C_p}{k}\left(\frac{K}{\rho}\right)^{\frac{2}{n+1}} x^{\frac{n-1}{2(n+1)}} [g\beta_0(T_{wL} - T_\infty)]^{\frac{3(n-1)}{2(n+1)}} \tag{27}$$

The expressions for Gr_{Lv} and Pr_{Lv} are obtained by substituting L in place of x in the preceding expressions. α_N and $f(\alpha_N)$ are given by Equations 11 and 12.

Presently, there is no experimental data available in order to support the above theoretical predictions.

Horizontal Cylinder

Acrivos [9] obtained the expression for Nusselt number during natural convection heat transfer to power fluid applicable to any two-dimensional surface at constant temperature as

$$Nu = -\theta'(0)\left(\frac{2n+1}{3n+1}\right)^{\frac{n}{3n+1}} Gr_c^{\frac{1}{2(n+1)}} Pr_c^{\frac{n}{3n+1}} \frac{(\sin\varepsilon)^{\frac{1}{2n+1}}}{[\int_0^{x1} (\sin\varepsilon)^{\frac{1}{2n+1}} dx_1]^{\frac{n}{3n+1}}} \tag{28}$$

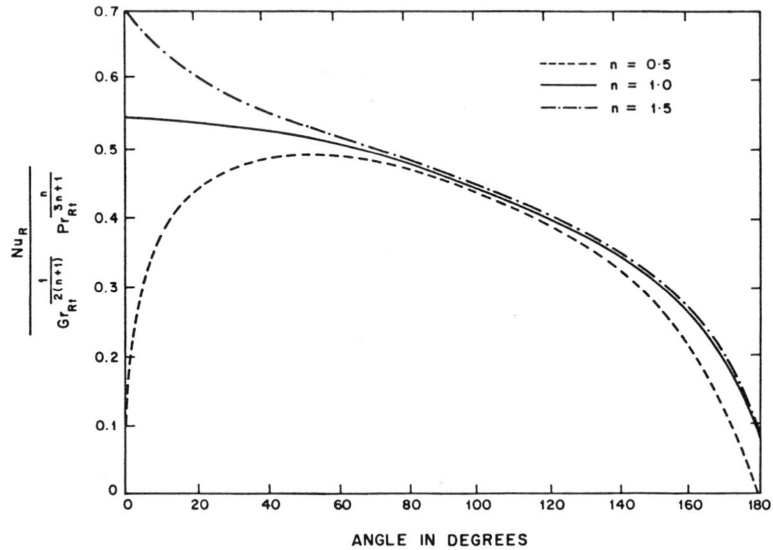

Figure 5. Variation of the local heat transfer rate along the surface of a horizontal cylinder [9].

where Gr_c and Pr_c are the characteristic Grashof number and Prandtl numbers defined as

$$Gr_c = \frac{\rho^2 l_c^{n+2}[g\beta_0(T_w - T_\infty)]^{2-n}}{K^2} \tag{29}$$

$$Pr_c = \frac{\rho C_p}{k}\left(\frac{K}{\rho}\right)^{\frac{n-1}{2(n+1)}} l_c^{\frac{n-1}{2(n+1)}} [g\beta_0(T_w - T_\infty)]^{\frac{3(n-1)}{2(n+1)}} \tag{30}$$

for a horizontal cylinder, $l_c = R$ (radius of the cylinder) and $\sin\varepsilon = \sin x_1$. A plot of the variation of Nusselt number for $n = 0.5$, 1.0, and 1.5 and for $0 < x_1 < \pi$ as given by Acrivos [9] is shown in Figure 5.

Gentry and Wollersheim [21] have provided the experimental data for natural convection heat transfer from an isothermal horizontal cylinder to power-law fluids. The experimental local Nusselt numbers were found to compare well with the theoretical predictions of Acrivos [9] as well as with their own.

Kim and Wollersheim [22] theoretically analyzed the natural convection heat transfer problem in the case of the horizontal cylinder maintained under constant heat flux conditions by the approximate integral method and gave the following expression:

$$Nu = \bar{K}(n)\, Gr_R^{\frac{3n+1}{2(n+1)(3n+2)}}\, Pr_R^{\frac{n}{3n+2}} \frac{(\sin x_1)^{\frac{1}{2n+1}}}{[\int_0^{x_1} (\sin x_1)^{\frac{1}{2n+2}} dx_1]^{\frac{n}{3n+2}}} \tag{31}$$

Their experimental data on dilatant (shear thickening) fluids for the isothermal as well as the constant heat flux case was found to agree with the theoretical predictions [21, 22] as has been shown in Figures 6 and 7.

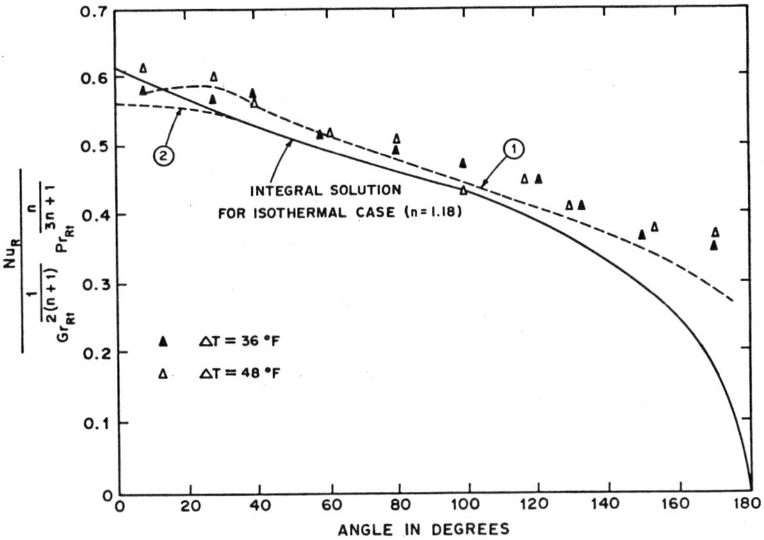

Figure 6. Variation of the local heat transfer rate along an isothermal surface of a horizontal cylinder for 38% corn starch in aqueous sucrose solution with power-law index n = 1.18. Curve 1 shows experimental results for Newtonian fluids; Curve 2 shows the integral solution for Newtonian fluids [22].

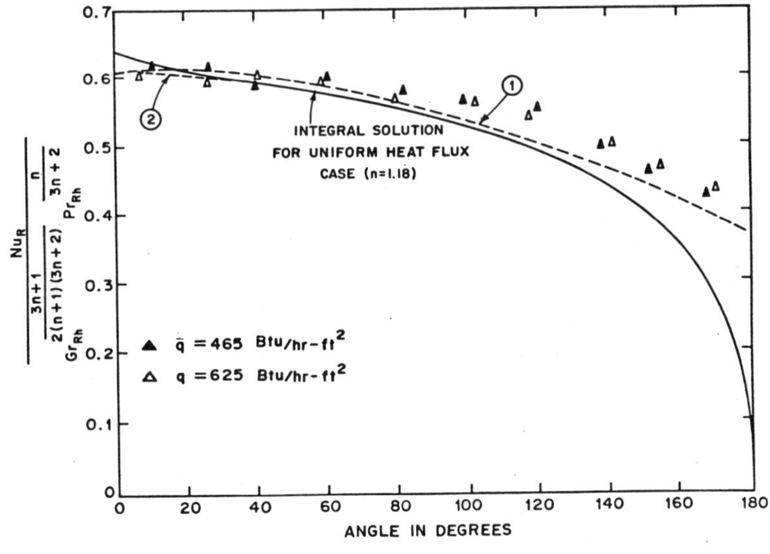

Figure 7. Variation of the local heat transfer rate along a uniform heat flux surface of a horizontal cylinder for the same system as in Figure 6 [22].

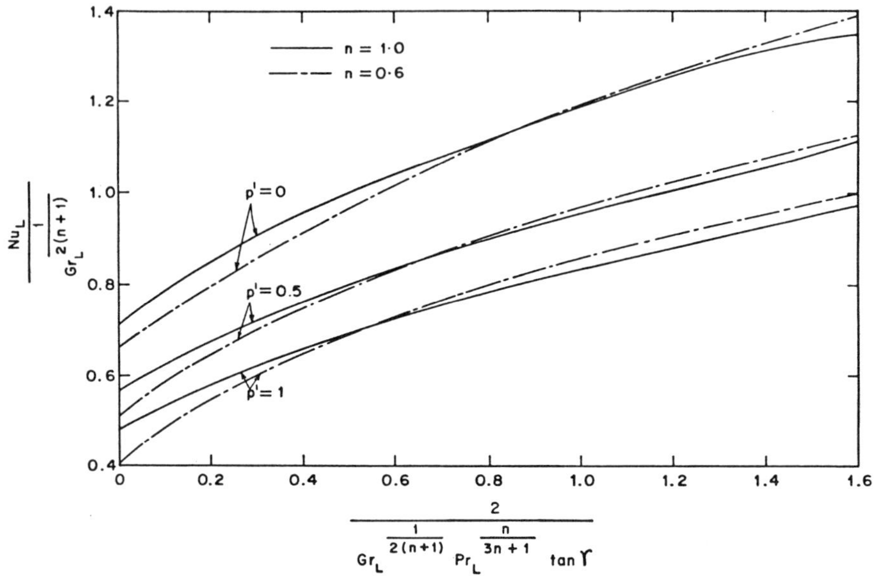

Figure 8. Variation of the heat transfer rate with curvature parameter along a slender vertical cone [12].

Table 2
Comparison of the Average Nusselt Number for the Constant Temperature Vertical Cone as Predicted by Acrivos [9] and Shenoy [12]

n	$Nu/Gr_L^{\frac{1}{2(n+1)}} Pr_L^{\frac{n}{3n+1}}$	
	Acrivos's Solution [9]	Shenoy's Solution [12]
0.5	0.65	0.623
1.0	0.71	0.718
1.5	0.75	0.767

Slender Vertical Cone

An expression for the mean Nusselt number for natural convection heat transfer from slender nonisothermal vertical cones has been provided as follows by Shenoy [12]:

$$Nu = \frac{4(3n+1)}{a_1[2n + (1+p)(3n+2)]} Gr_L^{\frac{1}{2(n+1)}} Pr_L^{\frac{n}{3n+1}} \tag{32}$$

where a_1 incorporates terms for the curvature effect. It was found that increasing the curvature gave rise to an increase in the Nusselt number and for power-law fluids this trend predominated at lower values of n. Figure 8 shows the relevant results for three different values of p relating to the constant temperature, constant heat flux, and linearly varying temperature case. The only other available information in the literature on natural

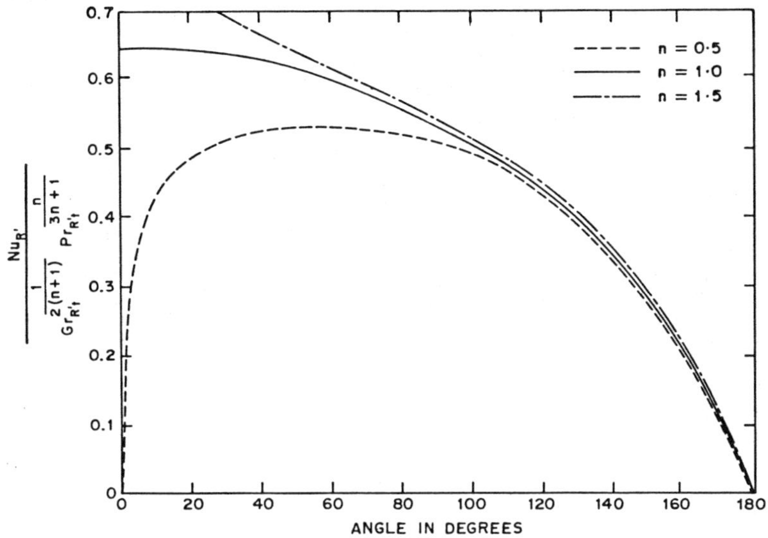

Figure 9. Variation of the local heat transfer rate along the surface of a sphere [9].

convection heat transfer from vertical cones to power-law fluid is that of Acrivos [9] but is limited to the constant temperature case. The results of Acrivos [9] when compared with Shenoy [12] show an error of less than ± 5% as can be seen from Table 2.

Sphere

Acrivos [9] obtained the Nusselt number for natural convection heat transfer to power-law fluids past three-dimensional bodies as

$$Nu = -\theta'(0)\left(\frac{2n+1}{3n+1}\right)^{\frac{n}{3n+1}} Gr_c^{\frac{1}{2(n+1)}} Pr_c^{\frac{n}{3n+1}} \frac{(r_1^n \sin\varepsilon)^{\frac{1}{2n+1}}}{\{\int_0^{x1} r_1^{\frac{3n+1}{2n+1}} (\sin\varepsilon)^{\frac{1}{2n+1}} dx_1\}^{\frac{n}{3n+1}}} \tag{33}$$

where Gr_c and Pr_c are as defined by Equations 29 and 30.

For a sphere, $l_c = R$ (radius of sphere) and $\sin\varepsilon = \sin x_i = r$ and the expression for the Nusselt number can be derived and plotted as shown in Figure 9 for $0 < x_1 < \pi$ and $n = 0.5$, 1.0, and 1.5.

Amato and Tien [23] have provided the experimental temperature and velocity profiles as well as the average Nusselt number values for flow of power-law fluids past a sphere. Experimental heat transfer variations were found to be in good agreement with the theory of Acrivos [9].

Horizontal Parallel Plates

Natural convection heat transfer to a power-law fluid between two horizontal parallel plates such that the bottom plate is heated poses a very interesting problem, popularly termed as the Benard convection problem, and has been the subject of a number of

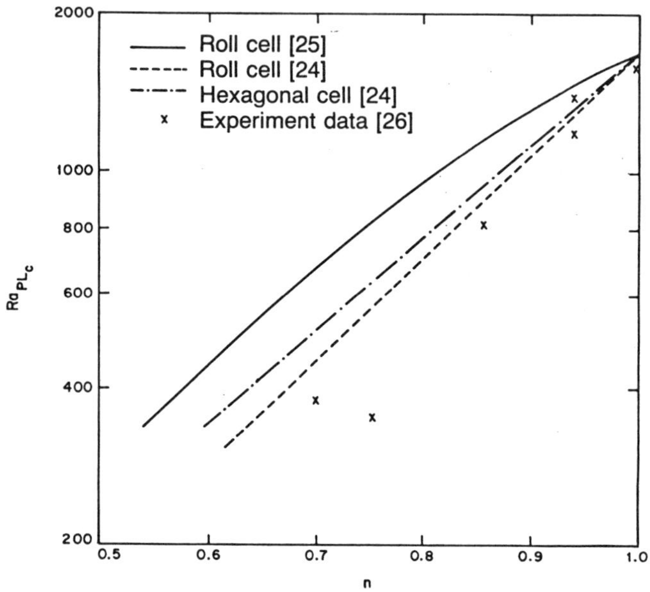

Figure 10. Comparison of the computed and experimental dependence of critical Rayleigh number on the pseudoplasticity index n [25].

investigations [24–28]. The onset of convection does not take place unless a critical value of the dimensionless Rayleigh number Ra_{PL} as defined below for power-law fluids is exceeded.

$$Ra_{PL} = \frac{\rho g \beta_0 (T_0 - T_1) d^{2n+1}}{K \alpha_0^n} \tag{34}$$

where T_1 and T_0 are the temperatures of the top and bottom plate respectively, d is the distance between the plates, and α_0 is the thermal diffusivity.

The resulting cellular convection that is set up after the critical Rayleigh number is exceeded leads to many-fold increases in the Nusselt numbers. Thus there exists a point of thermal instability when horizontal layers of fluids are heated from below. Tien et al. [24] and Ozoe and Churchill [25] studied the behavior of the natural convection between horizontal layers of a power-law fluid when heated from below. The critical Rayleigh number as predicted by the approximate analysis of Tien et al [24] was found to be lower than those of the exact predictions of Ozoe and Churchill [25] as can be seen from Figure 10. The experimental data of Tsuei and Tien [26] was, however, found to compare more closely with the approximate analysis of Tien et al [24]. This, of course, is by no means a devaluation of the exact solution of Ozoe and Churchill [25], as experiments of these kind are rather difficult and amenable to error. Both the approximate [24] and the exact [25] approaches predict that the critical Rayleigh number reduces with increasing pseudoplasticity, thus implying that thermal instability and Benard convection is easier set up in non-Newtonian fluid with values of n less than 1 than in Newtonian fluids. The work of Parmentier [27] and Parmentier et al. [28] shows that increasing pseudoplasticity results in changes in the flow patterns and leads to regions of fluid stagnation with very weak recirculating currents. The effect of the pseudoplasticity index and the Rayleigh number on

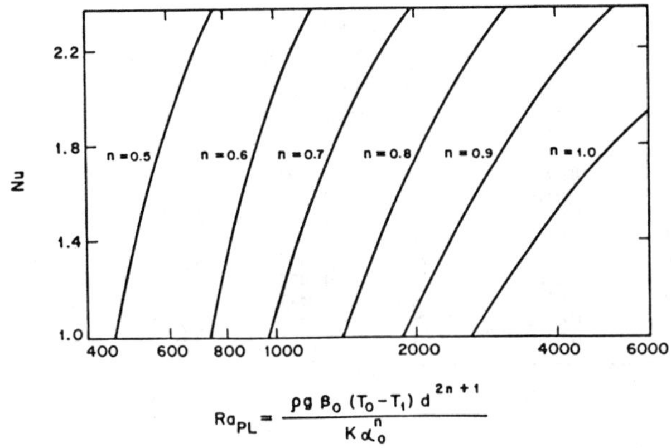

Figure 11. Variation of Nusselt number with Rayleigh number for varying power-law index n in the case of two horizontal plates with the bottom plate being at a temperature T_0 greater than the top plate temperature T_1 [25].

the Nusselt number during natural convection flow of a power-law fluid in horizontal layer is shown in Figure 11 as given by Ozoe and Churchill [25].

TURBULENT NATURAL CONVECTION

Turbulent natural convection heat transfer to power-law fluids has received very little attention in the literature. The reason for such a poor research interest in this area is the fact that for turbulence to set in, very large Grashof numbers are required, which are more difficult to achieve in high viscosity non-Newtonian fluids, and further, experiments to track turbulence are inherently tedious. The only available work on the turbulent natural convection heat transfer to power-law fluids is that of Shenoy and Mashelkar [10], which provides a theoretical analysis of the phenomena based on the approximate integral technique and the extended analogies of heat, mass, and momentum transfer for power-law fluids. In view of the difficulties involved in turbulent transport, their simplified approach, though not rigorous, is adequate to provide the predictions of the overall heat transfer coefficients and the effects that the non-Newtonian nature of the fluid is bound to have on the heat transfer. The Nusselt number expression as given by Shenoy and Mashelkar [10] for turbulent natural convection heat transfer to power-law fluids can be written as

$$Nu_x = c\, Gr_x^a\, Pr_x^b \tag{35}$$

where

$$Gr_x = \gamma_1^{-8\beta} \rho^{8\beta} x^{4\beta(2+n)} [g\beta_0(T_w - T_\infty)]^{4\beta(2-n)} \tag{36}$$

$$Pr_x = \frac{C_p}{k} \gamma_1^{4\beta} \rho^{1-4\beta} x^{\frac{3-4\beta(2+n)}{2}} [g\beta_0(T_w - T_\infty)]^{\frac{1-4\beta(2-n)}{2}} \tag{37}$$

$$a = \frac{[21 - 4\beta(8+n)]}{8[6 - \beta(10-n)]} \tag{38}$$

$$b = \frac{[3 - \beta(8+n)]}{[6 - \beta(10-n)]} \tag{39}$$

Table 3
Variation of the Various Parameters with Pseudoplasticity Index n

n	α	β	c	a	b
1.0	0.0790	0.250	0.0402	0.400	0.200
0.9	0.0770	0.255	0.0428	0.405	0.199
0.8	0.0760	0.263	0.0443	0.410	0.192
0.7	0.0752	0.270	0.0450	0.416	0.187
0.6	0.0740	0.281	0.0464	0.422	0.174
0.5	0.0723	0.290	0.0477	0.429	0.165
0.4	0.0710	0.307	0.0483	0.438	0.138
0.3	0.0683	0.325	0.0497	0.448	0.106
0.2	0.0646	0.349	0.0501	0.463	0.054

Values of c as well as of a and b for varying n are tabulated in Table 3. While c and a increase with increased pseudoplasticity, there is a continuous decrease in the value of b with reduced n values. The effect on the Nusselt number of pseudoplasticity is not obvious due to the uncertainty in predicting changes in the Grashof number and Prandtl number values with increasing shear-thinning behavior of the fluid. Equation 35 is, nevertheless, useful in heat transfer design whenever there is likelihood of turbulence in power-law fluids either because the heating surface in question is very large or the temperature difference between the heated surface and the surrounding fluid is very large.

LAMINAR MIXED CONVECTION

In most heat transfer situations, it is very difficult to identify clearly the conditions as being wholly in the forced convection regime or wholly in the natural convection regime since both mechanisms always operate simultaneously. Even though in some situations, if one mode of convection dominates the other, it is the combined effect of the two that actually determines the heat transfer characteristics. In fact, in any forced-convection circumstance, density differences are bound to give rise to natural convection currents, and depending upon whether the buoyancy forces assist or oppose the forced convection flow, the heat transfer coefficient would increase or decrease, respectively. The awareness that this could have a detrimental effect in the design of heat exchangers had led to the recent consideration of mixed-convection problems. Undoubtedly, mixed-convection problems are more difficult to analyze theoretically than the individual cases, and hence only the simple flow situations have been dealt with in the literature.

Vertical Flat Plate

Kubair and Pei [29] were the first to study the combined laminar forced and natural convective heat transfer from a vertical flat plate to a power-fluid, but their theoretical analysis does not appear to be correct as has been shown by Shenoy [30]. The discrepancy in Kubair and Pei [29] has also been pointed out by Lin and Shih [31] who argue that the similarity solution cannot exist when nonsimilarity arises due to non-Newtonian characteristics and buoyancy in mixed convection. With a view that there existed no correct solution for laminar mixed-convection heat transfer to power-law fluids from a vertical flat plate, Lin and Shih [31] and Shenoy [30] recently analyzed this problem. The approaches in the two cases are different. Lin and Shih [31] used the local similarity method to obtain a numerical solution, while Shenoy [30] provided a simple correlating equation by interpolating the two extremes of forced and free convection, considering each in terms of a boundary layer approximation. Though the solution obtained by Lin and Shih [31] is an exact one, it provides results for only specific values of the variables (as, for example, $Pr_{L,F} = 10$; $X/L = 0.01, 0.1, 1.0$; $Gr_x^{1/(2-n)}/Re_x^{2/(2-n)} = 0, 0.1, 1, 2$; $n = 0.5, 1.5$). In order to estimate the effect of Prandtl number, buoyancy, and flow index on heat transfer for any other values than those just mentioned, a numerical integration of their partial differential equations would have to be carried out. Contrary to this, the correlating equation of Shenoy [30], though approximate, provides a general solution to the problem of mixed convection to power-law fluids from a vertical flat plate and gives a good engineering estimate within reasonable accuracy as can be seen from Table 4. The correlating equation of Shenoy [30] for mixed-convection heat transfer from an isothermal vertical flat plate to a power-law fluid follows:

$$\frac{Nu_{x,M}}{Re_x^{\frac{1}{n+1}}} = \left\{ \left(\frac{1}{0.893}\right)^3 \frac{1}{18} \left[\frac{117}{560(n+1)}\right]^{\frac{1}{n+1}} \left[\frac{2n+1}{n+1}\right] Pr_{x,F} + 8\left[\frac{1}{2}\left(\frac{3}{10}\right)^{\frac{1}{n}} f(n) \left(\frac{2n+1}{3n+1}\right) Pr_{x,F}^{\frac{3n}{3n+1}} \left[\frac{Gr_x}{Re_x^2}\right]^{\frac{3}{(3n+1)(2-n)}}\right\}^{\frac{1}{3}} \tag{40}$$

where

$$Re_x = \frac{\rho U_\infty^{2-n} x^n}{K} \tag{41}$$

$$Pr_{x,F} = \frac{\rho C_p}{k}\left(\frac{K}{\rho}\right)^{\frac{2}{n+1}} x^{\frac{1-n}{n+1}} U_\infty^{\frac{3(n-1)}{n+1}} \tag{42}$$

$$f(n) = \frac{1}{15} - \frac{5}{126n} + \frac{1}{84n^2} - \frac{1}{486n^3} + \frac{1}{5103n^4} - \frac{1}{124740n^5} \tag{43}$$

The Grashof number Gr_x is as defined by Equation 13.

Shenoy [30] observed that the effect of buoyancy on forced-convection flow of power-law fluids was to increase the heat transfer rate and predominantly so for shear-thickening fluids

Table 4
Comparison of the Theoretical Results of Lin and Shih [31] Exact Analysis with the Approximate Approach of Shenoy [30]

n	$\frac{x}{L}$	$Nu_{x,M}/Re_x^{\frac{1}{n+1}}$							
		$Gr_x^{\frac{1}{2-n}}/Re_x^{\frac{2}{2-n}} = 0$		0.1		1		10	
		Shenoy	Lin and Shih	Shenoy	Lin and Shih	Shenoy	Lin and Shih	Shenoy	Lin and Shih
0.5	0.01	0.3920	0.4761	0.4142	0.5053	0.6143	0.6648	0.7688	0.7656
	0.1	0.5062	0.6183	0.5276	0.6512	0.7378	0.819	0.9103	0.9753
	1.0	0.6538	0.8007	0.6744	0.8365	0.8923	1.0570	1.0828	1.2205
1.5	0.01	0.8383	0.9493	1.0505	0.9799	1.3735	1.1417	1.5198	1.2437
	0.1	0.8383	0.9493	1.0505	0.9799	1.3735	1.1417	1.5198	1.2437
	1.0	0.6167	0.6973	0.7952	0.7251	1.0558	0.8646	1.1720	0.9475

(n > 1) in comparison to shear-thinning fluids (n < 1) as well as at higher Prandtl numbers. Presently, there is no information available in the literature for laminar mixed-convection past a vertical flat plate that is held at constant heat flux or variable temperature.

Inclined Flat Plate

The effect of natural convection on the forced convection heat transfer to power-law fluids past on inclined flat plate has been studied only for the isothermal case [32]. The approach of the analysis is the same as outlined earlier by Shenoy [30]. The final expression describing the combined effects of forced and natural convection is identical to that of an isothermal vertical flat plate except of an additional term of $\cos^{2-n}\gamma$. Thus, the controlling parameter Gr_x/Re_x^2 showing the relative importance of the natural and the forced convection effect would be modified to $Gr_x\cos^{2-n}\gamma/Re_x^2$ where γ is the acute angle of inclination of the plate from the vertical. Thus $\gamma = 0$ represents the vertical flat plate. Figure 12 shows the effect of the various parameters on the mixed convection heat transfer to an isothermal inclined flat plate. The discussion of the results based on Figure 12 is the same as that given earlier for the isothermal vertical flat plate. The effect of the angle of inclination was found by Shenoy [32] to decrease the Nusselt number with increasing departure from the vertical. There is no experimental data available in the literature to substantiate the theoretical findings both for the isothermal vertical flat plate or the inclined flat plate at the present time. The case of the constant heat flux as well as that of the variable temperature has yet to be studied theoretically and verified experimentally.

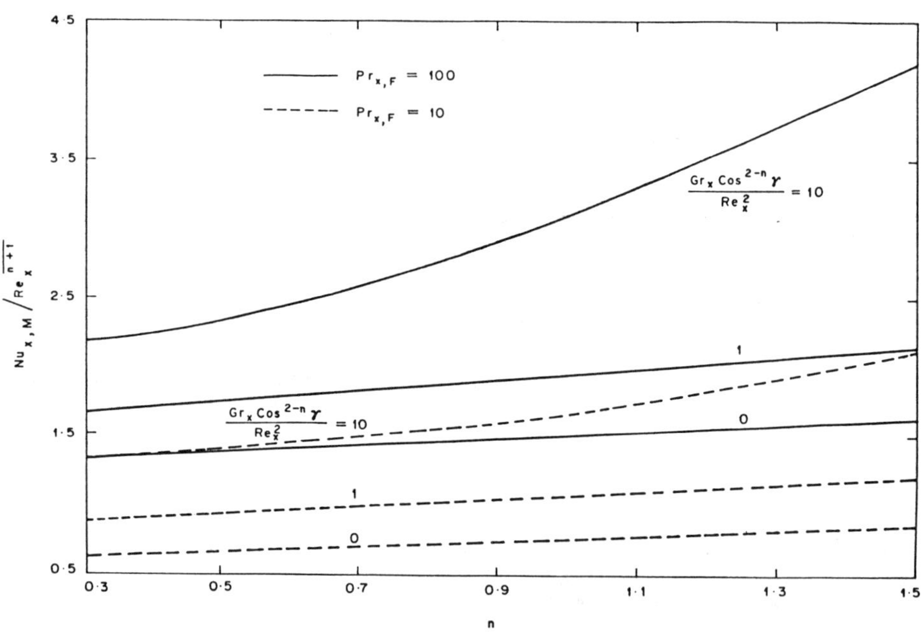

Figure 12. Variation of the local heat transfer rate with power-law index n for laminar mixed convection past an inclined vertical flat plate with an angle of inclination γ from the vertical [32].

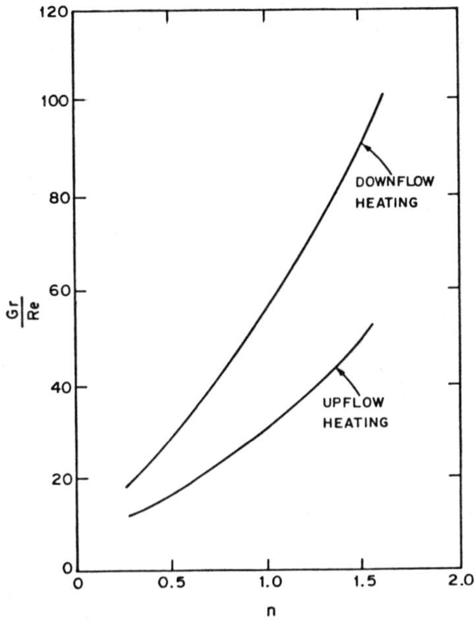

Figure 13. Variation of the mixed convection controlling parameter Gr/Re with power-law index for downflow and upflow heating in a vertical tube with constant wall heat flux [33].

Vertical Tubes

Laminar mixed convection heat transfer in vertical tubes is of immense pragmatic importance in heat transfer design. Different flow interactions between the forced and natural convection currents are encountered during the upward and the downward flow of the fluids and have to be dealt with separately. Available literature in this area includes the constant heat flux as well as the constant temperature case.

Constant Heat Flux–Upward Flow

The study of the effect of buoyancy on forced convection upward flow of a power-law fluid in a vertical tube under constant heat flux conditions was done by De Young and Scheele [33] and Marner and Rehfuss [34] through a theoretical numerical analysis. It was found that the controlling parameter to assess the relative effect of buoyancy on the forced flow was Gr/Re where the definitions of the respective dimensionless numbers have the conventional definition incorporating the radius of the tube as the length parameter and the average velocity in the tube as the velocity parameter. The value of Gr/Re has been found to be useful [33] for indicating the point of instability in the flow. In fact, the Gr/Re value at which the maximum velocity moved off the center was found to be a good criterion for assessing the stability of the flows [33]. Figure 13 shows the effect of the pseudoplasticity index on the stability of such a mixed-convection flow in a vertical tube and indicates that instabilities would be set up earlier in cases of decreasing n. The influence of the pseudoplasticity index on heat transfer can be seen from Figure 14 as given by Marner and

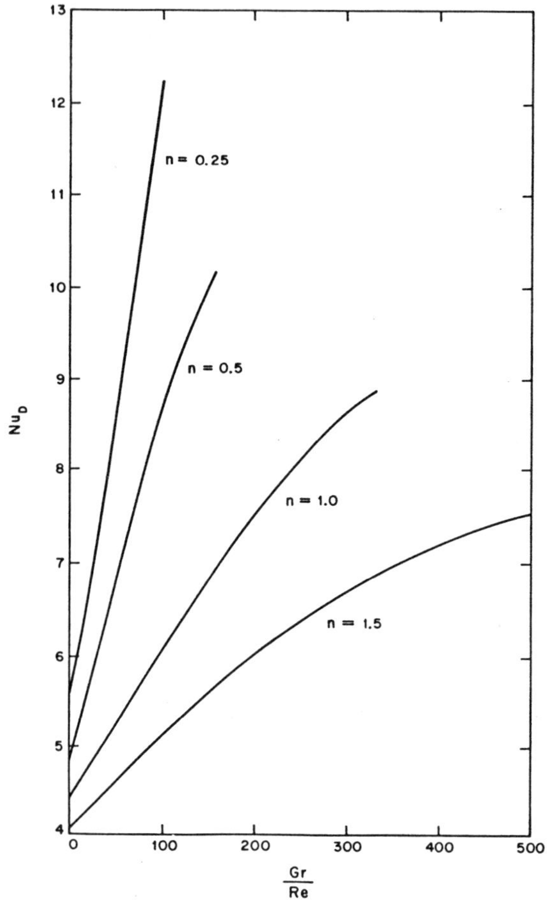

Figure 14. Variation of the Nusselt number with the controlling parameter Gr/Re for mixed convection flow in a vertical tube with constant wall heat flux [34].

Rehfuss [34]. Again, with decreasing n it is seen that the effect of buoyancy is to render an increase in the Nusselt number.

Constant Heat Flux-Downward Flow

De Young and Scheele [33] considered the case of downward flow, too, and showed that the buoyancy that results from increasing the flow in the upward direction is eventually capable of reversing the entire flow situation when its magnitude rises. Figure 15 shows that such a reversal of flow is likely to take place with increasing shear-thinning behavior. Again, the critical value of Gr/Re for flow stability is seen to decrease with reduction in n as shown in Figure 16, thus indicating that for both downflow as well as up-flow, the limiting Gr/Re values decrease with increasing pseudoplasticity.

Constant Wall Temperature–Upward Flow

Marner and McMillan [35] studied the effect of natural convection on the forced convection flow in an isothermal vertical tube, theoretically and experimentally and found that the essential conclusions regarding the criterion for stability and the Nusselt number trends were identical with those for the constant heat flux case. The experimental data on heat transfer to Carbopol solutions were found to agree with their theoretical predictions within ±15%. The literature presently is devoid of any theoretical or experimental study of the laminar mixed-convection heat transfer in isothermal vertical tubes with downward flow.

Horizontal Tubes

The effect of buoyancy on the forced-convection heat transfer to power-law fluids in horizontal tubes is an interesting and challenging problem from the fluid mechanics point

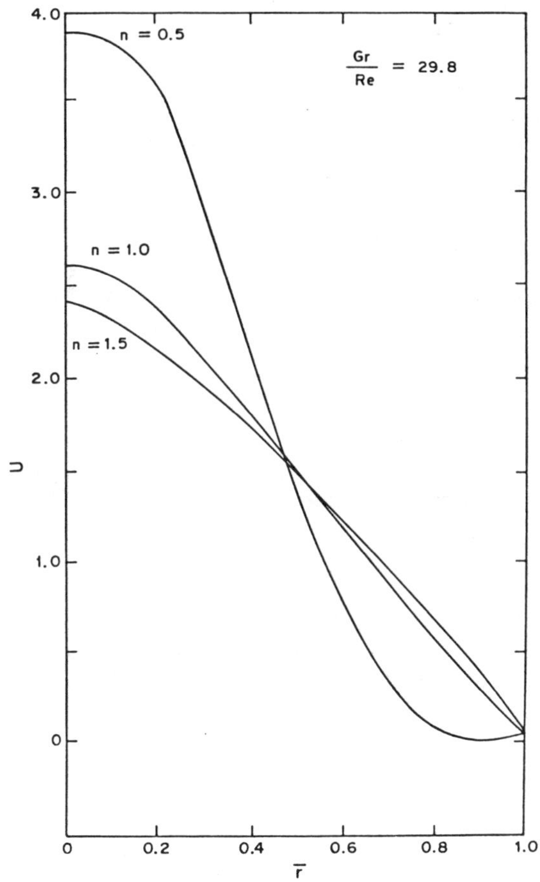

Figure 15. Variation of the dimensionless velocity with a dimensionless radial coordinate for downflow heating in a vertical tube with constant wall heat flux [33].

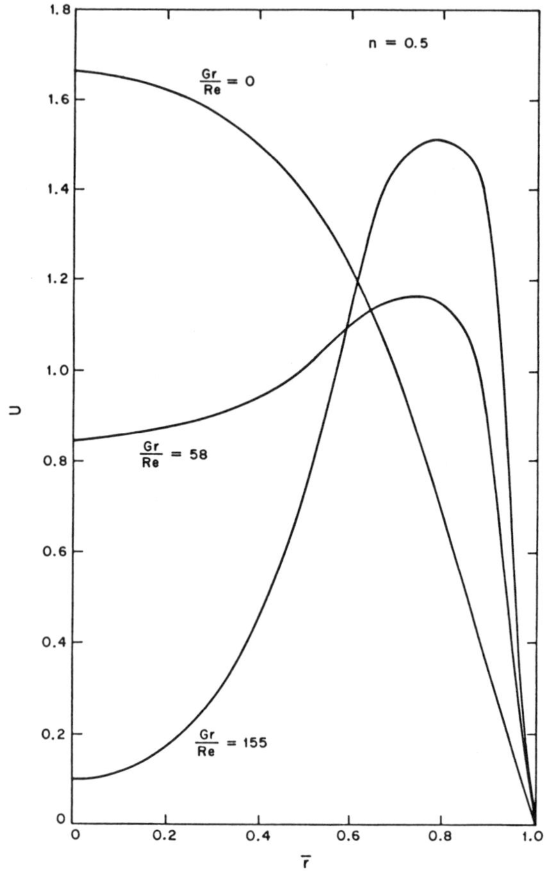

Figure 16. Variation of the dimensionless velocity with the dimensionless radial coordinate for mixed convection flow in a vertical tube with constant wall heat flux [34].

of view because of the secondary flow that sets into the system with increasing natural convection currents. When a horizontal tube is heated, the low-density fluid near the wall rises while the more dense fluid near the center moves downward until it meets the tube wall where sufficient heat transfer takes place and a continuous secondary circulation pattern is set up. This results in a significant increase in the heat transfer coefficient. For cooled tubes, the entire pattern will be reversed, and because the viscosity of the fluid increases as it cools, the pattern is more complex. However, a rigorous analysis of this problem has not yet been attempted for non-Newtonian fluids. The only available information for heat transfer in horizontal tubes is those well-known correlations of Metzner and Gluck [36] and Oliver and Jenson [37], namely

$$Nu_D = 1.75\left(\frac{3n+1}{4n}\right)^{\frac{1}{3}}\left(\frac{K_\infty}{K_w}\right)^{0.14}\left[Gz + 12.6\left(\frac{Gr_w Pr_w D}{L}\right)^{0.4}\right]^{1/3} \quad (44)$$

$$Nu_D = 1.75\left(\frac{K_\infty}{K_w}\right)^{0.14}[Gz + 0.0083\,(Gr_w Pr_w)^{0.75}]^{1/3} \tag{45}$$

The experimental data of Oliver and Jenson [37] did not fit the correlation of Metzner and Gluck [36], and hence they proposed a new correlation as given by Equation 45. There is insufficient evidence in the literature as to which of the Equations 44 or 45 should be used for design purpose. Hence, it would be recommended that both the correlations should be used to make a conservative estimate that would be useful in design of heat transfer equipment such as heat exchangers.

TURBULENT MIXED CONVECTION

Turbulent convection has not attracted the attention of research workers in the field due to a number of reasons that have been indicated earlier. The only flow situation that has been considered for theoretical analysis is the turbulent-flow combined forced- and natural-convection heat transfer to power-law fluids in vertical tubes [38]. The theoretical basis for the analysis is that the effect of buoyancy reduces the shear stress at the wall during upward flow and results in a situation similar to a buoyancy-free flow at some reduced value of Reynolds number. The effect on the Nusselt number due to the natural convection effects is given as

$$\frac{Nu'}{Nu} = \left\{1 - \left[\left(\frac{2}{\alpha}\right)^{\frac{2+n}{2}} e^{\frac{3.05n+3.8}{2.22}}\right]^{\frac{1}{n}} \frac{Gr}{Pr^{\frac{1}{2}}\, Re^{\frac{4n+2-\beta(2-n)}{2n}}}\left(\frac{K_w}{K_\infty}\right)^{\frac{1}{n}}\left(\frac{\rho_\infty}{\rho_w}\right)^{\frac{1}{2}}\right\}^{B} \tag{46}$$

where

$$Gr = \left(\frac{\rho_\infty(\rho_\infty - \bar{\rho})\,gD}{K_\infty^2}\right)^{2n+1} V_m^{2-2n} \tag{47}$$

$$Re = \frac{\rho_\infty V_m^{2-n} D^n}{K_\infty} \tag{48}$$

$$Pr = \frac{\rho_\infty C_p}{k}\left(\frac{K_\infty}{\rho_\infty}\right)^{\frac{2}{n+1}} D^{\frac{1-n}{n+1}}\, V_m^{\frac{3(n-1)}{n+1}} \tag{49}$$

The variation of B in Equation 46 with respect to changes in the pseudoplasticity index is shown in Table 5. Shenoy [38] found the criteria for a reduction of 5% or less in the heat transfer coefficient to be dependent on the following inequality.

$$\frac{Gr}{Pr^{\frac{1}{2}}\, Re^{\frac{4n+2+\beta(2-n)}{2n}}}\left(\frac{K_w}{K_\infty}\right)^{\frac{1}{n}}\left(\frac{\rho_\infty}{\rho_w}\right)^{\frac{1}{2}} < C \tag{50}$$

where

$$C = 0.1\left(\frac{\alpha}{2}\right)^{\frac{2+n}{2n}} e^{-\frac{(3.05n+3.8)}{2.22n}} \tag{51}$$

The values of C for varying pseudoplasticity indexes are also shown in Table 5, indicating that it is increasingly difficult to maintain the reduction in heat transfer due to the effect of

Table 5
Variation of B in Equation 46 with Respect to Changes in the Pseudoplasticity Index

n	α	β	B	C
1.0	0.0791	0.250	0.457	3.60×10^{-5}
0.9	0.0770	0.255	0.563	1.44×10^{-5}
0.8	0.0760	0.263	0.681	9.74×10^{-6}
0.7	0.0752	0.270	0.810	3.92×10^{-6}
0.6	0.0740	0.281	0.947	1.15×10^{-6}
0.5	0.0723	0.290	1.084	2.05×10^{-7}
0.4	0.0710	0.307	1.224	1.57×10^{-8}
0.3	0.0683	0.325	1.348	2.01×10^{-10}
0.2	0.0646	0.349	1.448	3.07×10^{-14}

natural convection to less than 5% during turbulent upward flow for increasing pseudoplasticity.

Though Equation 46 has been developed for upward flow, it can be used for downward flow by introducing a positive sign in place of the negative sign because the natural convection currents in such circumstances would cause an increase in stress across the buoyant layer and hence an increase in the Nusselt number.

CONCLUSIONS

The present chapter summarizes the developments in natural convection heat transfer to power-law fluids in external and internal flow situations under both laminar and turbulent conditions. The salient features of all the work in this area have been covered, and specifically, the relevant equations for estimating the heat transfer coefficient in each of the considered cases have been given so that they would be of direct use for design of heat transfer equipment. Special attention has been given to mixed-convection heat transfer to power-law fluids, as this is pragmatically more important in all real situations than the individual forced- and natural-convection cases. The entire coverage in this chapter is restricted to only inelastic non-Newtonian fluids which follow the power-law type shear stress versus shear rate behavior, and hence none of the presented design equations would be applicable to systems which are predominantly viscoelastic in nature.

NOTATION

- a exponent appearing in Equation 35 defined by Equation 38 with values for varying n tabulated in Table 3.
- a_1 parameter appearing in Equation 32 defined in Shenoy [12]
- b exponent appearing in Equation 35 defined by Equation 39 with values for varying n tabulated in Table 3
- B exponent in Equation 46 as defined by Shenoy [38]
- c coefficient appearing in Equation 35 defined in Shenoy and Mashkelkar [10] with values for varying n tabulated in Table 3
- d distance between two plates appearing in Equation 34
- D diameter of the tube
- $f(n)$ function of n defined by Equation 43
- $f(\alpha_N)$ function of α_N defined by Equation 12
- g acceleration due to gravity
- Gr Grashof number
- Gr_C power-law Grashof number based on the characteristic length of a two-dimensional

surface as defined by Equation 29

Gr_L Power-law Grashof number based on the slant height of a slender vertical cone defined in Shenoy [12]

Gr_R Grashof number based on the radius of the cylinder

Gr_w Grashof number evaluated at wall conditions of temperature and shear rate appearing in Equations 44 and 45 and defined in Metzner and Gluck [36]

Gr_x local distance-based Grashof number for turbulent natural convection of a power-law fluid defined by Equation 36

Gr_{Lt}, Gr_{Lh} power-law Grashof number based on length L of a constant temperature and constant heat flux vertical flat plate defined by Equations 7 and 16, respectively

Gr_{Rt}, Gr_{Rh} power-law Grashof numbers based on the radius of a constant temperature and a constant heat flux cylinder in Figures 5, 6, and 7 with forms similar to Equations 7 and 16, respectively

$Gr_{R't}$ power-law Grashof number based on the radius of a sphere in Figure 9 with a form similar to Equation 7

Gr_{xt}, Gr_{xh}, Gr_{xv} power-law Grashof numbers based on the local distance of a constant-temperature, constant heat flux and variable-temperature, vertical flat plate and defined in Equations 13, 20, and 26, respectively

Gz Graetz number in Equations 44 and 45, defined in Metzner and Gluck [36]

k thermal conductivity of fluid

K power-law model parameter in Equation 4 known as the consistency index

K_w, K_∞ Consistency indices evaluated at wall and bulk conditions, respectively, in Equations 44 and 45

$\bar{K}(n)$ Coefficient in Equation 31 defined in Kim and Wollersheim [22]

l_c Characteristic length of the heat transferring surface

n power-law model parameter in Equation 4

Nu, $\bar{Nu}$ local and average Nusselt numbers based on the length L of a vertical flat plate, respectively

Nu_D local Nusselt number based on the diameter of the tube

Nu_L local Nusselt number based on the slant height of a slender vertical cone

Nu_R, $\bar{Nu}_R$ local and average Nusselt numbers based on the radius of a cylinder, respectively

$Nu_{R'}$, $\bar{Nu}_{R'}$ local and average Nusselt numbers based on the radius of a sphere, respectively

Nu_x local Nusselt number based on the local distance x on the heat transferring surface

$Nu_{x,M}$ local Nusselt numbers based on the local distance x on the heat transferring surface for mixed convection

Pr_C power-law Prandtl number based on the characteristic length of a two-dimensional surface with a form similar to Equation 30

Pr_L power-law Prandtl number based on the slant height of a slender vertical cone defined in Shenoy [12]

Pr_w Prandtl number evaluated at wall conditions of temperature and shear rate appearing in Equations 44 and 45 and defined in Metzner and Gluck [36]

Pr_x local distance based Prandtl number for turbulent natural convection of a power-law fluid defined by Equation 37

Pr_{Lt}, Pr_{Lh} power-law Prandtl number based on length L of a constant-temperature and constant heat flux, vertical flat plate defined by Equations 8 and 17, respectively

Pr_{Rt}, Pr_{Rh} power-law Prandtl number based on the radius of a constant temperature and a constant heat flux cylinder in Figures 5, 6, and 7 with forms similar to Equations 8 and 17, respectively

$Pr_{R't}$ power-law Prandtl number based on the radius of the sphere in Figure 9 with a form similar to Equation 8

Pr_{xt}, Pr_{xh}, Pr_{xv} power-law Prandtl numbers based on the local distance of a constant temperature, constant heat flux and variable temperature vertical flat plate and defined in Equations 14, 21 and 27, respectively

$Pr_{x,F}$ power-law Prandtl number for forced convection based on the local distance and defined by Equation 42

q constant wall heat flux

$\bar{r}$ dimensionless radial coordinate in Figures 15 and 16

Ra_{PL} power-law Rayleigh number defined by Equation 34

Re Reynolds number

Re_x power-law Reynolds number based on local distance and defined by Equation 41

T temperature

T_0 temperature of bottom plate in Equation 34

T_1 temperature of top plate in Equation 34

T_w uniform wall temperature

$T_w(x)$ variable wall temperature as a function of x

T_wL wall temperature of length L of a variable temperature surface

T temperature of bulk of the fluid

u velocity component along x coordinate

U dimensionless axial velocity in Figure 15

U_{av} average velocity used as a scaling parameter for internal flow through pipes

v velocity component along y coordinate

x distance along a heated solid surface

y distance normal to the heated solid surface

Greek Symbols

α_N ratio of the termal to momentum boundary layer thickness for laminar natural convection to power-law fluids defined by Equation 11

α_0 thermal diffusivity of the fluid in Equation 34

α, β dimensionless functions of pseudoplasticity index n tabulated in Table 3

β_0 volume expansion coefficient of the fluid evaluated at T_∞

γ angle of inclination from the vertical

γ_1 parameter in Equations 36 and 37 defined in Shenoy and Mashelkar [10]

θ dimensionless temperature difference

ρ local density of the fluid at temperature T

ρ_∞ local density of the fluid at temperature T_∞

τ_{xy} shear stress in the $x - y$ direction

REFERENCES

1. Bird, R. B., Amstrong, R. C., and Hassager, O. *Dynamics of Polymeric Liquids, Vol. 1, Fluid Mechanics,* New York: John Wiley, 1977, pp. 205—213.
2. Rosen, M. R., "Characterization of non-Newtonian Flow" *Polym.-Plast. Technol. Eng.*, Vol. 12, No. 1 (1979), pp. 1—42.
3. Graessley, W. W., "Linear Viscoelasticity in Entangling Polymer Systems", *J. Chem. Phys.*, Vol. 54, No. 12 (1971), pp. 5143–5157.

4. Cramer, S. D., and Marchello, J. M., "Numerical Evaluation of Models describing non-Newtonian Behaviour," *AIChE J.*, Vol. 14, No. 6 (1968), pp. 980–983.
5. Houwink, R., *Elasticity, Plasticity and Structure of Matter*. New York: Dover, 1958.
6. Sherman, P., *Emulsion Science*, New York: Academic Press, 1968, Ch. 4, pp. 217–351.
7. Bird, R. B., Stewart, W. E., and Lightfoot, *Transport Phenomena*, New York: John Wiley, 1962; pp.
8. Shenoy, A. V., and Mashelkar, R. A., "Laminar Natural Convection Heat Transfer to a Viscoelastic Fluid," *Chem. Engg. Sci.*, Vol. 33 (1978), pp. 769–776.
9. Acrivos, A., "A Theoretical Analysis of Laminar Natural Convection Heat Transfer to non-Newtonian Fluids," *AIChE J.*, Vol. 6 (1960), 584–590.
10. Shenoy, A. V., and Mashelkar, R. A., "Turbulent Free Convection Heat Transfer from a Flat Vertical Plate to a Power-Law Fluid," *AIChE J.*, Vol. 24, No. 2 (1978), pp. 344–347.
11. Shenoy, A. V., and Ulbrecht, J. J., "Temperature Profiles for Laminar Natural Convection Flow of Dilute Polymer Solutions past on Isothermal Vertical Flat Plate," *Chem. Engg. Communications*, Vol. 3 (1979), pp. 303–324.
12. Shenoy, A. V., "Laminar Natural Convection Heat Transfer from a Slender Vertical Cone to a Power-Law Fluid," *Can. J. Chem. Engg.*, Vol. 61 (1983).
13. Tien, C., "Laminar Natural Convection Heat Transfer from Vertical Plate to Power-Law Fluid," *Appl. Sc. Res.*, Vol. 17 (1967), pp. 233–248.
14. Reilly, I. G., Tien, C., and Adelman, M., "Experimental Study of Natural Convective Heat Transfer from a Vertical Plate in a Non-Newtonian Fluid," *Can. J. Chem. Engg.* Vol. 43 (1965), pp. 157–160.
15. Chen, T. Y. W., and Wollersheim, D. E., "Free Convection at a Vertical Plate with Uniform Flux Condition in Non-Newtonian Power-Law Fluids," *J. Heat Transfer*, Vol. 95 (1973), pp. 123–124.
16. Shenoy, A. V., *Natural Convection Heat Transfer to Non-Newtonian Fluids*, Ph. D. thesis, University of Salford, England, 1977.
17. Dale, J. D., and Emery, A. F., "The Free Convection of Heat from a Vertical Plate to Several Non-Newtonian Pseudoplastic Fluids," *J. Heat Transfer*, Vol. 94, No. 1 (1972), pp. 64–72.
18. Dutta, A., and Mashelkar, R. A., "Taylor Diffusion in Polymer Solutions: Falsification by Slip Effects," *J. Appl. Polym. Sci.*, Vol. 27 (1982), pp. 2739–2742.
19. Dutta, A., and Mashelkar, R. A., "On Slip Effect in Free Coating of Non-Newtonian Fluids," *Rheol. Acta*, Vol. 21 (1982), pp. 52–61.
20. Mashelkar, R. A., and Dutta, A., "Convective Diffusion in Structural Fluids: Need for New Analysis and Design Strategies," *Chem. Engg. Sci.*, Vol. 37, No. 7 (1982), pp. 969–985.
21. Gentry, C. C., and Wollersheim, D. E., "Local Free Convection to Non-Newtonian Fluids from a Horizontal, Isothermal Cylinder," *J. Heat Transfer*, Vol. 96 (1974), pp. 3–8.
22. Kim, C. B., and Wollersheim, D. E., "Free Convection to Non-Newtonian Fluids from a Horizontal Isothermal Cylinder with Uniform Wall Heat Flux Conditions," *J. Heat Transfer*, Vol. 98 (1976), pp. 144–146.
23. Amato, W. S., and Tien, C., "Free Convection Heat Transfer from Isothermal Spheres in Polymer Solutions," *Int. J. Heat Mass. Transfer*, Vol. 19 (1976), pp. 1257–1266.
24. Tien, C., Tsuei, H. S., and Sun, Z. S., "Thermal Instability of a Horizontal Layer of Non-Newtonian Fluid Heated from Below," *Int. J. Heat Mass Transfer*, Vol. 12 (1969), pp. 1173–1178.
25. Ozoe, H., and Churchill, S. W., "Hydrodynamic Stability and Natural Convection in Ostwald–de Waele and Ellis Fluids: The Development of a Numerical Solution," *AIChE. J.* Vol. 18, No. 6 (1972), pp. 1196–1207.
26. Tsuei, H. S., and Tien, C., "Free Convection Heat Transfer in a Horizontal Layer of Non-Newtonian Fluid," *Can. J. Chem. Engg.*, Vol. 51 (1973), pp. 249–251.

27. Parmentier, E. M., "A Study of Thermal Convection in Non-Newtonian Fluids," *J. Fluid Mech.* Vol. 84 (1978), pp. 1–11.
28. Parmentier, E. M., Turcotte, D. L., and Torrance, K. E., "Studies of Finite Amplitude Non-Newtonian Thermal Convection with Application to Convection in the Earth's Mantle," *J. Geophysical Res.*, Vol. 81, No. 11 (1976), pp. 1839–1846.
29. Kubair, V. G., and Pei, D. C. T., "Combined Laminar Free and Forced Convection Heat Transfer to Non-Newtonian Fluids," *Intern. J. Heat Mass Transfer,* Vol. 11 (1968), pp. 855–869.
30. Shenoy, A. V., "A Correlating Equation for Combined Laminar Forced and Free Convection Heat Transfer to Power-Law Fluids," *AIChE J.*, Vol. 26, No. 3 (1980), pp. 505–507.
31. Lin, H. T., and Shih, Y. P., "Combined Laminar Free and Forced Convection from a Vertical Plate to Power-Law Fluids," *Chem. Engg. Communications,* Vol. 7 (1980).
32. Shenoy, A. V., "Laminar Mixed Convection Heat Transfer from an Isothermal Inclined Flat Plate to Power-Law Fluids," *AIChE J.*, Vol. 30 (1984).
33. De Young, S. H., and Scheele, G. F., "Natural Convection Distorted Non-Newtonian Flow in a Vertical Pipe," *AIChE J.,* Vol. 16, No. 5 (1970), pp. 712–717.
34. Marner, W. J., and Rehfuss, R. A., "Buoyancy Effects on Fully Developed Laminar Non-Newtonian Flow in a Vertical Tube," *Chem. Eng. J.,* Vol. 3 (1972), pp. 294–300.
35. Marner, W. J., and McMillan, H. K., "Combined Free and Forced Laminar Non-Newtonian Convection in a Vertical Tube with Constant Wall Temperature," *Chem. Engg. Sci.*, Vol. 27, (1972), pp. 473–488.
36. Metzner, A. B., and Gluck, D. F., "Heat Transfer to Non-Newtonian Fluids under Laminar-Flow Conditions," *Chem. Engg. Sci.*, Vol. 12 (1960), pp. 185–190.
37. Oliver, D. R., and Jensen, V. G., "Heat Transfer to Pseudoplastic Fluids in Laminar Flow in Horizontal Tubes," *Chem. Engg. Sci.*, Vol. 19 (1964), pp. 115–129.
38. Shenoy, A. V., "Natural Convection Effects on Heat Transfer to Power-Law Fluids flowing under Turbulent Conditions in Vertical Pipes," *Int. Comm. Heat and Mass Transfer,* Vol. II (1984), pp. 467–476.

CHAPTER 6

NATURAL CONVECTION IN EVAPORATING DROPLETS

Wen-Jei Yang

Department of Mechanical Engineering and Applied Mechanics
The University of Michigan
Ann Arbor, Michigan

CONTENTS

INTRODUCTION

All types of heat transfer phenomena known in a large volume of fluids are observed in a minute liquid drop (less than 3 mm in diameter) on a plate: natural convection followed by nucleate, transition and film boiling as the temperature difference between the plate surface and the liquid saturation point increases. Figure 1 depicts a schematic of droplet geometry at various heating regimes corresponding to those defined by the droplet lifetime curve, vaporization time versus plate temperature T_w, as shown in Figure 2 [1]. Notice that the lifetime curve resembles the boiling curve for a large liquid volume. As T_w is raised from room temperature at "a", the droplet experiences a vaporization process in the form of natural convection a–b, nucleate boiling b–c, transition boiling c–d, film boiling d–e, and finally spheroidal combustion e–f. Points c, d, and e correspond to the maximum boiling rate point, Leidenfrost point, and ignition point, respectively. As seen in Figure 1, the droplet forms a lens shape in the stage a–c, a film shape at point c, a main and several satellite fractions (some liquids only) in the stage c–d, and a sphere in the spheroidal state d–f. Vapor bubbles are generated in the stage b–c and at point c in larger droplets.

This chapter summarizes results of the research work which was supported by the National Science Foundation under Grant No. ENG-7816972.

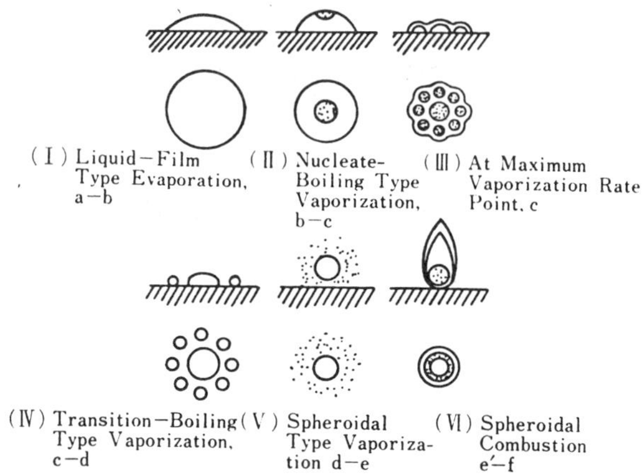

Figure 1. A schematic of geometry of minute drops at various heating regimes.

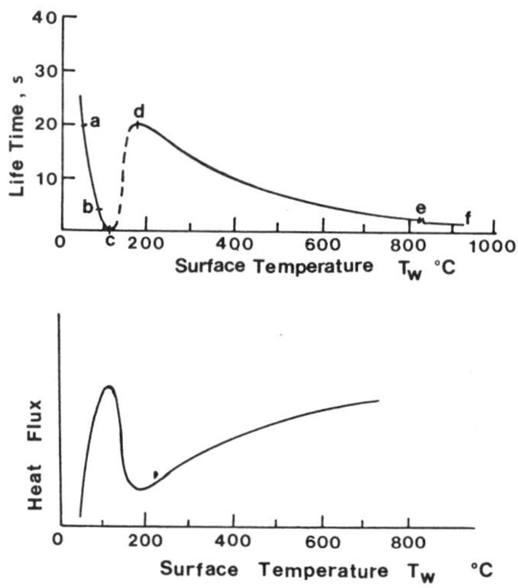

Figure 2. A typical curve for evaporation of benzene drops with initial diameter of 2.14 mm.

A comprehensive survey of the literature pertinent to the evaporation of liquid drops is available in Reference 2. The survey has revealed that the understanding of thermal instability and natural convection in droplets evaporating on a solid surface is very limited and that research on the subjects is still in its infancy. No analytical study exists due to difficult fluid dynamics and heat transfer problems presented by the droplet whose shape is constantly changing and whose interface properties are poorly defined. Its small and time-varying volume contribute extra complications and difficulties to experimental studies. This chapter is concerned with only natural convection in the stage a–b. Interfacial instability interacts with buoyancy-thermocapillary convection in addition to evaporation.

When a sessile droplet on a horizontal plate is evaporating in the air at room temperature, the absorption of latent heat to sustain the evaporation process causes temperature drops in both the air and liquid phases adjacent to the interface. This creates the formation of vertical thermal gradients in the droplet which is a typical unstable thermal state. In addition, there is a possibility that a disturbance such as a temperature change may, by chance, occur at certain spots on the droplet surface. The resulting nonhomogeneity in the interfacial temperature can cause longitudinal variations of interfacial tension. This interfacial traction may eventually induce fluid motion inside the droplet. The former is a classical Raleigh–Bénard (or thermal) stability problem, while the latter is commonly referred to as the Marangoni (or thermocapillary) instability. Thus, in the initial stage of droplet evaporation, one deals with a combined buoyancy-thermocapillary instability in a system of time-varying geometry.

The vapor evaporated at the interface diffuses into the air, while the dissolution of air into the droplet is considered negligible. However, water vapor in the air may be absorbed by the evaporating surface of liquid with high hydrophilicity, resulting in a combined interphase-intraphase mass transfer process.

In the case of a developing disturbance, the rate of droplet evaporation is strongly influenced by interfacial turbulence, a spontaneous agitation of the unequilibrated fluids. The origin of interfacial turbulence does not occur on the surface of every evaporating droplet. Depending on the extent of interfacial turbulence, droplet evaporating processes are classified into three categories: the unstable, substable, and stable type. The unstable-type evaporating process is characterized by vigorous interfacial turbulence, while the stable type has a very smooth droplet surface. The substable type has weak interfacial turbulence. The natural convection pattern inside the droplet varies depending on the type of evaporation process.

THERMOCAPILLARY INSTABILITY

To understand the mechanics of thermocapillary instability, let us consider an idealized case in which a two-dimensional disturbance is introduced in a sessile drop of very large curvature evaporating on a flat plate as illustrated in Figure 3. Initially, a uniform temperature T_a prevails in both the air and liquid phases. An evaporation process suddenly takes place, which causes the droplet surface temperature to fall to the saturated value T_s corresponding to the ambient pressure. The resulting temperature distribution is shown at the left side of Figure 3. If the disturbance in the form of infinitesimal roll-cells is, by chance, introduced in the process of evaporation, the roll cells carry the heat from both the air and liquid phases toward the interface at point 1. As a result, the temperature at point 1 increases, while point 2 remains unchanged. The temperature difference between these two points introduces a difference in the surface tension σ. For the majority of pure liquids, the surface tension-temperature gradient $d\sigma/dT$ is negative. Since T_1 is greater than T_2 and σ_1 is therefore less than σ_2, the difference $(\sigma_2 - \sigma_1)$ initiates an interfacial traction from 1 to 2. The traction is in the same direction as the rotation of the roll cells, resulting in an amplification of the disturbances. This explains the origin of Marangoni instability. Note that the disturbances in a sessile drop do not appear as two-dimensional roll cells. The surface curvature of the droplet forces cellular patterns to be unique and different from those observed in fluid layers. In addition to evaporation, heat and mass transfer, reaction processes, absorption-desorption phenomena, and deformation at interfaces can also induce surface tension inhomogeneities which lead to interfacial instability and convection.

NATURAL CONVECTION

Interfacial turbulence produced by surface tension gradients significantly augment both heat and mass transfer (for example see References 3 through 5). This flow phenomenon

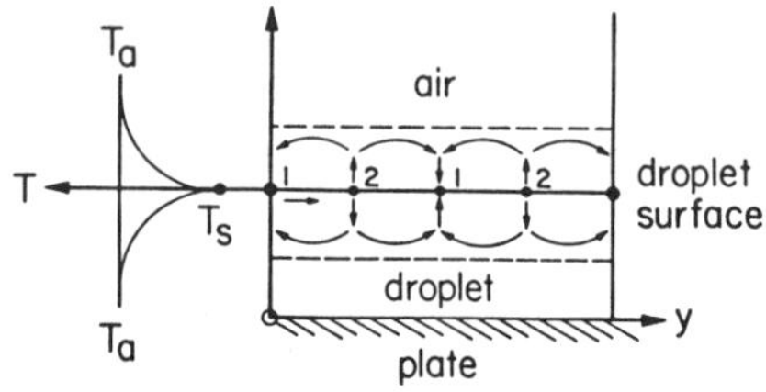

Figure 3. Mechanics of Marangoni instability in a droplet evaporating on a plate.

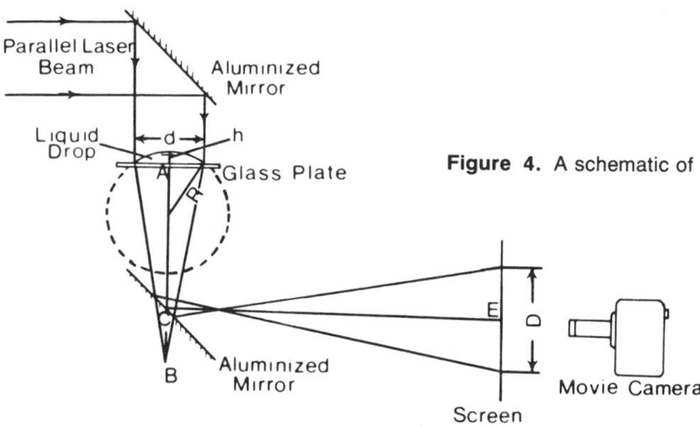

Figure 4. A schematic of a laser shadowgraphic system.

on a microscopic scale takes place on the surface of liquid droplets undergoing evaporation. Under this circumstance, only nonintrusive means should be employed in flow measurements or visualization.

A minute droplet placed on a flat plate takes the form of a spherical segment (at least very close in most cases) due to the action of surface tension. Since it is like a thin lens with a convex surface, parallel incident rays intersect at a single point (or at least in a very small region) after transversing the drop. This point is known as the principal focal point of the lens. When the drop reduces in size with evaporation, the focal point moves. The optical methods are the only nonintrusive means of measuring or visualizing flow phenomena in a droplet evaporating on a flat plate. However, both the Schlieren method and the Mach–Zehnder interferometer are unsuitable for this purpose because both techniques can not be applied to an object with an unsteady focal point. Currently, the only optical method for a continuous observation and recording is shadowgraphy, which is insensitive to a change in the focal point of the object.

Shadowgraphy is known to be the simplest of all optical methods. It requires only a collimating lens to make the light parallel. Figure 4 is a schematic diagram of the laser shadowgraphic system being used to study interfacial turbulence and internal convection [6, 7]. The various liquids of low boiling point listed in Table 1 were studied.

Interfacial Turbulence

The photographs by laser shadowgraphy exhibit two sets of rings: one appearing inside the drop image and the other outside the drop image, as schematically shown in Figure 5

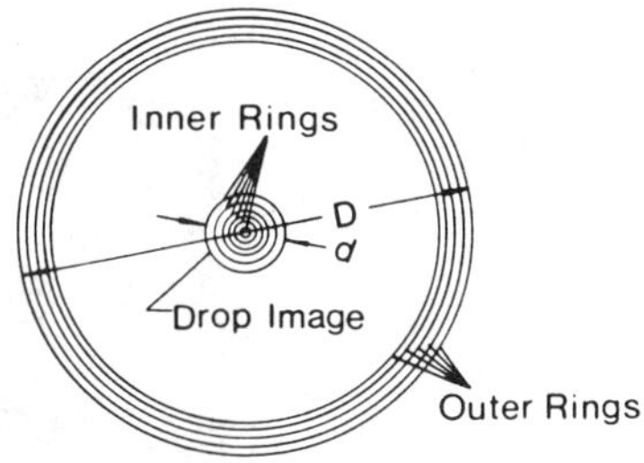

Figure 5. Two sets of rings in a drop image on a screen.

Table 1
Some Properties of Liquids Tested

Fluid	Boiling Point T_b, K	Dielectric Strength ε	Surface Tension σ, N/m	Water	Alcohol	Solubility Ether	Acetone	Benezene
Stable type								
Carbon tetrachloride	349.7	2.238	0.0269	i	s	inf	s	inf
Cyclohexane	353.9	2.023	0.0257	i	inf	inf	inf	inf
Ethyl ether	307.7	4.335	0.0172	del	inf	inf	v	inf
Substable type								
Ethyl acetone	350.3	6.02	0.024	s	inf	inf	v	v
Benezene	353.3	2.284	0.0292	del	inf	inf	inf	–
Chloroform	334.3	4.806	0.0273	del	inf	inf	s	inf
Methylene chloride	313	9.08	–	del	inf	inf	–	–
Unstable type								
Methanol	337.8	33.62	0.0224	inf	inf	inf	inf	inf
Ethanol	351.5	24.3	0.024	inf	inf	inf	inf	inf
Acetone	329.4	20.7	0.024	inf	inf	inf	inf	inf

[6]. The inner set of rings depicts the flow pattern at the liquid-solid interface, i.e., drop periphery, while the outer set of rings portrays the flow structure at the air-liquid and liquid-solid interfaces of the drop. In Figures 6A through 6E, the circle at the center of each photo represents the image of each drop. The inner rings in Figure 6F are similar in appearance for all drops. When a drop forms a spherical segment with a very smooth surface, the outer rings exhibit perfect concentric circles; Figure 6E shows this for cyclohexane. This type of phase change, classified as "stable-interface-type evaporation" occurs in the drops of cyclohexane, carbon tetrachloride, and ethyl ether. On the other hand, should a drop surface be rippled, the outer rings are dissolved to form irregular radiant stripes with a sawtooth-like circle (Figures 6A and 6B), for methanol and acetone, respectively. The drops of acetone, ethyl achohol and methanol exhibit the behavior, which is called "unstable-interface type evaporation." There is a group of liquids such as ethyl acetate, benzene, chloroform, and methylene chloride whose interfacial patterns are in between the two limits (Figures 6C and 6D) for ethyl acetate and methylene chloride, respectively. The outer rings are retained but their shape is distorted from a circular shape and spiked. This group is categorized into "substable-interface type evaporation."

An interfacial flow map is constructed in Figure 7 to define the domains of the three basic interfacial flow structures. The figure is a plot of the Marangoni number $Ma = (T_w - T_b)\, d_0\, (d\sigma/dT)/(\mu\alpha)$ versus the dielectric constant ε. Here, T_b denotes the liquid boiling temperature; d_0, equivalent diameter of the drop on the test plate at zero time; μ, kinematic viscosity; and α, thermal diffusivity. The unstable-interfacial-type liquids are characterized by high Ma and ε, whereas the liquids with low Ma and ε have stable interfaces during their evaporation. However, the addition of a small amount of stable-interfacial-type liquid can induce an interfacial turbulence in another stable-inter-

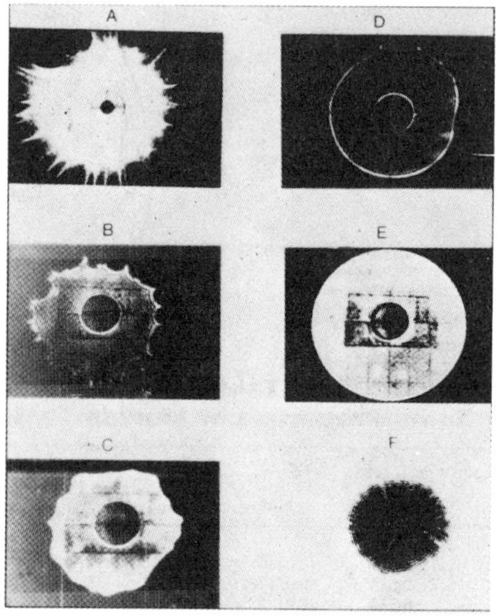

Figure 6. Instant laser shadowgraph of various drops evaporating on a plate: (A) methanol; (B) acetone; (C) ethyl acetate; (D) methylene chloride; (E) cyclohexane; (F) image of drop periphery.

facial-type liquid. For example, both carbon tetrachloride and ethyl ether take the form of a spherical segment with a perfectly smooth surface during evaporation. However, in the presence of one percent by volume of carbon tetrachloride, interfacial turbulence appears in the drop evaporation of ethyl ether.

Evaporation Rate

A method to determine the droplet volume-time history from the shadowgraph [6] has been developed. Results are presented in Figure 8 for the stable type and in Figure 9 for both substable- and unstable-interface-type evaporations. V and V_0 denote the instantaneous and initial drop volumes, respectively; t, time; and τ, droplet lifetime. In both the stable- and substable-evaporation cases, test results fall close to a single curve. In the unstable evaporation case, each liquid is represented by a specific curve. However, the time history can be divided into three stages. In the first stage, evaporation is controlled by the pure component (the original liquid); in stage 2, water is absorbed from the ambient and is transported to the edges (polarity); and in the final stage, evaporation is controlled by pure component-water. The broken lines show the vigorous evaporation stage in which determination of V is very difficult due to the irregularity in drop shapes.

Figure 10 correlates the dimensionless droplet lifetime against the dimensionless parameter E which is defined as $Ma^* \varepsilon^{1/4}$. Ma* denotes a modified Marangoni number defined as $\sigma d_0(T_w/T_b)/(\mu\alpha)$. In general, the larger ε, the higher is the hydrophilicity of the liquid. Figure 10 shows that τ decreases with an increase in E.

Flow Regimes

Traces of aluminium particles can be used to visualize natural convective flow patterns in an evaporating drop by the naked eye [6]. Convective currents in an unstable-type

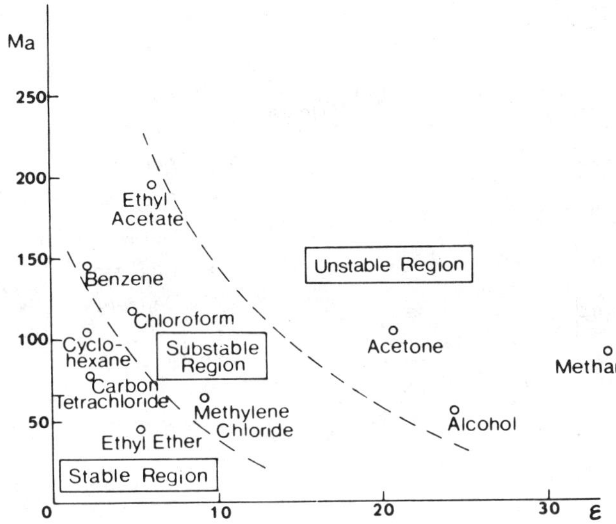

Figure 7. Interfacial flow map.

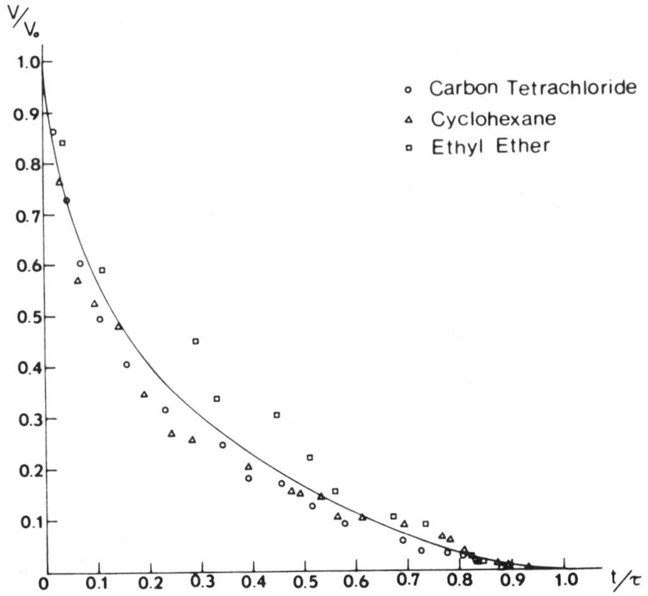

Figure 8. Time history of stable-type drop evaporation.

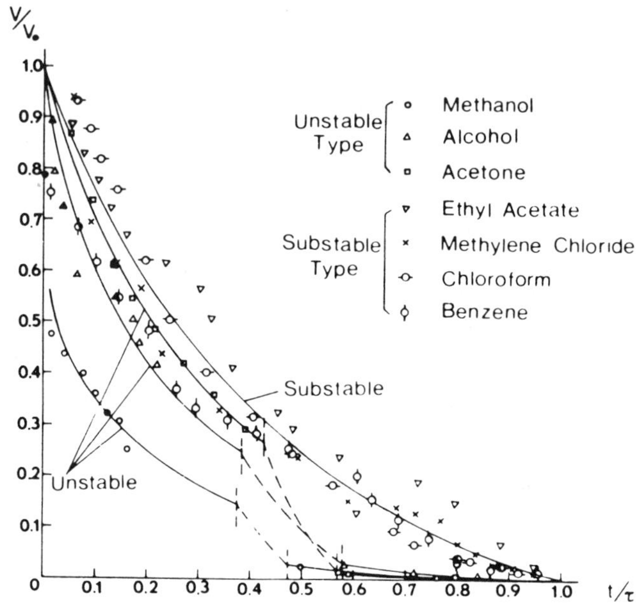

Figure 9. Time history of unstable- and substable-type drop evaporation.

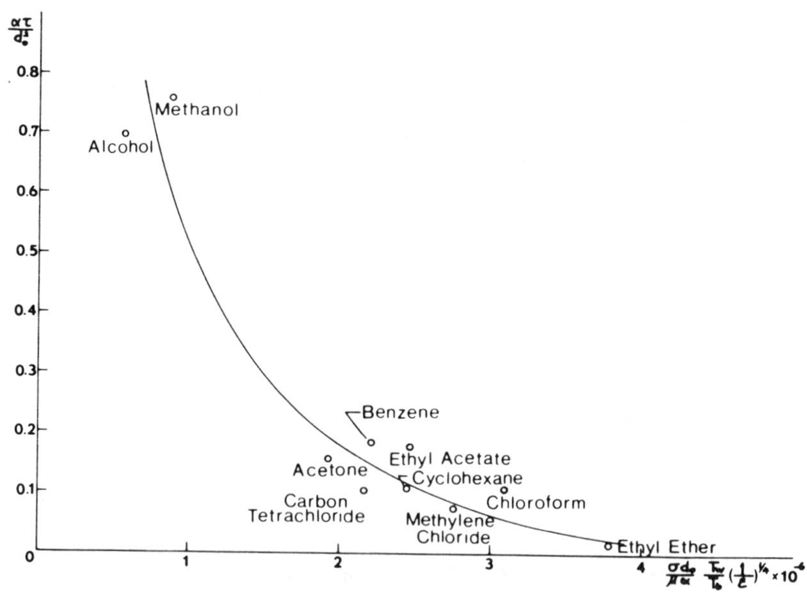

Figure 10. Correlation of drop lifetime of various liquids.

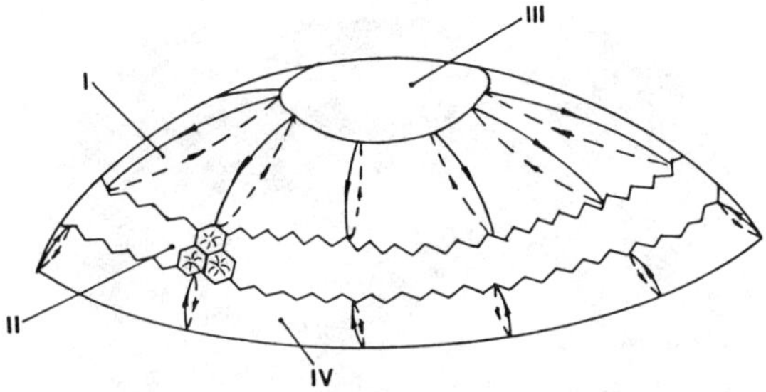

Figure 11. A schematic diagram of a flow pattern in a stationary drop evaporating on an isothermal plate.

evaporating droplet are too strong to identify any discernible flow pattern. Only in the stable- and substable-type evaporation, are reproducible flow regimes observed, as described in the following.

On an Isothermal Plate [2]

Figure 11 is a schematic diagram revealing the existence of four flow regimes: stagnant, buoyancy driven, surface-tension driven, and cellular motion. Region I is filled with intense natural convection currents, while the ring-shaped region II consists of a few layers of cellular structure. Located at the summit of the drop is the stagnation region III. A very weakly circulating region IV exists at the bottom of the drop. Figure 12 is a photocopy of flow patterns recorded using a coherent system [8]. It shows a top view of a chloroform drop about a few seconds after being placed on the plate. The heavy dark circle is the periphery of the drop. Two to three layers of the Bénard cells are visible. As evaporation proceeds, region I shrinks, while the cellular structure in region II moves toward region III.

As the plate temperature increases, the buoyancy effect is enhanced to promote the strength of natural convection currents, resulting in a reduction of the flow regimes in the stable- and substable-type drops from four to two, regions I and III.

On a Plate with a Temperature Gradient [9]

When a drop is placed on a plate with temperature gradient, it moves toward the colder side of the plate due to the Marangoni effect. Figure 13 shows a sequence of the change in the shape of an ethyl acetate drop. During its migration process, under the temperature gradient of approximately 2° C/mm, the drop produces a tail which breaks up and disappears through evaporation. The drop comes to a full stop when the balance of forces between the surface tension and gravity is achieved. The drop continues to evaporate thereafter until it completely disappears.

In stationary drops: The drop motion is prevented by raising the colder end of the plate to make a 7° angle of inclination. There exists four flow regions as schematically depicted in Figure 14. In region I, cold fluid descends along the drop surface, while warm fluid after being heated by the plate surface, ascends through the interior of the drop volume, thus forming a circulating route along the same longitude. The drop surface in this region is relatively placid, while active evaporation occurs because of exposure to the warm liquid stream. Region IV is small: it encompasses the area at the tip of the cold side where a pair of vortices are observed. The flow is governed by a surface tension force. In this region, the interface is appreciably agitated while the evaporative process is slow because of lower

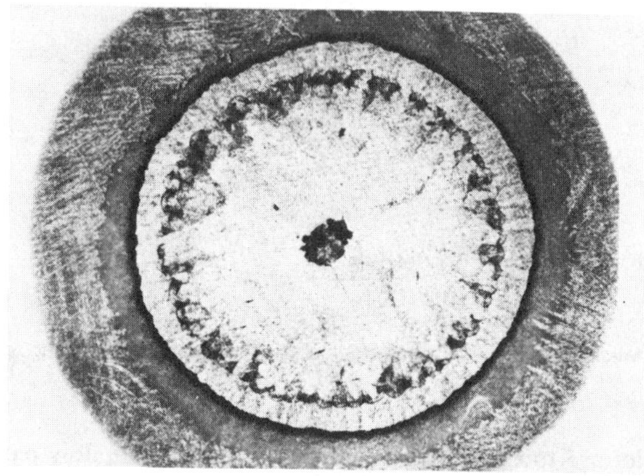

Figure 12. A photocopy of flow structure in a chloroform drop recorded by a coherent system.

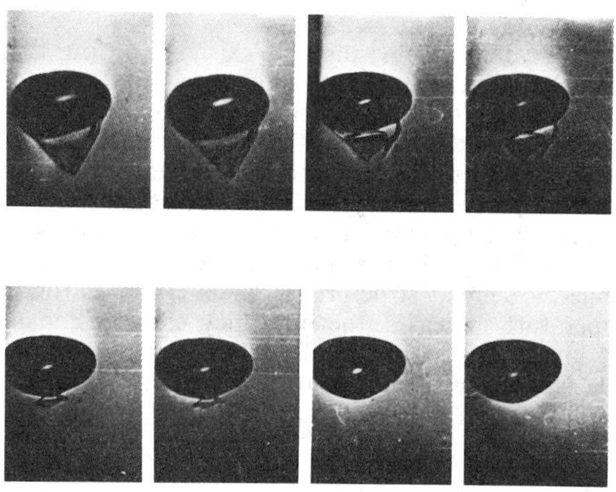

Figure 13. Change of shape of an ethyl acetate drop during its motion from warm to cold sides.

surface temperature and lack of internal circulation. Region III is located near the top of the drop where its two vortices run countercurrent with the vortex pair of region IV. These vortices serve to promote heat exchange between the warm and cold streams through mixing. Region II, which has a feather-like flow pattern, is observed on the hillside of the drop situated between the warm and cold regions. The flow stream carries the heat from the warm to cold sides. In view of their principal flow mechanisms, regions I, II, III, and IV can be referred to as the internal circulation, thermosiphoning, mixing, and surface circulation regions, respectively.

In moving drops: The drop is in transitory motion in the direction of a decreasing temperature gradient when the plate is set in a horizontal position. There are four flow regions as shown in Figure 15. The induction of drop motion causes a weakening of the flow strength in both regions I and II which are respectively controlled by buoyancy and

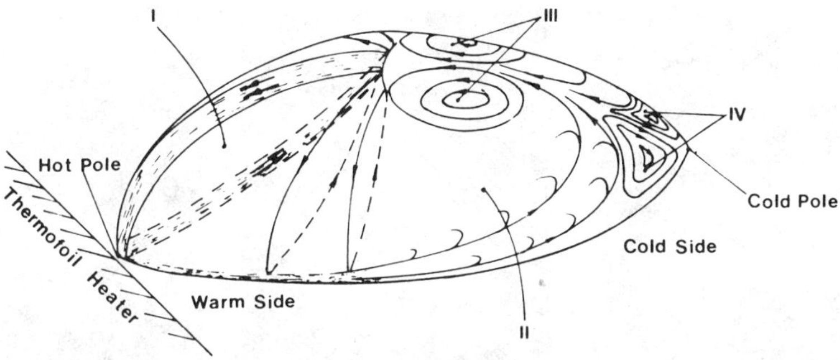

Figure 14. A schematic diagram of flow patterns in a stationary drop evaporating on an inclined plate (at a 7° angle) with a temperature gradient.

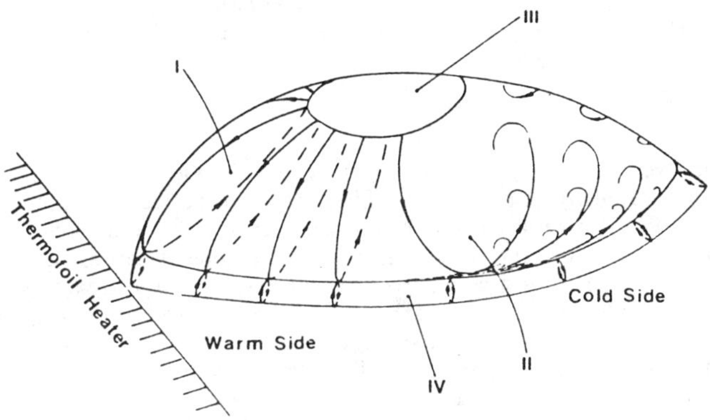

Figure 15. A schematic diagram of flow patterns in a moving drop evaporating on a horizontal plate with temperature gradient.

surface tension forces. The role of region III in mixing the warm and cold streams is diminished, forming a nearly stagnant space at the drop summit. However, the path of flow stream for transporting the heat from the warm to cold sides in region II is stretched. Eventually, two vortices in region IV of a stationary drop are replaced by a ring of small, two-dimensional roll cells for locomotion of the drop. Cold fluids of regions I and IV flow down into the partition between the regions. The roll cells are clearly visible on the warm side where flow circulation is strong. The visibility of these roll cells is reduced toward the cold side due to the weakening of flow strength. It is speculated that there exist concentric rings of roll cells near the solid-liquid interface in the drop interior which cause the drop to crawl on the plate. Due to the weakening of convective currents and mixing between the warm and cold streams, the evaporation rate is reduced in a moving drop.

Thermal Patterns

The shadowgraphic method is roughly sensitive to changes in the second derivative of density of the field. Due to the temperature difference between the ascending and

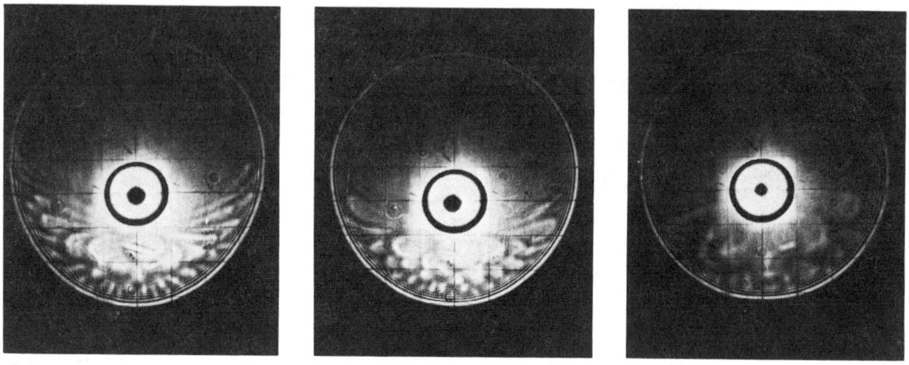

A B C

Figure 16. Shadowgraphs of a carbon tetrachloride drop at the (a) early stage, (b) intermediate stage, and (c) late stage of the evaporating process on an inclined plate with a temperature gradient.

descending fluid streams, the convection pattern behaves like lenses. The descending stream of cold fluid whose density is higher (in which the disturbance acts like a convex lens) produces a focusing effect on a parallel beam which traverses the fluid volume. Hence, it appears as a bright line. On the other hand, the ascending warm stream (in which the disturbance exerts the effect of a concave lens) appears as a dark area.

Figure 16 is the shadowgraphs of a carbon tetrachloride drop at the (a) early stage, (b) intermediate stage, and (c) late stage of the evaporating process on a plate with a temperature gradient. The images are magnified approximately 20 times. The initial drop volume at contact was 9 $\mu\ell$ with a diameter of approximately 4 mm on the plate surface.

The warm side (seen in the upper part of the photo) of the drop is almost uniform in temperature as a result of strong buoyancy-driven convection, as indicated by the nearly uniform dark image. On the other hand, the cold end of the drop (seen in the lower part of the photo) has a complex temperature variation which induces a surface tension-driven instability as evidenced by a feather-like pattern in the shadowgraphic image. Each bright line corresponds to a descending cold stream, while a dark line represents an ascending warm stream. This internal substructure continues to vary actively with time. In general, the cold side of the drop gradually becomes inactive, i.e. the nonuniformity of the temperature on the cold side diminishes gradually in the course of evaporation, as shown in Figures 16A through C. The warm and cold sides correspond to regions I and II in Figure 14, respectively, while the bright background cast by the laser beam on the photo wipes out the shadow of region III. The shadow of region IV can not be distinguished in the photo of Figures 16A through C.

Effects of Liquid Properties

Since both the Marangoni number and the dielectric constant tend to decrease with an increase in the liquid temperature, the characteristics of drop evaporation as shown in the flow map (Figure 7) may change from the unstable to substable type or from the substable to stable type as the heating of drops is increased. The effect of liquid properties is thus exemplified through different flow patterns in the surface-tension-driven convection region since the buoyancy-driven convection zone is generally well mixed in temperature. When the liquid temperature T_ℓ is appreciably below the saturation temperature T_b, a featherlike structure appears in the shadowgraphs. As T_ℓ is raised closer to T_b, the shadowgraphs exhibit cellular patterns near the image periphery. All of the flow patterns vary with time. In general, cells begin to appear at a certain value of T_ℓ and the cell size grows with an increase in T_ℓ toward T_b.

Effect of Air-Liquid Interfacial Geometry

A minute drop on a plate takes a lens shape of curvature R. As a result of the curvature, a surface tension force of $2\sigma/R$ is induced on the air-liquid interface. This surface force does not exist in the case of an evaporating liquid layer, where the interfacial curvature is infinity. Berg et al. [4] observed that in a pool of evaporating liquid, the metamorphosis varies with depth from the regular (Bénard) cells to the larger cells in which a substructure of ribs appeared followed by the dismemberment of the cells into a network of sluggishly moving hot and cold streamers. Beyond a depth of 1 cm, no further changes in flow form could be discerned. The basic structural forms (cells, ribs, streamers) appeared in all liquids except water. The properties of the liquid affected the flow structure to a much lesser extent compared with the depth of the evaporating liquid. In contrast, there exists four patterns of flow structures in a minute drop evaporating on a plate of uniform temperature, as shown in Figure 11; intensive convection current, cellular form, stagnation, and weak circulation in regions I–IV, respectively. In a liquid pool of the same depth as the diameter of these minute drops about 2–3 mm, one would observe only cellular flow. The difference between a cellular structure in a liquid layer and the flow pattern in a drop (Figure 11) may be attributed to the effect of the surface tension force $2\sigma/R$.

DROPLET MORPHOLOGY

When a minute drop is placed on a plate, it forms a lens shape. In the course of evaporation, both the stable and substable liquids retain their original shape as their volume diminishes. However, the unstable drop undergoes a series of complex changes in its shape. The drop morphology is intimately related to the speed of evaporation. Beside, the detailed profile of the liquid-gas interface near the contact line is important for determining wetting characteristics, apparent contact angle, and solid-liquid surface tension. The surface profile can be determined by an optical method, while the droplet structure of the unstable drop can be observed by direct photography.

Surface Profile

A droplet is placed on a transparent substrate which is in turn placed on the emulsion of a high-resolution photographic plate. The droplet is then illuminated from above by two plane waves of laser light which propagate with an angle 2θ between them and which are oriented symmetrically about the normal to the plate as shown in Figure 17.

Two sequential exposures of the plate are made: one before the drop is placed on the substrate and one after it is so placed. After this plate is developed it contains two superimposed, high spatial frequency, grating-like interference fringe patterns. One of these gratings consists of parallel, straight, equally-spaced fringes. The other grating is distorted because of the optical effects of the droplet. A set of moire fringes indicative of the difference of these two gratings can be observed over most of the droplet cross-section by illuminating the developed plate with ordinary white light.

Figure 18 depicts a schematic diagram showing various optical rays passing through the droplet and substrate. Those fringes which appear in the peripheral region a–a form a holographic interferogram which can be observed by illuminating the photographic plate

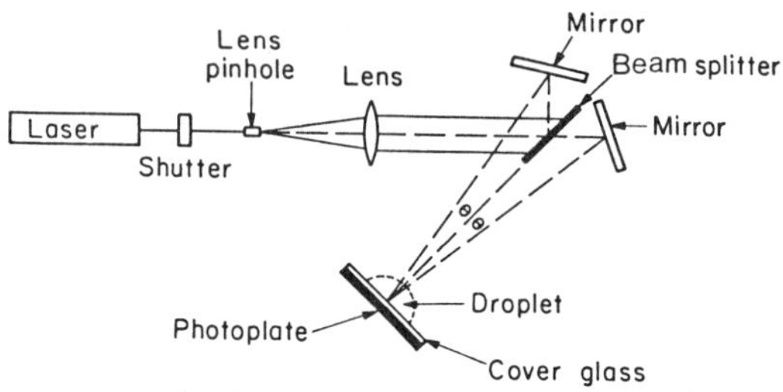

Figure 17. A schematic of an optical system for measurements of droplet profile.

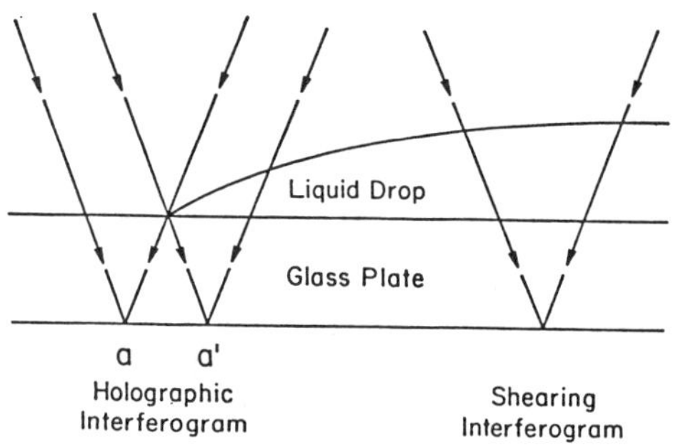

Figure 18. A schematic diagram of optical rays for producing shearing and holographic interferograms.

with white light traveling in the same direction as the left-most ray shown in Figure 18. The profile in this narrow peripheral region can approximately be determined using the relation

$$y \cong N\lambda / (n \cos \theta' - \cos \theta) \tag{1}$$

$$\cos \theta' = [1 - (\sin \theta/n)^2]^{1/2} \tag{2}$$

where y = the height of the droplet surface
λ = wavelength of laser light
n = refractive index of the liquid
N = fringe order number which is assigned integer values at the outer edge of the droplet

This analysis is approximate and is valid only for small values of θ and n nearly equal to the refractive index of the substrate.

The second set of fringes, which generally appears over most of the liquid cross-section, constitute a shearing interferogram. An approximate analysis of the droplet geometry can be made by

$$dy/dx \cong N\lambda/[2(y + d)(n \cos \theta' - \cos \theta) \tan \theta'] \tag{3}$$

Figure 19 is a photograph of the fringe pattern of a cyclohexane droplet (n = 1.44507) soon after being placed on the plate. As indicated in Figure 20, the closed fringes in the outer region of the droplet indicate that there is a reversal of curvature in microlayers near the droplet periphery in contact with the plate. In this photograph, the reversal occurs

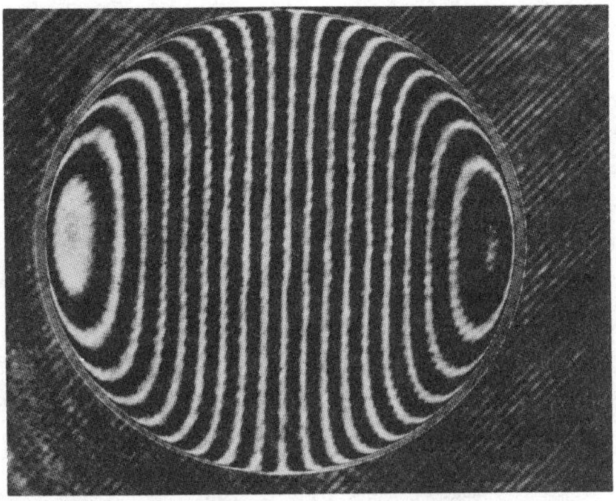

Figure 19. A photograph of the fringe pattern of a cyclohexane drop.

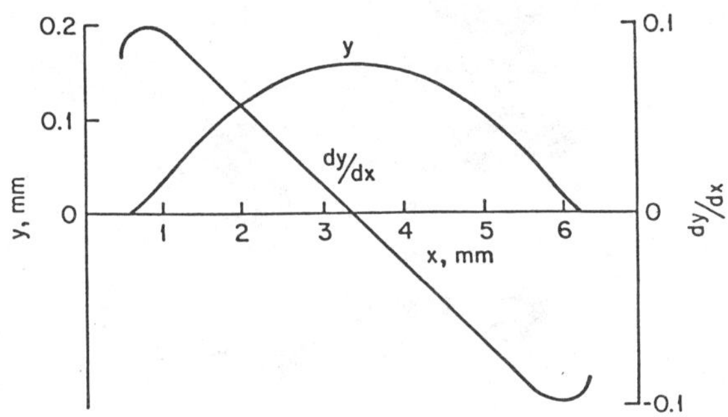

Figure 20. Surface profile and its variation of a cyclohexane drop calculated from its fringe pattern (Figure 19).

where the droplet height is 21 μm. As the difference in the refractive indices between the glass and the test fluid (for example, a light oil droplet) diminishes, the reversal of curvature reduces in magnitude and occurs at a location closer to the contact line. The reverse is also true.

Droplet Structure

The apparatus for direct photography consists of a white light source, a glass plate coated with developed emulsion on one side, and a movie camera, as shown in Figure 21. The developed emulsion has a specific density such that the photography density is approximately 1.7. When the light source and the camera are situated at proper positions, the process of the droplet evaporation can be recorded in photos with sharp contrast.

Two series of photos on an acetone drop with the camera at two different angles are shown in Figures 22 and 23. The unstable-interface-type evaporation process can be divided into three stages: quiescent, vigorous, and residual stages. The duration of these three stages

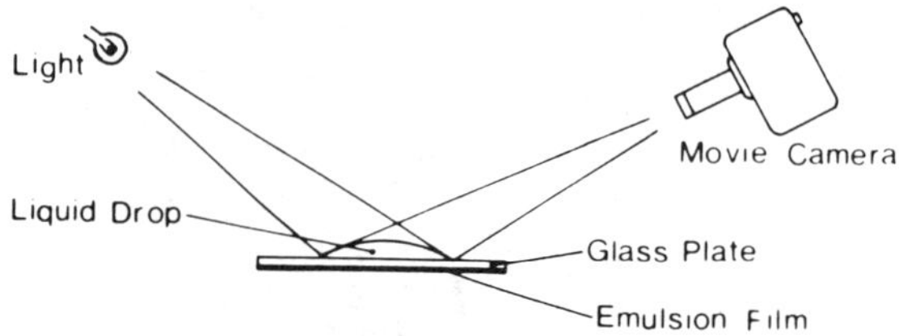

Figure 21. A schematic of direct photography of an evaporating drop.

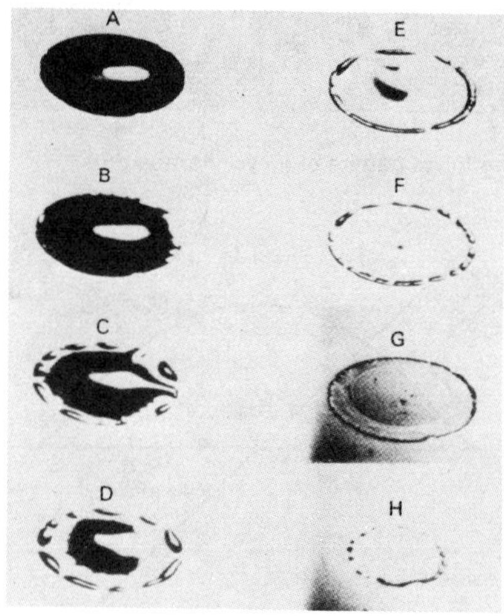

Figure 22. Direct photograph at various instants during evaporation process of an acetone drop on a glass plate.

varies with liquids. In the quiescent stage, the drop forms a spherical segment with a very smooth surface, as shown in Figure 22A. Then, tiny liquid blisters begin to appear at the drop periphery. The location and order of the blister formation are randomly formed until the entire periphery is filled with the blisters, Figures 22B through 22E and Figures 23A through 23C. Meanwhile, the spherical segment gradually deforms into a radial ridge but the air-liquid interface continues to sink in. In Figures 22F and 23D, the ridge disappears completely because the liquid has either evaporated or moved toward the drop periphery until it forms a very thin film surrounded by a circular torus-shaped periphery. Evaporation proceeds rapidly during this stage, particularly at the drop center, but more slowly at the torus surface because of high boiling point of the absorbed water. Due to a change in the surface tension/temperature gradient, the film then retracts and pulls the liquid in the torus toward the drop center, thus creating concentric ripples (Figures 22G and 23E). Finally, the torus retracts, attracting the remaining liquid to form a tiny hemisphere. Evaporation continues but at a very low rate in the residual stage. However, when the ambient temperature is raised by approximately 4° C, for example, due to heating by an illuminating lamp, the torus is not able to retract into a single-type drop due to insufficient surface tension force. The torus then breaks up into numerous tiny drops on a circle, as seen in Figures 22H and 23F.

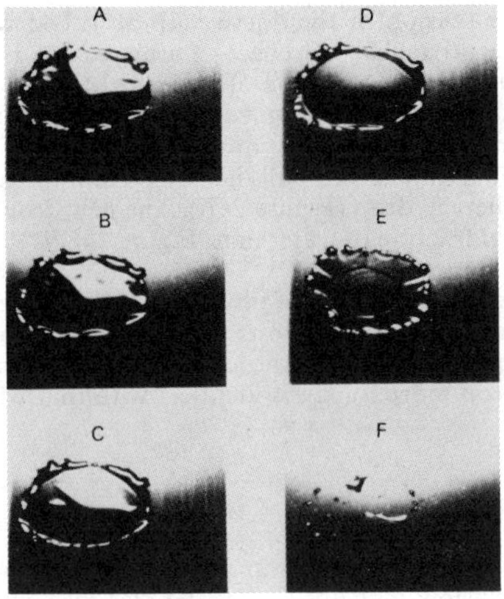

Figure 23. Direct photograph at various instants during evaporation process of an acetone drop on a glass plate.

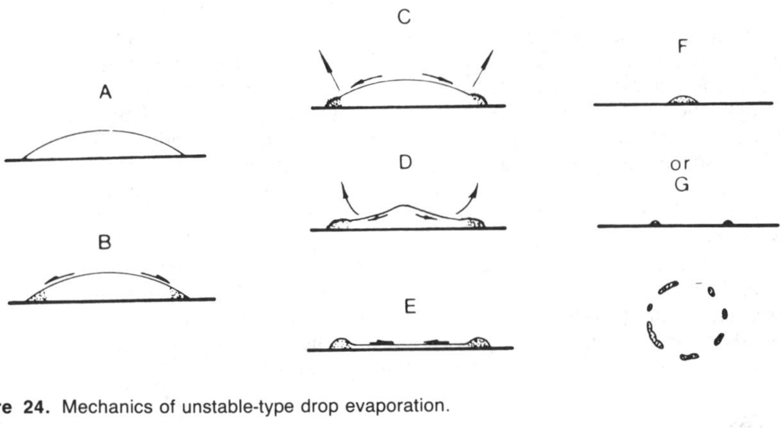

Figure 24. Mechanics of unstable-type drop evaporation.

A liquid with a larger dielectric constant is generally characterized by high hydrophilicity, as shown in Table 1. It is theorized that during evaporation, water molecules in the air are absorbed by the surface of the liquid drop, as shown in Figure 24A. The absorbed water molecules migrate down along the drop surface toward the periphery (Figure 24B). When the water concentration reaches a certain level, liquid blisters begin to appear in random order at the drop periphery (Figure 24C). This is because the repulsive pressure π (also referred to as "repulsive force") causes a reduction in surface tension from a value σ_0 for pure liquid:

$$\sigma = \sigma_0 - \pi \tag{4}$$

in which σ is the surface tension of the liquid with absorbed water molecules. When π exceeds σ_0, σ becomes negative. The net effect of a negative σ is a tendency for the drop surface to expand, leading to the formation of a torus at the periphery accompanied by a significant enhancement in the evaporation rate (Figure 24D). The drop surface recedes very rapidly until the central portion of the drop forms a very thin film (Figure 24E). The surface tension of the liquid film acts to pull the liquid from the torus toward the center to form a new, tiny hemispherical drop (Figure 24F). The new drop evaporates at a low rate as it contains a substantial fraction of water in it. Figure 24G is the alternate of Figure 24F when the ambient temperature is raised.

In order to prove there is no influence of physico-chemical characteristics of test surface on the nature of interfacial turbulence, tests on the evaporation of various liquids are repeated on a polished copper plate under the identical situation as when the glass plate was used. The variation of drop morphology is identical with that which has occurred on the glass plate [6].

CONCLUSIONS

The morphology of drop image, the drop volume-time history, and the internal flow structure lead to a conclusion that liquids can be classified into three types: unstable, substable, and stable evaporation. The interface of stable-type drops remains smooth and takes the shape of a spherical segment. The unstable-interface-type drop undergoes three stages of the evaporation process: quiet, rigorous, and residual. A theory is proposed and confirmed that the vigorous stage of an unstable interfacial-type drop is caused by the action of surface tension on the drop interface mixed with the water molecules that are absorbed from the ambient during the process of evaporation. An interfacial flow map is constructed which defines the domains. The physico-chemical characteristics of the test surface exert no influence on the nature of interfacial turbulence. When a drop is placed on a plate with a temperature gradient, the convective pattern on the warm side is induced by a buoyancy-driven instability, while that on the cold side is controlled by surface tension force. The major fraction of evaporation takes place on the warm side of the drop surface due to higher surface temperature and strong flow circulation. The flow regimes in between the warm and cold sides are governed by the combined effect of buoyancy and surface tension forces and are responsible for the exchange of heat between the warm and cold sides through mixing.

Drop motion alters some flow patterns in the four regimes and weakens both convective currents and stream mixing, resulting in a reduction in the evaporation rate. Liquid properties affect appreciably both the interfacial and internal flow structures and consequently alter the speed of drop evaporation. The contour of the air-liquid interface contributes to the complication of internal flow structures.

GLOSSARY

Marangoni effect – An interfacial movement induced by the temperature dependence of surface tension (more generally, a movement of an interface caused by longitudinal variations of interfacial tension).

Interfacial turbulence – a spontaneous agitation of the interface between two unequilibrated fluids.

Thermocapillary instability – interfacial tractions caused by temperature-dependent interfacial tension.

Interfacial flow map – a plot of the Marangoni number versus the dielectric constant which defines the domains of three basic interfacial structures.

Marangoni number – defined as $(-d\sigma/dT)(T_w - T_b)\,d_0/(\mu\alpha)$ where σ is the surface tension; T_w, plate temperature; T_b, liquid boiling point; d_0, equivalent drop diameter at zero time; α, thermal diffusivity; and μ, kinematic viscosity.

NOTATION

- D diameter of outermost ring on screen (Figures 4 and 5) (m)
- d diameter of drop (Figure 4) and its image on screen (Figure 5) (m)
- d_0 equivalent diameter of drop on test plate at zero time (m)
- E $(\sigma d_0/\mu\alpha)(T_w/T_b)/\varepsilon^{1/4}$
- f focal length of spherical-segmented drop (m)
- h height of spherical segment (Figure 4) (m)
- Ma Marangoni number $= (T_w - T_b)\, d_0\, (d\sigma/dT)/(\mu\alpha)$
- Ma* modified Marangoni number $= (\sigma d_0/\mu\alpha)(T_w/T_b)$
- N fringe order number
- n refractive index
- R radius of spherical segment (Figure 4) (m)
- T_ℓ drop temperature (° K)
- T_a ambient temperature (° K)
- T_w temperature of test plate (° K)
- T_b boiling temperature of liquid (° K)
- t time (s)
- V instantaneous volume of drop (m^3); V_0, initial value
- x distance measured from drop edge (Figure 20)
- y height of drop surface at x (Figure 20)

Greek Symbols

- α thermal diffusivity of liquid (m^2/s)
- ε dielectric constant
- λ wavelength of laser light (m)
- μ absolute viscosity (P · S)
- π repulsive pressure of thin layer of polar organic liquid over water (N/cm)
- σ surface tension at air-liquid interface of contaminated drop (N/cm); σ_0, of pure liquid
- τ drop lifetime (s)

REFERENCES

1. Yang, W.-J., ISAS Rep. No. 535 (Vol. 40, No. 15), University of Tokyo, Japan (Nov. 1975).
2. Zhang, N., and Yang, W.-J., *Experiments in Fluids,* 1: 101 (1983).
3. Sternling, C. V., and Scriven, L. E., *AIChE J.,* 5: 514 (1959).
4. Berg, J. C., Boudart, M., and Acrivos, A., *J. Fluid Mech.,* 24: 721 (1966).
5. Kenning, D. B. R., *Appl. Mech. Rev.,* 21: 1011 (1968).
6. Zhang, N., and Yang, W.-J., *J. Heat Transfer,* 104: 656 (1982).
7. Zhang, N., and Yang, W.-J., *Rev. Sci. Instrum.,* 54: 93 (1983).
8. Xu, Y., Zhang, N., and Yang, W.-J., *Preprint 3rd. Int. Symp. Flow Visualization,* 668 (1983).
9. Zhang, N., and Yang, W.-J., *Int. Heat Mass Transfer,* 26: 1479 (1983).

CHAPTER 7

PRINCIPLES OF HEAT AND MASS TRANSFER WITH LIQUID EVAPORATION

T. Kumada

Hokkaido University North 13, West 8
Sapporo 060, Japan

CONTENTS

INTRODUCTION

Heat and mass transfer with liquid evaporation is found everywhere in nature and is important in the functioning of many machines, such as wet-bulb hygrometers, air conditioning apparatuses, cooling towers, drying apparatuses, scrubbers, food storage systems, etc.

Early science and engineering activities on this subject, which can be traced as far back as about 1792 [1], concentrated on the use of the wet-bulb thermometer to measure humidity. The bulk of research on the wet-bulb thermometer was concerned with the determination of a psychrometric ratio, which is closely related to the theory of analogy between heat and mass transfer. The psychrometric ratio has been treated theoretically by many well-known scientists and engineers, and the results of these studies have stimulated disagreements concerning the significance of the data. The correlation of the psychrometric ratio has been gradually improved as more accurate information on the analogy between heat and mass transfer has been obtained. The final, simple, and practically important relationship between humidity and wet-bulb temperature is found from a very general starting point which should be applicable to mixtures other than air and water vapor.

Other activities on these subjects have been concerned with measurements of the heat and mass transfer coefficients of a wet surface (a free surface of liquid, a wetted-wall tower, and a droplet) for both natural and forced convection. The measurement of heat and mass transfer coefficients is considered to be of vital importance in the design and performance of the apparatuses concerned. Heat and mass transfer with liquid evaporation from a wet surface is, therefore, not only of practical importance in engineering, but is also an interesting subject for theoretical and basic science studies.

Most of the basic research has been concerned with heat and mass transfer from a flat plate, a cylinder, a liquid droplet, and a wetted column. Many experimental works have been conducted to measure the heat and mass transfer coefficients with evaporation from a flat plate. And the results of most studies have indicated clearly that not only is the Nusselt number much larger than the Sherwood number, but it also increases with an increasing evaporation rate, or with an increase in the gas temperature of the main stream. However, many studies report that if a transverse mass flow is in an opposite direction to heat flow, then the thickness of a boundary layer increases with blowing, thus leading to a decrease in heat transfer. Some confusion and disagreement, therefore, exist between the experimental data and the theoretical estimations.

In recent years the magnitude of the convective transport has been shown to depend upon the concentration gradient of vapor and the concentration of the nondiffusing gas. In the interpretation of gas phase mass transfer processes by molecular diffusion through a hypothetical stagnant film, the convective transport resulting from the diffusion process has usually been accounted for by the film pressure factor. For diffusion of one gas through a second stagnant gas, the film pressure factor becomes the log mean average partial pressure of the stagnant gas over the diffusion path, that is, P_{BM}. The role of the film pressure factor has been well confirmed for pure molecular diffusion, but only in recent years has attention been focused on verifying its role in heat and mass transfer in the turbulent flow of gas.

The study of the evaporation of water in superheated vapor instead of air is comparatively recent. Some experiments have shown that more water evaporates in superheated vapor than in air at the same temperature of the main stream for adiabatic evaporation and other experiments have shown a different conclusion. Some confusion and disagreement also exist concerning the results of these studies. Yoshida et al. [2] attempted to obtain heat and mass transfer data as a series on air and superheated vapor. Their conclusions are that the rate of evaporation of water has a point of limitation (inversion point) with respect to air and superheated vapor temperatures. The fact that water evaporates more rapidly into superheated vapor than into air at temperatures above this point, while the evaporation rate in humid air is less than in dry air, is true only in temperature ranges below this inversion point.

Although the prescribed experimental results are incomplete, they raise the following questions or curious points:

For low concentrations of vapor

- Is the Nusselt number much larger than the Sherwood number for simultaneous heat and mass transfer with evaporation, and is the analogy between heat and mass transfer not satisfied?
- Does the Nusselt number increase with an increasing rate of evaporation or an increase in the temperature of the main stream?

For higher concentrations of vapor

- How dependent is the rate of vapor diffusion and heat transfer through a hypothetical stagnant film on the effect of vapor concentration?
- Is it true that water evaporates more rapidly in superheated steam rather than in air at temperatures above the inversion point, as proposed by Yoshida et al.?

The conservation equations of heat and mass are simultaneously satisfied in the interface between gas and liquid phases for a system with liquid evaporation. In addition to this advantage, these two conservation equations are strictly connected and make the well-known relation that the heat consumption in the interface equals the heat obtained by multiplying an evaporation rate $\dot{m}$ with the latent heat of evaporation L_w. The relation gives a clue for determining the heat input, except for convection heat, onto the interface by assuming that the analogy between heat and mass transfer is well established. In the analysis of previously reported experimental data, this interconnected relation has not been sufficiently utilized.

The purpose of this text is to introduce the basic theory of the stagnant film model and the method of its application to evaluate the previously reported experimental data. The evaluation of the data based on the method gives new values of the Nusselt number which are free from erroneous heat input. Some suggestions for further development of the stagnant film model are given.

It is the author's hope that this brief examination will provide some basic information and give clear answers to the questions concerning heat and mass transfer with evaporation. The text also clarifies the following questions:

- What factors are important to measure exactly the heat and mass transfer coefficient with evaporation?
- How useful is it to adapt a wetted surface to measure the heat and mass transfer coefficient?

THEORY

A bulk of theoretical works concerning mass transfer cooling such as transpiration, ablation, and film cooling has been published. Most of these works must be applied to predict the transfer coefficients of heat and mass transfer with evaporation, if hydrodynamic and thermal conditions are equal with other mass transfer cooling

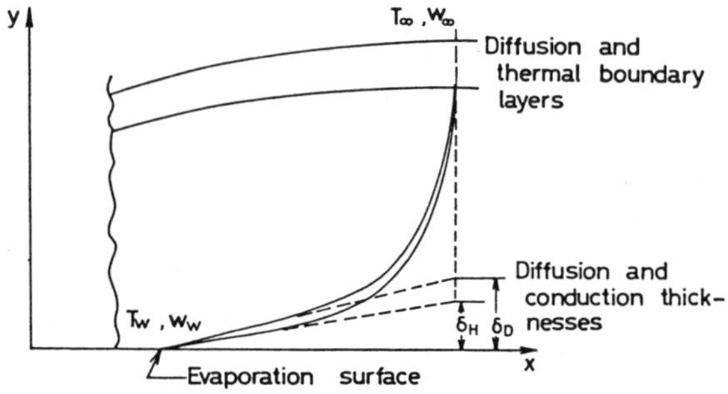

Figure 1. Stagnant film model and the thickness of diffusion and heat conduction layers.

conditions. Many methods have been presented in the literature which either utilize some integral relationship or attempt to solve the boundary layer equations themselves. These methods and the basic differential equations describing the heat and mass transfer process will not be introduced or discussed here as these subjects are covered extensively in many other texts.

The following is an introduction to the well-known theory of the stagnant film model. Based on the theory of the model, the author proposes a method to evaluate the previously reported experimental data. The method can be applied to correct error incurred in the heat transfer data.

STAGNANT FILM MODEL

Concentration and temperature gradients are confined to relatively narrow layers which are quite close to a wall or a wet surface, respectively, as shown in Figure 1. Simplified, the concentration and temperature curves can be replaced by the dashed lines. Then, the stagnant thickness of diffusion and thermal conduction layers are defined as δ_D and δ_H, respectively. The thicknesses, δ_D and δ_H, are usually different from each other because the Lewis number is not generally equal to unity and the blowing effect on δ_D and δ_H is different. This picture oversimplifies the actual process considerably, but it brings out great advantage when the effect of blowing on heat and mass transfer is considered.

Mass Transfer

It must be noted that this model has the possibility of being applied for every system of laminar or turbulent and natural or forced convection. As described on page 56 of Reference 3, the mass transfer takes place by diffusive and convective flux within the film for steady-state diffusion of only one of the two species in binary mixtures of ideal gases

$$\dot{m} = -D\,(d\rho_v/dy) + \rho_v v_v \dot{m} \tag{1}$$

where D is assumed to be constant through the film and physical properties c_p, D, λ, and ρ used in the following theory are evaluated by a method described later. For ideal gases, the volume fraction $\rho_v v_v$ is equal to the mole fraction of vapor Y. Equation 1 is rewritten as Equation 2.

$$\begin{aligned}\dot{m} &= -D/(1-Y)\,(d\rho_v/dy)\\ &= -D/(1-Y)\,(PM_v/RT)\,(dY/dy)\end{aligned} \tag{2}$$

where $\rho_v = Y \cdot PM_v/RT$ and T in the term PM_v/RT is a film temperature $(T_w + T_\infty)/2$. Integration for the case of a stagnant film of thickness δ_D gives

$$\dot{m} = (D/\delta_D)(PM_v/RT)\ln[(1 - Y_\infty)/(1 - Y_w)] = (D/\delta_D)(PM_v/RT)(Y_w - Y_\infty)/Y_{BM} \quad (3)$$

where $Y_{BM} = (Y_w - Y_\infty)/\ln[(1 - Y_\infty)/(1 - Y_w)]$. Equation 3 states that the transfer rate is proportional to the difference in the terminal values of the solute mole fraction across the diffusion path, and is inversely proportional to the logarithmic mean of Y_w and Y_∞ (Y_{BM}).

The author must introduce an additional correction factor β by Bedingfield and Drew [4], which accounts for some differences in the transport equations for heat and mass other than the replacement of mole fraction by temperature Sc by Pr. In this case D/δ_D in the previous equations is replaced by h_D, where $h_D = \beta h_{D0}$. β is given in Equations 20 and 21. The definition of the mass transfer coefficient gives a coefficient that is found to be independent of total pressure and only slightly dependent upon temperature:

$$h_D^* = \dot{m}/[(PM_v/RT)(Y_w - Y_\infty)] = \beta D/(\delta_D Y_{BM}) = \beta h_{D0}/Y_{BM} \quad (4)$$

The mass transfer coefficient in the limit of low solute concentration h_D (the transfer coefficient for the case of negligible convective flux, and hence also the coefficient for equimolar countercurrent diffusion) is then given by

$$h_{D0} = \lim_{\dot{m} \to 0} h_D^* = D/\delta_D \quad (5)$$

The ratio of the two coefficients h_D^* and h_{D0} is equal to the inverse of Y_{BM}

$$h_D^*/h_{D0} = \beta/Y_{BM} \quad (6)$$

For the case of a large concentration gradient, and hence large transfer rates, the ratio of the two coefficients may deviate from the simple expression of Equation 6. This deviation is the result of the interaction between the diffusion of vapor and the momentum transfer processes and the convection in the direction parallel to the surface.

Heat Transfer

As described on page 257 of Reference 3, it is easy to understand that the total flux of enthalpy into a differential element of thickness, dy, is made up of two parts: the flux of conduction $-\lambda(dT/dy)$, and the flux of enthalpy due to diffusion, $\dot{m}c_{pv}(T - T_w)$. When these quantities are evaluated for the flux vectors entering and leaving the differential element and their difference is set equal to zero, then the temperature distribution must satisfy Equation 7.

$$\lambda(d^2T/dy^2) - \dot{m}c_{pv}(dT/dy) = 0 \quad (7)$$

where it is assumed that the physical properties are constant through the stagnant film and are evaluated at the film temperature as a reference temperature. Equation 7 is integrated with the boundary condition $T = T_w$ and $\lambda(dT/dy)_w = \dot{m}_c L_w$ at the interface:

$$\lambda(dT/dy) - [\dot{m}_c L_w + \dot{m}c_{pv}(T - T_w)] = 0 \quad (8)$$

The first term on the left-hand side of Equation 8 represents the heat flow due to conduction, and the second term refers to the heat required to evaporate liquid from the surface plus the heat carried by the diffusing molecules. Rearrangement of Equation 8 gives:

$$(\dot{m}_c L_w/\lambda)\,dy = dT/[1 + c_{pv}(T - T_w)\dot{m}/(\dot{m}_c L_w)] \quad (9)$$

Equation 9 is integrated with the boundary condition that $T = T_\infty$ at the interface $y = \delta_H$:

$$\dot{m}_c L_w (\delta_H/\lambda) = (L_w/c_{pv})(\dot{m}_c/\dot{m}) \cdot \ln[1 + c_{pv}(T_\infty - T_w)\dot{m}/(\dot{m}_c L_w)] \quad (10)$$

Heat transfer can occur through not only convection but also through conduction and radiation, while mass transfer can occur only by convection. If heat is transferred by conduction through walls and radiation, except for convection, this heat must be substrated from $\dot{m}L_w$ when the heat transfer coefficient is defined in the following theory. Now, if the convective heat transfer coefficient is defined as:

$$h_H^* = \dot{m}_c L_w/(T_\infty - T_w) \quad (11)$$

where $\dot{m}_c$ is the evaporation rate due to only convective heat, the heat transfer coefficient h_H in the absence of mass transfer is defined as:

$$h_H = \lambda/\delta_H \quad (12)$$

Then, the left- and right-hand sides of Equation 10 are rewritten as

$$h_H^*/h_H = [h_H^*/h_H \ln(1 + c_w h_H/h_H^*)]/c_w \quad (13)$$

where $c_w = \dot{m}c_{pv}/h_H$. Rewritting Equation 13 gives

$$h_H^*/h_H = c_w/(\exp c_w - 1) \quad (14)$$

The function of c_w on the right-hand side Equation 14 is often called the Ackermann correction for mass transfer. As the Ackermann correction for evaporation is smaller than unity, a surface exposed to a hot gas can be partially protected from rapid heating by the bulk gas if the surface is kept wet with a volatile liquid, which evaporates and creates a positive, opposing mass transfer.

TEMPERATURE OF A WET SURFACE

Adiabatic Evaporation

One of the simplest applications of the theory of simultaneous transport of enthalpy and mass through the fluid near an interface is for the calculation of the steady-state temperature of a wet-bulb thermometer. The following theory can be applicable not only to a wet-bulb but also to a free surface of a pure liquid and a wet porous wick.

The solution of temperature distribution through a film can also be directly deduced by integration of Equation 7 with the boundary conditions that $T = T_{WB}$ at the interface of $y = 0$ and $T = T_\infty$ at the bulk-gas boundary of the film.

$$T = T_{WB} + (T_\infty - T_{WB}) \cdot [\exp(c_{WB} y/\delta_H) - 1]/(\exp c_{WB} - 1) \quad (15)$$

The conduction flux of heat at the interface is easily found using Equations 12 and 15

$$q_c = \lambda (dT/dy)_{WB} = h_H(T_\infty - T_{WB}) c_{WB}/(\exp c_{WB} - 1) \quad (16)$$

Note that q_c is not equal to the total enthalpy flux in the gas at the interface unless the standard state temperature is T_∞. If the enthalpies are computed relative to a standard state at the bulk gas temperature T_∞, the total enthalpy flux from the bulk gas stream at the interface δ_H between the bulk gas and the stagnant film is

$$q_t = q_c + \dot{m}_c c_{pv}(T_\infty - T_{WB}) = h_H(T_\infty - T_{WB}) c_{WB}/[1 - \exp(-c_{WB})] \quad (17)$$

The temperature of wet surfaces must be determined to the extent that at steady state the inward flux of heat is just equal to the enthalpy requirement at the surface:

$$h_H(T_\infty - T_{WB})c_{WB}/(\exp c_{WB} - 1) = L_{WB}h_{D,WB}(PM_v/RT)\ln[(1 - Y_\infty)/(1 - Y_{WB})] \quad (18)$$

where T in the term PM_v/RT is a film temperature $(T_{WB} + T_\infty)/2$. Equation 18 expresses the simple and practically important relationship between humidity and the wet-bulb temperature. It is noted that Equation 18 can be used to calculate the mole fraction of the diffusing vapor in the gas mixture from depression of the wet-bulb temperature if the ratio of the heat transfer coefficient to the mass transfer coefficient is known in the limit of a low solute concentration. Since Equation 18 contains both the heat and mass transfer coefficients, there is a relationship between the two through the analogy. The simplest form of the analogy is respresented by Colburn's relation,

$$h_{D,WB}(PM_v/RT)/h_H = (\beta_{WB}/c_p)(M_v/\bar{M})Le^{2/3} \quad (19)$$

in which the author has introduced a factor, β, following Bedingfield and Drew [4], to account for some differences in the transport equations for heat and mass transfer other than the replacement of mole fraction by temperature Sc by Pr. In particular, Bedingfield and Drew point to the difference in the meaning of "velocity" in the two equations, the velocity in the mass transfer equation being a molecule-mass average value. As shown by Mickley, Sherwood, and Reed [5], the velocities in both equations become the mass-average value if, instead of mole fraction, average molecular weight $\bar{M}$ is used as the dependent variable in the diffusion equation. According to Bedingfield and Drew, it follows that:

$$\beta_{WB} = \bar{M}_{WB}\ln(\bar{M}_{WB}/\bar{M}_\infty)/\{[(M_v - M_g)(1 - Y_{WB})] \cdot \ln[(1 - Y_\infty)/(1 - Y_{WB})]\} \quad (20)$$

The correction factor β is equal to unity in the limit of zero driving force for mass and seldom deviates more than 5%. In the limit of small driving force, $Y_W - Y_\infty$, Equation 20 becomes, approximately,

$$\beta_{WB} = 1 - [(M_v - M_g)/\bar{M}_{WB} + 1/(1 - Y_{WB})] \cdot (Y_\infty - Y_{WB})/2 \quad (21)$$

Introducing the analogy relationship between the heat and mass transfer coefficients according to Equation 19.

$$(T_\infty - T_{WB})c_{WB}/(\exp c_{WB} - 1) = L_{WB}(\beta_{WB}/c_p)(M_v/\bar{M})Le^{2/3} \cdot \ln[(1 - Y_\infty)/(1 - Y_{WB})] \quad (22)$$

in which Lewis number Le is equal to Pr/Sc or to $Dc_p\rho/\lambda$. Equation 22 is seen to be an equation determining T_{WB} for fixed Y_∞ and T_∞, because the interface mole fraction Y_{WB} is determined by the vapor-pressure curve of the evaporating liquid and the total pressure. If the temperature difference and the partial pressure difference are small, considerable simplification in Equation 22 is permitted. Introducing series approximations for both sides and replacing Y by w,

$$T_\infty - T_{WB} = (L_{WB}\beta_{WB}c_{pv})Le^{2/3}\rho(w_{WB} - w_\infty) \quad (23)$$

Equation 23 is often obtained by writing the heat and mass transfer rates as $h_H(T_\infty - T_{WB})$ and $h_D\rho(w_{WB} - w_\infty)$, respectively.

Radiative and Conductive Heat Contribution

If there is heat input q_s per unit area of wet surface and unit time, except for convective heat, into the surface, Equation 18 can be transformed as Equation 24

$$h_H(T_\infty - T_w)c_w/(\exp c_w - 1) + q_s = L_w h_D(PM_v/RT)\ln[(1 - Y_\infty)/(1 - Y_w)] \quad (24)$$

where $c_w = \dot{m}c_{pv}/h_H$ and $\dot{m}$ is the evaporation rate. Defining constant C as

$$C = 1 + q_s/[h_H(T_\infty - T_w)c_w/(\exp c_w - 1)] \quad (25)$$

Equation 24 can be transformed as Equation 26,

$$Ch_H(T_\infty - T_w)c_w(\exp c_w - 1) = L_w h_D(PM_v/RT)\ln[(1 - Y_\infty)/(1 - Y_w)] \quad (26)$$

where $h_D = \beta h_{D0}$ and the difference between β and β_{WB} can not be neglected when C is much different from unity. If the heat input q_s on a wet surface is known, then Equation 24 can be used to determine T_w for fixed Y_∞ and T_∞ or Y_∞ for fixed T_w and T_∞.

METHOD OF DETERMINATION OF C

Generally the heat input q_s contains both radiative and conductive heat transfer, and thus cannot be exactly determined. Then, neither T_w nor Y_∞ can be calculated by Equation 26, because C is an unknown parameter.

Small Blowing Factor

If $(h_D/h_H)(PM_v/RT)$ is constant in Equations 18 and 26, then C is expressed in the following relation by taking the ratio of Equations 18 and 26

$$\begin{aligned} C = {} & (\beta/\beta_{WB})(c_{WB}/c_w)[(\exp c_w - 1)/(\exp c_{WB} - 1)] \\ & \cdot (L_w/L_{WB})[(T_\infty - T_{WB})/(T_\infty - T_w)] \\ & \cdot \ln[(1 - Y_\infty)/(1 - Y_w)]/\ln[(1 - Y_\infty)/(1 - Y_{WB})] \end{aligned} \quad (27)$$

where T in Equations 18 or 26 is evaluated at a film temperature, $(T_{wB} + T_\infty)/2$ or $(T_w + T_\infty)/2$. The difference between T_{WB} and T_w is small enough to neglect the effect of change in physical properties. The temperature difference and the partial pressure difference are small, permitting considerable simplification in Equation 27. By introducing series approximations for the Ackermann factor on the left, neglecting the second order term of c_w and c_{WB} and assuming $L_w = L_{WB}$ and $\beta = \beta_{WB}$ Equation 27 becomes

$$\begin{aligned} C = {} & [1 + (c_w - c_{WB})/2][(T_\infty - T_{WB})/(T_\infty - T_w)] \\ & \cdot \ln[(1 - Y_\infty)/(1 - Y_w)]/\ln[(1 - Y_\infty)/(1 - Y_{WB})] \end{aligned} \quad (28)$$

If $(c_w - c_{WB})$ and $(Y_w - Y_\infty)$ are small enough, then Equation 28 can be approximated as follows,

$$\begin{aligned} C = {} & [(T_\infty - T_{wB})/(T_\infty - T_w)] \\ & \cdot [(Y_w - Y_\infty)/(Y_{WB} - Y_\infty)] \end{aligned} \quad (29)$$

If T_∞, $T_w(Y_w)$ and Y_∞ are known, the correction factor C can be readily calculated by Equations 27 through 29.

Relatively Large Blowing Factor

c_{WB} and c_w are expressed as follows:

$$c_{WB} = \dot{m}_c c_{pv}/h_H \quad (30)$$

$$c_w = \dot{m}c_{pv}/h_H \quad (31)$$

where

$$\dot{m}_c = h_{D,WB}(PM_v/RT)\ln[(1 - Y_\infty)(1 - Y_{wB})] \quad (32)$$

$$\dot{m} = h_D(PM_v/RT)\ln[(1 - Y_\infty)/(1 - Y_w)] \quad (33)$$

where for brevity the small difference of the temperature T is deleted in Equations 32 and 33. If the wet-bulb temperatures are not known, the value of C is also given by the following procedure. Substituting Equations 32 and 33 into Equations 30 and 31, respectively, c_{WB} and c_w are given by Colburn's relation, as follows,

$$c_{WB} = \beta_{WB}(c_{pv}/c_p)(M_v/\bar{M})\,Le^{2/3} \cdot \ln[(1 - Y_\infty)/(1 - Y_{WB})] \tag{34}$$

$$c_w = \beta(c_{pv}/c_p)(M_v/\bar{M})\,Le^{2/3} \cdot \ln[(1 - Y_\infty)/(1 - Y_w)] \tag{35}$$

C is already defined in Equation 25, and q_t is expressed as Equation 36

$$q_t = q_c + q_s = \dot{m}L_w \tag{36}$$

By substituting Equations 36 and 31 into Equation 25, then, C is expressed as follows:

$$C = (\dot{m}L_w/h_H(T_\infty - T_w)](\exp c_w - 1)/c_w \tag{37}$$

$$= (L_w/c_{pv})(T_\infty - T_w)(\exp c_w - 1) \tag{38}$$

If $(h_D/h_H)(PM_v/RT)$ is given by Colburn's relation, or in another form, C can be calculated from Equations 37 and 38.

Colburn's relation is not always valid for every system. For most natural convections, the exponent of the Lewis number in Equation 19 is given as the value of 0.2 to 0.33. It must, therefore, be calculated C and c_w with the relation of $(h_D/h_H)(PM_v/RT)$, appropriate to each system.

NUSSELT AND SHERWOOD NUMBERS WITH EVAPORATION

Adiabatic Evaporation

The Nusselt and Sherwood numbers can be defined on the basis of evaporation rate $\dot{m}_c$ for adiabatic evaporation:

$$Nu_e = \dot{m}_c \ell L_{WB}/[\lambda(T_\infty - T_{WB})] \tag{39}$$

$$Sh_e = \dot{m}_c \ell/[D(PM_v/RT)(Y_{WB} - Y_\infty)] \tag{40}$$

Substituting Equations 16 and 32 into Equations 39 and 40, respectively, then Nu_e and Sh_e are expressed by using Nu_{e0} and Sh_{e0} at the limit of $\dot{m} \rightarrow 0$ as follows:

$$Nu_e = h_H^* \ell/\lambda = (h_H \ell/\lambda)\,c_{WB}/(\exp c_{WB} - 1) = Nu_{e0} c_{WB}/(\exp c_{WB} - 1) \tag{41}$$

$$Sh_e = (h_{D,WB}\ell/D)\ln[(1 - Y_\infty)/(1 - Y_{WB})]/(Y_{WB} - Y_\infty) = Sh_{e0}\beta_{WB}/Y_{BM} \tag{42}$$

Where $Y_{BM} = (Y_{WB} - Y_\infty)/\ln[(1 - Y_\infty)/(1 - Y_{WB})]$. Equations 41 and 42 mean that Nusselt number Nu_e varies with $c_{WB}/(\exp c_{WB} - 1)$ and Sherwood number Sh_e with β_{WB}/Y_{BM}, respectively. If all physical quatities are accurately measured, and the stagnant film model is strictly valid, the values of Nu_{e0} and Sh_{e0} must agree well with those of Nu and Sh for the limit of $\dot{m} \rightarrow 0$.

Radiative and Conductive Heat Contribution

If there is heat input q_s per unit area of a surface and unit time, except convective heat, on the wet surface, then the Nusselt number cannot be calculated from Equation 39, because the value of the Nusselt number contains an error coming from q_s. The convective heat input $\dot{m}_c$ in $\dot{m}$ is given by

$$\dot{m}_c = \dot{m}/C, \tag{43}$$

then, the Nusselt number is expressed as:

$$Nu_e = \dot{m}\ell L_w/[C\lambda(T_\infty - T_w)] \tag{44}$$

and the corrected Nusselt number Nu_{e0} is expressed as

$$Nu_{e0} = Nu_e c_w/[(\exp c_w - 1)] \tag{45}$$

It is noticed that the Nusselt number with heat input q_s, expect convective heat, is corrected by dividing by C. The corrected Sherwood number is then free from the error coming from erroneous heat input q_s and is expressed as

$$Sh_{e0} = Sh_e \cdot Y_{BM}/\beta \tag{46}$$

where $Y_{BM} = (Y_w - Y_\infty)/\ln[(1 - Y_\infty)/(1 - Y_w)]$.

LIMIT OF APPLICABILITY OF A STAGNANT FILM MODEL

The stagnant film model was recognized very early to be an oversimplification of the actual conditions. The model can successfully predict the wet-bulb temperature or moisture content in the air by Equation 18, which is valid for a small evaporation rate. It is assumed in the model that the effective thickness of the film is constant and independent of evaporation rate $\dot{m}$ or blowing factor c. The applicability of the stagnant film model is limited mainly due to the change of the concentration or temperature profile. It is assumed that this change can be expressed as a variation of the stagnant film, which is herein called the apparent film thickness. The variation of the apparent film thickness occurs in the range of a large blowing factor c and is caused primarily by changes in the concentration or temperature profile, not by the real thickness of each profile. Many of the previously reported theories for transpiration or blowing have predicted accurately the heat and mass transfer by assuming no change in the hydrodynamic conditions for such a small blowing factor as that of the present case. If the apparent diffusion thickness changes with evaporation, then the correction of Sh_e based on the stagnant model does not give the same value of Sh for the limit of $\dot{m} \to 0$. If the difference between Sh_{e0} and Sh is assumed to be

Table 1
Classification of Previously Reported Experimental Works

Geometry	Type of Flow	Related Fields	Evaluation of Data	References
Horizontal flat surface	Forced convection parallel or perpendicular	Air-conditioning Drying Evaporation	Heat and Mass Transfer coefficients	[10], [12], [13], [14]
	Natural convection		Heat transfer Dependence of heat and mass transfer on evaporation rates	[19], [20]
Cylinder	Forced convection	Wet-bulb hygrometer	–	Not cited
Drop	Forced and natural convection	Spray evaporation Spray cooling Liquid fueled	–	Not cited
Tube (wet column)	Forced convection	Engine Chemical process	Heat and Mass Transfer coefficients	[2], [28]

due to the apparent change of the thickness of a stagnant film, then the ratio of Sh_{e0}/Sh gives the ratio of δ_D^* and δ_D;

$$Sh/(Sh_e \cdot Y_{BM}/\beta) = Sh/Sh_{e0} = \delta_D^*/\delta_D \tag{47}$$

where δ_D^* and δ_D are the apparent diffusion thicknesses for the evaporation rate $\dot{m}$ and the diffusion thickness for the limiting value of $\dot{m} \rightarrow 0$, respectively. It should be noted that the apparent change of the thickness of the stagnant film is not due to the real change of the stagnant film thickness, but that it expresses the difference between the thickness determined by the gross effect of blowing and that predicted with the stagnant film model. Then, the ratio of the two film thicknesses varies according to the following equation

$$\begin{aligned}\delta_D^*/\delta_D &= Sh/(Sh_e \cdot Y_{BM}/\beta)\\ &= (h_{D0}/\dot{m})(PM_v/RT)\ln[(1 - Y_\infty)(1 - Y_w)]\end{aligned} \tag{48}$$

In this equation, the numerator on the right-hand side is the evaporation rate predicted with the stagnant film model. The variation of the apparent thermal conduction thickness is similar to that of diffusion thickness, because the blowing factor c is not very large and the Lewis number not so much different from unity.

The variation of the ratio of two diffusion thicknesses is then approximately related to that of the thermal conduction layer as follows:

$$\begin{aligned}\delta_H^*/\delta_H = \delta_D^*/\delta_D = (h_{D0}/\dot{m})(PM_v/RT)\\ \cdot\ln[(1 - Y_\infty)/(1 - Y_w)]\end{aligned} \tag{49}$$

Then, Nu is related to Nu_{e0} by the following relation:

$$Nu = Nu_{e0}[c_w/(\exp c_w - 1)](\delta_H^*/\delta_H) = Nu_{e0}(\delta_H^*/\delta_H) \tag{50}$$

where δ_H^* is the apparent thickness of the thermal conduction layer at the blowing factor c_w, and c_w is the effective blowing factor which is not changed and expressed in $\dot{m}c_{pv}\delta_H/\lambda$, not in $\dot{m}c_{pv}\delta_H^*/\lambda$.

EVALUTION OF PREVIOUS EXPERIMENTAL DATA

Although a considerable amount of experimental work has been carried out in the past by various workers, their results differ appreciably and have been, in general, inconclusive and/or contradictory in regard to previous theoretical considerations. This may attest to the experimental difficulties involved in the accurate measurement of convective heat and mass transfer rates and the temperature and concentration differences over a wide range of inert gas composition in the various types of apparatuses employed. The previously reported experimental works are classified into four categories by the geometry of wetted surfaces, as shown in Table 1. Most of the literature does not report sufficient data to assess the methods described in the previous section. The author, therefore, evaluated the data of some of the reports, which are listed in the final column of Table 1. It is usually difficult to calculate the Nusselt number from the previously reported papers because of the insufficient data and the experimental conditions. The evaluation of the previous data teaches us that most of the curiosity described in the introduction comes from the erroneous measurements of evaporation rate, heat input, and surface temperature.

In the following section the author evaluates only the data obtained by the water-air system, because the physical properties related to evaluation of the data are known with higher accuracy.

PROPERTY EVALUATION

Physical properties c_p, D, λ, ρ, ν used in the present calculation are evaluated at the film temperature as a reference temperature. For the effect of vapor concentration, specific heat and density are evaluated at the weight of mole fraction. The diffusion constant is thought to be independent of vapor concentration. Thermal conductivity and viscosity are evaluated by the methods by Lindsay and Bromley [6] and Bromley and Wilke [7], respectively. Latent heat is evaluated at the temperature of a wetted surface.

HEAT AND MASS TRANSFER FROM A FLAT SURFACE

Heat and mass transfer with evaporation from a free or a wetted surface has attracted the attention of scientists and engineers for a long time [8–18]. Many workers have reported much enhancement of heat transfer as compared with that of a dry body under other equal hydrodynamic conditions. It is known that if a transverse mass flows in the opposite direction of heat flow, it leads to a decrease in heat transfer, providing that other physical mechanisms in the boundary layer do not change. Opinion is, therefore, divided on the subject of heat transfer with evaporation, and also on the qualitative effect of evaporation upon heat transfer.

Smolsky et al. [10] intensively studied heat and mass transfer from the free surface of a liquid (water, acetone, benzol, and butanol) into a hot air stream, and they obtained the following empirical relations from the treatment of their experimental data:

$$Nu_e = 0.086Re^{0.8}Pr^{0.33}Gu^{0.2} \text{ for heat transfer} \tag{51}$$

$$Sh_e = 0.094Re^{0.8}Sc^{0.33}Gu^{0.2} \text{ for mass transfer} \tag{52}$$

They also measured the heat transfer coefficient of a dry body under the same thermal, hydrodynamic, and other conditions, and obtained values which agreed satisfactorily with the well-known critical relation (Equation 53) within a scatter of 5%:

$$Nu = 0.037Re^{0.8}Pr^{0.33} \text{ for heat transfer} \tag{53}$$

$$Sh = 0.037Re^{0.8}Sc^{0.33} \text{ for mass transfer} \tag{54}$$

where Sh is derived from Nu based on the analogy between heat and mass transfer. Based on these results, Smolsky et al. concluded that the absolute values of Nu_e and Sh_e increase considerably with evaporation providing that other conditions are equal. Luikov [15] suggested that increased heat and mass transfer are due to the evaporation of fine drops which are carried away by air flow from the surface of a liquid into a boundary layer. Katto et al. [13, 14] also measured the heat transfer coefficient with water evaporation and confirmed that their data agreed with that predicted for Equation 51, within experimental error. Their results showed the occurrence of increased heat transfer in the case of mirror-like, smooth surfaced water, which led them to conclude that the hypothesis of volumetric evaporation is not always valid.

Mawatari et al. [12] measured concentration and temperature profiles on the wet surface of a porous plate in order to avoid the effect of erroneous heat gains and losses. Based on the measured profiles of concentration and temperature, local Sherwood and Nusselt numbers were calculated as follows:

$$Nu_{x,e} = [x/(T_\infty - T_w)]\,(\partial T/\partial y)_w \tag{55}$$

$$Sh_{x,e} = -\,x/[(1 - w_w)\cdot(w_w - w_\infty)]\,(\partial w/\partial y)_w \tag{56}$$

where $[1/(T_\infty - T_w)]\,(\partial T/\partial y)_w$ or $-\,1/[1 - w_w)\cdot(w_w - w_\infty)]\cdot(\partial w/\partial y)_w$ are obtained as a gradient at the origin of the curve of $(w_w - w)/(w_w - w_\infty)$ or $(T - T_w)/(T_\infty - T_w)$. They concluded that the mass transfer coefficient agreed well with the theoretical values, while the heat transfer coefficients were 5% to 40% larger than the theoretical values of no mass injection. The enhancement of the heat transfer coefficients is caused by inaccuracy in their method of measuring temperature. Their thermistor did not indicate the real gas temperature because of heat conduction from the stays and radiative heat transmission. On the contrary, the mixture of main flow air and vapor was inhaled into the pilot-like tube at the paper inlet velocity; therefore, the profiles of concentration were not disturbed by the tube. Mawatari et al. finally concluded that the Sherwood number $Sh_{x,e}$ is much more reliable than the Nusselt number $Nu_{x,e}$. Because they used a pitot-like gas sampling tube of 0.8 mm in diameter, and the pitot tube has been established as a reliabile means of measuring the profile of velocity in a boundary layer, this author agrees with the findings of Mawatari et al. The typical results of various works are presented in Figure 2 against the Reynolds number.

Temperature of a Wet Surface

The equilibrium temperature of a liquid surface is closely concerned with radiative and conductive heat gains and losses into the liquid vessel used in an experiment. Figure 3 shows a comparison between the wet-bulb temperature calculated by Equation 22 and the measured ones of water surfaces by Smolsky et al., Katto et al., and so on. The surface temperature by Mawatari et al. is not shown in Figure 3 because they conducted the evaporation experiment using a wetted sintered brass plate heated from the underside of the plate. If the temperature of a liquid surface is higher than the wet-bulb temperature corresponding to T_∞ and Y_∞, then the liquid vessel receives radiative and/or conductive heat. Figure 3 reveals that adiabatic evaporation was not realized in these experiments, and that a large amount of radiative and conductive heat entered their liquid vessels.

Calculation of Nusselt and Sherwood Numbers

If Nu_e (Sh_e), T_∞, T_w, and w_∞ (Y_∞) are given in the previously published reports, Nu_{e0} and Sh_{e0} can be calculated using Equations 45 and 46, based on the values of Nu_e and Sh_e. Figure 4 respresents Nu_{e0} and Sh_{e0} calculated with Equations 45 and 46. The overall results of Nu_{e0} are considerably smaller than the results of Nu_{e0}, and the analogy between heat and mass transfer is satisfied fairly well as shown in Figure 4. Some of the values of Nu_{e0} and Sh_{e0} are much larger than those predicted with Equations 53 and 54; these differences have come

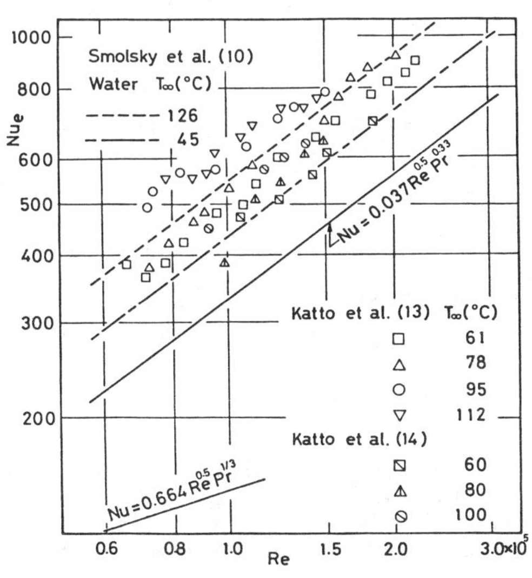

Figure 2. Typical results of previously reported Nusselt numbers, Nu_e.

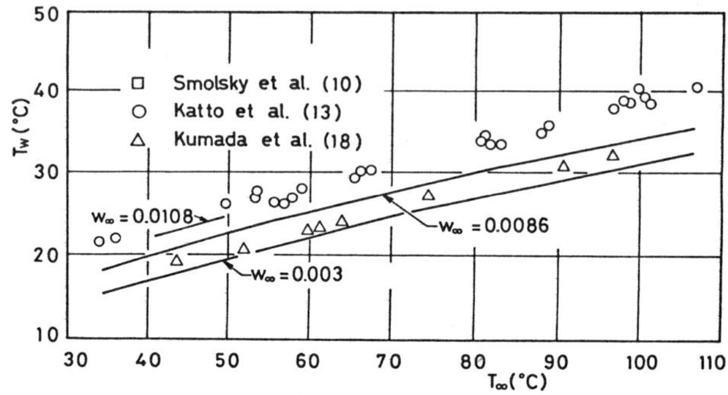

Figure 3. Comparison between temperatures measured at the surface of a liquid and the wet-bulb temperature for adiabatic evaporation.

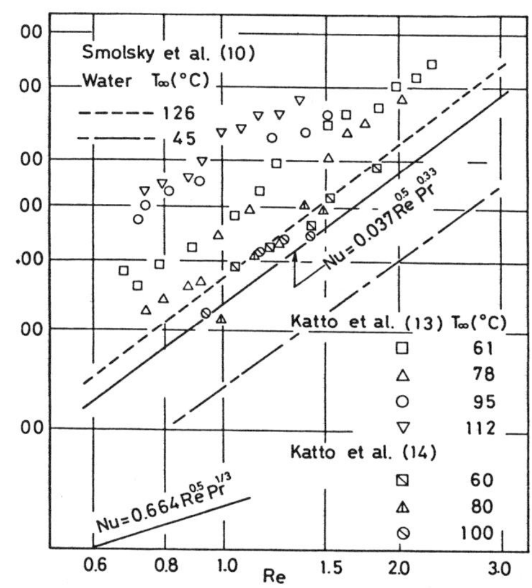

Figure 4. Typical results of Nusselt and Sherwood numbers Nu_{e0}, Sh_{e0} corrected with Equations 45 and 46.

mainly errors in the measuring of $\dot{m}$, $T_w(Y_w)$, and Y_∞ and from the formation of surface waves. The Nusselt number $Nu_{x,e}$ by Mawatari et al. was calculated with the following equation:

$$Nu_{x,e} = Sh_{x,e} \cdot D \cdot L_w (PM_v/RT)(Y_w - Y_\infty)/[C\lambda(T_\infty - T_w)] \tag{57}$$

Figure 5 shows the values of $Sh_{x,e}$ and $Nu_{x,e}$ calculated with Equation 57 against local Reynolds number Re_x, respectively. The value of $Sh_{x,e}$ in Figure 5 is slightly larger than that prescribed with Equation 54, and that of $Nu_{x,e}$ is slightly smaller than that of Equation 53. The analogy between heat and mass transfer does not appear to be sufficiently satisfied due to the effect of blowing. Figure 6 shows the values of $Sh_{x,e0}$ and $Nu_{x,e0}$ corrected for the blowing effect. Both the values of $Sh_{x,e0}$ and $Nu_{x,e0}$ are slightly smaller than these predicted with the following equations:

$$Nu = 0.0296Re^{0.8}Pr^{0.53}[1 - (x_0/x)^{0.9}]^{-1/9} \quad (58)$$

$$Sh = 0.0296Re^{0.8}Sc^{0.53}[1 - (x_0/x)^{0.9}]^{-1/9} \quad (59)$$

The slight differences of $Sh_{x,e0}$ and $Nu_{x,e0}$ from the values predicted with Equations 53 and 54 come from the increase of the apparent thickness of stagnant film at high evaporation rates. A detailed discussion of these differences is given later.

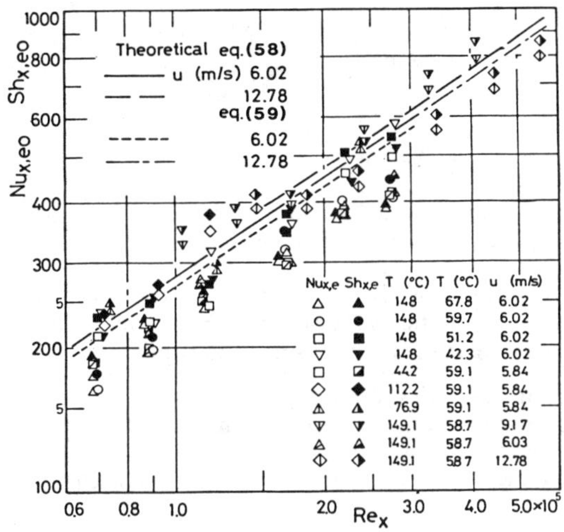

Figure 5. Local Nusselt and Sherwood numbers by Mawatari et al.

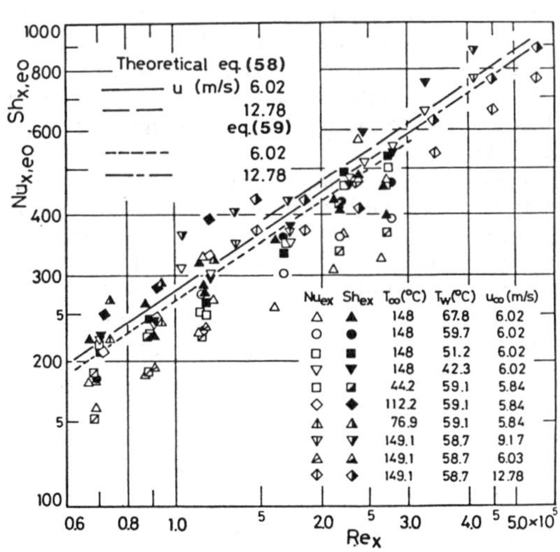

Figure 6. Corrected local Nusselt and Sherwood numbers by Mawatari et al.

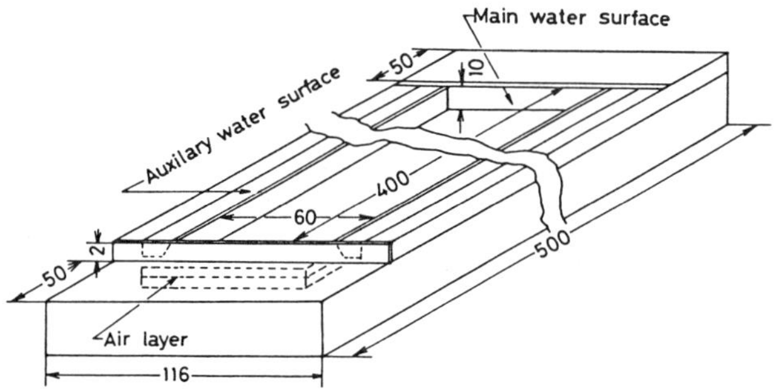

Figure 7. Details of the structure of the liquid pan.

Experimental Verification by the Author

The author et al. [16–18] conducted experiments for the purpose of obtaining more reliable data on the heat and mass transfer coefficients to verify the method of evaluating heat gains and losses into a liquid pan.

As pointed out in the foregoing discussion, it is very difficult to measure Nu_e and Sh_e under ideal conditions of adiabatic evaporation with no waves on a liquid surface. The following considerations were made in the design of the apparatus used in the experiments:

1. To minimize heat gains or losses into a liquid pan.
2. To design a simple structure liquid pan that will make possible exact calculations of the heat gains or losses into the pan.
3. To prevent the occurence of surface waves caused by the mechanical vibration induced by turbulent flow.

Figure 7 shows the details of the liquid pan. The water poured into the pan formed a free surface on the wet cloths covering the floor of the pan, which was rectangular and measured 60 mm wide and 400 mm long, beginning from a point 50 mm downstream of the entrance of the duct floor. The evaporation surface of the wet cloths was covered with a sheet of smooth and then silk cloth, which was completely wetted by capillary force. The pan had two auxiliary evaporation surfaces on its two sides, which served to prevent the edge effect of the evaporation surface and to cool both sides of the pan to prevent heat inputs through the lateral walls of both sides.

Figure 8 shows the values of the Nusselt and Sherwood numbers, which includes the error coming from radiative and conductive heat inputs and the effect of vapor blowing. The values of the Sherwood number are in fairly good agreement with the values predicted by Equation 54, except in the range of $2.0 \times 10^5 > Re > 1.5 \times 10^5$, where the transfer coefficients are enhanced due to the production of surface waves. The values of Sh_e obtained by the surface of the wet cloth are also in fairly good agreement with those predicted by Equation 54. Figure 9 shows the Nusselt and Sherwood numbers Nu_{e0} and Sh_{e0}, which are corrected for the effect of vapor blowing and erroneous heat inputs. The results showed clearly that there was no enhancement of the Nusselt number in the experiment.

Conclusions

The results obtained here suggested that the enhancement of Nu_e shown in the recently reported data was due to experimental errors which were produced mainly from erroneous heat inputs into the water pan used for evaporation. The other inconsistencies were

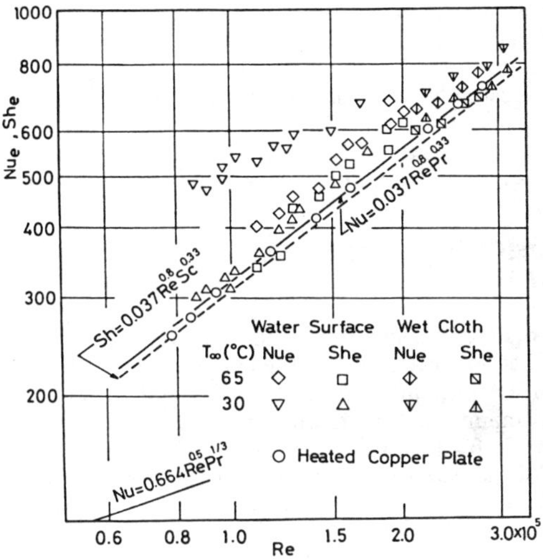

Figure 8. Nusselt and Sherwood numbers by the present author et al.

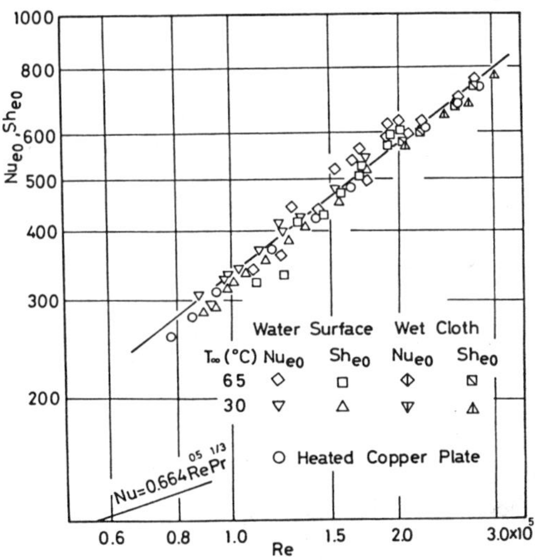

Figure 9. Corrected Nusselt and Sherwood numbers by the present author et al.

probably caused by the errors of T_w, $\dot{m}$, and by the enhancement in Nu_e by the waves of the liquid surface. If the measurement of Nu_e is performed under ideal conditions of adiabatic evaporation with no waves on the liquid surface, Nu_e may show a value close to that of a dry body for a small velocity of blowing, such as in the present case.

It is very curious that the problem of abnormal enhancement of heat transfer has not been solved up to now. This problem may be caused by the relatively small heat input by convection as compared with the heat inputs by conduction through the walls of a liquid pan. The condition of perfect adiabatic evaporation cannot be achieved easily due to the difficulty of preventing absolutely heat inputs, except for convection transfer of heat.

This problem is, interestingly, not encountered in the basic research on the wet-bulb hygrometer for the measurement of air humidity. In this case, the erroneous inputs are caused mainly by radiation from the surroundings and partly by heat conduction through a stem, because most of the experiments are performed for bluff-body objects.

NATURAL CONVECTION HEAT AND MASS TRANSFER FROM A FREE SURFACE

Sharpley and Boelter [19] confirmed the analogy between heat and mass transfer for the case of evaporation of distilled water, within the temperature limits of 53° F and 93° F, from a one-foot diameter surface into quiet air at 71° ± 1° F and 50% to 54% relative humidity. The water surface was maintained flush with the surrounding floor, except in the cases where the pans had rims. In further experiments to obtain supplemental data to support this report, Boelter et al. [20] used an apparatus which was essentially the same as that used by Sharpley et al., with the exception of the quieting chamber. The experimental conditions were within the temperature limits of water surfaces 63° and 200° F into quiet air at 65° to 80° F and 54% to 98% relative humidity. Sharpley et al. recommended Equation 60 for mass transfer of natural convection with evaporation.

$$Sh_e = 0.53\,(G\acute{r} \cdot Sr)^{0.26} \tag{60}$$

$$10^6 < G\acute{r} \cdot Sr < 10^8$$

Boelter et al. also recommended Equation 61

$$Sh_e = 0.643\,(G\acute{r} \cdot Sr)^{0.25} \tag{61}$$

$$10^8 < G\acute{r} \cdot Sr < 10^{11}$$

Sharpley et al. and Boelter et al. obtained the Nusselt number by multiplying the experimental values given by Bosch [21] for a vertical plate by 1.2. From Equations 60 and 61, the following ratios are computed, respectively:

Sharpley et al.

$$Nu/Sh_e = 1.29$$

Boelter et al.

$$Nu/Sh_e = 1.17$$

They concluded that these results indicated that the "equivalent film thickness" for nonisothermal mass transfer is 1.29, or 1.17 times the thickness of the "equivalent film thickness" for heat transfer for the system under consideration, within the range of experimental variables and the accuracy of measurement. Their Nusselt number, however, was estimated from the data of a vertical heated plate, and thus is not always applicable for these discussions. It is more desirable to calculate simultaneously the Nusselt and Sherwood numbers from the data given in References 19 and 20.

Balakhonova et al. [22] performed an experiment to confirm the effect of transverse mass flow on the heat transfer rate. Their experiment was carried out using vacuum equipment. They used two test sections, dry and liquid, which had the same horizontal level and heated surface area, and placed them into a vacuum chamber. The experiments were carried out at ambient pressure of 1.33×10^3 and 2.67×10^3 Newtons/m. The results showed that the ratio Nu_e/Nu rises to a maximum value larger than unity, then decreases and drops below

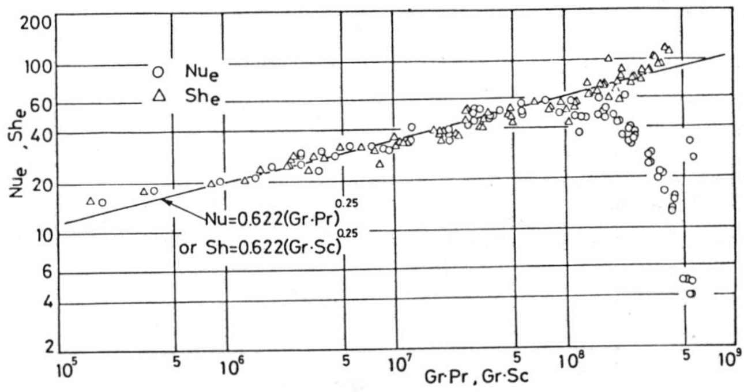

Figure 10. Nusselt and Sherwood numbers (Nu_e, Sh_e) for natural convection.

unity with an increasing of the stream flux density. It is unfortunate that Balakhonova et al. did not supply the data necessary to calculate the Nusselt and Sherwood numbers.

Calculation of Nusselt Number

Sharpley et al. and Boelter et al. gave all the values necessary to calculate Nu_e, Sh_e, Gr, and so on. The results of both Nu_e and Sh_e are shown in Figure 10. The overall results of the values of Nu_e are about equal to those of Sh_e in the range of lower concentrations. In the range of lower concentrations, each value of Nu_e or Sh_e is nearly equal to or a little lower than the following relations for a square, which was recently recommended by Yousel et al. [23].

$$Nu = 0.622\,(Gr \cdot Pr)^{0.25} \tag{62}$$

$$Sh = 0.622\,(Gr \cdot Sc)^{0.25} \tag{63}$$

where Sh is derived from Nu based on the analogy between heat and mass transfer. It is reasonable that the values of Nu by a circle are smaller than those by a square, because the edge region of higher heat transfer for the square is wider than that for the circle.

The measurement of the heat transfer coefficient in natural convection is particularly difficult in a gas, because the quantity of transfer heat by convection is not as large as the conduction and radiation gains and losses of heat. It is very difficult to measure accurately convective heat flux from the surface of a heated plate. On the contrary, it is necessary to measure only the decrease of liquid weight to determine the total heat consumption on a wetted surface if the erroneous heat inputs by conduction and radiation are corrected by the method described earlier.

In higher Rayleigh numbers, Nu_e decreases as compared with that predicted with Equation 62, and Sh_e increases as compared with the value obtained by Equation 63 as the Rayleigh number increases. This trend is closely related to the blowing factor, or to Y_{BM}, in the present case. Figure 11 shows the values of Nu_{e0} and Sh_{e0} corrected with Equations 45 and 46. The values of both Nu_{e0} and Sh_{e0} show similar variation, but the main differences between Sh_{e0} or Nu_{e0} and Sh or Nu are considered to come mainly from the change of diffusion thickness, the details of which will be discussed later.

Conclusions

There is no enhancement in the value of Nu_e for natural evaporation by Sharpley et al. and Boelter et al., if the method described earlier is followed. The values of Sh_e and Nu_e

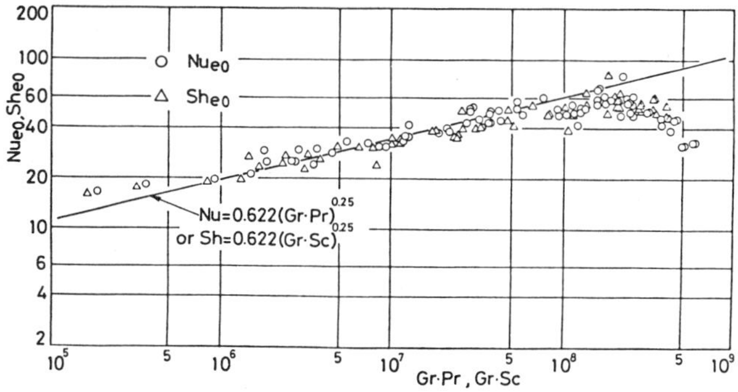

Figure 11. Corrected Nusselt and Sherwood numbers (Nu_{e0}, Sh_{e0}) for natural convection.

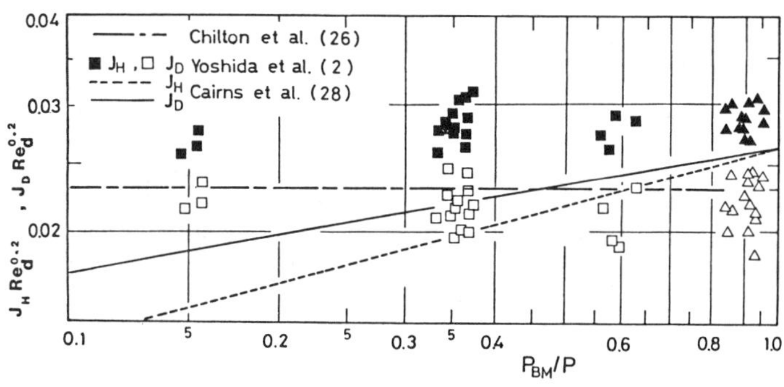

Figure 12. Comparison between the J_H and J_D factors by Cairns et al. and Yoshida et al.

agree fairly well with each other and with the values of the heat transfer experiment by Yousel et al. in the range of $Y_{BM} > 0.9$. The corrected values of both Sh_{e0} and Nu_{e0} vary similarly and are smaller than those predicted with Equations 62 and 63. In higher concentrations of vapor, the difference between the values of Sh_{e0} or Nu_{e0} and the values predicted by Equations 62 or 63 comes from the variation in the apparent thickness of the stagnant film assumed in the stagnant film model. It is necessary to take into account the change of the apparent stagnant film in order to understand the mechanism of the blowing effect on heat and mass transfer.

HEAT AND MASS TRANSFER OF A WETTED COLUMN

The heat and vapor transfer in a wetted wall tower is important in applications such as drying, air conditioning, and chemical processing. Heretofore, the effect of the concentration level on the transfer rate of this type of system has been considered in the design of equipment by following the predictions of the film theory based on the well-established effect of concentration level on mass transfer by molecular diffusion. Recent attempts [2, 27–29] to study the effect of the concentration level in turbulent gas streams, however, have yielded conflicting results with respect to the correct relationship between the concentration level and the dimensionless transfer coefficients, the J_H and J_D

factors, as shown in Figure 12. Gilliland and Sherwood [27] studied the evaporation of nine different liquids in a wetted-wall tower of 1.05 in. in inner diameter operating at pressures from 0.145 to 3.06 atm. They correlated their data on the basis of the Stefan equation (Equation 64) and the stagnant film theory and obtained:

$$\dot{m} = h_D (PM_v/RT)(Y_w - Y_\infty)/Y_{BM} \tag{64}$$

They also measured the heat transfer coefficients for heating air in upward and downward flow in a column without liquid. The experimental heat transfer data agreed closely with Equation 65 by McAdams [30, p. 219].

$$Nu = 0.023 Re^{0.8} Pr^{0.4} \tag{65}$$

Their mass transfer data are about 35% higher than that for heat transfer. Cairns et al. [28] and Yoshida et al. [2] also measured the heat and mass transfer coeficients in a wetted column. Their results did not agree with each other, particularly in the values of the heat transfer coefficients, which differed considerably. The heat transfer coefficients by Cairns et al. decreased, while those by Yoshida et al. increased as the concentration level of water vapor increased. The author doubts that the heat gains or losses into the surface of water were neglected in the previously reported data arrangement. The correction of the previously reported data was also made for this problem.

Method of Calculation

It is very difficult to determine the variation of temperature and concentration along a wetted column based on the inlet and outlet temperature and vapor concentration, because the flow is not developed, and the variation of the heat and mass transfer coefficients are complex and rather flat in the entrance region of a tube. The heat gains and losses into the surface from the bulk of water film are calculated on the basis of the method described earlier. The temperature of flowing liquid along an inner surface of the column is nearly constant in the down stream. Generally, the condition of an adiabatic evaporation is not satisfied for the entire wetted column, even if the adiabatic evaporation is realized at the inlet of the column. This means that there are erroneous heat gains or losses from the bulk of water into the surface of the liquid. The heat exchange between a main stream and the surface of water is mainly held in the entrance region of the columns, because the ratio of the length of the tube to its inner diameter is within 40. Then, the heat transfer coefficients are corrected by using the correction factor C and the blowing factor c_w at the inlet of the columns without a significant loss of accuracy.

T_{in} and Y_{in} are the mean temperature and mole fraction of the main stream at the inlet of a column, respectively. The correction factor C is calculated using T_{in}, Y_{in}, T_w and Y_∞, where T_w and Y_w are the temperature and the mole fraction of vapor corresponding to the temperature of the liquid film flowing on the inner surface of the column and the liquid film flowing on the inner surface of the column, respectively. The corrected J_H-factor is given as follows:

$$J_{He} = J_H/C \tag{66}$$

where $J_H = (h_{H,m}/c_{p,m} \cdot G_m) Pr^{2/3}$, which are given in their reports. The corrected J_H-factor is expressed as

$$J_{He0} = J_{He}(\exp c_w - 1)/c_w \tag{67}$$

The correction by Equations 66 and 67 is not, however, expected to be very accurate due to the complex distribution of temperature and vapor pressure along the columns and the difficulty of determining the real main stream condition. This correction gives only a rough approximation of the real value of the heat and mass transfer coefficient.

Water and Wet-Bulb Temperatures

If the temperature of water is higher than the wet-bulb temperature, then the water consumes its heat and evaporates itself. In this case the evaporation is enhanced as compared with that under the conditions of adiabatic evaporation. Conversely, if the temperature of water is lower than the wet-bulb temperature, then the water absorbs the convective heat from the gas stream and diminishes the evaporation from the surface of the water. Under these conditions, if the heat transfer coefficients are calculated based on the evaporation rates, they must include large errors originating from the difference between the wet-bulb and the measured temperatures. Figure 13 shows a comparison between the wet-bulb temperature calculated by Equation 23 and the measured ones of water by Cairns et al. and Yoshida et al. at the inlet and outlet of the columns. The temperatures of water by both Cairns et al. and Yoshida et al. are inconsistent with the wet-bulb temperatures not only at the inlet but also at the outlet of the columns, in spite of their insistence that their experiments were conducted at the condition of adiabatic evaporation. The temperatures of water by both Cairns et al. and Yoshida et al. approach the wet-bulb temperatures, which correspond to the temperature and the mole fraction of the main stream as P_{BM}/P decreases at both the inlet and outlet of the columns. This does not mean that the condition of an adiabatic evaporation is much more realized when P_{BM}/P is small, because the dependence of C on $T_w - T_{WB}$ significantly increases as P_{BM}/P decreases. Thus it is important that the mean temperatures of water is determined accurately when P_{BM}/P is small. The temperatures of water by Yoshida et al. are nearly equal to or higher than the wet-bulb temperature at the inlet of the columns, and those at the outlet significantly approach the wet-bulb temperatures. The variation of the wet-bulb temperature along the columns is strongly dependent on the evaporation rate, which makes it difficult to determine the mean temperatures which are appropriate for the data arrangement of the experiments.

Corrected J_H-factor and Its Explanation

Figure 14 shows the corrected J_H-factors in comparison with the original values by Cairns et al. and Yoshida et al. The corrected J_H-factors by both groups are in considerably good agreement with the values predicted by the relation of Chilton et al. [26] when P_{BM}/P is

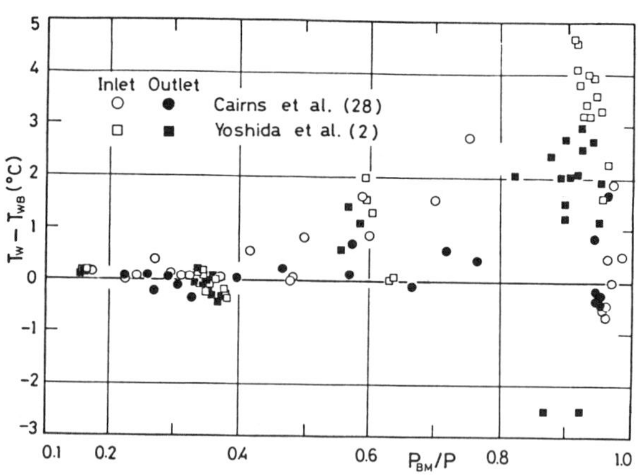

Figure 13. Comparison between the wet-bulb temperature calculated by Equation 23 and the measured ones of water by Cairns et al. and Yoshida et al. at the inlet and outlet of the columns.

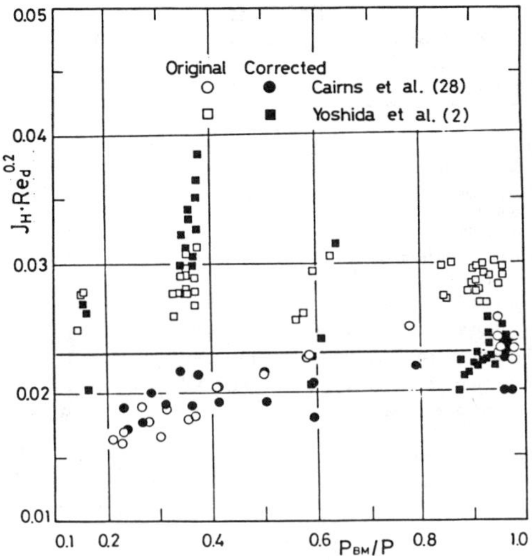

Figure 14. Comparison between the corrected J_H by the original values by Cairns et al. and Yoshida et al. Solid line shows value of Chilton et al.

nearly unity. The corrected J_H-factors by Cairns et al. are smaller than the predicted values in the range of low P_{BM}/P. The J_H-factors in Figure 14 and the J_D-factors in Figure 12 by Cairns et al. vary similarly as P_{BM}/P decreases. The variation of both the J_H- and J_D-factors by Cairns et al. against P_{BM}/P is explained by assuming the estimation error of the mean temperatures of water to be about 0.3° C. On the contrary, the values by Yoshida et al., though much more scattered, are in fairly good agreement with those predicted by the relation of Chilton et al., except for the case of $P_{BM}/P \approx 0.35$. The higher values of the J_H-factors come from small errors in the measured mean temperatures of water for the very small difference between the vapor pressures of the main stream and the surface of water. For the case of $P_{BM}/P \approx 0.35$ by Yoshida et al., the abnormal enhancement of the corrected J_H-factors can be explained by assuming that the measured mean temperatures are lower than the wet-bulb temperatures. The magnitude of these temperature differences may be estimated to be within only 0.3° C, on the average, between the measured and the wet-bulb temperatures when P_{BM}/P is about 0.35. It is very difficult to calculate the mean temperatures of water at such an accuracy of 0.3° C in such systems.

Conclusions

The heat gains and losses into a surface from the bulk of water film are calculated on the basis of the method discussed earlier. The conditions of adiabatic evaporation are not satisfied in either of the experiments by Cairns et al. and Yoshida et al. The corrected J_H-factors show fairly good agreement with those predicted by the relation of Chilton et al. over the entire range of P_{BM}/P, except for the humidity of 1.0 by Yoshida et al. It is postulated that the curious enhancement of the corrected J_H-factors for $P_{BM}/P \approx 0.35$ comes from the estimation error of the mean temperatures of the main stream and the surface of water. The corrected J_H-factors are very sensitive to the accuracy of water temperature for the case of the small difference between the vapor pressures at the main stream and the surface of water. This is the reason for the large difference between the J_H-factors by Cairns

Table 2
Effects of Partial Pressure of Vapor on Mass Transfer

Author	System		P_{BM}/P	n*	Geometry of a Wetted Surface
Film theory	–		–	1	–
Hinchley and Himus [8]	Water-air (evaporation)	natural forced	0.55–0.99	1 0	Flat plate
Wade [24]	Water, acetone, benzene-air (evaporation)	natural forced	0.68–0.99	1 0	Flat plate
Westkalmber et al. [25]	Tetrachloromethane-air (evaporation)		0.30–1.0	(0.83)	Flat plate
Boelter et al. [20]	Water-air (evaporation)		0.5–0.99	0.75	Circular flat plate
Gilliland and Sherwood [27]	Water, soluent-air (evaporation)		0.7–0.99	1	Wetted column
Yoshida et al. [2]	Water-air (evaporation)		0.356–0.955	1	Wetted column
Cairns and Roper [28]	Water-air (evaporation)		0.15–0.97	0.83	Wetted column
Shulman and Delany [29]	Tetrachloromethane-air (evaporation)		0.10–0.50	2/3	Column packed with Raschig rings
Tôei et al. [30]	Water-air (evaporation)		0.38–0.98	0.8	Liquid drop
Barnet and Kobe [31]	Water-air (evaporation)		–	1	Liquid drop

* Where n is an exponent in $(P_{BM}/P)^{-n}$

et al. and Yoshida et al. It is very difficult to realize the conditions of adiabatic evaporation in the experiments by a wetted column. This makes it difficult to determine the reasonable mean temperature of water to arrange the data. Many of the previous discussions are not meaningful, because the real experimental conditions are different from those assumed by Cairns et al., Yoshida et al., and others. Although the author has no data to explain the enchancement of evaporation rates for the case of superheated steam and speculates, he postulates that no realization of adiabatic evaporation and errors in measuring the water temperature cause the enhancement of evaporation rates.

HEAT AND MASS TRANSFER AT HIGHER CONCENTRATION OF VAPOR AND EVAPORATION RATES

A limited number of papers have reported experimental results for heat and mass transfer occurring at high concentrations. For a flat surface, Hincley and Himus [8] measured the mass transfer coefficient in the range of the ratio $P_{BM}/P\,(Y_{BM})$ from 0.55 to 0.99. The mass transfer coefficient is approximately inversely proportional to this ratio. Wade [29] also evaporated seven different liquids from a pan, 3.5-in. square, into a tunnel, 4.72 × 2.36 in. The evaporations were made into still air and into an air stream at 4.4 to 13.1 ft per sec. The ratio P_{BM}/P was varied between 0.68 and 0.99. The gas film coefficient, h_D, was independent of P_{BM}/P for forced convection and was inversely proportional to P_{BM}/P for natural convection. As described earlier, Boelter et al. [20] correlated their data by the dimensionless equation

$$h_D RTd/D = 0.643\,(G\acute{r} \cdot Sc)^{0.25} \tag{68}$$

in which the ratio of P_{BM}/P does not enter. Cairns et al. [28] correlated the data more satisfactorily by the dimensionless Equation 69

$$h_D RTd/D = 0.61\,(G\acute{r} \cdot Sc \cdot P_{BM}/P)^{0.25} \tag{69}$$

The ratio of P_{BM}/P was varied from 0.5 to 0.99 and the water temperature from 75° F to 202° F. The influence of P_{BM} on Equation 69 is not due to the effect predicted by the film theory and the Stefan equation. Table 2 shows the effects of partial pressure of the non-diffusional component on mass transfer coefficients according to the short review by Cairns et al.

The overall results show that Sh_e depends weakly on Y_{BM} for forced convection and strongly on it for natural convection.

Change of the Apparent Thickness of a Stagnant Film

The Sherwood number Sh_e changes according to the following two processes:

1. Change of a concentration profile by diffusion through an assumed stagnant film.
2. Change of the apparent film thickness due to the effect of blowing.

The apparent dependence of Sh_e on P_{BM}/P, as shown in Table 2, is the gross effect of the foregoing two processes. The former process enhances mass transfer proportional to $1/(P_{BM}/P)$, and the latter decreases it as δ_D/δ_D^* decreases. On the contrary, the Nusselt number Nu_e is depressed by both processes in proportion to both $c_w/(e^{cw} - 1)$ and δ_H/δ_H^*, where $c_w = \dot{m}c_{pv}\delta_H/\lambda$. Based on the theory described earlier, the change of the apparent film thickness can be calculated from the data previously reported. The apparent change of the stagnant film thickness is given by Equation 48 as follows:

$$\delta_D^*/\delta_D = Sh/(Sh_e \cdot Y_{BM}/\beta) \tag{48}$$

Figure 15 shows the various values of δ_D^*/δ_D against the blowing factor c_w. The apparent film thickness increases as c_w increases for both forced and natural convection. The results are very scattered from each other and do not show any trend or difference between natural and forced convection.

The author must point the theoretical conclusion that the stagnant model is satisfied even at a higher concentration level, if the evaporation rate or the blowing factor is small. In the experiments of the wetted columns, the concentration of vapor is rather high, but the differences in the vapor pressures and evaporation rates or the blowing factors are relatively small. This means that if the adiabatic evaporation is satisfied, the stagnant film model must be valid for the whole range of experimental conditions. The experimental data by the wetted columns show that the characteristics of the deviation are different from those predicted by the stagnant film model. The author speculates that the deviation comes mainly from the difficulty of conducting accurate experiments under the conditions of the small difference of vapor pressures and the difficulty of realizing adiabatic evaporation along the entire length of the columns.

On the contrary, in the experiments by Boelter et al., the concentration of vapor is relatively small but the blowing factor is rather large. The stagnant film model is not valid in the range of $c_w \gtrsim 0.3$. As mentioned previously section, there are only four experiments by Mawatari et al., Boelter et al., Cairns et al., and Yoshida et al., which were performed in the range of $c_w \gtrsim 0.3$. The author can find clear evidence that the film model is not valid in the range of the blowing factor $c_w \gtrsim 0.3$ in these experiments, as shown in Figure 15.

Conclusions

The author doubts that the stagnant film model is misunderstood in the application for the range of higher concentration and small blowing factors. The stagnant film model is strictly valid for the range of higher concentrations and smaller blowing factors. The blowing factor is only one parameter to determine that the stagnant film model is valid or not in the range of its larger values. The stagnant film model seems to be invalid at the blowing factor $c_w \gtrsim 0.3$ for both natural and forced convections in the papers discussed here. The dependence of vapor diffusion and heat transfer on the effect of vapor concentration must be expressed by the relations derived on the basis of the stagnant film model for such a small blowing factor as that of the wetted columns.

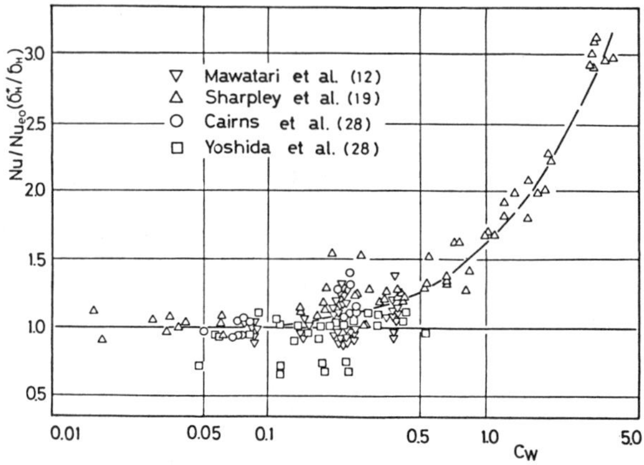

Figure 15. Various values of $\delta_D^*/\delta_D(\delta_H^*/\delta_H)$ against blowing factor c_w.

To understand the mechanism of varying the Sherwood and Nusselt numbers as c_w increases, it is necessary to take into account the effect of the changing of an apparent film thickness. The author proposes that the apparent thickness of the film be calculated by the ratio of Sh_{e0}/Sh, which gives the ratio of $\delta_D^*/\delta_D\,(\delta_H^*/\delta_H)$. The values of the Nusselt number corrected using δ_H^*/δ_H agree well with those of the Nusselt number Nu at the limit of $\dot{m} \rightarrow 0$.

However, in order to discuss the details of these problems, more reliable experimental data at higher concentrations of vapor and blowing factors are needed.

CONCLUDING REMARKS

In heat transfer experiments with evaporation, it is very important to measure exactly the temperature of a wet or a liquid surface. The temperature of the wet surface and main stream and the vapor concentration provide important information on heat gains and losses, except for convective heat. If the evaporation rate is known, then each quantity of these heats is identified based on the measured values of $T_w\,(Y_w)$, $\dot{m}$, Y_∞ and T_∞, using the analogy relation between heat and mass transfer. The detailed discussion presented herein concerning some of the previously reported data shows that the variation of the Sherwood and Nusselt numbers is due to experimental errors produced mainly from heat gains and losses from a liquid pan, or by the water used for evaporation. The other variations are probably caused by errors in measuring $T_w\,(Y_w)$, $\dot{m}$, and Y_∞.

In the higher range of evaporation rates, the change of the apparent thickness of thermal conduction and diffusion layers must be taken into account to understand the variation of the Sherwood and Nusselt numbers against the logarithmic mean of mole fraction Y_{BM} or the blowing factor c. The stagnant film model seems to be invalid at the blowing factor $c_w \gtrsim 0.3$ for both natural and forced convections. At smaller P_{BM}/P values, the psychometric oddity of wetted columns derives from the heat gains and losses from the surface of the liquid or the wetted surface, resulting in measure errors of the water temperature. The inverse point observed by Yoshida et al. is explained by considering these heat gains and/or the errors in measuring the temperatures of water, except for superheated steam. For the first approximation, the ratio of Sherwood number Sh_{e0} gives the ratio of film thickness δ_D^*/δ_D. The previously proposed correlation Sh_e against Y_{BM} must be improved by taking into account the change of apparent film thickness.

It is now possible to answer the six questions puised in the introduction.

For low evaporation rates

1. The analogy between heat and mass transfer is satisfied if the erroneous heat input into the interface is corrected and the physical quantities are measured accurately.
2. No enhancement in the Nusselt number is observed with an increase in the rate of evaporation or the temperature of the main stream.
3. The dependence of vapor diffusion and heat transfer on the effect of vapor concentration must be expressed by the relations derived on the basis of the stagnant film model if the evaporation rate or the blowing factor is small. The blowing factor is the only parameter which can assess the validity of the stagnant film model in the range of larger values.
4. It is speculated that water does not evaporate at higher concentrations of vapor more rapidly than at lower concentrations of vapor, if the conditions of adiabatic evaporation are satisfied.
5. It is very important to check whether adiabatic evaporation is realized and to measure correctly the temperature of a liquid surface in order to measure exactly the heat and mass transfer coefficient with evaporation.
6. It is possible to measure accurately the heat transfer coefficient for natural and forced convection by gas, using a wetted surface.

Further intensive research is necessary to make clear the effect of blowing and the mechanism of the variation of the apparent thickness of an assumed stagnant film.

NOTATION

C	correction factor for heat input
c_p	specific heat (kcal/(kg° C))
c	blowing factor for heat transfer ($\dot{m}c_{pv}/h_H$)
D	diffusion coefficient (m^2/h)
d	diameter (m)
G	mass velocity (kg/($m^2 \cdot$ h))
h_{D0}	mass transfer coefficient at $\dot{m} \to 0$ (m/h)
h_D	βh_{D0}, mass transfer coefficient (m/h)
h_D^*	$h_{D0}\beta/Y_{BM}$
h_H	heat transfer coefficient at $\dot{m} \to 0$ (kcal/(h $\cdot$ m^2 $\cdot$ ° C))
h_H^*	heat transfer coefficient (kcal/(h $\cdot$ m^2 $\cdot$ ° C))
L	latent heat (kcal/kg)
ℓ	length of a plate or a column (m)
M	molecular weight (kg)
$\dot{m}$	evaporation rate (kg/($m^2 \cdot$ h))
P	total pressure (atm)
P_{BM}	$(P_w - P_\infty)/\ln\{(1 - P_\infty)/(1 - P_w)\}$
q	heat coming into a wet surface (kcal/($m^2 \cdot$ h))
q_s	erroneous heat input a wet surface (kcal/($m^2 \cdot$ h))
q_t	total heat input a wet surface (kcal/($m^2 \cdot$ h))
R	universal gas constant (kcal/($m^2 \cdot$ h))
u	velocity along a surface (m/sec)
v	specific volume of vapor at total pressure (m^3/kg)
w	mass fraction of vapor
y	distance in direction perpendicular to a surface (m)
Y	mole fraction of vapor
Y_{BM}	$(Y_w - Y_\infty)/\ln[(1 - Y_\infty)/(1 - Y_w)]$

Greek Symbols

β	correction factor by Bedingfield and Drew
β'	thermal expansion coefficient
δ	thickness of a stagnant film (m)
δ^*	apparent thickness of a stagnant film (m)
λ	thermal conductivity (kcal/(m $\cdot$ h $\cdot$ ° C))
ρ	density (kg/m^3)
ν	dynamic viscosity (m^2/h)

Dimensionless Quantities

Gu	Gukhman number, $(T_\infty - T_w)/T_\infty$	J_D	mass transfer factor $(h_{D,m} \cdot P_{BM}/G_m)\, Sc^{2/3}$
Gr	Grashof number, $\ell^3 g\beta' (T_w - T_\infty)/(T_\infty \cdot \nu^2)$	Nu	Nusselt number $(h_H \ell/\lambda)$
Gŕ	Grashof number $\ell^3 g\, \{(\bar{M}_\infty T_w)/(\bar{M}_w T_\infty) - 1\}/\nu^2$	Pr	Prandtl number (ν/a)
Le	Lewis number (a/D)	Re	Reynolds number $(u\ell/\nu)$
J_H	heat transfer factor $(h_{H,m}/c_{p,m} G_m)\, Pr^{2/3}$	Sc	Schmidt number (ν/D)
		Sh	Sherwood number $(h_{D0}\ell/D)$

Subscripts

c	convection	v	vapor
D	mass transfer	w	wall or wet surface (including the free surface of liquid)
e	measured value corrected for erroneous heat input	WB	wet-bulb temperature for adiabatic evaporation
e0	corrected value for the effect of blowing and erroneous heat input	x	local
g	gas	∞	main stream
H	heat transfer	–	average value evaluated at the weight of mole fraction
m	mean value along a wetted column		

REFERENCES

1. Sherwood, T. K., "The Curious History of the Wet-Bulb Hygrometer," *Chem. in Canada* (June 1950), pp. 19–22.
2. Yoshida, T., and Hyōdō, T., "Evaporation of Water in Air, Humid Air, and Superheated Steam," *Ind. Eng. Chem. Process Des. Develop.*, Vol. 9, No. 2 (Feb. 1970), pp. 207–214.
3. Sherwood, T. K., Pigford, R. L., and Wilke, C. R., *Mass Transfer,* Int. Student ed., McGraw–Hill Kogakusha, LTD, 1975, pp. 56–57.
4. Bedingfield, C. H., and Drew, T. B., "Analogy between Heat Transfer and Mass Transfer–A Psychrometric Study," *Ind. Eng. Chem.,* Vol. 42 (June 1950), pp. 1164–1173.
5. Mickley, H., Sherwood, T. K., and Read, C. E., *Applied Mathematics in Chemical Engineering,* McGraw–Hill, New York, 1957.
6. Lindsay, A. L., and Bromley, L. A., "Thermal conductivity of Gas Mixtures," *Ind. Eng. Chem.,* Vol. 42, No. 8 (Aug. 1950), pp. 1508.
7. Bromley, L., and Wilke, C. R., "Viscosity Behavior of Gases," *Ind. Eng. Chem.,* Vol. 43, No. 7 (July 1951), pp. 1641–1648.
8. Hinchley, J. W., and Himus, G. W., "Evaporation in Currents of Air," *Trans. Inst. Chem. Engrs.,* Vol. 2 (1924), pp. 57–64.
9. Powell, R. W., and Griffiths, E., "The Evaporation of Water from Plane and Cylindrical Surfaces," *Trans. Inst. Chem. Engrs.,* Vol. 13 (1935), pp. 175–198.
10. Smolsky, B. M., and Sergeyev, G. T., "Heat and Mass Transfer with Liquid Evaporation," *Int. Heat and Mass Transfer,* Vol. 5 (1962), pp. 1011–1021.
11. Luikov, A. V., *Advances in Heat Transfer,* Vol. 1, Academic Press, New York, 1964, pp. 123–164.
12. Mawatari, K., Hiramoto, M., and Okazaki, T., "Measurements of the Concentration Boundary Layer in Heat and Mass Transfer," JSME 1967 Semi-Int. Symp. Papers, *Heat and Mass Transfer, Thermal Stress,* Vol. 1 (1967), pp. 249–257.

13. Katto, Y., and Aoki, H., "Peculiarity of Evaporating Liquid-Surface with Reference to Turbulent Heat Transfer," *Bulletin of JSME,* Vol. 12, No. 49 (1969), pp. 79–86.
14. Katto, Y., and Koizumi, H., "Turbulent Heat Transfer of a Gas Flow on an Evaporation Liquid Surface," *Bulletin of JSME,* Vol. 18, No. 122, 1975, pp. 866–873.
15. Luikov, A. V., "Heat and Mass Transfer with Transpiration Cooling," *Int. J. Heat Mass Transfer,* Vol. 6 (June 1963), pp. 559–570.
16. Kumada, T., et al., "Heat and Mass Transfer with Liquid Evaporation into a Turbulent Air Stream," *Letters in Heat and Mass Transfer,* Vol. 9, No. 1, 1982, pp. 1–9.
17. Kumada, T., et al., "Heat and Mass Transfer with Liquid Evaporation into a Turbulent Air Stream–1st report–Theoretical Discussions of Previously Reported Data," *Trans. JSME,* (Oct. 1984), pp. 2446–2453 (in Japanese).
18. Kumada, T., et al., "Heat and Mass Transfer with Liquid Evaporation into a Turbulent Air Stream–2nd report–Measurements of Heat and Mass Transfer Coefficients," *Trans. JSME,* (Oct. 1984), pp. 2454–2460 (in Japanese).
19. Skharpley, B. F., and Boelter, L. M. K., "Evaporation of Water into Quiet Air from a One-Foot Diameter Surface," *Ind. Eng. Chem.,* Vol. 30 (Oct. 1938), pp. 1125–1131.
20. Boelter, L. M. K., Gordon, H. S., and Griffin, J. K., "Free Evaporation into Air of Water from a Free Horizontal Quiet Surface," *Ind. Eng. Chem.,* Vol. 38, No. 6 (June 1946), pp. 596–167.
21. Bosch, T. M., "Die Wärmeübertragung," 3rd ed., Julius Springer, Berlin, 1936, pp. 164–167.
22. Balakhonova, V. I., Novikov, P. A., and Subach, V. M., "Effect of Mass Transfer on Heat Transfer for a Liquid Evaporating from a Free Surface in a Rarefied Gas Medium," Heat Transfer–Soviet Research, Vol. 3, No. 6 (Nov.–Dec. 1971), pp. 85–90.
23. Yousel, W. W., Tarasuk, J. D., and Mckeen, W. J., "Free Convection Heat Transfer from Upward-facing Isothermal Horizontal Surfaces," *ASME Journal of Heat Transfer,* Vol. 104 (Aug. 1982), pp. 493–500.
24. Wade, S. H., "Evaporation of Liquids in Currents of Air," *Trans. Inst. Chem. Engrs.,* Vol. 20, No. 1 (1942), pp. 1–13.
25. Westkalmper, L. E., and White, R. R., "Effect of Concentration Level on Mass Transfer Rates," *AIChE 5,* Vol. 1, No. 1 (Mar. 1957), pp. 69–74.
26. McAdams, W. H., "Heat Transmission," 3rd ed., McGraw–Hill, New York, 1954, pp. 219.
27. Gilliland, E. R., and Sherwood, T. K., "Diffusion of Vapors into Air Streams," *Ind. Eng. Chem.,* Vol. 26, No. 5 (May 1934), pp. 516–523.
28. Cairns, R. C., and Roper, G. H., "Heat and Mass Transfer at High Humidities in a Wetted Wall Column," *Chem. Eng. Sci.,* Vol. 3 (1954), pp. 97–100.
29. Shulman, H. L., and Delaney, L. J., "Performance of Packed Columns-5th. Effect of Solute Concentration on Gas-Phase Mass Transfer Rates," *AIChE 5,* Vol. 5, No. 3 (Sept. 1959), pp. 290–294.
30. Tõei, R., et al., "Evaporation from a Water Drop in the Steam of Steam-Air Mixtures," *J. Chem. Eng. Japan,* Vol. 30, No. 1 (Jan. 1966), pp. 43–49.
31. Barnet, W. I., and Kobe, K. A., "Heat and Vapor Transfer in a Wetted-Wall Tower," *Ind. Eng. Chem.,* Vol. 33, No. 4 (Apr. 1941), pp. 436–442.

CHAPTER 8

BUBBLE NUCLEATION, GROWTH AND DEPARTURE IN BOILING HEAT TRANSFER

Robert Cole

Department of Chemical Engineering
and
Institute of Colloid and Surface Science
Clarkson University
Potsdam, New York

CONTENTS

INTRODUCTION

Of all the problems in boiling heat transfer, the inability to properly characterize the boiling surface in terms of the distribution of nucleation sites is the most frustrating. Without this information it is simply not possible to predict, within reasonable accuracy, the heat flux density corresponding to a given wall superheat in the nucleate boiling regime.

The nucleation sites and their distribution are of importance because they are the source of the vapor bubbles which play such a dominant role in almost all of the proposed mechanisms of heat removal from the hot surface [1]. The identification of nucleation sites and the determination of the conditions under which sites may become active producers of bubbles represent the initial steps in the development of any physical model of the boiling process. Once produced, the bubbles grow and depart. Hence, the conditions necessary for initial growth, the rate of growth, and the forces resulting in departure must be defined. Lastly, the number density of active nucleation sites must somehow be specified.

Each of these topics will be considered in this chapter from the point of view that only by a sound understanding of the physics of the problem can a rational model of the boiling process be devised.

FUNDAMENTAL EQUATIONS OF SURFACE SCIENCE

Certain equations and/or conditions are fundamental to the science of surfaces and hence to a study of liquid-vapor interfaces. As it is inappropriate to develop these equations in a review of this type, they will simply be listed along with references to key derivations. Comments will be included where appropriate.

- Thermal equilibrium

$$T_L = T_V = T \tag{1}$$

- Mechanical Equilibrium

$$P_V - P_L = \sigma(1/r_1 + 1/r_2) \tag{2}$$

Equation 2 is the well known Laplace equation.

- Thermodynamic equilibrium

$$g_L(T, P_L) - g_V(T, P_V) = 0 \tag{3}$$

Equation 3 is the familiar condition that the chemical potential (Gibbs free energy per unit mass) be equal in each phase.

$$g_L(T, P_L) - g_V(T, P_L) = 2\sigma/(\rho_V r) \tag{4}$$

Equation 4 is developed [1] (pp. 114–116) using as a basis for thermodynamic equilibrium of the two phases the assumption that the availability function remain stationary, i.e., $dA = 0$. The equations of thermal and mechanical equilibrium are both employed in obtaining this result. A further condition used in the development of Equation 3, usually not stated, is that the vapor phase be incompressible.

At a flat interface ($r = \infty$), the pressure in both phases is identical (see Equation 2) and equal to the equilibrium vapor pressure P_∞. Thus

$$g_L(T, P_\infty) = g_V(T, P_\infty) = g_\infty \tag{5}$$

The equilibrium vapor pressure is used in "working" equations whenever possible as it is the pressure which appears in thermodynamic tables (e.g., the well known "steam tables").

$$[(\rho_L - \rho_v)/\rho_L](P_\infty - P_L) = 2\sigma/r \tag{6}$$

Equation 6 is called the Laplace–Kelvin equation. It is derived [1] (pp. 116–117) using the thermodynamic definition $(\partial g/\partial P)|_T = v$ and Equation 4 assuming the vapor phase to be incompressible. It can also be obtained directly from a simple force balance on the height of a liquid in a capillary tube immersed in a pool of liquid in equilibrium with its vapor (assuming the vapor phase to be incompressible). The Laplace–Kelvin equation is not a correction (as is sometimes stated) to the Laplace equation.

- Equilibrium superheat equation

$$T - T_S = 2\sigma T_S/(\rho_V h_{fg} r) \tag{7}$$

Table 1
Limiting Superheat of Some Organic Liquids

Substance	T_l, K	
	Theory (Equation 8) at $J = 1 \times 10^6 cm^{-3} \cdot s^{-1}$	Measured
Ethane	269.7	269.2
Propane	328.5	326.2
2-Methylpropane	360.8	361.0
n-Butane	378.3	376.9
Propene	323.5	325.6
2-Methylpropane	372.4	370.0
1-Butene	373.4	371.0
Propyne	361.4	356.8

Adapted from Porteous and Blander [3].

Equation 7 represents the equilibrium superheat required to maintain a finite-size vapor bubble having radius of curvature r. It is developed [1] (p. 117) by combining the Clausius–Clapeyron equation (which represents the equilibrium vapor pressure curve) and the Laplace–Kelvin equation. Because of the neglect of higher order terms in the expansion of a logarithmic term, the expression is valid only under the condition $(T - T_s)/T_S \ll 1$. This is usually not a serious problem, however, as absolute temperatures must be used. It is very important to recognize that Equation 7 is strictly applicable only to a system of uniform superheat. It is for this reason that no subscript appears on the system temperature.

BUBBLE NUCLEATION

Homogeneous Nucleation

At constant temperature and in the absence of a surface, the theory of homogeneous nucleation yields the following expression [1, 2] for either the pressure reduction or the temperature increase required to achieve nucleation.

$$\eta P_\infty - P_L = \left\{\frac{1.182 \cdot 10^5 \sigma^3}{T \ln[(3.73 \cdot 10^{35}/J)(\rho_L^2 \sigma/M^3 B)^{1/2}]}\right\}^{1/2} \tag{8}$$

where

$$\eta = \exp(v_L(P_L - P_\infty)/(RT)) \tag{9}$$

For determination of the temperature increase, thermodynamic tables will yield the desired liquid superheat temperature $T_L(P_\infty)$.

The liquid superheat temperatures predicted by Equation 8 are compared with experimental data [3] for organic liquids in Table 1. The agreement is obviously quite good. To date, however, no experiments have been able to achieve the superheat temperature of approximately 300°C predicted by Equation 8 for water. The closest experimental temperature, for the nucleation rate of 10^6 is 280°C reported by Apfel [4] using the exploding drop technique.

Equations similar to Equation 8 have been developed for homogeneous nucleation from binary solutions by Holden and Katz [5]. Additional studies include those by Pinnes and Mueller [6] and by Avedisian and Glassmann [7]. Forest and Ward [8, 9] and Mori et al.

[10] have investigated homogeneous nucleation in liquids containing dissolved gases. In both instances it is shown that the addition of a second component in solution can significantly affect the nucleation temperature or pressure. Blander [11] has presented a clear, detailed development of the theory of homogeneous nucleation which includes hydrodynamic constraints as developed by Kagan [12] following the method of Zel'dovich [13]. These latter, not-too-well-known developments are worth further study.

Heterogeneous Nucleation

When a new phase appears at an interface or boundary rather than in the bulk, the process is known as heterogeneous nucleation. For a smooth interface, as illustrated in Figure 1, a derivation similar to that of Equation 8 yields the expression

$$\eta P_\infty - P_L = \left\{\frac{1.182 \cdot 10^5 \sigma^3 f}{T \ln[(4.41 \cdot 10^{27} S/J)(\rho^{4/3}\sigma/M^{7/3}Bf)^{1/2}]}\right\}^{1/2} \tag{10}$$

where

$$f = (2 + 3\cos\theta - \cos^3\theta)/4 \tag{11}$$

$$S = (1 + \cos\theta)/2 \tag{12}$$

For $\theta = 0$, the liquid completely wets the surface and the superheat temperature is the same as the homogeneous case. (It is not possible to see that this is true by direct comparison of Equations 8 and 10 because the number density of nuclei ($N = 6.02 \cdot 10^{23} \rho_L/M$) used in developing Equation 8 has been replaced by $N^{2/3}$ in a similar development of Equation 10). Comparison of Equations 8 and 10 reveal that the effect of the surface is not felt until the contact angle is greater than 68°. The heterogeneous mechanism then yields the lower superheat temperature. For $0 = 90°$, the superheat is reduced by about 35%. For $\theta = 180°$, $f = 0$ and $A = 0$, and no superheat is possible (complete nonwetting).

Similar expressions have been developed for other geometries such as spherical and conical cavities, and spherical projections [14]. There (pp. 111–115) it is shown that poorly wetted surface cavities seemingly hold the greatest promise as nucleation sites for most engineering-type surfaces. Nucleation at liquid-liquid interfaces has been investigated by Jarvis et al. [15], Avedisian and Andres [16], and Avedisian and Glassman [17].

The kinetic limit of superheat as given by Equation 8 represents a temperature range (because it is dependent upon the nucleation frequency) over which homogeneous nucleation might be observed. The superheat limit may be lowered by the presence of dissolved gases, by the addition of a second component to form a solution, or by the presence of a smooth surface. In order for the presence of a smooth surface to lower the superheat limit to the range of observed values, it is necessary that the surface-liquid interface be a poorly wetted cavity.

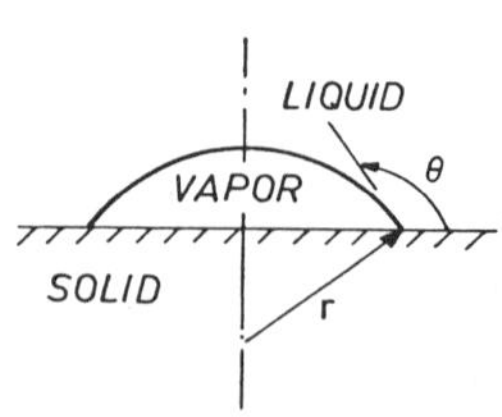

Figure 1. New vapor phase at a smooth solid-liquid interface [14].

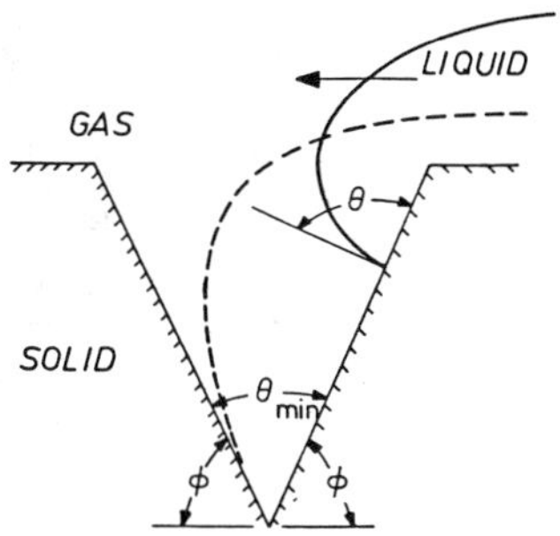

Figure 2. Advance of a liquid sheet over a gas-filled groove [18].

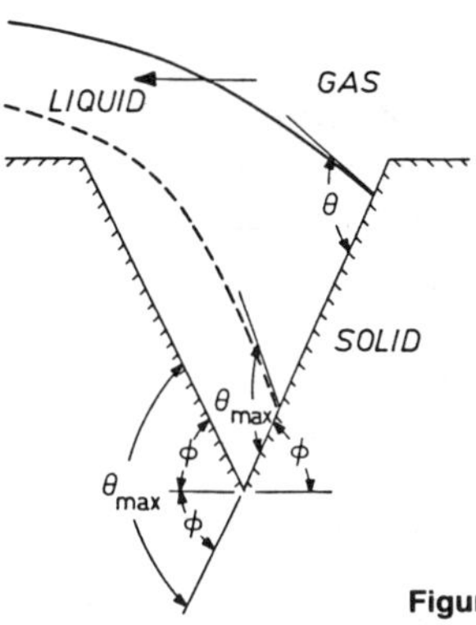

Figure 3. Advance of a gas-liquid interface over a liquid-filled groove [18].

Vapor Trapping

The spreading of liquid over a surface containing grooves and cavities has been considered by Bankoff [18]. Figure 2 illustrates the conditions for entrapment of gas in advance of a semi-infinite liquid sheet across a groove. It is apparent that if the contact angle θ is greater than the wedge angle $\pi - 2\varphi$, the advancing liquid will strike the opposite wall of the cavity before it has completed its advance down the near wall. Hence the condition for entrapment of gas (by the mechanism considered) is

$$\theta > \pi - 2\varphi \tag{13}$$

Figure 3 illustrates the condition for displacement of liquid from a groove by an advancing gas-liquid interface. It is apparent that if the contact angle θ is less than 2φ, the gas-liquid interface will contact the opposite side of the cavity before it has completed its advance down the near wall. Thus, the condition for failure to displace the liquid is

$$\theta < 2\varphi \tag{14}$$

By dividing the cavities into four classes (i.e., those that obey the first inequality but not the second, those that obey the second inequality but not the first, those that obey both inequalities, and those that obey neither), the limiting conditions as shown in Table 2 are obtained. From these limiting conditions, it is quite apparent that steep, narrow cracks and crevices that are poorly wetted by the liquid, maximize the tendency to entrap gases and vapors. Yet, as indicated, these are precisely the conditions which result in values of the superheat temperature sufficiently low that heterogeneous nucleation represents a potential mechanism.

Table 2
Limiting Cavity Conditions

Case	Type of Cavity	Condition	Trapping Ability
1	Steep	Poorly wetted	Traps gas only
2	Steep	Well wetted	Traps liquid only
3	Shallow	Poorly wetted	Traps gas and liquid
4	Shallow	Well wetted	Traps neither

From Cole 14.

Table 3
Initiation Heat-Transfer and Nucleation-site Data

Site	q, kW/m^2	$T_w - T_s$, K	R, μm	Type of Site
1	27.9	6.8	5	Pit, deep in parts
2	28.1	7.2		Largest cavity in crater of 80 μm
4	29.0	8.0	6	End of groove, 12 μm wide
8	29.0	8.3	4	Pit, deep central part of 4 μm diameter
18	29.6	9.0	5.5	Pit, elongated and deep, 16 × 18 μm
19	29.6	9.0	2.5	Pit, deep in parts
20	29.6	9.0	1.5	Site at end of groove, 20 μm wide
33	33.9	9.3	3	Pit, deep in parts
34	34.3	9.5	2	Pit, deep in parts
41	35.1	9.5	1.5	Elongated pit, deep at one end
42	35.1	9.5	1.5	Pit, largest in crater of 90 μm diameter
44	44.6	10.2	2.5	Pit, deep part of 2 μm diameter
45	44.6	10.5	1.5	Pit, deep part of scratch, 1 μm wide
50	44.6	10.6	1	Pit, shallow crater 25 μm wide

Adapted from Cornwell [19].

Nucleation from a Preexisting Gas or Vapor Phase

Most real surface contain natural or machine-formed pits, scratches, gouges, grooves, etc., of sizes ranging from the macroscopic to the microscopic. If the surface is poorly wetted by the liquid, it may be expected that many of the cavities will act as nucleation sites. Cornwell [19] identified the first 50 natural sites to nucleate on a prepared copper heat-transfer surface. Photographs of the sites at magnifications to 16,000 diameters, using a scanning electron microscope, showed them to be deep cavities with mean mouth radii in the range 1–6 μm. In some cases, the sites were seen to be situated in small craters. Photographs of some of the positively identified sites are shown in Fig. 4. Comments and supplemental data are presented in Table 3.

Substantial evidence has been obtained that nucleation from surface cavities is a result of a preexisting gas or vapor phase and not heterogeneous nucleation. A reasonably detailed description of the experiments leading to this conclusion has been presented in Reference 1 (pp. 119–122). In summary, the technique employed by most investigators is to subject the test system (liquid and surface) to high static pressures and then determine the superheat required to initiate boiling. Typical are the experiments of Sabersky and Gates [20]. Their experimental system, consisting of a 0.25 mm diameter wire immersed in water, was subjected to a pressure of 1,000 atm for periods in excess of 15 min. Upon reducing the pressure to atmospheric and heating the wire electrically, the superheat temperature required to initiate nucleation was found to be on the order of 100°F above the normal boiling point, i.e., a superheat of 100 F°. This is the magnitude of superheat which would

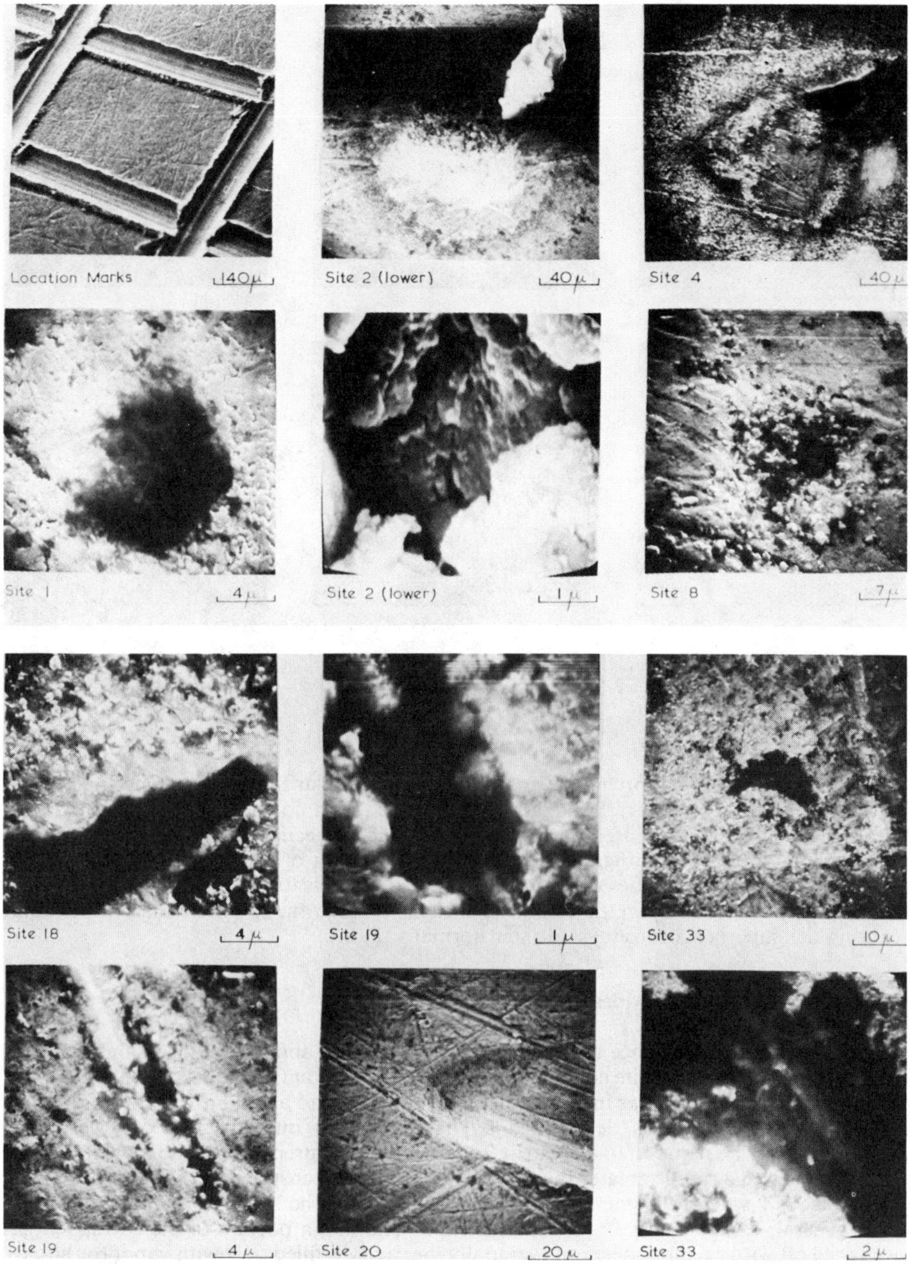

Figure 4. Scanning-electron-microscope photographs of selected natural nucleation sites. See Table 3 for comments and supplementary data [19].

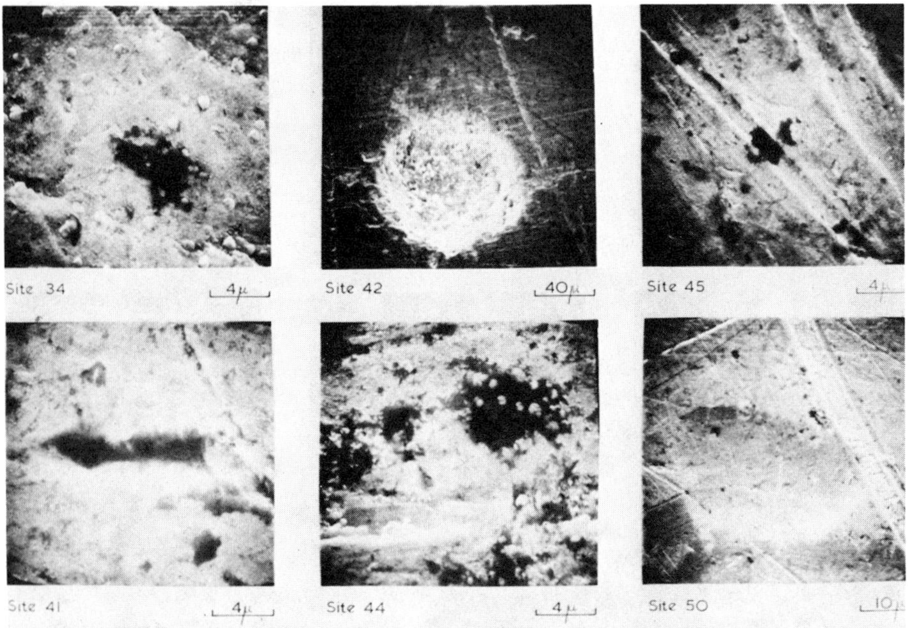

Figure 4. Continued.

be expected if the nucleation had been a result of a molecular mechanism. Once nucleation had been initiated, however, the superheat diminished, and no effects of the pressurization treatment remained evident. It was proposed that subjecting the liquid and wire to extremely high pressures either condensed entrapped vapor or forced undissolved gas into solution, so that large cavities were either eliminated as nucleation sites, or their gas content substantially reduced. Upon the commencement of boiling, the cavities refilled with vapor, causing the superheats to return to their normal values.

Behavior of Gas-Filled Cavities

The experimental evidence that nucleation results from undissolved gases entrapped in surface imperfections is quite convincing. If prepressurization techniques are employed and the amount of entrapped gas in any given cavity is reduced to a volume having a radius less than the critical value as determined by heterogeneous nucleation theory, very high superheats may be required to activate it. In the absence of prepressurization, an entrapped gas volume with a radius greater than the critical value will ensure that the site will activate spontaneously when the system temperature just exceeds the saturation temperature.

In general, once a cavity begins to emit vapor bubbles, a portion of the entrapped gas is carried off with each bubble until eventually the cavity is filled only with vapor (or liquid). The degree of difficulty of replacing the entrapped gas by vapor depends to a great extent on the geometry of the cavity. For example, a reservoir-type cavity, as illustrated in Fig. 5, is extremely stable and may bubble for days before its entrapped gas is completely replaced. Most cavities, however, are probably of such a geometry that over some finite period of time

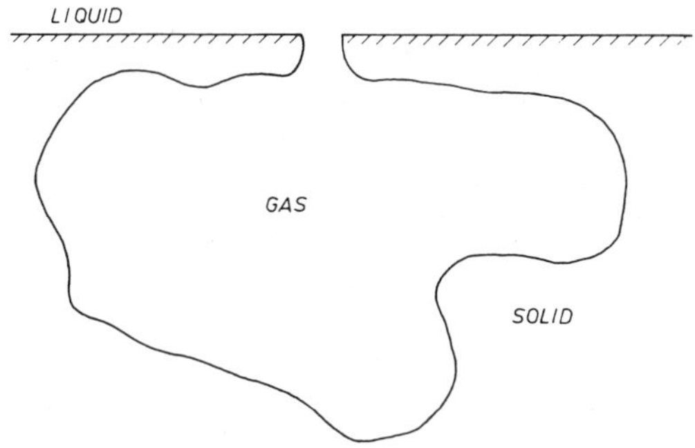

Figure 5. Reservoir-type surface cavity [14].

the noncondensable component will be completely carried away and replaced by vapor. Prior to this time, however, the pressure-radius relation may be expressed as

$$P_V + P_G - P_L = 2\sigma/r \tag{15}$$

Assuming that $P_V = P_\infty$ and substituting for $P_\infty - P_L$ from the Clausius–Clapeyron equation, the superheat is obtained as

$$T - T_s = \frac{T_s}{\rho_v h_{fg}}\left(\frac{2\sigma}{r} - P_G\right) \tag{16}$$

It should be expected then that nucleation from such a cavity will begin at a negligibly small superheat, increasing to some stable value as the noncondensable is swept out by dilution.

Behavior of Vapor-Filled Cavities

We now consider only cavities which have been swept free of entrapped gases as a result of bubble departure, or have been activated as a result of a growing bubble passing over them and displacing the liquid phase. For simplicity the cavities will be considered conical.

Forest [21] has considered the stability of a vapor nucleus in a conical shaped cavity such as illustrated in Figure 6. If for a given contact angle θ, the radius of curvature of the liquid-vapor interface is plotted as a function of axial position x, a cavity "characteristic curve" is obtained as shown in Figure 7. The characteristic curve in Figure 7 is typical of that for a cavity wetted by the liquid (hydrophilic cavity), for which $\theta < 90° + \beta$. A typical characteristic curve for a nonwetting (hydrophobic) cavity, for which $\theta > 90° + \beta$, is shown in Figure 8. For both figures, the portion of the curve from 0 to 1 corresponds to an interface with its 3-phase line entirely within the cavity. The portion of the curve from 1 to 7 corresponds to an interface with its 3-phase line within the corner region, i.e., between x_1 and x_2. Finally, the portion from 7 to 8 corresponds to an interface whose 3-phase line is outside of the cavity. In Figure 7, the corresponding free energy curves are shown as an aid to interpretation of the stability of the vapor volume at various axial positions. The reader is referred to the reference for a detailed discussion. Of importance to this review, the nucleus is shown to grow spontaneously when conditions, as given by

$$R^* = \frac{2\sigma}{\eta P + (C/C_s)P_L - P_L} \tag{17}$$

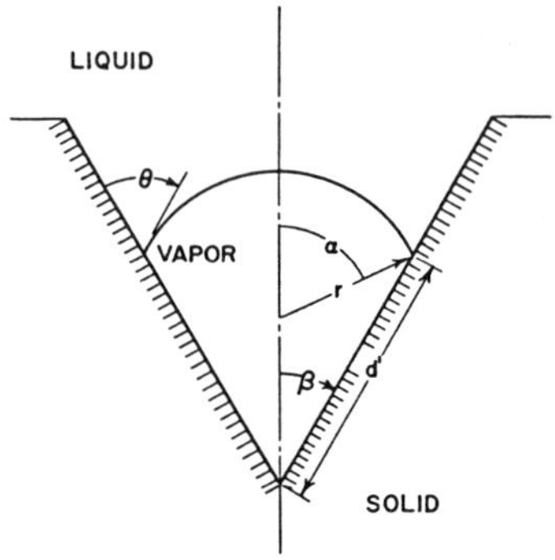

Figure 6. Cross-sectional view of a conical cavity with a vapor nucleus within the cavity [21].

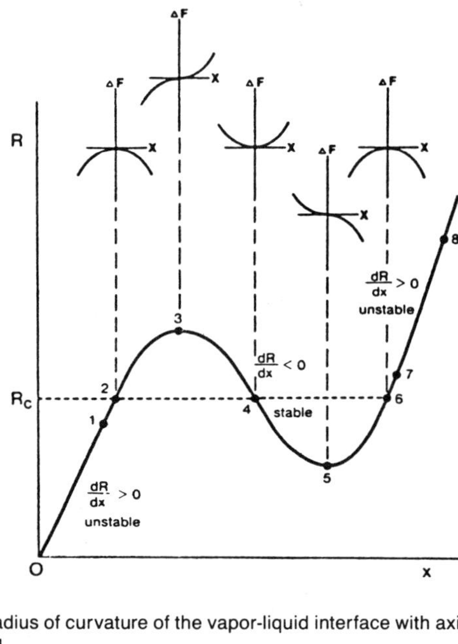

Figure 7. Variation of radius of curvature of the vapor-liquid interface with axial position for a wetted cavity where $\theta < 90° + \beta$ [21].

where

$$\eta = \exp[(v_L(P_L - P_\infty)/(RT)) - c] \tag{18}$$

are such that R*, the critical nucleus radius, equals the radius of curvature at R_5, the local minimum in the cavity characteristic curve (R_5 in Figure 8). Further, the minimum in the characteristic curve occurs at what might be considered the cavity mouth radius, so that it is this latter, measureable quantity which determines the superheat required for the nucleus to emerge from the cavity.

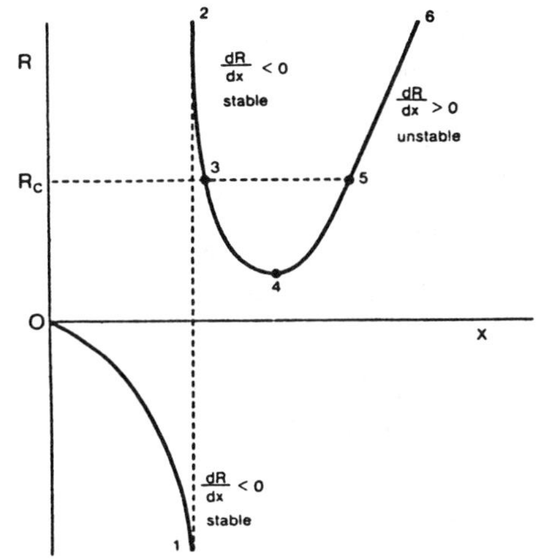

Figure 8. Variation of radius of curvature of the liquid-vapor interface with axial position for a non-wetting cavity where $\theta > 90° + \beta$ [21].

For negligible dissolved gas concentration and $\eta = 1$, Equation 17 may be combined with the Clausius–Clapeyron equation to obtain the equilibrium superheat equation describing spontaneous growth of the nucleus from the cavity.

$$T - T_S = 2\sigma T_S/(\rho_V h_{fg} R_c) \quad r/R_c \geqslant 1 \tag{19}$$

Experimental investigations of the deactivation characteristics of single cavities in uniformly superheated liquids have been conducted by Griffith and Wallis [22], Schultz et al. [23], and Jemison et al. [24]. Between the results of the first two references, a significant number of cavity geometries (conical, cylindrical, and reservoir) and cavity mouth radii (15–150 μm) were studied. The third reference presented preliminary studies with methanol as the boiling liquid. Most organic liquids wet the surface and tend to fill the nucleation cavities. Hence in Figure 9 which is a dimensionless representation of the data from References 23 and 24, the methanol data which lies outside of the ± 20% uncertainty limit was felt to be as result of premature deactivation of the nucleation sites. Further studies in this area using organic liquids and sites of varying depth-to-mouth ratio would be of value.

BUBBLE GROWTH

Once the vapor bubble has emerged from the nucleation cavity, it no longer exists in its reasonably sheltered world where temperature variations are probably not extreme, but instead finds itself in a hostile environment where it is exposed to a varying liquid temperature over most of its surface. The various thermal regions associated with the early stages of bubble growth, along with qualitative temperature profiles and approximate dimensions are illustrated in Figure 10 for a non-fully developed thermal boundary layer [25]. The "relaxation microlayer" is the layer of hot fluid lying directly over the nucleation cavity, which is pushed away by the vapor bubble in the process of emerging from the cavity and growing out into the adjacent liquid. At the moment that the vapor bubble emerges from or "caps" the cavity, the relaxation microlayer has the thickness of the developing thermal boundary layer. As bubble growth proceeds, the relaxation microlayer thins due

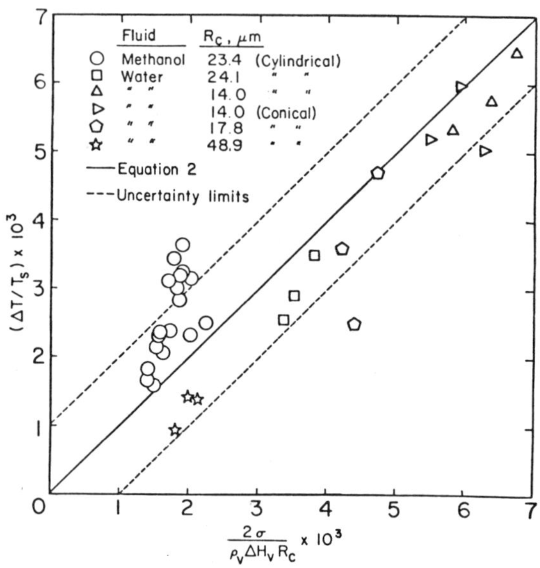

Figure 9. Dimensionless incipient superheat vs. dimensionless cavity mouth radius [24].

to liquid flow and to energy transport to the growing bubble and to the bulk liquid. Qualitative temperature profiles within the relaxation microlayer are shown for both subcooled and superheated bulk liquids. The "evaporation microlayer" is the layer of hot fluid left under a laterally growing vapor bubble. It differs from the relaxation microlayer in that it is in direct contact with the heating surface. As bubble growth proceeds, the evaporation microlayer thins out as a result of evaporation at the liquid-vapor interface, and may progressively disappear, producing an increasingly larger dry spot under the vapor bubble. The existence of the evaporation microlayer is dependent on the shape of the growing bubble; a spherically shaped bubble has no evaporation microlayer. For hemispherically shaped bubbles, the evaporation microlayer, is an important contributor to growth. The "developing thermal boundary layer" hereafter referred to simply as the thermal boundary layer is initially of zero thickness and will develop over a period of time to some steady-state thickness determined by the nature of the convective flow.

Initiation of Bubble Growth

Once the nucleus is exposed to the thermal field outside of the cavity, initiation of normal growth will depend upon the existence or development of a favorable thermal environment. Bergles and Rohsenow [26] expanded upon the initial growth hypothesis developed by Hsu [27] that the environment should be favorable for growth when the liquid temperature at the level of the top of the vapor cap be equivalent to that given by Equation 19. Their graphical approach is presented here because it conveys the physical sense of the model. Further, the result of the Bergles–Rohsenow model, whether it be graphical or analytical, has been very successful in predicting the wall superheat corresponding to the knee of the boiling curve, i.e., where the heat flux density begins to deviate significantly from that predicted by natural or forced-convection correlations.

Multiple Cavities

In a uniformly superheated system, the temperature of the vapor equals the temperature of the liquid equals the temperature of the wall. In a nonuniformly heated system, this is not true. Thus the meaning of T in Equation 19 is not clear in this more common, engineering type of system. Bergles and Rohsenow assume the equilibrium superheat equation to be a necessary condition for initiation of growth.

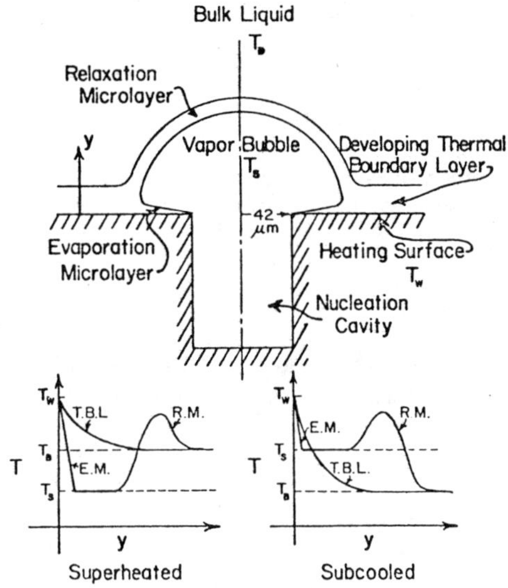

Figure 10. Thermal regions associated with a freshly nucleated bubble in transient boiling [25].

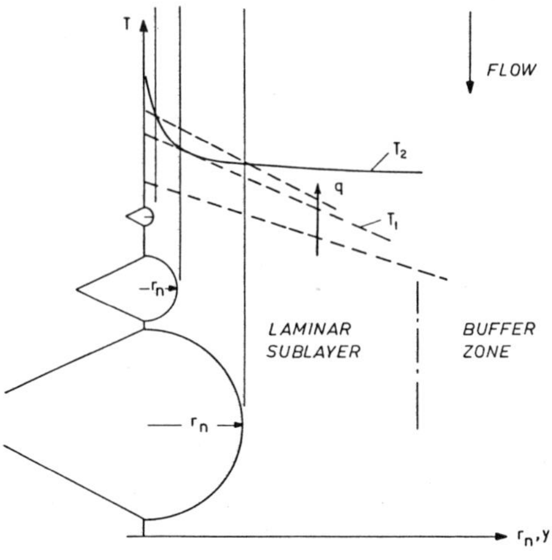

Figure 11. Initiation of bubble growth in forced convection [26].

$$\Delta T_{\infty} = T_V - T_S = \frac{2\sigma T_S}{\rho_V h_{fg} r} \tag{20}$$

As indicated, the system temperature T appearing in Equation 19 has been interpreted as the vapor temperature. Equation 20 also defines the quantity ΔT_{∞} for future use. With the physical properties in Equation 20 evaluated at the saturation temperature (T_S), and with the system pressure given, the uniform vapor temperature within a nucleus can be represented graphically as a function of nucleus radius, as shown in Figure 11. Because the bubble nucleus is generally smaller than the thickness of a fully developed thermal boundary layer, the temperature distribution near the wall can be approximated by the linear expression

$$T_L = T_W - \frac{qy}{k_L} \tag{21}$$

This distribution for three values of q is represented by the dashed lines in Figure 11. It is postulated that the hemispherical nucleus of radius r_n will grow if the liquid temperature at a distance $y = r_n$ from the wall is greater than the vapor temperature given by Equation 20. This criterion requires that all the liquid surrounding the vapor nucleus be able to transfer heat to the nucleus if it is to grow. It thus represents an upper limit for the wall superheat required to initiate growth and although it may overestimate the required superheat for an individual bubble, it may well be in agreement with the superheat required to increase the heat flux density; this being a result of the growth of many bubbles. The graphic procedure proposed by Bergles and Rohsenow for estimating the initiation of growth is illustrated in Figure 11. Once Equation 20 is plotted, the liquid temperature distribution is obtained from the relation

$$q = -k_L \left.\frac{dT}{dy}\right|_{y=0} = h(T_W - T_{L,\infty}) \tag{22}$$

It is presumed that $T_{L,\infty}$ is known and that the heat-transfer coefficient h can be estimated from standard forced-flow nonboiling correlations. Thus, for a given q, Equation 22 relates the wall temperature and the temperature gradient. According to the criterion, steady bubbling will occur at the heat flux density that brings the liquid and vapor temperature curves tangent to each other: i.e.

$$T_L = T_V \text{ at } y = r_n \tag{23}$$

and

$$\frac{dT}{dy} = \frac{dT}{dr} \text{ at } y = r_n \tag{24}$$

It is apparent that once tangency occurs, a slight increase in heat flux density (or wall temperature) results in the activation of a wide size range of additional cavities having radii both smaller and larger than that corresponding to the tangency conditions. Although agreement with experimental data is quite good, two additional conditions must be fulfilled for the method to work. First, a full spectrum of cavity sizes must exist; second, the cavities must contain entrapped gases or vapor. Fortunately, these conditions do exist for many commercially finished surfaces.

An analytical solution to the preceding equations yields for the necessary wall superheat

$$T_W - T_S = \frac{8\sigma T_S q}{k_L \rho_V h_{fg}} \tag{25}$$

When the heat transfer coefficient is low, or when the size range of cavities is limited, the tangency condition may not be satisfied. For this condition the necessary wall superheat is given by

$$T_W - T_S = \frac{qR_{max}}{k_L} + \frac{2\sigma T_S}{\rho_V h_{fg} R_{max}} \tag{26}$$

Single Cavities

Because of the assumption that all of the liquid surrounding the vapor cap be heated to the temperature determined by Equation 20 before initial growth may occur, Equation 26 will most likely overpredict the required superheat for a single bubble cap. The trends predicted by this equation have however been confirmed for a single cavity by Schultz et al. [23]. If a thermal boundary layer thickness is defined as

$$\delta = \frac{q}{k(T_W - T_S)} \tag{27}$$

Equation 26 can be rewritten in terms of δ and ΔT_∞ as

$$\frac{T_W - T_S}{\Delta T_\infty} = \frac{1}{(1 - R_C/\delta)} \tag{28}$$

No doubt because of the restrictive physical assumption, Equation 28 breaks down at R_c/δ equal to unity.

The Madejski model [28, 29] avoids the previous assumption by not requiring the bubble to be spherical and by interpreting the system temperature to be that of the liquid, which is then assumed to vary linearly within the boundary layer of thickness δ. The resulting solution yields the superheat as a continuous function of R_c/δ in terms of elliptic integrals. A graphical representation of the solution is shown in Figure 12 (the solid line) along with experimental initial growth data obtained by Schultz [23, 30]. The arrow in Figure 12 represents the trend of the data with increasing heat flux density, i.e., higher wall superheats and lower thermal-layer thickness at initial growth. The dashed lines represent uncertainty limits from the predicted curve; any data points lying outside the lines are presumed to disagree with the model predictions. Unfortunately, the data exhibit a marked effect of pressure unaccounted for in the model.

Growth Instabilities

Under steady-state conditions initial growth begins when the thermal environment in the surrounding liquid attains some necessary condition such as that proposed in the Bergles–Rohsenow model [26]. Under transient conditions however, it is possible that growth may be initiated prior to the establishment of a fully developed thermal boundary layer. Schultz and Cole [30, 31] have experimentally observed that under such conditions, once the bubble has successfully capped the cavity, it may then either collapse, remain stationary for a period of time prior to a slow but increasing rate of growth, or immediately begin its growth period. Further, depending upon the rate of growth relative to the state of development of the relaxation and evaporation microlayers, the bubbles may also exhibit subsequent instabilities characterized by dramatic changes in their rate of growth. An interesting example of the appearance, collapse, and subsequent growth of a vapor bubble is shown in Figure 13. The ordinate represents the radius in millimeters of the vapor cap (or bubble depending on the stage of growth) as determined from high speed (16,000 frames

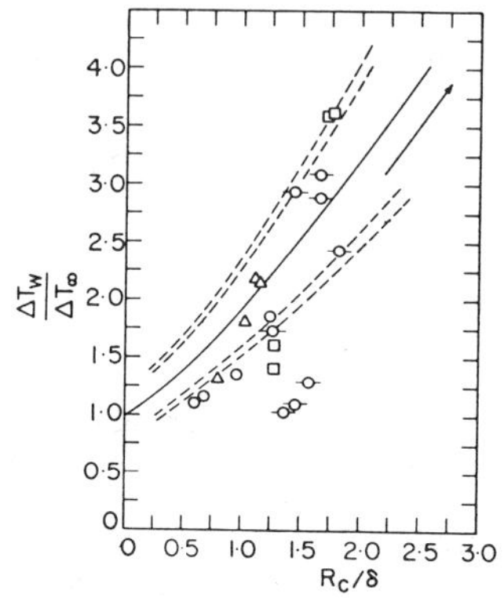

Figure 12. Comparison of experimental incipience data from single cavities with predictions of the Madejski model [1, 23, 30].

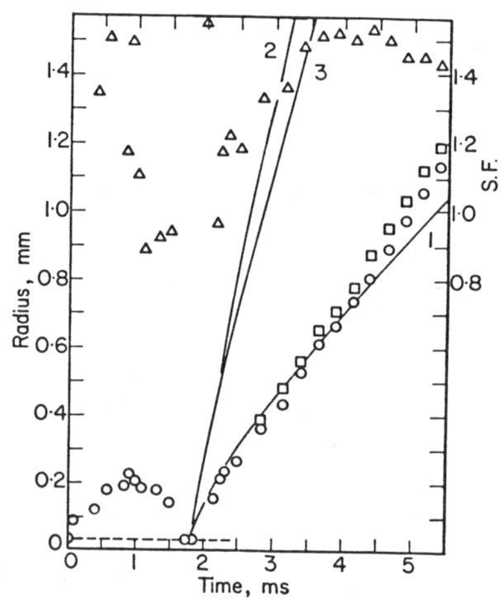

Figure 13. Collapse instability in initial bubble growth [30, 31]; P_L = 8.2 kPa, $T_B - T_S$ = 1.4 °C, t_w = 14.7 ms, $T_W - T_S$ = 9.0 °C, q = 111.9 kW/m^2, δ = 43.8 μm, R_C = 42 μm, N = 20 ms; (1) Mikic et al. [32] uniform superheat; (2) Mikic et al. [32] nonuniform superheat; (3) Cooper-Vijuk [39] microlayer evaporation.

per second) close-up motion picture photography. The abscissa is the time scale in milliseconds as determined from timing marks on the film. Information on the system pressure (P_L), bulk superheat ($T - T_S$), wall superheat ($T_W - T_S$), heat flux density (q), and developing thermal boundary layer thickness (δ), all at the moment at which the bubble first emerges from the single cylindrical nucleation site, are given in the figure caption along with the radius of the cavity. The horizontal dashed line represents the equivalent cavity radius. Analysis of the phenomenon utilized the nucleation criteria of Equation 20, the initial growth criteria of Madejski [28] as discussed in the previous section and the bubble radius which would be required to exist in metastable equilibrium at the uniform temperature of the bulk liquid as given by

$$R_{crit} = \frac{2\sigma}{P(T_B) - P(T_S)} \tag{29}$$

The bubble was able to emerge from and cap the cavity because the measured wall superheat of 9° C was greater than the superheat of 6.6° C required by Equation 20. After capping the cavity, the vapor cap began to grow using the available superheat stored in the existing, as yet undeveloped, thermal boundary layer of thickness 43.8 μm. That there was insufficient energy stored in the thermal layer at the moment of initial growth is indicated by the prediction of the Madejski analysis that a superheat of 12.5° C is required for normal, steady growth. The critical radius according to Equation 29, i.e., the radius below which the bubble will collapse and above which it will grow, is 0.22 mm. This appears to be just about the maximum in the growth curve. Thus the energy available in the underdeveloped thermal layer was insufficient to allow the vapor cap to achieve the critical radius and a subsequent surface tension induced collapse occurred. If the thermal layer had contained sufficient energy to allow the cap to achieve the critical radius, then growth would have continued. Instead, the bubble collapsed back to the cavity mouth, at which time the boundary layer had apparently become sufficiently developed to allow the process to begin again and this time achieve successful growth.

It is interesting to note that the growth rate is correctly predicted only when the bulk superheat of 1.4° C is used in the growth rate equation. The one used here (solid line 1) is that of Mikic et al. [32] which is a simple general relation for bubble growth in a uniformly superheated liquid and which is applicable in both the hydrodynamic (initial stage of growth) and diffusion (final stage of growth) controlled regions. The implication is that the energy available in the very thin relaxation microlayer (the thermal layer pushed away from the wall by the growing bubble) is depleted within the first fraction of a millisecond and the growth is subsequently controlled by the bulk superheat. Apparently the only effect of the thermal boundary layer is to influence the conditions necessary for the initiation of growth once the cavity has been capped.

A continuous growth curve starting from an underdeveloped thermal boundary layer is shown in Figure 14. If it is assumed that the thermal boundary layer of thickness 32.7 μm (at the moment of nucleation) is uniformly superheated to an average value $(T_W - T_S)/2$, the enthalpy contained in a cylindrical volume of radius 42 μm (equal to the mouth radius of the artificial nucleation site) is 1.7 μJ. This enthalpy, if used to transform saturated liquid to vapor, would yield a spherical volume of radius 0.11 mm. Since, by the end of two milliseconds the bubble radius was 0.2 mm, it seems reasonable to assume that the energy

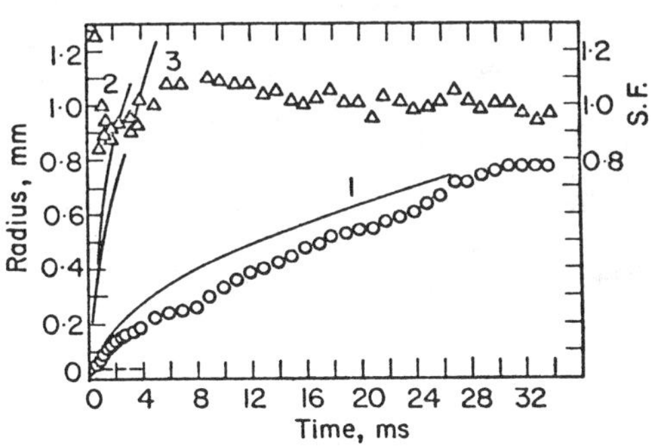

Figure 14. Continuous growth curve from an underdeveloped thermal boundary layer [25, 31]; P_L = 22.8 kPa, $T_B - T_S$ = 0.6 °C, t_w = 8.1 ms, $T_W - T_S$ = 4.1 °C, q = 71.1 kW/m², δ = 32.7 μm, R_C = 42 μm, N = 48 ms. (1) Mikic et al. [32] uniform superheat; (2) Mikic et al. [32] nonuniform superheat; (3) Cooper-Vijuk [39] microlayer evaporation.

in the displaced boundary layer (relaxation microlayer) had already been depleted and that most of the energy for further continuous growth would be supplied by the superheated bulk liquid. Microlayer vaporization was not a factor because the bubble was spherically shaped. Again, the resulting initial growth is well described by the Mikic et al. [32] model. The agreement between theory and experiment is especially notable because there are no adjustable constants in this equation.

A final interesting example of growth instabilities is shown in Figure 15. The first experimental zero growth plateau occurs near 0.2 mm and is apparently caused by an insufficiently developed thermal layer. Equation 20 predicts a minimum superheat of 4.8° C to cap the cavity. The experimental wall superheat was 6.2° C so this condition is satisfied. From Figure 12, for $R_c/\delta = 0.8$, the wall superheat required for bubble growth is 8.2° C. Again the thermal layer is apparently not sufficiently developed to allow the bubble to grow. From Equation 29 the critical radius corresponding to the bulk superheat of 1.4 °C is 0.19 mm. Judging from the observed fact that the bubble did not collapse, it must have just achieved the critical radius and paused. Observation of the high speed motion picture film shows that the bubble tends to become spherical as it approaches the first plateau and then flattens out during the period of zero growth. The impetus for further growth could then come either from a more fully developed thermal layer, or from microlayer evaporation. The appearance of a second plateau was unexpected. From the film, it was observed that the bubble was spherical during this entire zero growth period and in fact had appeared to be preparing for departure from the surface when it paused in its motion. Because of the spherical shape, microlayer effects cannot be responsible for its renewed growth, nor does it seem likely that the thermal boundary layer could be responsible. Thus it appears that the second plateau must be a result of hydrodynamic rather than thermal phenomena; possibly a perturbation in the system pressure.

Bubble Growth Rates in Nucleate Boiling

Experiment

In an extensive experimental study, Cole and Shulman [33] investigated bubble growth rates for a variety of fluids, at pressures ranging from 6.6 kPa (50 mm Hg) to 101 kPa (760 mm Hg) and Jakob numbers ranging from 24 to 792. At the lower pressures (higher Jakob numbers), discrepancies between experiment and existing theory (diffusion limited growth) were substantial, and it was suggested that the theory be reexamined to determine the relative importance of dynamic effects.

Stewart and Cole [34] continued the investigation, using a modified experimental system, and concentrated their efforts on studying the initial growth rate of vapor bubbles at a pressure of 4.9 kPa (37 mm Hg) and a nearly constant Jakob number of 1,030. A comparison of the experimental results with existing theoretical models confirmed the earlier findings. The data showed substantially lower initial bubble growth as compared to the predictions of the diffusion-controlled models, regardless of whether they were of the uniform superheat [35], nonuniform superheat [36, 37], or microlayer type [38]. Comparison with two models that included the effect of liquid inertia [32, 39] showed much closer agreement. As a consequence, the reduction in initial growth rate was attributed to the effect of liquid inertia.

Van Stralen et al. [40] developed a much larger and more sophisticated version of the boiling system used by Stewart and Cole, and obtained growth rate data (up to departure) in water boiling at pressures ranging from 2 kPa (15 mm Hg) to 26.7 kPa (201 mm Hg) and corresponding Jakob numbers ranging from 108 to 2,689. A feature of this investigation was the simultaneous use of two high-speed cameras; a Hitachi rotating-prism camera operated at 4,000 frames per second (showing a bubble during the entire lifetime), and a rotating-mirror Dynafax drum camera operated at 25,000 frames per second (showing

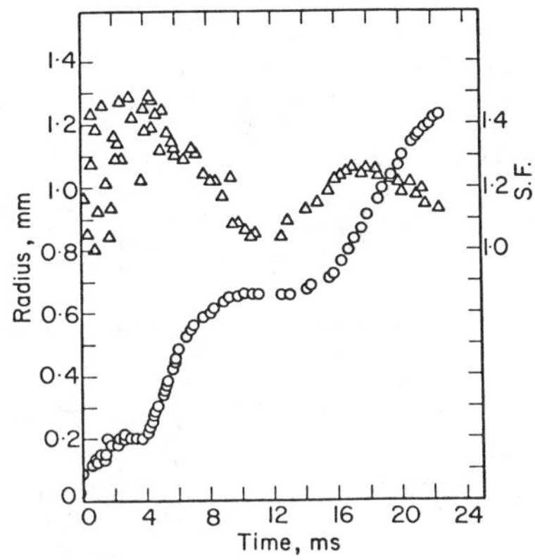

Figure 15. Plateau instabilities in initial bubble growth [31]; P_L = 12.3 kPa, $T_B - T_S$ = 1.4 °C, t_w = 20 ms, $T_W - T_S$ = 6.2 °C, q = 60.6 kW/m^2, δ = 52.1 μm, R_C = 42 μm, N = 47 ms.

initial bubble growth more accurately). The ignition of the flashlamp used in combination with the faster camera, was triggered to the instant of formation of an equilibrium bubble, by interruption of a laser beam projected across the artificial nucleation site. Experimental bubble growth curves R (t) are shown up to departure in Figures 16 through 18 for water boiling at pressures from 4 to 27 kPa. Detailed information on the values of the influential parameters is given in legends on the figures. The simultaneous bubble profiles (as a function of time) are drawn at the top of the figures. The most striking feature of the growth curves is that both the departure time (i.e., the time spent by the bubble on the surface) and the departure radius increase substantially with decreasing pressure. Also, in general, the slope of a growth curve (i.e., the bubble growth rate) at constant pressure decreases gradually with increasing time up to departure.

Experimental bubble growth curves for the aqueous binary system water-2-butanone [41] are shown up to departure in Figures 19 through 21 for 4, 10, and 15 wt% 2-butanone. Again values of the influential parameters are given below the figures and the time dependent, simultaneous bubble profiles are drawn at the top of the figures.

In general, directly after departure of a large bubble, an initially very rapidly growing thin vapor column occurs at the nucleation site as shown in Figures 22 through 25. This secondary bubble (vapor column) penetrates into the flattened lower boundary of the preceding large bubble. Apparently this curious event (which was observed only at the lower pressures) is a consequence of two originally different phenomena; the formation of a liquid jet in the wake of the departing large primary bubble and the existence of a dry area beneath the center of the growing primary bubble during adherence to the wall. The former is a phenomena that may occur at all pressures and is similar to cavitation where jet velocities up to 100 m/s have been observed in translating bubbles collapsing due to a pressure step [42]. The latter apparently occurs only at lower pressures or high heat flux densities where dry areas are most likely to form. The resulting pressure reduction in the wake of departing bubbles increases the superheating at the dry area substantially, and causes a very rapid growth of the vapor column. Due to its excess pressure, the secondary bubble follows the liquid jet immediately.

Figures 24 and 25 illustrate the phenomena at low subatmospheric pressures. Figure 26 shows a high-velocity liquid jet (in the absence of a secondary vapor column) penetrating through a large bubble. Jet velocities of approximately 5 m/s are estimated from the high-speed motion pictures.

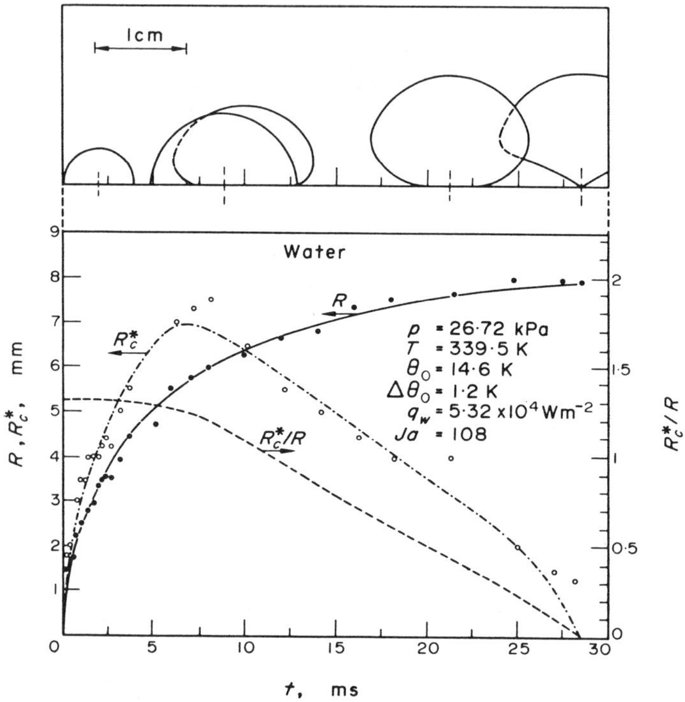

Figure 16. Water boiling at 26.72 kPa. Equivalent bubble radius R, contact radius R_C* and their ratio as a function of time. Corresponding bubble profile and limit of mode of hemispherical growth are shown in the upper drawing [40].

Theory

As a consequence of the inability of existing theoretical models of bubble growth to agree with the experimental results of Cole and Shulman [33] and Stewart and Cole [34], Van Stralen et al. [43] developed a comprehensive model for bubble growth in pure and binary systems, accounting for the presence of the surface, the nonuniformity of the temperature field, both inertial and diffusion controlled growth, and the influence of the bulk superheat (or subcooling). The model will be described in this section and compared with the results of Stewart and Cole [34] and the results of Van Stralen et al. [40, 41] presented in the previous section.

The model combines the Rayleigh solution for the hydrodynamically controlled initial phase of growth in an "in-parallel" fashion (as first suggested by Cooper and Vijuk [39]) with a diffusion-type solution which is dominant only at advanced times.

$$R = \frac{R_1 R_2}{R_1 + R_2} \tag{30}$$

Here, R_1 is the Rayleigh solution for bubble growth, given by

$$R_1 = \left\{\frac{2\,(P_{L,R} - P_{L,\infty})}{3\rho_L}\right\}^{1/2} t = \left\{\frac{2\rho_V h_{fg}\,(T_W - T_S)\,e^{-(t/td)^{1/2}}}{3\rho_L T_S}\right\}^{1/2} t \tag{31}$$

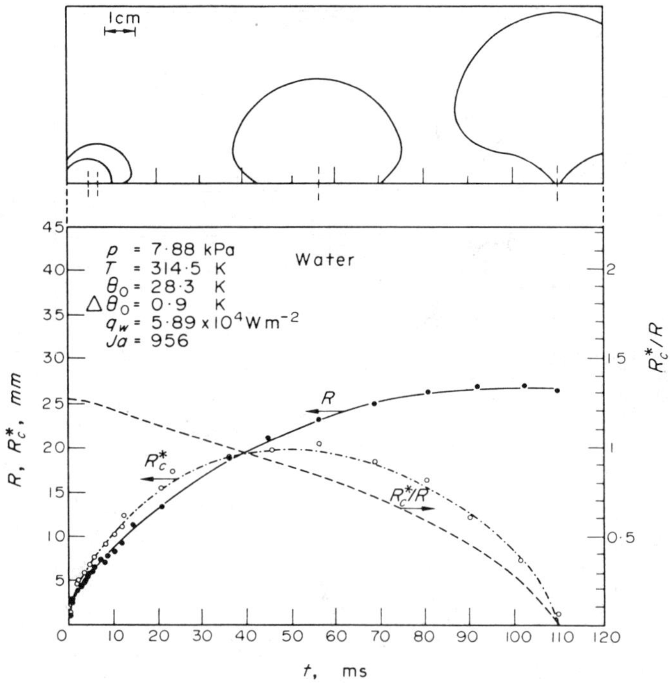

Figure 17. Water boiling at 7.88 kPa. Equivalent bubble radius R, contact radius $R_C{}^*$ and their ratio as a function of time. Corresponding bubble profile and limit of mode of hemispherical growth are shown in the upper drawing [40].

and R_2 is the diffusion solution; accounting for the relaxation microlayer (around the bubble dome), the evaporation microlayer (beneath the bubble), and the bulk liquid superheat.

$$R_2 = \left(\frac{12}{\pi}\right)^{1/2} \frac{C_{l,m}}{C_{l,p}} \left\{ be^{-(t/td)^{1/2}} + \frac{T_L - T_S}{T_W - T_S} \right\} N_{JaW} (\alpha_L t)^{1/2}$$

$$+ 0.3730 \frac{C_{l,m}}{C_{l,p}} (N_{PrL})^{1/6} e^{-(t/t0)^{1/2}} N_{Jaw} (\alpha_L t)^{1/2} \tag{32}$$

and

$$\frac{C_{l,m}}{C_{l,P}} = \frac{1}{1 + \dfrac{c}{h_{fg}} \left(\dfrac{\alpha}{D}\right)^{1/2} \dfrac{\Delta T}{G}} \tag{33}$$

In application the parameter "b" is determined from experimental data at departure conditions.

Thus from Equation 32

$$b = 1.3908 \frac{R_2(t_d) C_{l,p}}{N_{Jaw} (\alpha_L t_d)^{1/2} C_{l,m}} - 0.1908 \frac{C_{l,m}}{C_{l,P}} N_{Pr}^{-1/6} - 2.718 \frac{T_L - T_S}{T_W - T_S} \tag{34}$$

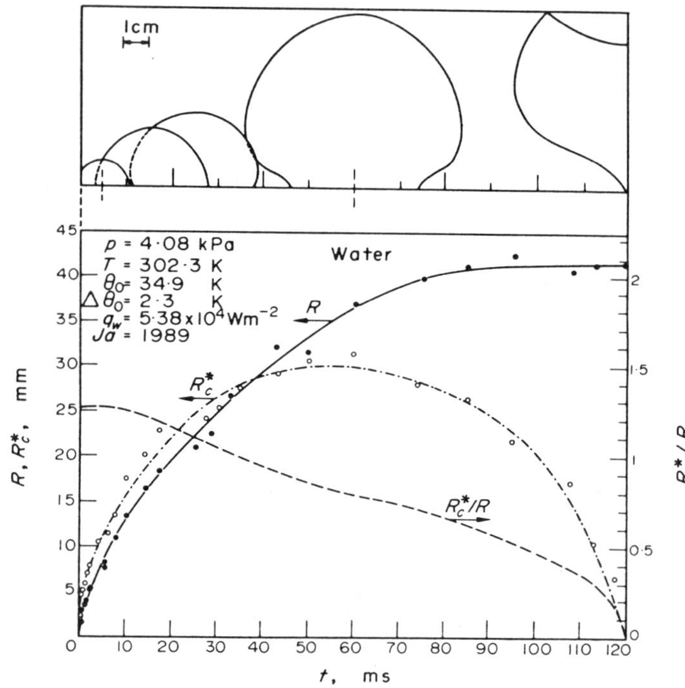

Figure 18. Water boiling at 4.08 kPa. Equivalent bubble radius R, contact radius $R_C{}^*$ and their ratio as a function of time. Corresponding bubble profile and limit of mode of hemispherical growth are shown in the upper drawing [40].

where $R_2(t_d)$ is calculated from Equation 30 using the experimentally determined departure radius $R(t_d) = R_d$, i. e.

$$R_2(t_d) = \frac{R(t_d)\,R_1(t_d)}{R_1(t_d) - R(t_d)} \tag{35}$$

and $R_1(t_d)$ is determined directly from Equation 31 using the experimentally determined departure time. For multicomponent systems, the Jakob number

$$N_{Jaw} = \frac{\rho_L c(T_W - T_S)}{\rho_v h_{fg}} \tag{36}$$

is evaluated in terms of the pure, less volatile component. For pure liquids, $C_{1,m}/C_{1,p} = 1$.

The bubble growth equation is compared with the experimental data of Stewart and Cole [34] in Figures 27 and 28, with the water data of Van Stralen et al. [40] in Figures 29 through 31, and with the aqueous binary data of Van Stralen et al. [41] in Figures 32 through 34. Additional comparisons are shown in Figures 35 through 37 for the slightly subcooled data of Schultz and Cole [25]. For all of the pure water results, comparisons are also made with other models. It is beyond the scope of this review to describe the details of the comparisons. The reader is referred to the individual papers for that purpose. It seems sufficient to state here that the agreement between the Van Stralen et al. model and experiment is excellent.

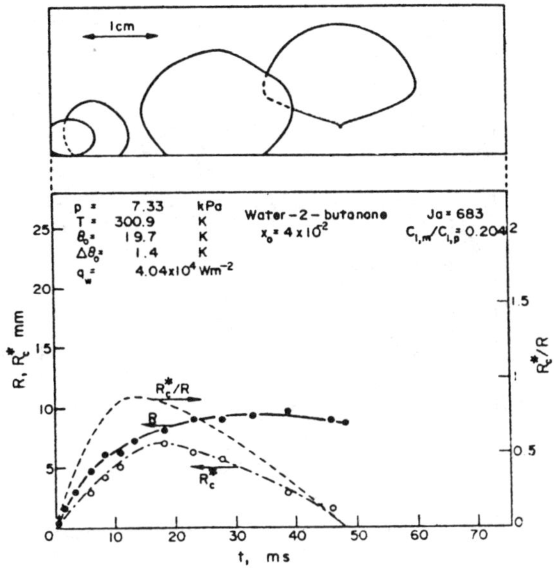

Figure 19. Water-2-butanone (4 wt% 2-butanone) boiling at 7.33 kPa. Equivalent bubble radius R, contact radius $R_C{}^*$ and their ratio as a function of time. Corresponding bubble profile is shown in the upper drawing. Initial mode of growth is ellipsoidal [41].

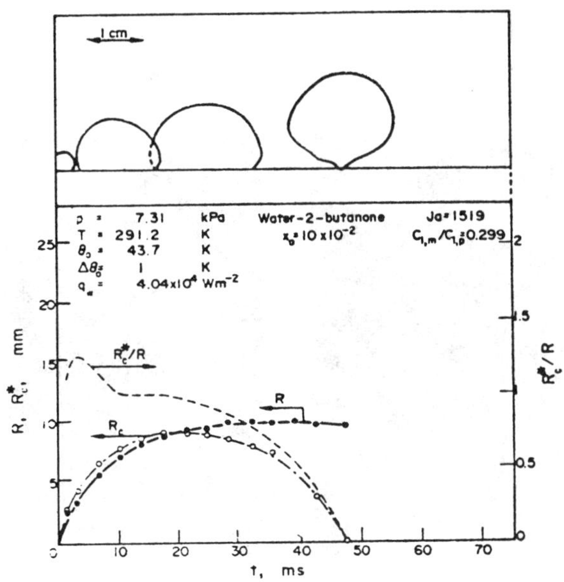

Figure 20. Water-2-butanone (10 wt% 2-butanone) boiling at 7.31 kPa. Equivalent bubble radius R, contact radius $R_C{}^*$ and their ratio as a function of time. Corresponding bubble profile is shown in the upper drawing. Initial mode of growth is ellipsoidal [41].

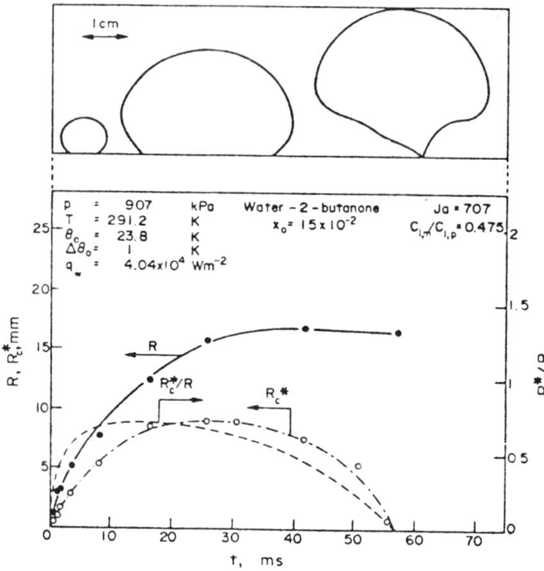

Figure 21. Water-2-butanone (15 wt% 2-butanone) boiling at 9.07 kPa. Equivalent bubble radius R, contact radius R_C^* and their ratio as a function of time. Corresponding bubble profile is shown in the upper drawing. Initial mode of growth is ellipsoidal [41].

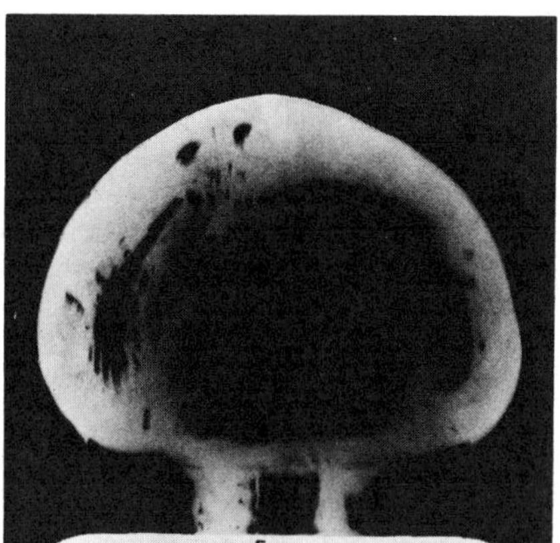

Figure 22. Water boiling at 2.04 kPa. A large "Rayleigh" vapor bubble shortly before departure [40].

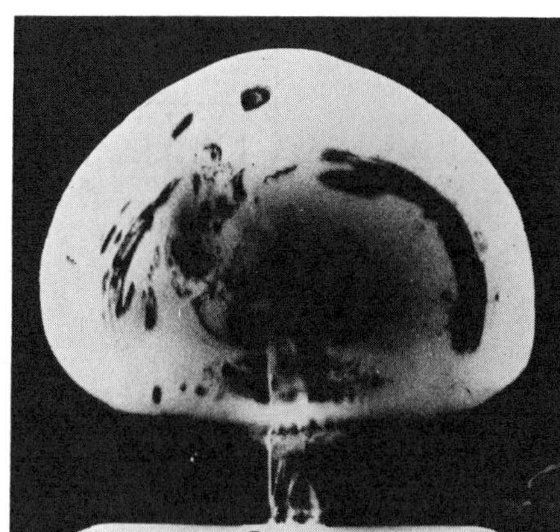

Figure 23. Water boiling at 2.04 kPa. A thin secondary vapor column succeeds a large "Rayleigh" vapor bubble [40].

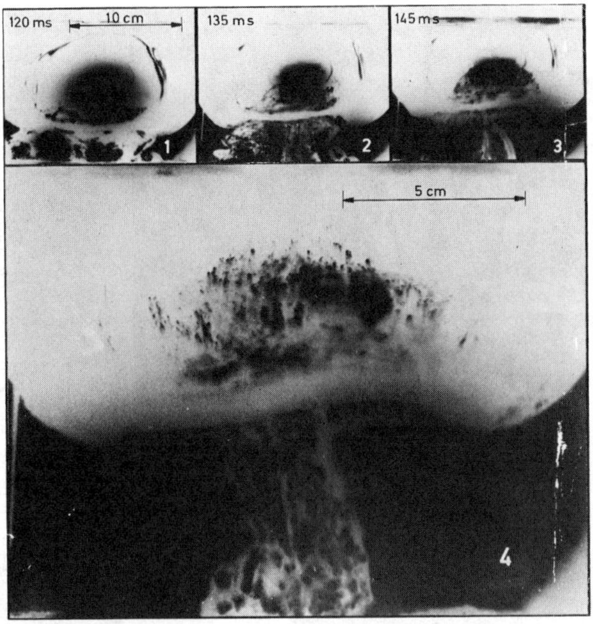

Figure 24. Water boiling at 4.08 kPa. A high velocity liquid jet followed by a hot vapor column is penetrating into a large vapor bubble. The lower side of the bubble shows an involution [40].

Subsequent studies by Van Stralen and his colleagues [44–47] have utilized numerical solutions of the equations of motion to approximate the flow and temperature fields around a growing bubble, to consider deviations from the spherical bubble shape (due to gravity), and to study the oscillations in the equivalent bubble radius observed experimentally.

BUBBLE DEPARTURE

The energy transported away from a heated surface by vapor bubbles may be expressed as

$$q_{LH} = \rho_V h_{fg} \sum_{s=1}^{N} f_s V_{ds}/A \tag{37}$$

where N refers to the number of simultaneously active sites, f_s is the frequency of bubble formation at a specific site and in terms of the individual bubbles formed at that site is given by

$$f_S = \frac{1}{(\tau_d + \tau_g)_s} = \frac{n}{\sum_{i=1}^{n} (\tau_d + \tau_g)_i} = \frac{n}{\sum_{i=1}^{n} 1/f_i} \tag{38}$$

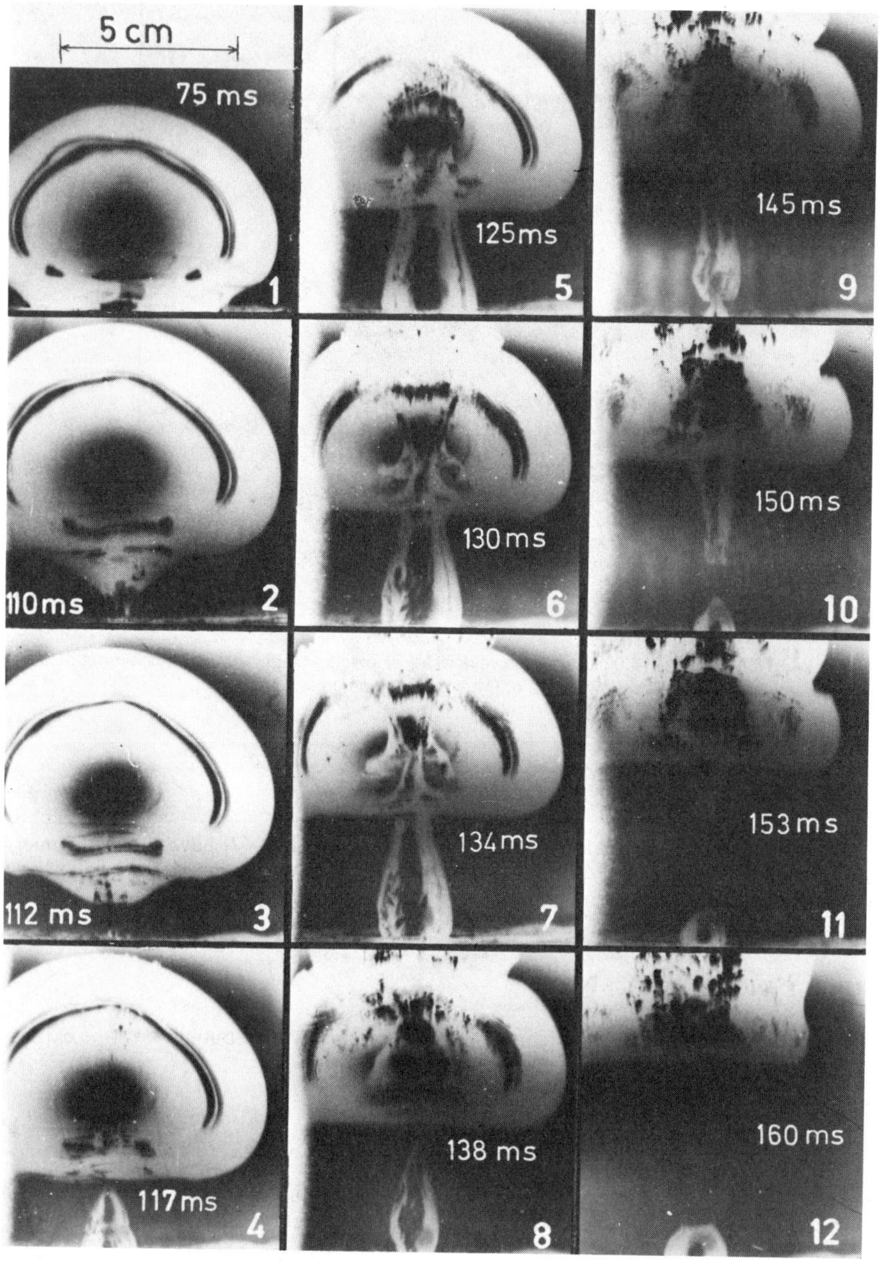

Figure 25. Water boiling at 7.88 kPa. A diverging high-velocity liquid jet followed by a hot vapor column is penetrating through a large vapor bubble [40].

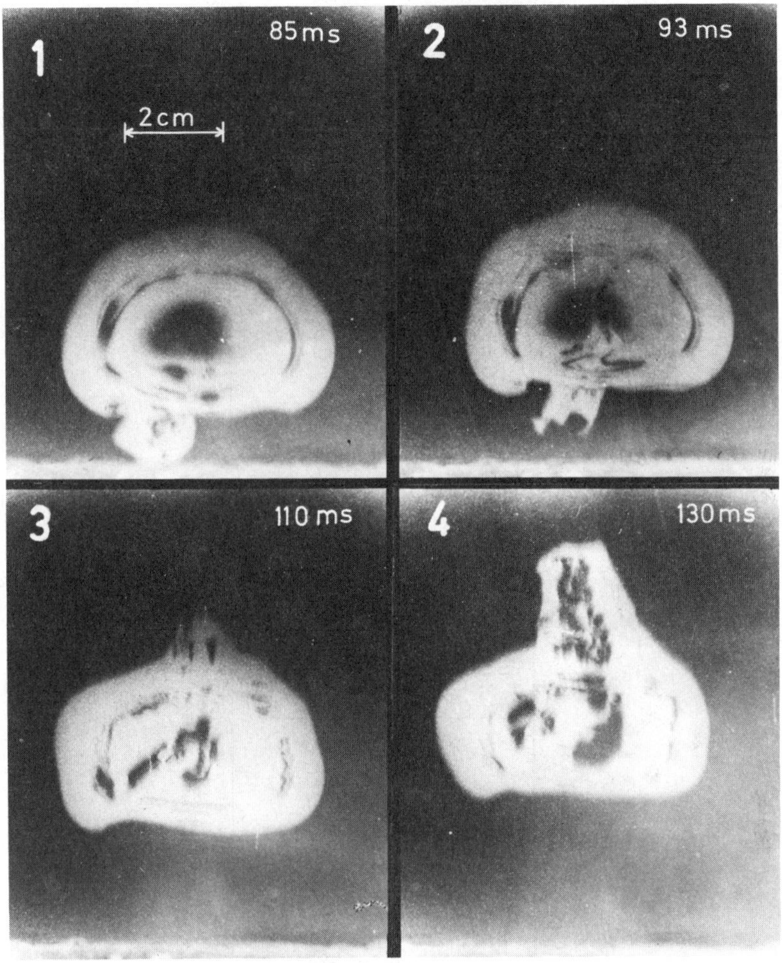

Figure 26. Water boiling at 13.21 kPa. A high velocity liquid jet (without secondary vapor column) is penetrating into a relatively large vapor bubble [40].

where n refers to the number of bubbles formed at a specific site. In similar fashion V_{ds}, the bubble volume at departure at a specific site, is given by

$$V_{ds} = \frac{\sum_{i=1}^{n} V_{di}}{n} \tag{39}$$

The departure diameter is often defined as that diameter corresponding to a sphere having the same volume as the bubble. It is apparent from Equation 37 that the important physical

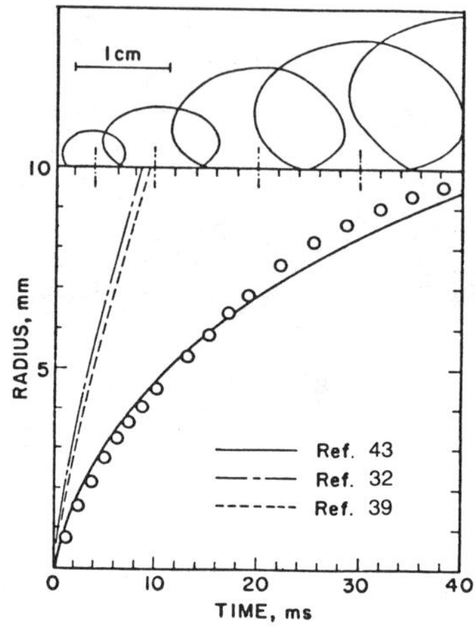

Figure 27. Water boiling at 4.9 kPa. Bubble growth up to departure in comparison with theoretical predictions [34]; $T_W - T_S = 20.8$ °C, $T_W - T_B = 19.7$ °C, $N_{JaW} = 1024$, $q = 171$ kW/m^2, $t_w = 78$ ms, $t_d = 58$ ms, $R_d = 11$ mm.

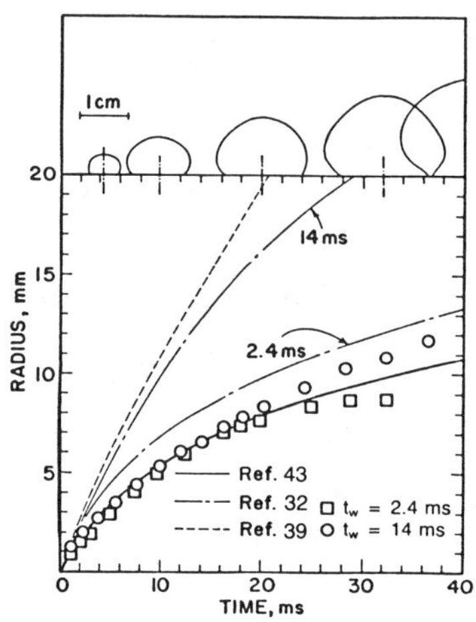

Figure 28. Water boiling at 4.9 kPa. Bubble growth up to departure in comparison with theoretical predictions [34]; $T_W - T_S = 21.5$ °C, $T_W - T_B = 20.5$ °C, $N_{JaW} = 1062$, $q = 152$ kW/m^2, $t_w = 2.4$, 14 ms, $t_d = 63$ ms, $R_d = 13$ mm.

parameters are the frequency of bubble formation and departure volume (or diameter) at given sites and the number of simultaneously active sites. Unfortunately these quantities are not independent of each other, both the bubble frequency and departure diameter being dependent to some extent on the bubble population density [48, 49]. Sultan and Judd [48], for example, in their study on the interaction of adjacent nucleating sites conclude that it will be necessary to think of area-averaged bubble emission frequencies in formulating boiling heat transfer models.

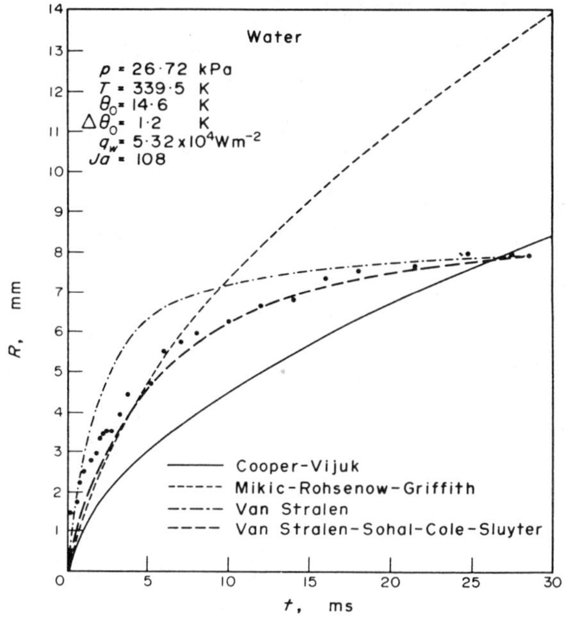

Figure 29. Water boiling at 26.72 kPa. Experimental bubble growth data up to departure in comparison with theoretical predictions [40], cf. Figure 16.

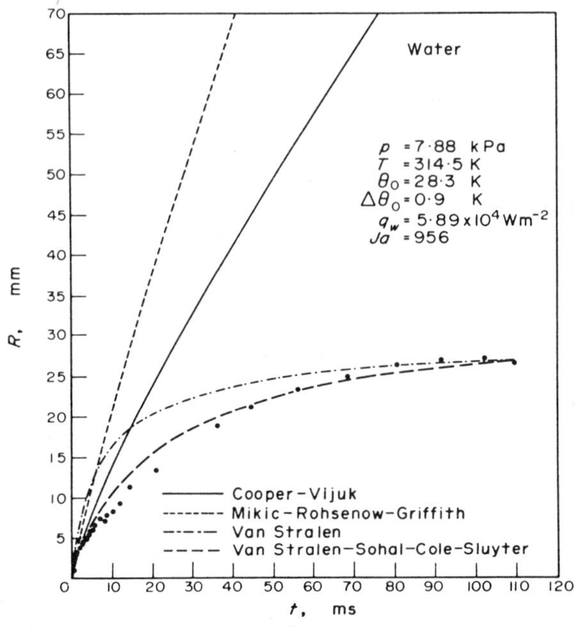

Figure 30. Water boiling at 7.88 kPa. Experimental bubble growth data up to departure in comparison with theoretical predictions [40], cf. Figure 17.

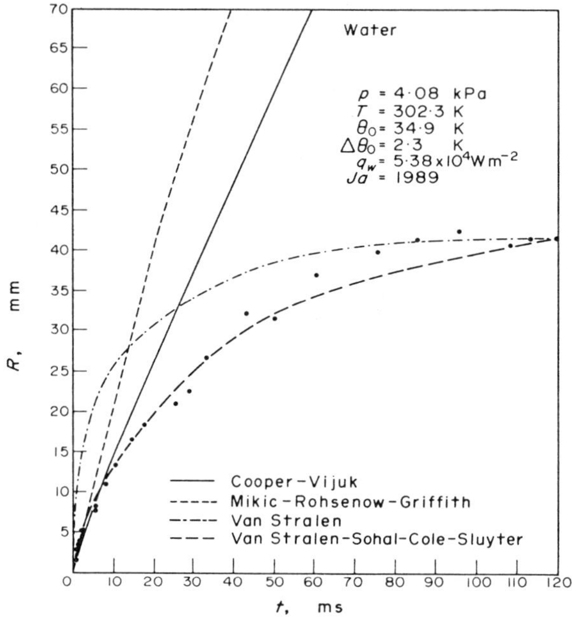

Figure 31. Water boiling at 4.08 kPa. Experimental bubble growth data up to departure in comparison with theoretical predictions [40], cf. Figure 18.

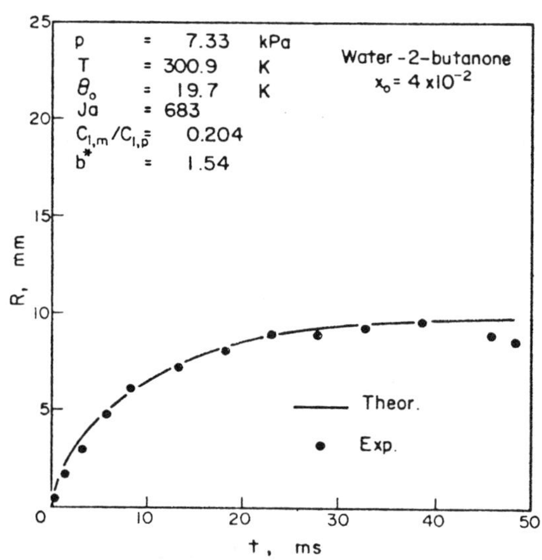

Figure 32. Water-2-butanone (4 wt% 2-butanone) boiling at 7.33 kPa. Experimental bubble growth data up to departure in comparison with the theory of Van Stralen et al. [43, 41], cf. Figure 19.

Bubble Departure Diameters

Cole and Rohsenow [49] developed a correlation for predicting bubble departure diameters which included data for such fluids as acetone, benzene, carbon tetrachloride, methanol, n- and isopentane, toluene, propane, n-butyl alcohol, Freon 12, nitrogen, and water, for pressures ranging from 0.066 to 140 atmospheres. The correlation is extremely

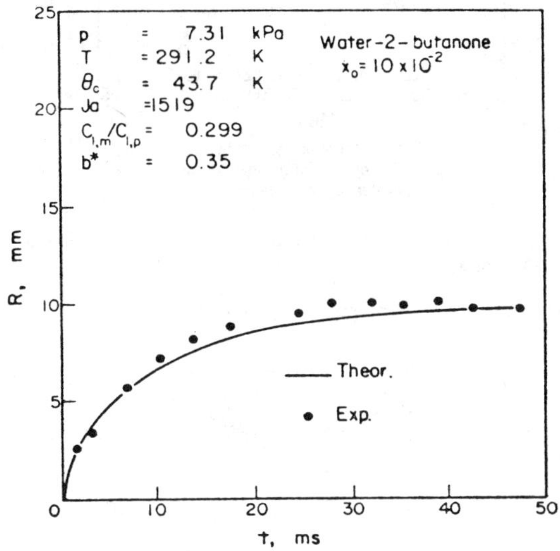

Figure 33. Water-2-butanone (10 wt% 2-butanone) boiling at 7.31 kPa. Experimental bubble growth data up to departure in comparison with the theory of Van Stralen et al. [43, 41], cf. Figure 20.

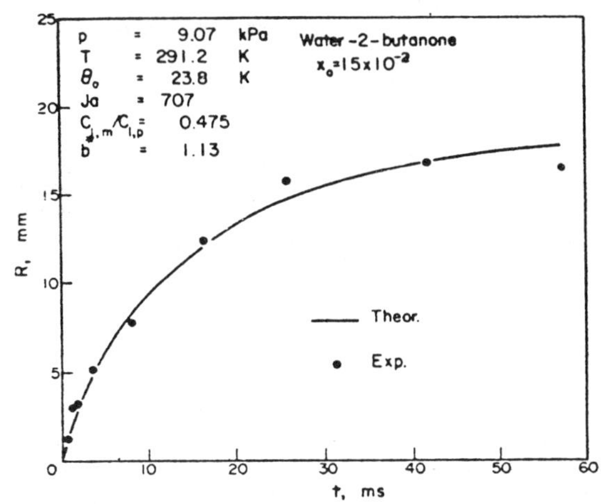

Figure 34. Water-2-butanone (15 wt% 2-butanone) boiling at 9.07 kPa. Experimental bubble growth data up to departure in comparison with the theory of Van Stralen et al. [43, 41], cf. Figure 21.

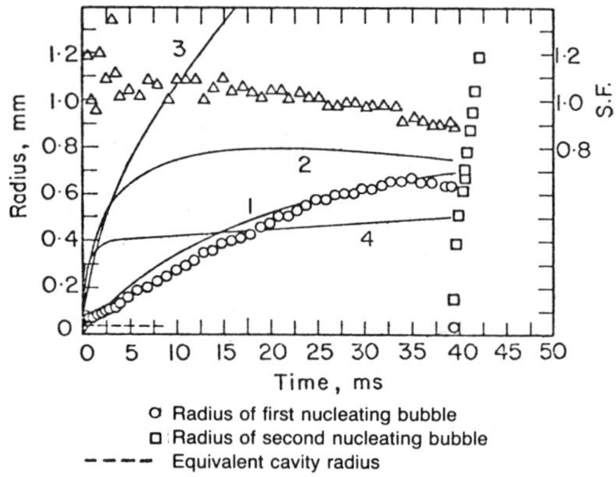

Figure 35. Water boiling at 39.7 kPa. Experimental bubble growth data up to departure (and growth of a second nucleating bubble) in comparison with theoretical predictions [25]; $T_B - T_S = -.5$ °C, $T_W - T_S = 3.1$ °C, $t_w = 8$ ms, $q = 73.1$ kW/m², $\delta = 33$ μm, $R_C = 42$ μm, N = 50 ms. (1) Van Stralen et al. [43]; (2) Mikic et al. [32]; (3) Cooper-Vijuk [39]; (4) Hsu-Graham [37].

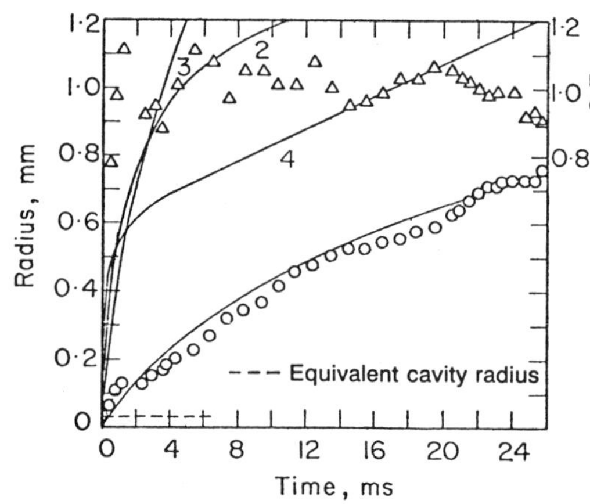

Figure 36. Water boiling at 39.7 kPa. Experimental bubble growth data up to departure in comparison with theoretical predictions [25]; $T_B - T_S = -.5$ °C, $T_W - T_S = 5.2$ °C, $t_w = 6.3$ ms, $q = 132$ kW/m², $\delta = 29.1$ μm, $R_C = 42$ μm, N = 50 ms. (1) Van Stralen et al. [43]; (2) Mikic et al. [32]; (3) Cooper-Vijuk [39]; (4) Hsu-Graham [37].

simple to use because all physical properties are evaluated at the saturation temperature and because the wall superheat is not required. Thus knowing the system pressure and the appropriate physical properties at the saturation temperature corresponding to that pressure, the average departure diameter can be calculated. The penalty to be paid is that two correlations are necessary as shown in Figures 38 and 39; one for water and one for all other fluids. Additionally, it is to be noted that there exists a minimum value of the modified Jakob number

$$N_{Ja}^* = \rho_L c T_S / \rho_V h_{fg} \tag{40}$$

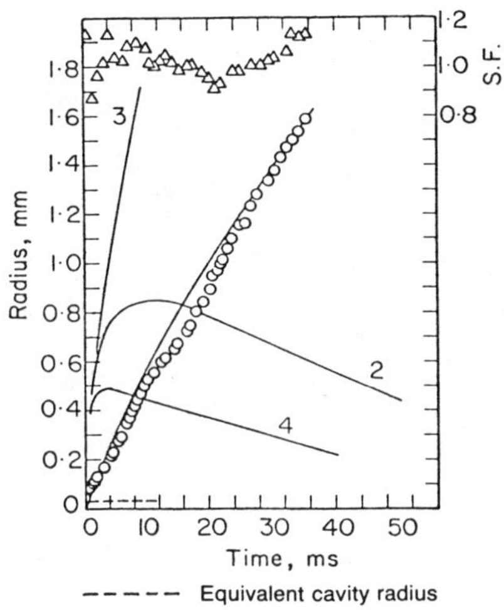

Figure 37. Water boiling at 39.7 kPa. Experimental bubble growth data up to departure in comparison with theoretical predictions [25]; $T_B - T_S = -1.1$ °C, $T_W - T_S = 5.2$ °C, $t_w = 4.4$ ms, $q = 166.5$ kW/m^2, $\delta = 25.2\ \mu$m, $R_C = 42\ \mu$m, N = 21 ms. (1) Van Stralen et al. [40]; (2) Mikic et al. [32]; (3) Cooper-Vijuk [39]; (4) Hsu-Graham [37].

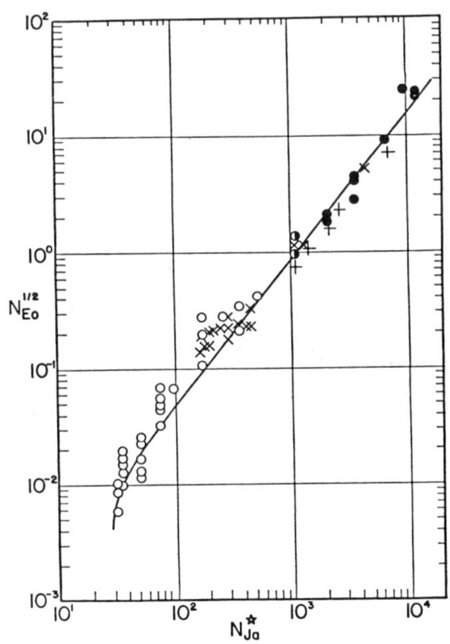

Figure 38. Cole-Rohsenow departure diameter correlation for water [49].

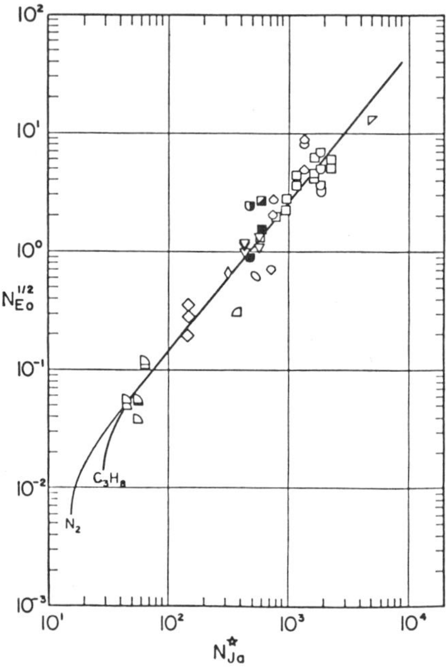

Figure 39. Cole-Rohsenow departure diameter correlation for liquid other than water [49].

for each fluid. This is a consequence of the existence of a minimum in the variation of the modified Jakob number with pressure as shown in Reference 49. The asymptotic expression which appears to best represent the water data is:

$$(N_{Eo})^{1/2} = (1.5 \times 10^{-4})(N_{Ja}^*)^{5/4} \quad P_r < 0.2 \tag{41}$$

where

$$N_{Eo} = g(\rho_L - \rho_V) D_d^2/\sigma \tag{42}$$

For all other fluids, the best asymptotic expression appears to be:

$$(N_{Eo})^{1/2} = (4.65 \times 10^{-4})(N_{Ja}^*)^{5/4} \quad P_r < 0.2 \tag{43}$$

Zijl et al. [47] have theoretically studied the growth and buoyancy-induced departure of vapor bubbles using global numerical methods. The results, following from prescribed initial temperature distributions are in agreement with the experimental results of Van Stralen et al. [40, 41] presented in the section on bubble growth.

Bubble Frequencies

At this time, correlations do not exist for bubble frequencies, which take into account the interaction between nucleation sites. Further, the expressions which have been proposed for single-nucleation sites all involve the departure diameter in one form or another.

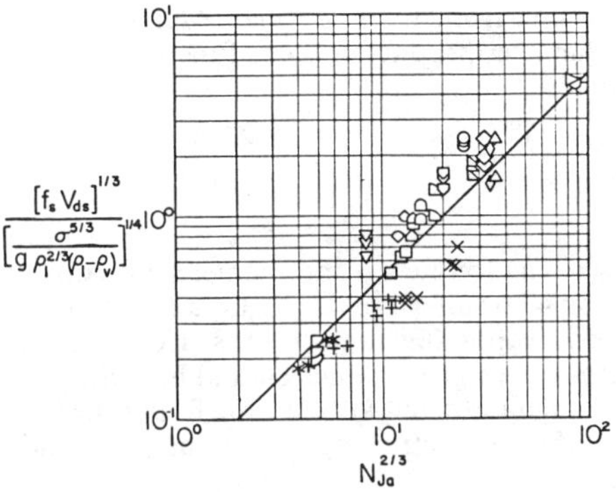

Figure 40. Correlation of dimensionless volumetric vapor flow rate with Jakob number [52]. Note that the data are plotted as the cube root of *both* the ordinate and the abscissa. This is because the data (both ordinate and abscissa) would otherwise vary over 6 orders of magnitude. Plotting as the cube root does not, in this instance, cause the correlation to appear better than it otherwise would.

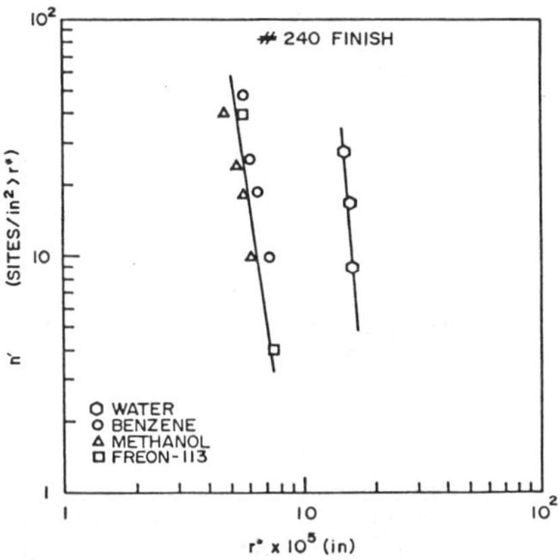

Figure 41. Cumulative site-density curves [53].

Frederking and Daniels [50] developed theoretical expressions for the bubble departure diameter and bubble frequency during film boiling on the basis of which the kinematic equation describing the vapor removal process by gravity (for film boiling) is

$$fD_d \sim [g(\rho_L - \rho_V)/\rho_L]^{1/2} \quad (44)$$

The volumetric vapor flow rate per cross section is

$$fD_d \sim [g(\rho_L - \rho_V)\sigma/\rho_L^2]^{1/4} \quad (45)$$

and the volumetric vapor flow rate is

$$fD_d^3 \sim \{\sigma^{5/3}/[g\rho_L^{2/3}(\rho_L - \rho_V)]\}^{3/4} \quad (46)$$

Extending these to surface boiling from *single cavities*, McFadden and Grassmann [51] proposed that the frequency-diameter product be some function of D_d, σ, ρ_L, and $(\rho_L - \rho_V)$. Dimensional analysis then yields

$$\frac{fD_d^2\rho_L^{1/2}}{(\sigma D_d)^{1/2}} \sim \left\{\frac{g(\rho_L - \rho_V)D_d^2}{\sigma}\right\}^m \tag{47}$$

as a possible relationship describing frequency and diameter in terms of liquid and vapor properties. If m = 1/4, Equation 45 results: if m = 1/2, Equation 44 results; and if m = 3/4, Equation 46 for the volumetric vapor flow rate results. Cole [52] proposed the Jakob number as an important dimensionless group which should be included in Equation 47 and using much of the experimental data existing at that time found the best representation to be as shown in Figure 40. For the volumetric vapor flow rate, this leads to the expression

$$fD_d^3 \sim \left\{\frac{\sigma^{5/3}}{g\rho_L^{2/3}(\rho_L - \rho_V)}\right\}^{3/4} N_{Jaw}^2 \tag{48}$$

When Equation 48 is combined with the *single cavity* correlation for the departure diameter using the same data

$$D_d^2 \sim \left\{\frac{\sigma}{g(\rho_L - \rho_V)}\right\} N_{Jaw}^2 \tag{49}$$

one obtains

$$fD_d \sim \left\{\frac{g(\rho_L - \rho_V)\sigma}{\rho_L^2}\right\}^{1/4} \tag{50}$$

which is identical to Equation 45, giving the interesting result that the volumetric vapor flow rate per cross section is independent of the Jakob number.

Bubble Population Density

The opening paragraph of this chapter stated that the inability to characterize the boiling surface in terms of the distribution of nucleation sites is perhaps the most frustrating problem in terms of predicting the boiling heat flux density. Perhaps it is appropriate then that a discussion of this topic be the last subject of this chapter.

From their early experiments, which indicated that a single dimension (the cavity mouth radius) was sufficient for characterizing a nucleation site, Griffith and Wallis [22] concluded that a knowledge of the size distribution of active cavities would be sufficient to fix the nucleation characteristics of a surface. More specifically, it was thought that a plot of the bubble population density n′ versus a characteristic radius r* such as given by Equation 7 should remain invariant for a given surface at differing pressures and with different fluids.

Although initial experiments [22] seemed to confirm this hypothesis, the later experiments of Lorenz et al. [53, 54] indicate that the plot of n′ vs. r* is not invariant, at least with respect to different fluids. In fact

$$r^*/R_c = f(\theta, 2\beta) \tag{51}$$

as shown by a more detailed analysis [1, 53] than that presented in this chapter. Thus for a given surface, r* is a function of contact angle θ. Cumulative nucleation population density data on a #240-finish copper boiling surface for four different liquids are shown in Fig. 41. When plotted in this fashion, the data are linear on a log-log plot. It is immediately apparent that the water data stand alone while the better-wetting organic liquids tend to group together; i.e., as predicted, the distribution is affected by differences in contact angle.

If the bubble population density is assumed to be of the form

$$n' = C_1 (R_s/R_C)^m \tag{52}$$

where R_s, C_1, and m are constants that depend on the surface, substitution of Equation 51 yields

$$n' = C_1 [R_s f(\theta, 2\beta)/r^*]^m \tag{53}$$

and from a plot such as Figure 41, the constant m and the product $C_1 [R_s f(\theta, 2\beta)]^m$ can be determined. Substituting for r* from Equation 19 gives

$$n' = C_1 [R_s f(\theta, 2\beta)]^m [\rho_V h_{fg} (T_w - T_s)/2\sigma T_s] \tag{54}$$

Equation 54 is used to describe the surface in the modified Mikic–Rohsenow model [55] for predicting the heat flux density in boiling systems.

NOTATION

b	constant, defined by Equation 34
B	constant, equals 1 for cavitation and 2/3 otherwise, dimensionless
c	liquid specific heat at constant pressure (kJ/(kg, K))
C	gas concentration (kg/m^3)
C_1	constant, dependent on the surface, see Equation 52
$C_{1,m}/C_{1,p}$	defined by Equation 33
D	mass diffusivity of more volatile component in less volatile component (mm^2/s)
D_d	departure diameter of a bubble (mm)
f	defined by Equation 11, dimensionless
f_i	frequency of bubble formation at a specific site (s^{-1})
f_s	average frequency of bubble formation at a specific site (s^{-1})
g	gravitational acceleration (m/s^2)
g	specific Gibbs free energy, (Equations 3 through 5 only) (kJ/kg)
G	vaporized mass fraction
h	heat transfer coefficient (kW/(m^2, K))
h_{fg}	latent heat of vaporization (kJ/kg)
J	nucleation frequency (nuclei/(s, cm^3))
k	thermal conductivity (kW/(m, K))
m	constant, dependent on the surface, see Equation 52
M	molecular weight (kg/kg-mole)
n	Number of bubbles formed at a specific site
n'	Bubble population density (mm^{-2})
N_{Eo}	Eotvos number, $g(\rho_L - \rho_V) D_d^2/\sigma$, dimensionless
N_{Ja}	Jakob number, $\rho_L c (T_W - T_S/(\rho_V h_{fg})$, dimensionless
N_{Ja}*	modified Jakob number, $\rho_L c T_S/(\rho_V h_{fg})$, dimensionless
N_{Pr}	Prandtl number, $\mu c/k$, dimensionless
P	pressure (kPa)
P_∞	equilibrium vapor pressure (kPa)
$P_{L,\infty}$	pressure of the liquid far removed from the bubble boundary, i.e. system pressure (kPa)
$P_{L,R}$	liquid pressure at the bubble boundary (kPa)
P_r	reduced pressure, P/P_{crit}, dimensionless
q	heat flux density (kW/m^2)
r	radius of curvature (μm)
r_n	radius of a vapor cap, see Figure 11 (μm)
r*	characteristic radius (μm)
R	bubble radius (mm)

R_C mouth radius of a surface cavity (μm)
R_s constant, dependent on the surface see Equation 52
R^* critical nucleus radius, defined by Equation 17 (μm)
R_{crit} defined by Equation 29, μm
R_{max} maximum mouth radius of a surface cavity (μm)
S defined by Equation 12, dimensionless
t time (ms)
t_d time spent by bubble on surface (ms)
T temperature (K or C)
ΔT temperature increase of liquid at bubble boundary due to vaporization of the more volatile component, with respect to original liquid. See References 1 and 43 for details (°K)
ΔT_∞ defined by Equation 20
v specific volume (m^3/kg)
V_{di} bubble volume at departure at a specific site (mm^3)
V_{ds} average bubble volume at departure at a specific site (mm^3)
x axial cavity position, see Figure 7
y distance from heated wall (μm)

Greek Symbol

α thermal diffusivity (mm^2/s)
β defined in Figure 6
δ thermal boundary layer thickness, defined by Equation 27 (μm)
η defined by Equation 9, dimensionless
θ contact angle, measured through the liquid phase
θ_0 wall superheat (° K)
ρ mass density (kg/m^3)
σ surface tension (N/m)
τ_d time spent by the bubble on the surface (t_d), s.
τ_g time between bubble departure and bubble initiation at a given site (s)
φ defined in Figure 2

Subscripts

B bulk
d departure
G gas
L liquid
R at bubble interface
S saturation
V vapor
W wall

REFERENCES

1. Van Stralen, S. J. D., and Cole, R., *Boiling Phenomena,* Vol. 1, Washington: Hemisphere Publishing Corporation, New York: McGraw–Hill Book Company, 1979, pp. 71–193.
2. Blander, M., and Katz, J. L., "Bubble Nucleation in Liquids," *AIChE Journal.* Vol. 21, No. 5 (Sept. 1975), pp. 833–848.
3. Porteous, W., and Blander, M., "Limits of Superheat and Explosive Boiling of Light Hydrocarbons, Halocarbons, and Hydrocarbon Mixtures," *AIChE Journal.* Vol. 21, No. 3 (May 1975), pp. 560–566.
4. Apfel, R. E., "Water Superheated to 279.5° C at Atmospheric Pressure," *Nature Physical Science,* Vol. 238 (July 24, 1972), pp. 63–64.
5. Holden, B. S., and Katz, J. L., "The Homogeneous Nucleation of Bubbles in Superheated Binary Liquid Mixtures," *AIChE Journal,* Vol. 24, No. 2 (March 1978), pp. 260–267.

6. Pinnes, E. L., and Mueller, W. K., "Homogeneous Vapor Nucleation and Superheat Limits of Liquid Mixtures," *J. Heat Transfer,* Vol. 101 (Nov. 1979), pp. 617–621.
7. Avedisian, C. T., and Glassman, I., "High Pressure Homogeneous Nucleation of Bubbles Within Superheated Binary Liquid Mixtures," *J. Heat Transfer,* Vol. 103 (May 1981), pp. 272–280.
8. Forest, T. W., and Ward, C. A., "Effect of a Dissolved Gas on the Homogeneous Nucleation Pressure of a Liquid," *J. Chem. Phys.,* Vol. 66, No. 6 (March 15, 1977), pp. 2322–2330.
9. Forest, T. W., and Ward, C. A., "Homogeneous Nucleation of Bubbles in Solutions at Pressures Above the Vapor Pressure of the Pure Liquid," *J. Chem. Phys.,* Vol. 69, No. 5 (Sept. 1, 1978), pp. 2221–2230.
10. Mori, Y., Hijikata, K., and Nagatani, T., "Effect of Dissolved Gas on Bubble Nucleation," *Int. J. Heat Mass Transfer,* Vol. 19 (1976), pp. 1153–1159.
11. Blander, M., "Bubble Nucleation in Liquids," in *Advances in Colloid and Interface Science,* Vol. 10, Amsterdam: Elsevier Scientific Publishing Company, 1979, pp. 1–32.
12. Kagan, Yu., "The Kinetics of Boiling of a Pure Liquid," *Russian Journal of Physical Chemistry.,* Vol. 34, No. 1 (January 1960), pp. 42–46. (English translation); *Zh. Fiz. Khim.,* Vol. 34 (1960), pp. 92–96.
13. Zel'dovich, Ya. B., *Acta Physiochim. U.S.S.R.,* Vol. 18 (1943), p. 1.
14. Cole, R., "Boiling Nucleation," in *Advances in Heat Transfer,* Vol. 10, J. P. Hartnett and T. F. Irvine, Jr. (Eds.) New York: Academic Press, 1974, pp. 85–166.
15. Jarvis, T. M., Donohue, M. D., and Katz, J. L., "Bubble Nucleation Mechanisms of Liquid Droplets Superheated in Other Liquids," *J. Colloid and Interface Science,* Vol. 50, No. 2 (Feb. 1975), pp. 359–368.
16. Avedisian, C. T., and Andres, R. P., "Bubble Nucleation in Superheated Liquid-Liquid Emulsions, "*J. Colloid and Interface Science, Vol. 64,* No. 3 (May 1978), pp. 438–453.
17. Avedisian, C. T., and Glassman, I., "Superheating and Boiling of Water in Hydrocarbons at High Pressures," *Int. J. Heat Mass Transfer,* Vol. 24, No. 4, 1981, pp. 695–706.
18. Bankoff, S. G., "Entrapment of Gas in the Spreading of a Liquid Over a Rough Surface," *AIChE Journal,* Vol. 4, No. 1, March 1958, pp. 24–26.
19. Cornwell, K., "Naturally Formed Site Cavities," *Letters in Heat Mass Transfer,* Vol. 4 (1977), pp. 63–72.
20. Sabersky, R. H., and Gates, C. W., Jr., "On the Start of Nucleation in Boiling Heat transfer," *Jet Propulsion* (Feb. 1955), pp. 67–70.
21. Forest, T. W., "The Stability of Gaseous Nuclei at Liquid-Solid Interface," *J. Appl. Phys.,* Vol. 53, No. 9 (Sept. 1982), pp. 6191–6201.
22. Griffith, P., and Wallis, J. D., "The Role of Surface Conditions in Nucleate Boiling," *CEP Symp. Ser. No. 30,* Vol. 56 (1960), pp. 49–63.
23. Schultz, R. R., Kasturirangan, S., and Cole, R., "Experimental Studies of Incipient Vapor Nucleation," *Can. J. Chem. Engr.* Vol. 53 (August 1975), pp. 408–413.
24. Jemison, T. R., Rivers, R. J., and Cole, R., "Incipient Vapor Nucleation of Methanol from an Artificial Site-Uniform Superheat," *J. Heat Transfer,* Vol. 104 (August 1982), pp. 567–568.
25. Schultz, R. R., and Cole, R., "Initial Bubble Growth in Slightly Subcooled Transient Boiling," *AIChE Symp. Ser. No. 199,* Vol. 76 (1980), pp. 310–317.
26. Bergles, A. E., and Rohsenow, W. M., "The Determination of Forced-Convection Surface-Boiling Heat Transfer," *J. Heat Transfer,* Vol. 86 (1964), pp. 365–372.
27. Hsu, Y. Y., "On the Size Range of Active Nucleation Cavities on a Heating Surface," *J. Heat Transfer,* Vol. 84 (August 1962), pp. 207–216.
28. Madejski, J., "Activation of Nucleation Cavities on a Heating Surface with Temperature Gradient in Superheated Liquid," *Int. J. Heat Mass Transfer,* Vol. 9 (1966), pp. 295–299.

29. Schmidt, R. J., and Cole, R., "Comments on Activation of Nucleation Cavities on a Heating Surface with Temperature Gradient in Superheated Liquid," *Int. J. Heat Mass Transfer*, Vol. 13 (1970), pp. 443–445.
30. Schultz, R. R., and Cole, R., "Transient Boiling from Artificial Sites at Low Pressures," *Proceedings of the 1974 Heat Transfer and Fluid Mechanics Institute*, L. R. Davis and R. E. Wilson (Eds.), Stanford: Stanford University Press (1974), pp. 38–56.
31. Schultz, R. R., and Cole, R., "Bubble Nucleation and Growth Instabilities in Transient Boiling," *Heat Transfer 1978*. Proc. of the Sixth International Heat Transfer Conference, Toronto, 1978, Vol. 1, Washington, D.C.: Hemisphere Publishing Corp. 1978, pp. 133–138.
32. Mikic, B. B., Rohsenow, M. W., and Griffith, P., "On Bubble Growth Rates," *Int. J. Heat Mass Transfer*, Vol. 13 (1970), pp. 657–666.
33. Cole, R., and Shulman, H. L., "Bubble Growth Rates at High Jakob Numbers," *Int. J. Heat Mass Transfer*, Vol. 9 (1966), pp. 1377–1390.
34. Stewart, J., and Cole, R., "Bubble Growth Rates During Nucleate Boiling at High Jakob Numbers," *Int. J. Heat Mass Transfer*, Vol. 15 (1972), pp. 655–663.
35. Scriven, L. E., "On the Dynamics of Phase Growth," *Chem. Engng. Sci*, Vol. 10 (1966), pp. 1–13.
36. Mikic, B. B., and Rohsenow, M. W., "Bubble Growth Rates in Non-Uniform Temperature Fields," *Progress in Heat and Mass Transfer*, Vol. 2, T. F. Irvine, et al., (Eds.), Oxford: Pergamon Press, 1969, pp. 283–293.
37. Hsu, Y. Y., and Graham, R. W., "An Analytical and Experimental Study of the Thermal Boundary Layer and Ebullition Cycle in Nucleate Boiling," *NASA TN D594*, May 1961.
38. Cooper, M. G., "The Microlayer and Bubble Growth in Nucleate Pool Boiling," *Int. J. Heat Mass Transfer*, Vol. 12 (1969), pp. 915–933.
39. Cooper, M. G., and Vijuk, T. M., "Bubble Growth in Nucleate Pool Boiling," *Heat Transfer 1970*, Vol. 5, U. Grigull and E. Hahne, (Eds.), Amsterdam: Elsevier Publishing Co., 1970, pp. B2.1, 1–7.
40. Van Stralen, S. J. D., et al., "Bubble Growth Rates in Nucleate Boiling of Water at Subatmospheric Pressures," *Int. J. Heat Mass Transfer*, Vol. 18 (1975), pp. 655–670.
41. Van Stralen, S. J. D., Sluyter, W. M., and Cole, R., "Bubble Growth Rates in Nucleate Boiling of Aqueous Binary Systems at Subatmospheric Pressures," *Int. J. Heat Mass Transfer*, Vol. 19 (1976), pp. 931–941.
42. Hermans, W. A. H. J., "On the Instability of a Translating Gas Bubble Under the Influence of a Pressure Step," doctoral thesis, Eindhoven University of Technology, Eindhoven, The Netherlands, 1973.
43. Van Stralen, S. J. D., et al., "Bubble Growth Rates in Pure and Binary Systems: Combined Effect of Relaxation and Evaporation Microlayer," *Int. J. Heat Mass Transfer*, Vol. 18 (1975), pp. 453–467.
44. Zijl, W., "Global Collocation Approximations of the Flow and Temperature Fields Around a Gas and a Vapour Bubble, "*Int. J. Heat Mass Transfer*, Vol. 20 (1977), pp. 487–498.
45. Van Stralen, S. J. D., Zijl W., and DeVries, D. A., "The Behaviour of Vapor Bubbles During Growth at Subatmospheric Pressures," *Chem. Engr. Sci.*, Vol. 32 (1977), pp. 1189–1195.
46. Joosten, J. G. H., Zijl, W., and Van Stralen, S. J. D., "Growth of a Vapour Bubble in Combined Gravitational and Non-Uniform Temperature Field," *Int. J. Heat Mass Transfer*, Vol. 21 (1978), pp. 15–23.
47. Zijl, W., Ramakers, F. J. M., and Van Stralen, S. J. D., "Global Numerical Solutions of Growth and Departure of a Vapor Bubble at a Horizontal Superheated Wall in a Pure Liquid and a Binary Mixture," *Int. J. Heat Mass Transfer*, Vol. 22 (1979), pp. 401–420.
48. Sultan, M., and Judd, R. L., "Interaction of the Nucleation Phenomena at Adjacent Sites in Nucleate Boiling," *J. Heat Transfer*, Vol. 65 (1969), pp. 211–213.

49. Cole, R., and Rohsenow, W. M., "Correlation of Bubble Departure Diameters for Boiling of Saturated Liquids," *AIChE Symp. Ser. No. 92,* Vol. 65 (1969), pp. 211–213.
50. Frederking, T. H. K., and Daniels, D. J., "The Relation Between Bubble Diameter and Frequency of Removal from a Sphere During Film Boiling," *J. Heat Transfer,* Vol. 88 (Feb. 1966), pp. 87–93.
51. McFadden, P. W., and Grassmann, P., "The Relation between Bubble Frequency and Diameter During Nucleate Pool Boiling," *Int. J. Heat Mass Transfer,* Vol. 5 (1962), pp. 169–173.
52. Cole, R., "Bubble Frequencies and Departure Volumes at Subatmospheric Pressures," *AIChE Journal,* Vol. 13, No. 4 (July 1967), pp. 779–783.
53. Lorenz, H. J., Mikic, B. B., and Rohsenow, W. M., "The Effect of Surface Conditions on Boiling Characteristics," *Heat Transfer 1974,* Proc. of the Fifth Int. Heat Transfer Conf., Tokyo, 1974, Vol. IV, Washington, D. C.: Hemisphere Publishing Corp., 1974 pp. 35–40.
54. Lorenz, H. J., Mikic, B. B., and Rohsenow, W. M., "Nucleation Site Activation in Saturated Boiling," *J. Heat Transfer,* Vol. 98, No. 2 (May 1976), pp. 340–341.
55. Mikic, B. B., and Rohsenow, W. M., "A New Correlation of Pool Boiling Data Including the Effect of Heating Surface Characteristics," *J. Heat Transfer,* Vol. 91, No. 2 (May 1969), pp. 245–250.

CHAPTER 9

FORCED-CONVECTION BOILING IN UNIFORMLY HEATED CHANNELS

Y. Katto

Department of Mechanical Engineering
University of Tokyo, Hongo, Bunkyo-ku
Tokyo, Japan

CONTENTS

INTRODUCTION

Forced convection boiling of the fluid flowing through a heated channel generally refers to nucleate boiling. As is well known, nucleate boiling is strongly affected by various characters of the heated wall such as the minute surface structure, affinity for the boiling fluid, etc. In addition, the behaviour of liquid and vapor in the vicinity of the heated surface is extremely complicated. Thus it is more difficult to make an accurate prediction of heat transfer in nucleate boiling than in the field of single-phase flow. One characteristic of nucleate boiling is that heat flux at the heated wall changes surprisingly with a slight difference in the wall temperature. Therefore, when a prediction of the local heat flux or the local heat transfer coefficient is made under the condition of fixed distribution of wall temperature, the magnitude of the error involved in the estimation of heat flux is generally large, and inaccuracies are also introduced to the prediction of the variation of flow configuration along the channel length.

In contrast, in the case of uniformly heated channels, the heat flux is known in most cases, and the main task is to predict the local wall temperature corresponding to the already-known heat flux. This case involves the same problem of error as in estimating heat flux, but it is lessened by the condition that the temperature of the heated wall causing nucleate boiling changes within a comparatively narrow range near saturation temperature. Hence, except for the special case when an extremely accurate prediction of the wall temperature is needed, the foregoing error does not become a serious problem as long as the heat flux remains below the critical heat flux, at which the excellent state of heat transfer of nucleate boiling abruptly deteriorates, resulting in an overheating of the wall.

For a uniformly heated channel, it is comparatively easy to grasp the variation of the flow configuration along the channel length. For example, when a uniformly heated round tube is fed with a liquid at an inlet subcooling enthalpy ΔH_i (i.e., saturated liquid enthalpy minus inlet liquid enthalpy), the equilibrium vapor quality χ at a distance z from the tube inlet is given as follows via the heat balance:

$$\chi = 4\frac{q}{GH_{fg}} \cdot \frac{z}{d} - \frac{\Delta H_i}{H_{fg}} \tag{1}$$

where q = the heat flux
G = the mean mass velocity of the fluid flowing in the tube
H_{fg} = the latent heat of evaporation
d = the tube diameter

For noncircular geometries, Equation 1 holds if d is replaced by the heated equivalent diameter d_{he} defined as:

$$d_{he} = \frac{4 \times \text{flow area}}{\text{heated perimeter}} \tag{2}$$

It must be noted that d_{he} is different from the hydraulic equivalent diameter in that it is defined by the heated perimeter instead of the wetted perimeter. Equation 1 indicates a linear relationship between the equilibrium quality χ and the axial distance z, as is illustrated in Figure 1. The equilibrium quality χ does not agree with true quality, but χ can be conveniently used in such cases as that of predicting the local void fraction.

The studies done on forced convection boiling in heated channels have been reviewed various places [1–6], and it is advisable to consult with them according to the need. The present chapter deals in particular with the forced-convection boiling in a uniformly heated vertical channel, where a single-component subcooled liquid is fed and the flow is stable with no oscillations. The explanations below will be given taking generalized forms as far as possible.

DEVELOPMENT OF FLOW STRUCTURE ALONG CHANNEL LENGTH

Even in the case of liquid flow without phase change in a heated channel, a change of velocity is caused by a change of density, but its effect is normally not noticeable. In contrast, the forced convection boiling in a heated channel is caracterized by the development of a flow configuration along the channel length because of the significant change in flow state due to phase change, and it must be noted that this character is also different from that of the adiabatic two-phase flow.

Figure 2 illustrates schematically the development of the flow configuration in a vertical tube heated uniformly at a comparatively low heat flux and fed with a liquid at an inlet subcooling enthalpy. When the mass velocity is high enough, the state of A appears, and then the flow state changes in the order of B, C, and D with decreasing mass velocity under the condition that the other factors are held fixed.

For the foregoing heated tube, Equation 1 suggests that the location where $\chi = 0$ moves toward the tube inlet with decreasing mass velocity G. Roughly speaking, the region of

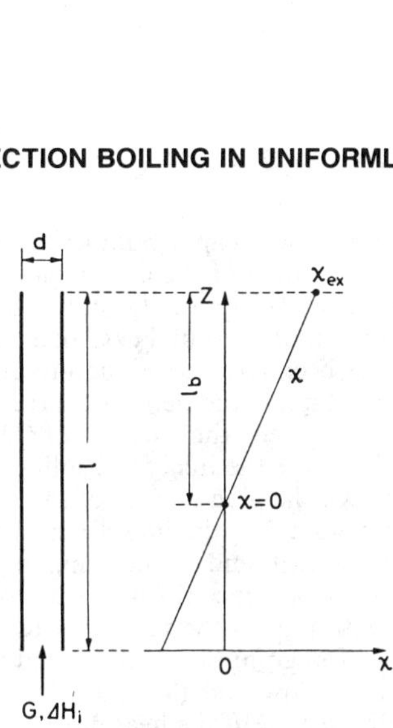

Figure 1. Variation of equilibrium quality X along tube length in a uniformly heated tube.

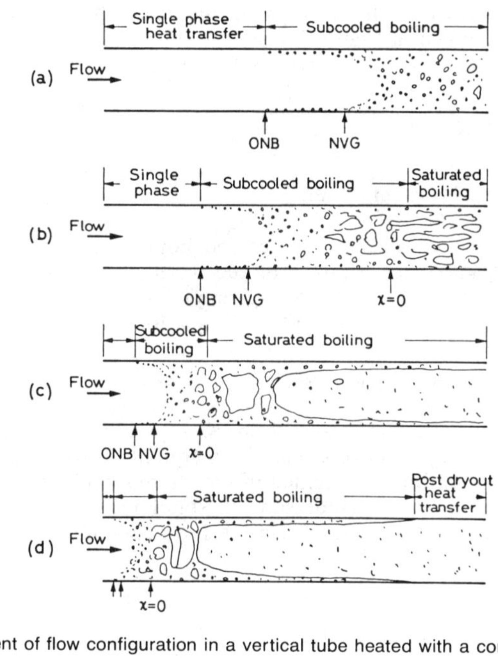

Figure 2. Development of flow configuration in a vertical tube heated with a comparatively low heat flux when the mass velocity is decreased in the order of A, B, C, and D.

$\chi < 0$ causes single-phase liquid heat transfer and subcooled boiling in succession, while the region of $\chi > 0$ causes saturated boiling. For nucleate boiling, however, the liquid in the vicinity of the tube wall must be superheated, so the liquid in the core flow is still subcooled to some extent at the location of $\chi = 0$. In other words, the region of subcooled boiling extends slightly downstream from the location of $\chi = 0$, as is shown in Figure 2.

The location where the incipient nucleate boiling takes place is represented by the point of ONB (onset of nucleate boiling) in Figure 2. It must be noted, however, that there is an area downstream from this location where vapor bubbles can exist only in the very vicinity of the wall due to a noticeable subcooling of the core region, and hence the void fraction does not increase, being kept at zero substantially along the tube length in this region. However, at a downstream location indicated by NVG (net vapor generation) in Figure 2, the subcooling of the core flow has been reduced enough to allow vapor bubbles departing from the tube wall to survive in the liquid, and hence void fraction begins to increase along the tube length. The axial distance between the location of ONB and that of NVG reduces with decreasing mass velocity G and/or the inlet subcooling enthalpy ΔH_i.

It will be noted that in Figure 2C, the greater part of saturated boiling is in the state of annular two-phase flow, and in Figure 2D, the liquid film flow on the wall dries out at a position upstream from the tube exit end, generating a special region of so-called post-dryout heat transfer. In this region, since the wall is not wet, its temperature becomes extraordinarily high compared with that in the upstream region.

It must be remembered that the foregoing state, illustrated in Figure 2D, is realized only when the heat flux is comparatively low and the tube wall material can still endure the temperature rise in the post-dryout region. If the heat flux is much higher, the phenomenon of critical heat flux takes place at the tube exit end under flow conditions such as those illustrated in Figure 2A or B, when the tube wall suffers damage due to physical burnout and hence has no chance to reach the state of Figure 2D. Usually, the critical heat flux of the type of Figure 2D is called dryout or high quality CHF, while the critical heat flux appearing in the state shown in Figure 2A or B is called DNB (departure from nucleate boiling), or subcooled and low-quality critical heat flux.

SUBCOOLED BOILING

Criterion for the Onset of Nucleate Boiling

In dealing with the heat transfer of subcooled boiling in a heated channel, it is requisite, first of all, to know the location of the onset of nucleate boiling. Bergles and Rohsenow [7] have presented the following empirical relationship between the heat flux q and the wall superheat at ONB $(T_w - T_s)_{ONB}$ for water in the pressure range of $p = 0.1 \times 10^6 - 13.6 \times 10^6$ Pa:

$$q = 1120\, p^{1.156} [1.8 (T_w - T_s)_{ONB}]^{2.16/p^{0.0234}} \tag{3}$$

where q is in W/m^2, p is in Pa, and $T_w - T_s$ is in °K. Equation 3 has been derived based on the relation between the superheat of the liquid in equilibrium with a bubble nucleus and the linear temperature distribution assumed in the viscous sublayer on the heated wall.

On the other hand, Davis and Anderson [8], standing on a principle similar to the foregoing gave the following generalized relationship between q and $(T_w - T_s)_{ONB}$:

$$q = \frac{\lambda_\ell \rho_v H_{fg}}{8\sigma T_s} (T_w - T_s)^2_{ONB} \tag{4}$$

where λ_ℓ = the thermal conductivity of liquid
ρ_v = the density of vapor
H_{fg} = the latent heat of evaporation
σ = the surface tension
T_s = the saturation temperature at system pressure

Davis and Anderson have shown that the prediction made by Equation 3 agrees fairly well with that made by Equation 4 for water over the ranges of $p = 0.7 \times 10^6 - 10.3 \times 10^6$ Pa and $q = 3.15 \times 10^4 - 3.15 \times 10^6$ W/m^2.

In the derivation of Equation 4, the critical radius r_{crit} of the bubble nucleus existing in the viscous sublayer is related to q and $(T_w - T_s)_{ONB}$ as:

$$q = \lambda_\ell \frac{(T_w - T_s)_{ONB}}{2r_{crit}}$$

Thus, eliminating $(T_w - T_s)_{ONB}$ between this equation and Equation 4 yields

$$r_{crit} = \sqrt{\frac{2\sigma\lambda_\ell T_s}{\rho_v H_{fg} q}} \tag{5}$$

For some fluid substances and conditions, there can be cases where the magnitude of r_{crit} predicted by Equation 5 is too high to drop in to the range allowed by the premise postulated in deriving Equation 4. Hence Butterworth and Shock [6] have recommended that Equation 3 be used for water at 0.1×10^6 to 13.6×10^6 Pa, while the following process is employed for water at $p < 0.1 \times 10^6$ Pa, organic liquids including refrigerants, and cryogenic liquids:

1. When r_{crit} evaluated by Equation 5 is less than 0.50 μm, Equation 4 is used to predict $(T_w - T_s)_{ONB}$.
2. When r_{crit} evaluated by Equation 5 is greater than 0.25 μm, the superheat ΔT_{sat} corresponding to a pressure difference $\Delta p = 2\sigma/r_{crit}$ is assumed to be equal to $(T_w - T_s)_{ONB}$.

Prediction of Local Wall Temperature

The Region Upstream from ONB

In this region, heat transfer is caused by a single-phase liquid flow. Thus, if it is a turbulent flow in a tube, the local heat transfer coefficient h_m is predicted by the Dittus–Boelter correlation as:

$$\frac{h_m d}{\lambda_\ell} = 0.023 \left(\frac{Gd}{\mu_\ell}\right)^{0.8} Pr_\ell^{0.4} \tag{6}$$

where μ_ℓ is the dynamic viscosity of liquid and Pr_ℓ is the Prandtl number of liquid. With this heat transfer coefficient h_m, the following equation holds:

$$q = h_m (T_w - T_b) \tag{7}$$

where T_b is the bulk temperature (mixed mean temperature) of liquid. If the specific heat of liquid $c_{p\ell}$ is assumed to be constant for simplicity's sake, the local value of T_b at an axial distance z from the tube inlet is given as follows via the heat balance:

$$T_b = T_{bi} + 4 \frac{q}{Gc_{p\ell}} \cdot \frac{z}{d} \tag{8}$$

where T_{bi} is the inlet temperature of the subcooled liquid fed to the heated tube. Equation 8 is also applicable to noncircular geometries if the diameter d is replaced by the heated equivalent diameter d_{he} defined by Equation 2.

The curves marked with single phase in Figure 3 represent the variation of T_b and that of T_w predicted by Equation 8 and by Equations 6 and 7, respectively, for a 0.01-m-diameter tube heated at $q = 2 \times 10^5$ W/m² and fed with water of $T_{bi} = 50°C$ at atmospheric pressure

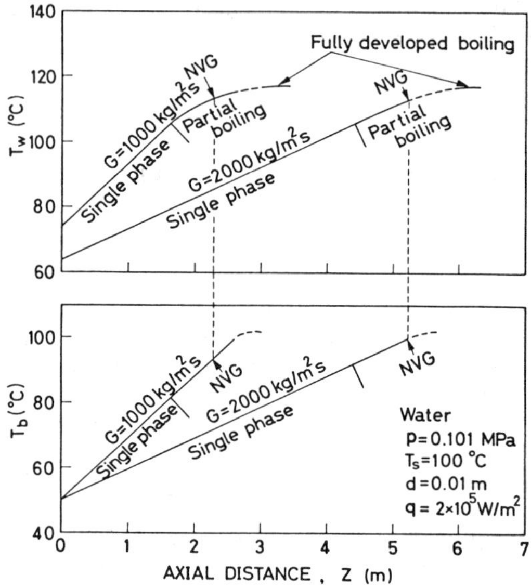

Figure 3. Variation of bulk liquid temperature T_b and wall temperature T_w along tube length for a 0.01-m-diameter tube fed with water of 50°C at atmospheric pressure and heated at a heat flux of $q = 2 \times 10^5 W/m^2$.

with mass velocity $G = 1{,}000$ or $2{,}000\ kg/m^2s$. The terminal point of the single-phase region has been determined through the wall temperature $(T_w)_{ONB}$ at the location of ONB predicted by Equation 3 for the foregoing conditions.

The Region Downstream from ONB

Heat transfer in this region is affected by nucleate boiling in such a way that the mechanism of heat transfer changes gradually along the tube length, tending toward a state where nucleate boiling prevails. Usually this final state is named "fully developed nucleate boiling," while the state in the region where the mechanism of heat transfer changes toward the final state is named "partial nucleate boiling."

Fully developed nucleate boiling is defined as the state dominated by nucleate boiling; accordingly it is generally postulated that heat transfer under this condition is no longer affected by the velocity of the forced flow as well as by subcooling, being expressed as:

$$q = B(T_w - T_s)^n \tag{9}$$

where B is a constant which depends on the combination of the fluid and the heated surface, and n is an index taking a value near 3 to 4 for commercially prepared surfaces. However, it must be noted that the foregoing concept of fully developed nucleate boiling is not necessarily realized in the strict sense of the word. Bergles and Rohsenow [7], examining the experimental data on heat transfer obtained for both pool and forced convection boiling on similar surfaces, have pointed out that Equation 9 cannot be based on a saturation pool boiling curve and that it must rather be based on actual forced convection data. The experimental data reported by Yilmaz and Westwater [9] for nucleate boiling on a heated cylinder in a crossflow suggests a tendency for the effect of forced convection velocity on nucleate boiling to be reduced noticeably with an increase in the heat flux when the other conditions are held fixed.

In regard to the local heat flux q at the position of the wall temperature T_w in the region of partial nucleate boiling, Bergles and Rohsenow [7] presented a relationship interposing between the single-phase forced convection heat transfer predicted by Equation 7 and the fully developed nucleate boiling predicted by Equation 9 as follows:

$$q = q_{FC}\left\{1 + \left[\frac{q_B}{q_{FC}}\left(1 - \frac{q_{Bi}}{q_B}\right)\right]^2\right\}^{1/2} \tag{10}$$

where q_{FC} and q_B are the heat fluxes predicted for the foregoing wall temperature T_w by Equation 7 and Equation 9, respectively, and q_{Bi} is the value of q estimated by Equation 9 for the wall temperature $(T_w)_{ONB}$ at the location of ONB as predicted by means of the previous subsection for a given heat flux q.

Now the relationship between q and T_w at the location of ONB, which is predicted by Equation 3, is represented by a broken curve in Figure 4 for water at atmospheric pressure where $T_s = 100°C$. Meanwhile, the thick curve in Figure 4 is the $q - T_w$ relationship for fully developed nucleate boiling, which is predicted by Equation 9; at present, the following Kutateladze correlation [10] for satured pool nucleate boiling is tentatively employed as Equation 9:

$$\frac{q\ell'}{(T_w - T_s)\lambda_\ell} = 7.0 \times 10^{-4}\,Pr_\ell^{-0.35}\left(\frac{q\ell'}{H_{fg}\rho_v\varkappa_\ell}\right)^{0.7}\left(\frac{p}{\sqrt{\sigma g(\rho_\ell - \rho_v)}}\right)^{0.7} \tag{11}$$

where $\ell' = \sqrt{\sigma/g(\rho_\ell - \rho_v)}$, and $\varkappa_\ell$ is the thermal diffusivity of liquid.

In Figure 4, thin curves shown in the region of single-phase liquid heat transfer, with T_b as a parameter, represent the $q - T_w$ relationship predicted by Equations 6 and 7, while those in the region of partial nucleate boiling represent the result predicted by Equation 10, under the same conditions as those in Figure 3 for $G = 1{,}000\,kg/m^2s$. Figure 5 represents similar results for $G = 2{,}000\,kg/m^2s$. It will be noted from these figures that as T_w increases under the condition of fixed T_b, the value of q predicted by Equation 10 approaches the state of fully developed boiling.

For the heat flux $q = 2 \times 10^5 W/m^2$ and the inlet liquid temperature $T_{bi} = 50°C$, as assumed in Figure 3, T_b and T_w start to change from point S toward point F in Figures 4 and 5, where the lines SI and IF correspond to single-phase heat transfer and partial nucleate boiling, respectively. Then, the relationship between T_b and T_w designated by line IF determines the variation of T_w along the tube length in the partial boiling region as shown in Figure 3.

Dimensionless Group Correlation Method

Other principles different in character from that of Equation 10 are used to predict the heat transfer in partial nucleate boiling region; one is a dimensionless group correlation method; Moles and Shaw [11] derived the following correlation based on the data for water, ethanol, iso-propanol, n-butanol, ammonia, aniline, and hydrazine:

$$q = q_{FC} \times 78.5\,(Pr_\ell)_f^{0.46}\left(\frac{q}{H_{fg}\rho_v u}\right)^{0.67}\left[\frac{H_{fg}}{c_{p\ell}(T_s - T_b)}\right]^{0.50}\left(\frac{\rho_v}{\rho_\ell}\right)^{0.70} \tag{12}$$

where q_{FC} and u are the heat flux and the bulk liquid velocity, respectively, appearing under conditions of nonboiling forced convection of liquid: q_{FC} is predicted by Equation 7, and $u = G/\rho_\ell$. Physical properties in Equation 12 are all evaluated at saturation temperature

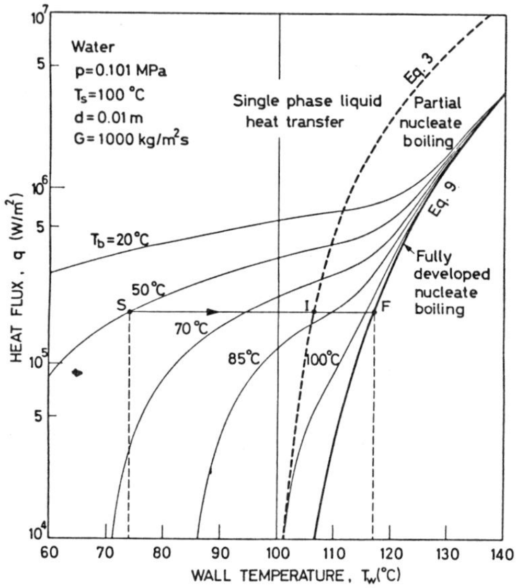

Figure 4. q – T_w relationship with bulk liquid temperature T_b as a parameter for the same condition as that in Figure 3 (G = 1,000 kg/m^{2}s).

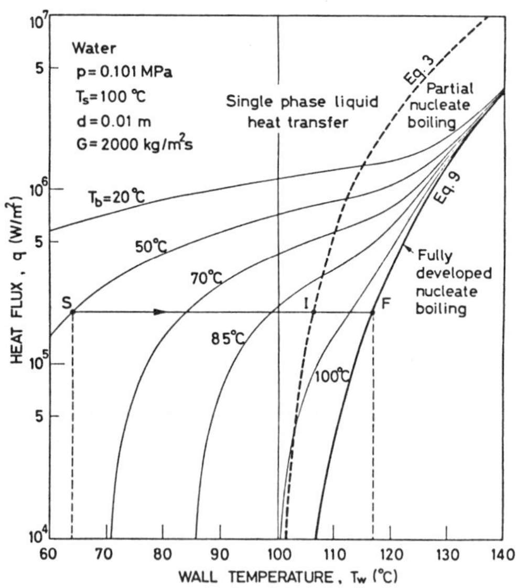

Figure 5. q – T_w relationship with bulk liquid temperature T_b as a parameter for the same condition as that in Figure 3 (G = 2,000 kg/m^{2}s).

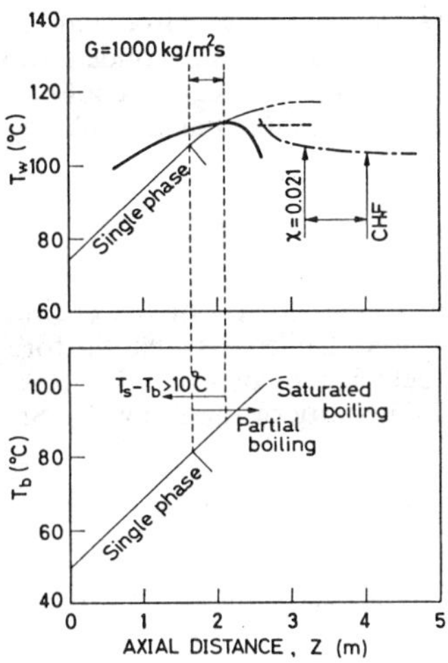

Figure 6. Variation of bulk liquid temperature T_b and wall temperature T_w along tube length for the same condition as that in Figure 3 (G = 1,000 kg/m²s).

except $(Pr_\ell)_f$ that is, the Prandtl number of liquid evaluated at the mean film temperature $(T_w + T_b)/2$.

Equation 12 applies in the region downstream from the point of ONB, but Moles and Shaw have recommended not applying it to the region where $T_s - T_b < 10°C$, because q predicted by Equation 12 tends to infinity as $(T_s - T_b) \rightarrow 0$. An approximate expression in the form of $(Pr_\ell)_f = a\,[b\,(T_w + T_b)/2 + c]^m$ is helpful in evaluating T_w from Equation 12 by computer.

The variations of T_b and T_w shown for $G = 1{,}000\ kg/m^2s$ in Figure 3 are reproduced by thin lines in Figure 6, where a thick continuous curve represents T_w predicted by Equation 12 together with its applicable range limited between two parallel vertical broken lines. The magnitude of T_w predicted by Equation 12 in the foregoing applicable range is close to that predicted by Equation 10, but a discontinuity of T_w appears at the point of ONB. Similar results to those above are shown in Figure 7 for $G = 2{,}000\ kg/m^2s$, but in this case no applicable range exists for Equation 12 if the condition of $T_s - T_b > 10°C$ is imposed, because the bulk temperature T_b at the point of ONB is higher than the value of $T_s - 10°C$.

Net Vapor Generation

The Starting Point of Net Vapor Generation

As has been mentioned earlier, vapor can be stored in the liquid flow in the region of net vapor generation, increasing void fraction along the channel length. Thus the premise on which the prediction of heat transfer made by Equations 10 and 12 relies begins to be violated, because these equations have been made based on Equations 6 and 7 relative to a single-phase flow; hence accurate prediction of the point where net vapor generation starts is important.

Saha and Zuber [12] assume that the point of initial net vapor generation depends on local conditions: at low mass flow rates, the phenomenon is dominated by thermal conditions, whereas at high mass flow rates it is hydrodynamically controlled. In the former case, the rate of evaporation at the wall and that of condensation in the liquid are presumed to be proportional to the heat flux q and the local subcooling $(T_s - T_b)_{NVG}$, respectively, so the following Nusselt number may be adopted as a similarity parameter:

$$Nu = \frac{qd_{hd}}{\lambda_\ell (T_s - T_b)_{NVG}}$$

where d_{hd} is the hydraulic equivalent diameter of the channel. On the other hand, it is presumed in the hydrodynamically controlled case that the foregoing thermal condition has already been satisfied and the detachment of vapor bubbles from the wall is controlled hydrodynamically, suggesting availability of the following Stanton number:

$$St = \frac{q}{Gc_{p\ell} (T_s - T_b)_{NVG}}$$

From the previous two equations, a Peclet number is also defined as:

$$Pe = Nu/St = Gc_{p\ell} d_{hd}/\lambda_\ell \qquad (13)$$

Then Saha and Zuber, analyzing the experimental data of boiling water, R-12 and R-114 in various channel geometries with wide ranges of system pressure, mass flow rate, and heat flux, derive the following criterion for the location of initial NVG:

$$\begin{aligned} &\text{For } Pe = Gc_{p\ell} d_{hd}/\lambda_\ell \leqq 70{,}000 \\ &\qquad Nu = qd_{hd}/[\lambda_\ell (T_s - T_b)_{NVG}] = 455 \\ &\text{For } Pe = Gc_{p\ell} d_{hd}/\lambda_\ell > 70{,}000 \\ &\qquad St = q/[Gc_{p\ell} (T_s - T_b)_{NVG}] = 0.0065 \end{aligned} \qquad (14)$$

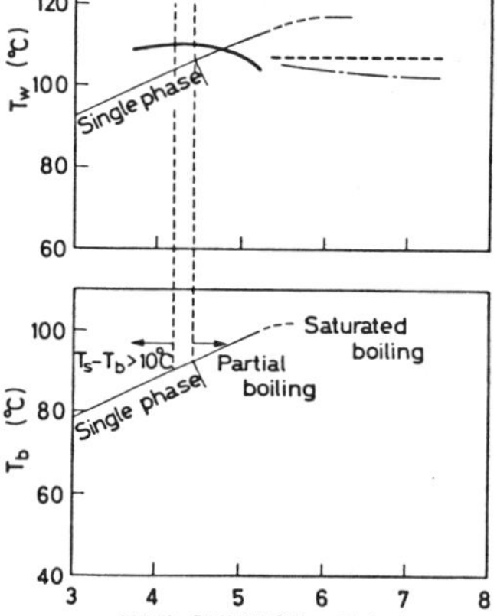

Figure 7. Variation of bulk liquid temperature T_b and wall temperature T_w along tube length for the same condition as that in Figure 3 (G = 2,000 kg/m²s).

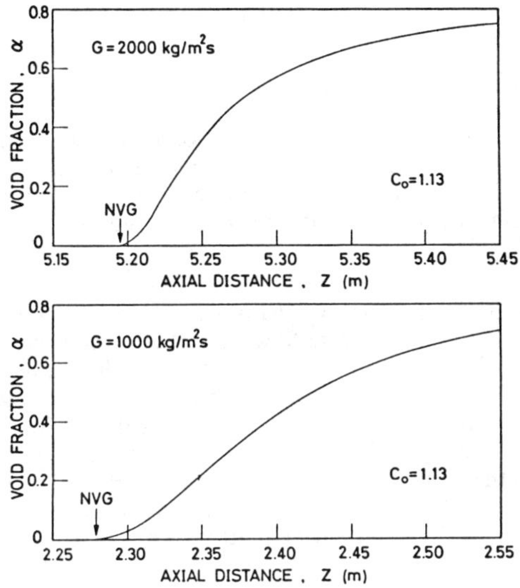

Figure 8. Variation of void fraction α along tube length for the same condition as that in Figure 3.

where T_b is given by Equation 8 for uniformly heated channels. The locations of initial NVG predicted by Equation 14 are shown in Figure 3.

Prediction of Vapor Void Fraction

For the variation of void fraction α which appears along the tube length after the net vapor generation commences, Saha and Zuber [12] have presented the following predictive equation:

$$\alpha = \frac{\chi^*}{C_0\left[\chi^*\left(1 - \frac{\rho_v}{\rho_\ell}\right) + \frac{\rho_v}{\rho_\ell}\right] + 1.41\,\frac{\rho_v}{G}\sqrt[4]{\frac{\sigma g(\rho_\ell - \rho_v)}{\rho_v^2}}} \tag{15}$$

with χ^*, the true vapor quality, given as:

$$\chi^* = \frac{\chi - \chi_{NVG}\exp\left[(\chi/\chi_{NVG}) - 1\right]}{1 - \chi_{NVG}\exp\left[(\chi/\chi_{NVG}) - 1\right]}$$

where χ is the equilibrium quality given by Equation 1 as a function of z, and χ_{NVG} is the value of χ at the location of initial net vapor generation. At $\chi = \chi_{NVG}$, Equation 15 gives $\chi^* = 0$, and hence $\alpha = 0$. Reference [12] presents three examples showing good agreement between the experimental data and the prediction made by Equation 15 with $C_0 = 1.13$. Figure 8 represents the results predicted by Equation 15 for the variation of α along the tube length under the conditions dealt with in Figure 3.

Hewitt [13] has recommended a method slightly different from the foregoing one to predict the void fraction, which includes processes that evaluate C_0 and another constant as well as χ^*.

SATURATED BOILING

Boiling in an Annular Flow Region

The most widely used correlation for saturated boiling in the annular flow region, such as that shown in Figure 2C and D, is the Chen correlation, where an additive-type mechanism composed of two-phase annular forced convection and nucleate boiling is assumed, and the total heat transfer coefficient h is written as:

$$h = h_\ell F + h_n S \tag{16a}$$

The first term on the RHS of Equation 16a is a forced convection term, where h_ℓ is the heat transfer coefficient if the liquid were flowing alone in the tube, and F is a multiplier to consider the effect of the increase in liquid velocity which is caused by the existence of vapor flow. The second term on the RHS of Equation 16a is a nucleate boiling term, where h_n is the heat transfer coefficient of ordinary type nucleate boiling, and S is a factor to consider the effect of the suppression of nucleate boiling due to the steepening of the temperature gradient in the liquid film on the wall, which is caused by the forced flow. Three heat transfer coefficients h, h_ℓ, and h_n included in Equation 16a are all defined with a temperature difference $T_w - T_s$, and hence equation 16a can be rewritten as:

$$q = q_\ell F + q_n S \tag{16b}$$

where $q = h(T_w - T_s)$, $q_\ell = h_\ell(T_w - T_s)$, and $q_n = h_n(T_w - T_s)$.

Based on the foregoing definition of h_ℓ, Chen [14] employes the Dittus-Boelter correlation, Equation 6, to predict q_ℓ in Equation 16b as follows:

$$\frac{q_\ell}{T_w - T_s}\frac{d}{\lambda_\ell} = 0.023\left[\frac{G(1-\chi)d}{\mu_l}\right]^{0.8} Pr_\ell^{0.4} \tag{17}$$

where χ is the local value of quality and hence $G(1-\chi)$ expresses the mass velocity if the liquid were flowing alone in the tube.

Then, if an analogy holds between momentum and heat transfer, the multiplier F can be assumed to be a function of $(dp/dz)/(dp/dz)_\ell$, where (dp/dz) is the pressure gradient of the two-phase flow, while $(dp/dz)_\ell$ is the pressure gradient if the liquid were flowing alone in the tube. Through analysis of the experimental data for water, methanol, cyclohexane, pentane, heptane, and benzene, Chen [14] succeeded in finding an empirical relationship of $F = [(dp/dz)/(dp/dz)_\ell]^{0.444}$. Recently, however, Bennett and Chen [15], based on the data for water, ethylene glycol, and aqueous mixture of ethylene glycol, have improved the foregoing equation by taking the effect of the Prandtl number into consideration as follows:

$$F = \left[\frac{Pr_\ell + 1}{2}\right]^{0.444}\left[\frac{dp/dz}{(dp/dz)_\ell}\right]^{0.444} \tag{18}$$

where Pr_ℓ is the Prandtl number of liquid. As for $(dp/dz)/(dp/dz)_\ell$ on the RHS of Equation 18, it is well known that when both liquid and vapor phases are in a turbulent state, it can be expressed as a function of the Lockhart-Martinelli parameter [16]:

$$X = \left(\frac{1-\chi}{\chi}\right)^{0.9}\left(\frac{\rho_v}{\rho_\ell}\right)^{0.5}\left(\frac{\mu_\ell}{\mu_v}\right)^{0.1} \tag{19}$$

For smooth tubes, Chisholm [17] has shown that the foregoing function is expressed approximately as:

$$\frac{dp/dz}{(dp/dz)_\ell} = 1 + \frac{20}{X} + \frac{1}{X^2} \tag{20}$$

Previously Chisholm and Laird [18] took the value of 21 instead of 20 in the second term on the RHS of Equation 20, but it seems to cause no noticeable difference in the result.

For q_n in Equation 17, Chen [14] employs the following generalized correlation proposed by Forster and Zuber [19] for saturated pool nucleate boiling:

$$\frac{q_n}{T_w - T_s} \frac{r_b}{\lambda_\ell} = 0.0015\, Re_b^{0.62}\, Pr_\ell^{0.33}$$

with

$$r_b = \frac{T_w - T_s}{H_{fg}\rho_v} \left(\frac{2\pi\lambda_\ell\rho_\ell c_{p\ell}\sigma}{\Delta p}\right)^{0.5} \left(\frac{\rho_\ell}{\Delta p}\right)^{0.25} \tag{21}$$

and

$$Re_b = \frac{\pi\lambda_\ell c_{p\ell}}{\mu_\ell} \left(\frac{\rho_\ell (T_w - T_s)}{\rho_v H_{fg}}\right)^2$$

where Δp is the difference in saturation vapor pressure corresponding to the temperature difference $T_w - T_s$. It is helpful if an approximate expression in the form of $\Delta p = k\,(T_w - T_s)^m$ is used in the region of interest. Figure 9 shows a comparison between the Forster-Zuber correlation, Equation 21, and the Kutateladze correlation, Equation 11, for boiling water at atmospheric pressure.

For the nucleate boiling suppression factor S, Chen assumed that since the temperature gradient at the wall is increased by the forced convective flow, the rate of bubble growth, and hence the subsequent heat transfer, is governed by an effective superheat less than the wall superheat $T_w - T_s$. Then, based on analysis of the experimental data, he presented the empirical value of S graphically as a function of $[G(1 - \chi)d/\mu_\ell]F^{1.25}$. Later, however, Bennett et al. [20] performed an analysis of this phenomenon to derive the following correlation:

$$S = \frac{\lambda_\ell}{h_\ell F y_0}\left[1 - \exp\left(\frac{-h_\ell F y_0}{\lambda_\ell}\right)\right] \tag{22}$$

where λ_ℓ is the thermal conductivity of liquid, $h_\ell = q_\ell/(T_w - T_s)$ is evaluated by Equation 17, F is evaluated by Equations 18 through 20, and y_0 is the distance from the wall defining the extent of the bubble growth region, which is given empirically by the following equation:

$$y_0 = 0.041 \sqrt{\frac{\sigma}{g(\rho_\ell - \rho_v)}} \tag{23}$$

After all, the first and the second terms on the RHS of Equation 16b are given by Equations 17 through 20 and Equations 21 through 23, respectively, as functions of χ and $T_w - T_s$. Accordingly, if a heat flux q is prescribed, the wall temperature T_w is calculated by computer as a function of χ, and hence of z; the thick chain line in the upper diagram of Figure 6 represents the results thus obtained for T_w under the same condition as that in Figure 3 with $G = 1{,}000\ kg/m^2s$.

The foregoing Chen correlation applies mainly to the annular flow region, so it is requisite to know the range of annular flow in the tube. According to the Hewitt-Roberts map [21] for vertical two-phase flow, it can be presumed that the annular flow appears in the range of $\chi > 0.021$ under the condition dealt with in Figure 6. On the other hand, the critical heat flux (CHF) also gives a limitation for the range of annular flow; the location of CHF indicated in Figure 6 is that predicted by the method described in the next section.

Boiling in the Region Preceding the Annular Flow Region

The Chen correlation described earlier is not necessarily applicable to the saturated boiling in the region preceding the annular flow region, such as is shown in Figure 2C and D. The satured boiling near the tube exit end in Figure 2B is also in the same situation. In fact, the Hewitt–Roberts map noted earlier suggests that under the conditions dealt with in Figure 7, the axial distance z at which the annular flow commences is excessively long, and hence saturated boiling can appear substantially in the form of Figure 2B only.

Shah [22] analyzed the experimental data for forced-flow saturated boiling in a wide range of conditions for water, freons, and cyclohexane, presenting a graphic correlation of heat transfer, which includes not only the region dealt with in the previous subsection but also a region termed "nucleate boiling regime," that is classified on the basis of the boiling heat transfer data substantially unaffected by the quality χ under the condition of low quality ($\chi \doteqdot 0$). Meanwhile, Lazarek and Black [23] experimented with R-113 boiling in a 3.15-mm-diameter tube, obtaining saturated boiling data unaffected by the quality χ, with which they derived the following correlation:

$$\frac{q}{T_w - T_s} \cdot \frac{d}{\lambda_\ell} = 30 \left(\frac{Gd}{\mu_\ell}\right)^{0.857} \left(\frac{q}{GH_{fg}}\right)^{0.714} \tag{24}$$

Although Equation 24 is based on the data of R-113 only in the range of $p = 1.3 \times 10^5 - 4.1 \times 10^5$ Pa, Lazarek and Black have presented a few results showing

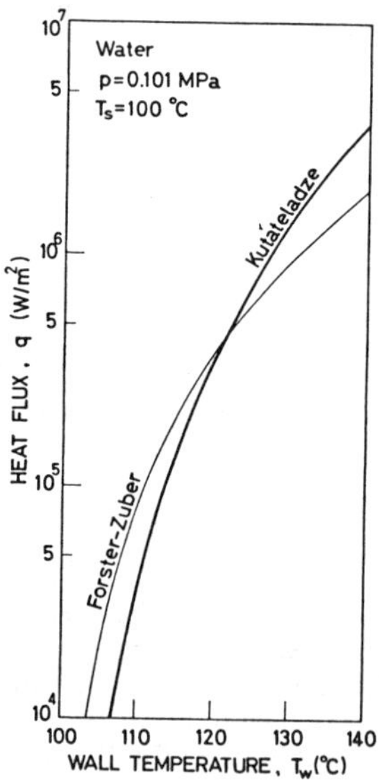

Figure 9. Comparison of Kutateladze and Forster-Zuber correlations for nucleate boiling of water at atmospheric pressure.

fairly good agreement of prediction between Equation 24 and the Shah correlation. The thick horizontal broken lines in the upper halves of Figures 6 and 7 represent the wall temperature T_w predicted by Equation 24. For comparison, the thin chain line in the upper diagram of Figure 7 represents T_w predicted by the Chen correlation (Equation 16b), though no annular flow pattern is expected to appear in this range.

CRITICAL HEAT FLUX

As noted earlier, when the critical heat flux condition takes place in the tube, the wall temperature T_w rises extraordinarily due to sudden deterioration of heat transfer. Particularly, when the CHF appears under the condition of subcooled or low-quality boiling, the heated tube suffers melting or physical burnout in most cases.

It is known that the CHF condition takes place normally at the channel exit end in the case of uniformly heated vertical channels. Therefore, it is important to examine whether or not the CHF condition appears at the exit end of the heated channel under given hydraulic and thermal conditions. There are many empirical correlations of CHF presented for water boiling in uniformly heated vertical channels (see References 24 and 25), but generalized correlations proposed recently based on the data for various kinds of fluids will be referred to in this section.

Critical Heat Flux in Tubes

Shah Correlation

Shah [26] analyzed existing data on CHF obtained for 11 kinds of fluids in wide ranges of conditions, presenting the following graphic correlation of CHF, which is applicable not only to the CHF in positive quality boiling but also to that in subcooled boiled. The dominant dimensionless parameter Y employed in this correlation is defined as:

$$Y = \left(\frac{Gc_{p\ell}d}{\lambda_\ell}\right)\left(\frac{G^2}{\rho_\ell^2 gd}\right)^{0.4}\left(\frac{\mu_\ell}{\mu_v}\right)^{0.6} \tag{25}$$

where g is the gravitational acceleration (the standard value at the earth's surface). Note that the magnitude of Y increases in proportion to $G^{[1.8]}$.

- In the case of $Y > 10^5$, the heat flux q_c to generate the critical heat flux condition at the tube exit end can be predicted by Figure 10 under given conditions of Y, ℓ/d, and $\Delta H_i/H_{fg}$, where ℓ is the length of the heated tube and ΔH_i is the subcooling enthalpy of the liquid at the tube inlet. When $Y \leqq 10^4$, read the value of q_c/GH_{fg} directly at the point (ℓ/d, $\Delta H_i/H_{fg}$). When $10^4 < Y \leqq 10^5$, find an intersecting point of the horizontal line passing through the point (ℓ/d, $\Delta H_i/H_{fg}$) and the slant straight line marked with $Y \leqq 10^4$, draw a vertical line through the foregoing intersecting point, and read the value of q_c/GH_{fg} at the point lying on the vertical line in proportion to Y between $Y = 10^4$ and $Y = 10^5$.
- In the case of $Y \leqslant 10^5$, the value of $(q_c/GH_{fg})_{\chi ex=0}$ is first estimated through Figure 11A for given conditions of ℓ/d, p_r and Y, where p_r is the reduced pressure (the ratio of the system pressure to the critical pressure). The expression refers to the critical heat flux caused under the condition of $\chi = 0$ at the tube exit. Usually, however, the critical heat flux takes place with $\chi_{ex} < 0$ (subcooled boiling) or $\chi_{ex} > 0$ (saturated boiling); so Figure 11B provides the correction value, $(q_c/GH_{fg})/(q_c/GH_{fg})_{\chi ex=0}$, through the height of the point (Y, χ_{ex}), which multiplied by the foregoing value of $(q_c/GH_{fg})_{\chi ex=0}$ determines the value of q_c/GH_{fg}. In this case, if $p_r > 0.6$, the height of the point (Y, χ_{ex}) must be modified before reading the value of $(q_c/GH_{fg})/(q_c/GH_{fg})_{\chi ex=0}$ by employing the right-hand

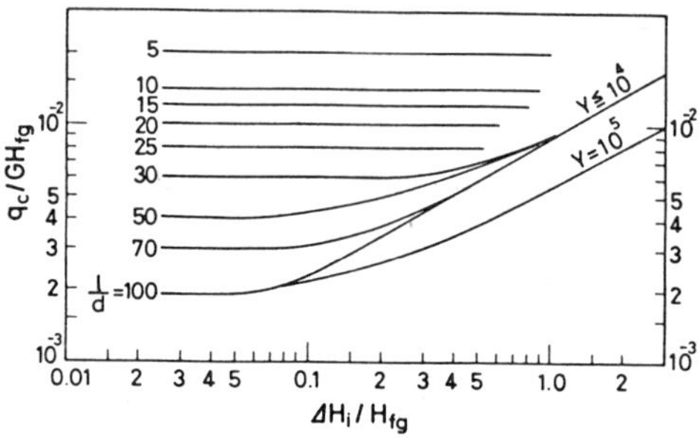

Figure 10. Shah graphic generalized correlation of CHF for Y ≦ 10^5 [24] (Reprinted with permission from Pergamon Press Ltd.).

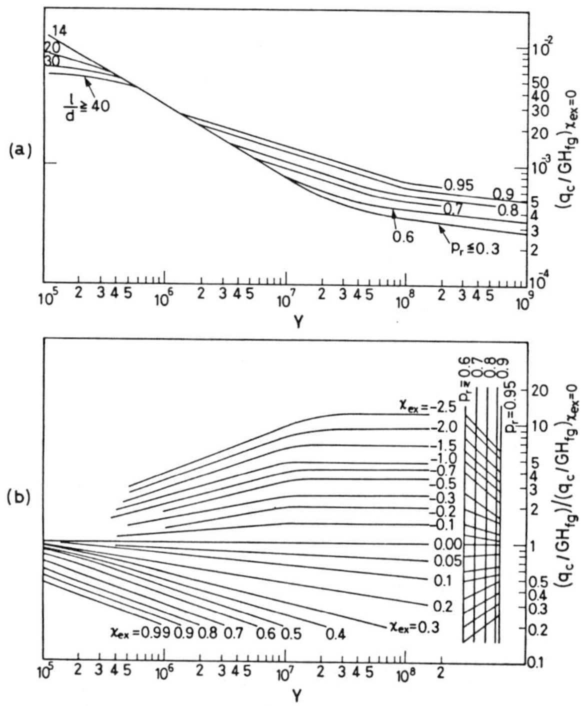

Figure 11. Shah graphic generalized correlation of CHF for Y > 10^5 [24] (Reprinted with permission from Pergamon Press Ltd.).

diagram in Figure 11B, which gives rise to a slight reduction of the height for $\chi_{ex} < 0$, while an increase for $\chi_{ex} > 0$ as will be readily noticed in the figure.

When a tube is heated with a heat flux q under given conditions of ℓ, d, G, and ΔH_i, the exit quality χ_{ex} is readily evaluated by Equation 1. If the critical heat flux q_c predicted by Figure 11 for this value of χ_{ex} is higher than the foregoing heat flux q, it means that this tube is free from critical heat flux.

When one needs the critical heat flux q_c for a tube under conditions of ℓ, d, G, and ΔH_i, the exit quality χ_{ex} is initially unknown. In this case, starting from an assumed value of χ_{ex}, the iterative calculation of q_c is repeated taking Equation 1 into account until the value of q_c converges. The relationships between q_c and G thus obtained by Figures 10 and 11 for water at various ρ_v/ρ_ℓ ratios under the conditions of $d = 0.008\,m$, $\ell_b/d = 25 \sim 500$, and $\Delta H_i = 0$ are represented in Figure 12 with a dimensionless form of q_c/GH_{fg} against $\sigma\rho_\ell/G^2\ell_b$, where ℓ_b is the length of the heated tube with $\Delta H_i = 0$ or $\chi = 0$ at the tube inlet (see Figure 1); this value is termed the boiling length. The discontinuity noted for the curves of $\ell_b/d = 25$, 50, and 100 in Figure 12 comes from the discontinuity existing between Figure 10 and 11. Meanwhile, Figure 13 represents the experimental data for water, potassium, nitrogen, R-12, R-113, R-114, and helium (those data employed in the analysis of CHF in Reference 27) in the same form as before. It will be noted that the results predicted by the Shah correlation in Figure 12 agree fairly well with the experimental results in Figure 13.

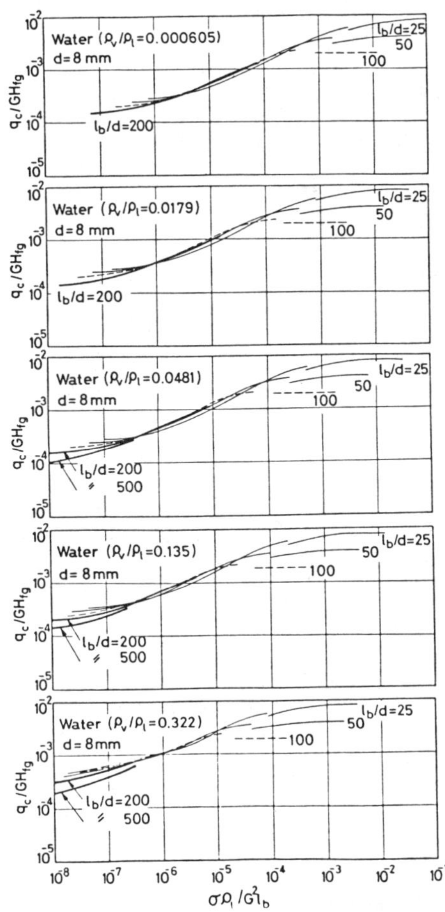

Figure 12. Relationship between q_c/GH_{fg} and $\sigma\rho_\ell/G^2\ell_b$ predicted by Shah correlation for water boiling in a 8-mm-diameter tube.

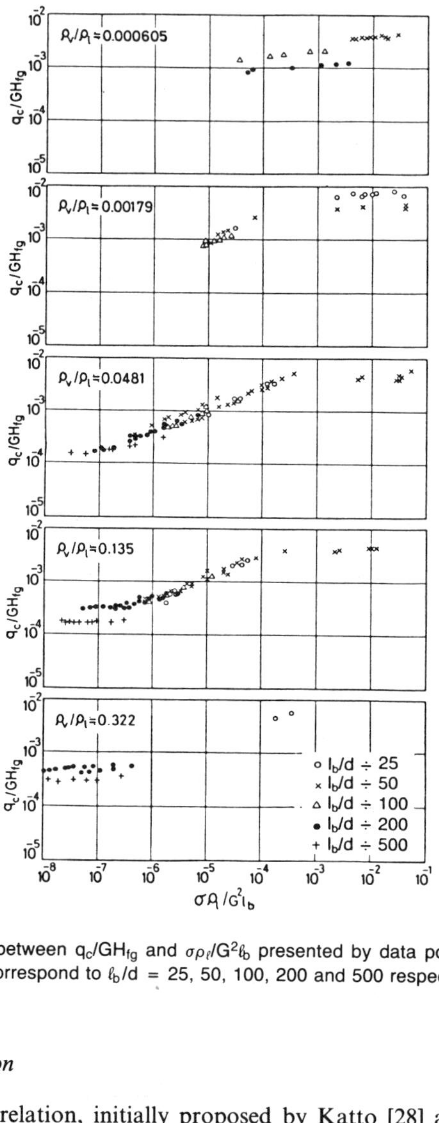

Figure 13. Relationship between q_c/GH_{fg} and $\sigma\rho_\ell/G^2\ell_b$ presented by data points of various fluids. Data points ○, X, Δ, ● and + correspond to ℓ_b/d = 25, 50, 100, 200 and 500 respectively.

Katto–Ohno Correlation

The generalized correlation, initially proposed by Katto [28] and recently refined by Katto and Ohno [29], is applicable to the CHF in positive quality boiling ($\chi_{ex} > 0$) based on the existing data for 14 kinds of fluids. This correlation is expressed in the following comparatively simple forms if the boiling length ℓ_b (Figure 1) is employed:

1. In the case of $\rho_v/\rho_\ell < 0.15$

$$q_c/GH_{fg} = C\,(\sigma\rho_\ell/G^2\ell_b)^{0.043}/(\ell_b/d) \tag{26}$$

$$q_c/GH_{fg} = 0.10\,(\rho_v/\rho_\ell)^{0.133}\,(\sigma\rho_\ell/G^2\ell_b)^{1/3} \cdot 1/(1 + 0.0031\ell_b/d) \tag{27}$$

$$q_c/GH_{fg} = 0.098\,(\rho_v/\rho_\ell)^{0.133}\,(\sigma\rho_\ell/G^2\ell_b)^{0.433}\,(\ell_b/d)^{0.27} \cdot 1/(1 + 0.0031/\ell_b/d) \tag{28}$$

where C is given as $C = 0.25$ for $\ell_b/d < 50$, $C = 0.25 + 0.0009\,[(\ell_b/d) - 50]$ for $\ell_b/d = 50 - 150$, and $C = 0.34$ for $\ell_b/d > 150$. Roughly speaking, Equations 26 and 27 correspond to the CHF in annular flow, and Equation 28 to the CHF in froth or bubbly flow (see Reference 29 for details). With increasing G (that is, with decreasing $\sigma\rho_\ell/G^2\ell_b$), Equations 26 through 28 are employed in the order of the first, second, and third equation so as to connect the value of q_c continuously. The upper diagrams in Figure 14 represent the CHF predicted by Equations 26 through 28 for $\rho_v/\rho_\ell = 0.000605, 0.0179$, and 0.0481.

2. In the case of $\rho_v/\rho_\ell > 0.15$,

$$q_c/GH_{fg} = C\,(\sigma\rho_\ell/G^2\ell_b)^{0.043}/(\ell_b/d) \tag{29}$$

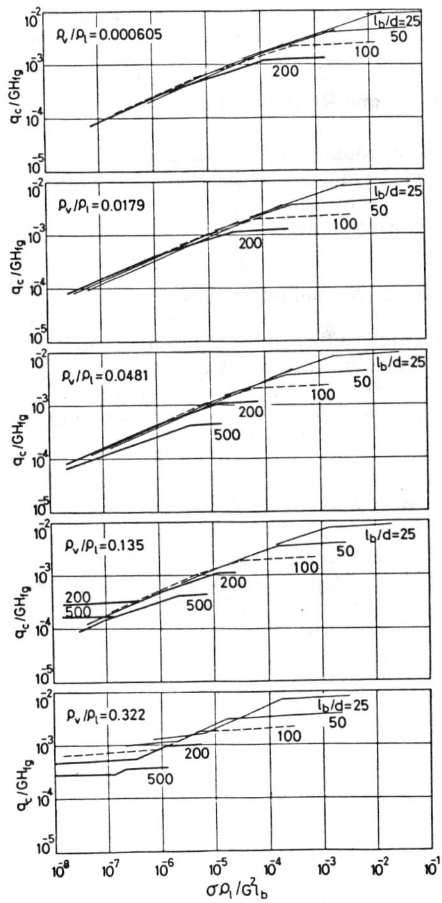

Figure 14. Relationship between q_c/GH_{fg} and $\sigma\rho_\ell/G^2\ell_b$ predicted by Katto-Ohno correlation.

$$q_c/GH_{fg} = 0.234\,(\rho_v/\rho_\ell)^{0.513}\,(\sigma\rho_\ell/G^2\ell_b)^{0.433}\,(\ell_b/d)^{0.27} \cdot 1/(1 + 0.0031\ell_b/d) \tag{30}$$

$$q_c/GH_{fg} = 0.0384\,(\rho_v/\rho_\ell)^{0.60}\,(\sigma\rho_\ell/G^2\ell_b)^{0.173} \cdot 1/[1 + 0.280\,(\sigma\rho_\ell/G^2\ell_b)^{0.233}\,\ell_b/d] \tag{31}$$

where C takes the same value as in Equation 26. The bottom diagram in Figure 14 represents the CHF predicted by Equations 29 through 31 for $\rho_v/\rho_\ell = 0.322$.

Strictly speaking, a transitional region between the foregoing state of 1 and that of 2 exists in the vicinity of $\rho_v/\rho_\ell = 0.15$. Accordingly, the second diagram from the bottom in Figure 14 represents the CHF predicted not only by Equations 26 through 28 but also by Equation 31. It will be seen that the result of Figure 14 also agrees fairly well with the trends of experimental data plotted in Figure 13.

The location of CHF indicated in Figure 6 is the result predicted by the Katto–Ohno correlation for the prescribed heat flux of $q = 2 \times 10^5\ \text{W/m}^2$, where the critical heat flux is evaluated by Equation 27, which corresponds to the annular flow pattern. In the case of Figure 7, no critical heat flux phenomenon is predicted in the region of interest.

Critical Heat Flux in Annular and Rectangular Channels

Compared with the CHF in round tubes, existing data are very few for CHF in annular or rectangular channels, and the ranges of experimental conditions are restricted. Thus, the following data correlations are in rather tentative style, so it is advisable to refer to the original papers for the applicable ranges.

Annular Channel with a Heated Shroud and an Unheated Rod

In this case, existing data are insufficient for a systematic analysis leading to a generalized correlation of these data. However, the existing data suggest that CHF can be predicted by Equations 26 to 31 in an approximate sense if the diameter d is replaced by the heated equivalent diameter d_{he}, that is, $d_{he} = (d_0^2 - d_i^2)/d_0$ from the definition of Equation 2, where d_0 is the diameter of the shroud and d_i is that of the rod.

Annular Channel with a Heated Rod and an Unheated Shroud

The existing CHF data for $\rho_v/\rho_\ell > 0.15$ are insufficient for a systematic analysis, while References 30 and 31 have presented the following correlations of CHF for $\rho_v/\rho_\ell < 0.15$ corresponding to Equations 26 through 28 for round tubes:

$$q_c/GH_{fg} = 0.25\,(\sigma\rho_\ell/G^2\ell_b)^{0.043}/(\ell_b/d_{he}) \tag{32}$$

$$q_c/GH_{fg} = 0.12\,(\rho_v/\rho_\ell)^{0.133}\,(\sigma\rho_\ell/G^2\ell_b)^{1/3} \cdot 1/(1 + 0.0081\,\ell_b/d_{he}) \tag{33}$$

$$q_c/GH_{fg} = 0.22\,(\rho_v/\rho_\ell)^{0.133}\,(\sigma\rho_\ell/G^2\ell_b)^{0.433}\,(\ell_b/d_{he})^{0.171} \cdot 1/(1 + 0.0081\,\ell_b/d_{he}) \tag{34}$$

where d_{he}, the heated equivalent diameter, is given from Equation 2 as $d_{he} = (d_0^2 - d_i^2)/d_i$, where d_0 is the diameter of the shroud and d_i is the diameter of the rod.

Rectangular Channel with Either One Side Wall or Two Facing Walls Heated Uniformly

Reference 32 has presented the following correlations of CHF for $\rho_v/\rho_\ell < 0.15$:

$$q_c/GH_{fg} = C\,(\sigma\rho_\ell/G^2\ell_b)^{0.043}/(\ell_b/d_{he}) \tag{35}$$

$$q_c/GH_{fg} = 0.15\,(\rho_v/\rho_\ell)^{0.133}\,(\sigma\rho_\ell/G^2\ell_b)^{1/3} \cdot 1/(1 + 0.0077\,\ell_b/d_{he}) \tag{36}$$

$$q_c/GH_{fg} = 0.26\,(\rho_v/\rho_\ell)^{0.133}\,(\sigma\rho_\ell/G^2\ell_b)^{0.433}\,(\ell_b/d_{he})^{0.171} \cdot 1/(1 + 0.0077\,\ell_b/d_{he}) \tag{37}$$

where C is given as $C = 0.25$ when ℓ_b/d_{he} is small (say, $\ell/d_{he} < 55$) and $C = 0.34$ when ℓ_b/d_{he} is large (say, $\ell_b/d_{he} > 110$) in a manner similar to Equation 26. The heated equivalent diameter d_{he} in Equation 35 through 37 had been defined by Equation 2.

FINAL REMARKS

All the examples employed in this chapter to explain the prediction of the variation of wall temperature and void fraction along the tube length are concerned with water at atmospheric pressure. This condition of fluid has been selected only because the physical states of steam and water at atmospheric pressure are so familiar to many people that readers may imagine the configuration of two-phase flow very easily. However, it must be remembered that water at atmospheric pressure is not necessarily a typical condition for boiling, because the vapor-to-liquid density ratio is exceedingly low ($\rho_v/\rho_\ell = 0.000624$) in this case.

NOTATION

C	constant (−), Equations 25, 28 and 34
C_0	constant (−), Equation 15
$c_{p\ell}$	specific heat of liquid at constant pressure ($Jkg^{-1}K^{-1}$)
d	diameter of tube (m)
d_{hd}	hydraulic equivalent diameter (m)
d_{he}	heated equivalent diameter (m), Equation 2
d_i	diameter of rod in annular channel (m)
d_0	diameter of shroud in annular channel (m)
F	multiplier (−), Equations 16b and 18
G	mass velocity ($kgm^{-2}s^{-1}$)
g	gravitational acceleration (ms^{-2})
H_{fg}	latent heat of evaporation (Jkg^{-1})
ΔH_i	inlet subcooling enthalpy (Jkg^{-1})
h_ℓ	heat transfer coefficient if the liquid were flowing alone in tube ($Wm^{-2}K^{-1}$), Equations 16 and 17
h_m	heat transfer coefficient of single phase liquid flow ($Wm^{-2}K^{-1}$)
ℓ	length of heated channel (m)
ℓ_b	boiling length (m), Figure 1
Pr_ℓ	Prandtl number of liquid (−)
p	system pressure (Pa)
p_r	reduced pressure (−)
q	heat flux (Wm^{-2})
q_c	critical heat flux (Wm^{-2})
q_ℓ	heat flux estimated by Equation 17 (Wm^{-2})
q_n	heat flux estimated by Equation 21 (Wm^{-2})
r_{crit}	critical radius of bubble nucleus (m)
S	nucleate boiling suppression factor (−), Equations 16b and 22
T_b	bulk temperature of liquid flow (K)
T_{bi}	value of T_b at tube inlet (K)
T_s	saturation temperature at system pressure (K)

T_w wall temperature (K)
u G/ρ_ℓ
X Lockhart–Martinelli parameter (−), Equation 19
Y dimensionless parameter (−), Equation 25
y_0 thickness of bubble growth region (m), Equation 23
z axial distance (m)

Greek Symbols

α void fraction (−)
$\varkappa_\ell$ thermal diffusivity of liquid (m^2s^{-1})
λ_ℓ thermal conductivity of liquid ($Wm^{-1}K^{-1}$)
μ_ℓ dynamic viscosity of liquid (Pa · s)
μ_v dynamic viscosity of vapor (Pa · s)
ρ_ℓ density of liquid (kgm^{-3})
ρ_v density of vapor (kgm^{-3})
σ surface tension (Nm^{-1})
χ equilibrium vapor quality (−)
χ^* true vapor quality (−)
χ_{ex} value of χ at the tube exit end (−)

Subscripts

NVG value at the starting point of net vapor generation
ONB value at the onset of nucleate boiling

REFERENCES

1. Collier, J. G., Convective Boiling and Condensation, 2nd ed., McGraw–Hill, New York, 1972, pp. 144–147.
2. Rohsenow, W. M., "Boiling," in *Handbook of Heat Transfer,* Rohsenow, W. M., and Hartnett, J. P., (Eds.), McGraw–Hill, New York, 1973, pp. 13.34–13.46.
3. Shock, R. A. W., "Estimation Methods for Forced Convective Boiling," in *Two-Phase Flow and Heat Transfer,* Butterworth, D., and Hewitt, G. F., (Eds.), Oxford Univ. Press, Oxford, 1977, pp. 200–222.
4. Mayinger, F., and Bucher, B., "Subcooled Boiling," in *Two-Phase Flows and Heat Transfer,* Kakaç, S., and Veziroglu, T. N., (Eds.), Vol. II. Hemisphere Pub. Corp., Washington, 1977, pp. 581–621.
5. Hsu, Y. Y., "Preburnot Convective Boiling," in *Handbook of Multiphase Systems,* Hetsroni, G., (Ed.). Hemisphere Pub. Corp., Washington, 1982, pp. 6.50–6.65.
6. Butterworth, D., and Shock, R. A. W., "Flow Boiling," in Proc. 7th Int. Heat Transfer Conference, Vol. 1. Hemisphere Pub. Corp., Washington, 1982, pp. 11–30.
7. Bergles, A. E., and Rohsenow, W. M., "The Determination of Forced-Convection Surface-Boiling Heat Transfer," *Trans ASME,* Ser. C, J. Heat Transfer, Vol. 86 (1984), pp. 365–372.
8. Davis, E. J., and Anderson, G. H., "The Incipience of Nucleate Boiling in Forced Convective Flow," *AIChE J.,* Vol. 12 (1966), pp. 774–780.
9. Yilmaz, S., and Westwater, J. W., "Effect of Velocity on Heat Transfer in Boiling Freon-113," *Trans. ASME,* Ser. C, J. Heat Transfer, Vol. 102 (1980), pp. 26–31.
10. Kutadeladze, S. S., "Heat Transfer in Condensation and Boiling," USAEC Rept-tr-3770, 1952.
11. Moles, F. D., and Shaw, J. F. G., "Boiling Heat Transfer to Sub-Cooled Liquids under Conditions of Forced Convection," *Trans. Inst. Chem. Eng.,* Vol. 50 (1972), pp. 76–84.
12. Saha, P., and Zuber, N., "Point of Net Vapor Generation and Vapor Void Fraction in Subcooled Boiling," in Proc. 5th Int. Heat Transfer Conference, Vol. IV, 1974, pp. 175–179.

13. Hewitt, G. F., "Void Fraction," in *Handbook of Multiphase Systems,* Hetsroni, G., (Ed.), Hemisphere Pub. Corp., Washington, 1982, pp. 2.76–2.94.
14. Chen, J. C., "Correlation for Boiling Heat Transfer to Saturated Fluids in Convective Flow," *Ind. Eng. Chem. Process Design Development,* Vol. 5 (1966), pp. 322–329.
15. Bennett, D. L., and Chen, J. C., "Forced Convective Boiling in Vertical Tubes for Saturated Pure Components and Binary Mixtures," *AIChE J.,* Vol. 26 (1980), pp. 454–461.
16. Lockhart, R. W., and Martinelli, R. C., "Proposed Correlation of Data for Isothermal Two-Phase, Two-Component Flow in Pipes," *Chem. Eng. Prog.,* Vol. 45 (1) (1949), pp. 39–48.
17. Chisholm, D., "A Theoretical Basis for the Lockhart-Martinelli Correlation For Two-Phase Flow," *Int. J. Heat Mass Transfer,* Vol. 10 (1967), pp. 1767–1778.
18. Chisholm, D., and Laird, A. D. K., "Two-Phase Flow in Rough Tubes," *Trans. ASME,* Vol. 80 (1958), pp. 276–286.
19. Forster, H. K., and Zuber, N., "Dynamics of Vapor Bubbles and Boiling Heat Transfer," *AIChE J.,* Vol. 1 (1955), pp. 531–535.
20. Bennett, D. L., Davis, M. W., and Hertzler, B. L., "The Suppression on Saturated Nucleate Boiling by Forced Convective Flow," *AIChE Symposium Ser.,* No. 199, Vol. 76 (1980), pp. 91–103.
21. Hewitt, G. F., and Roberts, D. N., "Studies of Two-Phase Flow Pattern by Simultaneous X-Ray and Flush Photography," UKAEA, AERE-M 2159, 1969.
22. Shah, M. M., "A New Correlation for Heat Transfer During Boiling Flow Through Pipes," *ASHRAE Trans.,* Vol. 82 (1976), pp. 66–86.
23. Lazarek, G. M., and Black, S. H., "Evaporative Heat Transfer, Pressure Drop and Critical Heat Flux in a Small Vertical Tube with R-113," *Int. J. Heat Mass Transfer,* Vol. 25 (1982), pp. 945–960.
24. Collier, J. G., *Convective Boiling and Condensation,* 2nd ed., McGraw–Hill, New York, 1972, pp. 259–269.
25. Lee, D. H., "Prediction of Burnout," in *Two-Phase Flow and Heat Transfer,* Butterworth, D., and Hewitt, G. F., (Eds.), Oxford Univ. Press, Oxford, 1977, pp. 295–322.
26. Shah, M. M., "A Generalized Graphical Method for Predicting CHF in Uniformly Heated Vertical Tubes," *Int. J. Heat Mass Transfer,* Vol. 22 (1979), pp. 557–568.
27. Katto, Y., "General Features of CHF of Forced Convection Boiling in Uniformly Heated Vertical Tubes with Zero Inlet Subcooling," *Int. J. Heat Mass Transfer,* Vol. 23 (1980), pp. 493–504.
28. Katto, Y., "A Generalized Correlation of Critical Heat Flux for the Forced Convection Boiling in Vertical Uniformly Heated Round Tubes," *Int. J. Heat Mass Transfer,* Vol. 21 (1978), pp. 1527–1542.
29. Katto, Y., and Ohno, H., "An Improved Version of the Generalized Correlation of Critical Heat Flux for the Forced Convective Boiling in Uniformly Heated Vertical Tubes," *Int. J. Heat Mass Transfer,* Vol. 26, No. 8 (1984).
30. Katto, Y., "Generalized Correlation of Critical Heat Flux for the Forced Convection Boiling in Vertical Uniformly Heated Annuli." *Int. J. Heat Mass Transfer,* Vol. 22 (1979), pp. 575–584.
31. Katto, Y., "General Features of CHF of Forced Convection Boiling in Vertical Concentric Annuli with a Uniformly Heated Rod and Zero Inlet Subcooling," *Int. J. Heat Mass Transfer,* Vol. 24 (1981), pp. 109–116.
32. Katto, Y., "General Features of CHF of Forced Convection Boiling in Uniformly Heated Rectangular Channels," *Int. J. Heat Mass Transfer,* Vol. 24 (1981), pp. 1413–1419.

CHAPTER 10

TRANSIENT BOILING HEAT TRANSFER UNDER FORCED CONVECTION

Isao Kataoka

Institute of Atomic Energy, Kyoto University
Uji, Kyoto 611, Japan

Akimi Serizawa

Department of Nuclear Engineering
Kyoto University, Sakyo-ku
Kyoto 606, Japan

CONTENTS

INTRODUCTION

Transient boiling heat transfer under forced convection is extremely important in analyzing thermohydrodynamic characteristics of various industrial equipment such as petrochemical plants and nuclear power plants. In normal operation of a plant, the transient boiling phenomenon sometimes occurs in the start-up or shut-down of the plant and also during a change of the operational conditions. It becomes more important particularly when the plants fall into off-normal conditions. An accurate knowledge of the transient boiling phenomenon is thus indispensable in the safety evaluation of the plants under accidental conditions.

There are three types of transients: power transients, flow transients, and pressure transients. The power transient is the phenomenon where the heat generation rate or heat input in a plant varies with time. In view of plant safety, a power increase is potentially disastrous and is thus more important than a power decrease. A typical example of the power transient is a reactivity accident in a nuclear reactor where the rate of heat generation in the reactor core increases exponentially with time as a result of an excessive reactivity insertion due to an accidental withdrawal of control rods.

The flow transient is caused by a reduction or loss of the pumping head or flow blockages with the flow decreasing with time. A pump-failure or loss of coolant accident in a nuclear power plant (LOCA) is one aspect of this mode of transient.

In pressure transients, the pressure of a plant varies with time. This transient is caused by a pipe rupture or misopening of relief valves and so on. The rate of pressure decrease depends on the size of the rupture or valve orifices. Since chemical and nuclear power plants are mostly operated at high pressure and high temperature, a sudden pressure decrease will result in a violent transient boiling accompanied by a large amount of the fluid rushing out. Coolability of the plant may be therefore deteriorated considerably. The LOCA is one of the most severe accidents in pressure transients because it may lead to a complete destruction of the reactor core if sufficient cooling is not secured.

In some plant accidents these three types of the transients take place at the same time, and the situation becomes more complicated. Therefore, accurate knowledge of the boiling phenomena during power, flow, and pressure transients is needed in an evaluation of the thermohydraulic characteristics of the plants under such accident conditions.

There are several important aspects involved in the transient boiling heat transfer under forced convection. They are the initiation of boiling, the boiling heat transfer coefficient, and the critical heat flux during postulated transients. Among these, the transient critical heat flux (transient CHF) is the most important parameter in practice. This is because once the heat flux exceeds the transient CHF, the temperature of a heated wall increases rapidly, leading to the melting or damaging of wall materials. Therefore, the transient CHF will be stressed in this text.

There are several mechanisms for CHF [1, 2] which are widely accepted. The fast burnout is encountered when the transition from nucleate boiling to film boiling occurs nearly instantaneously in a low steam quality flow (bubbly or slug flow). On the other hand, the slow burnout or dryout which corresponds to the transition from liquid film flow evaporation to vapor-phase heat transfer takes place in a high steam quality flow (annular or annular-dispersed flow). In general, the transient boiling phenomena are influenced by many factors. These are pressure, flow velocity, rate of transients, heat flux, flow geometry, etc. Among these relevant parameters, geometrical factors have crucial effects on the phenomena. The most important is a ratio between a heated length, ℓ, and a heated equivalent diameter, d_{he}, of the flow passage. In fact, experimental evidence indicates that the transient boiling phenomena depend very much on the ℓ/d_{he} ratio.

For large ℓ/d_{he}, the transient boiling aspects are greatly influenced by upstream conditions. The transient CHF which usually happens near the exit of a heated section is then determined by two-phase flow conditions at the exit. Since the local conditions of the two-phase flow there reflect the integrated upstream effects, the ℓ/d_{he} ratio has certainly a major influence on the transient CHF. The transient boiling characteristics under this circumstance are therefore basically determined by the transient two-phase flow characteristics. In view of this, accurate knowledges of the transient two-phase flow is also much needed. Practically, extensive progresses have been made recently on the transient two-phase flows [3], and the analysis of the transient boiling phenomena for large ℓ/d_{he} has been made possible using relevant two-phase flow models. The outline of the analysis will be described in the first part of this chapter, based on a simplified two-phase flow model.

When the ℓ/d_{he} ratio is small, the situation is quite different from that for large ℓ/d_{he}. The transient boiling phenomenon for small ℓ/d_{he} is supposed to be a localized phenomenon, and hence, it can not be predicted effectively by the transient two-phase flow models. It is well considered that the microscopic flow and heat transfer structure near the heated surface dominate the transient boiling mechanism for small ℓ/d_{he}. This interesting theme will be treated in the latter half of this chapter, referring to the experimental results which have been obtained so far.

TRANSIENT BOILING FOR LARGE ℓ/d_{he}

In this section, basic formulations of two-phase flow are described, and then, a simplified model is presented to predict the boiling phenomena in flow, power, and pressure transients with large ℓ/d_{he}. The experimental results on the transient boiling for large ℓ/d_{he} [4–28] are also reviewed.

Basic Formulation of Two-Phase Flow

An accurate formulation of gas-liquid two-phase flow is essential to the analysis of the transient boiling phenomena. A physically rigorous formulation of the gas-liquid two-phase flow is however a difficult task because both phases have complicated configurations and interact with each other. Therefore, appropriate approximations and time- and space-averagings are usually adopted in the practical formulations of two-phase flow. In fact, there have been proposed various formulations of two-phase flow with different approximations and averaging methods [29–36], i.e., homogeneous models, drift-flux models, and two-fluid models.

In a homogeneous model, a two-phase flow is treated as a mixture where gas and liquid phases are assumed to move at the same speed. For thermal equilibrium one-dimensional flow in a duct, continuity equations of mass, momentum, and energy for the two-phase mixture are given by

Mass

$$\frac{\partial \rho_m}{\partial t} + \frac{\partial}{\partial z}(\rho_m u) = 0 \tag{1}$$

Momentum

$$\rho_m \frac{\partial u}{\partial t} + \frac{\partial}{\partial z}(\rho_m u^2) = -\frac{\partial P}{\partial z} - \rho_m g \cos\theta - \frac{f}{2d_{he}} \rho_m u|u| \tag{2}$$

Energy

$$\frac{\partial}{\partial t}(\rho_m H_m) + \frac{\partial}{\partial z}(\rho_m u H_m) = \frac{\partial P}{\partial t} + \frac{4q}{d_{he}} \tag{3}$$

Here ρ_m is a mixture density defined as

$$\rho_m = \alpha\rho_g + (1-\alpha)\rho_\ell \tag{4}$$

and ρ_g, ρ_ℓ, and α are the densities of the gas and liquid phase and void fraction, respectively. The mixture enthalpy H_m is given in terms of the quality x as

$$H_m = xH_g + (1-x)H_\ell \tag{5}$$

where H_g and H_ℓ are the enthalpy of gas and liquid phases. u, P, f, d_{he} and q are velocity, pressure, friction factor, heated equivalent diameter, and heat flux, respectively. In homogeneous models, the void fraction and the quality are interrelated by

$$\alpha = x\rho_\ell/[x\rho_\ell + (1-x)\rho_g] \tag{6}$$

The drift flux model considers basically a two-phase flow as a mixture. However, in its formulation, the relative motion between the phases is taken into account in terms of the "drift flux" or the "drift velocity." The drift velocity V_{gj} and the drift flux j_{gj} are defined here as

$$V_{gj} = u_g - j \tag{7}$$

$$j_{gj} = \alpha V_{gj} \tag{8}$$

Here, u_g and j are the mass-weighted mean velocity of the gas-phase and mixture center-of-volume velocity which is given by

$$j = j_g + j_\ell \tag{9}$$

The volumetric flux of the gas and liquid phases, j_g and j_ℓ are

$$j_g = \alpha u_g \tag{10}$$

$$j_\ell = (1 - \alpha) u_\ell \tag{11}$$

The drift flux formulation of two-phase flows has been significantly developed by Zuber and Findlay [33], Wallis [32], and Ishii [30]. It consists of the continuity equations of mass, momentum, and energy for the mixture and gas phase mass continuity equation. For thermal equilibrium one-dimensional flow, they are written by

Mixture mass

$$\frac{\partial \rho_m}{\partial t} + \frac{\partial}{\partial z}(\rho_m u_m) = 0 \tag{12}$$

Gas phase mass

$$\frac{\partial}{\partial t}(\alpha \rho_g) + \frac{\partial}{\partial z}(\alpha \rho_g u_m) = \Gamma_g - \frac{\partial}{\partial z}\left(\frac{\alpha \rho_g \rho_\ell}{\rho_m} V_{gj}\right) \tag{13}$$

Mixture momentum

$$\frac{\partial}{\partial t}(\rho_m u_m) + \frac{\partial}{\partial z}(\rho_m u_m^2) = -\frac{\partial P_m}{\partial z} - \rho_m g \cos\theta - \frac{\partial}{\partial z}\left\{\frac{\alpha \rho_g \rho_\ell}{(1-\alpha)\rho_m} V_{gj}^2\right\} - \frac{f_m}{2d_{he}} \rho_m u_m |u_m| \tag{14}$$

Mixture energy

$$\frac{\partial}{\partial t}(\rho_m H_m) + \frac{\partial}{\partial z}(\rho_m u_m H_m) = \frac{4q}{d_{he}} - \frac{\partial P_m}{\partial t} - \frac{\partial}{\partial z}\left\{\frac{\alpha \rho_g \rho_\ell}{\rho_m} H_{fg} V_{gj}\right\} \tag{15}$$

Here, the mixture velocity u_m is defined as

$$u_m = [\alpha \rho_g u_g + (1 - \alpha) \rho_\ell u_\ell]/\rho_m \tag{16}$$

Γ_g appearing in Equation 13 is a vapor source term which is given by

$$\Gamma_g = q/H_{fg} \tag{17}$$

where H_{fg} is the latent heat of vaporization. With an assumption of no flow field gradients, the drift flux is given in terms of the bubble rise velocity u_∞ in an infinite medium as

$$j_{gj} = \alpha (1 - \alpha)^n u_\infty \ (n = 0 \sim 3) \tag{18}$$

An extensive work on the drift velocity has been conducted by Ishii [30], and the correlations for the drift velocity have been theoretically derived for various two-phase flow regimes [37]. Recent papers by Serizawa and Kataoka [38] and Žun and Serizawa [39] discuss the drift velocity in two-phase shear flows.

The two-fluid model is one of the most sophisticated and accurate two-phase flow formulations. It considers each phase separately. Therefore, the continuity equations of mass, momentum, and energy are necessary for each phase, and then, we have six continuity equations. Interactions between phases appear as interfacial terms in these equations [29, 31, 32, 35, 36]. The balance equations for thermal equilibrium one-dimensional flow in a duct are given by Ishii [29, 31] as

Mass

$$\frac{\partial \alpha_k}{\partial t} + \frac{\partial}{\partial z}(\alpha_k \rho_k u_k) = \Gamma_k \tag{19}$$

Momentum

$$\frac{\partial}{\partial t}(\rho_k \alpha_k u_k) + \frac{\partial}{\partial z}(\alpha_k \rho_k u_k^2) = M_{ik} - \frac{4\alpha_{kw}\tau_{iw}}{d_{he}} - \alpha_k \frac{\partial}{\partial z} P_k + \frac{4\alpha_{kw}\tau_{kw}}{d_{he}} + \alpha_k \rho_k g \cos\theta + u_{ki}\Gamma_k \tag{20}$$

Energy

$$\frac{\partial}{\partial t}(\alpha_k \rho_k H_k) + \frac{\partial}{\partial z}(\alpha_k \rho_k H_k u_k) = -\frac{4\alpha_{kw} q_w}{d_{he}} + \alpha_k \left(\frac{\partial}{\partial t} + u_k \frac{\partial}{\partial z}\right) P_k + H_{ik}\Gamma_k + \frac{q_{ki}}{L_s} \tag{21}$$

Here, the subscript k takes g for gas-phase and ℓ for liquid-phase, respectively. The subscripts i and w denote the values at gas-liquid interface and at the duct wall. τ and L_s are shear stress and a reciprocal of the interfacial area concentration a_i which is defined as the interfacial area per unit volume of the two-phase flow. That is

$$L_s = 1/a_i \tag{22}$$

The interfacial drag force M_{ik} satisfies

$$M_{ig} = -M_{i\ell} \tag{23}$$

In order to solve the previously mentioned conservation equations, we need the constitutive equations for the interfacial terms such as M_{ik}, τ_{iw}, u_{ki}, Γ_k, q_{ki} and H_{ik}. They are strongly dependent upon the two-phase flow regimes [3, 40–44].

As an example, we will now consider an annular-dispersed flow where the liquid phase flows as a liquid film on the duct wall and as small droplets entrained in the gas-phase core (Figure 1), since it is closely related to the burnout mechanism in transient boiling with large ℓ/d_{he}. For this flow situation the two-fluid model formulation has been applied recently by Kataoka and Ishii [43]. In this formulation, three sets of continuity equations for mass, momentum, and energy should be considered (i.e., two-fluid three-field formulation). According to Kataoka and Ishii [43], the mass continuity equations for droplets, liquid film, and gas phase are written as

Droplets

$$\frac{\partial}{\partial t}(\alpha_{\ell e}\rho_\ell) + \frac{\partial}{\partial z}(\alpha_{\ell e}\rho_\ell u_{\ell e}) = a_{i\ell f}(\dot{\varepsilon} - \dot{d}) + a_{i\ell e}\overline{m_{\ell e}} \tag{24}$$

Liquid film

$$\frac{\partial}{\partial t}(\alpha_{\ell f}\rho_\ell) + \frac{\partial}{\partial z}(\alpha_{\ell f}\rho_\ell u_{\ell f}) = -a_{i\ell f}(\dot{\varepsilon} - \dot{d}) + a_{i\ell f}\overline{m_{\ell f}} \tag{25}$$

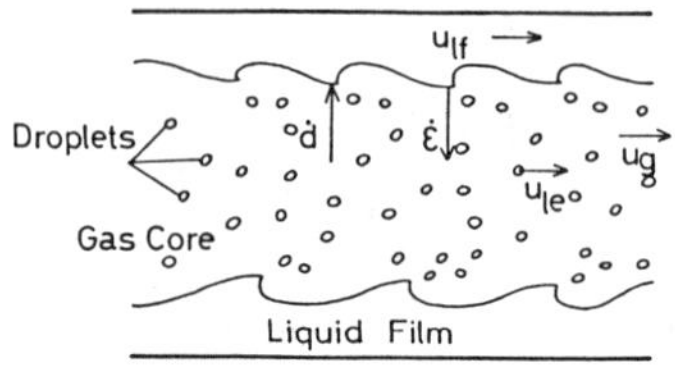

Figure 1. Annular dispersed flow model.

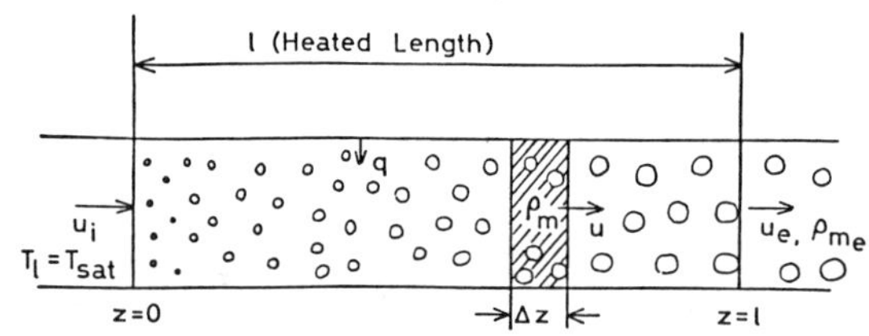

Figure 2. A simplified model of saturated flow boiling.

Gas

$$\frac{\partial}{\partial t}(\alpha\rho_g) + \frac{\partial}{\partial z}(\alpha\rho_g u_g) = -(a_{i\ell e}\overline{m_{\ell e}} + a_{i\ell f}\overline{m_{\ell f}}) \tag{26}$$

Here, a_i, u, and $\bar{m}$ are the interfacial area concentration, velocity, and the mass transfer rate between the droplets and the liquid film, for example, the evaporation and condensation rate for the flow with phase change. $\dot{\varepsilon}$ and $\dot{d}$ are the rate of entrainment and deposition, respectively. The subscripts ℓe, ℓf, and g represent liquid droplets, liquid film, and gas-phase, respectively. Constitutive equations for the previously mentioned parameters have been given also by Kataoka and Ishii [43, 44].

Among these three types of two-phase flow formulations, the two-fluid model reflects most accurately the physical aspects of two-phase flows. It is, however, noted that the two-fluid model formulation provides only numerical solutions for transient boiling phenomena at the present moment. The homogeneous model is, on the other hand, the most simplified model, but the accuracy of the prediction may be poorer than the others. The advantage of this model is that, under certain conditions, analytical solutions can be derived with this formulation, giving comprehensive overviews of the parameter-dependences of the transient boiling phenomena.

In the following sections will be described analyses based on the homogeneous model and some typical experimental results on the transient boiling phenomena with large ℓ/d_{he}.

Homogeneous Model Approximation for Transient Boiling Performance

In this section, we will consider general performances of the transient boiling including flow and power transients. We will start with a simplified homogeneous formulation of one-dimensional equilibrium two-phase flow in a duct. It is assumed here for the simplicity of analysis that the pressure drop along the flow passage and the external forces are neglected. We also assume a constant system pressure. Then, we may well consider only Equations 1 and 3 for the balance equations, which can be rewritten as

$$\frac{\partial \rho_m}{\partial t} + u\frac{\partial \rho_m}{\partial z} + \rho_m \frac{\partial u}{\partial z} = 0 \tag{27}$$

$$\rho_m \frac{\partial H_m}{\partial t} + \rho_m u \frac{\partial H_m}{\partial z} = \frac{4q}{d_{he}} \tag{28}$$

These equations can be solved using the Lagrangian coordinate [11, 18, 27, 45]. The Lagrangian derivative is given by

$$\frac{D}{Dt} = \frac{\partial}{\partial t} + u\frac{\partial}{\partial z} \tag{29}$$

Equations 27 and 28 are therefore replaced by

$$\frac{D\rho_m}{Dt} + \rho_m \frac{\partial u}{\partial z} = 0 \tag{30}$$

$$\rho_m \frac{DH_m}{Dt} = \frac{4q}{d_{he}} \tag{31}$$

Saturated Boiling

Saturated boiling is now considered first where the fluid enters a heated section at saturation temperature.

In order to solve Equations 30 and 31 for given initial and boundary conditions, we now consider, for convenience, fluid particles moving at a speed of u, as shown schematically in Figure 2. The basic concept here is that the fluid particles moving at a plane z are accelerated by an expansion of the fluid particles included in the region $z = 0 \sim z$ due to a change of phase during a postulated transient. The initial and boundary conditions are given by

$$u(0, t) = u_i(t) \tag{32}$$

$$\rho_m(0, t) = \rho_\ell \tag{33}$$

$$\begin{aligned} q(z, t) &= q(t) \text{ for } 0 \leqq z \leqq \ell \\ &= 0 \quad \text{for } z \leqq 0, z \geqq \ell \end{aligned} \tag{34}$$

For simplicity of the analysis, the heat flux is assumed uniform along the heated section. The amount of the heat which the fluid particles existing between $z(t) \sim z(t) + \Delta z(t)$ receive from the heated wall during a small time $t \sim t + \Delta t$ is given by

$$\Delta Q = p_h q(t) \Delta z(t) \Delta t \tag{35}$$

where p_h is a heated perimeter of the flow passage. Under thermal equilibrium conditions, this amount of heat is used to vaporize the liquid by $\Delta Q/H_{fg}$. Therefore, the volume of the fluid particles expands by ΔV which is given by

$$\Delta V = \Delta Q \left(\frac{1}{\rho_g} - \frac{1}{\rho_\ell}\right) / H_{fg} \tag{36}$$

The thickness, $\Delta z(t)$, of the fluid now considered becomes $\Delta z(t + \Delta t)$ at $t + \Delta t$.

$$\Delta z(t + \Delta t) - \Delta z(t) = \Delta V / A_s \tag{37}$$

where A_s is the flow area.

By substituting Equations 35 and 36 into Equation 37, we have

$$\frac{\Delta z(t+\Delta t)-\Delta z(t)}{\Delta t}=\frac{4}{d_{he}H_{fg}}\left(\frac{1}{\rho_g}-\frac{1}{\rho_\ell}\right)q(t)\,\Delta z(t) \tag{38}$$

Since the left-hand-side of Equation 38 represents the acceleration of the fluid particles $\Delta u(t)$, Equation 38 reduces to the following differential equation for $\Delta t \rightarrow 0$.

$$\frac{du}{dz}=\frac{4q(t)\,\Delta\rho}{d_{he}\rho_g\rho_\ell H_{fg}} \tag{39}$$

The density ρ_m of the mixture is, on the other hand, obtained as Equation 41 from a mass balance equation, Equation 40, and Equation 38 in the same manner, i.e., $\Delta t \rightarrow 0$.

$$A_s\,\Delta z(t+\Delta t)\,\rho_m(t+\Delta t)=A_s\,\Delta z(t)\,\rho_m(t) \tag{40}$$

$$\frac{d\rho_m(t)}{dt}=-\frac{4q(t)\,\Delta\rho}{d_{he}\rho_g\rho_\ell H_{fg}}\,\rho_m(t) \tag{41}$$

By solving Equations 39 and 41 with initial and boundary conditions given by Equations 32 ~ 34, we have thermohydraulic parameters of transient two-phase flow in a following manner.

First, by integrating Equation 39, we obtain the velocity of the fluid as

$$u=\frac{4q(t)\,\Delta\rho}{d_{he}\rho_g\rho_\ell H_{fg}}\,z+u_i(t) \tag{42}$$

By considering $u = dz/dt$, Equation 42 is rewritten by

$$\frac{dz}{dt}=\frac{4q(t)\,\Delta\rho}{d_{he}\rho_g\rho_\ell H_{fg}}\,z+u_i(t) \tag{43}$$

Integration of Equation 43 with the initial condition,

$$z = 0 \text{ at } t = t_i \tag{44}$$

yields

$$z=\int_{ti}^{t}u_i(t')\exp\left[-\frac{4\Delta\rho}{d_{he}\rho_g\rho_\ell H_{fg}}\int_{ti}^{t'}q(t'')\,dt''\right]dt' \cdot\exp\left[\frac{4\Delta\rho}{d_{he}\rho_g\rho_\ell H_{fg}}\int_{ti}^{t}q(t')\,dt'\right] \tag{45}$$

If $u_i(t)$ and $q(t)$ are known, the time t_i at which the fluid particles pass $z = 0$ is calculated in a form of

$$t_i = t_i(z, t) \tag{46}$$

The density of the two-phase mixture is obtained by integrating Equation 41 as

$$\rho_m(z,t)=\rho_\ell\exp\left(-\int_{ti(z,t)}^{t}\frac{4q(t')\,\Delta\rho}{d_{he}\rho_g\rho_\ell H_{fg}}\,dt'\right) \tag{47}$$

The void fraction $\alpha(z, t)$, steam quality $x(z, t)$, and the mixture enthalpy $H_m(z, t)$ are then calculated from Equations 4 ~ 6 and 47. Results are summarized in Table 1.

Once the initial and boundary conditions are specified, the thermohydraulic parameters of the transient two-phase flow can be thus known based on a simple homogeneous model. The solution scheme adopted here is called "method of characteristics," and Equation 43 is "characteristic equation" for partially differential equations (Equations 27 and 28). The method of characteristics has been applied earlier by Wallis [46], Zuber [47], and Lahey et al. [11, 45] to obtain the solutions of the transient two-phase flow based on the homogeneous formulations. Results developed here are more general since they include both power and flow transients.

Subcooled Boiling

When the fluid temperature at the inlet to the heated section is below saturation temperature, a nonboiling region appears initially. Therefore, the problem now is how to determine the location of boiling initiation. A simple assumption is that boiling commences when the fluid temperature exceeds saturation temperature. Although this approximation

Table 1
Thermohydraulic Parameters of Transient Two-Phase Flow Based on the Homogeneous Model (Saturated at Inlet)

Mixture velocity	$u(z,t) = \frac{4q(t)\Delta\rho}{d_{he}\rho_g\rho_\ell H_{fg}} z + u_i(t)$
Mixture density	$\rho_m(z,t) = \rho_\ell \exp\left[-\int_{t_i(z,t)}^{t} \frac{4q(t')\Delta\rho}{d_{he}\rho_g\rho_\ell H_{fg}} dt'\right]$ where $t_i(z,t)$ is given by $z = \int_{t_i(z,t)}^{t} u_i(t') \exp\left[-\frac{4\Delta\rho}{d_{he}\rho_g\rho_\ell H_{fg}} \int_{t_i(z,t)}^{t'} q(t'')\,dt''\right] dt'$ $\cdot \exp\left[\frac{4\Delta\rho}{d_{he}\rho_g\rho_\ell H_{fg}} \int_{t_i(z,t)}^{t} q(t')\,dt'\right]$
Void fraction	$\alpha(z,t) = \frac{\rho_m(z,t) - \rho_\ell}{\Delta\rho}$ $= \frac{\rho_\ell}{\Delta\rho}\left\{\exp\left[-\int_{t_i(z,t)}^{t} \frac{4q(t')\Delta\rho}{d_{he}\rho_g\rho_\ell H_{fg}} dt'\right] - 1\right\}$
Quality	$x = \frac{\rho_g\rho_\ell}{\Delta\rho}\left(\frac{1}{\rho_g} - \frac{1}{\rho_m}\right)$ $= \frac{\rho_g}{\Delta\rho}\left\{\exp\left[\int_{t_i(z,t)}^{t} \frac{4q(t')\Delta\rho}{d_{he}\rho_g\rho_\ell H_{fg}} dt'\right] - 1\right\}$
Mixture enthalpy	$H_m = H_\ell + x H_{fg}$ $= H_\ell + \frac{\rho_g H_{fg}}{\Delta\rho}\left[\exp\left(\int_{t_i(z,t)}^{t} \frac{4q(t')\Delta\rho}{d_{he}\rho_g\rho_\ell H_{fg}} dt'\right) - 1\right]$

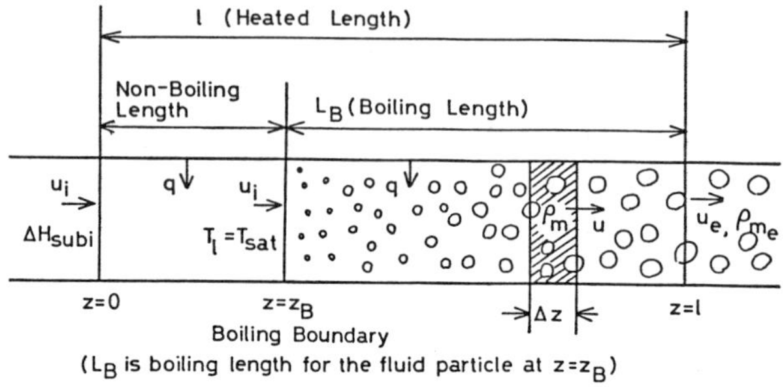

Figure 3. A simplified model of subcooled flow boiling.

is rough in some respect, it gives satisfactory results for the transient boiling with large ℓ/d_{he} [11, 46]. We therefore make this assumption, too. The boiling situation now considered is schematically shown in Figure 3.

Let us consider the motion of the fluid particles existing at a location z at time t. These fluid particles have passed z = 0, the inlet of the heated section, earlier at time $t_i(z, t)$, and the point of boiling initiation (boiling boundary) at time $t_{iB}(z, t)$.

Assuming thermal equilibrium, $t_{iB}(z, t)$ can be determined by Equation 48 in terms of the inlet subcooling enthalpy.

$$\int_{t_i(z,t)}^{t_{iB}(z,t)} q(t')\,dt' = \frac{d_{he}\rho_\ell \Delta H_{subi}}{4} \tag{48}$$

and

$$\Delta H_{sub} = C_{p\ell}(T_{sat} - T_{in}) \tag{49}$$

where T_{in} and T_{sat} are inlet liquid temperature and saturation temperature, respectively. If the thermal expansion of the liquid is neglected in the nonboiling region, the fluid particles are not accelerated there and their velocity is given by $u_i(t)$ (Equation 32). The mixture density is equal to ρ_ℓ. Therefore, Equations 42, 45, and 47 should be modified in the following manner to take into account the effect of nonboiling region.

Mixture velocity

$$\begin{aligned} u(z,t) &= u_i(t) && \text{for } t < t_{iB}(z,t) \\ &= \frac{4q(t)\Delta\rho}{d_{he}\rho_g\rho_\ell H_{fg}} L_B + u_i(t) && \text{for } t \geqq t_{iB}(z,t) \end{aligned} \tag{50}$$

Characteristic equation

$$\begin{aligned} z = {} & \int_{t_i(z,t)}^{t_{iB}(z,t)} u_i(t')\,dt' + \int_{t_{iB}(z,t)}^{t} u_i(t') \exp\left[-\frac{4\Delta\rho}{d_{he}\rho_g\rho_\ell H_{fg}} \int_{t_{iB}(z,t)}^{t'} q(t'')\,dt''\right] dt' \\ & \cdot \exp\left(\frac{4\Delta\rho}{d_{he}\rho_g\rho_\ell H_{fg}} \int_{t_{iB}(z,t)}^{t} q(t')\,dt'\right) \end{aligned} \tag{51}$$

Mixture density

$$\rho_m(z,t) = \rho_\ell \qquad \text{for } t < t_{iB}(z,t)$$
$$= \rho_\ell \exp\left(-\int_{tiB(z,t)}^{t} \frac{4q(t')\Delta\rho}{d_{he}\rho_g\rho_\ell H_{fg}} dt'\right) \qquad \text{for } t \geqslant t_{iB}(z,t) \tag{52}$$

Here, L_B appearing in Equation 50 is a boiling length which the fluid particles existing at (z, t) experience during their travels, and is given by

$$L_B(z,t) = z - \int_{ti(z,t)}^{tiB(z,t)} u_i(t')\,dt' \tag{53}$$

The location of the boiling boundary z_B can be obtained in terms of the boiling inception time at z_B, i.e., t_B, by putting

$$t_B = t_{iB}(z_B, t_B) \tag{54}$$

in Equations 48 and 49. The results obtained are

$$\int_{ti(zB,tB)}^{tB} q(t')\,dt' = \frac{d_{he}\rho_\ell \Delta H_{sub}}{4} \tag{55}$$

$$\int_{ti(zB,tB)}^{tB} u_i(t')\,dt' = z_B \tag{56}$$

For given t_B, z_B is calculated by eliminating $t_i(z_B, t_B)$ from Equations 55 and 56. On the other hand, if z_B is given, t_B is obtained as a function of z_B in the same way.

The relevant thermohydraulic parameters in the transient boiling under subcooled conditions can be thus estimated from the previous equations.

CHF-Prediction for Power and Flow Transients

In the previous section, the thermohydraulic characteristics of a transient two-phase flow have been shown to be estimated based on the homogeneous model formulations. One of the most important and interesting applications of the results is to predict the transient critical heat flux, CHF. To accomplish this, a CHF criterion based on the mechanisms of CHF is required. In case of slow burnout, the CHF occurs when the liquid film on a heated wall consumes or is disrupted by evaporation and entrainment of liquid droplets [1, 2]. The simplest assumption is that no liquid droplets are entrained in gas flow and, hence, the CHF is specified by the dryout condition, that is, all liquid evaporated. This assumption leads to $x_e = \alpha_e = 1$ and $\rho_{me} = \rho_g$, where the subscript e denotes the value at the exit of the heated section. However, many experimental evidences indicate that, even under steady-state conditions, the exit quality at dryout is not unity as shown schematically in Figure 4 [48], since a certain amount of the liquid flows as liquid droplets entrained in the gas flow. This experimental trend is predicted by using Katto's steady-state dryout correlation [49, 50], for example. This is shown in the following. Katto's steady-state dryout correlation is given by

$$\frac{q_c}{\rho_\ell u_i H_{fg}} = 0.25\left(\frac{\sigma\rho_\ell}{\rho_\ell^2 u_i^2 \ell}\right)^{0.043} \frac{d_{he}}{\ell} \tag{57}$$

This equation is applied to the case where the temperature of the fluid is at saturation temperature. For uniform heat flux distribution over the whole heated section, the exit quality under steady-state conditions is calculated as below by the heat-balance equation.

$$q_c p_h \ell = \rho_\ell u_i A_s x_e H_{fg} \tag{58}$$

By considering $d_{he} = 4A_s/p_h$, we have Equation 59 for the dryout quality x_e at the exit from Equations 57 and 58.

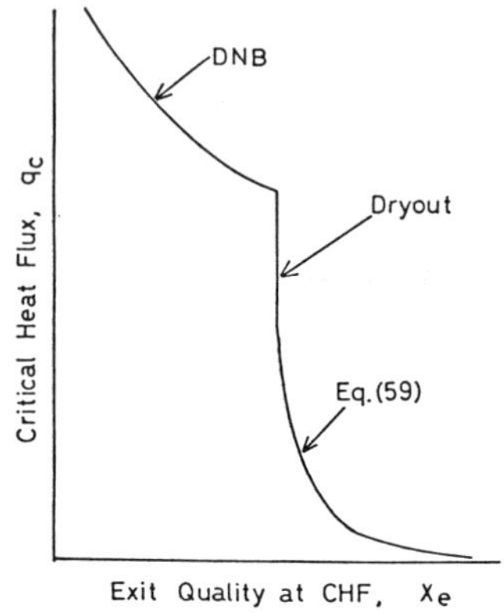

Figure 4. Critical heat flux vs. exit quality.

$$x_e = \left(\frac{\sigma\rho_\ell}{\rho_\ell^2 u_i^2 \ell}\right)^{0.043} = 0.88\left(\frac{d_{he}}{\ell}\right)^{0.094}\left[\frac{\sigma\rho_\ell}{(q_c/H_{fg})^2\ell}\right]^{0.047} \tag{59}$$

Equation 59 reflects the experimental trend of q_c vs. x_e for the dryout region shown in Figure 4.

For more general discussions, the mixture density at the exit where the CHF occurs can be derived from Equations 4, 6, and 58 in terms of the CHF q_c and inlet velocity u_i under steady-state condition. Namely,

$$\frac{\rho_{me}}{\rho_\ell} = \frac{1}{1 + \dfrac{4_{qc}\Delta\rho\ell}{d_{he}\rho_g\rho_\ell u_i H_{fg}}} \tag{60}$$

As has been described earlier, the CHF condition for steady-state is expressed in terms of two-phase flow parameters at the exit. Transient CHF criterion is not so much clear at the moment, but is supposed to be expressed also by two-phase flow parameters. A possible assumption is that the transient CHF occurs at the same two-phase flow situations as steady-state CHF. This hypothesis is supported by some experimental results [11, 14, 16] for large ℓ/d_{he}. The transient CHF is then given by

$$\left(\frac{\rho_{me}}{\rho_\ell}\right) \text{at transient CHF} = \left(\frac{\rho_{me}}{\rho_\ell}\right) \text{at steady-state CHF} \tag{61}$$

The transient CHF for large ℓ/d_{he} in power and flow transients can be therefore predicted by use of the analytical results of the transient two-phase flow described earlier and the foregoing CHF criterion. Some examples follow.

Power Transient – Stepwise Power Increase

$$q(t) = q_1 (< q_{cst}) \text{ for } t < 0, \quad = q_2 (> q_{cst}) \text{ for } t \geqq 0 \tag{62}$$

For $t < 0$, the steady-state boiling is prevailing at a heat flux q_1, which is lower than the steady-state CHF, q_{cst}, corresponding u_{i0}. At $t = 0$, the heat flux suddenly increases in a stepwise manner from q_1 to q_2. The transient CHF occurs when the surface temperature of the heater begins to increase abruptly, as shown in Figure 5. The time required from the initiation of transient to the occurrence of CHF is denoted by t_c (hereafter called "transient CHF time").

Assumptions:

1. Constant system pressure
2. Constant inlet velocity u_{i0}
3. Saturated fluid temperature at the inlet of a heated section.

The thermohydraulic parameters at the exit are calculated from the equations listed in Table 1 by putting

$$z = \ell \tag{63}$$

$$u_i(t) = u_{i0}\,(= \text{const.}) \tag{64}$$

The exit mixture density is then given by

$$\frac{\rho_{me}}{\rho_\ell} = \frac{q_2^*}{q_1^*(1 + q_2^*) + (q_2^* - q_1^*)\exp(q_2^* t^*)} \qquad \text{for } t^* \leqq \frac{1}{q_2^*}\ln(1 + q_2^*)$$

$$= \frac{1}{1 + q_2^*} \qquad \text{for } t^* \geqq \frac{1}{q_2^*}\ln(1 + q_2^*) \tag{65}$$

where t^*, q_1^*, and q_2^* are dimensionless time and heat fluxes defined as follows.

$$t^* \equiv t u_{i0}/\ell \tag{66}$$

$$q_j^* \equiv \frac{4\ell \Delta\rho q_j}{d_{he}\rho_g \rho_\ell H_{fg} u_{i0}} \quad (j = 1 \text{ or } 2) \tag{67}$$

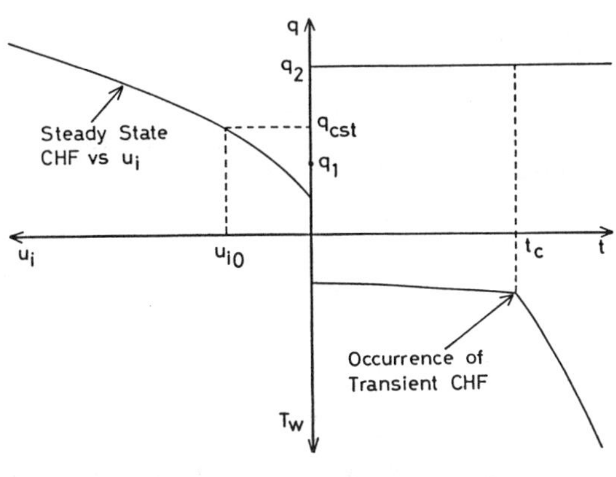

Figure 5. Performance characteristics in stepwise power increase.

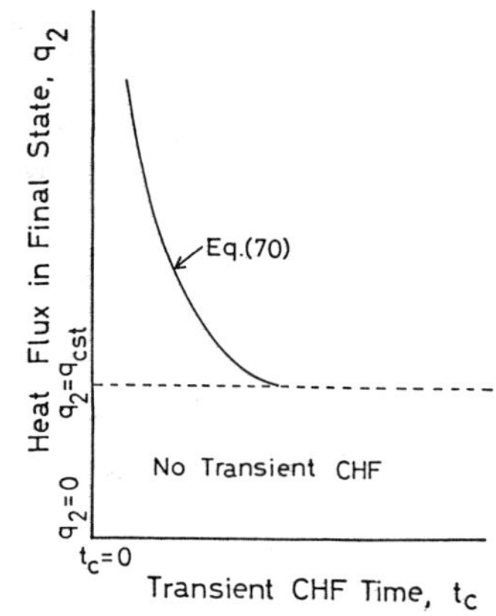

Figure 6. Transient CHF time vs. final heat flux in stepwise power increase.

If the transient CHF occurs at the exit, the mixture density at the transient CHF ($t = t_c$) is obtained straight from Equation 65 by replacing t with t_c that is

$$\left(\frac{\rho_{me}}{\rho_\ell}\right)_{\text{transient CHF}} = \frac{q_2^*}{q_1^*(1 + q_2^*) + (q_2^* - q_1^*)\exp(q_2^* t_c^*)} \tag{68}$$

On the other hand, from Equation 60, the mixture density at steady-state CHF, q_{cst}, is rewritten as

$$\left(\frac{\rho_{me}}{\rho_\ell}\right)_{\text{steady-state CHF}} = \frac{1}{1 + q_{cst}^*} \tag{69}$$

By applying the CHF criterion represented by Equation 61 to this case, Equations 68 and 69 yield the following CHF time t_c^*.

$$t_c^* = \frac{1}{q_2^*}\ln\left[\frac{q_2^*(1 + q_{cst}^*) - q_1^*(1 + q_2^*)}{(q_2^* - q_1^*)}\right] \tag{70}$$

Figure 6 represents Equation 70 in q_2 vs. t_c diagram. This picture shows that the transient CHF time t_c increases as the final heat flux q_2 is decreased.

Power Transient – Exponential Power Increase

$$\begin{aligned} q(t) &= 0 && \text{for } t < 0 \\ &= q_2 \exp(\omega t) && \text{for } t \geqq 0,\ (w > 0) \end{aligned} \tag{71}$$

The previously mentioned assumptions are applied here again. The present situation is schematically shown in Figure 7.

The thermohydraulic parameters at the exit can be calculated from Table 1 and Equations 63 and 64. The mixture density at the exit is thus given by

$$\frac{\rho_{me}}{\rho_\ell} \begin{cases} = \exp\left\{-\frac{q_2}{\omega^*}[\exp(\omega t)-1]\right\} & \text{for } 0 \leqq t \leqq t_0 \\ = \exp\left[-\frac{q_2}{\omega^*}\exp(\omega t)\right]\exp\left(-\mathrm{Ei}^{-1}\left\{\mathrm{Ei}\left[-\frac{q_2}{\omega^*}\exp(\omega t)\right]-\omega^*\exp\left[-\frac{q_2}{\omega^*}\exp(\omega t)\right]\right\}\right) & \text{for } t \geqq t_0 \end{cases} \tag{72}$$

where t_0 is specified by

$$\exp\left[-\frac{q_2^*}{\omega^*}\exp(\omega t_0)\right] = \frac{u_{i0}}{\ell}\int_0^{t0}\exp\left[-\frac{q_2^*}{\omega^*}\exp(\omega t')\right]dt' \tag{73}$$

The dimensionless increasing rate of the heat flux ω^* and the exponential integral $\mathrm{Ei}(-x)$ are defined by

$$\omega^* \equiv \omega\ell/u_{i,0} \tag{74}$$

$$\mathrm{Ei}(-x) = -\int_x^\infty \exp(-v)/v\,dv \tag{75}$$

Since the mixture density at the steady-state CHF is given by Equation 69, the transient CHF time t_c is calculated using the transient CHF criterion, i.e., Equation 61, in the same manner as in the case of stepwise power increase.

Referring to

$$q_{ctr} = q_2\exp(\omega t_c) \tag{76}$$

we have for the transient CHF q_{ctr}

$$\frac{q_{ctr}^*}{\omega^*} + \mathrm{Ei}^{-1}\left[\mathrm{Ei}\left(-\frac{q_{ctr}^*}{\omega^*}\right) - \omega^*\exp\left(-\frac{q_{ctr}^*}{\omega^*}\right)\right] = \ln(1+q_{cst}^*) \qquad \text{for } \omega \leqq \omega_0 \tag{77}$$

$$q_{ctr}^* = q_2^* + \omega^*\ln(1+q_{cst}^*) \qquad \text{for } \omega \geqq \omega_0 \tag{78}$$

where ω_0^* satisfies

$$\mathrm{Ei}\left(-\frac{q_2^*}{\omega_0^*}\right) = \mathrm{Ei}\left[-\frac{q_2^*}{\omega_0^*} - \ln(1+q_{cst}^*)\right] - \frac{\omega_0^*\exp(-q_2^*/\omega_0^*)}{1+q_{cst}^*} \tag{79}$$

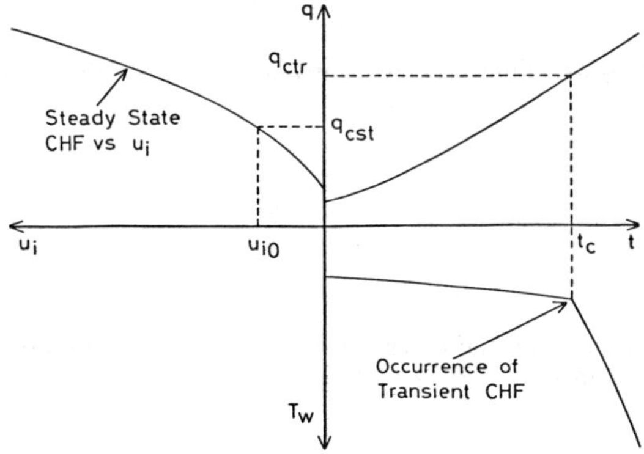

Figure 7. Performance characteristics in exponential power increase.

The variation of the transient CHF, q_{ctr}, with the rate of heat flux increase, ω, predicted by the theory is schematically given in Figure 8. As shown in this figure, the theory predicts that the transient CHF q_{ctr} is increasing with ω and higher than the steady-state CHF q_{cst}. It also indicates that q_{ctr} approaches q_{cst} for $\omega \rightarrow 0$. This means that for very slow increases of the heat flux, the transient process becomes almost "quasi-steady" as is anticipated from our intuition.

Flow Transient – Stepwise Velocity Decrease

$$u_i(t) = u_{i1} \text{ for } t < 0$$
$$= u_{i2} \text{ for } t \geqq 0, \ (u_{i2} < u_{i1}) \quad (80)$$

$$q_{cst2} < q_0 < q_{cst1} \quad (81)$$

Assumptions

4. Constant system pressure
5. Constant heat flux q_0
6. Saturated fluid temperature at the inlet of a heated section

For $t < 0$, the inlet velocity is kept at u_{i1}, and the steady-state boiling is prevailing. The steady-state CHF q_{cst1} corresponding u_{i1} is therefore larger than q_0. At $t = 0$, the inlet velocity changes suddenly from u_{i1} to u_{i2} ($u_{i1} > u_{i2}$). We now consider the case that the steady-state CHF q_{cst2} corresponding to u_{i2} is lower than q_0. Figure 9 shows the present situation.

The two-phase thermohydraulic parameters at the exit are calculated from Table 1 and Equation 80 in conjunction with

$$z = \ell \quad (82)$$

$$q(t) = q_0 \text{ (constant)} \quad (83)$$

The mixture density at the exit is thus obtained as

$$\frac{\rho_{me}}{\rho_\ell} = \frac{u_{i1}^+}{1 + u_{i2}^+ + (u_{i1}^+ - u_{i2}^+)\exp(t^+)} \qquad \text{for } t^+ < \ln\left(\frac{1 + u_{i2}^+}{u_{i2}^+}\right)$$
$$= \frac{u_{i2}^+}{1 + u_{i2}^+} \qquad \text{for } t \geqq \ln\left(\frac{1 + u_{i2}^+}{u_{i2}^+}\right) \quad (84)$$

where, t^+, u_{i1}^+, and u_{i2}^+ are dimensionless time and velocities defined as

$$t^+ \equiv \frac{4\Delta\rho q_0}{d_{he}\rho_g\rho_\ell H_{fg}} t \quad (85)$$

$$u_{ij}^+ \equiv \frac{d_{he}\rho_g\rho_\ell H_{fg}}{4\Delta\rho q_0 \ell} u_{ij} \ (j = 1 \text{ or } 2) \quad (86)$$

Assuming that the transient CHF takes place at the exit, we obtain the mixture density at the transient CHF from Equation 84 with $t = t_c$.

$$\left(\frac{\rho_{me}}{\rho_\ell}\right)_{\text{transient CHF}} = \frac{u_{i1}^+}{1 + u_{i2}^+ + (u_{i1}^+ - u_{i2}^+)\exp(t_c^+)} \quad (87)$$

The mixture density at the steady-state CHF is, on the other hand, determined by Equation 60 in terms of the steady-state CHF velocity u_{icst}. In the present case, this steady-state CHF velocity is defined as the inlet velocity supposed for the steady-state CHF value of q_0. If we define a dimensionless velocity u_{icst}^+ for u_{icst} similarly to Equation 86, we rewrite the mixture density at the steady-state CHF as follows.

$$\left(\frac{\rho_{me}}{\rho_\ell}\right)_{\text{steady-state CHF}} = \frac{u_{icst}^+}{1 + u_{icst}^+} \tag{88}$$

When the transient CHF criterion, Equation 61, is applied to Equations 87 and 88, we have the following equation for the transient CHF time t_c.

$$t_c^+ = \ln\left[1 + \frac{u_{i1}^+ - u_{icst}^+}{u_{icst}^+ (u_{i1}^+ - u_{i2}^+)}\right] \tag{89}$$

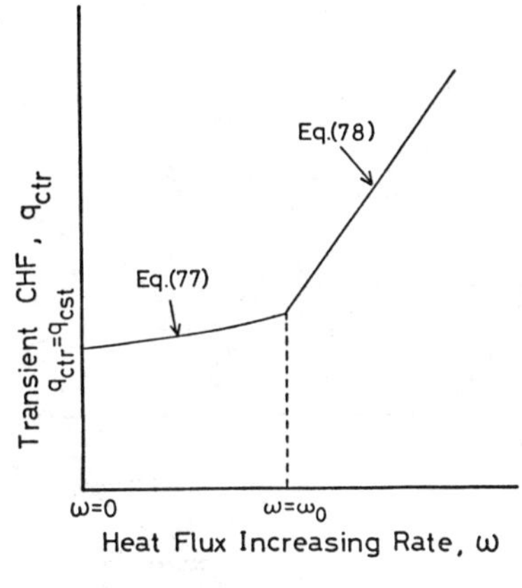

Figure 8. Transient CHF vs. exponential power increasing rate.

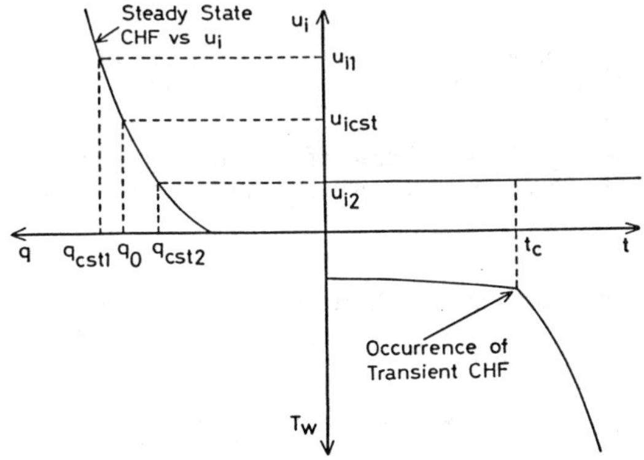

Figure 9. Performance characteristics in stepwise velocity decrease.

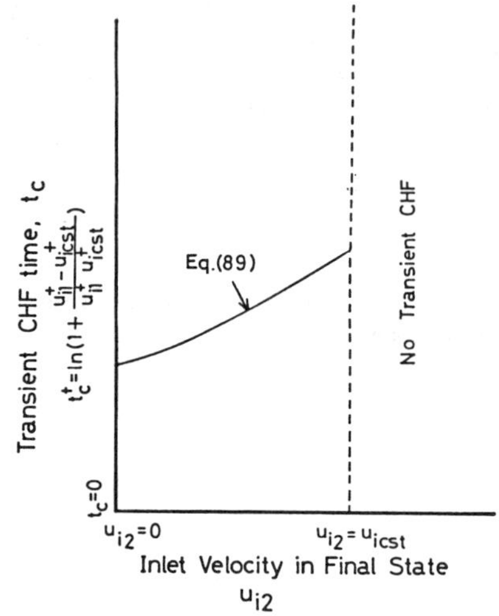

Figure 10. Transient CHF time vs. final velocity.

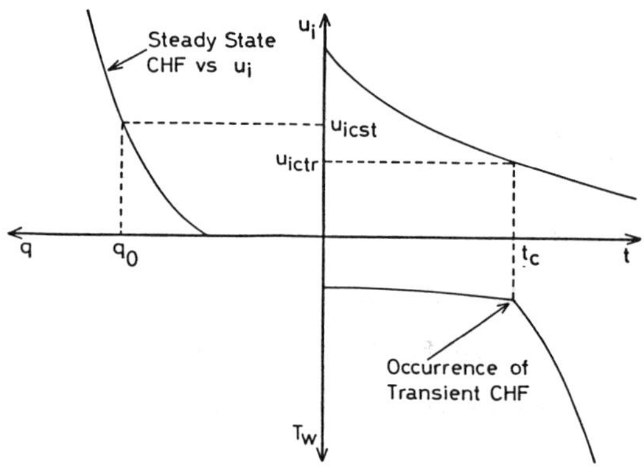

Figure 11. Performance characteristics in exponential velocity decrease.

Figure 10 shows schematically a variation of t_c with u_{i2} predicted by Equation 89 in a dimensionless form. This result indicates an increasing trend of the transient CHF time with the final velocity u_{i2}.

Flow Transient – Exponential Velocity Decrease

$$u_i(t) = u_{i1} \qquad \text{for } t < 0$$
$$= u_{i1} \exp(-\lambda t) \qquad \text{for } t \geqq 0, (\lambda > 0) \qquad (90)$$

For $t < 0$, the inlet velocity is kept constant at u_{i1}, and the steady-state boiling is maintained. At $t = 0$, the inlet velocity begins to decrease exponentially with time as expressed by Equation 90. The assumptions 4, 5, and 6 mentioned earlier are applied here again. The inlet velocity when the transient CHF occurs at $t = t_c$ is termed here as "transient CHF velocity" and denoted by u_{ictr}. The present situation is schematically given in Figure 11.

The thermohydraulic parameters at the exit are calculated from the results listed in Table 1, using Equations 82, 83, and 90. The mixture density is

$$\frac{\rho_{me}}{\rho_\ell} = u_{i1}^+\left[1 + \frac{\lambda^+}{1+\lambda^+}u_{i1}^+\exp(t^+) + \frac{u_{i1}^+}{1+\lambda^+}\exp(-\lambda t)\right] \quad \text{for } 0 \leqq t \leqq t_0$$

$$= \left[1 + \frac{1+\lambda^+}{u_{i1}^+}\exp(\lambda t)\right]^{-\frac{1}{1+\lambda+}} \quad \text{for } t \geqq t_0 \tag{91}$$

where t_0 is a solution to

$$\frac{4\Delta\rho q_0}{d_{he}\rho_g\rho_\ell H_{fg}}t_0 \equiv t_0^+ = \frac{1}{1+\lambda^+}\ln\left[1 + \frac{1+\lambda}{u_{i1}^+}\exp(\lambda^+ t_0^+)\right] \tag{92}$$

and the dimensionless rate of velocity decrease λ^+ is

$$\lambda^+ \equiv \frac{d_{he}\rho_g\rho_\ell H_{fg}}{4\Delta\rho q_0}\lambda \tag{93}$$

The definition of u_{i1}^+ has already appeared in Equation 86.

The mixture density at the transient CHF is therefore obtained from Equation 91 by placing $t = t_c$. Then, using Equations 61 and 88, we have a solution for the transient CHF time t_c. Substitution of thus calculated t_c into Equation 90 yields the transient CHF velocity in terms of u_{i1} and λ. The result is

$$u_{ictr}^+ = (1+\lambda^+)\Big/\left[\left(\frac{1+u_{icst}^+}{u_{icst}^+}\right)^{1+\lambda+} - 1\right] \text{ for } \lambda \leqq \lambda_0 \tag{94}$$

$$\left(\frac{u_{ictr}^+}{u_{i1}^+}\right)^{-\frac{1}{\lambda+}} + \frac{1}{\lambda^+}\left(\frac{u_{ictr}^+}{u_{i1}^+}\right) = \frac{(1+u_{icst}^+)(1+\lambda^+)}{\lambda^+ u_{icst}^+} - \frac{1+\lambda^+}{\lambda^+ u_{i1}^+} \text{ for } \lambda \geqq \lambda_0 \tag{95}$$

where λ_0 satisfies the following equation

$$\frac{\lambda_0^+ u_{i1}^+}{1+\lambda_0^+}\left\{(1+\lambda_0^+)\Big/\left[u_{i1}^+\left(\frac{u_{icst}^+ + 1}{u_{icst}^+}\right)^{1+\lambda_0^+} - u_{i1}^+\right]^{-\frac{1}{\lambda_0^+}}\right.$$

$$\left. + 1\Big/\left[\left(\frac{u_{icst}^+ + 1}{u_{icst}^+}\right)^{1+\lambda_0^+} - 1\right]\right\} = \frac{u_{i1}^+(1+u_{icst}^+)}{u_{icst}^+} - 1 \tag{96}$$

The transient CHF velocity predicted by Equations 94 and 95 is schematically shown in Figure 12 against the rate of velocity decrease. It is noted that the transient CHF velocity u_{ictr} is smaller than the steady-state CHF velocity u_{icst} and shows a decreasing tendency with increasing λ. For $\lambda \to 0$, Equation 94 indicates $u_{ictr} \to u_{icst}$, representing that the transient process is quasi-steady state. On the other hand, for $\lambda \to \infty$, Equation 95 calculates $u_{ictr} \to 0$.

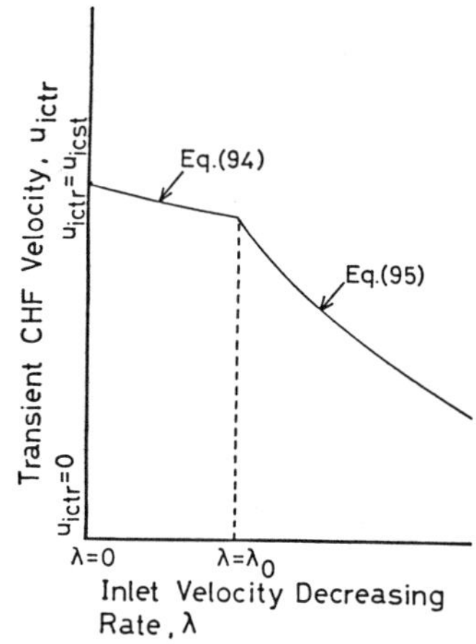

Figure 12. Transient CHF time vs. exponential velocity decreasing rate.

Experiments on Power and Flow Transients

Transient boiling during power and flow transients for large ℓ/d_{he} values has been extensively studied with primary emphasis given to safety analysis of nuclear reactors [4–18, 22–27]. A few extensive reviews have been also reported so far [51–53]. Along with these experiments, computer code developments have been stressed. Some works aimed at the simulation of nuclear reactor accidents, using large-scale experimental facilities which simulate nuclear reactors [54–58]. Some were devoted to studying basic phenomena involved in transient boiling, using laboratory-scale experimental facilities. In this section several experimental works stressing the fundamentals of the phenomena will be briefly reviewed.

Moxon and Edwards [4] carried out a series of experiments on forced convective transient boiling of water at a pressure of 6.9 MPa in stepwise power increase and gradual flow decay ($\ell/d_{he} = 200 \sim 400$). The test sections used are uniformly heated tubes (9 mm i.d., 1.83 m long, and 10.8 mm i.d., 3.66 m long), a nonuniformly heated tube (9.5 mm i.d., 3.66 m long), and a 3.66-m long uniformly heated 37-rods cluster. The initial inlet velocity ranged from 2.7 to 3.7 m/s. In flow transients, the inlet velocity was reduced to a half second, and the transient CHF time measured. Some of their experimental results are represented in Figures 13 and 14 for stepwise power increases and for flow reductions, respectively. In power transient tests, the results show that the transient CHF time increases with a decrease in the final heat flux q_2. Whereas in flow transients, the transient CHF time is shown to decrease with an increase in the heat flux q_0 for fixed inlet velocities, u_{i1} and u_{i2}. In both figures are also shown the predictions by a computor code "SLIP" [6], which uses practically a homogeneous two-phase flow formulation but which takes into consideration a slip velocity between the two phases in void fraction correlation. They reported that except for uniformly heated tubes the agreement between the predictions and experiments was not satisfactory because of the radial and axial mixing effects in momentum and heat transports.

Gaspari et al. [7] conducted an experiment in flow and power transients, using a 4-m long annulus (21-mm i.d. – 13.5-mm o.d.) with the outer tube heated ($\ell/d_{he} = 325$). Water was circulated at 5 MPa with the inlet velocity ranging from 0.5 to 3.3 m/s. In flow transients, the inlet flow was suddenly stopped by closing the valves. In power transients, the heat flux

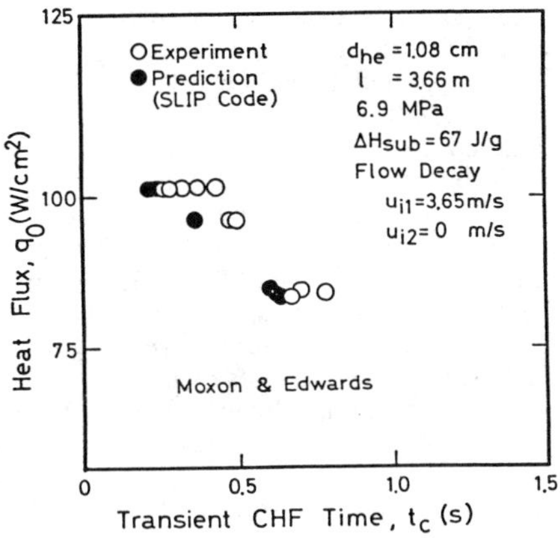

Figure 13. Transient CHF time vs. final heat flux in stepwise power increase [4].

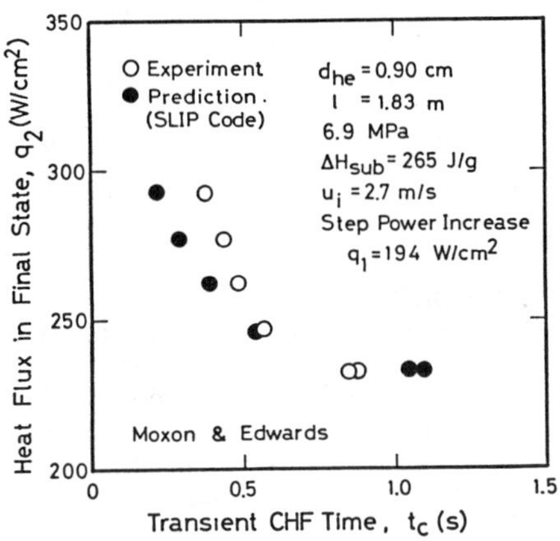

Figure 14. Transient CHF time vs. heat flux in flow decay [4].

was increased exponentially with time from a certain initial value q_1 to q_2. The exponential period of the power increase ranged from 0.5 to 1 s. In these transients, the CHF time and the time-varying average quality were measured. The results were compared with predictions by a homogeneous model, "TILT" code, which considers a slip velocity in the void fraction correlation and also the two-phase pressure drop along the flow direction. A satisfactory agreement was obtained between the theory and the experiment.

Shiralkar et al. [10] measured the transient CHF time in a flow transient test, using water at 6.9 MPa flowing in a 2.7-m long annulus of 31.4-mm i.d. – 13.7-mm o.d. with the inner tube heated ($\ell/d_{he} = 34$). The initial inlet velocity ranged from 0.4 to 1.8 m/s. In transient tests, the flow was reduced down to 10% of the initial velocity while the heat flux was kept constant at 67 ~ 179 W/cm². The results were compared with the Lahey et al. model [11], showing a good agreement. They also carried out 9- and 16-rods cluster tests under flow transients. The theoretical predictions compared generally well with the experiments except in a case of complete flow stoppage. In their calculation by the theory, they extended to zero velocity (flow stoppage) the steady-state CHF correlation originally derived empirically for forced-convective flows. They suggested therefore different CHF mechanisms between forced-convection and zero flow conditions.

Smirnov et al. [12, 13] studied experimentally the effect of the rate of inlet velocity decrease on the transient CHF velocity, using 1.4- and 0.5-m long, 4-mm i.d., single tubes ($\ell/d_{he} = 125, 350$) in water at 9.8 MPa. In their experiments, the inlet velocity was reduced to 40% ~ 70% of the initial velocity ranging 1.5 ~ 4.4 m/s. The heat flux, which was kept constant during the transients, was varied from 29 to 337 W/cm^2. It was shown in this experiment that the CHF velocity difference, $u_{icst} - u_{ictr}$, between the transient and the steady-state is proportional to a dimensionless rate of velocity decrease (du/dt)/u.

Iwamura et al. [15] carried out flow transient tests by linearly decreasing the inlet velocity of water flowing in a 0.8-m long, 10-mm i.d. tube at pressures 0.5 ~ 3.9 MPa ($\ell/d_{he} = 80$). The rate of velocity decrease ranged from 0.6 ~ 35%/s. The initial inlet velocity and the heat flux were 1,240 ~ 3,050 kg/m$^2\cdot$s, and 216 ~ 318 W/cm^2, respectively. Typical results obtained by them are given in Figure 15. They proposed the following empirical correlation to predict the transient CHF velocity.

$$\frac{u_{ictr}}{u_{icst}} = 1 \qquad \text{for } \lambda_r \leqq 11.14\left(\frac{\rho_g}{\rho_\ell}\right)^{0.3924}$$

$$= 1 - \left[0.05 + 6.426\cdot 10^{-5}\left(\frac{\rho_g}{\rho_\ell}\right)^{-1.315}\right]\log_{10}\left[0.08977\left(\frac{\rho_g}{\rho_\ell}\right)^{-0.3924}\lambda_r\right] \qquad (97)$$

$$\text{for } \lambda_r \geqq 11.14\left(\frac{\rho_g}{\rho_\ell}\right)^{0.3924}$$

where λ_r is a rate of flow decrease in %/s.

Cumo et al. [14] conducted a similar experiment with higher decreasing rate up to 100%/s, using Freon 12 flowing at pressures 1.1 ~ 1.8 MPa in a 2-m long 7.8-mm i.d. tube ($\ell/d_{he} = 256$). The heat flux was from 2.9 to 7.5 W/cm^2, and the initial velocity from 760 to 900 kg/m$^2\cdot$s. It was shown that the transient CHF is satisfactorily predicted by numerical calculations using two-phase flow code "PEPE" [59]. This code is based on a homogeneous two-phase flow model. They presented the following approximate correlation for the transient CHF velocity.

$$u_{ictr} = u_{icst} \qquad \text{for } \frac{t_{0.5}q_0}{q_{cst1}}\left(\frac{P_{cr}}{P}\right)^{2.2} \geqq 24\,\text{sec} \qquad (98)$$

$$u_{ictr} = u^q_{icst}(q_0 - \text{C.F.}) \qquad \text{for } \frac{t_{0.5}q_0}{q_{cst1}}\left(\frac{P_{cr}}{P}\right)^{2.2} \leqq 24\,\text{sec} \qquad (99)$$

where

$$\text{C.F.} = (q_{cst1} - q_0)\exp\left[-0.82\,(t_{0.5})^{0.15}\left(\frac{P_{cr}}{P}\right)t_c\right] \qquad (100)$$

$u^q_{icst}(q)$ is the inlet velocity corresponding to the steady-state CHF q, and was determined from their own data. From definitions, we have

$$u_{icst} = u^q_{icst}(q_0) \qquad (101)$$

$$u_{i1} = u^q_{icst}(q_{cst1}) \qquad (102)$$

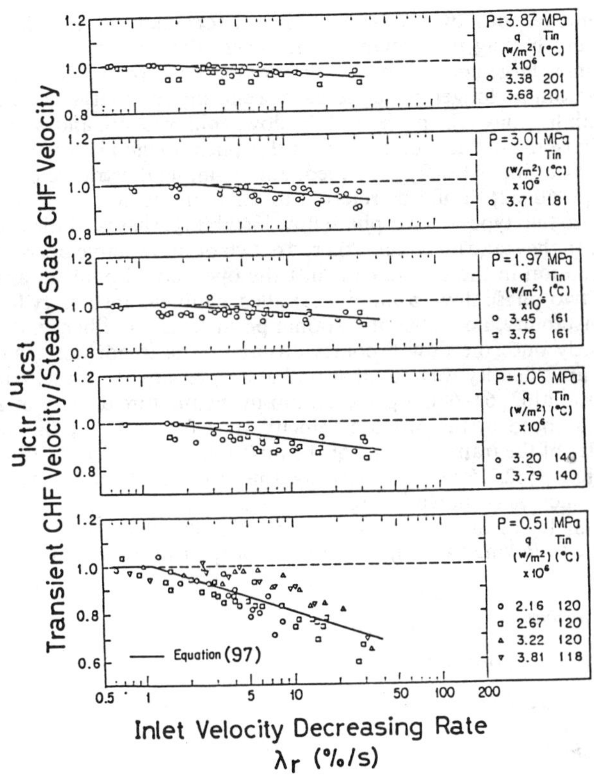

Figure 15. Effect of velocity decreasing rate on transient CHF velocity [15].

where u_{i1} is the initial inlet velocity, and q_0 the heat flux applied to the heated wall. $t_{0.5}$ appearing in Equations 98 and 99 is the time duration from the initiation of the transient till the time when the velocity attains the half of the initial velocity. P_{cr} is the critical pressure of Freon 12.

Leung [16, 17] measured the transient CHF time for flow transients with Freon 11 flowing in a uniformly or nonuniformly heated tube (11.7-mm i.d., 3.05 m in length and hence $\ell/d_{he} = 261$). The inlet velocity of Freon 11 was reduced at a rate of 4%/s or completely stopped in an abrupt manner. The initial velocity covered in this experiment was $0.5 \sim 2.2$ m/s at pressures $1.8 \sim 2.9$ MPa. Along with the experiment, he developed a computor code "CODA" based on a homogeneous flow model considering the pressure drop along the test section. The predicted transient CHF time by CODA compares well with the measured.

In summary, the results of the works reviewed here support that the homogeneous two-phase formulations are successfully applicable to predicting the transient CHF for both power and flow transients with large ℓ/d_{he}.

Power and Flow Transients Accompanied by Flow Instability

In the preceding paragraphs, the boiling during power and flow transients without flow instability has been exercised for given inlet velocities. In these cases, the momentum equation in the homogeneous two-phase flow model formulations can be neglected in analysing the transient boiling performances. However, when the pressure drop along a heated section becomes large comparably to the driving head (pump head and etc.), the momentum equation can no longer be neglected. Specifically for a large pressure drop, the two-phase flow instability may take place, and the flow, void fraction, and wall temperature begin to oscillate. Since the detailed discussion of this phenomenon is beyond the scope of this chapter, only its outline is briefly described, and a simple illustrative example will be given. The reviews [3, 60–62] should be referred to for more details.

In general, there are two types of two-phase flow instability, that is, static instability and dynamic instability. In the static instability, there are two or more stable operational points. Only a large perturbation in the system can shift the operational point to another stable point. This type of two-phase flow instability is called Ledinegg instability [63].

In the dynamic instability, no stable operational points exist, and hence, the oscillation continues permanently once the flow becomes unstable. The important examples of the dynamic instability are "density wave oscillation" and "pressure drop oscillation." In the density wave oscillation [62, 64–68], a perturbation in the mixture density of a two-phase flow propagates at a speed of the mixture velocity. The frequency of this oscillation is therefore on the order of the transport time of the mixture, i.e., ℓ/u. On the other hand, the pressure drop instability [62, 66, 69] is caused by a negative gradient in a two-phase flow pressure-drop curve such as is shown in Figure 16. The frequency range of this oscillation is related to the geometry of the system and also to two-phase compressibility.

Let us consider the following two important cases of the transient boiling accompanied by a flow instability.

- *Case 1: power transient*

$$\begin{aligned} &q(t)\text{: gradual increase with time} \\ &\Delta P_h(t) = \Delta P_{ho}\,(= \text{const.}) \end{aligned} \tag{103}$$

- *Case 2: flow transient*

$$\begin{aligned} &q(t) = q_0\,(= \text{const.}) \\ &\Delta P_h(t)\text{: gradual decrease with time} \end{aligned} \tag{104}$$

Here, ΔP_h is a pump driving head. It is assumed for both transients that the pressure drop ΔP_f along the heated section is comparable to the driving head ΔP_h, and that the transients are gradual enough to be regarded as quasi-steady state. The thermohydraulic parameters of the two-phase flows are therefore determined by the steady-state performances.

If we apply the homogeneous model to the present cases, the pressure drop ΔP_f along the heated section is calculated by integrating Equations 1 ~ 3 from $z = 0$ to $z = \ell$. This is given by References 70 and 71 as

$$\begin{aligned} \Delta P_f = \rho_\ell u_i^2\Bigg[K_{in} &+ \frac{f}{2d_{he}}(\ell - L_B) + \frac{fL_B}{2d_{he}}\left(1 + \frac{\Delta\rho}{\rho_g}\frac{x_e}{2}\right)\Bigg/\left(1 + \frac{\Delta u}{\mu_g}\frac{x_e}{2}\right)^{0.25} \\ &+ \frac{x_e}{\alpha_e}\frac{\rho_\ell}{\rho_g} + \frac{(1 - x_e)^2}{1 - \alpha_e} - 1 + K_e\left(1 + \frac{\Delta\rho}{\rho_g}x_e^{1.5}\right)\Bigg] + g\rho_\ell\ell - \Delta\rho g\langle\alpha\rangle L_B \end{aligned} \tag{105}$$

Here, K_{in} and K_e are the inlet and outlet orifice pressure drop coefficients, and $\langle\alpha\rangle$ the average void fraction over the boiling length L_B. As the heat flux increases, the void fraction, quality, and the boiling length increase also. Therefore, the pressure drop calculated from Equation 105 becomes larger. A variation of the pressure drop with the inlet velocity thus calculated shows an N-shaped curve as shown in Figure 16.

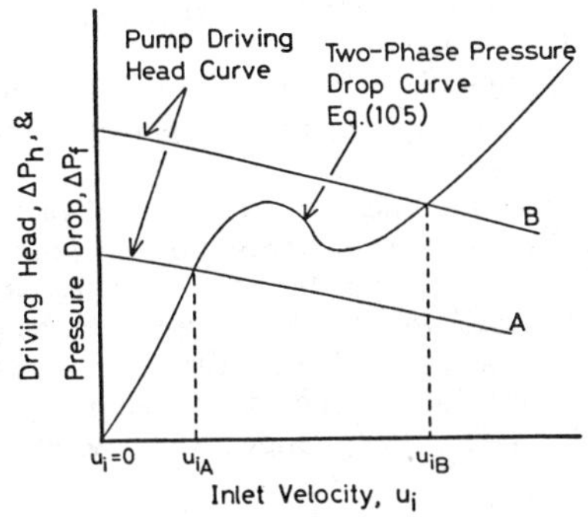

Figure 16. Flow characteristic of boiling two-phase flow.

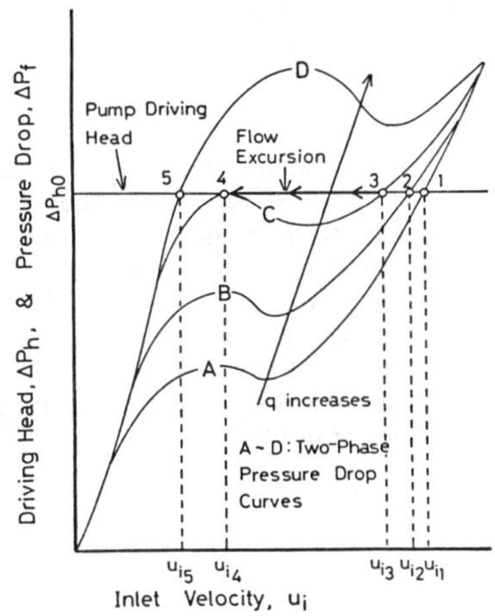

Figure 17. Flow characteristic in power increase.

For a given pump head ΔP_h, the inlet velocity u_i is determined so as to satisfy the following equation.

$$\Delta P_f = \Delta P_h \tag{106}$$

The inlet velocity u_i thus calculated is diagrammatically given in Figure 16 as the intersection between the pressure drop curve and pump head curve. The pump head ΔP_h is usually a decreasing function of the inlet velocity as represented in Figure 16 by curves A and B. However, it is assumed here for simplicity of the analysis that the pump head does not depend upon the inlet velocity (Figure 17).

- *Case 1: gradual power increase, Equation 103.* The variation of the inlet velocity with time is determined by the intersections of ΔP_f with ΔP_h, as shown in Figure 17. As the heat flux increases, the inlet velocity decreases gradually along $1 \rightarrow 2 \rightarrow 3$ due to an increase of the pressure drop. When the pressure drop curve contacts with a given ΔP_h line at a point 4 (curve C), a flow excursion occurs (Ledinegg instability), and the inlet velocity

u_i suddenly reduces in a great amount (3 → 4). A further increase of the heat flux induces a gradual decrease of the inlet velocity (4 → 5). The trace of the time-varying inlet velocity is shown in Figure 18, together with the heat flux variation q (t). Once the flow excursion occurs, the void fraction and the quality increase rapidly, often resulting in CHF.

- *Case 2: gradual pump head reduction, Equation 104.* The inlet velocity is determined again by intersections between the pressure drop curves and the pump head lines as shown in Figure 19. As the pump head is reduced, the inlet velocity changes gradually along 1 → 2 → 3. When a point is reached where ΔP_h line contacts the pressure drop curve at point 4, the flow excursion occurs, and, hence, the inlet velocity is much decreased suddenly from point 3 to 4. For a further decrease in the pump head, the inlet velocity changes rather gradually with a decreasing tendency. These variations are indicated in Figure 20.

Detailed analysis was undertaken by Ishii and Fauske [71] on the present subject in relation to the accident analysis of sodium-cooled fast breeder reactors. It is clear that the flow instability plays an important role in determining transient boiling phenomena, since it may lead to the transient CHF.

Pressure Transients

A serious pressure transient is normally accompanied by a violent boiling of the fluid due to a release of the energy stored in the liquid in addition to the heat flow through a heated surface. The boiling followed by a release of the stored energy of liquid in this way is called "flashing." In pressure transients, flashing is one important aspect in specifying thermohydraulic parameters during the transients.

A simple analytical treatment of a pressure transient performance is possible based on a homogeneous modeling of two-phase flow which is similar to that described for flow and power transients in previous paragraphs. The heat transferred to the fluid particles during the postulated transient (Equation 35) should be modified for the present case as

$$\Delta Q = p_h q(t) \Delta z(t) \Delta t - A_s \left(\frac{\rho_m}{\rho_\ell}\right) \Delta z(t) \frac{dH_\ell}{dt} \Delta t \tag{107}$$

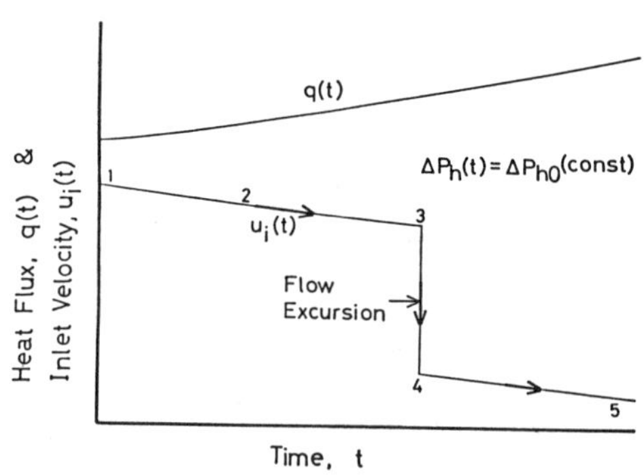

Figure 18. Time-variation of heat flux and inlet velocity during power increase followed by flow excursion.

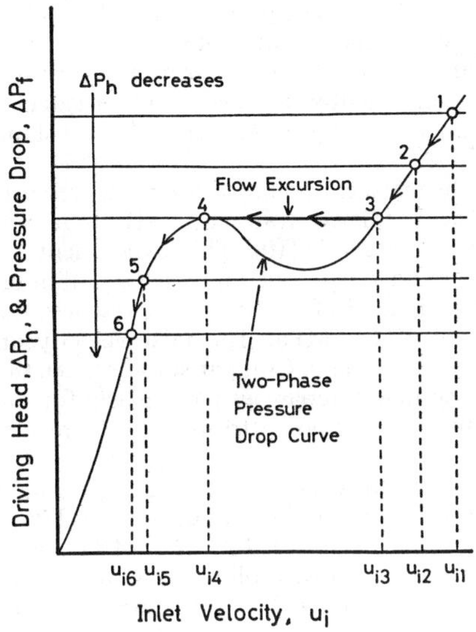

Figure 19. Flow characteristic in driving head decrease.

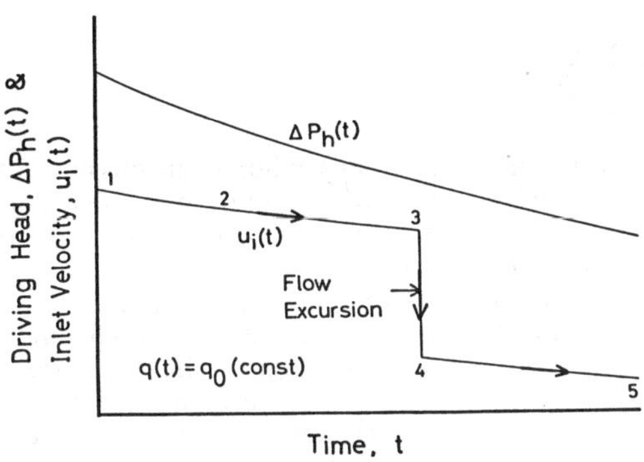

Figure 20. Time-variation of driving head and inlet velocity during driving head decrease followed by flow excursion.

The second term on the right-hand side of this equation represents a flashing term. The differential equation for the mixture density corresponding to Equation 41 is written by

$$\frac{d}{dt}\left(\frac{1}{\rho_m}\right) = \frac{4q(t)\Delta\rho}{d_{he}H_{fg}\rho_g\rho_\ell}\left(\frac{1}{\rho_m}\right) - \frac{\Delta\rho}{\rho_g\rho_\ell H_{fg}}\frac{dH_\ell}{dP}\frac{dP}{dt} \tag{108}$$

On the other hand, the characteristic equation, Equation 42, becomes

$$u = \frac{4q(t)\Delta\rho}{d_{he}H_{hg}\rho_g\rho_\ell}z - \int_0^z \frac{\rho_m\Delta\rho}{\rho_g\rho_\ell}\frac{dH_\ell}{dP}\frac{dP}{dt}dz + u_i(t) \tag{109}$$

By integrating Equations 108 and 109, we can obtain the thermohydraulic parameters, ρ_m, u, α, x, and H_m. However, these calculations are not so easy as in power and flow transients, since ρ_ℓ, ρ_g, H_{fg}, and H_ℓ change also with time during the postulated pressure transient. Numerical integrations are conveniently adopted in these calculations even though the previous thermodynamic properties are given in terms of the instantaneous pressure under thermal equilibrium conditions.

Lahey et al. [11] analyzed the transient boiling during a depressurization according to the method just mentioned. The predicted transient CHF time was compared with the experimental results by Shiralkar et al. [10]. This experiment by Shiralkar et al. was performed with water flowing in a 2.7-m long annulus (13.7-mm o.d. – 31.4-mm i.d.) and a 1.83-m long 9-rods cluster channel. The rate of depressurization was 0.345 and 0.69 MPa/s for an initial pressure of 6.9 MPa. The inlet velocity and the heat flux were 340 ~ 1,360 kg/m$^2\cdot$s and 63 ~ 158 W/cm^2. Comparisons between the theory by Lahey et al. and this experiment showed some discrepancies particularly for low velocity data. Lahey et al. concluded that these discrepancies are attributed to the reverse flow due to a violent flashing.

Leung [16, 17] studied experimentally the combined flow-pressure decay transients at low velocities, using Freon 11 flowing in a 11.7-mm i.d. tube of 3.06 m in length. The initial pressure and the initial inlet velocity were 2.2 ~ 2.9 MPa and 1,220 ~ 2,710 kg/m$^2\cdot$s, respectively. The rapid depressurization was followed by the inlet flow stoppage at a fixed heat flux ranging 4.4 ~ 6.0 W/cm^2. The results were compared with predictions by CODA code [17]. Good comparisons were obtained between them for the transient CHF time under low and reverse flow conditions. The homogeneous two-phase flow model adopted in this code uses the Griffith–Zuber correlation [72] for the steady-state CHF given as follows.

$$q_{cst} = 0.13\,(1-\alpha)\,\rho_g H_{fg}\left[\frac{\sigma\Delta\rho g}{\rho_g^2}\right]^{0.25} \tag{110}$$

The CODA code uses mass and energy equations in homogeneous model formulations. The measured inlet velocity was used as a boundary condition in the claculation. Leung [17] tried further to evaluate the inlet velocity according to the momentum equation based on the homogeneous model. But, the result was not succesful.

Gaspari et al. [7] conducted a series of experiments in a rapid depressurization followed by inlet flow stoppage. The test section used in their experiment is an annulus with the outer tube heated (13.5-mm o.d. – 21-mm i.d., 4-m in length). The initial pressure was 5.1 MPa, and the initial velocity of water ranged from 700 to 2,560 kg/m$^2\cdot$s. The heat flux was 34 ~ 83 W/cm^2. The measured transient CHF time was compared with the prediction by TILT code based on the homogeneous model. The agreement between the experiment and the prediction was less satisfactory than in the case of flow and power transients.

To conclude, mass and energy equations based on a homogeneous two-phase flow model show succesful results in analyzing the pressure transients. However, more advanced and comprehensive models are desired particularly for the momentum equation.

TRANSIENT BOILING FOR SMALL ℓ/d_{he}

When the ratio ℓ/d_{he} between the heated length and the heated equivalent diameter of the flow passage is small (order of unity or less), the transient boiling phenomenon becomes considerably different from that for large ℓ/d_{he}. The small ℓ/d_{he} ratio involves short heated-length tubes and small size heaters (wire, flat plate, sphere, and etc.) suspended in a flow with a large flow area. Typical boiling situations at near CHF for small ℓ/d_{he} are schematically shown in Figure 21.

Because of the short length of the heated section, the transient boiling phenomenon for small ℓ/d_{he} is not affected so much by the upstream conditions. In other words, the boiling

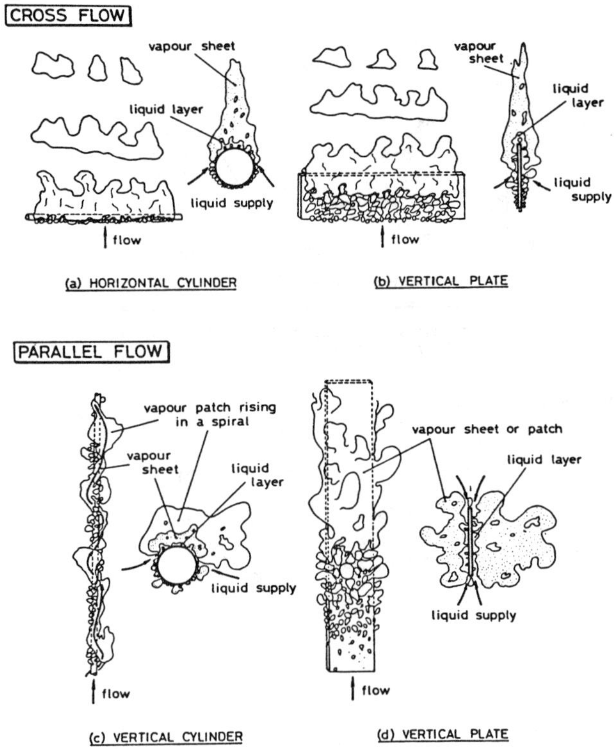

Figure 21. Vapor-liquid configuration at near CHF [136].

phenomenon can be regarded as "local phenomenon." The transient CHF usually occurs at a low steam quality, i.e., in bubbly or slug flow. The CHF mechanism is a transition from nucleate boiling to film boiling where the heater temperature increases very rapidly, often resulting in the melt-down of the heaters.

Unlike the transient boiling for large ℓ/d_{he}, it seems, at present, that there are few satisfactory analytical methods to predict the thermohydraulic characteristics of the transient boiling for small ℓ/d_{he}. Therefore, in this section, the transient boiling phenomenon for small ℓ/d_{he} is phenomenologically described based on the experimental findings obtained so far [73–95] including pool boiling data.

Steady-State Boiling Curve

The boiling processes for small ℓ/d_{he} and pools are conveniently expressed in a ΔT_{sat} vs. q coordinate (boiling curve) as shown in Figure 22. Here, q is the heat flux from a heater surface and ΔT_{sat} the excess temperature of the heater defined as

$$\Delta T_{sat} \equiv T_w - T_{sat} \tag{111}$$

where T_w is the surface temperature of the heater.

As the heat flux increases, the surface temperature of the heater and hence, the excess temperature ΔT_{sat}, increase also from the initial condition, point A. Boiling commences

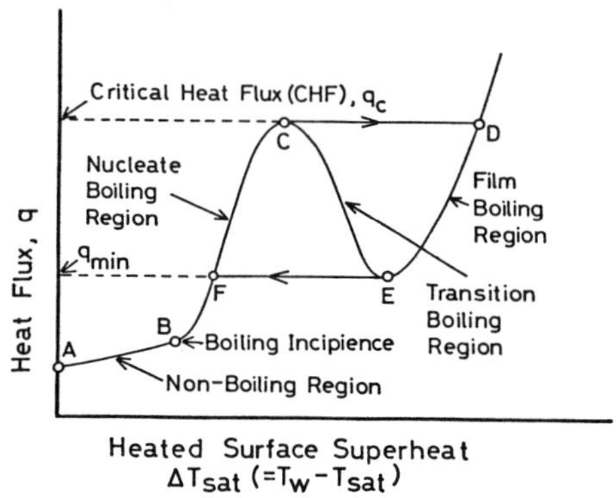

Figure 22. Boiling curve.

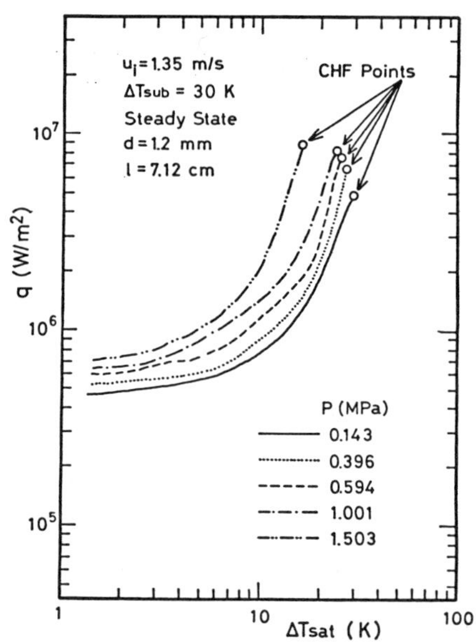

Figure 23. Steady-state boiling curve.

when a point B is reached, which is called "incipient boiling point" or "boiling incipience." The boiling mode in the region BC is called "nucleate boiling" because vapor bubbles are generated from nucleation sites (small cavities or pitts) distributed over the heated surface. At point C, the boiling mode suddenly changes from nucleate boiling to "film boiling" where vapor film covers the heater area as a vapor blanket. Since the heat transfer in film boiling is more deteriorated than that in nucleate boiling, the heater temperature jumps from point C to D (CHF) under heat-flux-controlled conditions. Therefore, the CHF under this condition is called also "DNB (departure from nucleate boiling) heat flux". When the heat flux is reduced from point D, the boiling process changes along the line DE, and, thus, the boiling transition occurs at point E from film to nucleate boiling. The heat flux at this point E is usually called "minimum heat flux." The region CE is realized in a cause of controlled heater temperature, and is called "transition boiling."

Figure 23 shows examples of the steady-state boiling curves obtained with small ℓ/d_{he} under forced convection of water [73, 74] (38-mm i.d. circular tube, $\ell/d_{he} = 0.024$). It is shown in this figure that the nucleate boiling region shifts towards smaller ΔT_{sat} with an increase of the pressure. However, as has been reported elsewhere, the nucleate boiling region is not affected so much by the liquid subcooling, velocity, and heater geometry. This result reflects that the nucleate boiling phenomenon is quite localized and limited to the very vicinity of the heater surface, and does not depend upon the bulk flow conditions (velocity, subcooling, and etc.).

The nucleate boiling heat transfer coefficient under forced convection

$$h_B \equiv q/\Delta T_{sat} \tag{112}$$

is therefore approximated by steady-state pool boiling correlations. One of such correlations for pool boiling is given by Kutateladze [96] as

$$h_B\left(\frac{\sigma}{g\Delta\rho}\right)^{1/2}/\lambda_\ell = C\left[\frac{q}{H_{fg}\rho_g\nu_\ell}\left(\frac{\sigma}{g\Delta\rho}\right)^{1/2}\right]^{0.7}[P/(\sigma g\Delta\rho)^{1/2}]^{0.7}\left(\frac{\nu_\ell}{a_\ell}\right)^{0.35} \tag{113}$$

The proportionality constant C in the Equation 113 depends on the surface condition of the heater. Equation 113 is rewritten in a form of

$$q \propto (\Delta T_{sat})^3 \tag{114}$$

Power Transients

Boiling patterns encountered in power transients where the power is increased in step-wise, rump-wise, exponential, or more sophisticated ways are illustrated in Figure 24 [136]. The method of classification is based on the charateristic behavior of surface temperature during a power increase.

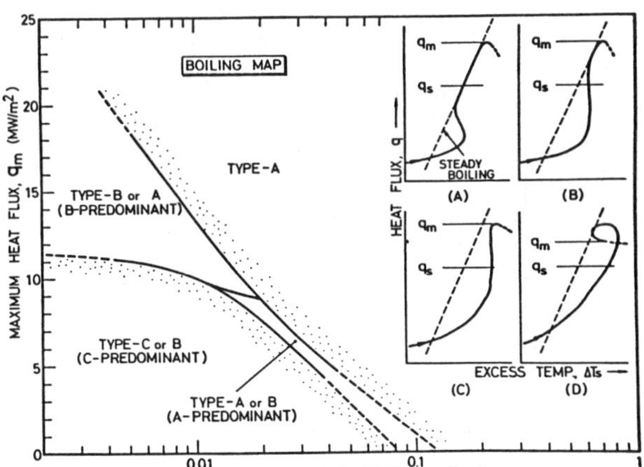

Figure 24. Boiling patterns observed in power transient (d = 1.2-mm o.d., ℓ = 10 cm).

A-type boiling occurs when the transient surface temperature begins to increase along the steady nucleate boiling curve and its extrapolation immediately after the initiation of boiling. B-type boiling is characterized by a long lasting temperature overshoot, but the surface temperature nevertheless increasses with the steady boiling curve at a heat flux beyond the steady-state CHF. In C-type boiling, the surface temperature continues to shift towards higher values than those in steady-state boiling until burnout. D-type boiling is

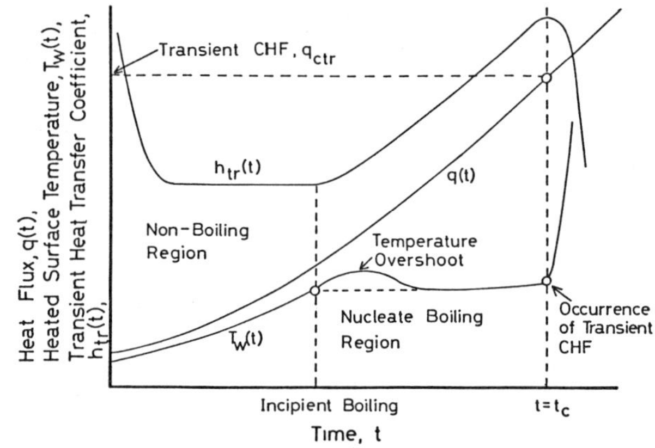

Figure 25. Time-varying parameters in exponential power increase.

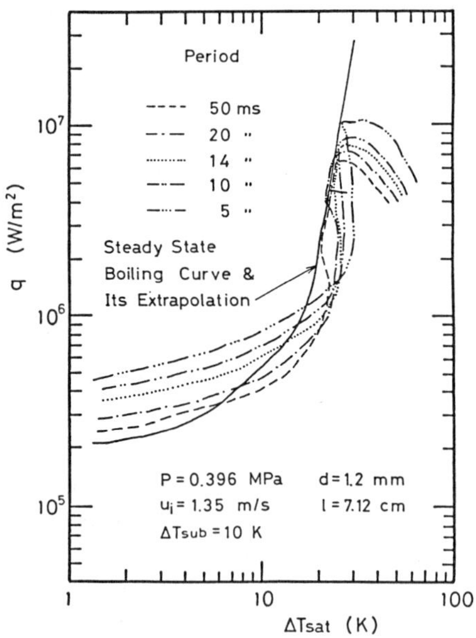

Figure 26. Transient boiling curve.

followed by a characteristic loop or loops in the boiling curve (temperature setback) which takes place at heat fluxes beyond the steady-state CHF.

Figure 24 shows also a map of previously mentioned boiling patterns based on the experimental results of Kataoka et al. [73] (water, pressure 0.1 ~ 1.5 MPa, velocity up to 4 m/s, subcooling up to 70° K, heater 1.2-mm o.d., 71-mm long). As shown in this figure, A- and B-type boilings are found to have been most often observed over a wide range of experimental variables. Therefore, these types of transient boiling are stressed in this section despite the practically important aspects of C-type boiling CHF.

Let us consider a case of an exponential power increase with time specified by the following equation.

$$\begin{aligned} q &= 0 && \text{for } t < 0 \\ &= q_2 \exp(t/\tau) && \text{for } t \geqq 0 \end{aligned} \tag{115}$$

Here, τ (> 0) is the "period" of the exponential power increase, and is the inverse of the power increasing rate ω (Equation 71).

$$\tau = 1/\omega \tag{116}$$

Figure 25 shows for the foregoing postulated transient the variation of the heat flux, surface temperature of the heater, and heat transfer coefficient with time. Typical examples of the experimental boiling curves for exponential power increase are given in Figure 26, showing the nonboiling region, the nucleate boiling region, and the occurrence of the transient CHF in a sequential manner (taken from Reference 73).

Nonboiling Heat Transfer

A large number of analytical and experimental works have been reported on the transient nonboiling heat transfer [73, 97–123]. This problem was analized for flows in a tube, annulus, and parallel plate by Siegel [98–101], Sparrow and Siegel [102], Perlmutter and Siegel [103, 104], Namatame [105], and Kawamura et al. [106–108]. Particularly, the works

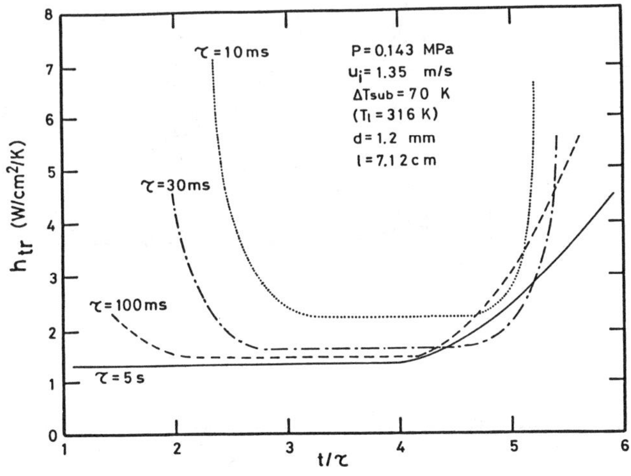

Figure 27. Transient nonboiling heat transfer coefficient.

by Cess [109], Goodman [110], Adams et al. [111], Soliman et al. [112–114], Chao et al. [115, 116], Hanawa [119–122] and Kataoka [73, 123] deal with the problem for small ℓ/d_{he}.

The experimental results obtained by Kataoka et al. [73] using a cylindrical heater in an exponential power increase will be briefly reviewed. The measured variation of the transient nonboiling heat transfer coefficient with time is represented in Figure 27, showing that the heat transfer coefficient h_{tr} initially decreases with time until it reaches an asymptotic value. The ratio of the asymptotic value of the transient nonboiling heat transfer coefficient, h_{tra}, to the steady-state nonboiling heat transfer coefficient h_{st} is calculated and shown in Figure 28. The data are well correlated by

$$h_{tra}/h_{st} = 1 + 0.1448\,\omega^{*0.97} \tag{117}$$

Here, the dimensionless parameter ω^* is defined as

$$\omega^* \equiv \omega\ell/u_i = \ell/(u_i\tau) \tag{118}$$

The steady-state nonboiling heat transfer coefficient is empirically given by

$$h_{st}\ell/\lambda_\ell = 0.2411\left(\frac{u_i\ell}{\nu_\ell}\right)^{0.7}\left(\frac{\nu_\ell}{a_\ell}\right)^{0.35} \tag{119}$$

Boiling Incipience

In order to estimate the initiation of transient boiling, we first consider a thermal equilibrium vapor bubble as shown in Figure 29. If we assume quasi-static bubble growth, then we have Equation 120 for the pressure balance between the vapor and the liquid.

$$P_g - P_\ell = \frac{2\sigma}{r} \tag{120}$$

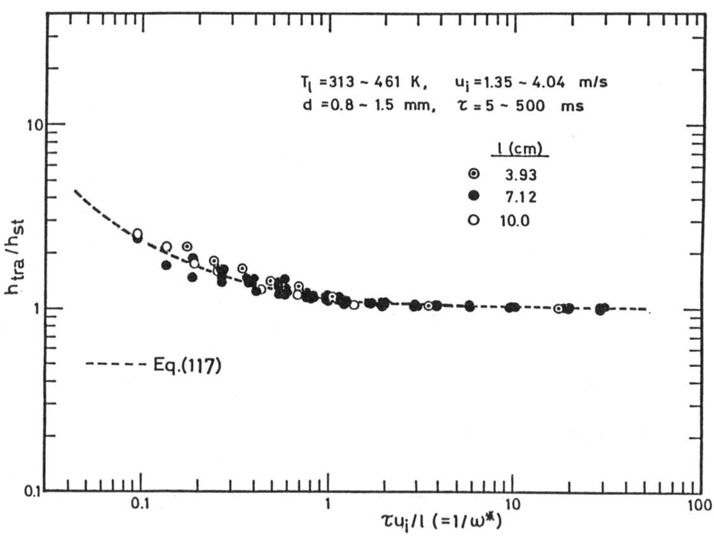

Figure 28. Asymptotic value of transient nonboiling heat transfer coefficient.

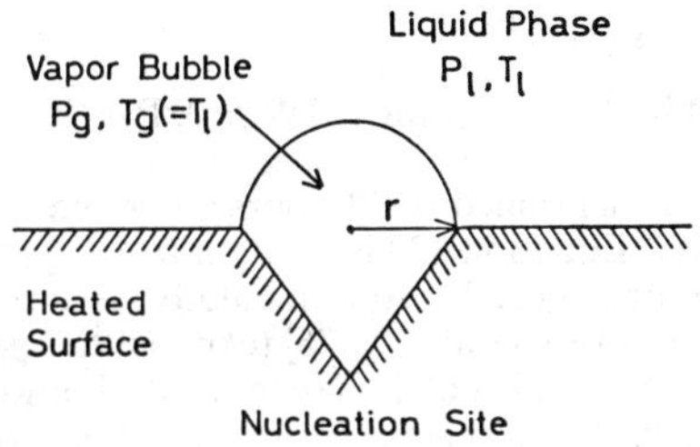

Figure 29. Vapor bubble at nucleation site.

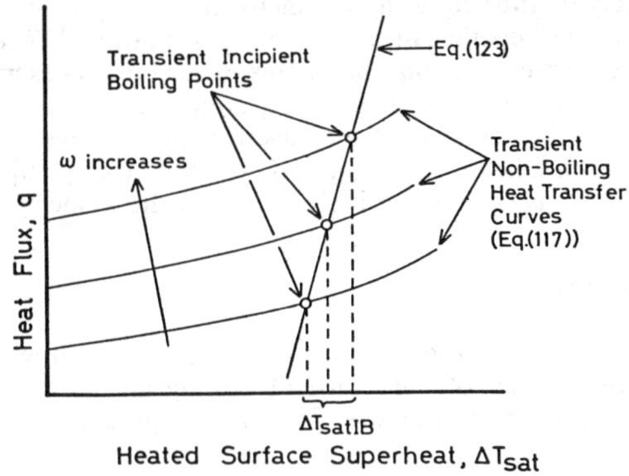

Figure 30. Determination of transient incipient boiling point.

where r is the radius of the nucleation site. Provided that the temperature of the vapor bubble is equal to the liquid temperature T_ℓ, the following equation is obtained from Clasius–Clapeyron's law

$$T_\ell - T_{sat} = \frac{R_v T_\ell^2}{P_\ell H_{fg}}(P_g - P_\ell) \tag{121}$$

where R_v is the gas constant. It is further assumed that boiling commences when the pressure difference given by Equation 121 exceeds the pressure difference due to surface tension, i.e., Equation 120. Then, we have

$$T_\ell - T_{sat} = \frac{R_v T_\ell^2}{P_\ell H_{fg}} \frac{2\sigma}{r} \tag{122}$$

It is noted here that Equation 122 was obtained for a single nucleation site with an assumption of uniform liquid temperature distribution near the heating surface. However, in practice, a number of nucleation sites with different sizes are distributed over the surface, and the liquid temperature distribution near the surface is not uniform. For such a case, Hsu and Graham [124] developed the theory to predict the incipient boiling point under steady-state conditions. Bergles and Rohsenow [125] extended the method and proposed the following approximate correlation of boiling inception for water at pressures from 0.1 to 1.4 MPa.

$$q_{IB} = 1.35\,P^{1.156}[1.8\,(T_w - T_{sat})]^{1.87/P^{0.234}} \tag{123}$$

where q_{IB} is the heat flux at boiling initiation (MW/m^2), P is the system pressure (MPa), and $T_w - T_{sat}$ in °K.

The boiling criterion just mentioned is, of course, derived initially for the steady-state boiling conditions. However, Equation 123 is thought to be applicable also to the transient boiling inception. This hypothesis can be experimentally verified by a detailed inspection of a series of experiments by Johnson et al. [75, 76] (exponential power increase with q small flat plate in water flow) and Sakurai et al. [92] (exponential power increase with cylindrical wires suspended in a water bath). In their experiments, a reasonable success was obtained in comparisons between the experiments and theories Fabic's model [126], Hsu–Graham model [124], and Bergles–Rohsenow model [125].

The surface temperature of the heater, T_{wIB}, at the transient boiling incipience can be obtained approximately by thus equating the incipient boiling criterion, Equation 123 and the nonboiling heat transfer coefficient, for example, Equation 117 for exponential power increase with cylindrical wires. The outline of this procedure is described in Figure 30. As shown schematically in this figure, the heat flux q_{IB} and the superheat of the heater surface ΔT_{satIB} at the transient boiling incipience increase for higher power increasing rate ω.

Schultz et al. [127, 128], Oker et al. [129], and Ngheim et al. [130] should be referred to for more details of the transient boiling incipience and the associated phenomena.

Nucleate Boiling Heat Transfer

The typical examples of the transient boiling curve have been shown previously in Figure 26. Figure 31 is another group of the transient boiling curves obtained for exponential power increases with a different heater size [73]. These figures indicate some characteristic

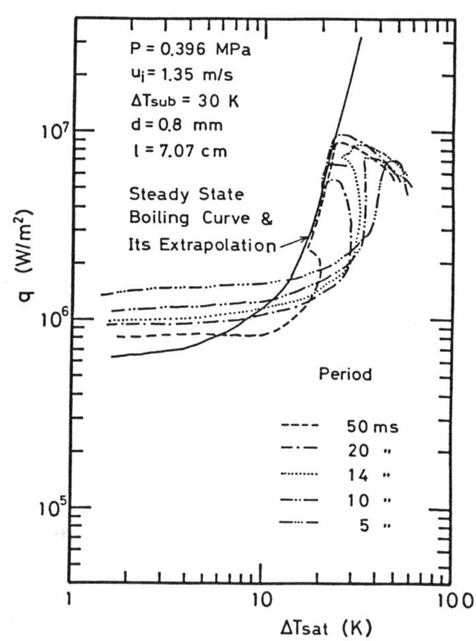

Figure 31. Transient boiling curve.

aspects of the transient boiling in addition to the types of boiling mentioned earlier, that is, "temperature overshoot" and the possible effect of the initial condition of the power transient.

The temperature overshoot in the transient boiling curves as shown in Figures 26 and 31 is considered to be caused by insufficient activation of the potential nucleation sites during the first stage of the nucleate boiling [131]. This phenomenon is reported in References 73 and 74 to have been observed over wide ranges of experimental parameters under forced convection of water in a pipe, i.e., velocity $0 \sim 4$ m/s, subcooling $0 \sim 70$ K, pressure $0.1 \sim 1.5$ MPa, and exponential period 5 ms ~ 10 s. A similar temperature overshoot has been reported earlier by Johnson et al. [75, 76] for flat plates and also by Aoki et al. [78] for an annulus.

We will now discuss this temperature overshoot in more detail. The maximum temperature overshoot ΔT_{ov}, which is defined as the point where the shift of the transient boiling curve is largest from the steady-state nucleate boiling curve as shown in Figure 32, is closely related to the fraction of the activated nucleation sites to all available sites. In this hypothetical view, the steady-state boiling is characterized by all the available nucleation sites being activated. On the other hand, the nonboiling heat transfer is characterized, of course, by no potential nucleation sites available. Therefore, the fraction of the activated nucleation sites during transients can be defined as [132]

$$\Phi \equiv \frac{\Delta T_{ov} - \Delta T_B}{\Delta T_N - \Delta T_B} \tag{124}$$

where ΔT_B and ΔT_N are defined in the figure. Based on the experimental results (some of them have been demonstrated in Figures 26 and 31), Sakurai et al. [132] found that this fraction takes nearly a constant value regardless of the pressure, velocity, and the exponential period, that is

$$\Phi = 0.68 \tag{125}$$

However, the subcooling-dependence of this parameter has not yet been made clear in their experiment. The experimental results and Equation 125 are shown in Figure 33. The maximum temperature overshoot can be thus determined by Equations 124 and 125 if ΔT_B

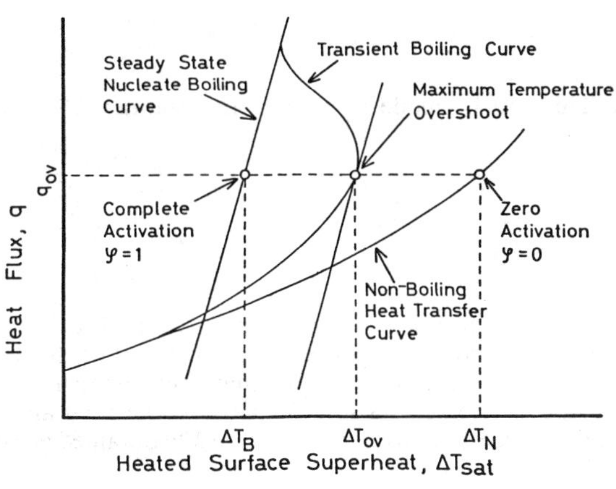

Figure 32. Temperature overshoot and nucleation site activation.

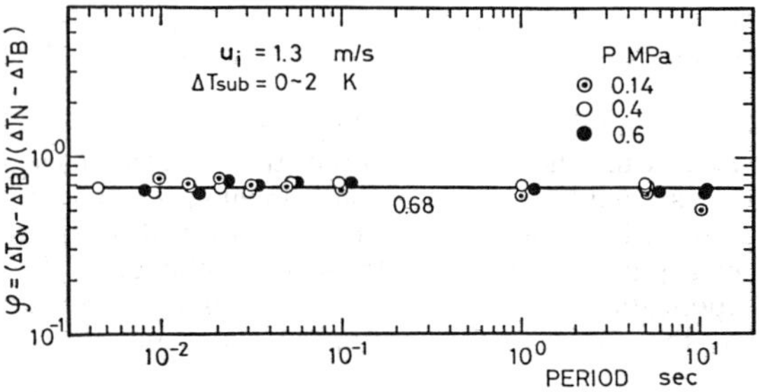

Figure 33. Fraction of activated nucleation sites [132].

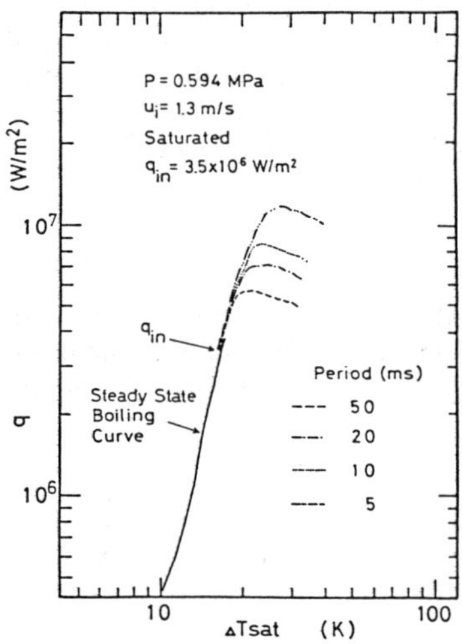

Figure 34. Effect of initial condition on transient boiling [133].

and ΔT_N are known. This can be made in a following manner. Here, the temperature differences are written as

$$\Delta T_N = q_{ov}/h_{tr} \tag{126}$$

$$\Delta T_B = q_{ov}/h_B \tag{127}$$

$$\Delta T_{ov} = q_{ov}/h_{ov} \tag{128}$$

where q_{ov} is the heat flux at the maximum temperature overshoot, and h_{tr}, h_B, and h_{ov} are the transient nonboiling heat transfer coefficient, the steady-state boiling heat transfer coefficient, and the heat transfer coefficient at the maximum temperature overshoot for constant heat flux q_{ov}. In order to solve a set of equations, we need another constitutive equation. This constitutive equation is given by Equation 129 obtained empirically from the experimental data of Kataoka et al. [73, 74].

$$h_{ov} = 0.6\,h_B \tag{129}$$

Since h_B is given by Equation 113 as a function of q_{ov}, and h_{tr} by Equation 119, the heat flux q_{ov}, and hence ΔT_{ov} are calculated from Equations 124 ~ 129.

Another interesting aspect of the transient boiling during a power increase is that, as shown in Figures 26 and 31, the transient boiling curve coincides after the temperature overshoot with the steady-state nucleate boiling curve and/or its extrapolation (A- and B-type boiling patterns). This phenomenon occurs even for a very rapid increase as shown in Figure 24. This way similarly observed earlier in pool transient boiling experiments [81–83]. For example, in Hall and Harrison's experiment [82], the exponential period of the power was less than 1 ms. Even in such extremely rapid transients, a short burst of the nucleate boiling was observed. It should be, however, noted that Figure 31 indicates also the transient boiling process which does not coincide with the steady-state nucleate boiling curve and its extrapolation during a transient (C-type boiling). In this mode of transient boiling, the available nucleation sites were not fully activated. A similar phenomenon was observed in the transient boiling experiment by Borishansky et al. [135], using fluids of low thermal conductivity. It is suspected that this type of boiling occurs in rather limited conditions.

Finally, the effect of the initial heat flux condition of the transient is discussed based on the experimental observations [133, 134]. Figures 34 and 35 are some examples showing the initial condition dependence of the transient boiling characteristics. The experimental results shown in Figure 34 were obtained by exponentially increasing the power to a heated wire from a heat flux q_{in} at which the steady-state nucleate boiling was initially maintained. In these cases, the transient boiling curve traces exactly the steady-state nucleate boiling curve and/or its extrapolation. This is followed by an idea that all the available nucleation sites have been fully activated at that initial condition, and, hence, no temperature overshoot was observed. Figure 35 shows transient boiling curves with different initial conditions for a constant exponential period of 10 ms. Here, q_{in0} is a nonboiling heat flux, q_{in1} and q_{in2} boiling heat fluxes under steady-state conditions. The temperature overshoot is clearly seen only in the case of q_{in0}. It is concluded from this fact that a certain time is needed for an available nucleation site to be activated, and that this time delay in the activation of nucleation sites is surely subjected to the time required for the formation of a thermal boundary layer near the heated surface.

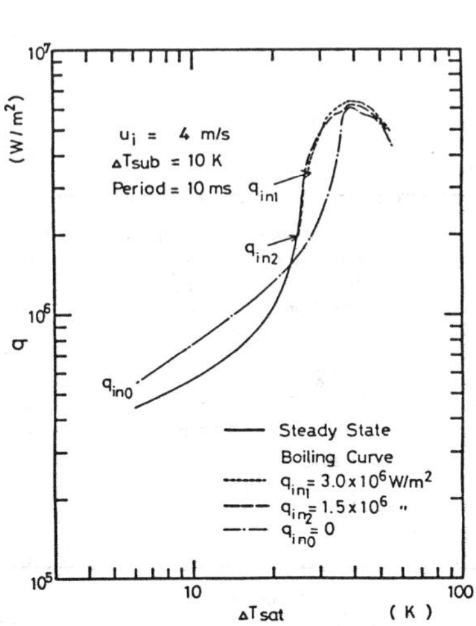

Figure 35. Effect of initial condition on transient boiling [132].

Transient CHF

The transient CHF for small ℓ/d_{he} is defined here as the heat flux where the surface temperature of the heater increases very rapidly in a low steam quality flow, i.e., bubbly or slug flow. The CHF mechanism is supposed to be a transition from nucleate boiling to film boiling, as has been stated earlier. In some cases such as shown in Figures 26 and 31, the maximum heat flux attained in the nucleate boiling region is regarded as the transient CHF.

Figure 36 shows some experimental transient CHF data [73, 74] for A-, B- and C-type boiling patterns (these boiling patterns are referred to Figure 24). As is clearly indicated in this figure, the transient CHF in A- and B-type boiling increases with a decrease in the exponential period. On the other hand, in C-type boiling, the experimental results indicate different trends, showing somewhat deteriorated CHF values. This could be an important and serious problem associated with safety evaluations of the plant. However, in the experiment of Kataoka et al. [73, 74], this was observed only in the limited range of experimental variables. Therefore, A- and B-type transient CHFs are discussed hereafter.

In general, the transient CHF, q_{ctr}, depends on the system pressure, flow velocity, liquid subcooling, heater geometry, the rate and the manner of power increase and so on. Figures 36 through 41 show the effects of these parameters on the transient CHF, showing increasing tendencies for higher velocity, subcooling, pressure, and thinner wire.

As is clear from previous discussions on the transient boiling curves and also from the experimental results shown in Figures 36 through 41, the transient CHF for A- and B-type boiling is thought to be expressed simply by the sum of the steady-state CHF, q_{cst}, and transient effects. By applying a dimensional analysis and using their own experimental data, Kataoka et al. [73] proposed the following empirical transient CHF correlation for exponential power increase with small wires.

$$\frac{q_{ctr} - q_{cst}}{GH_{fg}} = 0.2038\left(\frac{\rho_g}{\rho_\ell}\right)^{0.52}\left(\frac{\sigma\rho_\ell}{G^2\ell_0}\right)^{0.19}\left(\frac{\tau G}{\rho_\ell\ell_0}\right)^{-0.63} \tag{130}$$

where

$$G = \rho_\ell u_i \tag{131}$$

$$\ell_0 = \left(\frac{\sigma}{g\Delta\rho}\right)^{1/2} \tag{132}$$

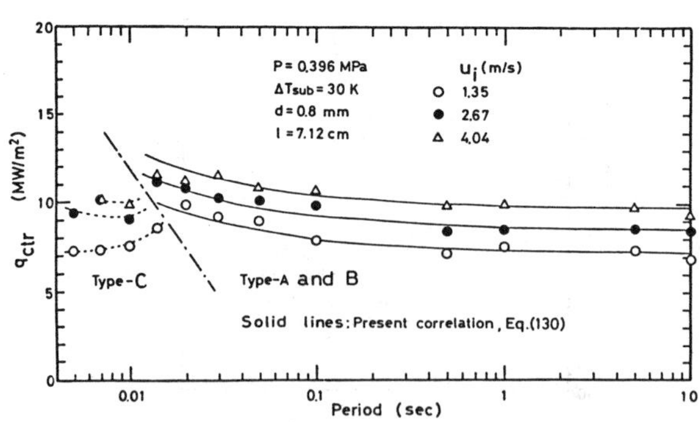

Figure 36. Transient CHF for A, B and C type.

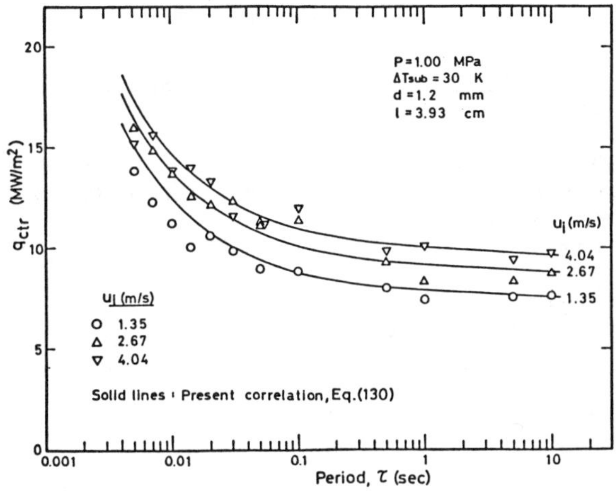

Figure 37. Effect of inlet velocity on transient CHF.

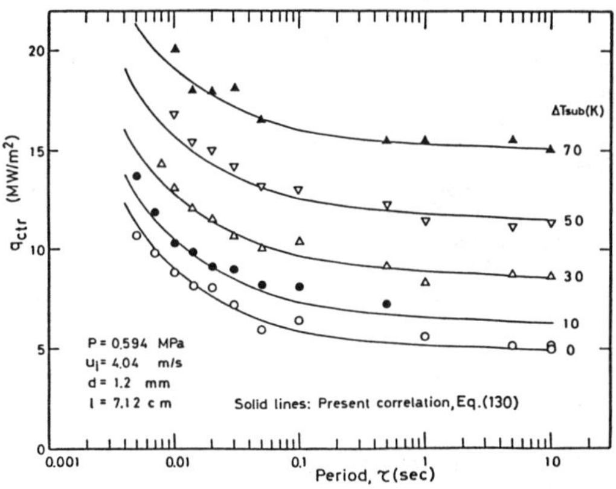

Figure 38. Effect of subcooling on transient CHF.

It should be noted in Equation 130 that the CHF difference, $(q_{ctr} - q_{cst})$, varies approximately as -0.6 power of the exponential period τ.

Here, the steady-state CHF q_{cst} in the previous correlation is given below also by Kataoka et al. [73] using their own experimental data under forced convection flows.

$$\frac{q_{cst}}{GH_{fg}} = 0.3740\left(\frac{\rho_g}{\rho_\ell}\right)^{0.66}\left(\frac{\sigma\rho_\ell}{G_2\ell_0}\right)^{0.40} \cdot\left\{1 + \left[0.03648\left(\frac{\ell_0}{d_{he}}\right)^{-0.20}\left(\frac{\rho_g}{\rho_\ell}\right)^{-0.79} + \varepsilon\right]\frac{\Delta H_{sub}}{H_{fg}}\right\} \qquad (133)$$

$$
\varepsilon = \begin{cases} 0 & (\Delta T_{sub} = 0 - 40°\,K) \\ 0.00808\left(\dfrac{\rho_g}{\rho_\ell}\right)^{-1.09} \dfrac{\Delta H_{sub} - \Delta H_{sub40}}{\Delta H_{sub}} & (\Delta T_{sub} = 40 - 70°\,K) \end{cases} \tag{134}
$$

Figures 42 through 44 represent comparisons of Equations 133 and 130 with their own experimental results of the steady-state CHF and the transient CHF, respectively, showing satisfactory agreement between them.

It is interesting here to see that Equation 130 can be rewritten as

$$
\frac{q_{ctr} - q_{cst}}{\rho_g H_{fg}\left(\dfrac{\sigma g \Delta\rho}{\rho_g^2}\right)^{1/4}} = 0.2038\left(\frac{\rho_g}{\rho_\ell}\right)^{-0.30}\left[\tau\left(\frac{\sigma g \Delta\rho}{\rho_g^2}\right)^{1/4}\Big/\ell_0\right]^{-0.63} \tag{135}
$$

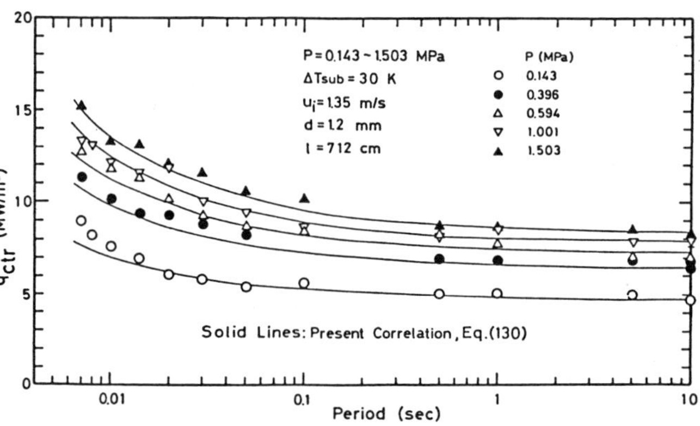

Figure 39. Effect of pressure on transient CHF.

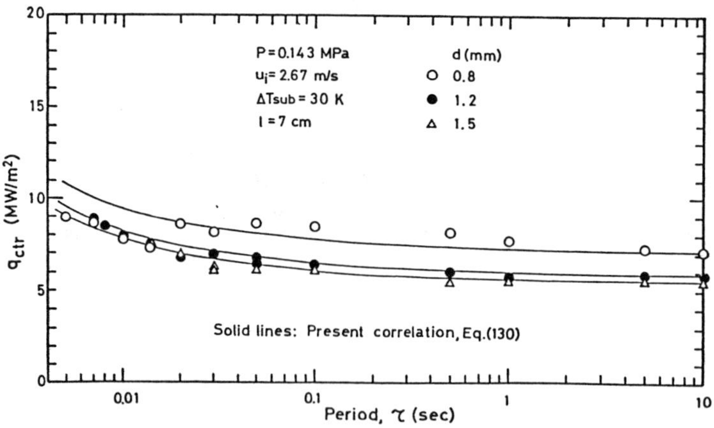

Figure 40. Effect of heater diameter on transient CHF.

This equation indicates that the difference between the transient CHF and the steady-state CHF, $(q_{ctr} - q_{cst})$ is independent of the flow velocity and subcooling. The effects of these parameters are included in the q_{cst} term. In view of this and close CHF mechanisms in forced-convective boiling for small ℓ/d_{he} and pool boiling, Equation 135 may well be successfully applied to the transient CHF prediction for pool boiling. In fact, a reasonable agreement is seen in Figure 45 between Equation 135 and the transient pool boiling CHF data obtained by Sakurai and Shiotsu [93] in water at pressures 0.143 – 2.0 MPa for an exponential power increase.

A detailed modeling of the transient CHF in power transients was carried out theoretically by Serizawa [136]. This model considers a thin liquid layer on the heated surface near CHF as shown in Figure 21, whose existence has been confirmed previously in experiments by many other investigators. A physical picture of this model is given in Figure 46. It is assumed that the transient CHF is caused by a deficiency of liquid within this thin liquid layer under the vapor blanket, with the liquid being supplied to the liquid

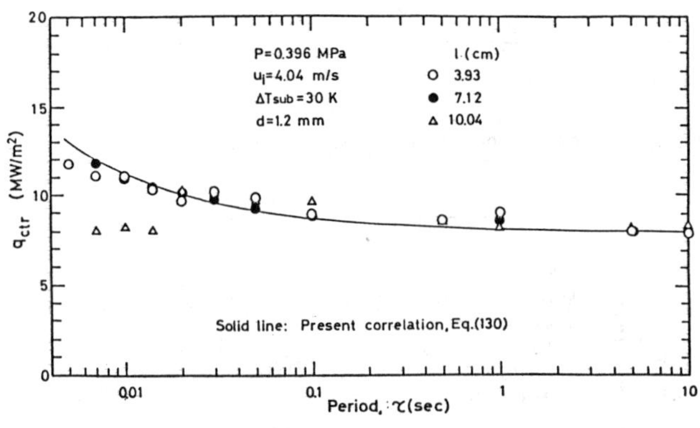

Figure 41. Effect of heater length on transient CHF.

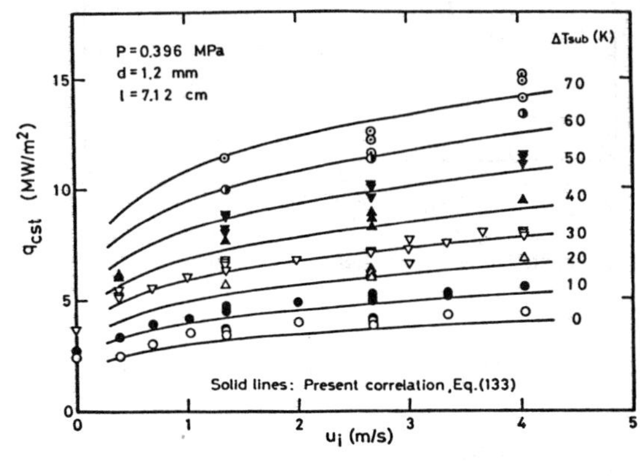

Figure 42. Steady-state CHF.

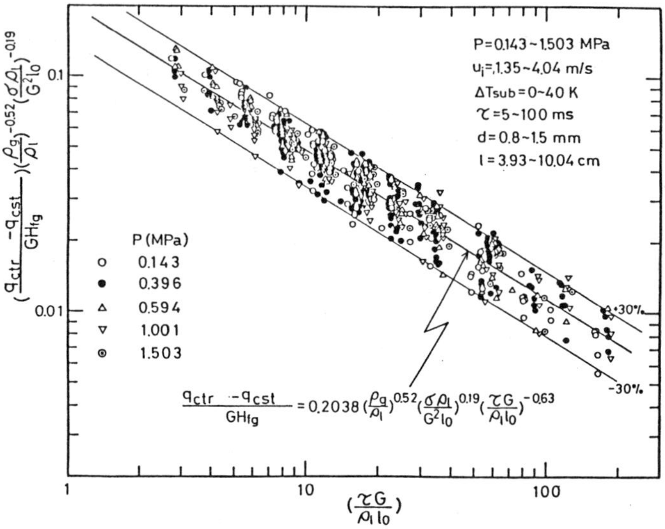

Figure 43. Difference between transient CHF and steady-state CHF [73].

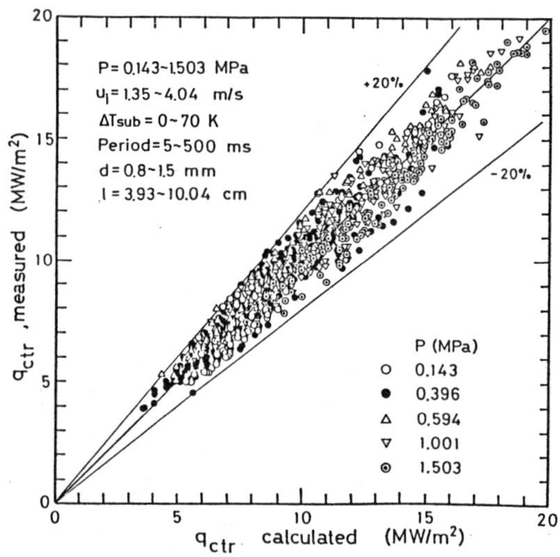

Figure 44. Correlative respresentation of the transient CHF correlation with experiment [73].

layer by $W_f(t)$ during the postulated transient. From the heat balance to the liquid layer, we have

$$\frac{d\delta(t)}{dt}=-\frac{1}{\rho_\ell H_{fg}}q(t)+\left(1+\frac{\Delta H_{sub}}{H_{fg}}\right)\frac{w_f(t)}{\rho_\ell A_f} \tag{136}$$

where A_f is heater area where the thin liquid layer exists, and $\delta(t)$ the thickness of the liquid film. Equation 136 can be integrated for given $q(t)$, $W_f(t)$, and $\delta(0)$, the initial value of the liquid film thickness.

After a detailed inspection of the transient CHF data reported elsewhere for various heater geometries and power increases, Serizawa determined $W_f(t)$ to be constant during the transient. He also derived theoretically the correlation for $\delta(0)$ in saturated pool boiling of water as

$$\frac{(\sigma g \Delta\rho/\rho_g^2)^{1/4}\,\delta(0)}{a_l} = 23\,(\rho_l/\rho_g)^{0.3}(1 - S_f)^{1/5} \tag{137}$$

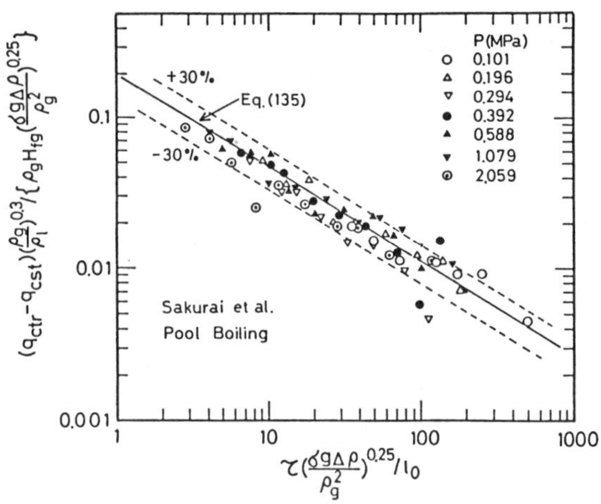

Figure 45. Correlative representation of the transient CHF correlation with pool boiling experiment [93].

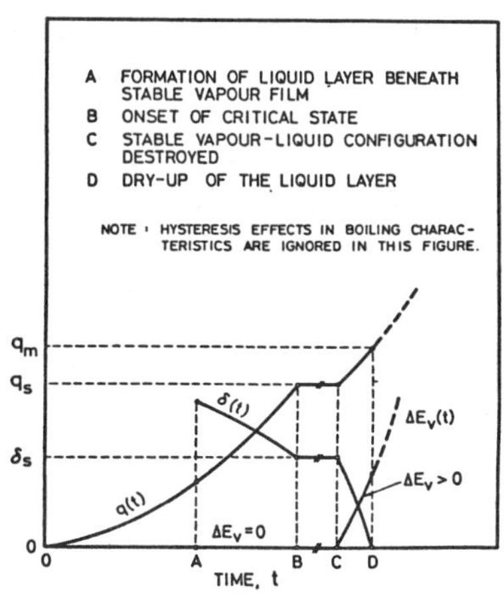

Figure 46. Physical model of CHF [136].

where the dry area fraction S_f is given by

$$S_f = 0.34\,(P/P_0)^{0.3} \exp[1.2\,(1 - P/P_0)]$$
$$P_0 = 0.1\,\text{MPa} \tag{138}$$

Figure 47 is a comparison between Equation 137 and some experimental data. This result was extended to forced-convection flows with and without subcooling. The time, t_c, required from the initiation of the transient until the CHF (point D in Figure 46) is evaluated by considering that the liquid film is consumed at that point, that is

$$\delta(t_c) = 0 \tag{139}$$

The transient CHF is then given by

$$q_{ctr} = q(t_c) \tag{140}$$

This model was compared satisfactorily with various transient CHF data for different geometries of the heater and flow and different ways of power increase [136].

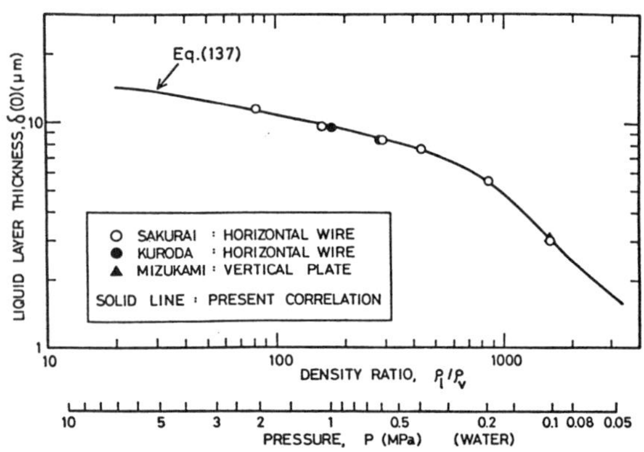

Figure 47. Initial liquid layer thickness in saturated pool boiling of water [136].

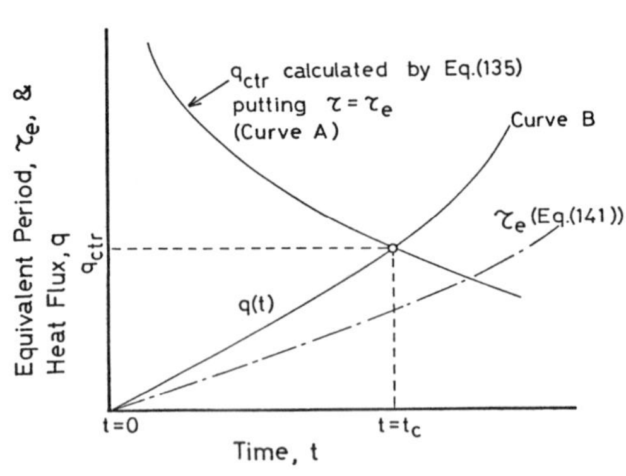

Figure 48. Determination of transient CHF using equivalent period.

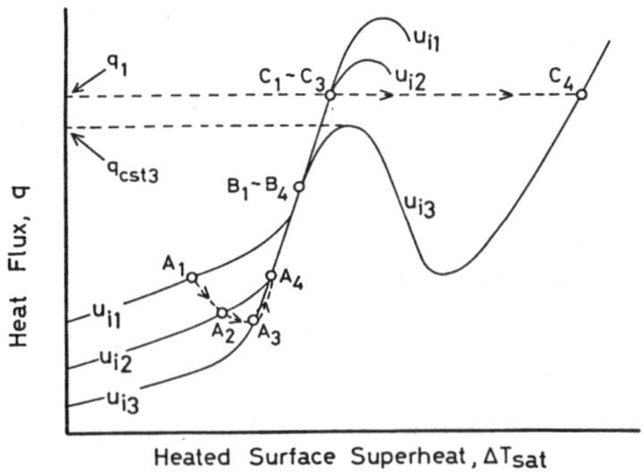

Figure 49. Heat transfer process in flow transient.

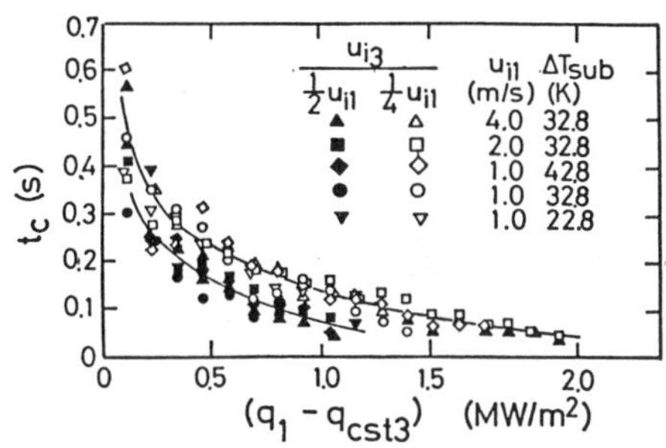

Figure 50. Transient CHF time vs. $(q_1 - q_{cst3})$ in flow transient [80].

Finally, a generalized method to predict the transient CHF behavior under a particular power increase from known correlations will be presented here. This method is a use of the equivalent period defined as

$$\tau_e \equiv \frac{\int_0^t q(t)\,dt}{q(t)} \tag{141}$$

The procedure to determine the transient CHF for particular types of power increases is schematically shown in Figure 48. By substituting the equivalent period τ_e into, for example, Equation 135 for the place of τ, we obtain q_{ctr} as a function of time (curve A in Figure 48). On the other hand, the heat flux applied to the heater is denoted by q (t) (curve B). The transient CHF occurs at a point of intersection between curve A and B. This method was partially verified by Kawamura et al. [11] for several types of power increases.

Flow and Pressure Transients

A quite limited number of experimental data are available now on the flow and pressure transients for small ℓ/d_{he} [79, 80, 94, 95]. The most important finding made by these works is that the transient heat transfer process follows the steady-state boiling performance, i.e., quasi-steady state in both flow and pressure transients. This is schematically shown in Figures 49 and 51 for flow and pressure transient, respectively.

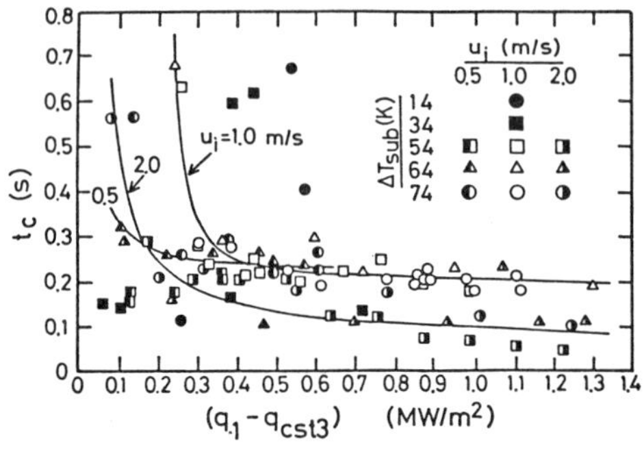

Figure 51. Heat transfer process in pressure transient.

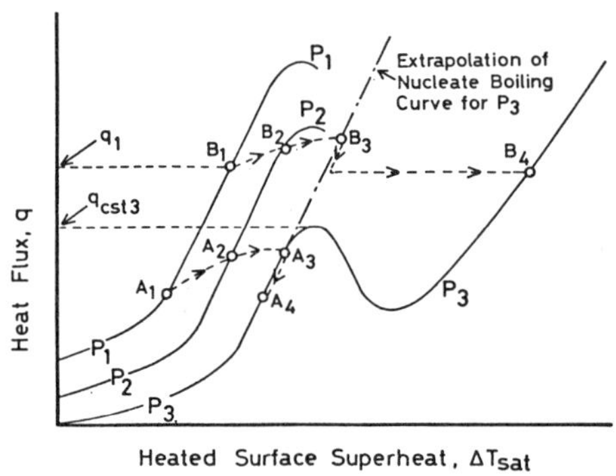

Figure 52. Transient CHF time vs. $(q_1 - q_{cst3})$ in pressure transient [79].

In Figure 49, the inlet velocity u_i decreases from $u_{i1} \rightarrow u_{i2} \rightarrow u_{i3}$. The steady-state boiling curves corresponding to $u_{i1} \sim u_{i3}$ are indicated in this figure by solid lines. When the velocity changes from a nonboiling condition (point A_1) along $u_{i1} \rightarrow u_{i2} \rightarrow u_{i3}$ for a constant heat-generation rate, the heat transfer process varies in a manner $A_1 \rightarrow A_2 \rightarrow A_3$ so as to attain the steady-state boiling condition A_4. It should be noted here that the heat flux decreases during this transient because a certain amount of heat generated in the heater is used to raise the heater temperature. If the transient process is initially at boiling condition (point B_1) where the heat flux is still lower than the steady-state CHF q_{cst3} in the final state, the heat transfer process remains almost unchanged ($B_1 \rightarrow B_4$). On the other hand, when the initial condition is point C_1 where the heat flux exceeds the steady-state CHF q_{cst3}, heat transfer process remains for a certain time ($C_1 \rightarrow C_3$ correspondingly to $u_{i1} \rightarrow u_{i3}$) on the steady-state nucleate boiling curve and/or its extrapolation until a point C_3 is reached where the transition from nucleate to film boiling occurs ($C_3 \rightarrow C_4$). These trends were experimentally observed by Aoki et al. [80]. The experimental data by Aoki et al. indicated also that the CHF time t_c decreases with an increase of the flux difference $(q_1 - q_{cst3})$ between the initial heat flux q_1 and the steady-state CHF in the final state q_{cst3}. This fact is clearly seen in Figure 50.

Figure 51 shows traces of the transient boiling process during a depressurization. The system pressure decreases along $P_1 \rightarrow P_2 \rightarrow P_3$. The steady-state boiling curves for $P_1 \sim P_3$ are also indicated in the figure by solid lines.

When the initial state is at point A_1 of which heat flux is lower than the steady-state CHF, q_{cst3}, in the final state, the heat transfer process changes along $A_1 \rightarrow A_2 \rightarrow A_3 \rightarrow A_4$ as the system pressure decreases like $P_1 \rightarrow P_2 \rightarrow P_3 \rightarrow P_4$ for a constant heat generation rate. At every moment during this transient, the transient boiling characteristics behave in quasi-steady state manners corresponding to the instantaneous system pressure. The heat flux during the transient becomes higher than the initial heat flux. This follows that the energy stored in the heater is released by a decrease in the heater temperature.

When the pressure transient starts from point B_1 where the heat flux is already higher than q_{cst3}, the heat transfer process moves along $B_1 \rightarrow B_2 \rightarrow B_3$ during the initial stage of the transient, staying at B_3 for a certain time. Point B_4 is finally attained where the transition from nucleate to film boiling takes place.

The phenomena just mentioned are reported to have been observed by Aoki et al. [79] and Toda et al. [94, 95]. Similarly in flow transients, the transient time t_c in pressure transients correlates also with the heat flux difference $(q_{1B} - q_{cs3})$, where q_{1B} and q_{cst3} are the initial heat flux and the steady-state CHF in the final state. Figure 52 shows variation of the transient CHF time with the heat flux difference just mentioned, obtained by using the experimental data of Aoki et al. [79]. The transient CHF time is thus decreasing for larger heat flux difference.

One of the important inferences obtained from the experimental evidences just mentioned is that the transient CHF in flow and pressure transients with small ℓ/d_{he} is closely related to the consumption of a thin liquid film by evaporation which is formed on the surface of the heater. Detailed analyses are still much needed to clarify the microscopic mechanisms involved in flow and pressure transients as well.

NOTATION

a_i	interfacial area concentration
$a_{i\ell e}$	interfacial area concentration of liquid droplets
$a_{i\ell f}$	interfacial area concentration of liquid film
a_ℓ	thermal diffusivity of liquid
A_f	heater area where thin liquid film exists
A_s	flow area
C	coefficient
$C_{p\ell}$	specific heat of liquid at constant pressure
$\dot{d}$	deposition rate of liquid droplets
d_{he}	heated equivalent diameter of heated section $(4A_s/p_h)$
Ei(-x)	exponential integral function
f	friction factor
f_m	friction factor for mixture
g	acceleration due to gravity
G	inlet mass velocity
h_B	boiling heat transfer coefficient
h_{st}	steady-state nonboiling heat transfer coefficient
h_{tr}	transient nonboiling heat transfer coefficient
h_{tra}	asymptotic value of transient nonboiling heat transfer coefficient for exponential heat flux increase
H_{fg}	latent heat of vaporization
H_g	enthalpy of gas
H_{ik}	enthalpy of k (gas or liquid) phase at interface
H_k	enthalpy of k phase
H_ℓ	enthalpy of liquid
j	total flux
j_g	gas flux (superficial velocity)
$j_{g\ell}$	drift flux
j_ℓ	liquid flux (superficial velocity)
K_{in}	inlet orifice pressure drop coefficient
K_e	exit orifice pressure drop coefficient
ℓ	heated length
ℓ_0	characteristic length scale
L_B	boiling length
L_s	length scale of gas liquid interface
$\overline{m_{\ell e}}$	mass transfer rate to droplets
$\overline{m_{\ell f}}$	mass transfer rate to liquid film
M_{ig}	interfacial drag for gas phase
M_{ik}	interfacial drag for k (gas or liquid) phase

Symbol	Definition
$M_{i\ell}$	interfacial drag for liquid phase
$n, n_1 - n_5$	exponents
p_h	heated perimeter
P	pressure
P_{cr}	critical pressure
P_g	pressure of gas phase
P_k	pressure of k (gas or liquid) phase
P_ℓ	pressure of liquid phase
P_m	pressure of mixture
q	heat flux
q^*	heat flux defined by Equation 67
q_0	heat flux that is kept constant in flow transient
q_1	heat flux in initial state
q_2	heat flux in final state (step increase) or initial value of exponential heat flux increase
q_c	critical heat flux (CHF)
q_{cst}	steady state CHF
q_{st1}	steady-state CHF corresponding to u_{i1}
q_{st2}	steady-state CHF corresponding to u_{i2}
q_{st3}	steady-state CHF corresponding to u_{i3} or P_3
q_{ctr}	transient CHF
q_{in}	heat flux in initial state
q_{in1}, q_{in2}	(steady-state boiling condition)
q_{in0}	heat flux in initial state (non-boiling condition)
q_{IB}	heat flux at boiling incipience
q_{ki}	heat flux of k (gas or liquid) phase at interface
q_{ov}	heat flux at maximum temperature overshoot
r	radius of nucleation site
R_v	gas constant
S_f	dry area fraction
t	time
t^*	dimensionless time defined by Equation 66
t^+	dimensionless time defined by Equation 85
t_0	time given by Equation 73 or Equation 92
$t_{0.5}$	time when initial velocity halves
t_B	time of boiling inception at $z = z_B$
t_c	transient CHF time
t_i	time when a fluid particle passes $z = 0$
t_{iB}	time when a fluid particle passes boiling boundary
T_{in}	inlet liquid temperature
T_ℓ	liquid temperature
T_{sat}	saturation temperature
u	velocity
u_∞	bubble rise velocity in an infinite medium
u_g	gas velocity
u_i	inlet velocity
u_i^+	dimensionless inlet velocity defined by Equation 86
u_{i0}	inlet velocity which is kept constant in power transient
u_{i1}	inlet velocity in initial state
u_{i2}	inlet velocity in final state (step velocity decrease)
u_{i3}	inlet velocity in final state (in Figure 49)
u_{icst}	steady-state CHF velocity corresponding to q_0
$u^q_{icst}(q)$	steady-state CHF velocity corresponding to q
u_{ictr}	transient CHF velocity
u_k	velocity of k (gas or liquid) phase
u_{ki}	velocity of k phase at interface
u_ℓ	liquid velocity
$u_{\ell e}$	velocity of liquid droplets
$u_{\ell f}$	velocity of liquid film
u_m	velocity of mixture
v	variable
V_{gj}	drift velocity
w_f	liquid supply to liquid film
x	quality (weight fraction of gas phase) or variable
x_e	exit quality
z	axial location
z_B	axial location of boiling boundary

Greek Symbols

Symbol	Definition
α	void fraction (volume fraction of gas phase)
$\langle\alpha\rangle$	average void fraction over heated section

α_e exit void fraction
α_k volume fraction of k (gas or liquid) phase
α_{kw} volume fraction of k phase at heated surface
$\alpha_{\ell e}$ volume fraction of liquid droplets
$\alpha_{\ell f}$ volume fraction of liquid film
Γ_g vapor source term
Γ_k source term of k (gas or liquid) phase
δ liquid film thickness
ΔE_v net evaporation rate
ΔH_{sub} inlet subcool enthalpy
ΔH_{sub40} inlet subcool enthalpy at 40° K subcooling
ΔP_h driving head of pump
ΔP_{h0} driving head of pump which is kept constant in power transient
ΔP_f pressure drop of heated section
ΔQ heat increment
Δt time increment
ΔT_B heated wall superheat for complete activation of available nucleation sites
ΔT_N heated wall superheat for zero activation of available nucleation sites
ΔT_{ov} maximum temperature overshoot
ΔT_{sat} heated wall superheat
ΔT_{sub} inlet liquid subcooling
ΔV volume increment
$\Delta \mu$ viscosity difference between gas and liquid
$\Delta \rho$ density difference between gas and liquid ($\rho_\ell - \rho_g$)
ε entrainment rate of liquid droplets
ε term in Equation 134
θ angle between flow direction and gravity
λ exponential velocity decreasing rate
λ^+ dimensionless exponential velocity decreasing rate defined by Equation 93
λ_0 exponential velocity decreasing rate given by Equation 96
λ_ℓ thermal conductivity of liquid
λ_r linear velocity decreasing rate
μ_g gas viscosity
ν_ℓ kinematic viscosity of liquid
ρ_g gas density
ρ_k density of k (gas or liquid) phase
ρ_ℓ liquid density
ρ_m mixture density
ρ_{me} exit mixture density
σ surface tension
τ exponential period for exponential heat flux increase
τ_e equivalent period given by Equation 141
τ_{iw} interfacial shear stress at heated wall
τ_{kw} shear stress of k (gas or liquid) phase at heated wall
Φ fraction of activation of available nucleation sites given by Equation 124
ω exponential heat flux increasing rate
ω^* dimensionless exponential heat flux increasing rate defined by Equation 74
ω_0 exponential heat flux increasing rate given by Equation 79

REFERENCES

1. Tong, L.S., and Hewitt, G.F., "Overall View Point of Flow Boiling CHF Mechanism," ASME paper, 72-HT-54 (1972).
2. Hewitt, G.F., "Critical Heat Flux in Flow Boiling," Proc. 6th Int. Heat Trans. Conf. Vol. 6, p. 143, KS-13, Toronto (1978).
3. Plesset, M.S., Zuber, N., and Catton, I., "*Transient Two-Phase Flow,*" Proc. 3rd CSNI Specialist Meeting on Two-Phase Flow, Hemisphere Pub. Co., New York (1983).
4. Moxon, D., and Edwards, P.A., "Dryout during Flow and Power Transients," UKAEA Report, AEEW-R-553 (1967).

5. Leslie, D. C., and Kirby, G. T., "Application of a Film Flow Model to Predicting Burnout under Transient Conditions," UKAEA Report, AEEW-R-542 (1967).
6. Moxon, D., "A Dynamic Programme for the Thermalhydraulic Behaviour of Boiling Loops," UKAEA Report, AEEW-R-448 (1968).
7. Gaspari, G. P., Granzini, R., and Hassid, A., "Dryout Onset in Flow Stoppage, Depressurization and Power Surge Transients," *Energia Nucleare*, 20 (10): 554 (1973).
8. Hassid, A., and Rychlichi, R., "An Attempt to Use Steady State Data for Determinig Critical Heat Flux in Transient Conditions," *Energia Nucleare*, 18 (6): 333 (1971).
9. Magni, A., "TILT: A Digital Simulation Programme for the Study of Hydrodynamic Processes and Core Heat Up of a Boiling Water Pressure Tube Reactor during Transient Conditions," *Energia Nucleare*, 20 (5): 293 (1973).
10. Shiralkar, B. S., et al., "Transient Critical Heat Flux-Experimental Results," General Electric Report, GEAP-13295 (1972).
11. Lahey, R. T., Jr., et al., "The Analysis of Transient Critical Heat Flux," General Electric Report, GEAP-13249 (1972).
12. Smirnov, O. K., Pashkov, L. T., and Zaitsev, V. N., "Investigation of Critical Heat Flux with Decrease in Flow through a Heated Tube," *Teploenergetika*, 19 (9): 83 (1972).
13. Smirnov, O. K., Zaitsev, V. N., and Serov, E. E., "Investigation of Burnout under Transient Hydrodynamic Conditions," *Teploenergetika*, 24 (5): 81 (1977).
14. Cumo, M., Fabrizi, F., and Palazzi, G., "Transient Critical Heat Flux in Loss-of-Flow Accident (L.O.F.A.)," *Int. J. Multiphase Flow*, 40: 497 (1978).
15. Iwamura, T., and Kuroyanagi, T., "Burnout Characteristics under Flow Reduction Condition," *J. Nucl. Sci. Technol.*, 19 (6): 438 (1982).
16. Leung, J. C. M., "Transient Critical Heat Flux and Blowdown Heat-Transfer Studies," Argonne National Laboratory Report, ANL-80-53, NUREG/CR-1559 (1980).
17. Leung, J. C. M., "Occurrence of Critical Heat Flux during Blowdown with Flow Reversal," Argonne National Laboratory Report, ANL-77-4 (1977).
18. Banerjee, S., and Hancox, W. T., "On the Development of Methods for Analysing Transient Flow Boiling," *Int. J. Multiphase Flow*, 4: 437 (1978).
19. Cermak, J. O., et al., "The Departure from Nucleate Boiling in Rod Bundles during a Pressure Blowdown," *Trans. ASME, J. Heat Transfer*, 92: 621 (1970).
20. Lawson, G. C., "Heat Transfer from Electrically Heated Rods during a Simulated Loss-of-Coolant Accidents," *Chem. Eng. Progr. Symp. Ser.*, 19 (119): 1 (1971).
21. Belda, W., and Mayinger, F., "Dryout Delay during Blowdown," *Chem. Eng. Progr. Symp. Ser.*, 73 (164): 21 (1977).
22. Hein, D., Kastner, W., and Mayinger, F., "Messungen der kritischen Heizflächenbelastung in Slabbünden bei stationärem und instationärem Betriebszustand," Proc. 4th Int. Heat Trans. Conf. B 6.5, Paris (1970).
23. Ishigai, S., et al., "Effect of Transient Flow on Premature Dryout in Tube," Proc. 4th Int. Heat Trans. Conf. B 6.13, Tokyo (1974).
24. MacDonald, P., Broughton, J. M., and Spore, Y. W., "An Evaluation of the Thermal-Hydraulic and Fuel Rod Thermal and Mechanical Deformation Behaviors during the First Power Burst Facility Nuclear Blowdown Tests," *Nuclear Technol.*, 44: 401 (1979).
25. LeTourneau, B. W., and Green, S. J., "Critical Heat Flux and Pressure Drop Tests with Parallel Upflow of High Pressure Water in Bundles of Twenty 1/2 in. Rods," *Nucl. Sci. Eng.*, 43 (1): 90 (1971).
26. Zielke, L. A., and Wilson, R. H., "Transient Critical Heat Flux and Spacer Grid Studies," *Nuclear Technol.*, 24: 13 (1974).
27. Marinelli, V., et al., "Use of the Lagrangian and Eulerian Points of View in the Transient Critical Heat Flux Calculations for BWR Rod Bundles and Experimental Verifications," *Nuclear Technol.*, 27: 207 (1975).

28. Leung, J. C. M., et al., "Critical Heat Flux Predictions during Blowdown Transients," *Int. J. Multiphase Flow,* 7 (6): 677 (1981).
29. Ishii, M., *Thermo-Fluid Dynamic Theory of Two-Phase Flow,* Eyrolles, Paris (1975).
30. Ishii, M., "One-Dimensional Drift Flux Model and Constitutive Equations for Relative Motion between Phases in a Various Two-Phase Flow Regimes," Argonne National Laboratory Report, ANL-77-47 (1977).
31. Ishii, M., and Mishima, K., "Study of Two-Fluid Model and Interfacial Area," Argonne National Laboratory Report, ANL-80-111, NUREG/CR-1873 (1980).
32. Wallis, G. B., *One-Dimensional Two-Phase Flow,* McGraw Hill Pub. Co., New York (1969).
33. Zuber, N., and Findlay, J. A., "Average Volumetric Concentration in Two-Phase Flow Systems," *Trans. ASME, J. Heat Transfer,* 87: 453 (1965).
34. Hewitt, G. F., and Hall-Taylor, N. S., *Annular Two-Phase Flow,* Pergamon Press, Oxford (1972).
35. Delhaye, J. M., "Equations Fondementales des Ecoulements Diphasiques, Part I and II," CEA-R-3429, France (1968).
36. Bouré, J., and Réocruex, M., "General Equations of Two-Phase Flows," Proc. 4th All Union Heat and Mass Transfer Conf., Minsk, USSR (1972).
37. Ishii, M., and Zuber, N., "Drag Coefficient and Relative Velocity in Bubbly, Droplet or Particulate Flows," *AIChE J.,* 25: 843 (1979).
38. Serizawa, A., and Kataoka, I., "Fundamental Aspects of the Drift Velocity in Turbulent Bubbly Flow," Tech. Repts Atom. Energy, Kyoto Univ. No. 182 (1980).
39. Žun, I., and Serizawa, A., "Analysis of Bubbly Flow in Vertical Tube (I) – Relative Velocity Profile," Proc. 4th Two-Phase Flow Symp. of Japan, p. 9, Kobe, Japan (1982).
40. Ishii, M., et al., "Two-Fluid Model and Importance of the Interfacial Area in Two-Phase Flow Analysis," Proc. 9th U.S. National Congress of Applied Mechanics, p. 73, Ithaca, New York (1982).
41. Ishii, M., and Kataoka, I., "Hydrodynamics of Annular Dispersed Flow," *Fluid Transient and Fluid Structure Interaction,* Y. W. Shin et al. (Eds.), p. 1, ASME Publication PVP-Vol. 64 (1982).
42. Ishii, M., and Katoka, I., "Interfacial Transfer in Annular Dispersed Flow," *Advances in Two-Phase Flow and Heat Transfer,* S. Kakac and M. Ishii (Eds.), Vol. 1, p. 93, Martinus Nijhoff Pub., Netherlands (1983). (Proc. NATO Advanced Research Workshop on Two-Phase Flow, Spitzingsee, FRG (1982).)
43. Kataoka, I., and Ishii, M., "Mechanism and Correlation of Droplet Entrainment and Deposition in Annular Two-Phase Flow," Argonne National Laboratory Report, ANL-82-44, NUREG/CR-2885 (1982).
44. Kataoka, I., and Ishii, M., "Entrainment and Deposition Rates of Droplets in Annular Two-Phase Flow," Proc. ASME/JSME Thermal Engineering Joint Conf., Vol. 1, p. 69, Honolulu, U.S.A. (1983).
45. Gonzales-Santalo, J. M., and Lahey, R. T., Jr., "An Exact Solution for Flow Transients in Two-Phase Systems by Methods of Characteristics," *Trans. ASME, J. Heat Transfer,* 95: 470 (1973).
46. Wallis, G. B., and Heasley, J. H., "Oscillation in Two-Phase Flow Systems," *Trans. ASME, J. Heat Transfer,* 83 (1961).
47. Zuber, N., and Staub, F. W., "The Propagation and Wave Form of the Vapor Volumetric Concentration in Boiling Forced Convection System under Oscillatory Condition," *Int. J. Heat Mass Transfer,* 9: 871 (1966).
48. Doroshuk, V. E., Levitan, L. L., and Lantzman, F. P., "Recommendations for Calculating Burnout in Uniformly Heated Tubes," *Teploenergetika,* 22 (12): 66 (1975).
49. Katto, Y., "Generalized Correlation of Critical Heat Flux for the Forced Convection Boiling in Vertical Uniformly Heated Round Tubes," *Int. J. Heat Mass Transfer,* 21: 1527 (1978).

50. Katto, Y., "Generalized Correlations of Critical Heat Flux for Uniformly Heated Annuli," *Int. J. Heat Mass Transfer,* 22: 575 (1979).
51. Morgan, R. P., "A Review and Discussion of Literature Concerning Transient Heat Transfer and Steam Formation," USAEC Report, TID-4500 (1967).
52. Hassid, A., "Critical Heat Flux (burnout) in Transients, Remarks on the Available Information," *Energia Nucleare,* 20 (12): p. 699 (1978).
53. Leung, J. C. M., "Critical Heat Flux under Transient Conditions; A Literature Survey," Argonne National Laboratory Report, ANL-78-39, NUREG/CR-0056 (1978).
54. Aerojet Nuclear Company, "RELAP4/MOD5: A Computor Programme for Transient Thermal Hydraulic Analysis of Nuclear Reactors and Related Systems, Vol. I–III," ANCR-NUREG-1335 (1976).
55. "TRAC-P1A, An Advanced Best Estimated Computor Programme for PWR LOCA Analysis," Los Alamos Scientific Laboratory Report, LA-7777-MS, Vol. 1, NUREG/CR-0665 (1979).
56. "K-FIX; A Computor Program for Transient Two-Dimensional, Two-Fluid Flow," Los Alamos Scientific Laboratory Report, LA-NUREG-6623 (1977).
57. Senglaub, M. E., "Low Heat Flux Sodium Voiding using the Drift Flux Formulation," Argonne National Laboratory Report, ANL/RAS-79-21 (1979).
58. Ferguson, D. R., et al., "The SUS 4A, LMFBR Accident Analysis Code System, a Progress Report," Proc. Int. Meeting on Fast Reactor Safety and Related Physics, p. 1225 Chicago (1976).
59. Cumo, M., Ferrari, G., and Urbani, G., "Prediction of Burnout Power with Freon up to the Critical Pressure," *Termotecnica,* 5: 269 (1977).
60. Tong, L. S., and Weisman, J., "*Thermal Analysis of Pressurized Water Reactors,*" 2nd ed., ANS Monograph, Americal Nuclear Society, Ill. (1979).
61. Lahey, R. T., Jr., and Moody, F. J., "*The Thermal Hydraulics of a Boiling Water Nuclear Reactor,*" ANS Monogrpah, American Nuclear Society, Ill. (1977).
62. Kakac, S., and Veziroglu, T. N., "A Review of Two-Phase Flow Instabilities," *Advances in Two-Phase Flow and Heat Transfer,* S. Kakac, and M. Ishii (Eds.), Vol. 2, p. 577, Martinus Nijhoff Pub., Netherlands (1983). (Proc. NATO Advanced Research Workshop on Two-Phase Flow, Spitzingsee, FRG (1982).)
63. Ledinegg, M., "Instability of Flow during Natural and Forced Circulation," USAEC Report, AEC-tr-1861 (*Die Wärme,* 61: 861 (1938).)
64. Ishii, M., and Zuber, N., "Thermally Induced Flow Instabilities in Two-Phase Mixture," Proc. 4th Int. Heat Transfer Conf., B 5–11, Paris (1970).
65. Saha, P., Ishii, M., and Zuber, N., "An Experimental Investigation of the Thermally Induced Flow Oscillations in Two-Phase Flow Systems," *Trans. ASME, J. of Heat Transfer,* 98: 616 (1976).
66. Stenning, A. H., and Veziroglu, T. N., "Flow Oscillation Modes in Forced Convection Boiling," Proc.. 1965. Heat Transfer and Fluid Mechanics Institute, Stanford Univ. Press, p. 301 (1965).
67. Yadigaroglu, G., and Bergles, A. E., "Fundamental and Higher Mode Density Oscillation in Two-Phase Flows," *Trans. ASME, J. Heat Transfer,* 94: 189 (1972).
68. Kakac, S., et al., "Sustained and Transient Boiling Flow Instabilities in Cross Connected for Parallel Channel Upflow System," 5th Int. Heat Transfer Conf., B 5–11, Tokyo (1974).
69. Ozawa, M., et al., "Oscillatory Flow Instability in Air-Water Two-Phase Flow System-1, Pressure Drop Oscillation," Bulletin of JSME, 22 (174): 1763 (1979).
70. Collier, J. G., "*Convective Boiling and Condensation,*" McGraw Hill Pub. Co., UK (1972).
71. Ishii, M., and Fauske, H. K., "Boiling and Dryout Behavior in a Liquid Metal Fast Breeder Reactor Subassembly Bundle under Low Heat Flux and Low Flow Condition," *Nucl. Sci. Eng.,* 84: 131 (1983).

72. Griffith, P., Pearson, J. F., and Lepkowski, R. J., "Critical Heat Flux during a Loss of Coolant Accident," *Nuclear Safety*, 18 (3): 298 (1977).
73. Kataoka, I., Serizawa, A., and Sakurai, A., "Transient Boiling Heat Transfer under Forced Convection," *Int. J. Heat Mass Transfer*, 26 (4): 583 (1983).
74. Kataoka, I., "Transient Boiling and Two-Phase Flow Problems Associated with Nuclear Reactor Safety," Ph. D. thesis, Kyoto University (1984).
75. Johnson, H. A., and Shrock, V. E., et al., USAEC TID Reports: SAN 1001 (1961) to SAN 1013 (1966).
76. Johnson, H. A., "Transient Boiling Heat Transfer to Water," *Int. J. Heat Mass Transfer*, 14: 67 (1971).
77. Martenson, A. J., "Transient Boiling in Small Rectangular Channels," WAPID-T-1290 (1963).
78. Aoki, S., Kozawa, Y., and Iwasaki, H., "Boiling and Burnout Phenomena under Transient Heat Input," Bulletin of JSME, 19 (132): 667 (1975).
79. Aoki, S., Inoue, A., and Kozawa, Y., "Transient Boiling Crisis during Rapid Depressurization," Proc. 4th Int. Heat Transfer Conf. B 6–3, Tokyo (1974).
80. Aoki, S., Kozawa, I., and Iwasaki, H., "A Study on Transient Burnout Phenomena during Loss of Flow," Proc. 12th National Heat Trans. Symp. Japan, B 116 (1975).
81. Rosenthal, M. W., "An Experimental Study of Transient Boiling," *Nucl. Sci. Eng.*, 2: 640 (1957), also Rosenthal, M. W., and Miller, R. L., ORNL-2294 (1957).
82. Hall, W. B., and Harisson, W. C., "Transient Boiling of Water at Atmospheric Pressure," Proc. 3rd Int. Heat Transfer Conf., Vol. 3, p. 186, Chicago (1966).
83. Sakurai, A., Mizukami, K., and Shiotsu, M., "Experimental Studies on Transient Boiling Heat Transfer and Burnout," Proc. 4th Int. Heat Transfer Conf. B 3.4, Paris (1970).
84. Cole, R., "Investigation of Transient Pool Boiling due to Sudden Large Power surge," NACA Technical Note 3885 (1956).
85. Hayashi, S., et al., "Experimental Study of the Temperature Overshoot and the Delay Time of the Transient Boiling," *J. of Atomic Energy Soc. Japan*, 2: 736 (1960).
86. Hayashi, S., Sakurai, A., and Iwazumi, T., "Transient Heat Transfer in Heterogeneous Water Reactor I," *J. of Atomic Soc. Japan*, 5: 403 (1963).
87. Lurie, H., and Johnson, H. A., "Transient Pool Boiling of Water on Vertical Surface with a Step in Heat Generation," *Trans. ASME, J. Heat Transfer*, 84: 217 (1962).
88. Sato, T., et al., "Experimental Studies of the Burnout under the Boiling I," *Trans. JSME*, 27: 722 (1961).
89. Sato, T., Michiyoshi, I., and Takeuchi, K., "Studies of the Burnout under the Boiling II," *Trans. JSME*, 27: 1817 (1961).
90. Tachibana, F., Akiyama, M., and Kawamura, H., "Heat Transfer and Critical Heat Flux in Transient Boiling I," *J. Nucl. Sci. Technol.*, 5: 117 (1968).
91. Kawamura, H., Tachibana, F., and Akiyama, M., "Heat Transfer and DNB Heat Flux in Transient Boiling," Proc. 4th Int. Heat Transfer Conf., B 3.3, Paris (1970).
92. Sakurai, A., and Shiotsu, M., "Transient Pool Boiling Heat Transfer, I," *Trans. ASME, J. Heat Transfer*, 99: 547 (1977).
93. Sakurai, A., and Shiotsu, M., "Transient Pool Boiling Heat Transfer, II," *Trans. ASME, J. Heat Transfer*, 99: 554 (1977).
94. Toda, S., Okubo, K., and Simamune, K., "Boiling Heat Transfer under Rapid Depressurization, I," 7th Nat. Heat Trans. Symp. Japan, II-3-2, p. 85 (1970).
95. Toda, S., and Okubo, K., "Boiling Heat Transfer under Rapid Depressurization, II," 8th Nat. Heat Trans. Symp. Japan, III-2-1, p. 365 (1971).
96. Kutateladze, S. S., "Heat Transfer in Condensation and Boiling," AEC-tr-3770 (1959).
97. Kalinin, E. K., and Dreister, G. A., "Unsteady Convective Heat Transfer and Hydrodynamics in Channels," *Advances in Heat Transfer*, 6: 367, Academic Press, New York (1970).

98. Siegel, R., "Transient Heat Transfer for Laminar Slug Flow in Ducts," *Trans. ASME, Ser. E,* 81: 140 (1959).
99. Siegel, R., "Forced Convection in Channel with Wall Heat Capacity and Wall Heating Variable with Axial Position and Time," *Int. J. Heat Mass Transfer,* 6: 607 (1963).
100. Siegel, R., and Sparrow, E. M., "Transient Heat Transfer for Laminar Forced Convection in the Thermal Entrance Region of Flat Ducts," *Trans. ASME, Ser. C,* 81: p. 29 (1959).
101. Siegel, R., "Heat Transfer for Laminar Flow in Ducts with Arbitrary Time Variations in Wall Temperature," *Trans. ASME, Ser. E,* 82: 241 (1960).
102. Sparrow, E. M., and Siegel, R., "Unsteady Turbulent Heat Transfer in Tubes," *Trans. ASME, Ser. C,* 82: 170 (1960).
103. Perlmutter, M., and Siegel, R., "Two-Dimensional Unsteady Incompressible Laminar Duct Flow with a Step Change in Wall Temperature," *Int. J. Heat Mass Transfer,* 3: 94 (1961).
104. Siegel, R., and Perlmutter, M., "Laminar Heat Transfer in a Channel with Unsteady Flow and Wall Heating Varying with Position and Time," *Trans. ASME, Ser. C,* 85: p. 358 (1963).
105. Namatame, K., "Transient Temperature Response of an Annular Flow with Step Change in Heat Generating Rod," *Nucl. Sci. Technol.,* 6: 591 (1969).
106. Kawamura, H., "Analysis of Transient Heat Transfer for Turbulent Flow in Annuli, Part 1, Heating Element with a Finite Thermal Capacity and No Thermal Resistance," *Trans. JSME,* 39: 2498 (1973).
107. Kawamura, H., "Analysis of Transient Heat Transfer for Turbulent Flow in Annuli, Part 2, Heating Element with a Finite Thermal Capacity and Finite Thermal Resistance," *Trans. JSME,* 42: p. 1207 (1976).
108. Kawamura, H., and Iwahori, K., "Experiment of Transient Heat Transfer for Turbulent Flow in a Circular Tube," *Trans. JSME,* 42: 2501 (1976).
109. Cess, R. D., "Heat Transfer to Laminar Flow across a Flat Plate with a Nonsteady Surface Temperature," *Trans. ASME, Ser. C,* 83: 274 (1961).
110. Goodman, T. R., "Effects of Arbitrary Nonsteady Wall Temperature on Incompressible Heat Transfer," *Trans. ASME, Ser. C,* 84: 347 (1962).
111. Adams, D. E., and Gebhart, B., "Transient Forced Convection from a Flat Plate Subjected to a Step Energy Input," *Trans. ASME, Ser. C,* 86: 253 (1964).
112. Soliman, M., and Johnson, H. A., "Transient Heat Transfer for Turbulent Flow over a Flat Plate of Appreciable Thermal Capacity and Containing Time-Dependent Heat Source," *Trans. ASME,* 89: p. 362 (1967).
113. Soliman, M., and Johnson, H. A., "Transient Heat Transfer for Forced Convection Flow over a Flat Plate of Appreciable Thermal Capacity and Containing an Exponential Time-Dependent Heat Source," *Int. J. Heat Mass Transfer,* 11: 39 (1968).
114. Soliman, M., "Analytical and Experimental Study of Transient Heat Transfer External Forced Convection," USAEC Report, SAN-1010 (1963).
115. Chao, B. T., and Jeng, D. R., "Unsteady Stagnation Point Heat Transfer," *Trans. ASME, Ser. C,* 87: 221 (1965).
116. Chao, B. T., and Cheema, L. S., "Unsteady Heat Transfer in Laminar Boundary Layer over a Flat Plate," *Int. J. Heat Mass Transfer,* 11: 1311 (1968).
117. Ishiguro, R., "A Study of Turbulent Heat Transfer for Thermally Unsteady System," Proc. of 9th National Heat Transfer Symposium of Japan, A 205, p. 315 (1972).
118. Mitazi, K., and Ishiguro, R., "Unsteady Forced Convective Heat Transfer," Proc. of 13th National Heat Transfer Symposium of Japan, A 209, p. 67 (1976).
119. Hanawa, J., "Transient Heat Transfer to Laminar Flow from a Flat Plate with Heat Capacity (Numerical Solution)," *Trans. JSME,* 41: 2119 (1975).
120. Hanawa, J., "Transient Convective Heat Transfer to Laminar Flow from a Flat Plate with Constant Heat Capacity," *Trans. JSME,* 46: 299 (1980).

121. Hanawa, J., "Transient Convective Heat Transfer to Turbulent Flow from a Flat Plate with Constant Heat Capacity," *Trans. JSME,* 46: 927 (1980).
122. Hanawa, J., "Transient Convective Heat Transfer to Turbulent Flow from a Flat Plate with Constant Heat Capacity," *Trans. JSME,* 46: 2348 (1980).
123. Kataoka, I., "Analysis of Transient Forced Convective Heat Transfer by Boundary Layer Approximation," Tech. Repts Inst. Atom. Energy, Kyoto Univ., No. 197 (1983).
124. Hsu, Y. Y., and Graham, R. W., "An Analytical and Experimental Study of the Thermal Boundary Layer and the Ebullition Cycle in Nucleate Boiling," NASA TN-D-594 (1961).
125. Bergles, A. E., and Rohsenow, W. M., "The Determination of Forced-Convection Surface-Boiling Heat Transfer," *Trans. ASME, J. Heat Transfer,* 86: 365 (1964).
126. Fabic, S., "Vapor Nucleation on Surfaces Subjected to Transient Heating," USAEC TID Report SAN 1008 (1964).
127. Schultz, R. R., and Cole, R., "Transient Boiling from Artificial Sites at Low Pressures," Proc. 1974 Heat Transfer and Fluid Mechanics Institute, p. 38 (1974).
128. Schultz, R. P., and Cole, R., "Bubble Nucleation and Growth Instabilities in Transient Boiling," Proc. 6th Int. Heat Transfer Conf. PB-4, Toronto (1978).
129. Oker, E., and Metre, H., Jr., "A Study of Transient Effects Leading up to Inception of Nucleate Boiling," Proc. 6th Int. Heat Transfer Conf. PB-5, Toronto (1978).
130. Ngheim, L., et al., "Prediction of Transient Inception of Boiling in Terms of a Heterogeneous Nucleation Theory," *Trans. ASME, J. Heat Transfer,* 103: 69 (1981).
131. Corty, C., and Foust, A. S., "Surface Variables in Nucleate Boiling," *Chem. Eng. Prog. Symp. Series,* Vol. 17: 51 (1955).
132. Sakurai, A., et al., "Transient Boiling Heat Transfer under Forced Convection, IV," paper presented at Ann. Meeting Atom. Energy Soc., Japan (1978).
133. Sakurai, A., et al., "Transient Boiling Heat Transfer under Forced Convection, II," 14th Nat. Heat Trans. Symp., Japan, B 114 (1977).
134. Sakurai, A., et al., "Transient Boiling Heat Transfer under Forced Convection, II," paper presented at Fall Meeting Atom. Energy Soc., Japan (1976).
135. Borishansky, V. M., and Fokin, B. S., "Onset of Heat Transfer Crisis with Unsteady Increase in Heat Flux," *Heat Transfer Soviet Research,* 1: 27 (1969).
136. Serizawa, A., "Theroretical Prediction of Maximum Heat Flux in Power Transients," *Int. J. Heat Mass Transfer,* 26 (9): 921 (1983).

CHAPTER 11

PREDICTION OF HEAT TRANSFER DURING FORCED-CONVECTION SUBCOOLED BOILING

M. Mohammed Shah

Consulting Engineer
15 Rush St.
Port Jefferson Station, New York

CONTENTS

INTRODUCTION

This paper is concerned with predicting heat transfer during forced-convection subcooled boiling prior to boiling crisis. Prediction of boiling crisis and heat transfer following boiling crisis are not covered.

In the following, some predictive techniques are given for calculating inception of boiling and heat transfer coefficients for flow in channels and during flow across cylinders. These predictive techniques are to be considered applicable only to single-component, nonmetallic, noncryogenic fluids, except where specifically noted otherwise. Separate sections of this paper have been devoted to metallic, cryogenic, and multicomponent fluids. These sections should be consulted for these special fluids.

It should be noted that subcooled boiling is also known as surface boiling.

INCEPTION OR ONSET OF NUCLEATE BOILING

Heat transfer due to convection on a wall is given by:

$$T_w = q/h_L + T_b \tag{1}$$

where h_L is the heat transfer coefficient due to single-phase convection.

The necessary condition for boiling to start is that $T_w > T_{SAT}$. However, this condition is not sufficient. In many cases, wall temperature can far exceed the saturation temperature without any boiling. For many practical problems, prediction of the onset of boiling is important.

Bergles and Rohsenow [1] presented an analytical model which has found wide acceptance. They postulated that a bubble of radius r can grow only if the temperature of the liquid at a distance r from the heating surface is equal to or greater than the temperature of vapor within the bubble. It was assumed that the heating surface contains cavities of the radius required by this condition. They further assumed that heat transfer through the liquid layer occurs by steady-state conduction. The equations resulting from this model were solved graphically and the following empirical equation fitted to the results:

$$q_{ib} = 15.6\, q^{1.156} (\Delta T_{SAT})_{ib}^{2.3/p^{0.0234}} \tag{2}$$

The subscript ib indicates incipient boiling. q is in Btu/hrft^2. p is in lb/in^2, and ΔT_{SAT} in °F. This equation covers the range p = 1 to 138 bars.

Davis and Anderson [2] solved the Bergles-Rohsenow equations analytically by making some simplifying assumptions. Their final result was

$$q_{ib} = \frac{k_\ell i_{fg} \rho_g}{8\sigma T_{SAT}} (\Delta T_{SAT})_{ib}^2 \tag{3}$$

Equation 3 agrees with Equation 2 as well as with the data of Bergles and Rohsenow [1] and Rohsenow and Clark [3], both for water at high pressure. However, the wall superheat was greatly underpredicted for the data of Backhurst [4] for benzene and Matzournis [5] for water, both at atmospheric pressure.

Frost and Dzakovic [6] also used the Bergles-Rohsenow model but arbitrarily included the Prandtl number to arrive at the following formula:

$$q_{ib} = \frac{k_\ell i_{fg}}{8\sigma T_{SAT} v_{fg}} \left[\frac{(\Delta T_{SAT})_{ib}}{Pr_\ell}\right]^2 \tag{4}$$

Frost and Dzakovic consider Equation 4 to be applicable to both forced- and natural-convection systems. The correlation was compared to data for water, hydrogen, freon, acetone, nitrogen, oxygen, neon, ethanol, n-pentane, carbon tetrachloride, benzene, and kerosene. Most of these data were for natural convection boiling (pool boiling).

If the heat flux is the independent variable, Equation 4 directly yields the ΔT_{SAT} at onset of boiling. If ΔT_{SAT} is the independent variable, Equation 4 has to be solved simultaneously with Equation 1. For forced-convection systems, h_L is calculated by a forced-convection correlation. For pool boiling, h_L is to be calculated by a correlation for natural convection.

If the liquid contains dissolved gases, onset of boiling occurs at superheats lower than for liquids free of dissolved gases. This was demonstrated experimentally by McAdams et al.

[7]. Heating the surface finish and microstructure can also affect incipient superheat. In view of the many variables involved, no predictive technique can be fully reliable. Subject to this limitation, the Frost and Dzakovic relation Equation 4, is recommended because of its agreement with a wide range of data.

HYSTERESIS

Heat transfer coefficients during subcooled boiling are often higher with decreasing heat flux than with increasing heat flux. At a constant heat flux, heat transfer coefficients are often higher with increasing subcooling than with decreasing subcooling. This phenomenon is known as hysteresis. The explanation is that with decreasing heat flux, the active bubble nucleation sites tend to remain active at lower heat flux as compared to the case of increasing heat flux. Similar is the situation with increasing subcooling as compared to decreasing subcooling.

Noncryogenic fluids exhibit no hysteresis under saturated conditions. Under subcooled conditions it occurs in many cases but not always. Hysteresis usually increases with increasing subcooling and heat flux and decreases with increasing mass flux. See for example the data of Hodgson [8] for water boiling in a tube.

No method for predicting the occurrence of hysteresis or its magnitude is available. For noncryogenic fluids, the wall temperature variations caused by hysteresis generally do not effect heat transfer coefficients by more than 20% to 30% which is within the accuracy of most predictive techniques.

HEAT TRANSFER DURING BOILING IN CHANNELS

Subcooled boiling in channels can be divided into two regimes, namely the "high subcooling" and the "low subcooling" regimes. In the high subcooling regime, bubble nucleation occurs only on a portion of surface, void fraction is low, bubbles grow and collapse while attached to the wall, and the wall temperature rises with decreasing subcooling. In the low subcooling regime, bubble nucleation occurs over almost the entire surface, void fraction is high, bubbles detach from the wall and are carried in the liquid core, and wall temperature remains constant with decreasing subcooling. The high subcooling regime is also known as the partial boiling regime and the low subcooling regime is also known as the fully developed boiling regime.

Some techniques for prediction of heat transfer in tubes and annuli are now presented. The most verified predictive technique is that of Shah [9, 10] which is discussed first.

Shah Correlation

The correlation of Shah was first presented in Reference 9. Further verification with data for annuli was given in Reference 10. This correlation uses the dimensionless parameters Bo, Ψ, and $\Delta T_{SC}/\Delta T_{SAT}$. The first two are defined as:

$$Bo = q/(G i_{fg}) \tag{5}$$

$$\Psi = q/(\Delta T_{SAT} h_L) \tag{6}$$

h_L is the heat transfer coefficient for single-phase liquid flow.

The value of Ψ at zero subcooling, Ψ_0, is given by the following relation:

$$\Psi_0 = 230\, Bo^{0.5} \tag{7}$$

If Equation 7 predicts $\Psi_0 < 1$, use $\Psi_0 = 1$.

In the high subcooling regime, it is assumed that the total heat flux is the sum of heat flux due to single-phase convection and the heat flux to nucleate boiling. Thus:

$$q = q_{SPC} + q_{nb} \tag{8}$$

At zero subcooling, the definition of Ψ_0 yields the following equation:

$$q = h_L (T_w - T_{SAT}) + h_L (\Psi_0 - 1)(T_w - T_{SAT}) \tag{9}$$

Comparison of Equations 8 and 9 shows that,

$$q_{nb} = h_L (\Psi_0 - 1)(T_w - T_{SAT}) \tag{10}$$

Thus for the high subcooling regime, the following relation results:

$$\Psi = \Psi_0 + \Delta T_{SC}/\Delta T_{SAT} \tag{11}$$

For the low subcooling regime,

$$\Psi = \Psi_0 \tag{12}$$

The demarcation between high and low subcooling regimes was found by trial and error and is as follows. When $\Delta T_{SC}/\Delta T_{SAT} > 2$, the regime is high subcooling. The regime is also that of high subcooling when,

$$\Delta T_{SC}/\Delta T_{SAT} > 6.3 \times 10^4 \, Bo^{1.25} \tag{13}$$

The regimes of high and low subcooling are shows graphically in Figure 1. The complete correlation is shown graphically in Figure 2.

For $Re_L > 2{,}300$ h_L is calculated by the Dittus-Boelter equation:

$$h_L = 0.023 (GD_{eq}/\mu_\ell)^{0.8} Pr_\ell^{0.4} k_\ell / D_{eq} \tag{14}$$

All properties are calculated at the bulk liquid temperature. The equivalent diameter D_{eq} for the circular tube is the inside tube diameter. For annuli, D_{eq} is defined as follows:

$$\delta > 4\,\text{mm}, \qquad D_{eq} = \frac{4 \times \text{flow area}}{\text{wetted perimeter}} \tag{15}$$

$$\delta < 4\,\text{mm}, \qquad D_{eq} = \frac{4 \times \text{flow area}}{\text{heated perimeter}} \tag{16}$$

Table 1
Complete Range of Parameters of Experimental Data with Which the Shah Correlation [9, 10] Has Been Verified

Geometry	Tubes: 2.4- to 27.1-mm ID annuli: δ = 1.0 to 6.4 mm, D_i = 4.5 to 42.3 mm. Internal, external, and two-sided heating.
Heating surface	Copper, nickel, inconel, glass, various stainless-steels
Fluids	Water, methanol, n-butyl alcohol, ammonia, R-11, R-12, R-113, K_2CO_3 solutions.
p_r	0.005 to 0.89
ΔT_{SC}, °K	0 to 153
$\Delta T_{SC}/T_{SAT}$	0 to 240
G, kg/m²s	200 to 87,000
$q \times 10^{-6}$ W/m²	0.01 to 22.9
$Re_L \times 10^{-3}$	1.4 to 360.0
$Bo \times 10^4$	0.1 to 54.0
Pr_ℓ	0.8 to 35.0

δ is the gap between the two tubes forming the annulus. All the data analyzed were for $Re_L > 2{,}300$ except one data set for an annulus which was for Re_L down to 1,400. This data set was better correlated when h_L was calculated by a laminar flow equation.

The complete range of data analyzed is given in Table 1. For the approximately 900 data points analyzed, the mean deviation of h_{TP} was found to be about 9% with over 97% of the data points within ± 30%. The two-phase heat transfer coefficients h_{TP} is defined as:

$$h_{TP} = q/(T_w - T_b) \tag{17}$$

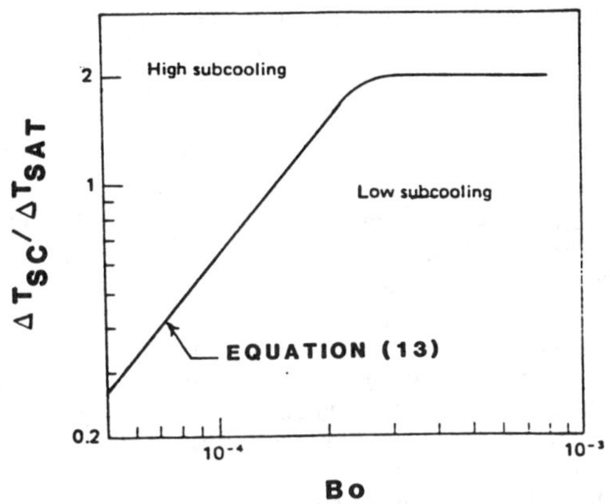

Figure 1. Demarcation between the regimes of high and low subcooled boiling according to the correlation of Shah [9] for boiling inside tubes and annuli.

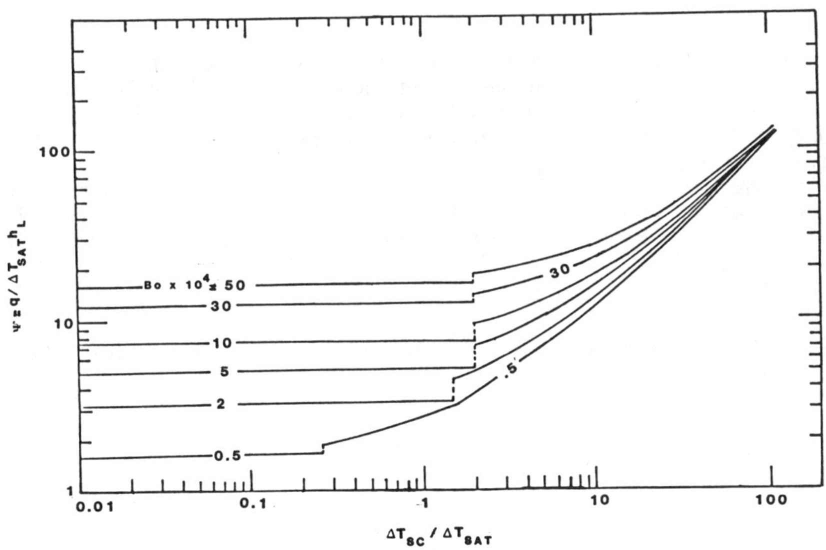

Figure 2. The correlation of Shah [9] for subcooled boiling inside tubes and annuli in graphical form.

Thus the accuracy of the Shah correlation for boiling heat transfer is comparable to that of correlations for single-phase heat transfer. The data analyzed are for a variety of heating surfaces and fluids and cover a very wide range of data. Hence it appears that this correlation can be applied to any commercially manufactured tube and a wide variety of fluids.

Techniques Based On Pool Boiling Correlations

Several predictive techniques have been proposed in which forced-convection heat transfer is calculated by superposing pool boiling and single-phase heat transfer. The best known among these is the correlation of Rohsenow [11] according to which:

$$q = q_{pb} + q_{SPC} \tag{18}$$

This can be written as:

$$h_{TP}\Delta T_b = h_{pb}\Delta T_{SAT} + h_L\Delta T_b \tag{19}$$

h_{pb} is calculated by a correlation for pool boiling and h_L by a single-phase heat transfer correlation such as the Dittus-Boelter equation. Thus it is assumed that forced convection does not affect pool boiling. Rohsenow used his own pool boiling correlation [12] to calculate h_{pb}. While Rohsenow was able to correlate several data sets, this agreement is not fully convincing as the value of factor C_{sf} in his pool boiling correlation could be adjusted to suit the data. The values of C_{sf} ranged from 0.0022 to 0.013. Bergles and Rohsenow [1] conducted tests in which forced convection and pool boiling tests were done on the same tube. Their measurements indicated that forced convection boiling could not be considered an extrapolation of pool boiling. They recommended that the forced-convection boiling curve be determined from forced-convection data, not pool boiling data. The Shah correlation is in accordance with this recommendation.

Another superposition technique has been proposed by Kutateladze [13]. It is expressed by the following equation:

$$\frac{h_{TP}}{h_L} = \left[1 + \left(\frac{h_{pb}}{h_L}\right)^n\right]^{1/n} \tag{20}$$

The exponent n = 2. While the Rohsenow correlation assumes that the nucleate boiling and single-phase convection effects are independent and additive, the Kutateladze approach assumes some suppression of nucleate boiling due to forced-convection effects. At n = 1, there is no suppression and the Rohsenow correlation is obtained. The increasing values of n represent increasing bubble suppression.

In the correlation of Chen [14] for saturated boiling, nucleate boiling component is calculated by a pool boiling correlation and then multiplied by a suppression factor which depends on two-phase Reynolds number. Collier [58] reports that it has been extended to subcooling boiling as follows:

$$q = h_{nb}\Delta T_{SAT} + h_L\Delta T_b \tag{21}$$

h_{nb} is calculated by the Chen correlation with the factor $F_c = 1$. Satisfactory agreement with subcooled boiling data for water, n-butyl alcohol, and ammonia is reported but no details are given.

Various Predictive Techniques

A dimensionless correlation presented by Hodgson [8] is as follows:

$$h_{TP} = 141\,Bo^{0.7}\left[\frac{\Delta T_{SC}C_{p\ell}}{i_{fg}}\right]^{-0.55}\left[\frac{Zi_{fg}}{XC_{p\ell}}\right]^{-0.08}\cdot h_L \tag{22}$$

where:

$$Z = -\frac{d(\rho_\ell/\rho_g)}{dp} \tag{23}$$

$$X = dT_{SAT}/dp \tag{24}$$

All properties are evaluated at the bulk fluid temperature. Satisfactory agreement was found with their own data for water and the data of Noel [15] for ammonia.

Numerous dimensional equations for water have been proposed. Among these, the best known is the following dimensional equation of Jens and Lottes [16]:

$$\Delta T_{SAT} = 25\,q^{0.25}e^{-p/62} \tag{25}$$

with ΔT_{SAT} in °C, q in MW/m^2, and p in bar (absolute). This equation was based on data for water flowing upward in nickel and stainless steel tubes of diameter 3.63 to 5.74 mm. Pressure varied from 7 to 172 bar, water temperature from 115° to 340°C, mass velocity from 11 to $1.04 \times 10^4\,kg/m^2s$ and heat flux up to $12.5\,MW/m^2$.

The methods proposed by Bergles and Rohsenow [1] is an extension of the Kutateladze method [13]. The total heat flux q is determined as follows:

$$q = q_{SPC}\left\{1 + \left[\frac{q_{nb}}{q_{SPC}}\left(1 - \frac{q_{ib}}{q_{SPC}}\right)\right]^2\right\}^{1/2} \tag{26}$$

q_{ib} is the heat flux for initiation of boiling which may be calculated for water by Equation 2. q_{nb} is the heat flux for fully developed nucleate boiling. It is to be calculated from data obtained with forced convection, for example the Jens-Lottes correlation. This method can be applied only if some forced convection boiling data (or correlation) are available for the liquid in consideration.

Percupile et al. [17] have given the following dimensional equation:

$$\Delta T_{SAT} = 2.28\,(\rho_\ell/\rho_g)^{-0.25}q^{0.3}D^{0.6} \tag{27}$$

ΔT_{SAT} is in °C, q in W/m^2, and D in m. This equation is based on their data for R-11, R-12, and R-113 flowing up through vertical tubes of 6-mm and 18.8-mm ID, pressure from 1.43 to 9.64 bar, q from 14.7 to $102\,kW/m^2$, and vapor quality from −0.06 to 0.21.

Design Recommendations

1. For prediction of boiling inception the correlation of Frost and Dzakovic, Equation 4, is recommended. It should be realized that boiling inception depends on so many factors that no predictive technique can be fully relied upon.
2. For calculation of heat transfer, the correlation of Shah is recommended for $Re_L > 2{,}300$. At lower Reynolds numbers, use the Rohsenow correlation Equation 19.

SINGLE CYLINDER WITH CROSSFLOW

Many heat exchangers, notably kettle reboilers and shell-and-tube evaporators, involve crossflow over tube bundles. For prediction of heat transfer in such tube bundles, it is desirable to be able to predict heat transfer to a single tube. See, for example, the computational method of Brisbane et al. [18]. For this reason, several experimental studies on heat transfer to single cylinders have been done and attempts made to develop predictive techniques. Some of these predictive techniques are now discussed.

Shah Correlation

Shah [19] has given a correlation which shows satisfactory agreement with available experimental data. The correlating parameters are the same as in his correlation for flow through tubes, presented earlier. The two correlations are similar but differ quantitatively.

At zero subcooling, the following equations apply:

$$\mathrm{Bo} > 2.5 \times 10^{-4}, \qquad \Psi_0 = 443\,\mathrm{Bo}^{0.65} \tag{28}$$

$$\mathrm{Bo} < 2.5 \times 10^{-4}, \qquad \Psi_0 = 19\,\mathrm{Bo}^{0.27} \tag{29}$$

If Equation 29 predicts $\Psi_0 < 1$, use $\Psi_0 = 1$.

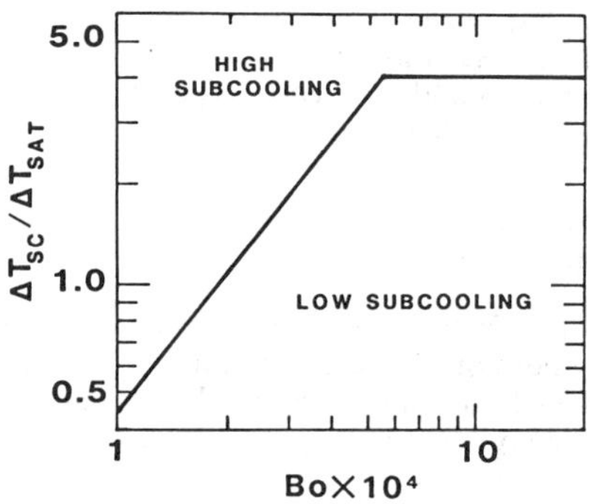

Figure 3. Demarcation between the regimes of high and low subcooled boiling regimes according to the correlation of Shah [19] for boiling with forced convection across cylinders.

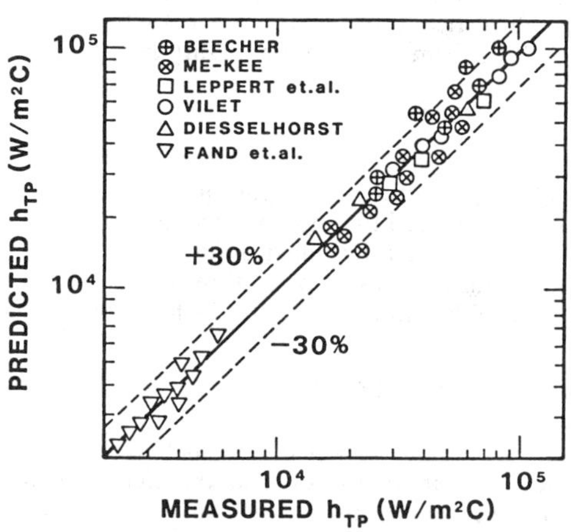

Figure 4. Comparison of the Shah correlation [19] with data for water boiling during flow across single cylinders.

There are two regimes of subcooling, namely the high subcooling regime and the low subcooling regime. In the low subcooling regime:

$$\Psi = \Psi_0 \tag{30}$$

In the high subcooling regime:

$$\Psi = \Psi_0 + \Delta T_{SC}/\Delta T_{SAT} \tag{31}$$

The demarcation between the two regimes of subcooling is shown graphically in Figure 3. It is seen that if $\Delta T_{SC}/\Delta T_{SAT} > 4$, the regime is that of high subcooling. The regime is that of high subcooling also when $Bo < 5.4 \times 10^{-4}$ and the following inequality is satisfied:

$$\Delta T_{SC}/\Delta T_{SAT} > 7.63 \times 10^4 \, Bo^{1.31} \tag{32}$$

For calculating the single-phase heat transfer, best results were found with the following equation:

$$\frac{h_L D}{k_\ell} = 0.21 \left(\frac{GD}{\mu_\ell}\right)^{0.62} Pr_\ell^{0.4} \tag{33}$$

All properties are calculated at the bulk fluid temperature. The mass velocity G is based on the minimum flow area, i.e., the clearance between the test cylinder and the flow channel in which it is located. The value of G used in calculating Bo in Equation 33 is the same as was used in Equation 5. Shah obtained Equation 33 by analyzing data for single-phase heat transfer.

The arguments used in developing this correlation are the same as those used in developing the Shah correlation for boiling in tubes, as described in the previous Section.

Table 2
Results of Comparison of Data for Crossflow Boiling with Shah Correlation [19]

Reference	D_t (mm)	Tube Material	Fluid	p_r	ΔT_{SC} (°C)	$\frac{\Delta T_{SC}}{\Delta T_{SAT}}$	u (m/s)	Re_L	q $\left(\frac{KW}{m^2}\right)$	Bo × 10^4	Dev. % ΔT_{SAT}	h_{TP}	No. of Data Points
Yilmaz [20]	6.4	copper	R-113	0.029	4 5	0.09 0.32	2.8 7.8	53,500 150,000	60 584	0.6 6.6	13.9	10.8	36
Beecher [26]	1.2	Stainless steel	water	0.005	0 80	0.0 12.0	0.9 1.5	1,940 4,400	13 850	4.8 27.4	26.1	13.1	14
Leppert et al. [23]	2.8	Stainless steel	water	0.005	10 23	0.5 1.5	0.17	1,480	86 286	23.3 77.9	11.0	5.9	10
McKee [25]	6.3 17.9	Stainless steel	water	0.005	7	0.14 1.10	1.14 2.17	25,000 135,300	22 286	0.7 8.0	18.8	14.2	50
Vliet [27]	3.2	Stainless steel	water	0.005	2 55	0.06 1.72	0.2 3.3	1,500 26,000	86 714	5.6 98.1	5.6	3.1	45
Fand et al. [21]	11.4	Stainless steel	water	0.014	32	1.12 2.92	0.02 0.12	970 4,830	166 366	6.6 66.2	42.3	16.8	13
	11.9	Titanium	water	0.009 0.018	39 52	1.3 173.0	0.02 0.12	700 4,710	160 323	6.0 64.0	24.5	5.8	61
Dieselhorst [28]	3.0	Stainless steel	water	0.005	0	0	0.24 0.44	2,500 4,600	150 1,000	2.9 10.4	8.2	8.2	15
Bitter [30, 31]	15.0	Porcelein, nickel coated	R-11	0.022 0.025	0	0	0.04 0.92	2,000 50,000	1 32	0.5 17.9	14.1	14.1	31
Lemmert & Chawla [24]	25.4	Copper	R-11	0.04	0 10	0 2	0.12 1.2	13,300 133,000	3 225	0.7 70.0	12.3	9.2	32
Fink et al. [29]	25.4	Copper	R-11	0.040	1 16	0.05 0.74	0.13 1.33	13,300 133,000	57 138	2.6 41.4	13.6	10.8	15
			R-113	0.051	1 16	0.05 0.87	0.13 1.33	11,600 116,000	57 138	4.3 51.7	19.2	16.0	12
All data	1.2 25.4			0.005 0.051	0 80	0.0 173.0	0.02 7.8	700 150,000	1 1,000	0.6 98.1	16.7	9.5	334

Shah compared this correlation with data from ten experimental studies. These were the only data in open literature. The range of parameters covered by these studies is listed in Table 2. He found that the use of Equation 33 gives satisfactory results with all data except those of Yilmaz [20] which are overpredicted. The Yilmaz data give satisfactory agreement when h_L is calculated by the following equation of Fand and Keshwani [21]:

$$Nu_f = (0.255 + 0.699\,Re_f^{0.5})\,Pr_f^{0.29} \tag{34}$$

All the properties in this equation are calculated at the film temperature. The Yilmaz data were analyzed using this equation while all other data were analyzed using Equation 33. The comparison between the data and correlation is shown in Table 2 and Figure 4. The heat transfer coefficients in all 334 data points were predicted with a mean deviation of 9.5%, with only 6 data points showing deviations greater than 30%. Thus the accuracy of the correlation is comparable to correlations for single-phase heat transfer. Better accuracy, though always desirable, will be difficult to achieve.

The data include four different tube materials and three fluids. Furthermore, the correlating parameters are the same as in the Shah correlation for boiling inside tubes which has been verified with many fluids and tube materials. Hence this correlation may be expected to be generally applicable to all common fluids and commercially prepared surfaces. The range of other parameters is very wide except that the reduced pressure ranged from 0.005 to 0.05. Application to higher pressures should be made with caution. As the shell side of heat exchangers is rarely operated at high pressures because of structural considerations, the verified pressure range covers most practical needs.

It should be remembered that this correlation is for forced convection boiling. At $G = 0$, heat transfer occurs by natural convection and pool boiling correlations apply. Hence the Shah correlation should not be used at velocities lower than those for which it has been verified.

Other Predictive Techniques

Several researchers have compared their measurements with the Rohsenow superposition correlation, Equation 19. Yilmaz and Westwater [22] and Leppert et al. [23] found satisfactory correlations by this method, using their own pool boiling data to calculate q_{pb}. McKee [25] using his own pool boiling, found that some of his data were satisfactorily correlated while some were under-predicted. Fand et al. [21] and Lemmert and Chawla [24], state that their data could not be correlated in terms of Equation 19. Hence the Rohsenow method has not been successful even where pool boiling data for the particular test section were available. The difficulties in accurately calculating pool boiling heat transfer are well-known.

Lemmert and Chawla [24] compared their data with the Kutateladze superposition correlation, Equation 20. Poor agreement was found. Fand et al. [21] found this equation unsatisfactory with $n = 2$, the value given by Kutateladze. However, their data were well-correlated with $n = 5.5$.

McKee [25] attempt to improve the Rohsenow method by distinguishing between areas with bubble nucleation and those without bubble nucleation. His equation is:

$$qA = q_{pb}A_{bn} + q_{SPC}(A - A_{bn}) \tag{35}$$

A_{bn} is the tube surface area with bubble nucleation and A is the total surface area. He determined A_{bn}/A by a dimensional equation which fits his data. Whether it fits other data, is unknown.

Lemmert and Chawla [24] also used the approach of Equation 35 but developed different expressions for determining A_{bn}/A and heat flux due to bubble nucleation. While their expressions fit their own data well, they have not been compared with any other data.

From the previous discussions, it is clear that none of these predictive techniques can be used with confidence.

Design Recommendations

Use the Shah correlation, Equations 30 and 31, with h_L calculated by Equation 33. If single-phase heat transfer data are available, adjust the constant in Equation 33 to fit them. The verified range of this correlation is given in Table 2. Use outside this range is not recommended. It is especially important that the correlation be not applied to u, G, and Re_L lower than the minimum listed in Table 2.

CRYOGENIC FLUIDS

Cryogenic fluids are Newtonian and their Prandtl numbers are comparable to that of water. Hence one would expect their behavior to be similar to that of ordinary fluids. Only two sets of analyzable data for subcooled boiling of cryogenic fluids are available. These are the data of Pappel and Hendricks [32] and Klimenko and Sudarchikov [33], both for nitrogen in electrically heated tubes. The present author found that data from both of these sources are in satisfactory agreement with the correlation of Shah presented earlier. Hence the available experimental data for subcooled boiling support the view that heat transfer to cryogenic fluids is no different from that of noncryogenic, nonmetallic fluids. However, there is some indication of differences which are now discussed.

One possible difference is in the influence of thermophysical properties of the heating surface. Stephan and Abdelsalam [34] developed correlations for pool boiling heat transfer. They found that for cryogenic fluids,

$$h_{pb} \propto S^{0.117} \tag{36}$$

where

$$S = \frac{(k\rho C_p) \text{ of heating surface}}{(k\rho C_p) \text{ of liquid}} \tag{37}$$

For noncryogenic fluids, they found no influence of S. This suggests that S may also have an influence on forced-convection boiling of cryogenic fluids.

Klimenko [35] analyzed the data for saturated boiling of nitrogen, neon, and hydrogen in tubes. He presented a correlation according to which:

$$h_{TP} \propto S^{0.2} \tag{38}$$

Shah [36] compared data for saturated boiling of helium, hydrogen, nitrogen, and argon in tubes with the correlations of Shah [37, 38] and Klimenko [35]. The Shah correlation has no factor for surface properties, and at zero vapor quality it reduces to Equation 7. Data were from 11 sources. Data from 9 sources showed adequate agreement with the Shah correlation. Data of Steiner et al. [39, 40] and Mohr and Runge [41] for nitrogen and neon, respectively, were underpredicted by a factor of up to 3. However, the deviations were unrelated to S. These data showed satisfactory agreement if h_{TP} at zero quality was calculated by the superposition correlations of Rohsenow or Kutateladze with h_{pb} by the Stephan–Abdelsalam correlation. However, this procedure resulted in overprediction of most other data. The Klimenko correlation showed satisfactory agreement with only 3 of the 11 data sets and the deviations were not related to S.

Hence the evidence from the experimental data from both saturated and subcooled boiling studies do not support the view that heat transfer to cryogenic fluid is influenced by S. Data from two sources are much higher than the Shah correlation and indicate little or no suppression of bubble nucleation by forced-convection. This anomaly remains unexplained.

Another difference is in hysteresis. Strong hysteresis has been reported during saturated boiling of helium [42, 43] and to a lesser extent in saturated boiling of nitrogen [39, 44] and neon [41]. Ordinary fluids shows no hysteresis during saturated boiling and thus this is a significant difference between the behaviour of cryogenic and noncryogenic fluids. In the data of Ogato and Sato [42] for saturated boiling of helium in tubes, the heat transfer coefficients with decreasing heat flux were up to four times higher than those with increasing heat flux. Jergel et al. [43] also noted strong hysteresis during saturated boiling of helium in rectangular channels. Hence one would expect strong hysteresis during subcooled boiling of helium. Klein [44] and Steiner and Schlünder [39] have reported hysteresis during saturated boiling of nitrogen in tubes but the difference between the higher and lower boiling curves did not exceed 20%. Similar results have been reported by Mohr and Runge [41] for saturated boiling of neon in tubes. Hence some hysteresis is likely in subcooled boiling of nitrogen and neon but is likely to be much less than with helium. It should however be noted that Klimenko and Sudarchekov [33] found no hysteresis during subcooled boiling of nitrogen and Papell and Hendricks [32] have made no mention of hysteresis during their experiments with subcooled boiling of nitrogen and neon.

Papell and Hendricks Correlations

Subcooled boiling of cryogenic fluids was the subject of studies by Papell and Hendricks [32]. In that study, nitrogen and neon flowed up through an electrically heated inconel tube of 19.8 mm ID. The parameters in nitrogen tests were: $p = 11.1$ to 28.7 bar, flow rate 2.7 to 6.5 g/s, $q = 4.1$ to 21.1 W/cm^2, and subcooling up to 26.5°K. The range of neon data was: $p = 11.2$ to 16.6 bar, flow rate from 4 to 10.8 g/s, $q = 1.8$ to 20.5 W/cm^2, and subcooling up to 11.6°K. All data were correlated within $\pm$ 30% by the following equation:

$$\frac{h_{TP}}{h_L} = 100\left[\mathrm{Bo}\left(\frac{\rho_b}{\rho_g}\right)\left(\frac{i_{fg}}{i_f - i}\right)^{1.2}\left(\frac{\rho_g}{\rho_\ell}\right)^{1.4}_{SAT}\right]^{0.7} \tag{39}$$

where i = the enthalpy of liquid at actual temperature
i_f = the enthalpy of saturated liquid
ρ_b = the density of liquid at bulk temperature

It should be noted that nitrogen data from this study show satisfactory agreement with the correlations of Shah [9] and Klimenko [35]. The data for neon were not given in an analyzable form.

Papell and Hendricks correlated the data for boiling inception by the following relation:

$$\mathrm{Bo}\,(\rho_b/\rho_g)\,(\rho_g/\rho_\ell)^{1.4}_{SAT}\left[\frac{i_{fg}}{i_f - i}\right]^{1.2} = 0.0014 \tag{40}$$

Design Recommendations

1. The correlation of Shah given in the section on "Heat Transfer during Boiling in Channels" is the best choice for calculating heat transfer in tubes and annuli. It should however be noted that some data are much higher than this correlation.
2. Very strong hysteresis may be expected with helium. It is expected that the Shah correlation will yield the lower boiling curve.
3. For inception of boiling, recommendations are the same as in the section mentioned in Recommendation 1. Equation 40 may be used in the range of data on which it is based.

LIQUID METALS

Boiling of liquid metals has been found to be considerably different from that of ordinary liquid. Correlations of heat transfer to nonmetallic fluids are usually inapplicable to metallic fluids. Ordinary liquids generally show incipient superheats of only a few degrees. In liquid metals superheats of over 100°K have been measured. Large fluctuations in wall temperatures and fluctuations in flow have often been experienced.

Boiling Inception And Instabilities

Several studies on potassium have shown that when liquid enters the heating tube in a subcooled state, boiling inception superheat is high and large fluctuations in wall temperatures occur. For example, Aladyev [45] reports the following typical behavior. Wall temperature first increases with time. When a certain value (greater that T_{SAT}) is reached, rapid boiling of potassium takes place and the wall temperature drops to T_{SAT} at the rate of 20° to 25°C/second. The process is then repeated. As the pressure increases, the superheat decreases, while the frequency increases. The observations of Hoffman and Krakoviak [46] are similar. With a plain stainless steel tube, they found superheats in excess of 160°C and wall temperature fluctuations at 0.025 Hertz. The superheat decreased sharply when the tube surface was provided with artificial nucleation sites. In the experiments of Noyes and Lurie [47] with sodium in an annulus, wall temperature fluctuations were also observed under subcooled conditions, but their amplitude and frequency were not as high.

Chen [48] measured incipient superheats in subcooled potassium at pressures between 0.55 and 0.96 bar flowing in tubes. Incipient superheats between 9° and 65°K were measured and those were found to be related to deactivation conditions preceding boiling tests and boiling pressure. Chen points out that analyses based on the Bergles–Rohsenow model, discussed earlier, will underpredict the incipient superheats for metals. This is because the cavity size corresponding to predicted inception conditions is so large that such cavities are likely to be flooded by liquid and hence inactive. As a consequence boiling will start only when superheat has increased sufficiently to activate smaller cavities containing trapped gases. Chen developed an analytical model which agrees with his own data.

Dodonov et al. [49] carried out an extensive study of incipient boiling superheat of potassium in straight and coiled tubes. The straight tubes were 6- to 6.1-mm ID stainless steel with their inner surface reamed on a lathe. The coiled section was made from a 10-mm ID stainless steel tube. The coil diameter was 162 mm, and coil pitch was 73 mm. The range of variables was: p from 0.6 to 3.6 bar, G from 0 to 277 kg/m^{2}s, q from 32 to 300×10^3 W/m^2, and inlet subcooling from 43° to 488°K. It was found that incipient boiling superheat was independent of G, q, and inlet subcooling. The unsteadiness in approach to boiling was not found to affect superheat. The superheat at the tube exit prior to vapor generation and the superheat in the center of tube with vapor generation were found to be the same. Superheats measured at different times and with different prior histories were found to be essentially the same. This contradicts the observations of Chen [47] discussed earlier. Superheat was found to depend on pressure. Their data were correlated by the following dimensional equation:

$$\Delta T_{ib} = 100 \exp(-50\, p_r) \tag{41}$$

ΔT_{ib} is in °K. This equation was found to also show reasonable agreement with data from several other sources.

Peppler [50] has reported experiments on incipient boiling superheat for flow of sodium in single channels and pin bundles. He states that every time boiling occurred in a new test section, no measurable superheat was observed. After repeated boiling under heat fluxes greater than 150 W/cm^2, the maximum superheat found was 25°C. This increase in incipient superheat may perhaps be explained in terms of surface deterioration caused by corrosion, as suggested by Schultheiss [51]. Peppler further reports that in boiling tests with pin bundles without gas injection, 5 Hz pulsations in outlet velocity were found. With gas injection, the pulsations were less evident.

Henry et al. [52] measured incipient superheats for sodium in an annulus 9.5-mm OD, 5.84-mm ID. Pressures varied from 0.14 to 0.68 bar. Superheats from 0° to 75°K were measured. The data were found to be consistent with the reentrant cavity model when history and gas solubility were properly accounted for.

Schmücker and Grigull [53] carried out tests on boiling of mercury in a tube. Boiling was found to be intermittent and the behaviour was generally similar to that of potassium described by Aladyev [45].

Tang et al. [54] conducted tests on potassium-mercury mixtures in tubes. Tests included transition from single-phase to subcooled boiling. They report no significant incipient superheats or instabilities. The tests of Mori et al. [55] on mixtures of mercury, bismuth, and lead also included transition from single-phase flow to subcooled boiling. They also do not report any significant incipient superheats or flow-temperature fluctuations.

Summarizing, boiling of pure subcooled liquid metals is often unsteady. Large amplitude cyclic variations in wall temperatures often occur. The variations are generally stronger for potassium than sodium. Mercury also shows behavior similar to potassium. Large incipient superheats have often been found with sodium and potassium. On the other hand some tests on sodium have shown negligible incipient superheat. Some data show incipient superheat to be related to previous history while some do not show such a relation. Finally mixtures of metals show no incipient superheat and no unsteadiness in boiling. Clearly, experimental evidence is conflicting and no method of predicting incipient superheat can be relied upon. However, Equation 41 for potassium shows reasonable agreement with a considerable amount of data.

Heat Transfer

Noyes and Lurie [47] studied subcooled boiling of sodium in an electrically heated annulus. Boiling heat transfer coefficients were found to be only slightly higher than those for single-phase forced convection.

Tang et al. [54] measured heat transfer to potassium-mercury mixtures in a 7.75-mm ID stainless steel tube. Pressure varied from 0.1 to 0.9 bar and subcooling was up to 167°K. Boiling heat transfer coefficients were found to be only slightly higher than those for single-phase convection.

Experiments on forced-convection boiling of rubidium and cesium have been reported by Fisher et al. [56, 57]. The tests were done in 7.1-mm ID tubes made of columbium-1 zirconium alloy. The range of parameters investigated for rubidium was: $G = 158$ to $1192\,kg/m^2s$, $T_{SAT} = 977$ to $1255°K$, $q = 0$ to $103\,KW/m^2$, inlet subcooling $= 111°C$ to $222°C$, and outlet quality from 0 to 0.8. The data for the nucleate boiling of rubidium were correlated by the dimensional equation:

$$h_{TP} = a\,q^n \tag{42}$$

Table 3
Value of a and n in the Correlation of Fisher (Equation 42) for Boiling of Rubidium

Temp. range (°F)	n	$\log_{10}a$
1,350–1,400	1.126	−2.046
1,400–1,450	0.917	−0.918
1,450–1,500	1.181	−2.294
1,500–1,550	1.325	−3.044
1,550–1,600	1.405	−3.414
1,600–1,650	1.333	−3.128
1,650–1,700	0.958	−1.268
1,700–1,750	1.528	−4.427

h_{TP} is in Btu/hr ft^2°F and q in Btu/hr ft^2. The value of a and n were determined at 50°F intervals and are given in Table 3.

Design Recommendations

1. No reliable method for calculating incipient superheat is available. For potassium, Equation 41 may be used in its verified range. For mixtures of mercury with other metals, incipient superheats is negligible.
2. No verified method for predicting the heat transfer rate is available. If no experimental data are available, the following conservative approach is suggested. Calculate pool boiling and single-phase-convection heat transfer coefficients separately. Use the larger of the two.

MULTICOMPONENT FLUIDS

In discussing the boiling of multicomponent mixtures, one must first distinguish between azeotropic and nonazeotropic mixtures. In azeotropic mixtures, the concentrations of components in the vapor and liquid phases are the same and remain the same during evaporation. Such mixtures behave like single-component fluids. Heat transfer to azeotropic mixtures can be calculated by the methods for single-component fluids, using the mean mixture thermophysical properties. Behaviour of nonazeotropic mixtures is much more complex and is the subject of this section.

Fink et al. [29] studied boiling of mixtures of R-11 and R-113 during forced flow across a heated cylinder. Heat transfer coefficients for the mixtures were always lower than what will be expected from linear interpolation of heat transfer coefficients for the pure fluids in proportion to concentration. Similar results have been reported during pool boiling of multi-component mixtures. The explanation is that the low boiling point component evaporates faster with the result that its concentration in the liquid surrounding the bubble is less than in the bulk of liquid. Thus the boiling point of the liquid surrounding the bubble is higher than that of the bulk liquid. As a consequence, the temperature difference causing bubble nucleation is lower than in the case of pure fluids.

Data for forced-convection subcooled boiling are scare and no verified method for predicting heat transfer is available. For rough estimates, the author's recommendation is to use the Kutateladze superposition method (Equation 20) with h_{pb} calculated by a correlation suitable for pool boiling of multicomponents mixtures. The single-phase heat transfer coefficients h_L is to be calculated using mixture mean properties. This recommendation is only for nonmetallic fluids. Recommendations for metallic fluids are given earlier.

NOTATION

- A area (m^2)
- Bo boiling number = $q/(G\,i_{fg})$, dimensionless
- C_p specific heat at constant pressure (J/kg°K)
- $C_{p\ell}$ specific heat of liquid at constant pressure (J/(kg°K)
- D diameter (m); when flow is inside the tube, D is the inside diameter; when flow is outside the tube, D is the outside diameter
- D_{eq} equivalent diameter of channel (m)
- D_i outside diameter of the inner tube of an annulus (m)
- D_0 inside diameter of the outside tube of an annulus (m)
- g acceleration due to gravity (m/s^2)
- G total mass flux (kg/m^{2}s)
- h heat transfer coefficient (W/m^2°K)

h_L heat transfer coefficient in single-phase flow assuming all mass (vapor + liquid) flowing in liquid form ($W/m^2\,°K$)
h_{nb} heat transfer coefficient due to nucleate boiling ($W/m^2\,°K$)
h_{pb} heat transfer coefficient calculated by a pool boiling correlation ($W/m^2\,°K$)
h_{TP} two-phase heat transfer coefficient $= q/(T_w - T_b)$ ($W/m^2\,°K$)
i enthalpy of liquid (J/kg)
i_f enthalpy of saturated liquid (J/kg)
i_{fg} latent heat of vaporization (J/kg)
k thermal conductivity ($W/m\,°K$)
Nu Nusselt number $= hD/k_\ell$, dimensionless
p pressure (N/m^2)
p_c critical pressure (N/m^2)
p_r reduced pressure $= p/p_c$, dimensionless
Pr Prandtl number of liquid $= C_{p\ell}\mu_\ell/k_\ell$, dimensionless
q heat flux (W/m^2)
q_{ib} heat flux at boiling inception (W/m^2)
q_{nb} heat flux due to nucleate boiling (W/m^2)
q_{pb} heat flux assuming pool boiling (W/m^2)
q_{SPC} heat flux due to single-phase convection (W/m^2)
r radius of bubble (m)
Re_L Reynolds number $= GD/\mu_\ell$, dimensionless
S parameter defined by Equation 37, dimensionless
T temperature (°K)
T_b bulk liquid temperature (°K)
T_{SAT} saturation temperature (°K)
ΔT_{SC} $= T_{SAT} - T_b$ (°K)
ΔT_b $= T_w - T_b$ (°K)
ΔT_{SAT} $= T_w - T_{SAT}$ (°K)
$(\Delta T_{SAT})_{ib}$ ΔT_{SAT} at boiling inception (°K)
u liquid velocity in the narrowest section (m/s)
v_{fg} specific volume of vapor minus specific volume of liquid both at saturation (m^3/kg)

Greek Symbols

Ψ $= h_{TP}/h_L$, dimensionless
Ψ_0 value of Ψ at zero subcooling and zero quality
ρ density (kg/m^2)
μ dynamic viscosity (Ns/m^2)
σ surface tension (N/m)
δ annular gap $= (D_0 - D_i)/2$ (m)

Subscripts

b bulk (i.e. mixing cup condition)
f with properties at film temperature $= (T_w + T_b)/2$
g of vapor
ℓ of liquid
SAT at saturation temperature
w at wall

Abbreviations

ID inside diameter
OD outside diameter
SS stainless steel

REFERENCES

1. Bergles, A. E., and Rohsenow, W. M., "The Determination of Forced-Convection Surface-Boiling Heat Transfer," *J. Heat Transfer* (Aug. 1964), pp. 365–372.
2. Davis, E. J., and Anderson, G. H., "The Incipience of Nucleate Boiling in Forced Convection Flow," *AIChE J.*, Vol. 12, No. 4 (July 1966), pp. 774–780.

3. Rohsenow, W. M., and Clark, J. A., "Heat Transfer and Pressure Drop for High Heat Flux Densities to Water at High Subcritical Pressure," *1951 Heat Transfer & Fluid Mechanics Institute Preprints,* Stanford Univ., Press, 1951; quoted by Bergles and Rohsenow [1].
4. Backhurst, J. E., Ph D. thesis, London Univ., 1965; quoted by Davis and Anderson [2].
5. Matzournis, B. G., Ph D. thesis, London Univ., 1958; quoted by Davis and Anderson [2].
6. Frost, W., and Dzakowic, G. S., "An Extension of the Method of Predicting Incipient Boiling on Commercially Finished Surfaces," ASME Paper 67-HT-61, 1967.
7. McAdams, W. H., et al., "Heat Transfer at High Rate to Water With Surface Boiling," *Ind. Eng. Chem.,* Vol. 41 (1949), pp. 1945–53.
8. Hodgson, A. S., "Forced Convection Subcooled Boiling Heat Transfer With Water in an Electrically Heated Tube at 100–550 lb./in.^{2}," *Trans. Instn. Chem. Engnrs.* Vol. 46 (1968), pp. T25–T31.
9. Shah, M. M., "A General Correlation for Heat Transfer During Subcooled Boiling in Pipes and Annuli," *ASHRAE Trans.,* Vol. 83, No. 1 (1977).
10. Shah, M. M., "Generalized Prediction of Heat Transfer During Subcooled Boiling in Annuli," *Heat Transfer Engng.,* Vol. 4, No. 1 (1983), pp. 24–31.
11. Rohsenow, W. M., "A Method of Correlating Heat Transfer Data for Surface Boiling of Liquids," *Trans. ASME.,* Vol. 75 (1952), pp. 969–976.
12. Rohsenow, W. M., "Heat Transfer With Evaporation," Lextures at the Symposium on Heat Transfer Univ. of Michigan, published by Univ. of Michigan, July 18–19 1952; quoted by Collier [58].
13. Kutateladze, S. S., "Boiling Heat Transfer," *Int. J. Heat Mass Transfer,* Vol. 4 (1961), pp. 31–45.
14. Chen, J. C., "Correlation for Boiling Heat Transfer to Saturated Fluids in Convection Flow," *Ind. Eng. Chem. Process Des. Dev.,* Vol. 5, No. 3 (1966), pp. 322–339.
15. Noel, M. B., "Experimental Investigation of The Forced-Convection and Nucleate Boiling Heat Transfer Characteristics of Liquid Ammonia," JPL, Calif. Inst. Techn., Report No. 32–125, July 1961.
16. Jens, W. H., and Lottes, P. A., "Analysis of Heat Transfer Burnout, Pressure Drop, and Density Data for High Pressure Water," ANL-4627, May 1951.
17. Percupile, J. C., Riedle, K., and Schmidt, F. K., "Experimental Investigation: Boiling Heat Transfer in Evaporator Tubes-Vertical Upflow," *AIChE Symp. Ser.,* Vol. 70, No. 70, No. 138, pp. 91–97.
18. Brisbane, T. W. C., Grant, I. D. R., and Whalley, P. B., "A Prediction Method for Kettle Reboiler Performance," ASHE Paper 80-HT-42, 1980.
19. Shah, M. M., "A Correlation For Heat Transfer During Subcooled Boiling on a Single Tube With Forced Crossflow," *Int. J. Heat & Fluid Flow,* Vol. 5, No. 1 (1984), pp. 13–20.
20. Yilmaz, S. B., "Effect of Velocity on Boiling Freon-113," doctorate thesis, Chem Eng. Dept., Univ. of Illinois, 1979.
21. Fand, R. M., et al., "Simultaneous Boiling and Forced Convection Heat Transfer from a Horizontal Cylinder to Water," *J. Heat Transfer,* Vol. 98 (1976), pp. 395–400.
22. Yilmaz, S. B., and Westwater, J. W., "Effect of Velocity On Heat Transfer to Boiling Freon-113" *J. Heat Transfer,* Vol. 102, 1980, pp. 26–31.
23. Leppert, G., Costello, C. P., and Hoglund, B. M., "Boiling Heat Transfer to Water Containing a Volatile Additive," *Trans. ASME.,* Vol. 80 (1958), pp. 1395–1403.
24. Lemmert, M., and Chawla, J. M., "Influence of Flow Velocity On Surface Boiling Heat Transfer Coefficients," in *Heat Transfer in Boiling,* E. Hahne and U. Grigull (Eds.), Academic Press, New York, 1977.
25. McKee, H. R., "Forced Convection Boiling From a Cylinder Normal to The Flow," PhD thesis, Chem. Eng. Dept., Oklahoma State Univ., 1967.
26. Beecher, N., "Heat Transfer to Water Flowing Normal to a Single Cylinder," M. S. thesis, Chem. Eng. Dept., Massachusetts Inst. Tech., 1948.

27. Vliet, G. C., "Local Boiling Peak Heat Flux For Water Flowing Normal to Cylinder," Ph. D thesis, Mech. Eng. Dept., Stanford Univ., 1962.
28. Dieselhorst, T., "Hydrodynamische und oberflächenspezifische Einflüsse auf die Siedekrisis beim Behältersieden," Dr. Ing. thesis, Techn. Univ. Munich, 1978.
29. Fink, J., Gaddis, E. S., and Vogelpohl, A., "Forced Convection Boiling of a Mixture of Freon-11 and Freon-113 Flowing Normal to a Cylinder," *Proc. Seventh Int. Heat Transfer Conf.,* Munich, paper FBS, 1982, pp. 207–211.
30. Bitter, R. C., "Heat Transfer From a Horizontal Tube With Transverse Flow of Evaporating Saturated R-11," Int. Inst. of Refrigeration, Commission B-1, B-2, and E-1, Freudenstadt, Fed. Rep. Germany, Supplement to Int. Inst. Refrig, Bulletin, 1972, pp. 97–107.
31. Bitter, R. C., "Zum Wärmeübergang von einem querangeströmten Rohr an Siedendes R-11 bei Ein- und Zweiphasenströmung," Dr. Ing. thesis, Tech. Univ., Clausthal, 1973.
32. Pappel, S. S., and Hendricks, R. C., "Boiling Incipience and Convective Boiling of Neon and Nitrogen," *Advances In Cryogenic Eng.,* Vol. 23 (1978), pp. 248–294.
33. Klimenko, V. V., and Sudarchikov, A. M., "Investigation of Forced Flow Boiling of Nitrogen in a Long Vertical Tube," *Cryogenics* (July 1983), pp. 379–385.
34. Stephan K., and Abdelsam, M., "Heat Transfer Correlations for Natural Convection Boiling," *Int. J. Heat Mass Transfer,* Vol. 23 (1980), pp. 73–87.
35. Kilmenko, V. V., "Heat Transfer Intensity at Forced Flow Boiling of Cryogenic Liquids in Tubes," *Cryogenics* (1982), pp. 569–576.
36. Shah, M. M., "Prediction of Heat Transfer to Cryogenic Liquids Boiling Inside Tubes," Cryogenics, May 1984, pp. 231–236.
37. Shah, M. M., "A New Correlation for Heat Transfer During Boiling Flow Through Pipes," *ASHRAE Trans.,* Vol. 82, No. 2 (1976) pp. 66–86.
38. Shah, M. M., "CHART Correlation for Saturated Boiling Heat Transfer: Equations and Further Study," *ASHRAE Trans.,* Vol. 88, No. 1 (1982).
39. Steiner, D., and Schlünder, E. U., "Heat Transfer for Boiling Nitrogen Flowing in a Horizontal Tube," in *Heat Transfer in Boiling,* E. Hahne and U. Grigull (Eds.), Academic Press, New York, 1977.
40. Müller, H., Bonn, W., and Steiner, D., "Heat Transfer and Critical Heat Flux at Flow Boiling of Nitrogen and Argon Within a Horizontal Tube," in *Heat Exchangers,* J. Taborek, G. F. Hewitt, and N. Afgan (Eds.), Washington, Hemisphere Publishing Co., 1983.
41. Mohr, V., and Runge, R., "Forced Convection Boiling of Neon in Horizontal Tubes," in *Heat Transfer in Boiling,* E. Hahne and U. Grigull (Eds.), Academic Press, New York, 1977.
42. Ogato, H., and Sato, S., "Forced Convection Heat Transfer to Boiling Helium in a Tube," *Cryogenics* (July 1980), pp. 375–380.
43. Jergel, M., Hechler, K., and Stevenson, R., "The Effect of Forced Circulation on Heat Transfer With Liquid Helium in Narrow Channels," *Cryogenics* (Oct. 1979), pp. 413–417.
44. Klein, G., "Heat Transfer for Evaporating Nitrogen Streaming in a Horizontal Tube," *Proc. Sixth Int. Cryogenic Eng. Conf.,* 1977, pp. 314–318.
45. Aladyev, I. T., "On The Nature of Liquid Superheat," *Progress In Heat & Mass Transfer,* Vol. 7 (1973), pp. 265–277.
46. Hoffman, H. W., and Krakoviak, A., "Convective Boiling With Liquid Potassium," *Proc. Heat Transfer & Fluid Mech. Inst.,* Stanford Univ. Press, 1964, pp. 19–37.
47. Noyes, R. C., and Lurie, H., "Boiling Sodium Heat Transfer," *Proc. Third Int. Heat Transfer Conf.,* Vol. 5, 1966, pp. 92–100.
48. Chen, J. C., "Incipient Boiling Superheats in Liquid Metal," *J. Heat Transfer* (Aug. 1968), pp. 303–312.
49. Dodonov, L. D., Koroleva, V. S., and Rzayev, A. I., "Experimental Study of Incipient Boiling Superheats for Potassium in Channel Flow," *Heat Transfer Sov. Res.,* Vol. 11, No. 4 (1979).

50. Peppler, W., "Sodium Boiling in Fast Reactors: A State of the Art Review," in *Thermal and Hydraulic Aspects of Nuclear Safety, Vol. 2, Liquid Metal Fast Breeder Reactors,* S. G. Bankoff (Ed.), ASME, New York, 1977.
51. Schultheiss, G. F., "Aspects of Liquid-Metal Superheat and the Effects on Dynamic Boiling," *Progress in Heat & Mass Transfer,* Vol. 7 (1973), pp. 279–298.
52. Henry, R. E., et al., "Incipient Superheat in a Convective Sodium System," *Proc. Fifth Int. Heat Transfer Conf.,* Vol. 4, 1974, pp. 305–309.
53. Schmucker, H., and Grigull, U., "Boiling of Mercury in a Vertical Tube Under Forced Flow Conditions," *Progress In Heat & Mass Transfer,* Vol. 7 (1973), pp. 363–376.
54. Tang, Y. S., et al., "Forced Convection Boiling of Potassium-Mercury Systems," *AIChE J.,* Vol. 10, No. 5 (1964), pp. 617–620.
55. Mori, Y., "Convective Boiling of a Binary Liquid Metal," in *Liquid Metal Heat Transfer & Fluid Dynamics,* J. C. Chen and A. A. Bishop (Eds.), ASME, New York, 1970.
56. Fisher, C. R., in ORNL 3605, Vol. 2, 1964, pp. 64–75.
57. Fisher, C. R., et al., "Forced Convection Boiling of Rubidium and Cesium Heat Transfer and Pressure Drop," AGN-8099, Feb. 1964, quoted by Noyes and Lurie [47].
58. Collier, J. G., "Convective Boiling and Condensation," 2nd ed., McGraw–Hill, London, 1982.

CHAPTER 12

LIQUID METAL HEAT TRANSFER IN TURBULENT PIPE FLOWS

Shong-Leih Lee*

Department of Power Mechanical Engineering
National Tsing Hua University
Hsinchu, Taiwan 300, Republic of China

CONTENTS

INTRODUCTION

Compared to nonmetallic liquids, liquid metals have many favorable thermal properties such as large thermal conductivities, high boiling points, high resistance to thermal decomposition, and extensive liquid-phase temperature ranges. Qualitatively, it has been recognized that the heat transfer coefficients are higher in liquid metals than in any other fluids for a given system and pumping power. Therefore, liquid metals are very desirable for certain important technological applications. The most familiar application is their use as coolants in nuclear reactors and their potential use as working media in space power plants in which small equipment size is essential.

In spite of the attractions and the potentialities of the uses of liquid metals in industrial applications, the mechanism of turbulent heat transfer in liquid metals is not well

* Presently at the Department of Mechanical and Aerospace Engineering, University of Missouri-Rolla, Rolla, Missouri, 65401.

The author would like to thank Kluwer Academic Publishers Group (*Applied Scientific Research*), Chemcan Publishers Limited (*Canadian Journal of Chemical Engineering*) and Pergamon Press (*International Journal of Heat and Mass Transfer and Chemical Engineering Science*) for their kind permissions to quote the work published in their journals.

understood. In the early literature [1–3] the turbulent Prandtl number, Pr_t, which is the ratio of eddy viscosity to eddy diffusivity, was taken to be a constant or unity. Based on this assumption, the well-known Martinelli–Lyon equation was proposed to predict the Nusselt numbers in pipe flows under uniform wall flux conditions. However, it has been found that experimental Nusselt numbers generally fall 30%–50% below the predicted values [4]. In order to investigate the invalidity of the Reynolds analogy ($Pr_t = 1$) in liquid metal pipe flows, various experiments [5, 6] have been conducted to determine the turbulent Prandtl number from measured temperature and velocity profiles. Wide differences are seen among the data, and it is difficult to get consistent values for the turbulent Prandtl number. Fortunately, it is generally agreed that Pr_t is larger than unity for liquid metals and is not a constant across the pipe.

In a theoretical analysis, Poppendiek and Palmer [7] assumed the eddy diffusivity to be negligible when compared with molecular diffusivity in liquid metals, and obtained predicted Nusselt numbers that are far below the experimental data. Deissler [8] modified the Reynolds analogy and proposed a model by allowing for heat transfer by conduction to or from a turbulent particle as it moves radially in the pipe. This also resulted in low Nusselt numbers. Later, Azer and Chao [8] proposed a mechanism based on a modification of Prandtl's mixing length hypothesis which assumes that there is a continuous change of momentum and energy during the movement of the eddy. Computations of the Nusselt number and the temperature profile in liquid metal were carried out under the wall condition of constant wall heat flux using the deduced expression for the turbulent Prandtl number. Comparison of the Nusselt number with the prediction of Lyon [2] was made and the improvement over Lyon's result was obvious. According to their results, the lowest possible Nusselt number is 7, but the bulk of experimental data [4] at low Peclet numbers indicate a value considerably less than 7. Azer and Chao [10] also used the theory to investigate liquid metal heat transfer in fully developed pipe flows with a constant wall temperature. Notter and Sleicher [11] solved the turbulent Graetz problem by a method of eigenfunction expansion. A new empirical turbulent Prandtl number expression was proposed and recommended for use together with values of eddy viscosity described in the literature [12] to predict values of eddy diffusivity for $Pr < 1$. Another model was proposed by Na and Habib [13] based on a modified form of the mixing length theory developed by Cebeci [14] for turbulent pipe flows. The model predicts the fully developed Nusselt number with $Pr = 0.02 - 14.3$ under a uniform wall flux condition and fits the available experimental data well for a range of the Peclet numbers. Both the eddy viscosity and eddy diffusivity were given in terms of the Nikuradse's [15] mixing length and damping factors, and the turbulent Prandtl number was given by the ratio of the two.

A model called the surface renewal model is found to be useful for low- to moderate-Prandtl-number fluids [16, 17]. This model is based on the concept that elements of fluid intermittently move from the turbulent core into direct contact with the wall. It gives predictions for ε_m and ε_h lying well above accepted correlations in the wall region due to no consideration given to the unreplenished fluid at the surface. The shortcoming of this model was later removed by the surface rejuvenation model [18]. Thomas [19] improved the mathematical complexities for the new model and gave a simplified formulation without losing any accuracy for practical purposes. However, this formula is applicable only in the wall region ($y^+ < 30$) because of the nature of the surface renewal model. It is interesting to note that the surface rejuvenation model suggests $Pr_t < 1$ for $Pr < 1$ within the wall region. This is opposite to the existing models [9–14].

Although axial conduction can be ignored in turbulent convection for nonmetallic fluids, this might not always be justified for liquid metals. In fact, the Peclet number can be as small as 5 in turbulent pipe flow for liquid metals. A number of prior works have investigated the effect of axial conduction on laminar heat transfer [20–23], but few have considered this effect in turbulent flow case. In turbulent flow, the effect was first considered by Schneider [24]. He considered the case of a liquid metal flowing through a round pipe under uniform wall temperature condition. The velocity was assumed to be uniform through out the pipe.

The inlet temperature was taken to be uniform (at $X = 0$), and the pipe is sufficiently long so that axial conduction is negligible before the pipe outlet is reached. Both the dimensionless axial and radial eddy diffusivities were assumed to be equal to unity ($\varepsilon_x = \varepsilon = 1$). Based on the assumptions, he concluded that for the purpose of computing the overall heat transfer rate, the axial conduction can be neglected when the Peclet number is larger than 100.

It is noted that previous publications have not given predictions or data at low Peclet numbers ($Pe < 100$) and low Reynolds numbers ($Re < 10^4$) for turbulent heat transfer. Under such conditions, the axial conduction effect may be significant in the entrance region and the influence of boundary sublayers must be taken into account. Recently, Lee [25] investigated the effect of axial conduction on a problem of liquid metal heat transfer in turbulent pipe flow with a uniform wall temperature. The Peclet numbers considered were less than 10^3 and Nikuradse's mixing length modified by Cebeci's [13] damping factor was adopted to approximate the eddy viscosity. The laminar sublayer and buffer zone were also considered in the flow field. To allow for the effect of upstream conduction, the fluid temperature was taken to be uniform at $X = -\infty$ and the pipe wall was adiabatic at $X \leqslant 0$. The results led to the conclusion that the effect is significant in the thermal entry-region for $Pe < 100$, but is negligible in the thermally fully developed region. Ignoring the axial conduction was shown to result in an error of about 45% (low) in the Nusselt number at a low Peclet number, say $Pe = 5$.

It is noteworthy that the bulk of experimentally determined Nusselt numbers are lower than those predicted by the existing theoretical analyses, especially in the low Peclet number range [4, 9, 10]. Lee [26] extended the work [25] to the boundary condition of uniform wall flux and provided detailed information for this case. Excellent agreement is observed upon comparing the resulting Nusselt numbers with the available experimental data in both the thermal entrance and fully developed regions. The solution indicates that the fully developed Nusselt number depends only on the Reynolds number at low Peclet numbers.

GOVERNING EQUATIONS

Consider a liquid metal flowing through a round pipe of infinite length. The flow is turbulent and fully developed. It is also assumed to be steady, Newtonian, and incompressible. The physical properties are constant. Viscosity dissipation, free convection, and tube wall thermal resistance are all assumed to be negligible.

After imposing the above assumptions and introducing the dimensionless transformation, $u = U/V$, $r = R/R_0$, $Pe = VD/\alpha$, $x = 4X/PeD$, $\varepsilon_x = 1 + (\varepsilon_h)_x/\alpha$, $\varepsilon = 1 + \varepsilon_h/\alpha$, $\sigma = \varepsilon_x/\varepsilon$, the energy equation becomes:

$$f\partial\theta/\partial x = g\partial\theta/\partial r + \partial^2\theta/\partial r^2 + (4\sigma/Pe^2)\,\partial^2\theta/\partial x^2 \qquad (1)$$

in which

$$f = u/\varepsilon \text{ and } g = 1/r + (d\varepsilon/dr)/\varepsilon. \qquad (2)$$

In the transformation, θ is defined as

$$\begin{aligned} \theta &= (T - T_W)/(T_0 - T_W) && \text{for uniform wall temperature} \\ \theta &= (T - T_0)/(q_W D/2k) && \text{for uniform wall flux} \qquad (3) \\ \theta &= (T - T_\infty)/(T_0 - T_\infty) && \text{for convective heat transfer wall condition} \end{aligned}$$

It is noted that fully developed flow idealization implies that both dimensionless velocity profile u and total eddy diffusivity ε are independent of x. The LHS of the energy Equation 1 is the forced-convection term; the first two terms on the RHS are the radial conduction terms; and the last term, $(4\sigma/Pe^2)(\partial^2\theta/\partial x^2)$, usually neglected for large Peclet number flows but rather important for low Peclet number flows, is the so-called axial conduction term. For liquid metals ($0.001 \leqslant Pr \leqslant 0.022$), the Peclet number can be as small as 5 in turbulent pipe flows. Therefore, this term is kept in the general governing equation here. As the Peclet number is sufficiently small, the effect of upstream conduction in the region $x \leqslant 0$ may be significant [25]. To consider this effect, the fluid temperature is taken to be uniform at $x = -\infty$ and the pipe wall is assumed to be adiabatic in $x \leqslant 0$ [25, 26]. The associated boundary conditions then are:

For uniform wall temperature condition

$$\begin{aligned} &\theta(-\infty, r) = 1,\ \theta(\infty, r) = 0,\ \partial\theta(x, 0)/\partial r = 0 \\ &\partial\theta(x, 1)/\partial r = 0 \qquad \text{at } x \leqslant 0 \\ &\theta(x, 1) = 0 \qquad \text{at } x \geqslant 0 \end{aligned} \tag{4}$$

For uniform wall flux condition

$$\begin{aligned} &\theta(-\infty, r) = 0,\ \theta(\infty, r) = \theta_{fd},\ \partial\theta(x, 0)/\partial r = 0 \\ &\partial\theta(x, 1)/\partial r = 0 \qquad \text{at } x \leqslant 0 \\ &\partial\theta(x, 1)/\partial r = 1 \qquad \text{at } x \geqslant 0 \end{aligned} \tag{5}$$

For convective heat transfer wall condition

$$\begin{aligned} &\theta(-\infty, r) = 1,\ \theta(\infty, r) = 0,\ \partial\theta(x, 0)/\partial r = 0 \\ &\partial\theta(x, 1)/\partial r = 0 \qquad \text{at } x \leqslant 0 \\ &\partial\theta(x, 1)/\partial r + (Bi/2)\,\theta(x, 1) = 0 \qquad \text{at } x \geqslant 0 \end{aligned} \tag{6}$$

It is noted that θ_{fd} in Equation 5 refers to the dimensionless fully developed temperature profile which includes the effect of axial conduction. If the Peclet number is large enough [25], say $Pe > 1{,}000$, so that the axial conduction term in the energy Equation 1 can be dropped, the boundary conditions in $x \leqslant 0$ previously listed may be replaced by a uniform temperature distribution at $x = 0$ [11, 12].

Before solving the energy Equations 1 and 2 with the associated boundary conditions 4, 5 or 6, the functions u, ε, and θ_{fd} must be determined. Many models to predict u and ε have been proposed in the literature, and the resulting Nusselt numbers have been found to vary with different models used. Some important models will be discussed in the following sections. For the boundary condition of a uniform wall flux, θ_{fd} can be determined by assuming that in the thermally fully developed region the bulk temperature rises linearly in the axial direction while the temperature profile keeps unchanged [27]. After substituting this assumption into the energy Equation 1 and 2 and then taking a heat balance over the region extending from $x = -\infty$ to an arbitrary axial position in the thermally fully developed region, one obtains [26]

$$\theta_{fd} = 2x + 2\int_0^r (1/r\varepsilon) \int_0^r ru\,(dr)^2 + (8/Pe^2)\int_0^1 2\varepsilon\sigma r dr - 4\int_0^1 ru \int_0^r (1/r\varepsilon) \int_0^r ru\,(dr)^3 \tag{7}$$

It is noted that the fully developed temperature profile θ_{fd} differs from that when axial conduction is neglected by the additional term

$$(8/Pe^2)\int_0^1 2\varepsilon\sigma r dr$$

where σ is the ratio of axial to radial eddy diffusivity ($\varepsilon_x/\varepsilon$) and is usually taken as unity [24–26]. For laminar pipe flow, $u = 2(1 - r^2)$, $\varepsilon = 1$ and $\sigma = 1$, and the fully developed temperature profile is then given by

$$\theta_{fd} = 2x + r^2 - r^4/4 - 7/24 + 8/Pe^2$$

SOME MODELS IN LIQUID METAL HEAT TRANSFER

Many models have been published for predicting the velocity profile u and eddy diffusivity ε in turbulent pipe flows. Among the models, the following are the most important ones for liquid metal heat transfer.

A Modified Mixing-Length Model by Azer and Chao

In 1925, Prandtl [28] proposed the well-known mixing-length hypothesis. It was presumed that the eddy preserves both its momentum and energy while traveling between layers of a fluid before mixing occurs. However, heat transfer at the surface of the eddy may become appreciable in fluids of high thermal conductivities. Azer and Chao [9] modified Prandtl's theory by considering this effect for liquid metals. The new model states that an eddy created at layer 1 with velocity u_1 and temperature t_1 will attain a velocity u_1' and a temperature t_1' before mixing takes place at layer 2. With this simple modification, one has

$$\begin{aligned} Pr_t = \frac{\varepsilon_m}{\varepsilon_h} &= \frac{(u_2 - u_1')/(u_2 - u_1)}{(t_2 - t_1')/(t_2 - t_1)} \\ &= \frac{1 + (t_1' - t_1)/(t_2 - t_1')}{1 + (u_1' - u_1)/(u_2 - u_1')} \end{aligned} \tag{8}$$

where $u_2 - u_1'$ and $t_2 - t_1'$ are the fluctuating axial component of velocity and temperature at layer 2, respectively. A similar event takes place for an eddy moving in the opposite direction.

To ascertain the influence of fluid drag and heat transfer at the surface of the eddy, it was assumed that the drag and heat transfer are both associated with the boundary layer on the surface of the eddy. For simplicity, the eddy was assumed to be spherical with radius a. As it moves from layer 1 to layer 2, its average axial velocity is $\frac{1}{2}(u_1 + u_2)$. Application of the boundary layer momentum principle yields,

$$\frac{u' - u_1}{u_2 - u_1'} = \frac{3}{2}\lambda\left(\frac{\ell}{d}\right)\left(\frac{u_2 - u_1'}{4v'}\right) \tag{9a}$$

where ℓ/v' is the time required for an eddy to travel from layer 1 to layer 2 and λ is the average friction coefficient defined by

$$\begin{aligned} \lambda &= \tau/\{\tfrac{1}{2}\rho[(u_2 + u_1)/2 - (u_1 + u_1')/2]^2\} \\ &= \tau/\{\tfrac{1}{2}\rho[(u_2 - u_1')/2]^2\} \end{aligned} \tag{9b}$$

To estimate the amount of heat transfer during the traveling time, the spherical eddy was conceived to have an average temperature $\frac{1}{2}(t_1 + t_1')$ as it is immersed in a fluid initially at a uniform temperature $\frac{1}{2}(t_1 + t_2)$. The film coefficient of heat transfer at the eddy surface is h_e. With this simplied model, a one-dimensional unsteady heat conduction equation was established. The amount of heat transfer then was evaluated by solving the equation. From the energy balance between the heat transfer at the eddy surface and a temperature increase in the eddy, the following formula was obtained

$$\frac{t_1' - t_1}{t_2 - t_1'} = \frac{3\alpha}{2a}\frac{h'}{1 + ah'}\left[\frac{\ell}{v'} + \frac{aH - 1}{\alpha H^2}F(Z)\right] \tag{10a}$$

where

$$F(Z) = \exp(Z^2)\,\mathrm{erfc}(Z) + 2Z/\sqrt{\pi} - 1 \tag{10b}$$

$$Z = H\sqrt{(\alpha \ell / v')} \tag{10c}$$

$$H = (1 + ah')/a \tag{10d}$$

$$h' = h_e/k \tag{10e}$$

Equation 10 contains the surface heat transfer coefficient of the eddy h_e. According to the proposed model, the heat transfer is characterized by the laminar boundary layer on the eddy surface. Possible influences of curvature behind it were ignored, and Pohlhausen's expression for flat plate was used to estimate the heat transfer coefficient h_e. For $Pr \ll 1$, hence,

$$Nu' \, Pr^{\frac{1}{2}}/Pe' = 1.64\,(\lambda/2) = 1.06/(Re')^{\frac{1}{2}} \tag{11}$$

where

$$Nu' = h_e(\pi a)/k = \pi a h'$$

$$Re' = [(u_2 + u_1)/2 - (u_1 + u_1')/2]\,\pi a/\nu = (u_2 - u_1')\,\pi a/2\nu$$

$$Pe' = Pr Re'$$

To consider the arbitrariness of the eddy shape and the nature of the heat transfer process in practice, Equation 11 was replaced by

$$Nu' Pr^{\frac{1}{2}}/Pe' = 1.64\,b/(Re')^{\frac{1}{2}} \tag{12}$$

where b is an empirical constant. Due to lack of available data in liquid metals, Sleicher's data [30] in air were adopted to determine the constant b. After a few calculations, Azer and Chao [9] assigned b = 2.5 for the model.

Introducing Equations 9, 10 and 12 into Equation 8, one obtains an expression for the turbulent Prandtl number in terms of $(u_2 - u'_1)/v'$, ℓ/a, and Re'. The quantities $u_2 - u_1'$ and v' are, respectively, the fluctuating velocities in the axial and radial direction. Both are complicated functions of space and time; their exact nature is not known. Customarily, they are treated on a statistical basis and expressed in terms of their root-mean square values. Numerous experiments were conducted to study the structure of turbulence in fully developed pipe flow. Laufer [29] carried out one of the most detailed investigations. His results may be closely approximated by

$$(u_2 - u_1')/v' = 1.018\,(1 + 1.778r)/(1 + 0.5625r) \tag{13}$$

The data were obtained for air, but were assumed to remain valid for liquid metals.

After a series of examinations, Azer and Chao [9] found that the turbulent Prandtl number is not sensitive to variations of ℓ/a in air. For simplicity, they selected $\ell/a = 2$ for their model of liquid metal heat transfer. Using the assumption and Nikuradse's mixing-length expression, there resulted [2]

$$Re'/Re^* = 0.276\,(1 + 1.778\,r)\,(0.14 - 0.08\,r^2 - 0.06\,r^4) \tag{14}$$

Substituting Equations 13 and 14 into the developing turbulent Prandtl number and Equation 8, and letting $\ell/a = 2$, it follows that

$$Pr_t = (1 + 0.75\,F_1/F_2)/(1 + 3.808\,\varphi/\sqrt{Re'}) \tag{15}$$

where $\varphi = (1 + 1.778\,r)/(1 + 0.5625\,r)$

$$F_1 = 2.031\,\varphi + [0.833\,(Pe')^{3/2}\,F(Z)]/[1 + 1.305\,(Pe')^{\frac{1}{2}}]^2$$

$$F_2 = 0.318\,Pe' + 0.244\,(Pe')^{\frac{1}{2}}$$

$$Z = 1.785\,\varphi^{\frac{1}{2}}[1.305 + (Pe')^{-\frac{1}{2}}]$$

For approximate calculations, the following equation may be used

$$Pr_t = \frac{1 + 380\,Pe^{-0.58}\exp(-y^{0.25})}{1 + 135\,Re^{-0.45}\exp(-y^{0.25})} \tag{16}$$

which gives a maximum deviation of less than 14%. More detailed information on the model can be found in the literature [9].

In predicting the fully developed Nusselt number for pipe flows, Azer and Chao [9] used Nikuradse's data [15] of eddy viscosity together with Equation 16 to approximate the eddy diffusivity. The data in the literature [15] were evaluated directly from the measured velocity gradient. To carry out an integration for radial heat flux distribution, they adopted von Karman's [31] logarithmic velocity profile in the calculation. Due to no consideration of the influences of the laminar sublayer and buffer zone, their results are valid only when the Reynolds numbers are sufficiently large.

Empirical Interpolation Model of Notter and Sleicher

Notter and Sleicher [11] constructed a purely empirical interpolation formula for the turbulent Prandtl number. Their model was based on observations from the available experimental data and theoretical analyses that the smaller the Pr the larger the Pr_t and that Pr_t approaches unity as the eddy viscosity ε^+ approaches infinity. These constraints were refined by Hill [32] who calculated the rate of heat transfer through an isotropically turbulent fluid with a uniform temperature gradient. The calculations with both the direct interaction and cumulate-discard approximations to the dynamics of turbulence were used to show that in the limit of zero Prandtl number, the turbulent Prandtl number is inversely proportional to $Pr\,\varepsilon^+$, and the turbulent Prandtl number is about 1.1 at $Pr = 0.1$. Notter and Sleicher assumed that Hill's refinement can be extended to shear flows and then constructed the empirical formula

$$Pr_t = (1 + B)/A \tag{17}$$

where $A = (0.025\,Pr\,\varepsilon^+ + B)\,[1 + 10/(35 + \varepsilon^+)]$

$$B = 90\,Pr^{1.5}(\varepsilon^+)^{0.25}$$

$$\varepsilon^+ = \varepsilon_m/\nu.$$

This formula fits the experimental data for Pr = 0.02–0.025 and 0.72. It is expected to remain good for liquid metals.

Notter and Sleicher further recommended Equation 17 together with a proposed empirical eddy viscosity formula [12] to predict the eddy diffusivity. The suggested eddy viscosity and its corresponding velocity profile are:

For $y^+ < 45$

$$\varepsilon^+ = 0.0083\,(y^+)^2$$

$$u^+ = 11\tan^{-1}(0.091\,y^+) \tag{18}$$

For $y^+ > 5$ with $y < 0.15$

$$\varepsilon^+ = 0.4\,(1 - y)\,y^+ - 1 \tag{19}$$

For $y > 0.15$

$$\varepsilon^+ = Re^*/[4\,dh/d\,(r^2)] - 1$$

$$u = u_{max} - h(y)\sqrt{(f_m/8)} \tag{20}$$

where the shear Reynolds number is defined as $Re^* = Re\sqrt{(f_m/8)}$ and the function $dh/d(r^2)$ is due to the velocity defect law. It is noted that Equations 18 is imprecise in the region $y^+ < 10$ where $\varepsilon_m \ll \nu$. All of the quantities, $dh/d(r^2)$, $h(y)$, u_{max}, and f_m can be found in Tables 1 and 2.

A Model with Damping Factor on the Wall

It is widely believed that the conventional mixing-length overstates the effect of turbulence in the immediate neighborhood of the wall. One method of accounting for the damping effect of the wall was proposed by van Driest [33]. The basis of this model can be stated as follows: An infinitely extended flat plate situated in an otherwise quiescent fluid environment oscillates harmonically in its own plane. Owing to the viscosity, the amplitude of the oscillations experienced by the fluid diminish exponentially with increasing distance from the plate; that is the damping factor is $\exp(-y^+/A^+)$, where A^+ is the damping constant. In the case in which the plate is stationary and the fluid is oscillating due to the dynamics of the turbulent eddies, the corresponding damping factor is $1 - \exp(-y^+/A^+)$.

Cebeci [14] followed the idea and obtained the expressions:

$$\varepsilon_m = \ell^2 (du/dy)$$

$$\varepsilon_h = \ell \ell_h (du/dy)$$

$$Pr_t = \ell/\ell_h \qquad (21)$$

Table 1
The Defect Law

y	h (y)	dh/d (r²)
0.15	5.60	9.70
0.20	4.80	8.80
0.30	3.55	7.65
0.40	2.60	7.30
0.50	1.80	7.20
0.60	1.15	7.20
0.70	0.65	7.20
0.80	0.30	7.20
0.90	0.07	7.20
1.00	0.00	7.20

From R. H. Notter, et al. [12]

Table 2
Friction Factor and Maximum Velocity

Re	f_m	u_{max}
5,000	0.0372	1.32
10,000	0.0305	1.27
20,000	0.0258	1.24
50,000	0.0209	1.22
100,000	0.0180	1.20
200,000	0.0157	1.19
500,000	0.0131	1.17
1,000,000	0.0118	1.16

From R. H. Notter, et al. [12]

where $\ell = [1 - \exp(-y^+/A^+)]\, k_m y$

$$\ell_h = [1 - \exp(-y^+\sqrt{Pr}/B^+)]\, k_h y$$

Note that ℓ and ℓ_h are mixing-lengths for momentum and heat, respectively. Both the mixing-length constants k_m and k_h can be assigned as 0.4 in the immediate neighborhood of the wall. To extend the model to be applicable from the wall to the center of the tube, Na and Habib [13] replaced both $k_m y$ and $k_h y$ by Nikuradse's mixing length and proposed the following expressions.

$$\varepsilon_m/\nu = (\ell^+)^2 (du^+/dy^+) \tag{22a}$$

$$\varepsilon_h/\nu = (\ell_0^+)^2\, DF\, DF_h (du^+/dy^+) \tag{22b}$$

where $\ell^+ = \ell_0^+ (DF)$

$$\ell_0^+ = V^* \ell/\nu = (Re^*/2)(\ell/R_0)$$

$$= (Re^*/2)(0.4\,y - 0.44\,y^2 + 0.24\,y^3 - 0.06\,y^4)$$

$$DF = 1 - \exp(-y^+/A^+)$$

$$DF_h = 1 - \exp(-y^+\sqrt{Pr}/B^+) \tag{22c}$$

The empirical constant A^+ was assigned the same numerical value used by van Driest [33], that is 26. Note that Cebeci [14] used a constant value of B^+ because he was interested mainly in air. For liquid metals, Na and Habib [13] found that the expression

$$B^+ = \sum_{n=1}^{5} C_n (\log Pr)^{n-1}$$

where $C_1 = 34.96$

$$C_2 = 28.79$$

$$C_3 = 33.95$$

$$C_4 = 6.33$$

$$C_5 = -1.186 \tag{23}$$

fits the available experimental data for Nusselt numbers very well. The turbulent Prandtl number then was given by

$$Pr_t = DF/DF_h \tag{24a}$$

At the wall, Equation 24a becomes

$$Pr_t = B^+/(A^+\sqrt{Pr}) \tag{24b}$$

Model for Low Peclet Number Flows

For low Reynolds number flows, say $Re < 10^4$, the influence of a laminar sublayer and buffer zone may be significant and must be taken into account in predicting the Nusselt numbers. The damping effect of the wall also is important. Under such a condition, Lee [25, 26] adopted the well-known three-region logarithmic velocity profile and the universal friction law in his calculations.

For low Peclet number flows, Lee [25, 26] suggested the use of a turbulent Prandtl number given by Azer and Chao [9] along with the eddy viscosity given by Na and Habib [13] to predict the total eddy diffusivity

$$\varepsilon = 1 + Pr\,\varepsilon^+/Pr_t \tag{25}$$

This model can be described as follows:

Velocity profile:

$$u^+ = y^+ \qquad \text{if } 0 \leqslant y^+ \leqslant 5$$

$$u^+ = 5.0\,\ell n\, y^+ - 3.05 \qquad \text{if } 5 \leqslant y^+ \leqslant 30$$

$$u^+ = 2.5\,\ell n\, y^+ + 5.5 \qquad \text{if } 30 \leqslant y^+ \tag{26}$$

In the preceding equations, $u^+ = U/V^*$ and $y^+ = V^*Y/\nu = (Re^*/2)y = Re\sqrt{(f_m/32)}\,y$. At a given Reynolds number, the friction factor f_m can be determined by the universal equation, namely,

$$f_m^{-0.5} = 2.0 \log(Re\, f_m^{0.5}) - 0.8$$

The dimensionless velocity u then can be taken as

$$u = u^+/u_m^+$$

where u_m^+ is the mean value of the dimensionless velocity u^+ over the cross-sectional area of the flow through the tube.

Eddy viscosity:

$$\varepsilon^+ = (\ell^+)^2(du^+/dy^+) \tag{27}$$

where $\ell^+ = (DF)(Re^*/2)(0.4y - 0.44y^2 + 0.24y^3 - 0.06y^4)$

$$DF = 1 - \exp(-y^+/26)$$

Turbulent Prandtl number:

$$Pr_t = \frac{1 + 380\,Pe^{-0.58}\exp(-y^{0.25})}{1 + 135\,Re^{-0.45}\exp(-y^{0.25})} \tag{28}$$

COMPARISONS AMONG MODELS

In this section, comparisons are made among the four models previously described, namely, the models by Azer and Chao [9], Notter and Sleicher [11], Na and Habib [13] and Lee [25]. As an illustration for comparisons, Figure 1 shows turbulent Prandtl numbers predicted by the various models for the case of $Pr = 0.02$ with $Re = 2.5 \times 10^5$. The experimental data from the literature [5, 6] are also plotted for comparisons. All of the predictions and data give large turbulent Prandtl numbers at the wall. It is apparent from Equation 25 that a large turbulent Prandtl number means a small eddy diffusity at given Prandtl and Reynolds numbers. The values by Azer and Chao [9] are seen to be greater than those by Notter and Sleicher [11] over the whole cross-sectional area of the pipe. Therefore, Azer and Chao's model [9] should predict a smaller heat transfer rate. The latter [11] seems to have a good agreement with the data of Brown et al. [6], but it does not necessarily give a better prediction because no consistent values for the turbulent Prandtl number can be found among the experimental data [5, 6]. The model of Na and Habib [13] gives a prediction close to the Reynolds analogy ($Pr_t = 1$) except for the region near the wall where a steep increase in the turbulent Prandtl number is seen. The Reynolds analogy gives the smallest turbulent Prandtl number. It is thought to overstate the eddy diffusivity in liquid metals. Both experimental data [5, 6] indicate that there exists a minimum turbulent Prandtl number near the buffer zone. This finding is different from the theoretical models, and more research is needed in this connection. If comparisons are made at a higher Reynolds number, a decrease in the turbulent Prandtl number will be seen in all of the experimental data and theoretical models.

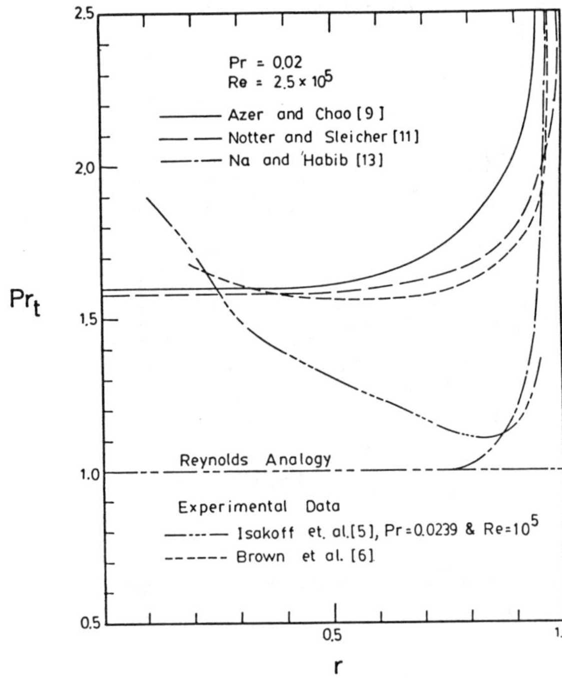

Figure 1. Comparison of various proposed turbulent Prandtl numbers for Pr = 0.02 and Re = 2.5 × 10^5. No consistent values can be found between the experimental data.

After the dimensionless velocity profile and total eddy diffusivity are predicted with the models, the functions f(r) and g(r) can be obtained from Equation 2 and the energy equation 1 may be solved. Lee [25] investigated the effect of axial conduction on a problem with uniform wall temperature. To allow for the axial conduction in the region $x \leqslant 0$, the fluid temperature was taken to be uniform at $x = -\infty$ and the pipe wall was adiabatic in $x \leqslant 0$. Both cases in which the axial conduction effect was included and neglected were solved by the method of separation of variables. The local Nusselt numbers were evaluated from

$$Nu(x) = -2[\partial\theta(x,1)/\partial r]/(\theta_b - \theta_w) \tag{29}$$

The error due to neglect of axial conduction was defined as

$$E = 1 - Nu_{na}/Nu \tag{30}$$

where the subscript *na* denotes "no axial conduction." Figure 2 illustrates the error for the cases of Pr = 0.001 and Pe = 40. It reveals that at small Peclet number, say Pe = 5, neglecting the axial conduction will result in an error of about 45% (low) in the entry region. This error should decrease with an increasing Peclet number; for example, only about a 7% error is obtained at Pe = 100. All the curves converge to zero as x approaches infinity, but to minus infinity as x approaches zero. The latter is due to the boundary condition $\theta(0, r) = 1$ set in the case of no axial conduction. At low Peclet numbers, the maximum error occurs about at X = 0.01 D. It is also seen that the error should increase gradually with increasing Prandtl number (or decreasing Reynolds number) at a certain

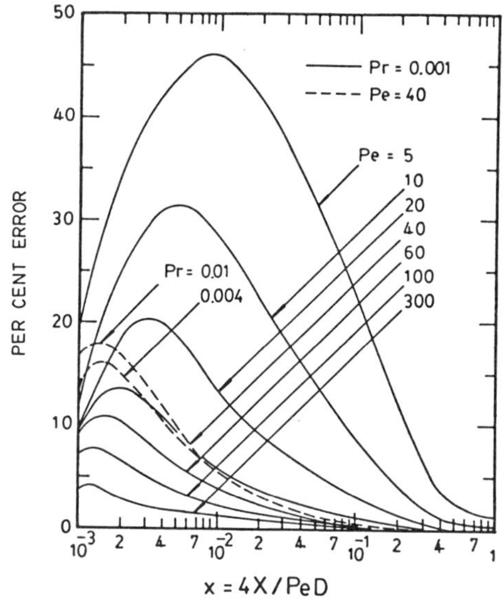

Figure 2. Errors in Nusselt number resulting from neglect of axial conduction. The maximum error is only about 7% at Pe = 100 [25].

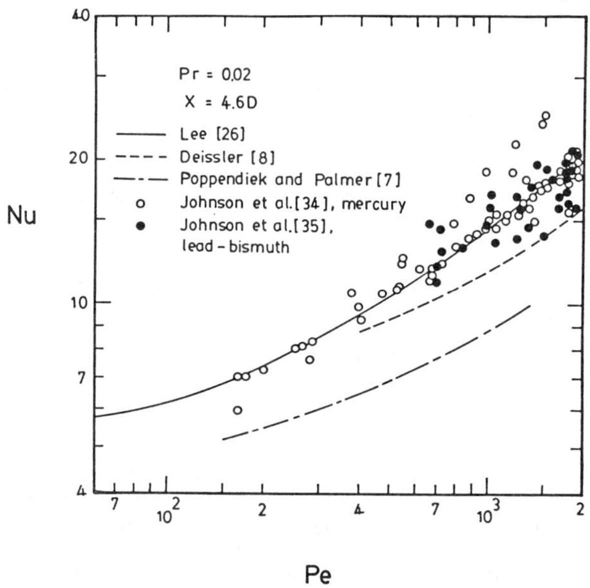

Figure 3. Comparisons of predicted Nusselt numbers with experimental data under a uniform wall flux condition at X = 4.6D [26].

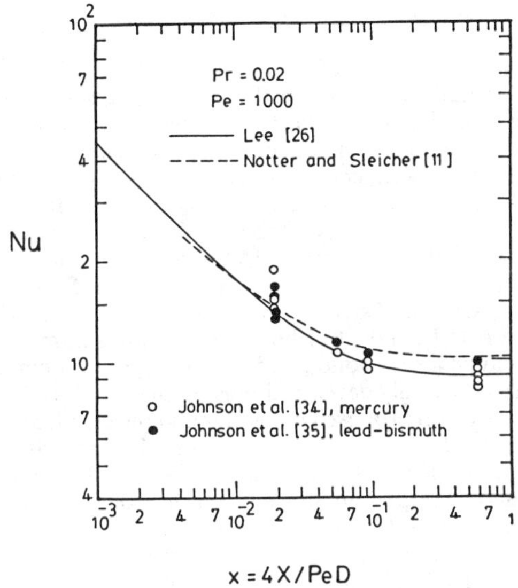

Figure 4. Comparisons of predicted Nu(x) curves with experimental data under a uniform wall flux condition for Pr = 0.02 and Pe = 1,000 [26].

Peclet number, say Pe = 40. Figure 2 indicates that the effect of axial conduction is significant at low Peclet numbers, but negligible when Pe > 100. It also verifies that axial conduction can be neglected in the thermally fully developed region for all Peclet numbers.

Figure 3 reveals comparisons of existing experimental data and theoretical predictions for various Peclet numbers at X = 4.6 D. The pipe wall is at the boundary condition of uniform heat flux. The experimental data for mercury and lead-bismuth were given by Johnson et al. [34, 35]. The fluid bulk temperatures were measured at the inlet and outlet of the test section and the wall temperature was measured at eight stations along the tube. Four of the stations were located at X/D = 4.6, 13.8, 23, and a position near the end of the tube whose length-to-diameter ratio is 73.6. The Prandtl number in both experiments was about 0.02. It is apparent from Figure 3 that the calculated Nusselt number by Lee [26] is in excellent agreement with the experimental data. The agreement is also found to be excellent if comparisons are made at other local stations. Because of the lack of other theoretical works on the local Nusselt number in the entrance region, only the predictions of Poppendiek-Palmer [7] and Deissler [8] are plotted for comparison. The prediction of Poppendiek and Palmer [7] was based on the assumption that the eddy diffusivity is negligible when compared with molecular diffusivity. Their solution is therefore independent of the Prandtl number and is seen to be too low compared with the experimental data. Deissler [8] obtained the solution numerically. To approximate the eddy diffusivity, he proposed a model based on a modification of the Reynolds analogy by allowing for heat transfer by conduction to or from a turbulent particle as it moves radially in the tube. This result is also seen to be low. Note that the Peclet numbers presented in both works are larger than 100 and thus the contribution of axial conduction is insignificant. Comparisons of the Nusselt numbers at various axial locations are illustrated in Figure 4 for the case of Pr = 0.02 and Pe = 1,000. An excellent agreement is observed between the experimental data [34, 35] and the computed result [26]. The prediction of Notter and Sleicher [11] is in good agreement with the results of Lee [26] in the entrance region, but is about 14% higher in the thermally fully developed region. Since the axial conduction is negligible throughout the pipe at Pe = 1,000, the difference is due to the different model used.

Figure 5 shows a comparison between the predicted Nusselt numbers and the experimental data in the thermally fully developed region for the wall condition of uniform heat flux. The available experimental data have Prandtl numbers in the range of 0.015 ~ 0.022, and all of the predictions are at Pr = 0.02. The agreement between the prediction by Lee [26] and the experimental data is excellent. The improvement of Lee's prediction over the early one by Azer and Chao [9] is obvious. In the thermally fully developed region, the axial conduction is negligible at all Peclet numbers. Therefore, this difference is due to the absence of the damping effect of a wall on the mixing length in the work of Azer and Chao [9]. In addition, to obtain a simplified closed form for fully developed Nusselt number, Azer and Chao [9] assumed the dimensionless radial heat flow rate, q/q_w, to be $r^{1.75}$ which is independent of both the Reynolds number and the Prandtl number. These errors should be more pronounced at low Reynolds number, say $Re < 10^5$. Note that according to Lee [26] a fully developed Nusselt number less than 7 at low Peclet numbers is possible. The prediction proposed by Notter and Sleicher [11] is seen to be about 10% to 15% higher than Lee's. This is due to a smaller turbulent Prandtl number predicted in the model of Notter and Sleicher [11]. The results of Na and Habib [13] seem to be in good agreement with those of the present study in a range of Peclet numbers, but their model is practical only when $Re > 10^4$ and $Pr > 0.02$.

For the problem with uniform wall temperature, comparisons between experimental data and theoretical predictions for the fully developed Nusselt numbers are given by Lee [25] as shown in Figure 6. The experimental data of Gilliland et al. [38] is the only one we can find which approximates the uniform wall temperature Graetz problem for liquid metals. In their system, mercury was taken as the working media. Heating was done by dropwise condensation of steam on the outside of a vertical nickel tube having a length-to-diameter ratio of 45. The Nusselt number reported was not for fully developed conditions, but rather

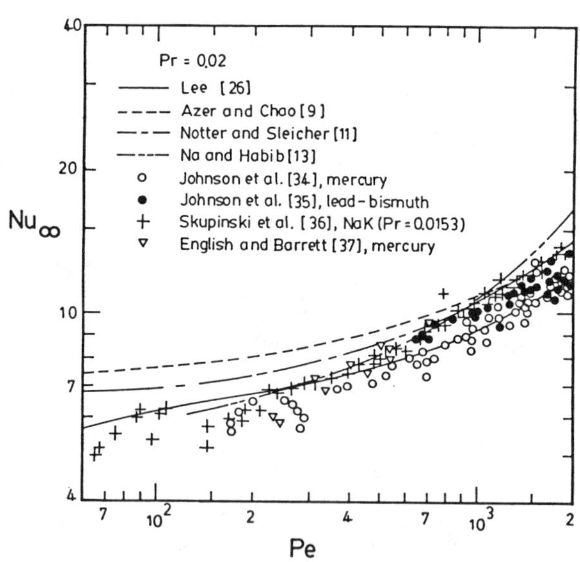

Figure 5. Comparisons of predicted fully developed Nusselt numbers with experimental data under uniform wall flux conditions for Pr = 0.02 [26].

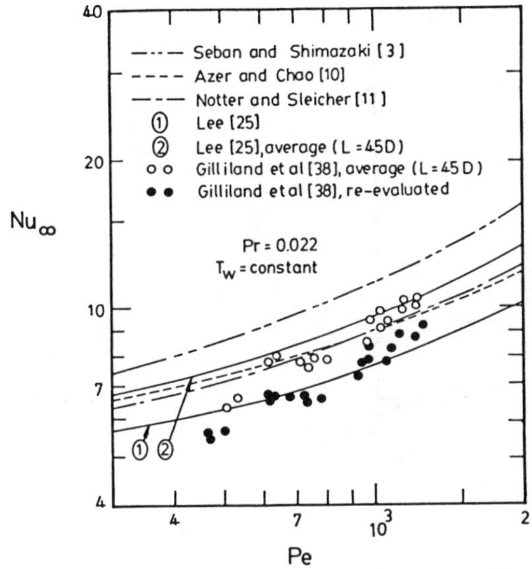

Figure 6. Comparisons of predicted fully developed Nusselt number with experimental data under uniform wall temperature conditions for Pr = 0.022 [25].

an overall average. For purposes of comparison, the prediction by Lee [25] was averaged in the region $0 \leqslant X \leqslant 45\,D$. This result is also plotted in Figure 6. It is apparent that the calculation is in agreement with the data of Gilliland et al. [38]. Nusselt numbers by Lee [25] are compared with the reevaluated experimental data due to Gilliland et al. [38] and obtained from the literature [10]. Again, the agreement is excellent. A recent experiment which was purported to be at uniform wall temperature was done by Sleicher et al. [39]. A high conductivity NaK mixture was used as the working medium and a water-cooled heat exchanger for removing heat was added in the test section in their experiment. This arrangement resulted in a wall condition somewhere between uniform wall temperature and uniform wall flux, especially at high Reynolds number flow. For these reasons, no attempt has been made to compare their data with the predicted results.

The other theoretical predictions of Seban and Shimazaki [3], Azer and Chao [10], and Notter and Sleicher [11] are also plotted in Figure 6 for comparisons. The prediction given by Seban and Shimazaki [3] results in the largest values of fully developed Nusselt numbers in comparison with the others depicted in Figure 6. In that early study, the turbulent Prandtl number was taken to be unity which is invalid for liquid metals. The prediction of Azer and Chao [10] is larger than that of Lee [25]. This difference may be due to the fact that the effects of wall damping and the boundary sublayers were not considered in the work of Azer and Chao [10]. The result of Notter and Sleicher [11] is also seen to be about 20% higher than Lee's. This difference is due to the different models used in the prediction.

RECOMMENDED DATA

Summarized in the following are some of the empirical and semiempirical relations for liquid metal heat transfer in fully developed turbulent flows inside smooth circular pipes under uniform wall heat flux and uniform wall temperature conditions.

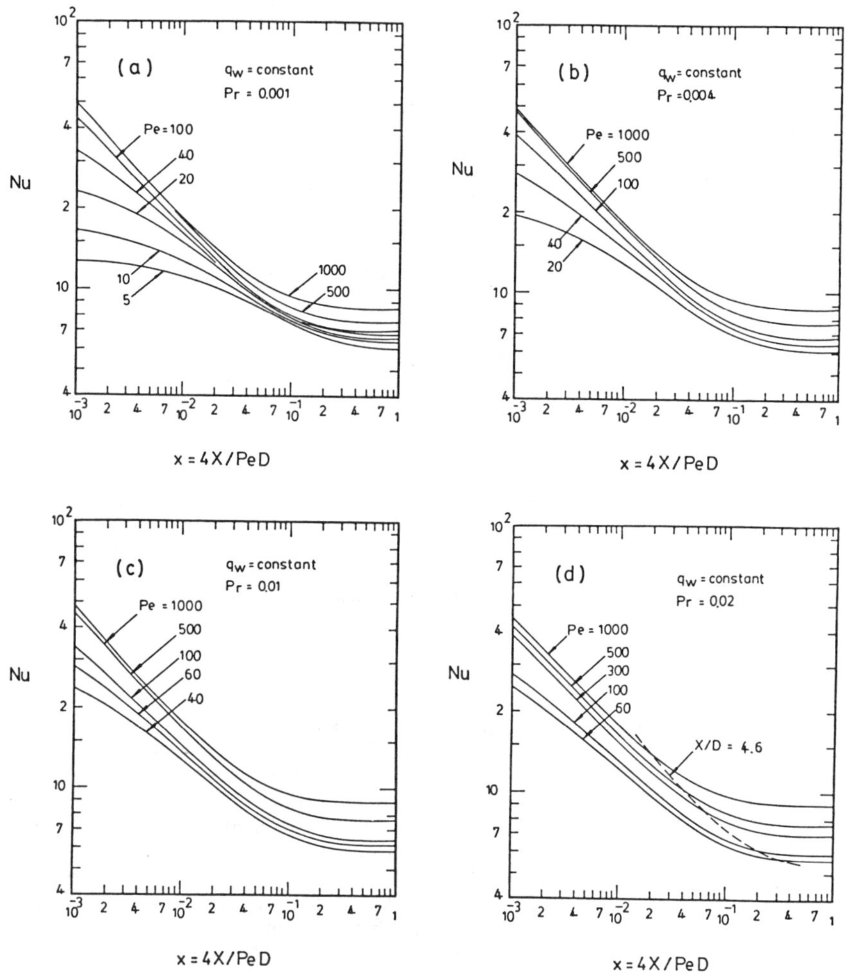

Figure 7. Local Nusselt number cuves Nu(x) under uniform wall flux. The larger the Pe, the higher the Nu value: (A) Pr = 0.001; (B) Pr = 0.004; (C) Pr = 0.01; (D) Pr = 0.02 [26].

Uniform Wall Heat Flux

By taking into account the effects of axial conduction, wall damping factor, and boundary sublayers, Lee [26] gave the correlations of the local Nusselt number vs. dimensionless axial coordinate x for the cases of Pr = 0.001, 0.004, 0.01, and 0.02 as shown on Figure 7A through D. The Peclet numbers considered were up to 1,000. It is seen from these figures that for each pair of Pr and Pe, the value of the Nusselt number approaches asymptotically to a certain constant as x approaches infinity. This value depends on the profile of θ_{fd}. For a given Prandtl number, the flow with larger Peclet numbers yields greater fully developed Nusselt numbers. In the thermal entrance-region, an increase in the Peclet number decreases the effect of axial conduction and the temperature profile tends to be more uniform at x = 0. This results in a higher Nusselt number as x approaches zero. Hence, for a given Prandtl number the Nusselt number curves do not cross each other as shown in each of the four figures. This is different from that of laminar heat transfer [20]. Note that the dimensionless axial coordinate x which is defined in terms of a Peclet number is different from the real axial position X/D. For example, the position X/D = 4.6 shown in Figure 7 D is not a vertical line but rather a curve. It is also noted that the contribution of axial conduction is rather important in the thermal entrance region when $Pe < 100$.

Figure 8 depicts the fully developed Nusselt number vs. Reynolds number with the Prandtl number as a parameter. The dashed lines indicate the variation of the Nusselt number with the Peclet number as a parameter. It is observed that the solid lines coincide with each other as the Reynolds number approaches a small value, say Re = 3,000, and the curves with larger Prandtl numbers have larger Nusselt numbers before they coincide with the other curves. The reason for this is that for a given Reynolds number, eddy diffusivity increases as the Prandtl number is increased, as can be seen from Equation 25, and this will give rise to the Nusselt number. It is noteworthy that for a small Reynolds number, say Re = 4,000, ε^+ is so small that eddy diffusivity is negligible compared to the molecular diffusivity for all Prandtl numbers in the range $0.001 \leqslant Pr \leqslant 0.02$ and therefore, the curves coincide in the low Reynolds number range. When the Reynolds number is less than 2,300, the flow field should be laminar and the resulting Nusselt number is 4.364 for all Prandtl numbers. At a given Peclet number below 450, say Pe = 200, an increase in the Prandtl number results in a decrease of the Reynolds number (since Re = Pe/Pr), which makes the decrease in ε^+ more pronounced than the increase in Pr/Pr_t. This should cause a decrease in eddy diffusivity and thus a decrease in the Nusselt number. Therefore, the curve with Pe = 200 shown in Figure 8 has a positive slope. It is noted that the influence of a laminar sublayer and buffer zone is significant when $Re > 10{,}000$. At high Peclet numbers, say Pe = 1,000, in both cases of Pr = 0.001 and 0.02, the influence of boundary sublayers is insignificant and the eddy viscosity ε^+ is rather large. It means that the dimensionless total eddy diffusivity ε is much larger than unity and eddy diffusivity should increase as the Prandtl number increases. Therefore, the curve with Pe = 1,000 has a negative slope.

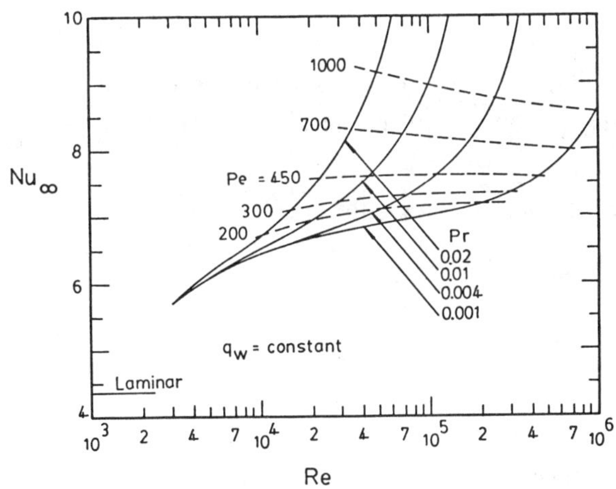

Figure 8. Fully developed Nusselt number vs. Reynolds number under a uniform wall flux condition. The curve with Pe = 100 approximately coincides with the curve for Pr = 0.001 [26].

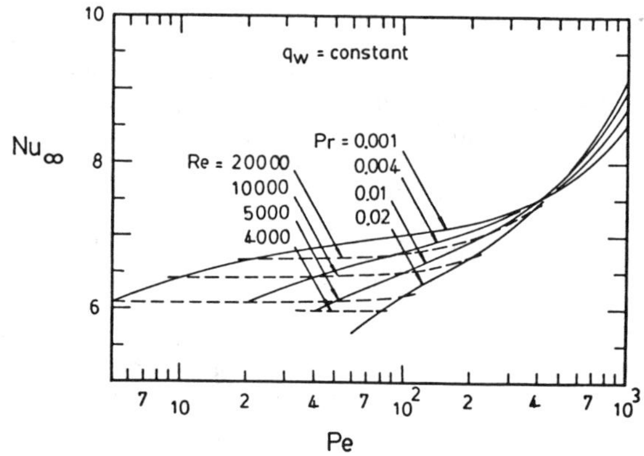

Figure 9. Fully developed Nusselt number vs. Peclet number under a uniform wall flux condition. Nu_∞ depends only on the Reynolds number for Pe < 100 [26].

On the basis of the foregoing discussion, there should exist a critical Peclet number near 450 whose slope is approximately zero. If the same results are depicted in fully developed Nusselt number vs. Peclet number correlations with the Prandtl number as a parameter as shown in Figure 9, a crossover near Pe = 450 (but not exactly) would be observed. The Nusselt number is larger for a larger Prandtl number when Pe > 450 and the difference diverges as the Peclet number approaches infinity. These findings are consistent with those reported by the existing predictions [9, 11]. The dashed lines indicate the variation of the Nusselt number with the Reynolds number as a parameter. It is seen that at a low Reynolds number, say Re = 10,000, an increase in the Prandtl number increases the Nusselt number slightly for Pe > 100. Below Pe = 100, the fully developed Nusselt numbers seem to depend on the Reynolds number only and the following interpolation formula

$$Nu_\infty = 3.01\, Re^{0.0833}$$

fits the calculated data well.

Other expressions developed by various investigators include the following:

- Notter and Sleicher [11, 39] predicted results

$$Nu = 6.3 + 0.0167\, Pe^{0.85}\, Pr^{0.08}$$

for $0.004 < Pr < 0.01$ and $10^4 < Re < 10^6$

$$Nu(X) = Nu_\infty[1 + 2/(X/D)]$$

for $X/D > 4$

- Azer and Chao [9] predicted result

$$Nu_\infty = 7.0 + 0.05\, Pe^{0.77}\, Pr^{0.25}$$

for $Pr < 0.1$ and $2 \times 10^3 < Pe < 1.5 \times 10^4$

- Skupinski [36] experimental data of NaK mixtures

$$Nu_\infty = 4.82 + 0.0185\, Pe^{0.827}$$

for $Pr = 0.0153$, $3.6 \times 10^3 < Re < 9.05 \times 10^5$ and $10^2 < Pe < 10^4$.

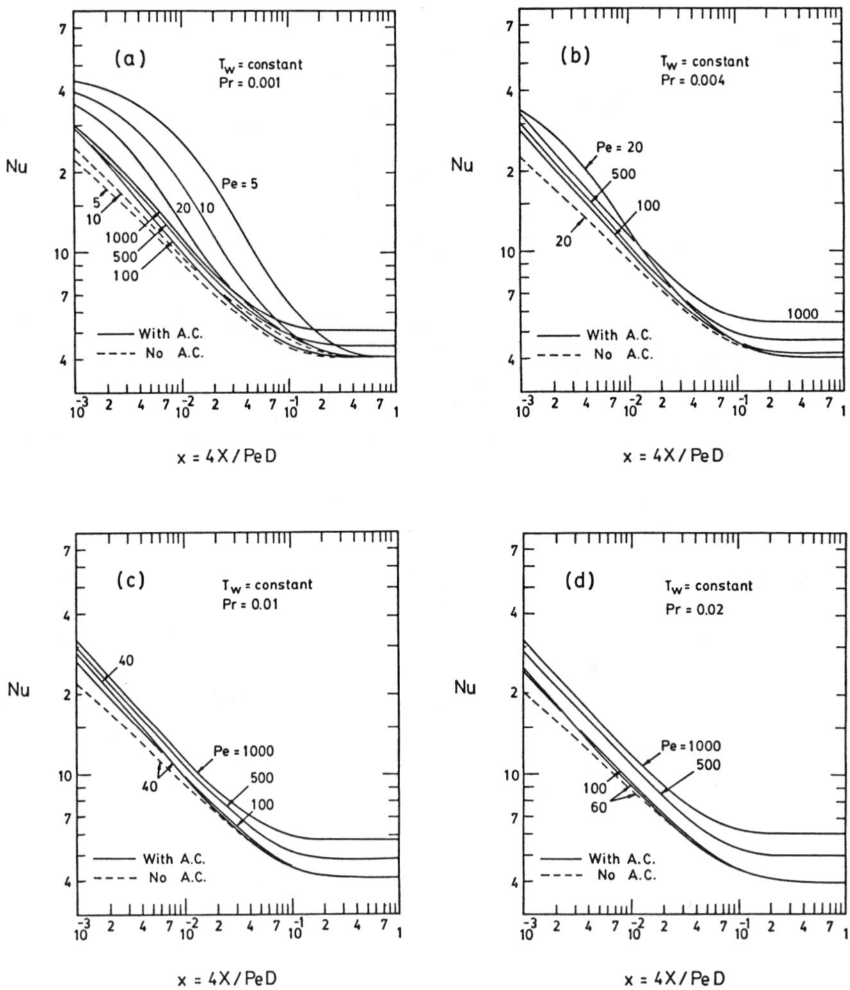

Figure 10. Local Nusselt number curves Nu(x) under uniform wall temperature. The effect of axial conduction is negligible in the thermally fully developed region: (A) Pr = 0.001; (B) Pr = 0.004; (C) Pr = 0.01; (D) Pr = 0.02.

Uniform Wall Temperature

Figures 10 A through D show the predictions by Lee [25] under a uniform wall temperature condition. In the figures, "No A.C." denotes no axial conduction. The ranges of Pr and Pe considered are $0.001 \leqslant \text{Pr} \leqslant 0.02$ and $5 < \text{Pe} < 10^3$. The corresponding thermally fully developed Nusselt numbers are plotted on Figure 11. The trend of the Nu_∞ vs. Pe curves is seen to be similar to the case of uniform wall heat flux Figure 9.

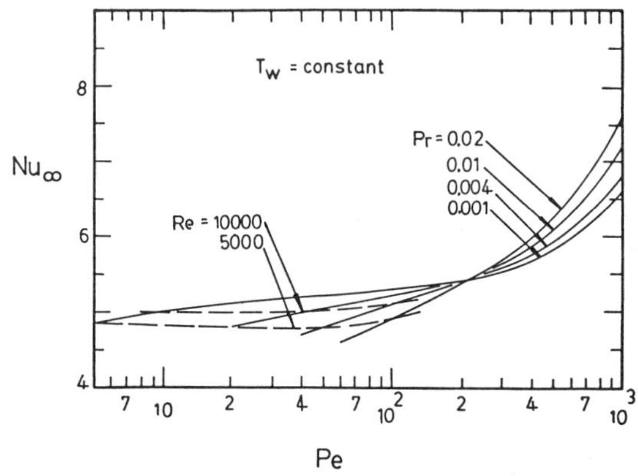

Figure 11. Fully developed Nusselt number vs. Peclet number under a uniform wall temperature condition. Nu depends only on the Reynolds number for Pe < 100 [25].

The available expressions for Nusselt numbers are listed in the following.

- Azer and Chao [10]

$$Nu_\infty = 5.0 + 0.05\,Pe^{0.77}\,Pr^{0.25}$$

for $Pr < 0.1$ and $2 \times 10^3 < Pe < 1.5 \times 10^4$

- Notter and Sleicher [11, 39]

$$Nu_\infty = 4.8 + 0.0156\,Pe^{0.85}\,Pr^{0.08}$$

for $0.004 < Pr < 0.1$ and $10^4 < Re < 10^6$.

$$Nu(X) = Nu_\infty[1 + 2/(X/D)]$$

for $X/D > 4$

- Lee [25]

$$Nu_\infty = 2.77\,Re^{0.0656}$$

for $Pe < 100$

NOTATION

a	radius of the eddy (m)
A^+, B^+	damping constants for eddy viscosity and eddy diffusivity, respectively
Bi	Biot number, h_0D/k
D	inner diameter of the pipe (m)
DF	damping factor for momentum mixing length
DF_h	damping factor for thermal mixing length
f	dimensionless quantity, Equation 2
f_m	friction factor in pipes
g	dimensionless quantity, Equation 2
h	heat transfer coefficient ($W/m\,^\circ C$)
h_0	heat transfer coefficient at the outer surface of the pipe ($W/m^2\,^\circ C$)

k	thermal conductivity of the fluid (W/m °C)
k_m, k_h	mixing-length constants for momentum and heat, respectively
ℓ	mixing-length for eddy viscosity (m)
ℓ^+, ℓ_0^+	dimensionless damped and undamped mixing lengths
Nu	Nusselt number (h D/k)
Pe	Peclet number (VD/α)
Pr, Pr_t	Prandtl and turbulent Prandtl numbers, $Pr=\nu/\alpha$, $Pr_t=\varepsilon_m/\varepsilon_h$
q	radial heat flow rate (W/m^2)
R	cylindrical coordinate in radial direction (m)
r	dimensionless cylindrical coordinate in radial direction, R/R_0
Re, Re*	Reynolds number and shear Reynolds number, $Re=VD/\nu$, and $Re^* = V^* D/\nu$
R_0	inner radius of the pipe (m)
T, T_0, T_∞	temperature, temperature at $X = -\infty$, and the ambient temperature (°C)
t_1, t_2	fluid temperature at layer 1 and 2, respectively (°C)
U	axial velocity (m/sec)
u	dimensionless axial velocity, $u = U/V = u^+/u_m^+$
u_1, u_2	axial velocity at layer 1 and 2, respectively (m/sec)
u^+	dimensionless velocity, U/V^*
u_m^+	mean value of u^+ over the cross-sectional area of the flow
V	mean velocity (m/sec)
V^*	shear velocity, $V\sqrt{(f_m/8)}$
v'	fluctuating velocity in radial direction (m/sec)
X	cylindrical coordinate in axial direction (m)
x	dimensionless cylindrical coordinate in axial direction, $x = 4X/PeD$
Y, y, y^+	defined as $Y = R_0 - R$, $y = Y/R_0$ and $y^+ = V^*Y/\nu$

Greek Symbols

α	thermal diffusivity (m^2/sec)
$\varepsilon, \varepsilon_x$	dimensionless total eddy diffusivity in r and x directions, respectively, $\varepsilon = 1 + \varepsilon_h/\alpha$ and $\varepsilon_x = 1 + (\varepsilon_h)_x/\alpha$
ε_m	radial eddy viscosity (m^2/sec)
ε^+	dimensionless radial eddy viscosity (ε_m/ν)
$\varepsilon_h, (\varepsilon_h)_x$	radial and axial eddy diffusivities (m^2/sec)
θ	dimensionless temperature, Equation 3
λ	friction factor at the eddy surface, Equation 9b
ν	kinematic viscosity (m^2/sec)
σ	ratio of axial to radial total eddy diffusivity, $\varepsilon_x/\varepsilon$

Subscripts and Superscripts

b	bulk
e	eddy
fd	fully developed
h	heat
m	momentum
max	maximum
w	wall
∞	fully developed region, $x = \infty$
′	prime, refer to eddy

REFERENCES

1. Martinelli, R. C., "Heat Transfer to Molten Metals," *J. Heat Transfer*, Vol. 69 (1947), pp. 947–959.
2. Lyon, R. N., "Liquid Metal Heat Transfer Coefficients," *Chem. Engng. Progr.*, Vol. 47 (1951), pp. 75–79.

3. Seban, R. A., and Shimazaki, T. T., "Heat Transfer to a Fluid Flowing Turbulently in a Smooth Pipe with Wall at Constant Temperature," *J. Heat Transfer*, Vol. 73 (1951), pp. 803.
4. Lubarsky, B., and Kaufman, S. J., *Review of Experimental Investigations of Liquid Metal Heat Transfer*, NACA TN 3336, 1955.
5. Isakoff, S. E., and Drew, T. B., "Heat and Momentum Transfer in Turbulent Flow of Mercury," *General Discussion on Heat Transfer*, London: Inst. Mech. Engrs., 1951, pp. 405–409.
6. Brown, H. E., *et al.*, "Temperature and Velocity Distribution of Heat in a Liquid Metal," *J. Heat Transfer*, Vol. 79 (1957), pp. 279–285.
7. Poppendiek, H. F., and Palmer, L. D., *Forced Convection Heat Transfer in Thermal Entrance Regions*, Part II, ORNL 914, Reactor Exp. Eng. Div., Oak Ridge Nat. Lab., May 1952.
8. Deissler, R. G., *Analysis of Turbulent Heat Transfer and Flow in Entrance Regions of Smooth Passages*, NACA TN 3016, 1953.
9. Azer, N. Z., and Chao, B. T., "A Mechanism of Turbulent Heat Transfer in Liquid Metals," *Int. J. Heat Mass Transfer*, Vol. 1 (1960), pp. 121–138.
10. Azer, N. Z., and Chao, B. T., "Turbulent Heat Transfer in Liquid Metals-Fully Developed Pipe Flow with Constant Wall Temperature," *Int. J. Heat Mass Transfer*, Vol. 3 (1961), pp. 77–83.
11. Notter, R. H., and Sleicher, C. A., "A Solution to the Turbulent Graetz Problem – III. Fully Developed and Entry Region Heat Transfer Rates," *Chem. Engng. Sci.*, Vol. 27 (1972), pp. 2073–2093.
12. Notter, R. H., and Sleicher, C. A., "The Eddy Diffusivity in the Turbulent Boundary Layer near a Wall," *Chem. Engng. Sci.*, Vol. 26 (1971), pp. 161–171.
13. Na, T. Y., and Habib, I. S., "Heat Transfer in Turbulent Pipe Flow Based on a New Mixing Length Model," *Appl. Sci. Res.*, Vol. 28 (1973), pp. 302–314.
14. Cebeci, T., *A Model for Eddy Conductivity and Turbulent Prandtl Number*, Rept. No. MDC-JO 747101. Long Beach, California: Douglas Aircraft Company, May 1970.
15. Nikuradse, J., *Gesetzmäßigkeit der Turbulenten Strömung in Glatten Röhren*, Forsch. Arb. Ing-Wes., No. 356, 1932.
16. Meek, R. L., "Mean Period of Fluctuation near the Wall in Turbulent Flow," *AIChE. J.*, Vol. 18 (1972), pp. 854.
17. Thomas, L. C., "Temperature Profiles for Liquid Metals and Moderate-Prandtl Number Fluids," *J. Heat Transfer*, Vol. 92 (1970), pp. 565.
18. Rajagopal, R., and Thomas, L. C., "The Formulation of Relationships for ε_M and ε_H Based on a Physically Realistic Model of Turbulence," *Int. Heat Transfer Conf.*, Tokyo, 1974.
19. Thomas, L. C., "A Formulation for ε_M and ε_H Based on the Surface Renewal Principle," *AIChE. J.*, Vol. 24 (1978), pp. 101–106.
20. Hsu, C. J., "An Exact Analysis of Low Peclet Number Thermal Entry Region Heat Transfer in Transversely Nonuniform Velocity Fields," *AIChE. J.*, Vol. 17 (1971), pp. 732–740.
21. Tan, C. W., and Hsu, C. J., "Low Peclet Number Mass Transfer in Laminar Flow Through Circular Tubes," *Int. J. Heat Mass Transfer*, Vol. 15 (1972), pp. 2187–2201.
22. Lee, S. L., and Hwang, G. J., "Solidification of a Low Peclet Number Fluid Flow in a Round Pipe with the Boundary Condition of the Third Kind," *Can. J. Chem. Engng.*, Vol. 58 (April 1980), pp. 177–184.
23. Lee, S. L., and Hwang, G. J., "Finite Element Solution of Low Peclet Number Fluid Flow in a Round Pipe with the Cauchy Boundary Condition," *Can. J. Chem. Engng.*, Vol. 59 (December 1981), pp. 760–765.
24. Schneider, P. J., "Effect of Axial Conduction on Heat Transfer in the Entrance Regions of Parallel Plates and Tubes," *Heat Transfer Fluid Mech. Institute*. Stanford: Stanford Univ. Press, 1956, pp. 41.

25. Lee, S. L., "Forced Convection Heat Transfer in Low Prandtl Number Turbulent Flows: Influence of Axial Conduction," *Can. J. Chem. Engng.*, Vol. 60 (August 1982), pp. 482–486.
26. Lee, S. L., "Liquid Metal Heat Transfer in Turbulent Pipe Flow with Uniform Wall Flux," *Int. J. Heat Mass Transfer*, Vol. 26 (1983), pp. 349–356.
27. Bird, R. B., *et al.*, *Transport Phenomena*, New York: John Wiley and Sons, 1960, pp. 295–296.
28. Prandtl, L., "Über die Ausgebildete Turbulenz," *Z. Angew. Math. u. Mech.*, Vol. 5 (1925), pp. 136. Also, translated as *Investigations on Turbulent Flow*, NACA TM 1231, 1949.
29. Laufer, J., *The Structure of Turbulence in Fully Developed Pipe Flow*, NACA TN 2954, 1953.
30. Sleicher, C. A., "Experimental Velocity and Temperature Profiles for Air in Turbulent Pipe Flow," *ASME Trans.*, Vol. 80 (1958), pp. 693.
31. von Kármán, T., "The Analogy Between Fluid Friction and Heat Transfer," *ASME Trans.*, Vol. 61 (1939), pp. 705.
32. Hill, J. C., "Estimation of Eddy Diffusivities in Isotropic, Turbulence," *64th Annual AIChE Meeting*, San Francisco, 1972.
33. van Driest, E. R., "On Turbulent Flow Near a Wall." *J. Aeronaut. Sci.*, Vol. 23 (1956), pp. 1007.
34. Johnson, H. A., *et al.*, *Heat Transfer to Mercury in Turbulent Pipe Flow*, California: Inst. Eng. Res., July 1953.
35. Johnson, H. A., *et al.*, "Heat Transfer to Molton Lead-Bismuth Eutectic in Turbulent Pipe Flow," *J. Heat Transfer*, Vol. 75 (1953), pp. 1191–1198.
36. Skupinski, E., *et al.*, "Determination des Coefficients de Convection d'un Alliage Sodium-Potassium dans un Tube Circulaire," *Int. J. Heat Mass Transfer*, Vol. 8 (1965), pp. 937–951.
37. English, D., and Barrett, T., "Heat Transfer Properties of Mercury," *J. Heat Transfer*, Vol. 73 (1951), pp. 458–460.
38. Gilliland, E. R., *et al.*, "Heat Transfer to Mercury." In *General Discussion on Heat Transfer*, London: Inst. Mech. Engrs., 1951, pp. 402–404.
39. Sleicher, C. A., "Temperature and Eddy Diffusivity Profiles in NaK," *Int. J. Heat Mass Transfer*, Vol. 16 (1973), pp. 1565–1575.

CHAPTER 13

MIXED CONVECTION IN BUOYANT PLUMES

Noor Afzal

Department of Mechanical Engineering
Aligarh Muslim University
Aligarh 202001, India

CONTENTS

INTRODUCTION

Classification of Buoyant Motion

The fluid motion driven by a source of buoyancy within a large boundary-free quiescent medium forms an important class of natural convection flows. The motions are primarily vertical, produced under the action of gravity by the density contrast between the source and its environment. Many types of flows can be identified depending on characteristics of the source and the environment into which they discharge.

The motion produced by a steady source of heat is called a plume; the buoyant region is continuous between the source and level of interest. When buoyancy is suddenly released the motion produced is known as thermal; the buoyancy is confined to a limited volume of fluid, which on rising, loses its connection with the source producing it. A starting plume is the flow generated when buoyancy is suddenly turned on and maintained steadily. The flow pattern at the front is similar to that of thermal, and well behind the front it is just similar to a steady plume. In a buoyant vortex ring the input of circulation at the source is also specified in addition to buoyancy. In a thermal the circulation is generated entirely by buoyancy: a thermal released from rest can be regarded as a special case of a buoyant vortex ring. In a plume the momentum flux is balanced by the buoyancy which produces it. If the momentum flux exceeds the buoyancy flux the flow is known as a buoyant jet, and it is known as a nonbuoyant jet if buoyancy is zero. As a buoyant jet develops, the momentum decreases more rapidly than in a plume; it will behave like a plume when an appropriate level of momentum is reached. The term "forced plume" is used for the motion generated by a finite source which emits fluxes of mass and momentum as well as buoyancy. This includes the plume and nonbuoyant jet as special cases. In this section, a brief description of natural convection laminar and turbulent plumes in a uniform environment is presented. They form one of the asymptotic regions in the mixed convection flow. The plume can be two-dimensional or axisymmetric. The idealized problem is that of the flow arising from a horizontal line or a point heat source in a uniform environment. The idealized flow will be approached from the natural convection arising from a long tube/wire or a sphere. For finite-shaped bodies in natural convection the heated fluid eventually leaves the body and rises freely in the environment and far away from the body the flow will approach that of a point heat source. Bodies of large aspect ratio give rise to a two-dimensional plume. In practice, the two-dimensional plume generated by a finite-sized heat source, sufficiently far away, is found to approach an axisymmetric plume.

The study of plumes is of interest in fire research, meteorology, industrial processes, discharge of technological waste, etc. In the industrial processes such as heat treatment, forging, foundry, and welding, the heat input gives rise to a buoyant plume. In the metallurgical industries there are open-heated surfaces (kettles of molten metal in the lead industry, electrolytic cells in zinc or copper refinery, etc.) at the floor level where plumes are formed that rise toward the roof. In technology, the discharge of the thermal energy and chemical waste products rejected to the atmosphere and ocean also gives rise to plumes. In natural convection heat transfer, the fluid particles in the boundary layer, after separating from the vicinity of the heated surface rise freely in the ambient medium to form a plume.

Mixed Convection Mechanism

In many situations an externally imposed flow exists in addition to natural convection from the source. In such flows both forced and free convection modes are present and the phenomena is known as mixed or combined convection. The mixed convection problems are of considerable interest in nature as well as in technology. These include, for example, thermal pollution, dispersion of pollutants, hot wire anemometry, hot film anemometry, flow past heated bodies, and spreading of fire. The dispersion of pollutants requires a study of buoyant motion in the moving evironment as the sources discharge their effluents at a temperature considerably higher than that of the environment. In transportation vehicles such as automobiles, aircrafts and submarines the source of buoyancy is movement through the ambient environment and the situation is again that of mixed convection.

If the external flow velocity is much larger than the characteristic natural convection velocity associated with the heat source, the flow may be termed as weakly buoyant plume. Here the basic problem is that of forced convection around the source where Oseen-linearization is valid and the buoyancy effects can be taken as perturbations on forced convection. If the external stream is weak and natural convection predominates, the flow is called strongly buoyant plume. Here the basic flow is that of a natural convection buoyant plume, and forced convection can be regarded as small perturbations. However, when the convective process is neither predominately forced nor predominately natural convection, both the modes are significant and one has to consider the full nonlinear equations. It is particularly interesting to know when a flow can be treated as forced or natural and or when neither of these modes can be neglected in analyzing for mixed or combined convection.

In mixed convection the direction of the forced stream with respect to the buoyant force vector plays a significant role. If the direction of the main stream and buoyant force vector are the same, the two convective processes assist each other, resulting in aiding flow, and the velocity in the plume increases. When the buoyant force vector and an oncoming stream are in opposite directions, the two convective processes oppose each other, resulting in an opposing flow situation, and the velocity in the plume decreases. In the opposing flow situation the buoyancy will bring the oncoming stream to rest at a certain distance from the heat source and reverse flow farther away. When incident velocity is inclined at an angle with respect to a buoyant force vector, the incident stream can be resolved into two components, parallel and normal to the buoyant force vector. The parallel component of oncoming streams acts on a buoyant plume very much in the spirit of aiding or opposing depending on whether the parallel component and buoyant force vector are in the same or opposite directions. The normal component of the velocity acts as cross flow on the plume which deflects the plume from its vertical position.

The mechanics of mixed convection from the heat source has come under active investigation relatively recently. Mixed convection in turbulent plumes has been almost exclusively studied for a point heat source subjected to horizontal stream (owing to its importance in dispersion of smoke emitted by a chimney). Since on this aspect several reviews are available [1–4], this chapter concerns itself mainly with laminar flows.

The equations governing the mixed convection in laminar buoyant plume under the Boussinesq approximation are

$$\operatorname{div} \underset{\sim}{u} = 0 \tag{1}$$

$$\underset{\sim}{u} \cdot \operatorname{grad} \underset{\sim}{u} = -\rho^{-1} \operatorname{grad} p + \nu \nabla^2 \underset{\sim}{u} + (e_x \cos\varphi + e_y \sin\varphi)\, g\beta (T - T_0) \tag{2}$$

$$C_p \underset{\sim}{u} \cdot \operatorname{grad} T = K \nabla^2 T \tag{3}$$

The coordinate x, measured from the heat source, is in the direction of the imposed forced stream that makes an angle φ in an anticlockwise direction with the vertical, and coordinate y is normal to it.

Here e_x and e_y = the unit vectors in x and y directions, respectively
$\underset{\sim}{u}$ = the velocity vector where components in x and y directions are u and v
p = the static pressure
T = the temperature
g = the gravitational acceleration
β = the volumetric expansion coefficient
ν = the molecular kinematic viscosity
K = the molecular thermal conductivity
C_p = the specific heat

The boundary conditions at infinity are those of prescribed free stream velocity U_∞ and temperature T_∞.

For $\varphi = 0$ and π, the imposed external stream is in the vertical direction. The external stream and buoyant force vector are in the same direction for $\varphi = 0$ and in opposite directions for $\varphi = \pi$. For the two cases the boundary-layer equations for mixed convection to a horizontal line or point source of heat subjected to a vertical free stream are given by

$$\frac{\partial u}{\partial x} + \frac{\partial v}{\partial y} + j\,\frac{v}{y} = 0 \tag{4}$$

$$\rho\left(u\,\frac{\partial u}{\partial x} + v\,\frac{\partial u}{\partial y}\right) = \frac{\partial \tau}{\partial y} + j\,\frac{\tau}{y} \pm \rho g \beta (T - T_\infty) \tag{5}$$

$$\rho C_p\left(u\,\frac{\partial T}{\partial x} + v\,\frac{\partial T}{\partial y}\right) = \frac{\partial q}{\partial y} + j\,\frac{q}{y} \tag{6}$$

where index $j = 0$ for a two-dimensional plume and $j = 1$ for an axisymmetric plume. The + sign corresponds to the case where buoyancy and forced convection effects are in the same direction (aiding flow situation) and the − sign to the case where buoyancy and forced convection are in opposite directions (opposing flow situation). The quantities τ and q are the appropriate shear stress and heat conduction at any point. For laminar flow τ and q are given by molecular transport as

$$\tau = \rho\nu \frac{\partial u}{\partial y}, \qquad q = K \frac{\partial T}{\partial y} \tag{7a}$$

The molecular Prandtl number Pr is defined as

$$Pr = \rho\nu C_p/K \tag{7b}$$

The velocity and temperature distributions must be symmetric with respect to the vertical axis

$$y = 0 \qquad v = \frac{\partial u}{\partial y} = \frac{\partial T}{\partial y} = 0 \tag{8}$$

Sufficiently far away ∞, from the heat source, the velocity and temperature are those of ambient fluid

$$y \to \infty, \qquad u \to U_\infty, \quad T - T_\infty \to 0 \tag{9}$$

The integral of the energy equation within a large control volume enclosing the heat source leads to

$$Q = 2\pi^j \rho C_p \int_0^\infty u(T - T_\infty) y^j dy \tag{10}$$

where Q is the heat released by the source. For a turbulent plume, τ and q are the appropriate Reynolds stress and heat flux. Based on the eddy viscosity closure model, their expressions are

$$\tau = \rho\nu_\tau \frac{\partial u}{\partial y}, \qquad q = K_\tau \frac{\partial T}{\partial y} \tag{11a}$$

where ν_τ and K_τ are eddy viscosity and conductivity and their values depend on flow rather than fluid. The turbulent Prandtl number Pr_t is defined as

$$Pr_t = \rho\nu_\tau C_p/K_\tau \tag{11b}$$

The boundary conditions at the center line and at the edge of the plume require τ and q to vanish there.

BUOYANT PLUME

Simple Analysis

Some insight into the structure of the free convection plume ($U_\infty = 0$) may be gained from the simple analysis now described. As plume develops it spreads outward in the downstream direction because the fluid particles are entrained in the plume owing to the friction at the boundaries. The equation of conservation of mass (Equation 4) applied to a control volume $y = 0$, $y = \delta(x)$, $x = x_0$ and $x = x_0 + \Delta x$ gives

$$V_E \delta^j = \frac{d}{dx} \int_0^\delta u y^j dy = \frac{d}{dx}(U\delta^{j+1}) \tag{12}$$

where V_E is the inward normal velocity component at the edge of the plume and U is the local mean streamwise velocity.

Laminar Plume

A dimensional argument relates U, δ, ν, x, and Pr as

$$\delta^2 = \frac{\nu x}{U} k(Pr) \tag{13}$$

Based on Equation 13 the relation in Equation 12 leads to $V_E \sim \nu/\delta$, and without loss of generality Equation 12 may be written as

$$\frac{d}{dx}(U\delta^{j+1}) = E(Pr)\nu\delta^{j-1} \tag{14}$$

where E (Pr) is the constant of proportionality known as the entrainment rate scale. For an axisymmetric plume (j = 1) Equation 14 was first proposed by Morton [5].

The equations for conservation of momentum (Equation 5) and buoyancy (Equation 6), using an idealized top hat profile (constant inside the plume and zero outside) can be reduced to

$$\frac{d}{dx} U^2\delta^{1+j} = B\delta^{1+j} \tag{15}$$

$$UB\delta^{1+j} = q \tag{16}$$

$$q = \frac{g\beta Q}{2\pi^j \rho C_p}, \quad B = g\beta(T_c - T_\infty) \tag{17}$$

Equation 16 expresses the fact that the total thermal energy in the plume remains constant. The conservation equations (Equations 14, 15 and 16) determine the gross parameters U, δ, and B for the two-dimensional (j = 0) and axisymmetric (j = 1) plumes.

For two-dimensional plume the solution is

$$\delta = \left(\frac{100}{27}\frac{E^3\nu^3}{q^3}\right)^{1/5} x^{2/5}$$

$$U = \left(\frac{15}{16}\frac{q^2}{\nu^2}\right)^{1/5} x^{1/5}$$

$$B = \left(\frac{36}{125}\frac{q^4}{E^2\nu^2}\right)^{1/5} x^{3/5} \tag{18a}$$

The mass flow rate in the plume

$$\dot{m} = U\delta = \left(\frac{125}{36} E^2\nu^2 q\right)^{1/5} x^{-3/5} \tag{18b}$$

increases with height above the line source as $x^{3/5}$ and also with heat input as $q^{1/5}$. As x increases the width of the plume increases as $x^{2/5}$, center line temperature decreases as $x^{-3/5}$ and velocity increases as $x^{1/5}$. The Grashof number of the plume

$$
\begin{aligned}
Gr_x^* &= \frac{g\beta\Delta T x^3}{\nu^2} \\
&= \left(\frac{36}{125} E^{-2} \frac{q x^3}{\nu^2}\right)^{1/5}
\end{aligned}
\tag{19}
$$

increases rapidly with x, and at a certain height from the source, the plume will eventually become turbulent.

For an axisymmetric plume (j = 1) the solution is

$$
\begin{aligned}
&\delta = (E^3\nu^3/q)^{1/4}x^{1/2} \\
&U = (q/\nu E)^{1/2}, \; B = q\,(E\nu x)^{-1}
\end{aligned}
\tag{20a}
$$

and mass flow rate in the plume is

$$\dot{m} = 2\pi U\delta^2 = 2\pi E\nu x \tag{20b}$$

The preceding relations give the variation of the physical parameters. The velocity in the axial direction is independent of height. The width of the plume increases with x and $x^{1/2}$ and the plume is paraboloid in shape. The numerical value of the entrainment constant E can be estimated by comparison with the exact solution described in the next section. Alternately, an additional relation for E may be obtained by considering the equation for kinetic energy. However, for present purposes it is sufficient to note the variation of plume characteristics with respect to x that forms the basis of the similarity analysis described in the next sections.

Turbulent Plume

When a plume becomes turbulent, eddy transport predominates over molecular transport. The simplest approach to describe the turbulent flow is to replace the molecular transport coefficient by a virtual eddy transport coefficient, and Equation 14 for entrainment can be expressed as

$$\frac{d}{dx}(U\delta^{j+1}) = 2eU\delta^j \tag{21}$$

$$e = E\,(Pr_t)/R_E, \; R_E = U\delta/\nu_\tau$$

where ν_τ = the eddy viscosity
Pr_t = the turbulent Prandtl number
R_E = the eddy Reynolds number

For a given free turbulent flow, the eddy Reynolds number and turbulent Prandtl number are universal numbers [6] and the entrainment coefficient e is a constant. The entrainment relation (Equation 21) implies that the mean lateral inflow velocity at the edge of the plume is proportional to U, and was first proposed by Morton, Taylor, and Turner [9].

For a turbulent plume the equations of conservation of momentum and buoyancy for the top hat profile remain the same as Equations 15 and 16, which along with Equation 21 determine the gross parameters δ, U and B for turbulent plumes. The solution to Equations 15, 16 and 21 for the two-dimensional (j = 0) turbulent plume is

$$\begin{aligned} \delta &= 2ex \\ U &= (qe^{-1}/2)^{1/3} \\ B &= (qe^{-1}/2)^{2/3} \\ \dot{m} &= U\delta = (4e)^{1/3}q^{1/3}x \end{aligned} \tag{22}$$

and for an axisymmetric (j = 1) turbulent plume

$$\begin{aligned} \delta &= 6ex \\ U &= \frac{5}{6e}\left(\frac{9}{10}eq\right)^{1/3}x^{-13} \\ B &= \frac{5q}{6e}\left(\frac{9}{10}eq\right)^{-1/3}x^{-5/3} \\ m &= U\delta^2 = 30\left(\frac{9}{10}\right)^{1/3}e^{4/3}q^{1/3}x^{5/3} \end{aligned} \tag{23}$$

is the solution of Morton, Taylor, and Turner [7] who adopted e = 0.085 [4]. It may be noted that under Equations 22 and 23 the two-dimensional and axisymmetric plumes are self-similar.

Two-Dimensional Plumes

Laminar Plume

The flow arising from the horizontal thermal source has been of interest to several investigators. Zeldovich [8] was the first to consider the similarity approach. Schuh [9] gave a similarity analysis for the flow and presented results for the Prandtl number Pr = 0.7. The closed-form solutions for Pr = 5/9 and 2 were given by Yih [10] and Fuji [11, 12]. Yih and Gupta [13] have developed an approximate solution for Prandtl numbers in the neighborhood of 5/9 that gives good results for air, Pr = 0.73. The solutions for a wide range of Prandtl numbers have been studied by Fuji [11], Fuji, Morioka, and Uehara [12], and Gebhart, Pera, and Schorr [14]. For large Prandtl numbers Spadding and Cruddace [15] and for small and large Prandtl numbers Kuiken and Rotem [16] have analyzed the flow.

For two-dimensional plumes the dependence of the velocity, temperature, and transverse-length scales on vertical coordinates, strength of the heat source, and fluid properties were determined earlier. Based on these scales the similarity variables may be taken as

$$\begin{aligned} \eta &= Gr_x^{1/5}y/x \\ \psi &= \nu Gr_x^{1/5}f(\eta),\ T - T_\infty = \theta_T Gr_x^{-1/5}h(\eta) \\ \theta_T &= \frac{Q}{\rho C_p \nu I},\ Gr_x = g\beta\theta_T x^3/\nu^2 \end{aligned} \tag{24}$$

where I is a constant number related to the fact that heat flux in the plume is constant, defined as

$$\int_{-\infty}^{\infty} f'h\,d\eta = I \tag{25}$$

In terms of the previous similarity variables, the two-dimensional laminar boundary layer equations (Equations 4 through 10) reduce to

$$5f''' + 3ff'' - f'^2 + h = 0 \tag{26}$$

$$5h'' + 3\Pr(fh)' = 0 \tag{27}$$

The symmetry of the flow along the centerline from Equation 6 requires

$$f(0) = f''(0) = h'(0) = 0 \tag{28}$$

The boundary conditions (Equations 28) of the quiescent ambient medium are,

$$f'(\infty) = h(\infty) = 0 \tag{29}$$

Equation 27 is solved subject to the second and third conditions of Equation 28 as

$$h = h(0)\exp\left(-\frac{3}{5}\Pr\int_0^{\eta} f\,d\eta\right) \tag{30}$$

As f is positive the solution of Equation 30 yields $h(\infty) = 0$. Thus the last of the conditions of Equation 29 given by physical consideration are not independent. Further, the homogeneous equations (Equations 26 and 27) are subjected to homogeneous boundary conditions (Equations 28 and 29). It is well-known that homogeneous equations under homogeneous boundary conditions possess the unique trivial solution and nontrivial eigen solutions.

The eigen solution can be determined but its multiplicative constant remains unspecified. In the literature, there has been some controversy over the manner in which the multiplicative constant is to be determined. In fact, the multiplicative constant can be determined by specifying at any desired location, any of the functions: velocity, temperature, heat flux, etc. It is more convenient to specify the functions at the boundaries: $f'(0)$, $h(0)$, $f(\infty)$, or I. Schuh [9]; Splading and Cruddace [15]; Fuji [11]; and Fuji et al. [12] have adopted $I = 1$, Gebhart et al. [14] and Jaluria [17] as $h(0) = 1$, and Brand and Lahey [18] as $f(\infty) = 1$. If desired, the condition $f'(0) = 1$ can as well be employed. All these normalization conditions are various versions of the eigen constant, and the solutions are interconvertible by a simple change of variables. However, when Equations 26 and 27 are numerically integrated some of the normalization conditions have a slight edge over the others. If $I = 1$ or $f(\infty) = 1$ is adopted, the solution of Equations 26–29 require the guessing of the two missing boundary conditions; namely, $f'(0)$ and $h(0)$. On the other hand when the normalization condition $f'(0) = 1$ or $h(0) = 1$ is adopted, then only one of the boundary conditions is missing that requires guessing. The solution obtained with any one of the normalization conditions can be transformed to other normalizations by a simple change of variables [13]. In the present review we adopt $I = 1$. The velocity and temperature profile for various values of the Prandtl number are presented in Figures 1 and 2, respectively.

The closed-form solutions for certain values of the Prandtl number can be obtained. For $\Pr = 2$ the solution of the Fuji [11] is

$$f' = a\,\mathrm{sech}^2\left[\left(\frac{3a}{10}\right)^{1/2}\eta\right], \quad h = \frac{4}{5}a^2\mathrm{sech}^4\left[\left(\frac{3a}{10}\right)^{1/2}\eta\right] \tag{31a}$$

where $a = f'(0)$ is given by

$$a = 1/4(15/2)^{3/5} \simeq 0.837 \tag{31b}$$

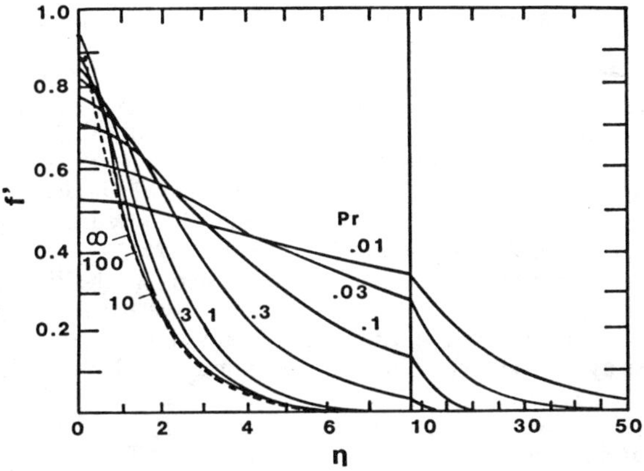

Figure 1. Dimensionless profiles of a vertical velocity component in a two-dimensional natural-convection laminar plume [12].

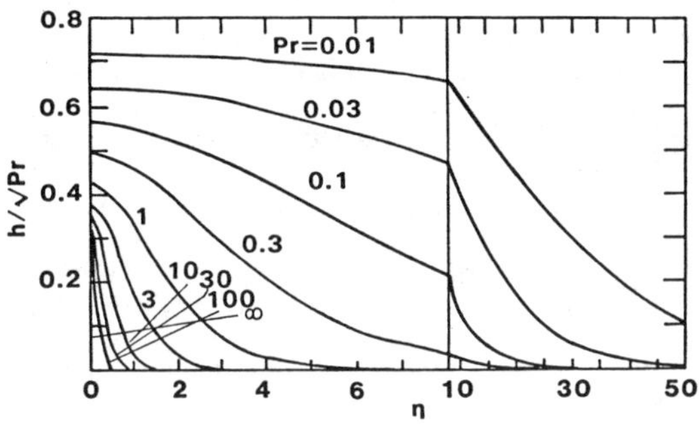

Figure 2. Dimensionless profile of temperature in a two-dimensional natural-convection laminar plume [12].

For Pr = 5/9 the closed-form solution of Fuji et al. [12] is

$$f' = \left(\frac{675}{2048}\right)^{1/5} \operatorname{sech}^2\left[\left(\frac{5}{768}\right)^{1/5}\eta\right], \quad h = \left(\frac{3}{640}\right)^{1/5} \operatorname{sech}^2\left[\left(\frac{5}{768}\right)^{1/5}\eta\right] \tag{32}$$

Based on different normalization conditions Yih [10] was the first to propose the solution for Pr = 2 and 5/9; slightly more accurate numerical coefficients in his solution have been reported in Yih and Gupta [13].

Yih and Gupta [13] have developed an approximate closed form solution for Pr in the neighborhood of 5/9 and their solution for air, Pr = 0.73, under present normalization, I = 1, is

$$f' = 0.821 \operatorname{sech}^2(0.424\eta), \qquad h = 0.387 \operatorname{sech}^2(0.424\eta) \tag{33}$$

This solution overestimates "centerline velocity" by 1.5 percent and centerline temperature by 3 percent and underestimates the entrainment rate in the plume by 5 percent, when compared with exact numerical solutions.

There have been several experimental investigations [13, 19–21] of the buoyant plume rising in air, water, silicone oil, and spindle oil, above a horizontal line source of heat. There is a discrepancy between theory and experiments. The observed centerline velocity is higher, the temperature is lower, and the plume width is more extensive than predicted by the theory. A comparison of the experimental results for temperature profile and centerline temperature with the theory is shown in Figures 3 and 4. The data on water and spindle oil are in fair agreement with the theory. The data for silicone oil show tendencies different from those for water in the range of small Gr_x, though the physical properties of both liquids are in the same order. All the data for air are about 15–20 percent lower than the theoretical predictions. Forstrom and Sparrow [19] suggested that the discrepancies may be explained by the introduction of virtual origin. Doubts have been cast in Reference 21 about whether such an explanation is entirely adequate. Fuji et al. [16] suggested that the difference is due to swaying and meandering of the plume. The theory deals with the flow in the region above the heat source ($x > 0$) only, which implies no entrainment from below the heat source ($x \leqslant 0$). However, in practice the entrainment from below the heat source does exist, which may be treated as a finite Grashof number effect. Riley [22] and Heiber and Nash [23] have analyzed the entrainment problem and associated first-order corrections to the leading order boundary-layer solution. These corrections are described later in the discussions on mixed convection. The first-order theory shows that at finite Grashof numbers the centerline velocity increases, the temperature decreases, the plume width is more extensive, and the trends are wholly consistent with the observations. However, the two-term result is not in complete quantitative agreement with the experiments. The virtual origin effects, although smaller in order of magnitude than the first-order corrections, would improve the results.

When a horizontal plate is placed to obstruct the entrainment from below the heat source, the first-order corrections to the plume boundary layer further improve the comparison, but there is no complete agreement with the observations [22]. It may be pointed out that the entrainment model being inviscid does not satisfy the no-slip condition on the horizontal plate. The correct entrainment model needs study of the viscous effects particularly on the horizontal plate, analogous to the work of Schneider [24] for axisymmetric plume. In the viscous entrainment model the velocity near the wall would decrease due to viscous effects and enhance the magnitude of induced axial component of velocity. The first-order corrections to the plume boundary layer corresponding to enhanced induced axial velocity would further improve the results.

The stability of laminar buoyant plume by employing the linear stability theory has been investigated by several workers [25]. Bill and Gebhart [26] have studied the transition domain where disturbances were found to be selectively amplified approximately as

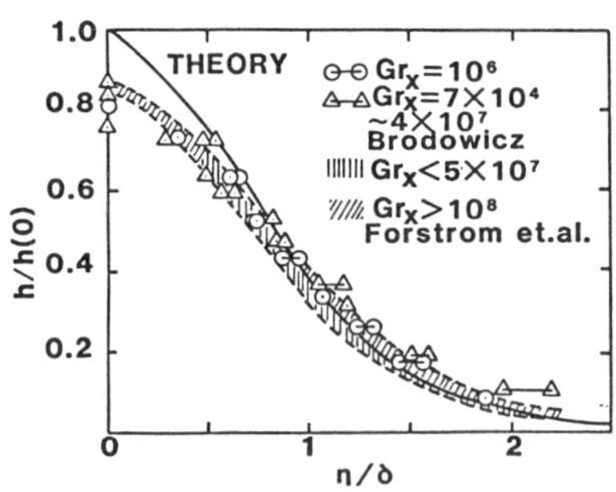

Figure 3. Comparison of theory and experiments for a dimensionless temperature profile in a two-dimensional natural-convection laminar plume in air [12].

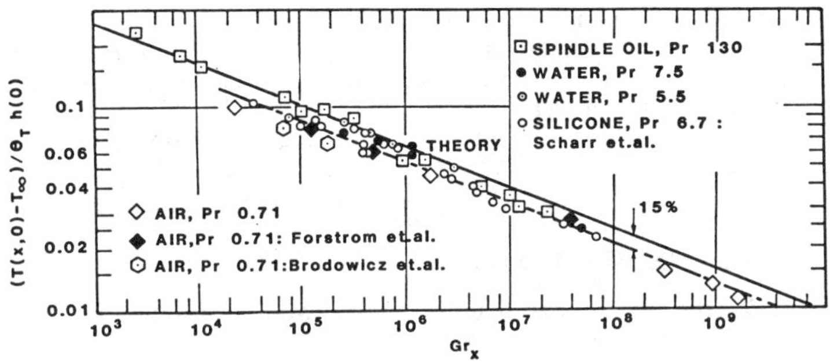

Figure 4. Comparison between theory and experiments for centerline temperature for a two-dimensional natural-convection laminar plume in fluids of different Prandtl numbers [12].

predicted by the linear stability theory. The beginning and the end of the transition were observed at $Gr_x = 6.4 \times 10^7$ and 2.95×10^8. Beyond this Grashof number the plume becomes turbulent.

Turbulent Plume

In a two-dimensional turbulent plume the dependence of centerline temperature on downstream distance and strength of heat source was studied by Zeldovich [8]. Rouse, Yih, and Humphreys [27] measured the velocity and temperature fields above a line fire. Their measurement gave

$$u = u_m \exp(-y^2/b^2)$$

$$T - T = T_m \exp(-y^2/b_T^2) \tag{34}$$

where

$$u_m = 1.8\,(g\beta Q/\rho C_p)^{1/3}, \quad T_m = 2.6\,(g\beta)^{1/3}(Q/\rho C_p)^{2/3}x^{-1}$$

$$b = x/\sqrt{32},\ b_t = x/\sqrt{42}$$

Lee and Emmons [28] measured only the temperature profile above a line fire. The detailed measurement for velocity, temperature, and turbulent structure of a two-dimensional plume have been reported by Kotsouinois [29, 30]. The measurements [30] obey these relations, but the numerical values are somewhat different [4].

The study of turbulent plumes based on the entrainment argument associated with corresponding integral equations has been considered by several workers, and a good review is available in Turner [31, 32], and Murgai [34]. Hence the present review emphasized a similarity analysis developed during, the last decade. The similarity analysis for a two-dimensional turbulent plume by employing the closure hypothesis of eddy viscosity, mixing length, and vorticity transfer theory has been presented by Rao [34] when the turbulent Prandtl number is unity and results are compared with the measurement of Rouse, Yih, and Humpreys [27]. The eddy viscosity hypothesis predicts good results near the axis of the plume but predicts a higher value of velocity and temperature towards the outer edge of the plume. On the other hand predictions based on mixing length are good towards the outer edge of the plume but give lower values of velocity and temperature near the axis of the plume. Tennekes and Lumley [35] have presented a detailed similarity analysis. Yih [36] considered the eddy viscosity model and presented closed-form solutions for the turbulent Prandtl number $Pr_t = 2/3$ and 2. Yih and Gupta [13] have developed an approximate solution, based on an exact solution of Yih for $Pr_t = 2/3$, that predicts results for $Pr_t = 1$, in good agreement with the measurement of Rouse, Yih, and Humphreys [27]. Teske, Lewellen, and Segur [31] employed a second-order closure model that reproduces most of the basic features qualitatively.

Based on the previous subsection, the similarity variable for a two-dimensional turbulent plume can be taken as

$$\psi = U_s \delta f(\eta),\ T - T_\infty = T_s h(\eta),\ \eta = y/x \tag{35}$$

$$U_s = (g\beta Q/\rho C_p)^{1/3},\ T_s = Q/(\rho C_p U_s \delta),\ \delta = x \tag{36}$$

In these variables Equations 4 through 6 and 11a for a two-dimensional turbulent plume with constant eddy viscosity yield

$$\lambda f''' + ff'' + h = 0 \tag{37}$$

$$\lambda h'' + Pr_t (fh)' = 0 \tag{38}$$

The boundary conditions (Equations 8 through 10) give

$$\eta = 0,\ f = f'' = h' = 0 \tag{39}$$

$$\eta \to \infty,\ f' \to 0,\ h \to 0 \tag{40}$$

$$\int_{-\infty}^{\infty} f' h d\eta = 1 \tag{41}$$

where λ is the inverse of the eddy Reynolds number defined as

$$\lambda = (U_s \delta/\nu)^{-1} \tag{42}$$

to be determined from comparison the solutions with the experimental data. The closed-form solutions for certain values of the turbulent Prandtl number can be obtained. For $Pr_t = 2$ the solution of Yih [36] is

$$f = a \tanh b\eta,\ \theta = c\,\mathrm{sech}^4 b\eta \tag{43}$$

$$a = 2b\lambda,\ c = (15/32)\,b\lambda,\ b = (128\lambda^3/15)^{-1/5}$$

For $Pr_t = 2/3$, the closed-form solution [36] is

$$f = A \tanh B\eta,\ \theta = C\,\mathrm{sech}^2 B\eta \tag{44}$$

$$A = 3B\lambda,\ C = 6B^4\lambda^2,\ B = (24\lambda^3)^{-1/5}$$

and by comparing the point of inflection with the experimental velocity profile (Equation 34), Yih obtained

$$\lambda = 0.01337,\ A = 0.282,\ AB = 1.99,\ C = 2.6$$

The solutions based on these constants are in reasonable agreement near the axis of the plume but give higher velocity and temperature near the outer part of the plume when compared with Equation 34 [27].

Yih and Gupta [13], basing their solutions on Equation 44 developed an approximate solution in the neighborhood of $Pr_t = 2/3$, and their values for A, B, and C are

$$A = 0.75/(\lambda^3 C) \tag{45a}$$

$$B = \{15\,Pr_t/[64\,(1 + Pr_t)\,\lambda^3]\}^{1/5} \tag{45b}$$

$$C = 1.6\,(1 + Pr_t)\,B^4/Pr_t^2 \tag{45c}$$

For $Pr_t = 1$ and $\lambda = 0.019$ the predictions based on Equations 44 and 45 shown in Figure 5 are in good agreement with the measurements [27].

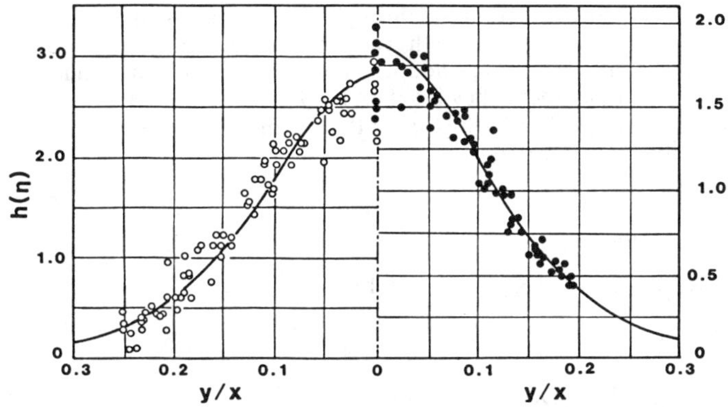

Figure 5. Comparison of the profiles of velocity and temperature between theory and experiments in a two-dimensional natural-convection turbulent plume [13].

Axisymmetric Plume

Laminar Plume

The buoyant plume arising from a point heat source was studied by Schuh [9] for Pr = 0.7. The closed-form solutions for Pr = 1 and 2 were given by Yih [10], Fuji [11] and Brand and Lahey [18]. Yih and Wu [38] have proposed an approximate solution for Prandtl numbers in the neighborhood of Pr = 1 that gives good results for Pr = 0.73. The solutions for a wide range of Prandtl numbers have been studied by Fuji [11] and Mollendorf and Gebhart [39]. The asymptotic analysis for low and high Prandtl numbers has been reported by Crane [40, 41].

Based on scales of velocity, temperature, and transverse length described earlier, the similarity varibles may be taken as

$$\eta = Gr_x^{1/4} y/x$$

$$\psi = \nu x f(\eta), \qquad T - T_\infty = \theta_p h(\eta) \tag{46}$$

$$\theta_p = \frac{Q}{2\pi\rho C_p \nu I x}, \qquad Gr_x = \frac{g\beta\theta p x^3}{\nu^2}$$

In terms of the foregoing similarity varibles, the axisymmetric boundary layer equations reduce to

$$f''' + (f - 1)(f'/\eta)' + \eta h = 0 \tag{47}$$

$$(\eta h')' + Pr(fh)' = 0 \tag{48}$$

The boundary conditions are

$$\text{at } \eta = 0 \qquad 2\eta^{-1} f - f' = 0, \quad (f'/\eta)' = 0, \quad h' = 0 \tag{49}$$

$$\text{as } \eta \to \infty \qquad \eta^{-1} f' \to 0, \quad h \to 0 \tag{50}$$

subject to the heat flux conditions

$$\int_0^\infty f' h \, d\eta = I \tag{51}$$

The problem governed by Equations 47 through 50 again forms an eigen value problem of the type described in connection with two-dimensional plume. For determining the multiplicative constant Fuji [11] has adopted I = 1, Jaluria [17] Mollendorf and Gebhart [39] $h(0) = 1$, and Brand and Lahey [18] $f(\infty) = 1$.

The closed-form solutions of Yih [10] for Pr = 1 and 2 may be conveniently written in terms of a variable X, defined as

$$X = 1 + A\eta^2 \tag{52}$$

For Pr = 1 the solution is

$$f = 6(1 - X^{-1}), \quad h = 2/3\,X^{-3}, \quad A = 1/12 \tag{53}$$

and for Pr = 2 is

$$f = 4(1 - X^{-1}), \quad h = \frac{5}{4}X^{-4}, \quad A = \sqrt{5}/16 \tag{54}$$

These solutions have also been discussed by Fuji [11] and Brand and Lahey [31]. Yih and Wu [38] have given two approximate solutions which become exact for Pr = 1 and 2. The first approximate solution for Pr in the neighborhood of unity, based on Equation 53 is

$$f = \frac{6}{Pr}(1 - X^{-1}), \quad h = \frac{2Pr}{3}X^{-3}, \quad A^2 = Pr^3/(50\,Pr + 54) \tag{55}$$

which gives good results for our Pr = 0.73, where maximum error in the value of A is about 7 percent. The second approximate solution based on Equation 54, is

$$f = \frac{8}{Pr}(1 - X^{-1}), \quad h = \frac{5Pr}{8}X^{-3}, \quad A^2 = 25\,Pr^3/[2048\,(2Pr + 1)] \tag{55}$$

predicts results for water Pr = 6.8, where the error in the value of A is about 12 percent.

The numerical solutions for velocity and temperature profiles from Fuji [11] for Pr values of 0.7, 1, 2.0, and 10 are shown in Figure 6 and 7. The general trend of the results is similar to that for two-dimensional plume described earlier. The plume thickness varies in accordance with the square root of the distance from the axis. Therefore, it may be anticipated that the paraboloidal coordinates (ω, χ) defined as

$$x = 1/2\,(\omega^2 - \chi^2), \; y = \omega\chi \tag{56}$$

should be optimal [42] in the sense of Kaplun [43]. The expressions for stream function and temperature in terms of paraboloidal coordinates are

$$\psi = 1/2\,\nu\omega^2 f\left[\left(\frac{2g\beta Q}{\pi\rho C_p\nu^3}\right)^{1/4}\chi\right] \tag{57a}$$

$$T - T_\infty = \frac{Q}{\pi\rho C_p\nu\omega^2}\,h\left[\left(\frac{2g\beta Q}{\pi\rho C_p\nu^3}\right)^{1/4}\chi\right] \tag{57b}$$

The solution in optimal coordinates contains the effects of displacement for $\chi \to \infty$ as

$$\psi = 1/2\,\nu\omega^2 f(\infty) \tag{58}$$

In terms of (x, y) coordinate system, this may also be written

$$\psi = 1/2\,\nu[x + \sqrt{x^2 + y^2}]\,f(\infty) \tag{59}$$

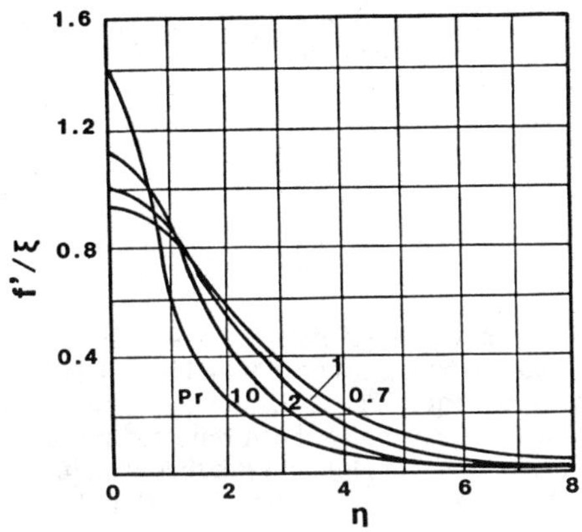

Figure 6. Dimensionless profile of a vertical velocity component in an axisymmetric natural-convection laminar plume [11].

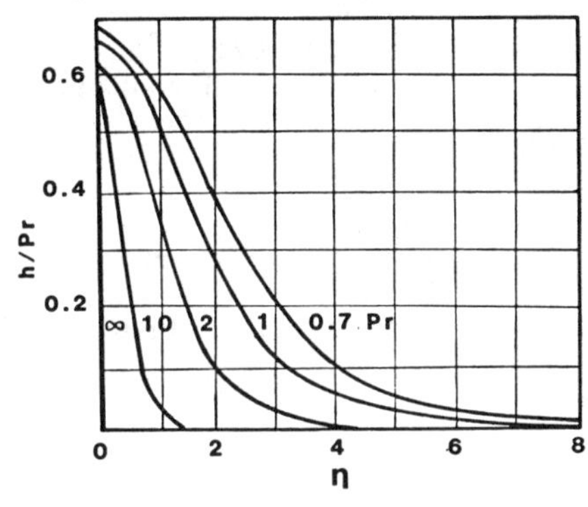

Figure 7. Dimensionless profile of temperature in an axisymmetric natural-convection laminar plume [11].

which represents the entrainment of the outer ambient fluid from all over the environment in the plume.

The stream function of outer flow has to satisfy the no-slip condition on the horizontal plate when it is placed just below the source to obstruct the entrainment from below. For this outer flow Schneider [24] has obtained an exact similarity solution of the Navier-Stokes equations such that both the no-slip condition at the horizontal plate and the entrainment condition at the axis are satisfied. It was found that the stream line of the inviscid solution is too close to the wall when compared with the viscous solution, which results in a decrease of velocity near the wall due to viscous effects and an increase of entrainment tangential velocity at the edge of the plume.

Turbulent Plume

The axisymmetric turbulent plume arising from a point heat source was first studied by Schmidt [44] experimentally as well as theoretically. Rouse, Yih, and Humphreys [27] also reported measurements on axisymmetric plume arising from a localized fire. The experimental results were fitted to empirical relations to get

$$u = 4.7 \left(\frac{g\beta Q}{\rho C_p x}\right)^{1/3} \exp\left[-1/2\left(\frac{y/x}{0.072}\right)^2\right] \quad (60)$$

$$g\beta(T - T_\infty) = 11.0 \left(\frac{g\beta Q}{\rho C_p y}\right)^{2/3} x^{-5/3} \exp\left[-1/2\left(\frac{y/x}{0.084}\right)^2\right] \quad (61)$$

whereas the measurements of Schmidt [44] gives 2.7, 13.7, 0.105, 0.099, in places of 4.7, 11.0, .072, and 0.084 in Equations 60 and 61. It has been noted by Yih [10] that the Schmidt data do not satisfy the momentum equation. The detailed measurements for velocity, temperature, and structure of turbulence have been reported recently by George, Alpert, and Tamanini [45], Nakagome and Hirata [46], and Beuther et al. [47]. These measurements suggest that the numerical coefficient 4.7 in Equation 60 within the scatter lies in the range 3.4–3.9.

Theoretically problem of axisymmetric turbulent plume based on the entrainment equation, described in the first section, and the integral momentum and energy equations have been studied extensively in the literature. Good reviews have been reported by Turner [31, 32], Murgai [33] and Baines [48]. Therefore, in this review we consider the similarity approach that will be helpful in the development of the turbulent modeling of plumes. Schmidt [44] considered the problem by employing the Prandtl's momentum transfer theory and Taylor's vorticity transfer theory as closure hypothesis and determined velocity and temperature distributions. Yih [36] considered the eddy viscosity model and gave closed-form solutions for the turbulent Prandtl number $Pr_t = 1.1$ and 2. Yih and Wu [38] have developed approximate solutions in the neighborhood of the two exact sulutions of Yih [36]. The second-order closure $K - \varepsilon$ model by Tennankore and Steward [49] and Chen and Nikitopoulos [50].

The similarity variables for an axisymmetric turbulent plume can be taken as

$$\begin{aligned} &\eta = y/\delta \\ &\psi = 3\,U_s\delta^2\lambda f(\eta) \\ &T - T_\infty = 3\,T_s\lambda h(\eta) \\ &U_s = (g\beta Q/\rho C_p\delta)^{1/3}, \quad T_s = Q/(\rho C_p U_s \delta^2), \quad \delta = x \end{aligned} \quad (62)$$

Using the eddy viscosity closure model the axisymmetric equations for turbulent flow, based on the similarity variables yield.

$$f''' + (5f - 1)(f'/\eta)' + f'^2/\eta + \eta h = 0 \quad (63)$$

$$(\eta h')' + 5\,Pr_t(fh)' = 0 \quad (64)$$

The boundary conditions are

$$\eta = 0;\ f = f' = h' = 0 \quad (65)$$

$$\eta \to \infty, \quad f' \to 0,\ h \to 0 \quad (66)$$

subject to the heat flux condition

$$18\,\pi\lambda^3 \int_0^\infty f'\theta d\eta = 1 \quad (67)$$

The closed-form solutions [36] for $Pr_t = 1.1$ and 2 are

$$f = B(1 - X^{-1}), \quad h = CX^{-m} \quad (68)$$

$$X = 1 + A\eta^2$$

where the constants for $Pr_t = 1.1$ are

$$m = 3,\ B = 12/11,\ C = 1{,}536\,A^2/121,\ \lambda^3A^3 = 1{,}331/829{,}942 \tag{69}$$

and for $Pr_t = 2$ are

$$m = 4,\ B = 4/5,\ C = 256\,A^2/25,\ \lambda^3A^2 = 625/18{,}432 \tag{70}$$

In the neighborhood of the preceding two values of Pr_t Yih and Wu [38] have developed approximate solutions. Their results in the neighborhood of $Pr_t = 1.1$ are

$$m = 3,\ B = \frac{6}{5Pr_t},\ C = \frac{5Pr_t}{27\lambda^3},\ A^2 = \frac{5}{27\lambda^3}\frac{Pr_t}{2.16 + 12Pr_t} \tag{71}$$

and in the neighborhood of $Pr_t = 2$ are

$$m = 4,\ B = \frac{8}{5Pr_t},\ C = \frac{25Pr_t}{144\pi\lambda^3},\ A^2 = \frac{625Pr_t^2}{73{,}728\pi\lambda^3} \tag{72}$$

The solutions Equation 68 based on Equation 72 were fitted for $Pr_t = 1$ to the data of Beuther, Capp, and George [51] and a comparison with $\lambda = 0.0156$, $A = 33$ is displayed in Figure 8. In the same Figure 8 solution (Equation 69) for $Pr_t = 1.1$ is also displayed by the dotted lines. The results describe the experimental data very well, though the results for $Pr_t = 1$ are in slightly better agreement than those for $Pr_t = 1.1$.

MIXED CONVECTION FROM LINE SOURCE

For weakly buoyant plume the problem has been studied by Wood [52] and Wesseling [53], and Nadai and Okazaki [54]. Of these, Reference 54 is also considered a perturbation on far field flow. Afzal [55] presented a complete solution describing the entire flow regime ranging from weakly to strongly buoyant plumes in aiding as well as opposing situations

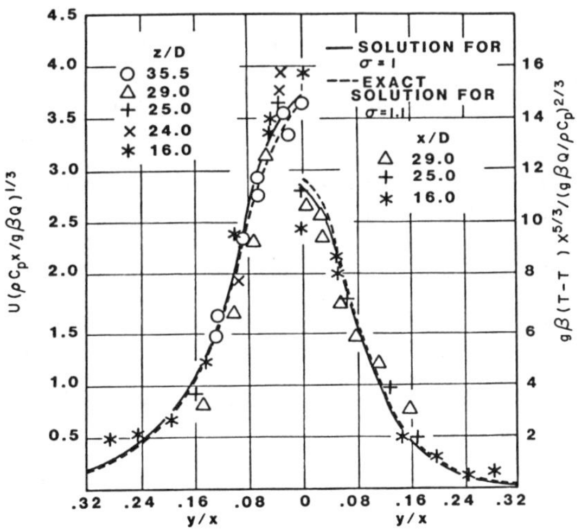

Figure 8. Comparison between theory and experiment in air for velocity and temperature profiles in an axisymmetric natural-convection turbulent plume [38].

for the Prandtl number 0.72. Haaland and Sparrow [56] have studied the effects of the Prandtl number in an aiding flow situation.

In the study of Wood [52] and Nakai and Okazaki [54] for weakly buoyant plume, the Oseen linearization of boundary-layer equations has been employed and explicit closed-form solutions have been obtained for the first-order terms for the velocity in the expansion near the source. For strongly buoyant plume, the first-order perturbation far flow field has also been studied by Nakai and Okazaki. Effects of the inclined free stream have been analyzed by Wood and the horizontal stream by Nakai and Okazaki [54]. For weakly buoyant plumes, Wesseling [53] has considered the leading terms in the Oseen linearization of the Navier-Stokes equations in the neighborhood of a horizontal line source of heat and free stream, directed at an arbitrary angle with respect to a buoyant free vector. It was shown that the Oseen-approximation does not admit solution with a uniform pressure at infinity, except for a vertically upward stream. Wesseling noted that the situation is analogous to the Stokes paradox. The paradox raised by Wesseling has already been resolved in Wood's earlier paper who has shown that the singularity in the pressure and the associated singularities in the velocity fit naturally into a wider picture of the flow which includes the nonlinear development of the buoyant plume at distance from the heat source larger than that at which Oseen linearized equations hold. Therefore, away from the heat source there is a region where buoyancy and forced-convection effects are of comparable magnitude, the Oseen linearization fails, and one has to consider the full nonlinear equations. At large distances from the heat source the problem is governed mainly by buoyancy.

Afzal [55] presented a complete solution of the mixed-convection problem for a vertical free stream by employing the boundary-layer equation for a Prandtl number of 0.72. The problem has been studied in terms of two coordinate expansions. The technique of direct coordinate expansion was used for a weakly buoyant plume valid for a near-source region. The closed-form solutions have been obtained for second-order terms for an arbitrary Prandtl number. The numerical solutions for the first eleven terms have been obtained for the Prandtl number of 0.72. In the inverse coordinate expansion for strongly buoyant plumes the first seven terms have been studied for the Prandtl number $Pr = 0.72$. The series for weakly buoyant plume was suitably transformed, and it describes very accurately the solution for the entire domain of flow in aiding as well as opposing flow situations. In the opposing flow situation the solutions are dual, corresponding to forward and reverse flow in the plume.

Haaland and Sparrow [56] have studied the aiding flow problem numerically by employing the Patankar and Spalding [57] finite difference scheme for Prandtl numbers of 0.72, 5, and ∞. For the case of the infinite Prandtl number, a method of inner and outer expansions has been used. For forced-convection-dominated flow their results compare well with the analytic solution of Afzal [55].

For a free-convection plume from a line source, the characteristic free-convection velocity, as shown earlier increases continuously as $x^{1/5}$. When an external mainstream with velocity $U_\infty(x)$ is imposed on the line source the problem is not amenable to similarity analysis except for the case $U_\infty(x) \sim x^{1/5}$. For a constant value of U_∞, the free stream velocity, the relative effects of buoyancy to forced convection increases as the boundary layer develops from the heat source. Thus, the mixed convection plume undergoes an evolution with an increasing vertical distance from the heat source. The evolution starts from the near source region where forced convection is dominant, passes through a region where forced convection and buoyancy are of comparable magnitude, and finally emerges as a far source region where natural convection is dominant. Therefore, based on the distance from the heat source the flow may be categorized as

1. *Weakly buoyant plume* The flow field near the heat source $x \ll x_0\,(x_0) = [U_\infty^5/\nu\,(g\beta\theta_T)]^2$ where forced convection is dominant and buoyancy effects can be taken as small perturbations. In the domain Oseen linearization may be fruitfully employed.

2. *Strongly buoyant plume* The region $x \gg x_0$ far away from the heat source where buoyancy is dominant and forced convection can be taken as perturbations. The asymptotic solutions correspond to natural-convection buoyant plume.
3. *Moderately buoyant plume* The region at moderate distances from the heat source $x \sim x_0$ where forced and free convections are of comparable magnitude.

In the presentation that follows the cases of weakly, strongly, and moderately buoyant plume are described for a vertical free stream. These results are based on boundary-layer equations. This is followed by the study of the entrainment of outer irrotational flow in the plume and the associated second-order corrections to the boundary layer for strongly buoyant plume, which forms the subject matter of the next section. The effects of inclination of the free stream with respect to the buoyant force vector are considered later.

Vertical Free Stream

Weakly Buoyant Plume

In the near-source region the forced convection is dominant and it is expected that the flow will behave analogous to a two-dimensional wake. In terms of the variables [55]

$$\psi = (\nu U_\infty x)^{1/2} F(\xi, \zeta),\ T - T_\infty = \theta_T (U_\infty x/\nu)^{-1/2} H(\xi, \zeta) \tag{73}$$

$$\zeta = y(\nu x/U_\infty)^{-1/2},\ \xi = \Gamma R_x^{1/2}$$

$$\theta_T = Q/\rho C_p \nu,\ \Gamma = Gr_x/R_x^3$$

$$Gr_x = g\beta\theta_T x^3/\nu^2,\ R_x = U_\infty x/\nu$$

the boundary-layer equations reduce to

$$F_{\zeta\zeta\zeta} + 1/2\, FF_{\zeta\zeta} \pm \xi H = 1/2\, \xi (F_\zeta F_{\zeta\xi} - F_\xi F_{\zeta\zeta}) \tag{74}$$

$$Pr^{-1} H_{\zeta\zeta} + 1/2\,(FH)_\zeta = 1/2\, \xi (F_\zeta H_\xi - F_\xi H_\zeta) \tag{75}$$

The boundary conditions at the vertical axis are of symmetry.

$$\zeta = 0:\ F = F_{\zeta\zeta} = H_\zeta = 0$$

For away from the source the velocity and temperature approache their ambient values

$$\zeta \to \infty,\ F_\zeta \to 1,\ H \to 0 \tag{77}$$

subject to normalized heat flux relation

$$\int_{-\infty}^{\infty} F_\zeta\, H d\zeta = 1 \tag{78}$$

Equations 74 through 78 describe the entire mixed convection regime. The plus and minus signs refer to upward (aiding) and downward (opposing) forced stream.

For weakly buoyant plumes the velocity and temperature can be expanded in the powers of ξ as

$$F = \sum_{n=0} (\pm\xi)^n F_n(\zeta),\ H = \sum_{n=0} (\pm\xi)^n H_n(\zeta) \tag{79}$$

The analysis then proceeds by substituting the foregoing expansions for F and H in the governing equations. Ordinary differential equations are obtained for successive-order approximations. The solution to the leading order terms

$$F_0 = \zeta,\ H_0 = \left(\frac{\sigma}{4\pi}\right)^{1/2} \exp\left(-\frac{Pr}{4}\zeta^2\right) \tag{80}$$

corresponds to Oseen linearization of boundary-layer equations. The closed-form solution for the first-order terms (F_1, H_1) for Pr = 1 is

$$F_1 = 1/2 \operatorname{erf}(1/2\zeta) \tag{81a}$$

$$H_1 = \frac{1}{16\pi}\left[\pi - 4\exp(-1/2\zeta^2) - \pi \operatorname{erf}^2(1/2\zeta) - 2\zeta\pi^{1/2}\exp(-1/4\zeta^2)\operatorname{erf}(1/2\zeta)\right] \tag{81b}$$

and for Pr ≠ 1 may be found [55]. The velocity and temperature at the axis of plume are given by

$$F_\zeta(\xi, 0) = 1 \pm \frac{(\mathrm{Pr}/\pi)^{1/2}}{1 + \mathrm{Pr}^{1/2}}\,\xi + 0(\xi) \tag{82a}$$

$$H(\xi, 0) = \left(\frac{\mathrm{Pr}}{4\pi}\right)^{1/2} \pm \frac{\mathrm{Pr}(\mathrm{Pr}+1)}{8\pi}\left[\tan^{-1}(\mathrm{Pr}^{-1/2}) + \frac{\mathrm{Pr} + \mathrm{Pr}^{1/2} + 2}{(1+\mathrm{Pr})(1+\mathrm{Pr}^{1/2})}\right]\xi + 0(\xi) \tag{82b}$$

As the plume develops it entrains the outer irrotational fluid at its edges. The stream function of outer irrotational flow determined from the solution of plume is

$$\begin{aligned}\Psi_1 &= \lim_{\zeta\to\infty}(\psi - \zeta\psi_\zeta)\\ &= (\nu U_\infty x)^{1/2}\Delta(\xi)\end{aligned} \tag{83a}$$

where Δ is the entrainment function defined as

$$\Delta = \lim_{\zeta\to\infty}(F - \zeta F_\zeta) \tag{83b}$$

The displacement thickness $\bar{\delta}^*$ for the plume may be defined as

$$U_\infty \bar{\delta}^* = \int_0^\infty u - U_\infty dy \tag{84}$$

Based on variables (Equation 73) the expression for $\bar{\delta}^*$ yields

$$\delta^* = \pm\,\xi\int_0^\infty (F_\zeta - 1)\,d\zeta \tag{85a}$$

where $\delta^* = \bar{\delta}^* g\beta\theta_T/U_\infty^2$ (85b)

is nondimensional displacement thickness. The entrainment function Δ is related to δ^* by

$$\Delta = \delta^*/(\pm\xi) \tag{86}$$

For weakly buoyant plume the expansion for δ^* is given by

$$\delta^* = \sum_{n=1}^{\infty} F_n(\infty)\,(\pm\xi)^{n+1} \tag{87}$$

For aiding flow (plus sign) the closed-form solutions (Equation 82) show that the centerline velocity increases and δ^* decreases with streamwise distance ξ. The effect of the increase of the Prandtl number for a fixed ξ is to increase the centerline velocity and decrease the centerline temperature. These trends are consistent with the requirement that the total convected energy in the plume remains constant.

The closed-form solution of further higher-order terms in expansions (Equation 79) is very laborious. Therefore the numerical solutions for the first eleven terms in the expansions (Equation 79) for the Prandtl number Pr = 0.72 have been determined by Afzal [55]. The results for the centerline velocity and temperature can be expressed as

$$F_\zeta(\xi, 0) = \sum_{n=0} a_n\,(\pm\xi)^n \tag{88a}$$

$$H(\xi, 0) = \sum_{n=0} b_n (\pm \xi)^n \tag{88b}$$

where coefficients a_n and b_n for Pr = 0.72 are given in Table 1. From the Domb-Sykes plots for series (Equation 88) it was concluded [55] that the series possess the nearest singularity as the square root on the real axis in complex ξ-plane at $\xi = \mp\xi_0$ where $\xi_0 = 1.6$ and the singularity may be represented as $(\xi_0 \pm \xi)^{1/2}$. This restricts the use of the series (Equation 88) for small values of ξ, the streamwise distance.

Moderately Buoyant Plume

The problem of combined forced and natural convection has been described earlier as two extreme cases: small buoyancy effects in a predominantly forced flow and small forced-convection flow in a predominantly natural-convection flow. However, in many situations the intermediate range where both convection processes are comparable must be considered. The partial differential equations in general require numerical integration by a finite-difference scheme. However, the convergence of the series for weakly buoyant plume can as well be accelerated by various methods [58]. It was found that the improved series describe the entire mixed-convection regime spanning from forced-convection to predominantly natural-convection plume [55].

A method to improve the series is to explore the possibility of completing it by finding the remainder. The remainder in the series (Equation 88) can be taken as proportional to $(\xi_0 \pm \xi)^{1/2}$, and the constant of proportionality is determined so that the coefficient of the last term ξ^{10} in the expansion for remainder and series (Equation 88) are equal. The completed series for centerline velocity and temperature are [55].

$$F_\zeta(\xi, 0) = a_{-1}(1 \pm \xi/\xi_0)^{1/2} + \sum_{m=0}^{9} a_m(\pm\xi)^m \tag{89}$$

$$H(\xi, 0) = b_{-1}(1 \pm \xi/\xi_0)^{1/2} + \sum_{m=0}^{9} b_m(\pm\xi)^m \tag{90}$$

the coefficients a_m and b_m are given in Table 1. The results for opposing flow (with minus sign) are considered first.

Table 1
Coefficients for Velocity and Temperature Series Expansion

n	a_n	b_n	a_n	b_n	$\mathscr{A}_n$	$\mathscr{B}_n$
−1			1.424909	-0.574973×10^{-1}	0.070814	−0.146915
0	1.0	0.239365	−0.425909	0.296863	0.757799	0.409872
1	0.258975	-0.125705×10^{-1}	−0.186617	0.539741×10^{-2}	0.402301×10^{-1}	-0.282694×10^{-1}
2	-0.51126×10^{-1}	0.301922×10^{-2}	0.184981×10^{-1}	0.211732×10^{-3}	0.662239×10^{-2}	-0.409804×10^{-2}
3	0.174952×10^{-1}	-0.101311×10^{-2}	-0.426239×10^{-2}	-0.135775×10^{-3}	0.216499×10^{-2}	-0.130641×10^{-2}
4	-0.722383×10^{-2}	0.397441×10^{-3}	0.127525×10^{-2}	0.547305×10^{-4}	0.886424×10^{-3}	-0.547620×10^{-3}
5	0.32893×10^{-2}	-0.170991×10^{-3}	-0.429007×10^{-3}	-0.210550×10^{-4}	0.408477×10^{-3}	-0.258411×10^{-3}
6	-0.159151×10^{-2}	0.782146×10^{-4}	0.151462×10^{-3}	0.793210×10^{-5}	0.200318×10^{-3}	-0.126495×10^{-3}
7	0.802681×10^{-3}	-0.373848×10^{-4}	-0.532445×10^{-4}	-0.287108×10^{-5}	0.100058×10^{-3}	-0.600005×10^{-4}
8	0.417275×10^{-3}	0.184713×10^{-4}	0.173742×10^{-4}	0.944801×10^{-6}	0.478534×10^{-4}	-0.250894×10^{-4}
9	0.221987×10^{-3}	-0.936619×10^{-5}	-0.439290×10^{-5}	-0.237803×10^{-6}	0.184331×10^{-4}	-0.734482×10^{-5}
10	0.120264×10^{-3}	0.484946×10^{-5}				

• Plane of symmetry for mixed convection in two-dimensional buoyant plume from low ξ series (a_n, b_n); completed low ξ series (a_n, b_n); and Eulerized series subsequently completed (A_n, B_n).
From Afzal [55].

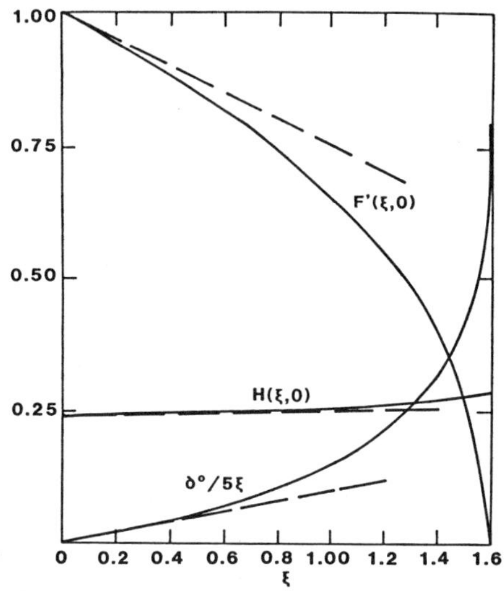

Figure 9. Nondimensional vertical velocity and temperature at the centerline and entrainment function in an opposing situation for laminar mixed convection due to a two-dimensional horizontal line heat source in a vertical stream for air Pr = 0.72 [55].

Opposing flow. The results for velocity and temperature in the plane of symmetry and entrainment function $\delta^* (= \Delta\xi)$ are displayed in Figure 9 against mixed convection parameter ξ. As ξ increases, the velocity at the axis decreases and becomes zero at $\xi = \xi_0$, where the buoyancy effect brings the oncoming stream to rest. The temperature at the axis increases with ξ. The entrainment function δ^* also increases with ξ indicating more extensive plume. The velocity and temperature profile obtained by completion of the series at various locations of η are displayed in Figures 10 and 11. These figures shows that as ξ increases the velocity decreases, temperature increases, and the plume becomes wider.

It may be pointed out that in opposing flow situations the nearest singularity in the complex ξ-plane is on the positive real axis at $\xi = \xi_0$. In general, this singularity cannot be eliminated, short of refining the whole theory. However, the square root on the positive real axis is not a real singularity, but an indication that the function is double valued. In the present case the dual solutions correspond to the forward and reverse flow in the plume. In between the forward and reverse flow cases there is a marginal case $\xi = \xi_0$ where centerline velocity zero. Therefore situations exist for $0 \leqslant \xi \leqslant \xi_0$ and for $\xi > \xi_0$ there is no solution to the problem. The solutions just described correspond to forward flow $[F_\zeta(\xi, 0) \geqslant 0]$. Some insight into the nature of reverse flow solutions $F_\zeta(\xi, 0) < 0$ in the domain $\xi \lesssim \xi_0$ may be gained by inverting the series (Equation 88) and subsequently summing by [N/N] Pade approximations. The application of the Pade approximation leads to reverse flow velocity, but unfortunately the successive Pade approximations are not quickly convergent. Finally as $\xi \rightarrow 0$, it may be anticipated from Equations 74 through 78 that the flow approaches a mixing layer of Chapman with a shift in origin. The mixing layer arises in the manner analogous to an axisymmetric buoyant plume (described in the next section).

Aiding flow. For aiding flow, the entire domain spanning from a purely forced-convection to a purely free-convection plume is of interest. The completed series (Equations 89 and 90) are valid in a very limited domain. In this case the nearest singularity is on a negative real

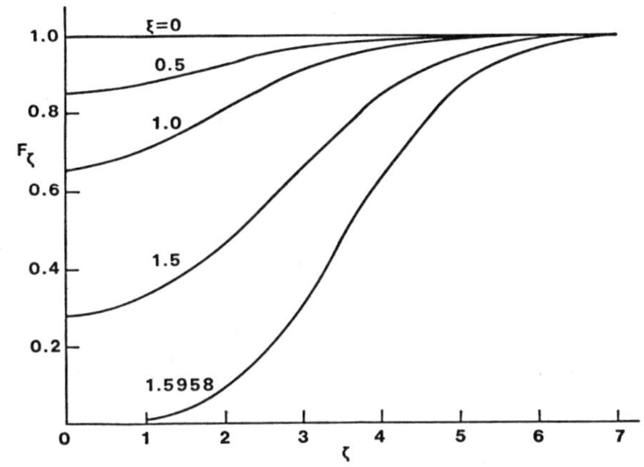

Figure 10. The profiles of vertical velocity in an opposing situation for laminar mixed convection due to a horizontal line heat source in a vertical forced stream for air Pr = 0.72 [55].

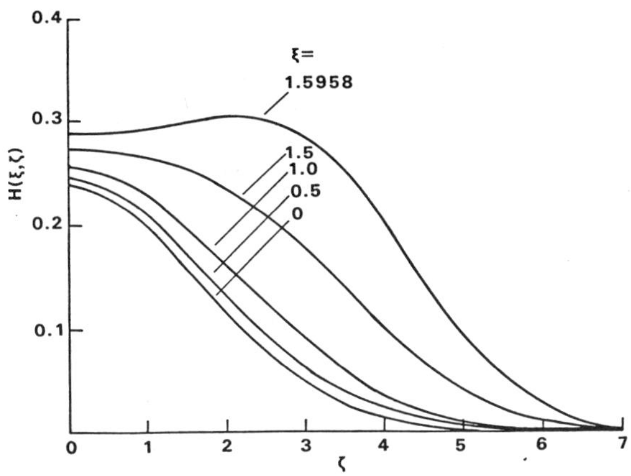

Figure 11. The profiles of temperature in an opposing situation for laminar mixed convection due to a horizontal line source in a vertical forced stream in air Pr = 0.72 [55].

axis at $\xi = -\xi_0$ in a complex ξ-plane. As the singularity off the positive axis is of no physical interest, it can be eliminated by mapping it away to infinity by the Euler transformation

$$Z = \xi/(\xi + \xi_0) \tag{91}$$

The series resulting from the Euler transformation is subsequently completed to get [55]

$$F_\zeta(\xi, 0) = (\xi/Z)^{2/5} [A_{-1}(1 - Z)^{2/5} + \sum_{m=0}^{9} A_m Z^m] \tag{92a}$$

$$H_\zeta(\xi, 0) = (Z/\xi)^{1/5} [B_{-1}(1 - Z)^{2/5} + \sum_{m=0}^{9} B_m Z^m] \tag{92b}$$

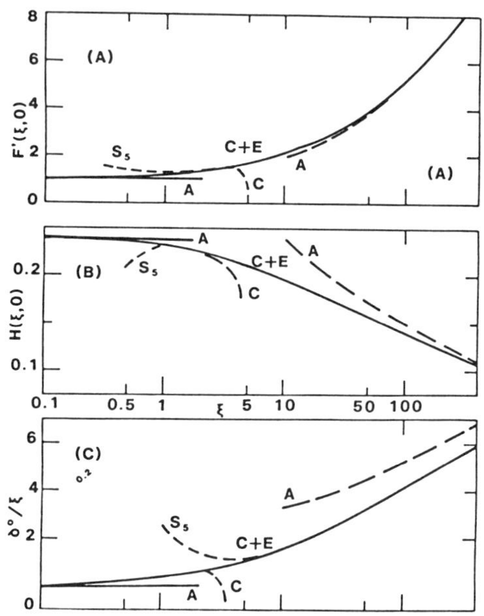

Figure 12. The characteristics of a laminar mixed-convection buoyant plume with a horizontal line source in a vertical free stream in an aiding the flow situation for air, Pr = 0.72: (A) centerline velocity; (B) centerline temperature; (C) entrainment function. A: Asymptotes for small and large, C: completed low series, C + E: completed Eulerized series, S_5: the sum of first five terms for large series [55].

where coefficients A_m and B_m are also given in Table 1. For purely free convection plume $Z \rightarrow 1\ (\xi \rightarrow \infty)$ the foregoing series yields

$$F_{\xi}(\xi, 0) = 0.8085\,\xi^{2/5},\quad H(\xi, 0) = 0.3752\,\xi^{-1/5} \tag{93}$$

whereas the corresponding exact results given earlier are 0.8096 and 0.3773. Thus, Equation 92 predicts results to two-digit accuracy even for the extreme case of a natural convection plume.

The results for velocity and temperature in the plane of symmetry and the entrainment function $\delta^* (= \xi\Delta)$ are displayed against a mixed convection parameter ξ in Figure 12a–c. In the respective figures describing the asymptotes for weakly and strongly buoyant plumes, completed Equations 89 and 90 for small ξ and the five term series Equation 120 for large ξ are also displayed. Figure 12a shows that the centerline velocity at $\xi = 0$ is equal to the forced-convection value and increases with increasing ξ, the distance from the source. This increase is due to the action of buoyancy which acts in the same direction as the forced-convection stream. Sufficient for down stream the velocity approaches a natural-convection buoyant plume. From Figures 12b it may be seen that the centerline temperature decreases with increasing ξ and that the decrease is more rapid as the plume evolves from forced convection to natural-convection dominance. Examination of Figure 12c shows that the magnitude of entrainment functions also increases with ξ. These figures show that the asymptotes for small and large ξ, underestimates the centerline velocity and overestimates the centerline temperature. The reliance on the asymptote for large ξ is good for velocity where $\xi \geqslant 100$ and for temperature when $\xi \geqslant 500$. The completed low ξ series (Equation 89) is good for $\xi \lesssim 3$ and five terms of the series (Equation 120) for $\xi \lesssim 1$. The

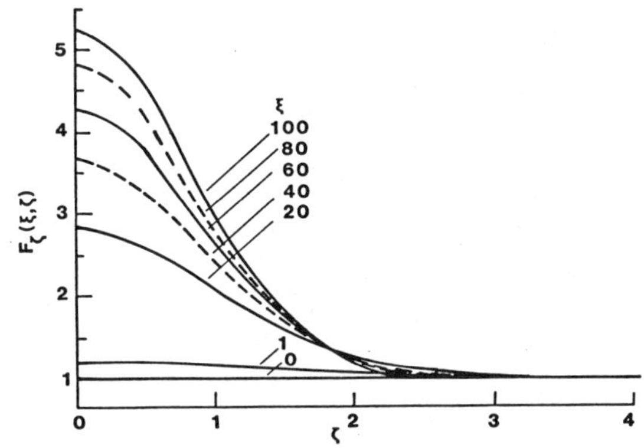

Figure 13. The profiles of velocity for laminar mixed convection in a horizontal line source of heat in a vertical stream for air Pr = 0.72 in an aiding situation [55].

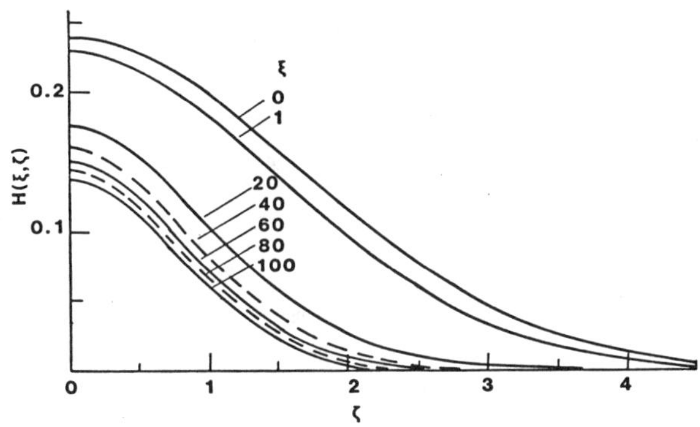

Figure 14. The profiles of temperature for laminar mixed convection in a horizontal line source of heat in a vertical forced stream for air in an aiding situation [55].

predictions based on the completed series (Equation 92) are extremely good throughout the domain spanning from forced to free convections. The velocity and temperature profile for $0 \leqslant \xi \leqslant 100$ obtained from the Euler transformation followed by the completion of the series at various values of ζ are shown in Figures 13 and 14. For further higher values of ξ velocity and temperature profiles are given in Figures 19 and 20. The figure shows that as ξ increases the centerline velocity increases, temperature decreases, and width of plume is less extensive.

The effects of the Prandtl number in the mixed-convection plume have been studied by Haaland and Sparrow [56] by numerically integrating the equations for $\text{Pr} = 0.72$, 5 and ∞, and their results for centerline velocity and temperature are shown in Figures 15 and 16. The purely free-convection asymptote (for $\text{Pr} = \infty$) and purely forced-convection asymptotes are also shown in these figures by dottes lines and their point of intersections are shown by arrows at that position. For $\text{Pr} = 0.72$ these authors have compared their results with Afzal [55] to show that the difference between two solutions for velocity is in the 0.3–0.4 percent range and for temperature is in 0–0.6 percent range. From Figure 15 it can be seen that for a fixed ξ the effect of the increase of the Prandtl number is to decrease the centerline velocity. On the other hand, the closed-form solution of Equation 82 for a weakly buoyant plume shows that for a fixed ξ the centerline velocity increases with Pr, a trend opposite to that shown in Figure 15. A close examination of their work (see Figure 5 [56]) reveals that the centerline velocity does increase with Pr, and therefore Figure 15 is likely to be due to oversight and probably typographic error; the curves labeled Pr = 0.72, 5, and ∞ should be changed respectively, to Pr = ∞, 5, 0.72. For the centerline temperature Figure 16 shows that for a fixed ξ the effect of an increase of the Prandtl number is a decrease in the temperature. This trend for temperature with respect to Pr is consistent with the closed-form solution (Equation 82) for weakly buoyant plumes. Haaland and Sparrow [56] have also proposed empirical expressions for centerline velocity and temperature

$$F_\zeta(\xi, 0) = \{1 + [\xi^{2/5} f_0'(0)]^{1/n}\}^n, \quad n = 0.53\text{–}0.09/Pr^{1/2} \tag{94}$$

$$H(\xi, 0) = CPr^{1/2}/\{1 + [CPr^{1/2}\xi^{1/5}/h_0(0)]^{1/m}\}^m \tag{95}$$

$$m = 0.26 + 0.04/Pr^{1/2}, \; C = (8\pi)^{-1/2} \simeq 0.282$$

where $f_0'(0)$ and $h_0(0)$ are the solution of the free convection asymptote (Equations 117 and 118). In the two extreme cases of $\xi = 0$ and ∞ these expression reduce to the correct asymptotes. In between the asymptotes the maximum error in Equation 94 for velocity is about 3 percent and in Equation 95 for temperature is 2 percent.

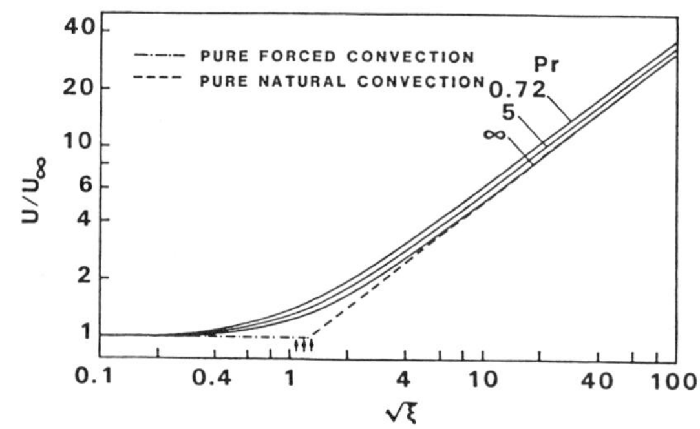

Figure 15. Effects of the Prandtl number on centerline velocity in laminar mixed convection due to a horizontal line source in a vertical stream in an aiding situation [56].

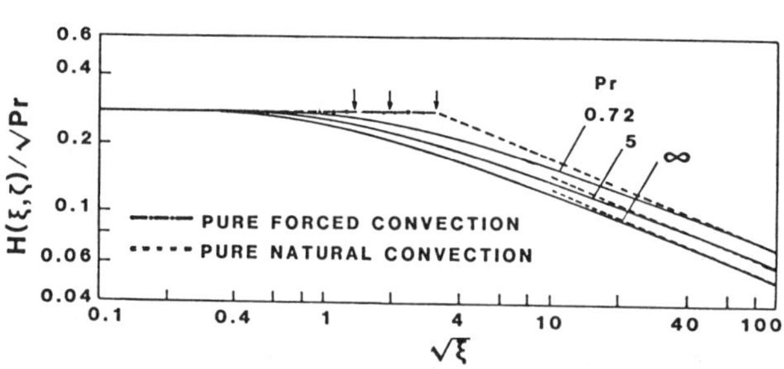

Figure 16. Effects of the Prandtl number on centerline temperature in laminar mixed convection due to a horizontal line source in a vertical stream in an aiding flow situation in a line source in a vertical stream [56].

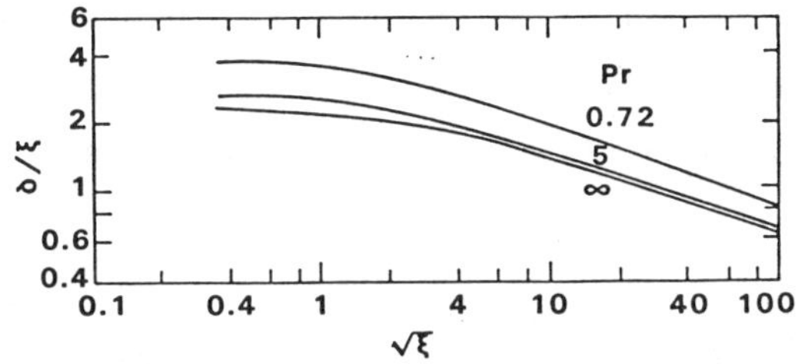

Figure 17. Half width of the plume from velocity profile in a laminar mixed convection due to horizontal line source in a vertical stream in an aiding situation [56].

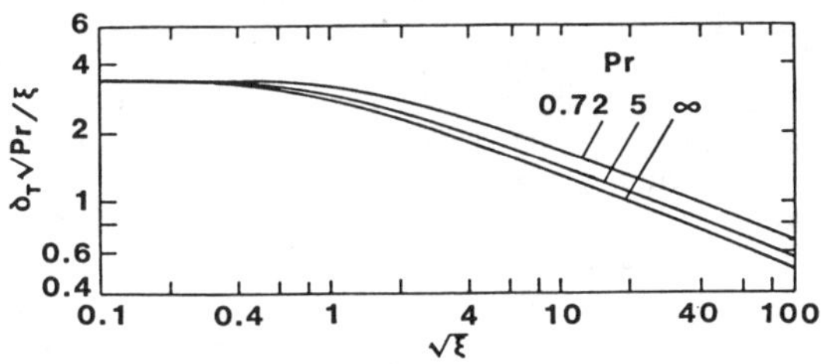

Figure 18. Half width of the plume from temperature profile in laminar mixed convection due to a horizontal line source in a vertical stream in an aiding situation [56].

The width of the plume is another important parameter. Based on velocity profile, width $\bar{\delta}$ may be defined as the distance from the axis where the free stream velocity defect is a small fraction (say k) of the velocity defect between the axis and free stream, i. e.,

$$\frac{u(x,\bar{\delta}) - U_\infty}{u(x,0) - U_\infty} = k \tag{96a}$$

Likewise the thermal width $\bar{\delta}_T$ may be defined as the distance from the axis of the plume where the temperature has dropped to a small fraction of temperature at the axis

$$\frac{T(x,\bar{\delta}_T) - T_\infty}{T(x,0) - T_\infty} = k \tag{96b}$$

As the plume is symmetric about the vertical axis, $\bar{\delta}$ or $\bar{\delta}_T$ represents half of the overall width of the plume. In these definitions the value of the fraction k is arbitrarily assigned.

For $k = 0.05$ the results [56] for δ/ξ and $\delta_T Pr^{1/2}/\xi$ are shown in Figures 17 and 18. Here δ and δ_T are related to dimensional $\bar{\delta}$ and $\bar{\delta}_T$ as

$$\delta = \bar{\delta} g\beta\theta_T/U_\infty^2, \quad \delta_T = \bar{\delta}_T g\beta\theta_T/U_\infty^2 \tag{97}$$

From Figure 17 it can be seen that the width δ from velocity increases with ξ and gives way to a somewhat slower increase at large ξ. The temperature width δ_T in Figure 18 shows that as $\xi \to 0$, $\delta_T Pr^{1/2}/\xi \to 3.46$, a value that can be determined from Equation 80 for weakly buoyant plume. At large ξ the nature of variation of δ_T is similar to δ. At large Pr the thermal width is very much smaller than the velocity's.

A more precise measure of the width is the displacement thickness δ^* defined by

$$\bar{U}\bar{\delta}^* = \int_0^\infty u - U_\infty dy, \qquad \delta^* = \bar{\delta}^* g\beta\theta_T/U_\infty^2 \tag{98}$$

where $\bar{U}$ is some characteristic velocity. The work of Afzal [55] adopted $\bar{U}$ as the free stream velocity U_∞, and the expression for δ^* is

$$\delta^* = \xi \int_0^\infty (1 - F_\zeta)\, d\zeta \tag{99}$$

Haaland and Sparrow [56] have raised objection to the use of U_∞ by Afzal [55] althrough their own dimensionalization of δ and δ_T [(see their equations [7]) are based on U_∞. Further, the results for δ^* are physically meaningful as they are directly related to the the entrainment of outer irrotational fluid in the plume. Reference 56 also suggested adopting $\bar{U} = u(x, 0)$ as the local centerline velocity. The local centerline velocity would work for an aiding flow case, but in the opposing flow situation its use is not desirable, since it decreases to become zero at certain locations.

The work of Afzal [55] as well as of Haaland and Sparrow [56] requires two different sets of equations for studying the entire mixed-convection regime from $\xi = 0$ to ∞: one set of equations in forced-convection-dominated variables, a second set in natural-convection-dominated variables, with a change over at certain values of ξ. The numerical solutions from either end generally become difficult and less accurate, as the mixed-convection parameter increases. For this reason, it is desirable to develop a single set of equations that remains valid for an entire mixed-convection domain. This objective may be achieved, for example, by recasting Equations 74 through 78 in terms of Z, the Euler variable (Equation 91). However, ξ_0 is unknown *ab initio* and without loss of generality we can set $\xi = 1$, in which case the variable Z becomes the conventional Euler transformation [59] defined as

$$Z_c = \xi/(1 + \xi) \tag{100}$$

In general, infinite domain of ξ can be mapped to a finite domain in terms of a parameter α defined as

$$\alpha = R_x^{n/2}/(R_x^{n/2} + Gr_x^{n/5}), \qquad n > 0 \tag{101}$$

that is related ξ or Z_c as

$$1 - \alpha = \frac{\xi^{n/5}}{1 + \xi^{n/5}} = \frac{Z_c^{n/5}}{(1 - Z_c)^{n/5} + Z_c^{n/5}} \tag{102}$$

The case of pure forced-convection plume corresponds to the condition when buoyancy is zero (i.e., $Gr_x = 0$, $\alpha = 1$ ($Z_c = 0, \xi = 0$), whereas pure free-convection plume corresponds to the condition $U_\infty = 0$; then $R_x = 0$, $\alpha = 0$ ($Z_c = 1, \xi = \infty$)).

In order to develop a single set of equations throughout the entire mixed-convection regime the following variables are utilized

$$\begin{aligned} \psi &= \nu(R_x^{n/2} + Gr_x^{n/5})^{1/n} F(\alpha, Y) \\ T - T_\infty &= \theta_T (R_x^{n/2} + Gr_x^{n/5})^{-1/n} H(\alpha, Y) \\ Y &= \frac{y}{x}(R_x^{n/2} + Gr_x^{n/5})^{1/n} \end{aligned} \tag{103}$$

which are related to the variables of weakly buoyant plume as

$$\zeta \to \alpha Y, \qquad F \to F/\alpha, \qquad H \to \alpha H \tag{104}$$

Based on prior transformations of variables the boundary-layer equations for mixed-convection plume yield

$$F_{YYY} + \frac{1}{10}(6-\alpha)FF_{YY} - \frac{1}{5}(1-\alpha)F_Y^2 + (1-\alpha)^{5/n}H$$

$$= \frac{n}{10}\alpha(1-\alpha)(F_\alpha F_{YY} - F_Y F_{Y\alpha}) \tag{105}$$

$$Pr^{-1}H_{YY} + \frac{1}{10}(6-\alpha)(FH)_Y = \frac{n}{10}\alpha(1-\alpha)(F_\alpha H_Y - F_Y H_\alpha) \tag{106}$$

with the boundary conditions

$$Y = 0,\ F = F_{YY} = H_Y = 0 \tag{107}$$

$$Y \to \infty, \quad F_Y \to \alpha^{2/n}, \quad H \to 0 \tag{108}$$

subject to the normalized heat flux relation

$$\int_0^\infty F_Y H d_Y = 1 \tag{109}$$

In the extreme cases of pure forced convection ($\alpha = 1$) and pure free convection ($\alpha = 0$) the nonsimilar terms on the right-hand side of Equations 105 and 106 vanish to yield the ordinary differential equations that are independent of n. In the mixed-convection regime $0 < \alpha < 1$ the problem is governed by partial differential equations. The solution may, in principle, be obtained for a given positive value of n, but it is desirable to choose it in a judicious manner. The positive number n, involved in Equation 101 represents the path in the mapping of an infinite domain in a ξ-plane to a finite domain in a α-plane. The mapping by the Euler transformation is more appropriate, and the value $n = 5$ follows from Equation 102. It may be noted that the behavior of partial differential equations in the neighborhood of the two asymptotes with $n = 5$ corresponds directly to that described in subsections dealing with weakly and strongly buoyant plume.

Strongly Buoyant Plume – Higher Order Effects

Far away from the heat source where the boundary layer is mainly governed by the buoyancy, the flow in the aiding case approaches natural-convection buoyant plume. For this case as well as for buoyant plume the appropriate variables depend on the nondimensional distance from the heat source as [55]

$$\psi = \nu Gr_x^{1/5} f(\varepsilon, \eta), \qquad T - T_\infty = \theta_T Gr_x^{-1/5} h(\varepsilon, \eta)$$

$$\eta = Gr_x^{1/5} y/x, \qquad \varepsilon = \xi^{-2/5} \tag{110}$$

These variables are related to the variables for weakly buoyant plume as

$$\eta = \zeta/\varepsilon^{1/2}, \quad f = \varepsilon^{1/2}F, \quad h = H/\varepsilon^{1/2} \tag{111}$$

Introducing these variables in the boundary-layer equations we get

$$f_{\eta\eta\eta} + \frac{3}{5}ff_{\eta\eta} - \frac{1}{5}f_\eta^2 + h = \frac{1}{5}\varepsilon(f_\varepsilon f_{\eta\eta} - f_\eta f_{\eta\varepsilon}) \tag{112}$$

$$Pr^{-1}h_{\eta\eta} + \frac{3}{5}(fh)_\eta = \frac{1}{5}\varepsilon(f_\varepsilon h_\eta - f_\eta h_\varepsilon) \tag{113}$$

The boundary conditions are

$$\eta = 0 \qquad f = f_{\eta\eta} = h_\eta = 0 \tag{114}$$

$$\eta \to \infty \qquad f_\eta \to \varepsilon,\ h \to 0 \tag{115}$$

subject to heat flux relation

$$\int_{-0}^{\infty} f_{\eta} h d\eta = 1 \tag{116}$$

For $\varepsilon = 0$ these are the equations for free-convection plume. For strongly buoyant plume $\varepsilon \ll 1$ the effects of forced convection can be studied by considering the perturbations expansion in terms of ε as

$$f = \sum_{m=0}^{4} \varepsilon^m f_m(\eta) + \varepsilon^5 \ln\varepsilon \tilde{f}_5(\eta) + \varepsilon^5 f_5(\eta) + \ldots \tag{117}$$

$$h = \sum_{m=0}^{4} \varepsilon^m h_m(\eta) + \varepsilon^5 \ln\varepsilon \tilde{h}_5(\eta) + \varepsilon^5 h_5(\eta) + \ldots \tag{118}$$

The expansion in ε is an inverse coordinate expansion as $\varepsilon \alpha x^{-1/5}$. It is well known that the inverse coordinate expansion suffers from indeterminacy giving to eigen-solutions. The indeterminacy physically represents the uncertainties about the details of the flow near the leading edge and arises from the fact that the initial conditions are not imposed on similarity solutions. If the eigen-solution is ignored, it gives rise to inconsistency in large η behavior of the solution when a term is of the same order as one of the eigen-solutions and that inconsistency would be resolved by inserting a log or log-log term in the series [60]. In the previous expansions the first eigen-solution is

$$\tilde{f}_5 = C_1(2\eta f_0' - 3f_0); \quad \tilde{h}_5 = C_1(2\eta h_0' + 3h_0) \tag{119}$$

occurs at a stage for which a particular integral in the sixth order is required, and to fulfill the condition of exponential decay at large η consisting of an eigen function multiplied by $\ln\varepsilon$, it is added to the expansion.

The ordinary differential equations governing the various order terms are given in Afzal [55]. The results for centerline velocity and temperature for $Pr = 0.72$ are

$$f'(\varepsilon, 0) = 0.8096 + 0.0631\varepsilon + 0.22356\varepsilon^2 + 0.1834\varepsilon^3 + 0.03886\varepsilon^4 + 0(\varepsilon^4) \tag{120a}$$

$$h(\varepsilon, 0) = 0.3773 - 0.1893\varepsilon - 0.008598\varepsilon^2 + 0.1132\varepsilon^3 - 0.06389\varepsilon^4 + 0(\varepsilon^4) \tag{120b}$$

The velocity and temperature profiles for $\varepsilon = 0$, 0.15, and 3.0 are shown in Figures 19 and 20 by solid lines. As ε increases, the centerline velocity increases, the temperature decreases, and the width of the plume is less extensive.

Higher order effects. The analyses considered earlier are based entirely on the classical boundary layer where the Grashof and Reynolds numbers are sufficiently high. When the Grashof or Reynolds numbers are moderately large the second-order corrections contribute significantly. Under this situation the plume becomes thick and interacts with outer irrotational fluid to entrain more ambient fluid in the plume. This entrainment gives rise to an induced flow in the outer layer that gives rise to additional second-order tangential components of velocity which in turn need second-order corrections to boundary layer associated with the plume. Further, the second-order effects when free stream velocity is nonzero also become significant. The stream function of outer flow may be taken as

$$\psi = U_{\infty} y + \Psi \tag{121a}$$

where the first term is due to incident stream and the second term represents the higher order corrections.

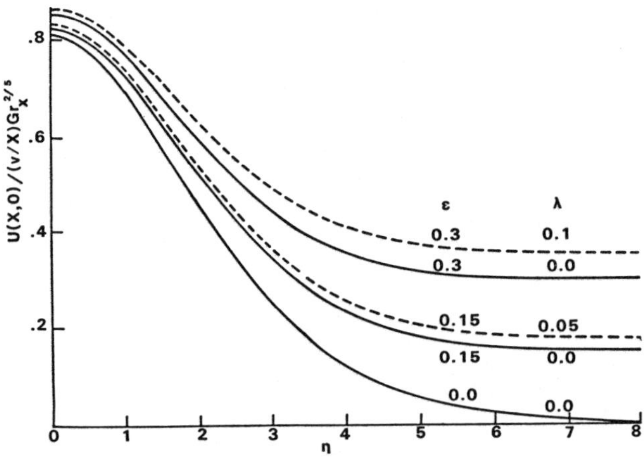

Figure 19. Effects of a moderately large Grashof number on velocity profiles due to a mixed convection in a strongly buoyant two-dimensional system for a vertical stream.

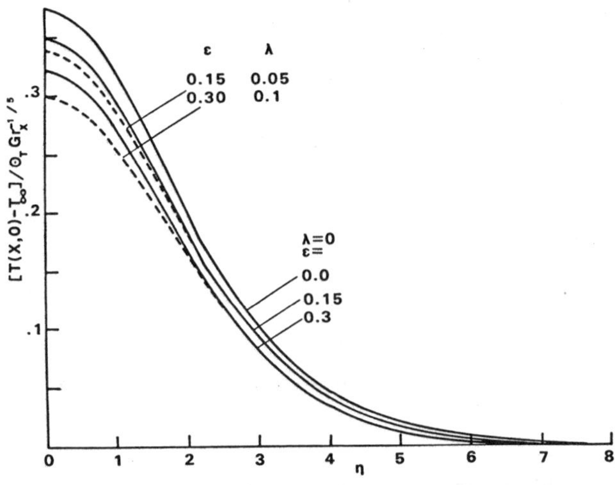

Figure 20. Effects of a moderately large Grashof number on temperature profiles due to mixed convection in strongly buoyant two-dimensional plume for a vertical stream.

As the plume develops it draws in fluid from the outside irrotational fluid and for the transverse velocity component at the edge of the plume the associated stream function is

$$\Psi'(x \geqslant 0,\ y = 0) = \nu Gr_x^{1/5} f(\varepsilon, \eta \rightarrow \infty) \tag{121b}$$

since the plume is thin and from the point of view of outer flow the condition may be transformed at the axis of the plume. The stream function Ψ' for the second correction is given by

$$\Psi'(x \geqslant 0,\ y = \delta) = \nu Gr_x^{1/5} A, \quad A = \lim_{n \rightarrow \infty} (f - \varepsilon\eta) \tag{122}$$

Based on Equation 117, Equation 122 yields

$$\Psi(x \geq 0, y = 0) = \nu Gr_x^{1/5} \left[\sum_{m=0}^{4} A_{0m}\varepsilon^m + 0(\varepsilon^4) \right] \tag{123}$$

$$A_{0m} = \lim_{\eta \to \infty} (f_{0m} - \eta f'_{0m}\delta_{1m}) \tag{124}$$

where δ_{1m} is the Kronecker delta.

In the outer layer the diffusive effects are negligible and the flow is governed by

$$\nabla^2\Psi = 0 \tag{125}$$

$$T - T_\infty = 0$$

The symmetry of the outer flow requires that

$$\Psi(x < 0; y = 0) = 0 \tag{126}$$

The solution to Equation 125 subject to the boundary conditions of Equations 124 and 126 in terms of complex potential W ($= \Phi + i\Psi$) in terms of complex variables $Z = x + iy$, is given by

$$W = \nu Gr_x^{-1/5} \sum_{m=0}^{4} A_{0m}\varepsilon^m \left[i - \cot\left(\frac{3-m}{5}\pi\right)\right] \left(\frac{Z}{x}\right)^{\frac{3-m}{5}} + 0(\varepsilon^4) \Bigg] \tag{127}$$

The velocity component in the vertical direction at y = 0 is given by

$$\Psi_Y(x, 0) = \left.\frac{dW}{dz}\right|_{y=0} = \frac{\nu Gr_x^{1/5}}{x} \left[\sum_{m=0}^{4} \frac{3-m}{5} \cot\left(\frac{3-m}{5}\pi\right) A_{0m}\varepsilon^m + 0(\varepsilon^4) \right] \tag{128}$$

The outer flow requires a second-order correction term to the variables of Equation 110 for velocity, and temperature can be written as

$$\psi = \nu Gr_x^{1/5} [f(\varepsilon, \eta) + \lambda F(\varepsilon, \eta) + \ldots] \tag{129a}$$

$$T - T_\infty = Gr_x^{-1/5}\theta_T [h(\varepsilon, \eta) + \lambda H(\varepsilon, \eta) + \ldots] \tag{129b}$$

where λ is a gauge function. The matching of the inner expansion Equation 129 with outer flow Equation 128 requires that

$$\lambda = Gr_x^{-1/5}$$

The second-order boundary layer equations are

$$F_{\eta\eta\eta} + \frac{3}{5} fF_{\eta\eta} + \frac{1}{5} f_\eta F_\eta + \frac{1}{5} \varepsilon(f_\eta F_{\eta\varepsilon} + f_{\eta\varepsilon}F_\eta - f_\varepsilon F_{\eta\eta} - f_{\eta\eta}F_\varepsilon) + H = D \tag{130}$$

$$Pr^{-1}H_{\eta\eta} + \frac{3}{5}(fH_\eta + 2f_\eta H + F_\eta h_\eta) + \frac{\varepsilon}{5}(f_\eta H_\varepsilon + F_\eta h_\varepsilon - f_\varepsilon H_\eta - F_\varepsilon h_\eta) = 0 \tag{131}$$

The boundary conditions are

$$\eta = 0 \qquad F = F_{\eta\eta} = H_\eta = 0 \tag{132}$$

$$\eta \to \infty \qquad F_\eta \to -\sum_{m=0}^{4} \frac{3-m}{5} \cot\left(\frac{3-m}{5}\pi\right) A_{0m}\varepsilon^m + 0(\varepsilon^4), \; H \to 0 \tag{133}$$

$$\int_0^\infty f_\eta H + F_\eta h d\eta = 0 \tag{134}$$

Here D is a constant of integration representing the effects of induced pressure gradient. The value of D may be estimated from Equation 130 for

$\eta \rightarrow \infty$ as

$$D = \frac{1}{5} \lim_{\eta \rightarrow \infty} (\varepsilon f_\eta F_\eta)_\varepsilon$$

$$= -\frac{\varepsilon}{25} \sum_{m=0}^{4} (2+m)(3-m) \cot\left(\frac{3-m}{5}\pi\right) A_{0m}\varepsilon^m + 0(\varepsilon^4) \tag{135}$$

The expansions for F and H are

$$F = F_0(\eta) + \varepsilon F_1(\eta) + 0(\varepsilon) \tag{136a}$$

$$H = H_0(\eta) + \varepsilon H_1(\eta) + 0(\varepsilon) \tag{136b}$$

The equations for (F_0, H_0) are

$$F_0''' + \frac{3}{5} f_0 F_0'' + \frac{1}{5} f_0' F_0' + H_0 = 0 \tag{137}$$

$$Pr^{-1} H_0'' + \frac{3}{5}(f_0 H_0' + 2f_0' H_0 + F_0' h_0') = 0 \tag{138}$$

$$\eta = 0 \qquad F_0 = F_0'' = H_0' = 0 \tag{139}$$

$$\eta \rightarrow \infty \qquad F_0 \rightarrow -\frac{3}{5} A_\infty \cot\left(\frac{3\pi}{5}\right)\eta + A_{10}, \quad H_0(\infty) = 0 \tag{140}$$

$$\int_0^\infty F_0' h_0 + f_0' H_0 d\eta = 0 \tag{141}$$

represents the correction to the first-order natural-convection plume at moderately large Grashof numbers studied by Riley [25] and Hieber & Nash [26]. The next order terms (F_1, H_1) are governed by the following equations

$$F_1''' + \frac{3}{5} f_0 F_1'' + \frac{2}{5} f_0' F_1' - \frac{1}{5} f_0'' F_1 + H_1 = -\frac{2}{5}(f_1' F_0')' - \frac{6}{25} A_\infty \cot\left(\frac{3\pi}{5}\right) \tag{142}$$

$$Pr^{-1} H_1'' + \frac{1}{5}(3f_0 H_1' + 7f_0' H_1 - F_1 h_0' + 3F_1' h_0) = -\frac{1}{5}(2f_1 H_0' + 6f_1' H_0 + 4F_0' h_1) \tag{143}$$

$$\eta = 0 \qquad F_1 = F_1'' = H_1' = 0 \tag{144}$$

$$F_1 \sim -\frac{2}{5} A_{01} \operatorname{Cot}\left(\frac{2\pi}{5}\right)\eta + A_{11} \quad \text{as } \eta \rightarrow \infty,\ H_1(\infty) = 0 \tag{145}$$

$$\int_0^\infty F_1' h_0 + F_0' h_1 + f_0' H_1 + f_1' H_0 d\eta = 0 \tag{146}$$

which represents the second-order corrections due to nonzero free stream velocity.

The numerical solutions to Equations 137 through 146 are obtained for Pr = 0.72, and the characteristic values are

$$
\begin{aligned}
&F'(\varepsilon,0) = 0.0944 + 0.07611\varepsilon + 0(\varepsilon)\\
&H(\varepsilon,0) = -\ 0.20177 + 0.07443\varepsilon + 0(\varepsilon)\\
&A_{\infty} = 2.0441 \qquad A_{01} = -\ 3.14397\\
&A_{10} = -\ .37765
\end{aligned}
\tag{147}
$$

The total centreline velocity from Equation 129a is

$$
u(x, y = 0) = \frac{\nu}{x} Gr_x^{2/5} \{f_0'(0) + \varepsilon f_1'(0) + \varepsilon^2 f_2'(0) + \varepsilon^3 f_3'(0)
$$

$$
+ \varepsilon^4 f_4'(0) + \ldots + \lambda[F_0'(0) + \varepsilon F_1'(0) + \ldots] + \ldots\} \tag{148}
$$

and a similar expression for temperature. As ε and Gr_x are two independent parameters, the relative magnitude of Gr_x with respect to ε decides the level of importance of the second-order effect. The relative effects of buoyancy to forced convection can also be studied in terms of a parameter Γ defined by Equation 73 that is related to ε and λ as

$$
\varepsilon^3 \Gamma = \lambda \tag{149}
$$

Eliminating λ in favor of Γ, Equation 148 becomes

$$
u(x, 0) = \frac{\nu}{x} Gr_x^{2/5} \{f_0'(0) + \varepsilon f_0'(0) + \varepsilon^2 f_2'(0)
$$

$$
+ \varepsilon^3 [f_3'(0) + \Gamma F_0'(0)] + \varepsilon^4 [f_4'(0) + \Gamma F_1'(0)] + (\varepsilon^4)] \tag{150a}
$$

Likewise the centerline temperature is given by

$$
T(x, 0) - T_{\infty} = \theta_2 Gr_x^{-1/5} \{h_0(0) + \varepsilon h_1(0) + \varepsilon^2 h_2(0)
$$

$$
+ \varepsilon^3 [h_3(0) + \Gamma H_0(0)] + \varepsilon^4 [h_4(0) + \Gamma H_1(0)] + 0(\varepsilon^4)] \tag{150b}
$$

The second-order effects for Γ are on the order of unity, which are on the same order as ε^3 in terms of the first-order boundary layer. When $\Gamma \gg 1$ (or more precisely $\Gamma\varepsilon^n \sim 0(1)$, $0 < n < 3$), the second-order effects become significant. For $\Gamma\varepsilon^3 \sim 0(1)$, the second-order effects become as dominant as first-order effects and the higher-order analysis breaks down. When $\Gamma \sim \varepsilon^m$, $m > 0$ the second-order effects become more and more insignificant as m increases.

The velocity and temperature profiles for various values of ε and λ are displaced in Figures 19 and 20. The solid lines for $\lambda = 0$ and $\varepsilon = 0$, 0.15, and 0.3 represents the solution for the first-order plume and the dotted lines represent the second-order corrections for these values of ε when $\lambda = 0.05$ and 0.1. The figure shows that for any fixed η the first-order effect of mixed convection and the second-order corrections increase the velocity and decrease the temperature as ε or λ increases. For given ε or λ the increase in the level of velocity is large towards the outer edge when compared to velocity at axis while the trend for temperature is just the opposite. A large η increase in the velocity at the edge of boundary layer is due to the mixed-convection effects associated with incident velocity and induced velocity. The increase in the level of velocity and the decrease in temperature is in accordance with the observations and the fact that the total heat flux remains constant in the plume.

Nonvertical Free Stream

If the free stream is externally imposed and inclined at an angle φ with respect to the buoyant force vector, the dynamics of the mixed-convection plume become complex. In the neighborhood of the heat source in the diffusive zone where Oseen linearization is valid, the problem has been considered by Wood [52] and Wesseling [53] for arbitrary φ and by Nakai and Okazaki [54] for $\varphi = \pi/2$. The particular solution for pressure has finite jumps, proportional to sin φ, across the plume, and Wesseling found that the general solution that gives uniform pressure at infinity does not exist, except when an imposed stream is in the vertical direction ($\varphi = 0$). This nonexistence was noted by Wesseling [53] as analogous to the Stokes paradox. The paradox raised by Wesseling has already been resolved in the earlier paper of Wood, who has shown that the singularity in pressure and associated singularity in velocity, across the plume, fits into a wider picture of flow which includes the nonlinear development of plume at distances from the heat source that are larger than that at which Oseen linearization holds. Wood has shown that a secondary irrotational motion is induced, outside the diffusive zone as well as in the nonlinear region, partly by virtue of the inflow at its edge and partly by virtue of the change of pressure that occurs across the plume. As the fluid in the plume is relatively lighter, the dynamic pressure beneath the plume is less than that above the plume and consequently the induced flow (in the irrotational zone) deflects the oncoming stream towards the low side of the plume.

When the incident stream is at angle φ, to the vertical, the location of plume is not known a priori. Let Ψ_0 be a reference streamline chosen arbitrarily from among the streamlines of the plume, and s and n, the orthogonal coordinates respectively, along the reference streamline and normal to it. If the plume is thin relative to the radius of curvature of the streamlines, the flow in the plume is governed approximately by

$$U\frac{\partial U}{\partial s} + V\frac{\partial U}{\partial n} = -\frac{\partial p}{\rho\partial s} + g\beta(T - T_\infty)\cos\varphi' + \nu\frac{\partial^2 U}{\partial n^2} \tag{151}$$

$$\frac{\partial p}{\rho\partial n} = g\beta(T - T_\infty)\sin\varphi' + \frac{d\varphi'}{ds}U^2 \tag{152}$$

$$U\frac{\partial T}{\partial s} + V\frac{\partial T}{\partial n} = \frac{\nu}{Pr}\frac{\partial^2 T}{\partial n^2} \tag{153}$$

$$\frac{\partial U}{\partial n} + \frac{\partial V}{\partial n} = 0 \tag{154}$$

where φ' denotes the inclination of the reference streamline from the vertical. The integral conditions from the momentum and energy equations are

$$\rho\int_0^\infty [U(U_\infty - U) + p - g\beta\int_0^s (T - T_\infty)\cos\varphi' ds]\,dn = 0 \tag{155}$$

$$\rho C_p\int_0^\infty u(T - T_\infty)\,dn = Q \tag{156}$$

The first condition expresses the fact that net drag is zero and the second condition the constancy of total heat flux.

The change in pressure across the entire width of the plume

$-\delta \leqslant n \leqslant \delta$ is

$$p(s,\delta) - p(s,-\delta) = g\beta\cos\varphi'\int_{-\delta}^{\delta}(T - T_\infty)\,dn + \frac{d\varphi'}{ds}\int_{-\delta}^{\delta}U^2 dn \tag{157}$$

The pressure change across the plume is contributed by two mechanisms: one associated directly with the buoyancy of the fluid in the plume and one with the plume curvature. When U is 0(1), the second contribution is plainly small. However, U may become large enough through the action of buoyancy and its order of magnitude may be anticipated from momentum Equation 151 as

$$U^2 = 0(1) + 0(p/\rho) + 0[g\beta \int_0^s (T - T_\infty)\cos\varphi' ds] \tag{158a}$$

$$p/\rho = 0[\int_{-\delta}^{\delta} g\beta (T - T_\infty)\sin\varphi' dy] + 0\left(\frac{d\varphi'}{ds}\right) \tag{158b}$$

Based on Equation 158 the second term on the right-hand side of Equation 152 may be expressed as

$$\frac{d\varphi'}{ds}\int_{-\delta}^{\delta} U^2 dn = \frac{d\varphi'}{ds}[0(\delta) + 0[\delta^2 g\beta(T - T_\infty)\sin\varphi'] + 0\left(\delta\frac{d\varphi'}{ds}\right)$$

$$+ 0[\delta g\beta(T - T_\infty)\cos\varphi']\,ds \tag{158c}$$

and small provided $sd\varphi'/ds$ is at most $0(1)$. Under this condition the plume is not far from the line of incident velocity, φ' may be regarded approximately equal to φ, and the normal momentum equation reduces to

$$\frac{\partial p}{\partial n} = \rho g\beta(T - T_\infty)\sin\varphi \tag{159}$$

Introducing the variables Equation 73 with x and y replaced by s and n, pressures can be expressed as

$$p = \rho U_\infty^2 P(\xi, \zeta) \tag{160}$$

Equations 151, 153, and 159 reduce to

$$2F_{\zeta\zeta\zeta} + FF_{\zeta\zeta} + (\zeta\Gamma\sin\varphi + 2\xi\cos\varphi)H = \xi(F_\zeta F_{\zeta\xi} - F_\xi F_{\zeta\zeta}) \tag{161}$$

$$2Pr^{-1}H_{\zeta\zeta} + (FH)_\zeta = \xi(F_\zeta H_\xi - F_\xi H_\zeta) \tag{162}$$

$$P_\zeta = \Gamma\sin\varphi H \tag{163}$$

The boundary conditions are

$$\zeta \to \pm\infty, \quad F_\zeta \to 1, \quad H \to 0 \tag{164}$$

$$\zeta = 0, \quad F = 0 \tag{165}$$

subject to the integral conditions

$$\int_{-\infty}^{\infty} F_\zeta(1 - F_\zeta) + \xi\cos\varphi H + Pd\zeta = 0 \tag{166a}$$

$$\int_{-\infty}^{\infty} F_\zeta H d\zeta = 1 \tag{166b}$$

When ξ and Γ are small, the solution to the momentum and energy equations

$$F_0 = \zeta$$

$$H_0 = e\exp\left(-\frac{Pr}{4}\zeta^2\right), \quad e = \left(\frac{Pr}{4\pi}\right)^{1/2} \tag{167}$$

is the same as Equation 80. For small values of Γ and ξ, varibles may be expanded

$$F = F_0(\zeta) + e[\Gamma\sin\varphi F_{10}(\zeta) + \cos\varphi F_{11}(\zeta)]$$

$$+ e^2[\Gamma^2\cos^2\varphi F_{20}(\zeta) + \xi\Gamma\cos\varphi\sin\varphi F_{21}(\zeta) + \xi^2\cos^2\varphi F_{22}(\zeta)] + \ldots \tag{168}$$

along with similar expansions for H/e and P. The ordinary differential equations govering the first-order terms ($F_{1j}, H_{1j}, j = 0, 1$) may be written as

$$2F_{1j}''' + \zeta F_{1j}'' - jF_{1j}' = -[\zeta(1-j) + 2j]H_0 \tag{169a}$$

$$2Pr^{-1}H_{1j}'' + \zeta H_{1j}' + (1-j)H_{1j} + F_{1j}H_0 + (1+j)F_{1j}H_0' = 0 \tag{169b}$$

$$\zeta \to \pm\infty, \qquad F_{1j}' \to 0, \quad H_{1j} \to 0$$

The closed-form solutions for F_{1j} are [52, 54]

$$F_{10} = \frac{1}{1-Pr}\int_0^\infty [\exp(-Pr\zeta^2/4) - Pr^{-1/2}\exp(-\zeta^2/4)]\,d\zeta \tag{170a}$$

$$F_{11} = \frac{|\zeta|}{1-Pr}\int_0^\infty \zeta^{-2}[Pr^{-1/2}\exp(-Pr\zeta^2/4) - \exp(-\zeta^2/4)]\,d\zeta \tag{170b}$$

The second-order terms in Equation 168 contains the cross product $\xi\Gamma\sin\varphi\cos\varphi$ of the two perturbation parameters. The two special cases where these cross product terms may be neglected is described in the following.

In the first case of horizontal stream studied by Nakai and Okazaki [54] $\varphi = \pi/2$, the buoyancy term multiplying $\xi\cos\varphi$ in Equation 161 vanishes and the expansions are in the powers of Γ. For this case the solutions for the first three terms in the expansions have been studied by Nakai and Okazaki [54]. The streamline $\psi = 0$ can be approximately expressed as

$$y = \zeta_*(\nu x/U_\infty)^{1/2} \tag{171}$$

$$\zeta_* = -[F_{10}(0)e\Gamma + \{1/2F_{20}''(0) - F_{10}(0)F_{10}'(0)\}(e\Gamma)^3 + \ldots]$$

where $F_{10}(0) = (Pr + Pr^{1/2})^{-1}$

and values of $F_{10}'(0)$ and $F_{20}'(0)$ from numerical solutions [54] are given in Figure 21.

The second case where $\Gamma\cos\varphi \ll \xi\sin\varphi$ has been studied is by Wood [52]. The ratio of $\Gamma\cos\varphi$ to $\xi\sin\varphi$ is $R_x^{1/2}\tan\varphi$, which is very small except for $\varphi = \pi/2$. In this case the term associated with $\Gamma\sin\varphi$ in the monumentum (Equation 161) may be neglected in comparison to $\xi\cos\varphi$, and the pressure change across the plume does not affect the solution. Therefore, the solution for the slanted plume can be obtained from those for the vertical case by putting $\xi\cos\varphi$ in place of ξ and s, n, U, and V respectively in place of x, y, u, and v. The reference streamline may be determined in the manner similar to Equation 171 or may be

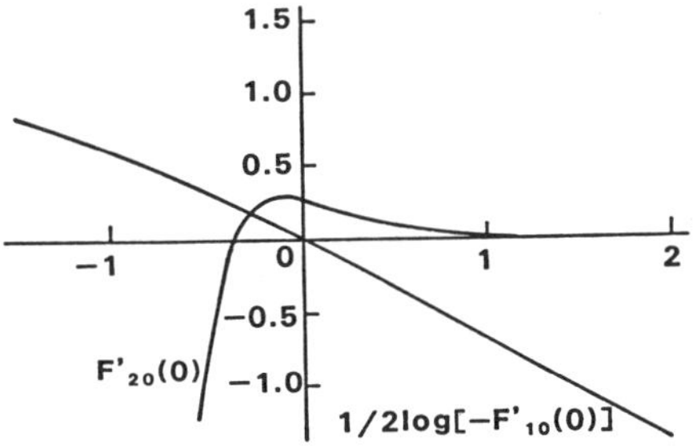

Figure 21. The characteristic values for a weakly two-dimensional laminar buoyant plume in a horizontal forced stream [54].

taken as the central streamline relative to which the streamwise velocity component and temperature are even functions of n. The slanted plume draws fluid from outside, and the transverse velocity component at the edge may be represented as

$$V_+ = \Gamma U_\infty \cos\varphi N(X) \quad (172a)$$

$$V_- = -\Gamma U_\infty \cos\varphi N(X) \quad (172b)$$

where $X = (\xi\cos\varphi)^2$ (173)

The net influx at the two edges of plume is

$$V_+ - V_- = 2\Gamma U_\infty \sin\varphi N(X) \quad (174)$$

The change in pressure across the plume obtained from Equation 163 is

$$\Delta p = \Gamma\rho U_\infty^2 \sin\varphi \int_{-\infty}^{\infty} H(X,\zeta)\, d\zeta = \Gamma\rho U_\infty^2 \sin\varphi M(X) \quad (175a)$$

This pressure jumps in terms of outer irrotational flow conditions may written as [52]

$$U_+ - U_- = -2\Gamma U_\infty \sin\varphi M(X) \quad (175b)$$

The secondary flow induced outside the plume by the net influx of Equation 175b across the width has been estimated by Wood [52] and his result for small $Z = X + iY$ is

$$u - iv = 1 - \frac{\Gamma}{\pi} e^{i\varphi}\log(-Z) + \Gamma C + 0(\Gamma Z^2) \quad (176)$$

$$= 1 + u_0 - iv_0 - \frac{\Gamma}{\pi}\log(x - iy) + 0(\Gamma Z^2) \quad (177)$$

where $u_0 - iv_0 = \Gamma[C - \frac{2e^{-i\alpha}}{\pi}\log(\Gamma\cos\varphi)]$

$$C = \frac{1}{\pi}\int_0^1 \{i[M(s) - M(0)]\sin\varphi - [N(s) - N(0)]\cos\varphi\}\frac{ds}{s}$$

$$+ \int_0^1 [iM(s)\sin\varphi - N(s)\cos\varphi]\frac{ds}{s}$$

Equation 177 shows that the plume is immersed in an outer flow whose complex velocity consists of a uniform contribution $1 + u_0 - iv_0$ together with a contribution varying as $\log(-x - iy)$. In the diffusive zone, where $x + iy = 0(1)$, the effect of outer flow on the convection zone has been treated by Wood using the Navier-Stokes equations.

The presence of term $\log Z$ in Equation 176 is because the influx in the plume and the pressure difference across the plume are initially nonzero. The expression shows that for small Z, irrotational velocity is less steeply inclined to the horizontal than the incident stream. The reason is that the transverse buoyancy force is balanced by a pressure gradient across the plume, and it is this counterforce to which irrotational flow responds. Therefore, in order to banlance the buoyancy the pressure p on the upper side of the plume has to be higher than on the lower side, and as a consequence the oncoming stream (in the irrotational zone) is deflected towards the lower side of the plume.

MIXED CONVECTION FROM POINT SOURCE

The mixed convection due to a stationary heat source in a uniform free stream has been studied by several workers. The weakly buoyant plume has been studied by Csanady [61], Hieber, and Gebhart [62], Wesseling [52], and Afzal [63] and the strongly buoyant plume by Appalaswamy and Jaluria [64]. Afzal [63] has also presented a complete solution for the entire mixed-convection regime for $Pr = 0.72$. Riley and Drake [65] have studied the problem for $Pr = 1$.

In the studies of weakly buoyant plume the Oseen linearization has been employed. Csanady [61] considered the horizontal stream and presented a solution for the weakly buoyant plume at large Reynolds numbers for a Prandtl number of unity. In the study of Wesseling [53] the forced stream is inclined at an arbitrary angle with respect to the buoyant force vector and a solution to the first-order perturbation for velocity based on the Oseen-Boussinesq linearization of the Navier-Stokes equation has been obtained. Hieber and Gebhart [62] in their study on mixed convection from a heat sphere in a vertical stream have also studied the weakly buoyant plume arising from the sphere for a Prandtl number of unity when the Reynolds number is large. For a vertical free stream, the weakly buoyant plume based on boundary-layer equations has been studied by Afzal [63]. In terms of perturbation expansion closed-form solutions for first-order perturbation in velocity and temperature have been obtained for general values of Prandtl numbers. Appalaswamy and Jaluria [64] studied the case of a strongly buoyant plume by considering a small perturbation to the free convection plume for Prandtl numbers of 0.7 and 7. A complete solution for the entire mixed convection regime has been given by Afzal [63] for the situation when aiding or opposing the main flow. It was shown that the solutions are unique, whereas in the opposing case solutions are dual corresponding to forward and reverse flows in the plume.

A moving heat source in a stagnant medium has been considered by Crane [66] and Miloh and Yahalom [67], whereby the observer's frame of reference is that of moving at the velocity of the heat source thus reducing it to a mixed-convection problem. The plume in a vertical direction studied by Crane [66] is oriented such that the plume descends with uniform velocity with the heat source at the origin (corresponding to the weakly buoyant plume of Afzal [63]). Miloh and Yahalom [67] studied the weakly buoyant plume arising from a source moving uniformly in the horizontal direction beneath a free surface or a horizontal rigid boundary by employing the Oseen linearization. When compared with Wesseling [53], their solution includes the correction term to accomodate the additional boundary condition at free surface.

Vertical Free Stream

Formulation

When a vertical main stream of velocity $U_\infty(X)$ is imposed on the point heat source the mixed convection flow is axisymmetric, that is, not amenable to the similarity analysis except for a uniform free stream, (i.e. U_∞ must be constant). This is because the characteristic free convection velocity associated with the source, as shown in the first section, is independent of x. Therefore in either of the forced- or free-convection-dominated variables the mixed convection for a uniform stream is governed by ordinary differential equations.

From the analysis of the previous section the similarity varibles may be defined as

$$\psi = \nu x F(\zeta) \tag{178a}$$

$$T - T_\infty = \theta_p H(\zeta) \tag{178b}$$

$$\theta_p = Q/(2\pi\rho C_p \nu x) \tag{178c}$$

$$\zeta = Jy^2/(2x^2) \tag{178d}$$

Afzal [63] considered two sets of equations: one set in forced-convection-dominated variables ($J = R_x$) and a second set in free-convection dominated variables ($J = Gr_x^{1/2}$).

However, if $J = R_x + Gr_x^{1/2}$ is adopted, a single set of equations will follow for the entire mixed-convection regime. A more general $J = (R_x^n + Gr_x^{n/2})^{1/n}$ also leads to a single set, and we shall return to these considerations later.

In the forced-convection-dominated variables $J = R_x$ and the axisymmetric boundary-layer equations yield

$$(\zeta F'')' + 1/2FF'' \pm \alpha H = 0 \tag{179}$$

$$Pr^{-1}(\zeta H')' + 1/2(FH)' = 0 \tag{180}$$

The boundary conditions at the axis are that of symmetry, and far away the flow approaches the uniform stream

$$\zeta = 0: \quad F = \zeta^{1/2}F'' = \zeta^{1/2}H' = 0 \tag{181}$$

$$\zeta \rightarrow \infty: \quad F' \rightarrow 1 \qquad H \rightarrow 0 \tag{182}$$

subject to normalized heat flux relation

$$\int_0^\infty F'H d\zeta = 1 \tag{183}$$

Here α is a parameter, independent of x, given by

$$\alpha = \frac{Gr_x}{2R_x^2} = \frac{g\beta Q}{4\pi\rho C_p \nu U_\infty^2} \tag{184}$$

which is a measure of the relative importance of the buoyancy effect with respect to the forced convection. The limiting cases of $\alpha \rightarrow 0$ and ∞ correspond respectively to weakly and strongly buoyant plumes. Furthermore the + and − sign of α in Equation 179 correspond to the cases when buoyancy is aiding and opposing the forced convection. As the plume develops, it entrains the outer irrotational fluid from its edge. The normal component of velocity at the edge of the plume is

$$yv|_{y\rightarrow\infty} = \nu\Delta \tag{185}$$

$$\Delta = \lim_{\zeta\rightarrow\infty} [\zeta - F(\zeta)]$$

where Δ may be termed as an entrainment coefficient.

The existence of the solutions for Equations 179 through 183 in a weaker sense can be analyzed by linearization where F in the nonlinear terms is replaced by ζ. With the linearization the solution to Equations 179 through 182 is [63]

$$F' = 1 - \frac{2A(\pm\alpha)}{Pr - 1}[E_1(Pr\zeta/2) - E_1(\zeta/2)] \tag{186}$$

$$H = A\exp(-Pr\zeta/2), \quad A > 0 \tag{187}$$

where $E_1(t)$ is the exponential integral defined as

$$E_1(t) = \int_t^\infty e^{-x}/x dx \tag{188}$$

$$\cong -\ln t + \gamma^* + 0(t) \quad \text{as } t \rightarrow 0$$

and $\gamma^* = 0.577$ is Euler's constant. Based on the preceding solution the heat flux relation (Equation 183) yields a quadratic equation

$$\hat{e}\alpha A^2 + \alpha_s(-2APr + Pr^2) = 0 \tag{189}$$

where $\hat{e} = +1$ for aiding flow and -1 for opposing flow, and α_s is given by

$$\alpha_s = (Pr - 1)\Big/\left[4Pr\ln\left(\frac{1 + Pr}{2}\right)\right] \tag{190}$$

The two roots of the equation are

$$A = -\Pr \frac{1 \pm [1 + \hat{e}\alpha/\alpha_s]^{1/2}}{\hat{e}\alpha/\alpha_s} \tag{191}$$

For the opposing case ($\hat{e} = -1$) the two roots are distinctly positive for $\alpha < \alpha_s$ and leading to dual solutions that correspond to forward and reverse flows in the plume. For $\alpha = \alpha_s$ the two roots coincide; the solutions is unique corresponding to a marginal state between the forward and reverse flow solutions. For $\alpha > \alpha_s$ the roots are complex implying the nonexistence of the solutions. For the aiding case ($\hat{e} = 1$) only one of the two roots is positive, implying that a unique solution exists for all σ. For strongly buoyant plume $\alpha \gg 1$, and it is appropriate to consider $J = Gr^{1/2}$ in Equation 178d or equivalently recast Equations 179 through 183 in terms of variable η defined as

$$\eta = \zeta/\hat{\beta}, \quad \hat{\beta} = \alpha^{-1/2} \tag{192}$$

to get the following equations

$$(\eta F'')' + 1/2\, FF'' \pm H = 0 \tag{193}$$

$$(2\eta H' + \Pr FH)' = 0 \tag{194}$$

$$\eta = 0, \quad F = \eta^{1/2}F'' = \eta^{1/2}H' = 0 \tag{195a}$$

$$\eta \to \infty, \qquad F' \to \hat{\beta}, \quad H \to \infty \tag{195b}$$

$$\int_0^\infty F'H d\eta = 1 \tag{196}$$

For aiding flow (+ sign), as $\hat{\beta} \to 0$ the flow approaches free-convection buoyant plume and a solution exists for all values of $\hat{\beta}$ between 0 and infinity. For opposing flow (− sign) the dual solutions exist only for $\hat{\beta} > \hat{\beta}_s$ (where $\hat{\beta}_s = \alpha_s^{-1/2}$), and $\hat{\beta} < \hat{\beta}_s$ the solutions are nonexistent.

It may be noted that for the case of opposing flow the present equations differ from Appalaswamy and Jaluria [64]. In their equations the buoyancy term occurs as $+H$ and the boundary condition as $f(\infty) = -\hat{\beta}$, whereas in Equations 193 and 195b these signs are opposite (i.e., the buoyancy term occurs as $-H$ and the boundary condition as $f'(\infty) = \hat{\beta}$). Further these signs in Appalaswamy and Juluria [64] cannot be interchanged by a transformation of variables to conform to Afzal [63]. Therefore, for the opposing flow case the formulation of Appalaswamy and Jaluria [64] is in error as these authors have adopted a positive sign with buoyancy terms instead of negative sign [63, 65].

Weakly Buoyant Plumes

When forced convection predominates over natural convection the variables F and H in Equations 179 through 183 are expanded by Afzal [63] in powers of α as

$$F = \sum_{n=0} F_n(\zeta)\,(\pm\alpha)^n, \quad H = \sum_{n=0} H_n(\zeta)\,(\pm\alpha)^n \tag{197}$$

The leading term F_0 satisfies the equation

$$(\zeta F_0'')' + 1/2\, F_0F_0'' = 0\,(\pm\alpha) \tag{198}$$

For the aiding case (+ sign) the solution is

$$F_0(\zeta) = \zeta \tag{199a}$$

In the opposing case (− sign) one of the solutions is Equation 199a, and the second solution, as described later, is that of a mixing layer, with a shift in the origin. Based on

Equation 199a, Equation 197 becomes the classical Oseen linearization, and the solution to lowest-order temperature distribution is

$$H_0 = 1/2\,\mathrm{Pr}\exp(-\mathrm{Pr}\zeta/2) \tag{199b}$$

The closed-form solutions for the first-order perturbations for velocity and temperature from Afzal [63] for Pr = 1 are

$$F_1 = 2(1 - e^{-\zeta/2}) \tag{200a}$$

$$H_1 = 1/2[2\ln 2 - 1 - \gamma^* - \ln\zeta - E_1(\zeta/2)]e^{-\zeta/2} \tag{200b}$$

and for $\mathrm{Pr} \neq 1$ are

$$F_1 = \frac{2\mathrm{Pr}}{\mathrm{Pr}-1}\left[\frac{\zeta}{2}E_1(\zeta/2) - \frac{\zeta}{2}E_1(\mathrm{Pr}\zeta/2) + 1 - e^{-\zeta/2} - \frac{1}{\mathrm{Pr}}(1 - e^{-\mathrm{Pr}\zeta/2})\right] \tag{201a}$$

$$H_1 = \frac{\mathrm{Pr}^3}{2(\mathrm{Pr}-1)}\left[\frac{1}{\mathrm{Pr}}\ln\frac{(1+\mathrm{Pr})^{\mathrm{Pr}}}{2\mathrm{Pr}^{\mathrm{Pr}}} + \frac{1}{\mathrm{Pr}}(\mathrm{Pr}\zeta/2 + 1)E_1(\mathrm{Pr}\zeta/2) - (\zeta/2 + 1)E_1(\zeta/2)\right.$$
$$\left. + \frac{1}{\mathrm{Pr}}(1 - e^{-\mathrm{Pr}\zeta/2}) - 1 + e^{-\zeta/2} + \frac{\mathrm{Pr}-1}{\mathrm{Pr}}(\gamma^* + \ln\zeta - \ln 2) + \frac{1}{\mathrm{Pr}}\ln\mathrm{Pr}\right]e^{-\mathrm{Pr}\zeta/2} \tag{201b}$$

The centerline velocity and temperature are given by

$$F'(0) = 1 + \frac{\mathrm{Pr}\ln\mathrm{Pr}}{\mathrm{Pr}-1}(\pm\alpha) + \dots \qquad \mathrm{Pr} \neq 1 \tag{202a}$$

$$= 1 + (\pm\alpha) + \dots \qquad \mathrm{Pr} = 1 \tag{202b}$$

$$H(0) = \frac{\mathrm{Pr}}{2} + \frac{\mathrm{Pr}^2}{2(\mathrm{Pr}-1)}\ln\left[\frac{(1+\mathrm{Pr})^{\mathrm{Pr}}}{2\mathrm{Pr}^{\mathrm{Pr}}}\right](\pm\alpha) + \dots \qquad \mathrm{Pr} \neq 1 \tag{203a}$$

$$H(0) = 1/2 + 1/4(2\ln 2 - 1)(\pm\alpha) + \dots \qquad \mathrm{Pr} = 1 \tag{203b}$$

At large Pr the validity of the solutions is restricted to $|\alpha| \ll (\ln \mathrm{Pr})^{-1}$. For $\alpha > 0$, these solutions have also been obtained by Crane [66] for a falling heat source, in a frame of reference fixed with the falling source.

Moderately Buoyant Plume

Opposing flow situation. For the case when buoyancy opposes the forced free stream the solutions are dual. For the computation of dual solutions it was found convenient to fix the values of $F'(0)$ at various levels and to determine the values of α and $H(0)$. From numerical results for the centerline velocity, temperature and the entrainment coefficient Δ are shown in Figures 22 and 23. The examination of Figure 22 shows that the centerline velocity, initially equal to the forced-convection value $F'(0) = 1$ decreases with increasing α and becomes zero at $\alpha = 0.6076$ (later denoted by α_s). This decrease is because the buoyancy opposes the forced convection and at $\alpha = \alpha_s$ the opposition of buoyancy is so strong that the oncoming stream becomes stagnant at the axis of the plume. At $\alpha = \alpha_s$, $F'(0) = 0$ and $dF'(0)/d\alpha$ is infinite, consequently the curve turns back implying the nonexistence of solutions for $\alpha > \alpha_s$. This is because the solution around $\alpha = \alpha_s$ possesses a square root singularity i.e. $F'(0) \sim \pm(\alpha_s - \alpha)^{1/2}$. Further, as α decreases from its value α_s the centerline velocity decreases further to become negative, implying the reverse flow. The fact that

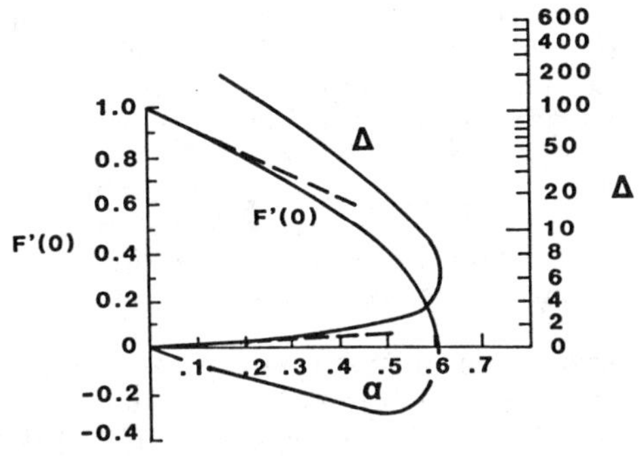

Figure 22. The centerline velocity and entrainment function in an opposing situation for laminar mixed convection due to a point heat source in a vertical free stream for air [63].

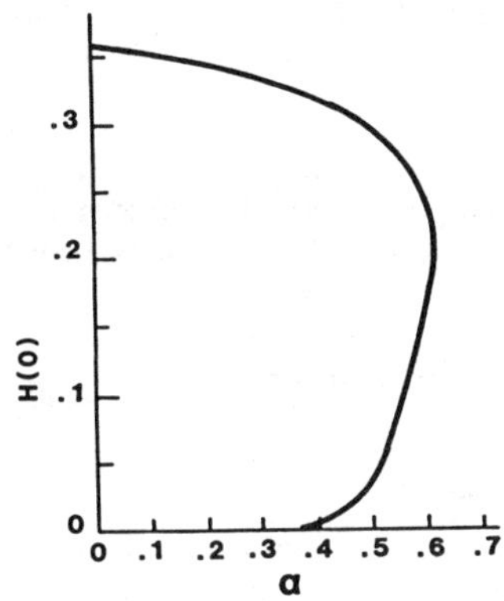

Figure 23. The centerline temperature for laminar mixed convection in an opposing situation due to a point heat source in a vertical free stream for air [63].

$F'(0) = 0$ at $\alpha = \alpha_5$ and $\alpha = 0$, in the reverse flow domain $F'(0)$ should be minimum at $\alpha = \alpha_m$, i.e., the magnitude of reverse flow velocity is maximum. Therefore for $\alpha < \alpha_s$ the solutions are dual, corresponding to forward and reverse flows in the plume, and $\alpha = \alpha_s$ is the marginal state (turning point) when the two solutions converge for a centerline velocity of zero.

The solution for the entrainment coefficient Δ, also displayed in Figure 22, shows clearly the duality of solutions. For a given α, the dual solutions for Δ are positive and the larger value corresponds to the reverse flow case. This is because in the reverse flow situation the thickness of the plume is much greater than in a forward flow situation, and consequently a large amount of fluid is entrained. The numerical results for the thickness of the plume δ (defined as a value of y where $u = 0.99\ U_\infty$) and the thickness of the reverse flow domain are given in Table 2.

The centerline temperature $H(0)$ displayed in Figure 23 again shows a duality of solutions for $\alpha < \alpha_s$ that predicts two positive values for the centerline temperature. In the forward flow situation $\alpha = 0$ and $H(0) = Pr/2$. The centerline temperature decreases as α increases and at $\alpha = \alpha_s$ $dH(0)/d\alpha$ is infinite, and consequently the curve turns back implying the nonexistence of solutions for $\alpha > \alpha_s$. In the reverse flow domain as α decreases from α_s to 0 the centerline temperature decreases continuously and at $\alpha = 0$ it approaches zero.

The velocity and temperature profiles for various values of α in the range $0 \leqslant \alpha \leqslant \alpha_s$ are displayed respectively in Figures 24 and 25 and the curves labeled 1, 2, 3, ... 11 correspond to values of α given in Table 2. The curve labeled 6 is for marginal state $\alpha = \alpha_s$, the curves 1–5 are for the forward flow domain and curves 7–11 are for the reverse flow domain. Figures 24 and 25 clearly show the duality of solutions corresponding to forward and reverse flow profiles. Curve 5 shows that the thickness δ_t of the reverse flow layer is zero at $\alpha = \alpha_5$ (curve 6) and increases as α decreases from α_s to 0. The magnitude of the reverse flow velocity $F'(0)$ is maximum in between $0 < \alpha < \alpha_s$, and zero at $\alpha = \alpha_s$ and 0. The temperature profile displayed in Figure 25 shows that for the forward flow case (curves 1–6) the temperature decreases monotonically from its value $H(0)$ at the axis to zero at infinity. However, in the reverse flow domain the temperature profile shows a maximum (say $H = H_{max}$ at $\zeta = \zeta_{max}$). As α decreases from its value α_s to 0, ζ_{max} increases and H_{max} decreases continuously and as $\alpha \to 0$, $H_{max} \to 0$ and $\zeta_{max} \to \infty$. The above behavior of the velocity and temperature profile in the neighborhood of $\alpha = 0$ suggests that velocity profile approaches a mixing layer, similar to that of Chapman, with a shift in origin. Here, the mixing layer arises in a manner analogous to that of Kennedy and Stewartson (see Reference 68, Chapter 4) where the pressure gradient pressure in a self-similar incompressible wake approaches zero.

In the reverse flow situation, when the boundary layer thickness (in terms of a transformed coordinate, say, ζ_∞) is appreciably larger than the forward flow situation, the utility of the boundary layer equations may be somewhat limited. If y_∞ is the thickness of the boundary layer in the physical coordinates, then from Equation 178d we have

$$(y_\infty/x)^2 = \zeta_\infty 2\nu/U_\infty x \tag{204a}$$

The boundary layer equations are valid for $y_\infty \ll x$ and the condition of Equation 204a is satisfied if

$$x \gg \nu\zeta_\infty/U_\infty \tag{204b}$$

Equation 204b implies that for small values of ζ_∞, the validity of the solutions extends to the near source region, whereas for large values of ζ_∞, the solution is valid relatively far

Table 2
Adverse Case: Characteristic Values of Buoyancy Layer for Various Values of Mixed Convection Parameter α

Legend in Figures 24 and 25	$F'(0)$	α	$H(0)$	$\delta[U_\infty/(2\nu x)]^{1/2}$	Δ	$\delta_t[U_\infty/(2\nu x)]^{1/2}$
1	1.0	0.0	0.36	0	0	
2	0.8	0.2141	0.3438	2.7823	0.5230	
3	0.6	0.3826	0.3234	3.2209	1.1817	
4	0.4	0.5052	0.2969	3.8932	2.0479	
5	0.2	0.5811	0.2608	4.0908	3.2695	
6	0.0	0.6076	0.2083	4.3204	5.2594	0.0
7	−0.2	0.5728	0.1188	5.0889	10.0744	1.633
8	−0.27	0.5112	0.4650E-1	6.1613	18.9371	2.750
9	−0.24	0.4054	0.9818E-2	7.7822	39.1690	4.485
10	−0.20	0.3609	0.3146E-3	9.3969	60.1973	5.820
11	−0.15	0.2675	0.5765E-4	12.1201	105.0362	8.116

From Afzal [63].

from the heat source. In the neighborhood of the region $x \sim \nu\zeta_\infty/U_\infty$ the boundary-layer equations fail and one has to consider the Navier-Stokes equations.

Aiding Flow Situation. When buoyancy and forced convection effects are in the same direction (+ sign) the situation results in aiding the two flows. For aiding flow, the solutions for the entire mixed-convection regime have been given by Afzal [63] and are based on two sets of equation. The solutions in the range $0 \leqslant \alpha \lesssim 1$ we obtained from Equations 179 through 183 in forced-convection dominated variables and in the range $1 \leqslant \alpha \leqslant \infty$ (or $0 \leqslant \hat{\beta} \leqslant 1$) from Equations 193 through 196 in free convection dominated variables with a

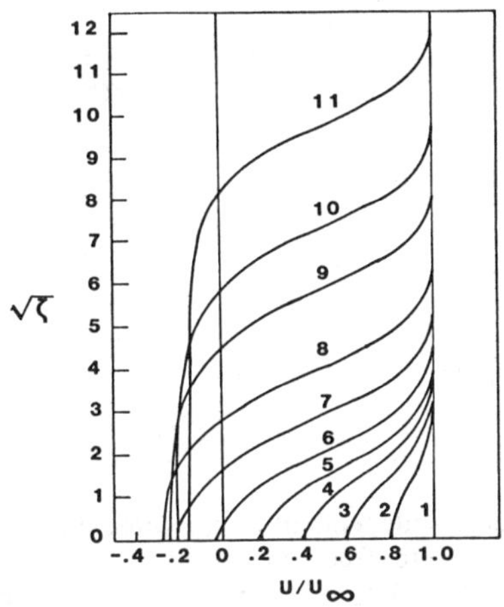

Figure 24. The profiles of vertical velocity for laminar mixed convection in an opposing situation due to a point heat source in a vertical forced stream for air [63].

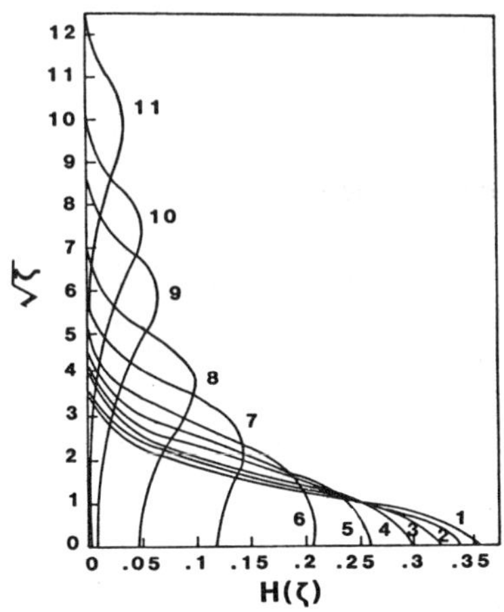

Figure 25. The profile of temperature for laminar mixed convection in an opposing situation due to a point heat source in a vertical forced stream for air [63].

change over at $\alpha = \tilde{\beta} = 1$ for Pr = 0.72. Appalaswamy and Jaluria [64] presented solutions for $\hat{\beta} \leqslant 0.1$, strongly buoyant flows for Pr = 0.7 and 7 and Riley and Drake for Pr = 1.

Here, we formulate a single set of equations, similar to that described in the section for mixed convection in a two-dimensional plume. A single set of equations for the entire mixed-convection regime may be obtained if in Equation 178 we set

$$J = (R_x^n + Gr_x^{n/2})^{1/n} \tag{205}$$

and the axisymmetric boundary-layer equations yield

$$2(\zeta F'')' + FF'' + \gamma H = 0 \tag{206}$$

$$2(\zeta H')' + Pr(FH)' = 0 \tag{207}$$

The boundary conditions are

$$\zeta = 0: \quad F = \zeta^{1/2}F'' = \zeta^{1/2}H' = 0 \tag{208}$$

$$\zeta \to \infty \quad F' \to \beta, \quad H \to 0 \tag{209}$$

subject to integral condition

$$\int_0^\infty F'Hd\zeta = 1 \tag{210}$$

Here γ and β are given by

$$\gamma = (1 - \Lambda)^{2/n} \qquad \beta = \Lambda^{1/n} \tag{211}$$

$$\Lambda = \frac{R_x^n}{R_x^n + Gr_x^{n/2}} \tag{212}$$

The mixed-convection parameter Λ, independent of x, represents the relative importance of forced convection when compared to free-convection. For the case of purely forced convection, the buoyancy is zero (i.e., $Gr_x = 0$ and $\Lambda = 1$) and for purely free-convection flow $U_\infty = 0$ and $\Lambda = 0$. Therefore Equation 212 maps the entire mixed-convection domain to $0 \leqslant \Lambda \leqslant 1$, for all values of n. The mapping constant n may be chosen such that the single set of equations in the asymptotic cases directly leads to the equations for weakly and strongly buoyant plume described earlier. Therefore we adopt

$$n = 2$$

$$\gamma = 1 - \beta^2 \qquad \beta = \Lambda^{1/2} \tag{213}$$

$$\Lambda = R_x^2/(R_x^2 + Gr_x)$$

For large values of Pr, the transformation

$$\zeta = \eta Pr, \quad f = FPr, \quad h = H/Pr \tag{214}$$

reduces Equations 206 through 210 to

$$2(\eta f'')' + Pr^{-1}ff'' + (1 - \beta^2)h' = 0 \tag{215}$$

$$2(\eta h')' + (fh)' = 0 \tag{216}$$

$$\eta = 0 \qquad f = \eta^{1/2}f'' = \eta^{1/2}h' = 0 \tag{217}$$

$$\eta \to \infty \qquad f' \to \beta, \quad h \to 0 \tag{218}$$

$$\int_0^\infty f'hd\eta = 1 \tag{219}$$

Equations 215 through 219 have been integrated numerically for various values of the Prandtl number Pr = 0.72, 6.8, 50, 100, 500, and ∞ for the entire mixed-convection regime $0 \leqslant \beta \leqslant 1$ and the result are described in the following.

The nondimensional axial velocity distribution $\beta u/U_\infty$ is presented in Figure 26 against η for various values of a mixed-convection parameter β and the Prandtl number of 0.72. The velocity is maximum at the axis of plume and away from the axis it decreases monotonically to approach the stream velocity. At $\beta = 1$ the velocity distribution is that of a forced-convection stream; as β decreases the velocity u/U_∞ increases due to the action of buoyancy, and at $\beta = 0$ the profile approaches free-convection buoyant plume. The effect of the Prandtl number on centerline velocity $f'(0) = \beta u(x,0)/U_\infty$ is displayed in Figure 27 against the mixed-convection parameter β. The centerline velocity $u(x,0)/U_\infty$ always increases with β. Further, the figure also shows that for a fixed β, the centerline velocity increases with an increase of the Prandtl number.

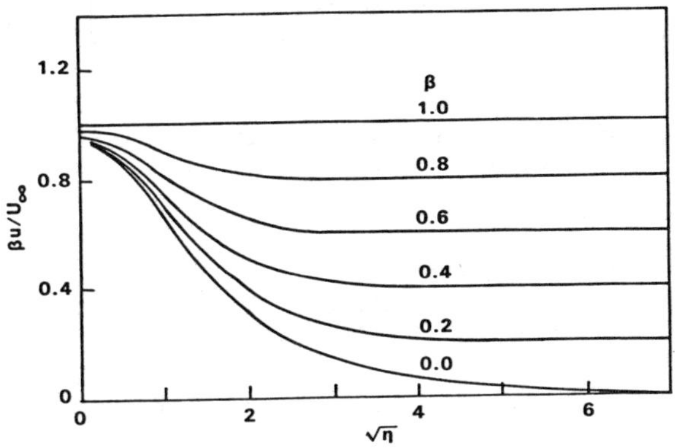

Figure 26. The profiles of velocity for laminar mixed convection in an aiding situation due to a point heat source in a vertical forced stream for air.

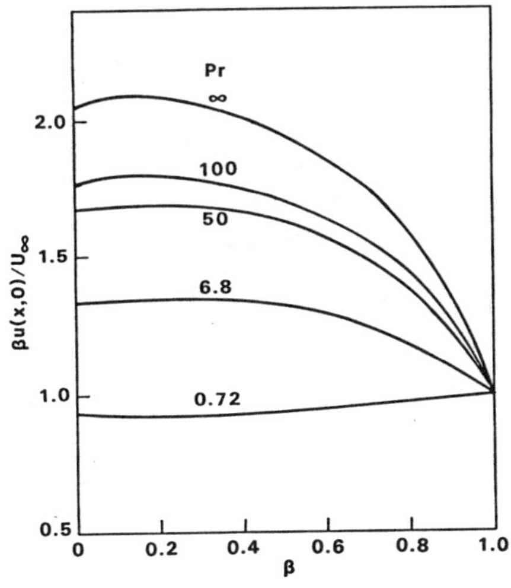

Figure 27. The centerline velocity for laminar mixed convection in an aiding situation due to a point heat source in a vertical forced stream for various Prandtl numbers.

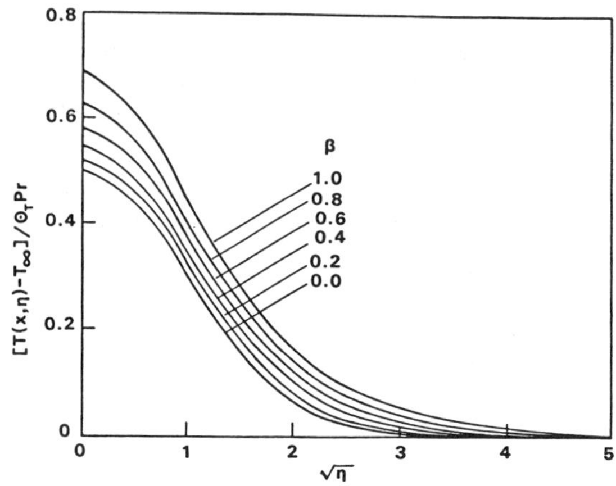

Figure 28. The profiles of temperature for laminar mixed convection in an aiding situation due to a point heat source in a vertical free stream for air.

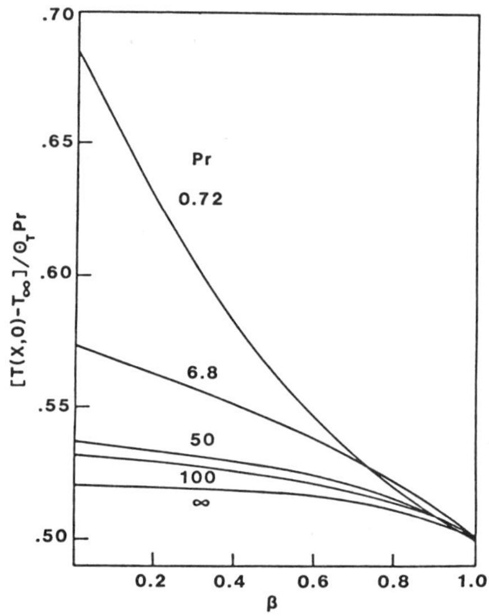

Figure 29. The centerline temperature for laminar mixed convection in an aiding situation due to a point heat source in a vertical forced stream for various Prandtl numbers.

The nondimensional temperature distribution $(T - T_\infty)/\theta_T Pr$ for $Pr = 0.72$ is displayed in Figure 28 against η for various values of the mixed-convection parameter β. The effect of the Prandtl number on centerline temperature $(T(x, 0) - T_\infty)/\theta_T Pr$ is presented against β in Figure 29. The figure shows that for $\beta \rightarrow 1$, the nondimensional centerline temperature approaches 0.5, and as $\beta \rightarrow 0$, the solution approaches the value for a free-convection plume. For fixed β, an increase in the Prandtl number results in an increase in the centerline temperature.

The half-width of the mixed-convection plume based on velocity δ and temperature δ_T is defined by Equation 96. The nondimensional velocity width $(\delta/x)\,(R_x/Pr\beta)^{1/2}$ is displayed in Figure 30 for various values of the Prandtl number against the mixed-convection parameter β. The width δ decreases as β decreases, and for $\beta \rightarrow 0$ it approaches forced-convection plume. Further for a fixed β, width δ decreases as Pr increases. The nondimensional temperature width $(\delta_T/x)\,(R_x/Pr\beta)^{1/2}$ is presented in Figure 31 against β. Near the forced-convection asymptote $\beta \rightarrow 1$, $(\delta_T/x)\,(R_x/Pr\beta)^{1/2} \rightarrow 3.46$, a value given by the closed-form Equation 199b. The qualitative results for δ are similar to δ_T and no additional comment is needed.

Nonvertical Free Stream

The weakly buoyant plume in an unbounded media where a forced stream is inclined at an angle φ with respect to the buoyant force vector has been analyzed by Wesseling [53].

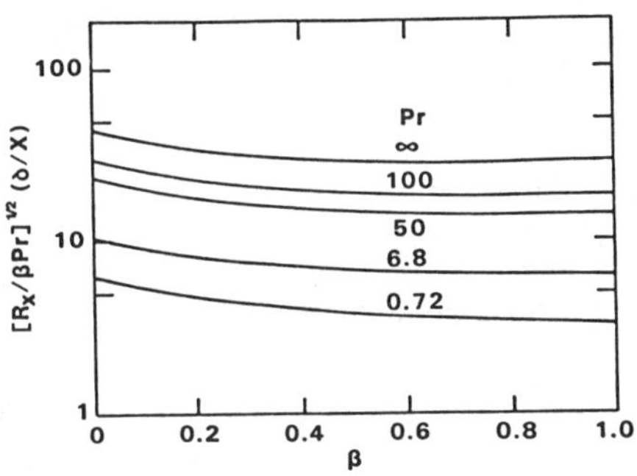

Figure 30. Half width of the plume from a velocity profile for laminar mixed convection in an aiding situation due to a point heat source in a vertical forced stream for various Prandtl numbers.

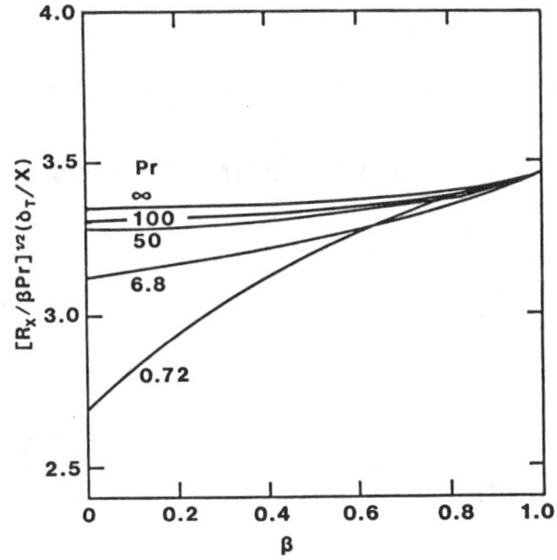

Figure 31. Half width of the plume from a temperature profile for laminar mixed convection in an aiding situation due to a point heat source in a vertical forced stream for various Prandtl numbers.

In nondimensional variables (with L, U_∞, ρU_∞^2, and $Q/(\rho CpU_\infty L^2)$ as reference length, velosity, pressure and temperature) the Oseen approximation

$$\begin{aligned} \underset{\sim}{u} &= e_x + \hat{\varepsilon}\underset{\sim}{u}_1 + \ldots \\ T - T_\infty &= \theta_0 + \hat{\varepsilon}\theta_1 + \ldots \\ p &= \hat{\varepsilon}p_1 + \hat{\varepsilon}^2 p_2 + \ldots \end{aligned} \tag{220}$$

of the Navier-Stokes Equations 1 through 3 yields

$$\frac{\partial\theta_0}{\partial x} = (2RePr)^{-1}\nabla^2\theta_0 \tag{221}$$

$$\frac{\partial \underset{\sim}{u}_1}{\partial x} = -\nabla p_1 + (2Re)^{-1}\nabla^2\underset{\sim}{u}_1 + \theta_0(e_x\cos\varphi + e_y\sin\varphi) \tag{222}$$

$$\nabla\cdot\underset{\sim}{u}_1 = 0 \tag{223}$$

The perturbation parameter $\hat{\varepsilon}$ is given by

$$\hat{\varepsilon} = g\beta Q/(\rho C_p U_\infty^3 L) \tag{224}$$

The boundary conditions at infinity are

$$\theta_0 = 0, \qquad \underset{\sim}{u}_1 = p_1 = 0 \tag{225}$$

The solution to Equation 221 for a point source is

$$\theta_0 = \frac{PrRe}{2\pi r}\exp[RePr(x - r)] \tag{226}$$

where $r^2 = x^2 + y^2 + Z^2$

Taking the divergence of Equation 222 and using Equation 223 one obtains

$$\nabla^2 p_1 = \frac{\partial\theta_0}{\partial x}\cos\varphi + \frac{\partial\theta_0}{\partial y}\sin\varphi \tag{227}$$

the equation for pressure distribution and its solution is

$$p_1 = \frac{1}{4\pi r}\{1 - \exp[RePr(x - r)]\}\left\{r\cos\varphi\frac{\partial}{\partial y}[\ln(r - x)] - \sin\varphi\right\} \tag{228}$$

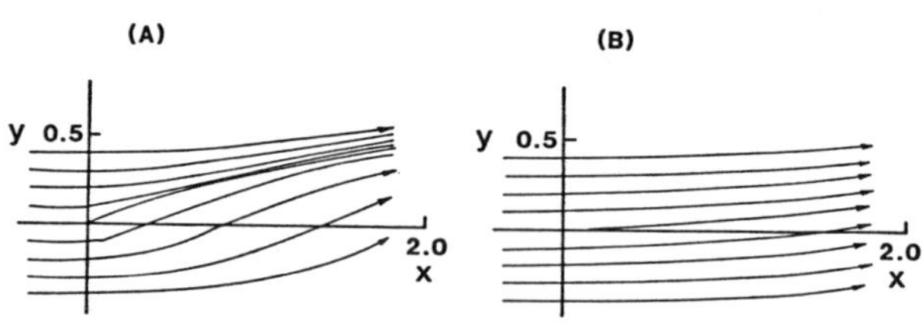

Figure 32. The streamlines in the plane of symmetry in a laminar weakly buoyant plume due to a point heat source in a horizontal forced stream ϵ = 0.1, Re = 10: (a) Pr = 3; (b) Pr = 1 [53].

The solution to the momentum (Equation 222), following the method of Lamb for the Oseen equation is given by

$$\underset{\sim}{u}_1 = \nabla\Phi_1 + \hat{e}_g\chi_1 \tag{229}$$

where $\hat{e}_g$ is the unit vector in the vertical direction and the function

χ_1 and Φ_1 for Pr = 1 are given by

$$\chi_1 = \frac{\text{Re}}{2}\exp\{\text{Re}(x - r)\} \tag{230a}$$

$$\Phi_1 = \frac{y\sin\varphi}{4\pi(r - x)}\{\exp[\text{Re}(x - r)] - 1\} - \frac{\cos\varphi}{2\pi}\{E_1[\text{Re}(x - r)] + \ln(r - x) + \exp[\text{Re}(x - r)]\} \tag{230b}$$

and $\Pr \neq 1$ the expressions may be found in Wesseling [54]. The special case Pr = 1, $\varphi = \pi/2$, in the limit of this solution as the Reynolds number goes to infinity has been obtained by Csanady [61]. As the potential of a horseshoe vortex is $y/(r - x)$, it is expected that the flow pattern will resemble a horseshoe-vortex flow pattern. Under favorable circumstances, the horseshoe-vortex type of behavior of smokestack plumes can be visually observed. For $\varphi = 0$, the streamline in the plane of symmetry is displayed in Figure 32 for Pr = 1 and 3. The examination of the figure shows that as Pr increases the heat remains more concentrated, resulting in a steep rise of streamlines that pass close to the heat source.

Miloh and Yahalom [67] studied the steady motion of a point heat source below and parallel to an interface, either a plane boundary or a free surface, in a medium which is otherwise at rest. The problem is considered in a frame of reference that is moving with a constant speed, the origin being at the free surface, at constant depth below the interface. In this frame of reference the problem is equivalent to mixed convection from weakly buoyant plume due to a stationary source subjected to a horizontal stream in a medium bounded by either a plane boundary or a free surface. In this situation the solution of Wesseling [53] for an infinite expanse of fluid needs correction due to the finite domain, i.e., horizontal boundary. The boundary conditions for a free surface require both a tangential and vertical component. The linearized version of free surface boundary conditions is

$$\left.\begin{aligned} &\frac{\partial u}{\partial y} + \frac{\partial v}{\partial x} = \frac{\partial w}{\partial y} + \frac{\partial v}{\partial z} = 0 \\ &p + \frac{1}{\text{Fr}}\hat{\eta} - \frac{1}{\text{Re}}\frac{\partial v}{\partial y} = 0 \end{aligned}\right\} \quad y = 0 \tag{231}$$

where $\hat{\eta} = \hat{\eta}(x, z)$ denotes the free surface elevation and $\text{Fr} = U_\infty^2/gL$ is the Froude number. The fact that the free surface is a stream surface requires

$$v = u\frac{\partial\hat{\eta}}{\partial x} + w\frac{\partial\hat{\eta}}{\partial z} \tag{232}$$

The temperature distribution on the interface is taken as a general Cauchy type

$$\frac{\partial\theta}{\partial n} + b\theta = 0 \quad \text{at } y = \eta \tag{233}$$

where n is the direction of normal at the free surface. The limiting cases $b = 0$ and $b \to -\infty$ correspond to the adiabatic and isothermal interface conditions, respectively. The expansions for the variables $\underset{\sim}{u}(u, v, w)$, p, and θ are the same (Equations 220) and the expansions for free surface elevations is

$$\hat{\eta} = \hat{\varepsilon}\eta_1 + \hat{\varepsilon}^2\eta_2 + \ldots \tag{234}$$

The Oseen-linearized equations (Equations 221 through 223) with $\varphi = \pi/2$ and $e_y = 1$ are considered. The solution is composed of two parts: a regular part and a singular part. The singular part is essentially identically with the solution in an infinite expanse of fluid given by Wesseling [53] for $\varphi = \pi/2$. The regular part is included so as to accomodate the additional boundary conditions (Equations 231 through 233) at the free surface. The solution to a zero-order temperature equation is given by

$$\theta_0 = \theta_{0s} + \theta_{0r} \tag{235}$$

where θ_{0s} is the singular part having the same solution as Equation 226 of Wesseling [54]

$$\theta_{0s} = \frac{\text{RePr}}{2\pi r_1} \exp[-\text{RePr}(r_1 - x)] \tag{236a}$$

where

$$r_1^2 = x^2 + (y - 1)^2 + Z^2 \tag{236b}$$

that accounts for the fact that the origin is not at the source. The equation satisfied by regular part θ_{0r} is the same as Equation 235 and the boundary condition at the interface is

$$\frac{\partial\theta_{0r}}{\partial y} + b\theta_{0r} = \frac{\partial\theta_{0s}}{\partial y} + b\theta_{0s} \quad \text{at } y = 0 \tag{237}$$

The solution for the regular part is determined by employing the double Fourier-transform. The total solution for θ_0 is given by

$$\theta_0 = \frac{\text{RePr}}{2}\left[\frac{1}{r_1} e^{-\text{RePr}(r1 - x)} + \frac{1}{r_2} e^{-\text{RePr}(r2 - x)} + 2b \int_t^\infty \frac{1}{r_2'} e^{-b(t - t') - \text{RePr}(r2 - x)} dt'\right] \tag{238}$$

where $r_2 = (x^2 + t^2 + Z^2)^2$

$r_2' = (x^2 + t'^2 + Z^2)^2, \qquad t = 1 + y$

For the two limiting cases of adiabatic interface ($b = 0$) and isothermal interface ($b \to \infty$) the Equation 238 yields

$$\theta_0(x, y, z) = \frac{\text{RePr}}{2}\left[\frac{1}{r_1} e^{-\text{RePr}(r1 - x)} \pm \frac{1}{r_2} e^{-\text{RePr}(r2 - x)}\right] \tag{239}$$

where upper (+ sign) corresponds to the adiabatic ($b = 0$) case and lower (− sign) to isothermal ($b \to -\infty$) case.

Likewise the solution of the Oseen-linearized momentum equation (Equation 222) may be expressed as a sum of regular and singular points

$$\underset{\sim}{u}_1 = \underset{\sim}{u}_{1r} + \underset{\sim}{u}_{1s} \tag{240}$$

$$p_1 = p_{1r} + p_{1s} \tag{241}$$

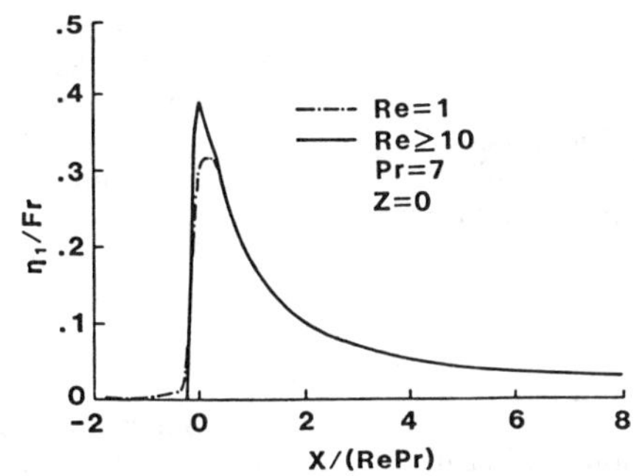

Figure 33. The free surface deflection (kinematic signature) due to a point heat source moving beneath the free surface [67].

For the singular parts p_{1s} and $\underset{\sim}{u}_{1s}$ the solution are given by Equations 228 and 229. The solution for the regular part that satisfies the free surface conditions has been considered by Miloh and Yahalom [67] by employing the double Fourier-transform. The closed-form solutions have been derived in the Fourier-transformed plane. The general case involves the numerical computations for inverse Fourier-transform. A special case of the small Froude number for the isothermal interface is considered [67] and the closed-form solution for free surface displacement of $\hat{\eta}$ is given by

$$\eta_1(x, Z) = \frac{Fr}{4\pi r_0(r_0 - x)}\left\{\frac{Pr}{2(Pr-1)}\left[e^{-RePr(r_0-x)} - e^{-RePr(r_0-x)}\right]\right. \\ \left. + \left[1 - e^{-RePr(r0-x)}\right] + \varphi(Fr^2)\right. \tag{242}$$

$$\text{where } r_0^2 = x^2 + Z^2 + 1 \tag{243}$$

The closed-form solution, Equation 242, for the free surface elevation $\hat{\eta}/Fr$ in the plane of symmetry ($Z = 0$) is depicted against x/RePr in Figure 33. The figure shows that the vertical coordinate is largely distorted with respect to the horizontal coordinate. For small values of Re the dependence x/RePr is observed for $x/RePr \ll 1$, and for $x/RePr \gg 1$ the solution does not vary with Re. Further the solution, Equation 242, shows that for large values of Re(> 500) a similar solution containing x/RePr is obtained which is also independent of Re.

NOTATION

C_p	specific heat
E	entrainment rate scale
e_x, e_y	unit vectors in x and y directions, respectively
F	function velocity
f	function
Gr	Grashof number
g	gravitational acceleration
H	temperature function
h	heat transfer coefficient
K	thermal conductivity
L	length
$\dot{m}$	mass flow rate
n	exponent
Pr	Prandtl number
p	pressure
Q	heat released
q	heat conduction
Re	Reynolds number
R_E	eddy Reynolds number

T temperature
U_∞ stream velocity
u velocity
V_+ transverse velocity component
v velocity
W complex potential defined by Equation 127
x coordinate
X parameter
y coordinate
Z_c Euler transformation defined by Equation 100

Greek Symbols

α parameter
β volumetric expansion coefficient or mixed convection parameter
Γ parameter defined in Equation 73
γ^* Euler's constant
Δ entrainment coefficient
δ thickness
δ^* displacement thickness
δ_{im} Kronecker delta
$\hat{\varepsilon}$ perturbation parameter
ξ parameter defined in Equation 73
η similarity variable
θ dimensionless temperature
Λ mixed convection parameter
λ inverse of eddy Reynolds number
ν kinematic viscosity
ν_τ eddy viscosity
ξ dimensionless parameter defined in Equation 73
ρ density
τ shear stress
φ' inclination of reference streamline from vertical
ψ stream function

REFERENCES

1. Briggs, G. A., *Plume Rise*. Oak Ridge, US Atomic Energy Commission, 1969.
2. Fay, J. A., "Buoyant Plumes and Wakes," in *Annual Rev. Fluid Mechanics,* Volume 5, M. Van Dyke and W. G. Vincenti (Eds.) California, Annual Reviews Inc. (1973), pp. 151–160.
3. Fischer, H. B., et al., Brooks, N. H., Mixing in Inland and Coastal Waters. New York: Academic Press, 1979.
4. List, E. J., "Turbulent Jets and Plumes," in *Annual Rev. Fluid Mechanics,* Volume 14, M. Van Dyke, et al. (Eds.) California: Annual Reviews Inc. (1982), pp. 189–212.
5. Morton, B. R., Entrainment Models for Laminar Jet, Plume and Wakes, *Physics of Fluids,* Vol. 10 (1967), pp. 2120–2127.
6. Townsend, A. A., *The Structure of Turbulent Shear Flows,* Cambridge University Press, 1975.
7. Morton, B. R., Taylor, G. I., and Turner, J. S., "Turbulent Gravitational Convection from Maintained and Instantaneous Sources," *Proc. Roy. Soc. A,* Vol. 234 (1956), pp. 1–23.
8. Zeldovich, Y. B., "Limiting Laws of Freely Rising Convection Currents," *Moscow, Zh. Eskp. Teoret, F12,* Vol. 7, 1937, p. 1463.
9. Schuh, H., "Boundary Layer of Temperature," in *Boundary Layer* W. Tollmien (Ed.), British Ministry of Supply, German Document Centre, Reference 3220T, 1948.
10. Yih, C. S., "Free Convection Due to Boundary Sources," Proc. of the First Symposium on the Use of Models in Geophysical Fluid Dynamics R. R. Long (Ed.), Washington Govt. Printing Source, 1956, pp. 117–133.
11. Fuji, T., Theory of the Steady Laminar Natural Convection Above a Horizontal Line Heat Source and a Point Heat Source," *Int J Heat Mass Transfer,* Vol. 6 (1963), pp. 597–606.
12. Fuji, T., Morioka, I., and Uehara, H., "Buoyant Plume Above a Horizontal Line Heat Source," *Int. J. Heat Mass Transfer,* Vol. 16 (1973), pp. 755–768.

13. Yih, C. S., and Gupta, A. S., "Plane Buoyant Plume," *Rev. Br C Mec. Rio de Janeiro,* Vol. 111, No. 2 (1981), pp. 49–56.
14. Gebhart, B., Pera, L., and Schorr, A. W., "Steady Laminar Natural Convection Plume Above a Horizontal Line Heat Source," *Int. J. Heat Mass Transfer,* Vol. 13, (1970), pp. 161–171.
15. Spalding, D. B., and Cruddace, R. G., "Theory of the Steady Laminar Buoyant Flow Above a Line Heat Source in a Fluid of Large Prandtl Number and Temperature Dependent Viscosity," *Int. J. Heat Mass Transfer,* Vol. 3, (1961), pp. 55–59.
16. Kuiken, H. K., and Rotem, Z., "Asymptotic Solution for Plume at Very Large and Small Prandtl Numbers," *J. Fluid Mechanics,* Vol. 45 (1971), pp. 586–600.
17. Jaluria, Y., *Natural Convection Heat and Mass Transfer,* Oxford, Pergamon Press.
18. Brand, R. S., and Lahey, F. J., "The Heated Laminar Jet," *J. Fluid Mechanics,* Vol. 29 (1967), pp. 305–315 (and Corrigendum Vol. 31 (1968), p. 206).
19. Forstrom, R. J., and Sparrow, E. M., "Experiments on Buoyant Plume Above a Heated Horizontal Wire." *Int. J. Heat Mass Transfer,* Vol. 10 (1967), pp. 321–331.
20. Brodowicz, K., and Kierkus, W. T., "Experimental Investigation of Laminar Free Convection Flow in Air Above Horizontal Wire with Constant Heat Flux," *Int. J. Heat Mass Transfer,* Vol. 13 (1970), pp. 1243–1248.
21. Schorr, A. W., and Gebhart, B., "An Experimental Investigation of Natural Convection Wakes Above a Line Heat Source, *Int. J. Heat Mass Transfer,* Vol. 13 (1970), pp. 557–571.
22. Riley, N., "Free Convection From a Horizontal Line Source of Heat," *Z. Angew. Math. Phys,* Vol. 25 (1974), 818–827.
23. Hieber, C. A., and Nash, E. J., "Natural Convection above a Line Heat Source: Higher-Order Effects and Stability," *Int. J. Heat Mass Transfer,* Vol. 18 (1975), pp. 1473–1479.
24. Schneider, W., "Flow Induced by Jets and Plumes," *J. Fluid Mech,* Vol. 108 (1981), pp. 55–65.
25. Gebhart, B., "Instability, Transition and Turbulence in Buoyancy Flows," *In Annual Reviews of Fluid Mechanics,* Vol. 5, M. Van Dyke and W. C. Vincenti (Eds.), California, Annual Reviews Inc., 1973, pp. 213–246.
26. Bill, R. G., and Gebhart, B., "The Transition of Plane Plumes," *Int. J. Heat Mass Transfer,* Vol. 18 (1975), pp. 513–526.
27. Rouse, H., Yih, C. S., and Humphrey, H. W., "Gravitational Convection from a Boundary Source," *Tellus,* Vol. 4 (1952), pp. 201 – 210.
28. Lee, S. L., and Emmons, H. W., "A Study of Natural Convection Above a Line Fire," *J. Fluid Mechanics,* Vol. 11 (1961), pp. 353–368.
29. Kotsovinois, N. E., "Plane Turbulent Buoyant Jet. Part 2: Turbulence Structure," *J. Fluid Mechanics,* Vol. 81 (1977), pp. 45–62.
30. Kotsovinois, N. E., and List, E. J., "Plane Turbulent Buoyant Jets. Part I: Integral Properties," *J. Fluid Mechanics,* Vol. 81 (1977), pp. 25–44.
31. Turner, J. S., *Buoyancy Effects in Fluids,* Cambridge University Press, 1973, pp. 165–206.
32. Turner, J. S., "Buoyant Plumes and Thermals," in *Annual Reviews of Fluid Mechanics,* Volume 1, W. R. Sears and M. Van Dyke (Eds.) Palo Alto: Ann. Rev. Inc., 1968, pp. 29–44.
33. Murgai, M. P., *Natural Convection from Combustion Sources,* New Delhi: Oxford & IBH, 1976.
34. Rao, V. K., "The Buoyant Plume Above a Line Source," *Atmospheric Environment,* Vol. 4 (1970), pp. 557–575.
35. Tennekes, H., and Lumley, J., *A First Cource in Turbulence,* Cambridge, M.I.T. Press.
36. Yih C. S., "Turbulent Buoyant Plumes," *Physics of Fluids,* Vol. 20 (1977), pp. 1231.
37. Teske, M. E., Lewellen, W. S., and Segur, H. S., "Turbulence Modeling Applied to Buoyant Plumes," *EPA Report* EPA-600/4-78-050, 1978, p. 47.

38. Yih, C. S., and Wu, F., "Round Buoyant Laminar and Turbulence Plumes," *Physics of Fluids,* Vol. 24 (1981), pp. 794–801.
39. Mollendorf, J. C., and Gebhart, B., "An Experimental and Numerical Study of the Viscous Stability of a Round Laminar Vertical jet with and without Thermal Buoyance for Symmetric and Asymmetric Disturbances," *J. Fluid Mechanics,* Vol. 61 (1973), pp. 367–399.
40. Crane, L. J., "Axially Symmetric Laminar Plume at Large Prandtl Numbers," *Z. Angew. Math. Phys.,* Vol. 25 (1974), pp. 635–641.
41. Crane, L. J., "Axially Symmetric Plume at Very Small Prandtl Numbers," *Z. Angew. Maths. Phys.,* Vol. 26 (1975), pp. 427–435.
42. Afzal, N., "Mixed Convection in an Axisymmetric Buoyant Plume," *Int. J. Heat Mass Transfer,* Vol. 26 (1983), pp. 381–388.
43. Kaplun, S., "Singular Perturbations and Fluid Mechanics," Lagrestrom, Liu, and Howard (Eds.) New York: Academic 1967.
44. Schmidt, W., "Turbulente Ausbreitung eines Stromes erhitzter Luft," I, II. Teil, *Z. Angew. Math. Mech.* Vol. 21 (1941), pp. 265–278 and 351–363.
45. George, W. K., Alpert, R. L., and Tamani, F., "Turbulence Measurements in an Axisymmetric Buoyant Plume," *Int. J. Heat Mass Transfer,* Vol. 20 (1977), pp. 1145–1154.
46. Nakagome, H., and Hirata, M., "The Structure of Turbulent Diffusion in an Axi-symmetric Thermal Plume," in *Heat Transfer and Turbulent Buoyant Convection,* Volume 1, D. B. Spalding and N. Afgan (Eds.), New York: McGraw Hill, 1977, pp. 361–372.
47. Beuther, P. D., "Experimental Investigation of a Turbulent Axisymmetric Plume," *Ph. D. thesis,* SUNY at Buffalo, 1980.
48. Baines, W. D., "Turbulent Buoyant Plumes," in *Heat Transfer and Turbulent Buoyant Convection,* Volume 1, D. B. Spalding and N. Afgan, (Eds.), New York: McGraw Hill, 1977, pp. 235–250.
49. Tennankore, K. N., and Steward, F. R., "Comparison of Several Turbulence Models for Predicting Flow Patterns within confined jets," *Cand. J. Chem. Engng,* Vol. 56 (1978), pp. 673–678.
50. Chen, C. J., and Nititopoulos, C. P., "On the Near Field Characteristics of Axisymmetric Turbulent Buoyant Jets in a Uniform Environment," *Int. J. Heat Mass Transfer,* Vol. 22 (1979), pp. 245–255.
51. Beuther, P. B., Capp, S. P., and George, W. K., Joint ASME/AIChE 18 National Heat Transfer Conference, San Diego, Calif., Aug 1979.
52. Woods, W. W., "Free and Forced Convection from Fine Hot Wires," *J. Fluid Mechanics,* Vol. 55 (1972), pp. 419–438.
53. Wesseling, P., "An Asymptotic Solution for Slightly Buoyant Plume," *J. Fluid Mechanics,* Vol. 70 (1974), pp. 81–87.
54. Nakai, S., and Okazaki, T., "Heat Transfer from a Horizontal Circular Wire at Small Reynolds and Grashof Numbers II Mixed Convection," *Int. J. Heat Mass Transfer,* Vol. 18 (1975), pp. 397–413.
55. Afzal, N., "Mixed Convection in a Two-dimensional Buoyant Plume," *J. Fluid Mechanics,* Vol. 105 (1981), pp. 317–368.
56. Haaland, S. E., and Sparrow, E. M., "Mixed Convection Plume Above a Horizontal Line Source Situated in a Forced Convection Approach Flow," *Int. J. Heat Mass Transfer,* Vol. 26 (Mar. 1983), pp. 433–444.
57. Spalding, D. B., *GENMIX: A General Computer Program for Two-Dimensional Parabolic Phenomena,* Oxford: Pergamon Press, 1977.
58. Van Dyke, M., "Analysis and Improvement of Perturbation Series," *Quart J. Mech. Appl. Math.,* Vol. 27 (1974), pp. 423–450.
59. Meksyn, D., *New Methods in Laminar Boundary Layer Theory,* Oxford Pergamon Press.

60. Van Dyke, M., "Perturbation Met ods in Fluids Mechanics," New York: Parabolic Press, 1975.
61. Csanady, G. T., "The Buoyant Motion within a Hot Gas Plume in a Horizontal Wind," *J. Fluid Mechanics,* Vol. 22 (1965), pp. 225–239.
62. Hieber, C. A., and Gebhart, B., "Mixed Convection from a Sphere at Small Reynolds and Grashof Numbers," *J. Fluid Mechanics,* Vol. 38 (1969), pp. 137–159.
63. Afzal, N., "Mixed Convection in an Axisymmetric Buoyant Plume," *Int. J. Heat Mass Transfer,* Vol. 26 (1983), pp. 381–388.
64. Appalaswamy, A. V., and Jaluria, Y., "Axisymmetric Flow in a vertical Uniform Free Stream," *J. Appl. Mech.,* Vol. 47 (1980), pp. 667–669.
65. Riley, D. S., and Drake, G. D., "Mixed Convection in an Axisymmetric Plume," *Quart. J. Mech. Appl. Maths.,* Vol. 36 (1983), pp. 43–54.
66. Crane, L. J., "The Round Plume Due to Falling Point Source of Heat," *Z. Angew. Math. Phys.,* Vol. 28 (1977).
67. Miloh, T., and Yahalom, Y., "On the Motion of Weakly Buoyant Heat Source Near an Interface," *J. Engng. Maths.,* Vol. 16 (1982), pp. 271–293.
68. Berger, S. A., *Laminar Wakes,* New York: Elsevier, 1971, pp. 76–95.

CHAPTER 14

NUCLEATION AND GROWTH IN THE DIFFUSION CLOUD CHAMBER

Richard H. Heist

Department of Chemical Engineering
University of Rochester
Rochester, New York

CONTENTS

INTRODUCTION

There are two general classes of nucleation phenomena: heterogeneous nucleation and homogeneous nucleation. The former involves a pre-existing nucleation site while the latter does not. Examples of heterogeneous nucleation are quite common and include rain drop formation, boiling, and most crystallization processes. Examples of the homogeneous nucleation, in the "real world" at least, are much less common and usually occur under carefully controlled laboratory conditions. Heterogeneous nucleation is a more difficult problem to handle because of the complicating effects of the pre-existing site; so much of the research carried out over the last sixty or so years has focused on homogeneous nucleation.

Much of the emphasis of this research has been on understanding the basis physics of the nucleation process and predicting the onset of nucleation and the dependence of the nucleation rate on temperature and other relevant parameters such as supersaturation and degree of undercooling or superheat. Vapor-to-liquid is probably the most commonly studied type of homogeneous nucleation, and it is vapor-to-liquid nucleation which will be of primary interest in this chapter.

There are a number of different experimental devices employed in the vapor-to-liquid nucleation research. The most common are the diffusion cloud chamber, the expansion cloud chamber [1–7], the supersonic converging-diverging nozzel [8–12], and the shock tube [13, 14]. Of these, the diffusion cloud chamber (DCC) is particularily attractive because of its steady-state operation and because it can be constructed and operated relatively easily. In this chapter we shall only be concerned with the diffusion cloud chamber.

In what follows we shall first briefly review the theory of homogeneous nucleation and then give a description of the diffusion cloud chamber. Next, we discuss nucleation experiments carried out with the diffusion cloud chamber at small nucleation rates. This will be followed by a discussion of the interesting behavior of the DCC at very large rates of nucleation; and finally by a discussion of work currently in progress at intermediate nucleation rates.

HOMOGENEOUS NUCLEATION

The formation of liquid from an existing vapor phase may be visualized as a two-step process: First, the appearance of small clusters or embryos of molecules of the new phase: and, second, growth of some of these clusters to form macroscopic quantities of the new phase. The first step of the process is the nucleation step and is of primary interest here. We shall review only the salient features of the theoretical description of the nucleation process since more complete treatments are available from a number of sources [15–27].

The thermodynamic aspects of nucleation may be treated by considering a simple system composed of a spherical liquid drop containing N_ℓ molecules surrounded by its vapor containing N_v molecules with the temperature, pressure, and total number of molecules held constant. The thermodynamic potential of the system, φ, can be written as

$$\varphi = N_v\mu_v + N_\ell\mu_\ell + 4\pi r^2\sigma \tag{1}$$

where μ_ℓ and μ_l = the (bulk) chemical potentials per molecule of the vapor and liquid, respectively
r = the radius of the liquid drop (assumed spherical)
σ = the surface tension of the bulk liquid

In writing Equation 1, we have made use of the capillarity approximation [28].

If we assume the liquid phase to be incompressible and the vapor phase ideal and then apply the equilibrium condition to Equation 1, we obtain

$$kT\ln\frac{P}{P_e} = \frac{2\sigma v_m}{r} \tag{2}$$

the well-known Kelvin equation, which expresses the dependence of the vapor pressure of a (spherical) liquid drop on its radius,

where P = the pressure of the vapor existing in the system
P_e = the equilibrium vapor pressure of the bulk liquid under its own pressure at temperature T
v_m = the molecular volume
k = the Boltzmann constant

The quantity P/P_e is called the supersaturation and is usually given the symbol S.

The thermodynamic potential of the system prior to formation of the liquid drop is

$$\varphi_0 = (N_v + N_\ell)\mu_v \tag{3}$$

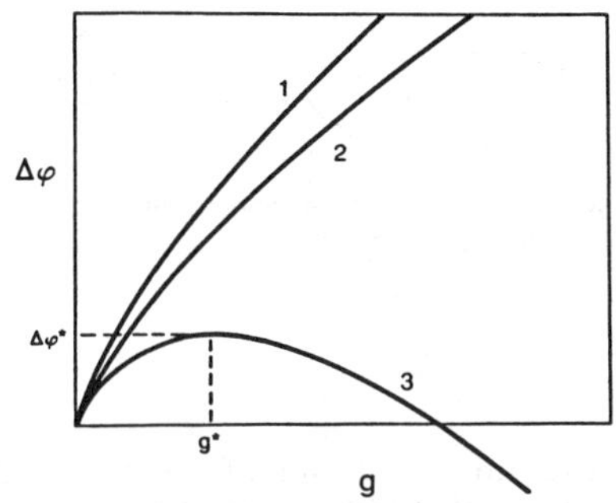

Figure 1. The variation of the thermodynamic potential of the droplet-vapor system as a function of droplet size as described in the text.

The difference between Equations 1 and 3 is a measure of the reversible work necessary to form the drop and can be written as

$$\Delta\varphi = g(\mu_\ell - \mu_v) + \xi g^{2/3} \tag{4}$$

where we let $N_\ell = g$ and $\xi = (4\pi)^{1/3}\sigma(3v_m)^{2/3}$. Equation 4 contains a term proportional to the droplet volume and a term proportional to the surface area of the drop. The relative effects of these two terms are illustrated in Figure 1. If $\mu_\ell > \mu_v$, the vapor is the more stable phase and is said to be undersaturated. Both terms in Equation 4 are positive and increase continuously with g. This is shown as curve 1 in Figure 1. If $\mu_\ell = \mu_v$, the vapor is saturated. The surface term is positive and $\Delta\varphi$ increases continuously with g. A liquid drop will still not form because of the work required to create the droplet interface. This is illustrated as curve 2 in Figure 1. If $\mu_\ell < \mu_v$, the liquid is the more stable phase and the vapor is said to be supersaturated. In this case $\Delta\varphi$ exhibits a maximum at the point the negative volume term begins to dominate the positive surface term. This is illustrated by curve 3 in Figure 1. This particular point is characterized by $g = g^*$ and $\Delta\varphi = \Delta\varphi^*$. A drop containing g^* molecules is called a critical size embryo or nucleus.

In the preceding discussion we have used the term "liquid drop" to refer to the various-size molecular clusters in the model. By convention those drops that are smaller than the critical size are called embryos, and those drops which are larger are called clusters.

By straightforward thermodynamic reasoning [22–24] it can be shown that at equilibrium the distribution of embryo sizes is given by the following expression [22, 25]:

$$N_g = N\exp(-\{[g(\mu_\ell - \mu_v) + \xi g^{2/3}]/kT\}) \tag{5}$$

where we have made use Equation 4 and N_g is the number of embryos containing g molecules and N is approximately the number of single (monomer) molecules. N is given by

$$N = F\left(\frac{N_a}{F}\right)^g \tag{6}$$

where N_a is the number of monomers and $F = N_a + \sum_g N_g$ is the total number of molecular "species" in the vapor. Under most conditions, $N_a \approx F$ so that $N \approx N_a$ [29].

In arriving at Equation 5 the embryo distribution was maintained at thermodynamic equilibrium. This is accomplished by not allowing formation of clusters significantly larger than the critical size. The conventional theory of nucleation is thus a "pretransition"

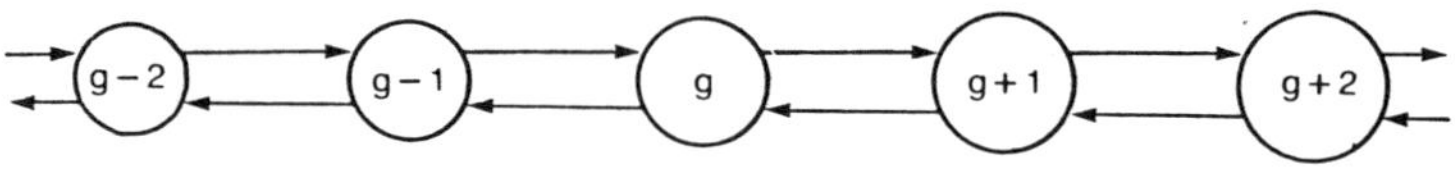

Figure 2. The "chain" of embryo sizes in a one-dimensional size space. The circles represent embryos containing g molecules, g + 1 molecules, g − 1 molecules, etc.

theory. Fortunately, the pretransition behavior of the system determines the essential aspects of the nucleation process so that little information is lost by imposing the artifical constraint upon the cluster size.

Although the Kelvin relation, Equation 2, provides the thermodynamic relationship between a liquid drop and its vapor, it provides no information as to how the drop was formed initially. The mechanism of droplet formation from its vapor is generally regarded as involving microscopic density fluctuations which give rise to subcritical embryos (so called heterophase fluctuations) [22]. The kinetic aspects of the nucleation process can be treated by viewing the growth and decay of these embryos as a sequence of reactions involving collisions between single molecules and embryos. "Higher order" events involving collisions between embryos are assumed to be rare events because of the low concentrations of the embryos relative to the single molecules. The series of reactions may be written as

$$\begin{aligned} A_1 + A_1 &= A_2 \\ A_2 + A_1 &= A_3 \ldots \\ &\cdot \quad \cdot \quad \cdot \\ A_{(g-1)} + A_1 &= A_g \ldots \\ &\cdot \quad \cdot \quad \cdot \end{aligned} \tag{7}$$

where A_1 represents a single molecule and A_g an embryo containing g molecules.

Alternatively, the growth and decay of an embryo may be pictured as the motion of an embryo along a "chain" in a one-dimensional size-space as illustrated in Figure 2. Here an embryo either moves forward one link of the chain by addition of a single molecule, or backward one link by the loss of a single molecule.

With this model for the nucleation kinetics, we can say something about the rate at which nucleation occurs. Normally, the nucleation rate is defined as the rate at which embryos pass through the critical size. Later we will link this definition to experimental observations by assuming the nucleation rate to be equal to the rate of appearance of liquid drops in the cloud chamber.

The rate of change of the number of embryos of size g is given by

$$\frac{\Delta n_g}{\Delta t} = (\alpha_{g+1} a_{g+1} n_{g+1} - \beta a_{g-1} n_{g-1}) - (\alpha_g a_g n_g - \beta a_g n_g) \tag{8}$$

where $\Delta n_g/\Delta t$ = the time rate of change of the number of g-size embryos
n_g = the number of g-size embryos
a_g = the surface area of a g-size embryo
α_g = the rate of loss of single molecules per unit surface area from a g-size embryo (generally an unknown function of embryo size)
β = the rate of addition of single molecules to an embryo per unit surface area

β is generally assumed to be independent of size and is usually approximated by the kinetic theory expression for the rate at which vapor molecules impinge upon a unit area of surface per unit time.

Equation 8 can be written as

$$\frac{\Delta n_g}{\Delta t} = J_{g-1} - J_g \tag{9}$$

where we have taken the net "flow" of embryos from size g to g + 1 to be

$$J_g = \beta a_g n_g - \alpha_{g+1} a_{g+1} n_{g+1} \tag{10}$$

If we consider the special equilibrium case, then the left-hand sides of both Equations 9 and 10 vanish and $n_g = N_g$ where N_g is given by Equation 5. Under these conditions we obtain the following relation

$$\frac{N_g}{N_{g+1}} = \frac{\alpha_{g+1} a_{g+1}}{\beta a_g} \tag{11}$$

which gives an expression for the evaporation rate, α_{g+1}, in terms of β, a_g, a_{g+1}, N_g, and N_{g+1}. Equation 11 is a statement of the principle of microscopic reversibility [22, 30].

Nucleation, however, does not occur in a system at equilibrium. During nucleation, the net "flow" of embryos given by Equation 10 will be nonzero; and, in addition, the distribution function, n_g, must be replaced with the nonequilibrium distribution function, f_g. Although f_g changes with time initially, a steady distribution of embryos is presumed to be established to that $J_g = J$, where J is a constant rate, independent of g. This is the rate which is identified as the nucleation rate and which is of interest in this analysis.

Utilizing Equation 11 and the nonequilibrium distribution function f_g, Equation 10 can be rearranged to give

$$J_g = \beta a_g N_g \left(\frac{f_g}{N_g} - \frac{f_{g+1}}{N_{g+1}} \right) \tag{12}$$

where we have assumed a steady distribution and a constant rate, J. Although it is possible to use Equation 12 and obtain an "exact" expression for J [25], it is usually more convenient to pass directly to the differential form,

$$J = -\beta a_g N_g \frac{d(f_g/N_g)}{d_g} \tag{13}$$

With suitable approximations [25], Equation 13 can be integrated giving

$$J = \frac{1}{d}\sqrt{\frac{2\sigma M}{\pi}} \left(\frac{P_e}{kT} \right)^2 S^2 \exp\left(-\Delta\varphi^*/kT\right) \tag{14}$$

which, when expressed in observable quantities, becomes

$$J = \sqrt{\frac{2}{\pi}} \frac{\sqrt{\sigma M}}{d} \left(\frac{P_e}{T} \right)^2 \frac{N_0^{3/2}}{R^2} S^2 \exp\left[-\frac{16\pi}{3R^3} \left(\frac{M}{d} \right)^2 \left(\frac{\sigma}{T} \right)^3 \frac{N_0}{(\ln S)^2} \right] \tag{15}$$

In Equation 14, $\Delta\varphi^*$ = the reversible work required to form the critical cluster (see Figure 1)
d = the liquid density

m = the molecular mass
σ = the bulk liquid surface tension
k = the Boltzmann constant
P_e = the equilibrium vapor pressure
S = the supersaturation

In Equation 15, N_0 = the Avogadro number
R = the gas constant
M = the molecular weight

In writing Equations 14 and 15 we have assumed a unit accommodation coefficient [25].

It is instructive, at this point, to examine the dependence of the nucleation rate on supersaturation. In Figure 3, we have plotted Equation 15 as a function of supersaturation for water vapor at 280° K. In the figure, the nucleation rate is seen to increase from immeasurably small values until a supersaturation (the so-called critical supersaturation) is reached above which the rate increases dramatically. This sudden, dramatic increase in the rate gives rise to the apparent discontinuous nature of a phase transition and has been termed by earlier investigators the "metastable limit" [19]. Also note that a variation of, say, ten drops/cc/sec in either direction in the neighborhood of the critical supersaturation has little impact on the value of the critical supersaturation. This has important bearing on experiments to be discussed later.

THE THERMAL DIFFUSION CLOUD CHAMBER

The thermal diffusion cloud chamber, or DCC for short, is a device used for creating and sustaining supersaturated vapors. It is a steady-state device and has been used extensively in atmospheric studies and for nucleation research. The DCC was originally developed by

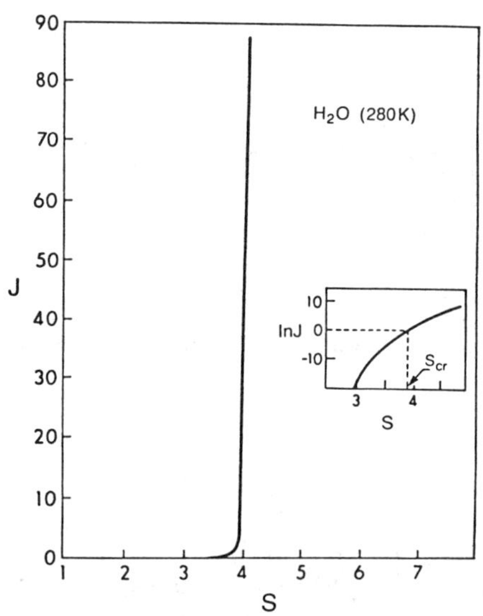

Figure 3. The variation of the nucleation rate of supersaturated water vapor as a function of supersaturation at 280°K. The inset illustrates how insensitive the value of the critical supersaturation is to small variations in the rate (about a rate of one drop/cc/sec).

Langsdorf [31]. Over the years it's operation has been refined by a number of investigators [32–36], and it has come to be used predominantly for nucleation research [36–49].

Basically, the DCC consists of two metal plates separated by a cylindrical, transparent ring. The space inside the ring, bounded by the plate surfaces is the actual working part of the chamber. The metal plates are usually made from a high thermal conductivity material, e. g., copper or aluminium, and machined to a cylindrical geometry. The transparent ring is usually Pyrex or quartz. The ring is separated from the metal plates by gaskets or O-rings usually of rubber, teflon, or soft metal in order to provide vacuum and pressure integrity inside the chamber. The plate surface can be the bare metal or coated, e.g., glass, teflon, or gold, for use with corrosive materials.

If the DCC is to be used as an upward diffusion chamber, the lower plate is heated during operation, and the upper plate is cooled. For use as a downward diffusion chamber, the upper plate is heated and the lower cooled. We shall be concerned only with the upward diffusion chamber here. A photograph of an upward DCC is shown in Figure 4. In the figure, a Pyrex ring is visible between the upper and lower plates (the exterior surfaces of both plates are covered with cork sheeting for thermal insulation), and the heating and cooling units are shown as the dark and light colored units, respectively, below and above the chamber plates. A shallow pool of the material under investigation is maintained on the lower plate (visible in Figure 4) as a source of vapor, and the region above the pool is filled with a light, inert carrier gas such as hydrogen or helium.

During operation the working fluid evaporates from the pool on the lower plate and diffuses through the carrier gas to the upper plate where it condenses, forming a thin film of condensate. The upper plate is machined with a small (one degree) bevel so the film slowly drains to the side and runs down the inside of the ring to the pool of working fluid on the lower plate. In order to prevent condensation from occurring on the inside of the ring, heater wires are wrapped around the outside of the ring (also visible in Figure 4).

Figure 4. A photograph of a typical diffusion cloud chamber.

In order to determine the partial pressure and temperature conditions inside the DCC the surface temperatures of the pool of working fluid on the lower plate and the film of condensate on the upper plate must be measuring along with the total pressure inside the chamber. The lower plate pool surface temperature is usually measured by passing thermocouples through the lower plate or through the ring and having the junction just dimple the pool surface. The upper plate film temperature is determined by imbedding thermocouples in the upper plate just above the plate surface and then calculating the temperature drop across the film [37]. In most cases, this temperature drop is negligible. The total pressure is measured by connecting the chamber to an external gas handling system. This connection is visible in Figure 4 on the chamber.

As long as the diameter-to-height ratio is sufficiently large (greater than about 5 or 6), the mass and energy transport inside the chamber may be considered one dimensional [37], and the resulting temperature and partial pressure profiles across the chamber can be obtained by solving the appropriate one-dimensional mass and energy transport equations. From this information the variation of the supersaturation between the chamber plates can be obtained. We shall return to this shortly.

A typical experiment begins with the DCC being properly cleaned and charged with the working fluid and carrier gas of interest and both plates at the same temperature. The supersaturation is unity (saturated) throughout the chamber. Gradually the lower plate is heated and the upper plate is cooled. As this process continues the supersaturation inside the chamber increases. During this phase of operation, foreign particulates present in the vapor will act as condensation centers and be "rained" out and bound by surface forces to the pool on the lower plate. Hence, the DCC is self cleaning. Eventually, the supersaturation and temperature at some level in the chamber, viz., the nucleation zone, is sufficient for homogeneous nucleation to begin, and small (10–20 microns) droplets are observed to form in this region and fall to the pool on the lower plate. The falling droplets are detected by light scattering using either white light or a laser beam as a light source. The scattered light is detected either by eye (white light) or by a photomultiplier tube. If the photomultiplier tube is used, the resulting signal is passed to an electronic counter or an oscilloscope to record the droplet count. In addition to the droplet count, droplet size can also be measured at any point in the chamber by time of flight measurement as the droplets pass through a set of closely spaced laser beams (see Appendix B for more details).

Since some substances, such as water and alcohol, undergo ion-induced nucleation at supersaturations lower than required for homogeneous nucleation, it is usually necessary to apply an electric field across the chamber plates to sweep out charged particles from the chamber interior. Typically, electric fields of the order of 35 VDC/cm are satisfactory.

Because the DCC must be diffusion controlled to function properly, it is important to suppress convection. There are several practical points to be made in connection with this requirement. First, it has been observed [37, 39] that the ratio of the total pressure inside the chamber to the equilibrium vapor pressure of the working fluid at the lower pool temperature must be greater than about 2 or 3. If this pressure ratio is too low, results from chamber experiments tend not to be reproducible. Second, if the carrier gas pressure (or molecular weight) is too great, the gas-phase density at the top of the chamber exceeds that at the bottom and natural convection results [39]. Third, if there is too much wall heat (from the heater wires) being added to the system, large convection cells are produced inside the chamber [42]. And, last, if the plate temperatures are not uniform across their surfaces, radial temperature gradients are set up and convection can result.

A serious limitation of the DCC has been the requirement that both the working fluid pool surface and the film of condensate on the upper plate remain liquid. With the wall thicknesses of the Pyrex or quartz rings generally used in DCC design (1/2 to 1 cm are typical values) and the need for the total pressure to be 2 or 3 times greater than the equilibrium vapor pressure at the working fluid pool surface (see the foregoing), the useful temperature range of the DCC was limited to between the triple point and (generally less than) the boiling point of the working fluid. This problem has been solved somewhat with the recent construction of a high-pressure DCC capable of sustained operation at temperatures in excess of 525° K and internal pressures in excess of 100 atm [50].

The temperature, partial pressure, supersaturation, and density profiles across the chamber can be determined in a relatively straightforward manner provided the one-dimensional transport approximation can be applied. Considering a two-component system and assuming no external forces, no net pressure gradient, no chemical reaction, and one component being stagnant, we can write the following expression for the diffusive flux:

$$L = -\frac{c\mathscr{D}_{12}}{(1-x)}\left[\frac{dx}{dz} + \alpha_T x(1-x)\frac{d\ln T}{dz}\right] \tag{16}$$

where L = a scaled molar flux equal to ℓh where ℓ is the molar flux of the diffusing material and h the height of the DCC (pool surface to upper plate condensate film, usually computed at one fourth the inside diameter of the chamber ring)
c = the total molar density
$\mathscr{D}_{12}$ = the binary diffusion coefficient
x = the mole fraction of diffusing material
α_T = the thermal diffusion ratio defined by $k_T/[x(1-x)]$

where k_T = the thermal diffusion factor
T = the absolute temperature
Z = the reduced chamber height, varying from zero at the lower pool surface to one at the upper film surface

The energy flux can be expressed as

$$Q = -\lambda\frac{dT}{dz} + LH + L\alpha_T RT(1-x) \tag{17}$$

where Q = a scaled heat flux equal to qh where q is the heat flux
λ = the mixture thermal conductivity
H = the vapor enthalpy
R = the gas constant

The remaining symbols have already been defined.

Equations 16 and 17 are solved numerically yielding the profile information necessary to evaluate the nucleation data. Typical profiles for supersaturated water vapor obtained using Equations 16 and 17 are shown in Figure 5. Here, the temperature, T, partial pressure,

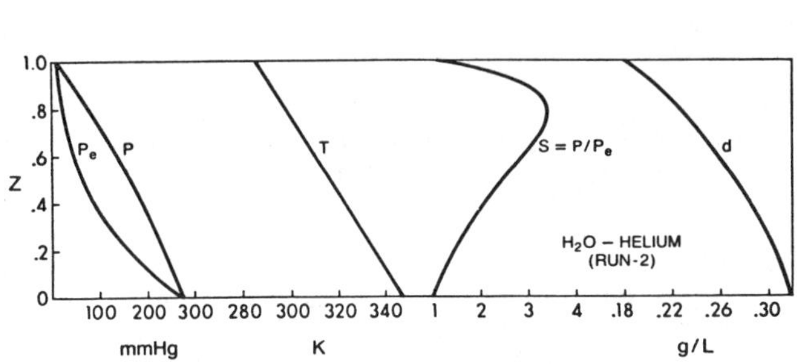

Figure 5. Typical conditions inside the diffusion cloud chamber during steady-state operation at zero or very low nucleation rates. The data is for water vapor with helium as a carrier gas and is taken from experimental run #2 in Reference 38 (reproduced in Figure 7).

P, supersaturation, S, and total density, d, profiles are shown. In addition, the equilibrium vapor pressure, P_e, at each reduced height is also plotted and compared with the actual partial pressure. The supersaturation profile is quite nonlinear due to the temperature dependence of the equilibrium vapor pressure. Also note that the total density throughout the chamber decreases from the lower to the upper surface. The behavior of these equations under various conditions and the sensitivity of the solutions to variations in property data has been discussed elsewhere [40, 44].

The most serious drawback of having to use Equations 16 and 17 is the need for a wide variety of accurate thermodynamic and hydrodynamic data. Indeed, this is one of the disadvantages of the DCC in general. Fortunately, in many cases, the solutions to Equations 16 and 17 are least sensitive to those properties which are most uncertain [44].

One other disadvantage of having to use Equations 16 and 17 is that since a numerical solution is required a mini- or mainframe computer is necessary. Online monitoring of chamber conditions is becoming an increasingly important capability in nucleation research, and it is very useful to be able to continually update chamber conditions based upon the values of the supersaturation and temperature in the chamber at any instant. This can be easily done by utilizing current microcomputer and digital technology provided a method of determining the partial pressure and temperature profiles amenable for use on a microcomputer is available. Although various simplifications for Equations 16 and 17 have been discussed in the literature [40], the results are either not sufficiently accurate (e.g., assuming linear profiles) or the solution still requires substantial computer horsepower. We provide here a simple method for calculating the temperature and partial pressure profiles which works reasonably well and is amenable for use on a microcomputer.

First, we ignore the Dufour and Soret terms (the terms containing α_T in Equations 16 and 17. Then we assume the mixture thermal conductivity is independent of temperature so the temperature profile is linear. Finally we approximate the binary diffusion coefficient as described in Reference 51. With these approximations the temperature and pressure profiles across the DCC become

$$T(Z) = [T(U) - T(L)]Z + T(L) \tag{18}$$

and

$$P(Z) = P_T - [P_T - P(L)]\left[\frac{P_T - P(U)}{P_T - P(L)}\right]^{\gamma} \tag{19}$$

respectively. In Equation 18, T (U) and T (L) are the upper and lower surface temperatures, respectively, and Z is the reduced chamber height. In Equation 19, P_T is the total pressure; P (U) and P (L) are the equilibrium vapor pressures at T (L) and T (U), respectively; and γ is given by

$$\gamma = \left[\frac{T(L)^{1-s} - T(Z)^{1-s}}{T(L)^{1-s} - T(U)^{1-s}}\right] \tag{20}$$

where s is the temperature dependence of the binary diffusion coefficient [51]. Representative solutions of Equations 18 and 19 are compared with "exact" solutions for supersaturated water vapor in Figure 6.

The linear temperature profiles (not shown) for the data shown in Figure 6 differ from the "exact" solutions by, at most, one degree. The partial pressure profiles (also not shown) agree to within 3 mm. The supersaturation profiles agree to within about 2% at the point where the nucleation rate is a maximum (indicated by the arrow in the figure). For comparison, the supersaturation profiles calculated assuming both temperature and partial pressure profiles to be linear for curves A and B differ by about 21% and 14%, respectively, at the point the supersaturation is maximum (indicated by the vertical bar in the figure). We have used Equations 18 and 19 with other working fluids and have found roughly the same agreement. It should be remembered that Equations 18 and 19 are not meant to replace Equations 16 and 17, only to supplement them for use with online monitoring and control systems.

THE DCC AT LOW NUCLEATION RATES

In describing nucleation experiments involving the DCC, it is convenient to divide the discussion into three sections corresponding to the range of nucleation rates of interest: the low nucleation rate regime; the high nucleation rate regime, and the intermediate nucleation rate regime. In the first section, the low nucleation rate regime, it is convenient to further break the discussion down into two subsections: critical supersaturation measurements; and nucleation rate measurements.

Critical Supersaturation Measurements

In nucleation research, the most common application of the DCC has been the measurement of the critical supersaturation. In Figure 3, the "threshold-like" dependence of the rate on supersaturation is clearly evident so it is reasonable to associate an onset of nucleation with a particular value of the supersaturation. The critical supersaturation can be calculated at a particular temperature, by setting the rate, J, in Equation 15 equal to one drop/cc/sec and solving the resulting expression for S. Here we have assumed the onset of nucleation to correspond to a rate of one drop/cc/sec. The result of this calculation for water is shown in Figure 7 as the solid line labeled BDZ (Becker, Döring, Zeldovitch).

Experimentally, the critical supersaturation for a particular working fluid vapor is obtained by configuring the chamber as described earlier and increasing the temperature difference between the chamber plates until a nucleation rate of approximately one to three drops/cc/sec is obtained. The reason for this less-than-precise specification of the rate will be discussed later. The resulting supersaturation and temperature profiles for that experiment (obtained from Equations 16 and 17) are then plotted against each other as illustrated by the numbered curves for supersaturated water vapor in Figure 7. After each experiment, the temperatures of both chamber plates are increased (or decreased) and adjusted until a steady rate of one to three drops/cc/sec is obtained again, this time at a slightly higher (or lower) temperature. The supersaturation profile is again plotted against the temperature profile giving another of the curves illustrated in Figure 7. Eventually the entire temperature range accessible to the DCC for a particular working fluid will have been covered and a family of curves, illustrated by the numbered curves in Figures 7, will have been generated. The envelope of all these curves is the experimentally determined variation of critical supersaturation with temperature. As can be seen in Figure 7 for water vapor, the experimentally determined envelope is about 7%–8% higher than the value predicted by nucleation theory. The actual envelope has not been drawn in for the sake of clarity.

Critical supersaturation measurements involving the DCC have been made, for the most part, on single-component systems. A partial list of these systems is given in Table 1 along with appropriate references. In addition to the single-component studies, there have also been a number of critical supersaturation experiments carried out on binary mixtures [52, 58] and on systems undergoing chemical reaction [43, 46, 49]. A partial list of these investigations is also provided in Table 1. Typical results of critical supersaturation measurements on the water-sulfuric acid and water-nitric acid systems are shown in Figure 8. Here, the activity of the water and the acid necessary to maintain a nucleation rate of one drop/cc/sec at a temperature of 298° K are plotted. Details of this experiment are given elsewhere [57] . In Figure 9 the critical supersaturations for two carboxylic acids are shown as functions of temperature. In each case the carboxylic acid vapor was undergoing a monomer-dimer reaction, $2A \rightleftarrows A_2$, during the nucleation process. The effect of the dimerization reaction is to stabilize the vapor with respect to condensation. Larger critical supersaturations are required in the presence of the reaction than without it. This effect is more pronounced with the lower molecular weight acids due to the larger reaction equilibrium constants. The different curves labeled BDZ, KSR, and FW represent different theoretical predictions of the critical supersaturation. Details of these and related experiments are available elsewhere [43, 46, 49].

Two points deserve mention at this juncture, both of which are related to the extremely strong dependence of the nucleation rate on supersaturation. The first can best be illustrated by considering Equation 15. The dependence of the nucleation rate on supersaturation is such that small variations of S produce large changes in J. This has already been illustrated in Figure 3, and is shown again in Table 2 where J, the nucleation rate of supersaturated water vapor, has been calculated for a variety of values of S at 300° K. In experimental measurements of the critical supersaturation, however, we set the rate at a constant value, say, one drop/cc/sec, and determine the corresponding supersaturation. If J were set to two or three drops/cc/sec instead, the corresponding change in S would be so small as to be beyond the experimental resolution of the DCC. This is the reason we need only to adjust the rate in the chamber to one to three drops/cc/sec in a critical supersaturation experiment.

The second point is concerned with the usefulness of the critical supersaturation experiment as a tool in nucleation research. The results of the water critical supersaturation experiments shown in Figure 7 (in addition to similar results for other substances in the literature, e.g., see references cited in Table 1) seem to suggest reasonable agreement with the classical nucleation model. If, however, the calculations shown in Table 2 are examined closely, it is seen that a 7% difference in supersaturation corresponds to a difference of roughly 10,000 in the nucleation rate as predicted by the classical model and that determined experimentally. When viewed in this fashion, the agreement is considerably less

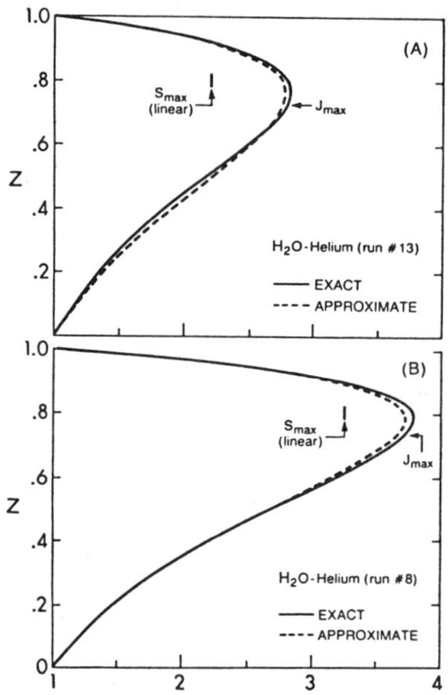

Figure 6. A comparison of the "exact" and an approximate method for calculating the supersaturation profile in the diffusion cloud chamber.

Table 1
A Partial List of Vapors Used in Homogeneous Nucleation Experiments with a Diffusion Cloud Chamber*

Vapor Type	Vapor Name	Reference
Single Component	Water	38
	Methanol†	36, 48
	Ethanol†	36, 48, 68
	n-Propanol†	58
	i-Propanol†	–
	Carbon tetrachloride	42
	Menthol	41
	Vinyl acetate‡	—
	m-Xylene	55
	Benzene	39
	1,2,2,2-tetrachloroethane	42
	Chloroform	42
	Freon II (CCl_3F)	42
	n-Hexane	37
	n-Heptane	37
	n-Octane	37
	n-Nonane	37
	Toluene	39
	O-Xylene	39
	n-Butylbenzene	39
Reacting Vapors	Formic acid	46
	Propanoic acid	46
	Decanoic acid	49
	Stearic acid§	–
	Acetic Acid	43
	Heptanoic Acid	49
	Myristic Acid	49
Binary Mixtures	Water-sulfuric acid	57
	Water-ethanol	55
	Water-methanol	58
	Water-nitric acid	57
	o-Xylene-m-xylene	55
	Water-n-propanol	58

* The corresponding references are indicated in parenthesis. The list is not comprehensive, and references cited may not denote all sources in the literature.
† A. Kacker and R. H. Heist, in preparation.
‡ M. A. Chowdhury, submitted to *J. Chem. Phys.*
§ R. H. Heist, unpublished data.

Table 2
The Homogeneous Nucleation Rate of Supersaturated Water Vapor at 300°K as a Function of Water Vapor Supersaturation*

Supersaturation	Rate
2.0	2.4×10^{-45}
2.2	3.3×10^{-29}
2.4	8.0×10^{-19}
2.6	1.2×10^{-11}
2.8	2.0×10^{-6}
3.0	1.8×10^{-2}
3.107	1.0
3.2	2.2×10^{1}
3.4	6.7×10^{3}
3.6	7.1×10^{5}
3.8	3.5×10^{7}
4.0	9.5×10^{8}

* The rates were calculated using Equation 15 and the necessary thermodynamic data provided in Reference 38.

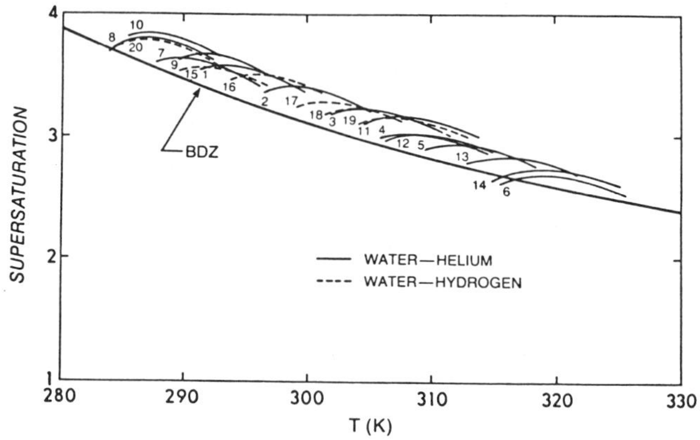

Figure 7. Variation of the critical supersaturation of water vapor with temperature. The envelope of the numbered curves is the experimentally determined variation of the critical supersaturation [38].

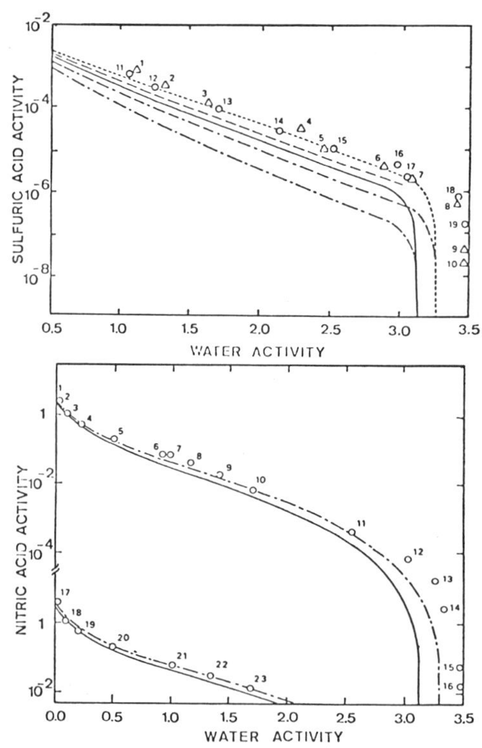

Figure 8. Homogeneous nucleation, theory, and experiment for the binary mixtures sulfuric acid-water (top plot) and nitric acid-water (lower plot) [57].

impressive. Based on this argument, it seems prudent to use critical supersaturation data only as qualitative support for any nucleation theory. The appropriate test of a nucleation model is the supersaturation and temperature dependence of the nucleation rate.

Nucleation Rate Measurements

The measurement of nucleation rate in the DCC is complicated by two factors: (1) the strong dependence of rate on supersaturation (see Figure 3 and Table 2); and (2) the deleterious effects of vapor depletion and latent heat of condensation associated with droplet growth on the supersaturation in the DCC. The former problem, the strong dependence of rate on supersaturation, requires careful control of chamber operating conditions, e.g., plate temperatures, total pressure, wall heat, etc., and a sufficiently accurate droplet counting system. The latter problem can generally be ignored (see below for details) in the range of small nucleation rates, e.g., less then 5–10 drops/cc/sec. In this range, the chamber temperature and supersaturation profiles are still accurately given by solutions to Equations 16 and 17.

The nucleation rate measuring system used in our laboratory is illustrated schematically in Figure 10. Here the DCC is shown (top view) with two laser beams passing through the chamber. The beam labeled "Radius" is used for measuring droplet radii and will be

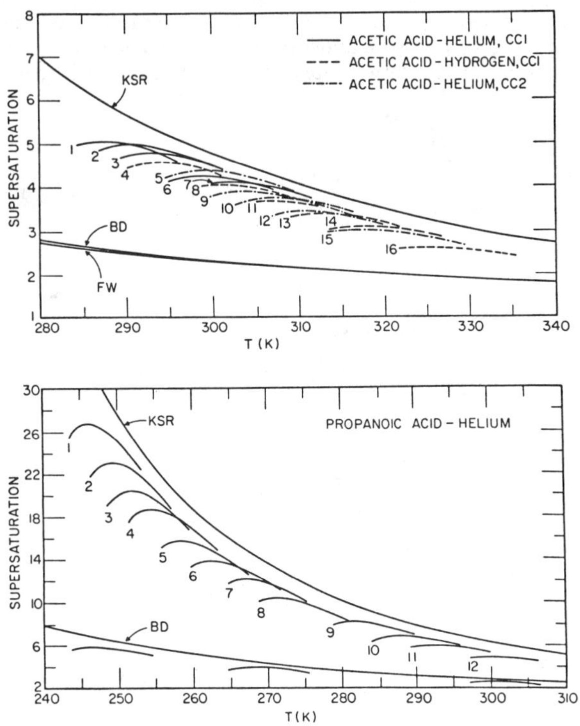

Figure 9. Variation of the critical supersaturation of acetic acid (top plot) and propanoic acid (lower plot) with temperature. The envelope of the numbered curves in each plot is the experimentally determined variation of the critical supersaturation [43,46].

discussed later. The beam labeled "Count" is used for droplet counting. It is directed through the chamber near the pool surface and is usually (though not necessarily) shaped into a rectangular geometry with a width of about 1 cm and a thickness of 0.1 cm or less being typical values. This beam can be translated throughout the chamber so that the droplet count can be measured at any point of interest in the chamber. Light scattered by a droplet passing throught this laser beam is collected by a lens assembly (not shown in Figure 10) and focussed on the photocathode of the photomultiplier tube assembly with the resulting signal being passed to droplet counter. Since the droplet count is related to the droplet flux by the counting area, it is important to know the actual counting area visible to the photomultiplier tube in any given experiment. We calibrate the viewing area so that for a droplet count measurement we can relate the distribution of scattered signal intensities to the counting area in the laser beam. Using a threshold setting on the counter allows us to record only those events corresponding to drops passing through a known area in the laser beam.

Once recorded by the counter, the counts are accumulated over a set time period, determined by a selectable time base, and at the end of that period the total count is passed to a microcomputer for storage and processing. Long counting periods are possible with the microcomputer-droplet counter assembly because the microcomputer also monitors the upper and lower plate temperatures in addition to the drop count to ensure data acquisition only during stable chamber operating conditions. This capability permits nucleation rate measurements to be made over the range of about 0.01 to 10 drops/cc/sec. Statistical correction for coincidence losses in this range of nucleation rates is generally not necessary.

Typical nucleation rate data obtained at several temperatures for ethanol vapor and n-propanol vapor are shown plotted against the nucleation zone supersaturation in Figure 11. The data in Figure 11 for nucleation rates less than about 5–10 drops/cc/sec correspond

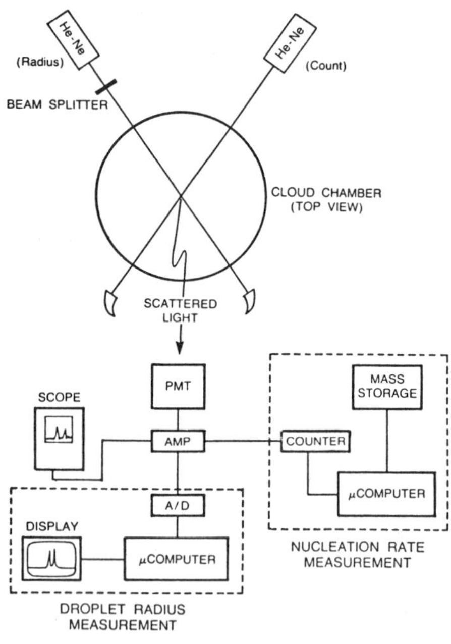

Figure 10. Schematic of the experimental set-up for nucleation rate and droplet radius measurements.

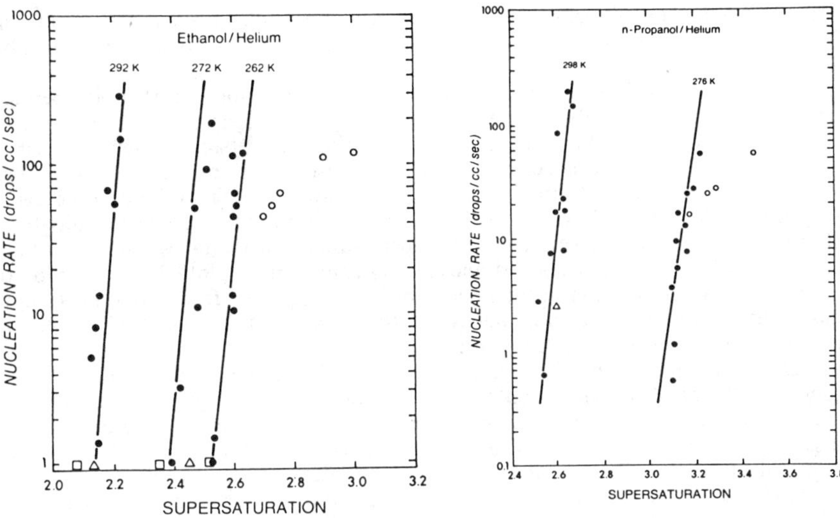

Figure 11. Variation of homogeneous nucleation rate of ethanol and n-propanol with nucleation zone supersaturation at several temperatures in the DCC. The solid circles in each plot are nucleation rate data corrected for vapor depletion and latent heat effects. See text for details.

to the low nucleation rate region just discussed. The solid circles represent the measured rate data. The open squares and triangles are critical supersaturation data from the literature [36, 48, 58]. The data represented by the open circles will be discussed later. The supersaturation and the temperature indicated in Figure 11 correspond to the conditions in the nucleation zone, the region of the DCC where nucleation is observed to occur. The zone supersaturation and temperature are calculating using Equations 16 and 17, and the rates are measured as described before. The rest of the data in Figure 11 will be discussed later.

The solid lines in Figure 11 were obtained using classical nucleation theory, Equation 15, and scaling by a (temperature dependent) factor to bring the classical theory into agreement with experiment. The scaling of the nucleation rate expression will be discussed later.

THE DCC AT HIGH NUCLEATION RATES

As a tool in nucleation research, the DCC is most commonly used at low and intermediate (see the following) rates of nucleation. However, when the chamber is operated under conditions producing extremely high nucleation rates an interesting oscillatory nucleation is observed. We discuss this phenomenon at this juncture both because it is important in its own right and because it will allow a more natural transition to the discussion of intermediate rates of nucleation later.

There is considerable interest in both physical and chemical systems that exhibit any kind of sustained oscillatory behavior. Typical examples of systems which can exhibit periodic behavior include: unstable chemical reactors in chemical engineering systems; biological "clocks" and competing populations in biological systems; fluid instabilities (e.g., Benard type) in hydrodynamic systems; certain structures in geological systems (e.g., oscillatory zoning in rock formation); and, oscillating reactions in chemical systems (e.g., Zhabotinskii reaction, Liesegang rings, etc.). The oscillatory behavior we describe here involves only nucleation, growth, and diffusion, no chemical reaction, and, as such, respresents an interesting addition to the relatively small group of physical systems known to exhibit oscillatory behavior.

During an experiment designed to observe oscillatory nucleation, the DCC is operated in much the same way as described earlier for nucleation rate measurements. The nucleated droplets form in a zone located in the upper portion of the chamber and are detected as they fall through a helium-neon laser beam located in the lower portion of the chamber. The light scattered by these falling droplets is detected by a photomultiplier tube, and the signal is displayed either as an oscilloscope trace or counted electronically. Typical results from a homogeneous nucleation experiment with supersaturated methanol vapor [59, 60] are shown in Figure 12. Here the measured nucleation droplet flux is plotted as a function of methanol vapor supersaturation in the nucleation zone at 283° K. It is important to note that this supersaturation is calculated neglecting the presence of the nucleated droplets in the cloud chamber, that is, using Equations 16 and 17.

For ease of description the data in Figure 12 are divided into three regions. The dependence of the data in region I on supersaturation is in good agreement with trends predicted using classical nucleation theory (Equation 15), although the magnitude differs substantially. The data in region II begin to deviate significantly from predictions of the theory and illustrate the effect of vapor depletion and latent heat on droplet formation in the nucleation zone. In region III the measured nucleation flux appears to be rather insensitive to the calculated supersaturation; and in addition, the character of the measured flux begins to change (as observed on the oscilloscope traces). No longer does there appear to be a random collection of individual droplets; but, rather, one begins to see organized collections of nucleation pulses appearing and disappearing where each pulse contains a number of individual droplets. These organized collections of nucleation pulses are short

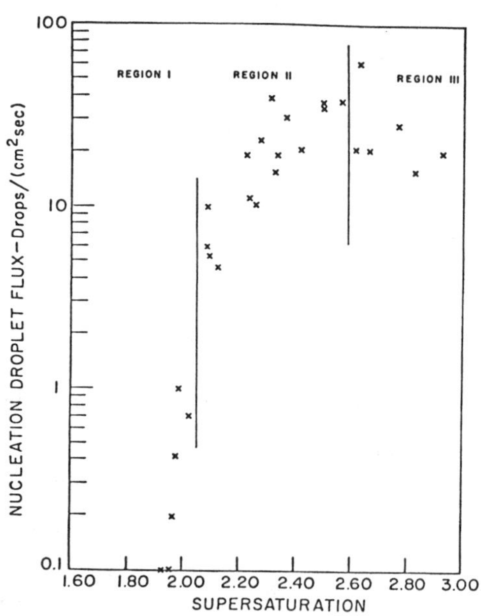

Figure 12. The variation of the nucleation droplet flux with nucleation zone supersaturation at 283°K for methanol vapor [59].

lived and occur relatively infrequently in the portion of region III shown in Figure 12. As the temperature difference between the two cloud chamber plates — and hence the supersaturation — is increased further, these collections of pulses occur more frequently and become longer lived. Increasing the vapor supersaturation still further results in a continuous, well-defined oscillatory nucleation mode. This behavior is best illustration by the photographs in Figure 13. Here the output from the photomultiplier was connected directly to a storage oscilloscope, and six traces were photographed using different oscilloscope sweep times [59]. The photographs shown in Figure 13 illustrate the highly regular nature of this oscillatory behavior. The total oscilloscope sweep time for each of these photographs is 1/2 second, 1 second, 2 seconds, 5 seconds, 10 seconds, and 20 seconds for photographs A–F, respectively. The higher frequency oscillation particularily evident in the first three photographs is constant and slightly less than 4 Hz. The individual pulses shown in Figure 13 do not represent an instantaneous picture of droplet formation in the nucleation zone. Rather, they reflect the shape of the nucleation pulse after it has fallen out

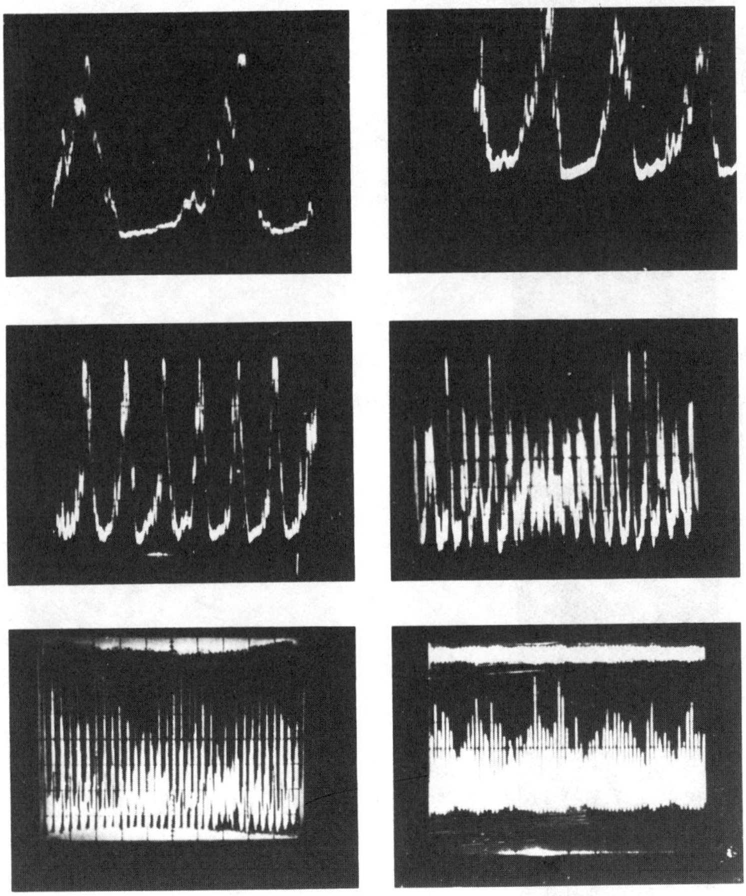

Figure 13. Photographs of oscilloscope traces illustrating the oscillatory nucleation described in the text [59].

of the nucleation zone and through the helium-neon laser beam. Careful examination of the individual pulses shown in photograph A or B reveals an asymmetry for each pulse and complete cessation of nucleation between the individual pulses [61]. The structure on the leading edge of a pulse represents those droplets nucleated first in that pulse. The scattered laser light intensity gradually builds to a maximum value and then falls quickly to zero at

Figure 14. Photograph of seven consecutive frames of a movie of the oscillatory nucleation. These seven frames represent one complete oscillation cycle. The view is through the side of the DCC [70].

the training edge of the pulse. The lower frequency components, evident in photographs D–F in Figure 13, will be discussed later.

In Figure 14 we have reproduced seven consecutive frames from a movie of the oscillatory nucleation taken during an experiment run under conditions similar to those used for Figure 14. In this figure the view is into the DCC from the side with a beam of white light directed in through the other side. The shiny regions at the top and bottom of each frame are the upper and lower liquid surfaces of the chamber, respectively. The dark, horizontal lines visible in each frame are the heater wires wrapped around the glass ring described earlier. The wave of droplets in one nucleation burst can be seen in the middle of the first frame. In the next four frames this wave can be seen falling through the chamber to the pool of methanol on the lower plate. In the second frame a new wave can be seen forming in the upper portion of the chamber. The frequency of these waves is approximately 4 Hz.

The following is a summation of the more important experimental observations [62, 63]: First, the frequency of the oscillations observed for the methanol-helium system ranges between 3 and 4 Hz, and both the frequency and the amplitude of the nucleation oscillations are relatively insensitive to the temperature differences between the upper and lower chamber plates once steady oscillatory nucleation is achieved. Second, the frequency of the nucleation oscillations increases only slightly with temperature in the nucleation zone. Third, using hydrogen instead of helium as the carrier gas increases the frequency of oscillations by approximately 25 percent. Fourth, the individual nucleation pulses are asymmetric with the leading edge of the pulse being more skewed than the trailing edge (see Figure 13). Fifth, nucleation stops completely between two nucleation pulses (see Figure 13). And, sixth, beginning with steady-state nucleation (no oscillations), as the temperature gradient is increased, the random nucleation characteristic of the steady state is momentarily interrupted and organized collections of pulses are seen for brief periods of time. As the temperature gradient is increased further, these transient excursions into the oscillatory mode occur more frequently; and as the temperature difference is increased still further, only the oscillatory mode is observed.

The following model has been proposed [63] to explain the oscillatory nucleation and to account, in general, for the effects of droplet nucleation and growth in a DCC. For convenience, we show a schematic of the DCC in Figure 15 with the nucleation zone and the region below the zone indicated along with the droplet and vapor fluxes. Typical supersaturation and temperature profiles have also been indicated.

During one oscillatory nucleation cycle, individual droplets, formed during the nucleation pulse, reduce the vapor concentration and perturb the local temperature with

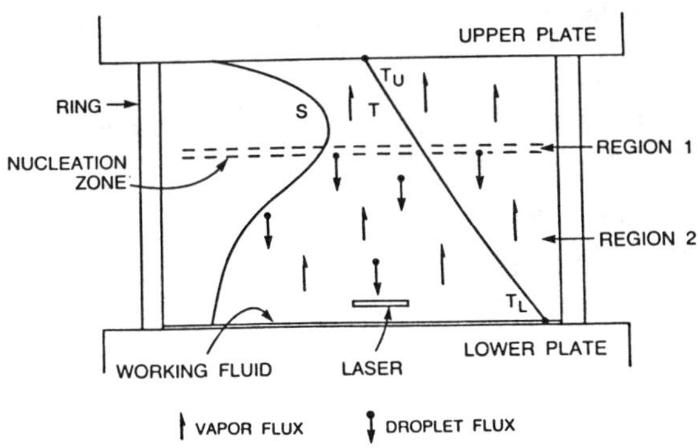

Figure 15. Diffusion cloud chamber schematic (not to scale) illustrating the regions 1 and 2 used in the vapor depletion and latent heat model described in the text.

latent heat of condensation as they grow in the nucleation zone. Since the dependence of nucleation rate on vapor supersaturation and temperature is extremely sensitive, these effects act to abruptly terminate nucleation thus completing a single pulse. After each complete pulse there is a finite time delay as the vapor and energy transport through the chamber act to refresh conditions in the nucleation zone. This transport process, however, is hindered because of the pulse of droplets falling through the chamber to the pool on the lower plate. The effect of these growing droplets on the mass and energy fluxes through the chamber gives rise to a kind of "delayed feedback" which is, in fact, crucial to the oscillation process. Without it, the nucleation oscillations can not be sustained. As the partial pressure and temperature in the zone return to their original values, nucleation begins again, and the cycle repeats.

Subject to certain assumptions, described in detail elsewhere [63], the following mass and energy conservation equations for the nucleation zone in the DCC and the region below the nucleation zone can be written down.

First, for the nucleation zone

$$dC_1/dt = K[C_2 - C_1]/\delta - R_1 \tag{21}$$

and

$$\begin{aligned} C_{T1}C_{P1}dT_1/dt = {} & K(C_2 - C_1)[\Delta H_v(T_1)]/\delta - F_{\ell 1}[\Delta H_\ell(T_1)] \\ & - U_1(T_1 - T_0) + U_1(T_2 - T_1). \end{aligned} \tag{22}$$

Next for region 2, below the zone

$$dC_2/dt = K(C_0 - C_2)/\ell - R_2 \tag{23}$$

and

$$\begin{aligned} C_{T2}C_{P2}dT_2/dt = {} & K(C_0 - C_2)[\Delta H_v(T_2)]/\ell - F_{\ell 2}[\Delta H_\ell(T_2)] \\ & - U_2(T_2 - T_1). \end{aligned} \tag{24}$$

In Equation 21 C_2 = the concentration of the diffusing vapor entering the nucleation zone
C_1 = the zone vapor concentration
K = a proportionality term which is, in general, concentration and temperature dependent and must be determined for each experiment
δ = the zone height
R_1 = the "reaction" term which expresses the volume specific rate of conversion of working fluid vapor to liquid by nucleation and growth

In general, R_1 is a complicated function of concentration and temperature and requires both a reliable nucleation rate expression and an appropriate droplet growth law or knowledge of the droplet size distribution in the nucleation zone as a function of time. The R_1 term is discussed in more detail in Appendix A.

In Equation 22 T_2 and T_1 = the input and the zone temperatures, respectively
C_{T1} = the total concentration (vapor plus carrier gas) in the zone
C_{P1} = the zone heat capacity
$\Delta H_v(T_1)$ and $\Delta H_\ell(T_2)$ = the vapor and liquid enthalpies, respectively, for the working fluid
$F_{\ell 1}$ = the liquid flux leaving the zone
T_0 = the zone temperature computed assuming no nucleation

U_1 and U_1' are related to the zone thermal conductivity. The first term on the right of the equality in Equation 22 represents the net enthalpy transport into the zone due to vapor transport. The temperature dependence of the vapor enthalpy has been suppressed. The second term is net enthalpy transport out of the zone due to liquid transport; and, again, the temperature dependence of the enthalpy has been suppressed. The last two terms in Equation 22 represent net energy transport from the zone by conduction.

In Equations 23 and 24, we have chosen the concentration and temperature "profiles" in the region below the nucleation zone (denoted region 2, for convenience) to be C_0 and T_0, the no nucleation values of the concentration and temperature in the system. The term, K, is the same as described earlier, and ℓ is the height of region 2. R_2 is the volume specific rate of vapor condensation in region 2. In Equation 24, C_{T2} and C_{P2} are the concentration and heat capacity for region 2; $F_{\ell 2}$ is the net liquid flux leaving region 2; and U_2 is related to the region 2 thermal conductivity.

The first term to the right of the equality in Equation 23 represents the net vapor transport to region 2, and the second term represents the vapor transport to the droplets as they pass through region 2. Similarly, in Equation 24, the first term is the net enthalpy transport to region 2, the second term represents the net enthalpy transport to the droplets as they pass through region 2 falling to the lower plate, and the last term represents net energy transport from region 2 by conduction.

A general expression for R_1 in Equation 21 is (see Appendix A of this chapter):

$$R_1 = \rho_\ell v_m [\dot{N}(t)\langle n(t)\rangle + N(t)\langle \dot{n}(t)\rangle + n_f N_d(t)] \tag{25}$$

In Equation 25
- ρ_ℓ = the liquid molar density
- v_m = the molecular volume
- $N(t)$ = the total number of droplets in the region of the nucleation zone defined as the system [64]
- $\langle n(t)\rangle$ = the average size of the distribution of droplets in the system
- n = the number of molecules in a droplet in excess of the critical radius
- n_f = the number of molecules in a droplet in excess of the critical radius as it falls from the nucleation zone
- $N_d(t)$ = the concentration of droplets leaving the nucleation zone each second

and the derivatives indicated are with respect to time.

$N(t)$ and $\langle n(t)\rangle$ are each defined in terms of $g(n, t)$, the distribution of droplet sizes in the system (see Appendix A for additional details).

$F_{\ell 1}$ in Equation 22 has the form

$$F_{\ell 1} = N_d(t)\, V_d \rho_\ell \tag{26}$$

where $N_d(t)$ has been previously defined, v_d is the droplet volume, and ρ_ℓ is the liquid molar density.

R_2 in Equation 23 has the form

$$R_2 = \frac{4\pi r_f \mathscr{D}_{12} N_2(t)}{\ell}[C_2 - C_{2e}(T_2)] \tag{27}$$

where
- r_f = the radius of the droplet as it falls from the nucleation zone
- $N_2(t)$ = the average number density of drops in region 2
- ℓ = the height (vertical dimension) of region 2
- $\mathscr{D}_{12}$ = the binary diffusion coefficient
- $C_{2e}(T_2)$ = the equilibrium concentration of working fluid vapor at temperature T

$F_{\ell 2}$ in Equation 24 has the same form as R_2 in Equation 27.

Before solving Equations 21 through 24, two additional points must be considered. First, to evaluate R_1 in Equations 21 and 22, we shall require an accurate expression for the nucleation rate as a function of vapor concentration and temperature. For convenience we choose Equation 15, the classical nucleation rate model. As discussed earlier, it will be necessary to scale Equation 15 to bring the predicted nucleation rate values in agreement with the measured ones. This scaling is done at low nucleation rates where accurate rate measurements are possible, and Equations 16 and 17 can be used to determine the zone supersaturation and temperature. For this analysis the scaling factor was assumed to be independent of concentration.

Second, we need to know δ, the nucleation zone thickness. The zone thickness can be measured only at low nucleation rates since at higher rates the falling droplets obscure the lower boundary of the zone. In order to obtain δ at higher rates we have adopted the following convention which predicts values of δ that agree reasonably well with those measured at low nucleation rates. Using Equation 15 and the no-nucleation concentration and temperature profiles (obtained from solution of Equations 16 and 17, a nucleation rate profile can be calculated for the entire chamber. Because of the strong dependence of nucleation rate on both supersaturation and temperature, a sharp peak in this rate profile is obtained in the upper half of the chamber corresponding to the region of maximum nucleation. The value of δ used in this analysis is set equal to the full width of this peak at half the maximum height.

We first present solutions to the isothermal version of the model, Equations 21 and 23. The reason being that the oscillatory behavior has been found to be due primarily to competing mass transfer effects and not to temperature or energy transfer considerations. The role of temperature is largely to modify the superficial features of the oscillations and will be discussed later.

Equations 21 and 23 have been solved numerically, and a typical solution is shown in Figure 16. In Figure 16, C_1, the concentration of the working fluid vapor in the nucleation zone is shown as a function of time. In Figure 17 the time evolution of the nucleation rate corresponding to this concentration is plotted. The oscillation frequency evident in Figures 16 and 17 is 3.4 Hz, in good agreement with the experimental observations discussed earlier. As is evident in Figure 17, the nucleation stops between the nucleation pulses in agreement with experimental results. With this isothermal version of the model, however, the calculated nucleation pulses tend to be rather symmetric; while, experimentally, they are observed to be somewhat asymmetric (see Figure 13). In addition, the magnitude of the calculated nucleation rate appears larger than is evident from the experimental data. It should be mentioned that when one does solve the nonisothermal version of the model (Equations 21–24), the calculated nucleation pulses are asymmetric and the nucleation rate is in better agreement with experimental observations. The isothermal results shown in Figures 16 and 17, however, suggest that the oscillatory nucleation is due primarily to vapor depletion effects and that latent heat contributions are of secondary importance.

We can examine the dependence of the solutions to Equations 21 and 23 on the value of C_0, the no-nucleation value of C. By varying only C_0, we obtain the results summarized in Figure 18. In Figure 18A the calculated frequency of the concentration oscillation is plotted as function of the no-nucleation value of the supersaturation, S_0, in the nucleation zone (S_0 is directly proportional to C_0). In the isothermal model, the frequencies of the concentration and the nucleation oscillations are identical. In Figure 18B the amplitude of the concentration oscillation is also plotted as a function of S_0. The data in figure 18A illustrate that the oscillation frequency increases slowly with increasing S_0, or what is roughly equivalent, the temperature difference between the upper and lower plates. Even when the temperature is permitted to change, the oscillation frequency still remains a relatively weak (increasing) function of S_0 which agrees with our experimental observations in the preceding.

The dependence of the oscillation amplitude on S_0 shown in Figure 18B also explains another experimental observation: The growth and decay of the "bunches" or groups of nucleated droplets as the DCC was being brought to the oscillatory mode of nucleation. If the temperature gradient across the chamber is such that our operating point is on part A

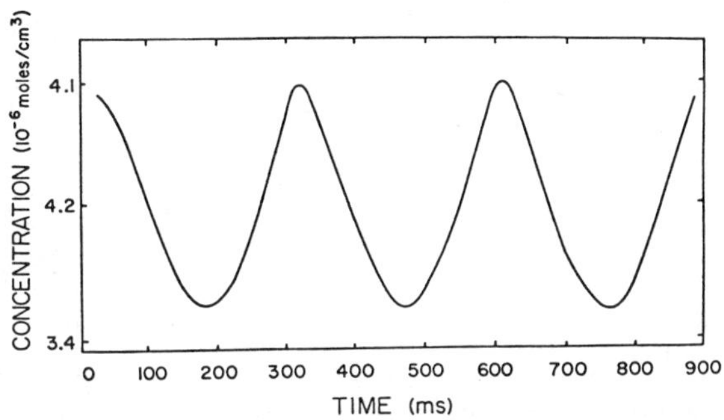

Figure 16. Typical solution of the isothermal nucleation and growth model, Equations 21 and 23. The figure shows the time variation of the methanol vapor concentration at 273°K in the nucleation zone (the time scale origin is relative) [63].

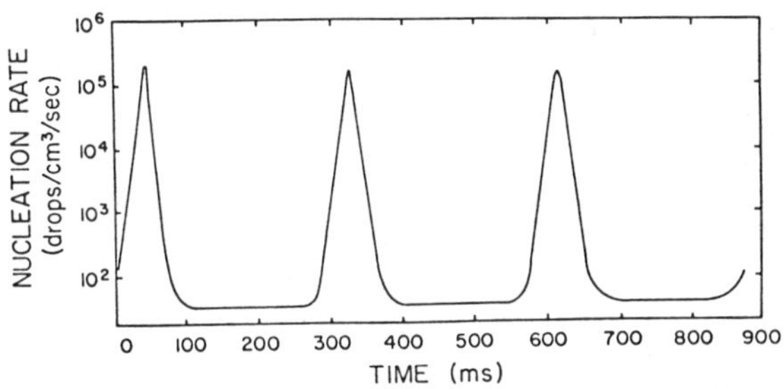

Figure 17. Predicted variation of the nucleation rate corresponding to the variation in concentration shown in Figure 16 [63].

of the curve in Figure 18B, then a variation in this gradient due to, say, changing operating conditions or variations in the temperature of the surface of the liquid pool on the lower plate could momentarily increase C_0 bringing the operating point closer to part B of the curve where oscillations are more developed. If these perturbations disappear, the operating point returns to part A of the curve and the oscillations disappear. Increasing the temperature gradient across the chamber (and consequently increasing the value of C_0) causes these transient excursions into the oscillatory mode to occur more easily and hence more often until a temperature gradient is reached (corresponding to part B of the curve in Figure 18B) for which the oscillatory mode is observed.

The dependence of the solution of the isothermal model on the value of K, the proportionality term in Equations 21 and 22, r_f, the falling radius of the droplet as it leaves the nucleation zone, and, the chamber height, have all been examined and the results are available elsewhere [63] so they will not be repeated here. It is sufficient to note that the temperature dependence of the nucleation oscillation and the carrier gas dependence can all be explained by the isothermal model.

Using both the mass and energy conservation equations, Equations 21–24, only three significant differences from the isothermal solutions are observed. First, there is the slight increase in oscillation frequency already mentioned; second, the shape of the nucleation pulses is asymmetric in better agreement with the observed shapes; and, third, the amplitude

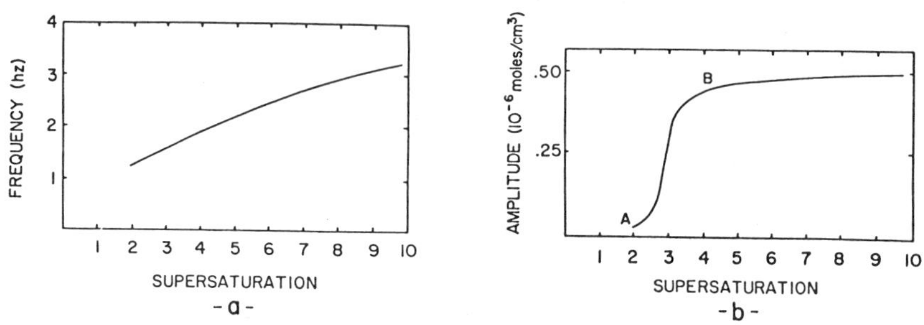

Figure 18. Variation of the (a) frequency and (b) amplitude of the concentration oscillations as a function of the no-nucleation value of the supersaturation, S_0, in the nucleation zone [63].

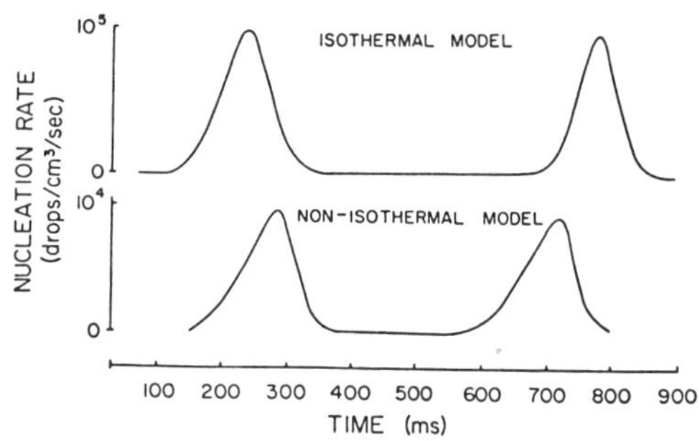

Figure 19. A comparison of the predicted nucleation rate variation as determined using both the isothermal (273°K) and the nonisothermal nucleation and growth models [63].

of the nucleation pulses decreases somewhat bringing the calculated number of nucleated droplets in any given pulse in closer agreement with experiment. These features are evident in Figure 19 where, for comparison, we have plotted the results of the isothermal and nonisothermal models for a set of identical chamber conditions.

Finally, the origin of the lower frequency oscillation which is evident in photographs D–F in Figure 13 is somewhat less certain. The interesting superposition of oscillatory modes appears to arise from the effects of hydrodynamic instabilities in the pool of working fluid on the lower plate of the chamber. We have monitored the surface temperature of the liquid pool at a point near our region of observation and have observed a complex oscillatory variation in surface temperature with time. The higher frequency components of these temperature oscillations have periods in the range of several seconds; while the lower frequency components have periods of the order of minutes. These higher frequency components appear somewhat regular and compare reasonably well with the period illustrated in photographs D–F in Figure 13. It should also be mentioned that with much longer records of the oscillatory nucleation, it is possible to observe lower frequency components of the nucleation oscillation which compare with the lower frequency temperature oscillations. These temperature oscillations at the surface of the liquid pool are consistent with the behavior of an unstable liquid layer which is being heated from below. It may be possible to extract information regarding the hydrodynamic instability in the liquid layer from the frequency spectrum of the nucleation oscillations.

THE DCC AT INTERMEDIATE NUCLEATION RATES

The primary motivation for the analysis in this rate regime is the desire to extend the utility of the diffusion cloud chamber for nucleation research. As mentioned earlier, the DCC is currently of primary use for measuring critical supersaturations, and for investigating various nucleation phenomena at low rates of nucleation [65]. Measurements involving nucleation rates much above a few drops/cc/sec are not feasible because of the deleterious effects of the growing droplets on the supersaturated vapor just described. An advantage of extending the range of nucleation rate measurements would be to allow direct comparison of nucleation rate data obtained with the DCC with rate data obtained using expansion cloud chambers. Ordinarily, the expansion cloud chamber is used to make nucleation rate measurements at high rates of nucleation, typically from several hundred to 1.0×10^8 drops/cc/sec [3,5,6]. If the TDCC could be used over the range of 0.01 to several hundred or a thousand drops/cc/sec, it would complement data obtained with the expansion cloud chambers and thus provide a range for nucleation rate data never before available in nucleation research.

During nucleation rate experiments, corresponding to the data shown in Figure 11, for example, the observed nucleation is steady (no oscillations) and uniform throughout the chamber (away from the walls). In order to account for the effects of vapor depletion and latent heat on the supersaturated vapor, we can apply the model described in the previous section [66].

At steady state the right-hand side of Equations 21 through 24 vanish giving rise to four algebraic equations relating the concentration and temperature in and below the nucleation zone. Under these conditions, Equation 25 can be simplified considerably since $g(n, t)$ is now time independent so the time derivatives of $N(t)$ and $\langle n(t)\rangle$ both vanish. With this simplification Equation 25 becomes

$$R_1 = \rho_\ell v_m [n_f J] \tag{28}$$

Where J is the steady-state nucleation rate. To use R_1 requires only the experimentally determined steady nucleation rate, J, and the droplet falling radius, both of which are measurable.

At steady-state, $F_{\ell 1}$ in Equation 22 becomes identical to R_1 in Equation 28, and R_2 and $F_{\ell 2}$ in Equations 23 and 24 assume the following form,

$$R_2 = F_{\ell 2} = \frac{4\pi r_f \mathscr{D}_{12} J \delta t_f}{\ell}[C_2 - C_{2e}(T_2)] \tag{29}$$

where t_f = the passage time of the droplet through region 2
R_f = the radius of the droplet as it falls from the nucleation zone
$\mathscr{D}_{12}$ = the binary diffusion coefficient
$C_{2e}(T_2)$ = the equilibrium vapor concentration at T_2

The other symbols have been defined previously. In deriving Equation 29 we made use of a conventional droplet growth law expression [67] and we assumed that the size of the droplet does not change as it falls through region 2. This last assumption is discussed in Appendix B along with a description of the droplet radius measurement. Substituting Equations 28 and 29 into the steady-state version of Equations 21 through 24 along with experimental values for J, r_f, lower, T(L), and upper, T(U), plate temperatures and total pressure, P_T, gives, upon solution, the zone supersaturation and temperature corrected for the effects of vapor depletion and latent heat in the DCC.

Solutions to the steady-state model for ethanol vapor and n-propanol vapor are shown in Figure 11 as solid circles. The rate data were obtained using the droplet-counting method described earlier. Coincidence corrections were always made on the rate data whenever appropriate. Notice, that for nucleation rates of about 5 drops/cc/sec and less at 262° K for ethanol and about 15 drops/cc/sec and less at 276° K for n-propanol, the value of the zone supersaturation calculated ignoring the presence of the nucleated droplets, agrees with the zone supersaturation predicted by the model. Notice further, that as the nucleation rate increases, the difference between the value of the zone supersaturation calculated ignoring the presence of the nucleated droplets (the open circles, shown only for the lowest temperature data in each plot) and the value predicted by the model equations increases rapidly. The effect of ignoring nucleation and growth throughout the chamber results in a significant overestimation of the vapor supersaturation in the nucleation zone and an underestimation of the zone temperature [66, 68, 69].

There is an upper limit to the useful range of nucleation rates accessible to the DCC in the intermediate regime. At nucleation rates approaches 1,000 drops/cc/sec or so the appearance of the "bunches" or groups of droplets described earlier produces significant undercounting errors which are difficult to correct for statistically. Although the nucleation is not oscillatory, this "coincidence" error effectively limits the upper range of nucleation rate measurement.

The corrected nucleation rate data are also compared to homogeneous nucleation theory in Figure 11. The solid lines represent the scaled nucleation theory predictions. As discussed earlier, the nucleation theory (Equation 15) works reasonably well in predicting the trend of nucleation rate data with supersaturation but in general, does not predict accurate absolute values of the nucleation rate. It is necessary to scale the predictions of the theory to force agreement with experiment. Equation 15 has been scaled at each of the temperatures shown in Figure 11 by multiplying by a term of the form $\exp(-\varphi'/kT)$ where $\varphi'/kT = A - B/T$. For ethanol, over the range of experimental conditions investigated, values of $A = 121.28$ and $B = 31{,}434.8$ give reasonable agreement with the measured rate data corrected for the effects of droplet growth. For n-propanol, the values of $A = 61.65$ and $B = 16{,}670.4$ work well [69].

SUMMARY

There are the number of points concerning nucleation and growth in the diffusion cloud chamber which are worthy of review. First, critical supersaturation measurements are, at best, only a qualitative measure of agreement between theory and experiment. The variation of nucleation rate with supersaturation and temperature is the preferred measurement. Second, only for small nucleation rates, e.g. < 5–15 drops/cc/sec, should the no-nucleation analysis of the partial pressure and temperature profiles, Equations 16 and 17, be used. Third, in describing the effects of nucleation and growth on a supersaturated vapor, it appears that vapor depletion due to droplet growth is more important than the accompanying latent heat of condensation. This seems particularily true at smaller nucleation rates but becomes less so at increasingly higher rates of nucleation. Fourth, vapor depletion and latent heat effects in the nucleation zone are important but not to the exclusion of these effects occurring below the nucleation zone. In fact, in the high nucleation rate regime, considering only vapor depletion and latent heat effects in the nucleation zone will not give rise to sustained oscillatory behavior [62]. These effects, occuring below the nucleation zone, must be included as well. The primary reason why oscillations are not observed when only Equations 21 and 22 are used (ignoring effects below the nucleation zone) is that, unlike the analogous CSTR (continuous stirred tank reactor) equations, the "reaction" term, R_1, decreases with increasing temperature. However, by including vapor depletion and latent heat effects below the zone, we are, in effect, adding a feedback loop to the process. Drops nucleated in the zone at a previous time are "fed back" into the input vapor stream so that the vapor feeding the nucleation zone is modified by growing droplets produced earlier in the zone. It is this delayed feedback process that sustains the oscillatory behavior [62, 63]. And, fifth, if the effects of droplet nucleation and growth on the supersaturated vapor are accounted for, steady-state nucleation rate measurements over a range of about 0.01 to a thousand or so drops/cc/sec are possible with the DCC thus extending its utility and making it a more general tool in nucleation research.

APPENDIX A

The "reaction" term, R_1, first introduced in Equations 21 and 22 represents the conversion, vapor $\Leftrightarrow$ liquid, occurring in the nucleation zone as droplets nucleate and grow. We provide here one method of obtaining an expression for R. This treatment follows that given in Reference 63.

Consider all the droplets in the system (a region of the nucleation zone in the center portion of the DCC of unit volume with thickness δ and area $1/\delta$) with radius equal to or greater than the critical radius, and let $g(n,t)dn$ be the number of these droplets in the range n to n + dn at time t, where

$$n = n_d - n_c \tag{30}$$

where n_d = the total number of molecules in a droplet of a given size
n_c = the total number of molecules in a droplet of critical radius, r_c
$g(n,t)$ = the distribution of droplet sizes in the system

The number of droplets in the system at any time is given by

$$N(t) = \int_0^f g(n,t)\,dn, \tag{31}$$

where the lower limit refers to the critical droplet size and the upper limit to the size of the droplet as it falls from the system.

The average droplet size can be expressed as

$$\langle n(t)\rangle = \frac{\int_0^f n g(n,t)\,dn}{N(t)} \tag{32}$$

where the average size, $\langle n(t)\rangle$, is, in general, a function of time.

The time variation of both the total number of droplets in the system and their average size can be expressed as

$$\dot{N}(t) = d/dt\left(\int_0^f g(n,t)\,dn\right) \tag{33}$$

and

$$d/dt[N(t)\langle n(t)\rangle] = d/dt\left(\int_0^f n g(n,t)\,dn\right) \tag{34}$$

which follows directly from Equations 31 and 32. Both Equations 34 and 35 can be evaluated directly provided $g(n,t)$ is known.

At any time the liquid mass in the system, m_ℓ, can be expressed as

$$m_\ell = \rho_\ell v_m \int_0^f n g(n,t)\,dn, \tag{35}$$

where ρ_ℓ is the liquid density and v_m the molecular volume. We have assumed the liquid to be incompressible. The time variation of the liquid mass is given as

$$\dot{m}_\ell = \rho_\ell v_m d/dt\left(\int_0^f n g(n,t)\,dn\right) \tag{36a}$$

$$= \rho_\ell v_m d/dt[N(t)\langle n(t)\rangle] \tag{36b}$$

$$= \rho_\ell v_m[\dot{N}(t)\langle n(t)\rangle + N(t)\langle \dot{n}(t)\rangle], \tag{36c}$$

which follows directly from Equations 33 through 35. In Equation 36c the change in liquid mass is shown to be due to a change in the number of droplets with time (nucleation and removal) and a change in the average size of the droplets (growth).

We can now relate the change in the liquid mass to be the vapor depletion through the following mass balance on the system:

$$\dot{m}_\ell = R_l - \rho_\ell v_m n_f N_d(t) \tag{37}$$

where R_l is the "reaction" term described earlier, $N_d(t)$ is the number of liquid droplets falling from the system per unit time per unit volume (see Equation 22) and the other terms have been defined. Equation 37 states that the rate of change of liquid mass in the system is equal to the total rate of change from vapor to liquid, R_l, minus the rate at which liquid leaves the system in the form of droplets of size r_f.

Using Equations 36 and 37 the rate of vapor depletion can now be written as

$$R_l = \rho_\ell v_m [d/dt(\textstyle\int_0^f ng(n,t)\,dn) + n_f N_d(t)] \tag{38a)]}$$

or

$$R_l = \rho_\ell v_m [\dot{N}(t)\langle n(t)\rangle + N(t)\langle \dot{n}(t)\rangle + n_f N_d(t)]. \tag{38b)}$$

R_l can now be evaluated provided $g(n,t)$ is known.

The problem, of course, is that $g(n,t)$ is not known. We can, however, obtain a distribution with the following numerical scheme which we will assume represents $g(n,t)$ closely enough to evaluate Equation 38 and, hence, solve Equations 21 through 24. If we consider a small increment of time and keep the vapor concentration and temperature in the system constant during this time, it is possible to estimate the number of droplets nucleated using the nucleation rate expression discussed earlier (Equation 15). In addition, during this same increment of time, all droplets nucleated in previous time increments are allowed to grow according to an appropriate growth law. This calculation is repeated and as the growing droplets reach the falling radius, they are removed from the system. This scheme allows one to compute both a $g(n,t)$ and a value for $N_d(t)$. The size distribution, $g(n,t)$, can be found at any time by sampling the number and the radii of all the droplets currently in the system; and the rate which droplets leave the system, $N_d(t)$, is just the rate at which droplets are passing the falling radius at a particular time.

Evaluating $g(n,t)$ and $N_d(t)$ using this method has the disadvantage of requiring considerable computational effort since the nucleation rate and the critical radius depend upon the vapor concentration and the temperature at the particular time they are computed; and the growth law is a function of vapor concentration, temperature, and the droplet size.

APPENDIX B

The experimental set-up used to measure the radius of the droplet as it falls from the nucleation zone is illustrated in Figure 10. The laser beam from the system labeled "Radius" is passed through a beam splitter which produces two closely spaces (0.05 to 0.1 cm separation is typical) beams and directed through the DCC. These beams can be translated throughout the chamber. The detection optics (not shown in Figure 10) are such that scattered light from a small region in the center of the chamber is passed to the PMT as a droplet falls through the two beams. The signal from the PMT is amplified and passed to a fast, 8-bit A/D converter. The digitized signal is acquired by the microcomputer and plotted in high resolution format on a display monitor. Typical examples of this screen display are shown in Figure 20. The oscilloscope trace, visible at the right of the top photograph, illustrates the same events shown on the display monitor on the left. The two groups of peaks shown in the top photograph represent two droplets falling through the two beams at two different (but close) instances in time. The separation of each set of peaks corresponds to the time of flight of a droplet as it passes through the beams. The lower photograph illustrates three such events. The separation, in time, of the two peaks in the center of the display in the lower photograph is 20.12 msec and is indicated in the upper left

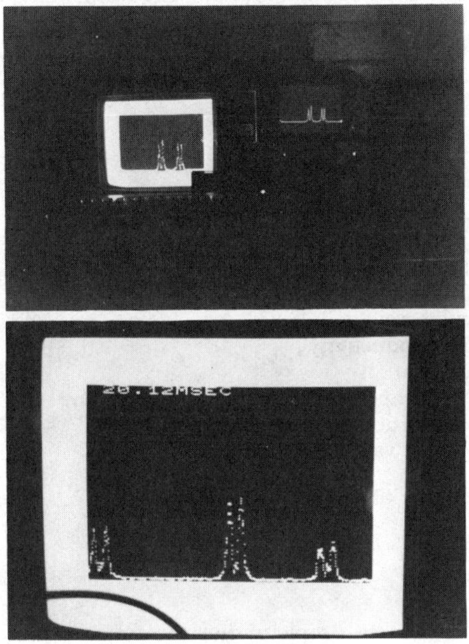

Figure 20. Photographs of typical computer screen displays of the digitized PMT output obtained as falling droplets pass through the dual laser beams (labeled Radius in Figure 10) and scatter light to the PMT. The oscilloscope trace, visible on the right in the top photograph, illustrates the same events shown on the computer display to the left. Additional details are given in Appendix B.

hand corner of the display. The crystal oscillator in the computer is used to measure the separation of the peaks. The droplet velocity can then be determined from the laser beam separation.

Once the droplet velocity is known, Stokes law,

$$v = dz/dt = \frac{2\rho_\ell r^2 G}{9\eta}, \tag{39}$$

is used to determine the radius.

In Equation 39 v = the falling velocity
ρ_ℓ = the liquid density
r = the droplet radius
G = the gravitational constant
η = the vapor-carrier gas mixture viscosity.

Once the droplet radius is known at a point in the chamber, a droplet growth law can be used to calculate the droplet radius as a function of chamber height. A droplet growth law typically used has the form [67]:

$$r(dr/dt) = \frac{S-1}{B+C} \tag{40}$$

where r is the radius and S the supersaturation. The forms for B and C are given by

$$B = \frac{\rho_\ell L_c^2 M}{\lambda R T^2} \tag{41}$$

and

$$C = \frac{\rho_l R T}{\mathscr{D}_{12} M P_e} \tag{42}$$

where ρ_ℓ = the liquid density
L_c = the heat of condensation
M = the molecular weight
λ = the thermal conductivity
R = the gas constant
T = the absolute temperature
$\mathscr{D}_{12}$ = the binary diffusion coefficient
P_e = the equilibrium vapor pressure

Dividing Equation 40 by Equation 39 gives

$$dr/dz = \frac{9\eta(S-1)}{2\rho_\ell G r^3 (B + C)} \tag{43}$$

which, when integrated, gives the droplet radius as a function of height (z) in the DCC. A typical solution to Equation 43 is shown in Figure 21. Here the experimental points represent the measured droplet radii at various positions in the chamber, and the solid curve

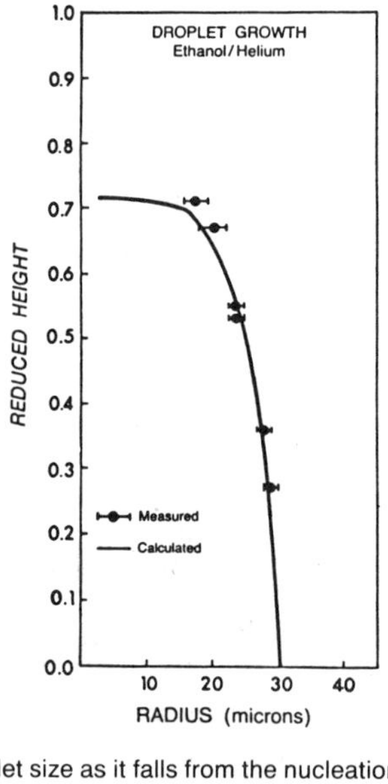

Figure 21. The variation of droplet size as it falls from the nucleation zone, through the DCC to the pool on the lower plate. Additional details are given in Appendix B.

is the solution to Equation 43 for the particular experimental conditions. The bars associated with each data point in Figure 21 represent the range (one standard deviation) of the droplet radii measured at that height. Similar curves, obtained using alternative methods of measuring the droplet radii and exhibiting similar behavior, have been reported in the literature [39, 68].

Using the solution to Equation 43, the falling radius of the droplet as it leaves the nucleation zone can be determined. For the conditions corresponding to the data shown in Figure 21, the lower boundary of the zone is located at about 0.65 reduced height and the zone thickness is 0.09 reduced height units. The chamber height in this experiment was 2.27 cm. As can be seen from the solid curve in the figure, the rate of change of the droplet size is not great from the lower boundary of the zone to the pool of working fluid on the lower plate. This is the justification for using a constant falling radius in the model for the region below the nucleation zone.

NOTATION

a_g	surface area of g-mer
C, C_T, C_0	vapor concentration
C_e	equilibrium vapor concentration
C_p	heat capacity
c	molar density
$\mathscr{D}_{12}$	binary diffusion coefficient
d	liquid density
F	total number of molecular species
$F_{\ell 1}, F_{\ell 2}$	liquid droplet fluxes in the DCC
f_g	number of clusters with g molecules
G	gravitational constant
g	number of molecules in liquid drop
g*	number of molecules in critical cluster
g (n,t)	droplet size distribution
$H, \Delta H_v(T)$	vapor enthalpy
$\Delta H_\ell(T)$	liquid enthalpy
h	cloud chamber height
J, J_g	nucleation rate
K	proportionality constant
k	Boltzmann constant
k_T	thermal diffusion factor
L	scaled vapor mass flux
L_c	heat of condensation
ℓ	vapor mass flux, height of region below zone
M	molecular weight
m_ℓ	liquid mass
N_ℓ	number of molecules in liquid drop
N_v	number of vapor molecules
N_a	number of vapor monomers
N_g	number of clusters with g molecules
N (t)	number of droplets in system at time t
$N_d(t)$	number density of drops falling from zone
$N_2(t)$	average number density of drops below zone
N_0	Avogadro number
n_c	number of molecules in critical cluster
n_d	number of molecules in liquid drop
n_g	number of clusters with g molecules
$\langle n(t) \rangle$	average droplet size
P, P (z)	chamber partial pressure
P_e	equilibrium vapor pressure
P_T	total pressure
Q	scaled energy flux
q	actual energy flux
R	gas constant
R_1, R_2	rate of conversion of vapor to liquid
r	radius
r_c	critical radius
r_f	falling radius
S, S_0	supersaturation
s	temperature dependence of binary diffusion coefficient
T, T_0	temperature (°K)
t	time
t_f	droplet falling time
U, U′	heat transfer coefficients
v	droplet fall velocity
v_m	molecular volume
x	mole fraction

Z	reduced cloud chamber coordinate	z	actual cloud chamber coordinate

Greek Symbols

α	evaporation rate from cluster	μ_ℓ	chemical potential (liquid)
α_T	thermal diffusion ratio	μ_v	chemical potential (vapor)
β	vapor condensation rate	η	viscosity
δ	nucleation zone thickness	$\varphi, \varphi_0, \varphi'$	thermodynamic potential
γ	exponent in Equation 19	ρ_ℓ	liquid density
λ	thermal conductivity	σ	surface tension

REFERENCES

1. Allard, E. F., and Kassner, J. L., Jr., *J. Chem. Phys.,* 42: 1401 (1965).
2. Kassner, J. L., Jr., and Schmitt, R. J., *J. Chem. Phys.,* 44: 4166 (1966).
3. Allen, L. B., and Kassner, J. L., Jr., *J. Colloid Interface Sci.*, 30: 81 (1969).
4. Schmitt, J. L., *Rev. Sci. Instrum.,* 52: 1749 (1981).
5. Wagner, P. E., and Strey, R. J., *Phys. Chem.,* 85: 2694 (1981).
6. Schmitt, J. L., Adams, G. W., and Zalabsky, R. A., *J. Chem. Phys.*, 77: 2089 (1982).
7. Miller, R., et al., *J. Chem. Phys.*, 78: 3204 (1983).
8. Wegener, P. P., and Lundquist, G., *J. Appl. Phys.*, 22: 233 (1951).
9. Hill, P. G., *J. Fluid Mech.,* 25: 593 (1966).
10. Stein, G. D., and Wegener, P. P., *J. Chem. Phys.,* 46: 3685 (1967).
11. Wegener, P. P., Clumpner, J. A., and Wu, B. J., *Phys. Fluids,* 15: 1869 (1972).
12. Wegener, P. P., and Wu, B. J., *Nucleation Phenomena,* A. C. Zettlemoyer (Ed.) (Elsevier, New York, 1977), pp. 325–417.
13. Barschdorff, D., *Phys. Fluids,* 18: 529 (1975).
14. Peters, F., *J. Chem. Phys.*, 77: 4788 (1982).
15. Volmer, M., and Weber, A., *Z. Phys. Chem.* (Leipzig), 119: 277 (1926).
16. Farkas, L., *Z. Phys. Chem.* (Leipzig), 125: 236 (1927).
17. Volmer, M., *Z. Phys. Chem.,* 25: 555 (1929).
18. Becker, R., and Doring, W., *Ann. Phys.* 24: 719 (1935).
19. Volmer, M., *Kinetics of Phase Formation,* (Theodor Steinkopff, Dresden, 1939).
20. Zeldovitch, J. B., *J. Exp. Theor. Phys.*, 12: 525 (1942).
21. Zeldovitch, J. B., *Acta. Phys. Chem. USSR,* 18: 1 (1943).
22. Frenkel, J., *The Kinetic Theory of Liquids* (Oxford University Press, London, 1946), Chap. 7.
23. Lothe, J., and Pound, G. M., *J. Chem. Phys.,* 36: 2080 (1962).
24. McDonald, J. E., *Am. J. Phys.,* 30 (12): 870 (1962).
25. McDonald, J. E., *Am. J. Phys.*, 31 (1): 31 (1963).
26. Reiss, H., Katz, J. L., and Cohen, E. R., *J. Chem. Phys.*, 48: 5553 (1968).
27. Abraham, F. F., *Homogeneous Nucleation Theory* (Academic Press, New York, 1974).
28. Abraham, F. F., *Homogeneous Nucleation Theory* (Academic Press, New York, 1974), pp. 55–58.
29. This condition may no longer apply in the critical region.
30. Abraham, F. F., Homogeneous Nucleation Theory (Academic Press, New York, 1974), p. 81.
31. Langsdorf, A., *Rev. Sci. Inst.,* 10: 91 (1939).
32. Nielson, C. E., Needels, T. S., and Weedle, O. H., *Rev. Sci. Inst.,* 22: 673 (1951).
33. Schaefer, V. J., *Ind. Eng. Chem.,* 44: 1381 (1952).
34. Franck, J. P., and Hertz, H. G., *Z. Physik,* 143: 559 (1956).
35. Hertz, H. G., *Z. Elektrochemie,* 60: 1196 (1956).

36. Katz, J. L., and Ostermier, B. J., *J. Chem. Phys.*, 47: 478 (1967).
37. Katz, J. L., *J. Chem. Phys.*, 52: 4733 (1970).
38. Heist, R. H., and Reiss, H. J., *Chem. Phys.*, 59: 665 (1973).
39. Katz, J. L., et al., *J. Chem. Phys.*, 62: 448 (1972).
40. Katz, J. L., and Mirabel, P. J., *Atmos. Sci.*, 32: 646 (1975).
41. Becker, C., Reiss, H., and Heist, R. H., *J. Chem. Phys.*, 68: 3585 (1975).
42. Katz, J. L., et al., *J. Chem. Phys.*, 65: 382 (1976).
43. Heist, R. H., Colling, K. M., and Dupois, C. S., *J. Chem. Phys.*, 65: 5147 (1976).
44. Colling, K. M., MS thesis, University of Rochester, 1976.
45. Reiss, H., Marvin, D. C., and Heist, R. H., *J. Colloid Interface Sci.*, 58: 125 (1977).
46. Russell, Y. G., and Heist, R. H., *J. Chem. Phys.*, 69: 3723 (1978).
47. Mirabel, P., and Clavelin, J., *J. Chem. Phys.*, 68: 5020 (1978).
48. Dupuis, C. S., MS thesis, University of Rochester, 1979.
49. Agarwal, G., and Heist, R. H., *J. Chem. Phys.*, 73: 902 (1980).
50. Heeks, M., MS thesis, University of Rochester, (1985).
51. Fuller, E. N., Schettler, P. D., and Giddings, J. C., *Ind. Eng. Chem.*, 58: 19 (1966).
52. Flood, H., *Z. Phys. Chem.*, A170: 286 (1934).
53. Reiss, H., Margolese, D. I., and Schelling, F. J., *J. Colloid Interface Sci.*, 56: 511 (1976).
54. Wegener, P. P., and Wu, B. J., *Faraday Discuss. Chem. Soc.*, 61: 77 (1976).
55. Mirabel, P., and Katz, J. L., *J. Chem. Phys.*, 67: 1697 (1977).
56. Boulaud, D., Madelaine, G., and Vigla, D., *J. Chem. Phys.*, 66: 4854 (1977).
57. Mirabel, P., and Clavelin, J. L., *J. Chem. Phys.*, 68: 5020 (1978).
58. Flageollet, C., Cao, M. D., and Mirabel, P., *J. Chem. Phys.*, 72: 544 (1980).
59. Heist, R. H., Fuchs, F., and Agarwal, G., *Chem. Eng. Commun.*, 5: 1 (1980).
60. Fuchs, A., MS thesis, University of Rochester, 1979.
61. A small, positive slope of the baseline signal clearly evident in photographs A and B in Figure 13 is an artifact of the measurement resulting from use of the AC-coupled input to the oscilloscope amplifier. With the DC-coupled input, the baseline is flat and the frequency unchanged.
62. Brito, J., MS thesis, University of Rochester, 1981.
63. Brito, J., and Heist, R. H., *Chem. Eng. Commun.*, 15: 133 (1982).
64. The system as used in this analysis is defined as a region of the nucleation zone in the center portion of the DCC of unit volume with thickness δ and area $1/\delta$.
65. Examples of such nucleation phenomena which are currently of considerable interest are (1) photo-induced nucleation (see, for example, Reiss, H., Marvin, D., and Heist, R. H., *J. Colloid Interface Sci.*, 58: 125 (1977); Katz, J. L., et al., *Science,* 196: 1203 (1977); Gertler, A. W., Berg, J. P., and El-Sayed, M. A., *Chem. Phys. Letts.*, 57: 349 (1978); Katz, J. L., et al., *Science,* 200: 769 (1978); Gertler, A. W., et al., *Chem. Phys. Letts.*, 42: 429 (1979); Kalisky, O., and Heist, R. H., *J. Chem. Phys.,* submitted for publication; and references cited within all of these papers.); and (2) gas phase polymerization studies (see, for example, Reiss, H., Marvin, D., and Heist, R. H., *J. Colloid Interface Sci.*, 58: 125 (1977); Reiss, H., and Chowdhury, M. A., in preparation; and Chowdhury, M. A., et al., in preparation).
66. Heist, R. H., Kacker, A., and Brito, J., *Chem. Eng. Commun.*, in press (1984).
67. See, for example, Friedlander, S. K., *Smoke, Dust and Haze, Fundamentals in Aerosol Behavior* (J. Wiley and Sons, New York, 1977) chap. 5; or Fukuta, N., and Walter, L. A., *J. Atmos. Sci.*, 27: 1160 (1970).
68. Flageollet-Daniel, C., Ehrhard, P., and Mirabel, P., *J. Chem. Phys.*, 75: 4615 (1981).
69. Kacker, A., and Heist, R. H., *J. Chem. Phys.*, 82: 2734 (1985). This method of scaling the nucleation rate has also been applied to expansion cloud chamber data (Wagner, P., and Strey, R., 56th Colloid and Surface Science Symposium, Blacksburg, Virginia, Juni 13–16, 1982).
70. Heist R. H., *IBM Perspectives,* in press (1984).

CHAPTER 15

CONVECTIVE AND RADIATIVE HEAT TRANSFER OF FLOWING GASEOUS-SOLID SUSPENSIONS

Shu Hasegawa, Ryozo Echigo, and Akihiko Shimizu

Department of Nuclear Engineering
Kyushu University
Fukuoka 812, Japan

CONTENTS

INTRODUCTION

Practicing engineers and researchers encounter the problems of gaseous/solid suspension flow in such areas as pneumatic conveying of granular or powdery materials, combustion of pulverized coal in boiler furnaces, various types of catalytic reactors in chemical engineering, solid propellant rocket fuel, and so on. Related research works have formed so far a vast bibliography the extent of which can not be covered here. Accordingly, the main focus of this chapter is on one particular aspect of the suspension flow phenomena, namely convective and radiative heat transfer of the suspension flows. Other comprehensive volumes [1–4] provide a general discussion of the topic.

Although engineers who deal with gas/solid suspension flows have different perspectives, in most cases of conventional application, suspended solid particles are always direct or indirect target materials for the industrial systems or equipment under consideration. For example, transported grain in the pneumatic conveying line is no doubt the direct target of the equipment. On the other hand, the suspended solid particles are sometimes regarded as unwelcome superfluity or a kind of contaminant material.

The engineer's task is then to separate and remove them from the loading gas stream. Even for the latter case, however, the existence of the solid is apparently a premise for the problem and the suspended solid is still a target material in a sense. In other cases, the solid particles are fed to the gas flow only to change the inherent characteristics of the loading gas and to acquire some favorable properties as an operating medium. The suspended solid of this case is therefore neither the target material nor the superfluity for the system. Any kind of solid will do for the suspended phase, so far as the necessary properties of the flowing media can be obtained. Probably the rocket engine exhaust which is sometimes fed with the suspended particulate phase in order to acquire the necessary thrust performance may be classified into this category.

Another outstanding example of such a case is the heat transfer improvement by use of the gas/solid suspension flows. This concept of making use of the flows as an effective means of heat transfer enhancement is comparatively new and may be traced back to the concept of a high-temperature nuclear reactor cooled by the suspension flow, proposed in 1954 by Schulderberg [5]. It may be easily understood that using the suspension flow as the heat transfer media instead of the usual gas flow requires troublesome and otherwise unnecessary devices which can efficiently feed the particles into and separate them from the gas stream as well as additional instrumentation for the measurement and control of the suspension. Also in the designing procedure, attention should be paid to erosion of the constituent materials of the flowing channels and to possible flow instability. The erosion of the duct materials inevitably results in the contamination of the suspended powder and this may aggravate the problems of induced radioactivity of the irradiated particles if used as the coolant of the nuclear reactors. Accordingly, the use of suspension flows as a heat transfer medium, offers several advantages over single-phase gas flow which compensate for the additional problems generated by their use. Suspension flows also show promise as a coolant for advanced applications.

Characteristic features of the gaseous solid suspension flow heat transfer may be compared with those of single-phase gas flow as follows.

- Heat capacity of the flowing media can be substantially and arbitrarily increased due to the addition of the solid phase. If extreme cases are excluded, the specific heats of the solid and loading gas are usually of the same order. If we use a gas and solid of precisely the same specific heats with the loading ratio of unity, it is said that we have acquired a new heat transfer media of doubled specific heat without any pressurization.
- Radiation heat transfer becomes the dominant heat transfer mode especially within a high-temperature region. The suspended solid particles emit and absorb the thermal radiation directly to or from the channel walls and exchange it with the surrounding gas through conduction and convection. Optical properties of the flowing media can be easily changed by selecting the appropriate particle size and solid loading ratio. A detailed mathematical description of the radiation heat transfer will be presented later.
- The turbulent structure of the loading gas stream might be substantially changed through the addition of the solid phase. This effect has both positive and negative sides depending on the situation. If sufficiently coarse particles are suspended, then the wake may be formed behind each particle and this in turn leads to the enhancement of the turbulent level of the flow and therefore of the convective heat transfer. On the other hand, finer particles tend to calm the fluctuation of the flow through viscous drag on their surfaces and also through their dividing the turbulent eddies into smaller ones, which leads to a higher dissipation rate of the turbulent energy. However, the effect of the particle size on the phenomena has not been fully understood and there exist experimental results which contradict each other. Although conclusions should be reserved until later, there seems to be more cases where the turbulent motion of the flow is substantially reduced below the level of the single-phase gas flow. It is true that there exist many research works that report the increase in heat transfer by use of the suspension flows. If the experimental data are precisely examined, however, the increase of the heat transfer is usually reported as the form of the increase in Nusselt numbers

or simply in heat transfer coefficients. Then if we have in mind the fact that the heat capacity of the flow is also increased, it is said in a sense that the improvement in heat transfer is quite natural. In order to examine the effects of the added solid on the turbulence level of the flow, the heat transfer comparison should be done between the suspension flow and the single-phase flow, the overall densities or the heat capacities of both being kept equal. (Due to the small size of the particles, heat transfer resistance between the particles and loading gas is usually minor and the two substances can be assumed to be in thermal equilibrium.) If the existing experimental data are reexamined based on this view, most of them are found to show the reduction in convective heat transfer. Therefore, if we have in mind the previously mentioned problems which inevitably accompany powdery or granular materials, and if the temperature of the system is not so high that we can not expect the radiative heat transfer to be dominant, then, the advantage of using the suspension flow as the heat transfer media in conventional industry is said to be less attractive.

An exception to this observation should be pointed out now, wherein the overall convective heat transfer is improved to a great extent even when the temperature of the system is not so high. The material density of the solid phase is usually higher than that of the loading gas phase by three or four orders of magnitude, so that the specific inertia of the former is larger than of the latter by the same order. Therefore, if the stream lines of the flow are not straight, as with the case of a curved tube, the solid particles which are of larger inertia are apt to flow outside the curvature and to become more populated there. As a result, they have more of a chance to penetrate into the viscous sublayer and to induce strong disturbances there. In the usual case of a moderate Prandtl number, most of the convective heat transfer resistance lies within the sublayer region. The overall convective heat transfer coefficient is then enhanced significantly, because there is no negative factor in heat transfer on the opposite (inner curvature) side of the wall. Two prominent examples of this phenomena will be presented later.

FLOW CHARACTERISTICS

Correlation of the Pressure-Drop Data

The designing procedure of any equipment which contains the fluid flow always begins with the estimation of pressure drop along the flowing line. In the field of the gaseous solid suspension flows, numerous investigations have been performed in relation to the estimation of the necessary pumping power of the pneumatic conveying lines. Besides this practical aspect, the pressure drop data are of interest because they give close insight into the turbulence mechanism of the suspension flows. Before some experimental data are examined, some remarks are presented here on the methods of correlating the pressure drop data of the suspension flows because several procedures are found in the literature. In referring to the existing data, attention should be given to which procedure is used.

Gasterstaedt [6] correlated his pressure drop data according to the following equation,

$$\Delta P_m/\Delta P_0 = 1 + \Gamma \cdot m \tag{1}$$

where Γ represents the ratio of the mass flow rate of the particulate phase to that of the gaseous phase, ususally referred to as the solid loading ratio, while Δp_m and Δp_0 are the pressure drops of the flowing line of specified length L with and without the suspended phase, respectively. Vogt and White [7], Farbar [8], and Mehta et al. [9] also adopted this approach. In this procedure, constant m must be determined experimentally and may be a function of geometric properties of the suspended particles, pipe roughness, gas Reynolds number, and so forth. As is easily imagined, universality and validity of this correlation is restricted, because a significant part of the measured pressure drop is due to the gravity force exerted on the suspended particles. The flow conditions would be quite varied between vertically upward and downward flows and both of them would also be different from the horizontal flows.

In other procedures which do not make use of Equation 1, the measured (total) pressure drop is divided into several contributions. According to Hariu and Molstad [10] it is expressed as,

$$\Delta P_m = \Delta P_{af} + \Delta P_{ap} + \Delta P_{pf} + \Delta P_{pp} + \Delta P_{ff} + \Delta P_{fp} \tag{2}$$

The first two terms of the right-hand side represent the pressure drops due to acceleration of the gaseous phase and the solid phase, respectively. Therefore both can be neglected when the flow is fully accelerated or developed. The next two terms of the equation are the pressure drops necessary to sustain the columns of both phases or simply static heads of both phases. Of course the signs before these terms should be changed in case of the vertically downward flows. The last two terms of the equation represent frictional pressure drops of both phases. A second method of correlating the pressure drop data is based on Equation 2 and is due to Ergun [11] who derived the following equation for the frictional pressure drop of the particulate phase in connection with the pressure drop of packed beds,

$$\frac{\Delta P_f}{L} = \frac{\alpha^2}{(1-\alpha)^3} \cdot \frac{\mu U_f}{d_p^2} + 1.75 \frac{\alpha}{(1-\alpha)^3} \frac{\rho_f U_f^2}{d_p} \tag{3}$$

where U_f, μ, d_p, and ρ_f denote gas velocity, gas viscosity, particle diameter, and gas density, respectively. The volume fraction of the suspended phase α $(= 1 - \varepsilon; \varepsilon =$ voidage) explicitly appears in this equation. For the packed beds, the estimation of α or ε is relatively easy, and it is usually fixed at a certain value. In contrast, the voidage of the suspension flow can not easily be obtained and Equation 3 is too sensitive to it. Therefore it was pointed out by Leung [12] that the correlation based on Equation 3 is not recommended.

The most widely used correlating method of the pressure drop data also takes advantage of Equation 2, but the frictional pressure drop of the particulate phase is expressed in the same form as for the gaseous phase.

$$\frac{\Delta P_{fp}}{L} = 2 f_s (1 - \varepsilon) U_p^2/D \tag{4}$$

Here, D denotes the diameter of the pipe and the friction coefficient of the particulate phase f_s must be determined experimentally. The superiority and flexibility of Equation 4 is apparent, and it is used in most of the recent experimental studies of the suspension flow pressure drop. However, the superiority is not perfect. In the first place, there appears again in the equation the voidage of the media which requires the estimation of the particles' average velocity within the duct. The particles' velocity is strongly affected by their collision with the duct wall and also by the interparticle collision. In practical use, the velocity data are not given a priori from the prescribed operating conditions. Therefore, Yang [13] proposed for this purpose the following equation based on the examination of the existing pressure drop data.

$$U_p = U_f - \sqrt{(1 + f_p U_p^2/2gD) \cdot \frac{4(\rho_p - \rho_f) d_p g \varepsilon^{4.7}}{3 \rho_f C_{DS}}} \tag{5}$$

where ρ_p and C_{DS} are material density of the particle and drag coefficient of a particle, respectively. Equation 5 contains again the voidage of the flowing media and another friction coefficient f_p is in turn related to the frictional pressure drop of the solid phase ΔP_{fp} as

$$f_p = \frac{2\,D\Delta P_{fp}}{\rho_m U_f^2 L} \qquad (\rho_m = \rho_p(1-\varepsilon) + \rho_f\varepsilon) \tag{6}$$

so that an iterative calculation is necessary to close the procedure. Equation 5 is reported to be able to correlate the existing particle velocity data within an accuracy of 20%.

A second disadvantage is more essential and related to the underlying assumption in using the original Equation 2. Even when the estimation of the accelerational and gravitational pressure drops ΔP_{ap}, ΔP_{pp}, etc. is appropriately made, residual frictional pressure drop cannot be divided into two portions that cannot be readily by theoretical equations. Therefore the friction coefficient of the clean gas flow of the same Reynolds number is usually applied for the evaluation of ΔP_{ff}, which apparently corresponds to the ideal that the flow structure is not changed from the unloading gas flow. There exist numerous reports which contradict this ideal. In Boothroyds paper [14], for example, some experimental data show even the reduction of a "total" pressure drop below that of the single-phase gas flow. If Equations 2 and 4 are applied to such data, a negative friction coefficient of the solid phase would be obtained. In pneumatic conveying problems, a major concern of the engineer is to transport as much material as possible within given operating conditions. Therefore the loading ratio of his interest is comparatively high and the ambiguity in the estimation of the gaseous-phase frictional pressure drop is trivial. On the other hand, it would become significant in the case of suspension flow with the purpose of heat transfer enhancement, because the loading ratio of interest would be in a comparatively lower region in order to assure smooth circulation. At the present stage, however, no appropriate alternative for Equations 2 and 4 exists. Therefore due caution should be retained in referring to the pressure-drop data and in inspecting the physical meanings.

Experimental Results

The pressure drop was measured by the authors [15] for dilute suspension flows of air and copper particles. Figure 1 depicts the schematic view of the experimental facilities. Acrylic resin pipes of a 28-mm inner diameter are smoothly connected to form the 3.8-m vertical downcomer and 6.2-m riser test sections. Typical results of the pressure variation along the test are shown in Figure 2 for both upward and downward flows. As is verified in Figure 2B, a considerably longer entrance length is needed for the vertical downflows of the suspension to be fully accelerated compared with the single-phase flow and also with the upward suspension flow. Also, the test tube of the downcomer was too short to realize the fully accelerated flows. Therefore, the pressure drop data was obtained only for the riser tube in the following. Recently, Kim and Seader [16] performed a similar measurement and obtained excellent results for the downward suspension flow. In Figure 2A, some arrows are indicated at the position where the flows are regarded tentatively to be fully developed and the data are summarized in Figure 3 in the form of dimensionless entrance length vs. *mixture Reynolds number*, defined as

$$Re_m = \frac{[\varepsilon\rho_f U_f + (1-\varepsilon)\rho_p U_p]\,D}{\mu} = (1+\Gamma)\,Re \tag{7}$$

According to Figure 3, the correlation of such an intuitive method is found to be useful for the simple evaluation of the entrance length.

The particle velocity of the riser test tube was obtained simply from the gas velocity minus the free falling terminal velocity of the particle because the voidage of the media in Equation 5 was found to be near-unity and the term containing f_p in Equation 5 was also found to be minor within parameter region examined. Regarding the former discussion on the frictional pressure drop, no attempt was made to divide the measured frictional pressure

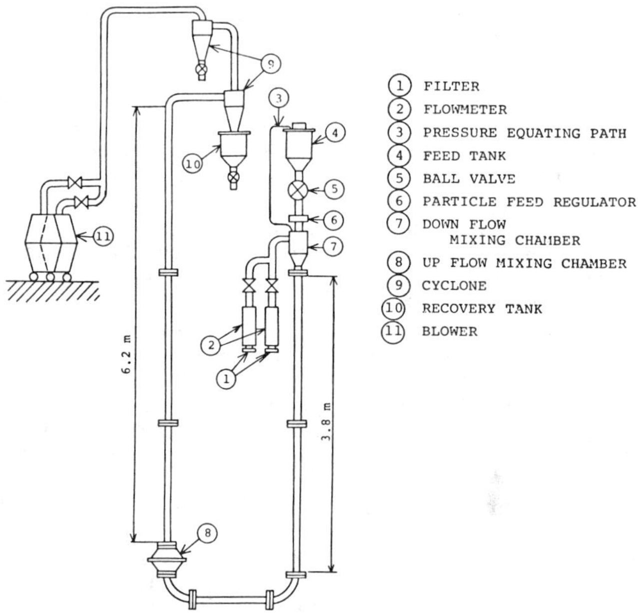

Figure 1. Experimental setup for pressure drop measurement.

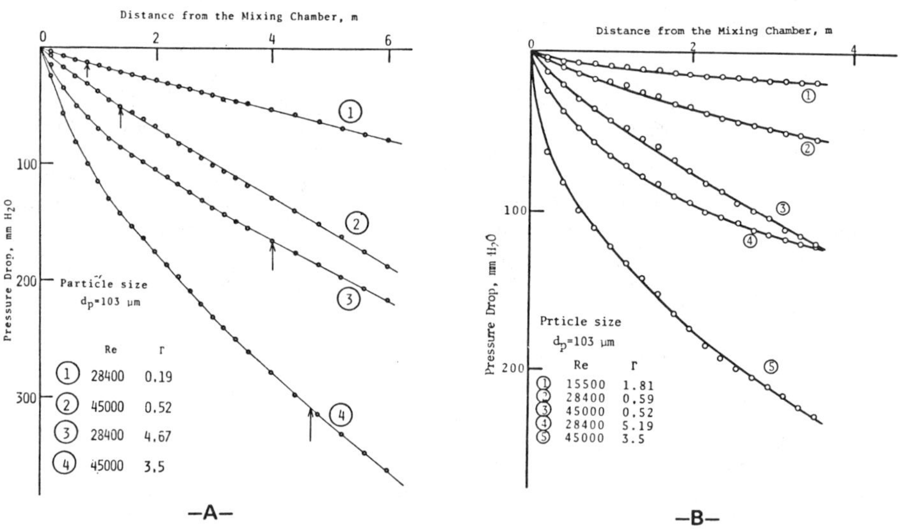

Figure 2. Typical variations of static pressure along the flowing lines: (A) vertically upward flow; (B) downward flow.

drop into two parts. Instead, the frictional pressure drops, including contributions from particles and gas, are treated as single-phase gas flow where solids have the same velocity as the gas. Results are summarized in Figure 4 along with those of literature studies. According to the figure, the frictional pressure drop ratio is found to be monotonously increased with the solid loading ratio for the prescribed particle size of 103 μm and no minimum is found in contrast with the results of finer particles by Boothroyd. Also it is found in the figure that an equation for the particulate phase friction coefficient proposed by Leung [12]

$$f_s = 0.05\, U_p^{-1} \tag{8}$$

U_p is in m/sec.

gives too large a value of the frictional-pressure ratio. The difference may be due to the difference in the loading ratio of interest, as was pointed out previously. In Figure 5, the same results are shown for four different sizes of particles. The abscissa of the figure is logarithmic in order to emphasize the smaller loading region. For the smaller particles, the frictional-pressure-drop ratio is found to be less than 1.0, implying an effective turbulence reduction due to the addition of solids.

RADIATIVE HEAT TRANSFER IN GASEOUS/SOLID SUSPENSIONS

In high and/or extremely high temperature heat transfer facilities with high thermal loads, radiative transfer is the most significant and prominent mode compared with others

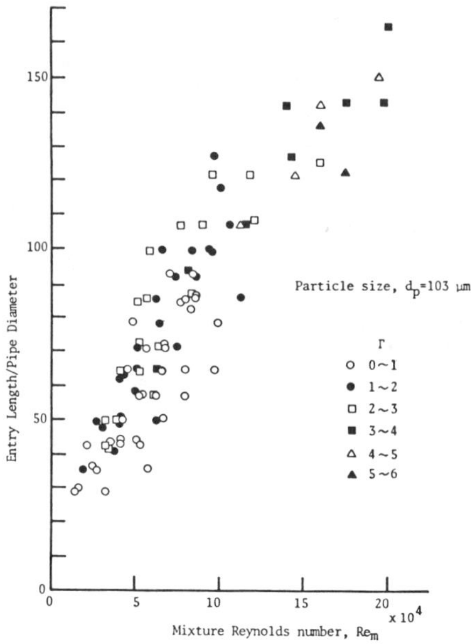

Figure 3. Change of the nondimensional entry length of the suspension flow with the mixture Reynolds number [15].

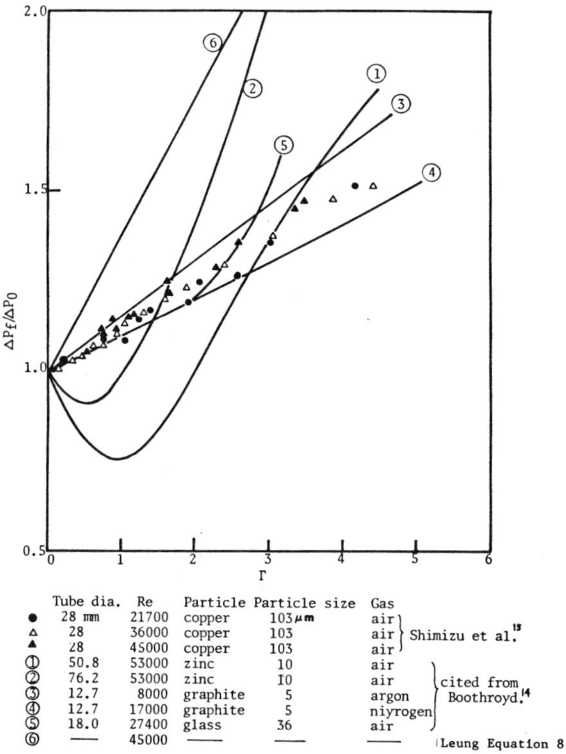

Figure 4. Frictional pressure drop ratio vs. solid loading ratio [14,15].

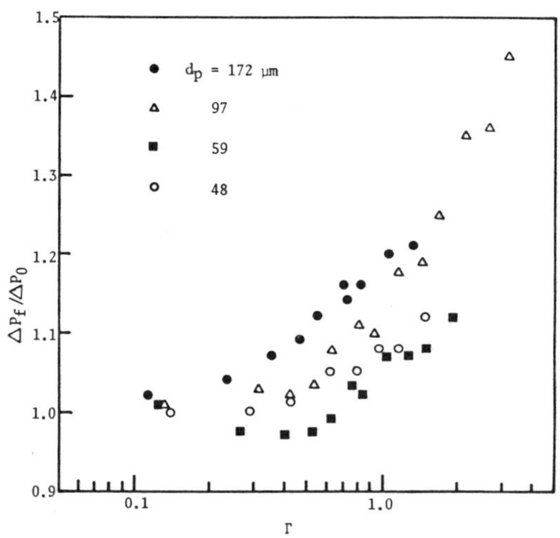

Figure 5. Frictional pressure drop ratio vs. solid loading ratio for different sizes of the particles. Emphasis is laid on the lower loading region [15].

such as conductive and convective heat transfer. However, the emissive power or absorption efficiency of gaseous bodies for thermal radiation is generally not high, so that the working gas must be pressurized, or the geometric dimension of the system must be large so as to improve the radiative heat transfer properties. Therefore, the gaseous solid suspensions take a priority over other heat transfer media regarding the heat transfer systems at high temperature and high heat flux because of the high emissive and absorptive powers of the media.

Hypothesizing, the major heat transfer mechanism in a gaseous solid suspension system may be depicted as in Figure 6, where the radiative heat exchange between the heating wall and the particulate phase and the radiative propagation among the particles are the dominant mode of heat transfer.

In the first part of this section, the analytical treatment of the radiative heat transfer of the gaseous suspension flow within circular tubes is presented [17, 18]. This geometry is chosen here for demonstrative purposes and the method can be equally extended to other geometries as well. In the next part, the experimental study on high-temperature heat transfer within circular tubes is presented, although detailed and quantitative comparison is not attempted.

Theoretical Analysis on Heat Transfer in a Circular Tube

Description of the Problem

The coordinate system of the flow field is shown in Figure 7. The flow regime is either laminar or turbulent, and the turbulent mechanism in flowing gaseous suspensions of fine particles is so complicated that no information on the eddy diffusivities for the heat and momentum of the particles and the gas, either experimentally or analytically, is available so far. In the present analysis, therefore, it is assumed as Tien et al. [19] did that the eddy diffusivity of the particles are equal to zero (that is, $\varepsilon_{H,P} = 0$) and the presence of particles has a negligible effect on the eddy diffusivity of the gas. It is readily seen that this is a close approximation to the actual system when the mean size of particles is very small and the loading ratio is less than or equal to unity.

The Nusselt number on the heat transfer between the particle and the gas Nu_p may be a function of the radial position r, due to the flow being turbulent near the particles so that the Nusselt number Nu_p is assumed in the following form

$$Nu_p = \frac{h_p d_p}{k_f} = 2 \times g_f \tag{9}$$

This relationship has no theoretical or experimental background except that this seems to be physically and intuitively acceptable. Here, $g_f = 1 + (\varepsilon_{H,P}/\nu)\,Pr$ and the value of $Nu_p = 2$ correspond to pure conduction for laminar flow or to the viscous sublayer in turbulent flow where the turbulent fluctuations vanish.

In addition to the subjects just discussed the following assumptions are introduced into the analysis.

- The pipe wall is isothermal (T_w = constant) and black for thermal radiation.
- The flow field of the two phases is hydrodynamically fully developed from the inlet and has a laminar or turbulent velocity distribution. The temperatures of the particles and the gas are the same at the inlet.
- The physical properties (such as ρ_f, c, μ, k_f, etc.) are constant.
- The particles are spheres of uniform size and thermal conductivity which are large enough to neglect radial variations of temperature within the sphere.
- The particles are uniformly distributed throughout the pipe cross section, and emit and absorb thermal radiation as a gray fluid in local thermodynamic equilibrium, but do not scatter. The gas is transparent for radiation.

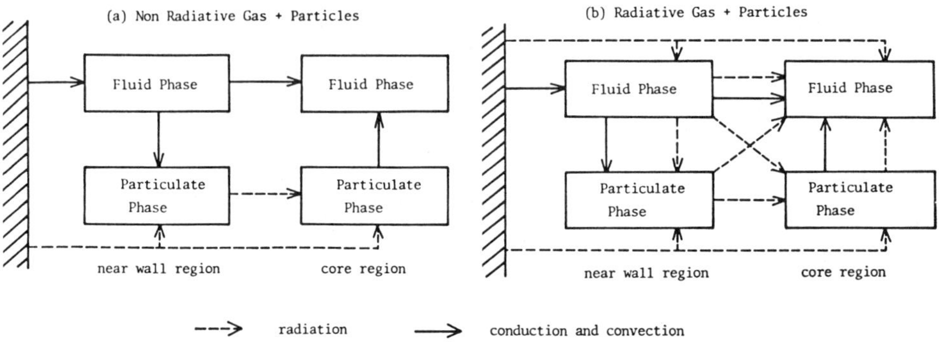

Figure 6. Heat transfer mechanism in gaseous suspension.

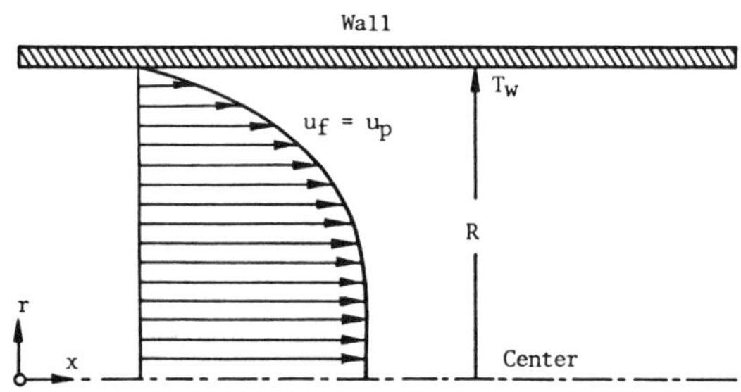

Figure 7. Cylindrical coordinate system for the analysis.

- The particles are sufficiently small and numerous to be considered as a continuum for thermal radiation propagation.
- The time-mean velocities of the two phases are equal and the presence of the particles has no effect on the velocity profile and the friction factor of the gas.
- The agglomeration, chemical reaction of the particles and viscous dissipation are not considered.
- The energy transports by collisions among particles or with the wall are negligible as compared with those by conduction and radiation.

Basic Equations and Numerical Method

In the foregoing postulation, a heat balance in pipe flow yields the basic equations governing the temperature field. For the gas,

$$C_f\rho_f U_f \frac{\partial T_f}{\partial x} + n_p h_p A_p (T_f - T_p) = \frac{C_f\rho_f}{r}\frac{\partial}{\partial r}\left[r\left(\frac{k_f}{C_f\rho_f} + \varepsilon_{H,f}\right)\frac{\partial T_f}{\partial r}\right] \tag{10}$$

and for the particles,

$$C_f\rho_{dp} U_p \frac{\partial T_p}{\partial x} + n_p h_p A_p (T_p - T_f) = \frac{C_p}{r}\frac{\partial}{\partial r}\left[r\varepsilon_{H,p}\rho_{dp}\frac{\partial T_p}{\partial r}\right] - \operatorname{div} q_R \tag{11}$$

The apparent density of the dispersed particles ρ_{dp} is assumed constant everywhere and ρ_{dp} is related to the density ρ_p with the following equation

$$\rho_{dp} = \rho_p n_p V_p \tag{12}$$

where V_p is the volume of a particle. The first term in both Equation 10 and Equation 11 describes heat convected along the pipe; the second term is the heat transfer rate between the phases (an identical term in both equations) while the first term on the right-hand side of the equations is the heat transferred radially due to eddy diffusion in case of turbulent flow. For the gas, molecular conduction is assumed to contribute additively to radial heat transfer whereas there is considered to be no similar term for the particles.

Let us consider the geometrical configuration for the radiative heat flux in Figure 8 where an arbitrary point is denoted by (x, r) and a dummy variable point by $(x_1, r_1, \emptyset)$. The distance S is then given by

$$S = [(x_1 - x)^2 + r^2 + r_1^2 - 2rr_1\cos\emptyset]^{1/2} \tag{13}$$

and S_w denotes the value of S when $r_1 = R$.

Assuming the fluid to be in thermodynamic equilibrium and not scattered, the divergence of the radiative heat flux vector is given by [20].

$$\begin{aligned} - \operatorname{div} q_R = \varkappa J_w \int_{-\infty}^{+\infty} \int_0^{2\pi} R(R - r)\exp[-\tau(S_w)]\frac{d\emptyset dx_1}{S_w^3} \\ + \varkappa^2 \int_0^R Br_1 \int_{-\infty}^{+\infty} \int_0^{2\pi} \exp[-\tau(S)] \times \frac{d\emptyset dr_1 dx_1}{S^2} - 4\pi\varkappa B \end{aligned} \tag{14}$$

where J_w is the intensity of radiation at the wall, B is the blackbody radiation, and $\tau(s)$ is the optical distance defined by

$$\tau(s) = \int_0^S \varkappa(S_1)\, dS_1$$

The absorption coefficient $\varkappa$ is given by the projected cross-section of particles normal to the radiative flux

$$\varkappa = \pi\left(\frac{d_p}{2}\right)^2 n_p \varepsilon_p \tag{15}$$

In deriving Equation 14, the approximation of the one-dimensional radiative heat flux is employed, based on the assumption that the radiative transfer in the axial direction is neglected as compared to the radial radiative heat transfer, i.e.

$$\frac{\partial q_{Rx}}{\partial x} \ll \frac{1}{r}\frac{\partial}{\partial r}(rq_{Rr})$$

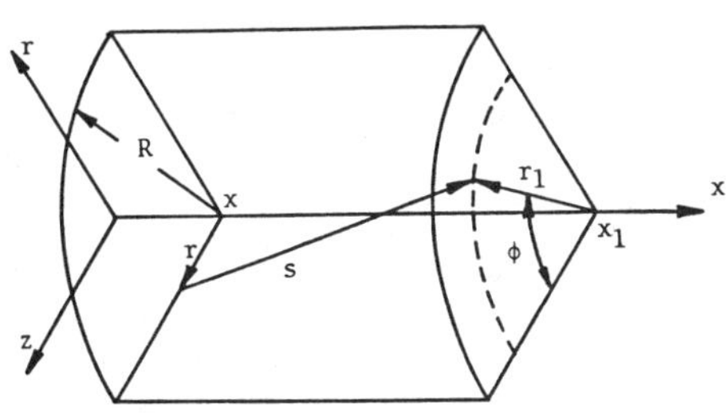

Figure 8. Geometrical configuration for radiative heat flux.

The one-dimensional approximation is valid if the temperature variation along the flow direction is slow compared to that of radial, that is,

$$\frac{\partial T}{\partial x} \ll \frac{1}{r}\frac{\partial}{\partial r}(rT)$$

which is the most prevalent for practical heat transfer facilities.

For the velocity distributions for a fully developed condition, it is assumed that the dimensionless profiles for both fluid and particles are the same so that for laminar flow profiles they are given by

$$U = \frac{U_f(r)}{U_{fm}} = \frac{U_p(r)}{U_{pm}} = 2\left[1 - \left(\frac{r}{R}\right)^2\right] \tag{16}$$

and for turbulent flow the velocity distribution is [21] expressed in the three subdivided regions.

For a viscous sublayer ($y^+ \leqq 5$),

$$u^+ = y^+, \varepsilon_{M,f} = 0 \tag{17a}$$

For a transition layer ($5 \leqq y^+ \leqq 30$),

$$u^+ = -3.05 + 5.0 \ln y^+, \frac{\varepsilon_{M,f}}{\nu} = \frac{y^+}{5}\left(1 - \frac{y^+}{R}\right) - 1 \tag{17b}$$

For a turbulent core ($y^+ \geqq 30$),

$$u^+ = 5.5 + 2.5 \ln\left[y^+ \frac{1.5(1 + r/R)}{1 + 2(r/R)^2}\right]$$

$$\frac{\varepsilon_{M,f}}{V} = \frac{0.4R^+}{6}\left[1 - \left(\frac{r}{R}\right)^2\right]\left[1 + 2\left(\frac{r}{R}\right)^2\right] \tag{17c}$$

where $u^+ = (u/u_m)/\sqrt{f/2}$
$y^+ = (y/R)(Re/2)\sqrt{f/2}$
$y^+ = R^+ - r^+$

and the eddy diffusivity for heat $\varepsilon_{H,f}$ is calculated by the use of the velocity distribution on the assumption that $\varepsilon_{H,f} \approx \varepsilon_{M,f}$. For brevity, the boundary conditions are taken as follows

$$\begin{aligned} r = R;&\quad T_f = T_p = T_w \\ r = 0;&\quad \partial T_f/\partial r = \partial T_p/\partial r = 0 \\ x = 0;&\quad T_f = T_p = T_0 \end{aligned} \tag{18}$$

Strictly speaking, however, the temperature discontinuity between the heated wall and the particles at the wall has to be included in the boundary condition, due to the absence of the conduction term in Equation 11. The results obtained with conditions of Equation 18 underestimate the temperature gradient at the wall as compared to those obtained with the so-called *radiation slip* condition [22], so that the heat transfer characteristics are also underestimated. This effect is, however, expected to be small, because when the solid particles are travelling in the immediate neighborhood of the wall before hitting it, the temperature of the particles due to the local condition is very close to that of the wall.

After transforming Equation 14 according to the method of Heaslet and Warming [23] and transforming Equations 10 and 11 into dimensionless forms with the assumption that $U_f = U_p = U$, it was found after some manipulation, for the gas

$$U\frac{\partial\theta_f}{\partial\xi} + M(\theta_f - \theta_p) = \frac{2}{\eta}\frac{\partial}{\partial\eta}\left(g_f\eta\frac{\partial\theta_f}{\partial\eta}\right) \tag{19}$$

and (when radial heat transfer by the particles neglected) for the particles

$$U\Gamma\frac{\partial\theta_p}{\partial\xi} + M(\theta_p - \theta_f) = \frac{2\tau_0^3}{N_R}H(\tau_0\eta) + \frac{2\tau_0^4}{N_R}\int_0^1 K(\tau_0\eta,\ \tau_0\eta_1)\,\eta_1\theta_p^4\,d\eta_1 - \frac{2\tau_0^2}{N_R}\theta_p^4 \tag{20}$$

where

$$H(\tau_0\eta) = \int_1^\infty K_1(\tau_0 y)\,I_0(\tau_0\eta y)\,\frac{dy}{y} \tag{21a}$$

$$K(\tau_0\eta, \tau_0\eta_1) = \begin{cases} \int_1^\infty K_0(\tau_0\eta y)\,I_0(\tau_0\eta_1 y)\,dy & (\eta_1 < \eta) \\ \int_1^\infty K_0(\tau_0\eta_1 y)\,I_0(\tau_0\eta y)\,dy & (\eta < \eta_1) \end{cases} \tag{21b}$$

$I_n(y)$ and $K_n(y)$ are modified Bessel functions of the first and second kind of n-th order, respectively. The dimensionless parameters are

$$g_f = 1 + \frac{\varepsilon_{H,f}}{\nu_f}Pr \qquad \Gamma = \frac{c_p\rho_{dp}u_{pm}}{c_f\rho_f u_{fm}}$$

$$N_R = \frac{\varkappa k_f}{4\sigma T_w^3} \qquad \tau_0 = \varkappa R \tag{22}$$

$$\varkappa = \pi\left(\frac{d_p}{2}\right)^2 n_p\varepsilon_p \qquad M = Nu_p\,\frac{2n_pA_pR^2}{d_p}$$

The function g_f is a characteristic parameter in turbulent flow and reduces to unity in laminar flow. The boundary conditions, then, become

$$\begin{aligned} &\eta = 1; \quad \theta_f = \theta_p = 1 \\ &\eta = 0; \quad \partial\theta_f/\partial\eta = \partial\theta_p/\partial\eta = 0 \\ &\xi = 0; \quad \theta_f = \theta_p = \theta_0 \end{aligned} \tag{23}$$

Equations 19 and 20 constitute simultaneous nonlinear integrodifferential equations, and it seems to be formidable to get the analytical solution by virtue of the high-order nonlinearity. However, it is amenable to the numerical calculations so that an iterative procedure is used for evaluating the nonlinear term based on a finite difference method, wherein care must be taken to subdivide the fine meshes in order to have a few nodal points in the viscous sublayer in turbulent flow.

Heat transfer evaluation. The total convective and radiative heat fluxes at the wall and the local Nusselt numbers are defined by the following equations.

$$q_T = q_C + q_R \tag{24}$$

$$q_C\frac{R}{k_fT_w} = \left.\frac{\partial\theta_f}{\partial\eta}\right|_{\eta=1} \tag{25}$$

$$F(\tau_0) = \int_1^\infty K_1(\tau_0 y)\,I_1(\tau_0 y)\,\frac{dy}{y^2} \tag{26}$$

where

$$q_R \frac{R}{k_f T_w} = \frac{\tau_0^2}{N_R} F(\tau_0) - \frac{\tau_0^3}{N_R} \int_0^1 \theta_p^4 \eta_1 H(\tau_0 \eta_1)\, d\eta_1 \tag{27}$$

$$Nu_{\zeta,T} = \frac{h_\zeta 2R}{k_f} = \frac{q_T \cdot 2R}{k_f(T_w - T_m)} = Nu_{\zeta,C} + Nu_{\zeta,R} \tag{28}$$

$$Nu_{\zeta,C} = \frac{q_C \cdot 2R}{k_f(T_w - T_m)} \tag{29}$$

$$Nu_{\zeta,R} = \frac{q_R \cdot 2R}{k_f(T_w - T_m)} \tag{30}$$

The dimensionless cup-mixing mean temperature of the gas, the particles, and the two phases are

$$\theta_{fm} = \int_0^1 U\theta_f \eta\, d\eta \Big/ \int_0^1 U\eta\, d\eta \tag{31}$$

$$\theta_{pm} = \int_0^1 U\theta_p \eta\, d\eta \Big/ \int_0^1 U\eta\, d\eta \tag{32}$$

$$\theta_m = \int_0^1 U(\theta_f + \Gamma\theta_p)\, \eta\, d\eta/(1 + \Gamma) \int_0^1 U\eta\, d\eta = \frac{\theta_{fm} + \Gamma\theta_{pm}}{1 + \Gamma} \tag{33}$$

respectively. Then, Equations 29 and 30 can be rewritten as follows

$$Nu_{\zeta,C} = 2\frac{\partial\theta_f}{\partial\eta}\bigg|_{\eta=1} \Big/ (1 - \theta_m) \tag{34}$$

$$Nu_{\zeta,R} = 2\left[\frac{\tau_0^2}{N_R} F(\tau_0) - \frac{\tau_0^3}{N_R} \int_0^1 \theta_p^4 \eta_1 H(\tau_0 \eta_1)\, d\eta_1\right] \Big/ (1 - \theta_m) \tag{35}$$

Analytical Results and Discussions

Temperature profiles: The typical results of the temperature profiles are illustrated for laminar and turbulent flows in Figures 9 and 10 respectively. In turbulent flow the temperature difference between gas and particles is so minute that the temperature differences are smeared out in the figure, while in laminar flow the temperature differences shown in Figure 9 are important for understanding a detailed heat transfer mechanism.

Examination of the temperature profiles show that the particle temperature exceeds to a certain extent depending mainly on parameter M for a wide range of radius except for a close vicinity of the heating wall. Consequently, the major heat transfer between the heating wall and particulate cloud and the presence of particles near the wall leads to an additional enhancement of conductive heat transfer from the wall to fluid.

Heat transfer characteristics. The heat transfer characteristics are summarized in Figure 11 which indicates the relationship between the convective and total Nusselt numbers against ξ. $Nu_{\xi,T}$ decreases to a certain minimum and turns to increase beyond this point. This behavior implies that there exists no asymptotic Nusselt number. In addition, if the comparison is made with the single-phase heat transfer, it is noted that the heat transfer characteristic is enhanced prominently by the flowing multiphase media, especially, at high temperatures (for small N_R).

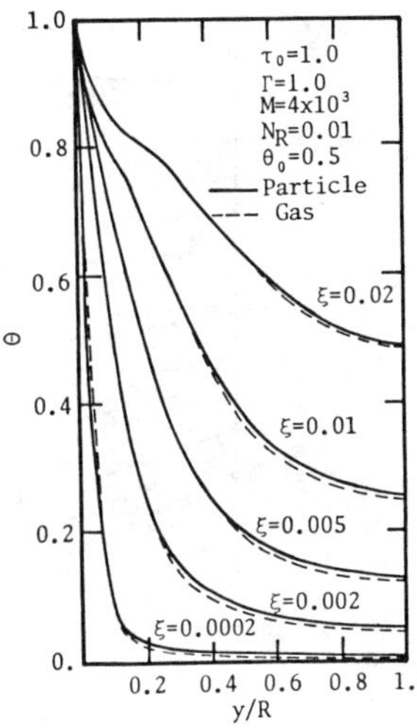

Figure 9. Results of the analysis. Temperature profiles for laminar flow [17].

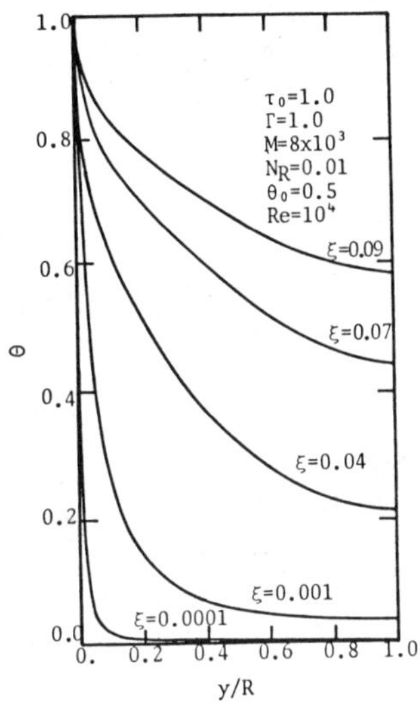

Figure 10. Results of the analysis. Temperature profiles for turbulent flow [18].

Experimental Study

There have been published a number of papers dealing with the experiments of gaseous solid suspensions heat transfer. However, most of the experimental studies have been executed at moderate temperature, wherein the effect of thermal radiation is not significant.

One of the most prominent features embodied in gaseous solid suspension heat transfer is radiative heat transfer so that in this article the experiment [24] carried at elevated temperature will be introduced in some detail.

Experimental Facility

A schematic diagram of HTL–GSS (heat transfer loop–gaseous solid suspension) is shown in Figure 12. This experimental closed loop mainly consists of three flow lines i.e. gas stream, particle stream, and gaseous solid suspension stream. Gas is circulated by a

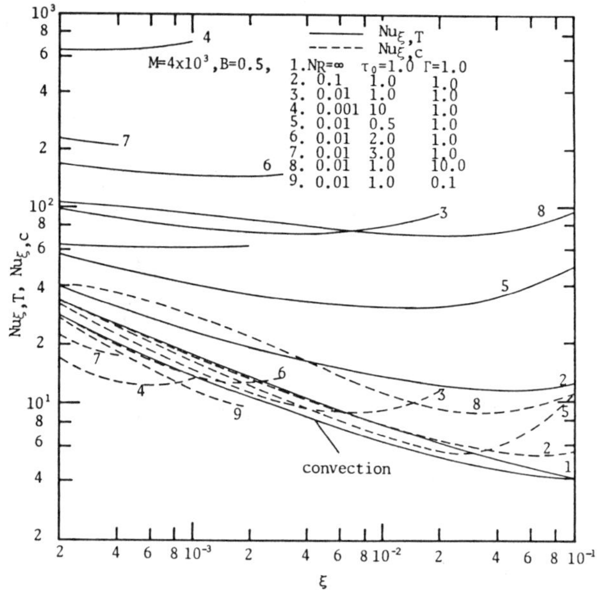

Figure 11. Results of the heat transfer analysis. Nusselt numbers vs. ξ for laminar case [17].

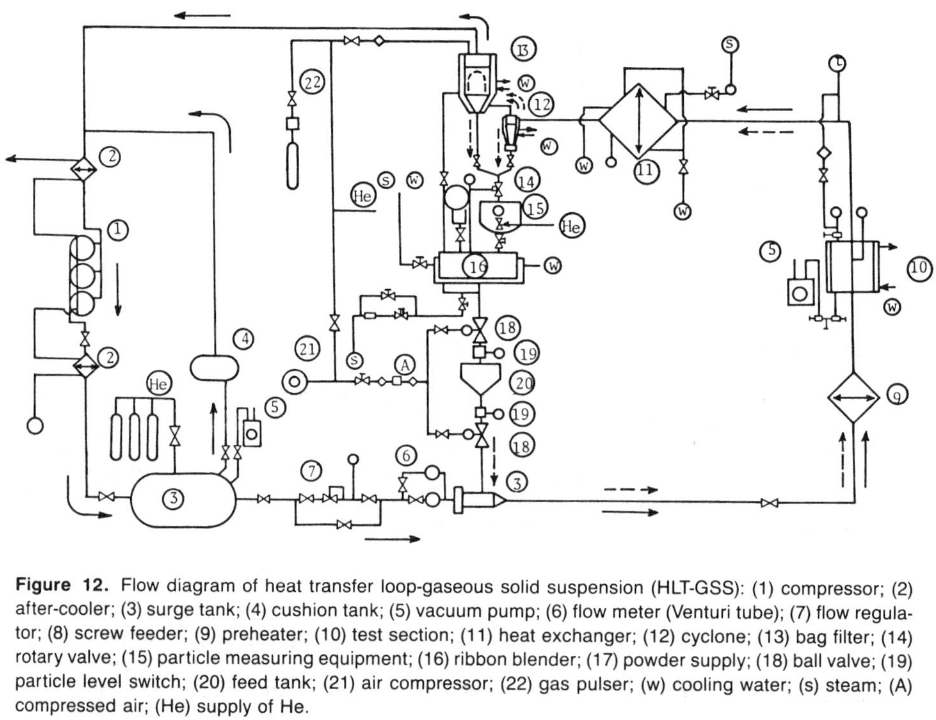

Figure 12. Flow diagram of heat transfer loop-gaseous solid suspension (HLT-GSS): (1) compressor; (2) after-cooler; (3) surge tank; (4) cushion tank; (5) vacuum pump; (6) flow meter (Venturi tube); (7) flow regulator; (8) screw feeder; (9) preheater; (10) test section; (11) heat exchanger; (12) cyclone; (13) bag filter; (14) rotary valve; (15) particle measuring equipment; (16) ribbon blender; (17) powder supply; (18) ball valve; (19) particle level switch; (20) feed tank; (21) air compressor; (22) gas pulser; (w) cooling water; (s) steam; (A) compressed air; (He) supply of He.

diaphragm-type compressor (1). The gas flow rate is adjusted by by-pass circuit valve and is measured by Venturi tube (6). A solid particle feed system consists of a ribbon blender (16), two ball valves (18), a feed tank (20) and a screw feeder (8). Leaving a heat transfer section (10), the suspension flow enters the gas-solid separator system, (cyclone (12) and bag filter (13). Particles are captured there and sent to the particle weight measuring equipment (15) through two rotary valves (14) and accumulated in a conical hopper within it. The time variation of the weight of the hopper is recorded automatically and the slope of the recorded curve gives the solid flow rate.

Various measures are taken in the heat transfer section to cope with the situation at high temperature and high thermal load. The main heater is made of an Inconel pipe of an 18-mm inner diameter (designated by D), a 0.75-mm thickness and a 1,000-mm length, which is equally divided into five segments (a 200-mm length regional pipe sections) by watercooled electrodes in order that regional variations of heating can be controlled. The pipe heater is wound by a fibrous insulator of an outer diameter of about 90 mm and is covered with asbestos. This heater assembly is contained in a vessel with a water jacket. A turbulence promotor (twisted tape) is inserted into the heating pipe in some cases.

This tape is made of a stainless steel plate of a 2,300-mm total length (2,000 mm twisted part), a 2.1-mm thickness, a 10-mm width and a 400-mm pitch. The tape end extends upstream by 500 mm from the starting point of the heating pipe. Asbestos blocks ($3 \times 10 \times 15\,mm^3$) are attached to the twisted part as the spacer libs for electrical insulation, the pitch of which is 100 mm.

Experimental Results and Discussions

The results of forced convection heat transfer of helium-graphite suspension (average particle diameter $d_p = 18$ (μm) in a straight tube without a turbulence promotor are shown in Figures 13 and 14 in case of a constant wall temperature condition up to 1173° K. The Reynolds number based on the gas phase at inlet Re_{fi} ranges from 1.0×10^4 to 2.0×10^4, and the particle loading ratio Γ reaches about 4. Prior to the experiments the heat losses are estimated through the preliminary measurements by realizing a similar thermal condition in the insulator vessel. The heat transfer coefficient and Nusselt number are locally evaluated as follows,

$$h_x = q_x/(T_w - T_b) \tag{36}$$

$$Nu_x = h_x \cdot D/k_{fbx} \tag{37}$$

where q_x is the net heat flux, k_{fb} is the thermal conductivity of the fluid phase at the suspension bulk temperature T_b and the suffix x means a length from the starting point of the heating pipe. The variation of Nu_b at the location $x/D = 50$ for $Re_{fi} = 1.0 \times 10^4$ are shown in Figure 13. It turns out that Nu_b increases only slightly with an increase in wall temperature T_w. However, it should be noted that high heat fluxes reduce the forced-convective Nusselt number for a single-phase flow as indicated by the correlation, for example, $Nu_b = 0.021 Re_b^{0.8} Pr_b^{0.4} (T_w/T_b)^{-0.5}$. Therefore, it is considered that both the radiative heat transfer mechanism of the particulate phase owing to the rise in wall temperature and fluid turbulence enhanced by fine particles might be compensated for by the foregoing reduction effect. As shown in Figure 13, Nu_b for $Re_{fi} = 2.0 \times 10^4$ decreases to a certain minimum with increasing Γ and beyond this point Nu_b becomes greater. It is also clear that its rising gradient becomes steeper with an increase in T_w.

In Figure 14 the heat transfer results are reproduced in the form of a convectional Nusselt ratio Nu_b/Nu_g, where Nu_g is the Nusselt number for a single gas flow obtained experimentally for the same x, T_w, and Re_{fi}. The effects of the wall temperature T_w on Nu_b/Nu_g is emphasized, because Nu_g is considerably higher for thermal load.

The results obtained in a straight tube might be explained by the suppressive effect on fluid turbulence by fine particles. To recover this defect and make use of larger inertia forces

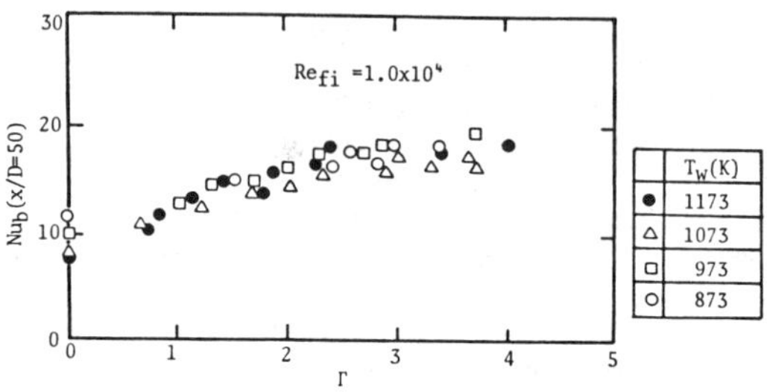

—A—

Figure 13. Variation of Nu_b with solid loading ratio at the specified location of the test section: (A) Re_{fi} = 1.0 × 10^4; (B) Re_{fi} = 2.0 × 10^4 [24].

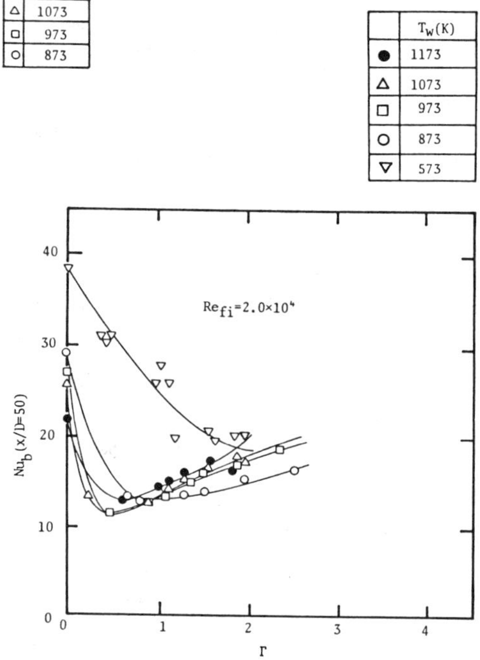

—B—

of particles, the previously mentioned turbulence promotor (twisted tape) is used in flowing gaseous-solid suspension systems, and its augmentation effect is experimentally examined as shown in Figure 15. It becomes evident that the local heat fluxes with the twisted tape increase remarkably with an increase in T_w.

AUGMENTATION OF CONVECTIVE HEAT TRANSFER OF GASEOUS SUSPENSIONS

As mentioned earlier, if the streamlines of the suspension flow are not straight but have a curvature, the difference in the inertia forces of both phases could possibly bring about unexpected heat transfer enhancement. An example of such phenomena has already been demonstrated in the former section. Namely, it was seen that the insertion of the twisted tape into the straight pipe resulted in a significant increase in the convective heat transfer rate, although the original objective was to make use of the radiative heat transfer mode. In this section, two more examples of such phenomena are presented: flows within helically coiled circular tubes, and impinging jets. In the former case, the analytical attempt is also presented. Direct comparison between the analysis and the experiment is impossible at the present stage, because the analysis is restricted to the laminar flow case. Nevertheless, the general flow characteristics may be grasped from it.

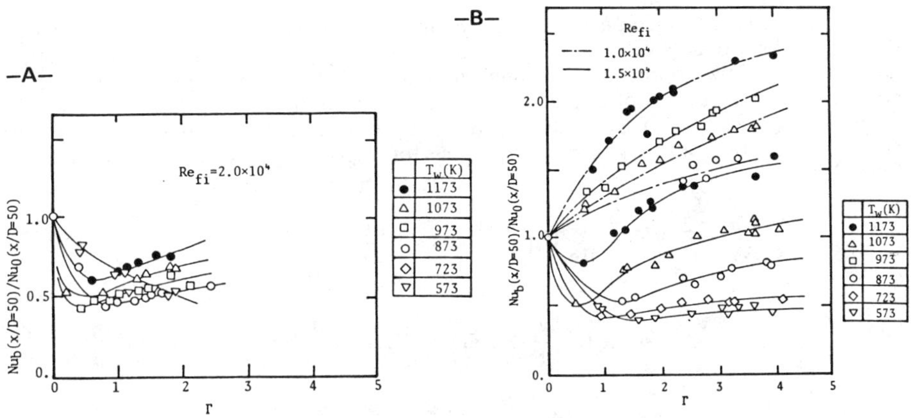

Figure 14. Variation of specific Nusselt number with solid loading ratio at the same position with Figure 13: (A) Re_{fi} = 2.0 × 10^4; (B) Re_{fi} = 1.0 × 10^4 and Re_{fi} = 1.5 × 10^4 [24].

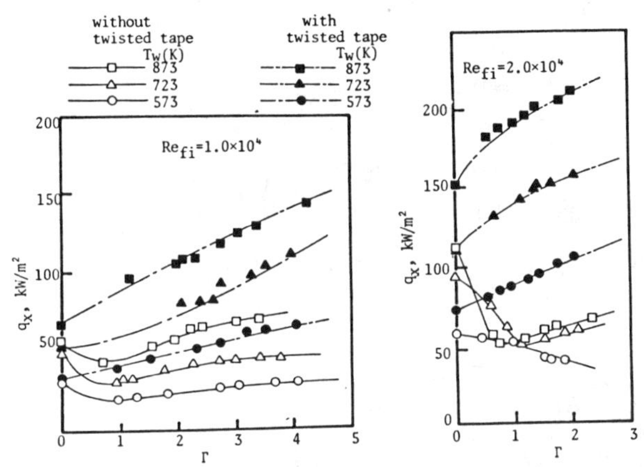

Figure 15. Comparison of the local heat flux with and without the inserted twisted tape into the test tube [24].

Gas/Solid Suspension Flows in Curved Tubes

Flow Characteristics

Single-phase gas or liquid flows within helically coiled circular tubes are simple means for heat transfer enhancement and are widely used in various types of heat exchangers. Therefore, there exist a lot of related research works at present. On the contrary, the counterpart of the gaseous solid suspension flows is extremely scarce and the only practical subject of research activity is that related to the problems of the erosion of the bend section of the pneumatic conveying line.

Contrasting to the single-phase flow [25], an inherent difference of inertia forces between the solid and gas phases will affect the flow structure to a great extent so that the so-called homogeneous flow approach (assuming equilibrium between both phases and taking account of an increment of apparent density) is not permissible. Figure 16 shows the coordinate system used in the analysis. After the prevalent postulations and assumptions are introduced, the coupled governing equations for respective phases are formulated in the following,

$$V_f\frac{\partial\bar{\omega}}{\partial R}+\frac{W_f}{R}\frac{\partial\bar{\omega}}{\partial\varphi}=\frac{2}{D_e}\nabla^2\bar{\omega}-\frac{2}{Re}U_f\left(\frac{1\partial U_f}{R\partial\varphi}\cos\varphi+\frac{\partial U_f}{\partial R}\sin\varphi\right)$$

$$-\frac{2\Gamma}{\Lambda D_e}\frac{1}{R}\left[G_p\left(\frac{\partial V_p}{\partial\varphi}-W_p-R\frac{\partial W_p}{\partial R}+R\bar{\omega}\right)+(V_p-V_f)\frac{\partial G_p}{\partial\varphi}-R(W_p-W_f)\frac{\partial G_p}{\partial R}\right] \quad (38)$$

$$\bar{\omega}=-\nabla^2\psi\left(\text{where }\nabla^2=\frac{\partial^2}{\partial R^2}+\frac{1}{R}\frac{\partial}{\partial R}+\frac{1}{R^2}\frac{\partial^2}{\partial\varphi^2};\ V_f=\frac{1}{R}\frac{\partial\psi}{\partial\varphi},W_f=-\frac{\partial\psi}{\partial R}\right) \quad (39)$$

$$V_f\frac{\partial U_f}{\partial R}+\frac{W_f}{R}\frac{\partial U_f}{\partial\varphi}=\frac{16}{D_e}\overline{DP}+\frac{2}{D_e}\nabla^2U_f+\frac{2\Gamma}{\Lambda D_e}G_p(U_p-U_f) \quad (40)$$

$$\frac{\partial}{\partial R}(G_pRV_p)+\frac{\partial}{\partial\varphi}(G_pW_p)=0 \quad (41)$$

$$V_p\frac{\partial V_p}{\partial R}+\frac{W_p}{R}\frac{\partial V_p}{\partial\varphi}-\frac{W_p^2}{R}-\frac{\cos\varphi}{R_c}U_p^2=\frac{2}{\Lambda D_e}(V_f-V_p) \quad (42)$$

$$V_p\frac{\partial W_p}{\partial R}+\frac{W_p}{R}\frac{\partial W_p}{\partial\varphi}+\frac{V_pW_p}{R}+\frac{\sin\varphi}{R_c}U_p^2=\frac{2}{\Lambda D_e}(W_f-W_p) \quad (43)$$

$$V_p\frac{\partial U_p}{\partial R}+\frac{W_p}{R}\frac{\partial U_p}{\partial\varphi}=\frac{2}{\Lambda D_e}(U_f-U_p) \quad (44)$$

where the dimensionless variables and parameters are defined as

$$\left.\begin{aligned}&U=u/u^*,V=v/u^*,W=w/u^*,R_c=r_0/r_c,R=r/r_0,R_p=d_p/2r_0\\&\overline{DP}=(\partial p/\partial x)/(8\mu u^*/r_0^2),Re=2\rho_f r_0 u_{fm}/\mu,D_e=Re\sqrt{r_0/r_c}\\&\Gamma=(\rho_{dp}U_p)/(\rho_f U_f),G_p=(\rho_{dp}/\rho_f)/\Gamma,\Lambda=(\rho_p/\rho_f)(d_p/r_0)^2/18\\&\zeta=(x/r_0)/D_ePr,u^*=u_{fm}\sqrt{r_0/r_c}\end{aligned}\right\} \quad (45)$$

Related to the boundary conditions, special attention is paid to the behavior of solid particles in close vicinity of the wall. If the radial force exerted on a particle contacting the wall is positive, the friction drag is taken into account for the equation of motion which constitutes one of the boundary conditions for the particulate phase.

$$\varphi=0,\pi:\partial U_f/\partial\varphi=\partial V_f/\partial\varphi=W_f=0,\partial U_p/\partial\varphi=\partial V_p/\partial\varphi=W_p=\frac{\partial G_p}{\partial\varphi}=0 \quad (46)$$

$$R=1:U_f=V_f=W_f=0 \quad (47)$$

$$\frac{W_p}{R}\frac{\partial W_p}{\partial \varphi}+\frac{\sin\varphi}{R_c}U_p^2$$
$$=\frac{2}{\Lambda D_e}(W_f-W_p)-fw\frac{W_p}{\sqrt{U_p^2+W_p^2}}\left(\frac{W_p^2}{R}+\cos\varphi\frac{U_p^2}{R_c}+\frac{2}{\Lambda D_e}V_f\right) \tag{48}$$

$$\frac{W_p}{R}\frac{\partial U_p}{\partial \varphi}=\frac{2}{\Lambda D_e}(U_f-U_p)-fw\frac{U_p}{\sqrt{U_p^2+W_p^2}}\left(\frac{W_p^2}{R}+\cos\varphi\frac{U_p^2}{R_c}+\frac{2}{\Lambda D_e}V_f\right)$$

Numerical solutions have been obtained by the S.O.R. method based on the finite difference scheme, in which the grid spacing is nonuniform with a higher density of nodal points near the wall. In Figure 17 typical results on the flow field are illustrated in which the parameters are set as $R_c = 100$, $R_p = 10^{-3}$, $\Lambda = 10^{-3}$, $f_w = 0.1$. Examination of the figures shows that particle loading deforms the velocity profile appreciably and enhances the secondary flow, and it is worthy to note that higher particle density is observed in the vicinity of the wall due to the centrifugal and Coriolis forces. The higher population of

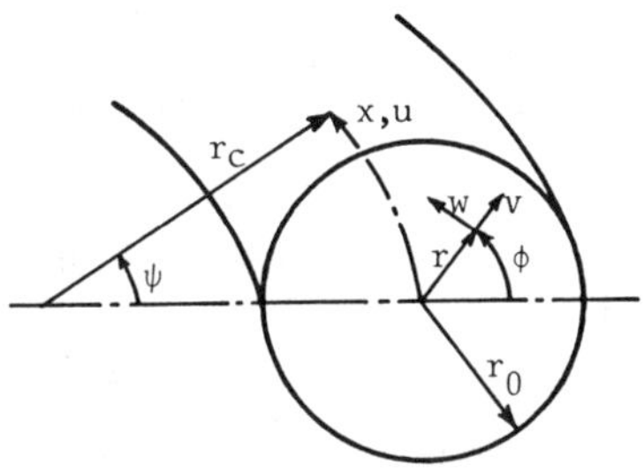

Figure 16. Coordinate system used in the analysis of the flow within the curved tube.

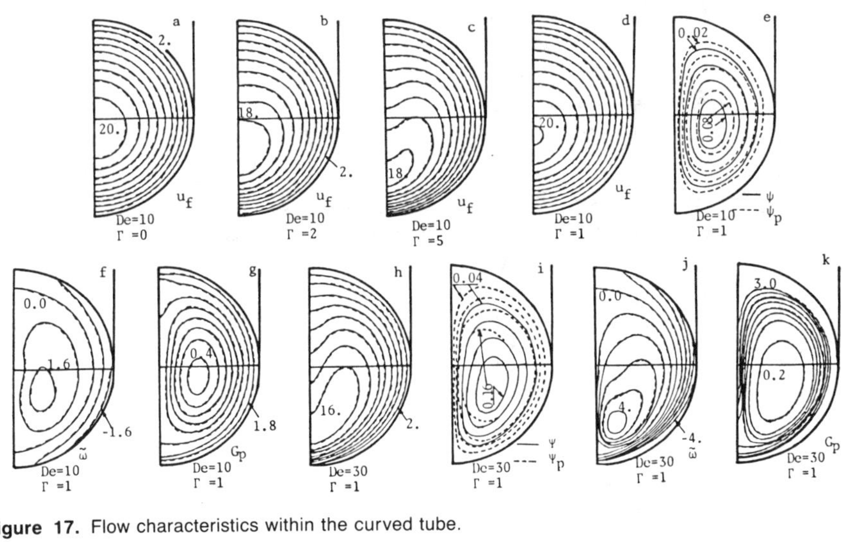

Figure 17. Flow characteristics within the curved tube.

particles near the wall will play an important role in the heat transfer mechanism which will be discussed briefly in the following part. Detailed behaviors of particulate phase in a close vicinity of the wall are depicted in Figure 18 for the Dean number De = 30 and particle loading ratio $\Gamma = 1$, respectively. For a set of parameters shown in the figure the particles contacting the wall ($R = 1 - R_p$) detach at 20 degrees due to the Stokes drag of the secondary flow and yield a steep increase of particle concentration G_p at $R = 1 - 2R_p$. The number density of particles at $R = 1 - 2R_p$ is gradually decreasing due to the dispersion (becomes zero at 170 degrees), while it comes to increase at $R = 1 - 4R_p$ ($\Delta R = 0.025$, $\Delta R_p = 0.001$). The limit of the most populated condition of attached particles to the wall in a single layer is expressed as

$$G_p \leqq (\rho_p/\rho_f)\,\pi/6\Gamma \tag{49}$$

The calculation results executed in this study clear the foregoing conditions. These behaviors are intuitive for understanding the flow and heat transfer mechanism in this system.

Heat Transfer Analysis and Discussions

Once a flow field is established it is comparatively easy to get a solution of energy equations for fluid and particle media. The energy equations in the dimensionless forms state

$$\mathrm{PrD_e}\left(V_f\frac{\partial\theta_f}{\partial R} + \frac{W_f}{R}\frac{\partial\theta_f}{\partial\psi}\right) + U_f\frac{\partial\theta_f}{\partial\zeta} = 2\nabla^2\theta_f + \frac{2}{3}\frac{Nu_p}{\Lambda}\Gamma G_p(\theta_p - \theta_f) \tag{50}$$

$$\mathrm{PrD_e}\left(V_p\frac{\partial\theta_p}{\partial R} + \frac{W_p}{R}\frac{\partial\theta_p}{\partial\varphi}\right) + U_p\frac{\partial\theta_p}{\partial\zeta} = \frac{2}{3}\frac{Nu_p}{\Lambda C^x}(\theta_f - \theta_p) \tag{51}$$

with the boundary conditions of

$$\varphi = 0, \pi : \partial\theta_f/\partial\varphi = \partial\theta_p/\partial\varphi = 0, \quad R = 1 : \theta_f = 1 \tag{52}$$

$$\theta = (T - T_0)/(T_w - T_0), \quad Pr = \mu c_f/k_f, \quad Nu_p = h_p d_p/k_f \tag{53}$$

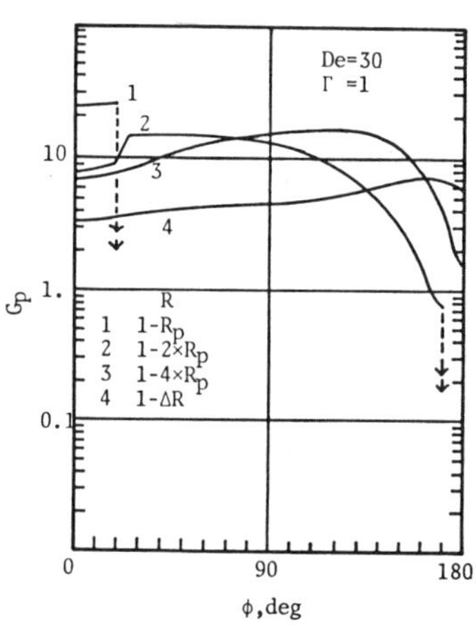

Figure 18. Number density of the particulate phase in the vicinity of the channel wall.

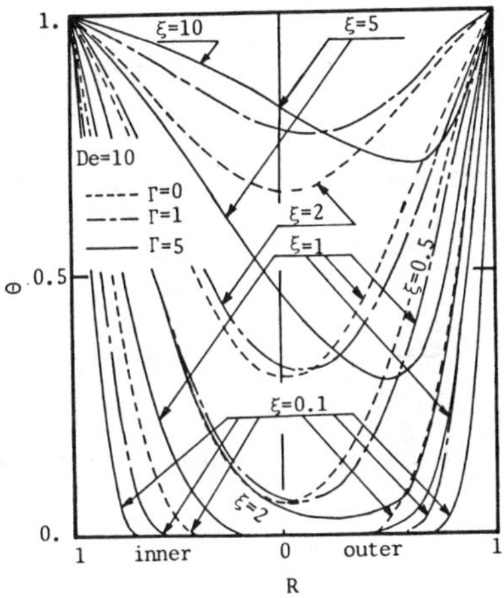

Figure 19. Temperature profiles along the $\phi = 0 - \pi$ axis.

For the numerical calculations some of the dimensionless parameters related to the energy equations are fixed as $Pr = 0.7$, $C^* = 1$, and $Nu_p = 2$. Figure 19 shows the typical temperature profiles along the $\varphi = 0 - \pi$ axis, where the temperature of particle is not drawn in the figures because the differences from the fluid are minor. In examining the particle temperature, we observe a wide variation over the flow cross section which is an important heat transfer characteristic. In order to discuss the heat transfer characteristics the mixing mean temperatures and Nusselt numbers are defined in the following,

$$\bar{\theta}_m = \frac{\int_A \theta_f W_f dA + \int_A C^* G_p \theta_p W_p dA}{\int_A W_f dA + \int_A C^* G_p W_p dA} \tag{54}$$

$$\bar{\theta}_f = \frac{\int_A \theta_f W_f dA}{\int_A W_f dA} \tag{55}$$

$$\bar{\theta}_p = \frac{\int_A \theta_p W_p dA}{\int_A W_p dA} \tag{56}$$

$$Nu(X, \varphi) = 2\left(\frac{\partial \theta_f}{\partial R}\right)_{R-1} / (1 - \bar{\theta}_m) \tag{57}$$

$$Nu(X) = \frac{1}{\pi}\int_0^{\pi} Nu(X, \varphi)\, d\varphi \tag{58}$$

$$\overline{Nu}(X) = -(1 + C^*\Gamma)\sqrt{R_c}\, \ln(1 - \bar{\theta}_m)/(2X) \tag{59}$$

The mixing mean temperatures ($\bar{\theta}_m$, $\bar{\theta}_f$, and $\bar{\theta}_p$) and mean Nusselt number Nu (x) against x are shown in Figure 20 for $D_e = 1$, 10, and 30 with loading ratio $\Gamma = 1$. Though the local temperature of fluid is always exceeding the particle temperature, the mixing mean temperature of fluid $\bar{\theta}_f$ is lower than that of particles $\bar{\theta}_p$. Noticing the local excursion of distribution of G_p, the physical meaning is selfexplanatory. It is important to gain insight

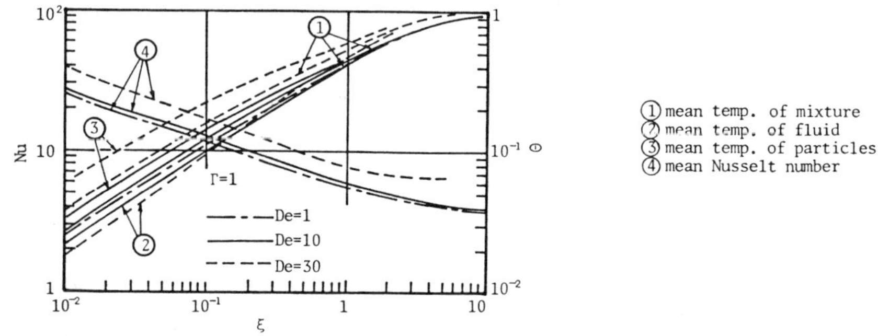

Figure 20. Mixing mean temperature and Nusselt number.

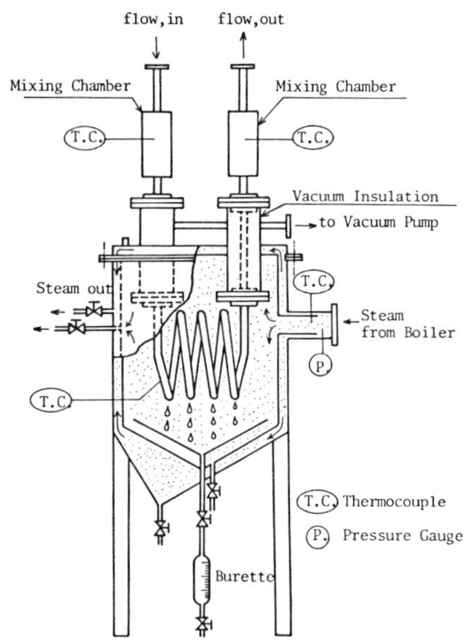

Figure 21. Experimental setup for the measurement of heat transfer within the curved channel.

into the circumferential distributions of the local Nusselt number Nu (x, ϕ). The difference between the outside $(\phi = 0)$ and inside $(\phi = \pi)$ of the curvature is prominent.

Experimental Study

To study particles in convective heat transfer within a curved duct, the experimental apparatus shown in Figure 21 was constructed. A 3.5-turn helically coiled circular tube with a 23.8-mm inner diameter and 308-mm helix diameter is set within the double-walled steam jacket. This test facility was set within the by-pass circuit of the HTL–GSS, and the suspension flow of nitrogen gas and graphite powder was fed into the test tube. The test tube was heated by saturated steam of atmospheric pressure which was introduced into the jacket from a boiler. The amount of added heat to the suspension flow was measured from the quantity of the collected condensate water, and also from the temperature difference between the inlet and outlet mixing chambers. Use was made of the vacuum insulation in order to minimize the heat transfer at the location other than the coiled test section.

Experimental results are summarized in Figure 22. As is expected, the Nusselt number of the suspension flow increases monotonously with the solid loading ratio. If examined more precisely, the slope of the increase is seen to take three stages. Namely, a slight increase is observed at the first stage followed by the steep increase with moderate loading. Finally, the slope again becomes smaller for higher loading. As was mentioned in earlier the increase in the Nusselt number is in a sense not strange, because the heat capacity of the media is elevated through the addition of the solid phase. Therefore another attempt was made to correlate the experimental results using the *mixture Reynolds number* defined in Equation 7 and the *mixture Stanton number*. The latter is defined as,

$$St_m = \frac{h}{U_f \rho_f C_f + U_p \rho_{dp} C_p} \doteq \frac{St}{1 + (C_p/C_f)\,\Gamma} \tag{60}$$

The numerator of the mixture Reynolds number (Equation 7), contains the added inertia due to the solid phase, while the denominator of the mixture Stanton number (Equation 60), represents the "mixture heat capacity flux." Therefore if the mixture Reynolds numbers are changed through one of two ways (namely by the change of gas velocity or through the change in solid loading), and both have the same effects on the mixture Stanton number, then it would be concluded that the increase in the Nusselt number is simply due to the increase in the heat capacity of the flowing media. Whole experimental results are replotted based on this consideration in the form of Re_m vs. St_m in Figure 23. The figure is a sort of *Moulie Diagram* with iso-Reynolds-number lines (solid lines) and iso-loading lines (broken lines). If the solid had only the capacity effect as previously mentioned, whole lines within Figure 23 would have been converged into the correlation of the single-phase flow within the curved tube. Corresponding to the change of the slope in Figure 22, the mixture Stanton

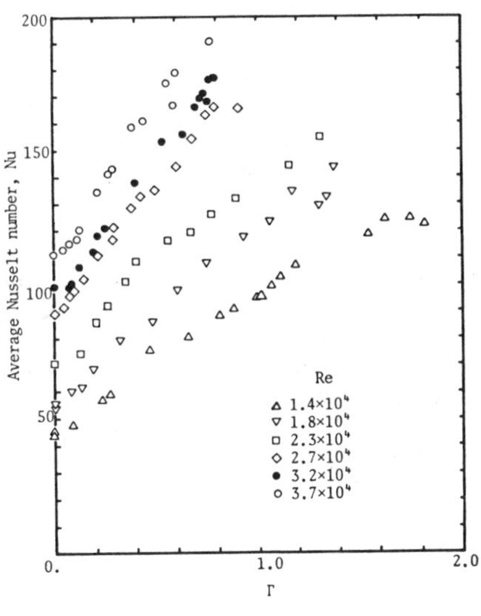

Figure 22. Average Nusselt number of the helically coiled tube vs. solid loading ratio.

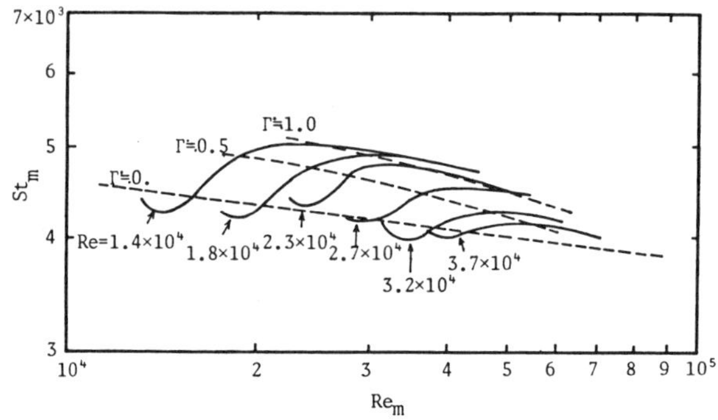

Figure 23. Mixture Reynolds number vs. mixture Stanton number. Effect of added solid on the convective heat transfer within the coiled tube.

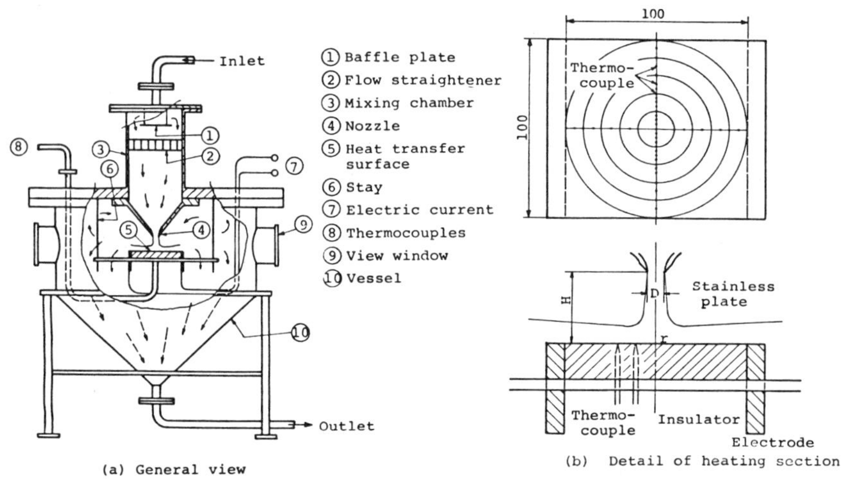

Figure 24. Impinging jet heat transfer test section [26].

number in Figure 23 displays somewhat interesting behavior. At the first stage, it decreases slightly with solid loading. It can be postulated that a turbulence-suppressing phenomenon exists even at light loading however above some critical loading value (about 0.2), the sublayer disturbing effect at the outer curvature region is surpassed, resulting in a steep increase in the mixture Stanton number. However, this effect seems to be saturated at the higher loading region, and the slope of the mixture Stanton number approaches that of the single-phase flow. Even if more solid is fed to the stream at this stage, the resulting effect on the heat transfer is said to be simply that of increased heat capacity of the media.

Impinging Jet Heat Transfer

One more typical example of the nonstraight streamline flow is the impinging jet. There exist many situations wherein free jets contain the dispersed phase and the impinging jets

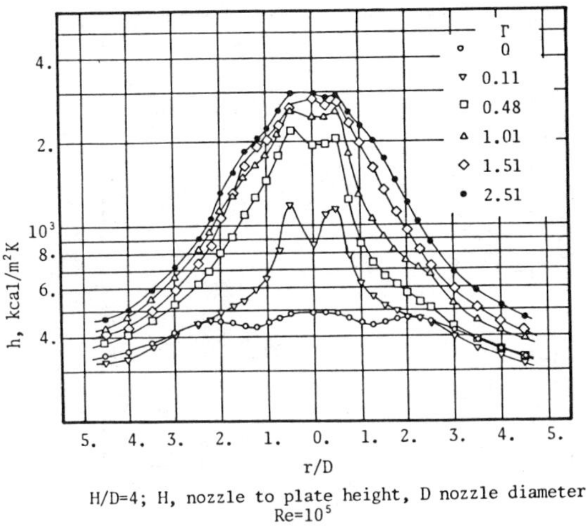

Figure 25. Radial profiles of the local heat transfer coefficients of the impinging jet, effect of the solid loading ratio [26].

of gas-liquid two-phase flow are also extensively investigated. Meanwhile, there seems to be no counterpart to the gas-solid impinging jet systems. Accordingly, no research work is found in the literature that deals with this subject. However, their potential advantage in heat transfer seems attractive as will be seen in the following.

Figure 24 is a brief sketch of the experimental setup [26] used in the heat transfer measurement of the impinging jet. This was also connected within the by-pass circuit of the HTL–GSS and fed with the same suspension media as described in the previous subsection. The heat transfer surface is made of a stainless steel plate of a 0.3-mm thickness and affixed on the asbestos insulator block. It is fed with the alternating current and cooled by the impinging jet of the suspension which issues from a round nozzle of 10 mm in diameter. The surface temperatures of the plate were measured by the CA thermocouples attached on the reverse side of the plate. Local heat transfer coefficients were obtained from the heat flux divided by the local temperature difference, the latter being defined between the surface temperature and jet temperature at the nozzle.

Typical results of the local heat transfer coefficients are shown in Figure 25 for several loading ratios including the clean gas case. A remarkable increase in heat transfer can be seen in the figure especially around the stagnation point. The nozzle-to-plate distance was fixed at four times the nozzle diameter. In the local heat transfer profile of a usual single-phase jet, a second peak other than that at the stagnation point appears as it is typically found in the figure. This peak is seen to become less clear with the increase in solid loading, because the whole heat transfer coefficient around the stagnation point is enhanced to a great extent due to the solid phase. Figure 26 is the specific Nusselt numbers obtained at the stagnation point. The maximum specific increase in the heat transfer coefficient at the stagnation point amounts to about 6.0 within the examined parameter region. Therefore the phenomena explained in the first section is typically realized in the impinging jet heat transfer. It is true that the heat transfer coefficient at the location far from the stagnation is not so high as those of the stagnation. However, if a concaved heat transfer surface is used, this effect of sublayer disturbance would be extended to the whole region. The maximum value of the stagnation point heat transfer coefficient within this investigation

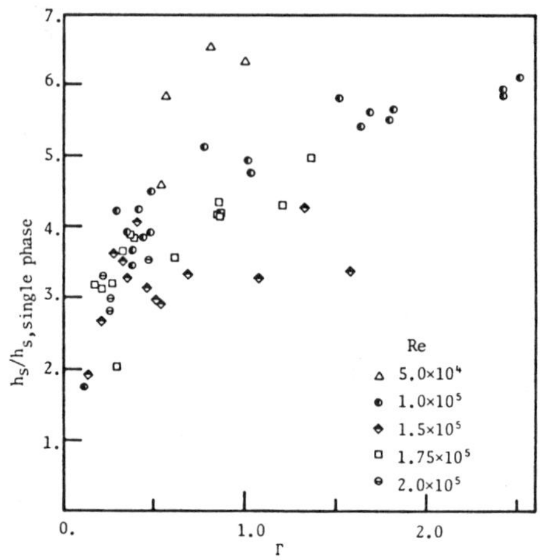

Figure 26. Specific Nusselt number at the stagnation point vs. solid loading [26].

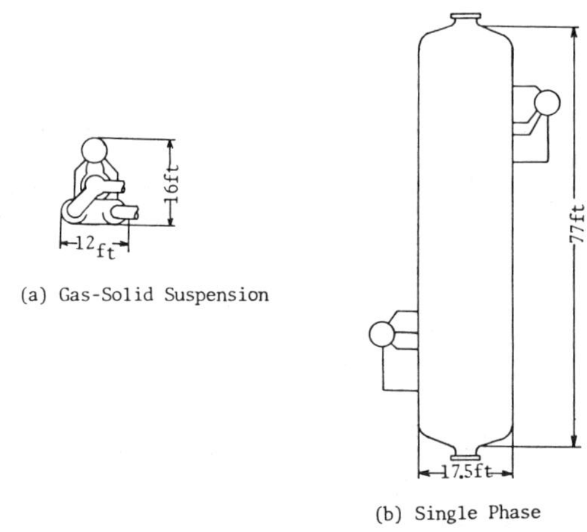

Figure 27. Comparison of the heat exchangers: (A) gas-solid suspension; (B) single phase flow [27].

was about 3,000 kcal/m^{2}hr K, which is compared favorably with the values of latent heat transfer.

ADVANCED APPLICATIONS

In connection with the preferable heat transfer characteristics of the gaseous solid suspension flows, several advanced applications can be conceived and some of them have already been proposed and examined. As was pointed out in the previous sections, the advantage of suspension heat transfer becomes clearer when the temperature and the thermal load of the system are high enough so that the radiation heat transfer becomes dominant. It is true that in order to realize those systems many problems remain to be solved in the future. Of these, powder handling technology, impurity control, material feasibility study, and so forth are included. However, the improvement of heat transfer is attractive from many points of view, and the authors conceive that more attention should be focused on the suspension flow heat transfer and on the related powder technology. In the present section, two advanced applications are introduced in order to demonstrate the potential utility of the gaseous suspension flows.

High-Temperature Heat Exchangers and Gas-Cooled Nuclear Reactor

An intensive application of gaseous solid suspension media has been examined by D. C. Schulderberg [5] to the coolant for nuclear reactors and also for heat exchangers because of the significant advantages in heat transfer performance, reduced pumping power, high specific power, reduced coolant volume fraction, lower coolant pressure, smaller containment structure, and so forth.

Extensive discussions based on the experiments concerning the pressure drop and heat transfer of gaseous solid suspensions have been reflected on the heat exchanger design with the Calder Hall boiler design, as illustrated in Figure 27. The overall heat transfer coefficient is the order of magnitude of 10 $W/m^2 K$ on a conservative basis. Other related features are the reduced coolant pressure and the reduced cost of the component, because a steam boiler for a gas-suspension system does not require the extended surface on the coolant side of the boiler tubes.

For the application to nuclear reactors, four basic thermal cycles as shown in Figure 28 have been proposed and the comparison of coolants have been examined in some detail, the results of which are tabulated in Table 1. It is understood that the gaseous solid suspension coolant is favorable not only for heat transfer characteristics but for the material compatibility.

Cooling Systems in Controlled Thermonuclear Reactors

Recent outstanding progress in the plasma confinement problem is now stimulating the advances in many aspects of related technical and engineering problems that are inevitable for the realization of thermonuclear fusion reactors. In most of these problems, researchers are compelled to face at first sight insurmountable critical conditions which they have never

Table 1
Comparison of Coolants

Coolant	Cost ($/lb)	Typical Maximum Loop Pressure (psia)	Typical Temp in Reactor (°F)	Material Compatibility	Th $\Sigma \times 10^4$ (cm^{-1})	Typical Film Coefficient, Core	Typical Average Heat Transport (Btu/ft^3)	Appreciable Radiation Damage Problems
Sodium	0.25	100	1,000	Poor	63	7,000	6,500(400)*	No
Fused salts	High†	100*	High†	Poor	100†	100	Good	–†
Organic fluids	0.65	400	600	Fair	100	1,200	1,100(50)	Yes
H_2O	0.25×10^{-3}	2,000	550	Fair	100	6,000	1,750(30)	No
D_2O	28.0	2,000	550	Fair	0.40	6,000	1,750(30)	No
CO_2	0.12	520	1,050	Good	0.04	400	217(700)	Yes
He	1.89	965	1,260	Excellent	0.00	1,250	234(685)	No
Gas suspension, N_2 (ws/wg = 20.0)	0.96	150	1,100	Very good	1.087	1,500	1,527(700)	?‡
Gas suspension, He (we/wg = 42.0)	1.02	350	1,260	Excellent	0.10	3,000	1,047(685)	Not expected

* Temperature rise 1°F through the core.
† Data not available or classified.
‡ Radiation damage to the coolant is not expected. Radiation catalyzed reactions with system material have not been explored.

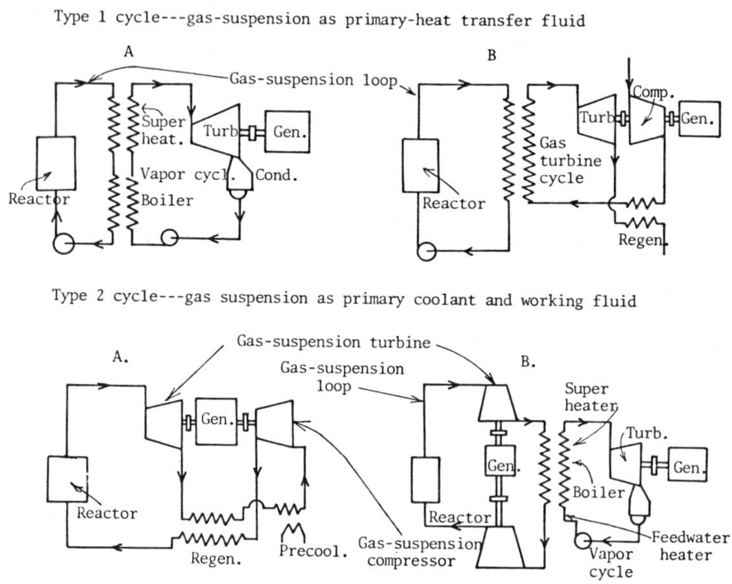

Figure 28. Proposed gas/solid thermal cycles for nuclear power plants [5].

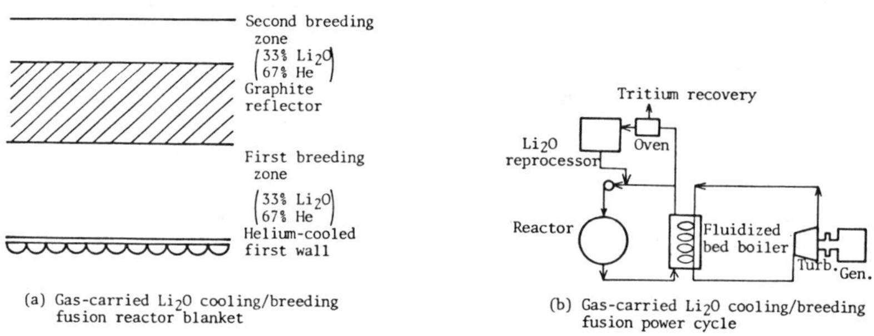

Figure 29. Blanket system of the gaseous solid suspension [28].

experienced in other conventional industrial equipment. The specification of the cooling system and tritium breeding method of the fusion reactor blanket is one such critical problem. As is well known, the roles of the fusion reactor blanket can be summarized as

1. The thermalization of high-energy neutrons and the conversion of their energy into heat.
2. Tritium breeding for fuel cycle.
3. A radiation shield for the protection of other instruments or structural materials as superconducting magnetic coils and also for human accessibility.

The volumetric heat generation rate within the blanket is of highest value at the first wall and decreases with the distance from there. Therefore in most of the tritium breeding region, it is believed that the heat removal can be achieved by the conventional coolants as duct flows of gas or liquid or two-phase combinations of them. The cooling of the first wall, on the contrary, belongs to such critical problems. The first wall is irradiated directly by high energy neutron flux and thermal radiation, the latter being originated from bremsstrahlung within the plasma. According to the recent design, estimated total energy flux amounts to the order of 1 MW/m^2, although the value tends to decrease from those of earlier designs. The material of the first wall should be of preferable nuclear properties, as higher resistance for sputtering, higher neutron multiplying ability for neutron economy and lower probability for other neutron exchange interactions particularly for helium-bubble forming reactions.

As well as these nuclear properties, it must have sufficient compatibility with the coolant materials and be of low cost. If the cooling of the first wall is not sufficient, then the thickness of the wall should be small enough to suppress the total heat generation within the wall through (n.γ) reaction, and this thin wall condition obviously conflicts with the required mechanical strength of the wall. As a result, the excellent heat transfer performance at the first wall is the key factor in order to make sure of a higher level freedom in the selection of wall materials and other designing procedures.

The candidate coolant materials proposed so far are roughly classified into liquid lithium, single-phase gas, molten salt, and two-phase flows of various combinations of them.

Liquid lithium is certainly a promising candidate, because of its excellent heat transfer properties. Furthermore, because it is the fertile material for the tritium breeding, the whole system can be extremely simplified. The main difficulties in using the liquid metal coolant lies in its high electric conductivity. When it flows within the strong magnetic field, generated by a plasma-confining magnetic coil, it receives a strong electro-magnetic deterring force in a direction which is perpendicular to both the flow and the magnetic field vectors. This force results in large flow resistance, and tends to consume the great part of the generated power as a form of pumping power. Furthermore, this force possibly suppresses the turbulent fluctuation of the flow and brings about the laminarization of the flow even when the Reynolds number of the flow is otherwise sufficient for the realization of the turbulent flow. As a result, the inherent advantage of excellent heat transfer with liquid metal can possibly be cancelled out as a whole.

A second candidate of the operating fluid is single-phase helium gas, which is of a low neutron-capture cross-section, of no phase change, stable, safe, and is an easily handled coolant material from physical, chemical, and nuclear aspects. Moreover, the designers are rescued from the troublesome problems of the electro-magnetic deterring force because it has no significant value of electric conductivity. However, the poor heat transfer properties of helium gas are the main abstacle that always afflicts the engineers as in HTGR.

Helium gas is made of single-atom molecule. Therefore, it has no mode of molecular vibration, which means that it is perfectly transparent for thermal radiation. It is true that the low density can theoretically be made up for through pressurization, but the excessive pressurization under the circumstance of high temperature and high thermal load inevitably brings about further restrictions into the selection of structural materials. Accordingly, the feasibility of first-wall cooling by helium gas seems doubtful.

A third group of candidate material is molten salts of a lithium compound such as FLIBE. Electric conductivities of the molten salts are generally by two or three orders of magnitude lower than those of liquid metals so that the problem of electro-magnetic flow resistance is significantly mitigated. It is said, however, that the compatibility of the molten salts with the existing structural materials is not necessarily favorable and their selection remains an important problem.

Having in mind these complicated situations, some researchers are making efforts to invent various kinds of other cooling systems. Of these new ideas, various types of two-phase flows exist and their common point is to avoid the electro-magnetic force while retaining the excellent heat transfer.

The concept of a gaseous-solid suspension-cooled fusion-blanket system is one such new candidate. The excellent heat transfer properties have been already explained in previous sections. In the following, the characteristic features of the suspension flow as the fusion-reactor blanket coolant are summarized.

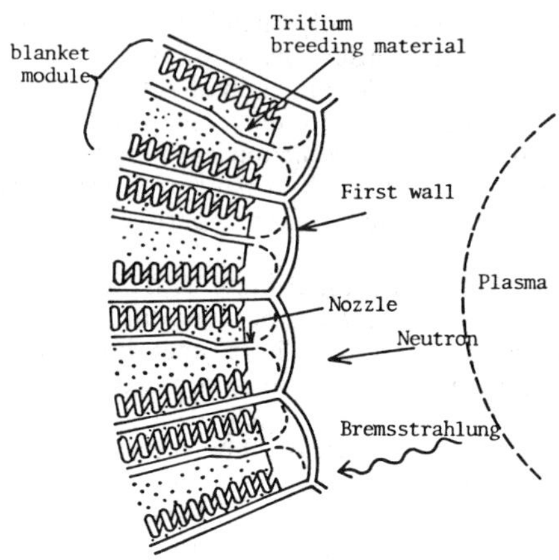

Figure 30. Blanket cooling by the gaseous solid suspension [26].

1. Estimation of the effects of suspended particles on the MHD properties of the media remains to further investigations. Unless the volume fraction of the solids is large enough for the particle-particle collision frequency to be siginificant, it is impossible to apply a high level of electric conductivity so that the MHD effect can be neglected.
2. It is not preferable to feed the turbine directly with a gas suspension from the viewpoint of blade integrity. Therefore, a heat exchanger that receives heat from primary suspension flow and transfers it to the secondary gas or liquid loop is necessary.
3. The selection of suspended phase materials is quite free, because they are fed to the flow only to enhance the heat transfer properties and the heat transfer enhancement depends only on the physical or geometrical qualifications of the suspended particles. Designers can choose the best powder suitable for surmounting the various restricting conditions just discussed. Only one exception to this is the case wherein the suspended powder serves as the fertile breeding material of tritium as well.
4. In order to keep the circulation of the coolant smooth, too dense a suspension flow should be avoided. Therefore the flowing media cannot be expected to have a function of softening the neutron energy spectra or of shielding (which is a situation similar to single-phase gas flow). However, it is possible to let the suspended particles take a role of tritium breeding by selecting some compound of lithium as the suspended solid. For example, Sze [28] proposed the system of He-Li_20 blanket concept, which is schematically shown in Figure 29. According to his proposal, produced tritium is recovered in the form of T_20 by continuously extracting 1% of the flowing Li_20 powder and introducing it into the oven of 1200° C for baking.
5. Impurity control for both helium gas and suspended powder may be a serious problem.
6. Depending on the kind of selected powder, the problem of duct erosion should be considered. According to the authors' experience obtained in the series of experiments using HTL–GSS, however, the use of graphite particles of fine size (less than 20 micron) is accompanied with no serious erosion with the duct of SUS-314. However, synthetic and systematic research which corresponds with the same condition as within the blanket is expected for practical use.
7. It is possible to set the blower within the closed loop of a primary suspension flowline. From the point of view of accidents and periodical inspection, systems containing particle flows, after receiving heat within the blanket and rejecting it to the secondary loop should be separated from the gas stream before allowing the flow to enter the blower and again fed to the gas at the downstream side of the blower. Therefore the

equipment which can efficiently and smoothly feed powder to the higher pressure side and recover it at the lower pressure side must be developed.

8. The heat exchanger can possibly be made as fluidized-bed type which averages the intermittently generated heat of the primary flow.
9. When the powder of nonfertile material is used for the suspended phase, the requirement of inventing the breeding and recovery system of tritium is similar as with the blanket cooled by the single-phase helium gas.

Figure 30 shows the conceptual idea of a gaseous suspension blanket proposed by the authors [26]. The key point of first-wall cooling is achieved by the impinging jet of the suspension flow which issues from the nozzle and impinges on the inner side of the first wall. The excellent heat transfer performance of the impinging jet has already been verified in the former section and is quite suitable for the severest condition of the first wall explained so far. Furthermore the cooling of the tritium breeding section can be achieved by the helically coiled tube flows of gaseous solid suspension flow.

NOTATION

C, C_f	specific heat of gas phase	n_p	number of particles
C_{Ds}	particle drag coefficient	p	pressure
D	pipe diameter	Pr	Prandtl number
D_e	equivalent diameter	q_R	heat flux
d_p	particle diameter	R	radius
f_s	friction factor	Re	Reynolds number
G_p	particle concentration	r	radius
g_f	parameter defined in Equation 22	S	distance
h_p	heat transfer coefficient	St	Stanton number
$I_n(y)$	n-th-order Bessel function of first kind	T	temperature
J_w	intensity of radiation	T_b	bulk temperature
$K_n(y)$	n-th order Bessel function of second kind	U_f	gas velocity
k_f	thermal conductivity of fluid	u^+	dimensionless velocity
L	length	V_p	particle volume
Nu	Nusselt number	x	distance
		y^+	dimensionless distance

Greek symbols

α	volume fraction of suspended phase	$\varkappa$	projected area of particles
Γ	particulate-to-gas-phase-mass ratio	μ	viscosity
Λ	parameter in Equation 45	ρ_f	fluid density
ε	void fraction or eddy diffusivity	$\tau(s)$	optical distance
θ_m	dimensionless mixing cup temperature	ω	radial velocity component

REFERENCES

1. Boothroyd, R. G., *Flowing Gas-Solid Suspension,* London: Chapmann and Hall, 1971.
2. Soo, S. L., *Fluid Dynamics of Multiphase Systems,* Waltham: Blaisdell Publishing Company, 1967.

3. Rudinger, G., *Handbook of Powder Technology,* Vol. 2, *Fundamentals of Gas-Particle Flow,* Williams, J. C. and Allen, T. (Eds.) Amsterdam: Elsevier Scientific Publishing Company, 1980.
4. Hetsroni, G., *Handbook of Multiphase Systems,* Washington: Hemisphere Publishing Corporation, 1982.
5. Schulderberg, D. C., et al., "Gaseous Suspension – A New Reactor Coolant," *Nucleonics,* Vol. 19, No. 8 (1961), pp. 67–76.
6. Gasterstaedt, J., "Die experimentelle Untersuchung des pneumatischen Fördervorganges," *Zeit. Vereines Deutscher Ingenieure,* Vol. 68 (1924), p. 617.
7. Vogt, E. G., and White, R. R., "Friction in the Flow of Suspensions," *Industrial and Engineering Chemistry,* Vol. 40, No. 9 (Sept. 1948), pp. 1731–1738.
8. Farbar, L., "Flow Characteristics of Solid-Gas Mixtures," *Industrial and Engineering Chemistry,* Vol. 41, No. 6 (June 1949), pp. 1184–1191.
9. Mehta, N. C., et al., "Pressure Drop in Air-Solid Flow Systems," *Industrial and Engineering Chemistry,* Vol. 49, No. 6 (June 1957), pp. 986–992.
10. Hariu, O. H., and Molstad, M. C., "Pressure Drop on Vertical Tubes in Transport of Solid by Gases," *Industrial and Engineering Chemistry,* Vol. 41, No. 6 (June 1949), pp. 1148–1160.
11. Ergun, S., "Fluid Flow Through Packed Columns," *Chemical Engineering Progress,* Vol. 48 (1952), p. 89.
12. Leung, S. L., and Wiles, R. J., "A Quantitative Design Procedure for Vertical Pneumatic Conveying Systems," *Industrial and Engineering Chemistry, Process Des. Dev.,* Vol. 15, No. 4 (1976), pp. 552–557.
13. Yang, W., "Estimating the Solid Particle Velocity in Vertical Pneumatic Conveying Lines," *Industrial and Engineering Chemistry, Fundamentals,* Vol. 12, No. 12 (1973), pp. 349–352.
14. Boothroyd, R. G., "Pressure Drop in Duct Flow of Gaseous Suspensions of Fine Particles," *Transactions of Institution of Chemical Engineers,* Vol. 44 (1966), pp. T306–T313.
15. Shimizu, A., et al., "Experimental Study on the Pressure Drop and the Entry Length of the Gas-Solid Suspension Flow in a Circular Tube," *International Journal of Multiphase Flow,* Vol. 4, No. 1 (1978), pp. 53–64.
16. Kim, J. M., and Seader, J. D., "Pressure Drop for Downflow of Gas-Solid Suspension," *AIChE Journal,* Vol. 29, No. 3 (May 1983), pp. 353–360.
17. Echigo, R., et al., "Radiative Heat Transfer by Flowing Multiphase Medium-Part II. An Analysis on Heat Transfer of Laminar Flow in an Entrance Region of Circular Tube," *International Journal of Heat and Mass Transfer,* Vol. 15 (1972), pp. 2595–2610.
18. Tamehiro, H., et al., "Radiative Heat Transfer by Flowing Multiphase Medium-Part III. An Analysis on Heat Transfer of Turbulent Flow in a Circular Tube," *International Journal of Heat and Mass Transfer,* Vol. 16 (1973), pp. 1199–1213.
19. Tien, C. L., "Heat Transfer by a Turbulent Flowing Fluid-Solid Mixture in a Pipe," *Journal of Heat Transfer,* Vol. 83-C (1961), pp. 183–188.
20. Pearce, B. E., and Emery, A. F., "Heat Transfer by Thermal Radiation and Laminar Forced Convection to an Absorbing Fluid in the Entry Region of a Pipe," *Journal of Heat Transfer,* Vol. 92-C (1970), pp. 221–230.
21. Reichardt, H., *Angew. Math. Mech.,* Vol. 31 (1951), p. 208.
22. Sparrow, E. M., and Cess, R. D., *Radiation Heat Transfer,* Belmont, Calf.: Brokks/Cole (1966), p. 267.
23. Heaslet, M. A., and Warming, R. F., "Theoretical Prediction of Radiative Transfer in a Homogeneous Cylindrical Medium," *Journal of Quantum, Spectroscopic and Radiative Transfer,* Vol. 6 (1966), pp. 751–774.
24. Hasegawa, S., et al., "Experimental Study on Forced Convective Heat Transfer of Flowing Gaseous Solid Suspensions at High Temperature," *International Journal of Multiphase Flow,* Vol. 9, No. 2 (1983), pp. 131–145.

25. Mori, Y., and Nakayama, W., "Study on Forced Convective Heat Transfer in Curved Pipes," *International Journal of Heat and Mass Transfer,* Vol. 8 (1965), pp. 67–82.
26. Shimizu, A., et al., "Impinging Jet Heat Transfer with Gaseous-Solid Suspension Medium," *Advances in Enhanced Heat Transfer,* Symposium Volume of the 18th National Heat Transfer Conference, San Diego: Aug. 1979, pp. 155–160.
27. Babcock and Wilcox Comp., "Final Report on the Gas-Suspension Project," AEC BAW-1159, 1959.
28. Sze, D. K., et al., "A Gas-Carried Li_20 Cooling/Breeding Fusion Reactor Blanket Concept," *Transactions of ANS,* Vol. 22 (Nov. 1975), p. 21.

CHAPTER 16

HEAT TRANSFER IN GAS-SOLID FLUIDIZED BEDS

W. E. Genetti

Chemical Engineering Department
University of Mississipi
University, Mississippi

CONTENTS

INTRODUCTION

The term "fluidization" was coined to described a particular mode of contacting granular solids with fluids. The response of the bed to upwards flows of gas through it is shown on Figure 1. When a gas is passed up through a bed of granular material, there is a flow rate at which the solids are suspended. This state is known as minimum fluidization. At this point the fluid/particle system begins to behave like a fluid and will flow under a hydrostatic head. The pressure drop across the bed will be equal to the weight of the bed. This pressure drop will be exceeded prior to minimum fluidization because residual packing and interlocking of particles within the bed must be overcome.

At higher flow rates, bubbles will be formed and at still higher flow-rates entrainment flow will occur. The bubbles are responsible for particle circulation within a fluidized bed.

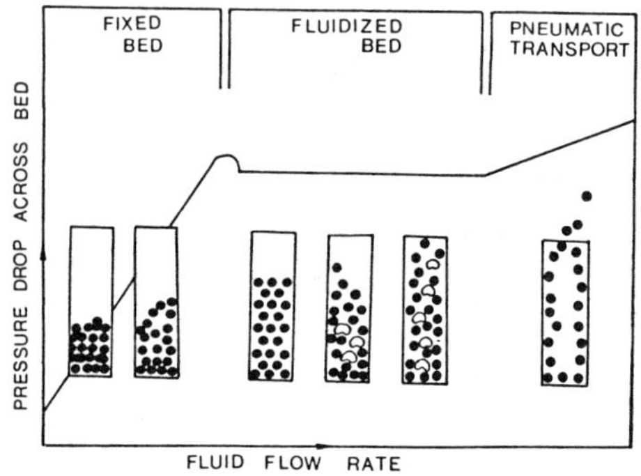

Figure 1. Stylized representation of bed response to gas flow (adapted from Botterill [49]).

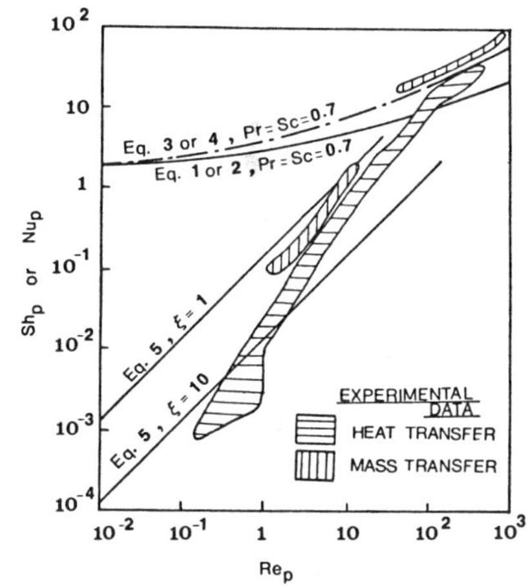

Figure 2. Gas-solid heat and mass transfer in fixed beds.

This circulation has a direct bearing on the heat transfer properties of gas-solid fluidized-bed systems.

However, in many large industrial reactors, where fluidizing velocities are considerably in excess of the minimum fluidization velocities, there are conditions of very violent agitation. The resemblence between the behavior of such beds and the description of the bed shown on Figure 1 may be minimal.

Fluidized beds have been used for heat exchange, in both physical operations and chemical processes, because rapid transport of heat can be obtained. Also because of rapid mixing, it may be possible that a uniform temperature can be maintained throughout the bed.

There are two aspects of fluidized-bed heat transfer–gas-particle heat transfer and surface-bed heat transfer. This chapter will deal primarily with gas-particle heat transfer.

GAS/PARTICLE HEAT TRANSFER

Applications of gas-particle heat transfer in fluidized beds have become numerous. These applications usually fall into three categories: heat transfer operations with no mass transfer; operations involving simultaneous heat and mass transfer, and processes requiring heat transfer to induce, sustain, or control a chemical reaction. The later category almost always involves simultaneous mass transfer.

The cooling of a granular material by direct contact with a gas may be a case of only heat recovery. Whereas the drying of feed grains involves simultaneous heat and mass transfer. The drying and devolatilization of coal in a fluidized bed will involve heat and mass transfer in addition to chemical kinetics.

The heat or mass transfer relationships for a single sphere moving through a fluid at a velocity of U_0 and gas-particle heat or mass transfer relationships for a fixed bed are often used as references in fluidized-bed systems.

Single Sphere

For a single sphere moving at a relative velocity U_0 through a fluid, the Sherwood Number, Sh_p, is given by Froessling [1] as

$$Sh_p = \frac{K_d d_p y}{\mathscr{D}} = 2.0 + 0.6\, Re_p^{1/2} Sc^{1/3}$$

$$Sc = \frac{\mu_g}{\rho_g \mathscr{D}} \tag{1}$$

$$Re_p = \frac{d_p U_0 \rho_g}{\mu_g}$$

where K_d = the mass transfer coefficient
d_p = the sphere diameter
y = the logarithmic mean fraction of nondiffusing component
$\mathscr{D}$ = the diffusivity
Sc = the Schmidt number
Re_p = the Reynolds number

The Ranz and Marshall [2] expression is a similar equation for heat transfer between a fluid and a single sphere; that is

$$Nu_p = \frac{h_p d_p}{k_g} = 2 + 0.6\, Re_p^{1/2} Pr^{1/3} \tag{2}$$

$$Pr = \frac{\mu_g C_g}{k_g}$$

where Nu_p is the Nusselt number and Pr is the Prandl number.

Equations 1 and 2 are shown on Figure 2. The limiting Nusselt or Sherwood number of 2 results from heat conduction or mass diffusion through a stagnant fluid.

FIXED BEDS

For fixed beds experimental Sherwood and Nusselt numbers are in good agreement for Reynolds numbers greater than about 100. Ranz [3] reports the following equation for the Sherwood number.

$$Sh = 2 + 1.8\, Re_p^{1/2}\, Sc^{1/3} \tag{3}$$

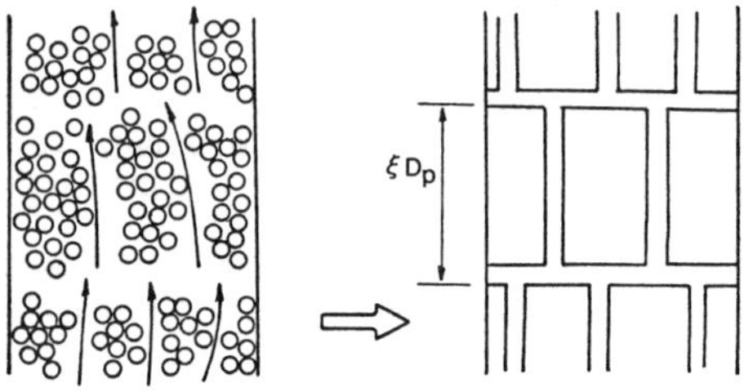

Figure 3. Channeling model [11].

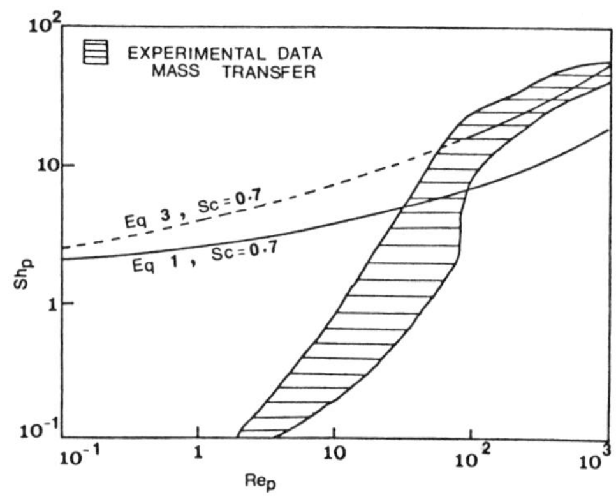

Figure 4. Gas-solid mass transfer in fluidized beds.

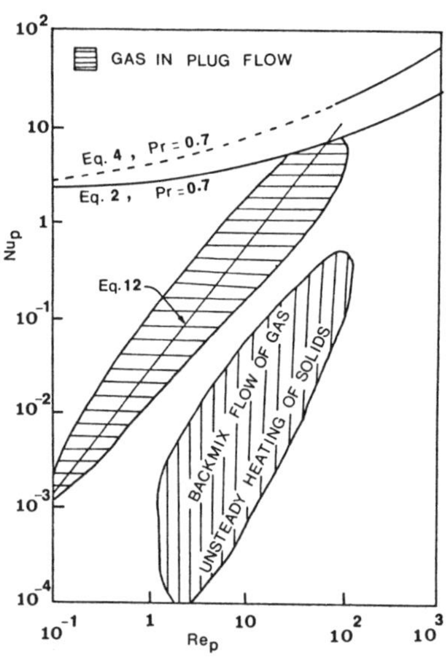

Figure 5. Gas-solid heat transfer in fluidized beds.

A similar expression is given by Kunii and Levenspiel [4] for heat transfer:

$$\begin{aligned} &\mathrm{Nu_p} = 2 + 1.8\,\mathrm{Re_p^{1/2}}\,\mathrm{Pr^{1/3}} \\ &\mathrm{Re_p} > 100 \end{aligned} \tag{4}$$

Equation 3 and 4 are also shown on Figure 2. For Reynolds numbers less than 30 in a fixed bed (usually with fine solids) it is difficult to determine a mass transfer coefficient for

Table 1
Experimental Investigation of Gas-Solid Heat Transfer in Fluidized Beds

Investigators	Measurement Technique	Assumed Gas Flow	Bed Dia. (cm)	Particle Dia. (mm)	Re_p	Pr	Nu_p
Kettenring et al. [12]	Steady-state	Plug	5.85	0.36–1.0	10–57	0.72	0.2–2.2
Walton et al. [15]	Steady-state	Plug	10.2	0.36–0.71	10–32	0.68	0.5–5
Heertjes & McKibbons [16]	Steady-state	Plug	6	0.36–1.1	8–52	0.73	0.5–5
Frantz [17]	Steady-state	Plug	10.2	0.23	1.3–2.8	0.65	0.02–0.06
Donnadieu [10]	Unsteady-state	Plug	20	0.09–1.1	0.1–100	0.70	2×10^{-3}–3
Wamsley & Johanson [19]	Unsteady-state	Backmix	5.93	0.14–0.91	1–40	0.70	10^{-3}–0.17
Fritz [20]	Unsteady-state	Backmix	40.6	0.06	1.2–13	0.70	6×10^{-5}–1.6×10^{-3}
Ferron [21]	Unsteady-state	Backmix	7.6–20	0.06	0.8–19	0.70	2×10^{-5}–8×10^{-3}
Yoshida [22]	Unsteady-state	Backmix	11	0.16–1.1	4–100	0.70	0.002–0.3
Sato et al. [23]	Steady-state with continuous feed and withdrawal of solids	Backmix	8	0.16–0.64	3–40	0.70	0.03–1.5

transfer between the gas and the particle because of the large specific surface area of fine particles. Reliable data is wanting. The data that does exist [5] would indicate that the Sherwood number is less than two for Reynolds numbers less than ten. For a long time it was thought physically impossible to have a Sherwood number less than the limiting value of two. That is it was thought that the most difficult situation is diffusion of mass through a stagnant gas.

Heat transfer data reported by numerous investigators [6–9] using a variety of techniques show the same trends as mass transfer. That is, as shown in Figure 2, Nusselt numbers are less than two for Reynolds numbers less than 30. A channeling model proposed to explain the apparent anomaly has been proposed by Kunii and Suzuki [11]. The concept shown on Figure 3 is the basis for the channeling model. The adjustable parameter in the model is the channel length factor, ξ, which is the channel length in particle diameters. The result of the model is Equation 5:

$$Nu_p = \frac{\varphi_s}{6(1-\varepsilon)\xi} Re_p Pr \tag{5}$$

where φ_s is the particle shape factor.

Equation 5 is also shown in Figure 2 for $\xi = 1$ and $\xi = 10$. This is the range of values reported for ξ by Grootenhuis [7] and Eichhorn [8] for particles less than 1 mm in diameter.

FLUIDIZED BEDS

Heat and mass transfer gas-particle coefficients are difficult to measure in gas fluidized beds for several reasons. Like fixed beds, large surface areas are involved which give rise to rapid attainment of thermal or concentration equilibrium between phases. The bubbling behavior of a fluidized bed gives rise to problems dealing with gas bypassing, sampling, representative temperature measurements, and temperature or concentration driving forces that approach zero.

Despite these difficulties data have been reported for both heat and mass transfer. Mass transfer data for several investigators is shown in Figure 4. For Reynolds numbers less than 100 the Sherwood numbers are less than those predicted by Equation 1 for single spheres. The data of Resnick and White [5] and Kettenring et al. [12] demonstrates the anomaly. The data [5, 13, 14] above a Reynolds number of about 100 show Sherwood numbers greater than the single-sphere curve and less than the fixed-bed results.

Reported [10, 12, 15–17, 19–21, 23] results for heat transfer as shown in Figure 5, vary several orders of magnitude. The large differences between investigators demonstrate the difficulty of the measurement as well as differences in assumed flow models. However, the trends are similar for heat and mass transfer. That is, the Nusselt number drops off rapidly for Reynolds numbers less than 100. Table 1 shows assumptions made by investigators.

In the calculation of the heat transfer coefficient, an accurate measurement of the temperature driving force is important. However, some investigators have assumed a bare thermocouple measures the solids temperature while others assume the bare thermocouple is the gas temperature.

Most experimental measurements of gas-particle heat transfer coefficients have involved either steady-state or unsteady-state experiments.

Steady-State Experiment

In this type of experiment, energy is added or removed from the bed so to keep the bed solids at a uniform temperature. Energy is added if the fluidizing gas is to be heated and removed if it is to be cooled.

The gas temperature is only different from the solids temperature in most cases near the distributor plate. Zabrodsky [24] gives examples showing that the fraction of the temperature difference retained after the gas has flowed past the first row of particles is usually small. Most investigators have assumed perfect backmixing of the solids and measure the solid temperature with bare thermocouples at the exit. At the exit the difference between the solid and gas temperatures is negligible. Both bare and suction thermocouples have been used to attempt to measure the gas temperatures at locations near the distributor. The bare thermocouple will give readings between the solids and gas temperatures. Suction thermocouples will at best give a time-average gas temperature. The bubble gas temperature will be higher than the time average temperature for a gas-cooling experiment. Figure 6 qualitatively shows the bare thermocouple reading, the suction thermocouple reading, the gas bubble temperature, and the solid temperature as a function of position. Backmixing of solids all the way back to the distributor is usually assumed. This may not be the case in reality.

Unsteady-State Experiment

With the unsteady-state experiments, the bed solids are heated with a hot gas. The inlet gas temperature is held constant. The outlet gas temperature is measured as a function of time. Complete gas backmixing in the bed has been assumed by most investigators using this technique. Therefore, the energy transferred from the gas to the particle can be expressed as

$$G_0C_g(T_{gi} - T_{g0}) = h_p a L_f (T_{g0} - T_s) \tag{6}$$

differentiating

$$\frac{dT_s}{dt} = \frac{h_p a L_f + G_0 C_g}{h_p a L_f} \frac{dT_{g0}}{dt} \tag{7}$$

the rate of energy transferred to the solids is

$$h_p a A L_f [T_{g0} - T_s] = W C_s \frac{dT_s}{dt} \tag{8}$$

substituting Equation 6 and Equation 7 into Equation 8

$$\frac{h_p a A L_f G_0 C_g (T_{gi} - T_{g0})}{h_p a L_f} = W C_s \frac{[h_p a L_f + G_0 C_g]}{h_p a L_f} \frac{dT_{g0}}{dt} \tag{9}$$

integrating

$$-\frac{[h_paAL_fG_0C_g]t}{WC_s[h_paL_f + G_0C_g]} = \ln\frac{\Delta T_g}{\Delta T_{ginitially}} \tag{10}$$

where $\Delta T_g = T_{g0} - T_{gi}$

At least one investigator [10] using the unsteady-state setup has used a plug flow assumption for the gas. This assumption results in an exponential gas temperature decay with distance from the distributor. It is possible to have a long flat tail of an exponential decay which would result in uniform gas temperature over much of the bed. In fact, this model is probably much closer to reality than complete backmixing of the gas. With the assumption of plug flow of the gas Equation 11 can be developed.

$$\ln\frac{\Delta T_g}{\Delta_{T_{ginitial}}} = -\frac{G_0C_gA}{WC_s}\left[1 - \exp\left(-\frac{h_paL_f}{G_0C_g}\right)\right]t \tag{11}$$

For either plug flow or complete backmixing h_p can be calculated from Equation 11 or 10 respectively if ΔT_g is measured as a function of time.

The backmixing and plug flow assumptions for the gas leads to different temperature driving forces and different calculated coefficients. Neither assumption fits real flow perfectly; however, the results from the plug flow assumption of the gas flow for both the steady- and unsteady-state techniques follow a consistent pattern that is independent of bed height. The backmixing assumption for the gas leads to coefficients that are functions of bed height, which suggests that the plug flow assumption for the gas is more meaningful.

Equation 12 correlates [4] Nusselt numbers determined with the assumption of plug flow of the gas for Reynolds numbers less than 100.

$$Nu_p = 0.03\,Re_p^{1.3} \tag{12}$$

Equation 12 was developed by Kunii and Levenspiel [4] from data reported by investigators who used small-scale fluidized beds (see Table 1). Kato and Wen [25] have stressed that such results are limited to small beds with small bubbles. That is, the use of Equation 12 may over predict heat transfer coefficients in industrial-scale beds where large bubbles may exist immediately above the distributor. The bubble size and bubble dynamics can greatly affect the apparent heat transfer coefficient.

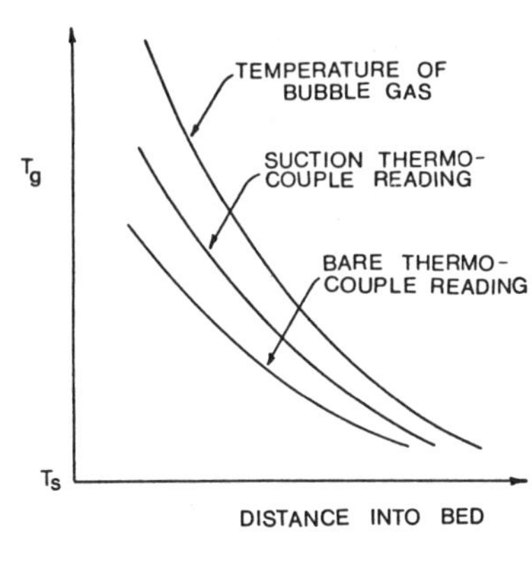

Figure 6. Interpretation of temperature measurements [4].

In order to predict gas solid heat transfer coefficients for large-scale equipment and to explain the anomaly as to why Nusselt numbers can be less than two, an understanding of the gas-solid heat transfer mechanism is needed.

PHENOMENA THAT AFFECT GAS-SOLID HEAT TRANSFER

Gas fluidized beds have been considered [25] to consist of a bubble and an emulsion phase. The bubble phase comprises regions of the bed with low solid density, and the emulsion phase is regions of higher solid density. To begin to have insight into the mechanisms of gas-solid heat transfer, it is necessary to have some insight into the gas and solid dynamics of both phases.

Bubble Size

Bubble size at the distributor is usually small, and it is a function of distributor type and orifice size. Bubbles coalesce, grow, and speed up as they move up the bed. It is well-known [26] that the mechanism of coalescence involves the capture of one bubble in the wake (low pressure region) of another from the rear. At the same bed level all bubbles are about the same size. Figure 7 shows [27] the bubble diameter as a function of U_0/U_{mf} and the distance from the distributor. The information shown in Figure 7 is for a bed diameter of 40 cm. In commercial beds with adequate internals, the bubble growth is controlled. In this case bubble size can be considered constant except near the distributor.

Mori and Wen [28] have reported that the maximum attainable bubble diameter, D_{bm}, due to total coalescence of bubbles is given by Equation 13:

$$D_{bm} = 0.64\,[A_t\,(U_0 - U_{mf})]^{2/5} \tag{13}$$

where A_t is in cm^2 and $U_0 - U_{mf}$ is in cm/sec and D_{bm} is in cm.

Mori and Wen [28] further express the bubble diameter D_b as an exponential function of distance from the distributor, x, and the bed diameter, D_t, as shown by Equation 14:

$$\frac{D_{bm} - D_b}{D_{bm} - D_{b0}} = \exp\left(-\frac{0.3\,h}{D_t}\right) \tag{14}$$

where D_{b0}, the initial bubble diameter, is given by Equation 15 and 16:

$$D_{b0} = 0.347\left[\frac{A_t\,(U_0 - U_{mf})}{n_d}\right]^{2/5} \tag{15}$$

for a perforated plate distributor or

$$D_{b0} = 0.00376\,(U_0 - U_{mf})^2 \tag{16}$$

for a porous plate distributor D_{b0} is in cm, A_t in cm^2, and $(U_0 - U_{mf})$ in cm/sec.

The constant 0.3 in Equation 14 was determined under the following experimental conditions:

$$0.5 \leqslant U_{mf} \leqslant 20\ \text{cm/sec}$$

$$0.06 \leqslant d_p \leqslant 0.045\ \text{cm}$$

$$U_0 - U_{mf} \leqslant 48\ \text{cm/sec}$$

$$D_t \leqslant 130\ \text{cm}$$

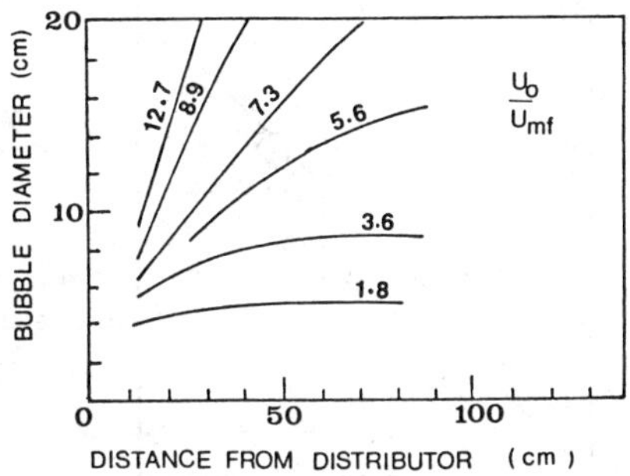

Figure 7. Bubble sizes at different bed heights (adapted from Kunii and Levenspiel [4]).

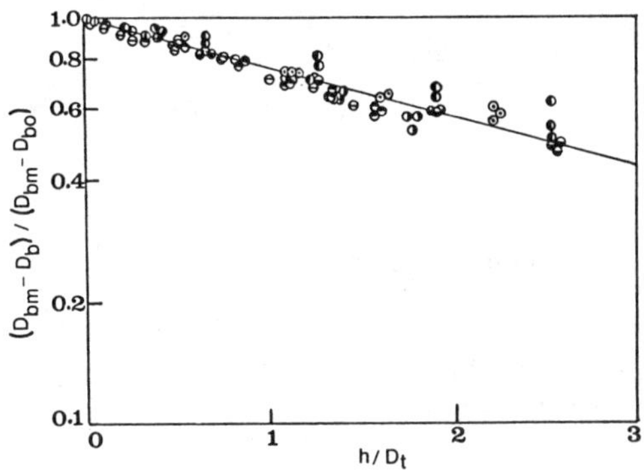

Figure 8. Dimensionless bubble diameter versus height-to-bed diameter [28].

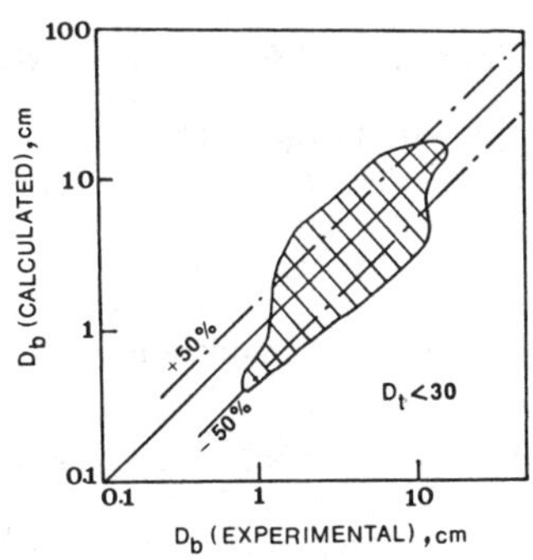

Figure 9. Comparison of calculated and experimental bubble diameters for beds less than 30 cm in diameter.

Figure 8 shows Equation 14 with the data of Rowe [29] and Werther [30] which cover a wide range of operating conditions. The value of 0.3 was determined statistically from more than 200 data points from various investigators as shown in Table 2. Figures 9 and 10 show that most of the experimental bubble diameters are within ±50 percent of values calculated from Equation 14.

Darton et al. [26] have also developed an expression for bubble diameter growth in a fluidized bed with a perforated distributor. Their expression is limited to nonslugging beds and is given by the following equation:

$$D_b = 0.54\,(U_0 - U_{mf})^{2/5}\,(h + \sqrt{A_0})^{4/5}/g^{1/5}$$

where A_0 is the catchment area for a bubble at the distributor plate.

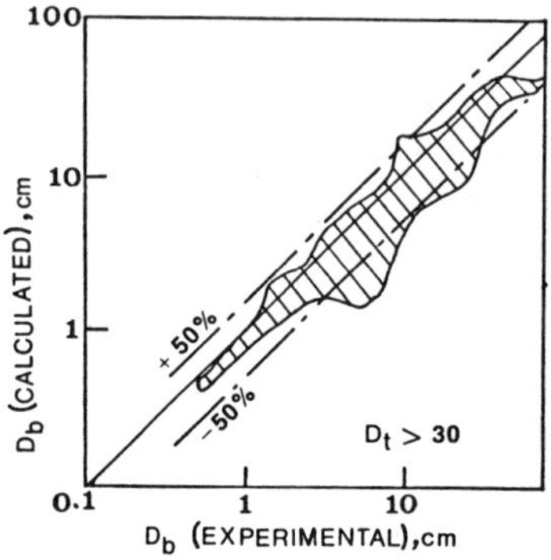

Figure 10. Comparison of calculated and experimental bubble diameters for beds greater than 30 cm in diameter.

Bubble Velocity

Davidson and Harrison [31] have reported the bubble rise velocity, U_B, as

$$U_B = U_0 - U_{mf} + 0.71\sqrt{gD_b} \tag{17}$$

Since the bubble diameter, D_b, increases with bed height (except with beds which contain internals), the bubble rise velocity, U_b, will increase with bed height.

Gas and Solid Circulation Within and Around a Bubble

Figure 11 shows Murray's model [32] of gas and solid circulation. The fraction of solids within the bubble has been considered negligible. The sketch shows the bubble, the cloud and wake, and the emulsion phase. Solids flow downward through the cloud and gas circulates upward through the bubble and then downward through the cloud. The wake is a low pressure region which will carry solids behind the bubble as the bubble rises. The wake is a major mechanism for solids mixing.

Voidage of Bubble and Emulsion

Experimental evidence [4] suggests that the solid fraction within the bubble is small and negligible except for heat and mass transfer calculations. If δ is the volume fraction of bubbles in the bed, then the average bed voidage, ε_f, is related to the voidage in the bubbles and emulsion by

$$\varepsilon_f = \delta\,\varepsilon_b + (1 - \delta)\,\varepsilon_e \tag{18}$$

The void fraction in the emulsion should be near the voidage at minimum fluidization; that is

$$\varepsilon_e \cong \varepsilon_{mf} \tag{19}$$

Substitution of Equation 19 into Equation 20 and rearranging gives

$$\varepsilon_{mf} = \frac{\varepsilon_f - \delta\varepsilon_b}{1 - \delta} \tag{20}$$

Table 2
Summary of Experimental Conditions for Bubble Size Measurement

Investigator	D_t (cm)	Solid Particles	d_p (cm)	U_{mf} (cm/s)	U_0/U_{mf}	Distributor, n_d
Werther [30]	20, 100	Quartz sand	0.0083	1.8	5	Porous plate
Geldart [35]	31	Sand	0.0128	1.2	2.6–7.7	Perforated Plate, 3,100
Rowe [29]	30 × 20	Alumina	0.021	2.54	1.25–2.5	Porous plate
		Carbon	0.0296	8.0	1.3–1.7	Porous plate
		Quartz	0.0135	2.75	2.2–6.6	Porous plate
		Ballotini	0.0325	8.0	1.6–2.4	Porous plate
		Glass powder	0.0268	5.5	1.7–2.7	Porous plate
Whitehead [36]	61 × 61	Silica sand	0.015	2.5	1.8–6.9	Tuyere, 4
	61 × 61	Silica sand	0.015	2.5	2.8–6.6	Tuyere, 16
	122 × 122	Silica sand	0.015	2.5	3.2–6.2	Tuyere, 64
	122 × 122	Silica sand	0.015	2.5	2.1–6.3	Tuyere, 16
Kunii [37]	20	M.S. cat.	0.015	2.0	9.5	Perforated plate, 79
	40	M.S. cat.	0.015	2.0	1.5–2.5	Perforated plate, 314
Yasui [38]	10.2	Glas beads	0.0242	7.56	1.5–2.5	Porous plate
	10.2	Glass beads	0.175	4.7	1.5–2.7	Porous plate
	10.2	U.O.P. cat.	0.006	0.418	2–10	Porous plate
	10.2	Coal	0.045	19.4	1.5–1.75	Porous plate
Toei [39]	10 × 10	Glass beads	0.0137	2.25	1.5–4.0	Porous plate
Kobayashi [40]	10	Crush silica	0.021	2.85	2–9.7	Perforated plate, 1,850
Miwa [41]	15	Sand	0.016	2.4	3.1–5.2	Perforated plate, 37
Tomita [42]	21	Sand	0.02	4	4.25	Perforated plate, 184
	38	Sand	0.02	4	4.25	Perforated plate, 575
	60	Sand	0.02	4	4.25	Perforated plate, 1,450
Baumgarten [43]	7.6	Glass beads	0.0074	0.72	2–84	Porous plate
Park [44]	10	Coke	0.0086	0.63	4–10	Porous plate
	10	Coke	0.0156	1.83	1.5–6	Porous plate
	10	Coke	0.034	6.8	1.5–3	Porous plate
Botton [45]	50	Sand	0.0071	1.0	2–15	Perforated plate, 78
Fryer [46]	23	Sand	0.0071	1.7	1.47	Bubble cap, 61

where the fraction emulsion is

$$1 - \delta = \frac{L_{mf}}{L_f} = \frac{\rho_f}{\rho_{mf}} = \frac{1 - \varepsilon_f}{1 - \varepsilon_{mf}} \tag{21}$$

now $\varepsilon_b \approx 1$ (solid fraction in the bubble is very small); therefore

$$\varepsilon_{mf} \approx \frac{\varepsilon_f - \delta}{1 - \delta} \tag{22}$$

For heat or mass transfer calculations the volume fraction of solids in the bubble, γ_b can be important. Experimental findings show that γ_b varies from 0.001 to 0.01

where $\gamma_b = \dfrac{\text{volume of solids in bubbles}}{\text{volume of bubbles}}$

$$\gamma_c = \frac{\text{volume of solids in clouds and wakes}}{\text{volume of bubbles}}$$

$$\gamma_e = \frac{\text{volume of solids in emulsion}}{\text{volume of bubbles}}$$

Kunii and Levelspiel [4] shows how γ_c and γ_e can be estimated.

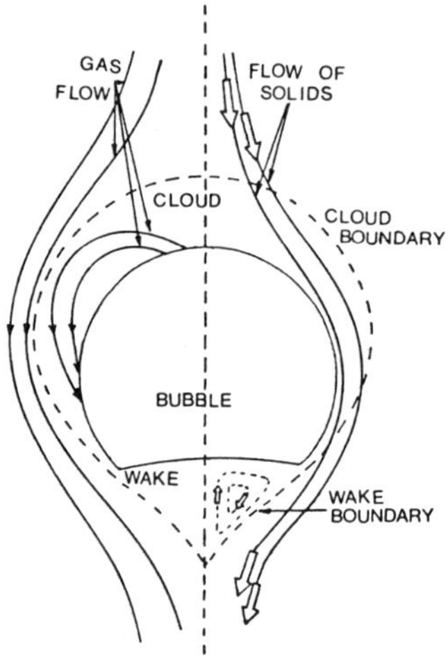

Figure 11. Murray's model of a fast two-dimensional bubble, cloud and wake [32].

GAS-SOLIDS HEAT TRANSFER COEFFICIENTS FROM HEAT TRANSFER MODELS

Kunii and Levenspiel's [4, 27] model was the first model which both quantitatively and qualitatively explained the heat and mass transfer anomaly (i.e. why Nusselt Numbers can be less than two) and also provided a method to predict heat or mass transfer coefficients in large beds with large bubbles. This model equates the energy transfer from the gas in the bubble to the energy taken up by the solids in the bubble and the energy transferred to the cloud. It was assumed that the heat transferred from the cloud to the emulsion is negligible. The gas flows upward through the bubble and downward through the cloud and then up through the wake as shown in Figure 11. Solids move downward through the cloud and bubble. The solids in the bubble are considered to be at a relative velocity between the gas and the solids equal to terminal velocity of the solids.

The model is as follows:

$$\text{heat loss by gas in bubble} = \text{heat taken up by solids in bubble} + \text{heat transferred to cloud where it is all absorbed}$$

$$U_b\rho_g C_g \frac{dT_{gb}}{dx} = \gamma_b h_{pt} a' (T_{gb} - T_s) + H_{bc}(T_{gb} - T_s) \tag{23}$$

$$T_{gapp} - T_s = \delta T_{gb} + (1 - \delta) T_{ge} - T_s \tag{24}$$

Gas temperature in the emulsion phase is approximately the solids temperature

$$T_s \approx T_{ge} \tag{25}$$

$$T_{gapp} - T_s = \delta [T_{gb} - T_s] \tag{26}$$

$$\frac{dT_{gapp}}{dx} = \delta \frac{dT_{gb}}{dx} \tag{27}$$

$$\frac{h_{app} d_p}{k_g} = Nu_{app} = \frac{d_p U_0 \rho_g C_g \dfrac{dT_{gapp}}{dx}}{k_g a (T_{gapp} - T_s)} = \frac{d_p U_0 \rho_g C_g \dfrac{dT_{gb}}{dx}}{a k_g (T_{gb} - T_s)} \tag{28}$$

substitution of Equation 23 into Equation 28 gives

$$Nu_{app} = \frac{U_0}{U_b}\frac{a'}{a}\left[\gamma_p Nu_{pt} + \frac{\varphi_s d_p^2 H_{bc}}{6\,k_g}\right] \tag{29}$$

where $a' = \frac{\varphi_s d_p}{6}$

$$a = \frac{\varphi_s d_p}{6(1-\varepsilon)}$$

if $U_0 \approx U_b$, then

$$Nu_{app} = \frac{1}{1-\varepsilon}\left[\gamma_p Nu_{pt} + \frac{\varphi_s d_p^2 H_{bc}}{6\,k_g}\right] \tag{30}$$

Nu_{pt} is calculated from Equation 2, the single-sphere correlation, using the terminal velocity of the solids in the Reynolds Number; that is

$$Re_{pt} = \frac{U_t d_p \rho_g}{\mu_g}$$

$$Nu_{pt} = 2 + 0.6\,Re_{pt}^{1/2}\,Pr^{1/3}$$

Equation 2 is justified since the particles are very dilute in the bubble phase.

The volumetric heat transfer coefficient, H_{Bc}, is given [4] by

$$H_{Bc} = 4.5\left(\frac{U_{mf}\rho_g C_g}{D_b}\right) + \frac{5.85\,(k_g \rho_g C_g)^{1/2} g^{1/4}}{D_b^{5/4}} \tag{31}$$

For the case of a commercial bed with internals, the bubble size is limited and can be considered constant (except in regions close to the distributor) and estimated from the geometry of the internals. For this situation the estimated bubble diameter can be substituted into Equation 31 and the gas-solid heat transfer coefficient can be determined from Equation 30.

For cases with no internals, an average gas-solid heat transfer coefficient can be estimated by using an appropriate average bubble diameter, such as

$$D_{bav} = D_{bm} - (D_{bm} - D_{b0})\,(1 - e^{-\frac{0.3\,Lf}{Dt}})\,\frac{D_t}{0.3\,L_f} \tag{32}$$

Local gas-solid heat transfer coefficients could be calculated by using Equation 14 to calculate local bubble diameters and then using Equation 31 to calculate local values of H_{Bc}. Equation 30 would then be used to calculate the local gas-solid heat transfer coefficient.

Figure 12 shows Equation 30 compared with the experimental data of Kettenring et al. [12] and Heertjes and McKibbons [16]. The details of the conditions of those experiments are shown in Table 1. Kunii and Levenspiel's model provides a method if estimating gas-solid heat transfer coefficients in large commercial beds. The use of Equation 12 for such beds may lead to large errors [28].

Kato and Wen [25, 55] have developed the bubble assemblage model which is shown in Figure 13, the bed is broken up into vertical steps with transfer from the bubble to the emulsion and from the emulsion to the bubble at each step. The volume of gas transferred

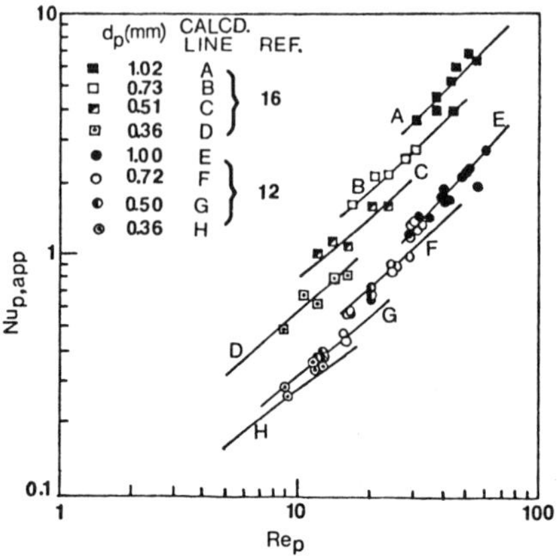

Figure 12. Comparison of calculated lines with experimental data using the bubbling-bed model [4].

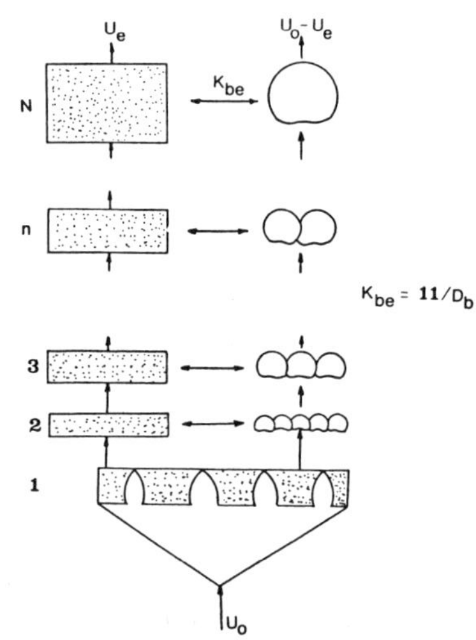

Figure 13. Bubble assemblage model [55].

from the bubble to the emulsion or from the emulsion to the bubble per unit time and volume of bubbles in the bed, K_{be}, is given [33] by

$$K_{be} = 11/D_b \tag{33}$$

Calculated gas-solid heat transfer coefficients from the bubble assemblage model are compared to experimental coefficients in Figure 14. Except for the data of Juveland et al. [34] agreement is within 100 percent.

PARTICLE-TO-PARTICLE HEAT TRANSFER

In some applications, such as fluidized-bed combustion of coal, all solid particles are not at the same temperature. During fluidized-bed coal combustion usually about 2 percent of the particles are coal. The remaining particles are limestone/dolomite or an inert material. Most of the coal particles at a given time have been devolatilized and are hot char particles. Heat transfer from these particles is by convection to gas phase, by particle-to-particle heat transfer, and by radiant transfer to other particles.

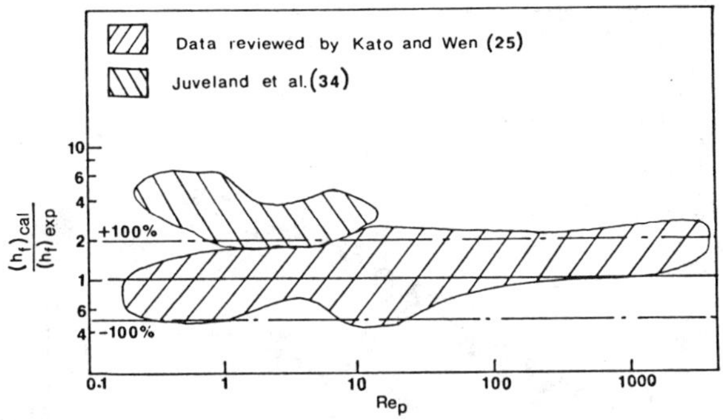

Figure 14. Comparison of experimental heat transfer coefficients with calculated values from the bubble assemblage model [25].

Wen and Chang [48] studied particle-to-particle heat transfer under conditions where radiant heat transfer could be neglected. In their experiments the bed was heated to a uniform temperature by an inlet air stream. A particle, with an embedded thermocouple, was inserted in the bed. The inserted particle temperature was followed as it came to equilibrium with the bed. The sum of the gas-solid and particle-to-particle heat transfer coefficient, h_{ba} was in this way determined.

These workers visualized a model in which the particles were classified into two groups–a and b. The two groups are at different temperatures. The fraction, f_b, is the average fraction of group b particles surrounding a group a particle. The particle-to-particle heat transfer coefficient is defined as

$$(q_p)_{ba} = (h_p)_{ba}(T_{pa} - T_{pb}) \tag{34}$$

and

$$(h_p)_{ba} = f_b(\bar{h}_p)_{ba} \tag{35}$$

where T_{pb} and T_{pa} are the temperatures of group b and a particles respectively and $(\bar{h})_{ba}$ is the particle-to-particle heat transfer coefficient when all the contacting particles are group b particles; that is $f_b = 1$.

The sum of the gas-solid and particles-to-particle components is

$$q_{ba} = q_f + (q_p)_{ba} = h_g(T_{pa} - T_g) + (h_p)_{ba}(T_{pa} - T_{pb}) \tag{36}$$

If the group a particles are dilute, as would be expected for coal particles in a fluidized-bed combustor, f_b will be nearly 1. Under these conditions Equation 36 will become

$$q_{ba} = h_g(T_{pa} - T_g) + (\bar{h}_p)_{ba}(T_{pa} - T_{pb}) \tag{37}$$

If the total heat transfer coefficient is defined as

$$h_{ba} = h_g + (\bar{h}_p)_{ba} \tag{38}$$

Then the relationship for a single a particle surrounded by b particles is obtained

$$q_{ba} = h_{ba}\left[T_{pa} - \frac{h_g}{h_{ba}}T_g - \left(1 - \frac{h_g}{h_{ba}}\right)T_{pb}\right] \tag{39}$$

Wen and Chang [48] estimated, depending on the gas flow rate and particle size, that from 10 to 35 percent of the total heat transfer was by a particle-to-particle mode. The combined

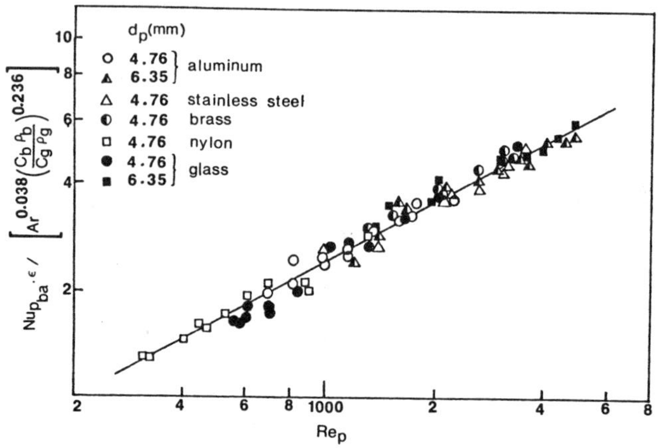

Figure 15. Combined gas-solid and particle-to-particle heat transfer in gas fluidized beds [48].

heat transfer coefficient, h_{ba}, is correlated in Figure 15. The data shown include aluminium, brass, and stainless steel particles of two sizes. The particle thermal conductivity is not a variable in the correlation. This raises the question: Is the enhancement due to particle-to-particle conduction? Botterill [49] suggests that the measurement is probably some measure of zones temporarily inaccessible to fluid pentration, since contact areas are too small for significant heat transfer by conduction.

At higher temperatures radiant heat transfer from hot particles to the colder particles in the bed can be significant. The radiant heat transfer from the group a particles to the group b particles can be estimated by

$$q_R = \varepsilon_a h_r (T_{pa} - T_{pb}) \tag{40}$$

where

$$h_r = \sigma (T_{pa}^4 - T_{pb}^4)/(T_{ba} - T_{pb}) \tag{41}$$

the total heat transfer is

$$q_T = q_R + q_{ba} \tag{42}$$

In the development of Equation 41 among other things is the assumption that the gas is transparent to radiation. This may not be the case in a combustor with high CO_2 and H_2O concentrations.

Fluidized bed combustors are usually operated [50] at about 800° C. The total emissivity of hot coal char is between [51] 0.8 and 0.95. The air temperature in the bed will be near the bed temperature except near the distributor. Pillai [52] has reported char temperatures during combustion are from 40° to 120° C higher than the bed temperature. Agarwal [51] and Ross and Davidson [53] have observed similar results.

Example Problem

For a fluidized-bed combustor operating at the following conditions

$$p = 1 \text{ atm}$$
$$T_{bed} = 800° \text{C}$$

$d_p = 4.75\,\text{mm}$, limestone particles
$G_0/G_{mf} = 2$

Fluidized with air

$\varepsilon_{char} = 0.9$ (emissivity)
$\varepsilon_{bed} = 0.45$ (void fraction)

calculate the rate of heat transfer from a burning char particle to the bed (gas and limestone particles) for a char surface temperatures of 840° C and 920° C. Also calculate the fraction of the heat transfer by radiation in both cases. Assume the char particle diameter is the same as the limestone diameter. From Leva [54]

$$G_{mf} = 0.00125 \frac{d_p^2 (\rho_s - \rho_g)^{0.9} \rho_g^{1.1} g}{\mu_g}$$

$$\rho_g = 0.32 \frac{\text{kg}}{\text{m}^3};\ \rho_s = 1680 \frac{\text{kg}}{\text{m}^3};\ \mu_g = 4.46 \times 10^{-5} \frac{\text{kg}}{\text{ms}}$$

$$G_{mf} = 0.00125 \frac{(.00475)^2 (1{,}680)^{0.9} (0.32)^{1.1} (9.8)}{4.46 \times 10^{-5}} = 1.45 \frac{\text{kg}}{\text{m}^2 - \text{s}}$$

$$G_0 = 2\,G_{mf} = 2.9 \frac{\text{kg}}{\text{m}^2\text{s}}$$

$$\text{Re}_p = \frac{g_0 d_p}{\mu_g^2} = \frac{(2.9)(.00475)}{4.46 \times 10^{-5}} = 309$$

$$\text{Ar} = \frac{G_0 d_p^3}{\mu_g} \rho_g (\rho_s - \rho_g) = \frac{(9.8)(.00475)^3 (0.32)(1680)}{(4.46 \times 10^{-5})^2} = 2.8 \times 10^5$$

$$C_b \approx \varepsilon_{bed} C_g + (1 - \varepsilon)_{bed} C_s = .45(0.22) + (0.55)(0.202) = 0.21 \frac{\text{KCal}}{\text{kg}\,^\circ\text{K}}$$

$$\rho_b \approx \varepsilon_{bed} \rho_g + (1 - \varepsilon_{bed})\,\rho_s = 0.45(0.32) + 0.55(1680) = 924 \frac{\text{kg}}{\text{m}^3}$$

$$\frac{\rho_b C_b}{\rho_g C_g} \approx \frac{(924)(0.21)}{(0.32)(0.22)} = 2{,}756$$

From Figure 15 at $\text{Re}_p = 309$

$$\text{Nu}_{pba}\,\varepsilon_{bed}/\text{Ar}^{0.038} \left(\frac{\rho_b C_b}{\rho_g C_g}\right)^{0.238} = 1.2$$

$$\text{Nu}_{pba} = \frac{(1.2)}{(0.45)} (2.8 \times 10^5)^{0.038} (2{,}756)^{0.238} = 28.6$$

$$h_{ba} = \frac{(28.6)(0.69)}{(.00475)} = 407 \frac{\text{watts}}{\text{m}^2\,^\circ\text{C}}$$

$$h_r = \sigma (T_{pa}^4 - T_{pb}^4)/(T_{ba} - T_{pb})$$

at $T_{pa} = 840° C = 1113° K$

$$h_r = (5.67 \times 10^{-8})[(1113)^4 - 1073^4]/40 = 296 \frac{\text{watts}}{\text{m}^2 °\text{C}}$$

$$q_T = \pi(.00475)^2[0.9(296) + 407]40 = 1.92 \text{ watts}$$

$$\eta = \text{fraction of heat transfer by radiation} = \frac{(296)(0.9)}{407 + (296)(0.9)} = 0.4$$

at $T_{pa} = 920°C = 1193°K$

$$h_r = (5.67 \times 10^{-8})(1{,}193^4 - 1{,}073^4)/120 = 331$$

$$q_T = \pi(.00475)^2[0.9(331) + 407]\,120 = 6.0 \text{ watts}$$

$$\eta = \frac{(331)(0.9)}{0.9(331) + 407} = .42$$

It is estimated that under the conditions of this illustration that from 40 to 42 percent of the heat transfer from the char particle is by radiation.

CLOSURE

Experimental data for both gas-solid heat and mass transfer have been presented for fixed and fluidized beds. The experimental data presented are for relatively small laboratory size beds. The bubble size is a very important parameter in the gas-solid heat or mass transfer mechanism. The bubble diameter is limited by the bed diameter or by the bed internals. Therefore, the experimental data presented may over predict the heat transfer coefficients in large commercial beds without internals.

The bubbling-bed model (as well as some other models) explains why fluidized bed gas-solid Nusselt or Sherwood numbers are less than the limiting value of two. The model was shown to quantitively predict mass and heat transfer coefficients for small-scale beds and to provide a method of estimating coefficients for large-scale beds. Furthermore, the bubbling-bed model and other models discussed give insight into the transport phenomena occurring.

NOTATION

a	surface area per unit volume of bed (m^{-1})
a'	surface-to-volume ratio for a particle (m^{-1})
A_0	catchment area for the bubble at the distributor (m^2)
A_t	total cross-sectional area of the bed (m^2)
Ar	Archimedes number, $gd_p^3\rho_g(\rho_s - \rho_g)/\mu_g^2$
C_g	gas heat capacity (J/Kg °C)
C_s	solids heat capacity (J/Kg °C)
d_p	particle diameter (mm)
D_b	bubble diameter (cm)
D_{b0}	initial bubble diameter (cm)
D_{bav}	average bubble diameter (cm)
D_{bm}	maximum bubble diameter (cm)
D_t	bed diameter (m)
$\mathcal{D}$	mass diffusivity (m^2/s)
f_b	fraction of group b particles surrounding a group a particle
g	acceleration due to gravity (m/s^2)
G_0	gas mass velocity (Kg/m^2s)
G_{mf}	gas mass velocity at minimum fluidization (Kg/m^2s)
h_p, h_g	gas-particle heat transfer coefficient (watts/m^2 °C)

$(h_p)_{ba}$ particle-to-particle heat transfer coefficient (watts/m^2 °C)
$(\bar{h}_p)_{ba}$ particle-to particle heat transfer coefficient when all contacting particles are group b particles (watts/m^2 °C)
h_{ba} total of gas-solid and particle-particle heat transfer coefficients (watts/m^2 °C)
h_r radiant heat transfer coefficient (watts/m^2 °C)
H_{bc} volumetric heat transfer coefficient (watts/m^3 °C)
k_g gas thermal conductivity (watts/m °C)
K_d mass transfer coefficient (m/s)
L_f fluidized bed height (m)
n_d number of orifices in the distributor
Nu_p particle Nusselt number
Nu_{pt} single particle Nusselt number at the terminal velocity
Nu_{app} apparent Nusselt number
P pressure (N/m^2)
Pr Prandtl number
q_f gas-solid heat flux (watts/m^2)
$(q_p)_{ba}$ particle-to-particle heat flux (watts/m^2)
q_{ba} combined gas-solid and particle-to-particle heat flux (watts/m^2)
q_T total heat flux (watts/m^2)
Re_p particle Reynolds number
Re_{pt} particle Reynolds number at terminal velocity
Sc Schmidt number
Sh_p particle Sherwood number
t time (s)
T_g gas temperature (°K)
T_{gb} gas bubble temperature (°K)
T_{gi} inlet gas temperature (°K)
T_{go} outlet gas temperature (°K)
T_{pa} surface temperature of group a particles (°K)
T_{pb} surface temperature of group "b" particles (°K)
T_{bed} bed temperature (°K)
U_B bubble velocity (m/s)
U_{mf} superficial velocity at minimum fluidization velocity (m/s)
U_0 superficial velocity (m/s)
W solids mass (Kg)
x vertical position above the bed (m)
y logarithmic mean fraction of nondiffusioning component

Greek Symbols

δ bubble fraction
ε void fraction
ε_e void fraction of emulsion phase
ε_f average bed voidage
ε_{bed} bed voidage
ε_{mf} bed voidage of minimum fluidization
$\varepsilon_{char}, \varepsilon_a$ emissivity for char or group "a" particles
ξ channel length factor
γ_B particle fraction in bubble
η fraction of heat transfer by radiation
φ_s shape factor
ρ_f bed density (Kg/m^3)
ρ_g gas density (Kg/m^3)
ρ_{mf} gas density at minimum fluidization (Kg/m^3)

REFERENCES

1. Froessling, N., *Gerland Beitr. Geophysics,* Vol. 52 (1938) pp. 170–179.
2. Ranz, W. E., and Marshall, W. R., "Evaporation from Drops," *Chem. Eng. Progr.,* Vol. 48 (Mar. 1952), pp. 141–146.
3. Ranz, W. E., "Friction and Transfer Coefficients for Single Particles and Packed Beds," *Chem. Eng. Progr.,* Vol. 48 (May 1952), pp. 247–260.
4. Kunii, D., and Levenspiel, O., *Fluidization Engineering,* Huntington: Krieger Publishing, 1977, pp. 108–137, pp. 195–223.
5. Resnick, W. E., and White, R. R., "Mass Transfer in Systems of Gas and Fluidized Solids," *Chem. Eng. Progr.,* Vol. 45 (June 1949), pp. 337–390.

6. Löp, C. O. G., and Hawley, R. W., "Unsteady-State Heat Transfer Between Air and Loose Solids," *Ind. Eng. Chem.*, Vol. 40 (June 1948), pp. 1061–1070.
7. Grootenhuis, R. C., Mackworth, A., and Saunders, O. A., *Proc. Inst. Mech. Eng.* (1951), pp. 363–370.
8. Eichhorn, J., and White, R. R., *Chem. Eng. Prog. Symp. Ser.*, Vol. 48 (1952), p. 11–19.
9. Kunii, D., and Smith, J. M., "Heat Transfer Characteristics of Porous Rocks: Thermal Conductivity of Unconsolidated Particles With Flowing Fluids," *AIChE J.*, Vol. 7 (Mar. 1961), pp. 29–34.
10. Donnadieu, G., *Rev. Inst. Franc. du Petrole,* Vol. 16 (1961), pp. 1330–1339.
11. Kunii, D., and Suzuki, M., "Particle-to-Fluid Heat and Mass Transfer in Packed Beds of Fine Particles," *Int. J. of Heat and Mass Transfer,* Vol. 10 (July 1967), pp. 845–853.
12. Kettenring, K. N., Manderfield, E. L., and Smith, J. M., "Heat and Mass Transfer in Fluidized Systems," *Chem. Eng. Progr.*, Vol. 46 (Mar. 1950), pp. 139–145.
13. Chu, J. C., Kalil, J., and Wetteroth, W. A., "Mass Transfer in a Fluidized Bed," *Chem. Eng. Progr.*, Vol. 49 (Mar. 1953), pp. 141–149.
14. Riccetti, R. E., and Thodos, G., "Mass Transfer in the Flow of Gases Through Fluidized Beds," *AIChE J.*, Vol. 7 (Sep. 1961), pp. 442–444.
15. Walton, J. S., Olson, R. L., and Levenspiel, O., "Gas-Solid Film Coefficients of Heat Transfer in Fluidized Coal Beds," *Ind. Eng. Chem.*, Vol. 44 (June 1952), pp. 1474–1480.
16. Heertjes, P. M., and Kibbins, S. W., "The Partial Coefficient of Heat Transfer in a Drying Fluidized Bed," *Chem. Eng. Sci.*, Vol. 5 (July 1956), pp. 161–167.
17. Frantz, J. F., "The Evaporation of Brine Solutions in a Fluidized Salt Bed," Louisiana State Univ.: Ph. D. thesis, 1958, pp. 25–36.
18. Genetti, W. E., Heat Transfer in a Fluidized Bed Tubular Heat Exchanger," Oregon State Univ.: Ph. D. thesis, 1968, p. 12.
19. Wamsley, W. W., and Johanson, L. N., "Fluidized Bed Heat Transfer," *Chem. Eng. Progr.,* Vol. 50 (July 1954), pp. 347–355.
20. Fritz, J. C., "Heat Transfer Between Air and Solid Particles in Sixteen-Inch Diameter Fluidized Beds," Univ. of Wisconsin: Ph. D. thesis, 1956, pp. 81–95.
21. Ferron, J. R., "Studies of Gas-Solid Heat Transfer and Gas Mixing in Fluidized Beds," Univ. of Wisconsin: Ph. D. thesis, 1958, pp. 153–169.
22. Yoshida, K., B. S. Thesis., Univ., of Tokyo, 1961, pp. 15–31.
23. Sato, K., Shirai, T., and Aizawa, M., *Annual Meeting of Soc. Chem. Engrs.,* Japan, 1950.
24. Zabrodsky, S. S., "Heat Transfer Between Solid Particles and a Gas in a Non-Uniformly Aggregated Fluidized Bed," *Int. J. Heat and Mass Transfer,* Vol. 6 (Jan. 1963), pp. 23–32.
25. Kato, K., and Wen, C. Y., "Gas-Particle Heat Transfer In Fixed and Fluidized Beds," *Chem. Eng. Progr. Symp. Series.*, Vol. 66, No. 105 (1970), pp. 100–108.
26. Darton, R. C., LaNauze, R. D., and Davidson, J. F., "Bubble Grwoth Due to Coalescence in Fluidized Beds," *Trans. I. Chem. E.*, Vol. 55 (1977), pp. 274–280.
27. Kunii, D., Yoshida, K., and Hiraki, I., *Proc. Intern. Symp. on Fluidization,* Amsterdam: Netherlands Univ. Press, 1967, pp. 147–159.
28. Mori, S., and Wen, C. Y., "Estimation of Bubble Diameter in Gaseous Fluidized Beds," *AIChE J.,* Vol. 21 (Jan. 1975), pp. 109–115.
29. Rowe, P. N., and Everett, D. J., "Fluidized Bed Bubbles Viewed by X-rays, Pt. III – Bubble Size and Number When Unrestrained Three Dimensional Growth Occurs," *Trans. Inst. Chem. Engrs.*, Vol. 50 (Jan. 1972), pp. 55–68.
30. Werther, J., "The Influence of the Bed Diameter on the Hydrodynamics of Gas Fluidized Beds," National AIChE Meeting, Detroit, 1973.
31. Davidson, J. F., and Harrison, D., *Fluidized Particles,* Cambridge: Cambridge Univ. Press, 1963, pp. 31–43.
32. Murray, J. D., Mathematical Aspects of Bubble Motion in Fluidized Beds," *Chem. Eng. Progr. Symp. Series,* Vol. 62, No. 66 (1966), pp. 71–82.

33. Kabayashi, H. F., Arai, F., and Snagawa, T., *Chem. Eng.*, Tokyo, Vol. 31 (1967), pp. 239–245.
34. Juveland, A. C., Deinken, H. P., and Dougherty, J. E., "Particle-to-Gas Heat Transfer in Fluidized Beds," *Ind. Eng. Chem. Fundamentals,* Vol. 3 (Nov. 1964), pp. 329–333.
35. Geldart, D., "The Size and Frequency of Bubbles in Two and Three Dimensional Gas-Fluidized Beds," *Powder Technol.*, Vol. 4 (1971), pp. 41–52.
36. Whitehead, A. B., and Young, A. D., *Proc. Intern. Symp. on Fluidization,* Endhoven, Netherlands, 1967, pp. 284–292.
37. Kunii, D., Yoshida, K., and Hiraki, I., "The Behavior of Freely Bubbling Fluidized Beds," *Proc. Intern. Symp. On Fluidization.,* Endhoven, Netherlands, 1967, pp. 243–254.
38. Yasui, G., and Johanson, L. N., "Characteristics of Gas Packets in Fluidized Beds," *AIChE J.*, Vol. 4 (Dec. 1958), pp. 445–452.
39. Toei, R., et al., "Behavior of Bubbles in Gas-Solid Fluidized Beds," *Chem. Eng.*, Tokyo, Vol. 29 (1965), pp. 851–858.
40. Kabayashi, H., and Arai, F., "Effects of Several Factors in Catalytic Reactions in a Fluidized Bed Reactor," *Chem. Eng.*, Tokyo, Vol. 29 (1965), pp. 888–891.
41. Miwa, K., et al., "Behavior of Bubbles in Gaseous Fluidized Bed," *Int. Chem. Eng.*, Vol. 12 (Jan. 1972), pp. 187–194.
42. Tomito, M., and Adachi, T., "Effect of the Bed Diameter on the Behavior of Bubbles in Gas Fluidized Beds," *Proc. 36th Annual Meeting of Soc. Chem. Eng. Japan.*, Vol. II, 1971, pp. 169–176.
43. Baumgarten, D. K., and Pigford, R. L., "Density Fluctuations in Fluidized Beds," *AIChE J.*, Vol. 6 (Mar. 1960), pp. 115–123.
44. Park, W. H., et al., "The Properties of Bubbles in Fluidized Beds of Conducting Particles as Measured by an Electroresistivity Probe," *Chem. Eng. Sci.*, Vol. 24 (May 1969), pp. 851–865.
45. Botton, R. J., "Gas-Solid Contacting in Fluidized Beds," *Chem. Eng. Prog. Symp. Series,* Vol. 101, No. 66 (1968), pp. 8–13.
46. Fryer, C., Ph. D. thesis, Australia: Monash Univ., 1974, pp. 34–39.
47. Kunii, D., and Levenspiel, O., "Bubbling Bed Model," *Ind. Eng. Chem. Fundamentals,* Vol. 7 (Aug. 1968), pp. 446–451.
48. Wen, C. Y., and Chang, T. M., "Particle-to-Particle Heat Transfer in Air-Fluidized Beds," *Proc. Int. Symp. On Fluidization,* Endhoven, Netherlands, 1967, pp. 491–506.
49. Botterill, J. S. M., *Fluid Bed Heat Transfer,* New York: Academic Press, 1975, pp. 1–4, pp. 145–156.
50. Agarwal, P. K., *Fluid Bed Combustion of Mississippi Lignite.* Univ. of Mississippi: Ph. D. dissertation, 1984, p. 43.
51. Incropera, F. P., and DeWitt, D. P., *Fundamentals of Heat Transfer.* New York: John Wiley and Sons, 1981, p. 572.
52. Pallai, K. K., "Devolatilization and Combustion of Large Coal Particles in a Fluidized Bed," unpublished personal communications, 1983.
53. Ross, I. B., and Davidson, J. F., "The Combustion of Carbon Particles in a Fluidized Bed," *Trans. I. Chem. E.,* Vol. 59 (1981), pp. 108–114.
54. Leva, M., *Fluidization,* New York: McGraw Hill, 1959, p. 69.
55. Wen, C. Y., and Genetti, W. E., "Fluidized Bed Heat Transfer," *AIChE Today Series* (1975), p. V-5.

CHAPTER 17

GAS CONVECTION AND UNSTEADY CONDUCTION IN FLUID-BED HEAT TRANSFER

Ronald L. Adams

Department of Mechanical Engineering
Oregon State University
Corvallis, Oregon

CONTENTS

INTRODUCTION

The purpose of this chapter is to present theoretical models of the conductive and convective components of heat transfer in gas fluidized beds, emphasizing the particular model developed by the author. In beds operating at high temperature (i.e. greater than about 700° K), the contribution of thermal radiation to heat transfer must also be considered, and information regarding this component can be found in Baskakov et al. [1], Kolar et al. [2], Chen and Chen [3], Brewster and Tien [4], and Brester [5] as well as work soon to be published by Mahbod and Adams [6] and Fatani and Adams [7]. The models to be discussed have been developed for the bubbling regime of fluidization as described by Kunii and Levenspiel [8] and Davison, Harrison, and de Carvalho [9]. The bubbling-bed assumption restricts the range of validity of the model, particularly for large particle fluidization as illustrated by Figure 1 [10]. As the figure indicates, the bubbling regime of fluidization is limited to a narrow range of fluidizing velocities at large particle diameters, while the turbulent regime becomes more prevalent. Though the mechanisms of heat transfer in turbulent fluidization (involving heat transfer from a particle-laden gas) are fundamentally different than in the bubbling regime, the theory to be presented has been found to yield reasonable results for a wide range of fluidizing velocities at large particle diameter. A model developed specifically for turbulent fluidization has been reported by Staub [11].

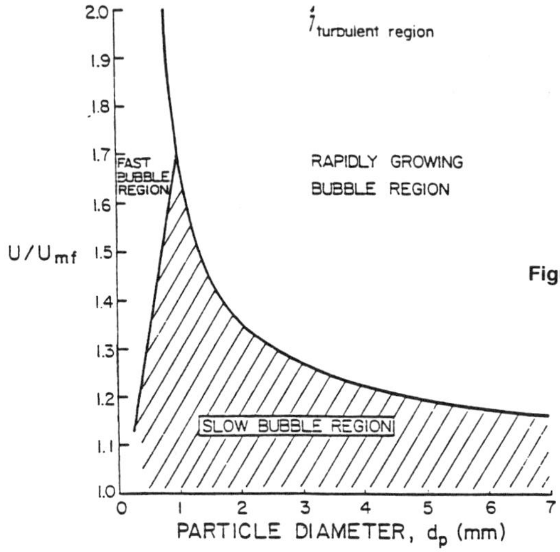

Figure 1. Regimes of bubbling fluidization [10].

The mechanisms of heat transfer in a bubbling fluidized bed have been described by a number of authors including Botterill [12] in his book on fluidized-bed heat transfer. In a bubbling fluidized bed, the heat transfer surface is alternately exposed to bubbles and emulsion (a mixture of particle and gas), and the heat transfer can be envisioned to occur by conduction and convection from the emulsion phase until it is displaced by a passing bubble rising through the bed. Note that this concept represents a significant simplification of the true situation involving constant particle motion as represented for a horizontal tube by Noack's experimental observations [13] and the calculations by Adams [14] in Figure 2. As a result of this interaction, the heat transfer is closely coupled to the hydrodynamic parameters, in this case emulsion-phase residence time and the fraction of time the heat transfer surface is exposed to bubbles. Fundamentally, the heat transfer mechanisms depend upon the phase (bubble or emulsion) which is in contact with the heat transfer surface. In the case of the bubble phase, the heat transfer is entirely due to gas convection and is strongly dependent upon the nature of the flow field within the bubble. While for the emulsion phase, heat transfer is determined from the combined effects of unsteady conduction within the solid particles and convection due the interstitial gas flow. Fortunately, the effects of gas convection and conduction can be uncoupled as many (e.g. Zabrodsky et al. [15], Catipovic et al. [16], Glicksman and Decker [17]) have assumed and will be shown theoretically in this chapter. Thus, the total time-averaged heat transfer can be viewed as the superposition of the effects of conduction and convection in the emulsion phase and gas convection in the bubble phase. This concept is expressed symbolically by Equation 1

$$\dot{q}'' = f_B \dot{q}''_B + (1 - f_B)(\dot{q}''_{gc} + \dot{q}''_{pc}) \tag{1}$$

where $\dot{q}''$ = the total heat transfer per unit area to the immersed surface

$\dot{q}''_B$ = the heat transfer due to convection from the bubble phase

$\dot{q}''_{gc}$ = the heat transfer due to convection from the emulsion phase

$\dot{q}''_{pc}$ = the particle convective heat transfer

f_B = the fraction of time the immersed surface is exposed to bubbles

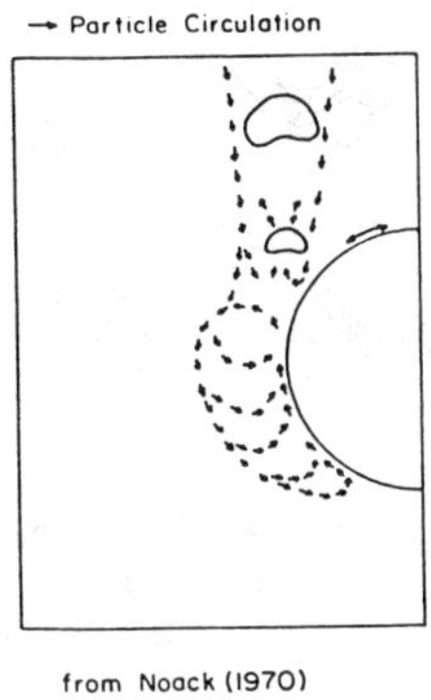

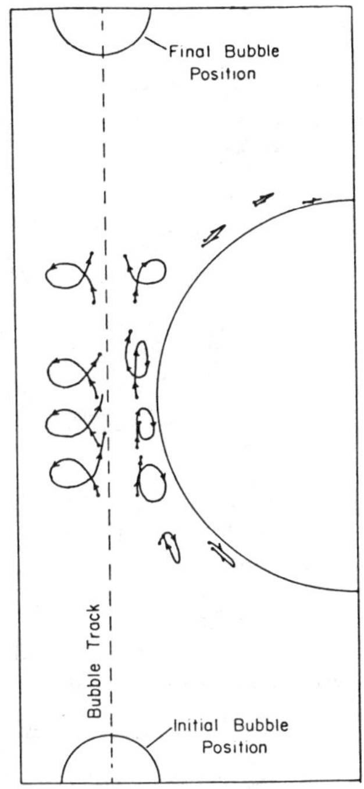

Figure 2. Particle motion near a horizontal surface, calculated trajectories and experimental observations [14].

In order for both gas convection and unsteady conduction effects to be significant it is necessary that the Reynolds number of the intersitial flow be large enough so that the thickness of the thermal boundary layer which forms along the heat transfer surface is smaller than the particle diameter, thus limiting the heat transfer influence to the first layer of particles adjacent to the heat transfer surface. Then the heat transfer phenomenon is conveniently described by the flow and unsteady conduction effects for the unit cell depicted in Figure 3. The elements of the cell include a single particle as well as the interstitial channel formed by the heat transfer surface and the adjacent particle surfaces. The characteristics of this cell will be described in detail later.

Each of the terms contributing to the heat transfer and represented in Equation 1 will be discussed separately in the following sections, then combined and used to determine heat transfer to a horizontal tube immersed in a two-dimensional bed for which data is available for comparison.

THE GAS-CONVECTION COMPONENT OF THE EMULSION-PHASE HEAT TRANSFER

Heat transfer by gas convection becomes important for large-particle and/or high-pressure fluidized beds. At a Reynolds number (based upon minimum fluidizing velocity, gas properties in the bed and particle diameter) greater than about 100, heat transfer by gas convection becomes significant. Investigation of the gas convection component of heat transfer has been motivated by the recent interest in fluidized-bed combustion of coal in which both large particles and high operating pressures have been considered.

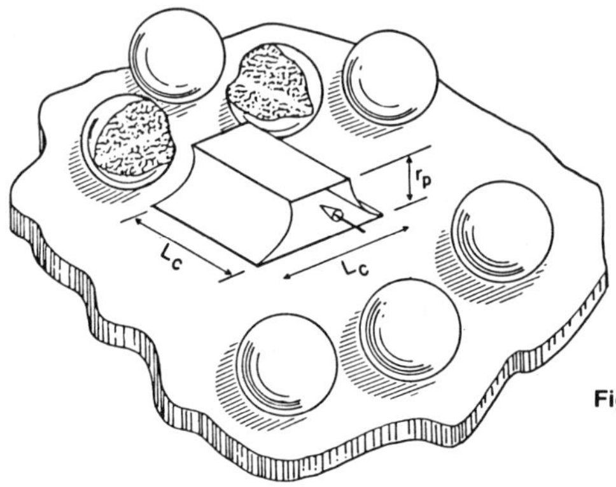

Figure 3. Geometry of the interstitial channel [30].

While a theoretical model of heat transfer by gas convection in a fluidized will be described here, many investigators (e.g. Botterill and Denloye [18], Xavier and Davidson [19], Captipovic et al. [16], and Baskakov et al. [1]) have assumed that the gas convection component of the emulsion-phase heat transfer is unaffected by fluidizing velocity and use results for heat transfer in packed beds to estimate this component. Among these is the expression for heat transfer coefficients for verticle tubes immersed in packed beds developed by Gabor [20] and adapted by Xavier and Davidson [19]

$$h_{gc} = \sqrt{4K_{mf}\rho_g C_{pg} u_{mf/\pi L}} + K_{mf/D} \tag{2}$$

where K_{mf} is the effective conductivity of the bed and L and D are the length and diameter, respectively, of the tube. The effective thermal conductivity of the bed, K_{mf}, can be estimated from Gelperin and Einstein [21] or the following from Xavier and Davidson [19] for particles less than 2 mm in diameter

$$K_{mf} = K_e^0 + 0.1\,\rho_g C_{pg} d_p u_{mf} \tag{3}$$

In Equation 3, the stagnant bed thermal conductivity, K_e^0, is determined from the correlations given in Kunii and Smith [22].

An alternate approach is to use an empirical correlation for the Nusselt number at minimum fluidizing conditions such as that of Baskakov and Suprum [23]

$$Nu_{pqc} = 0.0175\,Ar^{0.46}Pr^{0.33} \tag{4}$$

where $Ar = gd_p^3(\rho_s/\rho_g - 1)/\nu_g^2$, the Archimedes number, and Pr is the gas Prandtl number.

Though Equation 4 was originally developed from data for vertical tubes in beds operating at atmospheric pressure, Catipovic [10] has found that it produces equally accurate results for horizontal geometries when the overall heat transfer to the tube is determined. Also, Borodulya et al. [24] have successfully used the correlation at pressures up to 8.1 MPa.

Fundamentally, the convective heat transfer in a fluidized bed of large particles is determined by the nature of the flow of gas within the interstitial voids formed by the particle and heat transfer surfaces. The heat transfer will be enhanced by turbulence generated by the wakes of particles as well as the accelerations which occur as a result of blocked passages. The geometry of the interstitial flow is conveniently modeled as the channel configuration as described previously for spherical particles in Figure 3. The size of a representative channel, L_c, is determined by the average distance between particles in contact with the heat transfer surface and the boundaries include the heat transfer surface and cylindrical surfaces with radii equal to the average particle radius. Within these channels, the flow is partitioned into an inviscid core and surrounding viscous boundary

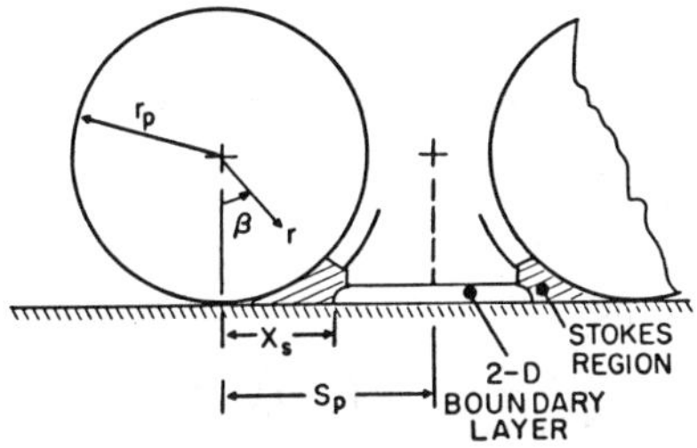

Figure 4. Two-region boundary layer flow model identifying the Stokes and two-dimensional regions [43].

layers. Furthermore, the influence of surrounding particles is assumed to be such that the flow within the channel is stagnation-like with the core velocity varying linearly from zero at the entrance to twice the average interstitial velocity at the exit.

The channel model provides a basis for determination of both particle-convective and gas-convective contributions to the emulsion-phase heat transfer. This follows by partitioning the boundary layer flow into a region near the center of the channel where the thermal effects of the particles can be neglected and into regions near the corners or particle contact points where the thermal effect of the particles is predominate. Furthermore, the flow near the corners is assumed to be Stokes like and the flow in the central region is assumed to be two-dimensional. This two-region model is illustrated in Figure 4. The Stokes region size parameter, X_s, is determined by requiring the Nusselt number to be continuous across the channel, a simple matching procedure, described below, which yields reasonable results.

The gas convection contribution to the emulsion-phase heat transfer is determined by the product of the Nusselt number for the two-dimensional boundary layer flow and the fraction of the heat transfer surface involved, i.e.

$$Nu_{pgc} = (1 - x_{s/sp})\,Nu_{p2D} \tag{5}$$

The convective heat transfer due to flow in the two-dimensional portion of the interstitial channels is obtained from an integral solution of the two-dimensional boundary layer equations. The effect of interstitial turbulence in this case is analogous to the effect of free-stream turbulence on the heat transfer to objects immersed in an air stream for which theoretical results have been obtained, by Smith and Kuethe [25] and Traci and Wilcox [26]. The effect is included here in an integral solution by introducing a transformation which effectively stretches the normal coordinate. Since gas property variations have little effect on heat transfer [27, 28], the calculation is performed using gas properties based upon the average temperature, $(T_B + T_W)/2$. The effect of turbulence is included through an eddy diffusivity term in the time-averaged equations. In dimensionless form, they are

$$\frac{\partial \upsilon}{\partial y} + \frac{\partial u}{\partial z} = 0 \tag{6}$$

$$\upsilon\frac{\partial u}{\partial y} + u\frac{\partial u}{\partial z} = -\frac{dp}{dz} + \frac{1}{Re}\frac{\partial}{\partial y}\left[(1 + \varepsilon_T)\frac{\partial u}{\partial y}\right] \tag{7}$$

$$\upsilon\,\partial\theta/\partial y + u\,\partial\theta/\partial z = 1/RePr\,\frac{\partial}{\partial y}\left[(1 + \varepsilon_T)\frac{\partial\theta}{\partial y}\right] \tag{8}$$

where ε_T is the dimensionless eddy diffusivity, and eddy diffusivities of heat and momentum are assumed to be equal and isotropic. In Equations 6 through 8, y is normal to the heat transfer surface and z is parallel to the interstitial flow. Also, the boundary conditions are at

$$y = o \text{ (heat transfer surface)} \quad (9)$$

$$u = v = \theta = o \quad (9a)$$

$$\varepsilon_T = o \quad (9b)$$

Thus

$$\frac{\partial}{\partial y}\left[(1 + \varepsilon_T)\frac{\partial u}{\partial y}\right] = \frac{dp}{dz} \quad (9c)$$

and

$$\frac{\partial}{\partial y}\left[(1 + \varepsilon_T)\frac{\partial \theta}{\partial y}\right] = o \quad (9d)$$

At

$$y = \delta \text{ (edge of boundary layer)} \quad (9e)$$

$$u = Q_g(z) \quad (9f)$$

$$\theta = 1 \quad (9g)$$

$$\frac{\partial u}{\partial y} = \frac{\partial^2 u}{\partial y^2} = \frac{\partial \theta}{\partial y} = \frac{\partial^2 \theta}{\partial y^2} = 0 \quad (9h)$$

$$\varepsilon_T = \beta_T \quad (9i)$$

where $Q_g(z)$ is the local interstitial gas velocity.

The integral transformation $d\xi = dy(1 + \varepsilon_T)$ is introduced to absorb the interstitial turbulence in the normal coordinate, and Equations 7 and 8 are integrated across the boundary layer following the von-Karman-Pohlhausen approach. The details of the analysis can be found in Adams [29] and Adams and Welty [30] and reveal the true dependent variables to be the velocity profile shape parameter, $\Omega_v = Re\delta_v^2 dQ_g/dz$ and the ratio of thermal and velocity boundary layer thicknesses, N_{HV}. Also the Nusselt number is expressed

$$Nu_p = \frac{\partial \theta}{\partial y}(0) = 2N_{HV}\sqrt{(\Omega_v/Re_p)\frac{dQg}{dz}} \quad (10)$$

The Nusselt number given by Equation 10 has been found to depend upon the turbulence parameter, β_T, which is related to the interstitial turbulence intensity, u′, through

$$\beta_T = K_T\frac{u' u_{ref} L_T}{\nu} \quad (11)$$

where K_T is assumed to be a universal constant, and L_T is the length scale for turbulence. Following Traci and Wilcox [26], the turbulence length scale is assumed to be

$$L_T \sim \frac{L}{\sqrt{Re_L}} \quad (12)$$

where L is an appropriate characteristic length (particle diameter in this case). Thus

$$\beta_T = K_T u' \sqrt{Re} \tag{13}$$

where Re_c is the Channel Reynolds number, and the universal constant K_T is obtained by matching predictions for stagnation-point heat transfer for a cylinder in an air stream with experimental results. Calculations performed using the integral analysis with $K_T = 0.13$ provide results which agree well with experimental data for stagnation-point flow.

The assumed linear variation in the interstitial velocity makes the interstitial flow equivalent to a two-dimensional stagnation point flow. Numerical results for this case show that N_{HV} is approximately independent of the intensity of interstitial turbulence (varying between 1.5 and 1.6 for a wide range of β_T and for $Pr = 0.7$) and Ω_v can be approximated by

$$\Omega_{vu'} = \Omega_{vu'=0}/(0.2 + 0.8\,e^{-0.0849\,u'\sqrt{Rec}}) \tag{14}$$

to within 5% of the numerical results. It is further suggested that the approximation

$$N_{HV} = 1.34\,Pr^{-0.4} \tag{15}$$

be used so that the proper variation in Nusselt number with Prandtl number is obtained (see, for example, Kays and Crawford [28]).

The above results can be combined to produce the approximation

$$Nu_{P2D} = 0.798\,Pr^{0.4}\sqrt{Re_c/[(S_p/r_p)\,(\Omega_{vu'}/\Omega_{vu'} = o)]} \tag{16}$$

where Re_c is the Reynolds number based upon local interstitial velocity and particle diameter with gas properties at the mean temperature and S_p is half the spacing between particles (approximately $0.75\,r_p/\sqrt{1 - \varepsilon}$) [30].

THE CONDUCTION COMPONENT OF THE EMULSION-PHASE HEAT TRANSFER

Heat transfer from the solids in a fluidized bed, often called the particle convective component, ultimately occurs by conduction across a gas layer residing between the solids and the heat transfer surface. This gas film can be viewed as a contact resistance across which the thermal energy within the solid particles is conducted. The details of this phenomenon are extremely complex due to the complex geometry of crushed solids, the influence of adjacent particles upon each other, and the influence of the gas flowing within the interstitial voids. As a result of this complicated nature, early attempts at modeling this component of the heat transfer involved treating the emulsion phase as a continuum with an effective thermal conductivity (largely influenced by the gas) and an effective thermal capacity (largely influenced by the properties of the solids), both determined experimentally. Most notable of these models is the packet model of Mickley and Fairbanks [31] in which a packet of emulsion material is envisioned to contact the heat transfer surface and undergo transient conduction of heat to the surface until it is renewed by a passing bubble according to the sequence illustrated in Figure 5. The packet model was later refined by others including the addition of contact resistance at the heat surface by Baskakov [32] and an alternate approach in which the effect of voidage variations near the heat transfer surface were included by Kubie and Broughton [33]. Taking the contact resistance to act in parallel with the resistance of the packet, Gelperin and Einstein [21] obtain the following expression for the particle convective heat transfer coefficient:

$$h_{pc} = \frac{1}{R_f + R_p} \tag{17}$$

with

$$R_p = \sqrt{4\,K_{mf}\,\rho_{mf}\,c_{ps}/\pi\bar{\tau}_e} \tag{18a}$$

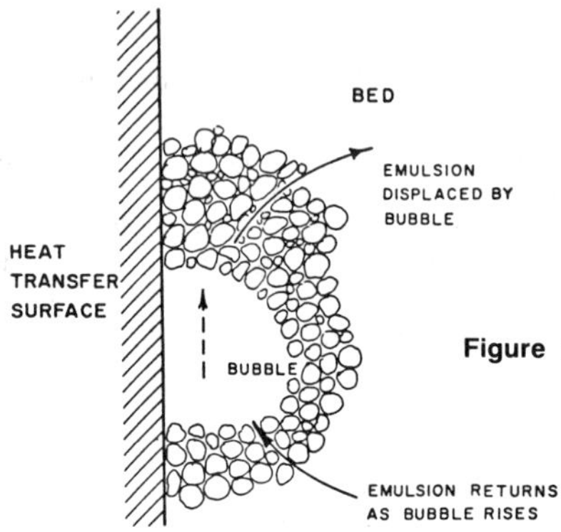

Figure 5. Cycle of particle replacement at a heat transfer surface.

where the emulsion phase thermal conductivity, K_{mf}, can be estimated using the expressions given in Equation 3. The contact resistance is determined by considering the heat transfer to occur by conduction across a gas layer of thickness δ so that

$$R_f = k_g/\delta \tag{18b}$$

The thickness of the equivalent gas layer, δ, appears to be in the range of dp/7 to dp/6 according to Zabrodsky et al. [15] and Catipovic et al. [16].

The model to be presented here involves an alternate approach to calculation of the conduction component in which the details of heat transfer to a single particle or representative cell are modeled. Such an approach was originally used by Botterill and Williams [34]. Since only the first layer of particles near the heat transfer surface is considered, the approach is limited to either very short contact times or beds of large particles. This limitation led Gabor [35] to an approximation in which the bed is modeled as a series of alternate slabs of solids and gas with the thickness of the solid slabs given as 2/3 of the particle diameter and the intervening gas layer thickness adjusted to obtain agreement between the theory and experimental results.

THE CONDUCTION COMPONENT IN BEDS OF LARGE PARTICLES

In beds of large particles, the influence of the heat transfer surface is confined to the first layer of particles in contact with the surface. The heat transfer from the particles then occurs by conduction across the gas lens between the particle and the heat transfer surface. The geometry of this region, called the Stokes region in the preceding analysis, was illustrated for spherical particles in Figure 4. As shown, the extent of the Stokes region along the heat transfer surface determines the portion of the heat transfer surface which is influenced by the particles. As the boundary layer becomes thinner, the size of this region diminishes and a smaller portion of the wall is influenced by the particles so that the heat transfer eventually becomes dominated by gas convection.

The heat transfer by conduction from the particle is determined by the product of the Nusselt number for the Stokes flow region and the fraction of the surface involved, i.e.

$$Nu_{ppc} = \frac{x_s}{S_p} Nu_{ps} \tag{19}$$

Since the thermal mass of the gas is being neglected, the Stokes region Nusselt number, Nu_{ps}, is determined from an analysis of the steady conduction within the gas layer between the particle surface and the heat transfer surface. The boundary condition for this analysis is determined by the instantaneous temperature distribution along the particle surface. The analysis is simplified by superposition of the contributions from the inviscid core and particle surface. Gas property variations are included by using Kirchoff's transformation (see, e.g. Aparci [36]) ($S = -\int_1^\theta k_g/k_g^* d\theta$) to obtain

$$\nabla^2 S = 0 \tag{20}$$

from the steady conduction equation.

As in Reference 30, the boundary conditions are as follows

$$S = S(T_B) \quad \text{along inner boundary of Stokes region} \tag{21a}$$

$$\frac{\partial S}{\partial n} = 0 \quad \text{along dividing boundary between 2-D and Stokes regions} \tag{21b}$$

$$S = S(T_W) = S_o \quad \text{along tube wall boundary} \tag{21c}$$

$$S = S(\beta) \quad \text{along particle surface} \tag{21d}$$

Note that the temperature parameter, S, can be related to dimensionless temperature, θ, for gases having $k_g \sim T^\alpha$, i.e.,

$$S(\theta) = -[(\beta_{c1} + \beta_{c2})^{\alpha+1} - (\beta_{c1}\theta + \beta_{c2})^{\alpha+1}]/\beta_{c1}(\alpha + 1) \tag{22}$$

with

$$\beta_{c1} = 2(1 - T_W/T_B)/(1 + T_W/T_B) \tag{23}$$

$$\beta_{c2} = 2(T_W/T_B)/(1 + T_W/T_B) \tag{24}$$

To facilitate the solution of Equation 20 for the Stokes region, two conformal mappings are used as illustrated in Figure 6 and symmetry about an axis through the contact point

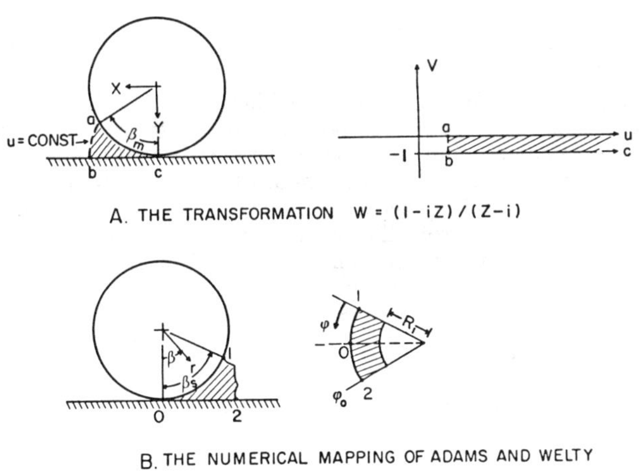

Figure 6. Conformal mappings used in evaluating the gas temperature distribution within the lens region [43].

is assumed. The contribution from the particle surface is obtained by first making the transformation suggested by Dunsky, et al. [37]

$$W = (1 - iZ)/(Z - i) = u\ iv \tag{25}$$

Then, because of the stretching of the gas layer produced by the transformation, the temperature along the particle surface (v = 0) is assumed to be slowly varying such that

$$S(u, v) \simeq S_0 + [S(u, 0) - S_o](v + 1) \tag{26}$$

Thus, the local heat transfer from the particle surface is given by the following Nusselt number:

$$Nu_{pss} = \frac{2}{\left|\frac{dz}{dw}\right|}[S(\beta) - S_o] = \frac{2[S(\beta) - S_o]}{(1 - \cos\beta)} \tag{27}$$

where β is the spherical coordinate from the contact point as shown in Figure 6B and $|dz/dw|$ is the scale factor for the transformation.

The contribution due to conduction from the inviscid core is obtained using the numerical conformal mapping of Adams and Welty [30] which reduces the Stokes region to a circular sector. On the circle plane, the particle surface temperature approximates a step profile so that the contribution from the gas is given by the Nusselt number

$$Nu_{psg} \simeq \left(\frac{S(\beta_s) + S_o}{h \ln R_1}\right) - \frac{4[S(\beta_s) - S_o]}{h\varphi_0} \cdot \cos\frac{\pi\varphi}{\varphi_0}\frac{(1 + R_1^{2\pi/\varphi 0})^2 - (1 - R_1^{2\pi/\varphi 0})^2}{(1 - R_1^{4\pi/\varphi 0})} \tag{28}$$

where $S(\beta_s)$ is the value of the temperature parameter at the Stokes region boundary on the particle, h is the scale factor for the trnsformation and φ, φ_0, and R_1 are as indicated in Figure 6B. Note that

$$R_1 = \left(1 - \frac{\varphi_0}{2}\right) \tag{29}$$

i.e., the Stokes region is assumed to form a curvilinear "square" on the circle plane as in Adams and Welty [30]. The Stokes region boundary on the sphere surface, β_s, of Figure 6B is determined from the intersection on the sphere of the constant u line from the boundary on the heat transfer surface. Then φ_0 is determined from the numerical mapping.

The lateral extent of the Stokes region is obtained by assuming that the heat transfer is continuous across the surface from the two-dimensional boundary layer flow near the center of the intenstitial channel to the contact point. Thus, x_s is obtained from

$$Nu_{pss}(x_s) + Nu_{psg}(x_s) = Nu_{p2D} \tag{30}$$

Using the approximations presented in Adams [38, 39], the edge location is determined from

$$\left(1 - \frac{x_s}{s_p}\right) \simeq \begin{cases} 0.76 + S_o k_1/Nu_{p2D} & Nu_{p2D} \geqslant Nu_{pm} \\ -k_2 Nu_{p2D}/S_o & Nu_{p2D} < Nu_{pm} \end{cases} \tag{31}$$

where k_1 and k_2 take on the packed bed values [39]

$$k_1 \simeq 5.26 - 7.72(s_p/r_p - 1) \tag{32}$$

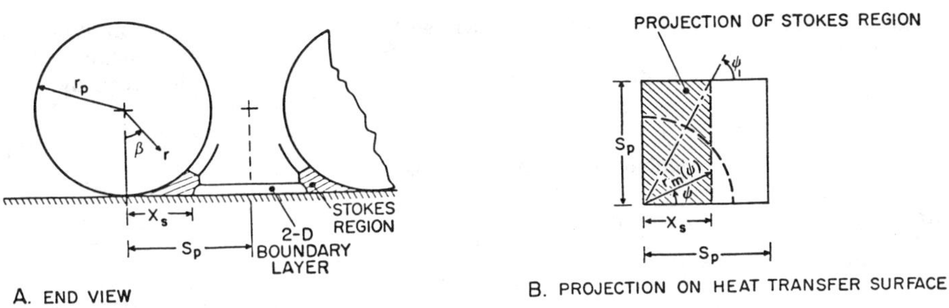

Figure 7. Stokes region projection on the heat transfer surface [43].

$$k_2 \simeq 0.0127 + 0.0222\,(s_p/r_p - 1) \tag{33}$$

and

$$Nu_{pm} = -\,2k_1S_o/(.76 - \sqrt{.578 - 4k_1k_2}) \tag{34}$$

Note that the two-dimensional Nusselt number was given previously as Equation 16.

The total heat transfer to the wall beneath the Stokes region is obtained by integrating the contribution given by Equation 27 over the portion of the particle surface influenced by the Stokes region and adding the inviscid core contribution, i.e.

$$Nu_{ps} = \{[S(\beta_s) + S_o]\,\beta_1 + [S(\beta_s) - S_o]\,\beta_2\}\,r_p/x_s + \frac{2r_p^2}{s_px_s}\int_0^{\pi/2}\int_0^{\beta m(\psi)} \frac{[S(\beta) - S_o]}{(1 - \cos\beta)}\sin\beta\, d\beta\, d\psi \tag{35}$$

where

$$\beta_1 = \varphi_0/2\ln R_1 \tag{36}$$

$$\beta_2 = 4/\pi\,[(1 + R_1^{2\pi/\varphi 0})^2 - (1 - R_1^{2\pi/\varphi 0})^2]/(1 - R_1^{4\pi/\varphi 0}) \tag{37}$$

and the integration limit $\beta_m(\psi)$ is determined from the intersection on the sphere of the constant u lines which originate at the boundaries of the Stokes region projected onto the heat transfer surface (see Figures 6B and 7). This procedure is consistent with the assumed temperature distribution given by Equation 26. When r_m is the projected radial coordinate from the contact point of this boundary as shown in Figure 7, the transformation Equation 25 gives

$$\beta_m(\psi) = \sin^{-1}\{r_m/r_p/[1 + (r_m/2r_p)^2]\} \tag{38}$$

while r_m follows from the projection of the Stokes region as shown in Figure 7.

$$r_m/r_p = \begin{cases} x_s/r_p\cos\psi & \psi \leqslant \psi_1 \\ s_p/r_p\sin\psi & \psi > \psi_1 \end{cases} \tag{39}$$

with

$$\psi_1 = \tan^{-1}(s_p/x_s) \tag{40}$$

Closure of the analysis is achieved once the particle surface temperature distribution (i.e. $S(\beta)$) is determined. This temperature distribution is established from an analysis of transient conduction within a spherical particle with convective boundary conditions as previously described. To simplify this analysis, the temperature distribution is assumed to be symmetric about an axis through the contact point and is therefore governed by

$$\frac{\partial \theta}{\partial Fo} = \frac{1}{r^2}\left[\frac{\partial}{\partial r}\left(r^2\frac{\partial \theta}{\partial r}\right) + \frac{1}{\sin\beta}\frac{\partial}{\partial \beta}\left(\sin\beta\frac{\partial \theta}{\partial \beta}\right)\right] \tag{41}$$

with initial and boundary conditions

$$\theta(r, \beta, Fo) = 1 \text{ at } Fo = 0^- \tag{42}$$

$$\theta(1, 0, Fo) = 0 \text{ at } Fo = 0^+ \tag{43}$$

$$\frac{\partial \theta}{\partial \beta} = \begin{cases} -\,Bi(\beta)\,\theta & \beta < \beta_s \\ -\,Bi(\beta)(\theta - 1) & \beta > \beta_s \end{cases} \tag{44}$$

Note that the condition given by Equation 43 sets the temperature of the contact point at the wall value while the convective boundary conditions (Equation 44) provide for cooling of the portion of the sphere influenced by the wall and heating of the upper portion by the bed.

The results of the calculation can be improved by including the effect of contact resistance for rough and nonspherical particles on the boundary conditions such as the model of Decker and Glicksman [40]. These effects are included in a recent analysis to be published by Mahbod and Adams [6].

The angular limit of the wall influence on the particle, β_s, indicated in Equation 44 is determined through Equation 30 and the numerical mapping with the geometry of Figure 6. This method of calculation results in an overestimation of the particle convective contribution because the surface area of the particle influenced by the wall for the temperature calculation will be smaller than that indicated by Figures 6A and 7.

The Biot number distribution for the lower portion of the sphere ($\beta < \beta_s$) is obtained from the Stokes region analysis just described and can be written

$$Bi(\beta) \simeq \frac{-k_g^*}{k_s\theta(\beta)}\{\alpha_1[S(\beta_s) + S_0] + \alpha_2[S(\beta_s) - S_0] - [S(\beta) - S_0]/(1 - \cos\beta)\} \tag{45}$$

where

$$\alpha_1 = \frac{1}{2h(\beta)\ln R_1} \tag{46}$$

$$\alpha_2 = \frac{-2}{h(\beta)/\varphi_0}\cos\pi\varphi/\varphi_0\left[\frac{(1 + R_1^{2\pi/\varphi_0})^2 - (1 - R_1^{2\pi/\varphi_0})^2}{(1 - R_1^{2\pi/\varphi_0})}\right] \tag{47}$$

For the upper portion of the sphere, the 2-D results are used as follows

$$Bi(\beta) = \frac{-k_g^*}{2k_s}(Nu_{p2D} + Nu_{p0}) \tag{48}$$

where $Nu_{p0} = 2/[1 - (1 - \varepsilon)]^{1/3}$, the minimum Nusselt number suggested by Galloway and Sage [41]. Note that the particle Nusselt numbers computed according to Equation 48 agree

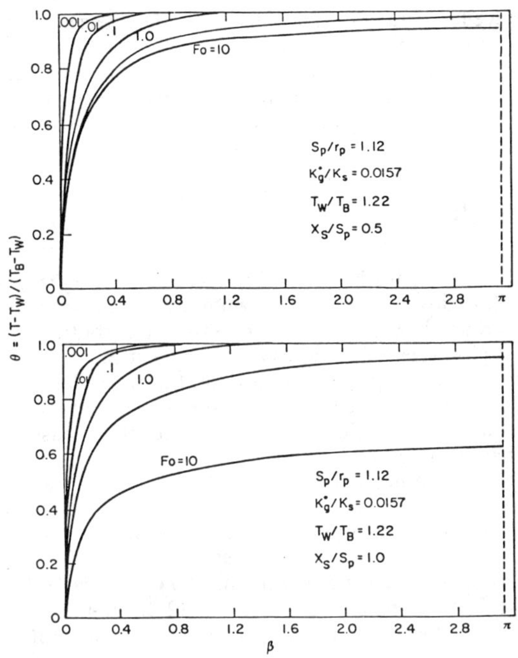

Figure 8. Distributions of dimensionless temperatures within a spherical particle for two values of the Stokes region size [43].

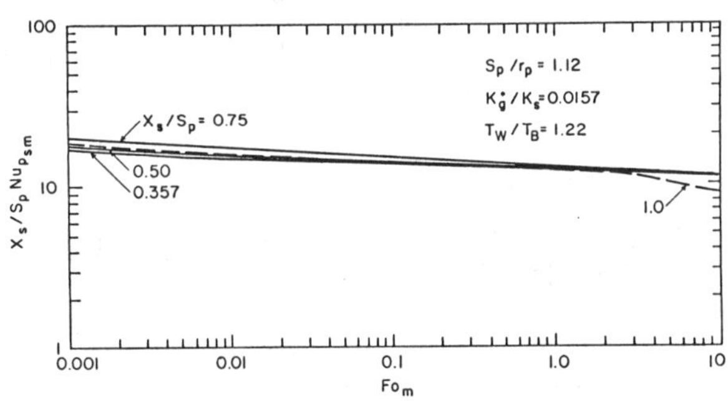

Figure 9. Particle convective contribution to the heat transfer [43].

closely with the empirical correlation for particle-gas heat transfer reported by Galloway and Sage [41].

A numerical solution of Equation 41 with the stated boundary conditions is obtained using an alternating direction implicit finite difference scheme (e.g. Reference 42). In the present context, the grid spacing is allowed to be fully variable to accommodate the large gradients in dimensionless temperature, θ, which occur near the contact point. The details of the approach can be found in Adams [43].

Typical particle surface temperature distributions are shown as a function of the Fourier number in Figure 8. The upper portion of the figure shows results obtained when half of the region between the particles is influenced (high Reynolds number) while the lower portion of the figure presents results for complete influence (very low Reynolds number).

Time-average Nusselt numbers were computed by assuming a uniform distribution of residence time so that

$$Nu_{psm} = \frac{1}{Fo_m} \int_0^{Fo_m} Nu_{ps} dFo \tag{49}$$

Typical values of this parameter for $k_g^*/k_s = 0.0157$ (corresponding to air-sand system) are shown in Figure 9 as a function of lens size. Shown in this figure are values of the product, $(x_s/s_p)\,Nu_{psm}$, which is primarily the particle convective contribution. Note the relatively weak influence of the region size on these results over most of the Fourier number range presented. Since the region size depends primarily on the Reynolds number, these results suggest that the particle convective contribution should depend mostly on the Fourier number and thermal conductivity ratio and thus be relatively insensitive to the Reynolds number. This result has been observed empirically by others and is taken as a basic assumption in many phenomenological models of fluid bed heat transfer [15, 16, 17].

Since it is impossible to distinguish experimentally the conductive and gas convective contributions to the heat transfer, the preceding result cannot be verified. However the total emulsion-phase heat transfer can be determined from analysis of measurements of instantaneous heat transfer. These data were obtained by Catipovic [10] in two-dimensional beds of sand (0.8 and 1.3 mm diameter) and dolomite (2, 2.85, 4.0, and 6.6-mm diameter) fluidized with air. Figure 10 illustrates this comparison when the computational parameters of Table 1 are used in the calculation. Shown in the figure are predicted time-averaged emulsion-phase Nusselt numbers along with the time-averaged maximum instantaneous Nusselt numbers obtained from the experiments. The best agreement between theory and experiment is obtained for the higher Reynolds number dolomite cases while the theory generally over predicts the lower Reynolds number sand cases except at the lower stagnation region where the theory is low. The discrepancy between theory and experiment for smaller particles is due to the single particle feature of the model, and Figure 10 provides an indication of the range of validity of the model.

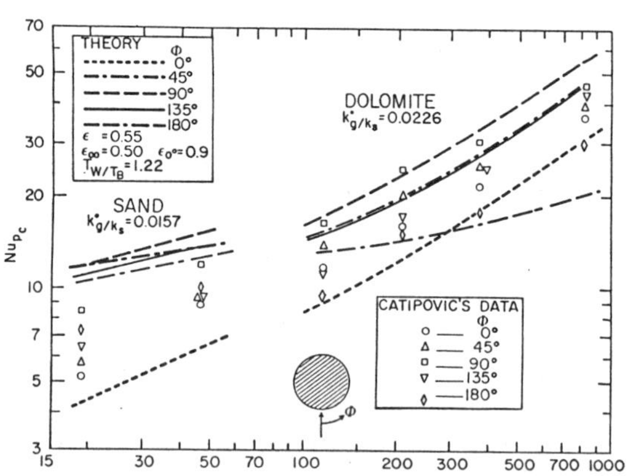

Figure 10. Calculated and measured values of emulsion-phase heat transfer to a horizontal tube, data of Catipovic [10] [43].

Table 1
Calculation Parameters

Bed Voidage	0.4	Packed
(ε_∞)	0.5	Fluidized
Surface voidage	0.44	Packed
(ε)	0.55	Fluidized
	0.90	$\theta = 0$
Temperature ratio (T_W/T_B)	1.22	
Interstitial turbulence intensity (u')	0.2	Galloway and Sage [41]
Prandtl number (Pr)	0.72	Air
Reynolds numbers at u_{mf}	46.2	1.3 mm sand
(Re_p)	116	2 mm dolomite
	208	2.85 mm dolomite
	371	4 mm dolomite
	817	6.6 mm dolomite
Fourier number range (Fo_m)	.0148 to 7.92	Catipovic* [10]
Bubble contact fraction (f_B)		Catipovic* [10]

* Interpolated from data for 1.3 mm sand and 4.0 dolomite.
From Adams [61].

HYDRODYNAMIC PARAMETERS

It is now clear that determination of the heat transfer by both unsteady conduction and by gas convection depends upon knowledge of the hydrodynamic characteristics of the bed. In the case of the conduction component, the emulsion-phase residence time must be known at the particular location along the heat transfer surface at which the heat transfer is to be evaluated. Though very little general information exists regarding this parameter the correlation suggested by Baskakov et al. [1] for verticle tubes can be used in beds of small particles (less than 1 mm diameter) as follows

$$\bar{\tau}_e \simeq 0.44\,\{d_p g/[u_{mf}^2(u_0/u_{mf} - A)^2]\}^{0.14}(d_p/D)^{0.225} \tag{50}$$

where $\bar{\tau}_e$ is in seconds, A is a constant between 0.7 and 0.9, D is the tube diameter, and d_p is particle diameter. Residence times for horizontal cylinders immersed in beds of intermediate to large particles (up to 6.6 mm diameter) have been reported by Catipovic [10] and Chandron and Chen [44]. Catipovic [10] also provides a compact correlation for bubble contact fraction for horizontal tubes in the form

$$1 - f_B = 0.45 + 0.061/(u_0 - u_{mf} + 0.125) \tag{51}$$

with u_0 and u_{mf} in m/s, while Baskakov, et al. [1] suggest

$$f_B = 0.33\left[u_{mf}^2\frac{(u_0/u_{mf} - A)^2}{d_p g}\right]^{0.14}$$

for vertical tubes.

The essential parameter in the evaluation of the gas convective heat transfer is the local gas velocity within interstitial passages of the emulsion phase, including the effect of bubbles. The interstitial velocity is obtained by reducing the momentum equations of Anderson and Jackson [45] and linearizing the Ergun equation for pressure drop in packed beds [8] to obtain

$$\vec{\nabla} p = -k_\beta(\vec{Q}_g - \vec{Q}_s) \tag{52}$$

where $\vec{Q}_g$ and $\vec{Q}_s$ are average gas and particle velocities, respectively, k_β is the drag parameter, and p is the gas pressure. In beds of large particles operating near minimum fluidizing conditions, the particle velocities induced by rising bubbles are much smaller than the interstitial gas velocities, and $\vec{Q}_s$ can be neglected.

When the voidage and k_β are constant, Equation 52 and conservation of mass reveal that $p/k\beta$ serves as a potential function for velocity difference, $\vec{Q}_g - \vec{Q}_s$. Also, when the potential flow model of Davidson [46] is used to determine the particle velocity field induced by rising bubbles (note that the solutions of the complete particle momentum equation reported by Martin-Gautier and Pyle [47] offer no substantial increase in accuracy), the interstitial gas velocity field is based completely on potential flow. These approximations were also used by Kilkis [48] to obtain a solution for the interstitial velocity field near surfaces immersed in a bubbling bed. Kilkis [48] used a finite element algorithm to carry out the analysis while the approach to be described here is due to Adams and Welty [49] and based upon analytical methods which employ complex variable theory and the method images for potential flow.

When Equation 52 is used to determine the local gas velocity at the heat transfer surface, the parameter, k_β, is based upon local surface voidage except in the high voidage lower stagnation region of horizontal tubes where the approximation breaks down. In this region, a correlation based upon numerical solutions by George et al. [50] is used to relate the potential flow, constant voidage pressure gradient to the actual variable voidage value.

For the particular case of a horizontal tube and two-dimensional bubbles, the velocity potentials can be constructed using complex variable theory and the method of images. The condition of flow tangency for the gas at a tube surface and for particles relative to a moving bubble (i.e. constant stream function) is satisfied by complex functions of the form

$$G(Z) = F(Z) + \bar{F}\left(\frac{b_i^2}{Z - Z_{0i}}\right) \tag{53}$$

in accordance with the circle theorem (see e.g. Batchelor [51], where b_i is the radius of the object and Z_{0i} is the location of the center of the object. While the condition of constant pressure along a bubble boundary (i.e. constant potential) is satisfied by functions of the form

$$H(Z) = F(Z) - \bar{F}\left(\frac{b_i^2}{Z - Z_{0i}}\right) \tag{54}$$

Equations 53 and 54 can be used to construct an algorithm similar to that reported by Halsey [52] to handle multiple reflections of images.

The emulsion-phase velocity field near a contacting bubble requires special consideration because the bubble and tube boundaries intersect and the images method cannot be used. However, it has been found that successive application of the Joukowski transformation,

$$\zeta = c\left(\frac{\tau + 1}{\tau - 1}\right) \tag{55}$$

where $\tau = [(Z + c)/(Z - c)]^{1/\alpha}$, $\alpha = 3/2$, and c is half of the chord length of the intersected region, will produce two pseudocircles with a negligible region of intersection. This transformed geometry is approximated as two tangent circles and the image method is applied on the transform plane.

Generally 5 image reflections are required when bubbles contact the tube while two or three reflections are sufficient for object separations greater than .25 tube diameter for boundary-condition accuracy to within 5%.

The analysis just described will provide instantaneous velocity fields provided the configuration of bubbles near the heat transfer surface is known. Results of calculations using this method have been reported by Adams and Welty [49].

In order to estimate the time-averaged heat transfer, the effect of bubbles could be assumed to be self-compensating so that the velocity distribution for the bubble-free case can be used. This approach provides the following for a horizontal tube

$$Q_g/u_{ref} = 2\,[k_\beta(\varepsilon_\infty)/k_\beta(\varepsilon)]\sin\theta/\varepsilon \tag{56}$$

with $u_{ref} \leqslant u_{mf}$

and where

$$k_\beta(\varepsilon) \simeq \frac{150}{Re_p}\left(\frac{1-\varepsilon}{\varepsilon\varphi_s}\right)^2 + \frac{1.75}{\varepsilon_\infty}\left(\frac{1-\varepsilon}{\varepsilon\varphi_s}\right) \tag{57}$$

from the linearized form of the Ergun equation. Also, finite difference solution of George et al. [50] for the interstitial flow field in the high voidage lower stagnation region of a horizontal tube can be combined with the gas convective heat transfer analysis to yield the approximation

$$Nu_{pZD}\varepsilon \geqslant 0.8 \simeq 1.1\left[\frac{k_\beta(\varepsilon_\infty)\,d_p/D\,Re_p\,Pr^{.7}(\Omega_v/\Omega_{v0})^{-1}}{\varepsilon_\infty k_\beta(\varepsilon)\left\{1+(\varepsilon/\varepsilon_\infty-1)^2\left[.065+d_p/D\left(12.13+\dfrac{991}{Re_p}\right)\right]\right\}}\right]^{1/2} \tag{58}$$

While the Stokes region contribution for the high voidage regions is estimated by taking $x_s = r_p$ so that

$$Nu_{pe}\varepsilon \geqslant 0.8 \simeq (1 - r_p/s_p)^2\,Nu_{p2D} + \left(\frac{r_p}{s_p}\right)^2\left[\frac{x_s}{s_p}Nu_{ps}(Fo_m)\right] \tag{59}$$

with $x_s/s_p Nu_{ps}(Fo_m)$ estimated from the transient conduction analysis for a particular gas/solid system.

BUBBLE-PHASE CONVECTIVE HEAT TRANSFER

The gas convection contribution from the bubble phase is dependent upon the flow field within the bubble which is produced as the gas flows into the bottom and out of the top of the bubble as shown in the inset of Figure 11. Thus, the bubble-phase heat transfer is highly dependent upon the hydrodynamic characteristics of the heat transfer surface/bubble geometry. An empirical correlation of the bubble-phase convective heat transfer to horizontal tubes has been obtained by Catipovic et al. [16] by assuming that the gas through-flow is equivalent to a particle-free flow at a "free stream" velocity of 3 U_{mf}. Reasonable agreement with experiment is obtained by using the single-phase correlation to obtain

$$Nu_{DB} = (0.88\,Re_{Dmf}^{0.5} + 0.0042\,Re_{Dmf})\,Pr^{0.33} \tag{60}$$

where the Reynolds number, Re_{Dmf}, is based upon u_{mf} and tube diameter.

In the case of a horizontal cylinder, the interstitial flow field model, previously described can be used to establish boundary conditions for a potential flow model for the bubble flow field. Then a complex potential describing the flow is assumed to be of the form

$$W_B = \sum_{n=0}^{N}\left[D_n Z^n + \bar{D}_n\left(\frac{a^2}{Z}\right)^{-n}\right] \tag{61}$$

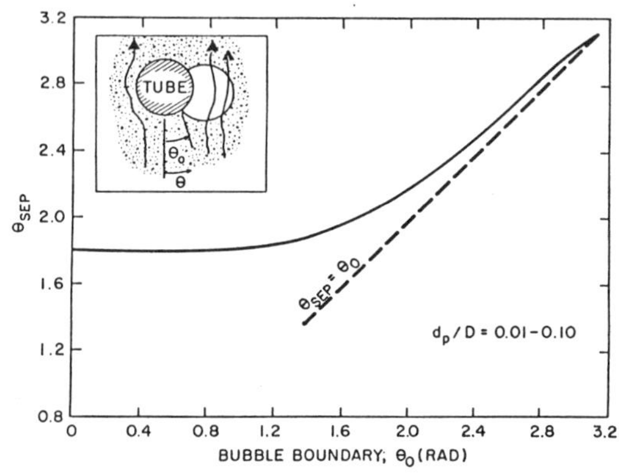

Figure 11. Location of boundary layer separation within a bubble contacting a horizontal tube [60].

where a is the tube radius and the complex coefficients, D_n, are obtained by making a least-square fit to the in-flow boundary condition at the bubble surface using the Householder transformation (see Ralston and Rabinowitz [53]). Note that a complex potential of this form automatically satisfies flow tangency at the tube surface, in accordance with the circle theorem. To obtain the heat transfer at the tube surface, the one-parameter method of Smith and Spalding [54] is used in place of the more complex model previously described, with the augmentation due to turbulence determined from the stagnation point flow results.

Accordingly, the local Nusselt number (here based upon tube diameter) is obtained from the following

$$Nu_D = \left(\frac{a}{Re_D} \frac{\int_{x0}^{x} U^{b-1}dx}{U^b} + \frac{U_0^b}{U^b {Nu_{D0}}^2}\right)^{-1/2} \tag{62}$$

where $a \simeq 9.07\,Pr^{-0.7}$, $b \simeq 2.95\,Pr^{.07}$, Pr is the gas Prandtl number, x is the dimensionless streamwise coordinate (relative to D), U is the dimensionless boundary layer edge velocity (relative to minimum fluidizing velocity), and U_0 and Nu_{D0} are evaluated at X_0, the leading edge of the bubble.

The analysis can be simplified for the case of a single slow bubble contacting a horizontal tube for which the boundary-layer edge velocity has been shown by Adams [55] to be

$$u/u_{mf} = 4\sin\theta \tag{63}$$

with θ measured from the lower stagnation point of the tube. Note that this result is independent of the size and position of the bubble.

For the contacting bubble, X_0 is the trailing edge location of the bubble so that Nu_{D0} is taken as the two-dimensional Nusselt number for the interstitial flow at this point. Thus, the local instantaneous Nusselt number within the contacting bubble becomes

$$Nu_D(\theta, \theta_0) \simeq \left[\frac{Pr^{.7}\,Re_D \sin^b\theta}{1.13 \int_{\theta_0}^{\theta} \sin^{b-1}\theta_1 d\theta_1 + \frac{0.297\,\varepsilon_\infty d_p}{\sqrt{1-\varepsilon_h}\,D} \sin^{b-1}\theta_0}\right]^{1/2} \tag{64}$$

where θ_0 locates the bubble trailing edge, d_p is the particle diameter, and ε_∞ is the bed voidage. Now, the time-averaged local bubble Nusselt number is obtained by assuming that the bubble trailing edge position, θ_0, is uniformly distributed between the lower stagnation point ($\theta_0 = 0$) and the position of interest ($\theta_0 = \theta$). Thus, the time-averaged local bubble Nusselt number is

$$\bar{N}u_D(\theta) = \frac{1}{\theta}\int_0^\theta Nu_D(\theta, \theta_0)\, d\theta_0 \tag{65}$$

The effect of free stream turbulence arising from the interstitial flow is included by assuming that the relative change is given by that for stagnation point flow as previously described, i.e.

$$\frac{(Nu_D)_{u'}}{(Nu_D)_{u'=0}} = \left(\frac{\Omega_{v0}}{\Omega_v}\right)^{1/2} \simeq \left(0.2 + 0.8\exp\left(-0.0849\,u'\,\frac{\sqrt{Re_D d_p}}{\varepsilon_\infty D}\right)\right)^{-1/2} \tag{66}$$

where Ω_v is the velocity profile shape parameter and u′ is the interstitial turbulence intensity.

When Equations 64 and 65 are combined and Equation 66 is used to account for the free-stream turbulence effect, the following expression is obtained for the time-averaged local bubble Nusselt number

$$\frac{\bar{N}u_D(\theta)}{\sqrt{Re_D Pr^{0.7}\Omega_{v0}/\Omega_v}} = \frac{\sin^{b/2}\theta}{\theta}\int_0^\theta \frac{d\theta_0}{\sqrt{1.1\dot{3}\int_{\theta_0}^{\theta}\sin^{b-1}\theta_1 d\theta_1 + \frac{0.297\,\varepsilon_\infty d_p}{\sqrt{1-\varepsilon_\infty}\,D}\sin^{b-1}\theta_0}} \tag{67}$$

Note that the R.H.S. of Equation 67 serves as an influence coefficient which depends primarily upon θ since b is a weak function of the Prandtl number and the second term under the radical is small.

In conjunction with Equation 67, it is also necessary to know the location of the boundary-layer separation within the contacting bubble so that the results can be properly applied. The separation location will depend upon the trailing edge location of the bubble (θ_0); however, estimates made using the Thwaites method (see, e.g., White [54]) give boundary-layer separation at $\theta = 1.8$ radians (103 deg) for θ_0 up to 1 radian (57.3 deg). For $\theta_0 > 1$ radian, the location of separation asymptotically approaches θ_0. These results are shown in Figure 11. From these results it would seem that contacting bubbles which do not surround the tube should not exist on the upper part of the tube, because the reverse flow produced would cause the bubble leading edge to collapse at the tube surface. This observation seems to be consistent witht he results of experimental studies of bubble/tube interaction performed by Hagar and Thomson [56].

The extension of the model to the case of a three-dimensional spherical bubble can be accomplished by increasing the bubble through-flow velocity by a factor of 1.5 in accordance with the isolated bubble fluid mechanical theory of Davidson [46]. This procedure results in an increase of about 22% in the bubble-phase Nusselt number over the two-dimensional case.

The time-averaged local bubble Nusselt number parameter given by Equation 65 is shown for particle sizes of 1.3 to 6.6 mm diameter and for a 50.8 mm diameter horizontal tube in Figure 12. For these calculations, the bed voidage was assumed to be 0.5, and the gas Prandtl number was taken as 0.72. The theoretical values of the heat transfer parameter are seen to be relatively insensitive to position (except near the lower stagnation point, $\theta = 0$) and particle size. Shown with calculated values are experimental results based upon Catipovic's [10] minimum local heat transfer coefficients obtained in a cold, two-dimensional bed and with an assumed turbulence intensity, u′, of 0.2 based upon single-bubble contact compare at least qualitatively with the experimental results with

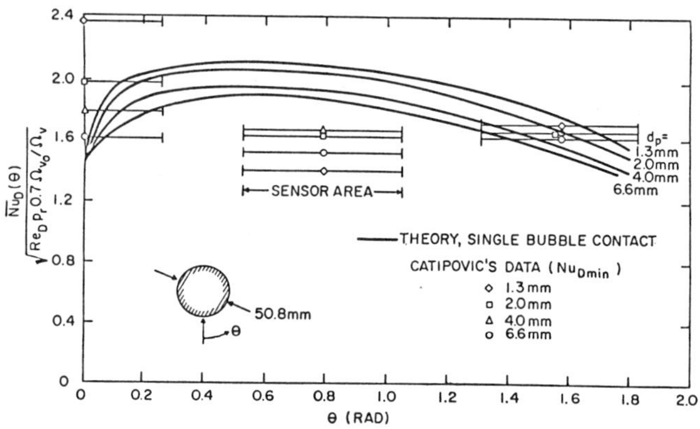

Figure 12. Calculated bubble phase convective heat transfer parameter compared to experimental data of Catipovic [10] for a horizontal tube [60].

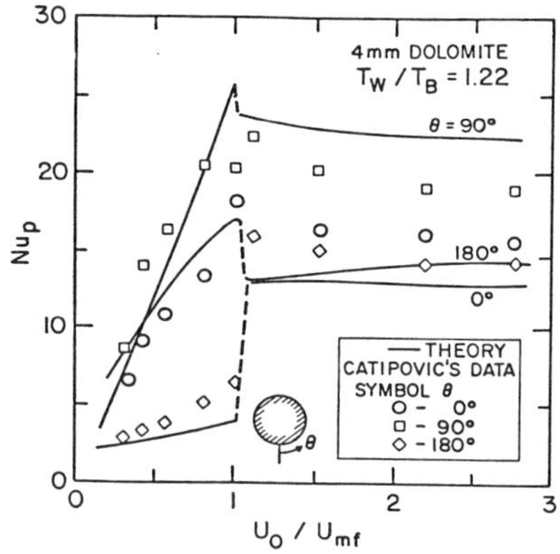

Figure 13. Calculated local Nusselt numbers for a horizontal tube immersed in a bed of 4-mm dolomite particles at low temperature compared to experimental data of Catipovic [10] [61].

reasonable agreement at $\theta = 0$ and $\pi/2$ (except for 1.3 mm diameter particles at $\theta = 0$). However, there are uncertainties in the comparison; the experimental results represent the effects of multiple bubble interaction, the bubbles are not circular, and minimum heat transfer levels may not always represent bubble levels.

HEAT TRANSFER TO A HORIZONTAL CYLINDER

The total heat transfer to an immersed surface can be determined by combining the conductive and convective components previously described. The gas-convection model can be used with any of the conduction models mentioned because of the independence of these components, though the results presented here are based upon the single-particle model previously described. In the case of immersed tubes, the review by Saxena et al. [57] provides a comprehensive summary of experimental correlations.

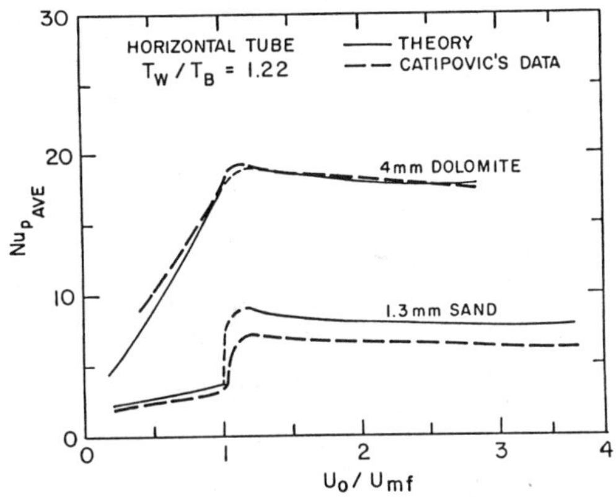

Figure 14. Calculated total heat transfer to a horizontal tube compared with data of Catipovic [10] [61].

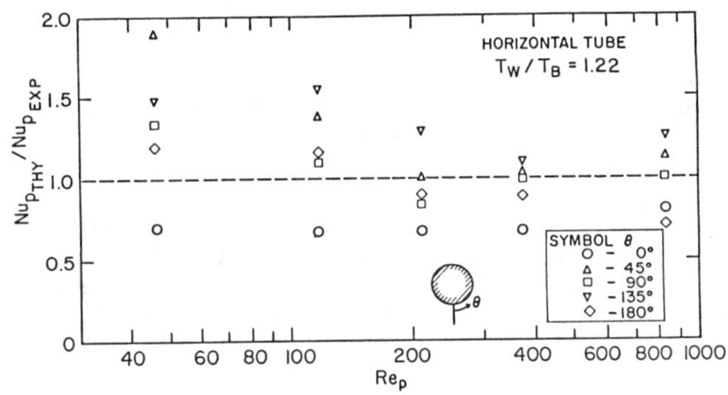

Figure 15. Comparison of measured and calculated values of local heat transfer at u_0/u_{mf} = 1.1, data from Catipovic [10] [61].

The detailed heat transfer and hydrodynamic parameters measured by Catipovic [10] provide sufficient information to both apply and test the model. Catipovic's experiments were conducted in low-temperature two-dimensional beds of sand and dolomite fluidized with air. Heat transfer coefficients, bubble contact fraction, and emulsion-phase residence times were determined locally for a horizontal tube of 50.8 mm diameter and with sand particles of 1.3 mm and dolomite particles of 2.0 mm, 2.8 mm, 4.0 mm, and 6.6 mm mean diameters. His hydrodynamic data obtained using capacitance probes provide the bubble contact fraction correlation given in Equation 51, while Catipovic's measured contact times are used directly. A summary of input parameters used in the calculation is provided in Table 1.

Local time-averaged Nusselt numbers are compared with data for 4 mm dolomite particles in Figure 13. Shown are results for the packed bed ($u_0 < u_{mf}$) as well as the fluidized state up to $u_0 = 2.8\,u_{mf}$. Generally, the trend of the theory is correct but predicted values are at best reasonable when compared to data. The overall time-average Nusselt number for the tube is shown with Catipovic's data for 4 mm and 1.3 mm diameter particles in Figure 14. Compensating errors in the theory (particularly at $\pi/4$ and $3\pi/4$ and not shown

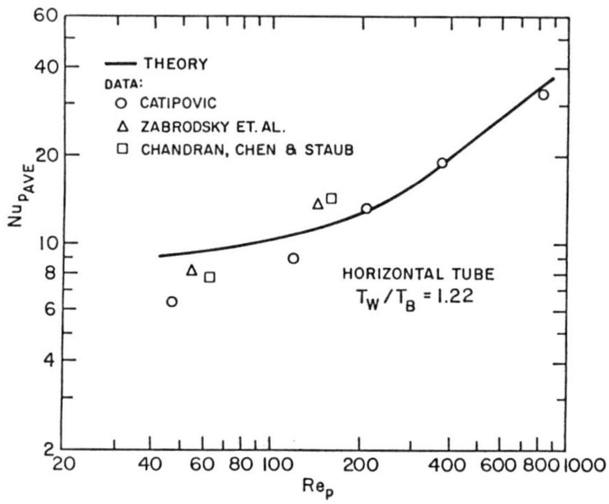

Figure 16. Total heat transfer to tubes immersed in fluidized beds at low temperature [61].

in Figure 13) result in excellent agreement between theory and experiment for 4 mm dolomite while the 1.3 mm sand case is overpredicted.

The overall success of the model in predicting local time-averaged Nusselt numbers at $u_0/u_{mf} = 1.1$ (so that significant bubbling occurs) is illustrated in Figure 13. Shown in this figure is the ratio of predicted and measured values over a range of particle Reynolds numbers from 46.2 to 817. These results show a trend toward over prediction at the lower Reynolds numbers (smaller particles), which is most likely due to the single-particle feature of the conduction model used in the calculation. Some improvement in these values can be obtained by using the method of Gabor [35] as demonstrated by the work of Fatani [58].

Though local values are not accurately predicted for all cases the errors seem to be self compensating so that reasonable agreement is obtained for the total heat transfer as Figure 16 shows. Included in this figure are results obtained by Zabrovsky et al. [15] as well as Chandron et al. [59].

NOTATION

A	constant appearing in formula for mean residence time
a	parameter appearing in Smith and Spalding Nusselt number formula and tube radius
Ar	Archimedes number
b	parameter appearing in Smith and Spalding Nusselt number formula
b_i	radius of bubble or tube
Bi	Biot number; $\dot{q}''r_p/k_s(T_W - T_B)$
C	coefficient in Joukowski transformation
C_{pg}	specific heat of gas
C_{ps}	specific heat of solids
D	tube diameter
$D_n, \bar{D}_n$	complex coefficient and conjugate
d_p	mean particle diameter
$F, \bar{F}$	complex potential function and conjugate
f_B	bubble contact fraction at immersed surface
Fo	Fourier number; $k_s\tau_e/(\rho_s C_{ps} r_p^2)$

Fo_m	Fourier number based upon $\bar{\tau}_e$
G, H	complex functions
g	gravitational acceleration
h	scale factor
h_{gc}	gas-convective heat transfer coefficient
h_{pc}	particle-convective heat transfer coefficient
i	index and unit complex number; $\sqrt{-1}$
k_e^0	effective thermal conductivity of quiescent bed
kg, kg*	gas thermal conductivity and gas thermal conductivity at average temperature; $(T_B + T_W/2$
K_{mf}	effective thermal conductivity of bed at minimum fluidizing conditions
K_T	interstitial turbulence parameter
K_β	pressure gradient parameter for interstitial flow
K_s	thermal conductivity of solids
K_1, K_2	constants
L	length
L_c	length dimension of interstitial channel
L_T	length scale of turbulence
n	index and normal coordinate
N_{HV}	ratio of boundary layer thickness; (δ_H/δ_V)
Nu_{DB}	Nusselt number for bubble-phase heat transfer; $\dot{q}_B'' D/k_g^* (T_B - T_W)$
Nu_{D0}	Nusselt number at bubble trailing edge
Nu_{pe}	Nusselt number for emulsion phase heat transfer; $\dot{q}_e'' d_p/k_g^* (T_B - T_W)$
Nu_{0m}	maximum Nusselt number
Nu_{ppc}	Nusselt number for particle convective heat transfer; $\dot{q}_{pc}'' d_p/k_g^* (T_B - T_W)$
Nu_{ps}, Nu_{psm}	Stokes region Nusselt number and mean value
Nu_{psg}	component of Nu_{ps} due to gas
Nu_{pss}	component of Nu_{ps} due to solids
Nu_{p0}	Nusselt number for gas-solid heat transfer in quiescent bed
Nu_{p2D}	Nusselt number for heat transfer due to two-dimensional portion of interstitial channel
p	gas pressure
Pr	Prandtl number
$\dot{q}''$	heat transfer rate per unit area
$\vec{Q}_g, \vec{Q}_s$	average velocities of gas and solid phase, respectively
$\dot{q}_B''$	bubble-phase heat transfer rate per unit area
$\dot{q}_e''$	emulsion-phase heat transfer rate per unit area
$\dot{q}_{gc}''$	gas-convective heat transfer rate per unit area
$\dot{q}_{pc}''$	particle-convective heat transfer rate per unit area
r	radial coordinate
Re	Reynolds number; $u_{ref}L/\nu_g$
Re_{Dmf}, Re_D	Reynolds number at minimum fluidizing condition based upon tube diameter; $u_{mf}d/\nu_g$
Re_c	channel Reynolds number; $Q_g d_p/\nu_g$
Re_L	Reynolds number based upon L
Re_p	Reynolds number at minimum fluidizing conditions based upon particle diameter; $u_{mf} d_p/\nu_g$
R_f	contact resistance for heat transfer

r_m	maximum value of projected radial coordinate
r_p	particle radius
R_s	solid phase resistance for heat transfer
R_1	dimension of transformed geometry of Stokes region
S, S_0	temperature-conductivity parameter defined by Kirchoff's transformation
S_p	half of distance between particles
T	temperature
T_B	bed temperature
T_W	wall temperature
T^*	average temperature; $(T_B + T_W)/2$
u, υ	real and imaginary parts of complex number and velocity components
u'	intensity of interstitial turbulence
u_{mf}	minimum fluidizing velocity
u_{ref}	reference velocity
U_0	velocity of gas at bubble trailing edge; also superficial velocity
W, W_B	complex potential
x, y, z	cartesian coordinates
x_s	width of Stokes region near particle contact point
x_0	location of bubble trailing edge
Z_{0i}	complex coordinate of center of bubble or tube

Greek Symbols

α	exponent for thermal conductivity variation with temperature and parameter in Joukowski transformation
α_1, α_2	parameters in expression for Biot number
β	spherical coordinate (colatitude)
β_1, β_2	Stokes region Nusselt number parameters
β_m	integration limit for Nusselt number calculation
β_{c1}, β_{c2}	conductivity-temperature function, S parameters
β_T	interstitial turbulence parameter
δ	boundary layer thickness
δ_H	thermal boundary layer thickness
δ_V	momentum boundary layer thickness
$\varepsilon, \varepsilon_\infty$	voidage, bed voidage
ε_T	turbulent diffusivity
θ	dimensionless temperature, $(T-T_w)/(T_B-T_W)$ and angular coordinate
θ_1	integration variable
ν_g	gas kinematic viscosity at mean temperature
ρ_g	gas density
ρ_s	solid density
τ	parameters in Joukowski transformation
τ_e	average emulsion-phase residence time
φ_0, φ	angular coordinate
φ_s	particle sphericity
$\Omega_v, \Omega_{vu'}, \Omega_{vu'=0}$	velocity profile shape parameter for 2-dimensional boundary layer flow
Ψ, Ψ_1	spherical coordinate (longitude)

REFERENCES

1. Baskakov, A. P., et al., "Heat Transfer to Objects Immersed in Fluidized Beds," *Power Technology*, Vol. 8 (1973), pp. 273–282.

2. Kolar, A. K., Grewal, N. S., and Saxena, S. C., "Investigation of Radiative Contribution in a High Temperature Fluidized-Bed Using the Alternative-Slab Model," *Int. J. Heat Mass Transfer*, Vol. 22 (1979), pp. 1695–1703.
3. Chen, J. C., and Chen, K. C., "Analysis of Simultaneous Radiative and Conductive Heat Transfer in Fluidized Beds," *Chem. Eng. Commun.*, Vol. 9 (1981), pp. 255–271.
4. Brewster, M. Q., and Tien, C. L., "Radiative Heat Transfer in Packed/Fluidized Beds: Dependent vs. Independent Scattering," *J. Heat Transfer*, Vol. 104, No. 4 (1982), pp. 573–579.
5. Brewster, M. W., "Radiative Heat Transfer in Fuidized Bed Combustors," ASME Paper 83-WA/HT-82 (1983).
6. Mahbod, B., and Adams, R. L., "A Model of the Radiative Contribution to Heat Transfer in a High-Temperature Large-Particle Gas Fluidized Bed," ASME/AIChE National Heat Transfer Conference, Aug. 1984.
7. Fantani, A., and Adams, R. L., "Heat Transfer in Beds of Small and Large Particles at High Temperatures," to be published.
8. Kunii, D., and Levenspiel, O., *Fluidization Engineering*, Wiley, New York, 1962.
9. Davidson, Harrison, and de Carvalho, "On the Liquidlike Behavior of Fluidized Beds," *Annual Review of Fluid Mechanics*, Vol. 9 (1977), pp. 55–86.
10. Catipovic, N. M., "Heat Transfer to Horizontal Tubes in Fluidized Beds: Experiment and Theory," Ph. D. thesis, Oregon State University, Corvallis, Oregon (1979).
11. Staub, F. W., "Solids Circulation in Turbulent Fluidized Beds and Heat Transfer to Immersed Tube Banks," Vol. 101, No. 3 (1979), pp. 391–396.
12. Botterill, J. S. M., *Fluid-Bed Heat Transfer*, Academic Press, New York, 1975.
13. R. Noack, *Chem. Ing. Tch.*, Vol. 42 (1970), p. 371. Also J. S. M. Botterill, *Fluid Bed Heat Transfer*, Academic Press, New York, 1975.
14. Adams, R. L., "An Unsteady Potential Flow Model of Bubble-Induced Particle Motion Near a Horizontal Tube in a Fluidized Bed," *Powder Technology*, Vol. 30 (1981), pp. 207–210.
15. Zabrodsky, S. S., et al., "Heat Transfer in a Large-Particle Fluidized Bed with Immersed In-Line and Staggered Bundles of Horizontal Smooth Tubes," *Int. J. Heat Mass Transfer*, Vol. 24, No. 4 (1981), pp. 571–579.
16. Catipovic, N. M., et al., "A Model for Heat Transfer to Horizontal Tubes Immersed in a Fluidized Bed of Large Particles," in *Fluidization – Proceedings of the 1980 Int. Fluidization Conf.* J. R. Grace and J. M. Matsen (Eds.), Plenum Press, New York, 1980, pp. 225–234.
17. Glicksman, L. R., and Decker, N., "Heat Transfer in Fluidized Beds with Large Particles," private communication.
18. Botterill, J. S. M., and Denloye, A. O. O., "Gas Convective Heat Transfer in Packed and Fluidized Beds – 1. A Theoretical Model," *AIChE Symposium Series No. 176*, Vol. 74 (1978), p. 194.
19. Xavier, A. M., and Davidson, J. F., "Heat Transfer to Surfaces Immersed in Fluidized Beds, Particularly Tube Arrays," *Fluidization*, Cambridge Univ. Press (1978), pp. 333–338.
20. Gabor, J. D., "Heat Transfer to Particle Beds with Gas Flows Less than or Equal to that Required for Incipient Fluidization," *Chem. Eng. Sci.*, Vol. 25 (1970), p. 979.
21. Gelperin, N. I., and Einstein, V. G., "Heat Transfer in Fluidized Bed," in *Fluidization*, J. F. Davidson and D. Harrison (Eds.), London, Academic (1971), p. 471.
22. Kunii, J., and Smith, J. M., "Heat Transfer Characteristics of Porous Rocks," *AIChE J.*, Vol. 6 (1960), p. 71.
23. Baskakov, A. P., and Suprum, V. M., "Determination of the Convective Component of the Heat Transfer Coefficient to a Gas in a Fluidized Bed," *Int. Chem. Eng.*, Vol. 12 (1972), pp. 324–326.
24. Borodulya, V. A., Ganzha, V. L., and Podberezkiy, A. I., "Heat Transfer Between a Fluidized Bed Operating Under Pressure and a Wall Immersed in it," *Heat Transfer – Soviet Research*, Vol. 14, No. 3 (1982), pp. 94–101.

25. Smith, M. C., and Kuethe, A. M., "Effect of Turbulence on Laminar Skin Friction and Heat Transfer," *Phys. Fluids*, Vol. 9 (1966), pp. 2337–2344.
26. Traci, R. M., and Wilcox, D. C., "Free Stream Turbulence Effects on Stagnation Point Heat Transfer," *AIAA J.*, Vol. 13 (1975), pp. 890–896.
27. Levy, S., "Effect of Large Temperature Changes (Including Viscous Heating) upon Laminar Boundary Layers with Variable Free Stream Velocity," *J. Aero. Sci.*, (1954), pp. 459–473.
28. Kays, W. M., and Crawford, M. E., *Convective Heat and Mass Transfer*, 2nd ed., New York, McGraw Hill, 1980.
29. Adams, R. L., "An Analytical Model of Heat Transfer to a Horizontal Cylinder Immersed in a Gas Fluidized Bed," Ph. D. thesis, Oregon State University, Corvallis, Oregon (1977).
30. Adams, R. L., and J. R. Welty, "A Gas Convection Model of Heat Transfer in Large Particle Fluidized Beds," *AIChE J.*, Vol. 25, No. 3 (May 1979), pp. 395–405.
31. Mickley, H. S., and D. F. Fairbanks, "Mechanism of Heat Transfer to Fluidized Beds," *AIChE J.* (1955), pp. 374–384.
32. Baskakov, A. P., "The Mechanism of Heat Transfer Between a Fluidized Bed and a Surface," *Int. Chem. Eng.*, Vol. 4 (1964), p. 320.
33. Kubie, J., and Broughton, J., "A Model of Heat Transfer in Gas Fluidized Beds," *Int. J. Heat Mass Transfer*, Vol. 18 (1975), pp. 289–299.
34. Botterill, J. S. M., and Williams, J. R., "The Mechanism of Heat Transfer to Gas-Fluidized Beds," *Trans. Instn. Chem. Engrs.*, Vol. 41 (1963), pp. 217–230.
35. Gabor, J. D., "Wall-to-Bed Heat Transfer in Fluidized Beds," *AIChE J.*, Vol. 18 (1972), p. 249.
36. Arpaci, V. S., *Conduction Heat Transfer*, Addison-Wesley, Reading, Mass., 1966.
37. Dunsky, V. D., Zabrodsky, S. S., and Tamarin, A. T., "On Mechanism of Heat Transfer Between a Surface and a Bed of Moving Particles," *Proc. 3rd Int'l Heat Transfer Conf.*, Chicago, Vol. 4, paper 144 (1966), pp. 293–297.
38. Adams, R. L., "An Approximate Formula for Gas Convection Dominant Heat Transfer in Large-Particle Fluidized Beds," *J. of Heat Transfer*, Vol. 103 (1981), pp. 395–397.
39. Adams, R. L., "Extension of the Adams-Welty Fluid Bed Heat Transfer Model to the Packed Bed Case," *J. of Heat Transfer*, Vol. 103 (1981), pp. 602–604.
40. Decker, N. A., and Glicksman, L. R., "Conduction Heat Transfer at the Surface of Bodies Immersed in Gas Fluidized Beds of Spherical Particles," *AIChE Symp. Ser.*, Vol. 77, No. 298 (1981), pp. 341–349.
41. Galloway, T. R., and Sage, B. H., "A Model of the Mechanism of Transport in Packed, Distended, and Fluidized Beds," *Chem. Engrg. Sci.*, Vol. 25 (1970), pp. 495–516.
42. Garnahan, B., Luther, H. A., and Wilkes, J. O., *Applied Numerical Methods*, Wiley, New York, 1969, Chap. 7.
43. Adams, R. L., "Coupled Gas Convection and Unsteady Conduction Effects in Fluid Bed Heat Transfer Based on a Single Particle Model," *Int. J. Heat Mass Trans.*, Vol. 25, No. 12 (1982), pp. 1819–1828.
44. Chandran, R., and Chen, J. C., "Bed-Surface Contact Dynamics for Horizontal Tubes in Fluidized Beds," presented at ASME/AIChE National Heat Transfer Conference, Milwaukee, Wis. (1981).
45. Davidson, J. F., and Harrison, D., *Fluidization*, New York, 1971.
46. Davidson, J. F., "Symposium on Fluidization – Discussion," *Trans. Inst. Chem. Eng.*, Vol. 39 (1961), p. 230.
47. Martin-Gautier, A. L. F., and Pyle, D. L., "The Fluid Mechanics of Single Bubbles," *Fluidization Technology*, Vol. 1, D. L. Keairns (Ed.), Hemisphere, Washington, D. C. (1976).
48. Kilkis, B., "A Two-Dimensional Finite Element Analysis of Hydrodynamical Behavior of Fluidized Beds," Lewis and Morgan (Eds.), Num. *Met. In Thermal Probs.* Swansea, UK, Pineridge Press (1979), pp. 840–849.

49. Adams, R. L., and Welty, J. R., "An Analytical Study of Bubble and Adjacent Tube Influence on Heat Transfer to a Horizontal Tube in a Gas Fluidized Bed," *J. of Heat Transfer*, Vol. 104 (1982), pp. 206–209.
50. George, A. H., Catipovic, N. M., and Welty, J. R., "An Analytical Study of Heat Transfer to a Horizontal Cylinder in a Large Particle Fluidized Bed," ASME Paper 79-HT-78 (1979).
51. Batchelor, G. K., *An Introduction to Fluid Dynamics*, Cambridge Univ. Press, England, 1967.
52. Halsey, N. D., "Potential Flow Analysis of Multielement Airfoils Using Conformal Mapping," AIAA Paper 79-0271, 1979.
53. Ralston, A., and Rabinowitz, P., *A First Course in Numerical Analysis*, 2nd ed., McGraw-Hill, New York, 1978.
54. White, F. M., *Viscous Fluid Flow*, McGraw-Hill, New York, 1974.
55. Adams, R. L., "An Exact Solution for the Inviscid Flow Field Within a Two-Dimensional Bubble Contacting a Horizontal Tube in a Gas Fluidized Bed," *Chem. Eng. Sci.*, to appear (1982).
56. Hager, W. R., and Thomson, W. J., "Bubble Behavior Around Immersed Tubes in a Fluidized Bed," *AIChE Symp. Ser.*, Vol. 69, No. 128 (1973), pp. 68–77.
57. Saxena, et al., "Heat Transfer Between a Gas Fluidized Bed and Immersed Tubes," *Advances in Heat Transfer*, Vol. 14, Irvine and Hartnett (Eds.), New York, Academic Press (1978), pp. 149–247.
58. Fatani, A. A., "An Analytical Model of Radiative and Convective Heat Transfer in Small and Large Particle Gas Fluidized Beds," Ph. D. thesis, Oregon State Univ., Corvallis, Oregon (1983).
59. Chandron, R., Chen, J. C., and Staub, F. W., "Local Heat Transfer Coefficients Around Horizontal Tubes in Fluidized Beds," *Journal of Heat Transfer*, Vol. 102, No. 1 (1980), pp. 152–157.
60. Adams, R. L., "An Approximate Model of Bubble Phase Heat Transfer to a Horizontal Tube in a Large Particle Fluid Bed," *Journal of Heat Transfer*, Vol. 104 (1982), pp. 565–567.
61. Adams, R. L., "Heat Transfer in Large Particle Bubbling Fluidized Beds," *J. of Heat*

CHAPTER 18

HEAT TRANSFER BETWEEN TUBES AND GAS-SOLID FLUID BEDS

N. S. Grewal

Mechanical Engineering Department
University of North Dakota
Grand Forks, North Dakota

CONTENTS

INTRODUCTION

Gas-solid fluid-bed heat exchangers are well known for significantly higher tube-to-bed heat transfer rates than conventional gas heat exchangers. Fluid-bed heat exchangers find applications in utilization of high-sulfur coal and low-rank coal, waste heat recovery systems, and calcination of radioactive waste. A good number of excellent reviews of studies of fluid-bed heat transfer are available in the literature [1–8].

The heat transfer rate between a tube and a gas-solid fluidized bed is influenced by a large number of variables such as particle size, particle size distribution, particle shape, particle and gas thermal properties, reactor geometry, type of gas distributor, fluidization conditions, radiative properties of bed material, heat transfer surface and fluidizing gas, and heat-transfer tube design. Before the presentation of predictive correlations for heat transfer between a tube and a fluid-bed, a brief review of theoretical models proposed to explain the mechanism of heat transfer in fluidized beds is given.

MECHANISMS OF HEAT TRANSFER BETWEEN TUBES AND GAS-SOLID FLUID BEDS

The total heat transfer coefficient, h_w, between a tube and a gas-solid fluidized bed can be taken approximately equal to three additive components [9];

$$h_w = h_{wc} + h_{wcv} + h_{wr} \tag{1}$$

The unsteady-state conduction component, h_{wc} (also called the particle convective component) is due to the heat transfer by unsteady-state conduction between the heat transfer surface and the solid particles directly adjacent to it and then the convection of the cooled particles to the bulk of the bed. For particle sizes between 40 to 800 μm, which fall in Geldart's Groups A and B [5], the contribution by the particle convection is predominant for nonpressurized systems at temperatures (< 900° K) such that radiation can be neglected. The gas convective component, h_{wvc}, is due to the augmentation of the heat transfer in the gas gaps between the particles and the heat transfer surface, and between neighboring particles by convective mixing of gas. Its contribution becomes increasingly important for denser and larger particles and at high operating pressures. The radiative component of the heat transfer coefficient, h_{wr}, becomes significant only for bed temperatures above 870° K [5]. The simple addition of these three components should be used with caution especially when two or more components are of comparable magnitude.

Unsteady State Conduction to Moving Solid Particles

A number of theoretical models have been proposed to model the tube-tobed heat transfer by unsteady-state conduction to moving solid particles. For vertical walls, they are essentially as follows:

1. The principal resistance to heat transfer from a heater surface to a bed is due to a fluid film, and the film resistance is decreased due to a reduction in film thickness caused by scouring action of moving solid particles [10–13]. However, experiments by various investigators [14, 15] showed that absorption of heat by solid particles rather than reduction of the film resistance was the main mechanism for heat removal from the surface in a fluid bed of small particles.
2. According to Mickley and Fairbanks [16], the heat is absorbed by "packet" of particles when resting close to the heater surface. The heated packet is then convected into the core of the bed and is replaced by a fresh packet of particles. The composition and effective properties of the packet are assumed to be the same as that of a bed at minimum fluidizing conditions. Figure 1 shows the main features of this model. If a packet at bulk bed temperature, T_b, comes in contact with a vertical heater at constant temperature, T_w, for a short time the local instantaneous heat transfer coefficient is given by the following equation.

$$h_{wiL} = (1/\sqrt{\pi})\,(k_{emf}\rho_{emf}C_{epmf})^{1/2}\tau^{-1/2} \tag{2}$$

here, k_{emf}, ρ_{emf}, C_{epmf} are the effective thermal conductivity, effective density, and effective specific heat of the packet, respectively. The local time-averaged heat transfer coefficient h_{wL} is then:

$$h_{wL} = (1/\sqrt{\pi})\,(k_{emf}\rho_{emf}C_{epmf})^{0.5}\int_0^{\infty}\tau^{-1/2}\psi(\tau)d\tau \tag{3}$$

where $\psi(\tau)$ denotes the frequency of occurrence of packets of age τ. The heat transfer rate between the surface and the gas when no packet is in contact with the surface is neglected. The predicted values of h_{wiL} for short residence times from Equation 2 are much higher than experimentally measured by Harakas and Beatty [17], Ernst [18], Botterill et al. [19], and Gabor [20]. Since thermal response is different in the gas and solid phases, the transient heat transfer in a gas-particle bed cannot be treated as heat transfer in a homogeneous medium as implied by using an effective thermal conductivity obtained from steady-state measurements.

The Mickley–Fairbanks model has been modified by several investigators [3, 9, 21–23] by the introduction of a contact resistance. The contact resistance at the wall-bed interface is attributed to various effects. For example, Kubie and Broughton [22] modeled voidage variations in the bed in the vicinity of the wall, and Wunschmann and Schlünder considered the reduced mean free path of the gas molecules. Antonshin et al. [24] later developed a

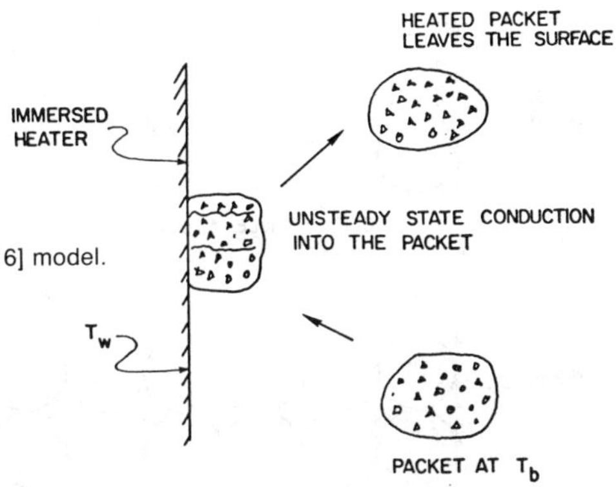

Figure 1. Schematic of Mickley and Fairbanks [16] model.

model without using a contact resistance. They concluded that if the temperature response of the gas and solid phases were each accounted for, rather than using average properties of the two-phase system, a contact resistance need not be assumed. Gelperin and Einshtein [6] have correlated various data based on the packet theory, but the use of their correlations require knowledge of parameters for each particular fluid-bed systems. It should be noted that since these are not models of fluidized bed-to-surface heat transfer but only of particle packet-to-surface heat transfer, an account of the bubble contact time is needed to calculate the average heat transfer coefficient.

Xavier and Davidson [25] used an approach similar to Baskakov et al. [9] and arrived at the following expression for h_{wc}:

$$h_{wc} = \frac{1}{(k_f/\delta_1) + [4\, k_{emf}\rho_{emf}C_{ps}(U - U_{mf})/(\pi \ell')]}(1 - \varepsilon_b) \tag{4a}$$

where

$$1 - \varepsilon_b = U_B/(U - U_{mf} + U_B) \tag{4b}$$

$$U_B = (U - U_{mf}) + 0.711(gD_{be})^{0.5} \tag{4c}$$

$$(\bar{d}_p/4) < \delta_1 < \bar{d}_p/10$$

The film thickness, δ_1 is taken equal to $\bar{d}_p/10$ and $\bar{d}_p/4$ for vertical and horizontal tubes, respectively. The characteristics dimension ℓ' is equal to the smaller of the vertical dimensions the tube and bubble length. The effective bubble diameter is calculated from the correlation proposed by Mori and Wen [26].

3. In another theoretical approach, heat transfer is modeled by unsteady-state conduction from the heater surface to a single particle [1, 5, 27], and to depths of two particles [19], and a particle chain of essentially unlimited length [28], (Figure 2). In this type of model, analysis is made of the heat as it is transferred to distinct particles, rather than by treating the bulk of the bed as a homogeneous medium using average bed properties and an effective thermal conductivity. In this model the primary resistance to heat flow is in the gas phase and the solid phase absorbs heat. After introducing a gas gap of 0.0075 d_p between the first row of particles and wall, and a gap of 0.015 d_p between particles, the model on a chain of particles of unlimited lengths [28] compared well with the data for both short and long particle residence times reported by several investigators [17, 19, 20].

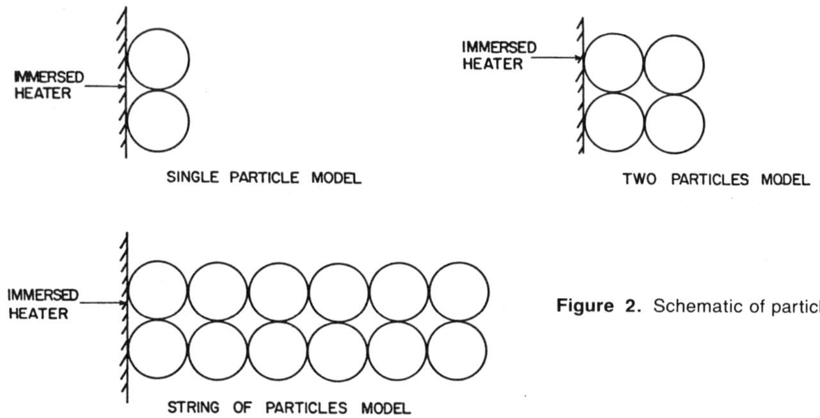

Figure 2. Schematic of particle models.

This chain-of-particles model coincides with the Mickley–Fairbanks model for long residence times.

Gabor [28] simplified the chain-of-particles model by approximating the gas-particle bed by a series of alternate gas and solid slabs (Figure 3). The alternate-slab model requires much less computing time. In all these models the knowledge of particle residence time and its distribution is needed to predict the heat transfer rates in fluidized beds.

If the characteristic conduction path through the gas between the particle and the surface is 1/10 the particle diameter, and if radiation, convection, and internal resistance of the particle can be neglected, the particle thermal time constant is [27]:

$$\tau_c = 1/15(\rho_s C_{ps} d_p^2/k_f) \tag{5}$$

If the time constant is much shorter than the packet residence time at the wall, then the packet renewal frequency plays an important role in the process of heat exchange between the wall and the bed. On the other hand, if the time constant is much larger than the packet residence time, the particle temperature can be assumed to be constant and consequently the heat transfer process can also be assumed to be steady state. For such a case, Glicksman and Decker [29] obtained the following expression for h_{wc} for spherical particles in contact with a surface when the contribution of bubble phase is neglected.

$$h_{wc} = 11.2(1-\delta)k_f/d_p \tag{6}$$

4. Recently Martin [30] has proposed a model which makes use of some basic ideas adopted from the kinetic theory of gases. The final expression to calculate h_{wc} is:

$$Nu_{wcp} = Nu_{max}\frac{(1-\varepsilon)}{C_k}\left(\frac{1-e^{-M}}{M}\right) \tag{7a}$$

where

$$M = \frac{6\, k_f Nu_{max}}{C_k \rho_s C_{ps}\{g\bar{d}_p^3\,(\varepsilon-\varepsilon_{mf})/[5\,(1-\varepsilon_{mf})\,(1-\varepsilon)]\}^{0.5}} \tag{7b}$$

$$Nu_{max} = 4\{[(1+2\sigma/\bar{d}_p)\ln(1+\bar{d}_p/2\sigma)]-1\} \tag{7c}$$

$$\sigma = 2(2/\gamma-1) \tag{7d}$$

$$C_k = 4\bar{d}_p/(U_p\tau) \tag{7e}$$

The values of Nu_{max} calculated from Equation 7c represent the maximum wall-to-particle heat transfer coefficient for short contact time [31]. Except for constant C_k, the foregoing equations contain only thermal and physical properties of solid particles and gas, particle diameter, bed void fraction, and void fraction at minimum fluidized conditions. The constant C_k was determined to be 3 from the data for glass beads. Using this value in Equation 7e, it is evident that the contact time of the particle is of the same order of magnitude as free-flight time for a particle to travel a distance equal to the particle diameter. The model is found to predict the effect of particle diameter, temperature, pressure, physical properties of the fluidizing gas and particles and gas velocity on the heat transfer coefficient quite well.

Gas Convection

Gas convective component of the heat transfer in a fluidized bed becomes important under the following conditions [1, 9, 32–44].

- Large particle diameter.
- High operating pressures.
- Dilute phase fluidization (very high velocities).
- Gas velocities near minimum fluidization conditions.
- Surface geometries which result in regions of either high voidage or stagnant particles.

Gabor [32] proposed a model to predict the heat transfer rate from an immersed vertical surface to bed for gas flows through the bed less than or equal to that required for minimum fluidization. The basis of the model is that all heat transfer is normal to the immersed surface and can be defined by an effective thermal conductivity for an orthorhombic array of spherical particles and that all heat is removed by gas flow. By considering the heat transfer coefficient at minimum fluidizing conditions obtained from the Gabor's model as a measure of gas convective component of the heat transfer coefficient, Xavier and Davidson [25] proposed the followig equation to predict h_{wcv}:

$$h_{wcv} = [4\, k_{emf} \rho_f C_{pf} U_{mf}/(\pi L_H)]^{0.5} + k_{emf}/D_T \tag{8a}$$

where the effective bed thermal conductivity, k_{emf}, at U_{mf} is given by:

$$k_{emf} = k_e + 0.1\, \rho_f C_{pf} \bar{d}_p U_{mf} \tag{8b}$$

The first term, k_e, in Equation 8b represents the effective thermal conductivity with no gas flow and is calculated following Kunii and Smith [45] or Krupiczka [46]. The second term represents the gas convective contribution.

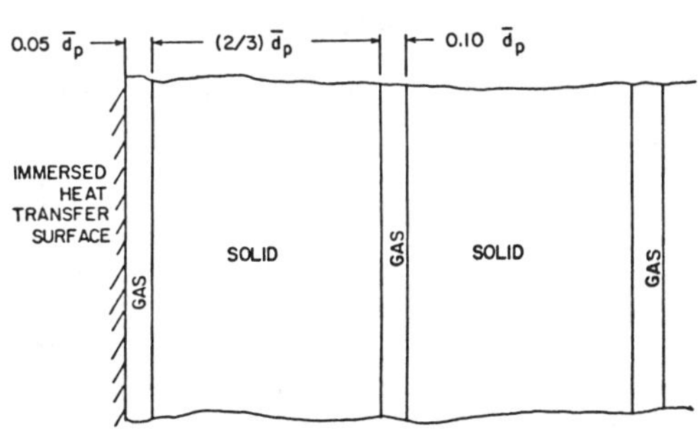

Figure 3. Gabor's alternate-slab model [28].

Botterill and Denloye [33] developed a model to describe heat transfer between a vertical cylindrical surface and packed beds and incipiently fluidized beds. They considered the bed as consisting of two regions:

- A region of increased voidage extending about half of a particle diameter from the cylindrical surface.
- The region comprising the rest of the bed.

The bed region near the wall was characterized by an effective thermal conductivity which was smaller than the effective thermal conductivity of the rest of the bed. The gas velocity and the voidage were assumed to be constant in each region. The first solution obtained was based on the steady-state heat transfer in the wall region, and it assumed that the temperature of the interface between the two bed regions is constant. The second solution was a numerical solution without the assumption of a constant interface temperature. By taking the gas connective component equal to the heat transfer coefficient at minimum fluidization conditions, this model can be used to predict h_{wcv}.

Based on the experimental data for an incipiently fluidized bed operating in the pressure range between 100 kPa to 1 MPa, Botterill and Denloye [33] proposed the following dimensional correlation for Nu_{wcvp}:

$$Nu_{wcvp} = 0.863\,Ar^{0.39}\bar{d}_p^{-1/2}, \qquad 10^3 < Ar < 2 \times 10^6 \tag{9}$$

By using the analogy between heat and mass transfer, Baskakov and Suprun [34] have developed a correlation for h_{wcv} based on data for mass transfer taken in a fluid bed operating at atmospheric pressure. For gas velocities below that necessary to give the maximum bed-to-surface heat transfer coefficient, U_{opt}, the proposed correlation is:

$$Nu_{wcvp} = 0.0175\,Ar^{0.46}Pr^{0.33}(U/U_{opt})^{0.3}, \qquad U_{mf} < U < U_{opt} \tag{10}$$

and for gas velocities higher than U_{opt} it is:

$$Nu_{wcvp} = 0.0175\,Ar^{0.46}Pr^{0.33} \tag{11}$$

The predictions from the preceding two correlations of Equations 9, 10 and 11 are good for data at atmospheric pressure. However, at high operating static pressure the Baskakov and Suprun correlation [34] does not predict the data well [33].

Adams and Welty [35] developed a theoretical model to calculate the instantaneous bubble and emulsion-phase gas convective heat transfer to a horizontal tube immersed in a fluidized bed. Because this model is too cumbersome to use for prediction purposes, this led Adams [36] to develop an approximate model for emulsion-phase gas convective heat transfer. According to this model [36], the Nusselt number is found to vary linearly with the Reynolds number.

Based on heat transfer studies done by Yagi and Kunii [47] in packed beds, Glicksman and Decker [27] recommended the following correlation for the estimation of the gas convective component:

$$Nu_{wcvp} = 0.042\,(1-\delta)Re_pPr \tag{12}$$

Catipovic et al. [37] proposed the following correlation for Nu_{wcvp}.

$$Nu_{wcvp} = (1-\beta)\,(0.0175\,Ar^{0.46}Pr^{0.33}) + \beta(\bar{d}_p/D_T)\,(0.88\,Re_{Tmf}^{0.5}+0.0042\,Re_{Tmf})Pr^{0.33} \tag{13}$$

In this equation the first term represents the gas convective heat transfer for the emulsion phase and is calculated with the help of the Baskakov and Suprun correlation [34]. The second term gives the gas convective heat transfer for the bubble phase.

According to Zabrodsky et al. [38] the gas convective component is represented by the following correlation:

$$h_{wcv} = 26.6\, U^{0.2} C_{pf} \rho_f \bar{d}_p \tag{14}$$

The above correlation is based on data for fluidized beds of large particles operating at atmospheric pressures.

Recently Ganzha et al. [39] have proposed the following correlation to estimate Nu_{wpcv} for vertical tubes:

$$Nu_{wcvp} = 0.12\, Re_p^{0.8} Pr^{0.43} (1-\varepsilon_w)^{0.133} \varepsilon_w^{-0.8} CT \tag{15a}$$

where

$$\varepsilon_w = \varepsilon_{wmf} + 1.65[(Re_p - Re_{pmf})\, Ar^{-0.5}]\,(1-\varepsilon_{mf}) \cdot \{1-\exp[-a\,Ar/(Re_p - Re_{pmf})^2]\} \tag{15b}$$

$$a = 0.367 \ln[(\varepsilon_{wmf} - \varepsilon_{mf})/(1-\varepsilon_{mf})] \tag{15c}$$

$$\varepsilon_{wmf} = 1 - \frac{(1-\varepsilon_{mf})\,[0.7293 + 0.5139\,(\bar{d}_p/D_T)]}{[1+(\bar{d}_p/D_T)]} \tag{15d}$$

$$CT = \frac{\{1-\exp[-0.1\, Ar^{0.1}(0.013/\bar{d}_p)]\}}{\{1-\exp[-0.1\, Ar^{0.1}(D_T/\bar{d}_p)]\}} \tag{15e}$$

The correlation is derived by considering that a turbulent boundary layer on the heat transfer surface is continuously disturbed by the front half of the solid particles and forms again in the wake and the dependence of Nu_{wcvp} on Re_p in fluidized beds is similar to that for turbulent gas flow over a flat plate.

Radiation

Zabrodsky [48] has discussed the rôle of radiative component of heat transfer in a gas fluidized bed. The radiation component, h_{wr}, can be considered insignificant at temperatures below 870° K [5]. At higher temperatures, the relative contribution by radiation becomes increasingly important with an increase in particle size, bed temperature, tube surface temperature, and radiative properties of bed material and tube surface.

A number of investigators [49–56] have attempted to develop theoretical models to predict the relative contribution due to radiation in high temperature fluidized beds.

Szekely and Fisher [49] using the Botterill and Williams [5] model for unsteady-state conduction from a single spherical particle in contact with a cold surface have obtained a simple expression for the radiative component. They concluded that the relative contribution due to radiation increases with an increase in bed temperature and particle residence time.

Yoshida et al. [50] have proposed an emulsion-layer model in which they have considered heat transfer by radiation through the bubble, and through the emulsion on the concept of effective thermal conductivity. They concluded that up to temperatures of 1,273° K the relative contribution due to radiation is insignificant for beds of small particles ($\bar{d}$ = 180 μm).

Vedamurthy and Sastri [51] calculated the radiation contribution in an air-ash system by assuming packets of particles 3 $\bar{d}_p$ thick which move from the core of the bed and reside at the heat transfer surface separated by a gas film 0.5 $\bar{d}_p$ thick for a given residence time. Heat is transferred from the packet to the wall by

- Conduction through the gas film.
- Radiation through the transparent film.

When the heat transfer surface is engulfed by bubble phase, the heat is transferred from the core of the bed to the surface by radiation. The bed and surface are considered as black bodies and the gas as radiatively transparent. They found that for a particle diameter of 500 μm, the radiative contribution is between 17 to 30 percent at a bed temperature of 1,173° K. Vedamurthy and Sastri [51] assumed a rather large gas film thickness (0.5 $\bar{d}_p$) between the packet and heat transfer surface which reduced both conductive and total heat transfer coefficients. For this reason as well as due to the assumption that the fluidized bed and heat transfer surface may be treated as a black body system, they overestimated the role of radiation. Wright et al. [52] made an additional assumption that the temperature of the radiating packet remains equal to the temperature of the bed core and this led to their estimation of radiation even larger than that of Vedamurthy and Sastri [51].

Bhattacharya and Harrison [53] have adopted the features of the Vedamurthy–Sastri model and by considering the gas to be radiatively absorbing and emitting and the bed to be gray, have computed the surface temperature dependence of radiative contribution for an air-sand system. They have found that for a particle diameter of 0.35 mm at T_b = 1,123° K, the contribution increases from about 10 percent at T_w = 473° K to about 20 percent at T_w = 1,073° K.

Thring [54] has examined the Vedamurthy–Sastri model with a smaller value of the surface film resistance. The model predictions are in good agreement with the experimental data for an air-sand system. He has also investigated the spherical particle and cubical particle models of Botterill and Williams [5] and concluded that the three models predicted widely different values for the radiative contribution but almost the same results for the maximum total heat transfer coefficient.

Basu [55] applied the single-particle model of Ziegler et al. [15] to investigate the effect of coal combustion on heat transfer rate to an immersed surface in a fluidized bed. The relative contribution of radiation was also investigated both theoretically and experimentally. The total heat transfer coefficient, in general, increased with an increase in carbon content of the bed. However, it decreased in some cases, depending on the relative size of the inert particles and coal particles. The relative contribution of radiation was experimentally determined to be between 5 to 10 percent for sand ($\bar{d}_p$ = 280 μm) at bed temperatures between 1,073 to 1,173° K, and wall temperature was between 293 to 303° K. However, the relative contribution of radiation as predicted from the model varied between 10 to 20 percent under similar conditions.

Kolar et al. [56] have extended the alternate-slab model of Gabor [28], as modified by Kolar et al. [57] to predict the radiative contribution in an air-sand fluidized bed. The model predicted that the percentage radiative contribution was significant for large particles, and high surface and bed temperature. The model generally over predicted the radiative contribution and the total heat transfer coefficients but within reasonable limits. The contact resistance at the wall surface was found to influence significantly the magnitude of relative contribution due to radiation.

All the preceding models are for fluidized beds of small particles. In order to estimate the radiative, as well as the total heat transfer coefficients, the knowledge of particle/packet residence time and its distribution is needed. Since this information is generally not available, the models, therefore, are not useful for prediction purposes.

Baskakov et al. [9] have experimentally investigated the relative contribution due to radiation in air fluidized at high temperatures. The results of this study are plotted in Figure 4. It is seen that the percentage radiative contribution increases with an increase in particle size and surface temperature. It is recommended that this figure be used to estimate the relative contribution due to radiation when the bed material and operating conditions are similar to this investigation.

The radiative component in fluidized beds of large particles can be estimated approximately from the following well-known equation:

$$h_{wr} = \frac{5.673\,[(T_b/100)^4 - (T_w/100)^4]}{(1/\varepsilon_T + 1/\varepsilon_B - 1)\,(T_b - T_w)} \tag{16}$$

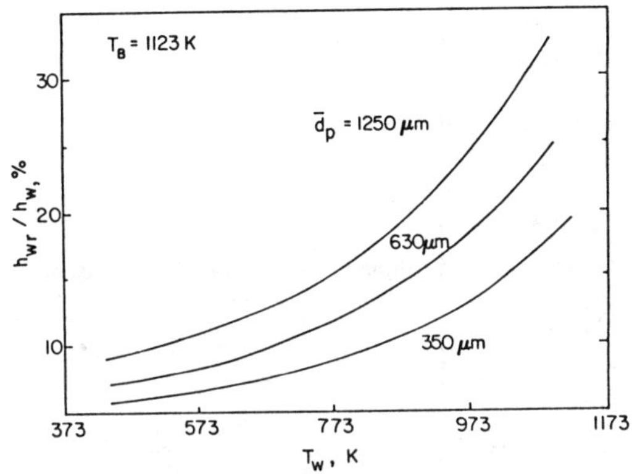

Figure 4. Relative contribution of radiation, h_{wr}/h_w, as a function of T_w for a bed of chamotte particles [9].

here ε_B is the effective emissivity of the bed and can be much higher than the emissivity of individual bed particles. The experimental values of ε_B for different materials are reported by Pikashov et al. [58]. For example, the effective emissivity of river sand (rounded particles; d_p = 1,000 − 1,500 μm; T_b = 773 to 1,373° K; U = 1.2 to 3.0 U_{mf}) is 0.85 as compared to the particle emissivity of 0.60. Based on these experimental data [58], the bed emissivity is approximately represented by:

$$\varepsilon_B = (1+\varepsilon_p)/2 \tag{17}$$

PREDICTIVE CORRELATIONS FOR HEAT TRANSFER BETWEEN TUBES AND GAS-SOLID FLUID BEDS

A large number of experimental studies to investigate the effect of various variables on h_w are reported in the literature [1–9].

The heat transfer coefficient, h_w, is shown plotted in Figure 5 as a function of superficial fluidizing velocity and particle diameter. The heat transfer coefficient increases with an increase in the value of U. The rate of increase is larger in the beginning but decreases as the fluidizing velocity is increased. The heat transfer coefficient attains a maximum value $h_{w\,max}$ at U_{opt}. With further increase in U, the value of h_w decreases. The dependence of h_w on U can be explained on the basis of "particle mode" of heat transfer [5]. The heat transfer rate between an immersed surface and a gas-solid fluidized bed of small particles mainly depends on particle concentration close to the heat transfer surface and particle residence time at the heat transfer surface. The particle residence time depends on solids mixing which is due to rising bubbles. The initial rise in h_w is mainly due to a lowering of particle residence time, resulting from greater particle mixing. The large degree of mixing is due to high bubble frequency resulting from a large U. However, the bulk particle packing density decreases with increase in U which causes a reduction in the heat transfer rate. At a particular fluidizing velocity, the value of h_w mainly depends upon the net contribution of these two opposing effects. With the increase in particle size, the contribution due to the unsteady-state heat conduction to moving solid particles, h_{wc}, becomes less important, therefore, the maximum for larger particles becomes less pronounced.

The dependence of $h_{w\,max}$ on particle diameter for a fluid bed operating at atmospheric conditions is shown in Figure 6. The heat transfer coefficient decreases with increase in

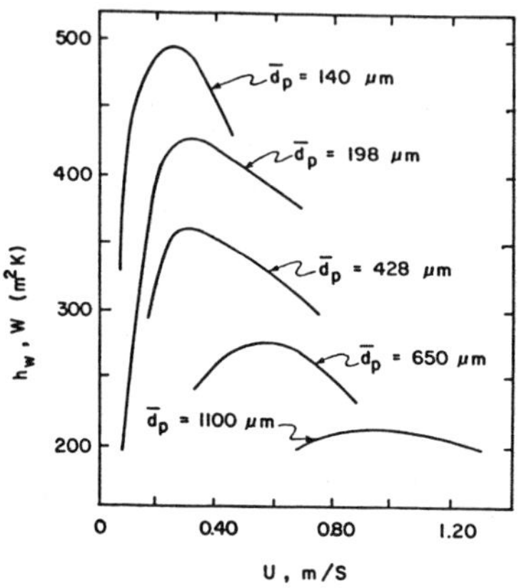

Figure 5. Heat transfer coefficient as a function of fluidizing velocity for different sizes of quartz sand [1].

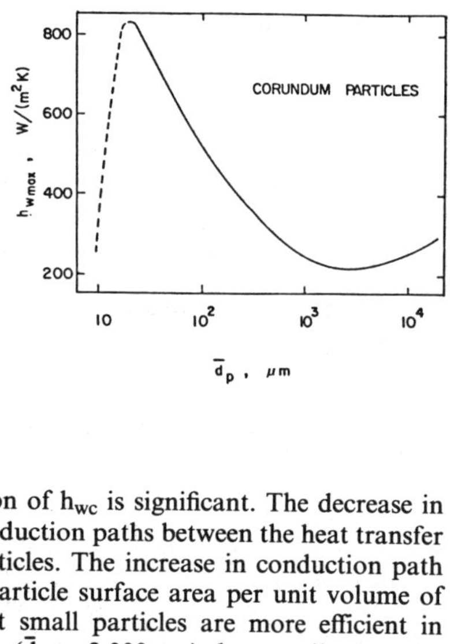

Figure 6. Maximum heat transfer coefficient as a function of particle diameter [9].

particle diameter in the range where the contribution of h_{wc} is significant. The decrease in h_w is primarily due to increase in the average gas conduction paths between the heat transfer tube and the first row of particles and between particles. The increase in conduction path increases the resistance to heat flow. Further, the particle surface area per unit volume of the bed is larger for smaller particles. As a result small particles are more efficient in exchanging heat with the surface. For large particles ($\bar{d}_p > 2{,}000$ μm) the contribution due to gas convection is significant. Therefore, h_w increases with an increase in particle diameter. The decrease in h_w for particles less than 20 μm, which fall in Group C of Geldart [59], is due to poor fluidization characteristics (agglomeration and Channeling) of such powders.

Local Heat Transfer Coefficients

The variation of the local heat transfer coefficient along the circumference of a horizontal tube is, generally, found to be the same in the reported investigations [9, 60–65]. At low gas velocities the maximum local heat transfer coefficients are observed at a place somewhat higher than the equator while minimum heat transfer coefficients are observed on the upstream and downstream sides. Figure 7 displays the local heat transfer coefficients, the intensity of solid particle movement close to the tube, the gas streamlines, and the movement of the solid particles for a cross section through the bed and horizontal tube for $U/U_{mf} = 1.34$ and $\bar{d}_p = 600$ μm. The maximum value of the local heat transfer coefficient occurs at 110° and this is due to intense particle circulation. The low values of the heat transfer coefficient on the downstream side are due to the presence of stagnant "caps" of defluidized solid particles. On the other hand, the low values of h_{wL} on the upstream side are due to gas pockets that are formed here from time to time. With an increase in fluidization velocity, the shape of the h_{wL} curve changes significantly. The absolute maximum moves from the equatorial lateral zones to the downstream side of the tube at higher velocities. The reason for this is that at higher velocities the bed porosity close to the tube increases significantly at the equatorial lateral zones of the tube, whereas at the downstream side of the tube, the bed density is relatively high because of the moving "cap" of almost defluidized solids.

The air pocket beneath an inclined tube is less stable than under a horizontal tube [61]. Therefore, for the same fluidizing velocity the values of h_{wL} at the upstream side of the slanted tube are higher than those for horizontal tubes.

The local gas velocity between tubes in an array is greater than a single tube for the same superficial fluidizing velocity. It has been observed that the cap of defluidized particles tends to be less stable on tubes in an array. This results in the maximum h_{wL} being displaced from the equatorial lateral zone toward the downstream side of the tube [60]. However, the total heat transfer coefficient for a tube in an array is generally lower than for a single tube.

From the studies [66–69] of the local time-averaged heat transfer coefficient for a vertical tube immersed in a fluidized bed, it may be concluded that h_{wL} tends to decrease with increasing height. For a bundle of vertical tubes, lowest coefficients in the dense phase are generally observed at the center location and higher values are observed at tube locations between the center and the wall of the column.

Average Heat Transfer Coefficients

For a fluid bed of small particles, the bed-to-surface heat transfer can be calculated with good reliability from theoretical models if particle residence times and their population near the immersed surface are known. However, the information about particle residence times is generally not known, and particle residence times cannot be calculated today a priori for practical fluidization. Therefore, empirical and semiempirical correlations are still used for

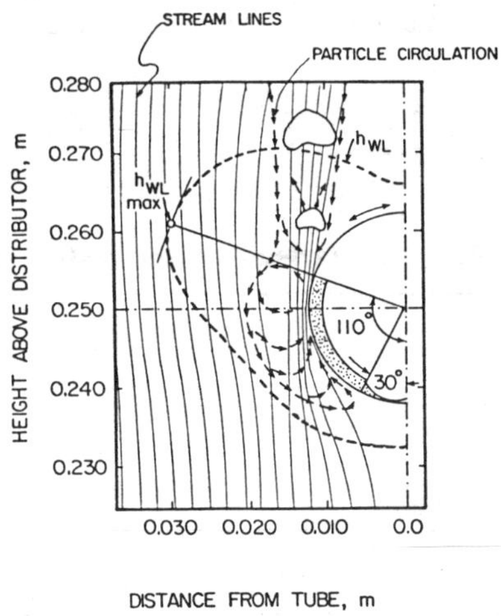

Figure 7. Variation of h_{wL} along the circumference of a horizontal tube [64].

predicting heat transfer rates. The use of these correlations should be limited to the experimental conditions for which they were derived. Extrapolation of these correlations to other conditions is risky because it can result in large discrepancies. All these correlations do not include the contribution due to radiation. Therefore, the h_{wr} should be estimated by the methods outlined in section [2–3].

Horizontal Smooth Tubes

Single tube. Vreedenberg [70] carried out extensive studies on heat transfer between a horizontal tube and an air-solid fluid bed. The tube was mounted 0.85 m above an orifice-type distributor in a 0.565-m-diameter column. The static bed height was 1.2 m. The effect of bed temperature (T_b = 313°–613° K), air mass velocity (G = 0.010–0.239 kg/m² s), particle shape (round and sharp), particle diameter ($\bar{d}_p$ = 64–316 μm), particle density (ρ_s = 1,600–5,150 kg/m³), and the tube diameter (D_T = 16.9–51.0 mm) on h_w was investigated. The experimental results for fine and light particles (a cracking catalyst, sand used for sand blasting, and dune sand) were correlated by

$$Nu_{wT} = 0.66\,[G\,D_T\rho_s\,(1-\varepsilon)/(\rho_f\mu\,\varepsilon)]^{0.44}Pr^{0.3} \tag{18}$$

for $[G\,\bar{d}_p\rho_s/(\mu\,\rho_f)] < 2{,}050$. The experimental results did not deviate by more than 42 percent than predicted from the Equation 18.

For coarser and heavier particles characterized by $[G\,\bar{d}_p\rho_s/(\mu\,\rho_f)] > 2{,}550$, the experimental results were correlated by

$$Nu_{wT} = 420\left(\frac{G\,D_T\rho_s}{\mu\,\rho_f}\cdot\frac{\mu^2}{\bar{d}_p^3\rho_s^2 g}\right)^{0.3} Pr^{0.3} \tag{19}$$

The experimental data were within ±29 percent from the correlation of Equation 19. For values of $[G\,\bar{d}_p\rho_s/(\mu\,\rho_f)]$ between 2,050 and 2,550, it was recommended that the average given by Equations 18 and 19 be used for prediction of h_w. Table 1 lists the range of variables over which Vreedenberg's correlations are applicable.

Vreedenberg's correlation of Equation 19, predicts a monotonic increase in h_w with an increase in G. However, the h_w, as discussed, passes through a maximum and then decreases with an increase in the bed porosity. Based on the data taken in a shallow fluidized bed of silica sand ($\bar{d}_p$ = 360–710 μm), Andeen and Glicksman [71] modified the correlation of Equation 19, by including the factor (1–ε), as follows

$$Nu_{WT} = 900(1-\varepsilon)\left(\frac{G\,D_T\rho_s}{\mu\,\rho_f}\cdot\frac{\mu^2}{\bar{d}_p^3\rho_s^2 g}\right)^{0.326} Pr^{0.3} \tag{20}$$

Table 1
Range of Variables Over Which Vreedenberg [70] Correlations are Applicable

Material	Particle Diameter $\bar{d}_p$ (μm)	Particle Density ρ_s (kg/m³)	Range of Mass Velocity G (kg/m²s)	Range of Bed Temperature T_b (°K)	Range of Reynold's Group $G\,\bar{d}_p\,\rho_s/(\rho_f\mu)$ (dimensionless)	Correlation Equation No.
Regenerated catalyst	64	1,600	0.010–0.149	313–533	49–740	18
Sand used for sand-blasting	136	2,600	0.046–0.077	313–473	670–1,290	18
Dune sand	213	2,660	0.056–0.096	313–513	1,270–2,550	18
New type of Maas sand	316	2,660	0.099–0.163	293–533	3,360–6,610	19
Sharp silver sand	215	2,660	0.089–0.151	313–533	2,050–4,060	19
Old type of Maas sand	353	2,660	0.148–0.242	313–593	5,620–10,900	19
Mixture of silver and old type of Maas sand	260	2,660	0.118–0.198	313–613	3,270–6,690	19
Round silver sand	259	2,660	0.091–0.150	313–553	2,520–5,500	19
Ground iron ore, wide sieve fraction	130	5,150	0.234	313–493	6,160–7,250	19
Ground iron ore, more homogeneous fraction	247	5,150	0.239	313–513	12,900–15,100	19

The modified Vreedenberg's correlation of Equation 20 predicted the experimental data of Vreedenberg [70] (excluding fine particles), Petrie et al [72], Ainshtein [73], and Andeen and Glicksman [71] very well. In this correlation, all the necessary properties of fluidizing gas are evaluated at the film temperature, T_f, where T_f is taken as the arithmetic average of the bed temperature and tube surface temperature. The bed porosity, ε, needed in the previous correlation is calculated with the use of the following correlation proposed by Goroshko et al [74].

$$\varepsilon = [(18\,\mathrm{Re_p} + 0.36\,\mathrm{Re_p^2})/\mathrm{Ar}]^{0.21} \tag{21}$$

In this correlation, the viscosity and density of the fluidizing gas are evaluated at the bed temperature.

For fluid bed of small particles, the dependence of h_w on the volumetric heat capacity of solid particles ($\rho_s C_{ps}$) is well known and has been verified experimentally [75, 76]. Therefore, Grewal and Saxena [77] proposed the following equation:

$$\mathrm{Nu_{wT}} = 47(1-\varepsilon)\left(\frac{G\,D_T\rho_s}{\mu\,\rho_f}\,\frac{\mu^2}{\bar{d}_p^3\rho_s^2 g}\right)^{0.325}\left(\frac{\rho_s C_{ps} D_T^{3/2} g^{1/2}}{k_f}\right)^{0.23}\mathrm{Pr}^{0.33} \tag{22}$$

This correlation is of the same form as the modified Vreedenberg's correlation except for an additional dimensionless factor, $(\rho_s C_{ps} D_T^{3/2}\, g^{1/2}/k_f)$, which accounts for the volumetric heat capacity of solid particles, the tube diameter, and the thermal conductivity of the fluidizing gas. This correlation is based on data for sand ($\bar{d}$ = 451,504 μm), silicon carbide ($\bar{d}_p$ = = 178,362 μm), alumina ($\bar{d}_p$ = 259 μm), dolomite ($\bar{d}_p$ = 312 μm), glass beads ($\bar{d}_p$ = = 265–427 μm) taken in a square air-solid fluid bed (.305 × .305 m) operating at atmospheric conditions, and Ziegler's data [75] on copper, nickel, and solder ($\bar{d}_p$ = 136 μm). All the properties of the fluidizing gas are to be evaluated at the film temperature. The bed porosity is calculated from the correlation of Equation 21. Tables 2 and 3 list the experimental conditions for various studies [60, 64, 70–73, 75, 77–90]. Generally the correlation predicted the data on small particles within ±25 percent. However, the experimental values of h_w of Cherrington et al. [89] were consistently larger than the predicted values of h_w.

Table 2
Range of Variables in the Experimental Data of Various Investigators

Parameter	Reference: Andeen and Glicksman [71]	Petrie et al. [72]	Ainshtein [73]	Grewal and Saxena [77]
Particle diameter, $\bar{d}_p$ (μm)	360–710	256–735	163–285	167–504
Particle density, ρ_s (kg/m^3)	2,660	1,490–3,012	2,651–2,659	2,490–4,450
Particle shape, φ_s	–	–	–	0.60–1.00
Particle specific heat, C_{ps} (kJ/(kg °K))	0.80	–	0.80	0.440–0.929
Bed material	Ottawa Sand	Sand, etc.	Sand	SiC, alumina silica sand, dolomite, glass beads
Tube diameter, D_T (mm)	19	12.7–35.6	22–30	12.7–28.6
Height of tube above distributor, H_d (m)	0.013	–	0.054–0.221	0.213
Bed size	0.61 × 0.61 m diameter	0.305–1.22 m diameter	.152–.275 m diameter	0.305 × 0.305 m diameter
Static Bed height, H_s (m)	0.02–0.04	–	0.35	0.237–0.367
Fluidizing gas	air	air	air	air
Fluidizing velocity, G; kg/(m^{2}s) (U, m/s)	(0.25–2.1)	0.08–0.384	(0.078–0.403)	0.086–0.735
Tube surface temperature, T_w (°K)	–	–	–	310–358
Bed temperature, T_b (°K)	–	400–772	295–406	298–315
Bed pressure (kPa)	100 kPa	100 kPa	100 kPa	100 kPa
Tube arrangement	Single row P/D_T = 2 Two rows of tubes in Staggered array	P = 127 mm, triangular array; P = 51 mm, in-line arrangement; P = 56 mm, triangular array	Single horizontal tube	Single horizontal tube

Table 3
Range of Variables Over Which Grewal and Saxena [77] Correction was Tested

Reference	Solid Particles: Material	Solid Particles: $\bar{d}_p$ (μm)	Tube Diameter D_T (m)	Bed Temperature T_b (°K)
Grewal and Saxena [77]	SiC	178, 362	0.0127	Room temperature
	Alumina	259	0.0127, 0.0286	Room temperature
	Silica sand	451	0.0127	Room temperature
	Dolomite	312	0.0127	Room temperature
	Silica sand	504	0.0127, 0.0191, 0.0286	Room temperature
	Silica sand	167	0.0127, 0.0286	Room temperature
	Glass beads	265	0.0191	Room temperature
	Glass beads	265	0.0286	Room temperature
	Lead-glass beads	241	0.0127	Room temperature
Grewal et al. [78]	Glass beads	265, 357, 427	0.0127	Room temperature
Chen [81]	Glass beads	240	0.025	Room temperature
Ziegler [75]	Nickel	136	0.0127	
	Copper	136	0.0127	Room temperature
	Solder	136	0.0127	
Antonshin [84]	Quartz sand	268	0.014, 0.021, 0.028	Room temperature
Cherrington et al. [89]	Glass beads	390, 1,000	0.0508	Room temperature
	Limestone	1,000	0.0508	
Howe et al. [87]	Dolomite	800	0.0508	1028, tube bundle
Gelperin et al. [60]	Quartz sand	350	0.020	Room temperature
Andeen and Glicksman [71]	Ottawa sand	360, 710	0.0191	Room temperature
Bartel and Genetti [86]	Glass beads	203, 279, 470	0.0159	Room temperature
Gelperin et al. [85]	Quartz sand	263	0.020	Room temperature
	Quartz sand	163	0.025	Room temperature
Noack [64]	Glass beads	600	0.025	Room temperature
Natusch and Blenke [80]	Glass beads	450	0.025	Room temperature
Janssen [90]	Quartz sand	1,000	0.018	473
Neukirchen and Blenke [88]	Glass beads	500	0.025	Room temperature, Tube bundle
	Glass beads	700	0.025	
	Quartz sand	720	0.025	
	Aluminum	750	0.025	
Vakhrushev et al. [79]	Coke	430	0.025	903
Tischchenko and Khvastukhin [82]	High alumina	970–1,450	0.036	1,273–1,373
Lese and Kermode [83]	Glass beads	295	0.0286	Room temperature

Glicksman and Decker [29] proposed the following dimensionless correlation to predict the average heat transfer coefficient for tubes immersed in fluidized beds of large particles ($\bar{d}_p > 1$ mm).

$$Nu_{wp} = (1-\delta)(9.42+0.042\,Re_p Pr) \tag{23}$$

The correlation is found to predict the recently reported data for large particles by a number of investigators quite well [38, 41, 43, 52, 91, 92]. All the necessary properties of the fluidizing gas are to be calculated at the film temperature. The values of the bubble fraction, δ, needed in this correlation, are calculated from the following relation:

$$\delta = 1-(L_{mf}/L_f) = 1-(1-\varepsilon)/(1-\varepsilon_{mf}) \tag{24}$$

The values of L_f/L_{mf} are obtained from the following correlation proposed by Babu et al. [93]:

$$L_f/L_{mf} = 1+[10.978(U-U_{mf})^{0.783}\bar{d}_p^{1.006}\rho_s^{0.376}U_{mf}^{-0.937}\rho_f^{-0.126}] \tag{25}$$

The value of δ can also be calculated with the use of the Equation 24 and by estimating ε from the following equation proposed by Staub and Canada [42]:

$$U/\varepsilon = 1.05\,U+(1-\varepsilon_{mf})\,U_{mf}/\varepsilon_{mf} \tag{26}$$

where ε_{mf} is obtained from the modified Ergun correlation proposed by Geldart and Cranfield [94].

$$1.75\,\rho_f U_{mf}^2+180[(1-\varepsilon_{mf})/\bar{d}_p]\,(\mu\,U_{mf})-\varepsilon_{mf}^3\bar{d}_p(\rho_s-\rho_f)g = 0 \tag{27}$$

The values of U_{mf} in Equation 27 can be estimated from the well-known correlation by Wen and Yu [95] or from the correlation proposed by Babu et al. [93].

Catipovic et al. [37] proposed the following correlation to predict heat transfer rate between horizontal tubes and fluidized beds of large particles ($\bar{d}_p > 1$ mm).

$$Nu_{wp} = 6\,(1-\beta) + (0.0175\,Ar^{0.46}Pr^{0.33})\,(1-\beta) + (\bar{d}_p/D_T)\,(0.88\,Re_{pmf}^{0.5} + 0.0042\,Re_{pmf})\,Pr^{0.33}\beta \tag{28a}$$

where the time fraction the tubes in a bundle are bathed by bubbles is given by:

$$1-\beta = 0.45 + (0.061)/[(U-U_{mf}) + 0.125] \tag{28b}$$

For the case of a single tube, the values of β are seven percent higher than its values for the tube bundles. In the correlation of Equation 28a, the first term represents the contribution due to particle convection, the second term represents the contribution due to gas convection for the emulsion phase, and the third term represents the contribution due to gas convection for the bubble phase. All the necessary properties of the fluidizing gas are to be evaluated at the bed temperature. The correlation predicted the data of various investigations quite well [37, 43, 91, 92]. However, the predictions for the data of Borodulya et al. [41] at high pressures (4.1–8.1 MPa) were rather poor.

Zabrodsky et al. [38] studied heat transfer between horizontal tube bundles and a rectangular fluidized bed (0.40×0.24 m) of large particles operating at atmospheric conditions. The effect of particle diameter ($\bar{d}_p = 2$–3 mm), shape, tube pitch, and fluidizing velocity ($U = 0.6$–2.8 m/s) was investigated. The experimental data for widely spaced tube bundles were correlated by the following semi-empirical equation:

$$h_w = [7.2\,k_f\,(1-\varepsilon)^{2/3}/\bar{d}_p] + 26.6\,U^{0.2}C_{pf}\rho_f\bar{d}_p \tag{29}$$

The correlation predicted the data taken in beds at atmospheric pressure [38, 91, 92] quite well. However, it greatly overpredicted the data of Borodulya et al. [41] at high pressures.

Recently Ganzha et al. [39] have proposed the following semi-empirical correlation to predict heat transfer between a fluidized bed of large particles ($\bar{d}_p > 1$ mm) and immersed tubes:

$$Nu_{wp} = 8.95\,(1-\varepsilon_w)^{0.667} + Nu_{wcvp} \tag{30}$$

here Nu_{wcvp} is given by Equation 15. The prediction from the correlation of Equation 30 are found to be in good agreement with the existing data [37, 38, 95–97] of large particles including high pressure data of Borodulya et al. [40, 41].

According to Martin, [30] the heat transfer coefficient between a tube and a fluidized bed of small and large particles is given by:

$$Nu_{wp} = Nu_{wpc} + 0.009\,Ar^{0.5}Pr^{0.33} \tag{31}$$

In this equation the first term represents the contribution due to particle convection and is calculated from Equation 7 while the second term represents the contribution due to gas convection and is taken from Baskakov et al. [9].

Xavier and Davidson [25] proposed that the heat transfer coefficient for a horizontal or vertical tube is given by:

$$Nu_{wp} = Nu_{wcp} + Nu_{wcvp} \tag{32}$$

where Nu_{wcp} and Nu_{wcvp} are estimated by using Equations 4 and 8, respectively. The validity of Equations 31 and 32 have not yet been tested in the light of existing data on large particles.

The maximum heat transfer coefficient between a tube and a fluidized bed of small particles ($\bar{d}_p < 1$ mm) can be estimated from the dimensional correlation proposed by Zabrodsky [1]:

$$h_{w\,max} = 35.7\,k_f^{0.6}\bar{d}_p^{-0.36}\rho_s^{0.2} \tag{33}$$

Table 4
Range of Variables Over Which Grewal [99] Correlation was Tested

Reference	Symbol	Solid Particles Material	d_p (μm)	Tube Diameter D_T (m)	Bed Temperature T_b (°K)
Howe et al. [87]	▲	Dolomite	800	0.0508	1028, tube bundle
Gelperin et al. [60]	⚠	Quartz sand	350	0.020	Room temperature
Andeen and Glicksman [71]	⦶	Ottawa sand	360, 710	0.0191	Room temperature
Bartel and Genetti [86]	◇	Glass beads	203, 279, 470	0.0159	Room temperature
Gelperin et al. [85]	⚠	Quartz sand	263	0.020	Room temperature
	⚠	Quartz sand	163	0.025	Room temperature
Noack [64]	⬘	Glass beads	600	0.025	Room temperature
Natusch and Blenke [80]	⟐	Glass beads	450	0.025	Room temperature
Janssen [90]	▼	Quartz sand	1,000	0.018	473
Neukirchen and Blenke [88]	◁	Glass beads	500	0.025	Room temperature
	◁	Glass beads	700	0.025	tube bundle
	◪	Quartz sand	720	0.025	
	⬓	Aluminum	750	0.025	
Vakhrushev et al. [79]	⛛	Coke	430	0.025	903
Lese and Kermode [83]	◈	Glass beads	295	0.0286	Room temperature
Grewal and Hajicek [91]	▽	Silica sand	971	0.025	1070, tube bundle
AECC [76]	○	Silica sand	172	0.0191	478, tube bundle
	●	Alumina	170, 266	0.0191	478, tube bundle
	△	Expanded alumina	163, 444	0.0191	478, tube bundle
	◓	Graphite	137	0.0191	478, tube bundle
	◒	Zircon sand	147	0.0191	478, tube bundle
	◇	Nickel	91	0.0191	394, tube bundle
	▲	Solder	88	0.0191	394, tube bundle
	□	Glass beads	164, 333, 494	0.0191	478, tube bundle
Grewal and Saxena [77, 100]	○	SiC	178, 362	0.0127	Room temperature
	◒	Alumina	259	0.0127, 0.0286	Room temperature
	◓	Silica sand	451	0.0127	Room temperature
	◑	Dolomite	312	0.0127	Room temperature
	●	Silica sand	504	0.0127, 0.0191, 0.0286	Room temperature
	⊕	Silica sand	167	0.0127, 0.0286	Room temperature
	◇	Glass beads	265, 357, 427	0.0127, 0.0191, 0.0286	Room temperature
	⬙	Lead-glass beads	241	0.0127	Room temperature
Chen [81]	⊖	Glass beads	240	0.025	Room temperature
Ziegler [75]	◪	Nickel	136	0.0127	Room temperature
	+	Copper	136	0.0127	Room temperature
	◐	Solder	136	0.0127	Room temperature
Antonshin [84]	△	Quartz sand	268	0.014, 0.021, 0.028	Room temperature
Cherrington et al. [89]	◈	Glass beads	390, 1,000	0.0508	Room temperature
	◇	Lime stone	1,000	0.0508	Room temperature

The predicted values of $h_{w\,max}$ from the Equation 33 are, generally, within ± 20 percent of the data. Zabrodsky et al. [98] expressed this equation in the dimensionless form as follows:

$$Nu_{wp\,max} = 0.88\,Ar^{0.213} \tag{34}$$

This correlation of Equation 34 is applicable for immersed bodies 20–60 mm in diameter, particle diameters between 280 and 1,000 μm and densities from 2,000–4,000 kg/m^3, and Ar 80–10^6. The properties of the fluidizing gas are to be evaluated at the film temperature. It may be noted that the preceding two correlations do not include the effect of the volumetric heat capacity of solid particles. This is one of the reasons why the correlation underpredicted the Aerojet Energy Conversion Company [76] data for solder and nickel by as much as 50 percent [99].

Recently Grewal [99] has proposed a correlation for $h_{w\,max}$ between a horizontal tube immersed in a fluidized bed of small particles.

$$Nu_{wp\,max} = 0.9(Ar\,D_{12.7}/D_T)^{0.21}\,(C_{ps}/C_{pf})^{0.2} \tag{35a}$$

or, if $(1-\varepsilon_{opt})$ is included:

$$Nu_{wp\,max} = 2(1-\varepsilon_{opt})\,(Ar\,D_{12.7}/D_T)^{0.21}\,(C_{ps}/C_{pf})^{0.2} \tag{35b}$$

The correlation of Equation 35 is based on data of Grewal and Saxena [77, 100], Ziegler [75], and AECC data [76]. All properties of the fluidizing gas are calculated at the film temperature. ε_{opt} is the bed porosity at superficial fluidizing velocity (U_{opt}), at which

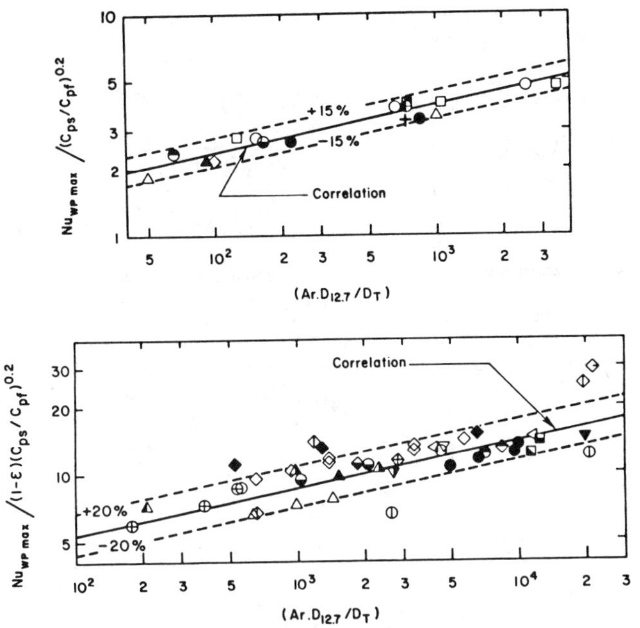

Figure 8. Comparison of experimental data with the correlation of Equation 35. Symbols and related details are given in Table 4 [99].

maximum value of the heat transfer coefficient occurs. The correlation of Todes et al. [101] is used for the prediction of the optimum fluidizing velocity for all bed material except smooth spherical spheres like glass beads [100].

$$Re_{p\,opt} = Ar/[18 + 5.22(Ar)^{0.5}] \tag{36}$$

The correlation of Varygin and Martyushin [102] is employed to calculate U_{opt} for glass beads.

$$Re_{p\,opt} = 0.118\,Ar^{0.5} \tag{37}$$

The value of ε_{opt} is calculated from the correlation [74] of Equation 21. It is seen from Figure 8 that the predicted values of $h_{w\,max}$ from the prior correlation of Equation 35 are, generally, within ± 20 percent of the data reported by 18 investigators. The correlation is applicable for immersed horizontal tubes 12.7–50 mm in diameter, solid particles with $\bar{d}_p$ = 80–800 μm, ρ_s = 1,000–8,900 kg/m^3, C_{ps} = 0.440–1.043 kJ/(kg) (° K), and $\emptyset_s$ = 0.60 to 1.0, bed temperature less than 870° K, and Ar between 75 and 20,000.

For large particles, $Nu_{wp\,max}$ is calculated from the following correlation proposed by Maskaev and Baskakov [103]:

$$Nu_{wp\,max} = 0.21\,Ar^{0.32} \tag{38}$$

The correlation is applicable to the fluidized bed of wide range of particles with $\bar{d}_p$ from 2,000–12,400 μm, ρ_s from 4,000–11,400 kg/m^3, and Ar between 1.4×10^5 and 3×10^8.

Tube bundles. The horizontal tubes in a bundle are generally arranged in an in-line or staggered array as shown in the Figure 9. The tube pitch is found to influence the h_w for

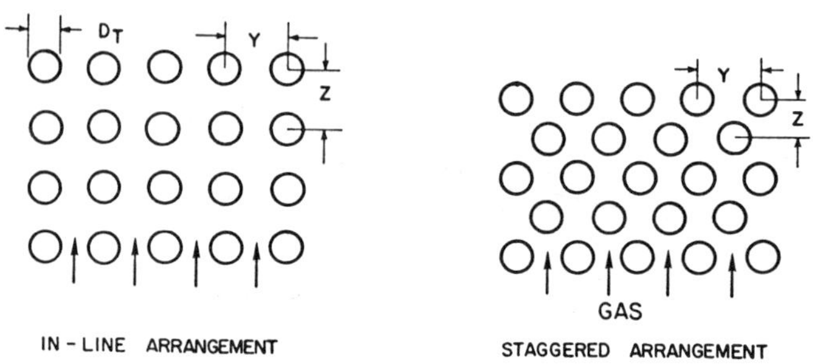

Figure 9. Arrangements of tube bundles.

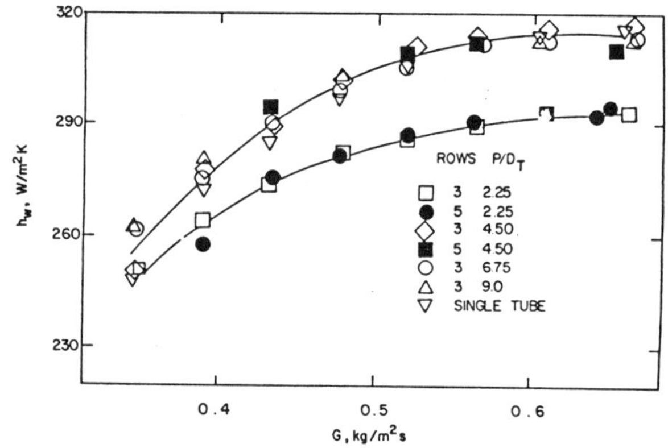

Figure 10. Dependence of heat transfer coefficient for horizontal tube bundles on G and P/D_T [8].

a tube bundle (Figure 10). It is generally agreed that the heat transfer coefficient decreases with a decrease in tube spacing [45, 83, 86, 104–108].

Bansal et al. [104] investigated the effect of relative tube pitch, p/D_T, on the heat transfer coefficient for a horizontal tube bundle ($D_T = 0.0254$ m) immersed in a fluidized bed of small particles ($\bar{d}_p = 230–715$ μm). The tubes in the bundle were located at the vertices of equilateral triangles. By using the modified Vreedenberg's correlation of Equation 20 for a single tube, and including a factor to account for the influence of tube pitch, the experimental data [104] were correlated by:

$$Nu_{WT} = 900(1-\varepsilon)\left(\frac{GD_T\rho_s}{\mu\rho_f}\,\frac{\mu^2}{\bar{d}_p^3\rho_s^2 g}\right)^{0.326} Pr^{0.3}[1-0.4(P/D_T)^{-2.5}] \tag{39}$$

for $P/D_T > 1.1$

Grewal [105] studied the heat transfer coefficient between a horizontal staggered tube bundle and an air-solid fluid bed. The tubes in the bundle were located at the vertices of equilateral triangles with pitch, P, varying between 1.75 and 9 times the tube diameter. Based on the experimental data, the following correlation was proposed:

$$Nu_{WT} = 47(1-\varepsilon)\left(\frac{GD_T\rho_s}{\mu\rho_f}\,\frac{\mu^2}{\bar{d}_p^3\rho_s^2 g}\right)^{0.325}\left(\frac{\rho_s C_{ps} D_T^{3/2} g^{1/2}}{k_f}\right)^{0.23} Pr^{0.33}$$

$$[1-0.21(P/D_T)^{-1.751}], \quad \text{for} \quad 1.75 \leqq P/D_T \leqq 9 \tag{40}$$

Table 5
Key to Figure 10

Reference	Symbol	Solid Particles	Experimental Conditions
Gelperin et al. [106]	△	Quartz sand, $\bar{d}_p$ = 350 μm, Φ_s = 0.84	D_T = 0.020; U = 0.15–0.85 m/s; P/D_T = 1.17–3.90
Howe and Aulisio [87]	▲	Dolomite, $\bar{d}_p$ = 688 to 1,370 μm	D_T = 0.0508; P/D_T = 3.0; T_b = 980°–1100° K; G = 0.89–1.5 kg/m²s
Petrie et al. [72]	○	Sand, etc., $\bar{d}_p$ = 326–735 μm	D_T = 0.0127–0.0355; P/D_T = 2.67–4.5; G = 0.080–0.384 kg/m²s; T_b = 405–772° K
McLaren and Williams [107]	●	Ash, $\bar{d}_p$ = 480 μm	D_T = 0.35, 0.060; P/D_T = 2.0–4.0; U = 0.3–1.2 m/s
Borodulya et al. [95]	◒	Sand, $\bar{d}_p$ = 250, 660 μm	D_T = 0.014; P/D_T = 2–6; U = 0.15–0.66 m/s
Andeen and Glicksman [71]	□	Ottawa sand, $\bar{d}_p$ = 360, 710 μm	D_T = 0.019; P/D_T = 2; U = 0.3–1.9 m/s
Bartel and Genetti [86]	◇	Glass beads, $\bar{d}_p$ = 203–470 μm	D_T = 0.0159; P/D_T = 1.7–8.5; G = 0.163–1.12 kg/m²s
Chandran et al. [96]	◆	Glass beads, $\bar{d}_p$ = 650 μm	D_T = 0.03175; P/D_T = 2.40; G = 0.37–1.7 kg/m²s
Staub and Canada [42]	⬘	Glass beads, $\bar{d}_p$ = 650 μm	D_T = 0.03175; P/D_T = 2.40; G = 0.14–0.86 kg/m²s
Canada and McLaughlin [43]	⊕	Glass beads, $\bar{d}_p$ = 650 μm	D_T = 0.03175; P/D_T = 2.40; G = 0.063–1.34 kg/m²s; Bed pressure = 0.1–1.0 MPa
Bansal et al. [104]	+	Glass beads, $\bar{d}_p$ = 230 μm; Silica sand, $\bar{d}_p$ = 715 μm; Ottawa sand, $\bar{d}_p$ = 514 μm	D_T = 0.0254; P/D_T = 1.3; G = 0.14–1.15 kg/m²s

Table 6
Correlations for Maximum Heat Transfer Coefficient Between a Horizontal Tube Bundle and a Fluidized Bed of Small Particles

Reference	Correlation	Experimental Conditions
Gelperin et al. [60]	$Nu_{wp\,max} = 0.79\,Ar^{0.22}[1-(D_T/y)]^{0.25}$	In-Line tube bundle; D_T = 0.020 m; quartz sand, $\bar{d}_p$ = 160–350 μm; Ar = 215–2,200; y/D_T = 1.5–10
Gelperin et al. [60]	$Nu_{wp\,max} = 0.74\,Ar^{0.22}\left[1-\frac{D_T}{y}\left(1+\frac{D_T}{Z+D_T}\right)\right]^{0.25}$	Staggered tube bundle; D_T = 0.020 m; quartz sand, $\bar{d}_p$ = 160–350 μm; Ar = 200–2,400; y/D_T = 2–9; Z/D_T = 1–10
Chekansky et al. [108]	$h_{w\,max} = 28.2\,\rho_s^{0.2}k_f^{0.6}\bar{d}_p^{-0.36}(\Delta/\bar{d}_p)^{0.04}(D_T/D_{20})^{-0.12}$	Staggered tube bundle; D_T = 0.05–0.1 m; porous corundum, $\bar{d}_p$ = 440 μm; quartz sand, $\bar{d}_p$ = 220 μm; experimental data of Gelperin et al. [60, 106]; Δ/d_p = 16–63
Grewal et Saxena [100]	$Nu_{wp\,max} = 0.9(Ar\,D_{12.7}/D_T)^{0.21}(C_{ps}/C_{pf})^{0.2}[1-0.21(P/D_T)^{-1.75}]$	Staggered tube bundle; D_T = 0.0127, 0,0286 m; silica sand, $\bar{d}_p$ = 167, 504 μm; alumina, $\bar{d}_p$ = 259 μm; experimental data of AECC [76]; Ar = 75–20,000; P/D_T = 1.75–9

The value of ε needed in the preceding equation is calculated from the correlation of Goroshko et al. [74]. For tube bundles with in-line arrangement, P is taken to be equal to the horizontal pitch, y. On the other hand, for tube bundle with staggered array P is taken to be equal to the minimum distance between the centers of the adjacent tubes. The correlation is found, in general, to predict the data of various investigators as shown in Table 5, as well as for small particles (Figure 11).

The heat transfer coefficient between a horizontal tube bundle and a fluidized bed of large particles can be estimated with the use of correlation of Glicksman and Decker [29] for a single horizontal tube or the Catipovic et al. [37] correlation for tube bundles.

The maximum heat transfer coefficient for a horizontal tube bundle immersed in a fluidized bed of small particles can be estimated from the correlations proposed by Gelperin et al. [60, 106], Chenkensky et al. [108], and Grewal and Saxena [100] (Table 6). A detailed comparison between the predicted values of $h_{w\,max}$ from these correlations and experimental data of twelve investigators is shown in Table 7. It is seen from Table 7 that the root-mean-square deviation of the predicted values of $h_{w\,max}$ using the Grewal and Saxena [100] correlation from the experimental data is generally smaller than the other correlations and

Table 7
Comparison of the Calculated Values of h_w from Different Correlations with Experimental Values of h_w

Reference	Solid Particles	D_T (m)	Experimental Conditions	Grewal et Saxena [100]	Gelperin et al. [60, 106]	Chen-kansky et al. [108]
Gelperin et al. [106]	Quartz sand, $\bar{d}_p$ = 350 μm	0.020	Y/D_T = 2–6; Z/D_T = 0.6–2.50; P/D_T = 1.17–3.90; staggered tube bundle; room temperature	7.7	9.7	10.3
Howe and Aulisio [87]	Limestone, $\bar{d}_p$ = 800 μm	0.0508	P/D_T = 3.0; Y/D_T = 3.0; Z/D_T = 1.49; staggered tube bundle; T_b = 1,000° K	7.0	17.5	38.1
Borodulya et al. [110]	Sand, $\bar{d}_p$ = 250, 660 μm	0.014	P/D_T = 2–6; in-line tube bundle; room temperature	12.2	21.0	13.9
Bartel and Genetti [86]	Glass beads, $\bar{d}_p$ = 203–407 μm	0.0159	P/D_T = 1.7–8.5; staggered tube bundle; room temperature	35.6	47.6	32.3
Chandran et al. [96]	Glass beads, $\bar{d}_p$ = 650 μm	0.03175	P/D_T = 2.40; Y/D_T = 3.2; Z/D_T = 1.80; staggered tube bundle; room temperature	10.9	9.2	19.7
Staub and Canada [42]	Glass beads, $\bar{d}_p$ = 650 μm	0.03175	P/D_T = 2.40; Y/D_T = 3.2; Z/D_T = 1.80; staggered tube bundle; room temperature	1.6	1.3	9.7
Canada and McLaughlin [43]	Glass beads, $\bar{d}_p$ = 650 μm	0.03175	P/D_T = 2.40; Y/D_T = 3.2; Z/D_T = 1.80; staggered tube bundle; room temperature bed pressure 0.1–1.0 MPa	19.6	24.6	19.7
Bansal et al. [104]	Glass beads, $\bar{d}_p$ = 230 μm; Silica sand, $\bar{d}_p$ = 715 μm; Ottawa sand, $\bar{d}_p$ = 514 μm	0.0254	P/D_T = 1.3; staggered tube bundle; room temperature	3.8	*	10.0
Xavier and Davidson [109]	Sand, $\bar{d}_p$ = 385, 885 μm	0.028	P/D_T = 2.71; staggered tube bundle; room temperature	17.1	14.2	23.9
Grewal and Hajicek [91]	Silica sand, $\bar{d}_p$ = 971 μm	0.0254	P/D_T = 1.90; Y/D_T = 2.11; Z/D_T = 1.58; staggered tube bundle; T_b = 1,070° K	4.7	11.7	19.8
Grewal and Saxena [100]	Silica sand, $\bar{d}_p$ = 167, 504 μm; alumina, $\bar{d}_p$ = 259 μm	0.0127, 0.0286	P/D_T = 1.75–9.0; staggered tube bundle; room temperature	3.4	18.6	6.6
Aerojet Energy Conversion Co. [76]	Details of solid particles are given in table 4	0.019	P/D_T = 3.77; Y/D_T = 5.33; Z/D_T = 2.67; staggered tube bundle; T_b = 394, 477° K	8.0	12.8	17.9

* *Imaginary Value of calculated* h_w

does not exceed 20 percent except for the data of Bartel and Genetti [86]. The correlation of Gelperin et al. [106] predicts imaginary values of $h_{w\,max}$ under the experimental conditions of Bansal et al. [104]. From this comparison, it may be concluded that the Grewal and Saxena [100] correlation predicts the existing data on small particles very well. For high temperature applications, the radiative component is important and is estimated following Baskakov et al. [9]. The correlation is applicable to Geldart's group B powder and the scale of the apparatus and operating conditions employed in the reported studies.

Vertical Smooth Tubes

Single tube. Vreedenberg [111] proposed two different correlations to predict the heat transfer coefficient between an axially located vertical tube and a fluid bed of small particles. One correlation was for fine and light particles, where viscous forces on the particles are predominant, for which an analogy between a fluidized bed and a flowing fluid was assumed and the other was for coarser and heavier particles where inertia effects were assumed significant. These correlations were based on the data of eleven investigators [16, 112–121].

For fine and light particles described by $G\,\bar{d}_p\rho_s/(\mu\rho_f) < 2{,}050$, the correlation obtained was

$$\frac{h_w(D_b-D_T)}{k_f}\left(\frac{D_T}{D_b}\right)^{1/3}\left(\frac{k_f}{\mu C_{ps}}\right)^{1/2} = 2.7\times10^{-16}\left(\frac{G(D_b-D_T)\,\rho_s}{\rho_f\mu}\right)^{3.4} \tag{41a}$$

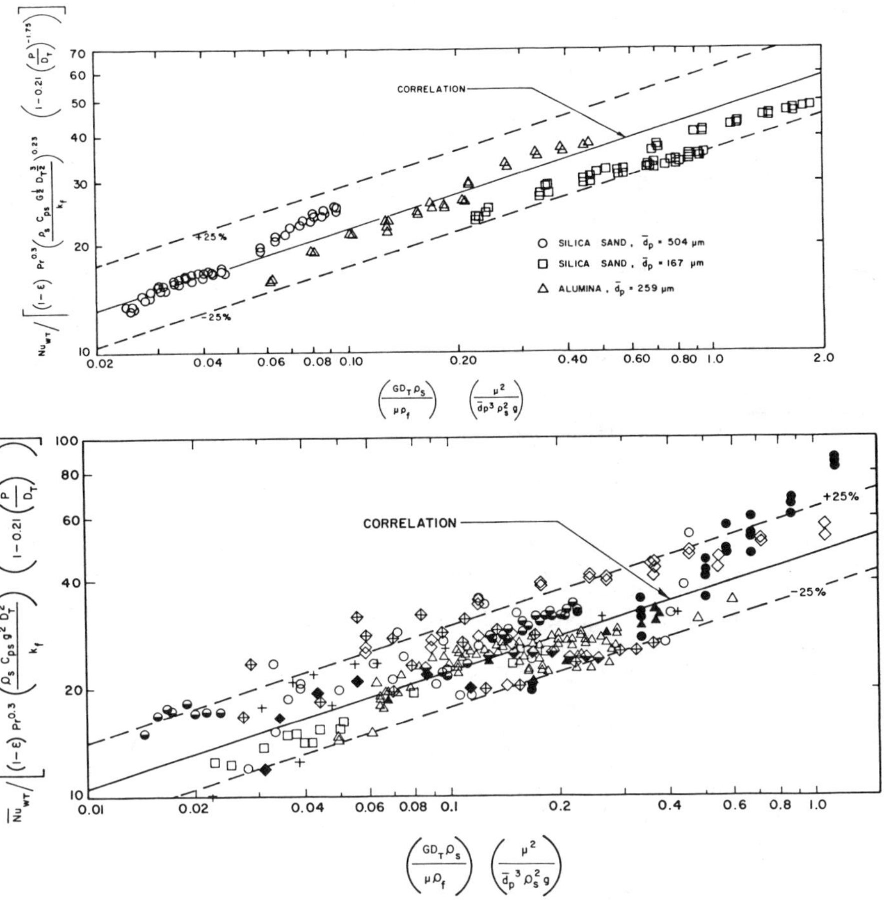

Figure 11. Comparison of raw data to regression and Equation 40. Symbols and related details are given in Table 5 [105].

for values of $G(D_b - D_T)\rho_s/\rho_f\mu < 0.237 \times 10^6$, and

$$\frac{h_w(D_b - D_T)}{k_f}\left(\frac{D_T}{D_b}\right)^{1/3}\left(\frac{k_f}{C_{ps}\mu}\right)^{1/2} = 2.2\left(\frac{G(D_b - D_T)\rho_s}{\rho_f\mu}\right)^{0.44} \tag{41b}$$

for $G(D_b - D_T)\rho_s/\rho_f\mu > 0.237 \times 10^6$.

For coarse and heavy particles represented by $G\bar{d}_p\rho_s/\rho_f\mu > 2{,}550$, the following correlation was recommended:

$$\frac{h_w(D_b - D_T)}{k_f}\left(\frac{D_T}{D_b}\,\frac{\bar{d}_p}{D_b - D_T}\,\frac{k_f}{C_{ps}\mu}\right)^{1/3} = 0.105 \times 10^{-3}\left(\frac{G(D_b - D_T)}{\rho_f\bar{d}_p^{3/2}g^{1/2}}\right)^2 \tag{42a}$$

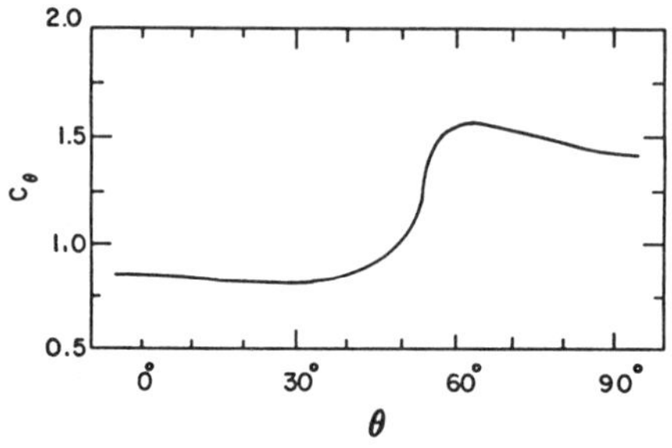

Figure 12. Dependence of C_θ on θ for a finned tube [123].

for $G(D_b - D_T)/\rho_f \bar{d}_p^{2/3} g^{1/2} < 1{,}070$, and

$$\frac{h_w(D_b - D_T)}{k_f}\left(\frac{D_T}{D_b}\frac{\bar{d}_p}{D_b - D_T}\frac{k_f}{C_{ps}\mu}\right)^{1/3} = 240\left(\frac{G(D_b - D_T)}{\rho_f d_p^{3/2} g^{1/2}}\right)^{-0.10} \tag{42b}$$

when $G(D_b - D_T)/\rho_f \bar{d}_p^{2/3} g^{1/2} > 1{,}070$.

For values of the modified Reynolds group $G\,\bar{d}_p \rho_s/\rho_f \mu$ between 2,050 and 2,550, it was recommended that the average of the values predicted by Equations 41 and 42 be used for the prediction of h_w. The range of applicability of the correlations of Equations 41 and 42 is given in Table 8. All the properties of the fluidizing gas and the bed particles needed in Equations 41 and 42 are evaluated at the bed temperature.

Wender and Cooper [122] correlated 323 experimental data points obtained by various investigators [16, 112–114, 117] by the following equation:

$$Nu_{wp} = 0.000351\, C_R(1-\varepsilon)\left(\frac{C_{pf}\rho_f}{k_f}\right)^{0.43}\left(\frac{G\,\bar{d}_p}{\mu}\right)^{0.23}\left(\frac{C_{ps}}{C_{pf}}\right)^{0.8}\left(\frac{\rho_s}{\rho_f}\right)^{0.66} \tag{43}$$

It may be pointed out that all the groups except $C_{pf}\rho_f/k_f$ in the previous equation are dimensionless. The factor C_R in the correlation is to allow for the effect of nonaxial-tube location. The value of C_R is 1, 1.76, and 1.57 when dimensionless radial location, $2r_b/D_b$, of the vertical tube is 0, 0.35, and 0.71, respectively. The correlation predicted the experimental data with a mean deviation of ±19.5 percent. Table 9 lists the range of variables over which the correlation is applicable.

The heat transfer coefficient, h_w, between a vertical tube immersed in a fluidized bed of large particles can be estimated by the correlations of Equations 23 and 30.

To estimate the maximum value of the heat transfer coefficient, $h_{w\,max}$, the correlation of Equations 33, 34, and 38 are recommended.

Tube bundles. Gelperin and Einshtein [3] reported the values of h_w for vertical tube bundles immersed in a fluidized bed of quartz sand. The tubes (D_T = 20–40 mm) were located at the vertices of equilateral triangles with relative pitch, P/D_T, between 1.25 to 5. For $U < U_{opt}$, the values of h_w for tubes in the bundle increased slightly as the radial distance from the centerline increased. However, for $U > U_{opt}$, the values of h_w were the same for all tubes except those located on the periphery. With the increase in the number of tubes ir a bundle, the heat transfer coefficient for the tubes tended to become equal due to the uniform gas flow. The bed height (H_s = 0.175–.500 m) and the height of the lower edge of the tubes from the distributor plate (H_d = 0.120–0.400 m) showed very small effects on h_w. When the tubes were loosely packed in the bundle, the relative pitch showed no influence on h_w. However, for tubes closely spaced in the bundle, the h_w decreased with a decrease in P/D_T. For example, a decrease of relative pitch from 5 to 2 resulted in a reduction of

Table 8
Range of Variables Over Which Correlations of Equations 41 and 42 are Applicable

Variable	Unit	Fine and Light Particles	Coarse and Heavy Particles
$G(D_b - D_T)\rho_s/\rho_f\mu$	Dimensionless	$.067 \times 10^6$–3.7×10^6	–
$G(D_b - D_T)\rho_f \bar{d}_p^{3/2} g^{1/2}$	Dimensionless	–	0.23×10^3–76×10^3
Mass fluidizing velocity (G)	kg/m^2s	0.005–0.62	0.032–2.6
Bed temperature (T_b)	°K	289–618	273–534
Thermal conductivity of gas (k_f)	W/K m	0.010–0.20	0.010–0.21
Viscosity of gas (μ)	NS/m^2	9.5×10^{-6}–31.5×10^{-6}	9.5×10^{-6}–28.2×10^{-6}
Density of gas (ρ_f)	kg/m^3	0.12–4.8	0.08–4.8
Density of particles (ρ_s)	kg/m^3	790–3,900	1,910–6,950
Specific heat of particles (C_{ps})	kJ/kg °K	0.675–1.060	0.470–0.970
Prandtl Number ($\mu C_{pf}/k_f$)	Dimensionless	0.097–1.00	0.039–0.98
Mean-volume-surface particle diameter ($\bar{d}_p$)	μm	38–178	40–900
Diameter of fluidized bed (D_b)	m	0.051–0.140	0.051–0.565
Outside diameter of vertical tube (D_T)	m	0.006–0.032	0.006–0.034
Ratio of diameter (D_b/D_T)	Dimensionless	4.40–16.0	4.40–20.0
Equivalent diameter of annulus ($D_b - D_T$)	m	0.041–0.108	0.041–0.531
Ratio of particle diameter to equivalent diameter of annulus $[\bar{d}_p/(D_b - D_T)]$	Dimensionless	0.41×10^{-3}–4.3×10^{-3}	0.40×10^{-3}–10.3×10^{-3}
Heated length of vertical tube	m	0.033–0.572	0.033–1.18
Static bed height (H_s)	m	0.1–1.0	0.1–1.8
Ratio of bed height to bed diameter	Dimensionless	1.0–13	0.7–14
Total heat transfer coefficient (h_w)	W/m^2 K	20–1,060	25–1,080

After Vreedenberg [111].

Table 9
Range of Variables Over Which the Correlation of Equation 43 Is Applicable

Variable	Unit	Range
Particle diameter, $\bar{d}_p$	μm	41–878
Specific heat of particles, C_{ps}	kJ/kg °K	0.84–1.32
Density of particles, ρ_s	kg/m^3	785–2,868
Particle fraction, $1-\varepsilon$	Dimensionless	0.55–0.603
Specific heat of gas, C_{pf}	kJ/kg °K	0.52–5.19
Density of gas, ρ_f	kg/m^3	0.176–7.770
Dynamic viscosity of gas, μ	N s/m^2	0.00105–0.00382
Thermal conductivity of gas, k_f	W/m °K	0.0255–0.3600
Diameter of fluidized bed, D_b	m	0.073–1.93
Length of heat transfer tube, L_H	m	0.100–5.334
Bed height at incipient fluidization or height of a freshly settled bed, L_{mf}	m	0.253–8.108
Heat transfer coefficient, h_w	W/m^2 K	57–1,028

After Wender and Copper [122].

the value of $h_{w\,max}$ by 5 to 7 percent and with a further decrease of P/D_T, the $h_{w\,max}$ decreased by 15 to 20 percent.

To estimate the h_w for loosely packed tube bundles, the correlation for an axially-located, single, vertical tube for example, Equations 41 to 43 are used.

The experimental data for $h_{w\,max}$ were correlated by the following dimensionless equation [3]:

$$Nu_{WP\,max} = 0.75\,Ar(1 - D_T/P)^{0.14} \tag{44}$$

The correlation is valid for small particles ($\bar{d}_p$ = 160–350 μm), P/D_T = 2–5, and Ar = 210–2,400.

Inclined Smooth Tubes

Genetti et al. [123] studied heat transfer coefficient between slanted tube (D_T = 16 mm) and air-fluidized beds of glass beads ($\bar{d}_p$ = 114–470 μm). The minimum heat transfer

Table 10
Summary of Heat Transfer Studies From Finned Tubes Immersed in a Fluidized Bed

Reference	Tube Base Diameter (mm)	Tube Orientation	Fin-Type	Fin-Length l (mm)	No. of Fins/m or Fin Spacing (mm)	Fin Thickness δ' (mm)	Fin Material	Air Fluidizing Velocity G (kg/m^2s)	Bed Material $\bar{d}_p$ (μm)	φ (dimensionless)	β (dimensionless)	Remarks
Petrie et al. [72]	19.0	Horizontal	Continuous transverse	10.0	197, 433	0.41	Aluminium	0.186–0.384	Silica sand 497, 530	0.1–0.42	–	Tube bundle P = 57 mm
Bartel et al. [124]	16.0	Horizontal	Serrated transverse	10.0–25.0	315	0.6–0.8	Carbon steel	0.089–1.09	Glass beads 114–470	0.2–0.5	–	Single tube
Genetti et al. [123]	16.0	Horizontal to vertical	Serrated transverse	19.1, 9.5	315, 236	0.64, 0.76	Carbon steel, Copper	0.081–1.090	Glass beads 114–470	–	–	Single tube
Bartel and Genetti [86]	16.0	Horizontal	Serrated transverse	3.0–22.0	315	0.50	Carbon steel	0.089–1.090	Glass beads 114–470	0.2–0.8	0.7–1.6 (G/G_{mf} = 4)	Tube bundle P = 27–122 mm
Priebe and Genetti [130]	13.0	Horizontal	Serrated transverse Spined tube	3.0–22.0 8.0–19.0	315	0.41–0.64	Carbon steel, Steel and Copper	0.163–1.120	Glass beads 203–470	–	0.8–1.80 0.4–1.60 (G/G_{mf} = 4)	Tube bundle P = 27–122 mm
Genetti and Kratovil [127]	11.5–19.0	Horizontal	Helical continuous transverse	6.0–10.5	197–709	0.4–0.64	Copper	0.136–0.949	Glass beads 173–551	–	1.3–2.9 (G/G_{mf} = 4)	Single tube
Zabrodsky et al. [131]	30.0	Horizontal	Triangular	5.0–22.0	29–100	–	Copper	0.58–1.87	Silica sand, 250–620; millet, 2,000; crushed fire clay, 975, 2,000, 3,000	0.72–1.02	–	Single tube; Tube bundle P = 60, 75, 100 mm
Natusch and Blenke [153]	25.0	Horizontal	Continuous longitudinal	10.0–40.0	(2–16)	2.0	Copper	–	Glass beads, 150–670	–	–	Single tube
Natusch and Blenke [80]	25.0	Horizontal	Continuous transverse	7.5–37.5	(2–10)	0.5–2.0	Copper	G > $6G_{mf}$	Glass beads 400–800	–	–	Single tube
Gelperin et al. [125]	24.0	Horizontal	Continuous transverse	4.0	(0–12)	4.0	Copper, aluminum	–	Quartz sand, 350	–	–	Single tube
Grewal et al. [132]	19.8–27.2	Horizontal	V-groove	0.68–	(0.79–5.08)	–	Bronze	0.1–0.67	Silica sand, 167, 504; Alumina 259	1.1–1.68	0.6–1.0	Single tube Tube bundle P = 1.75–3.50
Gelperin et al. [126]	24.0	Horizontal	Continuous transverse	up to 10.0	(6)	4.0	Copper, dur-aluminum	–	Quartz sand, 350	–	–	Single tube
Grewal and Saxena [128]	11.3–12.3	Horizontal	V-groove	0.21–0.68	(0.24–0.79)	–	Copper	G/G_{mf} = 1.2–8.5	Glass beads 265, 347, 427	–	1–1.4	Single tube
Krause and Peters [129]	19.2	Horizontal	Serrated transverse	4.76–17.46	300	–	Carbon steel	0.110–0.530	Glass beads 210, 430	0.23–0.65	–	Single tube
Staub and Canada [42]	25	Horizontal	Helical continuous transverse	13	98, 280	0.15–0.27	–	0.9–3.4	Glass beads 650	0.54–0.57	–	Tube bundle
Chen and Withers [134]	18.8–22.6	Vertical	Helical continuous	1.5–3.4	197–752	–	Copper	0.14–0.880	Glass beads 130–610	0.5–1.2	1.1–2.90 (G/G_{mf} = 3)	Single tube
Korolev and Syromvatnikov [138]	–	Vertical	Longitudinal and transverse	.40	(0.5)	–	Copper	0.080 0.620	Corundum 120; Chamotte 320; Polystyrene 720	–	–	Vertical tube
Vijayaraghavan and Sastri [136]	33.4–34.8	Vertical	Longitudinal and transverse V-groove	0.1, 0.8	1,250	–	Copper	0.060–1.060	Glass beads 160–360	0.56–0.83	–	Single tube
Genetti and Everly [37]	17.9–28.6	Vertical	Continuous groove	3.81–5.33	–	–	Copper	0.140–0.750	193–417	–	–	Spiral tube
Genetti et al. [135]	11.5–19.1	Vertical	Helical, segmental	0.32–0.87	(1.0–4.7)	0.4–0.79	–	0.170 0.780	Glass beads 190–410	–	–	Tube bundle

Table 11
Correlations for the Prediction of the Heat Transfer Coefficients for Finned Tubes

Reference	Correlation	Remarks
Bartel et al. [124]	$Nu_{wftp} = \dfrac{10(1-\varepsilon)^{0.5}}{[1+\lambda_1 Re_p \lambda_2]}$ where: $\lambda_1 = 58{,}000(\delta'/0.0254)^{2.93}(\ell/0.0254)^{(11570\delta' - 8.37)}$ $\lambda_2 = -0.000675(\delta'/0.0254)^{-1.872}(\ell/0.0254)^{(4220\delta' + 3.216)}$	Single horizontal tube
Genetti et al. [123]	$Nu_{wftp} = \dfrac{11(1-\varepsilon)^{1/2}}{1+\left[\dfrac{C_\Theta}{Re_p^{0.51}(\bar{d}_p/0.0002)^{0.35}}\right]^2}$ where: C_Θ is found from Figure 12	Single inclined tube
Natusch and Blenke [133]	$Nu_{wftp} = Nu_{wp} \exp[-0.0964(\ell/D_T)^{0.566}(U/U_{mf}+1.7)^{-0.592}]$ where: Nu_{wp} is the Nusselt number for a smooth tube	Single horizontal tube $190 < Ar < 30{,}000$
Natusch and Blenke [80]	$h_{wft} = h_{w\,exp}[(0.27\ell)/(s+\beta'\ell+3.8)]$ where: h_w is the heat transfer coefficient for a smooth tube, and $\beta' = 0.37/[0.36+(U/U_{mf})]+0.247$	Single horizontal tube $U/U_{mf} > 6$, $\bar{d}_p = 400$ to 800 μm
Bartel and Genetti [86]	$Nu_{wftp} = \dfrac{7.34\left(1-\dfrac{0.027+10.60\ell^{1.5}}{(P/0.0254)^{(1.112+29\ell^{0.6})}}\right)}{\left(1+\dfrac{0.00102+0.889\ell^{0.80}}{Re_p^{(0.33+3.1\times 10^{14}\ell^{9.33})}(\bar{d}_p/0.0254)^{(1.23-1.33^{0.23})}}\right)^2}$	Horizontal tubes bundle
Priebe and Genetti [130]	$Nu_{wftp} = \dfrac{(30{,}270\bar{d}_p-2.9)(Q/A_{wft})^{0.24}(\bar{d}_p/s)^{-(0.55+323\bar{d}_p)}}{\left(1+\dfrac{8.19+231\,s}{(G/G_{mf})^{0.164}}\right)^2}$	Horizontal tubes bundle
Priebe and Genetti [130]	$Nu_{wftp} = \dfrac{5.3(A'_{wft}/\ell)^{-0.1311}}{\left[1+\dfrac{(2.1-2.3\times 10^{-14}D_f^{-7.24})}{Re_p^{1.24}}\right]^2}$	Horizontal tubes bundle
Genetti and Kratovil [127]	$Nu_{wftp} = \dfrac{-0.26(\ell/s)+7.1}{\{1+0.074\,Re_p^{[-8.57(\ell/s)-3.4(\bar{d}_p/s)-0.685(\ell/s)+1.96]}(\bar{d}_p/s)^{-0.77}(\ell/s)^{0.75}\}^{0.24(\ell/s)+1}}$	Horizontal tubes bundle
Genetti and Everly [137]	$Nu_{wftp} = \dfrac{7.26}{[1+0.000235(G/G_{mf})^a(\bar{d}_p N_f/P_f)^b N_f^{2.87}]^2}$ where $a = 21.9(\bar{d}_p N_f/P_f)^{-0.428N_f}+0.52N_f^{-1.69}$ $b = -1.72N_f^{-0.173}$	Spiral tube
Zabrodsky et al. [131]	$Nu_{wfbp\,max} = Nu_{wp\,max}\,\eta\varphi'^{0.9}$ where $Nu_{wp\,max}$ = the maximum Nusselt number for a smooth tube	Single horizontal tube
Grewal et al. [132]	$Nu_{wftT} = Nu_{wT}[0.55+1.077(\bar{d}_p/D_T)^{-0.068}(P_f/\bar{d}_p)^{-0.171}]$ where Nu_{wT} is given by Equation 22	Single horizontal tube
Genetti et al. [135]	$Nu_{wftp} = 0.00366(1+\delta'/w)^{12}\left(\dfrac{\pi(D_{sT}+Z)(s+\delta')}{N_{pt}\delta' w}\right)^{3.11}(\bar{d}_p/s)^{0.75}(G/G_{mf})^{0.57}$	Vertical tube bundles
Zabrodsky et al. [131]	$Nu_{wfbp\,max} = Nu_{WP\,max}\,\eta\varphi'^{0.9}[(P-D_{ST})/P]^{0.16}$ where $Nu_{wp\,max}$ is the maximum Nusselt number for a smooth tube	Horizontal tube bundles
Grewal et al. [132]	$Nu_{wftT} = Nu_{wT}[0.55+1.077(\bar{d}_p/D_T)^{-0.068}(P_f/\bar{d}_p)^{-0.171}$ where Nu_{wT} is given by Equation 40	Horizontal tube bundles

coefficients were observed at an orientation angle of 45°. The experimental data were correlated by the following semiempirical equation based on the "particle mode" heat transfer mechanism.

$$Nu_{wp} = \frac{11(1-\varepsilon)^{0.5}}{1 + \dfrac{0.44 - [0.20(\theta-45)^2]/[(\theta-45)^2+120]}{Re_p^{0.24}\,(\bar{d}_p/0.0002)^{1.23}}} \tag{45}$$

There does not appear to be a particular advantage in using the inclined smooth tubes.

Finned Tubes

If the bed-side heat transfer coefficient is the limiting coefficient, then the heat transfer rates can be increased by increasing the surface area by the use of fins. Heat transfer from finned tube heat exchangers is represented by an effectiveness factor of the finned tube, Ø, defined by:

$$Ø = Q/h_{wb}A_{wft}(T_w - T_b) \tag{46}$$

Another measure of the thermal performance of a finned tube is a quantity known as the heat capacity function, β, which is defined as the ratio of the heat transfer coefficient for a finned tube based on the surface of a smooth tube with a outside diameter equal to a fin-tip diameter, h_{wfb}, to the heat transfer coefficient for a smooth, h_{wb}, both occupying the same superficial bed volume.

A number of investigations have been reported in the literature on the performance of transverse [42, 72, 86, 123–133] and longitudinal finned horizontal tubes [133], transverse [86, 134–136], and longitudinal finned [136], vertical tubes, spiral finned tube [137], and transverse and longitudinal V-groove vertical plates [138]. All these efforts are summarized in Table 10.

The fin effectiveness factor, Ø, increases with an increase in fluidizing gas velocity, decrease in fin height, increase in fin spacing, and increase in thermal conductivity of fin material.

An increase in tube-to-tube spacing in a bundle of short-finned tubes, which reduces the resistance to particle movement between the tubes, increases the heat transfer coefficient. However, an increase in tube-to-tube spacing for a bundle of long-finned tubes does not noticeably increase the heat transfer rate, implying that the resistance to particle movement between fins predominates.

A number of correlations have been proposed to predict the heat transfer coefficients for finned tubes and are listed in Table 11. The range of experimental conditions under which these correlations are valid are listed in Table 10.

Rough Tubes

The heat transfer coefficient between a rough tube and a gas-solid fluid bed has been found to depend on the type od surface roughness, and the roughness magnitude as compared to particle diameter [128, 136, 138, 139]. The heat transfer coefficient for rough tubes, h_{wfb}, is found to be smaller than that for the heat transfer coefficient for smooth tubes, h_{wb}, for fluidized beds of small particles. This is mainly due to an increase in resistance to heat flow because of longer heat conduction paths or larger values of bed porosity in the vicinity of a rough tube as compared to a smooth tube. The increase in particle residence time due to irregularities of a rough surface also reduces the heat transfer coefficient.

NOTATION

a	defined by Equation 15, dimensionless
A_w	surface area of a smooth tube (m^2)
A_{wfb}	surface area of a smooth tube with outside diameter equal to fin tip diameter (m^2)
A_{wft}	total surface area of a fin tube (m^2)
A'_{wft}	surface area of a spined tube per unit length (m)
Ar	Archimedes number; $= \bar{d}_p^3 \rho_f(\rho_s - \rho_f) g \mu^{-2}$; dimensionless
C_{epmf}	effective heat capacity of the packet; kJ/(kg °K)
C_k	constant; $= 4\,\bar{d}_p/(U_p\tau)$; dimensionless
C_R	factor to allow for the effect of nonaxial-tube location; dimensionless
C_θ	function of θ and is given in Figure 12; dimensionless
C_{pf}	specific heat of fluidizing gas at constant pressure (kJ/(kg °K))
C_{ps}	specific heat of solid particles at constant pressure (kJ/(kg °K))
CT	correction factor for tube diameter, Equation 15e; dimensionless
d_p	particle diameter (m)
d_{pi}	arithmetic average diameter of the successive screens (m)
$\bar{d}_p$	average particle diameter; $= 1/\Sigma\,(w/\bar{d}_p)_i$ (m)
D_b	bed diameter (m)
D_T	outside diameter of a smooth tube (m)
D_{be}	diameter of a sphere having same volume as a bubble (m)
D_f	diameter over spines (m)
D_{ST}	fin base tube diameter (m)
$D_{12.7}$	heat transfer tube 12.7 mm in diameter (m)
D_{20}	heat transfer tube 20 mm in diameter (m)
g	acceleration due to gravity (m/s^2)
G	superficial mass fluidizing velocity ($kg/(m^2 s)$)
G_{mf}	mass velocity of gas at minimum fluidization ($kg/(m^2 s)$)
h_w	total heat transfer coefficient for a smooth tube (W/(m^2 °K))
h_{wb}	total heat transfer coefficient for a smooth tube with outside tube diameter equal to fin-tip diameter (W/(m^2 °K))
h_{wc}	conductive component of the heat transfer coefficient (W/(m^2 °K))
h_{wcv}	gas convective component of the heat transfer coefficient (W/(m^2K))
h_{wfb}	total heat transfer coefficient for a finned tube based on the surface area of a smooth tube with outside diameter equal to fin-tip diameter (W/(m^2 °K))
h_{wiL}	instantaneous local heat transfer coefficient (W/(m^2 °K))
h_{wL}	time-averaged local heat transfer coefficient (W/(m^2 °K))
$h_{w,\max}$	maximum heat transfer coefficient for a tube (W/(m^2 °K))
h_{wr}	radiative component of heat transfer coefficient for a smooth tube (W/(m^2 °K))
H_d	height of the heat transfer tube above the distributor plate (m)
H_s	static bed height (m)
k_e	effective thermal conductivity of bed with no gas flow (W/(m °K))
k_{emf}	effective thermal conductivity of emulsion phase (W/(m °K))
k_f	thermal conductivity of gas (W/(m °K))
ℓ	fin length (m)
ℓ'	characteristics dimension and is equal to the smaller of the vertical dimension of the tube and bubble length (m)
L_f	expanded bed height (m)

L_{mf} bed height at minimum fluidization (m)

L_H length of heater (m)

m number of transfer units, defined by Equation 7b; dimensionless

N fluidization number; $= U/U_{mf}$; dimensionless

N_f number of flutes; dimensionless

N_{pt} number of fins per turn; dimensionless

Nu_{max} maximum Nusselt number for brief contact time;

$$4\left\{\left[\left(1+\frac{2\sigma}{d_p}\right)\ln\cdot\left(1+\frac{d_p}{2\sigma}\right)-1\right\};$$ dimensionless

Nu_{wcp} particle conductive component of the Nusselt number based on particle diameter; $= h_{wc}\bar{d}_p/k_f$; dimensionless

Nu_{wcvp} gas convective component of the Nusselt number based on particle diameter; $= h_{wcv}\bar{d}_p/k_f$; dimensionless

$Nu_{wfb\,max}$ maximum Nusselt number for finned tubes based on particle diameter; $(h_{wfb\,max}\bar{d}_p)/k_f$; dimensionless

Nu_{wftp} Nusselt number for finned tubes based on particle diameter, $(h_{wft}d_p)/k_f$; dimensionless

Nu_{wftT} total Nusselt number for finned tubes based on heat transfer tube diameter; $(h_{wft}D_T)/k_f$; dimensionless

Nu_{wp} Nusselt number based on particle diameter; $= h_w\bar{d}_p/k_f$; dimensionless

$Nu_{wp\,max}$ maximum value of Nusselt number based on particle diameter; $= h_{w\,max}\bar{d}_p/k_f$; dimensionless

Nu_{wrp} radiative component of the Nusselt number based on the particle diameter; $h_{wr}\bar{d}_p/k_f$; dimensionless

Nu_{wT} Nusselt number based on outside diameter of boiler tube; $= h_w D_T/k_f$; dimensionless

p center-to-center distance (pitch) between adjacent tubes in a bundle (m)

p_f fin pitch, defined as the distance between two identical points of the consecutive threads (m)

Pr Prandtl number; $= \mu\, C_{pf}/k_f$; dimensionless

Q total heat flow from tube to bed (W)

r_b radial location of vertical tube (m)

Re_p Reynolds number based on particle diameter at a given fluidizing velocity; $= U\rho_f\bar{d}_p/\mu$; dimensionless

Re_{pmf} Reynolds number based on particle diameter at minimum fluidizing velocity; $= U_{mf}\rho_f\bar{d}_p/\mu$; dimensionless

$Re_{p\,opt}$ Reynolds number based on particle diameter at optimum fluidizing velocity; $= U_{opt}\rho_f\bar{d}_p/\mu$; dimensionless

Re_{Tmf} Reynolds number based on tube diameter at minimum fluidizing velocity; $= U_{mf}\rho_f D_T/\mu$; dimensionless

s fin spacing (m)

T_b bed temperature (°K)

T_f film temperature; $= (T_w + T_b)/2$ (°K)

T_w outside tube wall temperature (°K)

U superficial gas fluidizing velocity (m/s)

U_B bubble velocity (m/s)

U_{mf} minimum fluidizing velocity (m/s)

U_{opt} superficial fluidizing velocity at which maximum value of h_w occurs (m/s)

U_p particle velocity (m/s)

w fin width (m)

w_i weight fraction of particles in a specific size range; dimensionless

Y horizontal pitch of tube bundle (m)

Z vertical pitch of tube bundle (m)

Greek Symbols

β time fraction that the tube is in contact with bubbles; dimensionless

β_1 heat capacity function; $= h_{wfb}/h_{wb}$; dimensionless

γ accommodation coefficient; dimensionless

δ bubble fraction in the bed; dimensionless

δ_1 gas film thickness (m)

δ' fin thickness (m)

Δ minimum distance between surfaces of adjacent tubes (m)

ε bulk bed porosity; dimensionless

ε_b void fraction; $= 1 - U_B/(U - U_{mf} + U_B)$; dimensionless

ε_B effective emissivity of the bed; dimensionless

ε_{mf} void fraction at minimum fluidization conditions; dimensionless

ε_{opt} void fraction at optimum fluidizing velocity; dimensionless

ε_p particle emissivity; dimensionless

ε_T emissivity of tube material; dimensionless

ε_W bed porosity in the vicinity of the immersed tube; dimensionless

ε_{Wmf} bed porosity in the vicinity of the immersed tube at minimum fluidizing velocity, dimensionless

η fin efficiency factor when fin thermal resistance is considered; dimensionless

θ angle of tube with respect to horizontal (degrees)

Λ mean free path of a gas molecule (m)

μ viscosity of fluidizing gas (Ns/m^2)

π constant; = 3.1416; dimensionless

ρ_{emf} particulate phase density (kg/m^3)

ρ_f fluidizing gas density (kg/m^3)

ρ_s density of solid particles (kg/m^3)

σ modified mean free path of a gas molecule; $= 2\Lambda\,(2/\gamma - 1)$ (m)

τ particle residence time (s)

τ_c particle thermal time constant (s)

φ fin effectiveness factor, defined by Equation 46; dimensionless

φ' finning factor; $= A_{wft}/A_w$; dimensionless

Φ_s particle sphericity; dimensionless

ψ_τ frequency of occurrence of packets of age τ (1/s)

REFERENCES

1. Zabrodsky, S. S., *Hydrodynamics and Heat Transfer in Fluidized Beds,* Cambridge: The M.I.T. Press, 1966, pp. 226–299.
2. Kunii, D., and Levenspiel, O., *Fluidization Engineering,* New York: John Wiley and Sons, 1969, pp. 265–301.
3. Gelperin, N. I., and Einshtein, V. G., "Heat Transfer in Fluidized Beds," in *Fluidization,* J. F. Davidson and D. Harrison (Eds.), New York: Academic Press, 1971, pp. 517–540.

4. Gutfinger, C., and Abuaf N., "Heat Transfer in Fluidized Beds," in *Advances in Heat Transfer*, Volume 10, J. P. Hartnett and T. F. Irvine, Jr. (Eds.), New York: Academic Press, 1974, pp. 167–218.
5. Botterill, J. S. M., *Fluid-Bed Heat Transfer*, New York: Academic Press, 1975, pp. 145–277.
6. Zabrodsky, S. S., Antonishin, N. V., and Parnas, A. L., "On Fluidized Bed-to-Surface Heat Transfer," *Canadian Journal of Chemical Engineering*, Vol. 54 (1976), pp. 52–58.
7. Saxena, S. C., et al., "Heat Transfer between a Gas Fluidized Bed and Immersed Tubes," in *Advances in Heat Transfer*, Volume 14, T. F. Irvine Jr. and J. P. Hartnett (Eds.), New York: Academic Press, 1978, pp. 149–247.
8. Grewal, N. S., "Experimental and Theoretical Investigations of Heat Transfer between a Gas-Solid Fluidized Bed and Immersed Tubes," Ph. D. thesis, University of Illinois at Chicago Circle, Chicago, Illinois, 1979, pp. 1–123.
9. Baskakov, A. P., et al., "Heat transfer to Objects Immersed in Fluidized Beds," *Powder Technology*, Vol. 8 (1973), pp. 273–282.
10. Leva, M., *Fluidization*, New York: McGraw—Hill, 1959, pp. 183–186.
11. Levenspiel, O., and Walton, J. S., "Wall Bed Heat Transfer in Fluidized Systems," *Chemical Engineering Progress Symposium Series*, Vol. 50, No. 9 (1954), pp. 1–13.
12. Dow, W. M., and Jacob, M., "Heat Transfer between a Vertical Tube and a Fluidized Air-Solid Mixture," *Chemical Engineering Progress*, Vol. 47 (1951), pp. 537–548.
13. Wasan, D. T., and Ahluwalia, M. S., "Consecutive Film and Surface Renewal Mechanism for Heat or Mass Transfer from a Wall," *Chemical Engineering Science*, Vol. 24 (1969), pp. 1535–1541.
14. Van Heerden, C., Nobel, P., and Van Krevelen, D. W., "Mechanism of Heat Transfer in Fluidized Beds," *Industrial Engineering Chemistry*, Vol. 45 (1953), pp. 1237–1242.
15. Ziegler, E. N., and Brazelton, W. T., "Mechanism of Heat Transfer to Fixed Surfaces in Fluidized Bed," *Industrial Engineering Chemistry Fundamentals*, Vol. 3 (1964), pp. 94–98.
16. Mickley, H. S., and Fairbanks, D. F., "Mechanism of Heat Transfer to Fluidized Beds," *AIChE Journal*, Vol. 1 (1955), pp. 374–384.
17. Harakas, N. K., and Beatty, K. O., Jr., "Moving Bed Heat Transfer: 1. Effect of Interstitial Gas with Fine Particles," *Chemical Engineering Progress Symposium Series*, Vol. 59, No. 41 (1963), pp. 122–128.
18. Ernst, R., "The Mechanism of Heat Transfer to Heat Exchangers in Fluidized Beds," *Chemie Ingenieur Technik*, Vol. 31 (1959), pp. 166–173.
19. Botterill, J. S. M., et al., "The Effect of Gas and Solids Thermal Properties on the Rate of Heat Transfer to Gas-Fluidized Beds," in *Proceedings International Symposium on Fluidization.*, A. A. H. Drinkenburg Ed. Amsterdam: Netherlands University Press, 1967, pp. 442–453.
20. Gabor, J. D., "Wall-to-Bed Heat Transfer in Fluidized Beds," *AIChE Journal*, Vol. 18 (1972), pp. 249–250.
21. Yoshida, K., Kunii, D., and Levenspiel, O., "Heat Transfer Mechanisms between Wall Surface and Fluidized Bed," *International Journal Heat Mass Transfer*, Vol. 12 (1969), pp. 529–536.
22. Kubie, J., and Broughton, J., "A Model of Heat Transfer in Gas Fluidized Beds," *International Journal Heat Mass Transfer*, Vol. 18 (1975), pp. 289–298.
23. Wunschmann, J., and Schlünder, E. V., "Heat Transfer from Heated Plates to Stagnant and Agitated Beds of Spherical Shaped Granules Under Normal Pressure and Vacuum," *Proceedings 5th International Heat Transfer Conference*, Vol. 5, 1975, pp. 49–54.
24. Antonishin, N. V., Geller, A., and Parnas, A. L., "Hyperbolic Equation of Heat Conduction for Dispersed Systems," *Journal of Engineering Physics* (English Translation), Vol. 26 (1974), pp. 353–356.

25. Xavier, A. M., and Davidson, J. F., "Heat Transfer to Surfaces Immersed in Fluidized Beds and in the Freeboard Region," *AIChE Symposium Series*, Vol. 77, No. 208 (1981), pp. 368–373.
26. Mori, S., and Wen, C. Y., "Estimation of Bubble Diameter i Gaseous Fluidized Beds," *AIChE Journal*, Vol. 21, No. 1 (1975), pp. 109–115.
27. Ziegler, E. N., Koppel, L. B., and Brazelton, W. T., "Effect of Solid Thermal Properties on Heat Transfer to Gas Fluidized Beds," *Industrial Engineering Chemistry, Fundamentals*, Vol. 3, No. 4 (1964), pp. 324–329.
28. Gabor, J. D., "Wall-to-Bed Heat Transfer in Fluidized and Packed Beds," *Chemical Engineering Progress Symposium Series*, Vol. 66, No. 105 (1970), pp. 76–86.
29. Glicksman, L. R., and Decker, N. A., "Design Relationships for Predicting Heat Transfer to Tube Bundles in Fluidized Bed Combustors," *Proceedings Sixth International Conference on Fluidized Bed Combustion*, Georgia, U.S.A., Vol. III. Technical Sessions, April 1980, pp. 1152–1157.
30. Martin H., "Heat and Mass Transfer in Fluidized Beds," International Chemical Engineering, Vol. 22, No. 1 (1982), pp. 30–43.
31. Schlünder, E. U., "Heat Transfer to Moving Spherical Packings at Short Contact Times," *International Chemical Engineering*, Vol. 20, No. 4 (1980), pp. 550–554.
32. Gabor, John D., "Heat Transfer to Particle Beds with Gas Flows Less Than or Equal to that Required for Incipient Fluidization," *Chemical Engineering Science*, Vol. 25 (1970), pp. 979–984.
33. Botterill, J. S. M., and Denloye, A. O. O., "Gas Convective Heat Transfer to Packed and Fluidized Beds – 1. A Theoretical Model," *AIChE Symposium Series*, Vol. 74, No. 176 (1978), pp. 194–202.
34. Baskakov, A. P., and Suprum, V. M., "Determination of the Convective Component of the Heat Transfer Coefficient to a Gas in a Fluidized Bed," *International Chemical Engineering*, Vol. 12, No. 2 (1972), pp. 324–326.
35. Adams, Ronald L., and Welty, James R., "A Gas Convection Model of Heat Transfer in Large Particle Fluidized Beds," *AIChE Journal*, Vol. 25, No. 3 (1979), pp. 395–405.
36. Adams, R. L., "An Approximate Formula for Gas Convection Dominant Heat Transfer in Large-Particle Fluidized Beds," *Journal of Heat Transfer*, Vol. 103 (1981), pp. 395–397.
37. Catipovic, N. M., et al. "A Model for Heat Transfer to Horizontal Tubes Immersed in a Fluidized Bed of Large Particles," in *Fluidization*, J. R. Grace and J. M. Matsen (Eds.), New York: Plenum Press, 1980, pp. 225–234.
38. Zabrodsky, S. S., et al. "Heat Transfer in a Large Particle Fluidized Bed with Immersed In-Line and Staggered Bundles of Horizontal Smooth Tubes," *International Journal of Heat Mass Transfer*, Vol. 24 (1981), pp. 571–579.
39. Ganzha, V. L., Upadhyay, S. N., and Saxena, S. C., "A Mechanistic Theory for Heat Transfer Between Fluidized Beds of Large Particles and Immersed Surfaces," *International Journal of Heat Mass Transfer*, Vol. 25, No. 10 (1982), pp. 1531–1540.
40. Borodulya, V. A., Ganzha, V. L., and Podberezsky, A. I., "Heat Transfer in a Fluidized Bed at High Pressures," in *Fluidization*, J. R. Grace and J. M. Matsen (Eds.), New York: Plenum Press, 1980, pp. 201–207.
41. Borodulya, V. A., et al., "High Pressure Heat Transfer Investigations for Fluidized Beds of Large Particles and Immersed Vertical Tube Bundles," *International Journal of Heat Mass Transfer*, Vol. 26, No. 11 (1983), pp. 1577–1584.
42. Staub, F. W., and Canada G. S., "Effect of Tube Bank and Gas Density on Flow Behavior and Heat Transfer in Fluidized Beds," in *Fluidization*, J. F. Davidson and D. L. Keairns (Eds.) Cambridge: Cambridge University Press, 1978, pp. 339–344.
43. Canada, G. S., and McLaughlin, M. H., "Large Particle Fluidization and Heat Transfer at High Pressures," *AIChE Symposium Series*, Vol. 74, No. 176 (1976), pp. 27–37.

44. Xavier, A. M., et al., "Surface-to-Bed Heat Transfer in a Fluidized Bed at High Pressures," in *Fluidization,* J. R. Grace and J. M. Matsen (Eds.) New York: Plenum Press, 1980, pp. 204–216.
45. Kunii, D., and Smith, J. M., "Heat Transfer Characteristics of Porous Rocks," *AIChE Journal,* Vol. 6, No. 1 (1960), pp. 71–78.
46. Krupiczka, R., "Analysis of Thermal Conductivity of Granular Materials," *International Chemical Engineering,* Vol. 7 (1967), pp. 122–124.
47. Yagi, S., and Kunii, D., "Studies on Heat Transfer Near Wall Surface in Packed Beds," *AIChE Journal,* Vol. 6, No. 1 (1960), pp. 97–104.
48. Zabrodsky, S. S., "Compound Heat Exchange Between a High Temperature Gas-Fluidized Bed and a Solid Surface," *International Journal of Heat Mass Transfer,* Vol. 16 (1973), pp. 241–248.
49. Szekely, J., and Fisher, R., "Bed to Wall Radiation Heat Transfer in a Gas-Solid Fluidized Bed," *Chemical Engineering Science,* Vol. 24, No. 5 (1969), pp. 833–848.
50. Yoshida, K., Ueno, T., and Kunii, D., "Mechanism of Bed-Wall Heat Transfer in a Fluidized Bed at High Temperatures," *Chemical Engineering Science,* Vol. 29 (1974), pp. 77–82.
51. Vedamurthy, V. N., and Sastri, V. M. K., "An Analysis of the Conductive and Radiative Heat Transfer to the Walls of Fluidized Bed Combustors," *International Journal of Heat Mass Transfer,* Vol. 17 (1974), pp. 1–9.
52. Wright, S. J., Ketley, H. C., and Hickman, R. G., "The Combustion of Coal in Fluidized Beds for Firing Shell Boilers," *Journal of Institute of Fuel,* Vol. 42 (1969), pp. 235–240.
53. Bhattacharya, S. C., and Harrison, D., "Heat Transfer in High Temperature Fluidized Beds," *European Congress on Particle Technology,* Nuremberg, 1977.
54. Thring, R. H., "Fluidized Bed Combustion for the Stirling Engine," *International Journal of Heat Mass Transfer,* Vol. 20 (1977), pp. 911–918.
55. Basu, P., "Bed-to-Wall Heat Transfer in a Fluidized Bed Coal Combustor," *AIChE Symposium Series,* Vol. 74, No. 176 (1978), pp. 187–193.
56. Kolar, A. K., Grewal, N. S., and Saxena, S. C., "Investigation of Radiative Contribution in a High Temperature Fluidized-Bed Using the Alternate-Slab Model," *International Journal of Heat Mass Transfer,* Vol. 22 (1979), pp. 1695–1703.
57. Kolar, A. K., et al., "On Gabor's Alternate-Slab Model," *Numerical Heat Transfer,* Vol. 1 (1978), pp. 425–430.
58. Pikashov, V. S., et al., "On the Experimental Study of the Fluidized-Bed Effective Emissivity," *Heat and Mass Transfer,* Vol. 5, Nauka i Tekhn, Minsk, 1968.
59. Geldart, D., "Types of Gas Fluidization," *Powder Technology,* Vol. 6 (1972), pp. 285–292.
60. Gelperin, N. I., Einshtein, V. G., and Zaikovskii, A. V., "Hydraulic and Heat Transfer Characteristics of a Fluidized Bed with a Horizontal Tube Bundle," *Journal of Engineering Physics,* Vol. 10 (1966), pp. 473–475.
61. Gelperin, N. I., et al., "Local Fluidized Bed-to-Surface Heat Transfer Coefficients," *Teor. Osnovy Khim. Tekhnol,* Vol. 2, No. 3 (1968), pp. 430–438.
62. Keairns, D. L., "Heat Transfer from Horizontal Tubes in a Rectangular Fluidized Bed," Research Report 68-9D3-273-R13, *Westinghouse Research Laboratory,* Pittsburgh, Pennsylvania, 1968.
63. Keairns, D. L., "Heat Transfer from a Horizontal Tube in a Fluidized Bed at 1600° F," Research Report 68-9D3-273-R13, *Westinghouse Research Laboratory,* Pittsburgh, Pennsylvania, 1968.
64. Noack, R., "Lokaler Wärmeübergang an Horizontalen Rohren in Wirbelschichten," *Chemie Ingenieur Technik,* Vol. 42, No. 6 (1970), pp. 371–376.
65. Berg, B. V., and Baskakov, A. P., "Investigation of Local Heat Between a Fixed Horizontal Cylinder and a Fluidized Bed," *International Chemical Engineering,* Vol. 14, No. 3 (1974), pp. 440–443.

66. Mickley, H. S., Fairbanks, D. F., and Hawthorn, R. D., "The Relation Between the Heat Transfer Coefficient and Thermal Fluctuations in Fluidized-Bed Heat Transfer," *Chemical Engineering Progress Symposium Series,* Vol. 57, No. 32 (1961), pp. 51–60.
67. Noë, A. R., and Knudsen, J. G., "Local and Average Heat Transfer Coefficients in a Fluidized Bed Heat Exchanger," *Chemical Engineering Progress Symposium Series,* Vol. 64, No. 82 (1968), pp. 202–211.
68. Genetti, W. E., and Knudsen, J. G., "Heat Transfer in a Dilute-Phase Fluidized-Bed Tabular Heat Exchanger," *Chemical Engineering Progress Symposium Series,* Vol. 30 (1968), pp. 147–156.
69. Selzer, V. W., and Thomson, W. J., "Fluidized Bed Heat Transfer – The Packet Theory Revisited," *AIChE Symposium Series,* Vol. 73, No. 161 (1977), pp. 29–37.
70. Vreedenberg, H. A., "Heat Transfer Between a Fluidized Bed and a Horizontal Tube," *Chemical Engineering Science,* Vol. 9 (1958), pp. 52–60.
71. Andeen, B. R., and Glicksman, L. R., "Heat Transfer to Horizontal Tubes in Shallow Fluidized Beds," *ASME–AIChE Heat Transfer Conference,* ASME Paper 76-HT-67, St. Louis, Missouri, 1976.
72. Petrie, J. C., Freeby, W. A., and Buckham, J. A., "In-Bed Heat Exchangers," *Chemical Engineering Progress,* Vol. 64, No. 67 (1968), pp. 45–51.
73. Ainshtein, V. G., "An Investigation of Heat Transfer Process Between Fluidized Beds and Single Tubes Submerged in the Bed" cited in *Hydrodynamics and Heat Transfer in Fluidized Beds,* by S. S. Zabrodsky, Cambridge: M. I. T. Press, Massachusetts, 1966, pp. 270–271.
74. Goroshko, V. D., Rozenbaum, R. B., and Todes, O. M., "Approximate Hydraulic Relationship for Suspended Beds and Hindered Falls," *Izvestiya Vuzov, Neft' i Gaz,* No. 1 (1958), pp. 125–131, cited in *Hydrodynamics and Heat Transfer in Fluidized Beds,* by S. S. Zabrodsky, Cambridge: M. I. T. Press, Massachusetts, 1966, pp. 71–73.
75. Ziegler, E. N., "Heat and Mass Transfer to a Gas-Fluidized Bed of Solid Particles," Report No. 6807, *Argonne National Laboratory,* Illinois, 1963.
76. "Evaluation of Fluid Bed Heat Exchanger Optimization Parameters," Final Technical Report submitted to the Department of Energy under Contract No. ACO3-78ET11343, *Aerojet Energy Conversion Company,* Sacramento, California, 1980.
77. Grewal, N. S., and Saxena, S. C., "Heat Transfer Between a Horizontal Tube and a Gas-Solid Fluidized Bed," *International Journal of Heat Mass Transfer,* Vol. 23 (1980), pp. 1505–1519.
78. Grewal, N. S., et al., "Effect of Distributor Design on Heat Transfer from an Immersed Horizontal Tube in a Fluidized Bed," *Chemical Engineering Journal,* Vol. 18 (1979), pp. 197–201.
79. Vakhrushev, I. A., Botnikov, Ya. A., and Zenchenkov, N. G., "Heat Transfer from a Fluidized Bed of Hot Coke to the Surface of Horizontal Tubes," *International Chemical Engineering,* Vol. 3, No. 2 (1963), pp. 207–210.
80. Natusch, H. J., and Blenke, H., "Warmeübertragung and Rippenrohren in Gas-Fliesbetten," *Chemie Ingenieur Technik,* Vol. 7, No. 10 (1973), pp. 293–296.
81. Chen, J. C., "Heat Transfer to Tubes in Fluidized Beds." *ASME–AIChE Heat Transfer Conference,* Paper No. 76-HT-75, St. Louis, Missouri, 1976.
82. Tishchenko, A. T., and Khvastukhin, Yu. I., *Fluidized Bed Furnaces and Heat Exchangers,* Kiev: Naukova Dumka, 1973.
83. Lese, H. K., and Kermode, R. I., "Heat Transfer from a Horizontal Tube to a Fluidized Bed in the Presence of Unheated Tubes," *Canadian Journal of Chemical Engineering,* Vol. 50 (1972), pp. 44–48.
84. Antonshin, N. V., cited in *Hydrodynamics and Heat Transfer in Fluidized Beds,* by S. S. Zabrodsky, Cambridge: M. I. T. Press, MA, 1966, pp. 289–291.
85. Gelperin, N. I., Einstein, V. G., and Romanova, N. A., *Khim Mashinostr,* No. 5, 1963, p. 13, cited in *Fluidization,* J. F. Davidson and D. Harrison (Eds.), New York: Academic Press, 1971, p. 489.

86. Bartel, W. J., and Genetti, W. E., "Heat Transfer from a Horizontal Bundle of Bare and Finned Tubes in an Air Fluidized Bed," *Chemical Engineering Progress Symposium Series,* Vol. 69. No. 128 (1973), pp. 85–93.
87. Howe, W. C., and Aulisio, C., "Control Variables in Fluidized Bed Steam Generation," *Chemical Engineering Progress,* Vol. 73, No. 7 (1977), pp. 69–73.
88. Neukirchen, B., and Blenke, H., "Gestaltung Horizontaler Rohrbundel in Gas-Wirbelschichtreaktoren nach Wametechnischen," *Chemie Ingenieur Technik,* Vol. 45, No. 5 (1973), pp. 307–312.
89. Cherrington, D. C., Golan, L. P., and Hammit, F. G., "Industrial Application of Fluidized Bed Combustion-Single Tube Heat Transfer Studies," *Proceedings of the Fifth International Conference on Fluidized Bed Combustion,* Washington, DC, 1977.
90. Janssen, K., Dr. Ing Thesis, Aachen, 1973.
91. Grewal, N. S., and Hajicek, D. R., "Experimental Studies of Heat Transfer from an Atmospheric-Fluidized-Bed Combustor to an Immersed Array of Horizontal Tubes," *7th International Heat Transfer Conference,* Paper No. HX23, Munchen, Germany, 1982.
92. Grewal, N. S., and Goblirsch, G., "Heat Transfer to Horizontal Tubes in a Pilot-Scale Fluidized-Bed Combustor Burning Low-Rank Coals," *ASME–AIChE International Heat Transfer Conference,* Paper No. 83-HT-91, Seattle, WA, 1983.
93. Babu, S. P., Shah, B., and Talwalker, A., "Fluidization Correlations for Coal Gasification Materials — Minimum Fluidization Velocity and Fluidized Bed Expansion Ratio," *AIChE Symposium Series,* Vol. 74, No. 176 (1978), pp. 176–186.
94. Geldart, D., and Cranfield, R. R., "The Gas Fluidization of Large Particles," *Chemical Engineering Journals,* Vol. 3 (1972), pp. 211–231.
95. Borodulya, V. A., et al., "Heat Transfer from In-Line and Staggered Horizontal Smooth Tube Bundles Immersed in a Fluidized Bed of Large Particles," *International Journal of Heat Mass Transfer,* Vol. 23 (1980), pp. 1602–1604.
96. Chandran, R., Chen, J. C., and Staub, F. W., "Local Heat Transfer Coefficient around Horizontal Tubes in Fluidized Beds," *Journal of Heat Transfer,* Vol. 102 (1980), pp. 152–157.
97. Golan, L. P., LaLonde, G. V., and Weiner, S. C., "High Temperature Heat Transfer Studies in a Fluidized Bed," *Proceedings 6th International Conference Fluidized Bed Combustion,* Atlanta, 1980, pp. 1773–1185.
98. Zabrodsky, S. S., et al., "The Choice of Design Correlation for the Estimation of the High-Temperature Fluidized-Bed-to-Immersed Body Heat Transfer Coefficient," *Vestn. Akad Nauk. BSSR,* Ser. Fiz-Energ Nauk No. 4, 1974, pp. 103–107.
99. Grewal, N. S., "A Correlation for Maximum Heat Transfer Coefficient Between a Horizontal Tube and a Gas-Solid Fluidized Bed of Small Particles," *Letters Heat Mass Transfer,* Vol. 9 (1982), pp. 377–385.
100. Grewal, N. S., and Saxena, S. C., "Experimental Studies of Heat Transfer Between a Bundle of Horizontal Tubes and a Gas-Solid Fluidized Bed of Small Particles," *Industrial Engineering Chemistry, Process Design and Development,* Vol. 22, 1983, pp. 367–376.
101. Todes, O. M., Bondareva, A. K., and Geltsiker, A. O., "*Doklady, No. 5–18 na 2-M Vsesovuzn, Soveshch, po Teoplo-i Massoobmenu,* Minsk, 1964, cited in Ainshtein, V. A., and Gelperin, N. I., "Heat Transfer Between a Fluidized Bed and a Surface," *International Chemical Engineering,* Vol. 6, No. 1 (1966), pp. 67–74.
102. Varygin, N. N., and Martyushin, I. G., "A Calculation of Heat Transfer Surface Area in Fluidized Bed Equipment," *Khim. Mashinostroenie,* No. 5 (1959), pp. 6–9.
103. Maskaev, V. K., and Baskakov, A. P., "Features of External Heat Transfer in Fluidization Beds of Coarse Particles," *International Chemical Engineering,* Vol. 14, No. 1 (1974), pp. 80–83.

104. Bansal, R. K., Kadaba, P. V., and Desai, P. V., "Heat Transfer from Horizontal Tubes in Gas Fluidized Beds," *ASME/AIChE National Heat Transfer Conference,* Paper No. 80-HT-115, Orlando, 1980.
105. Grewal, N. S., "A Generalized Correlation for Heat Transfer Between a Gas-Solid Fluidized Bed of Small Particles and an Immersed Staggered Array of Horizontal Tubes," *Powder Technology,* Vol. 30 (1981), pp. 145–154.
106. Gelperin, N. I., Einshtein, V. G., and Korotyanskaya, L. A., "Heat Transfer Between a Fluidized Bed and Staggered Bundles of Horizontal Tubes," *International Chemical Engineering,* Vol. 9, No. 1 (1969), pp. 137–142.
107. McLaren, J., and Williams, D. F., "Combustion Efficiency, Sulphur Retention and Heat Transfer in Pilot Plant Fluidized-Bed Combustors," *Journal of Institute of Fuel,* Vol. 42 (1969), pp. 303–308.
108. Chekansky, V. V., et al., "Heat Transfer from a Fluidized Bed to a Closely-Packed Staggered Bundle of Horizontal Tubes," Trudy SKB-3, Minavtoprom SSSR, Vyp. 4, ONTI, Minsk, 1970, pp. 143–152.
109. Xavier, A. M., and Davidson, J. F., "Heat Transfer to Surfaces Immersed in Fluidized Beds, Particularly Tube Arrays," in *Fluidization,* Davidson, J. F., and Keairns, D. L., (Eds.) Cambridge: Cambridge University Press, 1978, pp. 333–338.
110. Borodulya, V. A., et al., "Heat Transfer Between Gas-Solid Fluidized Beds and Horizontal Tube Bundles," *Letters Heat Mass Transfer,* Vol. 7 (1980), pp. 83–95.
111. Vreedenberg, H. A., "Heat Transfer Between a Fluidized Bed and a Vertical Tube," *Chemical Engineering Science,* Vol. 11 (1960), pp. 274–285.
112. Toomey, R. D., and Johnstone, H. F., "Heat Transfer Between Beds of Fluidized Solids and the Walls of the Container," *Chemical Engineering Progress Symposium Series,* Vol. 5, No. 5 (1953), pp. 51–63.
113. Mickley, H. S., and Trilling, C. A., "Heat Transfer Characteristics of Fluidized Beds," *Industrial Engineering Chemistry,* Vol. 41 (1949), pp. 1135–1147.
114. Baerg, A., Klassen, J., and Gishler, P. E., "Heat Transfer in a Fluidized Solids Bed," *Canadian Journal of Research Section F,* Vol. 28 (1950), pp. 287–307.
115. Miller, C. O., and Logwinuk, A. K., "Fluidization Studies of Solid Particles," *Industrial Engineering Chemistry,* Vol. 43, No. 5 (1951), pp. 1220–1226.
116. Vreedenberg, H. A., "Heat Transfer Between Fluidized Beds and Vertically Inserted Tubes," *Journal Applied Chemistry,* Vol. 2. Suppl. Issue No. 1 (1952), pp. 526–533.
117. Olin, H. L., and Dean, O. C., "Fluidization," Petroleum Engineer, Vol. 25, 23C (1953), pp. 5–12.
118. Wicke, E., and Hedden, K., "Strömungsformen und Wärmeübertragung in von Luft Aufgewirbelten Schuttgutschichten," *Chemie Ingenieur Technik,* Vol. 24, No. 2 (1952), pp. 82–91.
119. Brotz, W., "Untersuchungen über Transportvorgänge in Durchströmtem Gekorntem Gut," *Chemie Ingenieur Technik,* Vol. 28 (1956), pp. 165–174.
120. Wicks, E., and Fetting, F., "Wärmeübertragung in Gaswirbelschichten," *Chemie Ingenieur Technik,* Vol. 26, No. 6 (1954), pp. 301–309.
121. Fetting, F., and Wicke, E., "Strömungsformen und Wärmeübertragung in Gaswirbelschichten, *DECHEMA–Monogr.* Vol. 24 (1955), pp. 146–169.
122. Wender, L., and Cooper, G. T., "Heat Transfer Between Fluidized-Solids Beds and Boundary Surfaces—Correlation of Data," *AIChE Journal,* Vol. 4 (1958), pp. 15–23.
123. Genetti, W. E., Schmall, R. A., and Grimmet, E. S., "The Effect of Tube Orientation on Heat Transfer with Bare and Finned Tubes in a Fluidized Bed," *Chemical Engineering Progress Symposium Series,* Vol. 67, No. 116 (1971), pp. 90–96.
124. Bartel, W. J., Genetti, W. E., and Grimmet, E. S., "Heat Transfer from a Horizontal Discontinuous Finned Tube in a Fluidized Bed," *Chemical Engineering Progress Symposium Series,* Vol. 116, No. 67 (1971), pp. 85–89.

125. Gelperin, N. I., Einshtein, V. G., and Toskubaev, I. N., "Heat Transfer Between Finned Tubes and Fluidized Bed of Granular Material," *Khim. Tekhnolog,* Topliv i Masel, Vol. 2 (1972), pp. 42–43.
126. Gelperin, N. I., et al., "Fluidized Bed-to-Finned Tubes Heat Transfer," *Trudy TITChT,* Vyp. 1 (1974), pp. 165–168.
127. Genetti, W. E., and Kratovil, M. T., "Heat Transfer from Helical Finned Tubes in a Fluidized Bed," *The 83rd National AIChE Meeting,* AIChE Paper No. 46, Houston, TX, March 1977.
128. Grewal, N. S., and Saxena, S. C., "Effect of Surface Roughness on Heat Transfer from Horizontal Immersed Tubes in a Fluidized Bed," *Journal of Heat Transfer,* Vol. 101 (1979), pp. 397–403.
129. Krause, W. B., and Peters, A. R., "Heat Transfer from Horizontal Serrated Finned Tubes in an Air-Fluidized Bed of Uniformly-Sized Particles," *The Joint ASME/AIChE National Heat Transfer Conference,* Paper No. 80-HT-48, Orlando, FL, July 1980.
130. Priebe, S. J., and Genetti, W. E., "Heat Transfer from a Horizontal Bundle of Extended Surface Tubes to an Air Fluidized Bed," *AIChE Symposium Series,* Vol. 73, No. 161 (1977), pp. 38–43.
131. Zabrodsky, S. S., et al, "Heat Transfer of Single Horizontal Finned Tubes and Their Bundles in a Fluidized Bed of Large Particles," in *Fluidization,* J. R. Grace and J. M. Matsen (Eds.) New York: Plenum, 1980, pp. 195–200.
132. Grewal, N. S., Cheung, T. K., and Saxena, S. C., "Heat Transfer Between Horizontal Finned Tubes and a Gas-Solid Fluidized Bed," *Industrial Engineering Chemistry Process Design and Development,* 1985, in press.
133. Natusch, H. J. and Blenke, H. "Zur Wärmeübertragung and Horizontalen Längsrippenrohren in Gas-Fliesbetten," *Verfahrenstechnik,* Vol. 8, No. 10 (1974), pp. 286–293.
134. Chen, J. C., and Withers, J. G., "An Experimental Study of Heat Transfer from Plain and Finned Tubes in Fluidized Beds," *AIChE Symposium Series,* Vol. 74, No. 174 (1978), pp. 327–333.
135. Genetti, W. E., Yurich, S. P., and Vanderhoof, D. W., "Heat Transfer from Vertical Finned Tube Bundles in a Fluidized Bed," the joint *ASME/AIChE National Transfer Conference,* Paper No. 80-HT-118, New York, NY, 1980.
136. Vijayaraghavan, M. R., and Sastri, V. M. K., "Effect of Surface Roughness on Heat Transfer in Fluidized Beds," *Proceedings, Conference on Future Energy Production,* International Center Heat Mass Transfer, 1975, pp. 571–579.
137. Genetti, W. E., and Everly, D., "Heat Transfer from Spiral Tubing in an Air Fluidized Bed," *The 71st Annual Meeting of AIChE,* Miami Beach, FL, November 1978.
138. Korolev, V. N., and Syromyatnikov, N. I., "Heat Transfer from a Surface with Artificial Roughness to a Fluidized Bed," *Journal of Engineering Physics,* Vol. 28, No. 6 (1975), pp. 698–700.
139. Decker, N. A., and Glicksman, L. R., "Conduction Heat Transfer at the Surface of Bodies Immersed in Gas Fluidized Beds of Spherical Particles," *AIChE Symposium Series,* Vol. 77, No. 208 (1981), pp. 341–349.

CHAPTER 19

PERIODIC HEAT TRANSFER THROUGH INHOMOGENEOUS LAYERS

A. K. S. Thakur

Department of Physics
University of Sokoto
Sokoto, Nigeria

CONTENTS

INTRODUCTION

Over the last a few years heat transfer processes viz. steady state, transient, and periodic, in variety of isotropic materials have been studied with entire text books and journals devoted to the subject. But the heat transfer in nonisotropic materials received much less attention in spite of the fact that most of the materials exhibit change in their thermal properties with respect to temperature or position; only steady-state heat transfer has been described by Carslaw and Jaeger [1] in their book "Heat Conduction in Solids." In some cases heat transfer through nonisotropic materials has been studied by averaging out the thermal conductivity over the temperature or position domain. Thermal properties, namely thermal conductivity, density, and specific heat of several materials, demonstrate a marked spatial or temperature dependence. Many engineering materials, in particular building materials such as brick, concrete, asbestos, wood, and synthetic foam, possess porous, and nonuniform structure. As a rule of thumb, materials with high density exhibit high thermal conductivity, which of course greatly depends on porosity and moisture content; for instance asbestos shows an increase in thermal conductivity from 0.105 kJ/hm °C to 0.248 kJ/hm °C with an increase in density from 400 kg/m^3 to 800 kg/m^3. Thermal conductivity of some typical sample of dry soil corresponds to 0.51 kJ/hm °C and shows an almost threefold rise in wet conditions. Figure 1 illustrates thermal conductivities of some construction and insulating materials.

Similar to building materials, metals also show a change in thermal conductivity in different circumstances. A sharp decrease in the thermal conductivity of metals is recorded in presence of all kinds of admixtures. This decline can be explained on the basis of structural inhomogeneity, which brings about enhanced scattering of free electrons in metals. For example, thermal conductivity of pure copper is 396 kJ/hm °C but arsenic contamination brings it down to 142 kJ/hm °C. Thermal conductivity of alloys increases with rising temperatures (Figure 2), whereas thermal conductivity of pure metals decreases

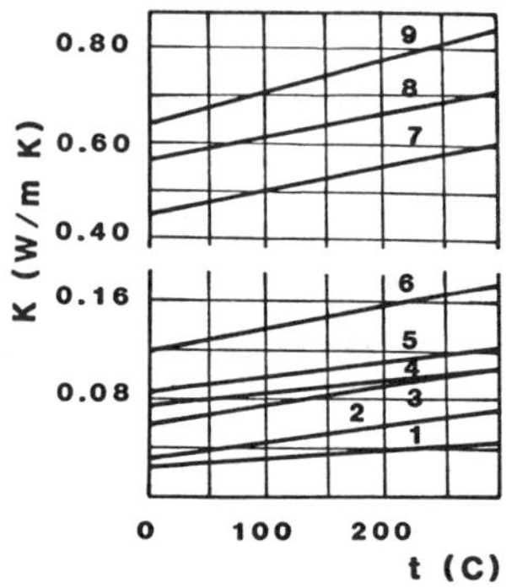

Figure 1. Thermal conductivity of building and heat insulating materials: 1—air; 2—mineral wool; 3—slag wool; 4—85% magnesia; 5—sovelite; 6—diatomaceous brick; 7—red brick; 8—slag concrete brick; 9—fire-clay brick [2].

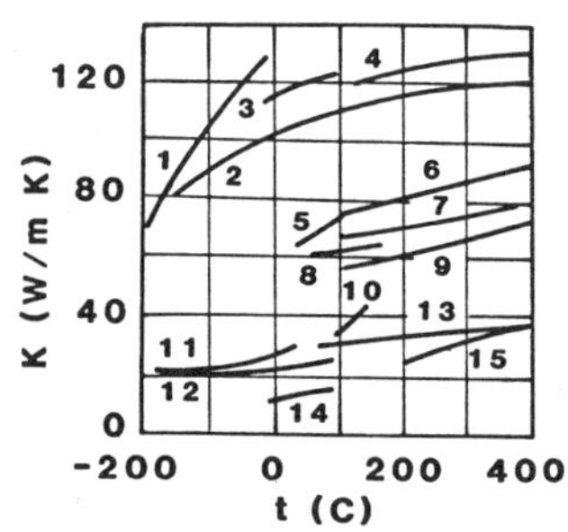

Figure 2. Thermal conductivity of alloys: 1—brass 18; 2—brass 30; 3—brass 12; 4—nichrome; 5—bronze; 6—manganese bronze; 7—gun bronze; 8—Sn-Zn alloy; 9—phosphorus bronze; 10—white metal; 11—constantan; 12—monel metal; 13—manganin; 14—nickel steel; 15—liquid Sn-Zn alloy [2].

with rising temperatures due to increase in collisional frequency and electron scattering; Figure 3 illustrates the thermal conductivity of various pure metals.

The thermal conductivity and density of liquids with the exception of water and glycerin decrease with temperature (Figure 4) and lies between 0.07 kJ/hm °C to 0.7 kJ/hm °C.

For gases, thermal conductivity and specific heat increase with rising temperature (Figures 5 and 6) and falls in the range 0.006 kJ/hm °C to 0.6 kJ/hm °C.

DIFFERENTIAL EQUATION OF HEAT CONDUCTION

The one-dimensional differential equation for heat conduction in a medium exhibiting spatial variations in its thermal properties may be expressed in cartesian coordinates as follows:

$$K \frac{\partial^2 \theta}{\partial x^2} + \frac{\partial K}{\partial x} \cdot \frac{\partial \theta}{\partial x} - \rho C \frac{\partial \theta}{\partial t} = H \tag{1}$$

where θ = temperature distribution inside the medium
C = specific heat of the medium

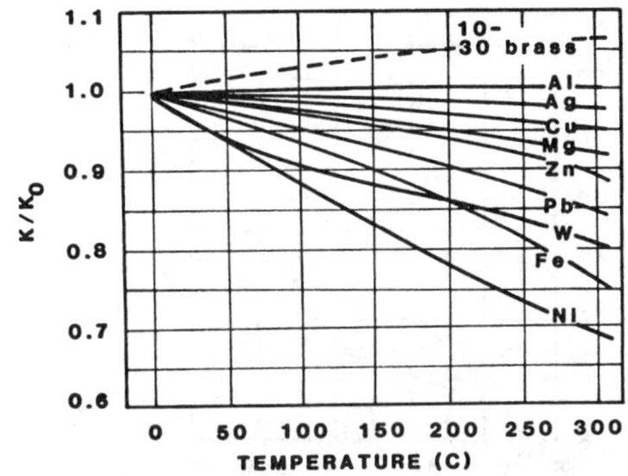

Figure 3. Dependence of thermal conductivity coefficient on temperature for some pure metals [2].

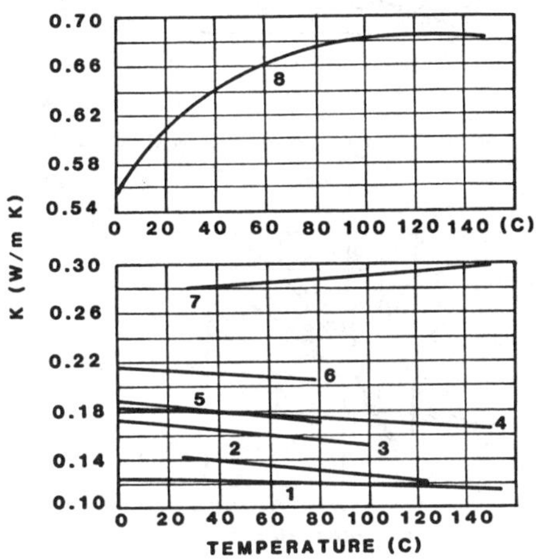

Figure 4. Thermal conductivity of liquids: 1—vaseline oil; 2—benzene; 3—acetone; 4—castor oil; 5—ethyl alcohol; 6—methyl alcohol; 7—glycerine; 8—water [2].

Term H accounts for the rate of heat production or absorption due to various chemical and physical reactions.

Likewise, the differential equation for heat conduction in a medium with temperature-dependent thermal properties is expressed as

$$K\frac{\partial^2\theta}{\partial x^2} + \frac{\partial K}{\partial \theta}\cdot\left(\frac{\partial\theta}{\partial x}\right)^2 - \rho C\frac{\partial\theta}{\partial t} = H \tag{2}$$

where K, ρ, and C all show temperature dependence.

BOUNDARY CONDITIONS

Boundary conditions for the periodic solution of differential equations for heat coduction in inhomogeneous medium discussed in the preceding section may be expressed as follows:

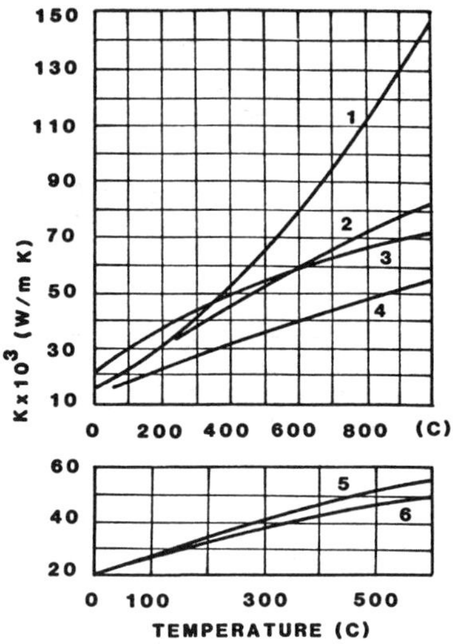

Figure 5. Thermal conductivity of gases: 1—steam; 2—carbon dioxide; 3—air; 4—argon; 5—oxygen; 6—nitrogen [2].

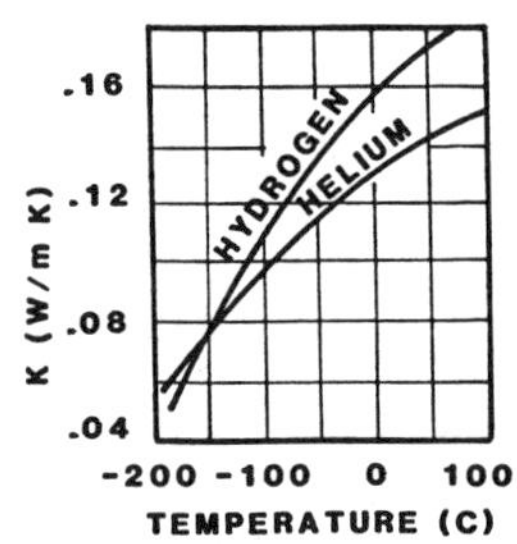

Figure 6. Thermal conductivity of helium and hydrogen [2].

1. *Periodic temperature.* We consider that the temperatures of the surroundings around the material are a periodic function of time and may be represented in terms of the Fourier series, thus

$$(\theta)_{x=0,L} = a_0 + \text{Real} \sum_{m=1}^{\infty} a_m \exp(im\omega t) \tag{3}$$

where a_0 = average temperature at the surface (°C)

$a_m = a'_m \exp(-i\sigma_m)$, where a'_m is the amplitude (°C) of the m-th harmonic in the Fourier expansion and σ_m is the corresponding phase factor (radians) in the Fourier representation of the prescribed surface temperature.

ω = angular frequency of temperature variation, e.g. for daily variation $\omega = 2\pi/24\ (h^{-1})$.

2. *Energy balance.* In some physical situations the surface temperature of the medium may not be equal to that of its surrounding environment and therefore the following energy

balance conditions are used in evaluating the periodic solution of the heat conduction equation.

$$\left(-K\frac{\partial\theta}{\partial x}\right)_{x=0,L} = \pm h[\theta_{SA} - (\theta)_{x=0,L}] \tag{4}$$

the positive sign is applicable to the x = 0 surface and the negative sign for the x = L surface, and h is the heat transfer coefficient between the environment and the medium (kJ/hm^2 °C).

In solar energy heat transfer problems temperature at the x = 0 surface exposed to the sun is known as solair temperature and is expressed as [3]

$$\theta_{SA} = [\alpha_S \cdot S(t) + h \cdot T_A - \varepsilon \cdot \Delta R]/h \tag{5}$$

where α_S = fraction of solar radiations absorbed at x = 0 surface
$S(t)$ = intensity of solar radiations (kJ/hm^2)
T_A = atmospheric air temperature (°C)
ε = long wave emissivity of x = 0 surface
ΔR = difference between the long-wave radiations, incident at x = 0 surface from the sky and surroundings, and the radiations emitted by black-body at atmospheric air temperature (kJ/hm^2)

3. *Semi-infinite medium.* In this case, the boundary condition at the x = 0 surface may be given by either of the two boundary conditions (1 and 2) discussed previously, depending upon the physical situation. The second necessary boundary condition for the region extending to infinity in the positive x-direction is

$$\theta \to 0 \quad \text{as} \quad x \to \infty \tag{6}$$

MEDIUM WITH SPATIALLY VARYING THERMAL PROPERTIES

Medium Without Heat Generation/Absorption: Exact Solutions

The one-dimensional heat conduction equation (Equation 1) for such a medium reduces to

$$K\frac{\partial^2\theta}{\partial x^2} + \frac{\partial K}{\partial x}\cdot\frac{\partial\theta}{\partial x} - \rho C\frac{\partial\theta}{\partial t} = 0 \tag{7}$$

The periodic solution of Equation 7 can be expressed in terms of the Fourier series as

$$\theta = \theta_0 + \text{Real}\sum_{m=1}^{\infty}\theta_m \exp(im\omega t) \tag{8}$$

In practice only five or six harmonics are sufficient for perfect convergence of the preceding series. Subsituting for θ in Equation 7 and writing the time-independent and time-dependent parts seperately, we have

$$K\frac{\partial^2\theta_0}{\partial x^2} + \frac{\partial K}{\partial x}\cdot\frac{\partial\theta_0}{\partial x} = 0 \tag{9}$$

and

$$K\frac{\partial^2\theta m}{\partial x^2} + \frac{\partial K}{\partial x}\cdot\frac{\partial\theta m}{\partial x} - im\omega\rho C\theta_m = 0 \tag{10}$$

The solution of Equation 9 is straight forward and is obtained by double integration as follows

$$\theta_0 = A_0 + B_0 \int K^{-1}\, dx \tag{11}$$

where A_0 and B_0 are constants of integration.

The time-dependent part, Equation 10, is nonhomogeneous in nature and thus cannot have a most general solution applicable to any arbitrary spatial profiles of thermal properties. However, the equation has been found to have closed-form-series solutions [4,5] for some of the following typical thermal conductivity profiles of interest; ρ and C are constant.

Linear Profile

Spatial variation in thermal conductivity may be expressed by the relation

$$K = K_0(1-\alpha_0 x)$$

and the solution of Equation 10 is expressible in the form of modified Bessel functions.

$$\theta_m = A_m J_0[\alpha_m(1-\alpha_0 x)^{1/2}] + B_m Y_0[\alpha_m(1-\alpha_0 x)^{1/2}] \tag{12}$$

where

$$\alpha_m = 2\exp\left(\frac{3}{4}\pi i\right)\left(\frac{m\omega\rho C}{\alpha_0^2 K_0}\right)^{1/2}$$

J_0 and Y_0 are the zero-order modified Bessel functions.

Exponential Profile

If thermal conductivity varies exponentially in space then

$$K = K_0 \exp(-\alpha_0 x)$$

and the temperature θ_m may be expressed in terms of first-order modified Bessel functions.

$$\theta_m = \left\{A_m J_1\left[2\beta_m^{1/2}\exp\left(\frac{\alpha_0 x}{2}\right)\right] + B_m Y_1\left[2\beta_m^{1/2}\exp\left(\frac{\alpha_0 x}{2}\right)\right]\right\}\cdot\exp\left(\frac{\alpha_0 x}{2}\right) \tag{13}$$

with

$$\beta_m = \exp\left(\frac{3}{2}\pi i\right)\left\{\frac{m\omega\rho C}{\alpha_0^2 K_0}\right\}$$

$K = K_0/(1-\alpha_0 x)$ *Profile*

In this case θ_m is given as follows

$$\theta_m = \left\{A_m J_{2/3}\left[\frac{2}{3}\alpha_m(1-\alpha_0 x)^{3/2}\right] + B_m J_{-2/3}\left[\frac{2}{3}\alpha_m(1-\alpha_0 x)^{3/2}\right]\right\}\cdot(1-\alpha_0 x)^{-1} \tag{14}$$

with

$$\alpha_m = \exp\left(\frac{3}{4}\pi i\right)\left(\frac{m\omega\rho C}{K_0\alpha_0^2}\right)^{1/2}$$

Parabolic Profile

The materials exhibiting thermal conductivity variation of the type

$$K = K_0(1-\alpha_0 x^2)$$

have temperature distribution θ_m in terms of extended power series.

$$\theta_m = A_m\left(1-\beta_m\left[\frac{\alpha_0 x^2}{2}+\sum_{n=1}^{\infty}\frac{(\alpha_0^{1/2}x)^{2n+2}}{(2n+2)!}\times\prod_{p=1}^{n}[2p(2p+1)-\beta_m]\right]\right)+B_m\left\{\alpha_0^{1/2}x+\sum_{n=1}^{\infty}\frac{(\alpha_0^{1/2}x)^{2n+1}}{(2n+1)!}\cdot\prod_{p=1}^{n}[2p(2p-1)-\beta_m]\right\} \quad (15)$$

with

$$\beta_m = \exp\left(\frac{3}{2}\pi i\right)\left(\frac{m\omega\rho c}{\alpha_0 k_0}\right)$$

Both power series that appeared in this solution are fast converging over a wide range of thermal conductivity, and in practice only four terms in each series are sufficient for fairly accurate results.

Various arbitrary constants A_0, B_0, A_m's and B_m's in Equations 11 through 15 may be evaluated with the help of appropriate boundary conditions applicable to the system.

Approximate Solutions

In the preceding section we have discussed various exact solutions of the heat conduction equation for inhomogeneous materials exhibiting different kinds of thermal conductivity variations. However, in practice, exact solutions for any arbitrary thermal conductivity profile may or may not exist, and at the same time a highly complicated exact solution is of little practical importance. Further, if we consider spatial variations in specific heat and density together with thermal conductivity, the heat conduction equation becomes extremely complex and the existence of a closed-form solution is doubtful. The method of approximate solutions is very popular in various branches of physics and provides fairly accurate and acceptable results even in extremely complex situations. The nonhomogeneous Equation 10 is also known to have approximate solutions [4] which may be applied to any system for which thermal conductivity, specific heat, and density exhibit spatial variation simultaneously. Interestingly, these solutions predict fairly accurate values of temperature and heat flux over a wide range of thermal parameters (Figure 7) and the accuracy of the method does not deteriorate with the complexity of the spatial variations in thermal properties of the material.

To develop the approximate solution of Equation 10 we consider θ_m to be the product of two unknown functions.

$$\theta_m = T\cdot G \quad (16)$$

Substitution of θ_m in Equation 10 results

$$KG\frac{\partial^2 T}{\partial x^2}+\left(2K\frac{\partial G}{\partial x}+G\frac{\partial K}{\partial x}\right)\frac{\partial T}{\partial x}+\left(K\frac{\partial^2 G}{\partial x^2}+\frac{\partial K}{\partial x}\cdot\frac{\partial G}{\partial x}-\mu m G\right)T = 0 \quad (17)$$

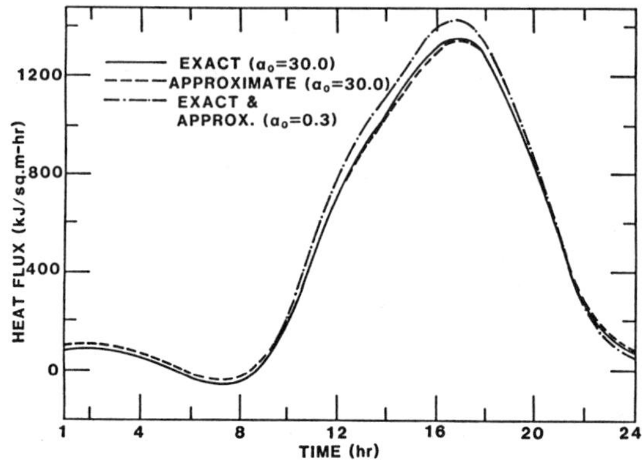

Figure 7. Comparison of heat flux through a concrete layer with parabolic variation in thermal conductivity: 1—exact method; 2—approximate method [4].

where

$$\mu_m = \exp\left(\frac{1}{2}\pi i\right)(m\omega\rho c)$$

Choosing the unknown function G to be such that the coefficient of $\frac{\partial T}{\partial x}$ equals zero; thus

$$G = \text{Constant} \cdot K^{-1/2} \tag{18}$$

and therefore Equation 17 reduces to:

$$\frac{\partial^2 T}{\partial x^2} + \psi^2 T = 0 \tag{19}$$

where

$$\psi^2 = \left(\frac{1}{2K} \cdot \frac{dk}{dx}\right)^2 - \frac{1}{2k} \cdot \frac{d^2 k}{dx^2} - \frac{\mu_m}{k} \tag{20}$$

Equation 19 is similar to the famous wave equation in quantum mechanics given by E. Schrödinger. It is evident that the solution of Equation 19 can be obtained readily only if thermal properties of the medium are constants. However, for media with spatially varying thermal properties the solution of Equation 19 may be obtained by following the approach of the W. K. B. approximation [6] (named after G. Wentzel, H. A. Kramers, and L. Brillouin) used for the solution of the Schrödinger wave equation. Later on the W. K. B. method became very popular and found wide applications in studies of electromagnetic waves propagation through stratified plasmas, nonlinear optics, and other branches of physics.

Consider that the function $\psi(x)$ expressed in Equation 20 to be a slowly varying function of the position coordinate so that the relation

$$\frac{1}{\psi^2} \cdot \frac{d\psi}{dx} \ll 1 \tag{21}$$

is satisfied. In such circumstances the approximate solution of Equation 19 may be expressed as

$$T = D\psi^{-1/2} \cdot \exp\left(\pm i \int_0^x \psi \, dx\right) \tag{22}$$

with D as some constant.

The condition given by Equation 21 has been found to be satisfied both analytically and numerically [4] over a wide range of variation in thermal properties of the medium.

Thus the time-dependent component θ_m of the temperature distribution in the inhomogeneous medium can be written in the following form after subsituting values for functions G and T from Equations 18 and 22 in Equation 16:

$$\theta_m = (k\psi)^{-1/2}\left[A_m \exp\left(i \int_0^x \psi \, dx\right) + B_m \exp\left(-i \int_0^x \psi \, dx\right)\right] \tag{23}$$

Constants A_m and B_m in Equation 23 are determined by appropriate boundary conditions.

Figure 7 illustrates the application of approximate methods in the study of conductive heat transfer through inhomogeneous layers. We consider a concrete slab exhibiting parabolic thermal conductivity variation which is exposed to the periodic temperature at one face while the other face is exposed to periodic ambient temperature. Various constants appeared in Equations 11, 15 and 23 are evaluated by means of appropriate boundary conditions viz. Equation 3 at $x = 0$ surface and Equation 4 at $x = L$ surface. Heat flux passing through the slab over a period of 24 hours computed both by means of exact solution and approximate solution is shown in Figure 7. Curve I shows the computed heat flux for $\alpha_0 = 30$ based on exact solution (i. e. Equation 15) and Curve II depicts heat flux for the same parameters computed from approximate solution. Heat flux for $\alpha = 0.3$ is represented by Curve III. It is evident from various curves that the approximate solution is in absolute agreement with the exact solution of the heat conduction equation for small values of α_0. Also, for very large variations in thermal conductivity (say $\alpha_0 = 30$) the approximate solution is in fine agreement with the exact solution and this fine agreement between the two establishes the utility of approximate solutions in the study of conductive heat transfer through inhomogeneous materials. Thus the approximate solution is highly versatile in heat transfer studies because it provides an easy method for computation of temperature distribution in any inhomogeneous material exhibiting any type of spatial variations in one or more of the thermal properties—specific heat, density, and thermal conductivity of the material.

Application of Exact and Approximate Solutions

Bounded Medium

Consider a concrete slab exhibiting parabolic spatial variation in thermal conductivity and satisfy periodic temperature boundary condition at $x = 0$ surface and energy balance

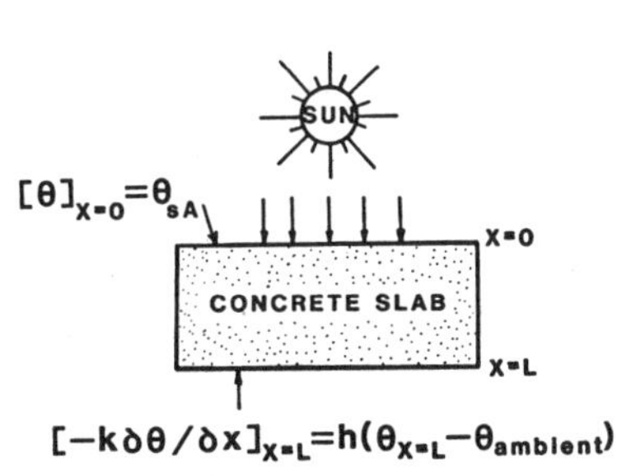

Figure 8. Ray diagram of an inhomogeneous layer subjected to periodic boundary conditions.

condition at x = L surface Figure 8. Therefore, the constants A_0, B_0, A_m and B_m in Equations 11 and 15 are evaluated as

$$A_0 = h(b_0 - a_0) \Big/ \left\{K_0 + \left[h \ln\left(\frac{1+\alpha_0^{1/2}L}{1-\alpha_0^{1/2}L}\right) \Big/ 2\alpha_0^{1/2}\right]\right\}$$

$$B_0 = a_0$$

$$A_m = a_m$$

$$B_m = \left\{hb_m - a_m\left[h\left(1-\beta_m\left\{\frac{\alpha_0 L^2}{2} + \sum_{n=1}^{\infty}\frac{(\alpha_0^{1/2}L)^{2n+2}}{(2n+2)!}\cdot\prod_{p=1}^{n}[2p(2p+1)-\beta_m]\right\}\right)\right.\right.$$
$$\left.\left. - K\beta_m\left\{\alpha_0 L + \sum_{n=1}^{\infty}\frac{\alpha_0^{1/2}(\alpha_0^{1/2}L)^{2n+1}}{(2n+1)!}\cdot\prod_{p=1}^{n}[2p(2p+1)-\beta_m]\right\}\right]\right\}$$
$$\times\left(h\left[\alpha_0^{1/2}L + \sum_{n=1}^{\infty}\frac{(\alpha_0^{1/2}L)^{2n+1}}{(2n+1)!}\cdot\prod_{p=1}^{n}[2p(2p-1)-\beta_m]\right]\right.$$
$$\left. + K\left\{\alpha_0^{1/2} + \sum_{n=1}^{\infty}\frac{\alpha_0^{1/2}(\alpha_0^{1/2}L)^{2n}}{(2n)!}\cdot\prod_{p=1}^{n}[2p(2p-1)-\beta_m]\right\}\right)^{-1} \tag{24}$$

The heat flux passing through the slab may be calculated by the following relation

$$\dot{Q} = -K_0(1-\alpha_0 L^2)\left(\frac{d\theta}{dx}\right)_{x=L}$$

Figure 9 shows variation of heat flux passing through 0.1 m thick layer over a period of 24 hours for different values of α_0. Also, heat flux computed for constant (average) thermal conductivity is shown in the same plot. For a large spatial variations difference between the heat flux calculated for average thermal conductivity and actual thermal conductivity profile increases significantly with increasing values of α_0.

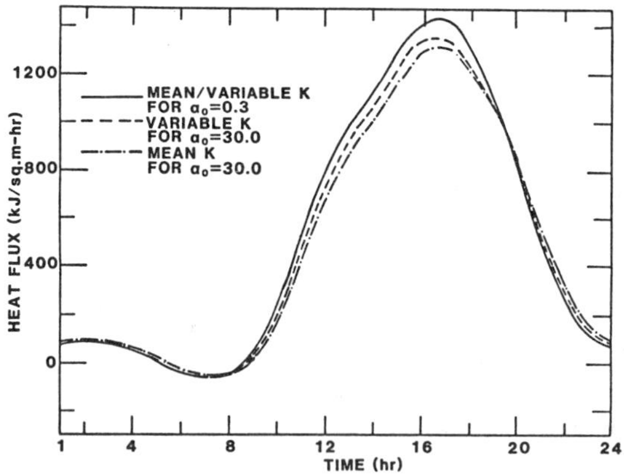

Figure 9. Heat flux through a concrete layer with parabolic variation in thermal conductivity: effect of α_0 [4].

Semi-infinite Medium

Let the solid be bounded by the x = 0 plane and extend infinitely in the positive x-direction. We consider periodic temperature at the x = 0 plane which remains finite as $x \to \infty$ and therefore the solutions obtained in the preceding sections should satisfy the condition $x \to 0$ as $x \to \infty$. Though the problem appears to be trivial, due to the extremely complex nature of exact solutions obtained earlier the boundary condition for a semi-infinite medium is not satisfied and hence these exact solutions fail to provide any physical solution for temperature distribution in a semi-infinite medium. However, the approximate solution (Equation 23) together with Equation 11 satisfy the appropriate boundary condition for a semi-infinite medium and provides good results with great computational ease [7]. A numerical study of Equation 23 by computer reveals that the term

$$\exp\{\textstyle\int_0^x \psi^* \, dx\}\,; \qquad \psi^* = i\psi$$

diverges with increasing x whereas the term

$$\exp\{-\textstyle\int_0^x \psi^* \, dx\}$$

converges; the only complex integral appearing in the solution may be solved explicitly for simple cases but can also be evaluated numerically by using Simpson's method with only 20 equispaced points. Since a diverging solution does not have any physical significance, the general solution for a semi-infinite medium may be expressed as

$$\theta = A_0 + \text{Real} \sum_{m=1}^{\infty} A_m[K\psi^*]^{-1/2} \cdot \exp\{im\omega t - \textstyle\int_0^x \psi^* \, dx\} \tag{25}$$

Constants A_0 and A_m are evaluated with the help of the boundary condition given by Equation 4.

$$A_0 = a_0$$

and

$$A_m = [ha_m] \cdot \left[h\{k\psi^*\}^{-1/2} - k\left\{\frac{\partial}{\partial x}(k\psi^*)^{-1/2}\right\} + \{k\psi^*\}^{1/2} \right]_{x=0}^{-1}$$

A temperature variation inside the semi-infinite medium (e.g., ground) has been shown in Figure 10 over a period of 24 hours. It can be seen that the temperature calculated for an average thermal conductivity deviates from the one evaluated for an actual thermal conductivity profile, presumably linear and this deviation increases with depth (Figure 11). Similar behaviour may also be seen for the heat flux [8] passing through different layers of the ground (Figure 12).

Medium with Internal Heat Generation/Absorption

We consider a particular case of heat transfer in salt gradient solar ponds which have been investigated at length for its suitability in electricity generation and interior space heating. A nonconvective zone of a solar pond (Figure 13) exhibits simultaneous spatial variations in thermal conductivity, specific heat, and the density along its depth. This zone absorbs solar radiations and conducts heat to the convective zone for heat extraction.

The differential equation of heat conduction in the nonconvective zone of the solar pond may be expressed as

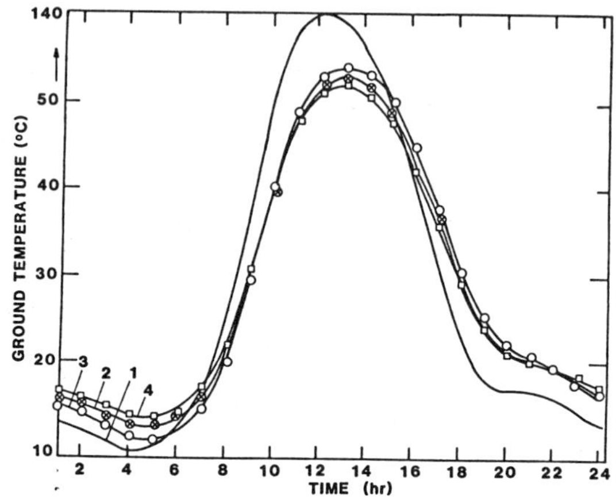

Figure 10. Daily variation in ground temperature: 1—surface temperature; 2—temperature at $x = 0.16$ m for average linear thermal conductivity variation; 3—temperature corresponding to linear variation in thermal conductivity; 4—temperature corresponding to parabolic variation in K [7].

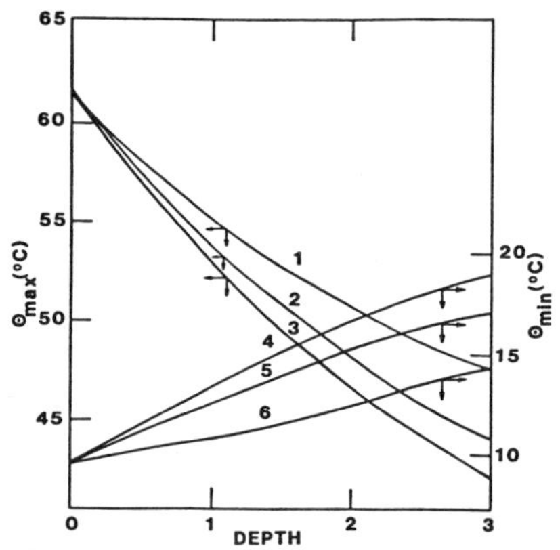

Figure 11. Variation of maximum and minimum ground temperature with depth: 1 and 6—linear variation; 2 and 5—average for linear profile; 3 and 4—parabolic variation in K [7].

$$K\frac{\partial^2\theta}{\partial x^2} + \frac{\partial K}{\partial x}\cdot\frac{\partial\theta}{\partial x} - \frac{\partial S}{\partial x} = \rho C\frac{\partial\theta}{\partial t} \tag{26}$$

The term $\frac{\partial S}{\partial x}$ in Equation 26 exhibits absorption of solar radiation through the depth of the nonconvective zone of the solar pond and may be expressed [9] in terms of the Fourier series.

$$S = [S_{00} + \sum_{m=1}^{\infty} S_{0m}\exp(im\omega t)]\cdot\sum_{j=1}^{\infty}\mu_j\exp(-n_jx)$$

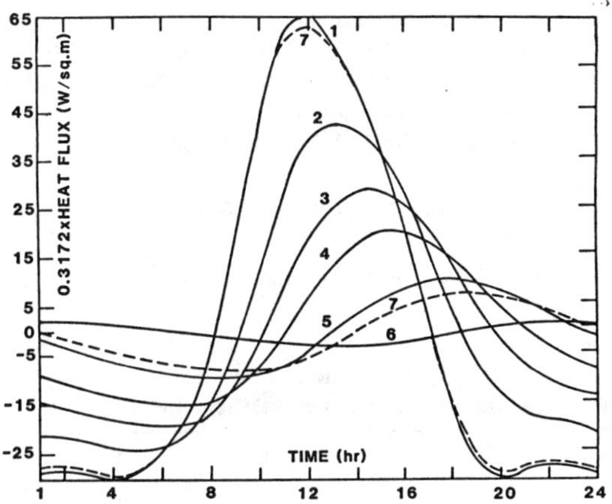

Figure 12. Daily variation of heat flux in the ground: 1—x = 0.025 m; 2—x = 0.05 m; 3—x = 0.075 m; 4—x = 0.10 m; 5—x = 0.15 m; 6—x = 0.30 m; and curves 7 and 8 correspond to the average value of linear thermal conductivity at x = 0.025 m and x = 0.30 m, respectively [8].

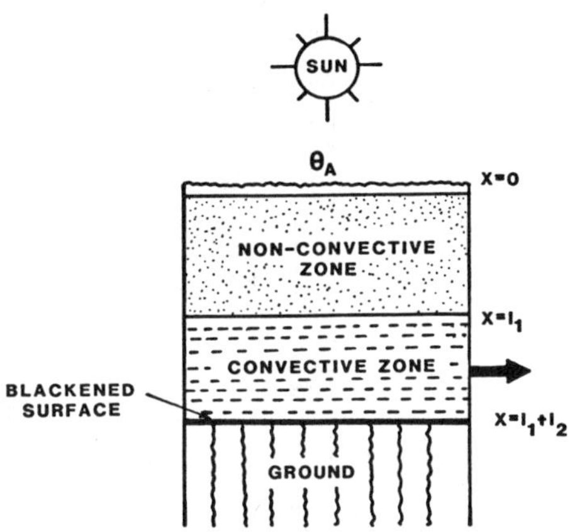

Figure 13. Ray diagram of a solar pond.

The coefficients μ_j and n_j are found by the curve-fitting method.

For the solution of Equation 26 we separate it into time-independent and time-dependent parts after substituting for θ from Equation 8.

$$\frac{\partial}{\partial x}\left\{K\frac{\partial\theta_0}{\partial x}\right\} + S_{00}\sum_{j=1}^{\infty}\mu_j n_j \exp(-n_j x) = 0 \tag{27}$$

and

$$\frac{\partial}{\partial x}\left(K\frac{\partial\theta_m}{\partial x}\right) + S_{0m}\sum_{j=1}^{\infty}\mu_j n_j \exp(-n_j x) = im\omega\rho C\theta_m \tag{28}$$

Equations 27 and 28 are nonhomogeneous and thus could be simplified by assuming constant thermal properties of the nonconvective zone brine [9]. However, we restrict ourselves to the most general solution [10] which accounts for all spatial variations in thermal properties of the brine. Thus following the method of variation of parameters due to Lagrange [11] the solution of Equation 27 may be given by

$$\theta_0 = A_0 + B_0 \int K^{-1}\,dx + \sum_{j=1}^{\infty} S_{00}\mu_j n_j \Big[\int K \exp(-n_j x) \cdot \Big(\int K^{-1}\,dx\Big)\,dx + \int K^{-1}\,dx \cdot \int K \exp(-n_j x)\,dx\Big] \tag{29}$$

Table 1
Thermal Properties of Some Insulating and Building Materials

Material	Average Temp. (°C)	K (W/m · K)	C (J/kg · K)	ρ (kg/m³)
Asbestos	0	0.150	1,046.7	576
	200	0.208	–	576
Cork	30	0.043	167.4	160
Cotton fabric	93	0.079	167.4	160
Diatoma ceous earth;				
powdered	38	0.052	879.2	224
	149	0.062	–	224
	315	0.079	–	224
Molded pipe	204	0.088	–	416
covering	870	0.152	–	416
Glass wool,	−7	0.038	–	24
fine	38	0.053	–	24
	93	0.074	–	24
packed	−7	0.027	–	96
	38	0.038	–	96
	93	0.050	–	96
Hair felt	38	0.046	–	131
Kaolin	500	0.260	–	432
insulaing brick	1,150	0.450	–	432
Kaolin insulaing	200	0.086	–	304
fire brick	760	0.190	–	304
85% magnesia	0	0.055	–	272
	93	0.064	–	272
Rock wool	−7	0.029	–	128
Rubber	0	0.151	2,009.7	1,201
Fire clay	200	1.004	837.4	2,306
	1,000	1.644	837.4	2,306
Masonary	21	0.657	837.4	1,698
Zirconia	200	1.454	–	4,869
	1,000	1.956	–	4,869
Chrome brick	200	1.419	–	3,940
	1,000	1.662	–	3,940
Concrete:				
stone	21	0.935	837.4	2,306
10% water	21	1.212	–	2,242
Glass window	21	0.779	837.4	2,723
Lime stone dry	21	0.692	921.1	1,682
Sand:				
Dry	20	0.346	–	1,522
10% water	20	1.038	–	1,602
Soil:				
Dry	21	0.346	1,842.2	–
Wet	21	2.596	–	–
Wood:				
Oak ⊥ to grain	21	0.208	2,386.6	817
∥ to grain	21	0.346	2,386.6	817
Pine ⊥ to grain	21	0.104	2,805.3	496
∥ to grain	21	0.242	2,805.3	496

Source: *Principles of Solar Engineering* by F. Kreith and J. F. Kreider; © 1978 Hemisphere Publishing Corporation, London.

The first two terms in this solution correspond to the homogeneous part of Equation 27.

The nature of Equation 28 is extremely complex because of various spatial variations in thermal properties of the medium and hence an explicit solution is not workable. We obtain a general solution of it by means of a method of variation of parameters and an approximate solution as developed earlier. The homogeneous part of Equation 28 which may be obtained by equating the second term to zero on the left is identical to Equation 10 and therefore has the solution of the type of Equation 23; ψ is again given by Equation 20. Once the solution for the homogeneous part is found, the particular solution of Equation 28 may be conveniently determined by means of variation of parameters technique.

$$\begin{aligned}\theta_{mp} = S_{0m} \sum_{j=1}^{\infty} \mu_j n_j (K\psi)^{-1/2} \cdot \{&\exp\left(-i \textstyle\int_0^x \psi\, dx\right) \\ &\cdot [\textstyle\int (K\psi)^{-1/2} \cdot \exp\left(i \textstyle\int_0^x \psi\, dx\right) \cdot \exp(-n_j x) \cdot M^{-1}\, dx] \\ &- \exp\left(i \textstyle\int_0^x \psi\, dx\right) \cdot \{\textstyle\int (K\psi)^{-1/2} \cdot \exp\left(-i \textstyle\int_0^x \psi\, dx\right) \\ &\cdot \exp(-n_j x) \cdot M^{-1}\, dx]\}\end{aligned} \tag{30}$$

with

$$\begin{aligned}M = (K\psi)^{-1/2} \cdot \{&\exp\left(-i \textstyle\int_0^x \psi\, dx\right) \cdot \frac{d}{dx}[(K\psi)^{-1/2} \\ &\cdot \exp\left(i \textstyle\int_0^x \psi\, dx\right)] - \exp\left(i \textstyle\int_0^x \psi .dx\right) \\ &\cdot \frac{d}{dx}[(K\psi)^{-1/2} \cdot \exp\left(-i \textstyle\int_0^x \psi\, dx\right)]\}\end{aligned}$$

Thus the most general solution for temperature distribution inside the nonconvective zone of a salt gradient solar pond may be written by summing up Equations 23, 29 and 30. Various constants are evaluated by using appropriate boundary conditions discussed in the thrid section of this chapter; however, for solar pond investigations the following boundary conditions are more appropriate:

$$-\left(K \frac{\partial \theta}{\partial x}\right)_{x=0} = h_1(\theta_{SA} - \theta_{x=0})$$

and

$$-\left(K \frac{\partial \theta}{\partial x}\right)_{x=l_1} = h_2(\theta_{x=l_1} - \theta_{extraction})$$

MEDIUM WITH TEMPERATURE DEPENDENT THERMAL CONDUCTIVITY

Consider a semi-infinite medium without heat generation or absorption exhibiting a temperature-dependent thermal conductivity. Let the temperature at $x = 0$ plane be periodic in time expressible in terms of the Fourier series.

$$T_{x=0} = A_0 + \text{Real} \sum_{n=1}^{\infty} a_n \exp(in\omega t) \tag{31}$$

where A_0 is the mean surface temperature and $T(x, t)$ is the temperature profile inside the medium.

In practice, thermal conductivity of most of the materials of interest may be conveniently expressed as a linear function of temperature.

$$K = K_0 + K_1\theta$$

where K_1 represents thermal conductivity per unit temperature and K_0 is the thermal conductivity at temperature A_0.

Subsituting for K in Equation 2, the differential equation for heat conduction may be re-written as

$$K_0 \frac{\partial^2\theta}{\partial x^2} + K_1\left(\frac{\partial\theta}{\partial x}\right)^2 + K_1\theta\frac{\partial^2\theta}{\partial x^2} = \rho C\frac{\partial\theta}{\partial t} \tag{32}$$

with

$$\theta = T(x, t) - A_0$$

Equation 32 may also be put in a more convenient form

$$\frac{\partial^2\theta}{\partial x^2} + \frac{K_1}{2K_0}\frac{\partial^2(\theta^2)}{\partial x^2} = \frac{\rho C}{K_0}\frac{\partial\theta}{\partial t} \tag{33}$$

Equation 33 is a nonlinear equation in temperature and can only be solved for an approximate solution to provide fairly accurate results. Let us consider that the thermal conductivity per unit temperature is very small, which is true for many insulating materials, i.e. $K_1/K_0 \ll 1$. In such conditions Equation 33 may be solved explicitly by treating K_1 as a perturbation [12]. The boundary condition at the surface is

$$\theta_{x=0} = T_{x=0} - A_0 = \text{Real}\sum_{n=1}^{\infty} a_n \exp(in\omega t) \tag{34}$$

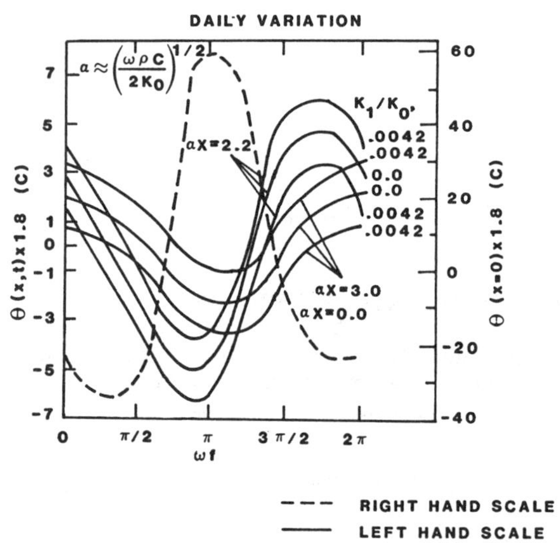

Figure 14. Daily temperature variation with time in polyurethane foam [12].

Following the procedure of the time-dependent perturbation theory [13] of quantum mechanics, Sodha et al. [12] obtained the solution of Equation 33 as

$$\theta = \text{Real} \sum_{n=1}^{\infty} a_n \exp(-\beta_n x) \cdot \exp(in\omega t) + \frac{K_1}{K_0} F(x, t)$$

with

$$\beta_n = n^{1/2}\left(\frac{\omega\rho C}{2\,K_0}\right)^{1/2} (1-i)$$

and the function F(x, t) is given by

$$\begin{aligned}
F(x, t) = {} & \frac{1}{2}\,\text{Real} \sum_{n=1}^{\infty} \frac{a_n a_n^*}{2}\{1-\exp[-(\beta_n+\beta_n^*)x]\} \\
& + \text{Real} \sum_{m=1}^{\infty} \frac{a_m a_{m+1}^*(\beta_m+\beta_{m+1}^*)^2}{[(\beta_m+\beta_{m+1}^*)^2-\beta_1^2]} \\
& \cdot \{\exp(-\beta_1 x)-\exp[-(\beta_m+\beta_{m+1}^*)x]\}\exp(i\omega t) \\
& + \left(\frac{2\,a_1^2\beta_1^2}{4\,\beta_1^2-\beta_2^2}\cdot[\exp(-\beta_2 x)-\exp(-2\,\beta_1 x)]\right. \\
& + \text{Real} \sum_{m=1}^{\infty} \frac{a_m a_{m+2}^*(\beta_m+\beta_{m+2}^*)^2}{[(\beta_m+\beta_{m+2}^*)^2-\beta_2^2]} \\
& \left. \cdot \{\exp(-\beta_2 x)-\exp[-(\beta_m+\beta_{m+2}^*)x]\}\right)\exp(i2\omega t) \\
& + \left(\frac{a_1 a_2(\beta_1+\beta_2)^2}{[(\beta_1+\beta_2)^2-\beta_3^2]}\cdot\{\exp(-\beta_3 x)-\exp[-(\beta_1+\beta_2)x]\}\right. \\
& + \text{Real} \sum_{m=1}^{\infty} \frac{a_m a_{m+3}^*(\beta_m+\beta_{m+3}^*)^2}{[(\beta_m+\beta_{m+3}^*)^2-\beta_3^2]} \\
& \left. \cdot \{\exp(-\beta_3 x)-\exp[-(\beta_m+\beta_{m+3}^*)x]\}\right)\exp(i3\omega t) \\
& + \text{higher harmonics}
\end{aligned}$$

It is interesting to observe that the first term on the right of Equation 35 represents a number of harmonics of the Fourier series which propagate linearly and independently of each other. The second term arises due to the interaction of various harmonics; the amplitude and phase of a particular harmonic at a point depends upon the amplitude and the phase of other harmonics at the surface. This phenomenon of harmonic mixing is not a unique case but is a well known characteristic feature of wave propagation in nonlinear media (e.g., the generation of harmonics by electromagnetic waves in nonlinear plasma and semiconductors [14]). Thus on account of temperature-dependent thermal conductivity the temperature inside the medium gets modified by an amount $K_1/K_0 \cdot F(x, t)$ corresponding to the temperature evaluated on the basis of constant thermal conductivity [15].

The temperature distribution in polyurethane foam, a good thermal insulator [16], has been illustrated in Figure 14 over a period of 24 hours for various depths. The magnitude of K_1/K_0 for rigid polyurethane varies from 0.002° C^{-1} to 0.0042° C^{-1}, depending upon the density and the method of preparation. K_0 corresponds to a temperature ≈24° C. In a low temperature region (<0° C) K_1/K_0 is negative and has a magnitude of the same order.

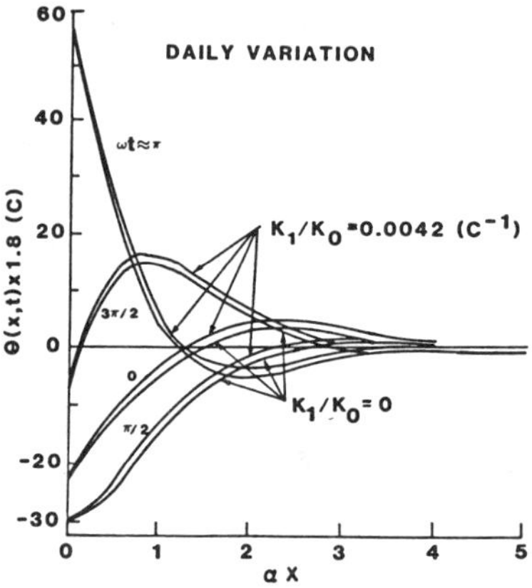

Figure 15. Temperature variation with depth in polyurethane foam [12].

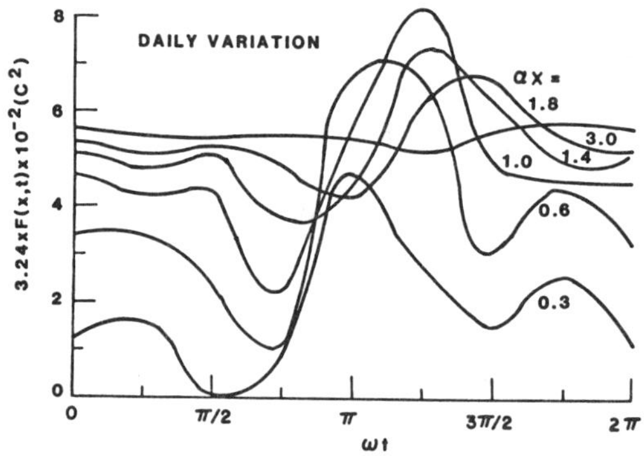

Figure 16. Daily variation of the function F(x,t) with time for polyurethane foam [12].

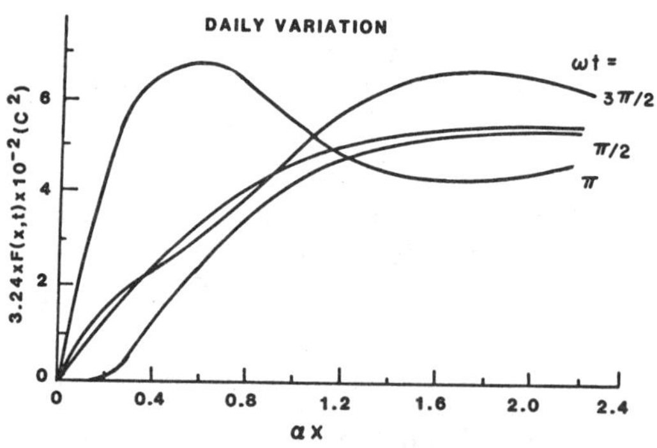

Figure 17. Daily variation of the function F(x,t) with depth for polyurethane foam [12].

Table 2
Thermal Properties of Some Metals and Alloys

Material	K (W/m·K) 0	100	300	500	C (J/kg·K) 0°C	ρ (kg/m³) 0°C
Aluminium	202.5	206.0	230.0	268.3	870.9	2,707
Bismuth	8.5	6.7	–	–	121.4	9,803
Copper, pure	387.7	377.3	367.0	358.3	38.1	8,938
Gold	292.5	294.0	–	–	125.6	19,269
Iron, pure	62.0	63.3	–	–	435.4	7,864
Lead	34.8	32.8	31.1	–	125.6	11,293
Magnesium	157.5	159.2	–	–	971.4	1,746
Mercury	7.8	–	–	–	138.2	13,599
Nickel	59.7	58.8	55.4	–	431.3	8,890
Silver	419.0	412.0	–	–	234.8	10,492
Tin	62.3	58.8	–	–	226.1	7,304
Zinc	112.5	110.8	102.1	–	381.0	7,144
Brass 70% Cu, 30% Zn	96.9	103.8	114.2	–	385.2	8,521
Bronze 75% Cu, 25% Sn	26.0	–	–	–	343.3	8,650
Cast iron						
Plain	57.1	55.0	48.0	43.0	460.6	7,592
Alloy	52.0	49.0	46.7	–	418.7	7,288
Stainless steel:						
Type 304	13.8	16.3	18.8	21.5	460.6	7,817
Type 347	13.8	16.1	19.0	22.2	460.6	7,817
Steel mild, 1% carbon	45.9	45.0	43.2	38.1	460.6	7,849
Constanton, 60% Cu, 40% Ni	21.5	22.1	–	–	418.7	8,922

Source: *Principles of Solar Engineering* by F. Kreith and J. F. Kreider; © 1978 Hemisphere Publishing Corporation, London.

K_1/K_0 increases significantly with moisture content [17]. The effect of temperature dependence on thermal conductivity becomes significant with increasing depths though the peak values get reduced (Figure 14). The temperature, in general, is higher for $K_1/K_0 > 0$ and lower for $K_1/K_0 < 0$ as compared to the case when $K_1/K_0 = 0$. For typical polyurethane foam $\alpha = 8.48\ m^{-1}$ for daily temperature variations.

The daily variation of temperature with depth (αx) for various values of time has been shown in Figure 15. For $\omega t = \pi$ the temperature decreases rapidly with increasing depth and then increases; however, for $\omega t = 0$, $\pi/2$ and $3\pi/2$ it first increases and then decreases slowly. Such typical variation is due to different phase legs introduced by the various harmonics which is more vivid if we examine the behavior of the function F(x, t). Figures 16 and 17 illustrate the variation of the function F(x, t) with time and depth respectively. It may be seen that the function F(x, t) is a periodic function of time. For $\omega t = 0$ and $\pi/2$ the function F(x, t) increases with increasing depth whereas for $\omega t = \pi$ and $3\pi/2$ it decreases with an initial increase in depth. The magnitude of F(x, t) diminishes with increasing depths and smoothes out at larger depths. This is due to the fact that the amplitude of various harmonics decays exponentially with depth, and for depths greater than $\alpha x = 3$ only the time-independent part predominates.

NOTATION

A_0	mean surface temperature
A_m, B_m	coefficients
a'_m	amplitude of m-th harmonic of Fourier expansion of temperature distribution
a_0	average temperature at surface
C	specific heat
D	constant in Equation 22
G	parameter in Equation 16

- H rate of heat production or absorption
- h heat transfer coefficient
- J_0 zero-order modified Bessel function
- K thermal conductivity
- L length
- M function defined in Equation 30
- n_j coefficient
- Q heat flux
- $S(t)$ intensity of solar radiation
- T temperature
- t time
- x coordinate
- Y_0 zero-order modified Bessel function

Greek Symbols

- α_s fraction of solar radiation absorbed
- ε emissivity
- θ temperature distribution
- κ thermal conductivity
- μ_j coefficient
- σ_m phase factor
- $\psi_{(x)}$ function by Equation 20
- ω angular frequency of temperature variation

REFERENCES

1. Carslaw, H. S., and Jaeger, J. C., *Conduction of Heat in Solids,* 2nd ed., University Press, Oxford, 1978.
2. Isachenko, V. P., Osipova, V. A., and Sukomel, A. S., *Heat Transfer,* 3rd ed., Semyonov, S. (Trans.) Mir Publishers Moscow, 1977, pp. 23–26.
3. Sodha, M. S., et al., "An Inexpensive Solar Collector/Storage System in the Ground," *J. Phys. D: Appl. Phys.,* Vol. 11 (1978), pp. 2375–2381.
4. Thakur, A. K. S., "Periodic Heat Transfer through an Inhomogeneous Medium," *Letters Heat Mass Transfer,* Vol. 9 (1982), pp. 385–394.
5. Watson, G. N., *Theory of Bessel Functions,* 2nd ed., University Press, Cambridge, 1966, p. 97.
6. Merzbacher, E., *Quantum Mechanics,* 2nd ed., John Wiley & Sons, Inc., New York, 1970, pp. 116–137.
7. Thakur, A. K. S., and Momoh, Musa M., "Temperature Variation in Upper Earth Crust Due to Periodic Nature of Solar Insulation," *Energy Convers. Mgmt.* (1983).
8. Thakur, Alok, "Heat Transfer in a Medium with a Density Gradient," *Appl. Energy,* Vol. 12 (1982), pp. 71–76.
9. Kaushik, N. D., Bansal, P. K., and Sodha, M. S., "Partitioned Solar Pond Collector/Storage System," *Appl. Energy,* Vol. 7 (1980), pp. 169–190.
10. Thakur, A. K. S., "Conductive Heat Transfer in Salt Gradient Stablized Solar Ponds," *Int. Commu. Heat Mass Transfer,* Vol. 10 (1983), pp. 335–340.
11. Kreyszig, E., *Advanced Engineering Mathematics,* 3rd ed., John Wiley and Sons, Inc., New York, 1976, pp. 99–101.
12. Sodha, M. S., et al., "Periodic Heat Transfer with Temperature Dependent Thermal Conductivity," *Int. J. Heat Mass Transfer,* Vol. 22 (1979), pp. 777–781.
13. Merzbacher, *Quantum Mechanics,* pp. 450–486.
14. Sodha, M. S., and Kaw, P. K., "Theory of Harmonics in Plasmas," *Adv. Electronics Electron Phys.,* Vol. 27 (1969), pp. 187–293.
15. Khatry, A. K., Sodha, M. S., and Malik, M. A. S., "Periodic Variation of Ground Temperature with Depth," *Solar Energy,* Vol. 20 (1978), pp. 425–427.
16. Healy, T. T., "Polyurethane Foams," in *Proceedings of the Symposium on Polyurethane Foams,* London, 1963, p. 125.
17. *ASHRAE Handbook of Fundamentals,* New York, 1967, p. 243.

CHAPTER 20

HEAT FLUX MEASUREMENTS

M. Zamora and A. Rey de Luna

Departamento de Termologia
Universidad de Sevilla, Spain

CONTENTS

INTRODUCTION

The study of physical far-from-equilibrium systems has revealed the need to develop new techniques to measure flux quantities in nonequilibrium situations. The most important flow quantity is undoubtedly heat flux, partly because of its energy properties used in the basic balances in nonequilibrium physics, and partly due to its own conceptual structure which is always related to temperature and to the evolutionary criteria of thermodynamics (both factors being hard to define in far-from-equilibrium systems). In fact this has been recognized theoretically by extended thermodynamics [1–4] which used fluxes, particularly heat flux, as independent variables, and by steady-state thermodynamics [5–6] which, from a macroscopic point of view, uses the same kind of variables. Experimentally, interest in the measurement of heat fluxes can be verified by the extensive proceedings of the many congresses about thermal conductivity and heat transfer, and by the many different methods proposed for the determination of heat flux in conductive systems. However, no method for the measurement of heat flow in a conductive system exists, and literature related to this aspect is limited [7–15].

This chapter is not an exhaustive study of the different methods of heat flux measurement, but rather we examines the use of a heat fluxmeter which has been developed and successfully applied to diverse physical problems. This heat fluxmeter behaves as a heat transducer in that its physical properties depend only on itself. This chapter is therefore dedicated to discussing the theoretical principles, the physical structure, the nomenclature, and a few of the applications of our heat transducer.

EXPERIMENTAL REQUIREMENTS AND NOMENCLATURE

A heat fluxmeter which is intended for general use should satisfy several experimental requirements. First, it should transform the heat flow which crosses it in other physical forms, preferably electrical and lineal. Secondly, its structure should be sufficiently rigid and versatile to permit its insertion between a system and an external thermal source or between different parts of a conductive system. In both cases, the introduction of the heat transducer should produce a minimal effect on the properties of the conduction of the systems under study. Finally. the heat fluxmeter should be a unidimensional heat conductor, limited by two isothermic surfaces imposed by the geometry of the problem. In this way, the heat fluxmeter will not introduce deformations in the isothermic surfaces of the physical problem. Also, the lower its thermal resistance and diffusivity, the smaller will be the temperature jump be between the two extremes.

A general method used to measure the heat flux in a unidirectional conductive system consists of introducing a thermal resistance as proof material in the system, and then determining the temperatures at the external surfaces (Figure 1). In steady state, the principles and limitations of this technique can be studied easily. Thus, using Fourier's law, it can be established that:

$$q = \frac{T_1 - T_2}{R_1 + R_2} \tag{1}$$

$$q' = \frac{T_1 - T_2}{R_1 + R_2 + R} \tag{2}$$

$$q' = \frac{\Delta T}{R} \tag{3}$$

where Equation 3 is the basic expression used to measure the heat flux in the system; but Equations 1 and 2 show that the heat flux measured with Equation 3 is not the correct value, due to the introduction of the thermal resistance, R, which modifies the heat flow crossing the total system. Therefore, the correct heat flow can be obtained by making an additional condition:

$$q = \lim_{R \to 0} q' = \frac{dT}{dR} \tag{4}$$

From Equation 4 several important results can be obtained. The thermal resistance of the proof material should be so small as to permit the resolution of the thermometers used. The use of a differential thermometer is preferable to the use of two absolute thermometers

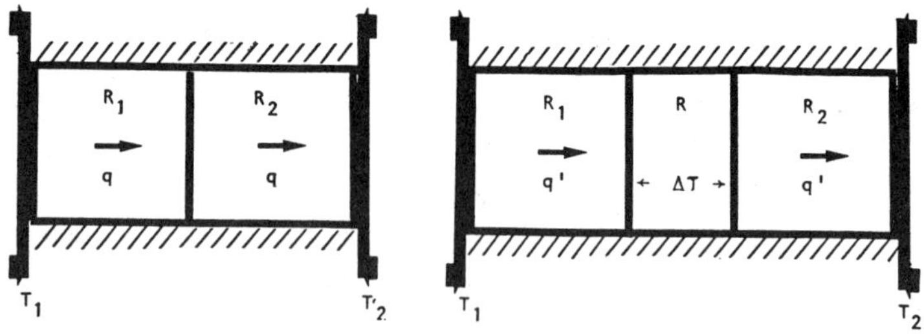

Figure 1. General method of measurement of heat flux.

on the external faces of the proof material, in that the relative error of measurement is less. Furthermore, in transient situations the heat capacity of the proof material should be approximately negligible.

On this basis, we have designed our heat fluxmeter. Essentially, it consists of a number N of thermocouples connected electrically in series in such a way that the consecutive junctions fit the external surfaces and the actual wires are the proof material. This device is generally known as a thermopile (Figure 2). A thermopile, properly chosen, satisfies all the conditions previously discussed, and falling within the experimental needs and permitted error margins.

First of all, the use of thermocouples connected in series allows an output to be obtained which is proportional to the difference in the temperature applied between its extremes and to the number of thermocouples used, secondly, the thermocouples should be constructed from wires, metals, alloys, or semiconductors, which can be selected to obtain a poor thermal resistance and a low heat capacity. Moreover, they should be situated between the external surfaces in such a way that the internal heat conduction is lineal, independent of the arrangement of its external surfaces, conditioned by the problem under study. Finally, the thermoelectric circuit may be constructed so as to obtain a rigid mechanically resistant structure.

Briefly, a thermopile is formed by 2N wires of different materials, situated alternately. Each pair of terminals are soldered to a small metallic piece at both ends, forming the junction. All the junctions at each end are glued together with a dielectric cement, which is a good heat conductor, thus assuring its electric isolation. In the same way, each group of junctions are joined together, and using the same cement, they are joined to external pieces of metal, which are also good heat conductors. These external pieces guarantee the isothermic properties of the limits of the thermopile. The position of the electric terminals in the thermopile can be arranged in contra-position or in referential positions (Figure 3).

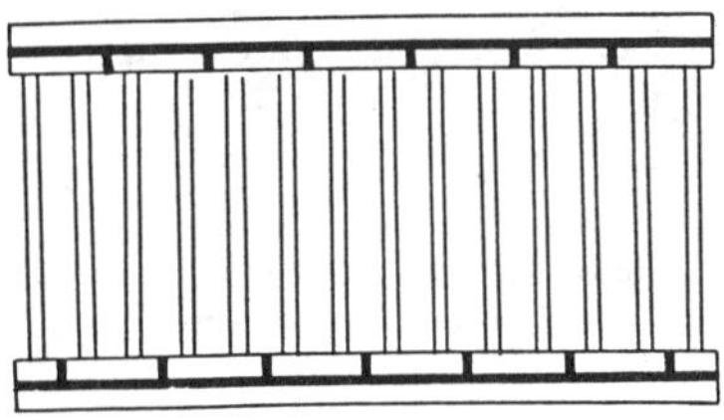

Figure 2. Diagram of a thermopile.

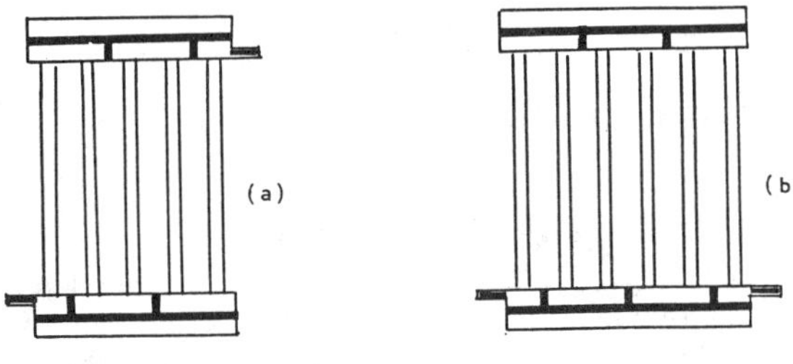

Figure 3. Arrangement of the electric terminals of a thermopile: (A) in contra-position; (B) in referential.

In order to measure heat fluxes, it is preferable to establish a referential arrangement. In this case, the side which supports both electric terminals is known as the referential face, while the opposite side is called the test face.

THEORETICAL BASIS

Seebeck's law is the physical principle that permits use of a thermopile as a heat fluxmeter [16–17]. As stated in Seebeck's law, in a steady state, the e. m. f. produced by a thermocouple is proportional to the increase in temperature applied to the junctions. In the case of a thermopile with N thermocouples, the e. m. f. is N times larger:

$$\varepsilon = N \cdot S_{AB} \cdot \Delta T \tag{5}$$

where ε is the e. m. f. output, S_{AB} the Seebeck coefficient of materials A and B, which constitute the thermocouples of the pile, and ΔT the temperature difference applied to the junctions.

Although Seebeck's law was established for steady states, it is frequently extended to transient states. We have used this application for several reasons, the first being the present lack of other laws which can be rigorously applied to these cases, the second corresponding to the real situation in which the temperature gradients applied to the heat fluxmeter are usually so small and their changes so light that they can be assumed to fall into the lineal approximation of the local equilibrium; and finally because the experimental results are in agreement with the calculations based on these hypotheses. For similar reasons, and because some orders of magnitude are inferior to the Seebeck effect, we have not considered the Thomson effect. On these assumptions, the behavior of the heat fluxmeter can be studied in different conditions.

Steady States

In steady states, Equation 5 is valid. Then ΔT can be eliminated using Fourier's law:

$$q = \frac{\Delta T}{R} \tag{6}$$

where q represents the total heat flux across the thermopile and R the total thermal resistance between the junctions. The result is:

$$q = \frac{1}{N \cdot S_{AB} \cdot R} \varepsilon \tag{7}$$

This expression permits the transformation of the thermopile in a heat fluxmeter by virtue of the definition of the "calibration constant," K:

$$K(T) = \frac{q}{\varepsilon} = \frac{1}{N \cdot S_{AB} \cdot R} \tag{8}$$

and the sensitivity of the heat fluxmeter

$$S(T) = \frac{\varepsilon}{q} = N \cdot S_{AB} \cdot R \tag{9}$$

Both quantities depend only on the parameters of the thermopile, which, in turn, are functions of the temperature. This is why the process of calibration of the heat fluxmeter should be done under different temperatures and repeated periodically, particularly if the materials used in the thermocouples undergo aging processes as in the case of semiconductors.

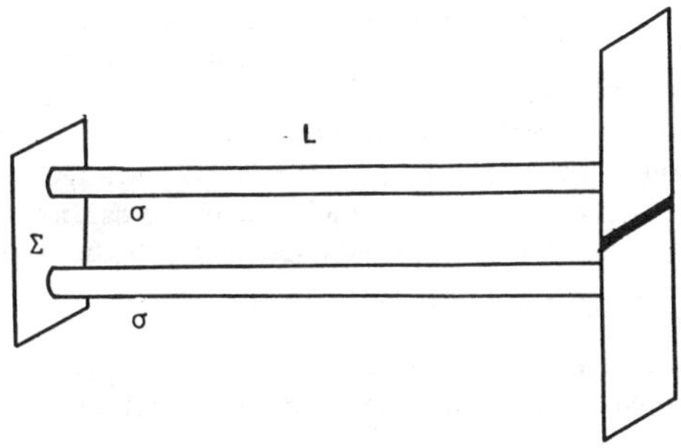

Figure 4. Definition of the "covering coefficient."

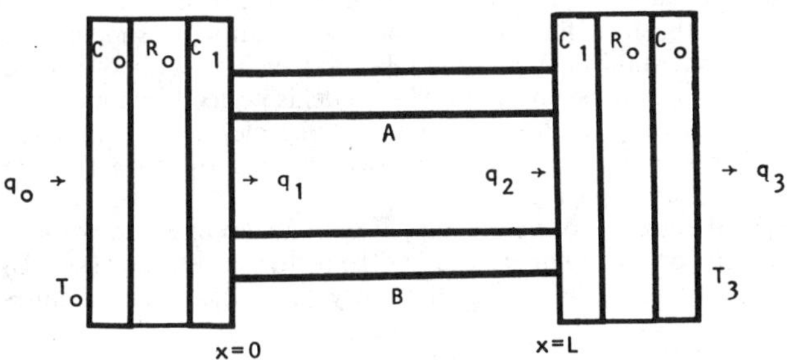

Figure 5. Simple model of a heat fluxmeter.

Henceforth, it will be assumed that the thermopile will be formed by N uniformly distributed thermocouples, by which only one thermocouple will be studied. Then the general results will be N times greater. Thus the thermal resistance per thermocouple, r, can be defined as:

$$r = N \cdot R \tag{10}$$

by means of which neither the calibration constant nor the heat fluxmeter sensitivity depend on the number of thermocouples included:

$$K(T) = \frac{1}{S(T)} = \frac{1}{r \cdot S_{AB}} \tag{11}$$

The "covering coefficient" is defined as the quotient between the sum of the different sections of the wires, σ, which are assumed to be equal, and the total section, Σ, corresponding to one junction (Figure 4). With this coefficient

$$\rho = \frac{2\sigma}{\Sigma}$$

the thermal resistance of a thermocouple is

$$r = \frac{L}{(\rho + (1-\rho)k_p/k) \cdot k\Sigma} \tag{12}$$

where $k = (k_A + k_B)/2$, k_A and k_B being the thermal conductivities of wires A and B respectively, and k_p the thermal conductivity of the material contained voluntarily or spuriously between the wires. Equations 11 and 12 show the dependence between the calibration constant and the covering coefficient. They also demonstrate the influence of the substance inserted between the wires. This substance is usually air in this case, assuming the wires are made of chromel and constantan and the quotient of conductivities k_p/k is lower than 12.5×10^{-3} for a wide range of temperatures around room temperature. However, this effect is included in the calibration. The upper limit of the application of this last law when air is included between the wires is approximately

$$\Delta T \cong 16 \cdot L^{-3}$$

where L is the length of the wires expressed in cm, which corresponds to the onset of convection, if this is not induced externally by other regions subject to different conditions.

Transient States

Let us now consider the transient states between two different well-defined steady states. In order to study the behavior of the heat fluxmeter in such a situation, a simple real, practical thermocouple model can be established, and the relation which exists between the input and output of the heat transducer can be analyzed. The model is perfectly symmetrical with respect to the central plane of wires. The four external metallic pieces are considered pure heat capacitors (thermal diffusivity zero), separated two by two by pure thermal resistances (thermal diffusivity infinite) (Figure 5).

The resolution of the differential equation of heat conduction in a slab or a rod without heat losses allows the Laplace transform of temperature and heat flow at the exit (x = L) to be written in function with the same values as at the entry (x = 0), all these values referring to the initial ones [18].

$$\begin{vmatrix} \bar{\theta}_{x=L} \\ \\ \bar{f}_{x=L} \end{vmatrix} = \begin{vmatrix} \cosh y & -\frac{r}{y}\,\text{senh}\, y \\ \\ \frac{y}{r}\,\text{senh}\, y & \cosh y \end{vmatrix} \begin{vmatrix} \bar{\theta}_{x=0} \\ \\ \bar{f}_{x=0} \end{vmatrix} \tag{13}$$

where $y = L \cdot \sqrt{p/\alpha} = \sqrt{prC}$, p being the Laplace variable and α, r, and C the thermal diffusivity, the thermal resistance, and the heat capacity, respectively, of the medium, and

$$\bar{\theta}_x = \bar{T}(x, p) - \frac{T(x, 0)}{p}$$

and

$$\bar{f}_x = \bar{q}(x, p) - \frac{q(x, 0)}{p}$$

Equation 14 can also be applied to media formed exclusively by pure thermal resistance, R_0, or pure heat capacity, C_0, but with matrices:

$$\begin{vmatrix} 1 & -R_0 \\ 0 & 1 \end{vmatrix} \quad \text{and} \quad \begin{vmatrix} 1 & 0 \\ -pC_0 & 1 \end{vmatrix}$$

respectively.

To obtain the Laplace transform of the mean instantaneous heat flux that crosses the heat fluxmeter:

$$\bar{q} = \frac{\bar{q}_0 + \bar{q}_3}{2} \tag{14}$$

as a function of the instantaneous temperature difference between the junctions:

$$\overline{\Delta T}(p) = \bar{T}(0, p) - \bar{T}(L, p) \tag{15}$$

the different matrices are adequately composed in agreement with the proposed model, and thus we obtain:

$$\bar{q} - \frac{q_0}{p} = Z(p)\left(\overline{\Delta T}(p) - \frac{\Delta T_0}{p}\right) \tag{16}$$

where

$$Z(p) = \frac{1}{2}\left[p(C_0 + C_1 + pR_0C_0C_1) + (1 + pR_0C_0) \sum_{A,B} \frac{y_i}{r_i} \coth \frac{y_i}{2}\right] \tag{17}$$

and q_0 and ΔT_0 are the heat flux and the temperature difference in the initial steady state.

By inverting Equation 16 we find:

$$q(t) - q_0 = \frac{\Delta T(t) - \Delta T_0}{r} + \left(\frac{C_0 + C_1}{2} + \frac{R_0C_0}{r}\right)\frac{d\Delta T}{dt} + \frac{R_0C_0C_1}{2}\frac{d^2\Delta T}{dt^2}$$

$$+ \sum_{A,B} \frac{2}{r_i} \sum_{n=1}^{\infty} \int_0^t \exp\left(-\frac{4\pi^2 n^2 \alpha_i (t - t')}{L^2}\right)\left(\frac{d\Delta T}{dt'} + R_0C_0 \frac{d^2\Delta T}{dt'^2}\right) dt' \tag{18}$$

and finally, integrating Equation 18 or applying the final value theorem to Equation 17, we obtain:

$$\int_0^{\infty} (q(t) - q_0)\, dt = \frac{1}{r}\int_0^{\infty} (\Delta T(t) - \Delta T_0)\, dt + \left(\frac{C_0 + C_1}{2} + \frac{R_0C_0}{r} + \frac{C}{12}\right)(\Delta T_{\infty} - \Delta T_0) \tag{19}$$

where C is the mean heat capacity of the wires and ΔT_{∞} the temperature difference between the junctions in the final steady state. The extension of these results to the complete heat fluxmeter is easy.

In order to investigate the results previously obtained, one should remember that the extension of Seebeck's law (Equation 5) can be introduced into Equations 18 and 19 to express the mean heat flux which crosses the heat fluxmeter as a function of the e.m.f. produced by it. Then Equation 18 will show that the instantaneous output e.m.f. of the heat fluxmeter is not a measure of the heat flux which crosses it at that moment. This expression is too complicated to be of practical use. Equation 19 is simpler than Equation 18 and much more important. Indeed, it shows that the total output or the area of the thermograph is proportional to the total heat transferred by the heat fluxmeter, allowing for a correction term. This term can be evaluated in that:

$$\Delta T_{\infty} - \Delta T_0 = \frac{1}{N \cdot S_{AB}}(\varepsilon_{\infty} - \varepsilon_0)$$

and $(\varepsilon_{\infty} - \varepsilon_0)$ represents the change in the zero line in the thermograph due to the change in the steady state. The upper limit of the integrals can be substituted by a sufficiently long time so as ensure that the final steady state is reached, otherwise the integrals tend towards infinite due to the change in the reference state.

The case in which the final steady state is the same as the initial steady state is particularly interesting. Here the corrective term in Equation 19 is cancelled out, as the temperature differences applied at the junctions are equal in the initial and final steady states. Thus the integral output or the total area of the thermograph, referred to the base line, is strictly proportional to the total heat transferred by the heat fluxmeter during the nonsteady situation, and the constant of proportionality is the calibration constant. This fact constitutes the central idea of dynamic or isoperibolic calorimetry [19]. In this type of calorimetry the sample is surrounded by a heat fluxmeter and adiabatic walls. The reference face of the heat fluxmeter is in thermal contact with a big block, preferably metallic and insulated, which has a heat capacity considerably greater than that of the sample and that of the fluxmeter itself. Thus the final state is closer to the initial state, both being equilibrium states. In this way, the total heat released or absorbed by the sample can be measured. This generalization has allowed the production of commercial calorimeters, which measure the total thermal effects of the sample, but it is not possible to know with sufficient accuracy

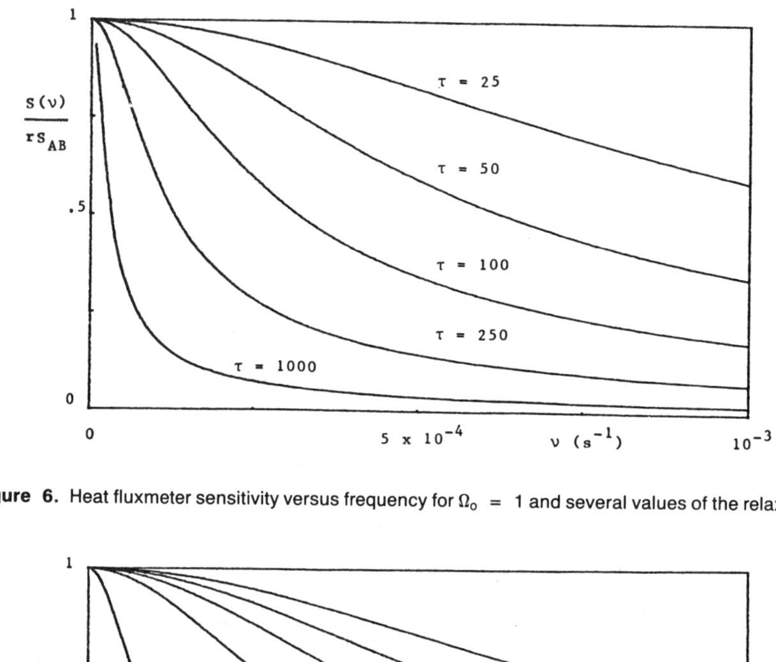

Figure 6. Heat fluxmeter sensitivity versus frequency for $\Omega_o = 1$ and several values of the relaxation time.

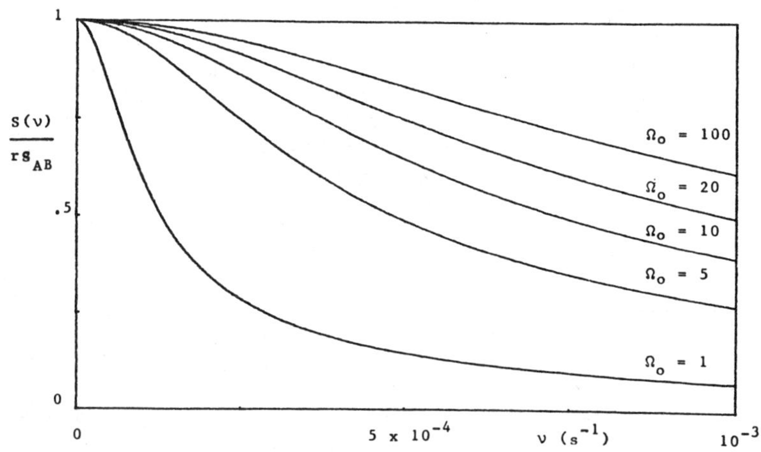

Figure 7. Heat fluxmeter sensitivity versus frequency for $\tau = 250$ s and several values of Ω_o.

the temporal generation of this heat, so that each problem would need a precalculated heat fluxmeter.

Oscillatory Regime

Equation 17 shows a relation between the input and output of the heat fluxmeter in Laplace space. Thus Z(p) can be recognized as its operational admittance, and the sensivity of the heat fluxmeter can be defined as a function of it:

$$S(\nu) = \frac{S_{AB}}{|Z(j\omega)|} \tag{20}$$

where ν is the frequency and $j = \sqrt{-1}$.

For simplicity of calculation, we can now make two approximations closely adjusted to the real situation: the first consisting of the assumption that $C_0 = C_1$ (i.e. that all the external slabs are made of the same material and are of the same thickness); the second taking $R_0 \sim 0$ (i.e. to make the resistances of contact between the slabs nil, this approximation being very close to reality if a high heat conductivity cement is used to join them). Then the sensitivity can be expressed by means of three nondimensional numbers:

$$x = \frac{r_A C_A}{r_A C_A + r_B C_B} = \frac{\alpha_B}{\alpha_A + \alpha_B}$$

$$s = \frac{r_A}{r_A + r_B}$$

$$\Omega_0 = \frac{r_A C_A + r_B C_B}{r C_0} \tag{21}$$

and the nondimensional frequency

$$\Omega = \omega(r_A C_A + r_B C_B) = 2\pi\nu(r_A C_A + r_B C_B)$$

where $r = r_A r_B/(r_A + r_B)$ and A and B are the materials forming the wires of the thermocouples.

If the two wires have similar thermal properties, $x \sim 1/2$, the sensitivity does not depend on s. In this case it can be demonstrate that:

$$\frac{S(\nu)}{rS_{AB}} = 2\left[\frac{\Omega}{2}\frac{\operatorname{senh}^2\frac{1}{2}\sqrt{\Omega} + \operatorname{sen}^2\frac{1}{2}\sqrt{\Omega}}{\left(\cosh\frac{1}{2}\sqrt{\Omega} - \cos\frac{1}{2}\sqrt{\Omega}\right)^2} + \frac{\Omega^2}{\Omega_0^2} + \frac{\Omega\sqrt{\Omega}}{\sqrt{\Omega_0}}\frac{\operatorname{senh}\frac{1}{2}\sqrt{\Omega} - \operatorname{sen}\frac{1}{2}\sqrt{\Omega}}{\cosh\frac{1}{2}\sqrt{\Omega} - \cos\frac{1}{2}\sqrt{\Omega}}\right]^{-1/2}$$

Figures 6 and 7 show $S(\nu)/rS_{AB}$ as a function of the relaxation time ($\tau = 2\pi(r_A C_A + r_B C_B)$) for $\Omega_0 = 1$ and as a function of Ω_0 for $\tau = 250$ s, respectively.

In conclusion, it must be pointed out that in order to design a heat fluxmeter, all the parameters defined in Equation 21 should be studied in detail together with the relaxation time of the fluxmeter itself and the quantities introduced in Equations 11. The design of a fluxmeter is, therefore, a delicate process in which the parameters previously defined and the natural limits of its mechanical properties may intervene.

CALIBRATION TECHNIQUES

Calibration of a fluxmeter consists of determining the quantities which permit its use as a heat transducer. As has already been shown, these quantities depend entirely on the heat fluxmeter itself, so that the process of calibration can be carried out outside the measuring circuit. Nevertheless, the dependence of the calibration quantities on the working temperature and the introduction of parasite thermal resistances between the heat fluxmeter and the measuring system, make it advisable to carry out the process of calibration in conditions which most resemble those of the measurement processes. Two kinds of calibration are possible, the stationary and the oscillatory. In the former, only the calibration constant is measured, while in the latter, this constant may be studied as a function of the frequency.

The calibration constant for the zero frequency can be obtained directly from Equation 8. For this reason, the reference face of the heat fluxmeter is placed in thermal contact with a heat source of a constant temperature, while the rest is kept adiabatically isolated. In such circumstances, a known power is released from the test face and the output is measured. The quotient of both quantities gives the desired calibration constant. In order to carry out this process more easily, all heat fluxmeters usually carry an electrical resistance uniformly distributed along the test surface. An alternative arrangement can be used in order to secure the adiabatic isolation. Two heat fluxmeters, with their reference faces in contact with two regulable heat sources, contain between their test faces an electrical resistance, supported by a narrow metal piece. It is thus possible to calibrate both heat fluxmeters at the same time by releasing different electric powers for different temperatures from each of the heat sources.

The previously mentioned methods can also be used to calibrate a heat fluxmeter in function with the frequency, but by releasing an alternative power in the electrical resistances. The frequency of the power will then be the independent variable. In this case, another method of calibration can be used. In fact, by using the first arrangement for a single fluxmeter, an alternative electric current is introduced into the thermoelectric circuit, and the Peltier effect will induce an alternative temperature difference at the junctions, and so consequently a heat flux which is also oscillatory. By filtering the output, it is possible to calibrate the heat fluxmeter. The supporting theory for this method is too vast to be described here.

Finally, the heat fluxmeter can be calibrated for nonoscillatory uses by means of Equation 19, when it works between two identical stationary states. In this case, the initial and the final temperature distributions are the same and the corrective term cancels out. With this method, there is no need to supply a continual electric power, a small discontinuous heat supply being enough to calibrate the transducer. This method is usually used in isoperibolic calorimetry.

HEAT FLUXMETER COMBINATIONS

Heat fluxmeters can be combined in the two classical forms of connection of electric batteries, in series and in opposition. In both cases, they measure the sum or the difference, respectively, of the heat fluxes that cross them. This property, particularly the possibility of connection in opposition, gives great versatility to this kind of measurement of heat flux, not found in other methods. This possibility has mainly been used between equilibrium states for dynamic calorimetry. But it is also possible to extend its use to the nonequilibrium phenomena in order to measure the energy accumulated by the sample between two steady states, as in differential thermal analysis, but in a more general way. For example, a sample is placed between two lineal heat fluxmeters and a temperature difference is applied to the whole. If both fluxmeters are mounted in opposition and the difference of temperature applied changes, by applying Equation 19, it is possible to obtain information about the energy accumulated by the sample, by means of a simple energy balance.

Another important aspect of connection in opposition is the elimination of external noise. This property (used for the first time by E. Calvet [19] in isoperibolic calorimetry] can be applied to heat fluxmeters whose reference faces are in thermal contact with a heat source, both in the steady state and the transitory state. As is well-known, all real thermal sources have an aleatory thermal noise due to external perturbation or circulatory fluid. This noise affects the reference junctions of the heat fluxmeters, producing an undesirable fluting effect in their output. To avoid this effect, an exact reply is set up in contact with the same heat source, so that the perturbations in both systems are in phase. The reply should be inert, that is, it should not produce any of the thermal effect problems. Then with both thermoelectric devices connected mutually in opposition, it is possible to decrease the fluting effect of the output of the heat fluxmeters.

Finally, another aspect of the heat fluxmeters must be pointed out. This element is a thermopile and, therefore, the inverse effect of Seebeck can be used, that is, the Peltier effect. Thus it is possible to insert a continuous electric current in a heat fluxmeter to maintain the temperature difference between its faces. This possibility can have many applications, from the calibration of the fluxmeter, as has already been seen, to the measurement of a heat flux without perturbation, using the heat fluxmeter as a "thermal potenciometer." All these possibilities should be taken into consideration when designing an experimental device, on the understanding that many of the applications of heat fluxmeters will be completely novel.

APPLICATIONS

The harbingers of the heat fluxmeters, in the version presented here, were dynamic calorimetry and differential thermal analysis. It would be impossible to enumerate all the applications of and the results obtained from these two important experimetal techniques here. However, to be fair, one must underline the important influence that both have had on the present technology of the heat transducer,

Various applications of the heat fluxmeters have been made, for both equilibrium and transport properties. A few of these follow:

- The determination of thermal and electric properties of dielectric in electric field [20–21]. Two fluxmeters of radial geometry have been used to measure the specific heat and the pyroelectric coefficient of a ferro-electric material in different electric fields round the Curie temperature. The heat fluxmeters were connected in opposition to two other fluxmeters set up in a passive symmetrical circuit.
- The measurement of the equivalent heat conductivity of a convective fluid [22–23]. A heat fluxmeter was used to measure the heat transfer of a liquid in convection. (See the Chapter 3, "Heat Flux in the Bénard–Rayleigh Problem" in this volume.)
- The measurement of the formation energy of a convective structure [24]. Two heat fluxmeters were used to measure the energy accumulated by a fluid as it passed from a conductive state to a convective one. In both situations the fluid is submitted to the same temperature difference. The necessary corrections were made. Both fluxmeters were connected in opposition to two others set up in another inert symmetrical device, to avoid the fluting effect at the output.
- Simultaneous measurements of the Seebeck and Peltier coefficients [25–26]. This is perhaps the best and the most original of the applications presented, from a technical point of view. Two conventional heat fluxmeters, connected to a common thermal source, support a sample thermopile. The whole forms a "thermoelectric bridge" which permits

the study of the thermoelectric coefficients of the sample thermopile, using the various properties described here. The results obtained show new possibilities for the use of heat fluxmeters.

NOTATION

C	heat capacity
K	calibration constant
$K_{A,B}$	thermal conductivities of materials A and B
L	length
q	heat flux
R	resistance
r	thermal resistance per thermocouple
S	sensitivity of heat fluxmeter
S_{AB}	Seebeck coefficient
T	temperature

Greek Symbols

α	thermal diffusivity
ε	electromotive force output potential
ν	frequency
σ	covering coefficient

REFERENCES

1. Muller, I., "Zum Paradoxen der Warmeleitungstheorie," *Zeitschrift fur Phys.*, 198: 329–335 (1967).
2. Lebon, G., "Derivation of Generalized Fourier and Stokes–Newton Equations Based of the Thermodynamics of Irreversible Processes," *Bull. Soc. Roy. Belgique*, Classe des Sciences, 64: 456–462 (1978).
3. Jou, D., Casas–Vazquez, J., and Lebon, G., "A Dynamical Interpretation of Second-Order Constitutive Equations of Hydrodynamics," *J. Non-equilib. Thermodyn.*, 4: 349–355 (1979).
4. Casas–Vazquez, J., "Thermodynamic Theory of Stability," in *Stability of Thermodynamic Systems* J. Casas–Vazquez and G. Lebon (Eds.), *Lecture Notes in Physics*, 164, Springer–Verlag, Berlin, 1982, pp 1–40.
5. Tykodi, R. J., *Thermodynamics of Steady States*, MacMillan, New York, 1967.
6. Benofy, S. J., and Quay, P. M., "The Thermodynamics of Systems in a Steady State," *J. Chem. Phys.*, 78: 3177–3190 (1983).
7. Godefroy, J. C., "Thin Film Transducers for Temperature and Heat Flux Measurements," *Rech. Aerosp.*, 2: 73–83 (1981).
8. Malcorps, H., "Frequency–Response of Heat Fluxmeters," *J. Phys.*, 14: 1054–1060 (1981).
9. Grishchenko, T. G., "Characteristics of the Use of a Heat-Flux Sensor in Instruments for Determining Thermal Conductivity," *J. Eng. Phys.*, 40: 640–645 (1981).
10. Woelk, G., "Fast Heat Flux Sensor for High Temperatures," Proceeding of Transducer Technology and Temperature Measurement Conference, Network, Buckingham (England), 1982, Vol. 3, pp 82–92.
11. Klems, J. H., and Dibartolomeo, D., "Large Area, Highsensitivity Heat-Flow Sensor," *Rev. Sci. Instrum.*, 53: 1609–1612 (1982).
12. Ravalitera, G., et al., "Characterization of New Heat Fluxmeters. Application to the Simultaneous Measurements of Thermal Fluxes and Temperature Variations," *Rev. Phys. Appl.*, 17: 177–185 (1982).
13. Leclercq, D., and Thery, P., "Apparatus for Simultaneous Temperature and Heat Flow Measurements Under Transient Conditions," *Rev. Sci. Instrum.*, 54: 374–380 (1983).
14. Prasad, B. V. S., and Mohanty, A. K., "Analysis and Calibration of a Foil Heat Flux Sensor for Convective Measurements," *J. Phys.*, 16: 1095–1099 (1983).

15. Atkinson, W. H., Strange, R. R., and Hobart, H., "Advanced High Temperature Heat Flux Sensors," *Adv. Instrum.*, 38: 1457–1479 (1983).
16. Callen, H. B., *Thermodynamics*, J. Wiley & Sons, New York, 1960.
17. De Groot, S. R., and Mazur, P., *Non-equilibrium Thermodynamics*, North-Holland, Amsterdam, 1962.
18. Carslaw, H. S., and Jaeger, J. C., *Conduction of Heat in Solids*, Oxford Univ. Press, London, 1959.
19. Calvet, E., and Prat, H., *Microcalorimétrie. Aplications Physico-Chimiques et Biologiques*, Masson & Cie, Paris, 1956.
20. Ramos, S., del Cerro, J., and Zamora, M., "Specific Heat of Triglicine Sulfate at Several Applied Electric Fields Near the Critical Temperature," *Phys. Stat. Sol.*, 61: 307–311 (1980).
21. del Cerro, J., Ramos, S., and Zamora, M., "Calorimetric Method for the Determination of the Pyroelectric Coefficient," *Phys. Stat. Sol.*, 62: 149–154 (1980).
22. Moreno, J., et al., "New apparatus for the Study of Rayleigh–Bénard Problem," *Rev. Sci. Instrum.*, 51: 82–85 (1980).
23. Zamora, M., Córdoba, A., and Moreno, J., "The Influence of Wavelength and Bimodal Transition on Heat Flux in the Bénard–Rayleigh Problem," *Physico Chemical Hydrodynamics*, 26: 4028–4034 (1983).
24. Zamora, M., and Rey de Luna, A., "Energy in the Bénard–Rayleigh Problem," in *Cellular Structures in Instabilities*, J. E. Wesfreid and S. Zalesky (Eds.), Lecture Notes in Physics, Springer-Verlag, Berlin, 1984, p. 342.
25. Jiménez, J., Rojas, E., and Zamora, M., "Design and Construction of Precision Heat Fluxmeter," *J. Appl. Phys.*, 56: 3353–3356 (1984).
26. Jiménez, J., Rojas, E., and Zamora, M., "Device for Simultaneous Measurement of the Peltier and Seebeck Coefficients. Verification of the Kelvin Relation," *J. Appl. Phys.*, 56: 3250–3255 (1984).

SECTION II

INDUSTRIAL OPERATIONS AND DESIGN

CONTENTS

CHAPTER 21

HEAT TRANSFER AND HYDRODYNAMICS OF TWO-PHASE ANNULAR FLOW

K. Suzuki

Department of Mechanical Engineering
Kyoto University
Kyoto 606, Japan

CONTENTS

INTRODUCTION

This chapter discusses the heat transfer and the hydrodynamics of two-phase annular flow. The annular flow is one of the various flow patterns appearing in the gas-liquid two-phase flows in pipes and it is encountered when the gas-to-liquid mass flow rate ratio is high. In annular two-phase flow, liquid flows wet the inner surface of the pipe and the gas-phase fluid flows occupy the central part of the pipe. Instability occuring at the interface of the two phases leads to the incipience of interfacial waves, and the large amplitude waves called disturbance waves are nonlinear in nature. Sometimes, the problem becomes more complicated by the large amount of droplets entrained from the waves. Gas-phase flow is turbulent in usual practical situations and the liquid flow also becomes turbulent when the liquid film Reynolds number is brought high enough.

The combined effects of these complicated aspects have prevented us so far from developing a good picture of the two-phase annular flow. Therefore, a large part of this chapter will be spent describing empirical relationships for the heat transfer and the hydrodynamics of the two-phase annular flow. The hydrodynamics will be discussed first in the following since it is the basis of understanding the heat transfer characteristics. In the sections for heat transfer, single-component boiling and condensation heat transfers and two-component evaporative heat transfer will be discussed.

GENERAL DESCRIPTION OF TWO-PHASE ANNULAR FLOW

Two-phase flow is characterized by the existence of the interface between the phases. The density difference between the two phases in gas-liquid two-phase flow is similar to that in gas-solid two-phase flow; in all other respects the flow are quite different (except for dispersed flows). In the case of gas-liquid two-phase flow, the interface can be deformed in various ways but not at all in the case of gas-solid two-phase flow. This deformability of the interface leads to the possible appearance of different flow patterns of gas-liquid two-phase flows in tubes and they are classified into several typical ones: bubbly flow, slug flow, annular flow, etc. Transition from one flow pattern to another is observed when the mass flow rates of the phases are changed. The observed flow pattern data are presented usually in terms of flow pattern maps. Detailed specification of each flow pattern and discussions of some typical flow pattern maps have been given by Delhaye [1] and Hewitt [2–4].

A schematic illustration of annular flow in a vertical tube is given in Figure 1. It is characterized by a gas flow in the core region of higher velocity than that of liquid film which flows clinging annularly to the inner surface of the tube. In downward vertical flows, the annular flow can be formed even at a rather low gas-to-liquid mass-flow-rate ratio, either artificially or in a tubular condensation device in practice. An important parameter of the flow in the liquid film is the liquid film Reynolds number defined by

$$Re_f = \frac{4M_f}{\pi D \mu_l} = \frac{m_f D}{\mu_l} \quad \left(\cong \frac{4\rho_l \bar{U}_f \delta}{\mu_l} \text{ for } \delta \ll D \right) \tag{1}$$

where M_f and $\bar{U}_f$ = the liquid film mass flow rate and the liquid averaged velocity in a film, respectively
m_f = the mass flux of liquid flowing in the film
D = the tube diameter
δ = the liquid film thickness
ρ_l and μ_l = the density and the viscosity of the liquid, respectively

Other important parameters to be used later follow:

Denoting the total mass flux by m, the liquid mass flux by m_l and the gas mass flux by m_g, the quality x is defined as follows:

$$x = \frac{m_g}{m} = \frac{m - m_l}{m} \tag{2}$$

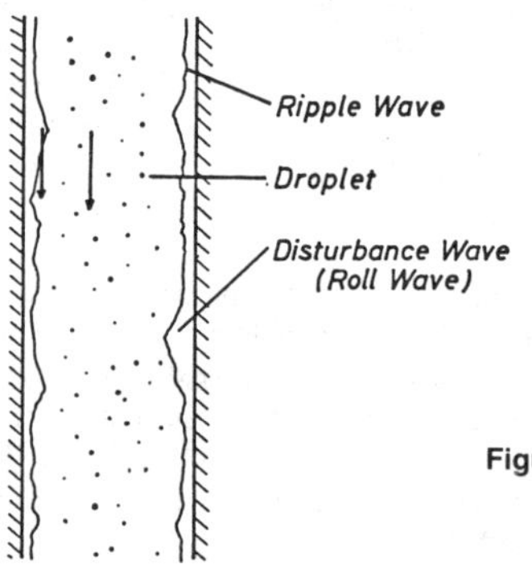

Figure 1. Schematic illustration of two-phase annular flow in a tube.

The void fraction α is the ratio of the cross-sectional area A_g occupied by gas-phase flow to the total tube cross-sectional area $A = \frac{\pi}{4}D^2$. Thus

$$\alpha = \frac{A_g}{A} = \frac{A - A_l}{A} \tag{3}$$

where A_l is the area occupied by the liquid flow. Introducing the averaged velocities of liquid and gas phases by $\bar{U}_l$ and $\bar{U}_g$,

$$M_l = m_l A = \rho_l A_l \bar{U}_l \tag{4}$$

$$M_g = m_g A = \rho_g A_g \bar{U}_g \tag{5}$$

$\bar{U}_l$ becomes equal to $\bar{U}_f$ when no dispersed phase exists so that $m_l = m_f$. In terms of $\bar{U}_l$ and $\bar{U}_g$,

$$\alpha = \frac{\dfrac{m_g}{\rho_g \bar{U}_g}}{\dfrac{m_l}{\rho_l \bar{U}_l} + \dfrac{m_g}{\rho_g \bar{U}_g}} \tag{6}$$

or with the slip ratio $S = \bar{U}_g/\bar{U}_l$,

$$\alpha = \frac{\dfrac{m_g}{\rho_g}}{S\dfrac{m_l}{\rho_l} + \dfrac{m_g}{\rho_g}} = \frac{\rho_l x}{S\rho_g(1-x) + \rho_l x} \tag{7}$$

Superficial velocities W_l and W_g of liquid and gas phases are also widely used. They are defined by:

$$W_l = \frac{m_l}{\rho_l} \tag{8}$$

$$W_g = \frac{m_g}{\rho_g} \tag{9}$$

and correspond to the phase-averaged velocities if each phase flows alone filling the entire tube. In terms of W_l and W_g, the quality x is expressed as follows:

$$x = \frac{\rho_g W_g}{\rho_g W_g + \rho_l W_l} \tag{10}$$

Assuming thermodynamic equilibrium between the phases, x can also be expressed for single-component flow in terms of enthalpies i_g, i and i_l:

$$x = \frac{i - i_l}{i_g - i_l} \tag{11}$$

where i_g = the enthalpy of the saturated vapor
i_l = the enthalpy of the saturated liquid
i = the local enthalpy of two-phase flow

However, difficulties often arise because the departure from the thermodynamic equilibrium exists actually. An example of this includes the supersaturation of vapor, subcooled boiling, superheated liquid, and the existence of droplets in superheated vapor.

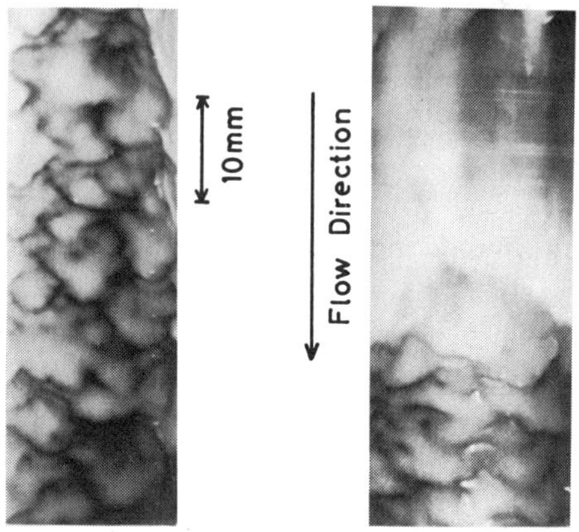

(a) Ripple wave regime (Re_l=210, Re_g=4.0×10^4)   (b) Disturbance wave regime (Re_l=500, Re_g=4.0×10^4)

Figure 2. Wave pattern visualized in an air-water downward annular flow [5].

Except at a low liquid-film Reynolds number, the surface of the liquid film is unstable and wavy in nature. Two different types of waves are observed, namely ripple waves and disturbance waves or roll waves. The former waves are of a ③-dimensional structure in contrast to the ②-dimensional structure of the disturbance waves. These are demonstrated in the two photographs of Figure 2A, B [5]. These photographs of scattered light from the TiO_2 fine particles dispersed in the liquid film of annular flow in a vertical tube were taken by a camera located outside the transparent tube. Therefore, the dark parts correspond to the troughs of the waves and the bright parts to the crests of the waves. While ripple waves are shaped like fish scales whose center is the crest as found in Figure 2A, the disturbance waves are almost peripherally uniform and much larger in size as seen in Figure 2B.

From the viewpoint of their effects on momentum, heat and mass transfer, the ripple waves are not so important but the disturbance waves play an important role. In the experiments on air-water annular flow heat transfer, the heat transfer coefficient is found to increase steeply with an increase of the air flow rate above its critical value for the inception of disturbance waves [6]. In other words, the disturbance waves cause the enhancement of heat transfer.

Figure 3 shows an example of simultaneous records of the wave height and the tube wall shear stress obtained in an air-water downward annular flow in a circular tube [7]. In this experiment, the appearance of disturbance waves was detected by a method paying attention to the fact that the wave height record shows a rapid increase for the disturbance waves. In Figure 3, two disturbance waves are detected and it is found that the wall shear stress can increase at the time of the detection of disturbance waves. This is the indication of enhanced momentum transfer due to the appearance of the disturbance waves. There may be several mechanisms for wave-induced enhancement of momentum and heat transfer including:

1. Non-linear wave effect.
2. Turbulence caused by the waves.
3. The alternate contraction and expansion of gas-phase flow cross-sectional area.

The non-linear effect has been modeled by Suzuki et al. [8] for a thin liquid film. It is commonly accepted that laminar-to-turbulence transition can occur at lower Reynolds numbers in the liquid film but some of the effects thought to have been turbulence effects might have been the nonlinear wave effect. The third effect is sometimes referred to as the roughness effect and the sinuous motion of gas-phase flow causes the momentum loss [7]. Sometimes, it is expected that the gas-phase flow separates at the crests of waves and recirculates at the front side of the waves. This has been discussed by Friedel [9].

Another important feature of the waves is the entrainment of droplets from the crests of the waves. Acceleration of the entrained droplets causes an additional pressure drop and the deposition of the entrained droplets plays an important role on heat transfer, especially, in postburnout heat transfer. Annular flow dispersed with a significant amount of liquid droplets is called annular-mist flow. The inception criterion for the droplet entrainment and the droplet entrainment fraction will be discussed later.

PRESSURE DROP

Pressure gradient (dp/dz) in two-phase flow may be divided into three parts corresponding to the gravitational term $(dp/dz)_g$, the acceleration term $(dp/dz)_a$ and the frictional term $(dp/dz)_f$, respectively. Thus,

$$\frac{dp}{dz} = \left(\frac{dp}{dz}\right)_g + \left(\frac{dp}{dz}\right)_a + \left(\frac{dp}{dz}\right)_f \tag{12}$$

In terms of the void fraction α, the first term can be written down as follows:

$$\left(\frac{dp}{dz}\right)_g = -[\rho_l(1-\alpha)+\rho_g\alpha]g \sin \gamma \tag{13}$$

where z is the axial distance measured upward, and γ the inclination angle of tube raised from the horizontal plane. To evaluate the gravitational term, the local void fraction is required to be known and this will be discussed in the next section. Friedel [9] presented an expression for the acceleration pressure drop Δp_a due to the change in quality x as follows:

$$\Delta p_a = \frac{\Delta J}{A} \tag{14}$$

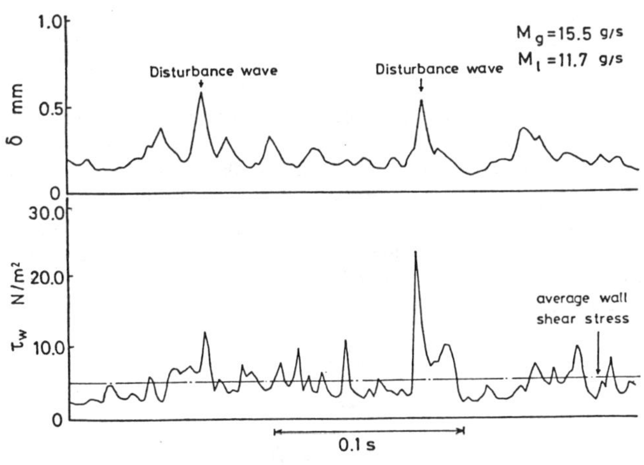

Figure 3. Fluctuation of wall shear stress and liquid film thickness [7].

where J is the total momentum flux defined by

$$J = \left(\frac{M_g^2}{\rho_g A_g} + \frac{M_l^2}{\rho_l A_l}\right) = \frac{M^2}{A}\left[\frac{x^2}{\alpha \rho_g} + \frac{(1-x)^2}{(1-\alpha)\,\rho_l}\right] \tag{15}$$

where M is the total mass flow rate.
Therefore, the acceleration pressure gradient is given by

$$\left(\frac{dp}{dz}\right)_a = \frac{-1}{A}\frac{dJ}{dz} \tag{16}$$

The frictional pressure gradient is most difficult to evaluate and has been discussed in many papers. It has been found convenient to use the following two-phase flow friction multipliers φ_g, φ_l, φ_{g0}, φ_{l0}

$$\varphi_g^2 = \frac{\left(\frac{dp}{dz}\right)_f}{\left(\frac{dp}{dz}\right)_{fg}} \tag{17}$$

$$\varphi_l^2 = \frac{\left(\frac{dp}{dz}\right)_f}{\left(\frac{dp}{dz}\right)_{fl}} \tag{18}$$

$$\varphi_{g0}^2 = \frac{\left(\frac{dp}{dz}\right)_f}{\left(\frac{dp}{dz}\right)_{fg0}} \tag{19}$$

$$\varphi_{l0}^2 = \frac{\left(\frac{dp}{dz}\right)_f}{\left(\frac{dp}{dz}\right)_{fl0}} \tag{20}$$

where $(dp/dz)_f$ is the two-phase frictional pressure gradient appearing in Equation 12. $(dp/dz)_{fg}$ is the single-phase flow frictional pressure gradient to be attained if the gas-phase fluid flows alone filling the entire tube at the actual flow rate M_g and $(dp/dz)_{fl}$ is the counterpart for liquid-phase flow at the flow rate M_l or $(M-M_g)$. $(dp/dz)_{fg0}$ and $(dp/dz)_{fl0}$ are the fictitious frictional pressure gradients for the gas-phase flow and for the liquid-phase flow, respectively, both flowing at the same flow rate equivalent to the total mass flow rate M. Introducing the Fanning friction factors f_g, f_l, f_{g0} and f_{l0}, the preceding single-phase pressure gradients can be expressed as follows:

$$-\left(\frac{dp}{dz}\right)_{fg} = f_g\frac{2\,\rho_g W_g^2}{D} = f_g\frac{2\,m_g^2}{\rho_g D} = f_g\frac{2\,m^2x^2}{\rho_g D} \tag{21}$$

$$-\left(\frac{dp}{dz}\right)_{fl} = f_l\frac{2\,\rho_l W_l^2}{D} = f_l\frac{2\,m_l^2}{\rho_l D} = f_l\frac{2\,m^2(1-x)^2}{\rho_l D} \tag{22}$$

$$-\left(\frac{dp}{dz}\right)_{fg0} = f_{g0}\frac{2\,m^2}{\rho_g D} \tag{23}$$

$$-\left(\frac{dp}{dz}\right)_{f10} = f_{10}\frac{2\,m^2}{\rho_l D} \tag{24}$$

The values of friction factors are obtained with the aid of usual friction formulae related to the Reynolds number. For example,

$$\text{Re} < 2{,}000; \quad f = \frac{16}{\text{Re}} \tag{25}$$

$$\text{Re} > 2{,}000; \quad f = 0.079\,\text{Re}^{-1/4} \tag{26}$$

The Reynolds number for each of Equation 21 through 24 can be defined as follows:

$$\text{Re}_g = \frac{\rho_g W_g D}{\mu_g} = \frac{m_g D}{\mu_g} \tag{27}$$

$$\text{Re}_l = \frac{\rho_l W_l D}{\mu_l} = \frac{m_l D}{\mu_l} \tag{28}$$

$$\text{Re}_{g0} = \frac{mD}{\mu_g} = \frac{1}{x}\text{Re}_g \tag{29}$$

$$\text{Re}_{l0} = \frac{mD}{\mu_l} = \frac{1}{1-x}\text{Re}_l \tag{30}$$

where μ_g and μ_l are the viscosities for the gas and liquid phases, respectively. In the homogeneous model to be mentioned later, the two-phase Reynolds number Re_{tp} is used.

$$\text{Re}_{tp} = \frac{mD}{\mu_{tp}} \tag{31}$$

where the two-phase viscosity is defined, for example, as follows:

$$\frac{1}{\mu_{tp}} = \frac{x}{\mu_g} + \frac{1-x}{\mu_l} \tag{32}$$

This is usually cited as suggested by McAdams et al. [10]. Except for some works including Beatty [11], the correlations for the frictional pressure gradient have been studied without specifying the flow pattern. Therefore, the discussions that follow may be used for the flow patterns other than the annular flow.

With the homogeneous model, the two-phase friction coefficient f_{tp} is introduced and the friction pressure gradient is expressed as follows:

$$-\left(\frac{dp}{dx}\right)_f = f_{tp}\frac{2\,m^2}{\rho_{tp} D} \tag{33}$$

where the homogeneous density ρ_{tp} is [10]:

$$\rho_{tp} = \frac{\rho_g \rho_l}{\rho_l x + (1-x)\rho_g} \tag{34}$$

and f_{tp} is evaluated using the two-phase Reynolds number Re_{tp} given by Equation 31 for Re in Equation 25 or 26. When Equation 26 is used for f_{tp} and f_{10}, the value of φ_{10}^2 may become:

$$\varphi_{l0}^2 = \left(1 + x\frac{\rho_l - \rho_g}{\rho_g}\right)\left(1 + x\frac{\mu_l - \mu_g}{\mu_g}\right)^{-1/4} \qquad (35)$$

The homogeneous model is simple to use but tends to underestimate the two-phase friction pressure gradient.

The most widely used method for the evaluation of the friction multipliers defined by Equations 17 through 20 originated from the works by Lockhart and Martinelli [12]. They introduced a parameter X defined as follows.

$$X^2 = \frac{\left(\frac{dp}{dz}\right)_{fl}}{\left(\frac{dp}{dz}\right)_{fg}} \qquad (36)$$

Their results for the correlation curve of multipliers φ_g and φ_l are indicated in graphical form and are reproduced in Figure 4. The correlation curve differs from one case to another depending on whether each phase flow is laminar ($Re_l < 2{,}000$ or $Re_g < 2{,}000$) or not. The first suffix of φ_g and φ_l designates the flow state (v for laminar (viscous) and t for turbulent) of the liquid flow having the frictional pressure gradient $(dp/dz)_{fl}$ and the second suffix denotes the same for the gas-phase flow of the frictional pressure gradient $(dp/dz)_{fg}$. Turbulent-turbulent is the most important flow regime and an approximate form of X_{tt} used by Lockhart and Martinelli is:

$$X_{tt} = \left(\frac{\mu_l}{\mu_g}\right)^{0.1}\left(\frac{\rho_g}{\rho_l}\right)^{0.5}\left(\frac{1-x}{x}\right)^{0.9} \qquad (37)$$

In this relation, $f \sim Re^{-0.2}$ has been used instead of Equation 26. As seen from Equation 37, X_{tt} is small for high quality or for annular flow and X_{tt} is large for low quality two-phase flow. Martinelli and Nelson [13] presented a correction factor of the multiplier curve for pressure variation.

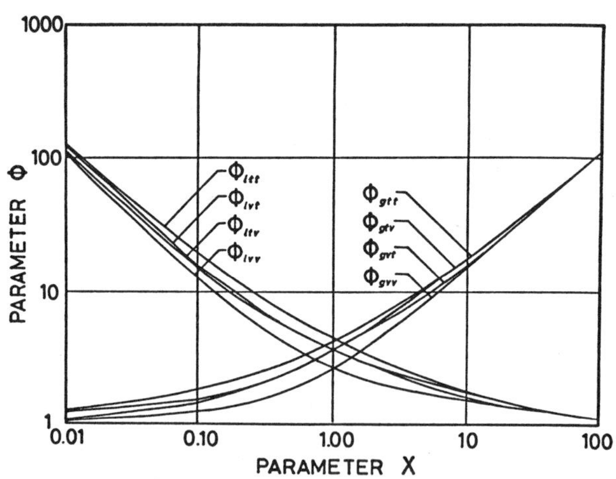

Figure 4. Correlation for pressure drop multiplier [12].

Chisholm and Laird [14] and Chisholm [15] suggested that the Lockhart–Martinelli correlation curves can approximately be expressed by the following type of simple equations.

$$\varphi_g^2 = 1 + CX + X^2 \tag{38}$$

$$\varphi_g^2 = 1 + \frac{C}{X} + X^2 \tag{39}$$

where C takes the following value:

$$\begin{aligned} C &= 20 \quad \text{for } X_{tt} \\ C &= 12 \quad \text{for } X_{vt} \\ C &= 10 \quad \text{for } X_{tv} \\ C &= 5 \quad \text{for } X_{vv} \end{aligned} \tag{40}$$

One of the deficiencies of the Lockhart–Martinelli method is that it does not include the known effect of mass flux on the friction multipliers. Baroczy [16] first succeeded in correlating the dependence of the friction multiplier φ_{l0}^2 on the mass flux. He has established the φ_{l0}^2 curve-versus-property index $(\mu_l/\mu_g)^{0.2}(\rho_g/\rho_l)$ for a standard mass flux = 1,356 Kg/m^{2}s as shown in Figure 5. For the previously mentioned property index, we have

$$\left(\frac{\mu_l}{\mu_g}\right)^{0.2}\frac{\rho_g}{\rho_l} = \frac{\left(\frac{dp}{dz}\right)_{fl0}}{\left(\frac{dp}{dz}\right)_{fg0}} \tag{41}$$

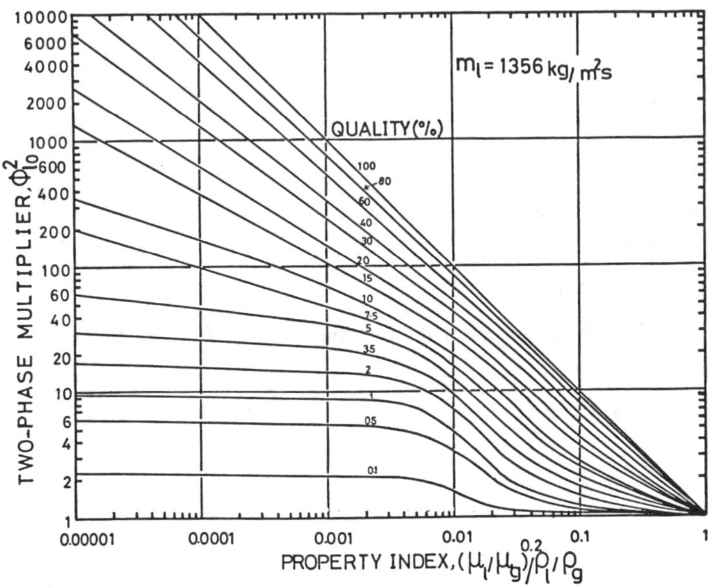

Figure 5. Friction multiplier ϕ_{l0}^2 versus property index [16].

if the friction law of $f \sim Re^{-0.2}$ is used again for each phase. This property index, therefore, corresponds to the Chisholm parameter Y to be referred to later:

$$Y^2 = \frac{\left(\frac{dp}{dz}\right)_{f l0}}{\left(\frac{dp}{dz}\right)_{f g0}} \tag{42}$$

Baroczy then deduced a mass velocity correction factor Ω which is graphically given as reproduced in Figure 6. The value of Ω must be multiplied by φ_{l0}^2 shown in Figure 5.

Chisholm [17] showed that the mass velocity effect can be expressed by the following relationship:

$$\varphi_{l0}^2 = 1+(Y^2-1)\,[Bx^{2-n/2}(1-x)^{2-n/2}+x^{2-n}] \tag{43}$$

where n = 0.2 and B is a parameter given by:

$0 < Y < 9.5$

$$B = \frac{55}{m^{1/2}} \qquad \text{for } m > 1{,}900 \text{ Kg/m}^2\text{s}$$

$$= \frac{2{,}400}{m} \qquad \text{for } 500 < m < 1{,}900 \text{ Kg/m}^2\text{s}$$

$$= 4.8 \qquad \text{for } m < 500 \text{ Kg/m}^2\text{s}$$

$9.5 < Y < 28$

$$B = \frac{520}{Ym^{1/2}} \qquad \text{for } m \leqq 600 \text{ Kg/m}^2\text{s}$$

$$= \frac{21}{Y} \qquad \text{for } m > 600 \text{ kg/m}^2\text{s}$$

$28 < Y$

$$B = \frac{15{,}000}{Y^2 m^{1/2}}$$

Friedel [9, 18] proposed the following relationship for φ_{l0}^2:

$$\varphi_{l0}^2 = 8.59\,\frac{x^{0.69}}{m^{0.29}D^{0.086}}\left(\frac{\rho_l}{\rho_g}\right)^{0.84} \tag{44}$$

This can be applied to the steam-water data in the range of the parameters,

$$344 \text{ Kg/m}^2\text{s} < m < 4{,}661 \text{ Kg/m}^2\text{s}$$

$$12.7 < (\rho_l/\rho_g) \text{ water} < 86.4$$

$$0.0032(m) < d_h < 0.0249(m)$$

$$0.0001 < x < 0.999$$

where d_h is the hydraulic diameter of the tube. Another form of the correlation proposed by Friedel has been discussed by Hewitt [19, 20].

Another empirical approach rather widely used is to correlate the interfacial friction factor f_i defined below as a function of film thickness. The interfacial friction factor is given by:

$$f_i = \frac{\tau_i}{\frac{1}{2}\rho_c W_g^2} \tag{45}$$

where τ_i is the shear stress acting on the interface and ρ_c is the mean density of the fluid flowing in the core region. The latter is given by:

$$\rho_c = \frac{m_l - m_f + m_g}{\frac{(m_l - m_f)}{\rho_l} + \frac{m_g}{\rho_g}} \tag{46}$$

where $(m_l - m_f)$ is the liquid mass flux entrained in the gas core. Empirical relationships for f_i are usually recommended in the following form:

$$f_i = f_{is} \cdot f\left(\frac{\delta}{D}\right) \tag{47}$$

where f_{is} is the friction factor for single-phase flow in a smooth tube like that given by Equation 26. The underlying concept for the above expression of f_i is that the interfacial wave whose height relates to the film thickness may act as the roughness element for the gas-phase flow in the core region [21, 22]. The functional form of $f(\delta/D)$ has been discussed

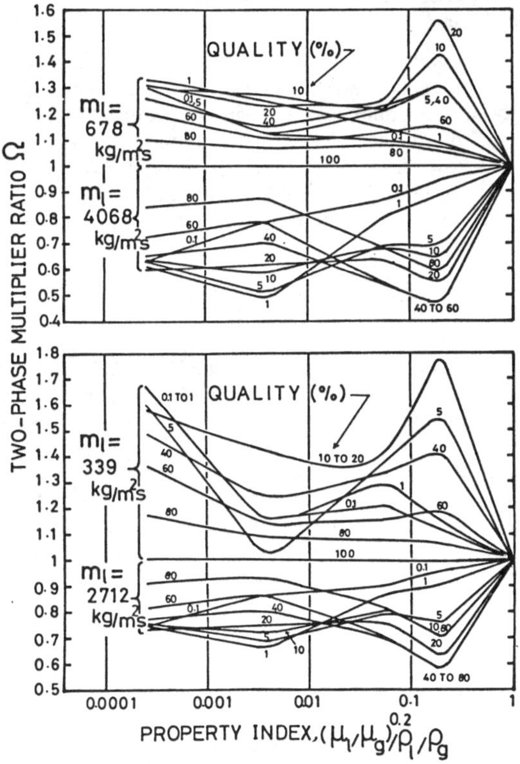

Figure 6. Correction factor for pressure drop multiplier accounting mass flux effect [16].

by many papers including those by Wallis [23] and Martindale and Smith [24]. One of the inconveniences of this approach is that the relationship between τ_i and $(dp/dz)_f$ is not simple unless the void fraction is known separately as will be discussed in the next section.

For practical purposes, the previously mentioned empirical formulae are used commonly but their physical background is not known well. Furthermore, we must admit that still discrepancies, which are sometimes quite large, exists between those formulae and the experimental data.

Apart from the empirical approaches already mentioned, there exist some analytical approaches. Connected with the present insufficient knowledge of two-phase annular flow, any of these analyses need the incorporation of some experimental information. Levy and Healzer [25] carried out an analysis for two-phase annular flow assuming a transition layer between the crest and trough of waves. Within this transition layer ($\delta_{min} \leqq y \leqq \delta_{max}$), it is assumed that

$$\frac{\rho}{\rho_l} = \left(\frac{y}{\delta_{min}}\right)^{-\xi} \tag{48}$$

$$u = U_w \tag{49}$$

where y = the distance from the tube wall surface
δ_{min} = the minimum thickness of the liquid film or the value of y at the through of waves
δ_{max} = the value of y at the crest of waves
ρ = the local apparent density of two-phase flow
U_w = the wave velocity

It is also assumed that the liquid flowing within this transition layer is the portion of liquid flow entrained into the core. The parameters ξ and U_w are chosen by comparing the computed wave velocity and entrainment rate with respective experimental counterparts. Inside the liquid film ($y \leqq \delta_{min}$), the velocity distribution is taken to be equal to that for the single-phase turbulent flow. Also in the core region ($y \geqq \delta_{max}$), the gas-phase velocity is assumed to yield to the logarithmic law of the wall, and its value at the position of the crest of the waves ($y = \delta_{max}$) is set by the following relationship. Denoting the wall shear stress by τ_w, that is

$$U^+ = \frac{U}{\sqrt{\tau_w/\rho}} = 5.5 \tag{50}$$

which corresponds to assume that the roughness at the interface is 3.32 times the minimum film thickness δ_{min}. They showed that the calculated values of the two-phase friction multiplier and minimum film thickness generally agree with existing empirical correlations and experimental data for annular flow regime.

Suzuki et al. [8] proposed a theoretical approach to a thin film two-phase annular flow based on a model for the effect of waves on interfacial momentum transfer. Their model introduces the apparent viscosities accounting for the wave effect, $\tilde{\mu}_l$ and $\tilde{\mu}_g$, for the liquid film and for the gas-phase flow near the interface, respectively. The final forms of the proposed expressions for the viscosities are as follows:

$$\tilde{\mu}_l = C\rho_l L^2 \omega \tag{51}$$

$$\tilde{\mu}_g = C\rho_g L^2 \omega \tag{52}$$

where L is the wave amplitude and ω the average number rate of waves passing a cross-sectional plane. Since the apparent viscosities are put proportional to L^2, the ripple waves contribute almost nothing to the interfacial momentum transfer, but the disturbance waves do significantly. With this model, they calculated the pressure gradient and the averaged film thickness for annular flow with thin liquid film. The computation of the averaged liquid-film thickness enables calculation of the void fraction as will be discussed in the next section. The computed film thickness is found to be larger for the flow rate conditions with the appearance of disturbance waves. In connection with this, the computed values of the pressure gradient for such conditions are smaller than the observed data, but are not so bad on the whole. The model proposed is equivalent to the empirical approach accounting for the roughness effect just mentioned, in the sense that the viscous sublayer of the gas-phase flow has been assumed broken up intermittently by the appearance of waves. As discussed by Friedel [9] and suggested from experimental work by Hagiwara et al. [7], the sinuous motion of gas-phase flow or the alternate contraction and expansion of the gas-phase flow area occurs when the gas flows over the crest of waves and causes additional momentum loss. Thus, the incorporation of this source of momentum loss into their model should be promising. This may become important for flows with thicker liquid film. In their theoretical treatment, the actually observed data were used for the values of L and ω so that their treatment is not still completely analytical. To devise a general means for the determination of such wave parameters is required.

VOID FRACTION AND FILM THICKNESS

Void fraction α is defined as the fraction of the tube cross-sectional area which is occupied by the gas-phase on an average and is an important parameter for the calculation of various two-phase quantities, as already seen by the inclusion of α in Equations 13 and 15. It is sometimes called the hold-up.

The homogeneous model assumes the local slip ratio to be unity everywhere (S = 1) so that for the void fraction it gives.

$$\alpha = \frac{\rho_l x}{\rho_g(1-x)+\rho_l x} \tag{53}$$

Sometimes, the drift flux model developed by Zuber and Findlay [26] is used as a tool for the determination of the void fraction. It is called a "model" but is actually a frame of generalized theory paying attention both to the relative velocity between the phases and to the nonuniformity of the flow and the local void fraction. The usual average of a quantity Q is defined as follows:

$$\langle Q \rangle = \frac{1}{A}\int_A Q\, dA \tag{54}$$

Against this, a weighted average of Q can also be introduced:

$$\bar{Q} = \frac{\frac{1}{A}\int_A \tilde{\alpha} Q\, dA}{\frac{1}{A}\int_A \tilde{\alpha}\, dA} \tag{55}$$

where $\tilde{\alpha}$ is the local void fraction. For simplicity, the averaged quantities in the numerator and denominator are expressed respectively as $\langle \tilde{\alpha} Q \rangle$ and $\langle \tilde{\alpha} \rangle$ in the following. The actual average velocities $\bar{U}_g$ for the gas phase introduced before have already been defined following the above general definition of $\bar{Q}$.

The drift velocities w_g and w_l are defined as the differences between the actual local velocities of the phases and the volumetric flux density of the mixture j by the relationships:

$$w_g = U_g - j \tag{56}$$

$$w_l = U_l - j \tag{57}$$

where

$$j = j_g + j_l \tag{58}$$

and

$$j_g = \tilde{\alpha} U_g \tag{59}$$

$$j_l = (1-\tilde{\alpha}) U_l \tag{60}$$

Defining the distribution parameter C_0 by:

$$C_0 = \frac{\langle \tilde{\alpha} j \rangle}{\langle \tilde{\alpha} \rangle \langle j \rangle} \tag{61}$$

the slip ratio S is described as follows:

$$S = \frac{\bar{U}_g}{\bar{U}_l} = \frac{\dfrac{\langle j_g \rangle}{\langle \tilde{\alpha} \rangle}}{\dfrac{\langle j_l \rangle}{\langle 1-\tilde{\alpha} \rangle}} = \frac{1-\langle \tilde{\alpha} \rangle}{\dfrac{1}{C_0 + \dfrac{\langle \tilde{\alpha} w_g \rangle}{\langle \tilde{\alpha} \rangle \langle j \rangle}} - \langle \tilde{\alpha} \rangle} = \frac{1-\alpha}{\dfrac{1}{C_0 + \dfrac{\langle \tilde{\alpha} w_g \rangle}{\alpha \langle j \rangle}} - \alpha} \tag{62}$$

where $\alpha = \langle \tilde{\alpha} \rangle$ has been used. Incidentally, when $w_g = 0$ everywhere, the previous relationship reduces to:

$$S = \frac{1-\alpha}{1/C_0 - \alpha} \tag{63}$$

which corresponds to the equation for the velocity slip ratio given originally by Bankoff [27].

Furthermore, the following relationship holds.

$$\langle j \rangle = \frac{m}{\rho_g \rho_l} [\rho_l x + \rho_g (1-x)] \tag{64}$$

Combining Equations 62 and 64 and using another form of the slip ratio definition by

$$S = \frac{\rho_l x (1-\alpha)}{\rho_g (1-x) \alpha} \tag{65}$$

the void fraction may be written as follows for the case of uniform distribution of w_g:

$$\alpha = \frac{\rho_l x}{C_0[\rho_l x + \rho_g(1-x)] + \rho_g \rho_l w_g / m} \tag{66}$$

The distribution parameter C_0 and the gas-phase drift velocity w_g are functions of flow pattern and other flow situation parameters. But their general expressions have been studied rather scarcely. For annular flow, C_0 takes a value close to unity [25] and w_g is not so large, at least, in the core region. Thus, for the annular flow, the value of α given by Equation 66 may roughly be estimated from Equation 53. Further discussion of this drift flux model has been made by a number of authors including Hewitt [4], Ishii [28, 29], and Rouhani [30].

An empirical relationship for the void fraction has also been presented in a graphical form by Lockhart and Martinelli in terms of their parameter X defined by Equation 36. Their result is reproduced in Figure 7. From a similar view point for the frictional pressure gradient, Martinelli and Nelson [13] introduce a pressure correction factor to the graphical relationship by Lockhart and Martinelli. Although this relationship was derived for steam-water two-phase flow, it has also been applied to the two-phase flow of other fluids.

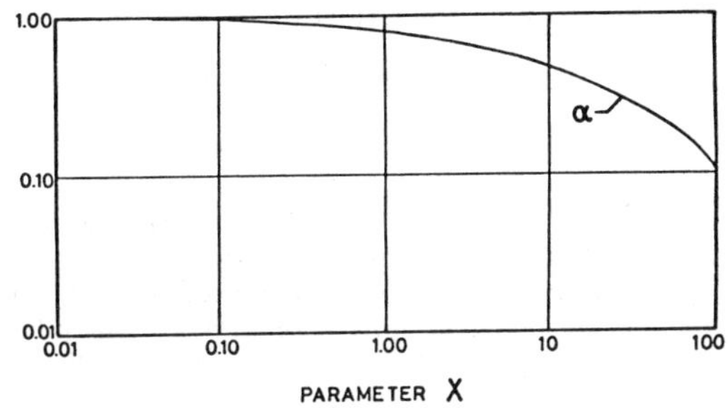

Figure 7. Void fraction [12].

This correlation is widely used but, as already mentioned, it cannot account for the effect of mass flux in an adequate way.

Some alternative methods have been researched, and Premoli et al. [31] present the correlation for the slip ratio taking account of the mass flux effect for wide range of data. Their correlation is

$$S = 1+E_1\left(\frac{\eta}{1+\eta E_2}-\eta E_2\right)^{1/2} \tag{67}$$

where η is a function of volumetric flow rate ratio ζ and is defined by:

$$\eta = \frac{\zeta}{1-\zeta}$$

$$\zeta = \frac{\dfrac{m_g}{\rho_g}}{\dfrac{m_g}{\rho_g}+\dfrac{m_l}{\rho_l}} = \frac{\alpha S}{\alpha S+(1-\alpha)}$$

and

$$E_1 = 1.578\ Re_{l0}^{-0.19}\left(\frac{\rho_l}{\rho_g}\right)^{0.22}$$

$$E_2 = 0.0273\ We\ Re_{l0}^{-0.51}\left(\frac{\rho_l}{\rho_g}\right)^{-0.08}$$

The Weber number We is defined by:

$$We = \frac{Dm^2}{\sigma\rho_l} \tag{68}$$

For annular flow without a significant amount of entrained droplets, the contribution from droplets flowing in the gas-phase core to the void fraction may be ignored with reasonable accuracy. Under this treatment, the void fraction can be related to the averaged film thickness δ as follows:

$$\alpha = \frac{(D-2\delta)^2}{D^2} \tag{69}$$

For thin film of $\delta/D \ll 1$, this may be approximated by

$$\alpha = \frac{D-4\delta}{D}$$

Thus, the methods for the calculation of film thickness are basically equivalent to the method for the evaluation of α. If the local liquid flow rate in the film is known, several methods are available for the calculation of δ [2, 3, 4, 19].

If the effects of flow acceleration in a tube can be ignored, the momentum equations for the liquid flow in the film and the gas-phase flow in the tube core may be written as follows:

$D/2 \geqq r \geqq r_i$;

$$0 = -\frac{dp}{dz} + \frac{1}{r}\frac{\partial}{\partial r}(r\tau_l) - \rho_l g \sin\gamma \tag{70}$$

$r_i \geqq r \geqq 0$;

$$0 = -\frac{dp}{dz} + \frac{1}{r}\frac{\partial}{\partial r}(r\tau_g) - \rho_g g \sin\gamma \tag{71}$$

where r is the radial coordinate, τ the local shear stress acting on the fluid element, and z is measured upward. Integration of Equation 70 once with respect to r gives the following differential equation for U_l:

$$r\mu_e\frac{dU_l}{dr} = -\frac{1}{2}\left(\frac{D^2}{4} - r^2\right)\frac{dp}{dz} + \frac{D}{2}\tau_w - \frac{1}{2}\left(\frac{D^2}{4} - r^2\right)\rho_l g \sin\gamma \tag{72}$$

where τ_w is the wall shear stress acting on the tube surface and the effective viscosity μ_e has been introduced:

$$\tau_l = \mu_e\frac{dU_l}{dr} \tag{73}$$

When the liquid film remains to be in laminar state, μ_e is just equal to the liquid viscosity μ_l. When the liquid film flow is turbulent, μ_e can be related to the local velocity distribution or its parameter following Dukler [32] as follows:

For $\dfrac{\rho_l u_\tau y}{\mu_l} < 20$

$$\mu_e = \mu_l + \rho_l n^2 U_l y\left[1 - \exp\left(-\frac{\rho_l n^2 U_l y}{\mu_l}\right)\right] \tag{74}$$

For $\dfrac{\rho_l u_\tau y}{\mu_l} > 20$

$$\mu_e = \frac{\kappa(dU_l/dy)^3}{(d^2U_l/dy^2} \tag{75}$$

where u_τ is the friction velocity $(\sqrt{\tau_w/\rho_l})$, y the distance from the tube wall, and n and κ the constants.

Equations 74 and 75 are respectively the Deissler expression and the von Kármán expression. They may be replaced by other alternative expressions of μ_e found in textbooks on fluid mechanics (e.g., Schlichting [33]).

Combining the integrated forms of Equation 70 and 71 and taking account of the interfacial shear stress matching condition $\tau_l = \tau_g = \tau_i$ at $r = D/2 - \delta$, the following relationship is obtained.

$$\frac{dp}{dz} = \frac{4\tau_w}{D} - [(1-\alpha)\rho_l + \alpha\rho_g]g \sin\gamma \tag{76}$$

The second term on the right-hand side of Equation 76 is the gravitational pressure gradient as seen from Equation 13. Then, it follows:

$$\frac{4\tau_w}{dz} = \left(\frac{dp}{dz}\right)_f \tag{77}$$

which can be calculated with a method mentioned earlier. Using this as a boundary condition for the velocity gradient at $r = D/2$, the velocity distribution inside the liquid film can be calculated. The liquid flow rate inside the film m_f is related to the integrated value of U_l as follows:

$$m_f = \int_{D/2-\delta}^{D/2} 2\pi\rho_l U_l r \, dr \tag{78}$$

If m_f is known, Equation 78 is used as a relationship to determine the liquid film thickness δ.

There is an alternative method using τ_i as a boundary condition for determining U_l. Integrating Equation 70 with respect to r from $D/2-\delta$ to $D/2$, the following relationship results in:

$$(1-\alpha)\frac{dp}{dz} = \frac{4\tau_w}{D} - \frac{4(D-2\delta)}{D^2}\tau_i - \rho_l g(1-\alpha)\sin\gamma \tag{79}$$

Combining this and Equation 76, the following result is obtained:

$$\frac{4\tau_i}{D} = \sqrt{\alpha}\left(\frac{dp}{dz}\right)_f - \sqrt{\alpha}\,(1-\alpha)\,(\rho_l - \rho_g)g \sin\gamma \tag{80}$$

For very thin film, the value of α on the right-hand side of the equation can be approximated by unity. Then, it follows

$$\frac{4\tau_i}{D} \cong \left(\frac{dp}{dz}\right)_f \tag{81}$$

If this relationship is used for determining the velocity gradient at the interface instead of giving the velocity gradient at $r = D/2$, the liquid velocity can also be calculated with a method similar to the preceding. However, if the approximation $\alpha \cong 1$ is not good, Equation 80 must be used. In that case, the calculation must be done with an iterative procedure because Equation 80 includes α which becomes known only after the value of δ has been determined.

Suzuki et al. [8] claims that the interfacial momentum transfer due to wave-induced motion of the fluids may be important for a thin liquid film. This leads them to add the wave effect viscosity $\tilde{\mu}_l$ given by Equation 51 into the expression of μ_e. The calculated results of δ for a wavy laminar liquid flow are a little overestimated but are generally reasonable.

As already mentioned some portion of liquid flow is entrained into the gas-phase flow in the form of droplets. Therefore, even if the local liquid flow rate m_l is known, the value of m_f in Equation 78 is still unknown. To make this known, we have to discuss the entrainment of droplets.

The entrainment of droplets can occur through several different mechanisms and causes a larger pressure loss in a tube. Ishii and Grolmes [34] studied a model for the shearing-off entrainment from role waves by the high-velocity gas-phase flow and presented a criterion for the inception of droplet entrainment. Unknown coefficients introduced into the criterion were determined referring to the data reported for different geometries of the tube and different flow direction, including their own data, and those by van Rossum [35], Ueda and Tanaka [36], Yablonik and Khaimov [37] and Zhivaikin [38].Their final comparison is cited in Figure 8. When $Re_f > 1{,}635$, the critical gas velocity is proposed to be independent of the value of Re_f. The criterion for $Re_f < 1{,}635$ shown by the solid line can be expressed as follows.

$$\frac{\mu_l W_g}{\sigma}\sqrt{\frac{\rho_g}{\rho_l}} = 11.78 N_\mu^{0.8} Re_l^{-1/3} \tag{82}$$

where N_μ is the viscosity number

$$N_\mu = \frac{\mu_l}{\left(\rho_l \sigma \sqrt{\dfrac{\sigma}{g\Delta\rho}}\right)^{1/2}} \tag{83}$$

and $\Delta\rho$ is the density difference between the phases and σ the surface tension. The inception criterion by Ishii and Grolmes are discussed also by Ishii and Kataoka [39] and Ishii [28].

The entrainment rate of droplets is also discussed by Ishii and Mishima [40], and they proposed the following correlation function for the droplet entrainment fraction E in the region far from the tube entrance.

$$E = \tanh\left(7.25 \times 10^{-7} j_g^{*\,5/2} D^{*5/4} Re_l^{1/4}\right) \tag{84}$$

where

$$E = (1 - m_f/m_l)$$

$$j_g^* = \frac{W_g}{\left[\dfrac{\sigma g \Delta\rho}{\rho_g^2}\left(\dfrac{\rho_g}{\Delta\rho}\right)^{2/3}\right]^{1/4}} \tag{85}$$

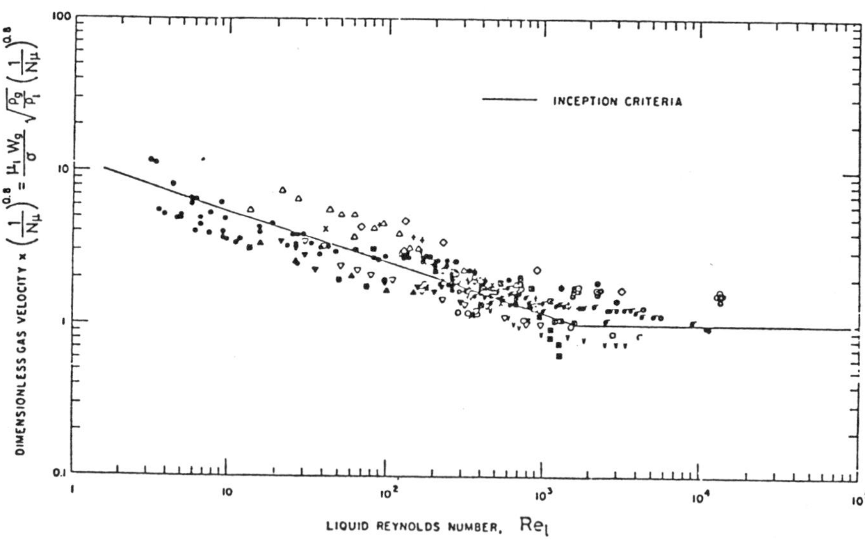

Figure 8. Inception criterion for droplet entrainment [34].

$$D^* = d_h \sqrt{\frac{g\Delta\rho}{\sigma}} \tag{86}$$

and d_h is the hydraulic diameter of the tube. Equation 84 holds for the axial region of nondimensional distance Z^* from the tube entrance

$$Z^* = 600\, d_h \sqrt{\frac{j_g^*}{Re_l}} \tag{87}$$

Equation 84 is compared with some experimental data in Figure 9. Incidentally, from the definition of the entrainment fraction E, the following relationship holds between Re_l, introduced in Equations 84 and 87 and defined by Equation 28, and Re_f defined by Equation 1.

$$Re_f = (1-E)\, Re_l \tag{88}$$

Discussions about the void fraction and the entrainment rate are also given by Hewitt [19, 41].

SINGLE-COMPONENT HEAT TRANSFER

In single-component two-phase flow heat transfer, interfacial heat transfer resistance may be negligibly small in usual circumstances. Thus, the heat transfer resistance in the liquid film is an important factor in that case.

Boiling Heat Transfer

For boiling heat transfer, the total tube wall heat flux q_w may be separated into two parts, the nucleate boiling heat flux q_{NB} and the convective heat flux q_c:

$$q_w = q_{NB} + q_c = h_{NB}\Delta T_{sat} + h_c \Delta T_b \tag{89}$$

where h_{NB} and h_c = the nucleate boiling heat transfer coefficient and the convective heat transfer coefficient, respectively and ΔT_{sat} the wall superheat or the temperature difference between the wall temperature T_w and the saturation temperature T_s given by:

$$\Delta T_{sat} = T_w - T_s \tag{90}$$

and ΔT_b is the difference between T_w and the fluid bulk mean temperature T_b defined by:

$$\Delta T_b = T_w - T_b \tag{91}$$

Except for subcooled boiling, $\Delta T_{sat} = \Delta T_b$ so that Chen [42] reduced the following relationship:

$$h = h_{NB} + h_c \tag{92}$$

where h is the coefficient of two-phase heat transfer with boiling for a high quality flow or for the annular two-phase flow. Heat transfer characteristics generally depend on the tube position (horizontal or vertical). In the following, the forced-convective boiling heat transfer in a vertical tube is mentioned first.

Figure 10 shows the two-phase heat transfer coefficient for steam-water annular flow given by Collier et al. [43]. The abscissa of the figure is the reciprocal of the Martinelli parameter X_{tt} for a turbulent-turbulent case, and from the left to the right, the quality increases. At low quality, the two-phase heat transfer coefficient depends on the heat flux q_w. This suggests that boiling is the important mechanism of heat transfer at low quality.

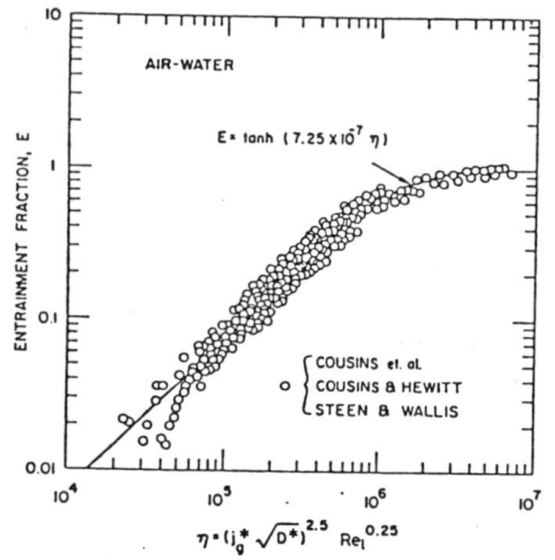

Figure 9. Correlation for entrainment fraction [40].

At high quality, however, the two-phase heat transfer coefficient solely depends on the Martinelli parameter and not on the heat flux. In this high quality condition, therefore, the convective heat transfer is believed to be predominant. This is because, on one hand, the higher vapor velocity brings the liquid velocity higher and the film thickness thinner so that the convective heat transfer resistance inside the film is brought smaller, and on the other hand the lower wall temperature caused by a higher convective heat transfer rate reduces the effective number of bubble nucleation sites.

The convective heat transfer coefficient h_c is usually expressed in the following form:

$$\frac{h_c}{h_{cl}} = F \tag{93}$$

where h_{cl} is the heat transfer coefficient to be obtained if the liquid is flowing alone in the tube and F accounts for the effect of the existing vapor flow. The correlation shown by the solid line running from the left bottom to the right top in Figure 10 for high quality region actually corresponds to F, so that the result of Collier et al. [43] shows

$$F = 2.9\left(\frac{1}{X_{tt}}\right)^{0.66} \tag{94}$$

at high quality. Chen suggested that F should be a flow parameter so that it is a function of the Martinelli parameter [42]. His correlation for F in a graphical form is reproduced in Figure 11. In his correlation, h_{cl} was evaluated with the Dittus–Boelter equation:

$$h_{cl} = 0.023\, Pr_l^{0.4}\, Re_l^{0.8}\, \frac{\lambda_l}{D} \tag{95}$$

where λ_l and Pr_l are the thermal conductivity and the Prandtl number of liquid, respectively. Bennett and Chen [44] extended the expressions of F to take into account the effect of the fluid Prandtl number. One of their correlation is:

$$F = Pr_l^{0.296}\varphi_l^{0.888} \tag{96}$$

which may be valid for the range of the Prandtl number from 1 to 6.

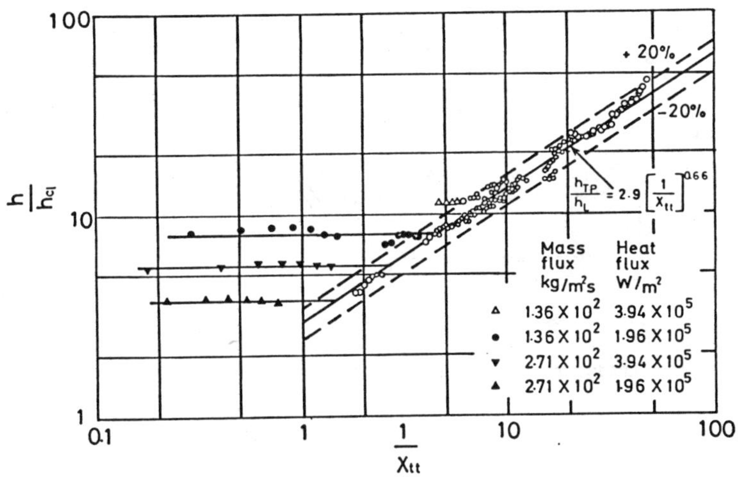

Figure 10. Variation of heat transfer coefficient ratio with Martinelli parameter [43].

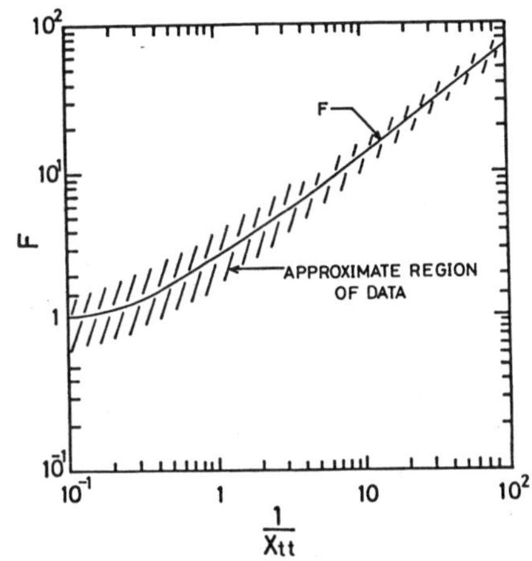

Figure 11. Convective heat transfer ratio [42].

Chen [42] introduced a suppression factor S_n of nucleate boiling in the expression of h_{NB} which is given by:

$$h_{NB} = S_n h_p \tag{97}$$

where h_p is the nucleate pool boiling heat transfer coefficient. Chen himself used the modified form of the Forster–Zuber equation [45] for h_p. But other correlations may be used as well. Chen also presented a correlation for the S_n factor in a graphical form which is shown in Figure 12. The abscissa of the figure is the two-phase Reynolds number. Inclusion of F in Re indicates that the suppression of nucleation is remarkable at high-quality or at thin-liquid film.

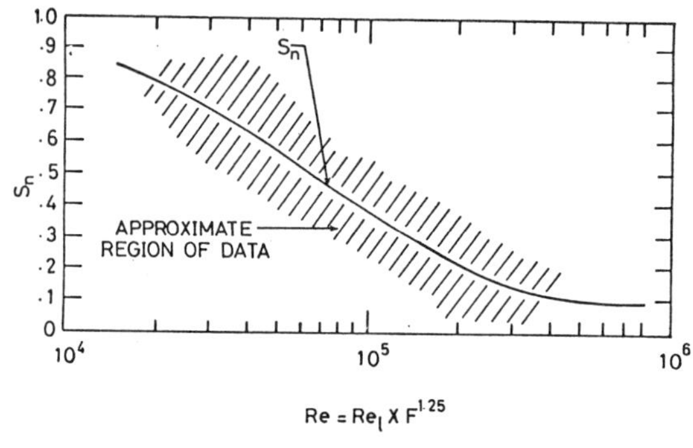

Figure 12. Suppression factor of nucleate boiling S [42].

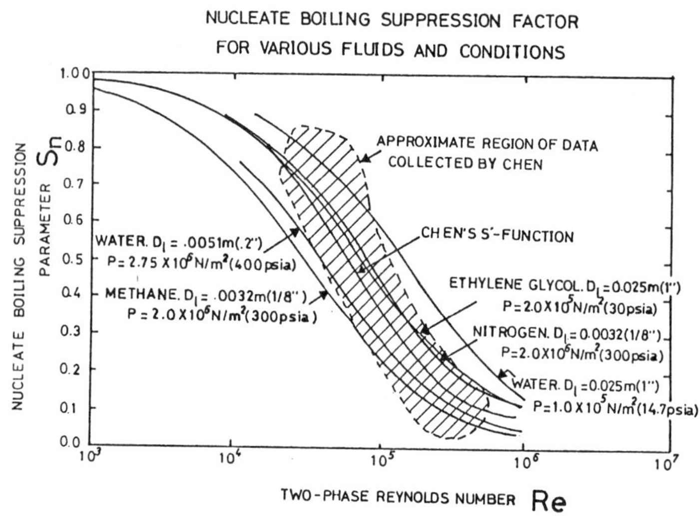

Figure 13. Suppression factor of nucleate boiling S [46].

Bennett et al. [46] derived the expression for S_n analytically, and it is given by:

$$S_n = \frac{\lambda_l}{h_{cl} F \delta_b}\left[1 - \exp\left(-\frac{h_{cl} F \delta_b}{\lambda_l}\right)\right] \tag{98}$$

where

$$\delta_b = 0.041\left[\frac{\sigma}{g\,(\rho_l - \rho_g)}\right]^{1/2}$$

They found that Equation 98 agrees well with the graphical presentation of S_n given by Chen. Their results are shown in Figure 13.

Other forms of the h correlation were proposed in many papers including those by Dengler and Addoms [47] and Guerrieri and Talty [48]. These are tabulated and reviewed by Rohsenow [49]. However, Chen's correlation is often used and may be best. Kihara et al. [50] checked the validity of his correlation and found very good agreement with their own experimental data of Freon-11.

In a horizontal tube, stratification of the flow is likely to occur at a low mass flow rate and, even at a high mass flow rate condition corresponding to the annular flow regime, peripheral nonuniformity of film thickness is often observed. Although the latter effect has not been clarified well enough, many correlations for heat transfer are proposed.

For refrigerant flows in horizontal tubes, many correlations of the following simple form have been proposed.

$$h = Am^{\alpha} q_w^{\beta} D^{\gamma} \tag{99}$$

Comparison of the values of α, β, and γ proposed by different authors are made by Butterworth and Shock [51] and large discrepancies can be found among the tabulated values. Slipcevic [52] proposed two sets of values for exponents α, β, and γ; one for a low mass flux condition and another for a high mass flux condition. For the low mass flux condition, $\beta = 0.7$ so that nucleate boiling was believed to be dominant. For the high mass flux condition, $\beta = 0$ so that the convective heat transfer must have been predominant. No explanation is given explicitly for flow pattern or even for stratification. Moreover, the proposed value 1.4 for α may be unreasonably large.

Figure 14 shows the experimental data of the two-phase heat transfer coefficient obtained by Bandel and Schlünder [53] for horizontal tube flow. Their results in the annular flow regime show that $h \propto m^{0.8}$ and the dependency of h on m is equivalent to that for single-phase turbulent flow heat transfer. They correlated the heat transfer data including some with boiling in the following form:

$$\frac{h}{h_{cl}} = C\left[\frac{\left(\frac{dp}{dz}\right)_f}{\left(\frac{dp}{dz}\right)_{fl}}\right]^{0.429} \tag{100}$$

Dembi et al. [54] compared several correlations including those of Chen [42] for vertical flows [55, 56]. They claim that inclusion of the Weber number in the correlation is important

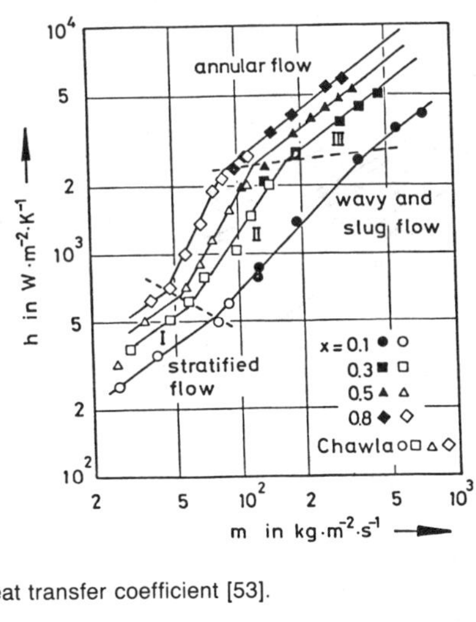

Figure 14. Two-phase heat transfer coefficient [53].

for horizontal flow heat transfer. They also concluded that the correlation by Bandel and Schlünder [53] and the graphic chart given by Shah [56] are good. Figure 15 shows the chart given by Shah. Abscissa C_0' is the convection number defined as:

$$C_0' = \left(\frac{1-x}{x}\right)^{0.8} \left(\frac{\rho_g}{\rho_l}\right)^{0.5} \tag{101}$$

and the parameter B_0 in the figure is the boiling number:

$$B_0 = \frac{q_w}{mH} \tag{102}$$

where H is the latent heat of vaporization. Another parameter Fr appearing for horizontal tube heat transfer is the Froude number $Fr = m^2/\rho^2 gD$.

Condensation Heat Transfer

Since it is impossible to maintain cooling surfaces in a state that maintains dropwise condensation for long periods of time, film condensation is of practical importance. This means the flow pattern more or less like the annular flow is formed in a case of vapor condensation inside a tube. The basis of theoretical approaches of film condensation has been established by Nusselt [57].

For film condensation on a vertical flat plate, the film thickness of condensate increases downward as shown in Figure 16. Ignoring the interfacial heat transfer resistance mentioned already and the convective transport of heat inside the film, the temperature distribution inside the film is given by a linear relationship:

$$\frac{T_s - T}{T_s - T_w} = 1 - \frac{y}{\delta} \tag{103}$$

where T_s = the vapor saturation temperature or the temperature of the film surface
T_w = the wall temperature
δ = the film thickness
y = the distance from the wall surface

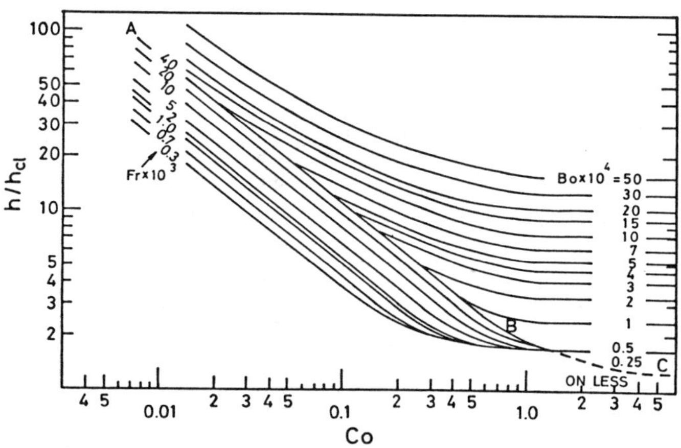

Figure 15. The chart correlation of two-phase heat transfer coefficient for annular flow [56].

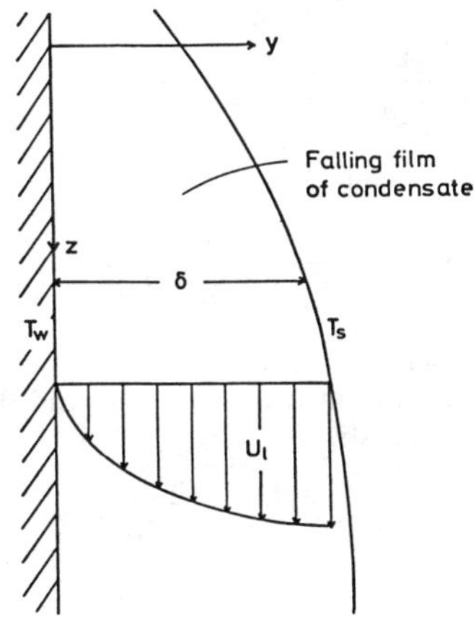

Figure 16. Film condensation on a vertical flat plate.

Then the wall heat flux is given by:

$$q_w = \frac{\lambda_l}{\delta}(T_s - T_w) \tag{104}$$

and the heat transfer coefficient by

$$h = \frac{\lambda_l}{\delta} \tag{105}$$

Considering that the vapor is stagnant, the static pressure gradient becomes:

$$\frac{dp}{dz} = \rho_v g \tag{106}$$

where ρ_v is the vapor density and z is measured downward. Thus, the equation of motion for condensate can be written as follows:

$$0 = \frac{d\tau_l}{dy} + (\rho_l - \rho_v)g \tag{107}$$

Assuming the laminar flow of condensate and the zero shear stress at the interface, the following relationship can be derived:

$$\mu_l \frac{d\,U_l}{dy} = g(\rho_l - \rho_v)(\delta - y) \tag{108}$$

With the boundary condition $U_l = 0$ at $y = 0$, the prior equation integrates to:

$$U_l = \frac{g(\rho_l - \rho_v)}{\mu_l}\left(\delta y - \frac{y^2}{2}\right) \tag{109}$$

Then, the condensate mass flow rate per unit spanwise width of film is given by:

$$\Gamma = \int_0^\delta \rho_l U_l\,dy = \frac{g\rho_l(\rho_l - \rho_v)\,\delta^3}{3\,\mu_l} \tag{110}$$

or

$$\frac{d\Gamma}{dz} = \frac{g\rho_l(\rho_l - \rho_v)\,\delta^2}{\mu_l}\frac{d\delta}{dz}$$

The wall heat flux q_w is related to $d\Gamma/dz$ as follows:

$$\frac{\lambda_l}{\delta}(T_s - T_w) = \frac{d\Gamma}{dz}\left[H + \frac{1}{\Gamma}\int_0^\delta \rho_l U_l C_l (T_s - T)\,dy\right] \tag{111}$$

where C_l is the specific heat of condensate. Substituting Equations 103 and 109 into the integral of Equation 111 and Equation 110 into $d\Gamma/dz$, the following relationship is obtained:

$$\delta^3 d\delta = \frac{\lambda_l \mu_l (T_s - T_w)}{g\rho_l(\rho_l - \rho_v)H'}\,dz \tag{112}$$

where

$$H' = H + \frac{3}{8}C_l(T_s - T_w) \tag{113}$$

Rohsenow [58] refined this expression taking into account nonlinear temperature distribution:

$$H' = H + 0.68\,C_l(T_s - T_w) \tag{114}$$

Solving Equation 112 with a condition $\delta = 0$ at $z = 0$, δ can be obtained:

$$\delta = \sqrt[4]{\frac{4\,\lambda_l \mu_l z (T_s - T_w)}{g\rho_l(\rho_l - \rho_v)H'}} \tag{115}$$

Thus, the local heat transfer coefficient becomes:

$$h = \sqrt[4]{\frac{g\rho_l(\rho_l - \rho_v)\lambda_l^3 H'}{4\,z\mu_l(T_s - T_w)}} \tag{116}$$

and the averaged heat transfer coefficient $\bar{h}$ between $z = 0$ and l is:

$$\bar{h} = 0.943\sqrt[4]{\frac{g\rho_l(\rho_l - \rho_v)\lambda_l^3 H'}{\mu_l(T_s - T_w)\,l}} \tag{117}$$

Equations 116 and 117 can be written in the following alternative forms:

$$\frac{h}{\lambda_l}\left[\frac{\mu_l^2}{\rho_l g(\rho_l - \rho_v)}\right]^{1/3} = 1.10\,Re_f^{-1/3} \tag{118}$$

$$\frac{\bar{h}}{\lambda_l}\left[\frac{\mu_l^2}{\rho_l g(\rho_l - \rho_v)}\right]^{1/3} = 1.47\,Re^{-1/3} \tag{119}$$

where the condensate film Reynolds number Re_f is given by:

$$Re_f = \frac{4\Gamma}{\mu_l} \tag{120}$$

When the subcooling of the condensate is considered to have been taken into account already in Equation 111, the influence of variable fluid properties must be accounted for.

This is usually done by introducing an effective film temperature T_{fm}:

$$T_{fm} = T_w + \chi(T_s - T_w) \tag{121}$$

Mincowycz and Sparrow [59] proposed 0.31 for χ.

The preceding simple analysis originating from Nusselt agrees well with the experimentally observed value of the heat transfer coefficient only at low Reynolds numbers, $Re_f < 10$. At higher Reynolds numbers, the assumption that the film flow is in laminar state may not be accurate enough. Colburn [60] attempted to take into account the effect of turbulence. His result reads:

$$\frac{h}{\lambda_l}\left[\frac{\mu_l^2}{\rho_l g(\rho_l - \rho_v)}\right]^{1/3} = 0.056\, Re_f^{0.2} Pr_l^{1/3} \tag{122}$$

He assumed that the film flow becomes turbulent when $Re_f > 2{,}000$. However, it is well known that the observed value of h is remarkably larger than the predicted value with a Nusselt-type analysis in the middle range of Reynolds numbers $10 < Re_f < 2{,}000$. This may be attributed partially to the effect of waves appearing on the interface [61]. Hirshburg and Florschuetz [61] treated this effect basically with a linear theory for waves and Suzuki et al. [63] with a wave effect model proposed in [8].

Condensation occurring in a tube differs from that on a vertical flat plate mentioned previously in a point that the vapor comes into the tube with some magnitude of velocity. This exerts the shear stress τ_l at the vapor-condensate interface. Due to Rohsenow et al. [64], the following relationship holds for laminar condensate film in downward flow instead of Equation 114 for $\tau_i = 0$.

$$\left[\frac{4\mu_l\lambda_l z(T_s - T_w)}{g\rho_l(\rho_l - \rho_v^*)H'}\right] = \delta^4 + \frac{4}{3}\left[\frac{\tau_i\delta^3}{(\rho_l - \rho_v^*)g}\right] \tag{123}$$

where ρ_v^* is the fictitious density of vapor defined by:

$$\left(\frac{dp}{dz}\right) = \rho_v^* g \tag{124}$$

ρ_v^* is equal to the actual vapor density ρ_v only when the vapor flow is stagnant. Otherwise, the pressure gradient is not solely determined from the vapor static head. Equation 124 clearly shows that δ becomes smaller at a higher shear rate. This leads to lower heat transfer resistance inside the film and hence to a higher heat transfer coefficient.

Figure 17 shows the theoretical results obtained by Dukler [32]. He assumed that the condensate velocity profile in the film is equal to that of single-phase flow at the same Reynolds number in a tube, and introduced the eddy viscosity defined by Equations 74 and 75. The parameter τ_i^* in the figure is the nondimensional magnitude of the interfacial shear stress τ_i. Rohsenow et al. [64] executed similar analysis before. They also derived a relationship for the dependency of the critical Reynolds number on the interfacial shear stress.

Recently, Blangetti [65] and Blangetti and Schlünder [66] carried out experimental and analytical works on condensation in a vertical tube. They took into account the suppression of turbulence in the liquid at the interface [67]. Their analytical results show the dependence

$$h \propto Re_f^{0.4} \tag{125}$$

They agree with their experimental data in the Reynolds number range of $Re_f > 900$ and and equivalent in form to the recommended curve by McAdams [68] for $\bar{h}$. They also suggests that the heat transfer at a lower value of Re_f may occur under the effect of interfacial waves.

As mentioned in the first part of this section, film condensation is important in a practical sense. This may also be applied to the condensation in horizontal tubes. This is different

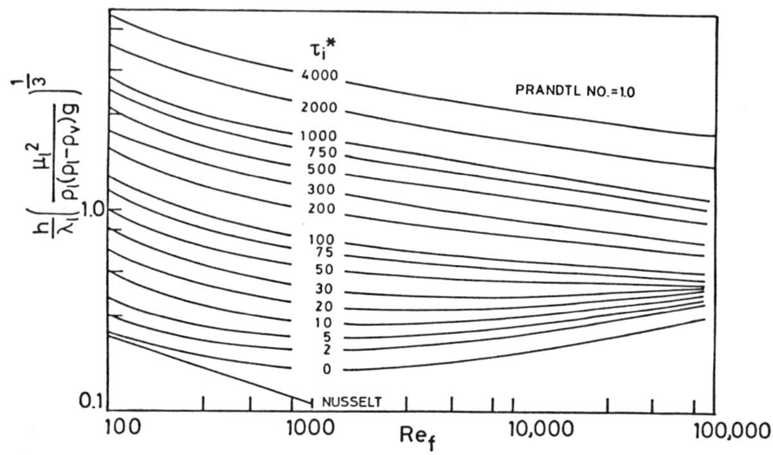

Figure 17. Local film-wise condensation heat transfer coefficient for condensate Prandtl number $Pr_1 = 1.0$ [32].

from the boiling heat transfer so dry-patches are unlikely to occur on the upper inner surface of a horizontal tube in the case of condensation. A part of the film formed there is likely to run down toward the lower part of the tube inner surface. Therfore, at a small vapor velocity or at a low interfacial shear stress, stratification of the flow may certainly occur. However, at a large vapor velocity or at a high interfacial shear stress, the effect of flow stratification on condensation heat transfer may be neglected. Thus, the relationships mentioned previously for the vertical tube may approximately be used for condensation in horizontal tubes at high vapor velocity.

TWO-COMPONENT HEAT TRANSFER

In single-component heat transfer, the interfacial heat transfer resistance is not important in usual situations as already mentioned. Against this, the gas-side heat transfer resistance is crucial in two-component heat transfer problems. Therefore, in two-component heat transfer problems, it is not enough to treat the flow and heat transport inside the liquid film only, but is required to treat the flow, heat, and mass transport in gas-phase flow simultaneously. In this section, attention will pe paid only to a case when evaporation proceeds at the interface at moderate heat flux. More detailed discussions of the same matter are given by Michiyoshi [68]. For condensation of a vapor mixture and boiling of a binary or multicomponent mixture, see References 69, 70, and 71.

For evaporative heat transfer in a two-phase two-component annular flow, several experiments have been conducted by Groothuis and Hendal [72], Kudirka et al. [73], Fried [74], Johnson and Abou–Sabe [75], and Pletcher and McManus [76]. Pletcher and McManus introduced an equilibrium temperature T_m to correlate the heat transfer data obtained within a horizontal tube. The local heat transfer coefficient defined by them is given by:

$$h = \frac{q_w}{T_w - T_m} \tag{126}$$

where the equilibrium temperature T_m may be calculated from the following heat balance equation if the change in flow kinetic energy is neglected:

$$\frac{4zq_w}{D} = m_{lz}(i_{lz}-i_{l0})+m_a(i_{az}-i_{a0}) + m_{v0}(i_{vz}-i_{v0})+\Delta m_v(i_{vz}-i_{l0}) \tag{127}$$

i_{lz}, i_{az}, and i_{vz} in the preceding equation are the enthalpy at the temperature T_m to be defined at a position z for liquid, noncondensable gas, and vapour, respectively. m_a is the mass flux of noncondensable gas, m_{lz} the local mass flux of liquid, m_{v0} the initial mass flux of vapor at the inlet to the tube z = 0, and Δm_v is the mean evaporating flux averaged between the position z and the inlet. i_{l0}, i_{a0}, and i_{v0} are the enthalpy at the inlet for the liquid, noncondensable gas, and vapor, respectively. Pletcher and McManus correlated the local heat transfer coefficient h and the averaged heat transfer coefficient $\bar{h}$ in the form divided by φ_l defined by Equation 18 against the Lockhart–Martinelli parameter X. They also show that the data by Johnson and Abou–Sabe [75] and Fried [74] group near the extended line of their correlation, although the data given by themselves are for the annular flow and the latter for the flow with lower quality.

A theoretical treatment for the same problem has been given by another paper by Pletcher and McManus [77]. Their analysis did not take into account the effect of gravity force so that it can also be applied to vertical annular flow at high shear rate. Their analysis assumes that the vapor is in saturated condition both at the interface and in the core region of a tube. They also uses the similarity between the heat transport and mass transport in gas-phase flow. The computed value of h/φ_l shows the same dependency on the parameter X as their experimental data but is higher than the latter by about 50%.

Recently, Vijay et al, [78] applied the Fried correlation of the heat transfer coefficient [74] to a vertical two-phase, two-component annular flow and proposed the following relationship for the annular flow regime:

$$\frac{h}{h_{cl}} = (\varphi_l^2)^{0.455} \tag{128}$$

Their correlation agrees rather well with the experimental data.

The previous forms of correlations are simple in form and convenient to use. However, a caution has been given by Suzuki et al. [8] about the usage of such correlations because they do not include the effect of temperature level. It is well established that the total heat flux at the interface q_i is given by the sum of the interfacial conduction heat flux q_{ci} and the removal rate of the latent heat of vaporization. The latter is given by the product of the latent heat of vaporization H and the evaporation flux $\dot{m}_v$. Thus,

$$q_i = q_{ci}+H\dot{m}_v \tag{129}$$

where the inward heat transport is taken to be positive. Dividing Equation 129 by a representative temperature difference ΔT, we have

$$\frac{q_i}{\Delta T} = \frac{q_{ci}}{\Delta T} + \frac{H\dot{m}_v}{\Delta T} \tag{130}$$

The left-hand side corresponds to the interfacial heat transfer coefficient. In a case when the heat transfer resistance inside the liquid film can be ignored, $q_i = q_w$. Then, the left-hand side of the preceding equation becomes identical with the overall heat transfer coefficient h. The first term on the right-hand side is the pure convective heat transfer coefficient h_c and the second term the heat transfer coefficient relating to the latent heat removal h_m. Then, in alternative form,

$$h = h_c+h_m \tag{131}$$

The value of $\dot{m}_v$ depends on the mass transfer rate of vapor, therefore, on the vapor pressure at the interface. The vapor saturation pressure is a nonlinear function of temperature so

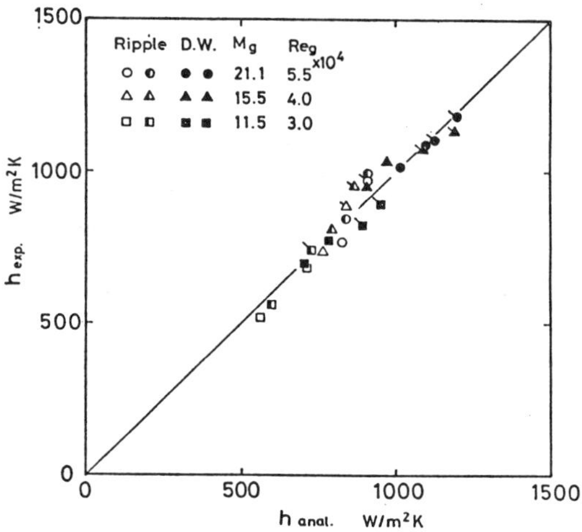

Figure 18. Comparison of heat transfer coefficient for two-component annular flow [8].

that the ratio between the first term and the second term of Equation 131 can vary with temperature. This indicates that the value of h depends not only on hydrodynamic parameters but also on thermodynamic parameters. The same argument can be applied to the cases when q_i differs from q_w. Suzuki et al. [8] proposed a nondimensional thermodynamic parameter β. It is given by:

$$\beta = \frac{H^2\tilde{M}_v}{\tilde{M}_a Cp_g P T_m V_{vm}} \tag{132}$$

where $\tilde{M}_v$ = the molecular weight of vapor,
$\tilde{M}_a$ = the molecular weight of noncondensable gas,
Cp_g = the specific heat of a mixture at constant pressure,
P = the total pressure,
T_m = the equilibrium temperature introduced by Pletcher and McManus [76],
V_{vm} = the specific volume of vapor at the temperature T_m.

V_{vm} is a sensitive function of temperature so that this parameter may be available for evaluating how the heat transfer coefficient depends on the temperature level. The higher the temperature level, the larger the evaporation effect or the heat transfer coefficient will be. The interfacial heat flux q_i is related to the interfacial conductive heat flux q_{ci} by the following equation with the above parameter β:

$$q_i = (1+\beta)q_{ci} \tag{133}$$

In the light of this equation, the overall heat transfer coefficient h may be expressed by:

$$h = (1+\beta)h_c \tag{134}$$

for such a thin liquid film whose heat transfer resistance can be ignored. Equation 134 may be used as a rule of thumb for the two-component annular flow heat transfer coefficient.

Suzuki et al. conducted a theoretical study of the evaporative heat transfer in a downward annular flow in a vertical tube. They showed that the evaporative heat transfer can rather easily be treated if the assumption that the vapor is in a saturated condition everywhere is adopted. They also showed that the energy and vapor mass transfer equations can be combined into a single equation for the gas temperature, which has the same form as the energy equation for single-phase flow. The parameter β has been introduced in the course of this derivation of the combined equation. The combined equation has been solved for the fully developed condition which is expected to exist at the low wall heat flux. Equation 133 was used as a matching condition for the energy equation at the interface between the liquid film and the gas-phase flow dispersed with droplets. They took into account the wave effect on the heat, mass, and momentum transfer by Equations 51 and 52. Their results are shown in Figure 18 in comparison with their experimental data obtained for air-water flow in a vertical tube of 26.4–mm ID [6]. The symbols specified as "Ripple" are for the flow conditions when only the ripple waves are observed on the interface and the symbols specified as "D.W." are for the conditions when the disturbance waves are observed as well. Their method would be used for the cases at low wall heat flux which was assumed in their calculation. Their experimental conditions compared in Figure 18 cover the ranges of the wall heat flux $0.93 \times 10^4\ W/m^2 \leqq q_w \leqq 2.87 \times 10^4\ W/m^2$, the tube wall temperature $313K \leqq T_w \leqq 343°\ K$, and the film Reynolds number $140 \leqq Re_f \leqq 500$.

NOTATION

- A flow cross-sectional area
- A_g cross-sectional area occupied by gas flow
- A_l cross-sectional area occupied by liquid flow
- B_0 boiling number defined by Equation 102
- C_l specific heat of condensate
- C_0' convection number defined by Equation 101
- C_0 coefficient introduced into Equation 61
- D tube diameter
- D^* non-dimensionalized tube hydraulic diameter defined by Equation 86
- d_h tube hydraulic diameter
- E droplet entrainment fraction $\left(=\frac{m_l - m_f}{m}\right)$
- F heat transfer coefficient factor defined by Equation 93
- Fr Froude number $\left(=\frac{m^2}{\rho^2 gD}\right)$
- f Fanning factor of single-phase flow
- f_g two-phase flow Fanning factor introduced in Equation 21
- f_{g0} two-phase flow Fanning factor introduced in Equation 23
- f_i interfacial friction factor introduced in Equation 33
- f_{is} single-phase flow friction factor
- f_l two-phase flow Fanning factor introduced in Equation 22
- f_{l0} two-phase flow Fanning factor introduced in Equation 24
- f_{tp} two-phase flow friction factor used in the homogeneous model and defined by Equation 33
- g gravitational acceleration
- H latent heat of vaporization
- H' apparent latent heat of vaporization defined by Equation 114
- h two-phase flow heat transfer coefficient
- $\bar{h}$ averaged heat transfer coefficient
- h_c convective heat transfer coefficient
- h_{cl} single-phase flow convective heat transfer coefficient
- h_{NB} nucleate boiling heat transfer coefficient
- h_p nucleate pool boiling heat transfer coefficient
- h_m equivalent heat transfer coefficient relating to latent heat removal from the interface $\left(=\frac{H\dot{m}_v}{\Delta T}\right)$
- i local two-phase flow enthalpy
- i_a enthalpy of noncondensable gas
- i_g saturated gas-phase enthalpy
- i_l saturated liquid-phase enthalpy
- i_v condensable gas enthalpy
- J total momentum flux

- j local volumetric flux density of two-phase mixture
- j_g gas phase volumetric flux density
- j_g^* nondimensional gas-phase volumetric flux density defined by Equation 85
- j_l liquid-phase volumetric flux density
- L averaged wave amplitude
- M_f liquid film mass flow rate
- M_g gas-phase mass flow rate
- M_l liquid-phase mass flow rate
- m total mass flux ($= m_l + m_g$)
- m_a mass flux of noncondensable gas
- m_f mass flux of liquid flowing inside film $\left(= \frac{M_f}{A}\right)$
- m_g gas-phase mass flux $\left(= \frac{M_g}{A}\right)$
- m_l liquid-phase mass flux $\left(= \frac{M_l}{A}\right)$
- m_v mass flux of condensable gas
- $\dot{m}_v$ evaporation flux at the interface
- N_μ viscosity number defined by Equation 83
- Pr_l Prandtl number of liquid
- p pressure
- Q arbitrary quantity
- q_c conductive heat flux
- q_{ci} interfacial conductive heat flux
- q_i interfacial total heat flux
- q_{NB} nucleate boiling heat flux
- q_w wall heat flux
- Re single-phase flow Reynolds number
- Re_f liquid film Reynolds number
- Re_g gas-phase flow Reynolds number defined by Equation 27
- Re_{g0} gas-phase flow Reynolds number defined by Equation 29
- Re_l liquid-phase flow Reynolds number defined by Equation 28
- Re_{l0} liquid-phase flow Reynolds number defined by Equation 30
- Re_{tp} two-phase flow Reynolds number defined by Equation 31
- τ_i radial position of interface
- S slip ratio
- S_n suppression factor of nucleate boiling
- T_b bulk mean fluid temperature
- T_{fm} effective film temperature
- T_m equilibrium temperature defined by Equation 127
- T_s vapor saturation temperature
- T_w wall temperature
- U streamwise velocity
- U^+ universal velocity defined by Equation 50
- $\bar{U}_f$ averaged liquid film velocity $\left(= \frac{M_f}{\pi \rho_l D \delta}\right)$
- $\bar{U}_g$ averaged gas-phase velocity ($= M_g/\rho_g A_g$)
- $\bar{U}_l$ averaged liquid-phase velocity ($= M_l/p_l A_l$)
- U_w wave velocity
- U_τ friction velocity
- V_{vm} specific volume of saturated vapor at temperature T_m
- We Weber number defined by Equation 68
- W_g superficial gas velocity
- W_l superficial liquid velocity
- w_g gas-phase drift velocity
- w_l liquid-phase drift velocity
- x vapor quality
- X Lockhart–Martinelli parameter defined by Equation 36
- Y Chisholm parameter defined by Equation 42
- y distance from the wall
- Z^* nondimensional streamwise distance defined by Equation 87
- z Streamwise distance

Greek Symbols

- α mean void fraction
- $\tilde{\alpha}$ local void fraction
- β nondimensional mass transfer effect parameter defined by Equation 132
- Γ condensate mass flow rate per unit film spanwise width
- γ inclination angle of tube raised from the horizontal plane

ΔJ	change of J
Δm_v	change of m_v
ΔP_a	pressure drop due to flow acceleration
ΔT	representative temperature difference
ΔT_b	temperature difference $(= T_w - T_b)$
ΔT_{sat}	wall superheat $(= T_w - T_s)$
$\Delta\rho$	density difference between two phases
δ	liquid film thickness
δ_b	thickness of bubble growth region introduced in Equation 98
δ_{max}	maximum film thickness
δ_{min}	minimum film thickness
ζ	volumetric flow rate ratio
η	function of $\zeta(= \zeta/(1-\zeta))$
λ_l	liquid thermal conductivity
μ_e	effective viscosity
μ_g	gas viscosity
$\tilde{\mu}_g$	gas-phase apparent wave viscosity
μ_l	liquid viscosity
$\tilde{\mu}_l$	liquid-phase apparent wave viscosity
μ_{tp}	two-phase flow viscosity defined by Equation 32
ρ_c	mean density of the fluid flowing in the tube core region defined by Equation 46
ρ_g	density of gas
ρ_l	density of liquid
ρ_{tp}	two-phase flow density defined by Equation 34
ρ_v^*	fictitious density of vapour defined by Equation 24
σ	surface tension
τ_g	shear stress in gas-phase flow
τ_l	shear stress in liquid-phase flow
τ_i	interfacial shear stress
τ_i^*	nondimensionalized interfacial shear stress
τ_w	wall shear stress
φ_g	two-phase flow friction multiplier defined by Equation 17
φ_{g0}	two-phase flow friction multiplier defined by Equation 19
φ_l	two-phase flow friction multiplier defined by Equation 18
φ_{l0}	two-phase flow friction multiplier defined by Equation 20
Ω	mass velocity correction factor for φ_{l0}^2
ω	averaged wave frequency

REFERENCES

1. Delhaye, J. M., "Two-phase Flow Patterns," in *Two-phase Flow and Heat Transfer in The Power and Process Industries* by A. E. Bergles et al., Washington D. C.: Hemisphere, 1981, pp. 1–36.
2. Hewitt, G. F., "Two-phase Flow Patterns and Their Relationship to Two-phase Heat Transfer," *Two-phase Flows and Heat Transfer,* Volume 1, S. Kakaç and F. Mayinger (Eds.), Washington D. C.: Hemisphere, 1977, pp. 11–35.
3. Hewitt, G. F., "Flow Patterns," in *Two-phase Flow and Heat Transfer,* D. Butterworth and G. F. Hewitt (Eds.), Oxford: Oxford University Press, 1977, pp. 18–36.
4. Hewitt, G. F., "Flow Regimes," in *Handbook of Multiphase Systems,* G. Hetsroni (Ed.), Washington D. C.: Hemisphere, 1981, pp. 2.3–2.43.
5. Hagiwara, Y., et al., "A Study on Liquid Film Flow Characteristics in Annular Two-phase Flow," in *Proc. 3rd Multi-phase Flow and Heat Transfer Symposium-Workshop,* T. N. Veziroglu (Ed.), 1983, pp. 43–46.
6. Hagiwara, Y., et al. "Studies on Thin Liquid Film of Annular-mist Two-phase Flow. I. Wave Characteristics and Heat Transfer," *Mem. Faculty Engng Kyoto Univ.,* Vol. 44, No. 2 (April 1982), pp. 309–328.
7. Hagiwara, Y., et al., "Simultaneous Measurements of Wall Shear Stress and Liquid Film Thickness in An Annular Mist Two Phase Flow in A Vertical Circular Tube," in *ASME/JSME Thermal Engineering Joint Conference Proceedings,* Volume 1, Y. Mori and W. J. Yang (Eds.), New York: The American Society fo Mechanical Engineers, 1983, pp. 55–62.

8. Suzuki, K., Hagiwara, Y., and Sato, T., "Heat Transfer anf Flow Characteristics of Two-Phase Two-Component Annular Flow," *Int. J. Heat Mass Transfer,* Vol. 26, No. 4 (1983), pp. 597–605.
9. Friedel, L., "Momentum Exchange and Pressure Drop in Two-phase Flow," in *Two-phase Flows and Heat Transfer,* Volume 1, S. Kakaç and F. Mayinger (Eds.) Washington D. C.: Hemisphere, 1977. pp. 239–312.
10. McAdams, W. H., "Vaporization Inside Horizontal Tubes. Part 2: Benzine-Oil Mixtures," *Trans. ASME,* Vol. 64 (1942), pp. 193–200.
11. Beattie, D. R. H., "A Note on The Calculation of Two-phase Pressure Losses," *Nuclear Engineering and Design,* Vol. 25 (1973), pp. 395–402.
12. Lockhart, R. W., and Martinelli, R. C., "Proposed Correlation of Data for Isothermal Two-phase, Two-Component Flow in Pipes," *Chemical Engineering Progress,* Vol. 45, No. 1 (1949), pp. 39–48.
13. Martinelli, R. C., and Nelson, D. B., "Prediction of Pressure Drop during Forced-Circulation Boiling of Water," *Trans. ASME,* Vol. 70 (1948), pp. 695–702.
14. Chisholm, D., and Laird, A. D. K., "Two-Phase Flow in Rough Tubes," *Trans. ASME,* Vol. 80, No. 2 (1958), pp. 276–283.
15. Chisholm, D., "A Theoretical Basis for The Lockhart–Martinelli Correlation of Data for Isothermal Two-phase, Two-Component Flow in Pipes," *Int. J. Heat Mass Transfer,* Vol. 10 (1967), pp. 1767–1778.
16. Baroczy, C. J., "A Systematic Correlation for Two-Phase Pressure Drop," *Chem. Engng. Prog. Symp. Ser.,* No. 44, Vol. 62 (1966), pp. 232–249.
17. Chisholm, D., "Pressure Gradients Due to Friction during The Flow of Evaporating Two-phase Mixtures in Smooth Tubes and Channels," *Int. J. Heat Mass Transfer,* Vol. 16 (1973), pp. 347–358.
18. Friedel, L., "Modellgesetz für Den Reibungs-druckverlust in Der Zweiphasenströmung," *VDI-Forschungs Heft* 572, 1975.
19. Hewitt, G. F., "Pressure Drop," *Handbook of Multiphase Systems,* G. Hetsroni (Ed.), Washington D. C.: Hemisphere, 1981, pp. 2.44–2.75.
20. Hewitt, G. F., "Multiphase Fluid Flow and Pressure Drop," *Heat Exchanger Hand Book,* E. U. Schlünder et al. (Eds.), Washington D. C.: Hemisphere, 1983, pp. 2.3.1.1–2.3.2.34.
21. Gill, L. E., et al., "Sampling Probe Studies of The Gas Core in Annular Two-phase Flow—I. The Effect of Length on Phase and Velocity Distribution," *Chem. Engng Sci.,* Vol. 18 (1963), pp. 525–535.
22. Hewitt, G. F., and Lacey, P. M. C., "The Breakdown of The Liquid Film in Annular Two-phase Flow," *Int. J. Heat Mass Transfer,* Vol. 8 (1965), pp. 781–791.
23. Wallis, G. B., "Annular Two-phase Flow," *Trans. ASME, J. Basic Engng,* Vol. 92–93 (1970), pp. 59–82.
24. Martindale, W. R., and Smith, R. V., "Two-Phase Two-Component Interface Drag Coefficients in Separated Phase Flows," *Int. J. Multiphase Flow,* Vol. 7 (1981), pp. 211–219.
25. Levy, S., and Healzer, J. M., "Application of Mixing Length Theory to Wavy Turbulent Liquid-gas Interface," *J. Heat Transfer,* Vol. 103 (1981), pp. 492–500.
26. Zuber, N., and Findlay, J. A., "Average Volumetric Concentration in Two-phase Flow Systems," *J. Heat Transfer,* Vol. 87 (1965), pp. 453–468.
27. Bankoff, S. G., "A Variable Density Single-fluid Model for Two-phase Flow with Particular Reference to Steam-water Flow," *J. Heat Transfer,* Vol. 82 (1960), pp. 265–272.
28. Ishii, M., "Wave Phenomena and Two-phase Flow Instabilities," in *Handbook of Multiphase Systems,* G. Hetsroni (Ed.), Washington D. C.: Hemisphere, 1982, pp. 2–95–2–128.
29. Ishii, M., "Drift Flux Model and Derivation of Kinematic Constituitive Laws," in *Two-phase Flows and Heat Transfer,* Volume 1, S. Kakaç and F. Mayinger (Eds.), Washington D. C.: Hemisphere, 1977, pp. 187–208.

30. Rouhani, Z., "Steady-State Void Fraction and Pressure Drop in Watercooled Reactors," in *Two-phase Flows and Heat Transfer*, J. J. Ginoux (Ed.), Washington D. C.: Hemisphere, 1978, pp. 241–273.
31. Premoli, A., et al., "Una Correlazionne Adimensionale per La Determinazione Della Densità di Miscele Bifasiche," La Termotecnica, Vol. 25, 1971, pp. 17–26.
32. Dukler, A. E., "Fluid Mechanics and Heat Transfer in Vertical Falling-Film Systems," *Chem. Engng Prog. Symp. Ser.*, No. 30, Vol. 56 (1960), pp. 1–10.
33. Schlichting, H., *Boundary Layer Theory*, 7th Ed., New York: McGraw–Hill, 1979, pp. 578–594.
34. Ishii, M., and Grolmes, M. A., "Inception Criteria for Droplet Entrainment in Two-phase Concurrent Film Flow," *AIChE Journal*, Vol. 21, No. 2 (March 1975), pp. 308–318.
35. van Rossum, J. J., "Experimental Investigation of Horizontal Liquid Films," *Chem. Engng Sci.*, Vol. 11 (1959), pp. 35–52.
36. Ueda, T., and Tanaka, T., "Studies of Liquid Film Flow in Two-phase Annular and Annular–Mist Flow Regions, Part 1 and 2," *Trans. JSME*, Vol. 39, No. 325 (1973), pp. 2842–2862.
37. Yablonik, R. M., and Khaimov, V. A., "Determination of The Velocity of Inception of Droplet Entrainment," *Fluid Mechanics Soviet Research*, Vol. 1, No. 1 (1972), pp. 130–134.
38. Zhivaikin, L. Ya., "Liquid Film Thickness in Film-type Units." *International Chem. Eng.*, Vol. 2, No. 3 (1962), pp. 337–341.
39. Ishii, M., and Kataoka, I., "Interfacial Transfer in Annular Dispersed Flow," in *Advances in Two-phase Flow and Heat Transfer*, Volume 1, S. Kakac and M. Ishii (Eds.), Boston: Martinus Nijhof, 1983, pp. 91–118.
40. Ishii, M., and Mishima, K., "Liquid Transfer and Entrainment Correlation for Droplet–Annular Flow," *Heat Transfer 1982, Proc. Seventh International Heat Transfer Conf.*, Volume 5, U. Grigull et al. (Eds.), Washington D. C.: Hemisphere, 1982, pp. 307–312.
41. Hewitt, G. F., "Void Fraction," in *Handbook of Multiphase Systems*, G. Hetsroni (Ed.), Washington D. C.: Hemisphere, 1982, pp. 2.76–2.94.
42. Chen, J. C., "Correlation for Boiling Heat Transfer to Saturated Fluids in Convective Flow," *Ind. Eng. Chem. Process Des. Development.*, Vol. 5, No. 3 (1966), pp. 322–329.
43. Collier, J. G., et al., "Heat Transfer to Two-phase Gas-liquid Systems Part II. Further Data on Steam/Water Mixtures in The Liquid Dispersed Region in An Annulus," *Trans. Instn. Chem. Engrs*, Vol. 42 (1964), pp. 127–139.
44. Bennett, D. L., and Chen, J. C., "Forced Convective Boiling in Vertical Tubes for Saturated Pure Components and Binary Mixtures," *AIChE Journal*, Vol. 26, No. 3 (1980), pp. 454–461.
45. Forster, H. K., and Zuber, N., "Dynamics of Vapour Bubbles and Booling Heat Transfer," *AIChE Journal*, Vol. 1. No. 4 (1955), pp. 531–535.
46. Bennett, D. L., et al., "The Suppression of Saturated Nucleate Boiling by Forced Convective Flow," *AIChE Symp. Ser.*, No. 199, Vol. 76 (1980), pp. 91–103.
47. Dengler, C. E., and Addoms, J. N., "Heat Transfer Mechanism for Vaporization of Water in A Vertical Tube," *AIChE Symp. Ser.*, No. 18, Vol. 52 (1956), pp. 95–103.
48. Guerrieri, S. A., and Talty, R. D., "A Study of Heat Transfer to Organic Liquids in Single-tube, Natural-circulation, Vertical-tube Boilers," *Chem. Engng. Prog. Symp. Ser.*, No. 18, Vol. 52 (1956), pp. 69–77.
49. Rohsenow, W. M., "Boiling," in *Handbook of Heat Transfer*, W. M. Rohsenow and J. P. Harnett (Eds.), New York: McGraw–Hill, pp. 13–1—13–75.
50. Kihara, D. H., et al., "Forced Convection Boiling Heat Transfer of Freon–11 in A Vertical Annular Passage," *Letters in Heat Mass Transfer*, Vol. 6 (1979), pp. 13–21.
51. Butterworth, D., and Shock, R. A. W., "Flow Boiling," *Heat Transfer 1982, Proc. Seventh International Heat Transfer Conf.*, Volume 1, U. Grigull et al. (Eds.) Washington D. C.: Hemisphere, 1982, pp. 11–30.

52. Slipcevic, B., "Heat Transfer to Boiling Fluorocarbon Refrigerants," *ASHRAE Journal*, Vol. 12 (June 1970), pp. 65–68.
53. Bandel, J., and Schlünder, E. U., "Frictional Pressure Drop and Convective Heat Transfer of Gas-liquid Flow in Horizontal Tubes," *Heat Transfer 1974, Proc. 5th International Heat Transfer Conf.*, Volume 4, 1974, pp. 190–194.
54. Dembi, N. J., et al., "Statistical Analysis of Heat Transfer Data for Convective Boiling of Refrigerants in A Horizontal Tube," *Letters in Heat Mass Transfer*, Vol. 5 (1978), pp. 287–296.
55. Chawla, J. M., "Wärmeübergang und Druckabfall in Waagerechten Rohren bei Der Strömung von Verdampfenden," *VDI–Forschungs Heft* 523, 1967.
56. Shah, M. M., "A New Correlation for Heat Transfer during Boiling Flow through Pipes," *ASHRAE Trans.*, Vol. 82, No. 2 (1976), pp. 66–86.
57. Nusselt, W., "Die Oberflachenkondensation des Wasserdampfes." *VDI Zeitschrift*, Vol. 60 (1916), pp. 541–546 and pp. 569–575.
58. Rohsenow, W. M., "Heat Transfer and Temperature Distribution in Laminar-film Condensation," *Trans. ASME*, Vol. 78 (1956), pp. 1645–1648.
59. Minkowycz, W. J., and Sparrow, E. M., "Condensation Heat Transfer in The Presence of Non-condensables, Interfacial Resistance, Superheating, Variable Properties and Diffusion," *Int. J. Heat Mass Transfer*, Vol. 9 (1966), pp. 1125–1144.
60. Colburn, A. P., "The Calculation of Condensation Where A Portion of The Condensate Layer Is in Turbulent Motion," *Trans. AIChE*, Vol. 30 (1933–34), pp. 187–193.
61. Kutatelatze, S. S., "Semi-empirical Theory of Film Condensation of Pure Vapours," *Int. J. Heat Mass Transfer*, Vol. 25, No. 2 (1982), pp. 653–660.
62. Hirshburg, R. I., and Florschuetz, L. W., "Liminar Wavy-film Flow: Part II, Condensation and Evaporation," *J. Heat Transfer*, Vol. 104 (1982), pp. 459–464.
63. Suzuki, K., et al., "An Analysis on Film Condensation Heat Transfer Taking Account of Interfacial Wave Effect," *Proc. 20th National Heat Transfer Symposium of Japan*, 1982, pp. 253–255.
64. Rohsenow, W. M., et al., "Effect of Vapor Velocity on Laminar and Turbulent-film Condensation," *Trans. ASME.*, Vol. 78 (1956), pp. 1637–1643.
65. Blangetti, F., "Lokaler Wärmeübergang bei Der Kondensation mit Überlagerter Konvektion im Vertikalen Rohr," Dr. Dissertation Universität Karlsruhe, 1979.
66. Blangetti, F., and Schlünder, E. U., "Local Heat Transfer Coefficients on Condensation in A Vertical Tube," *Heat Transfer 1978, Proc. Sixth International Heat Transfer Conf.*, Volume 2, Washington D. C.: Hemisphere, 1978, pp. 437–442.
67. Ueda, H., et al., "Eddy Diffusivity near The Free Surface of Open Channel Flow," *Int. J. Heat Mass Transfer*, Vol. 20 (1977), pp. 1127–1136.
68. Michiyoshi, I., "Two-phase Two-component Heat Transfer," *Heat Transfer 1978, Proc. Sixth International Heat Transfer Conf.*, Volume 6, Washington D. C.: Hemisphere, 1978, pp. 219–233.
69. Butterworth, D., "Condensation of Vapor Mixtures," In *Heat Exchangers Hand Book*, Volume 1, E. U. Schlünder et al. (Eds.) Washington D. C.: Hemisphere, 1983, pp. 2.6.3.1–2.6.3.10.
70. Collier, J. G., "Boiling of Binary and Multicomponent Mixtures." In *Heat Exchangers Hand Book*, Volume 1, E. U. Schlünder et al. (Eds.) Washington D. C.: Hemisphere, 1983, pp. 2.7.8.1–2.7.8.10.
71. Collier, J. G., *Convective Boiling and Condensation*, 2nd ed., New York: McGraw–Hill, 1981, pp. 394–426.
72. Groothuis, H., and Hendal, W. P., "Heat Transfer in Two-phase Flow," *Chem. Eng. Sci.*, Vol. 11, 1959, pp. 212–220.
73. Kudirka, A. A., et al., "Heat Transfer in Two-phase Flow of Gas-liquid Mixtures," *Ind. Engng. Chem. Fundamentals*, Vol. 4 (1965), pp. 339–344.
74. Fried, L., "Pressure Drop and Heat Transfer for Two-phase Two-component Flow," *Chem. Engng. Prog. Symp. Ser.*, Vol. 50 (1954), pp. 47–51.

75. Johnson, H. A., and Abou–Sabe, A. H., "Heat Transfer and Pressure Drop for Turbulent Flow of Air-water Mixtures in A Horizontal Pipe," *Trans. ASME,* Vol. 74 (1952), pp. 977–987.

76. Pletcher, R. H., and McManus, H. N., "Heat Transfer and Pressure Drop in Horizontal Annular Two-phase Two-component Flow," *Int. J. Heat Mass Transfer,* Vol. 11 (1968), pp. 1087–1104.

77. Pletcher, R. H., and McManus, H. N., "A Theory for Heat Transfer to Annular Two-Phase Two-Component Flow," *Int. J. Heat Mass Transfer,* Vol. 15 (1972), pp. 2091–2096.

78. Vijay, et al., "A Correlation of Mean Heat-Transfer Coefficient for Two-phase Two-component Flow in A Vertical Tube," *Heat Transfer 1982, Proc. Seventh International Heat Transfer Conf.,* Volume 5, Washington D. C.: Hemisphere, pp. 367–372.

CHAPTER 22

FLOW REGIME MAPPING FOR CONDENSATION INSIDE HORIZONTAL TUBES

C. P. Gupta and H. K. Varma

Department of Mechanical Engineering
University of Roorkee, India

T. N. Tandon

Department of Mechanical Engineering
M. M. M. Engineering College
Gorakhpur, India

CONTENTS

INTRODUCTION

An important difference between single-phase and two-phase flow is the existence of an interface between the two-phases which takes a wide variety of forms in two-phase flow. The different arrangements of the two-phases are called "flow regimes" or "flow patterns." The pressure drop and heat transfer behavior depend on the flow pattern, which exists for a given set of conditions.

The flow patterns during two-phase flow have been described differently by a large number of investigators. Equally large is the flow pattern terminology used by them. These flow patterns have been identified using various techniques and methods, viz. by direct observation through transparent pipes, by measuring the wall pressure fluctuations or void fraction fluctuations, by using electrical conductance probes, by studying the changes in local heat transfer values, by photography using X-rays or fluorescent light, etc. A critical review of the various measuring techniques used for identifying flow patterns during two-phase flow has been presented by Jones and Dalhaye [1].

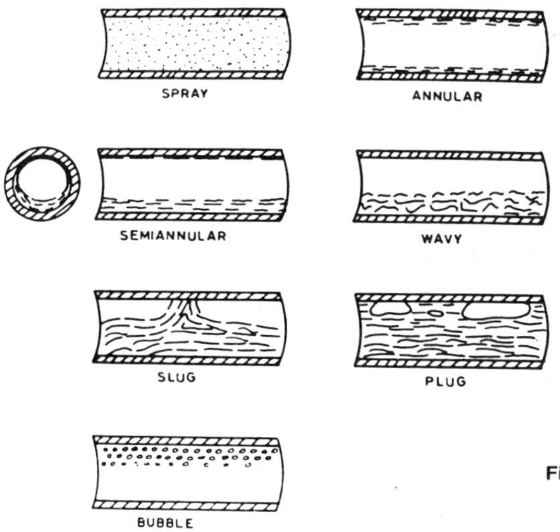

Figure 1. Depiction of various flow patterns.

From the thermal point of view, two-phase flows can be classified in the following categories.

1. Adiabatic.
2. Evaporation (boiling).
3. Condensation.

In this chapter, the flow patterns during condensation inside horizontal tubes are discussed.

IDENTIFICATION OF FLOW PATTERNS ON VISUAL BASIS

Flow visualization requires specific descriptions and characterizations of flow patterns. It appears from the literature that a large number of descriptions exist and that different investigators have employed somewhat varying terminology to describe the various flow patterns. It is, therefore, important to define clearly the classification most commonly employed for flow patterns during condensation inside horizontal tubes. These flow patterns are (Figure 1):

1. *Spray flow.* Flow in which no liquid film is present around the tube periphery and almost all of the liquid is dispersed as a mist in the vapor phase.
2. *Annular flow.* Flow in which the liquid forms an almost symmetrical film around the inside wall of the tube and the vapor flows as a central core.
3. *Semi-annular flow.* Basically the flow has the same appearance as annular flow but the liquid is asymmetrically distributed as a thin film around the top and sides of the tube and as a thick film at the bottom of the tube.
4. *Wavy flow.* Flow in which the liquid and vapor phases are completely separated by a disturbed interface caused by waves traveling in the direction of flow and with the liquid flowing in the bottom portion of the tube.
5. *Slug flow.* The liquid flowing in the lower portion of the tube and vapor flowing at a relatively higher velocity in the upper portion of the tube. The rapidly moving vapor causes frequent slugs of liquid to wet the upper wall of the tube.
6. *Plug flow.* Flow of a continuous liquid phase, with large plugs of vapor moving intermittently along the upper part of the tube.

7. *Bubble flow*. The liquid occupying most of the cross-section with vapor bubbles traveling in the top portion. The gas phase is discontinuous while the liquid phase is continuous.

FLOW REGIMES MAPS

The earliest generalized flow regimes map and perhaps most widely cited and used map was proposed by Baker [2]. He generated a generalized flow regimes map using the adiabatic gas-liquid flow data of Jenkins [3], Gazley [4], Kosterin [5], and Alves [6]. Although this map was originally proposed for adiabatic fluid flow, it has been widely used for single-component, two-phase flows, with heat transfer. The Baker's map is shown in Figure 2. He employed the dimensional group G_V/λ and the dimensionless group $G_L\lambda\Psi/G_V$ as ordinate and abscissa, respectively, for the map. The terms λ and Ψ are functions of fluid properties and were used to generalize the application of the flow regimes map for any gas-liquid combination since most of the data used in the development of the map were for the air-water system. The correction factor λ and Ψ are defined as

$$\lambda = [(\rho_V/\rho_A)(\rho_L/\rho_w)]^{1/2} \tag{1}$$

$$\Psi = (\sigma_w/\sigma)[(\mu_L/\mu_w)(\rho_w/\rho_L)^2]^{1/3} \tag{2}$$

Several investigators have pointed out the deficiencies of Baker's map which relate mainly to the use of a dimensional coordinate and repetition of mass velocity of the gas phase in both coordinates. Moreover, this map was developed from the experimental data obtained by using relatively larger diameter pipes (25 mm to 100 mm).

Later, when Scott [7] plotted the data of Hoogendoorn [8] and others for small-diameter pipes on the Baker's [2] map he found that many of the flow regime boundaries are not as distinct as the Baker's map indicated. He modified the Baker flow regimes map by depicting transition boundaries by relatively wide bands instead of sharp lines.

Recently, Taitel and Dukler [9] developed a purely theoretical model for transition between adjacent flow patterns for the case of steady cocurrent adiabatic gas-liquid flows without phase change in horizontal and near-horizontal pipes. Their approach represented a significant departure from the approach of previous investigators, because flow boundaries

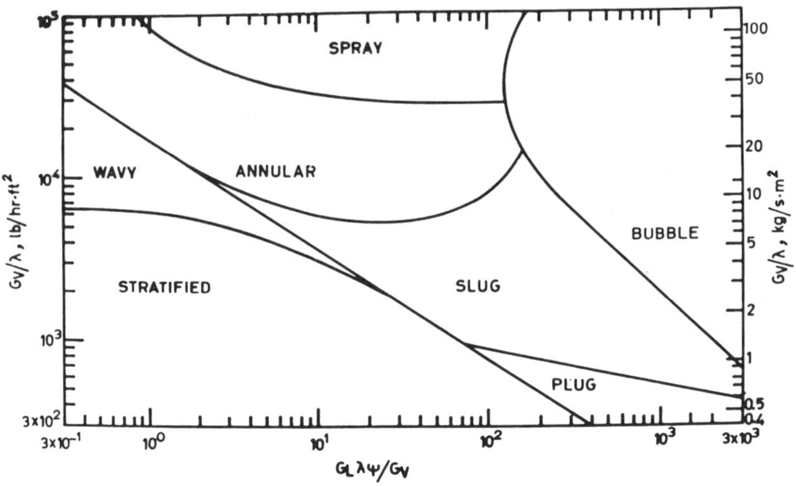

Figure 2. Baker's flow regime map [2].

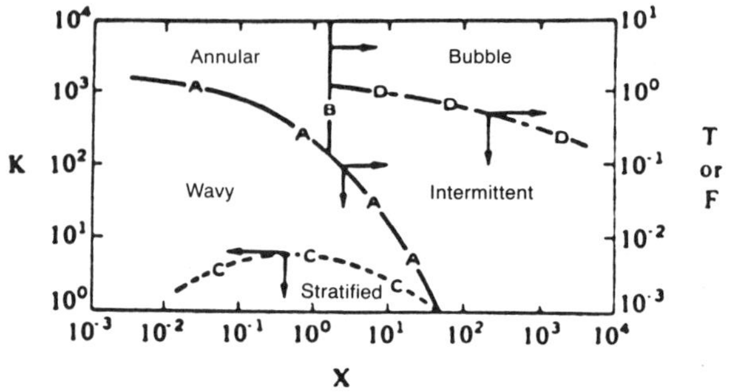

Figure 3. Taitel-Dukler flow regime map [9].

were predicted analytically and no flow regimes data were used. The generalized flow regimes map was actually a system of regime boundary criteria with different force balance ratios applicable to different boundaries. Taitel and Dukler expressed their results in terms of four dimensionless groups X, T, F, and K which are defined as

$$X = [(dp/dx)_{SL}/(dp/dx)_{sg}]^{1/2} \tag{3}$$

$$T = [|(dp/dx)_{SL}|/(\rho_L - \rho_g)g \cos \alpha] \tag{4}$$

$$F = \left[\frac{\rho_g v_{sg}^2}{(\rho_L - \rho_g)gD \cos \alpha}\right]^{1/2} \tag{5}$$

$$K = F\, Re_L^{1/2} \tag{6}$$

The prediction of the Taitel–Dukler theory for air-water flows at 25° C and 1 atm pressure in horizontal tubes of small sizes (upto 50 mm) were in reasonably good agreement with the empirical map of Mandhane et al. [10]. Taitel—Dukler, however, noted that larger sizes (such as 300 mm) can have a significant effect on the prediction of flow patterns. Their flow regimes map is shown in Figure 3.

FLOW PATTERN STUDIES OF CONDENSING FLUIDS

The problem of flow pattern prediction during condensation has not been given as much attention as that of flow pattern studies during boiling. Until the last decade no flow pattern studies during condensation were carried out, whereas several investigations were conducted pertaining to evaporative two-phase flow [11–19].

Bell et al. [20] in 1970 highlighted the interrelation of two-phase flow regimes and condensation heat transfer inside horizontal tubes. In the absence of any condensation flow regimes map they modified Baker's [2] map and suggested it for flow pattern prediction during condensation inside horizontal tubes.

A theoretical method of predicting flow patterns based on force balance criteria for cocurrent adiabatic gas-liquid mixture flowing upward in closed conduits was suggested by Quandt [21]. He used dimensionless criteria in terms of a mixture Froude number and a mixture Weber number to determine the dominant force acting on the flow which in turn determined which flow pattern would exist.

Soliman and Azer [22] were the first to develop flow pattern maps for condensation inside horizontal tubes. They observed the flow patterns during condensation of refrigerant 12 inside a 12.7 mm I.D. horizontal copper tube. They identified six major flow patterns, viz. spray, annular, semi-annular, wavy, slug, plug, and three transition flow patterns, viz. spray-annular flow, annular-wavy flow, and semi-annular-wavy flow. Soliman and Azer found poor agreement of their experimental data with Baker's [2] map and concluded that

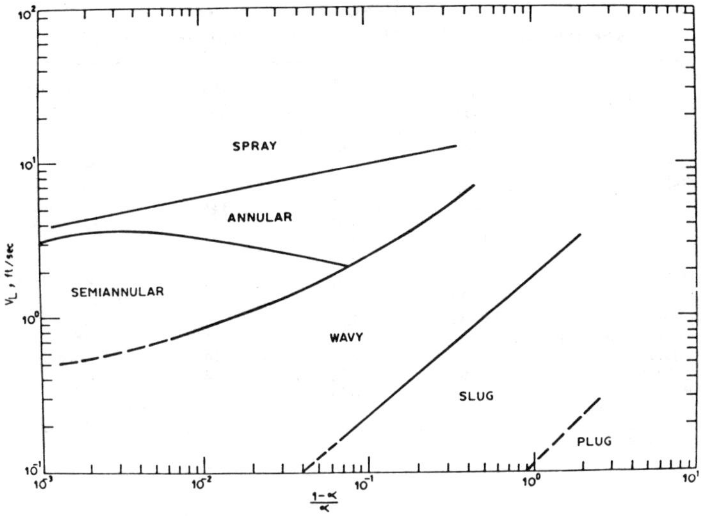

Figure 4. Soliman and Azer's flow map [26].

this map is inadequate to characterize different flow regimes for condensation. Out of the four flow maps generated by them they considered the flow regimes map with the Froude number and the gas volumetric ratio as coordinates to be the most general one.

Traviss and Rohsenow [23] observed the flow patterns during condensation of R–12 in a 8 mm I.D. horizontal tube. Their data could not be successfully represented on Soliman and Azer's [22] flow map with the Froude number and gas volumetric ratio as coordinates. They found their data to correlate fairly well when plotted on Baker's [2] original map and observed that Baker's map, as modified by Scott [7], agreed better with the flow regime data for small-diameter tubes. Traviss and Rohsenow proposed expressions for the annular to semi-annular flow regime transition and, based on their data, observed the transition to occur at a constant-film Froude number of 45. Later Rohsenow simplified these equations for flow transition.

In view of the opposite opinion of Traviss and Rohsenow [23] regarding the suitability of Baker's [2] map, Soliman and Azer [25, 26] conducted further flow pattern studies on three different tubes of inside diameters of 4.8 mm, 12.7 mm, and 15.9 mm with refrigerants R–12 and R–113 as condensing fluids. They again concluded that Baker's map was inadequate to characterize the twophase flow patterns during condensation. They also noted that their data of this investigation were not in satisfactory agreement with their earlier suggested flow map [22] with coordinates of the Froude number and gas volumetric ratio. Their flow pattern data of this investigation as well as that of their earlier investigation [22] correlated best on a map using V_L (average liquid velocity) and $(1-\alpha)/\alpha$ as coordinates, where void fraction α was calculated from Smith's [27] equation as follows. Their flow map is shown in Figure 4.

$$\alpha = \left\{1+(\rho_V/\rho_L)\left(\frac{1-x}{x}\right)\left[0.4+0.6\sqrt{\frac{(\rho_L/\rho_V)+0.4(1-x/x)}{1+0.4(1-x)/x}}\right]\right\}^{-1} \tag{7}$$

An effort has been made by some investigators [28–30] to select flow map coordinates based on force-balance criteria. Jastor and Kosky [29] performed experiments on condensing steam within a horizontal pipe, 12.5 mm I.D. and 2.20 m long. They suggested that the flow

Table 1
Condensation Flow Pattern Data Sources [28]

Fluid	Tube Inside Diameter (mm)	Test Section Length (m)	Number of Data Points	Symbol on Figures 5–8
R-12	8.0	4.42	83	○
R-12	12.7	0.61	91	●
R-12	4.8	0.61	66	○
	15.9	0.61	123	○
R-113	4.8	0.61	74	◇
	12.7	0.61	70	◇
	15.9	0.61	86	◇
R-12	12.7	0.61	25	▼
R-12	12.7	0.61	9	■
Steam	25.4	1.27	4	□
(no condensation)	50.8	1.27	4	□
Steam	22.0	6.70	14	▲
n-Pentane	22.0	6.70	74	▲
R-113	14.5	0.72	17	▽
R-11	16.0	4.40	13	△

From Breber, Palen, and Taborek [28].

transitions from the initial annular flow regime to a stratified regime are a function of the stress ratio, defined as the ratio of axial shear force to gravitational body force. Palen et al. [30] conducted visual investigations of condensation in a horizontal glass tube 22 mm I.D. and 6.7 m long with water and pentane as condensing fluids. They concluded that Baker's [2] map did not provide a realistic picture of the controlling regimes in condensation. Using their own flow regime data as well as that from other sources [22, 25, 23] they found that shear-controlled regimes (annular, mist-annular) and gravity-controlled regimes (wavy, stratified, slug) are quite well separated by dimensionless gas velocity $j_g^* = 1$. Generally, the transition regimes were within $j_g^* = 0.5$ and $j_g^* = 1.5$. The mist and mist-annular data were found to fall above $j_g^* = 5$.

The recent Taitel–Dukler [9] model, originally developed for transition between adjacent flow patterns for adiabatic flow without phase change, was tested by Breber et al. [28] for the case of condensation inside horizontal tubes. Using the experimental flow pattern data of References 31, 22, 23, 25 and some others, they modified the transition boundaries between different flow patterns and the Taitel–Dukler flow map with dimensionless gas velocity and Martinelli parameter, X, as coordinates. The data ranges are given in Table 1.

Sardesai et al. [32] delineated annular to stratified/wavy transition flow regimes for condensation inside a horizontal tube using local heat transfer coefficients. Their experimental data when plotted on the Taitel–Dukler [9] flow pattern map resulted in successful representation of annular to stratified/wavy transitions.

In a recent paper Soliman [33] examined the influence of tube diameter and fluid properties on the transition from annular to wavy and intermittent flow patterns during condensation inside horizontal tubes. While the vapor density had a noticeable effect, the influence of tube diameter was observed to be minor. He showed that the Traviss and Rohsenow [23] annular-to-semi-annular transition theory agreed well with small-diameter tubes (4.8 and 8 mm) but tended to overpredict the transition for large tube diameters (12.7 and 15.9 mm). Soliman developed a modified correlation based on the hypothesis that annular-to-wavy flow pattern transition occurs at a constant Froude number equal to 7, irrespective of tube diameter or condensing fluid.

Breber et al. [28] plotted the experimental flow pattern observations on the Taitel–Dukler map [9] as shown in Figures 5 through 8. Different symbols have been used for each data source and correspond to the designation given in Table 1. They made the following observations:

1. Relatively good agreement is found with data for tubes of larger diameters, ranging from 8 to 22 mm. For Soliman's [25] data with small-diameter tubes (4.8 mm), the agreement is poor; it can be seen that the wavy and slug flow pattern points are predicted as annular. Admitting that no definite reason could be given for this discrepancy, they conjectured that it may be due to surface tension effects which are not accounted for in their map and which may be important in small diameter tubes. They, however, noted that tubes of such small diameters are rarely used, particularly in process industries and therefore the anomaly is not serious insofar as practical condenser design is concerned.
2. The annular flow data are shown in Figure 5. Boundary A on the flow regime map corresponds to the theoretically derived criterion of Taitel and Dukler [9]. The degree of agreement is indeed remarkable. Figure 6 shows, similarly, the stratified wavy flow

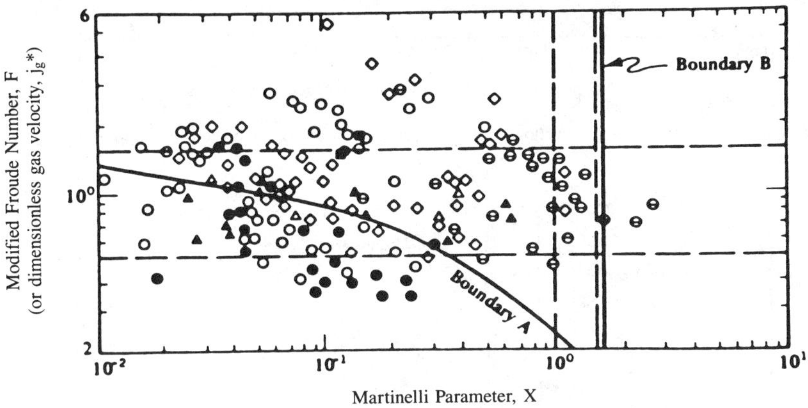

Figure 5. Comparison of Taitel-Dukler flow map with annular flow condensation [28].

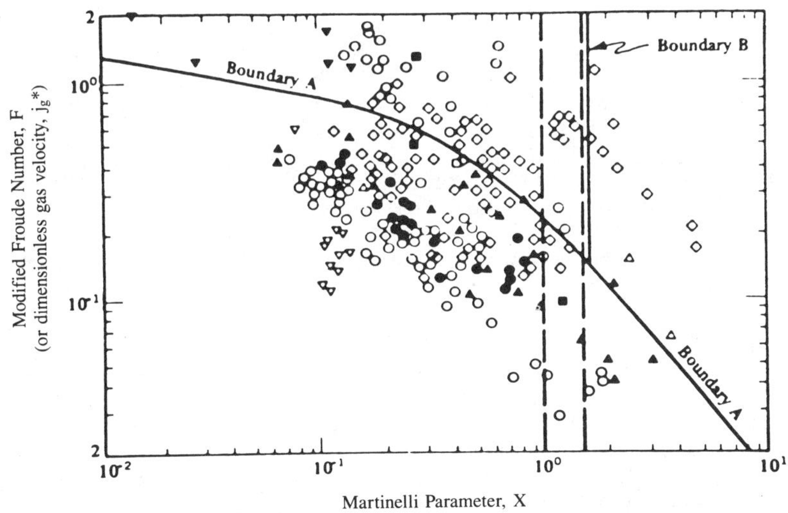

Figure 6. Comparison of Taitel-Dukler flow map with wavy flow condensation data [28].

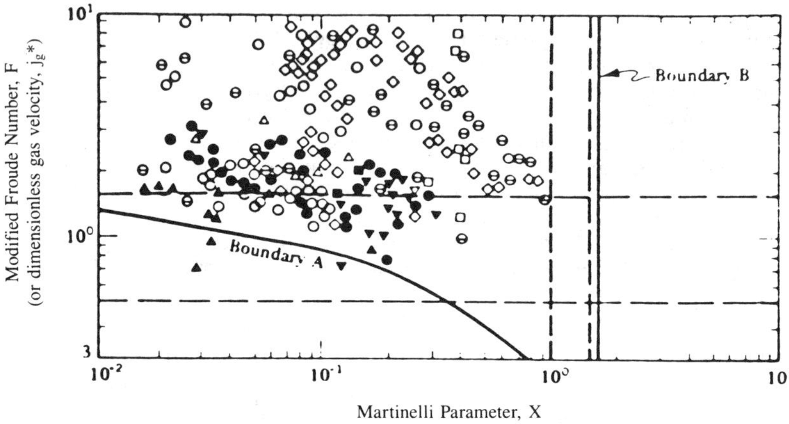

Figure 7. Comparison of Taitel-Dukler flow map with annular-wavy transition flow condensation data [28].

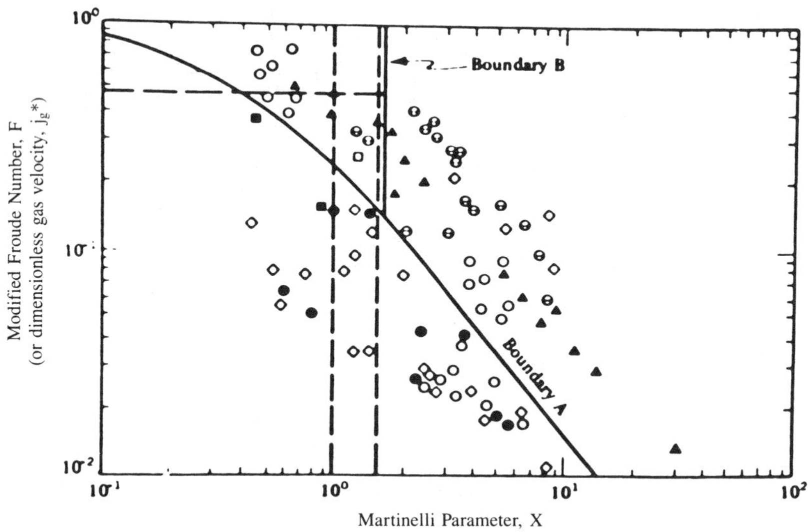

Figure 8. Comparison of Taitel-Dukler flow map with intermittent (slug-plug) condensation data [28].

pattern data. These points are, however, scattered both below and above the boundary line A of Taitel and Dukler. Thus, Breber et al. concluded that the transition between the annular and wavy flows is not abrupt and that a region rather than a single line defines the transition of flow patterns. Similarly, Figure 7 shows the points defined as wavy-annular or semi-annular and this also strenghtens the argument that a transition region is required. The flow data in Figure 6 are rather evenly divided by the line A.

3. The largest deviation is observed with respect to the slug flow as shown in Figure 8. It was found that at high mass velocities, some slug flow points were indicated in the annular-dispersed flow region on the Taitel–Dukler map, while some points at low mass velocities

were dispersed in the stratified-wavy region. Nevertheless, it was found that none of the slug flow data points fell to the left of $X = 0.4$ or above $F = 1$.

A set of simplified criteria in terms of X and F was recommended by Breber et al. [28] for more accurate prediction of horizontal tube-side condensation flow patterns. The following criteria were proposed by them:

- Zone I: Annular flow $j_g^* > 1.5,\ X < 1.0$
- Zone II: Stratified flow or wavy flow $j_g^* < 0.5,\ X < 1.0$
- Zone III: Slug or intermittent flow $j_g^* < 1.5,\ X > 1.5$
- Zone IV: Bubble flow $j_g^* > 1.5,\ X > 1.5$

The foregoing criteria are shown in Figure 9. Of course, no boundaries are exact. They contended that the only realistic way of defining the flow regimes on the map is based on the probability of finding a certain flow regime correctly on the map. Table 2 gives the percentage of the points for each flow pattern that are represented in the zones defined by the boundaries of Figure 9.

CRITICAL EVALUATION OF CONDENSING FLOW MAPS

It has been described above that the condensation flow maps which have been suggested for the prediction of flow patterns during condensation inside horizontal tubes are those of Soliman and Azer [25, 26] and Breber et al. [28]. It may, however, be noted that Soliman and Azer did not test their proposed map with the experimental flow pattern data of Traviss and Rohsenow [23]. Thus, to test the general applicability of the map of Soliman and Azer, the data of Traviss and Rohsenow have been plotted on their map and shown in Figure 10. It is apparent that a major shortcoming of the map is that nearly the entire semi-annular

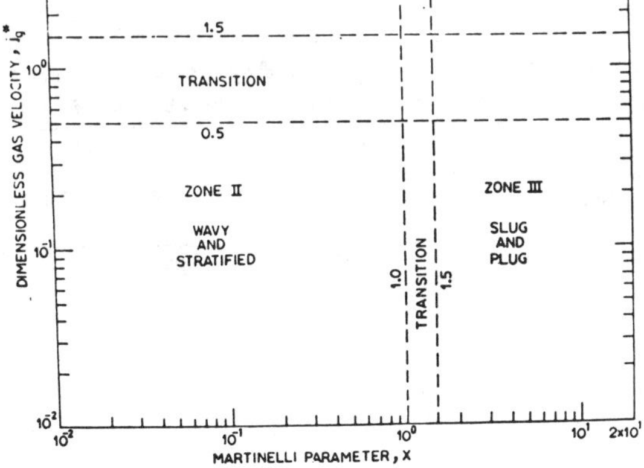

Figure 9. Breber et al. flow map [28].

Table 2
Distribution of Observed Flow Patterns on Breber et al. Map

Flow Pattern Zones	Percentage of Points of Each Type Observed in Each Zone: Annular and Mist I	Annular-Wavy Transition	Wavy and Stratified II	Slug and Plug III
Zone I: annular mist and annular	84	15	1	0
Zone II: wavy and stratified	0	9	85	6
Zone III: slug and plug	0	6	12	82
Transition between Zones I and II	24	55	20	1
Transition between Zones II and III	0	6	44	50

From Breber, Palen, Taborek [28].

Table 3
Range of Flow Conditions

Source	Refrigerant	Condensation Temperature (°C)	Inlet Superheat (°C)	Refrigerant Mass Velocity (kg/s-m^2)	Vapor Exit Quality	Average Cooling Heat Flux (W/m^2)
Soliman and Azer [22]	R-12	23–37	12–52	25.63–303.25	0.00–0.95	7,890–52,370
Traviss and Rohsenow [23]	R-12	10–40	–	100.8–988.67	0.010–0.96	6,310–63,100
	R-12	25–40	11–69	19.53–847.23	0.00–0.99	28,700–238,170
Soliman and Azer [25]	R-113	48–74	3–48	24.42–677.78	0.00–0.94	6,625–65,300

Table 4
Range of Physical Properties

Variable	Range of Variation
Liquid density (kg/m^3)	1,250–1,510
Vapor density (kg/m^3)	7.45–55.8
Liquid viscosity (N · s/m^2)	1.964×10^{-4}–5.043×10^{-4}
Vapor viscosity (N · s/m^2)	1.075×10^{-5}–1.34×10^{-5}
Surface tension (N/m)	0.0065–.0200

Table 5
Distribution of Flow Patterns Expressed as Percentage on the Flow Regimes Map of Tandon et al.

Observed Flow Pattern	Predicted Flow Pattern: Spray	Annular and Semi-annular	Wavy	Slug	Plug	Number of Points Observed
Spray	69	31	–	–	–	29
Annular and semi-annular	2	85	11	–	–	284*
Wavy	–	11	88	1	–	188
Slug	–	–	35	65	–	59
Plug	–	–	–	33	67	9
Spray-annular	64	36	–	–	–	11
Annular-wavy and semi-annular-wavy	–	49	51	–	–	84

* *Two percent of data falls above slug flow region which may be termed* bubble region *(not present for condensation except at high reduced pressure [28]).*

flow data points occupy the region of wavy flow. Also, for spray and slug flow patterns, the agreement is only moderate.

It has been reported that the flow map of Breber et al. [28] showed good agreement with the data of moderate and large tube diameters, but its agreement with the data of small tube diameters (4.8 mm) was unsatisfactory. Figures 11 and 12 which have been generated from the data of Soliman [25] for a tube diameter of 4.8 mm for condensing fluids R-12 and R-113, respectively, thus exhibit this lack of agreement. The following observations may be made from these figures:

- Wavy and slug flow data virtually occupy the transition zones and do not fall in their assigned regions.
- Semi-annular flow data during condensation occupy the spray-annular flow zone.

It is amply evident from this discussion that the two condensing flow maps show only limited agreement with the experimental data.

FLOW REGIMES MAP OF TANDON

Tandon, Varma, and Gupta proposed [34] a new flow regimes map on the basis of the flow pattern data of References 22, 23, and 25, which contain the bulk of the condensation flow pattern data available in open literature. The range of flow conditions covered are listed in Table 3, and the details of the physical properties are summarized in Table 4.

The dimensionless gas velocity, j_g^*, which was used as a map coordinate by Breber et al. [28] has also been used as one of the coordinates in the preceding map. The force-balance approach gives a dimensionless group that was termed by Wallis [35] as dimensionless gas velocity j_g^* defined as

$$j_g^* = xG/[gD\rho_V(\rho_L - \rho_V)]^{1/2} \tag{8}$$

The void fraction α, which significantly characterizes the two-phase flow pattern was selected as the other mapping parameter; α was calculated from Smith's [27] relation given

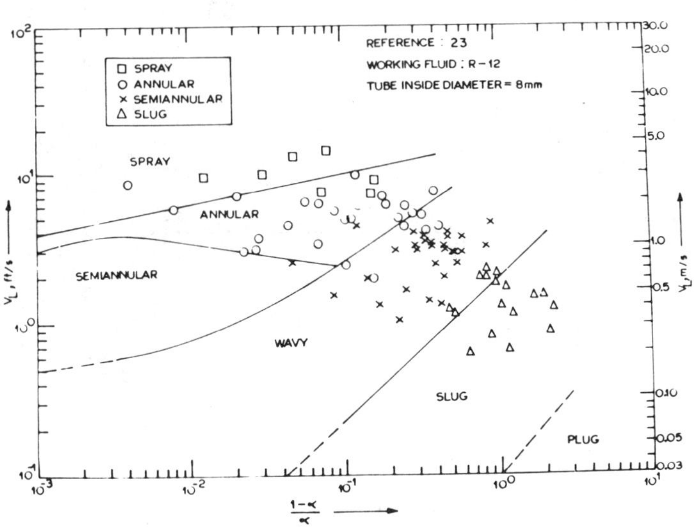

Figure 10. Comparison of Soliman et al. flow map with condensing data of Traviss and Rohsenow [23].

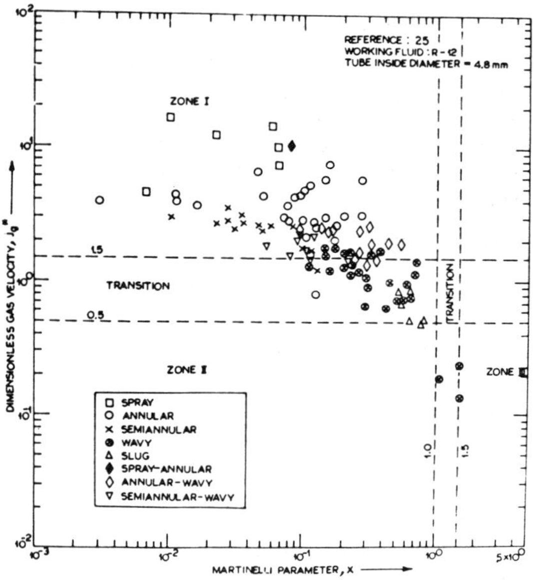

Figure 11. Comparison of Breber et al. [28] flow map with condensation data of Soliman [25] for 4.8-mm I.D. tube.

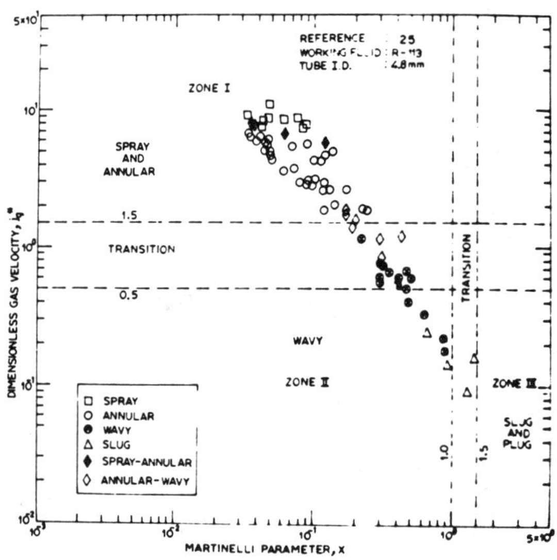

Figure 12. Comparison of Breber et al. [28] flow map with Soliman [25] flow pattern data for R-113 with 4.8-mm I.D. tube.

by Equation 7. Smith [27] claimed this equation to be valid for all flow conditions, irrespective of flow patterns, mass velocity, dryness fraction, or pressure. The preceding equation for void fraction was also used by Soliman and Azer [25, 26] in the development of their flow map.

In their flow regimes map Tandon et al. used the ratio $(1-\alpha)/\alpha$ instead of α as a coordinate because its use made the data spread much larger on the abscissa.

The selection of j_g^* and $(1-\alpha)/\alpha$ as map coordinates resulted in the best correlation of the data investigated. Figure 13 shows the data of References 22, 23, and 25 plotted on the proposed mapping coordinates and with the various transition boundaries drawn on the basis of these data delineating the various flow patterns. On the proposed flow regimes map, the transition lines drawn between the various flow patterns are sharp. Since annular and semi-annular flow data occupied nearly the same region on their map, these have been grouped together. Despite the wide range of flow conditions and different tube diameters it is seen that there is relatively good agreement for all tube diameters including the tube of small diameter (4.8 mm). This is an improvement over the map of Breber et al. [28]. It may be noted that the transition zones of the latter map are not identified in the map of Tandon et al.

Table 5 gives the agreement of the experimental data of References 22, 23, and 25 on their map. It is seen from Table 5 that the proposed map gives excellent agreement in cases of annular, semi-annular, and wavy flows which are predominant during condensation. It may be noted that annular flow exists over a major part of the condensing length [36] and wavy flow is another significant flow pattern during condensation [26]. For spray, slug, and plug flows the agreement is fair.

The results given in Table 5 may be appreciated better if the following facts are kept in view:

- 95 percent of annular flow data points fall in their specified region, an exceptional agreement.

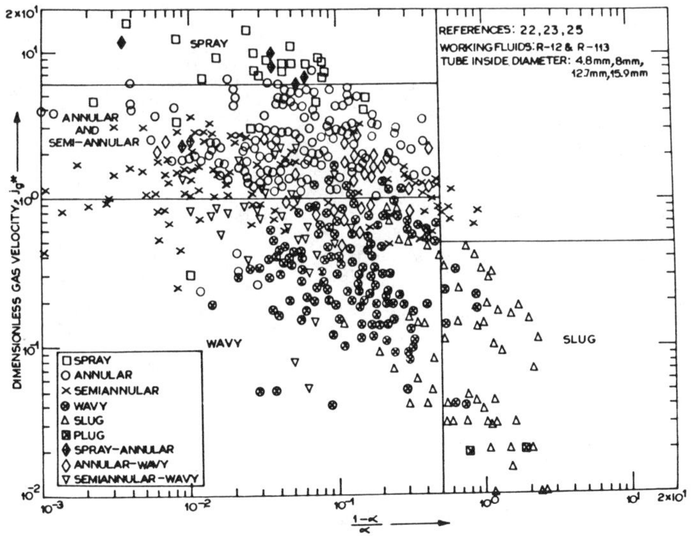

Figure 13. Proposed flow map showing all data points of references [22, 23, 25] on the Tandon et al. coordinates [34].

- 47 percent of semi-annular data points of Traviss and Rohsenow [23] fall in the wavy region on the proposed map. Traviss and Rohsenow did not observe any wavy flow which is an important flow pattern [26]. They considered semi-annular flow to be a transitional flow pattern. It is, therefore, likely that some of the semi-annular data might be belonging to the wavy region. If the semi-annular data of Reference 23 are neglected, the agreement for semi-annular data is 75 percent, and that for annular and semi-annular data taken together is 90 percent. It is, therefore, believed that prediction for annular and semi-annular flows would be still better than that given in Table 5.
- It was pointed out earlier that the wavy flow data for a 4.8-mm diameter tube did not fall in the appropriate zone on the map suggested by Breber et al. [28]. However, they made the following observations for this discrepancy: "... much of the discrepancy may just be due to the fact that it is extremely difficult to tell the difference between wavy and semi-annular flow for such a small tube."

Neglecting the 4.8-mm tube data, Breber et al. [28] found 80 percent of stratified-wavy flow data to fall in the proper region on the Taitel–Dukler [9] map. Considering the proposed map, if we ignore the 4.8-mm tube wavy flow data, the agreement for wavy flow data is 100 percent, a remarkable agreement. It is found that about 52 percent of wavy flow data for 4.8-mm diameter tube fall in the annular–semi-annular region. Therefore there is reason to agree with the apprehension of Breber et al. regarding the wavy flow data for the 4.8-mm diameter tube.

- Transition from one flow regime to other is not abrupt during the condensation process.

Figure 14 illustrates the three transitional flow patterns plotted on the proposed map. It is clear from this figure and Table 5 that the data points for the transitional flow patterns show reasonable scatter across their respective boundaries. This should be seen in the light of the fact that in the actual process there is a transition band between the two adjacent flow regimes.

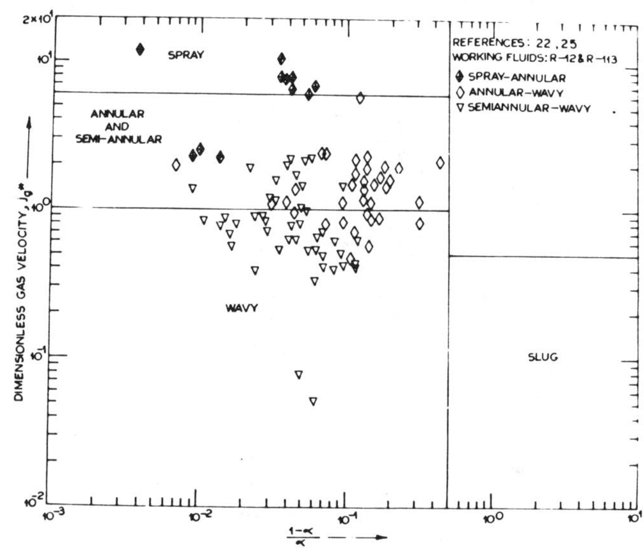

Figure 14. Transition flow pattern data of references [22, 25] on the Tandon et al. coordinates [34].

Tandon, Varma, and Gupta [34] have proposed the following criteria for delineating the various flow patterns during condensation inside horizontal tubes.

1. Spray: $j_g^* \geqq 6$ and $\frac{1-\alpha}{\alpha} \leqq 0.5$
2. Annular and semi-annular: $1 < j_g^* < 6$ and $\frac{1-\alpha}{\alpha} \leqq 0.5$
3. Wavy: $j_g^* \leqq 1$ and $\frac{1-\alpha}{\alpha} \leqq 0.5$
4. Slug: $0.01 \leqq j_g^* \leqq 0.5$ and $\frac{1-\alpha}{\alpha} > 0.5$
5. Plug: $j_g^* < 0.01$ and $\frac{1-\alpha}{\alpha} > 0.5$

Their flow regimes map is shown in Figure 15.

NEW EXPERIMENTAL DATA OF TANDON

In a recent investigation, Tandon et al. [37] have collected 208 flow pattern data points resulting from 60 test runs using R-12, and 146 flow pattern data points resulting from 39 test runs with R-22. These experimental results were employed to evaluate the prediction of the flow regimes maps of Baker [2], Soliman and Azer [26], Breber et al. [26], and Tandon et al. [34].

They have concluded, on the basis of their data, that Baker's [2] original map does not identify semi-annular flow. However, the semi-annular flow data of the present investigation

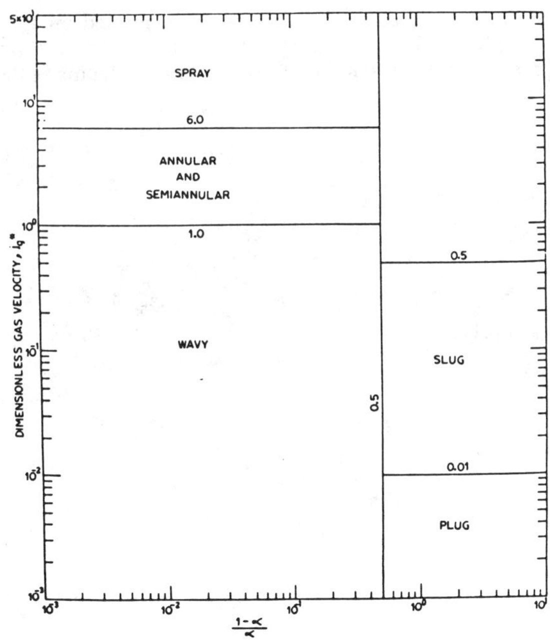

Figure 15. The Tandon et al. flow regime map [34].

occupied the annular region on Baker's map while most of the semi-annular flow data of Traviss and Rohsenow [23] occupied the slug region. Thus the Baker map has not been found to satisfactorily represent the two-phase flow patterns during condensation inside horizontal tubes. Soliman and Azer [22, 25, 26] and Palen et al. [30] after testing their condensation data on Baker's map also reached a similar inference. In the past, attempts have been made to reorganize the flow pattern boundaries of Baker's map so as to make it applicable to the condensation process. But it is apparent from the literature that such arrangements were satisfactory only for the test conditions of the investigation concerned. Therefore, such an attempt to modify the flow pattern boundaries of the original Baker's map will be of limited use only.

Figures 16 and 17 exhibit their observed flow patterns on Soliman and Azer's map [26]. An inspection of these figures reveals that:

- The agreement of the annular flow data during condensation is fairly good.
- The semi-annular flow data are poorly correlated on the map. These data points fall in the annular and wavy flow regions. Not even a single data point occupies the region assigned to it on the map.
- Spray, wavy, and slug flow data occupy the appropriate regions on the map. The plug flow data also occupies the assigned region (not shown).

Referring to Figures 18 and 19 which show the present data on the map of Breber et al. [28], it is seen that:

- The agreement of semi-annular flow data, which were assigned a transition zone between Zones I and II on the Breber et al. map, is unsatisfactory.
- The wavy flow data virtually occupy the transition zones.
- The spray, annular, and slug flow data mostly occupy the appropriate regions on the map. This is also true for the plug flow data (not shown).

Figures 20 and 21 illustrate the data of Tandon et al. plotted on their own map. It is apparent that except for wavy flow, the agreement for all other flow patterns is very good. Even for wavy flow the agreement is fair with 71 percent points falling in the assigned region.

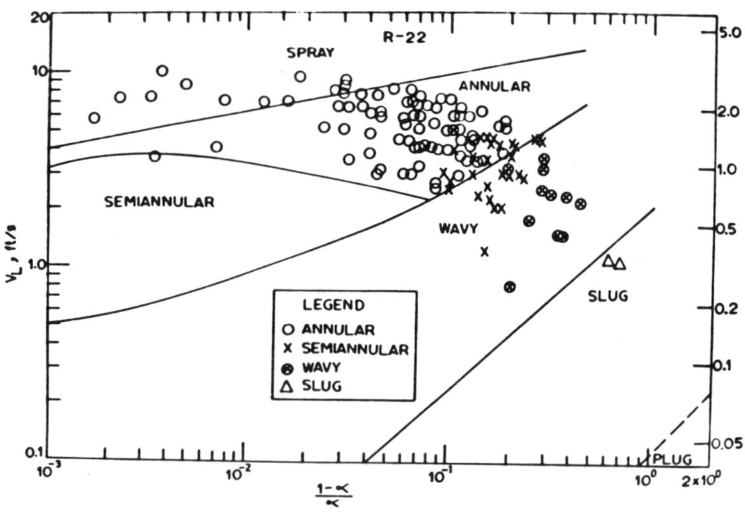

Figure 16. The Tandon et al. [34] flow patterns for R-22 plotted on the Soliman and Azer [26] map.

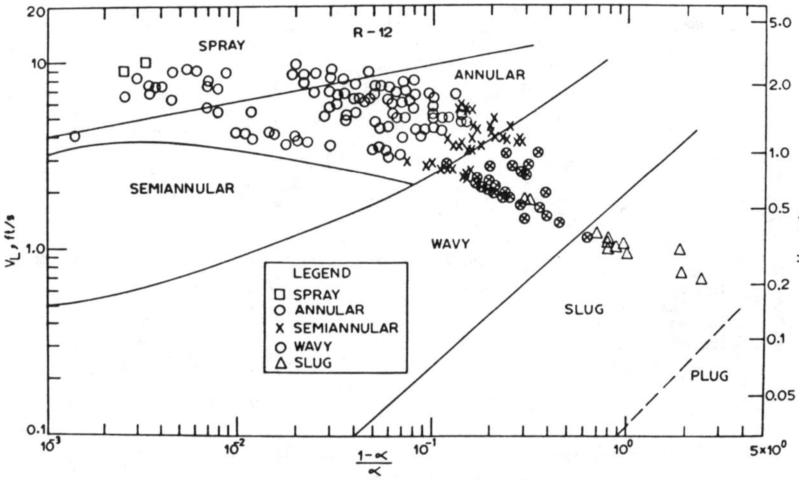

Figure 17. The Tandon et al. [34] flow patterns for R-22 plotted on the Soliman and Azer [26] map.

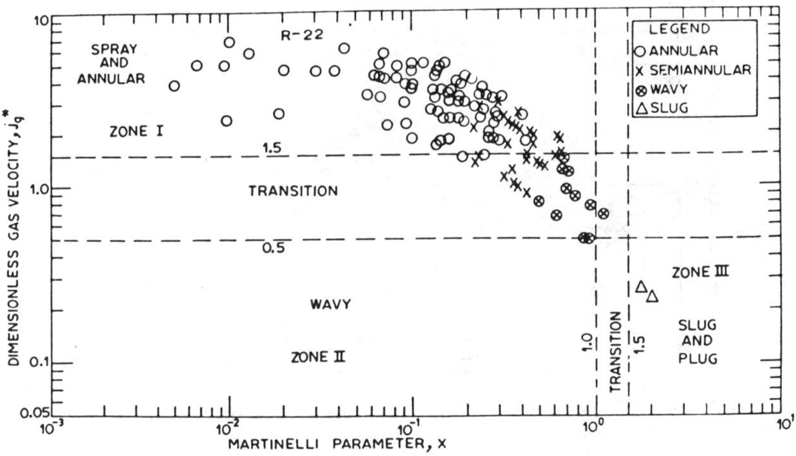

Figure 18. The Tandon et al. [34] flow patterns for R-12 plotted on the Breber et al. [28] map.

FLOW PATTERNS FOR BINARY MIXTURES

Tandon et al. [38] have also collected flow pattern data for three different mixtures of R-22 and R-12. They performed 109 steady-state runs and obtained 435 flow pattern data points. The range of operating conditions for mixtures is given in Table 6. The mixture flow pattern data were also tested against the flow maps of Baker [2], Soliman and Azer [26], Breber et al. [28], and Tandon et al. [34].

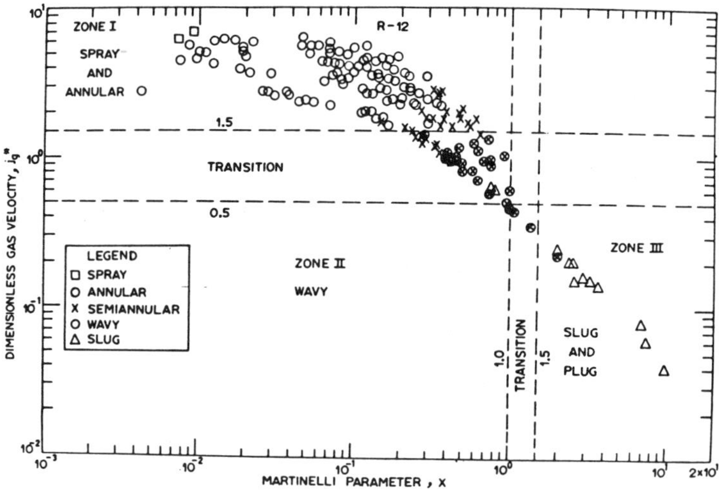

Figure 19. The Tandon et al. [34] flow patterns for R-12 plotted on the Breber et al. [28] map.

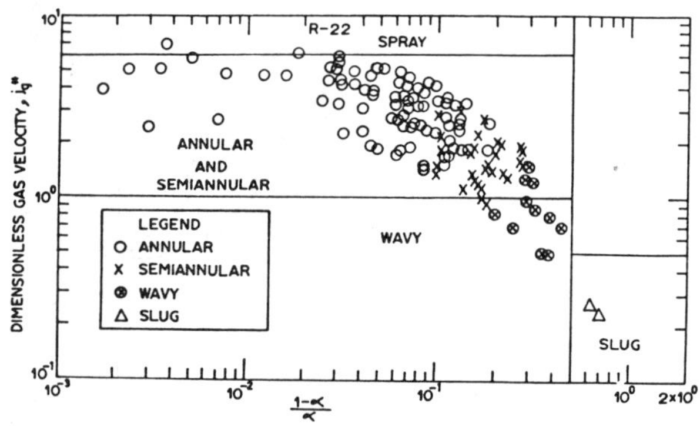

Figure 20. The Tandon et al. [34] flow patterns for R-22 plotted on the Tandon et al. flow map [34].

Table 6
Details of Test Runs and Range of Operating Conditions

Test Fluid	Number of Runs	Number of Flow Patterns Observed	Average Condensing Temperature (°C)	Refrigerant Mass Velocity	Vapor Quality at Visual Sections	Average Cooling Heat Flux in Test Sections (W/m²)
75% R-12 and 25% R-22	42	168	24–39	174.6–506.2	0.001–0.980	4,780–25,000
50% R-12 and 50% R-22	34	136	21–37	172.3–499.8	0.025–0.981	5,360–24,150
25% R-12 and 75% R-22	33	131	24.8–37	176.5–529.4	0.001–0.994	900–26,600

Figures 22 through 24 display the flow pattern data for the three mixtures on Baker's map. These figures lead to the following observations:

- The agreement of annular flow is good.
- None of the wavy flow pattern data fall in the specified region. Wavy flow data occupy annular and slug regions with the majority of data lying in the latter regime.
- Slug flow data are in agreement with the assigned region on the map, so is the case with plug flow data (not shown).
- Semi-annular flow data occupy the annular flow region on the Baker's map.

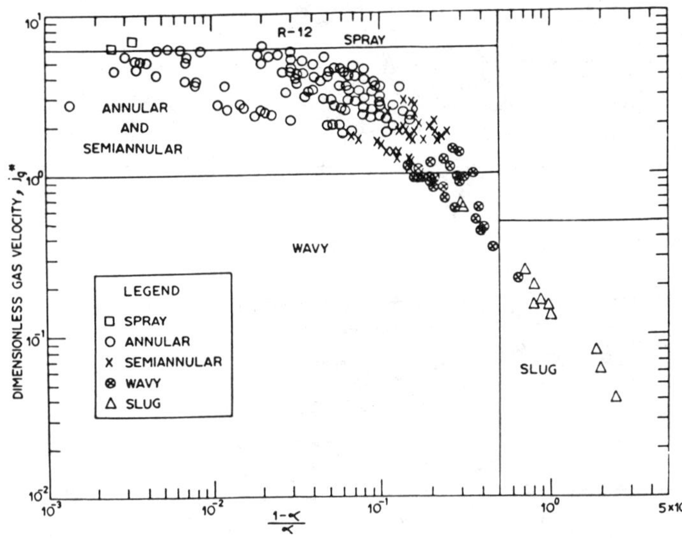

Figure 21. The Tandon et al. [34] flow patterns for R-12 plotted on the Tandon et al. flow map [34].

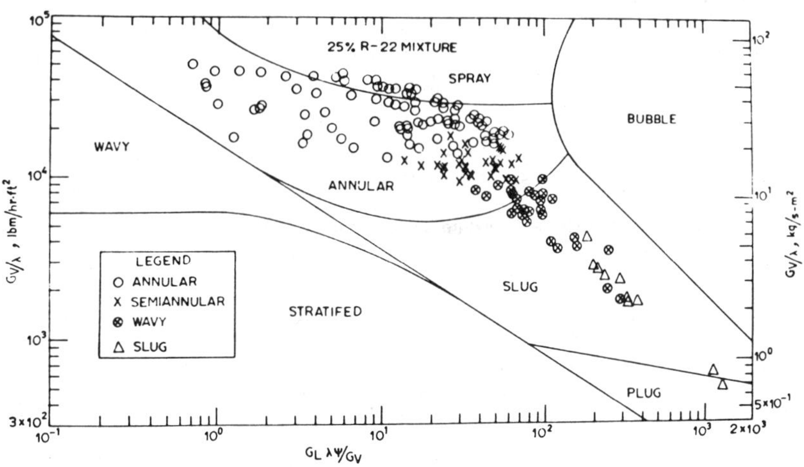

Figure 22. Flow patterns for 25% R-22 mixture plotted on the Baker map.

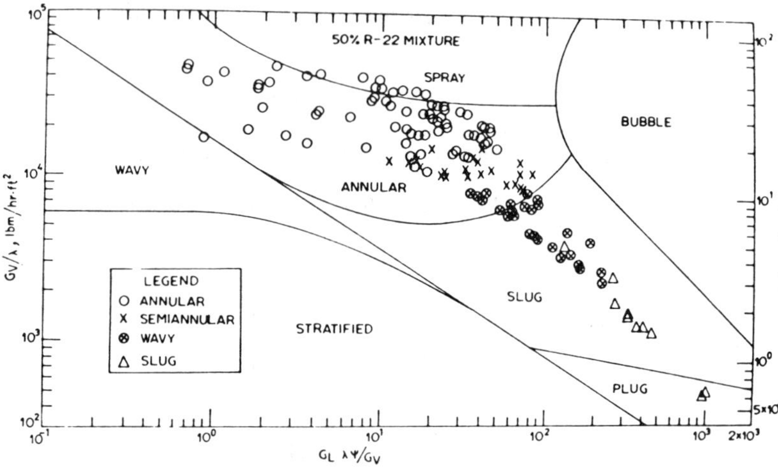

Figure 23. Flow patterns for 50% R-22 mixture plotted on the Baker map.

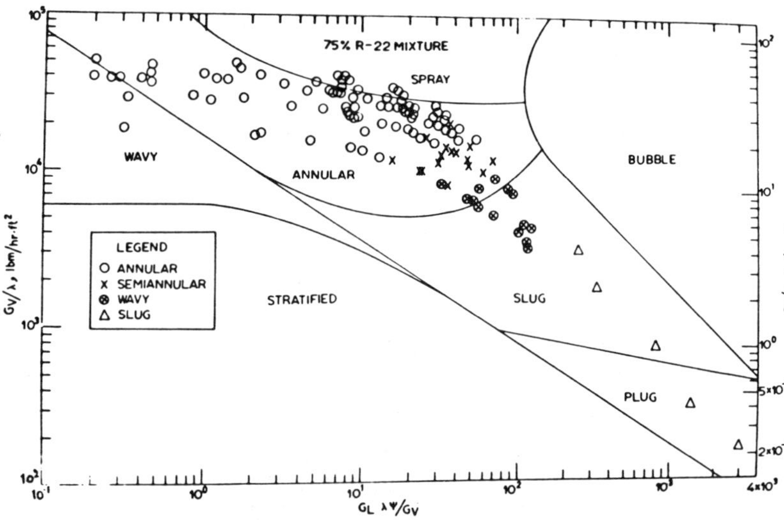

Figure 24. Flow patterns for 75% R-22 mixture plotted on the Baker map [2].

From the foregoing remarks it may be inferred that mixture flow pattern data also indicate that Baker's [2] map does not satisfactorily predict the flow patterns during condensation of binary mixtures inside horizontal tubes.

The comparison of mixture flow data on Soliman and Azer's [26] map is presented in Figures 25 through 27. An inspection of these figures discloses that

- Agreement for annular flow is moderate.
- None of the semi-annular flow data falls in the specified region. These data occupy the annular and wavy regions on the map.

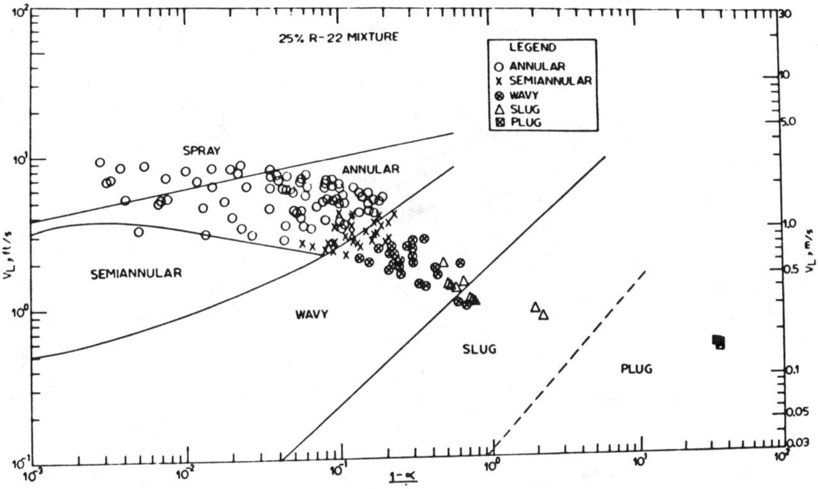

Figure 25. Flow patterns for 25% R-22 mixture plotted on the Soliman and Azer [26] flow map.

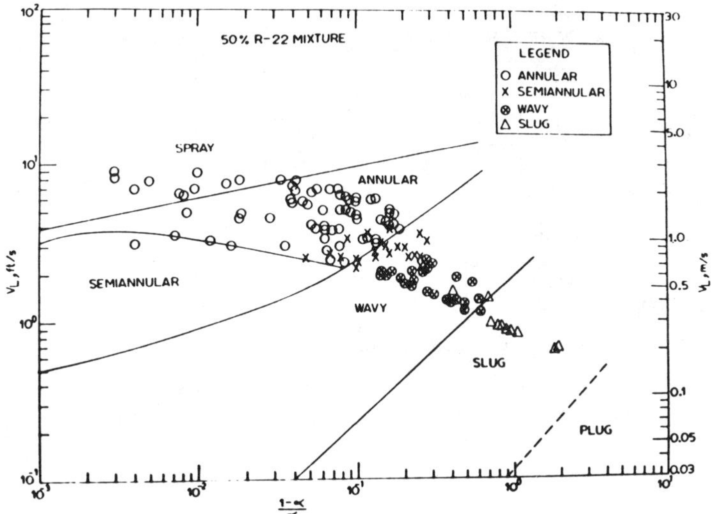

Figure 26. Flow patterns for 50% R-22 mixture plotted on the Soliman and Azer [26] flow map.

- The agreement with slug flow data is fair (not shown).
- Wavy and plug flow data occupy their specified regions on the map.

The mixture flow pattern data plotted on the Breber et al. [28] map are shown in Figures 28 through 30. These figures reveal that:

- The agreement of the semi-annular flow data is unsatisfactory. It may be noted that Breber et al. have shown the semi-annular flow pattern in the transition band between Zones I and II.

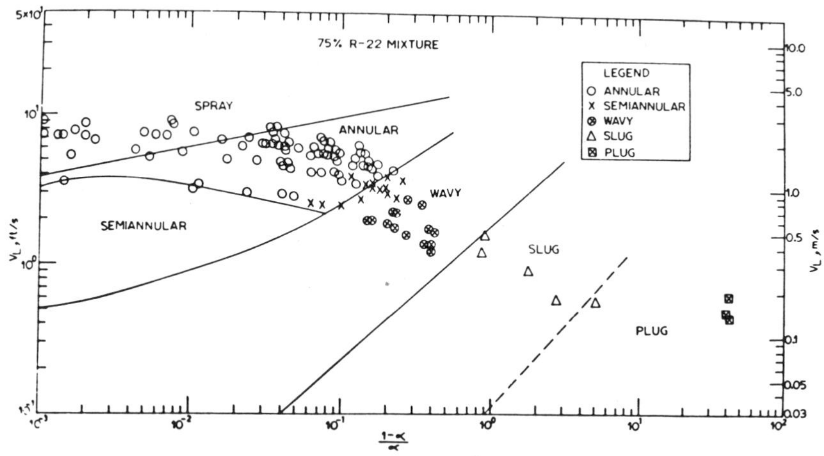

Figure 27. Flow patterns for 75% R-22 mixture plotted on the Soliman and Azer [26] flow map.

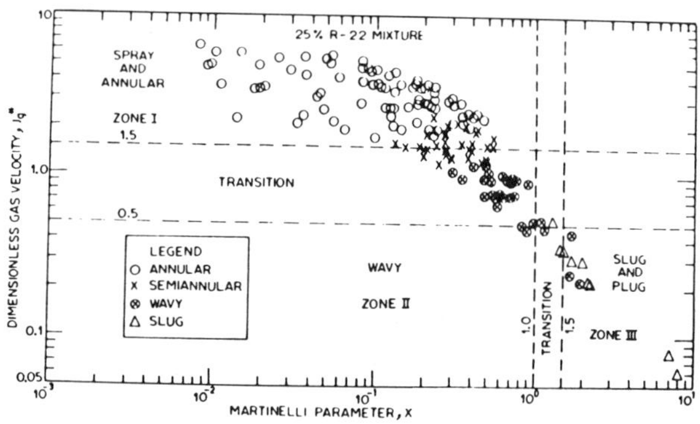

Figure 28. Flow patterns for 25% R-22 mixture plotted on the Breber et al. [28] map.

- The wavy flow data are poorly correlated on the map. These data virtually occupy the transition zones.
- Annular and slug flow data fall in appropriate zones on the map. The plug flow data also occupy the specified region (now shown).

Figures 31 through 33 exhibit the mixture flow pattern data represented on the flow regimes map suggested by Tandon et al. [34]. It is evident from these figures that:

- Annular, semi-annular, and slug flow data are in excellent agreement with predictions of the map. The same is the case with plug flow data (not shown).
- For wavy flow also, the agreement is good with 80 percent of the data occupying the assigned region.

COMPARISON OF VARIOUS FLOW MAPS

Tables 7 through 10 have been prepared to show the accuracy of the prediction of the four flow patterns maps viz. those of Baker, Soliman et al., Breber et al. and Tandon et al. for the experimental observations of binary mixtures of R-22 and R-12. A comparison of Tables 7 and 8 and Figures 22 through 24 and 31 through 33 show that the flow regimes map of Tandon et al. gives significantly improved predictions as compared to Baker's [2] map for annular, wavy, and slug flow patterns. The agreement of both maps with plug flow data is equally good.

Tables 7 and 8 and Figures 25 through 27 and 31 through 33 reveal that as compared to predictions of Soliman and Azer's [26] flow map, Tandon's map has substantially better

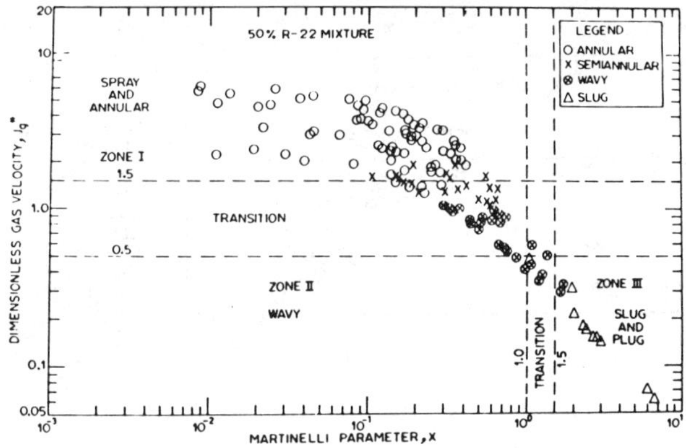

Figure 29. Flow patterns for 50% R-22 mixture plotted on the Breber et al. [28] map.

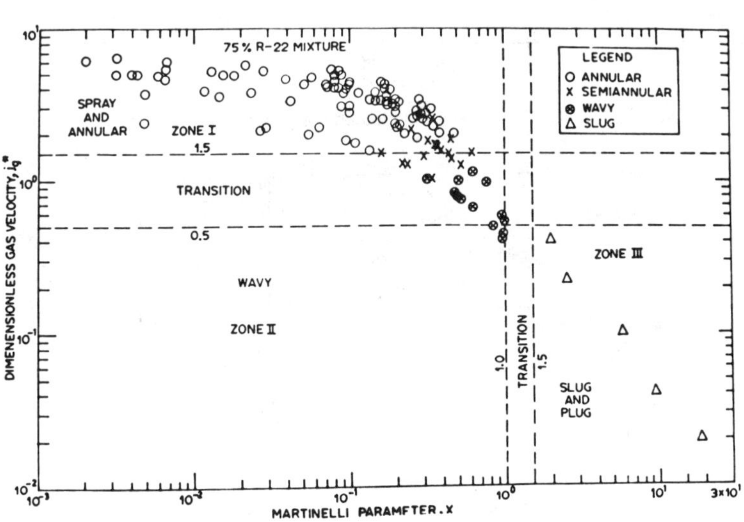

Figure 30. Flow patterns for 75% R-22 mixture plotted on the Breber et al. [28] map.

Table 7
Distribution of Observed Flow Patterns for Mixtures of R-22 and R-12 Expressed as Percentage on Tandon et al.

Observed Flow Pattern	Predicted Flow Pattern: Spray	Annular and Semi-annular	Wavy	Slug	Plug	Number of Data Points
Annular and semi-annular	1	99	–	–	–	330
Wavy	–	12	80	8	–	73
Slug	–	–	8	92	–	25
Plug	–	–	–	–	100	7

From Tandon, Varma, and Gupta [34].

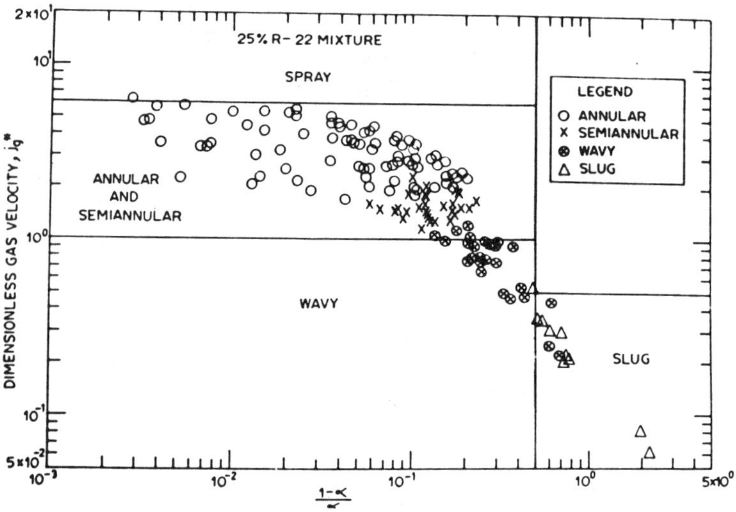

Figure 31. Flow patterns for 25% R-22 mixture plotted on the flow map of Tandon et al. [34].

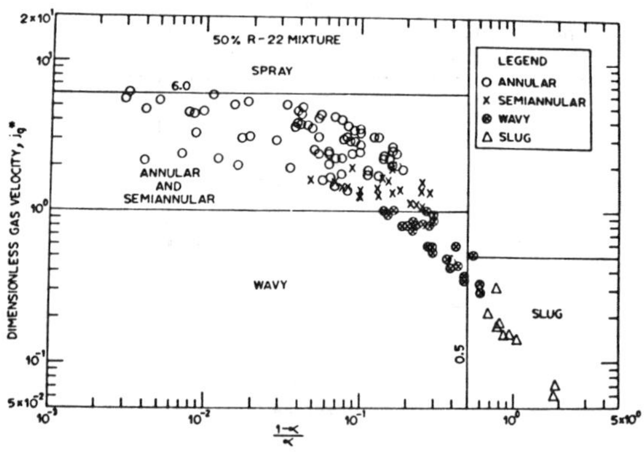

Figure 32. Flow patterns for 50% R-22 mixture plotted on the flow map of Tandon et al. [34].

Table 8
Distribution of Observed Flow Patterns for Mixtures of R-22 and R-12 Expressed as Percentage on Baker's [2] Map

Observed Flow Pattern	Predicted Flow Pattern: Spray	Annular	Wavy	Slug	Plug	Number of Data Points
Annular	14	84	2	–	–	255
Wavy	–	27	0	73	–	73
Slug	–	–	–	80	20	25
Plug	–	–	–	–	100	7

Note: Baker's map does not identify semiannular flow, hence these data are not included in Table 8. From Baker [2].

Table 9
Distribution of Observed Flow Patterns for Mixtures of R-22 and R-12 Expressed as Percentage on Soliman and Azer Map

Observed Flow Pattern	Predicted Flow Pattern: Spray	Annular	Semi-annular	Wavy	Slug	Plug	Number of Data Points
Annular	17	80	3	–	–	–	255
Semi-annular	–	41	0	59	–	–	75
Wavy	–	–	–	96	4	–	73
Slug	–	–	–	28	68	4	25
Plug	–	–	–	–	–	100	7

From Soliman and Azer [26].

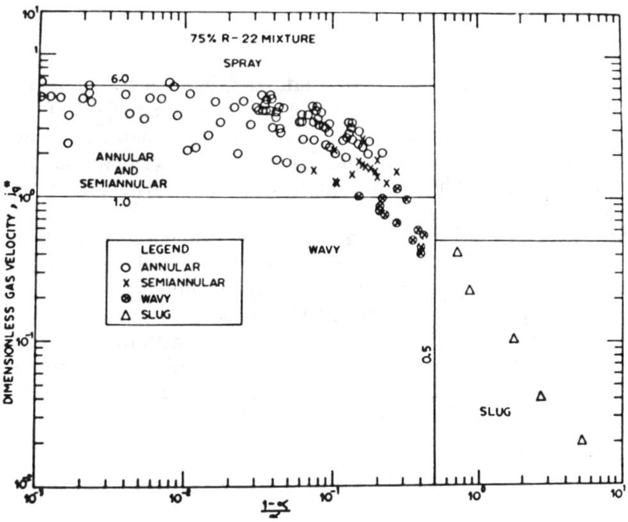

Figure 33. Flow patterns for 75% R-22 mixture plotted on the flow map of Tandon et al. [34].

Table 10
Distribution of Observed Flow Patterns for Mixtures of R-22 and R-12 Expressed as Percentage on Breber et al. Map

Observed Flow Pattern	Predicted Flow Pattern: Spray and Annular Zone I	Transition Between Zones I and II (Semi-annular)	Wavy Zone II	Transition Between Zones II and III	Slug and Plug Zone III	Number of Data Points
Spray and annular	99	1	–	–	–	255
Annular	45	55	–	–	–	75
Wavy	–	75	10	8	7	73
Slug and Plug	–	–	–	6	94	32

From Breber, Palen, Taborek [28].

Table 11
Overall Distribution of Condensing Flow Patterns on Proposed Flow Regimes Map

Observed Flow Pattern	Flow Patterns on Proposed Map: Spray	Annular and Semi-annular	Wavy	Slug	Plug	Number of Data Points
Spray	71	29	–	–	–	31
Annular and semi-annular	1	94	4	–	–	906
Wavy	–	7	83	10	–	302
Slug	–	–	25	75	–	98
Plug	–	–	–	14	86	21
Spray-annular	64	36	–	–	–	11
Annular-wavy and semi-annular wavy	–	49	51	–	–	84
Total points						**1,453**

* *One percent data occupied the region above the slug flow region, which has not been identified on the proposed map.*

agreement with semi-annular and slug flow data. Predictions for annular flow are also considerably better than that of Soliman and Azer's flow map. Even if the annular and semi-annular regions on their map are taken together, their map still shows much better agreement with the data. For wavy flow, the Soliman and Azer map had an edge over the predictions of Tandon's map. The agreement of both maps with plug flow data appears to be equally good.

The distribution of observed flow patterns on the Tandon map and Breber et al. [28] map is given in Tables 7 and 10, respectively. It is evident from these tables and Figures 28 through 33 that the former flow regimes map results in significantly improved predictions in respect to wavy flow pattern data. While only 10 percent of wavy flow data occupies the assigned region on the Breber et al. map, the agreement with the Tandon et al. [34] map is nearly 80 percent, a remarkable improvement. For semi-annular flow data also, the predictions on this map are substantially better than the Breber et al. map. For annular, slug, and plug flows the agreement of both maps with data is equally good.

OVERALL COMPARISON OF DIFFERENT FLOW REGIME MAPS

An overall comparison of the proposed flow regimes map with the entire condensing flow pattern data is shown in Table 11. In all over 1,450 data points resulting from over 555 test runs (which include over 205 runs of Tandon et al. [38] yielding over 780 flow pattern data points) have been considered. These data correspond to five tube diameters, six test fluids, and twelve data sets. It is apparent from Table 11 that the flow regimes map of Tandon et al. [34] shows excellent agreement for annular and semi-annular flows with 94 percent of the data falling in the assigned region. For wavy flow also the predictions are very good with 83 percent agreement. These flow patterns exist in the major part of the condensing length. Further, for spray and slug flow patterns the agreement is reasonably good (71 and 75 percent, respectively) and the agreement with plug flow data is 86 percent. Thus the predictions of their flow regimes map are more consistent as compared to the other flow maps including the one recently proposed by Breber et al. [28].

REFERENCES

1. Jones, O. C., and Delhaye, J. M., "Transient and Statistical Measurement Techniques for Two-Phase Flows: A Critical Review," *Int. J. Multiphase Flow,* Vol. 3, No. 2 (1976), pp. 89–116.
2. Baker, O., "Simultaneous Flow of Oil and Gas," *The Oil and Gas Journal,* Vol. 53 (1954), pp. 185–195.
3. Jenkins, R., "Two-Phase Two-Component Flow of Air and Water," M. S. thesis, University of Delaware, 1947.
4. Gazley, C., "Cocurrent Gas-Liquid Flow, III. Interfacial Shear and Stability," Heat Transfer and Fluid Mechanics Institute, Berkeley, California, 1949.
5. Kosterin, S. I., "An Investigation of the Influence of the Diameter and Inclination of a Tube on the Hydraulic Resistance and Flow Structure of Gas-Liquid Mixtures," *Izvest. Akad. Nauk. SSSR,* Otdel Tekh Nauk, No. 12 (1949).
6. Alves, G. E., "Cocurrent Liquid — Gas Flow in a Pipe Line Contactor," *Chem. Engg. Prog.,* Vol. 50 (1954), pp. 449–456.
7. Scott, D. S., "Properties of Co-current Gas-Liquid Flow," in Advances in Chem. Engg., Vol. 4, Academic Press, New York, (1963), p. 200.
8. Hoogendoorn, C. J., "Gas-Liquid Flow in Horizontal Pipes," *Chem. Engg. Science,* Vol. 9 (1959), pp. 205–218.
9. Taitel, Y., and Dukler, A. E., "A Model for Predicting Flow Regime Transitions in Horizontal and Near Horizontal Gas-Liquid Flow," *A.I.Ch.E. J.,* Vol. 22, No. 1 (January 1976), pp. 47–55.
10. Mandhane, J. M., Gregory, G. A., and Aziz, K., "A Flow Pattern Map for Gas-Liquid Flow in Horizontal Pipes," *Int. J. Multiphase Flow,* Vol. 1 (1974), pp. 537–553.
11. Hosler, E. R., "Visual Study of Boiling at High Pressure," *Chem. Engg. Prog. Symp. Ser. No. 57,* Vol. 61 (1965), pp. 269–279.
12. Hosler, E. R., "Flow Patterns in High Pressure Two-Phase (Steam-Water) Flow With Heat Addition," *Chem. Engg. Prog. Symp. Ser. No. 82,* Vol. 64 (1969), pp. 54–66.
13. Hsu, Y. Y., and Graham, R. W., "A Visual Study of Two-Phase Flow in a Vertical Tube with Heat Addition," NASA Technical Notes, NASA TN D-1564, 1964.
14. Staub, F. W., and Zuber, N., "Void Fraction Profiles, Flow Mechanisms, and Heat Transfer Coefficients for Refrigerant-22 Evaporating in a Vertical Tube," *ASHRAE Trans.,* Vol. 72, Part 1 (1966), pp. 130–146.
15. Suo, M., et al., "Investigation of Boiling Flow Regimes and Critical Heat Flux," U.S. Atomic Energy Commission, NYO-3304-3, 1965.
16. Vohr, J. H., "Flow Pattern of Two-Phase Flow — A Survey of Literature," U.S. Energy Commission, TID-11514, 1960.
17. Vohr, J. H., "A Photographic Study of Boiling Flow," U.S. Atomic Energy Commission, NYO-9650, 1963.
18. Zahn, W. R., "A Visual Study of Two-Phase Flow While Evaporating in Horizontal Tubes," *Trans. A.S.M.E.,* Series C, Vol. 86 (1946), pp. 417–429.

19. Zahn, W. R., "Flow Conditions When Evaporating Refrigerant-22 in Air Conditioning Coils," *ASHRAE Trans.,* Vol. 72, Part 1 (1966), pp. 82–89.
20. Bell, K. J., Taborek, J., and Fenoglio, F., "Interpretation of Horizontal In-Tube Condensation Heat Transfer Correlation with a Two-Phase Flow Regimes Map," *Chem. Engg. Prog. Symp. Ser. No. 102,* Vol. 66 (1970), pp. 150–163.
21. Quandt, E., Analysis of Gas-Liquid Flow Patterns, *Chem. Engg. Prog. Symp. Ser.* No. 57, Vol. 61 (1965), pp. 128–135.
22. Soliman, H. M., and Azer, N. Z., "Flow Patterns During Condensation Inside a Horizontal Tube," *ASHRAE Trans.,* Vol. 77, Part 1 (1971), pp. 210–224.
23. Traviss, D. P., and Rohsenow, W. M., "Flow Regimes in Horizontal Two-Phase Flow with Condensation," *ASHRAE Trans.,* Part 2 (1973), pp. 31–39.
24. Rohsenow, W. M. and Hartnett, I. P. (Eds), *Handbook of Heat Transfer,* McGraw-Hill Book Company, New York (1973).
25. Soliman, H. M., "Analytical and Experimental Studies of Flow Patterns During Condensation Inside Horizontal Tubes," Ph. D. dissertation, Department of Mechanical Engineering, Kansas State University, 1974.
26. Soliman, H. M., and Azer, N. I., "Visual Studies of Flow Patterns During Condensation Inside Horizontal Tubes," *Heat Transfer,* Vol. 3, (1974), pp. 241–245.
27. Smith, S. L., "Void Fraction in Two-Phase Flow: A Correlation Based Upon on Equal Velocity Head Model," *Proc. Instn. of Mech. Engrs.* (London), Vol. 184, Pt. 1, No. 36, pp. 647–657, 1969–1970.
28. Breber, G., Palen, J. W., and Taborek, J., "Prediction of Horizontal Tubeside Condensation of Pure Components Using Flow Regime Criteria," *Trans. ASME, J. of Heat Transfer,* Vol. 102, No. 3 (August 1980), pp. 471–476.
29. Jaster, H., and Kosky, P. G., "Condensation Heat Transfer in a Mixed Flow Regime," *Int. J. of Heat and Mass Transfer,* Vol. 19 (1976), pp. 95–99.
30. Palen, J. W., Breber, G., and Taborek, J., "Prediction of Flow Regimes in Horizontal Tube-Side Condensation," *Heat Transfer Engg.,* Vol. 1, No. 2 (1979), pp. 47–57.
31. Fujii, T., et al., "Condensation of R-11 Inside of Horizontal Tube," Report No. 1: Flow Regime Mapping and Pressure Drop, Trans of Japanese Society of Mechanical Engineers, Vol. 42, No. 363, pp. 3541–3550, November 1976.
32. Sardesai, R. G., Owen, R. G., and Pulling, D. J., "Flow Regimes for Condensation of a Vapour Inside a Horizontal Tube," *Chem. Engg. Science,* Vol. 36 (1981), pp. 1173–1180.
33. Soliman, H. M., "On the Annular-to-Wavy Flow Pattern Transition During Condensation Inside Horizontal Tubes," *Canadian J. of Chem. Engg.,* Vol. 60 (August 1982), pp. 475–481.
34. Tandon, T. N., Varma, H. K., and Gupta, C. P., "A New Flow Regimes Map for Condensation Inside Horizontal Tubes," *Trans. A.S.M.E., J. Heat Transfer,* Vol. 104, No. 4 (November 1982), pp. 763–768.
35. Wallis, G. B., *One-Dimensional Two-Phase Flow,* McGraw-Hill Book Co., 1969, p. 317.
36. Soliman, M., Schuster, J. R., and Bernson, P. J., "A General Heat Transfer Correlation for Annular Flow Condensation," *Trans. A.S.M.E., J. Heat Transfer* (1969), pp. 267–276.
37. Tandon, T. N., Varma, H. K., Gupta, C. P., "An Experimental Study of Flow Patterns During Condensation Inside a Horizontal Tube," *ASHRAE Trans.,* July 1983.
38. Tandon, T. N., "Forced Convection Condensation of Pure Refrigerants and Binary Mixtures Inside a Horizontal Tube," Ph. D. Thesis, Dept. of Mech. and Industr. Engg., Univ. of Roorkee, April 1983.

CHAPTER 23

HEAT TRANSFER AND PRESSURE DROP ACROSS FINNED TUBES

Kenichi Hashizume

Energy Science and Technology Laboratory
R&D Center, Toshiba Corporation
Ukishimacho 4-1
Kawasaki, Japan

CONTENTS

INTRODUCTION

Finned tubes find many applications for heat exchangers in various industrial processes such as boiler economizers, fired heater convection sections, waste heat boilers, and air cooled heat exchangers. Among various types of finned tubes, circular finned tubes are most commonly used because they are fabricated and available in a wide variety of fin geometries and materials, although studded, wired, plate, or some other types of finned tubes are also used. Therefore the characteristics of circular finned tubes are discussed mainly in this chapter.

Figure 1. Types of circular finned tubes: (A) spiral; (B) annular.

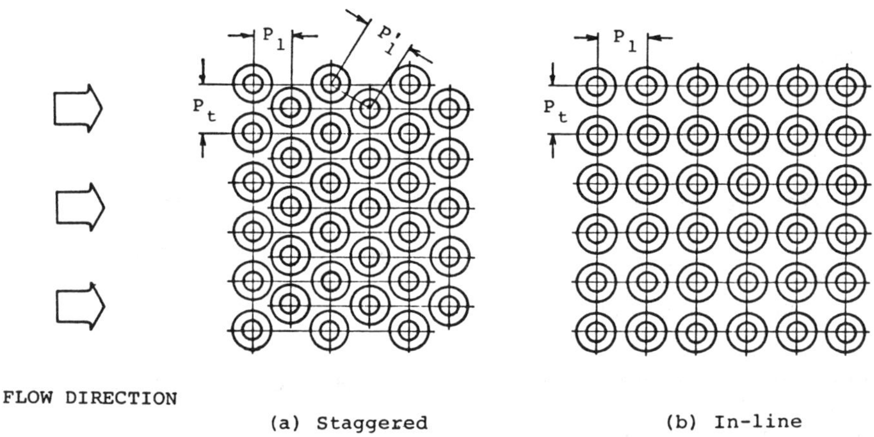

Figure 2. Tube layouts: (A) staggered; (B) in-line.

After the description of types of finned tubes, their layouts as a heat exchanger, and the parameters affecting the performance of finned tubes, correlations to predict the heat transfer coefficient and pressure drop for finned tube banks are reviewed. Then, remarks are given for designing finned tube heat exchangers, and lastly some special finned tubes and their characteristics are introduced.

TYPES OF FINNED TUBES AND TUBE LAYOUTS

Circular finned tubes can be classified as being of two different types, spiral or annular finned tubes, as illustrated in Figure 1. A spiral finned tube either has a metal tape wound around the tube to form fins, or is fabricated by extrusion (integral finned tube). Wound fins can be not only smooth but also segmented, corrugated, or slitted. Fins are attached to the tube by brazing, soldering, high frequency resistance welding, mechanical bonding, or by tension of the fins.

An annular finned tube is made from ring-shaped plates and installed on the tube. Fins are bonded by brazing or soldering onto the tube, or through expansion of the tube. This type of finned tube can bear various fin shapes such as square, rectangular, hexagonal, oval, or semicircular.

There is a variation of the annular finned tube with many tubes put together by many layers of multi-holed fin sheets. These are called plate finned tubes and used in mass-produced air coolers or air conditioning equipment.

In heat exchangers these finned tubes are arranged in two different layouts, staggered or in-line, as illustrated in Figure 2. Staggered layouts are mainly used in practical applications, because of their compactness and high heat transfer coefficient.

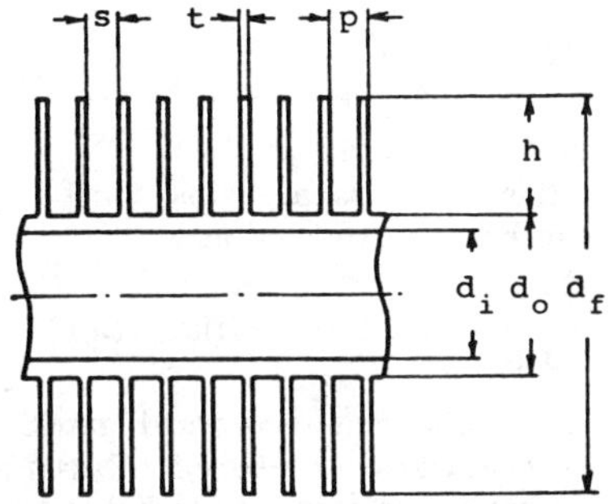

Figure 3. Fin geometries.

If the air or gas is very dirty involving much dust, in-line layouts are chosen. One reason for this is that cleaning lanes are required through the tube bank, and another reason is that the tubes in the second row in staggered layouts suffer more abrasion. (When dusty air flows through the first tube row, it is accelerated at the narrowest cross-section between the tubes causing the accelerated particles of dust to strike the surface of the tubes in the second row causing abrasion.) Some guidelines on selecting the type of finned tubes and fin geometries are given by Weierman [1] or Smittle [2].

PARAMETERS AFFECTING THE PERFORMANCE OF FINNED TUBES

The heat transfer and pressure drop across finned tubes are affected by flow conditions, fin geometries, tube layout, and the number of tube rows. The parameters for the flow conditions are the physical properties and the velocity of the fluid across the tube bank. They are presented in the dimensionless form of the Prandtl number Pr for heat transfer and the Reynolds number Re for heat transfer and pressure drop.

$$\mathrm{Re} = \frac{\rho u d}{\eta} \tag{1}$$

For the characteristic diameter d, several equivalent diameters are also chosen besides the outer diameter of the tube d_0.

Parameters for fin geometries as shown in Figure 3 are d_0, d_f, h, p, and t. The outer surface area of the tube A_0 is

$$A_0 = \pi d_0 L \tag{2}$$

and the total surface area of the finned tube A_t is

$$A_t = A_f + A_b \tag{3}$$

where

$$A_f = \left[\frac{\pi}{4}(d_f^2 - d_0^2)2x + \pi d_f t\right] nL \tag{4}$$

$$A_b = \pi d_0 (1 - nt) L \tag{5}$$

respectively. The finning factor σ is defined as

$$\sigma = \frac{A_t}{A_0} = 1 + \frac{d_f^2 - d_0^2}{2 d_0 p} + \frac{2ht}{d_0 p} \tag{6}$$

with

$$p = 1/n \tag{7}$$

and Equations 2 through 5.

If the fin thickness t is negligible compared to the fin pitch p (i.e. $t/p \ll 1$), then Equation 6 can be approximated as

$$\sigma \cong 1 + \frac{d_f^2 - d_0^2}{2d_0 p} = 1 + 2(h/p) + 2(h/d_0)(h/p) \tag{8}$$

Parameters for tube layouts shown in Figure 2 are P_t and P_l or P_l' for staggered layouts.

The average heat transfer coefficient α based on the outer surface area of the finned tube is presented either in the form of the Nusselt number Nu

$$Nu = \frac{\alpha d}{\lambda} \tag{9}$$

or in the form of the j-factor

$$j = \frac{\alpha}{\rho u c_p} Pr^{2/3} = Nu Re^{-1} Pr^{-1/3} \tag{10}$$

Besides the outer diameter of the tube d_0, the hydraulic equivalent diameter d_{eH} is presented as

$$d_{eH} = \frac{2A_t}{\pi L_p}$$

$$L_p = 2(2hn + 1)L$$

$$d_{eH} = \frac{(d_f^2 - d_0^2)/2p + d_f t/p + d_0(1 - t/p)}{(d_f - d_0)/p + 1} \tag{11}$$

or the equivalent diameter d_{eA}

$$d_{eA} = d_0 \frac{A_t}{A_0} = \sigma d_0 \tag{12}$$

are also used for the characteristic diameter d in Equation 9.

To calculate the overall heat transfer coefficient U from α, fin efficiency φ must be taken into account.

$$\frac{1}{U} = \frac{1}{\alpha} \frac{A_t}{\varphi A_f + A_b} + \frac{1}{\alpha_i} \frac{A_t}{A_i} \tag{13}$$

If the heat resistance due to conduction through the tube wall, contact resistance between the tube and fins, and fouling on the inside and outside cannot be neglected, these must also be added to the right-hand side of Equation 13.

The pressure drop across the finned tube bank ΔP is presented in the form of Euler number Eu using the mass flux at minimum flow area G_m

$$Eu = \frac{\Delta P \rho}{G_m^2 N} \tag{14}$$

or with the friction factor ξ

$$\Delta P = \xi \frac{G_m^2 N}{2\rho} \tag{15}$$

or in the American literature with Fanning factor f

$$\Delta P = 2f\frac{G_m^2 N}{\rho} \tag{16}$$

The relationship among Eu, ξ, and f is

$$Eu = \xi/2 = 2f \tag{17}$$

As in the case of using a characteristic diameter d in the Reynolds number to correlate pressure drop, the volumetric equivalent diameter d_{eV} is

$$d_{eV} = \frac{4V}{A_t}$$

$$V = P_t P_l L - \left[\frac{\pi}{4}d_0^2(1-nt)L + \frac{\pi}{4}d_f^2 nLt\right]$$

$$d_{eV} = \frac{\frac{4}{\pi}P_t P_l - [d_0^2(1-t/p) + d_f^2 t/p]}{(d_f^2 - d_0^2)/2p + d_f t/p + d_0(1-t/p)} \tag{18}$$

is also used besides the outer diameter d_0. Some other empirical equivalent diameters are also used sometimes. Equation 18 can be approximated as

$$d_{eV} \cong d_0\left(\frac{4}{\pi}\frac{P_t}{d_0}\frac{P_l}{d_0} - 1\right)/\sigma \tag{19}$$

if $t/p \ll 1$. This volumetric equivalent diameter d_{eV} represents the compactness of heat exchangers.

HEAT TRANSFER CORRELATIONS FOR SMOOTH CIRCULAR FINNED TUBES

Because of the complexity of the flow structure, theoretical treatment of flow across the finned tube bank is not easy. Therefore most investigations have been carried out experimentally to determine the heat transfer coefficient and pressure drop.

The experimental works that have been conducted and the empirical correlations that have been proposed are reviewed in this section for heat transfer and in the next section for pressure drop.

Each correlation has limitations for applicability, because they are derived purely experimentally. Some guidelines for selecting a correlation for designing a specific heat exchanger are described later.

Heat transfer correlations are presented in most cases by simple power law equations of dimensionless quantities involving the affecting parameters described in the previous section.

$$Nu = aRe^b Pr^{1/3}\varphi\psi C_N \tag{20}$$

or

$$j = aRe^{b-1}\varphi\psi C_N \tag{21}$$

where a and b are constant, φ is a function of fin geometries, ψ a function of tube layouts, and C_N a function of the number of tube rows.

Correlations for Staggered Layouts

Briggs–Young [3] Correlations

For low finned tubes

$$Nu_0 = 0.134\, Re_0^{0.681} Pr^{1/3} (s/h)^{0.200} (s/t)^{0.1134} \tag{22}$$

or

$$Nu_0 = 0.1507\, Re_0^{0.667} Pr^{1/3} (s/h)^{0.164} (s/t)^{0.075} \tag{23}$$

For high finned tubes

$$Nu_0 = 0.1378\, Re_0^{0.718} Pr^{1/3} (s/h)^{0.296} \tag{24}$$

These correlations were derived by Briggs and Young based on their own experimental data using six row tube banks in equilateral triangular pitches.

Schmidt [4] Correlation

$$Nu_A = 0.45\, Re_A^{0.625} Pr^{1/3} \tag{25}$$

or

$$Nu_0 = 0.45\, Re_0^{0.625} Pr^{1/3} \sigma^{-0.375} \tag{25$'$}$$

This correlation is based on data from various sources. Schmidt has proposed to introduce the Nusselt number and the Reynolds number based on the equivalent diameter d_{eA} to express the influence of fin geometries, especially of h/p on heat transfer.

Limitations given by Schmidt [5] $Re_0 = 2 \times 10^3 - 5 \times 10^5$, $\sigma = 5-12$ and $N \geqq 4$. But, Zukauskas [6] recommends this correlation for $Re_0 = 5 \times 10^2 - 10^4$.

Stasiulevicius–Survila [9] Correlations

For $Re_0 = 10^4 - 2 \times 10^5$

$$Nu_0 = 0.043\, Re_0^{0.8} (p/d_0)^{0.18} (h/d_0)^{-0.14} (P_t/P_l)^{0.2} \tag{26}$$

For $Re_0 = 2 \times 10^5 - 1.3 \times 10^6$

$$Nu_0 = 0.0069\, Re_0^{0.95} (p/d_0)^{0.18} (h/d_0)^{-0.14} (P_t/P_l)^{0.2} \tag{27}$$

Yudin and Tokhtarova [7] conducted an experiment in the higher range of Reynolds numbers up to $Re_0 = 10^6$ with finned tubes fabricated by lathe, and found that the tendency of the Nusselt number versus the Reynolds number changes discontinuously at about $Re_0 = 2 \times 10^5$, but they gave no generalized correlations. Stasiulevicius et al. [8] found the same tendency with their own experiment also in high Reynolds numbers using spirally wound trapezoidal finned tubes and derived heat transfer correlations. Stasiulevicius and Survila [9] added further data on tube banks in wider tube pitches and revised the previous correlations. These are Equations 26 and 27.

Zukauskas [6, 10] recommends these correlations for $p/d_0 = 0.125$–0.28, $h/d_0 = 0.125$–0.59, $P_t/d_0 = 2.17$–4.13 and $P_l/d_0 = 1.27$–2.14.

Zozulya–Khavin [11] Correlations

$$Nu_0 = 0.225\, Re_0^{0.62} Pr^{0.33} \left(\frac{P_t/d_0 - 1}{P_{tl}/d_0 - 1} \right)^{0.4} C_N \tag{28}$$

$$P_{tl} = \sqrt{0.25P_t^2 + P_l^2}$$

where $C_N = 1$ for $N \geqq 4$ and

$$C_N = \frac{1.70 + (N-2)}{N} \tag{29}$$

for N = 2 or 3, and

$$Nu_0 = 0.165\, Re_0^{0.6} Pr^{0.33} (P_t/d_0)^{0.5} \tag{30}$$

for N = 1. These correlations were obtained from an experiment on spirally wound, welded finned tubes with a large tube diameter (d_0 = 104 mm), which are used, for example in large capacity ammonia synthesis plants.
Limitations given are $P_t/d_0 = 1.54–2.50$ and $P_l/d_0 = 1.54–2.50$ for Equations 28 and 29, and $P_t/d_0 = 1.54–4.35$ for Equation 30.

Mirkovic [12] Correlation

$$Nu_H = 0.224\, Re_H^{0.662} Pr^{0.33} \left(\frac{1-t/p}{h/p}\right)^{-0,25} (P_t/d_0 - 1)^{0.1} \cdot (P_l/d_0 - 1)^{-0.15} \tag{31}$$

for $N \geqq 8$. Equation 31 is to be multiplied by the correction factor C_N according to Figure 4, if the number of rows is fewer than eight.

Mirkovic derived this correlation based on his own experiment, in which the fin geometries and the tube pitch were changed systematically.

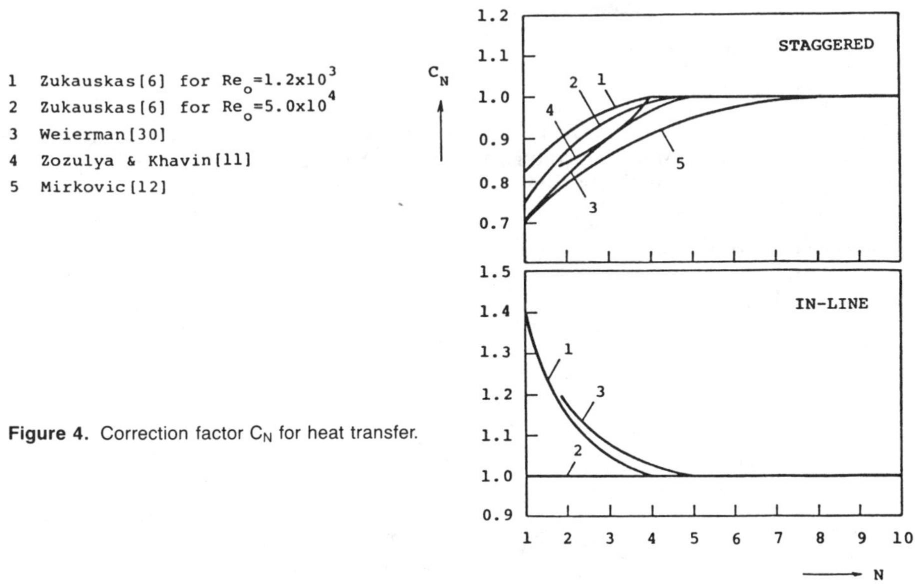

Figure 4. Correction factor C_N for heat transfer.

Correlations for In-line Layouts

Schmidt [4] Correlation

$$Nu_A = 0.30\, Re_A^{0.625} Pr^{1/3} \tag{32}$$

or

$$Nu_0 = 0.30\, Re_0^{0.625} Pr^{1/3} \sigma^{-0.375} \tag{32}$$

This correlation was derived from the same considerations as in Equation 25. Limitations given by Schmidt [5] are the same as for Equation 25, but the recommendation by Zukauskas [6] is $Re_0 = 5 \times 10^3 - 10^5$.

PRESSURE DROP CORRELATIONS FOR SMOOTH CIRCULAR FINNED TUBES

In the same manner as for heat transfer, pressure drop correlations are also presented by simple power law equations of dimensionless quantities involving the affecting parameters described earlier.

$$Eu,\ \xi,\ f = aRe^{-b} \varphi \psi C_N \tag{33}$$

where a and b are constant, φ is a function of fin geometries, ψ a function of tube layouts, and C_N a function of the number of tube rows.

Correlations for Staggered Layouts

Gunter–Shaw [13] Correlation

$$\frac{\Delta P d_{eV} \rho}{G_m^2 L_F} \left(\frac{\eta}{\eta_W}\right)^{0.14} \left(\frac{d_{eV}}{P_t}\right)^{-0.4} \left(\frac{P_1'}{P_t}\right)^{-0.6} = \zeta \tag{34}$$

$$\zeta = 90\, Re_V^{-1} \tag{35}$$

for $Re_V < 200$, and

$$\zeta = 0.96\, Re_V^{-0.145} \tag{36}$$

for $Re_V > 200$. Where L_F is the fluid flow length, η and η_W are the viscosity for an average fluid and at the surface wall temperature. Gunter and Shaw derived these correlations from an analysis of experimental data from various sources for air, water, and oil.

Jameson [14] Correlation

$$f = 1.532\, Re_J^{-0.25} \tag{37}$$

$$Re_J = \frac{G_m^2 d_{eJ}}{\eta}$$

$$d_{eJ} = \frac{d_{eH}}{\left[(h/2s)^{0.4}\left(\dfrac{1}{2\sqrt{P_t/d_{eH}-1}} + \dfrac{1}{2\sqrt{P_1'/d_{eH}-1}}\right)\right]^4} \tag{38}$$

Jameson experimented with finned tubes for commercial air coolers. By introducing the empirical equivalent diameter d_{eJ}, he was able to express the Fanning factor simply including the the tube pitch effects.

Robinson–Briggs [15] Correlation

$$Eu = 18.93\ Re_0^{-0.316}(P_t/d_0)^{-0.927}(P_t/P_1')^{0.515} \tag{39}$$

Robinson and Briggs conducted an experiment with smooth, integral, spiral finned tubes especially for designing air coolers, and found the effect of the ratio P_t/d_0 on pressure drop. But the effect of fin geometries was not investigated in this experiment.

Limitations given are $Re_0 = 2\times10^3$–5×10^4 and $P_t/d_0 = 1.8$–4.6.

Stasiulevicius–Survila [9] Correlations

$$Eu = 6.55\ Re_0^{-0.25}(1-p/d_0)^{1.8}(1-h/d_0)^{-1.4}(P_t/d_0)^{-0.55} \cdot (P_1/d_0)^{-0.5} \tag{40}$$

for $Re_0 = 10^4 - 10^5$, and

$$Eu = 0.37(1-p/d_0)^{1.8}(1-h/d_0)^{-1.4}(P_t/d_0)^{-0.55}(P_1/d_0)^{-0.5} \tag{41}$$

for $Re_0 = 10^5 - 1.3\times10^6$. Stasiulevicius and Survila derived these correlations in the same background as Equations 26 and 27 for heat transfer. They found that the Euler number becomes independent of the Reynolds number when $Re_0 \geqq 10^5$.

Zozulya–Khavin [11] Correlations

$$Eu = 2.8\ Re_0^{-0.25}\left(\frac{N+1}{N}\right) \tag{42}$$

for $N \geqq 2$, and

$$Eu = 19.2\ Re_0^{-0.33}(P_t/d_0)^{-1.3} \tag{43}$$

for $N = 1$. Zozulya and Khavin obtained these correlations based on the same experiment for Equations 28 to 30. Parameters on tube pitches are not included in Equation 42, because this effect could not be recognized in their experimental range.

Mirkovic [12] Correlation

$$Eu = 3.96\ Re_V^{-0.31}\left(\frac{1-t/p}{h/p}\right)^{-0.20}(P_t/d_0-1)^{0.14} \cdot (P_1/d_0-1)^{-0.18} \tag{44}$$

Mirkovic derived this correlation from the same experiment as for Equation 31. He chose the volumetric equivalent diameter d_{eV} as the characteristic diameter for pressure drop and the hydraulic equivalent diameter d_{eH} for heat transfer.

de Bruyn [16] Correlation

$$Eu = 104\ Re_0^{-0.58}(s/p)^{-7.47}(P_t/d_0)^{-1.45} \tag{45}$$

This correlation is based on de Bruyn's experiment and another experiment, where the effect of P_t on pressure drop was mainly studied. The finned tubes used had spirally wound L-type fins for air-cooled heat exchangers.

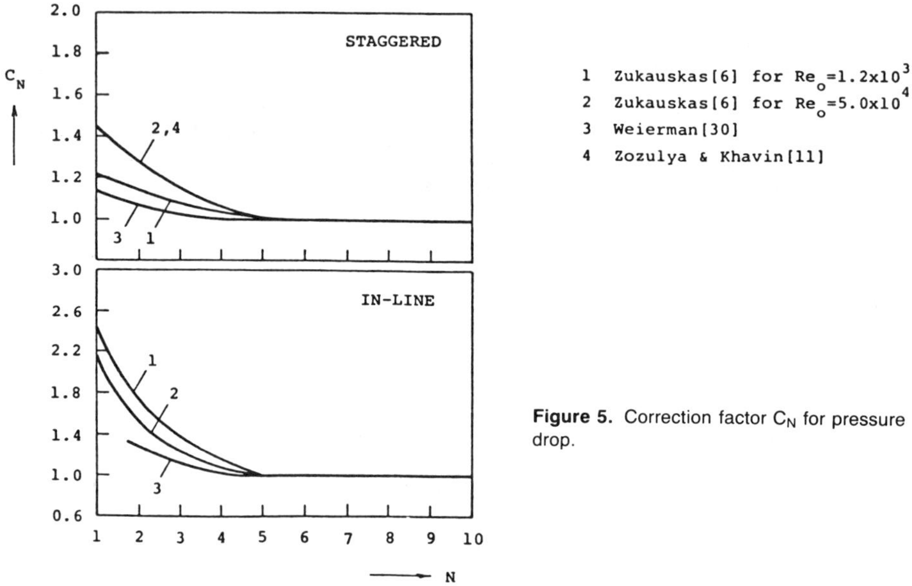

Figure 5. Correction factor C_N for pressure drop.

Lokshin–Fomina [17] Correlations

For $Re_0 = 10^3 - 10^5$

$$Eu = 3.0\, Re_0^{-0.25} \sigma^{0.5} (P_t/d_0)^{-0.55} (P_l/d_0)^{-0.5} C_N \tag{46}$$

For $Re_0 = 10^5\text{–}1.4 \times 10^6$

$$Eu = 0.18 \sigma^{0.5} (P_t/d_0)^{-0.55} (P_l/d_0)^{-0.5} C_N \tag{47}$$

Lokshin and Fomina compared data from several sources with Stasiulevicius–Survila correlations expressed in Equations 40 and 41, and found that these correlations do not agree with data for $\sigma \geqq 10$. They revised the Stasiulevicius–Survila correlations using the finning factor σ for the effect of fin geometries.

The correction factor for the tube row effect C_N can be read from the curves given by Zukauskas [6] in Figure 5. These correlations could be applied not only for spiral finned tubes, but also for annular finned tubes or bare tubes ($\sigma = 1$). Zukauskas [6] recommends these correlations for $\sigma = 1.9\text{–}16.3$, $P_t/d_0 = 1.6\text{–}4.13$, and $P_l/d_0 = 1.2\text{–}2.35$.

Schack [18] Correlations

Schack analyzed a total of 2,083 data points from various sources and derived these correlations.

For $Re_0 = 10^3 - 1.5 \times 10^5$

$$\xi = 10.45\, Re_0^{-0.22} \Pi \tag{48}$$

For $Re_0 > 1.5 \times 10^5$

$$\xi = 0.80 \Pi \tag{49}$$

where Π is

$$\Pi = \left(1 + \frac{d_f/d_0 + 1}{p/d_0 + t/d_0}\right)^{0.48} (P_t/d_0)^{-1.25} (A_m/A_d)^{0.53} \tag{50}$$

Correlations for In-line Layouts

Gunter–Shaw [13] Correlation

Gunter and Shaw stated that Equations 34 through 36 could be applied also for in-line layouts, if P_l is used instead of P_l'. But they did not show any comparison with experimental data.

Lokshin–Fomina [17] Correlation

$$Eu = 0.068\sigma^{0.5} \left(\frac{P_t/d_0 - 1}{P_l/d_0 - 1}\right)^{-0.4} C_N \tag{51}$$

for $Re_0 = 10^3 - 10^5$. This correlation was based on the same consideration as in Equations 46 and 47. The correction factor for the tube row effect C_N can be read from Zukauskas [6] shown in Figure 5.

Zukauskas [6] recommends this correlation for $Re_0 = 10^3$–10^5, $\sigma = 1.9$–16.3, $P_t/d_0 = 2.83$–3.13, and $P_l/d_0 = 1.2$–2.35.

Schack [18] Correlations

For $Re_0 = 10^3 - 10^5$

$$\xi = 0.565\, Re_0^{-0.16} \Pi \tag{52}$$

For $Re_0 \geqq 10^5$

$$\xi = 0.09\Pi \tag{53}$$

where

$$\Pi = (d_f/d_0 - 0.8)^{-0.32} (P_l/d_0)^{0.59} d_{eU}^{-0.67} \tag{54}$$

$$d_{eU} = 2(P_t/d_0)\left(\frac{d_f/d_0 - 1}{p/d_0 + t/d_0} + 1\right)^{-1} (A_m/A_d) \tag{55}$$

Schack gave these correlations from the same analysis as in Equations 48 through 50, based on 366 data points from various sources.

SOME REMARKS ON FINNED TUBE HEAT EXCHANGER DESIGN

Fin Efficiency

For the thermal design of finned tube heat exchangers using one of the correlations from the fourth section, fin efficiency φ in Equation 13 must be taken into account.

Gardner [19] derived generalized relationship theoretically for various fin shapes. The results were given as equations in terms of the Bessel function and also as diagrams. In one specific case of flat fins without a change of width in the fin height direction and with constant fin thickness, the equation becomes simple as follows.

$$\varphi = \tanh(mh)/(mh) \tag{56}$$

with

$$m = \sqrt{2\alpha/(\lambda_f t)} \tag{57}$$

Schmidt [20] showed that Equation 56 can be approximately applied also to circular fins, when the equivalent fin height (h_e)

$$h_e = h[1 + 0.35 \ln(d_f/d_0)] \tag{58}$$

is used instead of the actual fin height (h).

The theoretical treatment of the fin efficiency by Gardner is based on some idealized assumptions such as that the heat transfer coefficient is the same over all of the fin surface or the temperature of the surrounding fluid is uniform.

Actually in circular finned tubes, however, these assumptions are not valid. Large nonuniformity in the heat transfer coefficient has been observed over the fin surface in the circumferential and also in the fin height direction by Weiner et al. [21], Zukauskas et al. [22], Legkiy et al. [23, 24], Krückels and Kottke [25], or Jones and Russell [26].

Lymer and Ridal [27] compared the actual fin efficiencies of several finned tubes with the Gardner method and found that the Gardner method is unsatisfactory for surfaces with nonuniformly distributed heat transfer coefficients when fin efficiency is low.

To correct the theoretically obtained fin efficiency, some correction factors were proposed by Russian researchers in the form of

$$\varphi_{act} = \psi_u \psi_t \varphi_{th} \tag{59}$$

where φ_{act} = the actual fin efficiency
ψ_u = a correction factor for nonuniformity effect in the heat transfer coefficient
ψ_t = a correction factor for the trapezoidal profile of the fin
φ_{th} = the theoretically derived fin efficiency

For the correction factor ψ_u for the nonuniformity effect in the heat transfer coefficient

$$\psi_u = 0.97 - 0.056\, mh \tag{60}$$

was given by Zukauskas et al. [22] for mh = 0.3–3.0 and $Re_0 = 4.8 \times 10^4$–7.6×10^5,

$$\psi_u = 1 - 0.058\, mh \tag{61}$$

by Yudin and Tokhtarova [28] for σ = 2.4–18.2 and mh = 0.1–3.7, or a diagram by Kuznetsov and Pshenisnov [29].

For the correction factor ψ_t for the trapezoidal profile of the fin

$$\psi_t = 1 + (0.125 - 0.125\sqrt{t_1/t_2})mh \tag{62}$$

Table 1
Experimental Conditions for Comparison Between Spiral and Annular Finned Tubes

	d_o [mm]	d_f [mm]	p [mm]	t [mm]	σ	
Legkiy and Tupitsyn [32]	16.1	24.1	2.85	0.6	4.6	Single tube
Hashizume [33]	19.1	42.0	3.0	0.4	13.4	Tube banks in staggered and in-line layouts each with five rows

was given by Zukauskas [6], where t_1 and t_2 are the fin thickness at the tip and at the root, respectively. Equations 60 through 62 are valid only for staggered layouts. Weierman [30] gave the actual fin efficiency φ_{act} using data from various sources as follows.

$$h' = h + t/2$$

$$X = \tanh(mh')/(mh')$$

$$Y = X(0.7 + 0.3X)$$

$$\varphi_{act} = Y[0.45 \ln (d_f/d_0)(Y - 1) + 1] \tag{63}$$

Equation 63 can be applied for staggered and also for in-line layouts.

Tube Row Effect on Heat Exchanger Performance

Some heat transfer and pressure drop correlations include the correction factor C_N for the effect of tube rows, whereas in other correlations its effect is not considered. Figures 4 and 5 show the effect of the number of tube rows on the heat transfer coefficient and pressure drop from literature, in a representative form of

$$C_N = \frac{\alpha_N}{\alpha_\infty} \quad \text{or} \quad \frac{\Delta P_N}{\Delta P_\infty} \tag{64}$$

where the subscripts N and ∞ mean the average value of tube banks with N row tubes or a very large number of tube rows, respectively.

We can see from Figures 4 and 5 that the performance of tube banks with larger than four tube rows are not affected by the number of tube rows, and that the tube row effect in tube banks with fewer tube rows are stronger in in-line layouts than in staggered layouts.

The values of C_N for fewer than four tube rows, however, scatter widely. The most possible explanation of this may be the difference of the turbulence level of the approaching fluid.

Zozulya et al. [31] investigated the effect of the approaching flow turbulence level on heat transfer coefficients using a spiral finned tube bank in a six-row staggered layout. They measured the turbulence level in the longitudinal component and the heat transfer coefficient for each row and found that the heat transfer coefficient for the first row increases from 20% to 35% when the turbulence level increases from 2.5%–3% to 22%–25%. After the third or fourth row, however, this effect was not recognized.

Difference Between Spiral and Annular Finned Tubes

Experimental data for spiral and annular finned tubes are usually considered together. For example, Lokshin and Fomina [17] stated that Equations 46 and 47 can be applied both to spiral and annular finned tubes.

Legkiy and Tupitsyn [32] conducted an experiment on heat transfer with a single finned tube, and found that the heat transfer coefficients for the annular finned tube were somewhat higher than those for spiral finned tubes, but the discrepancy was within the experimental error.

Hashizume [33] experimented using both types of finned tubes in staggered and in-line layouts. He found that the heat transfer coefficient and pressure drop are from 4% to 14% and from 10% to 20% lower for annular finned tubes than for spiral finned tubes.

These two experiments do not show the same results. One of the reasons for this fact may lie in the difference of fin geometries and the tube layouts, as shown in Table 1. Further

Table 2
Heat Transfer Correlations

Author(s)	d_0 [mm]	h/d_0	h/p	σ	P_t/d_0	P_l/d_0	N	Re_0	Equation(s)
Staggered Briggs–Young [3]	13.6–15.9 11.1–40.9	0.09–0.34 0.35–0.84	1.1–1.9 1.6–6.3	3.2–6.1 6.2–22.0	1.5–2.0 1.8–2.6	Equilateral Triangular	≧ 6	1.2×10^3–2×10^4	23 24 22
Schmidt [4]	9.7–29.0	0.37–0.71	1.7–3.7	5–12	1.7–3.8	1.0–2.5	≧ 4	2×10^3–5×10^5	25
Stasiulevicius– Survila [9]	23.0–32.0	0.13–0.59	1.0–3.4	3.1–10.6	2.2–4.1	1.3–2.1	≧ 5	10^4–2×10^5 2×10^5–1.3×10^6	26 27
Zozulya– Khavin [11]	104.0	0.19	2.8	5.8	1.5–2.5 1.5–4.4	1.5–2.5 –	≧ 2 1	3.5×10^4–3×10^5	28 30
Mirkovic [12]	25.4–50.8	0.19–0.38	1.5–3.8	4.8–11.1	1.9–4.7	1.6–3.2	≧ 1	Re_H = 3×10^3–6×10^4	31 with Figure 4
In-line Schmidt [4]	15.0–29.0	0.36–0.70	1.4–4.3	5–12	1.7–3.0	1.9–4.0	≧ 4	2×10^3–5×10^5	32

Table 3
Pressure Drop Correlations

Author(s)	d_0 [mm]	h/d_0	h/p	σ	P_t/d_0	P_l/d_0	N	Re_0	Equation(s)
Staggered									
Gunter–Shaw [13]	0.5–50.8	–	–	–	1.3–5.0	1.3–5.0	≧ 3	$Re_V < 2 \times 10^2$ Re_V = 2×10^2–3×10^5	35 36
Jameson [14]	15.9–25.4	0.37–0.48	1.7–3.2	5.8–10.4	2.0–3.6	1.0–3.2	≧ 2	$Re_J = 5 \times 10^2$–10^4	37
Robinson– Briggs [15]	18.6–40.9	0.35–0.56	4.5–5.3	13.3–17.4	1.8–4.6	1.8–4.6	≧ 6	2×10^3–5×10^4	39
Stasiulevicius– Survila [9]	23.0–32.0	0.13–0.59	1.0–3.4	3.1–10.6	2.2–4.1	1.3–2.1	≧ 7	10^4–10^5 10^5–1.3×10^6	40 41
Zozulya– Khavin [11]	104.0	0.19	2.8	5.8	1.5–2.5 1.5–4.4	1.5–2.5 –	≧ 2 1	3.5×10^4–3×10^5	42 43
Mirkovic [12]	25.4–50.8	0.19–0.38	1.5–3.8	4.8–11.1	1.9–4.7	1.6–3.2	≧ 1	Re_V = 2×10^3–3×10^4	44
de Bruyn [16]	26.3	0.59	5.3–5.5	18.6	2.2–3.2	Equilateral Triangular	≧ 3	1.8×10^3–1.3×10^4	45
Lockshin– Fomina [17]	–	–	–	1.9–16.3	1.6–4.1	1.2–2.4	≧ 1	10^3–10^5 10^5–1.4×10^6	46 with Figure 5 47
Schack [18]	–	–	–	–	–	–	–	10^3–1.5×10^5 $> 1.5 \times 10^5$	48 49
In-line Lockshin– Fomina [17]	–	–	–	1.9–16.3	2.8–3.1	1.2–2.4	≧ 1	10^3–10^5	51 with Figure 5
Schack [18]	–	–	–	–	–	–	–	10^3–10^5 $> 10^5$	52 53

experiments are needed to clarify the difference in characteristics between spiral and annular finned tubes.

Guidelines for Selecting Correlations

Heat transfer and pressure drop correlations in the previous sections were obtained purely empirically based on data within some limited conditions. Therefore these correlations can, in principle, only be applied within the experimental conditions of data on which they were based. The best method to select one correlation for heat exchanger design is to find one for conditions matching those being designed. Tables 2 and 3 show the experimental conditions for each correlation.

When the conditions being designed cannot be perfectly matched to a set of conditions shown in Tables 2 and 3, then some correlations must be used with extrapolation. In such a case the important parameters to be taken into account are the ratio of fin height-to-tube diameter h/d_0, the ratio of fin height to fin pitch h/p, and the finning factor σ. For heat transfer and pressure drop characteristics of finned tubes with various fin geometries, we can consider two extreme cases. One extreme is the case when the fin height is too small compared to the tube diameter, i.e. $h/d_0 \ll 1$. Another one is the case when the tube can be considered to be negligible. In the former case the characteristics become approximated to that for bare tubes. In the latter case, the characteristics approach that for flow between parallel plates, where the dominant parameter is h/s or h/p when $t/p \ll 1$. The finning factor σ is a function of these two parameters h/d_0 and h/p as shown in Equation 8.

The characteristics of fin tubes used in practical situations can be related to those of bare tubes as tried by Biery [34] for heat transfer and by Kast and Eberhard [35] for pressure drop. For the performance of finned tube banks with tube rows fewer than four rows, the correlations listed in Tables 2 and 3 may be extrapolated.

SPECIAL FINNED TUBES AND THEIR CHARACTERISTICS

Segmented Finned Tubes

A segmented finned tube is fabricated by spirally winding and welding a fin strip on a tube. The fin strip is cut crosswise at certain intervals leaving uncut parts along one edge prior to being wound. As the fin strip is wound around the tube, the individual segments separate at the outer edge to form individual rectangular fins, as shown in Figure 6A.

This type of finned tube can have higher fins than the limitation in spirally wound smooth finned tubes, whose fin height is limited by the metal stretch at the fin tip. Another advantage in a segmented finned tube is the effect of heat transfer enhancement with a relatively small pressure drop increase. Figure 7 shows the performances of segmented finned tubes compared with those of smooth finned tubes with the same fin geometries and layouts, rewritten from Weierman [30]. From these figures it can be expected that

$$\frac{\alpha_{segmented}}{\alpha_{smooth}} > \frac{\Delta P_{segmented}}{\Delta P_{smooth}} \quad \text{for staggered layouts}$$

when $h/s \geqq 6$, although this is not comfirmed by experimental data.

Heat transfer and pressure drop data on some specific segmented finned tubes can be found in Rabas and Eckels [36], Weierman [37, 38], and Weierman et al. [39].

One important remark in using segmented finned tubes for heat exchangers, given by Smittle [2], was that this type of finned tubes should be used only in a clean gas stream, because soot can quickly accumulate in its segmentation, making cleaning real problem.

Corrugated Finned Tube

The corrugated finned tube shown in Figure 6B was born from consideration of cheap manufacturing. It is fabricated by spirally winding a corrugated fin strip on a tube. As the

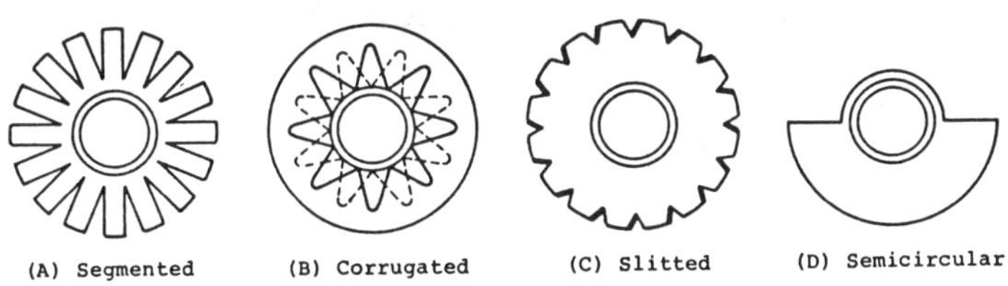

Figure 6. Special finned tubes: (A) segmented; (B) corrugated; (C) slitted; (D) semicircular.

fin strip is wound on the tube, the fin tip is stretched to becoming smooth, while the fin root remains corrugated.

Through corrugation of the fin strip, it becomes possible to form higher fins than the limitation in spirally wound smooth finned tubes, as in the case of segmented finned tubes. In addition to higher fin height, this type of finned tube has some advantages compared to smooth finned tubes.

The surface area A_f of corrugated fins per unit length is

$$A_f \cong 2\pi nh(d_0 + 2h) \tag{65}$$

whereas that of smooth fins is

$$A_f \cong 2\pi nh(d_0 + h) \tag{66}$$

from Equation 4 for $t/h \ll 1$. From Equations 65 and 66 it is evident that the surface area of the corrugated finned tube is larger than that of smooth finned tubes for the same fin height. The corrugation increases the contact length between the fins and tube, which is especially advantageous for tension wound finned tubes.

Furthermore the turbulence promotion effect can be expected from the corrugation. If the corrugation is made too deep, however, it will bring a pressure drop increase at the same time and also some air stagnation around the base of the corrugation, which do not contribute to heat transfer. Heat transfer and pressure drop data on some specific corrugated finned tubes can be found in Sherwin [40, 41] and Upmalis [42].

Slitted Finned Tube

By slitting fins the turbulent flow of air in the inter-fin spaces is promoted and heat transfer enhancement can be expected, which, however, in most cases is accompanied with an increase in pressure drop.

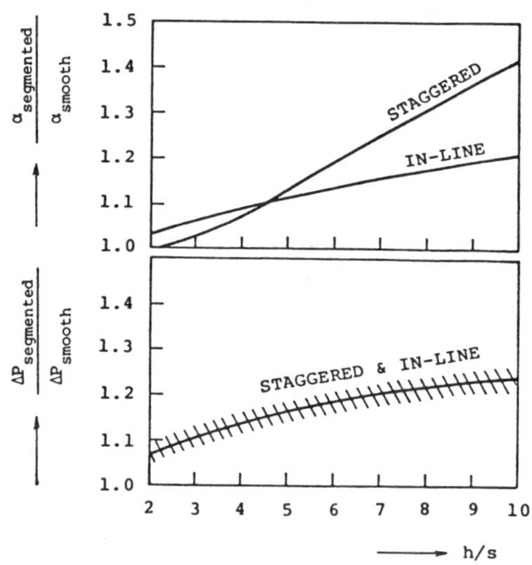

Figure 7. Comparison of performances between smooth and segmented finned tubes (after Weierman [30]).

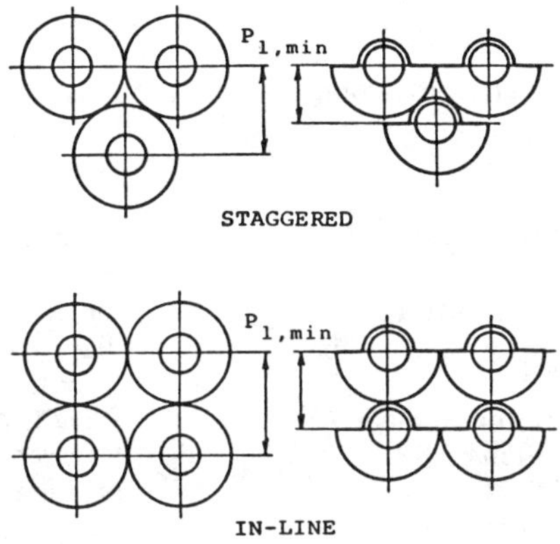

Figure 8. Minimum tube pitch $P_{1,min}$ of circular and semicircular finned tubes.

Kokorev et al. [43] developed a special slitted finned tube, whose fins are crosscut and deformed only in the tip section as shown in Figure 6C. Their experiment on this special finned tube with a slit depth of 2.5 to 4.0 mm from the fin tip and 11 to 25 slits over the circumference showed a heat transfer increase of 57% with a pressure drop increase of only 38% compared with smooth finned tubes with the same fin geometries.

This experiment suggests possibilities of developing other fin geometries that enhance heat transfer with a minimum amount of increase in pressure drop. But the accumulation of dust or soot must be taken into account for the same reason as in the case of segmented finned tubes.

Semicircular Finned Tube

Semicircular finned tube was developed especially for waste heat boilers, where the allowable pressure drop is limited, high compactness is required, and dryout risk should be reduced.

The downstream sides of the fins of this type of finned tubes are cut away, as shown in Figure 6D. The downstream side of the fins lies in the wake region, where heat transfer is poor. Therefore, the cutting away of this region of fins brings an increase of the average heat transfer coefficient compared with circular finned tubes.

This was first suggested by Weiner et al. [21]. Hashizume [33] conducted a flow pattern observation around the finned tube banks, and found that the flow pattern is unchanged by cutting away the downstream side of the fins. His experiment on heat transfer and pressure drop showed that the semicircular finned tubes have a 50% lower pressure drop for the same heat transfer coefficient than in circular finned tubes in staggered and also in-line layouts.

The semicircular finned tubes can be arranged more closely than the circular finned tubes, as shown in Figure 8, which realizes a compact heat exchanger. In addition, in case of waste heat boilers with horizontal evaporator tubes, there is a potential risk of dryout, causing severe corrosion if harmful chemicals are presented in the boiler water, as pointed out by Hinchley [44]. If the heat flux on the upper side of the evaporator tube is lower, which is the case for a horizontal semicircular finned tube with the heating gas flow upward, then the risk of dryout can be decreased. Thus, the semicircular finned tube seems to have particular application for use as evaporator tubes in waste heat boilers.

NOTATION

- A_b outside surface area excluding fins
- A_d frontal flow area
- A_f surface area of fins
- A_i inside surface area of tube
- A_m minimum flow area
- A_0 outside surface area of tube
- A_t total surface area of finned tube ($= A_f + A_b$)
- C_N correction factor for number of tube rows; Equation 64
- c_p specific heat of fluid
- d_{eA} area equivalent diameter; Equation 12
- d_{eH} hydraulic equivalent diameter; Equation 11
- d_{eV} volumetric equivalent diameter; Equation 18
- d_f outside diameter of fins
- d_0 outside diameter of tube
- Eu Euler number; Equation 14
- f Fanning factor; Equation 16
- G_m mass flux at minimum flow area ($= \rho u_m$)
- h fin height
- j Colburn j-factor; Equation 10
- L length of tube
- L_p perimeter of finned tube for d_{eH}
- N number of tube rows
- Nu_A Nusselt number based on d_{eA} ($= \alpha d_{eA}/\lambda$)
- Nu_H Nusselt number based on d_{eH} ($= \alpha d_{eH}/\lambda$)
- Nu_0 Nusselt number based on d_0 ($= \alpha d_0/\lambda$)
- n number of fins per unit tube length ($= 1/p$)
- P_l longitudinal tube pitch
- P_l' diagonal tube pitch in staggered layout
- Pr Prandtl number
- P_t transversal tube pitch
- ΔP pressure drop across tube bank
- Re_A Reynolds number based on d_{eA} ($= G_m d_{eA}/\eta$)
- Re_H Reynolds number based on d_{eH} ($= G_m d_{eH}/\eta$)
- Re_0 Reynolds number based on d_0 ($= G_m d_0/\eta$)
- Re_V Reynolds number based on d_{eV} ($= G_m d_{eV}/\eta$)
- s distance between adjacent fins ($= p - t$)
- t fin thickness
- U overall heat transfer coefficient based on A_t
- u_m fluid velocity at minimum flow area
- V net free volume for d_{eV}

Greek Symbols

- α outside heat transfer coefficient based on A_t
- α_i inside heat transfer coefficient based on A_i
- η viscosity of fluid
- λ thermal conductivity of fluid
- λ_f thermal conductivity of fin material
- ξ friction factor; Equation 15
- ρ density of fluid
- σ Finning factor ($= A_t/A_0$). Equation 6
- φ fin efficiency

REFERENCES

1. Weierman, C., *Oil Gas J.* (Nov. 1975), p. 64.
2. Smittle, D., *Power* (Aug. 1980), p. 76.
3. Briggs, D. E., and Young, E. H. *Chem. Eng. Prog. Symp. Ser., 59:* 1 (1963).
4. Schmidt, Th. E., *Kältetech., 15:* 98 and 370 (1963).
5. Schmidt, Th. E., in *Wärmepumpen,* Vulkan Verlag, Essen, 1978, p. 34.

6. Zukauskas, A., in *Heat Exchangers: Thermal–Hydraulic Fundamentals and Design*, S. Kakac, A. E. Bergles and F. Mayinger (Eds.), Hemisphere, Washington D. C., 1981, p. 49.
7. Yudin, V. F., and Tokhtarova, L. S., *Thermal Eng., 20* (2): 65 (1973).
8. Stasiulevicius, J., et al., *Heat transfer Sov. Res., 5* (6): 19 (1973).
9. Stasiulevicius, J., and Survila, V., *Proc. 5th Int. Heat Transfer Conf.*, Tokyo, Sept. 1974, Paper FC 6.5.
10. Zhukauskas, A. A., *Thermal Eng., 21* (5): 40 (1974).
11. Zozulya, N. V., and Khavin, A. A., *Heat Transfer Sov. Res., 6* (6): 82 (1974).
12. Mirkovic, Z., in *Heat Exchangers: Design and Theory Sourcebook*, N. F. Afgan and E. U. Schlünder (Eds.), Scripta, Washington D. C., 1974, p. 559.
13. Gunter, A. Y., and Shaw, W. A., *Trans. ASME,* Nov. 1945, p. 643.
14. Jameson, S. L., *Trans. ASME,* Nov. 1945, p. 633.
15. Robinson, K. K., and Briggs, D. E., *Chem. Eng. Prog. Symp. Ser., 62:* 177 (1966).
16. de Bruyn, R., *CSIR Rep. CENG 119* (PB 267660), 1976.
17. Lokshin, V. A., and Fomina, V. N., *Thermal Eng., 25* (6): 31 (1978).
18. Schack, K., *Chem.-Ing.-Tech., 51:* 986 (1979).
19. Gardner, K. A., *Trans. ASME,* Nov. 1945, p. 621.
20. Schmidt, Th. E., *Kältetech.-Klimatis., 18:* 135 (1966).
21. Weiner, J. H., Gross, D., and Paschkis, V., *Proc. General Discussion on Heat Transfer,* I. Mech, E., Sept. 1951, p. 154.
22. Zhukauskas, A., Stasiulevicius, J., and Skrinska, A., *Proc. 3rd Int. Heat transfer Conf.*, Chicago, 1966, Vol. 3, p. 299.
23. Legkiy, V. M., et al., *Heat Transfer Sov. Res., 6* (6): 101 (1974).
24. Legkii, V. M., Zholudov, Ya. S., and Gerashchenko, O. A., *J. Eng. Phys, 30:* 178 (1976).
25. Krückels, W., and Kottke, V., *Chem.-Ing.-Tech., 42:* 355 (1970).
26. Jones, T. V., and Russell, C. M. B., *ASME Paper,* 78–HT–30 (1978).
27. Lymer, A., and Ridal, B. F., *J. Brit. Nucl. Energy Conf.*, Oct. 1961, p. 307.
28. Yudin, V. F., and Tokhtarova, L. S., *Thermal Eng., 20* (3): 66 (1973).
29. Kuznetsov, N. V., and Pshenisnov, I. F., *Thermal Eng., 21* (8): 58 (1974).
30. Weierman, C., *Oil Gas J.,* Sept. 1976, p. 94.
31. Zozulya, N. V., Vorob'yev, Yu. P., and Khavin, A. A., *Heat Transfer Sov. Res., 5* (1): 154 (1973).
32. Legkiy, V. M., and Tupitsyn, Yu., K., *Heat Transfer Sov. Res., 10* (6): 112 (1978) and *Archiv f. Energiewirtsch.*, Aug. 1978, p. 612.
33. Hashizume, K., *Heat Transfer Eng., 3:* 15 (1981).
34. Biery, J. C., *J. Heat Transfer, 103:* 705 (1981).
35. Kast, W., and Eberhard, V., *Verfahrenstech., 16:* 342 (1982).
36. Rabas, T. J., and Eckels, P. W., *ASME Paper,* 75–HT–45 (1975).
37. Weierman, C., *16th Natl. Heat Transfer Conf.*, St. Louis, Aug. 1976, Paper No. 27.
38. Weierman, C., *Chem. Eng. Prog.*, Feb. 1977, p. 69.
39. Weierman, C., Taborek, J., and Marner, W. J., *AIChE Symp. Ser., 74:* 39 (1978).
40. Sherwin, K., *I. Mech. E. Symp. Compact heat Exchangers,* London, Oct. 1972, Paper C121/72.
41. Sherwin, K., *ASME HTD, 21:* 11 (1981).
42. Upmalis, A., *Wärme, 85:* 11 (1979).
43. Kokorev, B. I., et al., *Thermal Eng., 25* (2): 33 (1978).
44. Hinchley, P., *Chem. Eng. Prog.*, March 1977, p. 90.

CHAPTER 24

THE PERFORMANCE OF EXTENDED-SURFACE HEAT EXCHANGERS

B. M. Johnson

Pacific Northwest Laboratory
Richland, Washington

CONTENTS

INTRODUCTION

Extended surface heat exchangers are specified when heat is transferred between two gas-phase streams or between a gas-phase stream and a liquid stream (either single- or two-phase). In the first case the gas heat transfer coefficient is often so low that it is cost effective to incorporate extended surfaces into the design of both sides of the heat exchanger despite the increased cost of producing these extended surfaces. In the second case, the gas side heat transfer coefficient may be 10 to 50 times smaller than the liquid coefficient, and the use of a fin or extended surface will increase the gas-side conductance (hA) to more evenly proportion the thermal resistance on each side of the heat exchanger. In many applications, the gas-side extended surface can be added for a much lower unit cost than the base surface containing the liquid phase. For example, high-conductivity thin fins applied to tubes whose wall thickness is dictated by pressure requirements will be much less expensive than the equivalent area of base tube material.

The optimum design is ultimately determined by the lowest cost per unit of heat transferred, but the design may be constrained by a number of factors dictated by considerations outside the scope of the actual design of the heat exchanger. For example, the design may be constrained by allowable dimensions or the need for minimum frontal area within the constraints of allowable pressure drop; as for example when the cost of ancillary systems or structure may be more important than the cost of the heat exchanger.

This chapter constitutes work prepared for the U.S. Department of Energy under contract De–AC06–76RLO–1830.

Moreover, the lowest life cycle cost per unit of heat transfer must consider the cost of operations for the heat exchanger. Thus, pumping power and the effects of fouling must be taken into account. The former can be handled precisely; the latter generally requires a great deal of subjective judgment. But the judgment may dictate limiting the bounds of a design parameter (for example, the number of fins per inch) because of the potential for air-side fouling. In the process of optimization, the cost of fabrication is often difficult to put into algorithmic form, but the designer must be very aware that the exchanger design requiring either minimum surface area or minimum volume may not be the lowest cost because of differences in the fabrication cost of different designs.

This chapter includes a discussion of the types of extended surface heat exchangers and their relative performance. The chapter also suggests ways in which extended surface heat exchangers should be compared. A discussion of the enhancement of the extended surface heat exchanger performance by wet surface operation is included.

This chapter does not consider the design of compact heat exchangers or plate heat exchangers and is not a compendium of performance data, although it does reference sources of such data.

HEAT TRANSFER AUGMENTATION BY EXTENDED SURFACES

The use of extended surfaces is particularly cost effective in situations where the air-side heat transfer coefficient may be 10 to 50 times smaller than the liquid-side coefficient and the use of the fin surface will increase the air-side conductance to more evenly proportion the thermal resistance on each side of the heat exchanger. The degree to which the air-side conductance can be augmented depends primarily on three factors:

- The surface ratio (spacing and length of the extended surfaces).
- The fin efficiency (length, thickness, material of fins).
- Enhancement of the heat transfer coefficient on the extended surface through use of special fin configurations.

The significance of these factors on the augmentation that can be achieved is indicated by Equation 1, which defines the overall heat transfer coefficient, based on the internal surface area.

$$\frac{1}{U_i} = \frac{A_i}{\eta h_s A_s} + \frac{1}{h_i} \tag{1}$$

The air-side conductance is then

$$K_a = \eta h_s A_s / A_i \tag{2}$$

and increases in the surface ratio, A_s/A_i, can be achieved by increasing the number of fins per unit length or the length of the fins. The surface efficiency is influenced by the fin thickness, the thermal conductivity, and the fin length. The fin efficiency cannot generally be calculated in closed form, but for the special case of a straight fin with a rectangular cross section, the fin efficiency (η_f) is

$$\eta_f = \frac{\tanh m}{m} \quad \text{where} \quad m = L\sqrt{2h_s/k_t} \tag{3}$$

For other configurations the efficiency may be calculated from appropriate graphs or equations given in many heat transfer textbooks. The overall surface efficiency, which takes into consideration the existence of the outer tube surface not covered by fins, is calculated by

$$\eta = 1 - \frac{A_f}{A_s}(1 = \eta_f) \tag{4}$$

The fin-to-air heat transfer coefficient (h_s) will be influenced by the geometry of the fins and their spacing. The heat transfer coefficient in the region of the development of the thermal boundary layer is much higher than in the region where the boundary layer has had the opportunity to build. Enhancement of the heat transfer coefficient is often achieved by tripping the flow of air over the surface so as to force a redevelopment of the thermal boundary layer. Reducing the fin spacing may also increase the heat transfer coefficient by reducing the hydraulic diameter of the configuration.

It is evident that several tradeoffs exist among the design parameters. Increasing the surface ratio by increasing the number of fins per inch and their length will increase the air-side conductance but will also increase the pressure drop. Consequently, if the available power is kept constant, these measures for increasing surface ratio may decrease the air velocity sufficiently to offset the advantage of greater surface area. As the fin length increases, the fin efficiency will decrease and reduce the cost effectiveness of adding further surface. Air-side fouling will increase as the density of the fins increases, since airborne debris, dust, and seeds are more likely to be caught within the fin structure. The extent to which it is cost effective to increase the air-side conductance will depend upon the liquid-side conductance; the higher this term, the greater the incentive to increase the air-side conductance. Enhancing the surface heat transfer costs money and power, and the question of cost effectiveness must be addressed.

The design and optimization of air-cooled heat exchangers is a topic that has received considerable attention in the recent literature, for example, References 1 and 2. However, the wide variety of extended surface heat exchanger configurations is not simply an indication that the practice of optimization is just now coming into use, but attests to the fact that there are many constraints that can influence the design of extended surface heat exchangers. The optimum configuration can have a wide variety of shapes. New developments in manufacturing processes from time to time may greatly influence the selection of heat exchanger configuration.

Types of Extended Surfaces

Extended surface heat exchangers can be somewhat arbitrarily divided into three configurations:

- Finned tubes (individual tubes of circular or noncircular cross section).
- Plate fins.
- Plate exchangers with corrugated spacer surfaces.

The options available with the first two configurations will be briefly summarized. The third configuration has been extensively treated in numerous papers on compact heat exchangers (e.g. References 1 and 2.

Finned Tubes

This type of heat exchanger has a base tube to which additional surface is added by mechanical means. Most gas/liquid heat exchangers are operated in the crossflow, and the fins oriented so that their surfaces are parallel to the direction of gas flow.

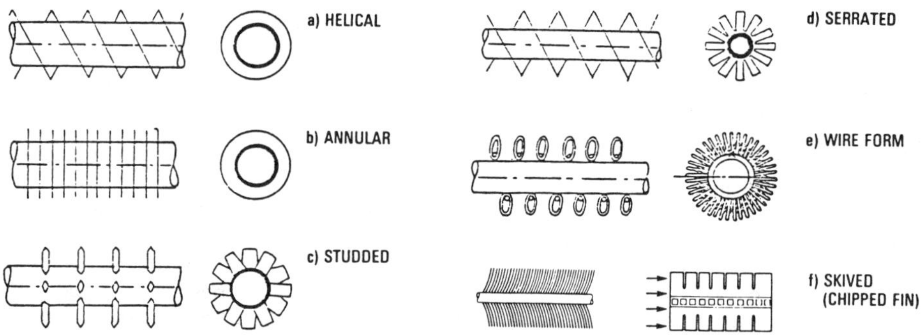

Figure 1. The variety of fin types includes: (A) helical; (B) annular; (C) studded; (D) serrated; (E) wire form; (F) skived.

These extended surfaces can have different shapes, forms, and sizes which affect the rate of heat transfer from the air to the surfaces. The surfaces can also differ by material and methods of manufacturing. The material will influence the fin efficiency. The method of manufacturing, in addition to influencing the cost, may affect the thermal resistance between the fin and the base tube.

Tube type: Round tubes with circular fins offer manufacturing advantages over other configurations, but the use of elliptical, flattened, and drop-shaped tubes can markedly reduce the form drag resistance. Brauer [4] compared the performance of round and oval tubes (8 fins/in.) in similar staggered tube banks and found the latter give 15% more heat tranfer and 18% less pressure drop. The tradeoff exists between improved heat transfer characteristics and higher manufacturing costs.

Fin type: Schematic diagrams of the common types of fins are given in Figure 1.

Helical fins, the most common fins applied on a tube, are normally manufactured by tension wrapping the fin material around the tube forming a continuous helical fin. Helical fins can also be formed by extrusion or by casting from a tube with a thick wall.

Annular fins are made of disks or plates mounted on the tube, individually separated from one another by spacers in the form of tube collars or protrusions stamped out of the fin's surface.

Studded fins are made by welding individual studs around the base of the tube. Serrated fin tubes are made by helically winding a continuous strip of metal that has been partially cut into narrow sections. Upon winding, the sections "separate" and form protrusions, similiar to the studded fins but connected at the base. Wire-form fins are formed by helically winding a continuous wire around the tube and along the tube. The purpose of the studded, serrated, and wire-form fins, as opposed to smooth circular fins, is to produce enchanced heat transfer by short flow lengths for boundary-layer development.

A variety of fin shapes may be made by skiving, or chipping up fins from a heavy walled, extruded tube that is either round or flattened (Figure 2). Cuts or slots may be incorporated by varying the extruded base shape.

Fin shape: The fins can be round, square, rectangular, or elliptical. The rectangular fins can be shared by more than one tube; when the extended surfaces are made of continuous annular fins attached to all the tubes in the bank, the fins are termed plate fins. This type of configuration is discussed subsequently.

Fin surface: In many cases, the surface of the fins is not smooth but may be artificially roughened, cut, corrugated, or perforated with the purpose of improving the heat transfer characteristics by interrupting the thermal boundary layer or directing the air flow to otherwise "blind" regions of the fins.

Figure 2. A skived- or chipped-finned tube.

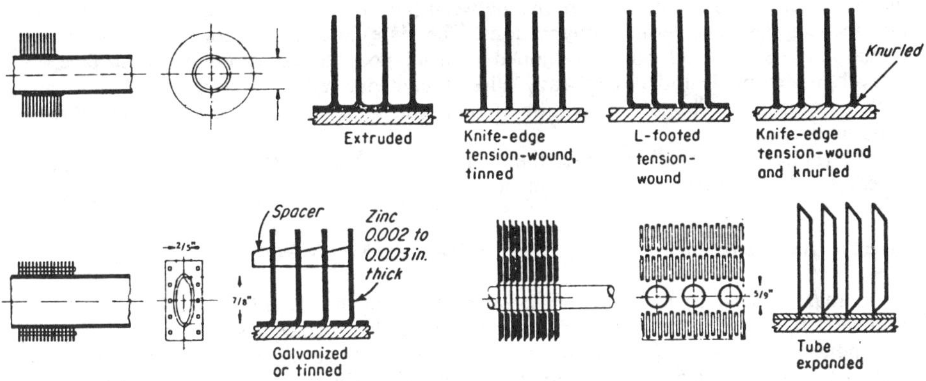

Figure 3. The variety of fin attachments to tubes.

The fin stock may become "roughened" when it is attached to the tube, depending on the manufacturing process. This will have an effect on the hydrodynamic and heat transfer characteristics of the tubes. An interrupted or slotted fin may have shapes punched out of the fin surface or partially stamped out. The fin surface of extruded fin tubes may be cut or slit by drawing the fin tube through a series of shears positioned around the fins. Wavy fins are manufactured by mechanically "crimping" the fin surface, generating a wavy pattern on the surface. This fin has more surface than a smooth tension-wound fin, but corrugations require a wider spacing of the fins on the tubes, offsetting the advantage of the increased area per fin.

Tube/fin attachment: Schematic diagrams of the most common ways of attaching the extended surfaces are shown in Figure 3. Integral-fin tubes are formed by an extrusion process starting with a thick solid-base tube or with a liner tube pressed into a aluminum or copper outer tube. A skiving or chipping process may also be used. For high-temperature applications, integral-fin tubes can also be made by machining or casting. Helically wound fins may be simply tension wound onto the base tube. However, the trend toward increasing the resistance to external corrosion and decreasing the thermal resistance between the fin and the base tube has led to L-footed fins, made by first forming the thin metal strip into an L shape and then tightly wrapping it around the surface of the tube. The foot of the L covers the base tube surface between adjacent fins. For some situations the L-foot may be made broader with a double step so that when wound onto the base tube, the fin overlaps onto the footing of the previous fin and thus more effectively covers the entire surface of the tube. For high-temperature applications, the fins are helically wrapped and fitted into a shallow groove on the tube. The groove is then closed on either side of the base fin, assuring mechanical tightness.

Most of the noncircular or continuous-plate fins are tightly fitted over the tube. The space between the fins is maintained by tube collars at the base or by small protrusions stamped out of the fin surface. A mechanical bond between the tubes and the fins can be produced by braising, dip galvanizing, or mechanically expanding the tubes into the fins by moving a ball through the inside of the tube. Serrated, studded, and wire-formed fins are attached by continuous arc welding.

Plate Fins

The plate-fin configuration can be regarded as a special case of fin tubes in which the fins are extended to become continuous between the tubes, and the previous description of fin tubes applies to plate-fin configurations. The plate-fin configurations are a special case of the annular fins and can be obtained with all possible variations of fin surface, i.e., smooth, interrupted, slotted, cut, wavy. Fin attachment is either by mechanical attachment such as braising or dip galvanizing, or by mechanical expansion of the tubes out into the plate-fin surfaces.

Web [3] has recently summarized the available literature on the performance of various types of extended surface heat exchangers.

Effect of Extended Surface Arrangement on Heat Exchanger Performance

Because the flow structure is very complex through the bundle of an extended surface heat exchanger, the performance of the air-side extended surface has remained an empirical art. Industry has made significant progress in developing proprietary fin configurations that have heat transfer coefficients 50% to 100% larger than those of flat fins. The local or point value of the transfer coefficient on the fin can vary over an order of magnitude for any one

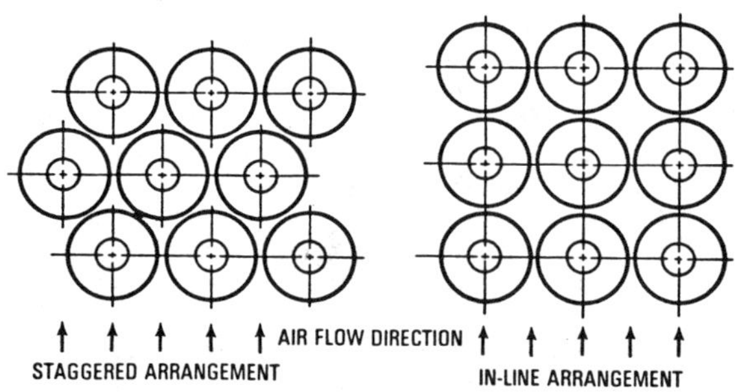

Figure 4. Tube bank layout: (A) staggered; (B) in-line.

given set of conditions. It is of value to understand these variations insofar as that knowledge can be used to modify the local or point conditions of flow over the surface and thus increase the local heat transfer coefficient and through it the average heat transfer coefficient over the fin. In this section, the influence of surface configuration on heat transfer performance is discussed only from the standpoint of the average coefficient without attempting to provide a mechanistic explanation of the localized influence of these configurations. Five aspects of the configuration are discussed.

- Tube bank layout (staggered versus in-line arrangement).
- Number of rows.
- Tube pitch.
- Finned tube variables (fin height, spacing, thickness).
- Fin tubes with surface modifications.

Effect of Tube Layout

The two basic arrangements of the tubes in a tube bank are staggered and in-line. These arrangements are shown in Figure 4. The only constructional difference between these two is that in the staggered arrangement, each alternate row is shifted half a transverse pitch. The finned tube itself can be of the various shapes and sizes as previously discussed. The two arrangements differ in the flow dynamics. The extent of transverse flow in the on-line arrangement is small in comparison to the staggered arrangement. The characteristics of heat transfer and the pressure drop of fin tubes in staggered and in-line layouts can be compared on the basis of tests reported by Brauer [5], Carnavos [6], and Rapas and Eckels [7]. For identical operating conditions, the heat transfer coefficients of the staggered arrangement are about 125% higher than for the in-line arrangement, whereas the pressure drop is only about 80% higher. Thus a staggered tube bank yields not only higher heat transfer coefficients, but more heat transfer per unit of friction. These conclusions are also supported by measurements of Weierman et al. [8] for spine fins. An explanation for this difference in heat transfer is that the rows of tubes behind the lead row of the in-line arrangement are in the low velocity wakes of the tubes. A strong bypass stream exists in any free area between the fins. Because of poor mixing between the wake region and the bypass stream, the actual temperature difference between the stream and the surface is much less than indicated by the overall mean temperature difference, based on the mixed outlet temperature. In contrast, with a staggered layout the temperature wakes are well mixed after each tube row. Thus the staggered arrangement benefits from higher local driving forces for heat transfer.

Effect of Number of Rows

The addition of a row of tubes in a crossflow direction is the common way of increasing heat transfer area when the face or approach area of the tube bank is fixed. Such an addition of area increases the length of flow path for the air and may affect the dynamics of flow with a possible effect on heat transfer pressure drop. Information on the effect of the number of rows is reported by Brauer [9], Cox [10], Jameson [11], Mirkovich [12], Ward and Young [13], and Weierman et al. [8].

A basic difference exists between the row effects for the in-line and a staggered arrangements. For an in-line arrangement, the heat transfer coefficient decreases with an increase in the number of rows. Brauer [9] showed that the one-row tube bank heat transfer coefficient is approximately 50%–80% higher than the average for 4 rows. For a staggered arrangement, the 4-row, tube bank has approximately a 30% higher heat transfer coefficient (average) than a 1-row tube bank. Ward and Young [13] and Mirkovic [12] measured individual coefficients of each row in an 8-row, staggered tube bank. Both reported that the heat transfer coefficient increased with the row number. At the third row the flow stabilizes and the coefficient remains essentially constant from the third row onwards. Ward and Young [13] found that the coefficient of the first row was about 65% to 85% of the coefficient of the last row. The coefficient of the second row was about 95% of the coefficient of the last row for a range of maximum velocities between 300 and 1,600 ft/min. Merkovic [12] reported the coefficients of the first and second rows were about 65% and 85% of the coefficient of the last row, respectively, for a maximum velocity of 1,200 ft/min. Figure 5 shows the effect of the number of rows on the average heat transfer coefficient as given in the cited references. The Ward and Young [13] lines apply for tube banks arranged in a tight pitch, with the clearance between the fins in the range of 10% to 20% of the pitch, between the tubes. The Merkovic [12] lines apply to wider pitches with clearances between the fins of approximately 50% of the pitch.

Strictly speaking, these graphs apply only for pure countercurrent flow with a capacity ratio of 1. The other extreme to this condition is a pure crossflow with isothermal heat transfer of the tube-side fluid, such as boiling or condensation. In this case, the mean temperature difference of the flows decreases with the number of rows, and the first row exhibits the highest heat flux even though it has the lowest coefficient. In this case the effect of the number of rows on the average coefficient is small and is evident from Brauer's results [9]; a bank of three or more rows has identical average outside coefficients.

Effect of Tube Pitch

Tube pitch is another variable commonly used in heat exchanger design to affect surface area and flow conditions. Jameson [11] carried out an extensive study on the effect of the tube pitch on heat transfer pressure drop. His results showed that pitch has no effect on the heat transfer coefficient on the basis of a Reynolds number defined with the base tube diameter and maximum flow velocities. A seeming effect of tube pitch can be shown if the heat transfer results are correlated with a Reynolds number using the hydraulic diameter as a characteristic dimension. As the pitch is increased, so the hydraulic diameter increases.

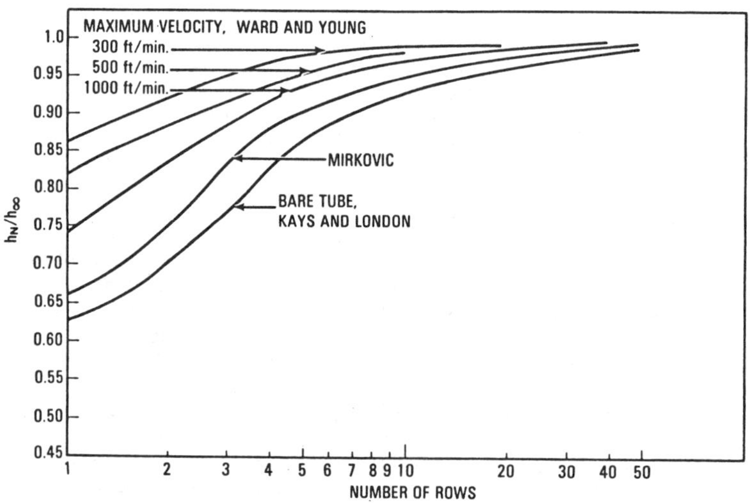

Figure 5. The effect of the number of rows on the average heat transfer coefficient: the ratio of the coefficient for N rows over an infinite number of rows.

But based on comparable air velocities, the pitch has no effect on the heat transfer coefficient. However, increasing the pitch does strongly affect (decreases) the pressure drop through the bundle.

Effect of Fin Tube Variables

The fin tube variables include fin height, spacing, and thickness. An increase in surface area of the tube bank can be achieved by employing fin tubes with higher fins and denser spacing. However, increasing the height of the fin will not only decrease its fin effectiveness but may result in lower convective coefficients than on shorter fins. Denser fins, i.e., more fins per unit tube length, may have lower heat transfer because of laminarization of the flow conditions between the fins.

Carnivose [6] investigated the effect of the fin spacing on both heat transfer and pressure drop. His results show that the heat transfer decreases and the pressure drop increases with an increase in the density of the fins, at least in the range of 16–30 fins per inch. His results show that the influence on performance decreases with an increase in the number of fins and may become independent of fin density. The fin height in these experiments was constant at 3/16 in.

Briggs and Young [14] and Kuntysh and Iokvedov [15] have determined from analysis of their data that the ratio of the fin height to the fin spacing ratio is a dominant variable affecting both the heat transfer and the pressure drop; that is, as this ratio increased, the heat transfer per unit area went down while pressure drop per row went up.

Rich [16] investigated systematically the effects of fin density on the performance of a continuous plate fin. The fin density varied from 2.92 to 20.6 fins per inch. The fin spacing appeared to have a small effect on heat transfer, about a 15% increase as the number of fins per inch increased from 2.92 to 20.6. This effect is smaller than the effect of fin spacing on round tubes; this difference which may be partially due to the fact that the plate fin retards the development of fully 3-dimensional turbulent flow.

Finned Tubes with Surface Modifications

The term surface modification refers to fin tubes which have extended surfaces that are mechanically deformed to various shapes by crimping, corrugating, cutting, slotting, or serrating the fins. In the case of plate fins, surface interruptions constitute modification. Taranyan et al. [17] and Kuntysh et al. [18] studied the effect of cut, round helical fins on the resultant heat transfer and pressure drop. The pressure drop of the cut fins was about 40% higher and the heat transfer about 25% higher than the pressure drop and heat transfer values for the equivalent smooth fins for identical maximum flow velocities. Preece et al. [19] compared the performance of a round, smooth L-footed fin with an identical fin tube that was modified by slotting and crimping the surface. They found that these surface modifications increased the pressure drop by an average of 70% while increasing the heat transfer only a nominal 10%. Data on other types of surface modifications are given by the previously mentioned investigators and also by Cox [10, 20, 21]. It can be concluded that any surface modification will increase the pressure drop as compared with the identical smooth fin. Some modifications such as serrated or cut fins will show an increase in heat transfer. Other types of surface modification will show only a nominal or no increase in heat transfer.

Heat Transfer and Pressure Drop Correlations

Because of the number of geometric variables in the complex flow conditions, the correlation of the performance of extended surfaces is a formidable task. In 1976 PFR

Engineering [22] carried out a very extensive review of the literature available on extended surfaces that could be used in very large air-cooled heat exchangers designed for dry cooling of power generation plants. All of the data which met certain criteria of reliability and precision were reduced to a common format in dimensionless form and a regression analysis of the data was performed. The results of scouting analysis of the form just reported served as a guideline for selecting the following correlation forms.

Heat Transfer

Round tubes with round smooth fins—staggered arrangement. The analysis showed that the heat transfer coefficient, expressed as a function of a Reynolds number based on the tube diameter, is independent of the tube pitch. The only geometrical variables affecting the heat transfer were the fin tube variables.

The fin tube variables were lumped into the area ratio variable (Ar), which expressed the ratio of the overall extended surface area to the area of the base tube. The value of Ar can be calculated from the following formula:

$$Ar = 1 + 2(N_f)(\ell_f)\left(1 + \frac{\ell_f + t_f}{d_r}\right) \tag{5}$$

Figure 6 shows the range of the Colburn j-factor as a function of Reynolds number for all of the finned tube data. A regression analysis of all the data yielded the following results:

$$j = 0.29\, Re_t^{-0.367} Ar^{-0.17} \tag{6}$$

This correlation predicts 80% of the data within ±10% and 95% of the data within ±18%.

Serrated or cut fins. The correlation for these fins is:

$$j = 0.195\, Re_t^{-0.3} Ar^{-.17} \tag{7}$$

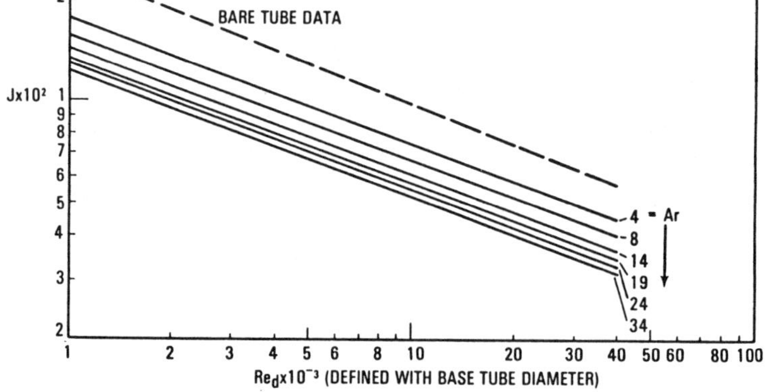

Figure 6. The variation of the j-factor with the area ratio Ar.

With this correlation 80% of the data was within ±15% and 95% of the data within ±20%. The correlation is valid for a module with six rows and with parameters in the following ranges:

$1{,}000 \leqq Re \leqq 40{,}000$
3/8 in. ≦ tube diameter ≦ 2 in.
300 (ft/min) ≦ flow velocity ≦ 1,500 (ft/min)
$4.0 \leqq Ar \leqq 34$
0.0175 cp ≦ viscosity ≦ 0.021 cp

A graphical presentation of the correlation for smooth and serrated fins is given in Figure 7.

Continuous plate fins: The data for continuous plate fins was little affected by the finned tube variables. The data by Rich [16], Cox [21], and the Trane Company data which was presented by Kays and London [1] were correlated with the following formula:

$$j = 0.13\, Re_t^{-.33} \tag{8}$$

With this correlation, 80% of the data were predicted within ±12%.

In-line arrangement. Existing data is not sufficient for a complete correlation. However, from the existing data the following is recommended:

- Smooth fins $j_{\text{in-line}} = 0.67\, j_{\text{staggered}}$.
- Serrated fins $j_{\text{in-line}} = 0.5\, j_{\text{staggered}}$.

The error for these formulas is estimated to be ±20%.

Pressure Drop

Round tube, helical round smooth fins—staggered. The pressure drop is both a function of the tube pitch (P) and the fin tube variables. The friction factor is not a simple function

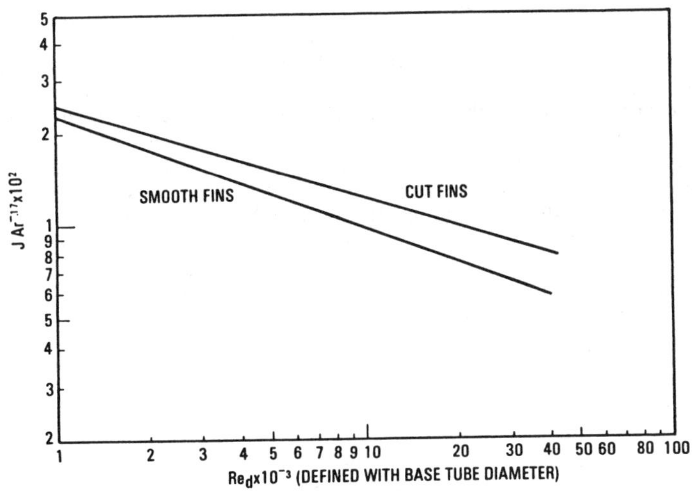

Figure 7. Heat transfer correlations for smooth and cut fins.

of the Reynolds number. Furthermore, it was found that the hydraulic diameter (D_h) is more suitable than the base diameter for correlation. The following correlation was derived:

$$F = (150\ Re^{-1.0} + 1.8\ Re^{-.2})\ (P/D_h)^{0.35} \tag{9}$$

for $(P/D_h) < 4.0$

F is the friction factor defined as follows:

$$F = \frac{\Delta P/N_t}{\rho avG_{max}^2/2g_c} \tag{10}$$

The range of the variables for this correlation is:

$$400 \leqq R_e \leqq 10{,}000$$

$$1.5 \leqq \frac{P}{D_h} \leqq 3.8$$

For all the cases when $(P/D_h) < 4.0$, a different correlation was developed:

$$F = 13.6\ Re^{-.3}(P/D_h)^{-.42} \tag{11}$$

Serrated or cut fins. Multiply the smooth fin correlation by a factor of 1.4.

Continuous plate fins.

$$F = (110\ Re_h^{-1.0} + 1.35\ Re_h^{-.2})\ (P/D_h)^{0.35} \tag{12}$$

In-line arrangement. Existing data is not sufficient for correlation. However, the in-line friction factor can be estimated as follows:

$$f_{in\text{-}line} = 0.6 f_{staggered}$$

This is valid for both smooth and serrated finned tubes.

COMPARATIVE EVALUATION OF EXTENDED SURFACES

The multiplicity of fin tube variables poses the problems of their relative merit of different kinds of fin tubes and the selection of a proper fin tube for a specific application. In any industrial application, the selection of the fin tubes is dictated by the combination of operating and capital costs, as well as a variety of technical constraints. For a fin tube module, the capital cost is primarily related to the size of the equipment, whereas the operating cost is primarily related to the fan and pumping power.

A measure of the efficiency of a module of a particular design is its heat duty under specified operating conditions. PFR Engineering [22] extended their assessment of extended surface data to consider the performance of several different arrangements of fin tubes in large installations for rejecting heat from power generating plants. Using the correlations just mentioned and a computer code to calculate the effectiveness of a particular heat exchanger design, they compared the performance of various modules designed to handle a fixed flow rate of water and observed the following:

- When the number of rows remains unchanged, an increase in the tube pitch will result in a decrease in the heat duty.
- As the number of rows increases, the heat duty increases. This is a consequence of the fact that more heat transfer area is added. However, the benefit of increasing the number of rows is reduced as the number of rows increases. This is because the tube-side water velocities decrease for a constant water flow rate.

Several approaches to characterizing the heat transfer performance of extended surface heat exchangers have been suggested in the literature. Cox and Jallouk [21] expressed the heat transfer rate at a unit temperature difference as a function of measured fan power. Both the heat transfer rate and the fan power were defined on the basis of a unit of heat exchanger surface area or volume. Cox and Jallouk then compared the efficiency of two surfaces by taking the ratio of the heat transfer rates evaluated at equivalent fan power. This ratio was defined as either the surface area efficiency or the volumetric efficiency, depending on the way the heat transfer rate and fan power were defined. However, the "standardized heat transfer rate" used by Cox and Jallouk [21] did not account for the influences of the total air flow on the log mean temperature difference in a heat exchanger of finite thickness. Consequently, the "standardized heat transfer rate" only partially reflected the incentives for high air flow at a given power expenditure. Cox's and Jallouk's line of reasoning results in the following expression of the relative performance of surface x as compared with surface A at the same fan power:

$$\frac{q_{f_x}}{q_{f_A}} = \left(\frac{D_{hA}}{D_{hx}}\right)^{2-4\left(\frac{1-n}{3-m}\right)} \left(\frac{\sigma_x L_x}{\sigma_A L_A}\right)^{1-\left(\frac{1-n}{3-m}\right)}$$

$$\cdot \left(\frac{j_{rx}}{j_{rA}}\right) \left(\frac{f_A}{f_x}\right)^{\left(\frac{1-n}{3-m}\right)} \mathrm{Re}^{\left[\left(\frac{1-n}{3-m}\right)(3-m_A)-(1-n_A)\right]} \tag{14}$$

Similarly, LaHaye et al. [23] developed a performance plot relating a dimensionless heat transfer factor (equivalent to the mean heat transfer coefficient) to a pumping power factor; the unique parameter was the ratio of the flow length between major boundary layer disturbances to the hydraulic diameter. In the fully turbulent region, this approach seems to result in the correlation of a number of surface geometries on the single-performance plot, with the nondimensional "flow length" being the important geometric characteristic of each surface. However, as with Cox and Jallouk [21], this approach considers that the air flow influences only the heat transfer coefficient; the approach neglects the influence of air flow on the air temperature range and consequently the effective temperature difference. LaHaye et al. presented a plot of dimensionless heat transfer coefficient represented by:

$$J = j\mathrm{Re} = \frac{hD_h}{k\mathrm{Pr}^{1/3}} \tag{15}$$

and a dimensionless fan power represented by:

$$F = f\mathrm{Re}^3 = \frac{p}{A_f L} \frac{g_c \rho^2}{2} \frac{D_h^4}{\mu^3} \tag{16}$$

The dimensionless parameter ℓ/D_h, the ratio of the distance between interruptions of the boundary layer to the hydraulic diameter, was the key parameter that appeared to correlate the performance of a wide variety of heat exchanger surface configurations. Figure 8 from LaHaye et al. [23] is the dimensionless plot on which all surfaces analyzed fall to within 20%. Figure 9 is the plot of $J = j\mathrm{Re}$, evaluated at $F = f\mathrm{Re}^3 = 10^9$, as a function of $/D_h$. Soland, Mack, and Rohsenow [3] utilized this approach to evaluate the performance of the plate fin surfaces summarized by Kays and London [1]. Results indicated that the method is universally applicable to any other heat exchanger surface as well.

Johnson, Kreid, and Hauser [24] presented a method of characterization that compares the total rate of heat transferred over the whole heat exchanger per unit of inlet temperature difference (rather than just the local rate of heat transfer per unit temperature difference) as a function of the required fan power. Thus, their characterization method includes the dependence of both the heat transfer coefficient and the effective temperature difference on the flow rate. Since the primary interest is in the air-side performance, and since the inside resistance is very much a function of the particular surface service and operating conditions, Johnson et al. based their development on zero resistance at the inside film and an isothermal fluid.

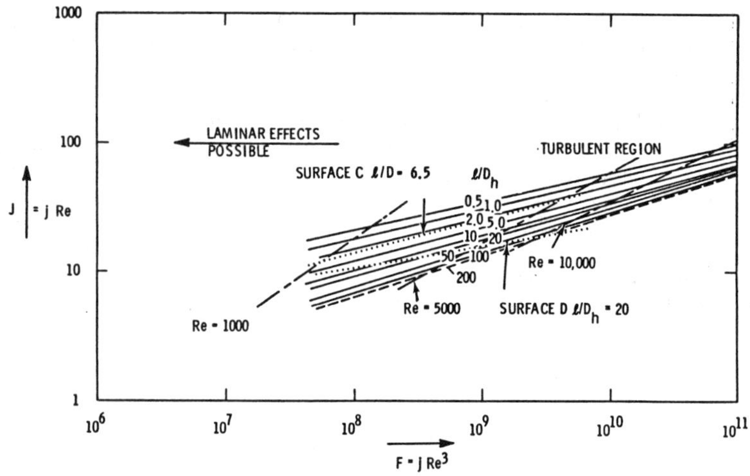

Figure 8. Idealized dimensionless performance plot [23].

Comparative Heat Duty

The comparison of heat transfer per unit of initial temperature difference can be made on several different bases:

- $Q/\Delta T_i$ per unit *frontal* area (designated as q_f) at the same fan power per unit frontal area.
- $Q/\Delta T_i$ per unit *surface* area (designated as q_s) at the same fan power per unit surface (or frontal) area.
- $Q/\Delta T_i$ per unit of heat exchanger *volume* (designated as q_v) at the same fan power per unit volume.
- Any of the above but at the same fan power per unit of heat transferred.

The pertinent equation for this relationship is the following for an air-cooled condenser:

$$\frac{Q}{\Delta T_i A_f} = q_f = C_1\left(\frac{P}{A_f}\right)^{1/(3-m)}\left\{1 - \exp\left[-C_2\left(\frac{P}{A_f}\right)^{-n/(3-m)}\right]\right\} \tag{17}$$

where $C_1 = \sigma C_p\left[\frac{g_c}{2Lv^2}\left(\frac{D_h}{f_r}\right)\left(\frac{D_h}{1{,}000\mu}\right)^m\right]^{1/(3-m)}$

$$C_2 = \left(\frac{C_i}{\sigma C_p}\right)^{-n}\left[4\,Pr^{-2/3}L\left(\frac{j_r}{D_h}\right)\left(\frac{1{,}000\mu}{D_h}\right)^n\right]$$

The equation for an air-cooled heat exchanger cooling a liquid through a temperature range is given Reference [24].

These expressions are obtained by combining the following equations.

Rate of heat transfer $\quad Q = \varphi C_{min}\Delta T_i \qquad (18)$

Required fan power $\quad P = \frac{\sigma A_f v^2 G^3}{2g_c}\left(K + \frac{fS}{\sigma A_f}\right) \qquad (19)$

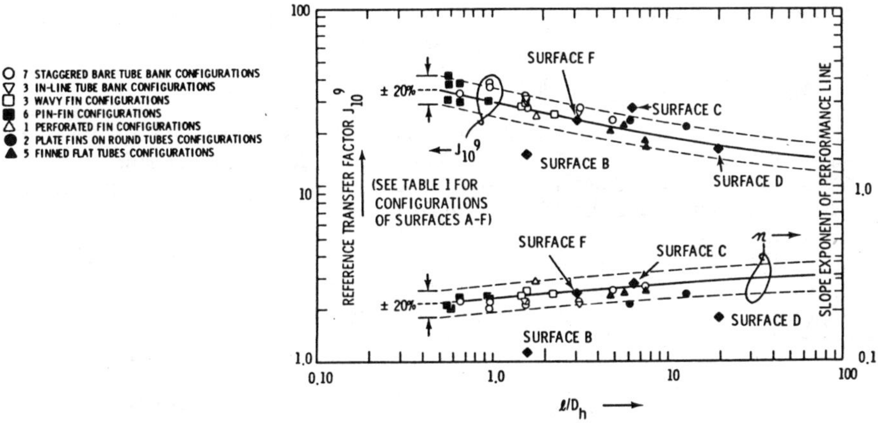

Figure 9. Characteristics of the dimensionless performance plot for 27 miscellaneous configurations [23].

Heat exchanger effectiveness for limiting case of a condenser

$$\varphi = 1 - \exp\left(-US/\sigma A_f G C_p\right) \tag{20}$$

Expressions for G and $Q/\Delta T_i A_f = q_f$ as a function of fan power are obtained by using the following assumptions:

- $K \ll fS/\sigma A_f$
- $S = sA_f L$
- $s = \dfrac{4\sigma}{D_h}$
- $f = f_r \left(\dfrac{Re}{1{,}000}\right)^{-m}$
- $j = \dfrac{h}{C_p G}(Pr)^{2/3} = j_r \left(\dfrac{Re}{1{,}000}\right)^{-n}$
- $U \approx h$ (i.e. no resistance to heat transfer on the inside surface of the heat exchanger)* (21)

* The resistance to heat transfer on the inside surface can be included through use of a comparable expression for

$$j_u = \frac{U}{C_p G}(P_r)^{2/3} = j_{ur}\left(\frac{Re}{1{,}000}\right)^{-n'}$$

where j_u = j-factor based on the overall heat transfer coefficient, u
j_{ur} = value of j_u at some reference value of the Reynolds number, Re_r, taken as $Re_r = 1{,}000$ in the above equation

j_u may be a straight line over only a limited range of Re. One can accommodate this in the following formulation by calculating the value of j_{ur} at a reference Reynolds number within the range of interest and then by using the appropriate quotient (Re/Re_r), where Re_r is the selected reference Reynolds number and n′ is the slope of the plot of log j_u versus log Re.

The expression for G is:

$$G = \left(\frac{P}{A_f}\right)^{1/(3-m)} \left[\frac{g_c(D_h)}{2\sigma L v^2 f_r}\left(\frac{D_h}{1{,}000\mu}\right)^m\right]^{1/(3-m)}$$

Equation 21 is then substituted into the expression for U as a function of the Reynolds number (from prior assumptions "e" and "f") and rearranged to give Equation 17.

The significance of the terms in Equation 17 can be seen by comparing it with the equation for the temperature rise of air through the heat exchanger.

$$Q = \sigma A_f G C_p \Delta T_i \beta_a$$

thus

$$C_1\left(\frac{P}{A_f}\right)^{1/(3-m)} = G(\sigma C_p)$$

This is also evident from Equation 20. Similarly the term in Equation 17 in the square brackets is seen to be equal to β_a. Then from Equations 18 and 19

$$C_2\left(\frac{P}{A_f}\right)^{-n/(3-m)} = \text{Ntu} = \frac{US}{\sigma A_f G C_p}$$

Heat Exchanger Performance Comparison

The importance of characterizing an extended surface (e.g., air-cooled) heat exchanger in terms of its actual performance rather than a heat transfer rate per unit temperature difference as was done by Cox and Jallouk [21] is shown in Figure 10. This figure compares the relative performance of several heat exchangers described in Table 1 [24, 25] with that of a reference surface Surface A (data from Reference 1). The heat rejection capabilities per unit frontal area and unit temperature difference are plotted against the fan power per

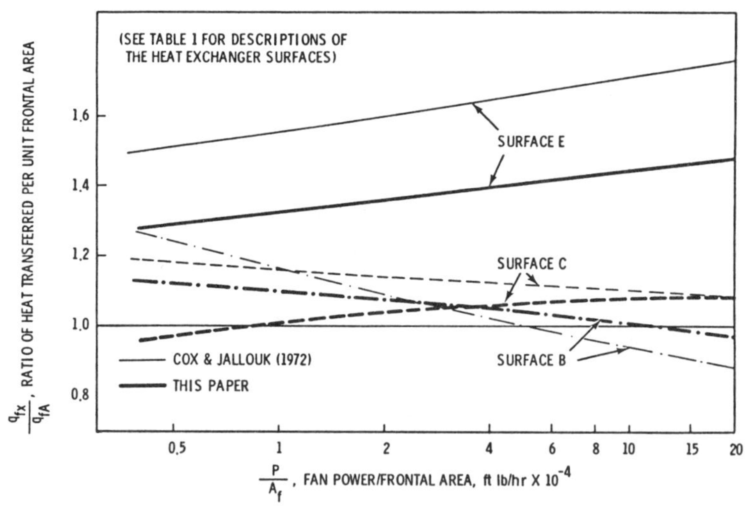

Figure 10. Heat exchanger relative performances compared on the basis of inlet conditions and whole exchanger conditions.

Table 1
Characteristic Dimensions of Heat Exchangers

	Surface A Helical Finned Tubes	Surface B Plate Fin	Surface C Plate Fin	Surface D Plain Chipped Fin	Surface E Slotted Chipped Fin	Surface F Slotted Chipped Fin
Description	Tube diameter = 0.42 in. Fin dia. (outside) = 0.861 in. Fin pitch: 8.72 in.$^{-1}$. Five rows staggered 0.975 in. × 0.80 in.	Plate fin slotted. Tube dia. 0.764 in. Fin pitch: 8.76 in.$^{-1}$. Six rows staggered 2.36 in. × 0.984 in.	Plate fin wavy. Tube dia. 0.625 in. Fin pitch: 10 in.$^{-1}$. Three rows staggered equilateral 1.5 in.	Extruded channel w/skived fins. Fin pitch: 9 in.$^{-1}$. Inline channel w/plain fins.	Extruded channel w/skived fins. Fin pitch: 12 in.$^{-1}$. Inline channel w/slotted fins.	Extruded channel w/skived fins. Fin pitch: 10 in.$^{-1}$. Inline channel w/slotted fins.
D_h ft	0.01452	0.0127	0.00972	0.01455	0.01088	0.0133
s, ft^2/ft^3	136.0	155.2	216.0	172.0	221.0	216.0
σ	0.494	0.493	0.534	0.629	0.603	0.719
L, ft	0.3384	0.492	0.375	0.575	0.575	0.4478
f_r	0.0531	0.032	0.0300	{0.0233* / 0.0249	0.0288	0.0327
m	0.23	0.0	0.600	{0.994* / 0.620	1.0	0.473
n	0.40	0.684	0.554	0.63	0.89	0.402
j_rE	0.0143	0.0107	0.0076	0.00741	0.00600	0.0105

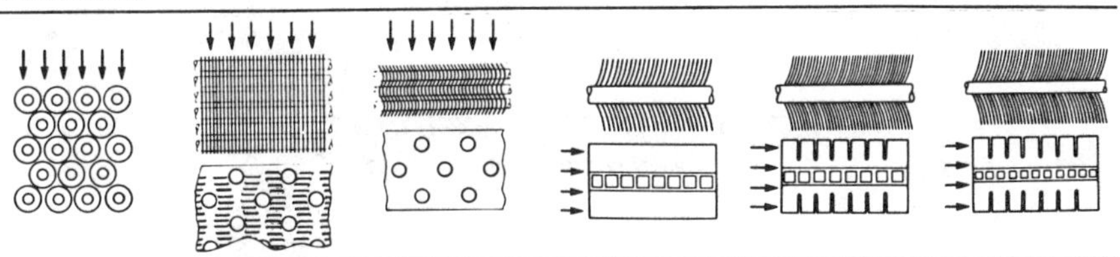

f_r *is the value of* f *at* Re = 1,000

* *For* Re < 1,500, $f = 0.0233\left(\frac{Re}{1,000}\right)^{-0.944}$; Re > 1,500, $f = 0.0249\left(\frac{Re}{1,000}\right)^{-0.620}$

unit frontal area. As previously stated for these calculations, the inside coefficient is assumed to be infinite and the inside surface isothermal (e.g., a condenser), so U is set equal to h. The relative performance (i.e., $q_f = Q/\Delta T_i A_f$ for Surface x divided by q_f for Surface A at the same fan power) predicted by Equation 17 is shown in bold lines in Figure 12. The relative performance predicted by the method used by Cox and Jallouk [21] (Equation 14) is shown in light lines.

The relation of fan power required at a corresponding Reynolds number through Surface A was calculated to provide the abscissa value.

For some surfaces, the difference in relative performance predicted by the two methods is very marked, showing both a different magnitude and a different trend with increased fan power (mass velocity of air). The difference between the two results, again, is that the present treatment accounts for the effects of higher air velocity on both the heat transfer coefficient and the effective temperature difference, while the former treatment includes only the first effect.

The calculated performance of several surfaces from Table 1 on two different bases is shown in Figure 11. Figure 11A depicts q_f, the heat rejected per unit frontal area; Figure 11B depicts q_v, the heat rejected per unit volume. The ranking of the heat exchangers changes with the basis of comparison. Figure 12 shows the relative rate of heat transfer compared to Surface A; i.e., q_{f_x}/q_{f_A} as a function of fan energy per unit of heat transferred, i.e., $P/(Q/\Delta T_i)$. This provides an even more appropriate comparison because it relates the frontal area required to transfer a stipulated heat load, if total available fan power is also stipulated.

Equation 17 does not provide complete insight into the predicted capabilities of particular heat exchanger surfaces. Although Equation 17 does provide a convenient means of

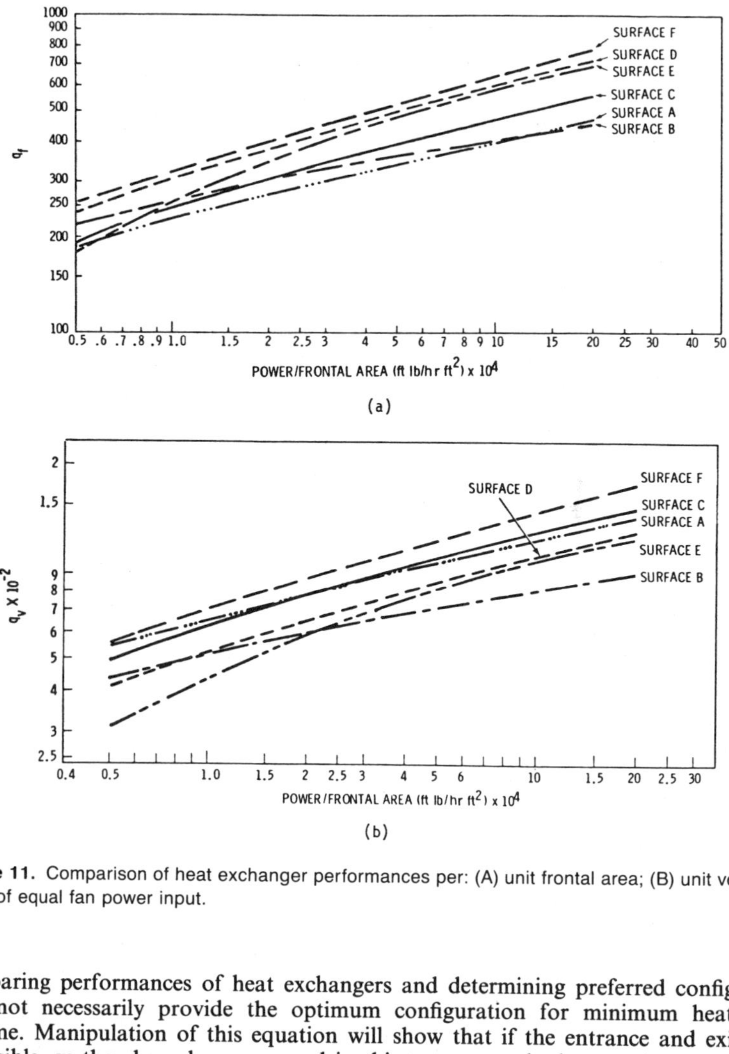

Figure 11. Comparison of heat exchanger performances per: (A) unit frontal area; (B) unit volume, on the basis of equal fan power input.

comparing performances of heat exchangers and determining preferred configurations, it will not necessarily provide the optimum configuration for minimum heat exchanger volume. Manipulation of this equation will show that if the entrance and exit losses are negligible, as they have been assumed in this treatment, the heat exchanger thickness for minimum heat exchanger volume approaches zero; i.e., a very wide, very thin heat exchanger has the minimum volume. However, the entrance and exit losses and expansion loss cannot be neglected in the analysis of such a configuration and if these losses are included, the "optimum" (i.e., minimum volume) configuration is one in which the air temperature range is limited to 0.2 to 0.3 times the initial temperature range [26]. Moreover, if detailed consideration of the costs for manifolds and tower structures is included, an optimum temperature rise of roughly 0.6 ΔT_i is usually indicated, as discussed by Fryer et al. [27].

Equation 17 is useful in visualizing the effects of changing design parameters such as tube pitch and tube spacing and fin spacing. These effects cannot be deduced from experimental values of j and f alone or from the simple ratio of heat transfer coefficients at equal power, as in Equation 14. For example, Figure 13 shows the data of Jameson [28], as reported by PFR Engineering [22]. If q_f versus P/a_f is calculated for these five configurations and then

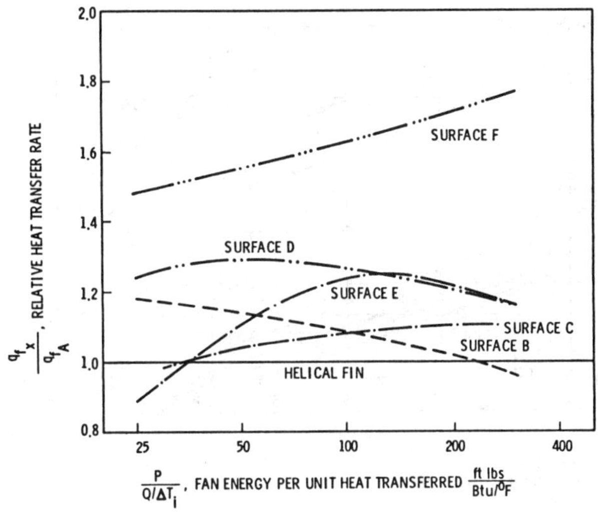

Figure 12. Heat exchanger relative performances at the same power per unit of heat transfer.

q_f versus $P/(Q/\Delta T_i)$ is plotted as in Figure 14, one would conclude that, depending on the value of $P/(Q/\Delta T_i)$ selected, either configuration "c" or "d" would be preferred. These configurations yield comparable values of q_f with about 40 percent fewer tubes (more open spacing) than configuration "a." However, if the number of rows is also allowed to be changed and the configurations are evaluated by Equation 17, the advantages of "c" and "d" are much less pronounced. For example, in Figure 15 the "poorest" configuration, "a," is compared with the "best" configuration, "d," by plotting $1/q_c = A_f/(Q/\Delta T_i)$ (frontal area required per unit of heat transferred for the respective heat exchangers at a select level of fan power) versus $A_s/(Q/\Delta T_i)$ (external surface area required per unit of heat transferred). Thus $1/q_c$ is a measure of the total number of tubes required. It is evident from this analysis that one should not compare performance on the basis of a fixed number of rows of tubes but on the optimum number of rows for each configuration and that fewer rows are required for the optimum with configuration "a." At the optimum number of rows the performance of configuration "a" would be within 6 to 7 percent of the performance of configuration "d."

Using a design optimization code for power plant dry cooling system, Fryer et al. [27] reported the results of varying tube spacing. The busbar cost of power was the optimized function of the code. In that study, a close-spaced, equilateral-pitched tube configuration optimized at four rows deep while more open configurations optimized at a greater number of rows. Costs included manifolding the heat exchanger tubes, connecting together the multitude of heat exchangers, piping coolant to the tower, providing all tower structures, as well as supplying power. Such a design optimization code is unwieldly to use for initial evaluation of surface configurations. Equation 17 is a convenient means of understanding situations such as this. The evaluation represented in Figure 15, for example, shows that air-side performance is generally better with an isosceles triangle pitch such that the flow areas between the rows are equal to the flow areas between the tubes in a particular row. This configuration would minimize the pressure drop that was due to expansion/contraction losses between the rows of the heat exchanger. The absence of this type of loss probably also contributes to the superiors performance of surfaces "B" through "F" in Table 1 as

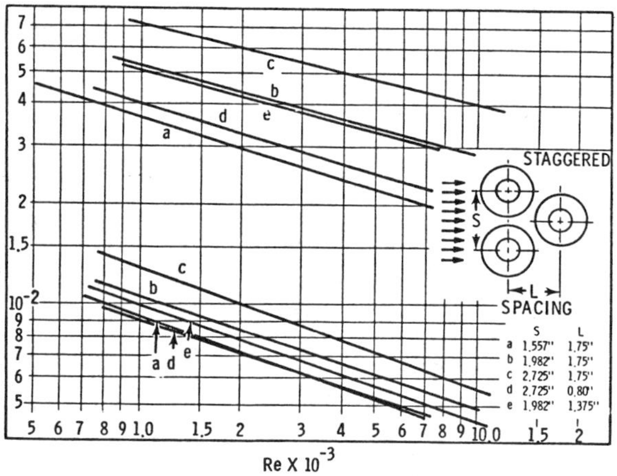

Reference: Jameson (1945)
Helically wound "L" type fins soldered to tube.
Tube outside diameter = 0.774 in.
Fin outside diameter = 1.463 in.
Fin pitch = 9.05 per in.
Fin thickness = 0.0120 in.
Number of rows in flow direction = 8.
Heat transfer area/bare tube area = 10.107.

		(a)	(b)	(c)	(d)	(e)
Flow passage hydraulic diameter	=	0.202 in.	0.322 in.	0.535 in.	0.21 in.	0.253 in.
Free flow area/ frontal area	=	0.455	0.572	0.688	0.537	0.572
Heat transfer area/ frontal area	=	122.07	95.068	68.145	68.145	95.07
Tube material/ Fin material	=	Admiralty/Copper				
Heat transfer area/ volume of exchanger	=	106.8	83.17	59.62	115.7	102.89
Depth of heat exchanger		1.143 ft.	1.143 ft.	1.143 ft.	0.589 ft.	0.924 ft.
Convective heat transfer coefficient						
Expressions for f and j as functions of Reynolds Number, $\underline{Re} = \frac{Re}{1000}$		$f = .0365\underline{Re}^{-.316}$ $j - .00925\underline{Re}^{-.387}$	$f = .052\underline{Re}^{-.278}$ $j = .0105\underline{Re}^{-.343}$	$f = .0715\underline{Re}^{-.264}$ $j = .0128\underline{Re}^{-.373}$	$f = .04\underline{Re}^{-.309}$ $j = .009\underline{Re}^{-.362}$	$f = .0506\underline{Re}^{-.275}$ $j = .0099\underline{Re}^{-.362}$

Figure 13. Effect of tube spacing on the heat transfer and pressure drop correlations.

compared with the conventional helicallywound round tubes arranged on close equilateral pitch in a heat exchanger.

When the surfaces listed in Table 1 are plotted on Figure 10 of LaHaye et al., the surfaces do not generally fall within the 20% variation band. One explanation is that the deviant data from this study are not in the turbulent region (for example, the slope of the f(Re) curve for the surface E is –1, indicating essentially laminar flow). A second explanation is that the correlations for Figures 9 and 10 do not include any consideration of fin efficiency.

The limitations of the correlation to turbulent flow stems from the variability of the laminar/turbulent transition zone as a function of the dimensionless parameter $F = fRe^3$. Within the turbulent range, the slope of the log-log plot of jRe versus fRe^3 in Figure 8 (and plotted as a function of ℓ/D_h in Figure 9) can be shown to be equal to $(1-n)/(3-m)$. The values of the slope calculated by LaHaye [23] range from about 0.2 to 0.3 as ℓ/D_h varies from about 0.5 to 70. If n is assumed equal to m, this range f values for the slope corresponds to values of m ranging from 0.5 down to 0.14, which is consistent with anticipated values of m in the turbulent range. The surfaces listed in Table 1 which do not

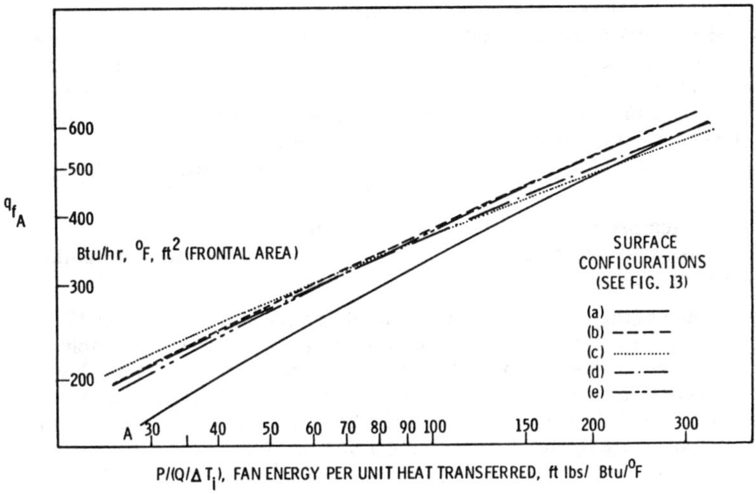

Figure 14. Rate of heat transfer (based on frontal area) as a function of fan power per unit of heat transferred for five different tube pitch and spacing configurations.

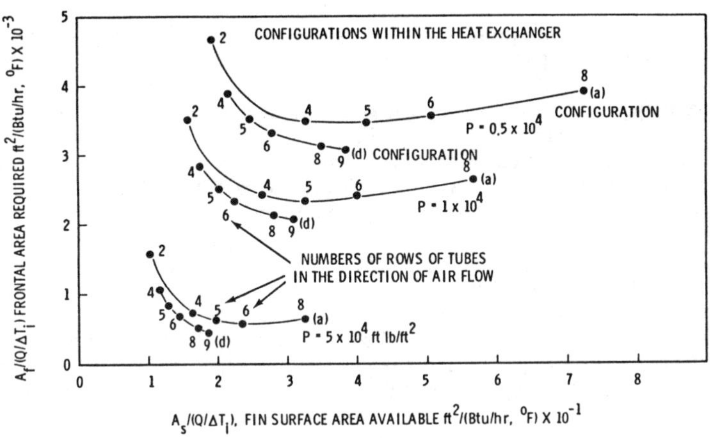

Figure 15. Relation between frontal area required and fin area available for two different configurations of fin tube bundles—arrangements A and D of Figure 13.

fit the correlation in Figure 9 had values of n or m greater than 0.5, indicating that the flow was predominantly either transition or laminar. The dimensionless group jRe is just $Nu/Pr^{1/3}$ and at a particular value of fRe^3 in the turbulent range, jRe reflects primarily the influence of developing flow from the disruption of the boundary layer, because of the nearly flat velocity profile in the main stream. In laminar and transition flow, the velocity profile is not nearly flat and disruption of the boundary layer has an effect that is very dependent on the local profiles and cannot be characterized by the simple relationship of ℓ/D_h. Unfortunately, the operating range of interest for most large extended-surface heat exchangers is not in the turbulent-flow range.

Bergles et al. [29–31] presented computer calculations of heat exchanger performance with enhanced surfaces in which nine different criteria were used to evaluate the relative performance. Performance curves for a number of these surfaces were presented. Shaw [2] presented an extended summary of over thirty different methods that have been proposed in the literature and provided recommendations on the various comparison methods. The differences involved a matrix of assumed conditions used in the comparison calculations. Three different assumptions were made relative to the temperature driving force in the comparisons:

1. No change in temperature through the heat exchangers.
2. One fluid isothermal, the other changing in temperature depending on the throughput.
3. Both fluid temperatures varying.

Eight different assumptions were used to account for the other operational parameters, which were kept constant between the comparison calculations. For example, assuming constant heat exchanger size, the heat duty was compared for constant air flow, constant pressure drop, constant air power requirement, etc. Similiar comparisons were made assuming constant heat duty and allowing the size of the heat exchanger to vary. These papers [29, 30] summarize the results of computer analyses but do not provide a convenient formulation to allow the reader to make his own comparisons on heat exchanger surfaces of interest.

ENHANCEMENT BY WET SURFACE OPERATION

Under certain operating situations both the heat duty required of the heat exchanger system and the ambient air temperature may be highly variable. To design an extended surface heat exchanger to meet the limiting conditions of maximum heat duty at maximum air temperature may represent a condition that occurs at a relatively small percentage of the time and may greatly increase the cost of the unit. Such a situation exists in the design of air-cooled heat rejections systems for power-generating plants. Although the design of such systems is dictated by the shortage or high cost of water available for evaporative cooling, it may be cost effective to incorporate a small amount of evaporative cooling into a system rejecting heat directly to the air through convective cooling. Wetting the extended surfaces of the heat exchanger would accomplish this, i.e. enhancing the heat transfer between the fins and the air by allowing heat and mass transfer to simultaneously occur from the surface. To avoid the deposition of scaling components on the heat transfer surface, it has been suggested the wetted surfaces must be kept wet while in this mode of operation, rather than permitting the water to evaporate to dryness on the surfaces and leave a film of dissolved solids. This type of operation has been called "deluge" cooling and has been extensively studied in connection with work to develop dry/wet cooling for power-generation plants. The enhancement of performance of extended surface heat exchangers by operating with a wet surface is discussed in this section.

Analytical Development

Relatively little work has been done on combined heat and mass transfer from extended surfaces. Threlkeld [31] discussed enhancement by wet surface operation as applied to finned air chillers with condensation of moisture on the surfaces. Tasnadi [32] briefly discussed this approach for enhancement of industrial heat exchangers. Kals [33] has analyzed the design of evaporative heat exchangers which use a configuration of serpentine coils of bare tubes over which water trickles from an overhead spray while air is passed up through the configuration. A recent paper by Leidenfrost and Korenic [34] provided a computer program for evaluation of evaporative cooling and heat transfer augmentation related to reduced condenser temperatures. The results discussed here are from a paper by Kreid,

Hauser, and Johnson [36] and more extensive reports on the development of deluge cooling for power plant applications [25, 36].

Surface Heat/Mass Flux Analogy

An approximate formulation of the heat transfer situation may be made on the basis of the physical similarity and resulting mathematical analogies between heat transfer and mass transfer. To highlight the similarities, an outline of the analogous steps in the parallel development of the equations for dry convective heat transfer and wet convective/evaporative heat transfer are shown in Tables 2 through 4. Figure 16 illustrates the situation of a wetted fin. In Table 2 the mass/energy balances are first applied to a differential surface area, resulting in analogous differential equations for the distribution of temperature and enthalpy. These equations are then extended to include effects of internal thermal resistance below the surface (Table 3). In Table 4 the local differential equations are integrated across the depth of a wet finned tube condenser to obtain expressions for the heat transfer and evaporation rates, and equations for the distribution of temperature and enthalpy in the system. Equations 22, 23, and 24 in Table 2 contain expressions for the energy and mass balances. Equation 23 is based on the assumption that the process of heat transfer and mass transfer are separable. The first term on the right in Equation 23 is the sensible heat transfer, which is assumed to be given by the equation for dry heat transfer at the same air-side conditions (i.e. by Equation 22). The second term on the right in Equation 23 is the latent heat term given by the product of the evaporation rate (assumed to be governed by diffusion as given by Fick's law) and the latent heat evaporation at the surface. Evaporation is assumed to be diffusion limited so that the surface air may be assumed saturated at the film temperature.

For water vapor at the temperature and pressures of interest in this situation, it may be assumed that the convective Lewis number (Le) is approximately equal to 1, where the convective Lewis number is defined by

$$Le = \frac{h_s}{\sigma_s C_a} \approx 1 \tag{26}$$

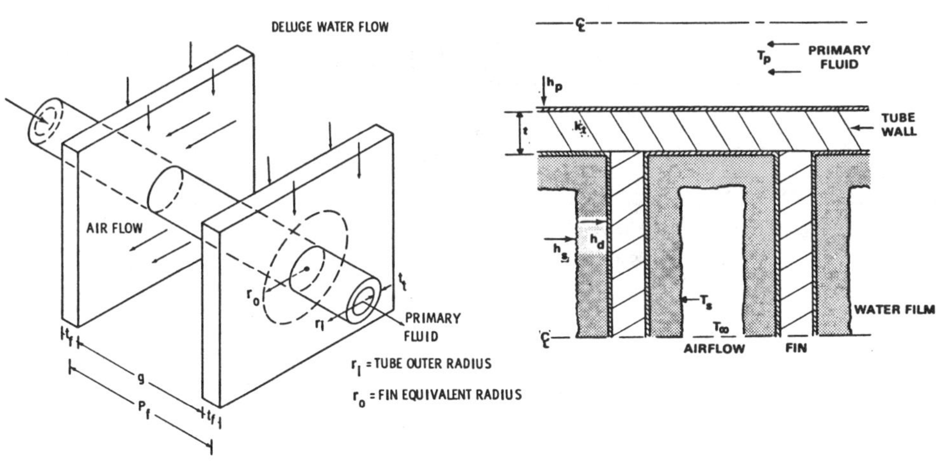

Figure 16. Representative schematic of a segment of a plate-fin assembly.

With this assumption, the term in brackets on the right in Equation 24 is very nearly equal to the difference between the enthalpy of moist air at the surface and at free stream conditions. Equation 25 then follows directly from Equations 23 and 24 with the assumption $Le = 1$.

The mass/energy balances for the air stream are given in Equations 27a and 27b. When combined with Equations 22 and 25 these yield analogous differential equations for the

Table 2
Summary of Equations for the Surface Mass/Energy Balance

Dry Heat Transfer		Dry/Wet Heat Transfer	
$\frac{dQ}{dA_s} = h_s(T_s - T_\infty)$	Equation 22	$\frac{dQ}{dA_s} \cong h_s(T_s - T_\infty) + \lambda_s\sigma_s(H'_s - H_\infty)$	Equation 23
		$\cong \sigma_s[C_a(T_s - T_\infty)Le + \lambda_s(H'_s - H_\infty)]$	Equation 24
		$\frac{dQ}{dA_s} \cong \frac{h_s}{C_a}(i'_s - i_\infty)$	Equation 25
$\frac{dQ}{dA_f} = G_a C_a\, dT_\infty$	Equation 27a	$\frac{dQ}{dA_f} \cong G_a\, di_\infty$	Equation 27b
$\frac{dT_\infty}{(T_s - T_\infty)} = \left(\frac{h_s D}{G_a C_a \delta}\right) d\chi$	Equation 28a	$\frac{di_\infty}{(i'_s - i_\infty)} = \left(\frac{h_s D}{C_a G_a \delta}\right) d\chi$	Equation 28b

Table 3
Summary of Model Equations for a Deluged Condenser in Crossflow

Dry Heat Transfer		Dry/Wet Heat Transfer	
$\frac{dQ}{dA_s} = U_0(T_p - T_\infty)$	Equation 30a	$\frac{dQ}{dA_s} = U_0^*\left(\frac{i'_p - i_\infty}{C_a}\right)$	Equation 30b
$U_0 = \left[\frac{1}{U_{pt}} + \frac{1}{h_s a_s}\right]^{-1}$	Equation 31a	$U_0^* = \left[\xi\left(\frac{1}{U_{pt}} + \frac{1}{h_d a_s^*}\right) + \frac{1}{h_s a_s^*}\right]^{-1}$	Equation 31b
$U_{pt} = \left[\frac{1}{h_p a_p} + \frac{t}{k a_t}\right]^{-1}$	Equation 32a	$\xi = \frac{(i'_p - i'_r)}{C_a(T_p - T_r)}$	Equation 32b
$\eta_f = f\left(\frac{r_i}{r_0}, Bi_f^{1/2}\right)$	Equation 33a	$\eta_f^* = f\left(\frac{r_i}{r_0}, Bi_f^{*1/2}\right)$	Equation 33b
$Bi_f = \frac{l_f^2}{k y_d} h_s$	Equation 34a	$Bi_f^* = \frac{l_f^2}{k y_d}\left[\frac{1}{h_d} + \frac{1}{\xi h_s}\right]^{-1}$	Equation 34b
$\frac{dT_\infty}{(T_p - T_\infty)} = \left(\frac{U_0 A_s}{\dot{m}_a C_a}\right) d\chi$	Equation 35a	$\frac{di_\infty}{(i'_p - i_\infty)} = \left(\frac{U_0^* A_s}{\dot{m}_a C_a}\right) d\chi$	Equation 35b

Table 4
Summary of Model Development for Extended Surface

Dry Heat Transfer		Dry/Wet Heat Transfer	
$\theta(\chi) = \frac{T_{pl} - T_\infty(\chi)}{T_{pl} - T_{\infty l}}$	Equation 41a	$\theta^*(\chi) = \frac{i'_{pl} - i_\infty(\chi)}{i'_{pl} - i_{\infty l}}$	Equation 41b
$= e^{-N\chi}$	Equation 42a	$= e^{-N^*\chi}$	Equation 42b
$N = \frac{U_0 A_s}{\dot{m}_a C_a}$	Equation 43a	$N^* = \frac{U_0^* A_s}{\dot{m}_a C_a}$	Equation 43b
$\dot{Q}_0 = \dot{m}_a C_a \varphi (T_{pl} - T_{\infty l})$	Equation 44a	$\dot{Q}_0 = \dot{m}_a \varphi^* (i'_{pl} - i_{\infty l})$	Equation 44b
$\varphi = 1 - e^{-N}$	Equation 45a	$\varphi^* = 1 - e^{-N^*}$	Equation 45b
$\dot{Q}_0 = U_0 A_s \Delta T_{lm}$	Equation 46a	$\dot{Q}_0 = U_0^* A_s \frac{\Delta i_{lm}}{C_a}$	Equation 46b
$\Delta T_{lm} = \frac{(T_p - T_\infty)_2 - (T_p - T_\infty)_1}{\ln\left\lvert\frac{(T_p - T_\infty)_2}{(T_p - T_\infty)_1}\right\rvert}$	Equation 47a	$\Delta i_{lm} = \frac{(i'_p - i_\infty)_2 - (i'_p - i_\infty)_1}{\ln\left\lvert\frac{(i'_p - i_\infty)_2}{(i'_p - i_\infty)_1}\right\rvert}$	Equation 47b

distribution of air temperature moving through the heat exchanger (Equation 28a) and the distribution of enthalpy of the air (Equation 28b). In principle, Equations 28a and 28b may be integrated if the relevant variables can be given as functions of the dimensionless distance.

Extensions to Finned Surface—The Deluge Model

The surface heat/mass transfer analysis has been extended to the treatment of heat transfer from finned surfaces, as illustrated by Figure 16. It is helpful to consider, for the moment, the enhancement to heat transfer from the finned surface brought on by wetting the surface with a water film. Comparing Equation 22 to Equation 25, one sees that heat transfer from a wetted surface is controlled by the difference in enthalpy of saturated air at the temperature of the wet surface and the air in the main stream. Thus, if it is assumed for the moment that the liquid film/air interface is at the same temperature as the dry surface, the heat transfer rate with wet film operation would be greater than that for a dry surface by the local augmenting factor (LAF) defined by

$$\mathrm{LAF} = \frac{i_s - i_\infty}{C_a(T_s - T_\infty)} \tag{29}$$

As a rough approximation, this factor may be set equal to the slope of the enthalpy-temperature curve for saturated air divided by the heat capacity of air. The value will depend on the temperature of the surface and the air, ranging, for example, from approximately 7 to 14 as the temperature changes from 50° F to 110° F.

The overall enhancement of performance will be considerably less than the LAF for the following reasons.

1. The other resistances to the heat transfer path are not enhanced, i.e., the resistance of the film and the tube wall.
2. As the rate of heat transfer from the fins is increased by wet-surface operation, the fin efficiency decreases and the average fin temperature drops.
3. An additional resistance is introduced into the heat transfer path, i.e., the resistance to heat transfer between the water film surface and the fin surface. This is convenently treated as a single resistance that includes the influence of possible incomplete wetting of the fin surface by the water film. The degree of incomplete wetting in a particular operating situation makes it difficult to anticipate this resistance without data on the heat exchanger itself.
4. The presence of the water film will decrease the air flow through the heat exchanger (assuming the same pressure drop), thus reducing surface heat transfer coefficient (i.e., increasing the air film resistance).

The interaction of the first three of these factors is brought out in the development of the deluge model, i.e., in the introduction of the overall heat and mass transfer coefficients. The prediction of the overall enhancement of performance when the heat exchanger is operated in the deluge mode is then compared with the measured performance of several heat exchangers.

The results of the deluge model development are summarized in Table 3. Equations 30a and 30b are the analogous equations for heat/mass transfer based on overall coefficients and overall driving potentials from the primary (tube) side to the free stream (air-side) conditions. The analogous expressions assumed for U_0 and U_0^* are given in Equations 31a and 31b, respectively.

The thermal resistance $(1/U_0)$ based on the total external surface area is given by the sum of the net inside thermal resistance $(1/U_{pt})$ plus the thermal resistance of the finned surface $(1/h_s a_s)$. The relative effective surface area a_s is given by

$$a_s = \frac{A_{st} + n_f A_{sf}}{A_{st} + A_{sf}} \tag{32}$$

The fin efficiency is defined in the conventional manner where the actual fin area (A_{sf}) is replaced by an equivalent area ($\sigma f A_{sf}$) which, when assumed to be uniformly at the fin root temperature T_r, will dissipate the same amount of heat as the actual finned surface. The fin efficiency can be obtained as a function of the heat exchanger geometry and operating conditions (as, for example, Equation 3).

With deluge operation, the comparable overall resistance term ($1/U_0^*$) has an additional term emanating from the resistance to heat flow between the deluge film and the fin (Equation 31b). The deluge film convective heat transfer coefficient (h_d^*) is an effective or "lumped" parameter that incorporates the effects of nonuniform water flow, air flow, and surface wetness. The relative effective surface area for wet operation (a_s^*) is analogous to a_s, but, lumped with h_d^*, a_s^* reflects the possibility of incomplete wetting of the fin surface as well as influence of lower fin efficiency because of the higher heat flow through the fins. The wet fin efficiency, needed to compute a_s^* by Equation 32, may be obtained from the equation or graph for the dry fin efficiency but using the transformation of variables given by Equations 34b and 35b. Detailed development and discussion of the methods described here for evaluation of h_d^* and the wet surface efficiency (n_f^*) are provided in Reference 37.

Since U_0^* applies to an overall enthalpy driving force, the additional thermal resistance terms (added to account for the resistance of the deluge water film ($1/h_d^* a_s^*$)) as well as the sum of the other resistances ($1/U_{pt}$) from the tube side to the air/water interface) must be multiplied by a resistance transformation parameter (ξ—Equation 32b). These resistance terms are computed for a temperature driving force, and multiplication by ξ converts their values consistent with the use of the enthalpy driving force. This transformation parameter (ξ) for the inside thermal resistances is calculated by matching the internal expression for Q, based on the temperature driving potential, to the surface heat flux, based on the enthalpy potential. The heat flux through the sum of the thermal resistances up to the fin root (including the effective resistance of the deluge water film) can be given by

$$\frac{dQ}{dA_s} = \frac{(T_p - T_r)}{\left(\frac{1}{h_p a_p} + \frac{t}{k a_t} + \frac{1}{h_d^* a_s^*}\right)} \tag{37}$$

$$= U_{pr}(T_p - T_r) \tag{38}$$

Converting this to the equivalent enthalpy difference, we obtain the needed transformation parameter as follows:

$$U_{pr}(T_p - T_r) = U_{pr}\left[\frac{C_a(T_p - T_r)}{(i'_p - i'_r)}\right]\left(\frac{i'_p - i'_r}{C_a}\right) \tag{39}$$

$$= \frac{U_{pr}}{\xi}\left(\frac{i'_p - i'_r}{C_a}\right) \tag{40}$$

where $$\xi = \frac{(i_p - i_r)}{C_a(T_p - T_r)} \tag{33b}$$

A means for evaluating this parameter is provided in the next section.

When the equations for dry surface heat flux (Equation 30a) and wet surface heat flux (Equation 30b) are equated to the respective heat/mass gains of the air (Equations 21, 25 and 26), expressions are obtained that describe the variations in air property with distance in the airflow direction (Equations 35a and 35b). For suitable conditions, these expressions can be integrated to obtain the distributions of temperature and enthalpy in a heat exchanger.

Analysis of a Wetted Plate-Fin Condenser

The principal assumptions made in the integration of Equations 36a and 36b for a plate-fin condenser in crossflow are the following:

- The primary or tube-side temperature is constant.
- The air mass flux (G) is uniform in any plane normal to the flow.
- The NTU parameters N and N* are constant.

The results of the integrations are summarized in Table 4.

Equations 42a and 42b give the distributions of the air temperature in a dry heat exchanger and the air enthalpy in a wet heat exchanger. In addition, by using psychrometric charts and the computed values of i_∞, the distribution of the dry bulb temperature in a wet core can also be predicted.

The total heat/mass transfer from the heat exchanger is obtained by a second integration where the results are given in terms of the effectiveness, Φ^*. The analogous expressions for heat and combined heat/mass transfer are given by Equations 44a and 44b where the corresponding expressions for the effectiveness are given by Equations 45a and 45b. The advantage to this approach is that the heat/mass transfer is given in terms of the inlet conditions without the need of the outlet properties.

Equations 35a and 35b can also be integrated to obtain alternate equivalent expressions for the dry and wet heat transfer, Equations 46a and 46b. A disadvantage of this approach is that both inlet and outlet conditions appear in the log mean property differences defined by Equations 47a and 47b. This generally requires an iterative solution technique. In addition, a crossflow correction would normally be required.

In principal, similar results could be obtained for other heat exchanger situations not involving a constant-temperature fluid, as in a condenser. However, in most cases, the results would be more complex and, in many cases, closed-form solutions could probably not be found. However, in these cases, existing numerical solutions or graphs of heat exchanger effectiveness, $\Phi(R, N)$ for dry operation, can be used to compute $\Phi^*(R^*, N^*)$ by transformation of variables. The capacity ratio $R = \dot{m}_a C_a / \dot{m}_p C_p$ can be transformed to R^* by multipliation with ξ. The solution for Φ in a condenser is a special case, where the dependence on R cancels out.

Evaluation of T_r, ξ, and h_d^*. To apply the deluge model for prediction of heat/mass transfer, values of ξ and h_d^* must be specified. An explicit expression (Equation 32b) has been derived for ξ. However, precise evaluation of ξ requires knowledge of the root temperature (T_r), which varies with core design and operating conditions in a manner that is not easily predicted.

When all the relevant operating and inlet conditions are known, including h_d^*, the value of the average or effective root temperature (T_r) can be estimated. Since the heat that is transferred from the system passes through a series resistance path, the "local" rate of heat transfer can be given by the following equivalent expressions:

$$\frac{dQ}{dA_s} = \frac{\left(\dfrac{i_p' - \bar{i}_r'}{C_a}\right)}{\xi\left(\dfrac{1}{h_p a_p} + \dfrac{t}{k a_t} + \dfrac{1}{h_d^* a_s^*}\right)} = U_{pr}(T_p - T_r) \tag{48}$$

$$\frac{dQ}{dA_s} = h_s a_s^* \left(\frac{i_r' - \bar{i}_\infty}{C_a}\right) \tag{49}$$

where U_{pr} is the inverse of the net internal resistance,

$$U_{pr} = \left(\frac{1}{h_p a_p} + \frac{t}{k a_t} + \frac{1}{h_d^* a_s^*}\right)^{-1} \tag{50}$$

Equation 48 gives the heat transfer from the tube side to the fin root, and Equation 49 gives the heat transfer from the root to the air. Equations 48 and 49 constitute a set of two simultaneous equations that can be solved for the two unknowns, dQ/dA_s and T_r. Unfortunately, these equations cannot be solved explicitly because of the complex interdependence of the parameters. Thus, an iterative solution is required.

Both h_d^* and T_r are generally unknown, but they can be determined from measurements of the heat transfer from the system Q_0. If we assume that the average value of dQ/dA_s is Q_0/A_s and use the measured average value of T_∞, Equations 48 and 49 can be solved simultaneously to obtain average values of T_r and h_d^* for each test. The value of ξ is then determined from the value for T_r. The results of this computation for the tests performed with Surface C, Table I (Reference 37) are shown in Figure 17.

Estimation of performance enhancement. Returning to the heat transfer performance of a wetted surface heat exchanger, it can be expressed in terms of the initial temperature difference (rather than initial enthalpy difference) by introducing the normalized driving potential Γ defined as:

$$\Gamma = \frac{(i_p' - i_\infty)_1}{C_a(T_p - T_\infty)_1} \tag{51}$$

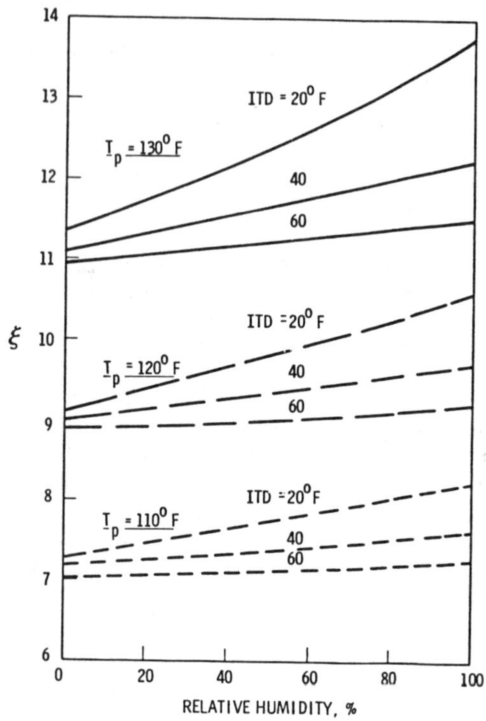

Figure 17. Variation of ξ with deluge operating conditions calculated for Surface C, Table II.

Thus, starting with Equation 44b,

$$\dot{Q}_0 = \dot{m}_a \varphi^* (i'_{p1} - i_{\infty 1}) \tag{44b}$$

and multiplying top and bottom by $C_a(T_p - T)_1$, we obtain for deluge operation.

$$Q = \dot{m}_a \Phi^* \frac{(i'_p - i_\infty)_1}{C_a(T_p - T_\infty)_1} \cdot C_a(T_p - T_\infty)_1$$

$$\frac{Q}{(T_p - T_\infty)_1} = \frac{Q}{ITD} = \dot{m}_a C_a \varphi^* \Gamma$$

The parallel expression for dry surface operation is

$$\frac{Q}{ITD} = \dot{m}_a C_a \Phi$$

There are a number of ways that the comparison between wet and dry performance can be made. One of the most useful is the performance ratio at the same core temperature, inlet air conditions, and air-side pressure drop. In this case the enhancement ratio is:

$$\frac{Q_{wet}}{Q_{dry}} = \frac{\dot{m}_{a\,wet}}{\dot{m}_{a\,dry}} \frac{\Phi^*}{\Phi} \Gamma \tag{53}$$

The parameter Γ is a purely thermodynamic variable that depends only on the inlet conditions (T_p and ITD). Figure 18 shows the range of this variable as a function of T_p

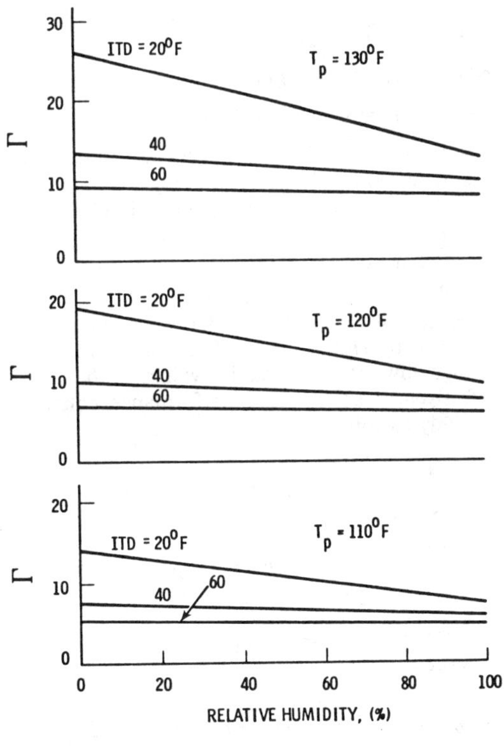

Figure 18. Variation of Γ with representative deluge operating conditions.

and ITD for conditions of interest in the rejection of heat from power plants. It is closely related to the Local Augmenting Factor previously noted (Equation 29).

The interaction of the three factors which reduce the surface enhancement or local augmentation factor are expressed in the ratio

$$\frac{\Phi^*}{\Phi} = \frac{1-\exp(-N^*)}{1-\exp(-N)} = \frac{1-\exp\left(\dfrac{U_0^* A_s}{\dot{m}^* C_A}\right)}{1-\exp\left(\dfrac{U_0^* A_s}{\dot{m}^* C_A}\right)} \tag{54}$$

The terms U_0 and U_0^* are expressed by Equations 31a and 31b.

$$U_0 = \left(\frac{1}{U_{pt}} + \frac{1}{h_s a_s}\right)^{-1} \tag{31a}$$

$$U_0^* = \left[\xi\left(\frac{1}{U_{pt}} + \frac{1}{h_d a_s^*}\right) + \frac{1}{h_s a_s^*}\right]^{-1} \tag{31b}$$

The overall transfer coefficient for deluge operation (U_0^*) appears to be smaller than the comparable coefficient for dry operation (U_0), because U_0^* applies to an overall enthalpy driving force while two of the resistance terms are applicable to a temperature driving force and must be multiplied by the transformation factor to compensate for the use of an overall enthalpy driving force. The relative value of U_0^*/U_0 can be approximated by relating all resistances to the outside film resistance ($1/h_s a_s$). Thus

$$\frac{U_0^*}{U_0} = \frac{1+f_1}{1+(f_1+f_2)} \tag{55}$$

where $f_1 = (1/U_{pt})/(1/h_s a_s)$ (56)

$f_2 = (1/h_d a_s^*)/(1/h_s a)$ (57)

The relative values of resistances would be unique to a particular heat exchanger. For the one reported by Hauser et al. [37], the values of these ratios were approximately $f_1 = 0.1$, $f_2 = 0.6$, and the value of ξ (Equation 32b) was approximately 9. For these values

$$\frac{U_0^*}{U_0} = 0.15$$

The ratio of the air flow under wet and dry conditions must also be known. In the studies reported by Hauser et al. [37], this ratio ranged from about 0.6 to 0.8, depending on the deluge flow rate. The relative values of the number of transfer units (N^*/N), heat exchange effectiveness (φ^*/φ), and performance enhancement (Q_{wet}/Q_{dry}) would consequently be roughly as follows:

$$\frac{N^*}{N} = \frac{U_0}{U}\frac{\dot{m}_a}{\dot{m}_a^*} = 0.15 \times 0.7 = 0.21$$

$$\frac{\Phi^*}{\Phi} = \frac{1-e^{-N^*}}{1-e^{-N}} = 0.42 \quad \text{for} \quad n \sim 2.4$$

$$\frac{Q_{wet}}{Q_{dry}} = 0.7 \times 0.4 \times \Gamma = 0.28\,\Gamma \tag{58}$$

Thus, the measured enhancement ratio would be about 0.3 times the value of local augmenting factor. Nevertheless, since this factor ranged from about 7 to 18, one could anticipate a measured enhancement from about 2 to 5, depending primarily on the operating temperature of the heat exchanger and the relative humidity.

Performance of Heat Exchangers Under Deluge Conditions

Hauser et al. [37, 38] evaluated three extended surfaces for wet/dry operation in the deluge mode in support of using this concept to develope lower cost dry cooling for power plants.

Two were plate-fin-type surfaces and the third was an integral spiral fin design. A brief summary of the results for dry and wet operation and the predictions of enhancement, computed with the deluge model, are given here. Additional data and more detailed discussion may be found in References 37 and 38.

The surfaces tested by Hauser et al. are described in Table 1 as Surfaces B and C. The third surface was four rows of 5/8-in. aluminium tubes with 1.375-in. extruded fins, eight per inch. The tubes were mounted on equilateral spacing of 1.4 in.; the surface was quite similar to Surface A in tube I and henceforth will be referred to as Surface A. All were operated in a slanting orientation to simulate one side of an A-frame configuration. Several different angles of inclination were tested.

For dry operation, the friction factor (f) and the surface heat transfer coefficient (h_s) were determined.

In wet operation the principal computed parameters were the friction factor (f^*) and the overall heat transfer coefficient for wet operation (U_0^*). The deluge model was then used to calculate an effective deluge film coefficient (h_d^*) and the enhancement due to wet-surface operation.

The dry heat transfer performances of the three surfaces appear very similiar for the range of air flow rates shown. Under deluge conditions the relative pressure drop characteristics of the three surfaces changed noticeably. The configuration of Surface A and Surface B caused the water film to break up into droplets, which tended to restrict the air flow. In Surface C the water flowed down the fins in a more-or-less continuous film. Thus, the pressure drop through Surface C was much less than the pressure drop through the other two surfaces at the same air velocity (Figure 19).

The primary derived parameter calculated from the data was the overall heat transfer coefficient (U_0^*). This parameter was calculated for each test using the following expression:

$$U_0^* = \frac{m_a C_a}{A_s} \ln \left| 1 - \frac{Q}{m_a(i_p' - i_\infty)_1} \right|$$

$$\varphi^* = 1 - \exp\left(-\frac{U_0^* A_s}{\dot{m}_a C_a}\right) \tag{59}$$

This expression is derived from Equations 44b and 45b

$$Q = m_a \varphi^* (i_p' - i_\infty)_1 \tag{44b}$$

where

$$\varphi^* = 1 - \exp\left(-\frac{U_0^* A_s}{\dot{m}_a C_a}\right) \tag{45b}$$

The other primary calculated variable is the effective deluge film heat transfer coefficient (h_d^*). It is based on the mathematical requirement that heat transfer to the surface from the inside must be equal to the heat rejected from the outer wet surface as defined by Equations 48 and 49. This procedure provides also the effective fin root temperature (T_r) needed to compute the thermal resistance transformation variable.

The heat transfer per unit inlet temperature difference (Q/ITD) is plotted vs. the dimensionless driving potential (Γ) in Figure 20. Data are shown for a range of Γ values

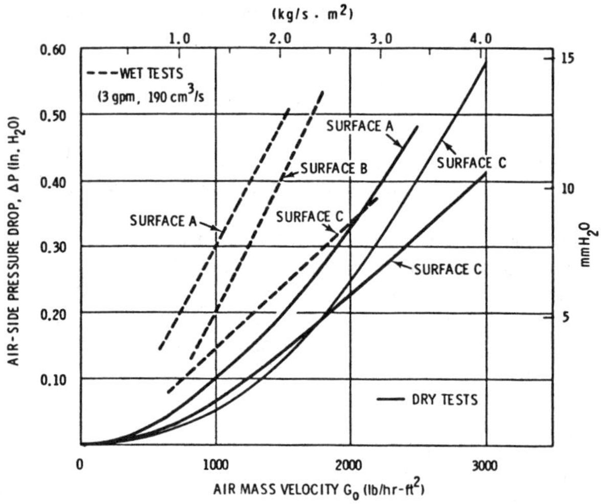

Figure 19. Air-side pressure drop through three heat exchangers (Surfaces A, B, C, Table II) operated dry and in the deluge mode.

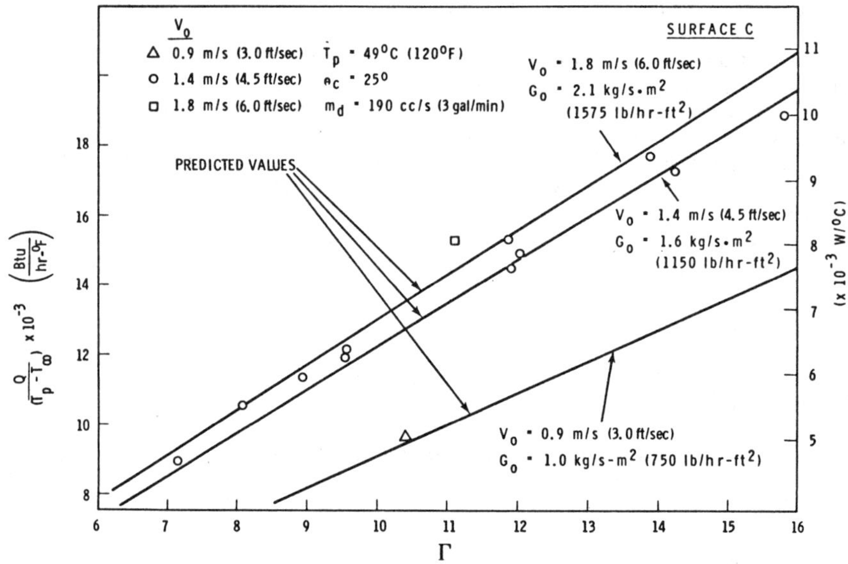

Figure 20. Heat transfer performance of Surface C (Table 1) as a function of psychrometric conditions.

corresponding roughly to ranges of relative humidity from 25% to 75% and inlet temperature differences from 5.6° C (10° F) to 27.8° C (50° F), all at V_0 = 1.4 m/s (4.5 ft/sec). Single data points are also shown for V_0 = 0.9 m/s (3 ft/sec) and 1.8 m/s (6 ft/sec). The solid line in each case shows the predicted correlation that results when an empirically determined value of the deluge film coefficient (h_d^*) is used in the heat transfer model. It is apparent that this data correlation technique is effective.

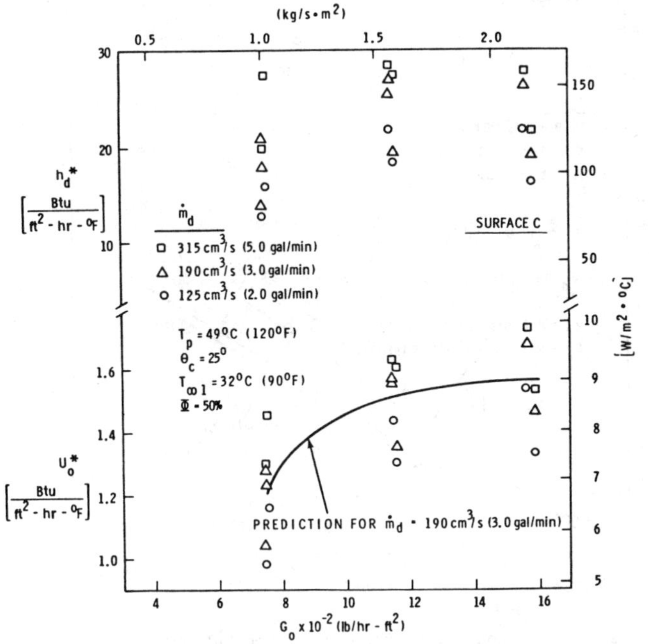

Figure 21. Variation of the deluge film heat transfer coefficient h_d^* and the overall deluge heat transfer coefficient U_o^* with deluge flow rate.

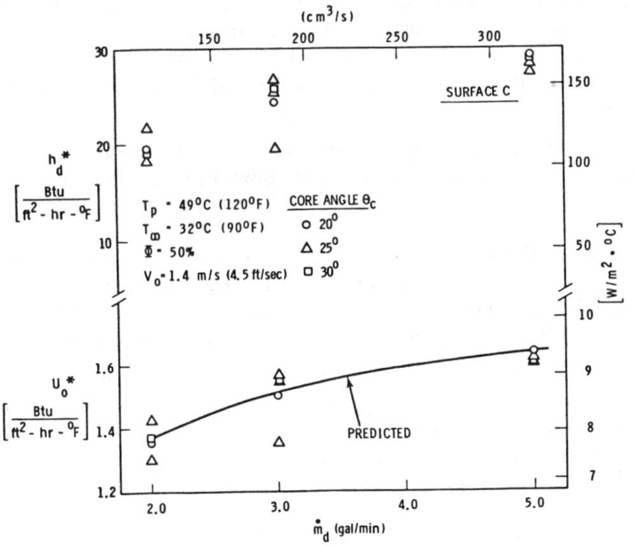

Figure 22. Variation of deluge film heat transfer coefficient h_d^* and the overall deluge heat transfer coefficient U_o^* with air flow rate.

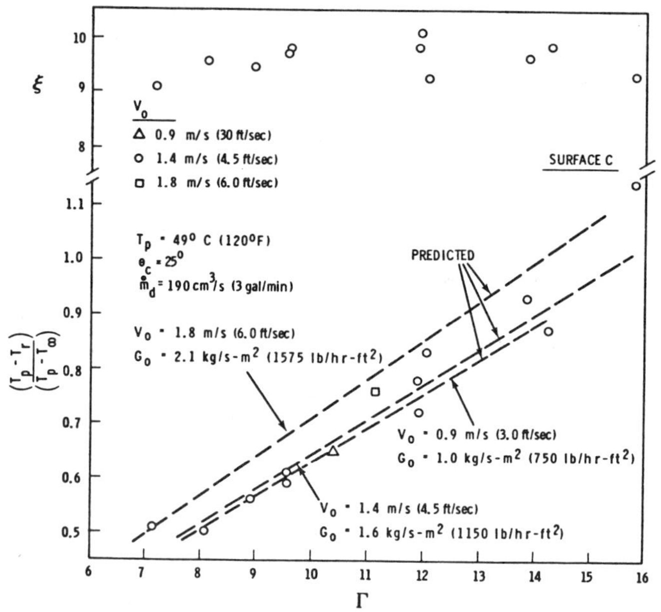

Figure 23. Calculated values of ξ and fin root temperature Tr (for Surface C, Table 1) as a function of psychrometric conditions.

Figures 21 and 22 show how U_0^* and h_d^* varied in these experiments with air flow rate, deluge water flow rate, and heat exchanger angle. In each case the solid line for U_0^* represents a prediction using the deluge model and an average value of h_d^*.

Although the scatter in the experimental values for both U_0^* and h_0^* is considerable, there are several apparent trends.

- From Figure 21 it appears that U_0^* increased slightly with increasing inclination (θ) at angles less than 25°. However, at 25° and 30°, U_0^* shows little (if any) apparent dependence on angle. There is no significant dependence on θ_c in the empiricially determined values for h_d^*.
- U_0^* and h_d^* increased with increasing deluge water flow rate $\dot{m}_d$ (Figure 21) and inceasing air flow rate G_0 (Figure 22). However, in both cases, the dependence on $\dot{m}_d$ and G_0 appeared to diminish with increasing flow rates. In fact, the data for h_d^* for G_0 1.6 kg/s-m^2 (1,200 lb/hr-ft^2) appeared to be essentially independent of air flow rate.

The root temperature (T_r) and corresponding values of the temperature-to-enthalpy transformation parameter (ξ) that were generated in the calculation of values of h_d^* are plotted in Figure 23 against the dimensionless driving potential (Γ). The significant result is that ξ is essentially independent of Γ. This greatly enhances the predictive use of the deluge heat transfer model because it implies that ξ can be assumed independent of meteorological conditions.

The wet-to-dry heat transfer ratio (Q_w/Q_d) evaluated at the same air-side pressure drop is plotted as a function of Γ in Figure 24. The equation for the correlation of Surface C is $Q_w/Q_d = 0.3\ \Gamma$, as described in the development of Equation 58.

Predictive Capabilities of the Deluge Model

The deluge model provides an excellent means for correlating data for wet heat transfer from a finned surface. The model requires knowledge of the deluge film coefficient (h_d^*), which at present can only be determined experimentally for each situation. This coefficient is essentially an empirical parameter in that it includes the effects of incomplete wetting of the surface by the deluge water. Consequently, existing correlations of heat transfer in falling films or other configurations are not applicable. Thus, it must be determined for each heat transfer surface and operating condition. From the results of Hauser et al. [37], h_d^* was insensitive to meteorological conditions and only moderately sensitive to operating conditions of core angle, air flow rate, and deluge water flow rate. In all cases, the use of a constant, average value of h_d^* resulted in predictions that were in good agreement with the data. The values of h_d^* for the three heat exchangers were similiar in spite of substantially different surface configurations, and consequently these data can probably be used with acceptable accuracy for estimating the performance of proposed wet/dry heat exchangers of similar design. As experience with various deluged surfaces increases, h_d^* values may be correlated to configurational parameters. Once the h_d^* values are known, heat exchanger performance in deluge operation can be predicted.

Dry/Wet Cooling Effectiveness

The potential effectiveness of dry/wet cooling is best illustrated by the results in Figure 24 from Hauser et al. [37]. For the two heat exchanger surfaces B and C, heat transfer enchancement factors ranging from 2 to 5 can be expected for the range of meteorological conditions for which deluge cooling would probably be used. Consequently, the size of a dry/wet cooling tower (as compared to an all-dry cooling tower) could be reduced to only 20% to 50% of the size of the all-dry tower. The tower would use water only in periods of peak cooling demand, and consequently the consumption of water would be reduced dramatically from that required for a wet cooling tower with equal capacity. Actually, water

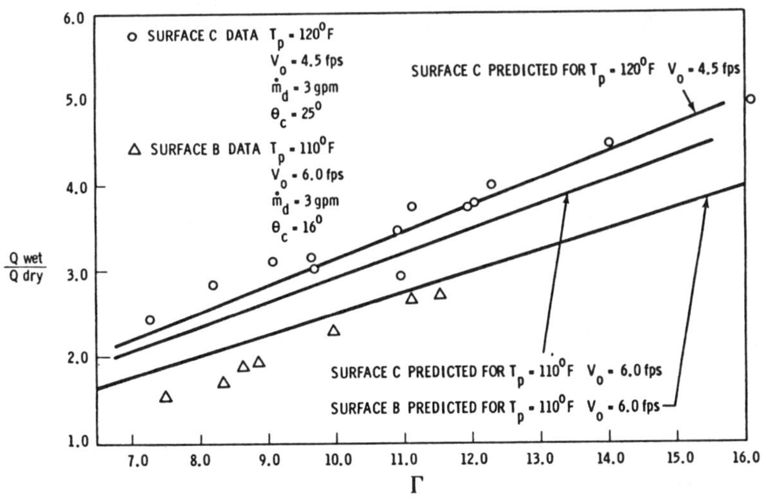

Figure 24. Enhancement ratio as a function of Γ, the psychrometric conditions for Surfaces B and C, Table 1.

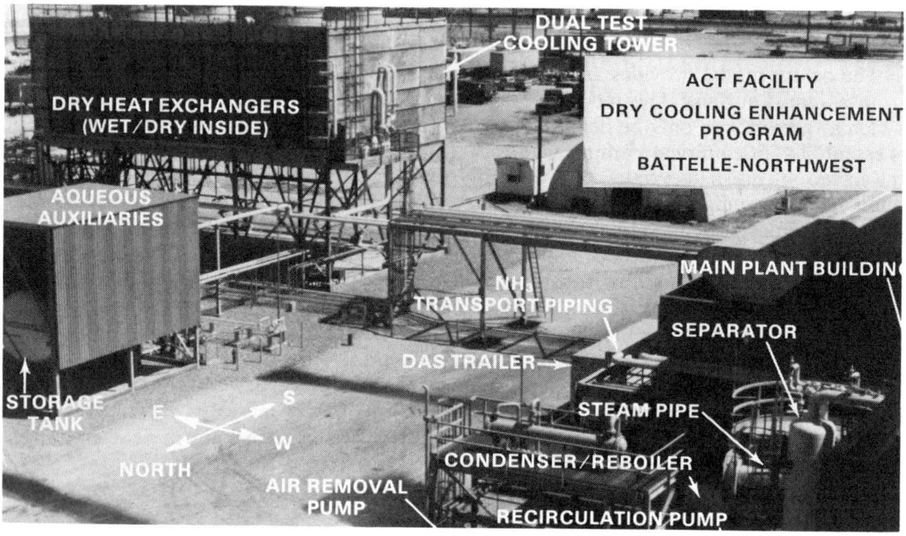

Figure 25. Advanced Concept Test Facility for dry/wet cooling at the Pacific Gas & Electric Kern Plant (heat duty—60 × 10^6 Btu/hr).

limitations and various options involved in minimizing power costs dictate a rather complex analysis in determining the best heat exchanger design and the optimum size of a dry/wet tower for a specific application. Nevertheless, if potential problems related to scaling and corrosion of the wetted heat transfer surfaces can be tolerated, deluge cooling appears to be an attractive alternative for power plant heat rejection where water availability is restricted. A test of this concept by the Electric Power Institute [39, 40] is currently under way at Pacific Gas & Electric's Kern Plant at Bakersfield, California (Figure 25). The facility, capable of reducing water consumption by 75%, is attached to a small house turbine and handles up to 60×10^6 Btu/hr of waste heat, equivalent to the heat rejection from a 15-MWe generation plant.

NOTATION

A area
A_c minimum cross-sectional area available to air flow
A_f total frontal area of heat exchanger
A_i inside surface area
Ar area ratio, total external surface area to base tube
A_s outside surface area
a_s relative surface area A_i/A_s
Bi Biot number for fins; defined by Equation 34a and b
C_{min} flow stream capacity rate; C_pW
C_p heat capacity of air
C_{pw} heat capacity of water
D_h hydraulic diameter ($= 4\sigma/s$)
d tube diameter
E overall fin efficiency based on air side conditions
F $f\,Re^3$, standardized power requirement for heat transfer
F friction factor defined by Equation 10
F_T ΔT effectiveness factor for cross-flow heat exchangers
f Fanning friction factor
f_r f evaluated at a reference Reynold's Number ($Re = 1{,}000$)
G mass velocity of air through heat exchanger of A_c
g_c gravitational constant

H moisture content of air
h effective heat transfer coefficient based on air side conditions and including the fin efficiency E
h_p product side heat transfer coefficient (Btu/hr-ft^2-°F)
ITD initial temperature difference
i enthalpy
J j Re; standardized heat transfer rate
j Colburn factor for heat transfer ($= h_a/GC_p(Pr)^{2/3}$)
j_r j evaluated at Re = 1,000
k thermal conductivity of air (or fin)
K entrance and exit loss factor; also air side conductance referred to inside surface area
L depth of heat exchanger module; also spacing between rows of tubes
Le convective Lewis number
ℓ flow length between major boundary layer disturbances
m slope of the log-log plot off is Re; i.e., the exponent in the expression $f = f_r(Re/1,000)^{-m}$
$\dot{m}$ mass flow rate of air
$\dot{m}_d$ flow rate of deluge water
N number of transfer units US/C_{min}
N_t number of tube rows in flow direction
n exponent in expression $j_r = j(Re/1,000)^{-n}$
P air power supplied by fan
ΔP pressure drop of air flowing through heat exchanger
P tube pitch—longitudinal
Pr Prandtl number ($= C_p\mu/k$)
Q heat transferred
Q_d heat transferred under dry conditions
Q_w heat transferred under wet (deluged) conditions
q_f $Q/A_f\Delta T_i$
q_s $Q/S\Delta T_i$
q_v $Q/A_fL\Delta T_i$
R capacity ratio, β_a/β_w
Re Reynold's number ($= GD_h/\mu$) (dimensionless)
Re_t Reynold's number referred to tube diameter Gd/μ
S total air side surface area of heat exchanger; also spacing between tubes
s specific surface per unit volume
T temperature
T_{ai} inlet temperature of air
T_{a0} outlet temperature of air
T_r effective average temperature at the root of the fins
T_{wi} inlet temperature of water
T_{w0} outlet temperature of water
ΔT temperature difference
ΔT_i $T_{wi} - T_{ai}$ = inlet, air/water temperature difference
t fin thickness
U overall heat transfer coefficient
v specific volume of air
W_a flow rate of air
W_w flow rate of water
y_d fin half-thickness

Greek Symbols

β_a temperature range of air ($= (T_{a0} - T_{ai})/\Delta T_i$) (dimensionless)
β_w temperature range of water ($= (T_{wi} - {}_{w0})/\Delta T_i$) dimensionless
Γ driving potential ratio
δ film thickness
η efficiency of finned surface
θ nondimensional temperature difference; also subtended angle between A-frame heat exchangers
λ heat of vaporization of water
ξ thermal resistance transfer parameter
ρ density of air
σ Ac/A_f = fraction of heat exchanger frontal area available for flow
Φ heat exchanger effectiveness; $Q/Q_{max} = Q/C_pW_aT_i$
μ dynamic viscosity

Subscripts

a air properties
d properties of deluge film
f properties of fin
i inside surface, also initial con-

	ditions
lm	log mean temperature or other property
m	mass transfer parameters
0	overall property or parameter
p	primary side property or parameter
r	root conditions
s	surface condition or parameter
t	tube wall property or parameter
v	vapor water property
w	liquid water property
∞	free stream condition
1	inlet conditions
2	outlet conditions

Superscripts

$(\ )^*$	denotes wet system parameter
$(\ )'$	denotes saturation condition

REFERENCES

1. Kays, W. M., and London, A. L., *Compact Heat Exchangers,* McGraw-Hill, New York, 1964.
2. Shah, R. K., "Compact Heat Exchanger Surface Selection Methods," in Heat Transfer 1978, Proceedings of the Sixth International Heat Transfer Conference, Vol. 4, pp. 193–199, Hemisphere, Washington, D. C., 1978.
3. Webb, R. L., "Air-Side Heat Transfer in Finned Tube Heat Exchangers," *Heat transfer Engineering,* Vol. 1, No. 3 (January–March 1980), pp. 33–49.
4. Brauer, H., "Compact Heat Exchangers," *Chemical and Process Engineering* (August 1964), pp. 451–460.
5. Brauer H., *Kaltetchnik,* 13 Jahrgang, Heft 8, 274–279 (1961).
6. Carnavos, T. C., U. S. Department of Commerce Clearing House, Report No. AECU–3970 (May 1958).
7. Rabas, T. J., and Eckels, P. W., "Heat Transfer and Pressure Drop Performance of Segmented Surface Tube Bundles," ASME 75–HT–45.
8. Weierman, C., Taborck, J., and Marner, W. J., "Comparison of the Performance of Inline and Staggered Banks of Tubes with Segmented Fins," 15th National Heat Transfer Conference, San Francisco, California, AICH Paper No. 71975.
9. Brauer, H., "Warme-und stromungstechniche anquer angestromten Reppenrokrundlen," *Chemie-Ing-Tech,* 33, Part II, (1961), pp. 431–438.
10. Cox, B., and Jallouk, P. A., *Experimental Determination of the Performance Characteristics of Eight Compact Heat Transfer Surfaces,* K–1832, December 1972, Union Carbide Corporation, Nuclear Division, Oak Ridge Gaseous Diffusion Plant, Oak Ridge, Tennessee.
11. Jameson, S. L., "Tube Spacing in Finned Tube Banks," *Trans. ASME,* No. 67 (November 1945), pp. 633–642.
12. Mirkovic, Z., Chapter 20 in *Heat Exchangers: Design and Theory Source Book,* by Afgan and Schlundr, McGraw-Hill Book Co., New York, 1974.
13. Ward, D. J., and Young, E. H., "Heat Transfer and Pressure Drop of Air in Forced Convection Across Triangular Pitch Banks of Finned Tubes," *CEP Symposium,* Series No. 29, Vol. 55 (1959), p. 37.
14. Briggs, D. E., and Young, E. H., "Convection Heat Transfer and Pressure Drop of Air Flowing Across Triangular Pitch Banks of Finned Tubes," *CEP Symposium,* Series No. 41, Vol. 59 (1963), p. 1.
15. Kuntysh, V. B., and Iokhevdov, F. M., "Effect of the Relative Interfin distance on the Thermal Efficiency Convective Heat Transfer in Finned Tube Bundles and on Augmenting of Heat Transfer," *Heat Transfer,* Soviet Research, Vol. 3 (2) (1971), pp. 50–61.

16. Rich, D. G., "The Effects of Fin Spacing on the Heat Transfer and Friction Performance of Multi-Row Smooth Plate Fin-and-Tube Heat Exchangers," *ASHRAE Trans. 79*, pt. 2 (1973), pp. 137–145.
17. Taranyan, I. G., Iokhvedon, F. M., and Duntysh, V. B., *High Temperature*, Vol. 10, No. 5 (1972), pp. 941–945.
18. Kuntysh, V. B., Iokhvedon, F. M., and Taranyan, I. G., "Effect of the Relative Depth of the Interfin Space on Heat Transfer from Bundles of Finned Tubes," *Teor. Asnovy Khimich Tekh*, Vol. 7, No. 4 (1973), pp. 616–621.
19. Preece, R. J., and Hitchcock, J. A., "Comparative Performance Characteristics of Some Extended Surfaces for Air-Cooled Applications," *Chem. Eng.*, London (June 1972), pp. 238–244.
20. Cox, B., *Performance of Compact Heat Transfer Surfaces–Finned & Bare Tube Banks*, Union Carbide Corporation, Nuclear Division, Oak Ridge Gaseous Diffusion Plant, Oak Ridge, Tennessee (November 1973).
21. Cox, B., and Jallouk, P. A., "Methods for Evaluating the Performances of Compact Heat Transfer Surfaces," ASME Paper 72WA/HT–56, Annual Meeting, New York (November 1972).
22. PFR Engineering Systems, Inc.," Heat Transfer and Pressure Drop Characteristics of Dry Tower Extended Surfaces, Part II: Data Analysis and Correlation." Report to Battelle Memorial Institute, Special Agreement BSA–988, June 1976. (Available from Pacific Northwest Laboratory.)
23. LaHaye, P. G., Neugebauer, F. J., and Sakhuja, R. K., "A Generalized Prediction of Heat Transfer Surface Performance and Exchanger Optimization," ASME Paper 72–WA/HT–55 Annual Meeting, New York (November 1972).
24. Johnson, B. M., Kreid, D. K., and Hauser, S. G., "A Method of Company Performance of Extended—Surface Heat Exchangers," *Heat Transfer Engineering*, Vol. 4, No. 1 (Jan.–March 1983).
25. Hauser, S. G., Kreid, D. K., and Johnson, B. M., *Dry/Wet Performance of a Plate-Fin Air-Cooled Heat Exchanger with Continuous Corrugated Fins*, PNL–3428, Pacific Northwest Laboratory, Richland, Washington (February 1981).
26. Johnson, B. M., and Dickinson, D. R., "On the Minimum Size for Forced Draft Dry-Cooling Towers for Power Generating Plants," ASME HTO, Vol. 6, Dry and Wet/Dry Cooling Towers for Power Plants Annual Meeting, Detroit (November 1973).
27. Fryer, B. C., et al., *An Engineering and Cost Comparison of Three Different All-Dry Cooling System*, BNWL–2121, Pacific Northwest Laboratory (September 1976).
28. Jameson, S. L., "Tube Spacing in Finned-Tube Banks," *Trans ASME 67* (1945), pp. 633–642.
29. Bergles, A. E., Bunn, R. L., and Junkhan, G. H., "Extended Performance Evaluation Criteria for Enhanced Heat Transfer Surfaces," *Lett. Heat Mass Transfer*, Vol. 1 (1974), pp. 113–120.
30. Bergles, A. E., Junkhan, G. H., and Bunn, R. L., "Performance Criteria for Cooling Systems on Agricultural and Industrial Machines," *SAE Trans*, Vol. 85 (1976), pp. 38–48.
31. Threlkeld, J. L., *Thermal Environmental Engineering*, Prentice-Hall, Inc., Englewood Cliffs, New Jersey, 1962, p. 253.
32. Tasnadi, C., "Calculation of Heat Transport Combined with Material Transfer by Analogy to Heat Exchanger Dimensioning," *Energy Es Atomtechnika*, Vol. 25, No. 6 (1972), pp. 252–257.
33. Kals, W., "Wet Surface Air Coolers: Characteristics and Usefulness," ASME Paper No. 72–HT–28 (1972).
34. Liendenfrost, W., and Korenic, B., "Analytical and Experimental Studies of Evaporative Cooling and Heat Transfer Augmentation Related to Reduced Condenser Temperature," *Heat Transfer Engineering*, Vol. 3, Nos. 3–4 1983, pp. 38–59.

35. Kreid, D. K., Hauser, S. G., and Johnson, B. M., "Formulation of an Approximate Heat/Mass Transfer Model for Calculating the Performance of Heat Exchange in Dry/Wet Operation," ASME/JSME Thermal Engineering Joint Conf. Paper Honolulu (March 1983). (Also submitted to *Heat Transfer Engineering*.)
36. Parry, H. L., et al., *Augmented Dry Cooling Surface Test Program: Analysis and Experimental Results,* PNL–2746, Pacific Northwest Laboratory, Richland, Washington (November 1978).
37. Hauser, S. G., Kreid, D. K., and Johnson, B. M. *Dry/Wet Performance of a Plate-Fin Air-Cooled Heat Exchanger with Continuous Corregated Fins,* PNL–3428, Pacific Northwest Laboratory, Richland, Washington (1981).
38. Hauser, S. G., et al., *A Progress Report on the Experimental Evaluation of Dry/Wet Air-Cooled Heat Exchangers,* PNL–4043, Pacific Northwest Laboratory, Richland, Washington (1981).
39. Johnson, B. M. (Ed.), *Development of An Advanced Concept Dry/Wet Cooling of Power Generating Plants,* Battelle-Northwest, Interim Report of Research Project RP422–3 with the Electric Power Research Institute, EPRI–CS–1668 (February 1981).
40. Allemann, R. T., et al., "Operation and Testing of a Phase-Change Wet/Dry Cooling System." in Proceedings of the 45th Annual American Power Conference, Vol. 45 (1983), p. 430.

CHAPTER 25

VIBRATION EFFECTS ON NATURAL CONVECTION HEAT TRANSFER FROM FINS

P. K. Nag

Department of Mechanical Engineering
Indian Institute of Technology
Kharagpur, India

CONTENTS

INTRODUCTION

The feasibility of rejecting waste heat by economical troublefree methods is the criterion upon which many proposed applications of electronic and thermoelectric devices depend. Vertical fins are extensively used to make better utilization of the heat rejection area. The performance of the thermal system can be substantially improved by a number of augmentative techniques to intensify heat transfer, and the most potent methods [1, 2] are:

1. Use of surface promoters to hasten the onset and intensity of turbulence.
2. Vortex flow.
3. Application of electrostatic or magnetic fields.
4. Fluid vibration by incorporating pulsations in the fluid.
5. Application of sound or mechanical vibration to the heated surface.

Although the selection of the particular method for augmenting natural convection heat transfer from fins depends on cost considerations as well as safety and reliability aspect, the most suitable method is to impart mechanical vibration to the heated surface.

EFFECTS OF FIN GEOMETRY AND FIN SPACING

Natural convection heat transfer from fins depends on the fin shape as well as the fin spacing. Rectangular fins are more common, and from a buoyancy consideration vertical fins are more effective. Starner and McManus [3] conducted experiments with rectangular vertical fin arrays and found that fins with wider spacings yielded higher coefficients of heat transfer, since these interfered the boundary layers to a lesser extent. Heat transfer was found to improved with a increase in fin length. Wellin and Woolridge [4] obtained an optimum aspect ratio S/H at a given temperature to yield a maximum heat transfer coefficient. Chaddock [5] correlated the dimensionless parameters involving fin height, fin length, fin spacing, etc. with the heat transfer coefficient. Aihara [6–9] through extensive experimentation determined the heat transfer coefficients for fin base, fin flats, and fin edges separately and developed empirical correlations. Sukhatme and Saikhedkar [10] theoretically analyzed the natural convection heat transfer from vertical fin arrays and compared their results with the published experimental data. The average Nusselt number computed with their data had been fitted to obtain the following two correlations:

1. For the range $0 < Ra_s < 250$,

$$\overline{Nu}_s = 0.079\,(S/H)^{0.579}\,(L/H)^{-0.261}\,(Gr_sPr)^{0.509} \tag{1}$$

2. For the range $250 < Ra_s < 3 \times 10^4$

$$\overline{Nu}_s = 1.619 + 11.0\,(H/S)^{-0.436}[1 - 3.8\,(L/H)^{0.068}\,(Gr_sPr)^{0.28}] \tag{2}$$

EFFECTS OF VIBRATION

It has long been recognized that transport processes can be significantly affected by inherent or induced oscillations. The most direct approach is to vibrate the heating surface mechanically, usually by an electrodynamic vibrator or a motor-driven eccentric. Frequencies are generally kept well below 1,000 Hz to achieve an adequate amplitude of vibration. Richardson [11] studied the vibration effects on heat transfer from plates, ducts, wires, and cylinders. Fand et al. [12, 13] and Fand [14] investigated the influence of horizontal and vertical mechanical vibrations on heat transfer by natural convection from a horizontal cylinder. The convection process in the vicinity of the cylinder is assumed to be quasi-steady, and the heat transfer may be described by conventional correlations for steady convection. For low values of amplitude-to-diameter ratios, there is no significant displacement of the cylinder through the fluid to enhance energy transport. The vibrational intensity defined as the product of the amplitude and frequency ($a \cdot f$) of vibration is low and natural convection dominates. However, when the vibrational intensity reaches a critical value, a secondary flow, called acoustic or thermoacoustic streaming, develops as a result of the interaction of the convection due to vibration with the convecton due to the buoyancy effect,

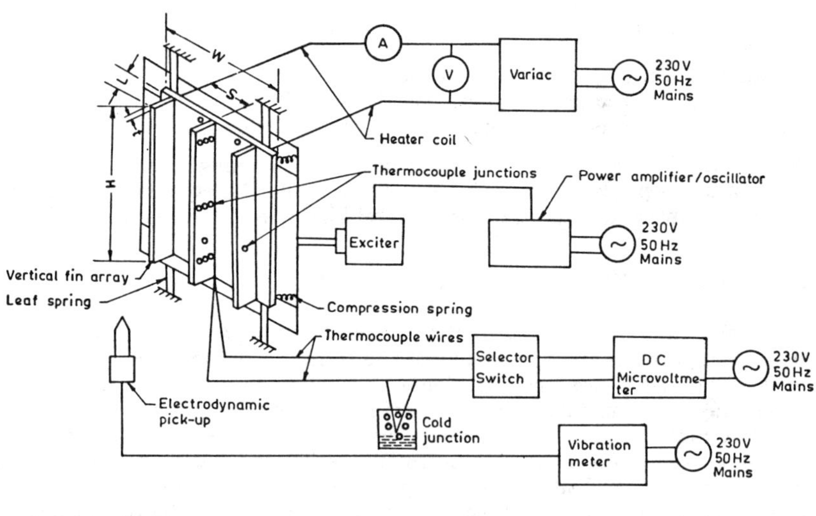

Figure 1. Schematic diagram of the experimental set-up.

enabling a net energy flux from the boundary layer. There are thus three distinct regions depending on the intensity of vibration as given by the vibrational Reynolds number, Re_v. In the region of low Re_v, natural convection dominates, which is followed by a transition region where natural convection and the "forced" convection due to vibration interact, and finally the region where forced convection due to vibration dominates. Dawood, Manocha, and Ali [15] showed that the relative increase of the heat transfer coefficient over that without vibration, i.e. $h_v - h_c/h_c$ depends on the vibrational Reynolds number and Grashof number, and obtained the following correlation

$$h_v - h_c = 12.9 + \frac{281.3\ af}{\left(\frac{\Delta T}{D}\right)^{0.1}} \tag{3}$$

for $0.015 < af < 0.15$ m/s and $22 \leqq \Delta T \leqq 85°$ C.

Blankenship and Clark [16, 17] studied both theoretically and experimentally the effect of induced transverse vibration on natural convection heat transfer from a vertical finite plate. It was found that the effect of vibration is negligible if the boundary layer remains laminar. But in the case of turbulent flow the heat transfer increases to a great extent under the influence of vibration. A smoke study was also made to note the value of $Gr \cdot Pr$ at which the flow changed from laminar to turbulent.

VIBRATION EFFECTS ON HEAT TRANSFER FROM FIN ARRAYS

Nag and Bhattacharya [18] studied the effect of vibration on heat transfer from vertical rectangular fin arrays of different fin spacings and fin lengths. The schematic diagram of the experimental set-up is shown in Figure 1. Nine sets of fin arrays, each consisting of three identical fins protruding from a base plate, all made from 13 mm thick aluminium plate, were used. At steady state, the heat dissipated from the electric heater was conducted through the base plate and then transferred to the air from the finned and flat surfaces of the test section by natural convection and radiation, and the rate of heat transfer was

$$Q = (h_c + h_r)\, A\, (\bar{T}_w - T_a) \tag{4}$$

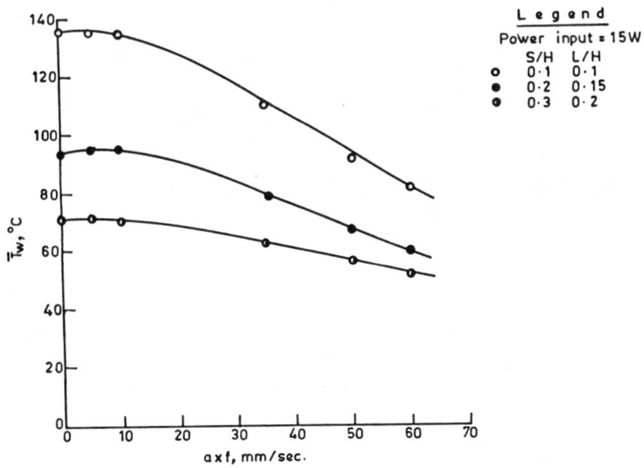

Figure 2. Variation of surface temperature (amplitude × frequency).

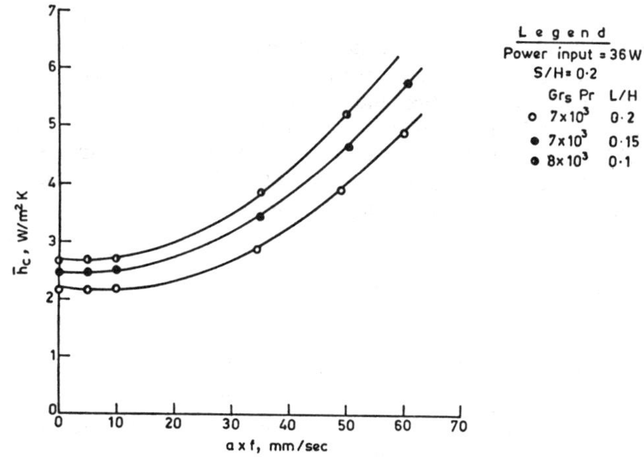

Figure 3. Variation of heat transfer coefficient (amplitude × frequency).

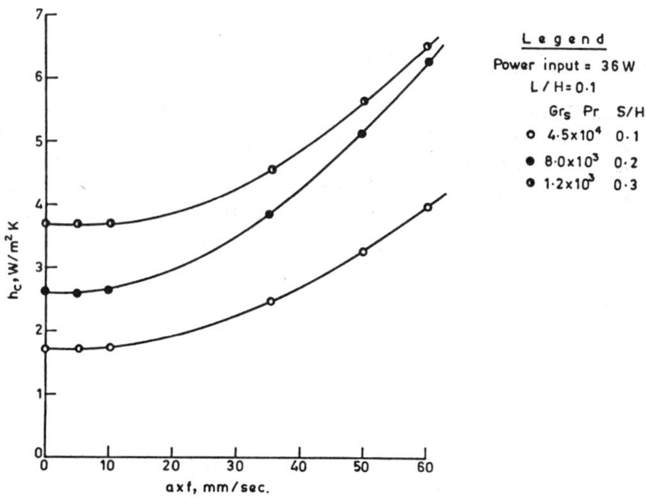

Figure 4. Variation of heat transfer coefficient (amplitude × frequency).

where $h_r = \sigma\varepsilon_1(\bar{T}_w + T_a)(\bar{T}_w^2 + T_a^2)$.

The average convective heat transfer coefficient h_c and Nusselt number $Nu_c = h_cS/K_f$ were evaluated and the gain in Nusselt number due to vibration $[(Nu_v - Nu_c)/Nu_c]100\%$ was computed. The Grashof number (Gr_s) and the vibrational Reynolds number (Re_v) were estimated using the expressions

$$Gr_s = \frac{g\beta(\bar{T}_w - T_a)S^3}{\nu^2}$$

$$Re_v = \frac{(af)S}{\nu}$$

Details of the results can be had in Reference 19. Typical plots are shown in Figures 2 through 6. It is observed that the change in surface temperature, the average heat transfer coefficient, or the Nusselt number was not significant up to a certain critical or threshold value of a · f which was found to be about 15 mm/s. The boundary layer was not disturbed by the low intensity vibration and remained laminar. Beyond this threshold value of a · f however, the rate of heat transfer considerably increased due to thermoacoustic streaming with the thermal boundary layer substantially agitated. The heat transfer coefficient decreased with the increase in L/H due to more conductive resistance as well as with the

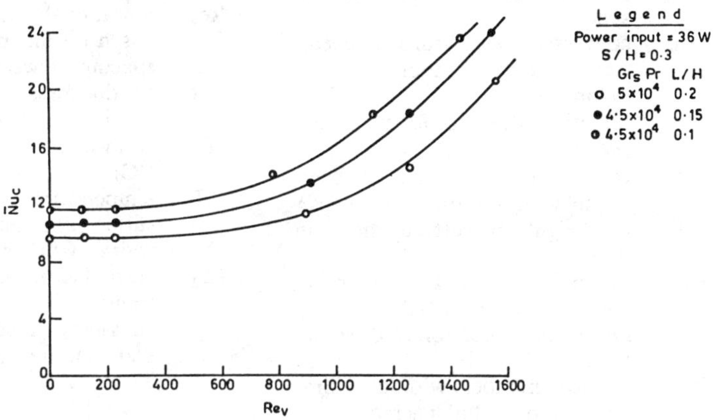

Figure 5. Variation of Nusselt number with vibrational Reynolds number.

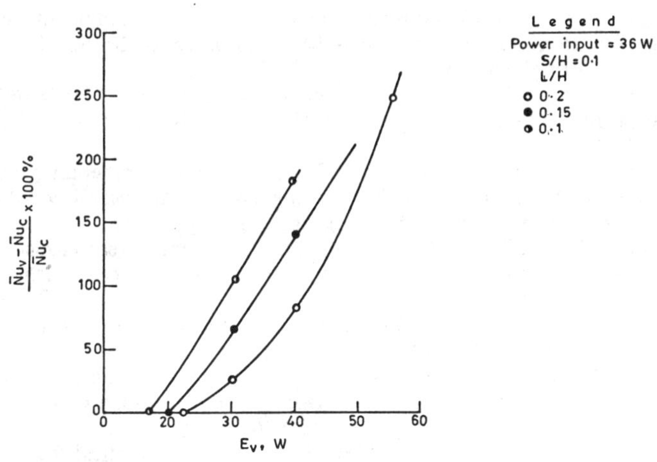

Figure 6. Percentage increase of Nusselt number varying with vibrational energy input.

decrease in S/H due to interference of the boundary layers. Figure 6 exhibits the percentage gain in Nusselt number due to vibration with the energy E_V imparted to vibrate a fin array. Up to a certain value of energy input E_V, the Nusselt number remained more or less constant, and beyond this value it increased rapidly with E_V. The maximum increase in heat transfer due to vibration was found to be about 250% against the energy input of 90 W for S/H and L/H values of 0.1 and 0.2, respectively.

NOTATION

a	amplitude of vibration (mm)
A	surface area exposed to air (m^2)
E_V	vibration energy input (W)
f	frequency of vibration (Hz)
Gr_s	Grashof number based on fin spacing
h_c	convective heat transfer coefficient without vibration ($W/m^2\,°K$)
h_r	radiation heat transfer coefficient ($W/m^2\,°K$)
h_v	convective heat transfer coefficient with vibration ($W/m^2\,°K$)
H	fin height in vertical direction (mm)
K_f	thermal conductivity of air at film temperature ($W/m\,°K$)
L	fin length in horizontal direction (mm)
Nu_c	Nusselt number without vibration based on fin spacing
Nu_v	Nusselt number with vibration based on fin spacing
Pr	Prandtl number of air
Q	heat input to heater (W)
Ra_s	Rayleigh number (Gr_sPr)
Re_v	vibrational Reynolds number based on fin spacing
S	spacing between two fins (mm)
t	fin thickness (mm)
T_a	ambient air temperature (°C)
T_w	surface temperature of fin array (°C)
ΔT	temperature difference between surface and ambient air (°C)
W	width of fin array (mm)
$\mathscr{F}_{1-2}$	view factor in radiation heat transfer
ε_1	emissivity of polished aluminium
σ	Stefan Boltzmann's constant

REFERENCES

1. Bergles, A. E., "Survey and Evaluation of Techniques to Augment Convective Heat and Mass Transfer," *Progress in Heat and Mass Transfer,* U. Grigull (Ed.), Vol. 1, Pergamon Press, Oxford, 1969, pp. 331–424.
2. Bergles, A. E., "Techniques to Augment Heat Transfer," *Handbook of Heat Transfer,* Warren M. Rohsenow and James P. Hartnett (Eds.), Section 10, McGraw–Hill Book Co., 1973
3. Starner, K. E., and McManus, H. N., Jr., "An Experimental Investigation of Free Convection Heat Transfer from Rectangular Fin Arrays," *Journal of Heat Transfer, Trans. ASME,* Vol. 65 (Aug. 1963), pp. 273–278.
4. Wellin, J. R., and Wooldridge, C. B., "Free Convection Heat Transfer Coefficients from Rectangular Vertical Fins," *Journal of Heat Transfer, Trans. ASME,* Vol. 87 (Nov. 1965), pp. 439–444.
5. Chaddock, J. B., "Free Convection Heat Transfer from Vertical Rectangular Fin Arrays," *ASHRAE Journal* (Aug. 1970), pp. 53–60.
6. Aihara, T., "Natural Convection Heat Transfer from Vertical Rectangular Fin Arrays, Part I—Heat Transfer from Base Plates of Vertical Open Channels," *Trans. ASME,* Vol. 34, No. 261 (May 1968), pp. 915–926.
7. Aihara, T., "Natural Convection Heat Transfer from Vertical Rectangular Fin Arrays, Part II—Heat Transfer from Fin Edges," *Bull JSME,* Vol. 13, No. 64 (Oct. 1970), pp. 1182–1191.

8. Aihara, T., "Natural Convection Heat Transfer from Vertical Rectangular Fin Arrays, Part III—Heat Transfer from Fin Flats," *Bull JSME,* Vol. 13, No. 64 (Oct. 1970), pp. 1192–1200.
9. Aihara, T., "Natural Convection Heat Transfer from Vertical Rectangular Fin Arrays, Part IV—Heat Transfer Characteristics of Non-isothermal Fin Arrays," *Bull JSME,* Vol. 14, No. 74 (Aug. 1971), pp. 818–828.
10. Sukhatme, S. P., and Saikhedkar, N. H., "Natural Convection Heat Transfer from Rectangular Cross Sectioned Vertical Fin Arrays," *Journal of Thermal Engineering,* Vol. 1, No. 3 (July–Sept. 1980), pp. 125–132.
11. Richardson, P. D., "Effects of Sound and Vibration on Heat Transfer," *Applied Mechanics Review,* Vol. 20, No. 3 (March 1967), p 201–211.
12. Fand, R. M., and Peables, E. M., "Heat Transfer from a Horizontal Cylinder Subjected to Horizontal Mechanical Vibrations," *Journal of Heat Transfer, Trans. ASME,* Vol. 84, No. 3 (1962).
13. Fand, R. M., et al., "Local Heat Transfer Coefficient Around a Horizontal Cylinder in the Presence of Sound," *Journal of Heat Transfer, Trans. ASME,* Vol. 84, No. 3 (1962).
14. Fand, R. M., "Mechanics of Interaction between Vibrations and Heat Transfer," *Journal of the Acoustical Society of America,* Vol. 34, No. 12 (1962), pp. 1887–1894.
15. Dawood, A. S., Manocha, B. J., and Ali, S. M. J., "The Effect of Vertical Vibrations on Natural Convection Heat Transfer from a Horizontal Cylinder," *Int. Journal of Heat and Mass Transfer,* Vol. 24, No. 3 (March 1981), pp. 491–496.
16. Blankenship, V. D., and Clark, J. A., "Effects of Oscillation on Free Convection from a Vertical Finite Plate," *Journal of Heat Transfer, Trans. ASME* (May 1964), pp. 149–158.
17. Blankenship, V. D., and Clark, J. A., "Experimental Effects of Transverse Oscillations on Free Convection of a Vertical Finite Plate," *Journal of Heat Transfer, Trans. ASME* (May 1964), pp. 159–165.
18. Nag, P. K., and Bhattacharya, A., "Effect of Vibration on Natural Convection Heat Transfer from Vertical Fin Arrays," *Letters in Heat and Mass Transfer,* Pergamon Press, Vol. 9, No. 6 (Nov. 1982), pp. 487–498.
19. Bhattacharya, Anjan, "Effect of Vibration on Natural Convection Heat Transfer from Vertical Fin Arrays," M. Tech thesis, Dept. of Mechanical Engg., I. I. T., Kharagpur, (June 1982).

CHAPTER 26

NUMERICAL MODELING OF HEAT EXCHANGERS

William T. Sha

Argonne National Laboratory
9700 South Cass Avenue
Argonne, Illinois

CONTENTS

This paper is based on work which was supported by the U.S. Department of Energy under ANL Contract No. 31–109–38–381, Foster Wheeler Contract Nos. 8–33–3088 and 8–51–3864, and the U.S. Nuclear Regulatory Commission under Contract A-2045.

INTRODUCTION

Accurate and detailed knowledge of the fluid flow field and thermal distribution inside a heat exchanger becomes invaluable as a large, efficient, and reliable unit is sought. This information is needed to provide proper evaluation of the thermal and structural performance characteristics of a heat exchanger. Thermal-hydraulically unbalanced design or operation of a heat exchanger often causes unequal thermal loadings among the heat-transfer tubes. This may result in excessive thermal stresses in the tube bundle, thus impacting upon the structural integrity and reliability of the unit. The structural integrity of the pressurebearing boundaries of heat exchangers employed in nuclear reactor systems has direct reactor safety implications.

The flow pattern and the thermal map of a shell-and-tube heat exchanger may be obtained either through actual testing of a scale model or by using an analytical prediction method. Scale-model testing is, in general, very expensive, and requires long lead time. Often instrumentation used in experiments must be able to withstand high temperature, high pressure, and a corrosive environment. Another problem associated with scale-model testing is that suitable similarity parameters must be obtained and matched between the scale model and the actual unit; otherwise all the conclusions and information drawn from the experimental data will be of limited value. Finally, both changing and upgrading the design of heat exchangers often are done routinely. If the design of a heat exchanger relies solely on experimental information, any significant deviation from the original design will require all tests to be repeated. This is a costly and time-consuming process.

It is to be noted that an analytical prediction method, when properly validated, will greatly reduce the need for model testing, facilitate the interpolating and extrapolating of test data, aid in optimizing heat-exchanger design and performance, and provide scaling capability. Thus tremendous savings of cost and time are realized. With the advent of large digital computers and advances in the development of computational fluid mechanics [1–3], it has become possible to predict analytically, through numerical solution the conservation equations of mass, momentum, and energy for both the shell-side and tube-side fluids. The numerical modeling technique will be a valuable, cost-effective design tool for the development of advanced heat exchangers.

In the past, investigations [4] have been made to obtain the flow distribution inside a heat exchanger through solutions of the continuum Navier–Stokes equations. The idea of using distributed resistance to simulate the presence of heat-transfer tubes and baffle plates on the shell side of a heat exchanger was first introduced by Patankar and Spalding [2]. These authors assumed that the space inside a heat exchanger is uniformly filled with fluid, throughout which, however, a resistance to fluid motion is distributed on a fine scale. But they did not calculate tube-side flow distribution. More recently, AbuRomia et al. [5] applied the distributed resistance concept to obtain the flow field between the typical tube support spans of the Clinch River Breeder Reactor Plant (CRBRP) intermediate heat exchangers. This analysis was limited to an isothermal flow field and included a turbulence model that is perhaps not appropriate for a tube bundle.

Recently a new porous-media formulation [6–10] was developed. It is rigorously derived through local volume averaging. The new porous-media formulation uses the concepts of volume porosity, directional surface permeabilities, distributed resistance, and distributed heat source or sink. All past analyses utilized the conventional porous-media formulation, in which only volume porosity, distributed resistance, and distributed heat source are used. Volume porosity is defined as the ratio of the volume occupied by fluid in a control volume to the total control volume. Directional surface permeabilities are similarly defined as the ratio of flow area of a control surface to the total control surface in each principal coordinate direction. The concept of *directional surface permeabilities* is new. In thermal-hydraulic analysis, resistance (friction factor) is, in general, not known precisely for most engineering applications; directional surface permeabilities are a function of geometry and are known precisely. The new porous-media formulation uses both distributed resistance and directional surface permeabilities for modeling velocity and temperature fields in anisotropic media. This is in contrast to the conventional porous media, in which only the distributed resistance is used. Thus, any error in estimating resistance will not be reflected entirely in the results using the new porous-media formulations, but will be in the conventional porous-media formulation. The concept of directional surface permeabilities greatly facilitates modeling velocity and temperature fields in anisotropic media and, in general, improves resolution and accuracy. It may be noted that, without the volumetric porosity and directional surface permeabilities, the calculated fluid momentum would be lower than the actual momentum; and without the distributed resistance, the calculated pressure drop would be less than the actual pressure drop.

POROUS MEDIA FORMULATION VIA LOCAL VOLUME AVERAGE

In general, fluid flow and heat transfer in a heat exchanger are complex. The complication arises from the fact that the flow domain of interest often contains both irregularly shaped fluids, such as bubbles, and solid structures. The cost of detailed thermal-hydraulic analysis with explicit treatment of these structures often is prohibitive, if not impossible. An alternative is to capture the essential features of the system and to express the flow and temperature field in terms of locally global quantities while sacrificing some of the details. Often, this is all that is required in most engineering applications. The present work is an attempt to achieve this goal by applying the local volume-averaging technique. In the course of derivations, fundamental averaging theorems and their basic premises are reconsidered.

Attention is focused on single-phase flows in a domain containing dispersed but stationary heat-generating or heat-absorbing solid structures. Local volume-averaged governing equations (i.e., conservation of mass, momentum, and energy) and interface relations between fluid and solid structures for such a system are derived. Extension of the resultant set of single-phase conservation equations to a multiphase system has been carried out and presented elsewhere [9, 10].

Averaging Relations

Consideration is hereby given to a domain consisting of a singlephase fluid and dispersed, stationary solid structures. Heat may be generated or absorbed by the structures. For an arbitrary point in the domain, we associate a *closed* surface A whose volume is V. A portion of V that contains the fluid is V_f and the total fluid-solid interface is A_{fs}. A portion of A through which the fluid may flow is A_e. A schematic of the system just described is illustrated in Figure 1.

Local Volume Average, Intrinsic Average, and Area Average

Let ψ be any intensive property *associated with the fluid.* It may be a scalar, vector, or second-order tensor. The local volume average of ψ is defined by

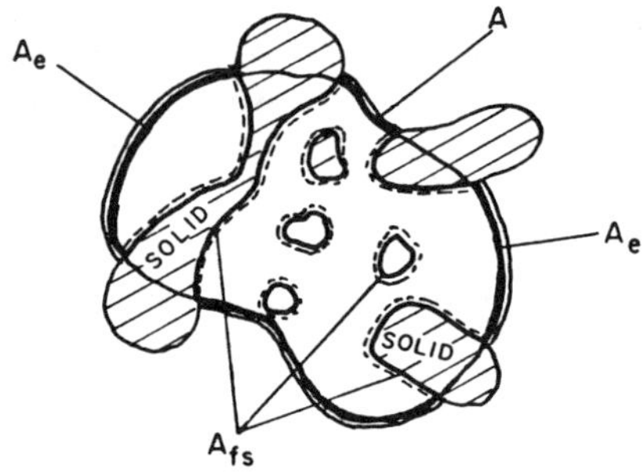

Figure 1. Local averaging volume of a physical system.

$$^{3}\langle\psi\rangle = \frac{1}{V}\int_{V} I(\mathbf{p})\psi \, dV \tag{1*}$$

where **p** is the position vector and the indicator function, I(**p**), is defined by

$$I(\mathbf{p}) = \begin{cases} 1, \text{ if the end point of } \mathbf{p} \text{ is in fluid} \\ 0, \text{ if the end point of } \mathbf{p} \text{ is in solid} \end{cases} \tag{2}$$

An equivalent form of Equation 1 is

$$^{3}\langle\psi\rangle = \frac{1}{V}\int_{V_f} \psi \, dV \tag{1a}$$

The intrinsic local volume average of ψ is

$$^{3i}\langle\psi\rangle = \frac{1}{V_f}\int_{V_f} \psi \, dV \tag{3**}$$

Likewise, the local area average of ψ is

$$^{2}\langle\psi\rangle = \frac{1}{A}\int_{A} I(\mathbf{p})\psi \, dA = \frac{1}{A}\int_{A_f} \psi \, dA \tag{4}$$

where A_f denotes the portion of A that is occupied by the fluid. The associated intrinsic area average is

$$^{2i}\langle\psi\rangle = \frac{1}{A_f}\int_{A_f} \psi \, dA \tag{5}$$

It is important to note that in Equation 4 the area A under consideration may *not* be the total enclosing surface. (If A is the enclosing surface in its entirety, then A_f in Equations 4 and 5 should be replaced by A_e; see Figure 1.) In fact, one is often concerned with a designated portion of it. For instance, in the Cartesian-coordinate system, the averaging

* The superscript 3 designates that the average is associated with volume; superscripts 2 and 1 are used for area and line (segment) averages.

** The superscript i denotes intrinsic average.

volume may be selected to be a parallelepiped $\Delta x \Delta y \Delta z$. The average mass flux through the surface ΔA_x (area $\Delta y \Delta z$) whose normal points in the direction of positive x axis are

$$^{2(x)}\langle \rho u \rangle = \frac{1}{\Delta A_x} \int_{\Delta A_x} I(\mathbf{p}) \rho u \, dA$$

$$= \frac{1}{\Delta A_x} \int_{\Delta A_{x,f}} \rho u \, dA \tag{4a}$$

where $\Delta A_{x,f}$ denotes the fluid portion of ΔA_x, ρ is the fluid density, and u is the fluid velocity in x direction. The corresponding intrinsic average is

$$^{2i(x)}\langle \rho u \rangle = \frac{1}{\Delta A_{x,f}} \int_{\Delta A_{x,f}} \rho u \, dA \tag{5a}$$

Volume Porosity and Directional Surface Permeabilities

The ratio of fluid volume V_f to the total volume V is defined to be the volume porosity, γ_v. Thus,

$$\gamma_v = \frac{V_f}{V} \tag{6}$$

Since $V_f = \int_V I(\mathbf{p})\, dV$, γ_v can also be written as

$$\gamma_v = \frac{1}{V} \int_V I(\mathbf{p})\, dV \tag{6a}$$

Furthermore,

$$^3\langle \psi \rangle = \gamma_v{}^{3i}\langle \psi \rangle \tag{7}$$

Analogously, we define the surface permeability γ_A associated with any surface (not necessarily closed) as

$$\gamma_A = \frac{A_f}{A} = \frac{1}{A} \int_A I(\mathbf{p})\, dA \tag{8}$$

where A_f is the portion of A that is occupied by the fluid. Consider, for example, the surface ΔA_x described earlier. Its directional surface permeability in the x direction is

$$\gamma_{A_x} = \frac{\Delta A_{x,f}}{\Delta A_x} = \frac{1}{\Delta A_x} \int_{\Delta A_x} I(\mathbf{p})\, dA \tag{9}$$

Clearly,

$$^2\langle \psi \rangle = \gamma_A{}^{2i}\langle \psi \rangle \tag{10}$$

Local Volume-Averaging Theorems

The volume averages of derivatives of ψ are given by the local volume-averaging theorems as [11, 12, 13, 14]

$$^3\langle \nabla \psi \rangle = \nabla^3\langle \psi \rangle + V^{-1} \int_{A_{fs}} \psi \mathbf{n}\, dA \tag{11}$$

$$^3\langle\nabla\cdot\psi\rangle = \nabla\cdot{}^3\langle\psi\rangle + V^{-1}\int_{A_{fs}}\psi\cdot\mathbf{n}\,dA \tag{12}$$

and

$$^3\langle\partial\psi/\partial t\rangle = \partial^3\langle\psi\rangle/\partial t - V^{-1}\int_{A_{fs}}\psi\mathbf{v}\cdot\mathbf{n}\,dA \tag{13}$$

Note that these averaging relations are subject to the restriction that [13]

(Characteristic length of pores and phases)

≪ (characteristic length of averaging volume)

≪ (characteristic length of the physical system). (14)

The averaging volume under consideration, therefore, cannot be arbitrarily small or become infinitesimal.

Local Volume-Averaged Conservation Equations

Conservation of Mass

The differential equation governing mass conservation for the fluid is

$$\frac{\partial\rho}{\partial t} + \nabla\cdot\rho\mathbf{v} = 0 \tag{15}$$

where ρ = fluid density
t = time
$\mathbf{v}$ = fluid velocity vector

Taking the local volume average of Equation 15 as defined by Equations 1 or 1a yields

$$^3\frac{\partial\rho}{\partial t} + {}^3\langle\nabla\cdot\rho\mathbf{v}\rangle = 0 \tag{16}$$

Application of the local volume-averaging theorem of the time derivative [11] to the first term in Equation 16 yields

$$^3\left\langle\frac{\partial\rho}{\partial t}\right\rangle = \frac{\partial^3\langle\rho\rangle}{\partial t} - \frac{1}{V}\int_{A_{fs}}\rho\mathbf{v}_{fs}\cdot\mathbf{n}\,dA \tag{17}^*$$

where $\mathbf{v}_{fs}$ is the velocity of A_{fs} and $\mathbf{n}$ is the unit outdrawn normal vector on A_{fs} and away from the fluid. For the case under study, $\mathbf{v}_{fs} = 0$ since the solids are stationary. It is interesting to note that the integral in Equation 17 may vanish even if $\mathbf{v}_{fs} \neq 0$ everywhere on A_{fs}. This would occur if the solids that are completely inside A are nondeformable and are in motion relative to the fluid, since for any interior solid of surface ΔA_{fs},

$$\int_{\Delta A_{fs}}\rho\mathbf{v}_{fs}\cdot\mathbf{n}\,dA = 0 \tag{18}$$

if the variation of fluid density over ΔA_{fs} is negligible.

In either case,

$$^3\frac{\partial\rho}{\partial t} = \frac{\partial^3\langle\rho\rangle}{\partial t} = \gamma_v\frac{\partial^{3i}\langle\rho\rangle}{\partial t} \tag{18a}$$

* In general, the velocity of the fluid-solid interface $\mathbf{v}_{fs}$ may not be identical to the fluid velocity $\mathbf{v}$ at the interface. This would occur when mass transfer takes place.

Applying the theorem of local volume averaging to a divergence due to Slattery [11] and Whitaker [12, 13] gives

$$^3\langle\nabla\cdot\rho\mathbf{v}\rangle = \nabla\cdot{}^3\langle\rho\mathbf{v}\rangle + \frac{1}{V}\int_{A_{fs}}\rho\mathbf{v}\cdot\mathbf{n}\,dA \tag{19}$$

Again, the second term on the right-hand side of Equation 19 vanishes because, for the problem under consideration, $\mathbf{v} = 0$ everywhere on A_{fs}. Consequently,

$$^3\langle\nabla\cdot\rho\mathbf{v}\rangle = \nabla\cdot{}^3\langle\rho\mathbf{v}\rangle = \frac{1}{V}\nabla\cdot\int_{V_f}\rho\mathbf{v}\,dV$$

Since, as has been shown by Slattery [11], for any quantity B that may be a scalar, vector, or tensor,

$$\nabla\int_{V_f}B\,dV = \int_{A_e}B\mathbf{n}\,dA \tag{20}$$

it follows then that

$$\langle\nabla\cdot\rho\mathbf{v}\rangle = \frac{1}{V}\int_{A_e}\rho\mathbf{v}\cdot\mathbf{n}\,dA \tag{21}$$

Substituting Equations 18a and 21 into Equation 16 yields

$$\gamma_v\frac{\partial^{3i}\langle\rho\rangle}{\partial t} + \frac{1}{V}\int_{A_e}\rho\mathbf{v}\cdot\mathbf{n}\,dA = 0 \tag{22}$$

which is the required local volume-averaged mass conservation equation. The need of introducing the concept of directional surface permeabilities is suggested by the second term of Equation 22 and others in the momentum and energy equations.

Conservation of Momentum

We begin with the dynamic equation of fluid motion:

$$\frac{\partial(\rho\mathbf{v})}{\partial t} + \nabla\cdot(\rho\mathbf{v}\mathbf{v}) = \rho\mathbf{g} - \nabla p + \nabla\cdot\tau \tag{23}$$

where $\mathbf{g}$ = body force per unit fluid mass; in the gravitational filed, it is simply the gravitational acceleration vector
p = static pressure
τ = stress tensor

For a Newtonian fluid

$$\tau = \left[\left(\lambda - \frac{2}{3}\mu\right)\nabla\cdot\mathbf{v}\right]\mathbf{I} + \mu\,[\nabla\mathbf{v} + (\nabla\mathbf{v})_c] \tag{24}$$

in which λ is the bulk viscosity, μ is the ordinary dynamic viscosity, $\mathbf{I}$ is the unit tensor, and $(\nabla\mathbf{v})_c$ is the conjugate of tensor $\nabla\mathbf{v}$.

Performing the local volume averaging of Equation 23 gives

$$^3\frac{\partial(\rho\mathbf{v})}{\partial t} + {}^3\langle\nabla\cdot(\rho\mathbf{v}\mathbf{v})\rangle = {}^3\langle\rho\rangle g - {}^3\langle\nabla p\rangle + {}^3\langle\nabla\cdot\tau\rangle \tag{25}$$

wherein $\mathbf{g}$ is taken as a constant. With the aid of the theorems of local volume averaging, Equation 20, and recognizing that $\mathbf{v}_{fs}$ and $\mathbf{v}$ vanish on A_{fs}, Equation 25 can be rewritten as

$$\gamma_V \frac{\partial}{\partial t} {}^{3i}\langle \rho \mathbf{v} \rangle + \frac{1}{V} \int_{A_e} \rho \mathbf{v}(\mathbf{v} \cdot \mathbf{n})\, dA$$

$$= \gamma_v {}^{3i}\langle \rho \rangle \mathbf{g} - \gamma_v \nabla {}^{3i}\langle p \rangle + \frac{1}{V} \int_{A_e} \tau \cdot \mathbf{n}\, dA - \gamma_v {}^{3i}\langle \mathbf{R} \rangle \tag{26}$$

in which **R** is the distributed resistance force exerted on the fluid by the dispersed solid per unit volume of the fluid. In writing Equation 26, use has been made of the following relationship

$$\int_{A_{fs}} (-p\mathbf{n} + \tau \cdot \mathbf{n})\, dA = -\int_{V_f} \mathbf{R}\, dV \tag{27}$$

which may be regarded as the defining equation for **R**. The negative sign is introduced to convey the notion that such force retards the motion. Equation 26 is the local volume-averaged linear momentum equation for the system under consideration.

Conservation of Energy

Energy equation in terms of internal energy. The differential transport equation for internal energy is

$$\frac{\partial(\rho e)}{\partial t} + \nabla \cdot (\rho e \mathbf{v}) = -p\nabla \cdot \mathbf{v} - \nabla \cdot \mathbf{q} + \dot{Q} + \Phi \tag{28}$$

where e = internal energy per unit fluid mass
$\mathbf{q}$ = heat-flux vector
$\dot{Q}$ = rate of internal heat generation per unit fluid volume due to extraneous heat source
Φ = dissipation rate of mechanical energy into heat

The local volume averaging of Equation 28 is

$${}^3\frac{\partial(\rho e)}{\partial t} + {}^3\langle \nabla \cdot (\rho e \mathbf{v}) \rangle = -{}^3\langle p\nabla \cdot \mathbf{v} \rangle - {}^3\langle \nabla \cdot \mathbf{q} \rangle + {}^3\langle \dot{Q} \rangle + {}^3\langle \Phi \rangle \tag{29}$$

Again, with the aid of the theorems of local volume averaging, etc., we obtain the local volume-averaged internal energy equation:

$$\gamma_V \frac{\partial}{\partial t} {}^{3i}\langle \rho e \rangle + \frac{1}{V} \int_{A_e} \rho e \mathbf{v} \cdot \mathbf{n}\, dA$$

$$= -\gamma_v {}^{3i}\langle p\nabla \cdot \mathbf{v} \rangle + \frac{1}{V} \int_{A} k\mathbf{n} \cdot \nabla T\, dA$$

$$+ \gamma_v {}^{3i}\langle \dot{Q}_{rb} \rangle + {}^{3i}\langle \dot{Q} \rangle + {}^{3i}\langle \Phi \rangle \tag{30}$$

in which the heat flux vector $\mathbf{q}$ is taken to be $-k\nabla T$, k being the fluid conductivity. $\dot{Q}_{rb}$ denotes the rate of heat liberation from the dispersed solids per unit volume of the fluid. Thus

$$\int_{V_f} \dot{Q}_{rb}\, dV = -\int_{A_{fs}} \mathbf{q} \cdot \mathbf{n}\, dA \tag{31}$$

Energy equation in terms of enthalpy. The differential equation for the transport of static enthalpy $h\left(= e + \frac{p}{\rho}\right)$ is

$$\frac{\partial(\rho h)}{\partial t} + \nabla \cdot (\rho h \mathbf{v}) = \frac{dp}{dt} - \nabla \cdot \mathbf{q} + \dot{Q} + \Phi \tag{32}$$

and the local volume-averaged enthalpy equation is

$$\begin{aligned} &\gamma_v \frac{\partial}{\partial t} {}^{3i}\langle \rho h \rangle + \frac{1}{V} \int_{A_e} \rho h \mathbf{v} \cdot \mathbf{n}\, dA \\ &= \gamma_v {}^{3i}\left\langle \frac{dp}{dt} \right\rangle + \frac{1}{V} \int_{A_e} k \mathbf{n} \cdot \nabla T\, dA \\ &\quad + \gamma_v ({}^{3i}\langle \dot{Q}_{rb} \rangle + {}^{3i}\langle \dot{Q} \rangle + {}^{3i}\langle \Phi \rangle) \end{aligned} \tag{33}$$

Since

$$\frac{dp}{dt} = \frac{\partial p}{\partial t} + \nabla \cdot (p\mathbf{v}) - p\nabla \cdot \mathbf{v}$$

$${}^{3i}\left\langle \frac{dp}{dt} \right\rangle = \frac{\partial {}^{3i}\langle p \rangle}{\partial t} + \nabla \cdot {}^{3i}\langle p\mathbf{v} \rangle - {}^{3i}\langle p\nabla \cdot \mathbf{v} \rangle$$

Reducing this New Formulation to Conventional Porous Media Formulation and Continuum

Equations 22, 26, and 30 or 33 form a set of local volume-averaged conservation equations and constitute a new porous media formulation. Directional surface permeabilities will be brought out through finite differencing the surface integrals of these equations. It can be shown readily that the conventional porous media formulation is a subset of the present formulation. Furthermore, the new porous media formulation can provide detailed and local velocity and temperature fields if the problems under consideration involve body shapes that do lend themselves to any one of the Cartesian, cylindrical, or spherical coordinate representations. Figure 2 shows that both the conventional porous media formulation and continuum are subsets of this new porous media formulation.

TURBULENCE MODELING

The turbulence model presented here is limited to single-phase applications. Two-phase or multiphase turbulence modeling is in its infancy and will not be discussed here. The turbulence model [15] developed and presented here is for quasi-continuum or porous media for the transport of momentum and heat in large tube bundles. This model utilizes the concepts of volume porosity, directional surface permeabilities, and distributed resistance. Although the model is relatively simple, it does take into account explicitly the effects of turbulent kinetic energy generation due to shear, viscosity, diffusion, geometric effects, buoyancy, and Reynolds number. This model is designed primarily for tube bundle analysis with a mesh size consisting of a number of tubes. The other models were developed for a continuum [16, 17, 18] and are more appropriate for use in the inlet and exit headers of a heat exchanger. They account for the transport effects of the turbulent kinetic energy k, turbulent kinetic energy dissipation rate ε, and scalar energy g. At any point in the flow, values for either k-ε-g of the three-equation model, or k-ε of the two-equation model are obtained from transport-type equations. For the three-equation model [16, 17], turbulent momentum fluxes (Reynolds stresses) and heat transport rates are obtained from an

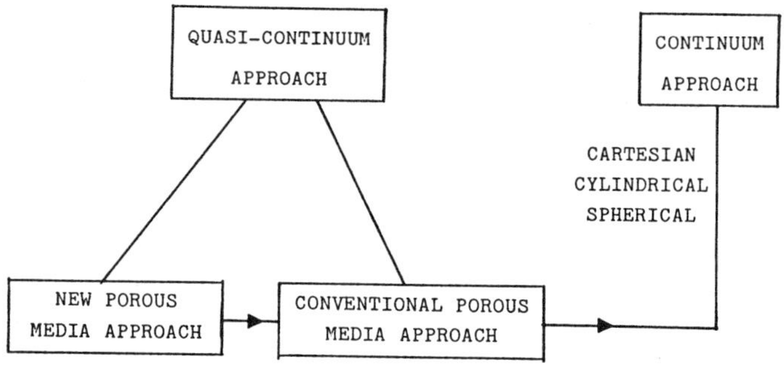

	New porous media approach	Conventional porous media approach	Continuum approach
VOLUME POROSITY	γ_v	γ_v	$\gamma_v = 1$
DISTRIBUTED RESISTANCE	R_x, R_y, R_z	R_x, R_y, R_z	$R_x = R_y = R_z = 0$
DISTRIBUTED HEAT SOURCE (SINK)	Q_v	Q_v	$\dot{Q}_{rb}$
SURFACE PERMEABILITY	$\gamma_x, \gamma_y, \gamma_z$	$\gamma_x = \gamma_y = \gamma_z = \gamma_v$	$\gamma_x = \gamma_y = \gamma_z = 1$

Figure 2. A unified approach to a fluid flow system.

algebraic formulation containing k-ε-g, the mean velocities, and temperatures, as well as their gradients. For the two-equation model [18], turbulent viscosity is calculated as a function of local k and ε.

Most heat exchangers are cylindrical in form; therefore, cylindrical coordinates r, θ, and z are used for all the formulations that follow.

Quasi-Continuum Model—Turbulent Kinetic Energy (k) Equation

When the flow is turbulent, the transport equation for turbulent kinetic energy may be expressed literally as:

$$\begin{array}{c}\text{net convection}\\\text{turbulent}\\\text{kinetic energy}\end{array} = \underbrace{\begin{array}{c}\text{production}\\\text{due to}\\\text{Shear}\end{array}}_{P_k} - \underbrace{\begin{array}{c}\text{dissipation}\\\text{due to}\\\text{viscosity}\end{array}}_{\varepsilon} + \begin{array}{c}\text{diffusion from}\\\text{neighboring}\\\text{computational}\\\text{cells}\end{array} + \begin{array}{c}\text{miscellaneous}\\\text{sources}\end{array} \tag{34}$$

Net Convection

This term can be written readily as

$$\gamma_v \frac{\partial \rho_k}{\partial t} + \nabla \cdot \gamma_A \rho \mathbf{v} k \tag{35}$$

Production Due to Shear

This process is represented as

$$P_k = \gamma_v \mu_t \frac{\partial U_i}{\partial x_j}\left(\frac{\partial U_j}{\partial x_i} + \frac{\partial U_i}{\partial x_j}\right) \tag{36}$$

where U_i is the velocity in i-th direction and μ_t is the turbulent viscosity.

An approximation of Equation 36 may be written as

$$\gamma_v c_s \mu_t \frac{(u-u_s)^2+(v-v_s)^2+(w-w_s)^2}{D_h^2} \tag{37}$$

where u, v, and w are velocity components in the x, y, and z directions, and u_s, v_s, and w_s are effective wall "slip" velocities designed to account for the fact that the flow in the viscous sublayer does not contribute to the production of turbulent kinetic energy. In practice, these slip velocities depend on the local Reynolds number. Initially, however, it appears advisable to use a constant value, which we take to be 50% of the corresponding velocities u, v, or w (implying that 50% of the velocity change occurs across the viscous sublayer—a reasonable estimate).

The coefficient c_s is included to account for the replacement of the x_i derivatives in Equation 36 by the hydraulic diameter. Its magnitude will be order unity. However, it probably will be somewhat larger than unity, since most of the energy production occurs near the wall, where velocity gradients are in fact steeper than implied by Equation 37.

Dissipation due to Viscosity

As in continuum models, we take the turbulent kinetic energy dissipation rate per unit volume ε to be equal to

$$\varepsilon = \gamma_v \rho k^{3/2}/\ell \tag{38}$$

In a sense, this may be regarded as defining the length scale ℓ.

Turbulence-Energy Diffusion

This will be represented by the gradient-type transport formulas used earlier to characterize the diffusion of heat and axial momentum:

$$\text{Diffusion rate} = \nabla \cdot \mu_{eff} \nabla \gamma_A k \tag{39}$$

where μ_{eff} is the effective viscosity.

Miscellaneous Effects

Baffle generation. The most reasonable simple proposition appears to be that a fixed proportion of the mean total energy lost in passing through the baffle is converted to turbulence (the remainder being directly dissipated to heat). Thus, we assume that the baffle loss should be expressed as

$$S_B = \gamma_v c_B \rho U^3 (A_x/A_0 - 1)^2 \text{ per unit cross-sectional area} \tag{40}$$

where U = the velocity perpendicular to the baffle,
A_x = the cross-sectional area of a computational cell upstream or downstream from the blockage,
A_0 = the open area of a computational cell at the blockage

This will produce a step change of turbulent kinetic energy. A little care is needed in transforming this equation to a quasi-continuum form for solution. The form loss conversion coefficient c_B should be of the order of 0.2, suggesting that about half the pressure loss goes into turbulent energy generation.

Buoyancy generation. For the case where the z axis is vertical with positive direction downward, the generation rate due to gravitational effects may be represented exactly as

$$S_G = -\gamma_v \frac{\dot{q}''_{z,t}}{c_p}\beta g \quad \text{per unit volume} \tag{41}$$

where $\dot{q}''_{z,t} = \dfrac{w_p^*}{\pi D_h}[\hat{\alpha}(T-T_w)]/[1+(\mu Pr_t)/(\mu_t Pr)]$

w_p^* = heat transfer perimeter
D_h = equivalent hydraulic diameter
$\hat{\alpha}$ = heat transfer coefficient
T, T_w = mean fluid and wall temperature
μ, μ_t = molecular and turbulent viscosity
Pr, Pr_t = molecular and turbulent Prandtl number
β = dimensionless volumetric expansion coefficient
g = gravitational acceleration
c_p = specific heat at constant pressure

Utilizing Equations 35 through 41, Equation 34 of turbulent kinetic energy becomes:

$$\begin{aligned}\gamma_v \frac{\partial \rho k}{\partial t} &+ \nabla \cdot \gamma_A \rho \mathbf{v} k \\ &= \gamma_v\{c_s\mu_t[(u-u_s)^2+(v-v_s)^2+(w-w_s)^2]/D_h^2 \\ &\quad -\rho k^{3/2}/\ell\} + \gamma_v c_B \rho\, |U|^3[(A_x/A_0)-1]^2/\delta x \\ &\quad -\gamma_v\{[\hat{\alpha}(T-T_w)\beta g/c_p]/[1+(\mu Pr_t)/(\mu_t Pr)]\}w_p^*/(\pi D_h) + \nabla \cdot \mu_{eff}\nabla\gamma_A k\end{aligned} \tag{42}$$

The effective viscosity μ_{eff} is given by

$$\mu_{eff} = \mu_t + \mu = c_\mu \rho k^{1/2}\ell + \mu \tag{43}$$

where

$c_\mu = c_{\mu\infty}$ for $Re_{max} > 2{,}000$
$c_\mu = c_{\mu\infty}(0.001\, Re_{max} - 1)$ for $10^3 \leqq Re_{max} \leqq 2\times 10^3$
$c_\mu = 0$ for $Re_{max} < 1{,}000$

and Re_{max} is the maximum of Re_x, Re_y, and Re_z.

The length scale appearing in Equation 43 and on the right-hand side of Equation 42 is taken as linearly proportional to the hydraulic diameter of the tube; i.e.,

$$\ell = c_\ell D_h \tag{44}$$

where c_ℓ is a coefficient.

The turbulent Prandtl number is obtained from [19]

$$Pr_t = 0.8[1-\exp(-6\times 10^{-5} Re_{max} Pr^{1/3})]^{-1} \tag{45}$$

The various empirical coefficients appearing in the model should be assigned initial values as follows:

Quantity	Initial Value	Quantity	Initial Value
c_B	0.2	$c_{\mu\infty}$	0.1
c_ℓ	0.4	u_s	0.5u
c_R	0.4	v_s	0.5v
c_s	2.0	w_s	0.5w

FINITE DIFFERENCE REPRESENTATION OF THE LOCAL VOLUME AVERAGED CONSERVATION EQUATIONS

The quasi-continuum-governing equations in cylindrical coordinates are finite-differenced in a staggered mesh system. The pressure p, temperature T, enthalpy h, density ρ, volume porosity γ_v, and turbulent kinetic energy k are defined at cell centers (i, j, k). The velocities, (u, v, and w) and directional surface permeabilities (γ_r, γ_θ, γ_z) are defined at cell surfaces $(i \pm 1/2, j, k)$, $(i, j \pm 1/2, k)$ and $(i, j, k \pm 1/2)$, respectively, as shown in Figure 3.

Finite Difference Equations

The finite difference equations for the local volume-averaged conservation of mass, momentum, and energy are presented for a variable mesh discretization. It is assumed here that the average of a product is equal to the product of averages.

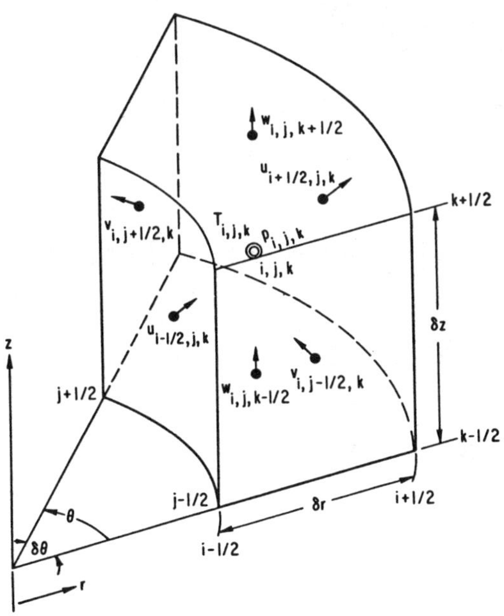

Figure 3. Locations of variables and indices about a computation cell.

Conservation of Mass

The conservation of mass demands that the following equation be satisfied:

$$(\gamma_v\rho)^{n+1}_{i,j,k} = (\gamma_v\rho)^n_{i,j,k} - \frac{\Delta t}{r_i\Delta r_1}(\langle\gamma_r\rho u r\rangle^{n+1}_{i+1/2,j,k} - \langle\gamma_r\rho u r\rangle^{n+1}_{i-1/2,j,k})$$

$$- \frac{\Delta t}{r_i\Delta\theta_1}(\langle\gamma_\theta\rho v\rangle^{n+1}_{i,j+1/2,k} - \langle\gamma_\theta\rho v\rangle^{n+1}_{i,j-1/2,k})$$

$$- \frac{\Delta t}{\Delta z_1}(\langle\gamma_z\rho w\rangle^{n+1}_{i,j,k+1/2} - \langle\gamma_z\rho w\rangle^{n+1}_{i,j,k-1/2}) + Sm_{i,j,k} \tag{46}$$

where Sm is the mass source term.

The cell differences in the various directions, e.g., Δr_1, $\Delta\theta_1$, Δz_1, etc. are defined in Appendix A. Convective terms are enclosed by the indicated brackets $\langle\,\rangle$ (note that brackets in this section do not indicate an average quantity); the operation indicated by these brackets provides for partial donor-cell differencing [20], e.g.,

$$\langle\gamma_r\rho u r\rangle^n_{i+1/2,j,k}$$

$$= (\gamma_r r u)^n_{i+1/2,j,k}\left[\left(\frac{1}{2}+\xi_x\right)(\rho)^n_{i,j,k} + \left(\frac{1}{2}-\xi_x\right)(\rho)^n_{i+1,j,k}\right] \tag{47}$$

where

$$\gamma_{r,i+1/2,j,k} = \frac{(r_{i+1/2}-r_i)\gamma_{r,i+1,j,k} + (r_{i+1}-r_{i+1/2})\gamma_{r,i,j,k}}{\Delta r_2}$$

$$\xi_x = a_0 \text{ sign } (u)^n_{i+1/2,j,k} + b_0\frac{(u)^n_{i+1/2,j,k}\Delta t}{\Delta r_2}$$

$$\langle\gamma_\theta\rho v\rangle^n_{i,j+1/2,k}$$

$$= (\gamma_\theta v)^n_{i,j+1/2,k}\left[\left(\frac{1}{2}+\xi_y\right)(\rho)^n_{i,j,k} + \left(\frac{1}{2}-\xi_y\right)(\rho)^n_{i,j+1,k}\right] \tag{48}$$

where

$$\gamma_{\theta,i,j+1/2,k} = \frac{(\theta_{j+1/2}-\theta_j)\gamma_{\theta,i,j+1,k} + (\theta_{j+1}-\theta_{j+1/2})\gamma_{\theta,i,j,k}}{\Delta r_2}$$

$$\xi_y = a_0 \text{ sign } (v)^n_{i,j+1/2,k} + b_0\frac{(v)^n_{i,j+1/2,k}\Delta t}{r_i\Delta\theta_2}$$

and

$$\langle\gamma_z\rho w\rangle^n_{i,j,k+1/2}$$

$$= (\gamma_z w)^n_{i,j,k+1/2}\left[\left(\frac{1}{2}+\xi_z\right)(\rho)^n_{i,j,k} + \left(\frac{1}{2}-\xi_z\right)(\rho)^n_{i,j,k+1}\right] \tag{49}$$

where

$$\gamma_{z,i,j,k+1/2} = \frac{(z_{k+1/2} - z_k)\gamma_{z,i,j,k+1} + (z_{k+1} - z_{k+1/2})\gamma_{z,i,j,k}}{\Delta z_2}$$

$$\xi_z = a_0 \text{ sign } (w)^n_{i,j,k+1/2} + b_0 \frac{(w)^n_{i,j,k+1/2}\Delta t}{\Delta z_2}$$

All other convective flux quantities in the following finitedifferenced equations are evaluated in a similar fashion.

In the equations for ξ_x, ξ_y, and ξ_z, a_0, and b_0 are input coefficients. When $a_0 = 1/2$ and $b_0 = 0$, the formulated differences constitute a full donor-cell difference and provide what is called upwind differencing. When $a_0 = b_0 = 0$, the usual central difference is obtained, whereas, when $a_0 = 0$ and $b_0 = 1/2$, an interpolated donor cell is formed. The value of ξ assumes that $\text{Max}\{|u|\Delta t/\Delta r, |v|\Delta t/r\Delta\theta, |w|\Delta t/\Delta z\} < 1/2$ everywhere. This criterion, in general, constitutes a reasonable upper bound on Δt for accuracy.

Since the mass is not identically conserved during each iteration, Equation 46 can be expressed in terms of the mass residue D, where:

$$\begin{aligned} D^{n+1}_{i,j,k} &= \gamma^n_{v,i,j,k}\left(\frac{\rho^{n+1}_{i,j,k} - \rho^n_{i,j,k}}{\Delta t}\right) \\ &+ \frac{1}{r_i \Delta r_1}\left(\langle\gamma_r \rho u r\rangle^{n+1}_{i+1/2,j,k} - \langle\gamma_r \rho u r\rangle^{n+1}_{i-1/2,j,k}\right) \\ &+ \frac{1}{r_i \Delta\theta_1}\left(\langle\gamma_\theta \rho v\rangle^{n+1}_{i,j+1/2,k} - \langle\gamma_\theta \rho v\rangle^{n+1}_{i,j-1/2,k}\right) \\ &+ \frac{1}{\Delta z_1}\left(\langle\gamma_z \rho w\rangle^{n+1}_{i,j,k+1/2} - \langle\gamma_z \rho w\rangle^{n+1}_{i,j,k-1/2}\right) + Sm_{i,j,k} \end{aligned} \tag{50}$$

The mass residue D requires that latest available values be used. The mass residue is used to calculate a pressure adjustment during the calculation. This point will be discussed in a later section.

Conservation of Momentum

The finite difference approximation to the momentum Equation 26, treats both the time-dependent and pressure-gradient terms implicitly, and the resistance-force terms semi-implicitly, whereas the remaining formulation is treated explicitly.

r-momentum equation. This is finite-differenced as:

$$\begin{aligned} (u)^{n+1}_{i+1/2,j,k} &= \left\{\frac{1}{(\gamma_v \rho)^n_{i+1/2,j,k}[1 + \Delta t (F_r)^n_{i+1/2,j,k}]}\right\} \\ &\cdot \Bigg[\overline{(\gamma_v \rho u)}_{i+1/2,j,k} \\ &- (\gamma_v u)^n_{i+1/2,j,k}\Delta t\left(\frac{\rho^n_{i+1/2,j,k} - \rho^{n-1}_{i+1/2,j,k}}{\Delta t}\right) \\ &- (\gamma_v)^n_{i+1/2,j,k}\Delta t\left(\frac{p^{n+1}_{i+1,j,k} - p^{n+1}_{i,j,k}}{\Delta r_2}\right)\Bigg] \end{aligned} \tag{51}$$

$$(F_r)^n_{i+1/2,j,k} = \frac{1}{2} f_r \mid u^n_{i+1/2,j,k} \mid \tag{52}$$

where f_r is the friction factor in the radial direction.

$$\begin{aligned}
\overline{(\gamma_v \rho u)}_{i+1/2,j,k} &= (\gamma_v \rho u)^n_{i+1/2,j,k} + (\gamma_v \rho)^n_{i+1/2,j,k} g_r \Delta t \\
&+ \frac{\Delta t}{r_{i+1/2}\Delta r_2}(\langle \gamma_r \rho r u^2 \rangle^n_{i,j,k} - \langle \gamma_r \rho r u^2 \rangle^n_{i+1,j,k}) \\
&+ \frac{\Delta t}{r_{i+1/2}\Delta \theta_1}(\langle \gamma_\theta \rho u v \rangle^n_{i+1/2,j-1/2,k} - \langle \gamma_\theta \rho u v \rangle^n_{i+1/2,j-1/2,k}) \\
&+ \frac{\Delta t}{\Delta z_1}(\langle \gamma_z \rho u w \rangle^n_{i+1/2,j,k-1/2} - \langle \gamma_z \rho u w \rangle^n_{i+1/2,j,k+1/2}) \\
&+ \Delta t \left[\frac{\langle \gamma_\theta \rho v^2 \rangle^n_{i+1/2,j,k}}{r_i} + (V_{\tau_r})^n_{i+1/2,j,k} + (Sm_r)_{i+1/2,j,k} \right]
\end{aligned} \tag{53}$$

and the radial viscous stress term is approximated by

$$\begin{aligned}
(V_{\tau_r})^n_{i+1/2,j,k} = &\frac{1}{r_{i+1/2}\Delta r_2}[(\mu_{eff}\gamma_r r)^n_{i+1,j,k}(u^n_{i+3/2,j,k} - u^n_{i+1/2,j,k})/\Delta r_3 \\
&- (\mu_{eff}\gamma_r r)^n_{i,j,k}(u^n_{i+1/2,j,k} - u^n_{i-1/2,j,k})/\Delta r_1] \\
&+ \frac{1}{r^2_{i+1/2}\Delta \theta_2}[(\mu_{eff}\gamma_\theta)^n_{i+1/2,j+1/2,k}(u^n_{i+1/2,j+1,k} - u^n_{i+1/2,j,k})/\Delta \theta_2 \\
&- (\mu_{eff}\gamma_\theta)^n_{i+1/2,j-1/2,k}(u^n_{i+1/2,j,k} - u^n_{i+1/2,j-1,k})/\Delta \theta_3] \\
&+ \frac{1}{\Delta z_1}[(\mu_{eff}\gamma_z)^n_{i+1/2,j,k+1/2}(u^n_{i+1/2,j,k+1} - u^n_{i+1/2,j,k})/\Delta z_2 \\
&- (\mu_{eff}\gamma_z)^n_{i+1/2,j,k-1/2}(u^n_{i+1/2,j,k} - u^n_{i+1/2,j,k-1})/\Delta z_3] \\
&- (\mu_{eff}\gamma_r)^n_{i+1/2,j,k}\left[\frac{2}{r^2_{i+1/2}}(v^n_{i+1/2,j+1,k} - v^n_{i+1/2,j-1/2,k})/\Delta \theta_1 \right. \\
&\left. + u^n_{i+1/2,j,k}/r^2_{i+1/2}\right]
\end{aligned} \tag{54}$$

The convective flux terms are evaluated as indicated in the preceding portion for the conservation of mass. For completeness, these will be shown for the radial momentum equation, but omitted for the other conservation of momentum equations. The convective flux terms are given by:

$$\begin{aligned}
\langle \gamma_r \rho r u^2 \rangle^n_{i,j,k} &= (\gamma_r r u)^n_{i,j,k}\left[\left(\frac{1}{2} + \xi_x\right)(\rho u)^n_{i-1/2,j,k} \right. \\
&\left. + \left(\frac{1}{2} - \xi_x\right)(\rho u)^n_{i+1/2,j,k}\right] \\
\langle \gamma_r \rho r u^2 \rangle^n_{i+1,j,k} &= (\gamma_r r u)^n_{i+1,j,k}\left[\left(\frac{1}{2} + \xi_x\right)(\rho u)^n_{i+1/2,j,k} \right. \\
&\left. + \left(\frac{1}{2} - \xi_x\right)(\rho u)^n_{i+3/2,j,k}\right]
\end{aligned}$$

$$\langle \gamma_\theta \rho u v \rangle^n_{i+1/2,j-1/2,k} = (\gamma_\theta v)^n_{i+1/2,j-1/2,k} \left[\left(\frac{1}{2} + \xi_y \right) (\rho u)^n_{i+1/2,j-1,k} + \left(\frac{1}{2} - \xi_y \right) (\rho u^n_{i+1/2,j,k} \right]$$

$$\langle \gamma_\theta \rho u v \rangle^n_{i+1/2,j+1/2,k} = (\gamma_\theta v)^n_{i+1/2,j+1/2,k} \left[\left(\frac{1}{2} + \xi_y \right) (\rho u)^n_{i+1/2,j,k} + \left(\frac{1}{2} - \xi_y \right) (\rho u)^n_{i+1/2,j+1,k} \right]$$

$$\langle \gamma_z \rho u w \rangle^n_{i+1/2,j,k-1/2} = (\gamma_z w)^n_{i+1/2,j,k-1/2} \left[\left(\frac{1}{2} + \xi_z \right) (\rho u)^n_{i+1/2,j,k-1} + \left(\frac{1}{2} - \xi_z \right) (\rho u)^n_{i+1/2,j,k} \right]$$

$$\langle \gamma_z \rho u w \rangle^n_{i+1/2,j,k+1/2} = (\gamma_z w)^n_{i+1/2,j,k+1/2} \left[\left(\frac{1}{2} + \xi_z \right) (\rho u)^n_{i+1/2,j,k} + \left(\frac{1}{2} - \xi_z \right) (\rho u)^n_{i+1/2,j,k+1} \right]$$

$$(u)^n_{i,j,k} = \frac{(r_i - r_{i-1/2})\,(u)^n_{i+1/2,j,k} + (r_{i+1/2} - r_i)\,(u)^n_{i-1/2,j,k}}{\Delta r_1}$$

$$(u)^n_{i+1,j,k} = \frac{(r_{i+3/2} - r_{i+1})\,(u)^n_{i+1/2,j,k} + (r_{i+1} - r_{i+1/2})\,(u)^n_{i-3/2,j,k}}{\Delta r_3}$$

$$(v)^n_{i+1/2,j-1/2,k} = \frac{(r_{i+1/2} - r_i)\,(v)^n_{i+1,j-1/2,k}}{\Delta r_2} + \frac{(r_{i+1} - r_{i+1/2})\,(v)^n_{i,j-1/2,k}}{\Delta r_2}$$

$$(v)^n_{i+1/2,j+1/2,k} = \frac{(r_{i+1/2} - r_i)\,(v)^n_{i+1,j+1/2,k}}{\Delta r_2} + \frac{(r_{i+1} - r_{i+1/2})\,(v)^n_{i,j+1/2,k}}{\Delta r_2}$$

$$(w)^n_{i+1/2,j,k-1/2} = \frac{(r_{i+1/2} - r_i)\,(w)^n_{i+1,j,k-1/2}}{\Delta r_2} + \frac{(r_{i+1} - r_{i+1/2})\,(w)^n_{i,j,k-1/2}}{\Delta r_2}$$

$$(w)^n_{i+1/2,j,k+1/2} = \frac{(r_{i+1/2} - r_i)\,(w)^n_{i+1,j,k+1/2}}{\Delta r_2} + \frac{(r_{i+1} - r_{i+1/2})\,(w)^n_{i,j,k+1/2}}{\Delta r_2}$$

All other variables at locations other than at the position where they are defined are calculated similarly. The ξ's are calculated as indicated in the equations for donor-cell differencing discussed earlier.

θ-momentum equation. This is finite differenced as

$$(v)^{n+1}_{i,j+1/2,k} = \left\{\frac{1}{(\gamma_v \rho)^n_{i,j+1/2,k}[1+\Delta t(F_\theta)^n_{i,j+1/2,k}]}\right\} \cdot \left[\overline{(\gamma_\theta \rho v)}_{i,j+1/2,k} - (\gamma_v v)^n_{i,j+1/2,k}\Delta t\left(\frac{\rho^n_{i,j+1/2,k} - \rho^{n-1}_{i,j+1/2,k}}{\Delta t}\right) - \frac{\gamma_{v,i,j+1/2,k}\Delta t}{r_i \Delta\theta_2}(p^{n+1}_{i,j+1,k} - p^{n+1}_{i,j,k})\right] \tag{55}$$

$$(F_\theta)^n_{i,j+1/2,k} = \frac{1}{2} f_\theta \mid v^n_{i,j+1/2,k} \mid \tag{56}$$

where f_θ is the friction factor in the θ direction.

$$\begin{aligned}\overline{(\gamma_\theta \rho v)}_{i,j+1/2,k} = {} & (\gamma_v \rho v)^n_{i,j+1/2,k} + (\gamma_v \rho)^n_{i,j+1/2,k} g_\theta \Delta t \\ & + \frac{\Delta t}{r_i \Delta r_1}(\langle\gamma_r \rho r u v\rangle^n_{i-1/2,j+1/2,k} - \langle\gamma_r \rho r u v\rangle^n_{i+1/2,j+1/2,k}) \\ & + \frac{\Delta t}{r_i \Delta\theta_2}(\langle\gamma_\theta \rho v^2\rangle^n_{i,j,k} - \langle\gamma_\theta \rho v^2\rangle^n_{i,j+1,k}) \\ & + \frac{\Delta t}{\Delta z_1}(\langle\gamma_z \rho v w\rangle^n_{i,j+1/2,k-1/2} - \langle\gamma_z \rho v w\rangle^n_{i,j+1/2,k+1/2}) \\ & + \Delta t\left[\frac{\langle\gamma_\theta \rho u v\rangle^n_{i,j+1/2,k}}{r_i} + (Sm_\theta)^n_{i,j,k} + (V_{\tau\theta})^n_{i,j+1/2,k}\right]\end{aligned} \tag{57}$$

and the azimuthal viscous stress term is approximated by

$$\begin{aligned}(V_{\tau\theta})^n_{i,j+1/2,k} = {} & \frac{1}{r_i \Delta r_1}[(\mu_{eff}\gamma_r r)^n_{i+1/2,j+1/2,k}(v^n_{i+1,j+1/2,k} - v^n_{i,j+1/2,k})/\Delta r_2 \\ & - (\mu_{eff}\gamma_r r)^n_{i-1/2,j+1/2,k}(v^n_{i,j+1/2,k} - v^n_{i-1,j+1/2,k})/\Delta r_4] \\ & + \frac{1}{r_i^2 \Delta\theta_2}[(\mu_{eff}\gamma_\theta)^n_{i,j+1,k}(v^n_{i,j+3/2,k} - v^n_{i,j+1/2,k})/\Delta\theta_4 \\ & - (\mu_{eff}\gamma_\theta)^n_{i,j,k}(v^n_{i,j+1/2,k} - v^n_{i,j-1/2,k})/\Delta\theta_1] \\ & \cdot \frac{1}{\Delta z_1}[(\mu_{eff}\gamma_z)^n_{i,j+1/2,k+1/2}(v^n_{i,j+1/2,k+1} - v^n_{i,j+1/2,k})/\Delta z_2 \\ & - (\mu_{eff}\gamma_z)^n_{i,j+1/2,k-1/2}(v^n_{i,j+1/2,k} - v^n_{i,j+1/2,k-1})/\Delta z_3] \\ & + (\mu_{eff}\gamma_\theta)^n_{i,j+1/2,k}\left[\frac{2}{r_i^2 \Delta\theta_2}(u^n_{i,j+1,k} - u^n_{i,j,k}) - v^n_{i,j+1/2,k}/r_i^2\right]\end{aligned} \tag{58}$$

The convective flux terms and the required donor-cell conditions are calculated in a similar fashion, as indicated earlier.

z-momentum equation. This is finite-differenced as

$$
\begin{aligned}
(w)^{n+1}_{i,j,k+1/2} = {} & \frac{1}{(\gamma_v \rho)^n_{i,j,k+1/2}[1 + \Delta t(F_z)^n_{i,j,k+1/2}]} \\
& \cdot \Bigg[\overline{(\gamma_z \rho u)}_{i,j,k+1/2} \\
& - (\gamma_v w)^n_{i,j,k+1/2} \Delta t \left(\frac{\rho^n_{i,j,k+1/2} - \rho^{n-1}_{i,j,k+1/2}}{\Delta t} \right) \\
& - \frac{\gamma_{v,i,j,k+1/2} \Delta t}{\Delta z_2} (p^{n+1}_{i,j,k+1} - p^{n+1}_{i,j,k}) \Bigg]
\end{aligned} \tag{59}
$$

$$
(F_z)^n_{i,j,k+1/2} = \frac{1}{2} f_z \mid w^n_{i,j,k+1/2} \mid \tag{60}
$$

where f_z is the friction factor in the z direction.

$$
\begin{aligned}
\overline{(\gamma_z \rho w)}_{i,j,k+1/2} = {} & (\gamma_v \rho w)^n_{i,j,k+1/2} + (\gamma_v \rho)_{i,j,k+1/2} g_z \Delta t \\
& + \frac{\Delta t}{r_i \Delta r_1} (\langle \gamma_r \rho r u w \rangle^n_{i-1/2,j,k+1/2} - \langle \gamma_r \rho r u w \rangle^n_{i+1/2,j,k+1/2}) \\
& + \frac{\Delta t}{r_i \Delta \theta_1} (\langle \gamma_\theta \rho v w \rangle^n_{i,j-1/2,k+1/2} - \langle \gamma_\theta \rho v w \rangle^n_{i,j+1/2,k+1/2}) \\
& + \frac{\Delta t}{\Delta z_2} (\langle \gamma_z \rho w^2 \rangle^n_{i,j,k} - \langle \gamma_z \rho w^2 \rangle^n_{i,j,k+1}) \\
& + \Delta t[(Sm_z)^n_{i,j,k+1/2} + (V_{\tau_z})^n_{i,j,k+1/2}]
\end{aligned} \tag{61}
$$

and the axial viscous stress term is approximated by

$$
\begin{aligned}
(V_{\tau_z})^n_{i,j,k+1/2} = {} & \frac{1}{r_i \Delta r_1} [(\mu_{eff} \gamma_r r)^n_{i+1/2,j,k+1/2} (w^n_{i+1,j,k+1/2} - w^n_{i,j,k+1/2})/\Delta r_2 \\
& - (\mu_{eff} \gamma_r r)^n_{i-1/2,j,k+1/2} (w^n_{i,j,k+1/2} - w^n_{i-1,j,k+1/2})/\Delta r_4] \\
& + \frac{1}{r_i^2 \Delta \theta_1} [(\mu_{eff} \gamma_\theta)^n_{i,j+1/2,k+1/2} (w^n_{i,j+1,k+1/2} - w^n_{i,j,k+1/2})/\Delta \theta_2 \\
& - (\mu_{eff} \gamma_\theta)^n_{i,j-1/2,k+1/2} (w^n_{i,j,k+1/2} - w^n_{i,j-1,k+1/2})/\Delta \theta_3] \\
& + \frac{1}{\Delta z_2} [(\mu_{eff} \gamma_z)^n_{i,j,k+1} (w^n_{i,j,k+3/2} - w^n_{i,j,k+1/2})/\Delta z_4 \\
& - (\mu_{eff} \gamma_z)^n_{i,j,k} (w^n_{i,j,k+1/2} - w^n_{i,j,k-1/2})/\Delta z_1]
\end{aligned} \tag{62}
$$

The convective flux terms and the required donor cell conditions are calculated in a similar fashion, as indicated in earlier.

Conservation of Energy

The conservation of energy is finite-differenced as**

$$
\begin{aligned}
h_{i,j,k}^{n+1} = h_{i,j,k}^{n} + \frac{\Delta_t}{(\gamma_v\rho)_{i,j,k}^{n}} \Bigg(& \frac{1}{r_i\Delta r_1}(\langle\gamma_r\rho urh\rangle_{i-1/2,j,k}^{n} - \langle\gamma_r\rho urh\rangle_{i+1/2,j,k}^{n}) \\
& + \frac{1}{r_i\Delta\theta_1}(\langle\gamma_\theta\rho vh\rangle_{i,j-1/2,k}^{n} - \langle\gamma_\theta\rho vh\rangle_{i,j+1/2,k}^{n}) \\
& + \frac{1}{\Delta z_1}(\langle\gamma_z\rho wh\rangle_{i,j,k-1/2}^{n} - \langle\gamma_z\rho wh\rangle_{i,j,k+1/2}^{n}) \\
& - \left[(\gamma_v h)_{i,j,k}^{n}\left(\frac{\rho_{i,j,k}^{n} - \rho_{i,j,k}^{n-1}}{\Delta t}\right)\right] + \left[\frac{(\gamma_v p)_{i,j,k}^{n+1} - (\gamma_v p)_{i,j,k}^{n}}{\Delta t}\right] \\
& + \frac{1}{r_i\Delta r_1}(\langle\gamma_v urp\rangle_{i+1/2,j,k}^{n} - \langle\gamma_v urp\rangle_{i-1/2,j,k}^{n}) \\
& + \frac{1}{r_i\Delta\theta_1}(\langle\gamma_v vp\rangle_{i,j+1/2,k}^{n} - \langle\gamma_v vp\rangle_{i,j-1/2,k}^{n}) \\
& + \frac{1}{\Delta z_1}(\langle\gamma_v wp\rangle_{i,j,k+1/2}^{n} - \langle\gamma_v wp\rangle_{i,j,k-1/2}^{n}) \\
& - (\gamma_v p)_{i,j,k}\Bigg\{\frac{1}{r_i\Delta r_1}[(ru)_{i+1/2,j,k}^{n} - (ru)_{i-1/2,j,k}^{n}] \\
& + \frac{1}{r_i\Delta\theta_1}[(v)_{i,j+1/2,k}^{n} - (v)_{i,j-1/2,k}^{n}] \\
& + \frac{1}{\Delta z_1}[(w)_{i,j,k+1/2}^{n} - (w)_{i,j,k-1/2}^{n}]\Bigg\} \\
& - \hat{\alpha}_{i,j,k}[(T)_{i,j,k}^{n} - (T_w)_{i,j,k}^{n}]\, W_p^{*}n \\
& + \frac{1}{r_i\Delta r_1}\Bigg\{\frac{[\gamma_r(\hat{k}+C)]_{i+1/2,j,k}^{n} r_{i+1/2,j,k}}{\Delta r_2}[(T)_{i+1,j,k}^{n} - (T)_{i,j,k}^{n}] \\
& - \frac{[\gamma_r(\hat{k}+C)]_{i-1/2,j,k}^{n} r_{i-1/2,j,k}}{\Delta r_4}[(T)_{i,j,k}^{n} - (T)_{i-1,j,k}^{n}]\Bigg\} \\
& \cdot \frac{1}{r_i^2\Delta\theta_1}\Bigg\{\frac{[\gamma_\theta(\hat{k}+C)]_{i,j+1/2,k}^{n}}{\Delta\theta_2}[(T)_{i,j+1,k}^{n} - (T)_{i,j,k}^{n}] \\
& - \frac{[\gamma_\theta(\hat{k}+C)]_{i,j-1/2,k}^{n}}{\Delta\theta_3}[(T)_{i,j,k}^{n} - (T)_{i,j-1,k}^{n}]\Bigg\} \\
& + \frac{1}{\Delta z_1}\Bigg\{\frac{[\gamma_z(\hat{k}+C)]_{i,j,k+1/2}^{n}}{\Delta z_2}[(T)_{i,j,k+1}^{n} - (T)_{i,j,k}^{n}]
\end{aligned}
$$

** On IBM computers with single precision, it is recommended that the density-time derivative be evaluated from the relationship $[(\partial\rho)/(\partial t)] = [(\partial h)/(\partial t)]\,(\partial\rho)/[\partial h(T)]$. For liquid sodium, $(\partial\rho)/[\partial h(T)]$ is a weak function of temperature for ranges of current interest [21].

$$- \frac{[\gamma_z(\hat{k}+C)]^n_{i,j,k-1/2}}{\Delta z_3}[(T)^n_{i,j,k}-(T)^n_{i,j,k-1}]\Big\}\Big) \tag{63}$$

$$\frac{\alpha_{i,j,k}W_P^*}{(1-\beta^n_{i,j,k})\pi(D_h)_{i,j,k}}\{[(T)^n_{i,j,k+1}-(T_w)^n_{i,j,k+1}] - [(T)^n_{i,j,k}-(T_w)^n_{i,j,k}]\}/\Delta t_2)$$

where $\hat{k}$ is the thermal conductivity, and n is the number of tubes per unit area.

$$B^n_{i,j,k} = \mu^n_{i,j,k}(Pr_t)^n_{i,j,k}/(\mu_t)^n_{i,j,k}(Pr)^n_{i,j,k}$$

$$C^n_{i+1/2,j,k} = (c_p)^n_{i+1/2,j,k}(\mu_t)^n_{i+1/2,j,k}/(Pr_t)^n_{i+1/2,j,k}$$

$$C^n_{i-1/2,j,k} = (c_p)^n_{i-1/2,j,k}(\mu_t)^n_{i-1/2,j,k}/(Pr_t)^n_{i-1/2,j,k}$$

All other quantities are defined in previous sections.

FINITE DIFFERENCE REPRESENTATION OF TURBULENT KINETIC ENERGY EQUATION

Following in a similar manner, the turbulent kinetic energy equation, (Equation 42), can be finite-differenced as

$$\begin{aligned}
k^{n+1}_{i,j,k} = {} & \frac{1}{(\gamma_v\rho)^n_{i,j,k}[1+\Delta t\,(k^n_{i,j,k})^{1/2}/\ell]}\Bigg[(\gamma_v\rho k)^n_{i,j,k} \\
& - \Delta t\left[(\gamma_v k)^n_{i,j,k}\left(\frac{\rho^n_{i,j,k}-\rho^{n-1}_{i,j,k}}{\Delta t}\right)\right] \\
& + \frac{\Delta t}{r_i\Delta r_1}(\langle\gamma_r\rho urk\rangle^n_{i-1/2,j,k}-\langle\gamma_r\rho urk\rangle^n_{i+1/2,j,k}) \\
& + \frac{\Delta t}{r_i\Delta\theta_1}(\langle\gamma_\theta\rho vk\rangle^n_{i,j-1/2,k}-\langle\gamma_\theta\rho vk\rangle^n_{i,j+1/2,k}) \\
& + \frac{\Delta t}{\Delta z_1}(\langle\gamma_z\rho wk\rangle^n_{i,j,k-1/2}-\langle\gamma_z\rho wk\rangle^n_{i,j,k+1/2}) \\
& + \gamma^n_{v,i,j,k}\Delta t\,(c_s\mu_t)^n_{i,j,k}\{[(u)^n_{i,j,k}-(u_s)^n_{i,j,k}]^2 \\
& + [(v)^n_{i,j,k}-(v_s)^n_{i,j,k}]^2+[(w)^n_{i,j,k}-(w_s)^n_{i,j,k}]^2\}/ \\
& (D^2_h)_{i,j,k}+\Delta t(BL)^n_{i,j,k}-\Delta t\,\gamma^n_{v,i,j,k}\{[\hat{\alpha}^n_{i,j,k}(T^n_{i,j,k} \\
& -(T_w)^n_{i,j,k})\,(W_P^*)_{i,j,k}\beta^n_{i,j,k}(g/c_p)^n_{i,j,k}]/[\pi(D_h)_{i,j,k} \\
& \cdot(1+B^n_{i,j,k})]\}+\Delta t\left(\frac{1}{r_i\Delta r_1}\left\{\frac{(\mu_{eff}\gamma_r r)^n_{i+1/2,j,k}}{\Delta r_2}\right.\right. \\
& \cdot[(k)^n_{i+1,j,k}-(k)^n_{i,j,k}] \\
& \left. - \frac{(\mu_{eff}\gamma_r r)^n_{i-1/2,j,k}}{\Delta r_4}[(k)^n_{i,j,k}-(k)^n_{i-1,j,k}]\right\} \\
& + \frac{1}{r_i^2\Delta\theta_1}\left\{\frac{(\mu_{eff}\gamma_\theta)^n_{i,j+1/2,k}}{\Delta\theta_5}[(k)^n_{i,j+1,k}-(k)^n_{i,j,k}]\right.
\end{aligned}$$

$$- \frac{(\mu_{eff}\gamma_\theta)^n_{i,j-1/2,k}}{\Delta\theta_3}[(k)^n_{i,j,k} - (k)^n_{i,j-1,k}]\Big\}$$

$$+ \frac{1}{\Delta z_1}\Big\{\frac{(\mu_{eff}\gamma_z)^n_{i,j,k+1/2}}{\Delta z_2}[(k)^n_{i,j,k+1} - (k)^n_{i,j,k}]$$

$$- \frac{(\mu_{eff}\gamma_z)^n_{i,j,k-1/2}}{\Delta z_3}[(k)^n_{i,j,k} - (k)^n_{i,j,k-1}]\Big\}\Big)\Big] \qquad (64)$$

where

$$(BL)^n_{i,j,k} = \gamma_{v,i,j,k} c_B \rho^n_{i,j,k+1/2} |w|^n_{i,j,k} (w^n_{i,j,k})^2 f_z$$

Here again, as for the resistance force terms in the momentum of turbulence, Equations 51, 55, and 59, the viscous dissipation rate term in the turbulent kinetic energy transport equation is treated semi-implicitly, i.e.,

$$\rho k^{3/2}/\ell = \rho^n_{i,j,k}(k^{n+1}_{i,j,k})(k^n_{i,j,k})^{1/2}/l_{i,j,k}$$

INITIAL AND BOUNDARY CONDITIONS

To solve the set of governing equations which consists of Equations 46, 51, 55, 59, 63, and 64, appropriate initial and boundary conditions must be specified. The following initial and boundary conditions are most likely to be used in engineering applications.

Initial Conditions

Initialization requires that values of all dependent variables be assigned before a solution sequence can begin. Values for the temperature, pressure, and three components of the velocity are specified for each interior point of the computational region. In addition, the density and enthalpy are calculated from the equation of state, utilizing the specified pressures and temperature. For optimum computer running time, an initial guess as close as possible to the expected solution should be used.

1. *Pressure Initialization.* The pressure is specified at the lowest level of cells. A pressure adjustment for gravity and the hydrostatic head is performed.
2. *Pressure Drop Initialization.* One available option for the initialization of pressure is a constant pressure gradient initialization in the axial direction only. With the constant pressure gradient along the same axis as gravity, the gravitational pressure gradient is included in the initialization of the constant pressure gradient.

Boundary Conditions

Boundary conditions for velocity, temperature, and pressure are specified for each boundary surface as well as wall functions for the turbulence near the boundary.

Velocity Boundary Conditions

- At all inlets and outlets, a predetermined normal velocity is specified. This specified velocity remains unchanged throughout the solution.
- As an option, a uniform transient velocity condition can be specified on both the tube and shell side. The normal velocity is set equal to a transient function at the beginning of each time step.
- A no-slip boundary condition specifies that the normal velocity as well as the tangential velocity at any solid surface be identically zero.

- A free-slip boundary condition specifies that the normal velocity at the wall be zero as well as the wall shear. This is accomplished by setting internal and external cell tangential velocities at the wall equal.
- For outlets, the so-called continuation boundary condition is specified. This simply requires the velocity to continue across the outlet plane with the same value as the adjacent internal value.

Temperature Boundary Conditions

- A uniform constant temperature boundary condition sets a predetermined temperature initially and remains unchanged throughout the solution.
- As an option, a uniform inlet temperature condition can be specified as a function of time on both the tube and shell side. The temperature is set equal to a transient function at the beginning of each time step.
- The adiabatic boundary condition is provided; i.e., boundary heat flux is set equal to zero.

Constant Pressure Boundary Conditions

In the implementation of this code, the specification of a pressure at an inlet or outlet cell, completely consistent with the continuative boundary condition, is tantamount to specifying a pressure level for the problem to be solved. Since the inlet or outlet cell cannot contain any residual mass from the specification of a continuative velocity boundary condition, the pressure in that cell cannot be adjusted by the scheme presented later; hence the initial pressure specified remains unchanged (constant) throughout execution.

Boundary Conditions for k

- *At solid wall.* k is set equal to zero. However, it is to be noted that near the solid surface, the turbulent quantities are strongly influenced by molecular viscosity. Turbulent quantities change extremely rapidly in the region next to the wall. This implies that an extremely fine grid system must be used if the turbulent transport equations are to be solved near the wall. Resolution of the velocity profile is important if the correct wall shear is to be computed. To avoid using large numbers of nodes near the wall, and to ensure correct shear stress near the wall, the logarithmic velocity profile will be prescribed between the computational cell next to the wall and the wall itself.
- At inlet. The measured value of k should be used if it is available. If there is no measured k, an estimated value of k is equal to $0.001\ U_0^2$ where U_0 is inlet velocity.
- At exit. Gradient of U_e is equal to zero where U_e is the exit velocity.

SHELL- AND TUBE-SIDE HEAT TRANSFER COUPLING

The development of a computational procedure or a computer code that would be used in the design of large heat exchangers requires a model for the heat transfer that takes place between the shell-side fluid and the tube-side fluid. Such a heat transfer model incorporates the tube-side fluid, tube wall, and heat transfer between the shell-side and tube-side fluids.

Shell-Side Energy Equation

The shell-side energy equation as described in Equation 63 will not be repeated here. It is sufficient to say that the energy equation contains a thermal sink (or source) term. This term accounts for the heat transfer either from the shell-side fluid to the tube wall, or from the tube wall to the shell-side fluid, and is a function of the number of tubes in the control volume, shell-side heat transfer coefficient, and surface heat transfer area.

Tube Wall Conduction Equation

For the tube wall, the temperature is given by

$$\rho_w c_{pw} \frac{dT_w}{dt} = \frac{T - T_w}{VR_0} A + \frac{T_t - T_w}{VR_i} A \tag{65}$$

where ρ_w = tube material density
c_{pw} = tube material specific heat
R_0 = heat transfer resistance between T and T_w
R_i = heat transfer resistance between T_t and T_w
T_t = tube-side fluid temperature
A = heat transfer area
V = total volume of tubes

The overall heat transfer resistances, R_0 and R_i, consist of the various resistances associated with conduction through the tube material, convective resistance due to fluid motion, and any resistance due to fouling. These resistances are in series with each other, and so, are additive. They are generally calculated in some ad hoc manner. In this case, half of the thickness of the wall material was associated with the tube side, while the other half thickness was associated with the shell side to simplify computation.

Continuum-Governing Equation on Tube Side

The flow distribution on the tube side of the heat exchanger is governed by the one-dimensional conservation equation. The assumptions employed in reducing the equation to one dimension and other specific simplifications are discussed in Reference 22. The equations are included here for completeness. The continuity and momentum equations, respectively, are

$$\frac{\partial \rho}{\partial t} + \frac{\partial}{\partial z}(\rho w) = 0 \tag{66}$$

and

$$\frac{\partial}{\partial t}(\rho w) + \frac{\partial}{\partial z}(\rho w^2) = -\frac{\partial p}{\partial z} + \rho g - \frac{f}{D}\left(\frac{1}{2}\rho w|w|\right) \tag{67}$$

where z = axial coordinate
t = time
w = axial velocity
ρ = density
p = pressure
g = axial body force
μ = molecular viscosity
f = pipe friction factor
D = tube inner diameter

The tube side energy equation is given as

$$\frac{\partial}{\partial t}(\rho h) + \frac{\partial}{\partial z}(\rho w h) = \frac{T_w - T_t}{R_i V} A + \frac{\partial}{\partial z}\left(\hat{k} \frac{\partial T_t}{\partial z}\right) \tag{68}$$

where h = enthalpy
A = heat transfer area
$\hat{k}$ = thermal conductivity

and subscript t denotes tube side.

Heat Transfer Coefficients

For a typical set of heat transfer coefficients for shell-tube counterflow heat exchangers with hot liquid sodium in the shell side and cold liquid sodium in the tube side, the following equations are recommended. It is to be noted that for different types of heat exchangers or different fluid/fluids, other sets of appropriate heat transfer coefficients must be provided.

Tube-Side Heat Transfer Coefficients

The low-temperature intermediate sodium flows down through the downcomer, makes a 180° turn, flows up into the heat transfer tubes, absorbs the heat from the primary side sodium, and exits from the top of the unit. For the turbulent liquid metal pipe flow with uniform heat flux, Dwyer [23] recommended the following equation:

$$\mathrm{Nu} = 5.0 + 0.025\,(\mathrm{Pe})^{0}$$

Nu and Pe are the Nusselt and Peclet numbers, respectively.

Shell-Side Heat Transfer Coefficients

The hot primary sodium flows down on the shell side of the tube bundle. Depending on the operating conditions and the design of the baffle plates, the flow direction could be oblique or axial, and the heat transfer mode could be forced convection or natural convection. The heat transfer coefficients are calculated from the following four equations and the largest value is used.

- Graber and Rieger [24] axial correlation—forced convection

$$\mathrm{Nu} = 0.025 + 6.2(P/D_0) + [0.032(P/D_0) - 0.007]\,(\mathrm{Pe})^{0.8-0.24(P/D_0)} \tag{69}$$

where P = pitch of tube bundle
D_0 = tube outside diameter

- Dwyer oblique flow correlation [23]—forced convection

$$\mathrm{Nu}_\beta = \left(\frac{\varphi_1}{D_0}\right)^{1/2}\left(\frac{P-D_0}{P}\right)^{1/2}\left(\frac{\sin\beta+\sin^2\beta}{1+\sin^2\beta}\right)^{1/2}[4.60+0.193(\mathrm{Pe})^{0.614}] \tag{70}$$

where β = oblique flow angle
φ_1 = unit hydrodynamic potential at the rod

- Laminar natural convection [23]

$$\mathrm{Nu}_x = 0.563\,\mathrm{Pr}^{0.46}(\mathrm{Gr}_x)^{1/4} \tag{71}$$

where Gr = Grashof number

- Turbulent natural convection [23]

$$Nu_x = 0.0295\left(\frac{Pr^{7/6}Gr_x}{1+0.444\,Pr^{2/3}}\right)^{2/5} \tag{72}$$

For the heat exchanger with the spacing between two adjacent tubes larger than 0.5 inch, the following two equations are also recommended for natural convection:

- Laminar natural convection [23]

$$Nu_x = 0.31(Gr_x\,Pr)^{1/4} \tag{73}$$

where the characteristic length x used in the Nusselt and Grashof numbers represents the fractional distance from the leading edge of the channel, not the actual distance.

- Turbulent natural convection [23]

$$Nu_d = 0.0280(Ra, d)^{0.355} \tag{74}$$

where the characteristic length in the Nusselt and Rayleigh numbers is the spacing between two adjacent tubes, d.

Finite Difference Representation and Solution Technique

The finite difference representation and solution technique for the coupled shell-side and tube-side heat transfer model is presented in Reference 23.

Heat Capacity Effects of Baffles and Walls during Transient

Baffle Thermal Capacity

For a baffle, the thermal capacity is obtained from

$$\rho_B c_{PB}\frac{dT_B}{dt} = \frac{T_f - T_B}{VR_u}A + \frac{T_f - T_B}{VR_d}A \tag{75}$$

where
ρ_B = baffle material density
c_{PB} = baffle material specific heat
T_B = baffle temperature
T_f = fluid temperature
R_u = heat transfer coefficient upstream of baffle
R_d = heat transfer coefficient downstream of baffle
V = heat transfer coefficient with material thickness

Thermal Capacity of Walls

The thermal capacity of walls is given by:

$$\rho_w c_{Pw}\frac{dT_w}{dt} = \frac{T_f - T_w}{VR_{int}}A + \frac{T_{ext} - T_w}{VR_{ext}}A \tag{76}$$

where
ρ_w = wall material density
c_{Pw} = wall material specific heat
T_f = fluid temperature
T_w = wall temperature
T_{ext} = external temperature (specified)
R_{int} = internal heat transfer coefficient

R_{ext} = external heat transfer coefficient (specified)
V = cell volume associated with material thickness

NUMERICAL SOLUTION TECHNIQUE

Mass Residue and Pressure Adjustment

The pressure is calculated by a scheme reported in References 25 and 26. This scheme is similar to that reported in Reference 21. The pressure is calculated from:

$$P^{n+1}_{i,j,k} = \tilde{P}_{i,j,k} + (\Delta p)_{i,j,k} \tag{77}$$

where the tilde ($\tilde{\ }$) as previously used indicates the latest iterative value available, and the pressure adjustment $(\Delta p)_{i,j,k}$ is determined from the mass residue $D_{i,j,k}$ (defined by Equation 50) as

$$(\Delta p)_{i,j,k} = -\frac{\omega}{(\partial D/\partial p)} D^{n}_{i,j,k} \tag{78}$$

ω is the relaxation factor, and $(\partial D/\partial p)$ is obtained once before the iteration sequence begins by determining the value of $\partial(D)_{i,j,k}/\partial(P)_{i,j,k}$ over all cells. This is accomplished by performing the differentiation of Equation 50 and accounting appropriately for the convective quantities by their definitions provided by the momentum equations.

The iteration for the velocity pressure adjustment continues until the maximum mass residue $|\tilde{D}_{i,j,k}|_{max}$ is less than a predetermined value DCONV, where DCONV for the quasicontinuum is given by

$$\mathrm{DCONV} = \varepsilon_1 \left[\left| \rho^{n}_{i,j,k} \left(\frac{|u^{n}_{i+1/2,j,k}| + |u^{n}_{i-1/2,j,k}|}{2\Delta x} + \frac{|v^{n}_{i,j+1/2,k}| + |v^{n}_{i,j-1/2,k}|}{2\Delta y} + \frac{|w^{n}_{i,j,k+1/2}| + |w^{n}_{i,j,k-1/2}|}{2\Delta z} \right) \right|_{max} + \varepsilon_2 \right]$$

where ε_1 and ε_2 are input convergence constants with preset values of 10^{-3} and 10^{-6}, respectively.

Solution Procedure

Explicit Time Advancement of Velocity

- Explicit portion of u-momentum equation and portion of resistance force term in x-direction invariant during time step.
- Explicit portion of v-momentum equation and portion of resistance force term in y-direction invariant during time step.
- Explicit portion of w-momentum equation and portion of resistance force term in z-direction invariant during time step.
- Explicit first approximation for the velocity components time advancement.

Mass—Momentum Iteration

- Calculate mass residue (DL) and adjust pressure to reduce mass residue.
- Adjust velocity components to account for new pressure distribution.

Reevaluate mass residue and adjust pressure.
Continue iteration sequence until either:

1. $|DL|_{max} < DCONV$, or
2. A specified number of iterations have been reached.

Explicit Time Advancement of Energy

- First enthalpy is advanced explicitly in time. After enthalpy has been evaluated at all internal cells, temperature and density are obtained from equations of state.
- First tube-side velocity field is implicitly advanced in time. Then tube-side temperature and tube-wall temperature are implicitly advanced in time.
- Turbulent kinetic energy is advanced explicitly in time.

Increment to new time step and evaluate explicit velocities. Continue to increment time until either:

1. A specified maximum time has been reached, or
2. A specified number of timesteps have been completed, or
3. A steady state has been reached.

The steady-state criteria are satisfied when:

- $\left(\frac{|\Delta h|}{h}\right)_{max} < \varepsilon_t^*$ and
- $\frac{|\Delta u|_{max}}{|U|_{max}} < \varepsilon_v^*$ and
- $\frac{|\Delta v|_{max}}{|U|_{max}} < \varepsilon_v^*$ and
- $\frac{|\Delta w|_{max}}{|U|_{max}} < \varepsilon_v^*$ and
- $|DL|_{max} < DCONV$ simultaneously

SOME NUMERICAL SIMULATIONS

Equations 46, 51, 55, 59, 63, and 64 constitute a set of governing equations coupled with appropriate initial and boundary conditions as outlined earlier. The coupling between shell side and tube side can also be viewed as a boundary condition either from the shell-side or tube-side point of view.

The following three numerical results presented were obtained from the COMMIX-IHX/SG code [3,27]. The COMMIX-IHX/SG code is a three-dimensional, steady-state/transient computer code for thermal-hydraulic analysis of heat exchangers and is developed primarily for LMFBR† Liquid Metal Fast Breeder Reactor applications. The shell side is capable of handling single-phase sodium and the tube side is capable of treating either single-phase sodium or two-phase flow water. The detailed description of COMMIS-IHX/SG can be found in References 3 and 27. A comparison between COMMIX results and experimental data is summarized below.

* The recommended values of ϵ_t^* and ϵ_v^* in the range of 10^{-2} to 10^{-3}.

† Liquid Metal Fast Breeder Reactor

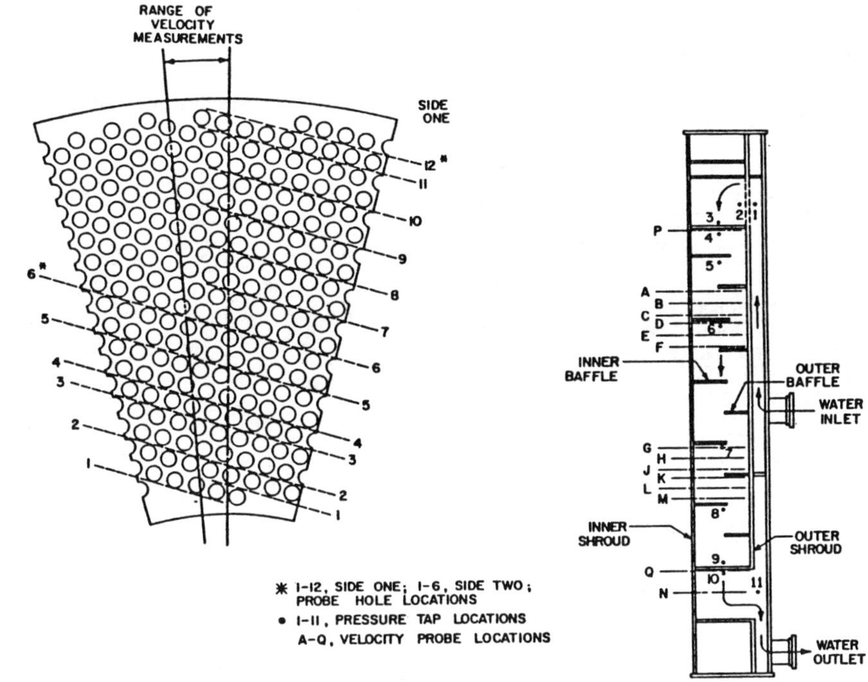

Figure 4. CRBRP-IHX tube bundle flow model.

30° Sector Full-size CRBRP Intermediate Heat Exchanger Flow Model Test

This is an isothermal flow model test conducted by Foster Wheeler Energy Corporation [4]. The 30° sector, full-scale test model duplicates the Clinch River Breeder Reactor Plant (CRBRP) intermediate-sodium-heat-exchanger-tube-bundle configuration, including the tubes, baffles, and inner and outer shrouds. The model length, however, is shorter by three baffle spans than the corresponding IHX bundle. Figure 4 shows the tube-bundle configuration and a vertical cross-sectional view of the experimental setup. The tube bundle has been assembled with 219 stainless steel tubes of 22-mm (7/8 in.) OD in an equilateral pitch of 33.3 mm (1.312 in.). The upper half of the bundle has 2/3 area overlapping baffles and the lower half contains 1/2 area overlapping baffles. Water at a temperature of 77° C (170° F) was used as a working fluid to simulate sodium viscosity and Reynolds number. The full flow rate for the model was 10.56 m^3/min (1,790 gpm), which corresponds to the full flow rate of the CRBRP-IHX.

The COMMIX-IHX/SG computer model of the foregoing 30° sector IHX test module is a two-dimensional, axisymmetric one, consisting of 9 radial nodes and 76 axial nodes. Several computer runs were made with different inlet flow rates ranging from 100 to 10 percent of full flow. The results for the axial velocity components at various elevations near the flow baffles are shown in the broken lines in Figures 5 and 6 for 100 and 40 percent of full flow conditions, respectively. Superimposed on these analytical velocity profiles are the corresponding measured velocity profiles. The agreement between analytical predictions

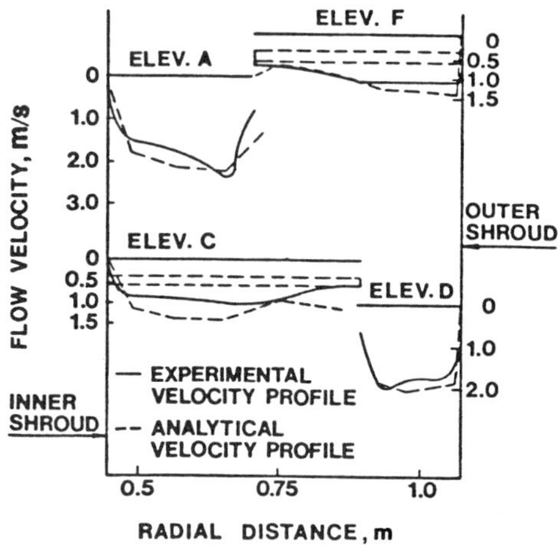

Figure 5. Axial velocity profile at 100 percent flow.

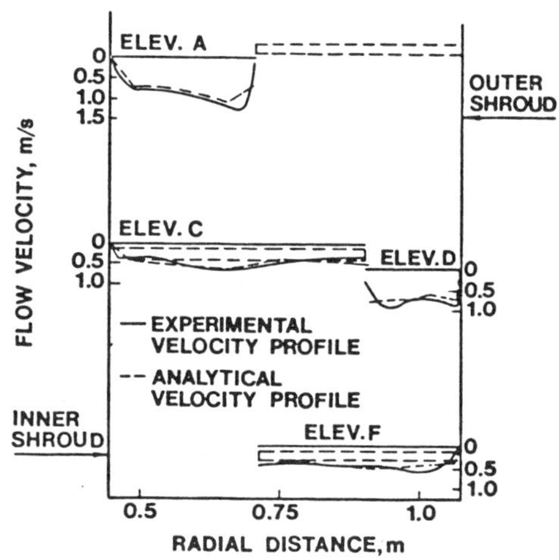

Figure 6. Axial velocity profile at 40 percent flow.

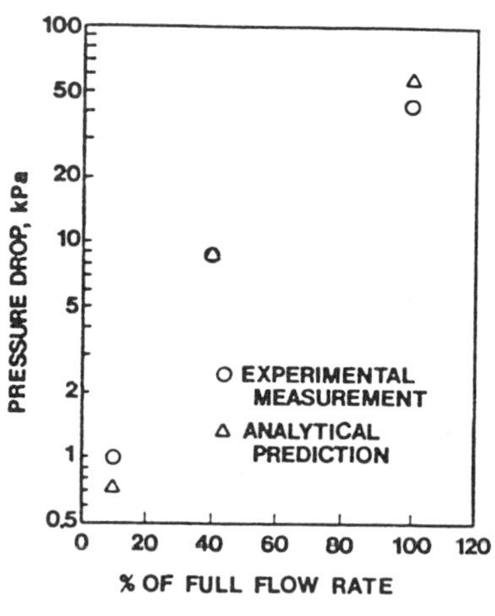

Figure 7. Thirty degree sector model pressure drop.

and experimental data appears satisfactory, considering experimental uncertainties and analytical complexities. At 100 percent flow rate, the analytical solution overestimates the axial velocity components, implying that the crossflow resistance correlation used in the analysis may be conservative. A comparison for the pressure drop across the test unit is shown in Figure 7. The agreement is quite good, even though the analytical predictions tend to be higher at 100 percent flow, and lower at 10 percent flow, than the measurements.

AI Modular Steam Generator Test

The Atomics International Modular Steam generator (AI-MSG) is a counterflow, once-through, sodium-to-water heat exchanger with an inverted hockey stick configuration. Sodium flows downward on the shell side, and water/steam flows upward inside 158 tubes. Sodium nozzles are located some distance away from water and steam nozzles, and semistagnant sodium regions exist in the tube bundle between the sodium outlet and the water inlet, and between the sodium inlet and the steam outlet. The physical dimensions of the MSG are summarized in Table 1. Further details of the MSG and its operation may be found in Reference 28.

The COMMIX-IHX/SG computer model for the AI-MSG unit, shown in Figure 8, is an axisymmetric model with 8 radial and 82 axial nodes, totaling 656 nodes. (See Figure 9

Table 1
AI-MSG Physical Dimensions

Tubes	
Outside diameter (mm)	15.875
Tube wall thickness (mm)	2.921
Tube length (tubesheet to tubesheet) (m)	
Shortest tube	20.88
Longest tube	21.47
Active heat-transfer length (m)	17.68
Active heat-transfer area (m^2)	139.0
Number of tubes	158.0
Average pitch-to-diameter ratio	1.885
Material	2 1/4 Cr-1 Mo steel
Tubesheets	
Thickness (mm)	152.4
Outside diameter (mm)	482.6
Shell	
Main shell outside diameter (mm)	457.2
Main shell thickness (mm)	19.05
Header shell outside diameter (mm)	609.6
Header shell thickness (mm)	50.8
Tube Spacers	
Thickness (mm)	19.05
Number of tube spacers	24.0
Support	8 tierods
Hole diameter (mm)	16.18

Table 2
AI-MSG Steady-State Operation Test Conditions (102.8 percent power)

Water flow (kg/hr)	57,800
Sodium flow (kg/hr)	735,000
Water inlet temperature (°C)	246
Sodium inlet temperature (°C)	464
Steam outlet pressure (MPa)	17.9
Steam outlet temperature (°C)	439
Sodium outlet temperature (°C)	341
Thermal duty (MW)	32.1

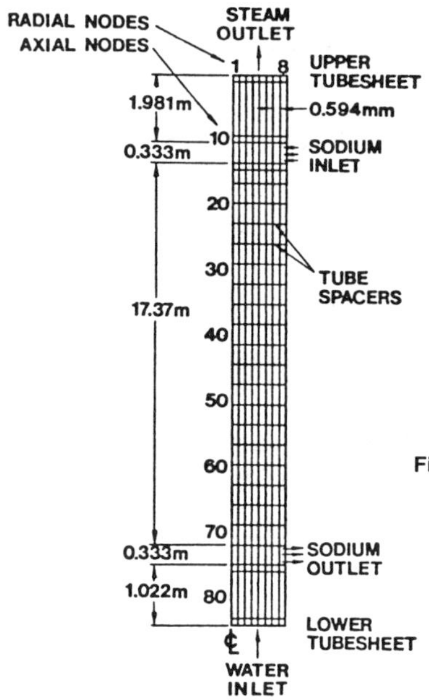

Figure 8. COMMIX model for modular steam generator.

for unit configuration.) The active heat-transfer region consists of axial nodes 11 through 73. The upper hockeystick bend area is simulated by a vertical cylindrical section (nodes 1 through 10) because the present version of the code does not have provisions to model a bend or elbow configuration. The lower stagnation region is modeled by nodes 74 through 82. In the radial direction, the tube bundle is modeled by nodes 1 through 6. Radial node 7 simulated the tubeless space between the bundle and the shell.

The thermal hydraulic operating conditions used for the COMMIX-IHX/SG simulation model are summarized in Table 2. This steady-state condition corresponds to 102.8 percent power condition.

In the MSG test, thermocouples were mounted on the shell wall and inside the tube bundle. The thermocouple data points for Table 3 test conditions are shown in Figure 9 as a function of the active heat-transfer length of the bundle. The solid line indicates the average temperature profile based on these thermocouple data. Superimposed on this, in broken lines, are the analytical results from the COMMIX-IHX/SG code. Fairly good agreement between the test data and predicted values is seen.

Also plotted in Figure 9 are the average water/steam temperature profiles. Note that, on the water/steam side, only the inlet water and outlet steam temperatures were measured by test facility thermocouples. The solid line water/steam temperature profile is computed based on the measured averaged sodium temperature along the steam generator.The discrepancy between the predicted exit steam temperature (435° C) and the measured value (439° C) is nearly within the experimental measurement error band of ±3° C that was estimated for the exit steam temperature. From a standpoint of enthalpy pickup by water/steam, the prior discrepancy of 4° C for the exit steam temperature amounts to less than a 0.3 percent difference.

Figure 10 shows the steam quality and the tube-wall temperature profiles for all radial tube rows that were simulated in the COMMIX-IHX/SG model. Generally, higher sodium temperatures exist around the outer tubes, rather than the inner tubes, so the heat input is greater for the outer tubes; therefore, the departure from the nucleate boiling (DNB) point occurs earlier (7.16 m from tube entrance) and at a lower quality (32 percent) for the outermost tubes. The average DNB quality for all tubes is 38 percent, which is close to the experimentally determined DNB quality of 40 percent. Due to the lack of local measurement data, a detailed comparison cannot be made.

The tube-side flow maldistribution, induced by thermal unbalance on the shell side, is predicted to be approximately ±2 percent. This relatively small variation is probably because the tube bundle diameter is relatively small (0.46 m). The shell-side entrance/exit crossflow effect is thus minimized. The overall tube-side pressure drop is computed to be 661 kPa, which compares favorably with the 669 kPa drop measured in the experiment.

Table 3
50-MWt SNR-Steam Generator Physical Dimensions

Tubes	
Outside diameter (mm)	17.2
Tubewall thickness (mm)	2.9
Heating surface based on OD (m^2)	104.2
Active heat transfer tube length (m)	13.87
Number of tubes	139
Tube arrangement	Triangular
Tube pitch (mm)	27.5
Material	2 1/4 Cr-1 Mo steel
Tube Spacer/Support	
Number of tube supports	20
Perforation	75%
Shell	
i.d. (mm)	468
Thickness (mm)	20

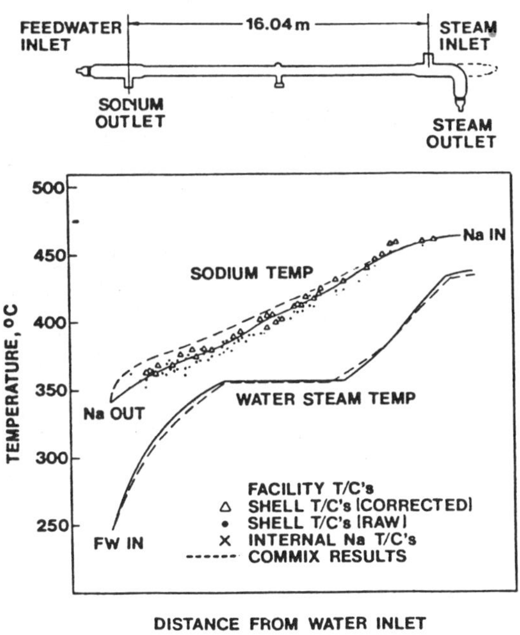

Figure 9. MSG performance data and COMMIX prediction.

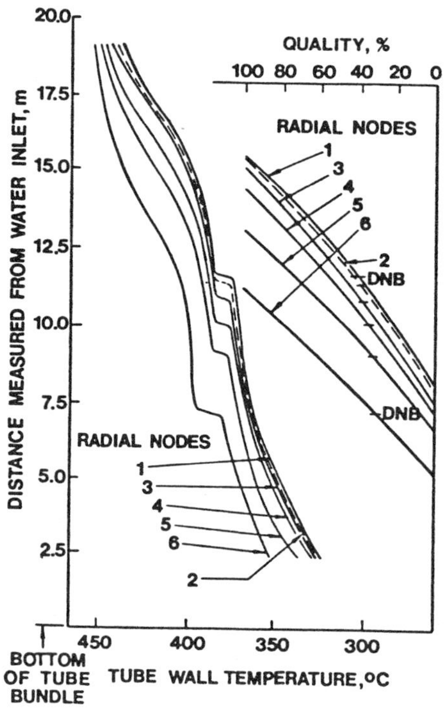

Figure 10. MSG steam quality and tube temperature profiles.

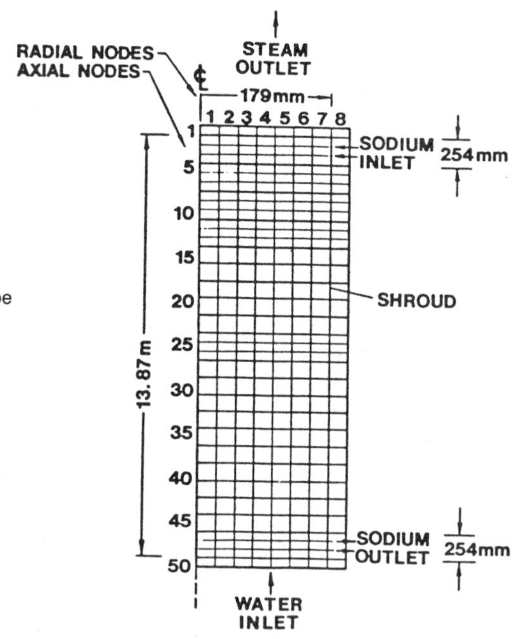

Figure 11. COMMIX model for SNR prototype steam generator.

50 MWt SNR-Steam Generator Test

The 50-MWt SNR-steam generator is a shell-and-tube, liquid sodium heated steam generator that was designed, constructed, and tested by NERATOOM for the Dutch LMFBR program [29]. Since actual design drawings for the steam generator were not available, a COMMIX computer model was constructed based on the limited information available in References 29–31.

The 50-MWt SNR-steam generator under consideration is a straight-tube, vertically mounted, shell-and-tube heat exchanger employing longitudinal flow on the shell side. The analytical computer model for the steam generator is an axisymmetric one with 8 radial nodes and 50 axial nodes as shown in Figures 11. In the radial direction the tube bundle is modeled by nodes 1 to 7. Since detailed drawings were not available, both the inlet and exit windows were assumed to be 154 mm high. This assumption is not expected to result in any significant errors. All support plates are assumed to have 75 percent open flow area. The physical dimensions of the unit are summarized in Table 3.

In the SNR steam generator test, seven thermocouples were placed at the steam exit ends of seven radially positioned heat transfer tubes (but only five thermocouple readings were reported); therefore, the computer model has 7 radial nodes to reflect this T/C arrangment and to provide proper comparisons between test data and predicted values. A comparison has been made in Figure 12 for the radial temperature distribution of the existing steam. In both experiment and prediction, the steam exit temperature is higher at the shroud location, probably due to the combined effects of the hotter shell-side fluid temperature at the entrance and of possible flow bypassing on the shell-side along the shroud. Agreement is excellent for the tubes located at both edges of the tube bundle, and deviations up to 11° C are realized in the middle of the bundle. On the average, the simulation underpredicts the exit steam temperature by approximately 6° C. However, in terms of an overall enthalpy (or thermal energy) pickup, this difference amounts to less than 0.5 percent difference. A few factors can cause this discrepancy. As indicated earlier, the physical dimensions used are only approximate at best because detailed drawings of the steam generator were not available at the time of this study. The heat-transfer and DNB correlations used in the analytical simulation also may contribute to the differences.

DISCUSSIONS AND CONCLUSIONS

The detailed development of a comprehensive, multidimensional, thermal-hydraulic, heat-exchanger analysis computer code is presented. The concept of porosity, directional

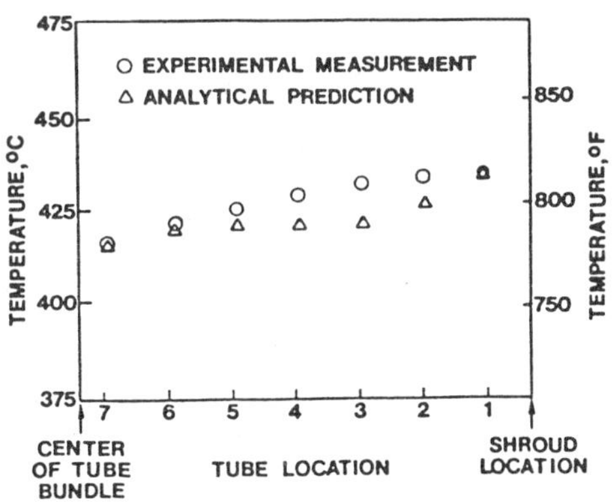

Figure 12. SNR radial steam exit temperature profile.

surface permeabilities, distributed flow resistance, and distributed heat source or sink has been employed on the shell-side fluid to modify the continuum Navier-Stokes and energy equations to properly account the blockage effects of the heat exchanger tube bundle. The tube-side flow is considered as a multichannel, parallel flow between two common inlet and exit plena. The conservation equations were put into finite-difference form using the implicit continuum-fluid Eulerian (ICE) method and solved on a digital computer.

The preceding analytical model was tested using three sets of experimental data. The first set of data was taken from isothermal water test results in a tube bundle, and the other two sets from the nonisothermal, sodium-heated steam generator tests. The comparisons between analytical predictions and experimental indications appear satisfactory, considering analytical complexities and experimental uncertainties, thus lending a credence to the analytical model. However, there definitely is room for further improvement in the analytical approach.

APPENDIX

The differentials in the r, θ, and z directions are approximated by the following difference formulas:

$$\Delta r_1 = r_{i+1/2} - r_{i-1/2}$$
$$\Delta r_2 = r_{i+1} - r_i$$
$$\Delta r_3 = r_{i+3/2} - r_{i+1/2}$$
$$\Delta r_4 = r_i - r_{i-1}$$

$$\Delta\theta_1 = \theta_{j+1/2} - \theta_{j-1/2}$$
$$\Delta\theta_2 = \theta_{j+1} - \theta_j$$
$$\Delta\theta_3 = \theta_j - \theta_{j-1}$$
$$\Delta\theta_4 = \theta_{j+3/2} - \theta_{j+1/2}$$

$$\Delta z_1 = z_{k+1/2} - z_{k-1/2}$$
$$\Delta z_2 = z_{k+1} - z_k$$
$$\Delta z_3 = z_k - z_{k-1}$$
$$\Delta z_4 = z_{k+3/2} - z_{k+1/2}$$

NOTATION

A	area
C_ℓ	coefficient in Equation 44
c_p	specific heat at constant pressure
c_{pw}	tube material specific heat
c_s	coefficient in Equation 37
D	tube diameter
D_h	equivalent hydraulic diameter
d	spacing between adjacent tubes
e	internal energy per unit fluid mass
f	friction factor
Gr	Grashof number
g	gravitational force
h	enthalpy
$\hat{k}$	thermal conductivity
k	thermal conductivity or turbulent kinetic energy
ℓ	length
Nu	Nusselt number
P	pitch of tube bundle
Pe	Peclet number
Pr	Prandtl number
$\underline{P}$	static pressure
$\dot{Q}$	rate of internal heat generation per unit fluid volume
q	heat flux
R	heat transfer resistance
Ra	Rayleigh number

Re Reynolds number
S_G buoyancy generation due to gravitational effects
S_m mass source term
T temperature
t time
U_i velocity in ith direction
u fluid velocity
u_s, v_s, w_s effective wall slip velocities
V volume
V heat transfer coefficient with material thickness in Equation 75
v fluid velocity vector
w_p^* heat transfer perimeter

Greek Symbols

$\hat{\alpha}$ heat transfer coefficient
β dimensionless volumetric expansion coefficient; angle
γ_v volume porosity
ε kinetic energy dissipation rate per unit volume
λ bulk viscosity
μ dynamic viscosity
μ_{eff} effective viscosity
μ_t turbulent viscosity
ρ density
τ shear stress
φ dissipation rate of mechanical energy into heat
ψ local volume average

REFERENCES

1. Harlow, F. H. (Ed.), *Computer Fluid Dynamics—Recent Advances,* AIAA Selected Reprint Series, *XV* (Feb. 1973).
2. Patankar, S. V., and Spalding, D. B., "A Calculation Procedure for the Transient and Steady-State Behavior of Shell-and-Tube Heat Exchangers," *Heat Exchangers: Design and Theory Sourcebook,* N. H. Afgan and E. V. Schlünder (Eds.), McGraw-Hill (1974), pp. 155–176.
3. Sha, W. T., et al., "Three-Dimensional Numerical Modeling of Heat Exchangers," *J. of Heat Transfer, 104:* 3: 397–573 (Aug. 1982).
4. AbuRomia, M. M., et al., "Flow Model Test Development of CRBRP Intermediate Heat Exchanger," ASME/IEEE Joint Power Generation Conference, ASME Paper No. 77–JPGC–NE–10, Los Angeles, Calif., (Sept. 18–21, 1977).
5. AbuRomia, M. M., Chao, B. C., and Cho, S. M., "Flow Distribution Analysis in Nuclear Heat Exchangers with Applications to CRBRP–IHX," *Proc. of 1976 Heat Transfer and Fluid Mechanics Institute,* Stanford U. Press (1976), pp. 468–482.
6. Sha, W. T., et al., "COMMIX-1: A Three-Dimensional Transient Single-Phase Component Computer Program for Thermal Hydraulic Analysis," Argonne National Laboratory, ANL–77–96, NUREG/CR–0785 (Sept. 1978).
7. Sha, W. T., et al., "A New Approach for Rod-Bundle Thermal-Hydraulic Analysis," Proc. of the Int. Meeting on Nuclear Power Reactor Safety, Brussels, Belgium (Oct. 16–19, 1978).
8. Sha, W. T., "An Overview of Rod Bundle Thermal-Hydraulic Analysis," Argonne National Laboratory, ANL–79–10, NUREG/CR–1825 (Nov. 1980).
9. Sha, W. T., Chao, B. T., and Soo, S. L., "Local Volume-Averaged Transport Equations for Multiphase Flow in Regions Containing Distributed Solid Structures," Argonne National Laboratory, ANL–81–69, NUREG/CR–2354 (Dec. 1981).
10. Sha, W. T., Chao, B. T., and Soo, S. L., "Porous Media Formulation for Multiphase Flow with Heat Transfer," Nuclear Engineering and Design, *75* (1983).
11. Slattery, J. C., "Flow of Viscoelastic Fluids through Porous Media," *AIChE J. 13:* 1066 (1967).

12. Whitaker, S., "Diffusion and Dispersion in Porous Media," *AIChE J. 13:* 420–427 (1967).
13. Whitaker, S., "Advances in Theory of Fluid Motion in Porous Media," *Ind. Eng. Chem., 61* (12): 14 (1969).
14. Gary, W. G., and Lee, P. C. Y., "On the Theorem for Local Volume Averaging of Multiphase System," *Int. J. of Multiphase Flow, 3:* 333–340 (1977).
15. Sha, W. T., and Launder, B. E., "A Model for Turbulent Momentum and Heat Transport in Large Rod Bundles," ANL–77–73 (January 1979).
16. Sha, W. T., and Launder, B. E., "A General Model for Turbulent Momentum and Heat Transport in Liquid Medals," ANL–77–78 (March 1979).
17. Yang, C. I., Sha, W. T., and "Numerical Modeling of Turbulent Flow in Pipes with Reynolds Stress Closure," Proc. of Joint ASME/ASCE Mechanics Conference, U. of Colorado, Boulder, Colorado (June 22–24, 1981).
18. Chen, F. F., et al., "Turbulence Modeling in the COMMIX Computer Code," ANL–83–65, NUREG/CR–3504 (to be published 1983).
19. Nijsing, R., and Eifler, W., "Temperature Fields in Liquid Metal Cooled Assemblies," Proc. Heat Mass Transfer, *7:* 115 (1973).
20. Harlow, F. H., and Amsden, A. A., "A Numerical Fluid Dynamics Calculation Method for All Flow Speeds," *J. Comp. Phy., 8:* 197–213 (1976).
21. Sha, W. T., et al., "COMMIX–1: A Three Dimensional Transient Single-Phase Component Computer Program for Thermal Hydraulic Analysis," Argonne National Laboratory Report ANL–77–96, NUREG/CR–0415 (1978).
22. Antonia, R. A., Dank, H. O., and Prabber, A., "Response of a Turbulent Boundary Layer to a Step Change in Surface Heat Flux," *J. of Fluid Mech., 80,* 1 (1977).
23. Dwyer, O. E., "Liquid Metal Heat Transfer," BNLj–11936R (1969).
24. Graber, H., and Riefer, M., "Experimental Study of Heat Transfer to Liquid Metals Flowing In-Line through Tube Bundles at Constant and Exponential Heat Flux Distribution," Atomkerhenergy, *19* (1972).
25. Amsden, A. A., and Harlow, F. H., "KACHINA: An Eulerian Computer Program for Multifield Fluid Flows," LA5680 (December 1974).
26. Harlow, F. H., and Amsden, A. A., "Numerical Calculation of Multiphase Fluid Flow," *J. Comp. Phys, 17:* 19–52 (1975).
27. Yang, C. I., Sha, W. T., and Kasza, K. E., "Thermal-Hydraulic Posttest Analysis for the ANL/MCTF 360° Model Heat-Exchanger Water Test under Mixed Convection," Proc. of 10th IMACS World Congress on Systems Simulation and Scientific Computation, Montreal, Canada (Aug. 1982).
28. McDonald, J. S., Harty, R. B., and DeBear, W. S., "Sodium Heated Steam Generator Test Performance," ASME–IEEE Joint Power Generation Conference, Paper 73–PWR–9, New Orleans, Louisiana (Sept. 16–19, 1983).
29. Ludwig, P. W. P. H., and Hus, B. M., "Some Results of the 50 MW Straight Tube Steam Generator Test in the TNO 50 MW SCTF at Hengelo," Study Group Meeting on Steam Generators for LMFBR's Bensberg, FRG (Oct. 14–17, 1974).
30. Van Westenbrugge, J. K., Tromp, T. J., and DeMaeijer, J. R. A., "Results of the Test and Inspection Program of the Prototype SNR–300 Steam Generator," Ferritic Steel for Fast Reactor Steam Generators, BNES, London (1978), pp. 16–21.
31. Knapp, M. H., Ratzel, W., and Van Westenbrugge, J. K., "Steam Generator Design and Experience in the SNR-Project," ASME Century 2 Nuclear Engineering Conference, Paper 80–C2/NE–31, San Francisco, California (August 18 19–21, 1980).

CHAPTER 27

NUMERICAL SIMULATION OF SHELL-SIDE FLOW AND TEMPERATURE DISTRIBUTION IN HEAT EXCHANGERS

H. de Bruijn

Group of Seismics and Acoustics
Department of Applied Physics
Delft University of Technology
The Netherlands

W. Zijl

Research & Development Group
Institute of Applied Geosciences TNO—D9V
(formerly Groundwater Survey TNO)
The Netherlands

CONTENTS

INTRODUCTION

The primary objective of a simulation of the thermal-hydraulic behavior of an intermediate heat exchanger in a sodium-cooled fast breeder reactor is to predict the temperature field under steady-state and transient conditions. Very often, the transient conditions arise from postulated hypothetical accidents, and the predicted temperature fields serve as a basis for stress analyses to prove the integrity of the component.

The Fluid-Structure Continuum

Only a couple of years ago, thermal-hydraulics dealt with heat exchangers on a gross average basis and could not adequately account for the variations in the heat exchanger and fluid parameters in space and time. In such an approach, the heat exchanger is divided into "stream tubes" based on "engineering intuition," and it is assumed that there is no cross flow between parallel stream tubes. Then, with some assumptions about the pressure distribution in the headers, the pressure drop and the flow and temperature distribution in the stream tubes can be calculated. Of course, for the tube-side of the heat exchanger such an approach is correct, but for the shell-side it represents a poor approximation making the results of the stress analysis based upon it doubtful.

Thermal-hydraulic simulation by computers allows a more detailed study by dividing the shell-side of the heat exchanger into a number of grid blocks (sometimes several thousands) and applying fundamental conservation equations for flow and heat transfer to each block.

Fluid motions at the shell-side of a tube bundle are governed by the fundamental laws based on the conservation of mass, momentum, and energy. However, from a practical standpoint, it is hopeless to try to apply these basic laws directly to the problems of heat exchangers. Instead, a semi-empirical approach is used where the concept of a porous medium is employed. An explanation of the equations describing this approach is the objective of the next major section.

Least Squares Numerical Approximation

Least squares numerical methods have been proposed by several authors as a generalization of the so-called penalty methods well-known in finite element theory. Most serious applications of finite elements in the field of heat and mass transfer, however, seem still to be based on techniques of the Bubnov–Galerkin kind. One can establish objectively that, up to now, least squares finite element methods (L. S. FEM) never became that popular. As it becomes clear now how L. S. FEM has to be practiced dealing with heat exchanger problems, this lack of popularity seems to be justified, indeed, by a number of misunderstandings.

First, properly speaking, L. S. FEM is *not* a finite element method at all. There is no classical variational principle; one has to find an approximation for the differential equations "as they stand." For this reason, the technique could probably be conceived much better as a kind of finite difference method.

Taking this view, it will be understandable that, for example, procedures of so-called "reduced integration" are merely standard for least squares finite elements. For this turns out to be the same as saying that there corresponds only *one* approximation to each equivalent finite difference cluster. Furthermore, F. D. techniques like "upwind differencing," which is the same as moving the integration point along a streamline, can be used to assure stability of least squares elements.

Dealing with a great number of different kinds of F. E. clusters in one mesh, as is the case with heat exchanger temperature calculations, another difficulty comes along. In the classical finite element approach, each approximation is, after its squaring, integrated over the element. Each approximation is thus weighted with the area of the element on which it is taken. As a rule, this results in a very bad conditioning of the global "stiffness" matrix. By discarding for a moment the finite element character of L. S. FEM, this phenomenon can be cured easyly and effectively; namely by weighting each approximation with its "length" in vector space.

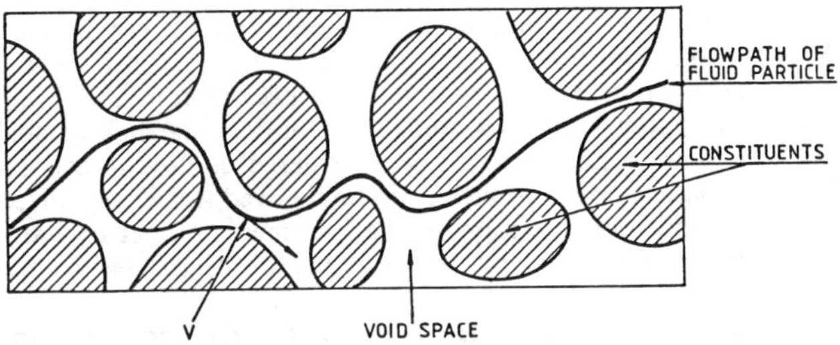

Figure 1. Liquid flow around the constituents of a porous medium.

On the other hand, least squares methods become rather clumsy, not to say completely unworkable, if one is not allowed to make use of some achievements of finite element theory. The "matrix method," for example, is in fact, nothing more than a modification of the classical "shape function method," a technique which is standard in current, F. E. applications. After squaring the approximations and differentiating to the unknowns, positive definite symmetric systems of equations result, which can also be handled and solved by standard finite element procedures (e.g. wave front methods).

We conclude that it apparently seems to be essential for the success of the least squares numerical method that finite differences and finite elements are treated on a uniform basis. A further explanation of this method is the objective of the third major section.

THE CONTINUUM APPROACH TO GENERALIZED POROUS MEDIA

The classical theory of porous media is devoted to the description of flow and transport through soils consisting of sand, clay, peat, etc. Typical applications are in the fields of petroleum reservoir engineering and groundwater hydrology. In these fields, a porous medium is defined as a solid structure containing a multiply connected void space through which a fluid can flow: (see Figure 1). The constituents of the solid structure (i.e. the sand, clay, peat) may be distributed randomly or in a regular way. The fluids (oil, water, gas) spread through the void space, thereby causing a pressure gradient which acts as a force on the constituents of the solid structure.

An important parameter in a porous medium is the porosity defined as a fraction of the control volume not occupied by the solid matrix, or solid structure. We can see that, if the control volume is of the size of a pore, the porosity would be either one or zero. As we increase the size of the control volume, the porosity value will fluctuate before reaching a representative value. The value of porosity associated with a point P is the representative value for a control volume of sufficiently large size containing P. Other physical properties are defined as a mean value at a point P in the porous medium for the same control volume. This is the socalled continuum approach, where the actual porous medium is replaced by a fictitious continuum. In text-books on classical porous media it is proved that the mean surface-porosity, that is the fraction of a sufficiently large control surface not covered by the solid structure, is equal to the earlier-defined (volume-) porosity. For a further discussion see Bear [8].

Generalized Porous Media

Looking around, we can see many nonclassical examples of fluid motion conforming to the earlier-presented definition. For instance, agricultural products are stored to be available over longer periods, and the container with these products can be considered as a porous medium for the cooling air flow. Other examples include filtration, chemical reactions using solid catalysts, adsorption, and mass transfer in packed columns. Finally, flow in the core structure of nuclear power reactors and their components such as shield rod arrays, heat exchangers, and steam generators can be considered as flow in a porous medium. In this presentation, nonclassical porous media will be referred to as generalized porous media, and we will mainly restrict our attention to the intermediate heat exchanger of the SNR-300 sodium-cooled fast breeder reactor. However, many of the results have a much broader applicability.

In Figure 2, a schematic example of a straight-tube shell-and-tube heat exchanger is presented. The flow distribution on the shell-side has a significant influence on the temperature distribution across the tube bundle and, consequently, on the thermal stresses caused by the temperature gradients. For that reason, this chapter will mainly be concerned with the equations describing flow.

The basic equations to be solved are well-established; they are the classical partial differential equations expressing conservation of mass, linear momentum, and energy for a Newtonian fluid, the so-called Navier-Stokes equations [1].

To obtain a well-posed partial differential problem, initial and boundary conditions must be prescribed.

The initial condition for the fluid velocity $\underline{v}$ is that $\underline{v}(\underline{r}, 0)$ must be prescribed at time $t = 0$, and the boundary condition at a boundary completely enclosing the fluid is that $\underline{v}$ must

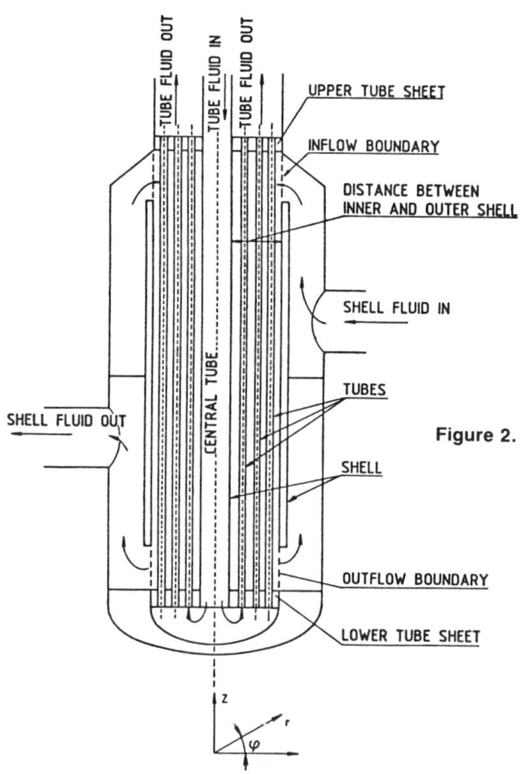

Figure 2. Straight tube shell-and-tube heat exchanger.

be prescribed for all times $t > 0$. For the (generalized) porous medium under consideration, this means that $v = 0$ must also be prescribed at the individual constituents of the solid structure (e.g. the tubes). If these constituents are in (vibrational) motion, this boundary condition becomes v = v-constituent.

Although the equations and boundary conditions needed for the prediction of the flow pattern in a porous medium are well-established, it is, however, also a well-known fact that the Navier-Stokes equations can hardly be solved. Even for simple geometries, analytical and numerical solutions can only be obtained for relatively low Reynolds numbers [2].

Furthermore, the requirement that $v = 0$ at all components of the solid structure is prohibitive even for low Reynolds number flow.

Consequently, from a practical standpoint, it is hopeless at this time to try to apply the basic Navier-Stokes equations directly to the problems of flow in porous media.

Several investigators have, therefore, attempted to describe the shell-side flow distribution approximately by partial differential equations describing a so-called fluid-structure continuum [3–7].

Such a continuum approach is well-known in petroleum reservoir engineering and groundwater hydrology, where the actual porous medium (sand, clay, peat, etc.) is replaced by a fictitious continuum to any point of which we can assign variables and parameters which are continuous functions of the space and time coordinates. In these classical fields, the equations describing flow are the continuity equation and Darcy's Law. Bear [8] provides a complete derivation and treatment of the dynamics and statics of fluids in porous media. However, most of the problems considered in the book by Bear are oriented towards groundwater hydrology. A derivation of the continuum equations for a tube bundle in heat exchangers has been presented by Zijl and De Bruijn [7].

Fluid-Structure Continuum Equations

Starting from the fundamental conservation equations, or Navier–Stokes equations, the fluid-structure continuum equations are derived and the resulting equations are given by:

$$\frac{\partial}{\partial t}(\rho\varphi)+\nabla\cdot(\rho\varphi\underline{\bar{v}}) = 0 \qquad \text{(continuity equation)} \tag{1}$$

$$\rho\dot{\underline{\bar{v}}} = -\nabla\bar{p}+\rho\underline{g}+\rho\underline{Q} \qquad \text{(momentum equation)} \tag{2}$$

In Equations 1 and 2, $\bar{v}$ is the mean fluid velocity in the void space between the constituents of the solid structure, $\dot{\underline{\bar{v}}}$ is the mean fluid motional acceleration given by $\dot{\underline{\bar{v}}} = \partial\underline{\bar{v}}/\partial t+\underline{\bar{v}}\cdot\nabla\underline{\bar{v}}$, $\bar{p}$ is the mean fluid pressure, ρ is the fluid density, φ is the porosity representing the ratio of the volume occupied by the fluid and the total volume, $\underline{Q}$ represents the frictional force distribution, and $\underline{g}$ is the gravitational acceleration. The force distribution $\underline{Q}$ is an unknown for which an additional expression must be found.

At low Reynolds number flow (based on the hydraulic diameter of the void space), the convective acceleration in Equation 2 may be neglected, and $\underline{Q}$ is linear in $|\underline{\bar{v}}|$. If, in addition, momentum transport may be considered as quasi-steady, the momentum Equation 2 simplifies to Darcy's Law:

$$\underline{u} = \varphi\underline{\bar{v}} = -\frac{k}{\mu}(\nabla\bar{p}-\rho\underline{g}) \tag{3}$$

where $\underline{u}$ is the volumetric flow rate in the space containing both the fluid and the solid structure (in the literature $\underline{u}$ is often denoted as the Darcy velocity), μ is the fluid dynamic viscosity, and k is the permeability depending only on the geometrical properties of the solid structure. In this way, the momentum (Equation 2) may be considered as a generalization of Darcy's Law (Equation 3).

The equations describing motion in the fluid-structure continuum have essentially the same character as the well-known Euler equations describing inviscid fluid dynamics; the only differences are the porosity φ representing the ratio of the volume occupied by the fluid and the total volume, and the continuously distributed force term $\underline{Q}$ (or permeability k) accounting for the flow resistance of the structure (the tubes).

The principal difference between the Navier–Stokes equations and the Euler equations is the absence of the second-order viscosity term $\mu\nabla^2\underline{v}$ in the Euler equations. One of the consequences of this absence is a simplification of the boundary conditions. Instead of boundary conditions for the three components of $\underline{v}$ for the Navier–Stokes equations, only the boundary condition for the normal component of the fluid velocity, $\underline{v} \cdot \underline{n} = 0$, holds for the Euler equations. This latter mathematical feature makes Euler-like equations especially well-suited for the description of a fluid-structure continuum, in contrast to Navier–Stokes-like equations, as will be shown in the following discussion.

From a physical point-of-view, the boundary conditions in a fluid-structure continuum are that no fluid is flowing out of the impermeable shell, i.e. $\bar{\underline{v}} \cdot \underline{n} = 0$ on the shell. Of course, in a fluid-structure continuum no boundary conditions may be prescribed on the individual constituents of the structure (e.g., the tubes). A mean volumetric flow rate of fluid is passing across a unit area containing the solid constituents, and the local fluid velocity, as it passes through the clearances between the individual solid constituents, will not be considered. Additional boundary conditions, e.g. for the velocity components parallel to the wall, for the so-called turbulent tangential stress, or for the vorticity, may not be prescribed since, if they are prescribed, there is no reason why the same conditions are not applied at the boundaries of the individual constituents of the solid structure (e.g., the tubes).

In conclusion, from a physical point-of-view, the boundary conditions in a fluid-structure continuum approximation should have a Euler-like character, and "improvement" of the momentum equation by adding a second-order term is nonphysical.

The Vorticity Equation

In classical inviscid fluid dynamics, it is customary to eliminate the pressure from Equation 2 by taking the curl of it and by using Equation 1, resulting in an expression describing the convection, attenuation, and production of vorticity divided by $\rho\varphi$, $\bar{\underline{\omega}} = (\nabla \times \bar{\underline{v}})/(\rho\varphi)$

$$\dot{\bar{\underline{\omega}}} = \bar{\underline{\omega}} \cdot \bar{\underline{\underline{D}}} + \frac{1}{\rho\varphi} \nabla \times \underline{Q} - \frac{\dot{\bar{\underline{v}}} - \underline{g} - \underline{Q}}{\varphi} \cdot \nabla\left(\frac{1}{\rho}\right) \tag{4}$$

where $\dot{\bar{\underline{\omega}}} = \partial\bar{\underline{\omega}}/\partial t + \bar{\underline{v}} \cdot \nabla\bar{\underline{\omega}}$, and $\bar{\underline{\underline{D}}} = (\nabla\bar{\underline{v}} + \bar{\underline{v}}\nabla)/2$ is the so-called deformation tensor. If $\nabla \times \underline{Q} = 0$, Equation 4 is exactly equivalent to the Helmholtz form of the Euler equations [9].

For the purpose of simplifying the discussion, let us assume that $\nabla\left(\frac{1}{\rho}\right) = \underline{0}$, i.e. natural convection will be neglected. In that case, the physical meaning of the so-called generalized Helmholtz (Equation 4) is that the vorticity of a piece of fluid (divided by $\rho\varphi$) changes with a rate $\bar{\underline{\omega}} \cdot \bar{\underline{\underline{D}}} + (\nabla \times \underline{Q})/(\rho\varphi)$ during the motion of this piece of fluid along its streamline. A consequence of this is that the vorticity must be prescribed both as an initial condition, and as a boundary condition on the inflow opening. If these latter conditions represent zero vorticity, and if $\nabla \cdot \underline{Q}\alpha - \bar{\underline{\omega}}$, it follows from Equation 4 that $\nabla \cdot \bar{\underline{v}} = \underline{0}$ in the complete flow field, which is a convenient simplification. In general, the condition $\nabla \times \underline{Q}\alpha - \bar{\underline{\omega}}$ does not hold and, as a consequence, vorticity is produced in the flow field. As a matter of course, this vorticity, $\rho\varphi\bar{\underline{\omega}}$, must be divergence-free, i.e. $\nabla \cdot (\rho\varphi\bar{\underline{\omega}} = 0$, since only in that case the vorticity can be interpreted as $\nabla \times \bar{\underline{v}}$.

From Equations 1 and 4 it can be derived that, if the condition $\nabla \cdot (\rho\varphi\underline{\bar{\omega}}) = 0$ is prescribed at the inflow boundary, then $\rho\varphi\underline{\bar{\omega}}$ is divergence-free in the complete flow field (provided, of course, that at $t = 0$, $\rho\varphi\underline{\bar{\omega}}(\underline{r}, 0)$ is divergence-free).

Furthermore, it can be derived that on the inflow boundary the condition $\underline{\bar{\omega}} = \underline{0}$ may be applied only if $\underline{n} \cdot \nabla \times \underline{Q} = 0$ at the inflow boundary [7].

The Permeability

First, as an example, low Reynolds number flow will be discussed. Low Reynolds number flow is generally applied in petroleum reservoir engineering and groundwater hydrology. For such a flow, the continuously distributed viscous drag force is in the opposite direction of flow, and it is given by:

$$\rho\underline{Q} = -\frac{\mu}{k}\varphi\underline{\bar{v}} = -\frac{\mu}{k}\underline{u} \tag{5}$$

where μ is the fluid dynamic viscosity, and k is the permeability of the solid structure. In general, the solid structure is anisotropic, which means that the permeability depends on the flow direction (i.e., $k = k(\underline{\bar{v}})$ with $\underline{\bar{v}} \cdot \partial k/\partial\underline{\bar{v}} = 0$). For instance, in the flow direction parallel to the tubes of a heat exchanger, the permeability is larger than in the flow direction normal to the tubes. For a given plane with θ = constant, the flow is in the r-direction normal to the tubes, and the following relationship holds:

$$\rho\underline{Q} = -\frac{\mu}{k_r}\varphi\underline{\bar{v}} \tag{6}$$

also, for flow in the z-direction parallel to the tubes the following relationship holds:

$$\rho\underline{Q} = -\frac{\mu}{k_z}\varphi\underline{\bar{v}} \tag{7}$$

where $k_z < k_r$.

One of the many possible combinations of Equations 6 and 7 is:

$$\rho\underline{Q} = \frac{\mu}{k_r\left(\frac{\bar{v}_r}{|\underline{\bar{v}}|}\right)^2 + k_z\left(\frac{\bar{v}_z}{|\underline{\bar{v}}|}\right)^2}\varphi\underline{\bar{v}}$$

or, equivalently,

$$k(\underline{\bar{v}}) = k_r\left(\frac{\bar{v}_r}{|\underline{\bar{v}}|}\right)^2 + k_z\left(\frac{\bar{v}_z}{|\underline{\bar{v}}|}\right)^2 \tag{8a}$$

Similarly, also the following choice is possible:

$$\frac{1}{k(\bar{v})} = \frac{1}{k_r}\left(\frac{\bar{v}_r}{|\bar{v}|}\right)^2 + \frac{1}{k_z}\left(\frac{\bar{v}_z}{|\bar{v}|}\right)^2 \tag{8b}$$

In fact, there is an infinity of ways in which Equations 6 and 7 can be satisfied, and only experiments can justify the ultimate choice of a friction model. For tube bundles some experimental work has been performed in this area [10].

For an isotropic medium, where $k_r = k_z = k$, Equations 8a and 8b simplify to $k(\underline{\bar{v}}) = k$ independent of $\underline{\bar{v}}$.

In Equation 5 it is assumed that the frictional force has always a direction opposite to the flow direction. However, in petroleum reservoir engineering and groundwater hydrology it is common practice to assume that Equations 6 and 7 hold simultaneously for the r- and z-components of the velocity, i.e.:

$$\underset{\sim}{Q} = (Q_r, 0, Q_z) = \mu\varphi\left(\frac{\bar{v}_r}{k_r}, 0, \frac{\bar{v}_z}{k_z}\right) \tag{9}$$

This latter expression leads to a mathematically simpler formulation, but the consequence is that the solid structure (e.g., the tube bundle) feels a force component normal to the direction of flow. This implies, for instance, that a solid structure dropped in a lake will sink to the bottom along a path not parallel with the direction of gravitational acceleration, but will instead follow a path with lateral displacement (see Figure 3) which seems to be nonphysical.

Also, in a tube bundle the permeability has different values for different values of the azimuthal orientation θ of the flow path in radial direction (see Figures 4 and 5), and it is not clear how this can be incorporated in a dyadic relationship such as Equation 9.

For these reasons we will only consider scalar permeabilities, as is customary in the field of thermal-hydraulics (see, however, the following discussion with respect to the existence of a lateral force).

As the flow velocity is increased, deviations from the linear relationship (5) are observed. The explanation is that, as the velocity is increased, deviation is due to inertial effects initially, followed later by turbulence effects. Hubbert [11] noted deviation from the linear relationship (Equation 5) at a low Reynolds number of flow about 1 (based on the grain diameter of unconsolidated media), whereas turbulence was not observed until the Reynolds number approached 600.

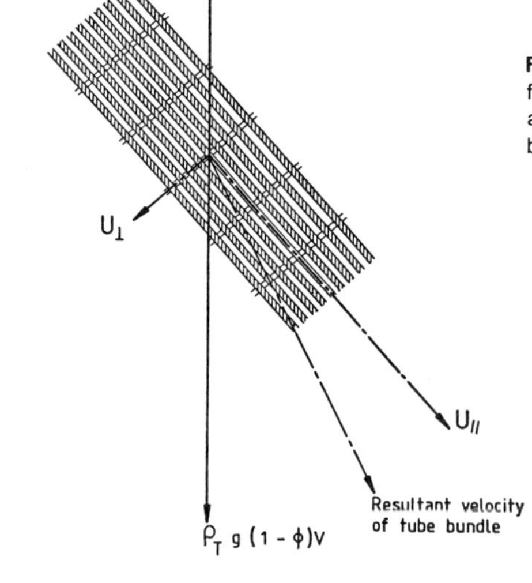

Figure 3. Generalized porous medium (tube bundle) falling downward in a pool of liquid. In this figure it is assumed that the permeability is a tensor, and equilibrium of forces then results in a lateral displacement.

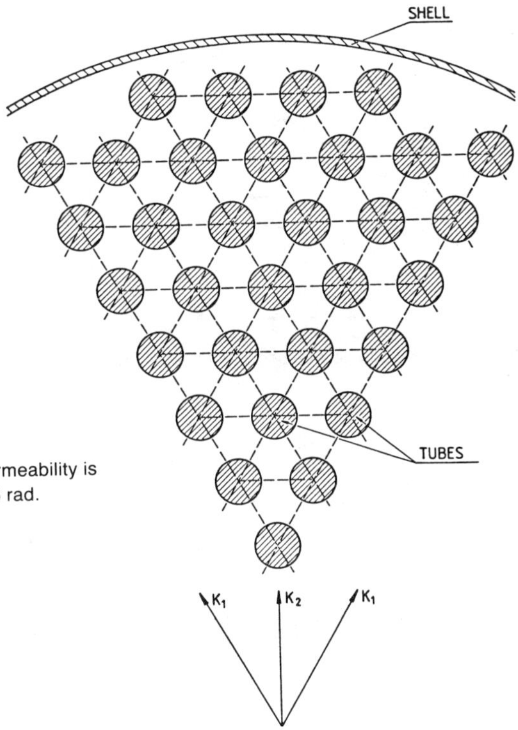

Figure 4. Tube bundle with $K_1 > K_2$; the permeability is different in directions making an angle of $\pi/6$ rad.

The transition from purely viscous flow, via laminar flow, to fully turbulent flow is a long one. This range of flow rates is adequately represented by an almost quadratic expression [10, 12].

In text-books on thermal-hydraulics the following relationship is given [7]:

$$\underline{Q} = -\frac{1}{2} K\, |\underline{\bar{v}}|\, \underline{\bar{v}}, \qquad k = \frac{2\mu\varphi}{lK\,|\bar{v}|} \tag{10}$$

where K is a positive friction factor accounting for the steady viscous drag and the pressure drop in the wakes at the constituents (e.g. the tubes), and M (depending on the direction of $\underline{\bar{v}}$ is the) virtual mass accounting for the fact that immersion of tubes in the liquid increases the kinetic energy without tubes (see Kelvin's theorem [13].

For non—Darcy flows a lateral force, perpendicular to the direction of flow may exist, similar to the lift acting on airfoils or to the Magnus effect caused by rotating-translating bodies. The phenomenon is due to inertial effects neglected in the Low–Reynolds number equations. The magnitude of this side force is given by [18]:

$$\underline{Q}_{lift} = -K'\underline{\bar{v}} \cdot (\nabla \cdot \underline{\bar{v}}) = -\frac{1}{2} K' \nabla |\underline{\bar{v}}|^2 + K' \underline{\bar{v}} \cdot \nabla \underline{\bar{v}}$$

Because of the uncertainty of the (small) value of K′, this term will be neglected.

Irrotational Flow

For low Reynolds number flow, the convective acceleration may be neglected, and substitution of Equation 5 into Equation 4 yields:

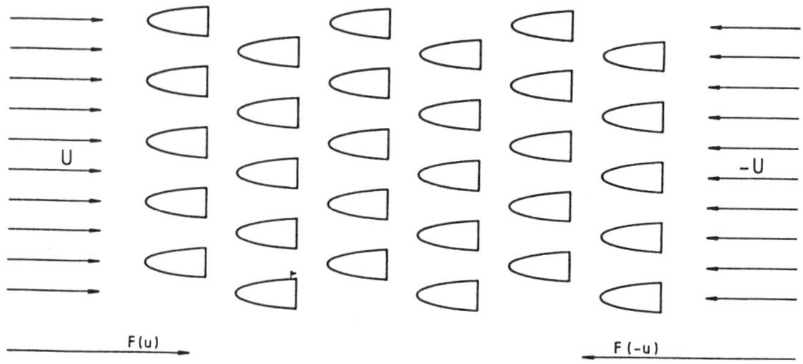

Figure 5. Tube bundle with peculiarly shaped tubes; the permeability is different in two opposite directions.

$$\frac{\partial \underline{\bar{\omega}}}{\partial t} = \frac{1}{\rho^2\varphi}\nabla\cdot\left(\frac{\mu}{k}\varphi\underline{\bar{v}}\right) - \frac{1}{\varphi}\left(\frac{\partial\underline{\bar{v}}}{\partial t} - \underline{g}\right)\cdot\nabla\left(\frac{1}{\rho}\right) \tag{11}$$

For a homogeneous ($\nabla\varphi = \underline{0}$, $\partial k/\partial \underline{r} = \underline{0}$) and isotropic ($\partial k/\partial\underline{\bar{v}} = \underline{0}$) medium under isothermal conditions ($\nabla\mu = (d\mu/dT)\nabla T = \underline{0}$, $\nabla\rho = (d\rho/dT)\nabla T = \underline{0}$), it follows from Equation 11 that, if $\underline{\bar{\omega}}(\underline{r}, 0) = \underline{0}$, $\underline{\bar{\omega}}(\underline{r}, t) = \underline{0}$ for $t > 0$.

The approximation of irrotational flow is very convenient from a mathematical point of view, and it is frequently applied in petroleum engineering and groundwater hydrology for different subregions of the petroleum or groundwater reservoir. The irrotational solutions in these regions are then coupled to each other by continuity conditions at the interfaces [14].

Also, for generalized porous media it would be very practical to find conditions where the flow is approximately irrotational in subregions. In the following discussion we will restrict our-selves to a tube bundle, but also for other generalized porous media similar arguments hold.

From the discussion just presented it is clear that the value of $\nabla\cdot\underline{Q}$ plays an important part, and it is given by:

$$\nabla\cdot\underline{Q} = -\frac{1}{2}K\,|\,\underline{\bar{v}}\,|\left[\gamma\underline{\bar{\omega}} - \frac{4}{5}\underline{\bar{v}}\cdot\nabla\ln|\,\underline{\bar{v}}\,| - \underline{\bar{v}}\left(\frac{\partial\ln R}{\partial\underline{\bar{v}}}\cdot\nabla\underline{\bar{v}}\right) - \underline{\bar{v}}\cdot\frac{\partial\ln R}{\partial\underline{r}}\right] + \frac{\partial\underline{\bar{\omega}}}{\partial t} - \nabla\cdot\left(\underline{\underline{M}}\cdot\frac{\partial\underline{\bar{v}}}{\partial t}\right) \tag{12}$$

where $\gamma = \rho\varphi$.

It has been assumed, in agreement with experimental data for turbulent flow in tube banks, that $K = R/|\,\underline{\bar{v}}\,|^{1/5}$ where $R = R(\underline{r}, \underline{\bar{v}})$ is a function of $\underline{r}$ and of the direction of the flow velocity only (i.e., $\underline{\bar{v}}.(\partial R/\partial\underline{\bar{v}}) = 0$) [15].

To simplify Equation 12, a steady-state situation will be considered first, i.e., $\partial\underline{\bar{v}}/\partial t = \underline{0}$. Furthermore, a homogeneous (but not isotropic) force distribution will be considered, i.e., $\partial R/\partial\underline{r} = \nabla R = \underline{0}$. Under these assumptions, Equation 12 simplifies to:

$$\frac{1}{\gamma}\nabla\cdot Q = -\frac{7}{10}K\,|\,\underline{\bar{v}}\,|\left\{\underline{\bar{\omega}} - \frac{2}{7}(\underline{e}\cdot\underline{\bar{\omega}})\underline{e} + \frac{5}{14}(\underline{\bar{v}}\cdot\underline{\bar{\omega}})\frac{\partial\ln R}{\partial\underline{\bar{v}}} - \frac{4}{7\gamma}\underline{e}\cdot(\underline{e}\cdot\underline{\underline{\bar{D}}}) - \frac{5}{7\gamma}\underline{\bar{v}}\cdot\left(\frac{\partial\ln R}{\partial\underline{\bar{v}}}\cdot\underline{\underline{\bar{D}}}\right)\right\} \tag{13}$$

In Equation 13 $\underline{e} = \underline{\bar{v}}/|\underline{\bar{v}}|$. Use has been made of the vector relation $\underline{\bar{v}} \cdot \nabla\underline{\bar{v}} = \nabla\frac{1}{2}|\underline{\bar{v}}|^2 + \gamma\underline{\bar{\omega}} \cdot \underline{\bar{v}}$ and of the Gibbs decomposition theorem for $\nabla\underline{\bar{v}}$: $\underline{a} \cdot \nabla\underline{\bar{v}} = \underline{a} \cdot \underline{\underline{\bar{D}}} + (\gamma\underline{\bar{\omega}} \cdot \underline{a})/2$, with $\underline{a} = \underline{e}$ and $\underline{a} = \partial \ln R/\partial\underline{\bar{v}}$, respectively.

It is important to note here that the deformation tensor, $\underline{\underline{\bar{D}}} = (\nabla\underline{\bar{v}} + \underline{\bar{v}}\nabla)/2$, at a point depends only on the irrotational part of the local velocity field, and not directly on the local value of the vorticity, $\gamma\underline{\bar{\omega}}$ [16]. Consequently, Equation 13 represents a splitting of $\nabla \cdot \underline{Q}$ into parts depending either on only irrotational or on only rotational modes of fluid motion.

When a two-dimensional flow or an axially symmetric flow without circulation around the axis of symmetry is considered, $\underline{\bar{v}} \cdot \underline{\bar{\omega}} = 0$ holds. In addition, $\underline{\bar{\omega}} \cdot \underline{\underline{\bar{D}}}$ in Equation 4 is equal to zero when a two-dimensional flow is considered and the term $\frac{D\underline{\bar{\omega}}}{Dt} - \underline{\bar{\omega}} \cdot \underline{\underline{\bar{D}}}$ in Equation 4 is equal to $r\frac{D}{Dt}\left(\frac{\underline{\omega}}{r}\right)$ when an axially symmetric flow without circulation around the axis of symmetry is considered. Under these conditions, a combination of Equations 4 and 13, neglecting natural convection, results in:

$$r^n \frac{D}{Dt}\left(\frac{\underline{\omega}}{r^n}\right) = -\frac{7}{10}K|\underline{\bar{v}}|\left[\underline{\bar{\omega}} - \frac{4}{7\gamma}\underline{e} \cdot (\underline{e} \cdot \underline{\underline{\bar{D}}}) - \frac{5}{7\gamma}\underline{\bar{v}} \cdot \left(\frac{\partial \ln R}{\partial \underline{\bar{v}}} \cdot \underline{\underline{\bar{D}}}\right)\right] \qquad (14)$$

where n = 0 for two-dimensional flow and n = 1 for axially symmetric flow without circulation around the axis of symmetry.

At the inflow boundary, $\underline{\bar{\omega}} = \underline{0}$, and moving with a piece of fluid along its streamline, it is observed from the relaxation character of Equation 14 that $\gamma\underline{\bar{\omega}}$ tends from zero to an asymptotic value of $\gamma\underline{\bar{\omega}} = \left[4\underline{e} \cdot (\underline{e} \cdot \underline{\underline{\bar{D}}}) + 5\underline{\bar{v}} \cdot \left(\frac{\partial \ln R}{\partial \underline{\bar{v}}} \cdot \underline{\underline{\bar{D}}}\right)\right]/7 = \gamma\underline{b}$ in a characteristic time interval $\Delta t \cong 2 \cdot 10/(7K\underline{\bar{v}})$. During that time interval, the piece of fluid under consideration has traveled a characteristic distance $\Delta L \cong |\underline{\bar{v}}|\Delta t = 20/(7K)$.

At a distance of order of magnitude L from the inflow boundary, where L is the distance between inner and outer shell, the flow becomes parallel to the shells. In a parallel flow region is $\underline{\bar{v}} \times \left(\frac{\partial \ln R}{\partial \underline{\bar{v}}} \cdot \underline{\underline{\bar{D}}}\right) = \underline{0}$, $\underline{e} \times (\underline{e} \cdot \underline{\underline{\bar{D}}}) = -\gamma\underline{\bar{\omega}}/2$ and $\underline{\omega} = (0, 0, -\partial v/\partial r)$. Substitution of these values into Equation 14 shows that in a parallel flow region where $\partial\gamma/\partial r = 0$ or $\partial\gamma/\partial z = 0$ the flow is irrotational.

An order-of-magnitude estimation results in:

$$|\underline{e} \times (\underline{e} \cdot \underline{\underline{\bar{D}}})| \cong \left(\frac{1}{2}\underline{\underline{\bar{D}}} : \underline{\underline{\bar{D}}}\right)^{1/2} \quad \text{and} \quad \left|\underline{\bar{v}} \times \left(\frac{\partial \ln R}{\partial \underline{\bar{v}}} \cdot \underline{\underline{\bar{D}}}\right)\right| \cong \left|\ln \frac{R_{perpendicular}}{R_{parallel}}\left(\frac{1}{2}\underline{\underline{\bar{D}}} : \underline{\underline{\bar{D}}}\right)^{1/2}\right|$$

near the inflow boundary. At a distance L from this boundary these terms are equal to zero, and at a distance L/2 these terms have a value of half the value at the inflow opening. Here $R_{perpendicular}$ and $R_{parallel}$ are the friction factors for flow normal and parallel to the tubes, respectively.

Summarizing: In a region near the inflow boundary, $\underline{\bar{\omega}}$ tends from zero at the inflow opening, via a maximum value at a distance of approximately L/2 from the inflow boundary to zero at a distance L.

At the location of maximum $|\underline{\bar{\omega}}|$, $|\underline{b}|$ has half the value of its maximum at the inflow opening and, due to the relaxation character of Equation 14, the maximum value of $|\underline{\bar{\omega}}|$ is equal to $\frac{1}{2}(L/\Delta L)|\underline{b}|$, or:

$$\gamma \mid \underline{\bar{\omega}} \mid_{max} \cong \frac{1}{80} KL \left(\frac{1}{2} \underline{\underline{\bar{D}}} : \underline{\underline{\bar{D}}} \right)^{1/2} \left(4^2 + 5^2 \left| \ln \frac{K_{perpendicular}}{K_{parallel}} \right|^2 \right)^{1/2} \quad \text{if} \quad KL < 80/7$$

$$\gamma \mid \underline{\bar{\omega}} \mid_{max} \cong \frac{1}{7} \left(\frac{1}{2} \underline{\underline{\bar{D}}} : \underline{\underline{\bar{D}}} \right)^{1/2} \left(4^2 + 5^2 \left| \ln \frac{K_{perpendicular}}{K_{parallel}} \right|^2 \right)^{1/2} \quad \text{if} \quad KL > 80/7$$

where the contributions of the two terms have been summed in a root mean square sense. In the following discussion, it is accepted that also for general three-dimensional flow this represents an acceptable approximation.

It follows from a Taylor series expansion and from the Gibbsian decomposition of $\nabla \underline{\bar{v}}$ that the velocity at a point $\underline{r} + \underline{h}$ can be expressed as:

$$\underline{\bar{v}}(\underline{r} + \underline{h}) = \underline{\bar{v}}(\underline{r}) + \underline{h} \cdot \nabla \underline{\bar{v}} + 0(h^2) = \underline{\bar{v}}(\underline{r}) + \underline{h} \cdot \underline{\underline{\bar{D}}} + \frac{1}{2} \gamma \underline{\bar{\omega}} \times \underline{h} + 0(h^2)$$

From this expression, it can be observed that the kinematical effect of vorticity with respect to irrotational modes of motion can be characterized by the ratio $\mid \gamma \underline{\bar{\omega}} \cdot \underline{h} \mid / \mid 2\underline{h} \cdot \underline{\underline{\bar{D}}} \mid$, or, since a characterization independent of $\underline{h}$ is preferable, by the ratio:

$$W_K = \frac{\gamma \mid \underline{\bar{\omega}} \mid}{2 \left(\frac{1}{2} \underline{\underline{\bar{D}}} : \underline{\underline{\bar{D}}} \right)^{1/2}}$$

where W_K is the so-called kinematical vorticity number [16]. Finally, it follows from the just-presented discussion that, at the place where $\mid \underline{\bar{\omega}} \mid$ has its maximum value, W_K has the following order of magnitude:

$$W_K \cong 0.025 KL \left(1 + 1.56 \left| \ln \frac{K_{perpendicular}}{K_{parallel}} \right|^2 \right)^{1/2} \quad \text{if} \quad KL < 11.4$$

$$W_K \cong 0.286 \left(1 + 1.56 \left| \ln \frac{K_{perpendicular}}{K_{parallel}} \right|^2 \right)^{1/2} \quad \text{if} \quad KL > 11.4$$

For a tube bank with values of $K = K_{perpendicular} = K_{parallel} = 40/m$ and $L = 0.25$ m, a value of $W_K = 25\%$ is found in the flow field at the place where maximum vorticity occurs. Also, if $K_{perpendicular} = 32/m$ and $K_{parallel} = 8/m$, or if $K_{perpendicular} = 20/m$ and $K_{parallel} = 0.4/m$, with a mean value for K of $(K_{perpendicular} + K_{parallel})/2$, a value of $W_K = 25\%$ is found for $L = 0.25$. Especially the latter two examples are representative for tube banks in actual heat exchangers. This means that, for practical purposes, the flow may be considered as irrotational, especially since the previously estimated contribution of vorticity to the complete flow field only applies in the relatively small region where vorticity has its maximum value.

The conclusion is that, for steady forced convection, where natural convection is negligible, the shell-side mean flow pattern in a homogeneous tube bank may be described by the following simple linear equations:

$$\frac{\partial}{\partial t}(\rho\varphi) + \nabla \cdot (\rho\varphi\underline{\bar{v}}) = 0 \tag{1}$$

$$\nabla \cdot \underline{\bar{v}} = \underline{0} \tag{15}$$

A more accurate calculation to account for the small amount of vorticity in the flow field is of hardly any use since it depends both on the prescribed velocity and vorticity at the inflow boundary, and on experimentally determined values of $K(\underline{\bar{v}})$. The accuracy of these values, in combination with the fluid structure continuum approximation, is certainly not so good to justify an accuracy better than 20% to 30% in the calculation of the mean flow velocity fields.

Transient Flow

As an example, to show the basic points, the continuity equation (Equation 1) will be simplified. On the length-scale of heat exchangers, the temperature-dependency of ρ may be excluded from Equation 1 (Boussinesq approximation, see Tritton [17]). Therefore, Equation 1 simplifies to:

$$\frac{\rho\varphi}{\kappa}\frac{\partial\bar{p}}{\partial t}+\nabla\cdot\underline{q}=0 \tag{16}$$

where $\underline{q} = \rho\varphi\bar{\underline{v}}$ is the mass flow density in the space containing both fluid and structure (tubes), and κ is the combined bulk modulus of fluid and structure defined by $\kappa = (\rho\varphi)\,(dp/\partial(\rho\varphi))$.

In a homogeneous medium, ρ, φ, and κ may be considered as constants, and it follows then from Equation 16 by taking the gradient that:

$$\frac{\rho\varphi}{\kappa}\frac{\partial}{\partial t}(\nabla\bar{p})+\nabla(\nabla\cdot\underline{q})=\underline{0} \tag{17}$$

Substitution of the momentum (Equation 2) into Equation 17 yields:

$$-\frac{\rho}{\kappa}\frac{\partial^2\underline{q}}{\partial t^2}+\frac{\rho^2\varphi}{\kappa}\frac{\partial\underline{Q}}{\partial t}-\frac{1}{\kappa\varphi}\frac{\partial}{\partial t}(\underline{q}\cdot\nabla\underline{q})+\nabla(\nabla\cdot\underline{q})=\underline{0} \tag{18}$$

where expressions quadratic in κ have been neglected. To further simplify, small Reynolds number flow in a homogeneous and isotropic medium will be considered. Substitution of the linear relationship (Equation 5) into Equation 18 and neglecting terms quadratic in κ and in the velocity yields:

$$\frac{\rho}{\kappa}\frac{\partial^2\underline{q}}{\partial t^2}+\frac{\mu\varphi}{\kappa k}\frac{\partial\underline{q}}{\partial t}-\nabla(\nabla\cdot\underline{q})=0 \tag{19}$$

From Equation 15 it follows that

$$\nabla\cdot\underline{q}=\nabla\rho\varphi\cdot\bar{\underline{v}}=\frac{\rho\varphi}{\kappa}\nabla\bar{p}\cdot\underline{q}=0(v^2)=\text{negligible,}$$

and consequently,

$$\nabla\cdot(\nabla\cdot\underline{q})=-\nabla^2\underline{q}+\nabla(\nabla\cdot\underline{q})=\underline{0} \tag{20}$$

Substitution of Equation 20 into Equation 19 results in the damped wave equation:

$$\frac{\rho}{\kappa}\frac{\partial^2\underline{q}}{\partial t^2}+\frac{\mu\varphi}{\kappa k}\frac{\partial\underline{q}}{\partial t}-\nabla^2\underline{q}=\underline{0} \tag{21}$$

The first and last terms of Equation 21 describe the propagation of acoustic waves through the porous medium. For the relatively slow thermal-hydraulic phenomena in heat exchangers, these effects are generally negligible. Also in petroleum reservoir engineering and groundwater hydrology these effects are negligible. The second and last terms of Equation 21 describe the viscous damping of changes in q, or the so-called storage. In petroleum reservoir engineering and groundwater hydrology the permeability is small and the dimensions of the domain involved are large, which means that storage may not always be neglected for the time-scales of interest. In large heat exchangers with a relatively compressible structure the storage term may eventually have some influence on very transient phenomena, but, it general, it may be neglected.

The most important point to be made here is that, for negligible wave propagation phenomena with respect to storage phenomena, i.e. for time-scales larger than $\rho k/(\mu\varphi)$, the transient terms in the momentum, Equation 2, and drag force Equation 10, may be neglected (leading to the neglect of the first term of Equation 21, and the first term of the continuity (Equation 1) may not be neglected (it leads to the second term of Equation 21).

This also justifies the use of the steady-state momentum equation for the derivation of irrotational motion (Equation 15) for transient phenomena.

The Temperature Equations

Conservation of energy for a representative heat exchanger control volume results in the following equation for the mean temperature $\bar{T}$ at the shell-side:

$$\rho\varphi C_p\left(\frac{\partial\bar{T}}{\partial t}+\underline{v}\cdot\nabla T\right)+\varphi' H(\bar{T}-\bar{\theta}) = 0 \tag{22}$$

and, for straight tubes parallel to the z-axis, the equation for the unsteady temperature at the tube-side, $\bar{\theta}$, is:

$$\rho C_p\left(\frac{\partial\bar{\theta}}{\partial t}+\bar{w}\frac{\partial\bar{\theta}}{\partial z}\right)-H(\bar{T}-\bar{\theta}) = 0 \tag{23}$$

In Equation 22 φ' is the volume fraction of liquid in the tubes. The value of H is given by $4h_{loc}/D$, where h_{loc} is the local heat transfer coefficient, and D is the inner diameter of the tube [12, 10] C_p = the liquid heat capacity.

An equation describing heat storage in the tube walls must be considered in addition:

$$\frac{\partial T_w}{\partial t} = a\frac{1}{r}\frac{\partial}{\partial r}\left(r\frac{\partial T_w}{\partial r}\right) \tag{24}$$

where $a = (\lambda/\rho C_p)_{tube}$ is the heat diffusity of the tubes.

In the case of steady state the terms with $\partial/\partial t$ will be deleted from Equations 22 and 23, while Equation 24 can be worked out to an analytical expression.

The boundary condition for the convection (Equation 22) is that $\bar{T}$ must be prescribed on the inflow opening only; this is similar to the earlier-discussed case for the vorticity convection (Equation 1). Also, for Equation 23, $\bar{\theta}$ must be prescribed at $z = 0$. From a physical point of view, it makes sense to include (turbulent) heat diffusion and dispersion terms; i.e., a term $\lambda\nabla^2 T$ to Equation 22 and a term $\lambda\partial^2\theta/\partial z^2$ to Equation 23; see Bear [8] for a thorough discussion of dispersion in porous media.

For example, if only thermally insulating tubes are present, heat transfer takes place by (turbulent) diffusion and dispersion only, i.e., not by heat transport from shell-side to tube-side liquid.

If these diffusion terms are present, either $\bar{T}$, or $\partial\bar{T}/\partial n$ or a combination of both must be prescribed at a boundary completely enclosing the temperature field; also, in that case, either $\bar{\theta}$, or $\partial\bar{\theta}/\partial z$ or a combination of both must be prescribed at $z = 0$ and at the end of the tubes. In the present study, however, diffusion terms are assumed to be sufficiently small with respect to the terms representing heat transport from shell-side to tube-side liquid.

Summary

Both for steady-state and transient thermal-hydraulic phenomena, the following flow equations should be solved:

$$\frac{\partial}{\partial t}(\rho\varphi)+\nabla\cdot(\rho\varphi\bar{\underline{v}}) = 0 \tag{1}$$

$$\nabla\cdot\bar{\underline{v}} = \rho\varphi\bar{\underline{\omega}} \tag{25}$$

$$\bar{\underline{v}}\cdot\nabla\bar{\underline{\omega}} = \bar{\underline{\omega}}\cdot\bar{\underline{\underline{D}}}+\frac{1}{\rho\varphi}\nabla\cdot\underline{Q}-\frac{\bar{\underline{v}}\cdot\nabla\bar{\underline{v}}-\underline{g}-\underline{Q}}{\varphi}\cdot\nabla\left(\frac{1}{\rho}\right) \tag{26}$$

where $\frac{1}{\rho} = \frac{1}{\rho_0}(1+\alpha\Delta T)$, α is the coefficient of thermal expansion.

If only forced convection is considered, the mean flow pattern in a homogeneous (part of the) tube bundle may often be considered as irrotational, and consequently, only a simple linear system of equations must be solved. Under the assumption that storage effects are negligible, these equations are:

$$\nabla\cdot\bar{\underline{v}} = 0 \tag{27}$$

$$\nabla\cdot\bar{\underline{v}} = \underline{0} \tag{28}$$

Finally, the temperature fields at the shell-side, $\bar{T}$, and tube-side, $\bar{\theta}$, are calculated from:

$$\rho\varphi C_p\left(\frac{\partial\bar{T}}{\partial t}+\bar{\underline{v}}\cdot\nabla T\right)+\varphi' H(\bar{T}-\bar{\theta}) = 0 \tag{22}$$

$$\rho C_p\left(\frac{\partial\bar{\theta}}{\partial t}+w\frac{\partial\bar{\theta}}{\partial z}\right)-H(\bar{T}-\bar{\theta}) = 0 \tag{23}$$

The equation for calculating the wall temperatures T_w is:

$$\frac{\partial T_w}{\partial t} = a\frac{1}{r}\frac{\partial}{\partial r}\left(r\frac{\partial T_w}{\partial r}\right) \tag{24}$$

LEAST SQUARES NUMERICAL APPROXIMATION

As stated enheir, the least squares finite element method is not a finite element method in the proper sense of the word. In contrast, there is much resemblance with classical finite differences. Therefore it appears to be essential for the succes of L. S. FEM that finite elements and finite differences be treated at a uniform mathematical basis [20].

Finite Difference Elements

Although finite differences and finite elements are distinct in origin, theory, and practice, they do have something in common [19]. Both methods use small clusters of nodal points, as their units for discretization. Within such a *finite cluster* (called finite difference star or finite element, respectively) function behavior is approximated by polynominal expansion, thereby interpolating the function values at the nodes.

As an example, take the four-node computational molecule of Figure 6, defined by:

$$(1) = (-1,-1); \quad (2) = (+1,-1); \quad (3) = (-1,+1); \quad (4) = (+1,+1) \tag{29}$$

Associated with this molecule is a vectorspace of dimension 4 spanned by a set of 4 independent polynomials.

By definition, the *finite difference* (F. D.) base of the four-node cluster is given by:

$$\ell_i = (1, \xi, \eta, \xi\eta); \quad i = 1, \ldots, 4.$$

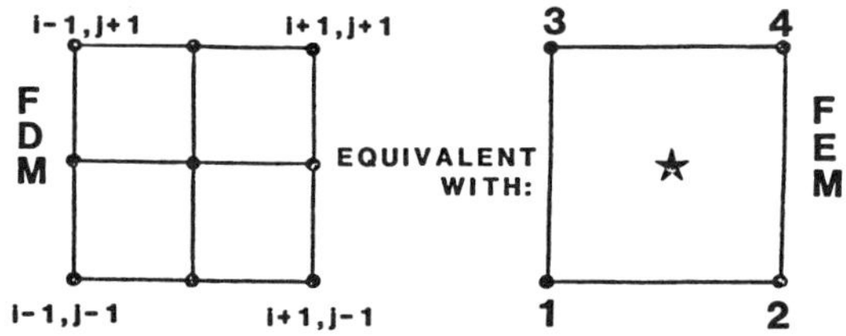

Figure 6. Four-node cluster.

An arbitrary approximation φ is defined upon this base as:

$$\varphi = a_1 + a_2\xi + a_3\eta + a_4\xi\eta; \qquad -1 \leqq \xi, \eta \leqq +1$$

or

$$\varphi = \ell_i a_i \tag{30}$$

The well-known summation convention is employed throughout this chapter. The polynomial approximation (Equation 30) is fully equivalent to a Taylor-series expansion around some origin 0:

$$a_1 = \varphi(0); \quad a_2 = \frac{\partial\varphi}{\partial\xi}(0); \quad a_3 = \frac{\partial\varphi}{\partial\eta}(0); \quad a_4 = \frac{\partial^2\varphi}{\partial\xi\partial\eta}(0)$$

By definition, the *finite element* (F. E.) base of the four-node cluster is given by the set of shape functions:

$$N_i = \left\{\frac{1}{4}(1\pm\xi)(1\pm\eta)\right\}; \qquad i = 1, \ldots, 4.$$

An arbitrary approximation φ is defined upon this base as:

$$\varphi = \frac{1}{4}(1-\xi)(1-\eta)\varphi_1 + \frac{1}{4}(1+\xi)(1-\eta)\varphi_2$$

$$+ \frac{1}{4}(1-\xi)(1+\eta)\varphi_3 + \frac{1}{4}(1+\xi)(1+\eta)\varphi_4 \tag{31}$$

The *transition matrix*, transforming the F. D. base into the F. E. base, is defined by $[c_{ij}]$, where:

$$N_i = \ell_i c_{ij} \quad \text{and} \quad a_i = c_{ij}\varphi_j \tag{32}$$

For many current molecule-forms, such as the cluster Equation 29, the transition matrix proves to have properties of *orthogonality*.

In our case:

$$[c_{ij}] = \frac{1}{4}\begin{bmatrix} 1 & 1 & 1 & 1 \\ -1 & 1 & -1 & 1 \\ -1 & -1 & 1 & 1 \\ 1 & -1 & -1 & 1 \end{bmatrix}$$

so that: $[c_{ij}]^T = \frac{1}{4}[c_{ij}]^{-1}$

Let us once more consider the finite *difference* star, depicted in Figure 6. Associated with this star are the following F. D. schemes:

$$\left.\frac{\partial\varphi}{\partial x}\right|_{i,j} = \frac{1}{4h}(\varphi_{i+1,j+1}+\varphi_{i+1,j-1}-\varphi_{i-1,j+1}-\varphi_{i-1,j-1})+0(h^2)$$

$$\left.\frac{\partial\varphi}{\partial y}\right|_{i,j} = \frac{1}{4h}(\varphi_{i+1,j+1}+\varphi_{i-1,j+1}-\varphi_{i+1,j-1}-\varphi_{i-1,j-1})+0(h^2)$$

$$\left.\frac{\partial^2\varphi}{\partial x\partial y}\right|_{i,j} = \frac{1}{4h^2}(\varphi_{i+1,j+1}+\varphi_{i-1,j-1}-\varphi_{i+1,j-1}-\varphi_{i-1,j+1})+0(h^2)$$

and, last but not least:

$$\varphi|_{i,j} = \frac{1}{4}(\varphi_{i-1,j-1}+\varphi_{i-1,j+1}+\varphi_{i+1,j-1}+\varphi_{i+1,j+1})+0(h^2)$$

See, for example, Ames [25] page 17. See also Forsythe and Wazor [29].

The associated quadrilateral *element*, with *reduced* integration [19], is depicted also in Figure 6.

Identify:

$$x/h = \xi, \quad y/h = \eta, \quad (i,j) \leftrightarrow (*)$$

$$(i-1,j-1) \leftrightarrow (1),\ (i-1,j+1) \leftrightarrow (3)$$
$$(i+1,j-1) \leftrightarrow (2),\ (i+1,j+1) \leftrightarrow (4)$$

The element is second order $0(h^2)$ accurate; which is obvious now.

The transition matrix between the element and the difference star is the orthogonal matrix Equation 32.

Generalization of the results above seems to be possible, and, in fact, proves to be appropriate. In N-dimensional space, a unit cube has 2^N nodal points. Accordingly, there are 2^N elementary polynomials in its F. D. base. Let the k-th coordinate in N-space be named $\xi(k)$.

The F. D. base of the unit cube can then be conceived as a direct product:

$$\{\ell_i\}_{i=1,2^N} = \bigotimes_{k=1}^{N} \{1,\xi(k)\}$$

The coefficients a_i in the Taylor-series expansion Equation 30 are, accordingly:

$$\{a_i\}_{i=1,2^N} = \bigotimes_{k=1}^{N} \{1,\partial/\partial\xi(k)\}\varphi(0)$$

The finite element (F. E.) base of the unit cube is given by the set of shape functions:

$$\{N_i\}_{i=1,2^N} = 1/2^N \bigotimes_{k=1}^{N} (1\pm\xi(k))$$

The transition-matrix, transforming the F. D. into the F. E. base, is orthogonal for all members of the N-cube family.

To be more specific, let n be the index of the n-th coordinate of the nodal point named (i). The coordinates themselves are then defined locally by the FORTRAN inline-function [27]:

```
kube (i, n) = 2 * mod ((i-1)/2**(n-1),2) - 1
```

where $i = 1, \ldots, 2^N$ and $n = 1, \ldots, N$; mod (k, ℓ) is the remainder of k divided by ℓ. The value of the i-th shape-function at an arbitrary point, with (local) coordinates p(n), is then calculated by the following algorithm:

```
shp      = 1./2**N
DO 1 n = 1,N
1 shp    = shp * (1.+kube (i, n) * p(n))
```

At last, the value of the approximation $\varphi = N_i \varphi_i$ at p can be calculated, taking the inner product of $(shp)_{i=1,N}$ and $(\varphi_i)_{i=1,N}$.

It is clear that the bilinear element Equation 29 can be considered as a special case of the N-dimensional cube, for N = 2.

In two-dimensional transient heat transfer analysis, the 3-cube can be used as an element for approximation. The third dimension is then interpreted as a time coordinate.

Other members of the N-cube family of course include the one-dimensional linear element, used in both space and time. And, last but not least, the zero-dimensional element; that is a single point in spacetime, used to describe known function values at the boundary.

A uniform treatment like the one given above can be seen to be advantageous now, because all those elements needed can be coded *at once;* thus saving a lot of FORTRAN writing, and ... errors!

Normed Approximations

At this stage, the four-node cluster could be completed with four so-called *integration points,* the coordinates of which are, in general:

$$(\xi_k, \eta_k) = (\pm p, \pm q)$$

where $k = 1, \ldots, 4$ and $p = q = 1/\sqrt{3}$. In this way, the finite cluster is turned into a conventional, numerically integrated, finite element fore (e.g. [19]). This case is worked out further in the paper of D.e Bruijn/Zijl [20]. Applying the least squares method, however, elements with only *one* integration point (reduced integration), as a rule, turn out to be far more appropriate. And the "matrix method" presented in [20, 21] can be simplified accordingly, without loss of generality.

The four-node cluster [24] is conceived as a parent-element for the general quadrilateral, which is derived from it by isoparametric mapping, (see [19] Chapter 8):

$$x = N_i x_i, \quad y = N_i y_i \tag{33}$$

The mapping is characterized by a Jacobian determinant:

$$J = \frac{\partial x}{\partial \xi}\frac{\partial y}{\partial \eta} - \frac{\partial x}{\partial \eta}\frac{\partial y}{\partial \xi}$$

which must be nonzero everywhere inside the element.

Applying a piece of standard finite element [19] theory we find:

$$\frac{\partial N_i}{\partial x} = \left(\frac{\partial y}{\partial \eta}\frac{\partial N_i}{\partial \xi} - \frac{\partial y}{\partial \xi}\frac{\partial N_i}{\partial \eta}\right) / J$$

$$\frac{\partial N_i}{\partial y} = \left(-\frac{\partial x}{\partial \eta}\frac{\partial N_i}{\partial \xi} + \frac{\partial x}{\partial \xi}\frac{\partial N_i}{\partial \eta}\right) / J \tag{34}$$

It can be shown that the functions $\partial(x, y)/\partial(\xi, \eta)$ and J are all linear in ξ, η.

Using the "shape function method," as described in Reference 19, the expressions of Equation 34 must be evaluated at an integration point (*).

Doing so, we construct, in fact, the following matrices:

$$DDX_j = \frac{\partial N_j}{\partial x}(*); \quad DDY_j = \frac{\partial N_j}{\partial y}(*)$$

Now $[DDX_j]$ and $[DDY_j]$ can be interpreted as matrix equivalents corresponding to the differentiations $\partial/\partial x$ and $\partial/\partial y$. They are called "differentiation-matrices," according to Csendes [22, 23], or "derivation matrices," according to Boisserie [24].

This set of elementary approximations is completed by adding the case of differentiation to the zero-th coordinate, previously [2, 3] called the projection matrix:

$$P_i = N_i(*) \tag{35}$$

The projection matrix (Equation 35) and differentiation matrices (Equation 34) may be called also the fundamental or universal matrices of the element. Nothing more seems to be needed, for consistent and effective application of finite element and finite difference methods.

For the family of N-cube elements, the fundamental approximations can be coded in a short-hand way, leaving free the number of dimensions actually implemented. This topic, however, will not be pursued further at the moment. The so-called „matrix-method," which is associated in a natural way with the use of $[DDX_i]$, $[DDY_i]$, and $[P_i]$ was demonstrated in earlier publications [21]. In the sequel, it will be referred to simply as the „approximation method."

Given an appropriate set of elementary approximations, it is very easy to write down the approximations for the partial differential *equations*, governing the problem at hand. This will be demonstrated now for the problem of shell-side flow in the SNR-300 heat exchanger. It was proven in earlier that this flow field can be assumed to be approximately irrotational. Furthermore, it is assumed that the problem as a whole is cylindrically symmetric. The differential (Equations 27 and 28) then simplify once again:

$$\frac{\partial(rv)}{\partial r} + r\frac{\partial u}{\partial z} = 0$$

$$\frac{\partial u}{\partial r} - \frac{\partial v}{\partial z} = 0$$

Where r, z = radial/axial coordinate; v, u = radial/axial velocity.

To apply now the approximation method conveniently, these equations must be presented first in so-called operator-form, that is:

$$\begin{bmatrix} \partial/\partial r \cdot r & r \cdot \partial/\partial z \\ -\partial/\partial z & \partial/\partial r \end{bmatrix} \begin{bmatrix} v \\ u \end{bmatrix} = 0 \tag{36}$$

As a unit for discretization of the Equation 36 the reduced quadrilateral (see Figure 6) proves to perform well. If the flow domain is assumed to be rectangular, built up from elements with sizes Δr and Δz, the elementary approximations can be constructed easily:

$$[P_i] = \frac{1}{4}[\ 1,\ \ 1,\ \ 1,\ \ 1]$$

$$[DDR_i] = \frac{1}{4}[-1,\ \ 1, -1,\ \ 1]/\Delta r$$

$$[DDZ_i] = \frac{1}{4}[-1, -1,\ \ 1,\ \ 1]/\Delta z$$

The numerical equivalents of Equation 36 are then FORTRAN-coded [14] according to the following algorithms:

```
  DO 1 I = 1,4
  A (2 * I - 1) = DDR(I) *R(I)
1 A(2 * I) = R(I) *DDZ(I)
  ----------------------------
  DO 2 I = 1,4
  A (2 * I - 1) = -DDZ(I)
2 A (2 * I) = DDR(I)
```

Where the radial velocities are associated with uneven indices, and the axial velocities with even indices, which is simply a matter of bookkeeping. R(I), I = 1, ..., 4 are the radial positions of the nodes of the element.

Note that this kind of making approximations is, so to speak, *structure preserving:* one can "read back" the original form of the P. D. E's from the FORTRAN text. This greatly facilitates such things as documentation, updating, and debugging of the computer program.

Given an appropriate set of universal matrices, it proves to be very easy to write down approximations for the partial differential equations, or variational principle, governing the problem. The matrix method always results in expressions of the form:

$$r = A_i \varphi_i \tag{37}$$

where r = residual of the equations, φ = vector of unknowns, A = so-called *approximation* matrix [20, 21] of the problem, specified for the element under consideration.

After constructing the cluster approximation matrices, some method must be devised to obtain a system of equations which is representative for the problem as a whole. So far, nothing has been said about the numerical method to be employed for that purpose. Assembling of the local approximations, such as Equation 36, could be accomplished in a finite element way by replacing all local indices I by global numbers, associated with the mesh as a whole. After that, simply *summing up* all the approximations, together with the boundary conditions, will result in a global set of F.D. equations.

Dealing with a great number of elements, of different sizes and kinds, this global system, however, could hardly be expected to behave like a well-conditioned one. In the worst case, some element contributions will be far too pronounced, while others, so to speak, will be almost invisible to the computer. Therefore, a kind of weighting procedure has to be devised to assure that all element approximations contribute with the same order of magnitude to the global scheme.

Insofar as the problem itself is not ill-conditioned, conditioning of the global matrix can be assured simply and effectively by the following procedure. Divide each approximation by its length in vector space. In other words: use *normed* approximations.

Let $[A_i]$ be a given approximation-matrix. The *norm* of $[A_i]$ is given by its length in vector space:

$$Sp(A) = \sqrt{\left(\sum_{i=1}^{N} A_i^2 \right)} \tag{38}$$

Divide each approximation by its norm:

$$A_i = A_i / Sp\,(A)$$

After that, Sp (A) = 1 for all approximations A.

In the sequel, all approximations will be assumed to be normed.

The L. S. FEM package

The *residual* r_E of the approximating equations, specified for a finite cluster E, is given by Equation 37:

$$r_E = A_i^E \varphi_i \tag{39}$$

where A_i^E is the normed approximation at E.

We shall consider, once again [20, 21], the general form of discretized equations, resulting from the *least squares* finite element principle of approximation (see [19] pg. 87–89). Applying the least squares criterion, the (weighted) sum of the squares of the residuals of the differential equations at all the integration points in the mesh should be a minimum:

$$\sum_E r_E^2 = \sum_E A_i^E \varphi_i \cdot A_j^E \varphi_j = \text{min.} \tag{40}$$

Here it is implicitly assumed that an element with several integration points can always be conceived as a superposition of reduced elements.

By analogy with the conventional finite element method [19], we can see that the following definition makes sense:

$$E_{ij} = A_i^E A_j^E \tag{41}$$

$[E_{ij}]$ is called *element matrix* or "local stiffness matrix." The algorithm of Equation 41 can be coded once and for all.

The least squares procedure of assembly (Equation 40) is completed by adding the element-matrices $[E_{ij}]$ to the *global* or *system matrix* $[S_{ij}]$ in the usual finite element way:

$$S_{ij} = \sum_E E_{ij} \tag{42}$$

Boundary conditions need not to be handled separately; they can be conceived as elements (= nodes) of *zero* dimension, which is quite as simple.

By differentiating to the unknowns φ_n, the minimum problem (Equation 40) is transformed into a system of linear equations:

$$S_{nj}\varphi_j = B_n \tag{43}$$

Where the *load* vector B is constructed from known contributions. Note that the S-matrix is symmetric, positive definite and, in many cases, banded. The global system of equations can be solved by straightforward Gaussian elimination, using the wave front method.

By application of the Schwartz-inequality at Equation 41 and 42 it can be shown that the following relations hold:

$$E_{ii} \cdot E_{jj} \geqq E_{ij}^2 \quad \text{and} \quad S_{ii} \cdot S_{jj} \geqq S_{ij}^2$$

The local and global matrices are diagonally dominant, in a mean square sense. Thus, for example, pivoting procedures will not be needed. Furthermore, due to norming of the approximations A_i, the following relations hold:

$$\sum_i E_{ii} = 1$$

$$\sum_i S_{ii} = \text{total number of elements}$$

It was pointed out already that least squares finite element methods have a strong relationship with conventional finite differences.

Let us now make this statement more precise. Let $A_i^E \varphi_i = 0$ with $i = 1, \ldots, N$, and let $E = 1, \ldots, m$ be a finite difference system of equations. Obviously, a series of numbers will be zero if and only if the sum of the squares of all those numbers is zero:

$$\sum_E (A_i^E \varphi_i)^2 = \sum_E \varphi_i A_i^E A_j^E \varphi_j = \text{minimum} = 0$$

It is clear now that this is nothing more than a special case of the least squares procedure (Equation 40).

For the minimum to be exactly zero, it is required that the number M of independent F.D. equations equals the number N of unknowns, so that M = N.

A "finite difference element" must therefore be chosen in such a way that the resulting number of discretized equations (= elements) exactly equals the number of unknowns (= nodes) in the grid.

This sounds trivial from a finite difference viewpoint. We think, however, it is, instead, a substantial problem, inherent to the F.D. method as such.

Using a conventional finite element procedure, on the contrary, no such problem arises. This is because the r_E are here, in fact, residuals of improper "equations," the square of which namely constitutes the integrand of the variational principle. And these residuals, of course, don't have to be zero.

In fact, the number of integration points (= reduced elements) is always *greater* than the number of unknowns: M > N. So that most F.E. methods give rise to *over*determined systems of F.D. equations, which are solved in a least squares sense.

In this respect, L.S. FEM shows some resemblance to conventional finite element methods. Indeed for the least squares numerical method to work, it is *not* demanded that its residuals be *exactly* zero. It must be assured, however, that they do *converge* to zero. Otherwise, no realistic solution can be expected. This enlarges somewhat the possibilities of least squares with respect to conventional finite differences.

At the same time, it is not clear at first sight why one should square working finite difference schemes, differentiate to the unknowns, and manipulate them as if they were finite elements. What should be the advantage of spending more computation time and computer memory?

The answer could be found in the fact that, once the F.D. schemes are squared, they can be handled as if they were conventional finite elements.

No further research is needed, therefore, to put the method to work. Moreover, F.E. modular concepts like "connectivity" and "isoparametrics" help to avoid the overhead, still common to F.D. programs [28] when attempting to get rid of inhomogeneous parts of the problem domain. At the other hand, some of the more robust F.D. concepts, such as "unconditional stability," have been entered for good in L.S. FEM.

The core of the package consists of a library of subroutines, invisible to the user. This library is "opened" by writing a user-defined MAIN program. MAIN has the following global structure [27]:

```
PROGRAM MAIN
Bookkeeping (TAPES, DIMENSION, COMMON)
Prepare data (using the LSFEM-library)
CALL SOLVER (..., FIELD, ...)
Evaluation of results (also using the library)
STOP & END
```

In fact, the MAIN program is built completely around the function SOLVER, with FIELD as an external in its parameter list. SOLVER is a library routine which solves a large set of positive definite symmetric equations, using mass storage. FIELD is a user-supplied procedure, with input parameter IG, being the nodal number, for which a row of matrix-coefficients ROW (NB1) is asked, as output, by SOLVER. FIELD is called somewhere in the body of SOLVER according to: CALL FIELD (IG, ROW, NB1); NB1 = bandwidth + 1.

Subroutine FIELD has the following global structure:

```
SUBROUTINE FIELD (IG, ROW, NB1)
Bookkeeping; DIMENSION MENU (max (LOOPS), 3)
DATA MENU /...; ...; .../
CALL LOGIKA (IG, LABEL, LOOPS)
DO 10 L = 1, LOOPS
```

```
   CALL TOPOL (IG, MENU (LABEL (L), 1))
   CALL PHYSIK (MENU (LABEL (L), 2))
   CALL PDE (MENU (LABEL (L), 3))
   CALL NORMAL
   CALL LSFEM
10 CALL INTEL (NB1)
   CALL UITEL (IG, ROW)
   RETURN & END
```

FIELD thus mainly consists of a number of calls to other modules.

Let us start with subroutine LOGIKA. The problem area is subdivided in a number of subsets, each of which is characterized by a logical predicate:

CASE (K) = true or false, as a function of IG.

Before returning to FIELD the indices (K) for which CASE (K)=.true. are stored into the array LABEL (LOOPS); so defining all the subsets of the mesh to which a current node IG belongs.

Each subset is characterized by the fact that it possesses a different topology, different physical properties and/or a different kind of "element" with respect to other subsets. LOGIKA is a user-made *classification procedure* for TOPOL, PHYSIK, and PDE.

The subroutines TOPOL, PHYSIK, and PDE are also, in principle, user-supplied. They define respectively: *connectivity,* "material *constants,*" and the kind of discretization *scheme* to be employed. These procedures are coded independently of each other and with no reference to the subset of LOGIKA. They possess the following general structure:

```
  SUBROUTINE PHYSIK (LABEL) for example.
  Bookkeeping
  GOTO (1, 2, 3, 4, ...), LABEL
1 CONTINUE
  calculation of constants type 1
  RETURN
2 CONTINUE
  calculation of constants type 2
  RETURN
  .........
  .........
  END
```

The interconnection between the LABEL's in LOGIKA and the LABEL's in TOPOL/PHYSIK and PDE is accomplished by a so-called MENU interface, consisting of three entries, to assure the modularity of this part of the coding. It is remarked further that, in contrast to the conventional F.E.M. packages generating the connectivity and material properties can be kept completely in-core.

A few words have to be said about the routine we called PDE. Although it is not impossible for the user to code such a routine himself, it is much more advantageous to supply the package with a number of preprogrammed modules, containing working distretization schemes for a number of commonly encountered partial differential equations. Reading through the following pragraphs, the reason for this will be still more clear. The task of routines like PDE is: to construct the approximation-matrix, according to the theory developed earlier in this section. Part of this task, again, will be accomplished elsewhere, by delegating the forming of elementary approximations and isoparametrics to specialized routines from the library.

After constructing the approximation, governing the problem corresponding to LABEL (L), the A_i's are normed according to Equation 38 with a call to NORMAL, and squared according to Equation 41 with a call to LSFEM. They are subsequently counted into the system matrix (Equation 42) by a call to INTEL; this routine uses the frontal method [19], it overwrites equations when they are no longer in use. Looping is performed through all the elements needed, to construct the complete matrix-row for the nodal point IG in question. At last, ROW itself is delivered by UITEL, thereby leaving FIELD and returning control to the SOLVER-program.

Steady-State Thermal Hydraulics

The numerical method has been worked out for three-dimensional, cylindrically symmetric flow and heat transfer in a heat exchanger [20, 21]. To describe the complex transport phenomena in the flow around the tubes, use has been made of the "liquid-structure continuum" approach with continuously distributed flow resistance and heat transfer mechanism, described earlier. This analysis leads to several different sets of partial diferential equations. They are all solved with the same least squares programming package, and with the same kind of finite elements.

The shell-side flow field was treated earlier. The equations are, once again:

$$\begin{matrix} [\partial/\partial r \cdot r & r \cdot \partial/\partial z] \\ [-\partial/\partial z & \partial/\partial r] \end{matrix} \begin{bmatrix} v \\ u \end{bmatrix} = 0 \tag{36}$$

Proper boundary conditions for this elliptic problem are: impermeability $u = 0$ or $v = 0$ at the solid boundaries, and prescribed (mutually consistent) velocity-profiles at the inflow and outflow openings. (Numerically, *one* point of the boundary can be left free.) As has been pointed out, the flow field can be modeled with a mesh, consisting of quadrilaterals with total reduction. In this case, the least squares method is fully equivalent with a volume method (finite differences). It was shown how Equation 36 is coded in FORTRAN, by using the matrix method.

The flow field, thus described, has to be calculated only once. Under transient conditions, multiplying it by a time-dependent scaling factor proves to be sufficient approximation.

After solving the Equation 36 for the velocities (v, u), which is a first-order linear problem, the streamlines at shell-side can be found, being the contour map of the stream function $\Psi(r, z)$:

$$\frac{\partial \Psi}{\partial r} = r \cdot u; \quad \frac{\partial \Psi}{\partial z} = -r \cdot v \tag{44}$$

One zero point value for Ψ must be prescribed. Also the pressure distribution $p(r, z)$ can be calculated, using the equations:

$$-\frac{1}{\rho}\frac{\partial p}{\partial r} = v\frac{\partial v}{\partial r} + u\frac{\partial v}{\partial z} + \frac{1}{2}\sqrt{K_z u^2 + K_r v^2} \cdot v$$

$$-\frac{1}{\rho}\frac{\partial p}{\partial z} = v\frac{\partial u}{\partial r} + u\frac{\partial u}{\partial z} + \frac{1}{2}\sqrt{K_z u^2 + K_r v^2} \cdot u \tag{45}$$

where it has been assumed that the flow is steady. Also, one zero point for the pressure must be given as a "boundary condition." A mesh of simple linear triangles can be used, to discretize the streamline and pressure (Equations 44 and 45) properly. The least squares procedure, in this case, has no finite difference equivalent. Undoubtedly, it is much more straightforward than designing marching schemes to any point in the pressure field; or solving a (second-order) Poisson equation with conditions for the pressure gradient at the boundary, often difficult to find. Setting up the least squares "variational principle" for streamlines:

$$\iint\left(\frac{\partial\Psi}{\partial r} - r\cdot u\right)^2 + \left(\frac{\partial\Psi}{\partial z} + r\cdot v\right)^2 rdrdz = \text{minimum} \tag{46}$$

We notice that the minimum will be zero only approximately.

The Euler equations, derived from Equation 46, are:

$$\frac{\partial}{\partial r} r \frac{\partial\Psi}{\partial r} + r\frac{\partial^2\Psi}{\partial z^2} = \frac{\partial}{\partial r}(r^2u) - \frac{\partial}{\partial z}(r^2v)$$

So that: $\nabla^2\Psi = 2u$

This is a Poisson equation for Ψ, with a nonzero right-hand side due to cylindrical symmetry. So Equation 46 establishes a link between least squares and conventional finite elements, for these kinds of "marching" problems.

As shown in section 2, using the continuum-approach, also the laws of energy conservation and heat transfer turn out to be, again, *first-order* partial differential equations of the hyperbolic type.

Assuming cylindrical symmetry and steady-state operating conditions, Equations 22 and 23 simplify to:

$$\text{PRM}\left(v\frac{\partial T_p}{\partial r} + u\frac{\partial T_p}{\partial z}\right) + \text{COEF}_p(T_p - T_s) = 0 \tag{47}$$

$$\text{SEK}\frac{\partial T_s}{\partial z} + \text{COEF}_s(T_s - T_p) = 0 \tag{48}$$

where r/z = cylindrical coordinates,
T = temperature of
p/s = shell-side/tube-side sodium, respectively.

As has been stated, the flow field (v, u) is calculated once. It is scaled up by: PRM/SEK = (time-dependent) velocity scaling factors; COEF = coefficients of heat transfer and -capacity, depending upon the constructive details of the apparatus.

Boundary conditions for the shell- and tube-side temperatures should be prescribed *only* at the respective inflow openings.

The rest of this paragraph is devoted to the question which element should be adopted for appropriate discretization of the temperature equations. It turns out that, if it is demanded that the numerical scheme be *stable* under all operating conditions, no compromise, as suggested in References 20 and 21 can be made.

The quadrilateral with reduced integration is adopted as an element to discretize the shell-side (Equation 46) at a rectangular mesh:

$$[\text{PRM}\,(v\partial/\partial r + u\partial/\partial z) + \text{COEF}_p]\cdot T_p - 1\cdot \text{COEF}_p\cdot T_s = 0 \tag{49}$$

The fundamental approximations are then given by:

$$\partial/\partial r = \frac{1}{2}[-(1-\eta) \qquad +(1-\eta) \qquad -(1+\eta) \qquad +(1+\eta)]/\Delta r$$

$$\partial/\partial z = \frac{1}{2}[-(1-\xi) \qquad -(1+\xi) \qquad +(1-\xi) \qquad +(1+\xi)]/\Delta z$$

$$1 = \frac{1}{4}[(1-\xi)(1-\eta) \qquad (1+\xi)(1-\eta) \qquad (1-\xi)(1+\eta) \qquad (1+\xi)(1+\eta)]$$

Where the position of the integration point (ξ, η) has to be chosen properly.

Let us define certain *dimensionless numbers* H, K, as follows:

$$H = \frac{PRM \cdot v}{\frac{1}{2}\Delta r \cdot COEF_p}, \quad K = \frac{PRM \cdot u}{\frac{1}{2}\Delta z \cdot COEF_p} \tag{50}$$

Working out now the approximation scheme Equation 49 yields:

$$\begin{aligned}
&[-H(1-\eta)-K(1-\xi)+(1-\xi)(1-\eta)]T_1^P \\
&+[+H(1-\eta)-K(1+\xi)+(1+\xi)(1-\eta)]T_2^P \\
&+[-H(1+\eta)+K(1-\xi)+(1-\xi)(1+\eta)]T_3^P \\
&+[+H(1+\eta)+K(1+\xi)+(1+\xi)(1+\eta)]T_4^P \\
&-(1\pm\xi)(1\pm\eta)T^s = 0
\end{aligned} \tag{51}$$

Let us assume, for simplicity, that:

$$H > 0, \quad K > 0 \quad \text{and} \quad H > K$$

Interpreting (H, K) as a dimensionless velocity vector, this situation can be depicted as in Figure 7.

A more serious restriction can be imposed by demanding that, in case of $H \to 0$ or $K \to 0$, the two-dimensional description automatically simplifies to one-dimensional theory.

This is accomplished most easily by positioning the integration point at the *characteristic* location going through a nodal point, in such a way, however, that the coordinates (ξ, η) lie *within* the element. In the situation sketched, only the characteristics 1 through 4 can be chosen.

Now in the very special case that $H = K = 0$, which corresponds to a total blockage of the shell-side flow, the differential equation (Equation 49) reduces to an algebraic one, namely:

$$T_p - T_s = 0$$

Such an equation is discretized best at a zero dimensional element. This must be one of the nodal points 2 or 4, if going continously from $H, K > 0$ to the situation of zero velocity. At the nodal point 2 however, the direction from which the flow comes from, a boundary condition could already be imposed. So, for reasons of consistency, an integration point can be placed only at the characteristic position for 1 through 4. To be more precise, the coordinates (ξ, η) must obey the relationship:

$$H(1-\eta) = K(1-\xi) \tag{52}$$

To proceed further, let us interpretete Equation 51 as a finite difference equation to calculate the shell-side temperature T_P at nodal point 4. Suppose we have a complete system of such equations, say:

$$A_{ij}T_j = 0 \quad \text{for} \quad i, j = 1, \ldots, N \tag{53}$$

Now a unique solution of the system (Equation 53) can be proven to exist, if it can be shown that the following conditions hold:

- $A_{ii} > 0$; $A_{ij} \leqq 0$ for all $j \neq i$.
- $A_{ii} \geqq \sum_{j=1,\ j\neq i}^{N} |A_{ij}|$ with strict inequality for some i.
- A is irreducible. (54)

For a proof of this theorem, see Ames [25] page 101; see also Varga [30].

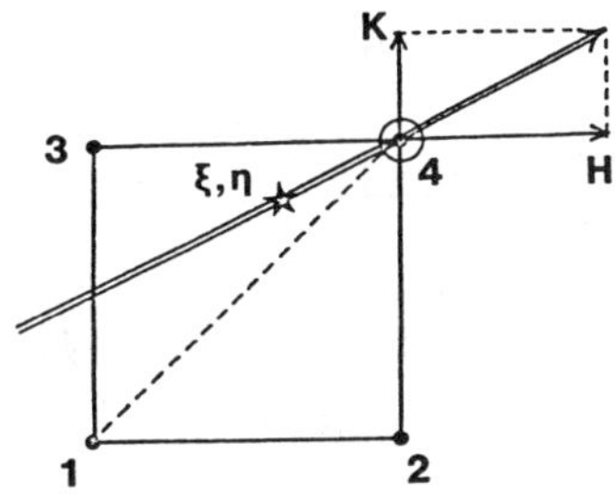

Figure 7. Reduced upwind integration.

It is more or less implicitly stated in this proof that a solution of Equations 53 and 54 will also be unconditionally *stable;* which is even of greater importance from a practical point of view.

Recalling Equation 51 we obtain the following set of inequality conditions:

$$A_{41} = -H(1-\eta)-K(1-\xi)+(1-\xi)(1-\eta) \leqq 0$$
$$A_{42} = +H(1-\eta)-K(1+\xi)+(1+\xi)(1-\eta) \leqq 0$$
$$A_{43} = -H(1+\eta)-K(1-\xi)+(1-\xi)(1+\eta) \leqq 0$$
$$A_{44} = +H(1+\eta)+K(1+\xi)+(1+\xi)(1+\eta) > 0$$
$$A_{4j} = -(1\pm\xi)(1\pm\eta) \leqq 0 \quad \text{for} \quad j = 5, \ldots, 8. \tag{55}$$

Because of $-1 \leqq \xi, \eta \leqq +1$ and H, K > 0, it is clear that the condition $A_{44} > 0$ is fulfilled without question.

The same thing can be asserted for $A_{4j} \leqq 0$, j = 5, ..., 8.

Looking at Figure 7, it is seen that the nodal characteristic is embedded in a "domain of influence," which is thought to be bounded by the lines 3–4 and 1–4. Therefore it is suggested that, physically speaking, 4 obtains temperature information from 1 and 3, but is not to be influenced by 2, which is outside the domain. For this reason, it would probably be a good idea to set the corresponding matrix coefficient to zero: $A_{42} = 0$ in Equation 55. Combining this with the characteristic equation Equation 52 we obtain for ξ:

$$-2H\xi+1-\xi^2 = 0$$

(Here it is tacitly assumed that $K \neq 0$.) The solution of which must be written as:

$$\xi = 1/(H+\sqrt{(1+H^2)}) \tag{56}$$

The reasoning is completed by asserting that, with this value for ξ:

$$A_{41} \leqq 0 \quad \text{and} \quad A_{43} \leqq 0$$

(Note that: $\xi > 1-H$).

Once the sign of the matrix coefficients has been established, it is a simple matter to show that the remainder of the conditions (Equation 54) is fulfilled almost automatically. So we have proved that, with Equations 52 and 56 determining the position of the integration points, the shell-side temperature problem as stated in Equation 49 has a unique solution. It can also be asserted that this solution is *unconditionally stable*.

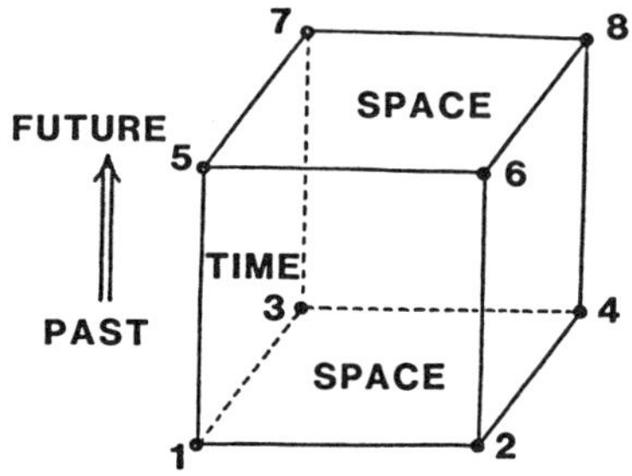

Figure 8. Space-time element.

Temperature Transients

The equations which describe temperature transients in the bundle of a shell and tube heat exchanger are given by Equations 22 through 24. Assuming cylindrical symmetry, they simplify as follows:

$$\frac{\partial T_p}{\partial t} + PRM\left(v\frac{\partial T_p}{\partial r} + u\frac{\partial T_p}{\partial z}\right) + COEF_p(T_p - T_w) = 0 \tag{57}$$

$$\frac{\partial T_s}{\partial t} + SEK \cdot \frac{\partial T_s}{\partial z} + COEF_s(T_s - T_w) = 0 \tag{58}$$

$$\frac{\partial T_w}{\partial t} = a\frac{1}{r}\frac{\partial}{\partial r}\left(r\frac{\partial T_w}{\partial r}\right) \tag{59}$$

where r, z, t = space-time coordinates
T = temperatures where p, s, w, refer to shell-side/tube-side, sodium, and tube-walls
(v, u) = irrotational flow field, calculated once
PRM/SEK = time-dependent velocity scaling factors
COEF, a = coefficients of heat transfer and capacity, respectively

Problems in space and time, such as temperature transients in a heat exchanger, will conveniently be described by space-time elements. In our opinion there is no reason why coordinates in space and the time coordinate should *not* be handled on an equal basis [28].

Let us consider for example the space-time cube, depicted in Figure 8. The F.D. base of this finite cluster is given by:

$$\ell_i = \{1, \xi, \eta, \zeta, \xi\eta, \xi\zeta, \eta\zeta, \xi\eta\zeta\}; \quad i = 1, \ldots, 8$$

Which expresses an arbitrary approximation in a Taylor-series expansion around the origin (0, 0, 0).

The F.E. base of the cluster is given by the shape-functions.

$$N_i = \left\{\frac{1}{8}(1 \pm \xi)(1 \pm \eta)(1 \pm \zeta)\right\}; \quad i = 1, \ldots, 8$$

The eight-node cluster is conceived as a parent element for the general cubic, which is derived from it by an isoparametric mapping (Equation 33). We can construct then differentiation-matrices such as Equation 34.

$$DDR_j = \frac{\partial N_j}{\partial r}, \quad DDZ_j = \frac{\partial N_j}{\partial z}, \quad DDT_j = \frac{\partial N_j}{\partial t}$$

For reasons of stability, elements with reduced upwind integration will be used for the temperature problem.

The eight-node cluster can be turned into a more conventional finite element, if we supply it with 8 integration points. This case is worked out in De Bruijn [20], and will not be considered further.

Let us consider now in more detail the differential equation for calculating the shell-side temperature field T_p, that is Equation 57:

$$[\partial/\partial t + PRM\,(v\,\partial/\partial r + u\,\partial/\partial z) + COEF_p] \cdot T_p - COEF_p \cdot T_w = 0$$

In analogy with Equation 50, let us define the dimensionless numbers H, K, L, according to:

$$H = \frac{PRM \cdot v}{\Delta r \cdot COEF_p}, \quad K = \frac{PRM \cdot u}{\Delta z \cdot COEF_p}, \quad L = \frac{1}{\Delta t \cdot COEF_p}$$

The integration point (ξ, η, ζ) must be chosen in such a way that the resulting F.D. scheme is unconditionally stable. Reasoning along the same lines as in Equation 32, the following recipe has been found:

$$(\xi, \eta, \zeta) = \mathrm{sign}\,(H, K, L) \cdot 1 - \lambda \cdot (H, K, L)/P$$

$$\text{where} \quad P = \max\,(|H|, |K|, |L|) \tag{60}$$

$$\lambda = 1 - 1/[P + \sqrt{(1+P^2)}]$$

In the case of $L = 0$, this reduces to the stability criterion of Equation 32.

Applying the matrix-method (Equation 30) now results in an approximation for each space-time element, of the form of Equation 39:

$$r_E = A_j^E \varphi_j$$

And the least squares principle (Equation 40) could, in principle, be applied.

A discrepancy with previous theory occurs, however, due to the fact that variables in a space-time element, when they belong to the *past,* are no longer *unknown.* Only those of the future are. The approximation matrix can thus be split up in two parts (see Figure 8):

$$\begin{bmatrix} \underset{j = 1, 2, 3, 4}{A_j} & \underset{j = 5, 6, 7, 8}{A_j} \end{bmatrix} \begin{bmatrix} \varphi_{1, \ldots, 4} \\ \varphi_{5, \ldots, 8} \end{bmatrix}$$

By differentiating only to the *future* variables $\varphi_{5, \ldots, 8}$, most of the procedures can be maintained. Element-matrix and -vector are constructed now as follows, moving known values to the right-hand side:

$$E_{ij} = A_i A_j; \quad b_i = -A_i A_j \varphi_j \tag{61}$$

where $5 \leqq i, j \leqq 8$; $5 \leqq i \leqq 8$, $1 \leqq j \leqq 4$, respectively.

These element-contributions are counted into the global matrix and load-vector (S_{ij}, B_i), a system which, in principle, must be solved at each time-step. The resulting time stepping scheme is thus an implicit one.

To discretize the tube wall equations (Equation 59) which are of second order in space, no use can be made of cubic elements, since the second derivative would then disappear. A proper element for discretization of the tube-walls is depicted in Figure 9.

Its F.D.-base is given by:

$$(1, \xi, \xi^2) \otimes (1, \eta)$$

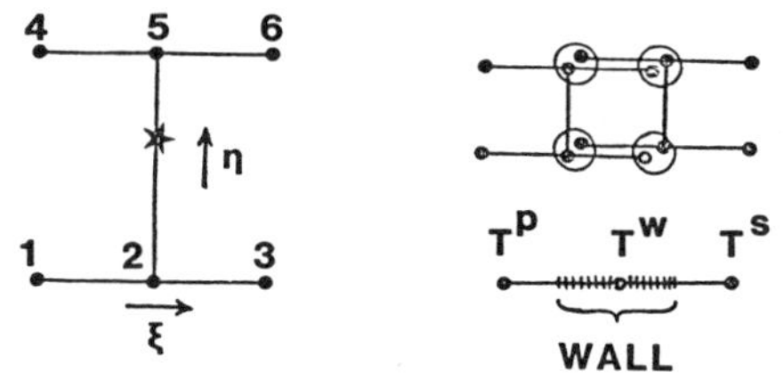

Figure 9. Tube wall element.

And its F.E.-base turns out to be:

$$\left(-\frac{1}{2}\xi+\frac{1}{2}\xi^2,\ 1-\xi^2,\ +\frac{1}{2}\xi+\frac{1}{2}\xi^2\right)\otimes\frac{1}{2}(1-\eta,\ 1+\eta)$$

The finite difference character of this element becomes still more clear if one looks at the way it is interconnected with other schemes: (see Figure 9). The integration point is chosen at $(0, \eta)$, where η is determined again by conditions of stability.

It can be justified that the tube walls can be discretized with sufficient accuracy by adopting just *one* element over the wall thickness, as depicted. The approximations for Equation 59, after time-stepping, then turn out to be of the following form:

$$[A_1^w A_2^w A_3^w]\begin{bmatrix}T^p\\T^w\\T^s\end{bmatrix}=A_4^w$$

Solving for T^w:

$$T^w=(A_4^w-A_1^w\cdot T^p-A_3^w\cdot T^s)/A_2^w \tag{62}$$

After discretization and time-stepping, also an approximation is found for the tube-side equations (Equation 58):

$$[A_1^s A_2^s A_3^s A_4^s]\begin{bmatrix}T_{1_w}^s\\T_{1_s}\\T_{2_w}\\T_2\end{bmatrix}=A_5^s \tag{63}$$

Substitution of Equation 62 into Equation 63 eliminates the tube-wall temperatures T^w, resulting in:

$$\begin{bmatrix}A_1^s-A_2^sA_3^w/A_2^w; & -A_2^sA_1^w/A_2^w;\\ A_3^s-A_4^sA_3^w/A_w^w; & -A_4^sA_1^w/A_2^w\end{bmatrix}\begin{bmatrix}T_{1_p}^s\\T_{1_s}\\T_{2_p}\\T_2\end{bmatrix}=A_5^s-A_2^sA_4^w/A_2^w-A_4^sA_4^w/A_2^w$$

In exactly the same way, the tube wall temperatures T^w can be eliminated from the shell-side approximations.

So the resulting global matrix needs only to be solved for the primary and secondary temperatures (T_p, T_s). This means a reasonable saving in computer memory and computation time. After solving for the temperatures of the fluid, the solid temperatures can be found using Equation 62.

The technique used here is known in finite element theory as *substructuring* or making *superelements*. Analogous procedures can be used to construct much more efficient (one-dimensional) temperature elements for the parallel flow region in a heat exchanger with straight tubes [26].

Modeling and Results

The SNR-300 is a sodium-cooled nuclear power plant, with a rated electrical output of 300 MW, situated at Kalkar in West Germany.

The heat exchanging system is looptype and consists of:

- The reactor.
- The primary pumps.
- The intermediate heat exchangers.
- The secondary sodium circuit with steam generators and secondary pumps.
- The tertiary circuit, a water/steam circuit, with turbines, condensors and feedwater pumps.

A picture of the intermediate heat exchanger (I.H.X.) is given in Figure 2.

At the shell-side hot primary sodium flows through a nozzle into the apparatus, and via an inlet perforation into the bundle. In this tube-bundle, heat exchanging with secondary sodium takes place. Next, the cooled primary sodium goes through an outlet perforation and streams out.

At the tube-side, cold secondary sodium flows down through a central tube. At the bottom of the apparatus, its speed is reversed, and via the lower tube-sheet it enters the bundle.

In the tube-bundle it warms up by exchanging heat with the primary sodium. Next, the heated fluid goes through the upper tube-sheet and streams out.

Geometrical data of the SNR-300 prototype are presented in the following table:

Inner shell diameter	0.460 m
Outer shell diameter	0.960 m
Length of tube bank	7.154 m
Height of inflow- and outflow openings	0.370 m
Outer tube diameter	0.0210 m
Inner tube diameter	0.0182 m
Pitch (triangular)	0.0270 m
Number of tubes	846

Process data are: shell-side mass flow rate 360 kg/s, tube-side mass flow rate 256 kg/s, shell-side inflow temperature 490° C, and tube-side inflow temperature 349° C.

Our heat exchanging components were tested at full scale, with true liquid sodium in it, at high temperatures, in a test facility owned by the Netherlands Organization for Applied Scientific Research (TNO). The experiments were carried out by people of TNO and Neratoom at the 50 MW test facilities in Hengelo, the Netherlands.

In the following short discussion, numerical results will be compared with experimental data.

These data have been obtained from the prototype model with the dimensions described in the previously presented table. Calculated and measured shell-side temperature profiles are shown in Figure 10 A and B.

The calculated temperature profiles, in axial direction, are almost straight lines. The measurements are plotted as separate points. Perhaps the agreement seems to be rather bad at first sight, but this is readily explained.

One always has to complete the picture with the radial temperature profiles, shown also. It is seen that a big spread in the measurements corresponds very precisely with steep gradients in the calculated lines.

Before the temperature calculations can start, however, at least the flow field at the shell-side must be calculated.

It was proved earlier that, under normal operating conditions, this flow field is virtually irrotational. Also, the original equations (Equations 1 and 2) were solved with the inflow boundary condition $\bar{\underline{\omega}} = \underline{0}$. Since this latter system is nonlinear, a generalized Newton iteration procedure was used.

From the numerical results, it turned out that the approximation of the flow field by irrotational flow theory compares to a calculation with the original equations within a maximum local error of 10% near the edge of the inflow boundary and outer shell. As a consequence of these two different flow models, the maximum difference in temperature for the two calculations is 7° C, or 5%. This small local difference is situated at the connection between lower tube sheet and central tube.

Of course, least squares finite difference elements were used to discretize the tube-bundle region of the I.H.X. The final mesh consists of 20 lines in radial direction, times 89 lines in axial direction, which gives us a total of 1,780 unknown temperatures and a bandwidth of 23.

As has been explained earlier a program MAIN must [27] be written by the user, mainly to define the sequence in which the calculations must be performed. In our case, first the flow field must be calculated, followed by a calculation of steady-state temperatures. Both serve as an input for the transient calculations:

```
   PROGRAM MAIN
   Bookkeeping
   CALL RASTER: generate grid.
   CALL SOLVER (..., FLOW, ...)
   Evaluate flow field/prepare other data
   CALL SOLVER (..., STEADY, ...)
   Evaluate steady-state temperature field
   DO 10 IT = 1, number of time-steps
   Define transient operating conditions
   CALL SOLVER (..., TRANSIENT, ...)
   Evaluate transient temperature field
10 CONTINUE
   STOP & END
```

The routines FLOW, STEADY, and TRANSIENT are foreseen with a structure analoguous to FIELD, as explained earlier. Assuming for a moment that a suitable library of approximation schemes PDE exists, then each of the routines FLOW, STEADY, and TRANSIENT must still be equipped with a classification LOGIKA, a connectivity TOPOL, physical properties PHYSIK, and a software interface defined by MENU data. The detailed structure of these subroutines is highly problem dependent.

To model the steady-state temperature field of a SNR-300 heat exchanger, 27 distinct classes were distinguished. The MENU card connected 15 different topologies, 15 kinds of physical properties, and 8 discretization schemes. Details of this modeling include boundary effects near the central tube and at the outer region of the bundle. These details, being one dimensional, are visualized as broken lines in Figure 10 B.

Summary

The numerical method developed in this chapter could be characterized by the following features:

1. The construction of elements in space and time, most of them being members of the isoparametric quadrilateral family with reduced integration.
2. Partially upwind integration of the elements describing convection, thus assuring numerical stability.
3. A matrix representation for the differential operators, which enables a very efficient coding of the differential equations govering the problem.

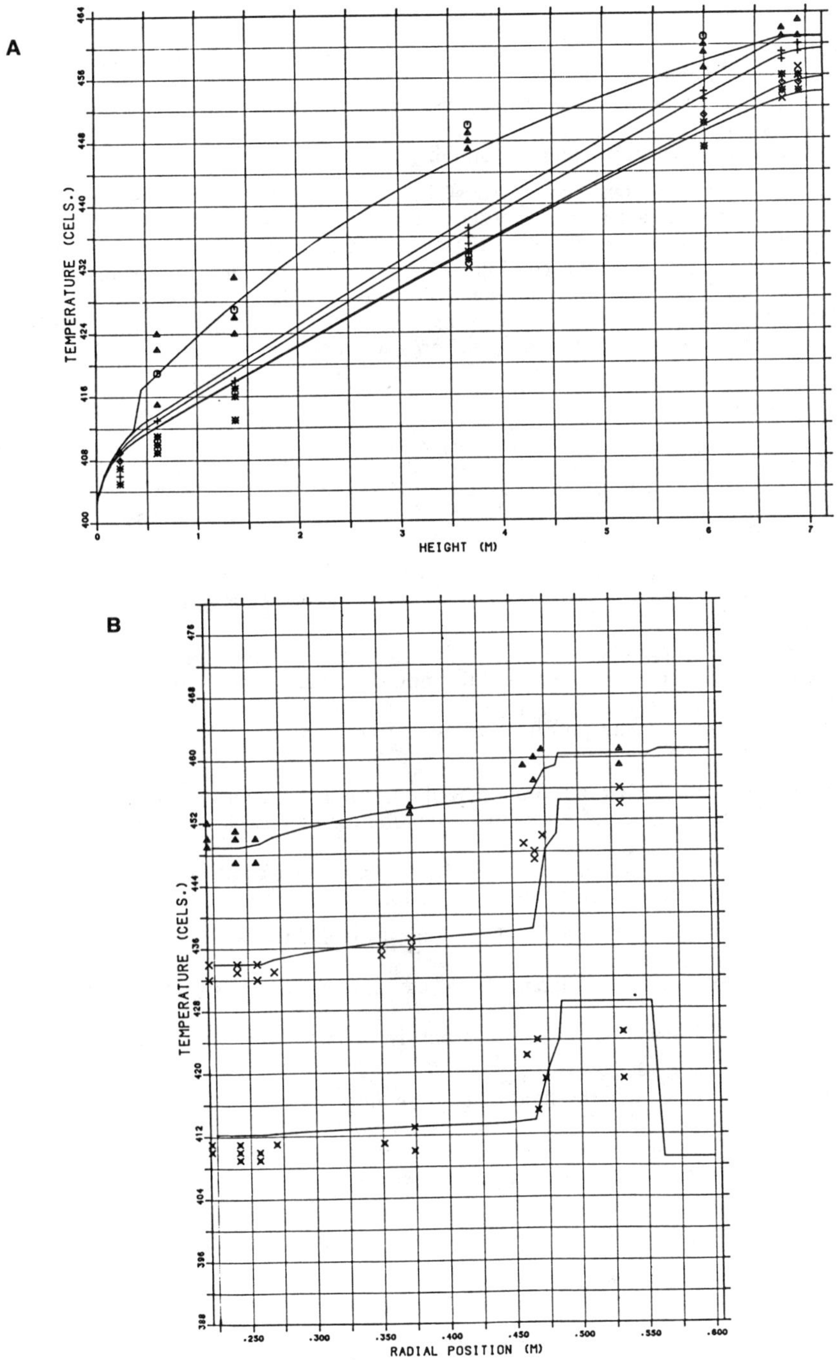

Figure 10. (A) Validation of steady-state analysis of IHX axial temperature profiles; (B) validation of steady-state analysis IHX radial temperature profiles.

4. An automatic scaling procedure, which ensures that the global equations of the problem will be optimally conditioned.
5. A least squares finite element procedure, resulting in a global matrix which is positive, definite, and symmetric.
6. Techniques of substructuring and repeatability.

The numerical method has been successfully applied to analyze three-dimensional, cylindrically symmetric flow and heat transfer in a heat exchanger. To describe the complex transport phenomena in the flow around the tubes, use has been made of the previously described section "liquid-structure continuum" approach, with continuously distributed flow resistance and heat transfer mechanism. This analysis leads to several different sets of first-order partial differential equations. They are all solved with the same kind of element, and with the same least squares (L. S. FEM) programming package.

CONCLUSIONS

The so-called fluid-structure continuum approach to described the thermal-hydraulic performance of heat exchangers had to be critically reviewed. A unification of this generalized theory with the classifical theory of flow in porous media was presented. Conclusions derived from the unification are:

- There is no difference between volume- and surface-porosity.
- A second-order "viscosity" term is not consistent with the fluid-structure continuum approximation.
- The permeability or drag coefficient is not a tensor but a scalar function of velocity.
- Under conditions of forced convection the flow is virtually irrotational both under steady-state and transient conditions.
- Under conditions of heat transport from shell-side to tube-side the temperature equations are of first order.

The resulting system of linear first-order partial differential equations was solved numerically by a finite element/finite difference approximation method based on the least squares minimum principle.

It proves to be essential for the success of least squares numerical methods that finite elements and finite differences are treated at a uniform mathematical basis:

- This enables a very efficient FORTRAN coding of the differential equations governing the problem.
- Numerical stability of finite difference elements can be assured by reduced upwind integration.
- The resulting set of algebraic equations is positive definite, symmetric, and optimally conditioned.

Our final conclusion is that the fluid-structure continuum approach presented in the second sectior together with the least squares numerical technique, presented in the subsequent section, provides a useful and accurate tool for the prediction of the thermal-hydraulic behavior of shell- and-tube heat exchangers.

EXPLANATION OF TENSOR NOTATION

Del Operator in Cartesian Coordinates

$$(\nabla\varphi)_i = (\text{grad}\ \varphi)_i = \frac{\partial\varphi}{\partial x_i} \qquad \text{gradient}$$

$$\nabla \cdot \underline{A} = \text{div}\ \underline{A} = \frac{\partial A_j^{(1)}}{\partial x_j} \qquad \text{divergence}$$

$$(\nabla \cdot \underset{=}{A})_i = (\text{curl } \underset{=}{A})_i = \varepsilon_{ijk} \frac{\partial A_k}{\partial x_j}^{(2)} \qquad \text{curl}$$

$$(\nabla \underset{=}{A})_{ij} = \frac{\partial A_j}{\partial x_i} \qquad \text{gradient tensor}$$

$$(\underset{=}{A}\nabla)_{ij} = \frac{\partial A_i}{\partial x_j} \qquad \text{transposed gradient tensor}$$

$$\underline{v} \cdot \nabla \varphi = v_j \frac{\partial \varphi}{\partial x_j}$$

convective time derivative

$$(\underline{v} \cdot \nabla \underset{=}{A})_i = v_j \frac{\partial A_i}{\partial x_j}$$

$$\frac{\partial \varphi}{\partial t}, \quad \left(\frac{\partial \underset{=}{A}}{\partial t}\right)_i = \frac{\partial A_i}{\partial t} \qquad \text{local time derivative}$$

$$\dot{\varphi} = \frac{D\varphi}{Dt} = \frac{\partial \varphi}{\partial t} + \underline{v} \cdot \nabla \varphi$$

motional time derivative

$$\dot{\underset{=}{A}} = \frac{D\underset{=}{A}}{Dt} = \frac{\partial \underset{=}{A}}{\partial t} + \underline{v} \cdot \nabla \underset{=}{A}$$

$$\nabla^2 \varphi = \nabla \cdot (\nabla \varphi) = \frac{\partial}{\partial x_j}\left(\frac{\partial \varphi}{\partial x_j}\right)$$

Laplacian

$$(\nabla^2 \underset{=}{A})_i = (\nabla(\nabla \cdot \underset{=}{A}) - \nabla \cdot (\nabla \cdot \underset{=}{A}))_i = \frac{\partial}{\partial x_j}\left(\frac{\partial A_i}{\partial x_j}\right)$$

$$\nabla \underset{=}{A} = \frac{1}{2}(\nabla \underset{=}{A} + \underset{=}{A}\nabla) + \frac{1}{2}\begin{vmatrix} 0 & (\nabla \cdot \underset{=}{A}) & -(\nabla \cdot \underset{=}{A})_2 \\ -(\nabla \cdot \underset{=}{A})_3 & 0 & (\nabla \cdot \underset{=}{A})_1 \\ (\nabla \cdot \underset{=}{A})_2 & -(\nabla \cdot \underset{=}{A})_1 & 0 \end{vmatrix} \qquad \text{Gibbsian decomposition}$$

(1) Two repeated indices denote summation over these indices.

$$\text{(2) } \varepsilon_{ijk} = \begin{cases} 0 \text{ if } i = j \text{ or } i = k \text{ or } j = k \\ +1 \text{ if } (i, j, k) = (1, 2, 3) \text{ or } (2, 3, 1) \text{ or } (3, 2, 1) \\ -1 \text{ if } (i, j, k) = (2, 1, 3) \text{ or } (1, 3, 2) \text{ or } (3, 2, 1) \end{cases} \qquad \text{(Levi-Civitatensor)}$$

REFERENCES

1. Bird, R. B., Stewart, W. E., and Lightfoot, E. N., Transport Phenomena, Wiley International, New York (1960).
2. Batchelor, G. K., *An Introduction to Fluid Dynamics,* Cambridge University Press, London (1967).
3. Patankar, S. V., and Spalding, D. B., "A Calculation Procedure for the Transient and Steady-State Behaviour of Shell-and-Tube Heat Exchangers, in *Heat Exchangers: Design and Theory Sourcebook,*" N. Afgan and E. U. Schlünder (Ed.), McGraw–Hill, New York, 1974.
4. Kao, T. T., and Cho, S. M., "A Numerical Solution Method for the Prediction of Flow and Thermal Distribution in Shell-and-Tube Heat Exchangers," Joint ASME/AICHE 18th Nat. Heat Transfer Conf., San Diego (1979).

5. Sha, W. T., "An Overview on Rod-Bundle Thermal-Hydraulic Analysis, *Nuclear Engineering and Design*," 62: 1–24 (1980).
6. Domanus, H. M., Shah, V. L., and Sha, W. T., "Applications of the Commix Code Using the Porous Medium Formulation," *Nuclear Engineering and Design*, 62: 81–100 (1980).
7. Zijl, W., and De Bruijn, H., "Continuum Equations for the Prediction of Shell-Side Flow and Temperature Patterns in Heat Exchangers," *Int. J. Heat Mass Transfer*, 26: (3) 411–424 (1983).
8. Bear, J., *Dynamics of Fluids in Porous Media*, American Elsevier, New York, 1972.
9. Saffman, P. G., "Dynamics of Vorticity," *J. Fluid Mech*, 106: 49–58 (1981).
10. Idel'Chik, I. E., "Handbook of Hydraulic Resistance: Coefficients of Local Resistance and Friction," The U. S. Atomic Energy Commission and The National Science Foundation, AEC-tr-6630, Israel Program for Scientific Translations Ltd. IPST Cat. No. 1505 (1966).
11. Hubbert, M. K., "Darcy's Law and the Field Equations of the Flow of Underground Fluids," *Trans. SPE of AIME*, 207: 222–239 (JPT) (1956).
12. Forschheimer, P., "Wasserbewegung durch Boden," *Z. Ver. Deut. Ing.*, 45: 1782–1788 (in German) (1901).
13. Lamb, H., *Hydrodynamics*, Cambridge University Press, London, 1974.
14. Zijl, W., "Finite Element Methods Based on a Transport Velocity Representation for Groundwater Motion," Accepted for publication in Water Resources Research (1983).
15. Wong, H. Y., *Handbook of Essential Formulae and Data on Heat Transfer for Engineers*, Longman, London, 1977.
16. Truesdell, C., *The Kinematics of Vorticity*, Indiana University Press, Bloomington, 1954.
17. Tritton, D. J., *Physical Fluid Dynamics*, Van Nostrand Reinhold Company, New York, 1977.
18. Happel, J., and Brenner, H., *Low Reynolds Number Hydrodynamics*, Prentice-Hall, Inc., Englewood Cliffs, N. J., 1965.
19. Zienkiewicz, O. C., *The Finite Element Method*, 3rd ed., McGraw–Hill, U. K., 1977.
20. De Bruijn, J. G. M., and Zijl, W., Least Squares Finite Element Alanysis of the Steady State and Transient Thermal Hydraulic Behaviour of L. M. F. B. R. Heat Exchangers, Proceedings of the Second International Conference on Numerical Methods in Thermal Problems, Venice, Italy, July 1981.
21. De Bruijn, J. G. M., Zijl, W., Least Squares Finite Element Solution of Several Thermal Hydraulic Problems in a Heat Exchanger, Proceedings of the Second International Conference on Numerical Methods in Thermal Problems, Venice, Italy, July 1981.
22. Csendes, Z. J., "A Finite Element Method for the General Solution of Ordinary Differential Equations," *Int. J. Num. Meth. Engng.*, 9: 551–561 (1975).
23. Csendes, Z. J., "A FORTRAN Program to Generate Difference Formulas," *Int. J. Num. Meth. Engng.*, 9: 581–599 (1975).
24. Boisserie, J. M., "Generation of Two- and Three-Dimensional Finite Elements," *Int. J. Num. Meth. Engng.* 3: 327–347 (1971).
25. Ames, W. F., *Numerical Methods for Partial Differential Equations*, 2nd ed. Acad. Press, 1977, Chapter 3–3.
26. Peters F. J., "A Novel Organization of Finite Element Calculations and Homogeneous Structures," *Int, J. Num. Meth. Engng.*, 17 (9): 1430–1435 (1981).
27. Metcalf, M., *FORTAIN Optimization*, Acad. Press, London, 1982.
28. Spalding, D. B., *Physical and Mathematical Basis of PHOENICS*, CHAM Limitad 1982, Technical Report.
29. Forsythe G. E., and Wazov, W. R., *Finite Difference Methods for Partical Differential Equations*, Wiley, New York, 1960.
30. Varga, R. S., "Matrix Iterative Numerical Analysis," Wiley, New York, 1962.

CHAPTER 28

THERMAL AND HYDRAULIC DESIGN OF HAIRPIN AND FINNED-BUNDLE HEAT EXCHANGERS

G. P. Purohit

Consulting Engineer
Parkview Lane
Irvine, California

CONTENTS

INTRODUCTION

Hairpin (or double-pipe) heat exchangers find application in a variety of services requiring small heat-transfer surfaces, particularly when one stream is a gas or viscous liquid, or its flow rate is small. Economics favor hairpin exchangers for high-pressure service because of their small shell diameter.

These exchangers are suited for handling dirty streams because of the ease of cleaning and maintenance. They are flexible, with their heat-transfer surfaces readily shifted for changing process conditions and heat loads. Spare parts are interchangeable, and available off-the-shelf (as is original equipment).

Longitudinally-finned-bundle exchangers, which are similar to the hairpin exchanger, are also available as standard designs. They are used extensively for heating viscous liquids, and can serve in applications involving fluids having significantly different heat-transfer characteristics and fouling resistances.

This chapter has been reprinted by special permission from *Chemical Engineering* (May 16, 1983; pp. 62–70), McGraw–Hill, Inc., New York, NY 10020.

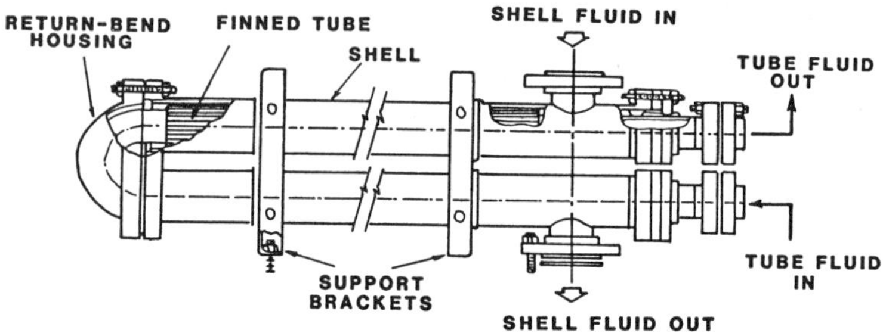

Figure 1. A hairpin (double-pipe) exchanger. These designs are easily stacked because inlets and outlets are close together.

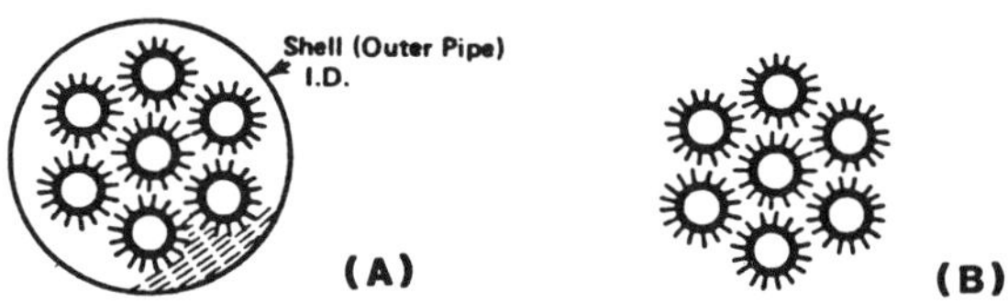

Figure 2. (A) Flow cross-section of a longitudinally finned multitube heat exchanger; (B) external heat transfer surface of a longitudinally finned multitube exchanger.

Table 1
Typical Dimensions of Hairpin Heat Exchangers

Parameter	Dimension range
Inside dia. of shell (outer pipe)	50–400 mm (2–16 in.)
Nominal length of section	1.5–12.2 m (5–40 ft)
Outside dia. of tube (inner pipe)	19–100 mm (0.75–4 in.)
Number of tubes	1–208
Number of fins/tube	0–72
Fin height	5.3–25 mm (0.21–1 in.)
Fin thickness	0.5–0.9 mm (0.02–0.035 in.)
Surface-area/section	0.25–150 m^2 (3–1,600 ft^2)

EXCHANGER CONFIGURATION AND FLOW DYNAMICS

A hairpin exchanger basically consists of concentric pipes (Figure 1). One fluid flows through the inner pipe (tube-side fluid), the other through the annular space (shell-side fluid). The inner pipes are connected by U-shaped return bends enclosed in a return-bend housing. Because the inlets and outlets are close together at one end, these exchangers can easily be stacked.

Hairpin exchangers are available in two basic types: single-tube and multitube. The first consists of a pipe within a pipe, and is generally called a double-pipe exchanger (Figure 1). In the multitube, a bundle of small-diameter tubes replaces the inner pipe. In either case, the tubes may be bare or have longitudinal fins (Figure 2A).

Fins increase heat-transfer surface per unit length (up to as much as 15 times), and reduce the size and number of exchangers required for a given service. Fins are usually welded to the tube but may be integrally formed (Figure 2B). They may be made of carbon steel or

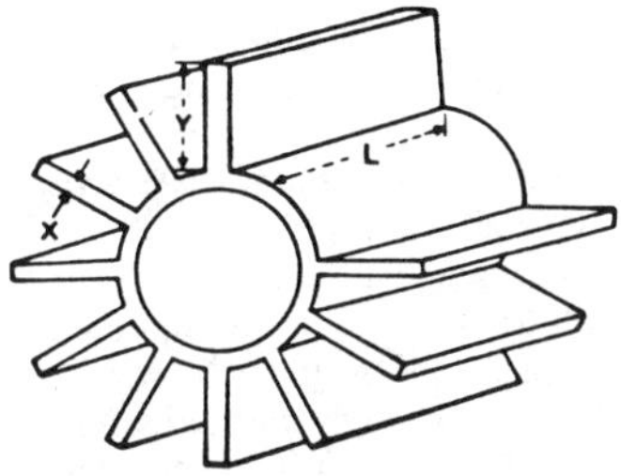

Figure 3. A continuous longitudinal fin.

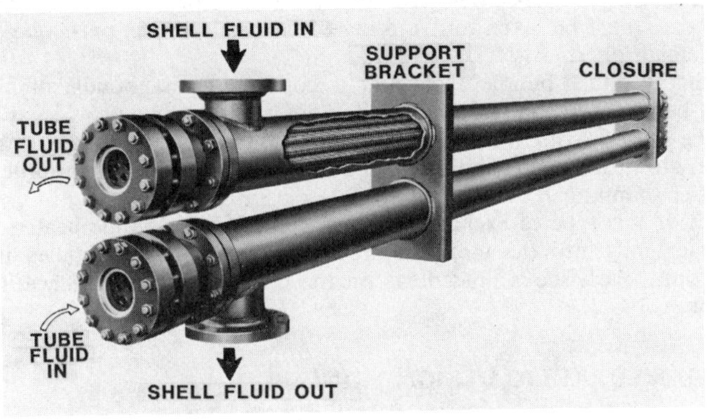

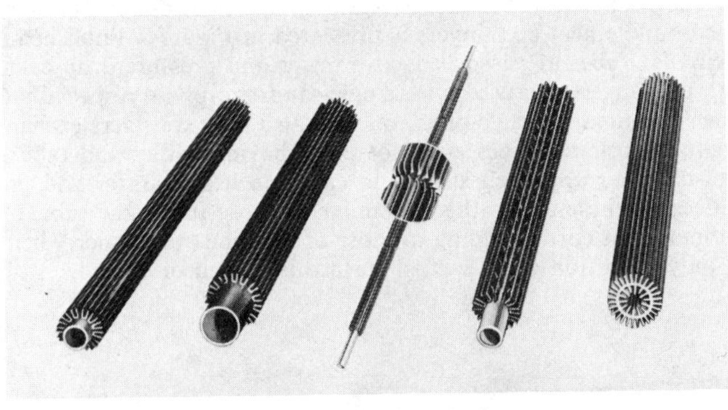

Figure 4. A multitube exchanger with longitudinally finned tubes.

of a variety of alloys, and may be cut and twisted to promote turbulent flow, or perforated for special applications. This discussion is limited to plain, continuous (uncut), longitudinal fins (see Figure 3 and 4).

Hairpin exchangers have shell diameters of 50–410 mm (2–16 in.) at nominal lengths of 1.5–12.0 m (5–40 ft). Surface area ranges from about 0.25–150 m^2 (3–1,615 ft^2) per section. Major dimensions are summarized in Table 1. Combinations of these dimensions yield a variety of design configurations available as standard equipment.

Flows in hairpin exchangers are countercurrent, which eliminates the F correction factor for the logarithmic mean temperature difference (LMTD) of the mixed countercurrent and cocurrent flows common in shell-and-tube exchangers. This results in up to 20% higher energy efficiency, correspondingly reducing the required heat-transfer surface. Pure countercurrent flows in a shell-and-tube exchanger would require a single-shell/single-tube pass configuration or a double-shell/double-tube pass configuration, at higher cost and greater complexity.

Pure countercurrent flow allows close temperature approaches and temperature crossovers. With conventional shell-and-tube exchangers of single-shell pass and multiple-tube passes, multiple shells in series would be required to eliminate crossovers.

In the shell-and-tube design, long temperature ranges (which usually result in temperature crosses) require either multiple shells in series or a single pass on both shell-side and tube-side. Using a single shell with a longitudinal baffle would result in high thermal stresses in the channel, tubesheet and shell, and in warpage of the baffle, causing leakage across it. An advantage of the hairpin exchanger is that its inner pipe can expand comparatively freely; however, care must be taken to avoid severe thermal stresses, particularly repetitive or cyclically reversing ones.

The longitudinally finned-bundle exchanger is composed of a bundle of finned tubes connected by U-bends and mounted in a shell. This configuration provides a single-shell pass and two tube passes. Available in shell diameters up to about 1,200 mm, this exchanger is used extensively to heat viscous liquids, such as asphalts and fuel oils, in order to reduce viscosity for easier pumping.

Two examples of this type of exchanger are tank-suction and inline heaters. The open end of the first extends into the tank (Figure 5A). The second is installed in pipelines (Figure 5B): In both, the viscous fluid flows on the finned side, and the heating medium through the tubes.

THERMAL AND HYDRAULIC DESIGN PRINCIPLES

A flowchart of the steps involved in the thermal and hydraulic design of longitudinal-flow hairpin and finned-bundle heat exchangers is presented in Figure 6. Published correlations for flow inside circular tubes are used in heat-transfer and pressure-drop calculations.

Although both exchangers are used with single- and two-phase (especially condensing) fluids, this chapter is limited to single-phase ones. It also only considers exchangers having continuous, longitudinal finned tubes, and not those having baffles and tube supports.

The equivalent-diameter approach is used to calculate heat transfer and pressure drop in the annulus, because these are in theory similar to those inside the tube, although the annulus has no dimensions corresponding to those of the tube (its geometry being distinctly dissimilar), and only a portion of its wetted surface is heated or cooled.

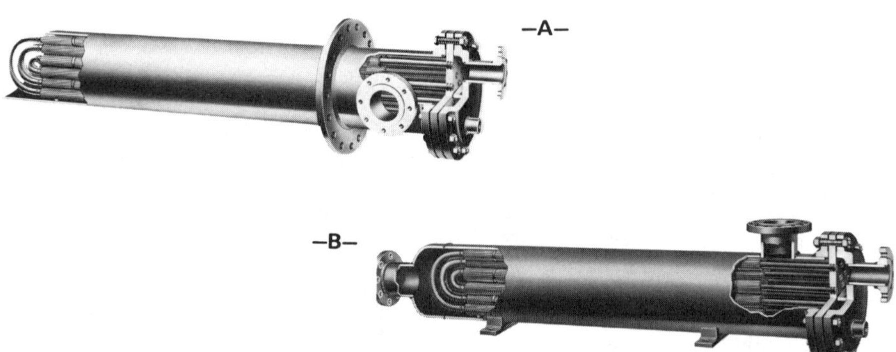

Figure 5. Suction and inline heaters which are often used to heat viscous fluids.

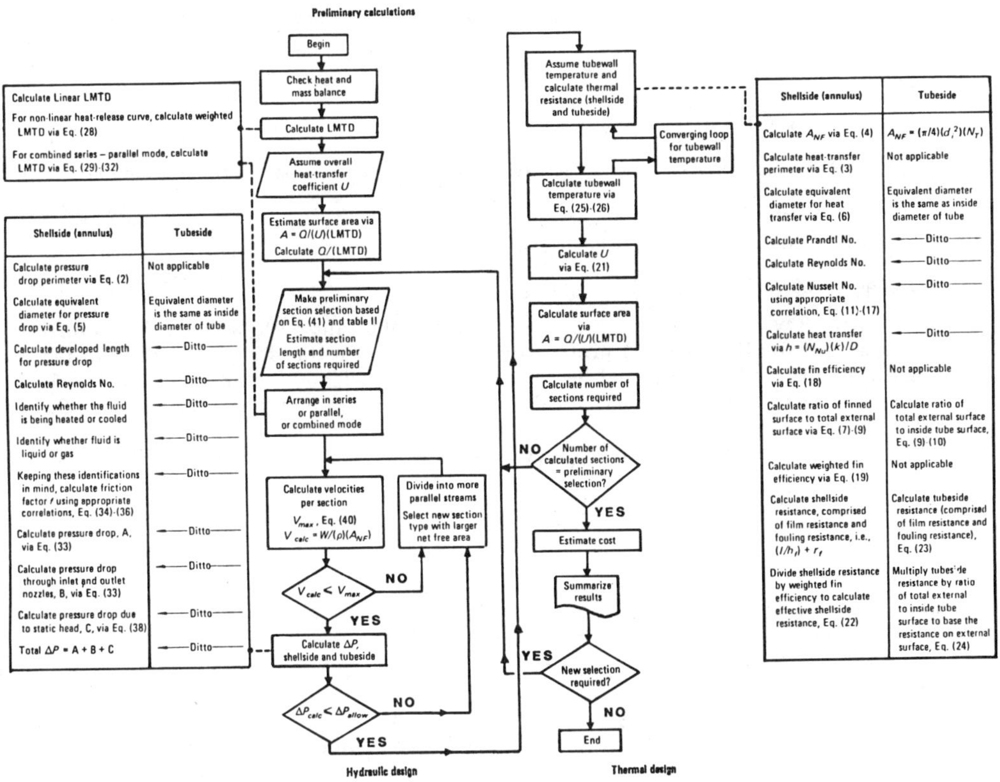

Figure 6. Flowchart showing logic sequence of steps involved in designing hairpin and bundle-type heat exchangers.

The simplest approach to applying inside-tube heat-transfer correlations to the annulus is to substitute equivalent diameter of the annulus for the inner-tube diameter:

$$d_e = 4r_h = \frac{4 \times \text{net free-flow area}}{\text{wetted (or heated or cooled) perimeter}} \tag{1}$$

This approximation is accepted for heat-transfer and pressure-drop calculations in annuli of small equivalent diameters. The validity of the equivalent-diameter approach has been substantiated by the results of experiments with finned annuli [1].

The total wetted perimeter of the annulus of a longitudinally finned heat exchanger (Figure 2A) is given by:

$$P_W = \pi(D_i + d_0 N_T) + 2(YN_F N_T) \tag{2}$$

And the heat-transfer perimeter of the annulus (Figure 2B) is given by:

$$P_{HT} = \pi d_0 N_T + 2(YN_F N_T) \tag{3}$$

The only difference between Equation 2 and 3 is the term D_i in Equation 2, which represents the inside diameter of the shell. This difference is due to the shell I.D. resistance to fluid flowing in the annulus. Such, however, is not the case for the heat-transfer perimeter, because heat transfer takes place from the total external surface and does not depend on the shell I.D.

The net free area in the annulus of a longitudinally finned exchanger (Figure 2A) is given by:

$$A_{NF} = (\pi/4)\,(D_i^2 - d_0^2 N_T) - (XYN_F N_T) \qquad (4)$$

Equivalent diameter based on wetted perimeter (also known as hydraulic diameter) is:

$$(d_e)_W = 4A_{NF}/P_W \qquad (5)$$

Equivalent diameter based on heat-transfer perimeter is:

$$(d_e)_{HT} = 4A_{NF}/P_{HT} \qquad (6)$$

Calculate Reynolds number, Graetz number, and the ratio D/L with the Equation 5 diameter. Use the Equation 6 diameter to calculate the heat-transfer coefficient from the Nusselt number and in evaluating the Grashof number. Slightly higher coefficients result from using the equivalent diameter based on wetted perimeter Equation 5 for heat-transfer calculations.

Total outside heat-transfer surface is the sum of the bare-tube and fin surfaces (Figure 3):

$$A_0 = 2N_T(\pi d_0 L - N_F LX) \qquad (7)$$

$$A_f = 2N_T N_F L(2Y + X) \qquad (8)$$

$$A_t = A_0 + A_f = 2N_T L(\pi d_0 + 2N_F Y) \qquad (9)$$

For bare tubes, $A_f = 0$, and $A_t = A_0$.

The inside heat-transfer surface (based on tube I.D.) is:

$$A_i = 2(\pi d_i L N_T) \qquad (10)$$

Area ratios A_f/A_t and A_0/A_t, or $1-(A_f/A_t)$, are needed for calculating weighted fin efficiency, and A_t/A_i for basing the inside (tube-side) heat-transfer coefficient on the total outside heat-transfer surface.

For standard equipment, calculated values for Equations 2 through 10 may be found tabulated in manufacturers' catalogs.

HEAT TRANSFER CORRELATIONS

For turbulent flow ($N_{Re} > 10{,}000$), with $L/D > 60$, and $0.67 < N_{Pr} < 120$, calculate N_{Nu} with the Sieder–Tate correlation [2]:

$$N_{Nu} = 0.023 N_{Re}^{0.8} N_{Pr}^{0.33} \varphi \qquad (11)$$

It is convenient to evaluate all fluid properties at the bulk temperature and apply a correction factor φ to account for the effects of property variations with respect to temperature. A common way of doing this is to lump the variant effects into a ratio of the dominant property evaluated at the surface temperature to that property evaluated at the bulk temperature.

For liquids, the dominant property is viscosity:

$$\varphi = (\mu_b/\mu_w)^n \qquad (12)$$

For both heating and cooling, $n = 0.14$, per experimental results [2]. Values of the correction factor n have also been derived analytically [3].

For gases, the key variable is absolute temperature:

$$\varphi = (T_b/T_w)^n \qquad (13)$$

For heating, n = 0.5, $T_b/T_w < 1$. For cooling, n = 0, $T_b/T_w > 1$. These values of n are based on gases for which temperature-dependent properties are similar to those of air between 310°–1,290° K [3].

For laminar flow ($N_{Re} < 2{,}100$), data are frequently correlated via three dimensionless numbers: N_{Nu}, based on arithmetic or logarithmic mean temperature difference; N_{Gz}, for the contribution due to forced convection; and N_{Gt}, for the contribution due to natural convection. At laminar flow, inside-tube heat transfer chiefly involves moderately to highly viscous liquids.

The Sieder–Tate correlation based on constant wall temperature can be used to calculate N_{Nu} for small tubes (I. D. up to 76 mm) and LMTD's (up to 315° K), and $N_{Gz} > 100$[2,4]:

$$N_{Nu} = 1.86[(N_{Re})(N_{Pr})(D/L)]^{0.33}\varphi \tag{14}$$

Equation 15 is a more general correlation, covering all diameters and ΔTs:

$$N_{Nu} = 1.86[(N_{Re})(N_{Pr})(D/L)]^{0.33}\varphi + 0.87[1 + 0.015(N_{Gr})]^{0.33} \tag{15}$$

For $N_{Gz} < 100$, use Hausen's correlation, developed on the basis of constant wall temperature [5]:

$$N_{Nu} = 3.66 + \left\{\frac{0.085[(N_{Re})(N_{Pr})(D/L)(A_{NF}/D^2)]}{1 + 0.047[(N_{Re})(N_{Pr})(D/L)(A_{NF}/D^2)]^{0.67}}\right\}\varphi \tag{16}$$

With liquids, φ is obtained as for turbulent flow, via Equation 12, for both heating and cooling [2]. With gases, the variation of fluid properties has only minor influence on inside-tube laminar-flow convective heat transfer, according to several investigations. For the temperature-ratio range of 0.5 – 2, n = 0 (φ = 1; T_x/T_b for heating, T_b/T_w for cooling) [6].

For transition flow ($2{,}100 < N_{Re} < 10{,}000$), heat transfer is unstable and difficult to define. Graphical methods relating laminar and turbulent correlations have been used. Hausen's equation is recommended [7]:

$$N_{Nu} = 0.116(N_{Re}^{0.67} - 125)N_{Pr}^{0.33}[1 + (D/L)^{0.67}]\varphi \tag{17}$$

The correction factor, φ, is determined via Equation 12.

Correlations for transition flow are discussed by Afgan and Schlunder [8].

DESIGN CORRELATIONS

Fin Performance

Fins present a special heat-transfer problem because ΔT cannot be specified independently. It is determined via thermal properties, fin geometry, and the heat-transfer coefficient, with the last being itself dependent on ΔT. Therefore, analyzing fins is difficult without resorting to simplified theory. Fin performance is gauged by fin efficiency, η_f, and fin effectiveness, ε_f.

Fin efficiency, η_f, is defined as the actual heat transferred by the fin divided by the heat that ideally would be transferred if the entire fin were at the base temperature. Based on simplifying assumptions, Equation 18 gives the efficiency of rectangular continuous longitudinal fins [9]:

$$\eta_f = \tanh(MY)/MY \tag{18}$$

Here, $M = (h_f P_f/k_f a_f)^{0.5}$; $a_f = XL$; and $P_f = 2(L + X) \cong 2L$. Substituting these values of a_f and P_f yields $M = (2h_f/k_f X)^{0.5}$.

Fin efficiency is inversely proportional to the finside film coefficient, h_f; for this reason, it is always advantageous to place the fluid having the lower heat-transfer coefficient or the higher fouling resistance on the finside of the exchanger. Equation 18 shows that fin efficiency decreases as fin height, Y, increases. Although efficiency provides some insight into how well a fin is contrived, it is not possible to design fins for a particular value of η_f.

Fin effectiveness, ε_f, is defined as the heat transferred through the fin, divided by the heat transferred through the same base surface having no fins. Effectiveness can be related to efficiency via: $\varepsilon_f = \eta_f$ (surface area of fin/cross-sectional area of fin). This indicates that increasing the fin height boosts fin effectiveness. Normally, the design objective is to achieve as high a fin effectiveness as possible. However, as noted previously, increasing the fin height adversely affects η_f, and thus, ε_f. Values of η_f and ε_f are useful in characterizing fins. Optimum fin performance depends on a tradeoff between the two.

Fin efficiency applies only to the finned portion of the total heat-transfer surface; the bare portion is regarded as having an efficiency of unity. Thus, efficiency of the total heat-transfer surface is expressed as a weighted fin efficiency:

$$(\eta_f)_{wtd} = \eta_f(A_f/A_t) + A_0/A_t \tag{19}$$

Here, $A_0/A_t = (1 - A_f/A_t)$.

Sometimes, it is necessary to know the tip temperature of the fin. The relationship between the fin-base and fin-tip temperatures is expressed by a tip temperature factor, E_T, defined as the ratio of the temperature driving force at the tip of the fin to that at its base [10]:

$$E_T = \frac{(T_b)_f - T_T}{(T_b)_f - T_w} = \frac{1}{\cosh(MY)} \tag{20}$$

Here, $(T_b)_f$ is the bulk temperature of the fluid on the finside, and T_w is the wall temperature at the bare surface (i.e., the fin base).

Temperature variation across the fin can sometimes be used to advantage in heating and cooling heat-sensitive fluids. In heating oil or asphalt with steam, for example, the average metal temperature on the finside is lower than at the bare surface. Therefore, heating takes place at a relatively low temperature, which reduces the tendency of the oil or asphalt to coke. If the fluid to be cooled is on the fin-side, cooling takes place at the higher temperature, and this helps prevent the solidification of viscous liquids at the tube wall.

Overall Heat-Transfer Coefficient

Based on total external surface, the overall heat-transfer coefficient is given by:

$$U = 1/(R_s + R_t + R_w) \tag{21}$$

The shell-side (fin-side) thermal resistance is:

$$R_s = [(1/h_f) + r_f]/(\eta_f)_{wtd} \tag{22}$$

The tubeside thermal resistance based on the inside tube surface is:

$$R_i = (1/h_i) + r_i \tag{23}$$

The foregoing resistance is referred to the total external surface as:

$$R_t = R_i(A_t/A_i) = [(1/h_i) + r_i]\,(A_t/A_i) \tag{24}$$

The tube wall resistance, R_w, equals the wall thickness divided by the thermal conductivity of the tube material; it is calculated only for bare tubes. In the case of finned tubes, the wall resistance is built into the fin efficiency, because the latter is based on uniform heat flux at the inside tube surface [9].

Tube Wall Temperature

As was explained previously, the tube wall temperature must be known to evaluate fluid properties at the surface temperature in order to calculate φ, the property-ratio correction factor. In the case of viscous liquid and laminar flow, the values of heat-transfer coefficients and pressure drops are greatly influenced by the tube wall temperature. Turbulent flow is also affected, but to a lesser extent.

For the purposes of this article, fluids of $\mu/_b > 1$ mPa (1 cP) are defined as viscous; for those of $\mu_b < 1$ mPa, μ_b/μ_w can be assumed to be unity and the effect of φ can be ignored.

The tube wall temperature depends on the bulk temperature of the two fluids and their film and fouling resistances. In calculating the wall temperature of bare tubes, the temperature difference across the metal wall is neglected, and the entire tube is considered to be at the temperature of the outside surface of the wall.

When the cold fluid is on the shellside, the wall temperature is given by:

$$\frac{T_w - (T_b)_C}{(T_b)_H - (T_b)_C} = \frac{R_s}{R_s + R_t} \tag{25}$$

When the cold fluid is on the tubeside, calculate the wall temperature via:

$$\frac{T_w - (T_b)_C}{(T_b)_H - (T_b)_C} = \frac{R_t}{R_s + R_t} \tag{26}$$

In Equation 25 and 26, the caloric temperatures of the hot and cold fluids may be used, instead of their bulk temperatures [4]. However, doing this involves additional calculations and does not significantly improve final results.

Tube wall temperature is computed via iteration. First, a wall temperature is assumed and the heat-transfer coefficients calculated. With these, the tube wall temperature is calculated by means of Equation 25 or 26. This temperature is used to recalculate the heat-transfer coefficients, with which another temperature is computed. The process is repeated until a fixed value for wall temperature is attained. This is shown in Figure 6 by a converging loop for tube wall temperature.

For finned tubes, no simple method is available for computing mean wall temperature for the combination of fins and base surfaces. A simplified approach is suggested for estimating the mean temperature of a finned tube as the arithmetic average of the fin-base (bare tube surface) and fin-tip temperatures (given by Equations 20, 25 and 26):

$$(T_w)_f = (T_w + T_T)/2 \tag{27}$$

Equation 27 will yield only approximate results, which may differ from those of more-precise methods, such as finite-difference schemes and the method in Reference 4. Discrepancies will be greater for high-fin tubes.

LMTD and True Temperature Difference

When the heat-transfer curve is linear with respect to temperature, LMTD is calculated conventionally. When the curve is not linear, it can be divided into zones that are approximately linear, and a weighted LMTD calculated via:

$$(\text{LMTD})_{wtd} = Q/\sum_{i=1}^{i=n} [Q_i/(\text{LMTD})_i] \tag{28}$$

Pure countercurrent flow offers the maximum thermal potential for heat transfer. When a large number of hairpin exchangers are required in a single service, it may not always be possible to connect both the shell sides and tube sides in series for pure countercurrent flow. An example would be when a large quantity of one fluid is to undergo a minor temperature change and a small quantity of the other a major change.

It may not be possible to circulate the large flow through the required number of hairpins connected in series, with the available pressure drop. In such a circumstance, the large flow may be manifolded into a number of parallel streams, and the small flow manifolded in series, in a series-parallel arrangement. The true overall mean temperature difference for such an arrangement is given by [9]:

$$(\Delta T)_{s-p} = S(T_{H_{in}} - T_{C_{in}}) \tag{29}$$

The dimensionless parameter S is defined as:

$$S = [W(C_p)_H(T_{H_{in}} - T_{H_{out}})]/UA(T_{H_{in}} - T_{C_{in}}) \tag{30}$$

The value of S depends on the number of hot and cold streams and their series-parallel arrangement:

For 1-series hot fluid and n_1-parallel cold streams:

$$S = \frac{(1-P_1)}{\left(\frac{n_1 R_1}{R_1 - 1}\right) \ln\left[\left(\frac{R_1 - 1}{R_1}\right)\left(\frac{1}{P_1}\right)^{1/n_1} + \frac{1}{R_1}\right]} \tag{31}$$

Here, $P_1 = (T_{H_{out}} - T_{C_{in}})/(T_{H_{in}} - T_{C_{in}})$, and $R_1 = (T_{H_{in}} - T_{H_{out}})/n_1(T_{C_{out}} - T_{C_{in}})$.

For 1-series cold stream and n_2-parallel hot streams:

$$S = \frac{(1-P_2)}{\left(\frac{n_2}{1 - R_2}\right) \ln\left[(1-R_2)\left(\frac{1}{P_2}\right)^{1/n_2} + R_2\right]} \tag{32}$$

Here, $P_2 = (T_{H_{in}} - T_{C_{out}})/(T_{H_{in}} - T_{C_{in}})$, and $R_2 = n_2(T_{H_{in}} - T_{H_{out}})/(T_{C_{out}} - T_{C_{in}})$.

The F correction factor must be applied to LMTD in the case of the longitudinal-flow bundle-type exchanger because the flow is not purely countercurrent. It is calculated using curves for single-shell pass, and for two, or multiples of two, tube passes, as given in TEMA for shell-and-tube exchangers [11].

Pressure Drop

Available pressure drop is usually a governing factor in heat-exchanger design. Normally, the higher it is, the smaller the exchanger size and cost. Total pressure drop includes fluid friction (the largest contributor), entrance and exit losses, static head, and moment-change effects.

Frictional pressure drop is calculated via the Fanning equation:

$$(\Delta P)_F = 4f\left(\frac{\rho v^2}{2}\right)\left(\frac{L_h}{D}\right) N = 4f\left(\frac{G^2}{2\rho}\right)\left(\frac{L_h}{D}\right) N \tag{33}$$

Here, L_h is the hydraulic length. For the shellside, $L_h = 2L$ for hairpin exchangers, and $L_h = L$ for bundle-type exchangers. For the tubeside, $L_h = 2L + L_U$ for both hairpin and bundle exchangers. Here, L_U is the equivalent straight length of the U-bend.

The friction factor, f, in Equation 33 depends on the Reynolds number.

In turbulent flow ($N_{Re} > 3{,}000$), f is strongly related to tube roughness.

With smooth pipe and $3{,}000 < N_{Re} < 3{,}000{,}000$, the empirical equation of Koo (with isothermal representation of heating and cooling) is used [12, 13]:

$$f = 0.0014 + (0.125/N_{Re}^{0.32})\psi \tag{34}$$

With clean commercial iron and steel pipes, the equation of Wilson et al. (with isothermal representation of heating and cooling) is used [14]:

$$f = 0.0035 + (0.264/N_{Re}^{0.42})\psi \tag{35}$$

For liquids, the viscosity correction factor in Equation 34 and 35 is taken to be: $\psi = (\mu_b/\mu_w)^m$, with $m = 0.25$ for both heating and cooling [15]. Values of m are also derived analytically and given as a function of N_{Pr} by Kays [3]. For nonviscous liquids, $\mu_b < 1$ mPa, and the effect of ψ can be ignored.

For gases, data on friction coefficients are limited. Kays recommends: $\psi = (T_b/T_w)^m$, with $m = 0.1$ for both heating and cooling [3]. This value of m is based on a gas for which temperature dependence on properties is similar to that of air in the temperature range 310°–1,920° K.

With fully developed laminar flow ($N_{Re} < 2{,}000$), the friction factor can be calculated via the Hagen-Poiseuille equation (with isothermal representation for heating and cooling):

$$f = (16/N_{Re})\psi \tag{36}$$

For liquids, the properties correction factor is used as analytically derived by Diessler [16]: $\psi = (\mu_b/\mu_w)^m$, with $m = -0.58$ for heating ($\mu_b/\mu_w > 1$), and $m = -0.5$ ($\mu_b/\mu_w < 1$).

For gases, the value of ψ given by Worsoe–Schmidt is used in the range 277° – 1,666° K: $\psi = (T_b/T_w)^m$, with $m = -1.0$ for both heating and cooling [17].

Transition flow ($2{,}000 < N_{Re} < 3{,}000$) is unstable and difficult to define. The pressure drop in this regime can be estimated by linear interpolation between laminar and turbulent Reynolds numbers.

Nozzle pressure drop is estimated by simply calculating kinetic energy:

$$(\Delta P)_n = K(\rho v_n^2/2) = K(G_n^2/2\rho) \tag{37}$$

Here, K is the velocity head, 1.0 at inlet and 0.5 at outlet.

Static head is calculated as:

$$(\Delta P)_s = \rho H \tag{38}$$

For horizontal shells, H is the total difference in elevation between the inlet and outlet nozzles.

Momentum-change pressure drop is calculated as:

$$(\Delta P)_m = G^2(1/\rho_{out} - 1/\rho_{in}) \tag{39}$$

With single-phase liquids and gases, the momentum-change pressure drop can be ignored.

GUIDELINES TO SELECTION

Maximum velocities in heat exchangers are limited by the effects of erosion, pressure drop, and vibration. These velocities are sometimes used as a guide during preliminary calculations for designing an exchanger within specified limits of pressure drop. Because of the effects of erosion, pressure drop, and vibration, the maximum velocity of water is limited to between 2.4–3.0 m/s for carbon steel tubes, and up to 4–5 m/s for some alloy materials. Copper alloy tubes permit lower velocities and titanium tubes higher velocities than these limits. A generalized equation based on these limits is used by some manufacturers to set maximum velocities for liquids and gases:

$$v_{max} = C/\rho^{1/2} \tag{40}$$

Here, C = 122 with SI units (v in m/s and ρ in kg/m^3); C = 100 with English units (v in ft/s and ρ in lb/ft^3), although some manufacturers assume this C to be 1.6 times higher.

A preliminary evaluation begins with assuming

1. An overall heat-transfer coefficient, with U dependent on fluid properties.
2. Allowable pressure drops.
3. Fouling resistances.

With these, the required heat-transfer surface is estimated.

A preliminary selection is next made of the type of hairpin section, depending on the ratio of duty/LMTD (also known as the UA product):

$$Q/(\text{LMTD}) = UA \cong 53{,}000 - 106{,}000 \text{ W/°K} \tag{41}$$

This ratio is used as a guide for limiting the size of the exchanger. However, other factors influence this limit. Criteria for the preliminary selection of the size of hairpin exchangers on the basis of this ratio are summarized in Table 2.

Having selected the section type and determined the necessary heat-transfer surface, the number of sections required for the chosen length can be calculated and arranged in series or in parallel, or in a combination—depending on flow rates, available pressure drops, and other considerations.

In the selection of section type, the choice of finned tubes over bare tubes depends on several factors. For the optimum design, the shell-side and tube-side effective film-coefficients referred to the same surfaces should be as high as possible, and be approximately of the same magnitude, within specified limits of pressure drops and fouling resistances. However, this is not always possible; the fluid film coefficients may differ significantly because of one or more of the following reasons:

1. The physical properties (density, viscosity, specific heat, and thermal conductivity) of the two fluids may differ greatly (as when one is a viscous liquid and the other is water).
2. The physical properties of the fluids may be similar, but the allowable pressure drops may differ greatly (i.e., one gas is at high and the other at low pressure).
3. Flow rates may differ greatly.
4. The fouling resistance of one of the fluids is high.

When the foregoing limitations are applicable, finned tubes are generally preferred. The fluid having the lower heat-transfer coefficient is placed on the fin side to keep the UA product of Equation 41 about equal by combining a high A with a low U.

Table 2
Preliminary Selection of Hairpin Exchanger Based on *Q*/LMTD

Q/LMTD, W/K [Btu/(h) (°F)]	Shell Dia. of Multitube Section (mm (in.))*
Greater than 79,000 [150,000]	305–406 (12–16)
Between 53,000 and 79,000 [100,000 and 150,000]	203–406 (8–16)
Between 26,500 and 53,000 [50,000 and 100,000]	152–254 (6–10)
Between 10,500 and 26,500 [20,000 and 50,000]	102–203 (4–8)†
Less than 10,500 [20,000]	51–102 (2–4)†

* *Length and number of sections required can be calculated based on total estimated surface area and known surface area per unit length of section*
† *Also consider double-pipe (single tube) unit.*

Table 3
Cost Breakdown of Hairpin Heat Exchangers

Component	Percent of Total Exchanger Cost			
	Double-pipe		Multitube	
	Bare	*Finned*	*Bare*	*Finned*
Tubes	15–25	20–40	30–40	35–60
Shell	40–55	40–50	35–50	25–40
Closure	20–35	20–35	15–30	15–30

Evaluation involves the following steps:

1. Calculate the velocities per section.
2. If the calculated velocity is larger than the maximum velocity Equation 40, divide the flow into parallel streams, or select a section type having a greater net free area.
3. When $v_{calc} \leqq v_{max}$, calculate the pressure drop. If the calculated ΔP is greater than the allowable ΔP, further split the flow into parallel streams, or select a section type having a larger net free area.
4. When $\Delta P_{calc} \leqq \Delta P_{max}$, proceed with the thermal calculations to determine the heat-transfer coefficients and the required surface area.
5. Depending on the percent over or under design, and on the difference between the calculated and allowable pressure drops, vary the exchanger length, number of sections, and other parameters to select a section type that will give the optimum design.

For cost estimating purposes, hairpin exchangers can be broken down into three major components: shell, tubes, and closure. The cost of the shell and tubes increases with longer exchanger length, whereas that of the closure remains fixed. This makes longer exchangers more economical.

In Table 3 cost breakdowns in percent of total exchanger cost are given for carbon steel at design pressures to 4,000 kPa (600 psi) and temperatures to 600° K, and for shell diameters to 200 mm in nominal lengths of 1.5–9.1 m. These percentages may vary with manufacturers due to differences in design patents and mechanical standards.

NOTATION

- A heat-transfer surface (m^2)
- A_{NF} net free area (m^2)
- a_f cross-section of fin (m^2)
- C numerical constant; Equation 40
- C_p specific heat (kJ/(kg) (°K))
- cosh hyperbolic cosine
- D characteristic diameter in heat-transfer and pressure-drop correlations (m)
- D_i inside diameter of shell (outer pipe) (m)
- d diameter of tube (inner pipe) (m)
- d_e equivalent diameter of annulus (m)
- E_T tip temperature factor; dimensionless
- F correction factor for LMTD; dimensionless
- f Fanning friction factor; dimensionless
- G mean mass velocity (kg/(m^2) (s))
- g acceleration due to gravity (9.81 m/s^2)
- H difference in elevation between inlet and outlet nozzles (m)
- h heat-transfer coefficient (W/(m^2) (°K))
- K velocity-head coefficient; dimensionless
- k thermal conductivity (W/(m) (°K))
- k_f thermal conductivity of fin (W/(m) (°K))
- L nominal length of exchanger section, also length of longitudinal fin (m)
- M parameter of fin efficiency; Equation 18 (m^{-1})
- m exponent of correction factor ψ for pressure-drop correlations
- N number of exchanger sections in series
- N_F number of fins per tube
- N_{Gt} Grashof number; $(\rho^2 g\beta\Delta TD^3)\mu^2$; dimensionless
- N_{Gz} Graetz number; $WC_p/kL = = (N_{Re}N_{Pr}(D/L))(A_{NF}/D^2)$; dimensionless
- N_{Nu} Nusselt number; hD/k; dimensionless
- N_{Pr} Prandtl number; $\mu C_p/k$; dimensionless
- N_{Re} Reynolds number; GD/μ; dimensionless
- N_T number of tubes in one leg of hairpin exchanger; number of tube holes in tubesheet of bundle exchanger (twice the number of U-bends), Equation 2, 3 and 4; or

	number of U-tubes in bundle exchanger, Equation 7, 8, 9 and 10.
n	exponent for correction factor φ in heat-transfer correlations; Equation 12
n_1, n_2	number of parallel cold and hot streams; Equations 31 and 32
P	perimeter (m)
P_1, P_2	temperature ratios, Equations 31 and 32
P	pressure drop (kPa)
Q	heat load or duty (W)
R	thermal resistance $((m^2)(°K)/W)$
R_1, R_2	temperature ratios; Equations 31 and 32
S	dimensionless parameter for true temperature difference of series-parallel arrangement; Equation 31
T	absolute temperature (°K)
ΔT	temperature difference (°K)
tanh	hyperbolic tangent
U	overall heat-transfer coefficient based on total external surface $(W/(m^2)(°K))$
v	mean stream velocity (m/s)
W	mass flow rate (kg/s)
X	fin thickness (m)
Y	fin height (m)

Greek Symbols

β	temperature coefficient of volume expansion or contraction for $N_{Gt}(1/\rho_{out} - 1/\rho_{in})/(1/\rho_b)(T_{out} - - T_{in})(°K^{-1})$
ε	fin effectiveness
η	fin efficiency
μ	dynamic viscosity (mPa)
ρ	mass density (kg/m^3)
τ	fouling resistance $((m^2)(°K)/W)$
τ_h	hydraulic radius (m)
φ	property-ratio correction factor for nonisothermal flow in heat-transfer correlations; Equation 11
ψ	property ratio correction factor for nonisothermal flow in pressure-drop correlations; Equation 34

Subscripts

allow	allowable
b	bulk
C	cold
calc	calculated
F	frictional
f	fins or finside
H	hot
HT	based on heat-transfer perimeter, Equations 3 and 6
h	hydraulic
i	inside of tubes
in	inlet
max	maximum
m	momentum
n	nozzles
0	outside of bare tubes
out	outlet
S	static head
s	shell-side or fin-side
s-p	series-parallel arrangement
T	fin tip
t	total external or tubeside coefficient based on total external surface
U	equivalent of U-bend
W	wetted; Equations 2 and 5
w	wall of bare tube
wtd	weighted

REFERENCES

1. DeLorenzo, B., and Anderson, E. D., "Heat Transfer and Pressure Drop of Liquids in Double Pipe Fintube Exchangers," *Trans. ASME,* Vol. 67 (1945), p. 697.
2. Sieder, E. N., and Tate, G. E., "Heat Transfer and Pressure Drop of Liquids in Tubes," *Ind. and Eng. Chem.,* Vol. 28, No. 12 (1936), pp. 1429–1434.

3. Kays, W. M., "Convective Heat and Mass Transfer," McGraw-Hill, Inc., New York, 1966.
4. Kern, D. Q., and Kraus, A. D., "Extended Surfaces Heat Transfer," McGraw-Hill, Inc., New York, 1972.
5. Perry, R. H., and Chilton, C. H., *Chemical Engineers' Handbook*, 5th ed., McGraw-Hill, Inc., New York, 1973.
6. Kays, W. M., and Nicoll, W. B., *Trans. ASME, J. of Heat Transfer*, Vol. 85 (1963), pp. 329–338.
7. Hausen, H. Z., *V. D. K. Beiheft Verfahrenstechnik*, Vol. 4 (1943), pp. 91–98.
8. Afgan, N., and Schlunder, E. U., *Heat Exchangers: Design and Theory Source Book*, McGraw-Hill, Inc., New York, 1974.
9. Kern, D. Q., *Process Heat Transfer*, McGraw-Hill, Inc., New York, 1950.
10. Lienhard, J. H., *A Heat Transfer Textbook*, Prentice-Hall, Inc., Englewood Cliffs, NJ, 1981.
11. "Standards of Tubular Exchanger Manufacturers Assn.," TEMA, Inc., New York, 1978.
12. Koo, E. C., Sc. D. thesis in chemical engineering, Mass. Inst. of Technology, 1932.
13. Drew, T. B., Koo, E. C., and McAdams, W. H., "The Friction Factor for Clean Round Pipe," *Trans. AIChE*, Vol. 28 (1932), pp. 56–72.
14. Wilson, R. E., McAdams, W. H., and Seltzer, M., "The Flow of Fluids Through Commercial Pipelines," *Ind. and Eng. Chem.*, Vol. 14 (1922), p. 105.
15. Rohsenow, W. M., and Hartnett, J. P., *Handbook of Heat Transfer*, McGraw-Hill, Inc., New York, 1973.
16. Diessler, R. C., Natl. Advisory Committee for Aeronautics, TN 2410, Washington, DC, July 1941.
17. Worsoe-Schmidt, P. M., Technical Report No. 247–8, Thermosciences Div., Dept. of Mech. Eng., Stanford University, Stanford, CA, Nov. 1964.

CHAPTER 29

ECONOMICS OF SHELL-AND-TUBE HEAT EXCHANGERS

G. P. Purohit

Consulting Engineer
Parkview Lane
Irvine, California

CONTENTS

INTRODUCTION

This cost-estimating method covers the fixed-tube-sheet, U-tube, splitring floating-head, and pull-through floating-head designs. For each of these types, it considers various configurations, as classified by the Tubular Exchanger Manufacturers Assn. (TEMA) (Figure 1) [1].

All the parameters are modeled as a function of shell inside diameter. The cost equations include the effects of shell-and-tube diameters, tube length, tube pitch and layout angle, tube construction (welded or seamless) and wall gage including average or minimum tolerances. Separate curves for shell- and tube-side take into account the effect of design pressures. Relative costs of various metals and alloys are extensively tabulated. A flowchart of the estimating procedure is presented in Figure 2.

ACKNOWLEDGEMENT

This chapter has been reprinted by special permission from *Chemical Engineering* (August 22, 1983, pp. 56–67), McGraw-Hill, Inc., New York, N. Y. 10020.

Using this method the various TEMA configurations, can be evaluated (e.g., shell diameters of 12 to 148 in., tube lengths of 8 to 36 ft, tube diameters of 3/4 to 2 in., tube-wall thicknesses of 10 to 20 Birmingham Wire Gage (BWG), tube passes of 1 to 8, shell design pressures of 100 to 2,800 psig, tube-side design pressures of 100 to 2,500 psig, construction of a wide range of alloys (also clad construction), and specialty exchangers with high-flux tubing).

The results of the cost-estimating method are compared against actual purchased prices, and found to be accurate to within ±10% in 60% of the cases, and to within ±15% in the remaining 40%. However, the overall accuracy for all the exchangers is within 2%.

DESIGN CONSIDERATIONS

Fixed-Tubesheet Design

Straight tubes in the fixed-tubesheet exchanger are secured at both ends in tubesheets welded to the shell. TEMA front-end stationary-head types A, B, or N and rear-end head types L, M, or N can be used with this type of exchanger (Figure 1). An example of a TEMA-type configuration is AEL. These exchangers are generally inexpensive. Besides having fewer parts, they are slightly smaller and of simpler construction than other types.

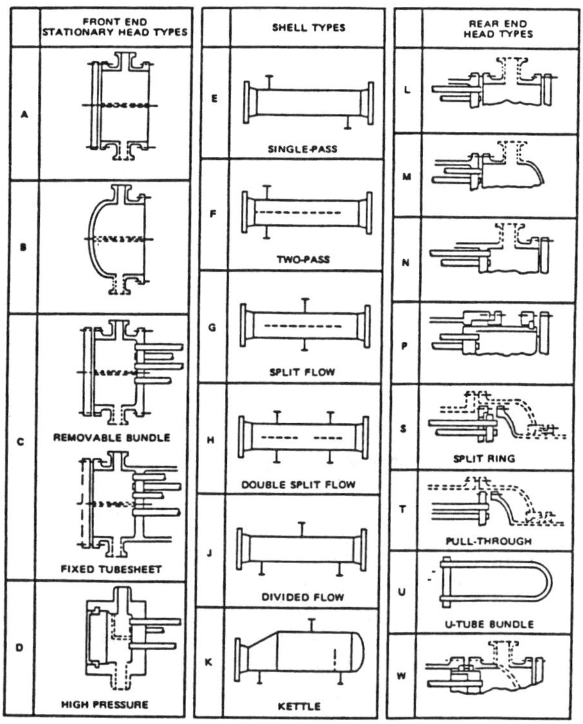

Figure 1. TEMA exchanger nomenclature along with cost factors for shell types and front-end and rear-end head types relative to respective designated baseline type.

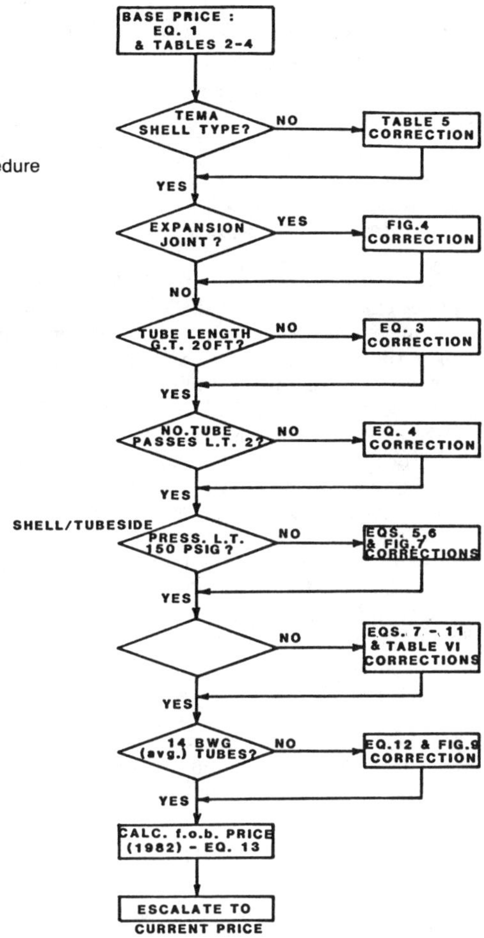

Figure 2. Flowchart for the cost-estimating procedure for shell-and-tube heat exchangers.

The tubes can be easily cleaned mechanically. However, because the tube bundle cannot be removed, the shell-side cannot be cleaned in this way. Chemical cleaning is possible but sometimes unsatisfactory. Because of the fixed tubesheet, an expansion joint in the shell may be necessary to prevent high stresses resulting from differential thermal expansion between the tubes and shell. The cost of the expansion joint may offset the price advantage of this type of exchanger over the others.

If corrosion occurs, fixed-tubesheet exchangers are difficult and expensive to maintain. Retubing may be expensive and time-consuming, and even impractical. Although bonnet or flat covers may be fitted at either end, flat covers are generally specified when frequent tube-cleaning is expected.

U-Tube Design

Both ends of U-shaped tubes are fastened to a single tubesheet in the U-tube design. This eliminates the problem of differential thermal expansion, because the tubes are free to expand and contract. This design may be fitted with any of the TEMA front-end stationary-head types (Figure 1). An example of a TEMA configuration is AEU. U-tube exchangers are moderate in cost.

Floating-Head Designs

Floating-head rear-end configurations (Figure 1) may be split-ring (TEMA type S), pull-through (TEMA type T), externally-sealed-tubesheet (TEMA type W), or outside-packed (TEMA type P). Types S and T are more common than types W and P. An example designation is BES.

The floating head eliminates the need for an expansion joint. Because the tube bundle can be removed, both the shell- and tube-side can be cleaned mechanically. Floating-head exchangers are more expensive than the fixed-tube and U-tube types.

The split-ring exchanger's shell-cover diameter is larger than the shell diameter, usually by about 3–4 in. In the pull-through design, the shell cover and shell are of the same diameter. Whereas the shell cover of the split-ring exchanger can be removed, this need not be the case with the pull-through design.

Split-ring construction permits tubes to be close to the inside of the shell. In the pull-through design, space must be provided around the periphery of the floating tubesheet to accommodate bolts. For the same number of tubes, therefore, the shell diameter of the pull-through exchanger is about 2 in. larger than that of the split-ring type. Because the seal is not visible externally in either type, a leak is not easily detected.

In the externally sealed design (TEMA type W), the shell- and tube-side streams are individually sealed and separated by a lantern ring. This type is the least expensive of the floating-head designs, and can be fitted with an A, B, or C type front head. It is limited to a maximum of two tube passes, 300 psi and 375° F.

In the outside-packed design (TEMA type P), a skirt attached to the floating tubesheet extends through the back of the shell. Several layers of packing in a packing gland seal the space between the skirt and shell. A packing failure is readily detected. This design permits high tube-side pressures and more than two tube passes. However, the packing tends to limit shell-side fluids to less than 150 psi and 300° F.

COST ESTIMATING PRINCIPLES

Costs of shell-and-tube exchangers vary according to: TEMA type; shell diameter; tube length, diameter, construction (welded or seamless), gage, pitch and layout; number of tube passes; shell- and tube-side design pressures; and materials of construction. Each of these factors is related to a baseline exchanger, which is defined in Table 1.

Unit labor costs are higher than unit material costs for small-diameter exchangers. As shell diameter increases, however, the unit material cost rises and the unit labor cost declines. For this reason, exchangers of large shell diameters are more economical on a $/ft^2 basis. This makes shell diameter a convenient parameter on which to base estimates of shell-and-tube exchanger costs.

Analysis of the purchase prices of many shell-and-tube exchangers received by means of competitive bids during the early part of 1982 resulted in the development of an exponential decay type Equation 1 for estimating the base cost of the exchanger defined in Table 1.

$$b = \left\{\frac{6.6}{1 - e^{[(7 - D_i)/27]}}\right\} pfr \qquad (1)$$

where D_i = shell I. D. or bundle diameter of a kettle reboiler (in.)
p = cost multiplier for tube O. D., pitch and layout angle (Table 2)
f = cost multiplier for TEMA-type front head (Table 3)
r = cost multiplier for TEMA-type rear head (Table 4)

Table 1
Definition of Baseline Shell-and-Tube Exchanger

Parameter	Base designation
Tubes	Welded c.s., 14 BWG, avg. wall
Nominal tube length	20 ft
Number of tube passes	1 or 2
Shellside design pressure, psig	≦ 150
Tubeside design pressure, psig	≦ 150
Material of construction	All carbon steel

Base price, which includes cost of ASME Sec. VIII, Div. 1 code stamp and of exterior paint, was developed via evaluation of bids by U.S. manufacturers.

Table 2
Cost Multiplier for Tube O.D., Pitch and Layout Angle

Tube O.D. in. × Pitch in.	Multiplier, P Triangular Pitch (30- or 60-deg.)	Square Pitch (45- or 90-deg.)
5/8 × 25/32	0.62	Not common
3/4 × 15/16	0.80	Not common
3/4 × 1	0.85	1.0 (base)
7/8 × 13/32	0.87	Not common
1 × 1 1/4	0.98	1.16
1 1/4 × 1 9/16	1.23	1.45
1 1/4 × 1 37/64	1.29	1.49
1 1/2 × 1 7/8	1.47	1.73
1 1/2 × 1 57/64	1.56	1.80
1 3/4 × 1 3/16	1.72	2.03
1 3/4 × 1 13/64	1.81	2.13
2 × 2 1/2	1.97	2.32
2 × 2 17/32	2.08	2.45

Table 3
Cost Multiplier for TEMA Front-end Stationary-Head Type

	TEMA Type	Multiplier, f
	No front end (special case)	0.95
B	Bonnet (integral cover) (base)	1.0
A	Channel and removable cover	1.02–1.03
N	Channel integral with tubesheet, and removable cover	1.05
C	Channel integral with tubesheet, and removable cover	1.06–1.07
D	Special high-pressure closure	1.5–1.7

Table 4
Cost Multiplier for TEMA Rear-end Head Type

	TEMA Type	Multiplier, r
S	Floating head with backing device (base)	1.0
M	Fixed tubesheet, like B stationary head	0.8
L	Fixed tubesheet, like A stationary head	0.83
N	Fixed tubesheet, like N stationary head	0.85
U	U-tube bundle	0.9
T	Pull-through floating head	1.05
P	Outside-packed floating head	1.04
W	Externally sealed floating tubesheet	1.02

Equation 1 is plotted in Figure 3 as a function of shell diameter for TEMA BES exchangers having 3/4-in. O. D. tubes × 1-in. pitch × 45-deg. layout. Figure 3 shows that as shell diameter increases, the price in $\$/ft^2$ decreases and tends to flatten out at a shell diameter of about 50 inch. As shell diameter increases, the $\$/ft^2$ rises exponentially, reaching infinity at a shell diameter of 7 in. Thus Equation 1 can theoretically be used for shell diameters as small as 8 in. However, at shell diameters of less than 12 in., double-pipe and multitube exchangers are more economical [2]. Equation 1 has been checked against purchase prices of exchangers ranging in diameter from 12 to 148 in. and found to be satisfactory.

Values for cost multipler p in Equation 1 are listed in Table 2 for commonly used tube diameters, pitches, and layouts. It can be seen from Table 2 that as the tube diameter increases, the cost multiplier p increases. This is due to the fact that the tube pitch is at least 1.25 times the tube O. D. This large diameter tube provides less surface area (for a given shell diameter) than small diameter tubes and increases cost almost proportionately. It can also be seen from Table 2 that triangular tube patterns (30 or 60 deg.) are more economical than square-pitch patterns (45 or 90 deg.). This is because square pitch patterns accomodate fewer numbers of tubes than triangular pitch in a given shell diameter. Values of p not given in Table 2 can be approximated by means of Equation 2:

$$p = 0.75(p_i)^2 (a/d_0) \tag{2}$$

In Equation 2, p_i = tube pitch, in., and a is a constant. For 45- or 90-deg. layouts, a = 1; for 30- or 60-deg. layouts, a = 0.85.

For a U-tube exchanger having a 35-in. shell diameter, 3/4-in. tube O. D., 1-in. tube pitch, 45-deg. tube layout, and a TEMA type A front end, p = 1.0 (Table 2), f = 1.02 (Table 3), and r = 0.9 (Table 4). Via Equation 1, b = $\$10/ft^2$. To arrive at a cost estimate for this exchanger, this base price must be corrected by means of other parameters, which are now discussed.

Correction for Shell Type

Equation 1 is based on TEMA shell type E. For shell types other than E, the base price calculated via Equation 1 must be adjusted by a costcorrection, C_s, from Table 5. For example, for a kettle reboiler (TEMA type K), C_s = 0.3. Therefore, the estimated price of a kettle reboiler is $\$13/ft^2$ – i.e. $\$10/ft^2 + (\$10/ft^2 \times 0.3)$.

Correction for Expansion Joints

If differential thermal expansion is expected, an expansion joint may be incorporated into the shell of a fixed-tubesheet exchanger. This, of course, increases the cost of an exchanger.

Two types of expansion joints are common: the flanged-and-flued and the bellows. The first is cheaper but limited to 3/16-in. expansion. For greater thermal expansion, the bellows type is usually selected.

Cost corrections for expansion joint as a fraction of base price are plotted in Figure 4 as a function of shell diameter. For example, consider a fixed-tubesheet exchanger with a 30-in. shell diameter at a base price of $\$10/ft^2$. A flanged-and-flued-type joint would add $\$1.4/ft^2$ to the base price, and a bellows type would boost the price by $\$2.9/ft^2$.

Table 5
Cost Corrections for TEMA Shell Type

	TEMA Shell Type	Correction, C_S
	Open-tube exchanger (no shell) (special case)	−0.2
E	One-pass (base)	0
J	Divided flow	0
X	Cross-flow	0
G	Split-flow	0.05–0.1
H	Double split-flow	0.1–0.15
F	Two-pass with longitudinal baffle	0.15–0.2
K	Kettle-type reboiler	0.25–0.35

Correction for Tube Length

On a $\$/ft^2$ basis, long, skinny exchangers are more economical than short, fat ones, because of the labor cost and material scrapped in cutting tubes to required lengths. However, the labor cost decreases somewhat as shell diameter increases, because of the larger number of tubes involved.

For the baseline exchanger, tubes are of 20-ft nominal length (Table 1). Cost-correction for tube lengths less than 20 ft can be calculated via Equation 3:

$$C_L = \left[1 - \frac{L}{20}\right]\left[1.5 - \frac{0.002083(D_i - 12)}{1 - (L/20)}\right] \tag{3}$$

If tube length, L, is 20 ft, $C_L = 0$. If $L > 20$ ft, according to Equation 3, C_L is negative, indicating that the unit cost of the exchanger ($\$/ft^2$) is less than for an exchanger having 20-ft tubes; however, no credit should be taken for tube lengths longer than 20 ft. If $L > 20$ ft, $C_L = 0$. Equation 3 is plotted in Figure 5. Although the curves only extend to shell diameters of 60 in., they are valid beyond this limit.

For an exchanger with a 35-in. shell diameter and 12-ft-long tubes, $C_L = 0.55$, via Equation 3. If the exchanger base price is $\$10/ft^2$, the shorter tubes would boost the price by $\$5.5/ft^2$.

Correction for Tube Passes

Larger numbers of tube passes increase exchanger cost by hiking labor costs and reducing the available heat-exchange surface. Exchangers can have as many as 20 tube passes.

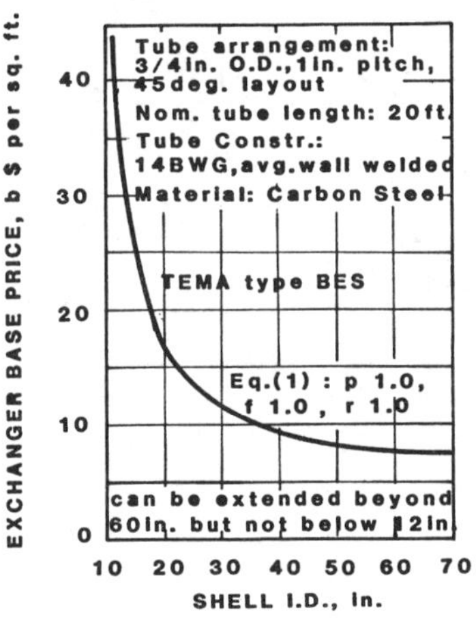

Figure 3. Base price of TEMA BES exchangers based on 1982 dollars.

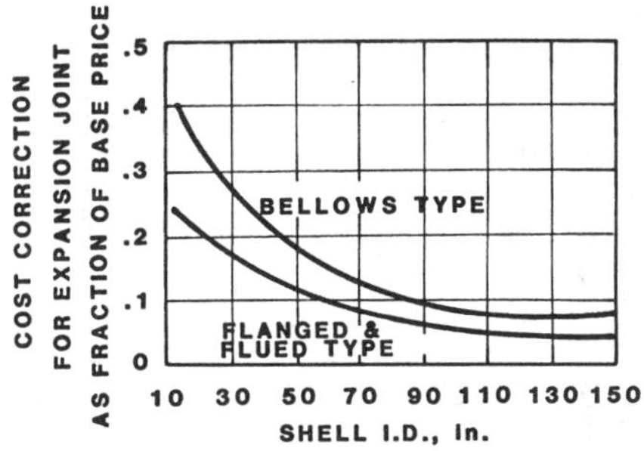

Figure 4. Cost corrections for shell expansion joints.

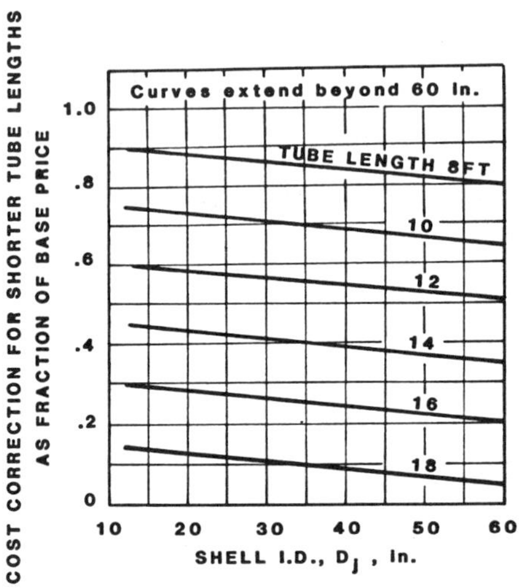

Figure 5. Cost corrections for tube lengths shorter than 20 ft.

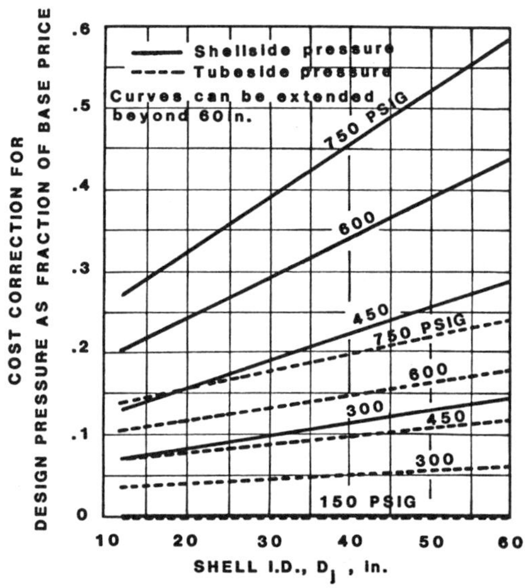

Figure 6. Cost corrections for shell- and tube-side design pressures (C_{PT}—tube-side; C_{PS}—shell-side).

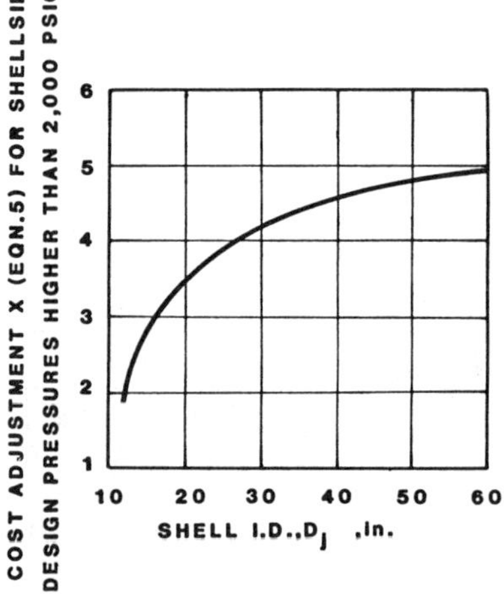

Figure 7. Additional cost correction for shell-side design pressures exceeding 2,000 psig.

Although based on limited data, Equation 4 is suggested for estimating the extra cost for more than two tube passes:

$$C_{N_{TP}} = (N_{TP} - 1)/100 \quad (4)$$

Here, N_{TP} = actual number of tube passes. For single- or two-tube passes, $C_{N_{TP}} = 0$.

For a $10/ft^2 base-price exchanger with 8 tube passes, the cost of the additional passes is $0.7/ft^2.

Correction for Design Pressures

Exchanger price rises with increasing design pressure. For shell-side design pressures higher than 150 psi, calculate the cost fraction to be added via Equation 5:

$$C_{PS} = [(P_{DS}/150) - 1]\,[0.07 + 0.0016(D_i - 12)] + X \quad (5)$$

For tube-side pressures greater than 150 psi, the cost fraction of base price to be added is calculated via:

$$C_{PT} = ((P_{DT}/150) - 1)\,(0.035 + 0.00056(D_i - 12)) \quad (6)$$

For shell-side and tube-side design pressures < 150 psig, $C_{PS} = 0.0$ and $C_{PT} = 0.0$.

Equation 5 (with X = 0) and Equation 6 are plotted in Figure 6. Although the curves extend to only 60-in. shell diameter, they can be extrapolated beyond this limit. As Figure 6 shows, the cost effect of shell-side pressure is twice that of tube-side, up to 2,000 psig, and even greater beyond 2,000 psig. The X term in Equation 5 is an adjustment factor for shell-side design pressures higher than 2,000 psig. Values of X (based on limited data) are plotted in Figure 7 as a function of shell diameter. For shell-side pressures equal to, and less than, 2,000 psig, X = 0.

For an exchanger with a 35-in. shell diameter and designed for a shell-side pressure of 450 psi and a tube-side pressure of 600 psi, $C_{PS} = 0.2136$ and $C_{PT} = 0.1436$. Therefore, the higher shell-side pressure adds $2.14/ft^2, and the tube-side pressure $1.44/ft^2, to the $10/ft^2 base cost of the exchanger.

MATERIALS OF CONSTRUCTION AND COST CORRECTIONS

Equation 1 is based on carbon-steel construction. Alloy construction can alter the price of an exchanger significantly, depending on the alloy's cost and its quantity in the exchanger. Sometimes various parts of the exchanger are made of different alloys, making cost estimating even more difficult.

Exchanger costs consist essentially of labor, profit and materials. The costs of materials can be broken down into those for shell, tubes, tube-sheets, and channels. These costs are related to shell diameter for a lowpressure (shell- and tubeside design pressures up to 150 psig), carbon-steel exchanger with a 3/4-in. tube O.D. × 1-in. tube pitch × 45- or 90-deg. layout in Figure 8.

Figure 8 shows that, with increasing shell diameter, unit labor cost declines, whereas the costs of the shell, channel, and tubesheet remain fairly constant—at approximately 10%, 6%, and 4%, respectively—relative to the total cost of the base carbon-steel exchanger. The unit tube cost, however, increases with shell diameter. The cost of carbon-steel tubes, y, as a fraction of the total price of the base carbon-steel exchanger can be calculated via Equation 7:

$$y = [0.129 + 0.0016(D_i - 12)]\,[d_0/0.75(p_i)^2 a] \quad (7)$$

In Equation 7, a = 0.85 for a triangular tube-pitch pattern (30 or 60 deg.), and a = 1.0 for a square tube-pitch pattern (45 or 90 deg.).

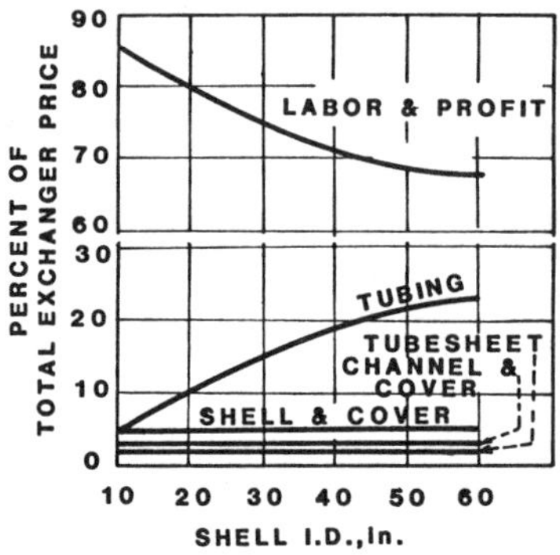

Figure 8. Typical price breakdown for low-pressure carbon-steel exchangers.

For a carbon-steel exchanger having a 35-in. shell diameter and tubes of 3/4-in. O.D. in a 1-in. square-pitch pattern, y = 0.166. In the case of a $10/ft^2$ base-price exchanger, this means that the cost contribution of the tubing is $10/ft^2 \times 0.166$, or $1.66/ft^2$. Similarly, the cost contribution of the shell is $10/ft^2 \times 0.1$, or $1/ft^2$; that of the channels is $10/ft^2 \times 0.06$, or $0.6/ft^2$; and that of the tubesheets is $10/ft^2 \times 0.04$, or $0.4/ft^2$. The total material cost is therefore: $1.66/ft^2 + 1/ft^2 + 0.6/ft^2 + 0.4/ft^2 = 3.66/ft^2$. The balance of $6.34/ft^2$ is attributed to labor costs and profit.

Cost corrections for tube material calculated from Equation 7 run higher—from 4% to 10%—in the shell diameter range of 20–12 in. The purpose of this is to provide conservative estimates for exchangers of small diameters with alloy tubes.

Knowing the cost relationships of the carbon-steel tubes, shell, channels, and tubesheets to the base carbon-steel exchanger, one can calculate the higher cost due to other-than-carbon-steel construction for these exchanger components by multiplying the relative cost of the other material to that of carbon steel.

Calculate the cost correction for tube material via:

$$C_{mt} = y(M_1 - 1) \tag{8}$$

The cost correction for shell material by:

$$C_{ms} = 0.1(M_2 - 1) \tag{9}$$

The cost correction for channel material from:

$$C_{mc} = 0.06(M_2 - 1) \tag{10}$$

And the cost correction for tubesheet material by:

$$C_{mts} = 0.04(M_2 - 1) \tag{11}$$

The value of y in Equation 8 is calculated via Equation 7. In Equation 8, M_1 is the ratio of tubing cost (welded or seamless) relative to welded carbonsteel tubes of the same diameter, wall thickness, and cut length (i.e., for welded carbon-steel tubes, $M_1 = 1$ and $C_{mt} = 0$). In Equations 9 through 11, M_2 is the relative cost of plate materials to carbon-steel plate of the same dimensions. For carbon steel, $M_2 = 1.0$ and C_{ms}, C_{mc}, and C_{mts} are all equal to zero. The cost ratios M_1 and M_2 for various materials used in heat exchangers listed in Table 6 are calculated from early 1982 prices.

Tubing prices depend on market conditions and such factors as required cut lengths and quantity of purchase. Price differences among manufacturers can differ by as much as 40%. Prices of low-alloy tubing vary significantly, depending on whether construction is welded

Table 6
Price Ratios, Related to Carbon Steel, of Various Materials for Shell-and-Tube Heat Exchangers

Material	M_1, Tubing-Price Ratio Relative to Welded Carbon-Steel Tubes		M_2, Price Ratio for Shell, Channel and Tubesheet Relative to Carbon Steel
	Welded	Seamless	
Carbon steel	1.0 (base)	2.50	1.0 (base)
Carbon steel, low alloys:			
1/2 Mo	1.04	2.60	1.04
1 Mo	1.05	2.70	1.05
2 1/2 Ni	1.15	2.90	1.15
3 1/2 Ni	1.20	3.10	1.20
2 Ni-1 Cu	—	3.30	1.30
Carbon steel, chromium-molybdenum alloys:			
1 Cr-1/2 Mo	Not standard	2.60	2.00
1 1/4 Cr-1/2 Mo	Not standard	2.70	2.10
2 1/4 Cr-1 Mo	Not standard	3.00	2.40
3 Cr-1 Mo	Not standard	3.20	2.50
5 Cr-1/2 Mo	Not standard	4.40	3.50
7 Cr-1/2 Mo	Not standard	5.50	Not standard
9 Cr-1 Mo	Not standard	6.10	Not standard
Stainless steels:			
304	2.80	6.50	3.70
304L	3.00	7.50	4.70
309	5.80	14.50	7.70
310	7.40	12.00	9.80
310L	7.60	12.40	10.10
316	4.70	10.10	6.20
316L	4.80	11.00	6.40
317	8.10	13.30	8.10
317L	8.30	13.60	8.30
321	4.20	9.50	5.60
329 (Carpenter 7 Mo)	10.50	17.20	10.50
330	7.90	12.90	9.50
347	5.50	13.70	7.30
405	6.00	15.00	6.90
410	6.90	17.20	7.90
430	5.40	10.60	6.20
439	5.00	11.20	5.80
444 (Alloy 18-2)	7.80	8.80	9.00
446	4.70	10.00	5.40
904L (Sandvik 2RK-65)	15.30	19.20	17.00
Sandvik 2RE-69	—	14.50	—
Sandvik 3RE-60	—	10.10	—
Sandvik 253 MA	—	12.70	—
Sandvik SAF 2205	—	11.80	—
Sanicro 28	16.10	20.20	18.20
E-brite-26-1 (XM-27)	9.00	—	10.00

Material	M_1, Tubing-Price Ratio Relative to Welded Carbon-Steel Tubes		M_2, Price Ratio for Shell, Channel and Tubesheet Relative to Carbon Steel
	Welded	Seamless	
Ferralium (Alloy 255)	12.00	23.90	14.00
Carpenter 20 Cb-3	15.10	—	16.00
Carpenter 20 Mo-6	18.90	—	—
AL-6-X	12.20	—	—
AL-29-4	12.00	—	—
AL-29-4-2	11.80	—	—
AL-29-4-C	5.0	—	—
Nickel 200	—	20.90	18.40
Monel 400 (Alloy 400)	—	15.50	14.50
Inconel 600 (Alloy 600)	19.40	—	15.30
Inconel 625 (Alloy 625)	—	32.70	27.40
Incoloy 800 (Alloy 800)	11.00	21.80	9.00
Incoloy 800H (Alloy 800H)	—	18.00	—
Incoloy 825	—	23.50	—
Hastelloy B-2	34.90	48.60	38.40
Hastelloy C-4	28.70	40.00	31.30
Hastelloy C-276	29.10	38.10	31.00
Hastelloy G	15.30	24.70	18.10
Hastelloy X	16.70	27.10	21.30
Titanium (Grade 2)	11.00	22.00	11.00
Titanium (Grade 7)	21.00	42.00	—
Titanium (Grade 12)	14.00	28.00	—
Zirconium 702	35.00	43.70	36.80
Zirconium 705	39.00	48.70	40.00
Aluminium	Not standard	1.60	1.60
Naval rolled brass	Not standard	3.50	3.50
Admiralty	Not standard	3.60	3.60
Aluminium brass	Not standard	3.70	3.70
Aluminium bronze (5%)	Not standard	4.10	4.10
Copper (arsenical or deoxidized	Not standard	4.20	4.20
90-10 cupro-nickel	3.50	4.60	4.60
70-30 cupro-nickel	4.20	5.50	5.50
Union Carbide high-flux tubing:			
SA-214, welded	4.40	—	—
SA-334-1, welded	4.70	—	—
SA-334-3, seamless	—	9.00	—
SA-214, fluted, welded	7.00	—	—
SA-334-1, fluted, welded	7.40	—	—
SA-210, fluted, seamless	—	8.40	—

or seamless. As can be seen in Table 6, the price of seamless carbon-steel tubing is 2.5 times that of welded carbon-steel tubing. As the material price increases, however, the cost difference between seamless and welded construction diminishes to as low as 10%. Therefore, depending upon whether tubes are welded or seamless, appropriate value of M_1 should be chosen from Table 6.

The M_2 ratios in table 6 for shell, channel and tubesheet materials are based on plate prices. The accuracy of M_1 and M_2 values is expected to be within $\pm 20\%$. The ratios should be updated periodically.

Exchanger components constructed of carbon steel or low alloys are sometimes clad with a more-expensive material, rather than making them entirely out of the more-expensive material. In such cases, the average of the material ratios of the base and clad materials should be used for costestimating purposes. If, for example, a carbon-steel ($M_2 = 1.0$) tubesheet is clad with Monel ($M_2 = 14.50$), the average $M_2 = 7.75$. For plate thicknesses $> 3/4$ inch clad construction is more economical. For plate thicknesses $\leqq 1/2$ inch, solid construction is generally more economical.

Correction for Tube Gage

The price of tubing depends on the tube-wall gage (because cost is proportional to the weight of the material), and on whether an average or minimum gage is specified. Because only positive tolerances are allowed, minimum-wall tubing will cost more than average-wall tubing by about 5%–10%.

This cost-estimating method is based on 14 Birmingham Wire Gage (BWG), average-gage wall. For gages other than this, the cost correction as a fraction of base price (C_g) is calculated from Equation 12:

$$C_g = y(g-1) \tag{12}$$

Values for y are calculated via Equation 7. Values for g, the cost multiplier for tube gage, are obtained from Figure 9, for average- or minimum-gage tubing walls.

Cost Premium for Special Cases

This estimating method is based on costs from competitive bids for a large number of shell-tube heat exchangers. When exchanger shops are very busy, or only a few exchangers are purchased, or rush delivery is required, exchanger prices will be higher than when bought via competitive bids. Although it is difficult to generalize about the size of this premium (because it will vary with circumstance and manufacturer), it cannot be ignored. A premium of 10% of the final estimated cost of the exchanger is suggested.

Estimating F.O.B. Exchanger Price

After the exchanger base price and all the corrections for alternatives have been calculated, the free-on-board (f.o.b.) January 1982 price of the particular exchanger can be estimated via Equation 13:

$$E_b = [b(1+C_T)A]N \tag{13}$$

In Equation 13, C_T represents the sum of all the applicable cost corrections for alternatives, as calculated via Equations 3 through 12 and obtained from Table 5 and Figure 4.

Because equipment costs generally change from the date that an order is placed to when it is shipped, estimators usually incorporate an escalation factor into cost estimates. When manufacturers submit bids, they usually include formulas for adjusting prices.

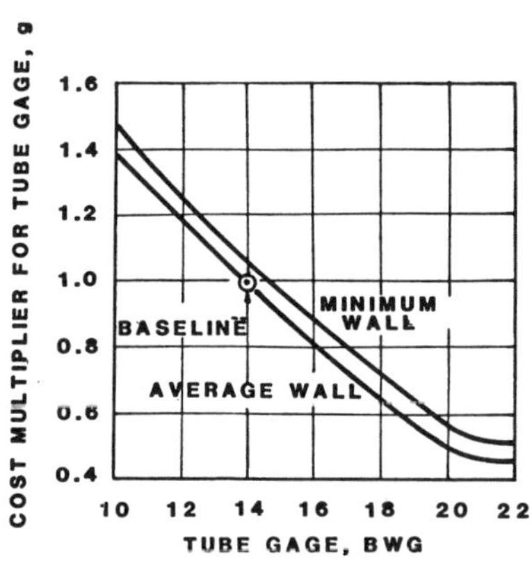

Figure 9. Tube-gage cost multiplier which depends on specifying average and minimum wall tolerances.

An exchanger price estimated by this method can be escalated to a price for a date beyond January 1982 by means of an escalation index. One such index that applies to process heat exchangers is the fabricated equipment component of the Chemical Engineering Plant Cost Index. This component is updated monthly.

Other possibilities are the Marshall & Swift Equipment Cost Index, the Nelson Refinery Cost Index (or the materials-cost component), and the U.S. Bureau of Labor Statistics General Purpose Machinery and Equipment Cost Index.

An estimated corrected base-price of an exchanger, E_b, is escalated from the January 1982 date to another date by multiplying it by the ratio of the values of the Fabricated Equipment component on the two dates.

When exchanger price quotations more current than those of the base date are readily available, an escalation index specific to exchangers can be determined. For comparable exchangers, the sum of the average of the lowest and second-lowest current quotations, divided by the sum of cost estimates for an equal number of exchangers, provides a highly reliable escalation index. An index determined in this manner should remain valid for about six months. Taking the average of lowest and second-lowest bids is recommended, because lowest bids are sometimes not realistic.

Estimating Delivered Cost

This method estimates exchanger cost f.o.b. the manufacturer's shop. The cost of the exchanger delivered to the plantsite depends on the distance between the shop and the site, the means of transportation, and the size and weight of the exchanger. Without knowledge of these details, the delivered cost cannot be accurately estimated. However, increasing the f.o.b. cost by 1%–4% propvides an approximate delivered cost. It is difficult to predict foreign prices based on the U.S. market because they are affected largely by the rate of exchange and the country's desire for the U.S. currency. For example, Italian prices are found to be 25%–35% lower than U.S. prices. The same is true for Japan and is expected to be true for Korean prices. The German prices on the other hand are found to be about 5% higher than U.S. prices.

A simple example that illustrates the method is presented in the worksheet shown in Figure 10. The problem is to estimate the f.o.b. cost of a pull-through floating-head, TEMA-type AJT, exchanger purchased in December 1982.

The single shell's diameter is 46 in. Nominal tube length is 16 ft, and surface area is 2,905 ft^2. The tubes are 1-in.-O.D., welded construction with a minimum wall gage of 12 BWG, arranged in a 1–1/4 in. rotated-square pitch and in four passes. Shell- and tube-side design pressures are 50 and 450 psig, respectively. The shell, channels, and tubesheets are constructed of 304 stainless stell, and the tubes of Sandvik 2RK-65.

These data are entered into Part I of the worksheet.

In Part II, the base price, b, is first calculated via Equation 1 and Tables 2, 3, and 4. Next, in order, are calculated: shell-type correction, C_s (Table 5); expansion-joint correction, C_x (Figure 4); tube-length correction, C_L via Equation 3; tube-pass correction, C_{NTP}, via Equation 4; shell-side-pressure correction, C_{PS}, via Equation 5 and Figure 7; tube-side-pressure correction, C_{PT}, via Equation 6; tube-material correction, C_{mt}, via Equation 8, with $y = 0.156$, from Equation 7; shell-material correction, C_{ms}, via Equation 9; channel-material correction, C_{mc}, via Equation 10; tubesheet-material correction, C_{mts}, via Equation 11; and tube-gage correction from Equation 12. All the foregoing corrections are then added to arrive at C_T.

Next, the estimate of the f.o.b. price, based on January 1982 prices, E_b, is calculated via Equation 13. Because the December value of the Fabricated Equipment component of the CE Plant Cost Index was 324.5—essentially its January 1982 value—the escalation index, E_I, is taken to be 1. Therefore, the December 1982 current-price estimate, E_c, is $131,000.

I. Exchanger Data Required For Cost Estimating:

TEMA Type AJT
Inside Diameter of Shell (Bundle Dia., D_j, inch) 46 (For Kettle Reboiler)
Nominal Tube Length, L, ft 16
Tube Outside Dia., d_o, inch 1.0
Tube Pitch, P_j, inch 1.25
Layout Angle, deg. 45
Tube Gage 12 Bwg/min
No. Tube Passes, N_{TP} 4
Type of Expansion Joint Not Applicable
Tube Construction Welded

Shell Side Design Pressure, PSIG 50
Tube Side Design Pressure, PSIG 410
Materials:
Tubes – SANVIK 2RK-65
Shell – 304 SS
Channel or Bonnet – 304 SS
Tube Sheet – 304 SS
Surface Area per Shell, A, ft² 2905
No. of Shells, N 1.0

Escalation Index EI – ____
Date – ____
Source – ____

II. Cost Estimating Calculations

1. Base Price, $b = \left[\frac{6.6}{1 - e^{(7 - D_j)/27}}\right] \times p \times f \times r = \_\_ \times \_\_ \times \_\_ \times \_\_ = \_\_ \frac{\$}{Ft^2}$ (p: Table II; f: Table III; r: Table IV)

2. **Corrections for Alternates as Fraction of Base Price, *b***

Correction for Shell Type (Table V)	C_s = ____		
Correction for Expansion Joint (Fig. 4)	C_x = ____		
Correction for Tube Length (Eq. (3))	C_L = ____		
Correction for Number of Tube Passes (Eq. (4))	C_{NTP} = ____		
Correction for Shell Side Design Pressure (Eq. (5) & Fig. 7)	C_{PS} = ____		
Correction for Tubeside Design Pressure (Eq. (6))	C_{PT} = ____		
Corrections for Materials (Table VI)			
– Tubes, Eq. (7) and Eq. (8), (y = 0.156)	C_{mt} = ____	M_1 =	15.3
– Shell, Eq. (9)	C_{ms} = ____	M_2 =	3.7
– Channel or bonnet, Eq. (10)	C_{mc} = ____	M_2 =	3.7
– Tubesheet, Eq. (11)	C_{mts} = ____	M_2 =	3.7
Correction for Tube Gage, Eq. (12) & Fig. 9	C_g = ____	g =	1.26
Total Correction	C_T = ____		

3. **Exchanger Price E_b (F.O.B., Based on January 1982 Prices)**

$E_b = b(1 + C_T) \times A \times M = \_\_ \times (\_\_ + \_\_) \times \_\_ \times \_\_ = \_\_ \$$

4. **Current Exchanger Price, E_c (F.O.B.)**

$E_c = E_b \times E_i = \_\_ \times \_\_ = \_\_ \$$

Figure 10. Example worksheet for cost-estimating shell-and-tube exchangers.

Regardless of the analytical approach taken, and the level of details incorporated into a cost-estimating method, exchanger price estimates based on it may vary widely, because of the many factors that affect exchanger costs. Indeed, an evaluation of a large number of competitive bids received from five manufacturers revealed differences between lowest and highest quoted prices of from 15% to as much as 60% (and, in some cases, even higher). Thus, it is important to always be aware that cost estimates are only "estimates."

Estimates derived by means of the method presented are compared against actual prices of 108 exchangers purchased during the first half of 1982. Included were exchangers of different TEMA types, geometry, and metallurgy with various design conditions. The results, summarized in Table 7, indicate that the estimates of all the individual exchangers

Table 7
Comparison of Purchase Prices (early 1982) and Estimated Prices of Shell-and-Tube Heat Exchangers

Exchanger Ab.	TEMA Type	Shell I.D., in. x Nominal Tube Length (in.)	Tube O.D. (in.) x Pitch, x Layout Angle (deg.)	BWG (Avg./Min.)	Tubes, Welded, Seamless	No. of Tube Passes, N_{TP}	Design Pressures, (psig) Shell-Tube	Metallurgy Tube/Shell/Channel or Bonnet/Tubesheet	No. of Shells	Surface Area per Shell, (ft²)	Exchanger FOB Price: Purchase ($)	Purchase ($/ft²)	Estimated ($)	Estimated ($/ft²)	Accuracy (%)
E-101	BEM (NO)*	12 × 144	3/4 × 15/16 × 30	14 (A)	W	4	180/130	All C.S.	1	185	7,700	41.62	7,600	41.08	− 1.3
E-102	BEM (NO)	13 × 144	3/4 × 15/16 × 30	14 (A)	W	2	100/100	All C.S.	1	310	9,600	30.97	10,500	33.87	+ 9.4
E-103	BEM (NO)	13 × 144	3/4 × 15/16 × 30	14 (A)	W	2	150/130	All C.S.	1	305	10,500	34.43	10,300	33.77	− 1.9
E-104	BEM (NO)	13 × 192	3/4 × 15/16 × 30	14 (A)	W	2	480/100	All C.S.	1	335	10,000	29.85	10,900	32.54	+ 9.0
E-105	BEM (FF)	15 × 96	3/4 × 15/16 × 30	14 (A)	W	1	200/100	All C.S.	1	305	11,600	38.03	10,700	35.08	− 7.8
E-106	BEM (NO)	15 × 120	3/4 × 15/16 × 30	14 (A)	W	1	130/100	All C.S.	1	370	11,300	30.54	10,600	28.65	− 6.2
E-107	BEM (FF)	17 × 192	3/4 × 15/16 × 30	14 (A)	W	4	130/100	All C.S.	1	590	13,700	23.22	12,200	20.74	−11.0
E-108	BEM (NO)	19 × 144	3/4 × 15/16 × 30	14 (A)	W	4	100/150	All C.S.	1	610	12,000	19.67	11,600	19.02	− 3.3
E-109	BEM (NO)	19 × 144	3/4 × 15/16 × 30	14 (A)	W	1	130/100	All C.S.	1	720	14,600	20.28	13,400	18.61	− 8.2
E-110	BEM (NO)	19 × 192	3/4 × 15/16 × 30	14 (A)	W	1	130/100	All C.S.	1	965	13,900	14.40	14,600	15.13	+ 5.0
E-111	BEM (BL)	21 × 240	3/4 × 15/16 × 30	14 (A)	W	4	200/100	All C.S.	1	1,300	17,000	13.08	18,100	13.92	+ 6.5
E-112	NKN (NO)	41/27 × 288	3/4 × 15/16 × 30	14 (M)	W	4	350/180	All C.S.	1	2,575	34,600	13.44	35,300	13.71	+ 2.0
E-113	AKL (NO)	46/28 × 192	3/4 × 1 × 30	14 (A)	W	1	230/230	All C.S.	1	1,660	23,200	13.98	24,200	14.58	+ 4.3
E-114	AEL (NO)	30 × 384	1 × 1 1/4 × 30	14 (A)	W	4	150/410	All C.S.	1	3,490	41,500	11.89	37,300	10.69	−10.1
E-115	BGM (NO)	30 × 420	3/4 × 15/16 × 30	14 (A)	W	2	300/1,000	All C.S.	1	5,125	65,700	12.82	57,200	11.16	−12.9
E-116	AKL (NO)	78/35 × 240	3/4 × 1 × 30	14 (A)	W	2	360/230	All C.S.	1	2,935	40,900	13.94	38,400	13.08	− 6.1
E-117	AFL (NO)	37 × 240	3/4 × 15/16 × 30	14 (A)	W	8	160/230	All C.S.	1	4,240	41,300	9.74	37,200	8.77	− 9.9
E-118	BEM (NO)	40 × 240	3/4 × 15/16 × 30	14 (A)	W	1	100/100	All C.S.	1	6,010	36,700	6.11	36,000	5.99	− 1.9
E-119	CEM (NO)	45 × 168	1 1/2 × 1 57/64 × 30	14 (A)	W	1	1,100/100	All C.S.	1	2,360	56,800	24.07	59,700	25.30	+ 5.1
E-120	BEM (NO)	58 × 144	3/4 × 15/16 × 30	14 (A)	W	1	100/100	All C.S.	1	7,325	52,600	7.18	54,700	7.47	+ 4.0
E-121	NEN (FF)	59 × 96	1 1/2 × 1 57/64 × 30	14 (A)	W	1	200/100	All C.S.	1	2,420	52,400	21.65	50,700	20.95	− 3.2
E-122	BKM (NO)	118/72 × 324	3/4 × 15/16 × 30	14 (M)	S	4	220/330	High-flux C.S. tube/C.S.→	1	24,650	469,000	19.03	454,100	18.42	− 3.2
E-123	NEN (FF)	83 × 96	1 1/4 × 1 37/64 × 30	14 (A)	W	1	100/100	All C.S.	1	5,800	79,400	13.69	85,400	14.72	+ 7.6
E-124	BEM (FF)	12 × 120	3/4 × 1 × 30	20 (M)	W	4	100/100	Ti (Gr2)/C.S./C.S./Ti (Gr2)	1	145	17,000	117.24	16,800	115.86	− 1.2
E-125	BEM (FF)	13 × 192	3/4 × 15/16 × 30	16 (A)	S	2	100/120	90-10 Cu-Ni/C.S./90-10 Cu-Ni→	1	340	20,700	60.88	17,900	52.65	−13.5
E-126	BEM (FF)	17 × 240	3/4 × 15/16 × 30	20 (M)	W	1	100/100	Ti (Gr2)/C.S./C.S./Ti (Gr2)	1	850	35,900	42.24	37,100	43.65	+ 3.3
E-127	BEM (FF)	18 × 120	3/4 × 15/16 × 30	16 (M)	S	2	100/150	Ni/C.S./Ni→	1	420	47,000	111.90	40,100	95.48	−14.7
E-128	BEM (NO)	19 × 192	3/4 × 15/16 × 30	16 (M)	W	1	130/100	304 S.S./C.S./304 S.S.→	1	960	23,200	24.17	21,000	21.88	− 9.5
E-129	BEM (FF)	27 × 192	3/4 × 15/16 × 30	16 (M)	S	1	100/100	Monel/C.S./C.S./Monel	1	1,985	69,700	35.11	77,800	39.19	+11.6
E-130	AJL (NO)	34 × 240	3/4 × 1 1/8 × 30	14 (M)	W	1	100/100	Ti (Gr2)/C.S./C.S./Ti (Gr2)	1	2,740	73,900	26.97	81,200	29.64	+ 9.9
E-131	–EM (NO)	36 × 144	3/4 × 1 1/4 × 30	20 (A)	W	1	130/100	Ti (Gr2)/C.S./C.S./Ti (Gr2)	1	2,400	65,000	27.08	74,800	31.17	+15.0
E-132	–EM (BL)	37 × 96	3/4 × 1 × 45	14 (A)	S	1	300/450	304-L S.S./C.S./304-L S.S.→	1	1,385	48,200	34.80	41,000	29.60	−14.9
E-133	BEN (BL)	37 × 132	3/4 × 1 × 30	14 (A)	S	1	300/300	All 304-L S.S.	1	2,350	68,700	29.23	67,600	28.77	− 1.6
E-134	–EM (NO)	42 × 120	3/4 × 1 1/4 × 30	20 (A)	W	1	130/100	Ti (Gr2)/C.S./C.S./Ti (Gr2)	1	2,650	72,000	27.51	78,300	29.55	+ 8.7
E-135	BXM (NO)	122 × 432	3/4 × 15/16 × 30	20 (M)	W	2	100/140	Ti (Gr2)/C.S.→	1	74,845	1,300,000	17.18	1,489,000	19.89	+14.5
E-136	NEN (FF)	128 × 336	1 1/2 × 1 57/64 × 30	14 (M)	S	1	100/100	All 1 Cr-1/2 Mo	1	40,550	669,000	16.50	600,000	14.80	−10.3
E-137	BEM (NO)	148 × 360	2 × 2 1/2 × 30	14 (M)	S	1	100/100	70-30 Cu-Ni/C.S./70-30 Cu-Ni→	1	45,565	1,214,000	26.64	1,070,000	23.48	−11.9
E-138	BEU	12 × 192	3/4 × 1 × 90	14 (A)	W	2	170/130	All C.S.	1	200	9,400	47.0	9,200	46.0	− 2.1
E-139	BEU	12 × 192	3/4 × 1 × 45	14 (A)	W	4	100/100	All C.S.	1	200	8,600	43.0	9,300	46.50	+ 8.1
E-140	BEU	13 × 96	1 × 1 1/4 × 30	14 (A)	W	2	100/120	All C.S.	1	140	8,700	62.14	7,800	55.71	−10.3
E-141	BEU	13 × 192	3/4 × 15/16 × 90	14 (A)	W	4	160/180	All C.S.	1	355	10,400	29.30	11,300	31.83	+ 8.6
E-142	AEU	15 × 240	1 × 1 1/4 × 30	10 (M)	W	2	2,520/200	All C.S.	1	305	31,500	103.28	36,200	118.69	+14.9
E-143	BEU	18 × 144	3/4 × 15/16 × 90	14 (A)	W	2	180/130	All C.S.	1	565	14,400	25.49	12,900	22.83	−10.4
E-144	BJU	23 × 192	3/4 × 1 × 45	14 (M)	S	2	100/100	All C.S.	1	900	16,200	18.0	18,100	20.11	+11.7
E-145	DEU	26 × 288	1 × 1 1/4 × 45	10 (M)	S	2	150/2,530	All C.S.	1	4,830	171,000	35.40	191,000	39.54	+11.7
E-146	BKU	49/29 × 192	3/4 × 1 × 90	14 (A)	W	2	650/740	All C.S.	1	1,780	47,100	26.46	40,000	22.47	−15.0
E-147	BKU	100/45 × 240	1 × 1 1/4 × 30	14 (A)	W	2	380/300	All C.S.	1	7,965	103,300	12.97	97,500	12.24	− 5.6
E-148	NFU	49 × 421	3/4 × 15/16 × 30	14 (A)	W	2	300/130	All C.S.	1	14,425	107,000	7.42	116,600	8.08	+ 9.0
E-149	BEU	52 × 192	3/4 × 1 1/4 × 30	14 (A)	W	2	350/650	All C.S.	1	3,040	55,300	18.21	47,100	15.49	−14.8
E-150	DEU	19 × 168	1 × 1 1/4 × 30	14 (M)	W	2	2,540/2,310	430 S.S./1¼Cr-½Mo→/430 S.S.	1	515	112,500	218.45	96,400	187.26	−14.3
E-151	BHU	21 × 144	1 × 1 1/4 × 45	13 (A)	W	2	300/650	E-brite-26-1/304-L S.S./304 S.S. clad on C.S.→	1	515	39,200	76.12	33,400	64.85	−14.8
E-152	BKU	42/21 × 192	3/4 × 1 × 90	16 (A)	S	2	150/150	Sandvik SAF 2205/C.S./304 S.S. clad on C.S.→	1	925	41,700	45.08	46,900	50.70	+12.5
E-153	BKU	33/22 × 144	3/4 × 1 × 90	16 (A)	W	2	300/650	Al-26-1/C.S.→/304 S.S.	1	585	31,100	53.16	29,800	50.94	− 4.2
E-154	AJU	23 × 288	1 × 1 1/4 × 30	16 (A)	S	8	350/100	304 S.S./304 S.S./C.S./304 S.S.	1	1,100	34,500	31.08	35,100	31.62	+ 1.7
E-155	BEU	24 × 120	3/4 × 15/16 × 30	16 (A)	S	4	210/100	All 304 S.S.	1	885	24,100	27.23	24,000	27.12	0.0
E-156	BEU	26 × 192	1 × 1 1/4 × 30	16 (A)	S	4	100/130	Sandvik 3 RE 60/304L S.S.→ Sandvik 3 RE-60	1	1,125	47,900	42.58	46,800	41.60	− 2.3
E-157 A, B	DEU	28 × 360	1 × 1 1/4 × 45	16 (A)	S	4	2,340/2,500	304 S.S./309 S.S. clad on 2¼Cr-½Mo→	2	1,950	753,000	193.08	649,800	168.62	−13.7

Table 7
Continued

Exchanger Ab.	TEMA Type	Shell I.D., in. x Nominal Tube Length (in.)	Tube O.D. (in.) x Pitch, x Layout Angle (deg.)	BWG (Avg./Min.)	Tubes, Welded, Seamless	No. of Tube Passes, N_{TP}	Design Pressures, (psig) Shell-Tube	Metallurgy Tube Shell/Channel or Bonnet/Tubesheet	No. of Shells	Surface Area per Shell, (ft^2)	Exchanger FOB Price — Purchase ($)	Purchase ($/$ft^2$)	Estimated ($)	Estimated ($/$ft^2$)	Accuracy (%)
E-158	DEU	32 × 132	1 × 1 1/4 × 45	12 (M)	S	2	2,770/2,540	321 S.S./304 S.S. clad on C.S./309 S.S. clad on 2¼Cr-½Mo →	2	1,290	499,000	193.41	436,400	169.15	−12.6
E-159 A, B	AEU	32 × 240	1 × 1 1/4 × 45	16 (A)	W	4	670/900	403 S.S./304 S.S. clad on C.S. →	1	1,815	60,700	33.44	51,700	28.48	−14.8
E-160	AEU	32 × 240	1 × 1 1/4 × 45	16 (A)	S	4	670/900	316L S.S./304 S.S. clad on C.S. →/316L S.S.	1	1,815	85,000	46.83	72,300	39.83	−14.9
E-161	BFU	34 × 180	3/4 × 15/16 × 30	16 (A)	W	2	720/350	304 S.S./C.S. 304 S.S. →	1	3,090	63,100	20.42	61,800	20.0	− 2.6
E-162	DEU	35 × 168	1 × 1 1/4 × 45	12 (M)	S	2	2,750/2,540	321 S.S./304 S.S. clad on C.S./309 S.S. clad on 2¼ Cr-½ Mo →	1	1,915	316,300	165.17	306,300	159.95	− 3.2
E-163 A, B	DEU	38 × 144	1 × 1 1/4 × 45	12 (M)	S	2	2,750/2,540	321 S.S./304 S.S. clad on C.S./309 S.S. clad on 2¼ Cr-½ Mo →	2	1,995	547,800	137.29	629,100	157.67	+14.8
E-164	AES	12 × 240	3/4 × 1 × 45	14 (M)	S	2	100/130	All C.S.	1	235	13,300	56.60	11,400	48.51	−14.3
E-165	AES	12 × 240	3/4 × 1 × 45	14 (M)	W	4	100/300	All C.S.	1	255	11,500	45.10	10,600	41.57	− 7.8
E-166	AES	15 × 192	3/4 × 1 × 45	14 (M)	W	4	230/100	All C.S.	1	400	12,100	30.25	13,900	35.25	+14.2
E-167	AES	15 × 240	3/4 × 1 × 45	14 (M)	W	4	150/130	All C.S.	1	470	15,200	32.34	13,000	27.66	−14.5
E-168	AES	17 × 192	3/4 × 1 × 45	14 (M)	W	4	290/100	All C.S.	1	465	14,000	30.11	14,200	30.54	+ 1.4
E-169 A, B	AES	19 × 144	3/4 × 1 × 45	12 (M)	S	4	100/100	All C.S.	2	415	34,500	41.57	29,400	35.42	−14.8
E-170 A, B	AES	19 × 192	3/4 × 1 × 45	14 (M)	W	2	150/130	All C.S.	2	645	28,400	22.02	31,500	24.42	+10.9
E-171 A, B, C	AES	19 × 192	3/4 × 1 × 45	14 (M)	W	2	230/170	All C.S.	3	660	44,800	22.63	50,200	25.35	+12.0
E-172	AES	19 × 240	3/4 × 1 × 45	14 (M)	W	4	300/130	All C.S.	1	735	17,800	24.22	15,500	21.09	−12.9
E-173	AES	21 × 144	3/4 × 1 × 45	14 (M)	W	2	150/300	All C.S.	1	650	15,700	24.15	17,800	27.38	+13.4
E-174	AES	21 × 192	3/4 × 1 × 45	14 (M)	W	2	150/420	All C.S.	1	855	18,300	21.40	19,600	22.92	+ 7.1
E-175	AES	21 × 192	3/4 × 1 × 45	14 (M)	S	6	250/130	All C.S.	1	740	21,700	29.32	20,000	27.03	− 7.8
E-176	AES	21 × 240	3/4 × 1 × 45	14 (M)	W	4	180/100	All C.S.	1	1,000	17,400	17.40	17,600	17.60	+ 1.1
E-177	AES	23 × 192	3/4 × 1 × 45	14 (M)	W	4	250/130	All C.S.	1	770	17,600	22.86	16,300	21.17	− 7.4
E-178	AES	23 × 192	1 × 1 1/4 × 45	12 (M)	W	2	180/500	All C.S.	1	830	20,200	24.34	21,000	25.30	+ 4.0
E-179 A, B	AES	23 × 192	3/4 × 1 × 45	14 (M)	W	2	180/100	All C.S.	2	910	34,400	18.90	35,900	19.73	+ 4.4
E-180	AES	23 × 192	3/4 × 1 × 45	14 (M)	S	6	250/130	All C.S.	1	920	23,800	25.87	22,600	24.57	− 5.0
E-181 A, B, C	AES	25 × 192	3/4 × 1 × 45	14 (M)	W	4	180/100	All C.S.	3	780	36,800	15.73	42,200	18.03	+14.7
E-182	AES	30 × 144	3/4 × 1 × 45	14 (M)	W	6	250/200	All C.S.	1	1,200	23,700	19.75	24,200	20.17	+ 2.1
E-183	AES	34 × 192	3/4 × 1 × 45	14 (M)	S	6	100/100	All C.S.	1	2,130	37,600	17.65	35,600	16.71	− 5.3
E-184	AJS	34 × 240	3/4 × 1 × 45	14 (M)	W	4	200/180	All C.S.	1	2,640	33,100	12.54	31,100	11.78	− 6.0
E-185	AES	37 × 240	3/4 × 1 × 45	14 (M)	S	4	200/130	All C.S.	1	3,350	48,100	14.36	45,200	13.49	− 6.0
E-186	BES	17 × 240	3/4 × 1 × 45	20 (M)	W	2	130/100	Ti (Gr2)/C.S. →/Ti (Gr2)	1	605	33,900	56.03	37,800	62.48	+11.5
E-187	BES	23 × 240	3/4 × 1 × 45	16 (M)	S	8	170/170	Monel/C.S. →/Monel	1	1,145	65,900	57.55	66,100	57.73	0.0
E-188	AET	30 × 240	3/4 × 1 × 45	14 (M)	W	2	150/100	All C.S.	1	1,760	25,900	14.72	22,200	12.61	−14.3
E-189	AJT	31 × 240	3/4 × 15/16 × 30	14 (M)	S	2	300/150	All C.S.	1	2,290	31,600	13.80	31,800	13.89	− 1.0
E-190	AET	33 × 192	3/4 × 1 × 45	12 (M)	W	2	180/500	All C.S.	1	1,935	34,900	18.04	34,000	17.57	− 2.6
E-191	AET	33 × 192	3/4 × 1 × 45	14 (M)	W	4	230/100	All C.S.	1	2,095	26,100	12.46	30,000	14.65	+14.9
E-192	AET	37 × 192	1 × 1 1/4 × 45	12 (M)	W	2	180/500	All C.S.	1	1,935	34,900	18.04	34,000	17.57	− 2.6
E-193	AJT	41 × 192	3/4 × 1 × 90	14 (M)	W	6	130/140	All C.S.	1	2,955	34,700	11.74	38,000	12.86	+ 9.5
E-194 A, B	AET	44 × 192	1 × 1 1/4 × 45	·14 (M)	W	2	180/410	All C.S.	2	2,830	83,000	14.66	85,000	15.02	+ 2.4
E-195	AET	44 × 192	1 × 1 1/4 × 45	12 (M)	W	2	290/410	All C.S.	1	2,830	44,000	15.55	47,600	16.82	+ 8.2
E-196	AJT	47 × 192	3/4 × 1 × 90	14 (M)	W	8	100/140	All C.S.	1	4,000	42,500	10.63	48,100	12.03	+13.2
E-197 A, B	AET	47 × 240	3/4 × 1 × 45	14 (M)	W	2	290/100	All C.S.	2	5,425	94,100	8.67	108,000	9.95	+14.8
E-198 A, B	AET	49 × 192	3/4 × 1 × 45	14 (M)	W	6	130/100	All C.S.	2	4,475	84,200	9.41	96,500	10.78	+14.6
E-199	AET	52 × 240	1 × 1 1/4 × 45	12 (M)	W	4	180/410	All C.S.	1	4,985	57,200	11.47	61,000	12.24	+ 6.6
E-200 A-H	AJT	54 × 240	1 × 1 1/4 × 45	12 (M)	W	4	100/410	All C.S.	8	4,950	522,000	13.18	495,000	12.50	− 5.2
E-201 A, B	AET	54 × 240	1 × 1 1/4 × 45	12 (M)	W	4	290/500	All C.S.	2	5,195	124,300	11.96	141,000	13.60	+13.4
E-202	AET	57 × 192	3/4 × 1 × 45	14 (M)	W	4	100/420	All C.S.	1	5,675	63,000	11.10	64,800	11.42	+ 2.9
E-203	AET	57 × 240	3/4 × 1 × 45	14 (A)	W	2	150/100	All C.S.	1	7,350	67,700	9.21	62,200	8.46	− 8.0
E-204	AET	59 × 240	3/4 × 1 × 45	14 (M)	W	4	150/420	All C.S.	1	7,905	152,600	9.65	150,700	9.53	− 1.2
E-205	AET	15 × 240	3/4 × 1 × 90	16 (A)	W	4	100/140	304L S.S./C.S. →/304L S.S.	1	970	36,200	37.32	37,800	38.97	+ 4.4
E-206	AET	32 × 240	1 × 1 1/4 × 45	14 (A)	S	2	290/500	All 5 Cr-½ Mo	1	1,730	62,000	37.05	52,700	30.46	−15.0
E-207	AET	46 × 240	1 × 1 1/4 × 45	16 (M)	S	4	100/410	Sandvik 2 RK-65/C.S. →	1	3,630	166,000	45.73	167,100	46.03	+ 1.0
E-208	AET	53 × 240	1 × 1 1/4 × 90	14 (A)	S	4	140/500	5 Cr-½ Mo/C.S. →	1	5,005	99,500	19.88	88,500	17.68	−11.1

* FF = Flanged-and-flued expansion joint, BL = bellows-type expansion joint, NO = No expansion joint

fall within ±15%. Of these 60% are found to be within ±10% and remaining 40% within ±15% of actual purchase prices. The average deviation is ±9.6%.

Accuracy is defined as: %-accuracy = ((estimated price–purchase price)/purchase price)100.

When the accuracies of the individual cost estimates for all of the exchangers fall within ±15%, the overall cost estimate for all of the exchangers purchased for a project will be more accurate, because positive and negative deviations will cancel each other out when added together. Thus, the total purchase price of all 108 exchangers in Table 7 is $10.615 million, and the estimated price of the total is $10.469 million—an overall accuracy of 1.4%.

Estimating Exchanger Weights

Along with exchanger costs, weight estimates are also required for shipping purposes, for foundation, structural, and support design calculations, and for lifting purposes. A correlation exists between the cost and weight of the exchanger. Thus, knowing the exchanger cost, the design engineer can calculate the *approximate weight* of a low pressure carbon steel exchanger using Equation 14:

$$W = 0.003(E_b)^{1.5} \tag{14}$$

where W = exchanger weight in lb
E_b = exchanger price in $ (1982 basis); calculated by Equation 13

The weight of an alloy exchanger can be estimated by first calculating the price E_b as if it were a carbon steel exchanger (i.e., by ignoring the cost corrections for metallurgy in Equations 8 to 11) and then using this price in Equation 14.

Equation 14 is limited to low pressure exchangers only and is not claimed to be exact, but gives the design engineer an idea about the exchanger weight.

NOTATION

- A surface area, ft^2/shell; Equation 13
- a cost multiplier for layout angle; Equation 2 and 7
- b base price of exchanger; Equation 1 (defined in Table 1) ($/$ft^2$)
- C_g cost correction for tube gage; Equation 12
- C_L cost correction for tube lenghts shorter than 20 ft; Equation 3
- C_{mc} cost correction for channel material; Equation 10
- C_{ms} cost correction for shell material; Equation 9
- C_{mt} cost correction for tube material; Equation 9
- C_{mts} cost correction for tubesheet material; Equation 8
- C_{NTP} cost correction for number of tube passes; Equation 11
- C_{PS} cost correction for shell-side design pressure > 150 psig; Equation 5
- C_{PT} cost correction for tube-side design pressure > 150 psig Equation 6
- C_S cost correction for TEMA shell type (Table 5)
- C_T sum of all applicable cost corrections; Equation 13
- C_x correction for expansion joint Figure 4
- D_i shell I.D. (bundle diameter for kettle reboiler); Equation 1
- d_0 tube outside diameter
- E_0 estimated price of exchanger (January 1982); Equation 13 ($)
- E_I escalating index
- f cost multiplier for type of TEMA front-end stationary head; Equation 1 (Table 3)
- g cost multiplier for tube gage; Equation 11
- L nominal tube length; Equation 3
- M_1 relative cost of alloy tubing to welded carbon-steel tubing;

Equation 8 (Table 6)

M_2 relative cost of plate materials to carbon-steel plate material of same dimensions for exchanger shells, channels, and tubesheets; Equations 9, 10, and 11 (Table 6)

N total number of shells per heat-exchange unit; Equation 13

N_{TP} number of tube passes; Equation 4

P cost multiplier for tube outside dia. × pitch × layout angle; Equations 1 and 2 (Table 2)

P_{DS} shell-side design pressure; Equation 5 (psig)

P_{DT} tubeside design pressure; Equation 6 (psig)

P_i tube pitch; Equation 2 (in.)

r cost multiplier for type of TEMA rear-end head (Table 4)

X adjustment parameter for shell-side design pressures greater than 2,000 psig; Equation 5 (Figure 7)

y cost of caron-steel tubes as a fraction of total price of base carbon-steel exchanger; Equations 7, 8, and 12

W weight of the exchanger; Equation 14 (lb)

REFERENCES

1. "Standards of Tubular Exchanger Manufacturers Assn.," Tubular Exchanger Manufacturers Assn., Tarrytown, N. Y.
2. Purohit, G. P., Thermal and Hydraulic Design of Hairpin and Finned-Bundle Exchangers," *Chem. Eng.* (May 16, 1983), p. 62.

CHAPTER 30

COUNTERCURRENT CONDENSATION DROBBLE COLUMN

B. Sudhoff, A. Steiff, and P. M. Weinspach

University of Dortmund
Federal Republic of Germany

CONTENTS

INTRODUCTION

In an indirect contact heat-exchanger heat is transported through a solid wall. Should the media, between which the heat is transported, be in direct contact with each other the heat transfer resistance of a separation wall is eliminated along with eventually appearing corrosion and incrustation.

These advantages are opposed by certain disadvantages: the media which come into contact may be polluted or may react with each other in an unwanted manner. Therefore the use of direct contact heat-exchangers is limited to particular systems. The drobble column is a special construction-type of a direct contact heat-exchanger. A flow of "two-phase" particles is evaporated or condensed in a continuous flow of liquid, which is not miscible with the phase-changing substance.

The direct cooling of hot gases and the heat transfer between two immiscible liquids in a drobble column are examples for heat transfer without phase change.

Just as in indirect heat transfer it is also possible in direct contact heat transfer to make use of the latent heat as well as of the sensible heat. In processes where the physical state of the substances changes, the combination of the occurring phases can be used as a distinctive feature. The assumed subdivision does not exclude that the phase-changing processes occur simultaneously.

Thus for example during the ice-crystallization in brines the released heat of crystallization can be transferred to a refrigerant, which is dispersed in the solution (freezing method). The liquefaction of the recirculated refrigerant can be conducted in a direct-contact condenser [1].

If the phase combination is limited to gas/liquid and therefore to evaporation and condensation processes, then processes using one or more substances are feasible. When water vapor is condensed into water then only one component appears. Several components take part when, for example, the heat released during an exothermic reaction, is taken out by an evaporating solvent.

Mass transfer coupled with heat transfer, as for example in a rectification column, will be excluded from further consideration.

Despite this, it is possible to differentiate among direct-contact heat exchangers with regard to interfacial area and phase contact in a way similar to the characterization of thermal separation apparatuses.

In a packed column a liquid film trickles down the surface of the packing countercurrent to a rising gas and forms a continuous interfacial area.

If one of the two phases occurs as fluid particles the interfacial area is discontinuous. In a tray column, for example, gas bubbles pass a liquid layer. After leaving the two-phase-layer on the plate, the bubbles join together to be redispersed on the next plate. The phase contact is thus discontinuous.

A continuous contact between drops and a continuous gas phase takes place in a spray column. The phase contact is also continuous in a drobble column. Here fluid particles are dispersed in a continuous liquid phase. These particles contain gas as well as liquid and thus are named "drobbles" because of their ambivalent character.

A literature bibliography about theoretical and experimental work on condensation and evaporation of drobbles is given in Reference 2. A synopsis about theoretical work on drobble columns and experimental research on evaporation of drobble swarms can be found in Reference 3. No experimental results about condensation in drobble columns are available so far, apart from what is described later in this chapter.

DROP, BUBBLE, DROBBLE

At the beginning of the heat transfer process the dispersed phase can be either liquid or gaseous. If the temperature of the continuous phase is lower than the dew point of the dispersed phase, a bubble condenses to a drop. On the other hand, if the temperature exceeds the boiling point of the dispersed phase, a drop evaporates to a bubble. This change of phase results in a considerable change of particle volume depending on the liquid and vapor densities of the dispersed phase

$$\frac{V_B}{V_T} = \frac{\rho_l}{\rho_g} \tag{1}$$

For spherical particles, the ratio of bubble and drop diameters is as follows:

$$\frac{d_B}{d_T} = \left(\frac{\rho_l}{\rho_g}\right)^{1/3} \tag{2}$$

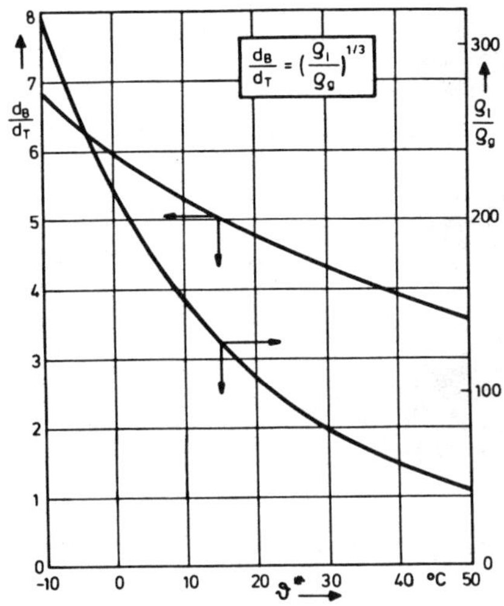

Figure 1. Ratios of liquid and vapor densities and of bubble and drop diameters for n-butane, at boiling point, as functions of phase-change temperature.

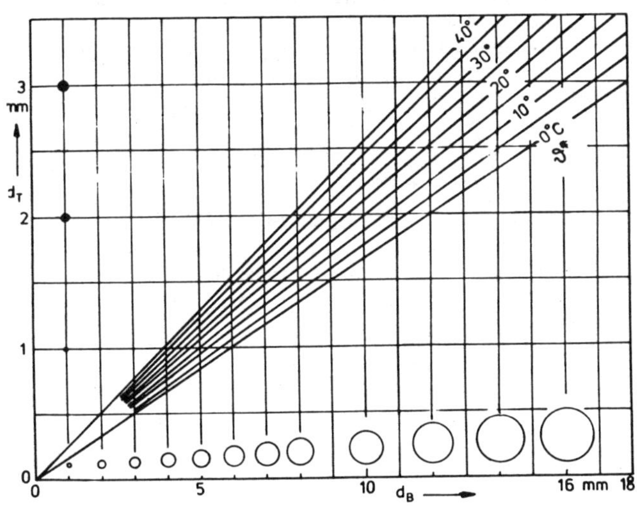

Figure 2. Relationship between the diameters of n-butane drops d_T and bubbles d_B of the same mass at boiling point. Phase change temperature T* is the parameter.

For n-butane at phase change temperatures from 0° C to 40° C, corresponding to equilibrium pressures of 1 to 4 bar, this ratio ranges between 4 and 6 (Figure 1). For example, at 15° C, a bubble of an equivalent spherical diameter of 14 mm condenses to a drop only 2.8 mm in diameter (Figure 2). While phase change is taking place, converting a drop to a bubble, or vice versa, a two-phase particle can be observed. Figure 3 shows such a butane particle in water. It contains both a liquid and a vapor phase, the vapor

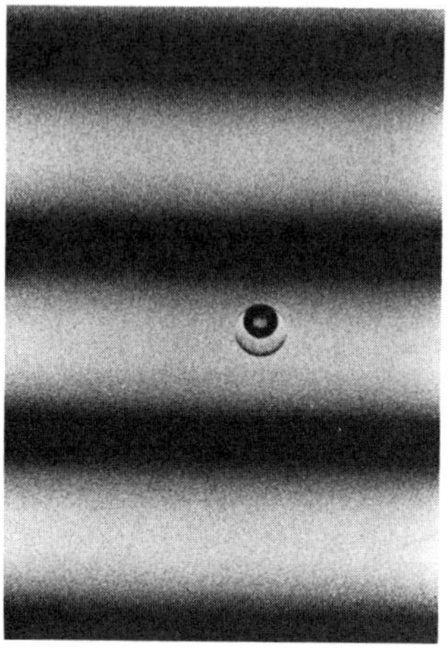

Figure 3. Photograph of a butane drobble in water (d = 1 mm).

accumulating in the upper part of the particle. Being neither a pure drop nor a pure bubble, this particle will be referred to as a drobble.

In Reference 4 it is designated as a drobble. While the mass of a drobble remains constant during the phase-change process, its mean density ρ_P varies with the changing volume or diameter (Figure 4):

$$\frac{m_P}{V_P} \equiv \rho_P = \rho_g \left(\frac{d_B}{d}\right)^3 = \rho_l \left(\frac{d_T}{d}\right)^3 \tag{3}$$

If the mass of vapor phase m_V (vapor fraction) or of the liquid phase m_K (condensate fraction) of a drobble is expressed as a fraction of its total mass m_P, the resulting evaporation ratio R_V or condensation ratio R_k, can be described as functions of the drobble diameter:

$$R_V = \frac{m_V}{m_P} = \frac{\left(\frac{d}{d_B}\right)^3 - \frac{\rho_g}{\rho_l}}{1 - \frac{\rho_g}{\rho_l}} = 1 - R_k, \tag{4}$$

$$R_K = \frac{m_K}{m_P} = \frac{1 - \frac{\rho_g}{\rho_P}}{1 - \frac{\rho_g}{\rho_l}} = \frac{1 - \left(\frac{d}{d_B}\right)^3}{1 - \frac{\rho_g}{\rho_l}} \tag{5}$$

It is worth noting that the condensation ratio of a butane drobble of half the bubble diameter is about 90%.

In order to illustrate the geometrical conditions, two models of a drobble are compared in Figure 5. In the first case, the condensate accumulates in the lower part of the drobble in the form of a spherical cap of height h_{Kk}. In the second, more realistic model, the gas phase forms a spherical bubble of diameter d_{Gk} within the drobble.

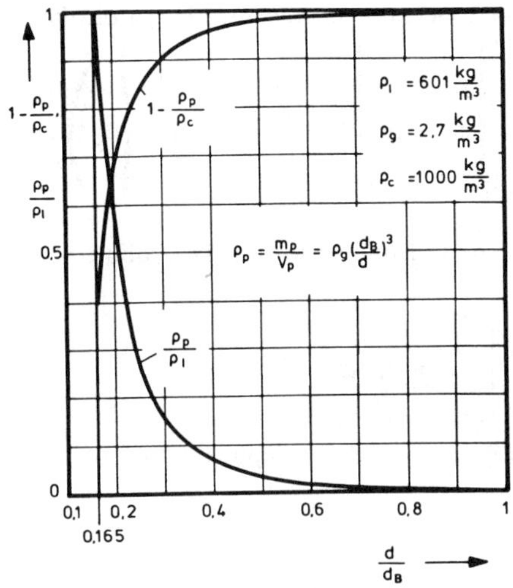

Figure 4. Density of a drobble ρ_p and deviation of the density ratio ρ_p/ρ_g from unity as a function of drobble diameter for n-butane at boiling point. If $d/d_B = 1$, the bubble is purely gaseous, for $d/d_B = 0.165$ the drop is purely liquid.

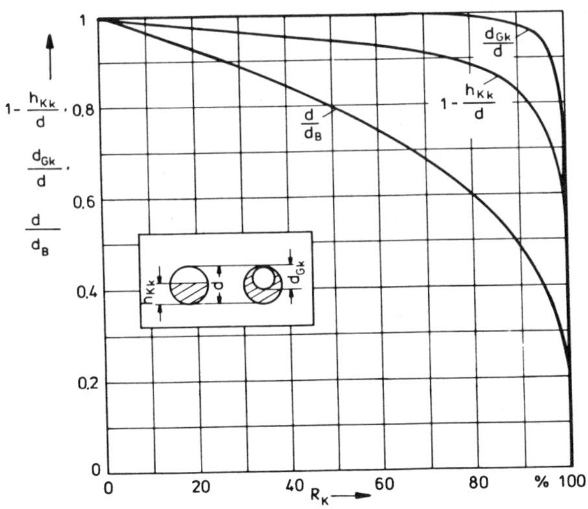

Figure 5. Diameter of drobble d, height of gaseous spherical cap ($d - h_{Kk}$), and diameter of the spherical bubble d_{Gk} inside the drobble as functions of condensation degree R_K (reference system n-butane, boiling point 20°C).

The height of the vapor spherical cap $(d-h_{Kk})$ based on the drobble diameter is:

$$1-\frac{b_{Kk}}{d}=0.5-\cos\left(\frac{1}{3}\left\{\pi+\arccos\left[2\frac{\left(\frac{d_B}{d}\right)^3-1}{\frac{\rho_l}{\rho_g}-1}-1\right]\right\}\right) \tag{6}$$

and the diameter of the spherical bubble within the drobble is:

$$\frac{d_{Gk}}{d}=\left[\frac{1-R_K}{1-R_k\left(1-\frac{\rho_g}{\rho_l}\right)}\right]^{1/3} \tag{7}$$

This ratio is equal to the cube root of the vapor fraction of the drobble:

$$\varepsilon_G=\frac{m_V}{\rho_g V_P}=\left(\frac{d_{Gk}}{d}\right)^3 \tag{8}$$

The graphs of Equations 6 and 7 are shown in Figure 5. Evidently the geometry of the condensate is of significance only at high condensation ratios. In other words, the gas-phase determines the geometry even at very low vaporization ratios.

This fact is made clear by the spherical drobble shown in Figure 3. The drobble diameter amounts to 1 mm and the gas sphere diameter reaches 0.65 mm which signifies a diameter ratio of 0.65.

Using Equation 7 the condensation degree can be calculated:

$$R_K=\frac{1-\left(\frac{d_{Gk}}{d}\right)^3}{1-\left(1-\frac{\rho_g}{\rho_l}\right)\left(\frac{d_{Gk}}{d}\right)^3} \tag{9}$$

The condensation degree is found to be 99.6% (compared to Figure 5). According to Equation 8 the gas however still takes up to 27.5% of the total drobble volume.

If a drobble consists of not just one component but also a second one which, with reference to temperature and pressure is not condensable, then the geometrical relations change. The drobble, in this case, is composed of the condensate and a gas vapor mixture. In the following m_P and also m_V will represent only the mass ratio of condensable mass in the drobble and in the gas vapor mixture respectively.

The alteration of the drobble diameter during the phase-change is dependent on the amount of inert gas:

$$\frac{d}{d_B}=\left[1-R_K(1-\Gamma)\left(1-\frac{\rho_g}{\rho_l}\right)\right]^{1/3} \tag{10}$$

If the molar fraction of the inert gas is exactly zero, then for one component there follows a relationship which is derived from Equation 5:

$$\frac{d}{d_B}=\left[1-R_K\left(1-\frac{\rho_g}{\rho_l}\right)\right]^{1/3} \tag{11}$$

Analogous to Figure 5, the diameter alterations of the drobble are shown in Figure 6.

Already small amounts of inert gas lead to noticeably larger dimensions. The described effect occurs most significantly when the total vapor is condensed. While in the presence

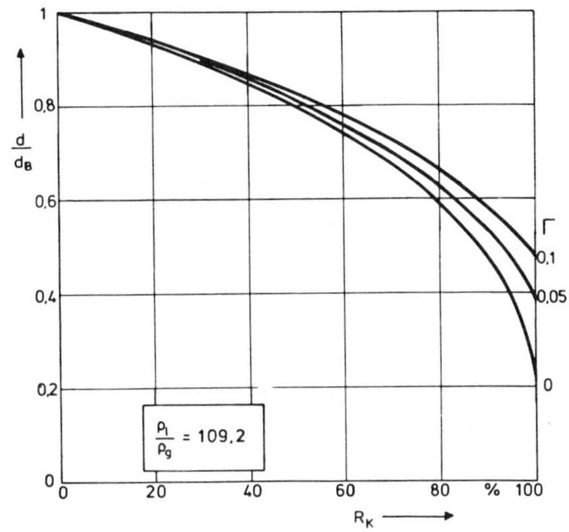

Figure 6. The drobble diameter, in relation to the bubble diameter as a function of the condensation degree and the molar fraction of inert gas for n-butane at 20°C.

of only one component a drop is finally built up, with an additional inert gas a drobble is obtained which is made up of the condensate and the noncondensable gas phase. Even with small amounts of inert gas the diameter of this drobble deviates substantially from the equivalent drop diameter, as seen in Figure 7.

$$\frac{d_f}{d_T} = \left[1 + \Gamma\left(\frac{\rho_l}{\rho_g} - 1\right)\right]^{1/3} \tag{12}$$

Here it is neglected that a complete condensation is not possible since the partial pressure of the condensing component, corresponding to its vapor pressure, does not disappear.

The inert gas effects not only a change of the geometrical relationships, but renders also an additional heat transfer resistance. This is valid in particular for condensation as it is for indirect heat transfer.

The technical use has not just to do with one particle but with particle collectives. If a drobble swarm flows through a liquid in a perpendicular cylindrical apparatus then this is called a drobble column in analogy to the bubble column and the drop column.

The results obtained about the fluid dynamics in the multi-phase columns without phase-change cannot be attributed to the drobble column because of the very large changes of all important parameters of the drobble swarm.

This will be demonstrated with the help of a simple model calculation for the drobble swarm condensation. The results of this calculation can be applied to the drobble swarm evaporation considering the exchanged initial and end conditions of the phase change. Through the lower part of the column flows a monodisperse bubble swarm which condenses in the middle section of the column and reaches the top part of the apparatus as a drop swarm. The two single-phase limiting states will now be compared with each other. Equation 2 is valid for the diameter ratio of the particles. The stationary rising velocity is calculated with the resistance laws for bubbles [5] and drops [6].

Figure 8 illustrates the velocity ratio for the system butane/water. Here the water velocity is assumed to be negligible. Further it is assumed that the particles also move in the swarm

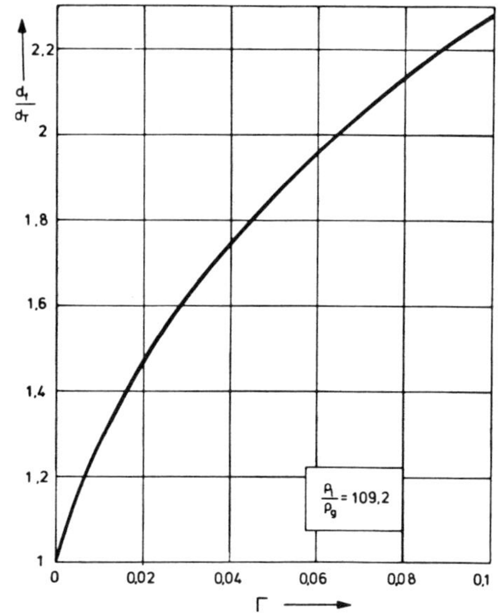

Figure 7. The drobble diameter in relation to the relevant drop diameter at complete condensation of the condensable vapor, as a function of the molar fraction of inert gas for n-butane at 20°C.

with the stationary rising velocity of the individual particles. If the flow through the column is continuous then the relation between the hold-up of bubbles and drops can be determined from their velocities. The quotient of the velocities of bubbles and drops is namely proportionally inverse to the ratio of the number of particles present in equally sized volume elements of the column:

$$\frac{n_T}{n_B} = \frac{w_B}{w_T}. \tag{13}$$

The hold-up is given by the ratio of dispersed-phase volume to the total dispersion volume. In the case of the bubble swarm this is the relative gas hold-up. The ratio of the hold-ups follows from Equation 13.

$$\frac{\varepsilon_T}{\varepsilon_B} = \frac{n_T d_T^3}{n_B d_B^3} = \frac{w_B}{w_T} \cdot \frac{\rho_g}{\rho_l}. \tag{14}$$

The volume-specific interfacial area results from the hold-up of the dispersed phase and the particle diameter:

$$a_P = 6\frac{\varepsilon_P}{d_P} \tag{15}$$

With Equations 2 and 14 the quotient of the interfacial areas of the bubble and drop swarm can now be determined.

$$\frac{a_T}{a_B} = \frac{w_B}{w_T}\left(\frac{\rho_g}{\rho_l}\right)^{2/3}. \tag{16}$$

If the particles are so small that their movement obeys Stokes law, they rise with the velocity:

$$w = \frac{gd^2}{18v_c}\left(1 - \frac{\rho_p}{\rho_c}\right). \tag{17}$$

The quotients of the velocities, volumetric hold-ups, and interfacial areas are only dependent, in this case, on the densities:

$$\frac{w_T}{w_B} = \frac{1 - \dfrac{\rho_l}{\rho_c}}{1 - \dfrac{\rho_g}{\rho_c}}\left(\frac{\rho_g}{\rho_l}\right)^{2/3} \tag{18}$$

$$\frac{\varepsilon_T}{\varepsilon_B} = \frac{1 - \dfrac{\rho_g}{\rho_c}}{1 - \dfrac{\rho_l}{\rho_c}}\left(\frac{\rho_g}{\rho_l}\right)^{1/3} \tag{19}$$

$$\frac{a_T}{a_B} = \frac{1 - \dfrac{\rho_g}{\rho_c}}{1 - \dfrac{\rho_l}{\rho_c}} \tag{20}$$

This special case cannot be expected in technical dispersion apparatuses. Bubbles of a few millimeters diameter or more show that the hold-up and the volume-specific interfacial area are much smaller for the dispersed liquid than for the dispersed gas (Figure 8).

MODELING OF THE DROBBLE-PHASE CHANGE

This section shows the mathematical description of phase change of a drobble using a simplified model.

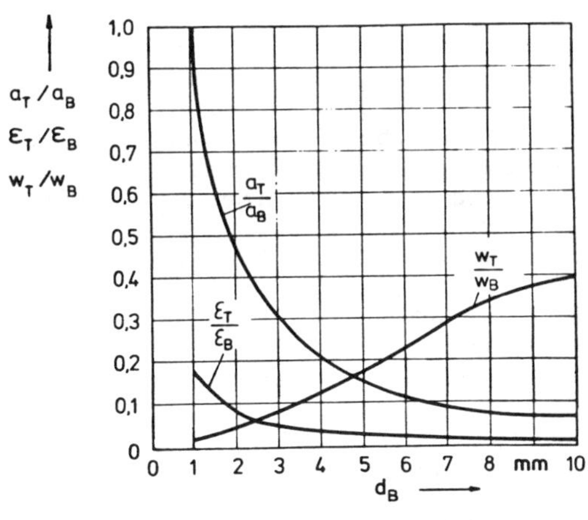

Figure 8. For comparison of bubble and drop swarms in the system, the following proportions are plotted: the rising velocities; the volume fractions of the dispersed phases; and the interfacial areas.

Assumptions

This model is based on the following assumptions and restrictions.

1. The dispersed phase is a pure substance. The continuous phase is a liquid.
2. No mass transfer occurs between the continuous and dispersed phases. They are immiscible and mutually insoluble.
3. The physical properties are constant.
4. The vapor pressure curve can be linearized in the observed temperature range.
5. The dispersed phase consists only of spherical drobbles.
6. The mean volume-specific mass of the dispersion is equal to the density ρ_c of the continuous phase, on account of the small hold-up of the dispersed phase.
7. Both the vapor density ρ_g and liquid density ρ_l of the dispersed phase are lower than the density ρ_c of the continuous phase.
8. The drobble moves parallel to the gravity vector (x-coordinate). The velocity vector of the dispersing agent is also vertical. Its velocity is negligible compared to that of the drobble. Therefore, the rise velocity of the drobble is equal to the drobble velocity in relation to a fixed reference system of coordinates:

$$w_d = w = \frac{dx}{dt} \tag{21}$$

9. The drobble motion is pseudo-steady so that the inertia term in the balance of forces is disregarded. Consequently, the drag coefficient depends only on the Reynolds number:

$$\psi = \frac{4}{3}\frac{gd}{w^2}\left(1 - \frac{\rho_p}{\rho_c}\right) \tag{22}$$

The quotient ρ_p/ρ_c is not negligible compared to unity.

10. The law of resistance, i.e., the relationship between the drag coefficient and the Reynolds number, is a power function:

$$\psi = b_1 Re^{b_2} \tag{23}$$

Using Equations 23 and 3 the Reynolds number can be expressed as a function of the drobble diameter:

$$Re = \left\{\frac{4}{3b_1} Ga\left[1 - \frac{\rho_g}{\rho_c}\left(\frac{d_B}{d}\right)^3\right]\left(\frac{d}{d_B}\right)^3\right\}^{1/(2+b_2)} \tag{24}$$

11. The particle swarms ascending in drobble columns are monodispersed with the corresponding diameter d at any column height x. The drobbles do not affect one another, so that a swarm can be regarded as independently ascending single drobbles.
12. Neither phase shows a back-mixing flow.
13. Break-up and coalescence effects do not occur. Heat transfer prior to and following the phase change is disregarded. At the beginning of phase change, the single-phase particle moves with uniform rise velocity.
14. The interface shows no resistance to transfer.
15. The phase change temperature T* is constant throughout the drobble, corresponding to the vapor pressure curve of the dispersed phase. Hence, the drobble has no inner transfer resistance. Therefore, all the heat transfer resistance acts in the continuous phase, so that the overall heat transfer coefficient can be equated to the continuous phase heat transfer coefficient α_c.
16. The heat transfer coefficient may be defined as a mean value over the total particle surface:

$$\alpha_c \equiv \frac{\dot{Q}}{(T_d - T_c)A_P} \tag{25}$$

It is calculated with the aid of relations between the Nusselt number and Reynolds and Prandtl numbers for heat transfer of submerged particles.

17. The heat transfer correlation is a power function:

$$Nu = a_1 Pr^{a_2} Re^{a_3} \tag{26}$$

It should be noted that for limiting case of Re = 0, the Nusselt number does not attain the limiting value of two, which, in theory, it should do.

Next, a general representation of the driving temperature difference should be deduced. With the inlet temperature T_{d0} of the dispersed phase at height x = 0 and assumptions 4 and 5, the temperature of the dispersed phase T_d at height x is as follows:

$$T_d = T_{d0} - \frac{dT^*}{dp} \rho_c g x \tag{27}$$

The temperature profile of the continuous phase depends on the method of operation (Figure 9). It can be expressed as a function of the condensation ratio. With inlet temperature T_{cE}, the energy balance for an adiabatic direct contact heat exchanger up to temperature T_c may be expressed by:

$$\dot{m}_c c_c (T_c - T_{cE}) = \dot{m}_d \Delta b_V R^* \tag{28}$$

Values of R^* vary according to the method of operation:

$$\text{Countercurrent condensation:} \quad R^* = 1 - R_k \tag{29}$$

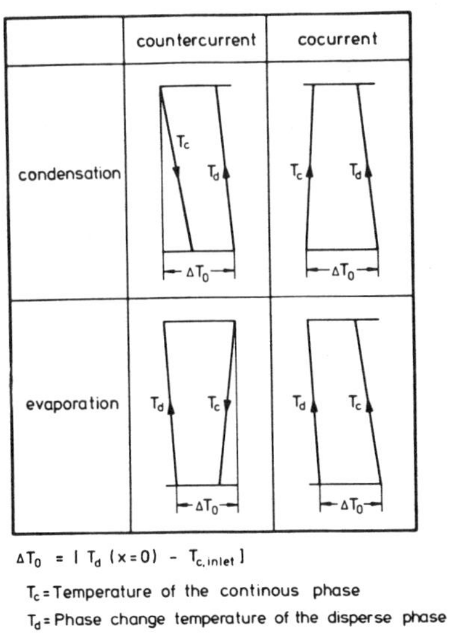

Figure 9. Temperature profiles in a direct contact heat exchanger with plug flow. $\Delta T_0 = T_d\ (x = 0) - T_{cE}$. T_c = temperature of continuous phase; T_d = phase-change temperature of dispersed phase.

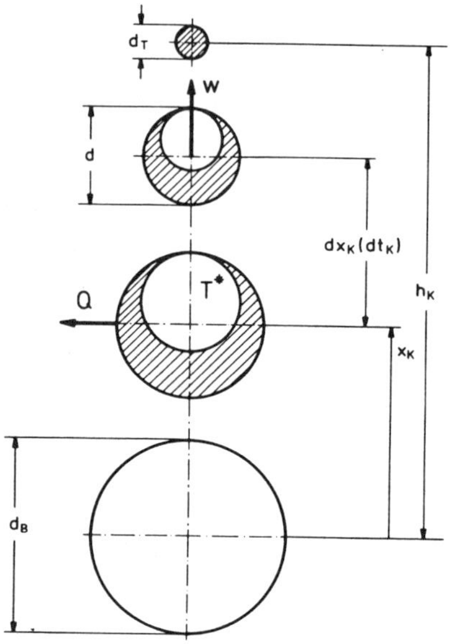

Figure 10. Model of condensation for a single drobble.

Cocurrent condensation: $R^* = R_k$ (30)

Countercurrent evaporation: $R^* = R_k = -(1-R_v)$ (31)

Cocurrent evaporation: $R^* = -(1-R_k) = -R_v$ (32)

The reference temperature difference for the dimensionless representation is defined by:

$$\Delta T_0 \equiv |T_{d0} - T_{cE}| \tag{33}$$

The relative temperature difference between the two phases:

$$\Delta T^* \equiv \frac{T_d - T_c}{\Delta T_0} \tag{34}$$

follows from Equations 27 and 28:

$$\Delta T^* = \frac{T_{d0} - T_{cE}}{\Delta T_0} - \frac{dT^*/dp}{\Delta T_0/(\rho_c g x)} - SR^* \tag{35}$$

ΔT^* is negative for condensation and positive for evaporation. Factor S in Equation 35

$$S \equiv \frac{\dot{m}_d \Delta b_V}{\dot{m}_c c_c \Delta T_0} = \frac{\dot{m}_d/\rho_g}{\dot{m}_c/\rho_c}\frac{1}{Ja} \tag{36}$$

represents the ratio of the phase-change enthalpy to the sensible heat produced by the temperature change ΔT_0 of the continuous liquid phase.

Model Equations

In order to establish the model equations, let us first look at a bubble with diameter d_B condensing to a drop with diameter d_T (Figure 10). During the condensation process, the drobble ascends to the condensation height h_K. At time t, a condensing bubble at position x has diameter d and transfers heat at flow rate $\dot{Q}$ to the supercooled liquid. For an infinitesimally short time interval dt_K or path dx_K, the transferred heat $\dot{Q}dt$ is equal to the product of the increase in condensate dm_K and latent heat of vaporation Δh_v:

$$\dot{Q}dt = \Delta h_v dm_K \tag{37}$$

On evaporation, heat is transferred to the drobble, while the condensate mass decreases and Equation 37 is also applicable. Using Equations 5 and 25 Equation 37 becomes

$$\alpha_c(T_d - T_c)\,dt = -\frac{1}{2}\frac{\Delta b_V \rho_g}{1-\dfrac{\rho_g}{\rho_l}}dd \tag{38}$$

Replacement of time increment by Equation 21 yields the second differential equation:

$$\alpha_c(T_d - T_c)\,dx = -\frac{1}{2}\frac{\Delta b_V \rho_g}{1-\dfrac{\rho_g}{\rho_l}}wdd \tag{39}$$

These equations can be expressed in dimensionless form using the bubble diameter d_B as a reference length and the relative temperature difference according to Equation 34 (see list of symbols):

$$\Delta T^* \mathrm{Fo}\, d\left(\frac{t}{\tau}\right) = -\frac{\dfrac{d}{d_B}}{2\left(1-\dfrac{\rho_g}{\rho_l}\right)\mathrm{Ja}\,\mathrm{Nu}}\, d\left(\frac{d}{d_B}\right) \tag{40}$$

$$\Delta T^* d\left(\frac{x}{d_B}\right) = -\frac{\mathrm{Pr}\,\mathrm{Re}}{2\left(1-\dfrac{\rho_g}{\rho_l}\right)\mathrm{Ja}\,\mathrm{Nu}}\, d\left(\frac{d}{d_B}\right) \tag{41}$$

For a numerical solution of these equations, the assumptions 3, 10, and 17 can be omitted.

Analytical solutions for the model differential Equations 40 and 41 are presented for different cases.

Analytical Solutions for Single Drobbles

First of all only single drobbles are considered. The factor S, defined by Equation 36 is in this case very small on account of low mass flow rate of the dispersed phase. Therefore, ΔT^* can be evaluated with Equation 35 where $S = 0$.

Solutions for Constant Temperature Difference

If the variation in the temperature of the drobble's phase change, caused by variation of pressure during ascent, can be neglected, the driving temperature difference remains constant throughout the entire process.

With Equations 35 and 33 the following is thus obtained:

$$\Delta T^* = \frac{T_{d0} - T_{cE}}{\Delta T_0} = \pm 1 \tag{42}$$

ΔT^* adopts the value of $+1$ for condensation, and -1 for evaporation.

Solution of the time-dependent differential equation. An analytical solution of Equation 40 is only possible at a constant value of ΔT^*. With Equations 24, 26, and 42 the following results from Equation 40

$$\mathrm{Fo}\, d\left(\frac{t}{\tau}\right) = \mp K_t \left\{\left[1-\frac{\rho_g}{\rho_c}\left(\frac{d_B}{d}\right)^3\right]\left(\frac{d}{d_B}\right)^3\right\}^{-E_t} \frac{d}{d_B}\, d\left(\frac{d}{d_B}\right) \tag{43}$$

with

$$E_t \equiv \frac{a_3}{2+b_2} \tag{44}$$

and

$$K_t \equiv \left[2\left(1-\frac{\rho_g}{\rho_l}\right)\mathrm{Ja}\, a_1\, \mathrm{Pr}^{a_2}\left(\frac{4\mathrm{Ga}}{3b_1}\right)^{E_t}\right]^{-1} \tag{45}$$

The upper sign in Equation 43 applies to condensation, and the lower one to evaporation. Integration proceeds via expansion of the power base in Equation 43 into a binomial series which, in this case, is convergent. With the boundary conditions for evaporation:

$$t_V = 0: \quad d = d_T$$
$$t_V = \tau: \quad d = d_B \tag{46}$$

and for condensation:

$$t_K = 0: \quad d = d_B$$
$$t_K = \tau: \quad d = d_T \tag{47}$$

the functional relationships between the drobble diameter and the time coordinate follow from Equation 43

$$\mathrm{Fo}\frac{t_V}{\tau} = K_t \sum_{j=0}^{\infty}\binom{-E_t}{j}\frac{(-\rho_g/\rho_c)^j}{2-3j-3E_t}\left[\left(\frac{d}{d_B}\right)^{2-3j-3E_t} - \left(\frac{\rho_g}{\rho_l}\right)^{2/3-j-E_t}\right]. \tag{48}$$

$$\mathrm{Fo}\frac{t_K}{\tau} = K_t \sum_{j=0}^{\infty}\binom{-E_t}{j}\frac{(-\rho_g/\rho_c)^j}{2-3j-3E_t}\left[1-\left(\frac{d}{d_B}\right)^{2-3j-3E_t}\right] \tag{49}$$

The overall time required for phase change is the same for both evaporation and condensation:

$$\mathrm{Fo} \equiv \frac{a_c\tau}{d_B^2} = K_t \sum_{j=0}^{\infty}\binom{-E_t}{j}\left(\frac{-\rho_g/\rho_c}{2-3j-3E_t}\right)^j \cdot\left[1-\left(\frac{\rho_g}{\rho_l}\right)^{2/3-j-E_t}\right] \tag{50}$$

The phase change profile is the same for both processes, with starting and terminal conditions merely being interchanged:

$$\tau_K = \tau_V = \tau = t_V + t_K \tag{51}$$

Numerical evaluation of the solution formulae presents no difficulties with a programmable pocket calculator, even in the presence of a very large number of terms. Various sample calculations have shown that, considering the greatly simplified model hypothesis, it is possible to limit the series to the first term without significant inaccuracies.

When j = 0, Equations 50, 48, and 51 reduce to the following:

$$\mathrm{Fo} = \frac{K_t}{(2-3E_t)}\left[1-\left(\frac{\rho_g}{\rho_l}\right)^{2/3-E_t}\right]$$

$$1-\frac{t_V}{\tau} = \frac{t_K}{\tau} = \frac{1-\left(\dfrac{d}{d_B}\right)^{2-3E_t}}{1-\left(\dfrac{\rho_g}{\rho_l}\right)^{2/3-E_t}} \tag{52}$$

Solutions of local differential equations. For an analytical solution, Equations 24, 26, and 42 are substituted into Equation 41

$$d\left(\frac{x}{d_B}\right) = \mp K\left\{\left[1-\frac{\rho_g}{\rho_c}\left(\frac{d_B}{d}\right)^3\right]\left(\frac{d}{d_B}\right)^3\right\}^E d\left(\frac{d}{d_B}\right) \tag{53}$$

with

$$E \equiv \frac{1-a_3}{2+b_2} \tag{54}$$

and

$$K \equiv \frac{\mathrm{Pr}^{1-a_2}}{2a_1\,\mathrm{Ja}\left(1-\dfrac{\rho_g}{\rho_l}\right)}\left(\frac{4\mathrm{Ga}}{3b_1}\right)^E \tag{55}$$

The minus sign applies to condensation and the plus sign to evaporation. With an expansion into series by analogy to the solution of Equation 43 and the boundary conditions for evaporation:

$$x_V = 0: \quad d = d_T$$

$$x_V = b_V: \quad d = d_B \tag{56}$$

and for condensation:

$$x_K = 0: \quad d = d_B$$

$$x_K = b_K: \quad d = d_T \tag{57}$$

the functional correlations between the drobble diameter and space coordinates are obtained from Equation 53

$$\frac{x_V}{d_B} = K\sum_{j=0}^{\infty}\binom{E}{j}\frac{(-\rho_g/\rho_c)^j}{1-3j+3E}\left[\left(\frac{d}{d_B}\right)^{1-3j+3E} - \left(\frac{\rho_g}{\rho_l}\right)^{1/3-j+E}\right] \tag{58}$$

$$\frac{x_K}{d_B} = K\sum_{j=0}^{\infty}\binom{E}{j}\frac{(-\rho_g/\rho_c)^j}{1-3j+3E}\left[1-\left(\frac{d}{d_B}\right)^{1-3j+3E}\right] \tag{59}$$

The total distance ascended by the rising drobbles during the phase-change process is the same for evaporation and condensation:

$$\frac{b_V}{d_B} = \frac{b_K}{d_B} = K\sum_{j=0}^{\infty}\binom{E}{j}\frac{(-\rho_g/\rho_c)^j}{1-3j+3E}\left[1-\left(\frac{\rho_g}{\rho_l}\right)^{1/3-j+E}\right] \tag{60}$$

The local dependence of phase change is similar in both cases with the starting and terminal conditions merely being interchanged:

$$b_V = b_K = b = x_V + x_K \tag{61}$$

In a practical application of the solution formulae, the summation must be discontinued at a finite value of "j." If a series is terminated after the first term, the following is obtained when j = 0.

$$\frac{b}{d_B} = \frac{K}{1+3E}\left[1-\left(\frac{\rho_g}{\rho_l}\right)^{1/3+E}\right] \tag{62}$$

$$1-\frac{x_V}{b} = \frac{x_K}{b} = \frac{1-\left(\dfrac{d}{d_B}\right)^{1+3E}}{1-\left(\dfrac{\rho_g}{\rho_l}\right)^{1/3+E}} \tag{63}$$

The model results for $\Delta T^* = \pm 1$ and j = 0 are illustrated in Figure 11 and Figure 12. They are valid for the density ratios ρ_l/ρ_g of n-butane at 20° C, a constant drag coefficient ($b_2 = 0$) and an exponent 0.5 on the Reynolds number in the heat transfer equation (Equation 61). The dependance of drobble diameter on instantaneous height can be calculated from Equation 63, and the corresponding condensation ratio and time coordinate can be determined from Equations 5 and 53 respectively.

$$\frac{x_K}{b} = \frac{1-\left\{1-\left[1-\left(\dfrac{\rho_g}{\rho_l}\right)^{2/3-E_t}\right]\dfrac{t_K}{\tau}\right\}^{(1+3E)/(2-3E_t)}}{1-\left(\dfrac{\rho_g}{\rho_l}\right)^{1/3+E}} \tag{64}$$

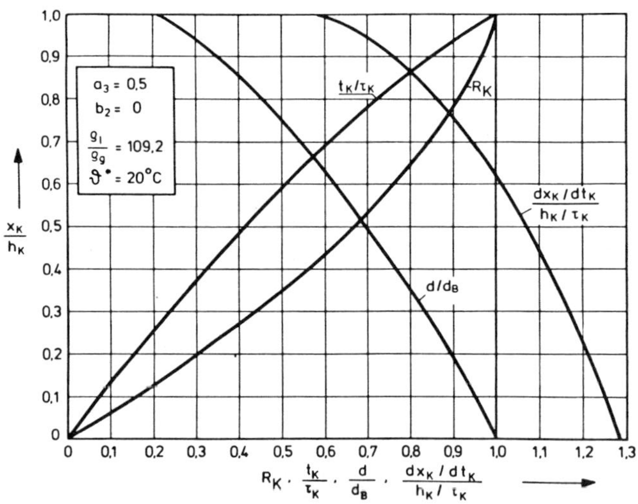

Figure 11. Model of single-drobble condensation (j = 0).

$$\frac{x_V}{b} = \frac{\left\{\left(\frac{\rho_g}{\rho_l}\right)^{2/3-E_t} + \left[1-\left(\frac{\rho_g}{\rho_l}\right)^{2/3-E_t}\right]\frac{t_V}{\tau}\right\}^{(1+3E)/(2-3E_t)} - \left(\frac{\rho_g}{\rho_l}\right)^{1/3+E}}{1-\left(\frac{\rho_g}{\rho_l}\right)^{1/3+E}} \tag{65}$$

Differentiation of this function produces the local drobble velocity, which is based here on the mean velocity:

$$\frac{dx_K/dt_K}{b/\tau} = \frac{1+3E}{2-3E_t}\,\frac{1-\left(\frac{\rho_g}{\rho_l}\right)^{2/3-E_t}}{1-\left(\frac{\rho_g}{\rho_l}\right)^{1/3+E}}\left\{1-\left[1-\left(\frac{\rho_g}{\rho_l}\right)^{2/3-E_t}\right]\frac{t_K}{\tau}\right\}^{(1+3E)/(2-3E_t-1)} \tag{66}$$

$$\frac{dx_V/dt_V}{b/\tau} = \frac{1+3E}{2-3E_t}\,\frac{1-\left(\frac{\rho_g}{\rho_l}\right)^{2/3-E_t}}{1-\left(\frac{\rho_g}{\rho_l}\right)^{1/3+E}}\left\{\left(\frac{\rho_g}{\rho_l}\right)^{2/3-E_t}+\left[1-\left(\frac{\rho_g}{\rho_l}\right)^{2/3-E_t}\right]\frac{t_V}{\tau}\right\}^{(1+3E)/(2-3E_t-1)} \tag{67}$$

Comparison with experimental values.: A heat transfer correlation, which was used by Sideman [7] for the phase change of drobbles, is used here for the comparison between the model calculation and the experimental data.

$$Nu = a_1 \cdot Pr^{a2} \cdot Re^{a3} = 0.27 \cdot Pe^{0.5} \tag{68}$$

Thus in the power product

$$a_1 = 0.27 \quad \text{and} \quad a_2 = a_3 = 0.5 \tag{69}$$

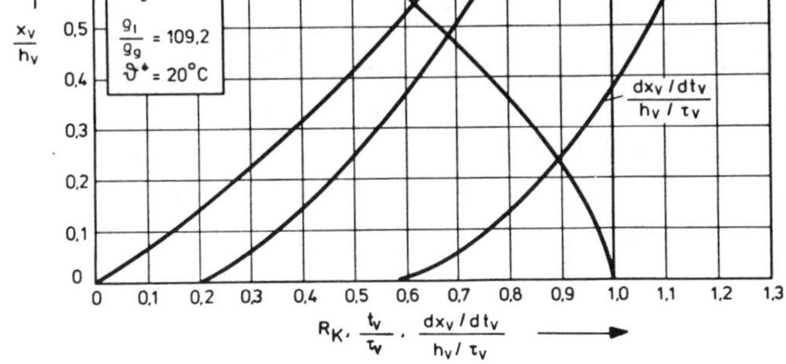

Figure 12. Model of single-drobble evaporation (j = 0).

The heat transfer coefficient calculated with Equation 68 increases with the condensation degree and hence with decreasing drobble diameter. At very high condensation degrees a decrease is, however, again observed, which is attributed to a very large fall in velocity under these conditions.

The drag coefficient $\Psi = 2.61$ valid for very large bubbles is employed for describing the drobble motion. In the resistance law (Equation 23) is $b_1 = 2.61$ and $b_2 = 0$. The latter is valid for all constant drag coefficients. Using experimental values of Nazir [8] comparisons were carried out with the time variation of drobble diameter in Figure 13 and in Figure 14 with the position line resulting from Equations 52, 62, and 65. In view of the greatly simplified model hypothesis the agreement is satifactory. In Figures 15 and 16 the required evaporation time and height according to Equations 52 and 62 respectively are compared with experimental values of Sideman [7]. A good agreement was obtained.

The most important parameter for equipment design, i.e. the overall height required, which can be calculated from Equation 62 is shown as a function of various parameters

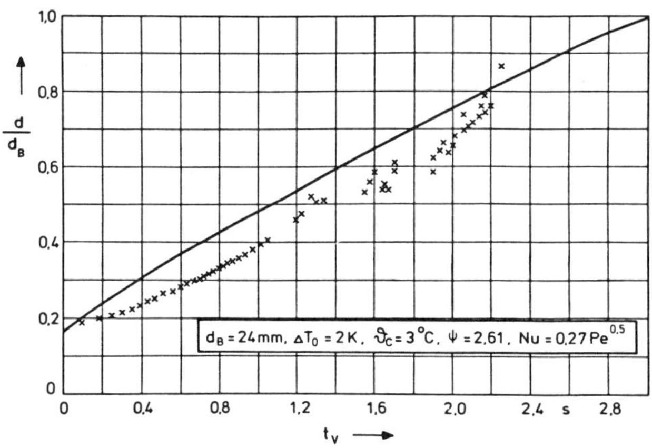

Figure 13. Diameter variation with time during single-drobble evaporation. x measurement points of Nazir.

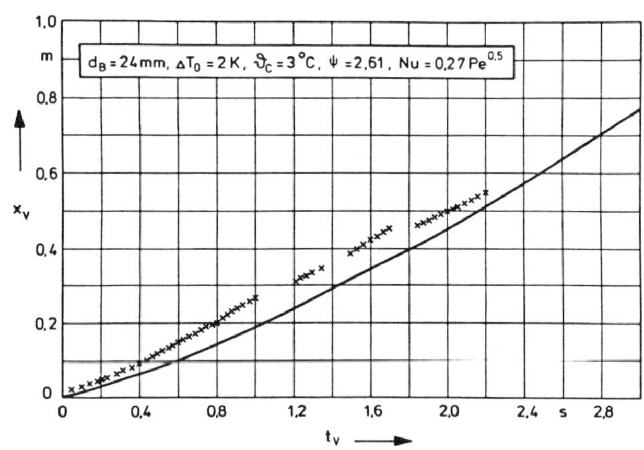

Figure 14. Positionline of an evaporating drobble. x measurement points of Nazir.

in Figure 17. A steep increase in this height with increasing bubble diameter and temperature is observed. However, the driving temperature difference, and the drag coefficient exert a strong reducing effect.

Solutions for a Variable Temperature Difference

With very small temperature differences, the dependence of the phase-change temperature on the height must be taken into consideration. In this case, Equation 35 is valid with S = 0. This, together with Equations 24 and 26 results in the following:

$$\left(1 \mp \frac{2}{q}\frac{x}{d_B}\right) d\left(\frac{x}{d_B}\right) = \mp K\left(\left\{1 - \frac{\rho_g}{\rho_c}\left[\left(\frac{d_B}{d}\right)^3\right]\frac{d}{d_B}\right\}^3\right)^E d\left(\frac{d}{d_B}\right) \quad (70)$$

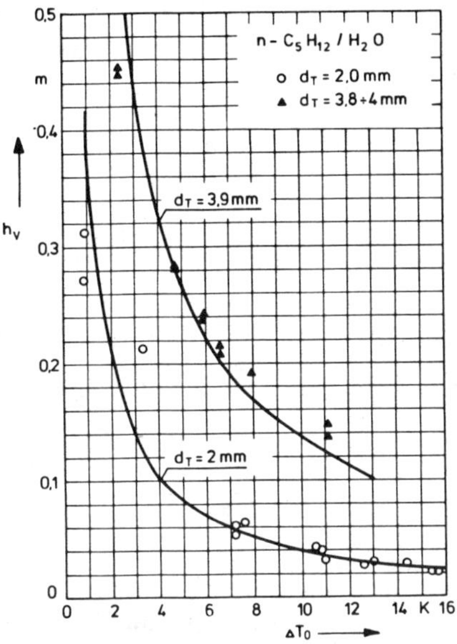

Figure 15. Comparison of evaporation heights, calculated with single-drobble model and experimental values of Sideman (▲ and 0).

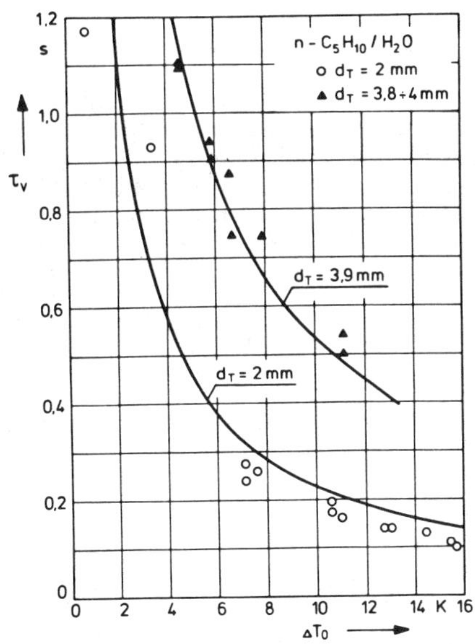

Figure 16. Comparison of evaporation time, calculated with single-drobble model and experimental values of Sideman (▲ and 0).

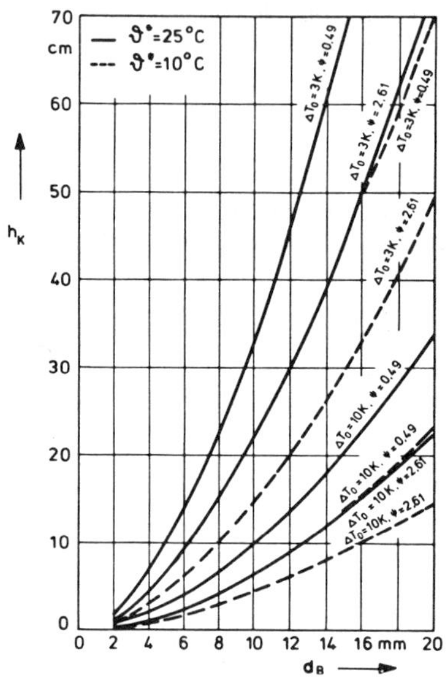

Figure 17. Condensation height calculated according to the drobble model, as a function of the bubble diameter. Parameters are two different coefficients ψ, two condensation temperatures T^*, and two temperature differences ΔT_0.

with

$$q \equiv 2\frac{\Delta T_0 dp}{\rho_c g d_B dT^*} \tag{71}$$

Denoting integral x/d_B for $\Delta T^* = \pm 1$ by I, according to Equations 58 or 59 the physically meaningful solution of Equation 70 is:

$$\frac{x}{d_B} = \frac{q}{2}\left[1 - \left(1 \mp \frac{4I}{q}\right)^{0.5}\right] \tag{72}$$

In Equations 70 and 72 the upper sign applies to condensation, while the lower one to evaporation. With condensation, it should be noted that $q > 1$, since otherwise the temperature difference vanishes before the condensation is completed. It follows from this, that it is possible to condense a bubble and to revaporize the resulting drop under a reduced static head. In contrast to Equation 60 for variable temperature differences, the overall height required for condensation is greater than for vaporization, although conditions are otherwise similar:

$$b_K - b_V = \frac{q}{2}\left[\left(1 + \frac{4I}{q}\right)^{0.5} - \left(1 - \frac{4I}{q}\right)^{0.5}\right] d_B \tag{73}$$

In Equation 73 $I = h/d_B$ according to Equation 60.

Analytical Solutions for Drobble Swarms

When examining drobble swarms, the mass flow rate $\dot{m}_d$ of the dispersed phase cannot be neglected but the temperature gradient of the vapor pressure curve can. Therefore the following results from Equation 35:

$$\Delta T^* = \pm 1 - SR^* \tag{74}$$

The plus sign applies to condensation and the minus sign to evaporation. Hence with Equations 29 through 32, the temperature gradient is obtained from Equation 74. This may be used together with Equations 24 and 26 in Equation 41. Both the numerator of the resulting expression, which has already appeared in Equation 53 and the reciprocal of the denominator (ΔT^*) are expanded into a binomial series. The Cauchy product of these two series can be integrated. Equations and solutions for the four methods of operation illustrated in Figure 9 are presented in the following.

Countercurrent Condensation (K_{gg}) *and Cocurrent Evaporation*

$$d\left(\frac{{}^{x}K_{gg}/V_{gl}}{d_B}\right) = \mp K \frac{\left\{\left[1-\frac{\rho_g}{\rho_c}\left(\frac{d_B}{d}\right)^3\right]\left(\frac{d}{d_B}\right)^3\right\}^E}{1-S(1-R_K)} d\left(\frac{d}{d_B}\right) \tag{75}$$

The upper sign applies to condensation, the lower to evaporation. With the corresponding boundary conditions Equation 56 or 57 the solution (Equations 76 through 78) valid for $0 \leqq S \leqq 1$, are obtained:

$$\frac{{}^{x}K_{gg}}{d_B} = \frac{K}{1+S\frac{\rho_g/\rho_l}{1-\rho_g/\rho_l}} \sum_{j=0}^{\infty}\sum_{k=0}^{j} \frac{\binom{E}{K}\left[\frac{S}{1-\frac{\rho_g}{\rho_l}(1-S)}\right]^{j-k}\left(\frac{\rho_g}{\rho_c}\right)^k}{1+3(j-2k+E)} \cdot \left[1-\left(\frac{d}{d_B}\right)^{1+3(j-2k+E)}\right] \tag{76}$$

$$\frac{{}^{x}V_{gl}}{d_B} = \frac{K}{1+S\frac{\rho_g/\rho_l}{1-\rho_g/\rho_l}} \cdot \sum_{j=0}^{\infty}\sum_{k=0}^{j} \frac{\binom{E}{K}\left[\frac{S}{1-\frac{\rho_g}{\rho_l}(1-S)}\right]^{j-k}\left(-\frac{\rho_g}{\rho_c}\right)^k}{1+3(j-2k+E)} \cdot \left[\left(\frac{d}{d_B}\right)^{1+3(j-2k+E)} - \left(\frac{\rho_g}{\rho_l}\right)^{1/3+j-2k+E}\right] \tag{77}$$

$$\frac{{}^{h}K_{gg}}{d_B} = \frac{{}^{h}V_{gl}}{d_B} = \frac{K}{1+S\frac{\rho_g/\rho_l}{1-\rho_g/\rho_l}} \cdot \sum_{j=0}^{\infty}\sum_{k=0}^{j} \frac{\binom{E}{K}\left[\frac{S}{1-\frac{\rho_g}{\rho_l}(1-S)}\right]^{j-k}\left(-\frac{\rho_{g90}}{\rho_c}\right)^k}{1+3(j-2k+E)} \cdot \left[1-\left(\frac{\rho_g}{\rho_l}\right)^{1/3+j-2k+E}\right] \tag{78}$$

$$\frac{{}^{h}K_{gg}}{d_B} = \frac{{}^{h}V_{gl}}{d_B} = \frac{K}{1+S\dfrac{\rho_g/\rho_l}{1-\rho_g/\rho_l}} \cdot \left\{ \sum_{j=0}^{\infty} \binom{E}{j} \frac{(-\rho_g/\rho_c)^j}{1-3j+3E}\left[1-\left(\frac{\rho_g}{\rho_l}\right)^{1/3-j+E}\right] \right.$$

$$+ \sum_{j=1}^{\infty}\sum_{k=0}^{j=1} \frac{\binom{E}{K}\left[\dfrac{S}{1-\rho_g/\rho_l(1-S)}\right]^{j-k}\left(-\dfrac{\rho_g}{\rho_c}\right)^k}{3\left(\dfrac{1}{3}+j-2k+E\right)}$$

$$\left. \cdot \left[1-\left(\frac{\rho_g}{\rho_l}\right)^{1/3+j-2k+E}\right]\right\} \tag{79}$$

The total height required is the same for both countercurrent condensation and cocurrent evaporation.

By numerical evaluation of the series, summation up to $j = 2$ should be carried out. Writing the infinite series in Equation 79 in a somewhat different form, when the density ratio ρ_g/ρ_l is negligible compared to unity, the overall height required can be regarded as composed of two independent terms i.e. one corresponding to single drobbles, and the other representing the swarm effect. Figure 18 shows the increase in the required height with the mass flow rate of the dispersed phase. For countercurrent condensation, the temperature variation of the continuous phase with height is shown in Figure 19. This temperature profile can be calculated with aid of Equations 74, 29, 34, and 76. As the semi-logarithmic representation shows, it is not an exponential process, occurring under the assumption of the volumetric heat transfer coefficient being constant.

Cocurrent Condensation (K_{gl}) *and Countercurrent Evaporation* (V_{gg})

$$d\left(\frac{{}^{x}K_{gl}/{}^{x}V_{gg}}{d_B}\right) = \mp K \frac{\left\{\left[1-\dfrac{\rho_g}{\rho_c}\left(\dfrac{d_B}{d}\right)^3\right]\left(\dfrac{d}{d_B}\right)^3\right\}^E}{1-SR_K} d\left(\frac{d}{d_B}\right) \tag{80}$$

The upper sign is valid for condensation, the lower for evaporation. The solutions, corresponding to the boundary conditions specified by Equation 56 or 57 depend on the value of the constant S. The following functions are defined in order to abbreviate the extensive solution equations and to provide a clear presentation:

$$F_I(a, b) = \frac{K}{1-\dfrac{S}{1-\dfrac{\rho_g}{\rho_l}}} \cdot \sum_{j=0}^{\infty} (-1)^j \sum_{k=0}^{j} \frac{\binom{E}{K}\left(\dfrac{\rho_g}{\rho_c}\right)^k\left(\dfrac{S}{1-S-\rho_g/\rho_l}\right)^{j-k}}{3\left(\dfrac{1}{3}+j-2k+E\right)}$$

$$\cdot \left[a^{3(1/3+j-2k+E)} - b^{3(1/3+j-2k+E)}\right] \tag{81}$$

$$F_{II}(a, b) \equiv \frac{K}{S}\left(1-\frac{\rho_g}{\rho_l}\right)\sum_{j=0}^{\infty}\sum_{k=0}^{j}\binom{E}{K}\left(\frac{\rho_g}{\rho_c}\right)^k\left(\frac{1-S-\rho_g/\rho_l}{S}\right)^{j-k}$$

$$\cdot \frac{(-1)^j}{3\left(E-j-\dfrac{2}{3}\right)}\left[a^{3(E-j-2/3)} - b^{3(E-j-2/3)}\right] \tag{82}$$

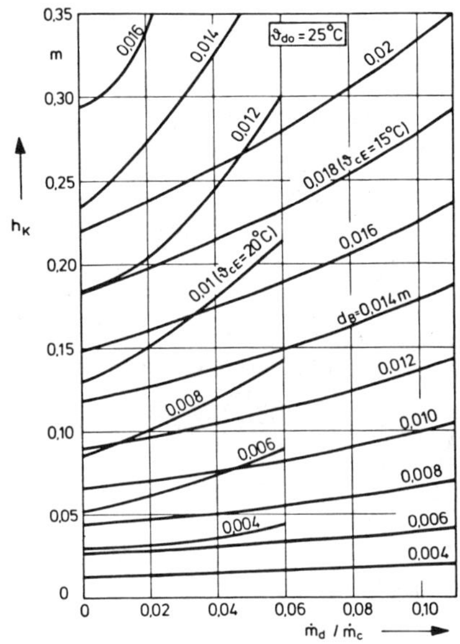

Figure 18. Condensation heights, calculated with drobble model (j = 2), for countercurrent condensation of n-butane in water, as a function of the mass flow ratio of the two phases for two different water inlet temperatures. Parameter is the bubble diameter.

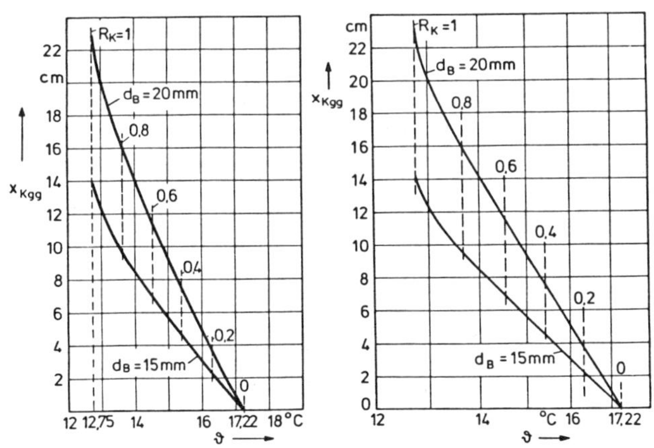

Figure 19. Heating profiles, calculated with the drobble model (j = 2), of the continuous phase with countercurrent condensation of n-butane in water for two different bubble diameters.

$Nu = 0.27\ Pe^{0.5}$

$\zeta^* = 24.3°C$

$\psi = 2.61$

$$\frac{\dot{m}_c}{\dot{m}_c} = 19.35$$

$$F_{III}(a, b) \equiv F_{II}\left\{a, \frac{d_S}{d_B}\right\} + F_I\left\{\frac{d_S}{d_B}, b\right\}$$

$$= \frac{K}{S}\left(1 - \frac{\rho_g}{\rho_l}\right) \sum_{j=0}^{\infty} (-1)^j \sum_{k=0}^{j} \binom{E}{K} \left(\frac{\rho_g}{\rho_c}\right)^k \frac{(d_S/d_B)^{3(j-k)}}{3\left(E - j - \frac{2}{3}\right)}$$

$$\cdot \left[a^{3(E-j-2/3)} - \left(\frac{d_S}{d_B}\right)^{3(E-j-2/3)}\right] + \frac{(d_S/d_B)^{3(k-j-1)}}{3\left(\frac{1}{3} + j - 2k + E\right)}$$

$$\cdot \left[\left(\frac{d_S}{d_B}\right)^{3(1/2+j-2k+E)} - b^{3(1/3+j-2k+E)}\right] \tag{83}$$

$$\frac{d_S}{d_B} \equiv \left(\frac{1 - S - \frac{\rho_g}{\rho_l}}{S}\right)^{1/3} \tag{84}$$

The solutions in the range $0 \leqq S \leqq \frac{1}{2}\left(1 - \frac{\rho_g}{\rho_l}\right)$ are:

$$\frac{{}^xK_{gl}}{d_B} = F_I\left(1, \frac{d}{d_B}\right) \tag{85}$$

$$\frac{{}^xV_{gg}}{d_B} = F_I\left[\frac{d}{d_B}, \left(\frac{\rho_g}{\rho_l}\right)^{1/3}\right] \tag{86}$$

$$\frac{{}^hV_{gg}}{d_B} = \frac{{}^hK_{gl}}{d_B} = F_I\left[1, \left(\frac{\rho_g}{\rho_l}\right)^{1/3}\right] \tag{87}$$

The solutions in the range $\frac{1 - \rho_g/\rho_l}{1 + \rho_g/\rho_l} \leqq S \leqq 1$ are:

$$\frac{{}^xK_{gl}}{d_B} = F_{II}\left[1, \frac{d}{d_B}\right] \tag{88}$$

$$\frac{{}^xV_{gg}}{d_B} = F_{II}\left[\frac{d}{d_B}, \left(\frac{\rho_g}{\rho_l}\right)^{1/3}\right] \tag{89}$$

$$\frac{{}^hV_{gg}}{d_B} = \frac{{}^hK_{gl}}{d_B} = F_{II}\left[1, \left(\frac{\rho_g}{\rho_l}\right)^{1/3}\right] \tag{90}$$

In the intermediate range $\frac{1}{2}\left(1 - \frac{\rho_g}{\rho_l}\right) < S < \frac{1 - \rho_g/\rho_l}{1 + \rho_g/\rho_l}$ the following is valid:

$$\frac{{}^hV_{gg}}{d_B} = \frac{{}^hK_{gl}}{d_B} = F_{III}\left[1, \left(\frac{\rho_g}{\rho_l}\right)^{1/3}\right] \tag{91}$$

The overall height required is the same in all ranges for countercurrent evaporation and cocurrent condensation. Depending on the drobble diameter, up to which they are integrated, the solutions read as follow:

$\frac{d}{d_B} \geqq \frac{d_S}{d_B}$:

Equation 88 for $\frac{{}^{x}K_{gl}}{d_B}$; Equation 86 for $\frac{{}^{x}V_{gg}}{d_B}$

$$\frac{d}{d_B} \leqq \frac{d_S}{d_B}$$

$$\frac{{}^{x}K_{gl}}{d_B} = F_{III}\left[1, \frac{d}{d_B}\right] \tag{92}$$

$$\frac{{}^{x}V_{gg}}{d_B} = F_{III}\left[\frac{d}{d_B}, \left(\frac{\rho_g}{\rho_l}\right)^{1/3}\right] \tag{93}$$

The maximum value of S is unity. In this case, the temperature of the continuous phase reaches the value of the phase change temperature. Thus heat transfer proceeds until thermal equilibrium is attained.

If S = 0, Equation 60 for single drobbles results from Equations 78 and 87, Equation 59 results from Equation 76 and Equation 85, and Equation 58 result from Equations 77 and 86.

EXPERIMENTAL SET-UP

The equipment illustrated in Figure 20 was used for the experimental investigations of direct-contact heat transfer with condensation in drobble columns. It consists of a stainless steel pressure tank, 40 cm in diameter and 1.5 m high.

The maximum gauge pressure was 2 bar. A visual observation of the process is made possible through parallel plane glass plates. A concentric perspex tube, 1 m in height, defines the volume for direct contact heat transfer. This tube is exchangeable so that the inner diameter d_S can be varied.

The continuous phase consists of water. n-Butane was selected as the dispersed phase. Water flows downwards through the column. Gaseous butane is dispersed by a sparger ring at the bottom of the column. Table 1 shows the characteristic data of the ring distributor. The water temperature is lower than the dew point of butane and, consequently, the gas bubbles condense to immiscible drops.

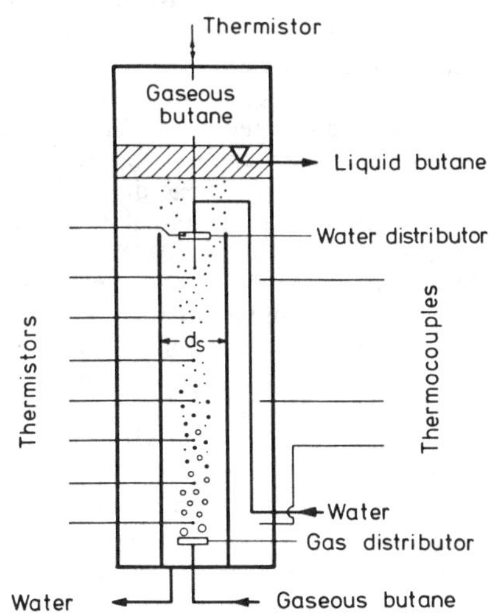

Figure 20. The direct-contact heat exchanger for the condensation of n-butane in water.

Table 1
Characteristic Data of the Gas Distributors

d_S mm	d_{R0} mm	d_o mm	n_0	t mm
70	50	2	26	6
70	50	1	52	3
140	100	2	52	6
140	100	1	124	2.5
190	138	2	72	6

Between the beginning and the end of this metamorphosis, the particles pass through a drobble state. The density of liquid butane being less than that of water the drops rise and coalesce to a liquid layer. Liquid butane flows out through a funnel.

In order to exclude inert gases, there is gaseous butane above the liquid layer. The pressure of this butane is regulated by means of a thermostated container, connected to the column. The height above the gas distributor, where the condensation of the drobble swarm is completed, can be determined visually and is referred to as condensation height h_{KV}.

The temperature distribution inside the heat exchanger tube can be measured by several thermistors, which can be displaced horizontally, and by one central thermistor, which can be moved in the vertical direction.

CONDENSATION HEIGHT AND TEMPERATURE PROFILES

Axial Temperature Profile

The temperature profile along the center axis of the condenser pipe was determined by an axially displaceable thermistor. In Figure 21 typical temperature profiles obtained at a constant flow rate of butane are presented for different water flow rates.

The axial temperature profile can be clearly divided into three different zones. In the first zone above the gas distributor, the measured temperature, after a short rise, is almost constant and acquires values which are higher than those of the measured temperature in the water outlet pipe. This difference in temperature results first from the fact that the thermistor also comes into contact with bubbles and drops and secondly from the fact that the here centrally measured temperature is higher than that at the periphery of the condensation column.

While the temperature increase measured in the first few centimeters above the gas distributor is independent of the flow rate, the nearly constant temperature measured in the subsequent zone increases with decreasing water flow rate. From the temperature constancy an intensive mixing of the dispersion can be concluded, which must be constricted to the central region, as the radial temperature distribution shows. The point of intersection between the vertical lines of the first zone and the regression lines through the measurement points of the second region can be seen as the range limit. The break in the temperature profiles for the different water flow rates can be well approximated by a line. The visual observation of the drobble condensation implies that the majority of the change in state takes place in the initial zone. It is not possible to measure this part exactly, but just to estimate it. It amounts to about 80% to 90%. The condensation reaches its termination in the second area above the gas distributor, where the temperature decreases constantly. The upper height of the second area, which is extractable from the temperature profile, shows the limit of the region in which the phase change takes place.

This limit is obtained from the point of intersection of the regression line through the experimental points of the second area with that through the third region or the lower part

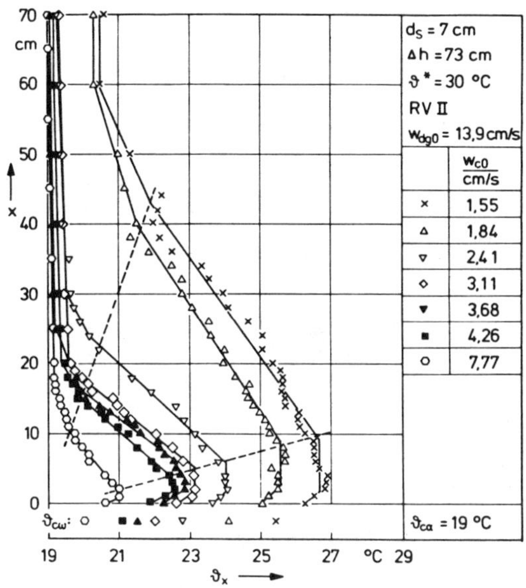

Figure 21. The axial temperature profile with constant butane flow rate and varying water flow rate.

of the third region. The break in the temperature profiles with different water flow rates can be approximated by a regression line.

In the highest third area of the heat exchanger pipe the axial temperature initially decreases more slowly than in the second area but also approximately linearly until it acquires almost constant values.

The supercooling of the drops after the total condensation only accounts for up to 6% of the water temperature change. The measured larger drop in temperature in the third area can be explained by two effects. To a small extent by the fact that the thermistor is touched by warmer drops and to a larger extent that the continuous phase does not flow through the drobble column free from back-mixing. This hypothesis is supported by the radial temperature profile (see Figure 25).

The most important information to be extracted according to the axial temperature profile is the height of the break point h_{KII} at the point of intersection between the regression lines of the second and third zones.

With butane flow rates other than those in Figure 21, small deviations from the described principal profile of the axial temperature can occur. Often, in the first zone above the gas distributor, only an increase in the temperature is observed and not a region of constancy (Figure 22). Also the measuring points of the third zone, can sometimes be approximated only by one regression line. Moreover in special cases with a small water rate it is also possible that the regression lines of the second and third areas acquire the same slope, and therefore there is no second break point to be calculated. If in this case a further reduction of the water flow rate is possible a point of intersection can be found again. Here the regression line of the third zone runs more horizontally than that of the second zone (Figures 23 through 25).

If the axial temperature profiles for different butane flow rates with equal water flow rates are plotted, a diagram (Figure 22) results which is very similar to Figure 21.

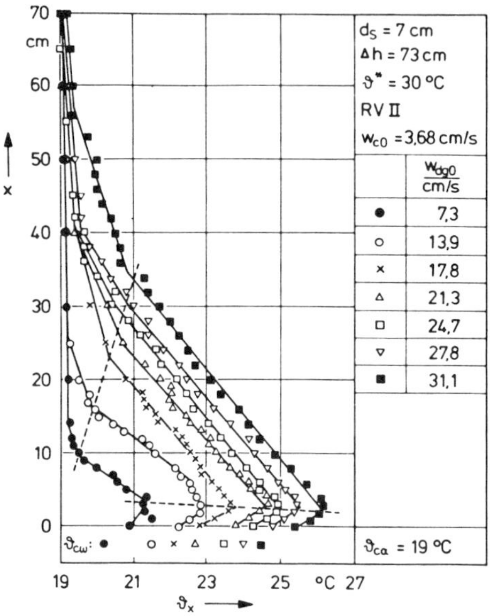

Figure 22. The axial temperature profile with constant water flow rate and varying butane flow rate.

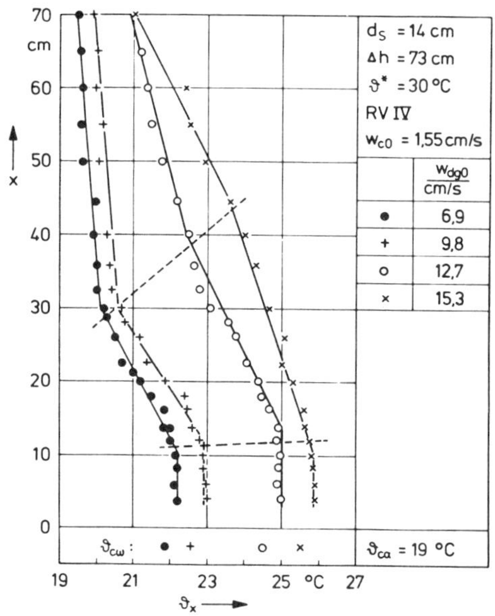

Figure 23. The axial temperature profile with constant water flow rate and varying butane flow rate.

Both the first break point between the first and second zone and the second break point between the second and third zone of the axial temperature profiles for different butane flow rates can be approximated by lines.

Radial Temperature Profiles

The temperature profile along a symmetrical axis of the column cross-section can be measured using horizontally displaceable thermistors. Figure 25 shows radial temperature profiles in which the measuring points are connected by polygone. In order to better compare the results which were obtained under different experimental conditions, the differences in temperature to the center value of the column are illustrated instead of absolute temperatures. The temperature differences measured at a constant height can show magnitudes which are half the size of the total temperature change that the water experiences during the flow through the direct contact heat exchanger.

Tha radial temperature profile becomes more horizontal with increasing distance from the gas distributor, with decreasing contents of gaseous dispersed component. The lower the water velocity, the later the reduction in the cross-sectional temperature variations occurs.

A very clear example of the effect of back-mixing of the water is shown in Figure 24 by the temperature profile for the very small superficial water velocity of 3.1 mm per second. The temperature, measured with the vertically displaceable thermistor in the center of the column only changes by 1° K along a distance of 65 cm while the total change in the water temperature is about 8° K and the largest part of the condensation takes place in the first 15 cm above the gas distributor.

Condensation Heights

While the value of visually determined condensation heights is very much dependent on the experimentator, the height of the second break point in the axial temperature profile is the result of objective measurement values and rather independent regression calculations. Figure 26 illustrates a comparison between both heights. In almost all cases, the visually-determined condensation height is lower than the break point height which likewise can be also called condensation height. Earlier it was shown from model calculations that the condensation height is a function of the mass flow ratio of both phases. Thus it is obvious to plot the experimentally determined condensation heights versus the ratio of the phase velocities as it is presented in Figure 27. The measurement values lie in a spread of 40%.

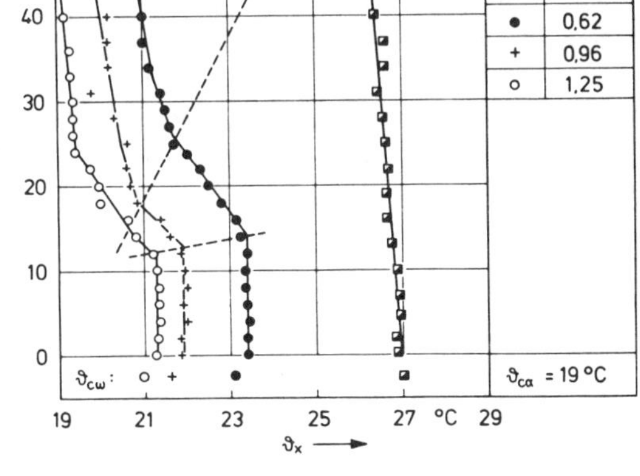

Figure 24. The axial temperature profile with varying water flow rate and constant butane flow rate.

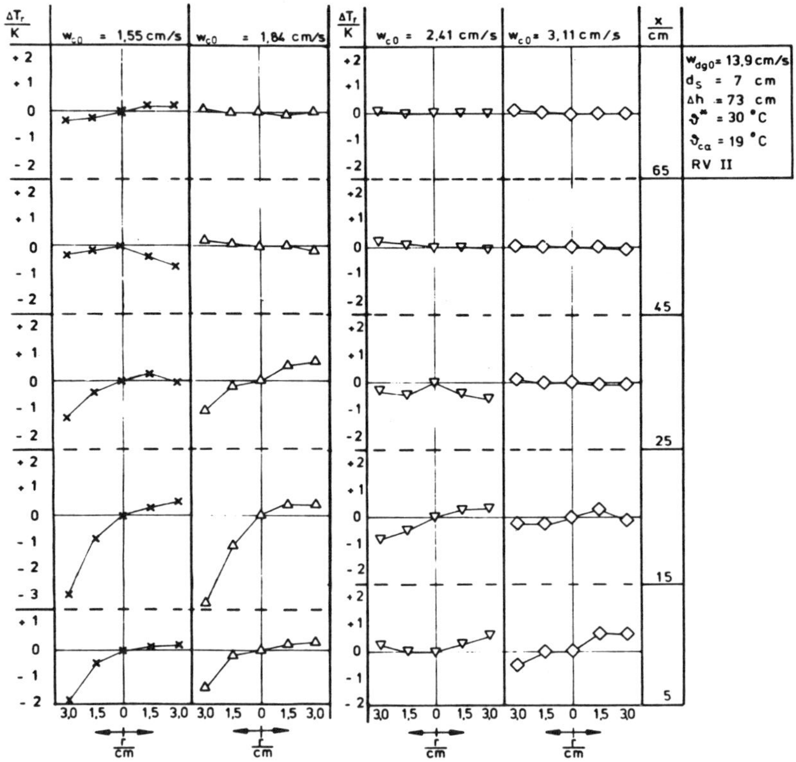

Figure 25. Radial temperature profiles for increasing water flow rate at constant butane flow rate.

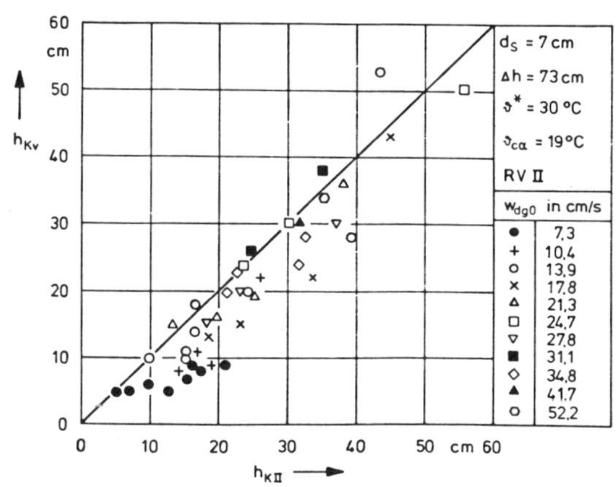

Figure 26. Comparison between visually obtained condensation heights and the height coordinates of the second break point.

Figure 28 shows the condensation height in relation to the drobble column diameter for different column diameters d_s.

It is seen here that smaller drobble columns show lower condensation heights for the parameter range considered.

Only those values of the 7-cm column are shown which lie in the gas velocity range of the larger drobble column. The condensation height of the 7-cm column diverges greatly from the illustrated line for superficial gas velocities above 15 cm/s.

Comparison between the Experimental Results and the Results of the Mathematical Model for Drobble Swarms.

The experimental results described for direct contact condensation of butane in water are now compared with the results of the mathematical model described earlier. The

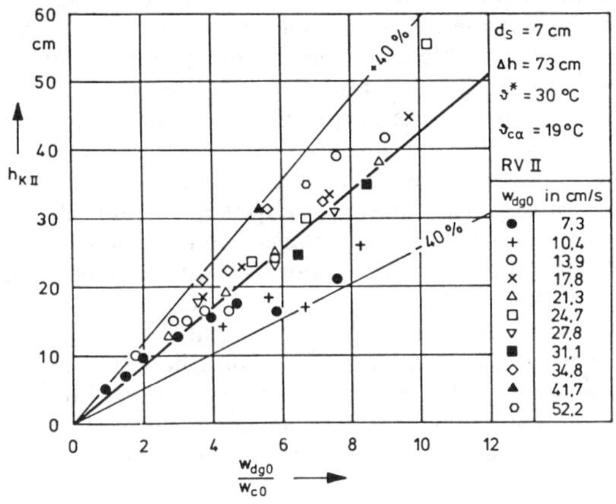

Figure 27. The condensation height as a function of the ratios of the superficial velocities.

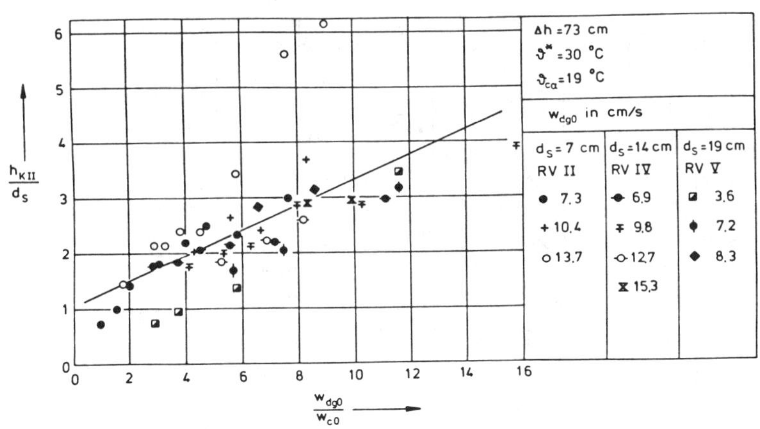

Figure 28. The condensation height in relation to the drobble column as a function of the ratio of the superficial velocities.

condensation height for countercurrent operation can be calculated using Equations 24 and 25. The infinite series of the equation can be terminated after the term for s = 2. From the resistance law, a relationship is obtained for middle-sized bubbles

$$\Psi = 7.03 \cdot 10^{-13}\,\mathrm{Re}^4 \tag{94}$$

The bubble diameter which is necessary for the calculation is obtained from Figure 32. (described later). The determined condensation heights are shown as a function of the ratio of superficial velocities in Figure 29 for the ring distributor RV II. The calculated values lie in a range between 4.3 cm and 7.5 cm.

If the bubble diameter or the flow rate of the dispersed phase increases, under otherwise unchanged conditions, the condensation height becomes larger. It decreases if the flow rate of the continuous phase increases. As Figure 32 shows, for Weber numbers greater than one, the bubble diameter decreases with increasing dispersed flow rate and increases again after running along the minimum.

The extremum lies at a Weber number of about seven (Figure 31). Until reaching this minimum the condensation height is affected by two counteractive influences. With constant water flow rate the height increase due to the increased butane flow rate prevails over the height decrease due to smaller bubble diameters. If the ratio of butane and water velocities remains constant, then with increased butane flow rate the condensation height decreases a little bit. Therefore in Figure 29 the curve calculated for a superficial gas velocity of 10.4 cm/s lies somewhat below the curve for 7.3 cm/sec.

The calculated curves shown in Figure 29 are valid for constant gas velocity. As the bubble diameter is also already determined thereby they are only dependent on the water velocity. The calculated curves show an almost linear connection between the condensation height and the inverse value of the superficial water velocity. Figure 29 contains the visually determined condensation heights beside these calculated heights. For very high water velocities, the measured and calculated values agree quite well.

This is because of the increased approximation of plug flow in the continuous phase with the increasing water flow rate, which is an important assumption of the mathematical model.

The deviation of the measured condensation heights from the calculated heights is very large as the water velocity decreases. In the calculated condensation height there only amounts to 15% of the measured heights. The following two phenomena should explain the causes of this discrepancy.

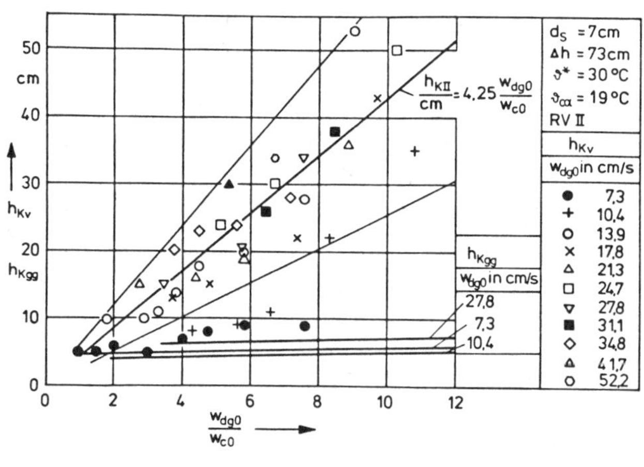

Figure 29. Comparison of the visually determined condensation heights with the calculated condensation heights for countercurrent condensation in a drobble column free from back-mixing.

While the bubble size and therefore the bubble rising velocity only slightly changes when the gas flow increases, the orifice velocity increases proportionally. This leads to a blocking of the bubbles in the upper zone near the gas distributer and finally to the formation of very large drobbles by the effects of coalescence which can be well observed in the experiment. Thus, first, higher condensation heights result due to the longer condensation time and secondly due to stronger back-mixing effects of the continuous phase.

As the radial temperature profiles show (Figure 25), the back-mixing of the continuous phase increases with decreasing water velocity. The resulting drop in the driving temperature difference, which can be observed in the axial temperature profiles (Figure 21), also leads to an important increase of the necessary condensation height in comparison to the case of plug flow.

RANGE OF OPERATION

The range of operation of a drobble column is limited by maximum and minimum flow rates of the dispersed and continuous phases. This range is illustrated in a performance diagram in which the superficial velocity of the gaseous dispersed phase is plotted versus that of the continuous phase. The performance diagram of a drobble column which was employed in the experimental investigation is shown in Figure 30. The symbols denote the operating points, where the temperature profiles were measured under steady-state conditions [3]. The range of operation is defined by limiting lines whose significance is explained in the following sections.

Minimum Gas Velocity

Sparger rings are used to distribute the gas in the drobble column. The minimum throughput in the distributor depends either on weeping in the case of large holes or on uniform gas distribution in the case of smaller holes. Mersmann [11] presents calculation equations for both cases. At a hole diameter of about 3 mm, weeping becomes significant for the system n-butane/water. With hole diameters of 1 and 2 mm the distributors employed in the present work should only be considered under conditions of uniform aeration.

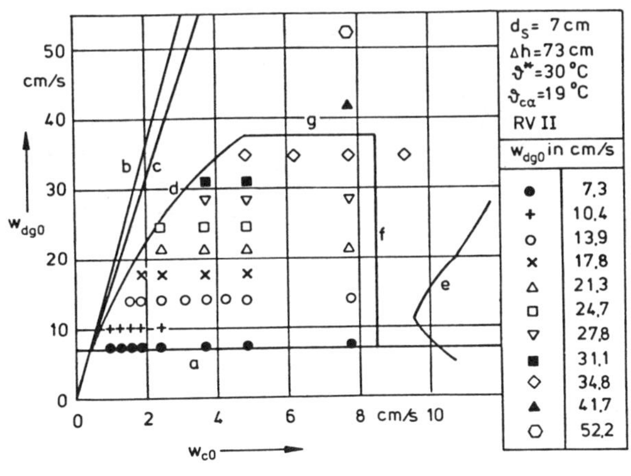

Figure 30. Performance diagram of a drobble column with countercurrent condensation.

Aeration by sparger rings operates steadily if flow occurs through all the holes. According to Mersmann [11] the lower limit is reached when the Weber number

$$\mathrm{We}_0 = \frac{\rho_g w_{dgD}^2 d_0}{\sigma} \tag{95}$$

reaches the value of two.

Hole velocity for uniform aeration by distributors with parallel openings can thus be calculated. The corresponding superficial velocity is obtained from

$$w_{dg0} = w_{dgD} \frac{n_0 d_0^2}{d_S^2} \tag{96}$$

The drobble column, whose performance diagram is shown in Figure 30 was operated with sparger ring RV II, which has $n_0 = 26$ holes diameter $d_0 = 2$ mm. For a pressure of 2.83 bar, a superficial velocity of 6.19 cm/s is obtained from Equations 95 and 96. To this value an uncertainty allowance of 15% is added [12], so that, finally, 7.12 cm/s is obtained as the minimum superficial velocity of the gaseous phase. The calculated value is identical to that obtained from the experiments and appears in Figure 30 as a horizontal straight line designated by a.

Minimum Liquid Velocity

Thermodynamic considerations permit the calculation of a minimum water flow rate. The water flows in at temperature $T_{c\alpha}$, and can reach the condensation temperature T^* under conditions of thermodynamic equilibrium. Since the water must remove the total condensation heat, increases in temperature are related to the flow rates via the energy balance.

$$\dot{m}_c c_c (T^* - T_{c\alpha}) = \dot{m}_d \Delta b_V \tag{97}$$

The result is a linear relationship between the minimum superficial velocity of water and that of the gas phase:

$$w_{c0} = \frac{\rho_g \Delta b_V}{\rho_c c_c (T^* - T_{c\alpha})} w_{dg0} \tag{98}$$

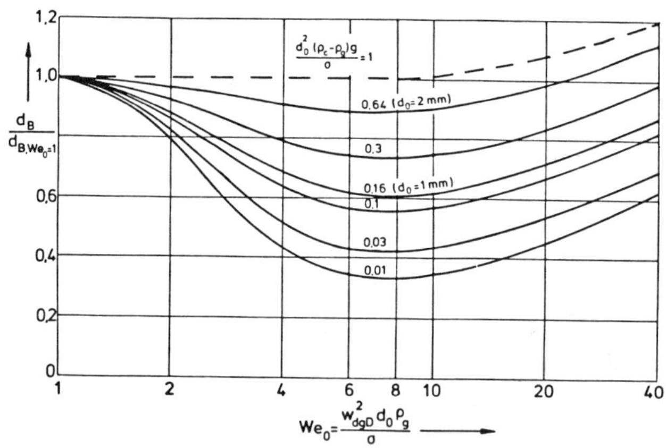

Figure 31. Bubble diameter versus Weber number, according to Mersmann [11].

Graphical representation of this equation is shown in Figure 30 (straight line b through the origin). Taking into account that the inlet gas is superheated to temperature $T_{d\alpha}$ and after condensation the liquefied gas is supercooled to the water inlet temperature $T_{c\alpha}$, the specific enthalpy change of the dispersed phase is

$$\Delta b_d = c_{pg}(T_{d\alpha} - T^*) + \Delta b_V + c_l(T^* - T_{c\alpha}) \tag{99}$$

On replacing Δh_V in Equation 98 by Δh_d, straight line c in Figure 30 is obtained.

Apart from the thermodynamic limiting lines b and c, another limiting line, characteristic of the equipment can be determined experimentally.

The condensation height h_{KV} increases with decreasing water flow rate. A reasonable operation of the drobble column is only possible if the condensation height does not exceed the distance Δh between the gas and water distributors. Therefore, the following must be valid:

$$b_{K\omega} \leqq \Delta b \tag{100}$$

With this condition, the obtained limiting velocity, characteristic of the equipment, is represented in Figure 30 by line d.

Maximum Liquid Velocity

After the occurrence of condensation, the dispersed phase consists of a swarm of drops which rise countercurrent to the continuous phase. If the downward velocity of the continuous phase exceeds the rise velocity w_T of a drop, the drop is entrained. With the admissible assumption that velocity of the continuous phase is identical to its superficial velocity, the condition

$$w_T = w_{c0} \tag{101}$$

represents an upper limit of the liquid velocity.

The drop diameter must be estimated since the volume of rising drops in the investigated drobble column were not measured. First, the diameter d_B of the dispersed bubbles is determined, and using the following equation:

$$\frac{d_B}{d_T} = \left(\frac{\rho_l}{\rho_g}\right)^{1/3} \tag{102}$$

the drop diameter d_T is obtained [13].

In order to predict the diameter of bubbles produced at the gas distributor, the method of Mersmann [11] is used. In the present case, calculations were performed with constant physical properties of water at 20° C and of butane at 30° C. The gas is assumed to be present as saturated vapor.

An important intermediate parameter in the course of the calculation is the Weber number of the hole, defined by Equation 95. Using Equation 96 the hole velocity can be eliminated so that the Weber number can be expressed as a function of the superficial velocity of the gas phase:

$$We_0 = \frac{\rho_g w_{dg0}^2 d_B^4}{\sigma n_0^2 d_0^3} \tag{103}$$

Hole velocity of the sparger ring is also a function of superficial velocity of the gaseous phase:

$$\dot{V}_{dgl} = \frac{\pi}{4} d_0^2 w_{dgD} = \frac{\pi}{4} \frac{d_S^2}{n_0} w_{dg0} \tag{104}$$

If the Weber number does not exceed unity, the bubble diameter is calculated, according to Reference 11 from:

$$d_B = \left[\frac{6}{\pi}\frac{V_B^* \dot{V}_{dgl}^{6/5}}{g^{3/5}}\right]^{1/3} \tag{105}$$

Here, the following equations are valid:

$$V_B^* = V_{Bl}^* + 4.7(\eta_c^*)^{3/4} + 1 \tag{106}$$

$$V_{Bl}^* = \pi(d_0\sigma)^* + 2.42\eta_c^*(V_{Bl}^*)^{-1/3} + 0.204(\eta_c^*)^{1/2}(V_{Bl}^*)^{-1/2}\left(1 - \frac{\rho_g}{\rho_c}\right)^{-1/2}$$
$$+ 0.0689(V_{Bl}^*)^{-2/3}\left(\frac{1.838\rho_c}{\rho_c - \rho_g} - 1\right) \tag{107}$$

$$\eta_c^* = \frac{\eta_c}{(\rho_c - \rho_g)g^{1/5}\dot{V}_{dgl}^{3/5}} \tag{108}$$

$$(d_0\sigma)^* = \frac{d_0\sigma}{(\rho_c - \rho_g)g^{2/5}\dot{V}_{dgl}^{6/5}} \tag{109}$$

Equation 107 must be solved iteratively. With Equation 103 through 109 it is possible to determine the bubble diameters, depending on superficial velocity of the gas phase.

If the Weber number is less than unity, only isolated bubbles are formed. Since the minimum gas velocity for a uniform distribution already requires a Weber number of two, smaller values are of no significance.

In the case of larger Weber numbers, the bubble diameter for a Weber number of one is calculated first with aid of Equations 103 through 109. The required bubble diameter can be determined from Figure 31. The diagram shows two lines for hole diameters of 1 and 2 mm, relating to the system n-butane/water, under given operating conditions. The bubble diameters calculated herewith are presented in Figure 32 for various employed sparger rings, as functions of the superficial gas velocity. The curves terminate at values which are obtained for a Weber number of forty, on account of the selected scale. At the relative maximum near the beginning of each curve, the Weber number is unity.

For sparger ring RV II, the bubble diameters depending on the superficial gas velocity were read from Figure 32 and the corresponding drop diameters were calculated, with the aid of Equation 102. The steady-state rise velocities of the drops are determined using the resistance law for rigid spheres put forward by Brauer [14].

$$\Psi_T = \frac{24}{Re_T} + \frac{3.73}{Re_T^{0.5}} - \frac{4.83 \times 10^{-3} Re_T^{0.5}}{1 + 3 \times 10^{-6} Re_T^{0.5}} + 0.49 \tag{110}$$

The rise velocity is set equal to the superficial velocity of the continuous phase so that curve e in Figure 30 could be constructed.

This predicted limiting curve cannot be verified experimentally for the following reasons:

The employed calculation method is based on measurement results, obtained in bubble and spray colums, and shows a deviation of 20% [11].

The break-up of a gas jet into single particles occurs in a system without a change of phase in a different manner from that at condensation point, where liquefaction can take place before the jet breaks up into bubbles.

The produced bubble or drobble swarm is not monodispersed.

The resultant size distribution of particles depends on the distributor geometry and operating conditions.

During the experimental investigations an entrainment of more than 5% of the total butane mass flow rate was regarded as unacceptable. The limiting curve for this arbitrary

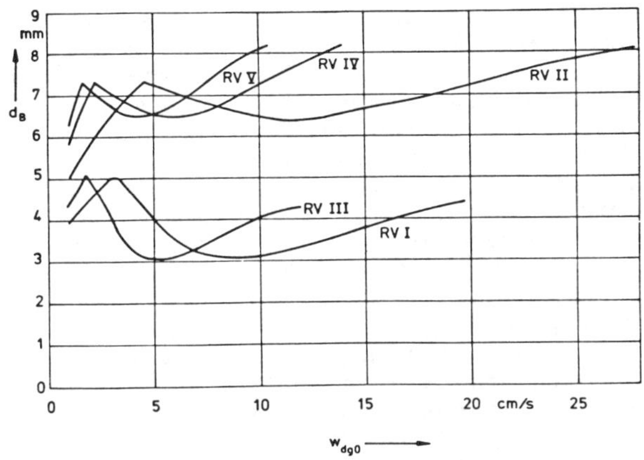

Figure 32. Bubble diameter versus superficial gas velocity for different sparger rings.

stipulation is shown in Figure 30 (curve f). The diagram contains three points which lie outside the range of operation. In spite of high butane entrainment the measurement program was completed. On account of measurement accuracy and the large span of operating points, line f cannot be regarded as the exact limit of the range of operation. Its position depends on the selected butane entrainment. Curve e represents the theoretical case of complete entrainment of a fictitious monodispersed drop swarm and comprises a useful approximation for the limiting curve f.

Maximum Gas Velocity

Entrainment of butane can be observed not only at high water velocities but also at high gas velocities. The relevant limiting line g is impossible to predict even approximately. The method, presented for determination of bubble diameter, cannot be employed in the region of line g (cf. Figure 32).

CONCLUSIONS

In this chapter, a new mathematical model of the phase change of drobbles and experimental results obtained at a drobble column with direct contact condensation were presented.

Herewith experimentally based guidances for the construction of a drobble column are available. As with countercurrent columns, for thermal separation processes, so also here the two main scales of a drobble column are initially determined by two almost separate paths and are then optimally harmonized considering the existing interdependances.

The diameter is mainly determined by fluid-dynamic principles, while the height is determined by the kinetics of the heat transfer process.

With reference to the phase velocities an operating point can be chosen with the help of a drobble column operation diagram and with consideration given to fluctuations of the flow rate. It is important to notice that some of the limiting lines of the diagram are dependent on the geometrical sizes of the apparatus. Superficial velocity and volume flow rate then lead to the column cross-section.

The condensation height can be determined for given flow rates. If the particular condensation height derived from the axial temperature profiles is used then a precaution is already included compared to the condensation height obtained visually.

The analytical solution of the mathematical model can be used if the gas velocities are lower than 10 cm/s and if the phase velocity ratio is no larger than four. As a precaution, a duplication of the calculated values is recommended.

It is to be expected that the model does not only lead to a usable solution for countercurrent condensation alone, which is verified experimentally but also in the case of cocurrent operations. Since in the cocurrent operation the flow is stabilized and the effects of back-mixing are minimized, at the same time the measurement values and the model calculations should even agree better.

While calculating the height of the apparatus it is assumed that the condensation height does not depend on the distance between the gas and the water distributor. This appears to be justified due to the comparative experiments with different distributor distances.

Due to the fact that the condensation height depends also on the column diameter it is important to coordinate both scales. The diameter should be small to achieve a small height, which implies that the superficial velocity increases and the operation limits are reached.

The gas distributor plays an important role in the interrelations between the condensation height and the operation range. If it produces small bubbles, the condensation level is reduced, however, the operation range is reduced too.

The volumetric heat transfer coefficient, which is a favorite characteristic number to describe direct-contact heat exchangers, should be used just as little as the concept of the transfer unit which is suggested in particular to describe the direct-contact heat transfer in drop columns. In both cases, values are deduced from integral measurements which characterize the total phase change, which however is not recognized at first sight. The volumetric heat transfer coefficient specifies the heat flow per unit volume. This heat flow is strongly dependent on location. It amounts directly above the gas distributor to a multiple of the value at the end of the condensation path. The reference volume of the volumetric heat transfer coefficient for the total condensation process is calculated together with the help of the condensation height. Since the latter is influenced by the value of the cross-sectional area of the drobble column care must be taken of a dimensioning using the volumetric heat transfer coefficient. It is therefore useful to immediately use the condensation height.

The practical uses for a drobble column are very limited because the substances involved may contaminate each other due to the direct contact. This is caused by a never completely expelled solubility and eventually by the entrainment of small particles.

A large problem in using the condensation drobble column is the presence of uncondensable gases. An inert gas prevents complete condensation if it is not soluble in the condensate. With increasing condensation degree, the molar fraction of the condensable components decreases. The partial pressure of the substance to be liquefied is in the case of equilibrium, as large as its vapour pressure at the operating temperature of the cooling liquid. Moreover the inert gas increases the gas volume of the drobble which therefore leads to higher rise velocities. Already small amounts of inert gas have a large influence on the condensation process.

A potential use of the condensation drobble column is the desalting of seawater using the freezing process. In the crystallization step of this process ice is frozen out by evaporating a substance like for example, n-butane which is not miscible with water. In a process with circulation of the refrigerant, the liquefaction can be conducted in a direct contact condenser, whereby the released condensation heat can be used to melt the crystals which have already been separated from the brine. In this case the drobble column has yet another phase namely the solid water ice. The apparatus should be cocurrently operated as the density of ice is smaller than that of water.

The water leaving the melting and condensation column must be refined due to the soluble refrigerant. Here, degasification and/or an adsorption process can be used.

NOTATION

a	thermal diffusivity (m^2/s)
a	value of the range limits in Equations 81 through 83
a	interfacial area per unit volume (m^2/m^3)
a_1, a_2, a_3	constants in Equation 26
A	area (m^2)
b	value of the range limits in Equations 81 through 83
b_1, b_2	constants in Equation 23
c	specific heat capacity (kJ/(kg°K))
d	diameter (m)
d_0	orifice diameter (mm)
d_{RO}	diameter of the distributor ring (mm)
d_s	diameter of the column (mm)
E	constant in Equation 54
E_t	constant in Equation 44
F_I, F_{II}, F_{III}	function in Equations 81 through 83
g	gravitational acceleration (m/s^2)
h	height (m)
Δh	difference in height (m)
h_K	condensation height (m)
Δh_V	specific heat of evaporation (kJ/kg)
I	integral x/d_B for $\Delta T^* = \pm 1$
j	index variable
k	index variable
K	constant in Equation 55
K_t	constant in Equation 45
m	mass (kg)
$\dot{m}$	mass flow (kg/s)
n	number of particles
n_0	number of orifices
p	pressure (bar)
q	constant in Equation 71
$\dot{Q}$	heat flow (kW)
r	radial coordinates (m)
R_K	condensation degree
R_V	evaporation degree
R^*	parameter in Equation 28
S	enthalpy ratio in Equation 36
t	orifice pitch (mm)
t	times coordiante (s)
T	temperature (°K)
T^*	boiling temperature (°K)
ΔT	difference in temperature (°K)
ΔT_0	difference in reference temperature (°K)
ΔT_r	radial difference in temperature (°K)
ΔT^*	dimensionless difference in temperature
V	volume (m^3)
w	particle velocity in relation to a fixed reference system (m/s)
w_{c0}	superficial velocity of the continuous phase (m/s)
w_{dg0}	superficial velocity of the gaseous dispersed phase (m/s)
x	height coordinate (m)

Greek Symbols

α	heat transfer coefficient $kW/(m^2K)$
Γ	molar fraction of the inert gas in the drobble (mole/mole)
ε_B	volume fraction of bubbles
ε_G	gas volume fraction of a drobble
ε_P	volume fraction of particles
ε_T	volume fraction of drops
ν	temperature (°C)
ν^*	boiling temperature (°C)
λ	thermal conductivity (W/(m°K)
ν	kinematic viscosity (m^2/s)
η	dynamic viscosity (kg/(ms))
$\Delta\rho$	density (kg/m^3)
τ	phase conversion time (s)

Subscripts

B	bubble
c	continuous phase
d	dispersed phase
E	entrance
f	end of the complete condensation

g gas phase of the dispersed medium
gg countercurrent
gl cocurrent
GK the gas phase of the drobble
K condensate, condensation
KK spherical calotte
KII second break point
l liquid phase of the dispersed medium
0 start of the phase conversion
p particle
s column
T drop
v visually
V vapor evaporation
α entrance
ω exit

Dimensionless Numbers

Fo Fourier number $\left(= \frac{a_c \cdot \tau}{d_{B2}}\right)$

Ga Galilei number $\left(= \frac{g d_{B3}}{\nu_{c2}}\right)$

Ja Jakob number $\left(= \frac{\rho_c c_c T_0}{\rho g \Delta h_v}\right)$

Nu Nusselt number $\left(= \frac{\alpha_c \cdot d}{\lambda_c}\right)$

Pe Péclet number (= Re · Pr)

Re Reynolds number $\left(= \frac{w_d \cdot d}{\nu_c}\right)$

ψ resistance coefficient $\left(= 4/3 \frac{gd}{w^2}\left(1 - \frac{\rho_g}{\rho_c}\right)\right)$

REFERENCES

1. Wiegandt, H. E., and von Berg, R. L., "Myths About Freeze Desalting," *Desalination,* Vol. 33 (1980), pp. 287–297.
2. Sudhoff, B., Plischke, M., and Weinspach, P. M., "Direct Contact Heat Transfer with Condensation or Evaporation of a Drobble," *German Chemical Engineering,* Vol. 5 (1982), pp. 24–43.
3. Sudhoff, B., "Direkter Wärmeübergang bei der Kondensation in Blapfensäulen," Dissertation, Universität Dortmund, 1982.
4. Melrath, C., "Darmstadt Drubbles," *Heat Transfer Engineering,* Vol. 1, No. 4 (Okt. 1980) pp. 20, 89.
5. Peebles, F. N., and Garber, H. J., "Studies on the Motion of Gas Bubbles in Liquids," *Chemical Engineering Progress,* Vol. 49, No. 2, (Febr. 1953), pp. 88–97.
6. Hu, S., and Kintner, R. C., "The Fall of Single Drops through Water," *AIChE Journal,* Vol. 1, No. 1 (March 1955), pp. 42–48.
7. Sideman, S., and Taitel, Y., "Direct Contact Heat Transfer with Change of Phase: Evaporation of Drops in an Immiscible Liquid Medium," *Int. J. Heat Mass Transfer,* Vol. 7 (1964) pp. 1273–1289.
8. Nazir, M., "Direct Contact Heat Transfer: Evaporation of Butane Drops in Brine," dissertation, University Strathclyde, Glasgow, 1972.
9. Simpson, H. C., Beggs, G. C., and Nazir, M., "Evaporation of Butane Drops in Brine," *Desalination,* Vol. 13 (1974) pp. 11–23.
10. Simpson, H. C., Beggs, G. C., and Nazir, M., "Evaporation of a Droplet of One Liquid Rising through a Second Immiscible Liquid," *J. Heat Transfer,* Vol. 5 (1974) pp. 59–63.
11. Mersmann, A., "Auslegung und Maßstabsvergrößerung von Blasen—und Tropfensäulen," *Chem.-Ing.-Tech.,* Vol. 49, No 9 (Sept. 1977) pp. 679–691.

12. Ruff, K., Pilhofer, Th., and Mersmann, A., "Vollständige Durchströmung von Lochböden bei der Fluiddispergierung," *Chem.-Ing.-Tech.,* Vol. 48, No 9 (Sept. 1976) pp. 759–764.
13. Sudhoff, B., Plischke, M., and Weinspach, P.-M., "Zum direkten Wärmeübergang bei der Kondensation oder Verdampfung eines Blapfens," *Chem.-Ing.-Tech.,* Vol. 53, No. 6 (June 1981), p. 471.
14. Brauer, H., "Impuls-, Stoff- und Wärmetransport durch die Grenzfläche kugelförmiger Partikeln," *Chem.-Ing.-Tech.,* Vol. 45, No. 18 (Sept. 1973) pp. 1099–1103.

CHAPTER 31

DESIGN AND SCALE-UP FOR HEAT TRANSFER TO COILS IN AGITATED VESSELS

M. Bruxelmane

Faculte Polytechnique de Mons
B-7000 Mons, Belgium

H. Desplanches

Ecole Supérieure de Chimie
Faculte de Saint Jerome
F-13397 Marseille Cedex 4, France

CONTENTS

INTRODUCTION

The geometrical and operating parameters characterizing the performance of a strirred tank are numerous. Consequently, a rational approach to a mixing system is difficult, considering the multiplicity of factors that must be taken into account and that can be modified. Before being able to specify the value for these factors, the design engineer must decide upon the two following items:

1. Selection of the impeller type: Circulation axial flow impellers such as propellers and pitched-blade turbines (PBT) or shear radial flow impellers such as hub mounted flat-blade turbines (FBT) and Rushton disc turbines (RDT).
2. Ascertainment of the optimum geometry and operating conditions of the stirred tank:

The achievement of these two objectives involves many interdependent considerations, as follows:

- First, the exact purposes of the mixing operation considered must be defined. This requires specification of the design parameters and the hydrodynamical actions required from the mixer.
- It is then necessary to determine the characteristic variables which control the value of the design parameters to achieve the desired process results. These variables could be, for example, the power per unit mass, the Froude number, the mean liquid circulation velocity, etc.
- Then, scale-up criteria and (at a given scale) design criteria can be specified.
- Finally, the following factors must be taken into account:
  1. Mechanical considerations (i.e. torque, shaft critical speed).
  2. Economical aspects (i.e. investment, power consumption).
  3. Operating constraints (corrosion problems, abrasion of particles, limited shear rates).

This section is concerned with heat transfer to a helical coil in an agitated Newtonian or non-Newtonian fluid in both the transitional and turbulent regions. In this case, the value of the process side heat transfer coefficient h or more exactly the product hA_H quantifies the purpose of the mixing operation considered. The characteristic variables which control the value of h, in the turbulent and transitional regions will be specified. General scale-up and design criteria will be defined.

Furthermore, this paper will also furnish the information needed for selection of the impeller type and for specification of the optimum geometry and operating conditions of the mixer, at any scale; mechanical, economical, and operating considerations being also taken into account.

GENERAL EQUATIONS OF HEAT TRANSFER TO COILS IN AGITATED TANKS

The overall heat transfer coefficient U between an agitated solution and a helical coil is related to the total rate of heat transfer q and the (process side) heat transfer area A_H by the following relation:

$$q = UA_H \mid \theta_i - \theta \mid_{lm} = UA_H \frac{\theta_{i1} - \theta_{i2}}{\ln [(\theta - \theta_{i1})/(\theta - \theta_{i2})]} \tag{1}$$

$$\cong UA_H \mid \theta_i - \theta \mid \tag{2}$$

with $A_H = \pi n_c d_c D_c[(\pi^2 + (p_c/D_c)^2]^{0.5}$ (2a)

where θ and θ_i = the agitated liquid bulk and the heat transport fluid temperatures, respectively
θ_{i1} and θ_{i2} = the inlet and outlet values of θ_i
n_c = the number of turns of the coil
p_c and d_c = the coil tube pitch and diameter, respectively
D_c = the mean coil diameter

Equation 2 must be used and is rigorous when θ_i is constant.

The overall resistance to heat transfer may be split into five components: the process-side film resistance, the eventual process-side fouling resistance r_s, the wall conduction term, the coil-side scale resistance r_{si}, and the coil inside film resistance. Consequently, one can write:

$$\frac{1}{U} = \frac{1}{h} + r_s + \frac{e_c}{\lambda_c}\frac{d_c}{d_{clm}} + r_{si} + \frac{1}{h_i}\frac{d_c}{d_{ci}} \tag{3}$$

where e_c and λ_c = the thickness and thermal conductivity of the coil tube, respectively
d_{ci} and d_c = the inside and outside coil tube diameters, respectively
d_{clm} = the logarithmic mean of d_c and d_{ci}
h_i = the coil-side heat transfer coefficient

The subscript i refers to the heat transfer fluid flowing inside the coil.

If the temperature of this fluid is varying ($\theta_{i1} \neq \theta_{i2}$), it is better to consider that the inside film resistance of the coil is equal to:

$$\frac{1}{h_{i\theta}}\frac{d_c}{d_{ci}}\left\{1 + \frac{\ln(h_{i2}/h_{i1})}{\ln[(\theta-\theta_{i1})/(\theta-\theta_{i2})]}\right\} \tag{4}$$

A linear relationship between h_i and the temperature θ_i of thermal fluid is then assumed. Heat transfer coefficients h_{i1}, h_{i2}, and $h_{i\theta}$ correspond respectively to the temperatures θ_{i1}, θ_{i2}, and θ.

INTERNAL HEAT TRANSFER COEFFICIENTS IN COILS

Ito [1] specified the initial point of turbulent flow of the heat transport liquid inside the coil tube in terms of a critical Reynolds number:

$$Re_{i,cr} = (4Q_i/\mu_i d_{ci})_{cr} = 2 \times 10^4 (d_{ci}/D_c)^{0.32}$$

When the flow is turbulent, the internal heat transfer coefficient h_i is usually determined by modified Dittus–Boelter equations:

$$\frac{h_i d_{ci}}{\lambda_i} = 0.0225 F_c \left(\frac{4Q_i\rho_i}{\pi\mu_i d_{ci}}\right)^{0.8}\left(\frac{\mu_i c_{pi}}{\lambda_i}\right)^{y_i}$$

i.e.

$$Nu_i = 0.0225 F_c\, Re_i^{0.8} Pr_i^{y_i} \tag{5}$$

y_i is equal to 0.3 when the fluid inside the coil is being cooled and at 0.4 when it is being heated. The corrective factor F_c takes into account of the coil tube curvature. λ_i, μ_i, c_{pi}, and Q_i are the thermal conductivity, the dynamic viscosity, the specific heat, and the flow rate of the heating or cooling medium at its mean temperature $\overline{\theta_i}$.

The following relationships of F_c were respectively proposed by Merkel [2] and Pratt [3]:

$$F_c = 1 + 3.54 d_{ci}/D_c \tag{6a}$$

$$F_c = 1 + 3.4 d_{ci}/D_c \quad (6b)$$

As for Rogers and Mayhew [4], they put forward:

$$Nu_i = 0.023\ Re_i^{0.85}\ Pr_i^{y_i}\ (d_{ci}/D_c)^{0.1} \quad (7)$$

Berg and Bonilla [5], on the other hand, found that the coil curvature did not affect h_i in the turbulent regime.

Marshall and Yazdani [6] recommended:

- For $d_{ci}/D_c < 0.005$, use Equation 2 with $F_c = 1$.
- For $0.005 \leqq d_{ci}/D_c \leqq 0.12$, use Equations 2 and 6b.
- For $d_{ci}/D_c > 0.12$, use Equation 7.

Seban and McLaughlin [7] considered the physico-chemical properties of the heat transport fluid at the film temperature θ_{if} which is the average of θ_i and tube inner wall temperature θ_{iW}. They correlated turbulent flow heat transfer data with the following equation:

$$Nu_{if} = 0.023\ Re_{if}^{0.85}\ Pr_{if}^{0.4} (d_{ci}/D_c)^{0.1} \quad (8)$$

According to Rogers and Mayhew [4], the constant in Equation 8 should be 0.021 instead of 0.023.

Finally, Mori and Nakayama [8] carefully investigated the heat transfer between gas in turbulent flow and inner wall of coils. They have obtained:

$$Nu_i = Re_i^{0.8} \{Pr_i/[26.2(Pr_i^{0.66} - 0.074)]\}\ (d_{ci}/D_c)^{0.1} \cdot \{1 + 0.098/[Re_i (d_{ci}/D_c)^2]\}^{0.2} \quad (9)$$

Mori and Nakayama also concluded that for coils entrance effects are negligible.

If the internal flow is not turbulent, see Seban and McLaughlin [7].

THE COIL PROCESS-SIDE HEAT TRANSFER COEFFICIENT—GENERALITIES

The published correlations are usually presented in a generalized form resulting from dimensional analysis as follows:

$$Nu = A\ Re^x\ Pr^y\ Vi^z\ MG_1^{\alpha 1}\ MG_2^{\alpha 2} \quad (10)$$

with $Nu = hd_c/\lambda$; $Re = \rho N d^2/\mu$; $Pr = \mu C_p/\lambda$; $Vi = \mu/\mu_w$

Nu, Re, and Pr are, respectively, the Nusselt number, the agitation Reynolds number, and the Prandtl number at the bulk temperature θ of the agitated liquid. λ, ρ, μ, C_p are the thermal conductivity, the density, the dynamic viscosity, and the specific heat of the process fluid at the temperature θ. N and d are the rotational speed and the diameter of the impeller. The Sieder-Tate number Vi allows the taking into account the opposite viscosity gradients existing at the external coil surface in cooling and in heating. μ_w is the dynamic viscosity of the agitated liquid at the mean temperature θ_w of the heat transfer surface. The MG_i terms are dimensionless geometrical factors.

Concerning the correlations available in the scientific literature, it must be noted that:

- The number of parameters actually investigated is often limited.
- Experimental studies at two scales have rarely been performed.
- This type of equation does not relate the heat transfer coefficient to a characteristic variable. For instance, the Reynolds number does not characterize perfectly the impeller action, because the ratio d/D is generally one of the corrective factors MG_i.
- Although several authors [16, 32, 33, 42, 9, 44] have pointed out the effect of some geometrical factors upon h, no explanation for their influence has usually been proposed.

- It must be noted that many authors did not determine their proposed value of y. Instead, they simply assumed the value y = 0.33, like in classical forced-convection heat transfer correlations.

THE COIL PROCESS-SIDE HEAT TRANSFER COEFFICIENT IN AGITATED NEWTONIAN FLUIDS IN THE TURBULENT REGION

Reference Correlation for the Different Types of Impeller

In order for an engineer to make use of the consequences from the definition of the main characteristic variable for h, here are the reference correlations which, in our opinion, are reliable for the different impeller types. The agitated tank contains a single coil. If several coils must be used, see Marshall [6] and Hicks [10].

Propellers [9]

$$Nu = 0.091\, Re^{0.67}\, Pr^{0.37}\, Vi^{c} (d/D)^{0.1} (d_c/D)^{0.5} \tag{11}$$

in heating and cooling; with $c = 0.174\,\mu^{-0.21}$, μ being expressed in $N \cdot s\, m^{-2}$ (in the turbulentregion we advise using c = 0.2); $H_L = D$; vessel with a flat bottom; $n_B = 4$; $B/D = 1/10$; $e_B = C_B = 0$; $D_c/D = 0.725$; $H_c/D = 0.66$; $C_c/D = 0.14$; $d_c/D = 0.0182$ and 0.0365; $e_c = d_c$; $400 \leqq Re \leqq 10^6$; $? \leqq d/D \leqq ?$; $n_p = 3$; p = d; pumping downwards; C = D/3; fluids = water and turbine oil; $\mu = 4.10^{-4}$–$0.4\, N \cdot s\, m^{-2}$; $2 < Pr < 6{,}330$.

Pitched-Blade Turbines Pumping Downwards [11]

$$Nu = 0.00752\, Re^{0.722}\, Pr^{0.4}\, Vi^{0.2} \tag{12a}$$

or

$$Nu = (0.0543\, Re^{0.5} + 0.000127\, Re)\, Pr^{0.4}\, Vi^{0.2} \tag{12b}$$

in cooling; with $H_L = D = 0.63$ or 1 m; vessel with a flat bottom; $n_B = 4$; $B/D = 1/12$; $e_B/D = 0.11$; $C_B = 0$; $D_c/D = 0.815$; $H_c/D = 0.808$; $C_{ci}/D = 0.1$; $d_c/D = 0.018$; $e_c/d_c = 2$; $22{,}400 < Re < 521{,}600$; $d/D = 1/3$; α = 45 deg; w/d = 1/5; $n_p = 4$; C = d; fluids = water and dilute aqueous solutions of sodium alginate; $4.2 < Pr < 41$; $0.55 < Vi < 0.95$.

$$Nu = 0.013\, Re^{0.684}\, Pr^{0.4}\, Vi^{0.2} \tag{13a}$$

or

$$Nu = (0.073\, Re^{0.5} + 0.000103\, Re)\, Pr^{0.4}\, Vi^{0.2} \tag{13b}$$

in cooling; with idem except d/D = 0.5.

The value of x appear to be somewhat high. However, similar values were proposed for propellers (x = 0.69 [12]), for FBT (x = 0.7 [13]) and for RDT (x ≅ 0.685 [77] and x = 0.69 [111]). The authors also observed that the values of x and y were interrelated. When y rises from 0.333 to 0.4, x is increased by at least 0.05. It should also be remarked that a lower value of x is obtained with the 4-PBT when d/D > 0.41. Generally, for a given impeller type (PBT, FBT, or RDT), it is noted that x was a little higher when the interactions of the circulation flow with the coil tubing became more important (e.g., a larger d_c and a smaller C_{ci} or d/D). Finally, it was observed that x increases with Re. This is the reason why Equations 12b, 13b, and 14b were utilized, which allowed this fact to be taken into account. As in the case of the classical Equation 10, Equations 12b, 13b, and 14b contain only two parameters but they generally improve the fitting of experimental results. Furthermore, this type of equation was also used in heat transfer between a single tube and a crossflowing liquid [14, 15].

Flat-Blade Turbines [11]

$$Nu = 0.0118\ Re^{0.684}\ Pr^{0.4}\ Vi^{0.2} \quad (14a)$$

or

$$Nu = (0.0583\ Re^{0.5} + 0.000115\ Re)\ Pr^{0.4}\ Vi^{0.2} \quad (14b)$$

in cooling; with $H_L = D$; vessel with a flat bottom; $n_B = 4$; $B/D = 1/12$; $e_B/D = 0.11$; $C_B = 0$; $D_c/D = 0.815$; $H_c/D = 0.808$; $C_{ci}/D = 0.1$; $d_c/D = 0.018$; $e_c/d_c = 2$; $12{,}900 < Re < 475{,}000$; $d/D = 1/3$; $w/d = 1/5$; $n_p = 4$; $C = H_L/3$; fluids = water and dilute aqueous solutions of sodium alginate; $4.6 < Pr < 49$; $0.51 < Vi < 0.93$.

The comments presented here for the PBT also apply to FBT. The value of x is nevertheless smaller.

Rushton Disc Turbines [16]

$$Nu = 0.17\ Re^{0.67}\ Pr^{0.37}\ Vi^{c}\ (d/D)^{0.1}\ (d_c/D)^{0.5} \quad (15)$$

in heating and cooling; with $c = 0.174\ \mu^{-0.35}$, μ being expressed in Nsm^{-2} (again, we advise $c = 0.2$ in the turbulent region); $H_L = D$; tank with a flat bottom; $n_B = 4$; $B/D = 1/10$; $e_B = C_B = 0$; $D_c/D = 0.725$; $H_c/D = 0.66$; $C_c/D = 0.15$; $d_c/D = 0.0182$ or 0.0365; $e_c = d_c$; $400 < Re < 1.5\ 10^6$; $d/D = 0.25, 0.33, 0.42, 0.5$, or 0.583; $d/l/w = 20/5/4$; $n_p = 6$; $C = D/3$; fluids = water and turbine oil; $\mu = 4.10^{-3}$–$0.4\ Nsm^{-2}$; $2 < Pr < 6{,}330$.

Effect of Geometrical Parameters Upon h

The effect upon h of the impeller geometrical parameters, which can be expressed by a power law (sometimes only in a limited range), are presented in the third column of Tables 1, 2, and 3. The corresponding geometries and operating conditions are briefly specified in the first two columns.

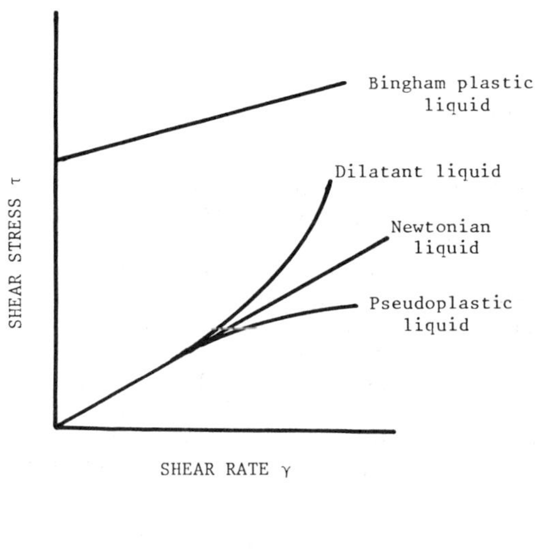

Figure 1. Flow curves for time-independent non-Newtonian fluids.

Table 1
Influence of Some Investigated Parameters upon the Coil Process Side Heat Transfer Coefficient h*

Experimental Conditions		Influence upon h of the Parameters Investigated						
Fixed Parameters	Investigated Parameter	Experimental {11}	From Circulation Hypothesis		From Turbulence Hypothesis		P/V_L (hA_H/V_L = const.)	
D = 1 m	d/D = 0.33; 0.41	$h \sim (Nd^2)^x$	$h \sim (Nd^2)^x$	{18[a]}				
$n_p = 4$	$2 \times 10^4 < Re < 5 \times 10^5$	x = 0.72	$h \sim (Nd^{2.2})^x$	{19[c]}			$P/V_L \sim d^{-1.36}$	{17}
			$h \sim (Nd^{2.23})^x$	{20[d]}				
D = 1 m	d/D = 0.33; 0.41; 0.5	$h \sim (Nd^{2.25})^x$	$h \sim (Nd^2)^x$	{17[d]}				
$n_p = 4$	$2 \times 10^4 < Re < 5 \times 10^5$	x = 0.72	$h \sim (Nd^{2.3})^x$	{21[c]}				
			$h \sim (Nd^{2.4})^x$	{17[c]}	$h \sim (Nd^{1.66})^x$	{23}		{17}
			$h \sim (Nd^{2.47})^x$	{22[a]}	$h \sim (Nd^{1.54})^x$	{17}	$P/V_L \sim d^{-2.1}$	
					$h \sim (Nd^{1.51})^x$	{24}		
D = 0.63 m	d/D = 0.333; 0.524							
$n_p = 6$								
Coil No. 5	Re ≅ 1.7×10^5	$h \sim (Nd^{2.22})^x$	$h \sim (Nd^{2.17})^x$	{25[c]}				
Coil No. 1	Re ≅ 6×10^4	$h \sim (Nd^{2.29})^x$	x = 0.72					
D = 1 m	n_p = 4; 6		$h \sim n_p^{0.19}$	{26[a]}	$h \sim n_p^{0.15}$	{28}		
N = const.		$h \sim n_p^{0.2}$			$h \sim n_p^{0.13}$	{24}	$P/V_L \sim n_p^{-0.25}$	{24, 29}
d = D/3	$2 \times 10^4 < Re < 5 \times 10^5$		$h \sim n_p^{0.22}$	{27[b]}	$h \sim n_p^{0.12}$	{29}		
D = 0.63 m	1 ≦ C/d ≦ 1.6†	$h \sim (C/d)^{-0.21}$	$h \sim (C/d)^{-0.13}$	{30[a]}				
N = const.	0.4 ≦ C/d ≦ 1.5	h_{max} for	h_{max} for	{26[a]}	$h \sim (C/d)^{-0.068}$	{24}		
d = D/3		C/d = 0.62	C/d = 0.68					
$n_p = 6$								
D = 0.63 m								
N = const.	d_c/D = 0.0190; 0.0286							
d = D/3								
$n_p = 4$	$2 \times 10^4 < Re < 7 \times 10^4$	$h \sim d_c^{-0.4}$	$h \sim d_c^{-0.27}$		$h \sim d_c^0$		$P/V_L \sim d_c^{1.7}$	
$n_p = 6$	$6 \times 10^4 < Re < 2.4 \times 10^5$	$h \sim d_c^{-0.27}$						
d = D/3	1,5 < p_c/d_c ≦ 3	$h \sim (p_c/d_c)^0$	$h \sim (p_c/d_c)^0$	{17}	$h \sim (p_c/d_c)^0$	{17}	$P/V_L \sim (p_c/d_c)^{4.5}$	
$n_p = 4$								
C = d								
Nd^2 = const.	D = 0.63; 1 m	$h \sim D^{-0.99}$	$h \sim D^{-1}$		$h \sim D^{-0.93}$		$P/V_L \sim D^{4.5}$	{17}
d = D/3	$2 \times 10^4 < Re < 2 \times 10^5$							
$n_p = 4$								
D = 0.63 m	0.46 ≦ H_L/D ≦ 1	$h \sim (H_L/D)^{-0.32}$	$h \sim (H_L/D)^{-0.31}$		$h \sim (H_L/D)^{-0.23}$	{24}	$P/V_L \sim (H_L/D)^{-0.6}$	
d = D/3			$h \sim (H_L/D)^{-0.26}$	{26[a]}			(V_L = const.;	
$n_p = 6$							H_L < 0.88D)	
C = d								

* *Verification of the validity of circulation and turbulence hypotheses* (PBT; Re > 2×10^4; C = D)
† *The effect of this parameter upon h is examined in detail in the text.*
a,b,c,d *Data derived from* Q_c, Q_p, t_M, N_{CH} *respectively.*
From Desplanches and Bruxelmane [11].

For the PBT, the effect of d/D, C/d, and D_c/D upon h are more complex and are discussed hereafter.

Effect of d/D (0.33 ≦ d/D ≦ 0.5)

When d/D is higher than 0.41, its effect upon h is more pronounced. This particular behavior is defined in terms of an exponential function in Table 1.

Effect of C/d (0.4 ≦ C/d ≦ 2.5)

Figure 1 illustrates the effect of C/d upon h, for a 6-PBT. The optimum value of C (for the coil geometry used) corresponds to C/d ≅ 0.62. The most unfavorable impeller positions

Table 2
Influence of Some Parameters Investigated in the Scientific Literature upon the Coil Process Side Heat Transfer Coefficient h*

Experimental Conditions		Influence upon h of the Parameters Investigated							
Fixed Parameters	Investigated Parameter	Experimental		From Circulation Hypothesis		From Turbulence Hypothesis		P/V_L, N (hA_H/V_L = const.)	
Six blade disk turbines ($Re > 10^4$; $H_L = D$; tank with flat bottom; $n_B = 4$)									
C = D/2	d/D = 0.17; 0.25;								
Vertical coils!	0.333; 0.5	$h \sim (Nd^{2.51})^{0.65}$	{31}						
D = 0.61; 1.22 m									
C = D/2	d/D = 0.33–0.5	$h \sim (Nd^{2.3})^{0.67}$	{77}	$h \sim (Nd^{2.57})^x$	{19[c]}				
				$\sim (Nd^{2.3})^x$	{35[c]}				
C = D/3	d/D = 0.213;	$h \sim (Nd^{2.27})^{0.67}$	{32}	$\sim (Nd^{2.29})^x$	{36[a]}	$h \sim (Nd^{1.66})^x$	{40}	$P/V_L \sim d^{-1.9}$	{40}
D = 0.36 m	0.33; 0.427			$\sim (Nd^2)^x$	{37[a]},			$N \sim d^{-2.3}$	
n_B = 2(!)					{38[c]}				
C = 0.27D	d/D = 0.25; 0.37	$h \sim (Nd^{2.3})^{0.64}$	{33}	$\sim (Nd^2)^x$	{37[d]}				
D = 1.22 m;				$\sim (Nd^2)^x$	{54[c]}				
H_L = 0.75D(!)									
C = D/3	d/D = 0.25; 0.33;	$h \sim (Nd^{2.15})^{0.67}$	{16}	$\sim (Nd^{1.8})^x$	{39[d]}				
Re > 400 (!)	0.42; 0.5								
C/D = ?	d/D = 0.25–0.33	$h \sim (Nd^2)^{0.65}$	{6}						
C = const.; $C_{ci} > C$	d/D = 0.39; 0.56	$h \sim (Nd^{1.88})^{0.63}$	{34}						
n_B = 2(!); d ≐ D/3	n_p = 2; 4; 6	$h \sim n_p^{0.28}$	{32}	$h \sim n_p^{0.35}$	{17[b]}	$h \sim n_p^{0.18}$	{40}	$P/V_L \sim n_p^{-0.4}$	{40}
N = const.; C = d						$h \sim n_p^{0.19}$	{29}	$N \sim n_p^{-0.42}$	
N = const.	d_c/D = 0.0268;	$h \sim (d_c/D)^{-0.45}$	{41}					$P/V_L \sim d_c^{2.2}$	{16}
d = D/3	0.0537	$h \sim (d_c/D)^{-0.48}$	{42}	$h \sim (d_c/D)^{-0.33}$		$h \sim (d_c/D)^0$	{16}		
C = d	d_c/D = 0.0182;	$h \sim (d_c/D)^{-0.5}$	{16}					$N \sim d_c^{0.73}$	
	0.0365								
N = const.	p_c/d_c = 2; 4	$h \sim (p_c/d_c)^0$	{16}	$h \sim (p_c/d_c)^0$		$h \sim (p_c/d_c)^0$		$P/V_L \sim (p_c/d_c)^{4.5}$	
d = D/3								$N \sim (p_c/d_c)^{1.5}$	
C = d									
Nd^2 = const.	D = 0.61; 1.22 m	$h \sim D^{-1}$	{31}	$h \sim D^{-1}$	{37[a]}	$h \sim D^{-0.86}$		$P/V_L \sim D^{5.23}$	
								$N \sim D^{1.08}$	
Propellers ($Re > 10^5$; $H_L = D$; tank with flat bottom, $n_B = 4$)									
C = d;	d/D = ?	$h \sim (Nd^{2.15})^{0.67}$	{9}	$h \sim (Nd^{2.3})^x$	{36[c]}				
Pumping downward									
non-Newtonian	d/D = 0.164; 0.329;	$h \sim (Nd^{2.15})^{0.62}$	{12}	$\sim (Nd^{2.29})^x$	{20[d]}	$h \sim (Nd^{1.66})^x$	{40}	$P/V_L \sim d^{-1.4}$	{40}
liq.; C = 0.3D	0.413			$\sim (Nd^{2.05})^x$	{19[c]}	$h \sim (Nd^{1.38})^x$		$P/V_L \sim d^{-2.3}$	{43}
Pumping up or down									
0.16 ≦ C/D ≦ 0.82	d/D = 0.164; 0.39;	$h \sim (Nd^2)^{0.67}$	{9}	$\sim (Nd^2)^x$	{35[c]}		{43}	$N \sim d^{-2.15}$	
D_c = 0.67D	0.493			$\sim (Nd^{1.71})^x$	{25[c]}				
Pumping upwards				$\sim (Nd^{1.7})^x$	{30[a]}				
N = const.	p/d = 1; 1.33	$h \sim (p/d)^{2.33}$	{44}	$h \sim (p/d)^{0.67}$	{45[c]}	$h \sim (p/d)^{0.38}$	{40}	$P/V_L \sim (p/d)^{-1.4}$	
Pumping upwards									
d = D/3; C = d	coils ≠ !!					$\sim (p/d)^{0.33}$	{43}	$N \sim (p/d)^{-1}$	(†)
Nd^2 = const.	0.16 ≦ C/D ≦ 0.82	$h \sim (C/D)^{-0.254}$	{44}	$h \sim (C/D)^{-0.024}$	{30[a]}	$h \sim (C/D)$	{40}		
Pumping upwards	d/D = 0.16; 0.39;								
	0.49								
N = const.	d_c/d = 0.0137;	$h \sim (d_c/D)^{-0.43}$	{44}			$h \sim (d_c/D)^a$		$P/V_L \sim d_c^{2.1+4.5a}$	
Pumping upwards	0.0274; 0.0548			$h \sim (d_c/D)^{-0.33}$					
d = D/3; C = d	d_c/d = 0.0182;	$h \sim (d_c/D)^{-0.5}$	{9}			a ≧ 0		$N \sim d_c^{0.7}$	
	0.0365								

* *Verification of circulation and turbulence hypotheses (Newtonian liquids; turbulent region). (In the estimations from circulation or turbulence hypothesis, x = 0.66.)*

† *From circulation hypothesis.*

a,b,c,d,e *Data derived from Q_c, Q_p, t_M, N_{CH}, N_{ma} respectively.*

From Desplanches and Bruxelmane [11].

Table 3
Influence of Some FBT Parameters Investigated by Nagata [46] upon the Coil Process Side Heat Transfer Coefficient h*

Experimental Conditions		Influence on h of the Parameters Investigated		
Fixed Parameters	Investigated Parameters	Experimental	From Circulation Hypothesis	
$n_p = 4$; $C = H_L/2$	$d/D = 0.25; 0.33; 0.41$	$h \sim (Nd^2)^x$ $x = 0.56!$	$h \sim (Nd^{1.9})^x$	{47[a]}
			$\sim (Nd^2)^x$	{37[a], 38[b]}
			$\sim (Nd^{2.29})^x$	{36[c]}
$d = D/2$; $n_p = 4$	$w/d = 0.666; 0.1; 0.133; 0.2$	$h \sim (w/d)^{0.3}$	$h \sim (w/d)^{0.33}$	{47[a]}
$d = D/2$	$n_p = 2; 3; 4; 6; 16$	$h \sim n_p^{0.2}$	$h \sim n_p^{0.17}$	{27[b]}
			$n_p^{0.36}$	{47[a]}
?	$H_L/D = ?$	$h \sim (H_L/D)^{-0.5}$	$h \sim (H_L/D)^{-0.22}$	{37[a]}

* *Test of the experimental results by the circulation hypothesis (Newtonian liquids; turbulent region;* $H_L/D = 1.333$; dished bottom; $n_B = 4$; $B = D/10$; $D_c/D = 0.75$; $H_c/D = 0.766$; $C_{ci}/D = 0.266$; $C/D = 0.66$.
The circulation hypothesis has been applied with $x = 0.66$.
[a,b,c,d] *Data derived from* Q_c, Q_p, t_M, N_{CH}, *respectively.*
From Desplanches and Bruxelmane [11].

ally inside the coil. For $D_c/D = 0.41$, the baffles had to be placed outside the coil. The effect of D_c/D can be characterized by multiplying the value of Nu corresponding to $D_c/D = 0.8$ by the corrective factor F_k defined as follows:

$$F_k = 2.766(D_c/D)^2 - 3.488D_c/D + 2.05 \tag{12c}$$

The minimum value of h was found to occur at $D_c/D \cong 0.63$.

Main Characteristic Variable for Nu

The definition of the main characteristic variable allows going beyond dimensional analysis. Then, we can draw very important practical conclusions especially concerning the definition of general scale-up rules, the estimation or the verification of the effects upon h are the mid-level of the liquid ($C/d = 1.5$ for the agitator used) and the proximity of the free surface.

Effect of D_c/D ($0.41 \leqq D_c/D \leqq 0.87$)

The effect of D_c/D upon h was investigated by Desplanches and Bruxelmane [11] in a tank equipped with a 6-PBT ($d/D = 0.33$). The four baffles were usually placed symmetric-of some geometrical parameters of the stirred tank, the selection of the impeller type, and of the geometry of the vessel/coil/impeller system.

Calderbank [48] using the Kolmogoroff theory of local isotropy, proposed the following general equation which would be valid irrespective of the geometry of the agitated system:

$$h/\rho C_p = 0.13\,[(P/V_L)\,(\mu/\rho^2)]^{1/4}\,Pr^{-2/3} \tag{16}$$

The term P/V_L is the power unit volume. The characteristic length considered is the scale of energy dissipating eddies. Other authors [49–52, 111] have also presented heat transfer correlations containing the term P/V_L. The use of P/V_L as characteristic variable is attractive considering the high value of the liquid turbulence intensity in agitated tanks.

However, the fifth column of Tables 1 and 2 clearly proved that the turbulence hypothesis using P/V_L as characteristic variable is not valid. Results arising from the turbulence hypothesis were obtained from power data, by considering $h \div (P/V_L)^{x/3}$.

On the other hand, the main characteristic variable which fixes the value of h could be the mean circulation velocity v_c in the vicinity of the heat transfer surface (circulation hypothesis). This variable is related (for a given value of D_c/D) to the total circulation flow rate Q_c of the liquid in the vessel by the following relation:

$$v_c = kQ_c/D^2 = k \cdot N_{qc} \cdot Nd^3/D^2 \tag{17}$$

where $Q_c = Q_p + Q_e$ (18)

Q_e = the liquid flow rate induced by Q_p, the pumping flow rate of the impeller.

The proportionality constant k depends upon the geometry of the agitated system. However, when the geometrical parameters do not appreciably affect the liquid flow pattern in the tank or the location of heat transfer area with respect to the liquid flow pattern, it may be assumed that for a given type of impeller, k is constant. The circulation number N_{qc} is independent of Re in the turbulent region; it is only a function of the geometry of the stirred tank. This has led us to propose the circulation Reynolds number Re_c as characteristic variable for Nu:

$$Re_c = \rho v_c d_c/\mu = \rho N_{qc} N d^3 d_c/\mu D^2 \tag{19}$$

The coil tube diameter d_c is selected as characteristic length of the heat transfer area. This is a similar situation to that of forced heat transfer between a single tube and a crossflowing liquid. Moreover, as a simplification, the proportionality constant k in Equation 17 was taken to be equal to 1.

The estimation corresponding to the circulation hypothesis are presented and compared with the experimental results in Tables 1 and 2. They were calculated from data obtained in baffled tanks without a helical coil. These data came from several sources:

- Total circulation flow rate Q_c.
- Agitator pumping rate Q_p.
- Mixing time t_M.
- Minimum stirrer speed N_{CH} for complete homogeneity of suspension.
- Minimum stirred speed N_{ma} required for bubble aspiration into the impeller, via vortices at the surface of the agitated liquid.

This is justified in a given tank, with impellers of a given type (strictly if d and C are constant), one observes that:

- The ratio Q_c/Q_p is practically constant.
- $t_M \sim Q_c^{-1}$ [53, 35, 54, 28, ...].
- $N_{CH} \sim Q_c^{-1}$ [17, 20, 37, ...].
- $N_{ma} \sim Q_c^{-1}$ [17, 55].

It should be noted that for PBT, in order to express the effect of H_L upon h according to the circulation hypothesis, the following assumptions were made:

- N_{qc} = constant regardless of the value of H_L/D.
- t_c = mean circulation time = V_L/Q_c.
- ℓ_c = length of the mean circulation path $\div (D+2H_L)$.
- Thus $v_c = \ell_c/t_c \div (1+2H_L/D)(H_L/D) \div (H_L/D)^{-0.43}$ for $0.46 \leq H_L/D \leq 1$.

Tables 1 and 2 reveal that the agreement between the experimental effects and those predicted from the circulation hypothesis is excellent. Consequently, contrary to what might be expected, it is the mean fluid velocity which controls the value of h and Re_c is the characteristic variable for Nu. The turbulence around the coil tubing appears to have limited effect upon h.

Another Characteristic Variable

It must be emphasized that although the bulk circulation velocity is effectively the main characteristic variable, the distribution curve of circulation velocities around the coil tubing should be taken into account for a rigorous approach to the problem. This is altogether an analogous situation to that of the mixing times, for which the Voncken model [35, 53, 54] for example, considers the mean circulation time and the distribution curve of circulation time. This corresponds, for a given tank, to the mean circulation velocity and the distribution curve of circulation velocity. Consequently, the effect upon h of parameters that appreciably affect the liquid flow pattern in the stirred tank, such as d and C, is more intricate than that estimated from the circulation model. The same is true for the case of factors, such as D_c, H_c, and C_{ci} that change the location of the heat transfer area in the liquid flow pattern, particularly with respect to the "circulation eye(s)" or the impeller pumping flow. In this connection, the flow patterns published in the scientific literature [e.g. 18, 56–58] are, in practice, very interesting. So, for the PBT's, the effect upon h of d/D when $d/D > 0.41$, is due to a translation of the circulation eye to the vessel wall and consequently to an enlargement of the circulation velocity in the coil zone. In the same way, it is normal to find a minimum value of h for $D_c/D = 0.63$ (d/D being equal to 0.33), since the coil then crosses the zone of the circulation eye. When the coil is moved away from that zone of reduced circulation, the value of h increases. As should be expected, Edney [33] found, for a 6–RDT, an unfavorable effect of D_c upon h, that is $h \div (D_c/D)^{-0.375}$. In the range considered ($0.458 \leqq D_c/D \leqq 0.792$), when D_c is increased, the coil is indeed brought nearer to the "circulation eyes" of the flow pattern. It is likely that an increase of D_c/D beyond 0.8–0.85 should be advantageous.

In any case, the observed effect of C and D_c upon h clearly indicate the danger of characterizing some geometrical effects by a power law and of using such correlations outside the range for which they have been established. This is especially the case when the effects of the considered parameter upon N_{qc} and/or the flow pattern around the coil tubes are intricate.

Conclusions

As all the experimental results indicate that for heat transfer between a helical coil and a stirred Newtonian liquid, in the turbulent region, the circulation Reynolds number is the main characteristic variable which fixes the value of the Nusselt number, we are able:

1. To increase greatly the prediction possibilities of the effects of at least the impeller geometrical parameters upon h.

   Indeed, in a given tank, v_c is also the main characteristic variable for the homogenization of miscible liquids, the production of a homogeneous liquid-solid suspension, and the aeration from the turbulent free surface of the tank contents. This is clearly proved by the calculations from circulation hypothesis in Tables 1 and 2. Thus, theoretically, information concerning the effect of a geometrical parameter on one of the variables Q_c, Q_p, t_M, N_{CH}, N_{ma}, or h can be used to forecast how the same geometrical parameter affects the others.

   Therefore, the authors applying the circulation hypothesis or using some experimental results recommend in Table 4 the effects of some geometrical parameters on h.
2. To test the correlations proposed in the scientific literature for h, Q_c, Q_p, t_M, N_{CH}, and N_{ma}.

   In Table 2, for instance, the effect of the propeller pitch p upon h, proposed by Skelland [44], seems to be exaggerated. This results from the fact that this author carried

Table 4
Effects of Some Geometrical Parameters upon the Process Side Heat Transfer Coefficient h

Impeller	Influence on h of the Parameter Investigated	Geometrical Conditions	References*
Every type	$h \sim (H_c/D)^0$	$C_{ci}/D \geq 0.1$	Desplanches [11]
		$C_{cs}/H_L \leq 0.9$	
	$h \sim (p_c/d_c)^0$	$1.5 < p_c/d_c \leq 3$	Desplanches [11]
	$h \sim n_B^0$	$2 \leq n_B \leq 6$	Skelland [12, 59]
		$10 < D/B < 12$	
	$h \sim (B/D)^0$	$10 < D/B < 12,$	Skelland [44, 59]
	e_B effect(‡)	$C_B = 0$	Oldshue [16]
			Desplanches [60]
PBT	effect of α (†)	$15 \leq \alpha < 90$	Medek [26a]
	effect of $w^{(x)}$ (†)	$0.15 \leq w/d \leq 0.3$	Medek [27b]
	$h \sim (2+D/H_L)^x$	$0.46 \leq H_L/D \leq 1$	Desplanches [11a]
	$\sim (H_L/D)^{-0.32}$		
FBT	$h \sim (w/d)^{0.33}$	$0.1 \leq w/d \leq 0.25$	Sasakura [17a]
	$h \sim n_p^{0.2}$	$2 < n_p \leq 16$	Nagata [46]
			Medek [26a]
RDT	$h \sim (w/d)^{0.46}$	$0.1 < w/d \leq 0.3$	Sato [61a]
	$h \sim (l/d)^0$	$0.25 \leq l/d \leq 0.31$	Bruxelmane [17a]
	$h \sim n_p^{0.33}$		

* a, b: *Data derived from* Q_c and Q_p *respectively.*
(†) $N_{qc} \sim N_{qp} \sim (\pi^2/6) \sin\alpha \cos\alpha\,(3z^2 - 2z^3) + \Pi^2 (w/d) z^2 \sin^2\alpha$ with $z = f(\alpha)$; $z \cong 0.66$ if $\alpha \geq 40$ deg.
(‡) *Wherever baffles were placed – either at the tank wall or inside the coil – little difference results in the heat transfer coefficient.*
From Bruxelmane and Desplanches.

Table 5
Comparison of Heat Transfer Performances of Two Pitched-Blade Turbines (PBT) and a Hub-Mounted Flat-Blade Turbine (FBT) and a Rushton Disk Turbine (RDT) in the same vessel

Experimental Conditions		P_1/P_2
Impeller 1	Impeller 2	(h = cst)
4-PBT	4-FBT	0.36 {26}
6-PBT	4-FBT	0.3 {26}
6-PBT	6-RDT	0.39 {26}
4-FBT	6-RDT	1.42 {26}

($d = D/3$; $5 \cdot 10^4 \leq Re \leq 5 \cdot 10^5$)
From Desplanches and Bruxelmane [11].

out his experiments with two different coils and did not consider the effects of D_c, H_c, and C_{ci}, upon h. In the same way, the results presented by Nagata [46] for the FBT are not quite reliable (see Table 3). First, the value of the exponent x for Re is abnormally low. Also, the effect of H_L/D upon h does not agree with the circulation hypothesis. For the 4–PBT, the same author [46] proposed $h \div (\sin\alpha)^{0.5}$ for $30 \leq \alpha \leq 90°$. The agreement with circulation hypothesis [27] is excellent for $\alpha \leq 60°$. On the other hand, for $\alpha \geq 60°$, the relationship is not valid.

3. It must be expected that the circulation impellers (i.e. the axial impellers: PBT or propellers) be the most efficient and that for a given impeller type, the geometries which increase the liquid circulation in the tank be recommended.

This is pointed out in the last columns of Tables 1, 2, and 5 which give the effect upon P/V_L of the parameter investigated, hA_H/V_L being kept constant. The most important parameters are d, n_p (as shown in the tables), as well as α and w/d. The unfavorable effect of an increase of d_c upon h and P/V_L must also be noted (Note: In a given tank, irrespective of d_c, the heat transfer area A_H is practically constant if D_c, H_c, and p_c/d_c are constant). However, whereas reduction of d_c involves an appreciable increase of h and h_i, it has the unfavorable effect of increasing the pressure drop in the coil [e.g., 62].

4. On the other hand, the scale-up rule for Nu is obviously to keep Re_c constant, whatever the impeller geometry may be. This corresponds to a constant Re in geometrically similar stirred systems.

   Consequently, the normal design procedure should be:
   - Utilization of a correlation or of an experimental result obtained from pilot scale in order to calculate the total rate of heat transfer in the geometrically similar industrial agitated system.
   - Utilization of the design rule, $Nu \div Re_c^x$, in order to estimate the effect upon h of a modification of the agitated system geometry or of the impeller type.

   For this purpose, the correlations for reference impellers as well as data and bibliographical references given in Tables 1, 2, 3, and 4 will be particularly useful. Nevertheless, in these calculations, the possible effect upon h of the distribution curve of circulation velocities around the coil tubing should always be taken into account.
5. The optimal geometry and operating conditions of the agitated system allowing performance at the heat transfer capacity required will be ascertained finally from mechanical and economical considerations. However, it should be recognized that other purposes of the mixing operation can also influence the impeller choice, as well as the optimal geometry and operating conditions of the stirred system.

THE COIL PROCESS SIDE HEAT TRANSFER COEFFICIENT IN AGITATED FLUIDS IN THE TRANSITION REGION

Introduction

In the transitional flow regime, as a result of an additional contribution of natural convection (NC), the logarithmic plot of $Nu/Pr^y\, Vi^z$ as function of the agitation Reynolds number Re can show significant deviations from a straight line when Re decreases. The existence of natural convection in stirred tanks was already pointed out by some research workers (31, 63, 64). However, only Desplanches and Bruxelmane [65, 66] proposed some correlations characterizing mixed (or combined) convection when free and forced convections simultaneously and appreciably exist.

According to Dunlap and Rushton [31], natural convection becomes actual under a certain level of turbulence intensity. They defined, (for RDT's, heat transfer device: vertical-tube baffles) a minimum rotational speed N_{c1} under which the heat transfer straight-line relationship breaks away from the forced-convection relationship and starts to approach the natural convection value. As for Petree and Small [64], they considered that natural convection heat transfer appeared when vertical gradients of temperature existed in the agitated liquid. So, they determined (for a two 6-FBT agitator; heat transfer device: vertical plate-coil baffles), a characteristic stirrer speed N_{c2}.

Those authors respectively proposed:

$$N_{c1} = 1.15(D/d)^{2.5}(n_B/2)^{0.3}\nu^{0.2} \tag{20a}$$

$$N_{c2} = 5.32\nu^{0.15} \tag{20b}$$

These expressions are directly related to the circulation capacity of the impeller (Desplanches and Bruxelmane [66]). Thus, contrary to Dunlap and Rushton, we can say that free convection is effective when the liquid circulation (and not its turbulence) is too weak.

On the other hand, Uhl [67] is of the opinion that the high values of the exponent c proposed by Oldshue and Gretton [16] in Equations 11 and 15 were probably due to strong natural convection effects for the water tests carried out by those authors in the transitional flow regime.

Desplanches and Bruxelmane [65] observed:

- Deviations from the forced-convection (FC) (straight-line relationship) in the early transition region ($Re \cong 100$) but also at the end of this region ($Re \cong 10^4$), although no temperature gradient was then detected in the agitated liquid.
- The exponents x and y of Re and Pr in the heat transfer correlation are not constant. So, the value of x passes from 0.45 in the early transition region to approximately 0.7 in the turbulent region. As for y, it ranges from 0.3 to 0.4. It must be noted that such variations were also found for simple geometries (single tube or plate) in a unidirectional flow.

We assume that the previous correlations are valid in both the turbulent region and in a wide range of the transition region.

Pure Natural Convection Correlation for a Submerged Helical Coil

Desplanches and Bruxelmane [66] have put forward the following pure ($N = 0$) natural convection correlation valid in cooling and heating.

$$Nu_{NC} = 0.53(Gr \cdot Pr)^{0.25}\, Vi^{0.11} \quad \text{if} \quad Ra = Gr \cdot Pr \leqq 10^6 \tag{21}$$

with $D_c/D \cong 0.8$; $C_{ci}/D = 0.11$; $p_c/d_c = 3$; $d_c/D = 0.018$ or 0.0286, impeller = 4-PBT.

$Gr = g\alpha d_c^3 \mid \overline{\Delta\theta_W} \mid \rho^2/\mu^2$ is the Grashof number in which the characteristic length considered is also the coil tube diameter. α, ρ, and μ are respectively the cubical expansion coefficient, the density, and the viscosity of the stirred liquid at its bulk temperature θ. $\overline{\Delta\theta_W} = \overline{\theta_W - \theta}$. Ra is the Rayleigh number.

Equation 21 yields values of Nu_{NC} near the ones calculated from Equations 22a and 22b published respectively by McAdams [68] as well as Churchill and Chu [69] for free-convection around a single horizontal cylinder:

$$Nu_{NC,1} = 0.53(Gr_f\, Pr_f)^{0.25} = 0.53\, Ra_f^{0.25} \tag{22a}$$

$$Nu_{NC,1} = 0.36 + 0.518\left(\frac{Gr_f\, Pr_f}{\left(1 + \left(\frac{0.559}{Pr_f}\right)^{9/16}\right)^{16/9}}\right)^{0.25} \tag{22b}$$

the physical properties of the agitated liquid being calculated at the film temperature $\theta_f = (\theta + \theta_W)/2$.

Finally, it should be remarked that:

- In a given stirred system or in geometrically similar stirred systems, the Grashof number (or better the Rayleigh number) is the characteristic variable which controls natural convection.
- The value of the proportionality constant in Equation 21 is a function of the geometry of the vessel/coil system especially of H_L/D, D_c/D, H_c/D, C_{ci}/D, and p_c/d_c. In this connection, Dapra et al. [71] performed trials with high $H_L/D \leqq 3(0.1 \leqq H_c/D \leqq 1)$ and showed that from the natural convection (h_{NC}) point of view, it was advisable to use.
  1. $D_c/D \cong 0.7$.
  2. Low value of C_{ci}, in heating. There is no heat transfer to the liquid located beneath the coil.
  3. A sufficient coil pitch ($p_c/d_c \geqq 1.5$).

4. A fin coil tube instead of a smooth tube (this is also interesting in forced convection considering the significant increase of heat transfer area).
5. A moderate fin-height $h_f(h_f/d_c \cong 0{,}18$ with $p_f/h_f \cong 0.66$; p_f being the fin pitch).

Moreover, h_{Nc} decreases when H_L and/or H_c increases.

Mixed Convection Correlations of the Fand Type

Initially, Desplanches and Bruxelmane [65] used the generalized mixed convection correlation of the Fand [71] type:

$$Nu = A_{MC}\,(Re_c^2 + \beta\,Gr)^{0.5x_{MC}}\,Pr^{y_{MC}}\,Vi^{z_{MC}} \tag{23a}$$

Its forms for natural and forced convection are

$$Nu_{NC} = C_{NC}\,Gr^{0.5x_{MC}}\,Pr^{y_{MC}}\,Vi^{z_{MC}} \tag{23b}$$

$$Nu_{FC} = A_{MC}\,Re_c^{x_{MC}}\,Pr^{y_{MC}}\,Vi^{z_{MC}} \tag{23c}$$

$$Nu/Nu_{FC} = (1 + \beta\,Gr/Re_c^2)^{0.5x_{MC}} \tag{23d}$$

The fittings of the experimental results obtained in cooling allow to draw the following conclusions:

- The natural convection limit forms (NCLF) obtained in the early transition region are very similar to the correlations proposed by McAdams[68], Morgan[72] and Fand[73] for free convection around a single horizontal cylinder.
- The NCLF are very clase to pure NC correlation (Equation 21).
- d_c is effectively the characteristic length in Gr.
- The NC is independent, which is normal, of the impeller geometry in a given vessel/coil/baffle system.
- In the early transition region, for a given impeller class (axial or radial impellers), Gr/Re_c^2 which is the ratio of buoyancy forces to inertial forces quantifies well the importance of natural convection with regard to forced convection (FC). The natural convection is significant when:

  $Gr/Re_c^2 > 1.5$–2 for the axial flow impellers, namely the PBT's

  $Gr/Re_c^2 > 0.8$ for the radial flow impellers

  The PBT's induce a liquid circulation more efficient around the whole coil tubing. This explains the different critical values of Gr/Re_c^2 just given.

However, a mixed-convection correlation of the Fand type presents some disadvantages:

- It imposes the same exponents of Vi and Pr for the NCLF and the FCLF though, as we will see, firstly $z_{NC} = 0.11$ and $z_{FC} = 0.2$ and secondly $y_{NC} = 0.25$ and y_{FC} varies from 0.3 to 0.4 over the whole transition region.
- The exponents of Gr and Re_c are coupled, whereas the exponent x_c of Re_c passes from 0.45 in the early transition region to about 0.7 in the turbulent region.
- In a given agitated system, β is a function of Re.
- In the same range of Re, in geometrically similar agitated systems or in a given vessel/coil/baffle system equipped with impellers of the same class, the values of β can be rather different.
- At constant Gr/Re_c^2, different values of Nu/Nu_{FC} may be observed. In Equation 23d, Gr/Re_c^2 characterizes only imperfectly the effects of natural and forced convection at high values of Re.
- At the end of the transition region ($Re > 5{,}000$), mixed convection Equation 23a may only be extrapolated to pure forced convection.

In conclusion, through the imbrication of the characteristic variables for free convection (Gr) and for forced convection (Re_c), the mixed-convection correlations of the Fand type are only valid over a limited range of Re.

Mixed Convection Correlations of the Churchill Type

The imperfections of Equation 23a have led us to test a mixed-convection correlation of the Churchill [74] type combining indepedently a natural-convection term and a forced-convection term:

$$Nu^m = Nu_{NC}^m + Nu_{FC}^m \tag{24}$$

with

$$Nu_{NC} = A_{NC}\, Gr^{x_{NC}}\, Pr^{y_{NC}}\, Vi^{z_{NC}} \tag{25a}$$

$$Nu_{FC} = A_{FC}\, Re_c^{x_{FC}}\, Pr^{y_{FC}}\, Vi^{z_{FC}} \tag{25b}$$

m characterizing the coupling of natural and forced convection.

Equation 24 may be directly extrapolated to the pure natural convection or to the pure forced convection whatever the value of Re may be.

It allows the quantification of the deviations with respect to the pure forced convection. Indeed, we have

$$\frac{Nu}{Nu_{FC}} = \left[1 + \left(\frac{Nu_{NC}}{Nu_{FC}}\right)^m\right]^{1/m} \tag{26}$$

Equation 24, we have supposed that the parameters A_{NC}, x_{NC}, y_{NC}, and z_{NC} are those determined in trials carried out in pure free convection (N = 0). As for A_{FC}, x_{FC}, y_{FC}, z_{FC}, and m, they were obtained from tests performed in mixed convection and in some cases in forced convection.

The values of the coefficients of the mixed convection Equation 24 in cooling at the beginning and at the end of the transition region [66] prove that:

- The value m = 3 appears to be the best choice in the different agitated systems investigated.
- y_{FC} varies from 0.3 to 0.4 over the transition region.
- $z_{FC} = 0.2$ in the transitional flow regime.
- Over the whole transition region, the use of the mixed-convection correlation [Equation 24] generally leads to a noticeable improvement of the data fitting with regard to the classical convection correlation [Equation 10].
- The influence of natural convection, characterized by the ratio Nu/Nu_{FC}, can become appreciable (higher than 1.15) even in the smaller tank and in spite of the moderate mean heat transfer driving force used ($\leq 15°$ C).
- The FCLF derived from the Fand or Churchill correlations are practically the same.

Thus, one value of m can characterize the coupling of the contributions of free and forced convection irrespective of the impeller type and the Re range considered. This value of m is identical with the one put forward by Churchill [74] and by Ruckinstein and Rajagopalan [75] for heat transfer by combined convection from a horizontal cylinder to a crossflowing fluid, forced flow being in the same direction as the natural-convection motion.

Consequently, the correlation

$$Nu = (Nu_{NC}^3 + Nu_{FC}^3)^{1/3} \tag{27}$$

provides a good representation of the heat transfer by mixed convection on the whole transition region, the expression of Nu_{FC} evolving according to the value of Re. So, we can see that heat transfer by mixed convection is effective for values of Nu_{NC}/Nu_{FC} comprised in the range 0.4 to 2.5. For lower (higher) values, we may consider that only forced (natural) convection occurs.

Finally, it must be added that trials performed by Desplanches and Bruxelmane [66] reveal weaker contributions of FC and NC in heating than in cooling. It seems that the circulation action of the impeller is opposed to the development of NC and inversely. However, the NC contribution is positive in cooling as well as in heating.

Characteristic Variable and Correlation for Forced Convection in the Transitional Flow Regime

Desplanches and Bruxelmane [65] verified the validity of the circulation hypothesis using Re_c as characteristic variable for Nu_{FC}. For this purpose, the effects on h of the following parameters: rotational speed N, diameter d, blade number n_p of 45 degree pitched-blade turbines pumping downwards, coil tube diameter d_c, and vessel diameter D were investigated. Some trials were also carried out with hub-mounted flat-blade turbines. The estimates of the effects were calculated from data of Q_c and of mixing time t_M. Attention must be drawn to the fact that the effects on h_{FC} of d and n_p (and probably the other geometrical parameters of the impeller) as well as of scale of the agitated system are practically the same as in the turbulent region. This corroborates, in a given vessel or in geometrically similar agitated systems, the following circulation correlations:

For PBT:

$$N_{qc} = \frac{Re}{87 + Re} N_{qc,turb} \qquad [22, 66]$$

For FBT and RDT:

$$N_{qc} = \frac{Re}{65 + Re} N_{qc,turb} \qquad [76]$$

$N_{qc,turb}$ being the circulation number in the turbulent region.

The forced convection correlations obtained in the transitional flow regime with large impellers (high d/D; [65]), the FCLF of the mixed-convection equations of the Fand or Churchill types and the utilization of the circulation hypothesis allow the proposal of the following forced-convection correlations valid in cooling on the whole transition region in agitated systems in which the liquid flow pattern is practically the same:

For Pitched-Blade Turbines Pumping downwards

$$Nu_{FC} = (0.992\, Re_c^{0.35} + 0.021\, Re_c)\, Pr^y\, Vi^{0.2} \qquad (28a)$$

with $Re_c = \rho N_{qc} N d^3 d_c / \mu D^2$ and $N_{qc} = 0.513(Re/(87 + Re))\,(D/d)^{0.6} n_p^{0.3} (d_c/D)^0$; $y = 0.4 - 0.1 \exp(-0.00024\, Re)$; $H_L = D$; vessel with a flat bottom; $n_B = 4$; $B/D = 1/12$; $e_B/D = 0.11$; $C_B = 0$; $D_c/D = 0.815$; $H_c/D = 0.808$; $C_{ci}/D = 0.1$; $d_c/D = 0.018$ or 0.0286; $e_c/d_c = 2$; $100 < Re < 20{,}000$; $0.27 < Re_c < 7700$; $0.33 \leq d/D \leq 0.524$; $w/d = 1/5$; $n_p = 4$ or 6; $\alpha = 45$ deg; $C = d$; fluids = glycerine, water or aqueous solutions of sodium alginate; $4.2 < Pr < 9{,}200$; $0.072 < Vi < 0.95$; $0.09 < Gr < 7.36\ 10^5$.

For Flat-Blade Turbines

$$Nu_{FC} = (1.019\, Re_c^{0.35} + 0.0241\, Re_c)\, Pr^y\, Vi^{0.2} \qquad (28b)$$

with idem except $N_{qc} = 0.47[Re/(65+Re)]\,(D/d)^{0.7}n_p^{0.24}(d_c/D)^0$; $180 < Re < 3\times10^4$; $n_p = 4$; $\alpha = 90$ deg; $C = H_L/3$.

These general FC correlations (Equation 28a and b) may be used in order to characterize the contribution of the forced convection in Equation 27.

Conclusions

In Newtonian agitated fluids, in the transitional flow regime, natural convection generally acts along with forced convection. The contribution of natural convection is positive and, at least in cooling, can be appreciable even at the end of the transition region though no temperature gradient is then detected in the agitated liquid.

Heat transfer by mixed convection can be characterized in cooling over the whole transition region by the equation of the Churchill type:

$$Nu = (Nu_{FC}^3 + Nu_{NC}^3)^{1/3}$$

combining independently a forced-convection term and a natural convection term. Equations 21 and 28 respectively give the expressions of Nu_{NC} and Nu_{FC} in vessel/coil/baffles systems geometrically similar to those investigated by the authors.

In a given vessel/coil/baffles system or in such systems geometrically similar:

- The Grashof number Gr (or better the Rayleigh number Ra) is the characteristic variable for Nu_{NC} whatever the impeller type.
- The circulation Reynolds number Re_c is the main characteristic variable which assesses the value of Nu_{FC} for impellers of a given class.

These conclusions directly lead to the scale-up criteria for Nu_{NC} and Nu_{FC} in geometrically similar vessel/coil/baffles systems:

$$Gr = \text{cst for } Nu_{NC}; \qquad Re_c = \text{cst for } Nu_{FC}$$

So, in geometrically similar agitated systems (vessel + impeller), we get the same value of Nu_{FC} if Re is kept constant.

Then, the following practical important conclusions can be drawn for heat transfer by forced convection from a helical coil to an agitated Newtonian liquid (at constant D_c/D, H_c/D and C_{ci}/D) in the transitional flow regime:

- Circulation impellers i.e. axial flow impellers (PBT and propellers, for instance) are the most efficient.
- It is recommended to enlarge economically the circulation action of the chosen impeller by an appropriate choice of its geometry.
- The utilization of the circulation hypothesis provides the estimation of the effects upon Nu_{FC} of geometrical parameters of the agitated system (d/D, n_p, w/d, α, d_c/D, p_c/d_c) which don't sensibly modify the distribution curve of circulation velocities around the coil tubing if the effect of those parameters upon N_{qc} are known.
- As a first approximation, Equation 28a is suitable to axial flow impellers (i.e. to propellers of any type specially to "marine" propellers) and the Equation 28b can be applied to radial flow impellers (rushton disc turbines, curved-blade turbines, etc.) if
    1. We have circulation data.
    2. C/d = cst for axial impellers.
    3. C/D = cst for radial impellers.
    4. The global coil geometry (D_c/D, H_c/D, C_c/D) remains identical.
- Any information concerning the effect of a geometrical parameter upon one of the variables h_{FC}, Q_c, Q_p, or t_M can be used to forecast how the same geometrical parameter affects the other variables. However, in the early transition region, the liquid flow pattern may be very complex. Closed circulation loops may exist and the use of a mixing-time

correlation to estimate the influence on h of some geometrical parameters may be dangerous.

- Use of the circulation hypothesis allows a critical examination of any experimental results about h_{Fc}, Q_c, Q_p, or t_M.
- Concerning the choice of the coil tube diameter d_c, it must first be recalled that, irrespective of d_c value, the heat transfer area A_H is almost constant if D_c, H_c, p_c/d_c are unchanged. At constant hA_H/V_L, reduction of d_c involves an increase of h_{FC}, of h_{NC} ($Nu_{NC}/Nu_{FC} \div d_c^{-0.25}$) and of pressure drop in the coil. So at constant Q_i, if the flow of the heating or cooling fluid is turbulent, the pressure drop in the coil is inversely proportional to $d_c^{5.8}$! [62].

When we extrapolate the experimental results from the pilot to the industrial scale, we must bear in mind that:

- $\overline{\Delta\theta}$ remains constant only if $Q_i \div hA_H$ at constant $\theta - \theta_{il}$.
- In geometrically similar agitated systems, A_H is proportional to the square of the scale factor F whereas the total rate of heat transfer q is proportional to F^3.
- The homogeneity of the temperature of the agitated liquid is mainly related to the mean circulation time $t_c = V_L/Q_c$. Accordingly, in the industrial prototype, if the scale-up criterion is less strict than t_c = cst, the local temperature gradients and thus the NC contribution will tend to be higher than in the pilot plant.
- The rate of energy dissipated by the impeller in the agitated liquid is not necessarily negligible compared with q, contrary to what is usually observed in the turbulent region. This is important when we are defining the optimal dimensions and working conditions of the agitated system.
- In geometrically similar agitated systems equipped with a PBT, for instance, at constant $h_{FC}A_H/V_L$:

  $$P/V_L \div D^{6.9} \quad \text{if } 200 < Re < 1{,}500$$

  $$P/V_L \div D^{4.6} \quad \text{if } 10^4 < Re < 2.10^4$$

  When extrapolating at constant P/V_L, we can keep q/V_L constant only by increasing the heat transfer driving force $\overline{\Delta\theta}$. Then, the contribution of free convection becomes more important too.
- As in mixed convection, two characteristic variables Gr and Re_c control the value of Nu; the contributions of NC and FC can be very different in geometrically similar agitated systems. In such systems, at constant Re_c, the group Gr/Re_c^2 which can quantify the importance of natural convection with regard to forced convection is proportional to the cube of the linear scale factor F. So, in scaling-up, a noticeable natural convection effect can take place for higher values of the agitation Reynolds number Re. Nevertheless, fortunately, the NC contribution always has a favorable effect on the heat transfer capacity of the agitated system.

THE COIL PROCESS SIDE HEAT TRANSFER COEFFICIENT IN AGITATED NON-NEWTONIAN FLUIDS

Introduction

Many engineers in chemical and allied industries are becoming increasingly concerned with the processing and mixing of moderate-to-high-viscosity non-Newtonian liquids. Such fluids are of considerable industrial importance as they include the following:

- Polymeric materials, such as synthetic fiber melts, molten plastics, rubbers, paints, varnishes, surface coatings, etc.

- Petroleum products, such as some lubricating oils, crude oils, greases, and waxes.
- Two-phase systems, such as the suspensions, slurries, dispersions, emulsions, pastes, creams, and gels encountered in the clay, paper, pottery, food, biochemical, pharmaceutical, photographic, and printing industries.

The viscosity of a Newtonian liquid is independent of the shear to which it may be subjected. Thus, although the shear stresses and strains vary greatly throughout an agitated liquid, the viscosity will be the same at all points in a Newtonian liquid. In contrast, in non-Newtonian fluids, the shear rate (velocity gradient) is not directly proportional to the applied shear stress. Thus, the apparent viscosity of such a liquid at any point in the agitated liquid depends on the magnitude of either the shear stress or the shear rate at that point and may also depend on the previous history of the liquid. The effect of shear rate on apparent viscosity will determine the viscosity profile throughout the stirred tank. Thus, difficulties arise in the development of heat transfer correlations for non-Newtonian liquids because the relationship (obtained in a rheometer) between shear stress τ and shear rate (or velocity gradient) γ in nonlinear. An apparent or effective viscosity μ_a of the fluid can be defined at any arbitrary shear stress as the viscosity of a Newtonian fluid which would flow at the same rate. Thus for a given shear rate γ, the apparent viscosity of the non-Newtonian fluid is given by the slope of line drawn from the corresponding point (γ,τ) of the flow curve $\tau = f(\gamma)$ to the origin (and not by the slope of the tangent to the flow curve in (γ,τ). i.e.

$$\mu_a = \tau/\gamma \tag{29}$$

As, this viscosity is not constant in the agitated solution, this leads to problems in specifying the Reynolds number, Prandtl number and viscosity ratio. Nevertheless, the first problem is to define the rheological properties of the agitated fluid.

Rheological Models

Various types of non-Newtonian behavior have been discussed in the literature [78, 79, 80...]. This paragraph will be concerned only with the simple time-independent fluids of the pseudoplastic (shear thinning), dilatant (shear thickening), and Bingham plastic (which

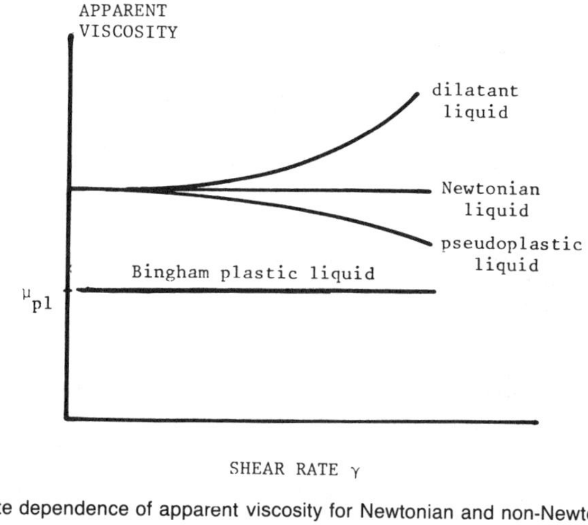

Figure 2. Shear rate dependence of apparent viscosity for Newtonian and non-Newtonian fluids.

exhibit a yield stress) types. Typical flow curves or rheograms, that is, plots of shear stress against shear rate, obtained by viscosimetric experiments are shown in Figure 1. The corresponding values of the apparent viscosity are given in Figure 2.

Pseudoplastic materials are those whose viscosity decreases with an increase in shear rate, whereas for dilatant fluids the viscosity increases with increasing shear rate. Pseudoplastic fluids are the most common. Examples of pseudoplastic fluids may usually be found in rubber solutions, polymer solutions, paints, fermentation broths, etc. Some aqueous suspensions of titanium dioxide and corn flour solutions are dilatant in character. Common examples of Bingham plastics are cement, rock, coal, chalk or detergent slurries, certain plastic melts, drilling muds, oil paints, toothpaste, and sewage sludges. The explanation of Bingham plastic behavior is that the fluid at rest contains a three-dimensional structure of sufficient rigidity to resist any stress less than the yield stress τ_y. If this stress is exceeded the structure completely disintegrates and the fluid behaves as a Newtonian liquid under a shear stress $\tau - \tau_y$.

The other more complex fluids which exibit time-dependent thixotropic and viscoelastic effects are not considered because no data on heat transfer in agitated vessels are available for these fluids at the present time.

The various types of behavior that have been presented in Figure 1 may be characterized by mathematical models of varying degrees of complexity. Some of these are straightforward attempts at curve fitting, giving an empirical relationship for the shape of the shear-stress/shear-rate curve for example, while others have some theoretical foundation.

Equations that have been used to describe these various types of fluid behavior follow. It must be noted that they are all approximations and appreciable deviations may occur in practice, particularly if a wide range of shear rate is considered. So, if an experiment is carried out over several decades of shear rate in a real pseudoplastic fluid, for instance, the logarithmic plot of the apparent viscosity as function of the shear rate is often of the type illustrated in Figure 3. These curves can frequently be approximated by the three linear regions A–B, C–D, and E–F and their extrapolations. At the ends of the shear rate range there are two regions in which the liquid is essentially Newtonian in character as the effective viscosities are uniform. It is difficult to provide valid generalization but many liquids meet these conditions for γ below about 0.1 s^{-1} and above $10^5 s^{-1}$. Data such as these are extremely difficult to obtain satisfactorily. However the low shear rate region is particularly

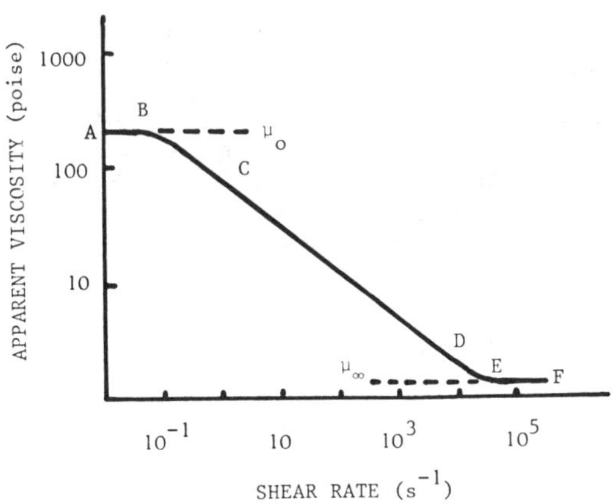

Figure 3. Apparent viscosity of a real pseudoplastic fluid.

important in agitation. Indeed, in stirred tanks, the shear rates are generally comprised between zero and one hundred s^{-1} and the zero-shear viscosity μ_0 may be useful in heat transfer problems.

Newtonian Liquid

For a Newtonian liquid, the viscosity is independent of shear rate and we have:

$$\mu = \frac{\tau}{\gamma} = \text{const.} \tag{30}$$

The Power-Law or Ostwald–De Waele Model

For pseudoplastic and for dilatant fluids, the relationships between shear stress and shear rate are often nearly straight lines on a log/log plot and may be approximated by an empirical functional relation known as the power law or the Ostwald–De Waele model.

$$\tau = K\,\gamma^n \tag{31}$$

In this relation which was originally proposed by Ostwald, the fluid consistency index K and the fluid behavior index n are constant for the particular fluid considered, over a limited range of shear rate. Thus, we must criticize the authors that propose the values of K and n without specifying the corresponding γ-range.

K has the dimensions of $ML^{-1}T^{n-2}$ and is a measure of the consistency of the fluid: the higher K the more viscous the fluid. As for n, it is dimensionless and is a measure of the degree of non-Newtonian behavior. It can be seen that if $n = 1$, then Equation 30 reduces to the Newtonian form. When n is between 0 and 1, Equation 31 describes the pseudoplastic behavior and when greater than 1, it is an appropriate form for the dilatant fluids. The greater the departure of n from unity the more pronounced are the non-Newtonian properties of the fluid. Besides, n is almost constant over a limited range of temperature.

The power-law model is widely applied and the apparent viscosity of the fluid is then defined by:

$$\mu_a = \tau/\gamma = K\,\gamma^{n-1} \tag{32}$$

Finally, it must be noted that for all values of n other than unity the apparent viscosity would approach an unrealistic limit for creeping flow, i.e. at very low rates of shear the pseudoplastic materials become very highly viscous and the dilatant fluids very mobile. Under such conditions, the power-law model breaks down, and it may well be useful to use more complex formulations than the simple power law. Of particular value for describing the behavior at relatively low shear rate ranges are the Ellis and Carreau equations. Both models are very similar in form; for the Carreau model, the functional dependence is on the shear rate γ, whereas it is on the shear stress τ for the Ellis model.

The Carreau Model [81, 82]

In the three-parameter Carreau rheological model:

$$\mu_a = \mu_0/(1 + t_k^2\gamma^2)^q \tag{33}$$

μ_0 is the zero-shear viscosity which could eventually be an estimation of the maximum fluid viscosity in the agitated solution. t_k is a time constant which is closely related to elastic time for most polymeric materials, and q is a dimensionless parameter.

The Ellis Model [82, 83]

$$\mu_a = \frac{\mu_0}{1+(\tau/\tau_{1/2})^{\delta-1}} \tag{34}$$

$\tau_{1/2}$ is the shear stress for which $\mu_a = \mu_0/2$, and δ is a dimensionless parameter.

The parameters q and δ are associated with the power-law behavior: $n \cong 1/\delta \cong 1-2q$ and obviously for pseudoplastic materials, for instance, q is greater than zero and δ greater than 1.

Bingham Plastic Equation.

The rheogram of a Bingham plastic fluid is a straight line having an intercept τ_y on the shear-stress axis. The yield stress τ_y is the stress which must be exceeded before any flow may take place.

The rheological equation for a Bingham plastic may be written:

$$\tau - \tau_y = \mu_{pl}\gamma \quad \text{if} \quad \tau > \tau_y \tag{35}$$

where μ_{pl} is the plastic viscosity or the coefficient of rigidity. It is the slope of the flow curve $\tau - \gamma$.

Though many real plastic fluids closely approximate this type of behavior, Casson [84] and Hershell–Bulkley [85], respectively, have proposed for these fluids:

$$\mu_a^{1/2} = \mu_C^{1/2} + (\tau_y/\gamma)^{1/2} \tag{36}$$

$$\mu_a = K_H\gamma^{nH-1} + \tau_y/\gamma \tag{37}$$

τ being greater than τ_y.

Characterization of the Viscosity of Agitated Non-Newtonian Liquids and Generalization of the Dimensionless Numbers

Dimensional Analysis

In order to characterize the heat transfer to helical coils in agitated non-Newtonian fluids, we may use the dimensional analysis. The viscosity μ is replaced by the parameters of the rheological model chosen to define the fluid behavior. So, Equations 38 and 40 are obtained for "Ostwaldian" and Bingham plastic fluids, respectively.

$$Nu = A_1 Re_1^{x1} Pr_1^{y1} n^{\beta 1} MG_1^{\alpha 1} MG_2^{\alpha 2} \ldots \tag{38}$$

with

$$Re_1 = \frac{\rho N^{2-n} d^2}{K} \quad \text{and} \quad Pr_1 = \frac{C_p K N^{n-1}}{\lambda} \tag{39}$$

$$Nu = A_2 Re_2^{x2} Pr_2^{y2} He^{\beta 2} MG_1^{\alpha 1} MG_2^{\alpha 2} \tag{40}$$

with

$$Re_2 = \frac{\rho N d^2}{\mu_{pl}}$$

$$Pr_2 = \frac{C_p \mu_{pl}}{\lambda}$$

$$He = \frac{\tau_y \rho \, d^2}{\mu_{pl}^2} = \text{Hedström number} \tag{41}$$

The expressions of the Reynolds number Re_1 (or Re_2) and of the Prandtl number Pr_1 (or Pr_2) are only empirical generalizations of the dimensionless Newtonian parameters. Their values may not be compared when the fluid behavior index n (or He) varies.

Definition of an Average Shear Rate [79, 88]

In an agitated vessel, the shear rate varies throughout the process liquid, thus producing a variation in the apparent viscosity if the fluid is non-Newtonian in character. This makes the specification of a viscosity, which is to be used in the Reynolds number, Prandtl number, and viscosity ratio extremely difficult, especially because there is little information on velocity profiles in the vessel. To overcome this difficulty, an appropriate average shear rate for the agitated liquid has to be defined.

Method 1. Magnusson [86] as well as Metzner and Otto [87] proposed a method of predicting such an average shear rate (that we shall note γ_{MO}) from measurements of power requirements for the agitation of Newtonian and non-Newtonian liquids in the laminar region. The average apparent viscosity is defined as the viscosity of the Newtonian fluid which requires the same power input as the non-Newtonian fluid, in the same vessel and the same rotational speed of the impeller. This value of the average apparent viscosity may be used to define the average shear rate γ_{MO} from the curve $\mu_a = f(\gamma)$ derived from the corresponding flow curve.

Metzner and Otto [87] next assumed that in a given stirred system or in geometrically similar stirred systems, the average shear rate γ_{MO} is directly proportional to the rotational speed of the impeller, that is:

$$\gamma_{MO} = k_{MO} N \tag{42}$$

where k_{MO} is a constant irrespective of the rheological properties of the agitated liquid. Theoretically, its value depends upon the geometry of the equipment and the type of impeller used. However, speaking practically, it is independent of d/D, n_p and the number of impellers on the agitator shaft [87, 89, 91, 92].

The validity of Equation 42 was subsequently established by Metzner and Taylor [89] in studies of flow patterns in baffled, agitated tanks. The Metzner–Otto constant k_{MO} can

Table 6
Typical Values of the Metzner–Otto Constant k_{MO} in Agitated Pseudoplastic Fluids

Impeller	k_{MO} Range	k_{MO} Recommended Values	References
Marine propeller*	9 to 11	10	90, 91, 79
PBT†	11 to 15	12	90, 91, 65, 92, 60, 79
FBT‡	10 to 14	11.5	91, 93, 76, 79, 60, 92
6-RDT	10 to 13	11.5	87, 90, 91, 94, 76, 95, 93, 79, 92, 105

* p = d or 2d ‡ w = d/5
† w = d/5; α = 45 deg.; *downwards pumping.*

be determined experimentally for a given equipment agitating a given non-Newtonian liquid as follows:

1. Select or experimentally determine, in the laminar region, the power number N_p vs, the agitation Reynolds number curve in the agitated system considered or in a geometrically similar agitated system to the latter using Newtonian liquids.
2. The power number is measured for a given impeller speed in the non-Newtonian fluid considered, in the laminar region.
3. The corresponding Reynolds number is read from the $N_p - Re$ plot obtained for the Newtonian fluids in the same geometrical configuration (step 1).
4. The value of μ_a under these agitation conditions is calculated from this Reynolds number.
5. The corresponding γ_{MO} is found from the curve $\mu_a = f(\gamma)$ or directly from the flow curve since $\mu_a = \tau/\gamma_{MO}$.
6. Substitution of γ_{MO} in Equation 42 results in the evaluation of k_{MO}. Repetition at various N (in the laminar region where the power number is most sensitive to viscosity changes) allows the best value of k_{MO} to be estimated. Typical values of k_{MO} in pseudoplastic fluids are presented in Table 6.

The Metzner–Otto shear-rate model allows us:

- To evaluate the range of shear rate prevailing under the working conditions of agitation considered and thus to choose the appropriate values of the parameters of the rheological equation chosen.
- To generalize the dimensionless numbers Re, Pr, and Vi whatever the rheological behavior of the fluid and without the assumption of any rheological model for describing the flow curve. For instance, if the power-law is valid, we have for μ_a and Re_a:

$$\mu_a = \mu_3 = K\,\gamma^{n-1} = K\,(k_{MO}N)^{n-1} \tag{43a}$$

$$Re_a = Re_3 = \rho\,Nd^2/\mu_3 = k_{MO}^{1-n}\,\rho\,N^{2-n}d^2/K \tag{43b}$$

If we use the Metzner–Otto model in combination with the three-parameter Carreau rheological model, we have:

$$\mu_a = \mu_4 = \mu_0(1 + t_k^2 k_{MO}^2 N^2)^{-q} \tag{44a}$$

$$Re_a = Re_4 = \frac{\rho N d^2}{\mu_0}(1 + t_k^2 k_{MO}^2 N^2)^q \tag{44b}$$

Still it must lastly be emphasized that the values of k_{MO} determined in the laminar region are supposed valid in the transition and in the turbulent regions. In this connection, in the early transition region (approximately $Re_a < 1{,}000$), for pseudoplastic fluids, the shear rate models of Metzner–Otto and Calderbank–Moo Young lead to non-Newtonian curves $N_p - Re_a$ which scatter below the Newtonian curves at Re_a greater than 30 [87, 91, 79, 105]. These discrepancies may be explained by the fact that local values of apparent viscosity increase with distance from the impeller because of the decline in shear rate with increasing radial distance. This retards the initiation of turbulence and extends the laminar region to Reynolds numbers well above the Newtonian limit of 10. In any case, this questions the adequacy of the Metzner–Otto shear rate model, at least in the early transition region.

Method 2. On the other hand, Metzner and Otto [87] also assumed that the power number in mixing and the friction factor in tube flow depend in the same way on n and by analogy proposed the following generalized Reynolds number for agitation of power-law fluids:

$$Re_5 = \frac{\rho N d^2}{K}(8\,N)^{1-n}\left(\frac{4\,n}{3\,n+1}\right)^n \tag{45}$$

We can show that:

$$\gamma = 8\,N\left(\frac{3\,n+1}{4\,n}\right)^{n/n-1} \tag{46}$$

$$C_1 = \frac{Re_3}{Re_5} = \frac{k_{MO}^{1-n}}{8^{1-n}}\left(\frac{3\,n+1}{4\,n}\right)^n$$

$$1 \leqq C_1 \leqq 1.25 \quad \text{if} \quad 1 \geqq n \geqq 0.6$$

Method 3. An alternative approach has been put forward by Calderbank and Moo-Young [91]. For power-law fluids, these authors have proposed the following expression for the average shear rate:

$$\gamma_{CM} = k_{CM} N\left(\frac{3\,n+1}{4\,n}\right)^{(n/n-1)} \tag{47}$$

The latter is identical to Equation 46 simply by replacing the factor 8 by the "generalizing" constant k_{CM}. Then:

$$Re_6 = \frac{\rho N d^2}{K}(k_{CM}N)^{1-n}\left(\frac{4\,n}{3\,n+1}\right)^n \tag{48}$$

For pseudoplastic fluids, with high speed impellers (propellers and turbines; $d/D < 0.66$; $n_B = 4$; $B/D = 0.1$), a value of $k_{CM} \cong 11$ is quoted. For dilatant materials with high speed impellers ($d/D > 1/3$; $n_B = 4$; $B/D = 0.1$):

$$k_{CM} = \frac{22\,(D/d)^2}{(D/d)^2 - 1} \tag{49}$$

For anchor agitation in pseudoplastic materials:

$$k_{CM} = 9.5 + \frac{9\,(D/d)^2}{((D/d)^2 - 1)} \tag{50}$$

It is worthwhile to compare Equations 45 (Metzner and Otto) and 47 (Calderbank and Moo-Young) for the prediction of the average shear rate γ in a vessel. For a wide range of n values ($0.05 < n < 1.68$), the term involving n in Equation 47 is $0.84 \pm 8.5\%$ and may thus be treated as a constant. Therefore both the methods of Metzner & Otto and Calderbank & Moo-Young predict a linear relationship between γ and N which is insensitive to changes in the behavior index n. So, the more complicated Calderbank–Moo Young model is in practice equivalent to this of Metzner–Otto and in addition it implies the use of the power-law model.

Extension of Newtonian Correlations

In studies of heat transfer to both Newtonian and non-Newtonian liquids (see Table 2A), Pollard and Kantyka [96] (coil and jacket), Suryanarayanan and Raja Rao [97] (coil) as well as Edney and Edwards [33] (coil) found that the correlations developed for Newtonian systems could be used to predict non-Newtonian data if the average apparent viscosity μ_a is used in place of the Newtonian viscosity. Furthermore, the use of Newtonian correlations in this way was suggested by Gluz and Pavlushenko [98], Skelland [79], Edwards and Wilkinson [88], etc. The similarity between Newtonian and non-Newtonian correlations has also been mentioned for heat transfer to coils by Krishnan and Pandya [99], Pandian and Raja Rao [100], Suryanarayanan and Raja Rao [97], Skelland and Dimmick [101], and for heat transfer to vessel wall by Carreau et al. [102], Hagedorn and Salamone [103] and

Martone and Sandall [104], for instance. All these authors indicated that it is possible to use the large number of Newtonian correlations in order to predict non-Newtonian behavior.

So, the non-Newtonian heat transfer correlation would be:

$$Nu = A_N(Re_a)^{x_N}(Pr_a)^{y_N}(Vi_a)^{z_N}MG_1^{\alpha_1}MG_2^{\alpha_2} \qquad (51)$$

A_N, x_N, y_N, z_N, α_1, and α_2 being the parameters of the Newtonian heat transfer correlation obtained in the same or in a geometrically similar agitated system. The apparent dimensionless groups being defined from the appropriate expression of the apparent viscosity.

The method of calculating h would be:

- Knowing the impeller rotational speed, calculate the average shear rate γ_{MO} from Equation 42. k_{MO} is evaluated from Table 6. For dilatant solutions use Equation 47 and evaluate k_{CM} from Equation 48.
- Determine the average apparent viscosity from the flow curve (shear-stress versus shear-rate plot) for the liquid used.
- Select the appropriate heat transfer correlations presented earlier.
- Replace the Newtonian viscosity in the agitation Reynolds number, Prandtl number, and the viscosity ratio by the average apparent viscosity μ_a. For many power-law fluids, the fluid behavior index n does not vary with temperature and we may therefore write:

$$Vi_a = K/K_W \qquad (52)$$

- Calculate h from Equation 51.

However, it must be noted that:

- It could be dangerous to use in the dimensionless groups characterizing heat transfer in the transition and turbulent regions, a viscosity which has been evaluated from information taken in the laminar region. Nevertheless, examination of the Newtonian correlations and the non-Newtonian correlations of Table 7A reveal that the heat transfer coefficient h is proportional to the liquid viscosity raised to the power of about −0.3. Thus in the previous procedure an error of 40% in the evaluation of the average apparent viscosity for a non-Newtonian liquid will result in an error of only 11% in the value of h. The described method (if it is applicable, that is to say, if the parameters A, x, y, and z are independent of the behavior indes), is therefore fairly insensitive to errors in the evaluation of the average shear rate and the average apparent viscosity.
- All the methods characterize the fluid viscosity in the region of the impeller and not near the coil though flow could be turbulent around the impeller and laminar or even nonexistent elsewhere.
- The viscosity ratio used, Vi_a, only takes into account the temperature differences near the heat transfer surface and not the variation of shear rate in that zone, though the latter factor should probably play a leading part with respect to the former.

Experimental Studies

General Review

In recent years, there have been several reviews (e.g. [88, 106]) pertaining to heat transfer from helical coils to agitated non-Newtonian liquids.

The available correlations for non-Newtonian fluids are presented in Tables 7A and 8. The geometrical characteristics and working conditions of the agitated systems correspond-

Table 7A
Coil Heat Transfer to Agitated Non-Newtonian Fluids*

Authors	Impeller	Re_a Range	Correlation
Skelland & Dimmick [101]	propeller ↑	$330–2.6 \times 10^5$	$Nu = 0.258\, Re_3^{0.62}\, Pr_3^{0.32}\, Vi_3^{0.2} \left(\frac{d}{D}\right)^{0.1} \left(\frac{d_c}{D}\right)^{0.5}$
Desplanches [107]	4-PBT ↓	83–4,855	$Nu = 0.01962\, Re_4^{0.71}\, Pr_4^{0.33}\, Vi_a^{0.17}$
Mizushina [108]	PBT; FBT & 4-RDT	$0.2–10^3$	$Nu = 36.7 n^{1.2}\, Re_7^{0.5}\, Pr_7^{0.3} \left(\frac{n_p w \sin \alpha}{H_c}\right)^{0.15} \left(\frac{d_c}{D}\right)^{1.7} Vi_a^{0.14}$
		$10^4–8 \times 10^4$	$Nu = 36.7\, Re_2^{0.5}\, Pr_2^{0.3} \left(\frac{n_p w \sin \alpha}{H_c}\right)^{0.15} \left(\frac{d_c}{D}\right)^{1.7} Vi_a^{0.14}$
Pandian & Raja Rao [100]	4-RDT	$100–10^4$	$Nu = 0.067\, Re_3^{0.65}\, Pr_3^{0.33}$
Suryanarayanan & Raja Rao [97]	4-RDT	200–21,700	$Nu = 0.21\, Re_6^{0.66}\, Pr_6^{0.33} \left(\frac{d}{D}\right)^{0.17} \left(\frac{C}{D}\right)^{0.13} \left(\frac{d_c}{D}\right)^{0.55} \left(\frac{D_c}{D}\right)^{-0.29}$
Edney et al. [33b]	6-RDT	$400–9.2 \times 10^5$	$Nu = 0.0675\, Re_3^{0.607}\, Pr_3^{0.345}\, Vi_3^{0.2}$
Edney et al. [33c]	6-RDT	$400–9.2 \times 10^5$	$Nu = 0.036\, Re_3^{0.641}\, Pr_3^{0.353}\, (D_c/D)^{-0.375}\, Vi_a^{0.2}$
Edney [33a]	6-RDT	$400–9.2 \times 10^5$	$Nu = 0.236\, Re_3^{0.641}\, Pr_3^{0.353}\, Vi_3^{0.2} \left(\frac{d}{D}\right)^{0.2} \left(\frac{d_c}{D}\right)^{0.5} \left(\frac{D_c}{D}\right)^{-0.375}$
Pollard & Kantyka [96]	Anchor	$200–6 \times 10^5$	$Nu = 0.077\, Re_6^{0.66}\, Pr_6^{0.33}\, Vi_a^{0.14} \left(\frac{d_c}{D}\right)^{0.52} \left(\frac{D_c}{D}\right)^{-0.27}$
Krishnan & Pandya [99]	Anchor	$100–5.5 \times 10^5$	$Nu = 0.0313 \delta^{9.5}\, Re_6^{0.66}\, Pr_6^{0.3}\, Vi_a^{0.18}$ with $\delta = (3n+1)/4n$
Blasinski et al. [113]	Anchor	$30–2.8 \times 10^4$	$Nu = 0.084\, Re_8^{0.55}\, Pr_8^{0.33}\, Vi_a^{0.14}\, (D_c/D)^{0.74}\, (p_c/d_c)^0$

* *See Table 7B for the corresponding working conditions and geometrical characteristics of agitated systems used.*
$Pr_8 = KC_p(4\pi N)^{n-1}/\lambda$; $Re_8 = \rho N d^2 (4\pi N)^{1-n}/K$; Vi_a *from Equation 52.*

ing to Table 7A are given in Table 7B. The form of the correlations listed in Tables 7A and 8, is similar to that for Newtonian liquids. The shear rate models of Metzner–Otto or Calderbank–Moo Young are generally used to define the Reynolds and Prandtl numbers as well as the viscosity ratio.

Mizushina, Ito, Murakami, and Kirii [108] used for pseudoplastic liquids:

$$Re_7 = Re_1/[1 - \exp(-1.4 \times 10^{-5} \times Re_1)] \tag{53a}$$

$$Pr_7 = Pr_1\, [1 - \exp(1.4 \times 10^{-5} \times Re_1)^{n-1}] \tag{53b}$$

So they tried an empirical definition of the shear rate in the coil region. For Bingham plastics, they suggested the plastic viscosity μ_{pl} cold be used as the average apparent viscosity in the Newtonian correlation. However, their corresponding correlation does not contain the Hedström number He.

Examination of Tables 7 and 8 reveals that relatively little work has been carried out on heat transfer to coils in non-Newtonian liquids. The fluids which have been used to study non-Newtonian heat transfer have generally been aqueous solutions of polymer or natural gum and slurries with low solids concentration, both of which exhibit pseudoplastic behavior and over moderate ranges of shear rate can be described by the power-law model. The latter is imprecise at low shear rates. Only Desplanches and Bruxelmane (Table 8) resort to a three-parameters rheological model. Moreover, the correlation corresponding to an extended n-range being much less accurate, these authors prefer to propose a separate correlation for each value of n. In addition, as the effects of Pr_a and Vi_a upon Nu are very connected, the values of x and y are also interrelated. Indeed, in non-Newtonian fluids, it is impossible to modify independently Rc_a and Pr_a by changing the impeller rotational speed. Consequently, these authors have chosen for the Prandtl number and the viscosity ratio the same exponents in Newtonian fluids i.e. y = 0.3 and z = 0.2.

There are no data on dilatant fluids and on time-dependent fluids. If the former are not common, on the other hand, the latter are worth considering. Indeed, viscoelastic fluids are frequently encountered and thixotropic fluids are of some importance.

Table 7B
Working Conditions and Geometrical Characteristics of Agitated Systems Corresponding to the Coil Heat Transfer Correlations Presented in Table 7A

Author	(•)	Tank				Coil			Impeller					Fluids	
		D (m)	H/D	n_B	D/B	D_c/D	d_c/D	e_c/d_c	Type	d/D	w/d	C/D	Type	n	Pr_a-range
Skelland [101]	±	0.46	1	4	9.6	0.671	0.014– 0.041	3–9	MP↑	0.164– 0.493	p = d	0.246– 0.302	Aq. carbopol sol.	0.53–0.91	11.8–1,110
Desplanches [107]	±	1.18	0.51!	4	14.7	0.805‖	0.0186	2.3	4-PBT↓	0.35	0.146	0.17!	Water; aq. SA sol.	0.77–1	720–10‖
Mizushina [108]	−	0.31	1.28	0	−	0.76	0.04; 0.054	1.14; 0.9	PBT† FBT‡ 4-RDT	0.4– 0.61	0.2	0.135– 0.57	Water; glycerine; castor oil; aq. CMC sol. Chalk/water slurries	? plastic	? ?
Pandian [100]	±	0.36	1	4	12	0.697	0.054	1	4-RDT§	1/3	1/4.37	1/3	Water; aq. CMC & PVA sol.	0.7–1	?
Suryanarayanan [97]	±	0.36	1	4	12	0.475– 0.781	0.045– 0.061	?	4-RDT§	0.22– 0.43	1/4.37	0.25– 0.6	Water; aq. CMC & SA sol.	0.47–1	49–1,220
Edney [33b]	−	1.22	0.754!	4	19	0.458	0.039	0.3!	6-RDT	0.25	0.2	0.27	Water; aq. CMC & polyacrylamid sol.	0.5–1	4.3–1,900
Edney [33c]	−	1.22	0.754!	4	19	0.458– 0.791	0.039	0.3!	6-RDT	0.25	0.2	0.27	Water; aq. CMC & polyacrylamid sol.	0.5–1	4.3–1,900
Edney [33a]	−	0.87; 1.22	0.754!	4	19	0.458– 0.791	0.018; 0.039	0.19; 0.25	6-RDT	0.35; 0.25	0.2	0.27	Water; aq. CMC & polyacrylamid sol.	0.5–1	4.3–1,900
Pollard [96]	±	0.3	1	0	−	0.49	0.044	0.89	Anchor	0.83	0.1	?	Water; aq. glycerol; chalk/water slurries	0.38–1	?
Krishnan [99]	−	0.51P	0.5	0	−	0.525	0.0375	0.167	Anchor	0.74	?	0.13	Water; aq. CMC & PVA sol.; $CaCO_3$/ glycerol slurries	0.66–1	95–4,950
Blasinski [113]	−	0.3	1	0	−	0.4– 0.73	0.04	0.083– 2.2	Anchor	0.95	0.05	0.167	Aq. CMC sol.	0.69–0.89	19–8,300

(•) + refers to heating and − to cooling
† n = 2.4; α = 45 deg.
‡ n = 2.4
§ 1/d = 1/3.5
‖ C_{ci}/D = 0.01!
P Dished bottom

Finally, let us point out that for the same effective viscosity and the same diameter and rotational speed of the impeller, the temperature gradients in agitated non-Newtonian fluids are higher than those observed in Newtonian liquids. This is due to the damping of the shear rate far from the impeller. Thus, the experimental correlations are less precise in non-Newtonian liquids, specially if Re_a is lower than 1,000.

Analysis of Experimental Results

The experimental results obtained on the one hand in the early transition region and on the other hand at the end of the transition region and in the turbulent region will be analyzed separately. However, generally speaking, the results of Edney [33] and Desplanches and Bruxelmane [65, 109] as well as Pollard and Kantyka [96] prove that in geometrically similar agitated systems, for the same non-Newtonian fluid, the Nusselt numbers are equal if the apparent agitation Reynolds numbers are kept constant and if natural convection is weak. In these conditions, we can say that the flow patterns are geometrically similar and the circulation Reynolds numbers equal. Thus, in spite of its imperfections, the Metzner–Otto shear-rate model allows this practical, important result.

Experimental Results in the Early transition Region. First, it must be noted that for successful heat transfer to non-Newtonian fluids, it is necessary to set in motion the whole contents of the vessel, i.e., to eliminate all stagnant regions in the vessel, including those at the surface. This condition has been detected visually by Metzner et al. [87, 90] and has been expressed, irrespective of the fluid behavior index, as a function of the ratio of impeller diameter to vessel diameter, d/D, the effective Reynolds number Re_3^* and the power input by unit volume. The results are summarized in Table 9.

Table 8
Correlations for Forced Heat Transfer to a Helical Coil in Agitated Non-Newtonian Fluids in the Transition Zone (Cooling Trials)

Vessel		Fluid				$Nu = A\,Re_4^x\,Pe_4^{0.3}\,Vi_4^{0.2}$			
Type(*)	D (m)	Nature	n	Re_4	Pr_4	A	x	$\bar{\Delta}$ (%)	$\frac{h_{NN}}{h_N}$
4-PBT pumping downwards (d/D = 1/3; w/d = 0.2; α = 45 deg.; C = d)									
STD	0.63	CMC 0.8%	0.74	278–2,494	710–1,270	0.0138	0.743	2.5	0.86
		CMC 1%	0.56	110–1,128	1,440–2,950	0.0026	1.014	3.8	1.08
		Guar 0.8%	0.46	29–541	1,780–5,600	0.00672	0.893	7	1.24
4-PBT pumping downwards (d/D = 0.524; w/d = 0.2; α = 45 deg.; C = d)									
STD	0.63	Glycerol	1	210–2,680	1,400–8,300	0.0513	0.586	0.837	1
		CMC 0.75%	0.807	1,150–4,520	475–760	0.0355	0.648	1.52	1.13
		CMC 0.8%	0.707	590–2,320	840–1,395	0.0212	0.721	1.07	1.1
		Guar 1%	0.383	63–700	3,420–9,730	0.00678	0.933	3.80	1.02
NSTD	1.18	SA 1%	0.82	760–4,600	766–1,280	0.0310	0.681	1.56	—
		SA 2%	0.73	230–1,730	2,130–3,520	0.0315	0.676	1.12	—
6-PBT pumping downwards (d/D = 1/3; w/d = 0.2; α = 45 deg.; C = d)									
STD	0.63	Guar 0.8%	0.46	29–990	1,465–5,691	0.0096	0.844	7	1.26
6-PBT pumping downwards (d/D = 0.524; w/d = 0.2; α = 45 deg.; C = d)									
STD	0.63	Glycerol	1	150–2,340	1,030–9,150	0.0555	0.585	0.797	1
		SA 2,5%	0.695	27–420	5,680–14,810	0.0233	0.751	4.21	1.04
		Guar 0.8%	0.451	67–1,370	1,730–6,010	0.0235	0.738	5.04	1.13
		Guar 1%	0.379	63–725	3,290–9,830	0.00795	0.933	5.58	1.15
		CMC 1.25%	0.597	60–530	4,460–11,260	0.0125	0.912	3.49	1.29
		CMC 0.8%	0.704	470–2,040	1,010–1,700	0.0259	0.708	2.12	1.11
		CMC 0.75%	0.774	900–3,300	600–1,020	0.0213	0.738	2.40	1.18
STD	1	Guar 1%	0.397	80–1,190	4,460–12,570	0.00819	0.912	6.55	—
4-FBT (d/D = 0.524; w/d = 0.2; C/D = 1/3)									
STD	0.63	Glycerol	1	200–2,410	1,470–5,450	0.0637	0.559	0.901	1
		SA 2.5%	0.705	34–420	3,610–11,940	0.0272	0.730	3.69	1.13
		Guar 0.8%	0.450	150–1,560	1,520–4,180	0.0460	0.646	2.59	1.28
		Guar 1%	0.376	110–740	3,230–8,070	0.0207	0.785	5.38	1.28
		CMC 1.25%	0.578	120–560	4,240–9,570	0.0164	0.864	5.28	1.35
6-RDT (d/D = 0.524; d/1/w = 20/5/4; C/D = 1/3)									
STD	0.63	Guar 0.3%	0.45	140–1,490	1,600–4,620	0.0547	0.614	2.53	—
		Guar 1%	0.38	110–745	3,190–7,730	0.0437	0.668	4.25	—

* *Geometry of standard vessels* (STD): H/D = 1; *flat bottom;* n_B = 4; D/B = 12; $D_c/D \cong 0.8$; $d_c/D \cong 0.0185$; $e_c/d = 2$. *For the non-standard vessel* (NTSD): *see Table 7B.* Guar 0.8% = Aq. guar (0.8% by wt) sol.; CMC = Aq. carboxymethylcellulose sol.; SA = Aq. sodium alginate sol.
From Desplanches and Bruxelmane [65, 109].

Table 9
Minimum Values of the Apparent Reynolds Number, Re_3^*, to Ensure Complete Fluid Motion in an Agitated Vessel

Impeller type	d/D 1/3	1/2	2/3	3/4
Propeller	500	290	200	–
Pitched blade turbine	120	50	40	30
Rushton disc turbine	270	110	90	50
Two FBT, D/2 apart	85	68	50	45

From Metzner et al. [87, 90].

As for Desplanches [60, 109], he observed that in the same agitated system, the value of Re_4^* under which the heat transfer relationship $Nu/Pr_a Vi_a = f(Re_a)$ breaks away from the forced-convection straight-line relationship is a function of the fluid behavior index. Besides, in these conditions, for the same d/D, like in Metzner's experiments, the value of Re_4^* is lower for PBT than for FBT and RDT.

However, we must observe that if in geometrically similar systems agitating the same fluid, we keep constant the apparent agitation Reynolds number ($N \div d^{-2/(2-n)}$ for Ostwaldian fluids), the flow patterns will be geometrically similar but the fluid velocities will not be equal in homologous points ($v_c \div N_{qc}Nd^3/D^2 \div Nd \div d^{n/n-2}$ at constant

Re_a!). Thus, a more conservative rule for a fluid-motion similarity in geometrically similar agitated systems would be:

$$N_{qc}Nd^3/D^2 = N_{qc}Nd = \text{constant}$$

or more simply but more coarsely:

$$Nd = \text{constant}$$

In addition, in these conditions, the mixing times will be longer and the temperature gradients and the natural convection will be more important in the large scale equipment.

Table 8 shows that in the early transition region ($Re_a < 1{,}000$), for pseudoplastic fluids, in a given agitated system or in geometrically similar agitated systems, at constant Re_a:

- If pseudoplasticity rises, i.e. if n becomes lower, A decreases and x increases strongly and y_{NN}/y_N deviates upwards from 1 ($y = Nu/Re^x Pr^y V_i^z$). The increase of x with the pseudoplasticity is also pointed out by Kipke [100] and Hagedorn and Salomone [103] for the heat transfer to the vessel wall.
- These effects are less marked:

  1. For radial (FBT and RDT) than for axial flow impellers (PBT).
  2. For RDT than for FBT.
  3. For a given impeller type, when Re_a increases.

Those observations question the generalization of dimensionless groups according to the Metzner–Otto shear-rate model and prevent to propose a correlation independent of rheological behavior of the fluid, at least in the early transition region.

Nevertheless, for pseudoplastic fluids, in a given agitated system or in geometrically similar agitated systems, the authors [109] propose the following empirical correlation:

$$Nu = A(n)\, Re_4^{x(n)}\, Pr_4^{0.3}\, Vi_4^{0.2} \tag{54}$$

with

$$A(n) = A_N n^{\alpha_1} \tag{55}$$

$$x(n) = x_N n^{\beta_1} \tag{56}$$

The values of the parameters α_1 and β_1 depend on Re_4. So, for PBT pumping downwards ($80 < Re_4 < 2 \times 10^4$; $0.38 < n \leqq 1$):

$$2.35 > \alpha_1 \geqq 0 \quad \text{and} \quad -0.5 < \beta_1 < 0$$

For radial flow impellers (FBT and RDT; $80 < Re_4 < 10^4$; $0.38 < n \leqq 1$):

$$2.5 > \alpha_1 \geqq 0 \quad \text{and} \quad -0.8 < \beta_1 < 0$$

For the range: $80 < Re_4 < 1{,}000$, the expressions of A(n) and x(n) are given in Table 10. They are less accurate for the CMC solutions and for the radial flow impellers.

On the other hand, on the base of these results, Desplanches and Bruxelmane [65, 109] have compared the heat transfer capacities and efficiencies of axial (PBT) and radial (FBT and RDT) impellers of the same diameter. Pronounced influences of the pseudoplasticity and agitation Reynolds number appear. The ratio of heat transfer capacities of axial impellers with respect to radial impellers (i.e. h_{PBT}/h_{4-FBT} and h_{PBT}/h_{6-RDT}) decrease with n at constant Re_4 and with Re_4 at constant n. Nevertheless axial impellers generally remain the most effective.

Finally, the influence upon h of some geometrical parameters of PBT that we investigated [65] are collected in Table 11. The influence upon h of impeller clearance C is also sensible and the optimum value of C/d is:

$$(C/d)_{opt} = 0.75$$

$h_{C/d = 0.75} = 1.13\, h_{c/d = 1}$ for n = 0.36 and $Re_4 = 700$. At the present time, the lack of hydrodynamical data (N_{qc} or N_{qp}) makes impossible any interpretation of those results from circulation theory. However, it must be noted that pseudoplasticity emphasizes the effect of d and n_p upon h. Thus, theoretically, we must cautiously use the effects of geometrical parameters on h, which are independent of the behavior index n. So, it seems odd that Edney [33], Suryanarayanan and Raja Rao [97] and Skelland and Dimmick [101] (see Table 7A) proposed, as also valid in the early transition region, effects of d/D, D_c/D and d_c/D upon h which do not take into account any influence of n.

Experimental Results at the End of the Transition Region and in the Turbulent region. At the end of the transition region and in the turbulent region, the phenomena seem less complex. Indeed, for FBT and RDT if $Re_a > 1{,}000–1{,}500$ and for PBT probably if $Re_a > 5{,}000–10{,}000$, we have:

$$A_{NN} \cong A_N \quad \text{and} \quad x_{NN} \cong x_N$$

This explains why Edney [33], Suryanarayanan and Raja Rao [97] and Skelland and Dimmick [101] who worked especially in the turbulent region and at the end of the transition region proposed correlations with exponents of Re_a and geometrical dimensionless groups which are constant and independent of n. In our opinion:

- In the transition region, above Re_a = 1,000, 1,500, 5,000, respectively for RDT, FBT, and PBT, one may use the Newtonian correlations by replacing the Newtonian viscosity μ by the average apparent viscosity μ_a.
- In the same way, in the turbulent region, the heat transfer correlations presented earlier will be used.

In addition, we must say that Bourne et al. [112] too consider that for heat transfer from anchor agitated non-Newtonian solutions to the vessel wall, a value of $Re_3 \cong 500–1{,}000$ must be attained in order to the Newtonian and non-Newtonian data converge.

In conclusion, we can say that even more in non-Newtonian fluids, the heat transfer correlations presented as valid at once in the turbulent region and in a wide range of the transition region are highly open to criticism.

Utilization of the Characteristic Variable Re_{ca}

Desplanches and Bruxelmane [11, 65, 66] have proved that the circulation Reynolds number is the main characteristic variable for Nu_{FC} in Newtonian fluids over the transitional and turbulent regions (for a given impeller type, in a given vessel/coil/baffles system or in geometrically similar vessel/coil/baffles systems). In order to get hydrodynamical information from heat transfer data, they consider [109] now that this is also true in non-Newtonian fluids whatever the behavior index may be. More precisely, they suppose that the characteristic variable for Nu was Re_{ca}, i.e., the circulation Reynolds number in the vicinity of the coil. Re_{ca} contains the average circulation velocity of the fluid v_{cNN} and the average apparent viscosity μ_{ac} in the coil region. Therefore, the detemination of μ_{ac} implies the definition of the relation existing between the average shear rate in the coil region, γ_c, and v_{cNN}. They propose the following expression of γ_c based on the dimensional analysis and the analogy with the flow of non-Newtonian liquids around immersed bodies (spheres, cylinders) or inside a tube:

$$\gamma_c = \left(\frac{3\,n+1}{4\,n}\right)^{1/n} \frac{v_{cNN}}{d_c} = F(n)\, \frac{v_{cNN}}{d_c} \tag{57}$$

The expression 1/n of the exponent of the proportionality constant is obtained from homogenization data in pseudoplastic liquids. We can see that at a given v_{cNN} an increase

Table 10
Expressions of A(n) and x(n)*

Impeller	A(n)	x(n)
6-PBT	$0.0555\ n^{2.06}$	$0.585\ n^{-0.48}$
4-FBT	$0.0637\ n^{1.35}$	$0.56\ n^{-0.35}$
	$0.0637\ n^{2.45}$*	$0.56\ n^{-0.77}$†
6-RDT	$0.075\ n^{0.54}$	$0.56\ n^{-0.18}$

* *Based on the Heat Transfer Correlation* $Nu = A(n)\ Re_a^{x(n)}\ Pr^{0.3}\ Vi^{0.2}$ *for Different types of Impellers* ($80 < Re_a < 1{,}000$; $d/D = 0.524$; $C = d$ for PBT; $C = H/3$ for FBT; $0.38 \leq n \leq 1$. *The other geometrical characteristic are given in Table 3*).
† $80 < Re_a < 500$.
From Desplanches and Bruxelmane [109].

Table 11
Influence of Some Geometrical Parameters upon the Coil Process Side Heat Transfer Coefficient h in Agitated Non-Newtonian Fluids (PBT; C = d)

Experimental Conditions: Fixed Parameters	Investig. Parameters	Fluid & Re_a Range	Infl. Investigated Parameter upon h
D = 0.63 m $n_p = 4$		Glycerine (n = 1) Re = 200–1,500	$h \sim (Re \cdot d^{0.42})^{0.55}$
	d/D = 0.333; 0.524	CMC 0.8% (n = 0.69) Re_a = 500–2,300	$h \sim (Re_a \cdot d^{0.8})^{0.75}$
D = 0.63 m $n_p = 6$	N	Guar 0.8% (n = 0.43) Re_a = 100–1,000	$h \sim (Re_a \cdot d^{0.85})^{0.79}$
D = 0.63 m d/D = 0.524 cst Re_a	n_p = 4; 6	Glycerine (n = 1) Re = 200–2,300	$h \sim (n_p^{0.34})^x$
		CMC 0.75% (n = 0.73) Re_a = 1,000–3,000	$h \sim (n_p^{0.88})^x$
		CMC 0.8% (n = 0.69) Re_a = 600–3,000	$h \sim (n_p^{0.54})^x$
		Guar 1% (n = 0.36) Re_a = 60–700	$h \sim (n_p^{0.78})^x$

From Desplanches and Bruxelmane [65]

of pseudoplasticity leads to an increase of γ_c. This is due to the higher damping of the fluid motion.

Then, the average apparent viscosity may be calculated from the power-law model, K and n corresponding to the shear rate range considered or from the Carreau model:

$$\mu_{ac} = K\gamma_c^{n-1} \tag{58a}$$

or

$$\mu_{ac} = \mu_0(1+t_k^2\gamma_c^2)^{-q} \tag{58b}$$

So, in the Re_4-range considered,

- By extension of the Newtonian heat transfer correlations using the circulation Reynolds number: $Nu = A_c Re_c^{xc} Pr^y Vi^z$ to non-Newtonian liquids by replacing the Newtonian viscosity by μ_{ac} that is to say by supposing for non-Newtonian fluids:

$$Nu = A_c Re_{ca}^{xc} Pr_{ca}^{y} Vi_{ca}^{z} \tag{59}$$

- By identification of Equation 59 to the corresponding pseudoplastic forced heat transfer correlations (cf. Table 8), it is possible to get dimensionless relationships of γ_c and v_{cNN} as functions of Re_4.

$$\frac{\gamma_c}{\gamma_{MO}} = k(n)\, Re_4^{b(n)} \div N^{(2-n)\,b(n)} \tag{60}$$

$$\frac{v_{cNN}}{Nd} = \frac{k(n)\, k_{MO}}{F(n)} \frac{d_c}{d} Re_4^{b(n)} N^{(2-n)\,b(n)} \tag{61}$$

with

$$k(n) = \left\{\frac{A(n)}{A_c}\left(\frac{d^2}{d_c^2}\frac{F(n)}{k_{MO}}\right)^{x_c}\right\}^{1/x_c(2-n)+y(n-1)} \tag{62}$$

$$(n) = \frac{x(n)-x_c}{x_c(2-n)+y(n-1)} \tag{63}$$

In summary, we can say:

- The model leads to Re_{ca} = cst, i.e., Nu = cst if Re_a and n are kept constant. This confirms previous results.
- The exponent b(n) which characterizes the deviation of the average shear rate (average fluid velocity) in the coil vicinity from the Metzner–Otto shear rate (from the impeller tip speed) increases with the fluid pseudoplasticity. So, γ_c increases with the impeller rotational speed faster than the Metzner–Otto model predicts, for instance, $\gamma_c \div N^2$ for $n \cong 0.4$. This agrees with the results of Metzner and Taylor [89] who determined by a photographic technique that the fluid velocities in the horizontal plane of the impeller increased exponentially with impeller speed in pseudoplastic systems.
- γ_c is a function of d/D, which is not the case of γ_{MO}.
- It seems that the model allows us to relate adequately v_{cNN} and v_{cN} irrespective of n. For instance, we have $v_{cNN} \div v_{cN}^2$ if n = 0.4. However, this should be confirmed by complementary experimental results. In any case, contrary to the Metzner–Otto shear rate model, the present model predicts in non-Newtonian fluids, effects of d/D and n_p on v_{cNN} and thus on h different from those observed in Newtonian fluids.

Another possibility of definition of the characteristic average shear rate in the transition region is to use a method proposed by Bourne et al. [112]. In their analysis, these authors consider Ostwaldian fluids and γ_B, an average of the shear rates within the volume V_L. As P is the rate of viscous energy dissipation, they write:

$$\frac{P}{V_L} \div \tau\gamma_B \div K\gamma_B^{n+1} \tag{64}$$

Then, in geometrically similar agitated systems, if $N_P \div Re_a^{a2}$ over the Re_a range considered, we have:

$$\gamma_B \div \{N^{3+2a2-a2n} d^{2(1+a2)}\}^{1/(1+n)} \tag{65}$$

$$\frac{\gamma_B}{\gamma_{MO}} \div Re_a^{1+a2/1+n} \tag{66}$$

We have observed that the effect of Re_a (or N) on the characteristic average shear rates resulting from their work and the effect from Bourne's analysis are very near.

Contribution of Natural Convection

As foreseen, at constant Re_a, Desplanches and Bruxelmane [65, 109] have found that the natural convection contribution is much more important in pseudoplastic fluids. It was specially true in highly viscous pseudoplastic fluids at low rotational speeds of a small impeller because then the liquid is practically motionless near the coil. For instance, for the 6-PBT if $28 < Re_a < 990$; $d/D = 0.33$; $n = 0.43$; Nu/Nu_{FC} can reach 1.8.

Pure Natural Convection Correlation in Non-Newtonian Fluids

For the immersed coil used in cooling, in their small vessel (cf. Table 8), Desplanches and Bruxelmane [109] put forward the following pure natural convection:

$$Nu_{CN} = 0.496\,(Gr_0\,Pr_0)^{0.25} \tag{67}$$

with $D = 0.63$ m; $N = 0$; $D_c/D = 0.81$; $d_c/D = 0.019$; $p_c/d_c = 3$; fluid = guar 0.8%; $n = 0.46$; $0.0118 < Gr_0 < 0.331$; $4{,}990 < Pr_0 < 10{,}860$; $128 < Ra_0 < 1{,}649$; $0.43 < Vi_0 <$ < 0.792.

The zero-shear viscosity μ_0 is used in the dimensionless groups. The use of Vi_0 does not improve the fitting of experimental results.

Though the investigated ranges of the dimensionless groups are not the same as in Newtonian fluids, Equation 67 is practically identical to Equation 21 obtained in Newtonian fluids.

Mixed Convection Correlations of the Churchill Type

Desplanches and Bruxelmane use the following mixed-convection correlation of the Churchill type:

$$Nu = \{(A\,Re_4^x Pr_4^{0.3} Vi_4^{0.2})^m + [0.496\,(Gr_0\,Pr_0)^{0.25}]^m\}^{1/m} \tag{68}$$

The best value of m is $m = 2$ in the different agitated systems considered (in cooling of pseudoplastic fluids). Thus again, one value of m can characterize the coupling of the contributions of free and forced convection. However, this value of m is lower than the one observed in Newtonian fluids ($m = 3$, see previous section). We can still say that:

- The contribution of natural convection becomes appreciable if $Nu_{NC}/Nu_{FC} > 0.2$. This value is lower than in Newtonian fluids and can be easily reached if the scale of the system is large.
- As in the Newtonian fluids, the contribution of natural convection is always positive.
- Equation 68 allows a better fitting of experimental results than the mixed-convection correlation of the Fand type.
- A better definition of the shear rate in the vicinity of the coil allows more accurate estimations of μ_a and natural-convection contribution and finally a better expression of the forced-convection limit form of the Churchill equation.

Conclusions

For forced convection heat transfer from an agitated non-Newtonian solution to a helical coil:

1. Circulation impellers are the most efficient.
2. It is recommended to enlarge economically the circulation action of the impeller by an appropriate choice of its geometry. The effects on Nu of the geometrical parameters are more important than in Newtonian liquids.
3. In geometrically similar agitated systems, the Metzner–Otto shear rate model allows the correlation of the experimental data and the scale-up criterion for Nu_{FC} which is to keep constant Re_a (for a given fluid, at least in the early transition region).
4. Specific heat transfer correlations have been proposed in the early transition region.

5. At the end of the transition region and in the turbulent region, the Newtonian correlations may be extended to non-Newtonian liquids by replacing the Newtonian viscosity by the average apparent viscosity μ_a.
6. Two models have been presented to estimate the average shear rate and fluid velocity in the coil vicinity.

In mixed convection, a correlation of the Churchill type has been presented. The natural-convection limit form allows an estimation of the contribution of this phenomenon. The latter is always positive in the cooling of pseudoplastic fluids and can be significant at large scale.

Finally, we must always keep in mind that in non-Newtonian fluids the heat dissipation corresponding to power consumption is generally nonnegligible.

NOTATION

A_H process side heat transfer area; Equation 2a (m^2)
B baffle width (m)
C height of impeller centerline above vessel bottom (m)
C_B clearance between baffles and vessel bottom (m)
C_c clearance between the center of the lower coil turn and vessel bottom (m)
C_{ci} coil clearance off vessel bottom (m)
C_{cs} higher level of the coil (m)
d impeller diameter (m)
d_c outside coil tube diameter (m)
D vessel diameter (m)
D_c mean helix diameter (m)
e_B clearance between baffles and vessel wall (m)
e_c coil spacing (m)
FBT hub-mounted flat-blade turbine
FC forced convection
FCLF forced convection limit form of Fand or Churchill equations
g gravitational acceleration (ms^{-2})
Gr $g\alpha d_c^3|\overline{\Delta\theta_w}|\rho^2/\mu^2$ = Grashof number (−)
h process side heat transfer coefficient ($Wm^{-2}\,°C^{-1}$)
H_c height of the coil (m)
H_L height of still liquid in tank (m)
He Hedström number defined in Equation 41
k_{CM} Calderbank–Moo Young constant; Equation 47
k_{MO} Metzner–Otto constant; Equation 42
K consistency index of an Ostwaldian fluid ($kgm^{-1}s^{n-2}$)
ℓ blade length (m)
m parameter in the mixed convection correlations of the Churchill type; Equation 24 and 68 (−)
MG_i dimensionless geometrical factor (−)
MP marine propeller
n fluid behavior index in rheological power-law model (−)
n_B number of baffles
n_c number of turns of the coil
N rotational speed of impeller (rev/s)
NC natural convection
NCLF natural convection limit form of Fand or Churchill equations
N_{CH} minimum stirrer speed for complete homogeneity of suspension (rev/s)
N_{ma} minimum stirrer speed required for bubble aspiration into the impeller, from the surface of the agitated liquid (rev/s)
n_p number of impeller blades
N_{qc} $Q_c/N \cdot d^3$ = circulation number (−)
Nu hd_c/λ = Nusselt number (−)
p propeller pitch (m)

- p_c coil tube pitch (m)
- P impeller power consumption (W)
- PBT pitched-blade turbine
- PP pseudoplastic
- Pr $\mu C_p/\lambda$ = Prandtl number (−)
- q total rate of heat transfer (J)
- Q_c total circulation flow rate (m^3s^{-1})
- Q_p impeller pumping rate (m^3s^{-1})
- Ra Gr · Pr = Rayleigh number (−)
- RDT Rushton disk turbine
- Re $\rho Nd^2/\mu$ = agitation Reynolds number (−)
- Re_a apparent Reynolds number (−)
- Re_c $\rho Nd^3/\mu D^2$ = circulation Reynolds number (−)
- Re_1 Re_a defined from dimensional analysis by Equation 39
- Re_2 Re_a defined from dimensional analysis by Equation 41 for Bingham plastic fluids
- Re_3 Re_a defined from the Metzner–Otto shear rate model and the power-law rheological model by Equation 43b
- Re_4 Re_3 but with the Carreau rheological model; Equation 44b
- Re_5 Re_a proposed by Metzner and Otto by analogy with the flow of non-Newtonian fluids in tubes; Equation 45
- Re_6 alternative of Re_5 proposed by Calderbank and Moo-Young; Equation 48
- Re_7 Re_a defined by Mizushina et al.; Equation 53a
- SA aqueous sodium alginate solution
- t_k characteristic time in Carreau rheological model (s)
- t_M mixing time (s)
- v_c mean circulation velocity of the liquid (ms^{-1})
- V_L liquid volume in the vessel (m^3)
- Vi μ/μ_w = viscosity ratio (−)
- w paddle width
- x Re exponent
- y Pr exponent
- z Vi exponent

Greek Symbols

- α pitch angle of blade to horizontal
- γ average shear rate (s^{-1})
- γ_{MO} average shear rate defined by Metzner and Otto in Equation 42 (s^{-1})
- $\bar{\Delta}$ average value of the absolute relative error (%)
- θ fluid bulk temperature (°C)
- λ fluid thermal conductivity at the temperature θ ($Wm^{-1}{}^\circ C^{-1}$)
- μ fluid dynamic viscosity at the temperature θ (Nsm^{-2})
- μ_0 zero-shear viscosity (Nsm^{-2})
- μ_{pl} plastic viscosity or coefficient of rigidity (Nsm^{-2})
- ϱ fluid density at the temperature θ (kgm^{-3})
- τ shear stress (Nm^{-2})
- τ_y yield stress of a Bingham plastic fluid (Nm^{-2})

Subscripts and Symbols

- a apparent
- f film
- FC forced convection
- i refers to the fluid flowing inside the coil
- lm logarithmic mean
- MC mixed convection
- N Newtonian fluid
- NC natural convection
- NN non-Newtonian fluid

W	refers to external surface of coil	$\sim$	approximately
1	inlet	$\neq$	different
2	outlet	$\uparrow(\downarrow)$	upwards (downwards)pumping
$\div$	proportional to		

REFERENCES

1. Ito, H., "Friction Factors for Turbulent Flow in Curved Pipes," *J: Basic Eng., Trans. A.S.M.E. Ser. D,* 81: 123–124 (1959).
2. Merkel, F., "Die Grundlagen der Warmeubertragung," Steinkopf, Leipzig, 1927, p. 51.
3. Pratt, N. H., "The Heat Transfer in a Reaction Tank Cooled by Means of a Coil," *Trans. Instn Chem. Engrs.,* 25: 163–180 (1947).
4. Rogers, G. F. C., and Mayhew, Y. R., "Heat Transfer and Pressure Loss in Helical Tubes with Turbulent Flow," *Int. J. Heat Mass Transfer,* 7(11): 1207–1216 (1964).
5. Berg, R. R., and Bonilla, C. F., "Heating of Fluids in Coils," *Trans. N. Y. Acad. Sci.,* 13: 12–18 (1950).
6. Marshall, V. C., and Yazdani, N., "Design of Agitated, Coil-in-Tank Coolers," *Chem. Proc. Eng.,* 51 (4): 89–97, 101 (1970).
7. Seban, R. A., and Mc Laughlin, E. F., "Heat Transfer in Tube Coils with Laminar and Turbulent Flow," *Int. J. Heat Mass Transfer,* 6: 387–95 (1963).
8. Mori, Y., and Nakayama, W., "Study on Forced Convective Heat Transfer in Curved Pipes. II. Turbulent Region," *Int. J. Heat Mass Transfer,* 10: 37–59 (1967).
9. Oldshue, J. Y., "Fluid Mixing, Heat Transfer and Scale-up," *Chem. Proc. Eng.,* 47: 183–188 (1966).
10. Hicks, R. W., and Gates, L. E., "Fluid Agitation in Polymer Reactors," *Chem. Eng. Prog.,* 71 (8): 74–79 (1975).
11. Desplanches H., et al., "Characteristic Variable, Prediction and Scale-up for Heat Transfer to Coils in Agitated Vessels. Part I: Newtonian Fluids in the Turbulent Region," *Chem. Eng. Res. Des.,* 61 (1): 3–12 (1983).
12. Skelland., A. H. P., and Dimmick, G. R., "Heat Transfer Between Coils and Non-Newtonian Fluids with Propeller Agitation," *Ind. Eng. Chem. Proc. Des. Dev.,* 8 (2): 267–274 (1969).
13. Appleton, W. T., and Brennan, W. C., "Some Observations on Heat Transfer to Agitated Liquids," *Can. J. Chem. Eng.,* 44: 276–80 (1966).
14. Douglas, W. J. M., and Churchill. S. W., "Recorrelation of Data for Convective Heat Transfer Between Gases and Single Cylinders with Large Temperature Differences," *Chem. Eng. Prog. Symp. Ser.,* 52 (18): 23–28 (1956).
15. Perkins, H. C. Jr., and Leppert, G., "Forced Convection Heat Transfer from a Uniformly Heated Cylinder," *J. Heat Transfer, Trans. ASME Ser. C,* 84: 257–263, (1962).
16. Oldshue, J. Y., and Gretton, A. T., "Helical Coil Heat Transfer in Mixing Vessels," *Chem. Eng. Prog.,* 50 (12): 615–621 (1954).
17. Bruxelmane, M., unpublished results.
18. Medek, J., and Fort, I., "Convective Flow in a Cylindrical System with Radial Baffles and Paddle Impeller with Inclined Plane Blades," Proc. 3rd European Conf. Mixing, Vol. 2, pp. 1–10, BHRA Fluid Engineering, Cranfield, 1980.
19. Prochazka, J., and Landau, J., "Studies on Mixing. 12—Homogenisation of Miscible Liquids in the Turbulent Region," *Coll. Czech. Chem. Comm.,* 26; 2961–73 (1961).
20. Kolar, V., "Studies on Mixing. 10—Suspending Solid Particles in Liquids by Means of Mechanical Agitation," *Coll. Czech. Chem. Comm.,* 26: 613–627 (1961).
21. Dickey, D. S., and Fenic, J. G., "Dimensional Analysis for Fluid Agitation Systems," *Chem. Eng.,* 83 (1): 139–145 (1976).

22. Hicks, R. W., Morton, J. R., and Fenic, J. G., "How to Design Agitators for Desired Process Response," *Chem. Eng.*, 83 (9): 102–110 (1976).
23. Landau, J., and Prochazka, J., "Studies on mixing, 17—Relation between the energy consumption and rate of homogenisation," *Coll. Czech. Chem. Comm.*, 29: 1866–1876 (1964).
24. Medek J., "Power Characteristics of Agitators with Flat Inclined Blades," *Int. Chem. Eng.*, 20 (4): 664–672 (1980).
25. Henzler, H. J., "Investigations on Mixing of Fluids," (Untersuchungen zum Homogenisieren Flüssigkeiten oder Gasen). VDI—Forschungsheft, 587, VDI—Verlag, Düsseldorf, 1978 (in German).
26. Medek, J., and Fort, I., "The Total Volumetric Flow Rate of Liquid Agitated by Pitched-Blade Turbines," *Chem. Prumysl*, 30: 622–627 (1980) (in Czech).
27. Medek, J., and Fort, I., "Pumping Effect of Impellers with Flat Inclined Blades," *Coll. Czech. Chem. Comm.*, 44: 3077–3089 (1979).
28. Fort, I., et al., "Studies on Mixing. 42—Hydraulic Characteristics of Paddle Impellers with Flat Inclined Blades," *Coll. Czech. Chem. Comm.*, 40: 3443–3458 (1975).
29. O'Kane, K., "The Effect of Geometric Parameters on the Power Consumption of Turbine Impellers Operating in Nonviscous Fluids," Proc. 1st European Conf. Mixing and Centrifugal Separation, Paper A3, pp. 23–31, BHRA Fluid Engineering, Cranfield, 1974.
30. Fort I., Valesova H., and Kudrna V., "Liquid Circulation in a System with Axial Mixer and Radial Baffles," *Coll. Czech. Chem. Comm.*, 36: 164–185 (1971).
31. Dunlap, I. R., and Rushton, J. H., "Heat Transfer Coefficients in Liquid Mixing Using Vertical-Tube Baffles," *Chem. Eng. Prog. Symp. Ser.*, 49 (5): 137–150 (1953).
32. Noorudin, and Raja, Rao, M.: "Heat Transfer Studies in Baffled Agitated Vessels," *Indian J. Technol.*, 4: 131–138 (1966).
33.a. Edney, H. G. S., "Heat Transfer in Agitated Vessels," Ph. D. thesis, Bradford University, 1973.
b. Edney, H. G. S., Edwards, M. F., and Marshall, V. C., "Heat Transfer to a Cooling Coil in an Agitated Vessel," *Trans. Instn Chem. Engrs.*, 51: 4–9 (1973).
c. Edney, H. G. S., and Edwards, M. F., "Heat Transfer to Non-Newtonian and Aerated Fluids in Stirred tanks," *Trans. Instn Chem. Engrs.*, 54: 160–166 (1976).
34. De Marteleire, E., "Heat Transfer in Turbine Agitated Gas-Liquid Dispersions," Paper XC2, pp. 7–35, Proc. Int. Symp. Mixing, Mons Polytech. Fac., Belgium, 1978.
35. Khang, S. J., and Levenspiel, O., "New Scale-up and Design Method for Stirrer Agitated Batch Mixing Vessels," *Chem. Eng. Sci.*, 31: 569–77 (1976).
36. Biggs, R. D., "Mixing Rates in Stirred Tanks," *A.I.Ch.E.J.*, 9: 636–40 (1963).
37. Schwartzberg, H. G., and Treybal, R. E., "Fluid and Particle Motion in Turbulent Stirred Tanks," *Ind. Eng. Chem. Fund.*, 7 (1): 1–12 (1967).
38. Holmes, D. B., Voncken, R. M., and Dekker, J. A., "Fluid Flow in Turbine-Stirred Baffled Tanks. I. Circulation Time," *Chem. Eng. Sci.*, 19: 201–208 (1964).
39. Rushton, J. H., and Fridley, D. L., "Suspension of Solids in Liquids in Cylindrical Vessels by Mixing Impeller," U. S. Atom. Energy Comm., TID 22073, 1965.
40. Rushton, J. H., Costich, E. W., and Everett, H. J., "Power Characteristics of Mixing Impellers," *Chem. Eng. Prog.*, 46: 395–404, 467–76 (1950).
41. Desplanches, H., recorrelated results of Oldshue [16].
42. Jha, R. K., and Raja, Rao, M.: "Heat Transfer Through Coiled Tubes in Agitated Vessels," *Int. J. Heat Mass Transfer.*, 10: 395–97 (1967).
43. Bates, R. L., Fondy, P. L., and Fenic, J. G., "Impeller Characteristics and Power," *Mixing—Theory and Practice* Uhl V. W. and Gray J. B. (Eds.) Vol. 1, Chap. 3, Academic Press, New York, 1967.
44. Skelland, A. H. P., et al., "Heat Transfer to Coils in Propeller-Agitated Vessels," *A.I.Ch.E.J.*, 11: 951–54 (1965).
45. Van De Vusse, J. G., "Mixing by Agitation of Miscible Fluids," *Chem. Eng. Sci.*, 4: 178–200, 208–220 (1955).

46. Nagata, S., Nishikawa, M., and Takimoto, T., "Turbulent Heat Transfer from Helical Cooling Coils in Agitated Vessels," *Heat Transf. Jap. Res.*, 1 (2): 49–56 (1972).
47. Sasakura, T., et al., "Mixing Process in a Stirred Vessel," *Int. Chem. Eng.*, 20 (2): 251–258 (1980).
48. Calderbank, P. H., "Mass Transfer" *Mixing—Theory and Practice,* Uhl V. W. and Gray J. B. (Eds.) Vol. 2, Chap. 6, Academic Press, New York, 1967.
49. Hiraoka, S., and Ito, R. J., "On the Relation of Power Input and Transport Phenomena at the Wall of Agitated Vessels," *Chem. Eng. Japan,* 6 (5): 464–467 (1973).
50. Sano, Y., Usui, H., and Saito, E., "The Correlation of Heat Transfer Coefficient at Outside Surface of Helical coil with Agitation Power," *Kagaku Kogaku Rombunshu,* 7 (3): 253–259 (1981) (in Japanese).
51. Blasinski, H., Heim, A., and Juncewicz, C., "Relation Between Mixing Power and Heat Transfer During the Mixing of Pseudoplastic Liquids with Ribbon Agitators," *Inzyneria Chem.,* 7 (3): 515–28 (1977).
52. Moo-Young, and Cross, J. V., "Heat Transfer Rates to Newtonian Fluids in an Anchor Agitated Kettle," *Can. J. Chem. Eng.*, 47: 363–373 (1969).
53.a. Voncken, R. M., "Circulation Flow and Mixing in Stirred Tanks," (Circulatiestroming and menging in geroerde vaten), doctoral thesis, Delft Technical University, 1966 (in Dutch).
b. Voncken, R. M., Holmes, D. B., and Den Hartog, H. W., "Fluid Flow in Turbine-Stirred, Baffled Tanks. II. Dispersion During Circulation," *Chem. Eng. Sci.,* 19: 209–213 (1964).
54. Bruxelmane, M., and Roustan, M., "Homogenization of Miscible Liquids," (Homogénéisation des liquides miscibles). A5910, pp. 1–11, Techniques de l'Ingénieur, Paris, 1983 (in French).
55. Van Dierendonck, L. L., Fortuin, J. M. H., and Verderbos, D., "Specific Contact Area in Gas-Liquid Reactors," Proc. 4th Eur. Symp. on Chemical Reaction Eng., Supplement to Chem. Eng. Sci., pp. 205–215, Pergamon Press, New York, 1971.
56.a. Fort, I., Grackova, Z., and Koza, V., "Studies on Mixing. 35—Flow Pattern in a System with Axial Mixer and Radial Baffles," *Coll. Czech. Chem. Comm.,* 37: 2371–2385 (1972).
b. Fort, I., Hruby, M., and Mosna, P., "Studies on Mixing. 36—Axial High Speed Rotary Mixer as an Axially Symmetrical Turbulent Jet," *Coll. Czech. Chem. Comm.,* 38: 1737–1749 (1973).
c. Fort, I., Koza, V., and Grackova, Z., "Studies on Mixing. 37—Liquid Flow at the Bottom of a Vessel with an Axial Mixer and Radial Baffles," *Coll. Czech. Chem. Comm.,* 38: 3074–93 (1973).
57. Salzman, R. N., and Weetman, R. J., "Effect of Side Flow on Mixing Impeller," *Chem. Eng. Prog.,* 77 (6): 71–75 (1981).
58. Nagata, S., et al., "Flow Patterns of Liquids in a Cylindrical Mixing Vessel with Baffles," *Mem. Fac. Eng.,* Kyoto Univ., 21: 260–74 (1959).
59. Skelland, A. H. P., Ulrich, J. A., and Mach, T. F., "Effects of Baffles on Heat Transfer to Coils in a Propeller-Agitated Vessel," *Chem. Eng. Prog. Symp. Ser.,* 61: 97–100 (1965).
60. Desplanches, H., "Heat Transfer from Submerged Coils in Stirred Tanks," (Analyse des transferts thermiques par serpentin immergé en cuve agitée), doctoral thesis, Université d'Aix-Marseille, 1983, (in French).
61. Sato, T., and Taniyama, I., "Discharge and Circulation—Flow Rate in Agitated Vessel," *Kagaku Kogaku,* 29 (3): 9–14 (1965).
62. Srinivasan, P. S., Nandapurkar, S. S., and Holland, F. A., "Pressure Drop and Heat Transfer in Coils," *Chem. Engr (London),* (213): 113–119 (1968).
63. Uhl, V. W., "Heat Transfer to Viscous Materials in Jacketed Agitated Vessels," *Chem. Eng. Prog. Symp. Ser.,* 51 (17): 93–108 (1955).

64. Petree, D. K. and Small, W. M., "Heat Transfer and Power Consumption for Agitated Vessels with Vertical Plate Coils," *A.I.Ch.E. Symp. Ser.*, 174 (74): 53–59 (1978).
65. Desplanches, H., Bruxelmane, M., and Chevalier, J. L., "Natural and Forced Heat Transfer Between Helical Coils and Agitated Newtonian or Non-Newtonian Fluids in the Transitional Flow Regime," Proc. 4th Eur. Mixing, Paper D3 pp. 141–156, BHRA Fluid Engineering, Cranfield, 1982.
66. Desplanches, H., and Bruxelmane, M., "Characteristic Variables, Prediction, Scale-up and Design for Heat Transfer to Coils in Agitated Vessels. Part II: Newtonian Fluids in the Transitional Flow Regime," to be published (in French).
67. Uhl, V. W., "Mechanically Aided Heat Transfer, *Mixing—Theory and Practice,* Uhl V. W. and Gray J. B. (Eds.), Vol. 1, Chap. 5, p. 295, Academic Press, New-York, 1967.
68. McAdams, W. H., *Heat Transmission* New-York, McGraw Hill Book Co. Inc., New York, 1954.
69. Churchill, S. W., and Chu, H. S., "Correlating Equation for Laminar and Turbulent Free Convection from a Horizontal Cylinder," *Int. J. Heat Mass Transfer,* 18 (9): 1049–53 (1975).
70. Dapra, L., and Henriette, J., "Heat Transfer from Immersed Coils in Continuous Flow-Heaters Suited for Hot Water Production. Working and Optimization," (Etude du transfert et du stockage de chaleur dans las accumulateurs à chauffage par serpentin immergé—Fonctionnement et optimalisation), personal communication, Mons Polytech. Fac., 1983.
71. Fand, R. M., and Keswani, K. K., "Combined Natural and Forced Convection Heat Transfer from Horizontal Cylinders to Water," *Int. J. Heat Mass Transfer,* 16: 1175–1191 (1973).
72. Morgan, V. T., "The Overall Convective Heat Transfer from Smooth Circular Cylinders," *Adv. Heat Transfer,* 11: 199–210 (1975).
73. Fand, R. M., Morris, E. W., and Lum, M., "Natural Convection Heat Transfer from Horizontal Cylinders to Air, Water and Silicone Oils for Rayleigh Number Between 300 and 2×10^7," *Int. J. Heat Mass Transfer,* 20: 1163–84 (1977).
74. Churchill, S. W., "A Comprehensive Correlating Equation for Laminar, Assisting, Forced, and Free Convection," *A.I.Ch.E.J.*, 23 (1): 10–16 (1977).
75. Ruckinstein, E., and Rajagopalan, R., "A Simple Algebraic Method for Obtaining the Heat and Mass Transfer Coefficients under Mixed Convection," *Chem. Eng. Commun.*, 4: 15–39 (1980).
76. Taniyama, I., and Sato, T., "Mixing Rates in Batch Agitated-Vessels," *Kagaku Kogaku,* 29 (9): 709–713 (1965) (in Japanese).
77.a. Havas, G., Deak, A., and Sawinsky, J., "Heat Transfer Coefficients in an Agitated Vessel Using Vertical Tube Baffles," *Chem. Eng. J.*, 23: 161–165 (1982).
b. Havas, G., Deak, A., and Sawinsky, J., "The effect of Impeller Diameter on the Heat Transfer in Agitated Vessels Provided with Vertical Tube Baffles," *Chem. Eng. J.*, 27: 197–198 (1983).
78. Wilkinson, W. L., *Non-Newtonian Fluids,* Pergamon Press, Oxford, 1960.
79. Skelland, A. H. P., *Non-Newtonian Flow and Heat Transfer,* John Wiley & Sons Inc., New York, 1967.
80. Coulson, J. M., and Richardson, J. F., "Non-Newtonian Technology," Chap. 6 in *Chemical Engineering,* Volume 3, Pergamon Press, Oxford, 1971.
81. Carreau, P. J., "Rheological Equations from Molecular Network Theories," *Trans. Soc. Rheol.*, 16 (1): 99–127 (1972).
82. Carreau, P. J., De Kee, D. and Daroux, M., "An Analysis of the Viscous Behaviour of Polymeric Solutions," *Can. J. Chem. Eng.*, 57: 135–140. (1979).
83. Bird, R. B., Armstrong, R. C., and Hassager, O., *Dynamics of Polymeric Liquids—Volume 1: Fluid Mechanics,* Wiley & Sons, New York, 1977.
84. Casson, N., *Rheology of Disperse Systems* Pergamon Press, New York, 1959.

85. Hershel, W. H., and Bulkley, R., *Kolloid Z.*, 39. 291 (1926), *Proc. Am. Soc. Test. Mat.*, 26: 621 (1926).
86. Magnusson, K., *"Effectbehovets beroende av den effektiva viscositeten vid omröring av strukturviskösa vätskor," Ingen. Vetensk. Akad. Tedsk. Tekn. Forsk. (IVA)*, 23 (2): 86–91 1952 (in Swedish).
87. Metzner, A. B., and Otto, R. E., "Agitation of Non-Newtonian Fluids," *A.I.Ch.E.J.*, 3 (1): 3–10 (1957).
88. Edwards, M. F., and Wilkinson, W. L., "Heat Transfer in Agitated Vessels. Part I —Newtonian Fluids. Part II—Non-Newtonian Fluids," *Chem. Engr (London)*, (264): 310–318 (1972) and (265): 328–335 (1972).
89. Metzner, A. B., and Taylor, J. S., "Flow Patterns in Agitated Vessels," *A.I.Ch.E.J.*, 6 (1): 109–114 (1960).
90. Metzner, A. B., et al., "Agitation of Viscous Newtonian and Non-Newtonian fluids," *A.I.Ch.E.J.*, 7 (1) 3–10 (1961).
91. Calderbank, P. H., and Moo-Young, M. B., "The Power Characteristics of Agitators for the mixing of Newtonian and Non-Newtonian Fluids," *Trans. Instn Chem. Engrs*, 39: 337–345 (1961).
92. Ducla, J. M., "Agitation of Hydrocolloid Solutions—Relationship Between Power Consumption and effective Viscosity," (Agitation mécanique de solutions d'hydrocolloides—Relation entre puissance consommée et viscosité effective), doctoral thesis, Université d'Aix-Marseille, 1981.
93. Nagata, S., et al., "Power Consumption of Mixing Impellers in Pseudoplastic Liquids," *J. Chem. Eng. Japan*, 4 (1): 72–76 (1971).
94. Godleski, E. S., and Smith, J. C., "Power Requirements and Blend Times in the Agitation of Pseudoplastic Fluids," *A.I.Ch.E.J.*, 8: 617–620 (1962).
95. Taguchi, H., and Miyamoto, S., "Power Requirements in non-Newtonian Fermentation Broth," *Biotech. Bioeng.*, 8: 43–54 (1966).
96. Pollard, J., Kantyka, T. A., "Heat Transfer to Agitated Non-Newtonian Fluids," *Trans. Instn Chem. Engrs*, 47: T21–27 (1969).
97. Suryanarayanan, S., Mujawar, B. A., and Raja Rao, M. R., "Heat Transfer to Pseudo-Plastic Fluids in an Agitated Vessel," *Ind. Eng. Chem. Proc. Des. Dev.*, 15 (4): 564–569 (1976).
98. Gluz, M. D., and Pavlushenko, I. S., "Experimental Investigation of Heat Transfer During Stirring of Non-Newtonian Liquids," *J. Appl. Chem. USSR*, 21: 2323–30 (1966).
99. Krishnan, R. M., and Pandya, S. B., "Heat Transfer to Non-Newtonian Fluids in Jacketed Agitated Vessels," *Indian Chem. Engr*, 8: 42–49 (1966).
100. Pandian, J. R., and Raja Rao, M., "Heat Transfer to Non-Newtonian Liquids in Agitated Vessels," *Indian Chem. Engr*, 12: 29–33 (1970).
101. Skelland, A. H. P., and Dimmick, G. R., "Heat Transfer Between Coilds and Non-Newtonian Fluids with Propeller Agitation," *Ind. Eng. Chem. Proc. Des. Dev.*, 8 (2): 267–274 (1969).
102. Carreau, P., Charest, G., and Corneille, J. L., "Heat Transfer to Agitated Non-Newtonian Fluids," *Can. J. Chem. Eng.*, 44 (2): 3–8, (1966).
103. Hagedorn, D., and Salamone, J. J., "Batch Heat Transfer Coefficients for Pseudo-plastic Fluids in Agitated Vessels," *Ind. Eng. Chem. Proc. Des. Dev.*, 6 (4): 469–475 (1967).
104. Martone, J. A., and Sandall, O. C., "Heat Transfer to Bingham Plastic Slurries in Agitated Vessels," *Ind. Eng. Chem. Proc. Des. Dev.*, 10 (1): 86–91 (1971).
105. Holland, F. A., and Chapman, F. S., "The Agitation of non-Newtonian Liquids," Chap. 10 in *"Liquid Mixing and Processing in Stirred Tanks,"* Reinhold Pub. Corp., New York, 1966.
106. Poggeman, R., Steiff, A., and Weinspach, P. M., Heat Transfer in Agitated Vessels with Single-Phase Liquids," *Chem. Ing. Tech.*, 51 (10): 948–959 (1979) (in German) or *Ger. Chem. Eng.*, 3: 163–174 (1980).

107. Desplanches, H., Llinas, J. R., and Chevalier, J. L., "Heat Transfer from a Helical Coil to Newtonian and Non-Newtonian Fluids Agitated by a Pitched-Blade Turbine," (Transferts thermiques entre un serpentin et des fluides newtoniens ou non-newtoniens agités par une turbine à pales inclinées dans une cuve). *Can. J. Chem. Eng.*, 58 (4): 160–170 (1980) (in French).
108. Mizushina, T., et al., "Experimental Study of the Heat Transfer to the Cooling Coil in an Agitated Vessel," *Kagaku Kogaku* (Abridg. Ed.), 5 (1): 93–97 (1967).
109. Desplanches, H., and Bruxelmane, M., "Scale-up and Design for Heat Transfer to Coils in Agitated Vessels. Part III: Non-Newtonian Fluids," to be published.
110. Kipke, K., "Heat Transfer to Aerated Non-Newtonian Liquids," (Wärmeübergang in begasten nicht-Newtonschen Flüssigkeiten) *Chem. Ing. Tech.*, 52 (12): 975–976 (1980) (in German) or paper presented at 2nd Eur. Cong. Biotech., Eastbourn, England, April 1981.
111. Bourne, J. R., Bürli, M., and Regenass, W., "Heat Transfer and Power Measurements in Stirred Tanks Using Heat Flow Calorimetry, *Chem. Eng. Sci.*, 36: 347–354 (1981).
112. Bourne, J. R., Bürli, M., and Regenass, W., "Power and Heat Transfer to Agitated Suspensions: Use of Heat Flow Calorimetry," *Chem. Eng. Sci.*, 36: 782–784 (1981).
113. Blasinski, H., Heim, A., and Kuncewicz, C., "Heat Transfer to a Spiral Coil during the Mixing of a Non-Newtonian Liquid by Anchor Mixers," *Int. Chem. Eng.*, 17 (3): 548–553 (1977).

CHAPTER 32

ESTIMATION OF HEAT LOSSES FROM FLOWS IN BURIED PIPES

Haim H. Bau

Department of Mechanical Engineering & Applied Mechanics
University of Pennsylvania
Philadelphia, Pennsylvania

CONTENTS

INTRODUCTION

In many industrial and natural processes, fluids are transported in buried pipes. Often, there is a temperature difference between the fluid inside the pipe and the surroundings outside the pipe. A few examples of such processes include steam distribution lines where power plant steam is distributed for district heating and/or for industrial purposes; oil and gas transmission lines where the fuel is heated or chilled in order to reduce pumping costs; underground electrical power transmission lines where fluids are used to enhance heat removal; water distribution lines in cold regions where one desires to avoid ice formation inside the pipe; heat removal from blood flowing in vessels beneath the skin and more.

The interest in the thermal interaction between the fluid inside the pipe and its surroundings arises for a variety of reasons. First, one is interested in estimating and optimizing the heat interaction between the fluid and the surroundings. For example, in the case of the oil transmission lines, one would like to estimate the magnitude of the heat losses and to minimize them as much as possible. Second, the heat interaction with the surroundings has a profound effect on the temperature distribution inside the pipe. Due to the temperature dependence of fluid properties (i.e. density and viscosity), the heat interaction will modify the internal flow field. The knowledge of the internal flow field is important for the estimation of the friction factor and pressure losses of the flow inside the pipe.

Third, the thermal interaction between the pipe and its surroundings may have an effect on the structural integrity of the pipeline. This last concern is mostly typical to cold regions. On the one hand, the introduction of hot pipelines into the permafrost will tend to lead to thawing around the pipe with a possibility of settlement or flow of the thawed soil. On the other hand, in regions which undergo seasonal thaws and freezings, there is a risk that a cold gasline may induce frostheave.

From the above, it is clear that when dealing with pipelines, one encounters a large number of phenomena. For lack of time, space, and expertise, we will not be able to review them all here. In particular, we shall not discuss here the transport process associated with cold climates. For a general review of the topic, however, we refer the reader to a text by Lunardini [1].

Furthermore, our discussion generally will be limited to only one part of the conjugate heat transfer problem. In actuality, the temperature and flow fields inside the pipe, in the soil outside the pipe, and in the atmosphere above the earth surface ought to be considered simultaneously. Obviously, this poses a very complex problem, and to the best of our knowledge, a general code which would facilitate such a simultaneous consideration is currently unavailable. Typically, one tends to decouple the various regions just mentioned through the introduction of some approximate boundary conditions. In this chapter, we generally focus on the transport processes in the soil outside the pipe. In the last section we present, however, a solution for the conjugate problem where the temperature fields inside and outside the pipe are calculated simultaneously.

The chapter is divided according to the various transport mechanisms in the medium outside the pipe. In the following section we consider a pipe buried in a conductive medium. This case is realistic when the pipe is buried in low permeability, dry soils. Otherwise, calculations based on this model would yield an underestimate of the actual heat transport. Later we consider a pipe buried in a fully saturated permeable medium. The results here are of particular importance when the saturating fluid is water. It is demonstrated that the fluid motion in the porous medium (soil) may enhance the heat transport by as much as an order of magnitude. In many cases, however, the soil is only partially saturated. In such cases, the estimates of the third section will provide an upper bound.

In the fourth section, we consider the more difficult problem of a partially saturated medium. Finally, we solve the conjugate heat transfer problem for a pipe buried in a conductive medium, and we also comment about the magnitude of the error introduced if the less rigorous techniques of the second section were to be used.

PIPES BURIED IN A CONDUCTIVE MEDIUM

In this section, we focus on conductive heat transport associated with pipes buried in a homogeneous, isotropic medium. Conductive heat losses will play a major role in situations where the pipe is buried in a low permeability, dry soil. The heat transfer estimates obtained here may also serve as a lower boundary of the actual heat losses in more complex situations.

Consider a pipe of radius r_1 buried at depth d beneath the surface of a porous medium (Figure 1). The bulk temperature of the fluid inside the pipe is T_b^* while the ambient temperature above the earth sur face is T_a^*. Superscript star (*) denotes dimensional quantities which later will be made nondimensional. In general, we assume boundary conditions of the third kind at both the pipe and the earth surfaces.

Because of the complicated geometry involved, it is convenient to use a bicylindrical coordinate system [2]. The connection between the bicylindrical coordinates (α, β) and the more familiar cartesian coordinates (x, y) is expressed by the formula:

$$x + iy = a \coth \frac{\alpha - i\beta}{2} \tag{1}$$

The bicylindrical coordinates are depicted in Figure 1, where constant α coordinates are the circles:

$$(x - a \coth \alpha)^2 + y^2 = \frac{a^2}{\sinh^2 \alpha} \tag{2}$$

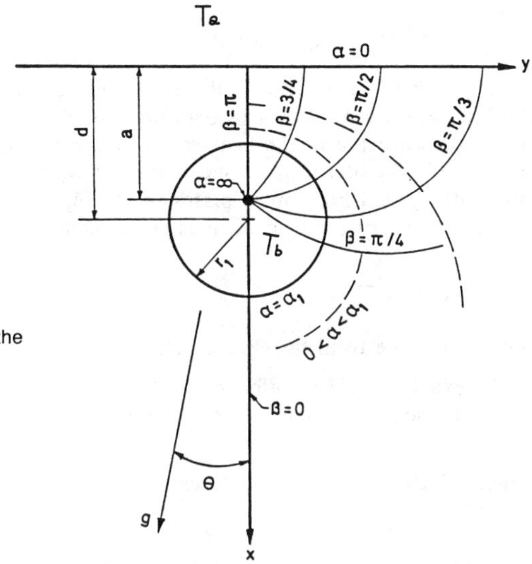

Figure 1. The geometrical configuration and the bicylindrical coordinate system.

Consequently, the pipe and the earth surfaces correspond to $\alpha = \alpha_1$ and $\alpha = 0$, respectively. The burial depth (d) and the scaling factor a are expressed, respectively, as $d/r_1 = \cosh \alpha_1$ and $a = r_1 \sinh \alpha_1$.

The bicylindrical coordinates are most natural for this geometry since the boundary conditions can be specified along constant value coordinates. The bicylindrical coordinates are also convenient for numerical computations since they map the semi-infinite physical domain into the rectangle $0 \leqq \alpha \leqq \alpha_1$ and $-\pi \leqq \beta \leqq \pi$.

The heat equation formulated in bicylindrical coordinates becomes

$$\frac{\partial T}{\partial t} = \frac{(\cosh \alpha - \cos \beta)^2}{a^2}\left(\frac{\partial^2 T}{\partial \alpha^2} + \frac{\partial^2 T}{\partial \beta^2}\right) \tag{3}$$

where

$$T = \frac{T^* - T_a^*}{T_b^* - T_a^*} \quad \text{and} \quad t = \frac{\kappa_2}{r_1^2} t^*$$

are the dimensionless temperature and time. κ_2 is the thermal diffusivity of the soil.

The boundary conditions of Equation 3 are:

$$\left.\begin{aligned} (\cosh \alpha_1 - \cos \beta)\frac{\partial T}{\partial \alpha} &= a\,Bi_1(1-T) \quad \text{at} \quad \alpha = \alpha_1 \\ (1-\cos \beta)\frac{\partial T}{\partial \alpha} &= a\,Bi_2 T \quad \text{at} \quad \alpha = 0 \end{aligned}\right\} \tag{4}$$

where $Bi_J = h_J r_1/k_2$ is the Biot number. Indexes $J = 1$ and $J = 2$ refer to the pipe and the earth surfaces, respectively. h_J is the heat transfer coefficient, and k_2 is the thermal conductivity of the soil.

In addition, continuity of temperature and heat flux requires that:

$$T(\alpha, \beta + 2\pi) = T(\alpha, \beta)$$

$$\frac{\partial T}{\partial \beta}(\alpha, \beta + 2\pi) = \frac{\partial T}{\partial \beta}(\alpha, \beta) \tag{4}$$

In general, the values of the heat transfer coefficients (h_J) at the pipe and the earth surfaces are not known a priori and should be obtained as part of the solution process, as we demonstrate in the last section. As a first approximation, one may use correlations available in the literature for heat transfer coefficients associated with pipe flow for h_1 [3, 4] and with heated horizontal plates for h_2 [3].

For engineering purposes, it is convenient to define a shape factor (S)

$$S = \frac{Q}{2\pi k_2 (T_b^* - T_a^*)} \tag{5}$$

where Q is the total heat losses from the pipe.

An exact general solution of Equations 3 and 4 is not known. Some special cases have, however, been considered in the literature.

Steady-State, Isothermal Surfaces

This is a classical case which is treated in many texts in applied mathematics and heat transfer (i.e. [5, 6]). Physically, this situation corresponds to very high values of the Biot number at both the pipe and the earth surfaces. Equations 3 and 4 become:

$$\frac{\partial^2 T}{\partial \alpha^2} = 0; \quad T(\alpha_1, \beta) = 1; \quad T(0, \beta) = 0 \tag{6}$$

The solution to this equation is:

$$T = \frac{\alpha}{\alpha_1} \tag{7}$$

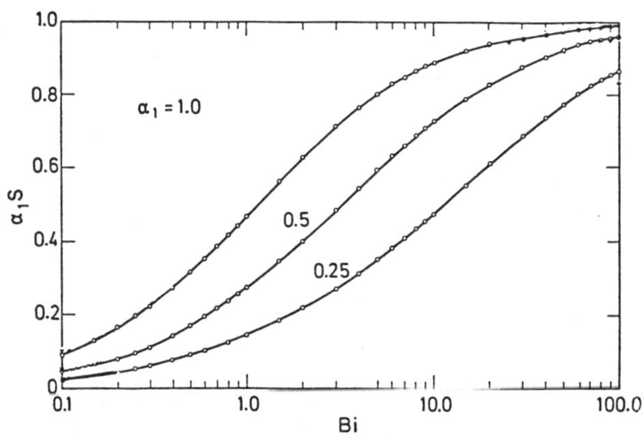

Figure 2. The shape factor (S) shown as a function of the Biot number (Bi_1) at the pipe surface, for various burial depths α_1 = 0.25, 0.5 and 1 (d/r_1 = 1.031, 1.128, and 3.762). $Bi_2 \rightarrow \infty$. The solid lines and the circles represent the exact (S) and approximate (S) solutions, respectively [7].

The shape factor is expressed as follows:

$$S = \frac{1}{\alpha_1} = \frac{1}{\cosh^{-1}(d/r_1)} \tag{8}$$

Steady-State, Boundary Conditions of the Third Kind

A series solution for this case has been provided by Bau and Sadhal [7] and DiFelice and Bau [8]. The temperature distribution is expressed in a Fourier series:

$$T = A_0\alpha + \sum_{n=0}^{\infty} (A_n \sinh n\alpha + B_n \cosh n\alpha) \cos n\beta \tag{9}$$

Note that the series Equation 9 satisfies the steady-state Equation 3 exactly. The coefficients A_n and B_n are evaluated by requiring that the series Equation 9 satisfy the boundary conditions Equation 4. Consequently, we obtain a set of linear algebraic equations for the coefficients A_n and B_n. The equations associated with the boundary condition Equation 4a are [8]:

$$\begin{aligned} &A_0(\cosh \alpha_1 + aBi_1\alpha_1) - \frac{1}{2}A_1 \cosh \alpha_1 + B_0 a\, Bi_1 \\ &\quad - \frac{1}{2}B_1 \sinh \alpha_1 = aBi_1 \qquad (n = 0) \end{aligned} \tag{10}$$

$$\begin{aligned} &A_0 + A_1(\cosh^2 \alpha_1 + aBi_1 \sinh \alpha_1) - A_2 \cosh 2\alpha_1 \\ &\quad + B_1(\sinh \alpha_1 \cosh \alpha_1 + aBi_1 \cosh \alpha_1) - B_2 \sinh 2\alpha_1 = 0 \qquad (n = 1) \end{aligned}$$

$$\begin{aligned} &-\frac{n-1}{2}A_{n-1} \cosh [(n-1)\alpha_1] + A_n[n \cosh n\alpha_1 \cosh \alpha_1 \\ &+ aBi_1 \sinh n\alpha_1] - \frac{n+1}{2}A_{n+1} \cosh (n+1)\alpha_1 \\ &-\frac{n-1}{2}B_{n-1} \sinh [(n-1)\alpha_1] + B_n[n \sinh n\alpha_1 \cosh \alpha_1 \\ &+ aBi_1 \cosh n\alpha_1] - \frac{n+1}{2}B_{n+1} \sinh [(n+1)\alpha_1] = 0 \qquad (n > 1) \end{aligned}$$

A similar set of equations is obtained from the boundary condition at $\alpha = 0$ (Equation 4b).

The preceding set of equations cannot be solved since any 2N equations involve 2(N+1) unknowns. However, A_n and B_n are decreasing sequences and can be truncated at some $n = N$. Should we set $A_{N+1} = B_{N+1} = 0$, the resulting 2N equations can be solved in a closed form for any N. The number of equations N is chosen in such a way as to satisfy the convergence criteria:

$$\frac{A_0^{N+1} - A_0^N}{A_0^N} < \varepsilon \tag{11}$$

where ε is a prescribed (small) number and the superscript N indicates the number of equations used.

The corresponding value of the shape factor is readily obtained as:

$$S = A_0 \tag{12}$$

In Figure 2, we depict (solid lines) the shape factor(S) as a function of Bi_1 for various burial depths and for $Bi_2 \gg 1$.

Since it may be inconvenient to carry out the exact calculations just presented, often for engineering purposes an approximate formula for the shape factor is desired. Lower (S_ℓ) and upper (S_u) bounds for the shape factor S are [8]:

$$S_\ell = aBi_1 \left[\left(\cosh \alpha_1 + \frac{h_1}{h_2} \cosh \alpha_2 \right) + aBi_1(\alpha_1 - \alpha_2)^2 - \left(1 + \frac{h_1}{h_2} \right)^2 \right]^{-1/2} \tag{13}$$

$$S_u = aBi_1 \left[aBi_1(\alpha_1 - \alpha_2) + \sinh \alpha_1 + \frac{h_1}{h_2} \sinh \alpha_2 \right]^{-1} \tag{14}$$

It is worth noting that for a variety of limiting cases the lower and upper bounds coincide. Clearly, the strategy to be adopted when one is solving a problem of this kind is first to calculate the lower and upper bounds for the shape factor. Only if those bounds are not sufficiently close need one seek an exact solution.

Often the lower bound (S_ℓ) is very close to the exact solution. We find that for $aBi_1 \, h_2/h_1 \geqq 1$, the deviation between the lower bound and the exact solution is well below 5 percent. For $aBi_1 \, h_2/h_1 \geqq 0.1$, this deviation is smaller than 20 percent. Thus, for most practical purposes, the lower bound (S_ℓ, Equation 13) can serve as an excellent approximation for the actual shape factor. In Figure 2, we denote by circles the lower bound (S_ℓ) for the shape factor. Note the excellent agreement between the exact and the approximate values.

Unsteady Problems

An exact solution for time-dependent problems is not available. Some approximate time-dependent solutions are given by Ioffe [9] and by Martin and Sadhal [10]; however, Ioffe [9], for instance, obtains his approximate solution in the form of integrals with oscillatory integrands which are difficult to compute numerically. On the other hand, Martin and Sadhal [10] use differential inequalities to determine lower and upper bounds on the transient temperature distribution around the pipe.

In the foregoing, we presented an analytical and a semi-analytical solution for the conduction problem in a homogeneous isotropic medium. For more general cases, when the medium is anisotropic or nonhomogeneous, it may be necessary to integrate the heat equation numerically. As we mentioned earlier, it is most convenient to carry out the numerical solution using the bicylindrical coordinates.

PIPES BURIED IN A FULLY SATURATED PERMEABLE MEDIUM

The soil outside the pipe is often permeable to fluid motion. The soil may be saturated with water or air. The fluid next to the pipe surface picks up heat from the pipe. As a result, it becomes hotter and thus lighter. Consequently, the fluid tends to rise and one expects that natural convection will take place in the medium outside the pipe. The heat transport mechanism should consist therefore of both conduction and thermal convection.

The convective effects are most significant when the saturating fluid is water as would be the case in offshore oil production when the pipe is buried beneath the ocean floor. In onshore applications, the soil is usually only partially saturated with water and the estimates given in this section may serve as an upper bound of the total heat losses. For more information about thermal convection in porous media, we refer the reader to an excellent review article by P. Cheng [11].

Here we use the Darcy–Oberbeck–Boussinesq equations to calculate the flow and temperature fields in the permeable medium outside the pipe. The equations are again formulated in bicylindrical coordinates (Figure 1).

The coordinate system was described at the beginning of the last section. The steady-state momentum and energy equations, written in terms of the stream function (ψ) and the temperature (T), are:

$$\frac{\partial^2\psi}{\partial\alpha^2}+\frac{\partial^2\psi}{\partial\beta^2}=aR\left\{[H(\alpha,\beta)\sin\theta+G(\alpha,\beta)\cos\theta]\frac{\partial T}{\partial\alpha}\right.$$
$$\left.+[H(\alpha,\beta)\cos\theta-G(\alpha,\beta)\sin\theta]\frac{\partial T}{\partial\beta}\right\} \tag{15}$$

$$\frac{\partial^2 T}{\partial\alpha^2}+\frac{\partial^2 T}{\partial\beta^2}=\frac{\partial\psi}{\partial\alpha}\frac{\partial T}{\partial\beta}-\frac{\partial\psi}{\partial\beta}\frac{\partial T}{\partial\alpha}$$

The functions H and G are:

$$H(\alpha,\beta)=\frac{1-\cosh\alpha\cos\beta}{(\cosh\alpha-\cos\beta)^2}$$

and

$$G(\alpha,\beta)=\frac{\sinh\alpha\sin\beta}{(\cosh\alpha-\cos\beta)^2}$$

We assume here that the soil surface is impermeable and maintained at a uniform temperature. Therefore, the boundary conditions are:

$$\psi=0;\quad T=1\quad \text{at}\quad \alpha=\alpha_1$$

and

$$\psi=0;\quad T=0\quad \text{at}\quad \alpha=0 \tag{16}$$

Additionally, we require

$$\psi, T(\alpha,\beta+2\pi)=\psi, T(\alpha,\beta)$$

The equations and the boundary conditions are written in nondimensional form, where the length scale is r_1, the velocity scale is κ_{eq}/r_1 and the temperature scale is $(T_b^*-T_a^*)$. κ_{eq}, the equivalent thermal diffusivity of the porous medium, is defined as:

$$\kappa_{eq}=\frac{k_2}{(\rho c_p)_f}$$

k_2 is the equivalent thermal conductivity of the saturated porous medium, and $(\rho c_p)_f$ is the product of the density (ρ) and the thermal capacity (c_p) of the saturating fluid. θ is the angle between the normal to the earth surface and the gravity vector (Figure 1). In most applications, k_2 can be estimated from

$$k_2=\varphi k_f+(1-\varphi)k_m \tag{17}$$

where φ is the medium porosity, and k_f and k_m are the thermal conductivities of the fluid and the solid, respectively. Since the fluid motion is very slow, we assume a thermal equilibrium between the fluid and the solid.

The Darcy–Rayleigh number (R), in Equation 15, is defined as

$$R=\frac{\gamma g\lambda(T_b^*-T_a^*)r_1}{\kappa_{eq}\nu_f} \tag{18}$$

where γ = the thermal expansion coefficient of the fluid
λ = the medium permeability
ν_f = the kinematic viscosity of the saturating fluid

The magnitude of R indicates the intensity of the convective motion.

Once the temperature field and the stream function are determined, the local heat flux (q) at the pipe ($\alpha = \alpha_1$) or the medium surface ($\alpha = 0$) can be readily obtained:

$$q = \left(\frac{\cosh\alpha - \cos\beta}{a}\frac{\partial T}{\partial \alpha}\right)_{\alpha = \alpha_1 \text{ or } \alpha = 0} \tag{19}$$

The total heat flow Q is

$$Q = \int_{-\pi}^{\pi} \left(\frac{\partial T}{\partial \alpha}\right)_{\alpha = \alpha_1 \text{ or } \alpha = 0} d\beta \tag{20}$$

and the Nusselt number (Nu) is

$$Nu = \frac{\alpha_1}{2\pi} Q \tag{21}$$

Equation 15 is nonlinear and an exact solution is not available. Bau [12] solved the prior equations using a regular perturbation expansion in terms of the Darcy–Rayleigh number. The procedure included expansion of the stream function and the temperature fields into a power series of the form:

$$\psi = \sum_{s=0}^{N} R^s \psi_s(\alpha, \beta) \quad \text{and} \quad T = \sum_{s=0}^{N} R^s T_s(\alpha, \beta) \tag{22}$$

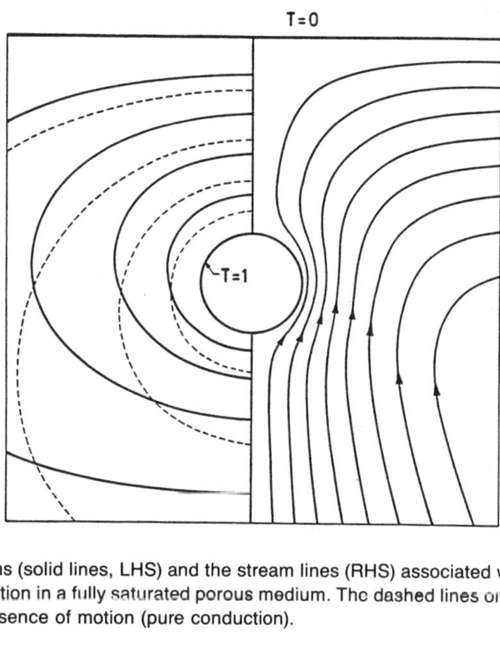

Figure 3. The isotherms (solid lines, LHS) and the stream lines (RHS) associated with low Darcy-Rayleigh number (R = 2) convection in a fully saturated porous medium. The dashed lines on the LHS correspond to the isotherms in the absence of motion (pure conduction).

We introduce the foregoing series into Equation 15 and compared terms of like power in R. As a result, we obtained a set of linear partial differential equations, which could be solved analytically. Such a procedure is adequate for small values of the Darcy–Rayleigh number.

The corresponding temperature and flow fields so obtained for R = 2 are depicted in Figure 3. The solid and dashed lines on the LHS of Figure 3 depict the isotherms with and without convection, respectively. The RHS of Figure 3 depicts the stream lines. The fluid rises next to the hot pipe, the heat is carried to the cold surface, and eventually the flow descends far from the pipe.

The analytical calculations were also used to construct a correlation for the Nusselt number in the form [12]

$$\mathrm{Nu} = 1 + \frac{c_1^2 R^2}{c_1 + c_2 R \cos\theta} \tag{23}$$

where

$$c_1 = 10^{-3}\left[-10.12 + 9.335\left(\frac{d}{r_1}\right) + 1.523\left(\frac{d}{r_1}\right)^2 - 0.034\left(\frac{d}{r_1}\right)^3\right]$$

$$c_2 = 10^{-3}\left[-1.599 + 1.865\left(\frac{d}{r_1}\right)^2 - 0.148\left(\frac{d}{r_1}\right)^3 + 0.005\left(\frac{d}{r_1}\right)^4\right]$$

The preceding correlation is accurate within 5 percent for $Rd/r_1 \leqq 60$. We emphasize again that this correlation is valid for a medium with impermeable and isothermal surfaces.

For higher values of the Darcy–Rayleigh number, one needs to integrate Equations 15 numerically. Typically, finite differences are used when the nonlinear terms in the equation are approximated using an upwind technique. Such a numerical simulation for the case of a pipe buried beneath a permeable surface has been carried out by Fernandez and Schrock [13]. Based on their numerical calculations and some experiments, they constructed a correlation for the heat transport which we reproduce here in our notation:

$$\mathrm{Nu} = \begin{cases} 1.6 - 0.6 \tanh\,[2.67 - 1.67c] & c < 1.6 \\ c & c \geqq 1.6 \end{cases}$$

$$\text{where} \quad c = \frac{1}{2}\alpha_1 R^{0.514}\left[\left(\frac{d}{r_1}\right)^2 - 1\right]^{0.257}(0.015 + 1.23e^{-\alpha_1}) \tag{24}$$

According to Reference 13, the preceding correlation is accurate within a standard deviation of 11.4 percent.

A rather interesting feature of the combined conductive-convective transport in the porous medium is the existence of an optimal burial depth for which the heat losses from the pipe are minimized. In Figure 4 we depict the total heat flow from the pipe (Q) as a function of the burial depth, as predicted by Equation 23 for various Darcy–Rayleigh numbers.

Clearly, for R > 0, the heat losses initially decrease with increasing burial depth, reach a minimum and then increase again. That is, there is an optimal depth for which the heat losses are minimized. The value of the optimal depth is a function of the Darcy–Rayleigh number. The magnitude of the optimal burial depth decreases as the Darcy–Rayleigh number (R) increases. This trend is described schematically by the dashed line in Figure 4. The physical explanation for this phenomena is rather simple. Recall that the heat transfer process consists of both conduction and natural convection. As the burial depth increases, the conductive heat losses are reduced as shown by the curve R = 0 in Figure 4. At the same time, however, the effective Rayleigh number ($R_{\mathrm{eff}} = Rd/r_1$) increases, which implies a more intense circulation and higher losses by convection. Consequently, there is some optimal value for which the total heat transfer is minimized. As the Darcy–Rayleigh number (R) increases, the convection plays a more important role and therefore the magnitude of the optimal value decreases. Clearly, one may achieve considerable energy savings by selecting the correct burial depth.

Finally, for demonstration purposes, we estimate the contribution of natural convection to the total heat losses from an isothermal pipe buried beneath an isothermal surface. Consider a pipe of radius 0.25 m, with a temperature difference of 60° C between the pipe and the earth surfaces. The soil outside the pipe is silica sand with average grain size of 2.54×10^{-4} m and permeability of $\lambda \approx 6 \times 10^{-11}$ m^2. The saturating fluid is water and its properties are evaluated at 40° C. The corresponding Darcy–Rayleigh number R $\approx$ 10. For a pipe buried at the depth of 2 m ($d/r_l \approx 8$), Nu = 2.8. That is, the thermal convection is almost twice as important as the conduction. For higher temperature differences and/or higher permeabilities, the convective effects will become even more important. Also note, that if the pipe were buried at a depth of 0.9 m ($d/r_l = 3.5$), the heat losses would be reduced by as much as 13 percent.

In this section, we considered transport processes associated with pipes buried in a fully saturated permeable medium. The transport process is formulated in terms of the Darcy–Oberbeck–Boussinesq equations. These nonlinear equations can be treated analytically for small values of the Darcy–Rayleigh number and for simple boundary conditions. For high values of the Darcy–Rayleigh number, one needs to apply numerical simulations. It seems that additional research is still needed to obtain correlations for the heat transfer under boundary conditions different that those described in this section, and more importantly, experiments are needed to verify the numerical results.

PIPES BURIED IN A PARTIALLY SATURATED POROUS MEDIUM

It is well-known that the equivalent thermal conductivity of soil varies strongly with the humidity content of the soil. For example, a dry Moon Valley soil has a thermal conductivity of about 0.25 W/m°K while the same soil with a moisture content of about 0.06 kg water/kg dry sand has a thermal conductivity of 1.0 W/m°K at 30° C [17]. Similar results are presented in Reference 20 for other types of sands.

In calculations of heat transfer in moist sands, as a first approximation, one may use the equivalent thermal conductivity of the soil, assume that conduction is the sole mode of heat transport and apply the methods described in the second section. In turns out that the preceding practice provides a quite satisfactory estimate for fine-mesh sands such as Fox loam [20]. However, for coarse sands, such as Mason sand [20], this practice may yield considerable errors.

The discrepancy between calculations based on the conductive model and the experimental results is due to the fact that conduction is not the sole mechanism of heat transport in humid soils. Indeed, in addition to conduction, one encounters also mass transport of water both in liquid and vapor phases. The transport mechanism is quite involved and for more details, we mention here a few references (e.g. [14–17, 22]).

Briefly, the voids in the sand are filled with liquid water and air. The air in the voids also contains water vapor. One can usually assume that the vapor is at the saturations pressure. Since the saturation pressure of the vapor is an increasing function of temperature, the vapor pressure next to the hot pipe is higher than that at the colder regions far away. Consequently, the vapor will diffuse from regions of high temperature to regions of cold temperature. Thus, we have a process of evaporation, mass diffusion, and condensation. Through these processes the vapor may transport a considerable amount of latent heat. As the liquid water evaporates, the liquid content next to the hot pipe decreases. We obtain a liquid concentration gradient which causes some of the liquid to be drawn back towards the hot region by capillary action. Since the capillary forces are proportional to the surface tension and since the surface tension decreases with increasing temperature, the region next

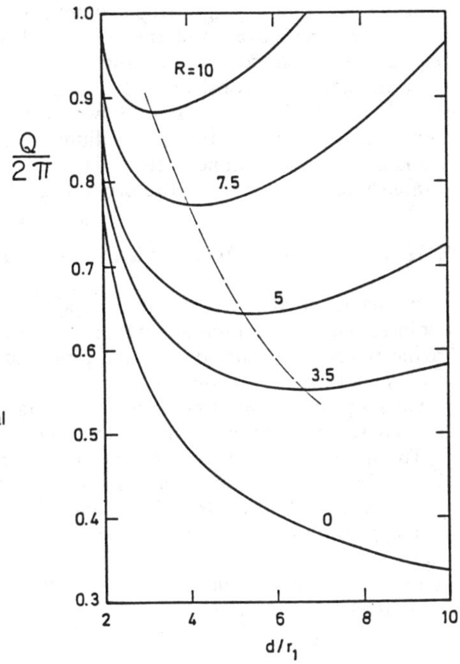

Figure 4. The heat losses as a function of the burial depth of various Darcy-Rayleigh numbers.

to the hot pipe will have a reduced humidity content. Two scenarios are possible. One includes a situation in which the moisture inflow balances the outflow and the region next to the pipe remains humid. The other possibility is that the hot region next to the pipe will dry out. The dry soil has a relatively low thermal conductivity and thus it may provide better thermal resistance.

The available literature contains a variety of models predicting heat transport in most soils. A simplified model for the diffusion process [17] consists of coupled conservation equations for the moisture and for the energy:

$$\frac{\partial w}{\partial t} = \nabla \cdot [D_w \nabla w + D_T \nabla T], \qquad (\rho c_p)_{eq} \frac{\partial T}{\partial t} = \nabla \cdot (k_{eq} \nabla T) + \varepsilon L \frac{\partial w}{\partial t} \tag{25}$$

In the prior equations, w is the humidity content of the soil and k_{eq} is the equivalent thermal conductivity of the solid, liquid, and vapor mixture. The latent heat carried by the vapor is included implicitly in k_{eq}. D_T and D_w are diffusion coefficients. $(\rho c_p)_{eq}$ is the heat capacity of the porous medium. ε is a phase conversion factor from liquid to vapor and L is the latent heat. It must be remembered that all the parameters in these equations are intricate functions of T and especially of w! Exact expressions for these coefficients are not known. Various models are suggested in the literature (e.g. [17, 20, 22]).

Additionally, Equation 25 is highly nonlinear, and typically one needs to use numerical techniques to solve it. Most of the numerical solutions available in the literature consider one-dimensional problems. For example, Hartley and Black [21] and Baladi et al. [22] consider cylindrical and spherical heat sources buried in an infinite medium. Both works pay special attention to the possibility of soil dryout next to the heat source. The numerical work compares favorably with experiments.

Another interesting application is described by Slegel and Davis [19] where wasted heat from, say, a power plant, is used to heat the soil through an underground piping system.

In order to avoid drying of the soil next to the pipe, leaky pipes are used. Thus irrigation and soil warming are simultaneously supplied. Consequently, in cold regions, the growing period can be extended during the spring and fall. This last work is of particular interest since the geometry considered is similar to the type we are discussing here.

To the best of our knowledge, however, general correlations for heat transport from pipes buried in a partially saturated medium are not currently available. It seems that a lot of work should still be done both in the area of mathematical modeling in general and in the establishment of the transport properties of the soil, in particular.

THE CONJUGATE HEAT TRANSFER PROBLEM

We have stated already in the introduction that an exact treatment of the heat transfer problem requires a simultaneous solution of the temperature and flow fields inside the pipe, in the porous medium around the pipe and in the ambient atmosphere above the earth surface. To the best of our knowledge solutions of this type are not currently available. A much simpler problem has been treated, however, by Bau and Sadhal [5], who considered laminar forced convection inside the pipe. The problem is set forth in the following.

The fluid properties are assumed to be temperature independent. The heat transfer in the soil consists of conduction alone, and the earth surface is maintained at a uniform temperature. Additionally, it is assumed that the temperatures vary linearly in the direction of the pipe axis. This would be the case if the pipe is heated by resistance heating, for example.

In what follows, indexes 1 and 2 will refer to conditions inside and outside the pipe, respectively. We assume

$$T_i^* = -\left(\frac{K\kappa_i}{2\bar{u}}\right) z + \theta_i(\alpha, \beta) \tag{26}$$

where z is a coordinate along the pipe axis, K is a constant, and $\bar{u}$ is the average velocity of the fluid inside the pipe. The local velocity u(r) is given by

$$\underset{\sim}{u}(r) = 2\bar{u}\left[1-\left(\frac{r}{r_1}\right)^2\right]\hat{e}_z \tag{27}$$

where r is a radial coordinate centered at the pipe axis and $\hat{e}_z$ is a unit vector parallel to the pipe axis.

Thus inside the pipe we solve

$$\frac{1}{\kappa_1}\underset{\sim}{u}\cdot\nabla T_1^* = \nabla^2 T_1^* \tag{28}$$

while outside the pipe in the soil we solve

$$\nabla^2 T_2^* = 0 \tag{29}$$

Substituting Equation 26 into Equations 28 and 29, we obtain

$$\begin{aligned}\nabla^2\theta_1 &= -K\left[1-\left(\frac{r}{r_1}\right)^2\right]\\ \nabla^2\theta_2 &= 0\end{aligned} \tag{30}$$

with the boundary conditions

$$\begin{aligned}\theta_1(\alpha_1, \beta) &= \theta_2(\alpha_2, \beta)\\ k_1\frac{\partial\theta_1(\alpha_1, \beta)}{\partial\alpha} &= k_2\frac{\partial\theta_2(\alpha_2, \beta)}{\partial\alpha}\end{aligned} \tag{31}$$

at the pipe surface (α_1) and

$$\theta_2(0, \beta) = 0 \tag{32}$$

at the soil surface ($\alpha = 0$).

Additionally, we require a finite temperature inside the pipe. That is,

$$\theta_1(\alpha, \beta) < \infty \text{ as } \alpha \to \infty.$$

The solutions for Equation 30 are:

$$\theta_1 = -K\left[\frac{1}{4}\left(\frac{r}{r_1}\right)^2 - \frac{1}{16}\left(\frac{r}{R}\right)^4 + A_0 + \sum_{n=1}^{\infty} A_n e^{-n\alpha} \cos n\beta\right]$$

$$\theta_2 = -K\left[B_0\alpha + \sum_{n=1}^{\infty} B_n \sinh n\alpha \cos n\beta\right] \tag{33}$$

with

$$A_0 = -\left[\frac{1}{4}\frac{k_1}{k_2}\alpha_0 + \frac{3}{16}\right]$$

$$A_n = -\frac{(-1)^n \sinh n\alpha_1}{2n[\sinh n\alpha_1 + (k_1/k_2)\cosh n\alpha_1]} \qquad n = 1, 2, \ldots$$

$$B_0 = -\frac{1}{4}\left(\frac{k_1}{k_2}\right)$$

and

$$B_n = \frac{(-1)^n e^{-n\alpha_1}}{2n[\sinh n\alpha_1 + (k_1/k_2)\cosh n\alpha_1]} \qquad n = 1, 2, \ldots$$

For the bulk temperature inside the pipe we obtain

$$\theta_b = K\left[\frac{11}{96} + \frac{1}{4}\frac{k_1}{k_2}\alpha_1 - \sum A_n e^{-2n\alpha_1}\right] \tag{34}$$

and the shape factor Equation 5 is

$$S = \left[\frac{11}{24}\frac{k_1}{k_2} + \alpha_1 + 2\sum_{n=1}^{\infty}\frac{k_2}{k_1}\frac{e^{-2n\alpha_1}}{n[\coth n\alpha_1 + k_2/k_1]}\right]^{-1} \tag{35}$$

In Figure 5, we depict the normalized temperature field (θ_1/θ_b) inside the pipe for various thermal conductivity ratios k_2/k_1. Note the asymmetry in the temperature distribution, resulting from uneven heat transport around the pipe surface. Unless the thermal conductivity of the fluid in the pipe is much larger than that of the medium outside the pipe, one may expect fairly high temperature differences inside the pipe, which may in turn give rise to secondary flows (the latter are excluded in the problem solved here).

Next, we examine how much of an error would be introduced if we ignore the temperature distribution inside the pipe and use the techniques presented in the second section. To facilitate comparison with the solution presented in this section, we assign for the heat transfer coefficient h_1 at the pipe surface the classical value $Nu = h_1 r_1/k_1 = 24/11$, which corresponds to pipe flow with uniform heating at the pipe surface. Using this value, we obtain a relationship between the two parameters, the Biot number at the pipe surface, Bi_1 and the thermal conductivity ratio k_2/k_1, in the form

$$Bi_1 = \frac{h_1 r_1}{k_2} = \frac{24}{11}\frac{k_2}{k_1} \tag{36}$$

In Figure 6 we present the shape factor (S) as a function of the burial depth for various values of k_2/k_1 (or Bi_1). The dashed and solid lines represent the exact solution and the approximate one. For high ratios k_2/k_1, the temperature distribution inside the pipe is almost symmetric (Figure 5) and the heat transfer coefficient on the pipe surface is almost uniform; thus the results of the exact and approximate solutions coincide. For low values

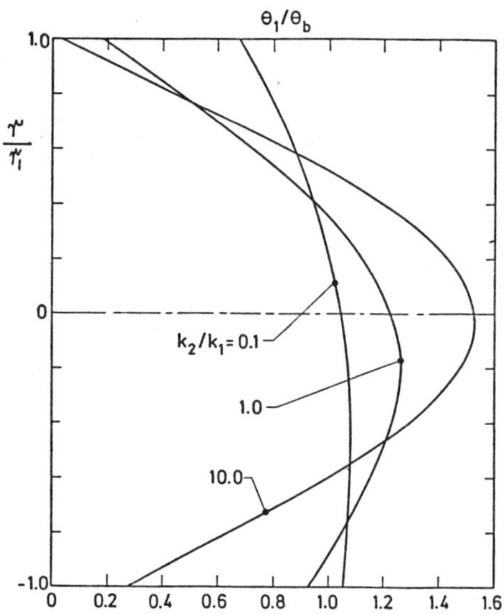

Figure 5. The temperature profile inside the pipe [7].

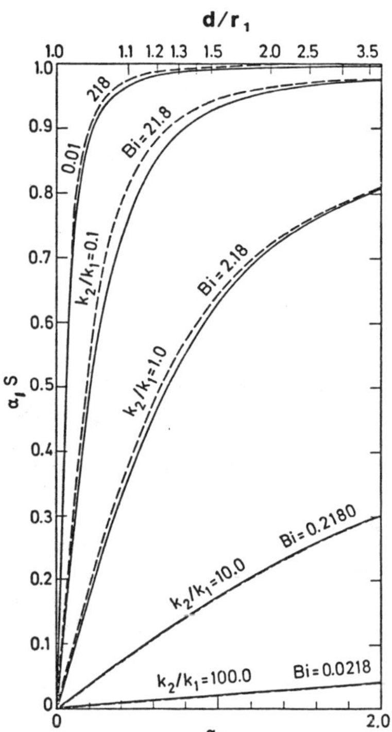

Figure 6. The shape factor as a function of the burial depth. The solid and dashed lines correspond to the exact and approximate solutions [7].

of k_2/k_1 and large burial depths, the temperature distribution inside the pipe is flat and again the exact and approximate models give similar results. It is interesting to note that the deviation between the two models is always smaller than 10% (Figure 6). This may give some indication as to the magnitude of the error incurred if the conjugate heat transfer problem is decoupled, assuming, of course, that reasonable boundary conditions are provided for that purpose.

In the foregoing, we presented an analytical solution for the conjugate heat transfer problem. In general, however, the conjugate problem is too complicated for analytical attack; thus one needs to use numerical techniques.

CONCLUSION

In this chapter we provide a brief survey of the methods of estimation of heat losses from buried pipes for three possible modes of heat transport: conduction (dry, low permeability soils), thermal convection (fully saturated soils), and mass diffusion (partially saturated soils). The order of presentation reflects an increasing order of difficulty. The conductive problem usually can be solved either analytically or numerically without major difficulties. In the fully saturated case, thermal convection must be taken into account. The Darcy–Oberbeck–Boussinesq equations usually provide an adequate description of the phenomena involved. For a very low Darcy–Rayleigh number one can use approximate analytical techniques. For moderate and high Darcy–Rayleigh numbers, numerical methods must be used. In the case of a fully saturated medium it is worth considering the possibility of minimizing the magnitude of the heat losses by selecting an optimal burial depth. The case of the partially saturated medium is the most complex one. It appears that only a small amount of research has been done in this area with regard to buried pipes.

In all these cases mentioned, there is a need for good data for the soil properties as well as for accurate measurement techniques. Finally, in order to achieve a faithful description of the real world a solution procedure to the conjugate heat transfer problem must be developed. Thus far, very little has been done in this respect.

NOTATION

A_n, B_n	coefficients
Bi	Biot number
c	function in Equation 24
c_p	specific heat at constant pressure
d	buried depth
$\hat{e}_z$	unit vector
H, G	functions
h	heat transfer coefficient
k	thermal conductivity
Nu	Nusselt number
Q	heat losses
q	local heat flux
R	Darcy–Rayleigh number
r	radius
S	shape factor
T	temperature
t	time
u	velocity
w	humidity content of soil
x, y	artesian coordinates
z	coordinate along pipe axis

Greek Symbols

α	thermal diffusivity
α, β	bicylindrical coordinates
γ	thermal expansion coefficient of fluid
θ	angle
θ_b	dimensionless bulk temperature
κ	thermal diffusivity
λ	medium permeability
ν_f	kinematic viscosity
ψ	stream function
ϱ	density

REFERENCES

1. Lunardini, V. J., *Heat Transfer in Cold Climates,* New York: Van Norstand, 1982.
2. Moon, P., and Spencer, D. E., *Field Theory Handbook,* Berlin: Springer Verlag, 1971, pp. 64 and 89.
3. Sieder, E. N., and Tate, G. E., "Heat Transfer and Pressure Drop of Liquids in Tubes," *Indust. Eng. Chem.,* Vol. 28 (1936), pp. 1429–1435.
4. Burmeister, L. C., *Convective Heat Transfer,* New York: Wiley, 1983. pp. 484–486.
5. Lebedev, N. N., Skalskaya, I. P., and Uflyand, Y. S., *Worked Problems in Applied Mathematics,* New York: Dover, 1965, p. 212.
6. Yovanovich, M. M., *Advanced Heat Conduction,* New York: Hemisphere, 1984.
7. Bau, H. H., Sadhal, S. S., "Heat Losses from a Fluid Flowing in a Buried Pipe," *Int. J. Heat Mass Transfer*, Vol. 25 (1982), pp. 1621–1629.
8. DiFelice, R. F., and Bau, H. H., "Conductive Heat Transfer between Eccentric Cylinders with Boundary Conditions of the Third Kind," *Trans ASME-J Heat Transfer,* Vol. 105 (1983), pp. 678–680.
9. Ioffe, I. A., "A problem of Transient Heat Conduction in a Semibounded Body with an Internal Cylindrical Heat Source," *J. Tech Physics,* Vol. 29 (1959), pp. 369–374.
10. Martin, W. W., and Sadhal, S. S., "Bounds on Transient Temperature Distribution Due to a Buried Cylindrical Heat Source," *Int J. Heat Mass Transfer,* Vol. 21 (1978), pp. 783–789.
11. Cheng, P., "Heat Transfer in Geothermal Systems," *Ad. in Heat Transfer,* Vol. 14, Hartnett, J. P., and Irvine, T. F. (Eds.) New York: Academic Press, 1978, pp. 1–105.
12. Bau, H. H., "Convective Heat Losses from a Pipe Buried in a Semi-Infinite Porous Medium," *Int. J. Heat Mass Transfer,* 27 (1984), pp. 2047–2056.
13. Fernandez, R. T., and Schrock, V. E., "Natural Convection from Cylinders Buried in a Liquid Saturated Porous Medium," *Proc of the Int. Heat Transfer Conf. Munich,* Vol. 2 (1982), pp. 335–340.
14. Hillel, D., *Soil and Water-Physical Principles and Processes,* New York: Academic Press, 1971, pp. 103–127.
15. Bear, T., *Dynamics of Fluids in Porous Media,* New York: American Elsevier, 1972.
16. Philip, J. R., and De Vries, D. A., Moisture Movement in Porous Materials under Temperature Gradients, *Trans Am. Geophysical Union,* Vol. 38 (1957), pp. 222–228.
17. Luikov, A. V., *Heat and Mass Transfer in Capillary Porous Bodies,* New York: Pergamon Press, 1966. Also *Int. J. Heat Mass Transfer*. Vol. 18 (1975), pp. 1–14.
18. Eckert, E. R. G., and Pfender, E., "Heat and Mass Transfer in Porous Media with Phase change," *Proc. of the 6th Int. Heat Transfer Conference,* Canada, 1978.
19. Slegel, D. L., and Davis, L. R., "Transient Heat and Mass Transfer in Soils in the Vicinity of Heated Porous Pipes," *Trans ASME-J. Heat Transfer, 99,* 1977, pp. 541–546.
20. Baladi, T. Y., Ayers, D. L., and Schoenhals, R. T., "Transient Heat and Mass Transfer in Soils," *Int. J. Heat Mass Transfer,* Vol. 24 (1981), pp. 449–458.
21. Hartley, T. G., and Black, W. Z., "Transient Simultaneous Heat and Mass Transfer in Moist, Unsaturated Soils," *Trans-ASME-J. Heat Transfer,* Vol. 103 (1981), pp. 376–382.
22. DeVries, D. A., "Heat Transfer in Soils," *Heat and Mass Transfer in the Biosphere,* DeVries, D. A., and Afgan, N. H., (Eds.), Washington: Scripta Book Co., 1975, pp. 5–28.

CHAPTER 33

PRINCIPLES OF EVAPORATIVE COOLING AND HEAT TRANSFER AUGMENTATION

W. Leidenfrost and B. Korenic*

School of Mechanical Engineering
Purdue University
West Lafayette, Indiana

CONTENTS

INTRODUCTION

For more than half a century evaporative cooling has been known to be a very effective method to enhance heat transfer in heat exchangers. This discussion will be limited to the analysis of processes which typically occur in heat exchangers, such as evaporative condensers or evaporative fluid coolers. In such devices a refrigerant is condensed at a constant temperature or a fluid is sensibly cooled inside the tubes of a bare tube of finned-tube heat exchanger. The outside surfaces of such exchangers are wetted by recirculated water which is in direct contact with an air stream. The evaporative cooling process occurs at the interface between the water film on the external heat exchanger surfaces and the air stream. By evaporating only a few percent of the water in the system, a significant cooling effect can be achieved. The theory behind evaporative cooling and the augmentation of heat transfer due to the presence of the water film will be discussed. Processes in common cooling towers which also take advantage of evaporative cooling, will not be treated, although there is no theoretical limitation in applying the same principles.

Evaporative cooling has been described and analyzed by many authors. Different methods were developed for different specific applications, but also different nomenclatures

* Present address: Baltimore Aircoil Co., P.O. Box 7322, Baltimore, MD.

have been employed. Although the evaporative cooling is a very old field in heat transfer, its mechanism was very often described in a simplified way. Modeling techniques were not perfected for practical use for the design of evaporative cooling equipment, and the ratings are typically provided in the form of charts and tables, based on extensive laboratory testings. Analytical tools were used mainly for data processing and for development of empirical equations.

With respect to heat transfer processes in evaporatively cooled devices, the driving potential in the first step of the heat flow process is the temperature difference between the process fluid and the surface of the external water film. The driving potential in the second step is a combination of the temperature and water vapor concentration difference between the water surface and the air. Sensible heat transfer between the water film and the air stream at the interface occurs because of the temperature gradients. The mass transfer, evaporation of the water from the interface to the air stream, occurs due to the water concentration gradient from the water interface to the air stream. Commonly, a single enthalpy driving force between the air saturated at the temperature of the water film interface and in the free stream is applied to simplify an analytical approach. This assumption was introduced by Merkel in his calculation of cooling towers. Merkel [1], as early as 1925, had demonstrated the analysis of the combined heat and mass transfer processes. Merkel's principal equations are still the basis of the treatment of evaporative cooling in handbooks such as *ASHRAE's Fundamentals [2]* and *Heat Exchangers: Design and Theory Sourcebook [3]*. The formulation of the principal equations however required six approximations as discussed by Yadigaroglu and Pastor [4], and the accuracy of the analysis is rather limited. Therefore, it can be used only for correlating experimental data obtained for a heat exchanger of a given geometry and operating conditions; but cannot be used for the accurate prediction of performance of the same heat exchanger operated under different conditions, or for heat exchangers of different geometry. Only recently, a more correct approach was shown for cooling towers by Sutherland [5] and by Mahumday [6]. Bošnaković [7] showed that graphical solutions are possible with the aid of Mollier $i-x$ diagrams and without Merkel's simplifying assumptions. However Bošnjaković's method could not easily be applied in a practical sense because of too many graphical integration steps, which could not be carried out without high-speed computers which were not available at that time.

It was not until recently, with the introduction of computers to aid engineering calculations, and with a broader use of the finite element techniques coupled with numerical methods, that the heat transfer field reached a new stage in its evolution. It seemed to be the right time to return again to the Mollier $i-x$ diagram for moist air in its original version, but to consider it in a different way, and to simulate the very complex evaporation processes with great speed and precision. Such a simulation provided an excellent basis for analytical treatment of evaporative cooling processes and heat transfer augmentation.

Since 1976 we have used this technique and have predicted the performance of evaporatively cooled condensers and compared them with air and water cooled devices [8, 9]. In the follow-up articles [10–12] these predictions were verified experimentally. A wider use of the same method is feasible and is expected to be utilized also in other applications of evaporative cooling for augmentation of heat transfer.

In the search for improving the efficiency of cyclic processes, attention was focused on the effect of condensing temperature on thermodynamic efficiencies of power-producing and power-consuming processes. It is well known that as a result of lowering the condensing temperature, a decrease of power input to refrigeration systems or an increase of power output of power producing systems is achieved. The first results in selection of smaller compressors, while the latter provides an increase of power output with a very small increase of feed pump power. By analyzing the evaporative condenser application for such processes it was proven that an evaporative condenser will provide economically the lowest possible condensing temperature.

The high rate of heat transfer augmentation in evaporative condensers is due to the fact that heat rejection from the condenser is mainly by means of latent heat of the evaporating water wetting the heat transfer surfaces. Under such conditions, the temperature difference between the condensing media and the water remains practically constant along the heat exchanging surfaces. In most instances that difference is much smaller than the logarithmic mean temperature difference between air and the condensing media under dry operations and is also smaller when compared with a water-cooled condenser where the water undergoes only a sensible heat transfer. For the sake of simplicity, some of the derivations made intially by Bošnjaković will be repeated here. Basic relationships and processes of moist air systems will be discussed first. Then the governing equations for the evaluation of transfer processes between air flowing over surfaces, covered by a water film, will be derived.

Computational procedures for plate-fin tube, bare tube, and vertical tube type condensers will be discussed.

Experiments will be described briefly and the results and the comparison with predictions will be presented. Unsteady-state performances of evaporative condensers during start-up or shut-down situations, or during changing climate or load conditions, will be dealt with. Finally, the computational procedures for the evaporative cooling of liquids and gases in coolers will be outlined.

BASIC RELATIONSHIPS OF MOIST AIR SYSTEMS

Dry air is a homogeneous mixture of approximately 79% nitrogen and 21% oxygen. When this dry air is in heat and mass transfer with water, steam will be added. For small pressures and when there is no chemical reaction, the resulting mixture of air and water vapor will follow Dalton's Law indicating that the total pressure P of the mixture is the sum of the partial pressures of the components.

$$P = p_A + p_W \tag{1}$$

where p_A is the partial pressure of dry air and p_W the partial pressure of water vapor. For each of the components the ideal gas law is valid, therefore

$$p_A V = m_A R_A T = m_A \frac{R_U}{M_A} T$$

and

$$p_W V = m_W R_W T = m_W \frac{R_U}{M_W} T \tag{2}$$

where R_A and R_W are the gas constants of air and water, respectively, R_U is the universal gas constant, and M_A = 29 kg/kmol and M_W = 18 kg/kmol and are the molecular weights.

In all moist air processes the mass of dry air m_A will not change when water vapor is added or subtracted; therefore, the water content x of moist air is given in relation to dry air and not to moist air, and is defined as

$$x = \frac{m_W}{m_A} \text{ [kg/kg]} \tag{3}$$

Therefore x is the mass of water in (1+x) mass of moist air or in 1 unit of mass of dry air.

From Equations 2 and 1 it follows that

$$x = \frac{M_W}{M_A} \frac{p_W}{p_A} = 0.622 \frac{p_W}{p_A}$$

$$= 0.622 \frac{p_W}{P - p_W} \tag{4}$$

Moist air can remain a homogeneous mixture only until it is saturated with water vapor. This condition, x_s, is reached whenever the partial pressure of the water vapor is equal to the saturation pressure, p_s, at the respective air temperature.

$$x_s = 0.622\frac{p_s}{P-p_s} \tag{4}$$

The saturation pressure p_s can be obtained from steam tables.

The relative humidity is defined as

$$\psi \equiv \frac{p_w}{p_s} \tag{5}$$

For technical computations it is more convenient to use the ratio x/x_s instead and

$$\varphi \equiv \frac{x}{x_s}$$

For low temperature processes under atmospheric pressure the partial vapor pressure is by comparison very small and

$$x \approx 0.622\frac{p_w}{P}$$

$$x_s \approx 0.622\frac{p_s}{P}$$

indicating that the relative humidity $\varphi = \psi$ and x can be given as

$$x = 0.622\frac{\varphi p_s}{100\,P-\varphi p_s} \tag{6}$$

Under supersaturated condition where $x > x_s$ the moist air becomes a heterogeneous mixture and water will be present in liquid form, as droplets or fog or in solid form as ice or snow. The water content in these forms is equal to $x-x_s$.

The heat and mass transfer under constant pressure will change the state of air, and the knowledge of its enthalpy or enthalpy changes is needed.

The enthalpy value of dry air is given by

$$i_A = c_{pA}(T-T_{Ref})$$

or

$$i_A = c_{pA}t \tag{7}$$

when the zero enthalpy value is selected to be at 0° C [T(K); t(°C)].

The enthalpy of the water vapor is

$$i_g = c_{pg}t+i_{fg} \tag{8}$$

where c_{pg} is the specific heat of water vapor and i_{fg} its latent heat at 0° C.

For moist air the enthalpy value becomes

$$i = i_A+xi_g = c_{pA}t+x(c_{pg}t+i_{fg}) \tag{9}$$

and for saturated air

$$i = c_{pA}t+x_s(c_{pg}t+i_{fg}) \tag{10}$$

Under supersaturated condition the moist air can also contain x_f liquid droplets and/or x_i ice crystals in the form of ice, fog, or snow flakes.

The total water content is then $x = x_g + x_f + x_i$ and the enthalpy becomes

$$i = c_{pA}t + x_g(c_{pg}t + i_{fg}) + x_f c_{pf}t - x_i(i_{fs} - c_i t) \tag{11}$$

where i_{fs} is the latent heat of fusion, c_{pf} and c_i are the specific heats of liquid water and ice respectively.

Moist air processes are usually studied with the help of the psychrometric chart [13], where the dry-bulb temperature is used as abscissa and the specific humidity (water content) as ordinate. Mollier [14] introduced another diagram using the enthalpy as ordinate and x as abscissa. This diagram has the advantage that processes of mixing and demixing and many others can be evaluated exactly by simple geometrical constructions, whereas the psychrometric chart will yield only approximate solutions. In addition such Mollier diagrams can be generalized for mixtures of two substances other than air and water.

This chapter deals with combined heat and mass transfer processes by computerized graphical methods, and therefore the Mollier i–x diagram will be used throughout.

Plotting i vs. x yields isotherms in the unsaturated region (see Equation 9) of constant slope according to the following equation

$$\left(\frac{di}{dx}\right)_t = c_{pg}t + i_{fg} \tag{12}$$

The slope increases slightly with temperature due to changing c_{pg}. For each of the isotherms the respective x_s value can be evaluated and the saturation line separating the unsaturated and foggy region is defined. The Mollier diagram so constructed will yield the unsaturated region confined in a very small area unsuitable for geometrical analysis. This drawback can be overcome by oblique-angled coordinates and making the 0° C isotherm horizontal.

The lines of constant enthalpy are parallel lines and have the same slope.

The isotherms in the foggy region have the slope $(\partial i/\partial x)_t = i_w = c_{pf}t$ which is almost identical to the slope of the enthalpy lines.

A Mollier i–x diagram for above 0° C and for P = atmospheric is shown in Figure 1.

This diagram can readily be used to determine enthalpy changes of air by heat or moisture addition or removal. Before this will be discussed in detail, a major advantage of enthalpy-composition diagrams (the so-called mixing line concept) will be introduced.

It is assumed that a mass of moist air m_1 of state condition t_1, x_1, i_1, is mixed with m_2 mass of air having state conditions t_2, x_2, i_2.

The conservation laws require that after mixing the mass of the mixture m is

$$m = m_1 + m_2 \tag{13}$$

and that the mass balance of water is satisfied by

$$mx = m_1x_1 + m_2x_2 \tag{14}$$

where x is the resulting moisture content of the mixture. The heat balance is given by

$$mi = m_1i_1 + m_2i_2 \tag{15}$$

Eliminating m with the help of Equation 13 yields

$$m_1(x - x_1) = m_2(x_2 - x)$$

and

$$m_1(i - i_1) = m_2(i_2 - i)$$

and by division

$$\frac{x - x_1}{i - i_1} = \frac{x_2 - x}{i_2 - i}$$

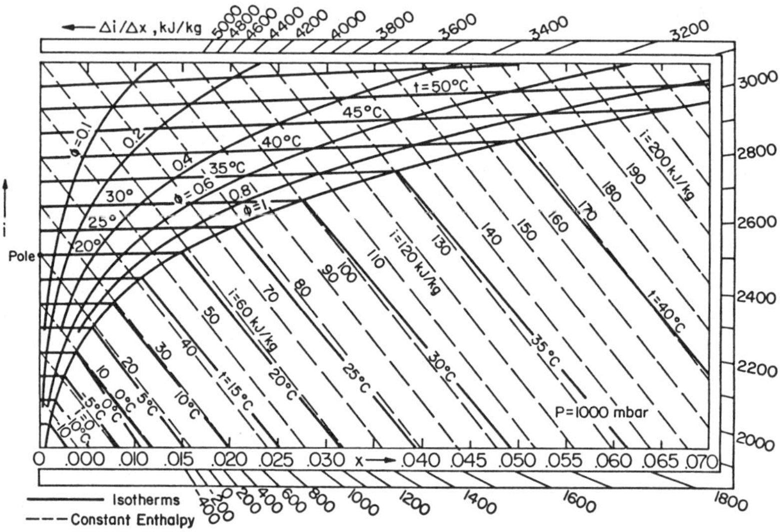

Figure 1. The Mollier i-x diagram.

This equation indicates that the state condition of the mixture must be situated on a straight line connecting the initial points in the i–x diagram. The location of the state point of the mixture follows from the equation

$$\frac{m_1}{m_2} = \frac{x_2 - x}{x - x_1} \tag{17}$$

The mixing line can also be found when water or pure water vapor both having $x = \infty$ is added to moist air. The mixing line is then a straight line through the state point of the moist air parallel to the line $di/dx = i$ or $di/dx = i_g$. These lines are lines from the "Pol" of the diagram to the respective i values on the margin of the diagram.

The final state x_2 on this line is determined by the amount of water or steam added.

$$x_2 - x_1 = \frac{m_W}{m_1}$$

or

$$x_2 - x_1 = \frac{m_g}{m_1} \tag{18}$$

The next important application of the i–x diagram is the determination of state changes of air resulting from heat and mass transfer processes with a water surface. It is assumed that the air enters a control volume at a mass flow rate $\dot{m}_A$ and at state condition x_A, t_A, and i_A and that heat and mass transfer takes place only at the boundary surface of the water. The air in the immediate vicinity must be in equilibrium with the boundary and $t_{Ab} = t_{Wb} = t_b$ and $x_{AB} = x_b$, where t_b and x_b are the state conditions on the saturation line. It is assumed that $t_b < t_A$ and $x_A < x_b$ and that the water remains at a constant temperature. Since $x_b > x_A$ and $t_A > t_b$ water will evaporate from the surface. The differential amount of water lost per time from a differential surface area dA_b is in accordance to Lewis

$$-d\dot{m}_W = \sigma(x_b - x_A)\, dA_b \tag{19}$$

where σ is a mass transfer coefficient in kg/m^2s. This amount is added to the air stream flowing through the control volume and changes its water content by

$$-d\dot{m}_W = \dot{m}_A\, dx_A$$

or

$$\dot{m}_A\, dx_A = \sigma(x_b - x_A)\, dA_b \tag{20}$$

The air experiences also heat transfer with the water surface and its enthalpy changes by

$$\dot{m}_A\, di = h_A(t_b - t_A)\, dA_b + i_{gb}\dot{m}_A dx_A \tag{21}$$

The first term on the right side of Equation 21 is due to convective heat transfer. The second term accounts for the enthalpy of the vapor transferred into the air stream by means of the mass transfer. Introducing dA_b from Equation 20 and multiplying the convective term by c_{pA}/h_A yields.

$$\frac{di}{dx} = \frac{c_{pA}(t_b - t_A)}{\dfrac{\sigma c_{pA}}{h_A}(x_b - x_A)} + i_{gb} \tag{22}$$

As shown in the i–x diagram of Figure 2 Equation 22 can be used to geometrically determine the state change direction of the air being in heat and mass transfer with an evaporating surface. This construction assumes that the so-called Lewis factor $\sigma c_{pA}/h_A > 1$. Therefore the boundary state B is displaced along the extension of the t_{Ab} isotherm yielding the so-called "pulling point" B_h. B_h represents an apparent state of air which is mixed with the air of state A, therefore the new states of air must be on the mixing line.

The Lewis factor in many instances is approximately equal to one and $B_h \equiv B$. For more correct evaluation its value can be computed from

$$\frac{\sigma c_{pA}}{h_A} = 1.1\frac{\ln \varepsilon}{\varepsilon - 1} \tag{23}$$

where

$$\varepsilon = (0.622 + x_b)/(0.622 + x_A)$$

The just-discussed determination of direction of state change of air is valid for any other heat and mass transfer processes with a water surface, no matter where A is located in relation to B.

The preceding derivations were based on the assumptions that the water was at a constant temperature and did not receive heat. In most real situations the water wetting a heat transfer surface will not only be in contact with the air but will also receive heat that is conducted through the wall due to heat transfer from a hotter fluid on the other side of the wall that is to be cooled. This fluid might undergo a phase change at a constant temperature t_c or might remain a single-phase fluid experiencing sensible heat transfer only.

The next treatment will consider first the case of an evaporative condenser; evaporative coolers will be treated later.

Considered is a wall over which water flows in a thin film at a flow rate $\dot{m}_W$. The free surface of the water is at temperature t_b and is in heat and mass exchange with air having a flow rate $\dot{m}_A$ and being at t_A, x_A, i_A. The water receives heat from the condensing media at t_c governed by an overall heat transfer coefficient U_W.

Along a differential surface area dA_c of the condenser the air will change in its properties by dt_A, dx_A and di_A, while its flow rate $\dot{m}_A$ = const. The water state changes by di_W and its mass flow rate by $d\dot{m}_W$.

The processes depend on nine parameters. The condenser heat load q_c, the mass flow rates of water and air, and their inlet and outlet state conditions. q_c commonly is given and the inlet flow rates and conditions can be selected or are known. The exit conditions however cannot be determined because not enough equations are available. However, with the aid of the Mollier i–x diagram, solutions are possible as was shown initially by Merkel and Busemann [15] and finally perfected by Bošnjaković's step-by-step integration procedure.

The heat received by the water from the condensing media minus the enthalpy loss of the water evaporating must equal the heat added to the water and the heat transferred to the air by a dry and wet process. The heat balance becomes

$$U_W\, dA_c(t_c - t_w) - i_{Wb}\, d\dot{m}_W = d(\dot{m}_W i_W) + h_A\, dA_c\,(t_b - t_A) + \sigma(x_b - x_A) i_{fg} \quad (24)$$

Solving $dA_c = dA_b$ from Equation 20 and introducing into Equation 24 yields

$$U_W(t_c - t_W) = \sigma(x_b - x_A)(i_W - i_{Wb}) + h_A(t_b - t_A) + \frac{\dot{m}_W}{\dot{m}_A}\frac{d_{iW}}{dx_A}\sigma(x_b - x_A) + \sigma(x_b - x_A) i_{fg} \quad (25)$$

Under normal operations of evaporatively cooled devices the water film is rather thin and $t_b \approx t_W$ therefore $i_W \approx i_{Wb}$. Introducing an overall heat transfer coefficient $U_{wb} = (A_W/A_b)U_W$ finally yields

$$U_{Wb}(t_c - t_b) = h_A(t_b - t_A) + \sigma(x_b - x_A)\left(i_{fg} + \frac{\dot{m}_W}{\dot{m}_A}\frac{di_W}{dx_A}\right) \quad (26)$$

Multiplying each term of Equation 26 by c_{pA}/h_A converts the heat fluxes to enthalpy values.

$$c_{pA}(t_c - t_b) = \frac{h_A}{U_{Wb}}\left[c_{pA}(t_b - t_A) + \frac{\sigma c_{pA}}{h_A}(x_b - x_A) \cdot \left(i_{fg} + \frac{\dot{m}_W di_W}{\dot{m}_A dx_A}\right)\right]$$

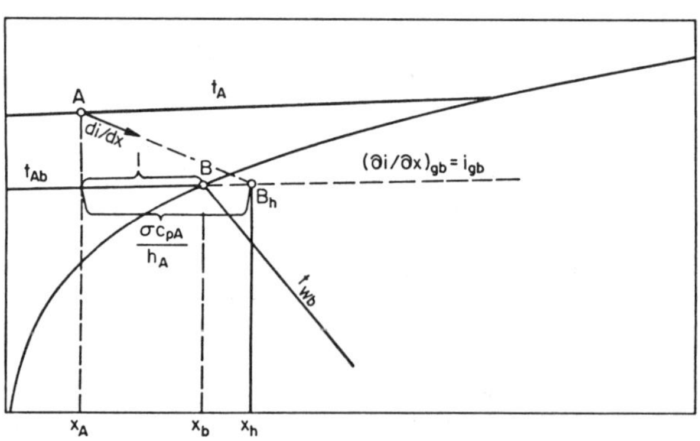

Figure 2. State change direction of air in heat and mass transfer with a water surface.

or

$$c_{pA}(t_c - t_b) = \frac{h_A}{h_A + U_{wb}}\left[c_{pA}(t_c - t_A) + \frac{\sigma c_{pA}}{h_A}(x_b - x_A)\left(i_{fg} + \frac{\dot{m}_W d_{iW}}{\dot{m}_A dx_A}\right)\right] \tag{27}$$

Note: c_{pA} is the specific heat of moist air to be evaluated from

$$c_{pA} = \frac{c_A + x_A c_g}{1 + x_A}$$

where c_A and c_g are the specific heats of dry air and water vapor, respectively. The enthalpy values of Equation 27 can be abbreviated, and the equation simplifies to

$$q_{bw} = q_h + q_\sigma + q_W \tag{28}$$

or:

$$\frac{U_{Wb}}{h_A} c_{pA}(t_c - t_b) = c_{pA}(t_b - t_A) + \frac{\sigma c_{pA}}{h_A}(x_b - x_A) i_{fg} + \frac{\sigma c_{pA}}{h_A}(x_b - x_A)\frac{\dot{m}_W}{\dot{m}_A}\frac{di_W}{dx_A}$$

These q values can be determined graphically in the i–x diagram and are therefore called the diagram values of the fluxes.

For the selected conditions of $t_c > t_A > t_g$ and $x_A < x_g$ the distribution of the heat input q_{bw}, into water heating q_W and air heating including evaporation is shown in Figure 3. The vertical line at x_A intersects t_b at state F and t_c at state M. As in Figure 2 the isotherm t_b along $\overline{FB}$ is extended by the Lewis factor yielding B_h. A line through B_h parallel to the wet isotherm t_W intersects the vertical x_A line at state H. A vertical line through x_h intersects the line parallel to t_b through M at state C_h, where $\overline{C_h B_h} = c_{pA}(t_c - t_b)$. The multiplication of $\overline{C_h B_h}$ by U_{Wb}/h_A equals q_{bW}. This value yields state D on x_h. A line through D parallel to t_W intersects the vertical x_A line at state P. All the diagram values are now given for the condition of moist air at state A in heat and mass transfer with the water surface at state B.

Then

$$\overline{HF} = q_\sigma;\ \overline{HA} = q_b;\ \overline{HP} = q_{bW};\ \overline{AP} = q_W;\ \overline{AF} = q_h$$

q_b is the diagram value of the heat flux at the water interface.

$$q_b = c_{pA}(t_b - t_A) + \frac{\sigma c_{pA}}{h_A}(x_b - x_A) i_{fg}$$

$$= i - i_A - c_{pW}(x_b - x_A) t_{Wb}$$

$$q_W = \frac{U_{Wb}}{h_A} c_{pA}(t_c - t_{Wb}) - q_b$$

Inspecting the figure indicates that the water-air interface receives heat both from the condensing process and the air. The sum of these heats is balanced by the heat needed for evaporation and heating the water. Therefore, for the conditions shown in Figure 3 q_h in Equation 28 is negative.

Obviously, there are many other heat and mass transfer situations where the relative conditions between the t's and x's are different from those just assumed. For such cases the same procedure of evaluating the diagram values holds true. The signs of the particular diagram values are as follows: $q_b > 0$ if H lies above A (heating on the water side); $q_h > 0$ if F lies abowe A (heat flow to air); $q_\sigma > 0$ if H lies above F (evaporation); $q_{bW} > 0$ if H lies above P (heating from outside); $q_W > 0$ if A lies above P (t_W increases in the direction of increasing x_A); $q_{bd} > 0$ if N lies above A (heat flow from wall to air).

This case can occur when part of the heat exchanger surfaces are not wetted and remaining dry. Under such conditions the state changes of the air will not be in the direction (di/dx) but in the direction $(di/dx)_R$ as shown by dotted lines in Figure 3. Any diagram value listed can be zero; this is indicated when the distance between the respective states on the x_A line is zero.

The diagram in Figure 3 shows the heat and mass transfer process between the fluids involved at a given location or at a given time within the heat exchanger and demonstrated the state changes which will occur for these conditions. Under unsteady-state operations or under changing external conditions different situations will occur at the same location, or for steady-state operations and constant external conditions different situations will be present at other locations. So for the analysis has not dealt with these facts.

STEADY STATE CHANGES OF AIR IN A HEAT EXCHANGER

It has been shown earlier that the state of air in heat and mass transfer with the water interface will change along the mixing line of slope di/dx. It will keep changing in this

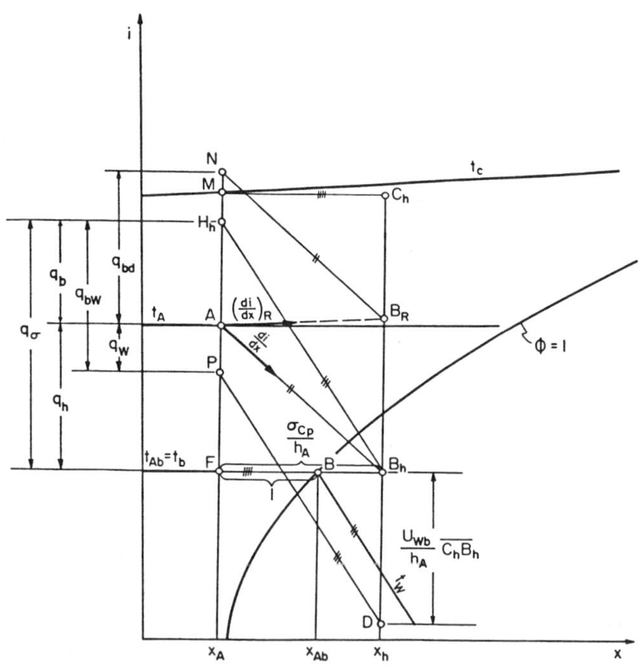

Figure 3. Distribution of heat fluxes for cooling wetted walls of condenser.

direction whenever the state of the "pulling point" does not change. Figure 3 however indicates that B or B_h must change for reasons that q_W is not zero and that the water temperature will increase. In an evaporatively cooled condenser the temperature t_c can be assumed to remain constant along the heat exchanger surfaces. Therefore the changes of t_W will remain small but still will not be zero, and even the sign of q_W can change which means that the B's are displaced in the opposite direction. Furthermore, the Lewis factor can change in magnitude which influences the location of B_h even when B remains at the same state condition. In a heat exchanger, where a liquid or a gas is cooled by evaporative means, the temperature of these fluids will change along their flow paths and t_W will change. These changes can be expected to be more noticeable than the possible changes of t_W in a condenser.

For all these reasons the final state of the air leaving a heat exchanger and therefore the overall heat load of the exchanger cannot be predicted unless the air enthalpy change between exit and inlet Δi_A and the q_W's can be calculated. This analysis can be performed and the changes of the A's and the B's can be determined by stepwise integration and by the geometrical means shown first by Bošnjaković. The integration can be carried out under all situations when $q_W \gtreqless O$ and Lewis factor $\geqq 1$ and also for the conditions when there is heat exchange with a dry surface. For reasons of brevity, the procedure will be demonstrated for the conditions where all surfaces are wetted and the Lewis factor is unity, therefore $B_R = B_h = B$ and the change of air state is directed towards B.

In Figure 4 it is assumed that the air state A_1, has moved to A and the initias distance $A_1 B$ taken as unity has diminished by a factor y. During this change B has not changed. A further change of A to A′ may be expressed by

$$\frac{dx_A}{x_b - x_A} = -\frac{dy}{y} \tag{29}$$

where the negative sign accounts for the fact that an increase of x_A corresponds to a decrease of y.

In accordance with Equation 20 and for a Lewis factor of unity

$$dx_A/(x_b - x_A) = dA_b\sigma/\dot{m}_A \cdot Fw$$

where $\sigma = h_A/c_{pA}$

$$dA_b = \frac{\dot{m}_A c_{pA}}{h_A}\,\frac{dx_A}{x_b - x_A} \tag{30}$$

$\dot{m}_A c_{pA}/h_A$ has the dimension of an evaporative surface area with respect to h_A and is called the unit surface area a_b, and Equation 29 becomes

$$-\frac{dy}{y} = \frac{dA_b}{a_b} \tag{31}$$

Integrating Equation 31 by assuming x_b = const. yields

$$\ln(1/y) = \Delta A_b/a_b = n' \tag{32}$$

ΔA_b is the evaporative surface required to change the air state A to A′ along the mixing line for which n′ numbers of unit surface areas a_b are needed.

For the stepwise integration of Equation 29 $(1-y)$ must be chosen small, otherwise B might be displaced and the assumption of x_b = const. is violated. Since high speed computers are available, $y = 0.99$ (indicating only 1% change of A towards B) can be selected. On the other hand, such fine steps are not always needed and $y \approx 0.95$ can be selected without markedly changing the results.

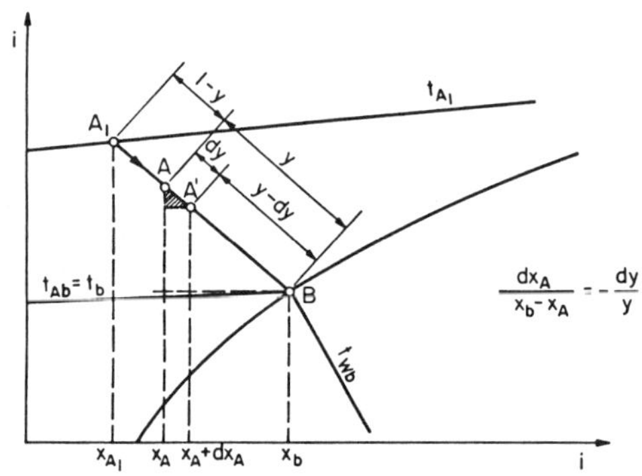

Figure 4. Determination of state changes by the partial value method.

The computer has been programmed to select smaller values of y when the state changes of A towards B indicate very small deviations from the mixing line, or to select larger y values for opposite conditions.

For the total wetted surface area A_b of the heat exchanger, $n\Delta A_b$ must be provided

$$A_b = nn'a_b = na_b \ln \frac{1}{y} = n\left[\frac{\dot{m}c_{pA}}{h_A}\right] \ln \frac{1}{y} \tag{33}$$

The increase of water enthalpy in accordance to Equation 28 is given by

$$di_{Wb} = q_W \frac{\dot{m}_A}{\dot{m}_W} \frac{dx}{(x_b - x_A)} \tag{34}$$

and depends on the magnitude and sign of q_W which can be determined for each pair of A and B as shown is Figure 3.

Integrating Equation 34 for the finite step Δx along ΔA_b yields

$$\Delta i_{Wb} = c_{pW} \Delta t_b = q_W \frac{\dot{m}_A}{\dot{m}_W} \ln \frac{1}{y} \tag{35}$$

The integration can be aided by introducing

$$\Delta i_F = c_{pA} \Delta t_b = (c_{pA}/c_{pW}) \Delta i_{Wb} \tag{36}$$

Equations 35 and 36 yield

$$\Delta i_F = q_W \left[\frac{\dot{m}_A c_{pA}}{\dot{m}_W c_{pW}} \ln \frac{1}{y}\right] = b \cdot q_W \tag{37}$$

where b is practically a constant for fixed flow rates and the step size y.

The determination of the new pulling point B_2 is shown graphically in the $i-x$ diagram of Figure 5 which is constructed in a way similar to Figure 3.

The preceding procedure is repeated for the required number of steps or until the required outlet state of the air is reached. For a fixed heat transfer area A_b the outlet state of the air will result.

The graphical solutions simulated analytically obviously require that the thermodynamic parameters of the moist air such as the saturation pressure of water vapor, humidity ratios, and enthalpies are given by mathematical relationships. Previously x was given in terms of p_s and i was given in terms of x. The relationship for $p_s = f(t)$ needs to be established. The most common temperature range for evaporative cooling processes will be in the temperature range 10°–42° C. For this range the partial pressure was established by a mathematical curve fit to literature data resulting in

$$p_s = \exp(0.0587t_A + 4.2866)$$

$$t_A = -73.00 + 17.03 \ln p_s \tag{38}$$

where p_s is given in kg/m^2 and t_A in °C. The coefficient of determination was $r^2 = 0.998768$ therefore, the equation represents the data excellently.

Under normal conditions the inlet air state to a heat exchanger will be unsaturated and the first or many of the following integration steps will yield new air states which still are in the unsaturated region. Each of these steps will yield changes of enthalpy and moisture content.

$$\Delta i_A = (1-y)(i_b - i_A) \tag{39}$$

and

$$\Delta x_A = (1-y)(x_b - x_s) \tag{40}$$

The new enthalpy and moisture contents are

$$i_{An+1} = i_{An} + \Delta i_A \tag{41}$$

and

$$x_{An+1} = x_{An} + \Delta x_A \tag{42}$$

For each integration step the new dry-bulb air temperature can be evaluated, as well as the respective saturation pressure and the relative humidity.

Inspecting Figure 5 demonstrates that the air flowing over the wetted surface area of a heat exchanger will reach or will be very close to saturated conditions providing the flow path is long enough or the air enters the passages at close to saturated condition. It might then happen that the next air state evaluated geometrically falls in the foggy region and its enthalpy and water content calculated as shown previously do not show the actual state which must lie on the saturation line with $\varphi = 100\%$.

This problem can be solved by evaluating the new air state by means of temperature increments

$$t_{A2} = t_{A1} + \Delta t_A \quad \text{for} \quad \varphi_{A2} = 100\% \tag{43}$$

And the state A_{2b} on the saturation line is evaluated from the moisture content

$$x_{As} = x_{A1} + \Delta x_A \tag{44}$$

where Δx_A is evaluated from Equation 40. The saturation temperature evaluated for x_{As} will differ from t_{A2} of Equation 43 and a mean temperature t_{A2m} is determined and used for the next integration step.

The increase of x_A by Δx_A for the stepwise processes changes the water flow rate by

$$\Delta \dot{m}_W = \pm(1-y)\dot{m}_A(x_A - x_b) \tag{45}$$

The positive sign holds for parallel flow while the negative sign holds for counterflow. Δi's are above A_1 for positive and below A_1 for negative values.

The sum of the $\Delta \dot{m}_W$ from inlet to exit of the heat exchanger yields the total mass of water evaporated. The part of the condenser heat load carried sensibly by the water, results in

$$\sum_1^n \Delta \dot{m}_{Wi} c_{pW} \Delta t_{Wi}$$

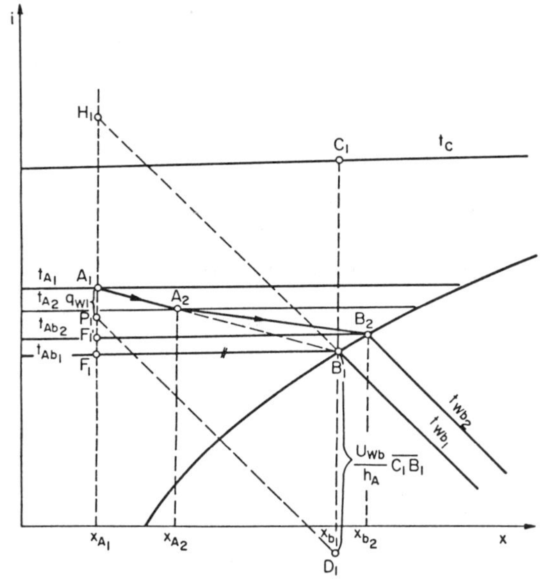

Figure 5. Successive changes of states of air and water.

Similarily to cooling towers, evaporatively cooled heat exchangers are commonly operated with recirculated water. The water is taken from a pond and is distributed over the heat exchanger surfaces from where it is returned by gravitation to the pond. When the water is sprayed, many of the droplets might not wet the surfaces but fall freely through the air passages back into the pond. During these processes part of the water evaporates and must be replenished.

When there is no heat exchange with the surroundings and no heat input from the recirculating pump, the water will reach under steady-state operation its cooling limit, i.e., its lowest possible temperature under given inlet conditions of the air.

In an evaporatively cooled device heat is added and the water will reach a temperature higher than the cooling limit. The water temperature will be practically the same at exit as at inlet, but at a different temperature level for different climate conditions.

For the situation of water recirculation one problem exists for the analysis, namely, the guessing of the first pulling point which must coincide with the last one.

This problem can be overcome, as was shown by Korenic [16] by selecting an initial water temperature t_{wi} which must be higher than a limited t_{WbL} where t_{WbL} can be calculated as

$$t_{WbL} = t_c - \frac{h_{AW}}{U_W c_{pbL}} (i_{bL} - i_{A1} - c_{PW}\, \Delta x_{bL} t_{WbL}) \tag{46}$$

where $\Delta x_{bL} = x_{bL} - x_{A1}$ and i_{bL} is the enthalpy of the pulling point B_L.

The selection of $t_{Wb1} > t_{WbL}$ must be checked by iteration which has to be repeated until B_1 and B_n coincide or deviate from each other by a small specified error.

Commonly the air flow is counter to the water flow but parallel flow arrangements can also be used. The computer programs can predict the state changes of air for both arrangements and also for water flows "once through," without recirculation. In the present article only counter flow with water recirculation is considered.

This section has demonstrated the computational procedures to determine heat and mass transfer processes in evaporatively cooled devices by making use of the basic relationships of moist air systems discussed earlier.

To apply the analysis to design heat exchangers of a desired performance requires in addition information for the evaluation of the heat transfer coefficients contained in the governing equations.

BASIC HEAT TRANSFER RELATIONSHIPS

The analytical procedures described in the previous chapters were used to predict performances of plate-fin tube and bare horizontal tube heat exchangers. In both, condensation occurred inside the tubes. Also studied was a vertical tube heat exchanger where the vapor was condensed on the outside of the tubes. The predicted performances of these type heat exchangers were checked experimentally with excellent results. There are many other types of heat exchangers of different geometry and different flow patterns of the fluids in heat exchange. The verification of the analysis by experiments for the previous three types of heat exchangers allows us to assume with confidence that all the other heat-exchanger types can be designed to perform as desired; expensive and time-consuming model tests on a wide scale are not needed. However, necessary adjustments must be made to account for differences in geometry and flow patterns, and for the resulting differences in heat transfer coefficients.

The heat transfer processes and coefficients of interest are those between the air and the water surface (h_A); between the water and the wetted solid surface (h_W); and between the condensing media or the fluid in a cooler (h_R); and also the conductive transport through the wall. The latter in most instances will represent only a very small heat transfer resistance and can be neglected. The mass transfer coefficient σ is also of interest. It can be concluded from the air-side heat transfer coefficient by means of the Lewis factor.

It has been shown that h_A in principle can be evaluated from literature data describing heat exchange with a solid wall. However, it must then be taken into account that the free exchange surface is that of the possibly wavy water film.

The heat transfer correlations depend on geometrical flow arrangement within a heat exchanger. Commonly a heat exchanger can be subdivided by lines of symmetry bordering so-called elemental channels. Those channels are shown in Figure 6 for a plate-fin-tube-type heat exchanger, and in Figure 7 for a bare tube heat exchanger. In both figures, in-line and staggered tube arrangements are considered. The concept of elemental channels assumes that the flows of air and water are the same in every channel.*

Therefore, heat and mass transfer processes need to be evaluated for one channel only. When the heat load of one channel has been determined the number of channels needed for the specified heat load of the condenser or cooler will result. Figure 6 and 7 show that the elemental channels are subdivided by modules of identical size.

The dimensions listed will be changed when the surfaces are wetted as shown in Figure 8 and the cross-sectional area for the air flow is decreased accordingly.

The cross-sectional area occupied by the water film A_{W0} in one module is for the plate-fin tube exchanger

$$(A_{w0})_F = 2(ST - DO)\delta_{W,\,av} + 2(SF - 2\delta_{W,\,av}) \tag{47}$$

For the bare tube exchanger this area is

$$(A_{w0})_B = 2\delta_{W,\,av}L \tag{48}$$

* This approximation can be eliminated in the computational procedure but will increase computing time.

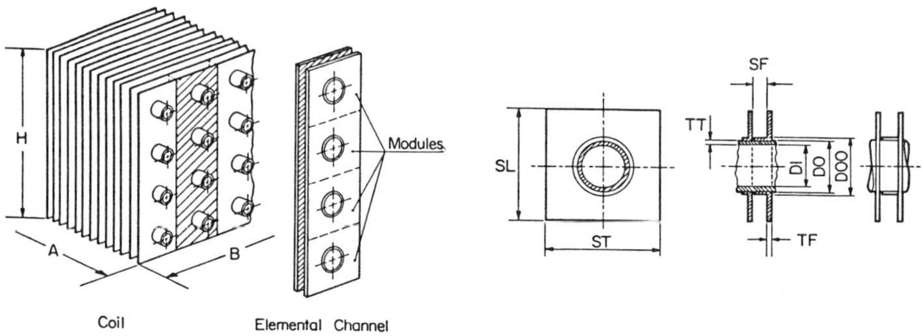

Figure 6. Elemental channel of a finned condenser coil and its modules.

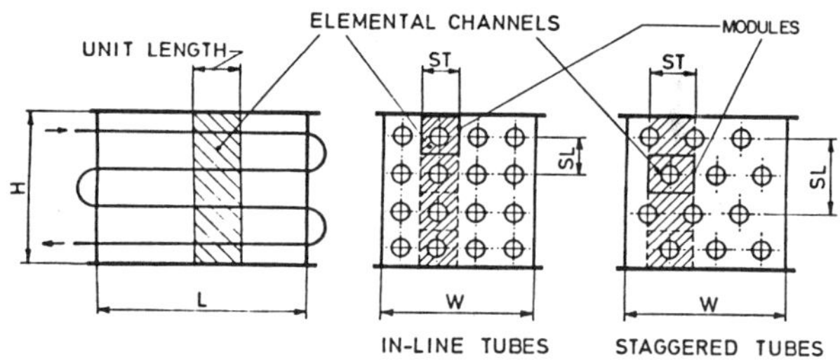

Figure 7. Elemental channels and modules in a bare tube heat exchanger.

The spacing theoretically available for the air flow is

$$A_{AW0} = A_{AM0} - A_{W0} \tag{49}$$

where A_{AM0} is the cross-sectional area in the absence of water film (dry case).

This area is for the finned exchanger

$$(A_{AM0})_F = (ST - DOO)SF \tag{50}$$

and for the bare tube exchanger

$$(A_{AM0})_B = (ST - DO)L \tag{51}$$

The hydraulic diameters of the air flow in the presence of a water film are for the finned exchanger

$$(D_{hAW})_F = \frac{4(A_{AW0})_F \cdot SL}{(Am)_F} \tag{52}$$

and for the bare tube exchanger

$$(D_{hAW})_B = \frac{4(A_{AW0})_B SL/2}{(A_m)_B} \tag{53}$$

The A_m's are the total surface areas of the respective modules.

$$(A_m)_F \cong A_f + \pi DOO \cdot SF \tag{54}$$

$$(A_m)_B \cong \pi DO \cdot L \tag{55}$$

The Reynolds number of the air is

$$Re_{AW} = \frac{G_{AW} D_{hAW}}{\mu_A} \tag{56}$$

where the mass flux of the air flow is

$$G_{Aw} = \rho_A v_A \tag{57}$$

v_A is the air velocity, ρ_A, μ_A are density and dynamic (absolute) viscosity of the air, respectively, at a given temperature. The heat transfer coefficient between the water interface and the air can be obtained from

$$h_{AW} = j_{AW} \frac{G_{AW} c_{pA}}{Pr_A^{2/3}} \tag{58}$$

where j_{AW} is the Colburn j-factor and can be obtained from experimental data [17] which can be replaced mathematically as

$$j = a\, Re_A^{-b} \tag{59}$$

The constants a and b must be given separately for each particular heat exchanger geometry. The heat transfer coefficient h_{AW} of Equation 58 must be used in the respective relationships of the first and second sections. When the evaporative heat exchanger is a finned coil, then $h_{AW} \cdot \eta_0$ must be used instead. This is needed to account for the fact that the fins are not at the same temperature as the tubes mainly for reasons that the conductive heat flow resistance of the fins is not zero. The latter is accounted for by the fin efficiency (η_f).

The fin effectiveness is

$$\eta_{0W} = 1 - \frac{A_{fw}}{A_{mw}} (1 - \eta_f) \tag{60}$$

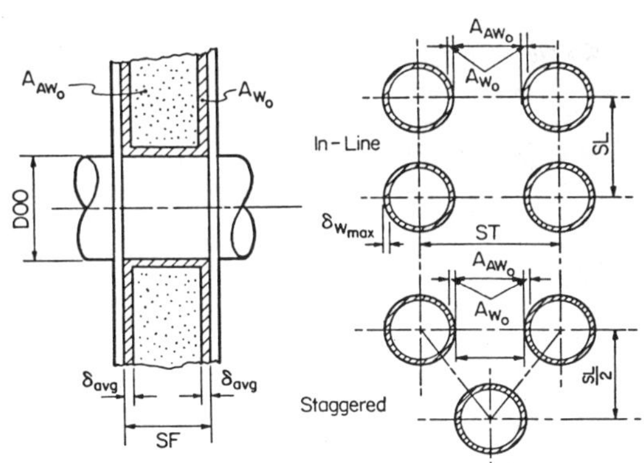

Figure 8. Cross-sectional air flow areas in wetted plate-fin-tube and bare-tube-type heat exchangers.

where η_{fW} = tanh mL/mL
A_{mw} = the total heat transfer area on the outer side of the module
A_{fw} = the total wet fin surface area on the same side
L = the radial height of the fin; = (ST − DOO)/2

$$m = \sqrt{\frac{2\, h_{Aw}}{k_f TT}}$$

where k_f is the thermal conductivity of the fin. For bare tube heat exchangers $\eta_0 = 1$.

The air flow rate through one elemental channel is $\dot{m}_{AwC} = V_A A_{Aw0} \cdot \rho_A$.

The evaluation of the foregoing air-side heat transfer coefficient required the knowledge of the water film thickness δ_W. This thickness will reach a maximum value in a plate-fin tube heat exchanger, where the water flows around the tube. Obviously the smallest cross-section for the air flow should not be blocked, and thus

$$2\,\delta_{W\,max} < SF \tag{61}$$

Water flows by gravity have been studied in the literature [18–21]. There it was shown that the flow remains laminar for $Re_w < 250$. Those conditions should be preferred, otherwise the film becomes too wavy and can interact with the air flow and dry spots might occur. In addition the pressure drop in the air will increase.

For laminar flow conditions of the water film the mass flow rate in the channel can be expressed as

$$\dot{m}_{Wc} = \frac{Re_W \mu_W}{2}[(ST - DOO) + SF] \tag{62}$$

The volumetric flow rate on one side is

$$\dot{V}_{Wc} = \frac{\dot{m}_{Wc}}{2\,\delta_W} \tag{63}$$

The thickness of the water film is given by

$$\delta_W = \left[\frac{3\,\nu_W \dot{V}_{Wc}}{ST\, g}\right]^{1/3} \tag{64}$$

and

$$\delta_{W\,max} = \left[\frac{3\,\nu_W \dot{V}_{Wc}}{(ST - DOO + SF)\cdot g}\right]^{1/3} \tag{65}$$

The mean velocities of the film are

$$v_{W,\,av} = \frac{g\delta_W^2}{3\,\nu_W} \tag{66}$$

and at critical cross-section

$$v_{W,\,av} = \frac{g\delta_{W\,max}^2}{3\,\nu_W} \tag{67}$$

The maximum velocity of a falling film is at its outer surface

$$v_{W\,max} = \frac{g\delta_W^2}{2\,\nu_W} = \frac{3}{2} v_{W,\,av} \tag{68}$$

Note: From case to case it must be investigated if the relative air velocity flowing counter or parallel to the falling film must be considered for h_A evaluation.

As stated Re_W should be less or equal to 250

$$Re_{W,\,in} = Re_{W,\,cr} = \frac{4\, v_{w,\,av,\,cr} \cdot \delta_{max}}{\nu_W} \tag{69}$$

and in areas above and below the tube

$$Re_w = \frac{4\, v_{W,\,av}\, \delta_W}{\nu_W} \tag{70}$$

This value should not be theoretically less than 10–30 otherwise the film will break up and dry spots will occur.

The Prandtl number of the falling water film having the temperature t_W can be obtained from property tables or can be computed from property data

$$Pr_W = \frac{\mu_W c_W}{k_W} \tag{71}$$

The film heat transfer coefficient between the water and the wetted fin h_W can be evaluated from the expression given by McAdams [18]

$$(h_W)_F = 0.01 \left[\frac{k_W^3 \rho_W^2 g}{\mu_W^2}\right]^{1/3} Pr_W^{1/3}\, Re_W^{1/3} \tag{72}$$

For bare tube heat exchangers Mizushina et al. [22, 23] list

$$(h_W)_B = 118 \left[\frac{3{,}600\, \dot{m}_{Wc}}{2\, DO}\right]^{1/3} \tag{73}$$

where $\dot{m}_{Wc}$ is the mass flow rate of water per unit length of the tube (kg/sm) and DO is the outside tube diameter. The quantity (3,600 $\dot{m}_{Wc}/2$) is commonly given by Γ representing the flow rate per unit breath.

$$(h_W)_B = 118 \left[\frac{\Gamma}{DO}\right]^{1/3} \tag{74}$$

In Reference 24 a similar correlation is reported

$$(h_W)_B = 87 \left[\frac{\Gamma}{DO}\right]^{1/3} \tag{75}$$

The difference of the constants listed possibly results from the fact that the water flow pattern or its characteristics were not the same for reasons that the geometry of the heat exchangers tested and the air flow rates were not identical.

The correlations shown in Equations 74 and 79 were based on experiments with staggered tube heat exchangers. For in-line tube measurements, no literature data could be found; however the authors measurements [10] gave the following correlations:

$$(h_W)_B = 262.17 \left[\frac{\Gamma}{DO}\right]^{0.252} \tag{76}$$

The constant in Equation 76 is much larger than the constants given in the two previous relationship but the exponent in Equation 76 is smaller. Values computed from Equation 76 are smaller than those obtained from Equation 74 and higher than those from Equation 75.

The differences are rather small, and in principle literature data can be used to evaluate the overall heat transfer coefficient between the two fluids flowing over the opposite surfaces of the separating wall.

To solve the relationships given previously the overall heat transfer coefficient U_W or U_{Wb} is needed. It is

$$U_W = \frac{1}{A_{mw}(R_W + R_{FTR})} \tag{77}$$

where

$$R_W = \frac{1}{\mu_0 h_W A_{mw}} \tag{78}$$

and R_{FTR} is the sum of resistances of the fin R_F the tube R_T and the convective resistance of the condensing media R_R or the convective resistance of a fluid in a cooler

$$R_F = TF/k_f A_{Fc}; \qquad A_{FC} = [\pi(DI + 2\,TT + TF)]/FPM$$

$$R_T = TT/k_T A_{To}; \qquad A_{To} = [\pi(DI + TT)]/FPM \text{ (with fins)}$$

$$R_R = 1/h_R A_{Ti}; \qquad A_{To} = \pi(DI + TT)L \text{ (bare tubes)}$$

$$A_{Ti} = (\pi DI)/FPM \text{ (with fins)}$$

$$A_{Ti} = \pi DL \text{ (bare tubes)}$$

FPM = number of fins per meter

The resistance R_R is not determined yet and h_R needs to be established. For a condensing medium

$$h_R = \frac{Nu_R k_R}{DI} \tag{79}$$

where for example in accordance to Chapman [25]

$$\text{For } Re_R < 50{,}000 \; Nu_R = 5.03\, Re_R^{0.3} Pr_R^{1/3} \tag{80}$$

and

$$\text{For } Re_R > 50{,}000 \; Nu_R = 0.02565\, Re_R^{0.8} Pr_R^{1/3} \tag{81}$$

The Reynolds number in the prior relationships is given by

$$Re_R = \frac{G_R D_{hR}}{\mu_R} \tag{82}$$

where G_R is the equivalent mass flux of the refrigerant defined as [25]

$$G_R = \frac{v_R \rho_{Rf}}{2}\left[1 + \left(\frac{\rho_{Rf}}{\rho_{Rg}}\right)^{0.5}\right] \tag{83}$$

The ρ's are the densities of saturated liquid and vapor, respectively at t_c, and v_R is the velocity of the refrigerant.

For single-phase fluids (in coolers) there are many correlations given in the literature. Only a few will be listed here.

For laminar flow and constant surface temperature (including entrance effects) [26, 27]

$$\overline{Nu}_D = 3.66 + \frac{0.0668\,(D/L)\,Re_D\,Pr}{1 + 0.04\,[(D/L)\,Re_D\,Pr]^{2/3}} \tag{84}$$

$$\overline{Nu}_D = 1.86\left(\frac{Re_D\,Pr}{L/D}\right)\left(\frac{\mu}{\mu_s}\right)^{0.14} \tag{85}$$

For fully developed turbulent flow [27, 28]

$$Nu_D = 0.023\,Re_D^{4/5}\,Pr^n \tag{86}$$

where n = 0.4 for heating and n = 0.3 for cooling.

$$Nu_D = 0.027\,Re_D^{4/5}\,Pr^{1/3}(\mu/\mu_s^{0.14}) \tag{87}$$

The properties are to be evaluated at the mean temperature but μ_s is the viscosity at wall temperature. The local Nusselt numbers in Equation 86 and 87 can be applied for both constant wall temperature or constant heat flux. Equations 84 and 85 give the averaged values over the length L. D in all the relationships is the hydraulic diameter.

The relationships given for turbulent flow are valid only for $L/D > 60$. For $L/D < 60$ entrance effects will be of influence and must be considered. This can be made by applying the correlation recommended by McAdams [18].

$$h = h_\infty(1 + C/(L/D)) \tag{88}$$

where the constant $C = f(L/D$, flow pattern) and h_∞ is the heat transfer coefficient for $L/D > 60$.

In accordance with the statements made in the introduction, vertical tubes wetted at their inside and heated at their outside by a condensing vapor were also considered. Those tubes can form an evaporative condenser cooling tower as described in Reference 12.

The air-side heat transfer coefficients for this type heat exchanger can be evaluated from the relationships given in Equations 84 through 88. The water film heat transfer coefficient can be given by

$$h_w = 724(\Gamma)^{1/6} \tag{89}$$

where Γ is given in lit/min-m. The tubes in such a cooling tower will be of rather large diameters and the condensate film formed will be very thin by comparison. Therefore curvature effects can be neglected and h_R follows from Nusselt film theory on a vertical wall [29],

$$h_R = 0.943\{(\rho^2 k^3 g i_{fg})/[L\mu^2(t_c - t_w)]\} \tag{90}$$

where the properties ρ, k, and μ are those of the liquid condensate. L is the tube length, and g is the acceleration due to gravity.

All the equations listed previously are sufficient to predict analytically the performance of an evaporatively cooled heat exchanger whenever the geometry of the design is specified, the flow rates of air, water, and fluid to be condensed (or cooled) are given, and when the inlet air state is known.

For the analysis, it is important that all the properties involved are represented by mathematical relationships as functions of pressure and temperature for the ranges of interest.

The statement that performances of evaporatively cooled heat exchangers can be predicted, obviously needs to be supported by experimental verification.

EXPERIMENTAL VERIFICATION OF THE NUMERICAL ANALYSIS

The analytical procedure developed in the previous chapters was checked initially by comparing experimental data of a manufacturer with the predicted values for identical flow conditions and geometry of a condenser heat exchanger.

The manufacturer provided data of two test runs for which the previously derived stepwise integration was used to predict the heat rejection rate from the change of state experienced by the air along its flow path and to predict the water temperature wetting the external surfaces. Furthermore the analysis was extended to determine the heat rejection rate of the condenser under dry operations. The condensing temperature of the refrigerant R–22 was $t_c = 40.8°$ C. The analysis assumed that all surfaces of the staggered bare-tube-type exchanger were completely wetted.

The total heat transfer area of the condenser was $A_t = 30.517\ m^2$. It was composed of 34 tubes having 12 passes. The steel tubes were of outside diameter $DO = 2.65 \times 10^{-2}$ m and of wall thickness $TT = 1.397 \times 10^{-3}$ m. The transverse tube spacing was ST = $= 5.334 \times 10^{-2}$ m; and the longitudinal tube spacing was $SL = 1.206 \times 10^{-1}$ m. The coil size was length TL = 7.92 m, width $W = 9.067 \times 10^{-1}$ m, and height $= 6.9 \times 10^{-1}$ m.

For the first test the following conditions were provided.

Total refrigerant flow rate	$\dot{m}_R = 2.92 \times 10^3$ kg/hr
Total air flow rate	$\dot{m}_A = 1.36 \times 10^4$ kg/hr
Total water flow rate	$\dot{m}_W = 1.25 \times 10^4$ kg/hr
$T_{Ain} = 32.03°$ C, $\varphi_{in} = 36.8\%$	(inlet air rate)

The heat rejection rates were

$$| q_c |_{predicted} = 153.4 \text{ kW}$$

$$| q_c |_{measured} = 160.1 \text{ kw}$$

(underpredicted by $\approx 4\%$)

The air at exit state $t_{AO} = 31.5°$ C, $\varphi_O = 85.8\%$. The water temperature from inlet to exit changed very little, and its average value was $t_{W,av} = 31.4°$ C. The air temperature from inlet to exit was higher than the water temperature. This means that the water evaporation rate or the latent heat transport was higher than the heat received from the condensing media (by approximately 1.3%).

In addition, a very important fact, was demonstrated, namely the heat transfer resistance between the water surface and the air was zero or the heat transfer coefficient is infinite. This obviously indicates the highest possible air-side heat transfer augmentation. As is demonstrated in Reference 8–10 an evaporatively cooled condenser can operate effectively even when the air dry bulb temperature entering the condenser is far above the condensing temperature.

Heat removal from the condenser however becomes impossible when the water temperature t_W or the inlet air wet-bulb temperature equals the condensing temperature t_c.

For the dry operation the analysis predicted that the coil could reject only

$$(q_c)_{dry} = 13.36 \text{ kW}$$

Comparing this value with $(q_c)_{pred}$ for the wet operation indicates an 11.5-fold increase of the heat rejection rate. This again demonstrates the heat transfer augmentation resulting from evaporative cooling. The computer also evaluated the amount of water evaporating during the process

$$(\dot{m}_W)_{lost} = 192 \text{ kg/hr}$$

The power for the water pump was $P_w = 0.943$ kW and for the fan $P_f = 0.432$ kW.

For the second test, the following conditions prevailed.

Total refrigerant flow rate: $\dot{m}_R = 3.46 \times 10^3$ kg/hr
Total air flow rate: $\dot{m}_A = 1.61 \times 10^4$ kg/hr
Total water flow rate: $\dot{m}_W = 1.48 \times 10^4$ kg/hr
$T_{A,in} = 30.7°$ C $\quad \varphi = 46\%$

The heat rejection rates were

$(q_c)_{predicted} = 164.5$ kW

$(q_c)_{measured} = 156.7$ kW

(overpredicted by $\approx 5\%$)

The other quantities computed were: $t_{Ao} = 31.2°$ C, $\varphi_o = 87.3\%$, $t_{w,av} = 31.4°$ C, $q_{dry} = 17.1$ kW (enhancement 9.59 fold), $(\dot{m}_W)_{lost} = 199$ kg/hr $p_W = 0.12$ kW, $p_f = 0.709$ kW.

For each of the test ≈ 75 integration steps were performed in computer simulation.

The comparison of the measured and predicted data indicate an excellent agreement especially when considering the vast numbers of parameters involved to analyze the complex combined heat and mass transfer processes. The correctness of the analytical procedure was also checked by experiments using a bench-type in-line bare tube heat exchanger and a bench-type vertical tube condenser. The results of this work [10, 12] did again yield excellent agreement between predicted and measured data. The studies described in these articles will not be repeated here for reason of brevity. However, one of the studies [10] will be briefly discussed because it addresses the computer search for the lowest possible condensing temperature reachable in an evaporatively cooled condenser, under given heat load and climate conditions. As stated in the introduction, lowering of condenser temperature is of benefit to both power-consuming and power-producing processes.

The computation of the air and water states within the evaporatively cooled commercial condenser coil as discussed did show that in both cases the air did not reach saturated condition. The heat exchanger could therefore be made larger to reach such conditions which would increase the heat rejection rate. The air and water temperatures changed only slightly along their flow paths and were very close to each other. Plotting t_c, t_A, and t_W versus the heat transfer exchanger area would yield a rather uncommon plot showing three lines practically parallel to each other. The temperature difference between air and water was of the order of one degree ($t_A > t_W$) and the temperature difference ($t_c - t_W$) was approximately 9° C. Finning this heat exchanger would increase the condenser load by approximately $\eta_0 A_f/A_b$ under otherwise identical conditions. The condenser would then be oversized for a refrigeration system of a specified constant cooling capacity. This can be compensated for by lowering the condensing temperature. This however will cause a lower quality of the condensate entering the evaporator and the cooling load would be increased. This increase in turn could be offset by lowering the refrigerant mass flow rate. Lowering t_c and $\dot{m}_R$ results in decreasing the condenser load, and t_c could be decreased further. By an iterative process [8–10] the computer can search for the lowest possible $t_c = t_c^*$.

For demonstration purposes, one of these studies [10] will be discussed here.

In this study t_c^* was considered only in respect to the constant condenser load; changes of $\dot{m}_R$ and compressor power were not included.

A small bench-type test condenser composed of in-line bare tubes was used for the studies. During the first part of the experiment the condenser was air-cooled. As shown in Figure 9 steady state was reached at a condensing temperature of 44.6° C under a measured heat input of 300 W (the computer predicted 308 W). The coil was then sprayed with water, the heat input remained constant but the condensing temperature dropped very rapidly and reached a final steady-state value of 24.4° C, which is below the inlet air temperature $t_{Ai} = 26.4°$ C. The computer predicted $t_c^* = 23.8°$ C. This value would have been reached

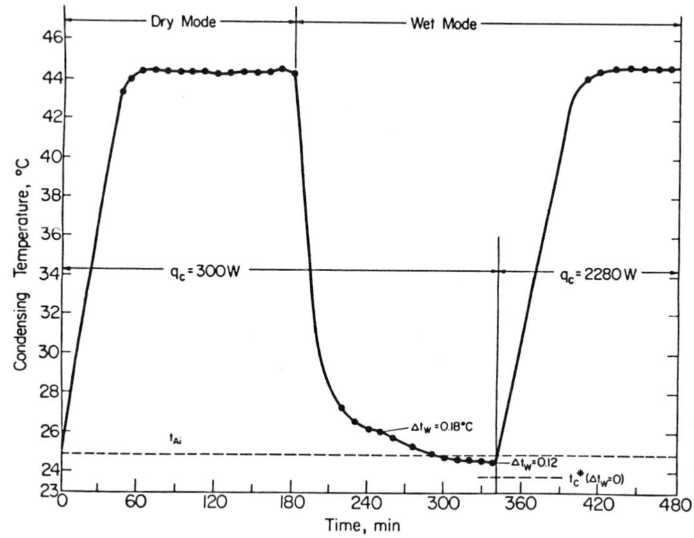

Figure 9. Heat removal rate from dry and wet condensers and condensing temperature.

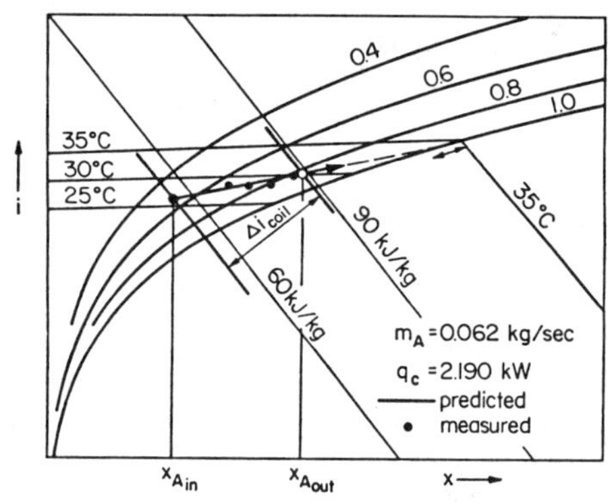

Figure 10. Predicted and measured state changes of air flowing through an evaporatively cooled condenser.

when the adiabatic condition assumed for the analysis would have been met by the experiment by making $\Delta t_W = 0$, i.e., by removing the water pump heat input. The condenser load was then raised until t_c under wet conditions reached the values observed under dry conditions. The heat input had to be increased to 2,280 W indicating an enhancement factor of 7.6. The computer predicted 2,180 W which again is in excellent agreement with the measured value.

In the i–x diagram of Figure 10 are shown the state changes of air and water predicted for the wet operation under steady heat input of 2,280 Watt, and constant inlet condition of the air, $t_{Ai} = 26.4°$ C, $\varphi = 53\%$. This line is slightly curved due to the changes epxerienced by the pulling point assumed to remain on the saturation line; Lewis factor = 1. During the test four air states at different locations in flow direction were determined by dry- and wet-bulb temperature measurements. These data practically fall on the predicted line. The small scatter possibly is due to the fact that the thermometers were difficult to maintain dry or wet as required. The measurements and the predicted data indicate that the air was not saturated at the exit ($t_{A0} = 30.4°$ C, $\varphi = 80.1\%$, $t_{W,av} = 35°$ C $\dot{m}_A = 0.0628$ kg/sec).

The agreement between experiments and predictions demonstrated in Figures 9 and 10 supports again the statements made before that a condenser (or cooler) can be designed properly under specified conditions. Expensive and time-consuming model tests on a large scale are not needed.

TRANSIENT RESPONSE OF EVAPORATIVE CONDENSERS

For the analysis, it is assumed that the system is initially in steady-state condition and exchanges a heat load q_{c1} at a condensing temperature t_{c1} by heat and mass transfer with air and water at fixed flow rates and constant state conditions. The water wetting the heat exchanger surfaces has reached an equilibrium temperature $t_{W,\,eq1}$ which is equal to the temperature of the water in the pond $t_{W,\,p1}$ and equal to the temperature of the sprayed water $t_{W,\,s1}$. These last two temperatures are equal, if there are no heat losses or gains from the surroundings and no water pumping heat input. At a reference time $\tau = 0^+$ the system is suddenly exposed to a change in operating conditions. This change is described as the step function of one independent variable recognizeable by the analytical model. This could be either the condensing temperature t_c or the heat load q_c, the flow rate of air $\dot{m}_A$, or of the refrigerant $\dot{m}_R$ or even the flow rate of water $\dot{m}_W$. Obviously, the change also could be due to a sudden change in climate condition. After the system has been exposed to any of these changes, it will asymptotically approach a new steady-state condition whenever no further changes take place. The condensing temperature will change to t_{c2} and the recirculated water will reach a new equilibrium temperature $t_{W,\,eq2}$. The analysis described in the previous chapters is used to calculate the new conditions and the new performance of the evaporative condenser, but it cannot determine the time period required by the system to reach the new equilibrium state. For this evaluation the analytical procedure must be extended. For the sake of simplicity it is assumed that the water in the system (pond, pump, lines, and coil) represents the largest heat capacity for which the time changes must be established. This assumption is correct for the evaporative condenser used is the bench-type experimental set-up described in Reference 10. Therefore the analysis can be checked by measurements. Obviously, in refrigeration or power-producing systems the heat capacities of all the other components must be considered. The analysis would then become rather complex and would be outside the scope of this chapter.

The system for the analysis is shown in Figure 11. Water at temperature t_W is recirculated. One part of the water flow $\dot{m}_{W,\,e}$ evaporates at t_W and is replenished by make-up water $\dot{m}_{W,\,mu}$ which is at $t_{W,\,mu}$. At initial time $\tau = 0$ the water in the system is at equilibrium condition $t_{Wi} = t_{W,\,eq1}$

The conservation of energy of the water flow can be stated as

$$\begin{bmatrix}\text{accumulation}\\ \text{of energy}\end{bmatrix} = \begin{bmatrix}\text{energy}\\ \text{input}\end{bmatrix} - \begin{bmatrix}\text{energy}\\ \text{output}\end{bmatrix} \tag{91}$$

The water receives the heat q_c from the condensing refrigerant and transfers the heat q_a to the air stream by means of a dry q_h and wet q_σ heat exchange. There is also heat input from the make-up water and heat output due to evaporation.

It is assumed that the internal resistance of the water volume is negligible in comparison with the external resistances (contact with the tube wall and heat transfer at the interface). The energy balance Equation 91 in accordance with Figure 11 can be given as

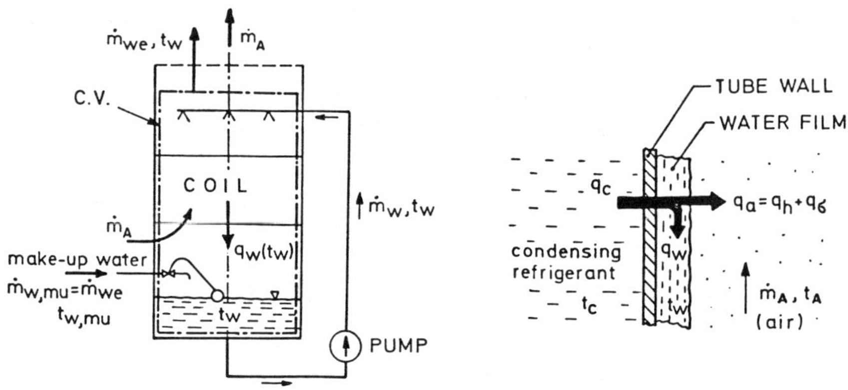

Figure 11. Energy balance on the water flow in an evaporative condenser.

$$\rho_W V_W c_W \frac{dt_W}{d\tau} = \dot{m}_{We} c_W t_{W,\,mu} - \dot{m}_{We} c_W t_w + q_c(t_W) - q_a(t_W) \tag{92}$$

Since

$$q_W(t_W) = q_c(t_W) - q_a(t_W) \tag{28}$$

Equation 92 becomes

$$\rho_W V_W c_W \frac{dt_W}{d\tau} = \dot{m}_{We} c_W (t_{W,\,mu} - t_W) + q_W(t_W) \tag{93}$$

V_W in the preceding relationships is the volume of water enclosed in the control volume of Figure 11.

Equation 93 can be solved only if $q_W(t_W)$ is known; $q_W(t_W)$ can be determined by following the procedure outlined earlier (see also Figure 3).

After applying the step function the new equilibrium water temperature $t_{W,\,eq2}$ is calculated first. In this case $t_{W,\,S2} = t_{W,\,p2} = t_{W,\,eq2}$. The computer program is then used to evaluate the system performances for conditions $t_{W,\,s} \neq t_{W,\,p}$. Different spray-water temperatures must be specified in the neighborhood of $t_{W,\,eq2}$ and the corresponding q_W is then determined.

The functional dependence $q_W(t_W)$ can be expressed by

$$q_W(t_W) = a + bt_W + ct_W^2 + dt_W^3 \tag{94}$$

where the coefficients a, b, c, and d are obtained by a least square method from the set of data points (t_{Wi}, q_{Wi}); i = 1, 2, 3 ... n. Introducing Equation 94 into Equation 93 yields

$$\frac{dt_W}{d\tau} = c_1 + c_2 t_W + c_3 t_W^2 + c_4 t_W^3 \tag{95}$$

where $c_1 = \dfrac{a + \dot{m}_{We} c_W t_{W,\,mu}}{\rho_W V_W c_W}$

$$c_2 = \frac{b - \dot{m}_{We} c_W}{\rho_W V_W c_W}$$

$$c_3 = \frac{c}{\rho_W V_W c_W}$$

$$c_4 = \frac{d}{\rho_W V_W c_W}$$

Equation 95 is a first-order nonlinear differential equation and can be solved by the fourth-order Runge–Kutta method to obtain

$$t_{Wi} \quad \text{for} \quad \tau_i; \quad \tau_i = \tau_0 + i\Delta\tau \quad (i = 1, 2, 3, \ldots)$$

where $\Delta\tau$ is a specified time increment.

Since Equation 95 is of the form $t_W = f(t_W)$ with initial value $t_W = t_{W0}$ for $\tau_0 = 0$ then

$$t_{Wi+1} = t_{Wi} + \frac{1}{6}(k_1 + 2k_2 + 2k_3 + k_4) \tag{96}$$

where
$$k_1 = \Delta\tau f(t_{Wi})$$
$$k_2 = \Delta\tau f(t_{Wi} + k_1/2)$$
$$k_3 = \Delta\tau f(t_{Wi} + k_2/2)$$
$$k_4 = \Delta\tau f(t_{wi} + k_3)$$

This allows a solution arbitrarily close to the equilibrium water temperature $t_{W,eq2}$ when $dt_W/d\tau = 0$. Since $t_W = t_{W,eq2}$ when $\tau \to \infty$ a desired tolerance must be specified to determine the "response time," that is, the time after application of a step input for the system to reach, within the specified tolerance in (°C or in %), the new equilibrium state condition.

If the water temperature does not change significantly from its initial value a linear curve fit can be applied to describe the functional relationship $q_W(t_W)$.

When $q_W(t_W)$ has been established the equation of the tangent line through the equilibrium point on the curve is determined by means of a Taylor-series expansion around the steady-state value $t_{W,eq2}$.

$$q_W \approx q_W(t_{W,eq2}) + q_W'(t_{W,eq2})\,(t_W - t_{W,eq2}) \tag{97}$$

where $q_W'(t_{W,eq2})$ is the first derivative of $q_W(t_W)$ evaluated at $t_{W,eq2}$.

In general Equation 97 has the form

$$q_W = a - bt_W \tag{98}$$

and Equation 93 becomes

$$\rho_W V_W c_W \frac{dt_W}{d\tau} = \dot{m}_{We} c_W (t_{W,mu} - t_W) + a - bt_W$$

or

$$\frac{dt_W}{d\tau} + At_W = B \tag{99}$$

where
$$A = \frac{\dot{m}_{We} c_W + b}{\rho_W V_W c_W}$$

$$B = \frac{\dot{m}_{Wc} c_W t_{W,mu} + a}{\rho_W V_W c_W}$$

The solution of Equation 99 is

$$t_W = C_1 e^{-A\tau} + C_2 \tag{100}$$

where $C_2 = \frac{B}{A}$

Introducing the initial condition

$t_W = t_{Wi}$ for $\tau = 0$ yields C_1

and Equation 100 is

$$t_W = \left(t_{W,mu} - \frac{B}{A}\right) e^{-A\tau} + \frac{B}{A} \tag{101}$$

For $\tau \rightarrow \infty$ the steady water temperature becomes $t_{W,eq2} = B/A$.

The described methodology for analyzing unsteady-state operations of evaporative condensers can be summarized by listing the five major steps to be taken.

1. Determine evaporative condenser performance at condition I at time $\tau \leqq 0$. Recirculated water is at $t_{W,eq1}$.
2. The system is exposed to a sudden change at time $\tau = 0^+$. Determine the steady-state evaporative condenser performance at this new condition II.
3. Determine the relationship between q_W and $t_W = t_{W,s}$, i.e., $q_W(t_W)$. Use the computer program to evaluate the system performance at $t_{W,s} \neq t_{W,p}$ at condition II.
4. Solve the differential Equation 95 and obtain $t_W(\tau)$.
5. Determine the evaporative condenser performance in the time domain. Since $t_W(\tau)$ is already known, it is assumed that the condensing temperature $t_c(\tau)$ is related in the time domain as $t_c[t_W(\tau)]$.

In general it is assumed that each analyzed case for the specified $t_{Wi}(\tau_i)$, where $i = 1, 2, 3, \ldots, n$, provides results which are valid at the given time τ_i.

If $t_W = t_{W2}$ at $\tau = \tau_2$ then $t_c = t_{c2}$ at $\tau = \tau_2$ as well as any other analyzed parameter. This implies that the system responds in the same way as the water temperature. When this temperature reaches equilibrium, the system is also in equilibrium.

The experiments [10] discussed earlier allowed us to verify the analytical technique.

The test were carried out continuously in the wet mode but the power input (and q_C) was gradually increased. For each power setting the system was allowed to reach equilibrium. Measurements were taken two minutes apart. Figure 12 shows for $\tau \leqq 0$ the equilibrium condition I reached under a constant heat input $q_h = 1{,}200$ W. At time $\tau = 0^+$ the power was suddenly changed to 1,400 W and maintained until new equilibrium state II was reached.

Thus the power input to the boiler (and also q_c) is recognized as the step function.

The computer predicted q_{c1} and $t_{W,eq1} = 24.7°$ C at measured t_{c1}. This represents the initial condition for the transient analysis. The computer predicted then q_{c2} and the new equilibrium water temperature $t_{W,eq2}$ for specified t_{c2} under constant power input of 1,400 W. In the next step the relationship between q_W and t_W was established for various spray-water temperatures different from $t_{W,eq2} = 26.0°$ C. In Figure 13 $q_W(t_{W,s})$, $q_a(t_{W,s})$ and the total predicted rejection rates $q_{c,p}(t_{W,s})$ are plotted. The latter are very little affected by $t_{W,s}$ because the heat input q_h was constant.

The measured heat rejection rate $q_{c,m}$ is smaller than q_h due to heat losses which are proportional to the temperature difference $(t_c - t_{amb})$.

In Figure 13 is also shown the change of condensing temperature with the spray-water temperature. It is also important to point out that $q_W(t_{W,s})$ intersects the abscissa at

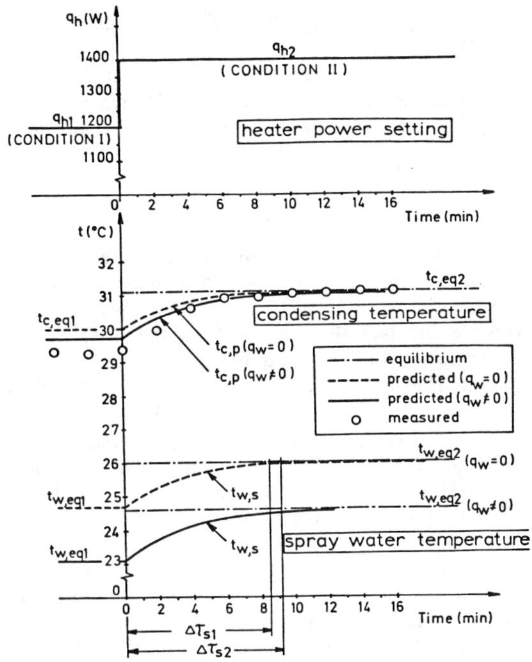

Figure 12. Transient response of evaporative condenser to the sudden change of power setting and condenser heat load.

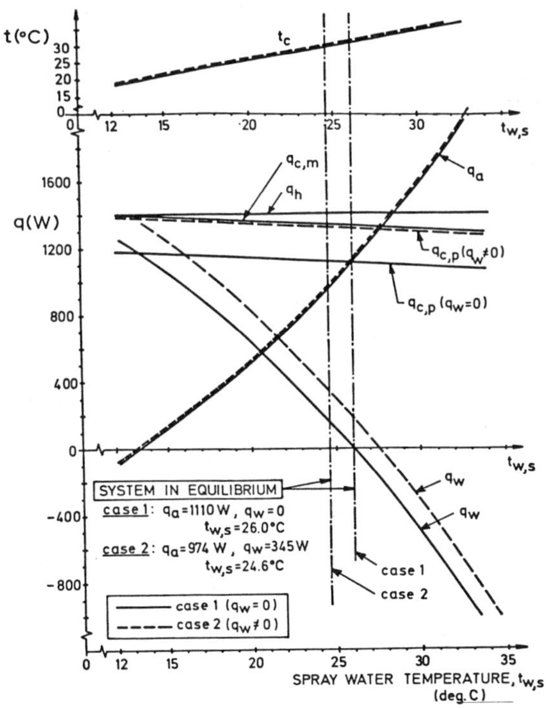

Figure 13. Effect of water spray temperature on q_W, q_a, q_c, and t_c.

$t_{W,s} = t_{W,eq2}$ where $q_W = 0$. (Theoretically in equilibrium there is no net heating or cooling of water.)

The results in Figure 13 indicate certain differences between predicted $q_{c,p}$ and measured $q_{c,m}$ heat rejection rates. They are shown as two parallel lines with the following values at equilibrium water temperature $t_{W,eq2} = t_{W,s2} = 26.0°$ C.

$q_{c,m} = 1{,}335$ W; $q_{c,p} = 1{,}100$ W

(underpredicted by 16.8%)

This difference results from the fact that q_W was assumed to be zero at equilibrium condition. This, however, was not the case because the water sprayed into the system to wet the coil was cooled by evaporation before it contacted the tubes of the heat exchanger. Despite the fact that $t_{W,s} = t_{W,p}$ the temperature difference between the water on the uppermost tubes and that on the lowermost tubes was not zero and therefore $q_W > 0$. Taking this into account makes $q_{c,p} = 1{,}319$ W which underpredicts in respect to the measured $q_{c,m}$ by only 1.2%. In addition $t_{W,eq1}$ changes from 24.7° to 23.1° C.

This clearly shows that in a spraying system the water will be precooled before it reaches the heat-exchanger surfaces. Thus the actual system showed a net water heating during steady-state operation. (For vertical tube condensers [12] the water was not sprayed and precooling did not occur.)

In order to demonstrate how temperatures of refrigerant, water, and air change in an evaporative condenser, three typical cases are shown in Figure 14. In case 1 the spray-water temperature $t_{W,s} = 23.6°$ C, and the temperature of the water in the pond is $t_{W,p} = 24.0°$ C. There is a net water heating in the system of $q_W = 265$ W. In case 2 $t_{W,s} = t_{W,p} = 26.0°$ C and $q_W = 0$. In the third case $t_{W,s} > t_{W,p}$. The water is cooled $q_W = -998$ W and a lower condensing temperature will result for the same power setting, which indicates a higher heat

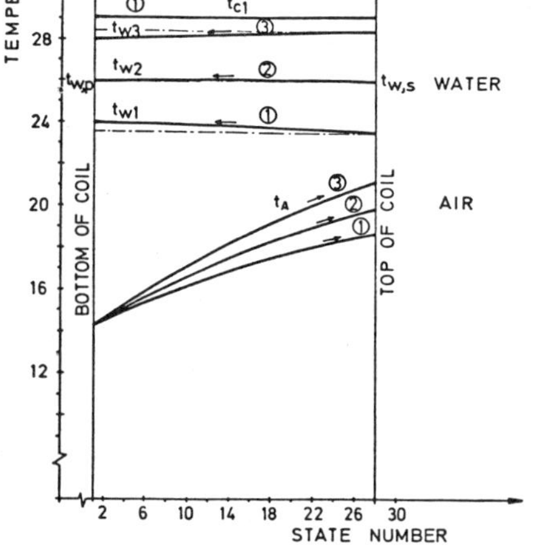

Figure 14. Temperature profiles of refrigerant, water, and air in the evaporative condenser for three different spray water temperatures according to case 1 of Figure 13.

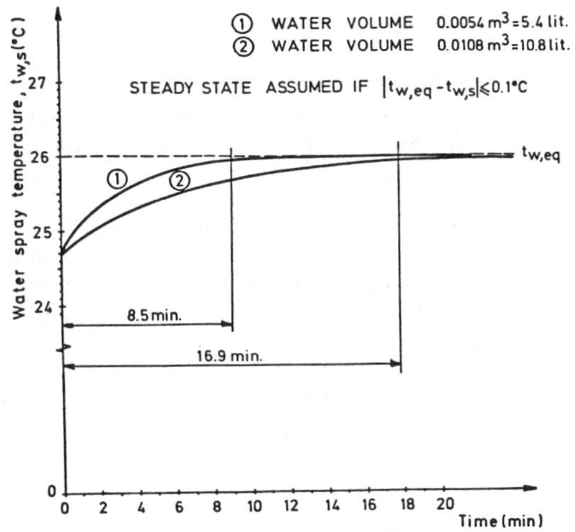

Figure 15. Effect of water volume in the system on transient response. (Solution given for analyzed case 1 where $q_W = 0$ during steady state.)

transfer augmentation. Obviously when the spray-water temperature is lower, the condensing temperature is lower.

The just-described experiments were conducted under the following conditions.

$$V_W = 0.0054\ m^3$$

$$(\dot{m}_W)_{1,200} = 9.793\ \text{lit/min}$$

$$(\dot{m}_W)_{1,400} = 9.893\ \text{lit/min}$$

$$(\dot{m}_R)_{1,200} = 22.64\ \text{kg/hr}$$

$$(\dot{m}_R)_{1,400} = 26.15\ \text{kg/hr}$$

$$(\dot{m}_A)_{1,200} = (\dot{m}_A)_{1,400} = 240.84\ \text{kg/hr}$$

$$t_{Ai} = 23.7°\ C$$

$$\varphi_{Ai} = 36\%$$

Figure 15 shows the settling time to be 8.5 min for the specified tolerance of $\leqq 0.1°$ C. This is the time needed to establish the new equilibrium condition after the heat input had been raised from 1,200 W to 1,400 W. Doubling the volume of the water in the system will increase under otherwise identical operating conditions the settling time to 16.9 minutes or $\approx$ twice as long as before.

This indicates that the fast response is obtained if evaporative condensers operate with least possible amounts of water in the pond or in the system as a whole.

EVAPORATIVE FLUID COOLERS

These devices are also known as indirect evaporative heat exchangers where a gas or a liquid is cooled by means of evaporative cooling from wetted walls separating the water film and the air stream from the fluid to be cooled. For liquid cooling these coolers quite often look like evaporative condensers. For cooling gases and in specific air, for air conditioning application, the cooler may be a combination of cross-flow plate heat exchanger and cooling tower.

Evaporative coolers were studied extensively by Mizushina [30, 31, 32] who also provided an approximate theory based on the enthalpy potential as a driving force. He obtained experimentally overall capacity coefficients of heat transfer between cooling water film and the air stream in terms of the respective Reynolds numbers. He also derived in a similar way an expression for the overall capacity coefficient of mass transfer.

Parker [24] reported laboratory tests and an approximate theory based on the enthalpy difference as the driving potential. The heat and mass transfer coefficients were evaluated as a function of process conditions. From all the various conditions investigated by Parker he concluded that the heat and mass transfer analogy cannot be used to obtain mass transfer coefficients for wetted surfaces from heat transfer data obtained on dry surfaces. The use of the analogy would yield uncorrect but "safe" approximations.

Pescod [33] studied cross-flow heat exchangers in dry and wet mode operations. He recognized that the heat and mass transfer processes are too complex and require the use of high speed computers. To avoid this approach he related the overall effectiveness in the wet mode to the performance of the dry heat exchanger as

$$\epsilon_W = \frac{t_{Ai} - t_{A0}}{t_{Ai} - t^*_{aWb,i}} \tag{102}$$

where t_{Ai} and t_{A0} are the inlet and outlet dry-bulb temperatures of the primary air conditioned air stream and $t^*_{aWb_i}$ is the inlet wet-bulb temperature of the wet secondary air stream in contact with the water film.

Elmahdy [34, 35] reported experimental and theoretical studies of finned dry and wet coils- and reported that the fin efficiency of the wetted fins is much smaller than that of the dry fins. He also used in his analysis the enthalpy as driving potential.

There are many more research results reported in the literature [36–40] dealing with various kinds of fluid coolers. All these studies also assume the enthalpy potential as the main driving force.

We previously have described and applied an analytical method to design evaporative condensers and to predict their performances and to verify the analysis by experiments. The analysis was based on the graphical procedure introduced by Bošnjaković, who also discussed briefly in his textbook how his step-by-step integration procedure can be applied for coolers of fluids whose temperatures do not remain constant along their flow paths. Therefore the methods described in detail in the second section, and partially in the first, can be used in an identical way for the analysis of coolers. This will be briefly outlined in this chapter.

Figure 16 shows a small finite elemental segment of a tube of length ΔL. The width of the element is equal to the transverse spacing ST between tubes of a heat exchanger, and its height is equal to the longitudinal spacing SL. The fluid to be cooled enters the tube segment at a flow rate $\dot{m}_F$ and at the temperature t_{Fi}. The fluid exits the element at a temperature $t_{F0} = t_{Fi} - \Delta t_F$. The water at flow rate $\dot{m}_W$ and at temperature t_{Wi} falls by gravity over the outside of the tube, spreads in a thin film, and leaves the element at temperature t_{W0}. The air flow rate $\dot{m}_A$ at temperature t_{Ai} enters the element from the bottom and leaves from the top at temperature t_{A2}. The processes between the air and water flows are similar to the processes described in previous chapters for the elemental channels of a condenser. The only difference is that the cooled fluid changes in temperature $t_{F0} < t_{Fi}$.

The stepwise integration is performed from the air inlet to the air outlet along the water film around the tube.

By selecting the step size y, the number of integration steps is calcuated as

$$n_s = \frac{A_{bt}}{a_b \ln 1/y} \tag{103}$$

where A_{bt} is the water interface area of the tube segment and $a_b = \dot{m}_A c_{pA}/h_A$ is the unit surface. (See Equation 30).

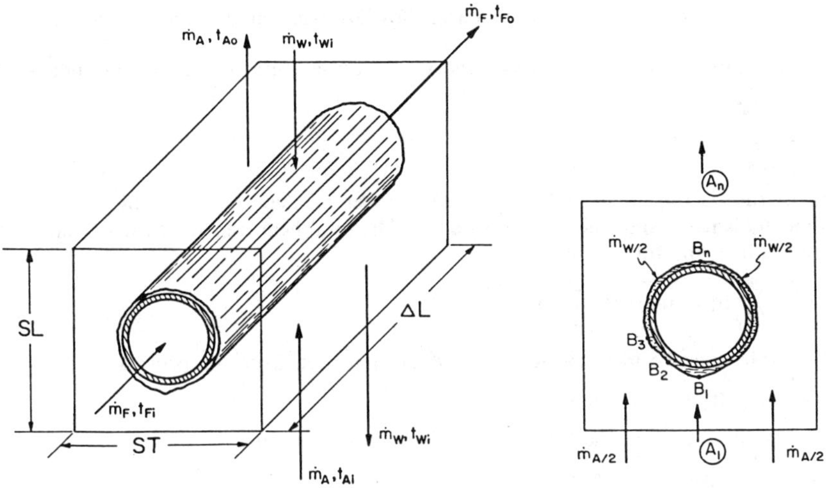

Figure 16. Evaporation process in the tube segment of a finite length ΔL for the analysis of fluid coolers.

The entering air state A_1 at temperature t_{A1} enthalpy i_{A1} and moisture content x_{a1} is completely defined. The temperature of the water leaving the tube t_{WO} (state B_1) must initially be assumed. The air state A_1 changes to A_2 in the direction of the line $\overline{A_1B_1}$ (see Figure 5). When A_2 is obtained the evaluation of the next boundary state B_2 is performed.

The relationship between the different diagram values of the heat fluxes are in accordance to Equation 28

$$q_F = q_{bW} = q_b + q_W = q_h + q_\sigma + q_W$$

q_F is the heat transferred from the fluid to be cooled which is equal the heat transferred through the tube wall to the water-air interface which serves as the reference heat transfer area.

The heat transferred through the water interface from the water film to the air stream $\dot{m}_A$ in the first integration step is

$$q_{b1} = q_{h1} + q_{\sigma 1} = i_{W1} - i_{A1} - c_{pW}(x_{b1} - x_{A1})t_{W1} \quad (104)$$

The diagram value of the heat flux transferred from the process fluid to the water-air interface is

$$q_{F1} = q_{bW1} = \frac{U_{Wb1}c_{pA1}}{h_{Aw1}}(t_{F1} - t_{W1}) \quad (105)$$

where U_{Wb1} is the overall heat transfer coefficient based on the film surface area.

The heat transferred to the water film in the first integration step is calculated as

$$q_{W1} = q_{bW1} - q_{b1} \quad (106)$$

The change of water enthalpy is

$$\Delta i_{W1} = q_{W1}\frac{c_{pA1}}{c_{pW1}}\frac{\dot{m}_{A1}}{(\pm \dot{m}_{W1})} \quad (107)$$

where the − sign stands for counter water/air flow arrangement and the + sign for parallel flow.

Once the enthalpy change is known the water temperature of the next boundary state B_2 is given by

$$t_{W2} = \frac{\Delta i_{W1}}{c_{pA1}} + t_{W1} \tag{108}$$

The procedure is repeated for the required number of steps n_s, i.e. until the outlet air condition A_n is reached at the temperature t_{Ao}. The total heat transferred from tube to the water and air in the tube element is

$$q_{bW} = \sum_{i=1}^{n_s} (q_{bi} + q_{Wi}) = \sum_{i=1}^{n_s} (q_{hi} + q_{\sigma i} + q_{wi}) \tag{109}$$

The heat rate must be identical to the heat rate removed from the process fluid

$$q_F = \dot{m}_F c_{pF}(t_{Fi} - t_{Fo}) = q_{bW} \tag{110}$$

The process fluid leaving the tube element has the temperature

$$t_{Fo} = t_{Fi} - \frac{q_{bW}}{\dot{m}_F c_{pF}} \tag{111}$$

The process fluid entering the next tube element is at the temperature t_{Fo} therefore $(t_{Fi})_{II} = (t_{Fo})_I$. The calculated t_{Fo} corresponds to the assumed temperature of the exit water t_{Wo}.

In an actual heat exchanger there are many such elements and the boundary conditions must be satisfied. This requires an extensive computing time, since the solutions must be obtained by an iterative technique. Typically the inlet temperature of the process fluid is known and also its flow rate. Commonly the air flow rate and its inlet state conditions are also known, the same is true for the water flow. With the assumption that the average spray water temperature is equal to the average leaving water temperature from the last tube row, one can obtain the temperatures of the leaving process fluid.

The demonstrated procedure can be applied to the bare-tube coil evaporative cooler, or to a cross-flow plate heat exchanger. In both cases one obtains the two-dimensional finite element analytical model, in which the process fluid temperature inside each considered finite elemental tube segment remains constant. The degree of complexity of this method depends very much on the geometry of an evaporative cooler and on the water and air flow direction which is either counterflow, parallel flow, or crossflow. For example, a cross-flow plate evaporative cooler can be subdivided into small finite rectangular elements, whose height is selected to correspond to a single step size in its direction, and whose length is arbitrarily selected to be of the same order of magnitude as its height. Each such element will be at constant water temperature, while the moist air flow and the process fluid flow would change stepwise from one side to another side of each elemental rectangular area. This method provides the two-dimensional temperature and enthalpy distribution of the fluids in the process.

The analysis is rather complex and quite involved. So was the analysis of evaporative condensers, which however was proven successful and it can be anticipated that any kind of evaporative cooler can be designed properly and their performances predicted reliably. This indicates again that expensive and very time consuming experiments on models are not really necessary on a large scale.

SUMMARY AND CLOSING REMARKS

Evaporative cooling for latent or sensible heat removal of fluids is a rather complex combined heat and mass transfer process, depending on a vast number of parameters Up to now, only approximate analysis has been widely in use. Such an analysis in most instances assists data processing and correlating numerous data, obtained by measurements on many heat exchangers of different design geometry, operating under various flow and state conditions of the fluids involved. The combination of the many parameters influencing the performance of evaporatively cooled heat exchangers is practically infinite and therefore, experimental data alone cannot provide a total understanding of the influence of the elements of the design because every elemental change requires new experimental verification. Computer aided design rapidly gains public interest in nearly every field, and its application in evaporatively cooled equipment design and development needs a better analytical basis. Data processing and development of empirical correlations, without solid and reliable analytical models, are still justified in the daily engineering practice, and provide satisfactory, but not optimized results. However, in the scientific world this practice is not justified because it has great limitations in the theoretical sense.

This article describes an exact analysis for the design of evaporatively cooled devices and verifies its correctness experimentally. However, the proper analytically supported design of heat exchangers requires iterative numerical methods and the use of high-speed computers. We have developed computer programs by an in-house effort and practically without external support. We have used these programs to predict heat exchanger performances and have searched for optimized design. The optimization was either directed towards the size, manufacturing, and operating costs, or in searching for the lowest possible operating condensing temperature under given climate and flow conditions. In order to improve a power consuming or producing thermodynamic process the lowest possible condensing temperature is necessary and can be obtained with augmentation of heat transfer by evaporative cooling.

We wish to make it very clear that the evaporatively cooled devices cannot be designed by analytical means only. This is also the case in even more advanced fields of science. Every new design must be also experimentally analyzed to verify the equipment reliability, which the analysis cannot provide. The experiments must be performed on a prototype to observe, for example the water film behavior, its stability and wetting characteristics, the thermal and mechanical characteristics during actual and prolonged operation, etc. However, the analytical knowledge is needed in the process of the optimization of the design of a new product to minimize repetitions and expensive data gathering, and to point at the strategic operating conditions which should be checked experimentally. It is hoped that the material provided in this article is sufficient enough to aid industry in developing similar computer programs for the analytically supported design of heat exchangers, which will help to decrease the loss of exergy and, therefore, the depletion rate of the finite conventional energy reserves.

NOTATION

Symbol	Definition
a	constant in Equation 59 and coefficient in Equation 94
a_b	unit vaporation surface (m^2)
A, A_1, A_2	air states in the Mollier i−x diagram
A_b, A_d, A_F, A_m	heat transfer areas of wetted and dry surfaces of fins and module (m^2)
A_{AWO}, A_{WO}	cross-sectional area for air flow and cross-sectional area occupied by water film (m^2)
b	exponent in Equation 59 and coefficient in Equation 94
B_1, B_2	boundary state of air-water interface
B_h	pulling point of completely wetted surface
B_R	pulling point of partially wetted surface
c	coefficient in Equation 94
c_{pA}, c_{PR}, c_{pW}	specific heats of air refrigerant water (J/kg-°K)
C	apparent air state in Figure 3
d	coefficient in Equation 94
D	apparent air state in Figure 3
D_h	hydraulic diameter (m)

- DI inside tube diameter (m)
- DO outside tube diameter (m)
- DOO diameter of tube with fin collar (m)
- F apparent air state in Figure 3
- FPM number of fins per unit length of tube (m^{-1})
- g gravitational acceleration (m/sec^2)
- G_A, G_R mass flux of air and of refrigerant in repective cross-sectional areas (kg/m^2-hr)
- h_A, h_R, h_W heat transfer coefficient of air, refrigerant and water (W/m^2-°K)
- h_∞ heat transfer coefficient inside a tube L/D > 60 ($Wm^{-2}K^{-1}$)
- H apparent air state in Figure 3 and heat exchanger heights shown in Figure 6 and 7
- i_A, i_W, i_i specific enthalpies of air, water, and ice (kJ/kg)
- i_{ig}, i_{fs}, latent heat of evaporation and fusion (kJ/kg)
- j_A Colburn j-factor for air; dimensionless
- k_A, k_W, k_R thermal conductivity of air, water and refrigerant (W/m-°K)
- K_1, K_2, K_3 coefficients of Equation 96
- L tube length and heat exchanger length (m)
- m mass of a mixture (kg)
- m_1, m_2 mass of component 1 and 2 in a mixture (kg)
- $\dot{m}_A$, $\dot{m}_W$ mass flow rates of air and water (kg/sec)
- M apparent air state in Figure 3
- M_A, M_W molecular weight of air and water (kg°Kmol)
- n number of parallel channels of the heat exchanger
- N apparent air state in Figure 3
- p_A, p_W partial pressure of air and water (Nm^{-2})
- p_s, p_{gb} saturation vapor pressure and partial pressure of water vapor at water-air outer (Nm^{-2})
- P total pressure of moist air (N/m^2) and apparent air state in Figure 3
- q_{bW}, q_d, q_h, q_W, q_σ diagram values for heat transfer at wetted and dry surfaces, dry heat transfer, water heating/cooling and wet heat transfer (kJ/kg)
- q_c condenser heat rejection rates (W)
- $q_{c,m}$, $q_{c,p}$, q_{cd}, q_{cw} measured and predicted condenser load, (W) and dry and wet mode condenser heat rejection rates (W)
- $q_{h,m}$ measured power input to boiler (W)
- SF fin spacing (m)
- SL longitudinal tube spacing (m)
- ST transverse tube spacing (m)
- t_A, t_W, t_c temperature of air water and condensate (°C)
- t_{amb} ambient temperature (°C)
- t_{gA}, t_{gW} temperature of air-water and water air interface (°C)
- t_c^* lowest possible condensing temperature (°C)
- $t_{c,m}$, $t_{c,p}$ measured and predicted condensing temperature (°C)
- T temperature (°K)
- TF fin thickness (m)
- TT tube wall thickness (m)
- U overall heat transfer coefficient ($Wm^{-2\circ}K^{-1}$)
- v velocity, m/s
- V_W volume of water in a system (m^3)
- W heat exchanger width shown in Figures 6 and 7
- x humidity ratio ($kgkg^{-1}$)
- x_A, x_b humidity ratio for moist air and at saturation at water state B
- y interval fraction (Figure 4)

Greek Symbols

- Γ flow rate per unit bredth (kg/m-hr)
- Δ differences of parameters
- δ_W water film thickness (m)
- η_t mean fin efficiency
- η_0 mean temperature effectiveness
- μ dynamic viscosity ($N s/m^2$)

ν	kinematic viscosity (m^2/sec)	τ	time (sec)
ρ	density (kg/m^3)	φ	air relative humidity (%)
σ	mass transfer coefficient (kg/m^2-hr)	ψ	air relative humidity (%)

Subscripts

A	air	m	surface of a module
b	air-water interface	mu	makeup
B	bare	O	for outlet conditions
C	condenser surface	R	refrigerant
d	dry surface	w	wetted surface
F	fin and fluid in a cooler	W	water
g	gas, vapor	W_b	water interfaces
i	for inlet conditions	W_e	water evaporating

Dimensionless Numbers

Nu	Nusselt number	Re	Reynolds number
Pr	Prandtl number		

REFERENCES

1. Merkel, F., "Verdunstungskühlung," *Forschungsheft 275, VDI,* Berlin 1925.
2. *ASHRAE Handbook of Fundamentals,* New York, 1977.
3. Mizushina, T., "Design of Cooler Condensers and Evaporative Coolers," in *Heat Exchanger Design and Theory Soucebook,* N. Afgan and E. U. Schlünder (Eds.), Hemisphere, Washington, DC, 1974.
4. Yadigaroglu, G., and Pastor, E. J., "An Investigation of the Accuracy of the Merkel Equation for Evaporative Cooling Tower Calculations," AIAA/ASME *Thermophysical and Heat Transfer Conference,* Boston, Massachusetts, July 1974.
5. Sutherland, J. W., "Analysis of Mechanical-Draught Counterflow Air/Water Cooling Towers" *Transactions of the ASME,* Vol. 105, August. 1983.
6. Mahumdar, A. K., Singal, A. K., and Spalding, D. B., "Numerical Modeling of Wet Cooling Towers: Part I—Mathematical and Physical Models, *Transactions of the ASME,* Vol. 105, Nov. 1983.
7. Bošnjaković, F., *Technical Thermodynamics,* Holt, Rinehart and Winston 1965 (Translation of *Technische Thermodynamik,* Part II. Theodor Steinkopff, Dresden und Leipzig, 1951).
8. Leidenfrost, W., and Korenic, B., "Analysis of Evaporative Cooling and Enhancement of Condenser Efficiency and of Coefficient of Performance," *Wärme- und Stoffübertragung* 12 (1979), pp. 5–23.
9. Leidenfrost, W., Lee, K. H., and Korenic, B., "Conservation of Energy Estimted by Second Law Analysis of a Power-Consuming Process," *Energy,* Vol. 5 (1980), 47–61.
10. Leidenfrost, W., and Korenic, B., "Evaporative Cooling and Heat Transfer Augmentation Related to Reduced Condenser Temperatures," *Heat Transfer Engineering,* Vol. 3 (1982), pp. 38–59.
11. Leidenfrost, W., and Korenic, B., "Experimentelle Überprufung einer Berechnungsmethode für die Leistungsvoraussage verdunstungsgekühlter Kondensatoren," *Brennstoff–Wärme–Kraft,* Vol. 34 (1982) pp. 9–14.
12. Fisher, U., Leidenfrost, W., and Li, J., "Hybrid Evaporative Condenser Cooling Tower," *Heat Transfer Engineering,* Vol. 4 (1983) pp. 28–41.

13. Carrier, W. H., "Rational Psychrometric Formulaes," *Transaction ASME,* (1911), pg. 1005–1053.
14. Mollier, R., Das i–x Diagramm für *Dampfluftgemische,*" *ZVDI* (1929).
15. Busemann, A., "Wärme- und Stoffaustausch dargestellt im Mollierschen Zustandsdiagramm für Zweistoffgemische," *VDI* Berlin (1933).
16. Korenic, B., "Augmentation of Heat Transfer by Evaporative Cooling *to Reduce Condensing Temperature,*" Ph. D. thesis, School of Mechanical Engineering, Purdue University, West Lafayette, Indiana, December 1980.
17. Kays, W., and London, A. L., *Compact Heat Exchangers,* 2nd ed., McGraw–Hill, New York, 1964.
18. McAdams, W. H., *Heat Transmission,* 3rd ed., McGraw–Hill, New York, 1954.
19. Sherwood, T. K., Pigford, R. L., Wilke, C. R., *Mass Transfer,* McGraw–Hill Chemical Enginering Series, New York, 1975.
20. Veniamin, L. G., *Physicochemical Hydrodynamics* Prentice-Hall, Inc., Englewood Cliffs, N. J., 1962.
21. Ceiro, R. L., and Whitaker, S., "Entrance Region Flows with a Free Surface: The Falling Liquid Film," *Chem. Eng. Sci.,* Vol. 26 (1971) p. 785.
22. Mizushina, T., Ito, R., and Miyashita, H., "Experimental Study of an Evaporative Cooler," *Int. Chem. Eng.,* Vol. 7 (1967) p. 727, 1967. (Translated from Kajaku-Kogaku, Vol. 31, (1967), p. 469.)
23. Mizushina, T., Ito, Z., and Miyaskita, H., "Characteristics of Method of Thermal Design of Evaporative Coolers." *Int. Chem. Eng.* Vol. 8 (1968), pp. 55, 532.
24. Parker, R. O., and Trybal, R. E., "The Heat Mass Transfer Characteristics of Evaporative Coolers," *Chem. Eng. Progr. Symp. Ser. Heat Transfer,* Buffalo, Vol. 57, No. 32 (1961), pp. 138–149.
25. Chapman, J. K., *Heat Transfer,* MacMillan Publishing Co., New York, 1974.
26. Hausen, H., *ZVDI* Beih. Verfahrenstechnik Vol. 4 (1943) p. 91.
27. Sieder, E. N., and Tate, G. E., *Ind. Eng. Chem.,* Vol. 28 (1936) p. 429.
28. Dittus, F. W., and Boelter, L. M. K., Univ. Calif. Berkeley, Publ. Eng. Vol. 2 (1930), p. 443.
29. Nusselt, W., "Die Oberflächenkondensation des Wasserdampfes," *Z. Ver. Deut. Ing.,* Vol. 60 (1916) p. 541.
30. Mizushina, T., "Design of Cooler Condensers and Evaporative Coolers," in *Heat Exchangers: Design and Theory Sourcebook,* N. Afgan and E. U. Schlunder (Eds.), MacGraw–Hill, New York, 1974, pp. 419–440.
31. Mizushina, T., Ito, R., and Miyashita, H., "Experimental Study of an Evaporative Cooler," *Int. Chem. Eng.,* Vol. 7 (1967), p. 727, (Translated from *Kagaku-Kogaku,* Vol. 32 (1968), p. 55.)
32. Mizushina, T., Ho, R., and Miyashita, H., "Characteristics and Methods of Thermal Design of Evaporative Coolers," *Int. Chem. Eng.,* Vol. 8 (1968) p. 532 (Translated from *Kagaku-Kogaku,* Vol. 32 (1968) p. 56.)
33. Pescod, D., *"A Heat Exchanger for Energy Serving in an Air-Conditioning Plant," Technical paper No. 2542, ASHRAE* Annual Meeting, Detroit, 1975.
34. Elmahdy, A. H., and Mitalas, G. P., "A Simple Model for Cooling and Dehumidifying Coils for Use in Calculating Energy Requirements for Buildings," *ASHRAE Transactions,* Vol. 83, Part 2, 1977.
35. Elmahdy, A. H., and Biggs, R. C., "Efficiency of Extended Surfaces with Simultaneous Heat and Mass Transfer," *ASHRAE technical paper No. 2744, ASHRAE Semiannual Meeting,* Atlantic City, Jan. 1983.
36. Ketteborough, C. F., and Hsieh, C. S., "The Thermal Performance of the Wet Surface Plastic Heat Exchanger Used as an Indirect Evaporative Cooler," ASME Paper No. 81–WA/Sol–11, 1981.
37. Maclaine-cross, I. L., and Banks, P. J., "A General Theory of Wet Surface Heat Exchangers and its Application to Regenerative Evaporative Cooling," *Journal of Heat Transfer,* Vol. 103 (August 1981), pp. 579–585.

38. Yang, Wen-jei, and Clar, D. W., "Spray Cooling of Air Cooled Compact Heat Exchangers," *Int. J. Heat Mass Transfer,* Vol. 18 (1975), pp. 311–317.
39. Kreid, D. K., Johnson, D. M., and Faletti, D. W., "Approximate Analysis of Evaporative Heat Transfer from a Finned Surface," *ASME Paper No. 78:HT–26,* AIAA/ASME Joint Thermodynamics and Heat Transfer Conference, Palo Alto, California, May 24–26, 1978.
40. Kreid, D. K., et al., "Performance of a Plate Fin Air-Cooled Heat Exchanger with Deluged Water Augmentation," *ASME Paper No. 79–WA/Ener.–4, 1979.*

CHAPTER 34

COOLING TOWER OPERATIONS

Nicholas P. Cheremisinoff

Exxon Chemical Company
Linden, New Jersey,

CONTENTS

INTRODUCTION

This chapter provides a review of standard design practices and operating guidelines for water cooling tower operations in industrial applications. Cooling towers are classified according to the method by which air is introduced to the tower, namely, atmospheric spray, natural-draft, mechanical-draft, deck-filled, spray-filled, coil shed, and hyperbolic towers.

In an atmospheric spray tower air movement depends on atmospheric conditions and the atomizing characteristics of the spray nozzles. Natural-draft cooling tower operations employ a chimney or stack to induce air movement. Mechanical-draft cooling towers utilize fans to move ambient air through the tower. Deck-filled towers contain tiers of splash bars or decks to assist in the breakup of water drops to increase the total water surface and subsequently the evaporation rate. Spray-filled towers depend only on spray nozzles for greater or larger interfacial areas for evaporation. Coil shed towers are composed of a combination structure of a cooling tower installed on top of a substructure that contains atmospheric section coils. Hyperbolic natural-draft cooling towers are typically large-capacity systems.

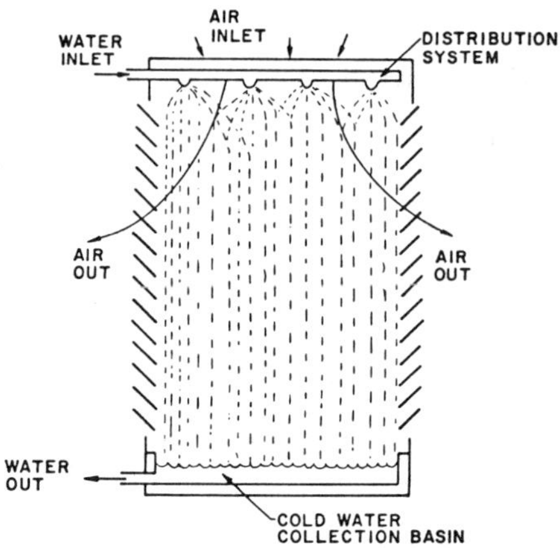

Figure 1. Atmospheric spray tower.

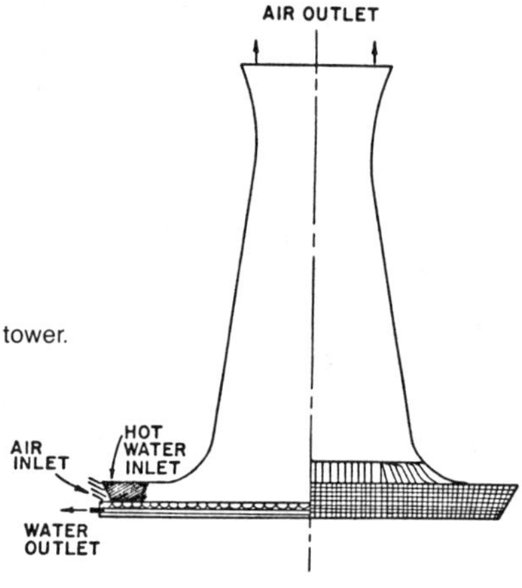

Figure 2. Hyperbolic natural-draft tower.

TOWER CONFIGURATIONS

In addition to the general class of draft, natural draft (natural convection), mechanical draft (forced convection), and mechanical and natural, distinction is made based on

1. The type of flow-crossflow, counterflow, cocurrent flow.
2. The type of heat dissipation—wet (evaporative cooling), dry, wet-dry.
3. The type of application-industrial or power plant.

Classifications of Cooling Towers

Details of the major configurations are shown in Figures 1 through 6 and a brief description of each type follows.

Atmospheric Spray Towers—Air movement depends on atmospheric conditions and the aspirating effects of the spray nozzles (refer to Figure 1).

Natural-Draft Cooling Towers—These depend on a chimney or stack to induce air movement through the tower (see Figure 2).

Mechanical-Draft Cooling Towers—These designs use a fan to move ambient air through the tower. They are subdivided into two different types—indirect and forced air towers, depending on whether the air is drawn or forced through the tower. At the same time, they

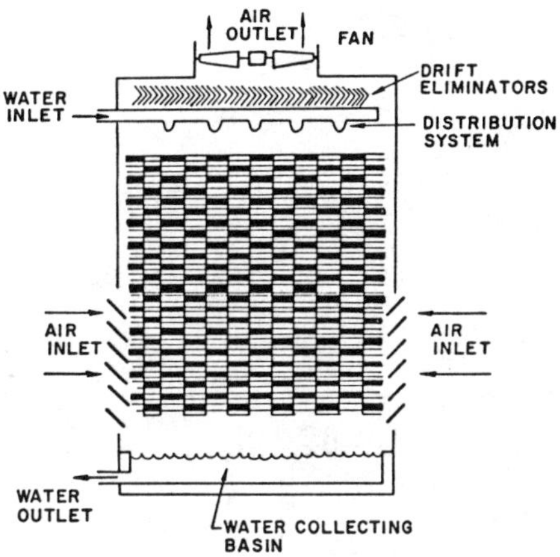

Figure 3. Counterflow tower.

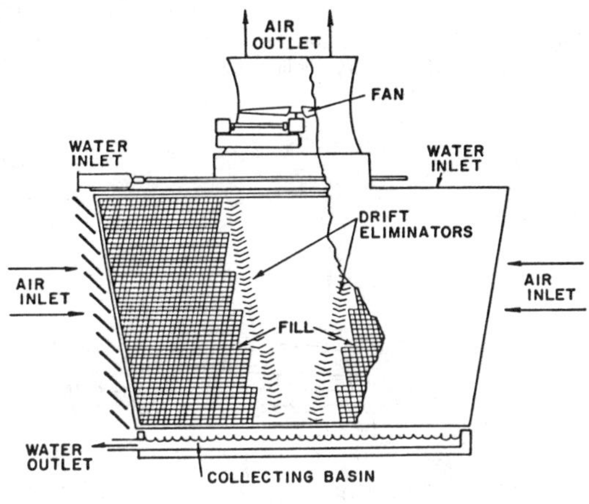

Figure 4. Crossflow tower.

can be grouped in crossflow or counterflow types, depending on the relative movement of air and water. In the crossflow type, air generally travels horizontally across the falling water, while in counterflow, it travels vertically upward through the falling water (see Figures 3 and 4). A more recent crossflow cooling tower is the wet-dry tower, which consists of a normal crossflow tower over which a few air coils are placed. The hot water is first cooled by an air-cooled heat exchanger and then drops to the wet cooling tower where more cooling is obtained by evaporation (refer to Figure 5).

Deck-Filled Towers—These contain tiers of splash bars or decks which assist in droplet breakup.

Spray-Filled Towers—These depend only on spray nozzles for water breakup.

Coil Shed Towers—These are composed of a combination structure of a cooling tower installed on top of a substructure that contains atmospheric section coils (see Figure 6).

Hyperbolic Towers—Natural-draft cooling towers with a hyperbolic configuration are usually constructed of concrete and can handle large capacities. They are generally used in power plants. Figure 2 illustrates this design.

Natural-Draft Towers—Figure 7 shows a cross section of a simple cooling tower formed by enclosing a spray pond with louvered walls. In this design, air changes direction in

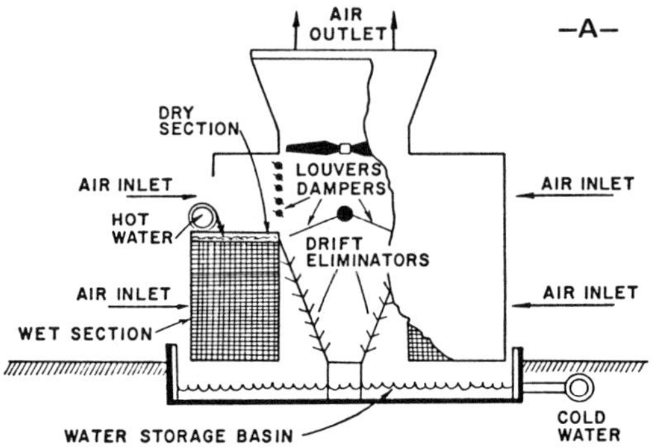

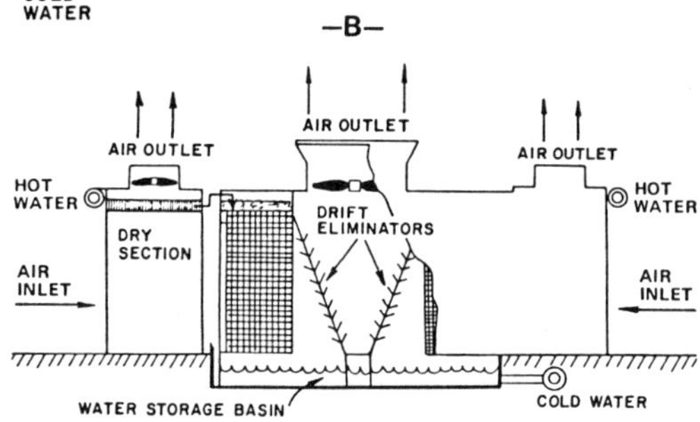

Figure 5. (A) Single structure-type wet/dry cooling tower; (B) wet/dry cooling tower.

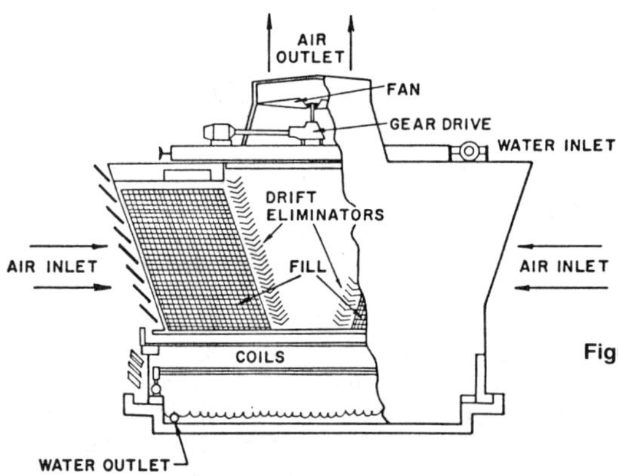

Figure 6. Coil shed cooling tower.

passing through the louvers as it leaves, and suspended water droplets impinge on the louver slats, deposit outward, and drain back into the tower basin.

This design is improved by utilizing low-pressure water sprays at the top, positioning horizontal decks in the tower, and arranging the walls so that the air enters horizontally and is discharged vertically (Figure 3). The vertical upward movement of air tends to slow the downward velocity of the water droplets, thus increasing the effective surface area for any given water loading. The fill material is usually staggered so that water droplets fall through a distance of only a few feet before striking another surface. Typical grids are constructed of redwood strips nailed to 1 × 2-inch stringers. Figure 8 shows some typical fill geometries.

Figure 7. Cross-section of a simple cooling tower formed by enclosing a spray pond with louvered walls.

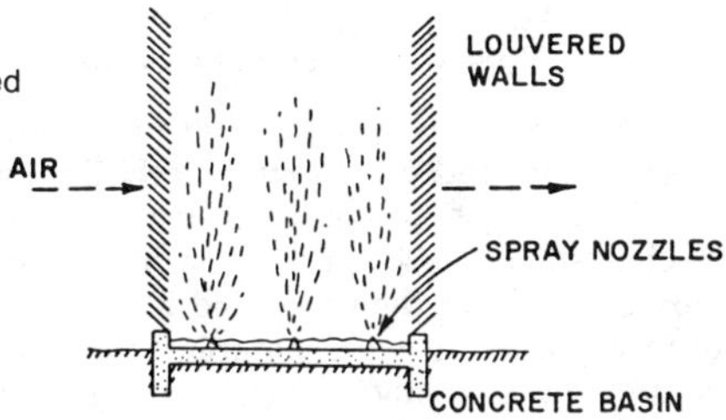

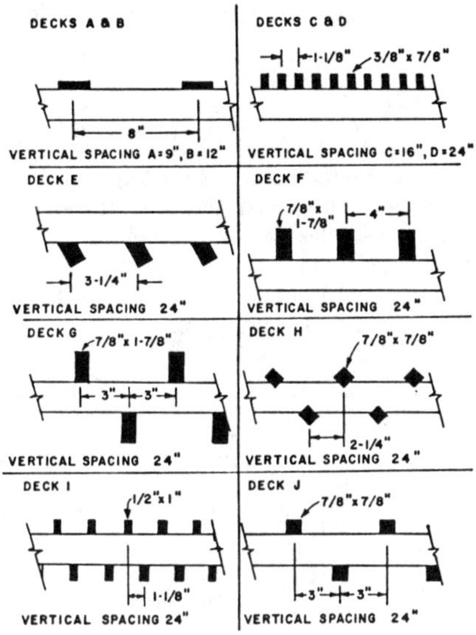

Figure 8. Redwood fill arrangements for cooling towers.

The increasing temperature of the upwardly moving air stream induces circulation by thermal convection. This is a favorable condition if the wind velocity is very low. Towers are generally placed in a side-by-side arrangement (i.e., in a row) at right angles to the direction of prevailing winds.

Hyperbolic Towers—The advantages of this configuration include:

1. Superior strength.
2. Close match to the natural flow of air through the tower shell.
3. The need for fewer materials of construction than for earlier shapes (less total volume is required because the hyperbolic shape of the shell provides good strength, so a thinner shell thickness can be used in comparison to older designs).

The performance of a natural-draft tower is characterized by a duty coefficient (C_D), which defines the overall capabilities of a tower under all operating conditions:

$$C_D = \frac{W_L}{90.59 \dfrac{\Delta Q}{\Delta T_W} \sqrt{\Delta T_a + 0.3124 \Delta Q}} \quad (1)$$

where W_L = water loading (lb/hr)
ΔQ = total heat of air through tower (Btu/lb dry air)

ΔT_W = change in temperature of water (°F)
ΔT_a = difference between dry-bulb and wet-bulb temperatures

Increases in loading, cooling range, and humidity improve cooling tower performance. Counterflow and crossflow hyperbolic towers are illustrated in Figure 9.

Of the two flow configurations, counterflow provides the more efficient heat transfer mechanism because the coolest water contacts the coolest air initially. In crossflow, air moves normal to the water and fill. These designs have fill in a ring outside the tower. This produces a lower water pumping heat than in counterflow. Fill inside a counterflow must be spread over a much larger area. Thus, crossflow units have shallower depths and vertical water risers have shorter pressure drops through the fill (much less than in a counterflow tower).

Mechanical-Draft Cooling Towers

Mechanical-draft cooling towers can be either field-constructed or factory-assembled. Installations that are constructed onsite are generally those employed at utility stations.

The growing size and numbers of central stations and new codes governing water qualities as well as dwindling water supplies have made once-through cooling increasingly difficult. Today's economics favor mechanical-draft towers over hyperbolics, except in very large applications. Figure 10 illustrates the various mechanical-draft tower types.

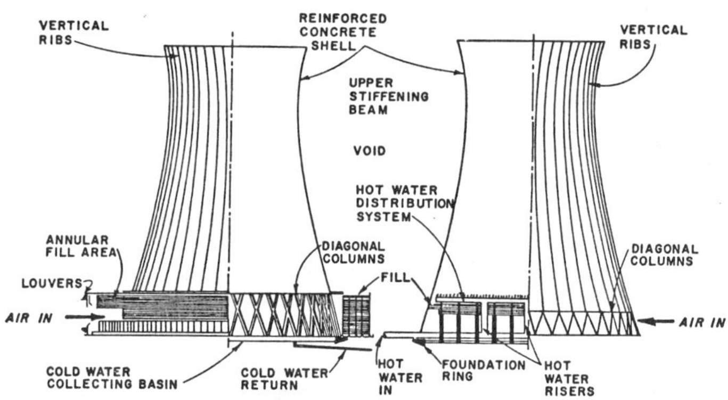

Figure 9. Crossflow and counterflow hyperbolic cooling towers.

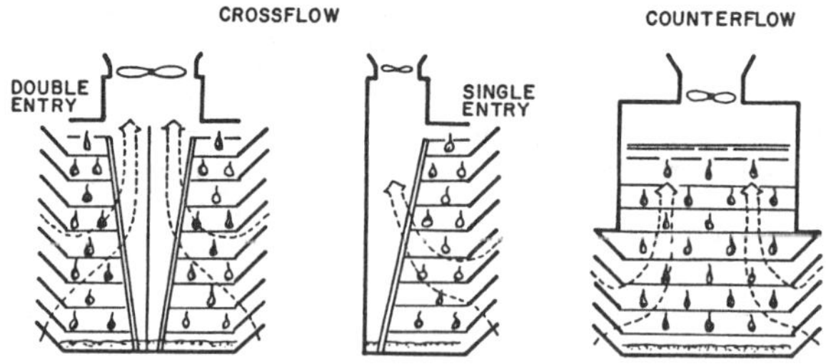

Figure 10. Large mechanical-draft cooling towers.

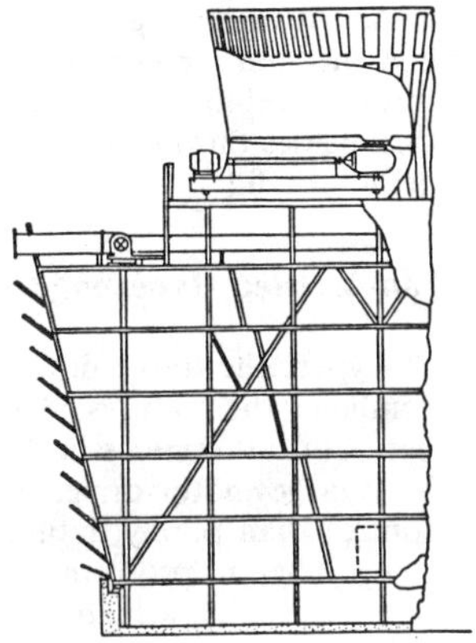

Figure 11. Design elements of a mechanical draft cooling tower.

Crossflow mechanical draft towers are thermally less efficient. They offer less resistance to air flow and can operate at higher velocities than counter-flow towers, which means less horesepower and smaller cell sizes for comparable outputs. In practice, however, both crossflow and counterflow designs have advantages and limitations, depending on the type of application.

A cross-section of a mechanical draft tower is shown in Figure 11. The primary design elements numbered in this figure are as follows:

1. *Electric motors*. These should be matched to power requirements of the fan.
2. *Fan blades*. These are fabricated of lightweight metal or plastics and are balanced.
3. *Speed reducer*. It is usually integrated with other elements, namely, motor, shaft and fan. Its job is to convert the high-rpm, horizontal rotation of the drive shaft into the vertical, low-rpm motion of the fan shaft.
4. *Header pipe*. This is usually made of redwood staves for very large towers.
5. *Interior supports*. These are an intricate cross-lacing of redwood beams arranged to give maximum support to the fill and fan deck. Glass-reinforced polyester connectors are positioned where diagonals cross upright. This permits horizontal members an even distribution of the load and a certain amount of expansion.
6. *Fill*. These are made from a number of materials; usually in a splash-type arrangement or wave-shaped bars supported in a close-knit grid.
7. *Louvers and casing*. These are constructed from asbestos-cement board, about 3/8-inch thick, in large towers. To eliminate splash out, louvers are sloped at an angle of about 40° with framework.
8. *Cold water basin*. This serves as a receptacle for cooled water and is usually made of concrete.

Factory-Assembled Towers

Elements of a factory-assembled tower are basically the same as field-erected towers. The main difference is that in factory-assembled units, the design is modular in concept, with units shipped as modules and arranged on site. Manufacturers offer considerable flexibility to permit best use of available space for their equipment.

Air paths may be crossflow or counterflow. Fan placement can be ahead of the fill section (forced draft) or behind it (induced draft). Manufacturers use different criteria in sizing units. Figure 12 presents one tower manufacturer's approach. Specific design details should be obtained from the tower manufacturer.

As shown in Figure 12, ambient air passes through the inlet cone of the centrifugal fan and through air vortex breakers to boost fan efficiency. During fan discharge, air passes through a diffuser, which equalizes flow and provides optimum static pressure regain. Air then passes through a series of serrated directional vanes, which ensure even air distribution below the fill.

Fan-Assisted Hyperbolic Towers

Hyperbolic tower designs also include fans for intermittent operation. This type tower combines the features of both natural-draft and mechanical-draft towers. These resemble natural-draft systems in that they employ a hyperbolic shell; however, they are smaller and have large motor-driven fans located at the periphery of the base. Fans provide greater control than purely natural-draft systems.

An inherent problem with mechanical-draft cooling towers is the potential for recirculation. Normally, a high air velocity discharge provided by fans will eliminate this problem. Table 1 compares mechanical and natural draft cooling towers.

Other Tower Configurations

A recent design developed by Baltimore Aircoil Co., Maryland, is shown in Figure 13. Large quantities of air are induced into the tower by cooling water as it is injected through spray nozzles at one end of a venturi plume. No fans are utilized in this design. Effective mixing of air and water in the plenum permits evaporative heat transfer to take place without the fill that is required in conventional towers. Cooling water collected in the sump

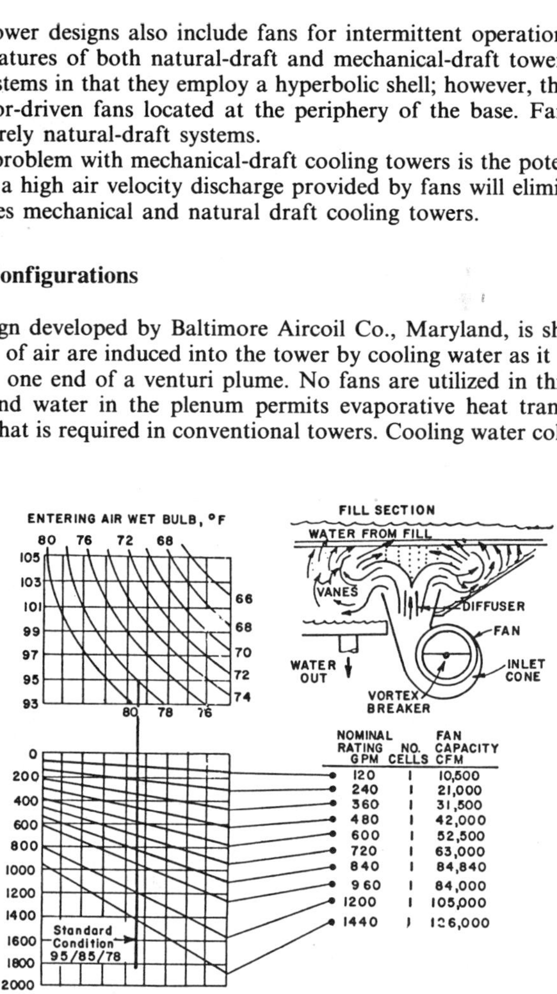

Figure 12. Upper right figure shows a forced-draft or blowthrough tower, which has a fan at the bottom for driving air through the fill above. Tower selection for smaller units can be made from the accompanying curves and table for a cold water temperature of 85°F (this is generally the water basin discharge temperature for small towers). As an example, enter at 104°F hot water temperature to a wet-bulb value of 75°F, then drop vertically to the water flow selected (580 gpm). This falls between curves that designate the manufacturer's distinct model size. Select the next larger size, i.e., the curve immediately below, and follow across to the recommended tower model.

is pumped through the cooling water circuit for the return cycle. Capacities generally range from 30 to 1,920 gpm for standard conditions. Since the tower has no moving parts, maintenance costs are low.

Under certain conditions, such as high water temperatures, insufficient water supplies and problems of blowdown disposal, systems that depend on convection and use air as the transport medium may be preferable. Two types of dry cooling towers are the direct and indirect systems. Figures 14 and 15 show these systems in operation for nuclear power plant cooling. Indirect units use a surface or jet condenser at the turbine to condense exhaust steam. Water from the condenser is pumped to the dry tower for cooling and recirculation

Table 1
Comparison Between Characteristics of Mechanical- and Natural-Draft Cooling Towers

Considerations	Mechanical Draft	Natural Draft
Location	Must be located in an area at some distance from the plant proper. This is necessary for air supply considerations and because of problems associated with fogging and drift from discharging air. Noise considerations also tend to dictate site selection.	Can be built adjacent to plant buildings, on centers 1.5 (d) where d = base diameter, so that piping costs are reduced. Their position must be chosen so as not to interfere with power plant exhaust plumes.
Materials of Construction	May be constructed of wood, metal or concrete. The structure must be designed to withstand wind or earthquake stress, dead loads such as weight of tower and circulating water and vibrations from mechanical equipment.	Constructed of thin concrete shells that have good wind resistance.
Initial Investment	Can be built with less expensive materials like wood, asbestos-cement board and plastic materials. Fan cost is higher.	Built with relatively expensive materials such as prestressed, precast and reinforced concrete and asbestos-cement for fill.
Operation & Maintenance Costs	Pumping head is less, so power cost for the circulating water pumps is less. Power cost for fans is considerable. Cost of maintaining fans and associated drives and transmissions is also significant.	Total operating cost will favor natural-draft towers.
Recirculation, Fogging	These are major problems. Design accommodation, restrictions on tower dimensions, orientation with prevailing winds, and added capacity for recirculation can boost tower cost.	Because of its elevated discharge, the natural-draft tower rarely has the trouble with recirculation and fogging.
Applications	Economics favor mechanical-draft towers over hyperbolics, except in very large installations.	Considering the climatic and load conditions, natural draft has its best application in the power industry. Selected when: (a) operating conditions consist of low wet-bulb temperature and high relative humidity; (b) a combination of low wet-bulb and high inlet and exit water temperature exists; and (c) heavy winter load is possible.

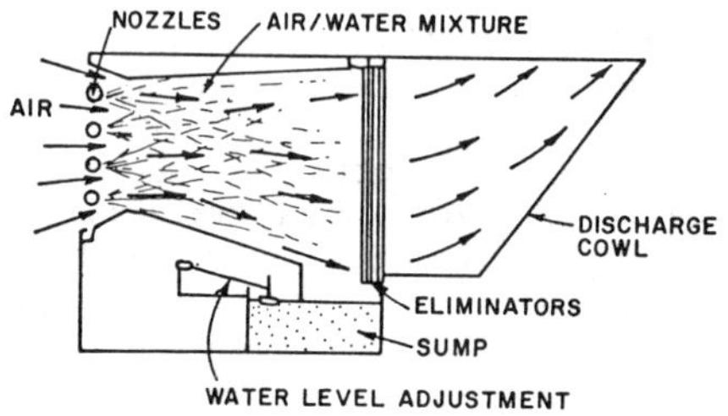

Figure 13. Cooling tower design developed by the Baltimore Aircoil Co.

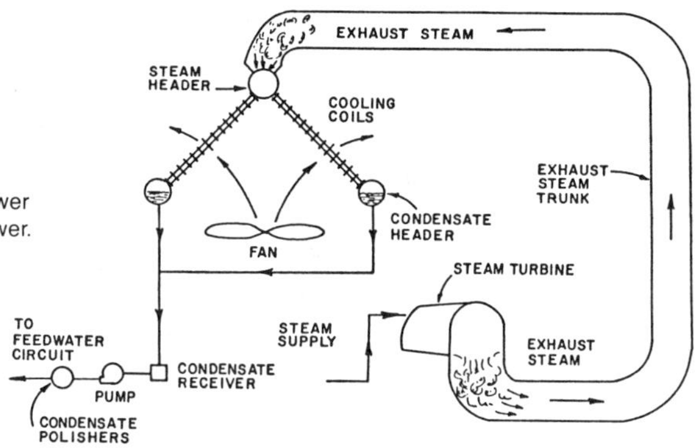

Figure 14. Direct, dry-type cooling tower condenser using a mechanical-draft tower.

back to the condenser. In the direct system, steam is condensed in cooling coils without interfacing with a condenser.

In the direct condensing system, a jet condenser is employed. Cool water from the dry tower is sprayed into the condenser, where it mixes with exhaust steam from the turbine. The mixture falls to the bottom of the condenser and is removed by circulating water pumps. A portion of this water flows to cooling coils, and an amount equal to the exhaust steam from the turbine is directed back to the feedwater circuit through condensate polishers for reevaporation in the steam generator. Because of the high degree of contacting when using a jet condenser, circulating water must be of condensate purity.

After passing through a water recovery turbine, circulating water is again introduced into the condenser and recycled through the system. The recovery turbine is coupled to the drive shaft of the circulating water pump to recover some of the pressure head between cooling coils and the condenser. Circulating water does not come into direct contact with cooling air; thus, no evaporative losses occur, as with the wet-type tower.

In direct condensing, turbine exhaust steam is conveyed through a trunk to the air-cooled coils, where cooling air passing over the finned coil surfaces condenses the steam. Steam enters the coil section and condenses as it travels downward, with steam and condensate flowing in the same direction, minimizing pressure loss and increasing heat transfer.

These designs have provisions for the removal of noncondensable vapors and air, for the prevention of freezing during cold weather. Excessive buildup of noncondensable vapors in the main condenser would prevent effective condensation. Protection agains ice formation is usually accomplished by warm air recirculation and/or fan control. Condensed steam from cooling coils flows by gravity to condensate receivers and is pumped back to the feedwater circuit by a condensate pump.

The major difference between the two systems is the large volume of exhaust steam that must be handled in the direct system, as compared to the smaller volume of circulating water

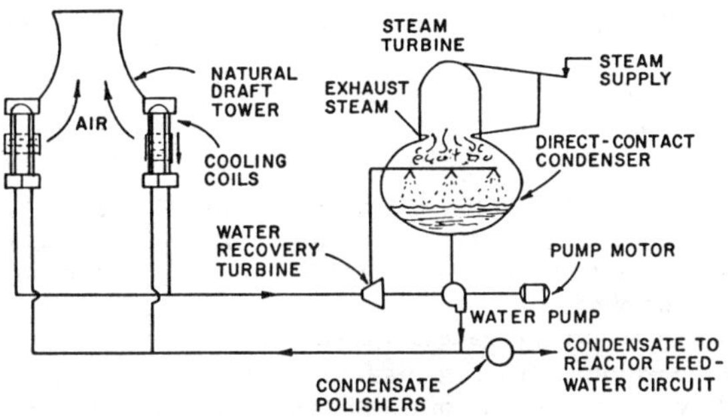

Figure 15. Indirect, dry-type cooling tower condensing system using a natural-draft tower.

in the indirect system. Cooling coils in the dry system operate under high vacuum, which is perhaps a disadvantage when compared to an indirect system that maintains positive water pressure in the cooling coils. Because of size requirement for steam pipes running from the turbine to the condenser, direct systems are limited to small- and intermediate-sized plants, while indirect systems are expected to be more economical and technically feasible for large plants.

The use of dry-closed water loop cooling systems eliminates the need for large water consumption in power plant cooling. It represents an attractive solution to many of the environmental problems associated with wet tower operations. It is, however, inherently costly. The low heat capacity of air and the low heat transfer coefficients of air-cooled heat exchangers make it necessary to have large air volumes, large surface areas and, consequently, large, costly towers. Mechanical-draft systems requiring large air volumes can result in high fan power requirements, which, in comparison to wet towers, result in significant replacement energy costs. Note also that since dry-bulb temperatures are greater than wet-bulb temperatures, dry cooling systems are forced to reject heat to a higher temperature sink. This results in higher turbine exhaust pressures and higher energy debits in comparison to wet cooling systems.

Recent developments promise significant cost savings over conventional dry cooling systems. McHale et al. [1] describe a proposed system that employs ammonia as the heat rejection fluid rather than water and uses high performance heat exchangers to further reduce costs. For the operation to be economical, the steam condenser ammonia exit quality must be maintained at or below 0.80 to ensure that the ammonia evaporator tubes are wetted at all times. At the separator, the ammonia vapor and liquid exiting the condenser are separated. The vapor is sent to the cooling tower, where it is condensed, and the liquid phase is returned to the steam condenser. Fans are used on the dry tower(s) to control the heat rejected as the system heat load varies. This control is provided by turning off cooling tower fans whenever the turbine exhaust pressure falls below the choke point. The reader is referred to the literature [1–5] for detailed discussions.

DESIGN BASICS

Cooling towers, along with air conditioning spray chambers, spray driers, spray towers, and spray ponds, are the largest users of applied diffusional heat transfer principles. The cooling water's available temperature is a major consideration, having direct impact on the economics of the design. This is especially true today for the chemical and power industries. Chemical plants establish their cooling water temperatures on the operating pressures on the condensers of distillation and evaporation operations and, consequently, on equipment preceding them. Power plants base it on turbine- or engine-discharge pressures and the ultimate heat recovery. To ensure that the proper water temperature is made available, careful analysis of process requirements, as well as careful review of cooling tower design principles, must be made.

Gas-Liquid Contacting

The available interfacial area for gas-liquid contacting establishes the thermal efficiency of the system. Consequently, the greater this interfacial contact area, the greater the degree of heat exchange. In spray ponds and spray towers this contact area is promoted by utilizing spray nozzles to generate small liquid droplets.

In addition to interfacial area, the time of contact between the two phases affects thermal efficiency. This is best illustrated by the example given in Figure 16. Consider a column Z feet tall and which is gravity fed with water in the form of droplets at a rate of 1 drop/sec. If there is no initial velocity imparted to the drops, then each drop descends through the column in accordance of free-fall:

$$\Delta Z = \frac{1}{2} g\theta^2 \tag{2}$$

where θ is the time it takes to fall through the column. If the column is 32 feet tall, then it takes $\sqrt{2}$ sec for a single drop to pass through. At the feed rate of 1 drop/sec then, there will always be 1.41 drops present in the column. At the discharge, 1 drop/sec will be removed continuously. Consequently, the interfacial or effective area in the column is that of $\sqrt{2}$ droplets.

By installing baffles or plates on which droplets can impinge or be deflected, an increase in the time it takes for each droplet to pass through the column is experienced. By interrupting the flow then, we might increase the droplet holdup time to 5 seconds. Then, as shown in Figure 16, although 1 drop/sec is fed to the column and 1 drop/sec discharges, there will always be 5 drops of water at any one time in the column for a continuous flow system. We thus have $5/\sqrt{2}$, or 3.5 times more effective surface than for the case of free-fall.

In gas absorption equipment, such as packed tower wet scrubbers, the interfacial gas-liquid contact area and the contact time are established in the form of a film surface. This film surface is generated by the exposed surface of the packing. The role of packing is to increase the turbulence and the contact surface between the two phases. It is also designed to provide a tortuous path to prevent bypassing of the phases.

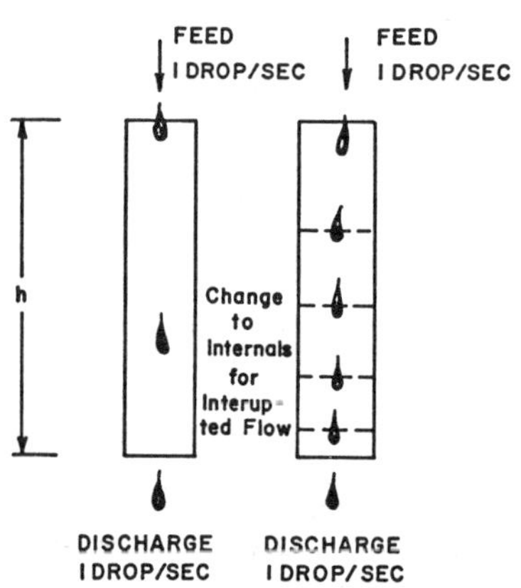

Figure 16. Free and interrupted flow through a column.

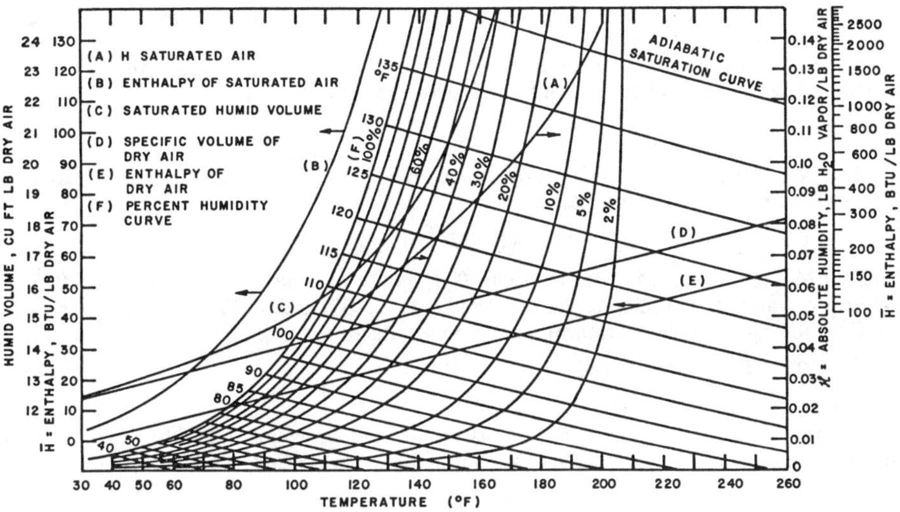

Figure 17. Psychrometric chart for the air-water system at 1 atm [6].

Cooling towers have different design restrictions than absorption columns. Because of large air volume and small allowable pressure drop requirements, common cooling tower designs employ wooden slats of either triangular or rectangular cross section at spaced intervals within the structure. This type of packing arrangement leaves the tower substantially unobstructed. The purpose of fill is to interrupt the downward flow of water. The free space between adjacent fill slats is relatively large; however, the horizontal projection of the fill is designed so that water droplets cannot fall through the tower without repeatedly striking successively lower slats. As the liquid descends through the tower it impinges on the top of each successive slat. This splits the flow, breaking it into turbulent flow as it continues to fall. This action forms new water droplets with each successive slat impinged on, thus exposing new water surface area for contact with the air. Cooling tower fill provides the interfacial contact by both film surface (as the liquid flows around the sides of the slats) and droplet surface. As the water strikes the top of the slats, droplets are created; however, more imporant, the draining action of falling water into turbulent flow creates droplets. Fill greatly increases the holdup time of liquid in the tower. Properly arranged fill will never allow liquid droplets to reach their free-fall velocity. Recall our simple example of water droplets falling freely in a column.

It is important that no bypassing of the two phases occurs in the tower. A uniform liquid distribution across the tower's cross section is essential. This is accomplished by spraying the water over the top of the tower. Spray nozzles also help to create droplet surface and increase contact time.

Psychrometric Chart and the Energy Balance

The standard psychrometric chart is shown in Figure 17. The basic chart consists of a humidity ($\mathscr{H}$)-temperature (dry-bulb) set of coordinates along with additional parameters (curves) of constant relative humidity, constant moist volume (humid volume), adiabatic cooling curves (which are the same as the wet-bulb or psychometric lines, for water vapor only) and the 100% relative humidity curve, also called the saturated-air curve. If any two values are known, we can determine the air-moisture condition on Figure 17 and evaluate all other required parameters.

The left-hand scale gives the enthalpy per pound of dry air of a saturated air-water vapor mixture. The enthalpy of the wet air (in Btu/1b dry air) is:

$$\Delta\bar{H} = \Delta\bar{H}_{air} + \Delta\bar{H}_{H_2O\,vapor}(\mathscr{H}) \tag{3}$$

There are alternate methods available for computing enthalpy values [7, 8]. Expanding Equation 3 for the air-water system:

$$\Delta H = 0.240(T-0°)+\mathscr{H}(1075+0.45(T-32°\ F)) \tag{4}$$

where $0.240(T-0)$ is $C_p(\Delta T)$ for air (with 0° F as the reference temperature, $\mathscr{H}(1{,}075)$ is the heat of vaporization of water at 32° F and $0.45(T-32)$ is $C_p(\Delta T)$ for water vapor (with 32° F as the reference temperature). Equation 4 reduces to:

$$\Delta\bar{H} = 0.240T+\mathscr{H}(1061+0.45T) \tag{5}$$

where T is in °F.

The use of these relationships in constructing and applying humidity charts is best illustrated by examining a simplified case, that of adiabatic cooling or humidification. Figure

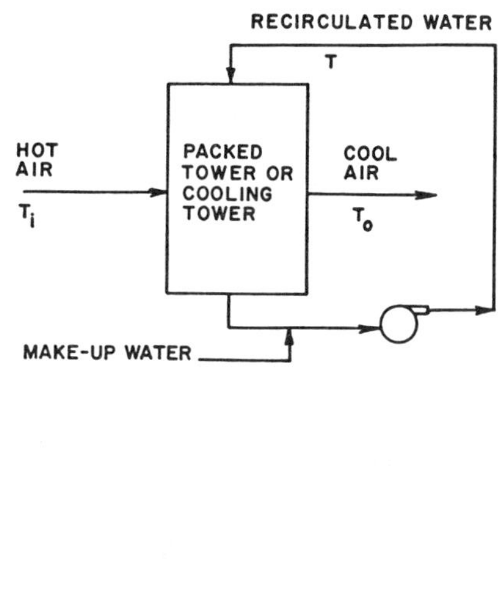

Figure 18. Example of adiabatic humidification.

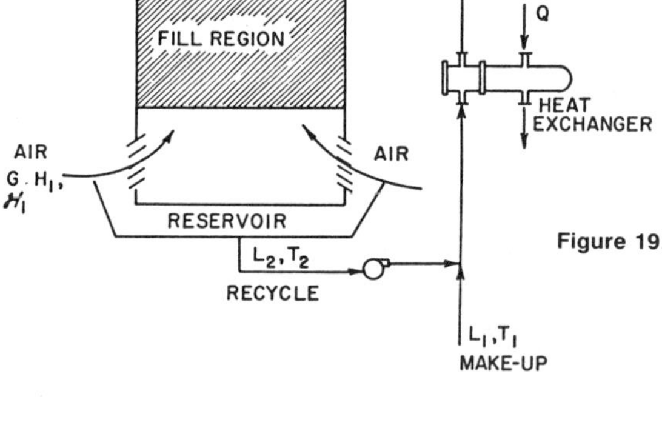

Figure 19. Countercurrent cooling tower operation.

18 illustrates this process between air and water that is recycled through the cooling tower. In this operation air is both cooled and humidified while a portion of the recirculated water undergoes evaporation. If equilibrium is reached the air becomes saturated and both the air and water temperatures are the same. If the operation occurs adiabatically, then by application of an energy balance we can develop an expression defining the adiabatic cooling line of the humidity chart:

$$\begin{pmatrix}\text{enthalpy of} \\ \text{air in}\end{pmatrix} + \begin{pmatrix}\text{enthalpy of } H_2O \text{ vapor} \\ \text{in air entering tower}\end{pmatrix} = \begin{pmatrix}\text{enthalpy of} \\ \text{exit air}\end{pmatrix} + \begin{pmatrix}\text{enthalpy of } H_2O \text{ vapor} \\ \text{in exit air}\end{pmatrix}$$

And defining T_s as the equilibrium temperature of the water,

$$0.240(T_i - T_s) + \mathscr{H}[\Delta\bar{H}_{\underset{T_s}{H_2O}} + 0.45(T_{air} - T_s)]$$
$$= 0.240(T - T_s) + \mathscr{H}_s[\Delta\bar{H}_{\underset{T_s}{H_2O}} + 0.45(T_s - T_s)] \tag{6}$$

where the subscript s refers to sautration conditions, which are actually the wet-bulb temperature (T_{wb}).

Equation 6 is rearranged to:

$$\frac{(\mathscr{H}_s - \mathscr{H})}{(T_{wb} - T_{air})} = -\frac{C_s}{\Delta\bar{H}_{vap}} \tag{7}$$

where C_s is the humid heat ($C_s = 0.240 + 0.45\mathscr{H}$).

Sample problems are given in Reference 5.

Material and energy balances directly applicable to cooling towers are now outlined. The idealized cooling tower operation is illustrated in Figure 19. The cooling tower operates with a heat source (a condenser for example) in a closed-loop arrangement. As shown, heated water along with some makeup to compensate for water losses through evaporation are sprayed over the top cross section of the tower.

It is convenient to develop energy and material balances on the basis of unit cross-sectional area of flow. The air loading is expressed as a superficial mass rate per unit cross-sectional area of flow, and thus has units of G 1b/(hr) (ft^2). The water loading at the top of the tower is expressed in the same manner as L 1b/(hr) (ft^2) of water. If Q is the heat rate (Btu/hr) through the condenser, then we can define the heat load per hr per square foot as q = Q/A, where A is the cross-sectional area of flow of the cooling tower. An energy balance about our system for air only yields the following:

$$q + L_1 C_{p_{H_2O}} T_1 = G(H_2 - H_1) \tag{8}$$

In writing Equation 8 we have used a reference temperature of 0° F for air. Note that $H_{1,2}$ is the enthalpy per pound of dry air and includes the heat of the vapor associated with a pound of dry air.

An energy balance for the water component alone gives:

$$q = LC_p(T_3 + T_2) + L_1 C_p(T_2 + T_1) \tag{9}$$

Equating both balances:

$$G(H_2 - H_1) = LC_p(T_3 - T_2) + L_1 C_p T_2 \tag{10}$$

where

$$L = L_1 + L_2$$

The amount of makeup water that is needed to compensate for evaporation losses is:

$$L_1 = G(\mathscr{H}_2 - \mathscr{H}_1) \tag{11}$$

Dividing Equation 10 by L_1 and using Equation 11, the balance becomes:

$$L_1\left(\frac{H_2-H_1}{\mathscr{H}_2-\mathscr{H}_1}\right) = LC_p(T_3-T_2)+L_1C_pT_2 \tag{12}$$

Combining Equations 12 and 9, a general expression for makup water requirements for any fixed set of conditions is obtained:

$$L_1 = \frac{q}{(H_2-H_1)/(\mathscr{H}_2-\mathscr{H}_1)-C_pT_1} \tag{13}$$

In cooling towers, cold air is used to cool hot water. As water descends through the tower, it is possible for the temperature of the water to become lower than the dry-bulb temperature of the incoming air; however, it will not fall below the inlet air's wet-bulb temperature. In the upper regions of a tower, hot water first contacts the discharging air which is still colder than the water. It should be noted that in this region the partial pressure of the water out of the liquid exceeds that which is in the exit air stream, while at the same time the water temperature is greater than the discharging air. Both these facts can be considered as potentials that tend to lower the water temperature via evaporation and sensible transfer to the air. The net result is an increase in the air enthalpy. Both potentials may operate adiabatically and in the same direction while saturating the air.

As water travels down through the lower regions of the cooling tower, it will have a temperature of, or less than, the dry-bulb temperature of the air. Under these conditions, sensible heat and mass transfer are in opposite directions. The wet-bulb temperature represents the limit to which the discharge water temperature can fall to in a cooling tower where adiabatic equilibrium exists between the water and inlet air.

Merkel [9] has developed the equations describing the performance of a cooling tower and Kern [10] has outlined the derivation in detail.

The total heat transfer in a cooling tower consists of two components: the transfer of heat by diffusion and by convection from the water to the air:

$$\begin{matrix}\text{total heat}\\\text{transfer}\\\text{per unit}\\\text{area}\end{matrix} = \begin{matrix}\text{diffusional}\\\text{heat transfer}\\\text{rate}\end{matrix} + \begin{matrix}\text{heat transfer}\\\text{by}\\\text{convection}\end{matrix}$$

$$q_t = q = q_D+q_c \tag{14}$$

where q's are in Btu/(hr) (ft^2).

The heat transferred by diffusion can be expressed in terms of the latent heat of vaporization of the water:

$$q_D = L_1\lambda_{avg} \tag{15}$$

where λ_{avg} is an average value for all the water vaporized in the tower. Combining this expression with Equation 9 and rearranging terms we get:

$$\frac{q_c}{q_D} = \frac{LC_p(T_3-T_2)+L_1C_p(T_2-T_1)-L_1\lambda}{L_1\lambda} \tag{16}$$

An expression for the makeup water, L_1 is given by Equation 11. Combining this and Equations 10 and 16 gives:

$$q_c = \frac{q_D}{\lambda}\left[\left(\frac{H_2-H_1}{\mathscr{H}_2-\mathscr{H}_1}\right)-\lambda\right] \tag{17}$$

Equation 17 provides the ratio of the heat transferred by convection to that transferred by diffusion.

The sensible heat transfer from the water to the air is:

$$dq_c = h(T_{H_2O} - T_{air})a\ dV \tag{18}$$

where a = effective surface area of water per unit volume of tower (as both droplet and film surface)
dV = differential tower volume

Interfacial area a consists of both droplet and film surface area. Film surface is independent of the thickness of the water film; however, the droplet surface is a function of both the liquid loading generating drops and the size distribution of droplets formed. We can bypass the difficulty in measuring "a" by measuring the product Ka for the entire tower at specific operating conditions.

The heat transfer area is, then:

$$dA_H = a\ dV \tag{19}$$

And

$$dq_c = GC_p\ dT_{air} \tag{20}$$

$$dq_D = \lambda\ dL \tag{21}$$

where dL represents the rate at which one component (water) diffuses. Using mass flux relationships, it can be shown that:

$$dL = K(\mathscr{H}_s - \mathscr{H})a\ dV \tag{22}$$

where K is the overall mass transfer coefficient (1b)/(hr) (ft^2) (1b/1b).

Substituting for dL into Equation 21, we obtain:

$$dq_D = K\lambda(\mathscr{H}_s - \mathscr{H})a\ dV \tag{23}$$

The total heat transfer, dq, is given by Equation 14, which in the differential form is:

$$dq = dq_c + dq_D = h(T_{H_2O} - T_{air})a\ dV + K\lambda(\mathscr{H}_s - H)a\ dV \tag{24}$$

and

$$dq = G\ dH \tag{25}$$

We can rewrite Equation 24 in a more usable form by defining average values for the humid heat, C_s, and the latent heat λ, and by neglecting superheat.

$$dq = Ka\ dV\left[(H_s - H) + C_s(T_{H_2O} - T_{air})\left\{\frac{h}{KC_s} - 1\right\}\right] \tag{26}$$

Note that dq can be expressed in terms of the enthalpy decrease of the total water quantity (or in terms of the enthalpy increase of the total air mixture). Both of these quantities are equal so that $dq = d(LC_pT_{H_2O}) = GdH$. The gas loading G is constant through the tower since it is expressed on a dry basis only. However, water loading is not constant due to evaporation losses. For normal operations, evaporation losses are generally less than 2% of the circulating water; therefore, a good estimate can be made by assuming constant, L. Hence:

$$d(LC_pT) = LC_p\ dT \tag{27}$$

and

$$LC_p\ dT = G\ dH \tag{28}$$

$$dq = d(LC_p\ dT) = G\ dH \tag{29}$$

In Equation 26 the quantity h/KC_s is the Lewis number, which is unity for the air-water system. Hence,

$$LC_p\,dT = G\,dH = K(H_s - H)a\,dV \tag{30}$$

From the overall energy balance (Equation 10), we introduce $L = L_1 + L_2$,

$$G(H_2 - H_1) = LT_3C_p - C_pL_2T_2 \tag{31}$$

Applying the assumption that negligible evaporation occurs in the system, then $L_2 \simeq L$, and,

$$G(H_2 - H_1) = LC_p(T_3 - T_2) \tag{32}$$

This enthalpy balance can be graphically represented by plotting the gas enthalpy $\bar{H}$ versus the water temperature T_{H_2O}, as in Figure 20. The line J–K represents Equation 32 and it passes through the coordinates, which represent the terminal conditons of the two fluids. Provided that $L - L_2$ is small, then JK is a straight line having a slope of LC_p/G. This is the operating line. Further, since $C_p = 1.0$ for water, its slope is the ratio of the liquid-to-air loading. Note that the equilibrium curve represents conditions of the gas at the gas-liquid interface; that is, it is the enthalpy of saturated gas at each temperature.

From diffusion theory, the number of transfer units ($Ntu = \int dy/(y - y'')$) represents a method of estimating the magnitude of effort needed to achieve a specified amount of mass transfer through diffusion. And if the height of a single transfer unit (Htu) is known, then we can compute the total height of the tower per unit cross section from the product Ntu (Htu). We can therefore rearrange the major cooling tower equation (Equation 30) to the following:

$$\int \frac{d\bar{H}}{\bar{H}_s - H} = Ka\frac{V}{G} \tag{33}$$

Multiplying by G/L and noting that C_p for water is 1.0:

$$Ntu' = \int \frac{dT}{\bar{H}_s - \bar{H}} = Ka\frac{V}{L} \tag{34}$$

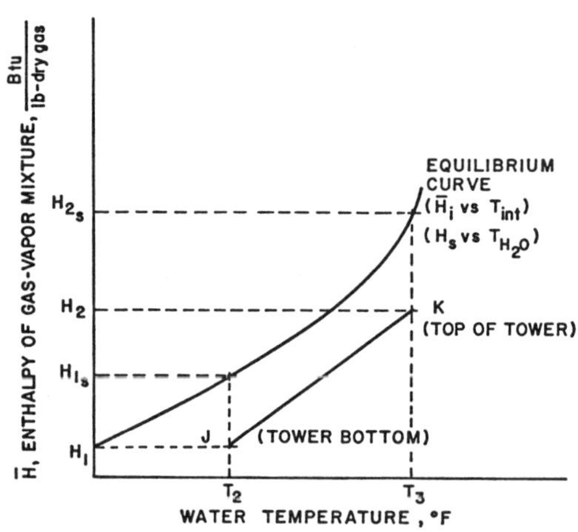

Figure 20. General operating diagram for a cooling tower.

Table 2

Heat Absorbed by Cooling Water for Various Operations

	Btu/min/ton	Btu/lb of steam	Btu/Bhp/hr
Air Compressors			
Single-stage			380
Single-stage with aftercooler			2545
Two-stage with intercooler			1530
Two-stage with intercooler and aftercooler			
Refrigeration			
Compression	250		2545
Absorption	500		
Steam Jet Refrigeration Condenser			
100 psi steam supply (dry)			
2 inch Hg condenser	900	1100	
Steam turbine condenser		1000	
Diesel Engine (Jacket Water)			
Four-cycle supercharged			2600
Four-cycle nonsupercharged			3000
Two-cycle, crankcase compressor			2000
Two-cycle, large unit			2300
Two-cycle, high speed			2100
Natural Gas Engines			
Four-cycle (250 psi compressor)			4500
Two-cycle (250 psi compressor)			3000

Note that Ntu′ is determined from process conditions imposed on the operation and is not based on the performance of the tower itself. The water temperature unfortunately is not a simple function of $\bar{H}_{vapor}$, hence Equation 34 must be solved either numerically or via graphic solution.

Note that the value of $\bar{H}$ at any point on the operating line in Figure 20 can be determined by:

$$H_2 = H_1 + \frac{L}{G}(T_3 - T_2) \tag{35}$$

The area of the plot bound between the equilibrium curve (saturation curve) and the operating line represents the potential that drives the total heat transfer. If the process conditions are changed such that the operating line JK in Figure 20 is shifted downward, then a larger area bound between the operating line and the saturation curve would be available. This means that fewer transfer units and less height of any type of tower are needed.

Design Considerations and Guidelines

The critical conditions that must be established before the design is initiated are the heat load, wet-bulb temperature, hot and cold water temperature, and water rate.

The determination of the heat to be dissipated by a cooling tower is an essential factor that not only affects the tower size, but also its effectiveness. If the heat load determination is not accurate, a larger or smaller size tower than is needed for a particular job could be selected.

The imposed heat load on a cooling tower is determined by the process operation. The degree of cooling is controlled by the normal operating temperature level of the process fluid. Sometimes a low temperature is desired to improve the quantity or quality of the final product or to increase process efficiency. For example, low operating temperatures are required in oil refineries to condense volatile vapors that otherwise would be lost. In other cases, high operating temperatures are desirable. It is important to determine the heat load as accurately as possible to select the proper size cooling tower.

Dependable information has been developed for the heat rejector requirements of different types of power equipment: air compressors, refrigeration equipment, steam condensers, diesel engines and natural gas engines. Table 2 gives information on the heat absorbed by cooling water for this type of power equipment.

The wet-bulb temperature is an important factor in sizing, selection and design. It is defined as the temperature to which air can be cooled adiabatically to saturation by the addition of water vapor and can be expressed as the lowest temperature to which water can be cooled by the evaporative method.

Theoretically, a cooling tower will cool water to the entering wet-bulb temperature when operating without a heat load; however, a thermal potential is required in all heat rejection processes, so it is not possible to cool water to the entering wet-bulb temperature when a heat load is applied. The wet-bulb temperature has a direct impact on the operating temperature of the plant and influences operating conditions, plant efficiencies, and operating cost.

The selection of the design wet-bulb temperature is made on the condition existing at the tower site and measured with a sling psychrometer. The selected temperature is normally close to the average maximum wet-bulb for the summer months.

Selection of wet-bulb temperatures that are not exceeded by more than 5% during a normal summer have given satisfactory performances for most industrial installations. The hours that wet-bulb temperatures exceed the average maximum by 5% need not be consecutive hours and may occur in periods of short duration. Examination of the daily wet-bulb temperature cycle as well as monthly and even yearly temperature cycles are required.

A distinction must be made between the ambient wet-bulb temperature and the cooling tower's inlet air wet-bulb temperature. The ambient wet-bulb temperature is the wet-bulb

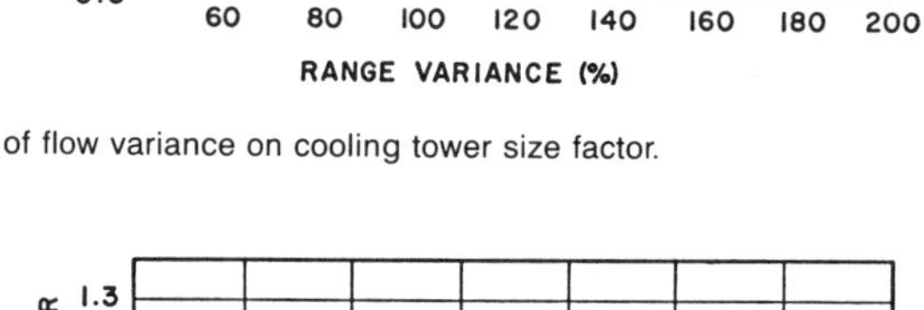

Figure 21. Effect of flow variance on cooling tower size factor.

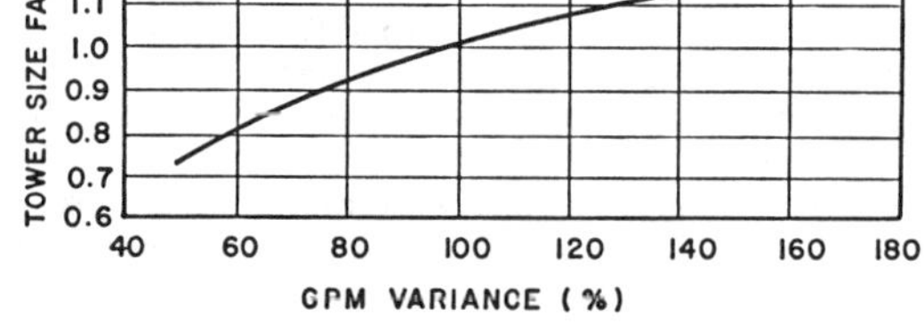

Figure 22. Effect of range variance on tower size factor.

temperature that exists around the outside of a cooling tower. The inlet wet-bulb temperature is the wet-bulb temperature of the air entering the tower. The former depends on the specific atmospheric conditions and the latter can be affected by the discharge of vapors being recirculated into the tower or heating equipment located nearby. The ambient wet-bulb temperature is the one that ASME Tower Test Code on Atmospheric Water Cooling Equipment recommends for official tower tests and normally is used in the Purchasing Specifications.

Wind velocities, wind directions, wet and dry-bulb temperatures are available from data collected by the U.S. Weather Bureau or the U.S. Military Forces covering a wide range of chronological periods. Sometimes information can be obtained from weather stations located at nearby airports.

Once the heat load is known and the wet-bulb temperature established, the selection of hot and cold water temperatures and gpm flow help to determine the size and, consequently, the cost of the cooling equipment. For a fixed heat load, the cold and hot water operating temperatures and the gpm are interrelated; that is, a change in one affects the other two. For example, if the cold water temperature increases slightly, the hot water temperature will decrease by a large quantity, providing a lower average cooling water temperature, all at the expense of increasing the gpm and, consequently, pumping costs. All three parameters must be properly evaluated to obtain the most favorable economic equipment.

The cooling tower range is defined as:

$$\frac{\text{heat load in Btu/min}}{\text{lb of water/min}} = \frac{\text{Btu/min}}{\text{gpm} \cdot 8.33} \tag{36}$$

The selection of the cooling range depends on process characteristics and requirements, as well as the type of cooling tower under consideration. In general, cooling ranges are divided into three categories:

Long range: 25–65° F
Medium range: 10–25° F
Short range: 5–10° F

Long ranges are often used in oil refineries and steel mill applications. Medium ranges are employed in power plants, while short ranges are used for refrigeration and air conditioning uses. The relation given by Equation 36 permits the range and/or gpm flow to be computed from knowledge of the heat load.

If we have a tower with a heat load, wet-bulb temperature, and cold water temperature already selected, the cooling range can be controlled by varying the quantity of water circulated, and vice versa. The net effect of varying the range (or the gpm), fixing the load, and varying the wet-bulb temperature and cold water temperature is a change in the tower size. Increasing the range will decrease the tower size. Increasing the gpm will increase the tower size. Figures 21 and 22 illustrate these points through the use of a dimensionless parameter called the tower size factor.

The term *approach* refers to the temperature difference of the cold water temperature leaving the tower and the wet-bulb temperature of the ambient air entering the tower. Once the wet-bulb temperature and range have been computed, the establishment of the approach fixes all the operating temperatures of the tower fluids.

The approach has a significant effect on the tower size, as shown in Figure 23. For a given heat load, gpm, and wet-bulb temperature, the cooling tower size increases as the approach decreases, and the closer the cold water temperature approaches the wet-bulb temperature, the greater the increase in the cooling tower size.

Approximately 80% of the transfer of heat is by evaporation and the remaining 20% by temperature gradient. Practically, the cold water temperature approaches the wet-bulb temperature of the air and the magnitude of this approach depends on the design of the cooling tower being a function of contact time between air and water, amount of fill surface, water distribution over the fill, and amount and size of water droplets formed. The cooling

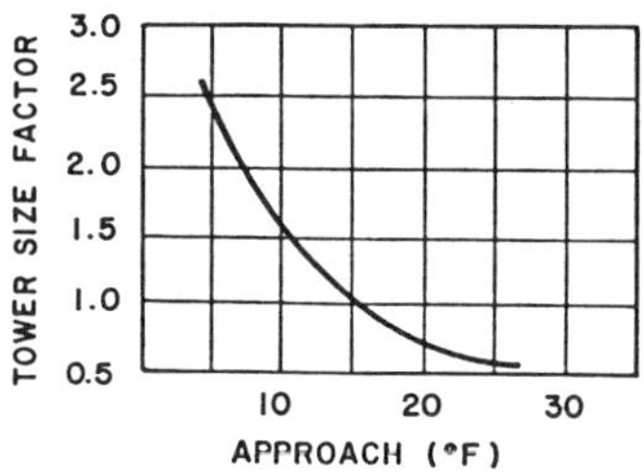

Figure 23. Shows variation in tower size factor with approach.

process involves both mass and heat transfer; the water surface existing on the tower packing contacts an air film, which is considered saturated at the water temperature. Heat is transferred by diffusion and convection between this film and the main body of air. The driving force is the difference of enthalpy between the film and the surrounding air.

The number of transfer units or tower characteristics is based on the overall heat and mass transfer expression given by Equation 34.

A more detailed examination of the equilibrium curve is warranted to better understand the design principles. The countercurrent system defined in Figure 19 is restated in Figure 24 in a slightly more simplified form to illustrate some important features on the enthalpy-temperature plot. In this figure, T denotes air temperature and t water temperature. The following curves are of importance:

Curve A–B has been discussed already. It represents the saturation curve. The points A, B are fixed by the outlet, t_2, and inlet, t_1, water temperatures.

Line A–B gives the enthalpy, $\bar{H}_s$, of air if saturated and at various temperatures included in the water cooling range; the points A and B are fixed by the outlet, t_2, and inlet, t_1, water temperatures, respectively.

Line C–D, the operating liner, gives the actual enthalpy, $\bar{H}$, of air plotted against water temperatures. The wet-bulb temperature at any point on C–D is found by projecting the point horizontally to the water operating line, then vertically to the abscissa of the diagram.

Line E–F represents actual enthalpy data of the air stream plotted against actual air temperatures. Should line E–F cross line A–B there is the possibility of fog formation.

Lines B–C and A–D represent the initial and final driving forces, respectively.

The area bound by A–B–C–D is proportional to the reciprocal of the value of the integral for Ntu'. This value defines the characteristics of the tower, and it is clear that the lower the area, the higher the value KaV/L, i.e., high contact area (a) and high contact volume (V); therefore, high investment cost. The reduction of the investment cost and, therefore, the increase of area A–B–C–D, can be achieved by increasing the approach or reducing the L/G ratio. By reducing the L/G ratio we obtain a high area mass velocity, i.e., large power consumption for the fans.

The approach and cooling range are also indicated in the diagram. Note that the smaller the approach, the smaller the area A–B–C–D and, consequently, the higher the investment cost.

Cooling tower performance is largely affected by internal arrangements. A large amount of information exists in the literature on the performance of different types of packing and fills [11, 12]. Diffusion calculations are usually presented in the form of a plot of K_G a versus G, where the gas film is diffusion controlling. Such plots are presented on the basis of pound-moles of water transferred with a driving potential expressed in atmospheres. In humidification problems common practice is to express pounds of water transferred and driving potential as humidity units. The relation between Htu' and Ka is:

$$\text{Htu}' = \frac{Z}{Ka}\frac{V}{L} \tag{37}$$

Table 3

Humidification Characteristics of Packing Materials

Packing	Size (in.)	Depth (ft)	L (lb)/(hr)(ft²)	G (lb)/(hr)(ft²)	γ	Ka (lb)(hr) (ft³)(lb/lb)	φ
Raschig Rings	1	2	500	250	0.5	226	14.3
			1500	250	0.5	468	29.6
			3000	250	0.5	635	40.2
	2	0.79	500	250	0.47	190	14.3
			1500	250	0.54	301	15.3
			3000	250	0.53	351	18.9
Berl Saddles	0.5	1.29	500	250	0.61	320	11.1
			1500	250	0.61	468	16.3
			3000	250	0.61	595	20.7
	1.5		500	250	0.52	200	11.4
			3000	250	0.52	383	21.8

Table 4

Guide to Packing Height Specification

Cooling Range (°F)	Approach (°F)	Packing Height (ft)
25-35	15-20	15-20
25-35	8-15	25-30
25-35	4-8	35-40

The relationship between K and K_G from absorption theory is:

$$K = K_G P_{B_{lm}} MW_B \tag{38}$$

where the subscript B refers to the air.

Htu′ or Ka represents the performance characteristics of a given fill or packing material. Recall that Ntu′ represents the degree of effort needed for meeting the process conditions. For packed towers the performance of packings over a wide range of conditions can be expressed by an equation having the following form:

$$Ka = \varphi G^{\gamma} \tag{39}$$

Table 3 gives some data for Raschig rings and Berl saddles. More data can be found in the literature [13, 14].

The Cooling Tower Institute has also recommended the following equation for the performance of commercially used cooling tower packings:

$$KaV/L = \varphi(L/G_a)^{-0.6} \tag{40}$$

Note that G_a is the air mass-velocity in lb dry air/(hr) and φ is a function of the type of packing and number of packing deck levels. This establishes the total height of the packing, and thus, the time of contact between the two phases. Packing height is selected on the basis of desired approach. As a rough guide, refer to Table 4.

In developing the design expressions, it was assumed that the overall mass transfer coefficient and the gas-side coefficient were the same. This assumption automatically implies that the liquid film does not offer any resistance to diffusion (i.e., the gas-side controls). This is not necessarily an accurate description, particularly if the humidifying liquid consists of an aqueous solution.

In the energy balance Equation 18, it was further assumed that h is the same as h_G, the heat transfer coefficient from the liquid-film/air-film interface to the air. If a significant resistance is associated with the liquid film, h is better expressed by the overall heat transfer coefficient (U), which consists of both h_G and h_L. Note that h_L is the convection coefficient from the liquid film to the interface [13]. When appreciable liquid film resistance exists, the energy balance is:

$$LC_p\, dT = h_L(T - T_i)a\, dV \tag{41}$$

where T_i is the gas-liquid interface temperature. Equation 41 correctly describes the rate of heat transfer from the liquid body through the liquid film to the interface. Similarly, we can write an expression for the rate of sensible heat transfer from the gas-liquid interface through the gas film to the gas body:

$$GC_s\, dT_{air} = h_G(T_i - T_{air})a\, dV \tag{42}$$

An expression for the material balance with interfacial values is:

$$G\, d\mathscr{H} = k(\mathscr{H}_i - \mathscr{H}) \tag{43}$$

By applying the Lewis number prediction ($h_G/KC_s = 1$), where k is the gas side mass transfer coefficient, and using Equation 30, we get:

$$LC_p\, dT = G\, dH = K(H_s - H)a\, dV \tag{44}$$

where the overall coefficient, K, is defined as h/C_s.

The number of diffusion units is computed from the relation $\int dT/(\bar{H}_s - \bar{H})$. Htu′ must be determined experimentally.

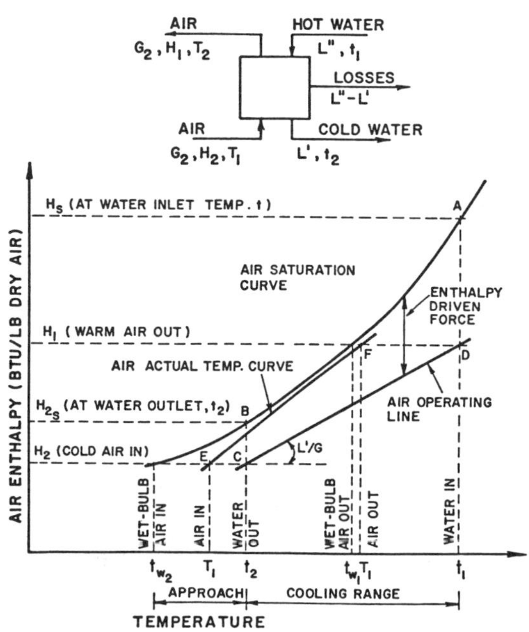

Figure 24. Shows design features for countercurrent cooling tower operation.

It should be noted that the minimum required coefficient (Ntu′) for a given temperature occurs at an L/G ratio of zero, which corresponds to an infinite air rate. That is, there is no increase in enthalpy. Therefore, the maximum driving force exists with least resistance to heat transfer.

If the air rate is lowered (i.e., increase L/G), then in effect the driving force is decreased and a greater degree of difficulty is reflected in the form of a larger value for Ntu′. This is illustrated by the enthalpy-temperature diagrams in Figure 25. Figure 25A shows the case of a counterflow cooling tower at constant conditions but variable L/G ratios.

The maximum L/G ratio corresponds to the case of minimum air rate. For a given temperature this occurs when the tower's operating line (line CA in Figure 25A) intersects the saturation curve. For this case the driving force is zero and Ntu′ becomes infinite. The point of zero driving force may occur at the outlet point. Figure 25B shows the effect of varying L/G for a crossflow tower. At zero L/G ratio the operating area becomes a single horizontal line identical to the counterflow case, and both coefficients become equal. Inreases in L/G result in an increase in the height of the operating area. This continues as the area extends to point A in Figure 25B as the limit.

Thus, both cross- and counterflow towers have the same minimum coefficient at an L/G ratio of zero, and both increase to infinity at $(L/G)_{max}$. A cooling tower can be designed to operate at any point between these limits; however, practical considerations limit the design further, which are determined by the air velocity.

A low air rate requires a large tower, while a high air rate in a smaller tower requires greater fan power. Limitations in air velocities are typically 300–500 fpm in counterflow towers, and 350–650 fpm in crossflow towers.

Higher velocities are obtainable in crossflow towers because of larger air inlet and greater eliminator area. Also, the discharging air does not pass through the water distribution system.

Cooling towers are capable of operating over a wide range of water rates, air rates, and heat loads. Variations are reflected in the approach of the cold water to the wet-bulb temperature. The available tower coefficient is not a constant but varies with operating conditions.

Cooling tower characteristics can only be deduced from actual field tests. Tower characteristics are generally presented in the form of an empirical correlation which establishes the relationship between the available coefficient and the operating conditions. Variations in the available coefficient occur because the calculations consider temperature change per

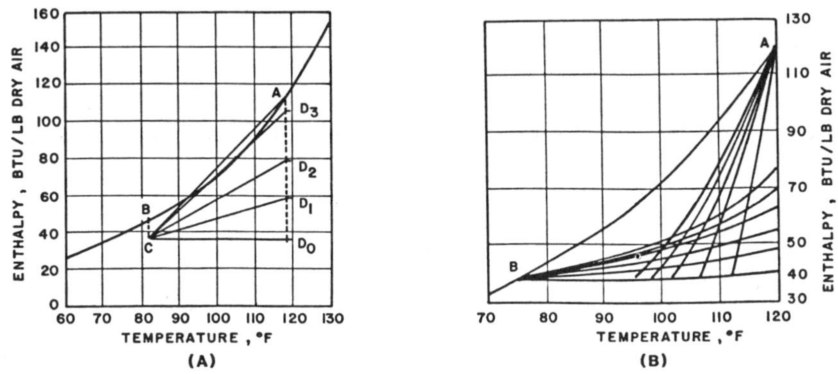

Figure 25. (A) Countercurrent cooling diagram for constant conditions and variable L/G; (B) crossflow tower cooling diagram.

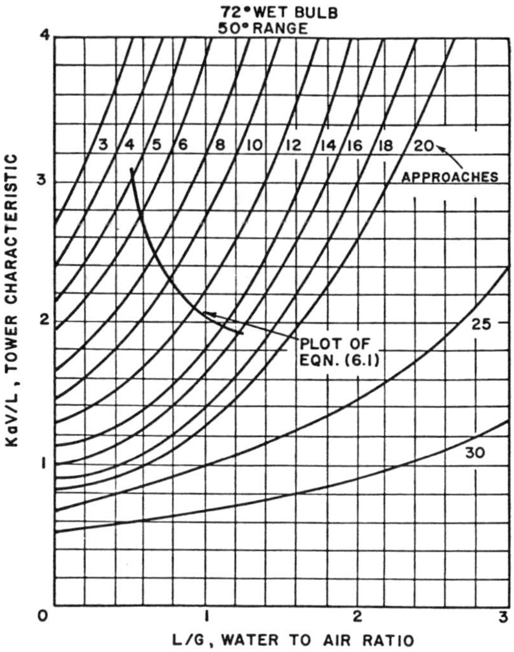

Figure 26. Generalized tower characteristic curves.

unit driving force without regard to tower height or water rate. Since the coefficients of mass and energy transfer increase with air velocity, this is reflected in the available coefficient.

Cooling tower data are most often plotted in the form of KaV/L versus L/G for various air wet-bulb temperatures and cooling ranges. This approach allows us to avoid solving the Ntu′ integral (Equation 34). These graphs have been published by the Cooling Tower Institute (CTI) and are similar to the one shown in Figure 26. Note that the validity of these graphs is good up to a 300-foot elevation above sea level.

Common practice is to neglect the effect of air velocity and develop the tower correlation in the form of a power law relation:

$$\frac{KaV}{L} \propto \left(\frac{L}{G}\right)^{n} \tag{45}$$

where the exponent n can vary between -0.35 and -1.1. Average values of n are between -0.55 and -0.65. A family of curves can also be generated with the following formula:

$$\frac{KaV}{L} \propto (L)^{m}(G)^{m} \tag{46}$$

where m typically varies between 0.6 and 1.1.

Tower performance is specified in terms of the cooling tower's range, approach, wet-bulb temperature, and water rate. The rating of a tower is established by developing a series of charts that relates these variables.

Rating factors correspond to tower units per gpm, which represents some functions of tower area in ft^2/gpm. Hence,

$$\text{rating factor (RF)} \times \text{gpm} = \text{tower units/area} \tag{47}$$

Rating charts are given by McDowell [15] for different temperature ranges. Figure 27 shows an example of a rating chart for one temperature range.

The preceding relations define the basic design theory of cooling towers. The following is an approximate procedure for obtaining a preliminary estimate of tower size for forced draft systems.

1. Specify operating conditions, i.e., water inlet and outlet temperatures and the inlet air wet-bulb temperature.
2. Prepare an enthalpy-temperature diagram. Select the exit air enthalpy so that the slope of the line for the air enthalpy is equal to the slope of the curve for the enthalpy of saturated air at the water outlet temperature.
3. Prepare a plot of $(1/(H_s - H_{air}))$ versus water temperature.
4. Using Simpson's rule or some other appropriate numerical integration technique, determine the area under the curve obtained from Step 3. From the area, the required value for the tower characteristic (KaV/L) can be determined.
5. Select a fill matrix geometry and obtain the data for its principal characteristics. Table 5 gives data for fill at 120° F water temperature.
6. Estimate the effect on the tower characteristic KaV/L on the deviation of the water inlet temperature from the 120° F for which the data of Table 5 were prepared.
7. Using the air enthalpy curve of Step 2, determine the L/G ratio from a heat balance.
8. The following equation can be used to estimate the number of decks of fill required and the height of the packed portion of the tower:

$$KaV/L = 0.07 + \varphi N(L/G)^{-\eta} \tag{48}$$

   where N is the number of decks, and φ and η are fill packing factors. Values to be used in this expression should be corrected for temperature deviations from the empirical constants in Table 5.
9. For the L/G of Step 6, determine the air flow rate for a water flow of 2,500 lb/(hr) (ft^2). If the resulting air flow exceeds 1,600 lb/(hr) (ft^2), determine the water flow that corresponds to an air flow of 1,600 lb/(hr) (ft^2).

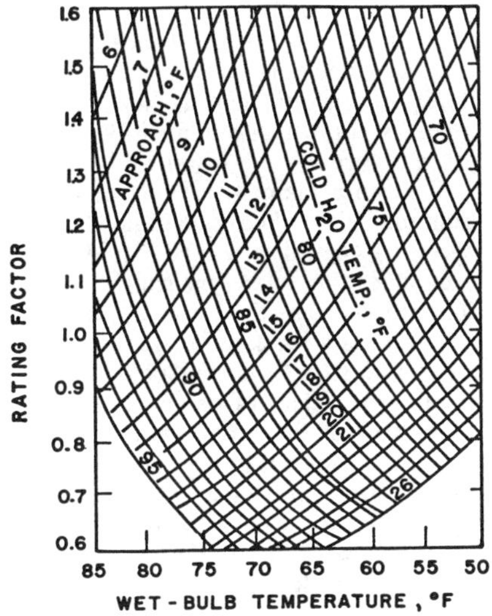

Figure 27. Countercurrent cooling tower rating chart for a 15° range [15].

Table 5
Fill Packing Factors at 120° F

Deck	ϕ	η
A	0.060	0.62
B	0.070	0.62
C	0.092	0.60
D	0.119	0.58
E	0.110	0.46
F	0.100	0.51
G	0.104	0.57
H	0.127	0.47
I	0.135	0.57
J	0.103	0.54

10. Determine the cross-sectional area of the tower using the water flow rate per unit area given by Step 8 and the total water flow rate established in Step 1.
11. Select the horizontal tower dimensions to establish an area defined by Step 9, maintaining the dimension in the direction of the prevailing wind to no less than the packed height and no more than 40 feet. The overall tower height is the packed height plus about 1.5 times the depth in the direction of the prevailing wind.

OPERATIONAL PROBLEMS AND GUIDELINES

The approach in a cooling tower is the difference between the water-exit temperature and the wet-bulb temperature. Basically, it is an index or gauge of how difficult the cooling operation is. This is a strong function of local climatological conditions (i.e., maximum temperatures and wind velocities). In the northeastern region of the U.S. the 5% wet-bulb temperature averages about 75° F. It is common practice in this region to cool the water in a cooling tower to a 10° approach, i.e., 85° F. The 5% wet-bulb temperature is about 80° F on the Gulf Coast, where typically a 5° approach is used (but again to 85° F). Note that when a cooling tower is designed for a definite approach to the 5% wet-bulb, there will be some time periods where the water discharged from the cooling tower will be warmer than originally specified in the process conditions. Cooling towers are normally designed to withstand wind speeds of up to 100 mph (or about 30 lb_f/ft^2).

A common misconception is that cooling towers cannot function when the inlet air is at its wet-bulb temperature. However, when air at the wet-bulb temperature enters the tower it acquires sensible heat from the hot water, thereby raising its temperature so that it is no longer saturated. Consequently, water evaporates continuously into the air phase as it ascends through the tower and cooling takes place.

There are many situations in which the tower's ground area may be too limited, an example being when a cooling tower is erected atop a building. In such a case, it may be necessary to utilize a high liquid loading without increasing the air rate. The maximum economical air velocity is roughly about 400 fpm. Referring back to Figure 25, we see that this will decrease the tower cross section but increase the slope of the operating line. The net result is a lower potential and a taller tower. This is the simplest observation made by an operator that the smaller the amount of air sent to the tower per pound of water, the smaller the extent of cooling.

In most of the situations described thus far it has been assumed that the incoming air is adiabatically saturated. This causes the operating line to drop from H_1-H_2 to H_1'-H_2' in Figure 28A. The shaded region on this plot represents the increase in potential. By not correcting for the enthalpy of the dry-bulb, we achieve results that are slightly conservative. For this reason it is common practice to specify only the wet-bulb in preparing designs.

By using a close approach, we arrive at the situation shown in Figure 28B. Both operating lines have the same L/G ratios. The operating line H_1'-H_2' attempts to perform the same cooling as H_1-H_2 and with the same inlet air but between two different temperatures (T_1', T_2'). The area between the saturation curve and the operating line is greatly reduced by H_1'-H_2'; hence, we obtain a lower potential.

Staging is a technique used for overcoming small L/G ratios. Staging in effect involves the use of two or more towers in series. Water introduced to the top of the first tower is hot and first contacts air of enthalpy, H_2, along the operating line, H_1-H_2 (Figure 28C). The water exits from the bottom at temperature, T_3, and is sent to the top of a second tower. The second tower will operate between H_1 and H_3 enthalpies. Both operating lines may have large slopes and not intersect the equilibrium curve. Water produced by a two-stage arrangement is generally regarded as chilled water. It is, however, a considerably expensive approach to cooling water. Both water cost and range are comparable to those of refrigerated water.

Elevation can also play a role. At a lower atmospheric pressure the saturation curve will be higher. This causes an increase in the potential and reduces the required tower size if operating conditions remain fixed. The reasons for this are that the humidity of saturated air at higher elevations is greater and although the partial pressure of the water is constant, by going to higher elevations the total pressure is reduced.

Since the atmosphere serves as the tower's cooling medium, operating performance is also affected by variations in atmospheric conditions. When the operation is at the design of the wet-bulb, the water should be introduced at the range and temperature specified in the guarantee. If, however, the wet-bulb temperature drops, in effect the potential in the tower increases. If L, G, and the heat load do not change, the water will still undergo the same degree of cooling in; however, the inlet and outlet temperatures will be lower than guaranteed. A tower can only continue removing the same heat load by automatically reducing the potential difference (i.e., the water temperatures reduce accordingly with the wet-bulb).

If either the water or air loadings to a specific tower are changed, then the number of diffusion units the system is capable of providing also changes. In general, the loadings on

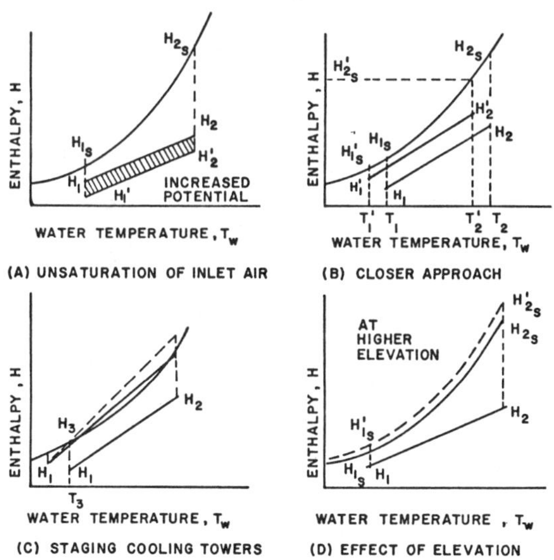

Figure 28. Shows the effects of varying process conditions on a cooling tower's enthalpy-temperature diagram.

a single unit do not change significantly. It is not uncommon for designs to anticipate a 25% deviation from the mean of the design water loading. The maximum discharge capacity will be roughly 120%–125% of the design. When less than 80% of the design water rate is used, the dispersion of droplets is reduced along with the total quantity of water. Cooling tower fabricators can specify the temperature range assumed for the cooling water at 80% and 120% of its design loading when the operating wet-bulb is below the design wet-bulb. Under this condition, it is desirable to use the increased potential to generate more cooling water at the original design temperature range.

Large cooling tower installations impose atmospheric discharge problems in the form of plumes, ground level fog, and icing. The tower exhaust (a saturated air-water vapor mixture) is warmer than outside ambient air, and hence, heat transfers from the plume to the atmosphere. At relatively high humidities, the surrounding air is too moisture laden to absorb the cooling tower vapor and the plume becomes supersaturated. A portion of the water vapor condenses into small droplets forming a ground fog, which limits visibility and, in extremely cold weather, causes icing. If the ambient temperature is high and the relative humidity low, water vapor is absorbed rapdily by the surrounding air and the plume extends only a very short distance from the tower outlet or does not form at all.

The interaction between the plume and atmospheric air is represented on the psychrometric chart in Figure 29. The saturation line (100% relative humidity) is the focus for saturation of air at various temperatures. At points above this line, the air is supersaturated and a plume is visible.

Point A in this plot represents the saturated cooling tower exhaust air. Point B represents a corresponding ambient atmospheric condition. The mixture of cooling tower effluent and atmospheric air follows the line from point A to point B if the mixing process is thorough.

The line from point A toward point B represents increased dilution of exhaust air. Fog will exist as long as the mixture of exhaust and ambient air is supersaturated (the region along line AC). In the unsaturated region (from point C to point B), the water droplets have evaporated and the plume is no longer visible. The formation of ground fog from a supersaturated plume depends on the cooling tower configuration and orientation, exhaust-air flow and temperature, and prevailing meteorological conditions (i.e., wind speed and direction, atmospheric temperature, humidity, and air stability).

Air approaching the tower is displaced, resulting in flow separation and wake formation. A strong negative pressure and secondary flow patterns in the wake are created, the naturally buoyant exhaust plume is drawn down and ground fog forms.

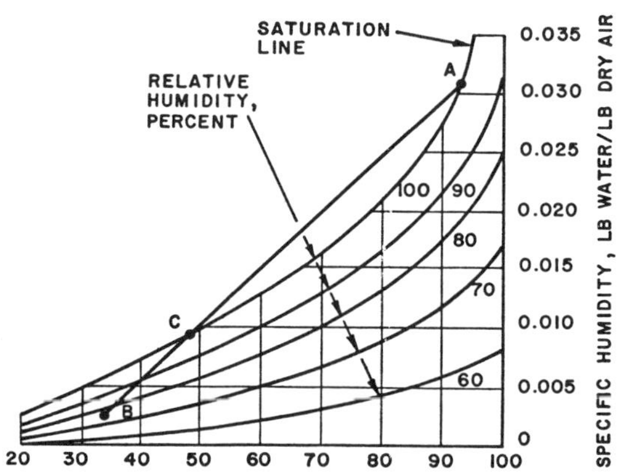

Figure 29. Shows portion of psychrometric chart pertaining to fog formation.

Relative humidity tends to be much higher at night than during the day. Cosequently, the critical period for the formation of cooling tower-induced ground fog is between 3 and 7 AM, especially during the fall and winter when ambient temperatures are lower.

For multicell mechanical-draft towers, the longitudinal axis must be oriented in the direction of the winter prevailing winds. This reduces the air pressure and individual stack exhausts combined together to form a large plume. The positive buoyant force of the individual plumes is combined and magnified, improving the plume rise and therefore diminishing the possibility of ground fog formation.

In places in which ground fog is undesirable, the dry-wet mechanical draft cooling tower is better suited. It will have its wet peaking tower out of service during the cooler months of the year, thereby substantially eliminating the fog problem and eliminating the icing problems of wet towers.

With regard to plume behavior, studies have shown that the atmospheric turbulence in the vicinity of natural-draft cooling towers can increase significantly due to the influence. Stack flue gases entering this turbulent region are more rapidly diluted. Unfortunately, this can lead to appreciable changes in the ground concentration of precipitated pollutants. As such, impairment is generally greater in the immediate vicinity of the plant and significantly lower at greater distances from the source. Perrinjaquet et al. [16] have studied the extent to which cooling tower plumes increase vertical diffusion to better estimate the environmental impacts of stack discharges.

Ground-level concentrations of pollutants in the vicinity of discharge sources can be estimated from diffusion models, which attempt to quantitatively describe the turbulent characteristics of the atmosphere. If operable cooling towers are located in the immediate vicinity of stacks, the large heat discharge from them can greatly intensify the turbulence within a fairly large volume. Under certain environmental conditions, this can result in dramatic increases in pollutant ground-level concentrations.

The studies by Perrinjaquet et al. and others [16, 17] showed that cooling tower plumes do not disperse in accordance with the prediction of common diffusion equations. Plumes were observed to consist of a throughly well-mixed core, which was screened from the undisturbed surrounding atmosphere by a band of relatively calm air. The mean square fluctuations of the vertical speed of the air inside and outside the plume centers have been estimated for different distances from the cooling towers and for different altitudes. Vogt and Geiss [17] have developed an empirical expression for determining the vertical turbulence intensity, i_u, which is applicable to the region between 200 and 500 meters above ground.

$$i_u = \bar{u}T \frac{17.3}{\frac{dT}{dZ} + 3.0} \tag{49}$$

where i_u = square root of the mean square fluctuation (turbulence intensity)
$\bar{u}$ = mean wind speed (m/sec)
$\frac{dT}{dZ}$ = vertical temperature gradient (°C/100 meters)

Note also that:

$$i_s^2 = i_{total}^2 - i_u^2 \tag{50}$$

where i_{total}^2 is the square of the fluctuations of the vertical speed of air at the center of the plume. Note that i_s^2 represents the inherent turbulence of the plume. It is found to decrease with increasing distance from the cooling tower. The inherent turbulence has a strong dependence on the thermal emission and the type of cooling tower. Equations 49 and 50 best represent natural-draft, wet cooling towers woth a thermal output in the range of 1,500 to 2,000 MW [16].

The increase in vertical turbulence intensity caused by cooling tower plumes can be estimated for each temperature gradient and increment of distance from the tower. This can be represented by well-known turbulence parameters developed for Gaussian plume models:

$$i_{total}^2 = i_u^2 + i_s^2 = i_u^2\left(1 + \frac{i_s^2}{i_u^2}\right) \equiv \beta i_u^2 \tag{51}$$

In this expression, β represents the increase factor of vertical diffusion due to the plume. Gaussian plume or dispersion models are based on standard deviations of the plume dimensions (σ_x, σ_y, σ_z). These represent a measure of the diffusive capacity of the atmosphere. They are dependent on the turbulence conditions of the atmosphere, the vertical temperature gradient (which helps to establish atmospheric turbulence in the vertical direction), and the transporting distance.

Classical theory of turbulent diffusion assumes some sort of constant mixing length, whereby:

$$\sigma \simeq \sqrt{i} \tag{52}$$

And the effective vertical dispersion coefficient in the plume is defined as:

$$\sigma_{z_{eff}} = \beta\sigma_z \tag{53}$$

where σ_z, in meters, is the vertical dispersion coefficient and is dependent on climatic conditions. If stable weather conditions exist with a temperature gradient of $+2°$ C/100 meters and wind speeds are mild (at or below 1 meter/sec), the β has a value between 30 (for 1-km distance from the tower) and 20 (for 6-km distance). This correction factor is much smaller if unstable weather conditions and high wind speeds exist. An advantage to using β is that it can be related to different meteorological classifications (or stability classes) once values for the vertical temperature gradients are known. A stability class comprises atmospheric conditions that possess approximately equal diluting characteristics.

When a stack is situated near the cooling tower, stack gases are very likely to enter the turbulent plume zone. Average yearly concentrations, which are normally computed from the frequency of various atmospheric stability classes at ground level, will not provide good estimates. The reason for these poor estimates is that for average situations stability class estimates assume conditions that are too stable and too low in turbulence. By employing the increase factor, β, in Gaussian dispersion models, it is possible to correctly estimate the extent to which ground level concentrations averaged over the year can change due to the influence of cooling tower discharges. Ground-level concentration predictions are best illustrated by a plot similar to Figure 30.

Blowdown

Blowdown is defined as the amount of water discharged from the system to control the concentration of salts or other impurities in the circulating water. Solids from dissolved chemicals and minerals in the source water accumulate in the circulating water, making blowdown a necessity operation. Blowdown depends on the allowable concentration of water constituents to meet water quality restrictions or for nonscaling requirements in the circulating water system. Varying the "bleed-off" controls the degree of concentration in the tower. This is measured in terms of cycles of concentration, π_c:

$$\pi_c = \frac{\text{dissolved solids in the cooling tower supply}}{\text{dissolved solids in the cooling tower makeup supply}} \tag{54}$$

For example, a cooling tower with water containing four times as much total dissolved solids as its makeup supply would be operating at four cycles of concentration. The cycles

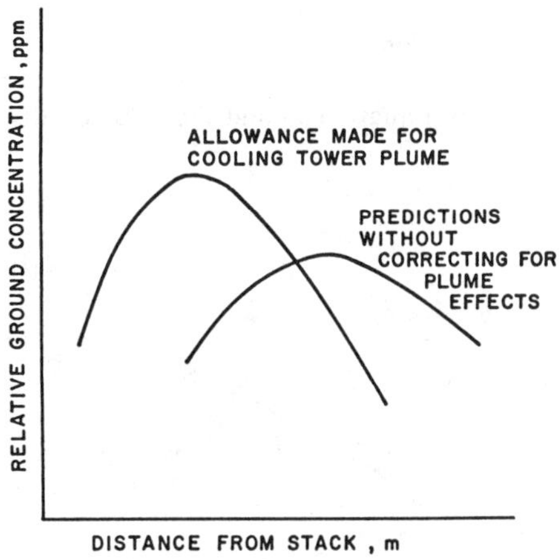

Figure 30. Typical plot comparing ground-level concentrations in cooling tower stack direction.

of concentration are determined by the cooling tower design, water characteristics, operating conditions and the type of treatment system employed.

Blowdown can be expressed as a percentage of the circulation rate:

$$M = E + W + B \tag{55}$$

where M = makeup water rate as a percentage of the circulation rate
E = evaporative water losses as a percentage of the circulation rate
W = windage and drift water losses as a percentage of the circulation rate
B = blowdown as a percentage of the circulation rate

Since the evaporation of 1 lb of water requires approximately 1,000 Btu, this heat will cool 200 lb of water by 10° F. Therefore, the evaporation losses are approximately 1% of the circulation water rates for each 10° F of cooling range:

$$E = \text{evaporation losses} = \frac{\text{cooling buty (Btu/lb)}}{1000} \tag{56}$$

Windage losses or drift vary with the type of tower and local conditions. Average estimates for normal tower operations are 0.3–1% of circulation for natural-draft towers and 0.1–0.3% of circulation for mechanical-draft towers.

Cycles of concentration compare the concentration of the dissolved salts in the circulating water with that of the makeup water. Usually, the circulating water salt content is limited to 3–7 times the makeup salt content (this is referred to the number of concentrations of 3–7).

The number of concentration is conveniently expressed as the ratio of the chlorides in the circulating water to chlorides in the makeup water. Chloride salts are soluble and will remain in solution; consequently, their concentration increases during the evaporative cooling process.

The basic items normally considered in a material balance of a cooling tower are the range, T, circulating water rate, C (in U.S. gpm), the ppm (parts per million of any soluble salts in C (X_c), and M.B, E, and W. X_c is normally expressed in terms of equivalent chlorides.

It must be noted that concentration of soluble salts in blowdown and windage losses equals that in circulating water. Thus, X_m = ppm of any soluble salts in M and

$$\pi_c = \frac{X_c}{X_m} = \text{cycle of concentration}$$

The overall balance is given by Equation 55 and the soluble salts balance is:

$$MX_m = BX_c + WX_c \tag{57}$$

or

$$MX_m = X_c(B+W)$$

$$\frac{X_c}{X_m} = \pi_c = \frac{M}{B+W}$$

and

$$B+W = M/\pi_c \tag{58}$$

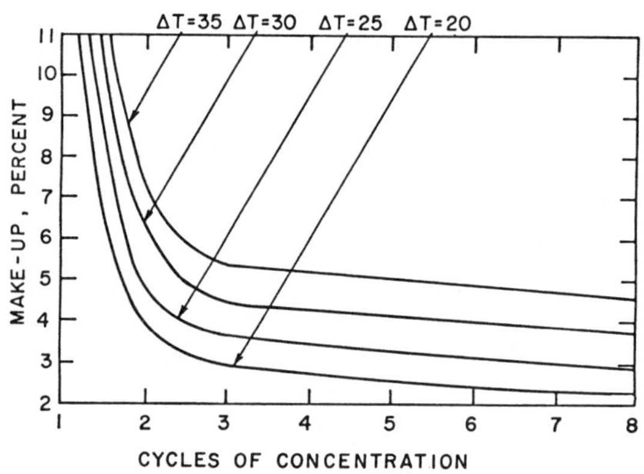

Figure 31. Chart for estimating cooling tower makeup requirements.

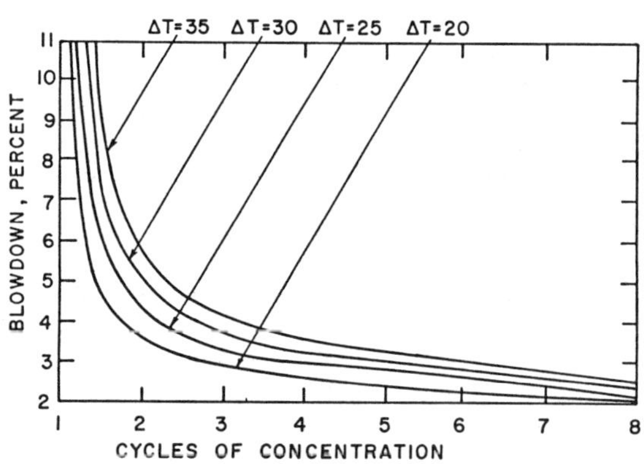

Figure 32. Chart for estimating cooling tower blowdown.

Substituting Equation 58 into Equation 55 we obtain:

$$M = E + \frac{M}{\pi_c} \tag{59}$$

Since $M(\pi_c - 1) = \pi_c E$, and $E = 0.1\ \Delta T$;

$$M = \frac{0.1\ \Delta T \pi_c}{\pi_c - 1} \tag{60}$$

and from Equation 58

$$B + W = \frac{0.1\ \Delta T}{\pi_c - 1}$$

and for $W \cong 0.3\%$,

$$B = \frac{0.1\ \Delta T}{\pi_c - 1} - 0.3 \tag{61}$$

Equations 60 and 61 are plotted in Figures 31 and 32. These plots can be used to obtain estimates of required water makeup and blowdown as functions of the number of concentration, π_c, respectively.

Retention time, τ, is defined as the theoretical time for a single water droplet to remain in the system:

$$\tau = \frac{V}{W + B}$$

where V is the total volume of the cooling tower.

Retention time is important as is affects the time a slug treatment will remain in the system. Lower bleed-off causes longer retention times.

The chemical composition of the cooling water makeup supply used in the plant determines the choice of the cycles of concentration. Some of the important constituents that must be controlled in the tower are calcium, magnesium, silica, carbonate, bicarbonate, and sulfate ions. Alkalinity levels are regulated by the addition of acid or alkali to achieve the desired pH. When adding H_2SO_4 (sulfuric acid) for pH control, it should be assured that calcium sulfate solubility limits are not exceeded.

Water Consumption and Recirculation Rates

Water consumption depends not only on the heat load but also on the ratio of the amounts of heat carried off by increasing the temperature of the air through evaporation. The amount of water lost in the form of suspended droplets in the exit air stream is usually less than 1% of the water consumption. Thus, the water consumption, L, of a cooling tower can be related to the air flow rate and the air inlet and outlet water vapor contents, x_1 and x_2, as follows:

$$L = W_a(x_2 - x_1) \tag{62}$$

If the air entering the tower is saturated, as much as one-third of the heat removed from the water may go into heating the air, while the balance will go into evaporating the water. Hence, water consumption will be only about two-thirds of what would be required if the entire heat load went into evaporation. On the other hand, at light loads, with a low temperature range and very dry air, evaporation of the water may actually reduce the air dry-bulb temperature so that heat is removed from rather than added to, the air, and the amount of heat going to evaporate the water actually exceeds the heat load on the tower.

Under some wind conditions, a portion of the warm moist air leaving the tower may recirculate back through the tower inlet and thus degrade performance. Forced-draft towers have recirculation rates that are about double those of induced-draft towers. Both water loading and tower height play the dominant role in recirculation. Correlations are given in the literature for defining the effects of these parameters, and corrections can be applied to the wet-bulb temperature [18, 19]. Cooling tower fabricators can supply data to estimate the severity of the problem.

NOTATION

Symbol	Definition
A	cross-sectional area
a	effective surface of water per unit volume of tower
B	blowdown
C_p	specific heat
C_s	humid heat
c	circulating water rate
E	evaporation losses
G	gas-mass rate loading
g	acceleration of gravity
H	enthalpy
Hdu	height of a diffusion unit
Htu′	height of a transfer unit
H	humidity, 1b water-vapor/1b-dry air
h	heat transfer coefficient in which one film controls
i_u	square root of mean square velocity fluctuations
K	overall mass transfer coefficient
k_G	gas film coefficient
L	liquid mass rate, loading
Le	Lewis number
MW	molecular weight
M	makeup water rate
Ntu′	number of transfer units
p	pressure (psia)
Q	heat load (Btu/hr)
q	heat transfer per unit area
R_c	recirculation rate
RF	rating factor
T	temperature
t	cold water temperature
U	overall heat transfer coefficient
$\bar{u}$	average wind speed
V	tower volume
W	windage or drift loss
Z	height

Greek Symbols

Symbol	Definition
β	increase factor of vertical dispersion
γ	exponent
θ	time
η	fill packing parameter
λ	latent heat of vaporization
ν	specific volume
π_c	cycles of concentration
σ	dispersion coefficient
τ	residence time
φ	packing factor

Subscripts

Symbol	Definition
A	diffusing component
AVG	average
B	the inert gas
C	convection
D	diffusion
G	gas
i	interfacial value
lm	log mean value
s	saturation conditions
t	total

REFERENCES

1. McHale, C. E., ct al., "New Developments in Dry Cooling of Power Plants," *Combustion,* 51 (11) (May 1980).

2. Van Der Walt, N. T., et al., "The Design and Operation of a Dry Cooling System for a 200 MW Turbo-Generator at Grootvlei Power Station—South Africa," paper presented at the IX World Energy Conference, Detroit, MI (1974).
3. Gottzmann, C. F., and O'Neill, P. S., "Field Experience with High Efficiency Heat Exchangers," paper presented at the AIChE 74th National Meeting (March 12–15, 1973).
4. Oleson, K. A., et al., "Dry Cooling Affects More Than Cost," *Elect. World* (July 1, 1972).
5. Cheremisinoff, N. P. and P. N. Cheremisinoff, *Cooling Towers: Selection, Design and Practice,* Ann Arbor Science Publishers, Ann Arbor, MI (1981).
6. Grosvenor, W. M., *Trans. AIChE.,* 1:184 (1908).
7. Bird, R. B., Stewart, W. E., and Lightfoot, E. N., *Transport Phenomena,* John Wiley & Sons, Inc., New York (1960).
8. Cheremisinoff, P. N., and Young, R. A., *Pollution Engineering Practice Handbook,* Ann Arbor Science Publishers, Ann Arbor, MI (1975).
9. Merkel, F., *Forschungsarb.,* 275:1–48 (1945).
10. Kern, D. Q., *Process Heat Transfer,* McGraw–Hill Book Co.; Inc., New York (1950).
11. Simpson, W. M., and Sherwood, T. K., *Refrig. Eng.,* 535 (1946).
12. Colburn, A. P., *Trans. AIChE,* 29:174 (1939).
13. McAdams, W. H., *Heat Transmission,* 2nd ed., McGraw–Hill Book Co., New York (1942).
14. Parekh, M., *Report, Chem. Eng.* (1941).
15. McDowell, D. W., Jr., "The Corrosion Engineer Looks at the Cooling Tower," *Cooling Tower Institute,* Houston, TX (June 1964).
16. Perrinjaquet, M., et al., "Increase in the Turbulent Diffusion of Pollutants by Cooling Towers," *Combustion,* 51 (11) (May 1980).
17. Vogt, K. J., and Geiss, H., "Tracer Experiments on the Dispersion of Plumes Over Terrain of Major Surface Roughness," KFA–Julich, Jul–1131–ST (Oct. 1974).
18. Lichtenstein, J., "Performance and Selection of Mechanical Draft Cooling Towers," *Trans. ASME,* 779 (1943).
19. Strauss, S., "Guide to Evalute Cooling Tower Performance," *Power,* (Oct. 1975).

CHAPTER 35

MECHANISMS OF QUENCHING SURFACES

R. A. Nelson

Los Alamos National Laboratory
Los Alamos, New Mexico

CONTENTS

INTRODUCTION

Quenching is the sudden cooling of an object caused by immersing it in a cooling fluid, such as water or oil. The quenching phenomenon exists in numerous industrial applications, such as metallurgical quenching [1, 2] or quenching of a container wall during filling with a cryogenic liquid. Recently, the quenching of uncovered nuclear reactor cores during postulated loss-of-coolant accidents has received much investigation [3, 4]. In this case, quenching is a transient cooling process taking place at the surface of a solid. For cooling to take place, the surface temperature must decrease. The qualitative value of these two temperatures is defined in the following paragraphs. The cooldown rate is proportional to the difference between the energy being transferred by convection from the surface and the energies being conducted to the surface and generated at the surface. The rate required to be classed as "sudden" is somewhat arbitrary.

From this brief description of quenching and a perusal of the quenching literature, it is obvious that several additional terms must be defined. These terms include cooldown, rewetting, and return to nucleate boiling and are sometimes used interchangeably with the term quenching. These definitions are presented with respect to a typical pool boiling curve, as shown in Figure 1.

A cooldown of the surface occurs when the surface temperature decreases. Surface cooldown can take place from any point on the boiling curve. This happens when the energy being transferred away from the surface by the fluid (through convection) is greater than the combined energy being conducted to the surface through the material and the energy generated at the surface. In the case of a body with no heat generation, cooldown is determined by the balance between conduction and convection.

The rewet of a point on the surface occurs when a point covered by vapor is once again contacted by liquid. The term rewet indicates a microscopic view of the heat transfer at a point on the surface and contains no statement as to the average surface temperature at which rewetting may occur. Thus, transient local rewetting of a point may occur in the film boiling region [5] for average surface superheats greater than the minimum surface superheat, ΔT_{min}.

A return to nucleate boiling (RNB) may be initiated from either the transition boiling region or the film boiling region, depending upon the original surface temperature. This occurs when the cooldown is sufficient to reach the nucleate boiling region. This process is also termed a quench, with the term quench principally being used in this chapter. The temperature at which quenching begins is therefore above the critical heat flux (CHF) superheat, ΔT_{CHF}, while the temperature to which the surface drops is below this value. Quenching and RNB relate to the macroscopic nature of the process.

In order to predict quenching, the analyst must adopt one of two extremes, either a simple quick calculation, using an available analytical model (which may be somewhat limited in capability), or a much longer and more difficult calculation, using numerical techniques to obtain an answer. This chapter will provide a brief summary of the more significant analytical and numerical models available to the analyst. It will also provide insight into the physics of the phenomena, for those who may either be required to address the process in greater detail or must investigate a problem not currently addressed by the literature.

In order to meet these objectives, this chapter will be divided into two major sections. The first will discuss conduction within the body being quenched. Based upon different assumptions involving the quench process, lump, one-dimensional, two-dimensional, and three-dimensional conduction models are obtained. An assumption of simple convective heat transfer models in this section allows many closed-form analytical models to be obtained. The second section will discuss the convective heat transfer process involved in quenching. It provides

1. The basis for understanding the more integral aspects of the process.
2. The information needed to represent the more complex convective heat transfer models required in the numerical approaches.
3. An understanding of the limitations of our current capabilities.

CONDUCTION

The general objective in the study of heat conduction is to predict temperature distribution within a body, or the history of temperature distributions within the body during a transient. Numerous textbooks [6, 7, 8] and reference books [9] study this aspect of heat transfer. Study of those portions of conduction necessary for the understanding of the process of quenching will be undertaken here.

The General Differential Equations

Fourier's law of heat conduction, in its ordinary constitutive vector form, provides the rate of energy transfer across a unit of area, to be given by

$$\vec{q}(\vec{x}, t) = -k\bar{\nabla}T(\vec{x}, t) \tag{1}$$

where $\vec{q}(\vec{x}, t)$ = the heat flux (W/m^2)
$\vec{x}$ = the position vector
t = time
k = the transport property known as thermal conductivity (W/m-K)

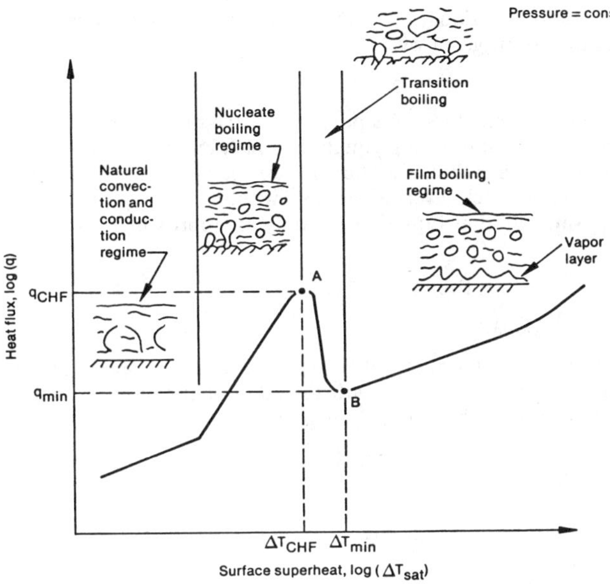

Figure 1. Conventional pool boiling curve.

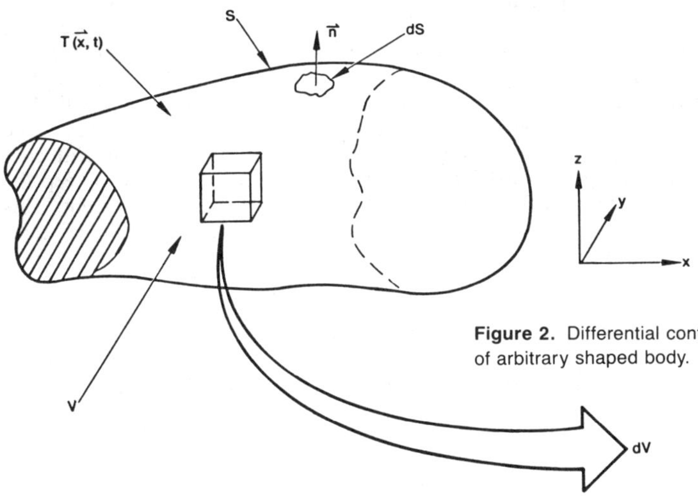

Figure 2. Differential control volume of volume dV of arbitrary shaped body.

The negative sign results from the fact that heat flows in the direction of decreasing temperature. Application of Fourier's law, and an energy balance on a differential control volume of volume dV of a body, as shown in Figure 2, can be used to obtain the parabolic form of the heat diffusion equation in the absence of density or pressure gradients given by

$$\vec{\nabla} \cdot \vec{q}(\vec{x}, t) + \rho c(\partial T(\vec{x}, t)/\partial t) = q' \tag{2}$$

where q' = the internal heat generation rate per unit volume (W/m^3)
ρ = the density (kg/m^3)
c = the specific heat (J/kg-K)

Substitution of Equation 1 into Equation 2 produces the diffusion equation in its most common form, given by

$$\vec{\nabla} \cdot (k\vec{\nabla}T(\vec{x}, t)) + q' = \varrho c(\partial T(\vec{x}, t)/\partial t)$$

Many investigators have pointed out the physical problem associated with the parabolic form of the heat diffusion equation, which implies an infinite speed of propagation for heat flow. This means that, if a temperature change occurs somewhere within a body, the change is felt immediately everywhere else within the body. This dilemma has been resolved by proposing [10] that Equation 1 is only an approximation of a more precise equation, given by

$$(\alpha/C^2)\,(\partial^2 T(x, t)/\partial t^2) + \varrho c(\partial T(\vec{x}, t)/\partial t) = q' + k\nabla^2 T(\vec{x}, t) \tag{3}$$

where α is the thermal diffusivity, and C is the speed of heat propagation. Employing Equation 3 in Equation 2, with the additional condition that q', k, ϱ, and c are constant, produces a hyperbolic equation given by

$$(\alpha/C^2)\,(\partial^2 T(x, t)/\partial t^2) + \varrho c(\partial T(\vec{x}, t)/\partial t) = q' + k\nabla^2 T(\vec{x}, t) \tag{4}$$

This equation is also known as the telephone equation.

Normally, the diffusivity in Equation 4 is 10 orders of magnitude less than C^2, making the effect of the finite speed of propagation negligible [11]. However, for solutions involving very short times, or at very low temperatures, the effect of finite propagation speed can become important. From a quenching standpoint, this limitation on the parabolic form of the heat diffusion equation affects study of local rewetting at a point, which may be fast at film boiling temperatures, and the quenching process which involves more exotic low temperature fluids such as liquid helium II. Baumeister and Hamill [12] studied the semi-infinite body problem, in order to investigate how short a time the parabolic diffusion equation could be applied. They obtained a relationship for the time necessary to keep the effect of the heat flux singularity in the parabolic solution to less than one percent. This relationship can be used as a guide to determine when the use of the parabolic equation might not be sufficient. It is given by

$$t > 2\alpha(10^4)/C^2$$

For the remainder of this chapter, Equations 1 and 2 will be the forms of Fourier's law and the diffusion equation used. Initial and boundary conditions are required, in order to solve the parabolic heat diffusion equation in any of its many expanded forms [9].

Initial and Boundary Conditions

Solution of the quenching problem involves the specification of a convective boundary condition at the surface of the body being quenched. This boundary condition can be expressed as

$$q_{conduction} = q_{convection} \quad \text{over S} \tag{5}$$

where S denotes the surface of the body as denoted in Figure 2. $q_{conduction}$ is obtained from the relationship

$$q_{conduction} = \vec{q}(\vec{x}, t) \cdot \vec{n} \tag{6}$$

where $\vec{n}$ is the unit vector normal to the surface S (see Figure 2), and $\vec{q}(\vec{x}, t)$ is given by Equation 1. The definition of $q_{convection}$ involves the definition of the boiling curve, as shown in Figure 1. It is the definition of this boiling curve which provides a nonlinear boundary condition, and its coupled effect with conduction, which makes the quenching process such a complex phenomenon to study. For this reason, numerous simplifications have been made in an effort to separate the two phenomena.

One of the major simplifications is to specify $q_{convection}$ through the appropiate rate equation known as Newton's law of cooling, given by

$$q_{convection} = h(T_s - T_f) \quad (7)$$

where h = the heat transfer coefficient ($W/m^2 - {}^\circ K$)
T_s = the surface temperature (K)
T_f = the fluid temperature (°K)

This reduces the problem primarily to specification of the heat transfer coefficient. The following sections will summarize the types of h's specified, and the physical problems which they represent. The initial conditions vary, depending upon the problem being solved. These will be discussed where appropriate. In general, it should be realized that this approach views quenching primarily as a conduction-dominated problem.

The Lumped Model

In terms of metallurgical quenching [1, 2] one of the common considerations involves immersion of a hot structure, such as an ingot or forging, at a uniform initial temperature, T_i, into a fluid, in order to quench it. The first step taken in analysis is the determination as to whether or not a simple lumped model can be applied. This is done by evaluating the Biot number. The Biot number is a dimensionless number and is given as

$$Bi = hL_c/k$$

where L_c is the characteristic length, given by L_c = volume (V)/surface area (S)

The Biot number represents a ratio of the resistance to conduction within the solid to the resistance of convection into the fluid. Hence, when $Bi \ll 1$ (usually taken as 0.1 or less), the resistance to heat flow through the solid is less than the resistance to heat flow from the body by convection. This allows a uniform temperature distribution to exist in the body during the quenching transient and eliminates the need to solve the diffusion equation. If $Bi > 1$, the diffusion equation must be solved to account for the temperature distribution within the body. The boundary condition for this solution is given by Equations 5 and 6. The initial condition, involved in either of these solutions, is that the body has a uniform temperature distribution previously denoted as T_i.

Incropera and Dewitt [6] provide an excellent summary of the lumped model and its validity. Figure 3 is a typical sketch of this situation and indicates that the problem has been reduced to an instantaneous balance, given by

rate of heat loss = rate of change of internal energy

or

$$-hS(T - T_f) = \varrho cV(dT/dt)$$

Separation of the variables and integration from the initial conditions leads to

$$(\varrho cV/hS) \int_{T_i}^{T} d(T - T_f)/(T - T_f) = \int_0^t dt \quad (8)$$

which leads to the solution given by

$$\theta = e^{-Bi\,Fo} \quad (9)$$

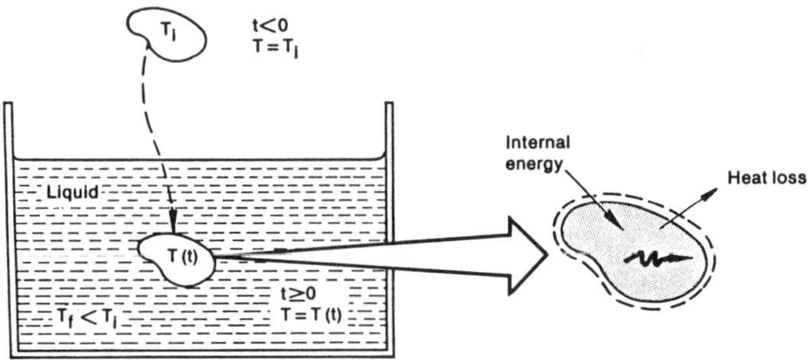

Figure 3. Cooling of a body showing lumped model conditions.

and introduces two additional dimensionless parameters. The first is the dimensionless temperature denoted by θ while the second is the dimensionless time, also known as the Fourier number, denoted by Fo. These are defined by

$$\theta = (T - T_f)/(T_i - T_f) \tag{10}$$

and

$$Fo = \alpha t/L_c^2$$

The dimensionless temperature, as shown in Equation 10, may change. It is however, always a ratio of temperatures which authors may choose slightly differently. Each different dimensionless temperature used within this chapter will be subscripted individually.

One assumption necessary to obtaining Equation 8 is that the heat transfer coefficient be constant. However, the heat transfer coefficient associated with the complete quench process is not a constant and may vary an order of magnitude or greater. Thus, obtaining Equation 8 and the final solution to the quenching process given by Equation 9 requires an approximation for h and provides only qualitative information.

A more exact solution to the problem including a nonlinear h requires leaving the h inside the right-hand integral of Equation 8. The most straightforward method by which to obtain a solution to the problem including a variable h is to use numerical integration techniques to evaluate the integrals of the resulting equation (including h within the integral). The definition and use of variable h's will be discussed in later sections.

Models Using The Heat Diffusion Equation

The lumped model with a closed-form solution not only requires that h be constant with respect to time but implies that, since the entire hot body is immersed in the fluid at time zero, a uniform heat transfer coefficient applies around the body. Orientation of the quenching surface with respect to gravity, and the manner in which the quenching fluid is applied, often make the assumption of a uniform heat transfer coefficient invalid. For example, the slow "end-first" immersion of a long hot rod into a quenching fluid produces a quench front which propagates up the rod as it is lowered into the fluid. This realization leads to the understanding that a quench front may exist during the quenching process.

The quench front is formed by quenching point loci on a body at any given instant of time. The quench front may or may not move, and the term implies a steep temperature gradient from the hot part of the body to the cold part. Just as the word "sudden," applied to quenching, was not exact, so is the term "steep," as applied to the gradient. Most often, a quench front does move. The velocity with which it propagates is called the quench front velocity, u. This concept of quench front velocity gives rise to a dimensionless velocity term called the Peclet number, which is given by

$$P = \rho c u L_c / k$$

The Peclet number occurs often in the solution of the diffusion equation associated with moving quench fronts.

Numerous simplifying assumptions have been made to represent the convection heat transfer process, in order to obtain solutions to the heat diffusion equation for quenching situations with moving quench fronts. Reviews of existing models can be found in the literature [4, 13–16].

To aid in understanding these models and their assumed initial and boundary conditions, it is useful to visualize the geometry associated with the most simple of these models. Figure 4 shows a sketch of a liquid film which quenches the wall as it flows down the wall. In this configuration, gravity pulls the liquid down the wall, resulting in a quenched surface covered by liquid and a hot unquenched surface ahead of the quench front.

An example of this type of behavior arises during the cooling phase of a hypothesized accident for a BWR (boiling water reactor), when sprays in the top of the reactor provide water to cool the hot, uncovered fuel rods. This water flows down the fuel rods from the top, producing a "top–down" quench. The rod's hot cladding surface throws the liquid at the quench front off the surface through vapor generation in the liquid film itself. This frees the liquid from its film layer and allows it to fall away from the clad surface. This rejection of the liquid from the surface at the quench front is called sputtering. Sputtering of fluid drops into the air-droplet or vapor-droplet mixture results in heat transfer ahead of the quench front (called precursory cooling), being small when compared to that behind the quench front. The quenching produced in this simple top-down quenching situation is primarily controlled by conduction. Heat energy ahead of the quench front is first transported by conduction to the lower temperature material behind the quench front. It is then transferred to the liquid film through the convection process.

Semeria and Martinet [17] and Yamanouchi [18] originally formulated the simple top-down quench problem. Their assumptions follow, with an indication of their effects on

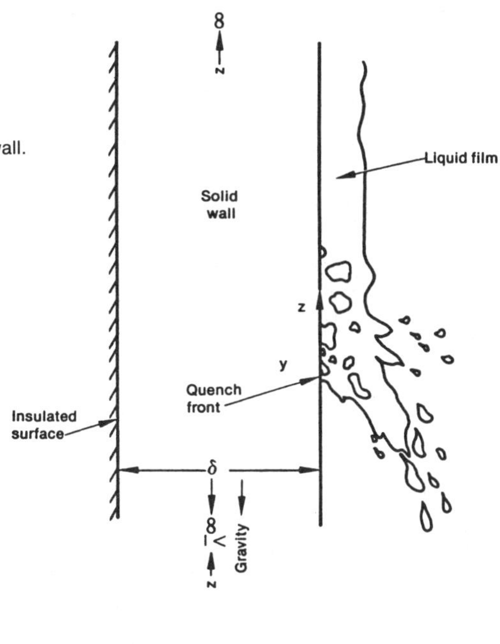

Figure 4. Falling liquid film quench of a hot wall.

the formulation. The discussion of later solutions to this and other more complex quenching situations will be done by relaxing these assumptions. The assumptions are as follows:

1. Constant physical and thermal properties in an infinite length wall of uniform thickness, δ.
2. Uniform behavior of the quench front with respect to depth of the wall. This eliminates temperature dependence in the x-direction, $\partial T/\partial x = 0$, and results in a two-dimensional problem.
3. One side of the wall is insulated. This is equivalent to the problem of a wall 2δ in thickness, with identical quench fronts on both sides.
4. No heat generation in the wall, $q' = 0$.
5. The coolant temperature is constant at the saturated value.
6. No transverse temperature gradient exists within the wall, $\partial T/\partial y = 0$. This assumption is valid for either thin walls or the case of low h's, with low quench front velocities. This latter case is represented by $Bi \ll 1$ and $P \ll 1$. Since the heat transfer coefficient varies with respect to position on the wall, the definition of the Biot number raises the question of which h to use. Assumption 8 will help clarify the question. This assumption, coupled with the second assumption, reduces the problem to a one-dimensional formulation.
7. A constant quench front velocity, u_0, produces an invariant temperature distribution, relative to the moving quench front. This reduces the rate of change of temperature with respect to time to

   $$\partial T/\partial t = (\partial T/\partial z)(dz/dt) = -u_0(\partial T/\partial z)$$

   and is obtained by transforming the z-axis, using the expression $z = z - u_0 t$, which produces a moving coordinate system, located at the quench front.
8. Generally, the local heat transfer coefficient at a point on the wall would be specified as being a function of x and z, i.e. $h_{local} = h(x, z)$. Assumption 2 eliminates the x dependence, to reduce the heat transfer coefficient to $h_{local} = h(z)$. Assumption of a two-region convective heat transfer model, with the wet region having a constant h and the dry region having an h of zero, is written in terms of the local heat transfer coefficient as

   $$h_{local} = \begin{cases} h_w & \text{if} \quad z \geqq 0, \\ 0 & \text{if} \quad z < 0. \end{cases}$$

   The very localized sputtering region is lumped into the wet region. This type convection model would have two Biot numbers associated with it. One is the wet region Biot number; the other, a dry region Biot number, with the dry region Biot number being zero. Thus, for assumption 6, the Biot number defined applies to the wet region. For more complex convective models, care must be exercised as to the manner in which Biot numbers are defined and used.

Use of these assumptions to reduce and solve the diffusion equation, as given by Equation 2, produced the solution [18] given in dimensionless form by

$$Bi^{1/2}/P = [\theta_q(\theta_q - 1)]^{1/2} \tag{11}$$

Here the dimensionless temperature has been redefined as

$$\theta_q = (T_i - T_{sat})/(T_q - T_{sat}) \tag{12}$$

where T_{sat} is the saturation temperature (°K), and T_q is the quench temperature (°K). T_i is the dry wall temperature at infinity, which is thought of as the initial wall temperature. Conceptually, the quench temperature should correspond to ΔT_{min} (Figure 1) and be that temperature at which the controlling mechanism at the wall changes from predominately vapor cover to a significant contribution, resulting from liquid-solid contact. In practice, due to the simple form of the solution and convective heat transfer model, the quench

Table 1
One-Dimensional Quenching Models

Reference	Experimental data correlated	Heat-transfer coefficients, $W/(m^2)(K)$	Quench temperature K	Comments and heat-transfer coefficient profile
Yamanouchi[18] Semeria and Martinet[17]	Yamanouchi[18]	$h_2 = 2\times10^5 - 10^6$ $h_3 = 0$	423	h_2 h_3
Yao[19]	No experimental data correlated			Constant internal heat generation included, assumes parabolic radial temperature profile h_2 h_3
Thompson[20]	Bennett et al[21]	$h_1, h_2 = 7\times10^6$ (peak values) $h_3 = 0$	T_{sat} + 373K	High-pressure data 6.9-69 bars abs. $h_2 = R\,\Delta T_s^3$ h_3
Sun et al[22]	Yamanouchi[18] Duffey and Porthouse[23] (only low-flow-rate data)	$h_1 = 570?$ $h_2 = 1.7\times10^4$ $h_3 = 0$	533	Sputtering region between location of incipience of boiling and quench front h_1 h_2 h_3
Ishir[24]	Bennett et al[21]	$h_1 = h_{CHF}$ $h_2 = 4\times10^5$ $h_3 = 0$	533-663	1. Sputtering region between CHF and quench front 2. Correlated a pressure range of 6.9-69 bars abs. 3. Defined thermal penetration length in sputtering region h_1 h_2 h_3
Sun et al[25]	Yamanouchi[18] Duffey and Porthouse[23]	$h_2 = 1.7\times10^4$ $h_3 = (h_2/N)_e 0.032$	533	Precursory cooling included h_2 h_3
Chun and Chon[26]	Case et al[27]	$h_2 = 2.56\times10^4$ $h_3 = 170$ $h_4 = 0$	533	Calculated length of dispersed flow region and mass carry-over to correlate h_3 h_2 h_3 h_4
Chan and Groimes[28]	Guerrero et al[32]			Used heat flux approximation 1. Area under boiling curve to CHF for sputtering 2. Radiation model for disposed flow $q_{sputter}$ rad
Andréoni[35]	Andréoni[35]	h_1, from Jens-Lottes correlation[36] h_2, extracted from rewetting data h_3, from experimental data	T_{sat} + 473K	h_1 h_2 h_3
Tan[37]	Tan[37]		Iloeje[38]	1. Boiling curve from steady state correlations 2. Thermal non-equilibrium included 3. Internal heat generation 4. Finite difference wall segments with varying h
Elias and Yadigaroglu[39]	Duffey and Porthouse[23] (only low-flow-rate data)	$h_1 = 170$ $h_2, h_3, h_4, \ldots$ = boiling curve approximation	533	1. Variable wall with temperature properties 2. Internal and/or external heat generation h_1 h_2 h_3 h_N
Kirchner[41]	FLECHT[42, 43]			1. Multiple region model 2. Boiling curves from steady state correlations
Fischer et al[44]	FLECHT[45] FLECHT-SET[46] Semiscale[47]			1. Miltiple region model 2. Boiling curves from steady state correlations

From Elias and Yadigaroglu [4].

temperature has been correlated with data and has been found to vary significantly. This results in the quench temperature being more like a correlation parameter than denoting the occurrence of altering physical phenomenon.

Table 1, which has been adapted from Elias and Yadigaroglu [4], provides a summary of some of the more significant one-dimensional models available. It presents them in order of increased complexity. As shown, Yamanouchi's results require a heat transfer coefficient of up to 10^6 W/m$^2 \cdot$ °K, which is very large to apply across the entire liquid wetted region. As observed by Carbajo and Siegel [16], the Yamanouchi model is conservative across most conditions and is accurate for only small values of Bi.

The effect of including internal heat generation, assumption 4, and nonzero wet side heat transfer, were investigated by Yao [19]. For this two-region heat transfer model, the temperature of the surrounding gas or vapor was assumed to be T_i, which is the initial temperature of the dry-side wall. This is different than the saturated conditions assumed by most models and represents an effort to more realistically model the dispersed flow region. The internal heat generation was simulated with a constant heat flux boundary condition replacing the insulated condition on the wall's dry side (assumption 3). Using a nonzero heat transfer coefficient on the wet-side also produces a nonzero wet-side Biot number (see assumption 8), so that the wet-side and dry-side quantities will be denoted with w and d subscripts (giving Bi_w and Bi_d as the wet-side and dry-side Biot numbers, respectively). A parabolic radial temperature profile was assumed, instead of using assumption 6. The solution obtained by use of an approximate integral technique is given by

$$P = \sqrt{2}\,(C_2^2A_d - C_3^2A_w)/\{8[C_1^2(C_3^2A_w + C_2^2A_d) + (C_2C_3)^2 \cdot (A_d + A_w) - (C_2^4A_d + C_3^4A_w)]\} \tag{13}$$

where
$$C_1 = Q[(1/Bi_w) - (1/Bi_d)] - 1$$
$$C_2 = [(\theta_q - 1)/\theta_q] + (Q/Bi_d)$$
$$C_3 = [(\theta_q - 1)/\theta_q] + (Q/Bi_d) - 1$$
$$Q = q'\delta/[k(T_i - T_{sat})]$$

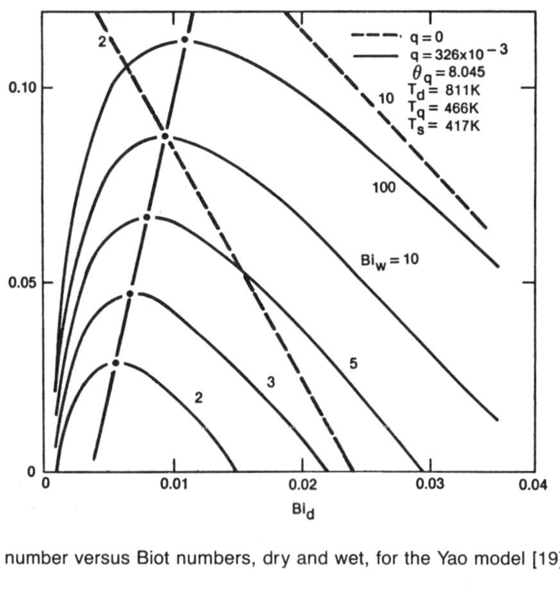

Figure 5. Peclet number versus Biot numbers, dry and wet, for the Yao model [19].

is the dimensionless internal heat source, and A_w and A_d are the effective wet and dry Biot numbers, given by

$$A_w = 3\,Bi_w/(3+Bi_w)$$

$$A_d = 3\,Bi_d/(3+Bi_d)$$

Figure 5 shows results [16] from Yao's model, Equation 13, for the two cases of no internal heat generation and a typical internal heat generation rate. In the case of no internal heat generation, the quench front rate is shown to decrease as Bi_d increases. This is caused by the assumption that the vapor or gas surrounding the dry wall is at T_i. As the dry side heat transfer improves, the vapor heats the wall near the quench front, slowing its progress. This characteristic of the model may or may not be acceptable to a user, depending upon the geometry of his problem. For example, with a large reservoir of vapor, having no close boundary in the negative y-direction associated with a quenching wall, this representation is realistic. However, for a limited, or small reservoir, having a close boundary in the negative y-direction, or a reservoir whose local temperature changes with the quench front motion, this model may not be realistic. Also shown in Figure 5 is a typical case of internal heat generation. Comparison with the unheated case shows that internal heat generation slows the progression of the quench front. Also for internal heat generation, there exists a maximum quench front velocity associated with each Bi_w. This characteristic is again associated with the T_i assumption for the vapor temperature. For $Bi_d < Bi_{d,\,optimum}$, internal heat generation heats the wall hotter than the vapor temperature so the increasing Bi_d increases the quench front velocity. For $Bi_d > Bi_{d,\,optimum}$, the dry wall ahead of the quench front is once again heated, instead of cooled, by the vapor. An expression for the heat generation rate required to fix the quench front location was obtained by setting $P = 0$ and is given by

$$Q_{fixed} = [\sqrt{A_w} - (\theta_q - 1)\,(\sqrt{A_w} + \sqrt{A_d})/\theta_q] \cdot (Bi_w + Bi_d)/(Bi_w\sqrt{A_d} + Bi_d\sqrt{A_w})$$

In an effort to account for the high heat transfer coefficients of Yamanouchi, Thompson [20] changed the wet region constant heat transfer coefficient of assumption 8 to a third power relationship, i.e. the heat transfer coefficient was taken as proportional to $(T_s - T_{sat})$ [3], more representative of nucleate boiling. Extremely high heat flux at the quench front was hypothesized as being potentially due to evaporation of a microlayer of liquid on the wall. The data from Bennett et al. [21] were used to correlate variables in the model.

In an attempt to improve the modeling of the sputtering region, Sun et al. [22] used a three-region model to replace the two-region model of assumption 8. The model replaced the single wet region constant heat transfer coefficient with two constant values, where the sputtering region was represented by a larger coefficient than the remainder of the wetted region. The sputtering region was defined as existing between the incipience of boiling and the quench front. For the range of validity of the one-dimensional theory, good agreement was found within the data of Yamanouchi [18] and Duffey and Porthouse [23].

In another effort to better represent the heat transfer in the sputtering region, Ishii [24] replaced the two-region heat transfer model with a three-region model, similar to that of Sun et al. [23]. The sputtering region was defined as existing between the CHF point and the quench front. From this factor, new dimensionless parameters were defined as

$$\theta^* = (T_i - T_q)/(T_q - T_{CHF})$$

$$P^* = \sqrt{\pi}\,(P/Bi)\,(T_q - T_{CHF})/(T_q - T_{sat})$$

$$Bi^* = (\pi/Bi)\,(T_q - T_{CHF})/(T_q - T_{sat})$$

to obtain a thin wall solution given by

$$Bi^{*1/2}/P^* = \theta^{*1/2} \quad \text{for} \quad \theta^* \leqq 1 \tag{14}$$

$$Bi^{*1/2}/P^* = [\theta^*(\theta^*+1)/2]^{1/2} \quad \text{for} \quad \theta^* > 1 \tag{15}$$

where, for thin walls, $Bi^* > (\theta^*+1)^{-2}$.

In an effort to relax the one-dimensional assumption given by 6 and to study thick walls ($Bi^* < (\theta^*+1)^{-2}$), Ishii implemented a penetration depth, λ, for which a uniform temperature distribution could be considered. This preserved the mathematical simplicity of the one-dimensional solution. The quench front velocity and sputtering length were obtained explicitly for both thin and thick walls, where λ was used as an effective wall thickness in the sputtering region for thick walls. Accuracy of the thick wall results depends upon the estimate of the penetration depth and were unrealistically short [4] for the data of Bennett et al.

Carbajo and Siegel [16] compared the results of the three-region models of Sun et al. [22] and Ishii's thin wall model [24] relative to Yamanouchi [18]. This comparison was done using the dimensionless temperature of Sun et al., given by

$$\theta_b = (T_b - T_{sat})/(T_q - T_{sat})$$

where T_b is the temperature (°K) at which boiling begins. Sun et al. define the sputtering region as being between the incipience of boiling and the quench front, while Ishii defines the sputtering region as beginning at the CHF point. To make this comparison, T_b was set equal to T_{CHF}. Figure 6 shows a comparison of these models for the case shown. For the Sun et al. model, the case of $\theta_b < .2$ which results when $T_b \rightarrow T_{sat}$, yields Peclet numbers the same as Yamanouchi. When $\theta_b \rightarrow 1$, which results as $T_b \rightarrow T_q$, two cases are shown, depending upon the ratio of wet region heat transfer coefficients. For the case of no heat transfer in the first wet region, $h_{wl} = 0$, no quench front motion results. When $h_{wl}/h_{w2} = 0.1$, the Peclet number is about 32% of Yamanouchi's. This generally shows the importance of modeling the sputtering region. If the quenching fluid has a $\theta_b < .2$, the results of a three-region model are the same as those of a two-region model.

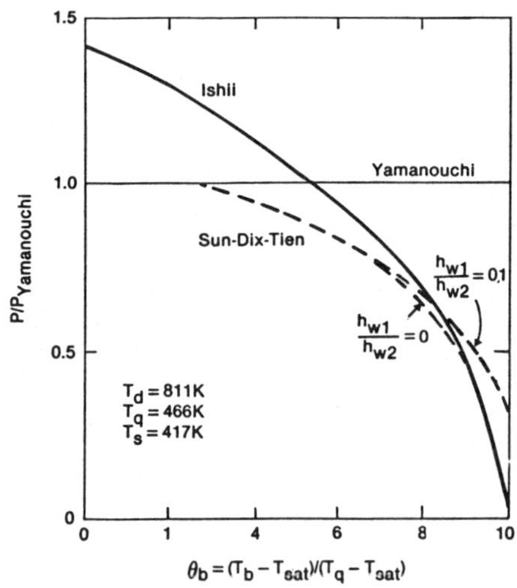

Figure 6. Comparison among Yamanouchi [18], Ishii [24], and Sun-Dix-Tien [22] models [16].

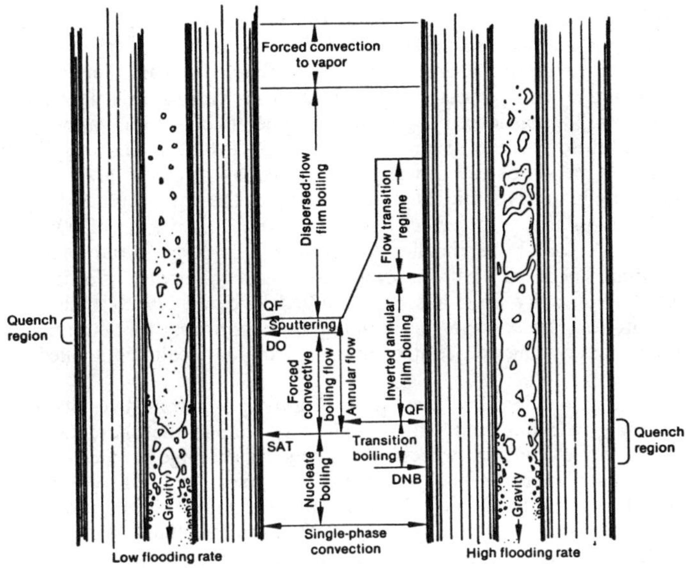

Figure 7. Flow and heat transfer regimes observed during bottom quenching: IB = incipience of boiling, DNB = departure from nucleate boiling, DO = dryout, QF = quench front, OE = onset of entrainment, SAT = slug annular transition.

Carbajo and Siegel also show that Equation 14 of Ishii's thin wall model differs from Yamanouchi by a factor of $1/\theta_b$, which is always greater than 1 and produces Peclet numbers greater than Yamanouchi's. Figure 6 shows the results of Equation 15. As $\theta_b \rightarrow 0$, Ishii's results differ by a factor of $\sqrt{2}$. While, as $\theta_b \rightarrow 1$, the results of Ishii and Sun tend to agree, if h_{w1}/h_{w2} is small. For fluid where θ_b and Bi are small, the Yamanouchi model is appropriate, and the Ishii model produces Peclet numbers which are too high.

To investigate the cooling effects of the droplets thrown off the wall at the quench front, Sun et al. [25] used a two-region model, but replaced the zero heat transfer coefficient for precursory cooling in the dispersed region with a coefficient decaying exponentially with respect to distance from the quench front. A solution containing Bessel functions was obtained. Quench front velocities, an order of magnitude higher than those without precursory cooling, were found possible, with the solution approaching Yamanouchi's for low precursory cooling. For low coolant flow rates, and consequent low quench front rates, the effect of precursory cooling is negligible. For high coolant flow rates, precursory cooling becomes significant. Duffey and Porthouse [23] data was correlated in order to obtain a representation for the heat transfer coefficient, with respect to flow rate. The resulting correlation is valid only within the data base, due to some subcooled data. This produces Peclet numbers as high as 10 resulting in questionable application of the one-dimensional theory.

Chun and Chon [26] also modeled the dispersed region. They used a three-region constant heat transfer coefficient to represent the wet region, dispersed flow region, and superheated vapor region. The hydraulic portion of the problem was represented by using the continuity equation for the coolant. A simple heat balance was used to determine the quantity of coolant in the dispersed region and the degree of vapor superheat in the vapor region. The data of Case et al. [27] was used during the model development.

The models just discussed have addressed the phenomenon of falling film quenching, which is a conduction-controlled process. This type quenching requires that the liquid be free to fall away from the plate once it is thrown from the wall. This requires either an unrestricted geometry, with respect to the liquid-vapor region, or a low liquid feed rate. Chan and Grolmes [28] considered the hydrodynamic effects of the vapor upon the liquid film. Countercurrent flow would normally occur for the geometry shown in Figure 3, where the liquid film and the drops fall downward (defined by direction gravity acts) while the vapor flows up. The hydrodynamic interaction between the liquid film and the vapor would be insignificant, unless the vapor velocity increased above a critical value referred to as the flooding velocity. At or above the flooding velocity, large-amplitude surface waves are created in the film. These may halt or reverse its flow significantly affecting the quenching rate. This flooding stability limit has been studied in detail [29–31]. Chan and Grolmes developed a criterion, based upon a flooding limit concept, to determine whether the rewetting process was conduction-controlled or hydrodynamically-controlled in a top-down quench.

Considering the case where the orientation of gravity is reversed results in what may be called bottom-up quenching, and may significantly change the characteristics associated with the convective heat transfer process. Generally, this change is reflected in the influence of precursory cooling upon the quench front. Two typical hydraulic configurations which may be encountered for this type of quenching are shown in Figure 7. This type of quenching requires some sort of confinement for the liquid, so that gravity may ensure liquid contact with the wall. Figure 7 shows two hot walls for two limiting cases of low and high liquid feed rates, called flooding rates. For low flooding rates, vapor generation below the quench front may throw some liquid droplets out of the bulk liquid volume. If the vapor velocity, relative to the size of the drop, is sufficient, the drop may be carried up the channel. This phenomenon is called entrainment and has received significant study [4, 33, 34]. For low flooding rates, entrainment provides some precursory cooling above the quench front and results in an annular flow regime. As shown, the annular flow regime has vapor flowing up the center of the channel, with a liquid film on the walls. For high flooding rates, the liquid is effectively driven past the quench front, resulting in an inverted annular flow film boiling regime. The inverted annular flow regime inverts the positions of the liquid and vapor, resulting in a liquid core, and provides significant precursory cooling above the quench front, when compared to the annular flow regime. The models discussed in Table 1 were developed to model the falling-film-type quenching and are not representative of the bottom-up type quenching. This is primarily due to the model of the convective heat transfer process assumed, i.e. assumption 8, and is not a limitation of the other assumptions.

Andreoni [35] used a three-region model to represent the convective heat transfer process in connection with the one-dimensional wall. Using data from the quenching of a tube with internal upflow, the quench front heat transfer coefficient was back-calculated using measured precursory cooling data and the Jens–Lottes nucleate boiling correlation [36]. Since Andreoni's tube was thick walled, the resulting heat transfer correlation is influenced by the application of the one-dimensional solution to the tube.

In order to study the relationship of conduction and precursory cooling, for bottom-up quenching with thin tubes, Tan [37] developed a one-dimensional conduction and hydraulic model, formulated in numerical form using a finite difference scheme. The process was assumed to be quasi-steady state, and a boiling curve, which included thermal nonequilibrium in the dispersed flow region, was used to obtain the convective heat flux from each wall segment. Thermal nonequilibrium exists when either the liquid or vapor, or both, in a two-phase mixture are not at the saturation temperature. In order to develop a criterion for determining when the effects of conduction should be included, a dimensionless parameter, β, was defined as

$$\beta = L(h\tilde{P}/kA)^{1/2}$$

where L was the quenching zone length, spanning the region from $\Delta T_{sat} = 0$ to ΔT_{min}, $\tilde{P}$ was the flow channel perimeter, and A was the channel flow area. The heat transfer

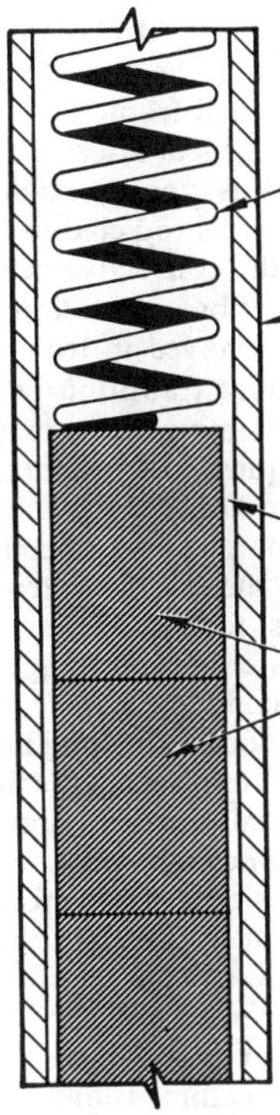

Figure 8. Side view of top of typical nuclear fuel rod.

coefficient, h, was defined as a length-averaged quantity over the distance L. For thin tube experimental data, also taken by Tan, the quenching process was found to be dominated by convective heat transfer when $\beta > 6$.

In an effort to generalize the heat transfer for either the falling film or bottom quenching cases, Elias and Yadigaroglu [39] developed a one-dimensional model, similar in concept to that of Tan [37], which divided the wall into an arbitrary number of segments. Variations of the physical properties of each segment, due to rapid temperature changes near the quench front, were considered (see assumption 1), as well as accounting for the variation in the heat transfer coefficient. Conventional steady-state heat transfer correlations were used with the model to compare to the low flow rate top-down quenching data of Duffey and Porthouse [24]. The effects of the decay heat associated with nuclear fuel rods was also studied, using the model [40].

The one-dimensional quenching models discussed so far have included conduction in the direction of the quench front motion, z-direction. As has been noted, this model applies to either thin walls or quenching processes where $Bi \ll 1$ and $P \ll 1$. A few investigators [41, 44] have used a one-dimensional model in the y-direction (see Figure 4). This is based upon the assumption that the convection process dominates the quenching process, with conduction in the z-direction moving little heat energy compared to that moved in the y-direction. In order for this assumption to be valid, a nonhomogeneous material with a higher resistance to heat flow in the z-direction than in the y-direction must be present. Application of this type of model has been primarily to the area of nuclear reactor safety and the modeling of nuclear fuel rods, as shown in Figure 8. This type of structure is nonhomogeneous and requires a numerical approach, such as the finite element or finite difference schemes, as opposed to an analytical solution. As shown by Tan's model, a thin wall may become convection dominated, justifying neglect of conduction effects in the z-direction. For the fuel rod cladding undergoing a rapid enough transient, the clad is effectively isolated from the stored energy of the interior fuel pellets [48], justifying the assumption at the quench front. The y-axis dependence is retained in the model to remove energy from the fuel pellets after the quench front has passed.

The problem with this type of one-dimensional model is the requirement that the transient be fast enough for the fuel rod clad to become conduction limited at the quench front. When this is not the case, the convective heat transfer used with the model must compensate for the lack of heat moved by conduction in the clad in the z-direction by artificially increasing the convective heat removed in front of the quench front. This generally involves raising ΔT_{min} in order to cause transition boiling sooner than it actually occurs.

In general, for a homogeneous material, application of such a one-dimensional model in the y-direction requires the instantaneous immersion of the hot body in the fluid, as has been discussed earlier. Otherwise, as noted by Tien and Yao [49], once a quench front enters the consideration for a homogeneous material, no matter how fast the quench front progresses ($P \gg 1$), the problem requires a two-dimensional representation. Thus, the practical problem is to determine what can be called instantaneous. This requires a comparison of the rate of quenching to the rate of coverage of the body by the quenching fluid. If a body is completely covered by the fluid, with very little of the body being quenched, or even before quenching begins, then instantaneous immersion can be assumed. The term "coverage" implies then a heat transfer model. With complete immersion, and no quench front, a uniform heat transfer coefficient is more likely to be justified. If the quenching rate is sufficiently high, or the coverage rate slow enough that a significant portion of the body is quenched before it is covered, an instantaneous immersion assumption is not valid. The heat transfer cannot be modeled with a single heat transfer coefficient in this situation.

Two-dimensional conduction models delete assumption 6 and eliminate the limitations that the wall be thin or that the quench front velocity be small. Semeria and Martinet [17], in addition to their one-dimensional model, also formulated the two-dimensional problem but did not obtain a solution. Table 2, again adapted from Elias and Yadigaroglu [4], summarizes some of the more significant two-dimensional models which have been de-

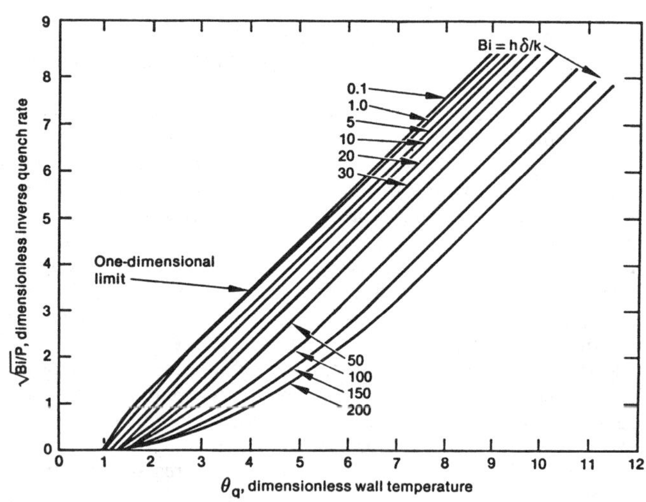

Figure 9. The calculated variation of $(Bi/P)^{1/2}$ with dimensionless wall temperature [56].

Table 2

Two-Dimensional Quenching Models

Reference	Experimental data correlated	Heat transfer coefficient (W/m^{2}K)	Quench temperature K	Comments and heat transfer coefficient profile
Duffey and Porthouse[50]	Yoshioka and Hasewaga[51] Yamanouchi[18] Duffey and Porthouse[23, 50] Andréoni[35] Martini and Premoli[52] Thompson[53] Campanile and Pozzi[54]	$h_2 = 10^4 - 2 \times 10^6$ $h_3 = 0$	463-523	h_2 h_3
Coney[56]	Bennett et al[21]	$h_2 = 0.94 - 1.3 \times 10^6$ $h_3 = 0$	T_{sat} + 341K	Pressure range 6.9-69 bars abs. h_2 h_3
Andersen and Hansen[58]	No experimental data correlated			h_2 h_3
Blair[55]	Thompson[59]	$h_2 = 1.7 \times 10^4$ $h_3 = 0$	533	Cylindrical geometry h_2 h_3
Yeh[60]	No experimental data correlated			Cylindrical geometry
Thompson[20]	Bennett et al.[21]	$h_1, h_2 = 4 - 8 \times 10^5$ (peak values) $h_3 = 0$	T_{sat} + 373K	Numerical solution $h_2 = R\,\Delta T_{sat}^3$ h_3
Tien and Yao[49]	No experimental data correlated			Wiener-Hopf technique h_2 h_3
Edwards and Mather[59]		$h_{max.} = 2 - 4 \times 10^5$		$q = q_0 e^{bz}$ $q = q_0 e^{az}$
Dua and Tien[65]	Duffey and Porthouse[23] Yamanouchi[18]	$h_2 = 1.7 \times 10^4$ $q_0 = q_0/N_e^{-az}$	533	Wiener-Hopf technique with precursory cooling h_2 q_3

From Elias and Yadigaroglu [4].

veloped. As with the one-dimensional models, these two-dimensional models are arranged with the simpler, top-down quench first and proceeding in order of increasing complexity.

Duffey and Porthouse [50] used a two-region heat transfer model and an infinite series representation of the temperature distribution for each region. The data of References 18, 23, 35, 50–54 were correlated to obtain the expression

$$\mathrm{Bi}/P = \pi[\theta_b(\theta_b - 4\mathrm{Bi}/\pi^2)]^{1/2}/2$$

for $1 \ll \mathrm{Bi} < \pi\theta_b/4$. This latter requirement is needed in order to keep the quantity in the radical positive and thus real. The solution reduced to the one-dimensional solution of Yamanouchi (Equation 11), for small Biot numbers. For large Biot numbers, a solution independent of wall thickness or thermal conductivity was found and is given by

$$\mathrm{Bi}/P = \pi\theta_b/2 \tag{16}$$

Blair [55] later showed the approximation to be inaccurate, due to use of only the first term of the series, since the series was a slowly converging one.

Coney [56] applied the separation of variables to the series solution in order to obtain a more accurate solution. Figure 9 plots the results, which can be approximated by

$$\mathrm{Bi}/P = 1.6\theta_b(\theta_b - 1) \tag{17}$$

Equation 17 applies when Bi/P < 1 and represents an approximate solution for higher values of Bi/P. Yu, Farmer, and Coney [57] presented the same solution in the form of tables and graphs.

Andersen and Hansen [58] used both finite difference and eigenfunction solution techniques to match the previously obtained results of Yamanouchi [18] and Duffey and Porthouse [50] for the two-dimensional, top-down quenching problem. Their results were presented in terms of modified Biot and Peclet numbers, as defined by

$$\overline{\mathrm{Bi}}_1 = \mathrm{Bi}[\theta_q(\theta_q - 1)]^{-\sqrt{\pi}/2}$$

$$\overline{\mathrm{P}}_1 = \mathrm{P}[\theta_q(\theta_q - 1)]^{(1-\sqrt{\pi}/2)/2}$$

The numerical results had two asymptotes, as represented by

$$\overline{\mathrm{P}}_1 = \overline{\mathrm{Bi}}_1^{1/2} \qquad \text{for small } \overline{\mathrm{Bi}}_1$$

$$\overline{\mathrm{P}}_1 = 2^{-(\sqrt{\pi}/4)}\,\overline{\mathrm{Bi}}_1 \qquad \text{for large } \overline{\mathrm{Bi}}_1$$

Blair [55] also used separation of variables and a series solution to solve the two-dimensional problem for a cylindrical rod. However, all the terms of the series were retained by deriving an accurate asymptotic representation for the terms, with the asymptotic solution given by

$$\theta_q = 1 + [2 + (\pi/2\mathrm{P}) + (2/\mathrm{P}^2)]\mathrm{Bi}/(\pi\mathrm{P}) + [(8/3) - \pi^2 - (13\pi/6\mathrm{P})]\mathrm{Bi}^2/[2(\pi\mathrm{P})^2] \qquad (18)$$

for P > 1 and Bi/P < 1. A first-order approximation for P ≫ 1 yields

$$\mathrm{Bi}/\mathrm{P} = \pi(\theta_q - 1)/2$$

which is different than the solution of Duffey and Porthouse as given by Equation 16. This difference is attributed to the use of only the first term of the series expansion by Duffey and Porthouse. A typical distribution of the temperature along the z-axis is shown in Figure 10. Yeh [60] later obtained similar results.

Thompson [20], in addition to his one-dimensional model using a third power representation of the wet front heat transfer, developed a similar two-dimensional model.

In an effort to generalize the two-dimensional solutions and to develop an understanding of the limiting cases, Tien and Yao [49] utilized the Wiener–Hopf technique [61, 62] and

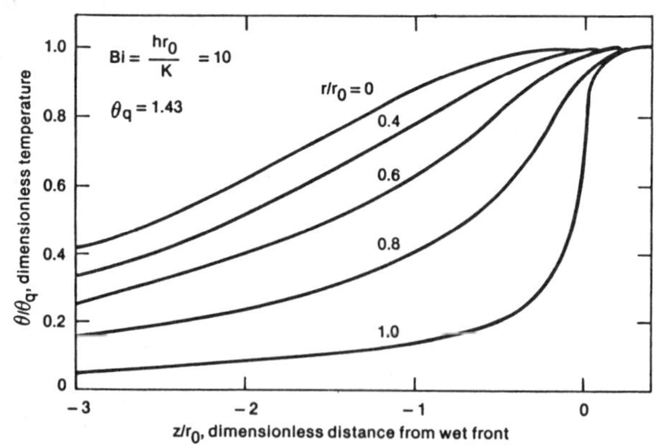

Figure 10. Axial temperature distributions for a rod or radius r_o [55].

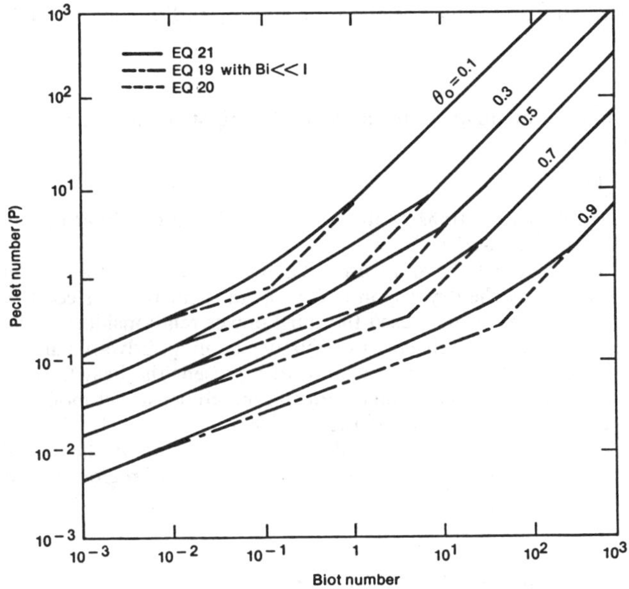

Figure 11. Variations of Peclet number in falling-film rewetting [49].

the kernel-substitution method [62] in order to obtain a two-dimensional solution of the two-region heat transfer model. The solution for small Peclet numbers reduces to

$$A/P^2 = [\theta_q(\theta_q - 1)]^{1/2} \tag{19}$$

where A is the effective Biot number as defined earlier for Equation 13. For small Biot numbers, $A \approx Bi$, so that Equation 19 reduces to Yamanouchi's result. For large Peclet numbers, the solution was

$$Bi/P = [(1+\sqrt{2})(\theta_q - 1)/\sqrt{2}] + [(1+\sqrt{2})^2(\theta_q - 1)^2/4] \tag{20}$$

The first term of Equation 20 is similar to Blair's asymptotic solution, given by Equation 18, with the constants being slightly different, i.e., 1.707, as compared to 1.571. Figure 11 shows a numerical solution obtained to fit the limiting cases, and is given by

$$\begin{aligned} Bi(1-\theta_0)/P\theta_0 = {} & [P/(1-\theta_0)]\,((1.707-0.25\theta_0)\,[3(1-\theta_0)^2 \\ & + P^2\theta_0]/\{[3(1-\theta_0)^2\,(1.707-0.25\theta_0)] \\ & + P[3(1-\theta_0)^2 + P^2\theta_0]\}) \end{aligned} \tag{21}$$

where $\theta_0 = (\theta_q + 1)/\theta_q$. Substitution of Equation 12 into the expression for θ_0 yields

$$\theta_0 = (T_i - T_q)/(T_i - T_{sat})$$

Dua and Tien [63] further simplified, and improved, the relationships to represent the limiting cases. Their relationships are given by

$$P = \overline{Bi}_2^{1/2} \quad \text{for} \quad \overline{Bi}_2 < 0.3 \tag{22}$$

$$P = 0.63\,\overline{Bi}_2 \quad \text{for} \quad \overline{Bi}_2 > 20.0$$

An improved semi-empirical matching relationship to replace Equation 21 was also found and is given by

$$P = [\overline{Bi}_2(1+.4\,\overline{Bi}_2)]^{1/2} \tag{23}$$

where $\overline{Bi}_2$ is a modified Biot number similar to that suggested by Andersen and Hansen [58] and is given by

$$\overline{Bi}_2 = Bi/[\theta_q(\theta_q-1)]$$

Figure 12 shows the plot of Equations 22 through 23, with the two limiting cases and the numerical results of Andersen and Hansen.

These two-dimensional models have all used the two-region heat transfer model, with zero heat transfer assumed in the dry region and constant heat transfer coefficients in the wet region, except for Thompson, who used the third power relationship in the wet region, as previously noted. To include the effects of precursory cooling, Edwards and Mather [64] used an exponential decay, with respect to length, to represent the heat transfer on either side of the quench front. As with the results from the one-dimensional models, precursory cooling was found to speed the advance of the quench.

Dua and Tien [65] also included the effects of precursory cooling on the two-dimensional model by assuming an exponentially decaying heat flux in the dry region. For $P \ll 1$, they obtained the relationship

$$P = Y - A_w/Y$$

where
$$Y = [\{A_w/[(\theta_q-1)Na\delta]\} + (\{A_w/[(\theta_q-1)Na\delta]\}^2 + 4A_w\theta_q/(\theta_q-1))^{1/2}]/2$$

and N and a are parameters used to define the precursory cooling or dry region heat transfer coefficient, h_d, as

$$h_d = (h_w/N)e^{-az}$$

For the other limiting case, $P \gg 1$,

$$P = 2\,Bi_w/(Z^2-1)$$

where Z is the real positive root of the cubic equation

$$Z^3 + \{\sqrt{2}\,Bi_w/(Na\delta) - [\theta_q(1+\sqrt{2})-1]/\sqrt{2}\}Z^2$$
$$+ [(N-1)(\sqrt{2}-1)Bi_w/(Na\delta)-1]Z + \{(N-1)Bi_w/(Na\delta)$$
$$+ \theta_q[(\theta_q-1)/(\sqrt{2}\theta_q) + 1 - (\theta_q-1)Bi_w/(\sqrt{2}\theta_q a\delta)]\} = 0$$

To bridge between the limits, they obtained the following semi-empirical relationship:

$$nP/(1-.01P^2) = n[(Z^2-1)/2Bi_w]\,[Y-(A_w/Y)] / \{[(Z^2-1)/2Bi_w] - .01[Y-(A_w/Y)]\}$$

with $n < 1$. For the data of Yamanouchi [18] and Duffey and Porthouse [23], $n = .01$ was found to provide the best agreement. Figure 13 provides results [16] obtained from the model for typical conditions, and uses $a = 0.05\ cm^{-1}$ as recommended. For values of $Bi_d > 0.0025$, it is shown that unrealistic results of decreasing quench velocity, with respect to increasing Bi_w are obtained. Comparing this result with the data used by Dua and Tien, Carbajo and Seigel [16] recommended that application of the model be limited to the range $0 < Bi_d < 0.0025$ and $a = 0.05\ cm^{-1}$.

Bonakdar and McAssey [66] combined the two-dimensional conduction model with the variable heat transfer coefficient model, as used by Tan [37] and Elias and Yadigaroglu [39]. This representation may be used to model both top-down and bottom-up quenching. A

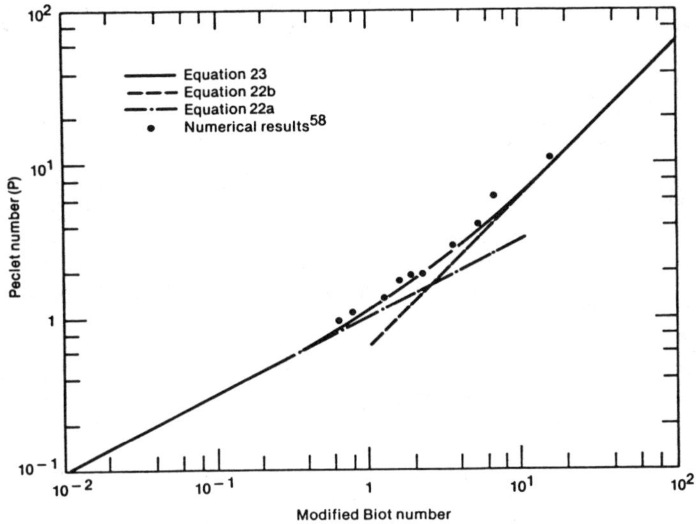

Figure 12. Variations of the modified Peclet number with respect to the modified Biot number [63].

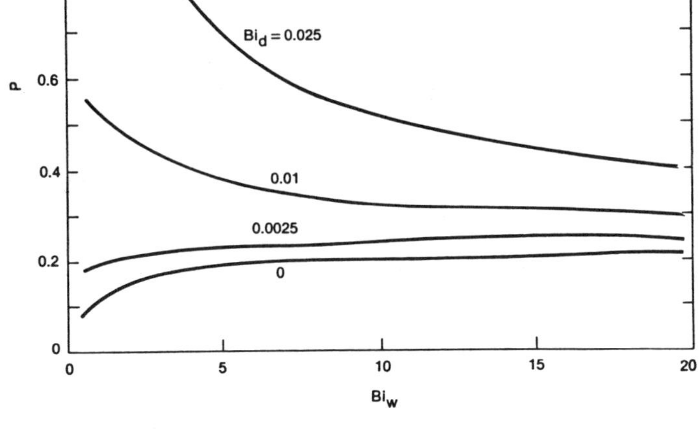

Figure 13. Effect of the dry side Biot number in the Dua and Tien [65] model [16].

numerical solution procedure was outlined, using a series expansion for each uniform heat transfer coefficient region and matching the temperature and heat flow (temperature gradient) at the interfaces between regions. The procedure was specialized to study a three-region heat transfer model similar to that obtained by Sun et al. [22]. Figure 14 presents a comparison of their results, given by

$$\bar{\theta}_s = (.551 + .228 \ln P) - \ln Bi(.111 - .032 \ln P) \tag{24}$$

to those of Sun et al. and the data of Yamanouchi [18] and Duffey and Porthouse [50]. In Equation 10, the fluid temperature, T_f, has been set equal to the saturation temperature, T_{sat} and denoted by θ_s. The bar over θ_s indicates an arithmetic average of θ_s at $z = 0$ for the sputtering region wall temperature and the dry wall temperature. Figure 14 is plotted as a function of θ_1, a dimensionless temperature used by Sun et al., where

$$\theta_1 = (1/\bar{\theta}_s) - 1 = (T_i - T_q)/(T_q - T_{sat})$$

and requires the assumption that the quench temperature, T_q, be equal to the arithmetic average of the sputtering region wall temperature and dry wall temperature. The two solutions agree well for $\theta_1 \geqq 0.2$. Equation 24 provides better results for $\theta_1 < 0.2$.

Considerable work has been done for bottom-up and top-down quenching situations, as has been discussed. Very little work has been done for other orientations of the quenching surface, with respect to gravity. The work of Chan and Banerjee [67–69] presents study of a horizontal system with an advancing quench front. The experimental work [67] was done on a thin walled tube with water fed from one end. A one-dimensional, two-fluid model was explicitly coupled with a three-dimensional conduction model of the tube wall and solved numerically. Steady-state heat transfer correlations were used, the boiling curve approach being somewhat like that employed by Tan [37] and others, but using correlations applicable to horizontal boiling. The model is very complex and is not easily applied.

CONVECTION

The preceding sections have discussed, in some detail, the numerous quenching models which have solved the diffusion equation, using different convective heat transfer boundary conditions and different assumptions concerning the quenching geometry and quenching surface orientation. The analytical solutions generally involve either complex conduction models, coupled with simple convective heat transfer models, or simple conduction models,

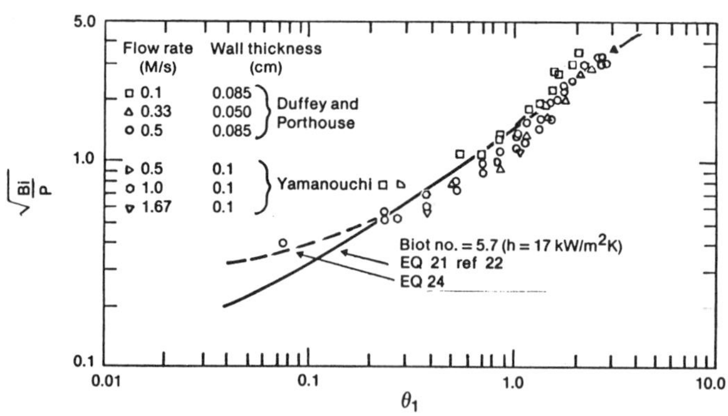

Figure 14. Comparison of predicted wet front velocity with experimental results [66].

coupled with more complex convection models. This arises naturally from the difficulties in obtaining an analytical solution. Many more complex convection models have been used in the numerical approaches to the problem; however, these models require a digital computer in order to implement them and are thus much more difficult to use. One fact remains consistent for all the models; that is the requirement for the definition of the convective heat transfer on the surface in order to predict its quenching behavior.

This section will address the question of convective heat transfer for the quenching process from a more conceptual viewpoint. This is done in order that

1. The understanding needed to determine the more complex convective heat transfer models in the numerical approaches be available.
2. An understanding of the more integral aspects of the process be developed.
3. An understanding of the limitations of our current capabilities be developed.

To meet these objectives, the topic of transient heat transfer will initially be discussed. This will be followed by a discussion of the ergodic assumption which is generally applied to steady-state convective heat transfer and how it relates to a transient situation. Then the effects of conduction upon transient convection are discussed. Finally, the boiling curve, with the multi-dimensional effects of forced convection upon the boiling curve, will be discussed, in order to determine how these influence the modeling of the quenching process.

Transient Conjugate Convection and The Quasi-Steady-State Analysis Approach

What is called, in the literature, the transient convection problem should not be confused with the transient conduction problem. The transient conduction problem solves the heat diffusion equation, given by Equation 2 with either constant heat transfer coefficients or heat transfer coefficients being known functions of time, temperature, and/or position. This is the type of problem which has been discussed in the preceding sections on conduction. The transient convection problem is a twofold one. It requires accounting for both the transient behavior of the convective fluid and the thermal heat capacity of the solid boundary. The solid-fluid interfacial behavior in this type problem is not known a priori. Thus, a simultaneous solution for both the temperature field in the solid and the temperature field in the fluid must be found with the two fields being coupled. This type coupled problem is referred to as a conjugate problem. A discussion of the transient conjugate convective heat transfer problem provides

1. The background needed to understand the physical coupling which exists between the fluid and the solid.
2. The information necessary to understand the limitations which exist in our current understanding of the quenching process.

Generally, the transient conjugate convection problem involves the solution of the energy equation for a single-phase fluid, coupled with the diffusion equation for the solid. The fluid is assumed incompressible, and the effects of axial diffusion and viscous dissipation are negligible, relative to the radial diffusion. Various boundary conditions simplify the problem. Often an approximate solution, using a simple quasi-steady-state approach, is adapted. In order to account for spatial variations in the fluid properties and/or wall temperatures, the quasi-steady-state approach uses steady-state heat transfer correlations, which are either constant with respect to time and location or which are applied in a "local condition" manner. The local condition model also assumes an isothermal surface for the solid, i.e., the solid can supply the energy required to hold the surface at a constant temperature. The numerical models, using a multiple-region representation for the heat transfer process in the conduction section, used this quasi-steady-state approach.

One of the early considerations undertaken in the investigation of the transient conjugate convective heat transfer problem was the effects of a step change in the wall temperature upon flow in a circular tube. This specification of the fluid temperature at the wall location eliminated the heat diffusion equation from the system of equations. Siegel [70] and Sparrow and Siegel [71] used this model to study the transient problem in laminar and turbulent flow, respectively. One of the major results, for small Prandtl number fluids, was the realization that, once enough time had passed for the fluid at the inlet to have progressed to a given location, the heat transfer at that location could be modeled using a local quasi-steady-state analysis. Before this minimal time, each fluid element passing a given location has seen a different heated length, causing different fluid conditions to pass the location, such that the local condition model does not apply. The time to reach this quasi-steady-state condition, t_{ss}, was estimated by the simple relationship given as

$$t_{ss} = z/u_\infty$$

where z is the distance measured from the entrance, and u_∞ denotes the fluid velocity a great distance away from the wall. Figure 15 presents a plot of the time to reach this steady-state flow condition, defined as the time for transient heat transfer to be within 5% of the steady-state value. As shown, this estimate works best for Prandtl numbers around unity and is also influenced by the Reynoldsnumber.

The Reynolds number and Prandtl number are two commonly used dimensionless parameters in heat and mass transfer considerations. The Reynolds number reflects the ratio of inertia and viscous forces and is given by

$$Re = u_\infty L_c/\nu$$

where ν is the kinematic viscosity. The Reynolds number is also used to reflect the speed of the fluid motion. As shown in Figure 15, for a step change in wall temperature, the faster the fluid moves, the quicker steady-state conditions are reached. The Prandtl number represents the ratio of the molecular momentum and thermal diffusivities and is given by

$$Pr = \nu/\alpha = C_p\mu/k,$$

where C_p is the specific heat at constant pressure, and μ is the dynamic viscosity. Thermal diffusivity dominates for small Prandtl numbers, Pr < 1. For the step change in wall

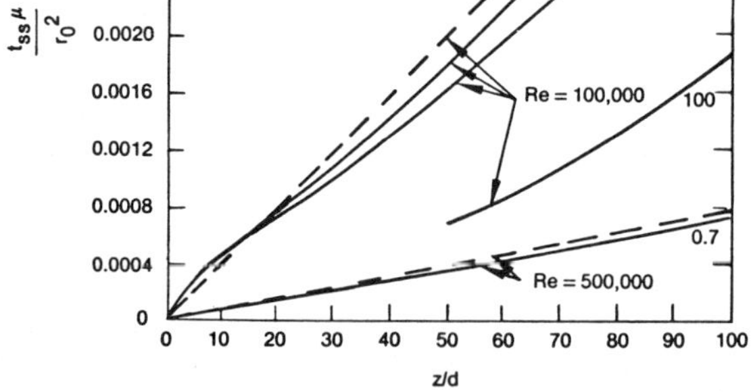

Figure 15. Time to reach steady state after a step jump in wall temperature [71].

temperature, approximate steady-state conditions are reached in the time z/u_∞. However, as the effect of molecular momentum increases and increases the Prandtl number, the transport of heat is sped up within the fluid and the time required to reach steady state decreased (Figure 15).

Kawamura [72, 73] included consideration of the storage capacity for the wall in his investigation of the transient conjugate problem for turbulent flow in a circular tube. A quasi-steady-state analysis was found to be valid when a nondimensional wall heat capacity term, β_k, was much much greater than 1. The heat capacity term was defined as

$$\beta_k = [h_{ss}(\rho c V)_w]/[A_{ht}(k\rho c)_f]$$

where h_{ss} is the steady-state heat transfer coefficient, A_{ht} is the heat transfer area, and the w and f subscripts represent the wall and fluid, respectively. An estimate for the time required for the heat transfer coefficient to decrease to within $1.1\,h_{ss}$ was given as

$$t_{ss} \approx 4\,(k\rho c)_f/h_{ss}^2$$

When β_k was not much much greater than 1, the quasi-steady-state solution was not valid.

Sucec [74] found a similar dependence on the solid properties, using a nondimensional number, a_1, defined as the ratio of the thermal capacity of the fluid to that of the solid. His improved quasi-steady-state approach yielded results that applied over a wider parametric range, i.e. $a_1 < 1$.

While the nondimensional parameters of Kawamura and Sucec are different, both indicated that a quasi-steady-state analysis can be applied when the thermal capacity of the solid sufficiently exceeds that of the fluid. This indicates that the wall responds more slowly to change than does the fluid. Thus, the fluid, when compared to the wall, adjusts more rapidly to any changes in either heat flux or temperature. This is the basic requirement in order to employ a quasi-steady-state analysis approach.

While the quenching of a surface is a transient process similar to the transient conjugate heat transfer problem, it is different in that:

1. The fluid is two-phase instead of single-phase.
2. The quenching surface is not necessarily isothermal.
3. Hydraulic conditions, other than those at the quenching front, may be changing.

However, insight into the conjugate problem also provides insight into the quenching process. This explains results such as those obtained in experiments by Veres and Florschuetz [75] and Ragheb, Cheng, and Groeneveld [76]. Both used large copper blocks, sufficient in size to meet thermal storage requirements relative to the fluid, in order to obtain transient quenching heat transfer data. This transient data compared well with steady-state data which were also obtained. Figure 16 shows typical results from Ragheb–Cheng–Groeneveld. Good agreement was found for the transition and film boiling regimes, with some problems at the lower temperature nucleate boiling regime.

An alternate method of approaching the transient convection problem is to look at the assumptions and limitations involved in obtaining steady-state convective heat transfer data. This understanding provides the background necessary to realize when these requirements are not met due to changing conditions in a transient situation. This approach to transient convection is discussed in the next section.

Steady-State Convective Heat Transfer, The Ergodic Assumption, and Transient Convection

Steady-state convective heat transfer has received extensive study. Such studies have investigated the effect of numerous factors including many geometries, internal and external boiling, forced and free convection, mixtures of fluids, and various hydraulic conditions.

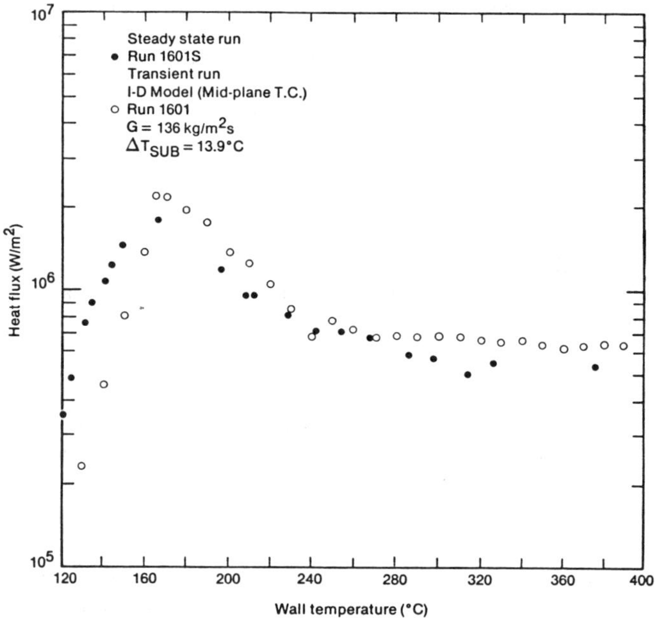

Figure 16. Comparison of boiling curves obtained from steady state and transient runs [76].

With this large data base, it is little wonder that the quasi-steady-state approach to transient analysis has evolved.

The previous section investigated the transient question from the conjugate problem's viewpoint. This section will address the transient convection question from an interfacial point of view, where the transient heat flux in functional form is given by

$$q = q(t, T_w, T_f, Pres, G, \alpha_v, \ldots) \tag{25}$$

with Pres representing the pressure, G representing the total mass flux, and α_v representing the void fraction. Newton's law of cooling, as given by Equation 7, is a special form of this more general expression. The steady-state form of Equation 25 is given by

$$q = q(T_w, T_f, Pres, G, \alpha_v, \ldots) \tag{26}$$

Other variables may be included, but these are sufficient to study the concepts involved.

In order to investigate the transient nature of the heat flux, the total derivative of Equation 25 with respect to time, can be written as

$$dq/dt = \partial q/\partial t + (\partial q/\partial T_w)(dT_w/dt) + (\partial q/\partial T_f)(dT_f/dt) + \ldots \tag{27}$$

Three types of transient conditions may be considered with respect to Equation 27. The first results when quasi-steady state convective heat transfer conditions exist for the transient, such that

$$\partial q/\partial t = 0 \tag{28}$$

For this condition, the transient nature of the heat flux, indicated by Equation 27, reduces to

$$dq/dt = (\partial q/\partial T_w)(dT_w/dt) + (\partial q/\partial T_f)(dT_f/dt) + \ldots \tag{29}$$

and can be determined using steady-state results ($\partial q/\partial T_w$, $\partial q/\partial T_f, \ldots$) obtained from Equation 26 and the transient wall and hydraulic conditions, as given by dT_w/dt, dT_f/dt, etc. This analysis is analogous to the local condition quasi-steady-state approach used with the conjugate convection problem.

The second case considers steady-state system conditions so that

$$dT_w/dt = dT_f/dt = \ldots = 0 \tag{30}$$

For this case, Equation 27 reduces to

$$dq/dt = \partial q/\partial t$$

Two further subcases exist. In the one, Equation 28 applies; in the other, it does not. These two subcases are dependent upon the time frame desired in the problem solution and will be discussed in greater detail later. However, the basic principle involves consideration of whether the time step desired is greater than the time associated with several cycles of the controlling mechanism in the convective process. If it is greater, Equation 28 applies, and what might be called the "trivial steady-state case" results. This result is the standard steady-state situation, which is independent of time. If Equation 28 does not apply, because of the time step desired, heat flux must be considered to be a function of time, and "steady-state boiling" becomes a transient consideration.

The third case occurs when the transient is such that neither Equations 28 nor 30 applies. This case is the true transient problem for which the relationships represented by Equation 25 must be determined. Each different transient will determine a different transient relationship. This presents a complex situation which is difficult to analyze.

From this simple perspective, it is easy to understand the desire to apply the quasi-steady-state analysis approach to the quenching process. To determine when this approach is valid requires determination of when Equation 28 is valid. Insight into this question can be obtained by looking at how the steady-state convective heat transfer relationships, represented by Equation 26, are obtained.

Steady-State Convective Heat Transfer and The Ergodic Assumption

Steady-state convective heat transfer information is obtained by analysis of experiments in which Equation 30 is valid. These experiments generally measure the power required to heat the wall or surface to a given temperature and hold it for the given hydraulic conditions. This power measurement is divided by the heated area to obtain an area average heat flux, which is denoted as q. The power may also be time-averaged, to minimize the effects of random variations in the power measurement. By running a number of different experiments, each having different hydraulic conditions, the relationship denoted by Equation 26 is obtained. Determination of a heat transfer coefficient relationship, using Newton's law of cooling as given by Equation 7 instead of Equation 26, is considered a simple extension.

Consider for a moment any arbitrary point (x, y) on a steady-state boiling surface undergoing uniform pool boiling (see Figure 17). The local instantaneous heat flux at this point will be denoted as $q^i_{x,y}$. An ensemble of $q^i_{x,y}$ is a collection of local instanteous heat fluxes from sample (x, y) points. An ensemble area average of the local instantaneous heat flux can be found and is given by

$$\langle q^i \rangle = \lim_{A \to \infty} (1/A) \iint_A q^i_{x,y}\, dA \tag{31}$$

This ensemble average is, in principle, the area-averaged heat flux normally measured on a large surface, i.e. $q = \langle q^i \rangle$, and is most often the quantity reported as steady-state data.

One practical question which arises in the consideration of this ensemble area average is how large an area is required, since an infinite area is not possible. This area, denoted

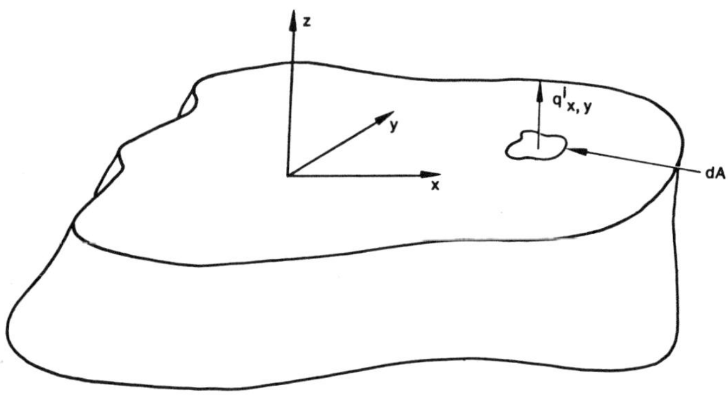

Figure 17. Local instantaneous heat flux at a point on a boiling surface.

as A_0, must be large enough to include a sufficient number of the basic phenomena which control the convective process. The type of phenomenon which is involved varies, depending upon the boiling regime. In nucleate boiling, for example, the process involves the growth and collapse or departure of bubbles from the surface. Thus, an area average must include a number of these nucleation sites. Further understanding of this space requirement can be obtained by investigating when does

$$\partial\langle q^i\rangle/\partial A = 0? \tag{32}$$

This derivative may be written in limit form as

$$\lim_{\Delta A \to 0} (\langle q^i\rangle|_{A_0+\Delta A} - \langle q^i\rangle|_{A_0})/\Delta A$$

where ΔA represents an arbitrary increment of area. Noting that $A = A_0 + \Delta A$, Equation 32 becomes

$$\partial\langle q^i\rangle/\partial A = \lim_{A \to A_0} (\langle q^i\rangle|_A - \langle q^i\rangle|_{A_0})/(A - A_0) = 0$$

Thus, in order to satisfy this equation for a finite consideration, where ΔA is any arbitrary size

$$\langle q^i\rangle|_A = \langle q^i\rangle|_{A_0}$$

Using the definition of an ensemble average, this result reduces to

$$(1/A_0) \iint_{A_0} q^i_{x,y}\, dA = [1/(A_0+\Delta A)] \iint_{A_0+\Delta A} q^i_{x,y}\, dA$$

Thus, as long as $A > A_0$, the boiling area is of sufficient size to include a statistically sufficient number of the basic phenomenon; including more area does not change the result. This argument for a minimum boiling area is analogous to the "grainy structure" argument often applied to a continuum in applying the fundamental laws of mechanics.

The minimum area requirement also gives rise to minimal dimension requirements associated with this area. This requirement arises when the area is broken into its basic components, as can be denoted by

$$dA = dx\, dy$$

and we realize that a sufficient number of the controlling phenomena should occur within either length over which dx or dy is integrated. This is required, since the controlling phenomena do interact with one another and require some minimum dimensions, x_0 and y_0, in order that the basic boiling phenomena remain unchanged. If λ_c were used to represent the size or dimension of this controlling phenomenon, x_0 would equal $n_0\lambda_c$ where n_0 represents the minimum number of phenomena required to get a meaningful statistical average. For the case of nucleate boiling, used earlier, n_0 would represent the minimum number of nucleation sites required. Similarly, $y_0 = n_0\lambda_c$. The requirement represented by Equation 32 can then be used to obtain the relationship that

$$\langle q^i \rangle|_{x_0, y_0} = \langle q^i \rangle|_{x_0+\Delta x, y_0+\Delta x}$$

or

$$[1/(x_0 y_0)] \int_{y_0} \int_{x_0} q^i_{x,y}\, dx\, dy$$
$$= [1/(x_0+\Delta x)(y_0+\Delta y)] \int_{y_0+\Delta y} \int_{x_0+\Delta x} q^i_{x,y}\, dx\, dy \qquad (33)$$

where Δx and Δy represent arbitrary increments of length. If one of these dimensions of the boiling surface decreases below its minimum, the heat flux becomes sensitive to that variable. For example, if the x-direction is assumed to be that dimension, an area-averaged heat flux can be obtained, but it is dependent upon the x dimension, as given by

$$\langle q^i \rangle (x_1) = 1/(x_1 y_0) \int_{x_1} \int_{y_0} q^i_{x,y}\, dy\, dx \qquad (34)$$

where $x_1 < x_0$.

This type of dimensional limit behavior was found by Bernath [77] in the study of the diameter effect of a horizontal heater on the critical heat flux in pools of saturated water at atmospheric pressure. As the heater diameter increased to, and then above, a value of 2.54 mm, the critical heat flux increased and then leveled off. This is also the principle utilized by Gunnerson and Cronenberg [78] in their prediction of the minimum film boiling conditions for spherical and horizontal flat plates. Figure 18 shows a plot of their theory for minimum heat flux, compared to data using copper spheres and nitrogen. As the diameter of the sphere increases, the analysis approaches the results of a flat plate.

For flowing conditions, where steady-state boiling may not remain uniform due to flow effects, this length criterion remains important. Boiling does vary with respect to the direction of travel. When the remaining dimension is large compared to its minimum requirement, an area-averaged relationship, similar to Equation 34, may be obtained, where x is the direction of travel and y is the remaining surface coordinate. However, another common practice under flow conditions is to obtain an average local heat flux relative to the position x. This quantity may be expressed as a line average, instead of an area average, and is given by

$$\langle q^i_x \rangle = (1/y_0) \int_{y_0} q^i_{x,y}\, dy \qquad (35)$$

When the energy equation for the fluid is used to transform the position x to quality X and an interfacial slip model is used, the functional relationship denoted by Equation 26 may be obtained. On occasion, the void fraction, α_v, is directly replaced with the quality. If the dimension in the y direction is decreased to $y_1 < y_0$, the resulting relationship would become

$$\langle q^i_x \rangle (y_1) = (1/y_1) \int_{y_1} q^i_{x,y}\, dy$$

This type dependence is seen for flow within a small tube. Similar considerations can be made for boiling on a vertical surface both with and without flow.

As long as the phenomena changing along the x-direction change slowly enough, with respect to position from one experiment to the next, the relationships derived from one case can be applied to another. This implies that $\partial\langle q^i_x \rangle/\partial x \approx 0$ or $\partial\langle q_r \rangle/\partial x$ are of approximately

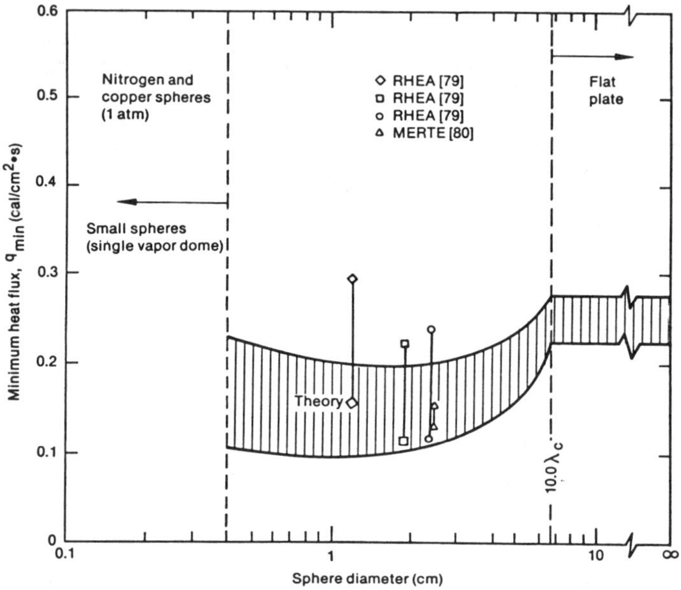

Figure 18. Minimum heat flux versus sphere diameter for large copper spheres in saturated liquid nitrogen [78].

equal value for the two experiments. However, as the phenomena change more rapidly with respect to position and become more "closely packed," their nature changes, producing changes in $q^i_{x,y}$. Thus, the relationship obtained with the boiling phenomena slowly changing over a great distance x will not apply to the case where conditions change rapidly over a very short distance x. This rapid change in local conditions happens frequently during the quenching process.

This discussion of the space averages, i.e., area or line averages, of the ensemble of local instantaneous heat flux has provided an understanding of space limitations in the determination and application of steady-state relationships. The part time plays in this consideration of steady-state relationships has not yet been investigated. If the assumption is made that the phenomenon occurring at each point on the surface is a random occurrence relative to all other points, the ergodic hypothesis [81] says that the ensemble average of any property is the same as the time average of that property for a single point. Thus, the assumption that the ergodic hypothesis holds true for the steady-state heat flux, shown in Figure 17, yields

$$q = \langle q^i \rangle = q_{x,y} \tag{36}$$

where

$$q_{x,y} = \lim_{t \to \infty} (1/t) \int_t q^i_{x,y} \, dt \tag{37}$$

This quantity $q_{x,y}$ is the local time-averaged heat flux occurring at the arbitrary point (x, y). Equation 36 indicates that the steady-state heat flux might be determined from either an instantaneous area-averaging or a local time-averaging operation. Practically speaking, the heat flux is generally determined from either an instantaneous area average, Equation 31, or from an instantaneous line integral (Equation 35). The use of an area or line average, as opposed to a time average, arises from the difficulty of measuring the local instantaneous heat flux at a point on the surface, required to determine $q_{x,y}$. The instantaneous area average (as noted earlier) and instantaneous line average are also sometimes time-averaged to minimize the effects of power measurement variations.

Just as with the question of how big an area was needed for $\langle q^i \rangle$, Equation 31, the question can be asked of how long a time, t, is needed for a valid $q_{x,y}$ to be obtained. This can be determined by finding when

$$\partial q_{x,y}/\partial t = 0 \tag{38}$$

As with the area consideration, the limit form of this derivative reduces to the requirement that

$$(1/t_0)\int_{t_0} q^i_{x,y}\,dt = (1/t_0+\Delta t)\int_{t_0+\Delta t} q^i_{x,y}\,dt \tag{39}$$

for any arbitrary time increment, Δt. If Δt_p represents the period of the controlling phenomenon and n_0 the minimal number of occurrences (cycles) of the phenomenon required, the minimum time for a statistically sufficient sample can be written as $t_0 = n_0 \Delta t_p$. If the time of interest is greater than this minimum time, t_0, the local time-averaged heat flux, $q_{x,y}$ is independent of time. This property is one of the characteristics of an ergodic process due to its stationary nature. However, if the time to be considered is less than t_0, then $q_{x,y}$ reflects a sensitivity to time. This lower limit on time is analogous to the dimensional requirement on $\langle q^i \rangle$. Thus, from Equations 36 and 38, an understanding is obtained of when the quasi-steady-state requirement, given by Equation 28, is satisfied for steady-state conditions. In order to analyze a steady-state situation for times shorter than t_0, the heat transfer must be known, with respect to its individual controlling phenomenon and on a time scale less than Δt_p. This is beyond the scope of most practical situations.

Therefore, for a standard steady-state situation, if the heat diffusion equation (reduced to its time-independent Poisson form) is solved using a steady-state convective boundary condition, where the heat flux is determined from standard techniques, indicated by Equation 36, the temperature at the surface may be thought of as either a time-averaged temperature or an area-averaged temperature. Special requirements in terms of size, orientation, and/or flow conditions on the boiling surface under consideration must be consistent with the type of ensemble spatial-averaging used in the determination of the steady-state convective relationships to be used for the convective boundary condition. The solution for a true local instantaneous point on the boiling surface at each instant of time, which implies that neither ensemble spatial-averaging or local time-averaging are to be used, requires measurement of $q^i_{x,y}$ even for steady-state conditions. To further study spatial and time-averaging operations, the work of Delhaye [82] provides an excellent base from which to begin.

Transient Convection and The Quasi-Steady-State Assumption

Having used the ergodic hypothesis in order to investigate both space and time constraints on steady-state convective heat flux relationships, as represented by Equation 26, it is possible to realize the limitations which apply when these same relationships are applied to a transient situation. As noted earlier, the real question arises as to when Equation 28 is satisfied. When Equation 28 is valid, a quasi-steady-state analysis approach, denoted by Equation 29, is valid. However, when Equation 28 is not satisfied, a determination of the transient relationship, as represented by Equation 25, must be undertaken.

The definition of when Equation 28 is valid for transients is not always clear. There are obvious situations in which the transient is slow and the equation is valid. For this case, little or no change occurs in either the hydraulic or thermal conditions during the minimum time period, t_0, denoted as the time required to obtain a valid time average for heat flux in the previous section. Thus, the phenomena which occur in this slow transient are the same as those which occur in the steady-state situation under the given hydraulic and thermal conditions. For this situation, the quasi-steady-state analysis procedure, represented by Equation 29, applies. This situation has been noted earlier by Veres and Florschuez [75] and applies at least during the film portion of the boiling curve, shown in Figure 16, obtained in the work of Regheb, Cheng, and Groeneveld [76].

The situation where the hydraulic or thermal changes are very rapid, such that a significant change takes place in these conditions within the period Δt_p, is the other extreme where the quasi-steady-state approach will not apply. The quantity Δt_p has been used to denote the time required for the phenomenon controlling the boiling process to undergo a single cycle. Thus a step change to the wall temperature when nucleate boiling is the controlling process will, at least initially, be a transient. This arises due to the fact that the bubble dynamics cannot instantly reflect the change in wall temperature. This is similar in nature to the step change in wall temperature for a flowing single-phase fluid, discussed earlier with respect to the conjugate problem. For those situations where the transient is extremely rapid, the heat diffusion equation, given by Equation 2, may also be inadequate. As noted earlier in the conduction section for this rare situation, the use of Equation 4 may be required.

Between these two extremes is obviously a transition region. The more rapid end of the transition region does not meet the requirements for a quasi-steady-state analysis approach, while for the other end, the quasi-steady-state approach can still provide an approximate solution.

It is possible during the quenching process to encounter all or just one of these transient regions. The film boiling region may satisfy the requirements in order that a quasi-steady-state approach is valid, but once cooling has dropped the temperature to the point that the transition boiling region has been reached, the quench may be rapid enough that the quasi-steady-state approach is not valid. While it is not known to be true, this may offer an explanation as to why the lower temperature discrepancy occurs in Figure 16. It is in this lower temperature region that cooling is the most rapid for the tests which were run by Regheb, Cheng, and Groeneveld.

Quenching and The Quasi-Steady-State Approach

Other than the simple models discussed in the conduction section, the quasi-steady-state analysis approach offers the only other practical method for analyzing the quenching process. The models developed by Tan [37], Kirchner [41], Fischer et al. [44], Bonakdar and McAssey [66], and Chan and Banerjee [69] which were discussed as part of the conduction section, all utilize this approach. The models are principally numerical models and do incorporate certain aspects of the conjugate problem. The principal portion of the conjugate problem, which they do incorporate, is the coupling between the hydraulics and the thermal structures. This coupling is generally accomplished in an explicit rather than implicit manner. This is accomplished numerically by solving for the hydraulic and thermal fields separately, where old time-step values of one are used to obtain new time-step values of the other. An implicit solution scheme would solve for both fields simultaneously. The models are different from the standard conjugate problem in that the hydraulic models do not represent the direction normal to the boiling surface and are thus one-dimensional along the boiling surface. This is principally due to the difficulty involved in using multi-dimensional models in two-phase flow. Since the dimension normal to the boiling surface is not available in the hydraulic model, convective heat transfer relationships as represented by Equation 26, must be used to couple the hydraulics with the thermal structures.

The spatial and time frame requirements for the steady-state convective relationships have been discussed, relative to the application of the quasi-steady-state models for the quenching process. If the quenching process is slow enough and the spacial region over which quenching takes place is wide (or long) enough, the application of steady-state relationships is acceptable, and the quasi-steady-state approach is valid. However, if the quenching process is very fast or takes place in a very restricted region compared to the region over which the data are taken, the application of steady-state relationships is not valid. With this consideration in mind, the determination of the heat flux or heat transfer coefficient, as undertaken by some of the transient conduction models in the conduction section, using simple convection models, represents just as valid a model as does the application of multi-region models, where the convective relationships are not really applicable around the quench front. The quantities obtained for these simple convective models are, therefore, a measure of a limited region heat flux, as denoted by Equation 34.

Another consideration, which has not been included in the discussion of transient convective heat transfer, is the effect the transport of thermal energy through the thermal structure may have upon the convective process. In the quasi-steady-state approach, it is assumed the thermal structure can supply the energy necessary to realize the steady-state heat flux. However, what happens in a transient when the thermal structure cannot transport the energy fast enough to supply this convective surface heat flux? The next section will investigate this influence.

The Effect of Conduction Upon The Transient Convective Process

The thermal response of a structure has been discussed and, as noted, is controlled by the heat diffusion equation, given by Equation 2, when coupled with the initial and boundary conditions. This thermal response is influenced by the ability of the structural material to both move thermal energy, as reflected in the thermal conductivity, k, and to store energy, as reflected by the ρc quantity. The energy which can be transported out of the "surface layer" of the structure to a fluid during a transient occurrence is, therefore, a combination of the energy stored and the energy which can be transported to replace the stored energy which was removed. Thus, it might be suspected that some combination of the quantities k, ρ, and c might reflect the ability of a thermal structure to react to a transient thermal occurrence. The quantity observed to be the key factor in early drop impingement studies [83], ΔT_{min} in flow boiling [84] and the Leidenfrost temperature [85, 86] is given by $\beta = (k\rho c)^{-1}$.

To develop an understanding of why this factor is important, it is beneficial to briefly review the work of Baumeister and Simon [86]. Their work investigated the role of the surface material and, to a lesser extent, that of surface conditions on the measurement of the Leidenfrost temperature. The Leidenfrost temperature, T_{Leid}, represents the temperature of a hot plate at which a liquid drop on the plate will begin to contact the plate surface. It is measured using the total evaporation time of the drop. For plate temperatures above T_{Leid}, vapor generation from the bottom of the drop supports it above the surface and produces long evaporation times. Hotter plate temperatures produce shorter evaporation times. As the plate temperature drops below T_{Leid}, the drop contacts the surface, speeding the evaporation process. The Leidenfrost temperature is defined as that temperature at which the drop has the longest total evaporation time in film boiling. Baumeister and Simon's investigation of the Leidenfrost temperature modeled the transient response of a semi-infinite solid at an initial temperature, T_o, to the application of a liquid drop of radius, R_o, gently placed upon the surface at time zero. Assuming the drop is symmetric about the origin, the heat diffusion equation can be written in dimensionless form as

$$\partial\theta/\partial t^* = N_r[(\partial^2\theta/\partial\eta^2) + (1/\eta)(\partial\theta/\partial\eta) + (\partial^2\theta/\partial z*^2)] \quad (40)$$

with the initial and boundary conditions given by

$$\theta = 1 \quad \text{for} \quad t^* = 0 \quad \text{and} \quad z^* \geqq 0$$

$$\partial\theta/\partial z^* = \theta \quad \text{for} \quad t^* > 0 \quad \text{and} \quad z^* = 0 \quad \text{and} \quad \eta \leqq \eta_0$$

$$\partial\theta/\partial z^* = 0 \quad \text{for} \quad t^* > 0 \quad \text{and} \quad z^* = 0 \quad \text{and} \quad \eta > \eta_0$$

$$\lim_{z^* \to \infty} = 1 \quad \text{for} \quad t^* > 0$$

where $\eta = hr/k$, $\eta_0 - hR_o/k$, $z^* = hz/k$, $t^* = t/t_\infty$, $N = h^2t_\infty/\beta$, θ is given by Equation 10, and t_∞ is a short characteristic which is of sufficient length for the drop to have entered steady-state film or nucleate boiling.

For $N_r \approx 0$, it can be seen from Equation 40 that the change in the surface temperature will be very small, and the surface is nearly isothermal. For very large N_r, the time-dependent temperature gradient will be large, and the surface will not be isothermal. For an infinite drop, given by $\eta_0 \to \infty$, Equation 40 reduces to

$$\partial\theta/\partial\tau = \partial^2\theta/\partial z*^2 \tag{41}$$

where τ was redefined as a new dimensionless time as given by

$$\tau = t^*N_r = h^2t/\beta$$

The solution of Equation 41 with the initial and boundary conditions as previously defined, was determined in Reference 8 and is shown in Figure 19 as the curve, denoted as η_∞. Radial conduction effects are shown to decrease the temperature drop for smaller drop sizes for finite η_0. The curves obtained for finite η_0 were obtained using a finite difference solution.

For $t \approx t_\infty$, sufficient time has occurred for the drop to enter its final boiling state, so that $\tau = N_r$. This reduces the abscissa of Figure 19 to N_r. Baumeister and Simon arbitrarily defined the region of $N_r < 0.001$ as the isothermal region, while $N_r > 0.01$ was defined as the nonisothermal region. The region between the isothermal and nonisothermal regions was called the intermediate region. Experimental results were obtained using water on several materials. The materials are denoted in their approximate region on Figure 19.

For the isothermal region, the temperature on the surface of the material under the drop, T_p, is approximately equal to the plate's initial temperature. This may be written in the limit as

$$\lim_{N_r \to 0} T_p = \lim_{\beta \to 0} T_p = T_o \tag{42}$$

for a finite h^2t_∞. Since Baumeister and Simon were investigating the influence of the material upon the measured Leidenfrost temperature, Equation 42 would be used to reach the conclusion that, for an isothermal material, the measured plate temperature was the Leidenfrost temperature, i.e.

$$\lim_{\beta \to 0} T_{Leid,\,meas} = T_{Leid,\,iso} = T_o \mid_{max\ evap\ time}$$

Comparison of pool boiling T_{min}'s with T_{Leid} leads to the conclusion that for saturated pools on isothermal materials the two quantities were equivalent, i.e.

$$\lim_{\substack{\beta \to 0 \\ T_f \to T_{sat}}} T_{min} = T_{min,\,iso} = T_{Leid,\,iso}$$

For materials other than isothermal materials, the surface temperature, and therefore the convective energy removed by the fluid during this short transient, is dependent upon the material. For the steady-state pool boiling curve, shown in Figure 1, this means that ΔT_{min} will be dependent upon the material, if the material is nonisothermal, causing the boiling curve to have differing characteristic dependent upon materials. This material dependence

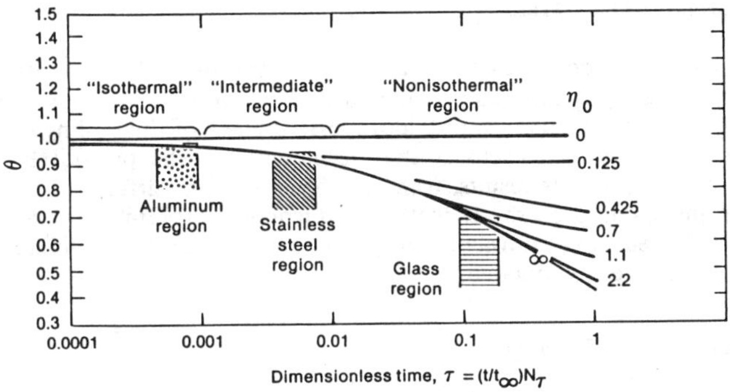

Figure 19. Effect of N_τ on surface temperature [86].

is most important in those post-CHF boiling regimes where significant liquid-solid contact may occur, i.e., in the transition boiling regime. Thus, nonisothermal materials will have different steady-state boiling curves, particularly in the transition boiling region, due to conduction limitations of the material.

Application of the quasi-steady-state approach to the quenching of a structure or body, as discussed earlier, still applies, with the added consideration of the material being quenched. For isothermal materials, the steady-state relationships, denoted by Equation 26, obtained for one material apply to all isothermal materials, ignoring for the moment the effects of surface conditions. However, for nonisothermal materials, the steady-state relationships become material-dependent for some regions. Thus, for nonisothermal materials, Equation 26 can be written as

$$q_{noniso} = q_{noniso}(T_w, (k\rho c)_w, T_f, Pres, G, \alpha_v, \ldots) \tag{43}$$

and the spatial and time frame requirements associated with the steady-state relationships must be met relative to the quasi-steady-state analysis. When these requirements are not met for a nonisothermal material, the transient relationship, as earlier denoted by Equation 25, also becomes material-dependent, as given by

$$q_{noniso} = q_{noniso}(t, T_w, (k\rho c)_w, T_f, Pres, G, \alpha_v, \ldots) \tag{44}$$

Surface conditions can change the boiling results obtained from a surface and may cause an isothermal material to appear to be a nonisothermal one. As pointed out by Yao and Henry [87], this arises when the properties of the surface material change, influencing the transient contact problem. This has been investigated by Baumeister and Simon [86]. Surface coatings, oxide layers, or contamination layers in the order of 10^{-6} m may thus have a significant effect. The work of Mowat et al. [88] provides an example of recent study on the effect of oxide films upon the quenching of metal tubes. Surface finish or roughness may produce similar results [86]. A rougher surface penetrates the vapor layer at a higher temperature, thereby producing liquid-solid contact at those higher temperatures. A recent model, developed by Witte, Chang, and Orozce [89] investigates this influence of liquid-solid contact and its enhancement due to surface roughness. Results indicate that this liquid-solid contact may represent a significant effect in the heat transfer process.

Thus, while an in-depth review of the effects of materials and surface conditions has not been possible, an understanding has been obtained which includes the influence of these factors. For the quenching process, these factors are significant, since the quenching structure or body involves this transition from a primarily vapor-covered environment to a liquid-covered one and must be influenced by the liquid-solid contact, or transient liquid contact, phenomenon.

Steady-State Convective Heat Transfer

This section qualitatively describes heat transfer mechanisms in the postcritical heat flux (post-CHF) region. These mechanisms provide an understanding and predictive capability for the quenching process. The section provides a brief review of the boiling curve and the heat transfer surface technique, discusses the mechanisms in the post-CHF region that produce quenching, and presents results, using the heat transfer surface, to investigate the forced-convective quenching process. Reduction of quenching experiment data is discussed, and the effects the quenching transient may have on the results of different types of quenching experiments are investigated.

The Heat Transfer Surface Technique

The heat transfer surface technique [90] was originally developed for RELAP4/MOD6 [44] and was initially presented at a Heat Transfer Workshop [91] sponsored by the U. S. Nuclear Regulatory Commission (NRC). Simultaneously, the technique was independently proposed by Collier [92]. Results utilizing this concept have been presented by Nelson and sullivan [93, 94] and Bjornard and Griffith [95, 96]. This section discusses the heat transfer surface concept and relates the surface to the classical pool boiling curve, as shown in Figure 1.

The function $z = f(x, y)$ may be viewed as a surface when z is plotted as a function of x and y. The insight gained from this visual approach can be quite helpful, particularly when a number of functions z_1, z_2, ..., z_n must be combined to cover a wide range for the independent variables x and y.

This mathematical concept of a surface has been termed "the heat transfer surface" when the functions z_1, z_2, ..., z_n are heat transfer correlations and the correlations are considered as functions of m variables. Thus, the heat flux defined by a number of different correlations is viewed as a function of wall superheat, quality, pressure, and mass flux. This has been defined in functional form in Equation 26 for boiling on an isothermal material, and by Equation 43 for a nonisothermal material. A family of heat transfer correlations used to build a surface is discussed in both References 44 and 97.

Within this chapter, the heat transfer surface represents the total wall heat flux being convected into the fluid from a given point on the wall. Also, for convenience thermodynamic equilibrium is assumed. Thermodynamic nonequilibrium, or individual liquid and vapor components from the wall, could be included; however, they add complexity to the discussion, while not changing the basic results.

The heat transfer surface is composed of two families of curves. One family is the family of "boiling curves." As shown in Figure 20, a boiling curve shows the dependence of heat flux upon wall superheat with all other variables held constant. This is more restrictive than required by the classical pool boiling curve shown in Figure 1. The other family of curves is termed the "isothermal curves." An isothermal curve shows heat flux as a function of void fraction for a given wall temperature. In most practical situations the void fraction is not available. Thus for the remainder of the chapter, the void fraction will be replaced with the thermodynamic equilibrium quality. While these two quantities are not equal [98], they do represent the same gross characteristic of the vapor and liquid combination. For a discussion involving thermodynamic nonequilibrium, the void fraction should be used.

Figure 21 constructs a heat transfer surface from these two families of curves with all other variables held constant. To aid in the visualization, different areas on the heat-transfer surface are denoted in terms of classical boiling regimes such as single-phase liquid heat transfer, nucleate boiling, transition boiling, film boiling, and single-phase vapor heat transfer. The slope of the heat transfer surface with respect to wall temperature, $\partial q/\partial T_w$, is positive in all the boiling regimes, except in the transition boiling region.

In order to relate the heat transfer surface approach to the classical pool boiling curve, as originally proposed by Nukiyama [99], assume Figure 21 can represent a pool boiling

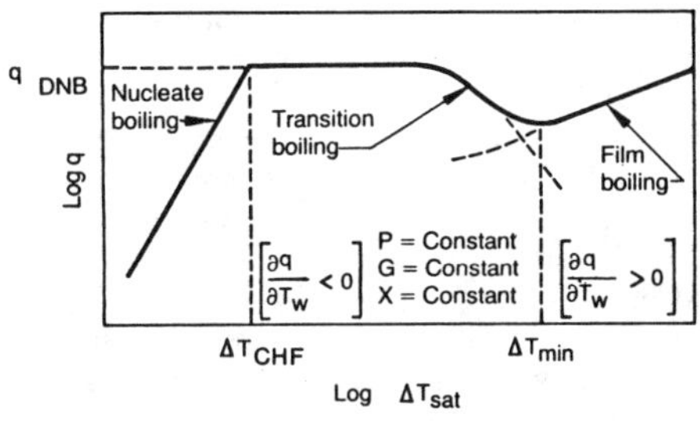

Figure 20. Boiling curve [90].

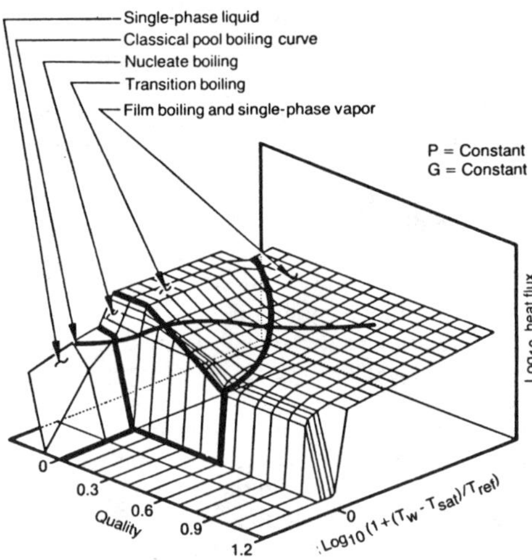

Figure 21. Heat transfer surface and regimes [90].

situation. The classical pool boiling curve can now be shown on the heat transfer surface. For this classical pool boiling curve, the "quality" used is the "time-averaged quality at a point on the wall" and changes as the wall temperature changes. In reality, for the pool boiling situation, there is only a narrow range of qualities, if not just one, for any given wall temperature, so that a full heat transfer surface as just assumed does not exist. Thus, a unique (or nearly unique) classical pool boiling curve exists for a given wall material, wall finish or condition, pressure, and geometry. For forced convective heat transfer, a unique classical boiling curve does not exist as it does for pool boiling. Instead, the heat transfer surface may be transversed in an infinite number of ways as determined by the coupled response of the wall, the local hydraulic conditions, and the hydraulic system. One aspect of this nonunique nature of forced-convective classical boiling curves will be fully discussed later.

Current Predictions of Quenching

Current predictions can be divided into two categories, depending upon whether the cooling is provided by pool boiling or forced convection. For the pool boiling situation, cooldown begins when a hot body or structure is covered by some cooling fluid. The convective energy removal provided by the fluid is characterized by the classical pool boiling curve, as shown by Figure 1. The boiling curve provides the necessary boundary condition for the solution of the heat diffusion equation. As just discussed, this boiling curve may or may not depend upon the material being quenched, or the quench transient itself, if the quench is very rapid.

Predictions of forced-convective quenching can be further divided into two subcases. The first case is the prediction of those experiments where quenching is controlled by conduction as a result of poor precursory cooling downstream of the quenching front. While work remains to be done on this case, it has been extensively studied and principal works on it were reviewed in the conduction section.

The second case, of predominant interest for this convection section, is the prediction of those experiments where precursory cooling is sufficient to cause cooling downstream of the conduction propagating quenching front. Several of the multi-region models reviewed in the conduction section were designed particularly for the modeling of this situation. Generally however, their emphasis was placed upon the solution of the heat diffusion equation to include multi-dimensional conduction, principally axial conduction, as opposed to a detailed description of the convective process.

For this case with significant precursory cooling, two axial locations are considered as an artificial means of envisioning the separation of conductive and convective effects. The first location is a point immediately downstream of the conduction propagating quenching front. With sufficient precursory cooling at this downstream point, the speed of the quenching front is increased [40]. Thus, the quenching of this downstream point is influenced by both the quenching front and convection, and the quench front will be termed a "conduction-convective propagating quench front." The second location is far downstream of the conduction-convective quenching front. If precursory cooling is sufficient, a "spontaneous" convective quench may occur at this point, prior to the arrival of the conduction-convection quenching front. Physically, for this type of quench to occur, some type of power distribution, such as cosine, must exist along the quenching surface. For the moment, axial conduction as this downstream point quenches is ignored, concentrating instead upon why quenching begins, in order to understand how forced-convective heat transfer influences this quench. It is the second location upon which most of the discussion in the remaining sections is concentrated. The integration of the effects of axial conduction upon this point is equivalent then to considering the first point and is briefly discussed later.

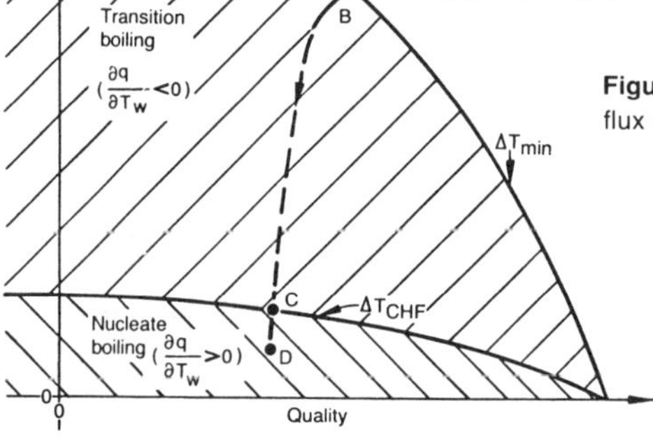

Figure 22. Loci of current maxima and minima of heat flux with respect to quality and superheat [90].

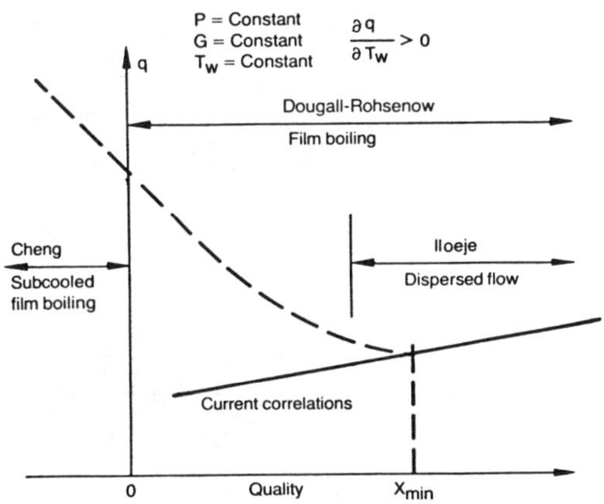

Figure 23. Synthesized low quality effect on film boiling [90].

To help envision how it has been presumed that forced-convective quenching takes place, it is helpful to first denote the relative maxima and minima of the heat transfer surface shown in Figure 21. The loci of these points are projected down onto the quality-wall superheat plane and denoted as ΔT_{CHF} and ΔT_{min}, as shown in Figure 22. Quenching has been thought of as being produced by two factors. First, the increased flow, downstream of the quench front, caused by fluid evaporation from the quenching fluid, causes a cooldown (point A to B) due to the classical Reynold's number effect in forced convective film boiling. Second, once transition boiling is reestablished at point B, a quench is guaranteed, without an increase in either internal heat generation or quality flowing by the point. This quench results from the feedback associated with the transition boiling region where a decrease in wall superheat produces an increase in heat flux, resulting in an increased cooldown rate. Movement on the heat transfer surface from point B to point C was though to be through phenomena quite similar to that occurring in the pool boiling transition boiling region.

Forced-Convection Post–CHF Heat Transfer and Quenching

It is the purpose of this section to study the forced-convection post–CHF region and to determine how convective quenching is produced. This section investigates that point experiencing spontaneous quench far downstream of the conduction-convection quench front and does not initially involve axial conduction effects.

The initial step in determining how forced-convective quenches are produced is to realize that the heat transfer mechanisms that occur during the low-void post–CHF region, where quenching generally occurs, are not well understood. Recalling the isothermal family of curves which constitute the heat transfer surface, shown in Figure 21, an isothermal line in the film boiling region ($\partial q/\partial T_w$) for typical film boiling correlations [100, 101] is shown as the solid line in Figure 23. The heat flux decreases as the quality decreases. The dashed line in Figure 23 represents a low-quality effect in the film boiling region, originally synthesized by Nelson [90] from the low-pressure work of Iloeje et al. [102], Dougall and Rohsenow [100] and Cheng et al. [103]. Current correlations do not have this synthesized characteristic.

The quality regions covered by References 100, 102, and 103, as shown in Figure 23, will be considered. Iloeje's three-step analytical model [102] for dispersed flow shows an X_{min}, as noted in Figure 23, to exist. The increase in film boiling heat flux for decreasing quality

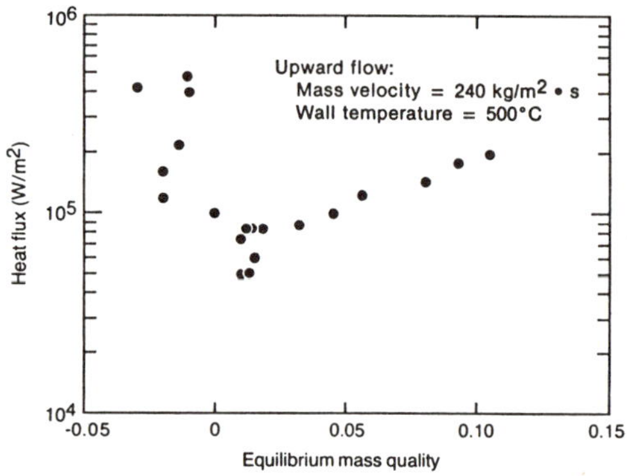

Figure 24. Post-dryout data [104].

is produced in the model by those drops which enter the thermal boundary layer but do not contact the wall. This increase in heat flux becomes more pronounced as more drops enter the thermal boundary layer. As quality continues to decrease, Dougall and Rohsenow [100] observed experimentally a continuing increase in the heat flux. This increase was attributed to the heat transfer in the inverted annular flow regime. Cheng et al. [103] studied experimentally the effects of subcooling on the boiling curves and found an increase in heat flux as the subcooling was increased.

Recent low-pressure data, taken by Barnard et al. [104] and Fung et al. [105] support the synthesized low-quality effect in the low-flow film boiling region. Typical results from Barnard et al. are shown in Figure 24. Both studies show these effects to be in the film boiling region, since the $\partial q/\partial T_w > 0$, and are not the result of entering the transition boiling region ($\partial q/\partial T_w < 0$) due to a quality change, as typically shown in Figure 22. It is this low-quality effect in the film boiling region that initiates low-pressure convection quenches. An example of this type of quenching is seen at the top of the heated core in the FLECHT forced-feed reflood experiments [45]. Once the top of the heated rods experience the convective quench, the upper quench front propagates down by conduction-convection to meet the conduction-convection front propagating up from the bottom.

High-pressure data supporting the low-quality effect was obtained by Polomik et al. [106] and has been utilized by Groeneveld [101] and others in the development of their film boiling correlations. However, the low-quality effect does not appear in the correlation due to the dominant effect, resulting from the mass flow produced by high-quality data at low flows and low-quality data at high flows. More recent high-pressure data (up to 9 MPa), obtained by Groeneveld et al. [107] further support the low-quality effect. Just as in the low-pressure convective quenches, the same basic low-quality effect should initiate convective quenches at high pressure. The Loss-of-Fluid-Test (LOFT) experiment L2–3 [108] provides an example of the influence of the type of convective-initiated quenching.

The preceding paragraphs have discussed film boiling characteristics which initiate convective high-temperature, low-quality film boiling quenches. While liquid may contact (rewet) the wall for high temperatures in the film boiling region, from a microscopic transient viewpoint [5], the question arises as to the wall temperature when this contact becomes significant from the macroscopic viewpoint and affects the heat transfer. Mathematically this translates into the question of what is the minimum wall superheat, as defined by Figure 20, or when is $\partial q/\partial T_w = 0$? As will be discussed in a moment, the methods used in the reduction of much of the current, forced-convection data make the direct answering of this question difficult. Thus, while more complex models can be constructed to address the question, only a simple model is proposed until sufficient properly analyzed data are obtained.

Much work has been done in determining how and when significant liquid contact begins [5, 87, 109, 110]. For reviews of the general work in this area, those by Comeau [111] and Fung [112] are useful. To more clearly define our question, recent publications by Gunnerson and Cronenburg [113] and Yao and Henry [87] prove quite useful. Gunnerson and Cronenburg present a generic approach to specifying the onset of stable film boiling and indicate the wall temperature is bounded by the expression

$$T_{sat} < T_i < T_{max,s}$$

where $$T_i = [T_w(k/\sqrt{\alpha})_w + T_f(k/\sqrt{\alpha})_f]/[(k/\sqrt{\alpha})_w + (k/\sqrt{\alpha})_f] \tag{45}$$

is the interfacial temperature as defined by Carslaw and Jaeger [8], and $T_{max,s}$ is the maximum liquid metastable temperature. Yao and Henry's work emphasizes the effects of pressure and materials upon the onset of stable film boiling. Their results indicate that, for pool boiling on a horizontal surface, the minimum film boiling point is determined by the mechanism that is stable to the lowest wall temperature of either the Taylor instability mechanism or the spontaneous nucleation mechanism. For water, the hydrodynamic instability can be assumed to control at pressures less than 1.5 MPa. The homogeneous nucleation temperature limit, T_{HN}, is used at higher pressures [87]. For the case of vertical boiling, the Helmhotz instability should replace the Taylor instability. However, due to the uncertainties in the effects of flows upon this instability limit and the fact that there is little practical difference between T_{HN} and $T_{max,s}$, the homogeneous nucleation temperature limit, for simplicity, is assumed to apply for all pressures. Thus, the wall temperature at which stable film boiling begins is assumed to result from the interfacial temperature equaling the homogeneous nucleation temperature and is given by

$$T_{w,HN} = \{T_{HN}[(k/\sqrt{\alpha})_w + (k/\sqrt{\alpha})_f] - T_f(k/\sqrt{\alpha})_f\}/(k/\sqrt{\alpha})_w \tag{46}$$

With respect to the wall superheat, this wall temperature for the onset of stable film boiling is denoted as $\Delta T_{HN} = T_{HN} - T_{sat}$. The assumption that the wall superheat, defined by ΔT_{HN}, indicates the onset of film boiling is utilized for all qualities at a given pressure. In the subcooled region ($X < 0$), this assumption is justified only for minimal subcooling, since a temperature distribution will exist in the inverted annular flow regime for the post–CHF region. In the two-phase regime, or with minimal subcooling as the vapor film on the wall is approached through the fluid, the liquid temperature will approach or reach the saturation condition. Thus, T_f in Equation 45 can be replaced by T_{sat}. Low-pressure, subcooled, forced-convective results presented by Cheng et al. [103] support this assumption as a first approximation. For higher subcooling, the bulk temperature of the fluid can be used for T_f in the model; however, the data show additional work is needed.

The effects of the film boiling low-quality changes upon the heat transfer surface can now be considered. This can be most easily done in terms of the relative maxima and minima for the new heat transfer surface, as shown in Figure 25. Figure 25 replaces the original single minimum, ΔT_{min}, of Figure 22 with two relative minima, X_{min} and ΔT_{HN}.

The X_{min} loci are those minimum heat flux points determined from isothermal curves, presented typically in Figure 23, as the wall temperature is parametrically varied. The curve denotes when $\partial q/\partial X$ changes sign and is controlled primarily by the hydraulic conditions adjacent to the wall. The ΔT_{HN} loci represent the wall temperatures below which liquid may contact the wall generally (macroscopically) and separate transition boiling ($\partial q/\partial T_w < 0$) and film boiling ($\partial q/\partial T_w > 0$). This line is controlled primarily by wall material properties, wall finish, and thermodynamic effects.

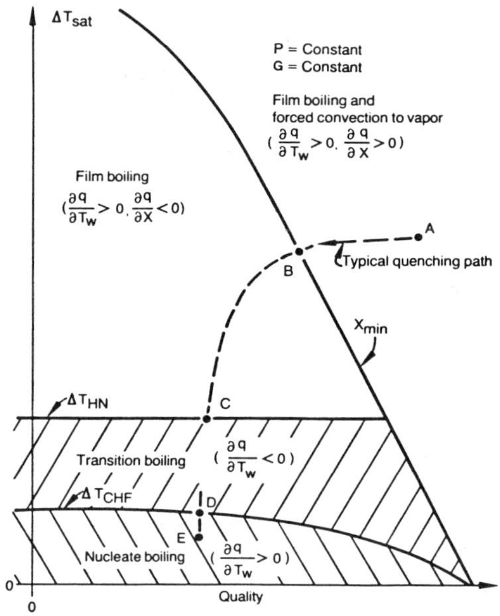

Figure 25. Loci of maxima and minima of heat flux with respect to quality and superheat showing effect of new mechanisms [90].

Effects of Forced-Convection Post–CHF Heat Transfer Upon Quenching

This section addresses the question of how the phenomena discussed in the previous section will influence forced-convective post–CHF data reduction and the quenching process. As noted earlier, this discussion is limited to that class of experiments where precursory cooling downstream of the conduction-convection propagating quench front causes cooldown.

The current literature on ΔT_{min} and transition boiling for forced convection provides, at best, a very confusing picture. The primary contributing factors in the confusion are believed to be:

- That the data reduction procedure may produce results dependent upon the quenching transient.
- That different types of experiments have been used.
- The effect of axial conduction upon ΔT_{min}.

The effect of axial conduction has undergone much discussion in the literature, while the effect of the other two factors has received little notice.

To understand how the data reduction procedure and the quenching transient influenced the results obtained for different experiments, two general types of experiments are discussed. Axial conduction is then factored into the discussion conceptually, after the effects of these two factors are understood. The first type of experiment considered is forced convective quenching experiments with flow inside long tubes and/or flow outside along bundles of tubes. The second is forced-convective quenching experiments, using short test sections with flow inside tubes having high thermal inertia.

Two types of data reduction procedures for a point far downstream of the quench front are considered. The quenching of this point will be slow enough and generally broad enough that a quasi-steady-state approach is valid. Thus, the heat flux can be thought of as being represented by Equation 26. To simplify this consideration, the fluid will be assumed to be saturated, $T_f = T_{sat}$, so that the wall superheat will be used as a variable instead of the wall temperature alone. This variable change represents only a rigid body translation of the origin on the temperature axis but is more consistent with standard practices. Thus, Equation 26 can be rewritten as

$$q = q(\Delta T_{sat}, \text{Pres}, G, X, \ldots) \tag{47}$$

The first data reduction procedure is implied by the classical pool boiling curve approach, where heat flux is considered a function of wall superheat, ignoring the effects of other changing variables. The second is that procedure inherent in the heat transfer surface approach, where the boiling curve represents the heat flux as a function of wall superheat, with all other variables held constant. For either case, the heat flux is assumed to be obtained from a one-dimensional (radial) inverse-conduction calculation, using thermocouple measurements at the elevation of interest. If the thermocouple is located on the wall, its measurement yields T_w directly; otherwise the inverse calculation must also yield this quantity.

In considering the classical pool boiling curve determined from the quenching of the point, the rate of change of heat flux is given by

$$dq/dt = [\partial q/\partial(\Delta T_{sat})]\,[d(\Delta T_{sat})/dt] + (\partial q/\partial \text{Pres})\,(d\text{Pres}/dt) + (\partial q/\partial G)\,(dG/dt) + \ldots \tag{48}$$

which is similar to Equation 29. Multiplying both sides of Equation 48 by $dt/d(\Delta T_{sat})$ then yields

$$dq/d(\Delta T_{sat}) = \partial q/\partial(\Delta T_{sat}) + \{(\partial q/\partial \text{Pres})\,(d\text{Pres}/dt) + (\partial q/\partial G)\,(dG/dt) + (\partial q/\partial X)\,(dX/dt) + \ldots\}/[d(\Delta T_{sat})/dt] \tag{49}$$

so that the slope of the classical boiling curve is given by the total derivative of heat flux with respect to wall superheat, $dq/d(\Delta T_{sat})$. Equation 49 indicates how different factors will affect the classical boiling curve when q is plotted against ΔT_{sat} for given times in the quenching transient. Thus, the classical boiling curve for quenching experiments with tubes and bundles is a complex relationship of the convective heat transfer $[\partial q/\partial(\Delta T_{sat}), \partial q/\partial X, \ldots]$, the hydraulic transient $(dX/dt, dP/dt, \ldots)$, and the conduction-convection transient of the wall $[d(\Delta T_{sat})/dt]$.

A special case of the general quenching experiments on tubes and bundles exists in those experiments that are run at constant pressure and inlet mass flux. For this case,

$$dP/dt = dG/dt = 0 \tag{50}$$

so that Equation 49 reduces to

$$dq/d(\Delta T_{sat}) = \partial q/\partial(\Delta T_{sat}) + (\partial q/\partial X)\,(dX/dt)/[d(\Delta T_{sat})/dt] \tag{51}$$

This classical data reduction procedure yields a minimum wall superheat, found from $dq/d(\Delta T_{sat}) = 0$, and a transition boiling region, where $dq/d(\Delta T_{sat}) < 0$. From Equation 51, it is observed that both the ΔT_{min} and transition boiling are dependent on the experimental transient. Transition boiling will thus appear to occur when, for example,

$$dX/dt < -\{[\partial q/(\partial T_{sat})]/[(\partial q/\partial X)\}\,\{d(\Delta T_{sat})/dt\} \tag{52}$$

Equation 52 can now be interpreted in terms of the new low-void film boiling effect for a typical quenching path, as shown in Figure 23. Quality is decreasing as the conduction-convection quench front progresses toward the point being considered. After X_{min} is crossed, the following relationships exist:

$$\partial q/\partial(\Delta T_{sat}) > 0, \quad \partial q/\partial X < 0, \quad d(\Delta T_{sat})/dt < 0$$

so that transition boiling appears to occur when dX/dt is decreasing at a sufficient rate. The apparent ΔT_{min} resulting from this interpretation, when Equation 52 is satisfied, is greater than ΔT_{HN} and, as shown, depends upon the experimental transient.

Another interesting result from this analysis is that the ΔT_{min} obtained from a quenching experiment, where Equation 52 is satisfied, will be hydraulically controlled by the transient. However, when the transient does not satisfy Equation 52, that is, does not decrease the quality fast enough, the resulting ΔT_{min} will be ΔT_{HN}, and the minimum will be thermodynamically controlled. Similar results are obtained for the more general quenching experiment, indicated by Equation 49; however, the relationships are more complex.

The heat transfer surface approach to data reduction to determine boiling curves yields only the heat transfer surface characteristic with respect to wall superheat, that is $\partial q/\partial(\Delta T_{sat})|_{X, G, Pres = constant}$. This characteristic must be obtained from the experiment by a transient analysis of the hydraulic conditions as a function of axial position, z, so that X = X(z, t), G = G(z, t), and Pres = Pres(z, t). Only in this manner can the values of heat flux and wall temperature be selected for constant values of X, G, and Pres. This analysis may require the use of a number of experimental runs in order to properly span the desired wall temperature range. If other characteristics of the heat transfer surface are desired, a similar procedure must be used. For example, the isothermal curves are obtained from $\partial q/\partial X|_{Tw, G, Pres = constant}$. In this manner, the experimentor can determine which parameters are necessary in the correlation of heat flux, since the partial derivatives of flux with respect to unimportant parameters will tend to be zero. The heat transfer surface procedure will ensure the proper separation of the variables. It is the improper separation of variables resulting from the classical approach, particularly quality and wall temperature, which has led to the confusion associated with ΔT_{min} and transition boiling for forced-convective heat transfer.

With this understanding of how the data reduction procedures influence the convective heat flux correlations obtained from the quenching of tubes or bundles for a point far downstream of the quenching front, the effect of axial conduction can be integrated into our thinking. Originally, the term $d(\Delta T_{sat})/dt$ in Equation 49 was discussed in terms of a one-dimensional conduction solution. However, this limitation to one dimension was not required by Equation 49. The limitation arises due to the problem caused by multi-dimensional conduction when the heat flux must be found from the wall thermocouples through an inverse-conduction solution. Without proper accounting of multi-dimension conduction, erroneous convective heat fluxes are found. For example, not including axial conduction in the data reduction when a quench front is near causes the ΔT_{min} to appear higher than it really is. For details of the mechanics of how axial conduction and convection combine to affect the results of quench front propagation, the work of Elias et al. [40] can be studied; however, their work does not include two-dimensional (radial and axial) conduction effects nor the low-void film boiling effect within its convection model.

This integration of multi-dimensional conduction into the situation yields a picture which applies to either of the two axial locations defined in the earlier section. Usually, the point located immediately downstream of the quenching front is the location that provides the majority of the data with respect to transition boiling, or the ΔT_{min}, when quenching of a tube or bundle is studied. Unfortunately, this is also the point where axial conduction and the rate of change of the hydraulic transient are at their maximum, making analysis of this type of experiment very difficult.

The short test section with high thermal inertia provides the advantage over a long thin tube in that it quenches, due to its shortness, with "constant" hydraulic conditions defined by the inlet conditions. Thus, the change in the hydraulic conditions over the short test section can be ignored, i.e.

$$dP/dt = dG/dt = dX/dt = 0$$

and Equation 49 reduces to

$$dq/d(\Delta T_{sat}) = \partial q/\partial(\Delta T_{sat}) \tag{53}$$

The high thermal inertia prolongs the quenching period so that sufficient data can be acquired during the true transition boiling period, $\partial q/\partial T_w < 0$.

Equation 53 indicates that the results obtained from this type of experiment will be the same for either of the data reduction procedures discussed earlier. This is an advantage, since the results from this type of experiment provide boiling curves that immediately separate the variables properly. The problem, if it can be termed that, created by this type of experiment has been that its results, $\Delta T_{min} = \Delta T_{HN}$, often disagree with tube results analyzed according to Equations 49 or 51, and confusion has arisen. Equation 53 also indicates why the results of approximation for defining the onset of film boiling earlier.

SUMMARY

This chapter has been divided into two major parts concerning conduction and convection. The conduction section reviews those quenching models which solve the heat diffusion equation for either constant heat transfer coefficients or ones known as functions of time, temperature, and/or position. The models are roughly grouped into lumped, one-dimensional, two-dimensional, and three-dimensional types. This section is meant to familiarize the analyst with those models which are available so that he may make simple and quick calculations.

The convection section reviews the quenching process from a more conceptual viewpoint, for those analysts who require a more fundamental understanding of the quenching process. This understanding should then provide the basis for development of the more complex multi-region heat transfer models discussed in the conduction section. This study was done by initially reviewing the transient convection heat transfer problem, which couples the temperature field solutions for both the solid and the fluid, where the solid-fluid interfacial behavior is not known a priori. This allows a clearer understanding of the quasi-steady-state approach, which is applied to the analysis of most quenching problems.

The quasi-steady-state approach is further investigated by reviewing transient and steady-state convective heat transfer, using the ergodic assumption. This allows a more precise definition of steady-state, quasi-steady-state, and transient convection. In order to further show the coupled nature of the process and the area and time constraints which apply to data obtained from steady-state experiments, the effects of conduction upon the steady-state convection process are also investigated. The influence of isothermal and nonisothermal materials and their importance to the quasi-steady-state approach are discussed. Finally, due to its control of the quenching process, the qualitative nature of the post-CHF region for steady-state boiling is considered. Data reduction of quenching experiments is discussed, and the manner in which the quenching transient may affect the results of different types of quenching experiments is investigated. This investigation provides an explanation of how minimum wall superheats greater than the homogeneous nucleation temperature result, as well as how these may be either hydrodynamically or thermodynamically controlled.

NOTATION

A_w, A_d effective wet and dry Biot numbers, respectively
Bi Biot number
C speed of heat propagation
C_P specific heat at constant pressure
c specific heat
Fo Fourier number

G total mass flux
h heat transfer coefficient
k thermal conductivity
L_c characteristic length
P Peclet number
Pr Prandtl number
Pres pressure
Q heat generation rate
q heat flux
q′ internal heat generation rate per unit volume
Re Reynolds number
S surface area
T temperature
t time
u_0 quench front velocity
V volume

Greek Symbols

α thermal diffusivity
α_v void fraction
β dimensionless parameter for conduction
δ thickness
λ penetration depth
θ dimensionless temperature
θ^* dimensionless temperature defined in Equation 14
μ dynamic viscosity
ν kinematic viscosity
ρ density
τ dimensionless time

REFERENCES

1. Van Vlack, L. H., *Elements of Materials Science,* Aldison–Wesley Publishing Company, Inc., Reading, Mass., 1960.
2. Dieter, G. E., *Mechanical Metallurgy,* McGraw–Hill Book Company, New York, New York, 1961.
3. Yadigaroglu, G., "The Reflooding Phase of the LOCA in PWRs. Part I: Core Heat Transfer and Fluid Flow," *Nuclear Safety,* Vol. 19, No. 1 (Jan.–Feb. 1978), pp. 20–36.
4. Elias, E., and Yadigaroglu, G., "The Reflooding Phase of the LOCA in PWRs. Part II: Rewetting and Liquid Entrainment," *Nuclear Safety,* Vol. 19, No. 2, (March–April 1978), pp. 160–175.
5. Bradfield, W. S., "Liquid-Solid Contact in Stable Film Boiling," ITEC Fundamentals, Vol. 5 (May 1966), pp. 200–204.
6. Incropera, F. P., and Dewitt, D. P., *Fundamentals of Heat Transfer,* John Wiley & Sons, New York, 1981.
7. Kreith, F., *Principles of Heat Transfer,* Intext Educational Publishers, 3rd ed., New York, 1976.
8. Carslaw, H. S., and Jaeger, J. C., *Conduction of Heat In Solids,* Oxford University Press, 2nd ed., London, 1959.
9. Rohsenow, W. M., and Hartnett, J. P., *Handbook of Heat Transfer,* McGraw–Hill Book Company, New York, 1973, pp. 3–1 to 3–134.
10. Chester, M., "Second Sound In Solids," *Physical Review,* Vol. 131, No. 5 (Sept. 1963), pp. 2013–2015.
11. Boley, B. A., *The Analysis of Problems of Heat Conduction and Melting: High Temperature Structures and Materials,* Pergamon Press, New York, 1964, pp. 260–315.
12. Baumeister, K. J., and Hamill, T. D., "Hyperbolic Heat Conduction Equation—A Solution for the Semi-Infinite Body Problem," *Journal of Heat Transfer* (November 1969), pp. 543–548.
13. Butterworth, K., and Owen, R. G., "The Quenching of Hot Surfaces by Top and Bottom Flooding—A Review," AERE–R7992, March 1975.
14. Sawan, M. E., and Carbon, M. W., "A Review of Spray–Cooling and Bottom–Flooding Work for LWR Cores," *Nuclear Engineering Design,* Vol. 32 (1975), pp. 191–207.

15. Fairbirm, S. A., Central Electricity Generating Board Report RD/B/N/4214, WR/HTSG/P (78) 15, 1978.
16. Carbajo, J. J., and Siegil, A. D., "Review and Comparison Among The Different Models For Rewetting in LWR's," *Nuclear Engineering Design,* Vol. 58 (1980), pp. 33–44.
17. Semeria, R., and Martinet, B., "Calefaction Spots on a Heating Wall: Temperature Distribution and Resorption," *Proc. Inst. Mech. Eng.,* Vol. 180, Part 3C (1965/1966), pp. 192–205.
18. Yamanouchi, A., "Effect of Core Spray Cooling in Transient State After Loss-of-Coolant Accident," *J. Nucl. Sci. Technol.,* Vol. 5 (1968), pp. 547–558.
19. Yao, L. S., "Rewetting of a Vertical Surface with Internal Heat Generation," AIChE paper 19, 16th National Heat Transfer Conference, Aug 1976, St. Louis, Mo., or AIChE Symposium, Vol. 73, 1976, pp. 46–50.
20. Thompson, T. S., "An Analysis of the Wet Side Heat Transfer Coefficient During Rewetting of a Hot Dry Patch," *Nuclear Engineering and Design,* Vol. 22 (1972), pp. 212–224.
21. Bennett, A. W., et al., "The Wetting of Hot Surfaces by Water in a Steam Environment at High Pressure," United Kingdom Atomic Energy Authority, AERE–R5146, 1966.
22. Sun, K. H., Dix, G. E., and Tien, C. L., "Cooling of a Very Hot Vertical Surface by a Falling Liquid Film," *Trans. ASME, Series C, Journal of Heat Transfer,* Vol. 96 (1974), pp. 126–131.
23. Duffey, R. B., and Porthouse, D. T. C., "Experiments on the Cooling of High-Temperature Surfaces by Water Jets and Drops," Proceedings of the CREST Specialist Meeting on Emergency Core Cooling for Light–Water Reactors, Report MRR–115, Vol. 1, Oct. 1972, Garching/Munich, Germany.
24. Ishii, M., "Study on Emergency Core Cooling," *Journal British Nuclear Energy Society,* Vol. 14, No. 3 (1975), pp. 237–242.
25. Sun, K. H., Dix, G. E., and Tien, C. L., "Effect of Precursory Cooling on Falling-Film Rewetting," ASME Winter Annual Meeting, ASME paper 74–WA/HT–52, 1974, New York, also *Journal of Heat Transfer,* Vol. 97, (1975), pp. 360–356.
26. Chun, M. H., and Chon, W. Y., "Analysis of Rewetting in Water Reactor Emergency Core Cooling Inclusive of Heat Transfer in the Unwetted Region," ASME Winter Annual Meeting, ASME paper 75–WA/HT–32, 1975, New York.
27. Case, P., et al., "The Rewetting Process of Hot Surfaces by Bottom Flooding," European Two-Phase Flow Group Meeting, June 1973, Brussels.
28. Chan, S. H., and Grolmes, M. A., "Hydrodynamically Controlled Rewetting," 975 National Heat Transfer Conference, ASME paper 75–HT–70, Aug. 1975, San Francisco, Calf.
29. Hewitt, G. F., and Wallis, G. B., "Flooding and Associated Phenomena in Falling Flow in a Vertical Tube," Multi-phase Flow Symp., 1974 ASME Winter Annual Meeting, Nov. 1974, Philadelphia, Pa.
30. Kutateladze, S. S., "Elements of the Hydrodynamics of Gas-Liquid Systems," *Fluid Mechanics–Soviet Research,* Vol. 1, No. 4 (1972), pp. 62–74.
31. Shires, G. L., and Pickering, A. R., "The Flooding Phenomenon in Countercurrent Two-Phase Flow," Proceedings on Symp. on Two-Phase Flow, Vol. 2, June 1965, Exeter, pp. B501–B538.
32. Guerrero, H. N., and Lowe, P. A., "Exploratory Single Tube, Top Flooding, Gravity Feed Heat Transfer Tests," 20th ANS Annual Meeting, June 1974, Philadelphia, Pa.
33. Wallis, G. B., *One-Dimensional Two-Phase Flow,* McGraw–Hill Book Company, New York, 1970.
34. Hewitt, G. F., and Hall-Taylor, N. S., *Annular Two-Phase Flow,* Pergamon Press, Inc., New York, 1970.
35. Andreoni, D., Exchanges Thermiques Lors due Renoyage d'un Coeur de Reacteur a Eau, Thesis from l'Universite Scientifique et Medicale de Grenoble et l'Institute National Polytechnique de Grenoble, Nov. 1975.

36. Jens, W. H., and Lottes, P. A., "Analysis of Heat Transfer Burnout, Pressure Drop and Density Data for High Pressure Water," ANL–4627, May 1951.
37. Tan, H. E., "A Model for Quenching with Vertical Flow Boiling," S. M. thesis, Department of Mechanical Engineering, M. I. T., Feb. 1975.
38. Iloeje, O. C., "A Study of Wall Rewet and Heat Transfer in Dispersed Vertical Flow," Ph. D. thesis, Department of Mechanical Engineering, M. I. T., July 1974.
39. Elias, E., and Yadigaroglu, G., "A General One-Dimensional Model For Conduction-Controlled Rewetting Of A Surface," *Nuclear Engineering and Design,* Vol. 42 (1977), pp. 185–194.
40. Elias, E., Arrieta, L., and Yadigaroglu G., "An Improved Model for the Rewetting of a Hot Fuel Rod," *Trans. ANS,* Vol. 24 (1976), pp. 299–300.
41. Kirchner, W. L., "Reflood Heat Transfer in a Light-Water Reactor," Ph. D., thesis, Department of Mechanical Engineering, M. I. T., Jan. 1976, also NRC report no. NUREG–0106 (Vols. 1 and 2).
42. Cermak, J. O., et al., "PWR Full Length Emergency Cooling Heat Transfer (FLECHT) Group I Test Report," WCAP–7435, Jan. 1970.
43. Cadek, F. F., et al., "PWR FLECHT (Full Length Emergency Cooling Heat Transfer) Group II Test Report," WCAP–7544, Sept. 1970.
44. Fischer, S. R., et al., "RELAP4/MOD6 A Computer Program For Transient Thermal-Hydraulic Analysis of Nuclear Reactors and Related Systems—User's Manual," E. G. & G. Idaho Inc. report CDAP TR 003, Jan. 1978.
45. Rosal, E. R., et al., "FLECHT-Low Flooding Rate Cosine Test Series Report," WCAP–8651, Dec. 1975, Westinghouse Electric Corporation.
46. Waring, J. P., and Hochreiter, L. E., "PWR FLECHT–SET Phase Bi Evaluation Report," WCAP–8583, 1975.
47. Peterson, A. C., et al., "Thermal and Hydraulic Response of the Semiscale MOD–1 Core During Forced-Feed Reflood Tests," TREE–NUREG–1001, EG&G Idaho, Inc., Oct. 1976.
48. Aksan, S. N., Nelson, R. A., and Tolman, E. L., "Application of an Assessed Analytical Capability to Predict Rapid Cladding Cooling and Quench Behavior During the Blowdown Phase of a Large Break LOCA," Second International Topical Meeting on Nuclear Reactor Thermal-Hydraulics, Vol. I, Jan. 1983, pp. 705–712.
49. Tien, C. L., and Yao, L. S., "Analysis of Conduction-Controlled Rewetting of a Vertical Surface," *Journal of Heat Transfer,* Vol. 97 (May 1975), pp. 161–165, also ASME paper 74–WA/HT–49.
50. Duffey, R. B., and Porthouse, D. T. C., "The Physics of Rewetting in Water Reactor Emergency Core Cooling," *Nucl. Eng. and Design,* Vol. 25 (1973), pp. 379–394.
51. Yoshioka, K., and Hasegawa, S., "A Correlation in Displacement Velocity of Liquid Film Boundary Formed on a Heated Vertical Surface in Emergency Cooling," *Journal of Nuclear Science Technology,* Vol. 7 (1970), pp. 418–425.
52. Martini, R., and Premoli, A., "A Simple Model for Predicting ECC Transients in Bottom Flooding Conditions," Proceedings of the CREST Specialist Meeting on Emergency Core Cooling for Light-Water Reactors, Oct. 1972, report MRR–115, Vol. 2, Garching/Munich, Germany.
53. Thompson, T. S., "Simulated Bottom Flooding Emergency Cooling of a Close-Spaced Rod Bundle," European Two-Phase Flow Meeting, 1972, Casaccia, Italy.
54. Campanile, A., and Pozzi, G. P., "Low Rate Emergency Reflooding Heat Transfer Tests in Rod Bundle," Proceedings of the CREST Specialist Meeting on Emergency Core Cooling for Light-Water Reactors, Oct. 1972, report MRR–115, Vol. 2, Garching/Munich, Germany.
55. Blair, J. M., "An Analytical Solution to a Two-Dimensional Model of the Rewetting of Top Spray Emergency Coolant," *Nuclear Engineering and Design,* Vol. 32 (1975), pp. 159–170.
56. Coney, M. W. E., "Calculations on the Rewetting of Hot Surfaces," *Nuclear Engineering and Design,* Vol. 31 (1974), pp. 246–259.

57. Yu, S. K. W., Farmer, P. R., and Coney, M. W. E., *International Journal of Multiphase Flow,* Vol. 3 (1977), pp. 415–443.
58. Andersen, J. G. M., and Hansen, P., "Two-Dimensional Heat Conduction in Rewetting Phenomenon," Danish Atomic Energy Commission Research Establishment, report # NORHAV–D–6, June 1974, Riso, Denmark.
59. Thompson, T. S., "Rewetting of a Hot Surface," Proceedings of 5th International Heat Transfer Conference, Vol. 4, Sept. 1974, Tokyo, Japan, pp. 139–143.
60. Yeh, H. C., "An Analysis of Rewetting of a Nuclear Fuel Rod in Water Reactor Emergency Core Cooling," *Nuclear Engineering and Design,* Vol. 34 (1975), pp. 317–322.
61. Carrier, G. F., Krook, M., and Pearson, C. E., *Functions of a Complex Variable,* McGraw-Hill, New York., 1966.
62. Moble, B., *Methods Based on the Wiener-Hopf Technique,* Pergamon Press, New York, 1958.
63. Dua, S. S., and Tien, C. L., "A Generalized Two-Parameter Relation For Conduction-Controlled Rewetting Of A Hot Surface," *International Journal of Heat Transfer,* Vol. 20 (1977), pp. 174–176.
64. Edwards, A. R., and Mather, D. J., "Some U. K. Studies Related to the Loss-of-Coolant Accident," Proceedings of the Topical Meeting on Water Reactor Safety, USAEC report CONF–730304, 1973, Salt Lake City, Utah, pp. 720–739, Aerojet Nuclear Company, NTIS.
65. Dua, S. S., and Tien, C. L., "Two-Dimensional Analysis of Conduction-Controlled Rewetting with Precursory Cooling," *Trans. of ASME, Series C, Journal of Heat Transfer,* Vol. 98 (1976), pp. 407–413.
66. Bonakdar, H., and McAssey, E. V., "A Method for Determining Rewetting Velocity Under Generalized Boiling Conditions," ASME 1980 Winter Annual Meeting, paper 80–WA/HT–40.
67. Chan, A. M. C., and Banerjee, S., "Refilling and Rewetting of a Hot Horizontal Tube Part I: Experiments," *Journal of Heat Transfer,* Vol. 103 (1981), pp. 281–286.
68. Chan, A. M. C., and Banerjee, S., "Refilling and Rewetting of a Hot Horizontal Tube Part II: Structure of a Two-Fluid Model," *Journal of Heat Transfer,* Vol. 103 (1981), pp. 287–292.
69. Chan, A. M. C., and Banerjee, S., "Refilling and Rewetting of a Hot Horizontal Tube Part III: Application of a Two-Fluid Model to Analyze Rewetting," *Journal of Heat Transfer,* Vol. 103 (1981), pp. 653–659.
70. Siegel, R., "Heat Transfer for Laminar Flow in Ducts With Arbitrary Time Variations in Wall Temperature," *Journal of Applied Mechanics, 27, Trans. ASME, Series E,* Vol. 82 (June 1960), pp. 241–249.
71. Sparrow, E. M., and Siegel, R., "Unsteady Turbulent Heat Transfer in Tubes," *Journal of Heat Transfer, Trans. ASME, Series E,* Vol. 82, No. 3 (Aug. 1960), pp. 170–180.
72. Kawamura, H., "Analysis of Transient Turbulent Heat Transfer in An Annulus: Part I: Heating Element With A Finite (Nonzero) Heat Capacity and No Thermal Resistance," *Heat Transfer-Japan Research,* Vol. 3, No. 1 (1974), pp. 45–68.
73. Kawamura, H., "Experimental and Analytical Study of Transient Heat Transfer For Turbulent Flow In A Circular Tube," *International Journal of Heat & Mass Transfer,* Vol. 20 (1977), pp. 443–449.
74. Sucec, J., "An Improved Quasi-Steady Approach For Transient Conjugated Forced Convection Problems," *International Journal of Heat & Mass Transfer,* Vol. 24, No. 10 (1981), pp. 1711–1722.
75. Veres, D. R., and Florschuetz, L. W., "A Comparison of Transient and Steady-State Pool-Boiling Data Obtained Using the Same Heating Surface," *Journal of Heat Transfer* (May 1971), pp. 229–232.
76. Regheb, H. S., Cheng, S. C., and Groeneveld, D. C., "Observations In Transition Boiling Of Subcooled Water Under Forced Convective Conditions," *International Journal of Heat & Mass Transfer,* Vol. 24, No. 7 (1981), pp. 1127–1136.

77. Bernath, L., "A Theory of Local-Boiling Burnout and Its Application to Existing Data," *Chem. Eng. Progr., Symp. Ser. 56,* No. 30 (1960), pp. 95–116.
78. Gunnerson, F. S., and Cronenberg, A. W., "A Prediction of the Minimum Film Boiling Conditions for Spherical and Horizontal Flat Plate Heaters," ASME paper 79–HT–45, 18th National Heat Transfer Conference, San Diego, Calf., Aug. 1979.
79. Rhea, L. G., and Nevins, R. G., "Film Boiling Heat Transfer from an Oscillating Sphere," *Journal of Heat Transfer* (May 1969), pp. 267–272.
80. Merte, H., and Clark, J. A., "Boiling Heat Transfer with Cryogenic Fluids at Standard, Fractional, and Near Zero Gravity," *Journal of Heat Transfer,* Vol. 86 (1964), pp. 351–359.
81. Tien, C. L., and Lienhard, J. H., *Statistical Thermodynamics, Revised Printing,* Series in Thermal and Fluids Engineering, Washington, 1979.
82. Bergles, A. E., et al., *Two-Phase Flow And Heat Transfer In The Power And Process Industries,* Hemisphere Publishing Corp. and McGraw-Hill Book Company, New York, 1981, pp. 40–97.
83. Wachters, L. H. J., et al., "The Heat Transfer from a Hot Wall to Impinging Mist Droplets in the Spheroidal State," *Chemical Engineering Science,* Vol. 21 (Dec. 1966), pp. 1231–1238.
84. Kalinin, F. K., et al., "Heat Transfer in Tubes with Rod Regime in the Case of Film Boiling of a Subcooled Liquid," *Concurrent Gas-Liquid Flow,* E. Rhodes and D. S. Scott (Eds.), Plenum, New York, 1969.
85. Baumeister, K. J., Simon, F. F., and Henry, R. E., "Role of the Surface in the Measurement of the Leidenfrost Temperature," *Augmentation of Convective Heat and Mass Transfer,* ASME, New York, pp. 91–101.
86. Baumeister, K. J., and Simon, F. F., "Leidenfrost Temperature—Its Correlation for Liquid Metals, Cryogens, Hydrocarbons, and Water," *Journal of Heat Transfer* (May 1973), pp. 166–173.
87. Yao, S. C., and Henry, R. E., "An Investigation of the Minimum Film Boiling Temperature on Horizontal Surfaces," *Journal of Heat Transfer,* Vol. 100 (May 1978), pp. 260–267.
88. Mowat, J. A. S., et al., "Investigations into the Effect of Surface Condition Upon Rewetting of Metal Tubes from High Temperature," AERE–R 10019, Dec. 1980.
89. Witte, L. C., Chang, K–H., and Orozce, J., "On the Role of Liquid-Solid Contact in Subcooled Flow Film Boiling From Submerged Bodies," Interfacial Transport Phenomena, ASME Heat Transfer Division Vol. 23, July 1983.
90. Nelson, R. A., "Forced-Convective Post–CHF Heat Transfer and Quenching," *Journal of Heat Transfer,* Vol. 104 (Feb. 1982), pp. 48–54.
91. Nelson, R. A., "The Heat-Transfer Surface Technique," LOCA Heat Transfer Workshop, Idaho Falls, Idaho, July 1975.
92. Collier, J. G., "Two-Phase Flow and Heat Transfer in Water Cooled Nuclear Reactors," Lecture Series Presented at Dartmouth College, Hanover, New Hampshire, Aug. 1975.
93. Nelson, R. A., and Sullivan, L. H., "RELAP4/MOD6 Reflood Heat Transfer and Data Comparison," CSNI Specialists Meeting on Transient Two-Phase Flow, Paris, France, June 1978.
94. Nelson, R. A., and Sullivan, L. H., "Blowdown Heat Transfer Surface in RELAP4/MOD6 and Data Comparison," ENS/ANS International Topical Meeting on Nuclear Power Reactor Safety, Vol. 2, Brussels, Belgium, Oct. 1978, pp. 1719–1728.
95. Bjornard, T. A., "Blowdown Heat Transfer In a Pressurized Water Reactor," Ph. D. thesis, M. I. T., Aug. 1977.
96. Bjornard, T. A., and Griffith, P., "PWR Blowdown Heat Transfer," ASME Symposium on the Thermal and Hydraulic Aspects of Nuclear Safety, Vol. 1, Nov. 1977, pp. 17–39.
97. Ransom, V. H., et al., "RELAP5/MOD1 Code Manual Volume 1: System Models and Numerical Methods," NUREG/CR–1826, EGG–207 Revision 1, March 1981.

98. Lahey, R. T., and Moody, F. J., *The Thermal-Hydraulices on a Boiling Water Nuclear Reactor,* American Nuclear Society, 1977.
99. Nukiyama, I., "Maximum and Minimum Values of Heat Transmitted from a Metal to Boiling Water Under Atmospheric Pressure," *Journal of the Society of Mechanical Engineers Japan,* Vol. 37, 1934.
100. Dougall, R. S., and Rohsenow, W. M., "Film Boiling in the Inside of Vertical Tubes with Upward Flow of the Fluid at Low Qualities," M. I. T. Technical Report 9079–26, 1963.
101. Groeneveld, D. C., "An Investigation Of Heat Transfer In The Liquid Deficient Regime," AECL–3281, Dec. 1969.
102. Iloeje, O. C., Rohsenow, W. M., and Griffith, P., "Three-Step Model of Dispersed Flow Heat Transfer (Post–CHF Vertical Flow)," ASME Paper 75–WA/HT–1, 1975.
103. Cheng, S. C., Ny, W. W. L., and Heng, K. T., "Measurements of Boiling Curves of Subcooled Water under Forced Convective Conditions," *International Journal of Heat & Mass Transfer,* Vol. 21 (1978), pp. 1385–1392.
104. Barnard, D. A., Glastonbury, A. G., and Ward, J. A., "The Measurement of Post-Dryout Heat Transfer at Low Pressure and Low Mass Quality under Steady State Conditions," Proceedings of the European Two-Phase Flow Group Meeting, Joint Research Center, ISPRE, June 1979.
105. Fung, K. K., Gardiner, S. R. M., and Groeneveld, D. C., "Subcooled and Low Quality Flow Film Boiling of Water at Atmospheric Pressure," a paper accepted for publication in Nuclear Engineering and Design.
106. Polomik, E. E., et al., "Heat Transfer Coefficients with Annular Flow During Once-Through Boiling of Water to 100 Percent Quality at 800, 1100 and 1400 psi," GEAP–3703, 1961.
107. Groeneveld, D. C., et al., "Chalk River Progress in Updating Best-Estimate Heat Transfer Correlations," Best-Estimate Heat Transfer Workshop, Washington, D. C., Jan. 1981.
108. Grush, W. H., et al., "Results and Predictions of Scaled, Nuclear Large Break Loss-of-Coolant Experiments," ASME Paper 80–WA/HT–49, ASME 1980 Winter Annual Meeting, Chicago, III., Nov. 1980.
109. Nishikawa, K., Fujii, T., and Honda, A., "Experimental Study on The Mechanism of Transition Boiling Heat Transfer," Bulletin of the JSME, Vol. 15 (1972), pp. 93–103.
110. Lee, L. Y. W., Chen, J. C., and Nelson, R. A., "Liquid-Solid Contact Measurements Using A Surface Thermocouple Temperature Probe In Atmospheric Pool Boiling Water," accepted to 22nd National Heat Transfer Conference, Niagara Falls, New York, 1984.
111. Comeau, J. G., "A Review of the Liedenfrost Phenomenon," Master of Mechanical Engineers Report, University of Ottowa, 1979.
112. Fung, K. K., "Post–CHF Heat Transfer during Steady-State and Transient Conditions," NUREG/CR–0195, ANL–78–55, June 1978.
113. Gunnerson, F. S., and Cronenburg, A. W., "On the Thermodynamic Superheat Limit for Liquid Metals and Its Relation to the Leidenfrost Temperature," *Journal of Heat Transfer,* Vol. 100, Nov. 1978, pp. 734–737.

CHAPTER 36

HEAT TRANSPORT IN NUCLEAR REACTOR CHANNELS

Jan Lach

Institute of Atomic Energy
Department of Reactor Engineering
Otwock-Swierk, Poland

Marian Kielkiewicz and Marek Kosinski

Warsaw Technical University
Institute of Heat Engineering
Warsaw, Poland

CONTENTS

INTRODUCTION

This chapter examines basic problems of heat transport in nuclear reactor channels during normal full-power reactor operation. Heat removal from the reactor corc is the same as in conventional thermal installations. However, the requirements concerning reactor calculations are critical because of safety issues concerning both fuel elements and the surrounding environment. The complexity of nuclear reactor systems must be emphasized, particularly in relation to safety along with the need for nonlinear coupling of neutronic and thermal reactor behavior.

The basic topics discussed below include: thermal energy generation, spatial power distribution, heat transfer with single-phase forced convection, heat transfer with boiling and boiling crisis. Since this is presented as an overview chapter, the reader is liberally referred to the literature for more in depth coverage.

NUCLEAR REACTORS

A neutron hitting a heavy nucleus may cause a fission reaction, i.e., disintegration of the original nucleus into new ones with the release of an enormous amount of energy and simultaneous emission of a few neutrons and gamma photons.

Two new nuclei, called fission fragments, usually have unequal masses and are left in an excited state.

Neutrons emitted in fission are called prompt neutrons and have an average kinetic energy about 2 MeV.

The average number of prompt neutrons per fission is about 2.5. The average energy released in fission is about 200 MeV. Most of it appears as a kinetic energy of fission fragments and prompt neutrons. This energy, appearing as heat, is millions of times greater than the energy released in most energetic chemical reactions.

Because more than one prompt neutron is emitted per fission, the possibility of achieving self-sustaining fission chain reaction in a device called a nuclear reactor is possible. In a nuclear reactor some of the neutrons emitted in fission are absorbed in different materials while some of them escape from the reactor.

A nuclear reactor is in the critical state when the number of neutrons produced in it in a unit of time is equal to the number of neutrons lost due to absorption and leakage in a unit of time. In other words, the critical state of a nuclear reactor is a state of balance between production and destruction of neutrons.

Obviously, this balance can be attained at different power levels. Note, however that, certain conditions have to be met in order to achieve a critical state.

For given reactor composition, the size of a reactor should not be too small, otherwise it would be impossible to account for the large leakage of neutrons. The minimal size of a nuclear reactor at which the critical state can be attained is called the critical size. The mass of nuclear fuel corresponding to it is called a critical mass.

Nuclear fuel is a composition of fissile isotopes with fertile isotopes and other materials. Fissile isotopes are those which quite easily undergoe fission when bombarded by low energy neutrons. To them belong U^{235}, U^{233}, and Pu^{239}. Only U^{235} can be found in nature. In natural uranium there is only 0.7% by weight of U^{235}; the others, U^{238}. U^{233} and Pu^{239}, are not found in nature. These are produced in nuclear reactors due to irradiation of so called fertile isotopes, namely U^{238} and Th^{232}. In most operating reactors, enriched uranium is used in which the content of U^{235} is higher than 0.7%.

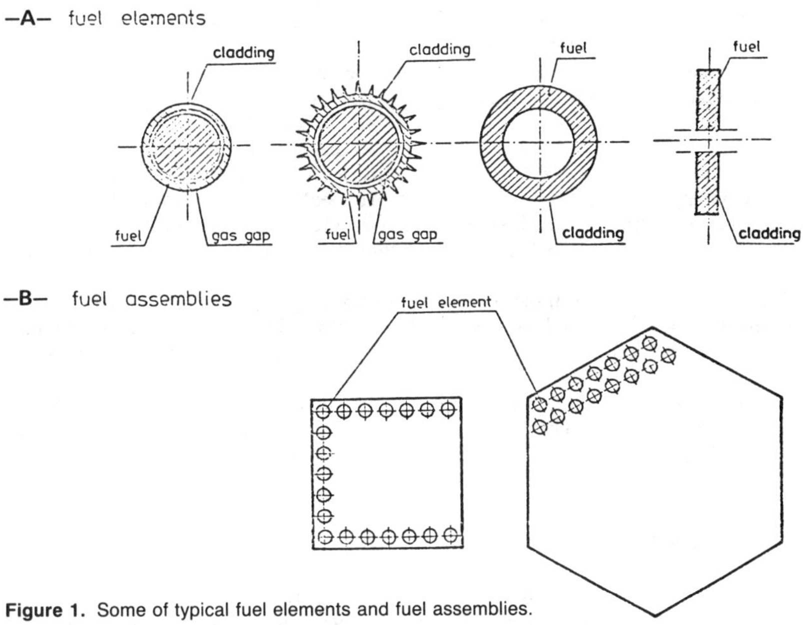

Figure 1. Some of typical fuel elements and fuel assemblies.

Fission of fissile isotopes may be induced by neutrons having different energies. Reactors in which the majority of fissions is caused by low energy neutrons ($\sim$ 0.025 eV) are called thermal reactors. Reactors in which the majority of fissions is caused by fast neutrons ($\sim$ tenths of MeV) are called fast reactors.

In thermal reactors neutrons must be slowed down from a high energy state (2 MeV) to a thermal energy state ($\sim$ 0.025 eV). Therefore a material of a low atomic mass, called a moderator, must be used. Neutrons lose their energy by collisions with moderator nuclei. The most frequently used moderators are: water, heavy water, and graphite (rarely beryllium).

A nuclear reactor in which nuclear fuel and other materials form a homogeneous mixture is called a homogeneous reactor. A nuclear reactor in which nuclear fuel is distinctly separated from other materials is called a heterogeneous reactor. Nearly all operating nuclear reactors, are of the heterogeneous type. Hence subsequent considerations will be limited to these.

In a heterogeneous reactor, nuclear fuel is contained in the fuel elements. These are usually in the form of cylindrical rods, although other geometries can be used. A cylindrical fuel element consists of a thin walled tube called cladding, inside which are stacked fuel pellets. The cladding must be tight to prevent the release of fission products.

Fuel elements are long and have a small diameter; hence they are susceptible to bending. In order to increase their stiffness and to ease handling, fuel elements are grouped into fuel assemblies. Some of fuel elements and fuel assemblies are shown in Figure 1. Fuel assemblies form the reactor core, which usually has an approximately cylindrical shape.

Heat generated in fuel elements by fissions is removed by coolant flowing through the core. Coolants may be gaseous or liquid. The most popular coolants are water, heavy water, carbon dioxide, helium and liquid sodium. In many cases the coolant also serves as a moderator.

Pressure in the cooling circuits of nuclear reactors usually falls in the range from a few tenths of MPa to about 16 MPa. Hence a reactor core must be housed in a pressure vessel.

Reactors where the entire core is placed in a pressure vessel are called vessel-type reactors and are the most popular in contemporary design.

Reactors whose core is divided into clusters of fuel elements positioned in separate pressure tubes are called channel-type reactors. This design is much less popular than the previous one. Therefore attention is restricted to vessel type reactors.

Other main components of a nuclear reactors are the control assemblies. These are usually clusters of rods or single rods (with circular, annular, and possibly other cross sections) made from material that can strongly absorb neutrons. Absorption of neutrons changes when control assemblies are inserted into a reactor core or removed from it. In this way a nuclear reactor can be controlled. A schematic of a vessel-type reactor is shown in Figure 2.

Coolant leaving a reactor in most cases passes through a heat exchanger (steam generator), where steam is generated to drive a steam turbine. Hence there are two circuits —primary and secondary (sodium cooled reactors have three circuits: primary, intermediate, and secondary).

There are also reactors whose coolant also serves as the working fluid passing through the turbine (steam or gaseous). There are several types of nuclear power reactors. The principal features and typical parameters of these are shown in Table 1. Basic information on nuclear power reactors can be found in the literature (e.g., [1–4]).

NEUTRON BALANCE EQUATIONS

The theory of nuclear reactors is based on neutron transport. Generally transport theory refers to the mathematical description of the transport of large numbers of identical particles like gas molecules, ions, neutrons, photons, electrons, etc. Therefore one can encounter transport equations in many fields of physics and engineering, for example in nuclear reactor theory, rarefied gas dynamics, plasma physics, and astrophysics.

Neutron Transport Equation

The dependent variable in the neutron transport equation is the neutron angular flux Φ, which in the most general case is a function of coordinates $\mathbf{r}$, direction unit vector $\boldsymbol{\Omega}$, energy E, and time t.

The neutron angular flux is defined as:

$$\Phi(\mathbf{r}, \boldsymbol{\Omega}, E, t) \overset{df}{=} vn(\mathbf{r}, \boldsymbol{\Omega}, E, t)$$

Table 1
Typical Average Parameters of Nuclear Reactors

		PWR	BWR	PHWR	GCR	AGR	HTGR	FBR
1	Fuel	UO_2	UO_2	UO_2	U	UO_2	UO_2, UC_2	UO_2+PuO_2
2	Fuel enrichment (%)	1.8–3.2	2–2.7	Nat.	Nat.	2–4	93	15–75
3	Cladding	Zircaloy	Zircaloy	Zircaloy	Magnox	SS	Graphite	SS
4	Moderator	H_2O	H_2O	D_2O	Graphite	Graphite	Graphite	—
5	Coolant	H_2O	H_2O	D_2O	CO_2	CO_2	He	Na
6	Core heat transfer area (m^2)	5,000	6,000	4,000	5,000	8,000	Not app.	500
7	Core power density (kW/dm^3)	100	50	10	1	2.5	10	500
8	Fuel power density (kW/kg)	30	20	20	3	10	600	1,000
9	Heat flux (MW/m^2)	0.7	0.5	0.5	0.2	0.1	Not app.	2
10	Cooling circuit pressure (MPa)	16	7	10	2.5	4	5	0.8
11	Coolant temperature rise in core (°C)	35	15	50	200	300	400	200
12	Coolant velocity (m/s)	4.5	2	4.5	25	25	30	7
13	Coolant mass flow rate (kg/s)	$2 \cdot 10^4$	$1.3 \cdot 10^4$	$1 \cdot 10^4$	$1 \cdot 10^4$	$5 \cdot 10^3$	$4 \cdot 10^2$	$6 \cdot 10^3$
14	Heat transfer coefficient ($W/m^2 \cdot K$)	$4 \cdot 10^4$	$8 \cdot 10^4$	$4 \cdot 10^4$	$6 \cdot 10^3$	$3 \cdot 10^3$	Not app.	$1 \cdot 10^5$

PWR — pressurized water reactor; BWR — boiling water reactor; PHWR — pressurized heavy water reactor; GCR — gas cooled reactor; AGR — advanced gas-cooled reactor; HTGR — high temperature gas-cooled reactor; FBR — fast breeder reactor.

where $\mathbf{v}$ = neutron speed = $v\mathbf{\Omega}$
n = neutron angular density
$n(\mathbf{r}, \mathbf{\Omega}, E, t)\, dV\, d\mathbf{\Omega}\, dE$ = the expected number of neutrons in volume dV about $\mathbf{r}$ with kinetic energy E in dE moving in direction $\mathbf{\Omega}$ in solid angle $d\mathbf{\Omega}$

The neutron transport equation has the form:

$$\frac{1}{v}\frac{\partial}{\partial t}\Phi(\mathbf{r}, \mathbf{\Omega}, E, t) = -\mathbf{\Omega}\nabla\Phi(\mathbf{r}, \mathbf{\Omega}, E, t) - \sum(\mathbf{r}, E, t)\Phi(\mathbf{r}, \mathbf{\Omega}, E, t)$$
$$+ \int \Phi(\mathbf{r}, \mathbf{\Omega}', E', t) \sum_s (\mathbf{r}, \mathbf{\Omega}' \to \mathbf{\Omega}, E' \to E)\, d\mathbf{\Omega}'\, dE'$$
$$+ S(\mathbf{r}, \mathbf{\Omega}, E, t) \tag{1}$$

where ∇ = nabla operator
$\sum$ = total macroscopic cross section
$\sum_s$ = macroscopic differential scattering cross section
S = neutron source

The initial condition is

$\Phi(\mathbf{r}, \mathbf{\Omega}, E, t) = \psi(\mathbf{r}, \mathbf{\Omega}, E)$ = given function

The boundary conditions are more complicated and depend on the specific problem of interest. For example, the boundary condition for a free surface follows. A free surface is defined such that neutrons can only escape a body through the surface; they cannot reenter it. Hence we would require that the angular neutron flux vanish on the surface for all inward directions:

$\Phi(\mathbf{r}, \mathbf{\Omega}, E, t) = 0$ on a free surface for all $\mathbf{\Omega}$ such that $\mathbf{\Omega} \cdot \mathbf{N} < 0$, where $\mathbf{N}$ = outward-bound unit vector normal to the surface.

The neutron transport equation is an integral-differential equation. The theory of it may be found for example in References 5 and 6.

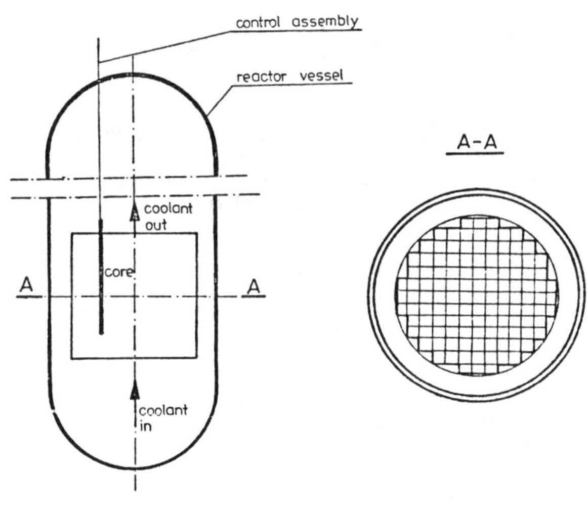

Figure 2. Vessel-type reactor.

Neutron Diffusion Equation

The solution of the neutron transport equation poses severe mathematical difficulties. Hence, there is a need for a development of a far simpler equation, which would be more amenable to mathematical treatment.

Such an expression, called the neutron diffusion equation, can be derived from transport theory.

It has the form:

$$\frac{1}{v}\frac{\partial}{\partial t}\Phi(\mathbf{r}, E, t) = \nabla D(\mathbf{r}, E, t)\nabla\Phi(\mathbf{r}, E, t) - \sum(\mathbf{r}, E, t)\Phi(\mathbf{r}, E, t)$$

$$+ \int \Phi(\mathbf{r}, E', t) \sum_{s}(\mathbf{r}, E' \to E)\, dE' + S(\mathbf{r}, E, t) \qquad (2)$$

where $\Phi(\mathbf{r}, E, t) = \int \Phi(\mathbf{r}, \mathbf{\Omega}, E, t)\, d\mathbf{\Omega}$

D = diffusion coefficient.

The initial condition has the form:

$$\Phi(\mathbf{r}, E, 0) = \psi(\mathbf{r}, E) = \text{given function}$$

The boundary conditions depend on the problem of interest. For a free surface it has the form:

$$\Phi(\mathbf{r}, E, t) + 2D(\mathbf{r}, E, t)N\nabla\Phi(\mathbf{r}, E, t) = 0$$

The Multigroup Form of Neutron Balance Equations

Before application of any approximate method of solution, the neutron balance equations must be recast in the so-called multigroup form. This consists of dividing the entire neutron energy range into a number of intervals which are called energy groups. Energy E belongs to the g-th energy group, if $E \in [E_{g-1}, E_g]$, $g = 1, 2, \ldots, G$, where G = total number of energy groups. One gets the set of G-coupled neutron balance equations. For example, from Equation 2 one obtains a set of equations for steady state in the form:

$$-\nabla D_g(\mathbf{r})\nabla\Phi_g(\mathbf{r}) + \sum_{g}(\mathbf{r})\Phi_g(\mathbf{r})$$

$$= \sum_{i \neq g}^{G} \Phi_g(\mathbf{r}) \sum_{ig}^{sl}(\mathbf{r}) + S_g(\mathbf{r}) \qquad g = 1, 2, \ldots, G \qquad (3)$$

$\sum_{ig}^{sl}$ is the macroscopic slowing down cross section from i-th to the g-th group. All other symbols are the same as before, and subscript g denotes that the quantity of interest is suitably averaged over the g-th energy group.

Computer-Based Approximate Methods in Neutron Transport Theory

The complexity of the neutron balance equations usually forces one to use computer-based approximate methods to obtain a solution. These are applied to the multigroup form of transport or diffusion equations.

There are many methods of approximate solution of the neutron transport equation, among the most popular are the method of discrete ordinates and the Monte Carlo method. Finite element methods have also gained popularity over the past few years.

The most popular methods of approximate solution of the diffusion equation are the finite difference method, the finite element method, and different coarse mesh methods. More information on approximate methods and computer codes can be found in the literature such as References 7 and 8.

Nuclear Data in Reactor Calculations

Nuclear reactor calculations require a knowledge of cross sections for different nuclear reactions induced by neutrons. Because of the enormous amount of nuclear data, storage on magnetic tapes is necessary. There exist many libraries of evaluated neutron cross sections. A good example is the ENDF library (Evaluated Nuclear Data File) developed in the United States and used in many countries around the world. Multigroup equations are used in reactor calculations, therefore it is necessary to generate group cross sections using libraries of evaluated neutron cross sections. Special computer codes are used for this purpose.

The group cross sections generated by these codes form multigroup cross-sections libraries in which a number of energy groups may extend up to several hundred. the basic information on these problems may be found for example in Reference 9.

Critical State of a Nuclear Reactor

The critical state is the normal operating state of a nuclear reactor. For the analysis of this state, in principle any of the neutron balance equations just described may be used with the source term to express the contribution from neutrons born in fissions. For example, the source term used in Equation 3 has the form:

$$S_g(\mathbf{r}) = X_g \sum_{g=1}^{G} \nu_g(\mathbf{r}) \sum_{g}^{f} (\mathbf{r}) \Phi_g(\mathbf{r})$$

where X_g = fraction of prompt neutrons belonging to g-th group
ν_g = average number of neutrons emitted per fission for g-th group
$\sum_{g}^{f}$ = macroscopic fission cross section for g-th group

The neutron balance equation for a critical state is a homogeneous equation, hence its solution gives only relative values of the neutron flux. The absolute value of the neutron flux can be determined when the power of the reactor is known.

A balance of neutrons in a reactor may be characterized by a neutron effective multiplication factor k. [10]. This is defined as the ratio of the number of neutrons produced in a reactor per unit time to the number of neutrons lost per unit time due to absorption and leakage. Hence for a critical state $k = 1$. When the number of neutrons produced exceeds the number of neutrons lost per unit time, then $k > 1$, and the reactor is supercritical. In the opposite case $k < 1$ and a reactor is subcritical. In normal operating conditions k does not differ too much from unity.

In many cases it is convenient to use related quantity called the reactivity ρ, defined below:

$$\rho = \frac{k-1}{k}$$

Obviously $\rho = 0$ for a critical state, $\rho > 0$ for a supercritical state, and $\rho < 0$ for a subcritical state.

HEAT GENERATION IN A REACTOR CORE

The energy of neutrons deposited in matter due to nuclear reactions caused by them appears as a heat. The main source of heat in nuclear reactors is a fission reaction taking place in the fuel elements, i.e., strictly speaking, in the fuel pellets. Heat sources in cladding, coolant, and other materials in a nuclear reactor result from other neutron-induced reactions such as scattering and absorption. There is also heating by gamma (γ) photons emitted by fission products and resulting from radiative capture of neutrons. In a nuclear reactor core a contribution from other than fission sources does not exceed a few percent and usually may be neglected. Obviously this contribution cannot be neglected in nonmultiplying media, e.g., in a biological shield of a nuclear reactor.

Heat Generation by Neutrons

Heating by neutrons at any spatial point can be expressed as:

$$q_V(\mathbf{r}) = \sum_{j=1}^{I} \int \Phi(\mathbf{r}, E) k_j(E) N_j(\mathbf{r})\, dE$$

where $q_V(\mathbf{r})\, dV$ = thermal power released in volume dV about $\mathbf{r}$
$N_j(\mathbf{r})$ = number density of isotope j at point $\mathbf{r}$
$k_j(E)$ = microscopic kerma factor for isotope j
J = number of isotopes

The term "kerma" is an acronym for kinetic energy released in material.

$$k_j(E) = \sum_i \sigma_{ij}(E) E_{ij}(E)$$

where $\sigma_{ij}(E)$ = microscopic neutron cross section for isotope j for reaction i at neutron energy E
$E_{ij}(E)$ = energy deposited per reaction i for isotope j at neutron energy E

In nuclear fuel the main heat source stems from the fission reaction, contributions from other reactions are negligible. Then

$$k_j(E) = \sigma_{fj}(E) E_{fj}$$

where σ_{fj} = microscopic fission cross section for isotope j
E_{fj} = fission energy released in isotope j

In fact $E_{fj} \approx E_f$ = const.; E_f is usually taken as equal about 180 MeV.

Hence in fuel

$$q_V(\mathbf{r}) = E_f \sum_j N_j(\mathbf{r}) \int \Phi(\mathbf{r}, E) \sigma_{fj}(E)\, dE \tag{4}$$

In thermal reactors, in which the majority of fissions is caused by thermal neutrons the preceding expression greatly simplifies to:

$$q_V(\mathbf{r}) = E_f \Phi_{th}(\mathbf{r}) \sum_j N_j(\mathbf{r}) \sigma_{fj}^{th} g_{fj} \tag{5}$$

where Φ_{th} = thermal neutron flux
σ_{fj}^{th} = microscopic fission cross section for thermal neutrons for isotope j
g_{fj} = Westcott coefficient for fission of isotope j

Hence, in thermal reactors heat generated in the fuel by fissions is proportional to the thermal neutron flux.

Heat Generation by Gamma Photons

Heating by gamma (γ) photons can be described by an expression similar to Equation 4. The microscopic kerma factor for photons takes into account three main effects: photoelectric, Compton, and pair production.

Heating by γ photons is important when decay heat is calculated in a reactor core after reactor shutdown or in nonmultiplying media (e.g., reactor shield).

Typical Heat Source Distributions in a Reactor Core

Heat sources in a reactor core are functionals of neutron flux as is seen from Equation 4. In thermal reactors heat sources are proportional to the thermal neutron flux.

Typical heat source distributions in a reactor core are shown in Figure 3.

The axial distribution is symmetrical with respect to half the core height where it reaches maximum value.

It can be approximated by a chopped cosine distribution:

$$q_v(z) = q_{v_{max}} \cos \frac{\pi z}{H_e} \tag{6}$$

where H_e = extrapolated height of a reactor core, $H_e = H + 2\delta$. The extrapolation length δ can be determined from reactor physics calculations.

The radial distribution cannot be described so simply. It would be well to have this as uniform as possible to attain uniform heat removal and fuel burnup throughout a core. It can be approximately achieved by a suitable distribution of fuel assemblies with different enrichments.

In physical calculations one considers a single fuel element with a coolant. This is referred to as an elementary cell. The local heat source distribution in an elementary cell is shown in Figure 4.

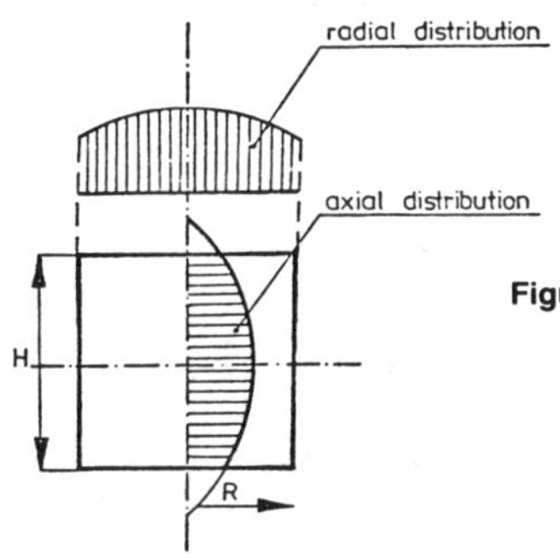

Figure 3. Typical heat source distribution in a cylindrical reactor core.

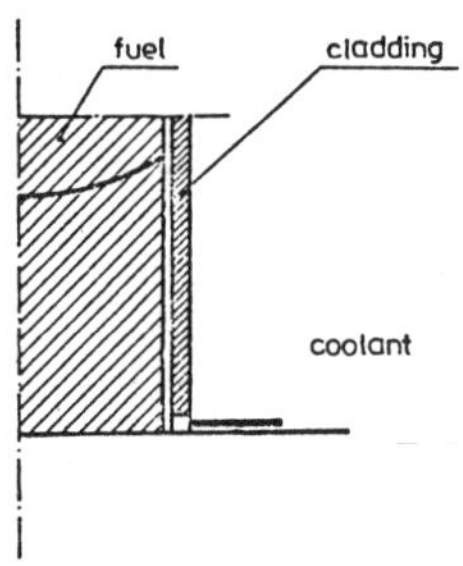

Figure 4. Heat source distribution in an elementary cell.

The heat source in fuel attains its minimal value at the center. The difference between the minimal and maximal value usually does not exceed 3%–4%. Hence it is customary to assume a constant radial heat source in the fuel. Heat sources in the gas gap, cladding, and coolant are usually neglected.

Decay Heat in a Nuclear Reactor Core

After shutdown of a nuclear reactor heat continues to be generated in the core. This heat, called decay or shutdown heat, is a result of radioactive decay of fission products.

The amount of decay heat depends on the time after shutdown and also on the operating time and power before shutdown. Results of calculations or measurements are presented in the form:

$$P/P_0 = f(t_0, t) \tag{7a}$$

where P = reactor decay power
P_0 = reactor power before shutdown
t_0 = reactor operating time at power P_0
t = time after reactor shutdown

When the operating time t_0 is sufficiently long (e.g. one year), then

$$P/P_0 = f(t) \tag{7b}$$

There are many expressions of the form of Equations 7a and b. For example

$$P/P_0 = 0.095\, t^{-0.26} \quad \text{for} \quad t > 200 \text{ s}$$

where t is in seconds.

Just after reactor shutdown P/P_0 is equal approximately 3% and later slowly decreases. Hence it is necessary to provide for suitable cooling of a reactor core after its shutdown.

HEAT REMOVAL FROM A REACTOR CORE

The maximum allowable thermal power of a nuclear reactor is established by the ability to effective remove heat from its core. The heat removal system should assure that the values of fuel temperatures, cladding temperatures, and heat fluxes in steady and transient states will be less than the maximal permissible levels.

Heat generated in fuel is transferred through fuel element components (fuel pellet, gas gap, cladding) to an external surface of cladding by conduction. From an external surface of cladding heat is transferred to the coolant by one- or two-phase forced convection.

In normal operating conditions the flow of coolant is turbulent. This results in high heat transfer coefficients and hence, enables effective heat transfer from fuel to coolant. Laminar flow takes place only in special cases (for example, removal of decay heat after reactor shut-down).

The choice of a suitable coolant is an important problem and depends on reactor type and the nuclear power plant thermal system. Values of thermal and hydraulic parameters encountered in nuclear reactor cores (see Table 1) clearly illustrate the complexity of the heat transfer analysis.

It is well known that analytical solutions of heat transfer problems for turbulent coolant flow through channels of complex geometries are impossible. Therefore, analysis of the heat removal from the core must rely strongly on experiments and requires the use of computer-based approximation methods.

Conservation Laws for Convective Heat Transfer in a Nuclear Reactor Core

The convective heat transfer from fuel elements to coolant is described by a set of equations based on the general laws of mechanics and thermodynamics. These equations express the principles of conservation of mass, momentum, and energy and are derived for an arbitrary control volume Ω with boundary Γ filled with flowing fluid.

These have the form [11–17]:

$$\iiint_{\Omega} (\partial \rho_c/\partial t) \cdot d\Omega + \iint_{\Gamma} \rho_c \cdot \mathbf{V} \cdot \mathbf{n} \cdot d\Gamma = 0$$

$$\frac{D}{Dt} \iiint_{\Omega} \rho_c \cdot \mathbf{V} \cdot d\Omega = \iint_{\Gamma} \mathbf{n} \cdot T_s \cdot d\Gamma + \iiint_{\Omega} \rho_c \cdot \mathbf{f} \cdot d\Omega$$

$$\iiint_{\Omega} [\partial(\rho_c e)/\partial t] \cdot d\Omega + \iint_{\Gamma} \rho_c \cdot e \cdot \mathbf{V} \cdot \mathbf{n} \cdot d\Gamma = \iint_{\Gamma} \lambda_c \cdot \text{grad}\, T_c \cdot \mathbf{n} \cdot d\Gamma$$

$$+ \iiint_{\Omega} q_{vc} \cdot d\Omega + \iiint_{\Omega} \rho_c \cdot \mathbf{f} \cdot \mathbf{V} \cdot d\Omega + \iint_{\Gamma} \mathbf{V} \cdot (T_s \cdot \mathbf{n}) \cdot d\Gamma \qquad (8)$$

The preceding integral equations may be transformed into partial differential equations by use of vector calculus (Gauss, Green, Stokes) and transport theorems. The following assumptions are applied: Newtonian fluid, Stokes' hypothesis, zero dilatational viscosity coefficient, and differentialibility of parameters. These allow the counterpart of Equation 8 in a differential form:

$$\partial \rho_c/\partial t + \text{div}\,(\rho_c \cdot \mathbf{V}) = 0$$

$$\rho_c \cdot D\mathbf{V}/Dt = \rho_c \cdot \mathbf{f} + \text{Div}\,[2\,\mu \cdot T_d - (p + 2/3\,\mu \cdot \text{div}\,\mathbf{V})\varepsilon]$$

$$\rho_c \cdot Du/Dt = \text{div}\,(\lambda_c\, \text{grad}\, T_c) - p \cdot \text{div}\,\mathbf{V} + q_{vc} + \mu \cdot \Phi \qquad (9)$$

or in terms of enthalpy

$$\rho_c \cdot Dh/Dt = \text{div}\,(\lambda_c\, \text{grad}\, T_c) + Dp/Dt + q_{vc} + \mu \cdot \Phi$$

The preceding equations are supplemented by additional equations and data such as:

- Equation of state $f(\rho_c, T, p) = 0$
- Properties of fluid e.g. $\mu = \mu(T, p)$, $\lambda_c = \lambda_c(T, p)$
- Mass force field $\mathbf{f} = \mathbf{f}(\mathbf{r}, t)$
- Heat source q_{vc}

Fourier's law and Newton–Stokes hypothesis have been taken into account in Equations 8 and 9.

Obviously it is necessary to formulate suitable initial and boundary conditions. The preceding equations are extremely complex, therefore simplifying assumptions must be introduced.

Usually in reactor channels one considers fully-developed flows where by the heat transfer intensity is the lowest. Assuming: $Dp/Dt = \Phi = \mathbf{f} = q_{vc} = 0$, $\lambda_c = \text{const}$, $\rho_c = \text{const}$, $\nu = \text{const}$ and neglecting axial heat conduction (usually $Pe \gg 10$) we obtain for turbulent single-phase, two-dimensional coolant flow the following momentum and energy conservation equations:

$$\partial/\partial x[(\nu + \varepsilon_m)\partial w/\partial x] + \partial/\partial_y[(\nu + \varepsilon_m)\partial w/\partial y] = -1/\rho \cdot \partial p/\partial z$$

$$\partial/\partial x[(\kappa + \varepsilon_h)\partial T_c/\partial x] + \partial/\partial y[(\kappa + \varepsilon_h)\partial T_c/\partial y] = w \cdot \partial T_c/\partial z \qquad (10)$$

For laminar flow $\varepsilon_m = \varepsilon_h = 0$.

A simple solution to this set of equations is possible only in the case of laminar coolant flow through channels of simple geometry with classic boundary conditions (T_w = const, q_w = const). For the case of turbulent flow which is typical for reactor channels, only approximate solutions are possible.

Useful and effective one-dimensional mass, momentum, and energy equations have been derived by J. E. Meyer [18], by suitable averaging on planes perpendicular to the coolant flow.

Introducing:

- Average density

$$\bar{\rho}_c = \frac{1}{A}\iint_A \rho_c \cdot dA$$

- Mass velocity

$$G = \frac{1}{A}\iint_A \rho_c \cdot w \cdot dA$$

- Effective specific volume

$$v_{ef} = 1/G^2\left[\frac{1}{A}\iint_A \rho_c \cdot w^2 \cdot dA\right]$$

- Average enthalpy over the channel cross section

$$\bar{h} = \frac{1}{\bar{\rho}_c A}\iint_A \rho_c \cdot h \cdot dA$$

- "Mixing cup" enthalpy

$$\tilde{h} = \frac{1}{G \cdot A}\iint_A \rho_c \cdot w \cdot h \cdot dA$$

Meyer obtained the following one-dimensional slice equations:

- Continuity equation

$$\partial(\bar{\rho}_c A)/\partial t + \partial(GA)/\partial z = 0 \tag{11}$$

- Momentum equation

$$\partial(GA)/\partial t + \partial(v_{ef}G^2A)/\partial z = -A \cdot \partial p/\partial z - \zeta \cdot G\,|G|\,A/(2\,\rho_c D_e) - g\bar{\rho}_c A \tag{12}$$

- Energy equation

$$\partial(\bar{\rho}_c \bar{h} A)/\partial t + \partial(G\tilde{h}A)/\partial z = \int_{\Pi_q} q_w \cdot d\Pi_q + \iint_A q_{vc} \cdot dA \tag{13}$$

Meyer's approach has found numerous and successful applications in many so called "channel codes" used in an analysis of a reactor core and primary cooling circuit.

Temperature fields in fuel elements are described by the heat conduction equation:

$$\operatorname{div}(\lambda_f \operatorname{grad} T_f) + q_v(\mathbf{r}, t) = \rho_f c_{pf} \cdot \partial T_f/\partial t \tag{14}$$

supplemented by suitable initial and boundary conditions. This is a quasilinear equation, since in general λ_f depends on temperature. The methods of solution of Equation 14 are presented in numerous publications [19–24].

Equations 11 through 13 for coolant, and Equation 14 for fuel elements are mutually connected since on cladding surface $q_w \neq$ const and $T_w \neq$ const. Simultaneous solution of the all differential equations is very complicated. Hence, in practice, these equations are separated.

Usually the rate of heat transferred by convection between a cladding surface and coolant is evaluated by means of the Newton cooling law:

$$q_w = \alpha(T_w - T_b) \tag{15}$$

where T_w, q_w, α are averaged quantities on the external surface of the cladding and T_b may be determined from Equation 13.

The simplicity of Equation 15 is a consequence of introducing the convective heat transfer coefficient which, unfortunately, is difficult to determine since the convective heat transfer is strictly related to the system hydrodynamics.

The nonuniform circumferential temperature distribution may be taken into account by means of the so-called conduction parameter ε [25, 26]. This parameter links geometrical and thermal parameters of a fuel element with the intensity of heat transfer on cladding surface. It is a measure of the ratio of heat flow along a fuel element and in coolant for the same temperature difference (if T_w = const then $\varepsilon \to \infty$ and if q_w = const then $\varepsilon \to 0$).

For the determination of temperatures, mass flow rates and pressures in the core, information on the heat transfer coefficient α and the friction factor ζ are needed. In practice, these can be determined using dimensional analysis and the theory of similarity combined with experiments.

Geometry of Cooling Channels

Geometry of cooling channels is strictly connected with geometry of fuel elements and different additional elements and their layout in a nuclear reactor core. It is convenient to

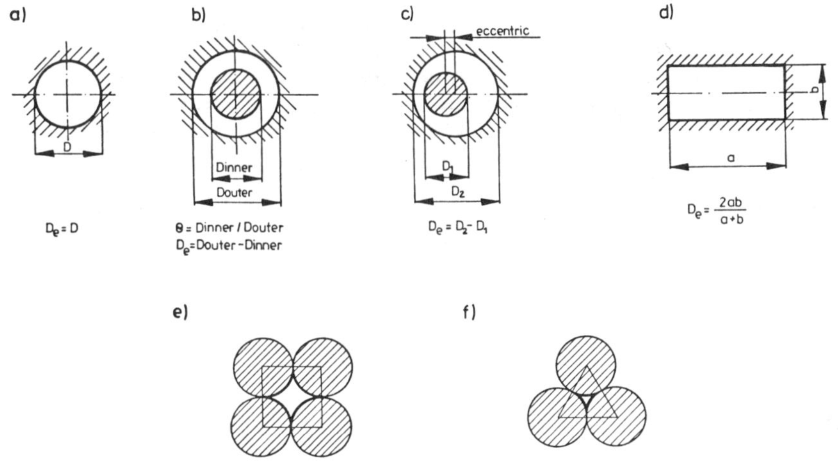

Figure 5. Various closed-channel geometries: (A) circular; (B) annular; (C) off-center annular; (D) thin rectangular; (E) square array with s/d = 1; (F) triangular array with s/d = 1.

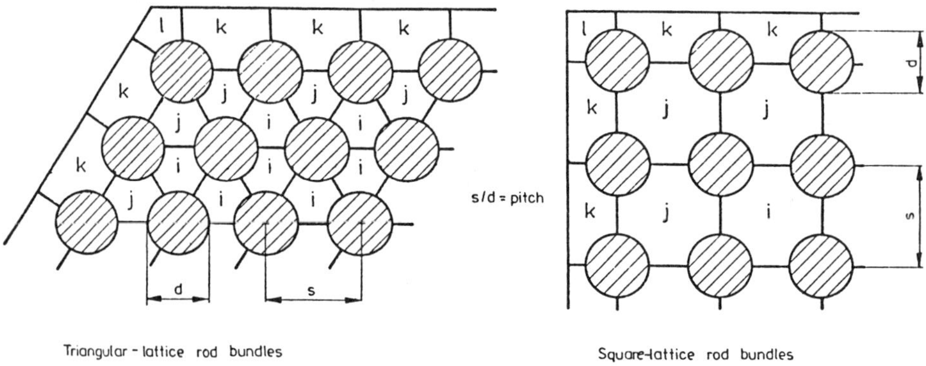

Figure 6. Noncircular open channels (subchannels) in two different fuel assemblies.

divide cooling channels into closed and open channels. Closed channels are bounded by a real wall. The open channels are partly bounded by a hypothetical wall.

Possible shapes of the closed and open channels are shown in Figure 5 and Figure 6. Open channels are also called subchannels. Hence, fuel assemblies are sets ("i"–"l" Figure 6) of parallel subchannels with mass, momentum, and energy transfer between them.

The complexity of subchannel cross-sectional shapes results from the presence of distancing elements, finned cladding (gas cooling), and special elements for increasing flow turbulence. These elements increase both heat removal intensity from fuel elements and, on the other hand, local friction factors.

Complex geometry of cooling channels and nonuniform axial and radial heat source distribution cause nonuniform heat loading of subchannels (particularly, the peripheral subchannels—"k," "l" (Figure 6)), nonuniform distribution of tangential stress, heat transfer coefficient, and temperature. For this reason, the behavior of peripheral subchannels has the significant influence (fast reactors) on permissible power of a fuel assembly.

In principle, it would be desirable to perform heat and mass transfer calculations for all subchannels in a fuel assembly. However, it is not always practical, possible, or suitable.

In practice, one can define (Figure 7) thermally and hydraulically isolated channels connected with a particular fuel element by neglecting interchannel influences and considering a cluster with an infinite number of fuel elements. This channel may be approximated by an equivalent annular channel (so-called the equivalent annulus model) with the hypothetical external wall, particularly when $s/d \geqq 1.3$ (thermal reactors). Such an approximation is not advisable in fast reactors, where $s/d \approx 1.1$–1.2. The isolated channel approach is also used in an analysis of a reactor primary circuit, particularly in transient states.

Nuclear Reactor Coolants

Nuclear reactor coolants should satisfy many, in general contradictory, thermal, nuclear, and economic requirements.

Thermal and economic requirements are the same as for coolants used in other applications. From the point of view of nuclear requirements, coolants should have small absorption and activation cross sections and high resistance to radiation damage.

If the coolant serves also as a moderator it should have a small mass number and high neutron scattering cross section. Therefore in selecting a reactor coolant certain criteria must be met. This problem is considered for example in References 27–29.

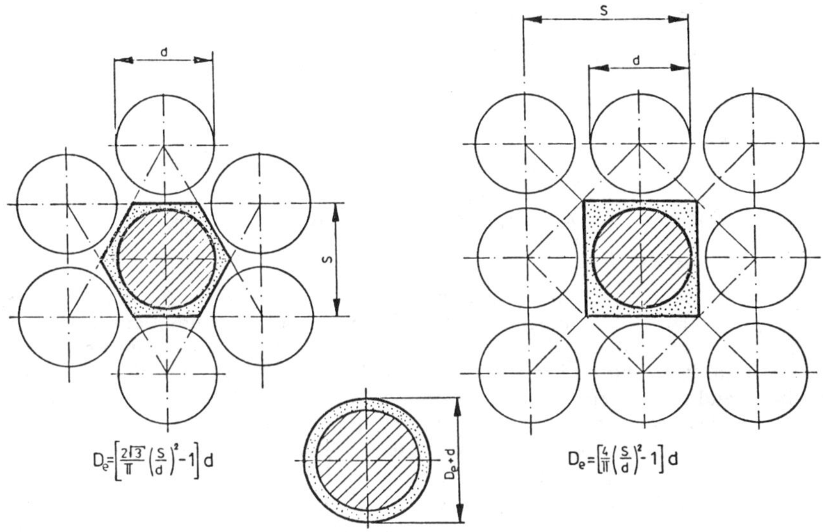

Figure 7. Isolated channel in triangular and square lattices and in equivalent annulus model.

Actually used coolants in nuclear power reactors are : water, heavy water, carbon dioxide, helium, and liquid sodium (in fast reactors). Thermal and physical properties of coolants depend on temperature and pressure and can be found in References 28–40.

Heat Transfer Through Fuel Element to Coolant

In general, cooling of fuel elements is described by the system of equations presented in the previous subsection. To determine the solution, a computer is needed. One may perform approximate analysis of heat transfer through an individual fuel element to coolant. The analysis enables an estimate of the maximum fuel and cladding temperatures both in "hot" and "average" channels, and provides functional relationships between parameters of fuel element components and coolant. In practice, these calculations are performed for the most thermally loaded fuel element in the most thermally loaded cluster ("hot" cluster). One of the simplest examples is heat flow through a cylindrical fuel rod of radius r_f surrounded by cladding of thickness δ_{cl} and cooled by constant mass-flow rate W of a single-phase coolant having bulk temperature $T_b(z)$ [27–29].

Making the following simplifying assumptions:

1. Steady state, $\partial/\partial t = 0$.
2. Distribution of heat generation is such as in reactor core without reflector, i.e., $q_v(r, z) = q_{v_0} \cdot \cos(\pi \cdot z/H_e)$, where q_{v_0} is the maximal value in a given fuel rod.
3. There is no heat generation in cladding and coolant.
4. Constant thermal conductivities λ_f and λ_{cl}.
5. Axial heat conduction in fuel, cladding, and coolant are negligible.
6. Lack of gas gap between fuel rod and cladding, perfect contact on the fuel-rod cladding interface.
7. Slug flow of coolant ($\tilde{h} = \bar{h} = h$), constant coolant properties.
8. Constant heat transfer coefficient

one obtains the conservation equations in the form:

$$\lambda_f/r \cdot \partial/\partial r\,(r \cdot \partial T_f/\partial r) + q_{v_0} \cdot \cos(\pi z/H_e) = 0$$

$$1/r \cdot \partial/\partial r\,(r \cdot \partial T_{cl}/\partial r) = 0$$

$$dh/dz = \pi_{heated}\, q_w/W \quad \text{where} \quad q_w = r_f^2 \cdot q_v/[2(r_f + \delta_{cl})]$$

with boundary conditions:

$$\partial T_f/\partial r|_{r=0} = 0, \qquad -\lambda_f \cdot \partial T_f/\partial r|_{r=r_f} = -\lambda_{cl} \cdot \partial T_{cl}/\partial r|_{r=r_f}$$

$$T_f|_{r=r_f} = T_{cl}|_{r=r_f}, \qquad -\lambda_{cl} \cdot \partial T_{cl}/\partial r\,|_{r=r_f+\delta_{cl}} = \alpha(T_{cl}|_{r=r_f} - T_b)$$

$$h|_{z=-H/2} = h_{in}$$

An analytical solution is shown in Figures 8 and 9. This treatment is useful in some preliminary calculations. In practice, most of the preceding assumptions do not hold. Physical properties (first of all, conductivities) of fuel element components depend on temperature and vary during irradiation [41–42]. Thus the heat conduction equations are quasi-linear, and consequently approximate solutions are available only. In addition, the distribution of heat generation differs largely from that of reactors without reflectors. A gas gap containing a gaseous mixture occurs usually between the fuel rod and cladding. Description of heat flow through the gap is complex since there are simultaneous processes of conduction, convection, and thermal radiation. The composition of the gas mixture varies continuously in view of diffusion of gaseous fission products. The problem of heat transfer through the fuel element to coolant is usually solved by use of computer codes (e.g. [50, 51]). A special code used to estimate heat flow through the gas gap [52] has lately been submitted for critical examination [53, 54]. Detailed description of the problems just disscused are given in Reference 49.

In thermal and hydraulic calculations a large number of simplifications are usually made, although their experimental verification is often difficult. Hence all unknows and uncertainties are estimated by the use of so-called hot-spot and hot-channel factors [15, 28, 49, 55].

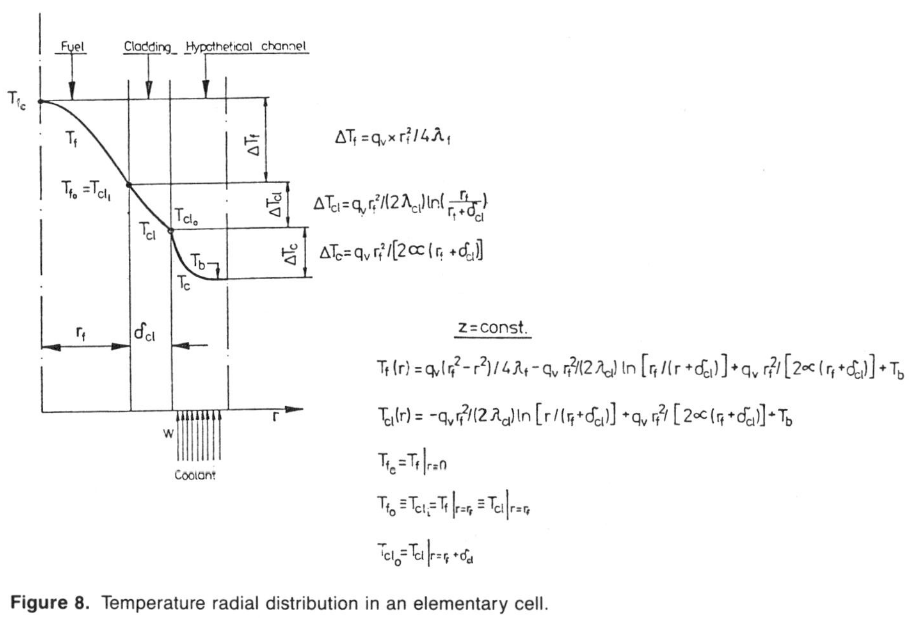

Figure 8. Temperature radial distribution in an elementary cell.

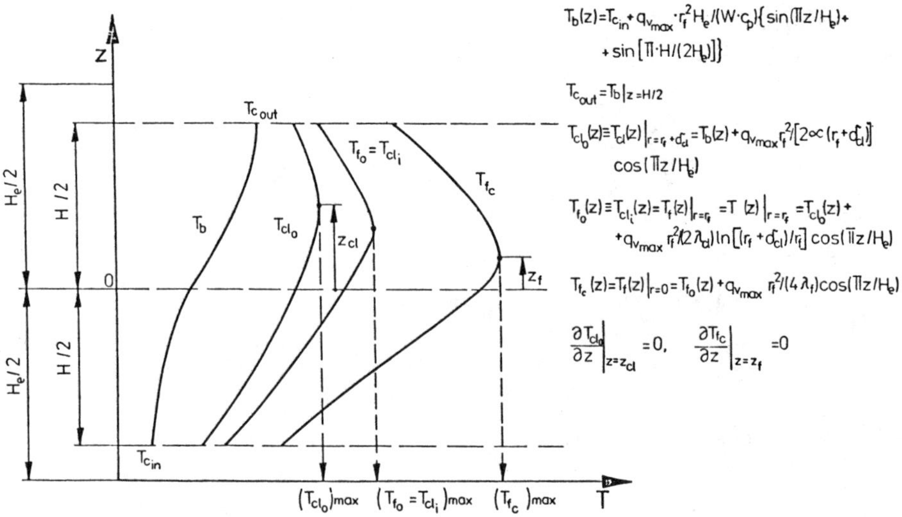

Figure 9. Temperature axial distribution in the fuel element and coolant.

Heat Transfer in Flow Through Rod Bundles

Analysis of thermal and hydraulic phenomena in both thermal and fast reactors requires consideration of thermal and hydraulic channel-interactions and possible cluster-interactions. Mass, momentum, and energy exchanges between subchannels and adjacent channels take place. The transverse flow is caused by flow distribution and rod spacers. Coolant flowing in intercluster spaces also takes place at part of the heat transfer between adjacent clusters. If cluster walls are perforated then intercluster hydrodynamic interactions occur.

Mathematical modeling of the preceding phenomena is based on mass, energy, and momentum (axial and transverse components) equations which are written for all subchannels in the rod bundle. In each fuel assembly, apart from a kind of lattice, four types of subchannels may be distinguished: central, side, adjacent to them, and corner ones (see Figure 6). Conservation equations are also formulated for the flow between clusters, or cluster interactions are considered via boundary conditions. Adding a heat transfer model to fuel elements gives the possibility of total analysis of rod bundles. The form of the equations depends on whether the coolant flow is single- or two-phase, and whether the state is steady or unsteady.

Heat transfer modeling in the flow through bundles was initiated by Rowe [56]. His work as well as that of other investigators [57–65] has greatly enhanced design capabilities, however further studies are necessary, particularly in the case of unsteady and two-phase flows. Thus, different versions of bundle codes, being the computer realization of the conservation equations, are still improved. The simplest system of mass, energy, and momentum equations for single-phase flow which served as the basis of COBRA [56] is given in the following (two-phase flow terms are neglected):

- Continuity equation

$$dm_m/dz = -\sum_{n=1}^{N} W_{mn}$$

- Energy equation

$$m_m \cdot dh_m/dz = (q_l)_m - \sum_{n=1}^{N} \begin{bmatrix} 0, & \text{if} & W_{mn} \geqq 0 \\ W_{mn} \cdot (h_n - h_m), & \text{if} & W_{mn} < 0 \end{bmatrix}$$

$$+ \sum_{n=1}^{N} W'_{mn} \cdot (h_n - h_m)$$

- Axial momentum equation

$$-dp/dz = (m_m/A_m)^2 \cdot \zeta_m/[2\rho_m(D_e)_m]$$

$$+1/A_m \cdot \sum_{n=1}^{N} f_T \cdot W'_{mn}(\bar{w}_m - \bar{w}_n)$$

$$+1/A_m \cdot \sum_{n=1}^{N} \begin{bmatrix} W_{mn} \cdot \bar{w}_m \cdot (f_D - 2), & W_{mn} \geqq 0 \\ W_{mn} \cdot (f_D \cdot \bar{w}_n - 2 \cdot \bar{w}_m), & W_{mn} < 0 \end{bmatrix}$$

- Transverse momentum equation

$$p_m - p_n = C_{mn} \cdot W_{mn} \cdot |W_{mn}|$$

where

$$W'_{mn} = 1/2 \cdot \beta \cdot S_{mn} \cdot (m_m/A_m + m_n/A_m)$$

and $m = 1, 2, \ldots, N_t$.

The coolant parameters are averaged over the channel cross section. Two-phase flow problems are described later in this chapter. Bundle code details are presented e.g. in References 66–73. Many of the mentioned codes require choosing from available correlations of the heat transfer coefficient, friction factor, and transverse flow parameters describing two-phase flow. In bundle codes the complex problems such as heat transfer in fuel rods, coolant flow distribution, form of the momentum equation, influence of rod spacers on transverse flow, accustic effects, heat transfer in two-phase flow, using steady-state empirical correlations in unsteady state should be solved. A large number of clusters in the core and subchannels in a cluster makes it impossible to use bundle programmes for computation of the entire reactor core. Therefore, as a first step, it is necessary to resign from a detailed heat transfer analysis by replacing each cluster by a single channel. Such calculations allow one to select the most heat overloated cluster ("hot" cluster). A detailed analysis of this cluster by use of a bundle programme constitutes the second step of calculations. If cluster walls are not perforated and the flow distribution to particular clusters is uniform, then the "hot" cluster is chosen according to the radial power distribution in the reactor core. In this case the first step in the calculation is superfluous. The second step allows one to determine the "hot" subchannel, which is next investigated as an individual fuel element (see the previous subsection).

SINGLE-PHASE FORCED CONVECTION IN REACTOR CHANNELS

Single-phase flow of different coolants takes place during normal full-power operation (steady state) of all nonhomogeneous reactors, except of BWR. In practice, the most important coolants are water (PWR) and liquid sodium (FBR).

Since the intensity of heat removal determines the temperature distribution in fuel elements and the latter establishes the maximum reactor power level, the coolant flow is highly turbulent and heat removal occurs under forced-convection conditions.

The analysis of single-phase forced convection leads to theoretical and experimental studies of nonisothermal, fully or nonfully developed turbulent or rare laminar flows through straight closed channels with complex geometry and, first of all, along rod bundles of triangular or square spacings (see Figures 5 through 7).

Additional difficulties are caused by: nonuniform heat flux distribution on channel walls, dependence of channel thermal load on its location in the reactor core as well as by dependence of coolant physical properties and various reactor materials on the temperature and, if need be, the pressure.

In this subsection, the analysis is confined to the steady state and includes determination of basic parameters (the heat transfer coefficient, the friction factor, the pressure drop) whose influence on the heat transport in nuclear reactor channels is decisive.

Convective Heat Transfer Coefficient

Heat removal from fuel elements by means of different cooling agents is a very complex phenomenon. Neglecting the usually small influence of radiation, heat flow between a cladding surface, and coolant takes place by conduction and mass transfer. In laminar flow, heat is transferred from fuel elements only by molecular conduction. A small effect of free convection may be neglected.

In turbulent flow, molecular conduction plays an unimportant role but there exist turbulent mixing currents and eddies which transport energy across streamlines. In addition, for a given fluid motion, the contribution of both of these heat transfer mechanisms depends on the type of coolant (ordinary fluids or liquid metals). The influence of cooling medium properties (in particular the Prandtl number) on the process of heat removal is essential in turbulent flow.

It is useful to distinguish between correlations used to calculate the heat transfer coefficient. Moreover, it should be taken into account the variety of reactor channel geometries.

Turbulent Flow

Turbulent flow is characteristic of a normal full-power reactor operation. It enables high values of the heat transfer coefficient, to be attained. In the case of fully-developed turbulent flow through straight channels with constant cross section, these coefficients to a small degree depend on boundary conditions (T_w = const, or q_w = const) given on the channel wall. Their variations in fluid flow direction are connected with dependence of coolant physical properties (mainly viscosity) on the temperature.

Difficulties relating to the theoretical approach of heat transfer in turbulent flow are commonly known.

Assuming that the channel has a simple geometry (circular tube), q_w = const, the flow is fully-developed, the fluid is incompressible with constant physical properties and neglecting the influence of conduction and turbulent heat transfer in the axial direction, one may obtain [74] an analytical solution of the energy Equation 10 in a from of the so-called Lyon integral [75]:

$$1/\mathrm{Nu} = 2\int_0^1 \{[\int_0^{\rho^*} (w/\bar{w}) \cdot \rho^* \cdot d\rho^*]^2/\{\rho^*[1+(\mathrm{Pr}/\mathrm{Pr}_t) \cdot (\varepsilon_m/\nu)]\}\} \, d\rho^* \tag{16}$$

where $\rho^* = r/R$.

In order to integrate Equation 16, one must know, the velocity profile and the relations of the turbulent eddy viscosity and the eddy diffusivity with physical and hydrodynamic fluid parameters.

Although hydrodynamic heat transfer theory has a number of remarkable achievements, in practice the heat transfer coefficients are calculated with the use of empirical correlations based on the similarity theory, the dimensional analysis, and generalization of many experimental data. The determination of quantities occurring in Equation 16 is difficult, especially in flows through complex shape channels commonly appearing in reactor cores.

Ordinary fluids. Determination of the forced-convection heat transfer coefficients is first of all based on empirical correlations obtained for flows through straight circular tubes with either q_w = const. or T_w = const. The same correlations may be used for noncircular closed-channels if the pipe diameter D_p is replaced by the equivalent (hydraulic) diameter D_e. This procedure sometimes leads to not so sufficiently correct results. The reason is underlined in the specific character of flow dynamics through some noncircular closed channels and in large temperature gradients occurring in reactor channel cross sections. Various correlations for different channel geometries are alluded to in the following.

Table 2 gives correlations which are usually recommended for the evaluation of the heat transfer coefficients in flow through circular closed channels with smooth walls.

The physical properties of coolants depend more or less on the temperature. Therefore there exists feedback between the velocity and temperature profiles. The viscosity variation is of most importance. This fact is taken into account either by an appropriate selection of a fluid reference temperature T_{ref} or by introducing the multiplier $(\mu_b/\mu_w)^n$ into the correlations. In selecting T_{ref} one should take into consideration the influence of the wall temperature T_w. This requirement is best satisfied by the temperature T_{ref} used in the Colburn correlation. However, application of the correlation requires a trial-and-error solution, because in practical problems both temperatures T_w and T_b are not known in advance. Two different types of correlations already discussed were suggested by Sieder-Tate and Mikheiev (see Table 2). Moreover, it is recommended in Reference 83 to set n = 0.11 for nonmetallic liquid coolants. This problem is more complicated as far as gases are concerned. The reguned data are available for example in References 84–89.

Apart from the prior empirical correlations, it is also known that the Petukhov formula 83, evaluated by use of the Lyon integral gives:

$$\alpha = \lambda_c/D_p \cdot (\zeta/8) \cdot Re \cdot Pr/[(1+900/Re)+12.7(\zeta/8)^{1/2} \cdot (Pr^{2/3}-1)]$$
$$10^4 < Re < 5 \cdot 10^6$$
$$0.5 < Pr\ 2{,}000 \tag{17}$$

The friction factor ζ is calculated from the Filonenko formula given in the next subsection.

One of the basic noncircular closed-channel geometries is an annular channel. Flows through a circular tube and between parallel plates constitute two extreme cases. In addition, there exists a strict relationship between the annular channels and the in-line cooled fuel rod bundles. This fact was mentioned when constructing isolated (equivalent) channels (see Figure 7).

Table 2
Empirical Forced Convection Correlations for Turbulent Flows through Circular Channels ($Nu = \alpha \cdot D_e/\lambda_c$, $D_e = D_p$)

Author(s)	Correlations	Range of Applications	References
Dittus–Boelter	$Nu \equiv Nu_{D-B} = C \cdot Re^{0.8} \cdot Pr^{0.4}$ $C = \begin{cases} 0.023 \text{ — Dittus–Boelter} \\ 0.019 \text{ — Rohsenow–Clark} \end{cases}$	$Re > 10^4$, $0.7 < Pr < 100$, $L/D_e > 60$ $T_{ref} = (T_b\|_{z=z1} - T_b\|_{z=z2})/2$	[76], [77]
Sieder–Tate	$Nu = Nu_{D-B} \cdot (\mu_b/\mu_w)^{0.14}$	$Re > 10^4$, $0.7 < Pr < 16{,}700$, $L/D_e > 60$ $\mu_b = \mu(T_b)$, $\mu_w = \mu = \mu(T_w)$	[78]
Colburn	$Nu = 0.023 \cdot Re^{0.8} \cdot Pr^{1/3}$	$Re > 10^4$, $0.7 < Pr < 16{,}700$, $L/D_e > 60$ $T_{ref} = (T_w + T_b)/2$ except for $c_p = c_p(T_b)$	[79], [80]
Mikheiev	$Nu \equiv Nu_M = 0.021 \cdot Re^{0.8} \cdot Pr^{0.43}$ $Nu = Nu_M \cdot (Pr_b/Pr_w)^{0.25}$	$10^4 < Re < 5 \cdot 10^6$, $0.6 < Pr < 2{,}500$ $T_{ref} = T_b$ except for $Pr_w = Pr(T_w)$	[81], [82]
Zhukauskas	$Nu = 0.0225 \cdot Re^{0.8} \cdot Pr^{0.6}$ (for gases only)	The same as in the Mikheiev correlation with $T_{ref} = T_b$	[74]

The most frequently cited empirical correlations are based on the following rule founded by Kays [90]. If both walls of an annular channel are heated independent of each other (i.e., with varied heat fluxes at the walls and along the axis, the Nusselt numbers (Nu_{inner} and Nu_{outer}) may be evaluated by the superposition principle. First, one should determine the Nusselt numbers (Nu^*_{inner} and Nu^*_{outer}) in two cases: heat transfer from the inner wall and outer wall. When heat is removed from one wall, the second one is considered as an adiabatic wall. In such a case, under the additional assumption that the q_{inner} and q_{outer} are constant, the Petukhov–Roizen correlation [84–86] is recommended in References 26 and 74:

$$Nu^*_{inner}/Nu = [1-\varphi(Pr)]\cdot(1/\Theta)^{n(Pr)}\cdot\delta, \qquad 0.03 \leqq \Theta \leqq 1$$

$$Nu^*_{outer}/Nu = 1-\varphi(Pr)\cdot\Theta^{0.6}, \qquad 0 \leqq \Theta \leqq 1 \tag{18}$$

where $\varphi(Pr) = 0.45/(2.4+Pr)$

$$n(Pr) = 0.16\cdot Pr^{-0.15}$$

$$\delta = \begin{cases} 1+7.5[(1/\Theta-5)/Re]^{0.6}, & \Theta < 0.2 \\ 1, & \Theta \geqq 0.2 \end{cases}$$

$$0.7 \leqq Pr \leqq 100, \qquad 10^4 \leqq Re \leqq 10^6, \qquad \Theta = D_{inner}/D_{outer}$$

and Nu is the Nusselt number for a circular tube ($D_e = D_{outer}-D_{inner}$) evaluated from either Equation 17 or its more general version:

$$Nu = (\zeta/8)\cdot Re\cdot Pr/[1.07+900/Re-0.63/(1+10\cdot Pr) + 12.7\cdot(\zeta/8)^{1/2}\cdot(Pr^{2/3}-1)]$$

The so-called adiabatic temperatures of nonheated walls T^{ad} are calculated using the formulas:

$$T^{ad}_{inner} = 22\cdot(0.27\cdot\Theta^2-1)\cdot Re^{-0.87}\cdot Pr^{-0.18}$$

$$T^{ad}_{outer} = \Theta\cdot T^{ad}_{inner}$$

Finally, the correlation takes the form:

$$Nu_{inner} = Nu^*_{inner}[1+(q_{outer}/q_{inner})\cdot T^{ad}_{inner}\cdot Nu^*_{inner}]$$

$$Nu_{outer} = Nu^*_{outer}[1+(q_{inner}/q_{outer})\cdot T^{ad}_{outer}\cdot Nu^*_{outer}] \tag{19}$$

In the case of gaseous coolants one may set $Pr \approx 0.7$ which simplifies the expressions adduced. Additional correlations are available in Reference 91. Heat transfer in a thin rectangular, square or triangular channels may be performed using the circular channel correlations given in Table 2. The reader should consult the literature for examples [92] since, in some cases, there exist more precise equations.

Heat transfer in in-line cooled rod bundles is far more complicated than in a circular channel or other closed channel cases. For $s/d < 1.15$ [74] circumferential variations of the heat transfer coefficient should be accounted for. One must then solve the differential equation conjugate system (10) and (14) with proper boundary conditions, which is extremely difficult. Difficulties are of a similar type when attempting to integrate Equation 16, although their scope is far larger because of the multidimensionality of the problem. For $s/d > 1.15$ [74] the circumferential wall temperature variations are insignificant and may be neglected. In such a case, the system of Equations 10 and 14 is separable. It suffices then to solve the conservation equations on the fluid, which, however, is still a nontrivial problem.

In practice, the heat transfer coefficient is calculated by using empirical correlations. Zhukauskas [74] recommends the following formula for nonmetallic coolants [93]:

$$Nu/Nu_p = 1 + 0.91 \cdot Re^{-0.1} \cdot Pr^{0.4} \cdot [1 - 2 \cdot \exp(-B)] \tag{20}$$

$$3 \cdot 10^3 < Re < 10^6,\ 0.66 < Pr < 5.0,\ 0.103 \leqq B \leqq 3.5,\ 1.02 < s/d < 2.5$$

where Nu_p is calculated based on the Dittus–Boelter correlation (see Table 2) and using an equivalent diameter $D_e = B \cdot d$ (see Figure 7). Hence, the coefficient B is given by

$$B = \begin{cases} 2\sqrt{3}/\pi \cdot (s/d)^2 - 1, & \text{for triangular array} \\ 4/\pi \cdot (s/d)^2 - 1, & \text{for square array} \end{cases}$$

In addition, there exists a number of correlations of the form

$$Nu = C(s/d) \cdot Re^m \cdot Pr^n \tag{21}$$

According to Weisman [94]:

$$C = \begin{cases} 0.026(s/d) - 0.006, & 1.1 \leqq s/d \leqq 1.5, \text{ for triangular array} \\ 0.042(s/d) - 0.024, & 1.1 \leqq s/d \leqq 1.3, \text{ for square array} \end{cases}$$

and $m = 0.8,\ n = 1/3,\ T_{ref} = T_b$

The correlation [26]:

$$C = 0.032 \cdot (s/d) - 0.0144$$

where $Re > 1.3 \cdot 10^4$ for $s/d = 1.1 \div 1.2$
$Re > 2 \cdot 10^4$ for $s/d = 1.2 \div 1.4$
$Re > 3 \cdot 10^4$ for $s/d = 1.4 \div 1.5$
and $m = 0.8,\ n = 1/3,\ T_{ref} = (T_b + T_w)/2$

verges upon the correlation above.
However, for a triangular array and liquid coolants, Subbotin–Ibragimov [91] recommend:

$$C = 0.0165 + 0.02 \cdot [1 - 0.91/(s/d)^2] \cdot (s/d)^{0.15}$$

where $5 \cdot 10^3 < Re < 5 \cdot 10^5,\quad 1.0 < Pr \leqq 20,\quad 1.1 \leqq s/d \leqq 1.8$

and $m = 0.8,\quad n = 0.4,\quad T_{ref} = T_b$

For gaseous coolants one may apply the correlation [95]:

$$Nu/Nu_p = 0.351 \cdot \lg(s/d - 1) + 0.0272, \quad 1.03 \leqq s/d \leqq 2.4 \tag{22}$$

where Nu_p is calculated with the help of the Dittus–Boelter formula (see Table 2).

The preceding adduced correlations (Equations 20 through 22) contain the characteristic array pitch s/d, the influence of which on heat transfer decreases together with its increase. Moreover, this parameter plays more of a role in liquid than gaseous coolants. The preceding correlations do not take into consideration the influence of disturbances caused by

- Rod spacers with a given geometry.
- Clad-finning.
- Special elements for increasing flow turbulence [96].

Liquid metals. Determination of the heat transfer coefficient in turbulent flow of liquid metals through closed or open channels is based on correlations different from those of ordinary fluids. It results from low values of Prandtl numbers for liquid metals ($Pr \sim 10^{-2}$). The turbulent eddy diffusivity is much less than molecular diffusivity. Therefore, the heat transfer correlations resemble those of laminar flow.

A classical correlation in the flow through pipes with q_w = const. is the Lyon–Martinelli Equations 75, 97.

$$Nu = 7.0 + 0.025 \cdot Pe^{0.8}, \quad Pe > 300 \tag{23}$$

obtained either by the use of the Lyon integral [75] with $\Psi \equiv \varepsilon_h/\varepsilon_m = 1$ or by an analogy between heat and momentum transfer [97]. Lyon suggested [98] the following extended form of Equation 23:

$$\mathrm{Nu} = 7.0 + 0.025 \cdot (\mathrm{Pe} \cdot \bar{\Psi})^{0.8} \tag{24}$$

and Dwyer [99, 100] recommended the average ratio of eddy diffusivities

$$\bar{\Psi} = 1 - 1.82/[\mathrm{Pr}\,(\varepsilon_m/\nu)_{max}^{1.4}] \tag{25}$$

where $(\varepsilon_m/\nu)_{max}$ is a function of the Reynolds number given graphically. Additional correlations may be found in numerous literature cited in review papers [40, 49, 101].

Dwyer [102–105] addressed himself to the problem of heat transfer in flows through annuli and between parallel plates. He generalized Equation 24 postulating:

$$\mathrm{Nu} = \alpha + \beta(\bar{\Psi} \cdot \mathrm{Pe})^{\gamma}, \qquad 10^2 \leqq \mathrm{Pe} \leqq 10^4 \tag{26}$$

where

- $\bar{\Psi}$ is given by Equation 25.
- $(\varepsilon_m/\nu)_{max}$ is given graphically for the mentioned channel geometries as a function of Reynolds number.
- α, β, γ are constants for flows through pipes (see Equation 24) and between parallel plates (if heat fluxes from both plates are equal then $\alpha = 0.49$, $\beta = 0.0596$, $\gamma = 0.688$ [106]), and functions of the ratio of inner-to-outer diameters Θ for flows through annuli (and function of

  $$\xi = \{(q_{w,\,inner}/q_{w,\,outer})/[(q_{w,\,inner}/q_{w,\,outer}) + 1/\Theta]\}$$

  for heat transfer from both walls).
- Pe and Nu are calculated using the channel equivalent diameter D_e

The values of the parameters α, β and γ for different geometries and heat transfer conditions are presented in above cited Dwyer's papers [102–106]. Some additional correlations for heat transfer in flows through annuli and between parallel plates are also presented in Reference 101. However, certain particular cases of heat transfer to liquid metals flowing in eccentric annuli are considered in References 107 and 108.

Equation 26 can be also applied for in-line flow between rod bundles with equilateral triangular or square spacing. The values of $(\varepsilon_m/\nu)_{max}$ are given, similarly as for the preceding geometries in Reference 100. However, the values of α, β, γ are functions of the pitch-to-diameter ratio (s/d). Assuming that the equivalent annulus model can be applied, Dwyer–Lyon–Maresca recommend [106, 109], $\alpha = 6.66 + 3.126 \cdot (s/d) + 1.184 \cdot (s/d)^2$, $\beta = 0.0155$, $\gamma = 0.86$ for triangular array with s/d = 1.3–3.0, $\alpha = 7.0 + 4.24 \cdot (s/d)^{1.52}$, $\beta = 0.0275 \cdot (s/d)^{0.27}$, $\gamma = 0.8$ for square array with s/d = 1.7–10.0.

In reactor cores cooled by liquid metals (first of all, Na), in view of the neutron balance, the pitch s/d is usually less than 1.3. In such a case, there occurs circumferential variations of the outside surface cladding temperature and heat transfer coefficient. Hence, it is necessary to calculate the local and average heat transfer coefficients. The equivalent annulus model cannot be applied. One should consider heat transfer in the real geometry (see Figure 7).

Dwyer–Berry and Subbotin–Ushakov [110–113] showed that for $s/d < 1.3$ the crucial influence on circumferential temperature and heat transfer coefficient distributions has the pitch s/d. If s/d tends to 1 then temperature variations rapidly increase and average heat transfer coefficients rapidly decrease. Moreover, some influence is also due to

- Fuel element geometrical parameters.
- Thermal conductivities of fuel element components and coolant.
- Low value of the Prandtl number.

In the Soviet literature, this influence is given by so-called thermal modeling of fuel elements [25–26, 114] by the help of the conduction parameter ε. Analysis of comprehensive experimental data obtained in the recent years in the Soviet Union and Czechoslovakia is presented in Reference 114, where the reader may also find a number of formulas concerning thermal and hydrodynamic calculations related to fast reactor cores cooled by liquid sodium. In addition, there is also given a comprehensive literature review. Special attention should be directed to reports of the FEI–Obninsk [115–124].

A few correlations recommended in [114] are given below. The following formula deals with triangular arrays with an infinite number of fuel elements:

$$\mathrm{Nu} = \mathrm{Nu}^{*}_{\mathrm{lam}} + \Psi(\varepsilon, \mathrm{s/d}) \cdot \mathrm{Pe}^{\gamma(\mathrm{s/d})}$$

$$1 \leqq \mathrm{Pe} \leqq 4{,}000, \quad 1.0 < \mathrm{s/d} < 2.0, \quad \varepsilon \geqq 0.01 \tag{27}$$

where $\Psi = 0.041/(\mathrm{s/d})^2 \cdot \{1 - \{[(\mathrm{s/d})^{30} - 1]/6 + (1.24\,\varepsilon + 1.15)^{1/2}\}^{-1}\}$

$\gamma = 0.56 + 0.19 \cdot (\mathrm{s/d}) - 0.1/(\mathrm{s/d})^{80}$

The Nusselt number $\mathrm{Nu}^{*}_{\mathrm{lam}}$ corresponding to laminar flow will be defined in the next section; similarly as $\mathrm{Nu}^{**}_{\mathrm{lam}}$ and $\mathrm{Nu}^{***}_{\mathrm{lam}}$. If (s/d) = 1 then

$$\mathrm{Nu} = \mathrm{Nu}^{**}_{\mathrm{lam}} + 0.041 \cdot \{1 - [(1.24\,\varepsilon + 1.15)^{1/2}]^{-1}\} \cdot \mathrm{Pe}^{0.65} \tag{28}$$

However, for $1.20 < \mathrm{s/d} < 2.0$ the formula holds:

$$\mathrm{Nu} = \mathrm{Nu}^{***}_{\mathrm{lam}} + 0.041/(\mathrm{s/d})^2 \cdot \mathrm{Pe}^{0.56 + 0.19(\mathrm{s/d})} \tag{29}$$

The accuracy of Equations 27 through 29 is about $\pm 15\%$.

In Reference 114, one may find formulas for the heat transfer coefficient in the subchannels of types "k" or "l" (see Figure 6). The average values of these coefficients on the peripheral element circumference are essentially less than those in the channels of types "i" and "j" (see Figure 6).

The maximal dimensionless circumferential temperature variation ΔT may be evaluated from the formula given in Reference 114:

$$\Delta T = \Delta T_{\mathrm{lam}}/[1 + \gamma(\varepsilon) \cdot \mathrm{Pe}^{\beta(\mathrm{s/d})}]$$

$$1.0 \leqq \mathrm{s/d} \leqq 1.15, \quad 1 \leqq \mathrm{Pe} \leqq 2{,}000, \quad \varepsilon \geqq 0.2 \tag{30}$$

where $\gamma(\varepsilon) = 0.008(1 + 0.03\,\varepsilon)$

$\beta(\mathrm{s/d}) = 0.65 + 51 \cdot \lg(\mathrm{s/d})/(\mathrm{s/d})^{20}$

and ΔT_{lam} corresponds to laminar flow (see next section). In the real cluster this variation is greater; in particularly, in case of peripheral rods, if $\mathrm{s/d} \rightarrow 1$ then the maximum cluster power can be determined. The preceding equations refer to smooth rod-bundle heat transfer in an equilateral triangular array. In practice, it is necessary to take into consideration the influence of different typs of rod spacers on the local and average heat transfer coefficients. A number of these correlations is demonstrated in Reference 114. Additional information with reference to heat transfer to liquid metals are discussed in References 125 through 132 and in Chapter 12.

Laminar Flow

In principle, the controlling heat transfer mechanism in laminar flow is conduction. Therefore it is not possible to obtain a high heat transfer coefficient α, although this coefficient is relatively high in most liquid metals. The result is that laminar flow of coolants cannot occur in reactor cores during normal full-power operation. This flow pattern can occur in various failures leading to reactor shut-downs. The removal of the residual heat should be intensive enough in order not to exceed admissible temperatures in all reactor core elements.

Ordinary fluids. If the coolant physical properties are constant then the fully developed laminar flow is described by the energy equation:

$$\lambda_c(\partial^2T_c/\partial x^2+\partial^2T_c/\partial y^2) = w \cdot \rho_c \cdot c_{p_c} \cdot \partial T_c/\partial z \tag{31}$$

Graetz [133] gave its solution for the case of flow through a straight circular tube with a uniform wall temperature T_w. The Nusselt number takes its minimal value $Nu_\infty = 3.656$ and the thermal entrance region length is differently estimated (e.g. $\delta_T = 0.055 \cdot D \cdot Re \cdot Pr$ [134]). In practical calculations Zhukauskas [74] recommends the empirical correlations given in References 83 and 135. In the case of uniform heat flux on the channel wall one has $Nu_\infty = 4.364$ and the appropriate convective correlations are listed in Reference 136. In addition, one may set $\delta_T = 0.07 \cdot D \cdot Pe$.

In laminar flow the equivalent-diameter concept may not be applied. Evaluating Nusselt numbers for noncircular closed channels is therefore based on correlations different than those of a circular tube. The determination of heat transfer between the wall and fluid is difficult in view of the complicated form of the velocity profile. Both for uniform wall temperature and uniform heat flux, Nu_∞ can be evaluated by using approximate methods only [134, 137]. For flows through annular channels with a uniform temperature of the heated wall, Nu_∞ may be estimated as follows

- $Nu_\infty = 3.96+0.9 \cdot \Theta^{-0.95}$ ($\Theta \geqq 0.2$) for heat flow between the inner wall and coolant.
- $Nu_\infty = 4.03 \cdot \exp(0.185 \cdot \Theta)$, where $\Theta \geqq 0.15$ for heat flow between the outer wall and coolant.

If both channel walls are cooled then the corresponding expressions may be found in [26, 91, 138]. In the cited literature, the values of Nu_∞ for uniform wall temperature and uniform heat flux are also adduced for the flow between parallel plates and through triangular and rectangular channels. The relation $Nu_\infty|_{T_w = const} < Nu_\infty|_{q_w = const}$ is independent on the closed-channel geometry.

Heat transfer in laminar flow through rod bundles differs to some extent from that in closed channels. This is a result of a lack of axial symmetry in different subchannels. If the pitch-to-diameter ratio $s/d < 1.3$, then the circumferential temperature and heat flux distributions are not uniform. These variations decrease when s/d increases. The values of Nu_∞ for in-line flow between rod bundle triangular and square spacings corresponding to uniform circumferential wall temperature and heat flux distributions are given in Tables 3 and 4. These results concerning both triangular- and square-lattice rod bundles have the following specific properties:

- $Nu_T > Nu_q$ with $\Delta Nu = Nu_T - Nu_q$ being essential when s/d tending to 1 [91]
- Both Nu_T and Nu_q increase when s/d does

To evaluate of Nu_∞ corresponding to a given value of the parameter ε the following relation is recommended [26]:

$$Nu(\varepsilon, s/d) = Nu_q(s/d)+f(\varepsilon, s/d) \cdot [Nu_T(s/d)-Nu_q(s/d)]$$

The function $f(\varepsilon, s/d)$ is given graphically and practically depends on ε only.

In the formulas-above there is not taken into account the dependence of coolant physical properties (first of all, viscosity) on temperature. In the contrary, the equations of energy (Equation 31) and motion (Equation 10) should be solved simultaneously. For flow between rod bundles with $s/d < 1.3$ thc preceding system of equations should be completed by the

Table 3
Values of Nu_∞ for Laminar Flow of Water through Fuel Rod Bundles with Equilateral Triangular Spacing

s/d	$Nu_\infty = Nu_T$ (T_w = const.)	$Nu_\infty = Nu_q$ (q_w = const.)	$Nu_T - Nu_q$	s/d	Nu_T	Nu_q	$Nu_T - Nu_q$
1.00	1.26	0.15	1.11	1.12	5.19	3.75	1.14
1.02	1.80	0.40	1.40	1.16	6.40	5.40	1.00
1.04	2.43	0.83	1.60	1.20	7.44	6.90	0.54
1.06	3.13	1.38	1.75	1.24	8.27	7.92	0.35
1.08	3.88	2.07	1.81	1.28	8.94	8.70	0.24
1.10	4.57	2.90	1.67	1.30	9.20	9.05	0.15

From Emelianov [26].

Table 4
Values of Nu_∞ for Laminar Flow of Water through Fuel Rod Bundles with Square Spacing

s/d	$Nu_\infty = Nu_T$ (T_w = const.)	$Nu_\infty = Nu_q$ (q_w = const.)	$Nu_T - Nu_q$	s/d	Nu_T	Nu_q	$Nu_T - Nu_q$
1.00	1.30	—	—	1.2	4.58	3.70	0.88
1.02	1.61	0.52	1.09	1.3	6.40	5.82	0.58
1.05	2.05	0.84	1.21	1.4	9.43	9.35	0.08
1.10	2.83	1.72	1.11	1.5	15.05	15.05	0

From Polanin, Ibragimov, and Sabelev [138].

heat conduction equation for the fuel element. The influence of the variation of physical properties due to the temperature gradient on heat-transfer correlations may be taken into consideration (multipliers: $(\mu_w/\mu_b)^n$ [134], $(Pr_w/Pr_b)^m$ [139]. One may also solve the system of conservation equations assuming some form of the function $\mu = \mu(T)$ [140–142]. Moreover, there are known relations that account for the effect of free convection.

Liquid metals. In principle, in laminar flow through closed channels the heat-transfer correlations for nonmetallic coolants may be used. The only obstacle is the range of the Prandtl number values. The axial heat conduction in metal is usually neglected when $Pe > 10$ [40]; i.e., $Re > \sim 10^3$. Otherwise, it should be taken into account. Empirical heat transfer correlations for closely packed rod bundles in equilateral triangular pitch only are given in the following.

For liquid-metal heat transfer in rod bundles (with an infinite number of fuel elements), the correlated result is [114]:

$$Nu = Nu^*_{lam} = \{7.55 - 6.3/(s/d)^{17(s/d)\cdot[(s/d)-0.81]}\} \cdot \{1 - 3.6/[(s/d)^{20} \cdot (1 + 2.5 \cdot \varepsilon^{0.86}) + 3.2]\}$$

for $1.0 \leqq s/d \leqq 2.0, \quad Pe \geqq 1, \quad \varepsilon \geqq 0.01$

In the case of tightly packed rod bundles, i.e. where $s/d = 1$, the preceding equation reduces to

$$Nu = Nu^{**}_{lam} = 1.25 \cdot [1 - 3.6/(4.2 + 2.5 \cdot \varepsilon^{0.86})]$$

If $1.2 \leqq s/d \leqq 2.0$ then the corresponding equation is

$$Nu = Nu^{***}_{lam} = 7.55 \cdot (s/d) - 20 \cdot (s/d)^{-13}$$

The circumferential variation of cladding surface temperature is correlated with the following equation [114]:

$$\Delta T_{lam} = 0.022\{(s/d)^3[(s/d)-1]^{0.4}-0.99\} \cdot [1-\tanh \cdot \{\{1.2\exp\{-26.4[(s/d)-1]\}+\ln\varepsilon\}\{0.84 + 0.2\{[(s/d)-1.06]0.06\}^2\}\}]$$

In clusters, these variations are somewhat more significant, first of all, on the surfaces of peripheral elements "k" and "l" (see Figure 6). Suitable correlations are reported in reference [114]. Some additional relavent information can be found in References 40, 105, 112–114, 132, 142–146.

Hydraulic Calculations of Reactor Channels

The goal of hydraulic calculations of a reactor core is the determination of pressure drop in reactor channels, pumping power, and flow distribution.

In the case of thermal reactors, where $s/d \sim 1.3$, it would be sufficient to use in these calculations a coolant velocity $\bar{w}$ averaged over the fuel assembly's cross section area.

More complicated situations arise in fast reactors, where $s/d \sim 1.1$–1.2 and detailed analysis of local hydrodynamic characteristics is necessary. This is mainly caused by specific interchannel interactions especially in peripheral channels.

Reactor Core Pressure Drop

The total pressure drop in a reactor core is the sum:

$$\Delta p_{total} = \Delta p_f + \Delta p_g + \Delta p_a + \sum \Delta p_{local} \tag{32}$$

In a case of liquid coolants, Δp_a may be neglected because of its low compressibility. Also Δp_g may be neglected because $\Delta p_g \ll \Delta p_f$. Changes of coolant density have to be taken into account for gaseous coolants since Δp_a is about 10% of Δp_{total}.

Friction loss. For laminar or turbulent flow through a circular tube, the frictional loss is defined as follows

$$\Delta p_f = \zeta \cdot (L/D_p) \cdot (\rho_c \cdot \bar{w}^2/2) \tag{33}$$

where ζ is the Darcy–Weisbach friction factor.

It can be determined theoretically only for the case of fully-developed isothermal laminar flow. For turbulent flow empirical and semi-empirical correlations exist. Some formulas for incompressible fully-developed isothermal laminar flow are given below [26, 138]:

- Circular tube

  $\zeta = \zeta_p = 64/Re.$

- Plane gap

  $\zeta = 1.5 \cdot \zeta_p.$

- Annular channel

  $\zeta = (1-\Theta^2)/[1+\Theta^2+(1-\Theta^2)/\ln\Theta]\zeta_p.$

For laminar coolant flow along a cluster of smooth fuel elements the friction factor ζ can be determined [114, 146, 147] from the empirical formula

$$\zeta = f(s/d) \cdot \zeta_p$$

where

$$
f(s/d) = \begin{cases} 0.41 + 1.9[(s/d) - 1]^{1/3}, & [146] \\ 1.369 - 0.709 \cdot (s/d)^{-10.7}, & [114,\ 147] \end{cases} \text{triangular lattice} \\
\begin{cases} 0.41 + 1.9[(s/d) - 1]^{1/2}, & [146] \\ 1.436 - 0.791 \cdot (s/d)^{-5.3}, & [114,\ 147] \end{cases} \text{square lattice}
$$

For noncircular channels the Reynolds number calculation is based on equivalent (hydraulic) diameter D_e. The average fluid velocity is calculated from the theoretical velocity distribution [74, 148, 149].

In the case of turbulent flow, which is typical for nuclear reactors, the friction factor can be determined from the formulas given below.

Friction factor ζ for fully-developed isothermal turbulent flow of incompressible fluid in circular tube (smooth or almost smooth):

$$
\zeta = \zeta_p = \begin{cases} 0.3164 \cdot Re^{-0.25}, \ Re \leqq 10^5, \ [150] \\ (1.82 \cdot \lg Re - 1.64)^{-2}, \ 10^4 \leqq Re \leqq 5 \cdot 10^6, \ [151] \\ 0.032 + 0.221 \cdot Re^{-0.237}, \ 10^5 \leqq Re \leqq 4 \cdot 10^6, \ [26] \end{cases} \tag{34}
$$

and

$$
\zeta_p^{-1/2} = \begin{cases} 2.0 \cdot \ln(Re \cdot \zeta_p^{1/2}) - 0.8, \ 10^4 \leqq Re \leqq 5 \cdot 10^6, \ [152] \\ -2 \cdot \lg[\delta/3.7 + 2.51/(Re \cdot \zeta_p^{1/2})], \ \text{turbulent flow}, \ [153] \\ 2 \cdot \lg(1/\delta) + 1.74, \ Re > 500/\delta, \ [138] \end{cases} \tag{35}
$$

Colebrook's correlation [153] is the most popular one in the western technical literature. It is plotted in Figure 10. This plot is known as Moody diagram [154] and is widely used in the whole range of turbulent flows.

The preceding formulas can be also used for noncircular channels when hydraulic diameter D_e is used instead of pipe diameter D_p. There exist also correlations for channels of different geometry [26, 91, 138]. For example, in a case of annular channel:

$$
\zeta = (1 + 0.04 \cdot \Theta) \cdot \{(1 - \Theta)/[1 + (1 - \Theta^2) \ln \Theta^2]\}^{0.62} \cdot \zeta_p
$$

where ζ_p is given by Equations 34 or 35.

The above formula is particularly recommended for large Θ values.

Determination of the friction factor in flow along closely packed rod bundles is more complicated. In this case, a certain role is played by anisotrop of flow. In addition, the transition region is much greater than that of closed channels.

The corresponding formulas containing the pitch-to-diameter ratio s/d can be found in References 26, 74, 91, 138, and 155. For example, the formula

$$
\zeta/\zeta_p = \begin{cases} 0.57 + 0.18[(s/d) - 1] + 0.53[1 - \exp(-a)], \\ a = 0.58\{1 - \exp\{-70[(s/d) - 1]\}\} + 9.2[(s/d) - 1] \\ \text{for triangular lattice,} \\ 0.59 + 0.19[(s/d) - 1] + 0.52\{1 - \exp\{-10[(s/d) - 1]\}\} \\ \text{for square lattice,} \end{cases}
$$

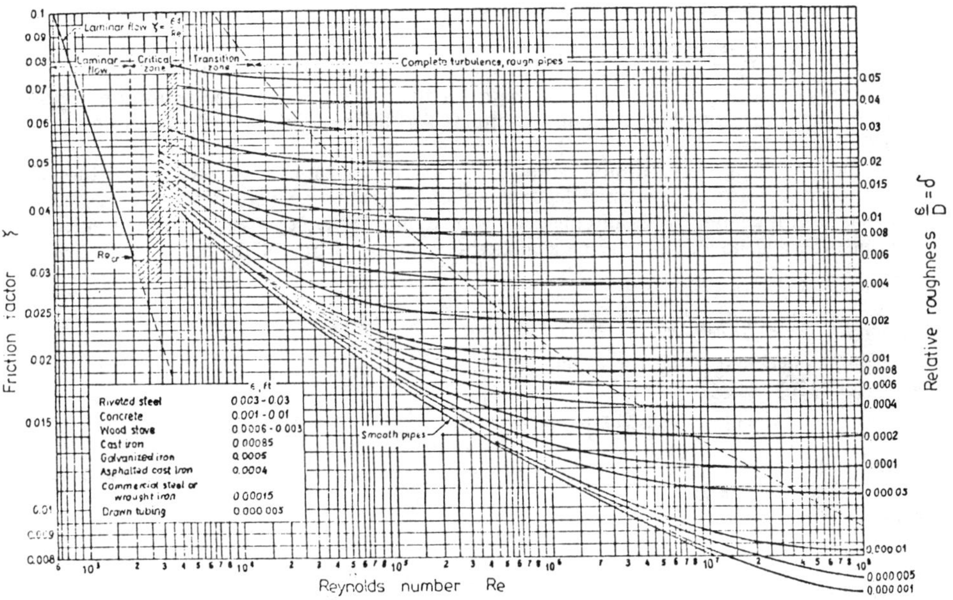

Figure 10. Moody friction factor chart. (Reprinted with permission from *Trans. ASME,* Vol. 66, Moody, L.F., "Friction Factors for Pipe Flows," Copyright 1944, ASME, pp. 671–684.)

for $1 < s/d < 10$, $2 \cdot 10^4 < Re < 5 \cdot 10^5$ (ζ_p is calculated from Equation 34 [150], is given in Reference 91. The correlation below

$$\zeta = C \cdot (\Pi_w/\Pi_f) \cdot (A_{ch}/A_f)^m \cdot Re^{-0.25}, \; 4 \cdot 10^3 < Re < 5 \cdot 10^4$$

where

$$C = 0.47, \; m = 0.35, \; 4 \cdot 10^3 \leqq Re \leqq 10^5; \text{ triangular lattice}$$

$$C = 0.38, \; m = 0.45, \; 10^3 \leqq Re \leqq 5 \cdot 10^4; \text{ square lattice}$$

is recommended in [138]. The following correlations are also useful:

- [114]: $(K/\zeta)^{1/2} = 2 \cdot \lg(Re \cdot \zeta^{1/2}/K^{1.5}) - 0.8$ with K being approximated for $1.0 \leqq s/d \leqq 1.4$ as follows:

$$K = \begin{cases} 1.277 - 0.532 \cdot (s/d)^{-10.7}; \text{ triangular lattice,} \\ 1.327 - 0.593 \cdot (s/d)^{-5.3}; \text{ square lattice,} \end{cases}$$

  remembering that, for $Re > 10^4$, one has

$$\zeta = 0.184 \cdot K^{1.2} \cdot Re^{-0.2};$$

- [147]: $\zeta = C \cdot a \cdot Re^{-m}$, $1 \leqq s/d \leqq 2$

  where

$$C = 1 + 0.316[1 + 2.83 \cdot \exp(-2 \cdot 10^3 \cdot Re)] \cdot [(s/d) - 1] + 0.93[1 + 2.95 \cdot \exp(-2 \cdot 10^{-3} \cdot Re)] \cdot \{1 - \exp\{-9.2[(s/d) - 1]\}\}$$

$$a = 0.19 + 20 \cdot \exp(-4 \cdot 10^3 \cdot Re) + 5.81 \cdot \exp(-1.6 \cdot 10^{-3} \cdot Re),$$

$$m = 0.25 + 0.75 \cdot \exp(-1.1 \cdot 10^{-3} \cdot Re).$$

The latter correlation is recommended both for turbulent flow and the transition region.

There are also a number of similar formulas for different subchannel-shapes occurring in a fuel assembly.

In the preceding correlations, isothermal flow of coolants was assumed. However, intensive heat transfer processes leading to considerable temperature gradients occur between reactor channel walls and the cooling medium. The Darcy–Weisbach friction factor in nonisothermal flow $\zeta_{\text{n-iso}}$ may be evaluated by means of one of the correlations below:

- Rohsenow–Clark [77]

  $$\zeta_{\text{n-iso}} = \zeta[\mu(T_w)/\mu(T_b)]^{0.60}$$

  recommended first of all for water.
- Taylor [156]

  $$\zeta_{\text{n-iso}} = (0.0028 + 0.25/Re_m^{0.32}) \cdot (T_b/T_w)^{0.5}$$

  where the Reynolds number Re_m is calculated based on $\rho_c(T_b)$ and $\mu(T_w)$ and temperatures are absolute.
- [83, 138]

  $$\zeta_{\text{n-iso}} = \zeta \cdot [1 - 0.5(1+M)^n \cdot \lg(1+M)]$$

  where

  $$M = [\mu(T_b)/\mu(T_w) - 1] \cdot [\mu(T_b)/\mu(T_w)]^{0.17}$$

  $$n = 0.17 - 2 \cdot 10^{-6} \cdot Re + 180/Re$$

  and $\quad 10^4 < Re < 3 \cdot 10^5, \quad 1.3 < Pr < 10.$
- [157]

  $$\zeta_{\text{n-iso}} = \zeta\{1 + \Pi_q/\Pi_w\{[\mu(T_w)/\mu(T_b)]^{0.6} - 1\}\}.$$

Zhukauskas [74] suggests more general expressions. However, a unique estimation of the influence of temperature gradient in the fluid on the friction factor value is not known. In the case of gaseous coolants, it is useful to account for the term Δp_a. Then Δp_{total} may be calculated from the following simplified formula [28, 29]:

$$\Delta p_{\text{total}} = \Delta p_f + \Delta p_a = (1/\bar{\rho}_c) \cdot G[\zeta \cdot L/D_e + (T_{c,\text{out}} - T_{c,\text{in}})/\bar{T}_c]$$

More detailed procedures are given in References 55 and 158.

Local losses. Additional pressure losses in straight reactor channels are caused by:

- Changes of channel cross-section, e.g., at inlet and outlet of reactor core.
- Rod spacers and finning of fuel elements.

The contraction and expansion coefficients may be found, e.g., in References 155 and 159. They depend on the Reynolds number and the ratio of channel cross-sectional areas. A variety of distancing elements and fuel element fins have produced the correlations cited, for example in References 26, 114, and 60, which are not sufficiently general. In each particular case, it is reasonable to examine the literature.

Pumping Power

The reactor is the basic element of the primary circuit. To ensure the required coolant flow through the reactor core, pumps (for liquid coolants) and gas-blowers are used.

Maintenance of continuous coolant circulation is due to the fact that the pumping power is equalized with the total pressure drop. If the pressure drop over reactor is taken into account only, then the pumping power $P_{pumping}$ is defined as follows:

$$P_{pumping} = \Delta p_{total} \times \text{total area of coolant flow} \times \text{average coolant speed/pump efficiency}.$$

In a reactor circuit this power is considerably greater since the total pressure drop is the sum of pressure drops in all elements of this circuit.

Flow Distribution

The spatial distribution of the heat generation requires readjustment of the mass flow rate to local power of fuel element. There exist a number of possibilities of realization of this postulate, which is often called the flow distribution [26–29 and 55]. One may assume a constant static pressure (or pressure gradient) in the channel cross-section. This method is usually termed as the equivalent diameter method. In more detailed calculations other methods are applied in which the reactor core is divided into a number of so-called hydraulic zones consisting of channels with similar thermal parameters (see e.g. [26]). For example, flow distribution is assumed to be proportional to the power of the most thermally loaded channel ("hot" channel) in a zone or such that the maximal fuel element temperature is the same in the all zones. Independent of the method, cooling of all fuel elements should be such that their maximal temperature values are less than the maximal permissible levels.

TWO-PHASE FLOW AND HEAT TRANSFER WITH BOILING

The thermal and hydraulic calculations of BWR and PWR reactor cores in steady and unsteady states require thorough analysis of coolant (mainly water) boiling.

In the case of two-phase flow, different nonhomogeneous phases may occur. The flow patterns for vertical circular tubes are shown in Figures 11 and 12 [161]. In the situation demonstrated in Figure 12 critical heat flux takes place (the fast increase of wall temperature and burnout).

In a case of complex geometry channels the following flow patterns can exist: bubble flow, slug flow, inverse annular flow (in the critical heat transfer conditions) and next when quality increases, complicated structures being the composition of the main flow patterns can occur.

More details about two-phase flow modes are described for example in References 162–166. Theoretical and experimental investigations of two-phase flow through reactor core are extremely complicated. It is connected with the presence of turbulent (in transient states also nonequilibrium) flow and nonuniform heat flux distribution.

In practice, correlations based on similarity theory and dimensional analysis for the case of q = constant are used, although correlations taking into account attributes of boiling in core channels exist (e.g. [166–170]).

Basic heat transfer phenomena and the predictive techniques which are particularly important in nuclear water reactor safety are described in [171–172].

In spite of great progress in numerical hydromechanics [64, 173–178] and in computer technology further experiments on boiling mechanisms during the flow are necessary. It results from the fact that reactor technology demands precise treatment of flow and heat transfer problems.

Void Fraction, Quality and Slip Relations

The description of two-phase flow requires three new parameters, not used in the single-phase case: void fraction φ, quality x, and slip ratio S.

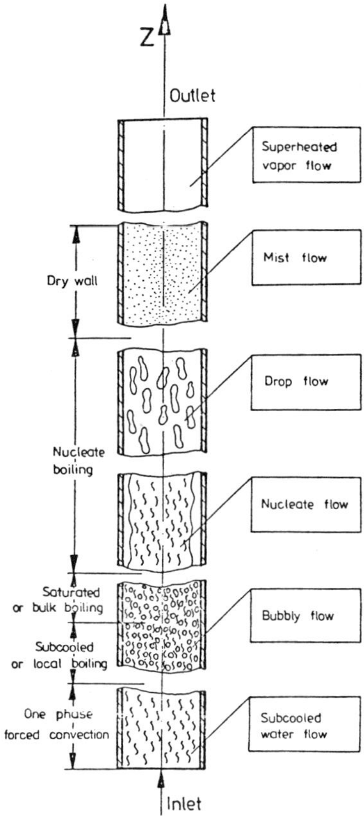

Figure 11. Flow regimes in a vertical heated pipe with upflow of coolant.

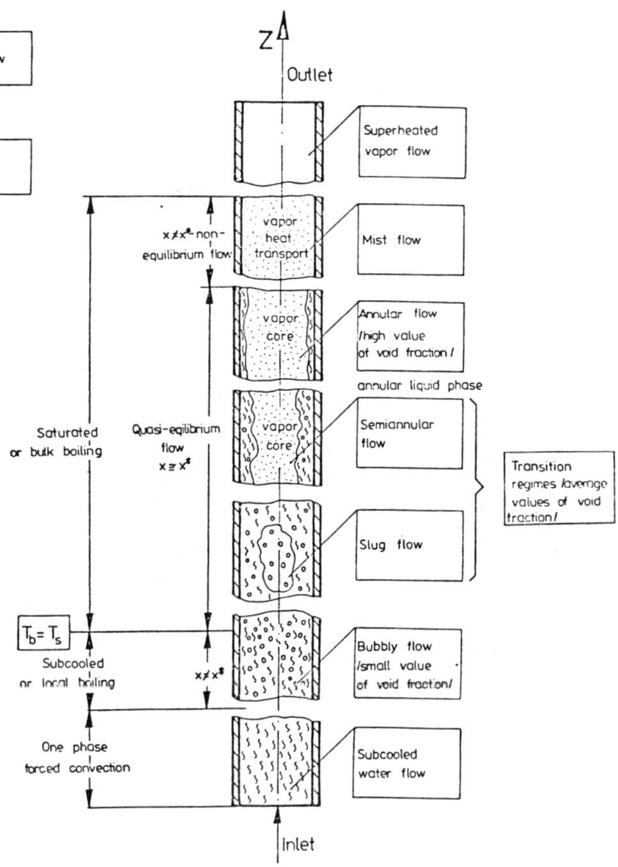

Figure 12. Flow regimes in a vertical heated pipe with upflow of coolant—case of the critical condition.

These are defined as follows:

$$\varphi = A_v/(A_v + A_l)$$

$$x = W_v/(W_v + W_l) = A_v G_v/(A_v G_v + A_l G_l)$$

$$S = w_v/w_l$$

These are quantities averaged over the channel cross section. It is possible to define average values of these parameters for the flow channel (A → V) as well as their local values for a given z.

Since φ can be measured [163, 179–181] then the average density of liquid-vapor mixture can be determined from the following relationship:

$$\rho_m = \rho_v \cdot \varphi + \rho_l \cdot (1 - \varphi)$$

The fundamental relationship between above parameters φ, x, S is:

$$(1 - \varphi)/\varphi = (1 - x) \cdot \rho_v \cdot S/(x \cdot \rho_l)$$

In homogeneous flow, $w_v = w_l$ and one has $S = 1$. Figure 13 shows φ versus x for liquid water at 6.89 MPa (1,000 psia) and several slip ratios.

Difficulties in the experimental and theoretical establishing of φ, x, S resulted in numerous correlations, most of which do not depend on two-phase flow patterns. Some useful correlations which can be used in reactor channel calculations are available (e.g. [55, 138, 182–191]). The most frequently used formula is that of Martinelli–Nelson (Figure 14 [182]).

More general definition of quality based on thermodynamic enthalpy is also used

$$x^* = (h_m - h_l)/\Delta h_{vl}$$

However, for complete thermodynamic equilibrium one has $x = x^*$. In reactor conditions, certain divergences between measured and calculated results must be taken into account.

For the typical boiling reactor channel the $\varphi(z)$ curve is shown in Figure 15. The vapor amount in the channel during subcooled boiling (region $L_1 L_2$) has an effect on the pressure gradient. For high subcoolings, after the onset of nucleation (region $L_1 L_1^*$) bubbles are attached to the heating surface and growing and collapsing take place. Voidage in this region is a wall effect.

For lower subcoolings, bubbles detach from the heating surface and condense as they are in motion through the slightly subcooled liquid core. Voidage in this region is a bulk fluid effect. Over the region $L_1 L_1^*$ the void fraction remains low and its influence on the pressure gradient may be neglected. Saha and Zuber [193] have proposed a method which

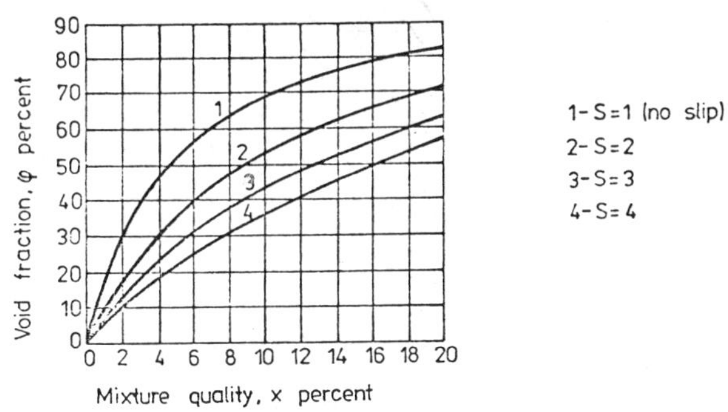

Figure 13. φ vs. x for water at 1,000 psia and various slip ratios in a flow system. (Reprinted with permission from EL Wakil, M.M., "Nuclear Power Engineering," Copyright 1962, McGraw Hill, p. 208.)

is recommended to calculate the point of net vapor generation. This point may be considered as the point of bubble detachement (L_1^*).

If $Pe_l \leqq 70{,}000$, then the value of $\Delta T^*_{L_l} = T_{sat} - T_c(L_l^*)$ is given by relationship

$$\Delta T^*_{L_l} = q_w \cdot D_e/(455 \cdot \lambda_l)$$

and if $P_{e_l} > 70{,}000$ then the value of $\Delta T^*_{L_l}$ is evaluated from the equation

$$\Delta T^*_{L_l} = q_w/(0.0065 \cdot G \cdot c_{pl})$$

For the region of detached voidage, Levy [195] assumed that the actual vapor mass fraction x(z) and the thermodynamic vapor mass fraction $x^*(z)$ are connected by formula

$$x(z) = x^*(z) - x^*(L_1^*) \cdot \exp[x^*(z)/x^*(L_1^*) - 1]$$

where $x^*(L_1^*)$ is thermodynamic vapor mass fraction at the point of bubble detachment, formed as

$$x^*(L_1^*) = -c_{pl} \cdot \Delta T_{sub}(L_1^*)/\Delta h_{vi}$$

In the preceding formula the following boundary conditions are fulfiled:

- If $x^*(z) = x^*(L_1)$ then $x(z) = 0$.
- $(dx/dz)|_{z = L_1^*} = 0$.
- $x(z) \rightarrow x^*(z)$ for $x^*(z) \gg |x^*(L_1^*)|$.

Two-Phase Pressure Drop

The total pressure drop in two-phase flow is expressed by the same formula (Equation 32) as in the case of single-phase flow. When determining the particular terms, various methods taking into consideration the two-phase flow specificity may be applied, for example, the homogeneous method, the Martinelli–Nelson method, the Thom method.

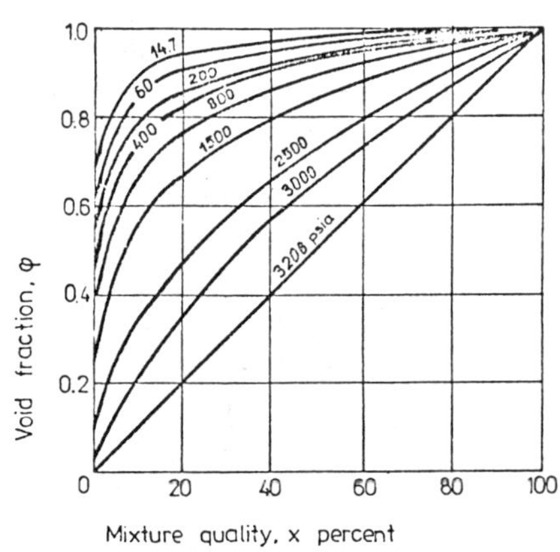

Figure 14. Martinelli-Nelson void fraction correlation (φ vs. x for steam-water mixture). (Reprinted with permission from *Trans. ASME,* Vol. 70, Martinelli, R.C., and Nelson, D.B., "Prediction of Pressure Drop During Forced Circulation Boiling of Water," Copyright 1948, ASME, pp. 695–702.)

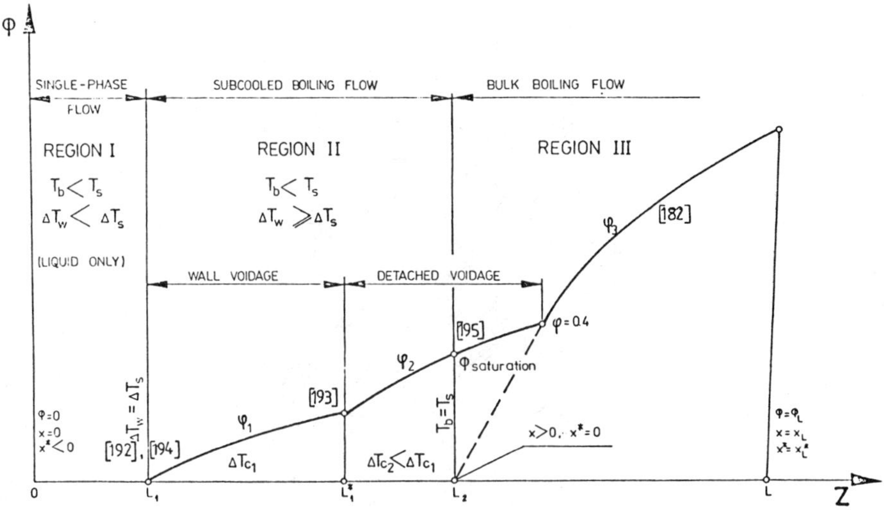

Figure 15. Characteristic regions of void fraction φ formation in boiling channels ($q_w < q_{w,crit}$ and $\Delta T_w = (T_b + q_w/\alpha_{convection}) - T_s$).

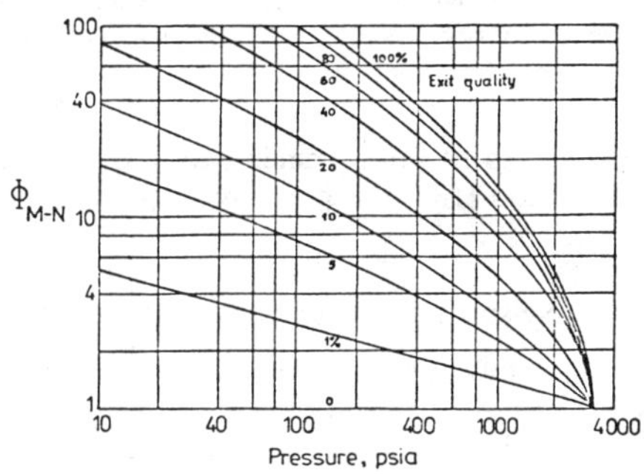

Figure 16. Martinelli-Nelson friction-multiplier as a function of pressure and exit quality. (Reprinted with permission from *Trans. ASME,* Vol. 70, Martinelli, R.C., and Nelson, D.B., "Prediction of Pressure Drop During Forced Circulation Boiling of Water," Copyright 1948, ASME, pp. 695–702.)

In the homogeneous method, the following relations are used:

$$\Delta p_f = \zeta(L/D) \cdot (G^2 v_m/2)$$

$$\Delta p_g = \rho_m \cdot g \cdot L \cdot \sin \omega$$

$$\Delta p_a = G^2 \cdot (v_{out} - v_{in})$$

where $v_m = 1/\rho_m = v_l + x \cdot \Delta v_{vl}$

$G = W/A$

ζ = the friction factor from Moody diagram (see Figure 10) for liquid single-phase flow of mass flow rate equal to the total mass flow rate

ω = the angle measured from horizontal

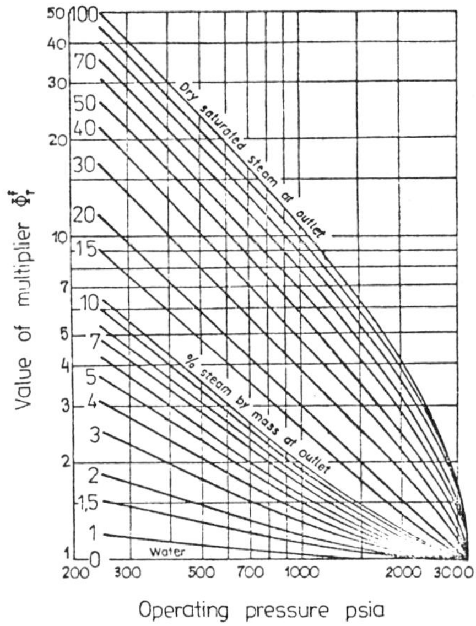

Figure 17. Thom frictional pressure-drop multiplier for boiling flow of water and steam. (Reprinted with permission from *Int. J. Heat Mass Transfer,* Vol. 7, Thom J.R.S., "Prediction of Pressure Drop During Forced Convection Circulation Boiling of Water," Copyright 1964, Pergamon Press Ltd., pp. 709–724.)

Other methods that follow represent improvement, over the homogeneous method.
In the Martinelli–Nelson method [182] the pressure drop is calculated according to:

$$(\Delta p_f)_{TP} = \Delta p_f \cdot \Phi_{M-N}$$

where Φ_{M-N} is the Martinelli–Nelson two-phase multiplier shown in Figure 16.

Although the friction multiplier Φ_{M-N} was evaluated for horizontal flow it is nevertheless often used for vertical reactor channel calculations.

Similar correlations are found by Thom [196]. They are based on experimental data obtained for vertical water-steam mixture flow through a pipe with uniform heat flux on its wall. The range of parameters examined by Thom is the nearest to those occuring in reactor channels (see Table 5). The Thom multiplier Φ_T^f given in Figure 17 is also a function of quality and pressure. The gravity contribution is evaluated assuming flow of saturated water at water-steam mixture mass velocity and using the multiplier Φ_T^g given in Figure 18, so

$$\Delta p_g = \rho_l \cdot g \cdot L \cdot \sin \omega \cdot \Phi_T^g$$

It is also known as the correlation developed by Baroczy in which the two-phase multiplier is a function of quality and so-called property index [197].

The method of the friction pressure drop evaluation was also elaborated by Armand [198].

More general treatment of the problem may be found in numerous references (e.g. [28, 29, 55, 92, 164, 166, and 190]).

Heat Transfer with Boiling

When the wall temperature remains below that necessary for nucleation, the mode of heat transfer is single-phase (liquid) convection. Nucleate boiling occurs when the wall

Table 5
Correlations for Heat Transfer in Two-Phase Forced Convection and Nucleate Boiling Regions

Author(s)	Correlations	Range of Application
Jens, Lottes [201]	$T_w - T_{sat} = 25.012(q_w/10^6)^{0.25} \exp(-0.016p)$	Subcooled boiling $p = 0.7\text{–}17.2$ [MPa] $G = 11\text{–}10{,}500$ [kg/m²/s¹] $q_w < 12.5 \cdot 10^6$ [W/m²]
Thom [189]	$T_w - T_{sat} = 22.52(q_w/10^6)^{0.25} \exp(-0.0115p)$	Subcooled boiling $p = 5.2\text{–}13.8$ [MPa] $G = 1{,}000\text{–}38{,}000$ [kg/m²/s¹] $q_w < 1.577 \cdot 10^6$ [W/m²] $x < 0.8$
Rasohin [138]	$\alpha_{NB} = 5.513p^{0.25}q_w^{2/3}$ $\alpha_{NB} = 0.048p^{4.33}q_w^{2/3}$	$p = 0.1\text{–}8$ [MPa] $p = 8\text{–}20$ [MPa]
Borishanski [138]	$\alpha_{NB} = 9q_w^{0.7}[3.3 - 0.0113(T_{sat} - 100)]$	$p = 3\text{–}20$ [MPa] $G = 100\text{–}4{,}000$ [kg/m²/s¹] $q_w < 2.3 \cdot 10^6$ [W/m²]
Labuncov [138]	$\alpha_{NB} = 3.4(10p)^{0.18}q_w^{2/3}/(1 - 0.045p)$	$p = 0.1\text{–}20$ [MPa]
Schrock–Grossman [202], [203]	$\alpha_{TP}/\alpha_l = Bq_w/(Gh_{vl}) + A(1/X_{tt})^n$ where $B = 7.39 \cdot 10^3 \quad A = 1.54 \cdot 10^{-4} \quad n = 0.66$ or $B = 6.7 \cdot 10^3 \quad A = 3.5 \cdot 10^{-4} \quad n = 0.66$ from [166] p. 248	Two-phase forced convection and nucleate boiling $p = 0.3\text{–}3.5$ [MPa] $G = 240\text{–}4{,}430$ [kg m⁻² s⁻¹] $x = 0.05\text{–}0.57$ $q_w = (0.19\text{–}4.57) \cdot 10^6$ [W m⁻²]
Chen [204]	$\alpha_{TP} = \alpha_{mic} + \alpha_{mac};\ \alpha_{mac} = \alpha_l F$ $\alpha_{mic} = 0.00122\left[\frac{\lambda_l^{0.79} \cdot c_{pl}^{0.45} \cdot \rho_l^{0.49}}{\sigma^{0.5} \cdot \mu_l^{0.29} \cdot \Delta h_{vl}^{0.24} \cdot \rho_v^{0.24}}\right] \cdot (\Delta T_{sat})^{0.24} \cdot (\Delta p_{sat})^{0.75} \cdot S$ with $\Delta p_{sat} = \frac{\Delta T_{sat}\Delta h_{vl}}{T_{sat}\Delta v_{vl}}; \quad \Delta T_{sat} = T_w - T_{sat}$ $F = f(1/X_{tt})$ and $S = f(Re_l F^{1.25})$ are given in Figures 20 and 21	Two-phase forced convection and nucleate boiling $p = 0.06\text{–}3.5$ [MPa] $x = 0\text{–}0.71$ $q_w = (0.0063\text{–}2.4) \cdot 10^6$ [W m⁻²]

Note: α in [W/m² · °C¹]; q_w in [W/m²]; p in [bars]; T_{sat} in [°C]; σ in [N/m¹]; $\alpha_l = 0.0023(\lambda_l/D_e) \cdot [G(1-x)D_e/\mu_l]^{0.8}(Pr_l)^{0.4}$

temperature is greater than the saturation temperature. This mode of heat transfer takes place in the presence of subcooled liquid in the liquid core. Qualitative transient from forced convection to nucleate boiling is shown in Figure 19. In reactor practice, the onset of subcooled boiling may be estimated from the intersection of the forced convection curve ABD′ with fully developed boiling curve C′EF (e.g. [199]). Other methods of onset subcooled boiling determination in flow through the tube are listed in References 163 and 166. For determination of the onset point in flow through rod bundles, Barulin et al. [170] suggest the following equation

$$\Delta h_{ONB} = (h_c - h_l)_{ONB} = -1.23 \cdot 10^6 \cdot q_w^{1.15} \cdot D^{0.3} \cdot (\rho_l/\rho_v)^{.2}/(3{,}600 \cdot G)$$

where
h = [J/kg]
q_w = [W/m²]
D = [m]
G = [kg m⁻²s⁻¹]

which is based on the experimental data.

At the same temperatures, the value of Δh_{ONB} is less than that for a tube. The difference results from the fact that the fluid bulk temperature in flow through rod bundles is greater than that one in flow through a tube. An appropriate correlation is given in Reference 200. From the point where the coolant temperature reaches the saturation temperature ($x = 0$) the saturated nucleate boiling mode appears. As the quality is increased through the

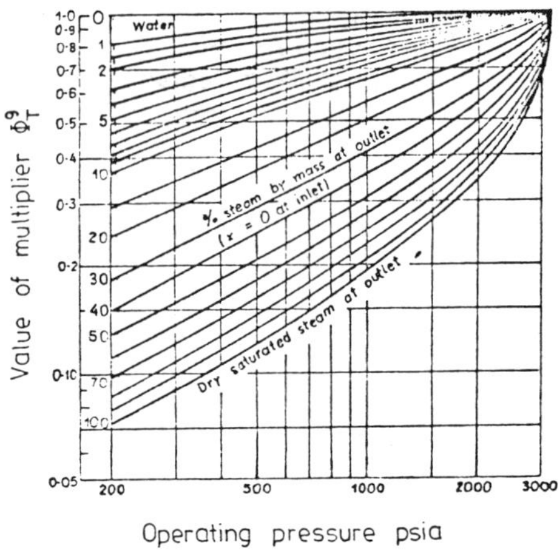

Figure 18. Thom gravitational pressure-drop multiplier for boiling flow of water and steam. (Reprinted with permission from *Int. J. Heat Mass Transfer,* Vol. 7, Thom J.R.S., "Prediction of Pressure Drop During Forced Convection Circulation Boiling of Water," Copyright 1964, Pergamon Press Ltd., pp. 709–724.)

saturated nucleate boiling region the process of "boiling" is replaced by the process of "evaporation." This transition is preceded by the flow pattern change from bubbly or slug flow to annular flow. Thus, in this region heat is mostly carried away from the wall by forced convection in the film to the liquid-vapor interface where evaporation takes place. In reactor technology, the correlations used to determine the heat transfer rates in subcooled and saturated nucleate boiling and forced convection vaporization are summarized in Table 5. Chen's correlation is generally accepted as one of the best available for saturated forced convective boiling region. Colier [166] gives approximative formulas for the functions shown in Figures 20 and 21 for F and S, respectively, as:

$$F = \begin{cases} 1 & \text{for} \quad 1/X_{tt} \leqq 0.1 \\ 2.35 \cdot (1/X_{tt} + 0.213)^{0.736} & \text{for} \quad 1/X_{tt} > 0.1 \end{cases}$$

$$S = 1/(1 + 2.53 \cdot 10^{-6} \cdot Re_l)^{1.17}$$

where $\quad Re_l = 10^{-5} G \cdot (1 - x) \cdot D_e/\mu_l$

These correlations are very useful and recommended for rapid calculations. Recently the correlation developed by Chen is used in the transient reactor analysis code TRAC [173].

Boiling Crisis

In forced convective boiling the boiling crisis occurs when the heat flux attains such a high level that the heat surface can no longer support continuous liquid contact. The peak heat flux is called critical heat flux (CHF).

The temperature difference between surface temperature and bulk temperature corresponding to point CHF is likewise called the maximum nucleate or the critical temperature difference.

The CHF is characterized by a sudden rise of surface temperature when the heated surface is covered by a stable vapor blanket or by small surface temperature spikes when dry spots appear and disappear.

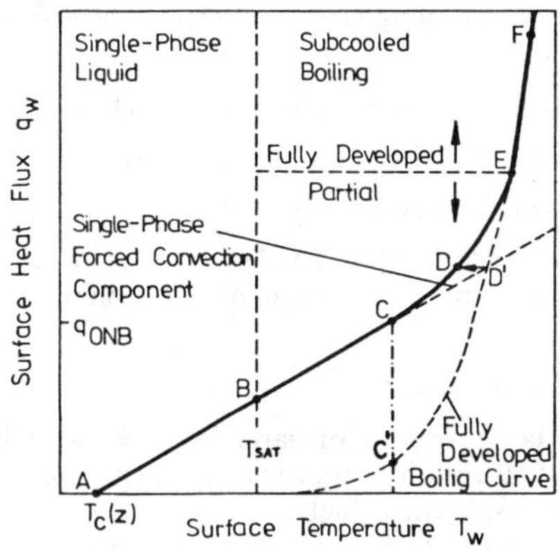

Figure 19. Subcooled boiling curve. (Reprinted with permission from Bergles, A.E., et al., "Two-Phase Flow and Heat Transfer in the Power and Process Industries," Copyright 1981, Hemisphere Publishing Corporation, p. 235.)

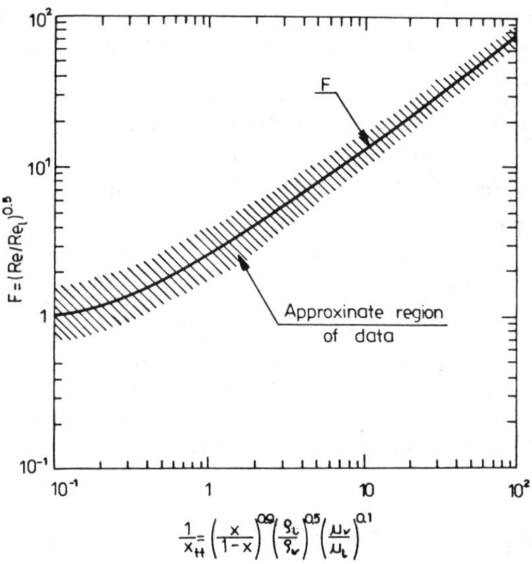

Figure 20. Reynolds number factor F for Chen correlation. (Reprint with permission from *Ind. Eng. Chem. Process Design and Developments,* Vol. 5, Chen, J.C., "Correlation for Boiling Heat Transfer to Saturated Liquids in Convective Flow," Copyright 1966, p. 322.)

A rapid increase of surface temperature of fuel elements during the boiling crisis is often sufficient to cause melting of the cladding, and releasing of considerable amount of fission products into the coolant flow.

Consequently, in water-cooled reactors CHF may be the most restrictive limit on the reactor power level.

CHF can occur either in a subcooled (low quality) bubbly flow or in saturated annular flow. The former is called departure from nucleate boiling (DNB) and the latter is called dryout. In general, the subject of boiling crisis and critical heat flux has been reviewed by Hewitt [205] and Tong [206].

In reactor practice, a minimum of the DNB ratio, which is defined by

$$\text{DNBR} = \frac{\text{DNB heat flux predicted by applicable correlation}}{\text{local reactor operating heat flux}}$$

is usually applied as a design criterion [49, 55, 190]. Experiments have shown that CHF is a function of the local coolant enthalpy, inlet subcooling, pressure, mass velocity, geometry, i.e.

$$q_{CHF} = q_{CHF}(h, h_{in}, p, G, D, L, \ldots)$$

Since CHF is affected by a large number of variables, it is difficult to obtain such CHF data which cover the range of all variables appearing in reactor design. The most interesting is CHF in rod bundle geometries, because fuel assemblies are widely used at present nuclear reactors. In particular, the annulus can be regarded as a single rod bundle (see Fig. 7).

It was confirmed that the critical heat flux value in rod bundles decreases as the vapor quality increases. This relation is approximately linear. The pressure increases in the case of G = constant, results generally in the decrease of the CHF value. Boiling crisis in cluster where rods are distributed compactly occurs usually in outside subchannels. The grids cause the flow redistribution and an increase of the critical heat flux. More details concerning the influence of different factors on the critical heat flux can be found for example, in References 168, 206 and 207.

In reactor technology, there are two methods which can be used for the calculation of the critical heat flux. In the first, different correlations based on experimental data of a given fuel cluster are used (e.g. Bezorukov correlation [208]):

$$q_{CHF} = 190 \cdot (1-x)^{-0.5+0.103 \cdot p} \cdot G^{-0.127+0.311(1-x)} \cdot (1-0.00182 \cdot p)$$

where $p = 75.5\text{–}167$ [bar]
$x = 0.07\text{–}0.4$
$G = 700\text{–}3800$ [$\text{kgm}^{-2}\text{s}^{-1}$]

used in the thermo-hydraulic calculation of VVER–1000

The second method is based on subchannels analysis. That is, the rod bundle is divided into subchannels. The thermo-hydraulic conditions are estimated for these subchannels, taking into consideration cross-flows generated by the pressure differences between subchannels and cross-mixing between adjacent channels. Therefore, the local value of critical heat

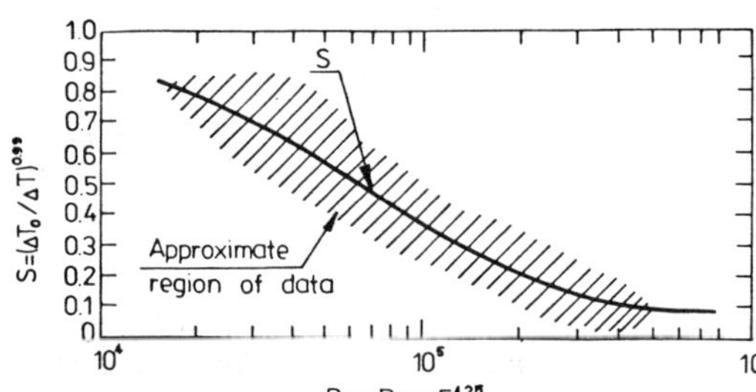

Figure 21. Suppression factor S for Chen correlation. (Reprint with permission for *Ind. Eng. Chem. Process Design and Developments*, Vol. 5, Chen, J.C., "Correlation for Boiling Heat Transfer to Saturated Liquids in Convective Flow," Copyright 1966, p. 322.)

Table 6
Critical Heat Flux Data

Name	Geometry	p [MPa]	Range of Application G [kg/m^{2}s^1]	x
Miropolski [210], [170]	Rod bundle	2.9–16.7	400–5,000	−0.1–0.3
Smolin–Poliakov [211]	Rod bundle	2.9–9.81	380–4,000	−0.2–0.25
Bezorukov [208]	Rod bundle	7.55–16.7	700–3,800	−0.07–0.4
Osmachkin [169]	Rod bundle	5–17	500–4,000	$\hat{x} = -1$–3
Babcock–Wilcox [212]	Rod bundle	1.66–13.8	1,020–5,400	−0.03–0.2
W-3 [213], [206]	Rod bundle Pipe	7–14	1,350–6,750	−0.25–0.15
Biasi [214]	Pipe	2.7–14	100–6,000	0–1
Bowring [215]	Pipe	2–19	136–18,600	0–1
Barnett [216]	Annulus	3–10	190–8,400	0–1
CISE-4 [209]	Pipe	4.5–15	1,000–4,000	0–1
Macbeth [217], [218]	Pipe	0.1–13.8	27–15,730	0–1
Griffith–Zuber [206], [219]	Short annulus	All range	Low	0–1

flux may be calculated using these local channel parameters in appropriate correlations. Application of the latter method is disscussed by Weisman and Bowring [71] and Gaspari [209]. Numerous correlations are currently available for the steady-state case. Reference to some of them are given in Table 6.

Many of the various correlations are adduced in Reference [206]. It is worth noting that some of these correlations describe DNB, some are valid for dryout and others deal with both kinds of boiling crisis. Recently, the necessity of verification of these correlations under depressurization transients (major LOCA and small breaks) has arisen. Comparison of experimental data obtained during LOCA simulations with those obtained by use of various correlations was made by Leung [220, 221].

Up to now, there are no correlations totally describing the problem of boiling crisis in reactor channels in steady- and unsteady-states. Therefore, in addition to reviewing existing correlations, further efforts for adopting recommended correlations must proceed.

FINAL REMARKS

This chapter is condensed and consequently does not include all problems connected with heat transfer in reactor channels. In reactors, feedbacks occur due to both thermal-hydraulic phenomena and nuclear processes. Unstable conditions arise from:

1. Variations in neutron flux generation over the volumetric thermal source strength distribution.
2. Distribution of temperatures followed by condition 1, which causes changes of properties of nuclear structural materials, temperature fields in the fuel elements, and heat transfer to coolants.
3. Coolant parameters resulting from condition 2 or other extraneous phenomena which affect the neutron balance.

In view of the complexity of these phenomena, experiments together with often subtle numerical methods have become a basic cognizance tool. Experiments are necessary as far as an analysis and modeling of two-phase flow and unsteady states are concerned. In recent years, experiments were related to reactor safety research in connection with LOCA (the loss-of-coolant accident) resulting from the rupture of the main pipe in a primary coolant circuit. When LOCA occurs almost all known heat transfer modes may appear in the reactor channel. The reader is referred to Reference 222 for detailed discussions.

In the first period of LOCA (5–100 ms) a rapid depressurization in the cooling system occurs as a result of subcooled water loss. In addition large dynamic loads inside the reactor core take place. As a rule, during the first second, the reactor is shutdown. Then its heat power decreases to the level of 6%–7% of the initial power. The energy released depends on the rate of fission product decay. After about 5 hours the power of the shutdown reactor is about 1% of the initial power (see the third section of this chapter). In the second period of blowdown the rate of depressurization decreases. A consequence of the heat generation and flashing process are large bubbles resulting in a fast growth of the void fraction. Reversal flow in the core may occur in the case of a break of the cold leg. Critical heat flux appears in reactor cooling channels. Leung [220] confirmed that it is dryout. Conditions of heat removal from fuel elements worsens considerably. There appears a transition regime and film boiling. An overview of various correlations and comparison with experimental works may be found in Groeneveld [223], Mayinger [224], and Tong [171]. One of the most important parameters, the cladding temperature, increases during the accident and therefore each analysis must take into consideration unsteady convection problems, boiling crisis and post-dryout heat transfer as well as problems of fuel rods rewetting. Water is injected to cold legs causing a refill of the lower plenum and then the entire core. It is characterized by steam-water interaction and quenching processes leading to pressure and flow oscillation [225, 226]. These processes are severly complicated. In principle, the resulting correlations are based on experiments. Research is conducted in many countries (the U. S., FRG, Japan, Soviet Union) in connection with the development of computer codes applying to numerical simulations of LOCA. The aim of this work is

- Development of needed correlations useful in blowdown analysis, transient heat transfer, behavior of fuel and fission products.
- Development of theoretical models such as two-phase flow models [227] and nonequilibrium processes [228].
- Development of multidimensionel hydraulic models and appropriate numerical methods [229, 176].
- Verification of codes by comparison with other ones and experimental tests.

One should underline that the research work is aimed to provide knowledge about secure and failure-free reactor operation.

NOTATION

A	coolant flow area, channel cross section area
A_{ch}	wetted perimeter of channels in rod cluster
A_f	perimeter of fuel elements in rod cluster
A_m	cross section area of the subchannel "m"
c_p	specific heat
C_{mn}	empirical function of pressure losses
d	fuel element diameter
D	diffusion coefficient, diameter
D_e	equivalent (hydraulic) diameter of non-circular channel, (= 4* flow area/wetted perimeter)
D/Dt	total derivative
Div	dyadic (tensor) divergence
e	specific total internal energy
E	energy
$\mathbf{f}$	unit mass force vector independent of flow motion
f_D	dimensionless coefficient used in the estimation of influence of molecular momentum exchange between subchannels on coolant enthalpy and mass flow rate distributions in subchannels
f_T	dimensionless coefficient used in the estimation of influence of turbulent momentum exchange between subchannels

	on coolant enthalpy and mass flow rate distributions in subchannels
F	Chen factor
g	Westcott coefficient; gravitational acceleration
G	number of energy groups; mass velocity
h	specific enthalpy
h_m, h_n	coolant enthalpies in the subchannels "m" and "n" (averaged over the subchannel cross section)
H	height of a reactor core
k	neutron effective multiplication factor; kerma factor
L	channel length
m_m	coolant flow through the subchannel "m"
n	angular neutron density; neutron density
$\mathbf{n}$ or $\mathbf{N}$	outward-bound unit vector normal to a surface
N	number density of isotope, total number of subchannels adjacent to the subchannel "m"
N_t	total number of subchannels in a rod cluster
Nu	Nusselt number
p	pressure
P	reactor power
Pe	Peclet number
Pr	Prandtl number
Pr_t	turbulent Prandtl number
q	heat flux at surface of channel
$(q_l)_m$	heat transferred to coolant per unit length of the subchannel "m"
q_v	volumetric heat internal generation, volumetric thermal source strength
$q_{w, crit}$, q_{crit}, q_{CHF}	critical heat flux
$\mathbf{r}$	radial coordinate
$\underline{r}$	position vector with respect to the fixed coordinate system
R	pipe radius
Re	Reynolds number
s	center-to-center distance (pitch)
S	neutron source, slip ratio, Chen suppression factor
S_{ij}	gap between adjacent fuel elements; gap through which cross-flow may occur
St	Stanton number
t	time
T	temperature
T_b	fluid-bulk temperature
$\bar{T}_c$	average coolant temperature in a channel
$T_{c, in}$	coolant temperature at the channel inlet
$T_{c, out}$	coolant temperature at the channel outlet
T_d	deformation velocity dyadic (tensor)
T_{ref}	some fluid reference temperature
T_s	symmetric stress dyadic (tensor)
u	internal energy per unit mass
v	neutron velocity, specific volume
V	volume
$\underline{V}$	coolant velocity vector
w	coolant velocity component in the axial direction
$\bar{w}$	average coolant velocity in the channel cross section
W	mass flow rate, coolant flow through channel
W_{mn}	coolant cross flow between the subchannel "m" and adjacent subchannel "n"
W'_{mn}	turbulent cross flow between the subchannel "m" and adjacent subchannel "n"
x	coordinate in channel cross section; mass quality
$\hat{x}$	generalized quality
X	fraction of prompt neutrons
X_{tt}	Lockhart–Martinelli factor

Greek Symbols

α	heat transfer coefficient
β	empirical turbulent mixing parameter
Γ	boundary-surface enclosing Ω

δ extrapolation length; relative roughness of the pipe inner surface
ε conduction parameter
ε_h turbulent, or eddy, diffusivity
ε_m turbulent, or eddy, diffusivity of momentum (turbulent kinematic viscosity)
ε unit dyadic (tensor)
ζ Darcy–Weisbach friction factor
Θ ratio of inner and outer diameters of annular channel
κ thermal diffusivity
λ thermal conductivity
μ dynamic viscosity
ν kinematic viscosity; average number of neutrons emitted per fission
Π_q, Π_{heated} heated perimeter of channel
Π_w wetted perimeter of channel
Π_f perimeter of fuel element
ρ reactivity, mass density
$\bar{\rho}_c$ average coolant density in the channel
σ microscopic cross-section, surface tension
Σ macroscopic cross-section
Φ angular neutron flux; neutron flux; Rayleigh dissipation function
Φ_{M-N} Martinelli–Nelson multiplier
Φ_T^f Thom frictional multiplier
Φ_T^g Thom gravitational multiplier
φ void fraction
Ω direction unit vector
$\bar{\Omega}$ control volume

Delta Quantities

Δh_{ONB} subcooling enthalpy at the onset point
Δh_{vl} latent heat of vaporization
Δp_a acceleration (or momentum) pressure drop
Δp_f friction pressure drop
Δp_g gravity pressure drop
Δp_{local} local losses
Δp_{total} total pressure drop
Δp_{TP} two-phase pressure drop
ΔT maximal circumferential temperature variation, dimensionless
$\Delta T_{L\dagger}$ subcooling at the point of net vapor generation
Δv_{vl} specific volume change for vaporization

Superscripts

ad adiabatic
f fission
sl slowing down
y coordinate in channel cross-section
Z axial coordinate

Subscripts

b at the fluid bulk temperature
c coolant
cl cladding
e extrapolated; equivalent
f fission; fuel
g number of energy group
i reaction
in inlet channel cross section
j number of isotope
l liquid
lv liquid-vapor mixture
m index of subchannel in rod cluster
max maximal
n index of subchannel adjacent to the subchannel "m"
noniso nonisothermal flow
0 before reactor shutdown

out	outlet channel cross section
ONB	onset of boiling
p	pipe
q	at constant heat flux on channel wall
s	scattering
sat	saturation
sub	subcooling
th	thermal
T	at constant wall temperature
TP	two-phase
tt	turbulent-turbulent; turbulent flow of both phases
v	vapor
w	wall; channel wall; at constant wall temperature
w, inner / inner	on the inner wall of annular channel
w_1 outer / outer	on the outer wall of annular channel

REFERENCES

1. Glasstone S., and Sesonske A., *Nuclear Reactor Engineering,* 3rd ed., Van Nostrand Reinhold Company, Princeton, 1981.
2. Connolly, T. J., *Foundations of Nuclear Engineering,* 1st ed., John Wiley § Sons, New York, 1978.
3. Smidt, D., *Reaktortechnik,* 1st ed., G. Braun, Karlsruhe, 1971.
4. "Directory of Nuclear Reactors," International Atomic Energy Agency, Vienna, 1959.
5. Duderstadt, J. J., and Martin, W. R., *Transport Theory,* 1st ed., John Wiley § Sons, New York, Chichester, Brisbane, London, 1979.
6. Bell, G. I., and Glasstone, S., *Nuclear Reactor Theory,* 1st ed., Van Nostrand Reinhold Company, Princeton, 1971.
7. Greenspan, M., Kelber, C. N., and Okrent, D., *Computing Methods in Reactor Physics,* 1st ed., Gordon and Breach, Science Publishers, Inc., New York, 1968.
8. Henry, A. F., *Nuclear Reactor Analysis,* 1st ed., M. I. T. Press, Cambridge, Mass., 1975.
9. Massimo, L., *Physics of High-Temperature Reactors,* 1st ed., Pergamon Press, Oxford, New York, Toronto, Sydney, Paris, Braunschweig, 1976.
10. Duderstadt, J. J., and Hamilton, L. J., *Nuclear Reactor Analysis,* 1st ed., John Wiley § Sons, Inc., New York, London, 1976.
11. Bird, R. B., Steward, W. E., and Lightfoot, E. N., *Transport Phenomena,* 1st ed., John Wiley § Sons, Inc., New York, 1960.
12. Loitsiansky, L. G., *Liquid and Gas Mechanics* (in Russian), 4th ed. Nauka, Moskva, 1973.
13. Landau, L. D., and Lifshits, E. M., *Fluid Mechanics* (in Russian), 1st ed., Gostekhizdat, Moskva, 1953.
14. Prosnak, W. J., *Fluid Mechanics* (in Polish), Vol. 1., 1st ed., PWN, Warsaw, 1970.
15. Phennigwerth, P. L., and Minkler, B., "Heat Transfer and Fluid Flow with Applications to Pressurized Water Reactors," WAPD–TM–1,000, Bettis Atomic Power Laboratory, Pittsburgh, Penn., January, 1971.
16. Staniszewski, B., "Heat Transfer–Theoretical Foundations," in Polish, PWN, Warsaw, 1963.
17. Madejski, J., *Theory of Heat Transfer,* (in Polish), PWN, Warsaw, Poznań, 1963.
18. Meyer, J. E., "Conservation Laws in One-Dimensional Hydrodynamics," Bettis Technical Review, Reactor Technology, WAPD–BT–20, September 1960, pp. 61–72.
19. Carslaw, H. S., and Jaeger, J. C., "Conduction of Heat in Solids," 2nd ed., Oxford University Press, Oxford, 1959.
20. Eckert, E. R. G., and Drake, R. M., *"Heat and Mass Transfer"* 2nd ed., McGraw Hill, New York, Toronto, London, 1959.
21. Jacob, M., *Heat Transfer,* Vol. I, John Wiley § Sons, New York, 1949.

22. Schneider, P. J., "Conduction Heat Transfer," Addison–Wesley Publishing Co., Reading, Mass., 1955.
23. Luikov, A. V., *"Analytical Heat Diffusion Theory"* Academic Press, New York, 1968.
24. Özisik, M. N., *Heat Conduction,* Wiley Interscience Publication, John Wiley § Sons, New York, Chichester, Brisbane, Toronto, 1980.
25. Ushakov, P. A., "Approximate Thermal Modelling of Cylindrical Fuel Elements," (in Russian), in *Liquid metals,* Atomizdat, Moskva, 1967.
26. Emelianov, I. Ia., et al., *Design of Nuclear Reactors* (in Russian), Energoizdat, Moskva, 1982.
27. Glasstone, S., *Principles of Nuclear Reactor Engineering,* D. Van Nostrand Company, Inc., Toronto, New York, London, 1955.
28. El Wakil, M. M., *Nuclear Power Engineering,* McGraw Hill, New York, Toronto, London, 1962.
29. El Wakil, M. M., *Nuclear Heat Transport,* International Textbook Company, Scranton, Toronto, London, 1971.
30. Cohen, P., *Water Coolant Technology of Power Reactors,* Gordon and Breach Science Publishers, Inc., New York, 1970.
31. Gebhardt, E., Thümmler, F., and Seghezzi, H. D., "Reaktorwerkstoffe. Teil 1. Metallische Werkstoffe 1964. Teil 2. Nichtmetallische Werkstoffe 1967," Teubner, Stuttgart.
32. Golden, G. H., and Tokar, J. V., "Thermophysical properties of sodium," USAEC Report ANL–7323, Lemont, Illinois, August 1967.
33. Hilsenrath, J., et al., *Tables of Thermodynamic and Transport Properties of Air, Argon, Carbon Dioxide, Carbon Monoxide, Hydrogen, Nitrogen, Oxygen, and Steam,* Pergamon Press, London, New York, Paris, Los Angeles, 1960.
34. Meyer, B., et al., *Thermodynamic and Transport Properties of Steam,* The American Society of Mechanical Engineers, New York, 1967.
35. Tipton, C. R. (Ed.), *Reactor Handbook,* Vol. 1: Materials, Interscience Publishers Inc., New York, 1960.
36. Ulybin, S. A., *Coolants of Nuclear Power Reactors* (in Russian), Energia, Moskva, 1966.
37. Vukalovich, M. P. et al., *Tables of Thermophysical Properties of Water and Water Steam* (in Russian), Izdatelstvo Standardov, Moskva, 1969.
38. Chirkin, V. S., *Thermophysical Properties of Materials in Nuclear Technology* (in Russian), Atomizdat, Moskva, 1968.
39. Vargaftik, N. B., *Handbook on Thermophysical Properties of Gases and Liquids* (in Russian), Nauka, Moskva, 1972.
40. Borishanski, B. M. et al., *Liquid metal coolants* 2nd ed., (in Russian), Atomizdat, Moskva, 1967.
41. Seddon, B. T., "Uranium Ceramics Data Manual: Properties of Interest in Reactor Design" British DEG–Report 120 and Addendum, March 7, 1960.
42. McIntosh, A., Heal, T. J. (Eds.), *Materials for Nuclear Engineers,* Temple Press Ltd., London, 1960.
43. Powers, A. E., "Fundamentals of Thermal Conductivity at High Temperatures," Rep. KAPL–2143, Knolls At. P. Lab., 1961.
44. Powers, A. E., "Conductivity in Aggregates," Report KAPL–2145, Knolls Atomic Power Laboratory, 1961.
45. Goldsmith, A., et al., *Handbook of Thermophysical Properties of Solid Materials,* Armour Research Foundation, The MacMillan Co., New York, 1961.
46. Smith, C. O., *Nuclear Reactor Materials,* Addison–Wesley Publishing Company, Reading, Palo Alto, London, 1967.
47. *Fundamental Properties of Materials in Nuclear Fuel Assemblies,* North-Holland Publishing Co., Amsterdam, 1969.
48. Christensen, J. A., et al., "Uranium dioxide thermal conductivity," *Trans. Am. Nucl. Soc.,* 7:2 (1964), p. 391.

49. Thompson, T. J., Beckerley, J. G. (Eds.), *The Technology of Nuclear Reactor Safety, Vol. 2: Reactor Materials and Engineering,* Chapter 16—"Heat Transfer" by H. Fenech, W. M. Rohsenow, MIT Press, Cambridge, Massachusetts, 1973.
50. Wagner, R. J., "HEAT 1–A One-Dimensional Time Dependent or Steady-State Heat Conduction Code for the IBM–650, IDO–16867, April 1963.
51. Hocevar, C. J., Wineinger, T. W., "THETA 1–B, A computer code for nuclear reactor core thermal analysis," IN–1445, February 1971.
52. Hann, C. R., et al., "GAPCON THERMAL–1: A Computer Program for Calculating the Gap Conductance in Oxide Fuel Pins," BNWL–1778 (UC–78), September 1973.
53. Loyalka, S. K., "A Model for Gap Conductance in Nuclear Fuel Rods," *Nuclear Technology,* Vol. 57, No. 2, 1982.
54. Chandola, V. K., and Loyalka, S. K., "Gap Conductance and Temperature Transients in Modified Pulse Design Experiments," *Nuclear Technology,* Vol. 56, No. 1, 1982.
55. Weisman, J. (Ed.), *Elements of Nuclear Reactor Design,* Elsevier Scientific Publishing Company, Amsterdam, Oxford, New York, 1977.
56. Rowe, D. S., "Cross-Flow Mixing Between Parallel Flow Channels During Boiling–Part 1–COBRA–Computer program for coolant boiling in rod arrays," BNWL–371, Pt. 1, UC–80, 1967.
57. Fajeau, M., "Programme FLICA–étude thermodynamique d'un reacteur ou d'une boucle d'essai," Rapport CEA–R–3716, CEN–Saclay, 1969.
58. Rowe, D. S., "COBRA–II: A Digital Computer Program for Thermal Hydraulic Subchannel Analysis of Rod Bundle Nuclear Fuel Elements," BNWL–1229, Pacific Northwest Laboratory, Richland, Washington, February 1970.
59. Vigassy, J., "COBRA–Computer Program for Coolant Boiling in Rod Arrays," KFKI–70–20 RPT, Hungarian Academy of Sciences, Central Research Institute for Physics, Budapest, June 1970.
60. Rowe, D. S., "Interim Report COBRA–III: A Digital Computer Program for Steady and Transient Thermal-Hydraulic Analysis of Rod Bundle Nuclear Fuel Elements," BNWL–B–82, Pacific Northwest Laboratory, Richland, Washington, 1971.
61. Rowe, D. S., "COBRA–IIIC: A Digital Computer Program for Steady State and Transient Thermal-Hydraulic Analysis of Rod Bundle Nuclear Fuel Elements," BNWL–1695, 1973.
62. Gosman, A., et al., "The SABRE–Code for Prediction of Coolant Flows and Temperatures in Pin Bundles Containing Blockages," AEEW–R–905, AEE Winfrith, October 1973.
63. Marr, W. W., "COBRA–3M: A Digital Computer Program for Analyzing Thermal-Hydraulic Behavior in Pin Bundles," ANL–8131, 1975.
64. Wnek, W. J., et al., "Transient Three-Dimensional Thermal-Hydraulic Analysis of Nuclear Reactor Fuel Rod Arrays: General Equations and Numerical Scheme (SCORE–EVET)," ANCR–1207, Aerojet Nuclear Company, Idaho Nat. Lab., Nov. 1975.
65. Stewart, C. W., et al., "COBRA–IV: The Model and the Method," BNWL–2214, Batelle Pacific Northwest Laboratories, Richland, Washington, July 1977.
66. Rogers, J. T., and Todreas, N. E., "Coolant Mixing in Reactore Fuel Rod Bundles–Single-Phase Coolants," *Heat Transfer in Rod Bundles, ASME,* 1968, pp. 1–56.
67. Ingesson, L. and Hedberg, S., "Heat Transfer Between Subchannels in a Rod Bundle," Paper No. FC 7.11, 4-th Int. Heat Transfer Conf., Versailles, France, 1970.
68. Stein, R. P., "Engineering Relationships for Turbulent Forced-Convection Heat Transfer in Duct with Flux Transient," ANL–7754, Argonne National Laboratory, Argonne, Illinois, June 1971.
69. Rogers, J. T., and Rosehart, R. G., "Mixing by Turbulent Inter-Change in Fuel Bundles. Correlations and Interferences," ASME Paper No. 72–HT–53, 1972.
70. Brown, W. D., et al., "Prediction of Cross Flow due to Coolant Channel Blockages," *Nucl. Sci. Engng.,* Vol. 57, No. 2, 1975, pp. 164–168.

71. Weisman, J., and Bowring, R. W., "Methods for Detailed Thermal and Hydraulic Analysis of Water Cooled Reactors," *Nucl. Sci. Engng.,* Vol. 57, No. 4, 1975, pp. 255–276.
72. Rowe, D. S., and Stewart, C. W., "Transient Hydraulic Analysis for Reactor Codes," ERDA Report CONF–750413, 1975, pp. II/17–II/26.
73. Wulf, W., "Development of a Computer Code for Thermal Hydraulic Analysis of Reactors," USAEC Report BNL–19978, Brookhaven National Laboratory, January 1975.
74. Zhukauskas, A. A., "Convective Heat Transfer in Heat Exchangers," (in Russian), Nauka, Moskva, 1982.
75. Lyon, R. N., "Forced Convection Heat Transfer Theory and Experiments with Liquid Metals," AECU–419, June 1949 and ORNL–361, August 1949.
76. Dittus, F. W., and Boelter, L. M. K., "Heat Transfer in Automobile Radiators of the Tabular Type," Univ. of California Publications, 2, 1930, pp. 443–461.
77. Rohsenow, W. M., and Clark J. A., "Heat Transfer and Pressure Drop Data for High Heat Flux Densities to Water and High Sub-critical Pressures," Heat Transfer and Fluid Mech. Inst., Stanford Univ. Press, Stanford, California, 1951.
78. Sieder, E. N., and Tate, C. E., "Heat Transfer and Pressure Drop of Liquids in Tubes," *Ind. Eng. Chem.,* Vol. 28 (1936), pp. 1429–1435.
79. Colburn, A. P., "A Method for Correlating Forced Convection Heat Transfer Data and a Comparison with Fluid Friction," *Trans. Am. Inst. Chem. Engrs.,* Vol. 29 (1933), pp. 174–210.
80. Colburn, A. P., "A Method of Correlating Forced Convection Heat Transfer Data and a Comparison with Fluid Friction," *Trans. A. I. Ch. E.,* Vol. 29, 1933, p. 174.
81. Mikheiev, M. A., "Foundations of Heat Transfer for Turbulent Liquid Flow in Pipes" (in Russian), *Izv. AN SSSR,* OTN, No. 10 (1952), pp. 1448–1454.
82. Mikheiev, M. A., *Fundamentals of Heat Transfer,* 2nd ed. (in Russian), Gosenergoizdat, Moskva, Leningrad, 1956.
83. Petukhov, B. S., *Heat Transfer in Nuclear Power Plants* (in Russian), Atomizdat, Moskva, 1974.
84. Petukhov, B. S., and Roizen, L. I., "Heat Transfer in Annular Pipes," (in Russian), *Inzh.-Fiz. Zhurn.,* Vol. 6, No. 3 (1963), pp. 3–11.
85. Petukhov, B. S., and Roizen, L. I., "Experimental Investigations of Heat Transfer in Turbulent Gas flow in Annular Pipes," (in Russian), *Teplofizika Vysokhikh Temperatur,* Vol. 1, No. 3 (1963), pp. 416–424.
86. Petukhov, B. S., and Roizen L. I., "Generalized Relationships for Heat Transfer in Annular Pipes," (in Russian), *Teplofizika Vysokikh Temperatur,* Vol. 12, No. 3 (1974), pp. 565–569.
87. Perkins, H. C., and Vorsce-Schmidt P., "Turbulent Heat and Momentum Transfer for Gases in a Circular Tube at Wall to Bulk Temperature Ratios to Seven," *Int. J. Heat and Mass Transfer,* Vol. 8, No. 7, 1965, pp. 1011–1031.
88. Taylor, M. F., "Correlation of Local Heat Transfer Coefficients for Single Phase Turbulent Flow of Hydrogen in Tubes with Temperatures Ratios to 23," NASA TND–4332, 1968.
89. Kurganov, V. A., and Petukhov, B. S., "Analysis and Generalization of Experimental Data on Heat Transfer in Pipes for Turbulent Flow of Gases with Variable Physical Properties," (in Russian), *Teplofizika Vysokikh Temperatur,* Vol. 12, No. 2, 1974, pp. 304–315.
90. Kays, W. M., *Convective Heat and Mass Transfer,* (Russian transl.), Energia, Moskva, 1972.
91. Subbotin, V. I., et al., *Hydrodynamics and Heat Transfer in Nuclear Power Plants,* (in Russian), Atomizdat, Moskva, 1975.
92. Rohsenow, W. M., and Hartnett, J. P. (Eds.), "Handbook of Heat Transfer," McGraw Hill Book Company, Inc., New York, 1973.

93. Markoczy, G., "Konvektive Wärmeübertragung in längsangeströmten Stabbündeln bei turbulenter Strömung," *Wärme und Stoffübertrag,* Vol. 5, No. 4 (1972), pp. 204–212.
94. Weisman, J., "Heat Transfer to Water Flowing Parallel to Tube Bundles," *Nucl. Sci. Engng.,* Vol. 6, No. 1 (July 1959), pp. 78–79.
95. Ain, E. M., and Puchkov, P. I., "Heat Transfer and Hydraulic Resistance in Gas-cooled Clusters of Smooth Rods," (in Russian), *Energomashinostroenie,* No. 11 (1964), pp. 21–22.
96. Sutherland, W. A., "Experimental Heat Transfer in Rod Bundles," *ASME,* New York, 1968, pp. 104–138.
97. Martinelli, R. C., "Heat Transfer to Molten Metals," *Trans. ASME,* Vol. 69, No. 8 (1947), pp. 947–959.
98. Lyon, R. N., "Liquid Metal Heat Transfer Coefficients," *Chem. Eng. Progr.,* Vol. 47, No. 2, 1951, p. 75.
99. Dwyer, O. E., "Eddy Transport in liquid-metal heat transfer," USAEC Report BNL–6149, Brookhaven National Laboratory, 1962.
100. Dwyer, O. E., "Eddy Transport in Liquid Metal Heat Transfer," *A. I. Ch. E. J.,* Vol. 9 (March 1963), pp. 261–268.
101. Dwyer, O. E., "Recent Developments in Liquid Metal Heat Transfer," *Atomic Energy Review 4,* No. 1 (1966), pp. 3–92.
102. Dwyer, O. E., "Equations for Bilateral Heat Transfer to a Fluid Flowing in a Concentric Annulus," *Nucl. Sci. Engng.,* 15 (1963), pp. 52–57.
103. Dwyer O. E., and Tu, P. S., "Unilateral Heat Transfer to Liquid Metals Flowing in Annuli, *Nucl. Sci. Engng.,* Vol. 15 (1963), pp. 58–68.
104. Dwyer, O. E., "On the Transfer of Heat to Fluids Flowing through Pipes, Annuli and Parallel plates," *Nucl. Sci. Engng.,* 17 (1963), pp. 336–344.
105. Dwyer, O. E., "Bilateral Heat Transfer in Annuli for Slug and Laminar Flows," *Nucl. Sci. Engng.* Vol. 19 (1964), pp. 48–57.
106. Dwyer, O. E., and Lyon, R. L., "Liquid Metal Heat Transfer," Proc. 3rd U. N. Int. Conf. PUAE, Geneva, Vol. 8, 1965, pp. 182–189.
107. Yu, W. S., and Dwyer, O. E., "Heat Transfer to Liquid Metals Flowing Turbulently in Eccentric Annuli," *Nucl. Sci. Engng.* 24 (1966), pp. 105–117.
108. Yu, W. S., and Dwyer, O. E., "Heat Transfer to Liquid Metals Flowing Turbulently in Eccentric Annuli," *Nucl. Sci. Engng.* 27 (1967), pp. 1–9.
109. Maresca, M. W., and Dwyer, O. E., "Heat Transfer to Mercury Flowing in Line Through a Bundle of Circular Rods," *J. Heat Transfer, Ser. C,* Vol. 86 (1964), p. 180.
110. Dwyer, O. E., "Analytical Study of Heat Transfer to Liquid Metals Flowing In-line Through Closely Packed Rod Bundles," *Nucl. Sci. Engng.* Vol. 25 (1966), pp. 343–358.
111. Dwyer, O. E., and Berry, H. C., "Effects of Cladding Thickness and Thermal Conductivity on Heat Transfer to Liquid Metals Flowing In-line Through Bundles of Closely Spaced Reactor Fuel Rods," *Nucl. Sci. Engng.,* 40 (1970), pp. 317–330.
112. Subbotin, W. I., et al., "Experimental Investigations on Models of Thermal Behavior of Reactor BOR–60 Fuel Elements," (in Russian), Rep. FEI–137, Obninsk, 1969.
113. Ushakov, P. A., et al., "Generalized Method of Determination of Temperature and Heat Removal in Triangular and Quadrangular Lattices of Cylindrical Fuel Elements," (in Russian), Rep. FEI–163, Obninsk, 1969.
114. "Recommendations on Thermal and Hydraulic Calculations of Fuel Assemblies in Fast Reactors," (in Russian), Sovet Ekonomicheskoi Vzaimopomoshchi, Postoianna Komisia po sotrudnichestvu v oblasti ispolzovania atomnoi energii v mirnykh celakh, Nauchno-tekhnicheski sovet po bystrym reaktoram, Moskva, 1981.
115. Zhukov, A. V., et al., "Calculation Method of Determination of Interchannel Mixing Influence on Temperature Field in Coolant Flowing Through Fuel Assemblies with Wire Spacers," (in Russian), Preprint FEI–512, Obninsk, 1974.
116. Zhukov, A. V., et al., "Experimental Investigation of Temperature Fields and Heat Removal in Triangular Lattices of Fuel Rod Simulators with Liquid Metal Cooling," (in Russian), Preprint FEI–800, Obninsk, 1978.

117. Zhukov, A. V., et al., "Temperature Fields in Deformed Fast Reactor Fuel Lattices for Uniform and Nonuniform Thermal Loads," (in Russian), Preprint FEI–909, Obninsk, 1979.
118. Zhukov, A. V., et al., "Influence of Lattice Deformation on Temperature Fields and Heat Removal for Fuel Elements in Characteristic Zones of Fast Reactor Fuel Assembly Model," (in Russian), Preprint FEI–979, Obninsk, 1980.
119. Zhukov, A. V., Kirillova, G. P., "Calculation of Temperature Fields in Inlet Part of Fuel Lattices in the Case of Turbulent Flow of Liquid Metal Coolant," (in Russian), Preprint FEI–715, Obninsk, 1976.
120. Sidelnikov, V. N., and Zhukov, A. V., "Calculations of Temperature Fields in Inlet Part of Fuel Lattices and Analysis of Influence of Variable Power (Flat Coolant Flow)," (in Russian), Preprint FEI–414, Obninsk, 1973.
121. Zhukov, A. V., et al., "Experimental Investigation of Temperature Fields in Inlet Part of Fuel Lattices in the Case of Turbulent Flow of Liquid Metal Coolant (for Simultaneous Hydrodynamic and Thermal Stabilization)," (in Russian), Preprint FEI–781, Obninsk, 1977.
122. Zhukov, A. V., et al., "Temperature Fields in Inlet Part and in Stabilized Heat Transfer Zone of Fuel Lattices for Simultaneous Hydrodynamic and Thermal Stabilization (liquid metal cooling)," (in Russian), Preprint FEI–883, Obninsk, 1978.
123. Vladimirov, M. A., et al., "Method of Calculation of Temperature Decrease of "Hot" Channels due to Heat Dissipation," (in Russian), Preprint FEI–437, Obninsk, 1973.
124. Kurbatov, I. M., "Calculation of Random Temperature Fluctuations in Reactor Core," (in Russian), Preprint FEI–1090, Obninsk, 1980.
125. Johannsen, K., "Analytical Investigation of the Effect of Variable Wall Heat Flux on Turbulent Liquid-Metal Heat Transfer Coefficients in Reactor Coolant Channels," *Atomkernenergie,* Vol. 1, 1968, pp. 21–24.
126. Nijsing, R., and Eifler, W., "Analysis of Liquid Metal Heat Transfer in Assemblies of Closely Spaced Fuel Rods," *Nucl. Engng. and Design,* Vol. 10 (1969), pp. 21–54.
127. Marchese, A. R., "Analytical Study of Heat Transfer to Liquid Metals Flowing Parallel Through Tightly Packed Fuel Rod Bundles," Proc. of the Symp. sponsored by the Nucleonic Heat Transfer Committee (K–13) of the ASME Transfer Division, 1970, pp. 15–29.
128. Oberjohn, W. J., "Turbulent Flow Thermal-Hydraulic Characteristics of Hexagonal Pitch Fuel Assemblies," Proc. of the Symp. sponsored by the Nucleonic Heat Transfer Committee (K–13) of the ASME Heat Transfer Division, 1970, pp. 30–40.
129. Ushakov, P. A., et al., "Heat Transfer to Liquid Metals in Regular Fuel Element Lattices," *Teplofizika Vysokikh Temperatur,* (in Russian), Vol. 15, No. 5 (1977), pp. 1027–1033.
130. Ushakov, P. A., et al., "Azimuthally Nonuniform Fuel Element Temperature Distributions for Regular Lattices and Turbulent Flow of Liquid Metals," (in Russian), *Teplofizika Vysokikh Temperatur,* Vol. 15, No. 1 (1977), pp. 76–83.
131. Minashik V. E., et al., "Temperature Calculation in Reactor core for Arbitrary Axial Thermal Power Distribution," (in Russian), *Atomnaia Energia,* Vol. 22, No. 5 (1967), pp. 362–366.
132. Proceedings of the ANS/ASME/NRC International Topical Meeting on Nuclear Reactor Thermal-Hydraulics, NUREG/CP–0014, Washington, D. C., 1980.
133. Graetz, L., "Über Wärmeleitungsfähigkeit von Flussigkeiten," *Ann. Phys. u. Chem.,* Bd. 25, 1885, pp. 773–778.
134. Petukhov, B. S., *Heat Transfer and Friction Losses for Laminar Liquid Flow in Pipes,* (in Russian), Energia, Moskva, 1967.
135. Grigull, U., and Tratz, H., "Thermischer Einlauf in ausgebildeter laminarer Röhrströmung," *Int. J. Heat and Mass Transfer,* Vol. 8, No. 5 (1965), pp. 669–678.
136. Sellers, J. R., Tribus M., and Klein J. S., "Heat Transfer to Laminar Flow in a Round Tube or Flat Conduit—the Graetz Problem Extended," *Trans. ASME,* Vol. 78, No. 2, (1956), pp. 441–448.

137. Clark, S. W., and Kays, W. M., "Laminar-Flow Forced Convection in Rectangular Tubes," *Trans. ASME,* Vol. 75, No. 7 (1953), pp. 859–866.
138. Polanin, L. N., Ibragimov, M. Kh., and Sabelev, G. I., "Heat Transfer in Nuclear Reactors," (in Russian), Energoizdat, Moskva, 1982.
139. Mikheev, M. A., and Mikheeva, I. M., "Foundations of Heat Transfer," (in Russian), Energia, Moskva, 1977.
140. Levêque, M. A., "Transmission de chaleur par convection," *Ann. des Mines,* Vol. 13 (1928), pp. 201–239.
141. Zhukauskas, A. A., and Zhiugzhda, I. I., "Heat Transfer in Laminar Liquid Flow," (in Russian), Mintis, Vilnius, 1969, pp. 261.
142. Liutikas, N. S., and Zhukauskas, A. A., "Investigation of Influence of Variable Viscosity on Laminar Heat Transfer in Flat Channel," (in Russian), *Tr. AN Lit. SSR, Ser. B,* No. 2 (49) (1967), pp. 97–114.
143. Dwyer, O. E., and Berry, H. C., "Laminar-Flow Heat Transfer for In-Line Flow Through Unbaffled Rod Bundles," *Nucl. Sci. Engng.,* Vol. 42 (1970), pp. 81–88.
144. Chia-Jung, Hsu, "Theoretical Solutions for Low Peclet Number Thermal Entry Region Heat Transfer in Laminar Flow Through Concentric Annuli," *Int. J. Heat and Mass Transfer,* Vol. 13, 1970, pp. 1907–1924.
145. Dwyer, O. E., and Berry, H. C., "Effects of Cladding Thickness and Thermal Conductivity on Heat Transfer for Laminar In-line Flow Through Rod Bundles," *Nucl. Sci. Engng.,* 42 (1970), pp. 69–80.
146. Subbotin, V. I., et al., "Heat Transfer and Hydrodynamics Calculations for Slug and Laminar Coolant Flow Through Regular Fuel Element Bundles," *Atomnaia Energia,* Vol. 33, No. 4 (1972), pp. 840–841, (in Russian).
147. Morozova, S. I., and Novofiliov, E. V., "Hydraulic Friction Coefficient in Transient Domain for Liquid Flow in Triangular Cell of Fuel Element Bundle," in *Heat Transfer and Hydrodynamics of One-Phase Flow in Fuel Element Bundles* (in Russian), Nauka, Leningrad, 1977, pp. 129–135.
148. Targ, S. M., *Main Problems of Laminar Flow Theory* (in Russian), Gostekhizdat, Moskva, 1948
149. Minashin, V. E., et al., *Heat Transfer in Liquid Metal Cooled Reactors* (in Russian), Atomizdat, Moskva, 1971.
150. Blasius, H., "Das Ahnlichkeitsgesetz bei Reibungsvorgängen in Flussigkeiten," *Forschg. Arb. Ing.-Wes.,* 131 (1913), pp. 1–34.
151. Filonenko, G. K., "Pipe Hydraulic Resistance" (in Russian), *Teploenergetika,* Vol. 4 (1954), pp. 40–44.
152. Prandtl, L., "Neuere Ergebnisse der Turbulenzforschung," *Zeitschrift VDI,* Vol. 77 (1933), pp. 105–114.
153. Colebrook, C. F., "Turbulent Flow in Pipes with Particular Reference to the Transition Region Between the Smooth and Rough Pipe Laws," *Proc. Inst. Civ. Eng.,* Vol. 11 (1939), p. 133.
154. Moody, L. F., "Friction Factors for Pipe Flows," *Trans. ASME,* Vol. 66 (1944), pp. 671–684.
155. Idelchik, I. E., *Handbook on Hydraulic Resistances* (in Russian), Mashinostrenie, 1975.
156. Taylor, M. F., "Correlation of Friction Coefficients for Laminar and Turbulent Flow with Ratios of Surface to Bulk Temperature From 0.35 to 7.35," NASA-TR-R-267, September 1967.
157. Tong, L. S., "Pressure Drop Performance of a Rod Bundle," *Heat Transfer in Rod Bundles, ASME,* 1968, pp. 57–69.
158. Waggener, J. P., "Friction Factors for Pressure-Drop Calculations," *Nucleonics,* Vol. 19, No. 11 (November 1961), pp. 145–147.
159. Kays, W. M., and London A. L., "Compact Heat Exchangers," 2nd ed., McGraw-Hill, New York, 1964.

160. Nowendstern, E. H., "Turbulent Flow Pressure Drop Model for Fuel Rod Assemblies Utilizing a Helical Wire-wrap Spacer System," *Nuclear Eng. and Design,* Vol. 22, 1972, pp. 19–27.
161. Madejski, J., "Boiling Heat Transfer and Two Phase Flow," (in Polish), Part 2, Nuclear Energy Information Center, Warsaw, 1973.
162. Tippets, F. E., "Critical Heat Fluxes and Flow Patterns in High Pressure Boiling Water Flows," ASME Paper 62-WA-162, 1962.
163. Bennet, A. W., et al., "Flow Visualisation Studies of Boiling at High Pressure," AERE R-4874, Harwell, England, 1965.
164. Wallis, G. B., "One-Dimensional Two-Phase Flow," McGraw-Hill Book Company, New York, 1969.
165. Butterworth, D., Hewitt, G. F. (Eds.), *Two-Phase Flow and Heat Transfer,* Oxford University Press, Oxford, 1977.
166. Bergles A. E., et al., *Two-Phase Flow and Heat Transfer in the Power and Process Industries,* Hemisphere Publishing Corporation, Washington, D.C., 1981.
167. Aden, V. G., et al., "The Study of Heat Exchange Enhancement in Models of Fuel Elements Bundles at the Coolant Boiling," paper presented at the 6th International Heat Transfer Conference, Toronto (August 7–11, 1978), Vol. 5, pp. 41–45.
168. Smolin, V. N., Poliakov, V. K., "Coolant Boiling Crisis in Rod Assemblies," paper presented at the 6th International Heat Transfer Conference, Toronto (August 7–11, 1978), Vol. 5, pp. 47–52.
169. Osmachkin V. S., "Problems of transient critical heat flux in rod bundles," paper presented at the 6th International Heat Transfer Conference, Toronto (August 7–11, 1978), Vol. 5, pp. 59–64.
170. Barulin Iu. D., et al., "Experimental Investigations of Heat Transfer and Hydrodynamics on Models of Fuel Assemblies of VVER Reactor," (in Russian), *Trudy VTI,* No. 11 (1977), pp. 214–220.
171. Tong, L. S., "Heat Transfer in Reactor Safety," paper presented at the 6th International Heat Transfer Conference, Toronto (August 7–11, 1978), Vol. 6, pp. 285–309.
172. Hsu, Y. Y., "Thermal Hydraulics Related to Reactor Safety," Nuclear Reactor Thermal-Hydraulic Topical Meeting, Saratoga, New York, October 6–8, 1980.
173. Vigil, J. C., and Pryor R. J., "Development and Assessment of the Transient Reactor Analysis Code (TRAC)," *Nuclear Safety,* Vol. 21, No. 2 (1980), pp. 171–183.
174. Burwell, H., "DRUFAN—An Advanced Best-Estimate Code for LWR Blowdown Analysis," Newsletter of the NEA, Bank Data Bulletin, No. 28, (Sept. 1982), pp. 103–142.
175. Pryor, R. J., "Computational Methods in Thermal Reactor Safety," NUREG/CR-0851, LA-7856-MS, Los Alamos Scientific Laboratory, Los Alamos, 1979.
176. Lyczkowski, R. W., "Numerical Techniques for the Computation of Transient Unequal Phase Velocity, Unequal Phase Temperature, Two-Phase Flow and Heat Transfer," paper presented at the Two-Phase Flow and Heat Transfer Symposium-Workshop, Florida, 1976.
177. Ranson, V. H., et al., "RELAP 5/MOD1 Code manual. Volume 1: System Models and Numerical Methods (draft)," NUREG/CR-1826, Idaho National Engineering Laboratory, Idaho Falls, Nov. 1980.
178. Agee, L. J., "RETRAN overview," *Nuclear Technology,* Vol. 61, No. 2 (May 1983), pp. 143–152.
179. Spigt, C. L., "The Hydraulic Characteristics of a Naturally Circulating Boiling Water System," *Atoomenergie Haar Toepass.,* Vol. 9, No. 1 (Jan. 1967), pp. 1–18.
180. Foglia, J. J., et al., "Boiling-Water Void Distribution and Slip Ratio in Heated Channels," USAEC Report BMI-1517, 1961.
181. Christensen, H., "Power-to-void Transfer Functions," USAEC Report ANL 6385, 1961.

182. Martinelli, R. C., and Nelson, D. B., "Prediction of Pressure Drop During Forced Circulation Boiling of Water," *Trans. ASME,* Vol. 70, (August 1948), pp. 695–702.
183. Levy, S., "Steam Slip—Theoretical Prediction from Momentum Model," *Trans. ASME, J. Heat Transfer,* Vol. 82 (1960), pp. 113–124.
184. Zivi, S. M., "Estimation of Steady-State Steam-Void-Fraction by Means of the Principle of Minimum Entropy Production," *Trans. ASME, J. Heat Transfer,* Vol. 86 (May 1964), p. 247.
185. Polomik, E. E., "Phase Velocities in Boiling Flow Systems by Total Energy and by Diffusion," ASME Paper No. 65-HT-34, 1965.
186. Kholodovski, G. E., "New Method of Experimental Data Generalization for Steam-Water Flow in Vertical Tubes," (in Russian), *Teploenergetika,* Vol. 4, No. 7 (1957), p. 68.
187. Bankoff, S. G., "A Variable Density Single-Fluid Model for Two-Phase Flow with Particular Reference to Steam-Water Flow," *Trans. ASME, J. Heat Transfer,* Vol. 82 (1960), pp. 265–272.
188. Hughmark, G. A., "Holdup in Gas-Liquid Flow," *Chem. Eng. Progr.,* Vol. 58 (1962), p. 62.
189. Thom, J. R. S., et al., "Boiling in Subcooled Water During Flow Up in Heated Tubes or Annuli," *Proc. Inst. Mech. Eng.,* 3C (1968), pp. 226–246.
190. Tong, L. S., and Weisman, J., "Thermal Analysis of Pressurized Water Reactors," *American Nuclear Society,* Hinsdale, Ill., 1970.
191. Osmachkin, V. S., and Borisov V. D., "Pressure Drop and Heat Transfer for Flow of Boiling Water in Vertical Rod Bundles," 4th Int. Heat Transfer Conf., Paper B 4.9, Paris–Versailles, 1970.
192. Bergles A. E., and Rohsenow W. M., "The Determination of Forced Convection Surface Boiling Heat Transfer," *Trans. ASME, J. Heat Transfer,* Vol. 86 (1964), p. 365.
193. Saha P., and Zuber N., "Point of Net Vapor Generation and Vapor Void Fraction in Subcooled Boiling," 5th Int. Heat Transfer Conf., Paper B 4.7, Tokyo, 1974.
194. Bowring, R. W., "Physical Model Based on Bubble Detachment and Calculation of Steam Voidage in the Subcooled Region of a Heated Channel," OECD Halden Reactor Project Report HPR–10, Halden, Norway, 1962.
195. Levy, S., "Forced Convection Subcooled Boiling Prediction of Vapor Volumetric Fraction," *Int. J. Heat Mass Transfer,* Vol. 10, 1967, pp. 951–965.
196. Thom, J. R. S., "Prediction of Pressure Drop During Forced Convection Circulation Boiling of Water," *Int. J. Heat Mass Transfer,* Vol. 7 (1964), pp. 709–724.
197. Baroczy, C. J., "A Systematic Correlation for Two-Phase Pressure Drop," NAA–SR–-Memo–11858, North American Aviation, March 1966.
198. Armand, A. A., "The Resistance During the Movement of a Two-Phase System in Horizontal Pipes," trans. by V. Beak, AERE Trans. 828. Izvestia Vsesoiuznogo Teplotekhnicheskogo Instituta (1), 1946, pp. 16–23.
199. Katama K. R., et al., "RELAP4/Mod5–A Computer Program for Transient-Hydraulic Analysis of Nuclear Reactors and Related Systems," Aerojet Nuclear Company, Report ANCR–NUREG–1335, September 1976.
200. Tarasova, N. V., and Orlov, V. M., Investigation of Hydraulic Resistance During Nucleate Boiling of Water in Pipe," (in Russian), *Teploenergetika,* Vol. 9, No. 6 (1962), pp. 48–51.
201. Jens W. H., and Lottes P. A., "Analysis of Heat Transfer Burnout, Pressure Drop and Density Data for High Pressure Water," Argonne National Laboratory, Ill., Report ANL–4627, 1951.
202. Schrock, V. E., and Grossman, L. M., "Forced Convection Boiling Studies–Forced Convection Vaporization Project," USAEC Rep. TID–14632, 1959.
203. Schrock, V. E., and Grossman, L. M., "Forced Convection Boiling in Tubes," *Nucl. Sci. Engng.,* Vol. 12 (1962), pp. 474–480.

204. Chen, J. C., "Correlation for Boiling Heat Transfer to Saturated Liquids in Convective Flow," *Ind. Eng. Chem. Process Design and Development,* Vol. 5 (1966), p. 322.
205. Hewitt, G. F., "Critical Heat Flux in Flow Boiling," 6th International Heat Transfer Conference, Vol. 6, Toronto, Canada, 1978, pp. 143–171.
206. Tong, L. S., "Boiling Crisis and Critical Heat Flux," U.S. Atomic Energy Commission, Office of Information Services, Oak Ridge, Tennessee, 1972.
207. Remizov, O. V., et al., "Some Features of Heat Transfer Burnout in Tubes with Non-uniform Axial Heat Flux Distribution," 6th International Heat Transfer Conference, Vol. 5, Toronto, Canada, 1978, pp. 53–58.
208. Bezorukov J. A., "Experimental Investigation and Statistical Analysis of Data Related to Critical Heat Flux in VVER–1000 Rod Bundles," (in Russian), *Teploenergetika,* No. 2 (1976), pp. 80–82.
209. Gaspari G. P., et al., "A Rod Centered Sub-channel Analysis With Turbulent (Enthalpy) Mixing for Critical Heat Flux Prediction in Rod Clusters Cooled by Boiling Water," 5th International Heat Transfer Conference, Vol. 44, Tokyo, Japan, 1974, pp. 295–299, CONF–740925 (1975).
210. Miropolski, Z. L., et al., "Statistical Correctness of the Critical Heat Flux Investigation," (in Russian), *Teploenergetika,* No. 4 (1969), pp. 49–52.
211. Smolin, V. N., and Poliakov, V. K., "Critical Heat Flux in Flow Through Rod Bundle," (in Russian), *Teploenergetika,* No. 7 (1967), pp. 54–58.
212. Gellerstedt, J. S., et al., "Correlation of Critical Heat Flux in a Bundle Cooled by Pressurized Water," The Winter Annual Meeting of the ASME, Los Angeles, 1969, p. 63.
213. Tong, L. S., "Prediction of Departure from Nucleate Boiling for an Axially Non-uniform Heat Distribution," *J. Nucl. Energy,* Vol. 6 (1967), p. 21.
214. Biasi, L., et al., "Studies on Burnout, Part 3," *Energ. Nucl.,* Vol. 14 (No. 9), 1967, p. 530.
215. Bowring, R. W., "Simple but Accurate Round Tube, Uniform Heat Flux Dryout Correlation over the Pressure Range 0.7 to 17 MPa," AEEW–R–789, U.K. Atomic Energy Authority, 1972.
216. Barnett, P. G., "A Correlation of Burnout Data for Uniformly Heated Annuli and Its Use for Predicting Burnout in Uniformly Heated Rod Bundles," AEEW–R–463, U.K. Atomic Energy Authority, 1966.
217. Macbeth, R. V., "Application of a Local Condition Hypothesis to World Data of Uniformly Heated Round Tubes and Rectangular Channels," AEEW–R–267, U.K. Atomic Energy Authority, 1963.
218. Thompson, B., and Macbeth, R. V., "Boiling Water Heat Transfer Burnout in Uniformly Heated Round Tubes: A Compilation of World Data with Accurate Correlations," AEEW–R–356, U.K. Atomic Energy Authority, 1964.
219. Griffith, P., and Smith, R. A., "A Simple Model for Estimating Time to CHF in a PWR LOCA," ASME Paper 76–HT–9, American Society of Mechanical Engineers, 1976.
220. Leung, J. C. M., "Critical Heat Flux Under Transient Conditions: a Literature Survey," Argonne National Laboratory, Argonne, Illinois, NUREG/CR–0056, ANL–78–39, 1978.
221. Leung, J. C. M., "Occurrence of Critical Heat Flux During Blowdown with Flow Reversal," Argonne National Laboratory, Argonne, Illinois, ANL–77–4, 1977.
222. Ybarrondo, L. J., et al., "The 'Calculated' Loss-of-Coolant Accident: a Review," AIChE Monograph Series, Vol. 68, No. 7, 1972.
223. Groeneveld, D. C., "Post-Dryout Heat Transfer: Physical Mechanisms and a Survey of Prediction Methods," *Nucl. Engng. Design,* Vol. 32 (1975), pp. 283–294.
224. Mayinger, F., and Longner, H., "Post-Dryout Heat Transfer," paper presented at the 6-th International Heat Transfer Conference, Toronto (August 7–11, 1978), Vol. 6, pp. 181–198.

225. Yadigaroglu, G., "The Reflooding Phase of the LOCA in PWRs. Part I: Core Heat Transfer and Fluid Flow," *Nuclear Safety,* Vol. 19, No 1 (1978), pp. 20–36.
226. Elias, E., and Yadigaroglu, G., "The Reflooding Phase of the LOCA in PWRs. Part II: Rewetting and Liquid Entrainment," *Nuclear Safety,* Vol. 19, No. 2 (1978), pp. 160–175.
227. Solbrig, C. W., et al., "Heat Transfer and Friction Correlations Required to Describe Steam-water Behavior in Nuclear Safety Studies," 15th National Heat Transfer Conference, San Francisco, California, August, 1975, CONF–750804–3.
228. Saha, P., "Review of Two-Phase Steam-Water Critical Flow Models with Emphasis on Thermal Nonequilibrium," NUREG/CR–0417, BNL–NUREG–50907, 1978.
229. Banerjee, S., and Hancox, W. T., "Transient Thermohydraulic Analysis for Nuclear Reactors," 6th International Heat Transfer Conference, Toronto, Canada, Vol. 6, 1978, pp. 311–337.

CHAPTER 37

THERMAL DRYING EQUIPMENT

Paul N. Cheremisinoff

Department of Civil and Environmental Engineering
New Jersey Institute of Technology
Newark, New Jersey

CONTENTS

INTRODUCTION

Thermal or heat drying is the process of evaporating water or other liquids from solids by thermal means. Water vapor is carried away by air and the equipment design establishes the conditions of the drying process (temperature, humidity, residence time, gas velocity, and flow direction of the gas stream across the drying surface).

There are basically three stages in heat or thermal drying:

1. *Initial drying* is where temperature and drying rates are increased to steady-state conditions. This stage is usually short with little actual drying taking place and accounts for the energy input to go past the latent heat of the vaporization state.
2. *Steady-state drying* is usually the longest stage; the solid's surface is saturated with water and surface moisture is replaced by water from the solid interior as fast as it evaporates. The drying mechanism is similar to moisture evaporating from a pool of water, the solid not significantly affecting the drying rate. During this stage the surface temperature of the solid-gas interface is kept at the wet-bulb gas temperature. So long as unbound surface moisture is present, the solid's temperature is at the wet-bulb temperature of the gas. Even though drying may be done with relatively hot gases, the solids themselves do not reach the elevated temperatures.
3. *Final drying* stage occurs when sufficient water has been evaporated so that the solids' surface is only partially saturated. Surface moisture is hence evaporated more quickly than it can be replaced by water from the solids' interior. Overall drying rates are lower in this stage and the solid/gas interface increases because the latent heat cannot be transferred from the solid to the gas phase as rapidly as sensible heat is received from the heating medium.

Solids moisture content is expressed in percent moisture, percent solids, or weight of water per weight of dry solids. Maximum drying or minimum moisture contents attainable with heat drying is dependent upon design and operation of the dryer and moisture content of the solids feed and its chemical composition. Chemical bonding can increase the amount of water retained in dried products.

Any drying process is concerned with materials transfer as well as heat transfer. Heat transfer, such as latent heat of vaporization and sensible heat are transmitted to the water through the pores of the solid. Mass transfer, the transfer of water from the inner portions of the paste or crystal mass to the surface, must be taken into account, indesign.

MASS TRANSFER AND HUMIDITY

The moisture content of the gas phase at a given temperature is humidity and must be considered in determining drying rates. Absolute humidity is a measure of the weight of water per unit weight of dry gas. In the thermal-drying of solids, water is transferred to the gas phase. The driving force for transfer is the difference between absolute humidity at the wetted solids/gas interface and the absolute humidity in the gas phase. The drying rate (transfer rate) can be expressed as:

$$W = K_y A(Y_s - Y_a)$$

where W = drying rate (pounds water per hr or kg/hr)
K_y = mass transfer coefficient of the gas phase (pounds water per hour per square foot per unit of humidity difference or kg/hr/m^2/unit of humidity difference)
A = wetted surface area exposed to drying mechanism (square feet or m^2)
Y_s = humidity at the solids/gas interface temperature (pounds water per pounds dry gas or kg/kg)
Y_a = humidity of the gas phase (pounds water per pounds dry gas or kg/kg)

HEAT TRANSFER AND TEMPERATURE

In thermal drying, the temperature difference between the heating medium and the solids/gas interface provides the driving force for heat transfer. Dryers are usually classified on the basis of the predominant heat transfer method to the wet solids being dried, and include convection (direct drying), conduction (indirect drying), and radiation (infrared or radiant heat drying).

Convection (direct drying) heat transfer is accomplished by direct contact between wet solids and hot gases. The sensible heat of the inlet gas provides the latent heat required for evaporating the water. Vaporized liquid is carried off by the hot gases. Direct dryers area common type used, and flash dryers, direct rotary dryers, spray dryers, and fluid bed dryers employ this method. Convective heat transfer can be expressed by:

$$q_{conv.} = h_c A(t_g - t_s)$$

where $q_{conv.}$ = convective heat transfer (Btu per hour or kJ/hr)
h_c = convective heat transfer coefficient (Btu per hour per square foot per °F, or kJ/hr/m^2/°C)
A = wetted surface area exposed to gas (square feet or m^2)
t_g = gas temperature (°F or °C)
t_s = solids/gas interface temperature (°F or °C)

Conduction (indirect drying), heat transfer is accomplished by contact of the wet solids with hot surfaces, as in the case of a retaining wall or partition between the wet solid and the heating medium. Vaporized liquid is removed independently of the heating medium; film dryers being an example employing this principle. Conductive heat transfer can be described by:

$$q_{cond.} = h_{cond.} A(t_m - t_s)$$

where $q_{cond.}$ = conductive heat transfer (Btu per hour or kJ/hr)
$h_{cond.}$ = conductive heat transfer coefficient (Btu per hour per °F or kJ/hr/°C)
A = area of heat transfer surface (square feet or m^2)
t_m = drying medium temperature (°F or °C)
t_s = solids temperature at drying surface (°F or °C)

The conductive heat transfer coefficient ($h_{cond.}$) is a term that includes heat transfer surface; solids' side and heating medium side film's effects. It is often difficult to determine appropriate values for heat and mass transfer coefficients for use in these equations and results obtained may be different from those in practice. The most usable design information is obtained by testing with actual process feeds simulating prototype operations. Additionally, methods for computing coefficients are described in the literature [1–4] and elsewhere in this book.

Radiation (infrared or radiant heat drying), heat transfer is effected by radiant energy supplied by electric resistance elements; by gas-heated incandescent refractories that also provide convective heating, or by infrared lamps. Radiation heat transfer can be expressed by:

$$q_{rad} = \varepsilon_S A\sigma(t_r^4 - t_s^4)$$

where q_{rad} = radiation heat transfer (Btu per hour or kJ/hr)
ε_s = emissivity of the drying surface (dimensionless)
A = solids' surface area exposed to radiant source, (square feet or m^2)
σ = Stefan–Boltzman constant (1.73×10^{-9} Btu per hour per square foot per °F or 4.88×10^{-8} kcal/m^2/hr/°k)
t_r = absolute temperature of the radiant source (°F)
t_s = absolute temperature of the solids' drying surface (°R)

ENERGY REQUIREMENTS

The principal cost in drying operations is heat. In all dryers with direct contact between the heating medium and material there is considerable heat loss, namely, all the heat that is required to raise the gases from ambient conditions to the temperature at which they are discharged from the dryer. Efforts to reduce this loss have led to recirculation and reheating of exhaust gas. When waste gases such as flue gases can be used, this loss is smaller, and compared to steam, the cost of raising the latter is eliminated, so that direct-heat dryers are economical.

Thermal drying (evaporation of water from solids) requires considerable energy. The amount of fuel or energy required for drying is dependent upon the amount of water evaporated. It is required that dewatering should preceed heat-drying to minimize energy requirements. Figure 1 shows a typical relationship between solids content and energy required to produce a product containing 10% moisture. Energy estimates in general, however, for heat-drying should be considered estimates, since values can vary considerably depending on dryer type, energy recovery, flow sheet of the process, and solids characteristics.

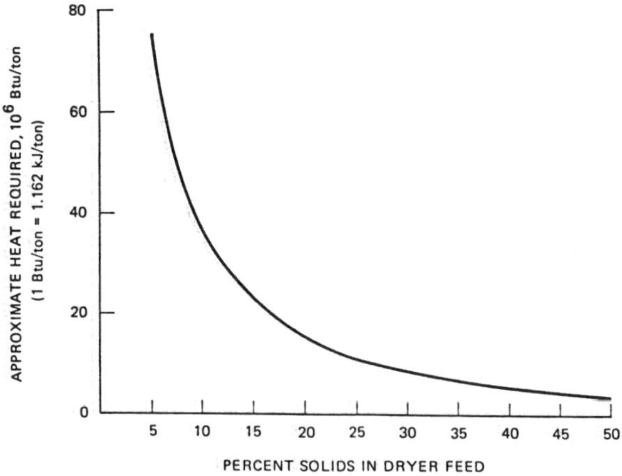

Figure 1. Estimate of energy needed to dry a wet product as a function of drying feed solids content. Assumptions are: 10% moisture in dried product; 2,000 Btu are required to evaporate one pound of water.

Table 1
Classification of Dryers Based on Materials Handling

Material	Type Dryer
Solids drying — Individual particles do not move in relation to each other.	Compartment dryers Tray dryers Tunnel dryers Belt dryers Vacuum shelf dryers
Solids drying — Individual particles move in relation to one another.	Drying towers Pan dryers Conveyor dryers Turbo dryers Rotary dryers
Drying solutions/slurries	Spray dryers Drum dryers Film dryers/wetted wall

Heat required to evaporate water from wet solids consists of:

1. Heat to raise wet solids to the temperature of the product as it leaves the dryer.
2. Heat to raise water temperature to the point of evaporation and then effect vaporization (latent heat).
3. Heat to raise the temperature of the exhaust gas, including water vapor, to the exhaust temperature.
4. Heat to offset heat losses.

CLASSIFICATION

A definition or classification of *drying* is almost impossible without overlapping the definitions of distillation or evaporation. Classification of various types of drying equipment which is given can be based on the method of heat transfer. An alternate classification may be based upon how the solids or material is handled and is shown in Table 1.

Dryers in which solids/individual particles do not move in relation to each other can be selected for those products where physical shape must be preserved, or carefully controlled conditions maintained for definite time intervals. Problems associated with this class of dryer are air conditioning, temperature and humidity control, and air circulation.

Dryers in which solid materials and their individual particles move in relation to each other include pan, conveyor, turbo, and rotary dryers of various types with revolving shells, agitators, or tube bundles. The problem in these dryers is largely one of agitation and material handling.

Drying liquids, slurries, or material of sufficiently low viscosity or fluid enough to be pumped can be processed in drum dryers, spray dryers, wetted wall evaporators or dryers.

OPERATIONAL AND SPECIAL CONSIDERATIONS

Size and number of dryers depends on process considerations and type of operation. If the operation is continuous, additional capacity may be required while maintenance is being performed. In situations where noncontinuous performance is contemplated evaporative capacity to handle all the solids must be provided, including when dryers are not on line.

Gas streams exhausted from dryers may contain odors and visible emissions. These can be significant in high-gas velocity processes where the product is subject to abrasion and dusting occurs. Effective control for such problems may include: afterburners, cyclones, wet scrubbers, baghouses, or electrostatic precipitators. Drying systems with heavy dusting have had problems with fires. Systems combining combustible particulate, elevated temperatures, oxygen, and high-gas velocities are susceptible to combustion.

Due to the large energy requirements for thermal-drying attention to the heat source for the drying medium must be given. Natural gas and fuel oil are most frequently used. Energy recovery within the system itself, such as use of heat exchangers on exhaust gas streams provide ways to reduce energy usage. Recovery of heat from a power source within the plant is another method.

In general, process operating temperatures will establish the heat transfer mode. Water and steam may generally be used below 350° F, petroleum derived oils to 550° F, synthetic organic fluids to 750° F, mixtures of inorganic salts to 1000° F, and certain liquified metals from 1000° to 2000° F. The choice of heat transfer mediums below 350° F and above 750° F depends on factors such as system design costs, thermal stability, heat transfer efficiency, corrosiveness, toxicity, flammability, etc.

Air flow is an important consideration in designing direct dryers. Air flows may be cocurrent, countercurrent, or crossflow. In direct drying, cocurrent flow has the advantage of higher thermal efficiency because of rapid cooling of the heating medium near the feed end with reduced heat losses through the dryer structure and the dried product is subject to lower gas temperatures near the discharge end. A further advantage of such flow is minimizing distillation of odors and increasing thermal efficiency by reducing heat losses with the dried product.

Air flow rates are a function of dryer design. Turbulent conditions are required to maintain and insure intimate contact between warm air and wet solids. A limiting factor in air flow rates may be dusting.

Thermal-drying in the process industries is usually preceeded by mechanical dewatering and may be followed by air pollution control systems as already mentioned. Mechanical dewatering is an important pretreatment, since it reduces volume of water to be removed in the dryer.

DRUM DRYERS

In drum drying, a liquid containing dissolved solids, or slurry containing suspended solids is spread on the surface of an internally heated horizontal drum. In its simplest arrangement the drum dips in a pan underneath where it is coated with the wet materials to be dried. As the drum revolves, liquid is vaporized, so that after 80%–90% of a revolution, the dried deposit can be scraped off by an adjustable/flexible blade. Rotation of the drum is such that all liquid portions are evaporated as the solid residue approaches the knife. Solids collect on an apron in front of the knife and fall into a container or screw conveyor.

Features of typical drum dryers are shown in Figures 2, 3, and 4. Drum dryers may be heated by steam entering through the trunnion and condensate discharged by means of a scoop or syphon through the second trunnion. Drum rotation will vary with application but generally is between 4–10 rpm with limits of between 1–20 rpm. Drums themselves for a single atmospheric drum dryer are fabricated from a single casting, the face turned on a lathe, ground, and polished. Drum dimensions for small units can be 24 inches in diameter by 24 to 36 inches long; 5–12 foot diameters for large size units.

Heat from condensing steam passes through condensate film, through the metal of the drum and then through the coating on the drum. The rate of evaporation is determined by the concentration at which the material is fed to the dryer, and equipment is designed and specified on the pounds of water evaporation per hour per square foot of drum surface. The capacity of a drum dryer depends upon its dimensions, speed of rotation, initial concentration of liquor or slurry, and residual allowable moisture in the product. Additional efficiency factors include heat resistance of the liquor film, steam pressure, and adhesion of drying coat to drum surface (the better the adhesion, the heavier the coating, the greater the amount dried in a given time for a given dryer). Capacity is proportional to surface

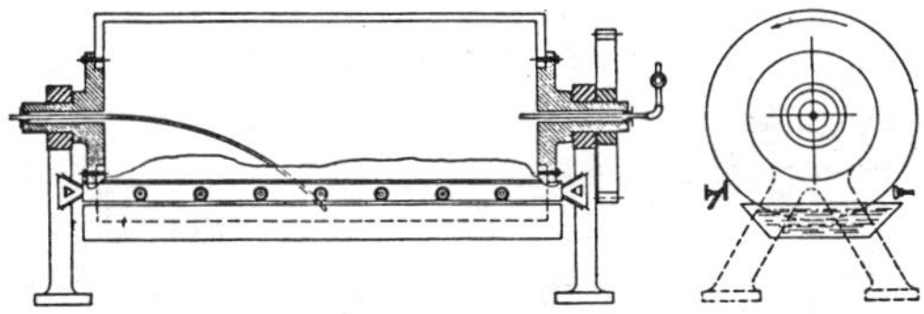

Figure 2. An atmospheric single-drum dryer with dip feed. Beside the knife, a spreader is shown that regularizes the coating.

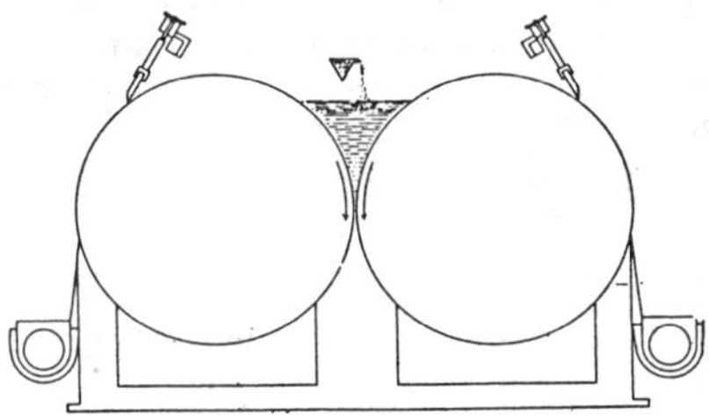

Figure 3. An atmospheric double-drum dryer, with center feed. The drums rotate at the rate of 4 rpm; the drum tops rotate toward each other. The dry material collects in the conveyors that run along the face of the drum (driving gear and vapor hood not shown). End boards form the two end walls that confine the liquid of slurry.

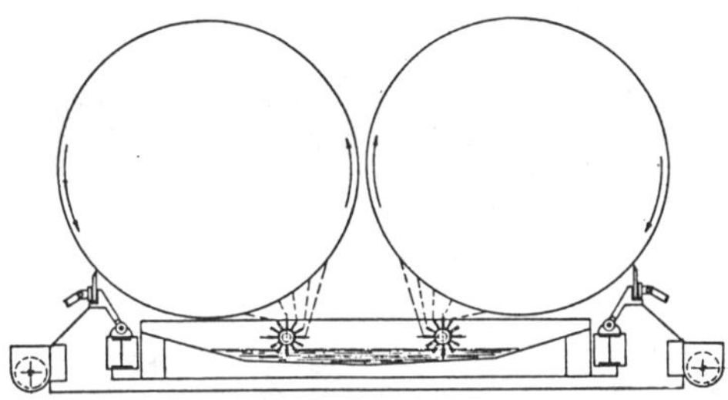

Figure 4. An atmospheric twin-drum dryer with splash feed. The rotation of the drums is opposite to that in the double-drum dryer.

area even though various products offer a wide range of properties particularly in adherence and liquid film resistance to heat.

Drum dryer heat economy tends to be very favorable because the heat of condensing steam is transferred directly to the material. In addition to theoretical amounts of heat required to bring the liquid to the boiling point, evaporative moisture must compensate for endothermic changes such as crystallization, as well as provide sufficient heat to compensate for radiation losses. Designs and equipment sizing should therefore be made on the basis of pounds of dry material produced and not on the basis of pounds of water evaporated.

Steam pressure is one of the factors that affect capacity and already mentioned. Higher steam pressure accelerates evaporation, however, there is a limit to the tensile strength of castings. In such cases it may be possible to choose a twin drum (double drum) dryer of smaller surface and still achieve the same production as on a single dryer of larger surface. For example a single drum 5 feet in diameter and 12 wide has a surface of 27,000 sq in., while for a double drum dryer with drums 42 inches in diameter and 100 inches wide each

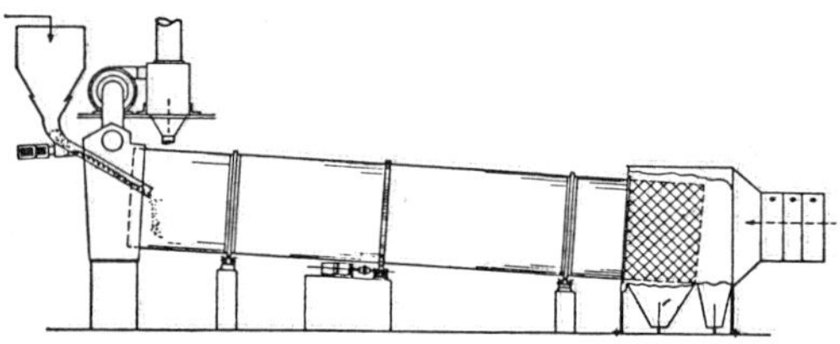

Figure 5. A direct-heat rotary dryer. The shell runs on two circular tails, carried by two sets of four idlers. The girth gear and its driving motor and connection are at the center of the shell. The wet material enters the hopper and the vibrating feeder; the dry material enters a delumper, which retains the lumps but permits the fines to drop to the hopper below, which feeds a conveyor. The combustion gasses enter at right, with secondary air regulated by inlet-baffles; the gases travel through the shell and leave through the exhauster at upper left.

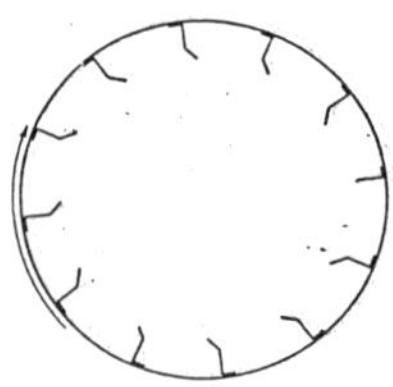

Figure 6. Cross-sectional view of the direct-heat rotary dryer, showing the lifting flights.

drum has a surface of 13,400 sq in. If the single drum receives 50 psig steam which is its maximum safe pressure, and the double drum may take 100 psig steam pressure, the capacity in dried material will be the same.

ROTARY DRYERS

Rotary dryers principally utilize a rotating cylinder, slightly inclined to the horizontal. The cylinder may be rotated as in revolving rotary dryers or the shell may be stationary and an agitator may revolve slowly (5–8 rpm), moving the material from one end to the other. Direct, indirect, and direct-indirect heat transfer may be used for effecting drying temperatures. The feed rate, speed of rotation or agitation, and volume of heated gases and their temperature are so regulated as to insure dry solids before discharge. Hot gas at temperatures of up to 1200° F can be added to the dryer, usually in a cocurrent flow pattern. Discharge temperatures are 180°–200° F; spent gases exit at about 300° F.

Figure 5 is a rotary dryer where the shell fits loosely into a stationary housing at each end. Wet material is brought to a chute running along the housing; the latter also carries an exhaust pipe. The revolving shell runs on two circular tracks, and is turned by a gear which meshes with a driven pinion. Inclination is typically 1:16 for high capacities and 1:30 for low ones. As the shell revolves the solids are carried upwards one-fourth the circumference and roll back exposing new surface. Efficiency is greatly improved by placing longitudinal shelves 3–4 inches wide on the inside of the cylinder. These longitudinal shelves are called lifting flights and carry part of the solids half way around the circumference and drop them through the whole of a diameter in the central part of the cylinder where the drying air is hottest and least moisture laden (see Figure 6). By bending the edge of the lifting flights slightly inward, a nearly uniform fall of material throughout the cross section of the cylinder can be effected.

Gas flow through rotary dryers may be either cocurrent or countercurrent and gas velocities will range from 2–12 ft/sec depending upon the application and to prevent dust from being entrained with the exhaust gas. The larger and denser the material being dried the greater drying air spatial velocities can be used. The mass of material in the dryer may be as high as 30%–40% of the volume of the shell so that the air has a narrower cross-sectional area to pass than the full diameter of the shell, and for the same volume of air the velocity is higher.

It has been stated that direction of drying gases may be either cocurrent or countercurrent to the solids flow. Cocurrent gas flow, entering at the upper or feed end and leaving at the discharge end, is favored when the material is heat sensitive although heat efficiency is somewhat lower. The governing factor in efficiency is the temperature drop of the air or gases. The higher the inlet temperature, the greater the efficiency.

More intimate contact between hot air or gases and material by the breakup of falling particles is further increased in the cellular type of rotary dryer whose lifting flights become cells filling the interior of the shell. As the shell rotates, material flows from one cell to another exposing wet surfaces.

The double-shell dryer with direct-indirect heat consists of an inner cylinder through which fire gases pass cocurrently to the lower end of the cylinder, there to turn and travel in the open shell space counter to the material until they leave at the point where wet feed material enters. Heat transfer from the inner cylinder is through the metal wall by conduction and radiation. The double-shell dryer becomes an indirect dryer by placing flues along the inside of the outer shell through which gases traveling countercurrent may pass on their way to the exit rather than in direct contact with the material.

The steam tube dryer has a revolving shell, and material drops through a bundle of steam tubes, which run the full length of the dryer parallel to the shell axis. Still another design of this type has a stationary shell and revolving bundle of tubes. There can be a number of modifications and combinations to these arrangements for special purposes.

INDIRECT DRYERS WITH AGITATION

Drying/heating/cooling can be achieved with an indirect heat exchanger with mechanical agitation of material. Examples of batch or continuous indirect heating are drying, cooling or reaction of solids, slurries, gels, filter cakes, powders, and viscous materials. One design for such operation consists of a stationary horizontal vessel containing a tubular rotor with verticallymounted double-walled discs. These discs provide approximately 85% of the heating surface. Other heating surfaces are the rotor shaft and inner wall of the jacketed vessel trough (see Figures 7 and 8).

During operation, a high relative velocity between the rotating discs and the product contributes to a high heat transfer coefficient. Agitator plows or scraper bars can be added to increase exposure of new surfaces and prevent material build-up. Additionally, rotor speeds can be set for optimum heat exchange and mixing. Residence times can range from minutes to several hours.

Another design available in this class of dryer is a hollow screw, jacketed-trough dryer that provides three to four times more heat transfer surface than simple jacket screw conveyors and six times more than water cooled drums, in either continuous or batch operation. Such models can operate in either a pressure or vacuum environment and can serve as a reactor for mixing liquid or gas reactants with product.

FLASH DRYING

The flash dryer is an application of the principal of spray drying to materials which are solid or semi-solid in the wet state. Wet material is dried while suspended in finely divided form in a flow of heated air. This is effected by dropping the wet solids into a high-temperature air stream which carries it to a hammer or cage mill or high speed agitator where exposed surface area is increased. Fine particles leave the mill through a duct sufficiently small to maintain transport velocities to reach a cyclone separator.

The flash dryer is an example of cocurrent gas flow operations. Solid particles may typically take 6–8 seconds to pass from entry point to the air stream of the collector. Air temperatures may typically drop from 1,300° F to 600° F in 2 seconds or 1,300° F to 350° F in 4 seconds. The drying time is therefore essentially 2–4 seconds and the material itself not raised in temperature by more than 100° F. Figures 9 and 10 show typical flash dryer arrangements and schematics. Heat source for drying gases may typically be an incineration process providing combustion air. Figure 11 shows a flash dryer installation drying potato starch.

SPRAY DRYERS

Spray dryers consist of a closed chamber in which a liquid solution of solids or low viscosity slurry carrying suspended solids is atomized. Hot air meets the droplets and travels with or past them, vaporizing the water so that at discharge only a dry solid remains in the drying air stream with water vapor. The dry particles travel at their transport velocity and follow the air currents. Moist air travels the same path and is discharged from the collector to the atmosphere.

Figure 12 shows the essentials of spray drying systems which consist of:

- The drying chamber.
- An atomizing device.
- Means of heating and moving the drying gases.
- A means of product recovery.

Drying chambers may be 10–20 feet in diameter and 10–30 feet high in conical or cylindrical shape. Drying air may be direct products of combustion (e.g., natural gas) or

Figure 7. Tubular rotor section can be divided into two sections for gradual heating and cooling processes using different temperatures. Two operating modes are possible with this unit as a dryer (courtesy Bepex Corp., Chicago, IL).

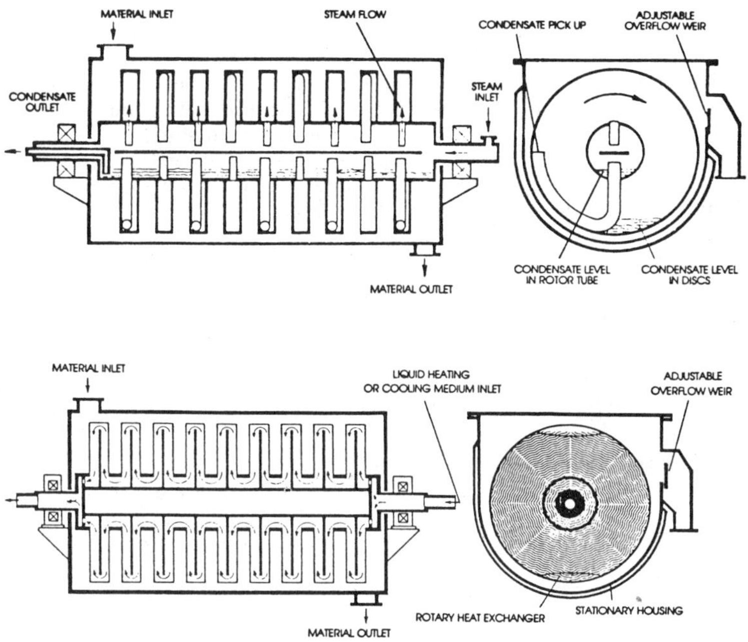

Figure 8. Heat transfer fluids flow through a tubular rotor and hollow discs mounted on the rotor. Fluids also flow through the jacket surrounding the rotor. Stationary agitator plows or scraper bars are located between the discs in order to achieve better mixing and to prevent material build-up on heat transfer surfaces. Transport of product is in an axial direction through the annular space between the discs and the vessel toward the discharge end, effecting high jacket heat transfer rates. Adjustable conveying vanes are fixed to the outer rim of the discs to control conveyance (courtesy of Bepex Corp., Chicago, IL).

heated indirectly by heat exchangers. Generally, however, flue gases are used as a heat source typically with indirect heat at 500° F and directly as high as 1,200° F, the temperature being regulated with dilution air. High inlet gas temperatures are favored, because the amount of applied heat transferred to wet feed is greater for maximum heat economy. Direct use of flue gases is not only more economical but in many cases reduces dust explosion danger due to the lower oxygen content of the drying gas.

Mechanisms for atomizing the liquid containing solids to be dried fall into three general categories:

- Pressure nozzles of special construction designed to impart a rotary motion to the fluid discharge spray liquor as a cone-shaped spray under pressures from 100–1,000 psi.
- A two-fluid-type nozzle in which feed liquor meets atomizing steam or air at 100 psi.
- A centrifugal atomizer consisting of a disc rotating up to 10,000 rpm.

Most units are designed on the basis of actual tests conducted in experimental spray dryers, tempered by the experience of the equipment manufacturer. The extent of testing will depend upon the material and desired capacity. A feasibility study in the laboratory will be followed by runs in a pilot plant spray dryer to obtain data for scaling up. For a given material and required characteristics, one set of operating conditions will produce the greatest economy in equipment selection and costs. A maximum drying air temperature difference consistent with product characteristics desired must be established.

FLUIDIZED-BED DRYERS

Fluid-bed dryers have been commercially used in the chemical process industries for about 30 years and are one of the newest of the unit operations. In addition to drying they have found use in combustion and reaction. An important limitation of fluidized beds is their limited range of drying air velocities, which must be sufficiently high to cause large particle fluidization without carrying off fines. Local air velocities through the bed plate openings may be quite high, depending upon available free area, but superficial velocities are typically 25–300 ft/min.

When a gas passes upward through a bed of finely divided solid particles, four general class conditions exist depending upon the characteristics of the particles, the geometry of the bed, and the velocity of the gas. These situations are

1. Fixed bed.
2. Dense phase fluidization.
3. Two-phase fluidization.
4. Pneumatic transport.

These conditions are shown in Figure 13.

In a fixed bed at low velocity, the passage of a gas up through the bed causes a relatively low pressure drop. The solid particles remain essentially stationary and the gas merely percolates through void spaces in the bed. The particles remain in place in the fixed bed

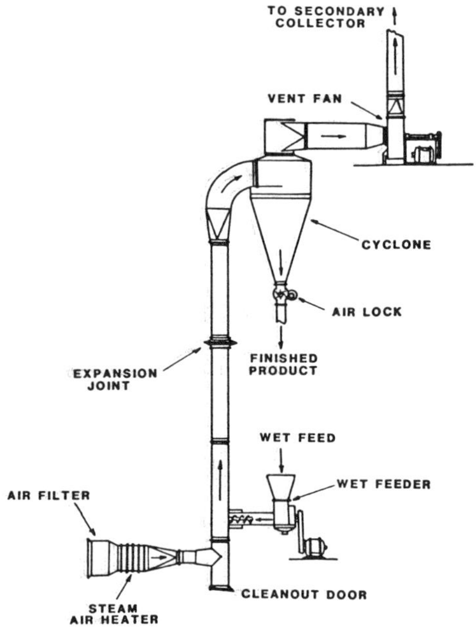

Figure 9. The principal elements of a simple flash drying system are an air heater, feeder to introduce the wet material into the system at a controlled rate, cyclone collector, vent fan, and necessary air piping. The wet material is dried by conveying it in the high velocity hot air stream. This type of system is particularly useful for drying free-flowing crystalline materials having a low surface moisture, such as sodium chloride, ammonium sulphate, potassium persulphate, and coal. The complete system is simple and compact and can be installed anywhere in the process line (courtesy C.E. Raymond Div. of Combustion Engineering).

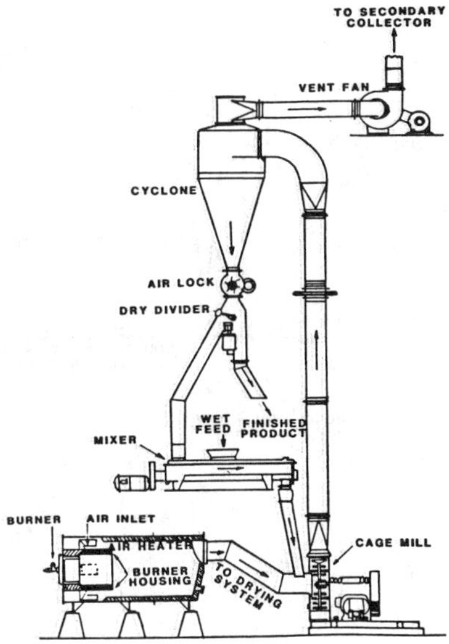

Figure 10. A system of flash drying with disintegration employs the use of a Raymond® Cage Mill. Any source of direct heat may be used. A mixer may be used for wet feed such as filter cakes, or omitted for non-sticky materials. This system handles a wide range of materials including acid-treated clay, spent grain, feed mixtures, calcium carbonate sludge, and sewage sludge. The unit is available in several standard sizes with capacities up to 36,000 lbs of water evaporation per hour (courtesy C.E. Raymond Div. of Combustion Engineering).

Figure 11. Photo of a flash dryer installation drying potato starch. This type of dryer is well suited for high evaporative loads as well as drying heat sensitive products because of its nearly adiabatic operation. Products with high moisture can be handled and dried uniformly by recycling part of the dry particles through a back-mixing unit (courtesy Bepex Corp., Chigaco, IL).

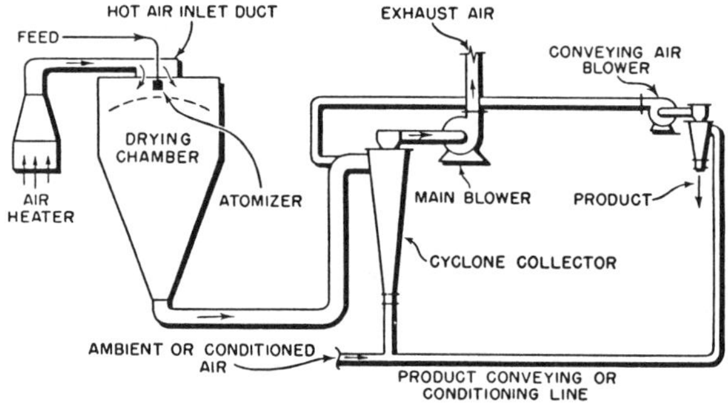

Figure 12. Typical spray-drying-system flowsheet.

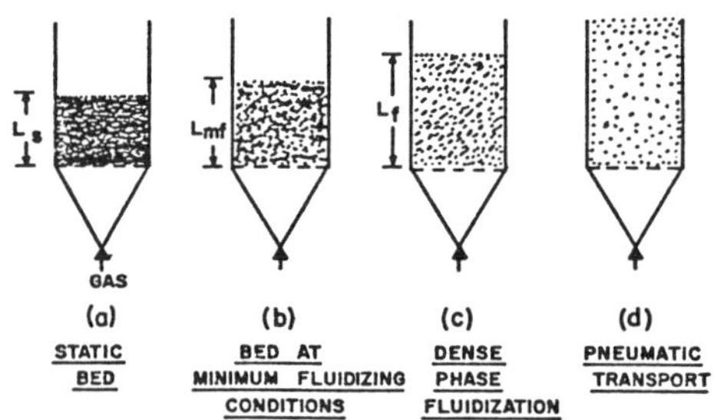

Figure 13. Fluid-bed situations.

as long as the pressure drop is less than the weight per unit area of the bed. This is represented in a graphical form as shown in Figure 14. Any further increase of the gas velocity causes the bed to expand slightly and particles gain freedom to move.

At a certain flow rate indicated by G_{mf}, the pressure drop becomes equal to the weight of the bed per unit area and the particles are suspended in the gas stream. At this point the bed is said to be in the fluidized state. This gas-solid system has liquid-like characteristics. Intense and violent mixing between the solids and gas occurs and results in completely uniform conditions of temperature, composition, and particle size distribution throughout the bed. The inception of fluidization, G_{mf}, is dependent on fluid and particle properties. The maximum gas velocity that can be allowed in a fluid bed is the terminal velocity G_t of the particle. These velocities are used in pneumatic transport. The limit of fluidizing velocity is $G_{mf} < G_f < G_t$. Depending on design, the actual gas velocity is chosen between the two extremes. It is because of this extensive combustion surface that the high thermal efficiency is possible in fluid-bed reactors. At operating equilibrium, the fluid bed acts as a boiling liquid.

A fluid-bed reactor or dryer system consists of the following:

1. The fluid-bed drying or combustion reactor, carbon steel lined with insulating and refractory to meet required incineration conditions.
2. The freeboard or disengagement zone designed so that there is a minimum of solid particulate carryover. Exiting gas may be brought in contact with feed and may be cooled by water content in the feed. It is also the largest section of the reactor. Proper design

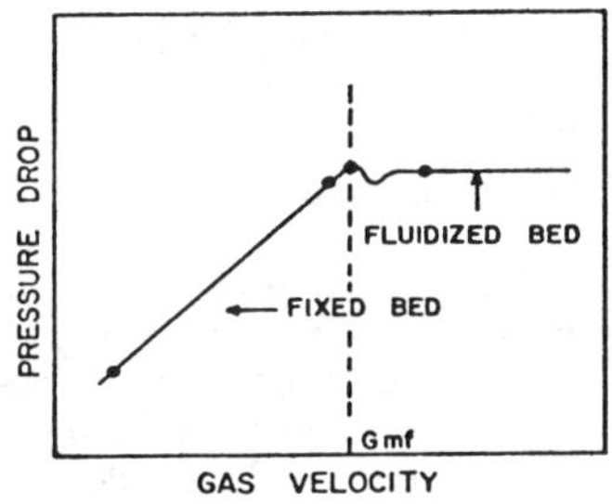

Figure 14. Variation of pressure drop with velocity in fixed and fluidized beds.

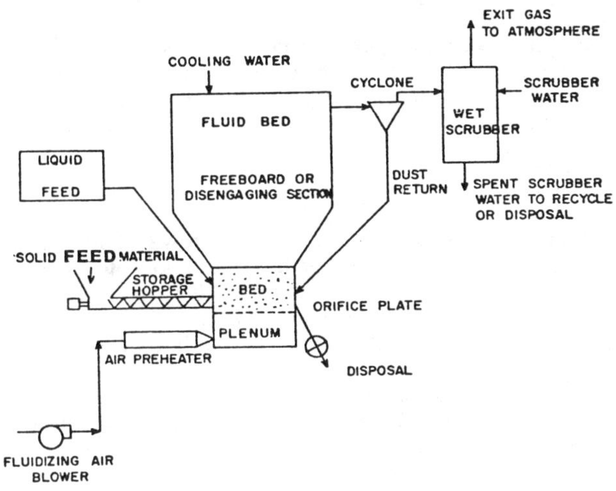

Figure 15. Typical fluid-bed incineration system.

of this section allows disengagement of solids from gases. This section may be designed with an increase in diameter to reduce exit gas velocity and ensure reduced dust loading.

3. The plenum chamber which ensures proper air distribution below the orifice plate.
4. The distributor or orifice plate that separates the plenum chamber from the fluid-bed zone. The fluidizing and combustion air passes through this plate which passes air but prevents the back-flow of bed material into the plenum chamber.
5. Air supply for fluidization and combustion in quantities and pressures (3–8 psig) supplied for the incineration and fluidizing requirement. The combustion system usually is forced draft. While a pre-heater may be desirable for start-up, secondary fuel also may be provided for injection directly into the fluid bed and/or freeboard section.
6. Dust collection systems provided to collect particulate matter from the exit gas. This may consist of both primary and secondary systems of low-pressure drop cyclones and wet scrubbers to control solid particulates in the submicron range. The latter would, of course, merit high-energy venturi scrubbers for consideration.
7. Feed and discharge systems to suit the particular materials handled. This would include such hardware as screw feeders, rotary valves, etc., depending on design and material requirements (type of solid or liquid).

8. Instrumentation to make the operation automatic and again to suit the individual problem's requirement. Fluid-bed systems lend themselves to a wide degree of instrumentation and imaginative engineering possibilities.

A schematic for a typical fluid-bed incineration system is shown in Figure 15, with interrelation between the components.

Careful design of the bedplate and occasional mechanical agitation of the bed is required in shallow beds to prevent air channeling. Hot air entering below the bed is uniformly distributed by pressure drop across the plate, and additional pressure drops resulting in the various area changes within the system require a pressure blower of 20–40 in. w.g. per bed. Fluid-bed dryers provide longer residence time, with particles attaining higher temperatures and absorb more sensible heat from the hot drying gas. Allowance should be made for this in the system's heat balance. Figure 16 shows a tube bundle for a fluid-bed internal heat source for indirect drying, and Figure 17 shows a fluidized-bed installation used for drying crystalline material.

Fluid-bed dryers can also be built for plug flow, in which the bed moves horizontally as well as to the random vertical motion created by the air flow. Plug flow residence time can be made uniform for all particles with preheating, cooling, and other operations in addition to drying, which can be performed at one or more temperatures in the same unit.

PURGE VESSELS

Purge vessel technology has developed as a direct outgrowth of storage and hopper dryer design. The typical hopper dryer consists of a vertically-oriented hopper with a large length-to-diameter ratio. Product is fed and discharged at equal rates with a constant product level being maintained in the hopper. Residence time then becomes a function of hopper volume and feed rate. Normally a gas purge is introduced into the bottom and travels countercurrent to product flow.

This concept can include a wide variety of vertical- and horizontal-oriented vessels which have countercurrent gas purge. These are involved in drying, cooling, heating, and reaction applications. They all feature good plug flow characteristics. The purge vessel itself may

Figure 16. Tube bundle for fluid-bed internal heat source (courtesy Bepex Corp., Chicago, IL).

Figure 17. Fluid-bed dryer installed in 1972 and used for drying crystalline material (courtesy Bepex Corp., Chicago, IL).

be designed as a plug flow hopper or may have an agitator installed when required. In general, the purge vessel minimizes mechanical requirements by using gravity and minimizing agitation requirements. Advantages include: absence of moving parts minimizes maintenance costs and assures reliability; operating costs are minimized because product is conveyed by gravity and/or with minimal agitation; plug flow characteristics are excellent when compared to most types of dryers; simple design offers minimal capital and installation costs. Applications include polyester, nylon, polyolefins, and many other granular chemicals, foods, and minerals. Figure 18 shows a schematic of a purge vessel with range of design parameters.

CASCADE-TYPE HEATERS

The vertical processor shown in Figure 19 operates by suspending a bed of solid particles in counterflow with a gas. Both streams—gas and material—are guided by a succession of sloped screens. This flow pattern insures complete heat transfer at low pressure drop and short, uniform residence time.

In addition to heating and cooling of granular materials, the cascade heater cooler can be used for drying or heat recovery from dirty gas streams. Gentle treatment of particles is provided by the aerodynamic forces supporting the bed of material. The unit handles a wide range of feed rates, and empties completely in seconds. Advantages include:

- Complete heat exchange due to counterflow approach, development of large heat transfer surface area, and turbulent flow.
- Space and energy requirements minimized.
- Offers excellent structural integrity with long life of internal components even when handling abrasive materials.
- Short residence time-requirements and a good approach to plug flow allows unit to be selfemptied in seconds.

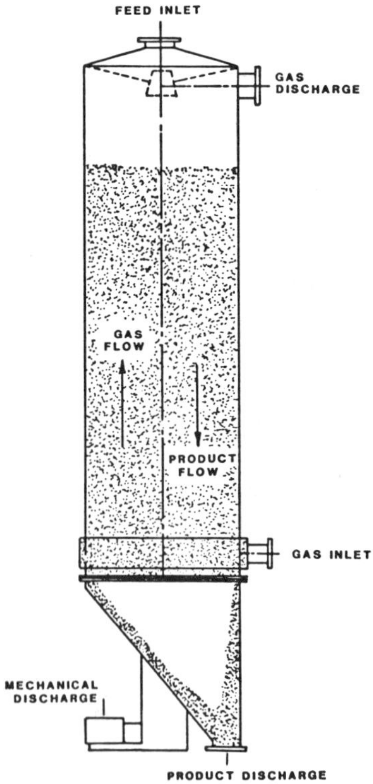

Figure 18. Purge vessel capabilities include a continuous drying process with direct or combination heat source. Temperatures range 0–1200°F. Material conveyance: mechanical or gravity; with holdup volume 0–1,000 cubic feet. Residence time can range from minutes to hours; the residence time distribution—plug flow with heat transfer surface 0–600 square feet (courtesy Bepex Corp., Chicago, IL).

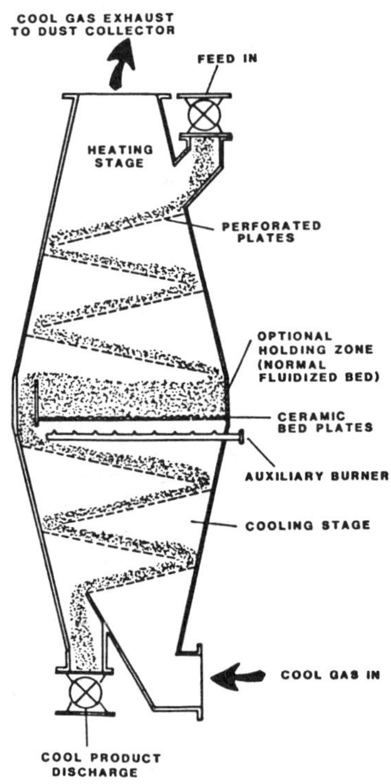

Figure 19. Cascade™ heater/cooler capabilities include continuous drying process with direct heat source in a temperature range from 0–2000°F. Materials are conveyed by pneumatic gravity and the holdup volume up to 20 cubic feet. Residence time distribution is plug flow with residence time ranging from seconds–minutes (courtesy Bepex Corp., Chicago, IL).

These systems should be designed as single-stage heaters or coolers only. Applications include:

- Heating.
- Cooling.
- Calcining of granular minerals and chemicals.
- Heat recovery from flue gases.

BELT DRYERS

These dryers generically are utilized in several forms depending upon the type of materials being dried. One such dryer consists of a set of hinged shelves between two endless chains, traveling on large sprocket wheels within a chamber where hot or warm gases circulate. The belt is long enough to keep the product within the heated chamber sufficiently long enough to effect drying. A *belt conveyor dryer* is a belt dryer made up of links with hooks, for example, traveling overhead. The material itself may form a belt as with cloth and yarn, which travels over rollers through a tunnel-like chamber countercurrent to the drying air. This type is known as a *festoon dryer*.

Belts may be made of coarse-mesh chain and material pressed into it in paste form. Such a belt is made to travel up and down the length of the drying chamber, festoon fashion. A number of such belts traveling horizontally for complete loops, mounted above one another travel back and forth in opposite directions. Material can be fed to the upper belt, dropping from one belt to the next.

Conveyor-type dryers consist of belts made of woven wire mesh on which material is spread; drying air passes through the meshes and bed of material. The belt travels on a chain, or is stretched on large pulleys. The whole assembly is enclosed in the drying cabinet.

DRYING ROOMS/TRAY, TRUCK, AND TUNNEL DRYERS

For drying materials which cannot be agitated, the drying room and the tray dryer, of which there are many varieties are largely used. In the tray dryer, shallow pans typically 2 inches deep by 2 × 3 ft are placed on a rack forming part of the dryer. In the truck dryer, the rack is on wheels and may be wheeled in and out of the dryer. Several tray or truck dryers may be typically set up side-by-side and air passed through them countercurrently and regulated by means of ducts, by-passes, and dampers. This latter arrangement is sometimes called a compartment dryer. The tunnel dryer is a variation of the truck dryer which has a long tunnel holding a number of trucks which travel in one direction. Air flow in such a dryer is countercurrent, being introduced at the dry end.

SHELF DRYERS

These dryers have hollow shelves, usually of welded steel with connections for incoming steam and outgoing condensate to each shelf. A number of shelves are superimposed in a square or rectangular box. The material is spread on the shelves directly, or on trays placed on the shelves and heat is through the metal wall of the shelf. In the atmospheric shelf dryer, an exhaust fan may be connected to the top of the drying cabinet.

The *rotary shelf dryer* has stationary, horizontal circular shelves and a rotating vertical central shaft with arms, carrying paddles or ploughs. Shelves are hollow, usually steel construction, with steam and condensate connections; they have alternating central and circumferential openings through which material makes its way downward, traveling the width of each shelf. Ploughs are set to sweep material as desired. A vertical turbo-dryer is a type of shelf dryer.

VACUUM DRYERS

Vacuum dryers offer a number of advantages over atmospheric drying. These are:

- Contamination by dirt in drying air is eliminated.
- Absence of oxygen in drying air which may cause oxidation in the product.
- Facilitates recovery of organic vapors if present.
- Lower working temperature which can be a distinct advantage for heat sensitive materials.

Disadvantages include greater first cost of equipment and greater operating costs with attending limits in production. However, in cases where the product is heat sensitive or solvent must be recovered such considerations are well justified. Many of the types of dryers described, such as pan, shelf, drum, and rotary dryers, can be designed for vacuum operation.

EVAPORATORS

Classification of drying is almost impossible without overlapping evaporation. Evaporators as a class of equipment are usually employed to evaporate water from a solution. The purpose is concentration of nonvolatile solutes such as inorganic salts, acids, bases, or organic compounds. Evaporation in many applications results in precipitation or crystallization products which are subsequently separated from solution by filtration, cyclones, settlers, wash columns, elutriation, with the product being precipitated solids, a concentrated solution, or both. Applications include use of the evaporator to recover a solvent such as potable water from saline water and the relatively pure condensed water in some cases is used for boiler feed makeup, dissolving, pump seals, instrument purges, equipment washing, etc.

A number of designs are available which basically include concentrating solutions where no precipitates are formed, and the others are used for salting or scaling applications. A variety of equipment described here is Swenson (a division of Whiting Corp.) evaporator designs which illustrate a broad range of evaporator designs.

In a single-effect, rising-film evaporator (see Figure 20), evaporation occurs inside the evaporator tubes, so it is used primarily for concentration nonsalting liquors. To provide for good heat transfer rates, the Δ between the heat transfer medium and the liquor should be greater than 15° F, and preferably greater than 20° F. Tubes are normally 3/4" to 2" in diameter and from 10 to 30 feet long.

Operation of the rising-film evaporator is straightforward. Liquor is fed into the bottom liquor chamber and then into the tubes. There it is heated with condensing steam or any other suitable heat transfer medium, or hot liquor. If the vapor pressure of the feed equals or exceeds the system pressure at the bottom tubesheet, vaporization will occur immediately. For colder feed, the lower portion of the tubes is used to preheat the liquor to its boiling point. Vaporization then begins at that height within the tubes where the vapor pressure of the feed liquor equals the system pressure. As the liquor climbs up the inside of the tubes, additional vapor is generated and the velocity of the liquid-vapor mixture increases to a maximum at the tube exit. The outlet mixture impinges upon a deflector, mounted above the top tubesheet of the heat exchanger, where gross, initial separation of the liquid from the vapor occurs.

Additional liquor is separated from the vapor by gravity as the vapor rises in the vapor body. A mesh-type or centrifugal entrainment separator can be installed near the top of the vapor body to remove most of the remaining traces of liquid from the vapor. The exit vapor is conducted to the next effect of a multiple-effect evaporator, to a compressor, or to a condenser. A Swenson vertical-tube surface condenser is shown in Figure 20. The concentrated liquor is discharged from a connection near the bottom of the vapor body.

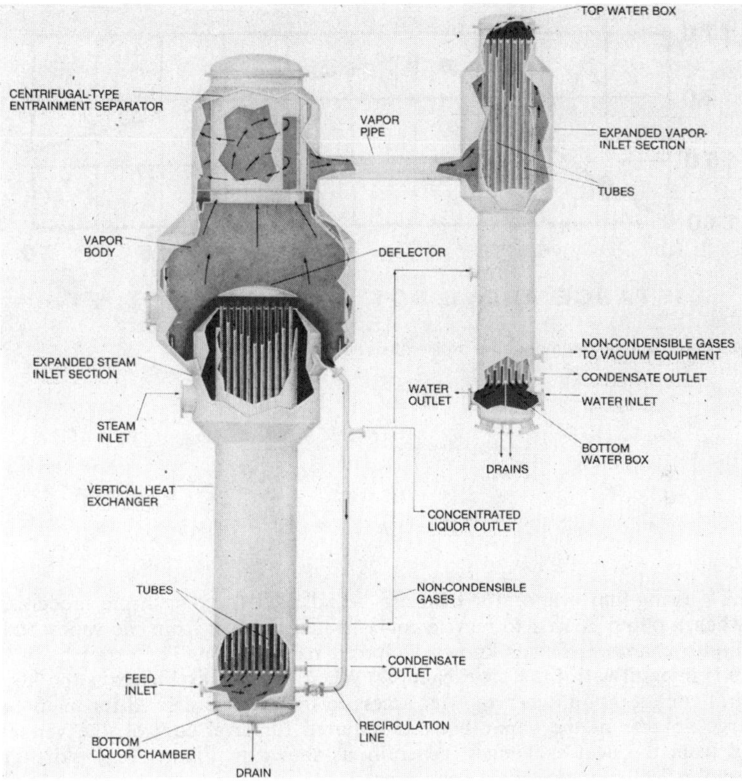

Figure 20. Swenson LTV rising-film evaporator with vertical tube surface condenser (courtesy Swenson Evaporators/Whiting Corp.).

Heat transfer rates are enhanced in the non-boiling section by surface or local boiling and in the boiling section by nucleate boiling. As expected, the heat transfer rates in the boiling zone are several times greater than those in the nonboiling zone, so it is important to reduce the nonboiling zone, to a minimum.

A typical fluid-temperature profile taken with a traveling thermocouple for a rising-film evaporator is shown in Figure 21. The maximum temperature occurred at a point eight feet above the bottom tubesheet where the boiling began.

Different two-phase flow schemes are created in the boiling zone, including slug flow, where a slug of liquor is followed by a slug of vapor, similar to the perking in a coffee percolator; annular flow, where a ring of liquor encases a center core of vapor and liquid mist; and mist flow, where vapor blankets the tube surface. Mist flow should be avoided because poor heat transfer results when there is not enough liquid present to wet the tube walls. To avoid mist flow, it is sometimes necessary to recycle concentrated product from the vapor body to the bottom liquor chamber so as to supplement the feed liquor. (This recirculation line is shown in Figure 20).

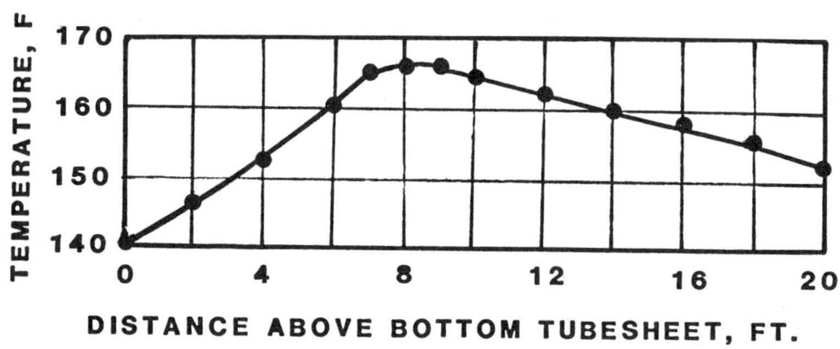

Figure 21. Typical fluid-temperature profile of rising-film evaporator.

The LTV rising-film evaporator can also be adapted to concentrate moderate scaling liquors when a pump is used to recycle concentrated product from the vapor body to the bottom liquor chamber. This is known as forced recirculation. The vapor body shown in Figure 20 is integral with the heat exchanger. When the heat exchanger is too large to use the integral configuration, when quicker access to the tubes is desired for maintenance, or when surge volume in the vapor body is required for level control, the vapor body is separated from the heat exchanger, generally as shown in Figure 22. A skirt-type baffle replaces the deflector as the initial separator.

FALLING FILM EVAPORATORS

A single-effect, falling-film evaporator with separate vapor body and heat exchanger is shown in Figure 22. Liquor is fed into the top liquor chamber of the heat exchanger where it is distributed to each tube. There are several different distribution devices for falling-film evaporators; a distribution plate is shown in Figure 22. The liquor accelerates in velocity as it descends inside the tubes because of the gravity and drag of the vapor generated by boiling. Liquid is separated from the vapor in the bottom liquor chamber of the heat exchanger and with a skirt-type baffle in the vapor body. A supplemental entrainment separator can be installed in the upper portion of the vapor body to reduce liquid entrained with the vapor to a minimum. The direct-contact condenser (shown in Figure 22) is used to condense the vapor with water. Concentrated liquor is discharged from the bottom liquor chamber and cone bottom of the vapor body.

The vapor body can also be provided as an integral component of the heat exchanger, similar to that shown in Figure 20, except the heat exchanger would be located above the vapor body in the falling-film configuration.

Evaporation also occurs inside the tubes of the falling-film evaporator. The unit can be used to concentrate the same nonsalting liquids concentrated in rising-film evaporators, and it is suitable for concentrating more viscous liquors. Tube sizes and lengths are comparable to those for rising-film evaporators.

The falling-film evaporator is particularly useful in applications where the driving force in temperature differences between the heat transfer medium and the liquid is small (ΔT's of less than 15° F). The retention time for liquor in this evaporator is less than that for a rising-film evaporator. The combination of short liquid retention time and the ability to operate at a low ΔT makes the falling-film evaporator ideal for concentrating the most heat-sensitive materials.

High heat transfer coefficients are attained in falling-film evaporators when a continuous film of liquid, preferably at its boiling point, flows down the inside tube wall with a vapor core in the tube center. For some applications, however, it is necessary to supplement an insufficient quantity of feed liquor with product liquor pumped to the top liquor chamber to avoid vapor blanketing of the inside tube surface.

A horizontal, cylindrical vapor-liquid separation device (see Figure 23) is sometimes used for multistage falling- and rising-film evaporators. The flow scheme (Figure 23) is for a multistage, falling-film waste liquor evaporator with mechanical recompression. The ability

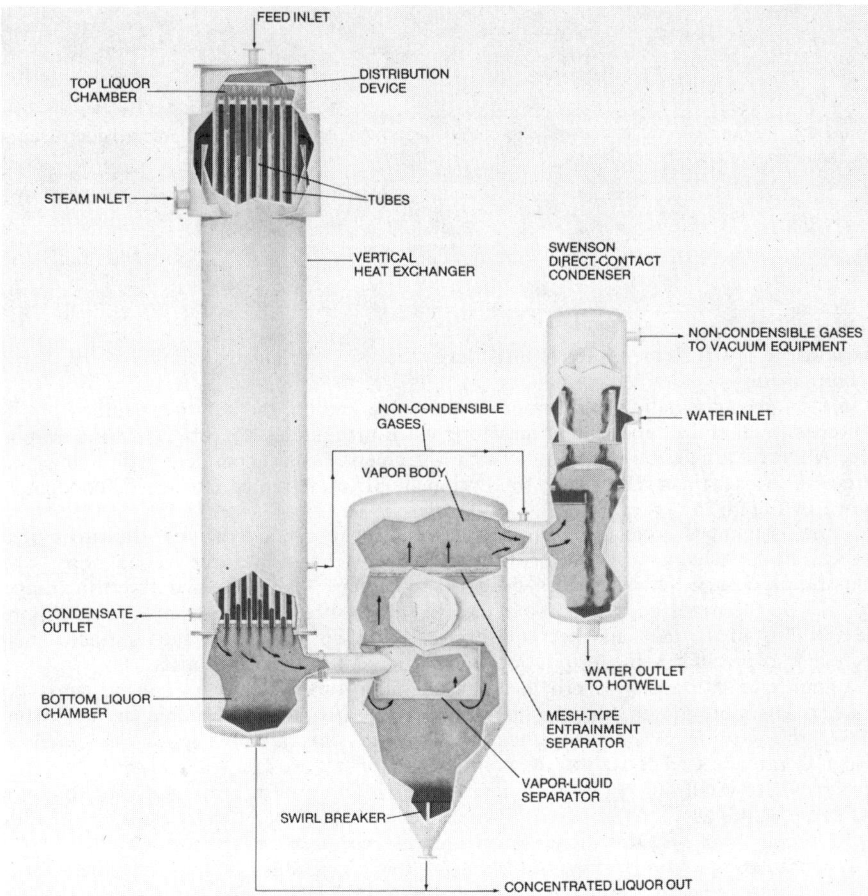

Figure 22. Swenson LTV falling film evaporator (courtesy Swenson Evaporators/Whiting Corp.).

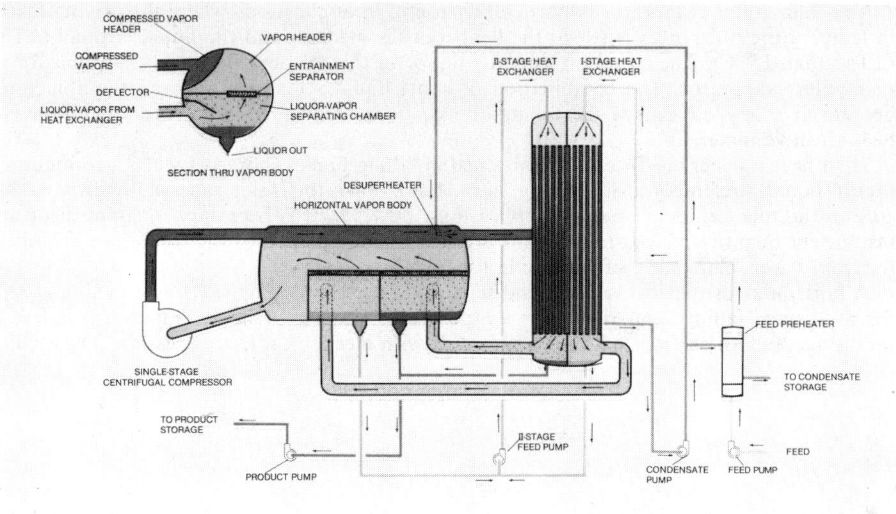

Figure 23. Falling film, multistage recompression evaporator for concentrating waste liquor (courtesy Swenson Evaporators/Whiting Corp.).

to operate at low ΔT's makes the falling-film evaporator particularly suitable for mechanical recompression.

In the forced-circulation evaporator, as the name implies, the liquor is pumped through the tubes to minimize tube scaling or salting when precipitates are formed during evaporation. A forced-circulation evaporator (with a submerged inlet) complete with a single-pass vertical heat exchanger, elutriating leg, cyclone, and top-mounted barometric condenser is shown in Figure 24.

Slurry is pumped from the bottom cone of the vapor body through the tubes of the vertical heat exchanger, where heat is added, and back into the vapor body where evaporation occurs. Sufficient slurry height (submergence) is maintained above the tangential inlet on the vapor body and above the top tubesheet of the heat exchanger to suppress mass boiling in the inlet and prevent surface (local) boiling on the tube surface. This is necessary to preclude salt precipitation on the tangential inlet and tubes.

A high circulation rate is provided for adequate tube velocity to achieve good heat transfer. Therefore, lower slurry temperature rises are assured which minimize supersaturation of the solution. A sufficient quantity of salt crystals is suspended in the circulating system to provide seed crystals in the boiling zone for salt growth. Adherence to these basic principles of crystallization results in coarse crystals and minimal wall and tube salting, so less equipment washing is required. This conserves energy because less steam is required to boil wash water and this increases on-stream time for the evaporator.

The circulating pump is usually of the axial-flow, singleelbow design, well-suited for the high flow rates and low pressure drops in Swenson-designed circulating systems. These heavy duty pumps operate at low speeds, which reduce maintenance and minimize mechanical attrition of the salt crystals.

Circulating piping interconnects the vapor body, the heat exchanger, and the pump. Conical liquor chambers provide gradual, low pressure drop transitions from the circulating piping to the tube bundle which is particularly important for the establishment of uniform feed to the tubes. To provide for thermal expansion without expansion joints, the circulating pump base is springmounted. As an alternative, the entire pump may be hung from the circulating piping.

The tangential inlet provides excellent mixing of slurry in the vapor body because of the circular motion it creates. Secondary vertical currents are also generated, mixing body slurry with the hotter slurry entering the vapor body to reduce the degrees of flash. This agitation minimizes salt buildup on the bottom cone of the vapor body. A swirl breaker is provided in the circulating slurry outlet.

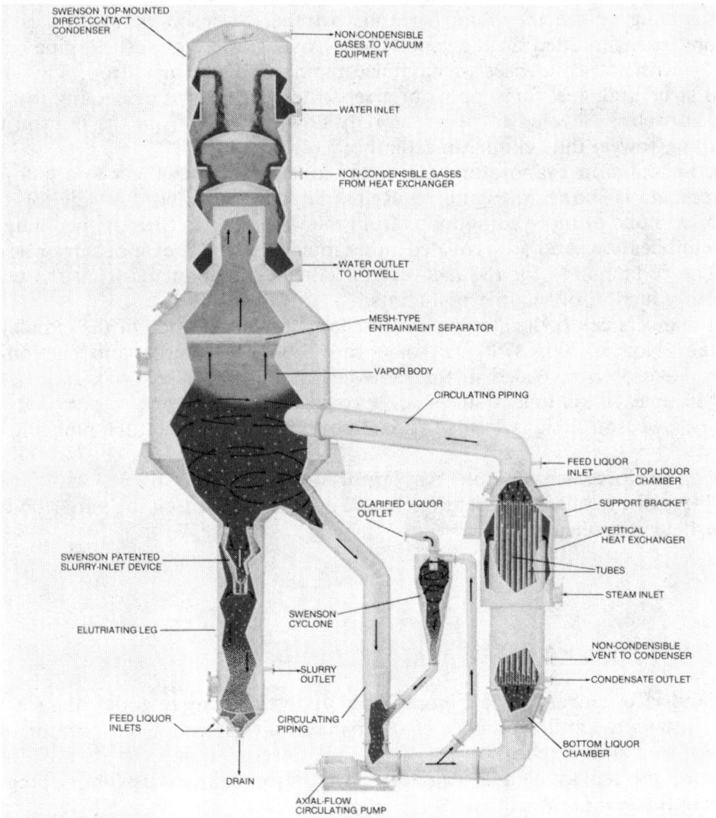

Figure 24. Forced circulation, submerged inlet, vertical tube evaporator (courtesy Swenson Evaporators/ Whiting Corp.).

The vapor body is conservatively designed both in diameter and height. It is important to have an adequate free space above the liquor level to allow the liquor droplets entrained in the vapor leaving the boiling surface to reach equilibrium and return by gravity to the circulating slurry. The large diameters result in low vapor valocities, which minimize entrainment and provide adequate retention time for salt growth.

A mesh-type entrainment separator may be installed in the upper portion of the vapor body to reduce solids carryover to normally less than 50 parts per million parts of vapor. Other types of entrainment separators are also available.

The elutriating leg, attached to the bottom cone of the vapor body, is a convenient device for thickening the slurry it receives from the vapor body and for salt crystal washing and classification. Slurry enters the top of the leg through a unique slurry-inlet device, which improves washing efficiency by reducing agitation in the leg (U.S. Patent No. 4, 113, 552). Salt crystals are fluidized and washed in the leg with a portion of the feed liquor which enters the bottom cone and is distributed with a perforated, dished head. Smaller crystals are washed into the vapor body for additional growth and the larger crystals are discharged from a connection near the bottom of the leg.

A low-pressure-drop liquid cyclone is sometimes used to clarify liquor discharged from the evaporator. The driving force is the pressure drop across the circulating pump. Thickened slurry is returned through a wide-open cyclone underflow connection to the circulating piping before the pump suction. Another Swenson innovation is the direct-contact condenser mounted on the vapor body. A short piece of vertical pipe connects the vapor body with the condenser to minimize piping and pressure drop. This design also eliminated structural steel for support of a separate condenser. For cooling tower applications, the hotwell is elevated to permit gravity flow of water from the hotwell to the top of the cooling tower, thus eliminating the need for a pump.

A forced-circulation evaporator with a horizontal heat exchanger and a top-mounted stripping column is shown in Figure 25. Reflux liquid is introduced on the top tray of the column to strip one or more compounds from the water vapor. Stripping columns are used for special applications and are provided either integral with the evaporator or as a separate column. The columns are for the recovery of valuable components from the water vapor and for the reduction of volatile pollutants.

Heat exchangers can be located on the discharge or suction side of the circulating pump (U.S. Patent No. 3, 300, 870). In some cases, both discharge and suction-side heat exchangers have been provided in the same circulating system to maximize heat-transfer surface. Multiple circulating systems are provided for the same vapor body on large evaporators. Swenson designs include special provisions to reduce tube plugging caused by salt pumps.

A variety of forced-circulation evaporator designs are available, with and without elutriating legs or cyclones. Different inlets and outlets are used on the vapor body, tailored for the particular application.

CALANDRIA EVAPORATOR

The calandria evaporator (see Figure 26) is applied less often today than it was years ago, though there are still a number of companies that prefer this evaporator for various applications. For most applications, however, the lower equipment costs for other designs has prompted the replacement of calandria evaporators with rising-film, falling-film, and forced-circulation evaporators.

The calandria evaporator has a heat exchanger (with tubes usually less than six feet long) integral with the vapor body. The level is maintained in the upper portion of the tubes or above the top tubesheet and the circulation pattern is up through the tubes and down through a central pipe called a "downcomer." Circulation is created by the difference in specific gravity between the body liquor and the heated liquor and vapor generated inside the tubes, plus a vapor lift effect.

The calandria evaporator can be used for salting-type applications; however, an agitator located inside or beneath the downcomer is recommended to suspend the salt crystals in the lower portion of the body. Although the agitator creates some flow through the tubes, most of the flow is still created by "thermo-syphon." The calandria is also used for batch-type concentration of liquors.

HEAT EXCHANGERS

Shell-and-tube heat exchangers are used for the majority of evaporator installations. The heating medium is almost always steam which is normally condensed on the outside of the

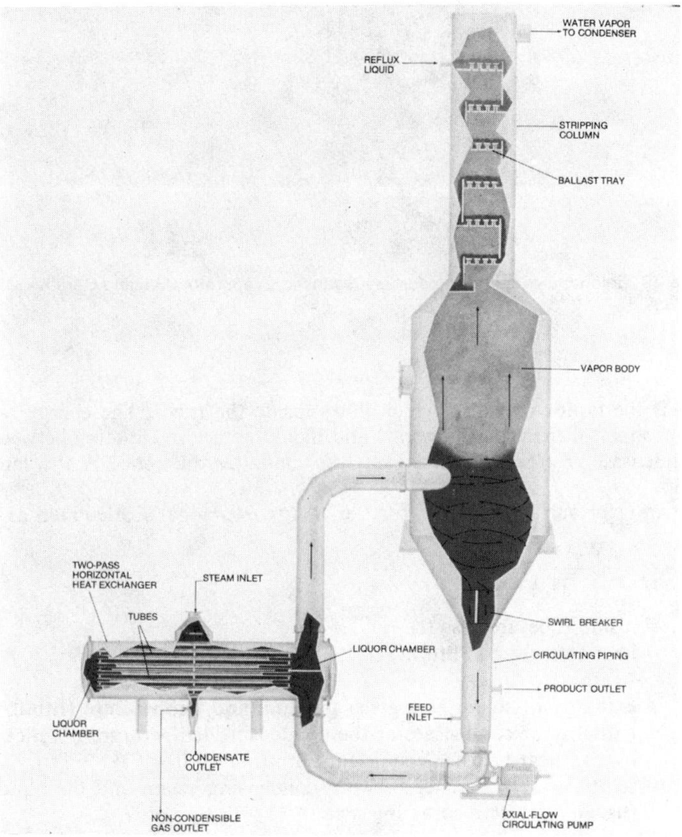

Figure 25. Forced circulation, submerged inlet, horizontal tube evaporator with stripping column (courtesy Swenson Evaporators/Whiting Corp.).

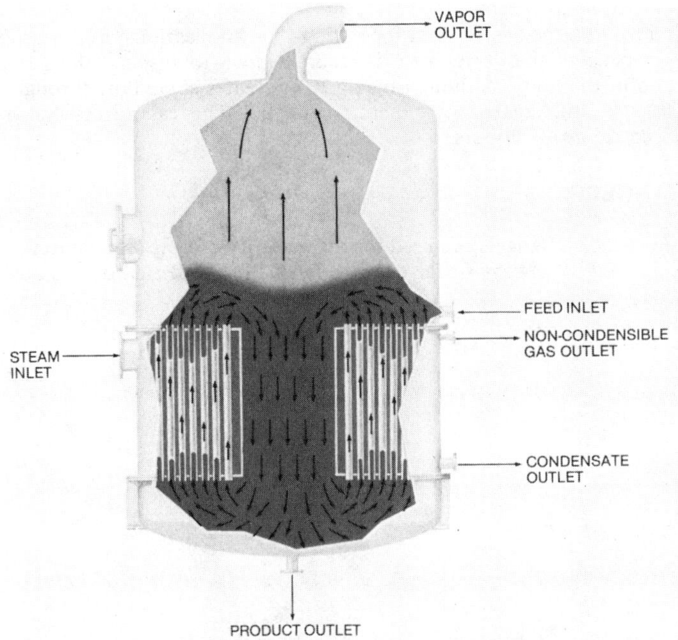

Figure 26. Calandria evaporator (courtesy Swenson Evaporators/Whiting Corp.).

tubes to heat the liquor or slurry which flows inside the tubes. The energy transferred is simply the product of the steam flow rate and the difference in enthalpy between the steam and its condensate. For saturated steam, the enthalpy difference is the latent heat of vaporization.

The heat-transfer surface required for the energy exchange is calculated as follows:

$$A_i = \frac{Q}{U_i \Delta T} = \frac{(W_s)(H_s - H_c)}{U_i \Delta T}$$

where A_i = inside tube area (sq ft)
Q = heat transferred (Btu/hr)
W_s = steam rate (lb/hr)
H_s = enthalpy of steam at a given pressure and temperature (Btu/lb)
H_c = enthalpy of condensate at the condensing temperature (Btu/lc)
U_i = overall heat-transfer coefficient (Btu/(hr) (sq ft) (°F))
ΔT = temperature difference between condensing steam and the liquor or slurry (usually determined as log mean), °F

Both vertical- and horizontal-tube heat exchangers are utilized. The horizontal exchanger is used for installations with limited headroom or where maximum liquor submergence is needed to prevent surface boiling and subsequent salt precipitation on the tubes. A typical horizontal heat exchanger is shown in Figure 25. Installations using vertical-tube heat exchangers are shown in Figures 20, 22, 23). Steam enters the enlarged shell section and flows around and up through the annulus formed between the inner and outer shells. The steam enters the tube bundle uniformly around the entire circumference beneath the top tubesheet. The steam inlet prevents condensate impingement on the tubes, reduces tube vibrations, provides for uniform steam distribution and, for some applications, acts as an entrainment separator. The noncondensable gases are swept downward and removed near the bottom tubesheet. Special vent baffles are provided for large diameter heat exchangers.

Internal tube supports are not normally required for vertical heat exchangers. The tubes are usually fastened mechanically to the top and bottom tubesheets by expansion of the tube into the tube holes.

PREHEATERS AND CONDENSATE FLASH SYSTEMS

Preheaters and condensate flash systems are often incorporated into process designs to reduce steam consumption. Preheaters are used to heat the feed to an evaporator with a heat-transfer medium, preferably of less energy value than that of the steam or vapor condensed in the evaporator heat exchanger. Typical heat-transfer fluids used in preheaters include evaporator condensate, evaporator vapor, and process liquor. Heat that is added in preheaters replaces the heat that would have been added in the evaporator heat exchangers; therefore, less heat-transfer surface area is required for the evaporator heat exchangers when preheaters are used.

A preheater is designed either as a separate exchanger or as an integral part of an evaporator heat exchanger or condenser. Preheaters can also improve heat transfer coefficients for rising- and falling-film evaporators because the nonboiling zone is reduced or eliminated. For some scaling applications, separate preheaters, with bypass piping, provide an additional benefit because scale forms in the smaller, easier-to-clean preheater rather than on the tubes of the larger evaporator heat exchanger.

Condensate flash systems are sometimes employed in place of, or to complement, preheaters. Condensate from a heat exchanger is flashed into a heat exchanger that is operated at a lower pressure to provide supplemental energy.

MULTIPLE EFFECTS

The multiple-effect configuration combines two or more evaporator bodies to conserve steam, which is condensed in the first-effect heat exchanger only. Water evaporated in the first-effect vapor body is condensed in the second-effect heat exchanger, which provides energy for evaporation in the second-effect vapor body (and so on for additional effects). Vapor from the last effect flows to a condenser. The last-effect vapor body is maintained at a high vacuum with a steam-jet air ejector or mechanical vacuum pump. The driving force is the pressure drop from the first to the last effect.

For the same overall pressure differential and process conditions, the evaporator with more effects will require less steam. A rough "rule of thumb" for comparison purposes is that the evaporation in each effect will be approximately 0.7 to 0.9 pounds for each pound of steam condensed in the first-effect heat exchanger. Thus, for the same evaporation rate, a quintuple-effect evaporator will require about 20% of the steam required for a single-effect evaporator. In actual practice, the steam economy can vary widely because of differences in feed temperature and other energy requirements such as heats of dilution and crystallization. For some applications, single-effect evaporators are used because multiple-effect evaporators cannot be economically justified. In particular, single-effect evaporators are utilized for small evaporation rates, or for liquids that boil at high temperatures (high boiling-point elevation liquors).

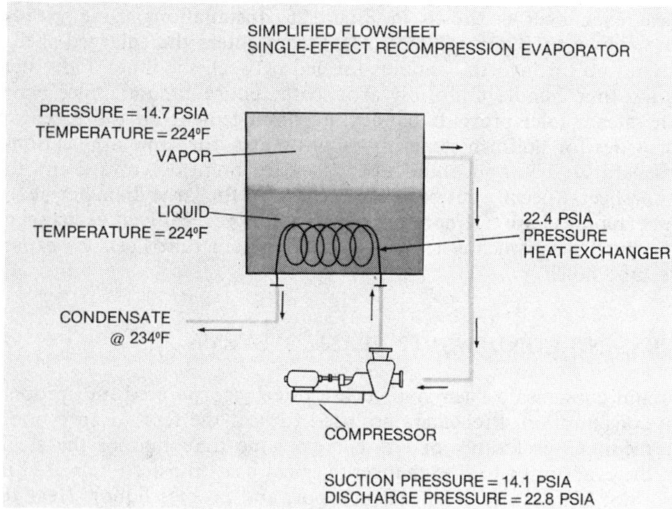

Figure 27. Simplified flowsheet of single-effect recompression evaporator (courtesy Swenson Evaporators/Whiting Corp.).

MECHANICAL RECOMPRESSION

Increasing energy costs have justified the increased use of mechanical recompression evaporators. The principle is simple. Vapor from an evaporator is compressed (with a positive-displacement, centrifugal, or axial-flow compressor) to a higher pressure so that it can be condensed in the evaporator heat exchanger. Various combinations are possible, including single-effect recompression, multiple-effect recompression, multiple-stage recompression, and single-effect recompression combined with a multiple-effect evaporator.

A simplified flowsheet of a single-effect recompression evaporator illustrates why mechanical recompression is energy efficient. All of the pertinent pressures and temperatures are given in Figure 27 for this example.

Based upon a 75% isentropic (adiabatic, reversible) compressor efficiency and a combined electric drive motor and gear reducer efficiency of 92%, the energy required to compress a single pound of vapor from 14.1 to 22.8 psia is only 49.3 Btu. Tor produce the equivalent steam from one pound of 234° F evaporator condensate requires 999 Btu. Therefore, the energy savings for a recompression evaporator are highly competitive with those of multiple-effect evaporators and depend upon the compression pressure ratio required and the relative cost of electric power and steam.

The compression ratio required comprises three components which are:

1. The boiling-point rise, i.e., the temperature of the boiling liquor minus the temperature of boiling water at the same pressure.
2. The ΔT required for heat transfer.
3. The pressure drop in the vapor pipe to and from the compressor.

Mechanical recompression is most practical for low ΔT's (larger heat-transfer areas) and low boiling-point elevations.

REFERENCES

1. Faust, A. S., et al., *Principles of Unit Operations,* John Wiley & Sons Inc., New York, 1962.
2. McCabe, W. L., and Smith, J. C., *Unit Operations of Chemical Engineering,* Third Edition, McGraw-Hill, NY, 1976.
3. Perry, R. H., and Chilton, C. H., *Chemical Engineers' Handbook,* Fifth Edition, McGraw-Hill, NY, 1973.
4. Treybal, R. E., *Mass Transfer Operations,* McGraw-Hill, NY, 1955.

CHAPTER 38

SPRAY DRYING

Paul N. Cheremisinoff

New Jersey Institute of Technology
Newark, New Jersey

John M. Phelps

Bowen Engineering Inc.
Somerville, New Jersey

CONTENTS

INTRODUCTION

Spray drying is a process in which a slurry of liquids and solids is atomized into droplets in a chamber through which heated gases are passing. The liquids are evaporated from the droplets and the solids are collected in a continuous fashion from the chamber.

Spray drying has been widely used in the pharmaceutical, chemical, and food industries. It has been used to a lesser extent in the metals-related industries. The purpose of this chapter is to review the spray drying process. The major topics to be discussed are: the basic concepts of spray drying, spray drying equipment design, atomization techniques/particle size distributions, and the economics of spray drying.

SPRAY DRYING EQUIPMENT DESIGN AND PROCESS

Figure 1 shows two basic designs for spray drying equipment. Figure 1A shows a means of atomizing a slurry using a rotary atomizer. Figure 1B shows the use of nozzles for the creation of the droplets. The spray drying process requires that the slurry be atomized into a mist in the presence of heated gases, usually air. The air can be heated using either direct or indirect gas or electric heaters.

The atomized droplets dry rapidly because of their high surface-to-volume ratio. The dried solids which remain are collected either from the chamber (coarse material) and from the cyclone (fines). Baghouse collectors or wet scrubbers can be used to clean the exhausted air. Baghouse collectors and electrostatic precipitators can sometimes also be substituted

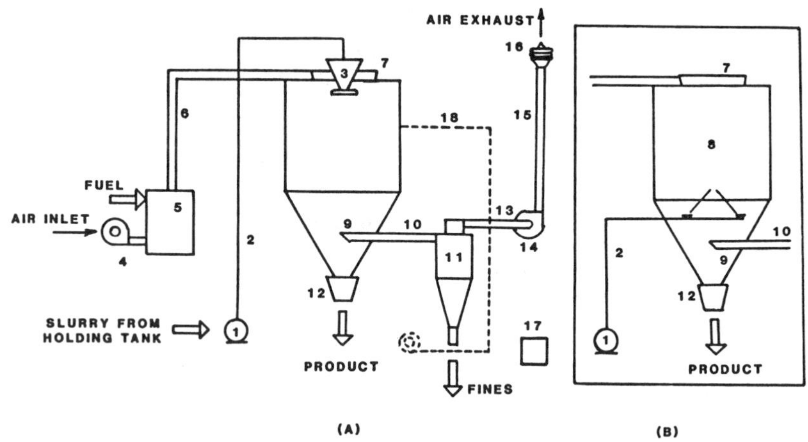

Figure 1. (A) Spray-drying installation with spinning disk atomizer; (B) installation with nozzle atomizer. (1) Feed pump, (2) feed pipe, (3) atomizer, (4) inlet fan, (5) air heater, (6) hot air duct, (7) air dispenser, (8) drying chamber, (9) outlet duct, (10) duct, (11) cyclone, (12) discharger, (13) duct, (14) exhaust fan, (15) exhaust duct, (16) air hood, (17) instrument panel, (18) fines recyling (optional).

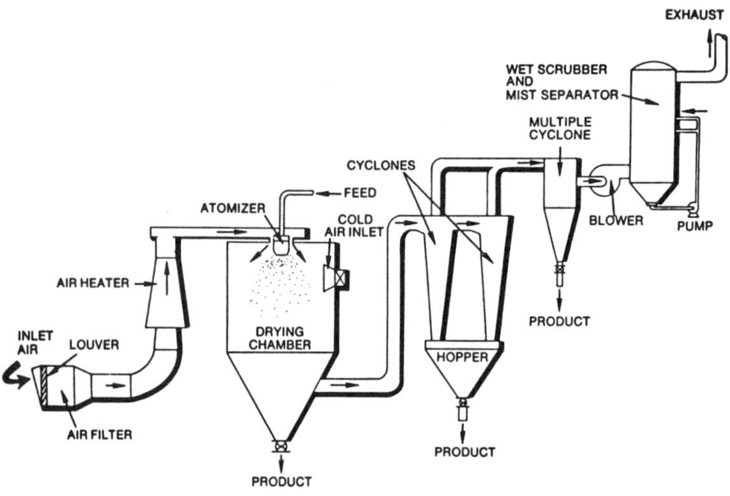

Figure 2. Total spray drying system.

for cyclone collectors. Figure 2 shows a schematic of the total spray drying system and supporting equipment. Figure 3 is a photo of typical pilot-plant spray dryer.

Other designs of spray dryers include the introduction of the air through a bustle at the cone cylinder interface or through ductwork centered above the cyclone exit. Also, the slurry can be introduced through a nozzle at the top of the spray dryer rather than at the bottom.

When the initial directions of the air and droplets are the same, drying is termed cocurrent. When they are opposite, it is termed countercurrent. Mixed flow or fountain

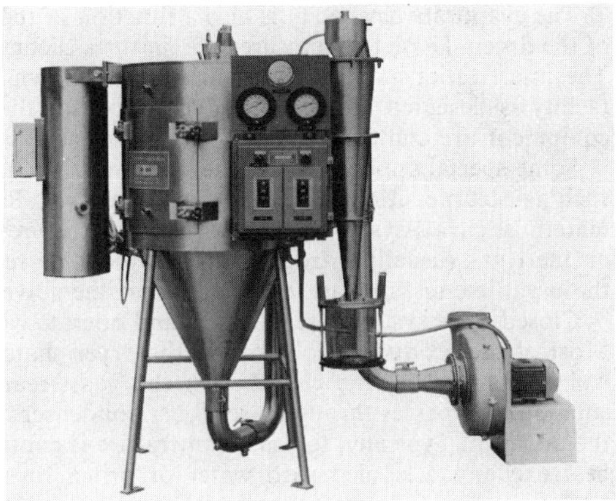

Figure 3. Typical small-scale pilot unit.

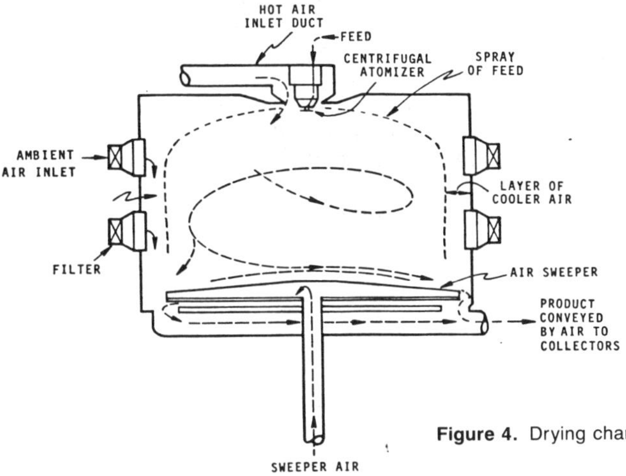

Figure 4. Drying chamber for heat-sensitive materials.

spray drying, shown in Figure 1B, allows the droplets to dry both in their upward and downward trajectories. Therefore, narrower, shorter dryers can be utilized. However, if partially dried product cannot be exposed to the high temperatures at the top of the dryer, this mode of spraying may not be applicable. A dryer design for heat-sensitive materials is shown in Figure 4.

In cocurrent, rotary atomization, the maximum temperature of a solid particle may approach within 10° C less than the outlet temperature of the dryer. This results from the evaporative cooling effect as the liquids leave the droplet as a dried particle is forming.

The size of the drying chamber is dictated by:

- The desired product's characteristics.
- The desired product throughput.
- The atomization technique.
- The solids content in the slurry.
- The evaporative capacity of the heating unit associated fans.

The evaporative capacity is also a function of the allowable inlet and outlet temperatures of the dryer. These temperatures are, in turn, dictated by the heat sensitivity of the product. The interrelationships of these factors are shown in Figure 5. Generally, a spray drying facility is designed as an integral unit so that the capacities of all the related pieces of equipment are compatible.

Some special applications in the hard metals industry require the use of organic solvents such as acetone, ethanol, hexane, and methanol. Because of the flammable nature of these materials, closed-system drying technology requires that the medium be spray dried using an inert gas (usually nitrogen is employed). By recycling the nitrogen-containing vapors, the organic emissions are eliminated and the solvent can be recycled.

Closed-cycle systems are gas tight and operate with pressures slightly above atmospheric. Most of the product is collected in the dryer chamber, though a small percentage of fines which pass out of the chamber in the gas stream is collected in the cyclone. The gas subsequently passes through a scrubber/condenser to recover the solvent and to recondition the nitrogen. Typically, the gas temperature is controlled using a recirculating scrubber and heat exchangers cooled with water or brine. In special cases, a cooling system may be necessary. The following is a list of suppliers of spray drying systems.

1. Anhydro
   165 John L. Dietsch Square
   Attleboro Falls, MA 02763
2. Bayliss Industries, Inc.
   P.O. Box 745
   Flemington, NJ 08822
3. Bowen Engineering, Inc.
   P.O. Box 898
   Somerville, NJ 08876
4. Damrow Company
   Div. of DEC International
   196 Western Ave.
   Font Du Lac, WI 54935
5. Komline Sanderson
   12 Holland Ave.
   P.O. Box 257
   Peapack, NJ 07977
6. Niro Atomizer, Inc.
   Oakland Ridge Industrial Center
   9165 Rumsey Rd.
   Columbia, MD 21045
7. Proctor Schwartz
   7th St. and Tabor Road
   Philadelphia, PA 19120
8. Swenson
   Div. of Whiting
   15700 Lathrop Ave.
   Harvey, IL 60426

CONTROL OF POWDER PROPERTIES

Spray drying can be used as either a means to remove water from a slurry or as a method for creating agglomerates of fine particles contained within the slurry. Agglomeration is realized by the inclusion of a binder.

Most of the applications of spray drying in the metals industry require the formation of free-flowing agglomerates. Many of these powders are for the production of a pressed part. Therefore, the particle size distribution and the bulk density are the two most important properties of the powders.

The particle size distribution is a function of the atomization conditions, and the solids content of the slurry. In general, a lower solids content will yield a finer average particle size. The attainable maximum solids content varies from material to material, but can usually be increased with the use of additives.

The bulk density may vary as a function of the solids content of the slurry, the inlet temperature, and slurry additives. Lower bulk densities are generally achieved from low solids content slurries. Also, excessive inlet temperatures can cause lower bulk densities. Rapid evaporation of the liquids causes the partially dried droplets to expand rapidly and therefore decrease density. Intentional introduction of frothing agents may cause air to be entrapped in the slurry also leads to lower bulk densities.

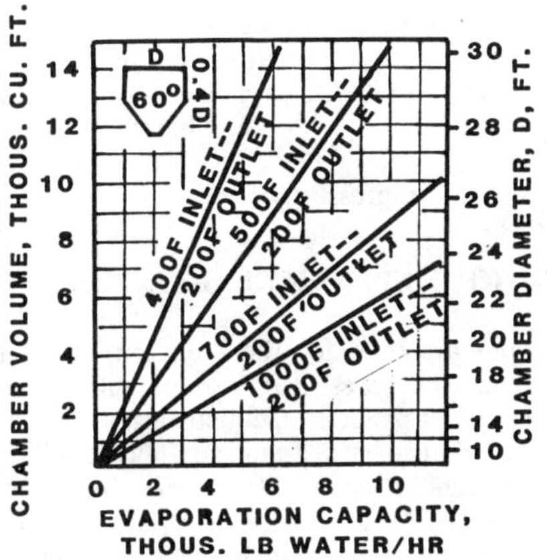

Figure 5. Relationship between chamber volume and evaporation capacity of spray dryers. (Data based on retention time of 20 seconds.)

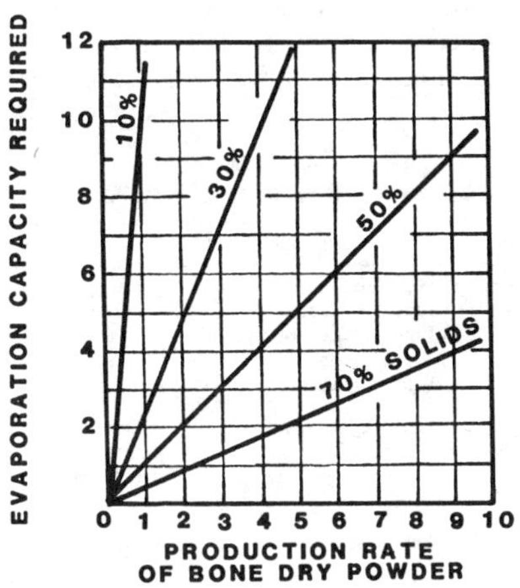

Figure 6. Curves for determining the evaporation capacity of spray dryers from the production rate and solids concentration. (Use consistent units for production and evaporation capacity.)

The moisture content of a powder is controlled by the outlet temperature of the spray dryer. For a given airflow and inlet temperature, the outlet temperature will be controlled by the rate of slurry feed and its percent solids. The higher the percentage solids, the less water that is required to be evaporated, and the higher the throughput of dry product. If a product can withstand higher inlet temperatures, its throughput can be increased. This relation is shown in Figure 6.

ADVANTAGES OF SPRAY DRYING

There are several advantages for the use of spray drying versus other processing techniques. Those listed below refer particularly to the ability to form agglomerates for subsequent pressing and sintering operations.

1. It is a continuous rather than a batch process.
2. The particle size, bulk density, and moisture content can be closely controlled.
3. The particle shape is spherical, allowing for excellent flowability.
4. The particles have very short exposure time. Therefore, heat sensitive materials can be spray dried.
5. Additives for lubrication in die pressing can be incorporated easily.

ATOMIZATION TECHNIQUES/PARTICLE SIZE DISTRIBUTIONS

Three standard techniques are used for the atomizing of slurries for the spray drying process. They include pneumatic (pressure), centrifugal (spinning disc), and two-fluid. Studies of the efficiency, resultant particle size distribution, and trajectory of particle droplets using these various techniques have been conducted. Table 1 shows the advantages and disadvantages of these techniques for the atomization of slurries. Table 2 shows the relative particle sizes produced by these techniques. Spinning atomizers use centrifugal force to atomize liquid fees. Typical peripheral speeds are 250 to 600 ft/sec at up to 60,000 lb/hr. Figures 7 to 17 show some atomizers and spray-dried products.

Note that, in general, the largest particle sizes (600 μm) are achieved by the pneumatic (or pressure) nozzle. The centrifugal atomizer (or the spinning disk) yields particles sizes up to 300 μm and the two-fluid nozzle particles only up to about 200 μm.

Centrifugal atomization entails the introduction of the slurry into a rotating device. This device is equipped with vanes or holes through which the slurry exits. The slurry is atomized

Table 1
Advantages and Disadvantages of Various Spray Drying Atomization Systems

Atomization System	Advantage	Disadvantage
Centrifugal (rotating disc)	High feed rates Less downtime Low pressure pumps can be used Seldom plugs	Larger diameter dryer required Coarse particle size not obtainable
Hydraulic (pressure)	Large agglomerate capability Smaller dryers can be used	Downtime due to part wear and plugging High pressure pumps required
Two-fluid (pneumatic)	Less part wear	Particle size distribution broad Compressed air needed

Table 2
Relative Particle sizes for Various Atomizing Systems

Atomizing System	Mean Particle Size
Centrifugal (spinning disc)	
High speed	25–100 μm
Medium speed	50–200 μm
Low speed	100–300 μm
Pneumatic (pressure nozzle)	
High pressure	25–100 μm
Medium pressure	50–200 μm
Low pressure	100–300 μm
Very low pressure	200–600 μm
Two-Fluid Nozzle	
High pressure	10– 50 μm
Medium pressure	25–100 μm
Low pressure	50–200 μm

into fine droplets, the average particle size of which is a function of the wheel design and the peripheral speed of the wheel. (See Figures 7–17). Each supplier of spray drying equipment has its own designs of wheels. Most of these incorporate a wear resistant material for extended life.

Centrifugal atomization has several advantages as listed in Table 2.

Because the openings in the wheel are very large, there is very little possibility for them to become plugged, therefore, ensuring continuity of atomization and very little downtime. Rotary atomizers are capable of very high feed rates. Rotary atomizers can typically develop up to 400 h.p. and even higher in certain applications.

Two major disadvantages of centrifugal atomization are:

1. Coarse particle sizes are not obtainable should they be desired.
2. A larger diameter dryer is generally required since most of the drying must take place in a horizontal plane.

Figure 7. Inverted-saucer disk gives high shearing action over sharp peripheral edge.

Figure 8. Slotted wheel is used where fine particles are desired or for high-viscosity feed.

Figure 9. Radial-hole disk provides high degree of particle size uniformity.

Figure 10. Multiple-tier design is used for high-capacity spray drying.

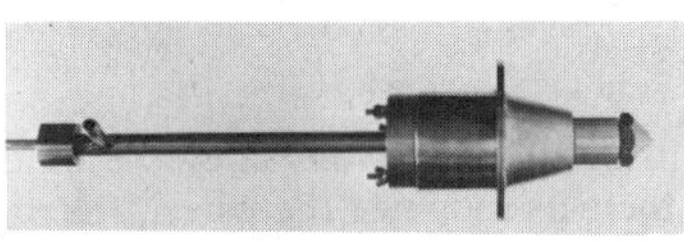

Figure 11. Two-fluid externally mixing nozzle atomizer for co-current spray drying.

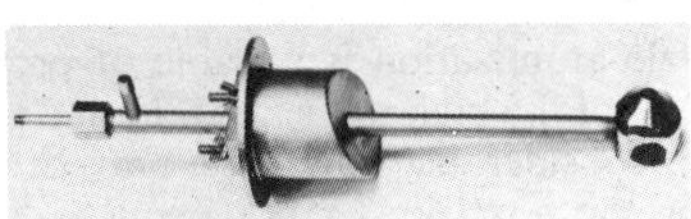

Figure 12. Two-fluid externally mixing nozzle atomizer for mixed-flow spray drying.

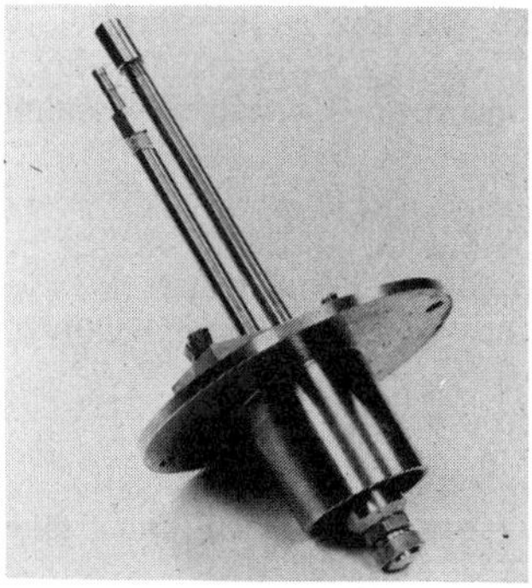

Figure 13. Two-fluid externally mixing nozzle atomizer for cocurrent spray drying.

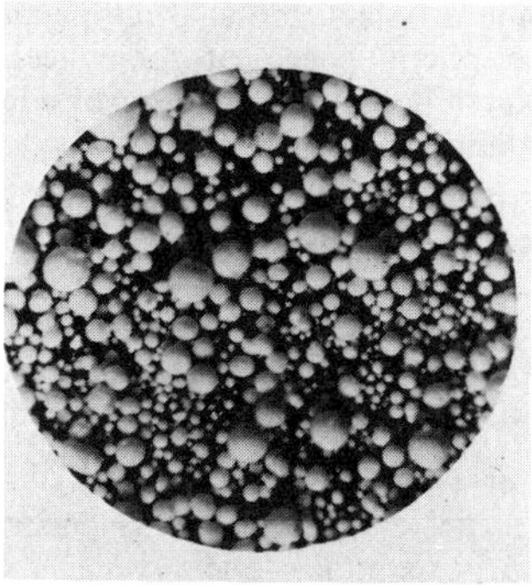

Figure 14. Two-mixed-flow spray-dried particle distribution from fluid nozzle.

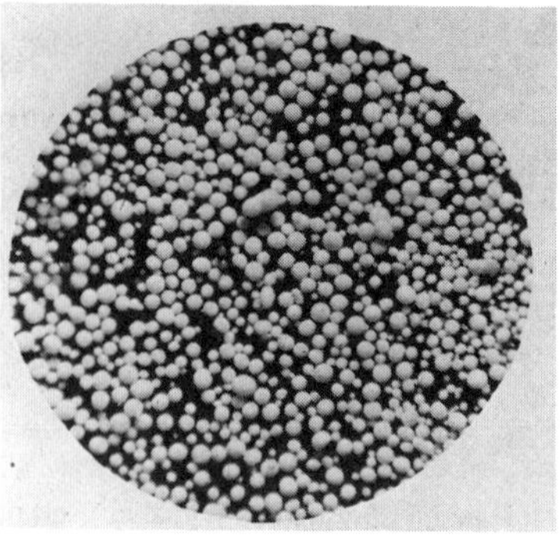

Figure 15. Mixed-flow spray-dried particle distribution from single-fluid nozzle.

Figure 16. One-half HP air-driven centrifugal spray machine 50,000 rpm for various atomizers.

The particles must be dried prior to hitting the sidewalls of the chamber. Because of the high momentum imparted to them as they exit the wheel, the dryer must be compatible with the required drying time.

Hydraulic or pressure nozzle atomization is capable of producing the largest diameter agglomerates. However, because of the high pressures required to atomize the slurry through a comparatively small orifice, considerable downtime may be experienced while replacing worn parts and in unplugging the nozzle.

Using atomizing nozzles in the fountain spray position extends the residence time of the droplets in the chamber. As a result, smaller dryers can be employed for nozzle atomization for materials which are not sensitive to heat.

Two-fluid atomization produces the broadest particle size distribution of all three techniques. However, it is a relatively easy technique for producing atomized droplets. Wear problems do not exist because high pressures are not employed. The orifices used are typically larger than those used for pressure (hydraulic) atomization, therefore, plugging is not as severe a problem. This technique does, however, require the use and additional

Figure 17. Bowen electric centrifugal spray machine with CSE atomizer.

expense of pressurized air. This mode of atomization finds ready acceptance where limited quantities of powder are required.

ECONOMICS OF SPRAY DRYING

Spray drying is one of the most economical ways of drying slurries. Even though the cost of capital equipment is high, the overall cost of spray drying is relatively low, compared to other processes.

The costs of spray drying is shown in Figure 18. Note that Figure 18 costs have been corrected to 1983 dollars using the ratios of the consumer price index for capital equipment for 1956 and 1981. The purchase price of a spray dryer may vary by a factor of two, depending upon the supplier and the completeness of the system supplied. The costs shown in Figure 18 are based on drying chambers of the proportions shown in the insert and include the following:

1. Drying chamber with insulation, door, double-hung inspection port, windows and floodlights, ductwork and air disperser, stainless-steel construction.
2. Stainless-steel cyclone dust collectors and carbon-steel stack.
3. Miscellaneous ductwork.
4. Oil or gas direct-fired air heater.
5. All motors and fans.
6. Air filters.
7. Centrifugal atomizer and drive.
8. Panel board and instrumentation for automatic control.
9. Engineering services for design and fabrication of items 1 to 8.
10. Erection in customer's plant of items 1 to 8.
11. Engineering starup service.

Total installed investment will range from 200 to 300 percent of the costs indicated when buildings, foundation, access platforms, piping, wiring, etc., are added.

The oil and power consumption for various-sized spray drying chambers is a function of the evaporative capacity. (See Figure 19). Using these data, one can calculate, as shown below, the estimated cost of spray drying per kilogram of product. For this example, an evaporative-capacity of 1,000 kilograms per hour has been chosen with a type 2 temperature profile within a 14-foot diameter dryer. It is assumed that the total installed cost of the spray

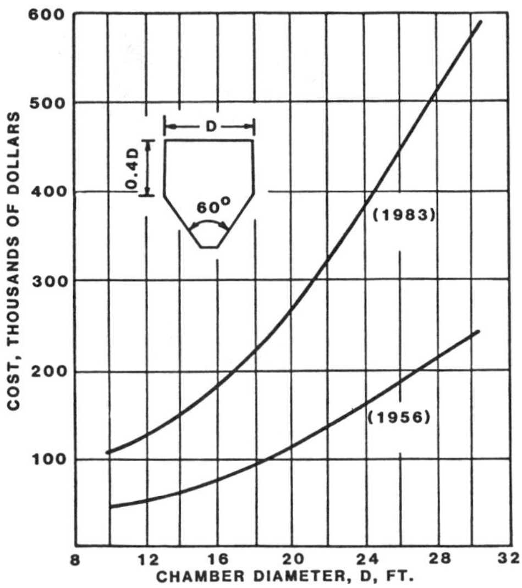

Figure 18. Cost of spray-drying equipment as a function of drying chamber size (from *Chemical Eng. Costs,* quarterly, 6:4 (1956), corrected to January 1983).

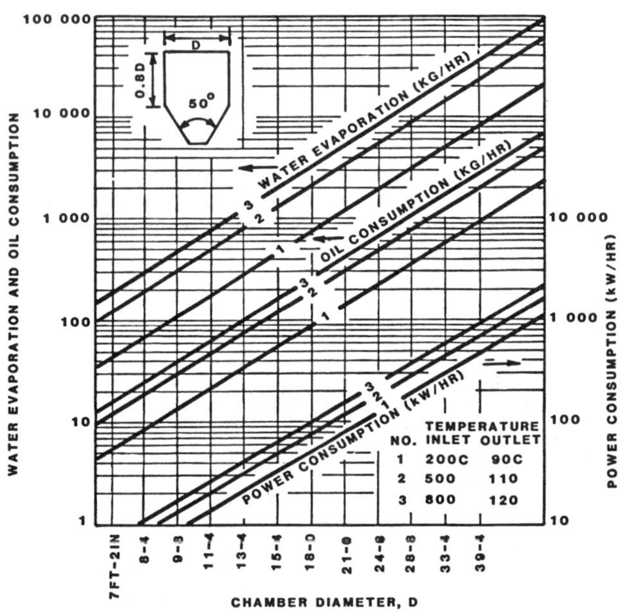

Figure 19. Water evaporation, oil consumption, and power consumption for spray dryers as a function of drying chambers and three different temperature profiles (assumed residence time of 22 seconds).

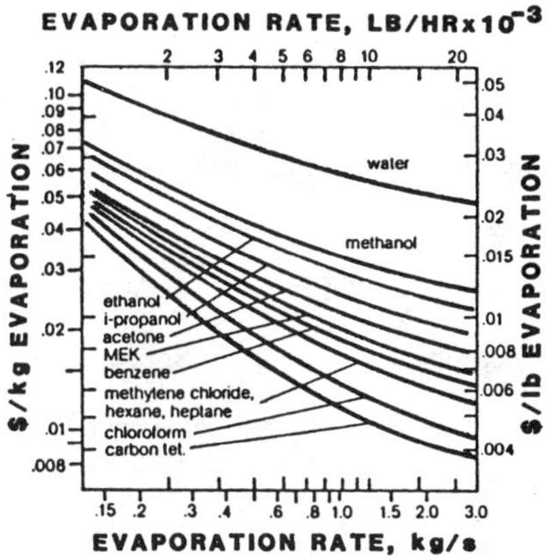

Figure 20. Approximate cost of evaporation of various liquids by spray drying.

dryer is \$450,000. and that the cost is written off at 20% per year. A 6,000-hour-per-year operating time is also assumed. In one hour of operation under these conditions, 100 kilograms of oil would be consumed and 50 kilowatt hours of power. The following calculation yields the cost of production of the power.

		Cost (\$/hr)
Equipment:	$\frac{\$450{,}000.}{6000\text{ h}} \times 0.2$	15.00
Labor:	\$20/manhour (including overhead)	20.00
Fuel: (Øil)	(100 kg/h) × (\$0.31/kg)	31.00
Power:	(50 kwh) × (\$0.05/kwh)	2.50
Total hourly cost		\$68.50

$$\text{Cost of evaporation: } \frac{\$68.50}{1000\text{ kg}} = \$0.0685 \text{ per kg of evaporation}$$

$$\text{Cost of production: } \$0.0685 \times \frac{\text{feed moisture (kg)}}{\text{feed solids (kg)}} \text{ per kg of product}$$

Therefore, for a solids content of 70%, the cost of production per kilogram is approximately \$0.03.

One advantage of spray drying with a volatile fluid rather than water is that the cost for evaporation is considerably diminished (see Figure 20). Normally, however, the cost of capital equipment is considerably higher because of the need for protection against explosions and fire and for gas and liquid recycling capability.

CUSTOM SPRAY DRYING SERVICE COMPANIES

Because of the initial high cost for installation of a spray drying facility, it is necessary to do extensive test work in order to determine the feasibility of the process for a given product. This testing can be performed by a number of sources. Most of the equipment manufacturers listed earlier will perform test work in order to aid in the specification of a dryer. Also, there are several custom spray drying companies throughout the United States which will spray dry quantities ranging from test lots through full production lots on a toll-fee basis. A list of some of these companies follows.

1. American Custom Drying
   109 Elbow Lane
   P.O. Box 309
   Burlington, NJ 08016
2. Cincinnati Chemical Processing
   458 Main St.
   Batavia, OH 45103
3. Custom Processing
   11 East 6th Ave.
   Trenton, NJ 08619
4. Estrone Chemical Co.
   Box 127
   Calvert City, KY 42029
5. General Spray Drying Services
   1001 Newark Ave.
   Elizabeth, NJ
6. Lea Manufacturing Co.
   Chemical Finishes Div.
   49 Waters Ave.
   Everett, MA 02149
7. Spray Drying Service Inc.
   505 North Ave.
   Garwood, NJ 07027
8. Spray-Tek, Inc.
   344 Cedar Ave.
   Middlesex, NJ 08046
9. Bowen Engineering, Inc.
   7 Finderne Ave.
   Bridgewater, NJ 08807

CHAPTER 39

CERAMIC SPRAY DRYING

John M. Phelps

Bowen Engineering, Inc.
Somerville, New Jersey

Olev Ratsep

Northeast Protech Inc.
Boonton, New Jersey

CONTENTS

INTRODUCTION

Spray drying is a liquid-solids separation process whereby virtually instantaneous evaporation occurs and dry, spherical particles are generated containing certain physical properties.

Spray drying, in brief, is accomplished by atomizing a pumpable liquid/solids formulation (i.e. emulsions, slurries/slips, solutions, or suspensions) into fairly uniform droplets inside a drying chamber through which a heated, dry gas is flowing. The gas provides the heat for evaporation to occur. The atomized droplets relinquish free (and to an extent bound) moisture to the gas primarily because of the high surface-area-to-volume ratio.

The total elapsed time for the entire spray drying process can range usually from 5 to 30 seconds, sometimes longer, depending upon the degree of residual moisture and the particle size, bulk density, required in the final product.

The dried solids, in the form of usually discrete spherical particles, remain suspended in the gas stream until they either settle out at the bottom of the drying chamber or are conveyed to, and recovered in, a secondary collection device (usually a cyclone or baghouse).

SPRAY DRYING AND CERAMICS

Today spray drying is a widely accepted method of processing ceramic slips in order to produce dry pressing grade powders.

While spray dryers were initially developed to process very heat-sensitive products (eggs, milk, and pharmaceuticals), by the mid 1960s more spray dryers were being used to process heat-insensitive inorganics (including ceramics) such as:

Aluminas	Graphites
Beryllias	Oxides
Carbides	PZT's
Catalysts	Steatites
Clays	Tile bodies
Ferrites	Titanates
Forsterites	Zirconias

Spray drying systems have specifically been developed and designed to satisfy the exacting requirements needed to be present in ceramic powders for subsequent processing. In general, using pressing grade ceramic powders as an example, the properties which remain important to favorable pressing operations are:

1. Flowability for rapid and complete filling of dies.
2. Accurately controlled high density.
3. Minimum use of additives.

The accurate control of these properties can result in decreased overall operating costs, especially from the standpoint of reducing the amount of rejected press powder.

Narrow, uniform particle size range (Figure 1) is thought to render improved flow properties by the premise that if all particles have the same spherical diameter they will exhibit the best flow characteristics. However, on the other hand, a wide particle-size distribution (Figure 2) will render higher density because the smaller, spherical particles will occupy the interstices between the larger ones.

In between the requirements of flowability and density a balance must exist. Occasionally, the desired particle-size, distribution can be obtained by thoughtful selection of the mode of atomization without the need for backmixing of fines or screening.

For pressing operations where flowability is the dominant requirement, the fines which are separated from the process airstream and collected in the secondary collector could be recycled. However, when density is the more important factor, part or all of the fines may be backmixed with the chamber bottom collected powder.

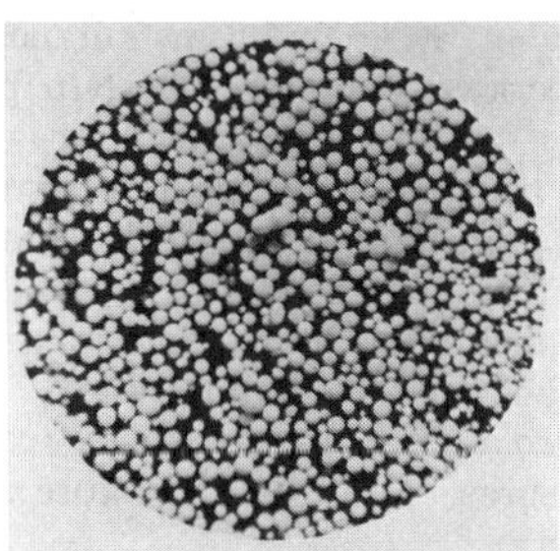

Figure 1. Spray-dried product showing narrow uniform particle size range.

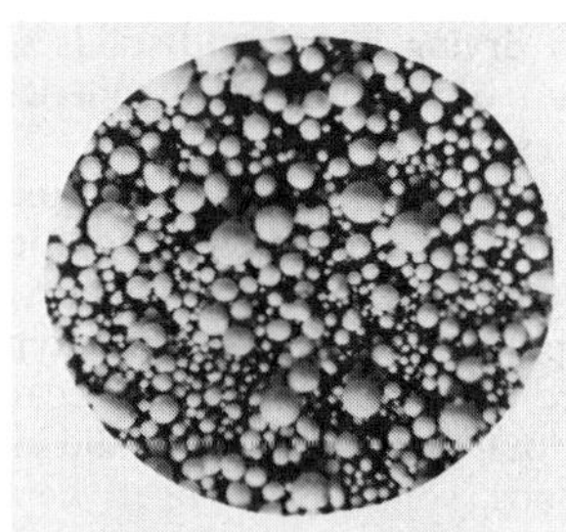

Figure 2. Spray-dried product showing a wide particle size distribution.

ECONOMICS OF SPRAY DRYING

Spray drying is one of the most economical process methods for drying ceramic slips. Though the initial capital investment for a spray drying system can be substantial, the overall cost of spray drying is relatively low compared to other processes.

Justification of spray drying requires more than an economic comparison of spray dryer prices. Labor, maintenance, and power costs are generally low for spray drying. Fuel costs may be greater, but they can be less for products where spray drying permits higher inlet drying air temperatures.

The most prevailing factor for many spray drying applications is the elimination of other processing steps which are cost prohibitive in terms of labor, maintenance, utilities, and product losses.

As an example, let us look at a spray drying application for a pressing grade alumina and see the economics on a cost per pound basis of product and evaporation.

Feedstock Definition

Feed type	Slurry
Feed identification	Alumina + Binder + Lubricant
Feed solvent	Water
Feed solids concentration	65%
Feed temperature	60° F
Feed rate	5,725 #/hr

Dryer Operation Parameters

Size of spray dryer	14′ diameter × 10′ cylinder height × 70° conical bottom
Mode of atomization	Centrifugal (rotary)
Dryer inlet temperature	650° F
Dryer outlet temperature	250° F
Dryer exhaust airflow	8,716 ACFM
Evaporation rate	2,000 #/hr
Product rate	2,980 #/hr
Residual product moisture	.10%
Utilities	
Natural gas	4,007 SCFH
Horsepower connected	110
Costs	
Spray drying system	$895,000
Natural gas	0.65 $/therm
Power	0.11 $/kw hr
Labor	$20/hr
Other	
Depreciation	10 years
Operation	4,000 hrs/year

Cost	Cost ($/HR)
Equipment	
$\frac{\$895{,}000}{4{,}000 \text{ hrs/yr} \times 10 \text{ yrs}}$	22.37

Natural gas

$\frac{4{,}007{,}000 \text{ BTU/hr}}{100{,}000 \text{ BTU/therm}} \times 0.65$ \$/therm	26.05

Power

110 HP × 0.746 kw/hp × 0.11 \$/kw hr	9.03

Labor

Assume 1 man at \$20/manhour (overhead included)	20.00
Total	77.45 \$/hr

Cost per pound evaporation

$\frac{77.45 \text{ \$/hr}}{2{,}000 \text{ \#/hr}}$	0.038 \$/#

Cost per pound product

$\frac{77.45 \text{ \#/hr}}{2{,}980 \text{ \#/hr}}$	0.025 \$/#

ADVANTAGES OF SPRAY DRYING

1. Single-step operation from liquid feed to dry product. Frequently eliminates such steps as filtering, grinding, and classifying.
2. The process is flexible in that it can be run on a continuous or intermittent basis.
3. Low maintenance costs because there are few moving parts (i.e. feed pump, fan/motor, rotary atomizer).
4. Most ceramic products spray dry to uniform spherical particles. Thus they are less dusty and more free flowing.
5. High purity is attainable because bickup of metallic contamination is minimized (usually 5 ppm or less).

Table 1
Design Parameters for Three Types of Spray Drying Systems

Design Parameter	Mode of Atomization: Two-Fluid Nozzle	Pressure Nozzle	Centrifugal
Chamber			
Diameter (feet)	4.5	8.0	16.0
Cylindrical height (feet)	15.0	10.0	12.0
Feed solids concentration (%)	68	68	68
Feed solids rate (lbs/hr)	180	445	1,500
Useful press powder rate (lbs/hr)	150	400	1,200
Press powder to fines ratio	5 : 1	9 : 1	4 : 1
Inlet/outlet air temperature profile (°F)	520 260	660 270	450 235
Heat duty to evaporate water (BTU/lb H_2O evaporated)	2,763	2,070	2,310
Overall heat duty (Btu/hr)	235,000	430,000	1,630,000
Connected horsepower	18	20	80
Estimated installed cost (\$) (less ancillary equipment)	186,000	226,000	670,000

6. A spray dryer designed for one specific application could be used for a number of other products with changes in operating parameters.
7. It is possible to control product qualities, such as final moisture and bulk density, by regulating temperatures. Particle size/distribution can be controlled by changing the mode of atomization or atomizing techniques.

ATOMIZATION AND SPRAY DRYER SELECTION

Comparing the efficacy of one atomization technique with another tends to become comparisons of biases. One technique that might be used is an evaluation of energy utilization along with systems costs to produce a given press body. As an example, this evaluation is predicated on a mean particle size of 100 microns for the basis of comparison; using two-fluid nozzle, pressure nozzle, and centrifugal atomization to generate the particle size.

Two-Fluid Nozzle

For this example, experience holds that a drying chamber 4-1/2 feet in diameter and 15 feet in cylindrical height is adequate to contain the spray pattern generated by an externally mixing two-fluid nozzle atomizer. A mixed-flow drying concept is used.

Further, experience has shown that a useful product rate in the order of 150 lbs/hr can be attained from a solids input of 180 lbs/hr. The rather tall form configuration of the drying chamber is dictated by the narrow spray pattern (about 20°) and the high velocity of the spray, which is typical of the externally mixing two-fluid nozzle.

Pressure Nozzle

As with the two-fluid nozzle, a mixed-flow drying concept is employed. Though in the case of using pressure nozzle atomization, a drying chamber of a squatter configuration is used. This is because the pressure nozzle generates a wider (40°–60°) included spray angle which requires a larger chamber diameter to permit the adequate drying of the larger particles produced. Since the velocity of the spray in the pressure nozzle is much less than that of the two-fluid nozzle, the drying chamber need not be as cylindrically tall in proportion. In this example a chamber of 8 feet in diameter and 10 feet in cylindrical height is used. A useful product rate of about 400 lbs/hr is readily attained in this size drying chamber, with a fines recycle rate of 45 lbs/hr or less. The fines recycle rate is greatly reduced because of the far narrower droplet distribution achieved with the pressure nozzle.

The pressure nozzle also enables a press body powder, with mean particle sizes ranging from about 105 to 150 microns, to be produced by employing a variety of atomizing pressures. These pressures would range from 450 to 150 psig respectively. The variance in atomizing pressures further allows for the selection of chamber sizing. For instance, the smaller mean particle size generated by the higher pressure might allow for the selection of a drying chamber of perhaps 6 feet in diameter. The chamber size for this example (8′ diameter × 10′ cylindrical height) allows for the proper drying of the larger particle size.

Centrifugal

The centrifugal atomizer essentially generates a horizontal spray pattern. This in turn dictates that the radial distance must be very wide in order to properly dry the largest droplets without impinging on the internal wall surfaces of the drying chamber. As part of this example, the drying chamber is 16 feet in diameter and 12 feet in cylindrical height. Typical design criteria then determine the remaining parameters, such as the useful product rate (1,200 lbs/hr) and a fines recycle rate of about 300 lbs/hr. Since a cocurrent drying concept is employed with centrifugal atomization, the inlet/outlet air temperature profile must be increased in order to produce a product density comparable to that produced using the mixed-flow concept associated with both types of nozzle atomization.

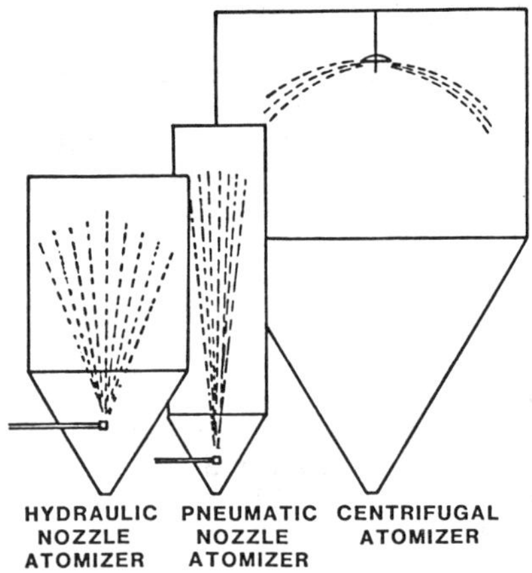

Figure 3. Relative chamber size to produce 100 microns mean particle size versus atomization mode.

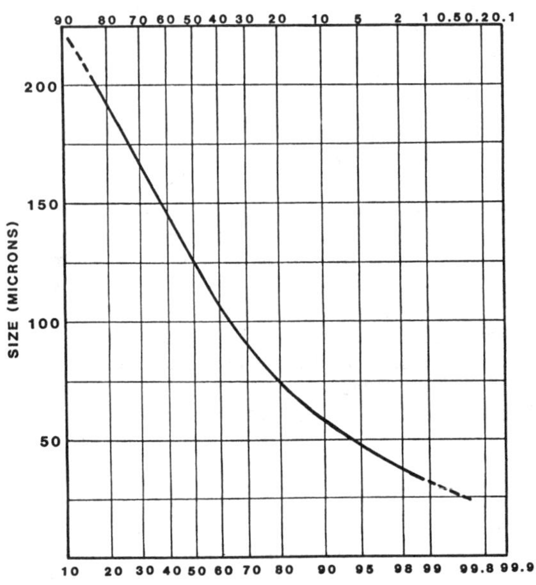

Figure 4. Pneumatic nozzle atomization, particle size distribution.

SPRAY DRYER SELECTION

In Table 1 is a tabulation of the pertinent design parameters for each of the three spray drying systems previously described. This data represents a ready comparison for producing a press body powder containing a mean particles size of 100 microns.

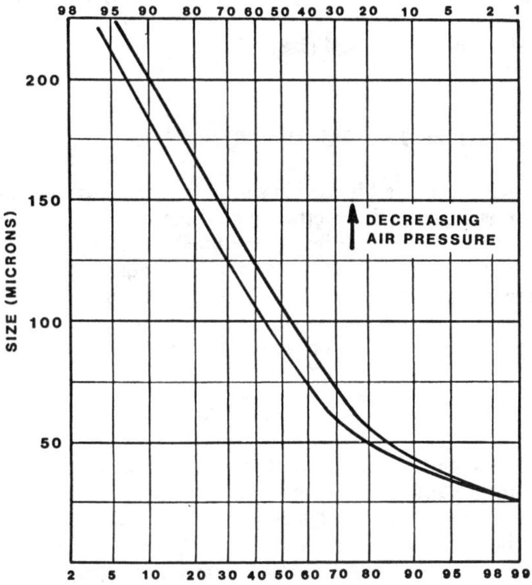

Figure 5. Pneumatic nozzle atomization with changing atomizing air pressure—cumulative particle size distribution.

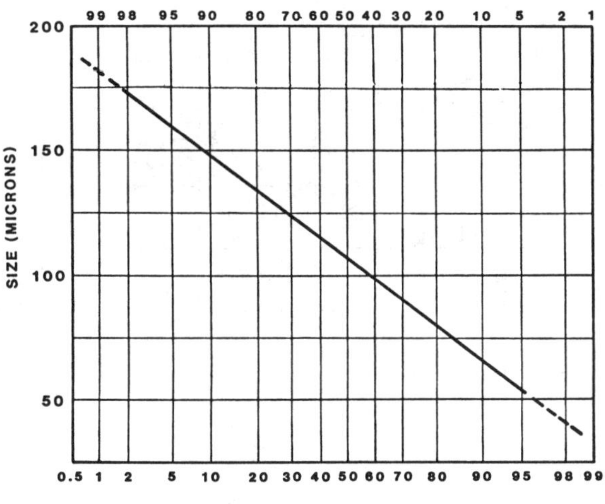

Figure 6. Hydraulic nozzle atomization—cumulative particle size distribution.

As can be seen in the tabulation, there are a significant number of alternatives which require careful consideration when selecting a spray drying system. The first question to be answered is the quantity of useful press powder required. In most instances this will amount to a few hundred pounds per hour. This virtually eliminates the centrifugal atomization system from consideration, with one exception. In the case of fluid-cracking catalysts (FCC) where a generally accepted average particle size of 60 microns is desired within a narrow distribution, the centrifugal atomization system would be the most appropriate selection.

Figure 3 illustrates the relative physical sizes of each spray drying chamber with regard to height/diameter ratios.

ATOMIZATION AND PARTICLE SIZE DISTRIBUTION

Another question which must be addressed in selecting a spray drying system when nozzle atomization is under consideration is whether the wider particle size distribution of the two-fluid nozzle or the narrower distribution of the pressure nozzle is the more acceptable.

Each method of atomization imparts its own unique stamp on the resultant press body powder by virtue of the mean particle size and the distribution that is generated.

Figures 4 through 8 illustrate the plots of some typical particle size distributions as obtained in spray dryers under actual product conditions.

Two-Fluid Nozzle

Very wide particle sizes and distributions are characteristics of the two-fluid nozzle atomizers-particularly the externally mixing type. Figure 4 illustrates the particle size distribution of a ferrite press powder produced with a two-fluid externally mixing nozzle. As Figure 5 confirms, the particle size distribution can be modified somewhat by changing the composition of the feed slip, selecting alternate nozzle metering parts (both compressed

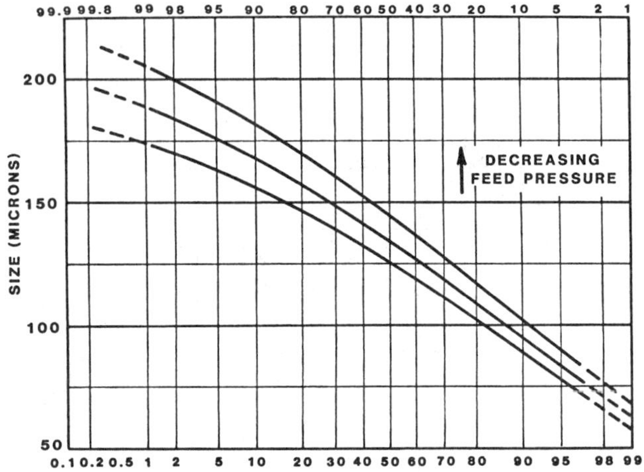

Figure 7. Hydraulic nozzle atomization with changing feed pressure—cumulative particle size distribution.

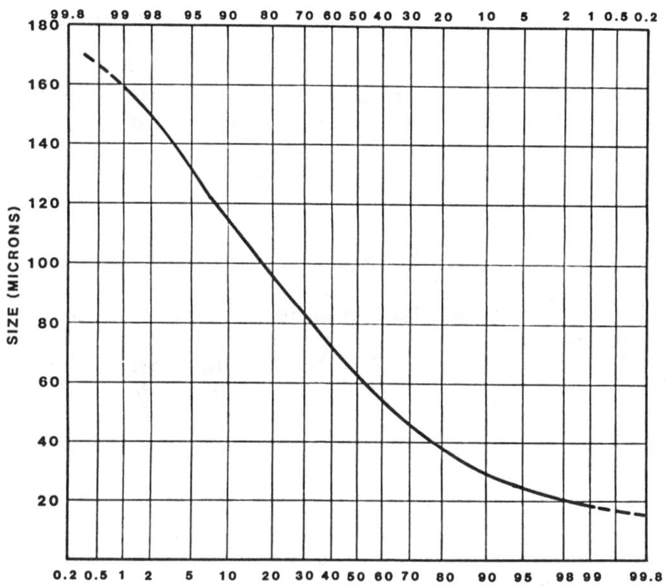

Figure 8. Centrifugal atomization—cumulative particle size distribution.

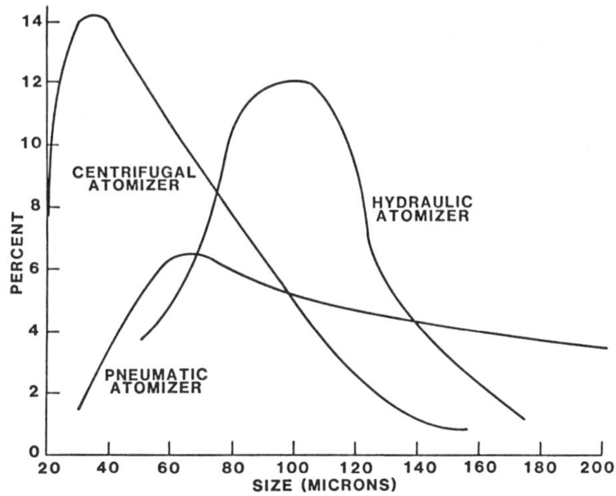

Figure 9. Histogram of particle size distribution generated by different atomizing modes.

and and fluid orifices) and varying the atomizing fluid (compressed air) pressure (volume). However, with few exceptions, the distribution will remain quite wide with a tendency to skew the curve more often towards the fines end. The press body tends to exhibit slightly higher bulk/green densities due to the finer particles filling in the voids or open areas between the larger coarse particles. This can be beneficial particularly where green density is important.

Pressure Nozzle

Figure 6 represents a truly normal distribution which is readily attainable using a low pressure nozzle atomizer (Delavan's SDX Series) with typical slip solids concentration. The use of other nozzle configurations (i.e. low slip solids and/or high atomizing pressures) will result in a wider distribution as depicted in Figure 7. More often than not, the powder obtained lacks the presence of dustiness. The lack of dust is desirable in a high speed pressing operation.

Centrifugal

Figure 8 illustrates that the centrifugal atomizer produces a particle size range which essentially follows a normal distribution; though it is somewhat skewed to the coarse end and slightly wide in this distribution. With some modification to the initial feed slip and the design of the centrifugal atomizer, the distribution can be altered to a certain extent.

CHAPTER 40

OPTIMIZING IN A CLASS OF DISPERSED-MATERIAL DRYING OPERATIONS

Stanislaw Sieniutycz

Institute of Chemical Engineering
Warsaw Technical University
00-645 Warsaw, Warynskiego 1, Poland

CONTENTS

INTRODUCTION

When considering optimization of various dryers, continuous or multistage (i.e. discrete) models may be used. In this work a general discrete approach is developed which has turned out to be very suitable for optimization of the class of dryers and other equipment in which the process state varies in a single distinguished direction of space or time. The processes investigated can be either continuous or discrete by physical nature. In the former case the discretization is the formal procedure serving for computational purposes.

Among processes occurring between gas and solid, Figure 1, there exist various processes with a dense suspension of granulated material, such as steady-state fluidization (in a single apparatus), fluidization in a cross-flow cascade, fluidization in a counterflow of multiplate apparatus, fluidization in a horizontal exchanger, and batch fluidization. In a batch process the state of a solid varies as a function of time, in others (steady-state) as a function of distance. The thermodynamical state of a solid in isobaric processes is described by enthalpy I_s and moisture content W_s, while a gas state by enthalpy i_g and humidity X_g. The initial

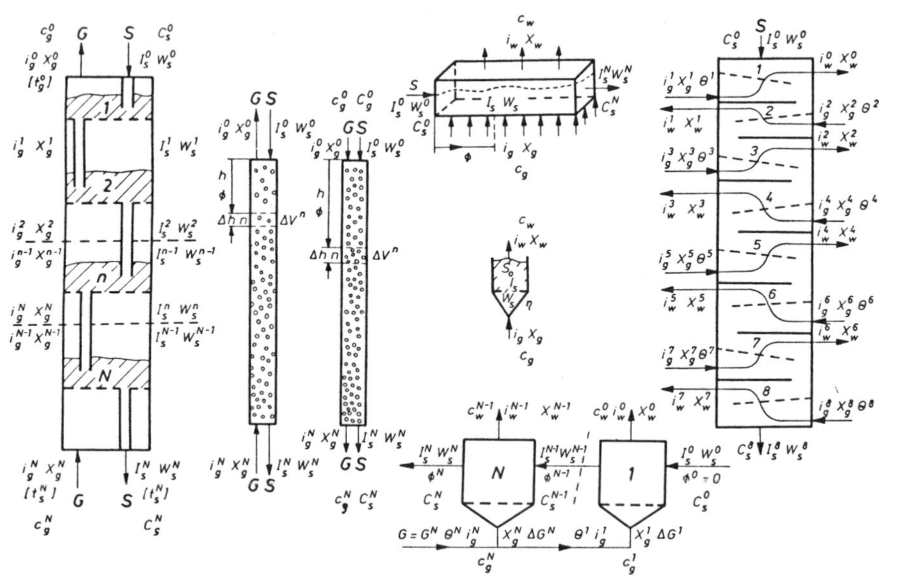

Figure 1. The processes under consideration.

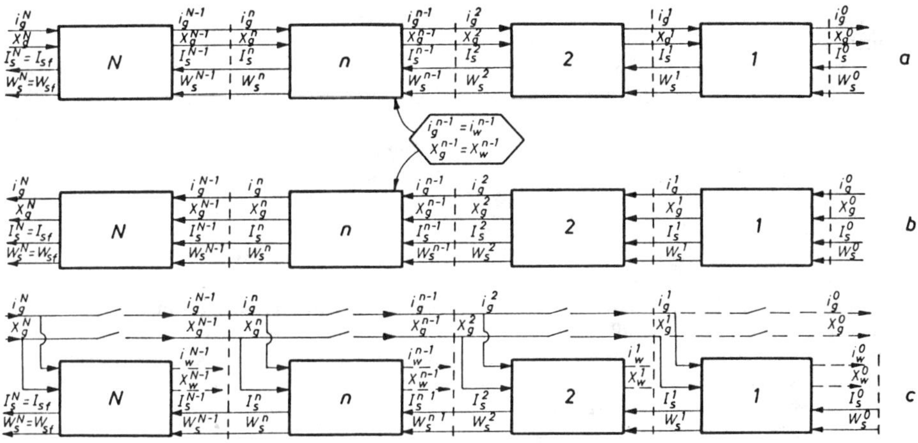

Figure 2. The block schemes of the processes under consideration: (A) countercurrent flow; (B) cocurrent flow; (C) crosscurrent flow and batch fluidization.

state of solid phase is assumed always as constant ($I_s^0 = \bar{I}_{si}$, $W_s^0 = \bar{W}_{si}$) while the final state $I_s^N = I_{sf}$, $W_s^N = W_{sf}$ may vary. Gas plays the role of the controlling phase, while the solid is the controlled phase.

The notations in Figures 1 and 2 correspond with "forward" algorithms in relation to the flow direction of the solid phase.

Apart from the previously mentioned fluidization processes Figure 1 also shows schemes of processes of pneumatic transport type with a dilute suspension of solid particles in counterflow and parallel-flow systems. There are moderate possibilities of controlling these processes in adiabatic systems and they refer to changes of inlet gas parametres.

We will consider the optimal transitions of a solid from one thermodynamical state I_s^0, W_s^0 to another I_s^N, W_s^N which occurs during the drying or wetting granulated material in all the just-described systems. With the assumption of the adiabatic character of every system as a whole, various optimization problems (based on a minimum economical cost criterion) will be formulated and solved (see text below as well as Reference 1).

Flow diagrams in Figure 2 describe *discrete* schemes of *all* the processes in question. According to designations in the diagrams, the examined processes can be described by suitable systems of ordinary difference equations. Such a description is quite exact for multistage processes which are discrete from their nature. On the other hand in the case of continuous processes, difference equations describe a discrete approximation of a model based originally on differential equations, this approximation being necessary for numerical calculation.

There exists a possibility of synthetic approach to optimization processes of Figure 1 with the use of difference schemes from Figure 2 in the sense of mathematical models and algorithms.

Owing to that, the processes are described together by a common generalized system of difference equations (Equations 8 through 10 in this paper), and they may be examined with the help of similar computational programs. Although the generalization has formal rather than physical character, it is very useful in optimization since it reduces considerably the number of programs and consumption of computer time.

In the event of fluidized processes (dense suspension of solid particles) the described difference equations system refers to the case when an outlet gas from the stage is in thermodynamical equilibrium with the solid phase leaving the stage* (equilibrium model). This case happens often in the industrial apparatus in which bed height is of about 0.5 m, and it occurs because of the large specific surface of a solid. In the instance of countercurrent and cocurrent flow processes with dilute particle suspension, the difference model is implied by both kinetical and balance considerations. Constancy of exchange coefficients of mass (K_g') and heat (α) and of global Lewis factor $Le = K_g'C_g/\alpha$ is assumed.

The difference model is valid for both hygroscopic and over-hygroscopic moisture contents of solids. It is a result of taking drying equilibrium curves into account when calculating enthalpy $i_s[I_s, W_s]$ and humidity $X_s[I_s, W_s]$ for a gas in equilibrium with solids. It is assumed that the tables $i_s[I_s, W_s]$ and $X_s[I_s, W_s]$ are given; the way to calculate them is discussed in References 1 and 2.

Performance index and state equations in this work contain some implicit relations. It should be pointed out that all these relations are of the same type namely, they are purely thermodynamical relations. This is why the determination of their detailed form does not depend on a mathematical problem, but on the examined physical system. Thanks to the use of implicit forms of the relations (e.g., $i_s[I_s, W_s]$), the derived equations are valid regardless of the materials used in a process, and moreover, they make rather complex mathematical descriptions clearer. In optimization programs, however, all thermodynamical functions have to be defined as analytical formulas or tables, depending on the examined physical system. Independent variables (e.g., W_s) and constant parameters (e.g., $\bar{X}_g$) are given most often in square brackets ([]) to stress the variables in relation to which a particular quantity is considered.

Recurrence optimization equations always operate with unit prices of a given phase, meant as functions of the enthalpy and absolute humidity. E.g., expression $c_g[i_g, X_g]$ denotes the price of gas mass unit, calculated for enthalpy i_g and humidity X_g.

PURPOSE OF OPTIMIZATION AND TYPES OF OPTIMIZATION PROBLEMS

The conventional method of carrying out fluidization cross-flow processes and a batch fluidization process (see Figures 1 and 2) consists in using constant policy (i_g, X_g) of inlet gas. In this case enthalpy of inlet gas i_g and its absolute humidity X_g do not vary with the residence time of solids for each of the discussed processes. It is known that driving forces of such processes are large at the beginning and small at the end of the process, and that this fact makes the discussed processes uneconomical (i.e., it causes large exploitation costs). Hence, it is useful to investigate cross-flow processes and batch fluidization in which the state of inlet gas varies, and it is required to find an optimal cost for changes of state with the residence time of a solid. Besides, it is advisable to examine a process with a constant but unknown state of inlet gas which can be made more economical by the selection of the gas parameter level.

This implies that there are two basic types of optimization problems for cross-flow fluidization processes and for batch fluidization.

In the simplest case of the inlet gas constant state the optimization problem is linked only with searching for *optimal level* of this state (function optimization).

In the more complex case, when the inlet gas state varies, we have the possibility to select this state for each stage of the process (functional optimization). From the mathematical point of view we then deal with a *variational problem* in a *continuous version* (continuous processes) or a *discrete version* (multistage processes). In this event both the optimal path of a process (trajectory) and the optimal final state of a solid are searched for. However,

* Bubbling has no influence on the form of the state equations, which are in fat just the balance equations.

in the precedent case we only look for the optimal final state of the solid phase (synonymously connected with the optimal state of an inlet gas).

It is obvious that for adiabatic countercurrent flow and cocurrent flow processes (schemes A and B, Figure 2) the economization increase can be carried out *only* by determining the optimal constant state of the inlet gas (or by determining the optimal final state of a solid). Although the possibilities of influencing the process are limited here, it should be kept in mind that the change of the inlet gas state (and the accompanying change of solid state) strongly affects the economical improvement of the discussed processes. Therefore, examining optimal conditions of carrying them out is also important.

REMARKS ON MATHEMATICAL METHODS

Two-point Boundary Conditions

In all the optimization cases discussed in this chapter we deal with a two-point boundary conditions problem; a part of the conditions $I_s^0 = \bar{I}_{si}$, $W_s^0 = \bar{W}_{si}$ is imposed at the beginning, another part (e.g., $I_s^N = \bar{I}_{sf}$, $W_s^N = \bar{W}_{sf}$) at the end of the process. Carrying out the optimization requires the prior finding of gas parameters i_g^N, X_g^N for all simpler processes (e.g. counterflow) with the trajectory determined *without* optimization. On the other hand, it is required to find a sequence $\{i_g^n, X_g^n\}$, n = 1, 2, ..., N, for more complex processes for which a trajectory is obtained *as a result* of optimization.

For simpler processes, in the case of dense phases the quantities i_g^N, X_g^N result from the solution of system of conservation equations. In the case of dilute suspensions, kinetic equations must also be included. Obtaining values i_g^N, X_g^N with the use of the classical method requires the assumption of two values, i_g^0, X_g^0, and solving the process equations system with initial conditions for I_s^0, W_s^0, i_g^0, X_g^0. Providing the resultant values I_s^N, W_s^N are equal to values of $\bar{I}_{sf}$, $\bar{W}_{sf}$ implied by final conditions, the point i_g^0, X_g^0 was chosen properly, otherwise the procedure should be repeated. This trial method as an optimization element would have to be repeated many times for various final conditions, so it would be very uneconomical and difficult to use. Difficulties of this sort occur both for continuous and multistage processes and, of course, they also appear for more complicated problems with a variational character of optimization problems.

Invariant Imbedding, [3.4]

It is known that for variational problems two-point boundary conditions do not constitute difficulty if the dynamical programming method [5] is used. Keeping this in mind and remembering the complicated character of analytical relations (drying equilibria), the dynamic programming method in "forward" version is used for variational problems (searching for the optimal sequence $\{i_g^n, X_g^n\}$, n = 1, ..., N). What should be the mathematical method in the case of nonvariational problems in which gas parameters i_g^N, X_g^N are searched for? It should not be a surprise that these quantities as well as the performance index may be obtained with the help of a (little known) method, being a counterpart to dynamic programming, called the invariant imbedding method [3, 4]. In this method (quite different from the previously discussed classic trial method) no extremization sequence is carried out on stages, as in dynamic programming. This is because decisions on stages 1, 2, ..., N − 1 do not appear for nonvariational problems. For the performance index (optimized only for n = N) the recursive equation, similar to that of dynamic programming [5], is valid but *without* the sign of extremizing. Gas parameters i_g^n, X_g^n are determined at successive stages from the operating line equations and not as result of a sequential optimization (occurring for dynamic programming). This implies that for nonvariational problems, for which the invariant imbedding method is used, the equations of the operating line must be always defined, even before optimization. For variational problems, however, using dynamic programming this line is determined by the optimization.

For cross-flow and batch fluidization processes (diagram C in Figure 2) both variational and nonvariational problems exist. It is pointed out by the discussed diagram which suggests either independence of states i_g^{n-1}, X_g^{n-1} and i_g^n, X_g^n, or their interdependence. Providing $i_g^{n-1} = i_g^n$, $X_g^{n-1} = X_g^n$, this relation corresponds with nonvariational optimization problems in which the *optimal level* of inlet gas constant parameters is searched for.

COMMON MATHEMATICAL MODEL

Example of Derivation of State Equations for Counterflow Processes

Not all process state equations from Figure 1 will be derived in this paper. Hereafter we limited ourselves to derive, as an example, counterflow process equations. For other cases the ready-and-easy-to-check results will be given.

In the case of the multistage counterflow (dense particle suspensions) the assumption of thermodynamical equilibrium of outlet phases is operative. Regarding the equilibrium relations $i_g^{n-1} = i_s[I_s^n, W_s^n]$, $X_g^{n-1} = X_s[I_s^n, W_s^n]$ (c.f. Figure 2, scheme a), in mass and energy balances of the n-th stage wo obtain the following relations:

$$I_s^n - I_s^{n-1} = G/S(i_g^n - i_s[I_s^n, W_s^n]) \tag{1}$$

$$W_s^n - W_s^{n-1} = G/S(X_g^n - X_s[I_s^n, W_s^n]) \tag{2}$$

They describe transformations of output state I_s^n, W_s^n into input state I_s^{n-1}, W_s^{n-1} depending on the values i_g^n, X_g^n. Hence, they are state equations corresponding with "forward" algorithm.

In the event of continuous counterflow the process stage is characterized by a small bed height $\Delta h = H/N$ (Figure 1) and small volume, ΔV^n. Therefore, it is necessary to consider kinetic equations. The energy exchange equation has the following form:

$$S(I_s^n - I_s^{n-1}) = \alpha a(t_g^n - t_s^n)\Delta V^n + i_{ps}^n K'_g a(X_g^n - X_s^n)\Delta V^n \tag{3}$$

where i_{ps}^n = partial enthalpy of moisture in a vapor state in equilibrium with solids ($i_{ps}^n = i_{ps}[I_s^n, W_s^n]$ and $X_s^n = X_s[I_s^n, W_s^n]$), while the mass exchange equation is as follows:

$$S(W_s^n - W_s^{n-1}) = K'_g a(X_g^n - X_s^n)\Delta V^n \tag{4}$$

Because enthalpies and concentrations are used for describing the examined processes, the right-hand side of Equation 3 should be expressed as a function of adequate enthalpies and humidities of a gas. In order to do this the following equation is employed:

$$i_g - i_s = C_g(t_g - t_s) + i_{ps}(X_g - X_s) \tag{5}$$

in Equation 3. This yields the equation:

$$S(I_s^n - I_s^{n-1}) = \alpha C_g^{-1} a\left[i_g^n - i_s^n + i_{ps}^n\left(\frac{K'_g C_g}{\alpha} - 1\right)(X_g^n - X_s^n)\right]\Delta V^n \tag{6}$$

The dimensionless time of continuous countercurrent process is defined by:

$$\varphi^n - \varphi^{n-1} = \frac{K'_g a \Delta V^n}{S} \tag{7}$$

which also determines the time increments $\Delta\varphi^n$ at the n-th stage (for multistage counterflow models Equations 1 and 2, an analogous formal definition: $\varphi^n - \varphi^{n-1} = G/S$ is not introduced (see Table 1) as it is neither a useful nor physical one.

Equations 4, 6 and 7 constitute state equations for continuous countercurrent processes. Although their physical origin is different from that of Equations 1 and 2, they also describe

Table 1
Definitions of Coefficients in Common Mathematical Model

No.	Name of Process and Type of Optimization Problem	$\theta^n = \Delta\Phi^n$	g^n	g_0	e	j	$\varphi^n = \sum_{l=1}^{n} \theta^l$
1	Steady-state one-stage fluidization (dense suspension, EM) nonvariational problem	$\frac{G}{S}$	$\frac{G}{S}$	$\frac{G}{S}$	1	0	$\frac{G}{S}$
2	Multistage countercurrent (dense suspension, EM) nonvariational problem	$\frac{G}{S}$	$\frac{G}{S}$	$\frac{G}{S}$	1	0	Not defined
3	Multistage crosscurrent (dense suspension, EM) or nonvariational problem	$\frac{\Delta G^n}{S}$	$\frac{\Delta G^n}{S}$	10^{10}	1	0	$\frac{G^n}{S}$
4	Batch fluidization (dense suspension, EM) variational or nonvariational problem	$\frac{\Delta\eta G}{S_0}$	$\frac{\Delta\eta G}{S_0}$	10^{10}	1	0	$\frac{G\eta}{S_0}$
5	Continuous countercurrent (dilute suspension, NM) nonvariational problem	$\frac{GN_g}{SN}$	$\frac{G}{S}$	$\frac{G}{S}$	$\frac{1}{Le}$	Le = 1	$\frac{GN_g}{S}$
6	Continuous cocurrent (dilute suspension, NM) nonvariational problem	$\frac{GN_g}{SN}$	$-\frac{G}{S}$	$-\frac{G}{S}$	$\frac{1}{Le}$	Le − 1	$\frac{GN_g}{S}$
7	Continuous crosscurrent in a horizontal exchanger (dense suspension, EM) variational or nonvariational problem	$\frac{G}{SN}$	$\frac{G}{SN}$	10^{10}	1	0	$\frac{G}{S}$

EM — equilibrium stage model
NM — nonequilibrium stage model

(as every state equation does) transformations of state coordinates at the stage depending on the controls. This stresses the mathematical character of the state equations (see Table 1).

Generalized State Equations, Operating Line Equations, and Balance Equations

Examining mathematical models of processes (Figure 1) leads to the conclusion that all these processes may be described by the following *generalized system of difference state equations:*

$$I_s^{n-1} = I_s^n - \theta^n e(i_g^n - i_s[I_s^n, W_s^n] + j i_{ps}[I_s^n, W_s^n](X_g^n - X_s[I_s^n, W_s^n])) \tag{8}$$

$$W_s^{n-1} = W_s^n - \theta^n (X_g^n - X_s[I_s^n, W_s^n]) \tag{9}$$

$$\varphi^{n-1} = \varphi^n - \theta^n \quad \text{(not used for multistage counterflow)} \tag{10}$$

referring to state transformations at the n-th stage. These equations cover also the already-examined counterflow processes.

In these equations the effect of solid state variables (enthalpy I_s^n and moisture content W_s^n) on enthalpy i_s and humidity X_s of gas and on equilibrium enthalpy of moisture i_{ps} (in vapor state) was pointed out.

Quantities θ^n, e, and j in Equations 8 through 10 are defined in relation to the type of process according to Table 1. The table also contains the definitions of other quantities discussed in the text below. They all constitute an input for a computer.

Symbols of state in Equations 8 through 10 and in Table 1 have the following meaning:

G = total dry gas mass flow in all systems in Figure 1

ΔG^n = gas mass flow at the n-th stage $\left(\sum_{k=1}^{n} \Delta G^k = G^n\right)$

S = dry solid mass flow

S_0 = mass of a solid in the system

η = residence time of a solid in the system; $\Delta\eta = \eta/N$

N = total number of stages

φ = total dimensionless residence time of solids, or a total dimensionless gas flow according to definitions in Table 1 (depending on a particular process, it is recommended to use one term or another).

$Le = K_g' C_g/\alpha$ = Lewis factor

$N_g = \dfrac{K_g' aV}{G}$ = total number of mass transfer units for the examined continuous system with a dilute suspension of solids. In this definition

V = volume,

a = specific surface

K_g' = mass exchange coefficient referring to the driving force $X_g - X_s[I_s, W_s]$ For fluidized systems (dense phases) quantity N_g does not appear in the model; the table describes in fact the case of $N_g \to \infty$.

As an example, a particular form of Equations 8 through 10 will be given for a crosscurrent cascade of fluidized-bed dryers. Regarding the appropriate definitions of the Table 1 we obtain:

$$I_s^n - I_s^{n-1} = \frac{\Delta G^n}{S}(i_g^n - i_s[I_s^n, W_s^n]) \tag{11}$$

$$W_s^n - W_s^{n-1} = \frac{\Delta G^n}{S}(X_g^n - X_s[I_s^n, W_s^n]) \tag{12}$$

$$\varphi^n - \varphi^{n-1} = \frac{\Delta G^n}{S} \tag{13}$$

which agrees with the mass and energy balance of the n-th stage. The quantity φ^n is a dimensionless total gas flow for an n-stages cascade while the ratio $g^n = \Delta G^n/S$ describes dimensionless gas flow at the n-th stage. Quantities connected with kinetics do not play any role in this case. The proceedings for other processes is similar.

The system of state Equations 8 through 10 deals generally with quantities of both a thermodynamic and kinetic nature. It is especially apparent in cases of dilute suspensions. The system contains:

State variables: I_s^n, W_s^n, φ^n.

Controlling variables: i_g^n, X_g^n, θ^n (these quantities determine decisions for variational problems).

Constant parameters: Le, N_g

Thermodynamical functions of state variables:

$$i_s[I_s, W_s], \quad X_s[I_s, W_s], \quad i_{ps}[I_s, W_s]$$

The explicit form of the thermodynamic functions depends on materials used in the process, and it is determined on the basic of drying equilibrium data (see Reference 2 for more information). Because of that Equations 8–10 describe the examined processes in a general form. For optimization calculations, analytical equations, or tables of the discussed functions for a particular material and gas are used.

Optimization of all processes (multistage and continuous) is based on discrete state (Equations 8 through 10). For multistage processes the number of stages N depends on a type of problem. In the case of continuous processes sufficiently large number of stages N is required in order to gain appropriate accuracy of approximation.

The system (Equations 8 through 10) will be used both for the problems in which optimal paths and optimal final states are searched for (variational problems) and for those in which gas parameters i_g^n, X_g^n are not selected and only optimal final states are searched for (nonvariational problems). In the latter problems which may refer to any of the processes in Figure 1, state Equations 8 through 10 must be accompanied by operating line equations.

Operating Line Equations

These equations (see Figure 2) are here defined as relations of the following form common for all the discussed processes:

$$\frac{I_s^n - I_s^{n-1}}{g_0} = i_g^n - i_g^{n-1} \tag{14}$$

$$\frac{W_s^n - W_s^{n-1}}{g_0} = X_g^n - X_g^{n-1} \tag{15}$$

The designations, according to Table 1, are:

$$g_0 = \begin{cases} G/S & \text{countercurrent flows} \\ -G/S & \text{concurrent flows} \\ \infty & \text{crosscurrent flows and batch fluidization} \end{cases} \tag{16}$$

The quantity g_0 is, then, a constant parameter defined according to the type of process.

Considering the condition $g_0 \rightarrow \infty$ for the last case in Equation 16, the approach consists of assigning a g_0 value in a computational program (hence numbers 10^{10} appear in appropriate positions of Table 1). It should be remembered that Equations 14 through 16 are valid *only* for nonvariational problems.

Operating line equations meant as Equations 14 through 16 are identical with material balance equations only for cases A and B in Figure 2 (counter- and cocurrent flow). For case C, however, (crosscurrent flow), they assume (according to the scheme in Figure 2) constant enthalpy and gas feed composition, and hence do not describe relations between parameters of inlet and outlet gas streams at each stage. These relations are described; for all cases, by balance equations.

Balance Equations

In order to determine changes of economical value of gas and solid streams at n-th stage it is necessary to know all input and output states. Energy and material balance at the n-th stage of any process yields (c.f. Figure 2):

$$g^n(i_g^n - i_w^{n-1}) = I_s^n - I_s^{n-1} \tag{17}$$

$$g^n(X_g^n - X_w^{n-1}) = W_s^n - W_s^{n-1} \tag{18}$$

where, according to Table 1:

$$g^n = \begin{cases} G/S & \text{for countercurrent flow} & \text{a} \\ -G/S & \text{for cocurrent flow} & \text{b} \\ \Delta G^n/S & \text{for crosscurrent flow} & \text{c} \\ \Delta\eta G/S_0 & \text{for batch process} & \text{c} \end{cases} \tag{19}$$

Equation 17 through 19 are employed for both variational and nonvariational problems.

Writing balance equations in a common form (Equations 17 through 19) requires remembering the meaning of index "W" in symbols i_W^{n-1}, X_W^{n-1} as well as in others (e.g., C_w^{n-1}), where this index is used. The index "W" means:

Gas outlet, for countercurrent flow	a
Gas inlet, for cocurrent flow	b
gas outlet, for crosscurrent flow and batch fluidization	c

Quantity g^n is called a dimensionless gas flow at the n-th stage.

Confrontation of Equations 14 through 16 and 17 through 19 leads to an obvious conclusion that for countercurrent flow (a) and parallel flow (b) : $i_g^{n-1} = i_W^{n-1}$ and $X_g^{n-1} = X_w^{n-1}$. For these processes the operating line gas parameters are at the same time the gas parameters resulting from the balance. It is, of course, not valid for crosscurrent flow and batch processes (c).

Additional Algebraic Restrictions

Optimization algorithms for processes in Figure 1 [1] are for a specified initial state $I_s^0 = \bar{I}_{si}$, $W_s^0 = \bar{W}_{si}$ and for the prescribed moisture content of a final state $W_s^N = \bar{W}_{sf}$ (technical requirement). Controlling gas parametres i_g^n, X_g^n have to lie within the range determined by inequalities:

$$t_{g*} \leqq t_g^n[i_g^n, X_g^n] \leqq t_g^* \tag{20a}$$

$$X_g^n \geqq 0 \tag{20b}$$

$$\beta_g[i_g^n, X_g^n] \leqq \beta_g^* \tag{20c}$$

$$\beta_g[i_w^{n-1}, X_w^{n-1}] \leqq \beta_g^* \tag{20d}$$

where t_{g*} and t_g^* = the least and the greatest allowed gas temperature respectively
β_g^* = the largest relative humidity

The same system of inequalities must be fulfilled by parameters i_w^{n-1}, X_w^{n-1}.

Significance of the Generalized Mathematical Model

The reader may be at first unfavorably impressed by a formal generalization in the mathematical model. In addition, the generalization is valid for various processes with various assumptions. For example, processes with dense solid suspensions (fluidized systems) are described at the condition $N_g \rightarrow \infty$, which is not used for processes with dilute suspensions (e.g., countercurrent settling).

Hence, it is worth noting that the purpose of the generalization is not to suggest physical resemblance of the examined processes but to utilize the formal similarity of suitable equations to reduce the number of optimization programs. Owing to the common model various processes may be examined with the use of the same, or similar, programs which are independent from the system structure and dependent only on the optimization problem.

For variational processes (crosscurrent flow and batch processes) disposing of state Equations 8 through 10 and balance Equations 17 through 19 is enough to formulate and solve an optimization problem. Optimization can be characterized by at most three decision variables at the n-th stage. These quantities are i_g^n, X_g^n, and θ^n (or g^n). For nonvariational problems formulating and solving the optimization problem requires additionally utilizing operating line Equations 14 through 16.

Advantages of dealing with one equations system for various processes are especially apparent for the case of nonvariational problems. Quantities θ^n and g^n are constant then, and—together with other coefficients (g_0, e, j) they constitute input data for only one computer program. Depending on the process type, the input varies only according to definitions in Table 1. One may imagine how ardous it would have been to write programs separately for each process if the discussed table had not been known.

It the case of variational problems some values of i_g^n, X_g^n, θ^n might be fixed for a particular optimization problem. Such a fixed value plays the role of a process parameter, not decision variable. Depending on the number of decisions, various specific cases of a state equations system (Equations 8 through 10) are dealt with. It appears advisable to employ various computational programs in these particular cases because of the possibility of specific simplifications. Hence, for variational problems, it is advisable to utilize several programs with their working form dependent not on a process type but on a type of optimization problem.

To sum up, thanks to employing the common mathematical model and the consequent use of definitions from Table 1, the number of optimization programs is lessened.

Limiting Transition to Continuous Processes Equations

In mathematical procedure the limiting transition from discrete to continuous models is carried out with the assumption that the number of process stages approaches infinity while the final and initial states, the total volume of a system, the total flows of both phases and the total residence time of solids remain fixed.

It should be noted (Table 1) that the limiting transition cannot be carried out for discrete equations of multistage countercurrent processes with dense particle suspensions because the increase of stage numbers N does not allow to hold the assumption of equilibrium between phases leaving stages ($N_g \to \infty$) to exist. However, the discrete model of continuous countercurrent flow processes with dilute particle suspensions can be effectively used (Table 1).

In a crosscurrent model the gas flow at every stage ΔG^n or the ratio G/N approaches zero when the number of stages grows. The discrete model of the batch fluidization process leads to a continuous model when the time interval of solid residence time $\Delta\eta$ decreases towards zero. For discrete models of continuous countercurrent and cocurrent processes (with dilute particle suspensions) the increase of stage numbers causes a diminishing of the number of mass transfer units N_g at the arbitrary stage ($\Delta N_g = N_g/N$ for N_g = const.). In all these cases the dimensionless time interval approaches zero. The system of differential state equations describing continuous processes in Figure 1 has the form:

$$\frac{dI_s}{d\varphi} = e(i_g - i_s[I_s, W_s] + ji_{ps}[I_s, W_s]\,(X_g - X_s[I_s, W_s])) \tag{21}$$

$$\frac{dW_s}{d\varphi} = X_g - X_s[I_s, W_s] \tag{22}$$

The particular form of energy Equation 21 for countercurrent flow and cocurrent continuous processes is (see also Equation 6)

$$\frac{dI_s}{d\varphi} = \frac{1}{Le}(i_g - i_s[I_s, W_s] + (Le - 1)i_{ps}[I_s, W_s]\,(X_g - X_s[I_s, W_s])) \tag{23}$$

while for crosscurrent continuous processes and for the batch fluidization process:

$$\frac{dI_s}{d\varphi} = i_g - i_s[I_s, W_s] \tag{24}$$

When variables i_g and X_g are not decisions, the differential version of the operating line equations system (Equations 14 and 15) should be added to the system of differential state Equations 21 and 22. Dividing both sides of Equations 14 and 15 by $0 = \Delta\varphi$ and calculating the limit for $\Delta\varphi \to 0$ we obtain:

$$\frac{1}{g_0}\frac{dI}{d\varphi} = \frac{di}{d\varphi} \tag{25}$$

$$\frac{1}{g_0}\frac{dW_s}{d\varphi} = \frac{dX}{d\varphi} \tag{26}$$

An interesting set of problems for the reader to address is:

1. Derive Equations 22 and 24 using the balance relations desribing the exchange of mass and heat in fluidization processes ($N_g \to \infty$). Prove that Equation 24 is valid in the case of the batch process with the omission of gas-state variations in an apparatus.
2. Using the definitions in Table 1 wtite down particular forms of state equations for all the processes in Figure 1 as they were performed for Equations 11 and 12. Explain the physical meaning of each result.

PERFORMANCE INDEX OF PROCESSES IN SPECIFIED APPARATUS

Economical Considerations

All optimization problems for the processes of Figure 1 will be considered for a definite apparatus (constant investment costs) and for fixed, total gas flow.* In order to find expressions describing performance indices, attention should be turned on the fact that for a given gas flow its pumping costs in any system may be assumed as constant and do not have to be taken into account (as the investment cost of a definite apparatus). Therefore, it is enough to investigate the flow balance of gas and solid economical values, which is similar to the mass and energy balance.

For any stage of a crosscurrent multistage apparatus (Figure 1) the economical balance of the inlet and outlet streams has the form:

$$P^n = \Delta G^n c_g^n + SC_s^{n-1} - \mu^{n-1}\Delta G^n c_w^{n-1} - SC_s^n \tag{27}$$

Quantities c_g and C_s in the previous formula are economical unit values of gas and solids referring to a mass unit of inert components (dry gas and dry material). The symbol μ^{n-1} denotes a so-called outlet gas utilization coefficient at the n-th stage. When $\mu^{n-1} = 0$, the outlet gas is released into the atmosphere, so its effective value equals zero. When $\mu^{n-1} = 1$, the outlet gas has a defined economical value, utilized in other operations. In the case of the partial usage of gas $0 < \mu^{n-1} < 1$. Quantity P^n is connected with that part of the total economic cost, which is the result of the process course. Costs P^n occur because of exploiting and creating streams of economic values in a system.

Equation 27 describes various levels of outlet gas utilization thanks to the introduced coefficient μ^{n-1}. An additional factor is then formally introduced to the discussed formula. The factor's aim is to fulfill another definition of costs [6] (appearing equally often in economical considerations), in which the costs are treated as an affect of the utilization of the economical values of the streams driving the process (inlet gas streams). If Equation 27 is generalized by employing the second coefficient u in the following way:

$$P^n = \Delta G^n c_g^n + SuC^{n-1} - \mu^{n-1}\Delta G^n c_w^{n-1} - SuC_s^n \tag{28}$$

* The exception will be gas allocation problems.

then in the case $\mu^{n-1} = 0$ and $u = 0$, the quantity P^n describes the effect of utilization of inlet gas economical values. Equation 28 common for the considered cases is easier to analyze (does not require repeating the derivation in each case) and its use in the computational program is not difficult thanks to the possibility of assigning coefficients μ^{n-1} and u as the computer input data. The mathematical, not physical, character of the generalization described by Equation 28 should be pointed out.

It is easier to deal with costs P^n linked with the unit solid flow S in the discussion. Equation 28 implies that unit costs P^n/S are described by the following relation:

$$P^n/S = c_g[i_g^n, X_g^n] - \mu^{n-1} c_g[i_w^{n-1} X_w^{n-1}]g^n + C_s[I_s^{n-1}, W_s^{n-1}] - C_s[I_s^n, W_s^n] \tag{29}$$

in which it is stressed that unit economical values of gas and material should be expressed as functions of the adequate state parameters. Forms of these functions can be determined with the use of exergy (available energy) tariff (c.f. [6] and [7])

$$c_g = e_g b_g; \quad C_s = e_s B_s \tag{30}$$

where b_g and B_s = thermal exergies of gas and solid referred to mass unit of dry gas and material

e_g and e_s = appropriate economical values of exergy unit* (computing b_g and B_s is described in References 2, 6, and 7).

As far as air is concerned this approach is justifiable by the fact that air production costs are always positive and they grow with the "distance" of the air state from the surroundings (state of zero economical value). The use of an exergetic tariff is not necessary, however. One can use the price scale operative at the definite factory.

Summing all the costs P^n/S for all process stages yields the following expression of performance index:

$$K = \sum_1^N (c_g[i_g^n X_g^n] - \mu^{n-1} c_g[i_w^{n-1}, X_w^{n-1}])g^n + u(C_s[I_s^0, W_s^0] - C_s[I_s^N, W_s^N]) \tag{31}$$

Equation 31 was derived for multistage crosscurrent processes. The appropriate expression for continuous crosscurrent processes (horizontal fluidized exchanger, Figure 1) is obtained with a limiting transition when the stage number approaches infinity:

$$K = \int_0^{\varphi f} (c_g[i_g, X_g] - \mu c_g[i_w, X_w])d\varphi + u(C_s[I_{si}, W_{si}] - C_s[I_{sf}, W_{sf}]) \tag{32}$$

Derivation of a performance index for the batch fluidization process (Figure 1) requires appropriate discussion. The adequate economical criterion for batch fluidization (in the case when $\mu = 1$ and $u = 1$) is equal to the difference between the total economic value of material entering to system and leaving it enlarged by the difference between the economic value in the system at the beginning and the end of the process. If this criterion is referred to mass units of a dry solids in the system, and if gas-state changes in the apparatus are neglected, coefficients μ and u can be introduced, and the dimensionless duration of batch fluidization can be defined as in Table 1 and Equation 32. In numerical applications this formula is replaced by its discrete counterpart, i.e. formula (31).

---

* The values of $e_s \cong e_g \cong 8.0$ \$/GJ have been evaluated from economic data for an air-water-silicagel system. Exergies b_g and B_s have included the effect of the chemical exergy and therefore, they differ from those of Reference 7.

Using flow diagrams of Figure 2 it is easy to note that Equation 31 can be generalized to countercurrent and cocurrent processes, providing an additional factor ψ^n is introduced (defined later) and the dimensionless gas flow definition g^n is extended according to rules of table 1. In the event of cocurrent flow the coefficient μ^{n-1} ceases being the outlet gas utilization coefficient and its role is taken over by coefficient ψ^n.

The performance index generalized for all processes of Figure 1 has therefore the common form:

$$K = \sum_1^N (\psi^n c_g[i_g^n, X_g^n] - \mu^{n-1} c_g[i_g^{n-1}, X_g^{n-1}])g^n + u(C_s[I_s^0, W_s^0] - C_s[I_s^N, W_s^N]) \tag{33}$$

In the *Case A*, when the gas streams utilization is complete, all coefficients ψ^n, μ^{n-1}, and u are equal to unity regardless of the number of the stage (see Table 2).

In the *Case B*, the process is examined which is characterized by the lack of outlet gas utilization at end stages. The coefficient in question take up following values:

$$\psi^n = 0 \text{ for } n = N \text{ and } \psi^n = 1 \text{ for } n \neq N \text{ and also } \mu^{n-1} = 1$$

for cocurrent flow, and

$$\mu^{n-1} = 0 \text{ for } n = 1 \text{ and } \mu^{n-1} = 1 \text{ for } n \neq 1 \text{ and also } \psi^n = 1$$

for countercurrent flow.

For crosscurrent flow all coefficients μ^{n-1} are equal zero and coefficients ψ^n are equal to unity. Coefficient u, however, is always equal to unity.

In the *Case C*, when total economical value of the inlet gas stream is examined, coefficients μ^{n-1} and ψ^n assume the same values as in the previous case, but coefficient u equals zero for all processes.

Numerical values of coefficients μ^{n-1}, ψ^n, and u are collected in Table 2 for Cases A, B, and C.

Using the table it is easy to prove that for countercurrent and cocurrent flows Equation 33 can be rewritten in the following way:

$$K = (\psi^N c_g^N - \mu^0 c_g^0)g + u(C_s^0 - C_s^N) \tag{34}$$

which results also directly from the stream balance of economical value for the whole system.

The Form of a Performance Index Serving to Derive Recursive Equations

To express performance index K in a form suitable for deriving the recursive equations of the considered problems, Equation 33 is used.

The price of gas $c_w^{n-1} = c_g[i_w^{n-1} X_w^{n-1}]$ should be expressed as a function of state variables I_s^n, W_s^n and gas parameters i_g^n, X_g^n being a solution of each optimization problem discussed here. The price arguments of c_w^{n-1} are variables i_w^{n-1}, X_w^{n-1}, which are expressed as functions of I_s^n, W_s^n, i_g^n, X_g^n by elimination of variables I_s^{n-1}, W_s^{n-1} from the generalized balance equations system of Equations 17 and 18 and the generalized state equations system of Equations 8 through 10. As a result we obtain for all considered processes.

$$i_w^{n-1} = i_g^n - \frac{e\theta^n}{g^n}(i_g^n - i_s[I_s^n, W_s^n] + ji_{ps}^n[I_s^n, W_s^n]\,(X_g^n - X_s[I_s^n, W_s^n])) \tag{35}$$

$$X_w^{n-1} = X_g^n - \frac{\theta^n}{g^n}(X_g^n - X_s[I_s^n, W_s^n]) \tag{36}$$

Equations 35, 36 and 33 allow for the presentation of the performance index in the form:

$$K = \sum_{1}^{N} \Bigg\{ \psi^n c_g[i_g^n, X_g^n] - \mu^{n-1} c_g \Bigg[i_g^n - \frac{e\theta^n}{g^n}(i_g^n - i_s[I_s^n, W_s^n] + ji_{ps}^n[I_s^n, W_s^n](X_g^n - X_s[I_s^n, W_s^n])), X_g^n - \frac{\theta^n}{g^n}(X_g^n - X_s[I_s^n, W_s^n]) \Bigg] \Bigg\} g^n + u(C_s[I_s^0, W_s^0] - C_s[I_s^N, W_s^N]) \tag{37}$$

The operating line equations (valid for nonvariational problems only) are

$$I_s^n - I_s^{n-1} = g_0(i_g^n - i_g^{n-1}) \tag{38}$$

$$W_s^n - W_s^{n-1} = g_0(X_g^n - X_g^{n-1}) \tag{39}$$

where $g_0 = \pm G/S$ for countercurrent and cocurrent processes, respectively, and $g_0 \to \infty$ for crosscurrent processes as well as for batch fluidization processes (in this latter case, a very large number instead of infinity is taken in computer program).

For variational problems Equations 35 through 37 will be applied to derive recursive equations of dynamic programming with decisions i_g^n, X_g^n, g^n. For nonvariational problems Equations 35 through 39 will be used to derive an invariant imbedding set describing functions $i_g^n[I_s^n, W_s^n]$ and $X_g^n[I_s^n, W_s^n]$ and index $K^n[I_s^n, W_s^n]$ in case $g^n = g = \text{const}$. For the index K^n the invariant imbedding equation will have the structure of the dynamic programming equation with the minimization sign omitted.

COMPUTATIONAL ALGORITHMS FOR NONVARIATIONAL PROBLEMS

Invariant Imbedding Equations for Gas Parameters

In the case of nonvariational problems (e.g., for countercurrent systems) operating line Equations 38 and 39 should be taken into consideration. If dimensionless time increments

Table 2
Values of Coefficients in a Performance Index

Case	Coefficient	Countercurrent Flow	Cocurrent Flow	Crosscurrent Flow
A	μ^0	1	1	1
B	μ^0	0	1	0
C	μ^0	0	1	0
A	ψ^N	1	1	1
B	ψ^N	1	0	1
C	ψ^N	1	0	1
A	μ^{n-1} $(n \neq 1)$	1	1	1
B	μ^{n-1} $(n \neq 1)$	1	1	0
C	μ^{n-1} $(n \neq 1)$	1	1	0
A	ψ^n $(n \neq N)$	1	1	1
B	ψ^n $(n \neq N)$	1	1	1
C	ψ^n $(n \neq N)$	1	1	1
A	u	1	1	1
B	u	1	1	1
C	u	0	0	0

g^n and θ^n are constant along the trajectory, then searching for optimal final states of definite processes is the only possible type of optimization. In the discussed case index n in symbols g^n and θ^n is omitted (i.e., symbols g and θ are dealt with).

In order to calculate the performance index K^n relations should be found in which the gas state i_g^n, X_g^n is expressed as a function of the solid state I_s^n, W_s^n, and of the stage number. Searching for functions $i_g^n[I_s^n, W_s^n]$ and $X_g^n[I_s^n, W_s^n]$ is connected with finding inlet gas parameters for n-stages apparatus in cases A and C in Figure 2 (countercurrent and crosscurrent flows). In the case B (cocurrent flows) the relations searched for do describe enthalpy and humidity of the outlet gas (at the n-th stage), but the inlet gas parameters i_g^0, X_g^0 can be very simply achieved as resulting from the material and energy balance of the n-stage cascade.

The crucial point of introducing invariant imbedding equations is recognizing the existence of functional relations:

$$i_g^{n-1} = i_g^{n-1}[I_s^{n-1}, W_s^{n-1}]; \tag{40}$$

$$X_g^{n-1} = X_g^{n-1}[I_s^{n-1}, W_s^{n-1}] \tag{41}$$

for the n−1-th stage cascade, which serves to find the discussed n-th stage equations.

Substituting variables I_s^{n-1} and W_s^{n-1} of state Equations 8 and 9 to operational line Equations 38 and 39 in which Equations 40 and 41 were exploited, we receive the system of two recursive invariant imbedding equations:

$$\begin{aligned} i_g^n[I_s^n, W_s^n] = {} & \frac{\theta e}{g_0}(i_g^n[I_s^n, W_s^n] - i_s[I_s^n, W_s^n] + ji_{ps}[I_s^n, W_s^n](X_g^n[I_s^n, W_s^n] \\ & - X_s[I_s^n, W_s^n])) + i_g^{n-1}[I_s^n - \theta e(i_g^n[I_s^n, W_s^n] - i_s[I_s^n, W_s^n] \\ & + ji_{ps}[I_s^n, W_s^n](X_g^n[I_s^n, W_s^n] - X_s[I_s^n, W_s^n])); \\ & W_s^n - \theta(X_g^n[I_s^n, W_s^n] - X_s[I_s^n, W_s^n])) \end{aligned} \tag{42}$$

$$\begin{aligned} X_g^n[I_s^n, W_s^n] = {} & \frac{\theta}{g_0}(X_g^n[I_s^n, W_s^n] - X_s[I_s^n, W_s^n]) + X_g^{n-1}[I_s^n - \theta e \cdot (i_g^n[I_s^n, W_s^n] \\ & - i_s[I_s^n, W_s^n] + ji_{ps}[I_s^n, W_s^n](X_g^n[I_s^n, W_s^n] - X_s[I_s^n, W_s^n])); \\ & W_s^n - \theta(X_g^n[I_s^n, W_s^n] - X_s[I_s^n\, W_s^n]))] \end{aligned} \tag{43}$$

Equations 42 and 43 should be accompanied by initial conditions. These conditions are obtained by transforming state Equations 8 and 9 into the following form (n = 1):

$$i_g^1[I_s^1, W_s^1] = \frac{I_s^1 - I_s^0}{\theta e} - ji_{ps}[I_s^1, W_s^1](X_g^1[I_s^1, W_s^1] - X_s[I_s^1, W_s^1]) + i_s[I_s^1, W_s^1] \tag{44}$$

$$X_g^1[I_s^1, W_s^1] = \frac{W_s^1 - W_s^0}{\theta} + X_s[I_s^1, W_s^1] \tag{45}$$

On the right-hand side of energy Equation 44 the function $X_g^1[I_s^1, W_s^1]$ described by Equation 45 appears. Hence

$$i_g^1[I_s^1, W_s^1] = \frac{I_s^1 - I_s^0}{\theta e} - ji_{ps}[I_s^1, W_s^1]\frac{W_s^1 - W_s^0}{\theta} + i_s[I_s^1, W_s^1] \tag{46}$$

Equations 45 and 46 constitute initial conditions of all of the examined processes. Note that, at the same time, they describe a one-stage steady-state fluidization process. The outline of the solution of Equations 42, 43, 45, and 46 is presented in a later section. The implicit character of Equations 42 and 43 in relation to i_g^n, X_g^n should be noted.

Limiting Form of Continuous Process Equations

The limiting transition to continuous processes is accomplished by a Taylor series expansion of the function i_g^{n-1}, X_g^{n-1} in Equations 42 and 43 about the point I_s^n, W_s^n and dividing both sides by $\theta = \Delta\varphi$. As a result we obtain the following system of partial differential equations and respective boundary conditions:

$$\frac{\partial i_g[I_s, W_s, \varphi]}{\partial \varphi} + \frac{\partial i_g[I_s, W_s, \varphi]}{\partial I_s} e(i_g[I_s, W_s, \varphi] - i_s[I_s, W_s] + ji_{ps}[I_s, W_s]\,(X_g[I_s, W_s, \varphi]$$
$$- X_s[I_s, W_s])) + \frac{\partial i_g[I_s, W_s, \varphi]}{\partial W_s}(X_g[I_s, W_s, \varphi] - X_s[I_s, W_s])$$
$$= \frac{e}{g_0}(i_g[I_s, W_s, \varphi] - i_s[I_s, W_s] + ji_{ps}[I_s, W_s]\,(X_g[I_s, W_s, \varphi] - X_s[I_s, W_s])) \quad (47)$$

$$i_g[I_s^0, W_s^0, \varphi] = i_s[I_s^0, W_s^0] \quad (48)$$

$$\frac{\partial X_g[I_s, W_s, \varphi]}{\partial \varphi} + \frac{\partial X_g[I_s, W_s, \varphi]}{\partial I_s} e(i_g[I_s, W_s, \varphi] - i_s[I_s, W_s] + ji_{ps}[I_s, W_s]\,(X_g[I_s, W_s, \varphi]$$
$$- X_s[I_s, W_s])) + \frac{\partial X_g[I_s, W_s, \varphi]}{\partial W_s}(X_g[I_s, W_s, \varphi] - X_s[I_s, W_s])$$
$$= \frac{1}{g_0}(X_g[I_s, W_s, \varphi] - X_s[I_s, W_s]) \quad (49)$$

$$X_g[I_s^0, W_s^0, \varphi] = X_s[I_s^0, W_s^0] \quad (50)$$

Boundary conditions, (i.e. Equations 48 and 50) describe the lack of the process in the case of the same initial and final states of solids. It is related to using gas in equilibrium with inlet solids. Equation System 47 through 50 is valid for continuous drying (or wetting) processes under consideration. In particular cases of the process in a horizontal exchanger and of the batch fluidization process, the right-hand sides of Equations 47 and 49 are equal to zero ($g_0 \to \infty$). As, in every case, the left-hand sides of Equations 47 and 49 describe total derivatives of functions $i_g[I_s, W_s, \varphi]$ and $X_g[I_s, W_s, \varphi]$, the discussed equations describe in our case the constancy of the inlet gas state (i_g, X_g) along the trajectory of a continuous process.

Equations of Invariant Imbedding Type for Performance Index

As a result of the invariant imbedding approach (which is just the dynamical programming approach in case variational problems) we receive cost-function data (nonvariational problems) or data of optimal performance index (variational problems) in the form of tables. These describe the data depending on the final solid state and the number of stages. In the case of variational problems, obtaining these data is accompanied by sequential minimizations at stages 1, 2, ..., n (in the case of nonvariational problems only the recursive computational procedure is used). So, in the following considerations the common notation will be used:

$$\tilde{K}^N[I_s^N, W_s^N] = \begin{cases} \min K \text{ expressed as function of } I_s^N, W_s^N, N \\ \text{for variational problems} \\ K \text{ expressed as function of } I_s^N, W_s^N, N \\ \text{for nonvariational problems} \end{cases} \quad (51)$$

Using state Equations 8 and 9 in the transformed form of Equation 37 in which the difference $C_s^0 - C_s^N$ was expressed as a sum of increments $C_s^{n-1} - C_s^n$ over all stages leads to the following recursive equation describing a performance index function $\tilde{K}^n[I_s^n, W_s^n]$ in nonvariational problems:

$$\begin{aligned}\tilde{K}^n[I_s^n, W_s^n] = {} & \Big\{\psi^n c_g[i_g^n, X_g^n] - \mu^{n-1} c_g\Big[i_g^n - \frac{e\theta}{g}(i_g^n - i_s[I_s^n, W_s^n] \\ & + ji_{ps}[I_s^n, W_s^n](X_g^n - X_s[I_s^n, W_s^n])), X_g^n - \frac{\theta}{g}(X_g^n - X_s[I_s^n, W_s^n])\Big]\Big\}g \\ & + u\{C_s[I_s^n - \theta e(i_g^n - i_s[I_s^n, W_s^n] + ji_{ps}[I_s^n, W_s^n](X_g^n - X_s[I_s^n, W_s^n])), W_s^n \\ & - \theta(X_g^n - X_s[I_s^n, W_s^n])] - C_s[I_s^n, W_s^n]\} + \tilde{K}^{n-1}[I_s^n - \theta e(i_g^n - i_s[I_s^n, W_s^n] \\ & + ji_{ps}[I_s^n, W_s^n](X_g^n - X_s[I_s^n, W_s^n])), W_s^n - \theta(X_g^n - X_s[I_s^n, W_s^n])]\end{aligned} \tag{52}$$

which is applied with condition $\tilde{K}^0 = 0$.

Quantities i_g^n, X_g^n in Equation 52 are defined as a result of the solution of recursive Equations 42 and 43 (tables of functions $i_g^n[I_s^n, W_s^n]$ and $X_g^n[I_s^n, W_s^n]$).

Equation 52 with the right-hand side modified by adding a minimization sign is valid for variational problems with decisions i_g^n, X_g^n when θ = const. It is an equation of dynamic programming.

For countercurrent and cocurrent flow processes (cases A and B, Figure 2) a simpler form of Equation 52 for a performance index is valid. In order to find this formula directly, quantities i_g^0, X_g^0 should be determined on the basis of the global cascade balance:

$$I_s^N - I_s^0 = g(i_g^N - i_g^0) \qquad (i_g^0 \equiv i_w^0) \tag{53}$$

$$W_s^N - W_s^0 = g(X_g^N - X_g^0) \qquad (X_g^0 \equiv X_w^0) \tag{54}$$

and the result should be substituted into Equation 34. This leads to the formula:

$$\begin{aligned}\tilde{K}^N[I_s^N, W_s^N] = {} & \psi^N c_g[i_g^N, X_g^N] - \mu^0 c_g\Big[i_g^N - \frac{I_s^N - I_s^0}{g}, \\ & X_g^N - \frac{W_s^N - W_s^0}{g}\Big]g + u(C_s[I_s^0, W_s^0] - C_s[I_s^N, W_s^N])\end{aligned} \tag{55}$$

A more formal way of obtaining Equation 55 consists in utilizing the data of Table 1 for countercurrent and cocurrent flow in the general Equation 52. For a one-stage fluidization process the functions $i_g^1[I_s^1, W_s^1]$ and $X_g^1[I_s^1, W_s^1]$ are known from Equations 45 and 46. Using this fact for Equation 55, regarding that $\theta = g$ and assigning: $\Psi^N = 1$ and $\mu^0 = \mu$, the following formula is obtained for the costs of the one-stage steady-state fluidization:

$$\begin{aligned}\tilde{K}^1[I_s^1, W_s^1] = {} & c_g\Big[\frac{I_s^1 - I_s^0}{\theta} + i_s[I_s^1, W_s^1], \\ & \frac{W_s^1 - W_s^0}{\theta} + X_s[I_s^1, W_s^1]\Big] - \mu C_g[i_s[I_s^1, W_s^1], X_s[I_s^1, W_s^1]] \\ & + u(C_s[I_s^0, W_s^0] - C_s[I_s^1, W_s^1])\end{aligned} \tag{56}$$

which allows for calculating costs of this process directly. The reader may check that simplifying Equation 52 with the help of the definitions of Table 1 leads also to Equation 56.

ALGORITHMS LEADING TO CALCULATION OF OPTIMAL COSTS AND DECISIONS IN VARIATIONAL PROBLEMS

Problems of Optimal Gas Allocation

In this section we consider variational problems for various fluidization processes using both the Equation 52 with the minimization sign and its simplifications and modifications. The latter refer to replacing constant flow θ by variable decision θ^n.

The simplest problem, concerned with the possibility of optimization at each process stage, is the problem of gas allocation in a crosscurrent multistage fluidized drying. In a more popular version of this problem the minimum utilization of a certain resource is discussed for a process with a specified initial state and specified some coordinates of a final state. Another (dual) formulation of this problem is also possible (e.g., searching for minimum resource allocations in a process with specified final state and with some free coordinates of the initial state). The latter formulation will be discussed later, as it leads here to the simplest computational procedure.

The considered problem appears for a cross-flow battery fed with a gas of given parameters i_g^0, X_g^0, independent from the stage number (see Figure 2 scheme C for the case where $i_g^n = i_g^0$ and $X_g^n = X_g^0$). The minimization of flux of total economical gas value is connected with assigning values $\mu^{n-1} = 0$ and $u = 0$ in Equation 2 as well as taking $\theta^n = g^n = \Delta G^n/S$ in this equation. Coefficient Ψ^n is equal to unity. Therefore, the minimum of the following sum is searched

$$K = \sum_1^N c_g[i_g^0, X_g^0]\theta^n \tag{57}$$

Regarding the constancy of inlet gas parameters, the minimization of the sum of Equation 57 is equivalent to the minimization of the total dimensionless gas flow $\varphi^N = \sum \theta^n$. The minimum of φ^N should be found for the two constraints imposed by state Equations 8 and 9, taken for variables I_s^{n-1} and W_s^{n-1}, with $e = 1$ and $j = 0$. An additional restriction is the condition of nonnegative values of the flow θ^n.

The "forward" dynamic programming algorithm searches for a single optimal trajectory family having various final states I_s^N, W_s^N, and uses a definite initial moisture content $W_s^0 = \bar{W}_{si}$, and free initial enthalpy I_s^0. Out of the family of trajectories using a given final moisture content $W_s^N = \bar{W}_{sf}$ and some final enthalpy I_s^N (which refers to a resulting enthalpy I_s^0 equal to initial process state enthalpy $\bar{I}_{si}$) a solution is found. The selection of the trajectory in question completes the solution of optimization problem. It is easily observed that the discussed problem exists only for multistage processes.

The examined optimization problem is two dimensional. Denoting

$$\min \sum_1^N \theta^n = f^N[I_s^N, W_s^N] \tag{58}$$

and assigning dimensionless gas flows θ^n for decision variables, we obtain, according to Bellman's principle of optimality [5], the following recursive equation:

$$f^n[I_s^n, W_s^n] = \min_{\theta n}. \{\theta^n + f^{n-1}[I_s^{n-1}, W_s^{n-1}]\}; \quad f^0[I_s^0, W_s^0] = 0 \tag{59}$$

in which variables I_s^{n-1}, W_s^{n-1} should be expressed as functions of state variables I_s^n, W_s^n and decision θ^n. Utilization for this purpose of state Equations 8 and 9 in which it is assumed that $i_g^n = i_g^0$ and $X_g^n = X_g^0$ and also $e = 1$ and $j = 0$ (see Table 1) yields:

$$\begin{aligned} f^n[I_s^n, W_s^n] = \min_{\theta n} \{&\theta^n + f^{n-1}[I_s^n - \theta^n(i_g^0 - i_s[I_s^n, W_s^n]), \\ &W_s^n - \theta^n(X_g^0 - X_s[I_s^n, W_s^n])]\} \end{aligned} \tag{60}$$

The reader may modify recursive Equation 60 to an alternative form in which the decision variables are the interstage moisture contents W_s^{n-1}.

Crosscurrent Decision Processes with Varying Temperature and Inlet Gas Flow

For crosscurrent processes (Figure 1) occurring in a given apparatus the important algorithms are those for varying enthalpy (or temperature) and constant humidity of the inlet gas. If the drying medium is air, humidity is most often equal to that of the environment X_e. Dimensionless gas flow $\theta^n(\theta^n \equiv g^n)$ may either be constant along the trajectory, $\theta^n = \theta$, or optimally chosen at each stage of the process. In the latter case, flow θ^n is a decision variable together with the inlet gas enthalpy i_g^n (or temperature t_g^n).

It might appear that two-dimensional recursive Equation 52 whose right-hand side was modified by the minimization sign in relation to decisions i_g^n and $\theta = \theta^n$, is quite enough as an optimization algorithm. In fact it is only so when a total inlet gas flow φ^N is not prescribed. Any restriction for φ^N and particularly the equality restriction $\varphi^N = \bar{\varphi}_f$ leads to a three-dimensional problem with state variables I_s^n, W_s^n, φ^n. Such a problem and the reduction of its dimensionality are described later.

In concrete calculations it is recommended to replace original decisions i_g^n and θ^n by other decision variables. These are: solid enthalpy before stage I_s^{n-1} and solid moisture content increment $\Delta^n \equiv W^{n-1} - W_s^n$.

To make quantities I_s^{n-1} and Δ^n decisions, the state Equations 8 through 10 [taken for $X_g^n = X_e$, e = 1 and j = 0] are transformed into following form:

$$I_s^n - I_s^{n-1} = -\left(\frac{i_g^n - i_s[I_s^n, W_s^n]}{X_e - X_s[I_s^n, W_s^n]}\right)\Delta^n \tag{61}$$

$$\varphi^n - \varphi^{n-1} = -\frac{\Delta^n}{X_e - X_s[I_s^n, W_s^n]} \tag{62}$$

$$W_s^n - W_s^{n-1} = -\Delta^n \tag{63}$$

obtained by division of the first and third expressions of Equations 8 through 10 by the second equation of this set and separation of decision Δ^n. The all variables in Equation 37 should be expressed as functions of state variables I_s^n, W_s^n, φ^n and of decisions I_s^{n-1} and Δ^n. It should also be assumed* that: $\psi^n = 1$, $X_g^n = X_e$, e = 1, j = 0, and $g^n = \theta^n$ in Equation 37. Using Equation 61 in simplified Equation 37 and considering definition $g^n = \theta^n$ described by Equation 62, we obtain:

$$K = \sum_1^N \left(c_g\left[\frac{(I_s^n - I_s^{n-1})(X_s^n - X_e)}{\Delta^n} + i_s^n, X_e\right] - \mu^{n-1} c_g[i_s^n, X_s^n]\right) \cdot \frac{\Delta^n}{X_s^n - X_e} + u(C_s^0 - C_s^N) \tag{64}$$

where $i_s^n = i_s[I_s^n, W_s^n]$ and $X_s^n = X_s[I_s^n, W_s^n]$.

Bellman's recursive equation for the preceding index is threedimensional. In order to reduce dimensionality it should be noted that variable φ^n does not explicitly appear on the left-hand side of state Equations 61 through 63 and of performance index (Equation 64). Hence, there exists a possibility of using the Lagrangian multiplier λ adjoint with variable

---

* see Tables 1 and 2.

φ^n to reduce dimensionality. Employing the Lagrangian multiplier relies on adding the following expression:

$$K_0 = \sum_1^N \frac{\lambda\Delta^n}{X_s[I_s^n, W_s^n] - X_e} \tag{65}$$

to the performance index (Equation 64).

Numerical value λ must be chosen so that the optimal value of the sum in Equation 65 was equal to the prescribed value $\varphi^N = \bar{\varphi}_f$. The use of the multiplier is connected with the minimization of sum K and K_0 for a given number of stages N and final moisture content $W_s^N = \bar{W}_{sf}$ with a specified, or free, final enthalpy I_s^N and which is essential with a free final time φ^N. Using the principle of optimality for the sum of costs K and K_0 we obtain, after utilizing Equation 63, the following recursive equation:

$$\tilde{F}^n[I_s^n, W_s^n, \lambda] = \min_{I_s^{n-1}, \Delta^n} \left\{ \left(c_g \left[\frac{(I_s^n - I_s^{n-1})(X_s^n - X_e)}{\Delta^n} + i_s^n, X_e \right] - \mu^{n-1} c_g[i_s^n, X_s^n] + \lambda \right) \right.$$
$$\cdot \frac{\Delta^n}{X_s^n - X_e} + u(C_s[I_s^{n-1}, W_s^n + \Delta^n] - X_s\ [I_s^n\ , \Psi_s^n\ ])$$
$$\left. + \tilde{F}^{n-1}[I_s^{n-1}, W_s^n + \Delta^n, \lambda] \right\} \tag{66}$$

in which symbol $\tilde{F}^n[I_s^n, W_s^n, \lambda]$ denotes optimal costs $K + K_0$ expressed as a function of state variables I_s^n, W_s^n and parameter λ.

Decision Problems of Batch Fluidization and of Process in Horizontal Fluidized Exchanger

The optimization problem previously discussed is simplified into the problem with one decision variable I_s^{n-1} in the case of continuous processes. This is connected with the transition of three difference state equations: Equations 61 through 63 into two differential equations describing derivatives dI_s/dW_s and $d\varphi/dW_s$. To solve the optimization problem in this case (horizontal exchanger and batch fluidization), it is enough to adopt a sufficiently small range of moisture content changes $\Delta^n = \Delta$ in state Equations 61 through 63 and recursive Equation 66. The problem becomes one-dimensional then, because solid moisture content W_s^n is at the same time connected with the stage number n:

$$W_s^n - \bar{W}_{sf} - n\Delta \tag{67}$$

Substituting Equation 67 into Equation 66 gives, for $\Delta^n = \Delta$, the following algorithm:

$$\tilde{F}^n[I_s^n, \lambda] = \min_{I_s^{n-1}} \left\{ \left(c_g \left[\frac{(I_s^n - I_s^{n-1})(X_s[I_s^n, \bar{W}_{si} - n\Delta] - X_e)}{\Delta} + i_s[I_s^n, \bar{W}_{si} - n\Delta], X_e \right] \right. \right.$$
$$\left. - \mu^{n-1} c_g[i_s^n[I_s^n, \bar{W}_{si} - n\Delta], X_s^n[I_s^n, \bar{W}_{si} - n\Delta]] + \lambda \right) \frac{\Delta}{X_s[I_s^n, \bar{W}_{si} - n\Delta] - X_e}$$
$$\left. + u(C_s[I_s^{n-1}, \bar{W}_{si} - (n-1)\Delta] - C_s[I_s^n, \bar{W}_{si} - n\Delta]) + \tilde{F}^{n-1}[I_s^{n-1}, \lambda] \right\} \tag{68}$$

In the case of continuous processes the optimization solution for varying inlet gas humidity has restricted practical significance. This is why the algorithm Equation 68 describing the situation where $X_g = X_e =$ const. is the most important for application in the discussed case. Considering the preceding, we shall discuss the numerical aspects of the solution of Equation 68 in a later section.

Crosscurrent Decision Processes for Varying Temperature and Humidity of Inlet Gas

In the present case discrete optimization for the crosscurrent multistage process with a small number of stages N is significant. It is often assumed in optimization problems that dimensionless gas flow $\theta^n = \theta$ is constant along the process trajectory. Fluidized cascades or crosscurrent multitray apparatus (Figure 1) with the constant quantity θ are frequently applied in practice.

The optimization problem for θ^n = const. consists in searching for an optimal trajectory and optimal decisions i_g^n, X_g^n, for specified values of one or both solid coordinates I_s^N, W_s^N as well as for a given number of process stages N.

The optimization algorithm is described by the general recursive Equation 52 modified by the presence of the minimization sign with respect to decisions i_g^n, X_g^n. It is easy to check, with the use of crosscurrent process data from Table 1 that the algorithm can be rewritten in the form:

$$\begin{aligned}\tilde{K}^n[I_s^n, W_s^n] = \min_{i_g^n, X_g^n} \{&(c_g[i_g^n, X_g^n] - \mu^{n-1} c_g[i_s^n, X_s^n])\theta + u(C_s[I_s^n - \theta(i_g^n - i_s^n),\\ &W_s^n - \theta(X_g^n - X_s^n)] - C_s[I_s^n, W_s^n]) + \tilde{K}_p^{n-1}[I_s^n - \theta(i_g^n - i_s^n),\\ &W_s^n - \theta(X_g^n - X_s^n)]\}\end{aligned} \tag{69}$$

The generalization of the just-discussed optimization problem is characterized by three decision variables. They are: inlet gas parameters i_g^n, X_g^n, and dimensionless gas flow θ^n

In the original version the generalized optimization problem is three-dimensional, with state variables: I_s^n, W_s^n, φ^n. (The presence of state variable φ^n is implied by the restriction for total gas flow). Reduction of dimensionality for such a problem is linked with introducing the Lagrangian multiplier λ. This corresponds with adding the term $\lambda\theta^n$ to the expression describing costs at the n-th stage. The two-dimensional recursive equation obtained as a result of this procedure is as follows:

$$\begin{aligned}\tilde{F}^n[I_s^n, W_s^n \lambda] = \min_{i_g^n, X_g^n, \theta^n} \{&(c_g[i_g^n, X_g^n] - \mu^{n-1} c_g[i_s^n, X_s^n] + \lambda)\theta^n\\ &+ u(C_s[I_s^n - \theta^n(i_g^n - i_g^n), W_s^n - \theta^n(X_g^n - X_s^n)] - C_s[I_s^n, W_s^n])\\ &\tilde{F}^{n-1}[I_s^n - \theta^n(i_g^n - i_s^n), W_s^n - \theta^n(X_g^n - X_s^n), \lambda]\}\end{aligned} \tag{70}$$

Economic considerations [1, 11] involving analysis of the investment costs show that λ is related to popular economic and technical quantities as follows

$$\lambda = \left(\frac{\tilde{z}}{\tau_p} + \beta\right)\frac{p}{\tau_u V_g \rho_g} + \frac{c_{el}\Delta P}{\eta' \rho_g} \tag{71}$$

where the first term on RHS pertains to the investment and the second to the gas pumping costs. The essential parameter in this equation is the fluidized apparatus price per unit area of sieve bottom, p. Thus λ is a measure of unit apparatus value in appropriate units. Iterating optimization computations for the same final state (I_{sf}, W_{sf}) and changing only λ, the tables $\varphi_f(\lambda)$ are obtained. The results show that φ_f is a decreasing function of λ. Since the state (I_{sf}, W_{sf}) is constant, the average process intensity, i.e. the vector $(I_{sc} - I_{sf})/\varphi_f$, $(W_{so} - W_{sf})/\varphi_f$ increases with unit apparatus price p or λ in an optimal process. Thus the value of λ, Equation 71, determines the optimal process intensity or, in other words, the degree of thermodynamic irreversibility which should be maintained in the optimal dryer. This general property of λ was confirmed for various dryers.

Optimizing of Crosscurrent Multistage Fluidized Drying by Discrete Maximum Principle with a Constant Hamiltonian

In References 1, 9, and 10 a new, very efficient, discrete optimization algorithm has been given, characterized by the constancy of a certain hamiltonian along the optimal process trajectory. This algorithm is here used (see also [11]) to solve the problem of the optimal sum of the investment and exploitation costs for the multistage crosscurrent drying (Figure 1) in which the final solid moisture content is preassigned whereas the final solid enthalpy I_s^N (or temperature) and total gas flow G^N are undetermined. The decision variables are temperatures t_g^n and humidities X_g^n of inlet gas and gas flows ΔG^n, for $n = 1, 2, \ldots, N$. In this problem $\mu = 0$ i.e. outlet gas is relased into the atmosphere and $u = 0$, (i.e., the change of initial and final solid values is ignored in the performance index). The performance index can be stated [11] in the simple form

$$K = \sum_{1}^{N} (c_g[t_g^n, X_g^n] + \lambda)\theta^n \tag{72}$$

containing the multiplier Equation 71 related to both investment and gas pumping costs. The λ is very suitable as a parameter of the optimization solution.

Applying the exergy tariff, $c_g = b_g \cdot e_g$, we exploit the convenient simplified form of exergy that is obtained after the Taylor expansion of the exact exergy function around the point (t_e, X_e) and neglecting the third order and higher terms. The unit gas price is then obtained as

$$c_g^n[t_g^n, X_g^n] = \frac{e_g}{2}\{A(t_g^n - t_e)^2 + B(X_g^n - X_e)^2\} \tag{73}$$

where $A = (c_p + X_e c_m)/T_e$ and $B = RT_e/(M_m + X_e M_g)X_e$ [1, 6]. Thus the final performance index, used in computations, is

$$K' = K/e_g = \sum_{1}^{N}\left\{\frac{1}{2}A(t_g^n - t_e)^2 + \frac{1}{2}B(X_g^n - X_e)^2 + \lambda'\right\}\theta^n \tag{74}$$

with $\lambda' = \lambda/e_g$. The state equations, under which K is minimized, are Equations 8 and 9 with $e = 1$ and $j = 0$ (Table 1). The minimum allowable gas humidity is equal to X_w. The working decisions are T_g^n, X_g^n, and θ^n (absolute temperatures of gas are used later).

The Optimization Algorithm with a Constant Discrete Hamiltonian

The mathematical model of the process (Equations 8 through 10) with $e = 1$ and $j = 0$ and Equation 74, is linear with respect to the unconstrained decisions θ^n. This fact proves the possibility of the use of the previously mentioned special discrete optimization algorithm of the maximum principle with a constant discrete hamiltonian.

The hamiltonian function is

$$H^{n-1}[W_s^n, I_s^n, z_1^{n-1}, z_2^{n-1}, T_g^n, X_g^n] = \frac{1}{2}A(T_g^n - T_e)^2 + \frac{1}{2}B(X_g^n - X_e)^2 + \lambda'$$

$$+ z^{n-1}((c_p + X_g^n c_m)(T_g^n - 273.15) + 2502.7X_g^n - i_s^n[W_s^n, I_s^n])$$

$$+ z_2^{n-1}(X_g^n - X_s^n[I_s^n, W_s^n]) = 0 \tag{75}$$

Table 3
Coefficients of Equations Describing Drying Equilibrium for Silica Gel-Water-Air System

i	a_i	b_i	c_i	d_i
1	7.1114×10^{-2}	1.3913×10^{-1}	1.6448×10^{1}	5.0111
2	5.2355	5.6037	4.5728×10^{2}	1.8365×10^{2}

Since the total gas flow is undetermined $\left(\text{free } G^N/S = \sum_{n=1}^{N} \theta^n\right)$ the constant value of the hamiltonian (Equation 75) is equal to zero.

The adjoint equations are (n = 1, ..., N):

$$\frac{z_1^n - z_1^{n-1}}{\theta^n} = \frac{\partial H^{n-1}}{\partial I_s^n} = z_1^{n-1} \frac{\partial i_s^n[I_s^n, W_s^n]}{\partial I_s^n} + z_2^{n-1} \frac{\partial X_s^n[I_s^n, W_s^n]}{\partial I_s^n} \tag{76}$$

$$\frac{z_2^n - z_2^{n-1}}{\theta^n} = \frac{\partial H^{n-1}}{\partial W_s^n} = z_1^{n-1} \frac{\partial i_s^n[I_s^n, W_s^n]}{\partial W_s^n} + z_2^{n-1} \frac{\partial X_s^n[I_s^n, W_s^n]}{\partial W_s^n} \tag{77}$$

As I_s^N is undetermined and W_s^N is prescribed we have the following final boundary conditions associated with Equations 76 and 77 respectively:

$$z_1^N = 0 \tag{78}$$

$$z_2^N \text{ — undetermined} \tag{79}$$

The partial derivatives appearing in Equations 76 and 77 are computed with the help of drying equilibrium equations [11] for the silicagel-water-air system:

$$i_s^n[I_s^n, W_s^n] = 4.19[a_1 W_s^n - b_1 + (c_1 W_s^n + d_1)\,(0.24 I_s^n + 32.6 W_s^n/(0.094 + W_s^n))^{-1}]^{-2} \tag{80}$$

$$X_s^n[I_s^n, W_s^n] = [a_2 W_s^n - b_2 + (c_2 W_s^n + d_2)\,(0.24 I_s^n + 32.6 W_s^n/(0.094 + W_s^n))^{-1}]^{-2} \tag{81}$$

where the coefficients a_i and b_i are given in Table 3.

Solving the transformed Equations 76 and 77 with respect to z_i^{n-1} (i = 1, 2) yields

$$z_1^{n-1} = \frac{z_1^n - \theta^n(z_1^n \beta_2^n + z_2^n \beta_1^n)}{1 + \theta^n(\alpha_1^n - \beta_2^n) + (\theta^n)^2\,(\alpha_2^n \beta_1^n - \alpha_1^n \beta_2^n)} \tag{82}$$

$$z_2^{n-1} = \frac{z_1^n \theta^n \alpha_2^n + z_2^n(\alpha_1^n \theta^n + 1)}{1 + \theta^n(\alpha_1^n - \beta_2^n) + (\theta^n)^2\,(\alpha_2^n \beta_1^n - \alpha_1^n \beta_2^n)} \tag{83}$$

where

$$\alpha_1^n = 2Y^n(c_1 W_s^n + d_1)E_1^n \tag{84}$$

$$\beta_1^n = 2Y^n(c_2 W_s^n + d_2)E_2^n \tag{85}$$

$$\alpha_2^n = 2Y^n[a_1(Y^n)^2 + c_1 Y^n - 3.064(c_1 W_s^n + d_1)/(0.094 + W_s^n)^2]E_1^n \tag{86}$$

$$\beta_2^n = 2Y^n[a_2(Y_n)^2 + c_2Y^n - 3.064(c_2W_s^n + d_2)/(0.094 + W_s^n)^2]E_2^n \tag{87}$$

$$Y^n = Y^n[I_s^n, W_s^n] = 0.24I_s^n + 32.6W_s^n/(0.094 + W_s^n) \tag{88}$$

$$E_i^n = ((a_iW_s^n - b_i)Y^n + c_1W_s^n + d_i)^{-3} \quad \text{for} \quad i = 1, 2 \tag{89}$$

The stationary optimal gas humidity X_g^n and gas temperature T_g^n are determined from the relations:

$$\frac{\partial H^{n-1}}{\partial X_g^n} = B(X_g^n - X_e) + z_1^{n-1}c_m(T_g^n - 273.15) + 2502.7 + z_2^{n-1} = 0 \tag{90}$$

$$\frac{\partial H^{n-1}}{\partial T_g^n} = A(T_g^n - T_e) + z_1^{n-1}(c_p + X_g^n c_m) = 0 \tag{91}$$

The solution of the system, Equations 90 and 91, with respect to X_g^n and T_g^n is

$$X_g^n = \frac{BX_e + 273.15z_1^{n-1}c_m - 2502.7 - z_2^{n-1} - z_1^{n-1}c_mT_e + (z_1^{n-1})^2c_mc_p/A}{B - (z_1^{n-1}c_m)^2/A} \tag{92}$$

$$T_g^n = T_ez_1^{n-1}(c_p + X_g^nc_m)/A \tag{93}$$

When stationary X_g^n computed from Equation 92 is less than X_n than optimal $X_g^n = X_w$ and in this case Equation 93 simplifies to

$$T_g^n = T_e - z_1^{n-1}(c_p + X_nc_m)A \tag{94}$$

The computational procedure is described in Reference 11.

OUTLINE OF COMPUTATIONAL PROBLEMS

Working State Variables and Preliminary Calculations

The first operation leading to obtaining numerical solutions is the determination of the grid of state variables I_s^n, W_s^n.

As the size of temperature changes is easier to perceive than the size of enthalpy changes, optimization calculations may be carried out for grid of variables t_s^n and W_s^n. Then in each node of the grid enthalpy I_s^n is calculated as well as other necessary functions of solid (see literature citations later). As a result, independently from the state variables applied, difference state equations and recursive equations (derived beforehand) may be used in an unchanged form.

The grid determination is connected with preassigning enthalpy and moisture content increments ΔI_s and ΔW_s (or temperature and moisture content increments Δt_s and ΔW_s) as well as prescribing the state variable values at the grid boundaries. Every node of the grid corresponds to a particular position in results table.

In preliminary calculations it is necessary to know the basic thermodynamical relations, e.g., $i_g[t_g, X_g]$, $I_s[t_s, W_s]$ etc. for the system gas-moisture-solid. Data for humid air are most common [6, 8]; thermodynamical functions of solid are discussed in Reference 2. For price calculations in the case when the exergy tariff of prices is used information in References 6, 7, and 11 may serve. Calculations of quantities depending only on the process state and independent of decision and number of the stages are common in all optimization algorithms (e.g., $i_s[t_s, W_s]$, $\beta_s[t_s, W_s]$). These calculations are performed once, even before the optimization procedure itself, and their results are stored as tables. Note that this fact is connected with the possibility of omitting index n in state variables (in recurrence equations).

At first, calculations of enthalpy i_s and humidity X_s of gas in equilibrium with a solid (t_s, W_s) are performed. For this purpose empirical formulas are applied which describe drying equilibrium (see, e.g., [2]). They allow the calculation of relative humidity $\beta_s[t_s, W_s]$ for gas in equilibrium with solids. Having this value, the absolute equilibrium humidity of gas $X_s[t_s, W_s]$ is then computed. Knowledge of humidity X_s and temperature t_s allows finding the gas equilibrium enthalpy $i_s[t_s, X_s]$ from the adequate thermodynamical formula. Then, other thermodynamic functions of a solid: enthalpy exergy, economic price, etc. (required for optimization computations) are calculated. These data are stored as tables.

Searching for Gas States by Invariant Imbedding

Gas parameters i_g^n, X_g^n should be determined sequentially for 1, 2, ..., N as functions of independent variables I_s^n, W_s^n belonging to a certain rectangle of enthalpy I_s^n and moisture content W_s^n (the grid t_s^n, W_s^n can also be used). At first, for n = 1, enthalpy $i_g^1[I_s^1, W_s^1]$ and humidity $X_g^1[I_s^1, W_s^1]$ are calculated from Equations 45 and 46. Then Equations 42 and 43 are solved for n = 2. This leads to obtaining functions $i_g^2[I_s^2, W_s^2]$ and $X_g^2[I_s^2, W_s^2]$ in tabularic form. Each position of the table contains two values i_g^2, and X_g^2 and refers to an appropriate node of computational grid.

Because of the implicit character of Equations 42 and 43 (the unknown variables i_g^n, X_g^n appear at both sides of the discussed equations) the solution is found with the help of the iterative method applied for all discrete values of I_s^n, W_s^n belonging to the examined rectangle. If, for instance, n = 2, then, for a particular grid node (definite values of I_s^2, W_s^2), two numerical values of i_g^2, X_g^2 are assumed and substituted into the right-hand sides of Equations 42 and 43. Calculation of the right-hand sides of these equations takes place on the basis of tables containing computational results of enthalpy $i_g^1[I_s^1, W_s^1]$ and humidity $X_g^1[I_s^1, W_s^1]$.

The form of recursive Equations 42 and 43 indicates that quantities i_g^1 and X_g^1 calculated for arguments $I_s^1[I_s^2, W_s^2]$ and $W_s^1[I_s^2, W_s^2]$ should be substituted into these equations. However numerical values of I_s^1, W_s^1 do not generally correspond with those discrete values independent variables I_s, W_s for which tables i_g^1, X_g^1 were computed. Hence, it is necessary to find the more accurate values of i_g^1, X_g^1 by interpolation in tables containing the result for the precedent (first) stage. After interpolation, the RHS of Equations 42 and 43 can be calculated.

If the numerical values of i_g^2, X_g^2 obtained in the prior manner are sufficiently close to the assumed ones (in the sense of a certain, positively determined error function which depends on difference between computed and assumed variables) then the assumption was proper. Then the numerical values of parameters i_g^2, X_g^2 and independent variables $I_s = I_s^2$, $W_s = W_s^2$ are stored and we procced to gas parameter calculations for another pair of values I_s^2, W_s^2 (another grid node). In the case of an inconsistency of the assumed and computed values of i_g^2 and X_g^2 the assumption has to be changed and the computation repeated. An optimization procedure was used here, for which the minimum of an error function, mentioned was searched for.

Finding of values i_g^2, X_g^2 for all discrete values of I_s^n, W_s^n (all grid nodes) yields the tables characterizing computational results for n = 2. These results serve to calculate the quantities i_g^3, X_g^3 characterizing the third stage of the process. Calculations for n = 3 are performed with the help of Equations 42 and 43 in the way analogous to the preciding for n = 2, etc.

The process performance index is computed with the use of Equation 16 on the basis of the obtained data i_g^n, X_g^n.

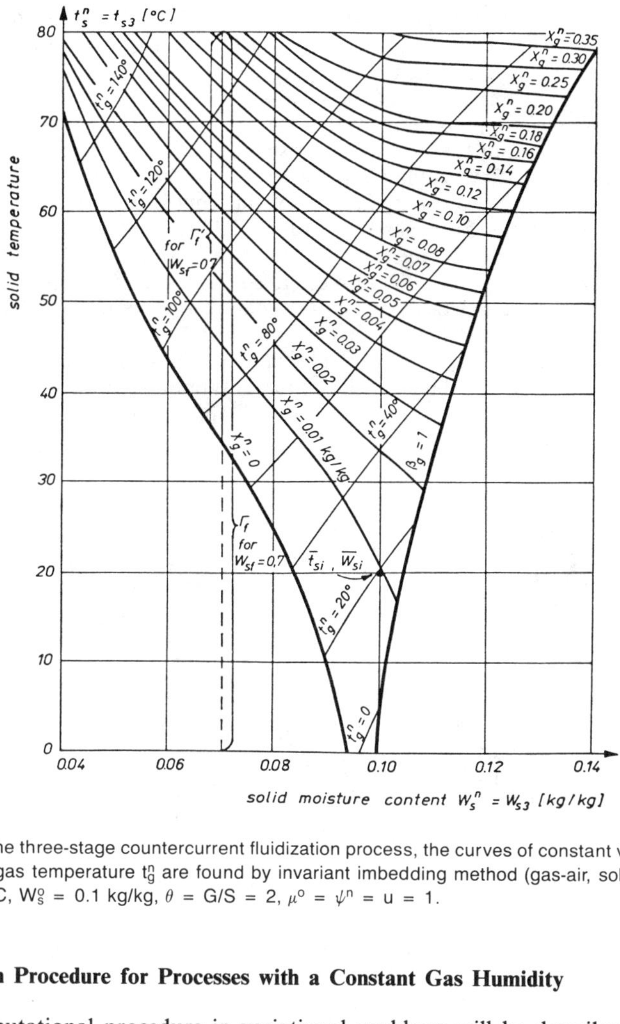

Figure 3. In the three-stage countercurrent fluidization process, the curves of constant values of gas humidities X_g^n and gas temperature t_g^n are found by invariant imbedding method (gas-air, solid-silica gel). Input data $t_s^o = 20°C$, $W_s^o = 0.1$ kg/kg, $\theta = G/S = 2$, $\mu^o = \psi^n = u = 1$.

Optimization Procedure for Processes with a Constant Gas Humidity

The computational procedure in variational problems will be described as an instance for the problem of a searching for optimal gas enthalpies i_g^n (or temperatures t_g^n) during drying a solid with a gas of a constant humidity X_e. Recursive Equation 68 should then be solved. The solution is found in such a region of variables I_s, W_s in which condition $X_s[I_s, W_s] \geq X_e$ is fulfilled. Equations 62 and 63 imply that in this case drying the material is possible (decrease of moisture content W_s with the increase of index n and time φ).

Optimization calculations are performed for discrete values of enthalpy I_s^n (increment $\Delta I_s = \text{const}$), fulfilling the condition $I_s^n > I_{s*}[n]$, where $I_{s*}[n]$ is a known lower-limit enthalpy corresponding with situation when $X_s[I_s, W_s] = X_e$.

According to recursive Equation 68, the optimization consists in searching for the value I_s^{n-1} which minimizes the sum of process costs at n-th stage and optimal cost of n − 1 stages. Optimal enthalpy I_s^{n-1} is found for each grid node of enthalpy I_s^n.

In the case when n = 1 there exists a free choice of just one numerical value of $I_s^{n-1} = \bar{I}_{si} = \text{const}$, and this is why optimization is trivial in this case. Function $F^1[I_s^1]$ is then obtained simply by substituting into the right-hand side of recursive Equation 68 the following values $I_s^{n-1} = \bar{I}_{si}$, $\tilde{F}^{n-1} = 0$ and $n = 1$. The discussed function is tabularized in the range $I_{s*} \leqq I_s \leqq I_s^*$, where I_s^* is solid enthalpy which, when achieved, signals the violation of the restriction for the temperature of the inlet gas.

The optimization procedure is nontrivial for $n \neq 1$. It consits in searching for the minimum of the expression in brackets of recursive Equation 52 versus enthalpy I_s^{n-1} and with constant enthalpy I_s^n. Both enthalpies must lie within the range in which $I_s > I_{s*}[n]$; discussion of the procedure ensuring fulfillment of this condition is omitted.

After finding the optimum for a particular value of I_s^n and calculating all the required data, calculations are repeated for the next discrete enthalpy value I_s^n. Finally it comes to so a large value of enthalpy I_s^n that any of discrete enthalpies I_s^{n-1} causes transgression of the condition for inlet gas temperature. Then the computer proceeds to find results of the next stage.

Apart from values of function $\tilde{F}^n[I_s^n]$, of optimal solid enthalpies $\hat{I}_s^{n-1}[I_s^n]$ and corresponding gas enthalpies $\hat{i}_g^n[I_s^n]$, dimensionless times of optimal process $\varphi^n[I_s^n]$ are also tabularized.

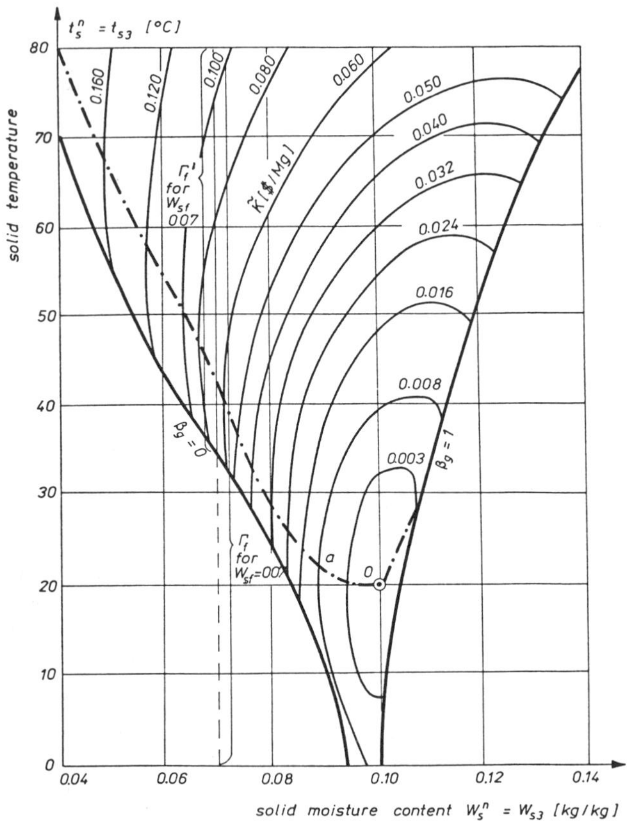

Figure 4. In the three-stage countercurrent fluidization process, the curves of constant values of economic costs $\tilde{K}^3$ $[t_s^3, W_s^3]$ are found by invariant imbedding. The input data is as in Figure 3.

Obtaining these results is based on the equation:

$$\varphi^n[I_s^n] = \frac{\Delta}{X_s[I_s^n, W_s^n - n\Delta] - X_e} + \varphi^{n-1}[\hat{I}_s^{n-1}[I_s^n]] \tag{95}$$

implied by difference Equation 62 for $\Delta^n = \Delta$.

Repeating computations for the some boundary of variables I_s^n, W_s^n and changing only the value of parameter λ, we obtain tables describing the relation between the multiplier λ and optimal final time φ^N for various final states I_s^N, W_s^N of a solid. With the use of these tables one may find (e.g., graphically on chart $\varphi^N - \lambda$) the numerical value of λ which solves the problem of transition from initial state $I_s^0 = \bar{I}_{si}$, $W_s^0 = \bar{W}_{si}$ to a prescribed final state $I_s^N = \bar{I}_{sf}$, $W_s^N = \bar{W}_{sf}$ in a given time $\varphi = \bar{\varphi}_f$. In order to obtain optimal trajectory and decisions, the discussed optimization computations should be performed once again for the preceding particular value of λ. This operation completes the computational procedure.

SELECTED RESULTS OF NUMERICAL COMPUTATIONS

Nonvariational Problems

Optimization results are presented in the form of tables or diagrams describing the optimal performance index and optimal decisions as functions of variables t_s^n, W_s^n, n (or variables I_s^n, W_s^n, n). All the prices appearing in the equations discussed were obtained on the basis of the exergy tariff (of [1, 6, 7]) (the economic values of the exergy unit were evaluated as $E_s = e_g = 8.0$ s/GJ for the silicagel-water-air system investigated).

Figure 3 illustrates computational results for temperatures t_g^n and humidities X_g^n for a three-stage countercurrent fluidized cascade in the case of an air-silicagel system. A series of the diagrams for n = 1, 2, ..., N describe thermodynamic parameters of inlet gas as functions of the thermodynamical final state of solid t_s^n, W_s^n and number of stages n. The existence of unattainable solid-final states t_s, W_s is observed (region for $X_g < 0$ in Figure 3). Results useful for practice are placed between the lines $X_g = 0$ and $\beta_g = 1$.

Data in Figure 3 allow for calculating economic costs of the process (Figure 4) as a function $\tilde{K}^n[t_s^n, W_s^n]$ and finding, on this basis, the optimal final temperature t_s^n along the prescribed line of constant final moisture content (see, for example, line Γ_f in Figures 3 and 4 for $\bar{W}_{sf} = 0.07$). The locus of optimal temperatures t_s^n is described by curve a in Figure 4.

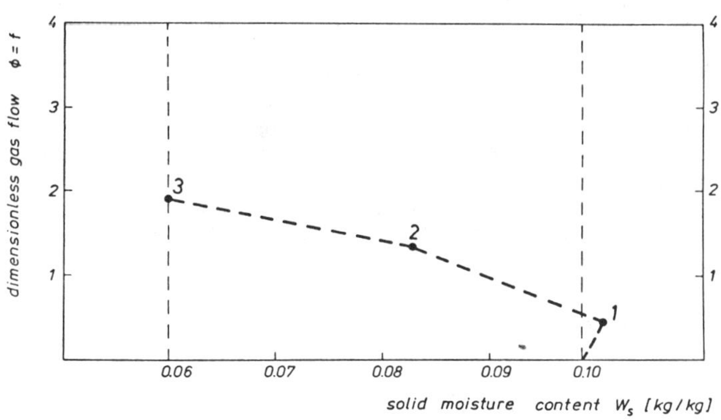

Figure 5. Wetting and drying of solids in optimal multistage fluidization process (gas allocation problem) (gas-air, solid-wood). Input data: $t_s^o = -8°C$, $W_s^o = 0.1$ kg/kg, $t_g^o = 130°C$, $X_g^o = 0.007$ kg/kg.

The diagram of Figure 3 is then applied again for finding *optimal* inlet gas parameters t_g^n, X_g^n. These parameters can be found on the admissible part of the straight line $W_s^n = \bar{W}_{sf}$ (line Γ_f' in Figure 3) for the optimal temperature t_s^n.

The minimum of economic costs (for W_s^n = const) can be determined also when gas temperature t_g^n is adopted as a decision. It is possible when the material state diagram contains complete data, i.e., curves t_g^n = const, X_g^n = const, and $\tilde{K}^n$ = const. (This was not done in the paper to keep the diagram readable). The readout of optimal gas parameters occurs then simultaneously with readout of optimal solid temperature t_s^n (also along the line W_s^n = const). The diagrams are then obtained which describe costs in relation to any variable characterizing thermodynamical gas or solid state (t_g^n, t_s^n, or X_g^n).

The example of the curve $\tilde{K}^n[t_g^n]$ for $n = 3$ is given in Reference 12. (There is a gas humidity scale, X_g^n marked on the curve.) The minimum of the curve corresponds with optimal conditions of the process.

Series of diagrams, analogous to the previously discussed, describe gas parameters and costs in nonvariational problems for crosscurrent and cocurrent contact.

Variational Problems

Gas Allocation Problem in Multistage Crosscurrent Fluidization

Figure 5 shows the example of optimal curve $\varphi^n[W_s^n]$. It is typical for processes in which cascade is fed with a cold solid with initial temperature lower than the dew point temperature of humid gas.

The curve in Figure 5 has a moisture content maximum which proves that wetting of the solid occurs at the first stage of cascade. This indicates an interesting fact—optimal decisions sometimes seem disadvantageous. Interpretation of this phenomenon is simplest if the equivalent problem of minimum final solid moisture content is considered for a specified total drying-gas flow.

For total gas flows that are not large the optimal decisions θ^n for the first stages of cascade (i.e., ratios $\Delta G^1/S$, $\Delta G^2/S$, etc.) may happen to be so small that the solid state will only slightly change in the first apparatus. Because of that the solid temperature and vapor pressure will only slightly increase. In view of low solid temperature in the first apparatus of cascade, condensation will take place up to transgressing the temperature of dew point. Only afterwards the solid body is dried and achieves minimum final moisture content for the required total gas flow G.

It is essential that apparently harmful wetting of the solid at the first stages of the battery is avoidable. If sufficiently large flow ΔG^1 exists then the solids will reach a moisture content lower than its initial state after every stage the of battery. However, in order to keep total dimensionless gas flow $G/S = f$ invariant, the remaining stages of the cascade should be fed with a smaller flow of a drying medium. Finally, it causes larger, nonoptimal, final moisture content of material.

The described fact indicates that the decision problem in the discussed case (small G/S) is not obvious, and apparently disadvantageous decisions are sometimes profitable.

Similar effects occur in multistage sorption of moisture on a hot solid (there is possibility of desorption in the first stages of the process).

Continuous Decision Processes with Constant Humidity of Inlet Gas

A dynamic programming algorithm for the considered case is described by recursive Equation 68. It refers to batch fluidization and a process in a horizontal fluidized exchanger.

Optimal decisions of the problem (achieved as a result of calculations performed in a computer) are presented in Figure 6. The diagram pertains to the case when the minimized

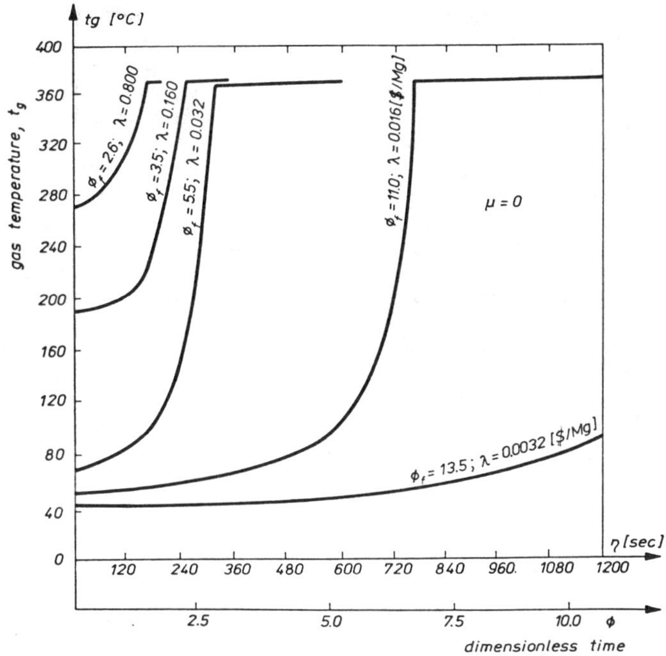

Figure 6. Optimal temperature policies t_g for various final process times ϕ_f in batch fluidization drying (the related Lagrangian multiplier values, λ, are also given). Input data: $t_s^o = 22.5°C$, $W_s^o = 0.1$ kg/kg, $\mu = 0$, $\psi = u = 1$, $t_{sf} = 115°C$, $W_{sf} = 0.03$ kg/kg, $t_g^* = 375°C$.

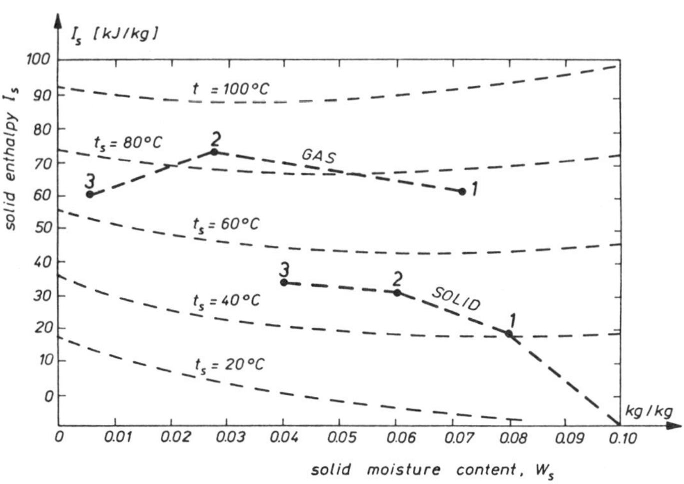

Figure 7. Three-stage crosscurrent fluidization drying with decisions t_g^n, X_g^n, $\theta^n = g^n$. The process representation on solid enthalpy diagram. Input data as for Figure 3.

criterion (taken with u = 1) is computed with the condition of no utilization of the outlet gas value ($\mu = 0$).

Figure 6 shows a diagram of optimal inlet-gas temperatures as a function of the current time φ and Lagrange multiplier λ (the values of total time φ_f, related to those of λ are also given). As the parameter λ increases, inlet gas temperatures become higher, which is caused by the increase of process intensity. Curves in Figure 6 indicate that optimal gas temperature t_g rises monotonically in time until it reaches maximum allowed gas temperature. Afterwards this temperature keeps constant.

The common method of carrying out the batch fluidization process consists in using the policy of constant inlet gas temperature. Optimization results show that the economical technique is linked with the policy of increasing gas temperature and that the policy of constant inlet gas temperature can be only a part of optimal policy, providing the upper limit of gas temperature is assumed.

The policy of optimal inlet gas temperature may be especially simply realized in the case of batch fluidization if the inlet gas stream is heated with intensity varying in time (variable flow of a heating medium or variable intensity of electric heating with the use of an autotransformer). For a horizontal fluidized exchanger the optimal policy can be realized only approximately, as in the case of other continuous steady-state processes (e.g., reactors). Inlet gas is then partitioned into several streams flowing to particular sections of the apparatus, and each stream is heated up to another temperature.

Decision Processes with Varying Temperature and Humidity of a Gas

The optimization problem for this considered process may refer to algorithm, Equation 69 (g = const.) or to algorithm, Equation 70 (g^n = var.). As the decision g^n has relatively small effect on the changes of the trajectories family when φ^N = const., it is

Table 4
Optimization Results for the Conventional and Unconventional Multistage Crosscurrent Fluidization Drying Gas-Air, Solid-Silica Gel

		λ' = 0.42 kJ/kg			λ' = 4.20 kJ/kg			λ' = 42.0 kJ/kg		
		Unconv. Proc.	Conv. Proc. Var. I	Conv. Proc. Var. II	Unconv. Proc.	Conv. Proc. Var. I	Conv. Proc. Var. II	Unconv. Proc.	Conv. Proc. Var. I	Conv. Proc. Var. II
T_g^1	K	303.31	302.48	304.47	330.83	328.77	337.30	441.32	434.49	446.02
T_g^2	K	302.46	302.48	304.47	330.03	328.77	337.30	433.54	434.49	446.02
T_g^3	K	301.83	302.48	304.47	323.21	328.77	337.30	336.62	434.49	446.02
X_g^1	kg/kg	0.0050	0.0046	0.0080	0.00001	0.00001	0.0080	0.00001	0.00001	0.0080
X_g^2	kg/kg	0.0047	0.0046	0.0080	0.00001	0.00001	0.0080	0.00001	0.00001	0.0080
X_g^3	kg/kg	0.0040	0.0046	0.0080	0.00001	0.00001	0.0080	0.00001	0.00001	0.0080
θ^1	kg/kg	8.14	7.02	8.61	3.39	2.64	2.85	1.51	0.90	0.88
θ^2	kg/kg	7.06	7.02	8.61	2.81	2.64	2.85	0.98	0.90	0.88
θ^3	kg/kg	5.68	7.02	8.61	1.67	2.64	2.85	0.20	0.90	0.88
W_s^0	kg/kg	0.2000	0.2000	0.2000	0.2000	0.2000	0.2000	0.2000	0.2000	0.2000
W_s^1	kg/kg	0.1589	0.1643	0.1654	0.1552	0.1663	0.1678	0.1443	0.1710	0.1723
W_s^2	kg/kg	0.1259	0.1311	0.1309	0.1188	0.1325	0.1331	0.1058	0.1355	0.1361
W_s^3	kg/kg	0.1000	0.1000	0.1000	0.1000	0.1000	0.1000	0.1000	0.1000	0.1000
I_s^0	kJ/kg	−11.20	−11.20	−11.20	−11.20	−11.20	−11.20	−11.20	−11.20	−11.20
I_s^1	kJ/kg	−12.91	−14.05	−8.66	−5.92	−7.08	3.93	21.22	18.49	24.53
I_s^2	kJ/kg	−11.97	−12.63	−6.81	−3.84	−5.47	5.66	23.72	22.87	28.61
I_s^3	kJ/kg	−9.87	−8.78	−3.51	−4.02	−2.18	8.20	17.84	24.44	29.57
K'	kJ/kg	13.952	13.977	16.486	54.484	54.593	64.443	205.02	207.21	216.68
$\sum \theta^n$	kg/kg	22.88	21.06	25.83	7.87	7.82	8.55	2.69	2.70	2.64
$\lambda'\tau$	kJ/kg	9.6096	8.8452	10.8486	33.054	31.8441	35.910	112.98	113.40	110.88
$\sum b_g^n \theta^n$	kJ/kgS	4.3424	5.1318	5.6374	21.430	22.7491	28.533	92.04	93.81	105.80

advisable to discuss solutions of both algorithms simultaneously. The obtained solution refers to crosscurrent multistage processes in Figure 1. It can also describe a discrete approximation of the solution for limiting continuous processes of the crosscurrent type, and the more stages N the more accurate the approximation is.

An appropriate representation of the changes of the thermodynamic states of the solid and gas along the optimal trajectory is an enthalpy-composition graph for the two-phase gas-moisture-solid system or an enthalpy plot of the moist solid. The latter is shown in Figure 7. This graph contains an example of a curve describing the optimal changes of a solid stream (lower line) and of a solid in thermodynamic equilibrium with the inlet gas (upper line). This sort of graph, with data obtained on the basis of the algorithm Equation 69, gives the best view of changes of the process driving force along the optimal trajectory. It indicates that in the optimal process, when $\mu = 1$, the driving force is nearly constant over, providing the restrictions for the thermodynamical state of the gas (t_g, X_g) are not operative.

As far as applications are concerned it is important to know trajectories of solid drying, along with which it is required to achiev low final temperatures of the material.

This requirement is often imposed by technology. In the discussed case trajectories are more complicated and the costs are high. This fact is a consequence of the operative restrictions for gas parameters. Obtaining a solid with a low final temperature and moisture content is bound with the necessity of using a cold dry inlet gas at at least one stage of the process. The price of a dry gas is high (so the operation is costly) and the optimal transition becomes more complex.

Optimization results show that achieving appropriately low final temperature and moisture contents of a solid often necessitates heating the solid, by dry gas, as a preliminary operation, and cooling the solid (with dry gas, too) as a final operation. This conclusion establishes a theoretical confirmation of the results known from drying practice which has presently been supported by quantitative results of optimization*.

It should be noted that the complex character of optimization solutions appears first as a result of a general requirement and that a concrete process must exist within a limited time frame. Furthermore the process must be characterized by a specified change of state, and hence also by a definite average intensity. A simple conclusion can be drawn in case of a free total drying time and a free final solid enthalpy, when the required final solid moisture content is larger than moisture content of a solid in equilibrium with the environment. The optimum drying process when associated with a lack of outlet gas utilization ($\mu = 0$) employs cold ambient air as the drying medium. In the case of utilizing the outlet gas ($\mu = 1$) and the free drying time, the tendency to quasistaticity is observed. Optimal solid residence time is very large while the optimal state of the inlet gas tends to arrive at equilibrium with a solid as much as process restrictions allow. When the number of operative restrictions increases, the more complex solutions, discussed earlier, are valid (see Table 4).

The results discussed so far have pertained to the case of the preassigned final state of a solid (I_s^N and W_s^N) and preassigned total gas flow, G^N = const, or φ^N = const. Frequently the constraints for I_s^N and G^N can be relaxed and then the optimal process costs are, of course, lower than, in the previous case. For free I_s^N or t_s^N and G^N the optimal results (the optimal values of t_g^n, X_g^n, θ^n) and the optimal trajectories (I_s^n, W_s^n) were obtained for the three-stage crosscurrent fluidized cascade with the help of the algorithm of discrete maximum principle with a constant hamiltonian. These results are stored in Table 4 as the

---

* It can be proved that optimal decisions and trajectories are the same for both u = 1 and u = 0 in cases when all the coordinates of the initial and final solid states are specified and a performance index has the Bolza form (Equation 37). It is easy to see that this conclusion has simplified our discussion of results.

data for unconventional control processes. They pertain to the following input data: N = 3, $W_s^0 = 0.2$ kg/kg, $I_s^0 = -11{,}2$ kJ/kg, $t_s^0 = 20°$ C, $W_s^N = 0{,}1$ kg/kg, $T_e = 293°$ K, $X_e = 0.008$ kg/kg, $X_w = 10^{-5}$ kg/kg. Various λ are used. From Table 4 and [11] it can be noted that:

- Along the optimal discrete trajectory (λ = constant, increasing n) the temperatures, humidities the and flows of the inlet gas decrease. When the apparatus price and/or gas pumping cost increases (n = constant, increasing λ) the optimal process intensity becomes larger, i.e., humidities and flows of gas decrease and gas temperature increases with λ. For λ = 0 the optimal air temperature is the ambient temperature T_e if W_{sf} exceeds the equilibrium value $W_e(T_e, X_e)$.
- The largest decrease of the solid moisture content is at the first stage of the cascade and the lowest at the third one. This property becomes more explicit when λ increases. For small λ the solid temperature decreases at the first stage and increases at the last one, whereas, for large λ, the reverse situation occurs.
- The general result is valid as follows: large energy consumption is reasonable only in sufficiently costly apparatus.

A comparison between the unconventional (control) processes and the two conventional processes (in which the optimal decisions t_g^n, X_g^n, θ^n are constrained to be the same at the every stage*) shows that when *all* possible parameters of the conventional process are optimized then its optimal cost can be only slightly higher than that found for the unconventional (control) processes. This is clearly seen for the variant I of the conventional process where the decisions (the same at every stage) are: temperature, humidity, and flow of the inlet gas. Whereas the optimal cost for the variant II of the conventional process, where the decisions are only temperature and flow of inlet gas, can be considerably higher than that for the unconventional control process. Hence the level of the all optimal decisions is a basic factor which makes it possible to improve the process economy. This conclusion is encountered also for other drying systems.

As the constant decision policies are most easy for the practical realization, the optimal conventional processes with the complete set of possible decisions may be recommended for many drying systems in the typical range of operational parameters. For some special range of the input data, especially for a low final solid moisture content W_{sf}, and when some unusual expensive drying agents are used the superiority of the unconventional processes, worked out here with details, should be more explicit. It should be remembered that, in the industry, even a cost decrease of several percent is important.

Drying Energy Savings Due to Optimization

In Reference 12 the performances of the various drying apparatus are assessed by defining the available energy savings or operational cost savings, attained among a class of nonoptimal processes under comparison, when the concrete nonoptimal drying process is being replaced by the best (optimal) process. When variations of investment terms are negligible the best process is characterized by the minimum available energy consumption. In the case where various apparatus are compared, i. e., when investment costs vary, the optimal process does not correspond with the minimum available energy consummed, i.e., the processes can be found, in the vicinity of the optimum, which have the energy consumption lower than the optimal (economical) process. There remains still, however, a subclass of

* The complex optimization precedures are not required when computing the conventional processes. Usually it is a differential calculus problem. Therefore the computational details for conventional processes are here omitted.

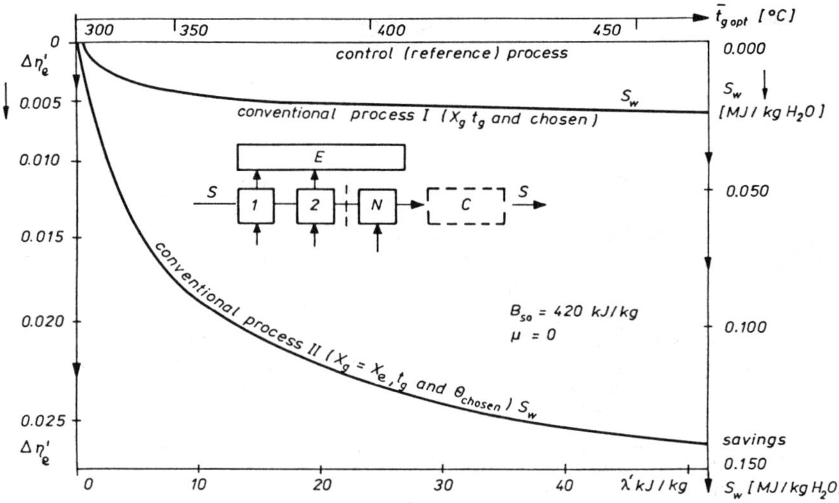

Figure 8. Increase of the saving ability with the intensity factor λ' for the two variants of the conventional processes of multistage fluidization drying. Variant I—decisions t_g, X_g and θ; variant II—decisions t_g and θ.

the processes characterized by an energy consumption larger than that in the optimal process, and for this subclass, the statement of the energy savings problem remains appropriate. The so-called energy saving ability function, introduced in Reference 12 determines quantitatively the savings of available energy, related to the unit mass of the moisture evaporated, obtained when the concrete nonoptimal drying process is replaced by the optimal one. The importance of the optimal process as the reference frame for the savings function should be understood. The quantitative relation between the reduction of the drying irreversibility, the increase of energy utilization efficiency η_e', and available energy savings can be studied by using this approach [12].

For example, in the case of multistage fluidized drying the saving ability function, expressed in available energy (exergy) terms,

$$S_w = \frac{\Delta\left[\sum_{1}^{N}(b_g^n - \mu^{n-1}b_w^{n-1})\theta^n + (B_s^0 - B_s^N)u\right]}{W_{s0} - W_{sf}} \tag{96}$$

where Δ pertains to the difference between the considered process and the optimal process.

Whereas the energy utilization efficiency

$$\eta_e' = \frac{B_s^N + \sum_{1}^{N}\mu^{n-1}\theta^n b_w^{n-1}}{B_s^0 + \sum_{1}^{N}\theta^n b_g^n} \tag{97}$$

where, in our case $\mu = u = 0$. With reference to the optimal process, the energy savings function s_w is presented in Figure 8 versus the investment factor λ', which is a measure of the optimal process intensity as well. The two variants of conventional processes, mentioned earlier, are considered. It is seen from Figure 8 that the saving ability increasses with λ', which means that replacement of the conventional process by the optimal control process is more resonable in the case of expensive drying equipment, where more intensive processes occur.

NOTATION

A	apparatus cross-sectional area (m^2)
a	solid specific surface (m^2/m^3)
B_s, b_g	energy of gas and solid per unit mass of absolute dry substance, respectively (kJ/kg)
C_g	heat capacity of gas per mass unit of dry gas (kJ/kgK)
C_s	solid price per unit mass of absolute dry substance ($/kg)
C_{el}	electric energy unit price ($/J)
C_s, C_s^n	solid price and its value after stage n, respectively ($/kg)
$C_{si} = C_s^o$, $C_{sf} = C_s^N$	initial and final solid price, respectively ($/kg)
c_g	inlet gas price (schemes a and c, Figure 2) or outlet gas price (scheme b) ($/kg)
c_s	price of gas at equilibrium with solid ($/kg)
c_w^n	outlet gas price (Schemes a and c, Figure 2) or inlet gas price (scheme b) ($/kg)
e, g, j	constant coefficients in state equations defined in Table 1
e_g, E_s	unit energy price for gas and solid phase, respectively ($/kJ)
$\tilde{F}^n[I_s^n, W_s^n]$	optimal overall cost modified by Lagrangian multiplier term ($/kg)
G	total dry gas mass flow rate for all systems in Figure 1 (kg/sec)
ΔG^n	gas flow rate trough stage n kg/sec)
g^n	dimensionless gas flow at stage n, defined in Table 1
g_o	parameter of operating line
H	total height of the system m
H^n	hamiltonian at stage n
I_s, I_s^n	solid enthalpy per unit mass of absolute dry substance and its value after stage n, respectively (kJ/kg)
I_s^o, I_s^N	initial and final solid enthalpy, respectively (kJ/kg)
I_{si}, I_{sf}	initial and final solid enthalpy, respectively (kJ/kg)
i_g, i_w	gas enthalpies, per unit mass of absolute dry gas taken before and/or after the stage in accordance with Figure 2 (kJ/kg)
i_s	enthalpy of gas in equilibrium with solid (kJ/kg)
i_{ps}	partial enthalpy of moisture in vapor state at equilibrium with solid (kJ/kg)
K	economic cost of the process defined in the text ($/kg)
Le	Lewis factor = kgCg/α
M	molar mass (kg/kmol)
N_g	overall gas number of mass transfer units/dilute suspensions)
N, n	final stage number and current stage number, respectively
P^n	unit cost of production at stage n ($/kg)
ΔP_g	pressure drop in fluidized layer (Pa)
p	unit apparatus price ($/m^2$)
S, S_o	dry solid flow rate and dry solid mass in a system, respectively (kg/sec, kg)
S_w	saving ability function (MJ/kgH_2O)
T	absolute temperature
t_g, t_g^*	inlet gas temperature and largest temperature, respectively (°C)
t_s	solid temperature (°C)
t_{si}, t_{sf}	solid initial and final temperature, respectively (°C)
u	coefficient modifying the performance index
$V, \Delta V^n$	total system volume and volume at stage n, respectively (m^3)
V_g	superficial gas velocity (m/s)
W_s	absolute moisture content of solid as mass of moisture per unit mass of dry material (kg/kg)

W_s^n solid moisture content after stage n (kg/kg)
$W_{si}=W_s^o$, $W_{sf}=W_s^o$ initial and final solid moisture content, respectively (kg/kg)
X_g, X_w gas absolute humidities (mass of moisture per mass unit of dry gas) taken before and/or after the stage in accordance with Figure 2 (kg/kg)
X_s humidity of gas at equilibrium with solid (kg/kg)
z adjoint variable
$\tilde{z}$ factor describing "freezing" of capital costs

Greek Symbols

α overall heat transfer coefficient ($kJ/m^2 - °C - h$)
β coefficient describing repairs ($year^{-1}$)
β_g, β_w inlet ant outlet gas relative humidities related to absolute humidities X_g and X_w
β_s equilibrium value of the relative gas humidity
μ,ψ gas utilization coefficients, defined in the text
λ Lagrangian multiplier, factor of investment costs (\$/kg, \$/Mg)
Δ increment
δ density (kg/m^3)
η,η' residence time of solid, energy utilization efficiency, respectively (S)
Φ dimensionless time defined in the text
τ_p, τ_u maximum acceptable payout time and utilization time of dryer, respectively (years, h/year)

Exact definitions of quantities: Φ^n, ϑ^n, g^n, g_o, e, j are given in Table 1 and quantities μ^n, ψ^n, u in Table 2. Parentheses ( ) in equations refer to multiplication and brackets [ ] in equations refer to functional dependence or minimization.

Subscripts

e environment
i, f initial and final, respectively
g gas (inlet and outlet, Figure 2)
m moisture
p vapor
s solid
w gas state defined by Figure 2
* least allowed value

Superscripts

n, N stages n or N, respectively
$-$ prescribed constant quantity
* largest allowed value
$\hat{}$ optimal final value
$\tilde{}$ cost optimum

REFERENCES

1. Sieniutycz, S., *Optimization in Process Engineering,* Warsaw, WNT, Poland, 1978.
2. Sieniutycz, S., "Computing of the Thermodynamic Functions in Gas-Solid Systems," *Reports of Inst. of Chem. Eng. Warsaw Tech. Univ.,* Vol. 2, No. 3. (1973), pp. 45–69.
3. Wing, E. M., "The Method of Invariant Imbedding and its Application to Transport Theory and Other Areas of Mathematical Physics," Dallas, Socony Mobil-Oil Co. Inc., 1965.
4. Lee, E. S., "Invariant Imbedding, a Versatile Computational Concept," *Ind. Eng. Chem.,* Vol. 60, No. 9 (1968), pp. 55–64.
5. Bellman, R., *Dynamic Programming,* Princeton Univ. Press, 1959.

6. Szargut J, Petela, R., *Exergy,* Warsaw, WNT Poland, 1965.
7. Sieniutycz, S., "The Thermodynamic Approach to Fluidized Drying and Moistering Optimization," *AIChE Journal,* Vol. 19, No. 3 (1973), pp. 277–286.
8. Bosniakovic, F., *Technical Thermodynamics,* Dresden-Leipzig T. Steinkopff, Germany, 1968.
9. Sieniutycz, S., "A Constant Hamiltonian Problem and an Introduction to Mechanics of Optimal Discrete Systems, "*Reports of Inst. of Chem. Eng. Warsaw Tech. Univ.,* Vol. 3, No. 1/2. (1974), pp. 27–55.
10. Sieniutycz, S., and Szwast, Z., "A Discrete Algorithm for Optimization with a Constant Hamiltonian and its Application in Chemical Engineering," *Intern. Chem. Eng.,* Vol. 23, No. 1. (1983), pp. 155–166.
11. Sieniutycz, S., and Szwast, Z., "Optimization of Multistage Fluidized Drying by Some Special Algorithm of Discrete Maximum Principle", *Chem. Eng. Journal,* Vol. 25, (1982), pp. 63–75.
12. Sieniutycz, S., "Development of Relation Between Drying Energy Savings and Thermodynamic Irreversibility, "Conference paper, International Meeting on Energy Savings in Drying Processes, Liege, Oct. 4–6, 1983. pp. A8.1–A8.13.

CHAPTER 41

THREE-DIMENSIONAL MODELING OF FLOW, HEAT TRANSFER, AND COMBUSTION

Bjorn H. Hjertager

Chr. Michelsen Institute
Dept. of Science and Technology
N-5036, Fantoft
Bergen, Norway

CONTENTS

The work presented in this chapter has been carried out partly while the author was affiliated at the Division of Thermodynamics, The Norwegian Institute of Technology, Trondheim, and partly at his present affiliation with the Chr. Michelsen Institute, Bergen. The author gratefully acknowledges financial support from the Royal Norwegian Council for Scientific and Industrial Research during his stay at The Norwegian Institute of Technology, and financial support from the oil companies BP Petroleum Development Ltd., Norway, Elf Aquitaine Norge A/S, Esso Exploration and Production Norway Inc., Mobil Exploration Norway Inc., Norsk Hydro and Statoil, during his present stay at Chr. Michelsen Institute. The author would also take the opportunity to acknowledge the support from his colleagues Professor B. F. Magnussen, The Norwegian Institute of Technology, and Dr. R. K. Eckhoff, Chr. Michelsen Institute.

INTRODUCTION

The Need for Three-Dimensional Modeling

Flow, heat transfer, and chemical reactions often take place simultaneously in various types of industrial equipment. The performance of such equipment is often strongly dependent on the interaction between these processes and the geometrical design parameters. Because of the complexity of such problems it has in most cases been impossible to solve these problems analytically. Design of industrial equipment has therefore mainly been based on experiments in full scale or in reduced scale by application of various empirical scaling techniques. These design techniques can be very expensive and may give insufficient and inaccurate information about the performance. The computer has made it possible to develop design techniques based on mathematical modeling of the basic processes that are more reliable and economical. To model these processes the basic equations of flow, heat transfer, and chemical reactions as well as the coupling between them must be solved in many cases for a three-dimensional space. The great variety and complexity of such problems calls for general and economical solution procedures.

Turbulence

One of the key problems in modeling three-dimensional flows is the ability to predict turbulent flows. Great advances have been made during later years in computing complex turbulent two-dimensional flows. Calculation of fully three-dimensional turbulent flows using advanced turbulence modeling has, however, only just begun.

The need for three-dimensional calculation procedures are obvious when we acknowledge that many processes of engineering importance take place within three-dimensional geometries. Examples are ventilation of rooms, flow in pumps and turbines, etc. To be able to calculate such cases with engineering accuracy, validations calculations of simpler three-dimensional cases must be performed. These calculations should be tested against detailed experimental data.

Combustion

Simulation of combustion processes in turbulent and compressible flows is a complex problem of both fundamental and practical importance. The nonlinearities and range of time scales in the problem makes it a demanding task from both computational and conceptual points of view. Adequate modeling of the coupling of fluid dynamics and reaction rate is the key to successful prediction of, for example, turbulent flame propagation, pollutant formation and overall efficiency of various combustion processes.

The shortage of fossil fuels has focused attention on more efficient use of these resources. To improve performance of and minimize pollution from combustion devices, there is a need for good predictive tools, with the objective to identify the optimum design or mode of operation. Furthermore, the increased production and thus transport and handling of large quantities of flammable gases have increased the need for adequate means of predicting flame and pressure development in possible accidental explosions, and thus enable adoption of realistic safety measures.

Some Prediction Methods

There are in the literature reported several procedures for numerical calculations of three-dimensional flows both for incompressible and compressible situations. The basic features of a procedure widely used have been presented by Patankar and Spalding [1] and applied to various types of incompressible steady-state flows, such as boundary layer flow, flow with combustion, and flow in heat exchangers.

This method employs several features of earlier procedures; the displaced grids for velocity and pressure by Harlow and Welch [2]; the guess and correct procedure for the coupling of velocity with pressure used by Chorin [3].

Several prediction methods capable of handling transient compressible flows with chemical reactions have also been published. All of these methods use explicit formulation of the momentum and scalar equations. The method of Butler and O'Rourke [4] uses the so-called ICE (Implicit Continuous-fluid-Eulerian) technique to eliminate the Courant sound speed criterion. This method together with the ALE (Arbitrary Lagrangian Eulerian) technique is used by Boni et al. [5] to study combustion in the stratified charge engine. Oran et al. [6] have used a fully explicit scheme to study the details of detonation waves. A special feature in their problem is to resolve and maintain steep gradients in the flow. This was accomplished by use of the nonlinear FCT (Flux Corrected Transport) technique. All methods mentioned above have a time-step limitation due to their explicit formulation. This may be a disadvantage if slow variations are studied or if a steady-state solution is sought. An alternative formulation is to use implicit methods. In principle these do not have any time-step limitations and may therefore also be used for steady-state incompressible as well as compressible turbulent flows.

Purpose and Contents of Chapter

This chapter will review a computation method based on the Patankar and Spalding [1] method in the form given by Hjertager and Magnussen [7]. This method was initially developed for incompressible reacting flows and has subsequently been extended by Hjertager [8] to take account of compressible flows. The method is therefore general, in a sense that it can handle both compressible and incompressible reacting or nonreacting multicomponent single-phase transient or steady-state turbulent flows in three-dimensional geometries.

The first part of this chapter formulates the basic equations to be solved. The following four sections outline the various models adopted, namely the turbulence, combustion, thermodynamics, and radiation models. Then a method of handling complex geometries is presented, followed by an exploration of the numerical calculation method. Several sample calculation of the numerical method including laminar and turbulent flows, combustion chamber processes and some compressible flows are then given. The last section is the conclusions.

GOVERNING EQUATIONS

The problem of turbulent transient flows can be handled by solving for the time evolution of time mean values of the dependent quantities, Φ, in the domain of interest. The time mean of a variable varying with time, t, may be expressed as:

$$\Phi(t) = \frac{1}{T}\int_{t}^{t+T} \hat{\Phi}(t+\tau)\, d\tau \tag{1}$$

where $\Phi(t)$ is the time mean value of instantaneous value $\hat{\Phi}(t)$ averaged over the time interval T. T must satisfy two competing demands. First, it must be small enough not to smear out the sought time dependence of the system under consideration. Secondly, it must be large enough to be able to produce sufficient information to enable relevant time mean values in the interval. This means time mean values of both the relevant quantities and their second-order correlations must be obtainable in the time interval T. This is often possible since conversely, turbulence has higher frequencies than the large-scale motion which generates turbulence. The equation of motion and energy may be expressed in tensor notation as:

$$\frac{\partial \rho}{\partial t} + \frac{\partial}{\partial x_i}(\rho U_i) = 0 \tag{2}$$

$$\frac{\partial}{\partial t}(\rho U_i) + \frac{\partial}{\partial x_j}(\rho U_j U_i) = -\frac{\partial p}{\partial x_i} + \frac{\partial}{\partial x_j}(\sigma_{ij}) + \rho g_i \tag{3}$$

$$\frac{\partial}{\partial t}(\rho h) + \frac{\partial}{\partial x_j}(\rho U_j h) = \frac{\partial}{\partial x_j}(J_{h,j}) + \frac{Dp}{Dt} + S_h \tag{4}$$

Here U_i = the velocity component in the x_i coordinate direction
g_i = the gravitation in x_i direction
p = the pressure
ρ = the density
h = the enthalpy
σ_{ij}, $J_{h,j}$ = the fluxes of momentum and energy, respectively

S_h is an additional source term which may include such sources as radiation influences and viscous dissipation. The conservation equation for a specie is given as

$$\frac{\partial}{\partial t}(\rho m_i) + \frac{\partial}{\partial x_j}(\rho U_j m_i) = -\frac{\partial}{\partial x_j}(J_{i,j}) + R_i \tag{5}$$

where R_i is the net reaction rate of the specie m_i.

TURBULENCE MODEL

Closure Assumptions

To solve the governing expressions Equations 2, 3, 4, and 5, the fluxes, $J_{\Phi,j}$, and the net rate of reaction R_i, have to be modeled together with specification of relevant boundary and intial conditions. Both the fluxes and the reaction rate are time-mean-averaged values of fluctuating quantities. The fluxes can, for a general scalar variable Φ, and a velocity component U_j, be expressed as:

$$J_{\Phi,j} = -\rho\overline{u_j\varphi} \tag{6}$$

$$\sigma_{ij} = -\rho\overline{u_i u_j} \tag{7}$$

where u_i and φ are the instantaneous fluctuations around the time mean values U_j and Φ, respectively. The overbar indicates time mean value over the time interval T as defined in Equation 1. When specifying the correlations given in Equations 6 and 7 it is usual to relate these to the product of time mean gradients of the relevant variables and an effective turbulent transport coefficient. For a general scalar variable Φ and a velocity component U_j the relations are:

$$J_{\Phi,j} = -\frac{\mu_{eff}}{\sigma_\Phi}\frac{\partial \Phi}{\partial x_j} \tag{8}$$

and

$$\sigma_{ij} = \mu_{eff}\left(\frac{\partial U_i}{\partial x_j} + \frac{\partial U_j}{\partial x_i}\right) - \frac{2}{3}\delta_{ij}\left(\rho k + \mu_{eff}\frac{\partial U_k}{\partial x_k}\right) \tag{9}$$

respectively.

Here $\delta_{ij} = 1$ if $i = j$ and $\delta_{ij} = 0$ if $i \neq j$. An effective turbulence viscosity μ_{eff} and the kinetic energy of turbulence have been introduced in the above expressions, together with an effective Prandtl/Schmidt number σ_Φ. The kinetic energy of turbulence, k, is related to the fluctuating turbulence velocity components in the three coordinate directions as:

$$k = \frac{1}{2}(\overline{u_1^2} + \overline{u_2^2} + \overline{u_3^2}) \tag{10}$$

The effective turbulence viscosity is given by the two turbulence parameters, the isotropic turbulence velocity u_t and a length scale, l, as:

$$\mu_{eff} = \mu_l + \rho u_t l \tag{11}$$

μ_l is the molecular viscosity.

Two-Parameter Turbulence Model

The determination of u_t and l are at present best done by application of the so-called k–ε model of turbulence given by Launder and Spalding [9]. The turbulence velocity is related to the kinetic energy of turbulence, k, as:

$$u_t = \left(\frac{2}{3}k\right)^{1/2} \tag{12}$$

and the length scale, l, is related to the kinetic energy of turbulence, k, and its rate of dissipation, ε, as:

$$l \sim \frac{k^{3/2}}{\varepsilon} \tag{13}$$

Inserting Equations 12 and 13 into Equation 11 gives as result:

$$\mu_{eff} = \mu_l + C_D\rho\frac{k^2}{\varepsilon} \tag{14}$$

C_D is a constant taken to be 0.09.

The conservation equations that determine the distribution of k and ε read:

$$\frac{\partial \rho k}{\partial t} + \frac{\partial}{\partial x_j}(\rho U_j k) = \frac{\partial}{\partial x_j}\left(\frac{\mu_{eff}}{\sigma_k}\frac{\partial k}{\partial x_j}\right) + G - \rho\varepsilon \tag{15}$$

$$\frac{\partial \rho \varepsilon}{\partial t} + \frac{\partial}{\partial x_j}(\rho U_j \varepsilon) = \frac{\partial}{\partial x_j}\left(\frac{\mu_{eff}}{\sigma_\varepsilon}\frac{\partial \varepsilon}{\partial x_j}\right) + C_1\frac{\varepsilon}{k}G - C_2\rho\frac{\varepsilon^2}{k} \tag{16}$$

The two new constants appearing above C_1 and C_2 are given the values 1.44 and 1.79, respectively. The Schmidt numbers σ_k and σ_ε are given the values 1.0 and 1.3, respectively, whereas the other Schmidt/Prandtl numbers are put equal to 0.7. The generation rate of turbulence is given by:

$$G = \sigma_{ij}\frac{\partial U_j}{\partial x_i} + \overline{\rho' u_i} g_i \tag{17}$$

The first part is related to turbulence production due to shear and expansion/compression, the second term is production due to buoyancy.

Treatment of Wall Regions

To be able to account for the large gradients of variables near the walls in a coarse grid computation, special assumptions must be made for these regions. The method used here is the wall function method given by Launder and Spalding [9]. In this method the variation of variables in near wall regions is taken to be similiar to that found in two-dimensional turbulent boundary layers. This means that the wall shear stresses and heat fluxes are given by:

$$\vec{\tau_w} = \frac{\rho(C_D^{1/2}k)^{1/2}\cdot\vec{V}\cdot\kappa}{\ln(Ey_{nw}^{+})} \tag{18}$$

and

$$q_w = \frac{c_{p,\,mix}\cdot\rho\cdot(C_D^{1/2}k)^{1/2}\cdot(T-T_w)\cdot\kappa}{\sigma_h(\ln(Ey_{nw}^{+})+\kappa\cdot P} \tag{19}$$

where

$$P = 9.24(\sigma_{h,\,1}/\sigma_h)^{0.75}-1.0) \tag{20}$$

$$y_{nw}^{+} = \frac{\rho(\sqrt{C_D}k_{nw})^{0.5}\cdot\Delta y_{nw}}{\mu_1} \tag{21}$$

In the equation for kinetic energy of turbulence the following assumptions are made:

- Diffusion towards the walls equals zero.
- Dissipation of k between the wall and near wall point is calculated from:

$$\int_0^{y_{nw}} \varepsilon\, dy = (\sqrt{C_D}k_{nw})^{1.5}\frac{1}{\kappa}\ln(Ey_{nw}^{+}) \tag{22}$$

In the equation for dissipation of kinetic energy of turbulence the following assumption for the near point is made:

$$\varepsilon_{nw} = C_D k_{nw}^{1.5}/(\kappa\Delta y_{nw}) \tag{23}$$

The constants in the wall functions are given the following values:

$\kappa = 0.42$

$E = 9.0$

which correspond to a smooth wall situation. If a rough wall situation exists in the problem, modifications to the constant E may be introduced as given by Jayatilleke [10].

COMBUSTION MODELING

The combustion process may often be treated as a single step irreversible reaction with finite reaction rate between fuel and oxydant. Hence, the reaction scheme may be written as:

$$1\text{ kg fuel} + s\text{ kg oxygen} \rightarrow (1+s)\text{ kg products} \tag{24}$$

Here s is the stoichiometric oxygen requirement to burn 1 kg of fuel. This simple reaction scheme results in mixture composition being determined by solving for only two variables, namely mass fraction of fuel, m_{fu}, and the mixture fraction, f:

$$\frac{\partial}{\partial t}(\rho m_{fu}) + \frac{\partial}{\partial x_j}(\rho U_j m_{fu}) = -\frac{\partial}{\partial x_j}(J_{fu,j}) + R_{fu} \quad (25)$$

$$\frac{\partial}{\partial t}(\rho f) + \frac{\partial}{\partial x_j}(\rho U_j f) = -\frac{\partial}{\partial x_j}(J_{f,j}) \quad (26)$$

Here R_{fu} is the time mean rate of combustion of fuel, whereas $J_{fu,j}$ and $J_{f,j}$ are the diffusive fluxes in the x_j-direction. The basis for this to be valid is that the effective turbulent Schmidt numbers are equal for all species, an approximation which is often found to be valid in turbulent flows.

The mixture fraction is defined as:

$$f = \frac{\beta - \beta_\infty}{\beta_0 - \beta_\infty} \quad (27)$$

where β is a conserved combined variable of, for example, mass fraction of fuel, m_{fu}, and mass fraction of oxygen, m_{O_2}, expressed as:

$$\beta = m_{fu} - m_{O_2}/s \quad (28)$$

β is the value of β at a fuel rich reference point, for example a fuel inlet point in the domain, and β_∞ is the value of β at an oxydant rich reference point, for example the ambient air condition. For a homogeneous premixed system the mixture fraction will be constant in the domain of interest and consequently only the m_{fu}-equation needs to be solved.

For the simple reaction scheme given in Equation 24 above an instantaneous rate of reaction can be written as:

$$\hat{R}_{fu} = \rho^2 \hat{k}_{fu} \hat{m}_{fu} \hat{m}_{O_2} \quad (29)$$

Here $\hat{k}_{fu}$ is the instantaneous value of the Arrhenius temperature dependence of the rate constant. If this rate constant and the mass fractions are written as the sum of a time mean and a fluctuating quantity ($\hat{m}_i = m_i + m_i'$), the reaction rate (ignoring density fluctuations) would be written as:

$$\hat{R}_{fu} = \rho^2 (k_{fu} + k_{fu}')(m_{fu} + m_{fu}')(m_{O_2} + m_{O_2}') \quad (30)$$

Multiplying out this expression and taking the time-mean value of the result gives:

$$R_{fu} = -\rho^2 (k_{fu} m_{fu} m_{O_2} + k_{fu} \overline{m_{fu}' m_{O_2}'} + m_{fu} \overline{k_{fu}' m_{O_2}'} + m_{O_2} \overline{k_{fu}' m_{fu}'} + \overline{k_{fu}' m_{fu}' m_{O_2}'}) \quad (31)$$

This relation shows that even the simplest reaction scheme gives rise to a rather complicated expression for the time mean reaction rate. This indicates that rigorous mathematical methods of attack of the reaction rate based on chemical kinetics in turbulent situations has immense obstacles prior to providing solutions. It is therefore necessary to seek, if possible, alternative and simpler methods.

Experiments have shown that the rate of combustion in flames is mainly dependent on hydrodynamic parameters. This implies that the combustion rate is limited by the rate of molecular mixing between the reactants. This mixing is directly linked to the rate at which turbulent eddies are dissipated. For a scalar variable this dissipation is formally expressed as:

$$\varepsilon_\Phi = D_\Phi \overline{\frac{\partial \Phi}{\partial x_k} \cdot \frac{\partial \Phi}{\partial x_k}} \quad (32)$$

D_Φ is the molecular diffusion coefficient. The hydrodynamic rate of dissipation of turbulent kinetic energy is denoted by ε. It is therefore assumed that combustion rate is proportional to the rate of dissipation of kinetic energy of turbulence

$$R_{fu} = -\rho\varepsilon\frac{g_\Phi^{1/2}}{k} \quad (33)$$

g_Φ is the variance of the fluctuations of the limiting specie in question. Magnussen and Hjertager [11] argue that the fuel, oxydant, and reaction products appear as intermittent fluctuating quantities. Consequently, the fluctuating species may be related to time-mean values of fuel, oxydant, or reaction products. Therefore:

$$R_{fu} = -A\rho\frac{\varepsilon}{k}m_{lim} \quad (34)$$

where m_{lim} is the smallest of the three mass fractions, namely fuel, m_{fu}, oxygen m_{02}/s or mass fraction of fuel already burnt, $m_{fu,\,b}$. A is a constant. In order for Equation 34 to be valid, the chemical kinetics of the system under consideration must be fast. In many cases this is not the case, especially in the fast transient combustion which for example occurs in high speed gas explosions.

A simple modification of the prior expression has therefore been proposed by Hjertager [12].

Based on the chemical kinetics of the system, a chemical time can be defined, τ_{ch}. Also, the lifetime of the turbulent eddies can be defined, τ_e. Ignition or extinction is assumed to occur when these two times are in a given ratio, namely:

$$\tau_{ch}/\tau_e = D_{ie} \quad (35)$$

The following modification to the rate expression in Equation 34 is therefore used:

- If $\tau_{ch}/\tau_e \geqq D_{ie}$ then $R_{fu} = 0.$
- If $\tau_{ch}/\tau_e < D_{ie}$ then $R_{fu} = -A\,\rho\frac{\varepsilon}{k}m_{lim}.$

(36)

This criteria is closely related to the models proposed by Radhakrishnan et al. [13] and Magnussen [14] for extinction phenomena. The eddy lifetime or mixing time is defined as:

$$\tau_e = \frac{1}{u_t} \approx \frac{k}{\varepsilon} \quad (37)$$

The chemical time is taken equal to the chemical induction time which is often expressed as:

$$\tau_{ch} = A_{ch}\exp(E/RT)(\rho m_{fu})^a(\rho m_{O_2})^b \quad (38)$$

Radhakrishnan et al. [13] proposed that the chemical time should be taken equal to the time for a laminar flame to propagate across a length equal to the Taylor micro scale, λ. Their application, however, was a correlation of blow-off velocity data in disc-stabilized premixed flames.

THERMODYNAMICS MODELING

For a single-step reaction of the form given in Equation 24 a total of five massfractions may be identified namely; fuel, nitrogen, oxygen, carbon dioxide, and water vapor.

The density of the gas mixture can be calculated from the ideal gas law as:

$$\rho = \frac{M_{av}\cdot p}{R_u\cdot T} \quad (39)$$

where the average molecular weight is calculated from:

$$\frac{1}{M_{av}} = \sum_{i=1}^{5} \frac{m_i}{M_i} \tag{40}$$

The enthalpy given in Equation 4 is defined as:

$$h = c_{pm, mix} T + m_{fu} \cdot H_{fu}$$

Here the mixture specific heat is calculated from each of the gases present in the mixture as:

$$c_{pm, mix} = \sum_{i=1}^{5} m_i \cdot c_{pm, i} \tag{41}$$

where

$$c_{pm, i} = \frac{1}{T - T_{ref}} \int_{T_{ref}}^{T} c_{p, i} \, dT \tag{42}$$

The specific heats are assumed to vary linearly with temperature:

$$c_{pm, i} = a_i + b_i \cdot T \tag{43}$$

Data for the a_i's and b_i's are taken from Khalil et al. [15] and given in Table 1 together with other data for the gas constituents. Data for natural gas is given as an example of a fuel gas.

RADIATION MODELING

Radiative heat transfer is different from convection and diffusion in such a way that in addition to the dependence on the three space coordinates it also depends on the space angle and optical wavelength. In this chapter we have retricted our attention to grey radiation.

The equation that governs the radiative transfer for an absorbing emitting nonscattering grey medium may be expressed as:

$$\frac{\partial I}{\partial s} = -K_R I + \frac{K_R \sigma T^4}{\pi} \tag{44}$$

Here s is the coordinate in a given direction.

The direct solution of this equation has been found to be cumbersome. It has therefore been the goal of many research workers within combustion to simplify the discretization of the angle. The work of Patankar and Spalding [16] introduces and extends the so called "flux-methods" for use in three-dimensional furnaces. This early, partly intuitive method, has been improved in later works. The works of De Marco and Lockwood [17] and Bartelds et al. [18] start with the basic equation of radiative transfer Equation 44 and make assumptions of the angular distribution of the intensity I. In the present work we have

Table 1
Thermodynamic Data

Gas Constituent	Chemical Formula	Molecular Weight (kg/kmol)	a_i (kJ/kg/K)	$b_i \times 10^{-3}$ (kJ/kg/K^2)
Fuel (natural gas)	—	18	1.0	2.055
Oxygen	O_2	32	0.8881	0.0977
Carbon dioxide	CO_2	44	1.0023	0.0865
Water vapor	H_2O	18	1.7402	0.3072
Nitrogen	N_2	28	0.8238	0.1983

adopted the method of De Marco and Lockwood [7]. They assumed a truncated Taylor expansion for the angular dependence as:

$$I = A_1\mu + A_2\eta + A_3\xi + B_1\mu^2 + B_2\eta^2 + B_3\xi^2 \tag{45}$$

Here μ, η, and ξ are the direction cosines of the unit vector and A_1, A_2, A_3, B_1, B_2, B_3 are coefficients which are only dependent on the three space coordinates. This angular dependence of the radiation intensity is substituted into Equation 44 and integrated over 6 different space angles. The result of this integration is a set of three flux equations:

$$\frac{\partial}{\partial x}\left[\frac{1}{K_R}\frac{\partial F_x}{\partial x}\right] = \frac{4}{3}K_R(3F_x - F_y - F_z - \sigma T^4)$$

$$\frac{\partial}{\partial y}\left[\frac{1}{K_R}\frac{\partial F_y}{\partial y}\right] = \frac{4}{3}K_R(3F_y - F_x - F_z - \sigma T^4)$$

$$\frac{\partial}{\partial z}\left[\frac{1}{K_R}\frac{\partial F_z}{\partial z}\right] = \frac{4}{3}K_R(3F_z - F_x - F_y - \sigma T^4) \tag{46}$$

Here the F_i's are related to the B_i's through:

$$F_x = \frac{\pi}{4}(2B_1 + B_2 + B_3)$$

$$F_y = \frac{\pi}{4}(2B_2 + B_1 + B_3)$$

$$F_z = \frac{\pi}{4}(2B_3 + B_1 + B_2) \tag{47}$$

Radiation influences the energy equation through absorption/emission of the radiative heat flux. This will include an additional source term in the energy equation of the form:

$$S_{h,\,rad} = -\frac{\partial}{\partial x_i} q_{i,\,rad} \tag{48}$$

This transforms to the following expression given by the preceding flux quantities F_i, as:

$$S_{h,\,rad} = \frac{16}{9}K_R(F_x + F_y + F_z - 3\sigma T^4) \tag{49}$$

The absorption coefficient K_R appearing in the prior equations may vary in the calculation domain and is dependent on the local concentration of the gas constituents. A simple relationship has been proposed by Khalil et al. [15] and reads:

$$K_R = 0.2 \cdot m_{fu} + 0.1 \cdot m_{pr} \tag{50}$$

MODELING OF COMPLEX GEOMETRIES

Many geometries found in industrial practice may contain a lot of geometrical details which may influence the process to be simulated. Examples of such geometries are heat exchangers with thousands of tubes and several baffles, and regenerators with a lot of internal heat absorbing obstructions etc. There are at least two routes for describing such complex geometries. We may choose to model every detail by use of very fine geometrical resolution, or we may describe the geometry by use of some suitable bulk parameters. Detailed descriptions will always need large computer resources both with regard to memory and calculation speed. It is not feasible with present or even with future computers to implement the detailed method for solving such complex problems. We are therefore forced to use the second line of approach, which incorporates the porosity/distributed resistance formulation of the governing equations. This method was proposed by Patankar and Spalding [19] and has been applied to the analysis of heat exchangers, regenerators, and nuclear reactors. Sha et al. [20] have extended the method to include advanced turbulence modeling.

The presence of geometrical details modifies the governing equations in two ways. First, only part of the total volume is available to flow and secondly solid objects offer additional resistance to flow and additional mixing in the flow. The modified equations for use in high density geometries may be expressed by

$$\frac{\partial}{\partial t}(\beta_v \rho \Phi) + \frac{\partial}{\partial x_i}(\beta_i \rho U_i \Phi) = \frac{\partial}{\partial x_i}\left[\beta_i \Gamma \frac{\partial \Phi}{\partial x_i}\right] + \beta_v (S_\Phi + R_\Phi) \tag{51}$$

Here Φ = a general variable.
β_v = the volume fraction occupied by the fluid
β_i = the area fraction available for flow in the x_i-direction
R_Φ = the additional resistance or additional mixing or heat transfer caused by solid obstructions in the flow

All the volume/area fractions (porosities) may take values between 0.0 (completely blocked), or 1.0 (completely open). Some R_Φ functions may be found in a report by Sha and Launder [21]. These functions depend on parameters like velocity, porosity, typical dimension, pitch between obstacles, obstacle shape, and orientation.

NUMERICAL PROCEDURES

Finite-Domain Equations

It is noted that all conservation equations just mentioned can be written in the following general form. (All porosities are set equal to unity for clarity):

$$\underset{\text{I}}{\frac{\partial}{\partial t}(\rho \Phi)} + \underset{\text{II}}{\frac{\partial}{\partial x_j}(\rho U_j \Phi)} = \underset{\text{III}}{\frac{\partial}{\partial x_j}\left(\Gamma_\Phi \frac{\partial \Phi}{\partial x_j}\right)} + \underset{\text{IV}}{S_\Phi}; \quad \Gamma_\Phi = \frac{\mu_{eff}}{\sigma_\Phi} \tag{52}$$

The equation has four distinct terms, namely: I transient, II convection, III diffusion, and IV source terms. A summary of all the equations needed for a typical combustion calculation are given in Table 2.

The calculation domain is divided into a discrete number of control volumes surrounding a grid point.

The staggered grid has been employed and is shown in Figure 1. The figure shows a central point P surrounded by its neighboring points H, L, N, S in the xz-plane and E, W normal to this plane (not shown). All dependent variables except velocity components are stored at the grid line intersections (nodes), whereas velocity components are stored midway between these nodes. The variables with reference to point P are enclosed in a "boomerang" as indicated in Figure 1. Locations between the main grid nodes are denoted by lower case letters. For example the location between node P and S will be denoted by s.

Finite-Domain Approximations

To solve the governing equations they are integrated over a control volume. If we, for instance, take the equation for a scalar quantity, the proper control volume is shown in Figure 1. The net diffusive and convective flux in the z-direction is expressed by:

$$A_N^\Phi(\Phi_P - \Phi_N) + A_S^\Phi(\Phi_P - \Phi_S) \tag{53}$$

Table 2
Conservation Equations

Conservation	Φ	σ_Φ	S_Φ
Mass	1.0	∞	0
Velocity in x-direction	U	1.0	$-\frac{\partial p}{\partial x}+\frac{\partial}{\partial x}\left(\mu_{eff}\frac{\partial U}{\partial x}\right)+\frac{\partial}{\partial y}\left(\mu_{eff}\frac{\partial V}{\partial x}\right)+\frac{\partial}{\partial z}\left(\mu_{eff}\frac{\partial W}{\partial x}\right)-\frac{2}{3}\frac{\partial}{\partial x}(\mu_{eff}\cdot\nabla\cdot\vec{U}+\rho k)$
Velocity in y-direction	V	1.0	$-\frac{\partial p}{\partial y}+\frac{\partial}{\partial x}\left(\mu_{eff}\frac{\partial U}{\partial y}\right)+\frac{\partial}{\partial y}\left(\mu_{eff}\frac{\partial V}{\partial y}\right)+\frac{\partial}{\partial z}\left(\mu_{eff}\frac{\partial W}{\partial y}\right)-\frac{2}{3}\frac{\partial}{\partial y}(\mu_{eff}\cdot\nabla\cdot\vec{U}+\rho k)$
Velocity in z-direction	W	1.0	$-\frac{\partial p}{\partial z}+\frac{\partial}{\partial x}\left(\mu_{eff}\frac{\partial U}{\partial z}\right)+\frac{\partial}{\partial y}\left(\mu_{eff}\frac{\partial V}{\partial z}\right)+\frac{\partial}{\partial z}\left(\mu_{eff}\frac{\partial W}{\partial z}\right)-\frac{2}{3}\frac{\partial}{\partial z}(\mu_{eff}\cdot\nabla\cdot\vec{U}+\rho k)$
Kinetic energy of turbulence	k	σ_k	$G-\rho\varepsilon$
Rate of dissipation	ε	σ_ε	$\frac{\varepsilon}{k}(C_1G-C_2\varepsilon)$
Enthalpy	h	σ_h	$16/9\cdot K_R(F_x+F_y+F_z-3\cdot\sigma T^4)+\frac{\partial p}{\partial t}+U\frac{\partial p}{\partial x}+V\frac{\partial p}{\partial y}+W\frac{\partial p}{\partial z}$
Mass fraction of fuel	m_{fu}	σ_{fu}	$-R_{fu}$
Mixture fraction	f	σ_f	0

$G=\mu_{eff}\left\{2\left[\left(\frac{\partial U}{\partial x}\right)^2+\left(\frac{\partial V}{\partial y}\right)^2+\left(\frac{\partial W}{\partial z}\right)^2\right]+\left(\frac{\partial U}{\partial y}+\frac{\partial V}{\partial x}\right)^2+\left(\frac{\partial U}{\partial z}+\frac{\partial W}{\partial x}\right)^2+\left(\frac{\partial W}{\partial y}+\frac{\partial V}{\partial z}\right)^2\right\}-\frac{2}{3}\nabla\cdot\vec{U}(\mu_{eff}\cdot\nabla\cdot\vec{U}+\rho k)$

$\nabla\cdot\vec{U}\equiv\frac{\partial U}{\partial x}+\frac{\partial V}{\partial y}+\frac{\partial W}{\partial z}$

The coefficients are:

$$A_N^\Phi = (\Gamma_{\Phi,n}/\Delta z_n+\langle-\rho_n W_N\rangle)\,\Delta x\,\Delta y \tag{54}$$

$$A_S^\Phi = (\Gamma_{\Phi,s}/\Delta z_s+\langle\rho_s W_P\rangle)\,\Delta x\,\Delta y \tag{55}$$

Here the expression in $\langle\ \rangle$ is the upwind differencing for the convection terms expressed by:

$$\langle e\rangle = 0.5(e+|e|) \tag{56}$$

This means that the value of Φ convected into or out of the control volume always has its upstream value.

The time derivative is expressed by:

$$b^\Phi(\Phi_P-\Phi_P^0) \tag{57}$$

where $b^\Phi = \rho_P^0\,\Delta x\,\Delta y\,\Delta z/\Delta t$

Φ_P^0 = the value of Φ at point P at the previous time level

Usually it is common to linearize the source term so that

$$S^\Phi = S_0^\Phi+S_P^\Phi\cdot\Phi_P \tag{58}$$

Here S^Φ, S_0^Φ, and S_P^Φ are the volume integral of the source term given in Equation 52.

Domain Equations

Expressions similar to Equation 53 can be derived for the net fluxes in the other two directions. This eventually gives the finite-domain equivalent to Equation 52:

$$A_P^\Phi\Phi_P = A_N^\Phi\Phi_N+A_S^\Phi\Phi_S+A_H^\Phi\Phi_H+A_L^\Phi\Phi_L + A_E^\Phi\Phi_E+A_W^\Phi\Phi_W+S_0^\Phi+b^\Phi\Phi_P^0 \tag{59}$$

Figure 1. Grid and control volume for scalar quantities.

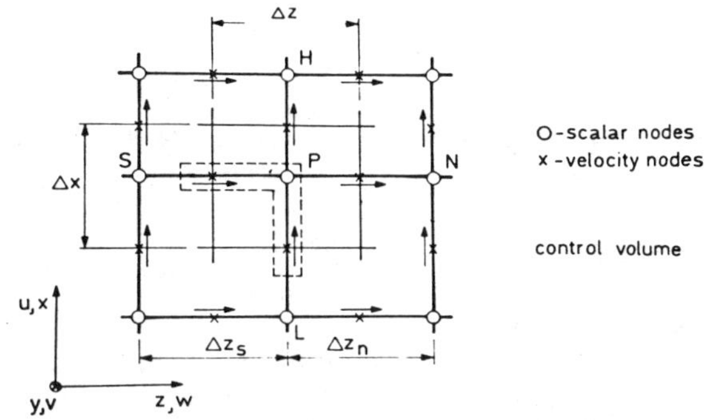

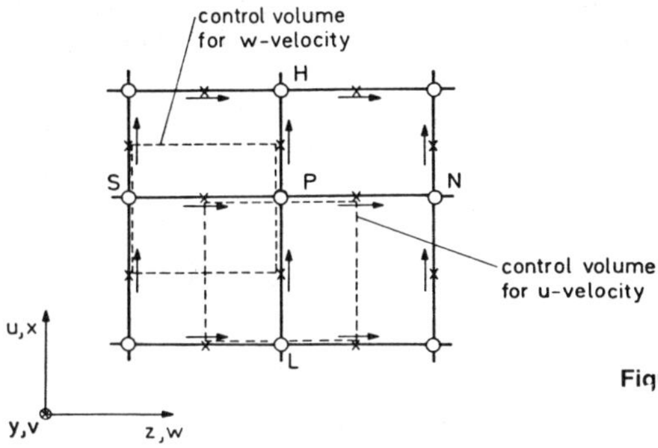

Fiqure 2. Control volumes for u- and w-velocities.

Here

$$A_P^\Phi = A_N^\Phi + A_S^\Phi + A_H^\Phi + A_L^\Phi + A_E^\Phi + A_W^\Phi + b^\Phi - S_P^\Phi \tag{60}$$

In order to secure stability of the numerical scheme, S_P^Φ should always be chosen negative.

The momentum Equation 3 can be integrated in way similar to the scalar equations. The only difference is that the control volume is different. For U- and W-velocities the proper volumes are shown in Figure 2. The finite domain equivalent for the momentum equations can thus be written as:

$$A_P^U U_P = \sum_i A_i^U U_i + b^U U_P^0 + \Delta y\, \Delta z\, (p_L - p_P) + S_0^U \tag{61}$$

$$A_P^V V_P = \sum_i A_i^V V_i + b^V V_P^0 + \Delta x\, \Delta z\, (p_W - p_P) + S_0^V \tag{62}$$

$$A_P^W W_P = \sum_i A_i^W W_i + b^W W_P^0 + \Delta x\, \Delta y\, (p_S - p_P) + S_0^W \tag{63}$$

The summation is performed around the neighboring points N, S, H, L, E, W as was shown in Equation 59. The constant A_i^i is also expressed in the same manner as in Equation 60.

At last the finite-domain approximation for the continuity equation is obtained by integrating it over the control volume in Figure 1. Thus:

$$(\rho_P - \rho_P^0)/\Delta t \cdot (\Delta x\, \Delta y\, \Delta z) + ((\rho U)_h - (\rho U)_l)\, \Delta y\, \Delta z + ((\rho V)_e - (\rho V)_w)\, \Delta x\, \Delta z + ((\rho W)_n - (\rho W)_s)\, \Delta x\, \Delta y = 0 \tag{64}$$

Solution Procedure

To solve Equation 59 through 64 the SIMPLE method (semi-implicit method for pressure linked equations) of Patankar and Spalding [1] has been adopted. This method has also been used in fully three-dimensional steady flows by Caretto et al. [22]. The procedure involves a series of guess and correct operations for the coupling of pressure with velocities. A description of the method shall now be given which also includes the handling of compressibility given by Hjertager [8].

To start in a forward time step we calculate all coefficients in the finite-domain Equations 59 to 64 and these are held constant during the step. Next a pressure field at the new time level has to be guessed. This or the result from the previous time level can be given. This field can be denoted by p^*. With this field we can solve the momentum Equations 61, 62 and 63 to give intermediate velocity fields at the new time level and from the ideal gas law Equation 39 we can calculate the intermediate density field. These fields can be denoted by U^*, V^*, W^*, and ρ^* and will in general not satisfy the continuity Equation 64. To get the velocity and density fields in accordance with continuity the following corrections are performed:

$$U_P = U_P^* + D_P^U(p_L' - p_P') \quad (65)$$

$$V_P = V_P^* + D_P^V(p_W' - p_P') \quad (66)$$

$$W_P = W_P^* + D_P^W(p_S' - p_P') \quad (67)$$

$$\rho_P = \rho_P^* + \left[\frac{\partial \rho}{\partial p}\right] I_P p_P' \quad (68)$$

$$p_P = p_P^* + p_P' \quad (69)$$

Here a new variable p′, the pressure correction, is introduced. The constants D come from the linearized momentum equations and are defined as:

$$D_P^U = \Delta y\,\Delta z / A_P^U; \quad D_P^V = \Delta x\,\Delta z / A_P^V \quad \text{and} \quad D_P^W = \Delta x \Delta y / A_P^W$$

Density variation with pressure is taken to be isentropic. This leads to

$$\left[\frac{\partial \rho}{\partial p}\right] I_P = \frac{1}{c_{s,P}^2}$$

where $c_{s,P}$ is the local sound speed at point P.

To determine the pressure correction p' the Equations 65 through 68 substituted into the continuity Equation 64 thus giving the following finite-domain equation for p':

$$A_P^p p_P' = \sum_k A_k^p p_k' + S^p \quad (70)$$

The summation is again performed around the 6 surrounding points. Further the coefficients are expressed as follows:

$$A_P^p = \sum_k A_k^p; \quad A_L^p = \Delta y\,\Delta z\left[\rho_l D_P^U + \frac{1}{c_{s,l}^2} \cdot \langle U_p \rangle\right] \quad \text{etc.} \quad (71)$$

The source term is:

$$S^p = [(\rho U^*)_l - (\rho U^*)_h]\,\Delta y\,\Delta z + [(\rho V^*)_w - (\rho V^*)_e]\,\Delta x\,\Delta z$$
$$+ [(\rho W^*)_s - (\rho W^*)_n]\,\Delta x\,\Delta y + (\rho_P^0 - \rho_P^*)/\Delta t \cdot \Delta x\,\Delta y\,\Delta z \quad (72)$$

We can see that the source term is the local unbalance in the continuity equation of the intermediate velocity fields.

When Equation 70 for the pressure corrections has been solved, the corrections (Equations 65 through 69) can be performed, and if the resulting mass unbalance is sufficiently small, then the values of the variables at the new time level have been found. If the mass unbalance is too large, repeated solution of Equation 70 and corrections Equations 65 to 69 must be performed.

Solution of the Algebraic Equations

In the preceding section we have referred to solving the equations for momentum and pressure correction. The actual method of solving is performed by the well-known tridiagonal matrix algorithm. This algorithm is used successively in each of the three coordinate directions. At each sweep the variables along a line are supposed unknown whereas the variables in the surrounding lines are supposed known. This practice yields that the variable at a point only is connected with variables at two neighboring points along the line in question. A triple sweep will thus consist of the three following stages.

Stage I:

$$A_P^\Phi \Phi_P^I = A_N^\Phi \Phi_N^I + A_S^\Phi \Phi_S^I + F_I(\Phi^0, S_0^\Phi)$$

Stage II:

$$A_P^\Phi \Phi_P^{II} = A_H^\Phi \Phi_H^{II} + A_L^\Phi \Phi_L^{II} + F_{II}(\Phi^o, \Phi^I, S_o^\Phi) \tag{73}$$

Stage III:

$$A_P^\Phi \Phi_P^* = A_E^\Phi \Phi_E^* + A^\Phi \Phi_W^* + F_{III}(\Phi^o, \Phi^{II}, S_o^\Phi)$$

One such triple sweep is performed for velocity components and Φ's, while three are done on the pressure correction (p'). In many cases the so-called block correction may also be included in the solution of the equations. Details of this technique may be found in the book by Patankar [31].

Boundary Conditions

Because we are working with the primitive variables U, V, W, and p, the boundary conditions are easily inserted. This means that at a wall, the normal velocity component is zero and tangential components equal the velocity of the wall (usually zero). When velocity components are imposed on a boundary the condition on pressure correction is the zero gradient type. For scalars (Φ's) the boundary condition is fixed value, fixed gradient, or a mixed condition, all dependent on the physical requirement.

Nonlinearities

The set of algebraic equations to be solved are only nominally linear. This means that special measures must be taken to secure numerical stability of the solution. This is done by under-relaxation in a steady-state problem and imposing limits to the time step taken in a transient problem. Under-relaxation means that only part of the implications of the algebraic equation is accepted. This means

$$\Phi_a = \alpha_\Phi \Phi_{ae} + (1-\alpha_\Phi)\,\Phi_{old} \tag{74}$$

Here α_Φ is the under-relaxation factor for the variable Φ in question. Typical values for a sought steady solution are $\alpha_U = \alpha_V = \alpha_W = 0.3$ to 0.5, $\alpha_p = 0.5$ to 0.8 and for other scalars $\alpha_\Phi = 0.5$ to 0.7. For a combusting flow with variable density, under-relaxation of the density of $\alpha_\varrho = 0.5$ may also be neccessary.

For a transient situation all α_Φ are set equal to 1.0 except α_P, and α_Q which are set to approximately 0.8. The time step are on the other hand limited by the Courant criteria, which says that the Courant number

$$N_c = \frac{\Delta t \cdot c_t}{\Delta x} \tag{75}$$

should be smaller than a given number usually on the order of unity. The typical velocity, c_t, to be used in the expression should be the absolute value of the local flow velocity for subsonic flows and the local sound speed for compressible supersonic flows.

SOME APPLICATIONS

Laminar Flows

We are concerned with the calculation of steady-state laminar flows presented by Hjertager and Magnussen [7]. To obtain steady-state solution we have gone through the time evolution and advanced in time until no significant change in the dependent variables were observed during a forward time step. To directize the space, 9 grid-points were used in each of the three coordinate directions. This rather coarse mesh was essentially chosen for economical reasons.

We have chosen three different flow cases for the present study. The first is the inlet flow in a square sectioned duct where one wall is allowed to move laterally. This flow falls into the category of three-dimensional boundary layer flows in ducts. It is characterized by accelerating positive central velocity in the axial direction and decreasing pressure.

The second flow case is the flow through a sudden expansion. The flow falls into the category of separated flows. The characteristics of the flow is a retarded central jet with regions of recirculating fluid. This type of flow is often encountered in technical devices, e.g., furnaces, the shell-side of heat exchangers, and flow past buildings or other obstacles.

The third and last flow case is the flow in a three-dimensional cavity with a moving wall. The practical occurence of this type of flow is, for instance, found in lubrication grooves in bearings.

The following paragraphs will now give some results from the computations of the cases just described.

The Inlet Flow in a Square Sectioned Duct with a Laterally Moving Wall

Figure 3 displays the flow geometry. A uniform distribution of the axial inlet velocity is assumed. At the outlet fully developed flow is assumed; thus zero gradients were imposed on all variables in the axial direction. Calulations were performed for two inlet Reynold's numbers 100 and 1,000 (based on inlet velocity, and ducts height) and the results were indentical when correlated as shown in Figures 4 through 9.

Figure 4 shows the axial increase in the maximum axial velocity for a moving wall and a wall at rest. Comparisons are shown with the boundary-layer solution of Patankar and Spalding [1] and the laser doppler anemometer measurement of Goldstein and Kreid [23]. As we can see the agreement is very good, taking into account the coarse grid that has been used. Although the boundary-layer solution should be less accurate than the present solution, it must be noted that the discrepancy between the present predictions and the measurements is probably due to the grid size. Figures 5 and 6 show the development of axial velocity profiles and the agreement with experiments is good. Figure 7 shows the pressure drop along the duct. The agreement between prediction and experimental data from Beavers et al. [24] is not so satisfactory. This again is probably due to the coarse mesh. But we can see that the axial pressure gradient tends to the experimental one as fully developed flow is approached. Figure 8 shows the fully developed cross-stream velocity along the

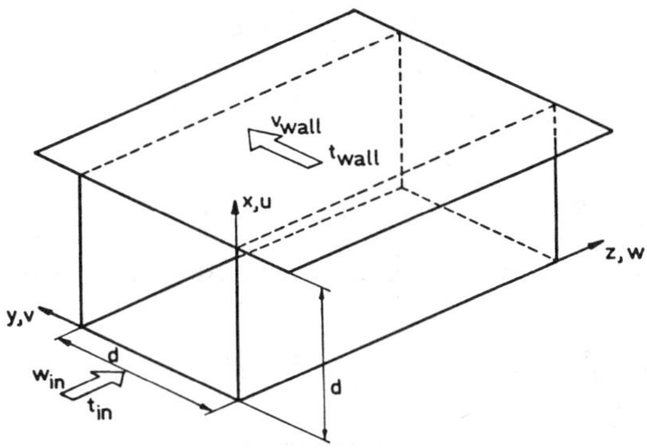

Figure 3. Flow geometry and notation.

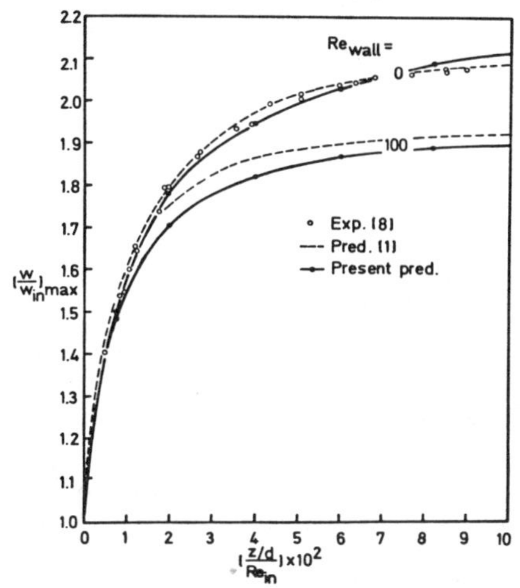

Figure 4. Axial variation of maximum axial velocity.

centerplane. Here the agreement with the corresponding boundary-layer solution is satisfactory. At last we will show an example of computation of a scalar. The problem is heat transfer from a warm wall to the cold fluid. The boundary conditions are uniform inlet temperature t_{in} and uniform wall temperature t_{wall}. The density was assumed constant, and the Prandtl number was unity. Figure 9 shows the results both for a moving wall and for a wall at rest. The agreement with the boundary-layer solution is good. The figure clearly shows how much a cross-stream velocity may influence heat transfer.

Flow Through a Sudden Expansion

Figure 10 shows the flow geometry for this case. The dimensions of the expansion are that reported by Durst et al. [25]. The upstream and downstream channel heights are respectively 4 and 12 mm. The width ratio B/h is 9.2, thus implying almost two-dimensional flow in the central region. The inlet profile was taken to be the experimental one. Three grid points in the x-directions were used to represent the inlet. Results of the calculation for a Reynolds number of 56, are shown in Figures 11 and 12. As we can see the agreement with experiments is good taking into account that 9 points in each direction were used.

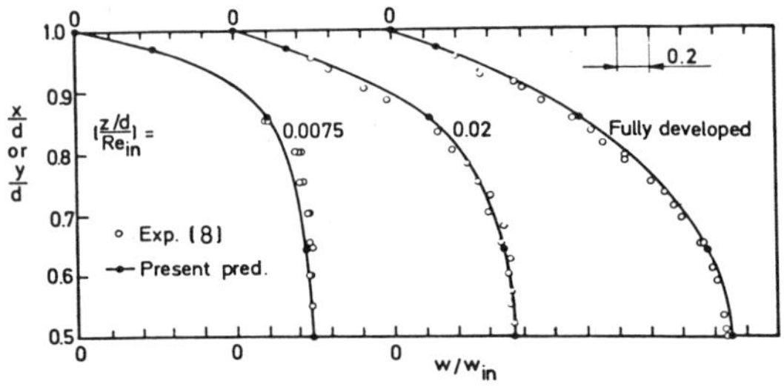

Figure 5. Development of axial velocity profile along transverse planes (Re_w = 0).

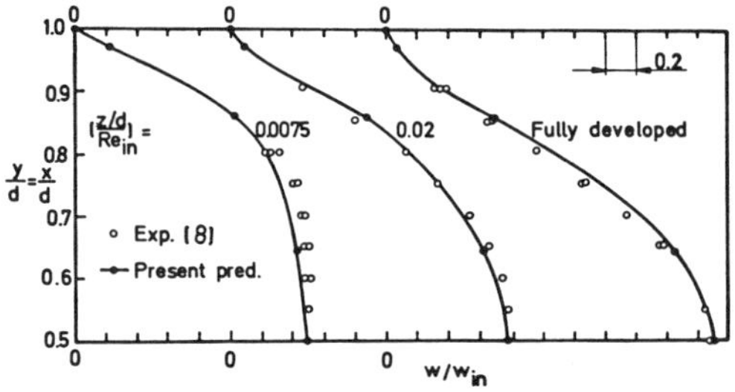

Figure 6. Development of axial velocity profile along diagonal planes (Re_w = 1).

Flow in a Three-Dimensional Cavity with a Moving Wall

Figure 13 shows the flow geometry of the last example in this study. The dimensions of the cavity were b/h = 1.0, b/l = 0.2, while the Reynolds number based on wall velocity and cavity height h was 100. From this computation we present the velocity field in the Figures 14 and 15. Although no measurements are available for comparison, the results are consistent with expectations.

Computation Details

All computations were done on a UNIVAC–1108 computer. To perform one time step for the hydrodynamic variables and one scalar variable, approximately 5 secs of CPU time were needed. To obtain steady-state solutions, 40 to 80 time steps had to be performed. The computer program and data took approximately 47 K of core memory.

Discussion

The preceding paragraphs have shown that computations of three different three-dimensional laminar flow cases were successfully computed. This is very encouraging taking account of the coarse mesh that was used. The results were obtained at a rather low cost, as calculation of one flow case took from 200 to 400 secs of CPU time.

Of course all the test cases in this study were laminar while most practical flows are turbulent. The effect of turbulence can be taken care of by mathematical modeling of the turbulence as indicated earlier.

Figure 7. Pressure drop in a square sectioned duct.

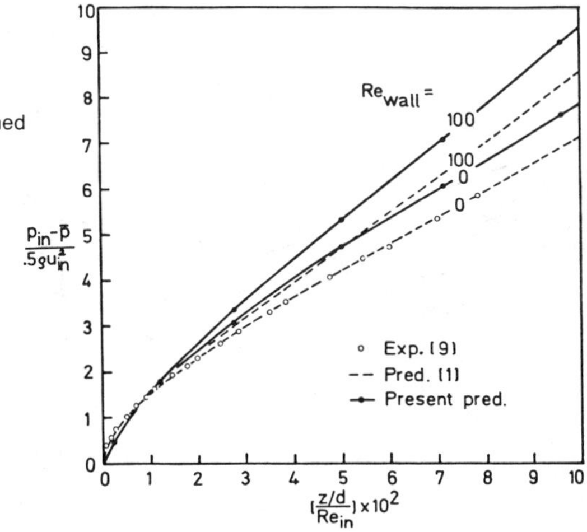

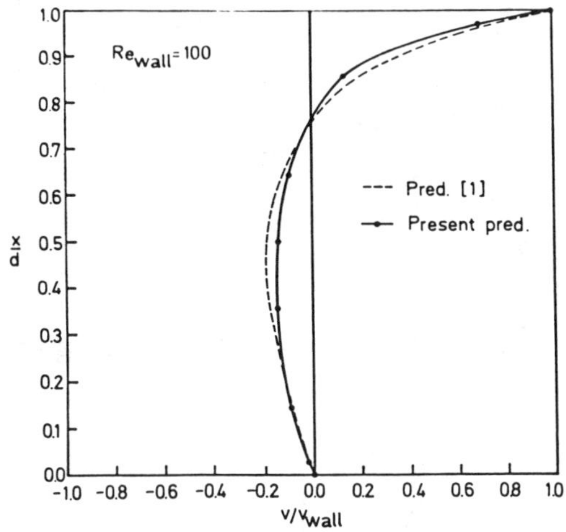

Figure 8. Fully developed v-velocity profile across vertical center plane.

Figure 9. Axial variation of dimensionless bulk temperature (Pr = 1).

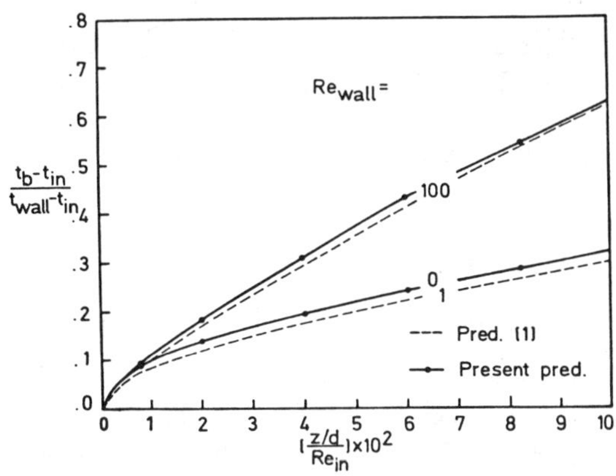

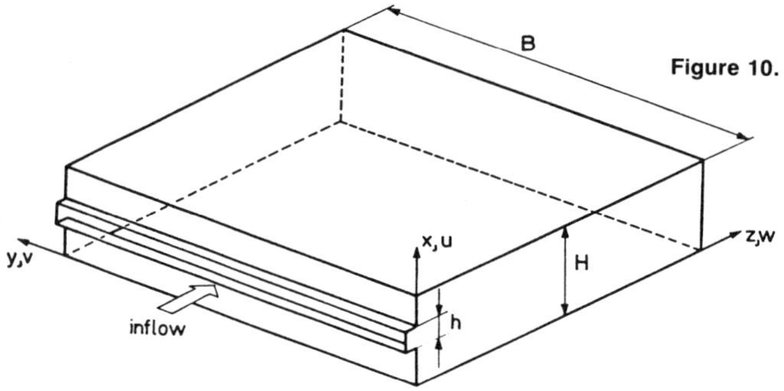

Figure 10. Flow geometry and notation.

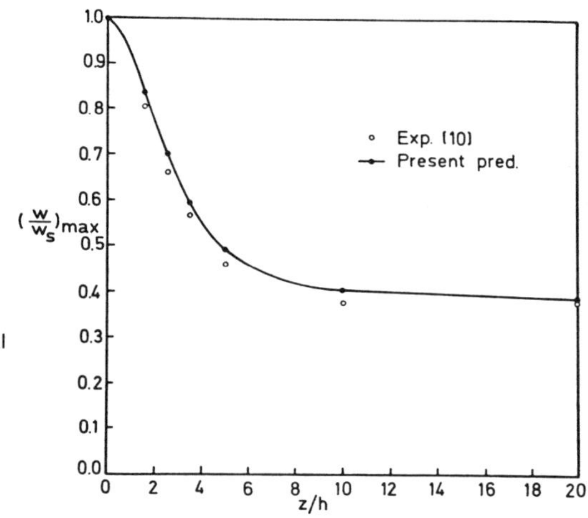

Figure 11. Axial variation of maximum axial velocity along center line (Re_s = 56).

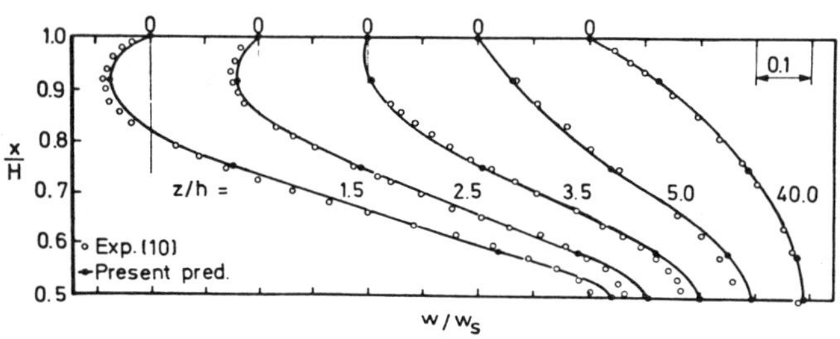

Figure 12. Development of axial velocity profile along transverse planes (Re_s = 56).

Turbulent Flows

Calculations have been performed by Hjertager and Magnussen [26] for two different isothermal flow cases. The geometries and notations are shown in Figure 16.

The NTH flow case shown in Figure 16 A, is given by Hjertager and Magnussen [27]. The figure shows one fourth of their flow domain. Because of symmetry properties, calculations are only performed for this domain. The flow is characterized by a jet issuing from a square duct centrally into a square sectioned enclosure of width D = 100 mm. Other geometrical data are:

d/D	0.1
L/D	3.0
d_o/D	0.2

The IC flow case shown in Figure 16 B is that set up by Nielsen et al. [28]. Only half of their flow domain is shown. Symmetry considerations make it possible to restrict the calculations to this domain. This flow is characterized by a jet injected through a square-

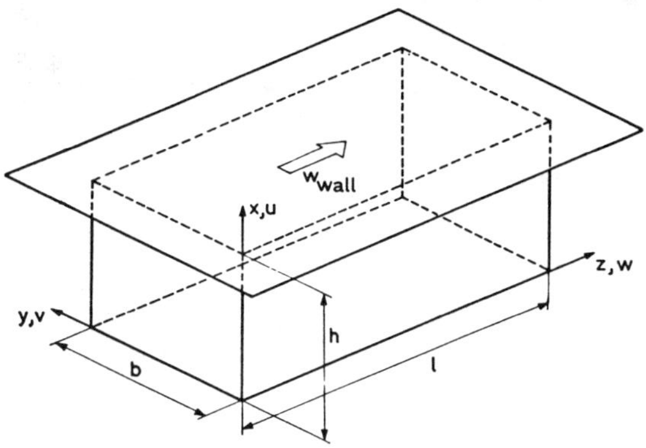

Figure 13. Flow geometry and notation.

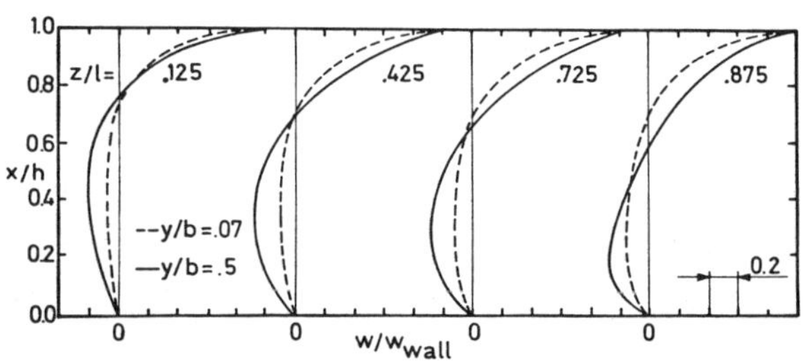

Figure 14. Distribution of w-velocity (Re_w = 100).

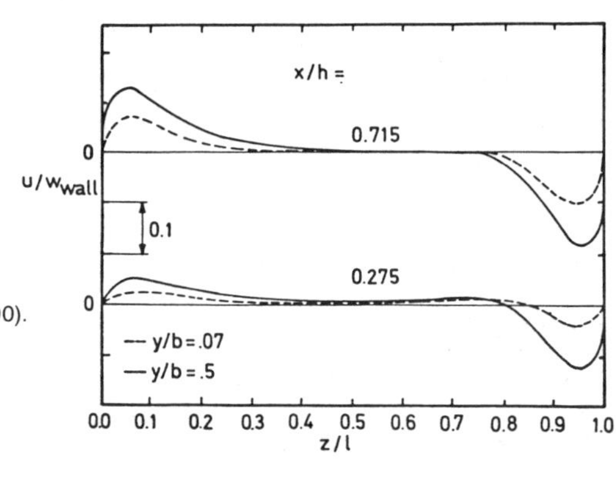

Figure 15. Distribution of u-velocity (Re_w = 100).

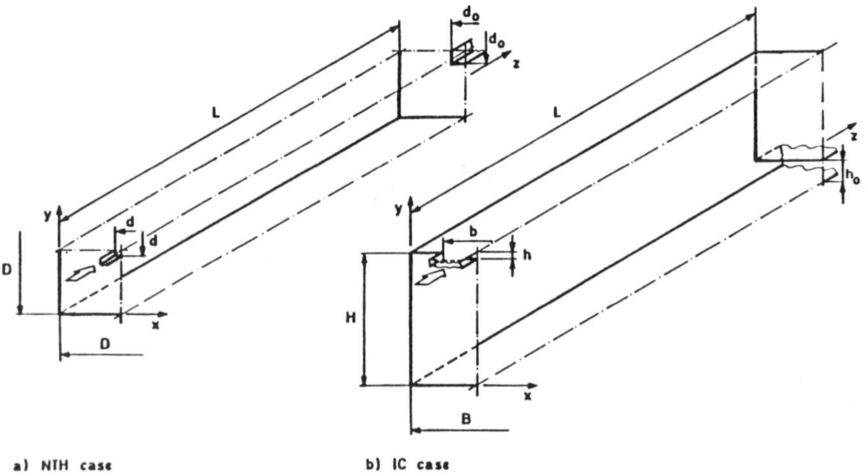

Figure 16. Flow geometry and notation for the two flow cases: (A) NTH case; (B) IC case.

sectioned slot aligned with the upper wall of the square-sectioned enclosure of height H = 89.3 mm. Other geometrical data for this case are:

h/H	0.056
B/H	0.5
L/H	3.0
B/H	1.0
h_o/H	0.16

The computational grid consisted of a $9 \times 9 \times 9$ and $7 \times 12 \times 12$ grid in the x, y, and z directions for the NTH and IC case, respectively. The inlet for both cases was approximated by four grid points. The distribution of the axial velocity across the inlet was taken from the experimental data. Kinetic energy of turbulence at the inlet was derived from the measured axial turbulence velocity of the inlet as:

$$k_{in} = \frac{3}{2} w'^2_{in} \tag{76}$$

Inlet values for the rate of dissipation of kinetic energy of turbulence are more difficult to find. We have chosen to estimate the inlet mixing length to be

$$l_{in} = 0.1\ h/2 \quad \text{and} \quad l_{in} = 0.1\ d/2$$

for the IC and NTH cases, respectively. The inlet values for the rate of dissipation are given by the following relation:

$$\varepsilon_{in} = C_D k_{in}^{1.5}/l_{in} \tag{77}$$

Test calculations showed that results were not strongly dependent on the estimates for the inlet conditions. Outlet boundary conditions for the axial velocity was computed using the nearby velocity distribution and overall continuity requirements. The other two velocity components were set equal to zero. The outlet conditions for kinetic energy of turbulence and its rate of dissipation were taken to be zero gradient in the z-direction. Boundary conditions for the walls have been treated previously and will not be elaborated here. Preliminary computations of the NTH case showed that it was impossible, with the limited number of grid points used, to be able to correctly calculate the potential core region of the jet. It was therefore found necessary to reject calculation in this region. Instead an ad hoc assumption was taken. The practice used was to insert the inlet boundary conditions at the end of the potential core. This means that the inlet conditions were moved forward z = 4.0 d and z = 8.0 h for the NTH and IC cases, respectively. The extent of the potential core was chosen according to the experimental findings. The inlet bulk velocity was W_b = 18.5 m/s for the NTH case and W_b = 15 m/s for the IC case.

Results

Figures 17 to 23 show the calculated result for the NTH case. The axial variation of axial mean and isotropic turbulence velocities are shown in Figure 17. The agreement with experimental data is good for both quantities. Figures 18 through 20 show calculated profiles of axial mean and isotropic turbulence velocities for the z/D = 1.0, 2.0, and 2.5 cross- sections. The agreement with experimental data is good for the z/D = 1.0 cross-section. We can see that for this cross-section there is a systematic overprediction in the recirculation region. This is in contrast to the results in Figure 19 and 20 where there is a systematic underprediction in the recirculation region. At these two stations there are also discrepancies between the measured axial turbulence velocity and the predicted axial isotropic $(2/3\ k)^{1/2}$ turbulence velocity. Especially near the corners of the enclosure there is some overprediction.

Figure 21 A and B shows comparison between measured and predicted contours of the axial mean velocity for the three cross sections x/D = 1.0, 2.0, and 2.5. The general features of this comparison are that the zero velocity regions are well predicted and that there is underprediction in the recirculation regions at z/D = 2.0 and z/D = 2.5. Figure 22 A through D displays the predicted velocity vectors in the xy planes. No comparative measurements were done for this case, but the results are consistent with expectations. At last Figure 23 A through E shows the predicted results of the isotropic turbulence velocity $(2/3\ k)^{1/2}$ for the yz planes. The general feature here is the large fluctuations in the jet-region and the low fluctuations in the recirculation regions. Another feature to note is that the isotropic turbulence velocity is fairly constant in the recirculation regions, ~ 1m/s, which corresponds to 5% of the inlet bulk mean velocity.

Figures 24 through 27 give the calculated results for the IC case. Comparisons of axial mean velocity profiles for two cross sections are shown in Figure 24. The agreement is in

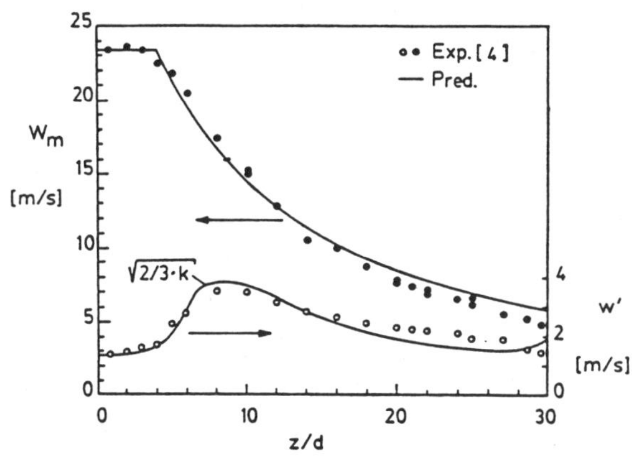

Figure 17. Axial variation of axial mean (•) and turbulence (○) velocity. Comparisons between experiments and prediction (NTH case).

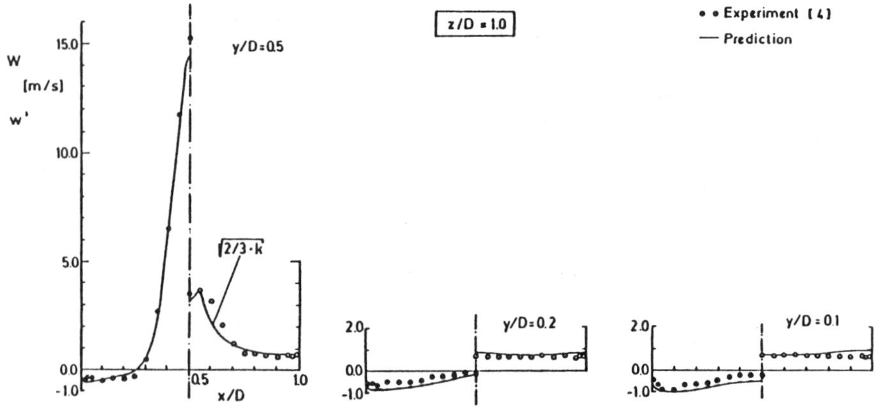

Figure 18. Profiles of the axial mean (•) and turbulence (○) velocities for the z/D = 1.0 plane. Comparisons between experiments and predictions (NTH case).

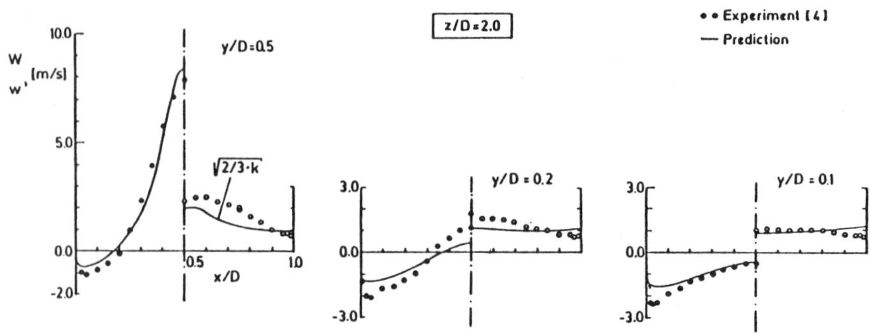

Figure 19. Profiles of the axial mean (•) and turbulence (○) velocities for the z/D = 2.0 plane. Comparisons between experiments and predictions (NTH case).

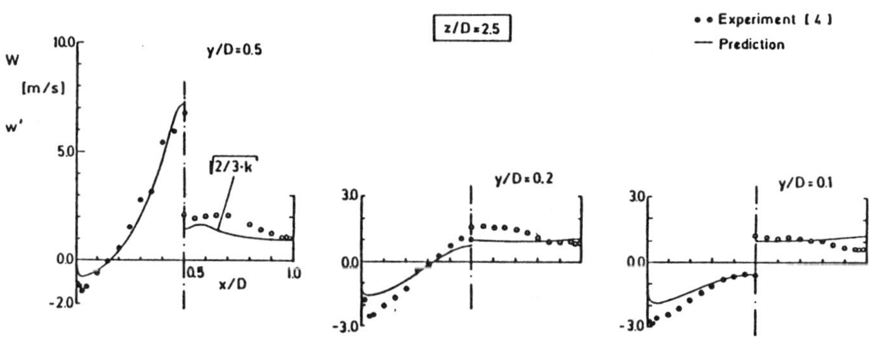

Figure 20. Profiles of the axial mean (•) and turbulence (○) velocities for the z/D = 2.5 plane. Comparisons between experiments and predictions (NTH case).

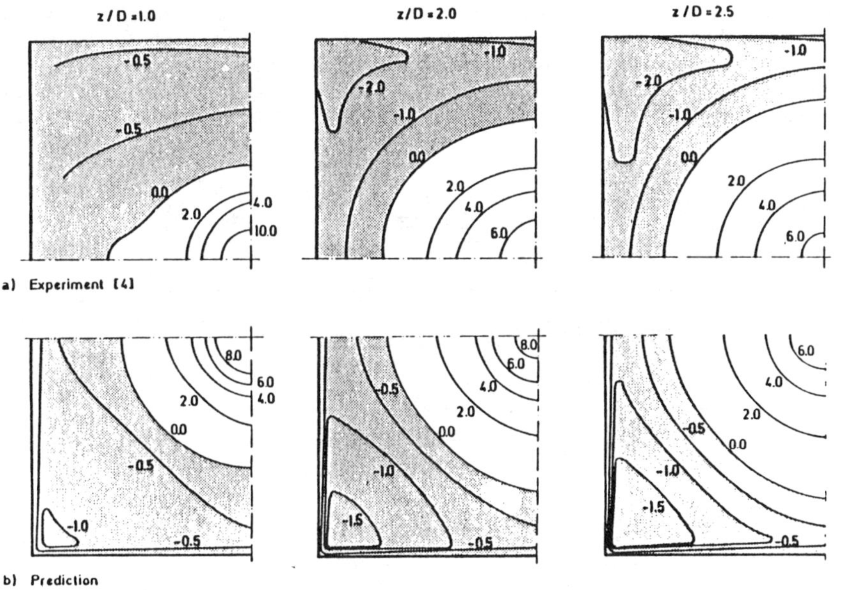

Figure 21. Comparisons between experimental (A) and predicted (B) contours of the axial mean velocity for the x-y planes (NTH case).

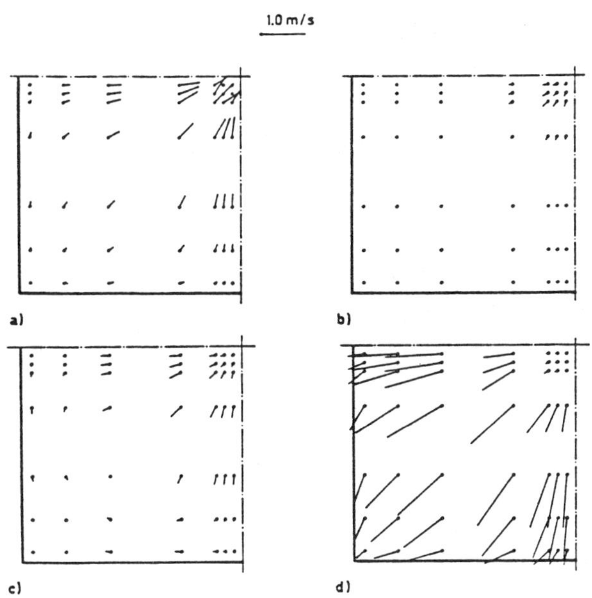

Figure 22. Predicted velocity vectors in the x-y planes (NTH case): (A) z/D = 0.1; (B) z/D = 1.7; (C) z/D = 2.7; (D) z/D = 2.95.

general good, but we can see that there are discrepancies especially in the jet flow region. Figure 25 A through F shows predicted contours of axial mean velocity for the xy planes. The maximum recirculation velocity normalized with the inlet bulk velocity is slightly in excess of 15%. The predicted mean velocity vectors in the xy planes are shown in Figure 26. Here we can note that a circulation flow pattern is set up in Figure 26 C through E.

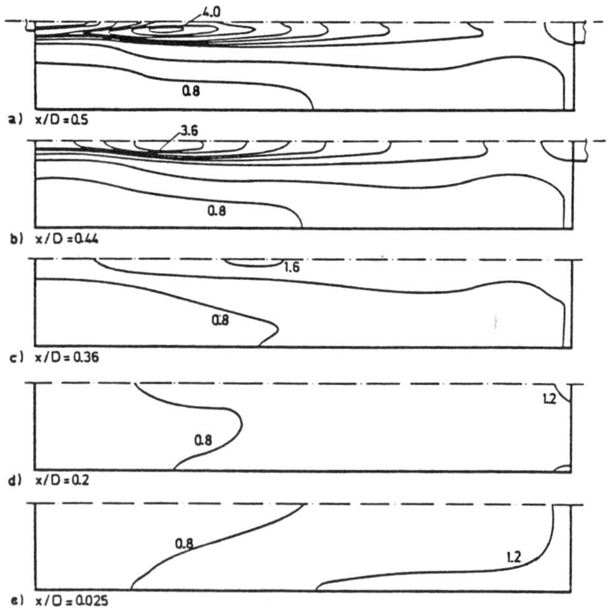

Figure 23. Predicted contours of the isotopic turbulence velocity ($\sqrt{2/3k}$) for the y-z planes. Numbers denote m/s. Spacing between contour lines equals 0.4 m/s (NTH case): (A) x/D = 0.5; (B) x/D = 0.44; (C) x/D = 0.36; (D) x/D = 0.2; (E) x/D = 0.025.

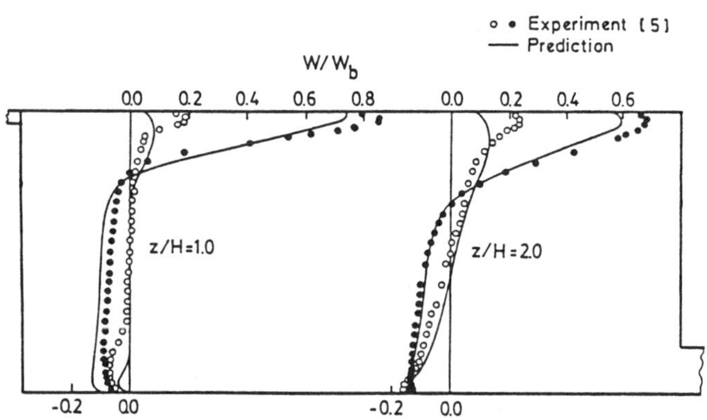

Figure 24. Comparisons between experimental and predicted profiles of the mean axial velocity at two stations z/H = 1.0 and z/H = 2.0. (•) x/B = 0.5, (○) x/B = 0.1 (IC case).

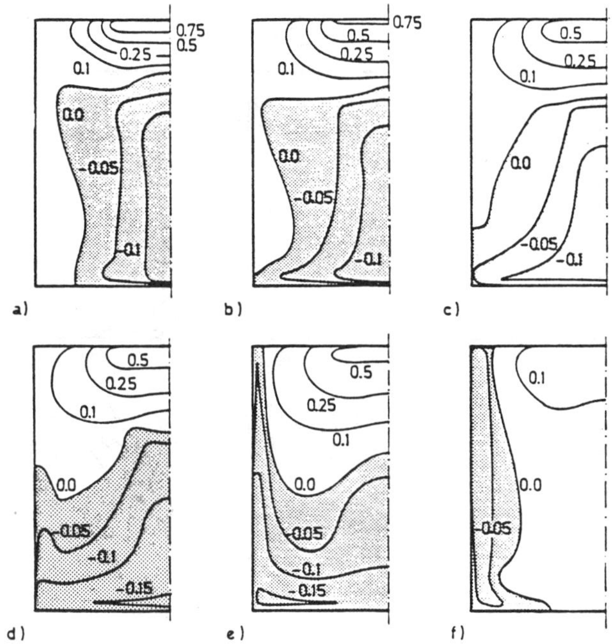

Figure 25. Predicted contours of the axial mean velocity for the x-y planes. Numbers denote velocity normalized with inlet mean velocity. W_b. (IC case): (A) z/h = 0.6; (B) z/h = 1.0; (C) z/h = 1.5; (D) z/h = 2.0; (E) z/h = 2.42; (F) z/h = 2.88.

This is in contrast to the NTH case where no such circulation was found. Figure 27 A through C gives the results of predicted contours of the isotropic turbulence velocity for the yz planes.

This flow case also displays the highest fluctuations in the wall jet region and the lowest in the recirculation regions. In the recirculation regions it is seen that this turbulence velocity lies between 1% and 7% of the inlet bulk mean velocity.

Computation Details

Converged solutions were obtained in approximately 150 iterations. When convergence was approached, the maximum change in the variables between iterations was 0.25% of a representative value of the variables in the calculation domain. Computations were performed on a UNIVAC–1108 computer, and the required time per iteration, point, and variable was approximately 1.5 m sec. Solutions were obtained by use of under-relaxation factors of 0.3 for velocity components and 0.5 for the other variables. The core memory requirements for these computations were approximately 45 K.

Discussion

The previous paragraphs have shown that computations of fully three-dimensional jet-induced flows in enclosures can be performed using a rather coarse grid lay-out. For both flowcases considered there are some discrepancies between experimental data and

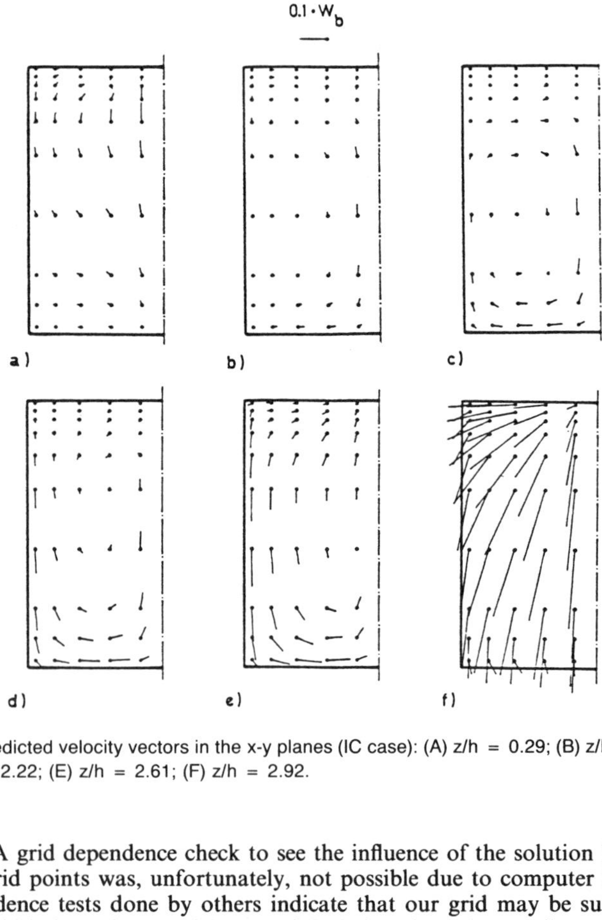

Figure 26. Predicted velocity vectors in the x-y planes (IC case): (A) z/h = 0.29; (B) z/h = 0.78; (C) z/h = 1.78; (D) z/h = 2.22; (E) z/h = 2.61; (F) z/h = 2.92.

predictions. A grid dependence check to see the influence of the solution by increasing the number of grid points was, unfortunately, not possible due to computer limitations. Such grid independence tests done by others indicate that our grid may be sufficient to obtain reasonable grid independence. Nielsen et al. [28] show in their two-dimensional calculation that an increase in the number of grid points from 17×20 to 39×39 only changes the solution by up to three percent. The calculations by Patankar et al. [29] use a grid of $10 \times 15 \times 15$ points and their comparison of predictions with experiments for time mean velocities are of similar quality as our data. They do not, unfortunately, show comparisons between predictions and experiments for turbulence quantities. Although both our prediction method and the methods of Nielsen et al. [28] and Patankar et al. [29] use similar types of upwind differencing schemes, which may introduce numerical smearing, this does not seem to upset the solutions.

Some explanations to this phenomena are offered by McCuirk and Rodi [30]. They have shown that the question of first- or second-order accurate finite-difference schemes are not always the proper indications of accuracy. They point out and show that it is more appropriate to look at the relative magnitude of the different terms occurring in the differential equations at different locations in the calculation domain to assess accuracy. A similar view on numerical diffusion is also expressed in the book of Patankar [31]. The most reasonable route to assessing the numerical accuracy is therefore to perform grid independence tests of the solutions. This was not possible in our case as stated previously. Although no general conclusions can be drawn on the cause of some of the discrepancies between

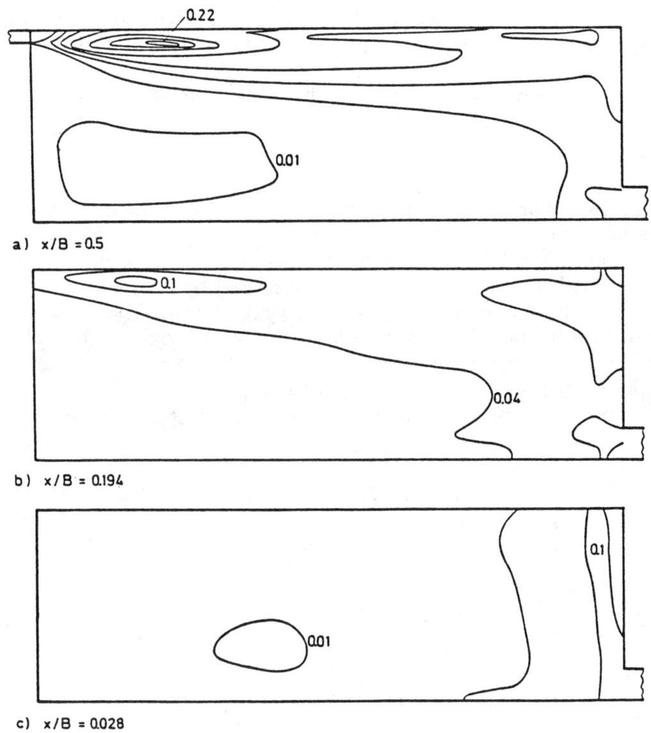

Figure 27. Predictea contours of the isotopic turbulence velocity ($\sqrt{2/3k}$) for the y-z planes. Numbers denote turbulence velocity normalized with inlet mean bulk velocity, W_b. Spacing between contour lines equals 0.03 W_b (IC case): (A) x/B = 0.5; (B) x/B = 0.194; (C) x/B = 0.028.

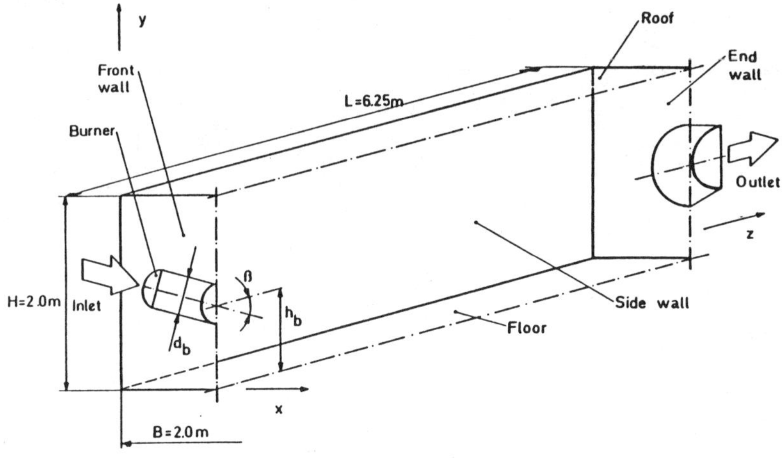

Figure 28. Geometry of furnace and notation.

predictions and experiments found in our work, we believe that the agreement is satisfactory and that useful information of engineering importance can be extracted. It should be emphasized that the constants appearing in the turbulence model has not been tuned to suit the present problems, but are basically the same as those used by others.

Combustion Chambers Flows

Detailed measurements of local properties in three-dimensional combustion chambers and furnaces are scarce. It seems that the work performed by the International Flame Research Foundation (IFRF) are the most extensive in this area. For this presentation we have chosen three different furnace configurations, used by Hjertager and Magnussen [32], as a basis for comparison between predictions and experiments. The two first cases are taken from the M3-trial program [33], and the last from the M2-trial program [34] performed at IFRF. The M3-trials are set up to collect data in a square-sectioned furnace with a watercooled floor while the M2-trials had cooling loops along the four walls within the same furnace geometry. The furnace geometry is schematically shown in Figure 28.

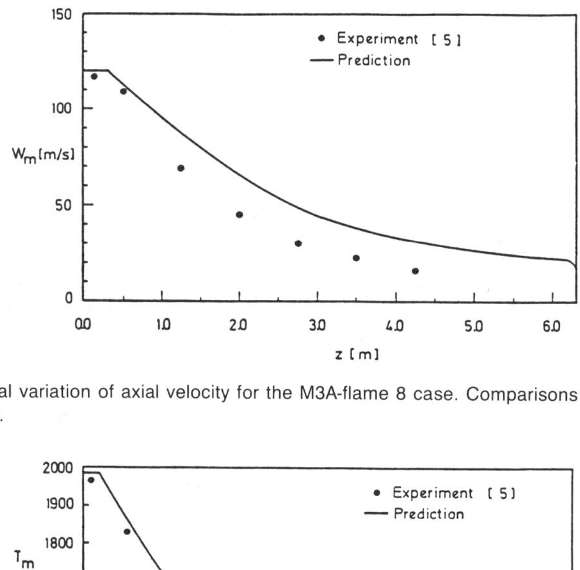

Figure 29. Axial variation of axial velocity for the M3A-flame 8 case. Comparisons between experiments and predictions.

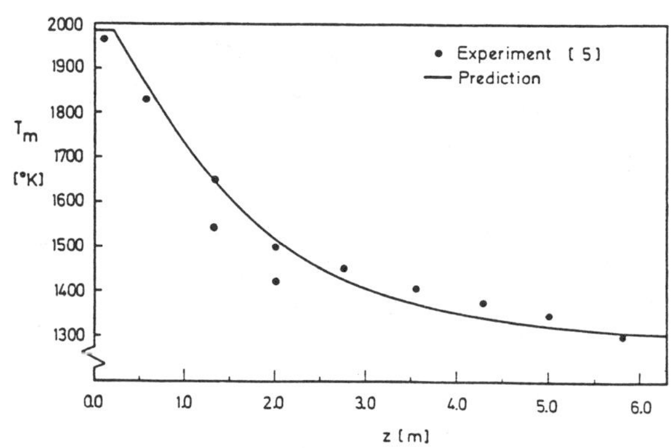

Figure 30. Axial variation of temperature for the M3A-flame 8 case. Comparisons between experiments and predictions.

Table 3
Geometric Data for the Furnaces

Case	d_b (m)	h_b (m)	Angle
M3A-8	0.263	0.88	0^0
M3A-10	0.273	0.72	25^0
M2-29	0.18	1.0	0^0

The two cases from the M3-program consisted of a tunnel burner where the combustion was completed before the gases entered the furnace.

In the case from the M2-program the fuel and air were injected through two concentric burner tubes with fuel in the center tube. The fuel was for all cases natural gas. Table 3 gives the different geometric data for the three cases. Due to symmetry considerations it was only necessary to perform calculations for one half of the furnace for the M3A cases and one fourth of the furnace for M2 case. The grid point distribution in x,y, and z directions was $12 \times 7 \times 12$ for the M3A cases and $9 \times 9 \times 11$ for the M2 case. Nonuniform grid distributions were used to concentrate points near the inlets and close to the walls where gradients in the dependent variables are large. Instead of evaluating the adsorption coefficient according to Equation 50 we chose to use a value equal to 0.2 m^{-1} in accordance with the work of Pai et al. [33]. The nominal heat input of all three furnace configurations was 3 MW.

Results

Axial variations of velocity and temperature along the burner axis are shown in Figures 29 and 30 for the M3A-flame 8 case. These figures show that the predicted axial velocity decay along the burner axis is somewhat slower than the experimental findings. The correlation between predicted and measured temperatures seems to be good.

Figure 31 displays the isotherms in the symmetry plane. We can observe that there is a difference of approximately 100° C between the gas near the adiabatic roof and the watercooled floor. The predicted isotherms are somewhat narrower along the burner axis than found in the experiments. The overall agreement is however satisfactory.

Figure 32 shows the predicted axial variation of the total heat flux towards the water-cooled floor. The agreement between prediction and experiment is satisfactory. The predictions of Pai et al. [33] are also shown.

Figures 33 and 34 display the predicted flow field for the M3-A flame 10 case. A characteristic feature is the rather complex flow field caused by the inclination of the burner towards the heat sink. Figure 35 shows a comparison between predicted and measured isotherms in the symmetry plane. A general agreement can be observed; however, significant descrepancies occur. These are approximately 100° C in the recirculation regions and temperatures on the order of 250° C in the jet flow regions.

The total heat flux distribution along the water-cooled floor is shown in Figure 36. Also shown together with the present prediction and experimental data are the predictions performed by Pai et al. [33]. Both predictions seem to underestimate the heat flux at the stagnation point. Different sources may have caused these discrepancies. The problems associated with the calculation of flows which are inclined to the computational mesh may have contributed significantly. In Figure 37 the predicted distribution of total heat flux to the floor is shown. There are great variations in the local heat fluxes ranging from approximately 200 kW/m^2 at the stagnation point to approximately 100 kW/m^2 in the corners. A comparison between the predicted heat fluxes in Figures 36 and 37 shows that the maximum local flux is almost 40% larger than the maximum transverse area averaged flux.

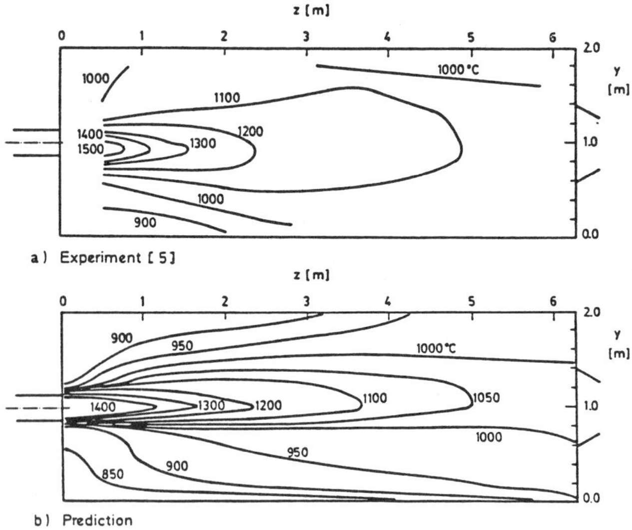

Figure 31. Comparisons between measured (A) and predicted (B) temperature distributions in the symmetry plane (M3A-flame 8 case).

The results for M2-flame 29 case are given in Figures 38 through 43. Predicted and experimental results for the axial variation along the burner axis of the mean temperature is shown in Figure 38. Figure 39 shows similar results for the combustion products CO_2 and H_2O whereas Figure 40 gives the results for the reactants, fuel and oxygen. The temperature variation is predicted excellently. The concentrations of fuel, oxygen, and combustion products are also well predicted. Some overpredictions of CO_2 and H_2O near the burner, Figure 39, may be due to the fact that the present treatment of the combustion does not take into account formation of intermediate reaction products such as CO and H_2.

The experimental data show that a significant amount of these products were present near the burner. The results in Figure 40 indicate that the flame length is correctly predicted. Figure 41 shows that the combustion model which has been applied in these predictions, allow fuel, oxygen, and products to be present at the same location. This is in agreement with experimental findings. The last two figures show the results of the heat transfer predictions. Figure 42 displays the axial variation of the area averaged total heat flux.

The agreement between experiments and calculations is satisfactory. Figure 43 shows that there are strong variations in the local heat transfer to the heat absorbing surfaces, even at the same axial position. The maximum local heat flux in Figure 43 is approximately 40% larger than the maximum area averaged flux in Figure 42.

Heat Transfer Characteristics

Table 4 summarizes experimental and predicted heat transfer efficiences. The table similarly gives the convective fraction of the total heat transfer for the two furnaces in the M3A. program. Apparently the increase in the efficiency from the M3A-8 to M3A-10 case is well predicted. This is also the case for the ratio Q_{con}/Q_{tot}.

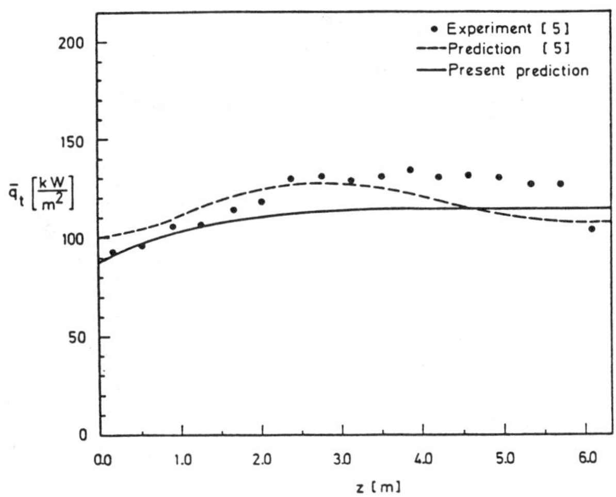

Figure 32. Axial variation of the transverse averaged total heat flux along the floor (M3A-flame 8 case).

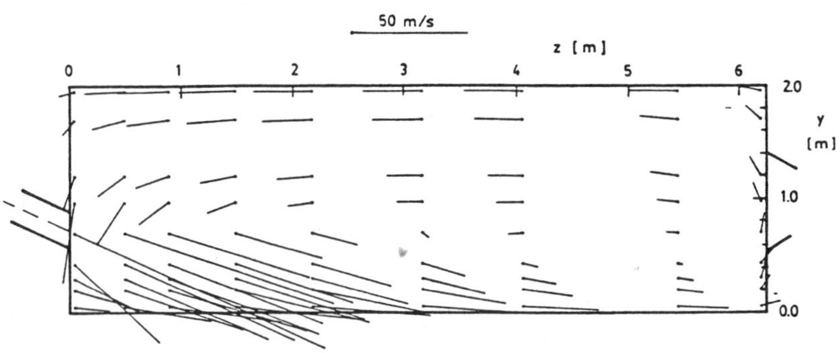

Figure 33. Predicted velocity vectors in the symmetry plane for the M3A-flame 10 case.

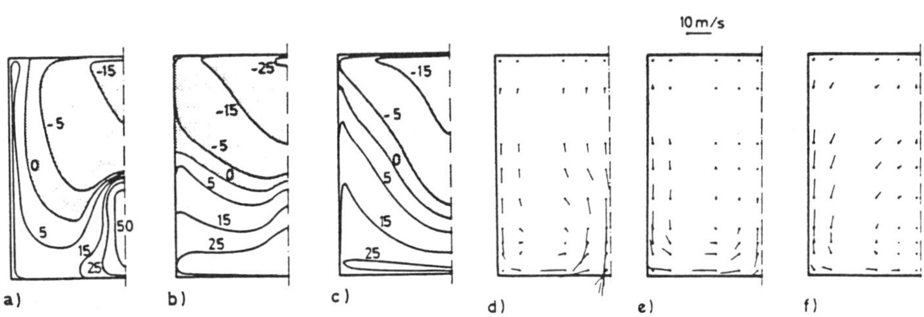

Figure 34. Predicted axial velocity contours—(A-C) and velocity vectors: (A) z = 0.7 m; (B) z = 2.7 m; (C) z = 4.8 m; (D) z = 0.9 m; (E) z = 3.2 m; (F) z = 5.5 m for x-y planes.

Table 4
Heat Transfer

Case	Efficiency (%) Present Prediction	Prediction [33]	Experiment [33]	Q_{con}/Q_{tot} (%) Present Prediction	Prediction [33]	Experiment [33]
M3A-8	45.5	48.5	51.8	23	21.4	17
M3A-10	54.6	52.5	58.4	51.5	28.4	40

Computation Details

The computaions were performed on a UNIVAC-1108 computer and the program and data took approximately 55 K of core memory. To obtain converged solutions, under-relaxation had to be used. A value of the relaxation factor of 0.3 was used for the velocity components and pressure, while the relaxation factor for the scalar variables was assigned a value of 0.5. The turbulent viscosity and density had relaxation factors of 0.3 and 0.1, respectively. Approximately 600 to 800 iterations were needed to obtain converged solutions. Due to computer limitations it was not possible to increase the number of grid points to test the grid dependence of the solutions. However, the agreement between numerical solutions and experimental data indicates that the grids used are satisfactory for many applications.

Discussion

The previous paragraphs have shown that quantitative calculations of three-dimensional furnaces can be performed. The accuracy of local temperatures is in the order of ±200° C, and the accuracy for local heat flux is approximately ±20%. The general trends are correctly predicted. This indicates that three-dimensional prediction methods of furnaces can be applied as a design tool for such furnace configurations as have been treated in this article. If on the other hand, more complex geometries are to be analyzed a more hybrid treatment may be used. We envisage the use of simple two-dimensional treatment near the burners to take account of the details, whereas three-dimensional treatment can be used for most of the furnace volume. This will incorporate iteration between two program systems. Although it might be more cumbersome to use, the gain in accuracy should be significant.

Compressible Flows

The method just described is incorporated into a two-dimensional computer code presented by Hjertager [8], and applied to study the following cases: The first is the shock tube problem that allows the model to be tested against known analytical solutions.

The second is calculation of flame and pressure development in the presence of turbulence-inducing obstacles, a problem of great importance in accidental gas explosions.

Shock Tube

The geometry and notation for this calculation are given in Figure 44.

A 1-m-long tube is initially divided in to a high-pressure and a low-pressure region separated by a diaphragm. At time equals zero the diaphragm is removed and two pressure waves are generated. A compression wave (shock) moving to the right and an expansion wave (rarefaction) moving to the left (Figure 44). To calculate this problem a one-dimensional grid layout is chosen. The gas is treated as an ideal nonreactive gas with molecular

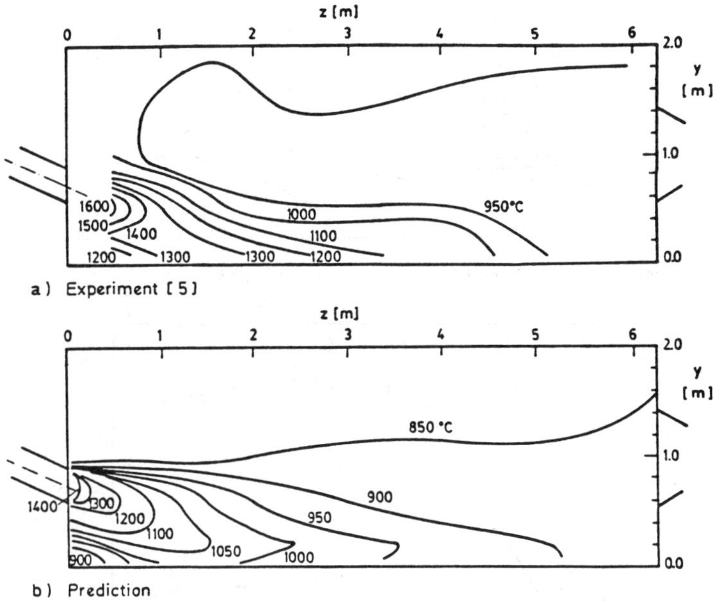

Figure 35. Comparisons between measured (A) and predicted (B) temperature distribution in the symmetry plane (M3A-flame 10 case).

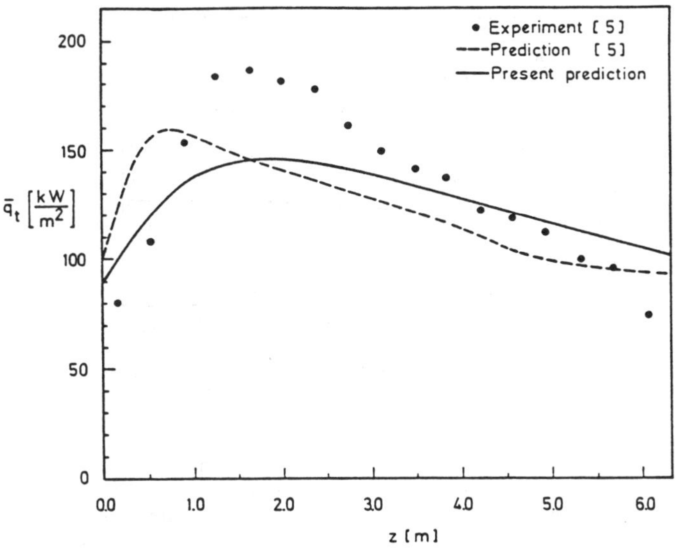

Figure 36. Axial variation of the transverse averaged total heat flux along the floor (M3A-flame 10 case).

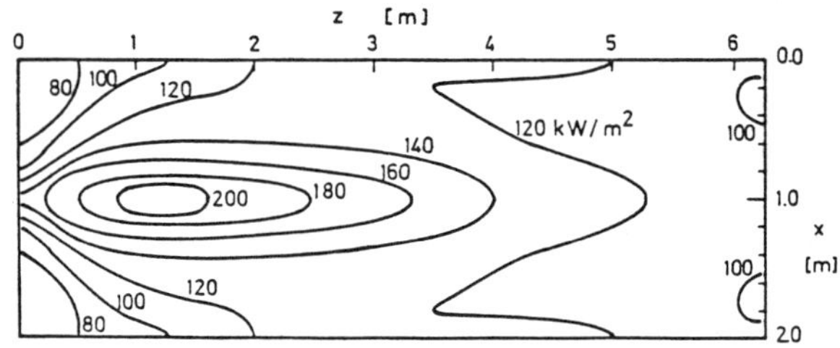

Figure 37. Predicted distribution of the total heat flux to the floor (M3A-flame 10 case).

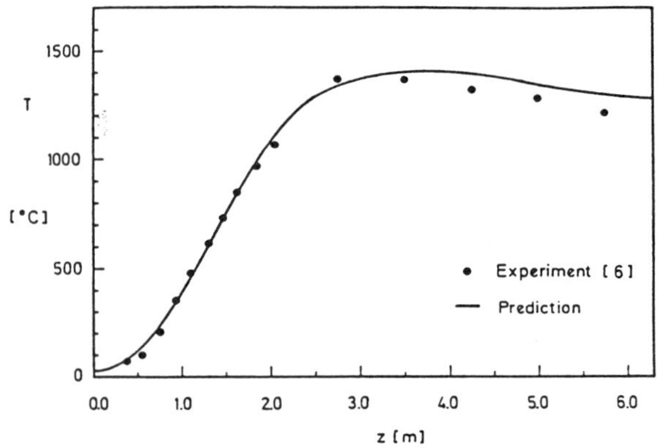

Figure 38. Axial temperature variation along the burner axis (M2-flame 29 case).

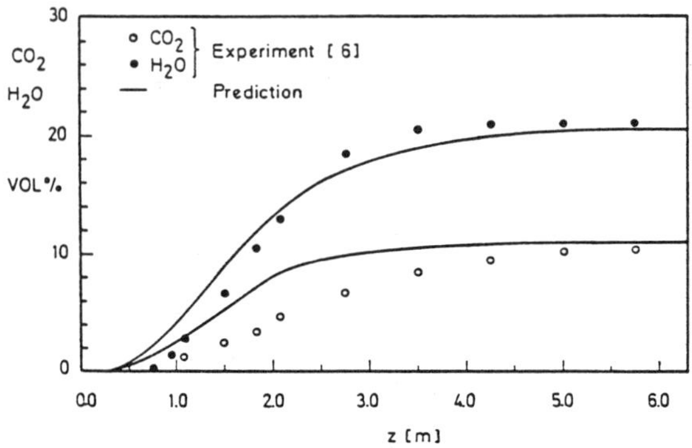

Figure 39. Axial variation of concentrations of CO_2 and H_2O along the burner axis (M2-flame 29 case).

weight equal to 29 and the ratio between specific heats equal to 1.4. Since this is a nonreacting situation, only equations for conservation of mass Equation 2, momentum Equation 3 and energy Equation 4 need to be solved. The Courant number, defined as $c_4 \cdot \Delta t/\Delta x$, is nominally chosen to be 0.2.

Figure 45 shows the grid number dependence of the pressure distribution at time equals 0.591 ms. It can be noted that the shock front position is well predicted even for coarse grids, whereas the steepness is more dependent on number of grid points. This is the usual feature for most methods which treats shock fronts in fixed-grid systems. The only simple remedy available for increased accuracy in such problems, is to increase number of points until acceptable grid independence is reached.

Figure 46 shows comparison between exact solutions from Harlow and Amsden [35] and predictions of shock pressure ratio and Mach number as function of initial pressure ratio. For pressure ratios tested, from 1 to 20, the agreement is satisfactory for both quantities. Calculations are also performed to test the stability of the numerical scheme. The Courant number is varied and converged solutions are obtained for values up to approximately 3.0. This is far beyond what any explicit method can perform.

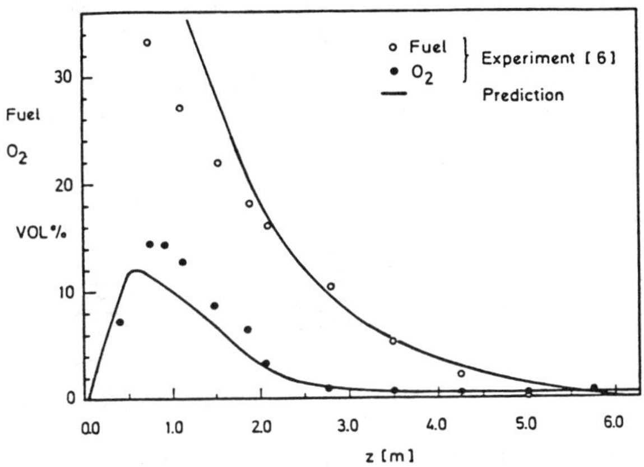

Figure 40. Axial variation of concentrations of fuel and O_2 along the burner axis (M2-flame 29 case).

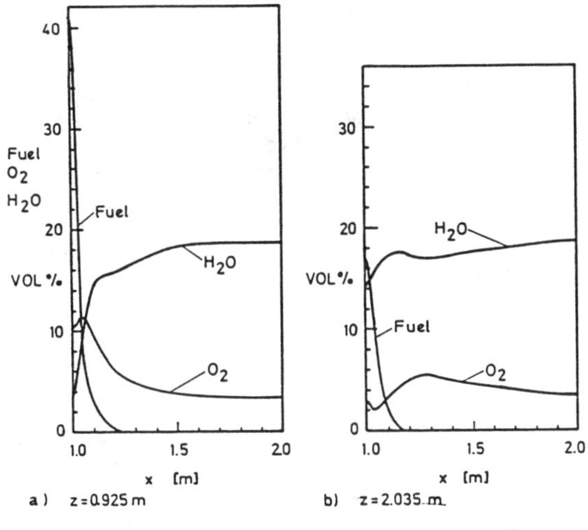

Figure 41. Transverse profiles of fuel, H_2O and O_2 at height y = 1.0 m (M2-flame 29 case): (A) z = 0.925 m; (B) z = 2.035 m.

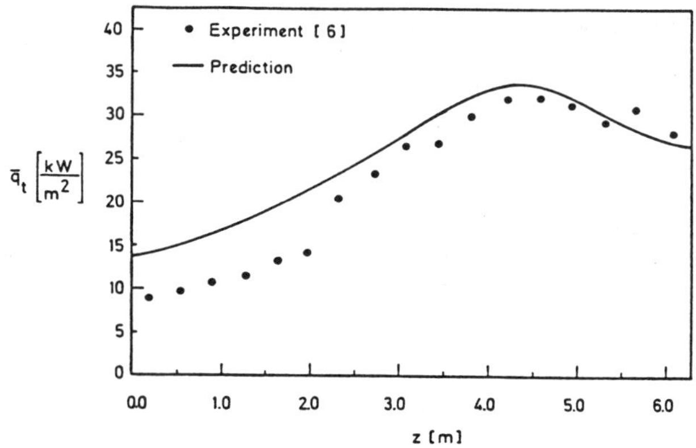

Figure 42. Axial variation of area averaged total heat flux towards the walls, floor and roof (M2-flame 29 case).

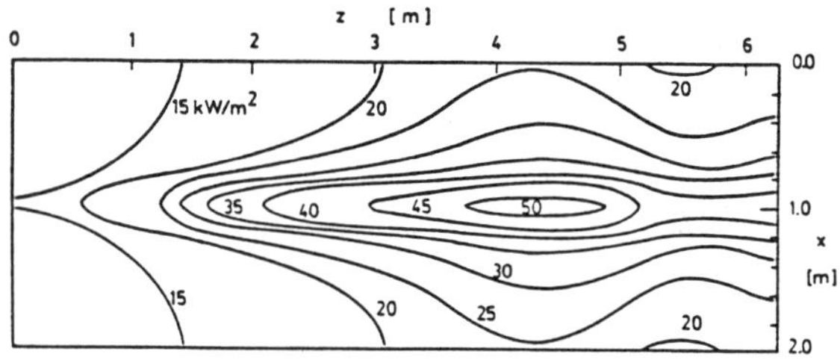

Figure 43. Predicted distribution of the total heat flux towards walls, floor, or roof (M2-flame 29 case).

Flame Acceleration by Turbulence

To be able to predict consequences of gas explosions in integrated industrial systems, the problem of flame and pressure development in turbulent environments must be addressed. Several experiments have recently been performed to investigate such problems.

The qualitative mechanisms for this flow, turbulence, and combustion interaction are fairly well established, but there is lack of quantitative methods to predict what happens in various geometries.

The present computation method is aimed at handling this complex interaction. The particular geometry that has been considered so far is a tube with axisymmetric repeated obstacles. This problem was studied by Moen et al. [36] in a large-scale tube facility situated at Sund close to Bergen, Norway. The tube was 10 m long and had a diameter of 2.5 m and thus a volume of 50 m^3. The ignition end was closed and the downstream end open. Turbulence inducing rings of various widths and number were inserted inside the tube. In the present calculation we focus only on one geometrical configuration, namely 5 equally spaced rings, each blocking off 30% of the free tube area. The calculation domain is shown in Figure 47 together with grid layout and dimensions. 30 grid points are used in the x-direction and 15 in the radial direction. The Courant number based on the unburnt condition was 0.5. The initial gas composition was a stoichiometric mixture of methane and air, and ignition was initiated by a planar source at the closed end (Figure 47). No attempt

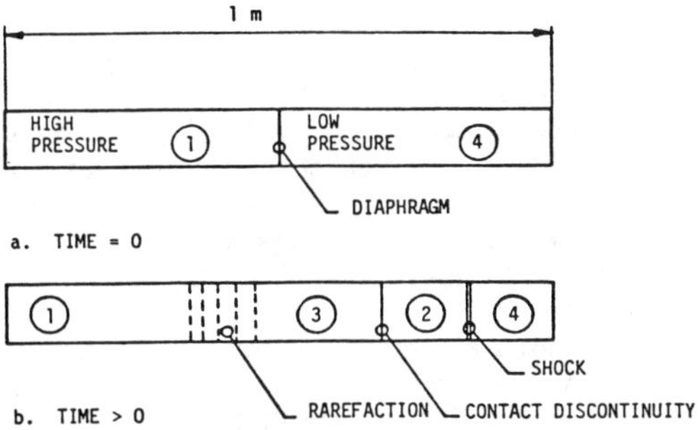

Figure 44. Geometry and notation for the shock tube problem.

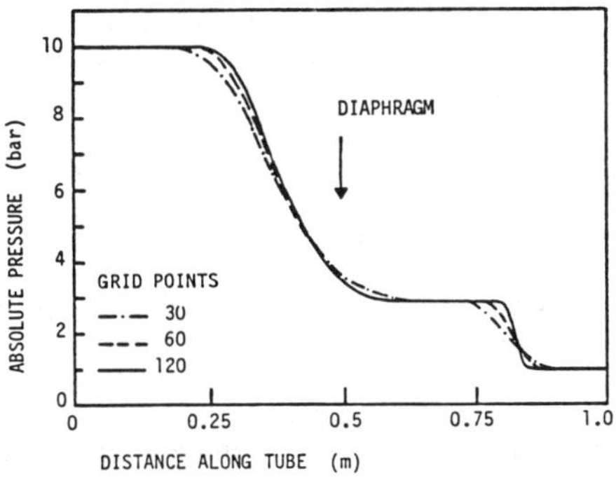

Figure 45. Grid dependence of pressure profile along the shock tube. Time equals 0.591 ms.

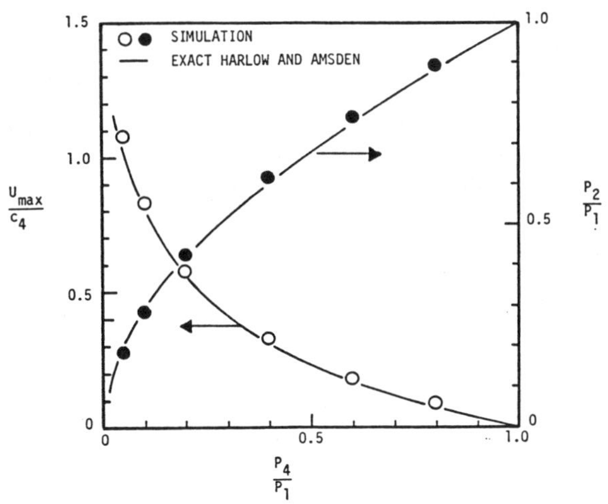

Figure 46. Variation of shock pressure ratio (p_2/p_1) and number (U_{max}/c_4) behind shock with initial pressure ratio (p_4/p_1).

was made to model details of the ignition process or the laminar to turbulent flow transition. In fact, the flow was treated as being in the turbulent regime from the start, and therefore, the turbulence and combustion models given in previous paragraphs were applicable. The ignition was accomplished by stating that 10% of fuel at the closed end of the tube was burnt and that the reaction rate and effective viscosity was dependent on the initial value of the dissipation rate ε_0. This value was varied until the resulting pressure and flame development at later times were almost independent of these initial conditions. The reason for this being valid, is the strong positive feedback mechanism which comes into play at the later stages of the process.

Figure 48 A and B shows a comparison between predictions and experiments for two typical pressure traces versus time, one recorded near the igniton end, and one near the downstream end of the tube. We can see that the general trends are reproduced in the calculations. This means increased pressure rise and peak pressure at the exit end of the tube as compared to the start end, and the later occurrence of the pressure peak at the start than at the exit. The largest discrepancies between experiment and simulation are found in the pressure decay period. A possible explanation to this may be the fact that Moen et al. [36] used a plywood ring at the exit end, whereas the other rings were made of steel. This means that the last ring was destroyed in each experiment, possibly giving rise to a faster pressure decay.

In Figure 49 A through J some details of various stages in the flame and pressure propagation are depicted. For each point of time stated, distributions in space of the following four quantitites are given. From top to bottom these are: velocity vectors, pressure, concentration of combustion product, and reaction rate. The meaning of the different vectors and contours is as follows: The largest velocity vectors displayed are 300 m/s and if higher vectors occur, only this maximum is shown. For the flame contours the full line represents 50% of fully burnt condition, whereas the other two contours represent 75% burnt (broken line) and 25% burnt (broken and dotted line).

The other two quantitites (pressure and reaction rate) are represented by contours scaled to the maximum value. This means that full lines represent 95%, broken lines 75%, broken dotted lines 50%, and dotted lines 25% of the maximum quantity.

Figure 49 A and B show the initial phases after ignition. Pressure waves are sent ahead of the flame, which is planar at these two times. In Figure 49 C and D distortion of the flame due to the influence of the first obstacle is observed. The peak reaction rate is located in the center in Figure 49 C, but in Figure 49 D the formation of large rates of reaction in the shear layer is seen. This again is confirmed by the pressure contours, which now has its maximum in the pocket (Figure 49 D).

Occurrence of highest pressure and rates of combustion in the shear layers of the pockets, are also evident in Figure 49 E, F, and G. The effect of turbulence enhancement on combustion and turbulent transport in the shear layer are depicted in Figure 49 F and G.

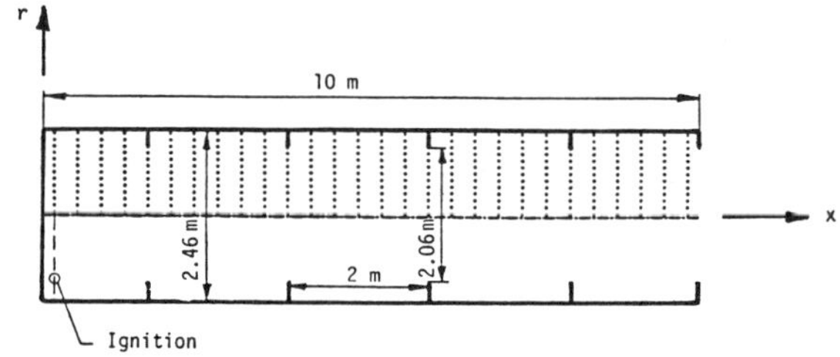

Figure 47. Geometry and grid layout for the flame acceleration problem.

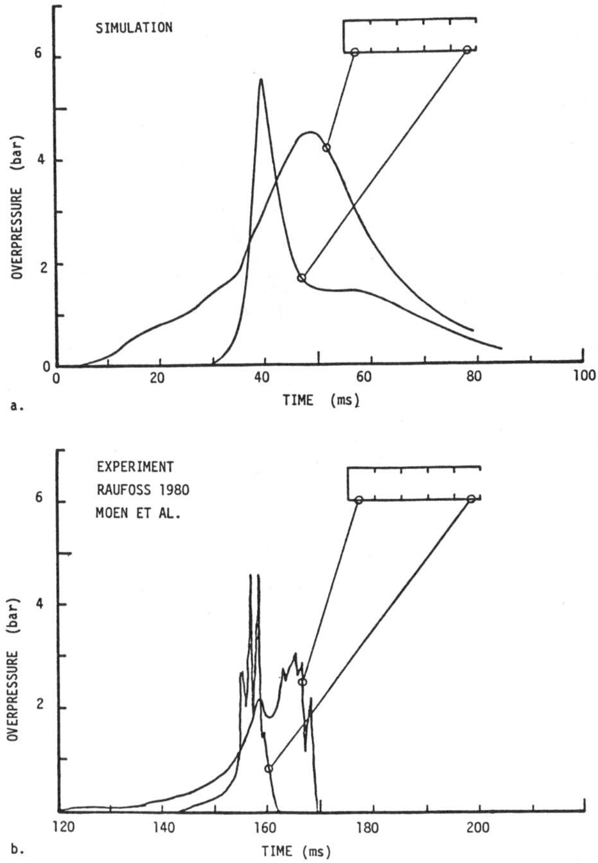

Figure 48. Variation of pressure with time for two positions along the tube: (A) prediction; (B) experiment.

In the first of these figures the leading flame is in the central part, whereas in Figure 49 G the flame is so distorted that the leading front is in the shear layer. Note that the peak rate of combustion in Figure 49 G is approximately 40 times larger than in the initial stages of propagation (Figure 49 A). After the flame has reached the outlet, the pressure becomes more constant in the various cross sections, and from Figure 49 I onwards the main process which occurs is venting and thus pressure relief. Although some combustion still occurs in the pockets (Figure 49 I and J) the rate is much lower than during the most vigorous periods (Figure 49 F to H).

Figure 50 shows comparisons between predicted and measured flame speeds along the center line of the tube. The general features of the flame propagation is reproduced, that is a wavy increase of speed. These waves are due to the alternating acceleration and deceleration of the flow through the ring orifices.

A severe test for any prediction method is to see the predicted effect in various scales of the geometry. This has been done for the present geometry and the result is shown in Figure 51. This figure shows the variations of the peak pressures at the closed and open ends of the tube with volume of the geometry. A strong increase of pressures form

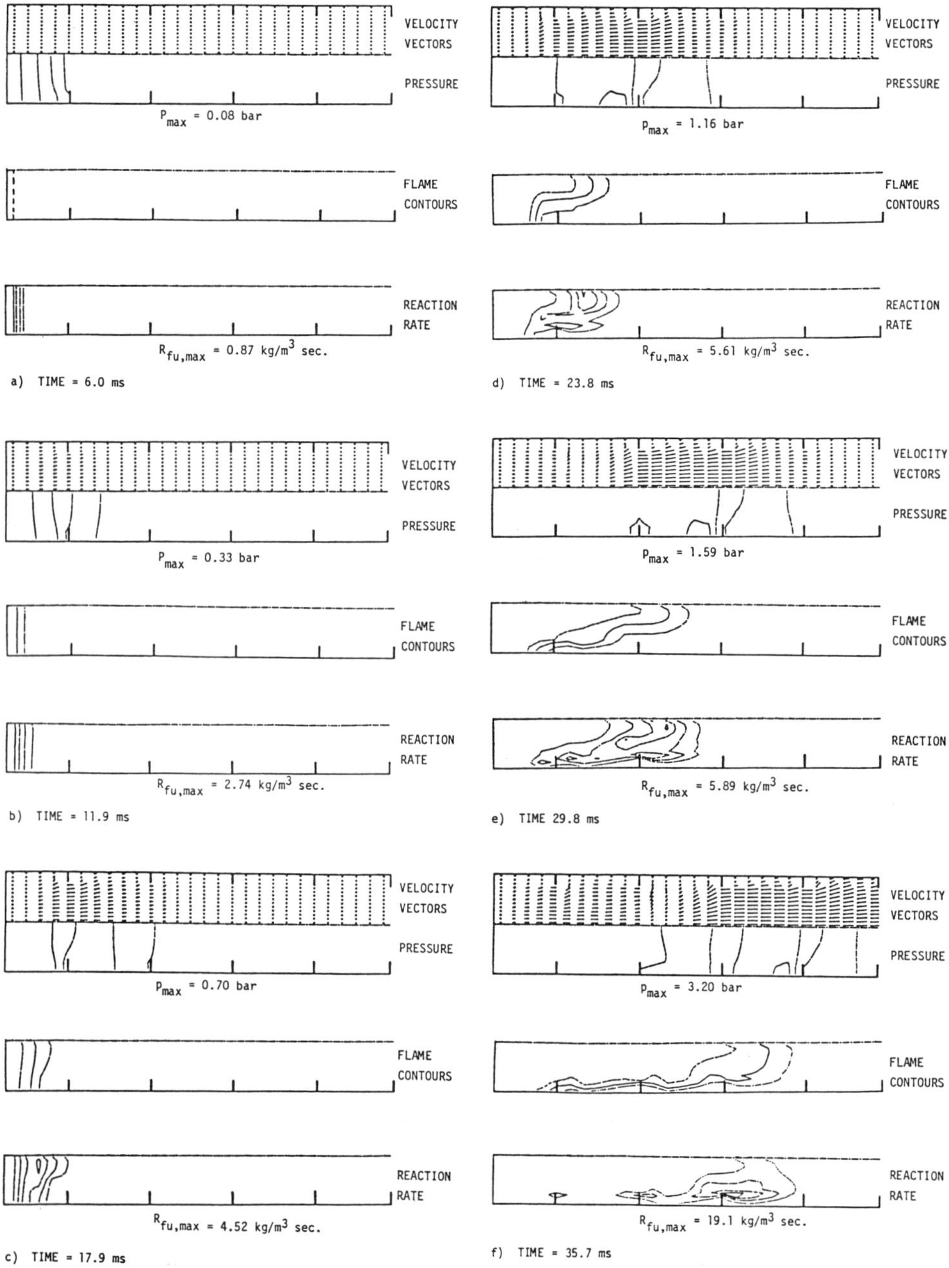

Figure 49. Velocity vectors, pressure-, flame- and reaction-rate contours for various time intervals (predictions): (A) time = 6.0 ms; (B) time = 11.9 ms; (C) time = 17.9 ms; (D) time = 23.8 ms; (E) time = 29.8 ms; (F) time = 35.7 ms; (G) time = 39.9 ms; (H) time = 41.1 ms; (I) time = 46.9 ms; (J) time = 52.2 ms.

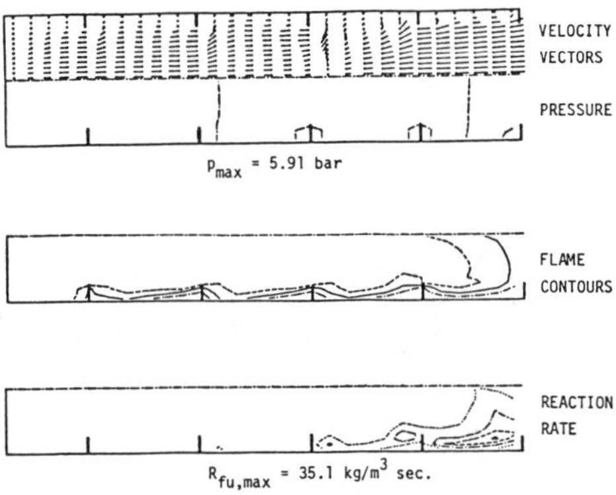

g) TIME = 39.9 ms

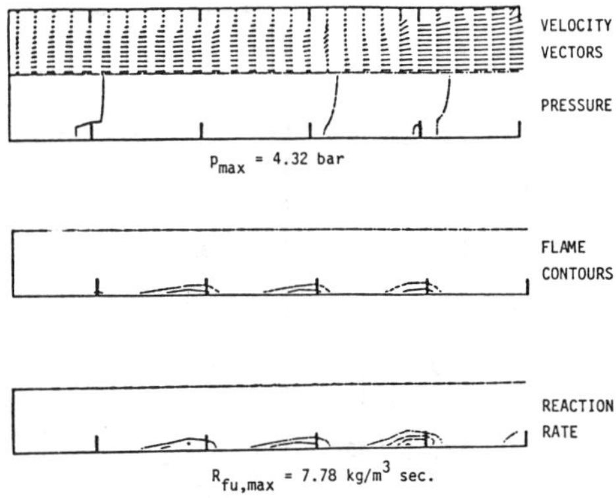

i) TIME = 46.9 ms

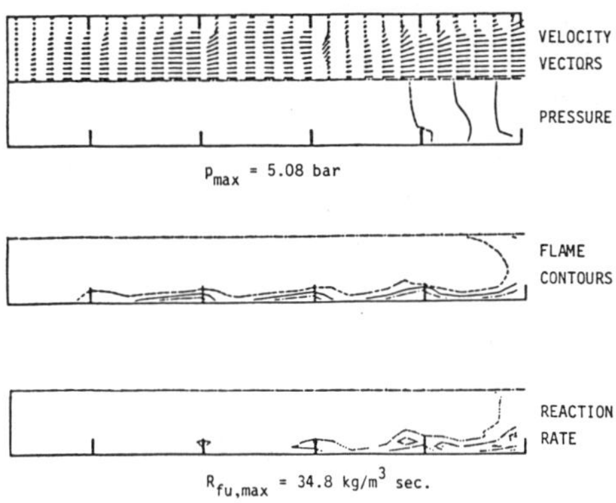

h) TIME = 41.1 ms

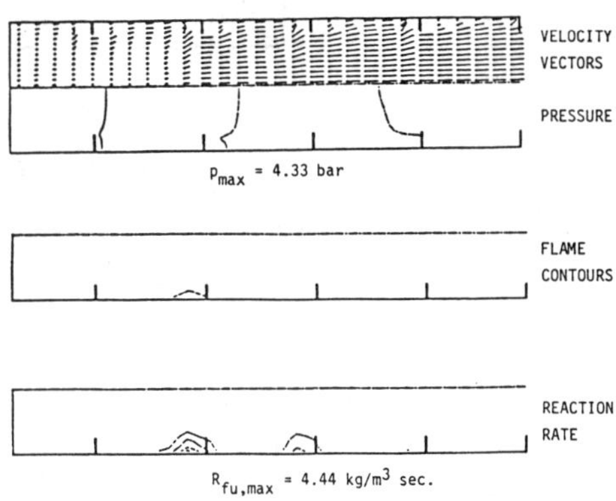

j) TIME = 52.2 ms

laboratory-scale (10^{-2} m^3, ~1.0 bar) to large scale (approx. 50 m^3, ~4.5 bar) is observed, whereas further pressure increase with larger scales beyond approximately 100 m^3 is moderate (up to ~5.5 bar). Moen [37] conducted a small-scale experiment (Scale 1/16) of the 50 m^3 Sund tube and found that the peak overpressures were in the order of 0.5 bar, which is comparable to what the present method predicts.

CONCLUSIONS

The present chapter has shown that predictions of flow, heat, and mass transfer processes in various multidimensional enclosures can be performed. Detailed comparisons with available experimental data are shown for various flow cases, ranging from laminar to turbulent flows involving combustion and radiative heat transfer and compressibility. For laminar flow the agreement is good taking into account that coarse grids are used. The results for turbulent isothermal flows indicate that the two-equation turbulence model together with the wall function method constitutes a satisfactory means of computing three-dimensional flows of engineering importance. This conclusion relates both to calculation of mean velocities and turbulence velocities.

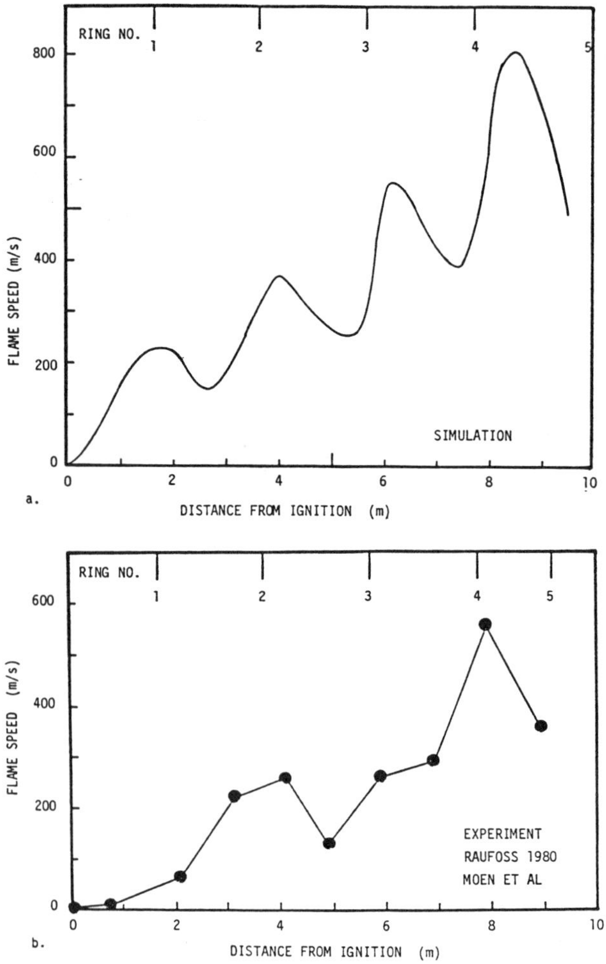

Figure 50. Variation of center line flame speed with distance from ignition: (A) prediction; (B) experiment.

The presented calculations of furnace flow processes reveal that satisfactory quantitative agreements between predictions and measurements have been found. The agreement for temperature calculations is in the order of $\pm 200°$ C and the agreement for heat flux predictions is approximately $\pm 20\%$. This is in many cases sufficient for engineering purposes.

The calculation method has also been tested for transient compressible reacting flows. The results of testing against available analytical solutions for the shock tube problem are good. Furthermore the problem of turbulent flame acceleration by obstacles has been modeled and comparison of predicted pressure pulses, flame speed, and pressure load scaling with available experimental data are good. However, the computation method for compressible reacting flows must be tested in various other geometrical configuration to further

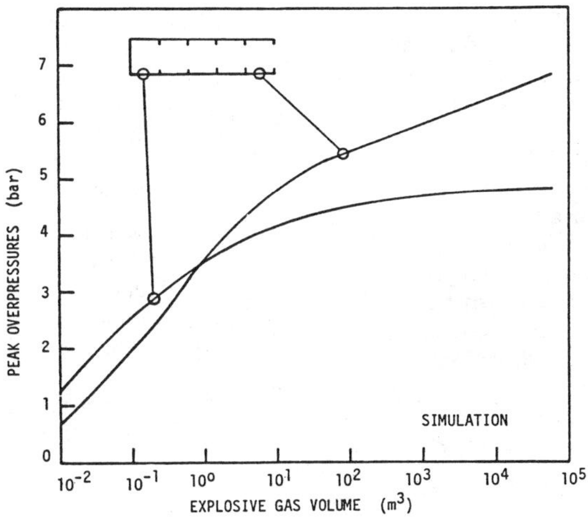

Figure 51. Variations of peak pressures with volume of explosive gas (scaling).

investigate the potential of the model to predict explosion effects in real-life geometries. Turbulence measurements of various quantities in transient combusting flows are highly needed to refine the various parts of the model.

NOTATION

Symbol	Meaning
A	Constant in combustion model or coefficient in finite-domain equation
A_1, A_2, A_3	coefficients in Taylor expansion for radiative intensity
a_i	contant in specific heat relation
B	width of furnace or width of enclosure or width of sudden expansion
B_1, B_2, B_3	coefficients in Taylor expansion for radiative intensity
b	width of injection nozzle or cavity or constant in finite-domain equation
b_i	contant in specific heat relation
C_D, C_1, C_2	constants in turbulence model
c	sound speed
$c_{pm,mix}$	specific heat of gas mixture
$c_{pm,i}$	specific heat of i-th component
$\frac{D}{Dt}$	substantial derivative
D	coefficient for pressure correction or width of enclosure
d	width of injection nozzle or dimension of duct side
E	constant in wall function
F	function
F_x, F_y, F_z	radiation flux quantities in x, y, and z directions
f	mixture fraction
G	generation of turbulent kinetic energy
H	height of furnace or channel height or enclosure height
H_{fu}	reaction enthalpy
h	enthalpy or channel height or injection nozzle height
h_b	vertical position of burner
I	radiation intensity
J	diffusive flux
K_R	absorption coefficient
k	kinetic energy of turbulence
L	length of furnace or enclosure
l	length scale of turbulence or length of square cavity
M	molecular weight

m	mass fraction
P	variable in wall function
P	pressure
p′	pressure correction
q	heat flux
Re	Reynolds number
R_u	universal gas constant
R_{fu}	reaction rate
r	radial direction
s	stoichiometric oxygen requirement to burn 1 kg of fuel
S_Φ	source term in Φ-equation
s	coordinate direction
T	temperature
Δt	time increment
t	time or temperature
U_i	time mean velocity components in x_i-direction
U, V, W	time mean velocity components in x, y, and z-directions
u_i	fluctuating velocity component in x_i-direction
$\vec{V}$	velocity vector parallel to walls
x_i	coordinate direction
x, y, z	coordinate directions
Δx, Δy, Δz	grid spacing
y^+	dimensionless, normal distance from walls

Greek Symbols

α	under-relaxation factor
β	angle of burner inclination or volume/area porosity or conserved variable in single-step combustion model
Γ_Φ	exchange coefficient for Φ
μ	viscosity, or direction cosine of the radiation intensity
η, ξ	direction cosines of the radiation intensity
ρ	density
ε	rate of dissipation of kinetic energy of turbulence
σ	Schmidt/Prandtl number or Stefan–Boltzmann constant
σ_{ij}	stress tensor
κ	constant in wall function
$\tau_{w,i}$	shear stress at the wall in x_i-direction
τ_w	wall shear stress
Φ	general variable
$\delta_{ij} = 0\ i \neq j$; $\delta_{ij} = i\ i = j$	Kronecker delta

Subscripts

av	average value
b	burner or bulk value
con	convective
E, e	east
eff	effective
fu	fuel
fu, b	fuel burnt
H, h	high
h	enthalpy
i	i-th specie
i, j, k	summation indicies
in	inlet
l	laminar
L, l	low
lim	limiting
m	maximum
max	maximum value
mix	mixture
N, n	north
nw	near wall value
O_2	oxygen
0	outlet
P	point
r	radiative
ref	reference value
S, s	south
s	peak value
t	turbulent
tot	total
U, V, W	reference to velocity
W, w	west
w, wall	wall values

Superscript

$-$	time mean value or area average	p	reference to pressure
*	intermediate values	o	reference to previous time level
$'$	correction or fluctuation	I, II, III	results of solution stages
U, V, W	reference to velocity	$\hat{}$	instantaneous value
Φ	reference to scalars		

REFERENCES

1. Patanker, S. V., and Spalding, D. B., "A Calculation Procedure for Heat, Mass, and Momentum Transfer in Three-Dimensional Parabolic Flows," *Int. Journal of Heat and Mass Transfer, 15:* 1787–1806 (1972).
2. Harlow, F. H., and Welch, S. E., "Numerical Calculation of Time-Dependent Viscous Incompressible Flow of Fluid with Free Surface," *Physics of Fluids,* 8 (12): 2182–2189 (1965).
3. Chorin, A. J., "Numerical Solution of Navier–Stokes Equations," *Maths. of Computation* 22 (104): 745–762 (1968).
4. Butler, T. D., and O'Rourke, P. J., "A Numerical Method for Two-dimensional Reacting Flows," 16th Symp. (Int.) on Combustion (1976), pp. 1503–1515.
5. Boni, A. A., Chapman, M., and Schneyer, G. P., "Computer Simulation of Combustion Processes in a Stratified Charge Engine," *Acta Astronautica,* 3: 293–307 (1976).
6. Oran, E., Young, T., and Boris, J., "Application of Time-Dependent Numerical Methods to the Description of Reactive Shocks," 17th Symp. (Int.) on Combustion (1978), pp. 43–53.
7. Hjertager, B. H., and Magnussen, B. F., "Computation of Some Three-Dimensional Laminar Incompressible Internal Flows," Proc. 1976 Heat Transfer and Fluid Mechanics Institute, Stanford University Press (1976), pp. 436–451.
8. Hjertager, B. H., "Simulation of Transient Compressible Turbulent Reactive Flows," *Combustion Science and Technology 27:* 159–170 (1982).
9. Launder, B. E., and Spalding, D. B., "The Numerical Computation of Turbulent Flows," *Computer Methods in Applied Mechanics and Engineering,* (3): 269–289 (1974).
10. Jayatilleke, C. L. V., "The Influence of Prandtl Number and Surface Roughness on the Resistance of the Laminar Sublayer to Momentum and Heat Transfer," in *Progress in Heat and Mass Transfer,* 1st ed., U. Grigull and E. Hahne, Pergamon Press, 1969.
11. Magnussen, B. F., and Hjertager, B. H., "On Mathematical Modeling of Turbulent Combustion with Special Emphasis on Soot Formation and Combustion," 16th Symp. (Int.) on Combustion, Combustion Institute, Pittsburg (1976), pp. 719–729.
12. Hjertager, B. H., "Numerical Simulation of Turbulent Flame and Pressure Development in Gas Explosions," *Fuel-air Explosions,* SM Study No. 16. University of Waterloo Press, Ontario, Canada (1982), pp. 407–426.
13. Radhkrishnan, K., Heywood, J. B., and Tabaczynski, R. J., "Premixed Turbulent Flame Blowoff Velocity Correlation Based on Coherent Structures in Turbulent Flows," *Combustion and Flame,* 12: 19–13 (1981).
14. Magnussen, B. F., "On the Structure of Turbulence and a Generalized Eddy Dissipation Concept for Chemical Reaction in Turbulent Flow," Proc. 19th AIAA Aerospace Science Meeting, St. Louis, Missouri, January 12–15 (1981).
15. Khalil, E. E., Spalding, D. B., and Whitelaw, J. H., "The Calculation of Local Flow Properties in Two-Dimensional Furnaces," *Int. J. Heat Mass Transfer,* 18: 775–791 (1975).
16. Patankar, S. V., and Spalding, D. B., "A Computer Model for Three-Dimensional Flow in Furnaces," 14th Symposium on Combustion, Combustion Institute, Pittsburgh (1973), pp. 605–614.

17. DeMarco, A. G., and Lockwood, F. C., "A New Flux Model for the Calculation of Radiation in Furnaces," Italian Flame Day, *La Revisita dei Combustiobili,* 29 (5–6): 184–196 (1975).
18. Bartelds, H., et al., "A Finite Difference Method for the Prediction of Radiant heat Exchange, Applicable to Various Furnace Geometries," Symp. Europeen Sur la Combustion, Vol. II (1975), pp. 757–762.
19. Patankar, S. V., and Spalding, D. B., "A Calculation Procedure for the Transient and Steady-State Behavior of Shell-and-Tube Heat Exchangers," *Heat Exchangers: Design and Theory Sourcebook,* N. H. Afgan and E. V. Schlünder (Eds.) McGraw-Hill, 1974, pp. 155–176.
20. Sha, W. T., et al., "Multi dimensional Numerical Modelling of Heat Exchangers," *Journal of Heat Transfer,* (104): 417–425 (1982).
21. Sha, W. T., and Launder, B. E., "A Model for Turbulent Momentum and Heat Transport in Large Rod Bundles," ANL–77–73, (1979).
22. Caretto, L. S., et al., "Two Calculation Procedures for Steady, Three-Dimensional Flows with Recirculation," Proc. of Third International Conference on Numerical Methods in Fluid Mechanics, Vol. II, (1973), pp. 60–68.
23. Goldstein, R. J., and Kreid, D. K., "Measurement of Laminar Flow Development in a Square Duct Using a Laser-Doppler Flowmeter," *Journal and Applied Mechanics,* 34: 813–818 (1967).
24. Beavers, G. S., Sparrow, E. M., and Magnuson, R. A., "Experiments on Hydrodynamically Developing Flow in Rectangular Ducts of Arbitrary Aspect Ratio," *Int. Journal of Heat and Mass Transfer, 13:* 689–702 (1970).
25. Durst, F., Melling, A., and Whitelaw, J. H., "Low Reynold's Number Flow over a Plane Symetric Sudden Expansion," *J. Fluid Mech.*, 64: 111–128 (1974).
26. Hjertager, B. H., and Magnussen, B. F., "Calculation of Turbulent Three-Dimensional Jet Induced Flow in Rectangular Enclosures," *Computer and Fluids,* 9: 395–407 (1981).
27. Hjertager, B. H., and Magnussen, B. F., "Laser-Doppler Velocity Measurements in the Three-Dimensional Flow of a Jet in a Square Enclosure," *Letters in Heat and Mass Transfer,* (8): 178–186 (1981).
28. Nielsen, P. V., Restivo, A., and Whitelaw, J. H., "The Velocity Characteristics of Ventilated Rooms," *Journal of Fluids Engineering,* 100: 291–298 (1978).
29. Patankar, S. V., Basu, D. K., and Alpay, S. A., "Prediction of the Three-Dimensional Velocity Field of a Deflected Turbulent Jet," *Journal of Fluids Engineering,* 758–762 (1977).
30. McGuirk, J. J., and Rodi, W., "A Depth-Averaged Mathematical Model for Near Field of Side Discharges into Open-Channel Flow," *Journal of Fluid Mechanics,* 68: 761–781 (1978).
31. Patankar, S. V., *Numerical Heat Transfer and Fluid Flow,* Chapter 5, McGraw-Hill, 1980.
32. Hjertager, B. H., and Magnussen, B. F., "Computer Simulation of Flow Heat Transfer and Combustion in Three-Dimensional Furnaces," *PCH Physico Chemical Hydrodynamics,* 3: 231–250 (1982).
33. Pai, B. R., Michelfelder, S., and Spalding, D. B., "Prediction of Furnace Heat Transfer with a Three-Dimensional Mathematical Model," *Int. J. Heat Mass Transfer,* 21: 571–580 (1978).
34. Michelfelder, S., and Lowes, T. M., Report on the M2-trials, Int. Flame Research Foundation, Document Number F36–a–4.
35. Harlow, F. H., and Amsden, A. A., "Fluid Dynamics," A LASL Monograph, LA–4700 (1971).
36. Moen, I. O., et al., "Pressure Development Due to Turbulent Flame Propagation in Large-Scale Methane-Air Explosions," *Combustion and Flame* 47: 31–52 (1982).
37. Moen, I. O., private communications (1980).

CHAPTER 42

DESIGN PRACTICES IN COMBUSTION AND INCINERATION

Kashi Banerjee and Paul N. Cheremisinoff

Department of Civil and Environmental Engineering
New Jersey Institute of Technology
Newark, New Jersey

and

Nicholas P. Cheremisinoff

Exxon Chemical Company
Linden, New Jersey

CONTENTS

INTRODUCTION

This chapter reviews design principles of incineration and combustion. Since the subject of combustion cannot be dealt with in a single chapter, emphasis is placed on one area of industrial application, namely incineration for the disposal of waste materials. Although application specific, the design basis for incinerators described herein is general.

Incineration can be defined as a unit operation that employs thermal decomposition via oxidation to reduce carbonaceous matter. The principal products of incineration are carbon dioxide, water, ash, and waste heat. In addition, various by-products having environmental importance are generated, such as sulfur and nitrogen compounds, halogens, and various heavy metals (e.g., lead, cadmium, mercury, and others).

Typical industrial incinerators are integrated systems composed of the incinerator, raw waste and auxiliary material handling equipment, exhaust gas cleaning devices, effluent liquid treatment, solids discharge control, and energy and by-product recovery operations. Basic methodology for scale-up and design of the integrated incineration process are described in the following. A brief glossary of combustion-related terms is included at the end of this chapter to facilitate discussions.

COMBUSTION PRINCIPLES

The principal products of combustion are carbon dioxide, water, sulfur dioxide, and nitrogen. These generally are in the most highly oxidized state that is stable for each element of the fuel. This definition of complete combustion is somewhat arbitrary. Some nitrogen is converted to oxides, in particular nitric oxide. Sulfur, if present, is also converted to its oxidized state. Metals such as iron and aluminum may also be converted to oxides.

In the application of incineration, combustion is a high-temperature process. Heat released by combustion is partly stored in the combustion products and partly transferred by conduction, convection, and radiation both to incinerator walls and to the incoming fuel. The portion of heat transferred to the incoming fuel is required for ignition.

Incinerators should be designed around the three "T's" of combustion [1, 2], namely, time, temperature, and turbulence in the presence of oxygen. Systems that do not embrace these factors in their design, experience operating and maintenance problems as well as pose enviromental hazards.

Time is accounted for in design through the space and volume of the combustion chambers of the incinerator. Volume must be sufficient to retain the gas flow long enough to allow complete combustion of fuel or solid wastes and volatile gases.

Temperature is a critical consideration. Heat is the driving force to sustain combustion, and in many instances supplemental heat must be supplied by auxiliary burners to preheat the incinerators or to support combustion of materials having high moisture levels and low heating values.

As noted in the previous chapter, the system hydrodynamics plays a major role in combustion efficiency. Turbulence can be designed into the incinerator by a series of baffles or constrictions. The purpose of turbulence is to promote mixing between the products of combustion and air (oxygen). Oxygen is provided to complete the combustion process. In the case of small refuse incinerators, application is aimed at the burning of cellulose (1 lb of cellulose releases 8,000 Btu's of heat). In this reaction, cellulose combines with oxygen to form CO_2 and H_2O. Municipal incinerators are designed on the basis of 4,500 to 5,000 Btu/lb of refuse burned because the heterogeneity of mixed refuse properties.

Fuel-to-Air Ratio

A basic requirement for efficient combustion is that fuel and air be mixed in proper proportions and under conditions that prompt ignition and maintain combustion. For gaseous, liquid, or pulverized solid fuels this criterion is readily met; however, with solid refuse the variability of physical and chemical properties often makes this a difficult condition to meet. Because of this variability and the prevalence of such materials as glass and metals, combustion is generally carried out in deep fuel beds.

We may consider an idealized system, such as a bed of double-screened coke. Depending on the relative movement of air and fuel and the resultant flow of heat and combustion materials, the fuel bed may be classified as follows:[3]

1. Underfeed beds in which fuel and air flow in the same direction.
2. Overfeed beds in which fuel and air flow in the opposite direction.
3. Crossfeed beds in which fuel and air flow in directions at some angle to each other so that the heat required for ignition neither flows directly against the air stream nor directly with the initial combustion products.

The underfeed bed arrangement is illustrated in Figure 1A. The flow of heat against the air stream produces a sharp temperature gradient at a level over the raw face designated as the ignition plane. Oxygen is rapidly consumed by the reaction:

$$O_2 + C \rightarrow CO_2 \tag{1}$$

as the gases move through the ignition plane and into the ignited fuel. As the oxygen is depleted the reaction becomes:

$$CO_2 + C \rightarrow 2\,CO \tag{2}$$

Reaction 2 shows that the CO_2 concentration decreases, producing carbon monoxide. The reaction absorbs heat so that the temperature falls as the gases flow toward the ash layer above the ignited fuel. The relative distribution of temperature and concentration of combustion products is shown in Figure 1B.

In the overfeed fuel bed (Figure 2A), the incoming air passes first through the ash layers, heat recuperation favors higher temperatures, and greater and more rapid conversion of CO_2 to CO takes place. The relative distribution of combustion products and temperature are shown in Figure 2B.

Figure 3 shows the crossfeed fuel bed where gas flow is from the raw fuel into the ignited fuel but heat flow for ignition does not flow directly against the gas stream. The relative direction of fuel and velocities are important in the fuel bed classification. Batch burning in a pot with top ignition and airfeed from the bottom is referred to as underfeed burning

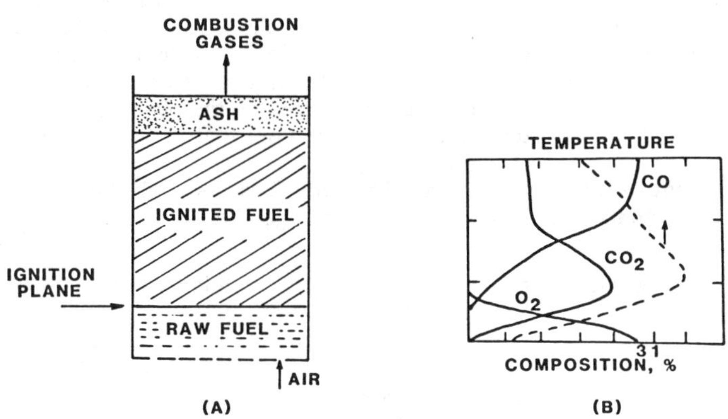

Figure 1. (A) Shows idealized underfeed fuel bed; (B) plot of relative distribution of temperature and combustion products.

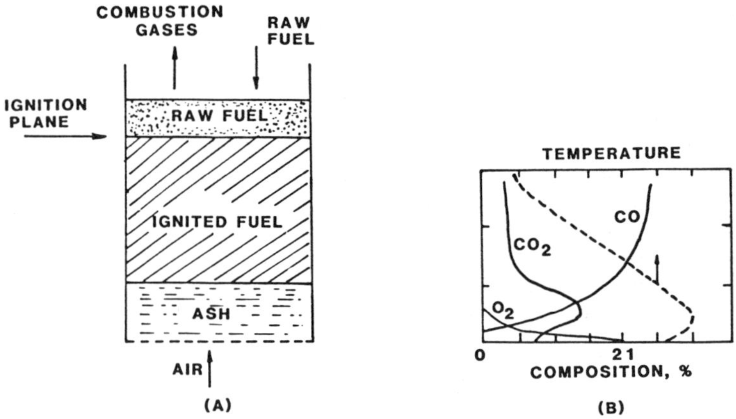

Figure 2. (A) Shows idealized overfeed fuel bed; (B) plot of relative distribution of temperature and combustion products.

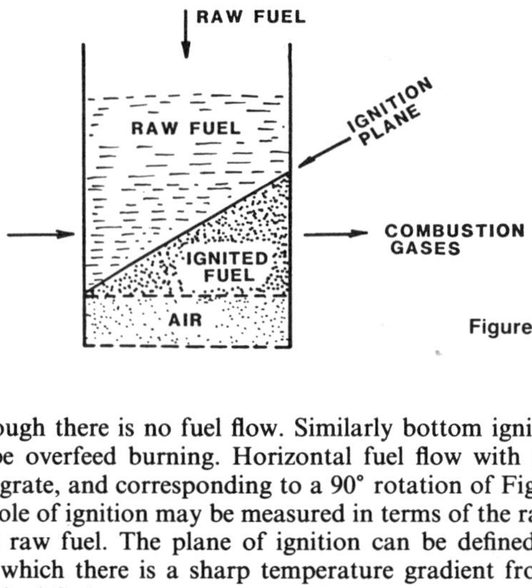

Figure 3. Shows idealized crossfeed fuel bed.

even though there is no fuel flow. Similarly bottom ignition with airfeed from the bottom would be overfeed burning. Horizontal fuel flow with airfeed from the bottom, as on a moving grate, and corresponding to a 90° rotation of Figure 3, would be crossfeed burning.

The role of ignition may be measured in terms of the rate of advance of the ignition plane into the raw fuel. The plane of ignition can be defined as the plane within the fuel bed beyond which there is a sharp temperature gradient from that of the raw fuel to that of fully ignited fuel.

The management of fuel and air can take many forms. Good management can be illustrated best by summarizing conditions that lead to good composition in deep fuel beds. These are as follows:

1. Raw fuel should be fed in such a manner as to provide underfeed or crossfeed burning. Large masses of raw fuel should never be dumped directly on top of the flames.
2. As in underfeed and crossfeed burning, gas flow should be from raw fuel into ignited fuel.
3. Overfire air must be mixed with the hot gases flowing out of the fuel bed. This is best accomplished by such devices as overfire-air jets to give turbulent mixing close to the intense-burning zone. That is, streams of gas from colder portions of the fuel bed must be directed into zones of intense burning.

Table 1
Classification of Refuse Materials

Refuse (Solid Wastes)	Composition	Source
Garbage	Wastes from the preparation, cooking, and serving of food Market refuse, waste from the handling, storage, and sale of produce and meat	
Rubbish	Combustible (primarily organic): Paper, cardboard, cartons; Wood, boxes, excelsior; Plastics; Rags, cloth, bedding; Leather, rubber; Grass, leaves, yard trimmings Noncombustible (primarily inorganic): Metals, tin cans, metal foils; Dirt; Stones, bricks, ceramics, crockery; Glass, bottles; Other mineral refuse	Households, institutions, and commercial concerns such as hotels, stores, restaurants, markets, etc.
Ashes	Residue from fires used for cooking and for heating buildings, cinders	
Bulky wastes	Large auto parts, tires Stoves, refrigerators, other large appliances Furniture, large crates Trees, branches, palm fronds, stumps, flotage	
Street refuse	Street sweepings, dirt Leaves Catch-basin dirt Contents of litter receptacles	Streets, sidewalks, alleys, vacant lots, etc.
Dead animals	Small animals: cats, dogs, poultry, etc. Large animals: horses, cows, etc.	
Abandoned vehicles	Automobiles, trucks	
Construction and demolition wastes	Lumber, roofing, and sheathing scraps Rubble, broken concrete, plaster, etc. Conduit, pipe, wire, insulation, etc.	
Industrial refuse	Solid wastes resulting from industrial processes and manufacturing operations, such as food-processing wastes, boiler house cinders, wood, plastic, and metal scraps and shavings, etc.	Factories, power plants, etc.
Special wastes	Hazardous wastes: pathological wastes, explosives, radioactive materials Security wastes: confidential documents, negotiable papers, etc.	Households, hospitals, institutions, stores, industry, etc.
Animal and agricultural wastes	Manures, crop residues	Farms, feed lots
Sewage-treatment residues	Coarse screenings, grit, septic-tank sludge, dewatered sludge	Sewage-treatment plants, septic tanks

From American Public Works Association [1].

4. Additional mixing with high amounts of excess air that may be desired to give lower gas temperatures must be delayed so as not to hinder good combustion.

More novel modes of fuel/air management are used in vortex [4] and fluidized-bed combustion [5]. In vortex combustion all of the air can be supplied as overfire air. The aerodynamics of the vortex is such that the air must follow a downward spiral to scrub the fuel bed and mix with rising volatile matter. In fluidized-bed combustion, refuse is fed into a turbulent bed of granular inert material. Through proper balance between heat input in fuel, heat losses to the surroundings, heat extraction from the bed and enthalpy of combustion products leaving the bed, the temperature can be maintained at a level that provides good combustion. The temperature also is practically constant throughout a

Table 2
Composition and Analysis of an Average Municipal Refuse
(Studies Made by Purdue University)

Component	Percent of All Refuse by Weight	Moisture (percent by weight)	Analysis (percent dry weight): Volatile Matter	Carbon	Hyd-rogen	Oxygen	Nitrogen	Sulfur	Non-com-busti-bles[b]	Calorific Value (Btu/lb)
				Rubbish, 64%						
Paper	42.0	10.2	84.6	43.4	5.8	44.3	0.3	0.20	6.0	7,572
Wood	2.4	20.0	84.9	50.5	6.0	42.4	0.2	0.05	1.0	8,613
Grass	4.0	65.0	—	43.3	6.0	41.7	2.2	0.05	6.8	7,693
Brush	1.5	40.0	—	42.5	5.9	41.2	2.0	0.05	8.3	7,900
Greens	1.5	62.0	70.3	40.3	5.6	39.0	2.0	0.05	13.0	7,077
Leaves	5.0	50.0	—	40.5	6.0	45.1	0.2	0.05	8.2	7,096
Leather	0.3	10.0	76.2	60.0	8.0	11.5	10.0	0.40	10.1	8,850
Rubber	0.6	1.2	85.0	77.7	10.4	—	—	2.0	10.0	11,330
Plastics	0.7	2.0	—	60.0	7.2	22.6	—	—	10.2	14,368
Oils, paints	0.8	0.0	—	66.9	9.7	5.2	2.0	—	16.3	13,400
Linoleum	0.1	2.1	65.8	48.1	5.3	18.7	0.1	0.40	27.4	8,310
Rags	0.6	10.0	93.6	55.0	6.6	31.2	4.6	0.13	2.5	7,652
Street sweepings	3.0	20.0	67.4	34.7	4.8	35.2	0.1	0.20	25.0	6,000
Dirt	1.0	3.2	21.2	20.6	2.6	4.0	0.5	0.01	72.3	3,790
Unclassified	0.5	4.0	—	16.6	2.5	18.4	0.05	0.05	62.5	3,000
				Food Wastes, 12%						
Garbage	10.0	72.0	53.3	45.0	6.4	28.8	3.3	0.52	16.0	8,484
Fats	2.0	0.0	—	76.7	12.1	11.2	0	0	0	16,700
				Noncombustibles, 24%						
Metals	8.0	3.0	0.5	0.8	0.04	0.2	—	—	99.0	124
Glass and ceramics	6.0	2.0	0.4	0.6	0.03	0.1	—	—	99.3	65
Ashes	10.0	10.0	3.0	28.0	0.5	0.8	—	0.5	70.2	4,172
				Composite Refuse, as Received						
All refuse	100	20.7	—	28.0	3.5	22.4	0.33	0.16	24.9	6,203

From American Public Works Association [1].

well-fluidized bed. Good heat transfer between hot granular material and incoming solid fuel causes prompt devolatilization and ignition.

Material and Thermal Balances

We shall consider the more complex application of solid waste incineration. Tables 1 and 2 provide general information and typical compositions of refuse. Since temperature, as well as the degree of contacting between fuel and air, is important to the combustion process, the effect of refuse composition on the air requirements, and the effect of excess air on the gas temperature produced must be examined. Information for the analysis of air, and the ultimate analysis and gross heating value of the refuse is first required.

The composition of air is approximately constant, except for its moisture content. The assumed composition on a basis of 70% relative humidity at 60° F is given in the footnote b of Table 3 [6]. Note that refuse composition varies considerably, depending on such factors as source (domestic, commercial, etc.), geography, and climatic conditions.

Material balances are performed on the basis of atomic weight units of the fuel elements and the moles of combustion products. This is a necessary simplification since each atomic weight unit contains the same number of atoms; and each volume of gas, under standard conditions of temperature and pressure, occupies the same volume. In other words, ideal gas behavior is assumed, which is a good approximation within the practical limits for air under atmospheric conditions as well as for combustion gases. The exception to this is when the moisture condensate is high at temperatures below the dewpoint. However, within these limits the gas composition in terms of percent by volume are the same as the mole percents.

The second column of Table 3 provides the weight percent in the particular refuse for each element as identified in the first column. Using a 100 lb refuse basis, the weight percents

Table 3
Calculation of Air Requirement and Gas Composition at Zero Excess Air

Component	Weight Percent	Atomic Weight	Atomic Weight Units	Moles of [a]Oxygen Required	Combustion Product	Moles of Combustion Products: From Combustion	From Air[b]	Total	Percent
Carbon	29.83	12.011	[c]2.484	[c]2.484	Carbon dioxide	2.484	[d]0.004	2.488	15.5
Hydrogen	[e]6.23	1.008	6.181	1.545	Water	3.090	[d]0.166	3.256	20.3
Oxygen	[e]43.45	16.000	2.716	−1.358	—	—	—	—	—
Nitrogen	0.37	14.008	0.026	—	Nitrogen	0.013	[d]10.283	10.296	64.2
Sulfur	0.12	32.066	0.004	0.004	Sulfur dioxide	0.004	—	00.004	00.02
Ash and metal	20.0	—	—	[f]0.052	—	—	—	—	—
Total	100.00			2.727				16.044	100.0

Moles of air required per 100 lb *fuel* = 2.727/.2069 = 13.18
Pounds of air required per pound of refuse = 13.18(28.7)/100 = 3.78
Moles of air per mole of gas = 13.18/16.044 = 0.8215
Moles of gas produced per pounds of refuse = 16.044/100 = 0.16044
a *Computation is based on* 100 lbs *of refuse.*
b *Assumed air composition, in volume fractions: carbon dioxide, 0.0003; nitrogen, 0.7802; oxygen, 0.2069; water, 0.0126. Assuming ideal gases, the volume fractions may be taken as mole fractions and are equal to the percentages by volume divided by 100. The composition as given is for rare gases included with the nitrogen and with moisture content corresponding to 70% relative humidity at* 60° F. *Air of this composition has a weight of* 28.7 lb *per mole of total gas.*
c *Per* 100 lb *of refuse.*
d 13.18(0.0126) = 0.166, *for instance.*
e *Includes hydrogen and oxygen from* 20% *moisture.*
f *An assumed value for partial burning of metals.*
From Orning [6].

then correspond to pounds of each element. Division by the atomic weight gives the atomic-weight units (fourth column) for each element. Each atomic weight unit of carbon requires 1 mole of oxygen to produce 1 mole of CO_2. These quantities are shown in the fifth and seventh columns of Table 3. An atomic-weight unit of hydrogen requires 1/4 mole of oxygen to produce 1/2 mole of H_2O. Each atomic weight unit of oxygen in refuse represents a 1/2 mole of oxygen that need not be supplied by air to form combustion products. It is assumed that the nitrogen of the refuse forms N_2 in the combustion products. This calculation ignores small percentages of nitrogen oxide that are typically found in combustion gases. The error is negligible in comparison to the percentages of other products. Thus each atomic weight unit produces 1/2 mole of N_2. Like the carbon, each atomic weight unit of sulfur requires a mole of O_2 to produce a mole of SO_2. Since the sulfur percentages are low and only a few percent of the SO_2 is oxidized to SO_3, the moles of SO_3 and the extra oxygen required for its formation can be neglected in determining gas composition and the oxygen requirement.

The total oxygen requirement, 2.727 moles, includes a negative item corresponding to oxygen in the refuse. The air required at zero excess, 13.18 moles per 100 lb. of refuse, is determined by dividing the oxygen requirement by its volume (or mole) fraction in air as shown at the bottom of Table 3.

The weight per mole of a mixed gas is calculated as the sum of volume fractions times its molecular weights. Thus, for air of the assumed composition (note b, Table 3) $0.0003(44.011)+0.7802(28.016)+0.2069(32)+0.0126(18.016) = 28.7$ lb/mole.

It now is apparent that 3.78 lb air are required per lb of refuse. The moles of combustion product derived from the air are found by multiplying the moles of air required by the volume fractions in air of CO_2, H_2O, and N_2. All the oxygen of the air, at zero excess, is included in the combustion product of the refuse. Since the percent by volume is practically equal to the mole percent, the composition of the product gas is obtained from the total moles for each gas as a percentage of the grand total.

The calculation of gas composition at various excess air levels is illustrated in Table 4 [5]. The composition is based on 100 moles of combustion products at zero excess air. Total moles per 100 moles of combustion products at zero excess air are given in column 3 for each excess air level. This will be used later for the thermal balance calculation. The gas composition is given on a moist basis. An approximate calculation of the theoretical air

Table 4
Effect of Excess Air on Gas Composition

Percent Excess Air[a]	Moles Excess Air[c]	Total Moles	Gas Composition[b]				
			CO_2	O_2	N_2	H_2O	SO_2
0	0.00	100.00	15.50	0.00	64.20	20.30	0.02
			15.50	0.00	64.20	20.30	0.02
50	[d]41.08	141.08	15.51	8.50	96.23	[e]20.82	0.02
			11.0	6.0	68.2	14.8	0.01
125	102.69	202.69	15.53	21.24	144.31	21.59	0.02
			7.7	10.5	71.2	10.6	0.01

a *Based on 100 moles of gas at zero excess air.*
b *Composition is given in moles in the first line and in percent in the second line.*

c *Percent of excess air* $= \dfrac{100\ (\text{air supplied} - \text{theoretical air requirement})}{\text{theoretical air requirement}}$

d *Excess-air requirement is calculated from the ratio of the theoretical air requirement to gas produced (from Table 2-1) as*

$$\left(\frac{\text{moles of air}}{\text{moles of gas}}\right)\ (\text{percent of excess air}) = 0.8215(50) = 41.08$$

moles of excess air per 100 moles of gas produced at zero excess air.
e *Calculation of the moles of product gas is illustrated for water at 50% excess air as (moles excess air)*

$$\left(\frac{\text{moles of water}}{\text{moles of air}}\right) + (\text{moles of water per 100 moles of product gas at zero excess air}) =$$

$41.08(0.0126) \times 20.30 = 20.82.$
From Bailie, Donner, and Galli [5].

Table 5
Air Requirement per 1,000 Btu for Various Compounds

Compounds	Pounds of air per 1,000 Btu
Methane	0.73
Methyl alcohol	0.67
Propane	0.73
Propylene	0.71
Benzene	0.74
Glucose	0.69
Glycol dipalmitate	0.74
Cellulose	0.68

From Bailie [5].

requirements can be based on the heating value of refuse and generally that appears in the fuel. Table 5 [5] gives the pounds of air requirement for releasing 1,000 Btu for complete combustion of the given fuel for a series of organic compounds containing carbon, hydrogen, and oxygen.

Figure 4 shows the air requirement as a function of refuse heating value. The plot is based on an air requirement of 0.7 lb/1,000 Btu of heat released. If the gross heating value of the refuse is known, the amount of air required per pound of refuse can be obtained directly from the plot for various excess air levels.

A modified Dulong formula is sometimes useful in estimating the gross heating value from the fuel analysis:

$$\text{gross Btu/lb} = 145.4\,C + 620\left(H - \frac{O}{8}\right) + 41\,S \tag{3}$$

where C, H, O, and S are the weight percentages of carbon, hydrogen, oxygen, and sulfur. The term $(H - O/8)$ represents hydrogen in the fuel that is combined with oxygen as moisture as well as hydrogen and oxygen that are combined with the fuel in some other form.

An approximate calculation of the excess air can be obtained from the product gas analysis expressed in percent by volume. The excess air is given as:

$$\% \text{ of excess air} = \frac{100\,(O_2)}{(20.69/78.02)\,(N_2) - O_2} \tag{4}$$

The denominator of this expression represents the oxygen requirement and is expressed as the difference between the oxygen equivalent to the nitrogen supplied and the oxygen remaining as such. Multiplying the ratio of oxygen remaining by the oxygen supplied provides the percent of excess air. Note that this computation is approximate because sulfur and nitrogen contents of the fuel are assumed.

Sulfur dioxide is included with CO_2 in the gas analysis. This involves the same oxygen consumption per mole. Error due to the approximation of the content is limited by the small retention of sulfur in the ash and by the small conversion to sulfur trioxide instead of the dioxide. Error due to nitrogen in the fuel is also small because of the relatively large portion of N_2 supplied by the air. Error in the present illustration is less than that due to the limited precision of the actual gas analysis. These errors may magnify with materials that are high in nitrogen content, such as proteins.

We now direct attention to the thermal balance. The enthalpy for a given amount of gas is the heat needed at constant pressure to raise the temperature from a standard state to some required level. A computation of the temperature of the product gas, with an assumed percentage of the heat derived from the fuel remaining in the product gas is needed. One approach is to perform such a calculation for an arbitrarily chosen fuel synthesis, however, for general purposes, it is more convenient to perform computations in an opposite direction. The enthalpy of that amount of product gas corresponding to 1 lb of a given fuel is calculated for a variety of excess air levels and temperatures. Comparison of these values to the percentage of the heating value of the fuel that may be assumed to remain in the product gas gives an estimate of the gas temperature.

Data required for this computation are given in Table 6. The enthalpy of the gas at a given temperature and excess-air level can be determined from the following relation.

Table 6
Enthalpies
(Btu per Pound Mole over Standard State*)

Temperature T (°F)	CO_2	O_2	N_2	H_2O
1,000	10,048	6,974	6,720	26,925
1,500	16,214	11,008	10,556	31,743
2,000	22,719	15,191	14,520	36,903
2,500	29,539	19,517	18,609	42,405

Enthalpy Equations

CO_2 $$H = 10{,}570\left(\frac{T+460}{1{,}000}\right) + 583.3\left(\frac{T+460}{1{,}000}\right)^2 + \frac{667.4}{(T+460)/1{,}000} - 7{,}085$$

O_2 $$H = 7{,}160\left(\frac{T+460}{1{,}000}\right) + 278.8\left(\frac{T+460}{1{,}000}\right)^2 + \frac{129.6}{(T+460)/1{,}000} - 4{,}163$$

N_2 $$H = 6{,}830\left(\frac{T+460}{1{,}000}\right) + 250.0\left(\frac{T+460}{1{,}000}\right)^2 + \frac{38.9}{(T+460)/1{,}000} - 3{,}811$$

H_2O $$H = 7{,}300\left(\frac{T+460}{1{,}000}\right) + 683.3\left(\frac{T+460}{1{,}000}\right)^2 + 14{,}810$$

* *Gas, except liquid water, at 1 atm pressure and 77° F.*
From Bailie [5].

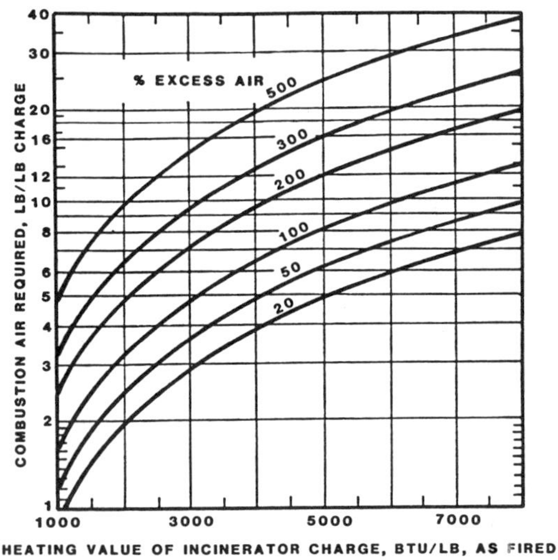

Figure 4. Plot of air requirement vs. heating value for refuse [5].

Table 7
Heat Content of Combustion Product Gases

Temperature (°F)	Heat Content of Gas from 1 lb of Refuse (Btu)	
	50% Excess Air	125% Excess Air
1,000	2,284	2,973
1,500	3,246	4,320
2,000	4,250	5,721
2,500	5,294	7,176

From Bailie [5].

$$\left(\frac{\text{theoretical moles of product gas}}{\text{pound of refuse}}\right)\left(\frac{\text{total moles of gas}}{\text{theoretical moles of gas}}\right)$$

$$\cdot\,[\text{Summation of (mole fraction of gas Component) Btu per mole of gas component}]$$

$$= \frac{\text{Btu in product gas}}{\text{pound of refuse}} \tag{5}$$

Illustrative data for such a calculation are given in Tables 3, 4, and 6 [5] Mole fractions are equal to the percentages divided by 100.

As an example, for 50% excess air and 1,500° F the result is-

$$0.16044\left(\frac{141.08}{100}\right)[0.110(16{,}214)+0.060(11{,}008)$$

$$+0.682(10{,}556)+0.148(31{,}743)]$$

$$= 3246 \text{ Btu/pound of refuse}$$

Data for two excess air levels and for four temperatures are given in Table 7. The gross heating value of the refuse in this example is 5,442 (Btu/lb). This heat must be distributed between the product gas and ash residues or be lost through the incincerator walls. If the loss through the walls and in ash residues are assumed to be 10% of the total, 4,898 Btu/lb of refuse must remain in the gas. An interpolation between the data of Table 7 [5] indicates that the gas temperature should be about 2,300° F at 50% excess air or 1,700° F at 125% excess air.

Combustion Calculations

Combustion of refuse involves the usual parameters but in a more complex manner because of the heterogeneous nature of the fuel. The most difficult factors to be accommodated are the amounts of moisture and the noncombustibles in the refuse. The average heat content of waste is actually consistent, but from instant to instant, the heat content can vary substantially. This implies that a degree of premixing is required to ensure that a minimum of moisture and maximum of combustibles are available to the process and that the resulting mixture is capable of supporting combustion.

Once mixing has been accomplished, the material must be dried and raised to their ignition temperature. At this point, air must be available in sufficient quantity to support combustion. The air and refuse must be mixed in a turbulent manner. This operation takes time, which means that certain furnace velocities and volume are required to properly complete the process and release the heat available in the refuse.

Drying is accomplished by convective and radiant heat transfer mechanisms. Convective transfer is derived from the air or gas flows, whereas radiant transfer occurs from the flames over furnace surface. The material can be either predried external to the furnace or dried within the unit itself as a result of the burning system's design.

Predrying can be accomplished in several ways, including flash drying, drying in rotary drums, trays, chute or fluid bed dryers. External drying can be accomplished with gases drawn from the furnace exit. These gases must be returned and reheated for deodorizing purposes. Drying can be done with preheated air from a gas to an air heat exchanger in the incinerator passage, a fired air heater, or in a steam coil air heater. If drying is accomplished with air, it must enter into the combustion process or the odors produced will have to be taken care of separately. Without mixing, the load on the drying system will be erratic.

A furnace can be designed so that drying is accomplished within the enclosure itself. Most modern incinerators incorporate furnace drying rather than predrying ahead of the furnace. An exception is where one is attempting to burn material such as sewage or industrial sludges.

If the drying is external to the surface, the refuse is brought to about 220° F and the rest of the sensible head addition required to bring the refuse to ignition temperature is added to the furnace. This requires an additional 1,200 Btu/lb to be added to the refuse in the external predrying system to evaporate residual moisture.

This 1,200 Btu was arrived at by comparing with the requirements for evaporating a pound of water and some of the factors that affect drying. It requires 1,122.3 Btu to evaporate 1 lb of water from a mass of refuse at an ambient temperature of 60° F. Since drying equipment are not 100% thermally effective, the heat input to accomplish the required evaporation must be greater. The effectiveness of a thermal drying process is governed by four fundamental factors:

1. Moisture dispersion within the refuse mass, allowing maximum exposure of moisture surfaces to the heated gas.
2. Large temperature differentials—high gas temperatures are required for rapid drying (i.e., a high rate of heat transfer).
3. Maximum refuse agitation to increase the rate of heat transfer.
4. Particle sizes must be at a minimum to increase the surface from which moisture can be evaporated and allow the moisture to reach the surface for drying.

This last requirement coupled with the fact that oxidation as well as evaporation are surface phenomena, indicates that a decrease in particle size of the refuse will improve not only its drying, but its combustion, once in the furnace. This argues for some form of preparation if maximum efficiency is to be obtained in both the drying and burning operations.

Internal drying, if the furnace is of the mass burning type, is usually accomplished on a stoker, the first grate system of which must dry and ignite the refuse. This is accomplished by designing the furnace so that the drying stoker is gas swept and exposed to the high-temperature furnace surfaces. This enables convective and radiant drying to be accomplished rapidly.

Once the mixed refuse has been dried, sensible heat from the furnace flame and enclosure must be added to bring the material to its ignition temperature and begin the rapid oxidation known as combustion, releasing the heat contained in the refuse fuel.

Burning With Sufficient Air

Although refuse appears highly variable, typical refuse is quite uniform in chemical analysis and fits into a family tree of fossil fuels as shown in Figure 5. Both industrial and municipal refuse bear close relation to wood, bark, and bagasse. Typical heating values of various wastes are listed in Table 8.

The variables in refuse are mostly the noncombustible residues and moisture, whereas the chemical composition of combustibles is uniform. Tests have shown that the heat of combustion on a moisture ash-free basis is very close to that of cellulose (8,000 Btu/lb for normal municipal and most industrial refuse fuel). This fact simplifies the calculations in the heat process and makes it possible to predict the behavior of the burning material. The most important variable in the moisture is the refuse, which is considered one of the main energy-consuming loads on the incinerator.

The heating values of the main constituents of municipal and industrial refuse on a moisture and ash-free basis are:

Wood — 8,420 Btu/lb
Brush — 8,600 Btu/lb
Plastic — 18,000 Btu/lb
Paper — 7,900 Btu/lb
Garbage — 7,280 Btu/lb

Although plastic could upset the situation as they grow in importance, the "family tree" of combustible materials in Figure 5 is based on a moisture and ash-free percentage of the material. Table 9 illustrates the relative quantities of air required and heat released from the various constituents in the combustion of refuse. To determine the necessary design parameters, the air and gas weight must be evaluated from an ultimate analysis. The ultimate analysis used in the sample calculations given below is as follows:

Carbon — 30%
Oxygen — 22%
Hydrogen — 4%
Moisture — 24%
Noncombustible — 20%

It may be of interest to compare these selected compositions with the ranges found by various investigators:

Carbon — 25% to 35%
Oxygen — 15% to 30%
Hydrogen — 3.5% to 4.5%
Moisture — 15% to 30%
Ash, metals, and glass (noncombustible) — 15% to 25%

Table 8
Refuse Heating Value

Type Refuse (%)	Moisture	BTU as fired
Domestic Refuse		
Paper, cardboard, cartons, bags	3	7,660
Wood crates, boxes, scrap	7	7,825
Brush, branches	17	7,140
Leaves	30	4,900
Grass	50	3,820
Garbage	75	1,820
Greenstuff	50	3,470
Greens	50	4,070
Rags, cotton, linen	10	6,440
Industrial Scrap Refuse		
Boot, shoe trim and scrap		8,500
Sponge waffle and scrap		8,500
Butyl soles scrap		11,500
Cement wet scrap		11,500
Rubber		12,420
Tire cord scrap		12,400
Rubber scorched scrap		19,700
Tires, bus and auto		18,000
Gum scrap		19,700
Latex coagulum		19,700
Latex waste, coagulum waste		12,000
Leather scrap		10,000
Waxed paper		12,000
Cork scrap		12,400
Paraffin		16,803
Oil waste, fuel oil residue		18,000
Plastic and Synthetic Refuse		
Cellophane plastic		12,000
Polyethylene		19,840
Polyvinyl chloride		17,500
Vinyl scrap		17,500
Aldehyde sludge		18,150
Solvent naptha		18,500
Carbon disulfite		8,000
Benzine		10,000
Miscellaneous		
Carbon to CO_2		14,093
Carbon to CO		4,347
Sulfur		3,983
Methane		23,879

From Combustion Engineering Inc. [8].

Table 9
Combustion Equations

Combustible	Molecular Weight	Equation	Heat Release (Btu/lb) (High)	Theoretical Air (lb/lb of Fuel Element)
Carbon	12	$C + O_2 \rightarrow CO_2$	14,100	11.52
Hydrogen	2	$H_2 + 0.5\,O_2 \rightarrow H_2O$	61,100	34.50
Sulfur	32	$S + O_2 \rightarrow SO_2$	4,000	4.32
Hydrogen sulfide	34	$H_2S + 1.5\,O_2 \rightarrow SO_2 + H_2O$	7,100	3.10
Methane	16	$CH_4 + 2\,O_2 \rightarrow CO_2 + 2\,H_2O$	23,880	17.28
Ethane	30	$C_2H_6 + 3.5\,O_2 \rightarrow 2\,CO_2 + 3\,H_2O$	23,320	16.12
Propane	44	$C_3H_8 + 5\,O_2 \rightarrow 3\,CO_2 + 4\,H_2O$	21,660	15.68
Butane	58	$C_4H_{10} + 6.5\,O_2 \rightarrow 4\,CO_2 + 5\,H_2O$	21,300	15.48
Pentane	72	$C_5H_{12} + 8\,O_2 \rightarrow 5\,CO_2 + 6\,H_2O$	21,090	15.38

From Fernandes [7].

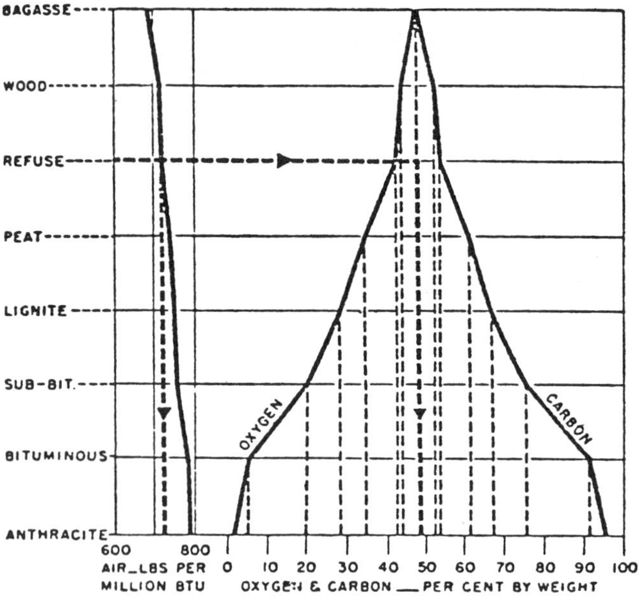

Figure 5. Family tree classification of combustible materials [7].

The final design should be based on the highest average heating value expected; otherwise the design could result in an undersized furnace volume or insufficient grate area.

Heat is usually expressed as million Btu/hr, which is the product of the fuel burned, in pound per hour times the heating value in Btu per lb (Btu/hr = lb/hr × Btu/lb).

In the following example the heating value is found to be approximately 5,000 Btu/lb as received, however, on a moisture and ash-free basis, it is 8,940 Btu/lb. If the ultimate analysis is adjusted to the moisture and ash-free basis, the following results are obtained:

Carbon — 53.5%
Oxygen — 39.3
Hydrogen — 7.2%

With this information, upon examining Figure 5, we note that 53.5% carbon and 39.3% oxygen come out exactly on the refuse line for a combustible material, requiring 720 lbs of air per million Btu of proper combustion.

Air and Gas Weight Determination

The refuse ultimate analysis chosen for the sample calculations requires the following theoretical air (see Table 9):

$$C \rightarrow 0.30 \times 11.54 = 3.46 \text{ lb air}$$

$$(H - 0/8) \rightarrow \left(4 - \frac{22}{8}\right) \times 34.56 = 0.43 \text{ lb air}$$

$$\text{Total} = 3.89 \text{ lb dry air}$$

Hence, the amount of theoretical atmospheric air to completely react with 1 lb of refuse is 3.89 lb. To ensure complete combustion, 50% excess air is assumed. This can vary with the type of firing and whether the furnace is water cooled or not. Greater amounts may be used to dilute the products of combustion and control furnace temperature, but this air is not actually part of the combustion process. The total combustion air to be used with 50% excess air, regardless of how it is admitted to the burning zone (under- or overfired air), is computed as $1.5 \times 3.89 = 5.84$ lb dry air per pound of refuse. This computation can be checked against Figure 6 if 5,000 Btu/lb refuse is converted to million Btu by dividing 5,000 into 1,000,000. It is then determined that there are 200 lbs of refuse per million Btu. Introducing this to the preceding calculation (5.84×200), 1,168 lb of air per million Btu are to be used with 50% excess air. This compares with 1,070 lb. air per 10^6 Btu from Table 9, indicating a required air quantity of 8.4% below the computed estimate.

Another quick method of estimating the theoretical air for a refuse fuel is determined by estimating heating value (HHV) and substituting it into the formula

$$T \cdot A = 7.7 \times 10^{-4} \times \text{HHV} = \text{pounds of air per pound of refuse} \tag{6}$$

The value obtained, 3.85 lb/lb, compares well with the computed value above. The constant in this equation (7.7×10^{-4}) varies somewhat with the variations in refuse composition. These results are good for rough design estimates.

The quantity of flue gas produced per unit mass of refuse in the above example can be determined from the foregoing results:

Carbon burned*	0.300
Hydrogen	0.040
Oxygen	0.220
Moisture	0.240
Dry combustion air (50% excess)	5.840
Weight of flue gas	6.640 lb gas per lb. of refuse (plus air moisture)

* Note that for the carbon burned value, if combustion is incomplete, assume that the combustion exhibits a hydrogen preferential and the percent unburned combustible is entirely attributable to the carbon in the refuse. Once the unburned carbon is computed, correct the carbon burned to reflect this effect.

Products of Combustion

The 6.64 lb of flue gas per pound of refuse indicated is the total product minus air moisture when this is added to the total, wet products of combustion are obtained. The quantity of moisture in the flue gas can be computed as follows:

Assuming 0.020 lb moisture per pound of dry air used, $0.020 \times 5.84 = 0.117$ lb/lb of refuse. So, total products of combustion for this example are $6.64 + 0.12 = 6.76$ lb/lb of refuse.

Moisture, 0.24	= 0.24
Hydrogen, 0.04×9	= 0.36
Air moisture from above	= 0.12
	0.72

Total moisture in flue gas 0.72 lb/lb of refuse

The dry product of combustion can now be computed:

$6.76 - 0.72 = 6.04$ lb/lb of refuse fired

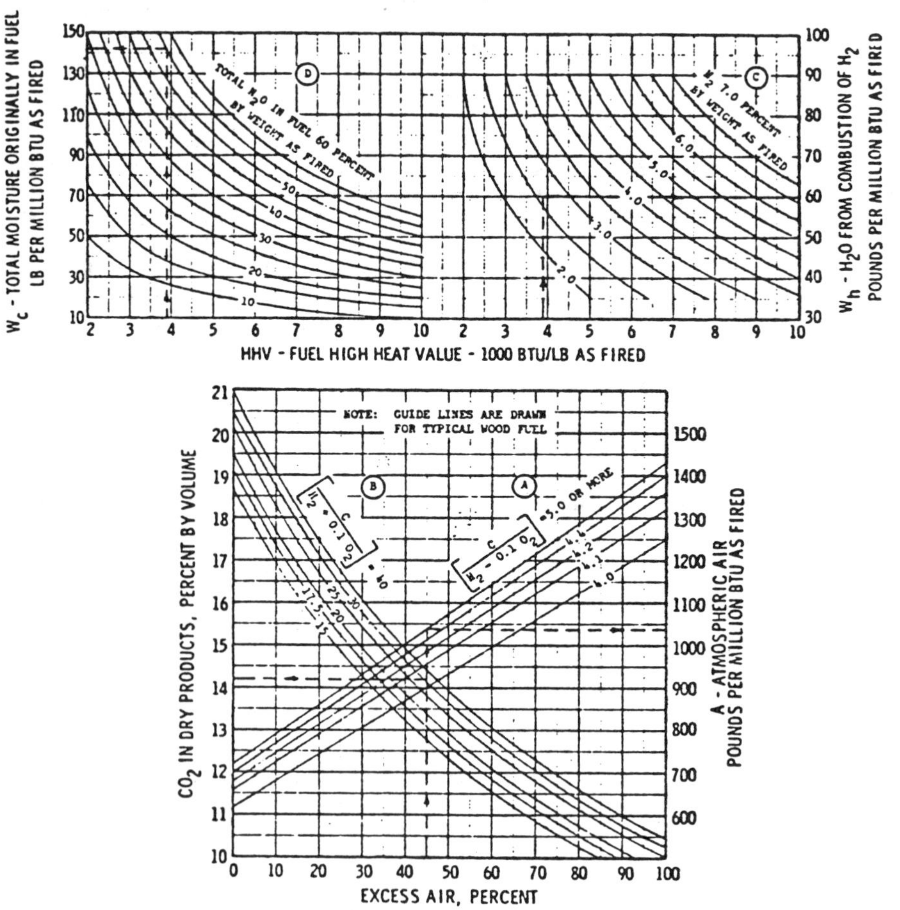

Figure 6. C, H_2 and O_2 on a combustible basis; plots are for wood and bagasse [7].

The combustion of the gaseous products of the refuse can be computed as follows from the assumed ultimate analysis:

$CO_2 = 0.30 \times 3.67 = 1.10$ lb/lb of refuse fired

H_2 (as computed above) $= 0.72$ lb/lb of refuse fired

$$O_2\left(\frac{3.89}{2} \times 0.232\right) = 0.45 \text{ lb/lb of refuse fired}$$

$N_2(3.89 \times 0.769 \times 1.5) = 4.49$ lb/lb of refuse fired

This assumption depends on the inlet air conditions. In this case temperature = 80° F and relative humidity = 90%. The moisture value is then read from the psychrometric chart.

Total wet products of combustion of 1 lb of the assumed refuse equals 6.76 lb/lb of refuse. Therefore:

$$CO_2 = \frac{1.10}{6.76} = 0.163 \text{ lb/lb of flue gas}$$

$$H_2O = \frac{0.72}{6.76} = 0.106 \text{ lb/lb of flue gas}$$

$$O_2 = \frac{0.45}{6.76} = 0.067 \text{ lb/lb of flue gas}$$

$$N_2 = \frac{4.49}{6.76} = 0.664 \text{ lb/lb of flue gas}$$

$$\text{Total} = 1.000 \text{ lb/lb of flue gas}$$

This can be expressed on a dry basis:

$$1.000 - 0.106 = 0.894 \text{ lb/lb of refuse (dry)}$$

$$CO_2 = \frac{0.163}{0.894} = 0.182 \text{ lb/lb of dry flue gas}$$

$$O_2 = \frac{0.067}{0.894} = 0.075 \text{ lb/lb of dry flue gas}$$

$$N_2 = \frac{0.664}{0.894} = 0.743 \text{ lb/lb of dry flue gas}$$

$$\text{Total} = 1.000$$

Volumetric composition of the dry products of combustion:

$$CO_2 = \frac{0.182}{44} = 0.00413 \qquad \frac{4.13}{32.87} = 0.126 = 12.6\%$$

$$O_2 = \frac{0.075}{32} = 0.00234 \qquad \frac{2.34}{32.87} = 0.071 = 7.1\%$$

$$N_2 = \frac{0.743}{28} = 0.02640 \qquad \frac{26.4}{32.87} = 0.803 = 80.3\%$$

$$\text{Moles of dry gas} = 0.03287 \qquad 100\%$$

If the CO_2 concentration is less than just indicated, the oxygen will be greater, indicating greater than 50% excess air should be used in the system. Conversely, a greater volume of CO_2 and less oxygen means that less than 50% excess air should be used.

Furnace temperature, an important parameter in the design of incinerators, is the temperature of the flue gases leaving the furnace. This can be approximated by determining an average specific heat and using it to compute the temperature.

The furnace exit temperature can be estimated from Figure 7. Using 24% moisture and 50% excess air used in the example, the temperature is found to be approximately 2,500° F.

The specific heat of CO_2 at constant pressure varies from 8.89 to 14.16 Btu/lm-mole°F between 80° F to 2,500° F. For an average of 11.53 Btu/lb-mole°F or 0.262 Btu/lb−°F.

Similarly for

Moisture	0.545 Btu/lb−°F
Oxygen	0.247
Nitrogen	0.274

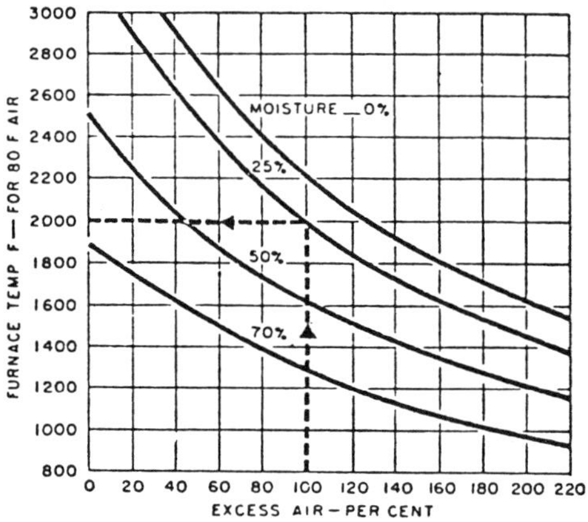

Figure 7. Plot of furnace gas temperature in a refractory furnace as a function of excess air and moisture level [7].

The temperature is computed as follows:

$$[\underset{CO_2}{(WC_p)} + \underset{H_2O}{(WC_p)} + \underset{O_2}{(WC_p)} + \underset{N_2}{(WC_p)}]\,(T_2 - T)$$

i.e., $\Sigma WC_p(T_2 - T)$

$$\begin{aligned} CO_2 &= 0.163 \times 0.262 = 0.0427 \\ H_2O &= 0.106 \times 0.545 = 0.0577 \\ O_2 &= 0.067 \times 0.247 = 0.0166 \\ N_2 &= 0.664 \times 0.274 = 0.1820 \\ \Sigma WC_p &= 0.2990 \end{aligned}$$

The furnace inlet temperature assumed here is 80° F.

The heat released and made available to the furnace per pound of refuse will be assumed as 4.70 Btu/lb. This assumes a 5% heat loss in incomplete combustion and through the furnace wall. Dividing by 6.76 lb of flue gas/lb of refuse, gives 703 Btu/lb of flue gas.

The estimated furnace exit temperature is:

$$(T_{exit} - 540) = \frac{703}{0.299} = 2350 \tag{7}$$

$$\begin{aligned} T_{exit} &= 2350 + 540 \\ &= 2890° \text{ R} \\ &= 2430° \text{ F} \end{aligned}$$

Additional discussions are given in Reference 9.

PARTICULATE CAPTURING MECHANISMS

Incineration operations produce a variety of emissions in the form of particulate matter and odorous and noxious gases. It is common practice therefore, to employ collection devices to decrease such emissions and to ensure maximum combustion efficiency. Because of the variable nature of the materials charged to incinerators and the changing modes of operation, the rate of particulate emissions can vary widely. For example, in the incineration of municipal wastes, typical emission values range from at 3 to 5 lb per ton to 10 to 25 lb per ton of refuse burned [10, 11]. There is general agreement, however, that the size of particulates is quite small. Flood [10] reports that the particles emitted from incinerators range from less than 5 micron (0.0002 in.) to about 120 microns (0.005 in.) in diameter. These fall into the category of fine dusts, as shown by the particle classification in Figure 8.

The removal of particles smaller than 50 microns is difficult and requires efficient collecting devices. Conventional particle capturing devices are based on gravity settling, inertia or momentum, filtration or electrostatic precipitation and agglomeration via sonic mechanical means to facilitate removal by increasing particle size. Both wet and dry media can be employed in most of these devices. To control or eliminate objectionable odors in circumstances where they cannot otherwise be readily handled, secondary combustion with or without catalysts, odor modification or masking and wet scrubbing are employed [13].

It is beyond the scope of this present chapter to discuss design and scale-up of particulate capturing and gas cleaning devices. We shall instead, concentrate on reviewing the fundamental mechanisms important to particle capture from gas streams and refer the reader to References 14 and 15 for design methodology of the actual unit operations.

Particles entrained in a gas stream are subject to several forces and actions (e.g., settling is induced due to the influence of gravity, particles collide with each other and agglomerate, impinge on obstacles such as walls, baffles, and liquid droplets, or attract small entities). By utilizing those forces and enhancing them with centrifugal, electrostatic, and sonic actions, separation of the particles from gas streams is achieved.

Particle dynamics is a complex subject, primarily due to the inability to adequately describe particle-particle interactions. We can, however, gain some insight on the dominant forces that must be accounted for in design by reviewing the behavior of single particles. Consider the limiting case of free fall of a single spherical particle. When there is relative motion between a fluid and a particle immersed in that fluid, such as with free-falling solids,

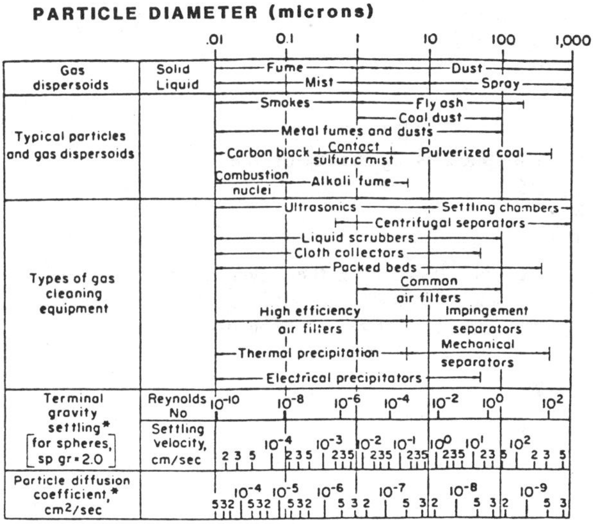

Figure 8. Particle size classification chart [12].

Table 10
Drag Coefficients for a Sphere

Reynolds Number Range	$10^{-4} \sim 10$ (Stokes Law)	$10 \sim 1{,}000$ (Transition)	$1{,}000 \sim 10^5$ (Newton Law)
Drag coefficient, C_D	24/Re	f(Re)	0.43
Setling Velcocity, u_t	$\frac{g\Delta\rho dp^2}{18\mu_f}$	$\left(\frac{4\Delta\rho d_p g}{3C_D\rho_f}\right)^{1/2}$	$1.7\left(\frac{\Delta\rho d_p g}{\rho_f}\right)^{1/2}$
Controlling resistance	Frictional	Friction plus Shape	Shape

From Cheremisinoff and Cheremisinoff [16].

drag forces are exerted on the particle. These forces are caused by skin friction and the differential pressure upstream and downstream of the particle resulting from the change in the streamlines of the gas flowing past the obstacle.

A free-falling particle accelerates until the frictional drag of the fluid surrounding the particle balances the gravitational force. From this point the particle continues to fall at a constant velocity, or at its terminal velocity. The terminal velocity of a free-falling sphere is expressed by the following force balance:

$$f = \frac{\pi}{g} C_D \rho_f d_p^2 U_t^2 V_p(\rho_p - \rho_f)_g \tag{8}$$

From when the terminal velocity is:

$$U_t^2 = \frac{8V_p(\rho_p - \rho_f)_g}{C_D \rho_f d_p^2 \pi} \tag{9}$$

where
- f = drag force
- C_D = drag coefficient
- ρ = fluid density
- d_p = particle diameter
- U_t = terminal velocity of particle
- V_p = particle volume
- ρ_p = particle density
- g = gravitational constant

Since the volume of a sphere is $V_p = (\pi d_p^3/6)$, then

$$U_t^2 = \frac{4}{3}\frac{d_p(\rho_p - \rho_f)_g}{\rho_f C_D} \tag{10}$$

The drag coefficient C_D is a function of the Reynolds number, which in turn is related to the diameter of the particle. The three ranges of Reynolds numbers considered in determining the drag coefficient are summarized in Table 10.

For settling in the laminar regime, frictional drag dominates, and from Table 10:

$$C_D = 24/Re = \frac{24\,\mu_f}{d_p U \rho_f} \tag{11}$$

Substituting for C_D in Equation 10 gives the terminal settling velocity for laminar conditions:

$$U_t = \frac{g d_p^2(\rho_p - \rho_f)}{18\,\mu_f} \tag{12}$$

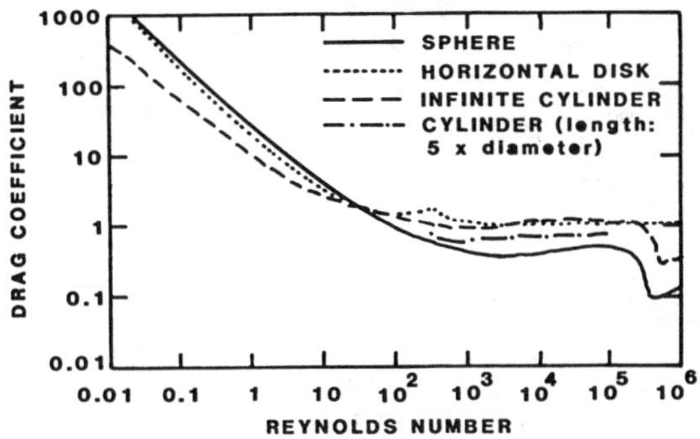

Figure 9. Plot of drag coefficients vs. Reynolds number.

Equation 11 is the Stokes law expression for the terminal velocity of a particle. The preceding relationship is based on a single sphere falling in a limitless gas expanse unhindered by other particles or boundary surfaces.

A plot of drag coefficient versus Reynolds number for spheres, disks, and cylinders is shown in Figure 9. Very small particles, whose diameters approach the mean free path of the gas molecules, fall faster than predicted by Stokes law. This occurs because the particles slip between the gas molecules. The Cunningham [17] correction factor is applied to Stokes law for such particles and results in the following expression for calculating terminal velocity:

$$U_t = \frac{g d_p^2 (\rho_p - \rho_f)}{18\, \mu_f}\left(1 + \frac{k\lambda}{d_p}\right) \tag{13}$$

where K is a constant ranging from 1.25 to 2.3 for different gases and particle sizes and λ is the mean free path of the fluid molecules. As shown in Table 11 corrections for small particles in air are significant, however, this correction approaches unity with increasing size.

Effect of Brownian Motion

Another factor that affects the settling of small particles is random Brownian motion resulting from particle bombardment by the gas molecules. The Brownian effect is significant for particles of 1 micron in size and smaller, and becomes increasingly important as the size of the particle decreases. A measure of the effect of Brownian motion is the displacement ratio, which is defined as the ratio of the random movement of a falling particle due to Brownian motion as compared with the distance it falls by gravity over the same time. The variation of the displacement ratio with size for particles falling in the air is shown in Table 12 [18]. A general formula expressing the displacement of the particles due to Brownian forces based on the theoretical derivation by Einstein is:

$$x = \left[\left(\frac{2RT}{N}\right)\left(\frac{t}{3\pi\mu_f d_p}\right)\left(1 + \frac{K\lambda}{d_p}\right)\right]^{1/2} \tag{14}$$

where
R = the gas constant
T = the absolute temperature
N = Avogadro's number

Table 11
Cunningham Correction Factors

Particle Diameter (microns)	Cunningham Correction
0.01	22.35
0.1	2.87
1.00	1.16
10.00	1.016
20.00	1.008

Mean free path of the fluid molecules, λ, is 6.53 × 10^{-6} cm.
From Zenz and Othmer [12].

Table 12
Displacement Ratio for Various Sizes of Particles in Air

Particle Size (Microns)	Displacement Ratio
0.1	17
0.5	0.45
1.00	0.09
5.00	0.002

From McCormic, Lucas, and Wells [18].

t = time
μ_f = gas viscosity
d_p = particle diameter
$(1+K\lambda/d_p)$ = the Cunningham correction factor

For settling in air at 20° C the equation is [20]:

$$x = 6.8\left[\frac{t}{d_p}\left(1+\frac{K\lambda}{d_p}\right)\right]^{1/2} \tag{15}$$

where x = average displacement (microns)
d_p = particle diameter (microns)
t = time (sec)

Particle Agglomeration

Agglomeration of micron and submicron particles into a cluster as a result of collision and adherence often can facilitate removal from a gas stream. The factors that effect agglomeration of small particles are particle size, the nature surface of the particles surfaces, the presence of absorbed gases on their surfaces, the rate of dispersion, humidity, and possibly the temperature and viscosity of the gas medium. As an approach to understanding the adhesion of particles, Bradley [21] and Hamaker [22] derived an expression for the adhesive force between two bodies based on London–van der Waals forces of attraction as follows:

$$f = \frac{(\pi^2 q_0^2 \lambda')(d_1 d_2)}{12x^2(d_1+d_2)} \tag{16}$$

where f = force of attraction (dynes)
q_0 = number of atoms per cubic centimeter of the substance
λ' = van der Waals constant

d_1, d_2 = diameter of particles (cm)
x = distance between particles (cm)

Brownian coagulation of smokes was studied intensively by Whytlaw et al. [23] where the following empirical expression was derived:

$$\frac{1}{n} - \frac{1}{n_0} = K_B t \tag{17}$$

where n = number of particles per c.c. at time t
n_0 = the original concentration
K_B = the agglomeration constant

Particle Adhesion

Adhesion of small particles to solid surfaces is another aspect of particle dynamics and is important in separation techniques such as filtration, impaction, and electrostatic precipitation. Factors that influence adhesion are particle size and shape, particle and surface material, surface roughness, temperature, and humidity of the ambient gas, time of particle-surface contact, and electrostatic forces. Corn [24] investigated the effects of several of these variables on the force of adhesion of Quartz and Pyrex-glass particles with diameters ranging from 20 to 90 microns on a flat Pyrex slide and on a glass microscope slide. His findings are summarized as:

1. Effect of particle size—The force of adhesion is proportional to particle size.
2. Effect of relative humidity—Adhesion increases with increasing relative humidity.
3. Effect of surface roughness—Adhesion decreases with increased surface roughness. Corn's experiments for adhesion of quartz particles to pyrex plates were correlated by the expression

$$S = 8.8 \times 10^3 e^{-0.053A} \tag{18}$$

   where S = average surface roughness in Angstroms
   A = percent of adhesion

4. Effect of particle-surface contact time—The force of adhesion was not significantly altered by the time of contact if the contact time was at least 5 minutes.

Impingement

Removal of dust particles by impingement is utilized principally in filtration devices, baffled chambers, and wet scrubbers. Separation of particles occurs because their inertia contained in a flowing gas exceeds that of the gas. Consider the example in Figure 10, where a gas moving at a velocity U flows past a cylinder of diameter D. Particles contained in

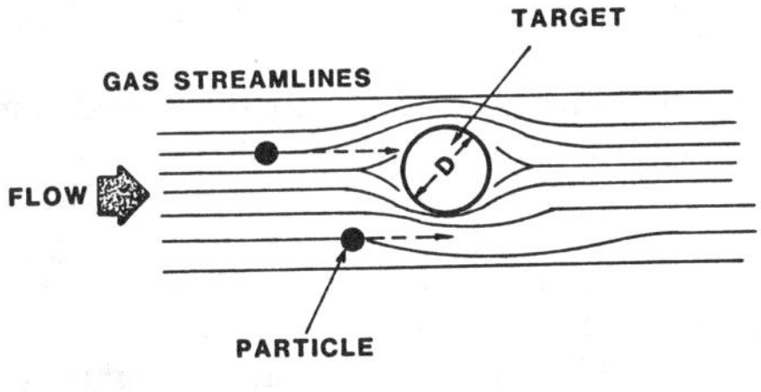

Figure 10. Illustrates impingement of particles onto a target.

the portion of gas stream will impinge onto the exposed target's surface area. The particle's inertia is defined as the change in motion and is equal to the product of the force which is attempting to create a change in the direction of motion and the time through which it acts. The generalized target efficiency equation for any single-stage impaction capture is:

$$\eta = \exp - \left\{ \frac{0.018}{D} \psi'^{0.5+R} - 0.6\, R'^{2} \right\} \tag{19a}$$

where η = effective target efficiency
D = collector surface diameter
R' = ratio of particle diameter to collector diameter
ψ' = dimensionless impaction parameter, defined as

$$\psi' = \left\{ \frac{C\rho_p v_r d_p^2}{18\mu D} \right\}^{1/2} \tag{19b}$$

where v_r is the relative velocity of the particles with respect to the target and C is a slip velocity correction factor for particles less than 1 μm in size:

$$C = 1 + \frac{3.45 \times 10^{-4} T}{d_p} \tag{20}$$

where T = absolute temperature (°R) and d_p is particle size in microns. Equation 20 is applicable over the Reynolds number range between 0.04 to 1.4.

Further discussions of particle dynamics are given by Lapple [25].

INCINERATION DESIGN PRINCIPLES

Incinerator Configurations

The empirical design methodology that has evolved from incinerator investigations has resulted in the development of two basic types of multiple-chamber incinerators, namely, the retort type and the in-line type. Other incinerator configurations include incinerators with vertically arranged chambers, L-shaped units, and units with separated chambers breeched together. These configurations are basically variations of the two basic designs. Each configuration has certain characteristics with respect to construction and operation, as well as limits to its applications. Figure 11 shows a cutaway view of a typical retort multiple-chamber incinerator. This type of unit derives its name from the return flow of effluent through the U-shaped gas path and the side-by-side arrangement of component chambers. Figure 12 shows a typical in-line design, so called because the various chambers follow one another in a line. In both types of multiple-chamber incinerators the combustion process proceeds in two stages, that is, primary or solid-phase, combustion in the ignition chamber, followed by secondary or gaseous-phase, combustion. The secondary combustion zone is composed of two parts: down draft or mixing chamber, and an up-pass expansion, or final combustion chamber. The gas flow and the combustion reactions in the two-stage process proceeds as follows.

The ignition-chamber reaction includes the drying, ignition and combustion of the solid refuse. As the burning proceeds the moisture and volatile component of the fuel are vaporized and partially oxidized in passing from the ignition chamber into the mixing chamber.

From the flame port the products of combustion and the volatile components of the refuse flow through the mixing chamber, at which point secondary air is introduced. The combination of elevated temperatures and the addition of combustion air, augmented by the mixing chamber, or secondary burners as necessary, assist in promoting the second stage

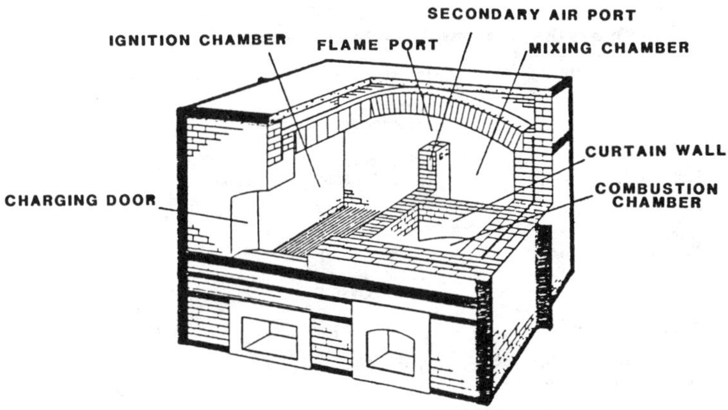

Figure 11. Retort-type multiple-chamber incinerator.

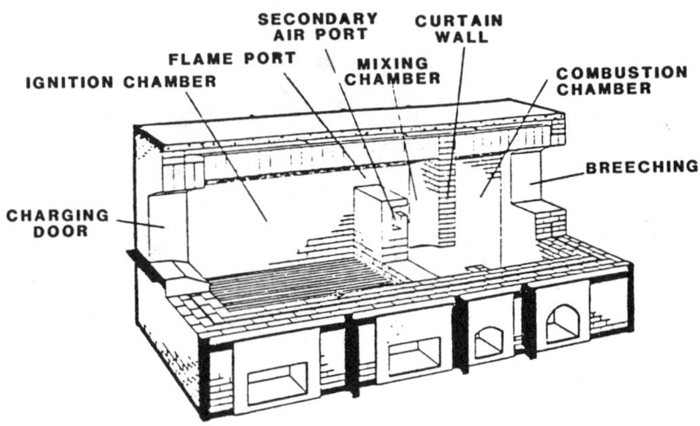

Figure 12. Shows in-line multiple-chamber incinerator.

of combustion process. Turbulent mixing, resulting from restricted flow areas and abrupt changes in flow direction, furthers the gaseous-phase reaction. In passing through the curtain-wall port from the mixing chamber to the final combustion chamber the gases undergo additional changes in direction, accompanied by expansion and final oxidation of combustible components. Fly ash and other solid particulate matter are collected in the combustion chamber by wall impingement and simple setting. The gases finally discharge through a stack or, in some installations, through a combination of a gas cooler (e.g., a water-spray chamber or scrubber) and induced draft system.

Other incinerator configurations in use for thermally decomposing chemical process industry wastes are as follows:

1. Multiple-hearth furnaces.
2. Fluidized-bed incinerators.
3. Liquid-waste incinerators.
4. Waste-gas flares.
5. Direct-flame incinerators.

Table 13
Operating Parameters for Incineration Process

Process	Temperature Range	Residence Time
Multiple-hearth	600–1,000° F (Drying zone) 1,400–1,800° F (Incineration)	0.25 to 1.5 hr
Fluidized-bed	1,400° F–1,800° F	Liquids and gases—seconds Solids—longer
Liquid incinerator	1,200–3,000° F	0.1–2 s
Direct flame	1,000–1,500° F	0.3–0.5 s
Catalytic combustor	600–1,000° F (1,500° F max)	1 s
Rotary kiln	1,500–3,000° F	Liquid and gases—seconds Solids—hours
Wet-air oxidation	300–550° F (1,500 psig)	10–30 min
Molten salt	1,500–1,800° F	0.75 s
Multiple chamber	1,500–1,800° F	Gases: seconds: solids—minutes
Pyrolysis	900–1,500° F	12–15 min

From Hitchcock [26].

Table 14
Incinerator/Waste Selection Chart

Waste Type	Rotary Kiln	Multiple Hearth	Fluidized Bed	Liquid Incinerator	Catalytic Combuster	Multiple Chamber Incinerator	Wet-Air Oxidation	Molten-Salt Incinerator
Granular homogenous	×	×	×					
Irregular bulky	×					×		
Low melting point (tar, etc.)	×		×	If material can be pumped				
Organic compounds with fusible ash consitients	×	×						×
Organic vapor laden				×	×			
High organic strength, toxic				×			×	
Organic liquids			×	×				×
Waste contains halogenated aromatic compounds (2,200° F)	×		×	If liquid				×
Aqueous organic sludges		×	×				×	

6. Catalytic combusters.
7. Rotary kilns.
8. Wet-air oxidation units.
9. Molten-salt incinerators.
10. Multiple-chamber incinerators.
11. Ship-mounted incinerators.

Each of these devices has advantages and disadvantages that must be evaluated prior to final process selection. Table 13 lists typical residence times and operating temperature-ranges for the various incineration processes. Classification of waste against the different incineration processes is presented in Table 14. The multiple-hearth, fluidized-bed, and liquid-waste incinerators can be operated under normal incineration and in a pyrolysis mode. A brief description of each major type follows.

Pyrolysis (air-starved) Incineration

Normal incineration requires 40%–100% excess air over the stoichiometric value. Pyrolysis is theoretically a zero-air indirect-heat process. However, in practical applications it is an air-starved process in that combustion is occurring with air levels less than stoichometric requirements for combustion. In pyrolysis waste organic compounds are distilled or vaporized to form combustible gas, which is discharged from the furnace. Heat for the process is provided by the partial combustion of the pyrolysis gas with the furnace and by the combustion of elemental carbon. The unoxidized portion of the combustible gas may be used as fuel in an external combustion chamber, with the resulting energy recovered by conventional-waste-heat-boiler technology. Fixed carbon levels in the furnace ash are higher for pyrolysis than for normal incineration. Pyrolysis is normally employed when the waste material has a high calorific content. Autogenous sludges with a high calorific value (Btu-to-moisture ratio greater than 3,500 Btu/lb of water) are best processed in oxygen-starved conditions. Normal incineration is used when a specific requirement exists for low level of fixed carbon in the furnace ash, and when the higher temperatures do not create ash fusion problems.

Multiple-Hearth Furnace

Figure 13 shows a multiple-hearth furnace incineration system. Such furnaces range from 6 to 25 ft dia, 12 to 75 ft high [27]. The diameter and number of hearths depend on the waste feed, the required processing time, and the type of thermal processing employed.

Normal incineration usually requires a minimum of six hearths, while pyrolysis applications require a greater number. Normally sludge or other waste material enters the furnace

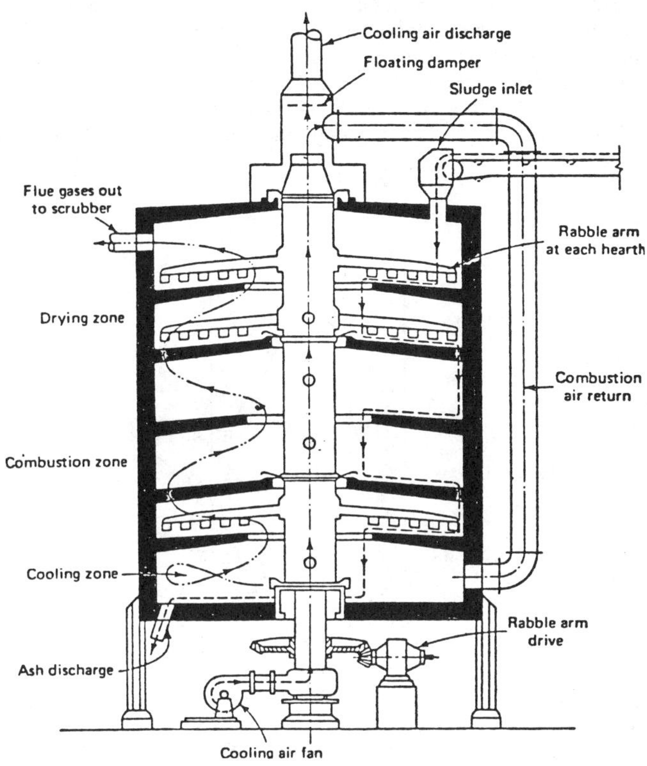

Figure 13. Shows multiple-hearth incinerator.

by dropping through a feed port located in the furnace top. Rabble arms and teeth, attached to a vertically positioned center shaft, rotate counterclockwise to spiral the sludge across the hearths and through the furnace. The waste drops from hearth to hearth through passages alternately located either along the periphery of the hearth or adjacent to the central shaft. Although the rabble arms and teeth all rotate in the same direction, additional agitation of the waste (back rabbing) is accomplished by reversing the angles of the rabble teeth. Waste retention time is controlled by the design of the rabble tooth pattern and the rotational speed of the central shaft.

Burners and combustion air ports are located in the walls of the furnace. Each hearth contains temperature sensors and controllers. The hearths are made of refractory, and the central shaft is cast iron and often insulated with castable refractories. The rabble arm and teeth are alloy castings. Materials vary in grade to suit waste requirements.

Cleaning of the multiple-hearth-furnace exhaust gas is usually accomplished by passing the hot gas from the furnace or from a waste heat reboiler through a precooler, where it is cooled to the adiabatic saturation temperature by spraying fine water droplets into the hot gas stream. Normally the adiabatic saturation temperature ranges from 170° to 190° F, depending on the water vapor content of the gas. The cool gas is then passed through a venturi throat into which additional water is sprayed. A 20- to 35-in. water column pressure-drop in the venturi throat provides the energy needed to collect the fine particulates on the water droplets in the gas stream.

The gas stream with entrained water then enters a particle disengagement unit, if additional cooling is required. The gas is now subcooled (approx. 120° F) and significantly stripped of water vapor, greatly reducing total volume. After passing through an induced draft fan, the effluent gas is discharged to the atmosphere.

Multiple-hearth furnaces operating in pyrolysis or oxygen-starved modes can handle feed materials with heat-release potentials greater than 25,000 Btu/lb of water and still maintain internal temperatures of 1,200° to 1,500° F. Temperatures inside the external combustion chamber can approach up to 3,000° F, greatly increasing the temperature driving force for energy recovery.

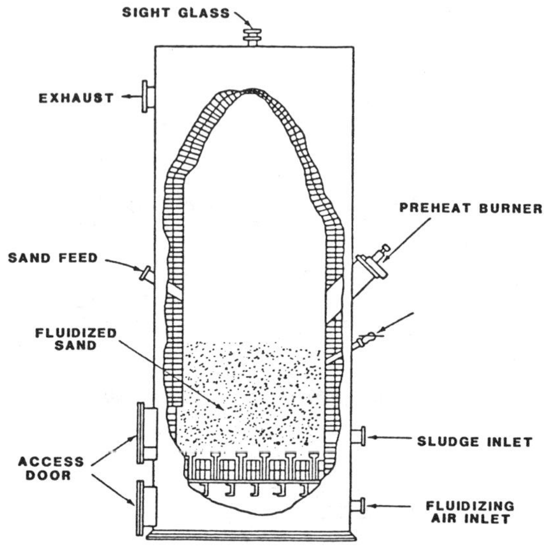

Figure 14. Shows a fluid-bed incinerator.

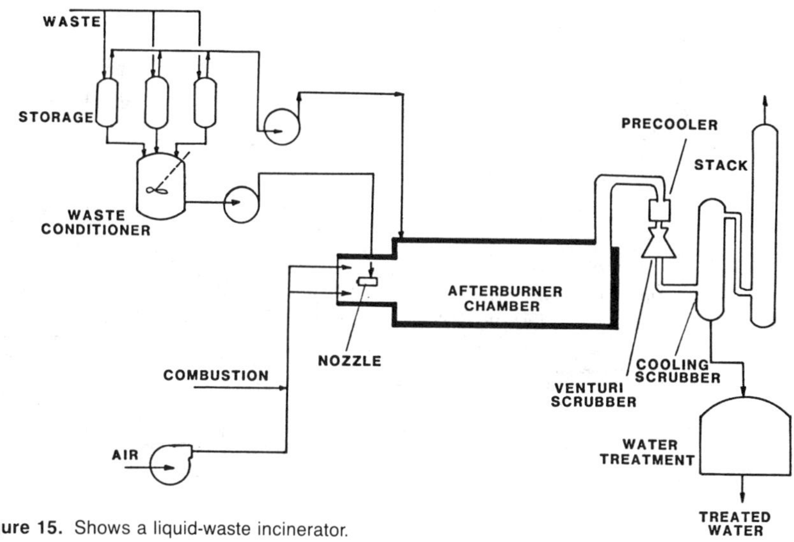

Figure 15. Shows a liquid-waste incinerator.

Fluidized bed

The fluidized-bed is a simple device consisting of a refractory lined vessel containing inert granular material (see Figure 14). Gases are blown through this material at a rate sufficiently high to cause the bed to expand and act as an ideal fluid. The fluidizing gases are injected through nozzles that permit flow up into the bed but restrict downflow of the material. Normally bed design restricts combustion to the immediate area of the bed. This maintains the "free board" area above the bed for separating the inert particles from the rising gases and for minor combustion of devolatilized components.

The hot gases leave the fluidized bed and enter heat-recovery or gas cleaning devices that are similar to those used with multiple hearth furnaces and other incinerators. Sludge of waste feed enters the bed through nozzles located either above or within the bed. Because of the intimate contact between combustion gases and the waste being burned, excess air for normal incineration is usually limited to approximately 40% above the stoichiometric air requirements for combustion of the waste.

Fluidized beds are subject to problems caused by low ash-fusion temperatures. These can be avoided by keeping the operating temperature below the ash fusion level or by adding chemicals that raise the fusion temperature of the ash to an acceptable level.

Liquid-Waste Incinerators

The liquid-waste incinerator is probably the most flexible, and certainly the most labor free. Here the feed waste acts as a liquid and exhibits a viscosity less than 10,000 SSU (see Figure 15). The heart of the system is a waste atomization device. Because a liquid combustion device is essentially a suspension burner, efficient and complete combustion is obtained only if the waste is adequately divided or atomized and mixed with the oxygen source. Atomization is usually achieved either mechanically using a rotary cup or a pressure atomization system. The burner nozzle is mounted at one end of the refractory-lined chamber and exhaust gases exit from the other end to gas-cleaning equipment. Liquid waste

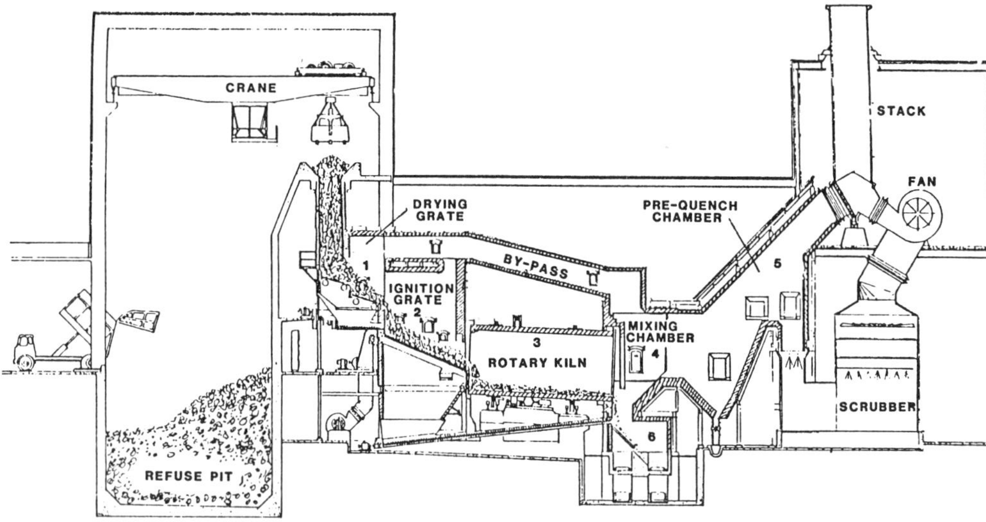

Figure 16. Shows a rotary kiln incinerator.

incinerator systems are equipped with waste storage and blending tanks to ensure a reasonably steady and homogeneous waste flow. The tank system is equipped to burn the liquid-waste fumes. See Reference 28 for further discussions.

Gas Incinerators

Waste gases are incinerated using direct-flame flare and catalytic combustors. The direct-flame incinerator is effective when the waste gas is not classified as a hazardous waste and has sufficient hydrocarbon content to act as a self-supporting fuel requiring only an oxygen source. Flares have reasonable combustion efficiencies but are unable to maintain hazardous waste. Flares are located on the ground or are elevated. Usually elevated flares are used due to their inherent safety benefits. It is essentially an open pipe through which the gases are passed and combusted using ambient oxygen. Steam is often used as an atomization medium to promote complete and smokeless combustion. Pilot burners are normally mounted on the flare for ignition purpose. These systems are described in detail in the next chapter.

Catalytic incinerators are considered for operation with waste containing hydrocarbon levels that are less than 25% of the lower explosive limit. When the waste gas contains sufficient heating value to cause concern about catalyst burnout, the gas may be diluted by atmospheric air to ensure operating temperatures within the operating limits of the catalyst. However, the waste gas usually contains combustible materials at levels far below those required to support autogenous combustion and usually must be preheated to the catalytic reaction temperatures. Catalytic-combustion systems often produce clean heated gas as a product and are well suited for waste heat recovery units. Such units significantly reduce the preheat fuel requirement. In addition to the temperature limitation, catalytic combustion incinerators are also sensitive to poisons, such as heavy metals, phosphates, arsenic compounds, and compounds of halogens and sulfur.

Rotary Kilns

The rotary kiln (Figure 16) is a cylindrical, horizontal, refractory-lined shell that is mounted at a slight incline. Rotation of the shell causes mixing of the waste with the combustion air. The length-to-diameter ratio of the combustion chamber normally varies between 2/1 and 10/1, and speed of the rotation is normally in the range of 1 to 5 rpm. Range of combustion temperature is from 1,500° to 3,000° F. Residence times vary from several seconds to hours, depending on the waste.

Rotary kilns are especially effective when the size or nature of the waste precludes the use of other types of incineration equipment. Special waste materials such as glass bottles, cardboard boxes, discarded packing cases, paper and other unmanageable solid wastes are often co-incinerated in rotary kilns. Kilns are usually designed for batch feeding, which allows for a highly flexible incineration. Continuous feeding is also possible if a reasonably homogenous feedstream is available. Rotary kilns are normally totally refractory lined with no exposed metallic materials of construction, so they resist higher incineration temperatures than others. Temperatures approaching 3,000° F, although certainly unusual, are within the capabilities of rotary equipment.

The unit is fired by conventional, as well as liquid-waste burner which enables the installation to double as a liquid-waste incinerator. Waste is delivered to the facility by dump trucks and deposited in a refuse storage area. The wastes are transferred to the shredder by a monorail clamshell and delivered to a ram feeder for introduction to the 1,600° F rotary incinerator. The exhaust gases from the facility pass through an afterburner chamber to ensure complete destruction of organic compounds, and then through a precooler, venturi scrubber, and packed column prior to discharge to the atmosphere. More detailed discussions are given in Reference 24.

Wet-Air Oxidation

This process (see Figure 17) operates on the principle that the rate of oxidation of organic compounds is significantly increased at higher pressure. Thus, by pressurizing an aqueous organic waste (pressures approach 1,500 psi), heating it to an appropriate temperature and

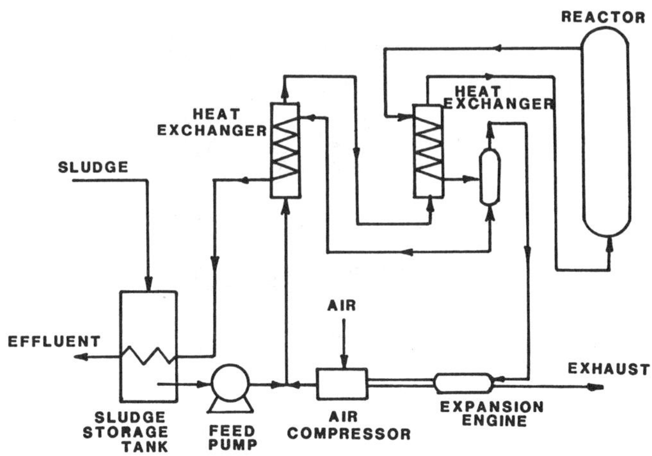

Figure 17. Illustrates the wet oxidation process.

then introducing atmospheric oxygen, an incomplete liquid-phase oxidation reaction is produced, which destroys most of the organic compounds. The process exhibits varying levels of combustion efficiency, depending upon the characteristics of the waste. It is often used as a pretreatment step to destroy toxic compounds before conventional biological wastewater treatment. Oxidation efficiency ranges between 600 to 100%. The recovery is done in a countercurrent heat exchanger. The process becomes thermally self-sufficient when the chemical oxygen demand of the influent waste reaches a level of 20,000 to 30,000 mg/L [30].

Molten-Salt Incinerators

This system is shown in Figure 18. Usually a molten-salt bath is composed of approximately 90% Na_2CO_3 and 10% Na_2SO_4 and is designed for operation in the range of 1,500° to 1,800° F. Sometimes K_2CO_3 is used for even lower incineration temperatures. The use of reactive salts, such as the eutectic mixtures NaOH-KOH and Li_2CO_3-K_2CO_3 produces the additional benefit of entrapping potentially toxic materials such as heavy metals (i.e., Hg, Pb, Cd, As, Se) [31]. This reduces the need of a pollution-control device. The spent salt often can be regenerated or may be land-disposed. Start-up and support fuels include gas, oil, and coal.

Waste such as free-flowing powders and shredded materials may be directly fed to molten-salt incinerators. Waste liquids may be sprayed into the combustion air and fed to the unit.

Multiple-Chamber Incinerators

Multiple-chamber incinerators are generally classified as retort types and in-line types. Design aspects of this system are described in the remainder of this chapter.

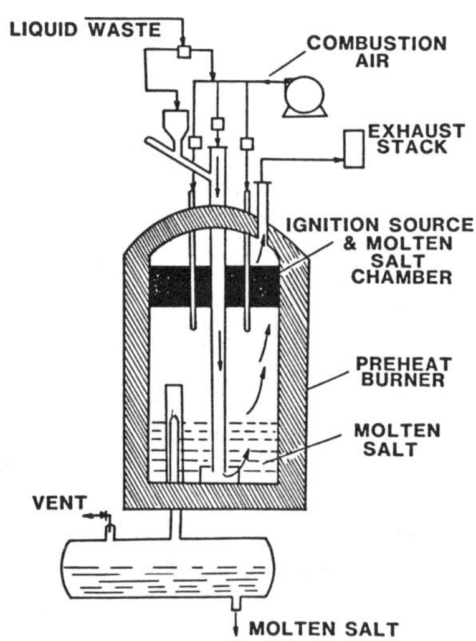

Figure 18. Shows features of a molten-salt incinerator.

Ship-Mounted Incinerators

To dispose of organics or organometallics that cannot be burned in conventional incinerators, ship-mounted incinerators are used. However, most of the companies that offer custom incineration are now faced with increasing U.S. EPA scrutiny of their effluents. One viable alternative may be incineration ships that burn hazardous and toxic wastes on the high seas.

The remainder of this chapter deals with specific design aspects of multiple-chamber incinerators for solid waste reduction.

In the case of refuse burning, the heterogeneous nature of the materials introduces a complication in that a large number of physical and operating variables exist and can therefore, only be effectively handled on an empirical basis. Theoretical treatment of the complex reactions taking place in some combustion process is, as yet, incomplete, but the empirical art of combustion engineering has developed to an advanced state. The principles

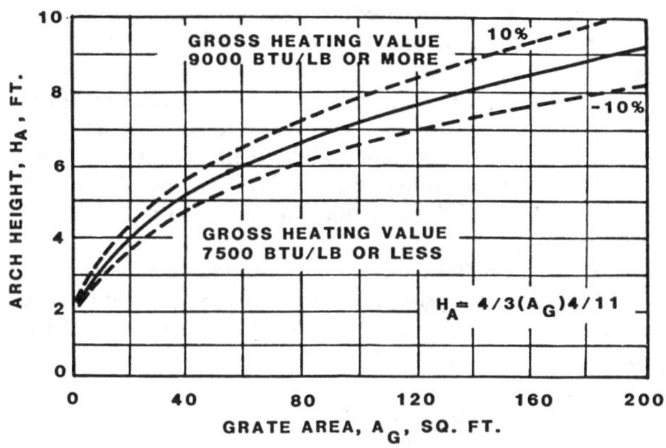

Figure 19. Plot of relationship between arc height and grate area for multiple-chamber incinerators.

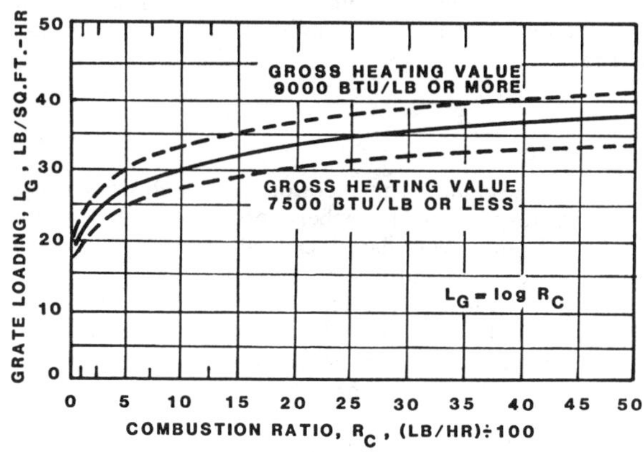

Figure 20. Plot of relationship between grate loading and combustion rate for multiple-chamber incinerators.

of solid-fuel combustion that in general apply to incineration and the basic prospects for combustion efficiency include the following [32]:

1. Air and fuel must be in proper proportion.
2. Air and fuel especially combustible gases, must be mixed adequately.
3. Temperature must be sufficient for ignition of both the solid fuel and the gaseous components.
4. Furnace volume must be large enough to provide the retention time needed for complete combustion.
5. Furnace proportions must be such that ignition temperatures are maintained and fly-ash entrainment is minimized.

Each of these is discussed in relation to the principal design components of incinerators.

Ignition Chamber

The ignition mechanism in a refuse-incineration process must be basically one of fuel-bed surface combustion. This is achieved by predominant use of overfire combustion air and minimum use of underfire air. In this context, overfire air is combustion air admitted into the ignition chamber at some point above the pile refuse. Such air is generally furnished through air-supply ports located in, or adjacent to, the charging door on the front wall of the incinerator. Underfire air is combustion air introduced into the ashpit beneath the fuel bed through airports located on, or adjacent to, the ashpit-cleanout doors.

By restricting the introduction of underfire air, relatively low fuel-bed temperatures are maintained, and entrainment of solid particulate matter in the effluent is minimized. If fuel-bed surface combustion through the use of overfire air is to be accomplished, the charging door must be located on the front wall of the ignition chamber or at the end of the chamber farthest from the flame port, because this is where the fresh charge of refuse is introduced. This method of introducing overfire air results in a movement of combustion air concurrent with the travel of the effluent, which has proved to be desirable for efficient combustion.

The emission of solid and liquid particulate materials in combustion effluents from multiple-chamber incinerators is principally a function of the mechanical and chemical processes taking place in the ignition chamber. The fundamental relationships to be considered in evaluating primary-combustion-chamber-parameters are length-to-width ratio, arch height, and grate loading. Formulas governing ignition-chamber design are postulated from data obtained from tests of units of varying proportions operated at maximum combustion rate.

Length-to Width-Ratios

In the retort type of multiple-chamber incinerator with rated design capacities of up to 500 lb/hr, satisfactory operating results have been obtained with length-to-width ratios varying from 2.0:1 to 2.5:1. In units with design capacities in excess of 500 lb/hr, optimum results are obtained with a length-to-width ratio of 1.75:1.

Although no sharp distinction has been established, optimum performance for burning rates ranging from 25 to 750 lb/hr has been obtained with the retort type of incinerator. Above this capacity it is difficult to obtain desirable combustion characteristics, proper flame travel, and combustion-air distribution and still retain the correct relationship of other critical design parameters.

In the in-line type optimum length-to-width ratios commence at 1.65:1 for the 750 lb/hr capacity unit and diminish linearly to 1.1:1 for incinerators with design capacities of 2,000 lb/hr or more.

Arch Height

Arch height has been observed to have an appreciate affect on contaminant discharges. Incinerators burning similar refuse and with similar grate areas but with different arch heights, have varying combustion rates and contaminant discharge characteristics.

Grate Loading

Acceptable grate loadings range from 15 to 25 lb/(ft^2) (hr) for incinerators with burning rates of 25 to 750 lb/hr. Practical considerations for charging and stoking in the smaller incinerators usually results in proportionally larger grate areas and lower grate loadings. For burning rates in excess of 750 lb/hr acceptable grate loadings range from 25 to 35 lb/(ft^2) (hr). A number of refuse incinerators have been designed to operate with grate loadings of 50 to 70 lb/(ft^2) (hr).

Optimum, values of arch heights and grate areas may be determined from Figures 19 and 20, respectively, by using the gross heating values of the refuse to be burned and hourly burning rates. The curves shown in each plot range over an upper gross heating value of 9,000 Btu/lb or more to a lower gross heating value of 7,500 Btu/lb or less. Interpolation between the upper and lower limits gives the correct arch height and grate area for refuse with a gross heat error between these values of $\pm 10\%$, which is considered to be reasonable. The relationship between arch height to grate area can be calculated from the empirical equation:

$$H_A = 4/3\,(A_G)^{4/11} \tag{21}$$

where H_A = arch height; defined as the average distance between the top of the grate to the arch in feet

A_G = grate area or horizontal cross-sectional area of the ignition chamber in square feet

Secondary Combustion Chamber

The gas-phase or secondary, combustion reactions in a multiple-chamber incinerator are controlled largely by the flame port, mixing chamber, and secondary combustion chamber. The relationship of these parameters is usually determined by certain limiting gas velocities or by the unit volume requirements for the type and quantity of refuse burned. This entails a determination of the combustion-air requirements, weight, and volume of combustion products to be handled, as well as combustion temperatures, heating values, moisture content, percent of combustibles, and ash content.

The primary effect of proper design of the secondary mixing and combustion chambers and ports has improved the combustion of volatile and solid components. The last chamber, or final combustion chamber, is intended to serve a dual purpose—it allows completion of the gas-phase combustion and also serves as a fly-ash settling chamber. The flame port is designed so as to provide a high-gas-velocity zone. Gases leaving the flame port make an abrupt change in the direction and are then expanded in the mixing chamber. The purpose is to promote turbulence and mixing of the effluent from the ignition chamber with secondary combustion air in a high temperature-flame zone.

Optimum design velocities determined for gas flows in the secondary combustion chambers and ports are summarized as follows:

- *Flame port*—Gas velocities at the flameport, or area provided above the bridge wall, should range from 45 to 65 ft/sec.
- *Mixing chamber*—Gas velocities in the mixing chamber may range from 20 to 35 ft/sec.

- *Curtain-wall port*—The gas passage beneath the curtain-wall port should be designed to give gas velocities ranging from 10 to 15 ft/sec.
- *Secondary combustion chamber*—Gas velocities in the secondary combustion chamber should not exceed 10 ft/sec.

Combustion Air

The final relationship to be considered in evaluating multiple-chamber design parameters affecting combustion efficiency and contaminant discharge are those of combustion-air supply and distribution, and the requirements for burners to supply auxiliary heat.

Sufficient air must be supplied to the incinerator to allow for the maximum combustion of the oxidizable materials charged and it must be introduced and distributed in such a manner as to reduce discharge contaminants to a minimum. This can be accomplished by providing air-port areas to allow for the introduction of at least 100 percent in excess of the amount of air theoretically required for complete combustion of the refuse. The air supply should be distributed so that not more than 10% of the total volume of air is introduced underfire or beneath the fuel bed. Approximately 70% of the air should be introduced overfire or above the fuel bed and 20% admitted through the secondary air-ports into the mixing chamber. Where excess air is required to control incinerator temperatures, the introduction of this air should be made through additional overfire or secondary air-ports. Secondary combustion air should be provided through controllable air-inlet ports located adjacent to the bridge wall.

It is not unusual for multiple-chamber incinerators to operate in a range of 100%–300% excess combustion air. Air-port areas ordinarily are sized to deliver about 50% of the total air required in the combustion process, that is, theoretical air plus about 100% excess air. The balance of excess air enters as leakage through expansion joints, through the charging door when refuse is introduced, and at other points of air leakage.

Auxiliary Heat

Often the inclusion of burners in the design of a unit is made in anticipation of human error. The use of burners to provide auxiliary heat in multiple-chamber incinerators is normally not required for the burning of Type 1 refuse.

The higher moisture content of Type 2 refuse makes it much more difficult to burn. This often requires the addition of a burner in the ignition chamber as well as a secondary burner in the mixing chamber. The function of a burner is of course to provide auxiliary heat, when and if needed. The higher the moisture content of the refuse, the greater are the auxiliary heat requirements. The determination of the size of burners required should be based on the highest moisture content of refuse expected to be burned in the unit. Auxiliary burners should be fired with either natural gas or manufactured gas.

Stack Draft

The normal method of producing a negative pressure within the ignition chamber is by the use of a natural-draft stack that utilizes the buoyancy of the hot flue gases. Draft produced in this manner is directly related to the height of the stack and to the difference in the reciprocals of the absolute temperature of the flue gas and ambient air. The theoretical draft requirements of a stack can be calculated from the following formula: [33]

$$D_t = 0.52\,PH\left(\frac{1}{T_0} - \frac{1}{T_a}\right) \qquad (22)$$

where D_t = theoretical draft in inches of water column
P = barometer pressure (psi)
H = height of the stack above breeching (ft)
T_0 = ambient temperature (°R)
T_a = average stack temperature (°R)

The velocity of the effluent in the stack and the cross-sectional area of the stack affect the usable or available draft. As the velocity within the stack increases or its cross-sectional area decreases, the losses due to friction increase proportionately. This reduces the available draft. Draft losses can be calculated from the following formulas [34]:

Frictional losses (round stacks):

$$F_s = \frac{0.008\ HV^2}{(D)\,(T)} \tag{23}$$

where F_s = friction loss (in. water column)
H = height of stack above breeching (feet)
V = velocity (ft/sec)
D = stack diameter (ft)
T = temperature (°R)

Frictional losses (rectangular stacks):

$$F_s = \frac{0.002\ HV^2}{(m)\,(T)} \tag{24}$$

where m = hydraulic radius (ft)

Expansion losses, which are usually negligible, can be estimated from

$$F_\ell = \frac{0.012\ V^2}{T} \tag{25}$$

where F_ℓ = headloss (in. water column)

The draft developed by the stack must be sufficient to overcome frictional losses and leave a net draft available of from 0.005 to 0.10 in. water column negative pressure for the inspiration of combustion air through the primary air-ports. The range of total available stack draft required is from 0.12 in. water column for a 50 lb/hr incinerator to 0.3 in. water column for a 2,000 lb/hr unit. Table 13 lists basic parameters, evaluation factors, and equations.

Design Calculations

In order to use the design factors given in Table 15, calculations are needed to organize incinerator data into a usable form. The calculations fall into three general categories:

1. Combustion calculations based on refuse composition, assumed air requirements, and estimated heat losses.
2. Flow calculations based on the properties of the products of combustion and assumed gas temperatures.
3. Dimensional calculations based on simple mensuration and empirical sizing equations.

The following assumptions are applied to the design calculations:

1. The burning rate and average refuse composition are assumed to be constant. The exception occurs when extremes in material quality and composition are encountered, in which case the most difficult burning condition is assumed.

2. The average temperature of the combustion products is determined through normal heat-balance calculations. Heat losses due to radiation, refractory heat storage and residue heat content may be assumed to average 20% to 30% of the gross heating value of the refuse during the first hour of operation.
3. The overall average gas temperature should be about 1,000° F when calculations are based on 300% excess combustion air and the heat-loss assumptions previously given.
4. The temperatures used in checking gas flow velocities are approximations of the actual temperature gradient in the incinerator since the products of combustion cool in passing from the flame port to the stack outlet.
5. In draft velocities in the combustion air-ports (overfire, underfire, and secondary) are assumed to be equal, with a velocity pressure of 0.1 in. water column (which is equivalent to 1,265 ft/min). Draft systems should be designed in such a way so that the available draft in the primary combustion chamber is about 0.1 in. water column. Oversizing of adjustable air-ports ensure maintenance of proper air induction.
6. Air-ports must be sized for admission of theoretical air plus 100% excess air. The remaining air enters the incinerator through the open charging door during batch operation and through expansion joints, cracks around doors, etc. Supplementary computations are usually required in determining necessary auxiliary-gas-burner sizes, stack draft control, and auxiliary-fuel-line piping. If the moisture content of refuse is less than 10% by weight, burners usually are not required. A moisture content of 10%–20% normally indicates the necessity for installation of mixing-chamber burners, and a

Table 15
Multiple-Chamber Incinerator Design Factors

Item and Symbol	Recommended Value
Primary combustion zone:	
Grate loading, L_G	10 log Rc [lb/(hr) (ft^2)]; Rc equals the refuse-combustion rate in lb/hr
Grate area, A_G	Rc ~ L_G (ft^2)
Average arch height, H_A	$\beta(A_G)^{4/11}$ (ft)
Length-to-width ratio (approximate): Retort; In line	Up to 500 lb/hr, 2.5 : 1 to 2 : 1. Over 500 lb/hr, 1.75 : 1, diminishing from about 1.7 : 1 for 750 lb/hr to about 1.2 : 1 for 2,000-lb/hr capacity. Oversquare acceptable in units of more than 11 ft ignition-chamber length.
Secondary combustion zone:	
Gas velocities:	
Flame port at 1,000° F, V_{FP}	55 ft/sec
Mixing chamber at 1,000° F, V_{MC}	25 ft/sec
Curtain-wall port at 950° F, V_{CWP}	About 0.7 of mixing-chamber velocity
Combustion chamber at 900° F, V_{CC}	5 to 6 ft/sec; always less than 10 ft/sec
Mixing-chamber downpass length, L_{MC}, from top of ignition-chamber arch to top of curtain-wall port	Average arch height (ft)
Length-to-width ratios of flow cross sections:	
Retort, mixing chamber, and combustion chamber	Range—1.3 : 1 to 1.5 : 1
In line	Fixed by gas velocities due to constant incinerator width
Combustion air:	
Air requirement batch-charging operation	Basis: 300% excess air: 50% air requirement admitted through adjustable ports: 50% air requirement met by open charge door and leakage
Combustion-air distribution:	
Overfire-air ports	70% of total air required
Underfire-air ports	10% of total air required
Mixing-chamber air ports	20% of total air required
Port sizing, nominal inlet-velocity pressure	0.1-in. water gage
Air-inlet ports oversize factors:	
Primary air inlet	1.2
Underfire-air inlet	1.5 for over 500 lb/hr to 2.5 for 50 lb/hr
Secondary air inlet	2.0 for over 500 lb/hr to 5.0 for 50 lb/hr
Furnace temperature (average temperature, combustion products)	1,000° F
Auxiliary burners (normal-duty requirements):	
Primary burner	3,000 to 10,000 Btu per lb of moisture in the refuse
Secondary burner	4,000 to 12,000
Draft requirements:	
Theoretical stack draft, D_T	0.15 for 50 lb/hr 0.30 for 1,000 lb/hr, uniformly increasing between sizes 0.35 for 2,000 lb/hr
Available primary-air-induction draft, D_A (assume equivalent to inletvelocity pressure)	0.1-in. water gage
Natural-draft stack velocity, V_S	Less than 30 ft/sec at 900° F

moisture content of more than 20% usually indicates that ignition-chamber burners also must be included.

The following example paraphrased from Williamson et al. [35] illustrates the necessary computations for incinerator designs.

Problem Design a multiple-chamber incinerator to burn 100 lb/hr of paper. The moisture content of the paper is 15%. Table 16 provides chemical properties and combustion data for cellulose-based materials.

Solution

1. *Combustion of refuse.*

   Dry combustibles

   (100 lb/hr) (0.85) = 85 lb/hr

   Moisture

   (100 lb/hr) (0.15) = 15 lb/hr

2. *Gross heat of combustion.* Gross heating value of dry paper is 7,590 Btu/lb, thus

   (85 lb/hr) (7,590 Btu/lb) = 645,200 Btu/hr

Table 16
Chemical Properties and Combustion Data for Paper, Wood, and Garbage

	Material							
	Sulfite Paper[a]		Average Wood[b]		Douglas Fir[c]		Garbage[d]	
Analysis								
Constituent:								
Carbon	44.34		49.56		52.30		52.78	
Hydrogen	6.27		6.11		6.30		6.27	
Nitrogen			0.07		0.10			
Oxygen	48.39		43.83		40.50		39.95	
Ash	1.00		0.42		0.80		1.00	
Gross heating value[e] (Btu/lb)	7,590		8,517		9,050		8,820	
Combustion Data								
	Cubic Feet	Pounds	Cubic Feet	Pounds	Cubic Feet	Pounds	Cubic Feet	Pounds
Constituent[f]:								
Theoretical air	67.58	5.165	77.30	5.909	84.16	6.433	85.12	6.507
40% saturation at 60° F	68.05	5.188	77.84	5.935	84.75	6.461	85.72	6.536
Flue gas with theoretical air:								
Carbon dioxide	13.993	1.625	15.641	1.816	16.51	1.917	16.668	1.935
Nitrogen	53.401	3.947	61.104	4.517	66.53	4.918	67.234	4.976
Water formed	11.787	0.560	11.487	0.546	11.84	0.563	11.880	0.564
Water (air)	0.471	0.023	0.539	0.026	0.587	0.028	0.593	0.029
Total	79.652	6.155	88.771	6.905	95.467	7.426	96.375	7.495
Flue gas with percent excess air as indicated:								
0	79.65	6.16	88.77	6.91	95.47	7.43	96.38	7.50
50.0	113.44	8.74	127.42	9.86	137.55	10.64	139.24	10.77
100.0	147.23	11.32	166.07	12.81	179.63	13.86	182.00	14.04
150.0	181.26	13.91	204.99	15.78	222.01	17.09	224.86	17.21
200.0	215.28	16.51	243.91	18.75	264.38	20.12	267.72	20.58
300.0	283.33	21.70	321.75	24.68	349.13	26.58	353.44	27.12

[a] Sulfite-paper constituents:
Cellulose ($C_6H_{10}O_5$) 84%
Hemicellulose ($C_5H_{10}O_5$) 8%
Lignin ($C_6H_{10}O_5$) 6%
Resin ($C_6H_{10}O_5$) 2%
Ash ($C_{20}H_{30}O_2$) 1%

[b] Kent, R. T., Mechanical Engineers Handbook, 11th. ed., Willey, N. Y. 1936, pp. 6–104.

[c] Kent, R. T., Mechanical Engineers Handbook, 12th. ed., Willey, N. Y. 1950, pp. 2–40.

[d] Estimated.

[e] Dry basis.

[f] Based on 1 lb.

3. *Heat losses.* From Table 16, 0.56 lb of water formed from the combustion of 1 lb of dry paper:

   Radiation, etc. (assume 20% loss)

   (645,200 Btu/hr) (0.20) = 129,040 Btu/hr

   Evaporation of contained moisture:

   (15 lb/hr) (1,060 Btu/lb) = 15,900 Btu/hr

   Evaporation of water from combustion:

   (0.56 lb/lb) (85 lb/hr) (1,060 Btu/lb) = 50,400 Btu/hr

   Total

   129,040 + 15,900 + 50,400 = 195,340 Btu/hr

4. *Net heat.*

   645,200 Btu/hr − 195,340 Btu/hr = 449,860 Btu/hr

5. *Weight of combustion products With 300% excess air.* From Table 16, 21.7 lb of combustion products from the combustion of 1 lb of paper with 300% excess air:

   Paper

   (85 lb/hr) (21.7 lb/lb) = 1,844 lb/hr

   Water

   (15 lb/hr)

   Total

   1,844 lb/hr + 15 lb/hr = 1,859 lb/hr

6. *Average gas temperature.* The specific heat of the products of combustion is approximately 0.26 Btu/(lb) (°F):

   $$T = Q/C_pM$$

   where T = temperature difference (°F)
   Q = net heat (Btu)
   C_p = specific heat (Btu/lb/°F)
   M = weight (lb)

   $$T = \frac{449{,}860 \text{ Btu/hr}}{(0.26 \text{ Btu/lb°F})\ (1{,}859 \text{ lb/hr})} = 930° \text{ F}$$

   $$T = T + 60° \text{ F}$$

   $$T = 930 + 60 = 990° \text{ F}$$

7. *Combustion-air requirements.* (Basis: use 300% excess air, 200% excess air is admitted through open charging door and leakage around doors, ports, expansion joints, etc.) From Table 16, 68.05 ft^3 of air is theoretically necessary to burn 1 lb of dry paper:

   $$\begin{aligned}(85 \text{ lb/hr})\ (68.05 \text{ ft}^3/\text{lb})\ (2) &= 11{,}680 \text{ ft}^3/\text{hr} \\ &= 192.3 \text{ ft}^3/\text{min} \\ &= 3.2 \text{ ft}^3/\text{sec}\end{aligned}$$

8. *Airport-opening requirements at 0.1 in. water column.* 1,265 ft/min is equivalent to a velocity pressure of 0.1 inch:

$$\text{Total} = \frac{(192.3\ \text{ft}^3/\text{min})\ (144\ \text{in}^2/\text{ft})}{1{,}265\ \text{ft/min}} = 22.0\ \text{in}^2.$$

Air supply from Table 15: Overfire 70%, underfire 10%, secondary 20%.

Overfire-air-port—air-port $(0.7)\ (22.0\ \text{in}^2) = 15.4\ \text{in}^2$.

Underfire-air-port—air-port $(0.1)\ (22.0\ \text{in}^2) = 2.2\ \text{in}^2$.

Secondary-air-port—air-port $(0.2)\ (22.0\ \text{in}^2) = 4.4\ \text{in}^2$.

9. *Volume of products of combustion.* From Table 16, 283.33 ft^3 of products of combustion are formed from the combustion of 1 lb of paper with 300% excess air. (Basis: 60° F and 300% excess air):

Paper

$$(85\ \text{lb/hr})\ (283.33\ \text{ft}^3/\text{lb}) = 24{,}080\ \text{ft}^3/\text{hr}$$

Water

$$\left(\frac{15\ \text{lb}}{\text{hr}}\right)\frac{379\ \text{ft}^3/(\text{lb})\ (\text{mole})}{18\ \text{lb/mole}} = 316\ \text{ft}^3/\text{hr}$$

Total

$$24{,}080 + 316 = 24{,}396\ \text{ft}^3/\text{hr} = 6.8\ \text{ft}^3/\text{sec}$$

10. *Volume of products of combustion through flame port.* Total volume minus secondary air:

$$6.8\ \text{ft}^3/\text{sec} - (3.2\ \text{ft}^3/\text{sec})\ (0.20) = 6.16\ \text{ft}^3/\text{sec}$$

11. *Flame-port area.* From Table 15, velocity is 55 ft/sec; thus:

$$\frac{(6.16\ \text{ft}^3/\text{sec})\ (1560°\ \text{R})}{(55\ \text{ft/sec})\ (520°\ \text{R})} = 0.34\ \text{ft}^2$$

12. *Mixing-chamber area.* From Table 15 velocity is 25 ft/sec; thus:

$$\frac{(6.8\ \text{ft}^3/\text{sec})\ (1460°\ \text{R})}{(25\ \text{ft/sec})\ (520°\ \text{R})} = 0.76\ \text{ft}^2$$

13. *Curtain-wall-port area.* From Table 15, velocity is 20 ft/sec; thus:

$$\frac{(6.8\ \text{ft}^3/\text{sec})\ (1410°\ \text{R})}{(20\ \text{ft/sec})\ (520°\ \text{R})} = 0.92\ \text{ft}^2$$

14. *Combustion-chamber area.* A velocity of 6 to 10 ft/sec is generally recommended for the combustion chamber (using 6 ft/sec); thus:

$$\frac{(6.8\ \text{ft}^3/\text{sec})\ (1360°\ \text{R})}{(6\ \text{ft}^3/\text{sec})\ (520°\ \text{R})} = 2.96\ \text{ft}^2$$

15. *Stack area.* A velocity on the order of 30 ft/sec is generally recommended for the stack (using 25 ft/sec); thus:

$$\frac{(6.8\ \text{ft}^3/\text{sec})\ (1360°\ \text{R})}{(25\ \text{ft/sec})\ (520°\ \text{R})} = 0.71\ \text{ft}^2$$

16. *Grate area.* From Figure 20, the grate loading for an average refuse rate is 18 lb/ft²/hr:

$$\frac{(100\ \text{lb/hr})}{18\ \text{lb/(ft}^2\text{) (hr)}} = 5.56\ \text{ft}^2$$

17. *Arch height.* From Figure 19,

    the arch height is 27 in.

18. *Stack height.* Examining Table 15, a theoretical draft value is obtained:

    $D_t = 0.17$ in. water column

    Applying Equation 22 to evaluate the height of the stack above breeching:

$$D_t = 0.52\ \text{PH}\left(\frac{1}{T_o} - \frac{1}{T_a}\right)$$

where D_t = draft (in. water column)
P = barometric pressure (psi)
H = height of stack above grates, (ft)
T_o = ambient temperature, (°R)
T_a = average stack temperature, (°R)

Hence,

$$H = \frac{D_t}{(0.52)\ (P)\ [(1/T_o) - (1/T_a)]}$$

$$= \frac{0.17}{(0.52)\ (14.7)\left(\frac{1}{520} - \frac{1}{1{,}360}\right)}$$

$$= 18.75\ \text{ft}$$

Problem. Calculate the size of a barometric damper to be installed in the breeching between the basement of a 100 lb/hr multiple-chamber incinerator and a 92 ft stack to limit the draft to 0.2 in. water column.

The following information is known: The stack is 8 in. square and has a cross-sectional area of 2.25 ft². The stack extends 92 ft above the breeching. The breeching itself is a 12-in.-diameter insulated straight duct that is 10 ft long.

Solution.

1. The theoretical draft in the breeching at the average gas temperatures is obtained from Equation 22:

$$D_t = 0.52\ \text{PH}\left(\frac{1}{T_0} - \frac{1}{T_a}\right)$$

where T_0 = ambient temperature (°R)
T_a = average stack temperature (°R)

For an average stack-gas temperature of 100° F:

$$D_t = (0.52)\ (14.7)\ (92)\left(\frac{1}{520} - \frac{1}{560}\right) = 0.092\ \text{in. water column}$$

The following tabulation of values is obtained from Equation 22:

Temperature (°F)	D_t (in. water column)	Temperature (°F)	D_t (in. water column)
100	0.092	400	0.53
200	0.29	500	0.62
300	0.43	600	0.69

2. The weight of air that must enter through the barometric damper to cool the combustion products from the multiple-chamber incinerator, neglecting heat losses is now evaluated. Assume a temperature of 300° F for the first calculation. Although neglecting heat losses will cause the damper to be somewhat oversized, the draft can still be regulated with the weights on the damper. However, with the damper undersized, the draft cannot always be controlled.

$$(W_A)(C_{p2})(T_2 - T_A) = (W_{pc})(C_{p1})(T_1 - T_2)$$

where W_a = weight of air entering through barometric damper (lb/sec)
W_{pc} = weight of stack gases (lb/sec)
C_{p1} = average specific heat of products of combustion from multiple-chamber incinerator over temperature range of T_1 to T_2 (Btu/(lb) (°F))
C_{p2} = average specific heat of air over temperature range T_2 to T_A (Btu/(lb) (°F))
T_2 = final temperature of stack gases (°F)
T_1 = average temperature of gases from multiple chamber incinerator (°F)
T_A = temperature of the air (°F)

The mass flow rate of the combustion products through the multiple chamber incinerator is 1,859 lb/hr, the temperature is 990° F, and the specific heat is 0.26 Btu/(lb) (°F). The specific heat of ambient air is 0.24 Btu/(lb) (°F).

$$(W_A \text{ lb/sec})(0.24 \text{ Btu/(lb) (°F)})(300° \text{ F} - 60° \text{ F})$$

$$= (1{,}859 \text{ lb/hr})\, 0.26 \left(\frac{\text{Btu}}{(\text{lb})\,(°\text{F})}\right)(990° \text{ F} - 300° \text{ F})$$

Hence,

$$W_A = 1.61 \text{ lb/sec}$$

3. Volume of air entering through the barometric damper (using the ideal gas law):

$$\text{Volume } (300° \text{ F}) = \frac{(1.61 \text{ lb/sec})\left(379 \dfrac{\text{ft}^3}{\text{mole}}\right)(760° \text{ R})}{(29 \text{ lb/mole})(520° \text{ R})}$$

$$= 30.8 \text{ ft}^3/\text{sec}$$

4. Volume of the products of combustion from the multiple-chamber incinerator at 300° F:

Volume of products of combustion is 6.8 ft^3/sec at 60° F

$$\text{Volume } (300° \text{ F}) = \frac{(6.8 \text{ ft/sec})(760° \text{ R})}{520° \text{ R}} = 9.93 \text{ ft}^3/\text{sec}$$

5. Total Volume flowing through the breeching and stack at 300° F:

$$\text{Total volume } (300° \text{ F}) = 30.8 + 9.93 \text{ ft}^3/\text{sec} = 40.73 \text{ ft}^3/\text{sec}$$

6. Velocity through breeching:

   Area of breeching $= 0.785$ ft

   $$\text{Velocity } (300^\circ \text{ F}) = \frac{40.73 \text{ ft}^3/\text{sec}}{0.785} = 51.9 \text{ ft/sec}$$

7. Frictional loss in the breeching (use Equation 23):

   $$F_B = \frac{0.008\,(H)\,(V)^2}{(D)\,(T)}$$

   where F_B = frictional loss (in. water column)
   H = length of breeching (ft)
   V = velocity (ft/sec)
   D = duct diameter (ft)
   T = temperature (°R)

   $$F_B = \frac{(0.008)\,(10)\,(519)^2}{(1)\,(760)} = 18.3 \text{ ft/sec}$$

8. Frictional loss in stack (use Equation 24):

   $$F_S = \frac{0.002\,(H)\,(V)^2}{(M)\,(T)}$$

   where F_S = loss (in. water column)
   M = hydraulic radius (ft)

   For rectangular cross section the hydraulic radius is

   $$M = \frac{(2.25 \text{ ft}^2)\,(12 \text{ in./ft})}{4(18 \text{ in.})} = 0.375 \text{ ft}$$

   $$F_S = \frac{(0.022)\,(92)\,(18.3)^2}{(0.375)\,(760)} = 0.216 \text{ in. water column}$$

9. Total frictional loss in breeching and stack:

   $= 0.283 + 0.216$ in. water column

   $= 0.499$ in. water column

10. Frictional losses (calculated by the preceding method with additional cooling air through the damper) for the assumed stack-gas temperatures, 400° and 500° F are given as follows:

Temperature (°F)	Frictional Loss (in. water column)
300	0.50
400	0.28
500	0.18

11. Stack gas temperature determination: A stack gas temperature of 375° F representing a difference of 0.2 in. water column between the theoretical draft and the frictional losses, is obtained from a plot of the preceding data as shown in Figure 21.

12. Weight of air entering through the barometric damper at 375° F:

$(W_A)(C_{P2})(T_2 - T_A) = (W_{PC})(C_{P1})(T_1 - T_2)$

where W_A = weight of air entering through barometric damper (lb/sec)
W_{PC} = weight of stack gases in (lb/sec)
C_{P1} = average specific heat of products of combustion from multiple-chamber incinerator over range of T_1 to T_2 (Btu/(lb) (°F))
C_{P2} = average specific heat of air over temperature range T_2 to T_A in (Btu/(lb) (°F))
T_2 = final temperature of stack gases (°F)
T_1 = average temperature of gases from multiple-chamber incinerator (°F)
T_A = temperature of air (°F)

$$(W_A \text{ lb/sec}) = [0.24 \text{ Btu/(lb) (°F)}]\,(375°\text{ F} - 60°\text{ F})$$
$$= (0.517 \text{ lb/sec})\,[0.26 \text{ Btu/(°F)}]\,(990°\text{ F} - 375°\text{ F})$$
$$W_A = 1.10 \text{ lb/sec}$$

13. Volume of air entering through the barometric damper:

$$\text{Volume (60° F)} = \frac{(1.10 \text{ lb/sec})\,(60 \text{ sec/min})\,(379 \text{ ft}^3/\text{mole})}{29 \text{ lb/mole}}$$
$$= 864 \text{ ft}^3/\text{min}$$

14. Area of barometric damper: The effective open area of a barometric damper is about 70% of its cross-sectional area. The area based on the calculated amount of air to be inspirated must therefore be increased accordingly. Velocity head at 0.2 in. water column and 60° F is 1,766 ft/min.

$$\text{Area} = \frac{(864 \text{ ft}^3/\text{min})\,(144 \text{ in.}^2/\text{ft}^2)\,(1.3)}{1{,}766 \text{ ft/min}} = 101 \text{ in.}^2$$

15. Diameter (sized to the nearest inch) of barometric damper:

$$\text{Area} = \frac{(\pi)\,(\text{diameter})^2}{4}$$
$$\text{Diameter} = \left[\frac{(4)\,(101)}{3.14}\right]^{1/2}$$
$$= 12 \text{ in.}$$

Gas Scrubbers

Gas Scrubbers employing water sprays are sometimes used in connection with incinerator operations in order to condition the effluent. Their purpose is to cool the effluent to a temperature low enough so that an induced-draft fan may be used to replace a stack or to remove large fly-ash particles. There are several basic considerations that are important in designing a gas scrubber. Perhaps the most important is that there be no carryover of water in the effluent discharged from the fan. In order to prevent this occurrence the scrubber design should provide sufficient residence time to completely vaporize the water entrained in the effluent. To accomplish this the water gas mixture should be retained within the scrubber for about 1 to 1 1/2 seconds and velocities should not exceed 15 ft/sec. The design parameters recommended for gas-scrubbers are summarized as follows:

1. The water rate to the scrubber should be about 1 GPM for every 100 lb/hr of rated incinerator capacity. This gives a water to gas ratio of about 1 GPM per 400 SCFM of effluent.

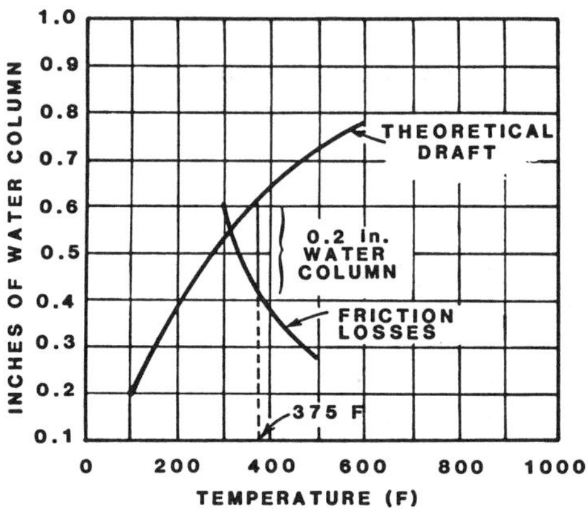

Figure 21. Draft at breeching of a multiple-chamber basement installation vs. average stack-gas temperature [4].

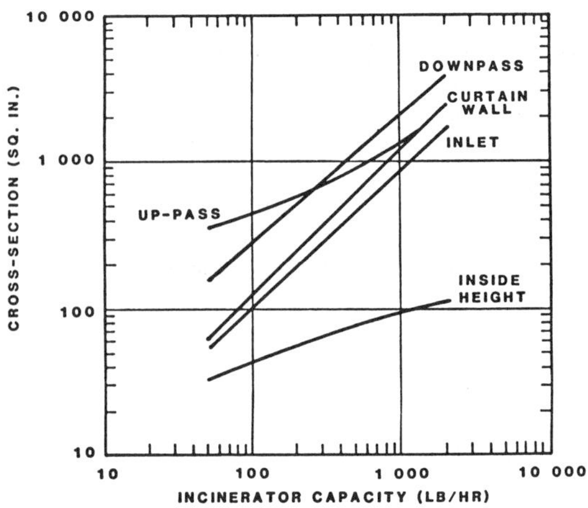

Figure 22. Plot for estimating the internal cross-section of scrubbers.

2. The exhaust fan should be designed to handle 700 CFM at 350° F for every 100 lb/hr incinerator capacity.
3. The fan should be designed to provide 0.5 in. static pressure for a 50 lb/hr incinerator, uniformly increasing to 1.5 in. for a 2,000 lb/hr incinerator. These static pressures should be developed with the fan operating at 350° F. The static pressure developed by fans operating at 350° F is approximately two-thirds of that developed when handling air at ambient temperatures. When selecting an induced-draft fan from the manufacturer's catalog, one should be chosen that will develop a static pressure 50% higher than those just given, or 3/4 inch for a 50 lb/hr incinerator to 2 1/4 inch for a 2,000 lb/hr incinerator.
4. The horsepower requirement of the fan should be based on the full capacity of the fan at ambient temperatures and not at 350° F.

The internal sizing of chambers and ports in a scrubber can be determined from the data given in Figure 22.

Refractory Walls, Linings, and Insulations

The minimum specifications for refractory materials used for lining the exterior walls of multiple-chamber incinerators are the following:

1. Type 1 and Type 2 refuse-burning service:
   - Firebrick, high heat duty—pyrometric cone equivalent not less than 32.5.
   - Castable refractory, not less than 120 lb/ft^3—pyrometric cone equivalent not less than 17.
2. Wood, sawdust, and other high temperature service:
   - Firebrick, superduty—pyrometric cone equivalent not less than 34.
   - Plastic—pyrometric cone equivalent not less than 34, not less than 130 lb/ft^3.

Minimum refractory thickness for lining exterior walls (including arches) of incinerators burning all classes of refuse are the following:

1. Up to and including 350 lb/hr capacity:
   Castable refractory or plastic—4 in, firebrick—4 1/2 in.
2. Above 350 lb/hr capacity: all refractory—9 in.

Stacks should be lined with refractory material with a minimum service temperature of 2,000° F. In low-capacity units the minimum lining thickness should be 2 1/2; in units larger than 350 lb/hr, 4 1/2 in.

Doors should be lined with refractory material with a minimum service temperature of 2,800° F. Units smaller than 100 lb/hr, should have door linings of 2-in. minimum thickness. In the size range of 100 to 350 lb/hr the linings should be increased to 3 inches. In the units with capacities from 350 to 1,000 lb/hr the doors should be lined with 4 inches of refractory. In the units of 1,000 lb/hr and more, linings should be 6 inches.

GLOSSARY

Auxiliary-fuel firing equipment—Equipment to supply additional heat, by the combustion of auxiliary fuel for the purpose of attaining temperatures sufficiently high (1) to dry and ignite the waste material (2) to maintain the ignition thereof, and (3) to promote complete combustion of combustible solids, vapors, and gases.

Breeching—The connection between incinerator and stack.

Bridge wall—A partition between the ignition and mixing chambers over which the product of combustion pass.

Btu—The amount of necessary heat to raise 1 lb. of water 1° F.

Damper—A manually or automatically controlled device used to regulate the rate of flow of air or other gases.

Draft—The pressure difference existing between the incinerator or any component part of the atmosphere, which causes a continuous flow of air and products of combustion through the gas passage of the incinerator to the atmosphere.

Drying hearth—A surface within the primary chamber upon which wet waste material is deposited for drying, prior to burning.

Excess air—The air remaining after a fuel has been completely burned, or that air supplied in addition to the theoretical quantity.

Flay ash—Suspended particulate matter in products of combustion.

Grate—Surface with suitable openings, to support the fuel bed and permit passage of air through the burning fuel.

Incinerator—Combustion apparatus in which solid, semi-solid or gaseous combustible wastes are ignited and burned to CO_2 and water vapor, and from which the solid residues contain little or no combustible material.

Mixing chamber—Chamber usually placed between the primary combustion chamber and the secondary combustion chamber where thorough mixing of the products of combustion is accomplished by turbulence created by increased velocities of the gases, checkerwork, and/or turns in direction of the gas flow.

Overfire air—Any air controlled with respect to quantity and direction, forced or induced, supplied through or adjacent to the fuel bed for the purpose of promoting combustion of the combustible materials in the fuel bed.

Primary combustion chamber—Chamber within an incinerator where primary ignition and burning of the waste occur.

Scrubber—Auxiliary equipment designed to remove pollutants in wet form from the products of combustion.

Secondary combustion chamber—Chamber within an incinerator where unburned combustible materials from the primary chamber are completely burned.

Settling chamber—Any chamber designed to reduce the velocity of the products of combustion to promote the settling of fly ash from the gas stream.

Stack—A vertical passage for conducting products of combustion to the atmosphere.

Standard conditions—Standard conditions shall mean a temperature of 70° F and a pressure of psia dry, and corrected to 12% CO_2.

Theoretical air—The exact amount of air required to supply oxygen for complete combustion of a given quantity of a specific fuel.

Underfire air—Any air controlled with respect to quantity and direction forced or induced, supplied beneath the grate, that passes through the fuel bed.

NOTATION

Symbol	Meaning
A	percent adhesion
A_G	grate cross section
C	slip velocity correction factor
C_D	drag coefficient
C_p	specific heat at constant pressure
D	diameter
D_t	theoretical draft
d_p	particle diameter
F_ℓ	head loss
F_s	friction loss
f	drag force
g	gravitational acceleration
H	stack height above breeching
H_A	arch height
K	Cunningham correction coefficient
K_B	agglomeration constant
m	hydraulic radius
N	Avagadro's number
n	particle number concentration
q_0	number atoms per unit volume
R'	particle-to-collector-diameter ratio
Re	Reynolds number
s	average surface roughness
T	temperature
TA	theoretical air
t	time
U	velocity
u_t	terminal settling velocity
V	stack gas velocity
V_p	particle volume
W	mass rate
x	particle displacement

Greek Symbols

Symbol	Meaning
η	effective target coefficient
λ	particle mean free path
λ'	van der Waals constant
μ	viscosity
ϱ	fluid density
ψ'	dimensionless impaction parameter

REFERENCES

1. American Public Works Association, Public Administration Service, *Refuse Collection Practice,* 3rd ed., Washington, DC (1966).
2. Cross, F., *Handbook of Incineration,* Technomic Publishing Co., Scranton, PA (1972).
3. Nicholls, P., U. S. Bureau of Mines Bull., 378 (1934).
4. Weintraub, M., A., Orning, A., and Schwartz, C. R., U. S Bureau of Mines Report, INV. 6908 (1967).
5. Bailie, R. C., Donner, D. M., and Galli, A. F., Proc. 1968 Nat. Incinerator Conference, New York, 12–17 (1968).
6. Orning, A. A., *Principles of Combustion,* Wiley-Interscience Pub. Co., NY, 1969.
7. Fernandes, J. H. "Incinerator Air Pollution Control," paper presented at the National Incinerator Conference, ASME, NY, (1968).
8. "Technical-Economic Study of Solid Waste Disposal Needs and Practices," Combustion Engineering Inc., for U. S. PHS.—PH86–66–133.
9. Stenburg, R. L., et al., *J. A. P. C. A.,* 297 (Feb. 1959).
10. Flood, L. P., *J. A. P. C. A.,* 9 (1): 63–68 (May 1959).
11. Lemhan, J. W., *J. A. P. C. A.,* 12 (9): 414–417 (Sept. 1962).
12. Zenz, F. A., and Othmer, D. F., *Fluidization and Fluid Particle Systems,* Reinhold Publishing Co., NY, 1960.
13. Turk, A., *Industrial Wastes,* 3: 9–14 (Jan–Feb 1958).
14. Cheremisinoff, P. N., and Young, R. A., *Pollution Engineering Practice Handbook,* Ann Arbor Science Pub., Ann Arbor, MI, 1975.
15. Cheremisinoff, N. P., and Azbel, D. S., *Fluid Mechanics and Unit Operations,* Ann Arbor Science Pub., Ann Arbor, MI, 1983.
16. Cheremisinoff, N. P., and Cheremisinoff, P. N., *Hydrodynamics of Gas-Solid Flows,* Gulf Pub. Co., Houston, TX, 1984.
17. Cunningham, E., *Proc. Roy. Soc. (London),* 83: 357 (1910).
18. McCormic, P. Y., Lucas, R. L., and Wells, D. F., "Gas-Solid Systems," Section 20, in *Chemical Engineer's Handbook,* 4th ed., R. H. Perry, C. H. Chilton and S. D. Kirkpatrick (Eds.), McGraw-Hill Book Co., NY, (1963) pp. 68–74.
19. Einstein, A., *Elemental Theory of Brownian Motion,* 41–42 (1907).
20. Loeb, L. B., *Kinetic Theory of Gases,* McGraw-Hill Book Co., NY, 1934.
21. Bradley, R. S., "The Cohesive Forces Between Solid Surfaces and the Surface Energy of Solids," *Phil. Mag.,* 13: 853–862 (1932).
22. Hamaker, H. C., "The London-van der Waals Attraction Between Spherical Particles," *Physica,* 4: 1058–1072 (1937).
23. Whytlaw–Gray, R., and Patterson, H. S., *Smoke,* Edward Arnold Co., London, 1962.
24. Corn, M., "The Adhesion of Solid Particles to Solid Surfaces II," *J. A. P. C. A.,* 11 (12): 566–584 (Dec. 1961).
25. Lapple, C. E., "Fine Particle Characteristics," *Standford Research Institute Journal,* 5: 95, (Third Quarter 1961).
26. Hitchcock, D., "Solid-Waste Disposal-Incineration," *Chemical Engineering* (May 21, 1979).
27. Sundstrom, D., and Klei, H. E. *Sludge Treatment Processes,* Prentice-Hall Inc., NJ, 1979.
28. "Hydrosystems for Liquid and Fume Incineration," BSP Div., Envirotech Corp., Belmont, CA, Bull. ESI 300–3–73–SM (1980).
29. Warner, A., Parker, C. H., and Baum, B., "Solid Waste Management of Plastics," Report of a Research Study for the Manufacturing Chemists Assoc., NY, 1970.
30. Stevens, J. L., Crumpler, E. P., and Shih, C. C., "Thermal Destruction of Chemical Wastes," 71st Annual AIChE meeting (Nov. 14, 1978).

31. Powers, P. W., *How to Dispose of Toxic Substances and Industrial Wastes,* Noyes Data Corp, Park Ridge, NJ (1976).
32. Williamson, J. E., MacKnight, R. J., and Chass, R. L., "Multiple-Chamber Incinerator Design Standards," Los Angeles County Air Pollution Control District (1960).
33. Kent, R. T., (Ed.) *Mechanical Engineer's Handbook,* 11th ed., John Wiley & Sons, Pub., NY, 1963. p. 6–104.
34. Griswold, J., *Fuels, Combustion and Furnaces,* McGraw-Hill Book Co., NY, 1976, p. 374.
35. Williamson, J. E., and George, R. E., *On-Site Incineration of Commercial and Industrial Wastes with Multiple Chamber Incinerators,* Wiley-Interscience Pub., NY, 1969.

CHAPTER 43

DESIGN GUIDELINES FOR FLARE GAS PRACTICES

K. Banerjee and P. N. Cheremisinoff

Department of Civil & Environmental Eng.
New Jersey Institute of Technology
Newark, New Jersey

and

N. P. Cheremisinoff

Exxon Chemical Co.
Linden, New Jersey

CONTENTS

INTRODUCTION

Direct discharge of waste or excess vapors to the atmosphere is unacceptable either because of restrictions imposed by local ordinances or plant practices; concentrations of the contaminants at ground or adjacent platform levels exceed permissible explosion or toxicological threshold limits; and/or meteorological conditions which promote high background concentrations of discharges. Nonhazardous vapors such as waste or low pressure steam are usually discharged directly to the atmosphere. In contrast, hydrocarbon vapors which are discharged on a continuous basis (for example, off-spec product or bypass streams generated during start-up) or intermittently and which cannot be directly discharged to the atmosphere must be disposed of through a closed system and burned in a flare.

The operation of gas flaring is aimed at converting flammable, toxic or corrosive vapors to either nontoxic emissions or to acceptable concentrations by means of combustion. Because flaring is an integral part of petroleum and petrochemical plant operations discussions on this subject are warranted. This chapter covers basic considerations for the

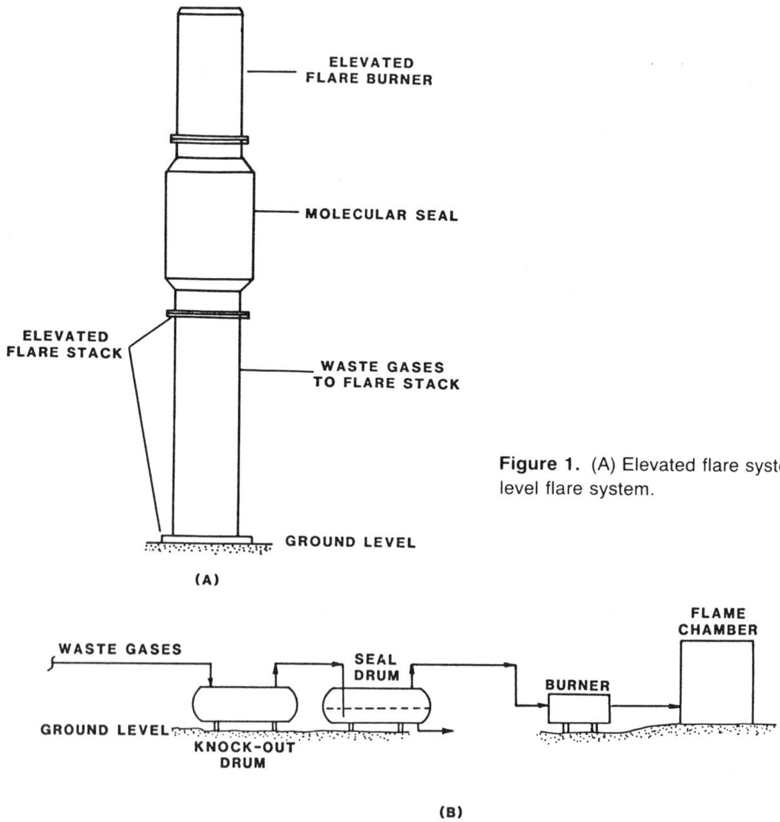

Figure 1. (A) Elevated flare system; (B) ground level flare system.

planning and design of flaring systems. Included are criteria for establishing safety facilities aimed at preventing overpressuring during the discharge of vapors and liquids. Among the standard pressure-relieving devices described are safety and relief valves, rupture disks, pressure control valves, and equipment blow down valves.

OVERVIEW OF FLARING PRACTICES

Flare systems are broadly categorized as elevated and ground-level designs. In elevated systems, combustion takes place at the discharge of a stack through a burner and igniter. Ground level flares are more elaborate in design, with basically three types of configurations. One arrangement uses a water spray to disperse the combustion gases; the venturi type takes advantage of the waste stream's kinetic energy to inspirate and mix combustion air with the gas; and finally, multi-jet designs distribute the gas through a cluster of burners.

Figure 1 illustrates the features of both elevated and ground level systems. In the elevated system (Figure 1A), the relieving gases are sent through an elevated stack from a closed collection system and burned off at the top. The generated flame is open in this case. Figure 1B illustrates a ground flare system, where principal components are a knock-out drum, multi-jet burners, a refractory lined rectangular flare box, and a seal drum. The flare flame is returned inside the flare chamber.

Designs often include several pilot burners which burn continuously. Systems must be continuously purged with an inert gas such as nitrogen or with natural gas to maintain an air-free line. During normal operations, if continuous venting is required, vapors may be routed back to a low pressure process stream, to a fuel system, or to an incinerator.

The major components of all flare systems are the relief, safety and depressuring valves; pressure relieving header(s) which convey discharges from safety and pressure control valves in the process unit to the flare; the knock-out drum located ahead of the flare stack; and the flare stack itself which consists of the riser structure, molecular seal, and burner tip.

The riser structure consists of two or more sections; the flare header enters at the bottom section which serves as a flare stack knock-out drum. Condensate carried over from the main knock-out drum is collected in the riser section.

The molecular seal is welded to the riser section. It provides a seal against air entering into the flare stack and minimizes the formation of explosive mixtures in the system. It resembles a bubble cap and creates a seal by taking advantage of the buoyancy of the purge gas to create a zone where the pressure exceeds atmospheric conditions.

The burner tip is a complete assembly which is connected to the molecular seal outlet. Accessories on the burner tips include about three or four gas pilots, a similar number of pilot gas/air mixture assemblies, and steam supply nozzles for steam injection.

Figure 2 shows a schematic diagram of the entire flare system. The relieving gases from safety relief valves are collected in a horizontal or vertical knock-out drum through a main header. Condensate carried in the gas is removed at this point. A constant liquid level is maintained in the boot of the drum. The liquid is pumped to a slop tank or reused in oil recovery facilities. Steam is normally used for winterizing. The gas from the knock-out drum is then sent to an elevated flare stack. At the bottom of the stack a seal of liquid is maintained.

The stack, comprised of a riser section, molecular seal, and burner tip, is automatically ignited through the line. A steam connection is provided for smokeless flares along with

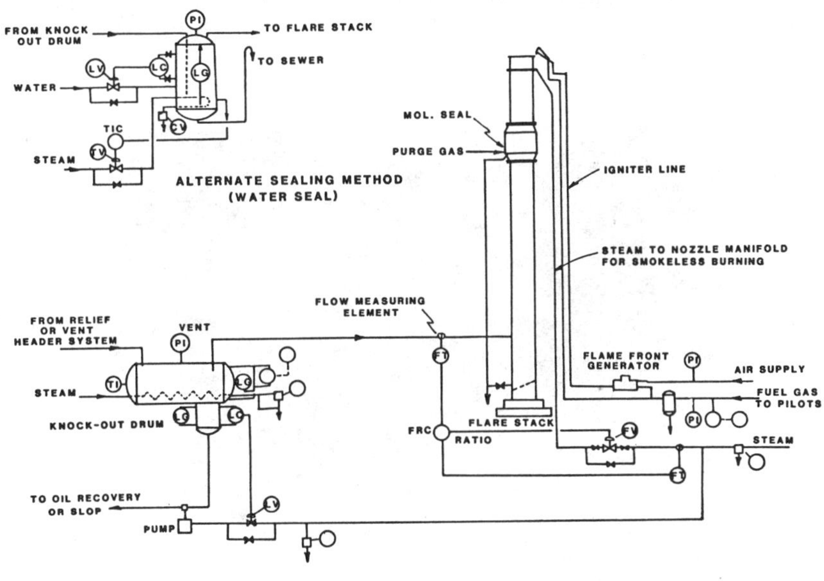

Figure 2. Schematic of typical flare gas system (adapted from API, RP 521).

a purge gas connection for maintaining an air-free system. The latter prevents flashback by maintaining the pressure at the molecular seal in excess of the atmosphere.

Note that a flame is referred to as being luminous when incandescent carbon particles are present. When these particles cool down, smoke is formed. Smoke formation mainly occurs in fuel-rich systems where a low hydrogen-atom concentration suppresses the smoke. Prevention of smoke in flares is normally accomplished by:

- By the addition of steam.
- By preparing a premixture of fuel and air before combustion (this provides sufficient oxygen levels for efficient combustion).
- By distribution of the flow of raw gases through a number of small burners.

Steam addition is most commonly used to generate a smokeless flare for economy and superior performance. In steam addition, the raw gas is preheated before entering the combustion zone. If the temperature is high enough, cracking of hydrocarbons occurs. This produces free hydrogen and carbon. When the cracked hydrocarbons enter the combustion zone, hydrogen reacts much faster than carbon. Unless the carbon particles are burned away, they cool down, thus forming smoke (see Figure 3). Consequently, to prevent smoke, either the hydrogen atom concentration must be decreased to ensure uniform burning of both hydrogen and carbon or sufficient oxygen must be provided for complete combustion. When steam is added, the hydrogen molecules also tend to separate which minimizes polymerization reactions and forms oxygen compounds which burn at a reduced rate and temperature. The net effect is to prevent cracking. Another explanation for the improved efficiency with steam addition is that the steam reacts with carbon particles forming carbon monoxide, carbon dioxide, and hydrogen, thereby removing the carbon which forms smoke after cooling.

CAUSES AND PREVENTION OF OVERPRESSURE

The initial design stage involves analysis of potential cases leading to discharges from pressure relief valves (i.e., establishing the maximum loading for emergency operations). The maximum load is composed of the individual contributions from the entire process. A

Figure 3. Smokeless steam flare-chemical plant-steam off. (Courtesy of National Air-Oil Burner Co., Inc., Philadelphia, PA.)

conservative design is one which assumes that all contributions are relieving simultaneously under emergency conditions.

It is preferable that relieving overpressures to the flare systems via the pressure relief valves be kept to a minimum, since these valves often do not reseat themselves. This leads to leakage, and consequently, reduced recovery of products. For minor operational upsets, and particularly during start-up, the use of pressure relief valves is undesirable. Therefore, overriding pressure control valves (PCV) strategically located on equipment are incorporated in the control strategy. Examples of such locations are the suction sides of compressors, overhead product lines of fractionating columns, at the beginning or at the end of a series of high pressure reactors, etc. The set point of the PCV is positioned above the operating pressure but below the set-point of the pressure relief valve. They are sized to handle about 40% to 100% of the flow of the safety valves. For about 90% of the emergency operations the system is depressurized through these valves so as to keep to operation of the pressure relief valves to a minimum.

Besides overriding pressure control valves, other remotely controlled valves (e.g., motor operated or solenoid operated valves) are also used.

Pressure vessels, heat exchangers, operating equipment, and piping are designed for a specified system pressure. This maximum allowable working pressure is about 10% higher than the normal operating pressure. Pressure relief valves are normally set at or below the maximum allowable working pressure in order to protect equipment. The relieving rate of a pressure relief valve depends upon the cause of system overpressure; principal ones being operational failures and plant fires [1].

Examples of operational failures are closed outlets on vessels, inadvertent valve opening, utility failure, and a variety of miscellaneous actions. In the case of an inadvertent closure of a block valve the pressure in a vessel can exceed the maximum allowable pressure. A pressure relief valve protects the vessel under this situation. If two vessels are in series and isolated by individual block valves, then each vessel must be protected by an individual pressure relief valve. Omission of the block valve in between the vessels or locking the same valve in the open position results in a common system. This arrangement may be protected by a single pressure relief valve. As in the case of block valves, each control valve should be considered as subject to inadvertent operation causing overpressure in the upstream section.

Another common error is the inadvertent opening of a valve from a higher pressure source (e.g., high pressure steam or process fluids connected to a low pressure system). This causes an overpressure exceeding the maximum allowable working pressure. If the block valve is intended for isolation only and normally remains closed, a pressure relief valve may be avoided by locking or sealing the same block valve.

Examples of miscellaneous actions/conditions leading to operational failures are:

- *Reflux failure of a fractionating column*—which can cause flooding of condensers resulting in an overpressure.
- *Heat exchanger tube failure*—when tube side pressure of an exchanger is much lower than the shell side, rupture of a tube can result from overpressure of the tube side.
- *Internal explosion*—a situation which is not predictable for conventional refinery installations. For some chemical reactions, it is possible to predict the probability of an explosion, in which case, special rupture disks are installed for quick disposal of vapors.
- *Chemical reaction*—where vessels may become overpressured because of an unbalanced reaction. Normally, sophisticated controls are used along with a safety valve in these cases.
- *Hydraulic expansion*—when a cold fluid is blocked in on hot exchanger surfaces, it will expand, causing a rise in pressure.
- *Accumulation of noncondensables*—these do not accumulate under normal conditions since they are released with the process streams; however, with certain piping configurations, it is possible for noncondensables to accumulate to the point that they may prevent condensation of a process stream in a condenser, thus resulting in overpressure.

In the case of uncontrolled fires, heat absorbed by exposed vessels or equipment causes vapor generation of the contents. A pressure relief valve is required to protect the vessel and relieve the generated vapor.

Each pressure relief valve should be individually analyzed for potential causes of overpressure. Valves should be sized for those situations that will require the maximum relieving rate. If a fire condition is controlling, two separate safety valves, one for a fire emergency and the other for an operational failure should be provided since the former is less likely to occur.

For fire emergencies an important consideration is the wetted surface area. The exposed surface area which is effective in generating vapor is that area wetted by its internal liquid level up to a maximum height of 25 ft above grade (this is the normal practice based on the average flame length). "Grade" is defined as any horizontal solid surface on which liquid could accumulate, i.e., roofs, solid platform, etc. The contents under variable level conditions would ordinarily be taken at the average inventory. Liquid-full vessels, horizontal or vertical (such as treaters), operate with no vapor space, and the wetted surface would be the total vessel area to a maximum height of 25 ft above grade. It should be noted that, in such a vessel, at the start of a fire the opening of the pressure relief may be due to thermal expansion of the liquid. The valve should, however, be sized based on the vapor generator at the relief pressure and the boiling point corresponding to that pressure.

The surface areas of typical vessels used in process operations are as follows:

- *Surge and reflux drums:* The wetted surface should be calculated using the high liquid level or 50% of the total vessel surface, whichever is greater, since 50% is the normal liquid level in these vessels.
- *Knock-out drums:* Knock-out drums usually operate with low liquid levels. If the normal level is not known, the inventory at the high level alarm should be used to estimate the wetted surface area.
- *Fractionating columns:* Fractionating columns usually operate with a normal liquid level in the bottom of the column plus a level on each tray. However, the entire wall of a fractionating column within the fire height limitation of 25 ft should be considered as wetted.
- *Working storage tanks:* Here the liquid level is independent of operation, and, therefore, the maximum level should be used for determining the wetted surface.

The wetted surface of spheres and spheroids are calculated as the area of the bottom half of the vessel or up to a height of 25 ft, whichever results in a greater surface area.

Where suitable drainage is provided to preclude an accumulation of flammable liquids directly beneath a vessel, the total heat input rate to the vessel may be computed from (as recommended by API publication RP 250):

$$Q = 21{,}000\ FA^{0.82} \tag{1}$$

where Q = total heat absorbed in Btu/hr
A = wetted surface in sq ft
F = environmental factor

Typical values of the environmental factor, F, for different insulation thicknesses are:

F = 1.0 for bare surface
F = 0.3 for 1 in. thickness of insulation
F = 0.15 for 2 in. insulation
F = 0.075 for 4 in. insulation

If insulation is required but the thickness is not known, an F-value of 0.3 is recommended. Figure 4 also provides a convenient plot of Equation 1 for graphical analysis.

If drainage is not provided for the area below the vessel (i.e., diked or curbed areas around a tank), then vapor relief for fire exposure should be computed from the following heat criteria:

- 20,000 Btu/hr/ft^2 for uninsulated vessel.
- 10,000 Btu/hr/ft^2 for 1 in. insulation.
- 6,000 Btu/hr/ft^2 for 2 in. insulation.
- 3,000 Btu/hr/ft^2 for 4 in. insulation.

These values are based on the wetted surface up to the normal liquid level, provided fireproof insulation is employed. With nonfireproof insulation it is best to assume that the vessel is bare.

For a fluid below the critical point (i.e., at relieving temperature and pressure), the vapor release rate can be computed from:

$$W = \frac{Q}{\lambda} \tag{2}$$

where W = vapor release rate lb/hr
Q = total heat input Btu/hr
λ = latent heat of fluid in vessel, evaluated at the relief valve inlet pressure Btu/lb

No credit is normally taken for the sensible heat capacity of the fluid.

For a fluid above its critical point, i.e., when pressure relief conditions are near or above the critical point, the vapor discharge rate depends on the rate at which the fluid will expand. The latent heat of vaporization at or near the critical point is almost zero in this case.

Once relieving loads of individual pressure relief valves are established, an analysis of the interactions between individual contributors is recommended. The simultaneous occurrence of two or more contingencies (i.e., double jeopardy) is highly improbable and is therefore often neglected when establishing the maximum system load. In determining the maximum load from a single contingency, all directly related contingencies which influence the load must be considered. For example, in a plant where a single boiler or source of steam is used for both process drives and electric power generation, the failure of a steam

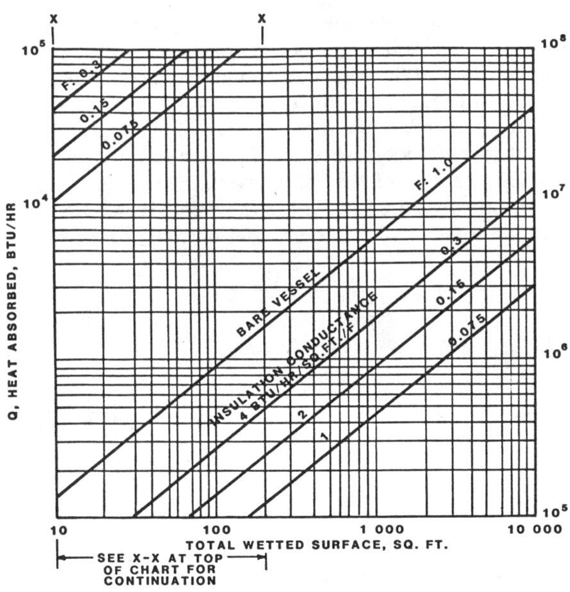

Figure 4. API formula for heat absorbed from fire on wetted surface of pressure vessel ($Q = 2{,}000\,FA^{0.82}$) [7].

source (a single contingency) can cause simultaneous loss of power (a directly related contingency). If the electrical system has an alternate source of supply, then only the loss of steam need be considered, provided the elapsed time for transfer switching is not long enough to be ineffective. In this situation, a power failure would not be a contingency directly related to the loss of steam.

If a certain contingency were to involve more than one unit, then the entire system must be evaluated. For example, if there is more than one reaction vessel in series, then in the event of a runaway reaction, all the reactors will be protected by a single pressure relief valve. The same situation can occur in the case of two or more fractionating columns in series. In the case of multi-stage compressors having individual pressure relief valves at each stage, the relieving rates from the pressure relief valves are not additive.

DESIGN METHODOLOGY FOR COLLECTION SYSTEMS

Vapors from different pressure relief valves and depressuring valves must first be collected in individual flare sub-headers located near each process area. Sub-headers must be interconnected to a main flare header which leads to a knock-out drum where condensates are removed.

The number of main flare headers and individual sub-headers depends upon the type of vapors being handled as well as the temperature and the back pressure limitations of the pressure relief valves.

The pressure level of flare headers depends on the type of pressure relief valves employed and the pressure levels of the equipment connected to the flare system.

The principal types of pressure relief valves are: conventional, balanced bellow type, piston type, and pilot operated.

Conventional pressure relief valves are those where the disk of the valve is held tight against the inlet nozzle by means of a spring. Figure 5A illustrates a conventional safety relief valve normally used in refineries and chemical plants. This type of valve is least expensive, but it is limited by a back pressure of 10% of the maximum allowable working pressure. The reason for the limited back pressure is as follows: Assume P_1 to be the maximum allowable working pressure in the vessel. Since this is also the set pressure of the safety valve, the spring is so loaded that the total downward force on the valve disk is exactly equal to the total upward force exerted on it by the process vapor when it reaches pressure P_1. That is, the spring force F_s is related to the inlet nozzle area A_N through applied pressure P_1 ($F_s = P_1A_N$). A slight pressure increase above P_1 lifts the valve disk up and relieves the vapor through the discharge nozzle of the valve. Accumulation or overpressure above the maximum allowable working pressure (P_1) within the vessel is safe up to about 10%, if the overpressure persists for a short period of time. Under normal operation when the operating

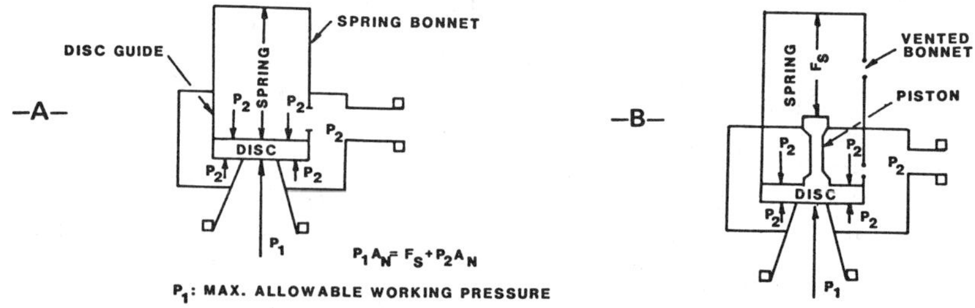

Figure 5. Shows effect of back pressure on (A) conventional safety valve; and (B) piston type valve.

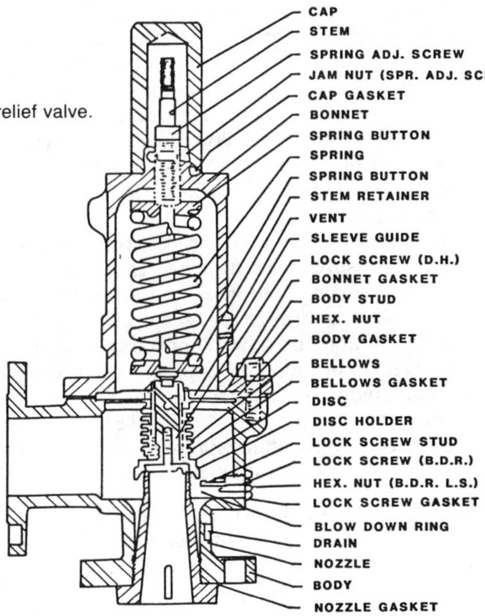

Figure 6. Shows a balanced bellows safety relief valve.

pressure is below P_1, the downward force exerted by the spring on the disk exceeds the upward force exerted by the vapor. Hence, the disk is held tight against the inlet nozzle under normal operating conditions. When the back pressure exceeds atmospheric, the combined downward force exerted by the spring and the force developed due to the back pressure is $F_s + P_2A_N$, where P_2 is the back pressure (see Figure 5A). In order to lift the valve disk against this combined downward force, the inlet vapor inside the vessel must be pressurized to a level greater than P_1. Hence, $P_1'A_N = F_s + P_2A_N$, where P_1' is the new pressure developed.

Since the maximum pressure is $P_1' = 1.1\ P_1$,

$$UF_s = P_1A_N$$

Hence

$$1.1P_1 \cdot A_N = P_1A_N$$

$$P_2 = 0.1P_1$$

Thus, the maximum allowable back pressure is 10% of the maximum allowable working pressure of the vessel.

The remaining three types of pressure relief valves do not depend upon the back pressure for performance. However, to ensure that the safety valves work at their maximum capacity, the back pressure is limited to 50% of the relief valve set pressure. With the balanced bellows valve (see Figure 6), the spring does not act directly on the disk. Instead, it serves on a bellows first, which in turn acts on the disk. The piston type works in a similar fashion. The cross-sectional area of both the piston and the bellows is the same as the inlet nozzle of the valve. The effect of the back pressure on the top and the bottom of the disk creates equal balancing forces (i.e., P_1A_N equals F_s).

Pilot-operated valves combine a pilot valve with the main valve. The spring of the main valve provides 75% loading on the disk and the remaining 25% is offered by the gas or vapor through the pilot valve. When the vessel reaches the maximum allowable working pressure, the pilot valve relieves the gas pressure which contributes to the disk load. Thus the safety valve becomes wide open [2, 3] (see Figure 7).

With nonconventional valves, the maximum allowable back pressure should be taken as 50% of the valve set pressure. This limit approaches the critical flow pressure. If the back pressure exceeds the critical flow pressure corresponding to the set pressure of the safety valve, the total pressure drop available for flow decreases. This state can potentially lead to overpressurization. The recommended back pressure is, therefore, a maximum 40% to 50% of the set pressure.

The design pressure levels of process equipment connected to a flare system also establish flare header pressure levels. In some cases, pressure fluctuations are significant. Hence, it is not always economical to interconnect the entire process to a common header. It is often more economical to arrange for flare headers of two pressure levels, one connecting the low pressure system and the other connecting a high pressure system.

The basis for collection philosophy of a flare system is based upon economics. A summary of the methodology is outlined below.

First consider the number of flare headers required. This depends on the economics, streamlined to accommodate minimum piping lengths and sizes. The following steps outline the procedure for comparative estimations:

1. Study of the plot plan layout: From the plot plan layout the number of safety valves, set pressures, individual relieving loads, relieving vapor temperatures, and properties of vapors must be noted.

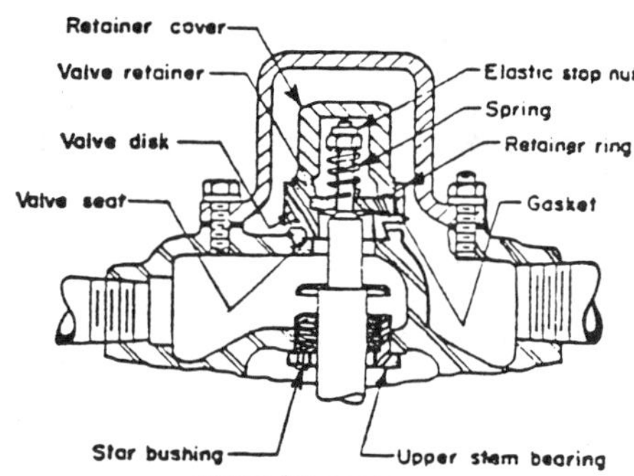

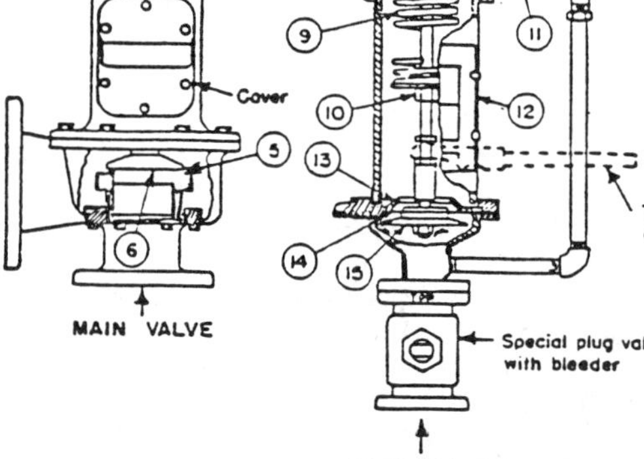

Figure 7. Pilot operated safety valve.

1 Lock nuts
2 Spring
3 Diaphragm
4 Diaphragm clamp nut
5 Seat
6 Pallet
7 Poppet assembly (see Detail "A")
8 Vent
9 Spring
10 Spring adjusting nut
11 Restriction orifice
12 Cover
13 Diaphragm ring
14 Pilot diaphragm
15 Safety piston

2. A single sub-header in each process area is drawn up connecting area pressure relief valves or depressuring valves.
3. The sub-headers are then connected to provide a single main flare header based on the shortest route.
4. The equivalent length of the main flare header is then calculated from the flare stack to the last safety valve, taking into consideration the straight length of the pipe and approximate equivalent lengths for bends, etc. If the actual location of the flare stack is not known a conservative estimate of 500 ft from the last piece of equipment can be assumed.
5. A trial estimate of the diameter of the flare header based on the maximum relieving flare load and a back pressure limitation of 10% for conventional valves and 40% for balanced type valves is made. Note, however, a single main header in most cases, turns out to be too large to be economical.
6. A second trial is required for two main flare headers, one collecting the low pressure flares (usually 5 to 10 psig), and the other collecting relatively high pressure flares (usually 15 to 20 psig). The two headers are connected to their individual knock-out drums. The vapor lines from the knock-out drums are combined into a single header which is connected to the flare stack.

   The maximum simultaneous load in each header must be calculated separately and the pressure drop is to be computed for the entire length of the pipe including the combined length from the knock-out drum to the stack.

   The sub-headers in each process area similarly will have two levels of flare headers. The line sizing of each level of sub-header in an individual area will depend on the maximum simultaneous flow in that area. Thus, the line sizing criterion of a sub-header may be the largest single flow due to a blocked outlet condition. This flow may not necessarily be the controlling load for the flare stack.
7. The materials of construction determine the final number of flare headers. Vapors that normally require expensive materials are: corrosive vapors (e.g., SO_2, H_2S), high temperature vapors (e.g., high temperature gases used for regeneration of catalysts in reactors), and low temperature vapors (e.g., vapors generated due to flashing across a control valve or a safety valve in a cryogenic system).
8. Wet and dry flares—sometimes relatively hot vapors carrying condensates may be separated from cold dry vapors. They do not run as separate headers, but either low pressure or high pressure flare headers may be associated with any one of them. Thus, a wet flare header may be, in fact, the low pressure header, and the dry flare header may be the high pressure flare or vice versa.
9. After the total number of flare headers has been established it may be necessary to recheck the vapor load in individual headers since introduction of a separate header may allow subtraction of the flow quantity from the low pressure header to which it was initially added.

Once the maximum vapor relieving requirement and the maximum allowable back pressure have been established, line sizing is performed with conventional calculations.

The flare piping system can be divided into the following sections:

- Individual discharge lines from the pressure relief valves.
- The sub-headers in each area connecting the discharge lines.
- The main flare header (or headers) connecting the sub-headers leading to the knock-out drum(s).
- The final header connecting the vapor line(s) from the knock-out drum(s) leading to the flare stack.

Since vapors in the flare headers are relieved from a high pressure system to atmospheric, there is an appreciable kinetic energy change throughout the line. Since compressible flow exists in the flare headers and lines are reasonably long, isothermal conditions may be assumed. When short, well-insulated vapor lines are employed, adiabatic flow is a better approximation. In general, all vapor flows that normally occur in process plants are somewhere in between adiabatic and isothermal. It has been observed that for the same flow rate and pressure drop, line sizing calculations based on compressible isothermal conditions provide an equal or a larger diameter pipe, and therefore, flare headers should be sized based on the more conservative predictions derived from isothermal conditions.

The following criteria are used in sizing flare headers:

- The back pressure developed at the downstream section of any pressure relief valve connected to the same headers should not exceed the allowable limit, i.e., 10% of the set pressure in psig for the conventional type and 40% to 50% of the set pressure in psig for the balanced type valve.
- Since the pressure drop is high, there is a possibility of approaching sonic conditions. This results in a potential noise problem. It is a good practice to limit the velocity to 60% of the sonic velocity or 0.6 Mach number.

The method of Lapple [4] provides good line-sizing estimates. The method employs a theoretical critical mass flow based on an ideal nozzle at isothermal conditions. For a pure gas, the mass flow can be computed from:

$$G_{ci} = 12.6P_0\left(\frac{M}{(2Z-1)\,t_0}\right)^{0.5} \tag{3}$$

where
G_{ci} = maximum mass flow or critical mass flow (lb/sec/ft^2)
P_0 = upstream pressure lb/in.^{2} abs.
M = molecular weight
T_0 = upstream temperature (°R)
Z = compressibility factor

The actual mass flow, G (lb/sec/ft^2) is a function of the critical mass flow G_{ci}, the line resistance N, and the ratio of the downstream to upstream pressures. These relationships are plotted in Figure 8. In the area below the dashed line, the ratio of G to G_{ci} remains constant, indicating that sonic flow exists. Thus, in sizing flare headers the condition must lie above the dashed line in Figure 8. The line resistance, N, is:

$$N = \frac{4fL}{D} + \Sigma K_i \tag{4}$$

Table 1
Resistance Coefficient K for Various Pipe Fittings

Fitting	K	Fitting	K
Globe valve, open	9.7	90° double-miter elbow	0.59
Typical depressuring valve, open	8.5	Screwed tee through run	0.50
Angle valve, open	4.6	Fabricated tee through run	0.50
Swing cheek valve, open	2.3	Lateral through run	0.50
180° close screwed return	1.95	90° triple-miter elbow	0.46
Screwed or fabricated tee through branch	1.72	45° single-miter elbow	0.46
90° single-miter elbow	1.72	180° welding return	0.43
Welding tee through branch	1.37	45° screwed elbow	0.43
90° standard screwed elbow	0.93	Welding tee through run	0.38
60° single-miter elbow	0.93	90° welding elbow	0.32
45° lateral through branch	0.76	45° welding elbow	0.21
90° long-sweep elbow	0.59	Gate valve, open	0.21

d/d':	0	0.2	0.4	0.6	0.8
Contractions (USASI)	—	—	0.21	0.135	0.039
Contractions (sudden)	0.5	0.46	0.38	0.29	0.12
Enlargements (USASI)	—	—	0.9	0.5	0.11
Enlargements (sudden)	0.0	0.95	0.74	0.41	0.11

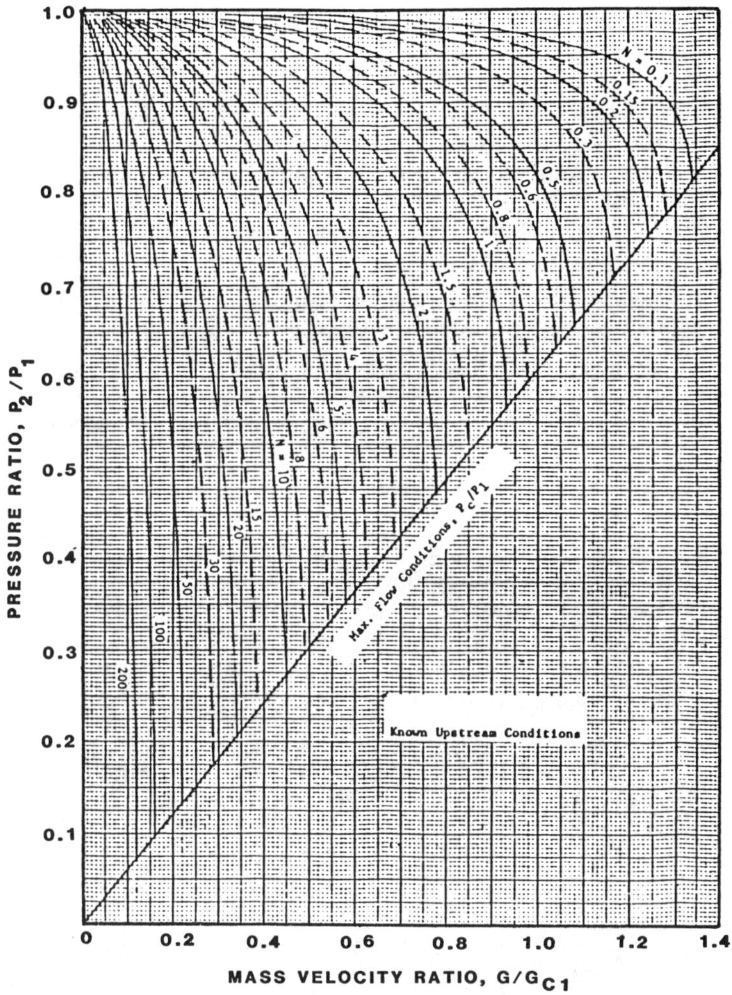

Figure 8. Pressure drop chart reported by Lapple [4].

where L = equivalent length of line (ft)
D = line diameter (ft)
f = Fanning friction factor
N = line resistance factor (dimensionless)
K_i = resistance coefficients for pipe fittings (refer to Table 1 for values) [6]

Lapple's method is useful when the upstream pressure of a header is known and the downstream pressure is to be evaluated. However, it is worthwhile to evaluate the pressure profile in the flare header as a function of distance from the stack. For this reason, one should calculate the pressure drop backwards, starting from the flare stack exit where the pressure is atmospheric. Figure 9 provides an alternate basis for evaluating pressure losses when the downstream pressure is known [5].

Although Figures 8 and 9 can be used for line sizing, it should be noted that the former requires more extensive trial and error.

The following steps summarize the procedure for sizing flare headers:

1. The pressure at the base of the flare stack is approximated as 2 psig. The pressure at the base may vary slightly depending on the type of seal used in the stack. Hence,

   Pressure at the base = Atm. pressure at the flare exit + 0.5 psi at the flare tip ΔP + 0.5 psi mol. seal ΔP + 1 psi
   ΔP due to flow through the stack height $\cong$ 2 psig

2. Compute the pressure in the knock-out drum:

   = 2 psig + ΔP required for the flow of the full load vapors from the knock-out drum to the stack + 0.5 psi ΔP assumed inside the knock-out drum.

3. For an initial trial, an inside pipe diameter is assumed based on 60% of the sonic velocity corresponding to the pressure and temperature at the base of the stack, i.e., at 2 psig and temperature = T_0 (upstream temperature since isothermality is assumed).

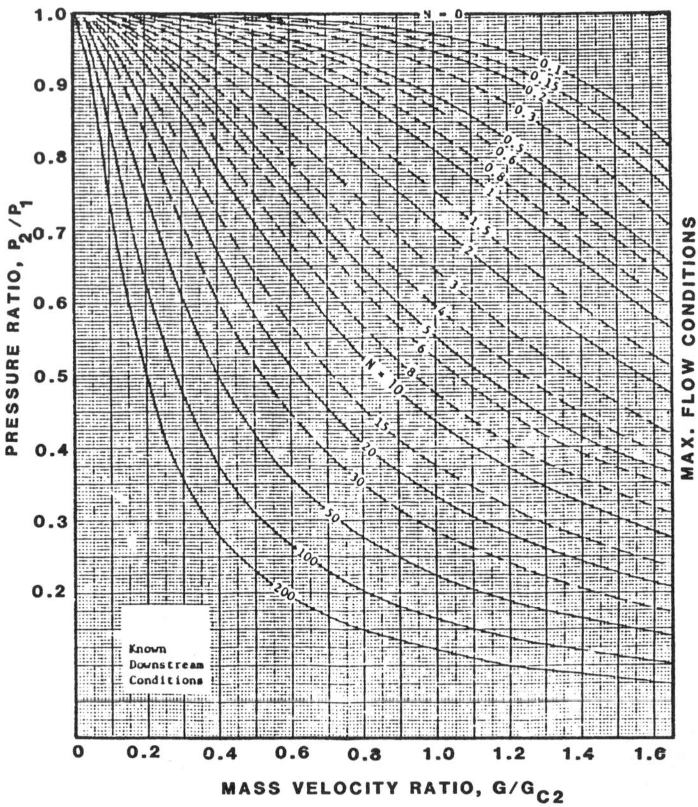

Figure 9. Pressure drop chart reported by Loeb [5].

The sonic velocity can be computed from:

$$V_s = 223\sqrt{\frac{\kappa T}{M}} \tag{5}$$

where V_s = sonic velocity (ft/sec)
κ = C_p/C_v of the gas normally between 1 to 1.8
T = temperature (°R)
M = molecular weight

The flare load, W, (lbs/sec) is known. The density of the vapor at 2 psig, T_0 (°R) is determined from the ideal gas law:

$$\rho = \frac{MP}{RT}$$

where M = mol. wt.
P = pressure in psia
T = °R
R = gas constant

Hence

$$\frac{4W}{\pi d^2} = \rho \times 223\sqrt{\frac{\kappa T}{M}} \tag{6}$$

where d = pipe I. D. (ft)

The inside diameter d can then be computed from Equation 6.

Once the diameter is known, the Reynolds number, Re, can be computed and the friction factor f obtained from Figure 10. Assuming a straight length of pipe for L = 500 ft, N (line resistance factor) can be calculated. Next G_{ci} is computed based on the downstream pressure and G/G_{c2} evaluated. From Figure 9 the ratio P_2/P_0 can be obtained. Since P_2 is known, p_0 can then be calculated. The pressure at the inlet of the knock-out drum is given by $P_0 + 0.5$ psi. Table 1 provides typical values of resistance coefficients for various pipe fittings.

4. From the konck-out drum, the individual flare headers can be sized. Based on a Mach number of 0.6 and the density corresponding to $(P_0 + 0.5)$ psia, a trial diameter can be established. The pressure at every intersection between the sub-header and the main header must be calculated, with the downstream pressure being $(P_0 + 0.5)$ psia. Knowing the pressure at the intersection of the sub-header and the main header, the pressure at the intersection of the sub-header and the discharge pipe of the safety valve is computed. Usually the discharge pipe of the safety valve is small and hence, a separate pressure drop calculation is not necessary. However, the velocity at the discharge pipe should be checked to ensure that it is below sonic conditions. If the discharge pipe runs a considerable distance before it ties in with the sub-header, a separate pressure drop calculation will be needed.

The sum of all the pressure losses starting from the flare stack up to the safety valve yields the total back pressure. This back pressure must be lower than the maximum back pressure allowed in the system and corresponding to the lowest set pressure of the safety valve.

DESIGN METHODOLOGY FOR CONDENSATE REMOVAL

Material entering the knock-out drum (or blow-down drum) will be a mixture of vapor and liquid. Liquid particle sizes less than 150 microns in size are readily burned in a flare,

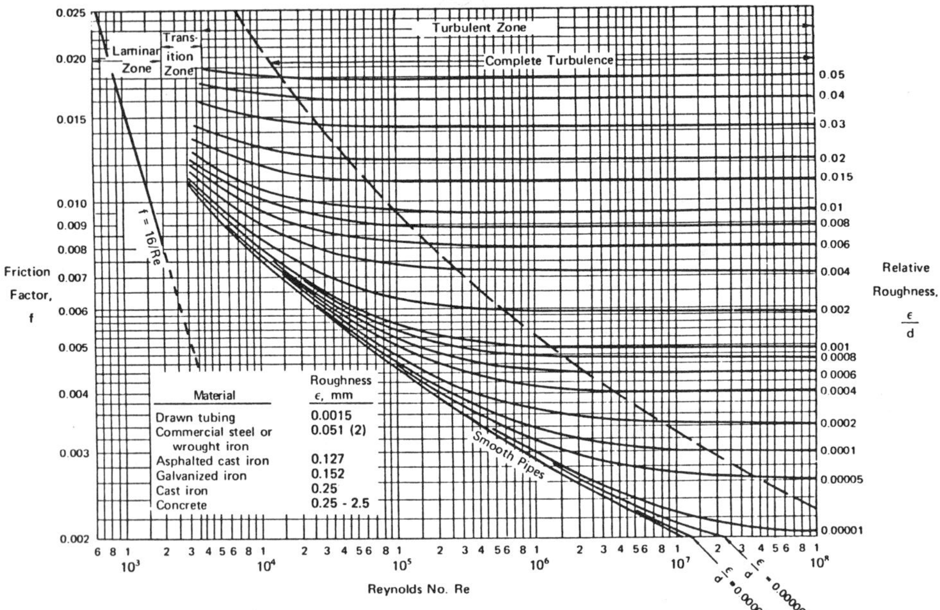

Figure 10. Generalized Moody chart for obtaining friction factor [6].

however, larger particles must be removed in the knock-out drum. This condensate is pumped out from the bottom of the knock-out drum either for reuse or disposal. (See Figure 11.)
pumped out from the bottom of the knock-out drum either for reuse or disposal.

In some process plants (e.g., ethylene production, coal gasification plant) hot vapors containing water are collected in a separate flare header (called a wet flare header). Liquid collected in the knock-out drum for the wet flare contains water and liquid hydrocarbons. In the same manner cold and dry hydrocarbon vapors are collected in a dry flare header. The hydrocarbon liquid collected in the knock-out drum of the dry flare is usually vaporized in a vaporizer located immediately below the knock-out drum and sent back to the flare. Figure 12 A shows a dry flare knock-out drum with a vaporizer at the bottom.

Knock-out drums are either horizontal or vertical types. They are also available in a variety of configurations/arrangements that include:

1. A horizontal drum with the vapor entering at one end of the vessel and exiting at the top of the opposite end (no internal baffling).
2. A horizontal drum with vapor inlets at each end of the horizontal axis, and a center outlet.
3. A horizontal drum with the vapor entering in the center and exiting at the two ends on the horizontal axis.
4. A vertical drum with the vapor entering at the top. These are equipped with a baffle that directs the flow downward. The outlet nozzle is located at the top of the vertical axis (as shown in Figure 12 B).
5. A vertical drum with a tangential nozzle.

A split entry or exit reduces the drum size for large throughputs. As a rule of thumb, when the drum diameter exceeds 12 ft, the split flow arrangement is normally more economical. Figure 12C shows a split flow horizontal drum with the recommended dimensions.

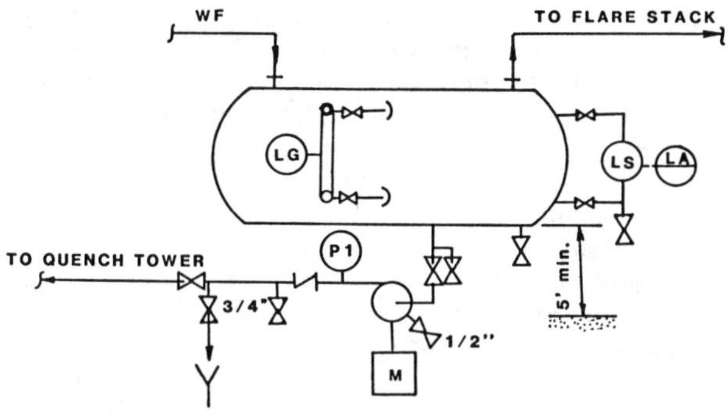

Figure 11. Typical knock-out drum with a level gauge, level switch, and level alarm. Also shows a two-inch utility connection for cleaning the drum with steam and a pump-out pump with pressure indicator and drain connection.

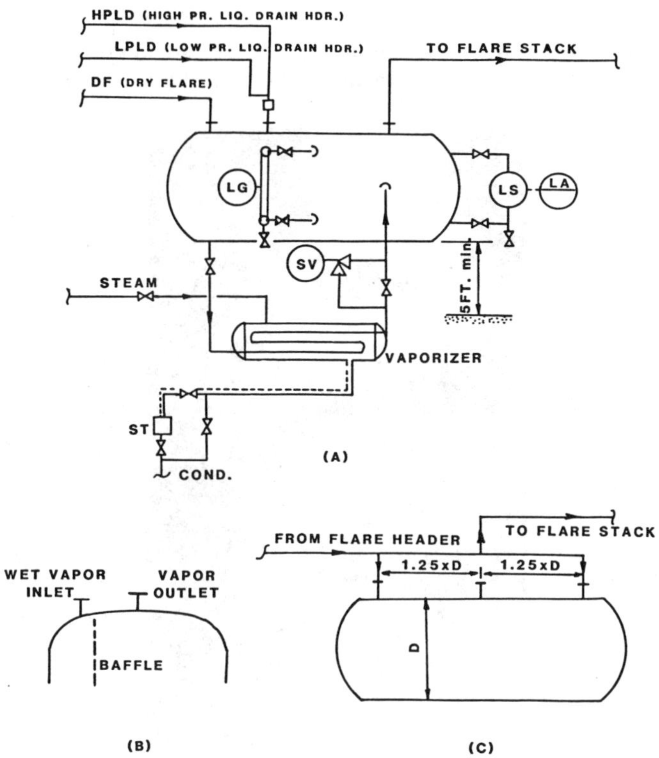

Figure 12. (A) Dry-flare knock-out drum; (B) shows vertical drum arrangement; (C) alternative split feed arrangement of knock-out drum.

Knock-out drums are usually sized by trial and error. Liquid particles drop out when the vapor velocity traveling through the drum is sufficiently low. In other words, the drum must be of sufficient diameter to effect the desired liquid-vapor separation.

Tan [8] gives the following formula for sizing horizontal drums:

$$W = 360D^2 \sqrt{(\rho_L - \rho_G)\, MP/T} \tag{7}$$

where W = lb/hr of vapor
ρ_L = liquid density (lb/ft^3)
ρ_G = gas density (lb/ft^3)
M = mol. wt. of the vapor
T = temperature of the vapor (°R)

P has units of psia, D in ft.

The formula is valid for particle sizes up to 400 microns.

Similar expressions are available for vertical knock-out drums. A formula for the vapor velocity is:

$$V = 0.4\sqrt{\frac{\rho_L - \rho_G}{\rho_G}} \text{ (ft/sec)} \tag{8}$$

SEAL SYSTEMS

Standard practice is to provide a seal at the base of the flare to prevent flashbacks from occurring. In the absence of a seal, a continuous quantity of gas must be bled to the flare

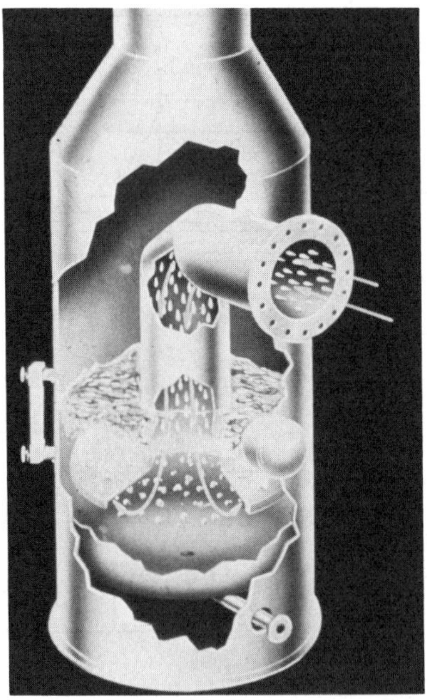

Figure 13. Water seal for flare staging or flashback protection. (Courtesy of National Air-Oil Burner Co., Inc., Philadelphia, PA.)

to maintain a net positive flow. Seals are of two main types: liquid and gas.

Liquid seals are further classified as seal drums and seal pipes (see Figure 13). In the former, a liquid seal is used in a seal drum located between the knock-out drum and the flare stack. Occasionally, a piping seal is used as a seal leg located at the bottom of the flare stack in place of a drum. This is often an integral part of the stack. Seal drums are either vertical or horizontal (see Figure 14A and 14B). The selection of the seal drum depends on the available space. Its purpose is to maintain a seal of several inches on the inlet flare header, preferably not exceeding 6 inches, otherwise it imposes back pressure to the knock-out drum. Water is normally used as a sealing liquid, and there is always a continuous flow of water with the overflow going to the sewer. If located in a cold climate either the water must be heated by a submerged steam heater or alternately the water may be replaced by liquids such as as alcohol, kerosene, etc., which do not reguire continuous flow.

The capacity of the seal drum is usually the volume corresponding to 8 to 10 ft of the vapor inlet line. In a vertical drum, the ratio of the inlet pipe cross-sectional area to the vessel free area for gas flow above the liquid should be at least 1 to 3 to prevent gas flow surges to the flare. The area for the gas above the liquid interface should be at least equal to that of a circle having a diameter D equal to 2d, where d is the inlet gas pipe diameter. This is derived as follows:

Assuming a vertical vessel of cross-sectional area $(\pi/4)D^2$ and inlet pipe $(\pi/4)d^2$, the annular area is $(\pi/4)(D^2-d^2)$. Since the suggested ratio is 1 : 3, $D^2-d^2 = 3d^2$ or $D = 2d$.

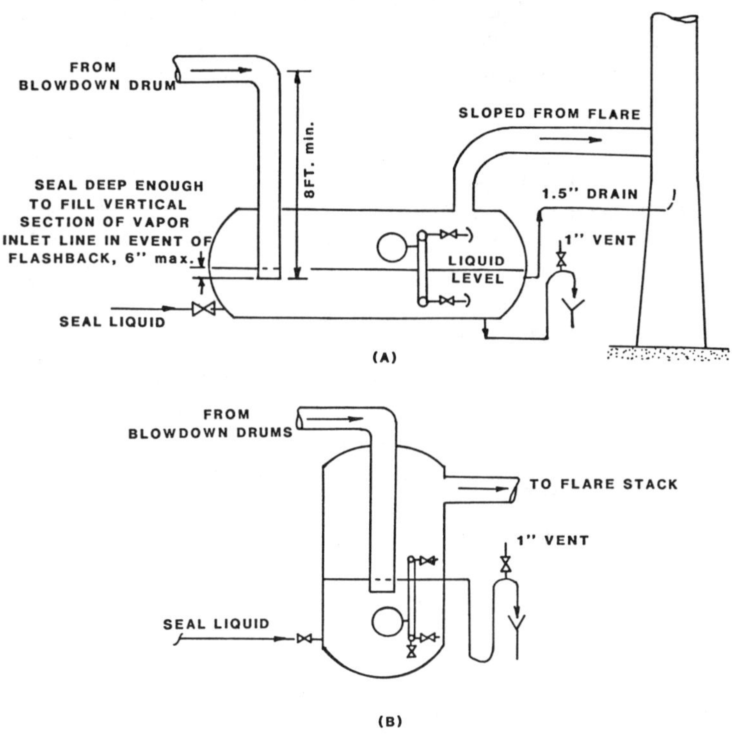

Figure 14. (A) Shows a horizontal seal drum; (B) a vertical seal drum.

The height of the vapor space above the liquid level in a vertical drum should be approximately 2 to 3 times the diameter (d) to provide disengaging space for entrained seal liquid. If a horizontal seal vessel is used, a minimum dimension of 3 ft between liquid level and top of the drum is recommended.

Seal pipes located at the base of the stack are less costly than drums. However, they can experience flow pulsations at low throughputs. Also, during large gas releases, the water seal may be blown out of the top to the flare stack. Some general guidelines for sizing seal legs are as follows:

- The slope of the inlet line is designed to provide a volume of water below the normal sealing water level equivalent to a volume of 10 ft of the inlet pipe.
- The depth of the water seal should not exceed 12 inches to prevent gas pulsation.
- The seal water level is maintained by a continuous flow of water at about 20 gallons per minute.
- Normal overflow is taken off the bottom of the seal through a seal leg. The height of the seal leg should be equivalent to about 175% of the pressure at the base of the stack during maximum vapor release so that gas release at the base is prevented. Figure 15 illustrates a typical seal leg arrangement.

An alternate *gas seal* is the "molecular" type seal which employs a purge gas having a molecular weight of 28 or less (e.g., N_2, CH_4, or natural gas). Because of the buoyancy of the purge gas, a zone having a pressure exceeding atmospheric is created. The molecular seal is located at the top of the flare stack immediately before the burner tip. Ambient air cannot enter the stack because of this high pressure. The recommended purge velocity through the molecular seal is about 0.1 ft/sec. If a molecular seal is not used, the recommended velocity is 1 ft/sec, thereby increasing the purge gas requirement. Figure 16 shows the features of the molecular seal.

FLARE BURNERS

Flare burners are located at the tip of the flare stack, with the top section typically about 12 ft in length. The burner diameter is sized to accommodate a specified velocity. Flame blow-out occurs when the exit velocity of vapors exceeds 20 to 30 percent sonic conditions

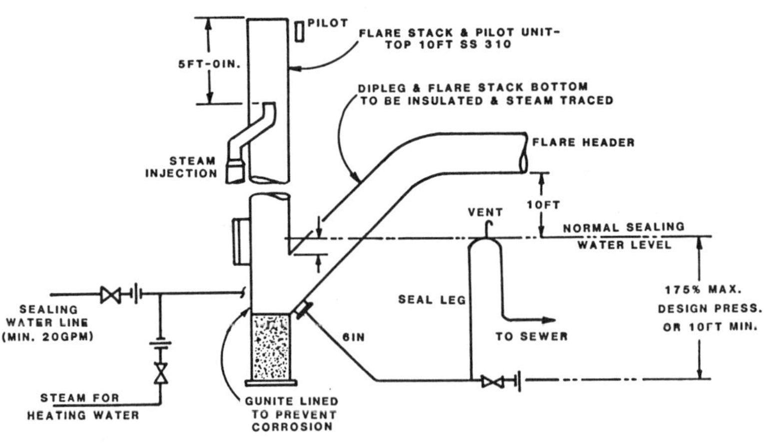

Figure 15. Seal-leg arrangement [9].

[8]. It is therefore good practice to size burners to accommodate discharge velocities less than 20 percent of the sonic velocity. The mass flow rate through the burner is:

$$W = 3{,}600\ \rho_G A_c \tag{9}$$

where W = mass flow rate (lb/hr)
ρ_G = density of the gas (lb/ft^3)
A_c = cross-sectional area (sqft)

The vapor density is:

$$\rho_G = \frac{MP}{10.73T}$$

and since the exit velocity equals 20 percent of sonic:

$$V = \frac{1}{5}\sqrt{\frac{g_c \kappa RT}{M}} \tag{10}$$

Hence, the flare tip cross-section is:

$$A_c = \frac{0.785}{144} = d^2 \tag{11}$$

where M = molecular weisht
P = absolute pressure of vapor (psia)
T = temperature (°F)
g_c = conversion factor, $32.17 \frac{(lb_m)\,(ft)}{lb_f(sec^2)}$
R = gas constant, $1{,}546 \frac{(ft - lb_f)}{(°R)\,(mole)}$
$\kappa = C_p/C_v = 1.2$ assumed
d = diameter of flare tip (in.)

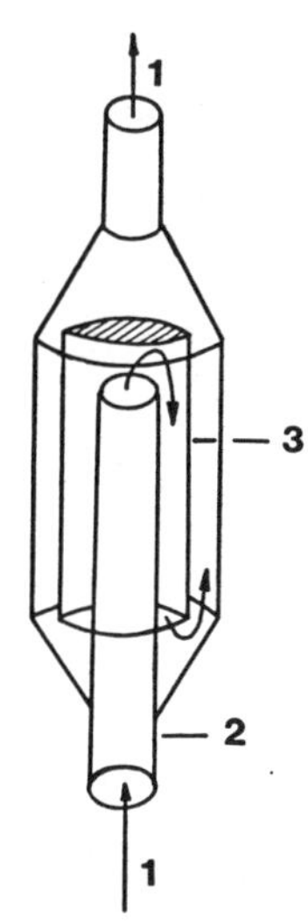

Figure 16. Shows a molecular seal. Note that the molecular seal is flange-connected to the burner tip at the top and to the flare stack section at the bottom: (1) flare gas (intermittent flow) and purge gas (continuous); (2) main section (having same diameter as stack); (3) inverted bucket.

Combining the above equations and using the values for g_c, κ, R, and P (where P = 14.7 psia):

$$d^2 = \frac{W}{1{,}370}\sqrt{\frac{T}{M}} \tag{12}$$

If based on the maximum rate, the diameter will be too large, then the normal flow must be used and checked for 40% of sonic velocity for the maximum load.

SAFETY CONSIDERATIONS FOR FLARING

The location of a flare stack is a safety related issue. Normally, it is located in areas which are on the leeward side of the plant (downstream of prevailing winds) and remote from operating and trafficked zones.

The height of the flare stack depends on several factors, namely:

- Heat released by the flare gas.
- Characteristics of the flame and the flame length.
- Emissivity of the flame.
- Radiation intensity of the flame.
- Ground level concentration of toxic gases present in the flare stream in the event of a flame blowout.

The burning characteristics of flames and the flame length are important parameters in sizing stacks. Burning characteristics are illustrated by the plots shown in Figures 17 through 19. The plot shows different zones of the flame spectrum in terms of dimensionless numbers which enables estimates of the critical flame points in each combustion zone. This helps to visualize how a flame profile may be superimposed on the loci of Figure 18.

Note that the flame height increases appreciably when combustible gas flow is sufficiently reduced so as to cause a shift into the laminar zone. By designing a flare tip which induces premixing of gas and air or selecting a "smokeless" design which induces partial premixing by agitation with steam, the increased peaking of the flare in the laminar zone may be avoided or reduced. This type of flare tip design also reduces the noise level.

Figure 16 should be used along with the following criteria:

- Peak at Reynolds number = 3,000.
- Valley at Reynolds number = 5,000.
- Blow off at Mach number = 0.2.

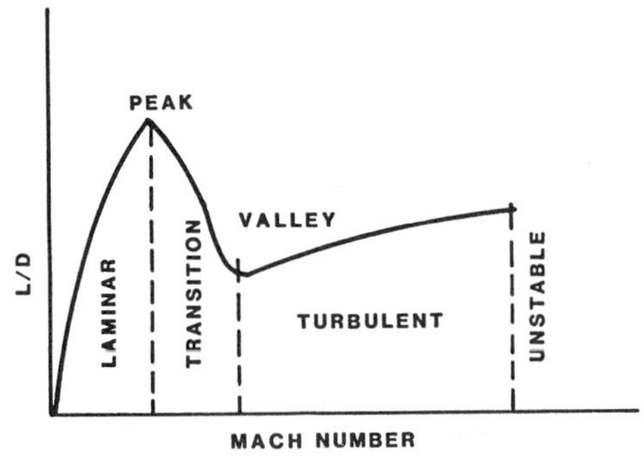

Figure 17. Burning characteristics of flames from circular ducts discharging vertically into quiescent air without premixing.

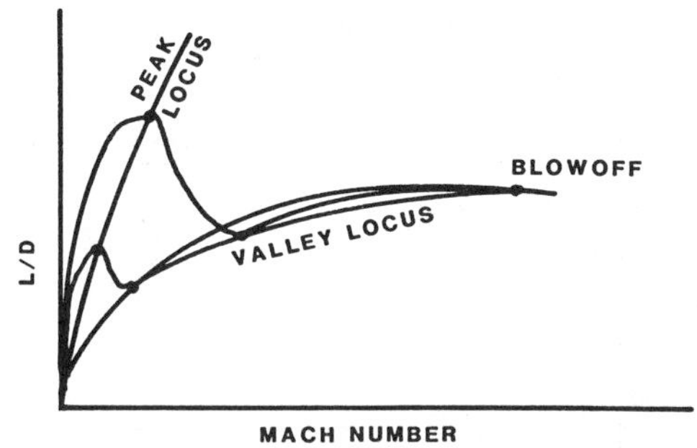

Figure 18. Shows superposition of typical flame characteristics on the locus curves.

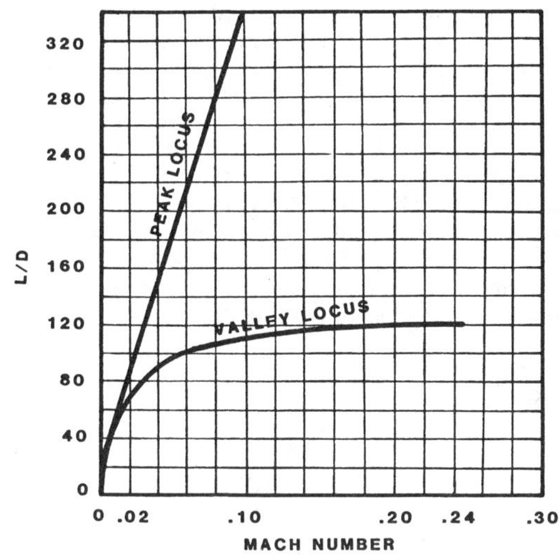

Figure 19. Plot of aspect ratio (L/D) versus Mach number.

Note that the Reynolds number is based on the stack diameter. Each of the above criteria refers to the gas state prior to combustion at the exit from the stack tip. A Reynolds number of 3,000 applies to the "peak loci curve," a Reynolds number of 5,000 applies to the "valley loci curve," and the blowoff Mach number aplies to the limit of the "valley loci curve." The blowoff point is reached when the velocity of gas leaving the stack causes the flame to separate from the tip at which point the flame becomes unstable.

For maximum stack discharge, a Mach number of 0.2 is recommended. From Figure 19 then, the corresponding L/D ratio is 118. From the stack diameter D, the flame length L can be determined.

The thermal radiation and escape time can be estimated from the data given in Table 2. Values are based on experimental data on the threshold limit of pain to the human body as a function of the radiation intensity in Btu/hr/ft^2, generated by a flame.

Figure 20 is a plot of the data reported in Table 1. A safe level of heat radiation intensity for unlimited time exposure has been found to be 440 Btu/hr/ft^2. It is apparent that a time

Table 2
Heat Radiation and Escape Time

Radiation Intensity (Btu/hr/ft^2)	Time to Pain Threshold (seconds)
440	Infinite
550	60
740	40
920	30
1,500	16
2,200	9
3,000	6
3,700	4
6,300	2

From American Petroleum Institute [12].

interval with varying radiation intensity must be allowed, to permit a human to escape from a suddenly released intense heat source. The varying radiation intensity results from an individual increasing his distance from the source of heat.

Assume an individual to be at the base of a flare stack when heat is suddenly being released. The average reaction time of an individual is between 3 and 5 seconds. Hence, during this short reaction time interval, the full radiated heat intensity will be absorbed. Then follows another short time interval (20 ft/sec is normally assumed to be the average escape velocity of a man) during which continually decreasing amounts of heat will be absorbed until a safe distance is reached (heat intensity for a safe location is 440 Btu/hr/sq.ft. and lower from Figure 17).

The maximum heat intensity which may be tolerated at the base of the stack corresponding to the limiting total heat absorbed may then be determined by

$$t_a \cdot q_a = t_r q_m + t_e \frac{q_M - q_m}{\ln \dfrac{(q_M)}{(q_m)}} \tag{13}$$

where $t_a = t_r + t_e$ (i.e., total time exposed = reaction time + escape time).
$t_a q_a$ = total heat flow/area for the exposure time
q_M = maximum radiation intensity
q_m = minimum radiation intensity

Figure 20 is the solution of the preceding equation. The escape time interval t_e depends on the stack height H. Therefore, the use of Figure 20 will not be fully understood until stack height criteria are established.

The following steps outline the approach to determining the flare stack height based on the radiation intensity:

- Calculate the radiation intensity using the following equation:

$$q = \frac{\varepsilon' Q}{4\pi X^2} \tag{14}$$

where q = radiation intensity (Btu/hr/sqft).
ε' = emissivity of the flame
Q = heat generated by the flame (Btu/hr)
X = distance from the center of flame, X_m feet above grade to point P (ft) (refer to Figure 20A)

Flame emissivity values for common gases are as follows [11]:

Gas	ε'
Hydrocarbon	0.4
Propane	0.33
Methane	0.2

A relationship between ε' and the net calorific value of a gas can be used in the absence of data [11]:

$$\varepsilon' = 0.2\left(\frac{h_c}{900}\right) 1/2 \tag{15}$$

where h_c = net heat value of a gas (NHV) in Btu/scf (60° F, 14.7 psia)

- Calculate the heat flow Q, Btu/hr

$$Q = Wh_c \frac{379}{M} \tag{16}$$

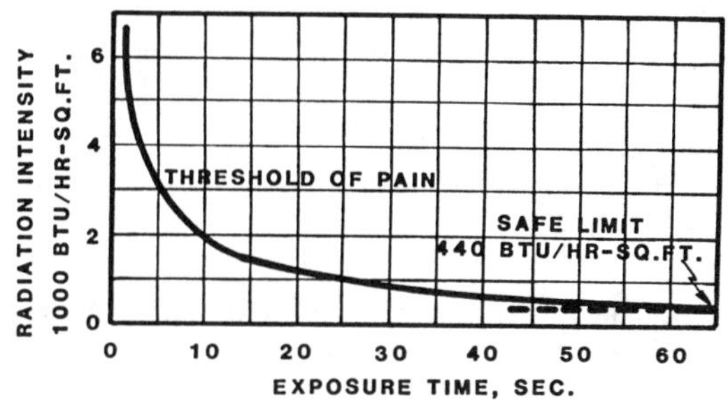

Figure 20. Plot of heat radiation vs exposure time for bare skin at the threshold of pain.

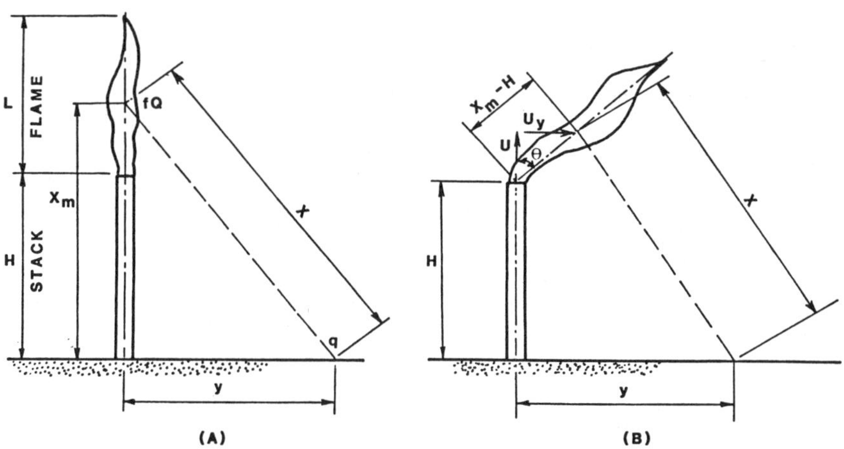

Figure 21. (A) Illustrates a flare stack and flame in stagnant surroundings; (B) shows flame in wind-blown environment.

where W = lb/hr of vapors released
h_c = net heating value of gas in Btu/scf (60° F, 14.7 psia)
M = molecular weight of the gas

- The formula for the stack height is first derived. Referring to Figure 21,

where $$X^2 = X_m^2 + y^2$$

$$X_m = \sqrt{H(H+L)}$$

where X_m = distance of the point of maximum intensity from grade (ft)
H = stack height (ft)
L = flame length (ft), L is 118D as shown

Hence

$$X^2 = H(H+L) + y^2$$

And from Equation 14

$$q_M = \frac{\varepsilon' Q}{4\pi H(H+L)}$$

where q_M is the maximum radiation intensity at the base of the flare (i.e., at $y = 0$ or $x = X_m$).

Hence, H is derived as:

$$H = \frac{1}{2}\left[\left(L^2 + \frac{\varepsilon' Q}{\pi q_m}\right)^{1/2} - L\right] \tag{17}$$

The shortest stack is obtained when q_M = 3,300 Btu/hr/sqft or from Figure 22 at $t_e = 0$. The limiting safe radial distance from the flame is:

$$X = \left[\frac{\varepsilon' Q}{4\pi 440}\right]^{1/2} \quad \text{and} \quad X^2 = \frac{\varepsilon' Q}{5{,}530}$$

Note that y = radial distance from the base of the stack = $[X^2 - H(H+L)]^{1/2}$. Allowing for the speed of escape (20 ft/sec) we have:

$$y = 20t_e = [X^2 - H(H+L)]^{1/2} \tag{18}$$

This defines the safety boundary, corresponding to quiescent ambient air.

Thus, the stack height H, the limiting heat radiation q_M, and the radial distance; y, can be evaluated by a trial and error procedure by assuming a value of t_e.

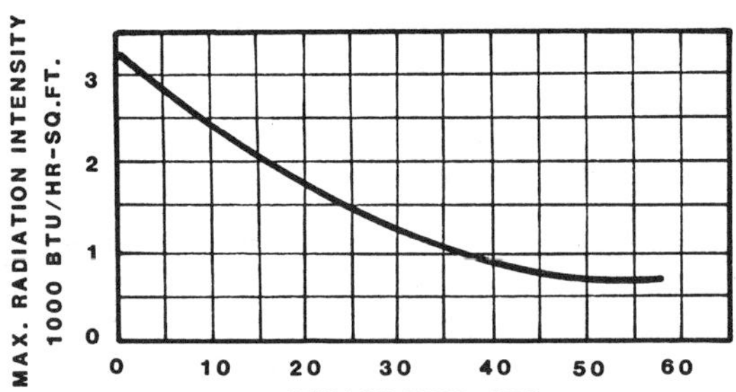

Figure 22. Plot of maximum radiation intensity vs. escape time, assuming a five-second reaction time.

The above analysis mut be extended to account for the more prevalent case of wind circulation in the vicinity of the flare. For locations where wind intensity is unknown, it is suggested that an average 20 mph wind be assumed acting in all directions which results in increasing the safe circular boundary by the resulting tilt of the flame. (This is illustrated in Figure 21B). The flame tilt and its effect on the safety boundary increase may be determined as follows:

$$\tan\theta = \frac{U_W}{U}$$

where U_W = wind velocity
U = flare exit velocity
$U_W = (X_m - H)\sin\theta$
$U_t = (X_m - H)\cos\theta$

$$y = \{X^2 - (H + (X_m - H)\cos\theta)^2\}^{1/2} + (X_m - H)\sin\theta \quad (19)$$

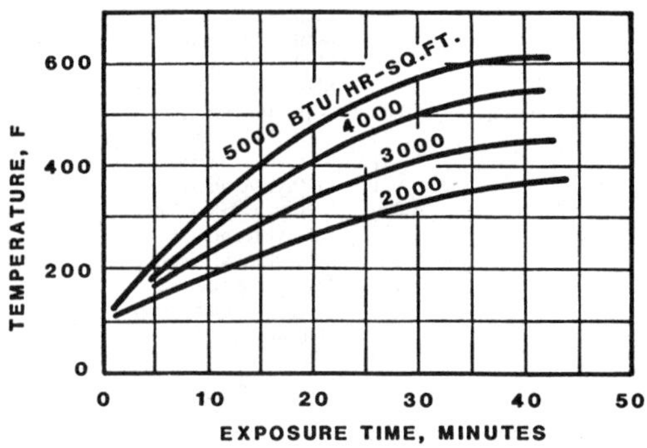

Figure 23. Plot of temperature of steel equipment vs. exposure time for different radiant heat intensities. Curves are based on 1/4 inch plate thickness with an effective emissivity of 10 and view factor of 0.5. Cooling caused by convection, etc., is neglected.

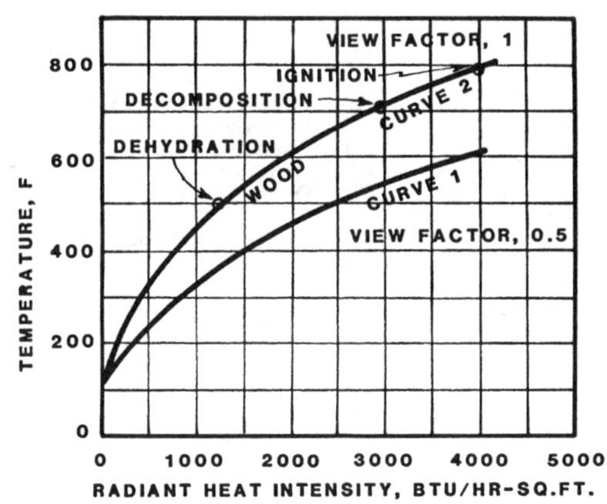

Figure 24. Plot of equilibrium temperature vs. radiant heat intensity: Curve 1—metal equipment; Curve 2—wood.

This formula establishes the limiting boundary for wind circulation. When evaluating wind effects on flame tilt, an average wind intensity should be used in the calculations.

The preceding analysis leads to a conservative design since calculations are based on thermal effects to bare skin. If protective clothing is worn by operating personnel along with proper shielding to reduce heat radiation, the required stack height can be greatly reduced. There is, however, a tradeoff in that the safety boundary increases. Since the head load of

Table 3
Threshold Limits for Certain Toxic Substances Gases and Vapors

Gas or Vapor	PPM	Gas or Vapor	PPM
Acetaldehyde	200	Hydrogen solenie	0.05
Acetic acid	10	Hydrogen sulfide	20
Acetic anhydride	5	Isodine	0.1
Acetone	1,000	Isophorene	25
Acrolein	0.5	Isopropylamine	5
Acrylonitire	20	Mesityl oxide	50
Ammonia	100	Methyl acetate	200
Amyl acetate	200	Methyl acetylene	1,000
Amyl alcohol	100	Methyl alcohol	200
Aniline	5	Methyl bromide	20
Arsinic	0.05	2-Methoxyethanol	25
Benxene	35	Methyl chloride	100
Benzyl chloride	1	Methylcyclohexane	500
Bromide	1	Methylcyclohexanol	100
Butadiene	1,000	Methylcyclohexanone	100
Butyl alcohol	100	Methyl formate	100
Butylamine	5	Methyl amyl alcohol	25
Carbon dioxide	5,000	Methylene chloride	
Carbon disulfide	20	(dichloromethane)	500
Carbon monoxide	100	Naptha (coal tar)	200
Carbon tetrachloride	25	Naptha (petroleum)	500
Chlorine	1	pNitroaniline	1
Chlorobenzene	75	Nickel carbonyl	0.001
Chloroform	100	Nitrobenzene	1
Cresol (all isomers)	5	Nitroethane	100
Cyclohexane	400	Nitrogen dioxide	5
Cyclohezanol	100	Nitromethane	100
Cyclohexanone	100	Nitrotoluene	5
Cyclohexene	400	Octane	500
Cyclopropane	400	Ozone	0.1
Diacetone alcohol	50	Pentane	1,000
0-Dichlorobenzene	50	Propyl ketone	200
1,1-Dichloroethane	100	Phenol	5
Diethylamine	25	Phenylhydrazine	5
Diisobutyl ketone	50	Phosgene (carbonyl chloride)	1
Dimethylaniline	5	Phosphine	0.05
Dimethylsulfate	1	Phosphorus trichloride	0.5
Diethylene dioxide	100	Propyl acetate	200
Ethyl acetate	400	Propyl alcohol	400
Ethyl alcohol (ethanol)	1,000	Propyl ether	500
Ethylamine	25	Propylene dichloride	75
Ethylbenzene	200	Pyridine	10
Ethyl bromide	200	Quinone	0.1
Ethyl chloride	1,000	Stibine	0.1
Ethyl ether	400	Styrene	200
Ethylene chlorohydrin	5	Sulphur dioxide	10
Ethylenedriamine	10	Sulphur hexafluoride	1,000
Ethylene dibromide	25	Sulphur monochloride	1
Ethylene dichloride	100	Sulphur pentafluoride	0.025
Ethylene oxide	100	1,1,2,2-Tetrachloroethane	5
Fluorine	0.1	Tetranitromethane	1
Formaldehyde	5	Toluene (toluol)	200
Gasoline	500	o-Toluidine	5
Hydrazine	1	Trichloroethylene	200
Hydrogen bromide	5	Trichloroethane	500

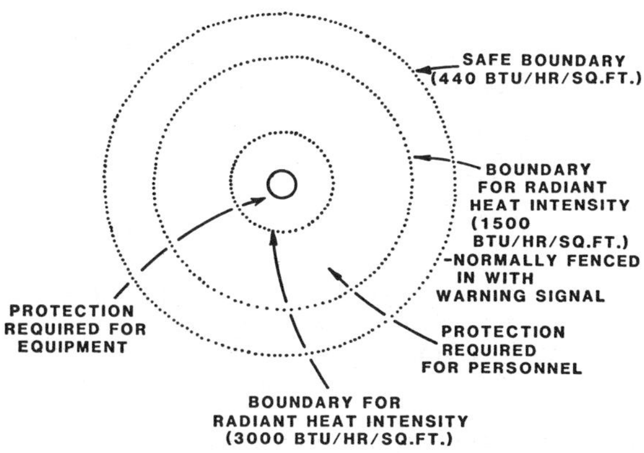

Figure 25. Shows circles of radiant heat intensity.

the flare, the flame length, and the safe radiation intensity (440 Btu/hr/sqft) remain the same, decreasing the stack height leads to an increase in the safety envelope.

An alternative method of stack height sizing is based on the allowable limit for radiation intensity. For operating personnel, the allowable intensity is 1,500 Btu/hr/sqft and for equipment 3,000 Btu/hr/sqft.

The 1,500 Btu/hr/ft^2 criterion is based on the following. In emergency releases, a reaction time of an operation of 3 to 5 seconds may be assumed. Perhaps 5 to 10 seconds more would elapse before the average individual could escape the area with an average escape velocity of 20 ft/sec. This would result in a total exposure period ranging from 8 to 15 seconds. The time to pain threshold corresponding to 1,500 Btu/hr/ft^2 is 16 seconds (from Table 1), before the individual could escape to a safe place. The effect of heat radiation on equipment is shown in Figure 23. [8]. The temperature of metal equipment increases with exposure time and the higher the radiant heat intensity, the greater the temperature. Curve 1 in Figure 24 shows the theoretical equilibrium temperatures based on a view factor of 0.5. The actual temperature on surfaces facing the flame will lie between curves 1 and 2.

The temperature of vessels containing liquid or flowing vapors may be lower because of cooling effects. Curve 2 applies to materials having a low heat conductivity coefficient (e.g., wood). In this case, equilibrium temperatures are reached within a shorter time as compared with metal objects. Dehydration of wood takes place at about 500° F, decomposition at 700° F, and ignition at around 800° F, corresponding to 1,300, 3,000, and 4,000 Btu/hr/ft^2, respectively. This means that wooden structures and vegetation exposed to heat intensities of 3,000 to 4,000 Btu/hr/ft^2 and higher may catch fire and burn. Paint on equipment may also be damaged. Therefore, it is recommended that equipment located in this area be protected by proper heat shielding or emergency water sprays.

The following steps outline design calculations by the alternate method:

- Compute the radial distance from the flame using Equation 14 for q = 15,000 Btu/hr/sqft.
- Compute the safe radial distance at q = 440 Btu/hr/sqft (Equation 14).
- A suitable value for q is assumed at the base of the stack. q = 3,000 Btu/hr/sqft is a good start since protective shielding will be provided at the stack.
- From Equation 17 compute H.

Figure 25 illustrates the different heat intensity loci that should be examined.

After the stack height has been established from radiation intensity values, the maximum permissible ground level concentrations of toxic gases in the event of a flame blowout should be evaluated. Table 3 provides data on toxicological threshold limits as reported by the Environmental Protection Agency (EPA).

Estimated ground level concentrations should be based on the emergency condition of a flame blowout. Calculations should be performed for a range of climatological conditions at the plant site. For a rough estimate the following empirical formula [13] can be used:

$$C_{max} = \frac{3{,}697\ vM\sigma_Z}{\mu H^2 \sigma_y}$$

$$X_{max} = (H/\sigma_Z)^{2/2-N} \tag{20}$$

where
- C = concentration at grade in ppm (volume)
- v = specific volume of toxic gas (ft^3/lb)
- M = weight discharge of pollutant component (ton/day)
- σ_Z = vertical diffusion coefficient
- μ = air velocity at grade
- H = stack height (ft)
- σ_y = horizontal diffusion coefficient
- X_{max} = distance from stack to the point of maximum conc. (ft)
- N = environmental factor

The following values are taken from the API Manual [14]:

$$\sigma_Z = 0.13$$

$$\sigma_y = 0.13$$

$$N = 0.25$$

Cheremisinoff et al. [15, 16] give detailed procedures for estimating pollution levels.

NOTATION

A	cross-sectional area	N	line resistance or environmental factor
$C_{p,v}$	specific heats at constant pressure and volume, respectively	P	pressure
C	concentration	Q	heat flow rate
d, D	diameter	q	heat flow per unit area
F	environmental factor	R	universal gas law constant
f	friction factor	Re	Reynolds number
G	mass flow rate per unit area	T	absolute temperature
g	gravitational acceleration	t	time
g_c	conversion factor	U	flare exit velocity
H	stack height	U_w	wind speed
h_c	heating value	V	velocity
K_i	resistance coefficient	W	vapor mass rate
L	length	X	distance from flame source
M	molecular weight	y	radial distance
Ma	Mach number	Z	compressibility factor

Greek Symbols

ε'	emmisivity	v	specific volume
θ	angle	σ_y, σ_z	horizontal and vertical diffusion coefficients
κ	ratio of specific heats		
μ	viscosity		

REFERENCES

1. American Petroleum Institute, *Refinery Practice,* 520: 8 (1969).
2. American Petroleum Institute, *Refinery Practice,* 520 (Part I): 18 (1973).
3. American Petroleum Institute, *Refinery Practice,* 520 (Part I): 26 (1973).
4. Lapple, C. E., *Trans. AIChE,* 39: 385 (1943).
5. Loeb, M. B., Report Tr–256–D, John F. Kennedy Space Center (1965).
6. Moody, LF., *Trans. ASME66* (1944).
7. American Petroleum Institute, *Refinery Practice,* 521: 45 (1969).
8. Tan, S. H., "Flare Design Simplified," in *Waste Treatment Flare Stack Design Handbook,* Hydrocarbon Processing, Houston (1968).
9. Perry, R., Cecil, H., and Chilton, H., (Eds.) *Chemical Engineers Handbook,* 5th ed., McGraw-Hill Book Co., NY, 1972.
10. American Petroleum Institute, *Refinery Practice,* 521: 50 (1969).
11. Kent, G. B., "Practical Design of Flare Stacks," in *Waste Treatment and Flare Stack Design Handbook,* Hydrocarbon Processing, Texas, (1968).
12. American Petroleum Institute, *Refinery Practice,* 521: 35 (1969).
13. American Petroleum Institute, *Refinery Practice,* 521: 64 (1969).
14. American Petroleum Institute, "Disposal on Refinery Wastes," 2:53 (1976), Ch. 9.
15. Cheremisinoff, P. N., and Young, R. A., *Pollution Engineering Practice* Handbook, Ann Arbor Science Pub., Ann Arbor, MI (1976).
16. Cheremisinoff, P. N., and Young, R. A., (Eds.) *Air Pollution Control and Design Handbook,* Part I, Marcel Dekker Inc., NY (1977).

INDEX

B

D

E

F

M

N

O

P

Q

S

U

V

W

Y

Z